# THE STRUCTURE OF EVOLUTIONARY THEORY

THE STRUCTURE OF EVOLUTIONARY THEORY

STEPHEN JAY GOULD

# The Structure Of Evolutionary Theory

THE BELKNAP PRESS OF
HARVARD UNIVERSITY PRESS
CAMBRIDGE, MASSACHUSETTS
AND LONDON, ENGLAND

Copyright © 2002 by the President and Fellows of Harvard College

Printed in the United States of America

*Library of Congress Cataloging-in-Publication Data*

Gould, Stephen Jay.
   The structure of evolutionary theory / Stephen Jay Gould.
     p.  cm.
   Includes bibliographical references (p. )
   ISBN 0-674-00613-5 (alk. paper)
   1. Evolution (Biology)   2. Punctuated equilibrium (Evolution)   I. Title.

QH366.2 .G663   2002
576.8—dc21         2001043556

Fifth printing, 2002

For Niles Eldredge and Elisabeth Vrba
May we always be the Three Musketeers
Prevailing with panache
From our manic and scrappy inception at Dijon
To our nonsatanic and happy reception at Doomsday
*All For One and One For All*

# Contents

# Expanded Contents

## Part II: Towards a Revised and Expanded
## Evolutionary Theory

### Chapter 8: Species as Individuals in the Hierarchical
### Theory of Selection

# THE STRUCTURE OF EVOLUTIONARY THEORY

# CHAPTER ONE

# Defining and Revising the Structure of Evolutionary Theory

## Theories Need Both Essences and Histories

In a famous passage added to later editions of the *Origin of Species,* Charles Darwin (1872, p. 134) generalized his opening statement on the apparent absurdity of evolving a complex eye through a long series of gradual steps by reminding his readers that they should always treat "obvious" truths with skepticism. In so doing, Darwin also challenged the celebrated definition of science as "organized common sense," as championed by his dear friend Thomas Henry Huxley. Darwin wrote: "When it was first said that the sun stood still and world turned round, the common sense of mankind declared the doctrine false; but the old saying of *Vox populi, vox Dei* [the voice of the people is the voice of God], as every philosopher knows, cannot be trusted in science."

Despite his firm residence within England's higher social classes, Darwin took a fully egalitarian approach towards sources of expertise, knowing full well that the most dependable data on behavior and breeding of domesticated and cultivated organisms would be obtained from active farmers and husbandmen, not from lords of their manors or authors of theoretical treatises. As Ghiselin (1969) so cogently stated, Darwin maintained an uncompromisingly "aristocratic" set of values towards judgment of his work—that is, he cared not a whit for the outpourings of *vox populi,* but fretted endlessly and fearfully about the opinions of a very few key people blessed with the rare mix of intelligence, zeal, and attentive practice that we call expertise (a democratic human property, respecting only the requisite mental skills and emotional toughness, and bearing no intrinsic correlation to class, profession or any other fortuity of social circumstance).

Darwin ranked Hugh Falconer, the Scottish surgeon, paleontologist, and Indian tea grower, within this most discriminating of all his social groups, a panel that included Hooker, Huxley and Lyell as the most prominent members. Thus, when Falconer wrote his important 1863 paper on American fossil elephants (see Chapter 9, pages 745–749, for full discussion of this incident), Darwin flooded himself with anticipatory fear, but then rejoiced in his critic's generally favorable reception of evolution, as embodied in the closing sentence of Falconer's key section: "Darwin has, beyond all his cotemporaries [sic], given an impulse to the philosophical investigation of the most backward and obscure branch of the Biological Sciences of his day; he has laid the foundations of a great edifice; but he need not be surprised if, in the progress of erection, the superstructure is altered by his successors, like the Duomo of Milan, from the roman to a different style of architecture."

In a letter to Falconer on October 1, 1862 (in F. Darwin, 1903, volume 1, p. 206), Darwin explicitly addressed this passage in Falconer's text. (Darwin had received an advance copy of the manuscript, along with Falconer's request for review and criticism—hence Darwin's reply, in 1862, to a text not printed until the following year): "To return to your concluding sentence: far from being surprised, I look at it as absolutely certain that very much in the *Origin* will be proved rubbish; but I expect and hope that the framework will stand."

The statement that God (or the Devil, in some versions) dwells in the details must rank among the most widely cited intellectual witticisms of our time. As with many clever epigrams that spark the reaction "I wish I'd said that!", attribution of authorship tends to drift towards appropriate famous sources. (Virtually any nifty evolutionary saying eventually migrates to T. H. Huxley, just as vernacular commentary about modern America moves towards Mr. Berra.) The apostle of modernism in architecture, Ludwig Mies van der Rohe, may, or may not, have said that "God dwells in the details," but the plethora of tiny and subtle choices that distinguish the elegance of his great buildings from the utter drabness of superficially similar glass boxes throughout the world surely validates his candidacy for an optimal linkage of word and deed.

Architecture may assert a more concrete claim, but nothing beats the extraordinary subtlety of language as a medium for expressing the importance of apparently trivial details. The architectural metaphors of Milan's cathedral, used by both Falconer and Darwin, may strike us as effectively identical at first read. Falconer says that the foundations will persist as Darwin's legacy, but that the superstructure will probably be reconstructed in a quite different style. Darwin responds by acknowledging Falconer's conjecture that the theory of natural selection will undergo substantial change; indeed, in his characteristically diffident way, Darwin even professes himself "absolutely certain" that much of the *Origin*'s content will be exposed as "rubbish." But he then states not only a hope, but also an expectation, that the "framework" will stand.

We might easily read this correspondence too casually as a polite dialogue between friends, airing a few unimportant disagreements amidst a commitment to mutual support. But I think that this exchange between Falconer and Darwin includes a far more "edgy" quality beneath its diplomacy. Consider the different predictions that flow from the disparate metaphors chosen by each author for the *Duomo* of Milan—Falconer's "foundation" *vs.* Darwin's "framework." After all, a foundation is an invisible system of support, sunk into the ground, and intended as protection against sinking or toppling of the overlying public structure. A framework, on the other hand, defines the basic form and outline of the public structure itself. Thus, the two men conjure up very different pictures in their crystal balls. Falconer expects that the underlying evolutionary principle of descent with modification will persist as a factual foundation for forthcoming theories devised to explain the genealogical tree of life. Darwin counters that the theory of natural selection will persist as a basic explanation of evolution, even though many details, and even some subsidiary generalities, cited within the *Origin* will later be rejected as false, or even illogical.

I stress this distinction, so verbally and disarmingly trivial at a first and superficial skim through Falconer's and Darwin's words, but so incisive and portentous as contrasting predictions about the history of evolutionary theory, because my own position—closer to Falconer than to Darwin, but in accord with Darwin on one key point—led me to write this book, while also supplying the organizing principle for the "one long argument" of its entirety. I do believe that the Darwinian framework, and not just the foundation, persists in the emerging structure of a more adequate evolutionary theory. But I also hold, with Falconer, that substantial changes, introduced during the last half of the 20th century, have built a structure so expanded beyond the original Darwinian core, and so enlarged by new principles of macroevolutionary explanation, that the full exposition, while remaining within the domain of Darwinian logic, must be construed as basically different from the canonical theory of natural selection, rather than simply extended.

A closer study of the material basis for Falconer and Darwin's metaphors— the *Duomo* (or Cathedral) of Milan—might help to clarify this important distinction. As with so many buildings of such size, expense, and centrality (both geographically and spiritually), the construction of the Duomo occupied several centuries and included an amalgam of radically changing styles and purposes. Construction began at the chevet, or eastern end, of the cathedral in the late 14th century. The tall windows of the chevet, with their glorious flamboyant tracery, strike me as the finest achievement of the entire structure, and as the greatest artistic expression of this highly ornamented latest Gothic style. (The term "flamboyant" literally refers to the flame-shaped element so extensively used in the tracery, but the word then came to mean "richly decorated" and "showy," initially as an apt description of the overall style, but then extended to the more general meaning used today.)

Coming now to the main point, construction then slowed considerably, and the main western facade and entrance way (Fig. 1-1) dates from the late 16th century, when stylistic preferences had changed drastically from the points, curves and traceries of Gothic to the orthogonal, low-angled or gently rounded lintels and pediments of classical Baroque preferences. Thus, the first two tiers of the main (western) entrance to the *Duomo* display a style that, in one sense, could not be more formally discordant with Gothic elements of design, but that somehow became integrated into an interesting coherence. (The third tier of the western facade, built much later, returned to a "retro" Gothic style, thus suggesting a metaphorical reversal of phylogenetic conventions, as

1-1. The west facade (main entrance) of Milan Cathedral, built in baroque style in the 16th century, with a retro-gothic third tier added later.

up leads to older—in style if not in actual time of emplacement!) Finally, in a distinctive and controversial icing upon the entire structure (Fig. 1-2), the "wedding cake," or row-upon-row of Gothic pinnacles festooning the tops of all walls and arches with their purely ornamental forms, did not crown the edifice until the beginning of the 19th century, when Napoleon conquered the city and ordered their construction to complete the *Duomo* after so many centuries of work. (These pinnacle forests may amuse or disgust architectural purists, but no one can deny their unintended role in making the *Duomo* so uniquely and immediately recognizable as the icon of the city.)

How, then, shall we state the most appropriate contrast between the *Duomo* of Milan and the building of evolutionary theory since Darwin's *Origin* in 1859? If we grant continuity to the intellectual edifice (as implied by comparison with a discrete building that continually grew but did not change its location or basic function), then how shall we conceive "the structure of evolutionary theory" (chosen, in large measure, as the title for this book because I wanted to address, at least in practical terms, this central question in the history and content of science)? Shall we accept Darwin's triumphalist stance and hold that the framework remains basically fixed, with all visually substantial change analogous to the non-structural, and literally superficial, icing of topmost pinnacles? Or shall we embrace Falconer's richer and more critical, but still fully positive, concept of a structure that has changed in radi-

1-2. The "wedding cake" pinnacles that festoon the top of Milan Cathedral, and that were not built until the first years of the 19th century after Napoleon conquered the city.

cal ways by incorporating entirely different styles into crucial parts of the building (even the front entrance!), while still managing to integrate all the differences into a coherent and functional whole, encompassing more and more territory in its continuing enlargement?

Darwin's version remains Gothic, and basically unchanged beyond the visual equivalent of lip service. Falconer's version retains the Gothic base as a positive constraint and director, but then branches out into novel forms that mesh with the base but convert the growing structure into a new entity, largely defined by the outlines of its history. (Note that no one has suggested the third alternative, often the fate of cathedrals—destruction, either total or partial, followed by a new building of contrary or oppositional form, erected over a different foundation.)

In order to enter such a discourse about "the structure of evolutionary theory" at all, we must accept the validity, or at least the intellectual coherence and potential definability, of some key postulates and assumptions that are often not spelled out at all (especially by scientists supposedly engaged in the work), and are, moreover, not always granted this form of intelligibility by philosophers and social critics who do engage such questions explicitly. Most importantly, I must be able to describe a construct like "evolutionary theory" as a genuine "thing"—an entity with discrete boundaries and a definable history—especially if I want to "cash out," as more than a confusingly poetic image, an analogy to the indubitable bricks and mortar of a cathedral.

In particular, and to formulate the general problem in terms of the specific example needed to justify the existence of this book, can "Darwinism" or "Darwinian theory" be treated as an entity with defining properties of "anatomical form" that permit us to specify a beginning and, most crucially for the analysis I wish to pursue, to judge the subsequent history of Darwinism with enough rigor to evaluate successes, failures and, especially, the degree and character of alterations? This book asserts, as its key premise and one long argument, that such an understanding of modern evolutionary theory places the subject in a particularly "happy" intellectual status—with the central core of Darwinian logic sufficiently intact to maintain continuity as the centerpiece of the entire field, but with enough important changes (to all major branches extending from this core) to alter the structure of evolutionary theory into something truly different by expansion, addition, and redefinition. In short, "The structure of evolutionary theory" combines enough stability for coherence with enough change to keep any keen mind in a perpetual mode of search and challenge.

The distinction between Falconer's and Darwin's predictions, a key ingredient in my analysis, rests upon our ability to define the central features of Darwinism (its autapomorphies, if you will), so that we may then discern whether the extent of alteration in our modern understanding of evolutionary mechanisms and causes remains within the central logic of this Darwinian foundation, or has now changed so profoundly that, by any fair criterion in vernacular understanding of language, or by any more formal account of departure from original premises, our current explanatory theory must be described as a different kind of mental "thing." How, in short, can such an intellectual entity be defined? And what degree of change can be tolerated or accommodated within the structure of such an entity before we must alter the name and declare the entity invalid or overthrown? Or do such questions just represent a fool's errand from the start, because intellectual positions can't be reified into sufficient equivalents of buildings or organisms to bear the weight of such an inquiry?

As arrogant as I may be in general, I am not sufficiently doltish or vainglorious to imagine that I can meaningfully address the deep philosophical questions embedded within this general inquiry of our intellectual ages—that is, fruitful modes of analysis for the history of human thought. I shall therefore take refuge in an escape route that has traditionally been granted to scientists: the liberty to act as a practical philistine. Instead of suggesting a principled and general solution, I shall ask whether I can specify an operational way to define "Darwinism" (and other intellectual entities) in a manner specific enough to win shared agreement and understanding among readers, but broad enough to avoid the doctrinal quarrels about membership and allegiance that always seem to arise when we define intellectual commitments as pledges of fealty to lists of dogmata (not to mention initiation rites, secret handshakes and membership cards—in short, the intellectual paraphernalia that led Karl Marx to make his famous comment to a French journalist: "je ne suis pas marxiste").

As a working proposal, and as so often in this book (and in human affairs in general), a "Goldilocks solution" embodies the blessedly practical kind of approach that permits contentious and self-serving human beings (God love us) to break intellectual bread together in pursuit of common goals rather than personal triumph. (For this reason, I have always preferred, as guides to human action, messy hypothetical imperatives like the Golden Rule, based on negotiation, compromise and general respect, to the Kantian categorical imperatives of absolute righteousness, in whose name we so often murder and maim until we decide that we had followed the wrong instantiation of the right generality.) We must, in short and in this case, steer between the "too little" of refusing to grant any kind of "essence," or hard anatomy of defining concepts, to a theory like Darwinism; and the "too much" of an identification so burdened with a long checklist of exigent criteria that we will either spend all our time debating the status of particular items (and never addressing the heart or central meaning of the theory), or we will waste our efforts, and poison our communities, with arguments about credentials and anathemata, applied to individual applicants for membership.

In his brilliant attempt to write a "living" history and philosophy of science about the contemporary restructuring of taxonomic theory by phenetic and cladistic approaches, Hull (1988) presents the most cogent argument I have ever read for "too little" on Goldilocks's continuum, as embodied in his defense of theories as "conceptual lineages" (1988, pp. 15–18). I enthusiastically support Hull's decision to treat theories as "things," or individuals in the crucial sense of coherent historical entities—and in opposition to the standard tactic, in conventional scholarship on the "history of ideas," of tracing the chronology of expression for entirely abstract concepts defined only by formal similarity of content, and not at all by ties of historical continuity, or even of mutual awareness among defenders across centuries and varied cultures. (For example, Hull points out that such a conventional history of the "chain of being" would treat this notion as an invariant and disembodied Platonic archetype, independently "borrowed" from the eternal storehouse of potential models for natural reality, and then altered by scholars to fit local contexts across millennia and cultures.)

But I believe that Hull's laudable desire to recast the history of ideas as a narrative of entities in historical continuity, rather than as a disconnected chronology of tidbits admitted into a class only by sufficient formal similarity with an abstract ideological archetype, then leads him to an undervaluation of actual content. Hull exemplifies his basic approach (1988, p. 17): "A consistent application of what Mayr has termed 'population thinking' requires that species be treated as lineages, spatiotemporally localized particulars, individuals. Hence, if conceptual change is to be viewed from an evolutionary perspective, concepts must be treated in the same way. In order to count as the 'same concept,' two term-tokens must be part of the same conceptual lineage. Population thinking must be applied to thinking itself."

So far, so good. But Hull now extends this good argument for the necessity of historical connectivity into a claim for sufficiency as well—thus springing a logical trap that leads him to debase, or even to ignore, the "morphology" (or idea content) of these conceptual lineages. He states that he wishes to "organize term-tokens into lineages, not into classes of similar term-types" (pp. 16–17). I can accept the necessity of such historical continuity, but neither I nor most scholars (including practicing scientists) will then follow Hull in his explicit and active rejection of similarity in content as an equally necessary criterion for continuing to apply the same name—Darwinian theory, for example—to a conceptual lineage.

At an extreme that generates a *reductio ad absurdum* for rejecting Hull's conclusion, but that Hull bravely owns as a logical entailment of his own prior decision, a pure criterion of continuity, imbued with no constraint of content, forces one to apply the same name to any conceptual lineage that has remained consciously intact and genealogically unbroken through several generations (of passage from teachers to students, for example), even if the current "morphology" of concepts directly inverts and contradicts the central arguments of the original theory. "A proposition can evolve into its contradictory," Hull allows (1988, p. 18). Thus, on this account, if the living intellectual descendants of Darwin, as defined by an unbroken chain of teaching, now believed that each species had been independently created within six days of 24 hours, this theory of biological order would legitimately bear the name of "Darwinism." And I guess that I may call myself kosher, even though I and all members of my household, by conscious choice and with great ideological fervor, eat cheeseburgers for lunch every day—because we made this dietary decision in a macromutational shift of content, but with no genealogical break in continuity, from ten previous generations of strict observers of kashrut.

The objections that most of us would raise to Hull's interesting proposition include both intellectual and moral components. Certain kinds of systems are, and should be, defined purely by genealogy and not at all by content. I am my father's son no matter how we interact. But such genealogical definitions, as validated by historical continuity, simply cannot adequately characterize a broad range of human groupings properly designated by similarity in content. When Cain mocked God's inquiry about Abel's whereabouts by exclaiming "Am I my brother's keeper" (Genesis 4:9), he illustrated the appropriateness of either genealogy by historical connection or fealty by moral responsibility as the proper criterion for "brotherhood" in different kinds of categories. Cain could not deny his genealogical status as brother in one sense, but he derided a conceptual meaning, generally accorded higher moral worth as a consequence of choice rather than necessity of birth, in disclaiming any responsibility as keeper. As a sign that we have generally privileged the conceptual meaning, and that Cain's story still haunts us, we need only remember Claudius's lament that his murder of his own brother (and Hamlet's father) "hath the primal eldest curse upon't."

Ordinary language, elementary logic, and a general sense of fairness all combine to favor such preeminence for a strong component of conceptual continuity in maintaining a name or label for a theory. Thus, if I wish to call myself a Darwinian in any just or generally accepted sense of such a claim, I do not qualify merely by documenting my residence within an unbroken lineage of teachers and students who have transmitted a set of changing ideas organized around a common core, and who have continued to study, augment and improve the theory that bears such a longstanding and honorable label. I must also understand the content of this label myself, and I must agree with a set of basic precepts defining the broad ideas of a view of natural reality that I have freely chosen to embrace as my own. In calling myself a Darwinian I accept these minimal obligations (from which I remain always and entirely free to extract myself should my opinions or judgments change); but I do not become a Darwinian by the mere default of accidental location within a familial or educational lineage.

Thus, if we agree that a purely historical, entirely content-free definition of allegiance to a theory represents "too little" commitment to qualify, and that we must buttress any genealogical criterion with a formal, logical, or anatomical definition framed in terms of a theory's intellectual content, then what kind or level of agreement shall we require as a criterion of allegiance for inclusion? We now must face the opposite side of Goldilocks's dilemma—for once we advocate criteria of content, we do not wish to impose such stringency and uniformity that membership becomes more like a sworn obedience to an unchanging religious creed than a freely chosen decision based on personal judgment and perception of intellectual merits. My allegiance to Darwinian theory, and my willingness to call myself a Darwinian biologist, must not depend on my subscription to all 95 articles that Martin Luther nailed to the Wittenburg church door in 1517; or to all 80 items in the *Syllabus of Errors* that Pio Nono (Pope Pius IX) proclaimed in 1864, including the "fallacy," so definitionally uncongenial to science, that "the Roman Pontiff can and should reconcile himself to and agree with progress, liberalism and modern civilization"; or to all 39 articles of the Church of England, adopted by Queen Elizabeth in 1571 as a replacement for Archbishop Thomas Cranmer's 42 articles of 1553.

Goldilocks's "just right" position between these extremes will strike nearly all cooperatively minded intellectuals, committed to the operationality and advance of their disciplines, as eminently sensible: shared content, not only historical continuity, must define the *structure* of a scientific theory; but this shared content should be expressed as a *minimal list* of the *few defining attributes* of the theory's *central logic*—in other words, only the absolutely essential statements, absent which the theory would either collapse into fallacy or operate so differently that the mechanism would have to be granted another name.

Now such a minimal list of such maximal centrality and importance bears a description in ordinary language—but its proper designation requires that evolutionary biologists utter a word rigorously expunged from our professional consciousness since day one of our preparatory course work: the concept that dare not speak its name—essence, essence, essence (say the word a few times out loud until the fear evaporates and the laughter recedes). It's high time that we repressed our aversion to this good and honorable word. Theories have essences. (So, by the way, and in a more restrictive and nuanced sense, do organisms—in their limitation and channeling by constraints of structure and history, expressed as *Baupläne* of higher taxa. My critique of the second theme of Darwinian central logic, Chapters 4–5 and 10–11, will treat this subject in depth. Moreover, my partial defense of organic essences, expressed as support for structuralist versions of evolutionary causality as potential partners with the more conventional Darwinian functionalism that understandably denies intelligibility to any notion of an essence, also underlies the double entendre of this book's title, which honors the intellectual *structure* of evolutionary theory within Darwinian traditions and their alternatives, and which also urges support for a limited version of *structuralist* theory, in opposition to certain strict Darwinian verities.)

Our unthinking rejection of essences can be muted, or even reversed into propensity for a sympathetic hearing, when we understand that an invocation of this word need not call down the full apparatus of an entirely abstract and eternal Platonic *eidos*—a reading of "essence" admittedly outside the logic of evolutionary theory, and historical modes of analysis in general. But the solution to a meaningful notion of essence in biology lies within an important episode in the history of emerging evolutionary views, a subject treated extensively in Chapter 4 of this book, with Goethe, Etienne Geoffroy St. Hilaire, and Richard Owen as chief protagonists.

After all, the notion of a general anatomical blueprint that contains all particular incarnations by acting as a fundamental building block (Goethe's leaf or Geoffroy's vertebra) moved long ago from conceptualization as a disembodied and nonmaterial archetype employed by a creator, to an actual structure (or inherited developmental pathway) present in a flesh and blood ancestor—a material basis for channeling, often in highly positive ways, the future history of diversity within particular phyletic lineages. This switch from archetype to ancestor permitted us to reformulate the idea of "essence" as broad and fruitful commonalities that unite a set of particulars into the most meaningful relationships of common causal structure and genesis. Our active use of this good word should not be hampered by a shyness and disquietude lacking any validity beyond the vestiges of suspicions originally set by battles won so long ago that no one can remember the original reasons for anathematization. Gracious (and confident) victors should always seek to revive the valid and important aspects of defeated but honorable systems. And the transcendental morphologists did understand the importance of designating a small but overarching set of defining architectural properties as legitimate essences of systems, both anatomical and conceptual.

Hull correctly defines theories as historical entities, properly subject to all the principles of narrative explanation—and I shall so treat Darwinian logic and its substantial improvements and changes throughout this book. But theories of range and power also feature inherent "essences," implicit in their logical structure, and operationally definable as minimal sets of propositions so crucial to the basic function of a system that their falsification must undermine the entire structure, and also so necessary as an ensemble of mutual implication that all essential components must work in concert to set the theory's mechanism into smooth operation as a generator and explanation of nature's order. In staking out this middle Goldilockean ground between (1) the "too little" of Hull's genealogical continuity without commitment to a shared content of intellectual morphology and (2) the "too much" of long lists of ideological fealty, superficially imbibed or memorized, and then invoked to define membership in ossified cults rather than thoughtful allegiance to developing theories, I will argue that a Darwinian essence can be minimally (and properly) defined by *three central principles* constituting a *tripod of necessary support,* and specifying the fundamental meaning of a powerful system that Darwin famously described as the "grandeur in this view of life."

I shall then show that this formulation of Darwinian minimal commitments proves its mettle on the most vital ground of maximal utility. For not only do these three commitments build, in their ensemble, the full frame of a comprehensive evolutionary worldview, but they have also defined the chief objections and alternatives motivating all the most interesting debate within evolutionary theory during its initial codification in the 19th century. Moreover, and continuing in our own time, these three themes continue to specify the major weaknesses, the places in need of expansion or shoring up, and the locus of unresolved issues that make evolutionary biology such a central and exciting subject within the ever changing and ever expanding world of modern science.

## The Structure of Evolutionary Theory: Revising the Three Central Features of Darwinian Logic

In the opening sentence of the *Origin*'s final chapter (1859, p. 459), Darwin famously wrote that "this whole volume is one long argument." The present book, on "the structure of evolutionary theory," despite its extravagant length, is also a brief for an explicit interpretation that may be portrayed as a single extended argument. Although I feel that our best current formulation of evolutionary theory includes modes of reasoning and a set of mechanisms substantially at variance with strict Darwinian natural selection, the logical structure of the Darwinian foundation remains remarkably intact—a fascinating historical observation in itself, and a stunning tribute to the intellectual power of our profession's founder. Thus, and not only to indulge my personal propensities for historical analysis, I believe that the best way to exemplify our modern understanding lies in an extensive analysis of Darwin's basic logical commitments, the reasons for his choices, and the subsequent manner in which these aspects of "the structure of evolutionary theory" have established and motivated all our major debates and substantial changes since Darwin's original publication in 1859. I regard such analysis not as an antiquarian indulgence, but as an optimal path to proper understanding of our *current* commitments, and the underlying reasons for our decisions about them.

As a primary theme for this one long argument, I claim that an "essence" of Darwinian logic can be defined by the practical strategy defended in the first section of this chapter: by specifying a set of minimal commitments, or broad statements so essential to the central logic of the enterprise that disproof of any item will effectively destroy the theory, whereas a substantial change to any item will convert the theory into something still recognizable as within the *Bauplan* of descent from its forebear, but as something sufficiently different to identify, if I may use the obvious taxonomic metaphor, as a new subclade within the monophyletic group. Using this premise, the long argument of this book then proceeds according to three sequential claims that set the structure and order of my subsequent chapters:

1. Darwin himself formulated his central argument under these three basic premises. He understood their necessity within his system, and the difficulty that he would experience in convincing his contemporaries about such unfamiliar and radical notions. He therefore presented careful and explicit defenses of all three propositions in the *Origin*. I devote the first substantive chapter (number 2) to an exegesis of the *Origin of Species* as an embodiment of Darwin's defense for this central logic.

2. As evolutionary theory experienced its growing pains and pursued its founding arguments in the late 19th and early 20th centuries (and also in

its pre-Darwinian struggles with more inchoate formulations before 1859), these three principles of central logic defined the themes of deepest and most persistent debate—as, in a sense, they must because they constitute the most interesting intellectual questions that any theory for causes of descent with modification must address. The historical chapters of this book's first half then treat the history of evolutionary theory as responses to the three central issues of Darwinian logic (Chapters 3–7).

3. As the strict Darwinism of the Modern Synthesis prevailed and "hardened," culminating in the overconfidences of the centennial celebrations of 1959, a new wave of discoveries and theoretical reformulations began to challenge aspects of the three central principles anew—thus leading to another fascinating round of development in basic evolutionary theory, extending throughout the last three decades of the 20th century and continuing today. But this second round has been pursued in an entirely different and more fruitful manner than the 19th century debates. The earlier questioning of Darwin's three central principles tried to disprove natural selection by offering alternative theories based on confutations of the three items of central logic. The modern versions accept the validity of the central logic as a foundation, and introduce their critiques as helpful auxiliaries or additions that enrich, or substantially alter, the original Darwinian formulation, but that leave the kernel of natural selection intact. Thus, the modern reformulations are helpful rather than destructive. For this reason, I regard our modern understanding of evolutionary theory as closer to Falconer's metaphor, than to Darwin's, for the Duomo of Milan—a structure with a firm foundation and a fascinatingly different superstructure. (Chapters 8–12, the second half of this book on modern developments in evolutionary theory, treat this third theme.)

Thus, one might say, this book cycles through the three central themes of Darwinian logic at three scales—by brief mention of a framework in this chapter, by full exegesis of Darwin's presentation in Chapter 2, and by lengthy analysis of the major differences and effects in historical (Part 1) and modern critiques (Part 2) of these three themes in the rest of the volume.

The basic formulation, or bare-bones mechanics, of natural selection is a disarmingly simple argument, based on three undeniable facts (overproduction of offspring, variation, and heritability)* and one syllogistic inference (natural selection, or the claim that organisms enjoying differential reproductive success will, on average, be those variants that are fortuitously better adapted to changing local environments, and that these variants will then pass their favored traits to offspring by inheritance). As Huxley famously, and ruefully, remarked (in self-reproach for failing to devise the theory himself), this argument must be deemed elementary (and had often been formu-

---

*Two of these three ranked as "folk wisdom" in Darwin's day and needed no further justification—variation and inheritance (the mechanism of inheritance remained unknown, but its factuality could scarcely be doubted). Only the principle that all organisms produce more offspring than can possibly survive—superfecundity, in Darwin's lovely term—ran counter to popular assumptions about nature's benevolence, and required Darwin's specific defense in the *Origin*.

lated before, but in negative contexts, and with no appreciation of its power—see p. 137), and can only specify the guts of the operating machine, not the three principles that established the range and power of Darwin's revolution in human thought. Rather, these three larger principles, in defining the Darwinian essence, take the guts of the machine, and declare its simple operation sufficient to generate the entire history of life in a philosophical manner that could not have been more contrary to all previous, and cherished, assumptions of Western life and science.

The three principles that elevated natural selection from the guts of a working machine to a radical explanation of the mechanism of life's history can best be exemplified under the general categories of agency, efficacy, and scope. I treat them in this specific order because the logic of Darwin's own development so proceeds (as I shall illustrate in Chapter 2), for the most radical claim comes first, with assertions of complete power and full range of applicability then following.

AGENCY. The abstract mechanism requires a locus of action in a hierarchical world, and Darwin insisted that the apparently intentional "benevolence" of nature (as embodied in the good design of organisms and the harmony of ecosystems) flowed entirely as side-consequences of this single causal locus, the most "reductionistic" account available to the biology of Darwin's time. Darwin insisted upon a virtually exceptionless, single-level theory, with *organisms* acting as the locus of selection, and all "higher" order emerging, by the analog of Adam Smith's invisible hand, from the (unconscious) "struggles" of organisms for their own personal advantages as expressed in differential reproductive success. One can hardly imagine a more radical reformulation of a domain that had unhesitatingly been viewed as the primary manifestation for action of higher power in nature—and Darwin's brave and single-minded insistence on the exclusivity of the organismic level, although rarely appreciated by his contemporaries, ranks as the most radical and most distinctive feature of his theory.

EFFICACY. Any reasonably honest and intelligent biologist could easily understand that Darwin had identified a *vera causa* (or true cause) in natural selection. Thus, the debate in his time (and, to some extent, in ours as well) never centered upon the existence of natural selection as a genuine causal force in nature. Virtually all anti-Darwinian biologists accepted the reality and action of natural selection, but branded Darwin's force as a minor and negative mechanism, capable only of the headsman's or executioner's role of removing the unfit, once the fit had arisen by some other route, as yet unidentified. This other route, they believed, would provide the centerpiece of a "real" evolutionary theory, capable of explaining the origin of novelties. Darwin insisted that his admittedly weak and negative force of natural selection could, nonetheless, under certain assumptions (later proved valid) about the nature of variation, act as the positive mechanism of evolutionary novelty—that is, could "create the fit" as well as eliminate the unfit—by slowly accumulating the positive effects of favorable variations through innumerable generations.

SCOPE. Even the most favorably minded of contemporaries often admitted that Darwin had developed a theory capable of building up small changes (of an admittedly and locally "positive" nature as adaptations to changing environments) within a "basic type"—the equivalent, for example, of making dogs from wolves or developing edible corn from teosinte. But these critics could not grasp how such a genuine microevolutionary process could be extended to produce the full panoply of taxonomic diversity and apparent "progress" in complexification of morphology through geological time. Darwin insisted on full sufficiency in extrapolation, arguing that his microevolutionary mechanism, extended through the immensity of geological time, would be fully capable of generating the entire pageant of life's history, both in anatomical complexity and taxonomic diversity—and that no further causal principles would be required.

Because primates are visual animals, complex arguments are best portrayed or epitomized in pictorial form. The search for an optimal icon to play such a role is therefore no trivial matter (although scholars rarely grant this issue the serious attention so richly merited)—especially since the dangers of confusion, misplaced metaphor, and replacement of rigor with misleading "intuition" stand so high. I knew from the beginning of this work that I needed a suitable image for conveying the central logic of Darwinian theory. As one of my humanistic conceits, I hoped to find a historically important scientific image, drawn for a different reason, that might fortuitously capture the argument in pictorial form. But I had no expectation of success, and assumed that I would need to commission an expressly designed figure drawn to a long list of specifications.

The specific form of the image—its central metaphorical content, if you will—plays an important role in channeling or misdirecting our thoughts, and therefore also requires careful consideration. In the text of this book, I speak most often of a "tripod" since central Darwinian logic embodies three major propositions that I have always visualized as supports—perhaps because I have never been utterly confident about this entire project, and I needed some pictorial encouragement to keep me going for twenty years. (And I much prefer tripods, which can hold up elegant objects, to buttresses, which may fly as they preserve great Gothic buildings, but which more often shore up crumbling edifices. Moreover, the image of a tripod suits my major claim particularly well—for I have argued, just above, that we should define the "essence" of a theory by an absolutely minimal set of truly necessary propositions. No structure, either of human building or of abstract form, captures this principle better than a tripod, based on its absolute minimum of three points for fully stable support in the dimensional world of our physical experience.)

But organic images have always appealed more strongly, and I preferred a biological icon. If the minimal logic can be represented by a tripod pointing downward, then the same topology can be inverted into a structure growing upward. Darwin's own favorite image of the tree of life immediately suggested itself, and I long assumed that I would eventually settle on a botanical icon. But I also remembered Darwin's first choice for an organic metaphor or picture of branching to capture his developing views about descent with modification and the causes of life's diversity—the "coral of life" of his "B Notebook" on transmutation, kept during the 1830's as he became an evolutionist and struggled towards the theory of natural selection (see Barrett *et al.*, 1987).

As I began to write this summary chapter, I therefore aimlessly searched through images of Cnidaria from my collection of antiquarian books in paleontology. I claim no general significance whatsoever for my good fortune, but after a lifetime of failure in similar quirky quests, I was simply stunned to find a preexisting image—not altered one iota from its original form, I promise you, to suit my metaphorical purposes—that so stunningly embodied my needs, not only for a general form (an easy task), but down to the smallest details of placement and potential excision of branches (the feature that I had no right or expectation to discover and then to exapt from so different an original intent).

The following figure comes from the 1747 Latin version of one of the seminal works in the history of paleontology—the 1670 Italian treatise of the Sicilian savant and painter Agostino Scilla, *La vana speculazione disingannata dal senso* ("Vain speculation undeceived by the senses"—Scilla's defense, at the outset of "the scientific revolution" of Newton's generation, for empirical methods in the study of nature, and specifically, in this treatise, for a scientific paleontology and the need to recognize fossils as remains of ancient organisms, not as independent products of the mineral kingdom). This work, famous not only for an incisive text, but also for its beautiful plates (see Fig. 1-3), engraved by an author known primarily as an artist of substantial eminence, includes this figure, labeled *Coralium articulatum quod copiosissimum in rupibus et collibus Messanae reperitur* ("Articulated coral, found in great abundance in the cliffs and hills of Messina").

This model, and its organic features, work uncommonly well as a metaphor for the Goldilockean position of definition by a barest minimum of truly fundamental postulates. For Scilla's coral, with its branching structure (see Fig. 1-4)—particularly as expressed in the lessening consequences of excising branches at ever higher levels nearer the top (the analogs of disconfirming theoretical features of ever more specialized and less fundamental import)— so beautifully captures the nature and operation of the intellectual structure that I defended above for specifying the essences of theories. The uncanny appropriateness of Scilla's coral lies in the fortuity that this particular specimen (accurately drawn from nature by Scilla, I assume, and not altered to assert any general point) just happens to include exactly the same number of branches (three) as my Darwinian essential structure. (They terminate at the same upper level, so I could even turn the specimen over into a tolerably unwobbly tripod!) Moreover, since this particular genus of corals grows in discrete segments, the joining points correspond ideally with my metaphor of chopping planes for excising parts of structures at various levels of importance in an intellectual entity. But, most incredibly, the segmental junctions of

1-3. The famous frontispiece from Scilla's treatise of 1670 defending the organic nature of fossils. The solid young man, representing the truth of sensory experience, shows a fossil sea urchin in his right hand to a wraithlike figure representing the former style of speculative thinking. With his left hand, the solid figure points to other fossils found in Sicily. The text proclaims: "Vain speculation undeceived by the senses."

this particular specimen just happen to occupy the exact places that I needed *a priori* to make my central point about lower choppings that destroy theories, middle choppings that change theories in a Falconerian way (major alterations in structure upon a preserved foundation), and upper choppings that change theories in the lesser manner of Darwin's Milanese metaphor (smaller excisions that leave the framework intact as well).

The central trunk (the theory of natural selection) cannot be severed, or the creature (the theory) dies. (The roots, if you will, represent sources of evidence; any one may be excised, if recognized as incorrect by later study, so long as enough remain to anchor the structure). This central trunk then divides into a limited number of major branches. These basic struts—the three

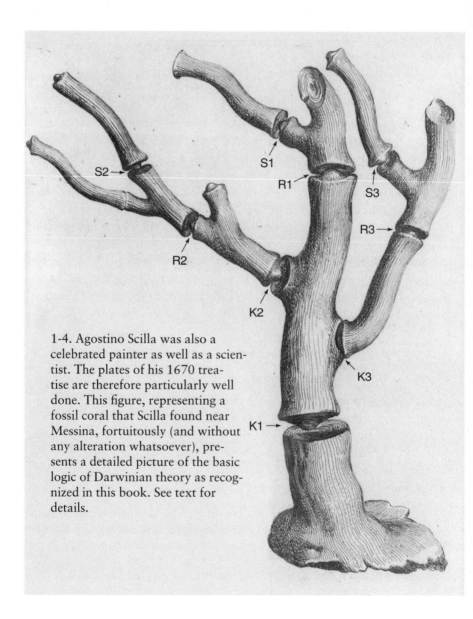

1-4. Agostino Scilla was also a celebrated painter as well as a scientist. The plates of his 1670 treatise are therefore particularly well done. This figure, representing a fossil coral that Scilla found near Messina, fortuitously (and without any alteration whatsoever), presents a detailed picture of the basic logic of Darwinian theory as recognized in this book. See text for details.

branches of the Darwinian essence in this particular picture—are also so essential that any severing of a complete branch either kills, or so seriously compromises, the entire theory that a new name and basic structure becomes essential.

We now reach the interesting point where excisions and regraftings preserve the essential nature of an intellectual structure, but with two quite different levels of change and revision, as characterized by Falconer's and Darwin's competing metaphors for the *Duomo* of Milan. I would argue that a severing low on any one of the three major branches corresponds with a revision profound enough to validate the more interesting Falconerian version of major revision upon a conserved foundation. (The Falconerian model is, in this sense, a Goldilockean solution itself, between the "too much" of full destruction and the "too little" of minor cosmetic revision.) On the other hand, the severing of a subbranch of one of the three branches symbolizes a less portentous change, closer to Darwinian models for the Milanese *Duomo*— an alteration of important visual elements, but without change in the basic framework.

My fascination with the current state of evolutionary theory, at least as I read current developments in both logic and empirics, lies in its close conformity to the Falconerian model—with enough continuity to make the past history of the field so informative (and so persistently, even emotionally, compelling), but with enough deep difference and intellectual fascination to stimulate anyone with a thirst for the intriguing mode of novelty that jars previous certainty, but does not throw a field into the total anarchy of complete rebuilding (not a bad thing either, but far from the actual circumstance in this case).

To summarize my views on the utility of such a model for the essence of Darwinian logic, I will designate three levels of potential cuts or excisions to this organic (and logical) coral of the structure of evolutionary theory, as originally formulated by Darwin in the *Origin of Species*, and as revised in a Falconerian way in recent decades. The most inclusive and most fundamental K-cuts (killing cuts) sever at least one of the three central principles of Darwinian logic and thereby destroy the theory *tout court*. The second level of R-cuts (revision cuts) removes enough of the original form on one of the three central branches to ensure that the new (and stronger or more arborescent) branch, in regrowing from the cut, will build a theory with an intact Darwinian foundation, but with a general form sufficiently expanded, revised or reconstructed to present an interestingly different structure of general explanation—the Falconerian model for the *Duomo* of Milan. Finally, the third level of S-cuts (subsidiary cuts) affects only a subbranch of one of the three major branches, and therefore reformulates the general theory in interesting ways, while leaving the basic structure of explanation intact—the Darwinian model for the *Duomo* of Milan.

I wrote this book because I believe that all three pillars, branches, or tripod legs, representing the three fundamental principles of Darwinian central logic, have been subjected to fascinating R-cuts that have given us at least the firm outlines—for the revised structure of evolutionary explanation remains a work vigorously in progress, as only befits the nature of its subject, after all!—of a far richer and fascinatingly different theory with a retained Darwinian core rooted in the principles of natural selection. In short, we live in the midst of a Falconerian remodelling of our growing and multiform, yet coherently grounded, intellectual mansion.

I will not, in this chapter, detail the nature of the K-cuts that failed (thus preserving the central logic of Darwinism), the R-cuts that have succeeded in changing the structure of evolutionary theory in such interesting ways, and the S-cuts that have refurbished major rooms in particular wings of the edifice—for these specifications set the subject matter of all following chapters. But to provide a better opening sense of this book's argument—and to clarify the nature of the three central claims of Darwinian logic—I shall at least distinguish, for each branch, the K-cuts that never prevailed (and therefore did not fell the structure) from the R-cuts that have affected each branch, and have therefore provoked our current process of building an enriched structure for evolutionary theory.

Returning to Scilla's coral (Fig. 1-4), consider the central branch as the first leg of the tripod (agency, or the claim for organismal selection as the causal locus of the basic mechanism), the left branch as the second leg (efficacy, or the claim that selection acts as the primary creative force in building evolutionary novelties), and the right branch as the third leg (scope, or the claim that these microevolutionary modes and processes can, by extrapolation through the vastness of geological time, explain the full panoply of life's changes in form and diversity).

The cut labeled K1 on Figure 1-4 would have severed the entire coral by disproving natural selection as an evolutionary force at all. The cut labeled K2 would have fully severed the second branch, leaving natural selection as a legitimate cause, but denying it any creative role, and thereby dethroning Darwinism as a major principle in explaining life's history. (We shall see, in Chapters 3–6, that such a denial of creativity underlay the most common anti-Darwinian argument in the first generations of debate.) The cut labeled K3 would have fully severed the third branch, allowing that natural selection might craft some minor changes legitimately called "creative" in a local sense, but denying that Darwin's mechanism could then be extended to explain the panoply of macroevolutionary processes, or the actual pageant of life's history. The success of *any one* of these K-cuts would have destroyed Darwinian theory, plain and simple. None of them succeeded, and the foundation of Darwinian central logic remains intact and strong.

In striking, and most positive, contrast, I believe that higher R-cuts—leaving the base of each major branch intact, but requiring a substantial regrowth and regrafting of an enlarged structure upon the retained foundation—have been successfully wielded against all three branches of Darwinian logic, as the structure of evolutionary theory developed in the last third of the 20th century (following too rigid a calcification of the original structure, a good adumbration of the coral metaphor!, in the hardening of the Modern Synthesis that culminated in the Darwinian centennial celebrations of 1959). On the first branch of agency, the cut labeled R1 (see Fig. 1-4) expanded Darwin's unilevel theory of organismal selection into a hierarchical model of selection acting simultaneously on several legitimate levels of Darwinian individuality (genes, cell-lineages, organisms, demes, species, and clades). I shall show in Chapters 3, 8, and 9 how the logic of this pronounced expansion builds a theory fascinatingly different from, and not just a smooth extension of, Darwin's single level mechanism of agency—my reason for portraying the hierarchical model as a deeply interesting R-cut rather than a more superficial S-cut.

On the second branch of efficacy, the cut labeled R2 accepts the validity of Darwin's argument for creativity (by leaving the base of the branch intact), but introduces a sufficient weight of formalist thinking—*via* renewed appreciation for the enormous importance of structural, historical, and developmental constraint in channeling the pathways of evolution, often in highly positive ways—that the pure functionalism of a strictly Darwinian (and externalist) approach to adaptation no longer suffices to explain the channeling of phyletic directions, and the clumping and inhomogeneous population of organic morphospace. The strict Darwinian form of explanation has thereby been greatly changed and enriched, but in no way defeated. I shall discuss the historical aspect of this branch in Chapters 4 and 5, and modern reformulations of this R2 cut in Chapters 10 and 11.

On the final branch of scope, the cut labeled R3 accepts the Darwinian contention that microevolutionary modes and principles can build grand patterns by cumulation through geological immensity, but rejects the argument that such extrapolations can render the entire panoply of phenomena in life's history without adding explicitly macroevolutionary modes for distinctive expression of these processes at higher tiers of time—as in the explanation of cladal trends by species sorting under punctuated equilibrium, rather than by extended adaptive anagenesis of purely organismal selection, and in the necessity of titrating adaptive microevolutionary accumulation with occasional resetting of rules and patterns by catastrophically triggered mass extinctions at time's highest tier. Chapters 6 and 12 discuss historical and modern critiques of Darwinian extrapolationism.

For now, I will say little about the even higher and more superficial S-cuts of subbranches, but I will at least indicate how I construe this category by stating a hypothetical example for each branch: an S1 cut, for example, might accept the selective basis of evolutionary change in a purely mechanical sense, but then deny full force to Darwin's deliciously radical philosophical claim that all apparent "higher level" harmony arises consequentially, through the invisible hand of lower levels acting for personal reproductive success. One might, in principle, propose such a revision by arguing that a higher force, operating by an overarching principle of order, "employs" natural selection as its mechanical agent. (I speak only hypothetically here, for no such defendable scientific hypothesis now exists, although the concept certainly remains intelligible. Explicitly theological versions don't count as science, whatever their kind or form of potential validity.) An S2 cut might be assayed by a developmental saltationist who accepted the selectionist basis of adaptive change but felt that, at a sufficient relative frequency to be counted as important, the initial steps of such changes may be larger than the pure continuationism of Darwinian selection can admit. And an S3 cut might accept the full validity of microevolutionary extrapolationism, but deny the subsidiary defense of progress that Darwin grafted onto this apparatus (see Chapter 6) with ecological arguments about plenitude and the priority of biotic over abiotic competition.

As a paleontologist and part-time historian of science by profession, my reading of these important R-cuts arose from a macroevolutionary perspective framed largely in terms of longstanding difficulties faced by Darwinism in extending its successes for explaining small changes in palpable time into equally adequate causal accounts for broader patterns and processes in geological history. I have, in this effort, also benefited from my personal study of Darwin's life and times, and especially the late 19th century debates on mechanisms of evolution (as promulgated largely by professionals who could neither fully understand nor accept the radical philosophical commitments underlying Darwin's view). This historical study allowed me to grasp the continuity in basic themes from Darwin's own formulation, through these foundational debates, right down to the major theoretical struggles of our own time. An appreciation of this continuity allowed me to discern and define the distinctively Darwinian view of life.

But I recognize only too well that every strength comes paired with weaknesses. In my case, a paleontological focus leads me into relative ignorance for an equally important locus of reform in the structure of Darwinism—increasing knowledge of the nature of genomes and the mechanics of development. (I try to cover the outlines of important theoretical critiques from this "opposite" realm of the smallest, but the relative weightings in my text reflect my own varying competencies far more than the merits of the cases. For example, although I do discuss, and perhaps even adequately outline, the importance of Kimura and King's neutralist theory in questioning previous assumptions of adaptationist hegemony, I surely do not give an appropriate volume of attention to this enormously important subject.)

Nonetheless, I hope that I have managed to present an adequate account of the coordinating themes that grant such interest and coherence to modern reformulations of the structure of evolutionary theory. Such thematic consistency in revision becomes possible largely because Darwin himself, in his characteristically brilliant way, tied the diverse threads of his initiating argument into an overall view with a similarly tight structure—thus granting clear definition to his own commitments, and also permitting their revision in the form of an equally coherent "package." I would argue, moreover, and without wishing to become extravagantly hagiographical (for I wrote this book, after all, primarily to discuss a critique and revision of strict Darwinism), that our modern sense of limitations in the canonical version arises from decisions that Darwin made for tough and correct reasons in the context of his initiating times—reasons that made his account the first operational theory of evolution in modern science. In particular, as Chapter 2 will discuss in detail, Darwin converted evolution from untestable speculation to doable science by breaking through the old paradox (as embedded most prominently in Lamarck's system) of contrasting a palpable force of small-scale change that could do little in extension, with a basically nonoperational (and orthogonal) mechanism of large-scale change putatively responsible for all the interesting patterns of life's history, but imperceptible and untestable from the uniformitarian study of modern organisms.

By claiming that the small-scale mechanics of modern change could, by extension, explain all of evolution, Darwin opened the entire field to empirical study. And yet, as Hegel and so many other students of change have noted, progress in human (and other) affairs tends to spiral upwards in cycles of proposal *(thesis)*, then countered by opposition *(antithesis)*, and finally leading to a new formulation combining the best aspects of both competitors *(synthesis)*. Darwin's thesis established evolution as a science, but his essential commitments, as expressed in the three legs of his necessary logical tripod (or the three branches of his conceptual tree or coral, as in the alternate metaphor of Fig. 1-4), eventually proved too narrow and confining, thus requiring an antithesis of extension and reformulation on each branch, and leading—or so this book maintains as a central thesis of its own—to a still newer and richer synthesis expressing our best current understanding of the structure of evolutionary theory.

In fact, and to repeat my summary in this different form, one might encapsulate the long argument of this book in such a Hegelian format. Pre-Darwinian concepts of evolution remained speculative and essentially nonoperational, largely because (see Chapter 3) they fell into the disabling paradox of contrasting an effectively unknowable large-scale force of cosmic progress against an orthogonal, palpable and testable small-scale force that could generate local adaptation and diversity, but that couldn't, in principle, explain the macroevolutionary pattern of life. Then Darwin, in his thesis (also an antithesis to these earlier sterile constructions), brilliantly argued that the putative large-scale force did not exist, and that all evolution could be explained by upward extrapolation from the small-scale force, now properly understood as natural selection. In a first stage of debate during the late 19th and early 20th centuries (Chapters 3–6), most critiques of Darwinism—one might designate them as a first round of ultimately destructive antitheses—simply denied sufficient agency, efficacy and range to natural selection, and reasserted the old claim of duality, with selection relegated to triviality, and some truly contrary force sought as *the* explanation for major features of evolution. Strict Darwinism eventually fended off these destructive critiques, reasserted itself in the triumphant, and initially (and generously) pluralistic form of the Modern Synthesis, but eventually calcified into a "hardened" version (Chapter 7).

Then, in a strikingly different, and ultimately fruitful, second round of antitheses, a renewed debate about central theoretical issues arose during the last three decades of the 20th century, and reshaped the field by recognizing that selection needed to be amplified, reformulated and invigorated by other, non-contrary (and, at most, orthogonal) causes, not rejected as wrong, or scorned as trivial (Chapters 8–12). The one long argument of this book holds that a synthesis (still much in progress) has now sufficiently coagulated from this debate to designate our best current understanding of the structure of evolutionary theory as something rich and new, with a firmly retained basis in Darwinian logic—in other words, and following the organizing and opening metaphor of this chapter, as a validation of Falconer's, rather than Darwin's, concept of the historical growth and change of Milan's cathedral.

Ariel's telling verse in Shakespeare's *The Tempest* proclaims in dense metaphor:

> Full fathom five thy father lies;
> Of his bones are coral made;
> Those are pearls that were his eyes:
> Nothing of him that doth fade
> But doth suffer a sea-change
> Into something rich and strange.

With the exception of one possible (and originally unintended) modern reading of these images, this famous and haunting verse provides a beautiful description of both the priceless worth and intriguing modern transformation of Darwin's original theory. (For the exception, several connotations of deep burial in the sea—full fathom five—might be viewed negatively, as in "deep sixing" or going to Davy Jones's locker. But, for natural historians who read this book, and coming from an invertebrate paleontologist as author, the seafloor could not represent a more positive resting place or point of origin— and I intend to evoke only these upbeat images in citing Ariel's lines.) Otherwise, Darwin's original structure has only yielded greater treasure in cascading implications and developments through the subsequent history of evolutionary thought—the conversion of the bones of an original outline into precious coral and pearls of current substance. Nothing of Darwin's central logic has faded or fully capsized, but his theory has been transformed, along his original lines, into something far different, far richer, and far more adequate to guide our understanding of nature.

The last three lines of Shakespeare's verse also appear on the tombstone of the great poet Percy Bysshe Shelley (also the author of the preface to his wife's novella, *Frankenstein*, which cites Erasmus Darwin in its first line of text). I believe that these words would suit, and honor, Charles Darwin just as well and just as rightly.

## Apologia Pro Vita Sua

### A TIME TO KEEP

The Preacher spoke ever so truly in writing his famous words (Ecclesiastes 3:1–7): "For every thing there is a season, and a time to every purpose . . . A time to break down, and a time to build up . . . A time to rend, and a time to sew: a time to keep silence, and a time to speak." Evolutionary theory now stands in the happier second state of these genuine dichotomies (in part because the first state had been mined to the limited extent of its utility): we live in a time for building up, for sewing together, and for speaking out.

Not all times are such good times, and not all scientists win the good fortune to live within these times of motion. For theories grow as organisms do, with periods of *Sturm und Drang,* long latencies of youth or ossifications of age, and some happy times of optimally productive motion in between (another Goldilockean phenomenon). I recently studied the life and career of E. Ray Lankester (Gould, 1999a), clearly the most talented evolutionary morphologist of the generation just after Darwin. He did "good" work and had a "good" career (see Chapter 10, pages 1069–1076 for his best theoretical efforts), but he never transcended the ordinary. Perhaps the limitation lay largely within his own abilities. However, I rather suspect that he did possess both the temperamental gumption and the requisite intellectual might—but that the tools of major empirical advance just didn't emerge in his generation, for he remained stuck in a relatively unproductive middle, as Darwin had seized the first-fruits from traditional data of natural history, and the second plucking required a resolution of genetic mechanisms.

I felt a similar kind of frustration in 1977, after writing my first technical book, *Ontogeny and Phylogeny* (see Chapter 10, pages 1061–1063). I had spent the best years of a young career on a subject that I knew to be relevant (at a time when most of the profession had forgotten). But then defeat snatched my prize from the jaws of victory. I am proud of the book, and I do believe that it helped to focus interest on a subject that became doable soon thereafter. But I ran up against a wall right at the end—for the genetics of development clearly held the key to any rapprochement of embryology and evolution, and we knew effectively nothing about eukaryotic regulation. Indeed, as we could then only characterize structural genes by electrophoretic techniques, our major "arguments" for regulatory effects (if they even merited such a positive designation) invoked such negative evidence as the virtual identity in structural genes between chimps and humans, coupled with a fair visceral sense of extensive phenotypic disparity in anatomy and behavior— with the differences then attributed to regulatory genes that we could not, at the time, either study or even identify.

By sheer good fortune (abetted in minuscule ways by my own pushes and those of my paleontological colleagues), the field moved fast and I lived long enough to witness a sea change (if I may cite Ariel yet again) towards potentiation on all three major intellectual and social substrates for converting a subject from great promise combined with even more frustrating inoperability, into a discipline bursting with new (and often utterly surprising) data that led directly to testable hypotheses about basic issues in the structure of evolutionary theory.

EMPIRICS. During the last third of the 20th century, new techniques and conceptualizations opened up important sources of data that challenged orthodox formulations for all three branches of essential Darwinian logic. To cite just one relevant example for each branch, theoretical development and accumulating data on punctuated equilibrium allowed us to reconceptualize species as genuine Darwinian individuals, fully competent to participate in processes of selection at their own supraorganismic (and suprademal) level— and then to rethink macroevolution as the differential success of species rather than the extended anagenesis of organismal adaptation (see Chapter 9). This validation of the species-individual aided the transformation of what had begun as a particular argument about group (or interdemic) selection into a fully generalized hierarchical theory, with good cases then documented from the genic to the cladal level (see Chapter 8).

On the second branch of full efficacy for natural selection as an externalist and functionalist process, the stunning discoveries of extensive deep homologies across phyla separated by more than 500 million years (particularly the vertebrate homologs of arthropod *Hox* genes)—against explicit statements by architects of the Modern Synthesis (see p. 539) that such homologies could not exist in principle, in a world dominated by their conception of natural selection—forced a rebalancing or leavening of Darwinian functionalism with previously neglected, or even vilified, formalist perspectives based on the role of historical and structural constraints in channeling directions of evolutionary change, and causing the great clumpings and inhomogeneities of morphospace—phenomena that had previously been attributed almost exclusively to functionalist forces of natural selection.

On the third branch of extrapolation, the discovery and relatively quick validation, beginning in 1980, of a truly catastrophic trigger for at least one great mass extinction (the K-T event of 65 million years ago), fractured the uniformitarian consensus, embraced by a century of paleontological complacency, that all apparent faunal overturns could be "spread out" into sufficient time for explanation by ordinary causes under plausible intensifications that would not alter conventional modes of evolution and extinction.

Moreover, as we shall see, these three apparently rather different kinds of data and their attendant critiques cohere into a revised general structure for evolutionary theory—thus marking our age as a time for building up and not only as a time for breaking down.

CONCEPTS. Following the Kantian dictum that percepts without concepts are blind, but concepts without percepts empty, these two categories interpenetrate as "pure" data suggest novel ideas (how can one *not* rethink the causes of mass extinction when evidence surfaces for a bolide, 7–10 km in diameter, and packing $10^4$ the megatonnage of all the earth's nuclear weapons combined), whereas "abstract" concepts then taxonomize the natural world in different ways, often "creating" data that had never been granted enough previous intellectual space even to be conceived (as when punctuated equilibrium made stasis a theoretically meaningful and interesting phenomenon, and not just an embarrassing failure to detect "evolution," in its traditional definition of gradual change—and paleontologists then began active studies of a subject that had previously been ignored as uninteresting, if conceptualized at all). But, speaking parochially as a student of the fossil record, I can at least say that the conceptual revolution in macroevolutionary thinking revitalized the field of paleobiology (even creating the name as a subdiscipline of paleontological endeavor). Whatever the varied value of different individual efforts in this burgeoning field, we may at least be confident that our profession will no longer be humiliated as a synecdoche for ossified boredom among the natural sciences—as *Nature* did in 1969 when editorializing about the salutary value of plate tectonics in revitalizing the geological sciences: "Scientists in general might be excused for assuming that most geologists are paleontologists and most paleontologists have staked out a square mile as their life's work. A revamping of the geologist's image is badly needed" (Anonymous, 1969, p. 903).

The intricate and multifaceted concepts that have nuanced and altered the central logic on all three branches of Darwinism's essential postulates represent ideas of broad ramification and often remarkably subtle complexity, as we vain scientists soon discovered in our fractured bubbles of burst pride— for we had been so accustomed to imagining that an evening in an armchair could conquer any merely conceptual issue, whereas we all acknowledge the substantial time and struggle that empirical problems, demanding collection and evaluation of data, often require. Yet, on these basic questions in formulating evolutionary theory, we often read and thought for months, and ended up more confused than when we began.

The general solution to such procedural dilemmas lies in a social and intellectual activity that scientists do tend to understand and practice better than colleagues in most other academic disciplines—collaboration. Far more than most colleagues, I have tended to work alone in my professional life and publication. But for each of the conceptually difficult and intellectually manifold issues of reevaluation for the central logic of the three essential Darwinian postulates, I desperately needed advice, different skills, and the give and take of argument, from colleagues who complemented my limited expertise with their equally centered specialties and aptitudes for other aspects of these large and various problems. Thus, on the first leg or branch of hierarchy theory, I worked with Niles Eldredge on punctuated equilibrium, and with Elisabeth Vrba on levels of selection and sorting. On the second leg of structuralist alternatives to adaptationist argument, I worked with Dick Lewontin on spandrels, Elisabeth Vrba on exaptation, David Woodruff on the functional and structural morphology of *Cerion,* and with "the gang of four" (increased to five with the later inclusion of Jack Sepkoski)—Dave Raup, Tom Schopf, Dan Simberloff, and me—on trying to specify how many aspects of apparently ordered phyletic patterns, heretofore confidently attributed to selection for little reason beyond the visual appearance of order itself, could plausibly be generated within purely random systems. And on the third leg of extrapolationism, my earliest interests in the logic and justification of uniformitarianism in philosophy, and of Lyellian perspectives in the history of science, could not have developed without advice and substantial aid (but not collaborative publication this time) with historians Martin Rudwick, Reijer Hooykaas, and Cecil

Schneer, and with philosophers Nelson Goodman, Bonnie Hubbard, and George Geiger. (Geiger, my mentor at Antioch College, was the last student of John Dewey and played with Lou Gehrig on the Columbia University base-ball team, thus embodying both my professional and avocational interests.)

In fact, and as a comment within the sociology of science, I would venture that future historians might judge the numerous seminal (and published) col-laborations between evolutionary biologists and professional philosophers of science as the most unusual and informative operational aspect of the recon-struction of evolutionary theory in the late 20th century. Research scientists tend to be a philistine lot, with organismic biologists perhaps at the head of this particular pack (for we work with "big things" that we can see and un-derstand at our own scale. Thus, we suppose that we can afford to be more purely empirical in our reliance on "direct" observation, and less worried about admittedly conceptual problems of evaluating things too small or too fast to see). Most of us would scoff at the prospect of working with a profes-sional philosopher, regarding such an enterprise as, at best, a pleasant waste of time and, at worst, an admission that our own clarity of thought had be-come addled (or at least as a fear that our colleagues would so regard our in-terdisciplinary collaboration).

And yet, the conceptual problems presented by theories based on causes operating at several levels simultaneously, of effects propagated up and down, of properties emerging (or not) at higher levels, of the interaction of random and deterministic processes, and of predictable and contingent influ-ence, have proven to be so complex, and so unfamiliar to people trained in the simpler models of causal flow that have served us well for centuries (see the next section on *Zeitgeist*), that we have had to reach out to colleagues ex-plicitly trained in rigorous thinking about such issues. Thus, we learned, to our humbling benefit, that conceptual muddles do not necessarily resolve themselves "automatically" just because a smart person—namely one of us, trained as a scientist—finally decides to apply some raw, naïve brain power to the problem. Professional training in philosophy does provide a set of tools, modes and approaches, not to mention a feeling for common dangers and fal-lacies, that few scientists (or few "smart folks" of any untrained persuasion) are likely to possess by the simple good fortune of superior raw brainpower. (We might analogize this silly and vainglorious, although regrettably com-mon, belief to the more popular idea that great athletes should be able to ex-cel at anything physical by reason of their general bodily virtue—a myth and chimaera that dramatically exploded several years ago when Michael Jordan discovered that he could not learn to hit a curve ball, just because he excelled so preeminently in basketball, and possessed the world's best athletic body in general—for he ended up barely hitting over 0.200 in a full season of minor league play. I do, however, honor and praise his persistence in staying the course and taking his lumps.)

Indeed, I know of no other substantial conceptual advance in recent science so abetted by the active collaboration of working scientists and professional philosophers (thus obviating, for once, the perennial, and justified, complaint of philosophers of science that no scientists read their journals or even encounter their analyses). Several key achievements in modern evolutionary theory, particularly the successful resolution of conceptual difficulties in formulating a workable theory of hierarchical selection (rooted in concepts like emergence and simultaneous selection at several levels that our minds, with their preferences for two-valued logics, don't handle either automatically, or well at all), have appeared as joint publications of biologists and philosophers, including the books of Sober and Wilson, 1998, and Eldredge and Grene, 1992; and articles of Sober and Lewontin, 1982, and Mayo and Gilinsky, 1987. My own understanding of how to formulate an operational theory of hierarchical selection, and my "rescue" from a crucial conceptual error that had stymied my previous thinking (see Chapter 8, pages 656–673), emerged from joint work with Elisabeth Lloyd, a professional philosopher of science. I take great pride in our two joint articles (Lloyd and Gould, 1993; Gould and Lloyd, 1999), which, in my partisan judgment, resolve what may have been the last important impediment to the codification of a conceptually coherent and truly operational theory of hierarchical selection.

ZEITGEIST. Although major revisions to the structure of evolutionary theory emerged mainly from the conventional substrates of novel data and clearer concepts, we should not neglect the admittedly fuzzier, but by no means unimportant, input from a distinctive social context, or intellectual "spirit of the times" (a literal meaning of *Zeitgeist*) that, at the dawn of a calendrical millennium, has suffused our general academic culture with a set of loosely coherent themes and concerns far more congenial with the broad revisions here proposed within evolutionary theory than any previous set of guiding concepts or presuppositions had been. Needless to say, *Zeitgeists* are two edged swords of special sharpness—for either they encourage sheeplike conformity with transient ghosts of time (another literal meaning of *Zeitgeist*) that will soon fade into oblivion, or they open up new paths to insights that previous ages could not even have conceptualized. Any intellectual would therefore be a fool to argue that conformity with a *Zeitgeist* manifests any preferential correlation with scientific veracity *ipso facto*. *Zeitgeists* can only suggest or facilitate.

Nonetheless, we would be equally foolish in our naïve empiricism if we claimed that major advances in science must be entirely data driven, and that social contexts can only act as barriers to our vision of nature's factuality. Both the social and scientific world were "ready" for evolution in the mid 19th century. People of equal intelligence could neither have formulated nor owned such a concept in Newton's generation, even if some hypothetical Darwin had then advanced such a claim (and probably ended up in Bedlam for his troubles). In Chapter 2, I shall document not only this general readiness of Western science within the *Zeitgeist* of Darwin's time, but also the specific social impetus that Darwin gained from studying the distinctive theories (also a product of the earlier Enlightenment *Zeitgeist*, and not accessible before) of Adam Smith and the Scottish economists. Thus, and by analogy a century later, the altered *Zeitgeist* of our own time may also facilitate a fruitful reconsideration of major evolutionary concepts that still bear the originating stamp of a Victorian scientific context strongly committed to unidirectional, single-level and deterministic views of natural causality—subtly controlling concepts that many scientists would now label as limiting and outmoded.

Although the next few paragraphs will be the most vague and impressionistic (I trust) of the entire book, I venture these ill-formulated statements about *Zeitgeist* because I feel that something important lurks behind my inability to express these inchoate thoughts with precision. I argue above (page 14) that the key concerns of the three essential branches of Darwinian logic might be identified as agency, efficacy and scope of natural selection. In each of these domains, I believe, the revised structure of evolutionary theory, as presented in this book, might be characterized as expansion and revision according to a set of coordinated principles, all consonant with our altered *Zeitgeist vs.* the scientific spirit of Darwin's own time. The modern revision seeks to replace Darwin's unifocal theory of organismic selection with a hierarchical account (leg one); his unidirectional theory of adaptational construction in the functionalist mode with a more balanced interaction of these external causes, treating internal (or structural) constraints primarily as positive channels, and not merely as limitations (leg two); and his unilevel theory of microevolutionary extrapolation with a model of distinctive but interacting modes of change, each characteristic for its tier of time. In short, a hierarchy of interacting levels, each important in a distinctive way, for Darwin's single locus; an interaction of environmental outsides with organic insides for Darwin's single direction of causal flow; and a set of distinctive temporal tiers for Darwin's attempt to situate all causality in the single microevolutionary world of our own palpable moments.

I do sense a common underlying vision behind all these proposed reforms. Strict Darwinism, although triumphant within mid 20th century evolutionary theory, embodied several broad commitments (philosophical or metatheoretical, in the technical sense of these terms), more characteristic of 19th than of 20th century thought (and, obviously, not necessarily wrong, or even to be discounted, for this reason—as nothing can be more dangerous to the progress of science than winds of fashion, and we do, after all, learn some things, and develop some fruitful approaches, with validity and staying power well beyond their time of origin and initial popularity). Some aspects of Darwin's formulation broke philosophical ground in a sense quite consonant with our modern *Zeitgeist* of emphasis upon complexity and interaction—particularly, Darwin's focus on the interplay of chance and necessity in sources of variation *vs.* mode of selection. Indeed, Darwin paid the usual price for such innovation in the failure of nearly all his colleagues, even the most intellectually acute, to grasp such a radical underlying philosophy. But, in many commanding respects, Darwinism follows the norms of favored scientific reasoning in his time.

The logic of Darwin's formulation rests upon several preferences in scientific reasoning more characteristic of his time than of ours—preferences that many scientists would now view as unduly restrictive in their designation of a privileged locus of causality, a single direction of causal flow, and a smooth continuity in resulting effects. Classical Darwinism follows standard reductionist preferences in designating the lowest level then available—the organism, for Darwin—as an effectively unique locus of causality (the first leg of agency). In this sense, the efforts of Williams and Dawkins (see Chapter 8) to reduce the privileged locus even further to the genic level (perforce unavailable to Darwin) should be read as a furthering and intensification of Darwin's intent—in other words, a basically conservative adumbration of Darwin's own spirit and arguments, and not the radical conceptual revision that some have imagined.

At this single level of causality, classical Darwinism then envisages a similarly privileged direction of causal flow, as information from the environment (broadly construed, of course, to include other organisms as well as physical surroundings) must impact the causal agent (organisms struggling for reproductive success) and be translated, by natural selection, into evolutionary change. The organism supplies raw material in the form of "random" variation, but does not "push back" to direct the flow of its own alteration from inside. Darwinism, in this sense, is a functionalist theory, leading to local adaptation as the environment proposes and natural selection disposes. Finally, classical Darwinism completes a trio of privileged causal places and consequently directional flows by postulating strict continuity in results, as local selection scales smoothly through the immensity of geological time to engender life's history by pure extrapolation of lowest-level modes and causes.

By contrast, the common themes behind the reformulations defended in this book all follow from serious engagement with complexity, interaction, multiple levels of causation, multidirectional flows of influence, and pluralistic approaches to explanation in general—a set of integrated approaches that strongly contribute to the *Zeitgeist* of our moment. To anticipate and make a preemptive strike against the obvious counterattack from Darwinian traditionalists, these alternative themes do not substitute a "laid back, laissez-faire, anything goes" kind of sloppy tolerance for contradiction and fuzziness in argument against the genuine rigor of old-line Darwinism. The social and psychological contributions of a *Zeitgeist* to the origin of hypotheses bear no logical relationship to any subsequent scientific defense and validation of the same hypotheses. Moreover, on this subject of test and confirmation, I espouse a rigorously conventional and rather old-fashioned "realist" view that an objective factual world exists "out there," and that science can access its ways and modes. Whatever the contribution of a Victorian *Zeitgeist* to Darwin's thinking, or of a contemporary *Zeitgeist* to our revisions, the differences are testable and subject to validation or disproof by the usual armamentarium of scientific methods. That is, either Darwin is right and effectively all natural selection occurs at the organismic level (despite the logical conceivability of other levels), or the hierarchical theory is right and several levels make interestingly different and vitally important simultaneous contributions to the overall pattern of evolution. The same ordinary form of testability can be applied to any other contrast between strict Darwinism and the revised and expanded formulations defended in this book.

As the most striking general contrast that might be illuminated by reference to the different *Zeitgeists* of Darwin's time and our own, modern revisions for each essential postulate of Darwinian logic substitute mechanics based on interaction for Darwin's single locus of causality and directional flow of effects. Thus, for Darwin's near exclusivity of organismic selection, we now propose a hierarchical theory with selection acting simultaneously on a rising set of levels, each characterized by distinctive, but equally well-defined, Darwinian individuals within a genealogical hierarchy of gene, cell-lineage, organism, deme, species, and clade. The results of evolution then emerge from complex, but eminently knowable, interactions among these potent levels, and do not simply flow out and up from a unique causal locus of organismal selection.

A similar substitution of interaction for directional flow then pervades the second branch of selection's efficacy, as Darwin's functionalist formulation—with unidirectional flow from an external environment to an isotropic organic substrate that supplies "random" raw material but imposes no directional vector of its own to "push back" from internal sources of constraint—yields to a truly interactive theory of balance between the functionalist Darwinian "outside" of natural selection generated by environmental pressures, and a formalist "inside" of strong, interesting and positive constraints generated by specific past histories and timeless structural principles. Finally, on the third and last branch of selection's range, the single and controlling microevolutionary locus of Darwinian causality yields to a multileveled model of tiers of time, with a unified set of processes working in distinctive and characteristic ways at each scale, from allelic substitution in observable years to catastrophic decimation of global biotas. Thus, and in summary, for the unifocal and noninteractive Darwinian models of exclusively organismal selection, causal flow from an environmental outside to an organismal inside, and a microevolutionary locus for mechanisms of change that smoothly extrapolate to all scales, we substitute a hierarchical selectionist theory of numerous interacting levels, a balanced and bidirectional flow of causality between external selection and internal constraint (interaction of functionalist and structuralist perspectives), and causal interaction among tiers of time.

Among the many consequences of these interactionist reformulations, punctuational rather than continuationist models of change (with stronger structuralist components inevitably buttressing the punctuational versions) may emerge as the most prominent and most interesting. The Darwinian mechanics of functionalism yield an expectation of continuously improving local adaptation, with longterm stability representing the achievement of an optimum. But interactionist and multi-leveled models of causality reconceptualize stasis as a balance, actively maintained among potentially competing forces at numerous levels, with change then regarded as exceptional rather than intrinsically ticking most of the time, and punctuational rather than smoothly continuous when it does occur (representing the relatively quick transition that often accompanies a rebalancing of forces).

To end this admittedly vague section with the punch of paradox (and even with a soundbite), I would simply note the almost delicious irony that the formulation of a hierarchical theory of selection—the central concept of this book, and invoking a non-vernacular meaning of hierarchy in the purely structural sense of rising levels of inclusivity—engenders, as its most important consequence, the destruction of a different and more familiar meaning of hierarchy: that is, the hierarchy of relative value and importance embodied in Darwin's privileging of organismic selection as the ultimate source of evolutionary change at all scales. Thus, a structural and descriptive hierarchy of equally effective causal levels undermines a more conventional hierarchy of relative importance rooted in Darwin's exclusive emphasis on the micro-evolutionary mechanics of organismal selection. And so, this structuralist view of nature's order enriches the structure of evolutionary theory—carrying the difference between strict Darwinism and our current understanding through more than enough metatheoretical space to fashion a Falconerian, not merely a Darwinian, rebuilding and extension for our edifice of coherent explanation.

## A PERSONAL ODYSSEY

For reasons beyond mere self-indulgence or egotism, I believe that defenders of such general theories about large realms of nature owe their readers some explanation for the personal bases and ontogeny of their choices—for at this level of abstraction, no theory can claim derivation by simple logical or empirical necessity from observed results, and all commitments, however well defended among alternative possibilities, will also be influenced by authorial preferences of a more contingent nature that must then be narrated in order to be understood. Moreover, and in this particular case, the structure of this book includes a set of vigorously idiosyncratic features that, if not acknowledged and justified, might obscure the far more important *raison d'être* for its composition: the presentation of a tight brief for substantial reformulation in the structure of evolutionary theory, with all threads of revision conceptually united into an argument of different thrust and form, but still sufficiently continuous with its original Darwinian base to remain within the same intellectual lineage and logic.

Two aspects of my idiosyncratic procedures require explicit commentary here because, at least as my intention, they should reinforce this book's central argument for coherence (logical, historical and empirical) of the revised and general structure of evolutionary theory, and not further the opposite, albeit customary, function of such "confessional" writing—namely, to slake authorial egos, fight old battles, and relate twice-told tales to one's own advantage (although I claim no immunity from these all too human foibles).

This book will be published in the Spring of 2002, an auspicious and palindromic year just one step out of the starting gate for a new millennium.

At the same time, and fortuitously, my 10th and last volume of monthly essays in *Natural History Magazine,* written without a single break from January 1974 to January 2001, will also appear in print. In an eerie coincidence (with no meaning that I can discern), my first technical book, *Ontogeny and Phylogeny,* appeared exactly 25 years before, in 1977, also at the same time as my first book of *Natural History* essays, *Ever Since Darwin.* This odd and twofold simultaneous appearance, 25 years apart, of my best youthful efforts in the contrasting (but not really conceptually different) realms of technical and popular science, and then of my best shots from years of greater maturity in the same two realms, has forced me to think long and hard about the meaning of continuity, commitment and personal perspective.

My popular volumes fall into the explicit and well recognized category of essays, a literary genre defined, ever since Montaigne's initiating 16th century efforts, as the presentation of general material from an explicitly personal and opinionated point of view—although the best essays (literally meaning "attempts," after all) tend to be forthright in their expression of opinions, generous (or at least fair) to other views, and honest in their effort to specify the basis of authorial preferences. On the other hand, technical treatises in science do not generally receive such a license for explicitly personal expression. I believe that this convention in technical writing has been both harmful and more than a bit deceptive. Science, done perforce by ordinary human beings, expressing ordinary motives and foibles of the species, cannot be grasped as an enterprise without some acknowledgment of personal dimensions in preferences and decisions—for, although a final product may display logical coherence, other decisions, leading to other formulations of equally tight structure, could have been followed, and we do need to know why an author proceeded as he did if we wish to achieve our best understanding of his accomplishments, including the general worth of his conclusions.

Logical coherence may remain formally separate from ontogenetic construction, or psychological origin, but a full understanding of form does require some insight into intention and working procedure. Perhaps some presentations of broad theories in the history of science—Newton's *Principia* comes immediately to mind—remain virtually free of personal statement (sometimes making them, as in this case, virtually unreadable thereby). But most comprehensive works, in all fields of science, from Galileo's *Dialogo* to Darwin's *Origin,* gain stylistic strength and logical power by their suffusion with honorable statements about authorial intents, purposes, prejudices, and preferences. I cannot think of a single major book in natural history—from Buffon's *Histoire naturelle* and Cuvier's *Ossemens fossiles* to Simpson's *Tempo and Mode,* and Mayr's *Animal Species*—that does not include such extensive personal information, either in explicit sections, or inserted by-the-by throughout. (Even so abstract a presentation as R. A. Fisher's *Genetical Theory of Natural Selection* gains greatly in comprehension through its long and final, if in retrospect regrettable, section on the author's idiosyncratic eugenical views about human improvement.) I have included personal discussion throughout this text, but let me also devote a few explicit pages to the two points that I regard as most crucial to understanding the general argument through (or despite) conscious idiosyncrasies in my presentation.

### History

Many technical treatises in science begin with a short section on previous history of work in the field—usually written in the hagiographical mode to depict prior history as a march towards final truths revealed in the current volume. Sometimes, authors get a bit carried away, and these historical sections expand into substantial parts of the final book. Lest anyone make the false inference that my full first half of history arose in this haphazard and initially unintended way, I hasten to assure readers that my final result was my intention from the start.

For several reasons, I always conceived this book as a smooth joining of two halves, roughly equal in length and importance. First, and ontogenetically, I had written my earlier technical book, *Ontogeny and Phylogeny*, in this admittedly unusual manner—and I remain pleased with both the distinctiveness and the efficacy of the result. Second, I believe that the history of evolutionary thought, and probably of any other subject imbued with such importance to our lives and to our understanding of nature, constitutes an epic tale of fascinating, and mostly honorable, people engaged in a great struggle to comprehend something very deep and very difficult. Thus, such histories capture a bit of the best in us (also of the worst, but all human endeavors so conspire)—a bit, moreover, that cannot be expressed in any other way. We really do need to honor the temporal substrate of our current understanding, not only as a guide to our continuing efforts, but also as a moral obligation to our forebears.

But a third and practical reason trumps all others. Although I would not state such a claim as a generality for all scientific analyses, in this particular case I do not see how the structure of evolutionary theory can be resolved and the appropriate weights of relative importance assigned to the different components thereof, absent such a historical perspective. (Would it not be odd to claim, in any case, that the quintessential science for resolving the nature of life's history can itself be understood as a pristine construction, a fully-formed conceptual entity drawn intact from some analog of Zeus's brow, rather than an "organic" structure of ideas with its own ontogeny and history?)

To give one example at the largest and at the smallest scales of my argument, I don't know how I could have properly defended my identification and explication of the threefold essence of Darwinian logic without documenting the history of theoretical debate in order to tease out the components that have always been most troubling, most central, and most directive. A pure description of the theory's abstract logic simply will not suffice. To epitomize, I have identified these essential components on three basic grounds: that logic compels (Chapter 2), that history validates (Chapters 3–7), and that current debate reaffirms (Chapters 8–12). The middle term of this epitome unites the end members; I cannot present a coherent or compelling defense without this linkage. The three issues of agency, efficacy and scope build the Darwinian essence both because the logical structure of the theory so dictates, and because the history and current utility of the theory so document.

To complement this most general statement with just one example of the utility of historical analysis at the smaller scale of details, I offer the following case as the strongest argument for my central claim that Darwin's brave attempt to construct a single-level, exclusively organismic theory of natural selection must fail in principle, and that all selectionists must eventually own a hierarchical model. What better evidence can we cite than the historical demonstration (see Chapters 3 and 5 for details) that each of the only three foundational thinkers who truly understood the logic of selectionism—August Weismann, Hugo de Vries, and Charles Darwin himself—tried mightily to make the single-level version work as a fully sufficient explanation for evolution. And each failed, after intense intellectual struggle, and for fascinatingly different reasons documented later in the book—Darwin for explaining diversity by reluctant resort to species selection; Weismann for a strongest initial commitment to a single level, and an eventual recognition of full hierarchy as the most important and distinctive conclusion of his later career (by his own judgment); and de Vries for reconciling his largely psychological fealty to Darwin as his intellectual hero, with his clearly non-Darwinian account of the origin of species and the explanation of trends (including an explicit coining of the term "species selection" for explaining cladal patterns).

One might cite various truisms telling us that people ignorant of history will be condemned to repeat its errors. But I would rather re-express this accurate and rueful observation in a more positive manner by illustrating the power of historical analysis to aid both our current understanding and the depth of our appreciation for the intellectual importance of our enterprise. Finally, and to loosen the rein on personal bravado that I usually try to hold at least somewhat in check, no scholar should impose a project of this length upon his colleagues unless he believes that some quirk of special skill or experience permits him to proceed in a unique manner that may offer some insight to others. In my case, and only by history's fortune of no immediate competition in a small field, I may be able to combine two areas of professional competence not otherwise conjoined among current evolutionists. I am not a credentialed historian of science, but I believe that I have done sufficient work in this field (with sufficient understanding of the difference between the Whiggish dilettantism of most enthusiastic amateurs, and the rigorous methods applied by serious scholars) to qualify as adequately knowledgeable. (At least I can read all the major works in their original languages, and I stay close to the "internalist" style of analysis that people who understand the logic and history of theories, but cannot claim truly professional expertise in the "externalist" factors of general social and historical context, can usefully pursue.) Meanwhile I am, for my sins, a lifelong and active professional paleontologist, a commitment that began at age five as love at first sight with a dinosaur skeleton.

Many historians possess deeper knowledge and understanding of their immediate subject than I could ever hope to acquire, but none enjoy enough intimacy with the world of science (knowing its norms in their bones, and its quirks and foibles in their daily experience) to link this expertise to contemporary debates about causes of evolution. Many more scientists hold superb credentials as participants in current debates, but do not know the historical background. As I hope I demonstrated by practical utility in *The Mismeasure of Man* (Gould, 1981a), a small and particular—but I think quite important—intellectual space exists, almost entirely unoccupied, for people who can use historical knowledge to enlighten (not merely to footnote or to prettify) current scientific debates, and who can then apply a professional's "feel" for the doing of science to grasp the technical complexities of past debates in a useful manner inaccessible to historians (who have therefore misinterpreted, in significant ways, some important incidents and trends in their subject). I only hope that I have not been wrong in believing that my devotion of a lifetime's enthusiasm to both pursuits might make my efforts useful, in a distinctive way, to my colleagues.

### Theory

I admire my friend Oliver Sacks extravagantly as a writer, and I could never hope to match him in general quality or human compassion. He once said something that touched me deeply, despite my continuing firm disagreement with his claim (while acknowledging the validity of the single statement relevant to the present context). Oliver said that he envied me because, although we had both staked out a large and generous subject for our writing (he on the human mind, me on evolution), I had enjoyed the privilege of devising and developing a general theory that allowed me to coordinate all my work into a coherent and distinctive body, whereas he had only written descriptively and aimlessly, albeit with some insight, because no similar central focus underlay his work. I replied that he had surely sold himself short, because he had been beguiled by conventional views about the nature and limits of what may legitimately be called a central scientific theory—and that he certainly held such an organizing concept in his attempt to reintroduce the venerable "case study method" of attention to irreducible peculiarities of individual patients in the practice of cure and healing in medicine. Thus, I argued, he held a central theory about the importance of individuality and contingency in general medical theory, just as I and others had stressed the centrality of historical contingency in any theoretical analysis and understanding of evolution and its actual results.

Oliver saw the theory of punctuated equilibrium itself, which I developed with Niles Eldredge and discuss at inordinate length in Chapter 9, as my coordinating centerpiece, and I would not deny this statement. But punctuated equilibrium stands for a larger and coherent set of mostly iconoclastic concerns, and I must present some intellectual autobiography to explain the reasons and the comings together, as best I understand them myself—hence my rip-off of Cardinal Newman's famous title for the best similar effort ever made, albeit in a maximally different domain. In his *Apologia Pro Vita Sua* (an apology for one's own life), Newman intends the operative word as I do, in its original and positive meaning, not in the currently more popular negative sense—"something said or written in defense or justification of what appears to others to be wrong or of what may be liable to disapprobation" (per *Webster's*).

As my first two scientific commitments, I fell in love with paleontology when I met *Tyrannosaurus* in the Museum of Natural History at age five, and with evolution at age 11, when I read G. G. Simpson's *The Meaning of Evolution,* with great excitement but minimal comprehension, after my parents, as members of a book club for folks with intellectual interests but little economic opportunity or formal credentials, forgot to send back the "we don't want anything this month" card, and received the book they would never have ordered (but that I begged them to keep because I saw the little stick figures of dinosaurs on the dust jacket). Thus, from day one, my developing professional interests united paleontology and evolution. For some reason still unclear to me, I always found the theory of how evolution works more fascinating than the realized pageant of its paleontological results, and my major interest therefore always focused upon principles of macroevolution.* I did come to understand the vague feelings of dissatisfaction (despite Simpson's attempt to resolve them in an orthodox way by incorporating paleontology within the Modern Synthesis) that some paleontologists have always felt with the Darwinian premise that microevolutionary mechanics could construct their entire show just by accumulating incremental results through geological immensity.

As I began my professional preparation for a career in paleontology, this vague dissatisfaction coagulated into two operational foci of discontent. First (and with Niles Eldredge, for we worried this subject virtually to death as graduate students), I became deeply troubled by the Darwinian convention that attributed all non-gradualistic literal appearances to imperfections of the geological record. This traditional argument contained no logical holes, but the practical consequences struck me as unacceptable (especially at the outset of a career, full of enthusiasm for empirical work, and trained in statistical techniques that would permit the discernment of small evolutionary

---

*As so much unnecessary rancor has been generated by simple verbal confusion among different meanings of this word, and not by meaningful conceptual disagreements, I should be clear that I intend only the purely descriptive definition when I write "macroevolution"—that is, a designation of evolutionary phenomenology from the origin of species on up, in contrast with evolutionary change *within* populations of a single species. In so doing, I follow Goldschmidt's own definitional preferences (1940) in the book that established his apostasy within the Modern Synthesis. Misunderstanding has arisen because, to some, the world "macroevolution" has implied a theoretical claim for distinct causes, particularly for nonstandard genetic mechanisms, that conflict with, or do not occur at, the microevolutionary level. But Goldschmidt—and I follow him here—urged a nonconfrontational definition that could stand as a neutral descriptor for a set of results that would then permit evolutionists to pose the tough question without prejudice: does macroevolutionary phenomenology demand unique macroevolutionary mechanics? Thus, in this book, "macroevolution" is descriptive higher-level phenomenology, not pugnacious anti-Darwinian interpretation.

changes). For, by the conventional rationale, the study of microevolution became virtually nonoperational in paleontology—as one almost never found this anticipated form of gradual change up geological sections, and one therefore had to interpret the vastly predominant signal of stasis and geologically abrupt appearance as a sign of the record's imperfection, and therefore as no empirical guide to the nature of evolution. Second, I became increasingly disturbed that, at the higher level of evolutionary trends within clades, the majority of well documented examples (reduction of stipe number in graptolites, increasing symmetry of crinoidal cups, growing complexity of ammonoid sutures, for example) had never been adequately explained in the terms demanded by Darwinian convention—that is, as adaptive improvements of constituent organisms in anagenetic sequences. Most so-called explanations amounted to little more than what Lewontin and I, following Kipling, would later call "just-so stories," or plausible claims without tested evidence, whereas other prominent trends couldn't even generate a plausible story in adaptationist terms at all.

As Eldredge and I devised punctuated equilibrium, I did use the theory to resolve these two puzzles to my satisfaction, and each resolution, when finally generalized and further developed, led to my two major critiques of the first two branches of the essential triad of Darwinian central logic—so Oliver Sacks's identification of punctuated equilibrium as central to my theoretical world holds, although more as a starting point than as a coordinating focus. By accepting the geologically abrupt appearance and subsequent extended stasis of species as a fair description of an evolutionary reality, and not only as a sign of the poverty of paleontological data, we soon recognized that species met all criteria for definition and operation as genuine Darwinian individuals in the higher-level domain of macroevolution—and this insight (by complex routes discussed in Chapter 9) led us to concepts of species selection in particular and, eventually, to the full hierarchical model of selection as an interesting theoretical challenge and contrast to Darwinian convictions about the exclusivity of organismal selection. In this way, punctuated equilibrium led to the reformulation proposed herein for the first branch of essential Darwinian logic.

Meanwhile, in trying to understand the nature of stasis, we initially focused (largely in error, I now believe) upon internal constraints, as vaguely represented by various concepts of "homeostasis," and as exemplified in the model of Galton's polyhedron (see Chapter 4). These thoughts led me to extend my doubts about adaptation and the sufficiency of functionalist mechanisms in general—especially in conjunction with my old worries about paleontological failures to explain cladal trends along traditional adaptationist lines. Thus, these aspects of punctuated equilibrium strongly contributed to my developing critiques of adaptationism and purely functional mechanics on the second branch of essential Darwinian logic (although other arguments struck me as even more important, as discussed below).

Nonetheless, and despite the centrality of punctuated equilibrium in developing a broader critique of conventional Darwinism, my sources extended outward into a diverse and quirky network of concerns that seemed, to me and at first, isolated and uncoordinated, and that only later congealed into a coherent critique. For this curious, almost paradoxical, reason, I have become even more convinced that the elements of my overall critique hang together, for I never sensed the connections when I initially identified the components as, individually, the most challenging and intriguing items I had encountered in my study of evolution. When one accumulates a set of things only for their independent appeals, with no inkling that any common intellectual ground underlies the apparent miscellany, then one can only gain confidence in the "reality" of a conceptual basis discerned only later for the cohesion. I would never argue that this critique of strict Darwinism gains any higher probability of truth value for initially infecting me in such an uncoordinated and mindless way. But I would assert that a genuinely coherent and general alternative formulation must exist "out there" in the philosophical universe of intellectual possibilities—whatever its empirical validity—if its isolated components could coagulate, and be discerned and selected, so unconsciously.

If I may make a somewhat far-fetched analogy to my favorite Victorian novel, *Daniel Deronda* (the last effort of Darwin's friend George Eliot), the hero of this story, a Jew raised in a Christian family with no knowledge of his ethnic origins, becomes, as an adult, drawn to a set of apparently independent activities with no coordinating theme beyond their relationship, entirely unknown to Deronda at the time of his initial fascination, to Jewish history and customs. Eventually, he recognizes the unifying theme behind such apparent diversity, and learns the truth of his own genetic background. (I forgive Eliot for this basically silly fable of genealogical determinism because her philosemitic motives, however naïve and a bit condescending, shine forth so clearly in the surrounding antisemitic darkness of her times.) But I do feel, to complete the analogy, rather like a modern, if only culturally or psychologically predisposed, Deronda who gathered the elements of a coherent critique solely because he loved each item individually—and only later sensed an underlying unity, which therefore cannot be chimaerical, but may claim some logical existence prior to any conscious formulation on my part.

In fact, the case for an external and objective coherence of this alternative view of evolution seems even stronger to me because I gathered the independent items not only in ignorance of their coordination, but also at a time when I held a conscious and conventional view of Darwinian evolution that would have actively denied their critical unity and meaning. I fledged in science as a firm adaptationist, utterly beguiled by the absolutist beauty (no doubt, my own simplistic reading of a more subtle, albeit truly hardened, Modern Synthesis) of asserting, à la Cain and other ecological geneticists of the British school, that all aspects of organismal phenotypes, even the most trivial nuances, could be fully explained as adaptations built by natural selection.

I remember two incidents of juvenilia with profound embarrassment today: First, an undergraduate evening bull session with the smartest physics major at Antioch College, as his skepticism evoked my stronger insistence that our science matched his in reductionistic rigor because "we" now knew for certain that natural selection built everything for optimal advantage, thus making evolution as quantifiable and predictive as classical physics. Second, as a somewhat more sophisticated, but still beguiled, assistant professor, I remember my profound feeling of sadness and disappointment, nearly amounting to an emotional sense of betrayal, upon learning that an anthropological colleague favored drift as the probable reason for apparently trivial genetic differences among isolated groups of Papua-New Guinea peoples. I remember remonstrating with him as follows: Of course your argument conforms to logic and empirical possibility, and I admit that we have no proof either way. But your results are also consistent with selection—and our panselectionist paradigm has forged a theory of such beauty and elegant simplicity that one should never favor exceptions for their mere plausibility, but only for documented necessity. (I recall this discussion with special force because my emotional feelings were so strong, and my disappointment in his "unnecessary apostasy" so keen, even though I knew that neither of us had the empirical "goods.") Finally, if I could, in a species of Devil's bargain, wipe any of my publications off the face of the earth and out of all memory, I would gladly nominate my unfortunately rather popular review article on "Evolutionary paleontology and the science of form" (Gould, 1970a)—a ringing paean to selectionist absolutism, buttressed by the literary barbarism that a "quanti-functional" paleontology, combining the best of biometric and mechanical analyses, could prove panadaptationism even for fossils that could not be run through the hoops of actual experiments.

Against this orthodox background—or, rather, within it and quite unconsciously for many years—I worked piecemeal, producing a set of separate and continually accreting revisionary items along each of the branches of Darwinian central logic, until I realized that a "Platonic" something "up there" in ideological space could coordinate all these critiques and fascinations into a revised general theory with a retained Darwinian base.

The first branch of levels in selection proceeded rather directly and linearly because the generality flowed so clearly from punctuated equilibrium itself, once Eldredge and I finally worked through the implications and extensions of our own formulations (Eldredge and Gould, 1972). Steve Stanley (1975) and Elisabeth Vrba (1980) helped to show us what we had missed in ramifications leading from the phenomenology of stasis and geologically abrupt appearance, to recognizing species as genuine Darwinian individuals, to designating species as, therefore and potentially, the basic individuals of macroevolution (comparable with the role of the organism in microevolution), to the validity of species selection, and eventually to the full hierarchical model and its profound departure from the exclusively organismal accounts of conventional Darwinism (or the even more reduced and equally monistic genic versions of Williams and Dawkins)—see Vrba and Gould, 1986. Finally, by adopting the interactor rather than the replicator approach to defining selection, and by recognizing emergent fitness, rather than emergent characters, as the proper criterion for identifying higher-level selection (Lloyd and Gould, 1993; Gould and Lloyd, 1999), I think that we finally reached, by a circuitous route around many stumbling blocks of my previous stupidity, a consistent and truly operational theory of hierarchical selection (see Chapter 8).

I must also confess to some preconditioning beyond punctuated equilibrium. I had admired Wynne-Edwards's pluck (1962) from the start, even though I agreed with Williams's (1966) trenchant criticisms of his particular defenses for group selection, rooted in the ability of populations to regulate their own numbers in the interests of group advantage. Still, I felt, for no reason beyond vague intuition, that group selection made logical sense and might well find other domains and formulations of greater validity—a feeling that has now been cashed out by modern reformulations of evolutionary theory (see especially Wilson and Sober, 1998, and Chapters 8 and 9 herein).

My odyssey on the second branch of balancing internal constraint with external adaptation in understanding the patterning and creative population of novel places in evolutionary morphospace followed a much more complex, meandering and diverse set of pathways. As an undergraduate, I loved D'Arcy Thompson's *Growth and Form* (1917; see Gould, 1971b, for my first "literary" paper), and wrote a senior thesis on his theory of morphology. But I thought that I admired the book only for its incomparable prose, and I attacked the anti-Darwinian (and structuralist) components of his theory unmercifully. I then took up allometry for my first empirical studies, somehow fascinated by structural constraint and correlation of growth, but thinking all the while that my task must center on a restoration of adaptationist themes to this "holdout" bastion of formalist thought—particularly the achievement of biomechanical optima consistent with the Galilean principle of decreasing surface/volume ratios with increasing size in isometric forms. I remain proud of my first review article, dedicated to this subject (Gould, 1966), written when I was still a graduate student, but I am now embarrassed by the fervor of my adaptationist convictions.

I emphasized allometric analysis, now in a directly multivariate reformulation, in my first set of empirical studies on the Bermudian pulmonate snail *Poecilozonites* (see especially Gould, 1969—the published version of my Ph.D. dissertation). And yet, of all the long and largely adaptationist treatises in this series, and for some reason that I could not identify at the time, the conclusion that I reached with most satisfaction, and that I somehow regarded as most theoretically innovative (without knowing why), resided in a short, and otherwise insignificant, article that I wrote for a specialized paleontological journal on a case of convergence produced by structural necessity, given modes of coiling and allometry in this genus, rather than by selectionist honing (for some cases rested upon ecophenotypic expression, others on paedomorphosis, and still others on gradual change that could be read as conventionally adaptive): "Precise but fortuitous convergence in Pleistocene land snails" (Gould, 1971c).

Five disparate reasons underlie my more explicit recognition, during the 1970's and early 1980's, of the importance and theoretical interest (and iconoclasm versus Darwinian traditions) of nonadaptationist themes rooted in structural and historical constraint. First, I stood under the dome of San Marco during a meeting in Venice and then wrote a notorious paper with Dick Lewontin on the subject of spandrels, or nonadaptive sequelae of prior structural decisions (Gould and Lewontin, 1979—see Chapter 11, pp. 1246–1258). Second, I recognized, with Elisabeth Vrba, that the lexicon of evolutionary biology possessed no term for the evidently important phenomenon of structures coopted for utility from different sources of origin (including nonadaptive spandrels), and not directly built as adaptations for their current function. We therefore devised the term "exaptation" (Gould and Vrba, 1982) and explored its implications for structuralist revisions to pure Darwinian functionalism. Third, I worked with a group of paleontological colleagues (Raup *et al.*, 1973; Raup and Gould, 1974; Gould *et al.*, 1977) to develop more rigorous criteria for identifying the signals that required selectionist, rather than stochastic, explanation of apparent order in phyletic patterns. This work left me humbled by the insight that our brains seek pattern, while our cultures favor particular kinds of stories for explaining these patterns—thus imposing a powerful bias for ascribing conventional deterministic causes, particularly adaptationist scenarios in our Darwinian traditions, to patterns well within the range of expected outcomes in purely stochastic systems. This work sobered me against such *a priori* preferences for adaptationist solutions, so often based upon plausible stories about results, rather than rigorous documentation of mechanisms.

Fourth, and most importantly, I read the great European structuralist literatures in writing my book on *Ontogeny and Phylogeny* (Gould, 1977b). I don't see how anyone could read, from Goethe and Geoffroy down through Severtzov, Remane and Riedl, without developing some appreciation for the plausibility, or at least for the sheer intellectual power, of morphological explanations outside the domain of Darwinian functionalism—although my resulting book, for the last time in my career, stuck closely to selectionist orthodoxy, while describing these alternatives in an accurate and sympathetic manner. Fifth, my growing unhappiness with the speculative character of many adaptationist scenarios increased when, starting in the mid 1970's, the growing vernacular (and some of the technical) literature on sociobiology touted conclusions that struck me as implausible, and that also (in some cases) ran counter to my political and social beliefs as well.

Personal distaste, needless to say, bears no necessary relationship to scientific validity. After all, what could be more unpleasant, but also more factually undeniable, than personal mortality? But when distasteful conclusions gain popularity by appealing to supposedly scientific support, and when this "support" rests upon little more than favored speculation in an orthodox mode of increasingly dubious status, then popular misuse can legitimately sharpen a scientist's sense of unhappiness with the flawed theoretical basis behind a particular misuse. In any case, I trust that this compendium of reasons will dispel Cain's (1979) hurtful assertion that Lewontin, I, and other evolutionists who questioned early forms of sociobiology by developing a general critique of adaptationism, had acted cynically, and even anti-scientifically, in opposing biological theories that we knew to be true because we disliked their political implications for explaining human behavior. My own growing doubts about adaptationism arose from several roots, mostly paleontological, with any displeasure about sociobiology serving as a late and minor spur to further examination and synthesis.

I then tried to apply my general critique of pure Darwinian functionalism, and my conviction that important and positive constraints could be actively identified by quantitative morphometric study (and not merely passively inferred from failures of adaptationist scenarios) in my work on "covariance sets" in the growth, variation, and evolution of the West Indian pulmonate *Cerion* (Gould, 1984b and c), a snail that encompasses its maximal diversity in overt form among populations within a constraining set of pervasive allometries in growth. I discuss some of this work in my text on the empirical validation of positive constraint (see Chapter 10, pages 1045–1051).

My doubts on the third branch of extrapolationism and uniformity began even earlier, and in a more inchoate way, but then gained expression in my efforts in the history of science, and not so much in my direct empirical work—hence, in part, the reduced attention devoted to this theme (Chapters 6 and 12) compared with the first two branches of selection's agency and efficacy. On a fieldtrip in my freshman geology course, my professor took us to a travertine mound and argued that the deposit must be about 11,000 years old because he had measured the current rate of accumulation and then extrapolated back to a beginning. When I asked how he could assume such constancy of rate, he replied that the fundamental rule of geological inference, something called "the principle of uniformitarianism" permitted such inferences because we must regard the laws of nature as constant if we wish to reach any scientific conclusions about the past. This argument struck me as logically incorrect, and I pledged myself to making a rigorous analysis of the reasons.

As a joint major in geology and philosophy, I studied this issue throughout my undergraduate years, producing a paper entitled "Hume and uniformitarianism" that eventually transmogrified into my first publication (Gould, 1965), "Is uniformitarianism necessary?" (Norman Newell, my graduate advisor, urged me to send the paper to *Science* where, as I learned to my amusement much later, my future "boss" at Harvard, the senior paleontology professor Bernie Kummel, rejected it roundly as a reviewer. Properly humbled—although I still regard his reasons as ill founded—I then sent the paper to a specialty journal in geology.)

May I share one shameful memory of this otherwise iconoclastic first paper, from which I still draw some pride? In my undergraduate work on this theme, I made a personal discovery (as others did independently) that became important in late 20th-century studies of the history of geology. I had been schooled in the conventional view that the catastrophists (*aka* "bad guys") had invoked supernatural sources of paroxysmal dynamics in order to compress the earth's history into the strictures of biblical chronology. I read and reread all the classical texts of late 18th and early 19th century catastrophism in their original languages—and I could find no claim for supernatural influences upon the history of the earth. In fact, the catastrophists seemed to be advancing the opposite claim that we should base our causal conclusions upon a literal reading of the empirical record, whereas the uniformitarians (*aka* "good guys") seemed to be arguing, in an opposing claim less congenial with the stereotypical empiricism of science, that we must make hypothetical inferences about the gradualistic mechanics that a woefully imperfect record does not permit us to observe directly.

But, although I had developed and presented an iconoclastic exegesis of Lyell, I simply lacked the courage to state so general a claim for inverting the standard view about uniformitarians and catastrophists. I assumed that I must be wrong, and that I must have misunderstood catastrophism because I had not read enough, or could not comprehend the subtleties at this fledgling state of a career. So I scoured the catastrophist literature again until I found a quote from William Buckland (both a leading divine and the first reader in geology at Oxford) that could be interpreted as a defense of supernaturalism. I cited the quotation (Gould, 1965, p. 223) and stuck to convention on this broader issue, while presenting an original analysis of multiple meanings— some valid (like the invariance of law) and some invalid (like my professor's claim for constancy in range of rates)—subsumed by Lyell under the singular description of "uniformity" in nature.

This work led me, partly from shame at my initial cowardice, and as others reassessed the scientific character of catastrophism, to a more general analysis of the potential validity of catastrophic claims, and particularly to an understanding of how assumptions of gradualism had so stymied and constrained our comprehension of the earth's much richer history. These ideas forced me to question the necessary basis for Darwin's key assumption that observable, small-scale processes of microevolution could, by extension through the immensity of geological time, explain all patterns in the history of life—namely, the Lyellian belief in uniformity of rate (one of the invalid meanings of the hybrid concept of uniformitarianism). This exegesis led to a technical book about concepts of time and direction in geology (Gould, 1987b), to an enlarged view that encouraged the development of punctuated equilibrium, and to a position of cautious favor towards such truly catastrophic proposals as the Alvarez theory of mass extinction by extraterrestrial impact—a concept ridiculed by nearly all other paleontologists when first proposed (Alvarez *et al.*, 1980), but now affirmed for the K-T event, and accepted as an empirical basis for expanding our range of scientifically legitimate hypotheses beyond the smooth extrapolationism demanded by this third branch of Darwinian central logic.

In addition to these disparate accretions of revisionism on the three branches of Darwinian central logic, one further domain—my studies in the history of evolutionary thought—served as a *sine qua non* for wresting a coherent critique from such an inchoate jumble of disparate items. Above all, if I had not studied Darwin's persona and social context so intensely, I doubt that I would ever have understood the motivations and consistencies—also the idiosyncrasies of time, place and manner—behind the abstract grandeur of his view of life. History, as I argued before (see p. 35), must not be dismissed as a humanistic frill upon the adamantine solidity of "real" science, but must be embraced as the coordinating context for any broad view of the logic and reasoning behind a subject so close to the bone of human concern as the science of life's nature and structure. (Of the two greatest revolutions in scientific thought, Darwin surely trumps Copernicus in raw emotional impact, if only because the older transition spoke mainly of real estate, and the later of essence.)

Some of my historical writing appeared in the standard professional literature, particularly my thesis about the "hardening" of the Modern Synthesis (Gould, 1980e, 1982a, 1983b), a trend (but also, in part, a drift) towards a stricter and less pluralistic Darwinism. Several full-time historians of science then affirmed this hypothesis (Provine, 1986; Beatty, 1988; Smocovitis, 1996). But much of the historical analysis behind the basic argument of this book had its roots (in my consciousness at least) in the 300 consecutive monthly essays that I wrote from 1974 to 2001 in the popular forum of *Natural History* magazine, where I tried to develop a distinctive style of "mini intellectual biography" in essay form—attempts to epitomize the key ideas of a professional career in a biographic context, and within the strictures of a few thousand words. By thus forcing myself to emphasize essentials and to discard peripherals (while always searching out the truly lovely details that best exemplify any abstraction), I think that I came to understand the major ideological contrasts between the defining features of Darwinian theory and the centerpieces of alternative views. In this format, I first studied such structuralist alternatives as Goethe's theory of the archetypal leaf, Geoffroy's hypothesis on the vertebral underpinning of all animals, and on dorso-ventral inversion of arthropods and vertebrates, and Owen's uncharacteristic English support for this continental view of life. I also developed immense sympathy for the beauty and raw intellectual power of various alternatives, even if I eventually found them wanting in empirical terms. And I came to understand the partial validity, and even the moral suasion, in certain proposals unfairly ridiculed by history's later victors—as in reconsidering the great hippocampus debate between Huxley and Owen, and recognizing how Owen used his (ultimately false) view in the service of racial egalitarianism, while Huxley misused his (ultimately correct) interpretation in a fallacious defense of traditional racial ranking.

Finally, my general love of history in the broadest sense spilled over into my empirical work as I began to explore the role of history's great theoretical theme in my empirical work as well—contingency, or the tendency of complex systems with substantial stochastic components, and intricate nonlinear interactions among components, to be unpredictable in principle from full knowledge of antecedent conditions, but fully explainable after time's actual unfoldings. This work led to two books on the pageant of life's history (Gould, 1989c; Gould, 1996a). Although this book, by contrast, treats general theory and its broad results (pattern *vs.* pageant in the terms of this text), rather than contingency and the explanation of life's particulars, the science of contingency must ultimately be integrated with the more conventional science of general theory as explored in this book—for we shall thus attain our best possible understanding of both pattern and pageant, and their different attributes and predictabilities. The closing section of the book (pp. 1332–1343 of Chapter 12) offers some suggestions for these future efforts.

When I ask myself how all these disparate thoughts and items fell together into the one long argument of this book, I can only cite—and I don't know how else to put this—my love of Darwin and the power of his genius. Only he could have presented such a fecund framework of a fully consistent theory, so radical in form, so complete in logic, and so expansive in implication. No other early evolutionary thinker ever developed such a rich and comprehensive starting point. From this inception, I only had to explicate the full original version, tease out the central elements and commitments, and discuss the subsequent history of debate and revision for these essential features, culminating in a consistent reformulation of the full corpus in a helpful way that leaves Darwin's foundation intact while constructing a larger edifice of interestingly different form thereupon. Clearly I do not honor Darwin by hagiography, if only because such obsequious efforts would make any honest character cringe (and would surely cause Darwin to spin in his grave, thus upsetting both the tourists in Westminster Abbey and the adjacent bones of Isaac Newton). I honor Darwin's struggles as much as his successes, and I focus on his few weaknesses as entry points for needed revision—his acknowledged failure to solve the "problem of diversity," or his special pleading for progress in the absence of any explicit rationale from the operation of his central mechanism of natural selection.

As a final comment, if this section has violated the norms of scientific discourse (at least in our contemporary world, although not in Darwin's age) by the liberty that I have taken in explicating personal motives, errors and corrections, at least I have shown how we all grope upward from initial stupidity, and how we would never be able to climb without the help and collaboration of innumerable colleagues, all engaged in the intensely social enterprise called modern science. I experienced no eureka moment in developing the long argument of this book. I forged the chain link by link, from initial possession of a few separate items that I didn't even appreciate as pieces of a single chain, or of any chain at all. I made my linkages one by one, and then often cut the segments apart, in order to refashion the totality in a different order. So many people helped me along the way—from long dead antecedents by their wise words to younger colleagues by their wisecracks—that I must view this outcome as a social project, even though I, the most arrogant of literati, insisted on writing every word. Perhaps I can best express my profound thanks to the members of such an intellectual collectivity by stating, in the most literal sense, that this book would not exist without their aid and sufferance. My formal dedication to my two dearest and closest paleontological collaborators in this effort to formulate macroevolutionary theory records the worthy apex of an extensive pyramid. Scientists fight and squabble as all folks do (and I have scarcely avoided a substantial documentation thereof in this book). But we are, in general, a reasonably honorable lot, and we do embrace a tendency to help each other because we really do revel in the understanding of nature's facts and ways—and most of us will even trade some personal acclaim for the goal of faster and firmer learning. For all the tensions and unhappinesses in any life, I can at least say, with all my heart, that I chose to work in the best of all enterprises at the best of all possible times. May our contingent future only improve this matrix for my successors.

## Epitomes for a Long Development

### LEVELS OF POTENTIAL ORIGINALITY

Most of this book can be described as extensive narration of work already done, and ideas already expounded elsewhere. But no one should write at such length merely to organize the conventional material of a field, and without an original structure, or a set of unconventional ideas, to propose. I wrote *The Structure of Evolutionary Theory* because I felt that I had followed a sufficiently idiosyncratic procedure to devise a sufficiently novel theoretical structure that then yielded a sufficient number of original insights on specific matters to qualify as a justification for spending so many years of a career, and daring to ask readers for such a non-trivial chunk of their attention.

As implied by the foregoing sentence, I think that whatever originality this work possesses might best be conceptualized at three levels of basic structure, primary justifications for the major components of theory, and specific insights or discoveries then developed under the aegis of this structure and theory. At the *first level of basic structure,* I believe that three features of organization set the novelty of presentation:

1. Developing an exegesis of essential components in the *logic* of Darwinian theory, as expressed in the agency, efficacy, and scope of selection as an evolutionary mechanism (Chapter 2).

2. Explicating the *history* of evolutionary thought as a complex and extended debate about these essential components, developed negatively at first by early evolutionists who sought alternative formulations to Darwinism (Chapters 3–6), and then positively in our times by scientists who recognized the need for extensive revisions and expansions that would build an enlarged structure upon a Darwinian foundation, rather than uproot the theoretical core of selectionism (Chapters 7–12).

3. Formulating an expanded *theory* that introduces substantial revisions on each branch of Darwinian central logic, but builds, in its ensemble, a coherently enlarged structure with a retained Darwinian base—moving from Darwin's single level of agency to a hierarchical theory of selection on the first branch; balancing positive sources of internal constraint (for both structural and historical reasons) with the conventional externalism of natural selection on the second branch; and recognizing the disparate inputs of various tiers

of time, rather than trying to explain all phylogenetic mechanics by uniformitarian extrapolation from microevolutionary processes, on the third branch.

At the *second level of validation for proposed revisions in the structure of evolutionary theory,* I have tried to develop broad arguments and empirical justifications for major changes and expansions on each of the three branches of Darwinian central logic. On the first branch of agency, the theory of punctuated equilibrium itself, initially formulated by Niles Eldedge and me, establishes the species as a true and potent Darwinian individual, and grants a minimal guarantee of descriptive independence to macroevolution by requiring a treatment of trends as the differential success of stable species rather than the adaptive anagenesis of lineages by accumulated and extrapolated organismal selection alone. Beyond punctuated equilibrium, the general rationale for a hierarchical theory of selection, as presented here through the interactor approach based on emergent fitnesses at higher levels, may establish a complete (and tolerably novel) framework not only for grasping the consistent logic of hierarchical selection, but also for viewing each level as potent in its own distinctive way, and for recognizing the totality of evolutionary outcomes as a realized balance among these potencies, and not as the achieved optimality of a single causal locus—a substantial difference from Darwinian traditions for conceiving the dynamics of evolutionary change. In working through the differences among levels—see Chapter 8, pp. 714–744—I was particularly struck by the surprising, but accurate and challenging, analogies (Lamarckian inheritance at the organismal level with adaptive anagenesis at the species level, for example); and by the different modes of equally effective change implied by disparate structural reasons for the establishment of individuality at various levels (particularly, the domination of selection over drift and drive at the organismal level *vs.* the potent balance among all three mechanisms at the species level).

On the second branch of efficacy, I have tried to make the most comprehensive case yet advanced for internal constraint as a positive director and channeler of evolutionary change, and not only as a negative brake upon pure Darwinian functionalism. I proceed by explicating two conceptually different forms of constraint—structural constraints as consequences of physical principles, and historical constraints as channels from particular pasts. I argue that each category challenges a different central tenet of Darwinism—structural constraint by establishing a substantial space for non-selectionist origin of important evolutionary features, and historical constraint for explaining the markedly inhomogeneous filling of morphospace as flow down ancient internal channels of deep homology, and not primarily as a mapping of adaptive design upon current ecological landscapes. Beyond any novelty in this general formulation, I have attempted to develop a conceptual space, and to establish practical criteria, for the identification of non-adaptive sequelae (spandrels), the evolutionary importance of their later cooptation for utility (exaptation), and the importance of such reservoirs of potential (exaptive pools) in explicating the important concept of "evolvability" in structural rather than purely adaptational terms.

On the third branch of scope, my contribution cannot claim much novelty, if only because I have not worked professionally in this area of paleontological research. But I do explicate, perhaps more fully than before, both the historical and conceptual reasons for regarding catastrophic mass extinction, and catastrophic mechanics in general (within their limited scope of validity), not as anti-selectionist *per se,* but rather as fracturing the extrapolationist premise of Darwinian central logic, and requiring that substantial aspects of phyletic pattern be explained as interaction between temporal extensions of microevolution and different processes that only become visible and effective at higher tiers of time. I try to resolve "the paradox of the first tier" (the empirical failure of Darwin's logically airtight argument for a vector of progress) by arguing that punctuated equilibrium at the second tier of phyletic trends, and mass extinction at the third tier of faunal overturn, impose enough of their own, distinctive and different, patterning to forestall the domination or pure imprint of extrapolated microevolutionary results upon the general pageant of life's history.

Finally, at the *third level of* those lovely *details* (where both God and the devil dwell, and where, ultimately, both the joy and power of science reside), I trust that any originality I have introduced at "higher" levels of theoretical structure gains primary expression and utility in the resolution of previously puzzling details, and in the identification of "little things" that had escaped previous notice or explicit examination.

For example, most original analyses and discoveries in the historical first half of this book flow directly from my organizing theme of identifying essential components in Darwinian logic, and then tracing both the early attempts to defeat, and our later efforts to modify and expand them through time. I was thus able to discover and identify Darwin's major encounter with higher level selection not in his recognized discussion of group selection for human altruism, but in his previously unexplicated admission of species selection to resolve the problem of diversity (see Chapter 3, pp. 246–250). In this case, I "lucked out" through an odd reason for previous ignorance of such an important textual revision—for Darwin omitted this material in his compressed and hasty discussion of diversity in Chapter 4 of the *Origin* (on this subject, the only Darwinian source generally known to professional biologists, who would immediately highlight the importance of any acknowledgment of species selection). But Darwin agonized over levels of selection at explicit length in the unpublished "long version" that only saw the light of printed day in 1975 (Stauffer, 1975), and that virtually no practicing biologist has ever read (whereas historians of science who do study this longer text usually lack sufficient knowledge of the technical debate about levels of selection to understand the meaning of Darwin's passages or to appreciate their import).

The same context led me to appreciate the previously unanalyzed development of a full hierarchical model by Weismann in his later works (Chapter 3, pp. 223–224), a formulation that Weismann himself identified as the most important theoretical achievement of his later career. Previous historians had written about his much longer and earlier explications of lower level selection (germinal selection in his terms), if only in the context of modern reductionistic breakdowns of Darwinism to selection among "selfish genes." But they had missed his later reversal and expansion to a full hierarchical model, despite Weismann's own emphasis. Similarly, de Vries's clear understanding of Darwinian logic had also been ignored because de Vries, as an opponent of the efficacy of Darwinian organismal selection (a painful decision for him, given his psychological fealty to Darwin, also explored herein), applied the logic to higher levels, and even devised the term "species selection" (Chapter 5, pp. 446–451)—a concept and coining previously entirely unremarked by historians (much to the embarrassment of scientists, including yours truly, who coined and explicated the same term much later in full expectation of pristine originality!).

Similarly, my sense of the logic in conflicts between constraint and adaptation (or internal *vs.* external, or formal *vs.* functional approaches) on the second branch helped me to pinpoint, or to make sense of, several important historical events and arguments that have not been properly treated or understood. Historians of science had not previously discussed orthogenetic theories in this fairest light, and had not distinguished the very different formulations of Hyatt, Eimer, and Whitman in terms of their increasingly greater willingness to accommodate Darwinian themes as well (see Chapter 5). The same framework allowed me to identify the crucial importance, and brilliant epitomization, of this issue in the final paragraphs of Chapter 6 ("Difficulties on Theory") in Darwin's *Origin,* a significance that had not been highlighted before.

I also traced the dichotomy of anglophonic preferences for functionalist accounts *vs.* continental leanings towards formalism back through the evolutionary reconstruction of the argument in the mid 19th century into the creationist formulations of Paley *vs.* Agassiz (Chapter 4), thus illustrating a pedigree for this fundamental issue in morphology that evolution may have recast in causal terms, but did not budge in basic commitments to the meaning of morphology. Among the little tidbits that emerge from such analyses, I even discovered that Darwin borrowed his clearest admission of co-opted utility from non-adaptive origins (unfused skull sutures in mammalian neonates, essential for passage through the birth canal, but also existing in birds and reptiles born from more capacious eggs) from the longer and more nuanced descriptions of Richard Owen, Britain's anomalous defender of formalism.

I also included some historical analyses in the book's second half on modern advances because I thought they could make an original contribution to arguments usually developed only in contemporary terms and findings. I have already mentioned my analysis of how the initial pluralism of the Modern Synthesis (embracing any mode of change consistent with known genetic mechanisms) hardened through subsequent editions of the founding volumes into pronounced preferences for adaptationist accounts framed only in terms of natural selection (Chapter 7). In addition, I think that my reexhumation of the debate between Falconer and Darwin on fossil elephants provides a good introduction to punctuated equilibrium (Chapter 9, pp. 745–749). The largely unknown paradox of Lankester's original definition of homoplasy as a category of homology, rather than in the opposite status held by the term today, provides the best entry I could devise for understanding the vital, but little appreciated and rarely acknowledged, theoretical differences between parallelism and convergence. In the absence of this context and distinction, the key importance of evo-devo and the discovery of deep homology among distant phyla cannot properly be grasped as a challenge and expansion of Darwinian expectations (Chapter 10).

I hope that my sympathetic portrayal of D'Arcy Thompson's theory of form (Chapter 11), despite my general disagreement with his argument, will help colleagues to understand the thrust and potential power of this unusual formulation of structuralist constraint on external grounds of universal physics. Although I am chagrined that I discovered Nietzsche's account of the distinction between current utility and historical origin so late in my work, I know no better introduction—from one of history's greatest philosophers to boot, and in his analysis of morality, not of any scientific subject—to the theoretical importance of spandrels and exaptation in the rebalancing of constraint and adaptation within evolutionary theory (Chapter 11, pp. 1214–1218). In a final historical analysis of the second part, I think that Darwin's own rationale for progress (Chapter 12, pp. 1296–1303), rooted not in the mechanics of natural selection itself, but in an ecological argument for extrapolation of biotic competition through time in a perpetually crowded world—an aspect of Darwin's thought that has very rarely been appreciated, formulated or discussed by historians—provided the best context I could devise for understanding why catastrophic mass extinction in particular, and non-extrapolation through tiers of time in general, play such havoc with Darwin's need for uniformity on the third branch of his essential logic.

The original claims in the book's second half on modern reformulations of evolutionary theory rest, necessarily and primarily, on theoretical insights and unusual conceptual parsings, rather than on novel data—if only because custom dictates that my extensive empirical documentation be presented in "review" format by collating published studies in support or refutation of general themes under discussion. But I have sometimes presented existing data in novel contexts—as in my analysis of the proper category for understanding the exaptive value of genes lost by founder drift in establishing the social cohesion (albeit transient) that has made the Argentine ant *Linepithema humile* such a successful invader of non-native Californian habitats (Chapter 11, pp. 1282–1284). I have also cited my own empirical studies, previously published but original in the more conventional sense, to support important pieces of more general arguments, including validation of punctuated equilibrium by dissection of a single bedding plane to reveal transition by absolute age dating of individual shells (Goodfriend and Gould, 1996), the "employment" of constraint by selection to yield several adaptive features by one heterochronic change in a case of neoteny in *Gryphaea* (Jones and Gould, 1999), and the explanation of most ordered geographic variation within a major subregion of *Cerion* as consequences of allometric correlations in growth (Gould, 1984b).

I tried (and utterly failed) to compose a selective listing, as provided above for the book's historical half, for original ideas about theoretical details developed in revising the three branches of Darwinian central logic in the book's second half on modern reformulations of evolutionary theory. I ripped up several attempts that read like the hodge-podge of a random laundry list rather than the ordered "sweet places" on a logical continuum. These highlights, I finally recognized, have little meaning outside the broader context of a linearly developing argument for each branch, and I will therefore make a second attempt, within the more detailed epitome of the next and final section of this chapter, to designate the points that struck me with the force of "aha," or that conveyed a hint of deeper, surprising, or more radical implications for reasons that I couldn't quite fathom directly, but that tickled my intuition at the edge of that wonderful, if elongate, German word: *Fingerspitzengefühl*, or feeling at the tip of one's finger. Most inchoate excitements of this sort lead to nowhere but foolishness and waste of time, but every once in a while, the following of one's nose catches a whiff of novelty. At least we must trust ourselves enough to try—and not take ourselves so seriously that we forget to laugh at our more frequent and inevitable stumbles.

## AN ABSTRACT OF ONE LONG ARGUMENT

I have insisted, borrowing Darwin's famous line in my arrogance, that this "whole volume is one long argument," flowing logically and sequentially from a clear beginning in Darwin's *Origin* to our current reformulations of evolutionary theory. But this structural thread of Ariadne can easily become lost in the labyrinth of my tendencies to expatiate on little factual gems, or to follow the thoughts of leading scientists into small, if lovely, byways of their mental complexities. Hence, I need to present summaries and epitomes as guidelines.

Long books, like large bureaucracies, can easily get bogged down in a baroque layering of summary within summary. The United States House of Representatives has a Committee on Committees (I kid you not), undoubtedly embellished with subcommittees thereof. And we must not forget Jonathan Swift's famous verse on the fractality of growing triviality in scholarly commentary:

> So, naturalists observe, a flea
> Hath smaller fleas that on him prey;
> And these have smaller still to bite 'em
> And so proceed *ad infinitum*.
> Thus every poet, in his kind,
> Is bit by him that comes behind.

I wrote, on page 13, that this book includes three levels of embedding for this long argument—the summary in this chapter, the epitome of Darwin in

Chapter 2, and the development of the totality. Now, and most sheepishly, I add two more, for a fractal total of five—the listed abstract, in pure "book order," of this section, and (God help us) the epitome of this epitome, presented now to introduce and guide the list.

I develop my argument throughout this book by asserting, first, that the central logic of Darwinism can be depicted as a branching tree with three major limbs devoted to selection's agency, efficacy and scope. Second, that Darwin, despite his heroic and explicit efforts, could not fully "cash out" his theory in terms of the stated commitments on each branch—and that he had to allow crucial exceptions, or at least express substantial fears, in each domain (admitting species selection to resolve the problem of diversity; permitting an uncomfortably large role for formalist correlations of growth as compromisers of strict adaptationism; expressing worry that mass extinction, if more than an artifact of an imperfect fossil record, would derail the extrapolationist premise of his system). Third, that the subsequent history of evolutionary debate has focused so strongly upon the key claims of these three essential branches that we may use engagement with them as a primary criterion for distinguishing the central from the secondary when we need to gauge the importance of challenges to the Darwinian consensus. Fourth, that we should not be surprised by the prominence of these three themes, for they embody (in their biological specificity) the broadest underlying issues in scientific explanation, and in the nature of change and history: levels of structure and causality, rates of alteration, directions of causal flow, the possibility of causal unification by reduction to the lowest level *vs.* autonomy and interaction of irreducible levels, punctuational *vs.* gradual change, causal and temporal tiering *vs.* smooth extrapolation. Fifth, that the most interesting and important debates in our contemporary science continue to engage the same three themes, thus requiring the vista of history to appreciate the continuity and logical ordering that extends right back to Darwinian beginnings. Sixth, that our best modern understanding of the structure of evolutionary theory has reversed the harmful dichotomization of earlier debates (Darwinian fealty *vs.* destructive attempts to trivialize or overturn the mechanism of selection) by confronting the same inadequacies of strict Darwinism, but this time introducing important additions and revised formulations that preserve the Darwinian foundation, but build a theory of substantial expansion and novelty upon a retained selectionist core.

This logic and development may be defended as tolerably impersonal and universal, but any book of this length and complexity, and of so idiosyncratic a style and structure, must also own its authorial singularities. *The Structure of Evolutionary Theory* emerges, first of all, from my professional focus as a paleontologist and a student of macroevolution, defined, as explained on page 38, as descriptive phenomenology prior to any decision about the need for distinctive theory (my view) or the possibility of full subsumption under microevolutionary principles (the view of Darwin and the Modern Synthesis). The contingency of history guarantees that any body of theory will underdetermine important details, and even general flows, in the realized pageant of life's phylogeny on Earth—and such a claim for nontheoretical in-dependence of macroevolution generates no dispute, even between rigid re-ductionists who grant no separate theoretical space to macroevolution, and biologists, like myself, who envisage an important role for distinctive macro-evolutionary theory within an expanded and reformulated Darwinian view of life.

In his description of the reductionist view of classical Darwinism—his own opinion in positive support, not a simplistic caricature in opposition—Hoffman (1989, p. 39) writes: "The neodarwinian paradigm therefore asserts that this history of life at all levels—including and even beyond the level of speciation and species extinction events, embracing all macroevolutionary phenomena—is fully accounted for by the processes that operate within pop-ulations and species." I dedicate my book to refuting this traditional claim, and to advocating a helpful role for an independent set of macroevolutionary principles that expand, reformulate, operate in harmony with, or (at most) work orthogonally as additions to, the extrapolated, and persistently relevant (but not exclusive, or even dominant) forces of Darwinian microevolution.

This perspective of synergy confutes the contrary, and ultimately destruc-tive, attempts by late 19th and early 20th century macroevolutionists to de-velop substitute mechanisms that would disprove or trivialize Darwinism, and that spread such a pall of suspicion over the important search for non-reductionistic and expansive evolutionary theories—a most unfortunate (if historically understandable) trend that stifled, for several generations, the unification and fruitful expansion of evolutionary theory to all levels and temporal tiers of biology. Thus, for example, my attempt to develop a specia-tional theory of macroevolution (Chapters 8 and 9), with species treated as irreducible Darwinian individuals playing causal roles analogous to those oc-cupied by organisms in Darwinian microevolution, represents an *extension* of Darwinian styles of explanation to another hierarchical level of analysis (with interestingly different causal twists and resulting patterns), not a refuta-tion of natural selection from an alien realm. (Such a speciational theory, however, does counter Hoffman's reductionistic claim of full theoretical suf-ficiency for "processes that operate within populations and species"—for, given the stasis of species under punctuated equilibrium, such macroevolu-tionary patterns originate by higher-order sorting among stable species, and not primarily by processes occurring anagenetically within the lifetime of these higher-level Darwinian individuals.) Similarly, the different rules of cat-astrophic mass extinctions require *additions* to the extrapolated Darwinian and microevolutionary causes of phyletic patterns, but do not refute or deny the relevance of conventional uniformitarian accretions through geological time. (In fact, a more comprehensive theory that seeks to integrate the rela-tive strengths, and interestingly disparate effects, of such different levels and forms of continuationist *vs.* catastrophic causality offers greater richness to Darwinian perspectives as both underpinnings and important contributors to a larger totality.)

A second authorial input must arise from the distinctive ontogeny of past work. *The Structure of Evolutionary Theory* occupies a much broader territory than my first lengthy technical book of an earlier career, *Ontogeny and Phylogeny* (1977b). The motivating conceit of the first book rested upon my choice of a much smaller compass defined by a much clearer tradition of definition and research. I thought—thus my designation of this strategy as a conceit—that I could quote, *in extenso* and from original sources, every important statement, from von Baer and before to de Beer and after, on the relationship between development and evolution. This potential for comprehensiveness brought me much pleasure and operational motivation.

In fact, I soon realized that I could not succeed, even within this limited sphere—and I therefore punted shamelessly in the final result. I did manage to quote every important passage on the *theoretical* relationship between these central subjects of biology, but I passed, nearly completely, on the actual use of these putative relationships in specific proposals for phylogenetic reconstructions. And, as all historians of science and practitioners of evolutionary biology know, this genre of "phylogenizing" represented by far (at least by weight) the dominant expression of this theoretical rubric in the technical literature. I would, by the way, defend my decision as entirely reasonable and proper, and not merely as practically necessary, because these specific phylogenetic invocations made effectively no contribution to the development of evolutionary theory—my central concern in the book—and remained both speculative and transient to boot. But I do remember the humbling experience of realizing that a truly full coverage could only represent a pipe dream, if applied to any important subject in a vigorous domain of research!

My personal love of such thoroughness (with the necessary trade-off of limitation in domain) posed a substantial problem when.I decided to expand my range from ontogeny and phylogeny to the structure of evolutionary theory. Of all genres in scholarship, I stand most strongly out of personal sympathy with broad-brush views that attempt to encompass entire fields (the history of philosophy from Plato to Pogo, or of transportation from Noah to NASA) in a breathless summary paragraph for each of many thousand incidents. Even the most honorable efforts by great scholars—former Librarian of Congress Daniel Boorstin's *The Explorers*, for instance—make me cringe for simplistic legends repeated and interesting complexities omitted. At some level, truly important and subtle themes can only be misrepresented by such a strategy.

But how then to treat the structure of evolutionary theory in a reputable, even an enlightening, way? Surely we cannot abandon all hope for writing honorably about such broad subjects simply because the genre of comprehensive listing by executive summary must propagate more mythology and misinformation than intrigue or understanding. As a personal solution to this crucial scholarly dilemma, and in developing the distinctive strategy of this book, I employed a device that I learned by doing, through many years of composing essays—a genre that I pursued by writing comprehensive personal treatments of small details, fully documentable in the space available, but also conveying important and general principles in their cascading implications. I vowed that I would try to encompass the structure of evolutionary theory in its proper intellectual richness, but that I would do so by exhaustive treatment of well-chosen exemplifying details, not by rapid summaries of inadequate bits and pieces catalogued for all relevant participants.

Under this premise, the central task then evolves (if I may use such a metaphor) into an extended exercise in discrimination. The solution may be labeled as élitist, but how else can selection in intellectual history be undertaken? One must choose the best and the brightest, the movers and shakers by the sieve of history's harsh judgment (and not by the transiency of immediate popularity)—and let their subtle and detailed formulations stand as a series of episodes, each conveyed by an essay of adequate coverage. Luckily, the history of evolutionary thought—as one of the truly thrilling and expansive subjects of our mental lives—has attracted some of the most brilliant and fascinating doers and thinkers of intellectual history. Thus, we are blessed with more than adequate material to light the pathway of this particular odyssey in science. Luckily too, the founding figure of Darwin himself established such a clear basis of brave commitment that I could characterize, and then trace down to our own times, an essential logic that has defined and directed one of the most important and wide-ranging debates in the history of science into a coherent structure, ripe for treatment by my favored method of full coverage for the few truly central items (by knowing them through their fruits and logics, and by leaving less important, if gaudy, swatches gently aside in order to devote adequate attention to essential threads).

A third, and final, authorial distinction—my treatment of history and my integration of the history of science with contemporary research on evolutionary theory—emerges directly from this strategy of coverage in depth for a small subset of essential items and episodes. My historical treatments tend to resolve themselves into a set of mini intellectual biographies (as exemplified and defended on page 46) for almost all the central players in the history of Darwinian traditions in evolutionary thought. I can only hope that this peculiar kind of intellectual comprehensiveness will strike some readers as enlightening for the "quick entrée" thus provided into the essential work of the people who led, and the concepts that defined, the history of the greatest and most consequential revolution in the history of biological science. (In most cases—a Goethe, Cuvier, Weismann, de Vries, Fisher or Simpson, for example—I chose people for their intrinsic and transcendent excellence. In fewer instances—an Eimer or Hyatt as proponents of orthogenesis, for example—I selected eminently worthy scientists not as great general thinkers, but as best exponents of a distinctive approach to an important subject in the history of debate on essentials of evolutionary theory.)

A few figures in history have been so prescient in their principal contribution, and so acute and broad-ranging in their general perceptions, that they define (or at least intrude upon) almost any major piece of a comprehensive discussion (A. N. Whitehead famously remarked, for example, that all philosophy might be regarded as a footnote to Plato). Evolutionary biology possesses the great good fortune to embrace such a figure—Charles Darwin, of course—at the center of its origin and subsequent history. Thus, Darwin emerges again and again, often controlling the logic of discussion, throughout this book—in his own full foundational exegesis (Chapter 2); but then, in later chapters, as the principal subject, and best possible exemplification, of other important subbranches on all three boughs of his essential logic (his reluctant acceptance of higher levels of selection in Chapter 3; his formalist contrast to his own functionalism in stressing "correlations of growth" in Chapter 4; his views on direction and progress in the history of life in Chapter 6, and, even in the book's second half on modern developments, for his discussion of discordance between historical origin and current utility as a point of departure for my treatment of exaptation in Chapter 11, and his attempt to underplay and undermine mass extinction as an introduction to my critique of uniformitarianism and extrapolationism in the final Chapter 12). Who could ask for a more attractive and effective coordinating "device" to tie the disparate strands of such an otherwise disorderly enterprise together than the genial and brilliant persona of the man who first gave real substance to the grandeur in this view of life?

Whatever my dubiety about the role and efficacy of abstracts (too often, as we would all admit in honest moments, our only contact with a work that we nonetheless then feel free to criticize in full assurance of our rectitude), I cannot deny that a work of this length, imbued moreover with a tendency to penetrate byways along a basic route that seems (at least to this author) adequately linear and logical, demands some attempt to list its principal claims in textual order. Hence, I now impose upon you the following abstract:

### Chapter 2: An exegesis of the origin

1. All major pre-Darwinian evolutionary theories, Lamarck's in particular, contrasted a primary force of linear progress with a distinctly secondary and disturbing force of adaptation that drew lineages off a main line into particular and specialized relationships with immediate environments. In his most radical intellectual move, expressing both the transforming depth and the conceptual originality of the theory of natural selection, Darwin denied the existence of a primary progressive force, while promoting the lateral force of adaptation to near exclusivity. In so privileging uniformitarian extrapolation as an explanatory device, Darwin imbued natural selection, the lateral force, with sufficient power to generate evolutionary change at all scales by accumulating tiny adaptive increments through the immensity of geological time.

2. The *Origin of Species* exceeds all other scientific "classics" of past centuries in immediate and continued relevance to the basic theoretical formulations and debates of current practitioners. Careful exegesis of Darwin's logic and intentions, through textual analysis of the *Origin,* therefore assumes unusual importance for the contemporary practice of science (not to mention its undeniable historical value *in se*).

3. Darwin famously characterized the *Origin* as "one long argument" without explicitly stating "for what?" Assumptions about the focus of this long argument have ranged from the restrictively narrow (for natural selection, or even for evolution) to the overly broad (for application of the most general hypothetico-deductive model in scientific argument, as Ghiselin has claimed). I take a middle position and characterize the "long argument" as an attempt to establish a methodological approach and intellectual foundation for rigorous analysis in historical science—a foundation that could then be used to validate evolution.

4. The "long argument" for historical science operates at two poles—methodological and theoretical. The methodological pole includes a set of procedures for making strong inferences about phyletic history from data of an imperfect record that cannot, in any case, "see" past causes directly, but can only draw conclusions from preserved results of these causes. Darwin develops four general procedures, all based on one of the three essential premises of his theory's central logic: the explanation of large-scale results by extrapolation from short-term processes. In order of decreasing information available for making the required inference, these four procedures include: (1) extrapolation to longer times and effects of evolutionary changes actually observed in historic times (usually by analogy to domestication and horticulture); (2) exemplification and ordering of several phenomena as sequential stages of a single historical process (fringing reefs, barrier reefs and atolls as stages in the formation of coral reefs by subsidence of central islands, for example); (3) inference of history as the only conceivable coordinating explanation for a large set of otherwise disparate observations (consilience); and (4) inference of history from single objects based on quirks, oddities and imperfections that must denote pathways of prior change.

5. The theoretical pole rests upon the three essential components of Darwinian logic: (1) *agency,* or organismal struggle as the appropriate (and nearly exclusive) level of operation for natural selection; (2) *efficacy,* or natural selection as the creative force of evolutionary change (with complexly coordinated sequelae of inferred principles about the nature of variation, and of commitments to gradualism and adaptationism as foci of evolutionary analysis); and (3) *scope,* or extrapolationism (as described in point 4 just above). The logical coordination of these commitments, and their establishment as a brilliantly coherent and intellectually radical theory of evolution, can best be understood by recognizing that Darwin transferred the paradoxical argument of Adam Smith's economics into biology (best organization for the general polity arising as a side consequence of permitting individuals to struggle for themselves alone) in order to devise a mechanism—natural selection—that would acknowledge Paley's phenomenology (the good design of organisms and harmony of ecosystems), while inverting its causal basis in the most radical of all conceivable ways (explaining the central phenomenon of adaptation by historical evolution rather than by immediate creation, and recognizing nature's sensible order as a side consequence of unfettered struggle among individuals, rather than a sign of divine intent and benevolence).

6. The first theme of agency: Darwin's commitment to the organismal level as the effectively exclusive locus of natural selection occupies a more central, and truly defining, role than most historians and evolutionists have recognized. Invocation of this most reductionistic locus then available (in ignorance of the mechanism of inheritance) embodies the intellectual radicalism of Darwin's theory—using Adam Smith to overturn Paley, and holding that all higher-order harmony, previously attributed to divine intention, arises only as a side-consequence of selfish "struggle" for personal advantage at the lowest organismal level. Darwin devoted far more of the *Origin* to defending this organismal locus than most exegetes have acknowledged, particularly in centering his only two chapters on specific difficulties in natural selection (7 on *Instinct* and 8 on *Hybridism*) to resolutions provided by insistence upon organismal agency—explaining the establishment of adaptive sterile castes in social insects by selection upon queens as individuals, and resolving sterility in interspecific crosses as an unselected sequel of differences accumulated by organismal selection in each of two isolated populations, rather than as a direct result of higher-level species selection, as Wallace affirmed and as Darwin strove mightily and consciously to avoid. We can also trace his struggle to affirm organismal exclusivity in his reluctances, underplayings and walling off (as unique and unrepeated elsewhere in nature) of the one exception (for human altruism) that the logic of his system forced upon his preferences.

7. For his defense of the second theme of efficacy—his assertion of natural selection as the only potent source of creative evolutionary change—Darwin recognized that his weak and negative force, although surely a *vera causa* (true cause), could only play this creative role if variation met three crucial requirements: copious in extent, small in range of departure from the mean, and isotropic (or undirected towards adaptive needs of the organism). I would argue that Darwin's most brilliant intellectual move lay in his accurate identification, through the logical needs of his theory and not from any actual knowledge of heredity's mechanism, of these three major attributes of variation—because he recognized that natural selection could not otherwise operate as a creative force in the evolution of novelties.

8. Gradualism enters Darwin's system as another deductive intellectual consequence of asserting that natural selection acts as the creative mechanism of evolutionary change. Gradualism has three distinct meanings in Darwinian traditions, with only the second (or intermediate) statement relevant to the central assertion of selection's creativity. First, gradualism as simple historical continuity of stuff or information underlies the basic factuality of evolution *vs.* creation, and does not validate any particular mechanism of evolutionary change. Second, gradualism as insensible intermediacy of transitional forms specifies the Goldilockean "middle position" required by the mechanism of natural selection to refute the possibility that saltational variation might engender creative change all at once, thus relegating selection to a negative role of removing the unfit. Third, gradualism as a geological claim for slowness and smoothness (but not constancy) of rate plays a crucial role in the third theme (see point 10 of this list) of selection's scope, or the extrapolatability of microevolution to explain all patterns in geological time—and is therefore the

aspect of gradualism that punctuated equilibrium refutes (for punctuated equilibrium questions Darwin's uniformitarian and continuationist beliefs, but not his mechanism of natural selection). This parsing of three distinctly different forms of gradualism, all embraced by Darwin for different reasons, alleviates the misunderstanding behind some unfortunate terminological wrangles without substance that have generated much heat (but little light) in recent debates.

9. The adaptationist program as a primary strategy of research emerges as the third major implication of advocating natural selection as the primary creative force in evolutionary change—for this Darwinian style of evolution must proceed step by step, with each tiny increment of change rendering organisms better adapted to alterations in local environments. To summarize all the key implications of this second theme of efficacy, the creativity of natural selection makes adaptation central, isotropy of variation necessary, and gradualism pervasive.

10. Restriction of agency to the organismal level, and assertions of selection's creativity, set a biological basis for the third essential claim of Darwinian logic—selection's scope, or the argument that this incremental and gradualistic style of microevolution can, by smooth extrapolation through the immensity of geological time, build the full extent of life's anatomical change and taxonomic diversity by simple accumulation. I focus my shorter discussion of this third essential theme not upon biological needs (already covered in the first two themes), but upon the requirement for similar gradualistic styles of change in the geological stage that must present the evolutionary play—particularly in Darwin's embrace of Lyellian uniformity, and his denial of catastrophism (through arguments about the imperfection of the fossil record to allay the literal appearance of such rapidity in geological data), for even a fully consistent, intellectually sound, and operationally potent theory will not regulate actual events if surrounding conditions debar its operation.

11. I use Kellogg's brilliant approach to the evaluation of Darwinian theory (published in 1907 in anticipation of centennial celebrations for Darwin's birth and the sesquicentenary of the *Origin*) to distinguish *alternatives* that deny the fundamental postulate of selection's creativity from *auxiliaries* that enlarge, adumbrate, or reformulate the theory of natural selection in basically helpful and consistent ways. I show that Darwinism may be epitomized by its three essential claims of agency, efficacy, and scope—and that the history of debate has always centered upon these themes, with critiques focusing upon destructive alternatives or constructive auxiliaries. I argue, as the major thesis of this book, that modern debates have developed important and coherent auxiliary critiques on all three branches of essential Darwinian logic, and that these debates may lead to a fundamentally revised evolutionary theory with a retained Darwinian core.

### Chapter 3: Seeds of hierarchy

1. Nearly all scientific revolutions originate as replacements and refutations of previous explanatory schemes, not as pure additions to a former state

of acknowledged ignorance. Lamarck's evolutionary theory, known to anglo-phonic readers as a first full account through the fair but critical descriptions of Lyell (in Volume 2, 1832, of the *Principles of Geology*), and from Chambers's promotion in the *Vestiges* of 1844, provided a context for Darwin's refutation. Darwin's single-level theory, based on the full efficacy of locally adaptive changes at the smallest scale, countered the only available alternative of Lamarckism by relocating the major phenomenon that generated change and required explanation (local adaptation for Darwin, general progress for Lamarck), and (far more radically) by reversing the conventional Paleyan explanation for the good design of organisms and the harmony of ecosystems (direct divine construction at the highest level *vs.* sequelae of natural selection working at the lowest level of organismal advantage).

2. Lamarck, a dedicated materialist with a two-factor theory of evolution as a contrast between linear progress up life's ladder and tangential deflections of diversity through local adaptation, has been widely misunderstood (and reviled), both in Darwin's time and today, as a vitalist and pure exponent of "soft" or Lamarckian inheritance (which he accepted as the "folk wisdom" of his day, and invoked primarily to explain the secondary process of lateral adaptation).

3. Darwin's theory of natural selection shared a functionalist basis with Lamarck in joint emphasis upon adaptation to external environment as the instigator of evolutionary change. But the two theories differ most radically in Darwin's citation of a single locus and mechanism of change—with the full range of evolutionary results proceeding by natural selection for local adaptation of populations to changing immediate environments, and all higher-level phenomenology emerging by sequential accumulation of such tiny increments through the immensity of geological time. By contrast, Lamarck advocated a two-factor theory, with local adaptation as a merely secondary and diverging process (and, as we all know of course, arising by soft inheritance of acquired features generated by adaptive effort during an organism's life, rather than by natural selection of fortuitous variation), set against a primary process of progressive complexification up the ladder of life. Thus, Darwin embraced Lamarck's secondary force (instantiated by a different mechanism), denied the existence of Lamarck's primary force, and argued that the secondary force of local adaptation also produced the large-scale results attributed by Lamarck to the primary force. Thus, this first major debate between evolutionary alternatives contrasted Lamarck's hierarchical theory with Darwin's single-level account. Hierarchy has been an important issue from the start (although, obviously, modern versions of hierarchical selection theory, advocated as the centerpiece of this book, bear no relationship, either genealogical or ideological, to this false, but fascinating, Lamarckian original).

4. Darwin explicitly rejected Lamarck's two-factor theory, correctly identifying the disabling paradox that rendered the theory nonoperational: "what is important cannot be observed or manipulated (the higher-level force of progress), and what can be observed and manipulated (the tangential force of local adaptation) cannot explain the most important phenomenon (progress

in complexification)." Darwin developed the first testable and operational theory of evolution by locating all causality in the palpable mechanism of natural selection.

5. In the first generation of Darwinian debate, August Weismann, clearly the most brilliant theorist of his time, and the only biologist (besides Darwin) who fully grasped the logic and implications of selection, wrestled with levels of selection throughout his career, and along an interesting path, finally developing a full hierarchical theory that he explicitly identified as the most important conclusion of his later work. He began by trying to refute Lamarckian inheritance (and Herbert Spencer's vigorous defense thereof) by advocating the *Allmacht* (omnipotence, or literally "all might" or complete sufficiency) of natural selection. He first attributed the degeneration of previously useful structures (a bigger problem for Darwinism than the explanation of adaptive features) to what he called "panmixia" (not the modern meaning of the term, but the effect of recombination, in sexual reproduction, between adaptive elements and inadaptive elements no longer subject to negative selection); then realized that this process could not explain complete elimination, thus leading him to propose a lower level of subcellular selection, potentially acting in opposition to organismal selection, and called "germinal selection"; and finally recognized that if levels of selection existed below the organismal, then the same logic implies the existence and potency of supraorganismal levels as well.

6. Darwin himself provides the best 19th century example—previously unrecognized because Darwin omitted this material, originally written for the unpublished "long version," from the *Origin*—of the need for a hierarchical theory of selection in any full account of the phenomenology of evolution. Entirely consistent single-level theories cannot be carried through to completion. Darwin admitted important components of species selection in capping his (still unsatisfactory) explanation for an issue that he ranked second in importance only to explaining the anagenesis of populations by natural selection: the resolution of organic variety and plenitude by a "principle of divergence" (his terminology). I document the largely unrecognized emphasis that he placed upon this principle of divergence (for example, the *Origin*'s famous single figure does not illustrate natural selection, as generally misinterpreted, but rather the principle of divergence). Darwin struggled to explain this descriptively higher-level phenomenon of taxonomic diversification as a fully predictable consequence of ordinary organismal selection, but he could not proceed beyond an argument that he himself finally recognized as forced, and even a bit hokey: the claim that natural selection will always favor extreme variants at the tails of a distribution for a local population in a particular ecology (the *Origin*'s diagram represents an exemplification of this claim). Eventually, Darwin realized that he needed to invoke species selection for a full explanation of the success of speciose clades—and this unknown argument, rather than his well-documented defense of group selection for human altruism, represents Darwin's most generalized invocation of selection at supraorganismal levels.

7. Hierarchical models of evolutionary processes (at least descriptively so, but causally as well) have been featured and defended by evolutionary theorists from the beginning of our science, although not always by good or valid arguments. This inadequately recognized theme explains the major contrast between Lamarck and Darwin, and coordinates the various disputes between Wallace and Darwin. Wallace simply didn't grasp the concept of levels at all, and remained so committed to adaptationism that he ranged up and down the hierarchy, oblivious of the conceptual problems thus entailed, until he found a level to justify his adaptationist bent. Darwin, by contrast, completely understood the problem of levels, and the reasons behind his strong preference for a reductionist and single-level theory of organismal agency—although he reluctantly admitted a need for species selection to resolve the problem of divergence. We can also understand why Wallace's 1858 Ternate paper, sent to Darwin and precipitating the "delicate arrangement," did not proceed as far to a resolution as later tradition holds, when we recognize Wallace's conceptual confusion about levels of selection.

### Chapter 4: Internalism and laws of form: pre-darwinian alternatives

1. In a brilliant closing section to his general chapter 6, entitled "difficulties on theory," Darwin summarized the logical structure of the most important challenge to his system, and organized his most cogent defense for his functionalist theory of selection, by explicating the classical dichotomy between "unity of type" and "conditions of existence"—or the formalism of Geoffroy *vs.* the functionalism of Cuvier—entirely in selectionist terms, and to his advantage. He attributed "conditions of existence" to immediate adaptation by natural selection, and then explicated "unity of type" as constraints of inheritance of homologous structures, originally evolved as adaptations in a distant ancestor. Thus, he identified natural selection as the underlying "higher law" for explaining all morphology as present adaptation or as constraint based on past adaptation. He also admitted, while cleverly restricting their range and frequency, a few other factors and forces in evolutionary explanation.

2. A fascinating, and previously unexplored, contrast may be drawn between the strikingly similar dichotomy, although rooted in creationist explanations, of Paley's functionalist and adaptationist theory of divine construction for individualized biomechanical optimality *vs.* Agassiz's formalist theory of divine ordination of taxonomic structure as an incarnation of God's thoughts according to "laws of form" reflecting modes and categories of eternal thought. Clearly, this ancient (and still continuing) contrast between structural and functional conceptions of morphology transcends and predates any particular mechanism, even the supposedly primary contrast of creation *vs.* evolution, proposed to explain the actual construction of organic diversity.

3. In the late 18th century, the great poet (and naturalist) Goethe developed a fascinating (and, in the light of modern discoveries in evo-devo, more than partly correct) archetypal theory in the structuralist or formalist mode—and explicitly critical of functionalist, teleological and adaptationist alterna-

tives—for the diversity of organs growing off the stems and roots of plants. He viewed cotyledons, and all the standard parts of flowers (sepals, petals, stamens and carpels), as modifications of a leaf archetype.

4. The famous early 19th century argument, culminating in the public debate of 1830 between Georges Cuvier and Etienne Geoffroy St. Hilaire (and analyzed by Goethe in his final paper before his death), did not, as commonly misinterpreted, pit evolutionary theories against creationist accounts (although Geoffroy favored a limited theory of evolution, while Cuvier remained strongly opposed), but rather represented the most striking and enduring incident in this older and persistent struggle between formalist (Geoffroy) and functionalist (Cuvier) explanations of morphology and taxonomic order. Geoffroy advocated the abstract vertebra as an archetype for all animals, beginning (largely successfully) with a common basis for anatomical differences between teleosts and tetrapods, moving to the putatively common design of insects and vertebrates (still with some success, partly confirmed by the Hoxology of modern evo-devo, but also including some "howlers" like the homology of arthropod limbs with vertebrate ribs), and crashing with the proposed homology of vertebrates and a cephalopod doubled back upon itself (the comparison that sufficiently aroused Cuvier's growing ire into a call for public debate). Geoffroy's theory of dorso-ventral inversion between insects and vertebrates was not a silly evolutionary conjecture about "the worm that turned" (as later caricatures often portray), and did not represent an evolutionary explanation at all, but rather expressed a formalist comparison based upon a common underlying structure, ecologically oriented one way in vertebrates (central nervous system up), and the other way in arthropods. The common impression of Cuvier's victory must be reassessed as a complex "draw," with Geoffroy's position abetted by the fortuity of his longer life and his courting of prominent literary friends as supporters (including Balzac and Georges Sand).

5. Adaptationist preferences have enjoyed a long anglophonic tradition, beginning with the treatises of Ray and Boyle, in Newton's founding generation, on final causes; then extending, in creationist terms, through Paley and the Bridgewater Treatises; and finally culminating in the radically reversed evolutionary explanations (but still retaining the same functionalist and adaptationist commitments) of Darwin, extending forward to Fisher and the Modern Synthesis. By contrast, continental traditions have favored formalist and structuralist explanations of morphology, from the creationist accounts of Agassiz, through the transitional systems of Goethe and Geoffroy, to the fully evolutionary accounts of Goldschmidt and Schindewolf in the mid 20th century. Interestingly, the complex views of Richard Owen, so widely misunderstood as an opponent of evolution (when he only rejected the predominant functionalism of traditional British approaches to morphology), may best be grasped when we understand him as a rare anglophonic exponent of a predominantly formalist theory. Owen, following Geoffroy, tried to explain the entire vertebrate skeleton, including the skull and limbs, as a set of modifications upon a vertebral archetype.

6. Darwin maintained a genuine interest in formalist constraints upon

adaptationist optimality for individualized features of anatomy—a theme that he epitomized as "correlations of growth." But he developed an explicit framework and rationale, most thoroughly discussed not in the *Origin* but in his longest 1868 book on *The Variation of Animals and Plants Under Domestication,* that relegated such formalist effects to a clearly subservient and secondary status, compared with natural selection and adaptation, in evolutionary causality.

### Chapter 5: Channels and saltations in post-Darwinian formalism

1. Galton's Polyhedron, the metaphor and model devised by Darwin's brilliant and eccentric cousin Francis Galton, and then fruitfully used by many evolutionary critics of Darwinism, including St. George Mivart, W. K. Brooks, Hugo de Vries, and Richard Goldschmidt, clearly expresses the two great, and both logically and historically conjoined, themes of formalist (or structuralist, or internalist, in other terminologies) challenges to functionalist (or adaptationist, or externalist) theories in the Darwinian tradition. This model of evolution by facet-flipping to limited possibilities of adjacent planes in inherited structure stresses the two themes—channels set by internal constraint, and evolutionary transition by discontinuous saltation—that structuralist alternatives tend to embrace and that pure Darwinism must combat as challenges to basic components of its essential logic (for channels direct the pathways of evolutionary change from the inside, albeit in potentially positive and adaptive ways, even though some external force, like natural selection, may be required as an initiating impulse; whereas saltational change violates the Darwinian requirement for selection's creativity by vesting the scope and direction of change in the nature and magnitude of internal jumps, and not in sequences of adaptive accumulations mediated by natural selection at each step).

2. Orthogenesis, as a general term for evolutionary directionality along channels of internal constraint, rather than external pathways of natural selection, existed in several versions, ranging from helpful auxiliaries to Darwinism, to outright alternatives that denied any creative potency to selection. Theodor Eimer, who coined the term orthogenesis, presented a middle version that tried to integrate internal channels of orthogenesis with external pathways of functionalist determination. But Eimer defended Lamarckian mechanics for his functionalism, thus leading him to oppose natural selection (he spoke of the *Ohnmacht,* or "without power," of selection, contrasted with Weismann's *Allmacht,* or "all power") despite his pluralistic linkage of formalist and functionalist explanations.

3. The orthogenetic theory of the late 19th century American paleontologist Alpheus Hyatt embodied maximal opposition to natural selection, and must be viewed as alternative, rather than auxiliary, to Darwinism. Hyatt conceived the pathway of ontogeny, modified only by heterochronic changes permitted under the biogenetic law, as the internal directing channel that natural selection could tweak, but not derail. Illustrating the influence of theory over perception, Hyatt found several parallel lineages of snails, running along

different segments of a common pathway, but all supposedly living in an identical environment—where others had reconstructed typical Darwinian monophyletic trees of phylogeny from the same stratigraphic section of freshwater planorbids. Hyatt, who engaged in a long and ultimately frustrating correspondence with Darwin on this subject, believed that lineages followed a preordained "ontogeny" of phyletic youth, maturity and old age, thus attributing the different internal responses of lineages living in the same environment to their residence in different stages of an ontogenetically fixed and shared phyletic pathway (a preset internal channel with a vengeance).

4. Charles Otis Whitman, a great early 20th century American naturalist, developed the most congenial auxiliary theory (to Darwinism) of orthogenesis in his extensive work on the evolution of color patterns in Darwin's own favorite organism, the domestic pigeon. Whitman argued that domestic pigeons in particular, and dove-like birds in general, followed a strong channel of internal predisposition leading in one direction from checkers to bars, and eventually to the obliteration of all color. (Darwin, by interesting contrast, argued for a reverse tendency from bars to checkers, but also held, as his basic theory obviously implies, that selection largely determines any particular event and that no internal predisposition can trump the dictates of immediate function.)

5. In his 1894 book on *Materials for the Study of Variation* (where he coined the term homeosis), William Bateson presented an extensive catalog of cases in discontinuous variation among individuals in a population and between populations of closely related organisms. He used these examples to develop a formalist theory of saltational evolution, strongly opposed to the adaptationist assumptions of Darwinian accounts. (Bateson's acerbic criticisms of adaptationist scenario-building and story-telling in the speculative mode emphasize a common linkage between structuralist preferences for mechanical explanation, and distaste for the adaptationist assumption that functional necessity leads and the evolution of form follows.) Although Bateson coined the term genetics, his personal commitment to a "vibratory" theory of heredity, based on physical laws of classical mechanics—an intuition that he could never "cash out" as a testable theory—prevented his allegiance to the growing influence of Mendelian principles.

6. Hugo de Vries, the brilliant Dutch botanist who understood the logic of selectionism so thoroughly and acutely (but largely in contrast with the only other biologists, Weismann and Darwin himself, who also grasped all the richness and range of implications, but with favor), developed a saltational theory of evolution, but explicitly denied any predisposition of lineages to follow internal channels of constraint. (He thus showed the potential independence of the frequently-linked formalist themes of channeling and saltation, a conjunction espoused by Bateson and Goldschmidt for example, but denied in the other direction by Whitman, who favored channeling but denied saltation by supporting a gradualist theory of orthogenetic change.) This fascinating scholar regarded Darwin as his intellectual hero and never forgot the kindness and encouragement conveyed by his mentor and guru during

their one personal meeting early in de Vries's career. But de Vries, who developed the theory of intracellular pangenesis (the ultimate source for the term "gene") in the late 19th century, and then (quite fortuitously and long after he had reached saltational conclusions for other reasons) became one of Mendel's rediscoverers, based his truly saltational theory of immediate macromutational origin of species on his work with the evening primrose, *Oenothera lamarckiana,* where he mistook an odd chromosomal organization that generates occasional saltations for a biological generality. De Vries, who understood the logic of selectionism so well, who knew that his macromutational theory refuted several essential components of Darwinian logic, but who could not bear (for largely psychological reasons) to forsake his intellectual and personal hero, insisted upon his larger fealty to Darwin, even though he had banned Darwinian mechanisms from the master's own realm of the origin of species. So de Vries developed a hierarchical theory that, while denying selection for the origin of species, restored selectionist logic at the higher level of phyletic trends by explicitly proposing "species selection" (his term) as a mechanism for generating broader phylogenetic patterns.

7. By proposing a comprehensive formalist theory in the heyday of developing Darwinian orthodoxy, Richard Goldschmidt became the whipping boy of the Modern Synthesis—and for entirely understandable reasons. Goldschmidt showed his grasp, and his keen ability to utilize, microevolutionary theory by supporting this approach and philosophy in his work on variation and intraspecific evolution within the gypsy moth, *Lymantria dispar.* But he then expressed his apostasy by advocating discontinuity of causality, and proposing a largely nonselectionist and formalist account for macroevolution from the origin of species to higher levels of phyletic pattern. Goldschmidt integrated both themes of saltation (in his concept of "systemic mutation" based on his increasingly lonely, and ultimately indefensible, battle to deny the corpuscular gene) and channeling (in his more famous, if ridiculed, idea of "hopeful monsters," or macromutants channeled along viable lines set by internal pathways of ontogeny, sexual differences, *etc.*). The developmental theme of the "hopeful monster" (despite its inappropriate name, virtually guaranteed to inspire ridicule and opposition), based on the important concept of "rate genes," came first in Goldschmidt's thought, and always occupied more of his attention and research. Unfortunately, he bound this interesting challenge from development, a partially valid concept that could have been incorporated into a Darwinian framework as an auxiliary hypothesis (and now has been accepted, to a large extent, if under different names), to his truly oppositional and ultimately incorrect theory of systemic mutation, therefore winning anathema for his entire system. Goldschmidt may have acted as the architect of his own undoing, but much of his work should evoke sympathetic attention today.

### Chapter 6: Pattern and progress on the geological stage

1. Darwin based his argument for a broad and general vector of progress in life's history not on the "bare bones" operation of natural selection (where he

had explicitly denied such an outcome as the most radical implication of his theory), but on subsidiary ecological claims for the predominance of biotic over abiotic competition, and for a geological history of plenitude in a persistently crowded ecological world, where one species must displace another to gain entry into ecosystems (the metaphor of the wedge). Darwin used these ecological sequelae, along with the gradualist and incrementalist logic of natural selection itself, as primary justifications for his third essential claim of selection's scope, or the uniformitarian extension of small-scale microevolution, in a smoothly continuationist manner, to explain all patterns of macroevolution by accumulation of increments through the immensity of geological time.

2. Such a claim requires that the geological stage operate in an appropriate, and "Goldilockean," manner—not too much change to debar the operation and domination of this slowly and smoothly accumulative biological mode, and not too little to provide insufficient impetus (within Darwin's externalist and functionalist theory) for attributing the amount of change actually observed to natural selection.

3. The primary claim of "too much" derived from the school of "catastrophism" in geology—a movement that has been unfairly stigmatized by later history, following Lyell's successful and largely rhetorical mischaracterization (he was a lawyer by profession), as an unscientific defense of supernaturalism to cram the observed results of geology into the strictures of biblical chronology, but that actually took the opposite position of strict empirical literalism (whereas uniformitarians argued that the numerous literal appearances of rapidity in the geological record must be "interpreted" as misleading consequences of how gradual change must be expressed in a woefully imperfect set of strata). The great catastrophist Cuvier, in particular, was an Enlightenment rationalist, not a theological apologist—and he based his defense of catastrophism upon his literalist reading of the paleontological and geological record.

4. The primary claim of "too little" geology followed Lord Kelvin's increasingly diminished estimates for the age of the earth (incorrectly made—although Kelvin accurately described the necessary, but (as it turned out) empirically false, logic required to validate his views—by assuming that heat now flowing from the earth represented a continuing loss from an originally molten state). Darwin worried intensely over Kelvin's claims, even referring to him as an "odious spectre" in a letter to Wallace. Darwin feared that Kelvin's low estimates would not permit enough time to generate the history of life under his slowly acting theory of gradualistic and accumulative change. Although this story has been told often, and has become familiar to scientists, an important (and decisive) aspect of the tale has rarely been exposed: Darwin fought this battle alone, and his strong distress illustrates the maximal, and unique, extent of his gradualistic and continuationist commitments. His closest colleagues, Wallace and Huxley, did not find Kelvin's low estimates unacceptable, but argued that we had only been led to expect such slow change from our previous conception of the earth's age, and that faster rates

of phyletic change, implied by Kelvin's dates, were entirely acceptable under their reading of evolution.

### Chapter 7: The modern synthesis as a limited consensus

1. From the anarchic situation that prevailed at the Darwinian centennial celebrations of 1909 (confidence in the factuality of evolution, linked with agnosticism about theories and mechanics, as the first fruits of Mendelism seemed, initially, to refute the gradualism and incrementalism of natural selection), the Modern Synthesis eventually emerged in two stages (following the union of Darwinian and Mendelian perspectives in the work of Fisher and others): first, by a welcome restriction that eliminated Kellogg's three alternatives in oppositional modes that would have destroyed Darwinism (Lamarckism as a substitute functionalism, and saltationism and orthogenesis as formalist alternatives), and reasserted, now in a context of Mendelian particulate inheritance, the adequacy of natural selection as a creative force; and second, by an increasingly dubious hardening, culminating in centennial celebrations for the *Origin* in 1959, that substituted an increasingly rigid adaptationism for an earlier pluralism that embraced all mechanisms (including genetic drift) consistent with known genetic principles, while favoring selection as a primary force.

2. In his founding book of 1930, *The Genetical Theory of Natural Selection*, R. A. Fisher showed how slow, gradualist evolution in large, panmictic populations (treated almost as an ahistorical system, analogous to effectively infinite populations of identical gas molecules free to move and diffuse by physical principles) could validate strict Darwinism under Mendelian particulate inheritance (with Darwin's own acceptance of blending inheritance exposed as a more serious impediment than Darwin himself had realized), and disprove saltational alternatives by the inverse correlation of frequency and magnitude in variation. To these mathematical and general chapters, Fisher appended a long closing section devoted to his eugenical theory that Western society had begun to degenerate seriously as a consequence of the social promotion of infertility (the rise in class level of "good" genetic stock, largely by their correlated tendency to have fewer children, thereby husbanding their economic resources to potentiate their social elevation). Fisher conceived this eugenical "blight" as entirely Darwinian in character—invisible in its gradual expression generation by generation, but ultimately more deadly than the explicit saltational degenerations stressed by most eugenicists.

3. In contrast with the initial pluralism of Haldane and Huxley (in the book that coined the Modern Synthesis), and of the first editions of founding documents for the second phase of the Synthesis (Dobzhansky's 1937 *Genetics and the Origin of Species*, Mayr's 1942 *Systematics and the Origin of Species*, and Simpson's 1944 *Tempo and Mode in Evolution*), later editions of these three documents encapsulated the hardening of this second phase, as initial pluralism yielded to an increasingly firm and exclusive commitment to adaptationist scenarios, and to natural selection as a virtually exclusive mechanism of change. Even Sewall Wright's views on genetic drift and shifting bal-

ance altered from initial stress upon stochastic alternatives to selection to an auxiliary role for drift (as an impetus for the exploration of new, and potentially higher, adaptive peaks) as one aspect of a more inclusive and basically adaptationist process. The complex reasons for this hardening include some empirical documentations of selection, but also involve a set of basically social and institutional factors not based on increasing factual adequacy.

4. If this hardening on the second Darwinian branch of selection's efficacy reflects a general trend within evolutionary theory, then we should find a similar Darwinian strengthening (and narrowing) on the other two branches of selection's agency (organismal *vs.* higher levels) and scope (adequacy to explain the entire geological record by extrapolated microevolution). The triumph (for good reasons at the time) of Williams over Wynne-Edwards affirms this trend for agency, although Williams's important clarification then unfortunately hardened (among epigones) into a dogmatic and *a priori* rejection of any hint of group selection. Similarly, the Synthesis's increasing confidence in the exclusivity of gradualistic microevolution deprived paleontology of any independent theoretical space, and relegated the field to documentation of an admittedly underdetermined pageant, built by the exclusive agency of microevolutionary principles. Several synthesists even denied the efficacy of differential speciation as an input to macroevolutionary pattern (branding the speciosity of some clades as a "luxury" rather than a crucial input to survival and flourishing), and attributed all higher-level change to extensions of gradualistic and adaptive anagenesis within unbranched lineages.

5. The trends to development, initial pluralism and later hardening of the Modern Synthesis win clearest expression in two sources of data: comparison of statements by leading scientists at the two contrasting centennial celebrations of 1909 and 1959 (for Darwin's birth and for the publication of the *Origin*); and by documentation of hardening in the summary statements (and increasingly dogmatic dismissal of alternatives) in leading textbooks for secondary and undergraduate courses in biology.

### Chapter 8: Species as individuals in the hierarchical theory of selection

1. Selectionist mechanics, in the most abstract and general formulation, work by interaction of individuals and environments (broadly construed to include all biotic and abiotic elements), such that some individuals secure differential reproductive success as a consequence of higher fitness conferred by some of their distinctive features, leading to differential plurifaction of individuals with these features (relative to other individuals with contrasting features), thus gradually transforming the population in adaptive ways. But the logic of this statement implies that organisms cannot be the only biological entities that manifest the requisite properties of Darwinian individuality—properties that include both vernacular criteria (definite birth and death points, sufficient stability during a lifetime, to distinguish true entities from unboundable segments of continua), and more specifically Darwinian criteria (production of daughters, and inheritance of parental traits by daughters). In

particular, and by these criteria, species must be construed not only as classes (as traditionally conceived), but also as distinct historical entities acting as good Darwinian individuals—and therefore potentially subject to selection. In fact, a full genealogical hierarchy of inclusion—with rising levels of genes, cell lineages, organisms, demes, species and clades—features clearly definable Darwinian individuals, subject to processes of selection, at each level, thus validating (in logic and theory, but not necessarily in the potency of actual practice in nature) an extension and reformulation of Darwin's exclusively organismal theory into a fully hierarchical theory of selection.

2. The validity of the "interactor approach" to defining the mechanics of selection, and the fallacy of the "replicator approach" expose, as logically invalid, all modern attempts to preserve Darwinian exclusivity of level, but to offer an even more reductionistic account in terms of genes, rather than organisms, as agents—with organisms construed as passive containers for the genes that operate as exclusive agents of natural selection. This false argument, based upon the true but irrelevant identification of genes as faithful replicators, must be replaced by the conceptually opposite formulation of a hierarchical theory of selection, with genes identified as only one valid, and lowest, level in a hierarchy of equally potent, and interestingly different, levels of Darwinian individuality: genes, cell lineages, organisms, demes, species and clades. Replication identifies a valid and important criterion for the crucial task of bookkeeping or tracing evolutionary change; but replicators cannot specify the causality of selectionist processes, which must be based upon the recognition and definition of interactors with environments. Even Williams and Dawkins, the two leading exponents of exclusive gene selectionism, have acknowledged and properly described the hierarchical causality of interaction (while proferring increasingly elaborate and implausible verbal defenses of gene selection in arguments about parallel hierarchies and Necker cubing of legitimate alternatives rooted in criteria of replication vs. interaction). Thus, Williams and Dawkins seem to grasp the validity of hierarchical selection through a glass darkly, while still trying explicitly to defend their increasingly indefensible preferences for exclusive gene selectionism.

3. The logic of hierarchical selection cannot be gainsaid, and even Fisher admitted the consistency, even the theoretical necessity, while denying the empirical potency, of species selection. Fisher based his interesting and powerful argument on his assumption that low $N$ for species in clades (relative to organisms in populations) must debar any efficacy for species selection in a world of continuous and gradualistic anagenesis rooted in organismal selection. However, Fisher's argument, although logically tight, fails empirically because species tend to be stable and directionally unchanging (however fluctuating) during their geological lifetimes, and the theoretically "weaker" force of species selection may therefore operate as the "only game in town" for macroevolution. The arguments for potency of species selection are stronger than corresponding assertions for interdemic selection (largely because species actively maintain their boundaries as Darwinian individuals, whereas demes remain subject to breakup and invasion). But, despite these intrinsic

weaknesses and problems, interdemic selection has now been empirically validated as an important force in evolution—thus strengthening a *prima facie* case for the even greater importance of species selection in macroevolution.

4. Two theoretical resolutions and clarifications have established both a sound theoretical basis, and a strong argument for the empirical potency, of species selection as an important component of macroevolution: first, the recognition of differential proliferation rather than downward effect as the most operational criterion for defining and recognizing species selection; second, the acknowledgment that emergent fitnesses under the interactor approach, rather than emergent features treated as active adaptations of the species, constitute the proper criterion for identifying species selection. The former insistence upon emergent features (by me and other researchers, and in error), while logically sound and properly identifying a small subset of best and most interesting cases, relegated the subject to infrequent operational utility, and thus to relative impotence. The proper criterion (under the interactor approach) of emergent fitness universalizes the subject by permitting general identification in the immediacy of the current mechanics of selection, and not requiring knowledge—often unavailable given the limits of historical archives—of adaptive construction and utility in ancestral states.

5. The six levels recognized for convenience, and not accompanied by any claim of completion or exclusivity—gene, cell lineage, organism, deme, species and clade—feature two important principles that make the theory of hierarchical selection so different from, while still in the lineage and tradition of, exclusivistic Darwinian organismal selection. First, adjacent levels may interact in the full range of conceivable ways—in synergy, orthogonally, or in opposition. Opposition has been stressed in the existing literature, but only because this mode is easier to recognize, and not for any argument of greater importance in principle. Second, the levels operate non-fractally, with fascinating and distinguishing differences in mode of functioning, and relative importance of components, for each level. For example, the different mechanisms by which organisms and species maintain their equally strong individuality dictate that selection should dominate at the organismal level, while selection, drift, and drive should all play important and balanced roles at the species level.

6. To cite just one difference (from conventions of the organismal level) for each nonstandard level, and to make the key point about distinctiveness of levels in an almost anecdotal manner: random change may be most prominent in relative frequency at the level of the gene-individual; true gene selection also plays an important, if limited, role (largely in the mode that has been given the unfortunate name—for its implication of opposition, almost in ethical terms, to the supposed standard of proper organismal selection—of "selfish DNA"); however, the Dawkinsian argument for exclusivity of genic selection only records the confusion of a preferred level of bookkeeping with an erroneous claim for a privileged locus of selection. Selection among cell-lineages, although ancestrally important in the evolution of multicellular organisms, has largely been suppressed by the organismal level in the interests

of its own integrity; failure of this suppression leads to the pyrrhic victory of cell-lineages that we call cancer. Interdemic selection, although once so widely rejected, probably plays an essential role in the evolution of social cooperation in general, and not only for such specific phenomena as human altruism. Species-level selection, combined with other species-level properties of drive and drift, establishes the independent basis for a distinctive speciational theory and reformulation of macroevolution. The highest level of clade selection, although sometimes operative, may be relatively weak by an extension of Fisher's argument about low $N$.

7. I explore the distinctive differences between levels of selection by trying to exemplify and "play out" the detailed disparities in a "grand analogy" between the conventional operation of organismic selection and the relative conceptual novelty of species selection. As an idiosyncratic sample of potential reforms and surprises, consider the following claims: First, the formulation of a general taxonomy for sources of change in hierarchically ordered systems, based on a primary distinction of "drive" for directed changes arising within an individual, based on change among lower-level individuals as constituent parts; and "sorting," with two causally distinct subcategories of "selection" and "drift" for change based on alterations of relative frequencies among individuals at the focal level itself. Second, the recognition, by following the logic of the analogy, of some strikingly counterintuitive comparisons that become both interesting and revealing upon subsequent reflection—including the likeness of Lamarckian change, construed as ontogenetic drive at the organismal level, with standard anagenetic transformation as organismal drive at the species level (transformation by directional change of constituent parts of a higher-level individual, in this case the organisms of a species); this similarity may also highlight the rather different reasons for general unimportance of both levels of drive—Lamarckism for the well-known reason of theoretical non-occurrence in a Mendelian world, and anagenesis based on the controversial claim for its evident plausibility in theory (as a basic Darwinian process), but rarity in fact, given the dominant relative frequency of punctuated equilibrium. Third, the establishment of a framework for distinguishing directional speciation as a form of reproductive drive (inherently biased differences in autapomorphies of descendant species *vs.* ancestral states) from true species selection as a higher order sorting among daughter species that arise with phenotypic differences randomly distributed about parental means. I believe that we have missed this crucial distinction because the analog of directional speciation at the organismal level—drives induced by mutation pressure—occur so rarely (for conventional reasons of organismal selection's power to suppress them) that we haven't considered the greater potency of analogous processes at other levels. Fourth, the importance of testing "Wright's Rule"—the claim that speciation is random with respect to the direction of evolutionary trends within clades—because the major alternative of directional speciation as the cause of trends holds such potential power at the species level, whereas its analog (drives of mutation pressure) assumes so little importance at the organismal level. Fifth, the potentially far greater im-

portance of drift (both species drift and founder drift) *vs.* selection as a mechanism of sorting at the species level, but not at the organismal level, where selection predominates in standard formulations. Sixth, the identification of an intrinsically, and probably unbreakable (in most cases), negative correlation between speciation and extinction propensities as the primary constraint operating to prevent the takeover of life by a few megaclades (which might dominate by enhancing speciation while retarding extinction among constituent species—or perhaps the Coleoptera have prevailed by this means). Seventh, the recognition that the organismal level operates uniquely in securing the integrity of its individuals by devices (physiological homeostasis among organs, and spatial bounding by an external surface) that "clear out" both drive from below and drift at its own level as mechanisms operating at high relative frequency—thus leaving selection in its most dominant position at this level. Perhaps our Darwinian prejudice for regarding selection as by far the most effective, or virtually the only important, process of evolutionary change arises more from the parochialism of our organismal focus (given our own personal residence in this category) than from any universal characterization of all levels in evolution.

### Chapter 9: Punctuated equilibrium and the validation of macroevolutionary theory

1. The clear predominance of an empirical pattern of stasis and abrupt geological appearance as the history of most fossil species has always been acknowledged by paleontologists, and remains the standard testimony (as documented herein) of the best specialists in nearly every taxonomic group. In Darwinian traditions, this pattern has been attributed to imperfections of the geological record that impose this false signal upon the norm of a truly gradualistic history. Darwin's argument may work in principle for punctuational origin, but stasis is data and cannot be so encompassed.

2. This traditional argument from imperfection has stymied the study of evolution by paleontologists because the record's primary (and operational) signal has been dismissed as misleading, or as "no data." Punctuated equilibrium, while not denying imperfection, regards this signal as a basically accurate record of evolution's standard mode at the level of the origin of species. In particular, before the formulation of punctuated equilibrium, stasis had been read as an embarrassing indication of absence of evidence for the desired subject of study—that is, of data for evolution itself, falsely defined as gradual change—and this eminently testable, fully operational, and intellectually fascinating (and positive) subject of stasis had never been subjected to quantitative empirical study, a situation that has changed dramatically during the last 25 years.

3. The key empirical ingredients of punctuated equilibrium—punctuation, stasis, and their relative frequencies—can be made testable and defined operationally. The theory only refers to the origin and development of species in geological time, and must not be misconstrued (as so often done) as a claim for true saltation at a lower organismal level, or for catastrophic mass extinc-

tion at a higher faunal level. Punctuation must be scaled relative to the later duration of species in stasis, and we suggest 1–2 percent (analogous to human gestation *vs.* the length of human life) as an upper bound. Punctuated equilibrium can be distinguished from other causes of rapid change (including anagenetic passage through bottlenecks and the traditional claim of imperfect preservation for a truly gradualistic event) by the criterion of ancestral survival following the branching of a descendant. Punctuations can be revealed by positive evidence (rather than inferred from compression on a single bedding plane) in admittedly rare situations, but not so infrequent in absolute number, of unusual fineness of stratigraphic resolution or ability to date the individual specimens of a single bedding plane. Stasis is not defined as absolute phenotypic immobility, but as fluctuation of means through time at a magnitude not statistically broader than the range of geographic variation among modern populations of similar species, and not directional in any preferred way, especially not towards the phenotype of descendants. Punctuated equilibrium will be validated, as all such theories in natural history must be (including natural selection itself), by predominant relative frequency, not by exclusivity. Gradualism certainly can and does occur, but at very low relative frequencies when all species of a fauna are tabulated, and when we overcome our conventional bias for studying only the small percentage of species qualitatively recognized beforehand as having changed through time.

4. Punctuated equilibrium emerges as the expected scaling of ordinary allopatric speciation into geological time, and does not suggest or imply radically different evolutionary mechanisms at the level of the origin of species. (Other proposed mechanisms of speciation, including most sympatric modes, envision rates of speciation even faster than conventional allopatry, and are therefore even more consistent with punctuated equilibrium.) The theoretically radical features of punctuated equilibrium flow from its proposals for macroevolution, with species treated as higher-level Darwinian individuals analogous to organisms in microevolution.

5. The difficulty of defining species in the fossil record does not threaten the validity of punctuated equilibrium for several reasons. First, in the few studies with adequate data for genetic and experimental resolution, paleospecies (even for such difficult and morphologically labile species as colonial cheilostome bryozoans) have been documented as excellent surrogates, comparable as units to conventional biospecies. Second, the potential underestimation of biospecies by paleospecies only imposes a bias that makes punctuated equilibrium harder to recognize. The fossil record's strongly positive signal for punctuated equilibrium, in the light of this bias, only increases the probability of the pattern's importance and high relative frequency. Third, the potential overestimation of biospecies by paleospecies is probably false in any case, and also of little practical concern because no paleontologist would assert punctuated equilibrium from the evidence of oversplit taxa in faunal lists, but only from direct biometric study of stasis and punctuation in actual data.

6. We originally, and probably wrongly, tried to validate punctuated equilibrium by asserting that, in principle, most evolutionary change should be

concentrated at events of speciation themselves. Subsequent work in evolutionary biology has not confirmed any *a priori* preference for concentration in such episodes. Futuyama's incisive macroevolutionary argument—that realized change will not become geologically stabilized and conserved unless such change can be "tied up" in the unalienable individuality of a new species—offers a far richer, far more interesting, and theoretically justified rationale for correlating episodes of evolutionary change with speciation.

7. Section III presents a wide-ranging discussion of why proposed empirical refutations of punctuated equilibrium either do not hold in fact, or do not bear the logical weight claimed in their presentation. Refutations for single cases are often valid, but do not challenge the general hypothesis because we anticipate a low relative frequency for gradualism, and these cases may reside in this minor category. Claims for predominant gradualism in the entire clade of planktonic forams may hold as exceptional (although, even here, the majority of lineages remain unstudied, in large part because they seem, at least subjectively, to remain in stasis, and have therefore not attracted the attention of traditional researchers, who wish to study evolution, but then equate evolution with gradualism). However, in these asexual forms with vast populations, gradualism at this level may just represent the expected higher-level expression of punctuational clone selection, as Lenski has affirmed in the most thorough study of evolution in a modern bacterial species—and just as gradual cladal trends in multicellular lineages emerge as the expected consequences of sequential punctuated equilibrium at the species level (trends as stairsteps rather than inclined planes, so to speak). Claims for genetic gradualism do not challenge punctuated equilibrium, and may well be anticipated as the proper expression at the genic level (especially given the high relative frequency of random nucleotide substitutions) of morphological stasis in the phenotypic history of species. Punctuated equilibrium has done well in tests of conformity with general models, particularly in the conclusion that extensive polytomy in cladistic models may arise not only (as usually interpreted) from insufficient data to resolve a sequence of close dichotomies, but also as the expectation of punctuated equilibrium for successive branching of daughter species from an unchanged parental form in stasis. In fact, the frequency of polytomy *vs.* dichotomy may be used as a test for the relative frequency of punctuated equilibrium in well resolved cladograms—a test well passed in data presented by Wagner and Erwin.

8. Section IV then summarizes the data on empirical affirmations of punctuated equilibrium, first on documented patterns of stasis in unbranched lineages; second on punctuational cladogenesis affirmed by the criterion of ancestral survival; third on predominant relative frequencies for punctuated equilibrium in entire biotas (with particularly impressive affirmations by Hallam, Kelley, and Stanley and Yang for mollusks; and by Prothero and Heaton for Oligocene Big Badlands mammals, where a study of all taxa yielded 177 species that followed the expectations of punctuated equilibrium and three cases of potential gradualism, only one significant); fourth on predominant relative frequencies for punctuated equilibrium in entire clades, with empha-

sis on Vrba's antelopes and, especially, Cheetham's rigorously quantitative and multivariate data of evolution in the bryozoan genus *Metrarabdotos,* perhaps the best documented and most impressive case of exclusive punctuated equilibrium ever developed. Finally, we can learn much from variation in relative frequencies among taxa, times, and environments—and interesting inferences have been drawn from recorded differences, particularly in Sheldon's counterintuitive linkage of stasis to rapidly changing, and gradualism to stable, environments.

9. Among many reasons proposed to explain the predominance of stasis, a phenomenon not even acknowledged as a "real" and positive aspect of evolution before punctuated equilibrium gave it some appropriate theoretical space, habitat tracking (favored by Eldredge), constraints imposed by the nature of subdivided populations (favored by Lieberman), and normalizing clade selection (proposed by Williams) represent the most novel and interesting proposals.

10. Among the implications of a predominantly punctuational origin of stable species-individuals for macroevolutionary theory, we must rethink trends (the primary phenomenon of macroevolution, at least in terms of dedicated discussion in existing literature) as products of the differential success of certain kinds of species, rather than as the adaptive anagenesis of lineages—a radical reformulation with consequences extending to a new set of explanations no longer rooted (as in all traditional resolutions) in the adaptive advantages conferred upon organisms, but potentially vested in such structural principles as sequelae (by hitchhiking or as spandrels) of fortuitous phenotypic linkage to higher speciation rates of certain taxa. In further extensions, macroevolution itself must be reconfigured in speciational terms, with attendant implications for a wide range of phenomena, including Cope's rule (structurally ordained biases of speciation away from a lower size limit occupied by founding members of the clade, rather than adaptive anagenesis towards organismal benefits of large size), living fossils (members of clades with persistently minimal rates of speciation, and therefore no capacity for ever generating much change in a speciational scheme, rather than forms that are either depauperate of variation, or have occupied morphological optima for untold ages), and reinterpretation of cladal trends long misinterpreted as triumphs of progressive evolution (and now reevaluated in terms of variational range in species numbers, rather than vectors of mean morphology across all species at any time—leading, for example, to a recognition that modern horses represent the single surviving twig of a once luxurious, and now depleted, clade, and not the apex of a continually progressing trend). By the same argument, generalized to all of life, we understand the stability and continued domination of bacteria as the outstanding feature of life's history, with the much vaunted progress of complexity towards mammalian elegance reinterpreted as a limited drift of a minor component of diversity into the only open space of complexity's theoretical distribution. But, to encompass this reformulation, we need to focus upon the diversity and variation among life's species, not upon the supposed vectors of its central tendencies, or even its pe-

ripheral superiorities. Hominid evolution must also be rethought as reduction of diversity to a single species of admittedly spectacular (but perhaps quite transient) current success. In addition, the last 50,000 years or more of human phenotypic stability becomes a theoretical expectation under punctuated equilibrium, and not the anomaly so often envisaged (and attributed to the suppression of natural selection by cultural evolution) both by the lay public and by many professionals as well.

11. Further extensions of punctuated equilibrium include the controversial phenomenon of "coordinated stasis," or the proposition that entire faunas, and not merely their component species, tend to remain surprisingly stable in composition over durations far longer than any model based on independent behavior of species (even under punctuated equilibrium) would allow, although other researchers attribute the same results to extended consequences of sudden external pulses and resulting faunal turnovers, while still others deny the empirics of coordination and continue to view species as more independent, one from the other, even in the classical faunas (like the Devonian Hamilton Group) that serve as "types" for coordinated stasis.

12. Punctuated equilibrium has inspired several attempts, of varying success in my limited judgment, to construct mathematical models (or to simulate its central phenomena in simple computer systems of evolving "artificial life") that may help us to identify the degree of generality in modes of change that this particular biological system, at this particular level of speciation, exemplifies and records. Punctuated equilibrium has also proved its utility in extension by meaningful analogy (based on common underlying principles of change) to the generation of punctuational hypotheses at other levels, and for other kinds of phenomena, where similar gradualistic biases had prevailed and had stymied new approaches to research. These extensions range from phyletic and ecological examples below the species level to interesting analogs of both stasis and punctuation above the species level. Non-trending, the analog of stasis in large clades, for example, had been previously disregarded—following the same fate as stasis in species—as a boring manifestation of non-evolution, but has now been recognized and documented as a real and fascinating phenomenon in itself. Punctuational analogs have proven their utility for understanding the differential pace of morphological innovation within large clades, and for resolving a variety of punctuational phenomena in ecological systems, including such issues of the immediate moment as rates of change in benthic faunas (previously the province of hypotheses about glacially slow and steady change in constantly depauperate environments), and such questions of broadest geological scale as the newly recognized stepped and punctuational "morphology" (correcting the hypothetical growth through substantial time of all previous gradualistic accounts) of mutual biomechanical improvement in competing clades involved in "arms race," and generating a pattern known as "escalation."

13. Punctuational models have also been useful, even innovative in breaking conceptual logjams, in nonbiological fields ranging from closely cognate studies of the history of human tools (including extended stasis in the *Homo*

*erectus* toolkit), and nontrending, despite classical (and false) claims to the contrary by both experts, the Abbé Breuil and André Leroi-Gourhan, for the 25,000 year history of elegance in parietal cave art of France and Spain—and extending into more distant fields like learning theory (plateaus and innovative punctuations), studies of the dynamics of human organizations, patterns of human history, and the evolution of technologies, including a fascinating account of the history of books, through punctuations of the clay tablet, the scroll, the codex, and our current electronic reformation (wherever it may lead), and long periods of morphological stasis (graced with such vital innovations as printing, imposed upon the unaltered phenotype of the codex, or standard "book").

14. In a long and final section, I indulge myself, and perhaps provide some useful primary source material for future historians of scientific conflicts, by recording the plethora of non-scientific citations, ranging from the absurd to the insightful, for punctuated equilibrium (including creationist misuses and their politically effective exposure by scientists in courtroom trials that defeated creationist legislative initiatives; and the treatment of punctuated equilibrium, often very good but sometimes very bad, by journalists and by authors of textbooks—the primary arenas of vernacular passage). I also trace and repudiate the "dark side" of non-scientific reactions by professional colleagues who emoted at challenges to their comfort, rather than reacting critically and sharply (as most others did, and as discussed extensively in the main body of the chapter) to the interesting novelty, accompanied by some prominent errors of inevitable and initial groping on our part, spawned by the basic hypothesis and cascading implications of punctuated equilibrium.

### Chapter 10: The integration of constraint and adaptation: historical constraint and the evolution of development

1. Although the directing of evolutionary change by forces other than natural selection has loosely been described as "constraint," the term, even while acknowledged as a domain for exceptions to standard Darwinian mechanisms, has almost always been conceived as a "negative" force or phenomenon, a mode of preventing (through lack of variation, for example) a population's attainment of greater adaptation. But constraint, both in our science (and in vernacular English as well), also has strongly positive meanings in two quite different senses: first, or empirically, as channeled directionality for reasons of past history (conserved as homology) or physical principles; and second, or conceptually, as an nonstandard force (therefore interesting *ipso facto*) acting differently from what orthodoxy would predict.

2. The classical and most familiar category of internal channeling (the first, or empirical, citation of constraint as a positive theme) resides in preferred directions for evolutionary change supplied by inherited allometries and their phylogenetic potentiation by heterochrony. As "place holders" for an extensive literature, I present two examples from my own work: first, the illustration of synergy with natural selection (to exemplify the positive, rather than oppositional, meaning), where an inherited internal channel builds two im-

portant adaptations by means of one heterochronic alteration, as neoteny in descendant *Gryphaea* species of the English Jurassic produces shells of both markedly increased size (by retention of juvenile growth rates over an unchanged lifetime) and stabilized shape to prevent foundering in muddy environments (achieved by "bringing forward" the proportions of attached juveniles into the unattached stage of adult ontogeny); second, an illustration of pervasiveness and equal (or greater) power than selective forces (to exemplify the strength and high relative frequency of such positive influences), as geographic variation of the type species, *Cerion uva,* on Aruba, Bonaire, and Curaçao, a subject of intense quantitative study and disagreement in the past, becomes resolved in multivariate terms, with clear distinction between local adaptive differences and the pervasive general pattern of an extensive suite of automatic sequelae, generated by nonadaptive variation in the geometry of coiling a continuous tube, under definite allometric regularities for the genus, around an axis.

3. For the second, or conceptually positive, meaning of constraint as a term for nonstandard causes of evolutionary change, I present a model that compares the conventional outcomes of direct natural selection, leading to local adaptation, with two sources that can also yield adaptive results, but for reasons of channeling by internal constraints rather than by direct construction under external forces of natural selection. In this triangular model for aptive structures, the *functional vertex* represents features conventionally built by natural selection for current utilities. At the *historical vertex,* currently aptive features probably originated for conventionally adaptive reasons in distant ancestors; but these features are now developmentally channeled as homologies that constrain and positively direct both patterns of immediate change and the inhomogeneous occupation of morphospace (especially as indicated by "deep homologies" of retained developmental patterns among phyla that diverged from common ancestry more than 500 million years ago). At the *structural vertex,* two very different reasons underlie the origin of potentially aptive features for initially nonadaptive reasons: physical principles that build "good" form by the direct action of physical laws upon plastic material (as in D'Arcy Thompson's theory of form), and architectural sequelae (spandrels) that arise as nonadaptive consequences of other features, and then become available for later cooptation (as exaptations) to aptive ends in descendant taxa. These two structural reasons differ strongly in the ahistoricist implications of direct physical production independent of phyletic context *vs.* the explicit historical analysis needed to identify the particular foundation for the origin of spandrels in any individual lineage.

4. As a conceptual basis for understanding the importance of recent advances in evo-devo (the study of the evolution of development), the largely unknown history of debate about categories of homology, particularly the distinction between convergence and parallelism, provides our best ordering device—for we then learn to recognize the key contrast between parallelism as a positive deep constraint of homology in underlying generators (and therefore as a structuralist theme in evolution) and convergence as the oppo-

site sign of domination for external natural selection upon a yielding internal substrate that imposes no constraint (and therefore as a functionalist theme in evolution). As a beginning paradox, we must grasp why E. Ray Lankester coined the term homoplasy as a category of homology, whereas today's terminology ranks the concepts as polar opposites. Lankester wanted to contrast homology of overt structure (homogeny in his terms, or homology *sensu stricto*) with homology of underlying generators (later called parallelism) building the same structure in two separate lineages (homoplasy, or homology *sensu lato*, in Lankester's terms). Because parallelism could not be cashed out in operational terms (as science had no way, until our current revolution in evo-devo, to characterize, or even to recognize, these underlying generators), proper conceptual distinctions between parallelism and convergence have generally not been made, and the two terms have even (and often) been united as subtypes of homoplasy (now defined in the current, and utterly non-Lankesterian sense, as opposite to homology). I trace the complex and confused history of this discussion, and show that structuralist thinkers, with doubts about panadaptationism, have always been most sensitive to this issue, and most insistent upon separating and distinguishing parallelism as the chief category of positive developmental constraint—a category that has now, for the first time, become scientifically operational.

5. I summarize the revolutionary empirics and conceptualizations of evo-devo in four themes, united by a common goal: to rebalance constraint and adaptation as causes and forces of evolution, and to acknowledge the pervasiveness and importance—also the synergy with natural selection, rather than opposition to Darwinian themes—of developmental constraint as a positive, structuralist, and internal force. The first theme explores the implications—for internally directed evolutionary pathways and consequent clumping of taxa in morphospace—of the remarkable and utterly unanticipated discovery of extensive "deep homology" among phyla separated at least since the Cambrian explosion, as expressed by shared and highly conserved genes regulating fundamental processes of development. I first discuss the role and action of some of these developmental systems—the *ABC* genes of *Arabidopsis* in regulating circlets of structures in floral morphology, the *Hox* genes of *Drosophila* in regulating differentiation of organs along the AP axis, and the role of the *Pax-6* system in the development of eyes—in validating (only partially, of course) the archetypal theories of 19th century transcendental morphology, long regarded as contrary to strictly selectionist views of life's history—particularly Goethe's theory of the leaf archetype, and Geoffroy's idea of the vertebral groundplan of AP differentiation. I then discuss the even more exciting subject of homologically conserved systems across distant phyla, as expressed in high sequence similarity of important regulators, common rules of development (particularly the "Hoxology" followed in both arthropod and vertebrate ontogeny), and similar action of homeotic mutations that impact Hoxological rules by loss or gain of function. Geoffroy was partially right in asserting segmental homology between arthropods and vertebrates, particularly for the comparison of insect metameres with rhom-

bomeric segments in the developing vertebrate brain (a small part, perhaps, of the AP axis of most modern vertebrates, but the major component of the earliest fossil vertebrates), where the segments themselves may form differently, but where rules of Hoxology then work in the same manner during later differentiation. I also defend the substantial validity of Geoffroy's other "crazy" comparison—the dorso-ventral inversion of the same basic body plan between arthropods and vertebrates.

6. The second theme stresses the even more positive role of parallelism, based on common action of regulators shared by deep homology, in directing the evolutionary pathways of distantly related phyla into similar channels of adaptations thus more easily generated (thereby defining this phenomenon as synergistic and consistent with an expanded Darwinian theory, and not confrontational or dismissive of selection). I discuss such broadscale examples as the stunning discovery of substantial parallelism in the supposedly classical, "poster boy" expression of the opposite phenomenon of convergence—the development of eyes in arthropods, vertebrates, and cephalopods. The overt adult phenotype, of course, remains largely convergent, but homology of the underlying regulators demonstrates the strong internal channeling of parallelism. The vertebrate and squid version of *Pax-6* can, in fact, both rescue the development of eyes in *Drosophila* and produce ectopic expression of eyes in such odd places as limbs. I also discuss smaller-scale examples of "convergence," reinterpreted as parallelism, for even more precise similarities among separate lineages within coherent clades—particularly the independent conversion of thoracic limbs to maxillipeds, by identical homeotic changes in the same *Hox* genes, in several groups of crustaceans. Finally, I caution against overextension and overenthusiasm by pointing out that genuine developmental homologies may be far too broad in design, and far too unspecific in morphology, to merit a designation as parallelism, as in the role of *distal-less* in regulating "outpouchings" so generalized in basic structure, yet so different in form, as annelid parapodia, tunicate ampullae and echinoderm tube feet. I designate these overly broad similarities (that should not be designated as parallelism, or used as evidence for constraint by internal channeling) as "Pharaonic bricks"—that is, building blocks of such generality and multipurpose utility that they cannot be labeled as constraints (with the obvious *reductio ad absurdum* of DNA as the homological basis of all life). By contrast, the "Corinthian columns" of more specific conservations define the proper category of important positive constraint by internal channelings of parallelism based on homology of underlying regulators (just as the specific form of a Corinthian column, with its acanthus-leafed capital, represents a tightly constrained historical lineage that strongly influences the particular shape and utility of the entire resulting building).

7. My third and shorter theme—for this subject, though "classical" throughout the history of evolutionary thought, holds, I believe, less validity and scope than the others—treats the role of homologous regulators in producing rapid, even truly saltational, changes channeled into limited possibilities of developmental pathways (as in Goldschmidt's defense of discontinuous

evolution based upon mutations in rate genes that control ontogenetic trajectories). I discuss the false arguments often invoked to infer such saltational changes, but then document some limited, but occasionally important, cases of such discontinuous, but strongly channeled, change in macroevolution.

8. The fourth theme of top-down channeling from full ancestral complements, rather than bottom-up accretion along effectively unconstrained pathways of local adaptation, explores the role of positive constraint in establishing the markedly non-random and inhomogeneous population of potential morphospace by actual organisms throughout the history of life. Ed Lewis, in brilliantly elucidating the action of *Hox* genes in the development of *Drosophila,* quite understandably assumed (albeit falsely, as we later discovered to our surprise) that evolution from initial homonomy to increasing complexity of AP differentiation had been achieved by addition of *Hox* genes, particularly to suppress abdominal legs and convert the second pair of wings to halteres. In fact, the opposite process of tinkering with established rules, primarily by increased localization of action and differentiation in timing (and also by duplication of sets, at least for vertebrate *Hox* genes), has largely established the increasing diversity and complexity of differentiation in bilaterian phyla. The (presumably quite homonomous) common ancestor of arthropods and vertebrates already possessed a full complement of *Hox* genes, and even the bilaterian common ancestor already possessed at least seven elements of the set. Moreover, the genomes of the most homonomous modern groups of onycophorans and myriapods also include a full set of *Hox* genes—so differentiation of phenotypic complexity must originate as a derived feature of *Hox* action, exapted from a different initial role. The Cambrian explosion remains a crucial and genuine phenomenon of phenotypic diversification, a conclusion unthreatened by a putatively earlier common ancestry of animal phyla in a strictly genealogical (not phenotypic) sense. The further evolution of admittedly luxuriant, even awesome, variety in major phyla of complex animals has followed definite pathways of internal channeling, positively abetted (as much as negatively constrained) by homologous developmental rules acting as potentiators for more rapid and effective selection (as in the loss of snake limbs and iteration of pre-pelvic segments), and not as brakes or limitations upon Darwinian efficacy.

### Chapter 11: The integration of constraint and adaptation: structural constraints, spandrels, and exaptation

1. D'Arcy Thompson's idiosyncratic, but brilliantly crafted and expressed, theory of form (1917, 1942) presents a 20th century prototype for the generalist, or ahistorical, form of structural constraint: adaptation produced not by a functionalist mechanism like natural selection (or Lamarckism), but directly and automatically impressed by physical forces operating under invariant laws of nature. This theory enjoyed some success in explaining the correlation of form and function in very simple and labile forms (particularly as influenced by scale-bound changes in surface/volume ratios). But similarly nongenetic (and nonphyletic) explanations do not apply to complex crea-

tures, and even D'Arcy Thompson admitted that his mechanism could not encompass, say, "hipponess," but, at most, only the smooth transformations of these basic designs among closely related forms of similar *Bauplan* (the true theoretical significance of his much misunderstood theory of transformed coordinates). In summary, D'Arcy Thompson, the great student of Aristotle, erred in mixing the master's modes of causality—by assuming that the adaptive value (or final cause) of well designed morphology could specify the physical forces (or efficient causes) that actually built the structures.

2. Stuart Kauffman and Brian Goodwin have presented the most cogent modern arguments in this tradition of direct physical causation. These arguments hold substantial power for explaining some features of relatively simple biological systems, say from life's beginnings to the origin of prokaryotic cells, where basic organic chemistry and the physics of self-organizing systems can play out their timeless and general rules. Such models also have substantial utility in describing very broad features of the ecology and energy dynamics of living systems in general terms that transcend any particular taxonomic composition. But this approach founders, as did D'Arcy Thompson's as well, when the contingent and phyletically bound histories of particular complex lineages fall under scrutiny—and such systems do constitute the "bread and butter" of macroevolution. Nonetheless, Kauffman's powerful notion of "order for free," or adaptive configurations that emerge from the ahistoric (even abiological) nature of systems, and need not be explained by particular invocations of some functional force like natural selection, should give us pause before we speculate about Darwinian causes only from evidence of functionality. This "order for free" aids, and does not confute, such functional forces as selection by providing easier (even automatic) pathways towards a common desideratum of adaptive biological systems.

3. I then turn to the second, and (in my judgment) far more important, theme of structural constraint in the fully historicist and phyletic context of aptive evolution by cooptation of structures already present for other reasons (often nonadaptive in their origin), rather than by direct adaptation for current function via natural selection. The central principle of a fundamental logical difference between reasons for historical origin and current functional utility—a vital component in all historical analysis, as clearly recognized but insufficiently emphasized by Darwin, and then unfortunately underplayed or forgotten by later acolytes—was brilliantly identified and dissected by Friedrich Nietzsche in his *Genealogy of Morals,* where he contrasted the origin of punishment in a primal will to power, with the (often very different) utility of punishment in our current social and political systems.

4. Darwin himself invoked this principle of disconnection between historical origin and current utility both in the *Origin*'s first edition, and particularly in later responses to St. George Mivart's critique (the basis for the only chapter that Darwin added to later editions of the *Origin*) on the supposed inability of natural selection to explain the incipient (and apparently useless) stages of adaptive structures. Darwin asserted the principle of functional shift to argue that, although incipient stages could not have functioned in the manner

of their final form, they might still have arisen by natural selection for a different initial utility (feathers first evolved for thermoregulation and later co-opted for flight, for example). Darwin used this principle of cooptation, or functional shift, in two important ways that enriched and expanded his theory away from a caricatured panselectionist version—as the primary ground of historical contingency in phyletic sequences (for one cannot predict the direction of subsequent cooptation from different primary utilities), and as a source of structural constraint upon evolutionary pathways. But these Darwinian invocations stopped short of a radical claim for frequent and important nonadaptive origins of structures coopted to later utility. That is, Darwin rarely proceeded beyond the principle of originally adaptive origin for different function, with later cooptation to altered utility.

5. This important principle of cooptation of preexisting structures originally built for different reasons has been so underemphasized in Darwinian traditions that the language of evolutionary theory does not even include a term for this central process—which Elisabeth Vrba and I called "exaptation" (Gould and Vrba, 1982). (The available, but generally disfavored, term "preadaptation" only speaks of potential before the fact, and has been widely rejected in any case for its unfortunate, but inevitable, linguistic implication of foreordination in evolution, the very opposite of the intended meaning!)

6. I present a list of criteria for recognizing exaptations and separating them from true adaptations. I also discuss some outstanding examples of exaptation from the recent literature, with particular emphasis on the multiple exaptation of lens crystallins (in part for their fortuitous transparency, but for many other cooptable characteristics as well) in so many vertebrates and from so many independent and different original functions.

7. The exaptation of structures that arose for different adaptive reasons remains within selectionist orthodoxy (while granting structural constraint a large influence over historical pathways, in contrast with crude panadaptationism) by confirming a Darwinian basis for the adaptive origin of structures, whatever their later history of exaptive shift. On the other hand, the theoretically radical version of this second, or historicist, style of structural constraint in evolution posits an important role for an additional phenomenon in macroevolution: the truly nonadaptive origin of structures that may later be exapted for subsequent utility. Many sources of such nonadaptive origin may be specified (see point 10 below), but inevitable architectural consequences of other features—the spandrels of Gould and Lewontin's terminology (1979)—probably rank as most frequent and most important in the history of lineages.

8. Spandrels (although unnamed and ungeneralized) have been acknowledged in Darwinian traditions, but relegated to insignificant relative frequencies by invalid arguments for their rarity, their structural inconsequentiality (the mold marks on an old bottle, for example), or their temporally subsequent status as sequelae—with the first two claims empirically false, and the last claim logically false as a further confusion between historical origin and current utility.

9. I affirm the importance and high relative frequency of spandrels, and therefore of nonadaptive origin, in evolutionary theory by two major arguments for ubiquity. First, for intrinsic structural reasons, the number of potential spandrels greatly increases as organisms and their traits become more complex. (The spandrels of the human brain must greatly outnumber the immediately adaptive reasons for increase in size; the spandrels of the cylindrical umbilical space of a gastropod shell, by contrast, may be far more limited, although exaptive use as a brooding chamber has been important in several lineages.) Second, under hierarchical models of selection, features evolved for any reason at one level generate automatic consequences at other levels—and these consequences can only be classified as cross-level spandrels (since they are "injected into" the new level, rather than actively evolved there).

10. The full classification of spandrels and modes of exaptation offers a resolving taxonomy and solution—primarily through the key concept of the "exaptive pool"—for the compelling and heretofore confusing (yet much discussed) problem of "evolvability." Former confusion has centered upon the apparent paradox that ordinary organismal selection, the supposed canonical mechanism of evolutionary change, would seem (at least as its primary overt effect) to restrict and limit future possibilities by specializing forms to complexities of immediate environments, and therefore to act against an "evolvability" that largely defines the future macroevolutionary prospects of any lineage. The solution lies in recognizing that spandrels, although architecturally consequential, are not doomed to a secondary or unimportant status thereby. Spandrels, and all other forms of exaptive potential, define the ground of evolvability, and play as important a role in macroevolutionary potential as conventional adaptation does for the immediacy of microevolutionary success. I emphasize the centrality of the exaptive pool for solving the problem of evolvability by presenting a full taxonomy of categories for the pool's richness, focusing on a primary distinction between "franklins" (or inherent potentials of structures evolved for other adaptive roles—that is, the classical Darwinian functional shifts that do not depart from adaptationism), and "miltons" (or true nonadaptations, arising from several sources, with spandrels as a primary category, and then available for later cooptation from the exaptive pool—that is, the class of nonadaptive origins that does challenge the dominant role of panadaptationism in evolutionary theory).

11. I argue that the concept of cross-level spandrels vastly increases the range, power and importance of nonadaptation in evolution, and also unites the two central themes of this book by showing how the hierarchically expanded theory of selection also implies a greatly increased scope for nonadaptive structural constraint as an important factor in the potentiation of macroevolution.

### Chapter 12: Tiers of time and trials of extrapolationism
1. Darwin clearly recognized the threat of catastrophic mass extinction to the extrapolationist and uniformitarian premises underlying his claim for full

explanation of macroevolutionary results by microevolutionary causes (and not as a challenge to the efficacy of natural selection itself). Darwin therefore employed his usual argument about the imperfection of geological records to "spread out" apparent mass extinction over sufficient time for resolution by ordinary processes working at maximal rates (and therefore only increasing the intensity of selection).

2. The transition of the impact scenario (as a catastrophic trigger for the K-T extinction) from apostasy at its proposal in 1980 to effective factuality (based on the consilience of disparate evidence from iridium layers, shocked quartz and, especially, the discovery of a crater of appropriate size and age at Chicxulub) has reinstated the global paroxysms of classical catastrophism (in its genuinely scientific form, not its dismissive Lyellian caricature) as a legitimate scientific mechanism outside the Darwinian paradigm, but operating in conjunction with Darwinian forces to generate the full pattern of life's history, and not, as previously (and unhelpfully) formulated, as an exclusive alternative to disprove or to trivialize Darwinian mechanisms.

3. If catastrophic causes and triggers for mass extinction prove to be general, or at least predominant in relative frequency (and not just peculiar to the K-T event), then this macroevolutionary phenomenon will challenge the crucial extrapolationist premise of Darwinism by being more frequent, more rapid, more intense and more different in effect than Darwinian biology (and Lyellian geology) can allow. Under truly catastrophic models, two sets of reasons, inconsistent with Darwinian extrapolationism by microevolutionary accumulation, become potentially important agents of macroevolutionary patterning: effectively random extinction (for clades of low N), and, more importantly, extinction under "different rules" from reasons regulating the adaptive origin and success of autapomorphic cladal features in normal times.

4. Catastrophic mass extinction, while breaking the extrapolationist credo, may suggest an overly simplified and dichotomous macroevolutionary model based on alternating regimes of "background" *vs.* "mass" extinction. Rather, we should expand this insight about distinctive mechanisms at different scales into a more general model of several rising tiers of time—with conventional Darwinian microevolution dominating at the ecological tier of short times and intraspecific dynamics; punctuated equilibrium dominating at the geological tier of phyletic trends based on interspecific dynamics (with species arising in geological moments, and then treated as stable "atoms," or basic units of macroevolution, analogous to organisms in microevolution); and mass extinction (perhaps often catastrophic) acting as a major force of overall macroevolutionary pattern in the global history of relative waxing and waning of clades. (I also contrast this preferred model of time's tiering with the other possible style of explanation, which I reject but find interesting nonetheless, for denying full generality to smooth Darwinian upward extrapolation from the lowest level—namely, an equally smooth and monistic downward extrapolation from catastrophic mortality in mass extinction to

diminishing, but equally random and sudden, effects at all scales, as proposed in Raup's "field of bullets" model.)

5. In a paradoxical epilogue, I argue (despite my role as a longtime champion of the importance and scientific respectability of unpredictable contingency in the explanation of historical patterns) that the enlargement and reformulation of Darwinism, as proposed in this book, will recapture for general theory (by adding a distinctive and irreducible set of macroevolutionary causes to our armamentarium of evolutionary principles) a large part of macroevolutionary pattern that Darwin himself, as an equally firm supporter of contingency, willingly granted to the realm of historical unpredictability because he could not encompass these results within his own limited causal structure of strict reliance upon smooth extrapolation from microevolutionary processes by accumulation through the immensity of geological time.

A FINAL THOUGHT. May I simply end by quoting the line that I wrote at the completion of a similar abstract (but vastly shorter, in a much less weighty book) for my first technical tome, *Ontogeny and Phylogeny* (1977b, p. 9): "This epitome is a pitiful abbreviation of a much longer and, I hope, more subtle development. Please read the book!"

PART I

THE HISTORY
OF
DARWINIAN
LOGIC
AND DEBATE

# The Essence of Darwinism and the Basis of Modern Orthodoxy: An Exegesis of the *Origin of Species*

## A Revolution in the Small

Our theatrical and literary standards recognize only a few basic types of heroes. Most are preeminently strong and brave; some, in an occasional bone thrown to the marginal world of intellectuals, may even be allowed to triumph by brilliance. But one small section of the pantheon has long been reserved for a sideshow of improbables: the meek, the mild, the foolish, the insignificant, the ornamental—in short, for characters so disdained that they pass beneath notice and become demons of effectiveness by their invisibility. Consider the secretaries or chauffeurs who learn essential secrets because patrician bosses scarcely acknowledge their personhood and say almost anything in their presence; or the pageboys and schoolgirls who walk unnoticed through enemy lines with essential messages to partisans in conquered territories.

Though few scholars have considered the issue in this light, I would argue that the intellectual agent of Darwin's victory falls into this anomalous category. To be sure, Darwin succeeded because he devised a mechanism, natural selection, that possessed an unbeatable combination of testability and truth. But, at a more general level, Darwin triumphed by allowing the formerly meek to inherit the entire world of evolutionary theory.

Darwin's theory explicitly rejected and overturned the two evolutionary systems well known in Britain during his time (see next chapter for details)—Lamarck's (via Lyell's exegesis in the *Principles of Geology*) and Chambers's (in the anonymously printed *Vestiges of the Natural History of Creation*). Both these theories sunk a deep root in the most powerful of cultural biases by describing evolution as an interaction of two opposing forces. The first—considered dominant, intrinsic and fundamental—yielded progress on the old euphonious (and sexist) theme of "the march from monad to man." The second—designated as secondary, diversionary and superimposed—interrupted the upward flow and produced lateral dead-ends of specialized adaptations, from eyeless moles to long-necked giraffes. Darwin, in his greatest stroke of genius, took this secondary force, proposed a new mechanism for its operation (natural selection), and then redefined this former source for superficial tinkering as fully sufficient to render *all* of evolution—thus branding the separate and more exalted force of progress as illusory.

Such an argument poses an obvious logical dilemma: how can such power be granted to a force formerly viewed as so inconsequential? After all, evolution must still construct the full pageant of life's history and the entire taxonomic panorama, even if we abandon the concept of linear order. Darwin's answer records the depth of his debt to Lyell, the man more responsible than any other for shaping Darwin's basic view of nature. Time, just time! (provided that the "inconsequential" force of adaptation can work without limit, accumulating its tiny effects through geological immensity). The theory's full richness cannot be exhausted by the common statement that Darwinism presents a biological version of the "uniformitarianism" championed by Lyell for geology, but I cannot think of a more accurate or more encompassing one-liner. (In a revealing letter to Leonard Horner, written in 1844, Darwin exclaimed: "I always feel as if my books came half out of Lyell's brains . . . for I have always thought that the great merit of the Principles [of Geology], was that it altered the whole tone of one's mind and therefore that when seeing a thing never seen by Lyell, one yet saw it partially through his eyes" (cited in Darwin, 1987, p. 55).)

Darwin, in his struggle to formulate an evolutionary mechanism during his *annus mirabilis* (actually a bit more than two years) between the docking of the *Beagle* and the Malthusian insight of late 1838, had embraced, but ultimately rejected, a variety of contrary theories—including saltation, inherently adaptive variation, and intrinsic senescence of species (see Gruber and Barrett, 1974; Kohn, 1980). A common thread unites all these abandoned approaches: for they all postulate an internal drive based either on large pushes from variation (saltationism) or on inherent directionality of change. Most use ontogenetic metaphors, and make evolution as inevitable and as purposeful as development. Natural selection, by contrast, relies entirely upon small, isotropic, nondirectional variation as raw material, and views extensive transformation as the accumulation of tiny changes wrought by struggle between organisms and their (largely biotic) environment. Trial and error, one step at a time, becomes the central metaphor of Darwinism.

This theme of relentless accumulation of tiny changes through immense time, the uniformitarian doctrine of Charles Lyell, served as Darwin's touchstone throughout his intellectual life. Uniformitarianism provides the key to his first scientific book (Darwin, 1842) on the formation of coral atolls by gradual subsidence of oceanic islands, long continued. And the same theme defines the central subject of his parting shot (1881), a book on the formation of vegetable mould by earthworms. Darwin, for lifelong reasons of personal style, did not choose to write a summary or confessional in lofty philosophical terms, but he did want to make an exit with guns blazing on his favorite topic. Ironically, Darwin's overt subject of worms has led to a common interpretation quite opposite to his own intent—his misrepresentation as a doddering old naturalist who couldn't judge the difference in importance between fishbait and fomenting revolution, and who, in recognizing evolution, just happened to be in the right place at the right time. In fact, Darwin's worm book presents an artfully chosen example of the deeper principle that underlay all his work, including the discovery of evolution—the uniformitarian power of small changes cumulated over great durations. What better example than the humble worm, working literally beneath our notice, but making, grain by grain, both our best soils and the topography of England. In the preface (1881, p. 6), Darwin explicitly draws the analogy to evolution by refuting the opinions of a certain Mr. Fish (wonderful name, given the context), who denied that worms could account for much "considering their weakness and their size": "Here we have an instance of that inability to sum up the effects of a continually recurrent cause, which has often retarded the progress of science, as formerly in the case of geology, and more recently in that of the principle of evolution."

Darwin waxed almost messianic in advancing this theme in the *Origin of Species*, for he understood that readers could not grasp his argument for evolution until they embraced this uniformitarian vision with their hearts. He confessed the *a priori* improbability of his assertion, given the norms and traditions of western thought: "Nothing at first can appear more difficult to believe than that the more complex organs and instincts should have been perfected, not by means superior to, though analogous with, human reason, but by the accumulation of innumerable slight variations, each good for the individual possessor" (1859, p. 459). In his short concluding section on our general reluctance to accept evolution, he did not—probably for diplomatic reasons—identify specific cultural or religious barriers; instead, he spoke of our unfamiliarity with the crucial uniformitarian postulate: "But the chief cause of our natural unwillingness to admit that one species has given birth to other and distinct species, is that we are always slow in admitting any great change of which we do not see the intermediate steps . . . the mind cannot possibly grasp the full meaning of the term of a hundred million years; it cannot add up and perceive the full effects of many slight variations, accumulated during an almost infinite number of generations" (1859, p. 481).

To impress readers with the power of natural selection, Darwin continually stressed the cumulative effect of small changes. He reserved his best literary lines, his finest metaphors, for this linchpin of his argument—as in this familiar passage: "It may be said that natural selection is daily and hourly scrutinizing, throughout the world, every variation, even the slightest; rejecting that which is bad, preserving and adding up all that is good; silently and insensibly working, whenever and wherever opportunity offers, at the improvement of each organic being in relation to its organic and inorganic conditions of life. We see nothing of these slow changes in progress, until the hand of time has marked the long lapse of ages" (1859, p. 84). Examine the smallest changes and variations, Darwin almost begs us. Let nothing pass beneath your notice. Cumulate, cumulate, cumulate:

Certainly no clear line of demarcation has as yet been drawn between species and sub-species . . . ; or, again, between sub-species and well-marked varieties, or between lesser varieties and individual differences. These differences blend into each other in an insensible series; and a series impresses the mind with the idea of an actual passage. Hence I look at individual differences, though of small interest to the systematist, as of high importance for us, as being the first step towards such slight varieties as are barely thought worth recording in works on natural history (1859, p. 51).

I need hardly stress Darwin's impact as one of the half dozen or so most revolutionary thinkers in western history. I want, instead, to emphasize a more curious aspect of his status—his continuing relevance, indeed his benevolent hovering over almost all our current proceedings. We may revere Newton and Lavoisier as men of equal impact, but do modern physicists and chemists actively engage the ideas of these founders, as they pursue their daily work? Darwin, on the other hand, continues to bestride our world like a colossus—so much so that I can only begin this book on the structure of evolutionary theory by laying out Darwin's detailed vision as a *modern* starting point, a *current* orthodoxy only lightly modified by more than a century of work. I do, in this book, advocate some major restructuring, in the light of new concepts and findings, and with the approbation of more and more colleagues as our understanding of evolution broadens. But Darwin remains our context—and my proposed restructuring represents an extension, not a replacement, of his vision. The hierarchical theory of selection builds a world different from Darwin's in many important respects, but we do so by extending his mechanism of selection to a larger realm than he acknowledged—that is, to levels both below and above his focus on the struggle among organisms.

When Cassius spoke his words about Caesar (paraphrased above), he added his puzzlement at Caesar's extraordinary success: "Upon what meat doth this our Caesar feed, that he is grown so great." I shall argue in this chapter that Darwin's continued, pervasive relevance arises from his capacity for revolutionary innovation at two opposite poles of scientific practice—the immediate strategy of formulating a methodology for everyday research, and the most general discussion of causes and phenomena in the natural world (the questions that will not go away, and that air continually from college bull sessions, to TV talk shows, to learned treatises on the nature of things). Darwin's residence at *both* poles of *immediate methodology* and *broadest theoretical generality* begins with his distinctive attitude towards the central importance of daily, palpable events in nature, and their power to account for all evolution by cumulation—hence my choice of an opening topic for this chapter (see Fig. 2-1).

Caesar voiced his suspicions of Cassius, fearing men who think too much (may all despots thus beware). But his grudging words of praise might well be invoked to epitomize the reasons for Darwin's unparalleled success: "He reads much; he is a great observer, and he looks quite through the deeds of men."

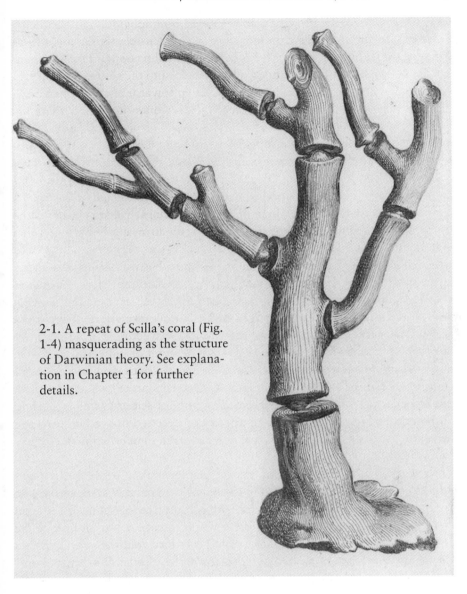

2-1. A repeat of Scilla's coral (Fig. 1-4) masquerading as the structure of Darwinian theory. See explanation in Chapter 1 for further details.

# Darwin as a Historical Methodologist

### ONE LONG ARGUMENT

An old quip, highlighting the intractability of philosophical dualism, proclaims: "what's matter? never mind; what's mind? doesn't matter." Predarwinian evolutionary systems embodied the same kind of Catch-22, this time in painful and practical terms, destined to ensnare any budding naturalist who hoped to study organisms by direct confrontation with testable hypotheses. Lamarck's system, for example, contrasted an intrinsic force of progress with a diversionary, and clearly secondary, force of adaptation to changing local environments. The secondary process worked in the immediate here and now, and might be engaged empirically by studies of adaptation and heredity. But the more important primary force, the source of natural order and the ultimate cause of human mentality, lurked in the background of time's immensity, and at the inaccessible interior of the very nature of matter. This characterization creates an intolerable dilemma for anyone who holds (as Darwin did) that science must be defined as testable doing, not just noble thinking. Recalling my opening quip, Lamarck's system virtually mocked the empirical approach to science, and forestalled any growing confidence in evolution: what is important cannot be seen; what can be seen is not important.

Darwin used a brilliant argument to cut through this dilemma, thus making the study of evolution a practical science. He acknowledged Lamarck's implied claim that small scale adaptation to local environment defines the tractable subject matter of evolution. But he refuted the disabling contention that adaptation in this mode only diverted the "real" force of evolution into side channels and dead ends. And he revised previous evolutionary thinking in the most radical way—by denying that Lamarck's "real" force existed at all, and by encompassing its supposed results as consequences of the "subsidiary" force accumulated to grandeur by the simple expedient of relentless action over sufficient time. Darwin established our profession not only by discovering a force—natural selection—that seems both powerful and true; he also, perhaps more importantly, made evolution accessible to science by granting to empiricists their most precious gifts of tractability and testability. The essence of Darwin's theory (specified in the next section) owes as much to his practical triumph at this immediate scale of daily work, as to his broadest perception that western views of nature had been seriously awry, and largely backwards.

Darwin, as we all know, began the last chapter of the *Origin* with a claim that "this whole volume is one long argument" (1859, p. 459). Fine, but an argument for what? For evolution itself? In part, of course, but such a general theme cannot mark the full intent of Darwin's statement, for the bulk of the *Origin* moves well beyond the basic arguments for evolution's factuality, as Darwin proceeds to craft a defense for natural selection and for the philosophy of nature so entailed. "One long argument" for natural selection, then? Again, in part; but we now confront the obverse of my last statement: too much of the *Origin* details basic evidence for evolution, independent of any particular mechanism of change. Instead, we must ask what deeper subject underlies both the defense of evolution as a fact and the proposal of a mechanism to explain its operation? How should we characterize the "one long argument" that pervades the entire book?

Ghiselin (1969) correctly identified the underlying theme as the construction, and defense by example, of a methodology—a mode of practice—for testing both the fact and mechanism of evolutionary change. But I cannot agree with Ghiselin that Darwin's consistent use of "hypothetico-deductive" reasoning constitutes his long argument (see Kitcher, 1985), for this style of scientific procedure, whatever its merits or problems, has been advocated as a general methodology for all scientific activity (see Hempel, 1965). Darwin, I believe, sought to construct and defend a working method for the special subject matter of evolutionary inquiry—that is, for the *data of history.*

Inferences about history, so crucial to any evolutionary work, had been plagued by problems of confidence that seemed to bar any truly scientific inquiry into the past. Darwin knew that evolution would not win respect until methods of historical inference could be established and illustrated with all the confidence of Galileo viewing the moons of Jupiter. He therefore set out to formulate rules for inference in history. I view the *Origin* as one long illustration of these rules. Historical inference sets the more general theme underlying both the establishment of evolution as a fact, and the defense of natural selection as its mechanism. The "one long argument" of the *Origin* presents a comprehensive strategy and compendium of modes for historical inference (see fuller exposition of this view in Gould, 1986). We must grasp Darwin's practical campaign on this battlefield in order to understand his radical philosophy, and to identify the features of his theory that count as essential to any definition of "Darwinism."

## THE PROBLEM OF HISTORY

Reading Darwin has been a persisting and central joy in my intellectual life. Lyell and Huxley may have been greater prose stylists, with more consistency in the ring and power of their words. Yet I give the nod to Darwin, and not only for the greater depth and power of his ideas. Darwin often wrote quite ordinary prose, page after page. But then, frequently enough to rivet the attention of any careful reader, his passion bursts through, and he makes a point with such insight and force (almost always by metaphor) that understanding breaks like sunrise. Every evolutionist can cite a list of favorite Darwinian passages, written on well-worn index cards for lectures (or, now, eternally embedded in Powerpoint files), posted on the office door or prominently displayed above the typewriter (now the computer terminal), or simply (and lovingly) committed to memory.

Several of my favorite passages celebrate the broadened understanding of nature that derives from recognizing organisms as products of history, rather than objects created in their present state. Darwin writes (1859, pp. 485–486):\*

---

\*I base this chapter on an exploration of the logic of argument in the first edition of the *Origin of Species* (1859). Provine (in lectures and personal communications) has argued that Darwinian historiography should focus on the definitive 6th edition of 1872, not only as Darwin's most considered and nuanced account, but primarily because this last edition has enjoyed such overwhelmingly greater influence through endless reprinting (continuing today) and translation into all major languages. The first edition had a print run of 1500 copies and sold out on the first day. I doubt that this original version ever reappeared in print before the facsimile edition edited by Mayr (1964), and this initial version remains rare relative to the ubiquitous sixth of almost every modern reprint. I agree with Provine's argument and, in fact, personally prefer the sixth edition for its subtleties on issues of macroevolution and adaptation. But I choose the first edition for this chapter as a necessary consequence of my idiosyncratic habits of historiographical work. I appreciate, and shame-

When we no longer look at an organic being as a savage looks at a ship, as at something wholly beyond his comprehension; when we regard every production of nature as one which has had a history; when we contemplate every complex structure and instinct as the summing up of many contrivances, each useful to the possessor, nearly in the same way as when we look at any great mechanical invention as the summing of the labor, the experience, the reason, and even the blunders of numerous workmen; when we thus view each organic being, how far more interesting, I speak from experience, will the study of natural history become!

By contrast, Darwin's chief quarrel with creationism resides not so much in its provable falseness, but in its bankrupt status as an intellectual argument— for a claim of creation teaches us nothing at all, but only states (in words that some people may consider exalted) that a particular creature or feature exists, a fact established well enough by a simple glance: "Nothing can be more hopeless than to attempt to explain the similarity of pattern in members of the same class, by utility or by the doctrine of final causes . . . On the ordinary view of the independent creation of each being, we can only say that so it is;—that it has so pleased the Creator to construct each animal and plant" (p. 435).

Moreover, and more negatively, creation marks the surrender of any attempt to understand connections and patterns. We express no causal insight whatever when we say that taxonomic order reflects the plan of a creator— for unless we can know the will of God, such a statement only stands as a redundant description of the order itself. (And God told us long ago, when he spoke to Job from the whirlwind, that we cannot know his will—"canst thou draw out leviathan with a hook?") Darwin, an ever genial man in the face of endless assaults upon his patience, directed several of his rare caustic comments against the ultimate idea-stopping claim that God so made it, praise his name. Darwin notes, for example, that horses are sometimes born with faint striping on their hides. A creationist can only assert that God made each equine species of zebras, horses, and asses alike, with such tendencies to vary and thereby to display, if only occasionally, the more comprehensive type.

---

lessly exploit, the historian's central concern for social context and the multifarious sources of intellectual arguments. But I am an internalist at heart, though wearing the sheep's clothing of my own Darwinian heritage with its emphasis on external adaptation, part by part. I love to follow the logic of argument, to treat a great text as Cuvier considered an organism—as an integrity, held together by sinews of logic (whatever the social or psychological origin of any particular item). I love to explore these connections, and to grasp the beauty of the totality. Thus, I prefer to practice the rather old-fashioned technique of *explication des textes* (see my longer rationale and attempt in Gould, 1987b, on Burnet, Hutton and Lyell). For this exercise, the first edition, despite its hurried composition as the scourge of Ternate breathed down Darwin's neck, represents the most coherent document, before all subsequent, externally-driven "adaptations" to critical commentary fixed the flaws and hedged the difficulties. Errors and inconsistencies build vital parts of integrity; I may share Cuvier's concern with necessary connections, but not his belief in optimal design. True integrity, in a messy world, implies rough edges, which not only have a beauty of their own, but also provide our best evidence for the logic of argument.

Evolution, on the other hand, supplies a true cause for an anomaly by positing community of descent with retention of ancestral states by heredity—something that might be tested in many ways, once we understand the mechanics of inheritance. (The following passage appears just before Darwin's summary to Chapter 5 on laws of variation.) Darwin lambastes the creationist alternative as causally meaningless: "To admit this view is, as it seems to me, to reject a real for an unreal, or at least for an unknown, cause. It makes the works of God a mere mockery and deception; I would almost as soon believe with the old and ignorant cosmogonists, that fossil shells had never lived, but had been created in stone so as to mock the shells now living on the sea-shore" (p. 167).

If we must locate our confidence about evolution in evidence for history—in part directly from the fossil record, but usually indirectly by inference from modern organisms—by what rules of reason, or canons of evidence, shall history then be established? Darwin's "long argument," in my view, can best be characterized as a complex solution to this question, illustrated with copious examples. We must first, however, specify the kinds of questions that *cannot* be answered. Many revealing statements in the *Origin* circumscribe the proper realm of historical inference by abjuring what cannot be known, or usefully comprehended under current limits. Darwin, for example, and following Hutton, Lyell and many other great thinkers, foreswore (as beyond the realm of science) all inquiry into the ultimate origins of things.* In the first paragraph of Chapter 7 on instincts, for example, Darwin writes (1859, p. 207): "I must premise, that I have nothing to do with the origin of the primary mental powers, any more than I have with that of life itself." Darwin invoked the same comparison in discussing the evolution of eyes, one of his greatest challenges (and firmest successes). He states that he will confine his attention to transitions in a structural sequence from simple to complex, and not engage the prior issue—answerable in principle, but beyond the range of knowledge in his day—of how sensitivity to light could arise within nervous tissue in the first place (1859, p. 187): "How a nerve comes to be sensitive to light, hardly concerns us more than how life itself first originated." Most crucially, and in a savvy argument that saved his entire system in the face of contemporary ignorance on a central issue, Darwin argues over and over again that we may bypass the vital question of how heredity works, and how variations arise—and only illustrate how evolution can occur, given the common-

---

*I have been both amused and infuriated that this issue still haunts us. I understand why American fundamentalists who call themselves "creation scientists," with their usual mixture of cynicism and ignorance, use the following argument for rhetorical advantage: (1) evolution treats the ultimate origin of life; (2) evolutionists can't resolve this issue; (3) the question is inherently religious; (4) therefore evolution is religion, and our brand deserves just as much time as theirs in science classrooms. We reply, although creationists do not choose to listen or understand, that we agree with points two and three, and therefore do not study the question of ultimate origins or view this issue as part of scientific inquiry at all (point one). I was surprised that Mr. Justice Scalia accepted this fundamentalist argument as the basis for his singularly inept dissent in the Louisiana creationism case, Edwards *v.* Aguillard (see Gould, 1991b).

place observation that sufficient variation *does* exist, and *is* inherited often enough:

> Whatever the cause may be of each slight difference in the offspring from their parents—and a cause for each must exist—it is the steady accumulation, through natural selection, of such differences, when beneficial to the individual, that gives rise to all the more important modifications of structure, by which the innumerable beings on the face of this earth are enabled to struggle with each other, and the best adapted to survive (p. 170—see also p. 131 for Darwin's argument that when we ascribe variation to "chance," we only mean to express our ignorance of causes).

Having established a domain of testability by exclusion, Darwin laid out his methodology for history—never explicitly to be sure, but with such accumulating force by example that the entire book becomes "one long argument" for the tractability of his new science. Those of us who practice the sciences of reconstructing specific events and unravelling temporal sequences have always fought a battle for appropriate status and respect, no less so today than in Darwin's time (see Gould, 1986), against those who would view such work as a "lesser" activity, or not part of science at all. History presents two special problems: (1) frequent absence of evidence, given imperfections of preservation; and (2) uniqueness of sequences, unrepeatable in their contingent complexity, and thereby distancing the data of history from such standard concepts as prediction, and experimentation.

We may epitomize the dilemma in the following way: many people define science as the study of causal processes. Past processes are, in principle, unobservable. We must therefore work by inference from results preserved in the historical record. We must study modern results produced by processes that can be directly observed and even manipulated by experiment—and we must then infer the causes of past results by their "sufficient similarity" (Steno's principle—see Gould, 1981c) with present results. This procedure requires, as Mill (1881) and other philosophers recognized long ago, a methodological assumption of temporal invariance for laws of nature. Historical study manifests its special character by placing primary emphasis upon comparison and degrees of similarity, rather than the canonical methods of simplification, manipulation, controlled experiment, and prediction.

Darwin had done some paleontological work, particularly in his treatises on barnacles (1851–1858), and his important discoveries of South American fossil vertebrates (formally named and described by Owen, at Darwin's invitation). But Darwin was not primarily a paleontologist, and he did not intend to base his argument for evolution on the evidence of fossils—especially since he viewed the stratigraphic record, with its vast preponderance of gaps over evidence, as more a hindrance than an aid to his theory (see chapters 9 and 10 of the *Origin*). Thus, of the two major sources for historical reconstruction—direct but imperfect information from fossils, and indirect but copious data from modern organisms—Darwin preferred the second as his wellspring of documentation. The *Origin* therefore focuses upon the establishment of a methodology for making inferences about history from features of modern organisms—and then using these multifarious inferences to prove both the fact of evolution and the probability of natural selection as a primary mechanism of change.

## A FOURFOLD CONTINUUM OF METHODS FOR THE INFERENCE OF HISTORY

Darwin, as a subtle and brilliant thinker, must be read on several levels. Consider just three, at decreasing domains of overt display, but increasing realms of generality: On the surface—a lovely, and not a pejorative, location for any student of nature—each book treats a particular puzzle: different forms of flowers on the same plant (1877), modes of formation for coral atolls (1842), formation of soil by worms (1881), styles of movement in climbing plants (1880a), the fertilization of orchids by insects (1862). At an intermediary level, as Ghiselin (1969) showed in his innovative study of the entire Darwinian corpus, each book forms part of a comprehensive argument for evolution itself. But I believe that we must also recognize a third, even deeper and more comprehensive layer of coordinating generality—Darwin's struggle to construct and apply a workable method for historical inference: a series of procedures offering sufficient confidence to place the sciences of history on a par with the finest experimental work in physics and chemistry. I have come to regard each of Darwin's books as, all at the same time, a treatment of a particular puzzle (level one), an argument for an evolutionary worldview (level two), and a treatise on historical methodology (level three). But the methodological focus of level three has usually been overlooked because Darwin chose to work by practice rather than proclamation.

Darwin recognized that several methods of historical inference must be developed, each tailored to the nature and quality of available evidence. We may order his procedures by decreasing density of available information. I recognize four waystations in the continuum and argue that each finds a primary illustration in one of Darwin's books on a specific puzzle in natural history. The *Origin of Species*, as his comprehensive view of nature, uses all four methods, and may therefore be read as a summation of his seminal contribution to the methodology of historical science. I shall list, and then illustrate with examples from the *Origin*, these four principles ordered by decreasing density of information.

UNIFORMITY. Or working up by extrapolation from direct observations on rates and modes of change in modern organisms. Call this, if you will, the worm principle to honor Darwin's last book (1881), which explains the topsoil and topography of England by extrapolating the measured work of worms through all scales of time, from the weight of castings left daily on a patch of sod to the historical and geological realms of millennia to millions of years.

SEQUENCING. Or the definition and ordering of various configurations, previously regarded as unrelated and independent, into stages of a single historical process. Here we cannot observe the changes between configurations directly and we must therefore work by recognizing them as temporally ordered products of a single underlying process of change. Call this, if you will, the coral reef principle to honor Darwin's first book (1842) on a scientific subject. His successful theory proposes a single historical process for the formation of coral atolls by recognizing three configurations of reefs—fringing reefs, barrier reefs, and atolls—as sequential stages in the foundering of oceanic islands.

CONSILIENCE (CONCORDANCE OF SEVERAL). We now reach a break in types of information. Methods 1 and 2 permit the reconstruction of historical sequences, either by extrapolating up from the most palpable and testable of daily changes (method 1), or by ordering a series of configurations as temporal stages (method 2). In many cases, however, we cannot reconstruct sequences, and must infer history from the configuration of a single object or circumstance. Of the two major methods for inferring history from single configurations, consilience calls upon a greater range of evidence. This word, coined by William Whewell in 1840, means "jumping together." By this term, Whewell referred to proof by coordination of so many otherwise unrelated consequences under a single causal explanation that no other organization of data seems conceivable. In a sense, consilience defines the larger method underlying all Darwin's inference from historical records. In a more specific context, I use consilience (see Gould, 1986) for Darwin's principal tactic of bringing so many different points of evidence to bear on a single subject, that history wins assent as an explanation by overwhelming confirmation and unique coordination. Call this, if you will, the different flowers principle to honor the extraordinary range of evidence that Darwin gathered (1877) to forge a historical explanation for why some taxa bear different forms of flowers on the same plant.

DISCORDANCE (DISSONANCE OF ONE). Here we reach a rock bottom of minimalism—unfortunately all too common in a world of limited information. We observe a single object, but not enough relevant items to forge consilience about its status as the product of history. How can we work from unique objects? How shall we infer history from a giraffe? Darwin tells us to search for a particular form of discordance—some imperfection or failure of coordination between an organism and its current circumstances. If such a quirk, oddity, or imperfection—making no sense as an optimal and immutable design in a current context—wins explanation as a holdover or vestige from a past state in different circumstances, then historical change may be inferred. Call this, if you will, the orchid principle (though I have also designated it as the panda principle for my own favorite example, perforce unknown to Darwin, of the panda's false thumb, Gould, 1980d), to honor Darwin's argument (1862) for orchids as products of history. Their intricate adaptations to attract insects for fertilization cannot be read as wonders of optimal design, specially created for current utilities, for they represent contraptions, jury-rigged from the available parts of ordinary flowers.

The *Origin of Species* presents an ingenious compendium of all four methods.

UNIFORMITY. People who do not understand science in their bones, and who think that revolutionary treatises must be presented as ideological manifestos at broadest scale, often express surprise and disappointment in reading the *Origin*, especially at Darwin's opening chapter. They expect fanfare, and they get fantails—pigeons, that is. But Darwin ordered his book by conscious intent and strategy. He knew that he had to demonstrate evolution with data, not simply proclaim his new view of life by rhetoric. Uniformitarianism embodied his best method based on maximal information—so he started from the smallest scale, change in domestication, and worked up to the history of life. As a member of two London pigeon fancying clubs (which he had joined, not from an abiding affection for this scourge of cities, but to gain practical information about evolution in the small), Darwin led from his acquired strength.

What better starting point, under method 1, than indubitable proof of historical change in domesticated plants and animals. The logic of the *Origin* employs one long analogy between artificial and natural selection, with uniformity as the joining point. Darwin writes in his introduction (p. 4): "At the commencement of my observations it seemed to me probable that a careful study of domesticated animals and of cultivated plants would offer the best chance of making out this obscure problem. Nor have I been disappointed; in this and in all other perplexing cases I have invariably found that our knowledge, imperfect though it be, of variation under domestication, afforded the best and safest clue."

Darwin continually drives home this analogy and extrapolation: if by artificial selection at small scale (as we know for certain), why not by natural selection at larger scale: "If it profit a plant to have its seeds more and more widely disseminated by the wind, I can see no greater difficulty in this being effected through natural selection, than in the cotton-planter increasing and improving by selection the down in the pods on his cotton-trees" (p. 86).

But this argument by uniformitarian extrapolation presents a serious difficulty (exploited by Fleeming Jenkin, 1867, in the famous critique that Darwin ranked so highly, and took so seriously in revising the *Origin*): change surely occurs in domestication, but suppose that species function like glass spheres with a modal configuration at the center and unbridgeable limits to variation representing the surface. Artificial selection could then bring morphology from the center to the surface, but no further—and the key argument for smooth extrapolation to *all* change over any time would fail. Darwin therefore staked a verbal claim for no limit. "What limit can be put to this power, acting during long ages and rigidly scrutinizing the whole constitution, structure, and habits of each creature—favoring the good and rejecting the bad? I can see no limit to this power, in slowly and beautifully adapting each form to the most complex relations of life" (p. 469).

Darwin then applied the full sequence of extrapolation to the natural world, beginning with individual variants as the source of subspecies, then moving to subspecies as incipient species, and finally to species as potential ancestors for branches of life's tree—a full range of scales from variation within a population to the entire pageant of life: "I look at individual differences, though of small interest to the systematist, as of high importance for us, as being the first step towards such slight varieties as are barely thought worth recording in works of natural history. And as I look at varieties which are in any degree more distinct and permanent, as steps leading to more strongly marked and more permanent varieties; and at these latter, as leading to subspecies, and to species" (p. 51).

Darwin invoked this first method, a strong argument based on maximal information at smallest scale, as his favored choice when available. To cite just three instances as a sampler: (1) the paleontological panorama may be read as a story of gradual evolution because species in adjacent strata show minimal differences, but these differences increase gradually as stratigraphic distance expands (p. 335). (2) When we find hints of the feather patterns of rock pigeon in highly modified breeds, we do not hesitate to interpret these designs as vestiges of an ancestral stock; therefore, the faint stripes that we sometimes observe in coats of young horses point to a common origin for all species in the clade of horses, asses and zebras (pp. 166–167). (3) Marine molluscs often exhibit brighter colors in warmer waters. We note this pattern both among varieties of a single species living in cold and warm waters, and among related species. A creationist explanation requires uncomfortable special pleading: God sometimes makes a species with bright shells in warm climates, but he allows other species to vary naturally, in the same geographic pattern, within a single created kind. An evolutionist, using method one, will recognize these phenomena as two stages in a single sequence of extrapolation from smaller to larger scale (p. 133).

SEQUENCING. We can use a second style of inference about temporal order when we cannot obtain adequate data about the nature of immediate changes at smallest scale. Since historical processes begin at different times and proceed at varying rates, all stages of a sequence may exist simultaneously (for example, stage one in case A, which began very recently; stage two in case B, which began at the same time, but has proceeded at an uncommonly rapid rate; and stage three in case C, which began long ago). Thus, fringing reefs, barrier reefs and atolls all exist now. When we recognize these forms as sequential stages of a single process, we may infer the pathway of history.

Darwin epitomizes method two in writing (p. 51): "A series impresses the mind with the idea of an actual passage." Invoking his usual starting point, Darwin presents a first example from breeds of domesticated pigeons. The more adequate data of method one—observed steps of passage, accumulating to greater and greater difference in time—no longer exist, for the transitional populations have died, and only a set of morphological "islands," representing a set of established breeds, remains. But these islands can be ordered as a plausible sequence of change between ancestral rock pigeons and the most aberrant of artificially produced breeds: "Although an English carrier or short-faced tumbler differs immensely in certain characters from the rock pigeon, yet by comparing the several sub-breeds of these breeds, more especially those brought from distant countries, we can make an almost perfect series between the extremes of structure" (p. 27).

Darwin uses method two in a special and crucial way throughout the *Origin*. Several of the most telling critiques against Darwin's style of evolution by gradualistic continuity—best represented in Mivart's famous argument (1871) about inviability of "incipient stages of useful structures" (see Chapter 11 for full treatment)—held that insensibly graded passages between putative ancestors and descendants could not even be conceptualized, much less documented. Charges of inconceivability took several forms, each reducible to the claim that you can't get from here to there, however well the beginning and end points may function. Consider the two most prominent formulations: (1) Early stages (when rudimentary) could provide no adaptive advantage, however valuable the final product (2) Major functional changes cannot occur because intermediary stages would fall into a never-never land of inviability, with the original (and essential) function lost, and the new operation not yet established.

Darwin offered a twofold response to these arguments, both using this second historical method of sequencing. He first presented theoretical arguments for the conceivability, even the likelihood, of intermediary stages in supposed cases of impossibility. He argued that early stages, too small to work in their eventual manner, could have performed different functions at the outset, and been coopted later for another style of life. (Incipient wings, originally used in thermoregulation, became organs of flight when they evolved to sufficiently large size to provide "fortuitous" aerodynamic benefits—see Kingsolver and Koehl, 1985, for an experimental validation of this scenario, and Gould, 1991b, for general discussion). As the misleadingly named principle of "preadaptation," this concept of functional shift became an important principle in evolutionary theory (see Chapter 11). Darwin writes, using a verbal intensifier rarely found in his prose: "In considering transitions of organs, it is so important to bear in mind the probability of conversion from one function to another" (p. 191).

As a response to charges of inviability for intermediary stages, Darwin invoked the important principle of redundancy as a norm for organic structures and functions. Most important functions can be performed by more than one organ; and most organs work in more than one way. By coupling these two aspects of redundancy, transitions in single organs can easily be conceived. An organ doesn't mysteriously invent a new function, but usually intensifies and specializes a previously minor use, while shedding an old primary operation. This previously major function can then be lost because other organs continue to do the same necessary job.

Ironically, we now recognize Darwin's favorite example of such redundancy as not only incorrect, but truly backwards (Gould, 1989b)—the evolution of lungs from swimbladders. (In fact, swimbladders evolved from lungs, see Liem, 1988). Darwin ran his transition in the wrong way, but his argument for redundancy as the key to viability for intermediary steps remains correct and crucially important, for the logic works equally well in either direction. Ancestral fishes maintained two systems for breathing—gills and lungs (as do modern lungfish, taxonomically called Dipnoi, or "two breathing"). The original lung probably played a subsidiary role in buoyancy; this function could be enhanced, and the original use in breathing deleted, because gills could adopt the entire respiratory burden. Darwin wrote (pp. 204–205): "For instance, a swim-bladder has apparently been converted into an air-breathing lung. The same organ having performed simultaneously very different functions, and then having been specialized for one function; and two very distinct organs having performed at the same time the same function, the one having been perfected whilst aided by the other, must often have largely facilitated transitions."

As a second response, Darwin proceeded beyond conceivability and tried to document actual sequences for supposedly impossible transitions—as in the evolution of a light-sensitive spot into an "organ of extreme perfection" like the vertebrate eye. These sequences cannot represent true phylogenies (since they consist solely of living species), but they do constitute structural series illustrating the conceivability of transitions. After admitting, for example, that the gradual evolution of such a miracle of workmanship as the eye "seems, I freely confess, absurd in the highest possible degree" (p. 186), Darwin presents a structural series of disparate animals, including working configurations proclaimed impossible by opponents: "Yet reason tells me, that if numerous gradations from a perfect and complex eye to one very imperfect and simple, each grade being useful to its possessor, can be shown to exist . . . then the difficulty of believing that a perfect and complex eye could be formed by natural selection, though insuperable by our imagination, can hardly be considered real" (p. 186).

Darwin applies this principle to behavior and its products, as well as to form. For the exquisite mathematical regularity of the honeycomb, he writes (p. 225): "Let us look to the great principle of gradation, and see whether Nature does not reveal to us her method of work." (See also page 210 on complex instincts and their explanation by the establishment of structural series.)

CONSILIENCE (CONCORDANCE OF SEVERAL). Darwin took great pride in his formulation of natural selection as a theory for the mechanism of phyletic change. But he granted even more importance to his relentless presentation of dense documentation for the factuality of change—for only such a cascade of data would force the scientific world to take evolution seriously. (The contrast between the *Origin* as a compendium of facts, and Lamarck's *Philosophie zoologique* as a purely theoretical treatise, strikes me as an even more distinguishing difference than the disparate causal mechanisms proposed by the two authors.) Facts literally pour from almost every page of the *Origin,* a feature that became even more apparent following Darwin's forced change of plans, and his decision to compress his projected longer work into the "abstract" that we call the *Origin of Species*—a revised strategy that led him to omit almost every reference and footnote, and almost all discursive discussion between bits of information. In some parts, the *Origin* reaches an almost frenetic pace in its cascading of facts, one upon the other. Only Darwin's meticulous sense of order and logic of argument save the work from disabling elision and overload.

Whenever he introduces a major subject, Darwin fires a volley of disparate facts, all related to the argument at hand—usually the claim that a particular phenomenon originated as a product of history. This style of organization virtually guarantees that Whewell's "consilience of inductions" must become the standard method of the *Origin*. Darwin's greatest intellectual strength lay in his ability to forge connections and perceive webs of implication (that more conventional thinking in linear order might miss). When Darwin could not cite direct evidence for actual stages in an evolutionary sequence, he relied upon consilience—and sunk enough roots in enough directions to provide adequate support for a single sturdy trunk of explanation.

Again, Darwin starts with pigeons, unleashing a cannonade of disparate arguments, all pointing to the conclusion that modern breeds of pigeons derive from a single ancestral stock. None of these facts permits the construction of an actual temporal series (methods one and two); but all identify the features of a current configuration that point to history as the underlying cause. Darwin, as usual, proceeds by particular example, but I doubt that a better general description of consilience could be formulated:

> From these several reasons, namely, the improbability of man having formerly got seven or eight supposed species of pigeons to breed freely under domestication; these supposed species being quite unknown in a wild state, and their becoming nowhere feral; these species having very abnormal characters in certain respects, as compared with all other Columbidae, though so like in most other respects to the rock pigeon; the blue color and various marks occasionally appearing in all the breeds, both when kept pure and when crossed; the mongrel offspring being perfectly fertile;—from these several reasons, taken together, I can feel no doubt that all our domestic breeds have descended from the *Columba livia* with its geographical subspecies (pp. 26–27).

Every scholar could cite a favorite case of Darwinian consilience. For my part, I especially admire Darwin's uncharacteristically long discussion (pp. 388–406) on transport from continental sources and subsequent evolution to explain the biotas of oceanic islands. Consider the main items in Darwin's own order of presentation:

(1) The general paucity of endemic species on islands, contrasted with comparable areas of continents; why should God put fewer species on islands?

(2) The frequent displacement of endemic island biotas by continental species introduced by human transport. If God created species for islands, why should species designed for continents so often prove superior in competition: "He who admits the doctrine of the creation of each separate species, will have to admit, that a sufficient number of the best adapted plants and animals have not been created on oceanic islands; for man has unintentionally stocked them from various sources far more fully and perfectly than has nature" (p. 390).

(3) Taxonomic disparity of endemic species within groups records ease of access, not created fit to oceanic environments: "Thus in the Galapagos Islands nearly every land bird, but only two out of the eleven marine birds, are peculiar; and it is obvious that marine birds could arrive at these islands more easily than land birds" (pp. 390–391).

(4) Biotas of oceanic islands often lack the characteristic groups of similar habitats on continents. On these islands, endemic members of other groups often assume the ecological roles almost always occupied by more appropriate or more competitive taxa in the richer faunas of continents—for example, reptiles on the Galapagos, or wingless birds on New Zealand, acting as surrogates for mammals.

(5) In endemic island species, features operating as adaptations in related species on continents often lose utility when their island residences do not feature the same environment: "For instance, in certain islands not tenanted by mammals, some of the endemic plants have beautifully hooked seeds; yet few relations are more striking than the adaptation of hooked seeds for transportal by the wool and fur of quadrupeds. This case presents no difficulty on my view, for a hooked seed might be transported to an island by some other means; and the plant then becoming slightly modified, but still retaining its hooked seeds, would form an endemic species, having as useless an appendage as any rudimentary organ" (p. 392).

(6) Peculiar morphological consequences often ensue when creatures seize places usually inhabited by other forms that could not reach an island. Many plants, herbaceous in habit on continents, become arboraceous on islands otherwise devoid of trees.

(7) Suitable organisms frequently fail to gain access to islands. Why do so many oceanic islands lack frogs, toads, and newts that seem so admirably adapted for such an environment? "But why, on the theory of creation, they should not have been created there, it would be very difficult to explain" (p. 393).

(8) Correlation of biota with distance. Darwin could find no report of terrestrial mammals on islands more than 300 miles from a continent. He presents the obvious evolutionary explanation for a disturbing creationist conundrum:

> It cannot be said, on the ordinary view of creation, that there has not been time for the creation of mammals; many volcanic islands are sufficiently ancient, as shown by the stupendous degradation which they have suffered and by their tertiary strata: there has also been time for the production of endemic species belonging to other classes . . . why, it may be asked, has the supposed creative force produced bats and no other mammals on remote islands? On my view this question can easily be answered; for no terrestrial mammal can be transported across a wide space of sea, but bats can fly across (p. 394).

(9) Correlation with ease of access. Creatures often manage to cross shallow water barriers between a continent and island, but fail to negotiate deepwater gaps of the same distance.

(10) Taxonomic affinity of island endemics—perhaps the most obvious point of all: why are the closest relatives of island endemics nearly always found on the nearest continent or on other adjacent islands?

Any honorable creationist, after suffering such a combination of blows, all implicating a history of evolution as the only sensible coordinating explanation, should throw in the towel and, like a beaten prizefighter, acknowledge Darwin as the Muhammad Ali of biology.

DISCORDANCE (DISSONANCE OF ONE). Consilience works as a cumulative argument for inferring history from objects and phenomena, rather than directly from sequences. You develop a line of attack, list numerous points, and then close in for the kill. But the empirical world often fails to provide such a bounty of evidence. Often, scientists must reason from a single object or situation—just the thing itself, not a network of arguments suitable for a broad consilience. Can history be inferred from such minimal information?

Thinkers, like soldiers, often show their true mettle in greatest adversity. I am particularly attracted by Darwin's approach to method 4, and have often cited his arguments in these "worst cases" as my primary illustration of his genius (Gould, 1986)—for Darwin met his greatest difficulty, and then not only devised a resolution, but also developed an argument of power and range. In other words, he turned potential trouble into one of his greatest strengths.

To infer history from a single object, Darwin asserts, one must locate features (preferably several, so the argument may shade into method three) that make no sense, or at least present striking anomalies, in the current life of the organism. One must then show that these features did fit into a clearly inferable past environment. In such cases, history—as expressed by preservation of signs from the past—provides the only sensible explanation for modern quirks, imperfections, oddities, and anomalies.

Darwin structured the *Origin of Species* as a trilogy. The first four chapters lay out the basic argument for natural selection. The middle five treat difficulties with the theory, and ancillary subjects that must be incorporated or explained away (rules of variation, nature of geological evidence, instincts, hybridism, and general objections). The final five chapters present the grand consilience by summarizing evidence for evolution itself—not so much for natural selection as a mechanism—from a broad range of disparate fields: geology*, geographic variation, morphology, taxonomy, embryology, and so forth.

The last part of the trilogy features method four. One might almost say that chapters 10–14 constitute one long list of examples for inferring history

---

*This tripartite structure of the *Origin* is masked by our tendency to treat the two geological chapters (9–10) as a unity. (Darwin even summarizes them together at the end of Chapter 10.) But Chapter 9, as the title proclaims ("On the imperfection of the geological record"), belongs to the discussion of difficulties in part 2 of the *Origin*—while Chapter 10 ("On the geological succession of organic beings") initiates part three on documentation of evolution as a fact. (Even the consolidated summary of Chapter 10 makes a clear break between these two disparate parts of Darwin's geological argument.)

from the oddities and imperfections of modern objects. (This arrangement of the last part struck me with particular force, as I reread the *Origin* before writing this book, and realized that the introductory paragraph for almost every new subject—from geographic variation to rudimentary organs—explicitly restates the general argument for method four.) Of course, the rest of the *Origin* also abounds with cases of method four, beginning as usual with examples from domestication. (Darwin argues that the chicks of wildfowl hide in grass and bushes to give their mother an opportunity for escape by flight. Domesticated chickens retain this habit, which no longer makes sense "for the mother-hen has almost lost by disuse the power of flight"—p. 216.)

Of subjects treated in this final part of the *Origin*'s trilogy, rudimentary organs represent, almost by definition, the "holotype" of method four. Darwin's definition, in the first sentence of his discussion, emphasizes this theme— "organs or parts in this strange condition, bearing the stamp of inutility" (p. 450). Nature tries to give us a history lesson, Darwin argues in some frustration, but we resist the message as inconsistent with received wisdom about natural harmony: "On the view of each organic being and each separate organ having been specially created, how utterly inexplicable it is that parts, like the teeth in the embryonic calf or like the shrivelled wings under the soldered wing-covers of some beetles, should thus so frequently bear the plain stamp of inutility! Nature may be said to have taken pains to reveal, by rudimentary organs and by homologous structures, her scheme of modification, which it seems that we wilfully will not understand" (p. 480). What else but imprints of history can explain rudimentary organs? Darwin ridicules the special pleading of creationist accounts as fancy ways of saying nothing at all. "In works on natural history rudimentary organs are generally said to have been created 'for the sake of symmetry,' or in order 'to complete the scheme of nature;' but this seems to me no explanation, merely a restatement of the fact. Would it be thought sufficient to say that because planets revolve in elliptic courses round the sun, satellites follow the same course round the planets, for the sake of symmetry, and to complete the scheme of nature?" (p. 453). Always searching for analogies with a short-term human history that we cannot deny, Darwin compares rudimentary organs with silent letters, once sounded, in the orthography of words: "Rudimentary organs may be compared with the letters in a word, still retained in the spelling, but become useless in the pronunciation, but which serve as a clue in seeking for its derivation" (p. 455).

Darwin continues the same argument as an underpinning for all discussions on other aspects of organic form. He introduces morphology as "the most interesting department of natural history, [which] may be said to be its very soul" (p. 434) and continues immediately with an example of method four: "What can be more curious than that the hand of a man, formed for grasping, that of a mole for digging, the leg of the horse, the paddle of the porpoise, and the wing of the bat, should all be constructed on the same pattern, and should include the same bones, in the same relative positions" (p. 434).

Similarly, the section on embryology begins with an example of method

four—the branchial circulation in young bird and mammalian embryos as indications of a "community of descent" with an aquatic past. This common
condition in embryonic frogs, birds, and mammals cannot reflect design for
*current* function: "We can not, for instance, suppose that in the embryos of
the vertebrata the peculiar loop-like course of the arteries near the branchial
slits are related to similar conditions,—in the young mammal which is nour-
ished in the womb of his mother, in the egg of the bird which is hatched in a
nest, and in the spawn of a frog under water" (p. 440).

The key argument of the section on taxonomy makes the same point in a
different form: if animals had experienced no history of change, and were created in accord with current needs and functions, then why should similar anatomical designs include creatures of such widely divergent styles of life? Darwin writes, in the opening paragraph of his discussion on taxonomy: "The
existence of groups would have been of simple signification, if one group had
been exclusively fitted to inhabit the land, and another the water; one to feed
on flesh, another on vegetable matter, and so on; but the case is widely different in nature; for it is notorious how commonly members of even the same
subgroup have different habits" (p. 411).

These arguments strike us as most familiar when based on organic form,
but fewer evolutionists recognize that method four also undergirds Darwin's
two chapters on biogeography (11 and 12). Darwin uses dissonance between
organism and dwelling place as the coordinating theme of these chapters: the
geographic distributions of organisms do not primarily suit their *current* climates and topographies, but seem to record more closely a history of opportunities for movement. Again, Darwin presents the basic argument in his first
paragraph (p. 346): "In considering the distribution of organic beings over
the face of the globe, the first great fact which strikes us is, that neither the
similarity nor the dissimilarity of the inhabitants of various regions can be accounted for by their climatal and other physical conditions."

Example tumbles upon example throughout these two chapters. Darwin
notes that northern hemisphere organisms of subarctic and north temperate
climes maintain far closer taxonomic similarity than the current geographic
separation of their continents would imply. He therefore interprets these likenesses as vestiges of history—preserved expressions of the glacial age, when
these climatic bands stood further to the north, near the arctic circle where all
northern continents virtually touch (p. 370). He also finds too much organic
similarity for the modern range of climatic differences along lines of longitude from north to south poles, and he again implicates the climax of glacial
ages as a time of formation (with modern persistence as a vestige), when even
a subarctic species might migrate in comfort, on a cold earth, across the equator from north to south along a single line of longitude. Invoking a complex
and graphic metaphor for history, Darwin writes of disjunct distributions on
opposite hemispheres, and of geographic refugia at high altitudes of lower
latitudes between these endpoints:

> The living waters may be said to have flowed during one short period
> from the north and from the south, and to have crossed the equator; but

to have flowed with greater force from the north so as to have freely inundated the south. As the tide leaves its drift in horizontal lines, . . . so have the living waters left their living drift on our mountain summits, in a line gently rising from the arctic lowlands to a great height under the equator. The various beings thus left stranded may be compared with savage races of man, driven up and surviving in the mountain fastnesses of almost every land, which serve as a record, full of interest to us, of the former inhabitants of the surrounding lowlands (p. 382).

Everyone cites the Galapagos in a virtual catechism about Darwin's evidence for evolution, but few biologists can state how he invokes these islands in the *Origin*. Most textbooks talk about a diversity of finches, each beautifully adapted to available resources on different islands, or of variation in tortoise carapaces from place to place. Both these stories exemplify both diversification and current adaptive value—but Darwin speaks not a word about either case in the *Origin!*

In fact, Darwin invokes the Galapagos primarily as an extended example of method four applied to biogeography: These islands house many endemic species, necessarily created *in situ* according to his opponents. But why then should all these endemics bear close relationship with species on the nearby American mainland? A creationist might say that God fits creatures to immediate circumstances, and that the Galapagos Islands, located so near America, must resemble America in environment, and therefore be best suited to house species of the same basic design. But now we grasp the beauty of the Galapagos as an almost uncannily decisive natural experiment for the influence of history. These islands do lie close to America, but could scarcely resemble the mainland *less* in climate, geology and topography—for the Galapagos are volcanic islands in the wake of a cool current that even permits access to the northernmost species of penguin! Therefore, if the Galapagos endemics resemble American species, they must be recording a history of accidental transport and subsequent evolutionary change—not similar creations for similar environments. Darwin's brilliant argument deserves citation *in extenso:*

> Here almost every product of the land and water bears the unmistakeable stamp of the American continent. There are 26 land birds, and 25 of these are ranked by Mr. Gould as distinct species, supposed to have been created here; yet the close affinity of most of these birds to American species in every character, in their habits, gestures, and tones of voice, was manifest. . . . why should this be so? Why should the species which are supposed to have been created in the Galapagos Archipelago, and nowhere else, bear so plain a stamp of affinity to those created in America? There is nothing in the conditions of life, in the geological nature of the islands, in their height or climate, or in the proportions in which the several classes are associated together, which resembles closely the conditions of the South American coast: in fact there is considerable dissimilarity in these respects. On the other hand, there is a considerable degree

of resemblance in the volcanic nature of the soil, in climate, height, and size of islands, between the Galapagos and Cape de Verde Archipelagos: but what an entire and absolute difference in their inhabitants! The inhabitants of the Cape de Verde Islands are related to those of Africa, like those of the Galapagos to America. I believe this grand fact can receive no sort of explanation on the ordinary view of independent creation; whereas on the view here maintained, it is obvious that the Galapagos Islands would be likely to receive colonists . . . from America; and the Cape de Verde Islands from Africa; and that such colonists would be liable to modifications—the principle of inheritance still betraying their original birth place (pp. 397–399).

Finally, in rereading the *Origin*, I was struck by another, quite different, use of the argument from imperfection—one that had entirely escaped my notice before. Darwin showed little sympathy for our traditional and venerable attempts to read moral messages from nature. He almost delighted in noting that natural selection unleashes a reign of terror that would threaten our moral values if we tried—as we most emphatically should not—to find ethical guidelines for human life in the affairs of nature. But I hadn't realized that he sometimes presents the apparent cruelties of nature as imperfections pointing to evolution by natural selection—imperfections relative to an inappropriate argument about morality to be sure, but imperfections that trouble our souls nonetheless, and may therefore operate with special force as suggestive arguments for evolution:

> Nor ought we to marvel if all the contrivances in nature be not, as far as we can judge, absolutely perfect; and if some of them be abhorrent to our ideas of fitness. We need not marvel at the sting of the bee causing the bee's own death; at drones being produced in such vast numbers for one single act, and being then slaughtered by their sterile sisters; at the astonishing waste of pollen by our fir trees; at the instinctive hatred of the queen bee for her own fertile daughters; at ichneumonidae feeding within the live bodies of caterpillars; and at other such cases. The wonder indeed is, on the theory of natural selection, that more cases of the want of absolute perfection have not been observed (p. 472).

I may have burdened readers with too much detail about Darwin's arguments for inferring history, but method inheres in this extended madness. My general argument holds that the *Origin* should be understood as a book encompassing two opposite, but complementary, poles of science at its best and most revolutionary—first, as a methodological treatise proving by example that evolution can be tested and studied fruitfully; and second, as an intellectual manifesto for a new view of life and nature. As a methodological treatise, the *Origin* focuses upon the palpable and the small—arguing that uniformitarian extrapolation into geological scales can render all evolution. We may therefore avoid any appeal to "higher" forces that cannot be studied directly because they work only in the untestable immensity of deep time, or occur so rarely that we can entertain little hope for direct observation during the short span of human history. The disabling Lamarckian paradox—what is important can't be studied; and what can be studied isn't important—therefore disappears, and evolution becomes, under Darwin's system, a working science for the first time. These features of methodology potentiate Darwin's theoretical overview (as we shall see in the next section), and therefore contribute indispensably to what may legitimately be called the essence of Darwinism, the *sine quibus non* for a Darwinian view of nature. This book argues that we can define such a set of basic commitments, but then maintains that these commitments have become inadequate in our times.

## Darwin as a Philosophical Revolutionary

### THE CAUSES OF NATURE'S HARMONY

#### Darwin and William Paley

In November 1859, just a week before the official publication date of the *Origin,* Darwin wrote to his neighbor John Lubbock* "I do not think I hardly ever admired a book more than Paley's 'Natural Theology.' I could almost formerly have said it by heart" (in F. Darwin, 1887, volume 2, p. 219).

The Reverend James McCosh receives my vote for the most interesting among a largely forgotten group of late 19th century thinkers who played a vital role in their own time—liberal theologians friendly to evolution (though not usually to Darwin's philosophy), and who prove that if any warring camps can be designated in this realm, the combatants surely cannot be labeled as science *vs.* religion (see Gould, 1999b), but rather as expressions of a much deeper struggle between tradition and reform, or dogmatics and openness to change. McCosh doesn't even merit a line in the *Encyclopedia Britannica,* though he did serve as president of Princeton University, where he had a major influence on the career of Henry Fairfield Osborn and other important American evolutionists.

In 1851, McCosh published an article entitled "Typical Forms" in the *North British Review.* Hugh Miller, the self-taught Scottish geologist and general thinker, called this article "at once the most suggestive and ingenious which we have almost ever perused," and urged McCosh to expand his argument to an entire volume. McCosh accepted this advice and, in collaboration with George Dickie, published *Typical Forms and Species Ends in Creation* in 1869. The Greek inscription on the title page—*typos kai telos* (type and pur-

---

*Later Lord Avebury and an author of many fine evolutionary works himself. But Lubbock's greatest contribution to human thought was probably indirect, a result of neighborly fellowship—for he sold to Darwin a corner of property that became the famous "sandwalk" where Darwin, perambulating and kicking aside a flint cobble for each circumnavigation, solved several riddles of life and human existence. Darwin graded the difficulty of his problems by the number of circuits required for solution—two-flint problems, five-flint problems, etc. I suspect that macroevolutionary theory must present us with at least a fifty-flint problem!

pose)—epitomizes the argument. McCosh holds that God's order and benevolence may be inferred from two almost contradictory properties that reside in tension within all natural objects—"the principle of order" and "the principle of special adaptation." (These two principles persist in Darwin's formulation under the names "Unity of Type" and "Conditions of Existence"—1859, p. 206, for example (see my extensive treatment of this passage on pp. 251–260), where their fundamental character merits upper case designations from Darwin.) McCosh defines his first principle as "a general plan, pattern, or type, to which every given object is made to conform"; and his second as a "particular end, by which each object, while constructed after a general model is, at the same time, accommodated to the situation which it has to occupy, and a purpose which it is intended to serve" (1869, p. 1). (If we call these two principles "anatomical ground plan" and "adaptation" we will be able to make the appropriate evolutionary translation without difficulty.)

McCosh argues that God's existence and benevolence can be inferred from either principle—from the first by the order of taxonomy, and the abstract beauty of bodily symmetry and structure; and from the second, by "adaptation,"* or the exquisite fit of form to function. McCosh also notes that the second, or functional, argument constitutes the "national signature" of British thought: "The arguments and illustrations adduced by British writers for the last age or two in behalf of the Divine existence, have been taken almost exclusively from the indications in nature of special adaptation of parts" (1869, p. 6).

The main lineage of this national tradition for "natural theology" based on the "argument from design" runs from Robert Boyle's *Disquisition About the Final Causes of Natural Things* (1688) and John Ray's *Wisdom of God Manifested in the Works of the Creation* (1691) in Newton's generation that promulgated what historians call "*the* scientific revolution"; to a grand culmination in William Paley's *Natural Theology* (1802), one of the most influential books of the 19th century; to an anticlimax, during the 1830's, in the eight "Bridgewater Treatises" (including volumes by Buckland and Whewell), established by a legacy from the deceased Earl of Bridgewater for a series of volumes "on the power, wisdom, and goodness of God, as manifested in the creation." Critics in Darwin's circle generally referred to this series as the "bilgewater treatises."

Revolutions usually begin as replacements for older certainties, and not as pristine discoveries in uncharted terrain. In understanding the second pole of Darwin's genius as the uncompromising radicalism of his new philosophy for life and history, we must first characterize the comfortable orthodoxy up-

---

*The word adaptation did not enter biology with the advent of evolutionary theory. The *Oxford English Dictionary* traces this term to the early 17th century in a variety of meanings, all designating the design or suitability of an object for a particular function, the fit of one thing to another. The British school of natural theology used "adaptation" as a standard word for illustrating God's wisdom by the exquisite fit of form to immediate function. Darwin, in borrowing this term, followed an established definition while radically revising the cause of the phenomenon.

rooted by the theory of natural selection. Darwin's essential argument begins with a definition of the dominant philosophy for natural history in his day—natural theology in the Paleyan mode.

At the outset of Chapter 4, I will say more about Paley and the alternative vision of continental natural theology (adaptationism vs. laws of form). For now, a simple statement of the two chief precepts of Paleyan biology will suffice:

NATURAL THEOLOGY IN GENERAL. The rational and harmonious construction of nature displays the character and benevolence of a creating God. In the last four chapters of his book, Paley tells us what we may infer about God from the works of creation. God's existence, of course, shines forth in his works, but this we know from many other sources. More specifically (and with a Paleyan chapter for each), nature instructs us about God's personality, his natural attributes, his unity, and (above all) his goodness.

PALEY'S PARTICULAR VERSION OF NATURAL THEOLOGY. Natural theology has been expressed in two basic modes (see Chapter 4), one primarily continental (laws of form), the other mainly British (adaptationism). Paley held that God manifests his creating power in the exquisite design of organisms for their immediate function. We all know Paley's famous opening metaphor: if I find a watch lying abandoned on an open field, I can conclude from the complex set of parts, all shaped to a common purpose and all well designed for a specific end, that some higher intelligence constructed the watch both directly and for a particular use. Since organisms show even more complexity and even more exquisite design, they must have been fashioned by an even greater intelligence. But fewer biologists know Paley's more specific argument *against* the alternative version of natural theology (laws of form), as presented in his chapter 15 on "relations." The parts of organisms exist in concert not because laws of form or symmetry demand one feature to balance another, but "from the relation which the parts bear to one another in the prosecution of a common purpose" (1803 edition, p. 296)—that is, to secure an optimal adaptation of the whole.

At the very outset of the *Origin*, Darwin tells us that his explanation of evolution will stress the Paleyan problem of exquisite adaptation. He writes, in the Introduction, that we could obtain sufficient confidence about evolution by "reflecting on the mutual affinities of organic beings, on their embryological relations, their geographical distribution, geological succession, and other such facts" (1859, p. 3). "Nevertheless," he continues, "such a conclusion, even if well founded, would be unsatisfactory, until it could be shown how the innumerable species inhabiting this world have been modified, so as to acquire that perfection of structure and coadaptation which most justly excites our admiration" (1859, p. 3). The explanation of adaptation therefore stands forth as the primary problem of evolution. Many lines of evidence prove *that* evolution occurred. But if we wish to learn *how* evolution works, we must study adaptation.

This basic Darwinian argument operates as a close copy of Paley's defense, recast in evolutionary language, for the English alternative in natural theology. We can infer, Paley often states, *that* God exists from innumerable aspects of nature. But if we wish to know any more about the creator—his nature, his attributes, his intentions—we must study the excellence of adaptation via the "argument from design." Paley writes (1803, p. 60): "When we are enquiring simply after the *existence* of an intelligent Creator, imperfection, inaccuracy, liability to disorder, occasional irregularities, may subsist, in a considerable degree, without inducing any doubt into the question."

On the other hand, adaptation in the fashioning of contrivances for definite ends reveals God's nature. Paley invokes this theme as a litany in developing his initial parable of the watch and watchmaker. He cites other possible explanations for the origin of the watch, and then intones, after each: "Contrivance is still unaccounted for. We still want a contriver" ("want," that is, in the old sense of "lack," not the modern "desire"—p. 13). "Contrivance must have had a contriver, design, a designer" (p. 14). Later, he tells us explicitly that nature can testify to God's character and goodness only by the phenomenon of adaptation (pp. 42–43): "It is only by the display of contrivance, that the existence, the agency, the wisdom of the Deity, *could* be testified to his rational creatures. This is the scale by which we ascend to all the knowledge of our Creator which we possess, so far as it depends upon the phenomena, or the works of nature . . . It is in the construction of instruments, in the choice and adaptation of means, that a creative intelligence is seen. It is this which constitutes the order and the beauty of the universe."

I had never read *Natural Theology* straight through before pursuing my research for this book. In so doing, I was struck by the correspondences between Paley's and Darwin's structure of argument (though Darwin, of course, inverts the explanation). Darwin did not exaggerate when stating to Lubbock that he had virtually committed Paley to memory. The style of Darwin's arguments, his choice of examples, even his rhythms and words, must often reflect (perhaps unconsciously) his memory of Paley. Consider just a few examples of this crucial linkage:

1. Paley, like Darwin, relies upon comparison and extrapolation from artificial to natural. Darwin moves from artificial to natural selection, Paley from human to animal machines. Both rely on the central argument that a common mechanism works much more powerfully in nature. Paley's words recall Darwin's argument that natural selection, working on all parts for so much time, must trump artificial selection, which only affects the few features we choose to emphasize in the short duration of human history. "For every indication of contrivance, every manifestation of design, which existed in the watch, exists in the works of nature; with the difference, on the side of nature, of being greater and more, and that in a degree which exceeds all computation" (1803, p. 19).

2. Both men invoke the same examples. Paley compares the eye and telescope; Darwin lauds the eye as the finest example of complex natural design, and then presents an evolutionary explanation. Paley cites the swimbladder as an independent device created for life in water; Darwin illustrates homology with the tetrapod lung and proposes an evolutionary passage.

3. Darwin often uses Paley's logic, sometimes against his predecessor. Paley, for example, dismisses arguments about "tendencies to order" or "principles of design" as empty verbiage, explaining nothing; a true cause must be identified, namely God himself. Darwin makes the same point, but cites evolution as the true cause, while branding statements about creation *ex nihilo* as empty verbiage. Paley writes (p. 76): "A principle of order is the word: but what is meant by a principle of order, as different from an intelligent Creator, has not been explained either by definition or example: and, without such explanation, it should seem to be a mere substitution of words for reasons, names for causes."

4. Paley discusses many themes of later and central importance to Darwin. He criticizes the major evolutionary conjectures of his day, including Buffon on "interior molds," and the idea of use and disuse. (Since I doubt that he had read Lamarck's earliest evolutionary work by 1802, Paley probably derived this aspect of Lamarck's theory from its status as folk wisdom in general culture.) Paley also states the following crisp epitome of the very argument from Malthus that so struck Darwin. (I am not claiming that this passage provided a covert source for Darwin's central insight. Darwin, after all, had also read Malthus.) "The order of generation proceeds by something like a geometrical progression. The increase of provision, under circumstances even the most advantageous, can only assume the form on an arithmetic series. Whence it follows, that the population will always overtake the provision, will pass beyond the line of plenty, and will continue to increase till checked by the difficulty of procuring subsistence" (p. 540).

This influence, and this desire to overturn Paley, persisted throughout Darwin's career. Ghiselin (1969), for example, regards Darwin's orchid book as a conscious satire on Paley's terminology and argument. Darwin called this work (1862), his next book after the *Origin of Species,* "On the various contrivances by which British and foreign orchids are fertilized by insects." Paley used the word "contrivance," as my previous quotations show, to designate an organic design obviously well-made by an intelligent designer. But Darwin argues that orchids must be explained as contraptions, not contrivances. Their vaunted adaptations are jury-rigged from ordinary parts of flowers, and must have evolved from such an ancestral source; the major adaptive features of orchids have not been expressly and uniquely designed for their current functions.

Now suppose, as a problem in abstract perversity, that one made a pledge to subvert Paley in the most radical way possible. What would one claim? I can imagine two basic refutations. One might label Paley's primary observation as simply wrong—by arguing that exquisite adaptation is relatively rare, and that the world is replete with error, imperfection, misery and caprice. If God made such a world, then we might want to reassess our decision to worship him. An upsetting argument indeed, but Darwin chose an even more radical alternative.

With even more perversity, one might judge Paley's observation as undoubtedly correct. Nature features exquisite adaptation at overwhelming relative frequency. But the unkindest cut of all then holds that this order, the very basis of Paley's inference about the nature of God, arises not directly from omnipotent benevolence, but only as a side-consequence of a causal principle of entirely opposite import—namely, as the incidental effect of organisms struggling for their own benefit, expressed as reproductive success. Could any argument be more subversive? One accepts the conventional observation, but then offers an explanation that not only inverts orthodoxy, but seems to mock the standard interpretation in a manner that could almost be called cruel. This more radical version lies at the core of Darwin's argument for natural selection. (Darwin actually employed both versions of the radical argument against Paley, but for different aspects of his full case. He invoked oddities and imperfections as his major evidence for the factuality of evolution (see pp. 111–116). But he used the more radical version—exquisite adaptation exists in abundance, but its cause inverts Paley's world—to construct his mechanism for evolutionary change, the theory of natural selection.)

We all understand, of course, that the force of Darwin's radicalism extends well beyond the inversion of an explanatory order; he also undercut a primary source of human comfort and solace. This book cannot address such a vital issue at any depth, but I must record the point—for this wrenching became so salient in subsequent human history. If the natural footprints of Paley's God—the source of our confidence in his character, his goodness and, incidentally, the only hint from nature that we should accept other revealed doctrines, in particular the idea of bodily resurrection (1803, pp. 580–581)—must be reconceived as epiphenomena of a struggle for personal success, then what becomes of nature's beauty, instruction and solace? What a bitter cup Darwin offers us, compared with Paley's sweet promise (1803, pp. 578–579): "The hinges in the wings of an earwig, and the joints of its antennae, are as highly wrought, as if the Creator had had nothing else to finish. We see no signs of diminution of care by multiplication of objects, or of distraction of thought by variety. We have no reason to fear, therefore, our being forgotten, or overlooked, or neglected."

But then, the man who served as the primary focus of Paley's veneration had also promised that the truth would make us free; and Darwin justly argued that nature cannot provide the source of morality or comfort in any case.

### Darwin and Adam Smith

Many scientists fail to recognize that all mental activity must occur in social contexts, and that a variety of cultural influences must therefore impact all scientific work. Those who do note the necessary link usually view cultural embeddedness as an invariably negative component of inquiry—a set of biases that can only distort scientific conclusions, and that should be identified for combat. But cultural influences can also facilitate scientific change, for incidental reasons to be sure, but with crucially positive results nonetheless—the exaptive principle that evolutionists, above all, should grasp and honor!

The origin of Darwin's concept of natural selection provides my favorite example of cultural context as a promoter.

The link of Darwin to Malthus has been recognized and accorded proper importance from the start, if only because Darwin himself had explicitly noted and honored this impetus. But if Darwin required Malthus to grasp the central role of continuous and severe struggle for existence, then he needed the related school of Scottish economists—the *laissez-faire* theorists, centered on Adam Smith and the *Wealth of Nations* (first published in the auspicious revolutionary year of 1776)—to formulate the even more fundamental principle of natural selection itself. But the impact of Adam Smith's economics did not strike Darwin with the force of eureka; the concepts crept upon him in the conventional fashion of most influences upon our lives. How many of us can specify a definite parental admonition, or a particular taunt of our peers, as central to the construction of our deepest convictions?

Silvan S. Schweber (1977), a physicist and historian of science, has traced the chain of influence upon Darwin from Adam Smith's school of Scottish economists—beginning in the early 1830's, and culminating in Darwin's intense study of these ideas as he tried to fathom the role of individual action during the weeks just preceding his "Malthusian" insight of September 1838. I believe that Schweber has found the key to the logic of natural selection and its appeal for Darwin in the dual role of portraying everyday and palpable events as the stuff of all evolution (the methodological pole), and in overturning Paley's comfortable world by invoking the most radical of possible arguments (the philosophical pole).

In fact, I would advance the even stronger claim that the theory of natural selection is, in essence, Adam Smith's economics transferred to nature. We must also note the delicious (and almost malicious) irony residing in such an assertion. Human beings are moral agents and we cannot abide the hecatomb*—the death through competition of nearly all participants—incurred by allowing individual competition to work in the untrammeled manner of pure *laissez-faire*. Thus, Adam Smith's economics doesn't work in economics. But nature need not operate by the norms of human morality. If the adaptation of one requires the deaths of thousands in amoral nature, then so be it. The process may be messy and wasteful, but nature enjoys time in abundance, and maximal efficiency need not mark her ways. (In one of his most famous letters, Darwin wrote to Joseph Hooker in 1856: "What a book a devil's chaplain might write on the clumsy, wasteful, blundering, low, and horribly cruel works of nature!") The analog of pure *laissez-faire* can and does operate in nature—and Adam Smith's mechanism therefore enjoys its

---

*"Hecatomb," an unfamiliar word in English, should enter the vocabulary of all evolutionists as a wonderfully appropriate description for this key aspect of Darwinism. A hecatomb is, literally, an offering of a hundred oxen in sacrifice. Yet, even in Homer, the word had come to designate any large number of deaths incurred as a sacrifice for some intended benefit—a good description of natural selection. And hecatomb trips so much more lightly off the tongue than "substitutional load."

finest, perhaps its only, full application in this analogous realm, not in the domain that elicited the original theory itself.

The primary argument of *laissez-faire* rests upon a paradox. One might suppose that the best path to a maximally ordered economy would emerge from an analysis conducted by the greatest experts all assembled, and given full power to execute their recommendations (the closest human analog to Paley's lone Deity), followed by the passage of laws to implement these rationally-derived, higher-level decisions. Yet Adam Smith argued that a society should follow the opposite path as a best approach to this desired end: law makers and regulators should step aside and allow each individual to struggle for personal profit in an untrammeled way—a procedure that would seem to guarantee the opposite result of chaos and disorder. In allowing the mechanism of personal struggle to run freely, good performers eliminate the less efficient and strike a dynamic balance among themselves. The "fallout," for society, yields a maximally ordered and prosperous economy (plus a hecatomb of dead businesses). The mechanism works by unbridled struggle for personal reward among individuals.

Schweber documents numerous sources in Darwin's wide readings for this central theme of political economy. In May 1840, for example, Darwin encountered the following passages in J. R. McCulloch's *Principles of Political Economy* (2nd edition of 1830—see Schweber, 1980, p. 268):

> Every individual is constantly exerting himself to find out the most advantageous methods of employing his capital and labor. It is true, that it is his own advantage, and not that of society, which he has in view; but a society being nothing more than a collection of individuals, it is plain that each, in steadily pursuing his own aggrandisement, is following that precise line of conduct which is most for the public advantage (p. 149). The true line of policy is to leave individuals to pursue their own interests in their own way, and never to lose sight of the maxim *pas trop gouverner* [not to govern too much]. It is by this spontaneous and unconstrained . . . effort of individuals to improve their conditions . . . and by them only, that nations become rich and powerful (p. 537).

The theory of natural selection lifts this entire explanatory structure, *virgo intacta,* and then applies the same causal scheme to nature—a tough customer who can bear the hecatomb of deaths required to produce the best polity as an epiphenomenon. Individual organisms engaged in the "struggle for existence" act as the analog of firms in competition. Reproductive success becomes the analog of profit—for, even more than in human economies, you truly cannot take it with you in nature.

Finally, continuing the analogy, Paley's dethronement follows the most radical path of supreme irony. For, in the ideal *laissez-faire* economy, all firms (purified in the unforgiving fires of competition) become sleek and well-designed, while the entire polity achieves optimal balance and coordination. But no laws explicitly operate to impose good design or overall balance by fiat—none at all. The struggle among firms represents the only causal process at work. Moreover, this cause operates at a lower level, and solely for the benefit of individual firms. Only as an incidental result, a side-consequence, does good design and overall balance emerge. Adam Smith, in coining one of the most memorable metaphors in our language, ascribed this process to the action of an "invisible hand." In the modern terms of hierarchy theory, we might say that overall order arises as an effect of upward causation from individual struggle. We may thus gain some clarity in definition, but we can't match the original prose. In his most famous words, Smith wrote in the *Wealth of Nations* (Book 4, Chapter 2): "He intends only his own gain, and he is in this, as in many other cases, led by an invisible hand to promote an end which was no part of his intention . . . I have never known much good done by those who affected to trade for the public good."

But Paley had assured us, in 500 closely-argued pages, that the analogous features of the natural world—good design of organisms and harmony of ecosystems—not only prove the existence of God, but also illustrate his nature, his personality, and his benevolence. In Darwin's importation of Adam Smith's argument, these features of nature become epiphenomena only, with no direct cause at all. The very observations that Paley had revered as the most glorious handiwork of God, the unquestionable proof of his benevolent concern, "just happen" as a consequence of causes operating at a lower level among struggling individuals. And, as the cruellest twist of all, this lower-level cause of pattern seems to suggest a moral reading exactly opposite to Paley's lofty hopes for the meaning of comprehensive order—for nature's individuals struggle for their own personal benefit, *and nothing else!* Paley's observations could not be faulted—organisms are well designed and ecosystems are harmonious. But his interpretations could not have been more askew— for these features do not arise as direct products of divine benevolence, but only as epiphenomena of an opposite process both in level of action and intent of outcome: individuals struggling for themselves alone.

I write this chapter with two aims in mind: first, to explicate the major sources and content of Darwin's argument; and second, to identify the truly essential claims of Darwinism, in order to separate them from a larger set of more peripheral assertions and misunderstandings—so that we can rank and evaluate the role of modern proposals and debates by the depth of their challenge to the central logic of our profession's orthodoxy. To fulfill this second goal, I try to identify a set of minimal commitments required of those who would call themselves "Darwinians." I argue that this minimal account features a set of three broad claims and their (quite extensive) corollaries. I then use this framework to organize the rest of this book, for I devote the historical chapters of this first part to pre- and post-Darwinian discussions of the three claims. Then, following a chapter on the construction of the Modern Synthesis as a Darwinian orthodoxy for the twentieth century, I revisit the three claims in the second part, this time by examining modern challenges to their exclusive sway.

By interpreting Darwin's radical theory as a response to Paley (actually an inversion), based on an importation of the central argument from Adam

Smith's *laissez-faire* economics, I believe that we achieve our best insight into the essential claims of Darwinism and natural selection. First, and foremost, we grasp the theoretical centrality of Darwin's conclusion that natural selection works through a struggle among *individual organisms* for reproductive success. Darwin's choice of levels, and his attempted restriction of causality to one level alone, then becomes neither capricious nor idiosyncratic, but, rather, central to the logic of an argument that renders the former "proof" of God's direct benevolence as an epiphenomenon of causal processes acting for apparently contrary reasons at a lower level. Second, we recognize the focal role of adaptation as the chief phenomenon requiring causal explanation—for good design had also set the central problem for English traditions in natural theology, the worldview that Darwin overturned by deriving the same result with an opposite mechanism.

These two principles—the operation of selection on struggling organisms as active agents, and the creativity of selection in constructing adaptive change—suffice to validate the theory in observational and microevolutionary expression. But Darwin nurtured far more ambitious goals (as the foregoing discussion of his methodology illustrates, see pages 97–116): he wished to promote natural selection, by extrapolation, as the preeminent source of evolutionary change at all scales and levels, from the origin of phyla to the ebb and flow of diversity through geological time. Thus, the third focal claim in the Darwinian tripod of essential postulates—the extrapolationist premise—holds that natural selection, working step by step at the organismic level, can construct the entire panoply of vast evolutionary change by cumulating its small increments through the fullness of geological time. With this third premise of extrapolation, Darwin transfers to biology the uniformitarian commitments that set the worldview of his guru, the geologist Charles Lyell.

## THE FIRST THEME: THE ORGANISM AS THE AGENT OF SELECTION

Once the syllogistic core,* the "bare bones" mechanism of natural selection, has been elucidated, two major questions—the foci of the next two sections

---

*By the "syllogistic core" of natural selection ("the bare-bones argument"), I refer to the standard pedagogical presentation of the abstract mechanism of the theory as a set of three undeniable factual statements followed by the inference of natural selection (the fourth statement) as a logical entailment of the three facts, *viz:*

1. Superfecundity: all organisms produce more offspring than can possibly survive.
2. Variation: all organisms vary from other conspecifics, so that each individual bears distinguishing features.
3. Heredity: at least some of this variation will be inherited by offspring (whatever the mechanism of hereditary transition—a mystery to Darwin, but the argument only requires *that* heredity exist, not that its mode of action be known).
4. Natural selection: if we accept these foregoing three statements as factual (2 and 3 ranked as "folk wisdom" in Darwin's time and could scarcely be doubted; while Darwin took great pains to validate 1 in early chapters of the *Origin*, showing, for example, that even the most slowly reproducing of all animals, the African elephant, would soon fill the

of this chapter—must be resolved before we can understand the theory's basic operation: the issues of *agency* and *efficacy*. The basic historical context of selection—its discovery and utilization by Darwin as a refutation of Paleyan natural theology through the imported causal structure of Adam Smith's invisible hand—grants primacy to the issue of *agency* (therefore treated here in the first of two sections on fundamental attributes). The rebuttal of the former centerpiece of natural history—the belief that organic designs record the intentions of an omnipotent creative power—rests upon the radical demotion of agency to a much lower level, devoid of any prospect for conscious intent, or any "view" beyond the immediate and personal. So Darwin reduced the locus of agency to the lowest level that the science of his day could treat in a testable and operational way—*the organism* (for ignorance of the mechanism of heredity precluded any possibility of still further reduction to cellular or genic levels). The purely abstract statement of natural selection (the syllogistic core) leaves the key question of agency entirely unanswered. Selection may be in control, but on what does selection act? On the subcellular components of heredity? on organisms? on populations? on species? or on all these levels simultaneously?

Darwin grasped with great clarity what most of his contemporaries never understood at all—that the question of agency, or levels of selection, lies at the heart of evolutionary causation. And he provided, from the depth of his personal convictions, the roots of his central premises, and the logic of his complete argument, a forthright answer that overturned a conceptual world—natural selection works on *organisms* engaged in a struggle for personal success, as assessed by the differential production of surviving offspring.

We all know that Darwin emphasized selection at the organismal level, but many evolutionists do not appreciate the centrality of this claim within his theory; nor do they recognize how actively he pursued its defense and illustra-

---

continent if all offspring survived and reproduced), then the principle of natural selection follows by syllogistic logic. If only some offspring can survive (statement 1), then, on average (as a statistical phenomenon, not a guarantee for any particular organism), survivors will be those individuals that, by their fortuity of varying in directions most suited for adaptation to changing local environments, will leave more surviving offspring than other members of the population (statement 2). Since these offspring will inherit those favorable traits (statement 3), the average composition of the population will change in the direction of phenotypes favored in the altered local environment.

As Darwin did himself in the *Introduction* to the *Origin*, nearly all textbooks and college courses present the "bare bones" of natural selection in this fashion (I have done so in more than 30 years of teaching). The device works well, but does not permit a teacher to go beyond the simplest elucidation of selection as a genuine force that can produce adaptive change in a population. In other words, the syllogistic core only guarantees that selection can work. By itself, the core says nothing about the locus, the agency, the efficacy, or the range of selection in a domain—the sciences of natural history—where all assessments of meaning rest upon such claims about mode, strength, and relative frequency, once the prior judgment of mere existence has been validated. Thus, an elucidation of this "syllogistic core" can only rebut charges of hokum or incoherence at the foundation. An analysis of the three key issues of the Darwinian essence, the subject of the rest of this chapter, then engages the guts of natural history.

tion. To explicate this issue, we must reemphasize the roles of William Paley and Adam Smith in the genesis of Darwin's system—using Smith to overturn Paley.

Adaptation and the "creativity of natural selection," as discussed in the next section, represent Darwin's evolutionary translation of Paley's chief concern with excellence in organic design. But the substitution of natural selection for God as creative agent, while disruptive enough to Western traditions, does not express the primary feature of Darwin's radicalism. To find this root, we must pursue a different inquiry about the locus of selection. After all, selection might operate at the highest level of species, even communities of species, for the direct production of order and harmony. We would then, to be sure, need to abandon God's role as an immediate creator, but what a gentle dispensation compared with Darwin's actual proposal: for if the agency of selection stood so high, God could be reconceptualized as the loving instigator of the rules. And the rules, by working directly for organic harmony, would then embody all that Paley sought to illustrate about God's nature.

Darwin's inversion of Paley therefore required a primary postulate about the locus of selection. Selection operates *on organisms*, not on any higher collectivity. Selection works directly for the benefit of organisms only, and not for any larger harmony that might embody God's benevolent intent. Ironically, through the action of Adam Smith's invisible hand, such "higher harmony" may arise as an epiphenomenal result of a process with apparently opposite import—the struggle of individuals for personal success. Darwin's revolution demands that features of higher-level phenomenology be explained as effects of lower-level causality—in particular, that the struggle among organisms yield order and harmony in the polity of nature.

Darwin's theory therefore presents, as the primary underpinning for its radical import in philosophy, a "reductionist" account of broadest-scale phenomena to a single causal locus at a low level accessible to direct observation and experimental manipulation: the struggle for existence among organisms. Moreover, this claim for organismal agency expresses Darwin's chief desideratum at each focus of his theory—at the methodological pole for tractability, and at the theoretical pole for reversal of received wisdom. Darwinians have often acknowledged the *descriptively* hierarchical character of nature—and some commentators have been misled to view Darwinism, for this reason, as hierarchical in mechanism of causal action as well. But Darwinism tries to explain all these levels by *one locus of causality*—selection among organisms. Strict Darwinism is a one-level causal theory for rendering nature's hierarchical richness. The major critique of our times, in advocating hierarchical levels of *causality*, therefore poses a fundamental challenge to an essential postulate of Darwin's system.

Consider four aspects and demonstrations of Darwin's conviction about the exclusivity of selection on organisms:

EXPLICIT STATEMENTS. Darwin did not passively "back in" to a claim for the organismic level as a nearly exclusive locus. He knew exactly what he had asserted and why—and he said so over and over again. Statements that selection works "for the good of individuals" recur, almost in catechistic form, throughout the *Origin:* "Natural selection will never produce in a being anything injurious to itself, for natural selection acts solely by and for the good of each (p. 201) . . . Natural selection acts only by the accumulation of slight modifications of structure or instinct, each profitable to the individual under its conditions of life" (p. 233). Even if higher-level order arises as a result, the causal locus must be recognized as individual benefit: "In social animals [natural selection] will adapt the structure of each individual for the benefit of the community; if each in consequence profits by the selected change" (p. 87).

Several other statements illustrate Darwin's emphasis on struggle among organisms, and his desire to avoid all implication that members of a species might amalgamate to collectivities functioning as units of selection in themselves. He continually stresses, for example, that competition tends to be more intense among members of a single species than between individuals of different species—thus emphasizing the difficulty of forming such collectivities. Moreover, Darwin's development of the theory of sexual selection, and his increasing reliance on this mechanism as his views matured, also forestalls any temptation to advocate group selection—as no form of intraspecific competition can be more intense than struggle among similar individuals for personal success in mating.

RESPONSE TO CHALLENGES IN THE *ORIGIN*. The primary commitments of a theory lie best revealed, not so much in the initial exposition of their logic, but in their later employment to resolve difficulties and paradoxes. Darwin devotes much more of the *Origin* than most readers have generally realized to defending his single-level theory of selection on organisms.

Darwin structured the *Origin* as a trilogy—a first part (4 chapters) on the exposition of natural selection, a last section (5 chapters) on the evidence for evolution, and a middle series of 5 chapters on difficulties and responses. Two chapters of this middle section treat a broad range of potential challenges to the creativity of selection and its sequelae—chapter 9 on the geological record (to defend gradualism in the face of apparently contradictory evidence), and chapter 5 on laws of variation (to assert the isotropy of variation—see pp. 144–146). A third (chapter 6) treats general "Difficulties on Theory," mostly centered on gradualism.

Darwin therefore devotes only two of these five chapters, 7 on "Instinct" and 8 on "Hybridism," to specific difficulties—that is, to issues of sufficient import in his mind to merit such extensive and exclusive treatment. Readers have not always discerned the common thread between these two chapters—Darwin's defense of struggle among organisms as the locus of selection. The chapter on hybridism presents, as its central theme, an argument against species selection as the cause of sterility in interspecific crosses. The chapter on instinct treats the more general subject of selection's application to behavior as well as to form, but Darwin devotes more than half of this chapter to social insects, and he presents his primary examples of differentiation among castes and sterility of workers as threats to the principle of selection on organisms.

Darwin raises two separate challenges to natural selection for the case of sterile castes in the Hymenoptera. How, first of all, can sterile castes evolve adaptive differences from queens (and from each other), when individuals of these castes cannot reproduce? If non-reproductive organisms can evolve adaptations, mustn't selection then be working at the higher level of colonies as wholes? Darwin answers, by analogy to domesticated animals once again, that differential survival of non-reproductives may still record selection on fertile members of the population. After all, a breeder can improve the distinct form of castrated animals (raised for food or labor), by mating only those fertile individuals that sire non-reproductives with the most advantageous traits (as recognized by the correlation of selectable features in parents with different traits in their castrated offspring):

> I have such faith in the powers of selection, that I do not doubt that a breed of cattle, always yielding oxen with extraordinarily long horns, could be slowly formed by carefully watching which individual bulls and cows, when matched, produced oxen with the longest horns; and yet no one ox could ever have propagated its kind. Thus I believe it has been with social insects: a slight modification of structure, or instinct, correlated with the sterile condition of certain members of the community, has been advantageous to the community: consequently the fertile males and females of the same community flourished, and transmitted to their fertile offspring a tendency to produce sterile members having the same modification. And I believe that this process has been repeated, until that prodigious amount of difference between the fertile and sterile females of the same species has been produced, which we see in many social insects (p. 238).

(This quotation illustrates a common source of misunderstanding. Darwin does often use such phrases as "advantageous to the community." By our later linguistic conventions, such a statement might seem to signify a leaning to group selectionist arguments. But these conventions did not exist in Darwin's generation. Note how he uses this phrase only as a description of a result. Darwin identifies the causal process yielding this result, in this case and almost every other time he invokes such language, as selection on organisms, with benefit to communities as an epiphenomenal effect.)

The second challenge, the origin of sterility itself, seems more serious—for how could selection, especially in its necessarily gradualistic mode, promote the diminution of reproductive power in individuals? Clearly, the increasingly sterile workers cannot be promoting their own fitness; but their labor may aid their entire nest or hive. Must not the evolution of sterility therefore provide *prima facie* evidence for group selection, and for the failure of Darwin's argument about the exclusivity of selection on organisms?

Darwin does indeed refer to sterility as "one special difficulty, which at first appeared to me insuperable, and actually fatal to my whole theory" (p. 236). He then offers an explanation, based exclusively on organismal selection and similar to his argument about differences in form between workers and reproductives (p. 236): "How the workers have been rendered sterile is a difficulty; but not much greater than that of any other striking modification of structure; for it can be shown that some insects and other articulate animals in a state of nature occasionally become sterile; and if such insects had been social, and it had been profitable to the community that a number should have been annually born capable of work, but incapable of procreation, I can see no very great difficulty in this being effected by natural selection."

The phrase "profitable to the community" seems to imply group selection but, as argued above, this modern interpretation need not reflect Darwin's intent. He did not, after all, know about haplodiploidy, different degrees of relatedness, or parent-offspring conflict. He does not argue here at the *locus classicus* for modern theories of group selection—altruism defined as the rendering of aid (at personal peril or expense) to non-relatives. Rather, he views the hive as a group of cooperating bodies, all tightly related and all generated by the queen. Anything beneficial to the hive fosters the reproductive success of the queen in ordinary natural selection upon her as an individual. The sterility of a worker does not differ in principle from the horns of an ox—a trait not found in parents, but produced by selection *on* parents. A queen that can generate more sterile workers might be favored by selection just as a breeder picks cows that yield castrated oxen with longer horns.

At most, one might hold that Darwin treats the entire hive as an entity—a statement about higher-level selection on the "superorganism" model (see D. S. Wilson and Sober, 1989, and Sober and Wilson, 1998). But here we meet an issue that must be regarded as more linguistic than substantive. Just as Janzen (1977) wishes to identify a clone as a single EI (for "evolutionary individual"), and to treat single bodies of rotifers or aphids as parts, so too might Darwin view the bodies in a hive as iterated organs of the whole. Nonetheless, selection acts on the queen as an individual reproducer. The determinants of her success undoubtedly include the form and function of her sterile offspring. Natural selection can "get at" a beaver through the form of its dam, or at a bird through the shape of its nest—and we do not talk about selection on the higher-level entity of organism plus product. Why should selection not "get at" the queen ant or bee through the conformation of the hive and the function of its members? (See Ruse, 1980, for a parallel argument, in agreement with mine, on Darwin's explanation of hymenopteran castes by organismic selection.)

Darwin takes up a different challenge to the exclusivity of organismic selection in the next chapter on "Hybridism." Crosses between varieties of a species are usually fertile, but crosses between species are generally sterile, or at least greatly impaired in fecundity. Under the guiding precepts of gradualism and uniformitarian methodology, we must view species as former varieties promoted by selection to the greater difference of true distinctness. But natural selection could not have built sterility in gradual degrees from an original fertility between parent and offspring—for sterility cannot benefit the hybrid individual: "On the theory of natural selection the case is especially important, inasmuch as the sterility of hybrids could not possibly be of any advantage to them, and therefore could not have been acquired by the continued preservation of successive profitable degrees of sterility. I hope, however, to be able to show that sterility is not a specially acquired or endowed quality but is incidental on other acquired differences" (p. 245).

Darwin considers two possible explanations. He constructs his entire chapter on hybridism as a defense of natural selection in its ordinary, organismal mode through the rejection of one explanation based on species selection and the advocacy of another rooted in selection on organisms with an interesting twist. Darwin admits that species selection, at first glance, seems to provide a simple and attractive solution: interspecific sterility must originate as an adaptation of species, built and promoted to preserve integrity by preventing introgression and subsequent dissolution. (A. R. Wallace strongly promoted this view. Darwin's firm rejection led to a protracted argument that strongly colored their relationship—see Kottler, 1985; Ruse, 1980.)

But Darwin rejected this explanation because he could not conceive how a species might act as an entity in this manner. Nonetheless, he could not possibly argue in response that hybrid sterility arose by direct selection for the trait itself. He therefore proposed a subtle argument, almost surely correct in our current judgment, for the origin of hybrid sterility as an incidental consequence of other differences established by organismal selection. A. R. Wallace, in striking contrast, remained so committed to viewing every natural phenomenon as a direct adaptation that he willingly roamed up and down among levels of selection (quite unaware of the logical difficulties thus entailed) until he found a locus that could support a direct adaptive explanation.

Darwin argued that any population, in diverging far enough from an ancestor to rank as a separate species, must undergo a series of changes (usually extensive), mediated by natural selection and leading to a set of unique features. Any two species will therefore come to differ in a series of traits directly built by natural selection. These disparities will probably render the two species sufficiently unlike, particularly in rates and modes of reproduction and development, that any hybrids between them will probably be stunted or infertile—not because selection acted directly for sterility, but only as an incidental effect of differences evolved by natural selection for other reasons. Although interspecific sterility cannot be *built* directly by selection for its advantages to organisms, this feature can and will originate *as a consequence* of ordinary selection on organisms. Darwin contrasts his proposal with Wallace's alternative based on direct adaptation *via* species selection:

> Now do these complex and singular rules indicate that species have been endowed with sterility simply to prevent their becoming confounded in nature? I think not. For why should the sterility be so extremely different in degree, when various species are crossed, all of which we must suppose it would be equally important to keep from blending together? . . . The foregoing rules and facts, on the other hand, appear to me clearly to indicate that the sterility both of first crosses and of hybrids is simply in-

cidental or dependent on unknown differences, chiefly in the reproductive systems, of the species which are crossed (p. 260).

In what I regard as Darwin's most brilliant use of his favorite device—argument by analogy—he then compares hybrid sterility with incompatibility in hybrid grafts (whereas grafts between varieties of the same species usually "take"). I find this comparison particularly compelling because we would not be tempted to construct an argument about species selection to explain the incompatibility of grafts—as no advantage for the integrity of species accrues thereby, especially since the "experiment" of grafting between two species almost never occurs in nature. Yet the logical structures of these two arguments about grafting and sterility, as well as the attendant results, share an identical logic—joining within species, and maintenance of separation between species, based upon incidental effects wrought by increasing degrees of difference evolved for other reasons:

> It will be advisable to explain a little more fully by an example what I mean by sterility being incidental on other differences, and not a specially endowed quality. As the capacity of one plant to be grafted or budded on another is so entirely unimportant for its welfare in a state of nature, I presume that no one will suppose that this capacity is a *specially* endowed quality, but will admit that it is incidental on differences in the laws of growth of the two plants . . . The facts by no means seem to me to indicate that the greater or lesser difficulty of either grafting or crossing together various species has been a special endowment; although in the case of crossing, the difficulty is as important for the endurance and stability of specific forms, as in the case of grafting it is unimportant for their welfare (pp. 261–263).

Darwin then drives the point home with a lovely prose flourish (and a memorable visual image!) in explicitly rejecting an appeal to supraorganismal selection. Nature knows no explicit principle of higher-level order. "There is no more reason to think that species have been specially endowed with various degrees of sterility to prevent them crossing and blending in nature, than to think that trees have been specially endowed with various and somewhat analogous degrees of difficulty in being grafted together in order to prevent them becoming inarched in our forests" (p. 276).

THE DEVELOPMENT OF DARWIN'S VIEWS ON ORGANISMIC SELECTION. If the first edition of the *Origin* only marked a waystation in fluctuation or degree of commitment, then Darwin's stand on organismic selection, however strongly expressed in this initiating volume, might not be deemed so central to his worldview. But Ruse (1980) has documented Darwin's continuing and increasing attention to this issue—particularly as he argued with Wallace (see also Kottler, 1985) about the principle of incidental effects to explain hybrid sterility as a side consequence of natural selection rather than a direct product of species selection. Ruse writes (1980, p. 620): "By the end of the decade [the 1860's] with respect to the animal and plant worlds, there was nothing implicit about Darwin's commitment to individual selection. He had looked long and hard at group selection and rejected it."

HOW DARWIN STRUGGLES WITH, AND "WALLS OFF," EXCEPTIONS. The exegetical literature on Darwin usually states that he allowed only two exceptions, in the entire corpus of his writing, to the exclusivity of natural selection on organisms—first, in permitting some form of group selection for the neuter castes of social insects, and second, for the origin of human moral behavior. I agree with Ruse (see point 2 just above) that Darwin did not stray from his orthodoxy for social insects, though some of his terminological choices invite misinterpretation today. For human morality, on the other hand, Darwin did throw in the towel after long struggle—for he could not render altruism towards non-relatives by organismal selection. Nonetheless, a theory often becomes sharpened (not destroyed or even much compromised in a world of relative frequencies) by specifying a domain of exceptions—provided that the exceptions be rare in occurrence, and peculiar in form. As humans, we surely have a legitimate personal interest in our moral behavior, but we cannot enshrine this property as occupying more than a tiny corner of nature (whatever its eventual impact upon our planet, and whatever our parochial concern for its uniqueness).

In the *Descent of Man*, Darwin presents his most interesting and extensive discussion of supraorganismal selection. As an example of his clarity on the issue of levels of selection, consider the following passage on why natural selection could not foster altruistic behavior within a tribe—with an explicit final statement that differential success among distinct tribes should not be called natural selection:

> But it may be asked, how within the limits of the same tribe did a large number of members first become endowed with these social and moral qualities, and how was the standard of excellence raised? It is extremely doubtful whether the offspring of the more sympathetic and benevolent parents, or of those who were the most faithful to their comrades, would be reared in greater number than the children of selfish and treacherous parents of the same tribe. He who was ready to sacrifice his life, as many a savage has been, rather than betray his comrades, would often leave no offspring to inherit his noble nature . . . Therefore it seems scarcely possible (bearing in mind that we are not here speaking of one tribe being victorious over another) that the number of men gifted with such virtues, or that the standard of their excellence, would be increased through natural selection, that is, by the survival of the fittest (1871, vol. 1, p. 163).

In the light of this conundrum, and as part of his resolution, Darwin does allow for selection at the tribal level defined as differential success of groups with more altruists: "It must not be forgotten that although a high standard of morality gives but a slight or no advantage to each individual man and his children over the other men of the same tribe, yet that an advancement in the standard of morality, and an increase in the number of well-endowed men will certainly give an immense advantage to one tribe over another" (1871, p. 166).

This passage has often been quoted, but without its surrounding context of contrary alternatives and restrictive caveats, as a clean example of Darwin's move to a higher level of selection when required. But such an interpretation seriously misrepresents Darwin's motives and logic. He did make the move, but only as one factor in a surrounding context of mitigation. I regard these mitigations and restrictions to hold the line of organismal selection (expressed in three distinct arguments, discussed below) as far more interesting than the move itself, for Darwin's extreme reluctance to address selection at any level other than the organismic lies so well exposed in the totality.

1. The *Descent*, as a whole, rests upon the strongest mode of argument for organismal selection. Darwin did not write a separate book on human evolution; his ideas (mostly speculative) on this subject occupy the first, and shorter, part of a two volume treatise entitled, in full: *The Descent of Man, and Selection in Relation to Sex*. In other words, Darwin wrote the *Descent* as an introduction to his general exposition of sexual selection. We might regard the two parts as oddly juxtaposed until we realize that many of Darwin's major arguments about human evolution—in the establishment of secondary sexual characters, and in differentiation among races, for example—invoke sexual selection by intraspecific competition, rather than ordinary natural selection as adaptation to external environments. As Ruse (1980) notes, Darwin viewed sexual selection as the strongest general argument against group selection, for its theme of relentless struggle in mating among members of a population guarantees that individualism must reign, largely by precluding the formation of alliances that higher-level selection could exploit. (Modern notions of sexual selection do envision the formation of such alliances, so the argument may strike us as incorrect today—but Darwin conceived sexual selection as a hyperindividual mode.)

2. Darwin does not present his argument for tribal selection as a happy solution to the problem of morality, but only as one potential factor among others. He also devises an argument based on organismal selection—in the form that would be called "reciprocal altruism" today: "As the reasoning powers and foresight of the members became improved, each man would soon learn from experience that if he aided his fellow-men, he would commonly receive aid in return. From this low motive he might acquire the habit of aiding his fellows" (1871, p. 163).

3. Darwin presents tribal selection as a peculiarity based on the uniqueness of human consciousness, and thus as a strictly circumscribed exception to the generality of organismal selection throughout living nature. As conscious beings, we become especially sensitive to the "praise and blame" of our fellows. If altruistic behavior gains a status as virtuous, then we might be persuaded—against our deeper biological drive for seeking personal advantage—to engage in such behaviors in order to foster praise or avoid calumny. In other words, a form of "cultural evolution," rooted in our unique level of consciousness, could overcome the behaviors driven by organismal selection, and

could establish a preference for altruistic acts that might then serve as a basis for tribal selection. But such an argument cannot enjoy wide application in nature, as all other species lack this special mental mechanism for spreading abstract ideas against the thrust of natural selection:

> We may therefore conclude that primeval man, at a very remote period, would have been influenced by the praise and blame of his fellows. It is obvious, that the members of the same tribe would approve of conduct which appeared to them to be for the general good, and would reprobate that which appeared evil . . . A man who was not impelled by any deep, instinctive feeling, to sacrifice his life for the good of others, yet was roused to such actions by a sense of glory, would by his example excite the same wish for glory in other men, and would strengthen by exercise the noble feeling of admiration. He might thus do far more good to his tribe than by begetting offspring with a tendency to inherit his own high character (1871, p. 165).

Note also how Darwin, in this passage, explicitly limits within tribal boundaries the extent of such spread against organismal selection. If some form of group selection had to be acknowledged for a special case, Darwin sought to confine its operation to the smallest aggregation within the species—and then to let these small collectivities struggle with others in a minimal context of groupiness.

Thus, in permitting a true exception to organismal selection, Darwin's primary attitude exudes extreme reluctance—restriction to minimal groupiness, provision of other explanations in the ordinary organismal mode, limitation to a unique circumstance in a single species (human consciousness for the spread of an idea against the force of organismal selection), and placement within a more general argument for sexual selection, the strongest form of the orthodox mode.

In my researches for this book, I made a discovery that strongly supports this view of Darwin's attitude towards supraorganismal selection. I found that the traditional sources (Ruse, Kottler and others) did not identify Darwin's major, explicit struggle to contain an apparent need for higher-level selection, and to assert exclusivity for the organismal mode. He fought a far more important battle with himself on an issue well beyond particular problems raised by single taxa (sterility of worker castes or human morality): the explanation of the principle that he ranked second only to natural selection itself as a component of evolutionary theory—the "principle of divergence." (Evolutionists have not recognized this important component of Darwin's developing ideas about selection because he excised this discussion as he abstracted his longer work to compose the *Origin*. But the full version exists in the uncompleted manuscript of his intended larger work—edited and published by Stauffer, 1975, but not widely read by practicing biologists.) Moreover, in his long version, Darwin wrestles not with the lowest interdemic level of tribal selection, but with species selection itself. I will present a full exposition in Chapter 3 (pp. 224–250), but should mention for now that Darwin's tactic closely follows his argument about human morality, and therefore emphasizes his extreme reluctance to embrace supraorganismal selection, and his almost desperate effort to confine explanation to the organismal mode. The recognition that Darwin, despite such strong reluctance, could not avoid some role for species selection, builds a strong historical argument for the ineluctability of a hierarchical theory of selection. (I shall show in Chapter 3 that none of the few 19th century scientists who truly grasped the full range and subtlety of selectionist theory could avoid important roles for levels other than the organismic.)

As with the next topic of creativity for natural selection (pp. 137–159), the issue of levels in selection has resounded through the entire history of evolutionary theory, and continues to set a major part of the agenda for modern debate—as it must, for the subject lies (with only a few others) at the very heart of Darwinian logic. Wallace never comprehended the question of levels at all, as he searched for adaptation wherever he could find it, oblivious to any problems raised by the locus of its action; Kropotkin, in asserting mutual aid, never grasped the problem either; Weismann shared Darwin's insight about the problem's fundamental nature, but also came to understand, after a long and explicit intellectual struggle with his own strong reluctance, that exclusivity must yield to hierarchy (pp. 197–224).

In our generation, Wynne-Edwards (1962) riled an entire profession by defending the classical form of group selection as a generality, while Williams (1966) penned a powerful rebuttal, urging us all to toe the Darwinian line (see Chapter 7 for a full account). The classical ethologists invoked various forms of group selection (often by default); the sociobiologists proclaimed a revolution by reaction and return to the pure Darwinism of individual advantage. Dawkins (1976) attempted an even stronger reduction to exclusivity for genic selection, but his false argument rests on a confusion of bookkeeping with causality, and his own later work (1982) negates his original claim, though Dawkins seems unaware of his own contradictions (see Chapter 8). Supporters of hierarchy theory—I am one, and this is a partisan book—are revising Darwinism into a multilevel theory of selection.

This issue will not go away, and must excite both interest and passion. Nothing else lies so close to the raw nerve of Darwin's radicalism. The exclusivity of organismal selection, after all, provides the punch line that allowed the vision of Adam Smith to destroy the explicit beauty and harmony of William Paley's world.

Viewed in this light, the *Origin*'s very few statements about solace become particularly revealing. Darwin had just overturned a system that provided the philosophical basis of human comfort for millennia. What could he supply in return, as we continue to yearn for solace in this vale of tears? One might be tempted to read the few Darwinian statements about solace as peculiar, exceptional, even "soft" or illogical. But we should note another feature of these statements as well: they yield no ground whatever on the key issue of organismal struggle. Solace must be found in other guises; the linchpin of selection as struggle among organisms cannot be compromised.

Darwin offers two sources for solace. First, the struggle, however fierce, usually brings no pain or distress to organisms (humans, with their intrusive consciousness, have introduced a tragic exception into nature). "When we reflect on this struggle, we may console ourselves with the full belief, that the war of nature is not incessant, that no fear is felt, that death is generally prompt, and that the vigorous, the healthy, and the happy survive and multiply" (p. 79).

Second, this struggle does lead to general improvement, if only as an epiphenomenon, and whatever the cost: "As natural selection works solely by and for the good of each being, all corporeal and mental endowments will tend to progress towards perfection" (p. 489). Darwin could never compromise his central logic; for even this "softest" of all his statements explicitly asserts that selection can only work on organisms—"for the good of each being." And why not? The logic of organismal struggle includes both fierce beauty and empirical adequacy—whatever the psychic costs. And, since roses by other names smell just as sweet, then beauty, even as an epiphenomenon, becomes no less pleasing, and no less a balm for the soul.

## THE SECOND THEME: NATURAL SELECTION AS A CREATIVE FORCE

The following kind of incident has occurred over and over again, ever since Darwin. An evolutionist, browsing through some pre-Darwinian tome in natural history, comes upon a description of natural selection. Aha, he says; I have found something important, a proof that Darwin wasn't original. Perhaps I have even discovered a source of direct and nefarious pilfering by Darwin! In the most notorious of these claims, the great anthropologist and writer Loren Eiseley thought that he had detected such an anticipation in the writings of Edward Blyth. Eiseley laboriously worked through the evidence that Darwin had read (and used) Blyth's work and, making a crucial etymological mistake along the way (Gould, 1987c), finally charged that Darwin may have pinched the central idea for his theory from Blyth. He published his case in a long article (Eiseley, 1959), later expanded by his executors into a posthumous volume entitled "Darwin and the Mysterious Mr. X" (1979).

Yes, Blyth had discussed natural selection, but Eiseley didn't realize—thus committing the usual and fateful error in this common line of argument—that all good biologists did so in the generations before Darwin. Natural selection ranked as a standard item in biological discourse—but with a crucial difference from Darwin's version: the usual interpretation invoked natural selection as part of a larger argument for created permanency.* Natural selection,

---

*Only two exceptions have been noted to this generality—both in the domain of anomalies that prove the rule. The Scottish fruit grower Patrick Matthew (in 1831) and the Scottish-American physician William Charles Wells (in 1813, published in 1818) spoke of natural selection as a positive force for evolutionary change, but neither recognized the significance of his speculation. Matthew buried his views in the appendix to a work entitled "Naval Timber and Arboriculture"; Wells published his conjecture in a concluding section,

treating the origin of human races, to a paper on the medical case of a piebald woman. He presented this paper to the Royal Society in 1813, but only published it as he lay dying in 1818—as a subsidiary to his two famous essays on the origin of dew, and on why we see but one image with two eyes.

Matthew, still alive and vigorously kicking when Darwin published the *Origin,* wrote to express his frustration at Darwin's non-citation. Darwin offered some diplomatic palliation in the historical introduction added to later editions of the *Origin,* while professing, with ample justice, that he had meant no malice, but had simply never encountered Matthew's totally forgotten and inauspiciously located speculation. He responded to Matthew's ire in the *Gardener's Chronicle* for April 21, 1860: "I freely acknowledge that Mr. Matthew has anticipated by many years the explanation which I have offered of the origin of species, under the name of natural selection. I think that no one will feel surprised that neither I, nor apparently any other naturalist, has heard of Mr. Matthew's views, considering how briefly they are given, and that they appeared in the Appendix to a work on Naval Timber and Arboriculture."

Wells' article is particularly intriguing, if only for an antiquarian footnote, in the context of this book's focus on supraorganismal levels of selection. Although Wells has often been cited as a precursor, very few citationists have read his paper, and have therefore simply assumed that he spoke of natural selection by Darwin's route of advantages to individuals within populations. In fact, as I discovered (Gould, 1983a), Wells attributes racial differentiation in skin color to group selection among populations.

I do not wish to make overly much of this point, as "precursoritis" is the bane of historiography; yet I am tickled by the ironic tidbit, in the light of later orthodoxy, that the first formulation of natural selection went forward in the supraorganismic mode. The point should not be overstressed, if only because Wells reached this alternative by the fallacious argument that favorable variants could not spread within populations. Echoing Jenkins' later criticism of Darwin, Wells held that blending inheritance prevents the transformation of populations from within because advantageous variants "quickly disappear from the intermarriages of different families. Thus, if a very tall man be produced, he very commonly marries a woman much less than himself, and their progeny scarcely differs in size from their countrymen" (1818, pp. 434–435).

Populations must therefore be transformed by fortuitous spread and propagation within small and isolated groups: "In districts, however, of very small extent, and having little intercourse with other countries, an accidental difference in the appearance of the inhabitants will often descend to their late posterity" (p. 435). Change may then occur within an entire species by group selection among these differentiated populations:

> Of the accidental varieties of man, which would occur among the first few and scattered inhabitants of the middle regions of Africa, some would be better fitted than the others to bear the diseases of the country. This race would consequently multiply, while the others would decrease, not only from their inability to sustain the attacks of disease, but from their incapacity of contending with their more vigorous neighbors. The color of this vigorous race I take for granted . . . would be dark. But the same disposition to form varieties still existing, a darker and a darker race would in the course of time occur, and as the darkest would be the best fitted for the climate, this would at length become the most prevalent, if not the only race, in the particular country in which it had originated (pp. 435–436).

Note Wells' unquestioned assumption that our original color must have been white, and that dark skin could only arise as a modification of the type. As a final interesting footnote, Wells denied (probably wrongly) that dark skin could be adaptive in itself, and argued for its establishment in Africa as a result of noncausal correlation with unknown physiological mechanisms for protection against tropical disease. Thus, Wells presents an "internalist" explanation based on what Darwin would later call "correlation of growth." With this argument about channels, and his basic claim for group selection, Wells' departure from Darwin's later preferences lie very much in the spirit of modern critiques, though for reasons that we would now reject (as if our anachronistic judgment mattered).

in this negative formulation, acted only to preserve the type, constant and inviolate, by eliminating extreme variants and unfit individuals who threatened to degrade the essence of created form. Paley himself presents the following variant of this argument, doing so to refute (in later pages) a claim that modern species preserve the good designs winnowed from a much broader range of initial creations after natural selection had eliminated the less viable forms: "The hypothesis teaches, that every possible variety of being hath, at one time or other, found its way into existence (by what cause or in what manner is not said), and that those which were badly formed, perished" (Paley, 1803, pp. 70–71).

Darwin's theory therefore cannot be equated with the simple claim that natural selection operates. Nearly all his colleagues and predecessors accepted this postulate. Darwin, in his characteristic and radical way, grasped that this standard mechanism for preserving the type could be inverted, and then converted into the primary cause of evolutionary *change*. Natural selection obviously lies at the center of Darwin's theory, but we must recognize, as Darwin's second key postulate, the claim that natural selection acts as *the creative force* of evolutionary *change*. The essence of Darwinism cannot reside in the mere observation that natural selection operates—for everyone had long accepted a *negative* role for natural selection in eliminating the unfit and preserving the type.

We have lost this context and distinction today, and our current perspective often hampers an understanding of the late 19th century literature and its preoccupations. Anyone who has read deeply in this literature knows that no argument inspired more discussion, while no Darwinian claim seemed more vulnerable to critics, than the proposition that natural selection should be viewed as a *positive* force, and therefore as the primary cause of evolutionary change. The "creativity of natural selection"—the phrase generally used in Darwin's time as a shorthand description of the problem—set the cardinal subject for debate about evolutionary mechanisms during Darwin's lifetime and throughout the late 19th century.

Non-Darwinian evolutionists did not deny the reality, or the operationality, of natural selection as a genuine cause stated in the most basic or abstract manner—in the form that I called the "syllogistic core" on page 125 (still used as the standard pedagogical device for teaching the "bare bones" logic of Darwinism in general and introductory college courses). They held, rather, that natural selection, as a headsman or executioner, could only eliminate the unfit, while some other cause must play the positive role of constructing the fit.

For example, Charles Lyell—whom Darwin convinced about the factuality of evolution but who never (much to Darwin's sadness and frustration) accepted the mechanism of natural selection—admitted that he had become stymied on the issue of creativity. He could understand, he wrote in his fifth journal on the "species question" in March, 1860, how natural selection might act like two members of the "Hindoo Triad"—like Vishnu the preserver and Siva the destroyer, but he simply could not grasp how such a force could also work like Brahma, the creator (in Wilson, 1970, p. 369).

E. D. Cope, chief American critic and exponent of neo-Lamarckism, chose a sardonic title to highlight Darwin's supposedly fatal weakness in claiming a creative role for natural selection. He called his book *The Origin of the Fittest* (1887)—a parody on Darwin's "survival of the fittest," and a motto for what natural selection could *not* accomplish. Cope wrote: "The doctrines of 'selection' and 'survival' plainly do not reach the kernel of evolution, which is, as I have long since pointed out, the question of 'the origin of the fittest.' This omission of this problem from the discussion of evolution is to leave Hamlet out of the play to which he has given the name. The law by which structures originate is one thing; those by which they are restricted, directed, or destroyed, is another thing" (1887, p. 226).

We can understand the trouble that Darwin's contemporaries experienced in comprehending how selection could work as a creative force when we confront the central paradox of Darwin's crucial argument: natural selection makes nothing; it can only choose among variants originating by other means. How then can selection possibly be conceived as a "progressive," or "creative," or "positive" force?

In resolving this paradox, Darwin recognized his logical need, within the basic structure of his argument, to explicate the three main requirements and implications of an argument for selection's creativity: (1) the nature of variation; (2) the rate and continuity of change; (3) the meaning of adaptation. This interrelated set of assertions promotes natural selection from mere existence as a genuine, but secondary and negative, mechanism to domination as the primary cause of evolutionary change and pattern. This set of defenses for selection's creativity therefore ranks as the second of three essential postulates, or "minimal commitments" of Darwinian logic.

As the epitome of his own solution, Darwin admitted that his favored mechanism "made" nothing, but held that natural selection must be deemed "creative" (in any acceptable vernacular sense of the term) if its focal action of differential preservation and death could be construed as the primary cause for imparting direction to the process of evolutionary change. Darwin reasoned that natural selection can only play such a role if evolution obeys two crucial conditions: (1) if nothing about the provision of raw materials—that is, the sources of variation—imparts direction to evolutionary change; and (2) if change occurs by a long and insensible series of intermediary steps, each superintended by natural selection—so that "creativity" or "direction" can arise by the summation of increments.

Under these provisos, variation becomes raw material only—an isotropic sphere of potential about the modal form of a species. Natural selection, by superintending the differential preservation of a biassed region from this sphere in each generation, and by summing up (over countless repetitions) the tiny changes thus produced in each episode, can manufacture substantial, directional change. What else but natural selection could be called "creative," or direction-giving, in such a process? As long as variation only supplies raw material; as long as change accretes in an insensibly gradual manner; and as long as the reproductive advantages of certain individuals provide the statistical source of change; then natural selection must be construed as the directional cause of evolutionary modification.

These conditions are stringent; and they cannot be construed as vague, unconstraining, or too far in the distance to matter. In fact, I would argue that the single most brilliant (and daring) stroke in Darwin's entire theory lay in his willingness to assert a set of precise and stringent requirements for variation—all in complete ignorance of the actual mechanics of heredity. Darwin understood that if any of these claims failed, natural selection could not be a creative force, and the theory of natural selection would collapse. We pay our highest tribute to the power of natural selection in recognizing how Darwin used the theory to deduce a set of necessary properties for variation, well before science understood the mechanism of heredity—and in noting that these properties then turned out to be both basically correct and also entailed by the causes later discovered!

### The requirements for variation

In order to act as raw material only, variation must walk a tightrope between two unacceptable alternatives. First and foremost, variation must exist in sufficient amounts, for natural selection can make nothing, and must rely upon the bounty thus provided; but variation must not be too florid or showy either, lest it become the creative agent of change all by itself. Variation, in short, must be copious, small in extent, and undirected. A full taxonomy of non-Darwinian evolutionary theories may be elaborated by their denials of one or more of these central assumptions.

COPIOUS. Since natural selection makes nothing and can only work with raw material presented to its stringent review, variation must be generated in copious and dependable amounts (especially given the hecatomb of selective deaths accompanying the establishment of each favorable feature). Darwin's scenario for selective modification always includes the postulate, usually stated explicitly, that all structures vary, and can therefore evolve. He argues, for example, that if a short beak were favored on a full-grown pigeon "for the bird's own advantage" (p. 87), then selection would also work within the egg for sufficient beak strength to break the shell despite diminution in overall size of the beak—unless evolution followed an alternate route of selection for thinner shells, "the thickness of the shell being known to vary like any other structure" (p. 87).

Darwin's faith in the copiousness of variation can be gauged most clearly by his response to the two most serious potential challenges of his time. First, he acknowledges the folk wisdom that some domestic species (dogs, for example) have developed great variety, while others (cats, for example) differ far less among populations. If these universally recognized distinctions arise as consequences of differences in the intrinsic capacity of species to vary, then Darwin's key postulate of copiousness would be compromised—for failure of sufficient raw material would then be setting a primary limit upon the rate and style of evolutionary change, and selection would not occupy the driver's seat.

Darwin responds by denying this interpretation, and arguing that differing intensities of selection, rather than intrinsically distinct capacities for variation, generally cause the greater or lesser differentiation observed among domestic species. I regard this argument as among the most forced and uncomfortable in the *Origin*—a rare example of Darwinian special pleading. But Darwin realizes the centrality of copiousness to his argument for the creativity of natural selection, and he must therefore face the issue directly:

> Although I do not doubt that some domestic animals vary less than others, yet the rarity or absence of distinct breeds of the cat, the donkey, peacock, goose, etc., may be attributed in main part to selection not having been brought into play: in cats, from the difficulty in pairing them; in donkeys, from only a few being kept by poor people and little attention paid to their breeding; in peacocks, from not being very easily reared and a large stock not kept; in geese, from being valuable only for two purposes, food and feathers, and more especially from no pleasure having been felt in the display of distinct breeds (p. 42).

Second, copiousness must also be asserted in the face of a powerful argument about limits to variation following modal departure from "type." To use Fleeming Jenkin's (1867) famous analogy: a species may be compared to a rigid sphere, with modal morphology of individuals at the center, and limits to variation defined by the surface. So long as individuals lie near the center, variation will be copious in all directions. But if selection brings the mode to the surface, then further variation in the same direction will cease—and evolution will be stymied by an intrinsic limitation upon raw material, even when selection would favor further movement. Evolution, in other words, might consume its own fuel and bring itself to an eventual halt thereby. This potential refutation stood out as especially serious—not only for threatening the creativity of natural selection, but also for challenging the validity of uniformitarian extrapolation as a methodology of research. Darwin responded, as required by logical necessity, that such limits do not exist, and that new spheres of equal radius can be reconstituted around new modes: "No case is on record of a variable being ceasing to be variable under cultivation. Our oldest cultivated plants, such as wheat, still often yield new varieties: our oldest domesticated animals are still capable of rapid improvement or modification" (p. 8).

I cannot here provide a full history for the subsequent odysseys of these key Darwinian precepts. But a few cursory comments indicate how these claims have remained central and contentious throughout the history of post-Darwinian thought, and how they continue to underlie important debates within Darwinism today.

The argument about copiousness, particularly as expressed in the claim for limits to further variability after intense selection, dogged the 19th century literature and emerged as a key issue in the biometrician *vs.* Mendelian debates early in our century (see Provine, 1971). Castle (1916, 1919) pursued his famous experiments on selection in hooded rats in order to test the hypothesis of limits imposed by variability upon continued change. One of the most appealing features of Mendelism—and a strong reason for acceptance following its "rediscovery" in 1900—lay in the argument that mutation could restore variation "used up" by selection. Nor has the issue abated today. In another form, copiousness underlay the great debate between Dobzhansky and Muller (see Lewontin, 1974)—the classical *vs.* the balance view in Dobzhansky's terminology. Kimura's (1963, 1983) modern theory of neutralism may be invoked to acknowledge the fact of copiousness while avoiding the pitfalls of genetic load—and therefore becomes "neoclassical" in Lewontin's terminology.

SMALL IN EXTENT. If the variations that yielded evolutionary change were large—producing new major features, or even new taxa in a single step—then natural selection would not disappear as an evolutionary force. Selection would still function in an auxiliary and negative role as headsman—to heap up the hecatomb of the unfit, permit the new saltation to spread among organisms in subsequent generations, and eventually to take over the population. But Darwinism, as a theory of evolutionary change, would perish—for selection would become both subsidiary and negative, and variation itself would emerge as the primary, and truly creative, force of evolution, the source of occasionally lucky saltation. For this reason, and quite properly, saltationist (or macromutational) theories have always been viewed as anti-Darwinian—despite the protestations of de Vries (see Chapter 5), who tried to retain the Darwinian label for his continued support of selection as a negative force. The unthinking, knee-jerk response of many orthodox Darwinians whenever they hear the word "rapid" or the name "Goldschmidt," testifies to the conceptual power of saltation as a cardinal danger to an entire theoretical edifice.

Darwin held firmly to the credo of small-scale variability as raw material because both poles of his great accomplishment required this proviso. At the methodological pole of using the present and palpable as a basis, by extrapolation, for all evolution, Darwin longed to locate the source of all change in the most ordinary and pervasive phenomenon of small-scale variation among members of a population—Lyell's fundamental uniformitarian principle, recast for biology, that all scales of history must be explained by currently observable causes acting within their current ranges of magnitude and intensity. "I believe mere individual differences suffice for the work," Darwin writes (p. 102). At the theoretical pole, natural selection can only operate in a creative manner if its cumulating force builds adaptation step by step from an isotropic pool of small-scale variability. If the primary source of evolutionary innovation must be sought in the occasional luck of fortuitous saltations, then internal forces of variation become the creative agents of change, and natural selection can only help to eliminate the unfit after the fit arise by some other process. Darwin, again using domestication as an analog, passionately defends the central role of variation so small as to pass beneath nearly everyone's notice (p. 32):

> If selection consisted merely in separating some very distinct variety, and breeding from it, the principle would be so obvious as hardly to be worth notice; but its importance consists in the great effect produced by the accumulation in one direction, during successive generations, of differences absolutely inappreciable by an uneducated eye—differences which I for one have vainly attempted to appreciate. Not one man in a thousand has accuracy of eye and judgment sufficient to become an eminent breeder. If gifted with these qualities, and he studies his subject for years, and devotes his lifetime to it with indomitable perseverance, he will succeed, and may make great improvements; if he wants [that is, lacks] any of these qualities, he will assuredly fail.

Saltational variation has always served as a rallying point for non-Darwinian evolutionary argument (see Chapters 4 and 5 for a full discussion). T. H. Huxley centered his own doubts about natural selection firmly upon Darwin's preference for change by insensible steps. Bateson (1894), in developing the concept of homeosis, and D'Arcy Thompson (1917), in his ideas on noncontinuity in certain geometrical transformations, advanced saltation as an explicitly anti-Darwinian argument. The early mutationists read Mendel as a warrant for discontinuous change, and a disproof of strict Darwinism as espoused by the "biometricians." Goldschmidt (1940; see Gould, 1982a) joined some interesting views on developmental discontinuity to an untenable genetic theory, all the better to espouse a saltationist view that made him the chief whipping boy of the Modern Synthesis.

Reciprocally, Darwinians countered with strong and explicit support. R. A. Fisher began his great book (1930) by rooting a defense of Darwin in a linkage of copiousness with small-scale variation—specifically, by arguing for an inverse correlation of frequency and effect, and then claiming that variations of large effect therefore become too rare to serve as evolution's raw material.

UNDIRECTED. Textbooks of evolution still often refer to variation as "random." We all recognize this designation as a misnomer, but continue to use the phrase by force of habit. Darwinians have never argued for "random" mutation in the restricted and technical sense of "equally likely in all directions," as in tossing a die. But our sloppy use of "random" (see Eble, 1999) does capture, at least in a vernacular sense, the essence of the important claim that we do wish to convey—namely, that variation must be *unrelated to the direction of evolutionary change;* or, more strongly, that nothing about the process of creating raw material biases the pathway of subsequent change in adaptive directions. This fundamental postulate gives Darwinism its "two step" character, the "chance" and "necessity" of Monod's famous formulation—the separation of a source of raw material (mutation, recombination, etc.) from a force of change (natural selection).

In a sense, the specter of directed variability threatens Darwinism even more seriously than any putative failure of the other two postulates. Insufficient variation stalls natural selection; saltation deprives selection of a creative role but still calls upon Darwin's mechanism as a negative force. With directed variation, however, natural selection can be bypassed entirely. If adaptive pressures automatically trigger heritable variation in favored directions, then trends can proceed under regimes of random mortality; natural selection, acting as a negative force, can, at most, accelerate the change.

Lamarckism (defined in the modern sense of "soft" heredity) represents the quintessential theory of directed variability. Variation arises with intrinsic bias in adaptive directions either because organisms respond creatively to "felt needs" and pass acquired features directly to their offspring, or because environments induce heritable variation along favored pathways. Other directional theories differ in viewing intrinsic variation as unrelated to adaptation, but still capable of overwhelming any counteracting selection, and therefore setting the path of evolutionary change. Historically important theories in this mode include various notions of orthogenesis that postulate the inevitable origin of hypertrophied and inadaptive structures; and theories of "racial life cycles" that envision an ineluctably aging protoplasm doomed to extinction despite any effort at "rejuvenation" by natural selection. (I shall discuss such ideas in Chapter 5.)

Darwin clearly understood the threat of directed variability to his cardinal postulate of creativity for natural selection. He explicitly restricted the sources of variation to auxiliary roles as providers of raw material, and granted all power over the direction of evolutionary change to natural selection. Drawing his customary analogy to artificial selection, Darwin writes (p. 30): "The key is man's power of accumulative selection: nature gives successive variations; man adds them up in certain directions useful to him. In this sense he may be said to make for himself useful breeds."

Darwin also understood that variation could not be construed as truly random in the mathematical sense—and that history did not imply or require this strict form of randomness. He recognized biased tendencies to certain states of variation, particularly reversions toward ancestral features. But he viewed such tendencies as weak and easily overcome by selection. Thus, by the proper criterion of relative power and frequency, selection controls the direction of change: "When under nature the conditions of life do change, variations and reversions of character probably do occur; but natural selection, as will hereafter be explained, will determine how far the new characters thus arising shall be preserved" (p. 15).

We may summarize Darwin's third requirement for variation under the rubric of *isotropy,* a common term in mineralogy (and other sciences) for the concept of a structure or system that exhibits no preferred pathway as a consequence of construction with equal properties in all directions. Darwinian variation must be *copious in amount, small in extent, and effectively isotropic.* (Think again of a dynamic sphere, with all radii accessible. The modal form lies at the center and may move by selection along any radius. At any new location, a sphere of comparable size may be reconstituted about the altered modal form.) Only under these stringent conditions can natural selection—a force that makes nothing directly, and must rely upon variation for all raw material—be legitimately regarded as creative.

### Gradualism

Darwinism, like most comprehensive and complex concepts, defies easy definition. Darwinism cannot be analogized to an object, like the Parthenon, with a clear criterion of membership for each potential slab (whether now resident in the British Museum or in Athens). Moreover, the various propositions of Darwinism cannot be regarded as either independent or of equal force. Darwinism cannot be construed as a deductive system, with some defining axioms and a set of logical entailments tied together like a classical proof in plane geometry. But neither can Darwinism be viewed as a set of separate stones, all of similar size, and each ejectable from a bag without great disturbance to the others.

As discussed at length in Chapter 1 (pp. 12–24), I view the conceptual structure of Darwinism much like the metaphor that Darwin himself first used (see Barrett et al., 1987) for depicting evolution (in the "B Notebook" on transmutation kept during the 1830's)—the "coral of life" (later superseded, in Chapter 4 of the Origin, and in other writings, by the tree of life). The central trunk (the theory of natural selection) cannot be severed, or the creature dies (see Fig. 1-4, p. 18). The first-order branches are also so fundamental that any severing of a complete branch converts the theory into something essentially different that must be newly named. (I have suggested that the theory of natural selection includes three major branches, discussed in sections B-D of this subchapter.) Each major branch then divides into smaller sub-branches. (In the present section C, I argue that the second major branch, the claim for "creativity of natural selection," divides into three important sub-branches of "requirements for variation," "gradualism," and "the adaptationist program.")

As further argued in Chapter 1, this model allows us to address the important question of dispensability. At some level above the base, we may excise a sub-branch, deny its premises, and still consider ourselves Darwinians. I envision the central trunk and first-order branches as indispensable. Along the continuum from necessary to avoidable, we may begin to make selective negations at the level of sub-branches, but not without severe stress to the entire structure. Thus, T. H. Huxley could oppose gradualism and still consider himself a supporter of natural selection (though his approbation remained ambiguous and indifferent at best, and his role as "Darwin's bulldog" rested upon his defense of evolution itself, not his explication of natural selection). And a modern developmental saltationist might call himself a Darwinian, though not without an array of "buts" and qualifications.

One other feature of the model requires explicit commentary. I have chosen a coral in preference to the more conventional tree, because the branches of many corals form a network by lateral anastomoses (while each limb of a tree stands free, and may be chopped off without necessarily affecting the others).

The premises of Darwin's theory (the branches and sub-branches of the coral model) are organically connected. One might be able to excise a single branch without killing the others, but some pain and readjustment will certainly be felt throughout the entire structure. The three sub-branches of the "creativity" limb, for example, are strongly conjoined in this manner. If variation forms an isotropic sphere (the expectation of sub-branch one), then change by natural selection can only occur a short step at a time (as predicted by the gradualism of sub-branch two). And if variation imposes no constraint upon the direction of change (an inference from isotropy), then natural selection works freely and adaptation prevails (as required by sub-branch three).

Finally, as so often emphasized throughout this book, we must recognize and embrace natural history as a science of relative frequencies. None of these basic Darwinian premises operates without exception throughout nature. Darwin insisted*—explicitly and vociferously—that natural selection only enjoyed a predominant relative frequency, not exclusivity: "the main but not exclusive means of modification," as he writes at the close of the introduction (p. 6). Darwin then extended his claim for a predominant relative frequency, but not for exclusivity, to all other sub-branches of his essential argument as well. Failure of raw material might occasionally explain a puzzling absence of evolutionary modification—but lack of selective pressure for change surely represents the more likely explanation for stasis by far. Substantial change might occur as a very rare event, but *most* alteration must be insensible, even on geological scales: "We see nothing of these slow changes in progress, until the hand of time has marked the long lapse of ages" (p. 84).

Understanding Darwin's mode of justification by relative frequency be-

---

*Charles Darwin surely ranks as the most genial of history's geniuses—possessing none of those bristling quirks and arrogances that usually mark the type. Yet, one subject invariably aroused his closest approach to fury—the straw-man claim, so often advanced by his adversaries, that he regarded natural selection as an exclusive mode of change in evolution. Darwin, who understood so well that natural history works by relative frequency, explicitly denied exclusivity and argued only for dominance. So frustrated did he become at the almost willful misunderstanding of a point so clearly made, that he added this rueful line to the 6th edition of the *Origin* (1872b, p. 395): "As my conclusions have lately been much misrepresented, and it has been stated that I attribute the modification of species exclusively to natural selection, I may be permitted to remark that in the first edition of this work, and subsequently, I placed in a most conspicuous position—namely at the close of the Introduction—the following words: 'I am convinced that natural selection has been the main, but not the exclusive means of modification.' This has been of no avail. Great is the power of steady misinterpretation."

Darwin's good friend G. J. Romanes, author of a famous essay on Darwin's pluralism vs. the panselectionism of Wallace and Weismann, wrote of this statement (1900, p. 5): "In the whole range of Darwin's writings there cannot be found a passage so strongly worded as this: it presents the only note of bitterness in all the thousands of pages which he has published." But Darwin wrote other bristling statements on the same sensitive subject. In 1880, for example, he castigated Sir Wyville Thomson for caricaturing him as a panselectionist: "This is a standard of criticism not uncommonly reached by theologians and metaphysicians when they write on scientific subjects, but is something new as coming from a naturalist . . . Can Sir Wyville Thomson name any one who has said that the evolution of species depends only on natural selection?" (1880b, p. 32).

comes vitally important because selective quotation represents the most common error made by evolutionists in interpreting his work and theory. The *Origin*, as a volume of single authorship, maintains a stronger plot line and features fewer inconsistencies than the Bible; but Darwin and the Good Lord do share the common trait of saying something about nearly everything. Wrenched from context and divorced from a crucial assessment by relative frequency, a Darwinian statement can be found to support almost any position, even the most un-Darwinian.

Since Darwin prevails as the patron saint of our profession, and since everyone wants such a preeminent authority on his side, a lamentable tradition has arisen for appropriating single Darwinian statements as defenses for particular views that either bear no relation to Darwin's own concerns, or that even confute the general tenor of his work. Thus, for example, Darwin wrote extensively about variational constraint, and he maintained great interest in this topic (see Chapter 4). But the logic of his work entails adaptive control of evolutionary change and isotropy of variation as generally prevalent—and Darwin ultimately comes down (as he must) on the side of these necessary underpinnings for natural selection. Proper textual analysis requires that general tenor, not selective statement, be presented. Two basic procedural modes, each with distinctive criteria, set the framework for such textual analysis. The empirical mode makes its judgments of importance by relative frequency and interconnectedness of statements. Meanwhile, and simultaneously, the logical mode employs theoretical consistency as an arbiter for judging the validity and power of the structure of argument. We revere Darwin because he unfailingly manifested the two key traits of brilliance and honesty. He knew where his arguments led, and he followed them relentlessly, however unpleasant the consequences. We do him the greatest possible disservice when we approach his work as a superficial grazer, searching for some particular item of personal sustenance, while ignoring the beauty and power of *general tenor and logical entailment*.

I raise this point here because abuse of selective quotation has been particularly notable in discussions of Darwin's views on gradualism. Of course Darwin acknowledged great variation in rates of change, and even episodes of rapidity that might be labelled catastrophic (at least on a local scale); for how could such an excellent naturalist deny nature's multifariousness on such a key issue as *the* character of change itself? But these occasional statements do not make Darwin the godfather of punctuated equilibrium, or a cryptic supporter of saltation (as de Vries actually claimed, thus earning a unique and official rebuke from the organizers of the Darwinian centenary celebration at Cambridge—see p. 416).

Gradualism may represent the most central conviction residing both within and behind all Darwin's thought. Gradualism far antedates natural selection among his guiding concerns, and casts a far wider net over his choice of subjects for study. Gradualism sets the explanatory framework for his first substantive book on coral reefs (1842) and for his last on the formation of topography and topsoil by earthworms (1881)—two works largely devoid of reference to natural selection. Gradualism had been equated with rationality itself by Darwin's chief guru, Charles Lyell (see Chapter 6). All scholars have noted the centrality of gradualism, both in the ontogeny (Gruber and Barrett, 1974) and logic (Mayr, 1991) of Darwin's thought.

I will not play "duelling quotations" with "citation grazers," though a full tabulation of relative frequencies could easily bury their claims under a mountain of statements. For the present assessment of branch two ("creativity of natural selection") on the coral of essential Darwinian logic, the necessity of gradualism will suffice. Selection becomes creative only if it can impart direction to evolution by superintending the slow and steady accumulation of favored subsets from an isotropic pool of variation. If gradualism does not accompany this process of change, selection must relinquish this creative role and Darwinism then fails as a creative source of evolutionary novelty. If important new features, or entire new taxa, arise as large and discontinuous variations, then creativity lies in production of the variation itself. Natural selection no longer causes evolution, and can only act as a headsman for the unfit, thus promoting changes that originated in other ways. Gradualism therefore becomes a logical consequence of the operation of natural selection in Darwin's creative mode. Gradualism also pervades the methodological pole of Darwin's greatness because the uniformitarian argument of extrapolation will not work unless change at the grandest scale arises by the summation through time of small, immediate, and palpable variations.

Gradualism, for Darwin, represents a complex doctrine with several layers of meaning, all interconnected, while remaining independent in some important senses. I shall consider three increasing levels of specificity, arguing, on the Goldilocks model, that one meaning is too nebulous, another overly wrought, but the third (in the middle) "just right" as the crucial validator of natural selection (whereas the other two meanings play equally crucial roles for other aspects of Darwin's view of life).

HISTORICAL CONTINUITY OF STUFF AND INFORMATION. At the broadest level, gradualism merely asserts unbroken historical connectedness between putative ancestor and descendant, without characterizing the mode or rate of transition. If new species originate as creations *ex nihilo* by a divine power, then connectivity fails. The assertion of gradualism in this broadest meaning encapsulates the chief defense for the factuality of evolution. Such a contention could not be more vital to Darwin's revolution of course, but this sense of gradualism only asserts *that* evolution occurred, while telling us nothing about *how* evolution happens; the logical tie of gradualism to natural selection cannot reside here.* Thus, this first, or "too big," sense of gradual-

---

*Some modern evolutionists have made the error of assuming that contemporary debates about gradualism engage this now obvious and entirely uncontroversial meaning. Thus Gingerich (1984a), abandoning his earlier and properly empirical approach to gradualism (sense iii of p. 152) vs. punctuation (1976), argues that gradualism must be true *a priori*, as equivalent to "empiricism" in paleontology. He then provides a curious definition of stasis as "gradualism at zero rate"—an oxymoron with respect to the definition of gradualism that punctuated equilibrium opposes with a prediction of stasis. I was, at first, deeply

ism validates evolution itself (*vs.* creationism), but not Darwin's, or anyone else's, proposed mechanism of evolutionary change.

INSENSIBILITY OF INTERMEDIACY. We now come to the heart of what natural selection requires. This second, "just right," statement does not advance a claim about how much time a transition must take, or how variable a rate of change might be. The second meaning simply asserts that, in going from A to a substantially different B, evolution must pass through a long and insensible sequence of intermediary steps—in other words, that ancestor and descendant must be linked by a series of changes, each within the range of what natural selection might construct from ordinary variability. Without gradualism in this form, large variations of discontinuous morphological import—rather than natural selection—might provide the creative force of evolutionary change. But if the tiny increment of each step remains inconsequential in itself, then creativity must reside in the summation of these steps into something substantial—and natural selection, in Darwin's theory, acts as the agent of accumulation.

This meaning of gradualism underlies Darwin's frequent invocation of the old Leibnizian and Linnaean aphorism, *Natura non facit saltum* (nature does not proceed by leaps). Darwin's commitment to this postulate can only strike us as fierce and, by modern standards, overly drawn. Thus, Darwin writes (p. 189): "If it could be demonstrated that any complex organ existed, which could not possibly have been formed by numerous, successive, slight modifications, my theory would absolutely break down." And lest we doubt that "my theory" refers specifically to the mechanism of natural selection (and not simply to the assertion of evolution), Darwin often draws an explicit link between selection as a creative force and gradualism as an implied necessity: "Undoubtedly nothing can be effected through Natural Selection except by the addition of infinitesimally small changes; and if it could be shown that . . . transitional states were impossible, the theory would be overthrown" (in *Natural Selection*—see Stauffer, 1975, p. 250). And in the concluding chapter of the *Origin*: "As natural selection acts solely by accumulating slight, successive, favorable variations, it can produce no great or sudden modification; it can act only by very short and slow steps. Hence the canon of 'Natura non facit saltum' . . . is on this theory simply intelligible" (p. 471).

But would the theory of natural selection "absolutely break down" if even a single organ—not to mention an entire organism—could arise by large and discontinuous changes? Does Darwinism truly require the following extreme

---

puzzled by Gingerich's definition until I realized the source of his confusion. He had switched definitions from the empirical issue of rates (meaning iii of this discussion)—a lively and testable argument opposing stasis to gradualism defined as a rate of *change*—to the completely settled question of historical continuity. Does anyone seriously think that supporters of punctuated equilibrium, or any scientist for that matter, would deny historical continuity? His argument therefore dissolves into the empty linguistic effort of trying to win a debate by shifting a definition. The question of punctuated equilibrium will be resolved by empirical testing under the third definition of gradualism. (See Chapter 9 for a full discussion of this issue.)

formulation: "Natural selection can only act by the preservation and accumulation of infinitesimally small inherited modifications" (p. 95). At some level of discontinuity, of course, Darwin's strong statement must prevail. If the altered morphology of new species often arose in single steps by fortuitous macromutation, then selection would lose its creative role and could act only as a secondary and auxiliary force to spread the sudden blessing through a population. But can we justify Darwin's application of the same claim to single organs? Suppose (as must often happen) that developmental heterochrony produces a major shift in form and function by two or three steps without intermediary stages. The size of these steps may lie outside the "normal" variation of most populations at most moments, but not beyond the potential of an inherited developmental program. (Incidentally, these types of changes represent the concept that Goldschmidt embodied in the legitimate meaning of "hopeful monster," before he made his unfortunate decision to tie this interesting concept to his fallacious genetics of "systemic mutation"—see Chapter 5 and Gould, 1982a.)

Would natural selection perish if change in this mode were common? I don't think so. Darwinian theory would require some adjustments and compromises—particularly a toning down of assertions about the isotropy of variation, and a more vigorous study of internal constraint in genetics and development (see Chapter 10 for advocacy of this theoretical shift)—but natural selection would still enjoy a status far higher than that of a mere executioner. A new organ does not make a new species; and a new morphology must be brought into functional integration—a process that requires secondary adaptation and fine tuning, presumably by natural selection, whatever the extent of the initial step.

I believe, therefore, that Darwin's strong, even pugnacious, defense of strict gradualism reflects a much more pervasive commitment, extending far beyond the simple recognition of a logical entailment implied by natural selection—and that this stronger conviction must record such general influences as Darwin's attraction to Lyell's conflation of gradualism with rationality itself, and the cultural appeal of gradualism during Britain's greatest age of industrial expansion and imperial conquest (Gould, 1984a). Huxley's savvy assessment of the *Origin* still rings true, for while he offered, in his famous letter to Darwin, written just as the *Origin* rolled off the presses, to "go to the stake" for Darwin's view, he also stated his major criticism: "You have loaded yourself with an unnecessary difficulty in adopting *Natura non facit saltum* so unreservedly" (in L. Huxley, 1901, p. 189).

Darwin persevered nonetheless. We often fail to recognize how much of the *Origin* presents an exposition of gradualism, rather than a defense of natural selection. As a striking example, the famous (and virtually only) statement about human evolution asserts the pedagogical value of gradualism—not natural selection—in our Socratic quest to know ourselves: "Psychology will be based on a new foundation, that of the necessary acquirement of each mental power and capacity by gradation. Light will be thrown on the origin of man and his history" (p. 488).

Chapter 9 on geological evidence, where the uninitiated might expect to find a strong defense for evolution from the most direct source of evidence in the fossil record, reads instead as a long (and legitimate) apologia for a threatening discordance between data and logical entailment—a fossil record dominated by gaps and discontinuities when read literally *vs.* the insensible transitions required by natural selection as a creative agent. Darwin, with his characteristic honesty, states the dilemma baldly in succinct deference to his methodological need for equating temporal steps of change with differences noted among varieties of contemporary species: "By the theory of natural selection all living species have been connected with the parent-species of each genus, by differences not greater than we see between the varieties of the same species at the present day" (p. 281).

Darwin, as we all know, resolved this discordance by branding the fossil record as so imperfect—like a book with few pages present and only a few letters preserved on each page—that truly insensible continuity becomes degraded to a series of abrupt leaps in surviving evidence:

> Why then is not every geological formation and every stratum full of such intermediate links? Geology assuredly does not reveal any such finely graduated organic chain; and this, perhaps, is the most obvious and gravest objection which can be urged against my theory. This explanation lies, as I believe, in the extreme imperfection of the geological record (p. 280).
>
> He who rejects these views on the nature of the geological record, will rightly reject my whole theory (p. 342).

SLOWNESS AND SMOOTHNESS (BUT NOT CONSTANCY) OF RATE. Darwin also championed the most stringent version of gradualism—not mere continuity of information, and not just insensibility of innumerable transitional steps; but also the additional claim that change must be insensibly gradual even at the broadest temporal scale of geological durations, and that continuous flux (at variable rates to be sure) represents the usual state of nature.

This broadest version of gradualism does not hold strong logical ties to natural selection as an evolutionary mechanism. Change might be episodic and abrupt in geological perspective, but still proceed by insensible intermediacy at a generational perspective—given the crucial scaling principle that thousands of generations make a geological moment. For this reason, Eldredge and I have never viewed punctuated equilibrium, which does refute Darwinian gradualism in this third sense, as an attack on the creativity of natural selection itself (Eldredge and Gould, 1972; Gould and Eldredge, 1977, 1993). The challenge of punctuated equilibrium to natural selection rests upon two entirely different issues of support provided by punctuational geometry for the explanation of cladal trends by differential species success and not by extrapolated anagenesis, and for the high relative frequency of species selection, as opposed to the exclusivity of Darwinian selection on organisms (see Chapters 8 and 9).

Some *fidei defensores* of the Darwinian citadel have sensed the weakness of this third version of gradualism, and have either pointed out that the creativity of natural selection cannot be compromised thereby (quite correct, but then no one ever raised such a challenge, at least within the legitimate debate on punctuated equilibrium); or have argued either that Darwin meant no such thing, or that, if he really did, the claim has no importance (see Dawkins, 1986). This last effort in apologetics provides a striking illustration of the retrospective fallacy in historiography. Whatever the current status of this third formulation within modern Darwinism, this broadest style of gradualism was vitally important to Darwin; for belief in slow change in geological perspective lies at the heart of his more inclusive view about nature and science, an issue even larger than the mechanics of natural selection.

Darwin often states his convictions about extreme slowness and continuous flux in geological time—as something quite apart from gradualism's second meaning of insensible intermediacy in microevolutionary perspective. Evolutionary change, Darwin asserts, usually occurs so slowly that even the immense length of an *average* geological formation may not reach the mean time of transformation between species. Thus, apparent stasis may actually represent change at average rates, but to an imperceptible degree even through such an extensive stretch of geological time! "Although each formation may mark a very long lapse of years, each perhaps is short compared with the period requisite to change one species into another" (p. 293).

Change not only occurs with geological slowness on this largest scale; but most transformations also proceed in sufficient continuity and limited variation in rate that elapsed time may be roughly measured by degree of accumulated difference: "The amount of organic change in the fossils of consecutive formations probably serves as a fair measure of the lapse of actual time" (p. 488).

Darwin presents his credo in crisp epitome: "Nature acts uniformly and slowly during vast periods of time on the whole organization, in any way which may be for each creature's own good" (p. 269). Note how Darwin concentrates so many of his central beliefs into so few words: gradualism, adaptationism, locus of selection on organisms.

But the most striking testimony to Darwin's conviction about gradualism in this third sense of slow and continuous flux lies in several errors prominently highlighted in the *Origin*—all based on convictions about steady rate (gradualism in the third sense), not on the insensible intermediacy genuinely demanded by natural selection (gradualism in the second sense), or on the simple continuity of historical information required to validate the factuality of evolution itself (gradualism in the first sense). For example, Darwin makes a famous calculation (dropped from later editions) on the "denudation of the Weald"—the erosion of the anticlinal valley located between the North and South Chalk Downs of southern England (pp. 285–287). He tries to determine an average value for yearly erosion of seacliffs today, and then extrapolates his figure as a constant rate into the past. His date of some 300 million years for the denudation of the Weald overestimated the true duration by five times or more. (The deposition of the Chalk, an Upper Cretaceous formation, persisted nearly to the period's end 65 million years ago.)

Moving to a biological example that underscores Darwin's hostility to episodes of "explosive" evolutionary diversification (he used his usual argument about the imperfection of the fossil record to deny their literal appearance and to spread them out in time), Darwin predicted that the Cambrian explosion would be exposed as an artifact, and that complex multicellular creatures must have thrived for vast Precambrian durations, gradually reaching the complexity of basal Cambrian forms. (When Darwin published in 1859, the Cambrian had not yet been recognized, and his text therefore speaks of the base of the Silurian, meaning lower Cambrian in modern terminology): "If my theory be true, it is indisputable that before the lowest Silurian stratum was deposited, long periods elapsed, as long as, or probably far longer than, the whole interval from the Silurian age to the present day; and that during these vast, yet quite unknown periods of time, the world swarmed with living creatures" (p. 307).

Paleontologists have now established a good record of Precambrian life. The world did swarm indeed, but only with single-celled forms and multicellular algae, until the latest Precambrian fauna of the Ediacara beds (beginning about 600 million years ago). The explosion of multicellular life now seems as abrupt as ever—even more so since the argument now rests on copious documentation of Precambrian life, rather than a paucity of evidence that could be attributed to imperfections of the geological record (see Chapter 10, pp. 1155–1161). Darwin on the other hand, predicted that complex, multicellular creatures must extend far into the Precambrian. He wrote: "I cannot doubt that all the Silurian [= Cambrian] trilobites have descended from some one crustacean, which must have lived long before the Silurian [= Cambrian] age" (p. 306). Darwin also conjectured, again incorrectly, that the ancestral verterbrate, an animal with an adult phenotype resembling the common embryological *Bauplan* of all modern verterbates, must have lived long before the dawn of Cambrian times: "It would be vain to look for [adult] animals having the common embryological character of the Vertebrata, until beds far beneath the lowest Silurian strata are discovered" (p. 338).

Darwin struggled for clarity and consistency. He did not always succeed. (How can an honest person so prevail in our complex and confusing world? I shall, for example, examine Darwin's ambivalences on progress in Chapter 6.) Darwin did not always keep the different senses of gradualism distinct. He frequently conflated meanings, arguing (for example) that the validity of natural selection (sense 2) required an acceptance of slow and continuous flux (sense 3). Consider once again the following familiar passage: "It may be said that natural selection is daily and hourly scrutinizing, throughout the world, every variation, even the slightest . . . We see nothing of these slow changes in progress, until the hand of time has marked the long lapse of ages" (p. 84).

This conflation came easily (and probably unconsciously) to Darwin, in large part because gradualism stood prior to natural selection in the core of his beliefs about the nature of things. Natural selection exemplified gradualism, not vice versa—and the various forms of gradualism converged to a single, coordinated view of life that extended its compass far beyond natural selection and even evolution itself. This situation inspired Huxley's frustration as he remonstrated with Darwin (see the famous quote on p. 151): you will have enough trouble convincing people about natural selection; why do you insist upon uniting this theory with an unnecessary and, by the way, false claim for gradualism?

We can best sense this overarching Darwinian conviction in a lovely passage that conflates all three senses of gradualism—the rationalist argument against creationism, the validation of natural selection by insensible intermediacy, and the slow pace of change at geological scales—all in the context of Darwin's homage to his guru Lyell, and his aesthetic and ethical convictions about the superiority of these "noble views" about natural causation and the nature of change:

> I am well aware that this doctrine of natural selection . . . is open to the same objections which were at first urged against Sir Charles Lyell's noble views on "the modern changes of the earth, as illustrative of geology;" but we now very seldom hear the action, for instance, of the coast-waves, called a trifling and insignificant cause, when applied to the excavation of gigantic valleys or to the formation of the longest lines of inland cliffs. Natural selection can act only by the preservation and accumulation of infinitesimally small inherited modifications, each profitable to the preserved being; and as modern geology has almost banished such views as the excavation of a great valley by a single diluvial wave, so will natural selection, if it be a true principle, banish the belief in the continued creation of new organic beings, or of any great and sudden modification in their structure (pp. 95–96).

### The adaptationist program

Darwin's three constraints on the nature of variation form a single conceptual thrust: variation only serves as a prerequisite, a source of raw material incapable of imparting direction or generating evolutionary change by itself. Gradualism, in the second meaning of insensible intermediacy, then guarantees that the positive force of modification proceeds step by tiny step. Therefore, the explanation of evolution must reside in specifying the causes of change under two conditions that logically entail a primary focus on adaptation as a canonical result: we know the general nature of change (gradualism), and we have eliminated an internal source from variation itself (the argument for isotropy). Change must therefore arise by interaction between external conditions (both biotic and abiotic) and the equipotential raw material of variation. Such gradual adjustment of one to the other must yield adaptation as a primary outcome.

Adaptational results flow logically from the mechanisms defining all other subbranches on the limb of Darwinism designated here as the "creativity of natural selection." But Darwin constructed this limb in reverse order in the psychological development of his theory. For he had long viewed an explanation of adaptation as the chief requirement of evolutionary theory. He sought the causes of evolution within his patrimony—the English tradition in natural theology—and he attempted to subvert this patrimony from within by accepting its chief empirical postulate of good design and then providing an inverted theoretical explanation (see p. 125).

When Darwin permits himself to make one of his rare forays into lyrical prose, we can grasp more fully (and dramatically) the extent of his feelings and the depth of his conviction. Consider the following passage on why the basic results of evolution and variation teach us so little about the origin of species, and why an understanding of mechanism requires an explanation of adaptation:

> But the mere existence of individual variability and of some few well-marked varieties, though necessary as the foundation for the work, helps us but little in understanding how species arise in nature. How have all those exquisite adaptations of one part of the organization to another part, and to the conditions of life, and of one distinct organic being to another being, been perfected? We see these beautiful co-adaptations most plainly in the woodpecker and missletoe; and only a little less plainly in the humblest parasite which clings to the hairs of a quadruped or feathers of a bird; in the structure of the beetle which dives through the water; in the plumed seed which is wafted by the gentlest breeze; in short, we see beautiful adaptations everywhere and in every part of the organic world (pp. 60–61).

Pursuing the theme of rare Darwinian lyricism as a guide to what he viewed as essential, consider his convictions about the overwhelming power of natural selection—a point that he usually conveyed by comparison with the limitations of artificial selection in breeding and agriculture:

> Man can act only on external and visible characters: nature cares nothing for appearances, except in so far as they may be useful to any being. She can act on every internal organ, on every shade of constitutional difference, on the whole machinery of life. Man selects only for his own good; Nature only for that of the being which she tends. Every selected character is fully exercised by her; and the being is placed under well-suited conditions of life. Man keeps the natives of many climates in the same country; he seldom exercises each selected character in some peculiar and fitting manner; he feeds a long and a short beaked pigeon on the same food; he does not exercise a long-backed or long-legged quadruped in any peculiar manner; he exposes sheep with long and short wool to the same climate. He does not allow the most vigorous males to struggle for the females. He does not rigidly destroy all inferior animals, but protects during each varying season, as far as lies in his power, all his productions . . . Under nature, the slightest difference of structure or constitution may well turn the nicely-balanced scale in the struggle for life

and so be preserved. How fleeting are the wishes and efforts of man! how short his time! and consequently how poor will his products be, compared with those accumulated by nature during whole geological periods. Can we wonder, then, that nature's productions should be far "truer" in character than man's productions; that they should be infinitely better adapted to the most complex conditions of life, and should plainly bear the stamp of far higher workmanship? (pp. 83–84).

But Darwin's world also differs strongly from Paley's, and the outcome of natural selection, however great the power of Darwin's mechanism, cannot be perfection, but only improvement to a point of competitive superiority in local circumstances. Natural selection operates as a principle of "better than," not as a doctrine of perfection: "Natural selection tends only to make each organic being as perfect as, or slightly more perfect than, the other inhabitants of the same country with which it has to struggle for existence" (p. 201). Thus, the signs of history will not be erased; creatures will retain signatures of their past as quirks, oddities and imperfections (see pp. 111–116 on methodology). Natural selection will fashion the organic world, while leaving enough signs of her previous handiwork to reveal a forming presence.

I have called this section "the adaptationist program," rather than, simply, "adaptation" because Darwin presents a protocol for actual research, not just an abstract conceptual structure. The relevant arguments may be ordered in various ways, but consider this sequence:

- Adaptation is the central phenomenon of evolution, and the key to any understanding of mechanisms.
- Natural selection crafts adaptation.
- Natural selection maintains an overwhelmingly predominant relative frequency as a cause of adaptation. Variation only provides raw material, and cannot do the work unaided.

Adaptation may be viewed as a problem of transforming environmental (external) information into internal changes of form, physiology and behavior. Two forces other than natural selection might play such a role—the creative response of organisms to felt needs with inheritance of acquired characters (Lamarck's system), or direct impress of environments upon organisms, also with inheritance of traits thus acquired (a system often associated with Geoffroy Saint-Hilaire). Darwin regards both alternatives as true causes, and he explicitly contrasts them with natural selection in several passages within the *Origin*. But, in these statements and elsewhere, he always grants natural selection the cardinal role by virtue of relative frequency—"by far the predominant Power," he writes on page 43, in upper case for emphasis. "Over all other causes of change, I am convinced that Natural Selection is paramount" (in *Natural Selection*, 1975 edition, p. 223).

In this light, how should evolutionists proceed if they wish to discover the mechanisms of change? Should they study the causes of variation (a vitally important issue, but unresolvable in Darwin's time, and not the cause of change in any case)? Or should they examine the large-scale phenomena of taxonomic order or geographic distribution (issues of great import again, but lying too far from immediate causation)? Instead, the best strategy, Darwin asserts, lies in the study of adaptation, for adaptation is the direct and primary result of natural selection; and the relative frequency of selection stands so high that almost any adaptation will record its forming power.

Adaptation therefore becomes, for Darwin, the primary subject for practical study of evolutionary mechanisms. Recall the basic methodological problem of a science of history (see p. 102): science aims, above all, to understand causal processes; past processes cannot be observed in principle; we must therefore learn about past causes by making inferences from preserved results. Adaptation is the common and coordinating result of nearly any episode of non-trivial evolutionary change. Adaptation not only pervades nature with an overwhelming relative frequency, but also embodies the immediate action of the primary cause of change—natural selection. The adaptations of organisms therefore constitute the bread and butter objects of study in evolutionary biology. Our first order approach to change must pose the following question in any particular case: what adaptive value can we assign; how did natural selection work in this instance? In a revealing statement, Darwin rolls all exceptions, all ifs and buts, into a set of subsidiaries to adaptation forged by natural selection—as either consequences of adaptation, inherited marks of older adaptations, or rare products of other processes: "Hence every detail of structure in every living creature (making some little allowance for the direct action of physical conditions) may be viewed, either as having been of special use to some ancestral form, or as being now of special use to the descendants of this form—either directly, or indirectly through the complex laws of growth" (p. 200).

The primary anti-Darwinian argument of late 19th century biology proceeded by denying a creative role to natural selection—but Darwin countered with a strong riposte. If adaptation pervades nature and must be constructed by natural selection, and if the steps of evolutionary sequences are generally so tiny that we may seek their source in palpable events subject to our direct view and manipulation, then we not only gain a theoretical explanation for evolutionary change. We also obtain the practical gift of a workable research program rooted in the observable and the resolvable.

But nothing so precious comes without a price, or without consequences. Darwin's argument works; no logical holes remain. But the research program thus entailed must embody attitudes and assumptions not necessarily true—or at least not necessarily valid at sufficiently high relative frequency to make the world exclusively, or even primarily, Darwinian. To accept Darwin's full argument about the creativity of natural selection, one must buy into an entire conceptual world—a world where externalities direct, and internalities supply raw material but impose no serious constraint upon change; a world where the functional impetus for change comes first and the structural alteration of form can only follow. *The creativity of natural selection makes adaptation central, isotropy of variation necessary, and gradualism pervasive.*

But suppose these precepts do not govern a commanding relative frequency of cases? What if adaptation does not always record the primacy of natural selection, but often arises as secondary fine tuning of structures arising in other ways? What if variation imposes strong constraints and supplies powerful channels of preferred direction for change? What if the nature of variation (particularly as expressed in development) often produces change without insensible intermediacy?

All these arguments merge into a structuralist critique that seriously challenges the predominant functionalism of classical Darwinism. As a common thread, these challenges deny exclusivity to natural selection as the agent of creativity, and claim a high relative frequency of control by internal factors. McCosh was right in establishing his pre-evolutionary contrast of a "principle of order" and a "principle of special adaptation" (see p. 116). Darwin was right in translating this distinction into evolutionary terms as "Unity of Type" and "Conditions of Existence," though he was probably wrong in his fateful decision—the basis of Darwinian functionalism—to yoke the two categories together under a common cause by defining unity of type as the historical legacy of previous adaptation, thus asserting the domination of natural selection (1859, p. 206—see extensive commentary in chapter 4). And E. S. Russell (1916) was also right in contrasting the "formal or transcendental" with the "functional or synthetic" approach to morphology.

We are children of Darwin, and an English school of adaptation and functionalism far older than evolutionary theory. Darwin's key claim for the creativity of natural selection—and the resulting sequelae of gradualism, adaptationism, and the isotropy of variation—builds the main line of defense for this powerful and venerable attitude towards nature and change. For many of us, these claims lie too close to the core of our deeply assimilated and now largely unconscious beliefs to be challenged, or even overtly recognized as something potentially disputable. Yet a coherent alternative has been proposed, and now provides one of the three most trenchant modern critiques of strict Darwinism. I believe that these critiques, taken together, will reorient evolutionary theory into a richer structure with a Darwinian core. But we cannot appreciate the alternatives until we grasp the basis of orthodoxy as an argument of compelling brilliance and power. Important critiques can only operate against great orthodoxies.

### THE THIRD THEME: THE UNIFORMITARIAN NEED TO EXTRAPOLATE; ENVIRONMENT AS ENABLER OF CHANGE

The first two themes—causal focus on organisms as agents of selection and creativity of selection in crafting adaptation—establish the biological core of Darwinian theory. That is, they perform the *biological* "work" needed to assure the third and last essential component of a Darwinian worldview: the uniformitarian argument for full application in extrapolation to all scales and times in the history of life. Mere operation in the microevolutionary here and now cannot suffice. Natural selection must also assert a vigorous claim for preeminence throughout the 3.5 billion years of phylogeny, lest the theory be reduced to an ornamental device, imposing only a fillip of immediate adaptive detail upon a grand pageant generated by other causes and forces.

Darwin, who fledged professionally as a geologist (the subject of his first three scientific books in the 1840's, on coral reefs, volcanic islands, and the geology of South America), and who regarded Charles Lyell as his intellectual hero, while embracing his mentor's doctrine of uniformitarianism as the core of his own philosophy as well, fully understood that his revolution would succeed only if he could show how natural selection might act as architect for the *full panoply* of life's history throughout geological time. The "methodological pole," one of the two foci of Darwin's revolution (see Section II of this chapter), brilliantly develops a set of procedures for defending extrapolation in various contexts of limited evidence.

The link of the first two themes (agency and efficacy) to this third theme of extended scope or capacity—thus forming in their threefold ensemble a minimally complete statement of revolution—received succinct expression in Ernst Mayr's (1963, p. 586) epitome of Darwinism as preached by the Modern Synthesis: "All evolution is due to the accumulation of small genetic changes, guided by natural selection [the first two themes of agency and efficacy], and that transpecific evolution [the third theme of scope, or uniformitarian extension] is nothing but an extrapolation and magnification of the events that take place within populations and species."

In this book, my explicit discussion for this third theme of extrapolation (Chapters 6 and 12) shall be shorter than my treatment of the first theme of agency (Chapters 3 and 8–9), leading from Darwin's nearly exclusive focus on the organismal level to the modern revision of hierarchical selection theory, and the second theme of efficacy (Chapters 4–5 and 10–11) on older and modern critiques of panadaptationism, with an emphasis on structural principles and constraints. I allocate my attention in this unequal manner because the first two themes already include, within themselves, the *biological* arguments for extrapolation, as embodied in Darwin's uniformitarian beliefs and practices. For my explicit and separate treatment of the crucial extrapolationist theme in this work, I therefore follow a different strategy, if only to avoid redundancy in a book that we all undoubtedly regard, author and readers alike, as quite long enough already! I will not rehearse Darwin's biological arguments for extrapolation, but will rather, as a "place holder" of sorts, concentrate upon the nature of the geological stage that must welcome Darwin's biological play.

I proceed in this way for a principled reason, and not merely as a convenience. All major evolutionary theories before Darwin, and nearly all important versions that followed his enunciation of natural selection as well, retained fealty to an ancient Western tradition, dating to Plato and other classical authors, by presenting a fundamentally "internalist" account, based upon intrinsic and predictable patterns set by the nature of living systems, for development or "unfolding" through time. (Ironically, such internalist theories follow the literal meaning of "evolution" (unfolding) far better than the Darwinian system that eventually absorbed the term. Darwin understood this etymological point perfectly well, and he initially declined to use the word "evolution"—preferring "descent with modification"—probably because he recognized the difference between the literal meaning of "evolution" and his own concept of life's history and change by natural selection—see Gould, 2000a.)

Darwin's theory, in strong and revolutionary contrast, presents a first "externalist" account of evolution, in which contingent change (the summation of unpredictable local adaptations rather than a deterministic unfolding of inherent potential under internal, biological principles) proceeds by an interaction between organic raw material (undirected variation) and environmental guidance (natural selection). Darwin overturned all previous traditions by thus granting the external environment a causal and controlling role in the direction of evolutionary change (with "environment" construed as the ensemble of biotic and abiotic factors of course, but still external to the organism, however intrinsically locked to, and even largely defined by, the presence of the organism itself). Thus, and finally, in considering the validity of extrapolation to complete the roster of essential Darwinian claims, the role of the geological stage becomes an appropriate focus as a surrogate for more overtly biological discussion.

If the uniqueness of Darwinism, and its revolutionary character as well, inheres largely in the formulation of natural selection as a theory of interaction between biological insides and environmental outsides—and not as a theory of *evolutio*, or intrinsic unfolding—then "outsides" must receive explicit discussion as well, a need best fulfilled within this treatment of extrapolation. Under internalist theories of *evolution*, environment, at most, holds power to derail the process by not behaving properly—drying up, as on Mars, or freezing over, as nearly occurred on Earth more than once during our planet's geological history. Under Darwinian functionalism, however, environment becomes an active partner in both the modes and directions of evolutionary change.

As the Utopian tradition recognizes, we can often devise lovely and optimal systems in abstract principle, but then be utterly unable to apply them in practice because an imperfect world precludes their operation. The central logic of Darwinism faces an issue of this kind. The two essential biological postulates of natural selection—its operation at the organismal level, and its creativity in crafting adaptations—build a sufficient theoretical apparatus to fuel the system. The play of evolution can run with such a minimal cast, but we do not know whether the drama can actually unfold on our planet until we also examine and specify the character of the theater—the geological and environmental stage for the play of natural selection. The geological stage therefore becomes a major actor in the drama set on its own premises.

Moreover, and reinforcing my argument that Darwin's strength lies in his brave specificity, Darwin places a great burden on geology and environment by devising such stringent conditions for the nature of this external setting. Again, we encounter the Goldilocks problem—environment cannot impose too much or provide too little, but must be "just right" in the middle.

Environment, as an active Darwinian agent, cannot underperform. In particular, an absence of environmental change would probably bring evolution to an eventual halt, as selective pressures for adaptive alteration diminished (see Stenseth and Maynard Smith, 1984). Purely biotic interaction might drive evolution for some time following a cessation of environmental change, but probably not indefinitely.

The possibility of too little change has rarely been viewed as a threat to Darwinism, largely because the geological record seems so clearly to emphasize potential dangers in the other direction (though see pp. 492–502 on Lord Kelvin). The specter of "too much" change, on the other hand, has haunted Darwinism from the start. In particular, if the theory of geological catastrophism were generally true, or even just sufficiently important in relative frequency, then Darwinism would be compromised as the primary agent of pattern in the history of life.

By catastrophism, I mean to designate the classical theory of global paroxysm as a primary agent of geological change—in particular, the idea that mass extinctions thus engendered might lie largely outside the domain of traditional Darwinism. Of course, mass extinctions cannot be construed as "undarwinian" *per se*. If environment changes so rapidly that organisms cannot adapt fast enough by natural selection, then many species will die. But, in a conceptual world of relative frequency, where Darwinism must not only operate, but also dominate as the creator of change, such formative power for mass extinction constitutes a serious challenge. If we survey the entire history of life, and find that catastrophic mass extinction, with non-Darwinian fortuity in causes of change (on either the "random" or the "different rules" model—see Chapter 12, and Gould, 1985a, 1993c), establishes more features of overall pattern than the ordinary interplay of taxa during normal times (between such episodes of coordinated death) can build and maintain, then Darwin's view of life lacks the generality once accorded. In particular, the key uniformitarian argument will then fail. The adaptive struggles of immediate moments will not extrapolate to explain the patterns of life's history. Moreover, if these undarwinian components of fortuity in extinction, and success for reasons unrelated to the original adaptive basis of traits, also maintain strong influence at lesser scales of smaller mass extinctions (Raup and Sepkoski, 1984), and even, in a fractal manner, for some ordinary extinctions in normal times (Raup, 1991), then the challenge may become truly pervasive.

These characterizations of Darwinian requirements cannot be dismissed or downgraded as conjectures or reconstructions, only inferentially based on deductions from premises stated by Darwin for different reasons. Darwin devoted an entire chapter of the *Origin,* number 10 "on the geological succession of organic beings," to an exploration of the geological stage and its requirements for natural selection. He argues that biotic competition, gradualistically expressed through time as coordinated waxing and waning of interacting clades, marks the overall pattern of life—and that the apparent fossil evidence for more rapid change, set by physical environments and leading to mass extinctions, must generally be read as artifacts of an imperfect record (see Chapter 12 for detailed exegesis of Darwin's arguments on this subject).

This issue exposes another essential Darwinian theme not yet discussed (but receiving full treatment in Chapter 6)—the nature of competition; the prevalence of biotic over abiotic effects; the metaphor of the wedge; and the fundamental role of Darwinian ecology as a validator of progress (in the absence of any available defense from the bare-bones mechanism of natural selection itself). Thus, the argument for uniformitarian change in geology undergirds a central conviction of the Darwinian corpus.

We cannot overestimate the depth of Darwin's debt to his intellectual hero, Charles Lyell. The uniformitarianism of his mentor not only provided, by transfer into biology, a theory of evolutionary change. The doctrine of uniformity also supplied, on its original geological turf, a world that could grant enough slow and continuous environmental change to fuel natural selection—but not so much, or so quickly, that selection would be overcome, and the rein of pattern seized by environment in its own right. In natural selection, environment proposes and organisms dispose; this subtle balance of inside and outside must be maintained. But in a world of too much environmental change, the external component does not only propose, but can also dispose of organisms and species without much backtalk. Darwinism does not run well on such a one-way street.

## Judgments of Importance

In the difficult genre of comprehensive historical reviews, a few special books stand out as so fair in their judgments and so lucid in their characterizations that they set the conceptual boundaries of disciplines for generations. In morphology, E. S. Russell's *Form and Function* (1916) occupies this role for the brilliance and justice of its characterizations, even though Russell, as an avowed Lamarckian, made no secret about his own preferences (and made the wrong choice by modern standards). In evolutionary biology, similar plaudits may be granted to Vernon L. Kellogg's *Darwinism Today* (1907). Kellogg, a great educator and entomologist from Stanford, had collaborated with David Starr Jordan on the best textbooks of his generation. He also played an ironic role in the history of evolution by serving a term (while America maintained her early neutrality) as chief agent for Belgian relief, posted to the German General Staff in Berlin during World War I. There, he listened in horror to German leaders perverting Darwinism as a justification for war and conquest—and he exposed these distortions in his fascinating volume, *Headquarters Nights* (1917). William Jennings Bryan read this book and, understanding the abuse but blaming the victims of misinterpretation rather than the perpetrators, launched his campaign to ban the teaching of evolution as a result (see Gould, 1991b).

As the Darwinian centennial of 1909 neared, Kellogg decided to write a volume providing a fair hearing for all varieties of Darwinism, and all alternative views in a decade of maximal agnosticism and diversity in evolutionary theories. Kellogg's book adopts the same premise as this treatise—that

Darwinism embodies a meaningful central logic, or "essence," and that other proposals about evolutionary mechanisms can be classified with reference to their consonance or dissonance with these basic Darwinian commitments.

I was particularly pleased to learn that Kellogg's categories, though differently named and parsed, are identical with those recognized here. He divides the plethora of proposals under discussion in his time into those "auxiliary to" and those "alternative to" natural selection. Among auxiliaries that aid, expand, improve, or lie within the spirit of Darwinism, Kellogg highlights two principal themes: studies of Wagner, Jordan, and Gulick on the role of isolation in the formation of species; and hierarchical models of selection as espoused by Roux and Weismann (discussed in detail in Chapter 3). I noted with special gratification that Kellogg recognized hierarchy as an auxiliary, not a confutation, to Darwinism, for this same contention sets a principal theme of this book.

In his second category of confutations, Kellogg identified "three general theories, or groups of theories, which are offered more as alternative and substitutionary theories for natural selection than as auxiliary or supporting theories" (1907, p. 262): Lamarckism (inheritance of acquired characters in the form advocated by late 19th century neo-Lamarckians), orthogenesis, and heterogenesis (Kellogg's designation for saltationism).

Kellogg's taxonomy works particularly well in evaluating the central principles of Darwinism. His "auxiliaries" aid selection (by addition of other principles that do not challenge or diminish selection, or by expansion of selection to other levels); but his "alternatives" confute particular maxims of the minimal commitments for Darwinian logic. The Kelloggian "alternatives" all deny the fundamental postulate of creativity for selection by designating other causes as originators of evolutionary novelties, and by relegating selection to a diminished status as a negative force. Each alternative rejects a necessary Darwinian postulate about the nature of variation (see pp. 141–146): Lamarckism and orthogenesis deny the principle of undirected variability; saltationism refutes the claim that variation must be small in extent.

I warmly endorse Kellogg's approach. As practicing scientists, we often do not pay enough respect to the logical structure of an argument—to its rigors and its entailments. We tend to assume that conclusions flow unambiguously from data, and that if we observe nature closely enough, and experiment with sufficient care and cleverness, the right ideas will somehow coalesce or flow into place by themselves. But scholars should know, from the bones and guts of their practice, that all great theories originate by intense and explicit mental struggle as well. We should not castigate such efforts as "speculation" or "armchair theorizing"—for mental struggle deserves this designation only when the thinker opposes or disparages our shared conviction that, ultimately, empirical work or testing must accompany and validate such exercises in thought (and then all scientists would agree to let the calumny fall). Great theories emerge by titration of this basically lonely mental struggle with the more public, empirical acts of fieldwork and benchwork.

One need look no further than Charles Darwin for proper inspiration. He rooted his theory in practical testability, and he continually devised and performed clever experiments, despite limited resources (of available equipment and personnel at Down, not of funds; for Darwin was a wealthy man and did not need to spend his time seeking patronage, his generation's equivalent of modern grant swinging). But natural selection did not flow from the external world into a *tabula rasa* of Darwin's mind. He carried out with himself, as recorded in his copious notebooks (Barrett *et al.*, 1987), one of the great mental struggles of human history—proposing and rejecting numerous theories along his slow and almost painful journey by inches, accompanied by lateral feints and backward plunges, towards the theory of natural selection. That theory, when fully formulated in the 1850's, emerged as an intricately devised amalgam of logically connected parts, each with a necessary function—and not as a simple message from nature. We must treat this theory, as Kellogg does, with respect for its integrity.

With the coalescence and hardening of the Modern Synthesis (Gould, 1983b), culminating in the Darwinian celebrations of 1959, an orthodoxy descended over evolutionary theory, and a generation of unprecedented agreement ensued (often for reasons of complacency or authority). However, the press of new concepts and discoveries has since fractured this shaky consensus, and we now face a range of options and alternatives fully as broad as those available in the contentious decade of Kellogg's review. In this renewed context, I recommend Kellogg's procedure as both intellectually admirable and maximally useful—namely, *to arrange and evaluate various views and challenges by classification according to their attitudes towards the minimal commitments of Darwinism.* I say "admirable" because such an approach pays proper respect to the intellectual power of Darwin's synthesis, and "useful" because a taxonomy by minimal commitments of an essential logic allows us to rank, assess, and interconnect an otherwise confusing array of proposals and counterproposals. And just as the widespread debate of Kellogg's time led to the Modern Synthesis of the next generation, I believe that the renewed arguments of our day will pay dividends in the form of a richer and more adequate consensus for our new millennium. Kellogg's characterization of his own era therefore becomes relevant to our current situation:

> The present time is one of unprecedented activity and fertility both in the discovery of facts and in attempts to perceive their significance in relation to the great problems of bionomics. Both destructive criticism of old, and synthesis of new hypotheses and theories, are being so energetically carried forward that the scientific layman and educated reader, if he stand but ever so little outside of the actual working ranks of biology, is likely to lose his orientation as to the trends of evolutionary advance. Precisely at the present moment is this modification of the general point of view and attitude of philosophical biologists unusually important and far-reaching in its relation to certain long-held general conceptions of biology and evolution (1907, p. ii).

I have therefore followed Kellogg's lead and attempted, in this introductory chapter, to characterize the central logic and minimal commitments of Darwinism—an essence, if you will, to invoke a good word and concept that has become taboo in our profession. I will then use this characterization as a foundation for classifying various challenges and controversies—just as Kellogg did—according to their stance towards the essential concepts of Darwinism. The most interesting and far-reaching challenges directly engage these essential concepts, either as alternatives to refute them in part, or as auxiliaries to expand and reinterpret them in fundamental ways. *This book presents, as its primary thesis, the notion that (i) Darwinism may be viewed as a platform with a tripod of essential support; (ii) each leg of the tripod now faces a serious reforming critique acting more as an auxiliary than an alternative formulation; and (iii) the three critiques hold strong elements in common, and may lead to a fundamentally revised evolutionary theory with a retained Darwinian core.*

We must rank challenges by their degree of engagement with the Darwinian core; we cannot follow a strategy of mindless "raw empiricism" towards the *Origin* and simply compile a list of Darwin's mistakes. All great works are bursting with error; how else could true creativity be achieved? Could anyone possibly reformulate a universe of thought and get every detail right the first time? We should not simply count Darwin's errors, but rather assess their importance relative to his essential postulates. (Consider, for example, the standard rhetorical, and deeply anti-intellectual, ploy of politically motivated and destructive critics, American creationists in particular. They just list the mistakes, envelop each in a cloud of verbal mockery, and pretend that the whole system has drowned in this tiny puddle of inconsequential error.)

I suggest that we use the list of minimal commitments to gauge the status of Darwin's errors. Very few faults of simple fact can, as individual items, be of much consequence unless they confute a core commitment. Darwin argued, for example, that swimbladders evolved into lungs (see p. 107) though exactly the opposite occurred—but no premise of the general theory suffers any injury by this mistake, however embarrassing. What about more important theoretical claims like Darwin's hypothesis of "pangenesis" as a mechanism of heredity (Darwin, 1868)? Again, Darwin's view of life would have been easier to vindicate if the theory had been affirmed, but none of his three essential postulates about the nature of variation fell with the disproof of pangenesis, and the core commitments remained intact, if unproven. What about the impact of major claims that turn out to be basically true, Mendelism for example? We must make our judgment by assessing their engagement with the core commitments. In the first decade of the 20th century, most evolutionists invoked Mendelism as a saltational theory of macromutation against the Darwinian core commitment to small-scale variation (see Chapter 5). Later, largely through R. A. Fisher's analysis and the resolution of the Mendelian *vs.* biometrician debate, macromutations were rejected, "ordinary" small-scale variation granted a Mendelian basis, and Mendelism comfortably reinterpreted as support for the same core commitment. Again, challenges and new proposals must be judged and ranked by their engagement with the essence of a reigning theory. Darwinism embodies a definable set of minimal commitments; all great theories do and must.

We should use this perspective of engagement with the core commitments to assess the relative theoretical importance of issues now commanding attention among evolutionists. For example, Kimura's theory of neutralism (1983) ranks as fundamental and reformative for proposing a new domain of causation at high relative frequency. I regard as unfair, and disrespectful of Darwin's clear commitments, the common rhetorical strategy of arguing, as Stebbins and Ayala did for example (1981), that selection and neutralism should be judged as competing paradigms comfortably embraced within the Modern Synthesis. The Synthesis, as an intellectual structure, has always been understood as Darwinism strengthened by modern knowledge about genetics and heredity. The Synthesis must therefore assert a dominant relative frequency for selection. Of course such a theory allows for neutrality—one could scarcely deny either the mathematics or the conditions of potential operation—but only at a low relative frequency, so that the preeminence of selection will remain unchallenged.

Kimura's claim for high, even dominant, relative frequency of neutral *change* at the nucleotide level introduces a world different from Darwin's. At most, one can say that this world, largely invisible at the organismic level, does not subvert Darwin's proposal that selection dominates the phenotypic realm of overt form, function, and physiology of organisms. But in so saying, we must admit that a large part of reality, though unaddressed by Darwin himself, cannot be explained on Darwinian principles if Kimura's theory holds. Darwinism does not fall thereby, but a new and distinct domain, primarily regulated by a different style of causality, has been added to evolutionary explanation. How can one deny that evolutionary theory becomes substantially reformulated and enriched thereby? Why would one want to issue such a denial, unless psychic health depends upon the continued assertion of comfortable orthodoxy, whatever the required twist of logic?

My own expertise lies in paleontology, and this book shall emphasize critiques from the attendant domain of macroevolution, descriptively defined as patterns and causes of evolution at and above the species level. (I acknowledge, of course, the fascination and transforming power of work at the molecular level. I also recognize that macroevolution must shake hands with molecular genetics in order to forge the new consensus. If this book slights the molecular side, my own ignorance stands as the only cause, and this work necessarily suffers thereby.)

Basically, I shall defend the view that each leg of Darwin's essential tripod, as explicated in this chapter, now faces a serious critique from the domain of macroevolutionary change. These critiques rank as *auxiliaries* to Darwinism in Kellogg's sense; for they either expand or add to the core commitments. But the expansions are large and the additions substantial—so the resulting revision can no longer be called ordinary Darwinism in any conventional meaning. I am convinced that the three critiques intertwine in a potentially unified way. But consensus is premature and we can only see the resulting shape of the revised and unified theory through a glass darkly—though in the future, no doubt, face to face.

Proceeding in reverse order through critiques of Darwinism's three core claims, catastrophic mass extinction, and more general views about fortuity in abiotically driven extinction at all levels, challenge Darwin's essential notion of a dominant relative frequency for biotic struggle in a crowded world—the third leg of the tripod, as represented by the geological stage required for an evolutionary play based entirely on extrapolation of microevolutionary principles (Chapters 6 and 12). The general idea of constraint—more in the positive sense of internally biased channels for change, rather than the negative meaning of limited variation for potentially useful alterations (see Gould, 1989a)—rejects the key Darwinian notion of isotropy in raw material, and consequent control of evolutionary direction by natural selection. Constraint therefore challenges the second leg of the tripod—the "creativity of natural selection"—not by confuting the proposition that natural selection acts as a creative force, but by insisting on diminished relative frequency and a sharing of control. Moreover, by reasserting the structuralist side of the old dichotomy between structure and function in biology—an issue far predating evolution, and inherent in the struggle between continental vs. Paleyan approaches to natural theology—the idea of constraint reengages one of the deepest issues in all the life sciences (Chapters 4–5 and 10–11).

Most importantly, and as the best integrator of all three critiques, the hierarchical theory of natural selection, by asserting both the existence and relative importance of selection at all levels from genes to species, challenges the first leg of the tripod—the insistence, so crucial to Darwin's radical overthrow of Paley via Adam Smith, that selection works almost exclusively on organisms (Chapters 3 and 8–9). I believe that this hierarchical theory provides the most fundamental, and potentially unifying, of all critiques—for I suspect that many constraints will be explained as effects of lower level selection indirectly expressed in phenotypes; while the contribution of mass extinction to repatterning life's history will include a crucial component of selection at levels above the organismic. Moreover, the attendant need to reconceptualize trends and stabilities not as optimalities of selection upon organisms alone, but as outcomes of interactions among numerous levels of selection, implies an evolutionary world sufficiently at variance from Darwin's own conception that the resulting theory, although still "selectionist" at its core, must be recognized as substantially different from current orthodoxy—and not just as a dash of spice on an underflavored dish. I therefore devote the largest section of this book's second half (Chapters 8 and 9) to defining and defending this hierarchical theory of selection.

If the next generation of evolutionists follows and extends this protocol at the outset of our new millennium, as presaged by the tentative work and exploration of so many scientists at the close of the last millennium, then we shall honor, all the more, the vitality of the tight definitions and firm commitments proposed by Darwin himself at the foundation of our discipline. Few theories hold the range of power, and the intricacy of logic, necessary to generate an intellectual structure of such continuing fascination and relevance. We do not pay our proper respect to Darwin by bowing before the icons of his central propositions, but by engaging these focal precepts as living presences, ripe for reformulation, almost 150 years after their initial presentation. In Darwin's own world of continuous flux, anything that lasts so long becomes a many-splendored thing. In a revised world of structuralism, we might say that Darwin first located and embellished one of the few brilliant and coherent positions in an intellectual universe with few nucleating places. Either formulation engenders the same result of abiding respect for Darwin's view of life—leading to proper thanks owed by all of us for the good fortune of such an interesting founder. What greater pleasure can we know than to engage Darwin in dialogue—as we can and must do, because his theory rests upon a powerful and defining essence. Darwin, in short, is the extraordinary man who, all by himself, embodied the only three beings proclaimed worthy of respect by Baudelaire—for he pulled down an old order, and came to know a large part of the new world that he created. *Il n'existe que trois êtres respectables: le prêtre, le guerrier, le poète. Savoir, tuer, et créer.* There exist only three beings worthy of our respect: the priest, the warrior, and the poet. Know, kill, and create.

# CHAPTER THREE

# Seeds of Hierarchy

## Lamarck and the Birth of Modern Evolutionism in Two-Factor Theories

### THE MYTHS OF LAMARCK

In 1793, the French revolutionary government, having expunged the past by executing a monarch, proclaimed a new beginning of time. They renamed the months, and started the calendar all over again with the foundation of the Republic in September 1792. The old months had honored emperors and gods, but the new months would celebrate the passing of seasons by weather and activity—*Brumaire* (the foggy month in fall), *Thermidor* (the hot times of mid summer), and *Nivose* (for the depth of a frosty winter), for example.

Jean-Baptiste-Pierre-Antoine de Monet, Chevalier de Lamarck (1744–1829)—now redesignated, with democratic brevity, as Citoyen Lamarck—became professor of "lower" animals (the old Linnaean classes Insecta and Vermes, later renamed "invertebrates" by Lamarck himself) at the newly founded *Muséum d'Histoire Naturelle* in 1793. (His previous work, nearly all in botany, had not prepared him for this new role, though he had long been an avid shell collector and student of conchology.) Until 1797, he had supported the conventional idea of species as fixed entities. But he then became an evolutionist, first expressing this new view of life in his inaugural lecture for the *Muséum* course of 1800, and then in three major works—the *Recherches sur l'organisation des corps vivans* of 1802; his most famous work, the *Philosophie zoologique* of 1809; and the *Histoire naturelle des animaux sans vertèbres* of 1815–1822.

In an ironic symbol, Lamarck first presented his evolutionary theories in an inaugural lecture pronounced on the 21st day of *Floréal,* year VIII (May 11, 1800)—in the month of flowering. For Lamarck's theory suffered the opposite fate of withering, and the scorn of inattention. We all know the image of Lamarck—an impression carefully nurtured, for different reasons, by friends and foes alike—as a lonely man (a prophet before his time to some, a kook to others), penniless, friendless, and, finally, blind; living out the last days of a long and sad life, supported only by his devoted daughters.

This image of a forgotten failure was fostered by the two greatest figures of 19th century natural history—first by Cuvier, and later by Darwin. Darwin said little about Lamarck (see pp. 192–197), but his derision still permeates our view. Cuvier did far more damage. I don't know what lip service Cuvier gave to the ancient maxim *de mortuis nil nisi bonum* (say only good of the dead), but he violated this precept with avidity in writing *éloges* (eulogies) of deceased colleagues. Cuvier, the consummate politician, understood the power thus granted to shape history in his own favor. For what forum could be less subject to rebuttal, and therefore more suited for easy passage into received truth. As master of *éloges*, Cuvier held enormous power over his colleagues, as long as he could outlive them! (see pp. 309–312 on Geoffroy's revenge for the same reason). His official *éloge* of Lamarck is a masterful, if repugnant, document of propaganda directed against a close colleague and former friend who had (in Cuvier's view) gone beyond the pale in both methodology of research and content of belief. Cuvier used his *éloge* as an opportunity to castigate Lamarck, and thus provide a lesson in proper procedure for aspiring scientists.

Cuvier began with cloying praise, and then portrayed his need to criticize as a sad duty: "In sketching the life of one of our most celebrated naturalists, we have conceived it to be our duty, while bestowing the commendation they deserve on the great and useful works which science owes to him, likewise to give prominence to such of his productions in which too great indulgence of a lively imagination had led to results of a more questionable kind, and to indicate, as far as we can, the cause or, if it may be so expressed, the genealogy of his deviations" (1832, 1984 edition, p. 435).

Cuvier then dismembered Lamarck on two grounds. First, with justice in the claim (however unkind the rhetoric), he castigated Lamarck for reaching too far without foundation, and for building all-encompassing systems in the speculative mode. (This criticism reflected Cuvier's main unhappiness with Lamarck's science. Cuvier viewed himself as a modernist, committed to rigorous empirical documentation, and no extension beyond direct evidence in the search for explanations—as opposed to Lamarck's unfruitful, comprehensive speculation in the antiquated *esprit de système,* or spirit of system): "He had meditated on the general laws of physics and chemistry, on the phenomena of the atmosphere, on those of living bodies, and on the origin of the globe and its revolutions. Psychology, and the higher branches of metaphysics, were not beyond the range of his contemplations; and on all these subjects he had formed a number of definite ideas . . . calculated to place every branch of knowledge on a new foundation" (1832, 1984 edition, p. 442).

Cuvier acknowledged Lamarck's excellent efforts in morphology and taxonomy, but then damned him for denigrating this admirable work as a trifle compared with all-embracing and useless theories. What a sorry spectacle: Lamarck in his armchair, challenging the great Lavoisier, the icon and martyr of true science. (Lavoisier was beheaded during the Reign of Terror.) "So intimately did he identify himself with his systems, and such was his desire that they should be propagated, that all other objects seemed to him subordinate, and even his greatest and most useful works appeared in his own eyes merely as the slight accessories of lofty speculations. Thus, while Lavoisier was creating in his laboratory a new chemistry, founded on a beautiful and methodical series of experiments, M. de Lamarck, without attempting experiment, and destitute of the means of doing so, imagined that he had discovered another" (1832, 1984 edition, p. 442).

After ridiculing Lamarck's general method of system building, Cuvier mounted his second attack and dismembered the particular content of Lamarck's system, especially his evolutionary views. Cuvier did his former colleague a lasting disservice by caricaturing Lamarckian evolution as the outcome of organic will, based on desires, translated into phyletic progress. Cuvier's rhetoric was brilliant, his characterization grossly distorted:

> Wants and desires, produced by circumstances, will lead to other efforts, which will produce other organs. . . . It is the desire and the attempt to swim that produces membranes in the feet of aquatic birds; wading in the water, and at the same time the desire to avoid wet, has lengthened the legs of such as frequent the sides of rivers . . . These principles once admitted, it will easily be perceived that nothing is wanting but time and circumstances to enable a monad or a polypus gradually and indifferently to transform themselves into a frog, a stork, or an elephant (1832, 1984 edition, p. 446).

Finally, in an ultimate dismissal from a "hard" scientist (and with the tone of the Yahoo), Cuvier concludes: "A system established on such foundations may amuse the imagination of a poet; a metaphysician may derive from it an entirely new series of systems; but it cannot for a moment bear the examination of anyone who has dissected a hand . . . or even a feather" (1832, 1984 edition, p. 447).

Cuvier's caricature remains potent in our worst modern misunderstanding of Lamarck as a mystical vitalist, advancing the idea of an ineffable organic will against the ordinary physical causality of science. (Tit for tat, and however unfairly, Lyell hurt Cuvier even more in return by caricaturing him as a theologically tainted, antiscientific catastrophist in geology.) But Lamarck, schooled (along with Cuvier) in the ideals of the French Enlightenment, was an ardent materialist. His idiosyncratic and unfruitful views about the nature of matter (arising primarily from his anti-Lavoisierian chemistry) led to predictions of odd behavior for living bodies, but his basic notions of reduction and causality remained in the scientific mainstream. In his last great work, and in the context of his evolutionary theory, Lamarck defended a conventional view of mechanistic causality, and derided all teleological interpretations. Goals, he argued, are false appearances reflecting an underlying causal necessity:

> It is chiefly among the living, and most notably among Animals, that some have claimed to glimpse a purpose in nature's operations. Even in this case the purpose is mere appearance, not reality. Indeed, in every type of animal organism, there subsists an order of things . . . whose only effect is to lead to what seems to us to be a goal, but is essentially a necessity. The order achieves this necessity through the progressive develop-

ment of parts, which are [also] shaped by environmental conditions (1815, in Corsi, 1988, p. 190).

Since watchdogs tend to be more vigilant than publicists, Lamarck's opponents among the natural theologians often noted (and deplored) his materialism. The pious Reverend William Kirby, one of Britain's greatest entomologists, made a statement that I regard as both trenchant and descriptively accurate (in Burkhardt, 1977, p. 189): "Lamarck's great error, and that of many other of his compatriots, is materialism; he seems to have no faith in anything but *body*, attributing every thing to a physical, and scarcely anything to a metaphysical cause. Even when, in words, he admits the being of a God, he employs the whole strength of his intellect to prove that he had nothing to do with the works of creation. Thus he excludes the Deity from the government of the world that he has created, putting nature in his place."

Curiously, each generation of historians and biological commentators has to discover, anew and for itself—and by reading original sources rather than imbibing mythology—this general and mainstream scientific position held by a man with such idiosyncratic views on specific subjects (see Mayr, 1972, and Simpson, 1961a, for the scientists; Gillispie, 1959; Burkhardt, 1977; and Corsi, 1988, for historians). For example, Gillispie wrote in his classic article for the Darwinian centenary (1959, p. 275): "Life is a purely physical phenomenon in Lamarck, and it is only because science has (quite rightly) left behind his conception of the physical that he has been systematically misunderstood and assimilated to a theistic or vitalistic tradition which in fact he held in abhorrence."

This correction allows us to see Lamarck as a key figure in and of his time—an age as rife with intellectual, as with political, ferment—and not as a painfully peripheral, and actively marginalized, oddball. In a meticulous analysis of French scientific thought, Corsi (1988) has placed Lamarck's views firmly amidst the debates of his age, and also demonstrated that his theories were not so ignored or ridiculed as tradition maintains.

I am not arguing that Lamarck was popular in his day, only that he was contemporary. In many ways, Lamarck became his own worst enemy, and he owed his fall from favor towards obscurity as much to his own unfortunate habits as to the peculiarity of his ideas. He possessed no political skills, and could only fare badly in any match with the masterful Cuvier (in an age that must rank as the best and the worst of all political times). He continued to practice the old style of speculative system building in an increasingly empirical climate. He was combative, and so self-assured, that affirmation without any documentation became his principal style of argument. Consider this claim for use and disuse from the *Philosophie zoologique* (1809, 1984 edition, p. 108): "Nothing of all this can be considered as hypothesis or private opinion; on the contrary, they are truths which, in order to be made clear, only require attention and the observation of facts." Lamarck's certainty extended even to the maximally dubious subject of weather forecasting: "I am not submitting an opinion, but announcing a fact. I am indicating an order of things that anyone can verify through observation" (in Corsi, 1988, p. 59). The old story that Napoleon refused a copy of Lamarck's *Philosophie zoologique* is apparently true (unlike most legends in the history of science). But Napoleon's motive has not generally been recognized: he mistook the nature of the gift, thinking that he had been offered one of Lamarck's discredited volumes of weather predictions for the coming year!

### LAMARCK AS A SOURCE

The preceding section on Lamarck as a man of his time may seem peripheral, if not wholly out of place, in a chapter on hierarchical causation in evolutionary theory, but this theme holds a definite place in the logic of my presentation. Such a diffuse and comprehensive idea as evolution can claim no single initiator or unique starting point. The search for precursors in ancient Greece, while overextended (Osborn, 1894), rests upon a legitimate foundation. But Lamarck holds a special place as the first to transcend footnote, peripheral commentary, and partial commitment, and to formulate a consistent and comprehensive evolutionary theory—in Corsi's words (1988, p. xi) "the first major evolutionist synthesis in modern biology."

Moreover, even in a book parochially skewed to British and American evolutionary theory, Lamarck still merits the status of an ultimate source. German and French biologists could cite a variety of references from their indigenous movements of *Naturphilosophie* (Oken, Meckel, or Goethe himself, for example) and the revolutionary times of the Age of Reason (Buffon, Maupertuis, Diderot, and a host of largely forgotten Enlightenment figures). England could boast a few precursors (including Darwin's grandfather Erasmus), but no strong movements. Ironically, as Darwin, Wallace, and all the great mid-century evolutionists acknowledged, Lamarck instigated *both* major treatments of evolutionary thought in English before 1859: first, the accurate and extensive, if negative, presentation of Lamarck's system by Charles Lyell in the first four chapters of *The Principles of Geology*, Volume 2 (1832); and second, the anonymously published (1844) *Vestiges of the Natural History of Creation*. The author of that scandalous and widely debated book, the Scottish publisher Robert Chambers, acknowledged Lamarck, via Lyell, as his major source of inspiration.

I have (see Chapter 1) rejected Hull's genealogical approach to the definition of theories, but I certainly defend this criterion (almost tautologically, I suppose) for the tracing of influences. Of Lamarck's foundational impact on English evolutionary thought, Hull (1985, p. 803) writes: "Darwin first confronted a detailed explication of the species problem in the context of Lyell's refutation of Lamarck in his *Principles of Geology* . . . Others like Spencer and Chambers were converted by reading Lyell's refutation of Lamarck; still others like Wallace and Powell were led to entertain the possibility of evolution by reading Chambers." We cannot, in short, view Lamarck as an oddity, cast aside by his own contemporaries, and irrelevant except as a whipping boy ever since. And we must acknowledge that Lamarckism, properly defined, forms a coherent and innovative system in the context of its own time.

Lamarck's active centrality provides a foundation to my historical argument because his theory, as presented in the next section, rests upon the concept of hierarchy, with distinct causes at two primary levels. Lamarck's hierarchy differs radically in form and logic from any acceptable modern version; indeed, I shall reject the basis of Lamarck's notion as an important component in developing the modern interpretation.

Lamarck's concept became Darwin's context. In perhaps the most widely quoted of all his letters, Darwin wrote to Hooker on January 11, 1844 (in F. Darwin, 1887, volume 2, p. 23): "I am almost convinced (quite contrary to opinion I started with) that species are not (it is like confessing a murder) immutable. Heaven forfend me from Lamarck nonsense of a 'tendency to progression,' 'adaptations from the slow willing of animals,' etc.! But the conclusions I am led to are not widely different from his; though the means of change are wholly so. I think I have found out (here's presumption!) the simple way by which species become exquisitely adapted to various ends."

Hierarchy has resided at the heart of evolutionary theory from the very beginning, despite a temporary eclipse during the rally-round-the-flag period of strict Darwinism at the middle of the 20th century. When the *Beagle* docked at Montevideo, Darwin received his most precious item of mail—volume two of Lyell's *Principles of Geology*. His joy at this gift, and his careful study of the contents, are well attested. This volume began with a long and careful exposition of Lamarck's theory, fairly but negatively described by Lyell. Darwin formulated his focal concept of small-scale change, based on organismal advantage as the mechanism (by extrapolation) for all evolution, as an explicit denial of Lamarck's hierarchy of causes. I believe that Darwin correctly rejected an untenable theory of hierarchy based on distinct causes for different levels, but that (in a historically portentous example of the cliché about babies and bathwater) he carried a good thing too far by dismissing the general concept entirely. I conceived this book—*The Structure of Evolutionary Theory*—both as a celebration of Darwin's exemplary toughness, and as a call for the reinstitution of causal hierarchy, properly reformulated.

## LAMARCK'S TWO-FACTOR THEORY: SOURCES FOR THE TWO PARTS

In the short period of 1797 to 1800, beginning with the Directory in power and culminating in Napoleon's coup d'état of 18th Brumaire year VIII (November 9, 1799), Lamarck became an evolutionist and constructed the major features of his theory. Scholars have identified many sources as Lamarck's primary impetus—his developing views on spontaneous generation, his work on living and fossil shells (Burkhardt, 1977), the implications of his unconventional theories in physics and chemistry (Corsi, 1988). But I wish to present the logic, rather than the ontogeny, of his final and completed argument.

Lamarck's evolutionary system attempts to marry two sets of ideas, each embodying a primary module of his conceptual world. These two sets commingle at their edges, but their distinction establishes the basis of Lamarckism, and provides a hierarchical context for this first comprehensive attempt to formulate an evolutionary theory.

### The first set: environment and adaptation

The first set focuses on adaptation, and links this key attribute of organisms to the history of environments, the general pace of change, and the intimate relationship between physical and biological worlds through time. (Corsi, 1988, grants primacy to this set in the ontogeny of Lamarck's developing ideas; I accept this assertion but note that the same set, curiously, becomes secondary in the logic of Lamarck's fully formulated argument.) The framework can be entered in several places, with the rest of the edifice following by implication from a few basic premises. Lamarck's views on extinction provide a good beginning.

In opposition to his colleague Cuvier, and acting as a major source of their estrangement, Lamarck denied that true extinction (defined as termination of genealogical lines) could occur—though he allowed an exception for large quadrupeds wiped out by human predation. (Cuvier, on the other hand, embraced extinction both as the foundation of geological ordering, and as a cardinal indication that animals cannot evolve to match changing environments.) Yet, as a molluscan paleontologist, Lamarck knew that the morphologies of organisms within major taxonomic groups changed in an orderly manner through time. Evolution of outward form, with consequent preservation of lineages from extinction, represents the only alternative to termination of lineages followed by creation of new and different morphologies.

Lamarck far out-Lyelled Lyell in his commitment to uniformitarian geology (an ironic fact, given Lyell's lambasting of Lamarckian biology in his own treatise on geological uniformity). Lamarck's geological volume, the *Hydrogéologie* of Year X (1802), may strike us as bizarre in several particular assertions; but his general commitment to uniformity cannot be denied as a primary feature. Lamarck would admit no causes not now observable in operation; in particular, no paroxysms or catastrophes beyond the range of modern effects. He adopted Hutton's rigidly ahistorical vision (see Gould, 1987b) and postulated a geological history ruled by aqueous erosion (hence the title of his work). Cycles of construction and erosion unfold so many times, and in so similar a manner, that individual moments lose any distinctness, given past and future repetition of their features. Ocean waters carve mountains and continents (though Lamarck made an exception for volcanoes built by magmas). Currents tend to flow from east to west, and continents therefore erode on their eastern borders and accrete by deposition at their western edges. In a sense, therefore, continents undergo a slow westward march around the globe. This curious circumnavigation has occurred several times during the earth's extended history. But why doesn't the process yield directionality as erosion wears continents down to flat plains permanently below water? Lamarck countered with his distinctive mineralogical thesis: all rocks arise as ultimate products of organic deposition. Erosion may break continents into tiny comminuted grains, bits, and dissolved material; but as long as organisms maintain their steady state of abundance, these raw materials will be taken up again and redeposited as new rocks fashioned from the products of life.

In summary, if lineages could not become extinct, if climate and geology changed in a continuous and insensibly gradual manner throughout geological time, and if the forms and functions of organisms always matched the features of their local environments, then gradual, adaptive evolution becomes a logical necessity.

But by what mechanism will this ineluctable evolution occur? In particular, since steady, continuous alteration of environment provides the impetus for organic change, how does information flow from new environments to modify the old forms of organisms? Lamarck's answer to this riddle—building only one corner of his complete system—invokes the familiar ideas that later generations would call "Lamarckism" when the rest of his edifice had been forgotten. Lamarck begins by formulating the central principle of his functionalist credo—the counterintuitive statement, later embraced by Darwin as well, that form follows function as the order of life's history. When we contemplate any adaptation of an organism, and consider the intricate correlation of form with function, we naturally assume (or so Lamarck asserts) that form comes first, and that function can only follow. (God makes a wing, and a bird can then fly, to cite a nonevolutionary example.) But Lamarck advanced the paradoxical reverse order as his key premise: new habits lead to altered structures.*

In Lamarck's proposed mechanisms, environment changes first. Indeed, environment changes slowly and continuously on our uniformitarian planet. "Every locality," Lamarck writes (1809, p. 111), "itself changes in time as to exposure, climate, character and quality, although with such extreme slowness, according to our notions, that we ascribe it to complete stability." Organisms must accommodate to these changes by alterations in their habits—chewing with greater strength if the food gets tougher, moving more vigorously if the temperature gets colder. These altered habits, if long sustained, must feed back upon the organism in the guise of altered morphology or physiology—a thicker beak to crack the harder seeds, longer hair on a tougher skin to resist the cold.

At this point in the argument, the famous "Lamarckian" theory of inheritance finally enters. As many scholars have documented, "soft," or "Lamarckian" inheritance represented the folk wisdom of Lamarck's time, and cannot be regarded as an innovation of the *Philosophie zoologique*. Therefore, the restriction of "Lamarckism" to this relatively small and non-distinctive corner of Lamarck's thought must be labeled as more than a misnomer,

*This reversed order does not constitute a general claim for evolution against previous creationist models. This reversal represents, rather, one major style of evolutionary argument—the functionalist response. The alternative, structuralist stance—the evolutionary version of function following form—sets a major theme of this book, both in the historical precedents of Goethe, Geoffroy, Bateson and others, and in modern notions of constraint and exaptation.

and truly a discredit to the memory of a man and his much more comprehensive system. In any case, the changes wrought by new habits during an organism's lifetime can be passed directly to offspring in the form of altered heredity. Soft inheritance may have been the standard belief of the time, but Lamarck certainly recognized its crucial and particular role in his system. He wrote with his characteristic lack of doubt (1815, in Burkhardt, 1984, p. xxix): "The law of nature by which new individuals receive all that has been acquired in organization during the lifetime of their parents is so true, so striking, so much attested by the facts, that there is no observer who has been unable to convince himself of its reality." Lamarck abstracts his idea of inheritance as two principles, usually printed in italics in his texts to emphasize their importance, and known ever since as:

- use and disuse
- the inheritance of acquired characters (1809, volume 1, p. 113)

Even if this theory of inheritance ranked as folk wisdom of the day, Lamarck's revolutionary statement, one of the great transforming insights in the history of human thought, resides in the preceding principle that translates this mode of inheritance into a theory of evolution—the triggering of change in form by prior alterations in behavior. Lamarck clearly recognized the central role of this claim, for he always cited this counterintuitive sequence of causality—from altered environments to changed habits to modified form—as the linchpin of his entire system. In the *Philosophie zoologique*, he quotes his own earlier statement from the *Recherches* of 1802: "It is not the organs, that is to say, the nature and shape of the parts of an animal's body, that have given rise to its special habits and faculties; but it is, on the contrary, its habits, mode of life and environment that have in course of time controlled the shape of its body, the number and state of its organs and, lastly, the faculties which it possesses" (1809, p. 114). Lamarck then makes his threefold causal chain—environment to habits to form—even more explicit (1809, p. 126): "This is a fact that can never be disputed; since nature shows us in innumerable other instances the power of environment over habit and that of habit over the shape, arrangement and proportions of the parts of animals."

Causality might run from altered environment to changed organism, but Lamarck insisted that he did not view organisms as passive writing slates, ripe for inscription by the modifying hand of environment. Environmental change translates to adaptation of form only through the intermediary of organic action expressed, in higher creatures at least, as altered habits: "Whatever the environment may do, it does not work any direct modification whatever in the shape and organization of animals. But great alterations in the environment of animals lead to great alterations in their needs, and these alterations in their needs necessarily lead to others in their activities. Now if the new needs become permanent, the animals then adopt new habits which last as long as the needs that evoked them" (1809, p. 107).

These statements about the responses of animals to "felt needs" (Lamarck used the word *besoins*) left Lamarck open to charges of mystical vitalism when distorted for rhetorical purposes by Cuvier, or approached with over-wrought caution by Darwin. One might caricature this part of Lamarck's system by saying that a giraffe felt a need for a long neck, stretched ever so hard, and then passed the results of these successful efforts directly to offspring. But a fair assessment of Lamarck's actual words shows that he advocated no ineffable willing, but only the commonplace idea that a change in environment can, in an almost mechanical way, elicit an organic response in terms of altered habits: "Variations in the environment induce changes in the needs, habits, and mode of life of living beings . . . these changes give rise to modifications or developments in their organs and the shape of their parts" (1809, p. 45).

This first set of Lamarckian ideas contains nothing that should have offended Darwin, while several points embody the deeper functionalist and adaptationist spirit of the Darwinian view of life. Darwin did not grant such crucial emphasis to soft inheritance, although he accepted the principles of use and disuse and inheritance of acquired characters, and he awarded them a subsidiary role in his own theory. But two key items in this first set might be designated as decidedly Darwinian in spirit, if only because they advance and presage two of the half dozen most important ideas in Darwin's theory: the uniformity of environmental change,* and the functionalist first principle that change of habit sets the pathway to altered form. The mechanisms of change differ to be sure—altered habits establish new selection pressures for Darwin, but induce heritable modifications more directly for Lamarck—but both thinkers share a functionalist commitment.

I would argue that the structuralist-functionalist dichotomy precedes any particular theory of mechanism within either camp. Thus, we may view Lamarck and Darwin as occupying the common ground of functionalism, with their differing mechanisms of natural selection and soft inheritance as versions of the same deeper commitment. Therefore, if Lamarckism only encompassed this first set of ideas, we might interpret Lamarck as the inception of a smooth transition to Darwin. But Lamarckism also includes a second set of concepts, which, when combined with the first set into Lamarck's full system, builds an evolutionary theory truly opposed to Darwin's chief theoretical concept and operational principle as well.

### The second set: progress and taxonomy

The first set, by itself, leads to a logical dilemma for Lamarck's view of life and his professional commitments. Adaptation to changing local environments may be well explained, but Lamarck's truly ahistorical uniformitarianism implies that life can manifest no progress, or no linear order at all, if adaptation matches creatures to an environmental history without direction.

---

*Lamarck's commitment to gradualism as a general philosophy matched Darwin's in centrality and strength, thus forging a connection deeper than a shared attitude toward environment alone. Lamarck wrote (1809, p. 46): "Consider . . . that in all nature's works nothing is done abruptly, but that she acts everywhere slowly and by successive stages."

(This issue arose for many environmental determinists in both creationist and evolutionist camps. Buckland, and most of his catastrophist colleagues, maintained their allegiance to life's increasing perfection by positing a directional history of environmental change—increasing inclemency, for example, requiring improvement in organic design to meet the growing challenge. This option was not open to Lamarck, who espoused a steady-state, nondirectionalist geology.)

Yet Lamarck firmly advocated a taxonomic ordering of organisms by the conventional scheme of increasing perfection in organization. This subject greeted him on a daily basis, for he held the post of curator for invertebrates at the *Muséum* in Paris, and his yearly courses featured this organizing theme of linear order. (As a pedagogic device, Lamarck usually started with humans, as the "highest" creature, and then discussed the rest of nature as a degradation from maximal complexity. He defended this procedure, even in his evolutionary writings, as a method for teaching, even though historical order had actually moved from simple to complex—for he argued that one must understand the full and final possibilities before grasping the imperfect and incipient beginnings.)

Lamarck argued that a second set of forces, distinct from the causal flow of environment to organism, produced nature's other primary pattern of advancing complexity. But this claim for an efficient and universal cause of progress engendered another dilemma: why, on our present and ancient earth, do some organisms still maintain the simplest anatomies? Why were these forms not pushed up the ladder of complexity ages ago? Lamarck resolved this problem with the last major argument of his full system—continuous spontaneous generation. New life continues to arise from chemical constituents; these simple forms begin their march up the ladder, while replacements at their lowly status continue to form anew. (Thus, in a curious sense, as Simpson and others have noted, Lamarck's evolutionary system operates as a grand steady state, even as any particular bit of protoplasm moves on a historical path up the sequence. The ladder of life really operates as a continuous escalator, with all steps occupied at all moments. The simplest forms continue to arise by spontaneous generation from chemical constituents formed by the breakdown of higher creatures upon their individual deaths.)

Lamarck argued that his unconventional chemistry, emphasizing the role of fire and the motions of subtle fluids, engendered these two central phenomena—spontaneous generation and progress up the ladder—as consequences of deeper physical principles. Lavoisier had destroyed the old quadripartite taxonomy of air, water, earth, and fire in developing his theory of chemical elements. Lamarck opposed the "new chemistry" by asserting the old taxonomy, and his own distinctive claim for the primacy of fire. Much of Cuvier's disdain focused not on Lamarck's biology, but on his allegiance to this antiquated chemistry.

Lamarck, who excelled in crisp assertion but not in clear exposition, never fully specified why chemicals should aggregate to life, or what subtle motions of physical fluids would build the increasing complexity of anatomy. He held that the products of spontaneous generation arose as small, soft and un-differentiated primal forms. The complexifying force—which Lamarck usually called *le pouvoir de la vie* or *la force qui tend sans cesse à composer l'organisation*—resides in the motion of fluids and their inevitable tendency to carve channels, sacs, and passageways in soft tissues. This process, extended through time, gradually builds ever greater complexity. In the *Histoire naturelle* of 1815, Lamarck offers his most explicit statement about this process: "As the movement of fluids . . . accelerates, the vital forces would grow proportionately, and so will their power. The rapid motion of fluids will etch canals between delicate tissues. Soon their flow will begin to vary, leading to the emergence of distinct organs. The fluids themselves, now more elaborate, will become more complex, engendering a greater variety of secretions and substances composing the organs" (1815, in Corsi, 1988, p. 189).

Lamarck did clearly assert that these internal carvings of complexity maintained a relentless and intrinsic causal basis separate from the apparatus of response to "felt needs" used in building adaptations to changing local environments. He contrasted the two sets of forces in writing: "There exists a variety of environmental factors which induces a corresponding variety in the shapes and structure of animals, independent of that special variety which necessarily results from the progress of the complexity of organization in each animal" (1809, p. 112). He also stated that the entire escalator of complexity could run a full course in a constant environment: "If nature had given existence to none but aquatic animals and if all these animals had always lived in the same climate, the same kind of water, the same depth, etc., etc., we should then no doubt have found a regular and even continuous gradation in the organization of these animals" (1809, p. 69).

Lamarck therefore proposes two distinct sets of forces to construct what he regarded as the two preeminent features of life—progress in linear order, and adaptation to environment. The interactions of these sets—not the causes or properties of either one—establish the foundation of Lamarckism, properly defined in his own expansive terms.

### Distinctness of the two sets

I shall argue in the next section that these two sets of concepts must be regarded as both logically distinct and opposed in Lamarck's system. My basis for regarding Lamarckism as a theory of hierarchy lies in this division. Lamarck, as we shall see, always presents the two sets as separate in his later evolutionary writing, and scholars of Lamarckism have accepted this contrast as crucial (Burkhardt, 1977; Mayr, 1972; Simpson, 1961). But Lamarck, as noted several times above, remains a frustrating figure for historians. His assertions are bold, even dogmatic; but his arguments tend to be sketchy, full of elisions, or even self-contradictory. These frustrations become most apparent in Lamarck's treatment of his two primary forces (as Corsi, 1988, has discussed with great insight). The explicit assertions of his later works rank the two forces as distinct and opposed, but both the ontogeny and logic of

Lamarck's argument show more "leakage" than his words would suggest. Consider the following, as stressed by Corsi and Burkhardt:

ONTOGENY. Although Lamarck presents the forces of adaptation as deviations from, and therefore secondary to, the primary causes that build complexity, he apparently developed his mechanism for progress from his previously formulated ideas about adaptation (Corsi, 1988; and newly discovered evidence in Gould, 2000d). Still, the psychological source of a theory needn't map its eventual logical structure, and this point, while interesting, scarcely compromises the distinctness and ranking of the two sets.

CAUSATION. At several points, Lamarck breaches the boundaries between his sets in discussing causation.

(1) Soft inheritance works in both sets. Whether an organism becomes more complex because fluids carve channels by intrinsic chemistry, or becomes better adapted because habits change in response to altered environments, the acquired features must still be passed to offspring by direct inheritance. Still, a common mechanism may work in two modes, and this linkage does not compromise Lamarck's claim for distinctness.

(2) The style of action for soft inheritance in adaptation depends upon the state of complexity engendered by opposing forces of progress. Lamarck divided organisms into three ascending groups designated, in the old Aristotelian terms, as insensitive, sensitive, and rational. The first group, too simple to mount a creative response to external change, reacts to environment not by altered habits, but by direct influence. The capacity for active response, Lamarck's famous *sentiment intérieur,* only arises in the second group and unleashes the tripartite causal sequence of changed environment to altered habit to modified form.

(3) The real blurring occurs when we try to make sense of Lamarck's claim that forces of progress can build the entire sequence from infusorian to complex vertebrate without any environmental change. Lamarck surely makes this assertion explicitly, without hesitation (see citation on p. 187), and the distinctness of his two forces depends upon this potential independence. But Lamarck does not work out a consistent justification, and several frustrated historians have even argued that he could not have done so without contradiction—that his system, in other words, suffers from a true defect in logic of argument on this point.

The simplest organisms, Lamarck states, are carved out and complexified by "subtle" and "imponderable" fluids—caloric and electricity in his system. These fluids work in their intrinsic way to produce increased complexity. But as animals differentiate and harden, fluids must flow in preset channels; the weak imponderables then lose their power to mold, and the body's own ponderable fluids must assume this role. (Lamarck locates this transition at about the echinoderm grade of organization.) At this level, the "power of life" should become inoperable without an impetus from environmental change—and the two sets of forces should therefore commingle. Protected inside a rigid body, and constrained to flow in preexisting channels, how can the ponderable fluids produce further advances in complexity unless changed environments elicit altered habits, thus modifying form and permitting the ponderables to flow in new ways (see Corsi, 1988, p. 200; and Burkhardt, 1977, p. 147).

PATTERN. The "pure" distinction of progress and adaptation should produce a single linear chain (for progress) with lateral deflections (for adaptation). Lamarck tried to construct such a topology, but could not carry his scheme to completion at two important places—the top and bottom of the ladder—where environment intruded upon the chain to blur the distinction of forces.

(1) Two sequences of spontaneous generation. Lamarck first proposed a single linear series of animals, starting with the spontaneous generation of infusorians (protistans) as free-living creatures in water. These unicells then aggregated to polyps and their relatives, and then to simple, bilaterally symmetrical worms (see Fig. 3-1). However, Lamarck later discovered worms (the acoelous platyhelminths in modern terminology) without nerve cords. These worms ranked "higher" than polyps on grounds of their mobility, but could not be the descendants of polyps, unless the nerve cords of polyps had degenerated and disappeared—an impossibility under the "force that tends incessantly to complicate organization." Thus, worms without nerve cords must represent part of a second and separate sequence of progress. Lamarck proposed an origin for this second sequence in the spontaneous generation of even simpler worms as parasites within the bodies of other organisms. If dif-

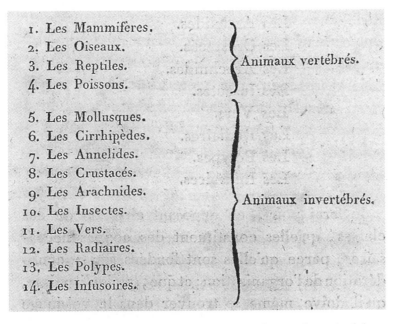

1. Les Mammifères.
2. Les Oiseaux.
3. Les Reptiles.
4. Les Poissons.
} Animaux vertébrés.

5. Les Mollusques.
6. Les Cirrhipèdes.
7. Les Annelides.
8. Les Crustacés.
9. Les Arachnides.
10. Les Insectes.
11. Les Vers.
12. Les Radiaires.
13. Les Polypes.
14. Les Infusoires.
} Animaux invertébrés.

3-1. Lamarck's linear series of animal organization, from volume 1 of the *Philosophie zoologique* of 1809. (Author's collection.)

ferent environments—a pond and the body of a complex creature—encourage disparate inceptions for sequences of progress, then the two forces commingle (Fig. 3-2).

(2) Ramification at the top. Lamarck could not rank the vertebrates in linear order. He followed the conventional path of fish to reptile, but could not convince himself that birds fell between reptiles and mammals in a genealogical sense. He therefore permitted a fork, provoked by the environmental set of forces, at the very top of a ladder supposedly built by the unilinear impetus of progress (Fig. 3-3): "We cannot doubt," he wrote with characteristic certainty (1809, p. 176), "that the reptiles by means of two distinct branches,

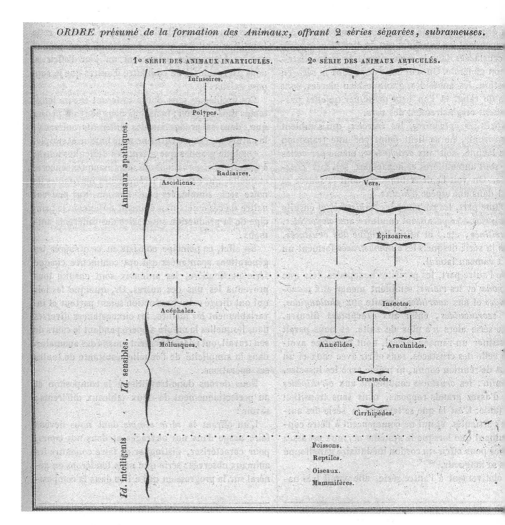

*ORDRE présumé de la formation des Animaux, offrant 2 séries séparées, subrameuses.*

3-2. Lamarck's later conception of two chains of being with different starting points, the first (to the left) from free-living single-celled infusorians, the second (to the right) beginning with parasitic worms spontaneously generated within the bodies of higher organisms. From Lamarck, 1815. (Author's collection.)

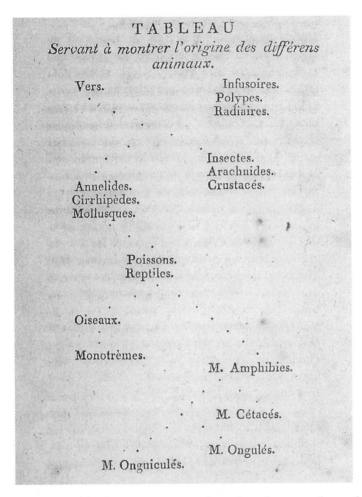

TABLEAU
*Servant à montrer l'origine des différens animaux.*

Vers.

Infusoires.
Polypes.
Radiaires.

Insectes.
Arachnides.
Crustacés.

Annelides.
Cirrhipèdes.
Mollusques.

Poissons.
Reptiles.

Oiseaux.

Monotrèmes.

M. Amphibies.

M. Cétacés.

M. Ongulés.

M. Onguiculés.

3-3. Lamarck allowed his linear sequence to branch at the apex of complexity (shown as the bottom in this figure because he begins with the lowest forms and works down by descent). Lamarck could not rank birds and mammals as part of the single sequence, and therefore allowed a branch after reptiles with birds on one side (left, culminating in egg-laying monotremes), and mammals to the right. From additional material added to the end of Volume 2 of the *Philosophie zoologique* of 1809. (Author's collection.)

caused by the environment, have given rise, on the one hand, to the formation of birds, and, on the other hand, to the . . . mammals."

I doubt that we can take this analysis much further. Historians often err in trying to wrest consistency from great thinkers at all costs. Some issues are too difficult, too encompassing, too important, too socially embedded, or just too devoid of evidence, for resolution even by the finest scientists. Darwin never consolidated his contradictory ideas about progess (see Chapter 6), and Lamarck never found a thoroughly consistent way to fulfill his desired argument for a full separation between two forces pulling evolution in orthogonal directions—up the ladder of progress, and sideways into lateral paths of adaptation. Lamarck may never have completed his scheme with success and consistency, but he made his desires clear to the point of redundancy. Lamarck's two-factor theory holds the distinction of being both the first evolutionary system in modern Western thought, and a strong argument for causal hierarchy. The two levels—in strong contrast with modern theories of hierarchy—are both causally distinct and contradictory for Lamarck, thus inspiring Darwin's legitimate disparagement. Lamarck's distinction of levels, as discussed in the next section, unites hierarchy and evolution at the starting gate of the subject's modern history.

## LAMARCK'S TWO FACTOR THEORY: THE HIERARCHY OF PROGRESS AND DEVIATION

Lamarck had separated his sets of forces in order to account for the two primary attributes of natural order—features that seemed to play off against, or even to contradict, each other. First, organisms form a progressive sequence from monad to man, but the sequence abounds with gaps and deviations—so some other force must be disrupting a potentially smooth gradation. Second, organisms are well adapted to their environments, but most adaptations, from the tiny eyes of moles to the legendary necks of giraffes, represent particular specializations and departures from type (with many adaptations counting as losses or degenerations); therefore, adaptation cannot account for the sequence of progress.

Lamarck joined the two sets in a discordant union that operated more like a tug of war than a harmony. This partnership made no pretense to equality. A primary and dominating force—the march of progress—struggled to order organisms in a simple and sensible way; while a secondary and disrupting force—*l'influence des circonstances,* or adaptation to local environments—tore this order apart by pushing individual lineages into lateral deviations from the main track, thereby making the order of life rich, messy, and replete with clumps and gaps. This clear distinction of merit—the regular vs. the deviant, the progressive vs. the merely fit—imparts the character of hierarchy to Lamarck's uneasy marriage of forces, with a primary factor doing its inexorable, underlying work at a higher level, while a secondary but more immediate factor of disruption plays upon the products of this higher level, pushing some forms into the side-channels of its influence. Burkhardt (1977, p. 87) captures both the hierarchy and conflict of forces in his epitome of Lamarck's system as an attempt to explain "how organisms would develop *naturally*" along a chain of progress "were it not for the constraining accidents of history" pushing lineages into side channels of adaptation.

Lamarck worked his way slowly towards this final system of hierarchy and relative importance. The *Floréal* lecture of 1800 states that the "principal masses" of major taxonomic units "are almost regularly spaced" (1800, 1984 edition, p. 416), but designates some peculiarly adapted species as "lateral ramifications" and "truly isolated points." But this lecture cites only environment as a trigger of change. The *Recherches* of 1802 adds the theme of "organic movements" forming new organs and faculties in an intrinsic sequence of advance. By the time of his most famous work, the *Philosophie zoologique* of 1809, Lamarck "was explicit in portraying the diversity of animal form as a result of two separate processes" (Burkhardt, 1977, p. 145), and he had formulated the arguments of hierarchy and relative importance as well. The *Histoire naturelle* of 1815–1822 then consolidates and advocates the hierarchical two-factor theory even more strongly.

In the *Philosophie zoologique,* Lamarck begins by claiming that, in an ideally simple world, a single sequence of progress would regulate all taxonomic order:

> It may then be truly said that in each kingdom of living bodies the groups are arranged in a single graduated series, in conformity with the increasing complexity of organization and the affinities of the object. This series in the animal and vegetable kingdoms should contain the simplest and least organized of living beings at its anterior extremity, and ends with those whose organization and faculties are most perfect. Such appears to be the true order of nature, and such indeed is the order clearly disclosed to us by the most careful observation and an extended study of all her modes of procedure (1809, p. 59).

But this principle of progress remains insufficient in our actual world, where environmental change elicits adaptations off the main sequence: "It does not show us why the increasing complexity of the organization of animals from the most imperfect to the most perfect exhibits only an *irregular gradation,* in the course of which there occur numerous anomalies or deviations with a variety in which no order is apparent" (1809, p. 107).

These "anomalies and deviations" are produced by a second, and clearly subsidiary, force—a "special factor" that thwarts the "incessantly working" source of general progress, and riddles the chain with gaps and lateral branches: "If the factor which is incessantly working toward complicating organization were the only one which had any influence on the shape and organs of animals, the growing complexity of organization would everywhere be very regular. But it is not; nature is forced to submit her works to the influence of their environment, and this environment everywhere produces variations in them. This is the special factor which occasionally produces . . . the often curious deviations that may be observed in the progression" (1809, p. 69).

This special factor may be identified as environmental adaptation, initiated by changed habits and abetted by soft inheritance in the principles of use and disuse and the hereditary passage of acquired characters: "The environment exercises a great influence over the activities of animals, and as a result of this influence the increased and sustained use or disuse of any organ are causes of modification of the organization and shape of animals and give rise to the anomalies observed in the progress of the complexity of animal organization" (1809, p. 105).

As a primary sign of our estrangement from Lamarck's world, and our lack of understanding for his system, all "standard" text-book examples of Lamarckian evolution ignore his fundamental, higher-level principle of progress, and only cite instances of lateral twigs built as highly specialized adaptations. We do this, I suppose, because adaptation and specialization constitute the major theme in our modern evolutionary vocabulary (in the altered guise of Darwinian causation), while the bulk of Lamarck's system has passed beyond our notice into cognitive dissonance. In any case, every classical example—from eyeless moles, to webbed feet of water birds, to long legs of shore birds, to the blacksmith's strong right arm—ranks as a lateral deviation, not a stage on the main sequence. As for the greatest cliché and exemplar of all, the ubiquitous giraffe of our text-books, happily munching leaves at the tops of acacia trees, Lamarck provides only one paragraph of speculation—with no elaboration, no measurements, no data at all. An example can become a knee-jerk standard for many reasons, with cogent, complete documentation not always prominent among them (see Gould, 1991b, on the evolution of horses and the size of *Hyracotherium*). Nor does simple repetition enhance the probability of truth! Lamarck wrote this and only this about giraffes (even repeating a common error about differential lengths of fore and hind limbs):

> It is interesting to observe the result of habit in the peculiar shape and size of the giraffe *(Camelo-pardalis)*: this animal, the tallest of the mammals, is known to live in the interior of Africa in places where the soil is nearly always arid and barren, so that it is obliged to browse on the leaves of trees and to make constant efforts to reach them. From this habit long maintained in all its race, it has resulted that the animal's forelegs have become longer than its hind-legs, and that its neck is lengthened to such a degree that the giraffe, without standing up on its hind-legs, attains a height of six meters (1809, p. 122).

The final, complex order of life arises from an interplay of the two forces in conflict, with progress driving lineages up the ladder and adaptation forcing them aside into channels set by peculiarities of local environments: "The state in which we find any animal, is, on the one hand, the result of the increasing complexity of organization tending to form a regular gradation; and, on the other hand, of the influence of a multitude of very various conditions ever tending to destroy the regularity in the gradation of the increasing complexity of organization" (1809, p. 107). In his strongest characterization of the two forces as conflicting, Lamarck tells us in another passage that "nature's work [of progress] has often been modified, thwarted and even reversed by the influence exercised by very different and indeed conflicting conditions of life upon animals exposed to them throughout a long succession of generations" (1809, p. 81).

Two additional statements in the *Philosophie zoologique* give dramatic expression to the absolute distinction of the forces, and to their hierarchical character, with progress as primary and regular, and diversity as secondary and disturbing. The first provides a vivid iconography of the two-factor theory: progress builds the regular and rising trunk; diversity snatches some items off this upward highway and pulls them into orthogonal blind alleys—"lateral ramifications" that peter out into "isolated points": "These irregularities . . . are found in those organs which are the most exposed to the influence of the environment; this influence involves similar irregularities in the shape and condition of the external parts, and gives rise to so great and singular a diversity of species that, instead of being arranged like the main groups in a single linear series as a regularly graduated scale, these species often constitute lateral ramifications around the groups to which they belong, and their extremities are in reality isolated points" (1809, p. 59). The second features Lamarck's explicit statement about hierarchy, translated into differential taxonomic levels of attention. The broad forces of progress set relations among orders and classes; smaller and more immediate episodes of adaptation establish species and genera:

> Nature . . . has really formed a true scale . . . as regards the increasing complexity of organization; but the gradations in this scale . . . are only perceptible in the main groups of the general series, and not in the species or even in the genera. This fact arises from the extreme diversity of conditions in which the various races of animals and plants exist; for these conditions have no relation to the increasing complexity of organization; but they produce anomalies or deviations in the external shape and characters which could not have been brought about solely by the growing complexity of organization (1809, p. 58).

Lamarck's increasing conviction about the distinctness, hierarchical character, and conflicting nature of the two forces culminates in his last major work, the *Histoire naturelle* of 1815–1822. The force of progress has now become a "predominant prime cause," while adaptation ranks below as occasional and foreign, a disturbance strong enough to disrupt but not to efface nature's deeper law:

> The plan followed by nature in producing animals clearly comprises a predominant prime cause. This endows animal life with the power to make organization gradually more complex . . . Occasionally a foreign, accidental, and therefore variable cause has interfered with the execution of the plan, without, however, destroying it. This has created gaps in the series, in the form either of terminal branches that depart from the series in several points and alter its simplicity, or of anomalies observable in specific apparatuses of various organisms (1815, in Corsi, 1988, p. 189).

### ANTINOMIES OF THE TWO-FACTOR THEORY

We have seen how Lamarck formulated and intensified the two-factor theory from his first exposition of evolution in 1800 to his last and most comprehensive works. One might even say, following a developmental metaphor, that the two sets of forces differentiated from an originally more inchoate conglomeration of evolutionary ideas, gradually becoming more different and more sharply defined. As the two sets grew along their orthogonal pathways, they became alternate centers of nucleation for the full realm of evolutionary ideas. All major items of this new conceptual world fell to one side or the other—as though two suns had entered an originally homogeneous universe, and all particles had to enter either one or the other gravitational system. Lamarck's two-factor theory separated the universe of evolutionary concepts into a set of dichotomies best characterized as antinomies, an unfamiliar word designating contradictions between two equally binding principles (and originally used to specify differences between ecclesiastical and secular law when both vied for domination in medieval states). The precipitation of ideas about Lamarck's two axes established a long list of antinomies that, in an important way, has set the agenda of evolutionary biology ever since. Darwin opposed this structure of antinomies; others have advanced strong defenses, in whole or in part. The modern theory of hierarchy depends upon a selective defense, but in a manner radically different from Lamarck's formulation. Consider a few key items:

IDEAL ORDER (REAL) *vs.* DISRUPTION (DISTURBING). The interpretation of diversification by adaptation as lateral to, and disruptive of, an underlying lawful regularity marks an old tradition that Darwin fought fiercely by elevating the supposed "disturbing" force to the cause, by extrapolation, of all evolution. Curiously, since old traditions die hard, this antinomy remains potent (even under a Darwinian rubric) in the common claim that anagenesis, or evolutionary trends in lineages, should be viewed as distinct from cladogenesis, or diversification—and that speciation is, in Julian Huxley's words (1942, p. 389), "a biological luxury, without bearing upon the major and continuing trends of the evolutionary process."

Behind this issue, of course, and particularly well expressed in this first antinomy, stands the ancient credo of essentialism. Just as the essence or type never becomes fully incarnate in an actual object (because any material being must be subject to all the slings and arrows of outrageous fortune in our palpable world of accidents), so does the outside world of changing environments deny full expression to the ideal march of progress.

PROGRESS *vs.* DIVERSITY. Lamarck's expression of the fundamental vs. the disruptive; note Huxley's words above for a modern expression.

INTERNAL *vs.* EXTERNAL, OR INTRINSIC *vs.* INTERACTIVE. The march of progress is intrinsically and internally generated as a consequence of the chemical properties of matter; this march represents the "essential" process of life, moving ever forward in the absence of any push or disturbance from external forces. Lamarck makes this contrast explicit in arguing that the march of progress would proceed to smooth completion even in an absolutely constant environment.

IMMANENCE *vs.* UNPREDICTABILITY. The forces of progress, arising as consequences of chemical laws, generate a set of predictable products inherent in the constitution of nature. Since the chain is constantly replenished by spontaneous generation, all stages exist at all times, and the entire sequence constitutes a permanent part of nature. But the disturbing force of environmental change introduces the oddness and unpredictability of contingency. We can never know exactly what climatic change will occur, which lineage will be diverted into its channel, and how the resulting adaptation will form. Thus, our actual world becomes filled up with unique particulars. Giraffe necks do not arise by first principles of natural law, but as a contingency of dry climates and acacia trees at a particular time and place.

TIMELESS *vs.* HISTORICAL. History requires distinctive moments that tell a story as a sequence of events. The force of progress may confer history upon any particular bolus of protoplasm as it mounts the ladder. But, in a larger sense, this force also cancels the usual meaning of history. Each step becomes predictable and repeatable; and each exists at every time (since spontaneous generation continually replenishes the base). Thus, Lamarck's perfecting force becomes essentially ahistorical. The rungs of the ladder are permanent and always occupied; items pass up and through, but the forms are timeless. Genuine history enters via the disturbing force of environmental adaptation instead. A mole without eyes, a stork with long legs, a duck with webbed feet—all originate as nonrepeatable objects of a historical moment, triggered by a particular change of environment in a unique time and place.

HIGHER TAXA *vs.* SPECIES AND GENERA. Lamarck's two-factor theory is hierarchical. The force of progress—paramount, primary, and underlying—produces patterns of nature at the broadest scale, and therefore forges the relationships among higher taxa in our classifications. The force of adaptation is secondary, disruptive, and subsidiary. It seizes individual lineages and pulls them from the main sequence into side channels that always peter out as dead ends. This lower-level force produces the smaller units, the species and genera of our taxonomies.

ELUSIVE *vs.* PALPABLE. This last antinomy does not form part of Lamarck's scheme, but becomes important in later interpretations, particularly in Darwin's refutation. The force of progress lies deeper within and operates at a higher level; the force of adaptation works palpably at the surface of things. One can, at least in principle, observe climates getting colder and elephants growing thick coats of fur in direct response; but advance up the ladder lies further from our view in an abstract distance. Lamarck might have denied that the causes of progress posed any greater difficulty in recognition and observation; in his conceptual world, these forces arose from the chemical nature of matter—and therefore became just as accessible as the immediate causes of adaptation. But when Lamarck's theories of physical causation collapsed, the force of progress became elusive—something operating so slowly, and at such high taxonomic levels, as to be effectively invisible in the here and now of testable science. Darwin based his theory upon a reformulation of this seventh antinomy—by arguing that palpable and immediate forces of adaptation did not oppose an inscrutable and untestable force of progress—but rather *became* the source of progress as well (and hence the only primary cause of all evolution) when extrapolated, by principles of uniformitarianism, into the immensity of time. All evolution therefore entered the realm of the testable.

Modern evolutionists may read this list with an odd feeling of *déjà vu*—in the backward sense that we have already encountered all these issues in the modern debates of our own professional careers, but didn't know that our forebears had struggled over the same themes. Doesn't the late 20th century debate about micro and macroevolution raise the same questions about different causes at higher and lower taxonomic levels (the basis of Goldschmidt's argument, for example—see pp. 451–466); and don't extrapolationists still charge the defenders of higher level causality with proposing untestable theories of evolutionary change?

I do not find this persistence surprising at all (see Gould, 1977a). I have already cited (p. 58) A. N. Whitehead's famous remark that all later philosophy is a footnote to Plato. He did not mean to argue, by this statement, that no one (including himself) had thought anything new for more than two thousand years. Rather, he wished to defend the proposition that truly deep issues are few and not so obscure. The first great thinker should be able to lay out a framework and specify the primary questions. Later history must recycle the same issues, while offering new explanations in abundance (especially in empirical realms where truly novel information becomes available). Lamarck, for all his stubbornness and for all the idiosyncrasies of his theory, was a great thinker, and he did find a location for all major questions within his system. His theory therefore becomes a starting point, and later debate must engage the same issues. Given the central theorem of this book, I am especially gratified that Lamarck based the initiating system of our profession upon a theory of hierarchy—in a form that did not work, based on causes that we must reject for Darwin's good reasons; but a theory of hierarchy nonetheless. Evolutionary theory therefore set its roots, and cut its teeth, in the concept of hierarchical levels of causality.

## An Interlude on Darwin's Reaction

In the flood of Darwinian scholarship unleashed after the centennial celebrations of 1959 and continuing unabated today, I regard no reform as more important than the thorough debunking of the romantic myth that Darwin, alone and at sea, separated from the constraints of his culture, apprehended evolution as an objective raw truth of nature. This Galapagos myth, rooted in tortoises and finches, is demonstrably false in Darwin's particular case, and surely bankrupt as a general statement about human psychology and the sociology of knowledge.

Darwin saw many wonderful things on the *Beagle;* nature challenged him, broadened his view, and instilled flexibility. Darwin returned to England with the tools of conversion, but still as a creationist, however suffused with doubts and questions (Sulloway, 1982a; Gruber and Barrett, 1974; Schweber, 1977; Kohn, 1980; Desmond and Moore, 1991; Browne, 1995). As for the Galapagos, he had missed the story of the finches entirely, because he had been fooled by their convergences and had not recognized the underlying taxonomic unity (Sulloway, 1982b; Gould, 1985c). Darwin had been informed that the tortoises differed from island to island, but had failed to appreciate the significance of this claim. Naturalists then believed, falsely, that the Galapagos tortoises were not indigenous but had been transported in recent memory by Spanish buccaneers and placed on the Galapagos as a source of food for revictualizing ships. Thus, Darwin must have reasoned, if the tortoises had only been on the Galapagos for two or three centuries, differences among islands could not be consistent or meaningful. The *Beagle* had provisioned itself with several tortoises, stored live in the hold as meat on the hoof, so to speak. In a modern version of fiddling while Rome burned, Darwin partook of the feasts but made no plea for conservation when his shipmates then pitched the carapaces overboard.

Darwin became an evolutionist by returning to England and immersing himself in the scientific culture of London—by arguing with colleagues, by reading and pondering (mostly in the library of the Atheneum Club), by seeking good advice (learning from ornithologist John Gould, for example, that those diverse Galapagos birds were all finches). He exploited a broad range of humanistic Western culture in pursuing his struggle for intellectual reform in natural history. He read Plato, Milton, and Wordsworth. He constructed the theory of natural selection, as argued in the last chapter, in conscious analogy with the laissez-faire theories of Adam Smith and the Scottish economic school. Darwin, without the impetus and challenge of this intellectual environment, might have become a country parson, with a beetle collection maintained by an ecclesiastical sinecure as the remnant of a childhood passion for natural history.

In this enlarged perspective on the origin of Darwin's evolutionary views, the importance of his precursors becomes greatly enhanced. Lamarck and Chambers* do not figure as irrelevancies to be ignored, or (even worse) as im-

---

*Darwin, in distancing himself from precursors as he formulated and refined his own theory in the years before 1859, usually drew a primary contrast with Lamarck. But he sometimes added the anonymous author of the *Vestiges of the Natural History of Creation* (written by the Scottish publisher Robert Chambers in 1844, though his authorship did not become officially known until the last edition of 1884, two years after Darwin's death), pairing "Mr. Vestiges" with Lamarck as the entirety of a background to be rejected with vigor on the old principle that the enemy within can be more distressing than the enemy without. Darwin wrote to Hooker in an undated letter between 1849 and 1853: "Lamarck . . . in his absurd though clever work has done the subject harm, as has Mr. Vestiges." Darwin then adds, with his endearing capacity for self-deprecation: ". . . and, as (some future naturalist attempting the same speculations will perhaps say) has Mr. D . . ." (in F. Darwin, 1887, vol. 2, p. 39).

The publication of the *Vestiges* in 1844 unleashed a firestorm of criticism from all sides. "From the bottom of my soul," wrote the dour creationist (and Darwin's teacher in geology) Adam Sedgwick, "I loathe and detest the Vestiges." Following the prejudices of his age, Sedgwick conjectured that anything so stupid must have been written by a woman. But serious evolutionists also took offense at Chambers's rank amateur ignorance of natural history, and at the purely speculative character of his assertions (including the claim that birds evolved to mammals in two steps via the intermediary of a duck-billed platypus). Nonetheless, *Vestiges* became a *succès de scandale*, going through 12 editions in 40 years,

pediments to be overcome; they represent an essential part of his context for study, and they played a major role in shaping the radical and different character of Darwin's distinctive theory.

I believe that Lamarck had a far greater influence on Darwin than tradition has allowed (a point advanced by other historians of science as well—see Corsi, 1978; Mayr, 1972, p. 90). I base this claim on Darwin's own contact with Lamarck's works, his private reactions as recorded in letters, and the eventual content of his theory. I don't claim that Darwin devised natural se-

---

and surely doing some service in making the subject of evolution discussible. The pairing of Lamarck with Chambers does, therefore, represent the totality of highly noticed, fully evolutionary systems available in England as Darwin formulated his theory.

This pairing becomes an important footnote to my argument because Chambers's theory is also hierarchical in the same general sense as Lamarck's—that is, in advocating separate realms of causes for progress and deviation, with progress at a higher level and difficult to discern, and deviation as immediate and palpable, but incapable of generating the full taxonomic order of life.

Chambers's particular theory, however, differed radically from Lamarck's, and "Mr. Vestiges" attacked his French predecessor and inspiration (via Lyell) as a man "whose notion is obviously so inadequate to account for the rise of the organic kingdoms, that we only can place it with pity among the follies of the wise" (1844, p. 231). But Chambers thoroughly misunderstood Lamarck, accepting the usual caricature of mystical will leading to adaptation as Lamarck's complete system. Somehow Chambers missed the hierarchical character of Lamarck's theory; he understood that the force of adaptation could not yield progress, but, unaware that Lamarck had ascribed life's ladder to different causes acting at another level, he concluded that Lamarck had erred in trying to explain all evolution by adaptation alone.

Chambers remedied this misperception by devising a theory every bit as hierarchical as Lamarck's actual proposal, but radically different in mechanism. Chambers's system works by extended analogy to von Baer's laws of embryonic differentiation (Gould, 1977b, pp. 109–112). Linear progress follows the embryonic sequence of the highest organism. The higher-level cause of progress pushes creatures up the sequence. Meanwhile, the lower-level force of deviation pulls organisms into adaptive configurations at various plateaus of design. A small impetus from the cause of progress may bring a developing animal to the plateau of fishes; a greater push will lead to reptilian, mammalian, or even human grade. Progress depends upon an ability "to protract the straightforward part of the gestation over a small space" (1844, p. 213), thereby resisting the lateral force of adaptation at lower plateaus. Chambers even introduces the cogent idea (still arousing modern debate) that higher-level causes may be hard to ascertain because they operate so rarely. They may work quite regularly and in an absolutely lawlike way, but only once every million years or so—and our chance of observing them then becomes vanishingly small. (This attack on a dogmatic uniformity, based only on observed modern causes, emerges in such recent ideas as the bolide impact theory of mass extinction.) Thus, for Chambers, embryological transcendence to the next stage of progress may occur regularly and rapidly, but very rarely—while the lateral forces of adaptation operate around us all the time.

Darwin therefore faced only two widely-discussed evolutionary systems as he formulated his own. Both were hierarchical, and both proposed an elusive higher-level force of progress, paired with a palpable but limited lower-level force of adaptation. Darwin must have been struck by the enigma that what he could see didn't matter (in the long run of evolutionary advance), and what mattered couldn't be seen. How much of his distinctive single-level theory of extrapolation arose in reaction to this intractable dilemma of hierarchical theories posed in the old and invalid style by Lamarck and Chambers? Hierarchy, as this chapter holds for its primary theme, has been a crucial ingredient of evolutionary theories from the start, and may be more responsible than we have recognized for the eventual character of Darwin's system.

lection as a conscious point-by-point contrast or refutation of Lamarck, but I suspect that Darwin clearly recognized what he liked least in Lamarck and strove to formulate a theory of opposite import.

Darwin said little about Lamarck in his published works, with no explicit reference to Lamarck's evolutionary views in the first edition of the *Origin*, and grudging praise in the historical preface added to later editions. But we know that he studied Lamarck intensely, and didn't like what he read. Darwin owned a copy of the 1830 printing of the *Philosophie zoologique* (see Hull, 1985, p. 802), and read the book while making heavy annotations at least twice. More important, perhaps, Lamarck had provided Darwin's introduction to evolution via Lyell's fair but critical exegesis in the *Principles of Geology*.

Lyell's characterization becomes particularly interesting because he emphasizes, in his masterful prose, the very two points that Darwin would strive most mightily to correct. First, Lyell castigates Lamarck for making assertions without a shred of direct evidence. Note that Lyell directs his scorn not at the palpable forces of lateral adaptation, but at claims for the origin of new organs as increments of complexity wrought by the forces of progress:

> We point out to the reader this important chasm in the chain of the evidence, because he might otherwise imagine that we had merely omitted the illustrations for the sake of brevity, but the plain truth is, that there were no examples to be found; and when Lamarck talks "of the effects of internal sentiment," "the influence of subtle fluids," and the "acts of organization," as causes whereby animals and plants may acquire *new organs,* he gives us names for things, and with a disregard to the strict rules of induction, resorts to fictions, as ideal as the "plastic virtue," and other phantoms of the middle ages (Lyell, 1832, p. 8).

Second, and more important for my argument, Lyell gives a crisp and accurate account of Lamarck's hierarchical view of evolutionary causality, emphasizing the contrast between the regular cause of progress, and the disrupting force of adaptation. The passage, worth quoting *in extenso*, probably represents Darwin's first contact with this invalid style of hierarchical theory:

> Nature is daily engaged in the formation of the elementary rudiments of animal and vegetable existence, which correspond to what the ancients termed spontaneous generations . . . These are gradually developed into the higher and more perfect classes by the slow, but unceasing agency of two influential principles: first, the *tendency* to *progressive advancement* in organization, accompanied by greater dignity in instinct, intelligence, etc.; secondly, the *force of external circumstances,* or of variations in the physical condition of the earth, or the mutual relations of plants and animals . . . Now, if the first of these principles, the tendency to progressive development, were left to exert itself with perfect freedom, it would give rise, says Lamarck, in the course of ages, to a graduated scale of being, where the most insensible transition might be traced from the simplest to the most compound structure, from the humblest to the most exalted de-

gree of intelligence. But in consequence of the perpetual interference of the *external causes* before mentioned, this regular order is greatly interfered with, and an approximation only to such a state of things is exhibited by the animate creation, the progress of some races being retarded by unfavorable, and that of others accelerated by favorable, combinations of circumstances. Hence, all kinds of anomalies interrupt the continuity of the plan, and chasms, into which whole genera or families might be inserted, are seen to separate the nearest existing portions of the series (Lyell, 1832, pp. 13–14).

Darwin's public silence (or mild approbation) is belied by his consistently negative attitude towards Lamarck, as recorded in private letters extending from the 1840's to post-*Origin* years. In 1844, he wrote to Hooker on the dearth of available evolutionary writing: "With respect to books on this subject, I do not know of any systematical ones except Lamarck's, which is veritable rubbish" (in F. Darwin, 1887, volume 2, p. 29). The most interesting post-*Origin* references occur in letters to Lyell, who criticized Darwin for not giving Lamarck sufficient credit. In responding to Lyell's first reaction to the *Origin,* Darwin wrote on October 11, 1859 (in F. Darwin, 1887, volume 2, p. 215): "You often allude to Lamarck's work; I do not know what you think about it, but it appeared to me extremely poor; I got not a fact or idea from it." Perhaps, but I suspect that Darwin got many a concept *against* it.

Darwin's longest statement, a testy comment directed against Lyell's repeated designation of Lamarck as a source (though mitigated at the end by Darwin's usual humor), conveys special insight in Darwin's stated rationale for rejecting Lamarck's theory so firmly:

Lastly, you refer repeatedly to my view as a modification of Lamarck's doctrine of development and progression. If this is your deliberate opinion there is nothing to be said, but it does not seem so to me. Plato, Buffon, my grandfather before Lamarck, and others, propounded the obvious view that if species were not created separately they must have descended from other species, and I can see nothing else in common between the "Origin" and Lamarck. I believe this way of putting the case is very injurious to its acceptance, as it implies necessary progression, and closely connects Wallace's and my views with what I consider, after two deliberate readings as a wretched book, and one from which (I well remember my surprise) I gained nothing. But I know you rank it higher, which is curious, as it did not in the least shake your belief. But enough, and more than enough. Please remember you have brought it down on yourself!! (in F. Darwin, 1887, volume 2, pp. 198–199).

Note the basis of Darwin's critique—"very injurious . . . as it implies necessary progression." In other words, Darwin dismisses the higher-level cause of Lamarck's hierarchy. Darwin's own theory, of course, rested on the complete sufficiency, by extrapolation, of the lower-level force of adaptation in

Lamarck's system (as produced by the different mechanism of natural se-lection).

Lamarck's hierarchical theory set a context (in opposition) for Darwin's distinctive single-level theory of extrapolation, based on uniformitarian as-sumptions, from the palpable and small-scale cases of adaptation that sur-round us to all evolutionary changes at all scales of time and magnitude. Nat-ural selection does not emerge from the raw observation of nature, but as a complex idea embedded both in observation and in Darwin's voracious study and trenchant analysis of contemporary ideas in biology and general culture. Lamarck's hierarchical theory formed an important, though not widely rec-ognized, part of the mix, distilled by Darwin to extract a theory that would change the world.

Darwin directed his anti-hierarchical theory against Lamarck's old and in-valid concept of hierarchy—different and opposed causes at distinct levels. Darwin labored mightily to encompass the entire domain of evolutionary causation within a single level—natural selection working on organisms. He knew what he wanted to do, and he pursued and extended the logic of his ar-gument relentlessly. Most of his supporters (including Wallace and Huxley) never understood the subtle logic of the single-level theory. Among the few who did, Weismann also made a strenuous effort to bring the system to com-pletion. I find the strongest historical support for modern versions of hierar-chy (same causes working in different ways at various levels, in direct con-trast with Lamarck's notion of disparate causes in opposition), in the intense intellectual struggle carried out by the two greatest selectionists of the 19th century—Weismann for lower levels, and Darwin himself for upper levels—to bring the nonhierarchical theory of selection to completion and sufficiency. Both men, as we shall see in the subsequent sections of this chapter, struggled valiantly, but could not prevail. (Chapter 5 will then discuss (in the salta-tionist context of his version) the other major hierarchical system of evolu-tionary thought before the Modern Synthesis—the fascinating and subtle early 20th century theory of de Vries on reintroducing selection at the species level after denying its central importance at Darwin's own level of the organ-ism.) Call it the bad penny that keeps cropping up, or the pearl of great price always found within, but hierarchy seems unavoidable. Could the basic rea-son for this persistence find an explanation in something so lovely, and so beautifully simple, as truth value?

# No *Allmacht* without Hierarchy: Weismann on Germinal Selection

## THE *ALLMACHT* OF SELECTION

In 1893, Herbert Spencer, who had a word (many of them) and a thought for nearly everything,* published a long critique in the *Contemporary Re-*

---

*Spencer's star has fallen dramatically. He was once renowned as a polymathic philoso-pher; he is now generally regarded as an unread eminent Victorian with acute logorrhea.

*view*—"The Inadequacy of 'Natural Selection.'" He strongly supported the Lamarckian principle of use and disuse with inheritance of acquired characters and, while not denying the importance of Darwin's principle, railed against the exclusivity claimed for natural selection by August Weismann and his school, variously labelled as "strict," or "ultra" or "neo" Darwinism. Weismann quickly rose to the challenge, choosing for his title a phrase that would become a motto for his approach. He called his rebuttal, in German, and parrying Spencer directly: *Die Allmacht der Naturzüchtung*—a title rendered by the English translator as "The All-Sufficiency of Natural Selection" (although I would prefer "Omnipotence," or the literal "All-Might").

This exchange (Spencer, 1893a and b; Weismann, 1893) became the focal point and most widely cited set of documents in the great debate between "neo-Darwinism" and "neo-Lamarckism," perhaps the hottest subject in evolutionary theory at the end of the 19th century (see Kellogg, who wrote, 1907, p. 134—"The best known part of the general debate was that carried on directly by Weismann and Spencer in the *Contemporary Review*." These terms, as so often noted, bear little relation to the chief concerns of the name-bearers. Neo-Lamarckians bypassed Lamarck's central concept of materialistic progress and focused on a theory of heredity that Lamarck espoused as the folk wisdom of his day, not as anything distinctive in his system. Neo-Darwinism referred to the panselectionism of Weismann and Wallace, an attitude explicitly and pointedly rejected by Darwin, who gave selection pride of place (hence the association), but granted other forces (including "lamarckism") important, if lesser, roles in evolutionary change).

Passions ran high; I own Weismann's annotated copies of Spencer's articles, and his anger drips off the pages (Fig. 3-4). The two warriors thrusted and parried on both high and low roads, mixing some good arguments about the structure of evolutionary explanation with *ad hominem* charges of incompetence. Weismann (1893, p. 317) disparaged Spencer for being merely a philosopher, and not a true scientist: "[I] can only explain Mr. Spencer's ignoring such cogent instances by supposing that, as a philosopher, he is unacquainted with the facts by personal observation, and that therefore they appear less weighty to him than to a naturalist; for I would not for a moment suppose that he purposely evades the difficulties which face his opinion, as is the manner of popular orators and advocates—and alas! even of some scientists." Spencer, in his touché (1893, p. 23), replied, not entirely without justice as we shall see, that Weismann had hidden poor arguments under the cloak of authority as a practicing scientist: "Now it is doubtless true that as a naturalist he may claim for his 'opinion' a relatively great weight. Still, in pursuance of the method of science, it seems to me that something more than an opinion is required as the basis for a far-reaching theory."

---

But he can lay claim to at least one undoubted legacy in our parish; he popularized and won acceptance for the name of our subject, a word that Darwin initially rejected and adopted with resignation only late in life after Spencer's usage had triumphed (Gould, 1977b)—evolution.

This chapter explores the hints, inklings, and tentative formulations of hierarchical selection theory developed and published during Darwin's century. As a primary goal, I wish to bring to light the buried and forgotten discovery made by all strict Darwinians: that they could not carry through the logic of organismal selection to render all evolution without a crucial assist from selection at other levels. Hierarchy theory only became a major and explicit theme in evolutionary thought during the late 20th century, but a secret of history reveals that none of the great thinkers who struggled, with uncompromising respect for logic, to establish a general theory based on organismal selection alone could ever make the argument work without an appeal, sometimes in frustration, to hierarchy.

Alfred Russel Wallace and August Weismann stand out as the two principal "neo-Darwinians" of the late 19th century, the men most strongly dedicated to the *Allmacht* of selection. They therefore become the test cases for my assertion that hierarchy cannot be avoided. I shall bypass Wallace, though he fits my claim that no pure selectionist could avoid hierarchy, because I find no

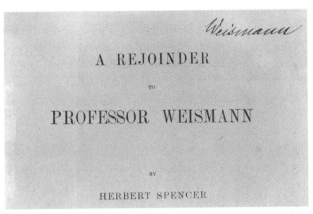

3-4. Weismann's personal copy (see his signature in upper right hand corner) of Spencer's reply to their first round of polemics. The two marginal comments on page 12 read (in translation of Weismann's German): "Unpermissibly weak!" And (ironically): "As if that were certain!" (Author's collection.)

evidence that he ever clearly conceptualized the issue of levels in selection. Wallace felt entirely comfortable with selection on all levels (see Kottler, 1985) and never seemed to grasp either the logic of Darwin's central commitment to the organismal level, or the problems involved in claiming that selection on other units (particularly higher "individuals") could be effective in the face of strong selection at the organismal level. Wallace maintained such an unshakable and primary commitment to the ubiquity of good design that he unhesitatingly invoked higher levels to preserve an argument for active selection whenever a focus on organisms raised the specter of nonadaptation (notably in his uncritical advocacy of species selection for sterility in interspecific crosses, rather than accepting Darwin's argument for infertility as a side consequence of accumulated differences in two diverging populations—see pp. 131–132).

But August Weismann represents the ideal test case for my assertion. Once he had declared war on "Lamarckian" inheritance, Weismann dedicated his professional life to promoting the *Allmacht* of selection. He grasped the logic of Darwin's argument in all its details and extensions. He recognized the centrality of selection on organisms, and he struggled to make Darwin's single-level theory work for all phenomena of evolution. His famous 1893 paper on the *Allmacht* of selection presents, as its central theme, an explicit defense for exclusivity of the organismal level—or "personal selection" in his terms.* Later, and largely in response to strong arguments made by Spencer, Weismann admitted that he could not rely on personal selection alone. He could continue to promote *Allmacht* only by recognizing another level of "germinal selection" for subcellular components of the germ-plasm.

Moreover, Weismann gradually extended the theory of germinal selection, from an *ad hoc* aid for personal selection (in the original formulation of 1895 and 1896) to a fully articulated theory of hierarchy replete with notions of independence and conflict between levels (1904 version). Finally, Weismann came to regard hierarchical selection as the linchpin and completion of his entire theory (see pp. 221–224)—though we have forgotten his cogent arguments, and usually depict him as the champion of conventional, organismic selection. Weismann's intellectual journey, his relentless probings and frequent reformulations, leading finally (and perhaps inexorably) to a full theory of hierarchy, provide an object lesson in the logic of evolutionary ar-

---

*Meaning, of course, not subjectivity in argument, but selection on organisms, or persons. This term became popular in Germany via Haeckel's theory of structural hierarchy (see pp. 208–210, this chapter), in which the body of an organism—*Eine Person*—enjoyed no special status, but merely represented one level of a six-tiered system ranging from "plastids" (subcellular parts) to "corms" (colonies). I rather wish that we could use this strong, jargon-free term today—for I would gladly adopt "personal selection" in preference to "organismal selection." But "personal" encompasses, alas, too wide a range of different meanings in the American vernacular, and I therefore desist. This charming term of ordinary language has held fast in at least one area, however. The parts of siphonophores (entire organisms by homology) are called "persons"—even in technical literature, where we can read about "polyp persons" and "mudusa persons" as "organs" of the differentiated colony.

gument, and the needs imposed by completeness and coherence once we abandon the myopia of regarding "organized adaptive complexity" (Dawkins, 1986) as the only focus for evolutionary explanation (with all else arising by extrapolation therefrom).

## WEISMANN'S ARGUMENT ON LAMARCK AND THE *ALLMACHT* OF SELECTION

I first learned about August Weismann in high school biology as the man who "disproved" Lamarckism by cutting off mouse tails for numerous generations and noting the fully retained tails of all offspring (a good example of terrible teaching based upon the myth of crucial experiments as the source of all insight in science). Weismann did perform these experiments (1888, in 1891, pp. 431–461), but they (by his own admission) did little to combat Lamarckism, which is, as supporters parried, a theory about the inheritance of functional adaptations, not of sudden and accidental mutilations.

Weismann's strong anti-Lamarckian argument does not rest upon an experiment, or an empirical observation at all. The rejection of soft inheritance arises as a logical deduction from Weismann's most distinctive contribution—his theory of inheritance and the continuity of germ-plasm (1885, in 1891, pp. 163–256). If germ-plasm is "immortal" (by passage across generations) and soma-plasm limited in existence by the death of each multicellular organism; and if germ-plasm is sequestered early in ontogeny ("locked away" as the guardian of posterity, and protected from all somatic influence); then Lamarckian inheritance becomes structurally impossible because acquired somatic adaptations cannot affect the protected germ plasm. Weismann wrote in his *Allmacht* paper (1893, p. 608): "Nature has carefully enclosed the germ-plasm of all germ-cells in a capsule, and it is only yielded up for the formation of daughter-cells, under most complicated precautionary conditions."

Once Lamarckian inheritance becomes impossible, Weismann's argument for the *Allmacht* of selection proceeds in four logical steps. This fourfold development will strike most modern scientists as curious and unsatisfactory, for the sequence not only requires no empirical contribution, but actively denies the possibility of effective input from this conventional source of scientific affirmation. The argument breaks no rules of logic, but several of its premises are (to say the least) not self-evidently true.

1. Adaptation is ubiquitous in nature; explaining adaptation therefore becomes the chief goal of evolutionary theory. As "the greatest riddle that living Nature presents to us" (1909, p. 18), Weismann identified "the purposiveness of every living form relative to the conditions of its life, and its marvelously exact adaptation to these" *(loc. cit.)*.

> I believe it can be clearly proved that the wing of a butterfly is a tablet on which Nature has inscribed everything she has deemed advantageous to the preservation and welfare of her creatures, and nothing else (1896, p. 5).

Everything we see in animals is adaptation, whether of today, or of yesterday . . . Every kind of cell . . . is adapted to absolutely definite and specific functions, and every organ which is composed of these different kinds of cells contains them in the proper proportions, and in a particular arrangement which best serves the function of the organ . . . The organism as a whole is adapted to the conditions of its life, and it is so at every stage of its evolution" (1909, pp. 64–65. This statement comes from Weismann's contribution to the "official" centennial celebration of Darwin's birth. Thus, Weismann chose to honor Darwin by stressing panselection.)

2. Adaptation must be attributed either to some materialistic cause, or to teleology (in the classic sense of spiritually directed purpose). The validity of science depends upon our ability to supply explanations in the former mode.

3. Among materialistic proposals, only Lamarckism and natural selection can explain adaptation—for adaptation is ubiquitous and clearly too complex to ascribe to chance or to render as a side consequence of any process serving unrelated ends.

4. Since Lamarkism is logically impossible (under the doctrine of continuity of germ-plasm), selection must be correct. To assert the *Allmacht* of selection, we need no evidence beyond the disproof of Lamarckism. In fact, given the complexities of nature, and our inability to reconstruct past conditions in detail, we probably could not supply adequate direct evidence in any single case.

We accept it, not because we are able to demonstrate the process in detail, not even because we can with more or less ease imagine it, but simply *because we must, because it is the only possible explanation* that we can conceive. For there are only two possible *a priori* explanations of adaptations for the naturalist—namely, the transmission of functional adaptations [i.e. Lamarckism] and natural selection; but as the first of these can be excluded, only the second remains . . . We are thus able to prove by exclusion the reality of natural selection, and once that is done, the general objections which are based on our inability to demonstrate selection-value in individual cases, must collapse, as being of no weight . . . It does not matter whether I am able to do so or not, or whether I could do it well or ill; once it is established that natural selection is the only principle which has to be considered, it necessarily follows that the facts can be correctly explained by natural selection (1893, pp. 336–337).

In 1893, when he made this bold assertion to counter Spencer's claim for the "inadequacy of natural selection," Weismann advocated a kind of double exclusivity—for natural selection over Lamarckism, and for selection upon organisms as the only mode of Darwinian action. As a terminological matter, Weismann equated the general phrase "natural selection" with selection upon organisms alone ("personal selection" in his words. For example, he wrote (1903, vol. 2, p. 126): "It is upon this that the operation of natural selection, that is, personal selection, must depend"). However, spurred by Spencer's critique, he soon expanded the boundaries of selection to include other levels of nature's hierarchy.

## THE PROBLEM OF DEGENERATION AND WEISMANN'S IMPETUS FOR GERMINAL SELECTION

As discussed in the last chapter, the primary and standard refutation of Darwinism by late 19th century evolutionists held that natural selection could eliminate, but not create—and that some other factor must therefore be identified to explain the *origin* of adaptations and species. For example, T. H. Morgan wrote in 1905, before he became a Darwinian: "It appears that new species are born; they are not made by Darwinian methods, and the theory of natural selection has nothing to do with the origin of species, but with the survival of already formed species" (in Kellogg, 1907, p. 95).

Darwinians, of course, understood this challenge, and responded with the argument that differential survivals, long cumulated, produce gradual and substantial changes meriting the designation "creative." Weismann himself, for example (1896, p. 1), spoke of "the opposition of our own day, which contends that selection cannot create but only reject, and which fails to see that precisely through this rejection its creative efficiency is asserted."

On this contentious question of creativity, several standard anti-Darwinian arguments invoked the earliest stages of features easily recognized as adaptive in their perfected form, for selection can preserve and accentuate a feature fully in place, but how can an organism move from an initial "there" to a fully functional "here"? Two claims predominated (see Mivart, 1871, for the classic statement that provoked Darwin's own response in later editions of the *Origin*): first, that initial steps are too small to provide any conceivable benefit in selection; and second, that earliest stages cannot initiate the final function in any sense (a bird cannot fly with 5 percent of a wing).

Darwinians developed satisfactory responses to both arguments about incipient stages of useful structures—the palpable value of tiny benefits for the first, and the principle of functional shift (preadaptation) for the second (see extensive discussion in Chapter 11, pp. 1218–1246). But the same problem seemed far more acute for the opposite dilemma of degeneration. Incipient stages of useful structures posed enough difficulties, although ultimate adaptiveness did suggest a Darwinian solution. But what conceivable pressure of natural selection could account for gradual stages in the *disappearance* of a *functionless* organ—for loss of function should remove a structure from the domain of selection entirely, and knowledge about an eventually adaptive state could not be invoked to guide an explanation for intermediary stages along such a functionless path.

(We might designate this problem by its classic example—the complete disappearance of eyes in some cave fishes. Despite a century of adequate Darwinian explanation, this issue continues to provide a rallying point for vernacular Lamarckism. I can testify to this in a personal way. As a result of writing more than 300 popular monthly essays on evolutionary topics during the past 25 years, I have become a statistically adequate sampling point, through thousands of letters received from lay readers, for both the frequency and intensity of standard confusions about our profession. I can testify that three items top the list of puzzlement: (1) evolution seen as anagenesis rather than branching ("if humans evolved from apes, why are apes still around"); (2) panselectionism ("what is the adaptive significance of male nipples"); and (3) Lamarckism and the failure of natural selection ("doesn't the blindness of cave fishes imply a necessary space for Lamarckian evolution by disuse").)

The problem of incipiency in degeneration poses more difficulty than the opposite issue of construction—for what can mediate the sequence if selection does not regulate the final outcome? Weismann struggled to encompass this issue with his favored apparatus of *Allmacht* for selection—and he failed. Degeneration acted as the lever that pried Weismann from his panselectionism, and led him through a chronological series of honorable changes that must be read, in one sense, as retreats from a former pugnacious insistence on *Allmacht*, but that also represents a complexification and strengthening of his original views.

Consider the example that Spencer raised with such effectiveness against Weismann, and that eventually prompted the theory of germinal selection—reduction of hind limbs in some whales to tiny vestiges with no exterior expression at all. Two classical explanations had been invoked by panselectionists: (1) the limbs became so reduced by ordinary negative selection, as a consequence of the hindrances they imposed upon efficient, streamlined swimming; (2) the limbs are not, in themselves, harmful, but energy invested in any useless structure must handicap a creature relative to conspecifics with fewer vestiges and neutral organs.

Weismann invoked these standard arguments, but he became convinced (long before his debate with Spencer) that only part of the puzzle could be resolved thereby. Selection would reduce the limbs to some degree (perhaps considerably), but surely the increments of further reduction soon become too small for granting a continuing, believable role to selection. Consider the figures that Spencer presents (1893b, p. 25), based on the efforts of a Dr. Struthers of Aberdeen, who had "kindly taken much trouble in furnishing the needful data, based upon direct weighing and measuring and estimation of specific gravity." Spencer cites a Greenland Right Whale, weight 44,800 pounds, femur weight, 3-1/2 ounces; and a Razorback at 56,000 pounds, with a femur weight of 1 ounce—"so that these vanishing remnants of hind limbs weighed but 1/896,000th part of the animal." Could one possibly believe that a profound relative, but inconsequential absolute, reduction—from a two ounce to a one ounce femur, for example—might materially aid streamlining (especially since external expression had disappeared long before) or conserve meaningful energy? Weismann accepted the implausibility of such a claim and recognized that he would have to seek an explanation beyond organismal selection for such late stages in the reduction of degenerate organs. "To use Herbert Spencer's striking illustration, how could the balance between life and death, in the case of a colossus like the Greenland whale, be turned one way or another by the difference of a few inches in the length of the hind-leg, as compared with his fellows, in whom the reduction of the hind-limb may not have gone quite so far? . . . Further reduction to their modern state of great degeneration and absolute concealment within the flesh of the animal cannot be referred even to negative selection" (Weismann, 1903, vol. 2, p. 114).

This example, and the general phenomenon of degeneration, deeply troubled Weismann because common sense seemed to demand that his Lamarckian bugbear and bogeyman—so recently and, as he thought, finally and effectively buried—be disinterred to explain reduction as inheritance of features shriveled by disuse. Spencer himself raised this example in order to defend a Lamarckian explanation *prima facie*:

> Thus, the only reasonable interpretation is the inheritance of acquired characters. If the effects of use and disuse, which are known causes of change in each individual, influence succeeding individuals . . . then this reduction of the whale's hind limbs to minute rudiments is accounted for. The cause has been unceasingly operative on all individuals of the species ever since the transformation began. In one case see all. If this cause has thus operated on the limbs of the whale, it has thus operated in all creatures on all parts having active functions (Spencer, 1893b, p. 26).

Weismann first attempted to resolve the difficulties posed by degeneration with his hypothesis of *panmixia* (not the later Fisherian definition now familiar to evolutionists). By panmixia, Weismann referred to the effect of recombination in sexual reproduction (amphimixis in his vocabulary) upon organs no longer subject to selection. When selection operates, Weismann argued, organs will be actively maintained, with constant vigilance and no relaxation, at the peak of their potential size and complexity by elimination of individuals bearing substandard parts. But as soon as selection ceases to act, formerly "substandard" attributes will no longer be eliminated; they now mix freely with "good" parts, and the organ slides, by continuous dilution, down an inclined plane towards total elimination. In a poignant example (since poor eyesight plagued his own career), Weismann wrote (1903, pp. 114–115). "If this conservative action of natural selection secures the maintenance of the parts and organs of a species at their maximum of perfection, it follows that these will *fall below this maximum as soon as the selection ceases to operate* . . . Those with inferior organs of vision will, *ceteris paribus*, produce as good offspring as those with better eyes, and the consequence of this must be that there will be a general deterioration of eyes, because the bad ones can be transmitted as well as the good, and thus the selection of good eyes is made impossible."

By his own admission and explicit defense (see p. 201), Weismann's argument for *Allmacht,* and against Lamarckian inheritance, rested upon a logical structure of inferences from premises, not upon observation—for an empirical approach, Weismann held, could not achieve resolution, given the impossibility of "seeing," at their minute sizes, the material bearers of heredity. Panmixia did compromise *Allmacht* in a sense, for this process yielded evolutionary change without selection. But following Kellogg's key distinction of auxiliary from contradictory hypotheses (see pp. 163–169), panmixia worked as an adjunct and aid—a mopping-up operation for organs fallen below the purview of selection, and, more importantly, a moat to prevent the incursion of a true enemy, the antiselectionist forces of Neo-Lamarckism. (Lamarck battled against Darwin for the common ground of universal adaptation, while panmixia only worked to finish what selection had started, and only in the limited domain of degeneration.)

But Weismann's panmixia, having no support beyond the internal logic of the argument itself, could not survive the detection and exposure of crucial flaws. Spencer was not the first writer to illustrate the weaknesses of panmixia, but the debate of 1893 does mark Weismann's last attempt to explain degeneration by panmixia alone, and therefore contains the seeds for his next and final attempt—the theory of germinal selection.

Spencer, referring to "the vexed question of panmixia" (1893b, p. 22), offered three major rebuttals. "When from the abstract statement of it we pass to a concrete test, in the case of the whale, we find that it necessitates an unproved and improbable assumption respecting plus and minus variation; that it ignores the unceasing tendency to reversion; and that it implies an effect out of all proportion to the cause" (1893b, pp. 28–29). The second point, based on Galton's principle of regression to the mean, denies that "minus" variations can continue to accumulate differentially; the third brands panmixia as too weak a force to secure the total elimination of a useless organ. The first argument, however, proved to be not only decisive in itself, but unusual in scientific discourse by accusing Weismann (correctly) of conflating linguistic usage with biological reality.

Weismann continually argued that selection maintained an organ "at its highest level." Relaxation of selection might then impel an accumulation of previously eliminated variation in the minus direction only. But, as Spencer and others protested, why should selective optimization hold an organ at the summit of its potential size and complexity. Shouldn't optimality lie somewhere in the middle of a possible range, with selective elimination of both plus *and* minus variations? "Take the case of the tongue," Spencer argued (1893b, pages 23–24). "Certainly there are tongues inconveniently large, and probably tongues inconveniently small. What reason have we for assuming that the inconveniently small tongues occur more frequently than the inconveniently large ones?" Without the invalid metaphor of selective summits, panmixia cannot reduce an organ to oblivion, for release from selection does not impart an inexorably downward trend to preserved variation.

All these objections can be combined into a single claim, which Weismann found so compelling that he eventually surrendered panmixia as a fully adequate explanation of degeneration. Panmixia is a genuine, but weak, force; it can reduce the average value of an organ to a state somewhat below its former functional size. But panmixia cannot solve the central question of degeneration: what propels a useless organ all the way down the slide and into history's dumpster? Weismann admitted his failure (1896, p. 22), and later summarized this ultimately unsuccessful episode in his quest to understand degeneration:

> As my doubts regarding the Lamarckian principle grew greater and greater, I was obliged to seek for some other factor in modification, which should be sufficient to effect the degeneration of a disused part, and for a time I thought I found this in panmixia, that is, in the mingling of all together, well and less well equipped alike. This factor does certainly operate, but the more I thought over it the clearer it became to me that there must be some other factor at work as well, for while panmixia might explain the deterioration of an organ, it could not explain its decrease in size, its gradual wearing away, and ultimate total disappearance. Yet this is the path followed, slowly indeed, but quite surely, by all organs which have become useless (1903, vol. 2, p. 115).

Weismann therefore needed another kind of auxiliary hypothesis to preserve the *Allmacht* of selection against resurgent Lamarckism. He had tried the mechanics of inheritance as expressed in the doctrine of panmixia; now he would expand the domain of selection itself. He would depart from Darwin's distinctive focus on struggle among organisms, and attempt to identify a source of directional variation in an analogous competition among determinants of heredity within germ cells—a "germinal selection." Weismann devised a truly ingenious argument: if natural selection can produce trends in the morphology of phenotypes, then an intracellular, germinal selection might yield directionality in the variation presented to conventional selection upon organisms. If the determinants of a useless organ predictably lose in an intracellular struggle for existence, then a trend to complete elimination—an apparent example of Lamarckian inheritance by the principle of disuse—might still be attributed to selection. This new mechanism could not be equated with Darwinian selection upon struggling organisms, but "germinal selection" did represent a process of the same form and logic, but applied to replicating objects at a subcellular scale rather than to entire organisms.

Weismann first proposed the theory of germinal selection as a brief note in his last rebuttal to Herbert Spencer, thus marking Britain's Victorian pundit as a chief source (in reaction) to the first explicit theory of hierarchical selection. (*Neue Gedanken zur Vererbungsfrage, eine Antwort an Herbert Spencer,* Jena, 1895). Weismann then elaborated the theory in 1896 (presented to the International Congress of Zoology at Leiden on September 16, 1895; first published in *The Monist* in January, 1896, then as a separate pamphlet, translated into English later that year). Weismann's fullest development, with some remarkable changes by extension, appeared in his most important book, *Vorträge über Descendenztheorie* (1902), translated into English by J. Arthur and Margaret R. Thomson as *The Evolution Theory* in 1903. A comparison of the original 1896 version with the fullest exposition of 1902 provides a fascinating exercise in itself, and also becomes a crucial argument for this book—for Weismann moved from a limited hypothesis proposed only as an adjunct to natural selection to a fully articulated theory of hierarchy, including concepts of independence and conflict between levels.

## SOME ANTECEDENTS TO HIERARCHY IN GERMAN EVOLUTIONARY THOUGHT

Germinal selection certainly finds its immediate source in Weismann's war with Lamarckism, his debate with Spencer, and his severe, longstanding difficulty with the problem of degeneration. But Weismann's eventual embrace of hierarchy as an ultimate argument against Lamarckism also grew from a deeper foundation in German evolutionary thought. This lineage of argument is virtually unknown to English-speaking evolutionists, for the roots lie in the two most important untranslated documents of 19th century German evolutionary biology—the *Generelle Morphologie* (1866) of Ernst Haeckel and the *Jugendwerk* of a man who eventually made his considerable mark in another area of biology, Wilhelm Roux's (1881) *Der Kampf der Theile im Organismus (The Battle of Parts in the Organism)*. Neither Haeckel nor Roux proposed a theory of causal hierarchy across levels of selection; both, in fact, spoke in the name of reductionism. Yet by denying, in very different ways, the exclusivity, or even the privileged status, of the organism as a causal agent in evolution, and by focusing attention on a structural hierarchy of levels, both Haeckel and Roux provided central ingredients to Weismann's theory of evolutionary hierarchy.

### Haeckel's descriptive hierarchy in levels of organization

*Generelle Morphologie der Organismen* (1866), Haeckel's first book, represents an eclectic mixture of militant reductionism and old-fashioned idealistic morphology, all united to an evolutionary theory every bit as idiosyncratic. (Haeckel dedicated volume two, jointly, to Darwin, Lamarck, and Goethe—and its central argument represents an odd fusion of their disparate ideas.) Haeckel's later notoriety rested almost entirely on the second volume, with its celebrated evolutionary trees (so often reproduced in modern textbooks), based largely on his "biogenetic law," ontogeny recapitulates phylogeny (Gould, 1977b). The first volume, entitled "Allgemeine Anatomie" and dedicated to Carl Gegenbaur, has largely been forgotten. This first volume consists of two major parts, each attempting to establish a formal science for morphological study and each, following Haeckel's invariable practice, studded with a baroque terminology of his own construction. (Haeckel, with a sure sense of what R. K. Merton (1965) would later call the eponymous strategy for renown, coined new terms shamelessly, recognizing (I suspect) that a few would probably hang on to bear his legacy (an *r*-selection approach to the courting of fame). The vast majority quickly succumbed to the negative selection of incomprehensibility, but survivors include ontogeny, phylogeny, heterochrony, ecology, and Monera.)

The second science, "promorphology," tried to establish a physical, or crystallographic basis for organic form. Haeckel created a forest of terms illustrated in two complex plates, but never established any useful connections with physical or chemical principles. (Haeckel promoted his much vaunted mechanistic reductionism more by verbal proclamation than by deed, but the influence of a well-articulated philosophy consonant with social trends of an age must never be doubted.) The first science, "tectology," tried a different approach to reductionism—not subsumption under physical laws, but breakdown to component parts.

As the "basic principle" of tectology, Haeckel stated that all organic objects must be built from components in a structural hierarchy of six ascending levels. But, in applying this pronouncement to actual cases, Haeckel makes a fascinating intellectual move, proving that his allegiance lay as much with holistic traditions of an older idealistic morphology, as with the militant physical reductionism that won his lip service and fit with many of his social and political goals (Gasman, 1971). For Haeckel did not argue, in the manner of most 19th century reductionists, that his first and lowest level stands as fundamental and basic (also "closer" to physics), with subsequent levels only treated as amalgamations based on principles of joining. Instead, Haeckel proclaimed a form of equality among the six levels (while not denying the compositional theme that lower units join to build higher entities). He referred to tectology as the "doctrine of organic individuality" *(Lehre von der organischen Individualität)*, and insisted that the objects at each of the six levels be designated as "individuals" in their own right—"individuals of the first order," "individuals of the second order," etc. He placed "plastids" (cells and cell components) on the first rung, organs (including tissues and organ systems) on the second, antimeres or *Gegenstücke* (symmetrical parts, including rays or body halves of bilateral creatures, literally "counterparts") on the third, metameres or *Folgestücke* (body segments, literally "following pieces") on the fourth, persons (or vernacular "individuals") on the fifth as *"morphologische Individuen fünfter Ordnung,"* and colonies or "corms" on the sixth and last plane.

This equalization of status prompted the interesting consequence, with reference to natural selection, of denying to organisms their privileged Darwinian role as exclusive evolutionary agents. Natural selection surely ascribed evolutionary change to a struggle among *individuals* for reproductive success. But Haeckel insisted that objects at *all* six levels counted as "individuals," and that no level could claim any special status as evolutionary agent. Organisms represent only one waystation in the ascending hierarchy. Perched on the fifth rung, they are made of metameres and aggregate into corms—just as organs are made of plastids and aggregate into antimeres. In an insightful statement on the role of language in prejudicing thought, Haeckel wrote of his fifth level (1866, vol. 1, pp. 318–319):

An unbiassed and more deeply probing conception of organic individuality shows that these "true" or absolute individuals are, in fact, only relative . . . Although these "true" individuals are, in most higher plants and coelenterates, only the subordinate components of a higher-standing unity (the colony), nonetheless the individuality of humans and higher animals leads us to the erroneous conception that morphological individuals of the fifth order are the "true" organic individuals. This concept has become so general, and has been so strongly fixed in both scientific and vernacular consciousness, that we must mark it as the major source of the numerous and varied interpretations and debates that prevail on the subject of organic individuality.

Haeckel's concept of structural levels and the non-distinctive status of organisms entered Weismann's argument in two crucial places—first, very generally, when Weismann used the same style of thinking to establish a hierarchy of (hypothetical) entities as the physical bearers of heredity within germ cells (see p. 214); and, second, quite specifically, when Weismann invoked Haeckel's six-part hierarchy (1896, p. 42) to argue that the struggle for existence starts within germ cells, but then extends up through all Haeckelian categories to colonies at the top.

### Roux's theory of intracorporeal struggle

Wilhelm Roux's *Der Kampf der Theile im Organismus* evoked a wide range of reactions. Roux's teacher, Gustav Schwalbe, warned him against ever publishing such a "philosophical" book again. Haeckel, another teacher, liked the work for its consonance with his own ideas, while Darwin himself, during the last year of his life, became greatly intrigued, writing to G. J. Romanes on April 16, 1881:

> Dr. Roux has sent me a book just published by him . . . It is full of reasoning, and this in German is very difficult for me, so that I have only skimmed through each page; here and there reading with a little more care. As far as I can imperfectly judge, it is the most important book on Evolution which has appeared for some time . . . Roux argues that there is a struggle going on within every organism between the organic molecules, the cells and the organs . . . If you read it, and are struck with it (but I may be wholly mistaken about its value), you would do a public service by analyzing and criticizing it in 'Nature' (in F. Darwin, 1887, vol. 3, p. 244).

(Note how, contrary to the prevalent historical myth of the aged Darwin as the reclusive "sage of Down," he actually (and actively) kept his ears alert, and his fingers right on the pulse of evolutionary debate. Romanes represented just one among several younger colleagues and supporters whom Darwin often recruited, both overtly and nonsubtly, to carry forth his interests in both the public and the professional arena.)

Obviously, Roux had adopted Darwinian language for his title. Just as obviously, he hoped to apply the Darwinian apparatus at a level below its conventional locus of organisms. Roux's book surely occupies a place in the history of hierarchy theory, if only because its verbal image of struggling parts led many evolutionists, notably Weismann himself, to consider multiple levels of selection. But, curiously, as several critics soon noted (Plate, 1905; Kellogg, 1907; and even, with some ambiguity, Weismann, 1904, as well), Roux's theory does not really treat descent at all. Weismann's germinal selection, as we shall see, is a true theory of suborganismal selection and inheritance; but Roux's battle of the parts includes no statement about heredity, and ranks instead as a theory of functional adjustment in development.

Roux argued that the construction of a harmonious and well-designed organism emerges from a struggle among parts competing for limited nutriment. Lung cells compete with liver cells, and bone cells battle with other bone cells for best locations in the flow of nutriment. To cite Roux's favorite example, made even more famous by D'Arcy Thompson's later analysis (1917, and see Chapter 11, pp. 1195–1196), the bony trabeculae in the head of a human femur form a virtual diagram of forces imposed on the bone during locomotion, and must therefore be optimally designed to counter stress. But no one can argue that details of the arrangement in any single bone represent an evolutionary adaptation, if only because the trabeculae of a broken, and improperly mended, femur reform along the new stress lines of a limping walk.

Roux argued that stresses establish lines of preferred flow for nutriment. Bone cells that happen to lie in the stream prosper and proliferate; others in less advantageous positions wither and die—leading to a functional honeycomb of struts and empty spaces. Roux used this argument to explain the functional design of tissues and organs in general, but he focused upon such complex and exquisite examples of optimal form as the barbules on bird feathers, the hairs that cover the spiracles of many insects, the arrangement of muscle fibers in the walls of blood vessels, and the bony trabeculae discussed above.

This "battle of the parts" may account for the flexible construction of optimal form in each organism. Indeed, such a principle, appropriately modernized, remains essential for a developmental biology that cannot invoke a specially-tailored gene for each villus on an intestinal surface. But Roux's proposal cannot operate as a theory of *evolutionary* change for two reasons. First, the struggling parts do not vary in heritable ways, and victory cannot lead to beneficial changes in future generations. Bone cells that prosper on the growing trabeculae cannot be designated as superior to, or even in any sense intrinsically different from, the losing cells that die for lack of nutriment in the spaces between. The winners owe their success only to the good fortune of a favorable location. Kellogg (1907, p. 207) wrote: "This competition depends chiefly on the hazard of position . . . Not the best qualified but the best situated fibers have vanquished the others by robbing them of food and thus finally destroying them." Second, Roux's *Kampf der Theile* includes no theory about inheritance. No matter how exquisite or optimal the outcome for any one organism, the results of the struggle cannot be imparted to offspring. (The capacity for functional adaptation might, of course, be heritable and might evolve by ordinary natural selection, but Roux never discusses this quite different issue.)

Weismann reacted to Roux's theory in a complex and ambiguous manner. He always credited Roux as an antecedent of germinal selection (a reasonable attribution, if only because an explicit metaphor of struggling parts can direct another scientist's thinking to a truly selectionist theory, even if the original proposal operated in a different domain). Weismann, particularly in his early work, seems to credit Roux—incorrectly—as a true suborganismal selectionist: "Functional adaptation is itself nothing else than the efflux of intrabiontic selective processes" (1896, p. 15). Roux's theory, he argues in several passages, rests upon a variational base, and is therefore Darwinian.

But, by 1904, Weismann had recognized that Roux's suborganismal struggle could not operate as a theory of evolutionary change: "There is an essential difference between personal and histonal selection, inasmuch as the latter can give rise to adaptive structural modifications corresponding to the needs of the tissue at the moment, but not to permanent and cumulative changes in the individual elements of the tissue" (1904, volume 1, p. 248). "No one will be likely to suppose that the distorted position of the spongiosa of a badly healed fracture could reappear in the straight bone of a descendant" (*ibid.*, p. 251).

Moreover, Weismann added, even the metaphorical linkages of Roux to Darwin cannot be logically sustained. Most of Roux's examples do not include competition among members of the same cell population (as in bone cells within the developing femur), but between entirely different organs: liver cell with lung, or kidney or heart. This process cannot be viewed as a struggle for existence at all, but only as a sorting out of different "species" into their appropriate places: "The struggle for existence and for descendants, in this case, is between two kinds of cell which were different from the beginning, and of which one has the advantage at one spot, another at another" (1904, volume 1, p. 248). Weismann then drew a striking analogy* between differ-

---

*This remarkable passage anticipates our modern debates about the efficacy of species selection. Weismann's analogy surely holds: this particular case involves no directional selection, but only a sorting out and consequent balance among three species, each in its proper place—just as lung, kidney, and heart develop where they should, and to their appropriate size. However, if such a competition led not to balance and stability, but to differential birth and death of the entities involved, then we could speak of directional species selection. This argument cannot apply at a lower level within an organism, for lung cannot defeat liver without destroying the entire system (but consider cancer as an event of this type). Nonetheless, the higher level version remains potentially valid for competition among species, for the success of one bird species over another will not cause an island to founder into the ocean. Species selection does operate in this manner—and some critics (e.g. Maynard Smith, 1988) have denied a creative status to this higher level because species selection only sorts entities already shaped by organismal selection, a position that I shall challenge in Chapter 8). After all, organismal selection also works only by sorting

ent tissues in an organism, and different species of birds in a broad geographical region:

> The case may be compared to that of a flock of nearly allied species of bird, of which one species thrives best in the plains, another among the hills, and a third among the mountain forests, all mingled together in a vast new territory to which they had migrated, and in which all three kinds of conditions were represented. A struggle would arise among the different species, in which in every case the particular species would be victorious which was best adapted to the local conditions . . . This would be the result of a struggle between the three species, not between individuals within each species, and it could not therefore bring about an improvement of a single species, but only the local prevalence of one or another (1904, volume 1, pp. 248–249).

Weismann's strong and valid critique of Roux leaves us with a puzzle: why did Darwin, who understood the nature of selection so much better than anyone else (see next section), become so intrigued with Roux's book, if *Kampf der Theile* does not really develop a selectionist, or even a truly evolutionary, theory at all? Several resolutions may be suggested. Most mundanely, Darwin was no German scholar, and he may not, as he himself suggested to Romanes, have properly understood the theory in his cursory reading. Secondly, Darwin was not a strict selectionist, and may simply have valued Roux's insights on functional adaptation, including the Lamarckian implications for a theory of heredity by extension. But, in a third and intellectually more intriguing hypothesis, perhaps Darwin valued *Kampf der Theile* for two genuine benefits or consonances that Roux's book granted to natural selection—one practical, the other metaphorical.

In a practical sense, Roux explicitly provided the resolution of a paradox that had plagued natural selection—the problem of too much adaptation ("organs of extreme perfection" in Darwin's designation in Chapter 6 of the *Origin*). Can we really believe that organismal selection constructs each barbule on every feather—even with the immensity of geological time and the hecatomb of deaths in each generation? Roux offered Darwin a sensible exit from such an untenable implication: selection builds the capacity for an automatic functional response that can directly shape each organism in minutely adaptive ways during growth: "Through the capacity of the struggle of parts, a much higher perfection, the purposefulness of the functioning part down to the last molecule, can arise, and occur much more rapidly, than if it had to originate, by the Darwin-Wallace principle, through selection of variation in the struggle for existence among individuals" (Roux, 1881, p. 239).

But Roux offered even more, by way of metaphor, to Darwin's cardinal vision—the paradox of higher stability arising through struggle among lower elements. Functional adaptation might not rank as an evolutionary theory,

---

variants produced at a lower level—yet we rightly deem such a process creative in the building of adaptations.

but such a process does produce *internal* order within a body by struggle—just as natural selection engenders the *external* harmonies of adaptive design and ecological balance: "As the struggle of parts [*Kampf der Theile*] yields purposefulness within an organism . . . so does the analogous struggle for existence [*Kampf um's Dasein*] among individuals yield purposefulness with respect to external conditions of existence" (Roux, 1881, p. 238). Roux also echoed Darwin's most general and most important philosophical principle:

> To many, the direction of this book may well seem very strange—for it holds that, in an animal, in which everything is so exquisitely ordered, in which all the different parts interlock with such excellence, and work together in such perfected coordination, that a struggle of parts occurs, so that in one place, where everything works together according to firm principles, a conflict among the individual parts exists. But how can an entity [*ein Ganzes*] exist, whose parts are at variance? . . . How shall the good and the stable arise from struggle and battle? . . . All good can only arise from struggle [*alles Gute nur aus dem Kampfe entspringt*] (1881, p. 64).

Darwin himself could not have penned a better epitome for his most radical claim.

### GERMINAL SELECTION AS A HELPMATE TO PERSONAL SELECTION

Weismann proposed the theory of germinal selection as a logical solution to the problem of degeneration in a non-Lamarckian world. But germinal selection only makes sense under Weismann's concept of inheritance—yet another theory of structural hierarchy, and explicitly linked by Weismann to Haeckel's 6-fold sequence as a further break-down and elaboration (for germ plasm) of categories within Haeckel's lowest unit of "plastids," or cellular constitutents (1896, p. 42).

In Weismann's admittedly hypothetical system, the fundamental sub-microscopic particles of heredity are called "biophors." Biophors aggregate to "determinants," the key unit for the theory of germinal selection. The logic of panselectionism requires a high degree of easy dissociability among genetic "particles" responsible for "traits" that can be individually optimized to construct well-adapted organisms—for if "particles" become too tightly linked or coordinated, then each change entails too many consequences for other traits, and constraints begin to prevail over adaptation. "Determinants" play this necessary role in Weismann's panselectionist theory of heredity. Each determinant builds an organ or a particular part of the body—in other words, an "item" of the phenotype that selection can mold independently.

Determinants, like their constituent biophors, are invisible and hypothetical. They aggregate into the first observable unit, the "id"—an earlier use of a term that Freud coopted for a much different role and purpose (just as paleontologists had coined and developed a meaning for "mutation" (Waagen, 1869) before the new science of genetics outcompeted us with a later and altogether different definition). Each id contains a determinant for every trait, and can therefore build a body. Weismann identified ids with the disk-like microsomes, recently observed as linearly ordered on chromosomes. The chromosomes themselves, or "idants" in Weismann's system, carry ids in rows, and stand atop the hierarchy of hereditary units.

Germinal selection rests upon the notion that determinants within germ cells may be analogized to organisms within habitats. Just as organisms struggle for limited resources (and not all can survive), determinants battle for the restricted flow of nutriment available to any cell. The winners grow and proliferate; the losers wither or disappear entirely. The strength of determinants governs the phenotypic expression of their resulting trait. Thus, if the determinants of a particular trait decrease or wither by germinal selection within cells, the trait will suffer in expression by exhaustion of its molecular base.

Weismann viewed germinal selection as an analog to interspecific struggle, rather than to competition among members of a single population—that is, determinants for one organ battle for limited nutriment with determinants for other parts of the body. But, unlike Roux's *Kampf der Theile*, Weismann's germinal selection does operate as a theory of altered heredity, and therefore as a potential evolutionary mechanism—for determinants weakened by germinal selection not only build a diminished body part; they also pass fewer (and debilitated) offspring determinants to germ cells of the next generation.

The ingenuity of this argument lies in its capacity to take the most serious potential challenges to the *Allmacht* of selection—a group of overt phenomena that seem to lie outside the possible control of organismal selection, and therefore within the domain of Lamarckism or orthogenesis—and to reinterpret them as consequences of selection acting *at a lower level*. For if organismal selection can produce directional trends in phenotypes during geological time, then germinal selection can forge trends in strengthening or weakening determinants (and their phenotypic expressions) across generations. The gradual and unidirectional shriveling to oblivion of an organ not subject to personal selection certainly suggests Lamarckian inheritance and evolutionary loss by disuse; but if we descend a level and peer within the germ cells, we may envision (though we cannot directly see) a constant competition and selection among determinants, with losers paying the usual price of gradual and inexorable elimination. The *Allmacht* of strict Darwinism may be sacrificed, as organismal selection loses its exclusivity; but selection itself remains preeminent by expansion:

> Powerful determinants in the germ cell will absorb nutriment more rapidly than weaker determinants. The latter, accordingly, will grow more slowly and will produce weaker descendants than the former . . . Since every determinant battles stoutly with its neighbors for food, that is, takes to itself as much as it can, consonantly with its power of assimilation and proportionately to the nutriment supply, therefore the unimpoverished neighbors of this minus determinant will deprive it of its nutriment more rapidly than was the case with its more robust ancestors (1896, p. 24).

Weismann offered the same pugnacious defense for germinal selection that he had long championed for the *Allmacht* of organismal selection; the argument must hold, lest we be driven either to mysticism or to the patently false Lamarckian mechanism:

> No one who is unwilling to accept germinal selection can be compelled to do so, as he might be to accept the Pythagorean propositions. It is not built up from beneath upon axioms, but is an attempt at an explanation of a fact established by observation—the disappearance of disused parts. But when once the inheritance of functional modifications has been demonstrated to be a fallacy . . . he who rejects germinal selection must renounce all attempt at explanation. It is the same as in the case of personal selection. No one can demonstrate mathematically that any variation possesses selection value, but whoever rejects personal selection gives up hope of explaining adaptations, for these cannot be referred to purely internal forces of development (1903, volume 2, p. 121).

Weismann, at least for public consumption, insisted that germinal selection represented the advancing wave of triumphant Neo-Darwinism. (In the late 19th century, "Neo-Darwinism," a term coined by Romanes, referred to the panselectionist school of Wallace and Weismann, not to the pluralism of Darwin himself. The modern meaning, associated with the evolutionary synthesis of the 1930's and onward, is not genealogically linked to this earlier definition.) Many critics responded by charging that this invisible process represented little more than an *ad hoc* hypothesis invented to save selection from the otherwise unexplainable phenomenon of degeneration. Kellogg (1907) provided, I believe, the most balanced perspective. He labelled germinal selection as "a new and radically un-Darwinian theory" (1907, p. 134)—recognizing that Darwin's own theory of "natural selection" specified organisms as the locus of selection. But he respected the theory, recognized its similarity with the selectionist logic of classical Darwinism, and regarded germinal selection as a credible attempt to explain, in expanded Darwinian terms, the apparently un-Darwinian property of directional variation. He wrote: "Obviously Weismann in his theory of germinal selection has preserved the actuality of the struggle and the selection, but with a 'rehabilitation' of natural selection in the real Darwinian meaning and only fair application of the phrase, the new theory has nothing to do. It is, much more, a distinct admission of the inadequacy of natural selection to do what has long been claimed for it. It is the first serious attempt at a causo-mechanical explanation of a theory of orthogenesis, that is, variation along determined lines" (Kellogg, 1907, p. 199).

I particularly value Kellogg's interpretation because the logic of his argument correctly represents, in my view, the relationship of modern hierarchical selection theory to classical Darwinism and to the Modern Synthesis as well. Hierarchy should be viewed as an expansion of Darwinism in its continuing reliance on selection as the primary mechanism of evolutionary change. But, at the same time, hierarchical selection does not merely extend Darwin's exclusively organismal version in a simple, comfortable, or inexorable manner. Rather, in fracturing Darwin's reliance on organisms as evolutionary agents, and in rendering evolutionary variation and change by interaction among levels, the theory of hierarchical selection introduces enough conceptual novelty, and disperses enough inadequate orthodoxy, to rank as a new, and in some respects radical, formulation, rather than a fully comfortable expansion.

Kellogg also grasped the weaknesses of germinal selection (which, of course, would soon become irrelevant when the codification of Mendelism disproved Weismann's conjecture about the physical mechanism of heredity). He asked (1907, pp. 200–201) (1) why measured variation so often conformed to a normal distribution if germinal selection could act so powerfully to promote directional variation; (2) why species generally displayed such geological stability if germinal selection provided such a strong mechanism for orthogenesis; and (3) why, if determinants wage such constant battle, each against all others, severe deprivation of food often produced a proportionately dwarfed organism, rather than a creature lacking phenotypic expression for particular losing determinants in such intensified struggle.

Weismann originally developed germinal selection to explain the disappearance of degenerate organs, once he recognized that panmixia could only yield a partial reduction. But he soon expanded the scope of his new theory into the domain of positive selection as well, hoping to resolve thereby most of the remaining dilemmas in classical Darwinism. The main promise of germinal selection lay in its capacity to explain a phenomenon that could scarcely be more inconvenient for Darwinism—directed variation. Darwin had emphasized the "random" or undirected character of variational raw material as a prerequisite for advocating natural selection as the cause of evolutionary change (see previous discussion of this crucial principle on pp. 144–146). For directed variation, if sufficiently powerful, would demote natural selection to a negative force for eliminating the unfit (while the fit arise automatically by differential variation), and a minor accelerator of trends originating for internal reasons (since random mortality can sustain an evolutionary direction imparted by biased variation).

But germinal selection could now explain directed variation by differential survival of struggling components within germ cells. Weismann therefore made a sweep through evolutionary problems that might be resolved by his new and selectionist theory of directed variation. In his original formulation of 1896, Weismann identified three possible roles for germinal selection in positive adaptation.

1. Fleeming Jenkin (1867) had troubled Darwin with his analogy of potential variation within a species to a rigid glass sphere. Selection could be effective along any radius, but the movement of a modal form from center to surface exhausted all possible variation, and positive selection must therefore damp itself to oblivion before achieving any substantial change. But, Weismann argued, germinal selection of successful determinants establishes a

highway for new variation, and any positive natural selection can automatically engender, by germinal selection, a wave of increasing variation ahead of the advancing mean.

2. Spencer (1893a and b) had invested his major defense of Lamarck in the phenomenon of co-adaptation. How could natural selection, working separately on each trait, produce an intricate coordination of numerous parts, all changing in the same direction? But Weismann now argued that any positive organismal selection will strengthen the determinants of all traits involved, thereby triggering a coordinated trend in germinal selection. Co-adaptation becomes less puzzling if all positively selected traits achieve such a strong boost from germinal selection: "As soon as utility itself is supposed to exercise a determinative influence on the direction of variation, we get an insight into the entire process and into much else besides that has hitherto been regarded as a stumbling block to the theory of selection . . . as, for example, the like directed variation of a large number of already existing similar parts, seen in the origin of feathers from the scales of reptiles" (1896, p. 39).

3. In his furthest extension—a remarkable claim given his previous faith in the *Allmacht* of purely organismal selection—Weismann now argued that any intricately precise adaptation probably requires a boost from germinal selection to reach a pinnacle of exquisite design. Writing of mimicry in Lepidoptera, he argued:

> It would have been impossible for such a minute similarity in the design, and particularly in the shades of the coloration, ever to have arisen, if the process of adaptation rested entirely on personal selection . . . In such cases there can be no question of accident, but the variations presented to personal selection must themselves have been produced by the principle of the survival of the fit! And this is effected, as I am inclined to believe, through such profound processes of selection in the interior of the germ plasm as I have endeavored to sketch . . . under the title of germinal selection (1896, p. 32).

This list, while granting broad scope and power to the newly formulated domain of suborganismal struggle, also illustrates the strictly limited conceptual role that Weismann envisaged for germinal selection when he first formulated the concept in 1895 and 1896. The action of germinal selection must *always* be synergistic with conventional organismal selection. In degeneration, the "type" example if you will, germinal selection "finishes off" what natural selection began but could not complete. In all other cases of evolutionary construction (maintenance of variation, co-adaptation, and "extreme perfection"), germinal selection works hand in hand with organismal selection, either to supply positive variation, or to accelerate change by supplying an impetus in the same direction from another level.

Over and over again, Weismann emphasizes the purely synergistic action of germinal with organismal selection. Thus, for example, ordinary natural selection can initiate the decline of an organ rendered useless by environmental change. In fishes that have migrated into dark caves, organismal selection against eyes must lead to differential survival of animals with weakened determinants for these organs. Once debilitated, these determinants continue to lose battles of germinal selection, and phenotypic expression sinks below the threshold that organismal selection could regulate: "In short, only by this means are the determinants of the useless organ brought upon the inclined plane, down which they are destined slowly but incessantly to slide towards their complete extinction" (1896, p. 25).

The case for synergism becomes even clearer when selection acts for the increasing complexity of an organ—for the two levels need not engage in a relay, but may now work continuously together in the same direction. Organismal selection favors a stronger part, thereby preserving organisms with more powerful determinants for the part, and unleashing a necessarily synergistic germinal selection: "As soon as personal selection favors the more powerful variations of the determinant . . . at once the tendency must arise for them to vary still more strongly in the plus direction . . . From the relative vigor or dynamical status of the particles of the germ plasm, thus, will issue spontaneously an ascending line of variation, precisely as the facts of evolution require" (1896, p. 27).

Weismann explicitly identified this inherent and automatic synergism as a major insight, and a logical completion of his argument for the *Allmacht* of selection. Noting "the harmony of the direction of variation with the requirements of the conditions of life" (1896, p. 38), Weismann continues: "The degree of adaptiveness which a part possesses itself evokes the direction of variation of that part. This proposition seems to me to round off the whole theory of selection and to give to it that degree of inner perfection and completeness which is necessary to protect it against the many doubts which have gathered around it on all sides like so many lowering thunderclouds . . . The principal and fundamental objection that selection is unable to create the variations with which it works, is removed by the apprehension that a germinal selection exists." And, more succinctly for the centennial of Darwin's birth, Weismann reduced this theme to a celebratory aphorism (1909, p. 39): "Germinal selection supplies the stones out of which personal selection builds her temples and palaces: adaptations"—thus amalgamating the synergistic link of levels, the *Allmacht* of selection, and the primacy of adaptation (as recorded in a metaphorical linkage with the noblest of buildings, both spiritual and temporal).

## GERMINAL SELECTION AS A FULL THEORY OF HIERARCHY

Weismann had granted an important role to germinal selection in his initial formulation, but he had not yet developed a full theory of hierarchy, for germinal selection could only walk the same paths established by ordinary natural selection—at most accelerating or intensifying the journey. But when Weismann wrote the major book and summation of his later career, *Vorträge über Descendenztheorie* (1902; English translation, 1903), he devoted two full chapters to germinal selection and, without explicitly acknowledging the change or extension, radically altered his conception into a full selectionist theory of hierarchy, the first such proposal in the history of evolutionary thought.

Taking an opposite tack from his original formulation, Weismann now proposed to define the scope of germinal selection by what such a process could accomplish *without* and *beyond* organismal selection: "We shall attempt to gain clearness as to what it can do, and how far the sphere of its influence extends, and, in particular, whether it can effect lasting transformations of species without the cooperation of personal selection, and what kind of variations we may ascribe to it alone" (1903, vol. 2, p. 126).

Weismann now recognized that he could use germinal selection to escape the straitjacket of Panglossian adaptation. He could finally admit the nonadaptive character of some phenotypic features without straying from the *Allmacht* of selection, or giving comfort to Lamarckism—for traits arising by germinal selection may be neutral or even harmful to survival in the Darwinian world of competition among organisms. Weismann recognized two classes of nonadaptive features potentially ascribable to germinal selection.

NEUTRAL FEATURES OF SMALL OR NO IMPORTANCE. Since germinal selection can promote changes invisible to organismal selection, Weismann wondered whether such minor variations as human racial differences—attributed by Darwin to sexual selection based on disparate standards of beauty arising for capricious reasons in various societies—might actually arise as effects of germinal selection: "It cannot be denied that there are characters which have no special biological significance [although] . . . it is difficult and often impossible to point these out with certainty. The shape of the human nose and the human ear, the color of the hair and the iris, may be such indifferent characters whose peculiarities are to be referred solely to germinal selection" (1903, vol. 2, p. 134).

ORTHOGENETIC DRIVES THAT MAY YIELD HARMFUL FEATURES AND EVEN LEAD TO EXTINCTION. In a remarkable departure from the almost strident panselectionism of his earlier years, Weismann now approved certain claims for orthogenesis, and admitted the existence of harmful trends (based on directionality in variation) that organismal selection could not reverse. He even accepted the classic examples of the anti-Darwinian schools as orthogenetic and harmful *prima facie*—the antlers of Irish Elks, and the massive canines of saber-toothed cats (1903, vol. 2, p. 139). He embraced the best cases of his opponents because germinal selection—once he reversed his original view and grasped its power to work *against* organismal selection— could convert these enemy troops to the doctrine of *Allmacht*. For when germinal selection acts with sufficient power, then all the determinants favored by organismal selection may be eliminated entirely, leaving only the vigorous determinants of harmful orthogenetic features, and rendering conventional selection impotent for lack of raw material: "In this case the variation-direction which had gained the mastery in all ids could no longer be sufficiently held in check by personal selection, because the variations in the contrary direction would be much too slight to attain to selective value" (1903, vol. 2, p. 139).

Following the early 20th century vogue for eugenics, Weismann used this argument to promote a positive program of breeding to save the human race. Germinal selection must be responsible for any decline of the human stock engendered by relaxation of natural selection—for panmixia lacks sufficient strength to produce such an effect, while Lamarckian inheritance does not exist. This orthogenetic deterioration by germinal selection can only be reversed by a reimposition of Darwinian competition in the organismal mode, with reproductive success to the victors. Arguing that military service might operate as a good filter for identifying bodies well suited for success in organismal selection, Weismann suggested (1903, vol. 2, p. 147): "It would indeed be well if only those who had gone through a term of military service were allowed to beget children" (thus adding another example to the compendium of social nonsense advanced by prominent evolutionists in the name of Darwinism—see Chapter 7, and Fisher, 1930).

But, far more important than merely extending the domain of germinal selection to explain potentially nonadaptive organismal traits, Weismann also enlarged the conceptual realm of levels in selection from a narrow mechanism for synergism to a full theory of hierarchy. Weismann had now worked his way through the logic of multi-level selection theory, and had recognized the two key ingredients of any full account.

INDEPENDENCE OF LEVELS AND POTENTIAL FOR CONFLICT. The attribution of orthogenesis to germinal selection implies that suborganismal selection can act separately from conventional Darwinian selection, and even work in a contrary direction to decrease phenotypic adaptation. Thus, the process that Weismann had originally promoted to make natural selection even more effective had become, by honest probing into all corners and implications of the argument, a separate cause that could work either with or against the canonical Darwinian mode. In fact, Weismann now argued that potential independence from the Darwinian level of organismal selection establishes the primary significance for germinal selection in evolutionary theory (1903, vol. 2, p. 119): "In this fact lies the great importance of this play of forces within the germ-plasm, that it gives rise to variations quite independently of the relations of the organism to the external world. In many cases, of course, personal selection intervenes, but even then it cannot directly effect [*sic*, and correctly; he means 'cause,' not just 'affect'] the rising or falling of the individual determinants—these are processes quite outside of its influence."

In his most revealing change, Weismann even reinterprets his type case of degeneration. He had previously tried to explain degeneration as a result of germinal selection completing a process that natural selection had started but could not finish. Now, without any change in evolutionary mechanics, he speaks of germinal selection working differently and independently. He even claims that degeneration offers the purest case for potential independence of levels—for germinal and organismal selection usually act together, thus rendering their individual contributions operationally inseparable. But we know that organismal selection has disappeared in the final stages of degeneration, and we can therefore observe the unsullied action of germinal selection alone!

"In one direction variation can be proved to go on without limit, and that is downwards, as is proved by the fact of the disappearance of disused organs, for here we have a variation-direction, which has been followed to its utmost limit, and which is completely independent of personal selection; it proceeds quite *uninterfered with* by personal selection, and is left entirely to itself" [Weismann's italics] (1903, vol. 2, p. 129). At England's major centennial celebration of Darwin's birth, Weismann presented an even stronger statement (1909, p. 38): "Useless organs are the only ones which are not helped to ascend again by personal selection, and therefore in their case alone can we form any idea of how the primary constitutents behave, when they are subject solely to intragerminal forces."

Such independence of levels implies potential conflict, with stability achieved through balance, or by the victory of one level. Natural selection, as a powerful force operating directly on phenotypes, usually prevails in its own domain of visible form. If germinal selection weakens a useful organ, natural selection can intervene in an antagonistic mode (except in degeneration, where natural selection ceases to act, and germinal selection reigns unchecked). "If a struggle for food and space actually takes place, then every passive weakening must lead to a permanent condition of weakness and a lasting and irretrievable diminution in the size and strength of the primary constituent concerned, unless personal selection intervenes, and choosing out the strongest among these weakened primary constituents, raises them again to their former level. But this never happens when the organ has become useless" (1903, vol. 2, p. 122).

Similarly, if germinal selection initiates an orthogenetic trend, natural selection can impose a halt by eliminating organisms with traits exaggerated beyond utility, thereby removing their positive determinants from the germinal population. (This basic antagonism, Weismann finally concluded in his strongest recognition of potential conflict between levels, may be virtually omnipresent in nature, and therefore fundamental to evolution, because positive organismal selection almost always elicits an upward trend in determinants by Weismann's earlier argument for synergism. Most stable species may not be quiescent with respect to selection, but balanced by a policing of germinal selection with opportune removals based upon Darwinian organismal competition.) "In the majority of cases the self-regulation which is afforded by personal selection will be enough to force back an organ which is in the act of increasing out of due proportion to within its proper limits. The bearers of such excessively increased determinants succumb in the struggle for existence, and the determinants are thus removed from the genealogical lineage of the species" (1903, vol. 2, p. 130).

But germinal selection can also triumph, and such victories may not be infrequent in nature. All orthogenetic and nonadaptive traits may record the potency of suborganismal processes in conflicts between levels of selection: "All excessive or defective hereditary malformations may be referred to germinal selection alone, that is, to the long-continued progressive or regressive variation of particular determinant-groups in a majority of ids" (1903, vol. 2, p. 138).

EXPANSION OF THE HIERARCHY BOTH UP AND DOWN FROM A DARWINIAN FOCUS ON ORGANISMS. Weismann devised germinal selection as an *ad hoc* hypothesis to resolve his longstanding embarrassment over the problem of degeneration. In 1896, he applied selection throughout Haeckel's hierarchy of "individuals," extending from cell constitutents to clonal colonies (1896, p. 42), and recognizing three primary levels—Darwin's conventional struggle for existence among organisms, Roux's histonal selection, and his own germinal selection.

But by 1903, in a statement that I regard as wonderfully prophetic of current concerns, Weismann had proceeded beyond his immediate theoretical needs to full generality. He had used germinal selection to break through the *Allmacht* of exclusivity for Darwin's level. But now he recognized the inexorable logic of a fully developed and extended theory of hierarchy—reaching right up to species selection at the top.

> I have called these processes which are ceaselessly going on within the germ-plasm, Germinal Selection, because they are analogous to those processes of selection which we already know in connection with the larger vital units, cells, cell-groups and persons. If the germ-plasm be a system of determinants, then the same laws of struggle for existence in regard to food and multiplication must hold sway among its parts which hold sway between all systems of vital units—among the biophors which form the protoplasm of the cell-body, among the cells of tissue, among the tissues of an organ, among the organs themselves, as well as among the individuals of a species and between species which compete with one another (1903, vol. 2, p. 119).

If Weismann had presented this full elaboration of hierarchy only as a footnote, a flash of insight in a book devoted to other goals, I could not claim him as an intellectual forebear of our modern excitement. Good ideas originate in fair abundance, and we must look to development and application for our main criteria of sustained scientific worth. If Weismann had even devoted an isolated chapter to hierarchy, I would note his insight with praise, but grant him limited success for failing to recognize the power of this theme as an organizing framework for evolutionary mechanisms. But, in fact, Weismann did fully grasp the fundamental difference between classical Darwinism and the expanded theory of interacting levels of selection—and he regarded his exposition of hierarchy as both the central feature of his mature thinking, and the unifying concept of all evolution: "This extension of the principle of selection to all grades of vital units is the characteristic feature of my theories; it is to this idea that these lectures lead, and it is this—in my own opinion—which gives this book its importance. This idea will endure even if everything else in the book should prove transient" (1903, in preface, vol. 1, p. ix).

How ironic that a man who so explicitly promoted the centrality of hierarchy should be remembered today primarily for his earlier advocacy of *Allmacht* at the traditional level of organismal struggle alone! A study of Weismann's intellectual ontogeny should lead us to respect the logic and power of hierarchy, whatever our eventual judgment about merit and utility.

Weismann, by far the most thoughtful of Darwinians in the first generation after the founder himself, could not "cash out" the exclusivity of organismal selection when he pushed the theory to the edges of its necessary application. He then invented an auxiliary theory of suborganismal selection to rescue himself from an uncomfortable corner; but he eventually followed the relentless logic of his own argument to a full theory of hierarchy.

The theory of hierarchical selection does not constitute either a small and merely incremental nuance, or a modern concoction and exaggeration of bored Darwinians trying to stir up some trouble. Hierarchy has accompanied the theory of selection from its very inception—if only because no truly tenacious and thoughtful Darwinian could ever avoid its appeal and logic, while at least one of the wisest and the most committed adherents, August Weismann, came to regard hierarchy as the implied and necessary centerpiece of any evolutionary theory fully rooted in selectionist principles, and truly comprehensive in explanatory range. We have largely forgotten Weismann's intellectual journey today. But we should recover his chain of argument—for his motives and insights retain full validity, even if later discoveries about the physical basis of heredity invalidated his particular form of suborganismal selection.

## Hints of Hierarchy in Supraorganismal Selection: Darwin on the Principle of Divergence

### DIVERGENCE AND THE COMPLETION OF DARWIN'S SYSTEM

Charles Darwin cannot be judged as a consistently felicitous writer, but he could turn a phrase with the very best of craftsmen. As noted before, many of his lines, particularly his wonderful metaphors, have become parts of our culture—the image of the "entangled bank" at the very end of the *Origin*, or the "tree of life" that closes Chapter 4 on natural selection. His posthumously published autobiography contains many memorable and oft-quoted statements, including his description of intellectual eureka: "I can remember the very spot in the road, whilst in my carriage, when to my joy the solution occurred to me" (in F. Darwin, 1887, vol. 1, p. 84).

Poll our biological colleagues, and most will tell you that this horse-drawn epiphany describes Darwin's Malthusian insight of September 1838, and the resulting formulation of natural selection. But the passage refers to a much later event, and this error of attribution may be the most common in all Darwinian exegesis. The statement recounts an insight—the "principle of divergence" in his own description—that Darwin ranked as equal in importance with natural selection itself, an idea whose formulation sometime in the early to mid 1850's (the true date of the carriage ride) allowed him to complete his theoretical structure and begin writing his *magnum opus*.

Darwin describes the phenomenon that a principle of divergence must resolve, and states his surprise at his own obtuseness before the fateful carriage ride:

But at that time [after the Malthusian insight of 1838 and his composition of the sketches of 1842 and 1844] I overlooked one problem of great importance; and it is astonishing to me, except on the principle of Columbus and his egg,* how I could have overlooked it and its solution. The problem is the tendency in organic beings descended from the same stock to diverge in character as they become modified. That they have diverged greatly is obvious from the manner in which species of all kinds can be classed under genera, genera under families, families under suborders, and so forth (in F. Darwin, 1887, vol. 1, p. 84; the carriage statement directly follows).

Darwin (*loc. cit.*) then epitomizes the solution that he named "the principle of divergence" and ranked with natural selection as a foundation of his theory: "The solution, as I believe, is that the modified offspring of all dominant and increasing forms tend to become adapted to many and highly diversified places in the economy of nature."

Darwin's principle of divergence has puzzled many biologists: why did Darwin rank the concept so highly, and as a principle separate from natural selection? May we not view divergence as a logical consequence or simple spin-off from natural selection itself? Yet when one considers the issue in Darwin's terms, both his separation of divergence from natural selection, and his joy in resolution, make excellent sense. Natural selection, as formulated under the Malthusian insight of 1838, states a principle of anagenetic change within phyletic lines—an argument about adaptation to local circumstances (biotic and abiotic). This principle says nothing, by itself, about diversification, or splitting of one lineage into two or several descendant

---

*The reference to Columbus and his egg puzzled me. The line is often quoted by Darwin scholars, but never explained—so either everyone (but me) knows the old tale, or else most people share my ignorance, and pass the issue by in embarrassed silence. Darwin apparently cites the story as a standard motto, or a schoolboy tale taught to everyone. But modern Americans (of my generation and younger) do not know the story—at least in my informal, but reasonably obsessive, survey. I asked several older Europeans, and caught a ray of light because some recalled such a tale from their distant educational pasts, but couldn't dredge up the details. Finally, the letters column of the *New York Times* came through. The old chestnut—this one I do know—about balancing an egg at the Spring equinox received an ample airing in 1989, both in editorials and letters. A Mr. Louis Marck unwittingly submitted this lovely resolution of Darwin's puzzling line in a letter of March 26 entitled "Columbus also had a way with eggs":

> In "It's Spring. Go Balance an Egg" (editorial March 19), you say that cheaters "crack the shell to create a flat bottom." According to a tradition strangely unknown in this country, one person who did that very thing, not as a cheater, but to prove a point, was Christopher Columbus.
>
> My German dictionary of quotations places the apocryphal incident in 1493, at a banquet given in honor of Columbus by Cardinal Mendoza. When the difficulty of his voyage of discovery was put into question, Columbus challenged his interlocutors to balance an egg. When they failed, he did it by cracking the shell.
>
> In German, as well as Spanish, "the egg of Columbus" has become proverbial for solving a difficult problem by a surprisingly simple knack or expedient.

taxa. So much of what Darwin needed to explain—plenitude in ecology, branching models in phylogeny, the hierarchical structure of taxonomy, to name just a few items of obvious centrality—rested upon the fact of diversification, not adaptation (see Mayr, 1992, on Darwin's *several* theories of evolution).

One might say—indeed many of us do say, thus leading us to downgrade and misinterpret Darwin's explanation of diversity—that "divergence of character" requires no separate principle beyond adaptation, natural selection, and historical contingency. After all, the earthly stage of evolution provides ecological and biogeographical prerequisites for diversification. Climates alter; topography changes; populations become isolated, and some, adapting to modified environments, form new species. What more do we need? Insofar as Darwin considered the issue at all between 1838 and the early 1850's, his thinking followed this general line (Sulloway, 1979; Ospovat, 1981). But Darwin grew dissatisfied with a theory that featured a general principle to explain adaptation, but then relied upon historical accidents of changing environments to resolve diversity. He decided that a fully adequate theory of evolution required an equally strong principle of diversity, one that acted intrinsically and predictably. If adaptation and diversification specify the central phenomena of evolution, each must have its principle, and their union would then define his complete theory.

(In modern evolutionary parlance, we may relate the growing intensity of Darwin's search for a general, "law-of-nature" explanation of divergence to his changing views about allopatric and sympatric speciation. During the 1840's, when diversity did not greatly trouble him as a theoretical issue, Darwin tended to view speciation as allopatric, and therefore as a consequence of historical accidents in geography and ecology. When a population becomes spatially isolated, he reasoned, natural selection can act independently upon it, and eventually accumulate enough divergence from the ancestral form to establish a new species. But Darwin's preferences then shifted to sympatric views of speciation—and he therefore developed a conviction that some general law, and not just historical accidents of isolation, must promote the multiplication of species. A complete theory of natural selection required that this elusive "law" of speciation or divergence also be based on the predictable operation of organismic selection. In the light of our current preferences for allopatric speciation, Darwin's shift may seem ironic, but our opinions and certainties, as presently defined, must be deemed irrelevant to such historical analysis.)

In the context of this book and its principal theme of hierarchical selection, I stress the centrality of Darwin's changing views on divergence because I think that I have made a small discovery about the structure of his argument. I shall try to show that this most brilliant of all theorists, this rigorously honest thinker who worked so diligently to explain all evolution as a consequence of organismic struggle, tried mightily to render his second touchstone, his "principle of divergence," by ordinary natural selection—and he failed. He could not succeed because the logic of his argument demands a major role for sorting and selection at the species level—and Darwin, with characteristic honesty, faced his distress head on.

I am fascinated that the exegetical literature on Darwin's attitude to supraorganismal selection (Ghiselin, 1974; Ruse, 1980, for example) has focused entirely on putative cases of group selection (sterility in plant hybrids; neuter castes in Hymenoptera) and has quite properly concluded that Darwin, with a possible exception for invoking family or clan selection to explain human moral traits, doggedly and consistently carried through his program for the exclusivity of organismic selection (see Chapter 2, pp. 125–137). In so doing, these scholars have missed the one area—the heart of Darwin's argument about diversity—where his logic falters because he needs (but hesitates to embrace in his distress) the apparatus of species selection. I suspect that the internal problems in this centerpiece of Darwin's thought have not been addressed, or even recognized, because species selection itself did not become a subject of importance (or even an acknowledged subject at all) until recently, while debate about conventional group selection has long raged. Darwin has therefore been well combed for comments about interdemic selection, while his main engagement with supraorganismal selection on species went unnoticed.

In any case, whatever our attitude or ignorance today, Darwin clearly regarded his solution to the problem of divergence as his second great achievement (after natural selection), and as the capstone to his theory. As Ernst Mayr notes (1985, pp. 759–760): "He referred to it always with great excitement, as if it had been a major departure from his previous thinking." On June 8, 1858, Darwin wrote to Hooker after completing his extended discussion of the principle of divergence for Chapter 6 of *Natural Selection* (the "big species book" that would never be completed because Wallace's paper, arriving within 10 days of this letter to Hooker, derailed his leisurely plans and led him to compose the "abstract" (in his own description) that we call *The Origin of Species*). "I am confined to the sofa with boils," he begins, "so you must let me write in pencil." He then goes on to describe "the 'Principle of Divergence,'" which, with 'Natural Selection,' is the keystone of my book" (in F. Darwin and Seward, 1903, volume 1, p. 109).

A year before, in September 1857, Darwin wrote his first complete account of the principle of divergence in a famous letter to Asa Gray at Harvard University. Gray had explicitly asked Darwin for an epitome of his evolutionary theory (previously revealed only to Darwin's closest confidants, Hooker and Lyell in particular): "It is just such sort of people as I that you have to satisfy and convince and I am a very good subject for you to operate on, as I have no prejudice nor prepossessions in favor of any theory at all" (quote in Kohn, 1981, p. 1107). Darwin responded positively with a lucid summary of his theory in six points:

- The power and effect of artificial selection.
- The even greater power of natural selection working on all characters at once and over vastly longer spans of time.

- The operation of natural selection at the organismal level, powered by the Malthusian principle that all species produce far more offspring than can possibly survive.
- A description of how natural selection works in nature.
- A defense of gradualism as the solution to standard problems in accepting the factuality of evolution.
- An explication of the principle of divergence.

This account of the principle of divergence also became the first published version because Lyell and Hooker included this letter to Gray among the documents published in the Linnaean Society's journal for 1858—the "delicate arrangement" that presented Darwin and Wallace jointly, stressing Darwin's priority but publishing Wallace's paper on the independent discovery of natural selection *in toto*. Darwin's sixth point neatly summarizes his ideas on divergence:

> Another principle, which may be called the principle of divergence, plays, I believe, an important part in the origin of species. The same spot will support more life if occupied by very diverse forms. We see this in the many generic forms in a square yard of turf, and in the plants or insects on any little uniform islet, belonging almost invariably to as many genera and families as species. We can understand the meaning of this fact amongst the higher animals, whose habits we understand. We know that it has been experimentally shown that a plot of land will yield a greater rate if sown with several species and genera of grasses than if sown with only two or three species. Now, every organic being, by propagating so rapidly, may be said to be striving its utmost to increase its numbers. So it will be with the offspring of any species after it has become diversified into varieties, or sub-species, or true species. And it follows, I think, from the foregoing facts, that the varying offspring of each species will try (only a few will succeed) to seize on as many and as diverse places in the economy of Nature as possible. Each new variety or species, when formed, will generally take the place of, and thus exterminate its less well-fitted parent. This I believe to be the origin of the classification and affinities of organic beings at all times; for organic beings always seem to branch and sub-branch like the limbs of a tree from a common trunk, the flourishing and diverging twigs, destroying the less vigorous—the dead and lost branches rudely representing extinct genera and families (from the 1858 published version, often reprinted, as, for example, in Barrett *et al.*, 1987).

Darwin continually awarded his principle of divergence the central role specified in this letter to Gray. Nearly half of the key chapter in the *Origin of Species* (number 4 on "Natural Selection") treats the principle of divergence, closing with the celebrated metaphor of the tree of life, sketched out at the end of point 6 to Gray. The only figure in the entire *Origin of Species* occurs in chapter 4. The intent of this famous diagram (reproduced here as Fig. 3-5 on p. 242) has almost always been misunderstood by later commentators.

Darwin did not draw this unique diagram simply to illustrate the generality of evolutionary branching, but primarily to explicate the principle of divergence.

After more than a century in limbo, Darwin's principle of divergence has been exhumed and subjected to careful scrutiny by historians of science. No subject in Darwinian studies has been more actively pursued during the past 25 years, and many excellent analyses have been published on the genesis and utility of the principle of divergence (Limoges, 1968; Sulloway, 1979; Browne, 1980; Schweber, 1980, 1985, 1988; Ospovat, 1981; Kohn, 1981, 1985). Therefore, the importance of this principle has finally been recognized. Ospovat, for example, writes (1981, pp. 170–171): "Darwin's 'principle of divergence' [is] the most important addition to his theory between 1838 and 1959 [*sic,* for 1859] and the one most intimately associated with the transformation of his theory after 1844." In all this literature, however, only Schweber has grasped Darwin's difficulties with divergence as an unresolved struggle between levels of explanation. Yet this theme, particularly Darwin's inability to "cash out" his usual argument about organismal struggle at the level of species birth and death, holds, I believe, the key to Darwin's treatment.

Darwin's argument about divergence begins with an unquestioned premise that strikes us as curious today (for we are immediately tempted to mount a challenge), but resonates with a central theme of Darwin's century—the clear and inherent "good" of maximizing the amount of life in any given region, and the consequent necessity for a cause to insure this natural goal. Maximization, Darwin argues, arises by diversification: the more taxa in a given area (and the more different), the greater the total quantity of life. This theme can be traced to Darwin's earliest "transmutation notebooks" of the 1830's, the primary documents of his quest to formulate evolution: "The end [that is, goal] of formation of species and genera is probably to add to quantum of life possible with certain pre-existing laws—if only one kind of plant not so many" (C Notebook, p. 146—in Barrett *et al.,* 1987).

In the fullest discussion within *Natural Selection* (written in early 1858), Darwin firmly links maximization of life to diversification of taxa: "I consider it as of the utmost importance fully to recognize that the amount of life in any country, and still more that the number of modified descendants from a common parent, will in chief part depend on the amount of diversification which they have undergone, so as best to fill as many and as widely different places as possible in the great scheme of nature" (p. 234 of Stauffer edition, 1975).

Darwin proposes that the vague concept of "amount" or "maximization" of life be quantified chemically as total metabolic flow through a given area in a given time—and he illustrates the primary dependence of this quantity on diversification:

> The fairest measure of the amount of life [is] probably the amount of chemical composition and decomposition within a given period. Imagine the case of an island, peopled with only three or four plants of the same order all well adapted to their conditions of life, and by three or four in-

sects of the same order; the surface of the island would no doubt be pretty well clothed with plants and there would be many individuals of these species and of the few well adapted insects; but assuredly there would be seasons of the year, peculiar and intermediate stations and depths of the soil, decaying organic matter etc., which would not be well searched for food, and the amount of life would be consequently less, than if our island had been stocked with hundreds of forms, belonging to the most diversified orders (*ibid.*, p. 228).

Darwin then provides examples from agriculture and domestication. Several varieties of wheat, sown together on a plot, will yield more grain per acre than a monoculture. In one experiment, two species of grass yielded 470 plants per square foot, while a plot of 8 to 20 species produced a thousand plants per square foot (*ibid.*, p. 229): "I presume that it will not be disputed that on a large farm, a greater weight of flesh, bones, and blood could be raised within a given time by keeping cattle, sheep, goats, horses, asses, pigs, rabbits and poultry, than if only cattle had been kept" (*ibid.*, p. 229).

But why was Darwin so wedded to a principle of maximization that would strike most of us today as both metaphysical and indefensible (ecosystems, after all, can work perfectly well with far fewer species and lower chemical "yield" per spot)? Schweber (1980), I think, has provided the correct answer by stressing Darwin's allegiance to one of the most popular philosophical approaches of his day—the "Benthamite optimization calculus" promoted by Jeremy Bentham, and many other prominent thinkers in several disciplines, as the utilitarian principle in philosophy and political economy, the "greatest good for the greatest number." William Paley, the intellectual hero of Darwin's youth (see p. 116), spoke for a utilitarian consensus in writing (quoted in Schweber, 1980, p. 263): "The final view of all rational politics is, to promote the greatest quantity of happiness in a given tract of country . . . and the quantity of happiness can only be augmented by increasing the number of percipients or the pleasure of their perceptions." In other words, make more objects and make them better. Nature achieves this desired maximization and progress by diversifying the number of species in each region of the globe.

Darwin explicates and defends the maximization of life with his favorite rhetorical device—analogy—and by invoking another fundamental tenet in the political economy of his era: the division of labor. As taxa specialize ever more precisely to definite and restricted roles in local ecologies, more species can be supported (leading to maximization of life as measured by chemical throughput). In a note of September 23, 1856, Darwin drew a direct parallel between diversification in nature and the economic principle of division of labor: "The advantage in each group becoming as different as possible, may be compared to the fact that by division of labor most people can be supported in each country." For the public presentation in the *Origin* three years later, Darwin retained the centrality of division of labor, but chose a biological analogy drawn from the French zoologist Henri Milne Edwards (who had, himself, credited Adam Smith and the political economists, and who characterized his own view as an extension of their principle):

The advantage of diversification in the inhabitants of the same region is, in fact, the same as that of the physiological division of labor in the organs of the same individual body—a subject so well elucidated by Milne Edwards. No physiologist doubts that a stomach by being adapted to digest vegetable matter alone, or flesh alone, draws most nutriment from these substances. So in the general economy of any land, the more widely and perfectly the animals and plants are diversified for different habits of life, so will a greater number of individuals be capable of there supporting themselves (Darwin, 1859, pp. 115–116).

Consider the form of the classic argument in Darwin's two analogical sources. Adam Smith began the *Wealth of Nations* by discussing pinmaking to illustrate the advantages of division of labor. Smith states the basic argument in the very first words of his classic book: "The greatest improvement in the productive powers of labor, and the greater part of the skill, dexterity, and judgment with which it is anywhere directed or applied, seem to have been the effects of the division of labor." Pinmaking, Smith tells us, may be "a trifling manufacture," but "18 distinct operations" are still needed to make the final product. If a single worker performed all these tasks, he "certainly could not make twenty" pins in a day, but allocation of separate tasks to 10 people (with some individuals continuing to perform 2 or 3 of the 18 operations) allowed one small factory to make 48,000 pins per day, or 4800 per man. Now who benefits from this division of labor? In part, the workers who hone their skills and participate in the resulting prosperity. But primarily the larger polity—the factory through profits, or society itself in the availability of moderately priced goods. Similarly, in Milne Edwards' physiological division of labor, the prime beneficiary cannot be the organ (an omnivore's stomach works perfectly well *qua* stomach), but again the larger polity, in this case the organism.

Applying the same logic to Darwin's analog, the beneficiary of life's diversification through division of labor is not the individual, or even the species, but the larger polity—or life itself through the principle of maximization. Thus we can grasp the link between division of labor as a pervasive structural principle, and Darwin's goal in application—the *summum bonum* of maximization of life, achieved through division of labor with the *larger polity,* or life itself, as the beneficiary.

The logic in this chain of reasoning also illustrates why the coupling of maximization with division of labor cannot validate Darwin's "principle of divergence." These arguments may indicate why maximization should occur, but do not explain *how* such a plenteous state of nature arises. In other words, this chain of reasoning does not propose a *cause* for maximization. Above all else, Darwin clearly understood that his distinctive style of evolutionary argument demanded an explanation for any higher level phenomenon as a consequence of struggle among *individual organisms* for reproductive success. Maximization and division of labor represent phenomenological statements about the constitution of life and ecology, not claims about the efficient causes of diversity. Such statements provide, in other words, a basic description of diversity, but emphatically not a "principle of divergence." Darwin's insight in his carriage did not merely systematize the notions of maximization and division of labor; Darwin had known and used these concepts for years. The transforming insight—the argument worthy of being called a "principle of divergence" and becoming a keystone of his entire evolutionary theory—occurred when Darwin recognized how he could apply his distinctive style of argument, based on organismal selection, to the higher-level phenomenology of diversity. In other words, the "principle of divergence" embodies Darwin's argument for how and why *ordinary natural selection* must, as a predictable consequence, yield divergence of character, leading to multiplication of successful taxa, extinction of others, ecologic plenitude, maximization of life, and the hierarchical structure of taxonomy.

### THE GENESIS OF DIVERGENCE

This perspective on the jelling of Darwin's principle of divergence resolves a recent debate among historians about the timing and reason for Darwin's formulation. Proposals for timing have ranged from the late 1840's to 1858, with cardinal inspirations from biogeography (Sulloway), systematics (Limoges, Ospovat), the "botanical arithmetic" of chapter 2 in the *Origin*, leading Darwin to defend the greater evolutionary potential for diversification in large genera (Browne), inspiration from the arguments of political economy (Schweber) and switch from allopatric to sympatric models of speciation (Kohn). All these influences surely played their parts, for the principle of divergence calls upon a wide and complex range of convictions, spanning many years and much turmoil in Darwin's mind. But Darwin's formulation and formalization of *the* "principle of divergence" records his conviction, and his great pleasure, that he could encompass all these ideas as predictable consequences of *natural selection* working by struggle *among organisms*—that he could, in other words, bring all the higher-level phenomenology of maximization, division of labor, and so forth, into his own distinctive explanatory framework.

Schweber dates the first full formulation to 1856 and writes (1988, p. 135): "That Darwin had the 'keystone' of the argument by January 1855 is probably correct, but I would also suggest that the argument was still not complete in an important way—at least insofar as an explicit presentation is concerned. All the arguments up to that point referred to levels of descriptions above individuals: varieties, species, and higher taxa. Natural selection operated on individuals, and the linkage by which diversity is accomplished had to be explicitly stated." Using this insight, Schweber regards the following note of September 23, 1856, as the first explicit formulation.* Just

---

*Since these words directly follow the statement, quoted on p. 230, about division of labor in political economy, Schweber locates a primary influence in this interdisciplinary transfer—not only, of course, via the specific linkage in this particular quotation, but primarily because the dominant political economy of individualism, the philosophy of Adam Smith, Jeremy Bentham and scores of followers, had always been a central inspiration for Darwin from the Malthusian insight onward—see pp. 121–125 and Schweber, 1977.

as unfettered individual competition yields the best social order in Adam Smith's world, so too will natural selection among organisms lead to maximization by division of labor: "The advantage in each group becoming as different as possible, may be compared to the fact that by division of labor most people can be supported in each country.—Not only do the individuals in each group strive against the others, but each group itself with all its members, some more numerous, some less, are struggling against all other groups, as indeed follows from each individual struggling" (Darwin, September 23, 1856, cited in Schweber, 1988). (This consistent stress on the role of individuals as primary causal agents also characterizes the writings in political economy that so influenced Darwin. For example, I omitted by ellipsis an intermediary passage in the statement from Smith quoted above on p. 124. It reads: "Although we speak of communities as sentient beings; although we ascribe to them happiness and misery, desires, interests, and passions; nothing really exists or feels but individuals. The happiness of a people is made up of the happiness of single persons.")

This proper characterization of Darwin's argument also overturned the most sensational charge ever based on the principle of divergence, and made with such attention in the public arena by Brackman (1980)—the claim that Darwin received Wallace's paper from Ternate earlier than the "official" date (June 18, 1858), and then proceeded to steal the principle of divergence from him, thus formulating his complete theory by ripping off Wallace and covering up the evidence. This charge, which can only be supported by ignorance of detail (see the analysis of Kohn, 1981), falls apart once we recognize Darwin's full principle of divergence as an explanation of maximization by natural selection through division of labor. Darwin clearly formulated this complete argument in 1856, and sent a lucid epitome to Asa Gray in 1857. Thus, a possible receipt of Wallace's paper earlier in June, or even in late May of 1858, cannot affect this chronology.

Brackman, of course, does not deny these facts. He must therefore claim that Darwin had been spooked by Wallace for years, that he pinched the initial idea of diversification from Wallace's 1855 paper, and that he then moved faster (and stealthily) when the firmer statement arrived in 1858. But if we turn to Wallace's 1855 paper, we note that this article contains nothing relevant to a principle of divergence properly defined as a set of complex arguments for linking natural selection on organisms with the phenomenology of higher levels of biological organization. At most, Wallace's 1855 paper includes a passing description of the simple property of divergence itself—a fact well recognized by Darwin, who had been noting the centrality of this theme since the transmutation notebooks of the late 1830's (see quote on p. 229). (We shall also see, at the close of this section (p. 248), that even Wallace's 1858 paper contains only a cursory statement about divergence with no hint of the central feature of bridging levels.) Brackman has confused the noting of a fact with the development of an explanation. He has also failed to recognize Darwin's long awareness of the fact and its importance.

### DIVERGENCE AS A CONSEQUENCE OF NATURAL SELECTION

In Chapter 2, I noted the radical character and intellectual power of Darwin's primary argument as embedded in the Malthusian insight about natural selection. In an irony that overturned the entire tradition of natural theology, Darwin held that all the higher order "harmonies" of good design and ecological balance arose as side consequences of a process—struggle among organisms for personal reproductive success—that would demand an opposite interpretation if we sought moral messages in nature. Now, in the mid 1850's, Darwin attempted the same philosophical coup to accomplish for diversity exactly what he had done for adaptation in the initial formulation of natural selection—that is, to render a higher level "good," the maximization of life through division of labor, as a side consequence of organismic struggle. In January 1855 (in the note that Schweber views as the genesis of the principle of divergence), Darwin takes this fateful repeat step into the philosophical radicalism of rendering higher harmonies by individual struggle (quoted in Schweber, 1988): "On Theory of Descent, a divergence is implied and I think diversity of structure supporting more life is thus implied . . . I have been led to this by looking at heath thickly clothed by heather and a fertile meadow both crowded, yet one cannot doubt more life supported in second than in the first; and hence (in part) more animals are supported. This is the final cause but mere result from struggle (I must think out last proposition)."*

---

*Darwin's abysmal handwriting causes endless trouble for scholars. All historians have recognized the crucial status of this final sentence, but each major exegete reads the line in a different way. Ospovat (1981, pp. 180–181) offers the following interpretation: "This is not final cause, but mere results from struggle, (I must think out this last proposition.)" Browne's version (1980, p. 71) reads: "This is not final cause but mere result from struggle (I must think out this last proposition.)" And Kohn (1985, p. 256) offers this deciphering: "This is not final cause, but more [a] result from struggle, (I must think out this last proposition.)" Aside from minor points of articles and punctuation, two disagreements are potentially substantial: First, did Darwin write "mere" or "more" with respect to struggle? "Mere" would be stronger, for then the higher order harmony of ecosystems becomes nothing but a consequence. But "more" still conveys the same sense—for higher order maximization of life would still represent *more* a consequence of individual struggle than anything else. Second, did Darwin say that maximization is *not* a final cause (Ospovat, Browne, Kohn), or does he choose to view such abundance of life as the final cause of struggle (Schweber)? These different readings seem to suggest a serious discrepancy, but, in fact, the meaning will be much the same in either case. For Darwin tells us, one way or the other, that individual struggle provides the generating cause, with maximization of life arising as a consequence. Thus, Darwin argues either that Aristotle's notion of "final cause" ("purpose" in the vernacular) has no place in science (since maximization of life only represents a result of struggle); or he states that we may continue to use the term "final cause" in an informal sense, so long as we acknowledge the underlying mechanism, or efficient cause, producing the phenomenon. (We may, in this case, view maximization as a "final cause" so long as we recognize its origin in struggle, and not in created harmony.) The same terminological ambiguity continues today in evolutionary theory. We use the language of final cause, or purpose, in describing adaptation, if only because we do not wish to abandon ordinary language. We say that giraffes evolved long necks "in order to" eat high foliage. But we recognize the causal basis of such adaptive "purpose" in natural selection by unconscious organismic struggle.

So Darwin recognized in early 1855 that maximization would have to be explained by natural selection (a "mere result from struggle"); he also stated that development of such an argument would be complex and difficult ("I must think out last proposition"). The "principle of divergence of character," or more succinctly the "principle of divergence," emerged as the result of this intellectual labor. How, then, did Darwin finally render maximization of life as a consequence of struggle, or ordinary natural selection?

Darwin's solution, embedded as Kohn notes (1985) in his increasing willingness to accept sympatric speciation, holds that natural selection will generally favor the most extreme, the most different, the most divergent forms in a spectrum of variation emanating from any common parental stock. Thus, each vigorous and successful stock produces a cone of varying forms about its own modal design (see Fig. 3-5 on p. 242). If natural selection generally favors extreme variants in such arrays—the core claim of the "principle of divergence of character"—then vigorous ancestors will generate two or more descendant taxa fanning out towards maximally different form and adaptation. Two sequelae now complete the argument by drawing both ecological plenitude and taxonomic structure from the principle of divergence: First, the process of divergence must continue (see Fig. 3-5), impelling each vigorous descendant to produce still more advantageous extremes—thereby entraining phyletic trends of constantly increasing specialization. (The full extension elevates subspecies to species, species to genera, etc.—as extreme variants proliferate and diversify. The taxonomic tree of life emerges as an ultimate result.) Second, descendants will, in general, be competitively superior to parents, and must therefore tend to exterminate them in competition—for the number of species cannot increase indefinitely, and some ecological mechanism for replacement of ancestors must exist.

In the *Origin*, Darwin begins, in his characteristic fashion, by analogy to artificial selection. Breeders, he argues, tend to favor extreme variants when trying to improve a stock; nature must follow suit (1859, p. 112). For the breeder's conscious aim, Darwin substitutes the natural advantages of extreme variants: "The more diversified the descendants from any one species become in structure, constitution, and habits, by so much will they be better enabled to seize on many and widely diversified places in the polity of nature, and so be enabled to increase in numbers" (1859, p. 112). With a botanical example, Darwin then strongly argues that divergence occurs because natural selection tends to favor extreme variants:

We well know that each species and each variety of grass is annually sowing almost countless seeds; and thus, as it may be said, is striving its utmost to increase its numbers. Consequently, I cannot doubt that in the course of many thousands of generations, the most distinct varieties of any one species of grass would always have the best chance of succeeding and of increasing in numbers, and thus of supplanting the less distinct varieties; and varieties, when rendered very distinct from each other, take the rank of species (1859, pp. 113–114).

Parental forms will then tend to succumb because natural selection favors their extreme and divergent descendants in competition: "As in each fully stocked country natural selection necessarily acts by the selected form having some advantage in the struggle for life over other forms, there will be a constant tendency in the improved descendants of any one species to supplant and exterminate in each stage of descent their predecessors and their original parent" (1859, p. 121).

All evolutionists know that the *Origin of Species* contains only a single figure. This statement has been endlessly repeated in textbooks and lectures, but the true significance of this figure remains obscure, because we nearly always misinterpret the diagram (Fig. 3-5). We read this sole figure as Darwin's basic illustration of evolution as a branching process. But Darwin did not construct his diagram for such a general purpose. Rather, he devised this unique figure to provide a surgically precise description of the principle of divergence, accompanied by several pages of explanatory text (pp. 116–126). Note how only two species of the original array (A–L) ultimately leave descendants—the left extreme A and the near right extreme I. Note how each diversifying species first generates an upward fan of variants about its modal form, and how only the peripheral populations of the fan survive to diversify further. Note that the total morphospace (horizontal axis) expands by divergence, even though only two of the original species leave descendants. Darwin writes (1859, p. 121): "In each genus, the species, which are already extremely different in character, will generally tend to produce the greatest number of modified descendants; for these will have the best chance of filling new and widely different places in the polity of nature: hence in the diagram I have chosen the extreme species (A) and the nearly extreme species (I), as those which have largely varied, and have given rise to new varieties and species." Darwin also states that the success of extremes records the action of natural selection in its usual mode of organismic struggle: "And here the importance of the principle of benefit being derived from divergence of character comes in; for this will generally lead to the most different or divergent variations (represented by the outer dotted lines) being preserved and accumulated by natural selection" (1859, p. 117).

### THE FAILURE OF DARWIN'S ARGUMENT AND THE NEED FOR SPECIES SELECTION

I apologize to readers for this laborious (though, I trust, not uninteresting) exposition of Darwin on divergence, but I have now reached the crux both of this argument and, in one sense, of this entire book as well. I am advocating both the necessity and importance of a hierarchical expansion of the theory of natural selection, defending this position by combining the standard techniques of validation in science and scholarship: empirical example, the logic of argument, and historical illustration. Why, then, should so much space be accorded to Darwin's views on divergence of character—especially since I have just documented Darwin's attempt to render this second keystone of his theory as a consequence of ordinary natural selection at the organismic level?

I have proceeded in this way for a simple reason: Darwin's argument doesn't work, and he came to recognize this failure in the face of his brave attempt. Darwin struggled mightily to render this second keystone of his full theory by natural selection alone, but he could not carry the logic to completion. He failed because his full argument demands a major contribution from species level selection (or, at the very least, strong attention to explicit sorting at the species level). I don't know that Darwin ever grasped this need in a fully explicit way, committed as he was to the exclusivity of selection on organisms. But he recognized the crucial difficulty at several places in his exposition; and, with his usual honesty, he made his distress palpable again and again.

I perceive his discomfort in the labored description of divergence given in the *Origin*—15 pages for a few points repeated many times, first in one way, then in another, all in a book so compressed that Darwin wanted to include the word "abstract" in the title. (John Murray, his publisher, demurred on obvious practical grounds). I sense Darwin's malaise in the fact that for this concept alone (among all the complex ideas developed in the *Origin*) he supplied both a figure and a meticulous "caption" (as it were), running for nearly ten pages. I note his dissatisfaction in the frequent shifting of attribution within his text—from consequences of organismic struggle (his usual and distinctive argument) on the one hand, to advantages for higher level units (usually species) on the other. In my reading, these shifts cannot be interpreted as comfortable transitions rooted in a confident reduction of higher level phenomenology to lower level causality (as he had achieved in explaining adaptation by natural selection), but must instead be regarded as genuine gropings and confusions. Finally, Darwin recorded his distress in explicit exclamations of doubt—from his "I must think out [this] last proposition" of the 1855 note, to his description of divergence as "this rather perplexing subject" in the *Origin* (1859, p. 116).

If the founder of the non-hierarchical organismal view, this doggedly persistent, fiercely honest and brilliant thinker, tried so hard to make the canonical argument of natural selection work for the central higher-level phenomenon of species diversity—and could not bring the logic of his argument to a satisfactory completion—then perhaps his failure tells us something about the necessity of hierarchical selection. Darwin, by his own formulation, faced two great issues—adaptation and diversity. He tried to render both by natural selection based on struggle among organisms—adaptation by the Malthusian insight of 1838, diversity by the principle of divergence formulated in the mid 1850's. His explanation worked well, or at least sprung no logical holes, for adaptation. But he could not carry the same argument through, despite extensive and valiant attempts, for diversity—the primary domain of species selection, as all modern advocates hold (see, for example, Gould and Eldredge, 1988, replying to Maynard Smith's 1988 misconception).

Schweber (1980, 1985, 1988) noted Darwin's trouble with discordance between levels, though he does not provide the technical arguments detailed below. In Schweber's view, Darwin was driven to formulate an argument that "does not cohere" (Schweber, personal communication) because his century's ignorance of hereditary mechanisms drove him to describe variation within species and varieties treated as units, while the causal structure of natural selection rested upon individual organisms. Arguments about organisms and species are not comfortably intertwined or mutually supporting within Darwin's conceptual structure: rather, the two levels remain discordant and inadequately (if not illogically) bridged. "This difference in the 'units' used is important. It accounts for the fact that at times levels of description were interchanged and some confusion necessarily crept in" (Schweber, 1980, p. 240). "There was no link between adaptation and speciation, except whatever could be supplied by a quasi-historical developmental idea of optimizing the amount of life" (ibid., pp. 287–288). "The problem of the different levels of descriptions was confined to how the properties of variations in individuals . . . were responsible for the assumed variability characteristic of varieties and species. This problem Darwin never solved" (ibid., p. 288).

We can exemplify Schweber's perceptions about Darwin's incoherence of argument by dissecting the logic of Darwin's attempt to use ordinary natural selection as the basis of divergence. For three basic reasons, his attempt to invoke selection among organisms as an explanation for patterns in speciation and extinction—the heart of the "principle of divergence," and the *primum desideratum* for a complete theory of natural selection—fails because the level of species must be addressed both directly and causally, while Darwin's rationale for explanation from below includes gaps and fatal weaknesses.

### The calculus of individual success

Darwin treats the principle of divergence in two extensive discussions—the long and even labored account of chapter 4 in the *Origin of Species* (1859, pp. 111–126), and the even more detailed exposition intended for the "big species book" that Wallace interrupted and Darwin never published because he rushed to compose the *Origin* instead. The manuscript for most of this larger project survives, including the full discussion of divergence intended for chapter 6 "On Natural Selection." This text, published under R. C. Stauffer's editorship in 1975, treats the principle of divergence on pages 227–251.

When, in the 1970's, I first read the *Origin* with the notion of hierarchical selection in mind, I was fascinated by Darwin's struggle to bridge the levels, and his ultimate lack of success. Schweber speaks of "incoherence"; I would rather describe Darwin's "moves" of argument as an oscillation between one mode and the other. In some passages (including those cited above), Darwin speaks of ordinary natural selection and the advantages enjoyed by extreme variants. In others, he judges the success of a parental form not by the vigor or competitive prowess of offspring, but by the number of descendant species emanating from a root stock.

These themes could, of course, be complementary. One perspective might imply and grade into the other; the two levels could represent alternate solutions of the Necker cube (see Dawkins, 1982)—that is, views of the same configuration from different vantage points. The individual success of extreme organisms may simply imply, *ipso facto* and necessarily, the ultimate multiplication of species, and success measured by number of descendant taxa might therefore act as a surrogate for, or simple extension of, natural selection. (For this reason, I assume, Darwin includes the principle of divergence as the largest section of chapter 4, entitled natural selection.)

But this argument does not cohere, and I think that any careful reader must be struck by Darwin's discomfort as he mixes and juggles the argument for success of extreme organismal variants with the calculus of advantage mapped by number of descendant taxa. Darwin's argument falters because the use of a lower level (success of extreme variants, in this case) to explain a phenomenon at a higher level (multiplication of species) can only work if "perfect transfer" can be defended—that is, if the lower level entails the higher as a direct consequence without any intervention (even a synergistic boost, not to mention a contrary force) from causes at the higher level itself. Darwin understood this principle perfectly well. Indeed, he was probably the only man who, in this infancy of evolutionary science, had carefully and consistently thought the logic of selection through to this correct interpretation. Thus, Darwin tried to construct an argument for perfect transfer—but he failed. Darwin advanced these claims in *Natural Selection,* but ultimately dropped them from the *Origin,* because (I suspect) he recognized their weaknesses. (I criticize the arguments for perfect transfer in points two and three below; in this subsection, I document Darwin's bold claim for a higher level calculus of individual success.)

Consider Darwin's first hypothetical example of the principle of divergence in the *Origin* (1859, p. 113). He speaks of a "carnivorous quadruped" that, by ordinary natural selection, has expanded in population to the limits of local environments. To do even better in the struggle for life, this form must now diversify into several descendant taxa. But how can the canonical argument for natural selection be cashed out in terms of multiplicity of descendant species? The logic of individual struggle carries no implications about the splitting of populations (especially for an *in situ* sympatric splitting that implies general predictability rather than the simple good fortune of geographic isolation).

Take the case of a carnivorous quadruped, of which the number that can be supported in any country has long ago arrived at its full average. If its natural powers of increase be allowed to act, it can succeed in increasing (the country not undergoing any change in its conditions) only by its varying descendants seizing on places at present occupied by other animals: some of them, for instance, being enabled to feed on new kinds of prey, either dead or alive; some inhabiting new stations, climbing trees, frequenting water, and some perhaps becoming less carnivorous. The more diversified in habits and structure the descendants of our carnivorous animals became, the more places they would be enabled to occupy (Darwin, 1859, p. 113).

Darwin continually invokes the calculus of individual success by *number* of descendant taxa (see also 1859, p. 116, and Stauffer, ed., 1975, p. 228): "As a general rule, the more diversified in structure the descendants from any one species can be rendered, the more places they will be enabled to seize on, and the more their modified progeny will be increased" (Darwin, 1859, p. 119).

Often, he mixes both criteria—the adaptive success of extreme variants in struggle, and the calculus of descendant taxa—in a single statement.

> Here in one way comes in the importance of our so-called principle of divergence: as in the long run, more descendants from a common parent will survive, the more widely they become diversified in habits, constitution and structure so as to fill as many places as possible in the polity of nature [organismic level], the extreme varieties and the extreme species will have a better chance of surviving or escaping extinction, than the intermediate and less modified varieties or species [taxon level]. But if in a large genus we destroy all the intermediate species, the remaining forms will constitute sub-genera or distinct genera, according to the almost arbitrary value put on these terms (in Stauffer, ed., 1975, p. 238).

But are these two statements really equivalent? Does the lower level claim for organismic advantage imply the higher level phenomenon of species proliferation without reference to any higher level causes?

In his most striking passage, indicating that he did grasp the need for higher level sorting based upon such group properties as the range of variation, Darwin attributes the success of introduced placentals in Australia not, as we might anticipate from ordinary natural selection, to the adaptive biomechanical superiority of placental design (honed in the refiner's fire of more severe competition in Eurasia and America), but to a greater range of placental variation across taxa, produced by their later stage in the historical process of divergence.

> A set of animals, with their organization but little diversified, could hardly compete with a set more perfectly diversified in structure. It may be doubted, for instance, whether the Australian marsupials, which are divided into groups differing but little from each other, and feebly representing, as Mr. Waterhouse and others have remarked, our carnivorous, ruminant, and rodent mammals, could successfully compete with these well-pronounced orders. In the Australian mammals, we see the process of diversification in an early and incomplete stage of development (Darwin, 1859, p. 116).

### The causes of trends

Trends represent the primary phenomenon of evolution at higher levels and longer time scales. Trends therefore pose the key challenge, the ultimate making or breaking point, for extrapolationist theories that seek the causes of macroevolution in microevolutionary processes centered upon organismic selection. Darwin understood and accepted this challenge; his principle of divergence marks his attempt to depict trends as extrapolated results of natural selection. (The principle of divergence attempts to explain morphological trends by specialization and progressive departure from ancestral form, and also to account for numerical trends by multiplication of some taxa at the expense of others within a clade.)

Advocates of species selection hold that trends must be described as the differential birth and death of species (not the simple anagenetic extrapolation of change within a population), and that the causes for such differentials must be sought, at least in part, in irreducible species-level fitness (see Chapter 8). The standard extrapolationist rejoinder invokes two arguments: (1) Differential death and survival rather than differential birth (of species) usually fuels trends. The death and persistence of groups can be reduced more easily to organismic competition, while differential production of species more often demands irreducible causes, for an organism cannot speciate by itself, while the death of a population may represent no more than the accumulated demise of all organisms. (2) The cause of differential survival or death must be reducible to ordinary natural selection.

Darwin did not offer his principle of divergence as a rejoinder to any explicitly developed theory of species selection, for no such formulation existed when he wrote. But he understood the logical requirements of his theory so well that he provided the necessary rationale without the spur of a formally stated alternative. He also, and uniquely, reinforced his argument with an illustration of the need for differential survival *of certain kinds of variants* within random arrays. In *Natural Selection,* Darwin presents his case as a *second figure* (reproduced here as Fig. 3-6) that he did not include in the abridged *Origin of Species* in 1859. (Virtually no one knew about the existence of this figure or argument until Stauffer published the manuscript of *Natural Selection* in 1975.)

The basic figure of both *Natural Selection* and the *Origin* illustrates Darwin's claim that only a few vigorous species will produce the variants leading to the "recruitment" of new species. (These vigorous species are the extreme forms favored by ordinary natural selection—A and M in *Natural Selection,* A and I in the *Origin,* see Figs. 3-5 and 3-6.) The variants of these vigorous species radiate in an even fan, or random array, about the modal form of their ancestor. A trend then arises because ordinary natural selection favors extremes within the fan. Darwin recognizes that a trend to specialization and diversity cannot be generated only by the greater vigor of extreme species in the initial array; he must also defend a second proposition about differential survival among offspring of these favored extremes.

The second figure of *Natural Selection* now comes into play (Fig. 3-6). Darwin shows us what happens under a regime of random survival within the fan of variants generated by the favored vigorous species at ecological extremes, as opposed to a regime of selection positively directed towards extreme mem-

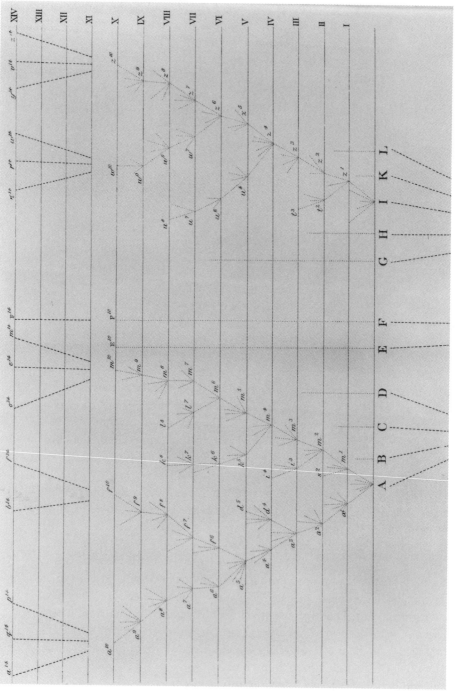

3-5. The famous and only diagram published by Darwin in the *Origin of Species*, 1859. Darwin did not construct this much misunderstood diagram as a simple and general model of branching phylogeny, but quite specifically as an illustration of his

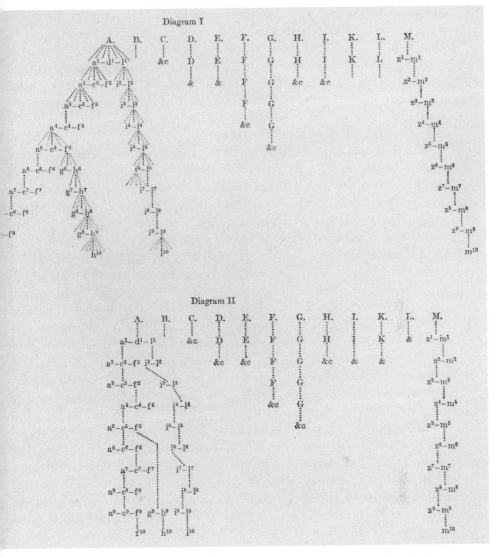

3-6. The expanded version of Darwin's figure, drawn in preparation for the long version, *Natural Selection*, that was never published because Darwin hurried to complete the shorter *Origin of Species* after receiving Wallace's manuscript. This version was not published until 1975 (see Stauffer, 1975). Here, Darwin shows us how the expectation of enhanced success for extreme variants within each fan, as predicted by his Principle of Divergence, will lead to trends (upper part of diagram); whereas random survival of variants (lower part of diagram) yields no trend and no ecological expansion. See text for details.

bers of each fan. The trend to diversification halts in the random regime. The final products need not become any more distinct than the initial parents ($f^{10}$, $h^{10}$, and $l^{10}$ lie right under ancestor A, while $m^{10}$ does not differ from parental M). On the other hand, differential survival of extremes *within* the fans will produce trends (also seen in Fig. 3-6). Darwin writes of the second diagram (the lower half of Fig. 3-6), where variation to the left and right of A represents greater or lesser adaptation to drought in plants: "Everything is the same as in diagram I . . . except that it is left to mere chance in each stage of descent, whether the more or less moisture loving varieties are preserved; and the result is, as graphically shown, that $a^{10}$ and $l^{10}$ [sic, he has no $a^{10}$ in the drawing, but represents the leftward extreme as $f^{10}$] differ in this respect; and so in other respects, hardly more than did the first varieties ($a^1$ $l^1$) which were produced" (in Stauffer, ed., 1975, p. 244).

The argument for a trend that can be reduced to natural selection therefore hinges upon reasons for differential survival of extremes *within* the fans of varying species; for the trend cannot emerge simply from the greater evolutionary vigor of the ancestral extremes themselves. (Interestingly, Darwin never considers the alternative, more congenial to species selection, of greater *production* of variants at the extremes, with random survival within fans.) Darwin now makes his crucial move for ordinary natural selection, using his principle of divergence: extremes enjoy differential survival within the fans of variants, because natural selection favors adaptation to peripheral, over adaptation to central, "stations" in any region. (We must remember that all members of the fan are well adapted to their own local bits of the environment). Now, at the crux of his development, Darwin tries to defend his position on the differential value of extreme stations, and his argument falls apart—to be rescued only with a forced and self-contradictory *ad hoc* hypothesis (explicitly stated in *Natural Selection*, but wisely omitted from the *Origin*).

Darwin provides two potential reasons for differential success of organisms adapted to extreme environmental stations. The first remains perfectly acceptable, and would pass muster today as a standard ecological argument featured in all textbooks—reduced competition in less "crowded" extreme environments: "From our principle of divergence, the extreme varieties of any of the species, and more especially of those species which are now extreme in some characters, will have the best chance, after a vast lapse of time, of surviving; for they will tend to occupy new places in the economy of our imaginary country" (in Stauffer, ed., 1975, p. 239).

If Darwin had stopped here, his argument would have remained consistent, if dangerously weak. But his relentless probing would not permit such a course—for he knew that a key problem remained unsolved:* extreme variants may be favored in their own extreme environments, but why should they

---

*If historians and historically-minded scientists, myself included, often develop an admiration bordering on reverence for Darwin, our judgment arises from his persistent thoroughness, his insistence on following a train of complex thought into all ramifications and difficulties, and his internal need to resolve each and every little puzzle before achieving satisfaction. Darwin therefore, over and over again, provides resolutions to puzzles that none of his contemporaries even considered or conceptualized. In this sense, no other evolutionist of his generation came close to rivaling Darwin in sophistication—and extensive logical sloppiness permeates the work of many other thinkers. Darwin never resolved several difficult issues (progress, divergence), but he thought about them with almost chilling clarity and integrity.

prevail over other variants in more central environments? After all, the more central ancestor continues to survive after the extreme variant buds off. The ancestor should therefore be favored in its own central environment, and the descendant in its new peripheral station. Why, then, should the descendant ever replace the ancestor—so long as central environments persist along with marginal places? In all his writing on divergence, Darwin recognized that trends to specialization could not occur unless extreme descendants tended to wipe out more central ancestors in competition: trends, in other words, required a pattern of differential extinction as well, for the number of species in a region cannot increase indefinitely.

And here, after so much effort and careful development, Darwin bogged down. For this most resolutely higher level phenomenon of the supposed differential success of extreme *vs.* central species, Darwin could not provide a tenable argument based upon natural selection. With evident discomfort, Darwin resorted to an *ad hoc* assumption: he argued that while extreme variants adapt to their marginal stations, they also retain *all the adaptations* of their parents for the original central habitats. Thus, the descendant extremes remain as good as their parents in the ancestral environment, while adding a capacity for survival in marginal habitats.

But how can such a proposition be defended? Why should a species that has left one environment, and explicitly adapted to another, still retain all its prowess in an environment no longer inhabited (and from which it has actively diverged)? Not only does this proposition make no sense *prima facie;* such a claim also contradicts the canonical argument (often embraced by Darwin and his contemporaries) that specialization leads to "locking in" and decreased flexibility. In short, Darwin knows that he has run into a severe logical problem in trying to justify a central implication of his general argument: the differential survival of extreme taxa with a consequently preferential extinction of central species. How can such a pattern be explained—for central and marginal species should not, after all, be in overt competition, and central environments cannot be regarded as generally more evanescent than extremes? Darwin therefore invokes his *ad hoc* argument (described just above) for an expanded range of adaptation in extreme species, in order to place the organisms from these extremes into competition with their parents, thus generating a hypothetical explanation for differential parental death in terms of natural selection.

Darwin begins by stating his *ad hoc* assumption: "As $m^1$ tends to inherit all the advantages of its parent M [see right side of Fig. 3-6], with the additional advantage of enduring somewhat more drought, it will have an advantage over it, and will probably first be a thriving local variety, which will spread and become extremely common and ultimately, supplant its own parent" (in Stauffer, ed., 1975, p. 239).

But Darwin immediately senses a problem and recognizes that descendants might not retain the parental range, and that ancestors might survive the onslaught of their phyletic children by living in a different station, thereby avoiding competition. "If $m^{1-10}$ had been produced, capable of enduring more drought, but not at the same time enduring an equal amount of moisture with the parent M, both parent and modified offspring might co-exist: the parent (with perhaps a more restricted range) in the dryer stations, and $m^{1-10}$ in the very dryest stations" (ibid., p. 240).

Darwin then rejects this reasonable but debilitating scenario with his *ad hoc* assumption, though he senses the weakness of his proposal and salvages his argument almost apologetically, especially at the end:

> In the imaginary case of the varieties $m^{1-10}$ which are supposed to inherit all the characters of M, with the addition of enduring more drought; these varieties would inhabit stations, where M could not exist, but in the less dry stations $m^{1-10}$ would have very little power of supplanting their parent M; nevertheless during unusually dry seasons $m^{1-10}$ would have a great advantage over M and would spread; but in damper seasons M would not have a corresponding advantage over $m^{1-10}$ for these latter varieties are supposed to inherit all the characters of their parent. So there would be a tendency in $m^{1-10}$ to supplant M, but at an excessively slow rate. It would be easy to show that the same thing might occur in the case of many other new characters thus acquired; but the subject is far too doubtful and speculative to be worth pursuing (ibid., pp. 241–242).

### Species selection based on propensity for extinction

Relentlessly probing as usual, Darwin now identifies another weak point in his argument. If divergence follows this predictable and necessary pattern, given the propensity of natural selection to favor extreme variants in all directions, then what prevents this inexorable process from reaching absurdity in a final state of such precise and extended diversification that each species contains but a single individual? (This issue became important for Darwin when he moved from his earlier allopatric view of speciation to embrace a largely sympatric model with an intrinsic and predictable "motor" for the generation of species by selection of extreme variants): "But if the time has not yet arrived, may it not at some epoch come, when there will be almost as many specific forms as individuals? I think we can clearly see that this would never be the case" (in Stauffer, ed., 1975, p. 247).

Darwin proposes three reasons for nature's avoidance of such an absurd outcome, the first conventional, but the second and third invoking species selection on population properties of size and variability. For the first reason, Darwin cites ecological notions that have since become standard—limiting similarity and a restricted number of "addresses" in the economy of nature. Diversity does beget more diversity, and the physical environment sets no strict *a priori* upper bound. But limits imposed by "inorganic conditions" will eventually cause selection to rein in the intrinsic process of ever finer diversification:

> Firstly, there would be no apparent benefit in a greater amount of modification than would adapt organic beings to different places in the polity

of nature; for although the structure of each organism stands in the most direct and important relation to many other organic beings, and as these latter increase in number and diversity of organization, the conditions of the one will tend to become more and more complex, and its descendants might well profit by a further division of labor; yet all organisms are fundamentally related to the inorganic conditions of the world, which do not tend to become infinitely more varied (ibid., p. 247).

But Darwin also recognizes that ecological limits may not be sufficient to restrain diversification, and he advances two other mechanisms, based on species selection against traits of small populations. First, Darwin argues that long before diversity reaches a physical limit (as quoted just above), species selection against finely divided taxa with consequently small populations will balance ordinary natural selection for further diversification: "If there exist in any country, a vast number of species (although a greater amount of life could be supported) the average number of individuals of each species must be somewhat less than if there were not so many species; and any species, represented by but few individuals, during the fluctuation in number to which all species must be subject from fluctuations in seasons, number of enemies, etc., would be extremely liable to total extinction" (ibid., pp. 247–248).

Finally, Darwin cites another reason for species selection against populations of unusually small size. Such populations are not only more prone to extinction; they are also less subject to further speciation because such a restricted number of individuals per population implies an insufficiency of opportunity for the origin of rare favorable variants: "Lastly we have seen . . . that the amount of variations, and consequently of variation in a right or beneficial direction for natural selection to seize on and preserve, will bear some relation within any given period, to the number of individuals living and liable to variation during such period: consequently when the descendants from any one species have become modified into very many species, without all becoming numerous in individuals, . . . there will be a check amongst the less common species to their further modification" (ibid., p. 248).

In a lovely closing metaphor, Darwin provides a fine description of the central hierarchical concept of balance produced by negative feedback between levels. Number of species will equilibrate at a stable level of diversification when positive selection at the organismal level (Darwin's argument for the advantages of extreme variants) becomes balanced by negative selection at the species level (disadvantages of small population size): "the lesser number of the individuals," Darwin writes, "serving as a regulator or fly-wheel to the increasing rate of further modification, or the production of new specific forms" (ibid., p. 248).

In summary, I have documented both Darwin's discomfort with his forced attempt to explain the primary species-level phenomenon of diversity by natural selection of extreme organismal variants, and his inability to complete the argument without an explicit invocation of species selection against taxa with small populations. His attempt to render his "principle of divergence" by organismal selection alone founders on three bases, the first two negative in logical barriers forced by the premise of organismal exclusivity, and the third positive in a potential "rescue" sought by acknowledging a necessary role for species selection: (1) He promotes a calculus of ultimate organismal success in terms of number of descendant taxa, but cannot extend his argument for diversification by selection of extreme variants to achieve the required perfect transfer to the species level. (2) In order to explain trends, he backs himself into a contradictory and *ad hoc* explanation for the elimination of ancestral forms by natural selection in competition with descendants. (3) Faced with the logical dilemma of runaway diversification under natural selection alone, he advocates negative species selection based upon small population sizes to bring the process of divergence into equilibrium.

## POSTSCRIPT: SOLUTION TO THE PROBLEM OF THE "DELICATE ARRANGEMENT"

The power of a new framework often becomes most apparent in its capacity to solve small and persistent puzzles. I therefore end this section with a solution to an old conundrum, and with another refutation for Brackman's (1980) claim that Darwin pinched the principle of divergence from Wallace, and then lied to cover up the theft. What did Wallace say about divergence? Did he really develop the principle in useful ways that Darwin had not anticipated, and might have coveted as his own? When we turn to Wallace's celebrated Ternate paper (sent to Darwin in 1858), we find only a cursory statement about divergence. Wallace only discusses anagenetic trends of descendants away from ancestors. He does not even consider the production of multiple taxa from single sources: "But this new, improved, and populous race by itself, in course of time, gives rise to new varieties, exhibiting several diverging modifications of form, any of which, tending to increase the facilities for preserving existence, must, by the same general law, in their turn become predominant. Here, then, we have *progression* and *continued divergence* deduced from the general laws which regulate the existence of animals in a state of Nature, and from the undisputed fact that varieties do frequently occur" (Wallace, 1858, in Barrett et al., 1987).

We must conclude that Wallace regarded a principle of divergence as "no big deal." He grasped the idea in outline and apparently found no problem therein. His short statement could not possibly have taught Darwin anything useful, for Darwin had already carried the argument far beyond this basic comment. Why then did Wallace fail to share Darwin's puzzlement, excitement and sense of complexity about the principle of divergence? I can imagine two explanations. Either Wallace simply didn't think the issue through to all the difficulties and implications that Darwin recognized. (After all, malarial fits on Ternate are less conducive to deep thought than years of protracted strolling around the sandwalk at Down.) Or he did think the issue through and, finding nothing problematical, therefore devoted little attention to the subject. If the second alternative is correct, then my framework for considering the principle of divergence as a problem in levels of selection can resolve Wallace's puzzling lack of appreciation.

As Kottler (1985) has shown, Wallace and Darwin were not identical peas from the pod of natural selection. They battled long and hard on several crucial issues, mostly involving Wallace's panselectionism vs. Darwin's more subtle view of adaptation (Gould, 1980d). One key area of disagreement centered upon the target of natural selection. Darwin labored to work out a consistent theory that virtually restricted selection to struggles among organisms (see Chapter 2 for his interesting reasons). Wallace, as Kottler shows, never grasped the centrality, or even the importance, of the issue of levels and agency for a theory of natural selection. He moved from level to level as the situation seemed to imply, choosing whatever target of selection would best support his panselectionist leanings. (For example, he ascribed hybrid sterility to species selection in order to preserve his conviction that features of such importance must originate as active adaptations; whereas Darwin, committed to a consistent theory of organismal selection, regarded hybrid sterility as an incidental side consequence of accumulated differences arising by ordinary selection in two initially isolated lineages—see p. 131.) Thus, if Wallace ever pondered the principle of divergence to the point of recognizing an issue in levels of selection, he would not have responded, as Darwin did, with such a sustained, almost impassioned, quest for resolution. Wallace would not have identified any problem at all, for he never grasped the thorny issue of a need to specify levels in the first place. A simple statement about divergence would have sufficed—as Wallace indeed provided. Darwin, in his overreliance upon organismal selection, may never have reached the finish line in explicating the principle of divergence; but Wallace scarcely got off the blocks.

## Coda

No one would argue that persistence in history makes anything right or even worthy—lest cruelty, murder and mayhem win our imprimatur by a misplaced criterion of longevity. Still, in the world of ideas, long pedigrees through disparate systems, and recurrence in the face of attempted avoidance, usually signify something about the power of an idea, or its necessary place in the logic of a larger enterprise.

Causal—and not merely descriptive—accounts of hierarchy have infused evolutionary biology in this way from the beginning of our subject. Lamarck initiated this conceptual rubric with a version in the invalid mode of causal differences based on opposition between levels. Darwin knew what he didn't like about this style, and his theory—preserved unchanged to our own orthodox commitments today—sinks a strong foundation in an active rejection of hierarchy in this inherently combative mode. Weismann and Darwin himself—the two greatest evolutionists and deepest thinkers with an explicit commitment to the single-level theory of natural selection—tried to extend the logic of this idea to encompass every important issue in evolution. Both became stymied, and eventually surrendered, with pleasure in Weismann's case at least, to the need for hierarchy in devising any complete system for the logic of evolutionary explanation in selectionist terms—Weismann on subcellular selection for explaining trends, Darwin on species selection for encompassing diversity. Both men accepted causal hierarchy in the modern and valid sense of similar forces working in distinct ways at different levels.

Very few evolutionists know anything about this history, and may therefore doubt the importance of the subject. But downgrading on this criterion would represent a great mistake rooted in the conservative premise that anything vital must be easily visible in all contexts. Anyone tempted to accept such a basis for dismissal should consider the conventional tales of rulers and conquerors—virtually the only subject matter of so many secondary school and undergraduate history courses—and recognize what modern scholars have taught us by probing the hitherto invisible pathways of daily life among ordinary people. Compare the overt and conventional history of diplomacy with the often more potent, but academically invisible, history of technology.

Evolutionary thought began with hierarchy, wrongly conceived. Our canonical theory of natural selection arose as an attempted rebuttal. The most brilliant practitioners of that theory could not bring the argument to completion; so both Weismann and Darwin brought hierarchy back, in a valid style this time, to render the theory of selection both coherent and comprehensive. I previously offered a choice of proverbs: you may view hierarchy as a bad penny or a pearl of great price. But hierarchy, like the poor, has always been with us (and, perhaps, shall inherit the earth as well!). This situation can only recall James Boswell's famous statement about one of Dr. Johnson's colleagues who lamented that he had tried to be a philosopher, but had failed because cheerfulness always broke through. Too many of us have tried to be good Darwinian evolutionists, and have felt discouraged because hierarchy always breaks in. I suggest that we rejoice—with good cheer—and welcome an underappreciated and truly indispensable old friend.

# CHAPTER FOUR

# Internalism and Laws of Form: Pre-Darwinian Alternatives to Functionalism

## Prologue: Darwin's Fateful Decision

Thinking in dichotomies may be the most venerable (and ineluctable) of all human mental habits. In his *Lives and Opinions of Eminent Philosophers* (circa A.D. 200), Diogenes Laertius wrote: "Protagoras asserted that there were two sides to every question, exactly opposite to each other."

Darwin follows this tradition of dichotomy in a passage that he earmarked for special impact as the concluding paragraph of his crucial Chapter 6, "Difficulties on Theory." I regard this passage as among the most important and portentous in the entire *Origin,* for these words embody Darwin's ultimate decision to construct a functionalist theory based on adaptation as primary, and to relegate the effects of constraint (a subject that also commanded his considerable interest—see Section IV of this chapter) to a periphery of low relative frequency and subsidiary importance. Yet this passage, which should be emblazoned into the consciousness of all evolutionary biologists, has rarely been acknowledged or quoted. Darwin begins (1859, p. 206), expressing his alternatives in upper case (and using the categories of the great debate between Cuvier and Geoffroy—see Section III of this chapter): "It is generally acknowledged that all organic beings have been formed on two great laws—Unity of Type, and the Conditions of Existence."

Conditions of Existence, of course, expresses the principle of adaptation—final cause or teleology to pre-evolutionists. Organisms are well designed for their immediate modes of life—and intricate adaptation implies an agent of design, either an intelligent creator who made organisms by fiat as an expression of his wisdom and benevolence, or a natural principle of evolution that yields such adjustment between organism and environment as a primary result of its operation. (Both Darwinian natural selection and Lamarckian response to perceived needs, for example, build adaptation as the most general consequence of their basic mode of action.)

Darwin then continues by defining the other side of the classical dichotomy: Unity of Type (1859, p. 206): "By unity of type is meant that fundamental agreement in structure, which we see in organic beings of the same class, and which is quite independent of their habits of life." In another critically placed passage, introducing the subject of "Morphology" in Chapter 13, Darwin waxes almost poetic about unity of type (p. 434): "This is the most interesting department of natural history, and may be said to be its very soul. What can be more curious than that the hand of a man, formed for grasping, that of a mole for digging, the leg of the horse, the paddle of the porpoise, and the wing of the bat, should all be constructed on the same pattern, and should include the same bones in the same relative proportions."

These two principles have always dwelled together in exquisite tension. Any complete account of morphology must call upon both phenomena, for most organisms are well adapted to their immediate environments, but also built on anatomical ground plans that transcend any particular circumstance. Yet the two principles seem opposed in a curious sense—for why should structures adapted for particular ends root their basic structure in homologies that do not now express any common function (as in Darwin's example of mammalian forelimbs)?

The designation of one principle or the other as the causal foundation of biology virtually defines the position of any scientist towards the organic world and its causes of order (see, especially, Russell's superb 1916 book on this dichotomy). Shall we regard the plan of high-level taxonomic order as primary, with local adaptation viewed as a set of minor wrinkles (often confusing) upon an abstract majesty? Or do local adaptations build the entire system from the bottom up? This dichotomy set the major debate of pre-Darwinian biology: does God reveal himself in nature primarily by the harmony of taxonomic structure, or by the intricacies of particular adaptations (see Section II, this chapter)? This dichotomy continues to define a major issue in modern evolutionary debates: does functional adaptation or structural constraint maintain priority in setting evolutionary pathways and directions (see Chapters 10–11)?

This issue of primacy between the two principles has held the central stage of natural history for so long that national traditions have developed, with continental preferences usually emphasizing unity of type (despite important exceptions like Georges Cuvier), and mainstream anglophonic science generally favoring adaptation (with exceptions for a few important pluralists like Richard Owen, or dissenters like William Bateson or D'Arcy Thompson). We often blunder in our historical understanding by assuming that evolution must be an ultimate watershed, marking a complete break between a bad before and an enlightened after. In fact, many continuities pass right through Darwin's rupture of history, with evolution only providing a different explanation for unaltered principles and phenomena. The good ship Dichotomy—Unity of Type vs. Conditions of Existence—entered the Darwinian current by converting its terms from a debate about God's primary mode of self-expression in nature to an argument about constraint and adaptation in evolution.

We cannot understand Darwin without grasping this fundamental continuity in national styles. As a young man, Darwin adored Paley's *Natural Theology* (see p. 116); later, in a courageous act of intellectual parricide, he constructed a theory that subverted Paley's mode of explanation. But Darwin never abandoned Paley's conviction that adaptation must be designated as the primary phenomenon of natural history. Darwin remained true to an English tradition stretching at least as far back as Robert Boyle and John Ray in the late 17th century of Newton's founding generation for modern science, running through Paley, the Bridgewater Treatises, Wallace, and Poulton in Darwin's own time, on to R. A. Fisher and finally to E. B. Ford, A. J. Cain, and R. Dawkins in the later 20th century.

When we properly place Darwin in this lineage, a genealogy unfractured by evolutionary theory, we can make sense of his fateful decision for resolving Unity of Type vs. Conditions of Existence at the end of Chapter 6—a choice faithful to Paley and the English tradition in reaffirming the primacy of adaptation. Darwin writes, in words that define the causal basis of his theory (and continuing from the previous quotation on p. 25):

> On my theory, unity of type is explained by unity of descent. The expression of conditions of existence, so often insisted on by the illustrious Cuvier, is fully embraced by the principle of natural selection. For natural selection acts by either now adapting the varying parts of each being to its organic and inorganic conditions of life; or by having adapted them during long-past periods of time: the adaptations being aided in some cases by use and disuse, being slightly affected by the direct action of the external conditions of life, and being in all cases subjected to the several laws of growth. Hence, in fact, the law of the Conditions of Existence is the higher law; as it includes, through the inheritance of former adaptations, that of Unity of Type (1859, p. 206).

Darwin's brilliant intellectual move clearly expresses the revolutionary impact of evolutionary explanations against the previous range of creationist paradigms. Creationist biology saw Unity of Type and Conditions of Existence, homology and adaptation, as opposite, but equally contemporary (or timeless), poles in a dichotomy of originating forces. Darwin literally added a new dimension to the debate—the axis of history. (And no intellectual expansion can be more profound than the introduction of a new dimension, orthogonal to previous modes of explanation.)

Thus, in this passage, Darwin makes a stunningly simple suggestion to break the impasse between Unity of Type and Conditions of Existence. (And yet, to be able to see anything at all in this clear and simple light, one must first grasp the revolutionary implications of evolution itself—the truly difficult intellectual transition out of Paley's world!) To be sure, the homologies of Unity of Type do not embody, and seem actively to oppose, current functions. Must Unity of Type therefore represent a principle of order dichotomously contrary to adaptation? In a world without history, where all features of organisms express their initially created state, the answer must be "yes." But the addition of history, by a theory of genealogical connection, permits (and even privileges) another interpretation. Suppose that Unity of Type records no mysterious groundplan of created design, but only the actual, retained form of a common ancestor at the base of a bush of descent? Then homology can be simply explained as passive retention in the genealogy of diversified descendants—not an archetype of intelligent design, but only the signature of history.

In this context of evolutionary reform, we may then inquire about the causes of these common ancestral structures in distant pasts. And Darwin now makes his fundamental choice by affirming fealty to the English lineage of adaptationist thought. He argues that ancestral structures, forming the great homologies of Unity of Type, initially arose, by natural selection, as adaptations to "organic and inorganic conditions of life" in ancestral environments. Thus, the dichotomous poles of Unity of Type and Conditions of Existence achieve a single and unified explanation under natural selection—as immediate adaptations to present environments (Conditions of Existence), or as adaptations to ancient environments, transmitted by inheritance to diversified descendants (Unity of Type). The old dichotomy, in fact, expresses no clash of opposites at all, but only marks the temporally sequential representations of one dominant principle in evolution—adaptation by natural selection. Thus, since adaptation embodies the principle of Conditions of Existence, and since adaptation builds both ends of the old dichotomy, Conditions of Existence becomes the victorious pole of the old contrast, in Darwin's own words the "higher law; as it includes, through the inheritance of former adaptations, that of Unity of Type."

Yet Darwin, far too sophisticated a thinker to embrace extreme positions, could not claim that natural selection and adaptation—though responsible for *both* poles of the old dichotomy—reigned exclusively in nature. Darwin knew that primary judgments in natural history must be rendered in terms of relative frequencies. Indeed, he had written as his last line in the Introduction to the *Origin of Species,* first edition: "I am convinced that Natural Selection has been the main but not exclusive means of modification" (1859, p. 6). He also reacted as strongly as his genial temperament ever permitted against those who charged him with false claims of exclusivity. In such cases, he usually cited this line from the *Origin* in vindication—as in his famous, almost rueful statement (*Origin,* 6th ed., 1872b, p. 395): "As my conclusions have lately been much misrepresented, and it has been stated that I attribute the modification of species exclusively to natural selection, I may be permitted to remark that in the first edition of this work, and subsequently, I placed in a most conspicuous position—namely, at the close of the Introduction the following words: 'I am convinced that natural selection has been the main but not the exclusive means of modification.' This has been of no avail. Great is the power of steady misrepresentation."

Thus, while extending natural selection to cover both poles of the old dichotomy between unity of type and conditions of existence, Darwin also listed the main supplements to selection among causes of evolutionary change: use and disuse, direct action of external conditions, and laws of growth. We reject the first two today, and Darwin also grants them little space by his qualifiers: "in some cases" and "being slightly affected." But Darwin put more store by the third—laws of growth—as indicated by his only positive qualifier: "being in all cases subjected to the several laws of growth." And we would offer the same judgment today, since laws of growth, under the more fashionable designation of "developmental constraints," have become a "hot topic" in evolutionary biology once again (see Chapter 10). And now we come to the Darwinian trope of argument, the ploy that makes this chapter (and, to a large extent, this entire book) necessary.

Darwin wrote his crucial closing paragraph of Chapter 6 to argue that Unity of Type should be subsumed under Conditions of Existence—for Unity of Type, he asserted, only expresses past episodes of ordinary adaptation and natural selection, subsequently inherited by numerous modern descendants. Unity of Type has always defined the main arena for naturalists who view adaptation as secondary, and some principle of morphological order (for many versions exist) as primary. Darwin removed the rationale for a separate principle of Unity of Type by noting that ancient adaptations would, if inherited throughout a subsequent lineage, become sources of deep homology. Yet he could not deny—and had no desire to subvert—the idea of morphological principles working separately from natural selection, and building exceptions to adaptation. In this sense, Darwin supported the concept of constraint, but only if this principle could be carefully circumscribed within a category distinctly subsidiary to natural selection in relative frequency and biological importance. Darwin fully understood the crucial role of relative frequency in evolutionary arguments, and he rested his case for natural selection squarely upon such a judgment of quantitative importance. In so doing, he pursued the following strategy: take the old dichotomy, and show that both poles arise as products of natural selection. Then, having removed constraint as the primary cause of one pole (where a high relative frequency could not have been denied), allow constraint to reenter as a subsidiary force to natural selection (with a consequent guarantee of low relative frequency). Natural selection then becomes the primary force of evolution. Recall the full title of Darwin's book: *On the Origin of Species by Means of Natural Selection, or the Preservation of Favored Races in the Struggle for Life.*

The classical form of a relative frequency argument upholds a favored position and then degrades an alternative by two strategies, both used by Darwin in making constraint subservient to natural selection.

NOOKS AND CRANNIES. Argue that your principle works nearly all the time, while the alternative occupies just a few subordinate holes of absence. By attributing both poles of the classic dichotomy (unity of type and conditions of existence) to natural selection as a primary cause, thereby robbing constraint of its potentially largest domain, Darwin granted dominance to adaptation.

SEQUELAE. Argue that your principle works as a prior and primary cause (in both temporal order and effect), and that the alternative only produces secondary modifications upon this fundamental action. In the closing paragraph of Chapter 6, all forces other than selection become sequelae to its primary action.*

Thus, Darwin invoked relative frequency to uphold his evolutionary world view: a theory of trial-and-error externalism, with natural selection as the only major creative force for change, and with internal variation restricted to the role of generating raw material for selection's perusal, and not of supplying important or consistent direction. Why, then, do many evolutionary biologists continue to demur? Darwin's basic argument in closing Chapter 6 can only be judged both brilliant and undoubtedly correct. Most homologies of Unity of Type are, indeed, adaptations inherited from a distant past, not fodder for constraint theorists who wish to demote the relative frequency and importance of natural selection. (Homologies of Unity of Type do act as phyletic constraints upon present possibilities—elephants will never fly—but such current limitations exist as consequences of initial adaptations, and therefore cannot stand against natural selection in any toting of relative frequencies.)

Modern constraint theorists, myself included, balk at Darwin's resolution because his argument demotes a large chunk of biology to a chink in a corner. The old Unity of Type theorists, lacking the alternative of "just history," did falsely assume that deep homology must stand against adaptation. But much validity still attends their cardinal insight that principles of design, laws of growth, rules of architecture, nature of materials—generalities transcending the particulars of specific genealogical pathways—work as important interior channels of constraint in the positive sense of that undervalued word: for constraints not only prevent evolutionary motion by failing to supply variation; they also act positively to set preferred channels of change. Internal forces do not only present isotropic raw material to the fully creative externality of natural selection. Constraint does not exist in subservience to adaptation under the nooks-and-crannies and sequelae arguments of relative frequency. Constraint may never again (and rightly so) be able to claim pri-

---

*I don't wish to jump the gun towards modern incarnations in this historical chapter, but I can't resist noting that this style of treating adaptation vs. constraint has been upheld by Darwinians ever since, thus generating the extreme frustrations of iconoclasts and reformers who wish to assert the importance of constraint in setting—not merely limiting—evolutionary pathways. Later Darwinians (often much more extreme than Darwin himself, and much more inclined to demote constraint even further) would frequently deny that they ignored constraint, pointing to a footnote or side comment that made concession by lip service to the possibility of such influence. But such purposefully restricted acknowledgment cannot count as balance, pluralism, or fairness. Rather, such treatment amounts to classical dismissal by the proper criterion of relative frequency!—and the frustrations of a C. H. Waddington or an E. C. Olson (at the Chicago centennial meetings of 1959) must be understood as deep and justified (see Chapter 7). Most people don't appreciate the style and power of relative frequency, and they will misread the strategy of dismissal by footnote as adequate pluralism rather than sharp rejection. I will never forget George Oster's reply to John Maynard Smith at the Chicago macroevolution meeting in 1980 (cited in full on p. 1023), when John insisted that he had (as an adaptationist) always acknowledged constraint, and George, recognizing footnote and side commentary as dismissal, replied: "Yes John, you may have had the bicycle, but you didn't ride it."

nacy, as the old Unity of Type theorists held, but partnership with adaptation remains a reasonable and minimal demand.

We may epitomize Darwin's brilliant reconstruction of causation in natural history, the new dimension that he added, and the one that we need to reinsert, in diagrammatic form. Pre-evolutionary theorists, entirely lacking the concept of historical change, attributed created form to a dichotomous distinction of causes: immediate and functional vs. deeper and architectural (Fig. 4-1). Darwin literally added the dimension of history, but removed a previous axis of explanation by redefining constraints of Unity of Type as consequences of past adaptation (ancient Conditions of Existence)—Figure 4-2. Yet Darwin understood that he had not abolished the concept of constraint in undermining the primary example—homologies of Unity of Type—by reallocation to the opposite camp. He therefore, in the same passage, established a different domain for constraint, as a category subservient to adaptation by the two standard arguments of relative frequency: the spatial claim of limited room (nooks and crannies) and the temporal claim of secondary status (sequelae to adaptation)—see Figure 4-3.

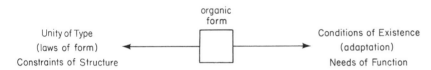

4-1. The standard pre-evolutionary and dichotomous conception of the causes of form as working either by adaptation to immediate conditions of existence, or by manifestation of laws of form that reflect unity of type.

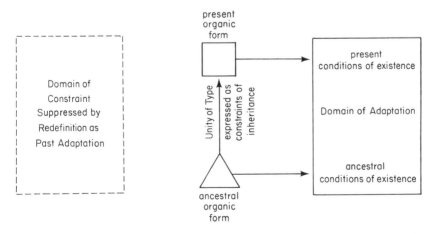

4-2. Darwin literally adds a third dimension of history for the explanation of form. But he greatly devalues the domain previously ascribed to unity of type, admitting constraints of laws of form only by redefining such similarities as homologies based on the inheritance of past adaptations, and therefore adaptational in their origin and primarily due to the other (and now predominant) domain of conditions of existence.

But many 19th century biologists, and many evolutionists in our day again feel that Darwin demoted constraint too far, and that the two domains—constraint and adaptation—must again share potential partnership, as expressed by the important relative frequency of each component. We would therefore restore the strength of the dimension that Darwin first eliminated (when he reinterpreted Unity of Type as a consequence of adaptation) and then reintroduced in weakened form (when he allowed laws of growth to fill nooks and crannies in a domain ruled by natural selection)—see Figure 4-4.

This full model of Figure 4-4 shows three dimensions of form and their interactions: adaptation, constraint, and history. A current trait of an organism may arise as an immediate adaptation to surrounding environments, as a constraint not particular to the contingent history of its lineage (architectural or structural principles, correlations to current adaptations), or by inheritance of an ancestral form (often called historical or phylogenetic constraint, but quite different in principle from nonhistorical styles of constraint). This distinction suggests a recursion, because contributions from the axis of "his-

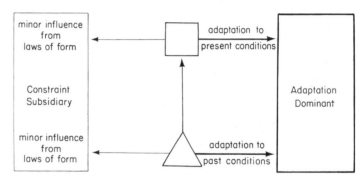

4-3. Darwin does allow minor influence for constraint apart from mere inheritance of past adaptations. See text for details.

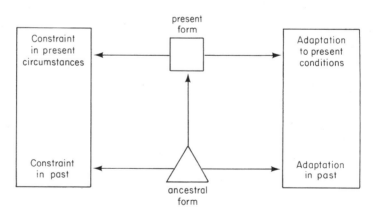

4-4. Constraint reestablished as equal in importance to adaptation as an immediate cause of form. See text for details.

tory" represent traits that, at their origin in an ancestor, arose as either adaptations or constraints. Nonetheless, the immediate form of an organism can still be meaningfully parsed into three major contributions of current adaptation, current constraint, and historical inheritance—Figure 4-5. This insight has generated the various "triangular" models of evolutionary causation that have gained vogue in recent years (see Fig. 4-6).

These issues and parsings have pervaded natural history since Plato and Aristotle argued about abstract form vs. teleology. Darwin made a seminal

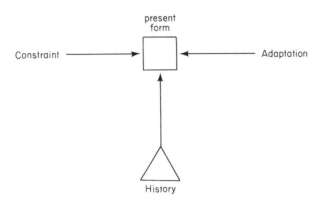

4-5. Because constraint and adaptation act either from the past or in the present, we may envisage three primary determinants of present form: present constraint, present adaptation, and inheritance due to past history of either constraint or adaptation.

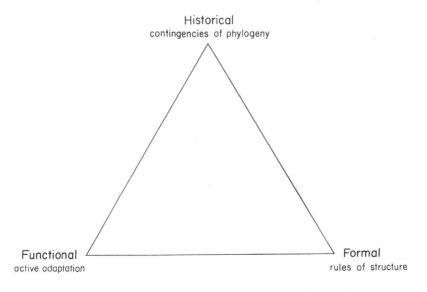

4-6. One of a group of "triangular" models constructed to express the major influences upon the genesis of form. The three vertices of this triangle refer to the three influences depicted in Figure 4-5.

contribution by adding a dimension for history, and by formulating a theory that granted controlling relative frequency to adaptation. But he did not invent the issues, or the scheme of classification. The triumph of Darwinian functionalism did, however, erase much historical memory for the old alternative of constraint. The next two chapters sink their rationale in a simple premise: Our current need to reinvigorate constraint as a vital topic in evolutionary explanation (see Stearns, 1986; Maynard Smith *et al.*, 1985; Gould, 1989a, 1992b)—based upon advancing knowledge of genetic architecture, development and macroevolution—requires that we rediscover this legacy of structuralist thought,* and recognize that the entire history of evolutionary theory has been pervaded by an issue that simply would not disappear, if only because the dialectic of inside and outside, structure and function, design and adaptation, must be resolved at some fascinating interplay and synthesis, not as a victory for either pole in a debate without true sides.

## Two Ways to Glorify God in Nature

We cannot comprehend the past from the vantage point of a newly-constructed present reality. Once the 19th century had discovered evolution as the primary cause of relationships among organisms, the historical axis not only sprang into being as a pole of explanation, but quickly assumed a primary status (Figs. 4-2 to 4-6). More than a century later, we can hardly imagine a biology without this theme. What kind of questions could be posed before history became an option for resolution? What kinds of explanations could be rendered when a biologist couldn't ask (or even conceptualize): "How has this feature changed from an ancestral state; what do its differing forms in various species tell us about phyletic relationships; what are the causal bases for both the origin and later alterations of this feature?"

Immediate appearance in a fully formed state provides the only alternative to history—whether such "creation" be achieved by the direct hand of a divine agent, or by spontaneous organization from elements according to some unknown law or principle of nature. If basic taxa originated as we find them now, then the range of theoretical explanation remains wide. Species might be purposely ill-designed to suit the black humor of a diabolical creator; or they might be cobbled together with no rhyme or reason by forces of universal randomness. The list of possibilities continues *ad infinitum*.

But, in fact, Western cultural traditions greatly limited the range of accept-

---

*I shall discuss, in subsequent sections of this and the next chapter, the functionalist (Paley) vs. structuralist (Agassiz) versions of natural theology, the central role of laws of form in the pre-Darwinian evolutionary debates (particularly the struggle between the structuralist Geoffroy Saint-Hilaire and the functionalist Cuvier), Darwin's own treatment of constraint and correlation of growth, alternatives to Darwinism based upon the centrality of constraint (Galton's polyhedron, various theories of orthogenesis, Bateson on homeosis, de Vries on saltation, Goldschmidt on jumps within channels). The structuralist alternative has always been pursued as an active option by some of the finest thinkers in our field.

ble alternatives. Very few creationists could imagine that species might be purposely ill-formed, or constructed in a disorganized fashion. With these attributes—purpose and order—as part of a cultural heritage, the basic explanations for organic form could be reduced to two major alternatives, expressing the primacy of one or the other overarching principle for a rational and benevolent world. These principles have been called structuralism and functionalism, order and teleology, laws of form and adaptation, Unity of Type and Conditions of Existence. These poles set the dichotomy that Darwin expanded by introducing history (see last section), but never really fractured because the new axis of time could also be divided into structural vs. functional explanations for ancestral forms. This dichotomy continues to set an important agenda for evolutionary theory at the opening of a new millennium, especially since the overly adaptationist Modern Synthesis (representing a temporary triumph of the functionalist pole—see Chapter 7) has yielded to a pluralism of structuralist alternatives as partners rather than subsidiary forces (Chapters 10 and 11).

In this light, I find it fascinating that the oldest tradition in modern natural history—the natural theology of so many pre-Darwinian biologists*—also existed in two primary versions, expressing the two poles of the same dichotomy. Since Darwin built his evolutionary theory in continuity with the pole favored by a long English heritage—the adaptationism of William Paley—this subject cannot be dismissed as an arcane issue from a forgotten past, but remains a vital presence in our daily concerns (by our own fundamental evolutionary criterion of genealogy and phyletic heritage!). For we still struggle with adaptation and constraint just as Paley and Agassiz contrasted the comparable positions in natural theology: "the creator foresaw the needs of each species and created just those organs that were necessary to carry them out" vs. "God had in the beginning established laws, and nature was left to unfold in accordance with them" (characterizations of Appel, 1987, p. 7). Do not Fisher vs. Wright, or Cain and Maynard Smith vs. Goodwin and Kauffman carry on the same debate, evolutionarily transmogrified of course?

Natural theology held, as a central premise, that the works of nature not only demonstrated God's presence, but could also reveal his character as well. We could learn *about* him, not only persuade ourselves *that* he exists. Paley's full title (1802) reads: *Natural Theology: or, Evidences of the Existence and Attributes of the Deity, Collected from the Appearances of Nature.* From this shared premise, two traditions proceeded, both "preadapted" to a later evolutionary transformation.

---

*Not all biologists, by any means, favored the arguments of natural theology. Our anglophone parochialism leads us to emphasize this attitude, which held greater sway in Protestant Britain than elsewhere (and had much less influence in Catholic France). Many of the continental formalists, for example, maintained little enthusiasm for such direct providentialism, and tended either towards a pantheism of uncaring (if pervasive) divine presence, sometimes even to materialism (Geoffroy as a child of the Enlightenment), or at least towards the less radical notion that God made nature's laws and then bowed out of her affairs.

In this section I shall contrast the two great texts of these alternative tradi-tions—Paley's *Natural Theology* (1802) with Agassiz's *Essay on Classifica-tion* (1857). The two works dovetail with remarkable symmetry in their op-position: Paley the British adaptationist vs. Agassiz the continental formalist. One might almost believe that the two works were explicitly written to flesh out (and fully clothe) the central dichotomy of form, with each awarded ex-actly half the totality. In a curious sense, this lack of contact almost allows the two texts to speak to each other—as if they formed a sand painting with one (Paley for temporal priority) filling in half the area up to an elaborate and jag-ged boundary, and the other then pouring sand of a different color right up to the previous boundary, leaving no space between at the contact. I am puzzled that these two texts have not been explicitly contrasted before.

### WILLIAM PALEY AND BRITISH FUNCTIONALISM: PRAISING GOD IN THE DETAILS OF DESIGN

Just a few years before Paley wrote his *Natural Theology* in 1802, Coleridge's Ancient Mariner (1798) proclaimed his hard-won message to a wedding feast, and to the world:

> He prayeth best, who loveth best
> All things both great and small;
> For the dear God who loveth us,
> He made and loveth all.

Paley probably appreciated the sentiments and surely longed to extend the ar-gument. He entertained no doubt that all things proclaimed God's *existence*. But he believed that we must be able to learn more if we hope to use natural theology as a strategy of exegesis. That is, we must also be able to infer im-portant aspects of God's *nature* and *character* from the works of creation.

The search to infer God's attributes from general features of natural objects led Paley to open his book with one of the most famous images in all English literature—a strong competitor with Adam Smith's "invisible hand" (a line also found in Paley, 1803, p. 344) and Darwin's tangled bank or tree of life. The good Reverend, crossing a heath on shank's mare, bumps his foot against a stone, feels the pain, but learns nothing about the origin of rocks because the object is too simple and too disordered to reveal a source of production. But if he should then kick a watch, he would surely know that the timepiece had been fashioned by a purposeful agent:

> When we come to inspect the watch, we perceive (what we could not dis-cover in the stone) that its several parts are framed and put together for a purpose, e.g. that they are formed and adjusted as to produce motion, and that motion so regulated as to point out the hour of the day . . . The inference, we think, is inevitable; that the watch must have had a maker; that there must have existed, at some time and at some place or other, an artificer or artificers who formed it for the purpose which we find it ac-

tually to answer; who comprehended its construction, and designed its use (Paley, 1803, p. 203—I am using my personal copy of the widely read 1803 edition for all quotes).

Two features of the watch compel this conclusion. First, and less important, its *complexity*—for chance could not make anything so intricate: "What does chance ever do for us? In the human body, for instance, chance, i.e. the operation of causes without design, may produce a wen, or a wart, a mole, a pimple, but never an eye" (1803, pp. 67–68). Second, and far more important, the watch's *design, its adaptation to a clearly perceived end.*\* A high degree of order might arise from laws of nature with no reference to final cause, but complexity for a clear purpose implies a designer. "There cannot be design without a designer; contrivance without a contriver; arrangement, without anything capable of arranging" (p. 12). Thus does Paley attack his hypothetical opponent and partial straw man throughout his work.† "Nor would any man in his senses think the existence of the watch, with its various machinery, accounted for, by being told that it was one out of several possible combinations of material forms; that whatever he had found in the place where he found the watch, must have contained some internal configuration or other; and that this configuration might be the structure now exhibited" (p. 6).

The watch implies, by its utility, a mind capable of forethought, design and construction: "In the watch which we are examining, are seen contrivance, design; an end, a purpose; means for the end, adaptation to the purpose. And the question, which irresistibly presses upon our thoughts, is, whence this contrivance and design. The thing required is the intending mind, the adapting hand, the intelligence by which that hand was directed" (p. 16).

But organisms surely display more complexity and more purposeful design than any watch. Just as Darwin would exalt natural selection as vastly more powerful than artificial human selection in breeding or agriculture, so does

---

\*The word adaptation did not enter biology with the advent of evolutionary theory. The *Oxford English Dictionary* traces it to the early 17th century in a variety of meanings, all referring to the designing or suitability of an object for a particular function, or the fit of one thing to another. The British school of natural theology used "adaptation" as its standard word for illustrating God's wisdom by the exquisite fit of form to immediate function. Darwin, in borrowing this term, simply followed an established definition while completely revising the cause of the phenomenon.

†Paley frames his hypothetical opponent as a somewhat caricatured workbench materialist who believes that all natural order arises from physical laws. For Paley, this opponent exists in two versions, one more dangerous—the true atheist who denies God outright; and the theist who has abandoned a directly caring and providential God for a deity who set up the laws of nature at the beginning and then bowed out (or the deist who sees spirit in everything, but calls this directing force physical law, and owns no caring, personal God). Apparently, Paley never conceptualized, as another potential opponent worthy of explicit refutation, the possibility of a principle of selection, in Darwin's version or otherwise. That is, his caricature depicts order as arising from laws of nature, but he never imagines that *good* order could also emerge as a residue of trying many things out and rejecting most. Such selectionism represents, to us today, an obvious potential alternative to Paley's only conceptual model for order without apparent purpose: direct construction by the action of physical laws.

Paley identify God's work as incomparably superior to any human art. If the existence of the watch implies a skilled craftsman, how can we even conceive the more awesome skill of he who made all living things: "For every indication of contrivance, every manifestation of design, which existed in the watch, exists in the work of nature; with the difference, on the side of nature, of being greater and more, and that in a degree which exceeds all computation" (p. 19).

In succinct epitome of the entire argument, Paley writes (p. 473): "The marks of design are too strong to be got over. Design must have a designer. That designer must have been a person. That person is God."

Since we often misuse the past for ridicule, Paley has emerged as everybody's favorite whipping boy from the bad old days of creationism. As a lively writer, he is, to be sure, eminently quotable. And he does sometimes stray into the kind of Panglossian perfectionism (or, rather, far-fetched rationalization for beneficence within apparent evil) that Voltaire savaged with such glee in *Candide*.

Paley, for example, does engage in "just-so" storytelling to support adaptationist explanation, though he presumably read this account of *Babyrussa* in a fallacious traveller's report, and can only be charged with insufficient skepticism, not fabrication (Fig. 4-7):

> I shall add one more example for the sake of its novelty. It is always an agreeable discovery, when, having remarked in an animal an extraordinary structure, we come at length to find out an unexpected use for it. The following narrative furnishes an instance of this kind. The babyrouessa, or Indian hog, a species of wild boar found in the East Indies, has two bent teeth, more than half a yard long, growing upwards, and (which is the singularity) from the upper jaw. These instruments are not wanted for defense, that service being provided for by two tusks issuing from the under jaw, and resembling those of the common boar. Nor does the animal use them for defense. They might seem therefore to be both

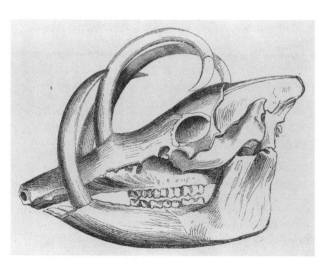

4-7. Paley does not include a drawing of the skull of *Babyrussa*, but this figure comes from an equally interesting source —P. H. Gosse's *Omphalos* of 1857, his treatise arguing for the sudden and recent creation of the Earth, including all its fossils, which therefore only display an appearance of great age.

a superfluity and an incumbrance. But observe the event. The animal sleeps standing; and, in order to support its head, hooks its upper tusks on the branches of trees (pp. 270–271).

More in the Panglossian mode, pain (an adaptation, Paley argues, for signalling distress to the mind so that we may care for our bodies) also shows God's benevolence on the theme of the old moron joke—we feel so good when the suffering stops! (On the subject of good in apparent noxiousness, compare John Ray (1735, p. 309) on why God made lice: "I cannot but look upon the strange instinct of this noisome and troublesome creature a louse, of searching out foul and nasty clothes to harbor and breed in, as an effect of divine providence, designed to deter men and women from sluttishness and sordidness, and to provoke them to cleanliness and neatness. God Himself hateth uncleanliness, and turns away from it." Or, as Robert Burns would later generalize the lesson in "To a Louse": "Oh wad some power the giftie gie us / To see oursels as ithers see us!") "A man resting from a fit of the stone or gout, is, for the time, in possession of feelings which undisturbed health cannot impart. They may be dearly bought, but still they are to be set against the price. And, indeed, it depends upon the duration and urgency of the pain, whether they be dearly bought by suffering a moderate interruption of bodily ease for a couple of hours out of the four and twenty" (pp. 523–533).

To complete the picture of joyous nature made by a loving God, signs of non-utility in sheer behavioral exuberance, particularly in the play of young creatures, testify to the sheer pleasure of being alive on such a wondrous planet:

> Swarms of newborn flies are trying their pinions in the air. Their sportive motions, their wanton mazes, their gratuitous activity, their continual change of place without use or purpose, testify their joy, and the exultation which they feel in their lately discovered faculties . . . Other species are running about with an alacrity in their motions which carries with it every mark of pleasure. Large patches of ground are sometimes half covered with these brisk and sprightly natures. If we look to what the waters produce, shoals of the fry of fish frequent the marshes of rivers, of lakes, and of the sea itself. These are so happy, that they know not what to do with themselves (pp. 490–491).

(Paley's prose may be purple, but his purpose is sanguine. He argues, in stating his primary case, that organic adaptation proves the personhood of God. But we want to know more. God could, after all, be a consummate craftsman, but a crabby character. Paley's arguments on pain and natural happiness indicate that God is not only skillful, but also benevolent as well.)

These statements, taken out of context (as usually done), promote an unfair caricature of a subtle argument. Paley cannot be dismissed as an intellectual slouch. His *Evidences of Christianity* (1794) remained a required text for entrance to Cambridge University until the 20th century, and Darwin would never have chosen a cardboard dogmatist for a hero or, later, for an opponent worthy of overturning as the essential thrust of a revolutionary theory (see pp. 116–125). Paley's totality presents a subtle, coherently reasoned brief for an adaptionist natural theology.

First of all, Paley cannot be caricatured as a Panglossian perfectionist. He states explicitly that we cannot use perfection as a criterion for identifying good design, or even as the necessary mark of divinity in craftsmanship: "It is not necessary that a machine be perfect, in order to show with what design it was made: still less necessary, where the only question is, whether it were made with any design at all" (p. 5).

Paley also provides, if only occasionally, positive arguments for imperfection, as in feathers of an ostrich's wing. "The filaments hang loose and separate from one another, forming only a kind of down; which constitution of the feathers, however it may fit them for the flowing honors of a lady's headdress, may be reckoned an imperfection in the bird, inasmuch as wings, composed of these feathers, although they may greatly assist in running, do not serve for flight" (p. 236). And he acknowledges that the creator's preference for utility lies revealed in the overwhelming relative frequency, not the ubiquity, of adaptation (but adding the conventional rider, still commonly advanced today, that, if we look hard enough, we will discover uses for traits now judged "nonadaptive"). "Instances . . . where the part appears to be totally useless, I believe to be extremely rare: compared with the number of those, of which the use is evident, they are beneath any assignable proportions; and perhaps, have never been submitted to a trial and examination sufficiently accurate, long enough continued, or often enough repeated" (Paley, 1803, p. 64).

In fact, Paley uses adaptationism primarily as a theoretical argument about depth of causality, not as an excuse to rhapsodize about happy nature. Opponents who wish to see "physical law" as the source of form might cite sexual generation and embryology as leading examples. But these processes only provide the immediate physical continuity of efficient causation: "The truth is, generation is not a principle but a process" (p. 453). We need a deeper reason, a true principle, for the evident adaptation of form to function—in short, a final cause. Even if watches gave birth to new watches, Paley argues, we would not identify ontogeny as the ultimate source of timekeeping. Neither can embryology be the cause of optical excellence in the human eye, if only because "things generated possess a clear relation to things not generated" (p. 455)—the eye to external light and to the objects we need to see in this case. (We now recognize this otherwise persuasive argument as wrong only because life, unbeknownst to Paley, possesses history and mutability.)

But the main case for taking Paley seriously lies in his formulation and refutation of opposing visions. Anyone can spin out a rationale for an *idée fixe,* but a well-crafted system requires both full analysis and principled denial of alternatives. *Natural Theology* merits our respect, and becomes a key document for this chapter on the history of functionalism vs. formalism, because Paley recognized the structuralist alternative and provided a coherent defense. His arguments span two chapters (15 on relations and 16 on compensations), treating the phenomenon always viewed as crucial and primary by advocates of structural constraint—stable correlations among parts of the body.

Since Paley's main argument holds that intricate contrivance implies a contriver, two main rebuttals might be offered in principle: (1) the adaptations exist, but they originated by a natural process of evolution, not by creative acts of a deity; (2) organisms were created, but adaptation does not permeate or even dominate their form.

Since Paley never imagined the alternative of natural change by selection or weeding out, he confines his refutation of adaptive evolution to the "Lamarckian" principle of use and disuse. (I doubt that Paley, writing in 1802, knew Lamarck's work directly, since his French colleague had just begun to publish evolutionary views. But use and disuse, as an item of folk wisdom, frequently entered the arguments of evolutionists.) Paley begins empirically by pointing out that centuries of disuse do not cause organs to disappear, though modesty leads him to cloak a classic case entirely in untranslated Latin: "The mammae of the male have not vanished by inusitation; *nec curtorum, per multa saecula, Judaeorum propagini deest praeputium*" [nor has the foreskin of Jews become any shorter in offspring through many centuries of circumcision] (p. 446).

Paley then asks, more theoretically, how any natural evolution of useful structures could be attributed to a stimulus structurally unrelated to biological form, and often inorganic. The eye is a contrivance for perceiving light, but light cannot make an eye. "Yet the element of light and the organ of vision, however related in their office and use, have no connection whatever in their original. The action of rays of light upon the surfaces of animals has no tendency to breed eyes in their heads. The sun might shine forever upon living bodies without the smallest approach towards producing the sense of sight" (p. 317).

When two structures have been similarly fashioned for a common purpose by a strengthening of one and a weakening of the other (the subject of Paley's "compensations" in chapter 16), natural adjustment by evolution might be defended (as when an elephant elongates its trunk to compensate a shortness of neck). But Paley denies this "best case" by the standard argument that intermediary stages could not be well designed: "If it be suggested, that this proboscis may have been produced in a long course of generations, by the constant endeavor of the elephant to thrust out his nose, (which is the general hypothesis by which it has lately been attempted to account for the forms of animated nature), I would ask, how was the animal to subsist in the meantime, during the process, until this elongation of snout were completed? What was to become of the individual, whilst the species was perfecting?" (p. 299).

If the first alternative (adaptation, but by evolution) can be thus refuted, how can the second possibility (creation, but with adaptation secondary or absent) be dismissed as well? Paley now meets the formalist alternative face-to-face—and rejects this last challenge with three arguments that, taken together, develop his strongest case for adaptationism (the first two remain in prominent use today):

1. Formalists do not deny the evident utility of most organic structures. The focus of their argument, rather, rests upon a claim for temporal and causal *primacy* (homology based upon historical order for evolutionists, or similarity based upon repeated themes in manufacture for creationists). Adaptationists hold that structures must evolve or be fashioned for utility: functional needs come first, and form follows. Formalists argue, on the other hand, that morphology may arise for reasons other than use, with later "uptake" of function as subsidiary: that is, form comes first, and organisms may then discover usages. In a remarkable passage, showing his grasp of this fundamental alternative (now being reasserted as the basis for revival of interest in constraint among modern evolutionists), Paley admits that the formalist argument must be acknowledged as "intelligible": "To the marks of contrivance discoverable in animal bodies, and to the argument deduced from them, in proof of design, and of a designing Creator, this turn is sometimes attempted to be given, viz. that the parts were not intended for the use, but that the use arose out of the parts. This distinction is intelligible. A cabinet-maker rubs his mahogany with fish-skin; yet it would be too much to assert that the skin of the dogfish was made rough and granulated on purpose for the polishing of wood, and the use of cabinet makers" (p. 72).

Paley's refutation invokes the classic response: the formalist argument will work for simple structures like fish-skin, but not for complex organs, composed of multiple parts, all apparently adjusted for current function. "Is it possible to believe that the eye was formed without any regard to vision; that it was the animal itself which found out, that, though formed with no such intention, it would serve to see with; and that the use of the eye, as an organ of sight, resulted from this discovery, and the animal's application of it?" (p. 73).

2. The first argument epitomizes a conceptual mainstay of formalism, but the empirical foundation of structuralist morphology has always depended more strongly upon correlation among parts of an organism, buttressed by the inference that structural relations, rather than utility, establish the linkage. Again Paley provides the classic functionalist refutation, still prominently in use. The correlations, he argues, do not arise by formal necessity, or "laws of growth," but as coordinated adaptations, each separately useful and required for good design. Swans have long necks and webbed feet for reasons of common function, not "necessary connection": "The long neck, without the web foot, would have been an incumbrance to the bird; yet there is no necessary connection between a long neck and a webfoot. In fact they do not usually go together. How happens it, therefore, that they meet, only when a particular design demands the aid of both?" (p. 293).

Paley then discusses a favorite example of British adaptationists since John Ray, and a pest in British gardens from time immemorial: the mole. "From soils of all kinds the little pioneer comes forth bright and clean. Inhabiting dirt, it is, of all animals, the neatest" (p. 294). Paley defends adaptation with an explicit rejection of the strongest argument for constraint (what Darwin would later call "correlation of growth"). Recalling his opening metaphor, Paley writes: "Observe then, in this structure, that which we call relation. There is no natural connection between a small sunk eye and a shovel pal-mated foot. Palmated feet might have been joined with goggle eyes; or small eyes might have been joined with feet of any other form. What was it there-fore which brought them together in the mole? That which brought together the barrel, the chain, and the fusee, in a watch: design; and design, in both cases, inferred, from the relation which the parts bear to one another in the prosecution of a common purpose" (p. 296).

3. But what can an adaptationist say about the overarching homologies of broad taxonomic structure? Are these widespread properties not formal con-straints, logically prior to any subsequent utility forged by specific tinkering with such common elements? (We certainly acknowledge such priority today, but we also recognize Darwin's incisive argument that these "phyletic" con-straints may have arisen as ancestral adaptations—see last section. Paley en-joyed no conceptual access to this legitimate adaptationist exit from the di-lemma.)

In a clever twist of argument, Paley turns homology to the cause of adapta-tion in two steps:

(1) God devised general plans with foreknowledge of their requisite modi-fication for specific purposes in individual species. For if these grand homologies had been generated automatically by abstract laws of nature, with no reference to final causality, how could such widespread structures be so subtly subject to such varied adaptation in the service of so many particu-lar modes of life? "Whenever we find a general plan pursued, yet with such variations in it, as are, in each case, required by the particular exigency of the subject to which it is applied, we possess, in such a plan and such adaptation, the strongest evidence, that can be afforded, of intelligence and design . . . If the general plan proceeded from any fixed necessity in the nature of things, how could it accommodate itself to the various wants and uses which it had to serve, under different circumstances, and on different occasions?" (Paley, 1803, p. 227).

(2) Yet Paley recognized the potential circularity in this claim, if taken by it-self. To be sure, once such homologies have been established, they may be ex-amined for susceptibility to adaptive modification. But why did God proceed in this manner at all? Why didn't he just make each species from scratch, opti-mally suited for its own peculiar mode of life? Why bother with common plans at all, when creatures sharing the plans work so differently? Here, at the crux of his difficulty, Paley invokes a venerable solution that has always (both then and now) struck critics as at least slightly sophistic (in the sense that any potential refutation could be so "accommodated," thus making the theory irrefutable, untestable, and therefore useless): God shows his greatness by limiting his own power with ordered principles (secondary causes based on natural laws) and structural designs (grand homologies):

> God, therefore, has been pleased to prescribe limits to his own power, and to work his ends within those limits. The general laws of matter have perhaps the nature of these limits . . . These are general laws; and when a particular purpose is to be effected, it is not by making a new law, nor by the suspension of the old ones, nor by making them wind and bend and yield to the occasion (for nature with great steadiness adheres to, and supports them), but it is, . . . by the interposition of an apparatus corresponding with these laws, and suited to the exigency which results from them, that the purpose is at length attained. As we have said, therefore, God prescribes limits to his power, that he may let in the exercise, and thereby exhibit demonstrations of his wisdom (p. 43).

After all, adaptationism only requires that organic designs be complex and work well, not that they embody perfection: "Contrivance, by its very definition and nature, is the refuge of imperfection. To have recourse to expedients, implies difficulty, impediment, restraint, defect of power" (pp. 41–42).

Paley's closing paean, following this last statement, exalts adaptation as logically necessary, quite apart from any factual validation. Contrivance not only sets the dominant pattern of empirical nature. Such good design also represents the only way that God *could* proclaim his existence in principle! To quote the passage of page 119 once again:

> It is only by the display of contrivance, that the existence, the agency, the wisdom of the Deity, *could* be testified to his rational creatures. This is the scale by which we ascend to all the knowledge of our Creator which we possess, so far as it depends upon the phenomena, or the works of nature. Take away this, and you take away from us every subject of observation, and ground of reasoning . . . Whatever is done, God could have done, without the intervention of instruments or means: but it is in the construction of instruments, in the choice and adaptation of means, that a Creative Intelligence is seen. It is this which constitutes the order and beauty of the universe (p. 42).

Paley's argument coheres, yet sounds a peculiarly limited range of notes—the reason for my "sand painting" metaphor of page 262. Paley does mention the grand homologies that underlie all taxonomy—but only in a paragraph or two, and only to offer an adaptationist riposte. He does formulate the structuralist argument based on correlation—but only in passing reference, and only for refutation. We might be tempted to offer the Philistine's retort—"oh well, Paley was just a philosopher; what did he know about real biology?" But modern disciplinary boundaries did not exist in 1800, and great biologists, including Darwin, valued Paley above all other books in natural history. Moreover, as I shall show in the next section, a fine working biologist like Agassiz could present the other side with equally uncompromising exclusivity.

We must therefore grasp Paley's restricted compass as a consciously-chosen vision of life's substance and meaning. As such, we may utilize, for our own instruction, a position so unsullied by nature's real complexities. We know that life cannot work at such a conceptual extreme, but any consistent and well-argued defense of such an edge remains fascinating—at least in illustrating a set of mental habits that still motivates scientists. Just as we learn to grasp nature through controlled and simplified cases (the experimental method), so may we also comprehend mind by its defense of coherence at the philosophical endpoint of a continuum.

## LOUIS AGASSIZ AND CONTINENTAL FORMALISM: PRAISING GOD IN THE GRANDEUR OF TAXONOMIC ORDER

Louis Agassiz, as the first permanent immigrant among great European biological theorists, became the symbol and actuality of maturation and prestige for American natural history in the mid 19th century. Romantic mythology proclaims that he ventured forth as an intrepid pioneer in a quest for pristine knowledge and uncharted species. In fact, Agassiz's primary reasons for resettlement were far more mundane—escape from trouble and hope for a new beginning. He had suffered the two classic reversals of personal misfortune after years of intellectual triumph: bankruptcy (when his lithographic press, initially established to print the plates for his *Poissons fossiles,* failed) and familial strife (when his wife moved out after he had turned their home into a factory and boarding house for workers at his press). In any case, whatever the complex motives, Agassiz's decision to settle permanently at Harvard established a happy incongruity within an expanding and accepting culture—a great francophone theorist, with traditional continental attitudes, living in Yankee Boston.

Agassiz (1807–1873) came to America with grand plans to invest his boundless energy in systematic work on undescribed native faunas, following his own maxim: "study nature, not books." But, as a consummate academic politician and promoter, he became sidetracked over the years (an old story, as deep as human nature itself), and published little technical work during his last two decades. The frustration in this familiar tale of good intentions lies best exposed in Agassiz's grandest project and its failure.

Early in the 1850's, he announced plans for a lavish 10-volume work to be called *The Natural History of the United States.* He gathered more than 2,000 paid subscriptions in advance, and began collection (for an initial monograph on turtles) with his old and characteristic zeal. But he soon bogged down—permanently. Only four volumes ever appeared (with the descriptive and taxonomic work largely done by others), and he talked less and less about his grand design as the years ticked away. Nonetheless, while still imbued with initial enthusiasm, he wrote, as a book-length introduction to volume 1, his finest theoretical work, the *Essay on Classification.* Published in 1857, and revised in 1859 (ironically just 3 months before publication of Darwin's *Origin,* the book that would undermine the central premise of Agassiz's work), the *Essay on Classification* stands as a unique and incongruous document—a statement of natural theology in the highest tradition of continental formalism, published in the most English of American cities. Agassiz never mentions Paley by name, but his volume presents an almost perfect counterpoise to Paley's *Natural Theology* from the other pole of the great dichotomy in approaches to form—particularly, in this case, to the question of how an omnipotent God would manifest his glory in nature.

Modern supporters of systematics, in a world increasingly dominated by trendier forms of biological research, often feel beleaguered, and therefore impelled to provide a wider rationale for pursuing classification, an enterprise unfairly burdened with such epithets as "stamp collecting" by a miscomprehending public. Today, the rationale for systematics tends to be given—quite legitimately of course—in terms of our current crises in environmental deterioration and declining biodiversity. Yet if any systematist ever yearned for a maximally grand rationale for his chosen profession, he could not find, or even imagine, a more audacious document than Agassiz's *Essay on Classification*. (Unfortunately, changing philosophies and increasing knowledge have rendered Agassiz's argument obsolete, but we may still sense, and should still admire, the style and grandeur of his claim.)

In baldest terms, and from a Platonic perspective (with organisms construed as temporary, material incarnations, representing the permanent and transcendent mental structures of an overarching creative force), Agassiz argues that taxonomy should be regarded, in principle, as the highest of the sciences. For species embody ideas in God's mind; and actual organisms then became transient configurations that represent, or incarnate, these ideas. Relationships among species, as expressed in classification, therefore reveal the structure of God's thought, for if each species denotes a divine idea, then their interconnections in taxonomy display the order of God's mentality.

Agassiz poses the key question: "Are these divisions artificial or natural? Are they the devices of the human mind to classify and arrange our knowledge in such a manner as to bring it more readily within our grasp and facilitate further investigations, or have they been instituted by the Divine Intelligence as the categories of his mode of thinking?" (1857, pp. 7–8). He then provides his firm answer: "To me it appears indisputable, that this order and arrangement of our studies are based upon the natural primitive relations of animal life,—those systems [of classification] . . . being in truth but translations, into human language, of the thoughts of the Creator."

With this vision, Agassiz cuts through an old argument about the differential "reality" of categories in a Linnaean hierarchy: Are species real and higher levels artificial? Are all categories real or do they only express the practical needs of human convenience? If, as Agassiz argues, the entire taxonomic system, when properly "discovered," records the structure of God's thoughts, then all categories must be objective segments of this divine totality. Only organisms have material existence, but taxonomic categories embody higher reality as direct expressions of the divine mind:

> Is not this in itself evidence enough that genera, families, orders, classes, and types have the same foundation in nature as species, and that indi-

viduals living at the same time have alone a material existence, they being the bearers, not only of all these different categories of structure upon which the natural system of animals is founded, but also of all the relations which animals sustain to the surrounding world,—thus showing that species do not exist in nature in a different way from the higher groups, as is so generally believed? (1857, p. 7).

Agassiz shares Paley's primary goal, the fundamental "research program" of "natural theology"—to infer, from the organic works of nature, not only God's existence, but as much as possible about his intellect and goodness. Yet, despite this common aim, Paley and Agassiz could not have advocated more disparate constructions of divine presence in nature.

Every good debater, following the principle of dichotomy, knows that arguments fare best by contrast with alternatives. Moreover, the more caricatured and cardboard the alternative, the better for your side (so long as you don't depict your opponent as so much of a strawman that he becomes unbelievable). Agassiz presents his vision of classification by contrast with a "materialist" alternative of his own construction. He defines a materialist as a naturalist who attributes the forms and properties of organisms to the shaping power of constant physical laws (secondary, efficient causes), and not to direct decisions of divine will. A materialist may escape the charge of godlessness by arguing for divine establishment of natural laws at the beginning of time. But if God then absconds forevermore, and lets nature work in such an automatic and heartless mode, what practical difference could we discern between outright materialism and such a divine clockwinder? "I allude here," Agassiz writes (p. 9) in defining his opponents, "only to the doctrines of materialists." The issue reduces to a simple dichotomy (given the inconceivability of other alternatives, including randomness): are taxa fashioned by laws of nature (and therefore in harmony with physical order), or by God as incarnations of His categories of thought? Agassiz states the contrast, and announces his own allegiance: "Others believe that there exist laws in nature which were established by the Deity in the beginning, to the action of which the origin of organized beings may be ascribed; while according to others, they owe their existence to the immediate intervention of an Intelligent Creator. It is the object of the following paragraphs to show that there are neither agents nor laws in nature known to physicists under the influence and by the action of which these beings could have originated" (1857, p. 13).

In a grand verbal flourish, Agassiz then upholds taxonomy as the highest science, while branding the materialist alternative both dreary and soul destroying (as well as wrong). Taxonomic order records divine mentality:

I confess that this question as to the nature and foundation of our scientific classifications appears to me to have the deepest importance, an importance far greater indeed than is usually attached to it. If it can be proved that man has not invented, but only traced the systematic arrangement in nature, that these relations and proportions which exist throughout the animal and vegetable world have an intellectual, and ideal connection in the mind of the Creator, that this plan of creation, which so commends itself to our highest wisdom, has not grown out of the necessary action of physical laws, but was the free conception of the Almighty Intellect, matured in his thought, before it was manifested in tangible external forms,—if, in short, we can prove premeditation prior to the act of creation, we have done, once and forever, with the desolate theory which refers us to the laws of matter as accounting for all the wonders of the universe, and leaves us with no guard but the monotonous, unvarying action of physical forces, binding all things to their inevitable destiny (1857, p. 9).

By setting up his argument in this manner, Agassiz immerses himself directly into the formalist-functionalist debate—with his own version of natural theology as a strictly, almost excessively, formalist proposal: taxonomic order at all levels, not the behavior and function of individual creatures, records God's nature and intent. But by characterizing (or caricaturing) his opposition as a claim for the direct production of form by physical forces, he places the chief category of putative evidence against his vision—correlation between morphology and physical conditions of life—into the functionalist camp. (One might object in principle that such a functionalist conclusion need not follow from Agassiz's version of "materialism." After all, morphology might be fashioned by laws of nature, but without functional excellence. Still, Agassiz's chosen definition should not be dismissed as self-serving because theorists who have espoused direct production of form by physical laws—D'Arcy Thompson (1917, 1942) in particular (see pp. 1179–1208)—have indeed used mechanical optimality as the criterion for their claim).

Thus, Agassiz commits himself to a "two-fisted" argument within the formalist-functionalist dichotomy: to demonstrate that taxonomic structure is a product of divine thought, he must show that classification records an anatomical order independent of external conditions of life (the positive argument for formalism), and also that a fit of form to immediate function cannot represent the generating principle of organic order (the negative argument against functionalism).

Agassiz, of course, does not deny that organisms tend to be well adapted; no formalist has ever made so strong a claim against the Paleyan alternative. He argues, rather—as formalists have done throughout history, no less so today than in Agassiz's time—that adaptation only expresses a secondary tinkering and minor adjustment of prior and fundamental *Baupläne* built by formalist principles. In its strongest version, Agassiz's brand of formalism labels adaptation as a delusion because good fit only confuses our search for a deeper order by imposing a superficial overlay of specific and immediate adaptation upon a *Bauplan*, thereby obscuring the more important underlying structure.

Agassiz's chief positive argument rests upon his unswerving allegiance to Cuvier's establishment of four anatomical ground plans as the foci of animal design: Radiata, Mollusca, Articulata, and Vertebrata.* Agassiz was particularly impressed that von Baer, the century's greatest embryologist, had, independently of Cuvier, recognized the same system by developmental standards. If morphology and embryology coincided so well, and if the greatest students of both subjects had reached agreement by such different criteria, then the fundamental principle of natural order must lie revealed:

> If we remember how completely independent the investigations of K. E. von Baer were from those of Cuvier, how different the point of view was from which they treated their subject, the one considering chiefly the mode of development of animals, while the other looked mainly to their structure; if we further consider how closely the general results at which they have arrived agree throughout, it is impossible not to be deeply impressed with confidence in the opinion they both advocate, that the animal kingdom exhibits four primary divisions, the representatives of which are organized upon four different plans of structure, and grow up according to four different modes of development (1857, p. 231).

But how shall taxonomists characterize the basis for this primary division into four? As God's mind lies so far beyond our poor faculties, we cannot identify his intent (though we can certainly record his decisions); but we may surely specify the criteria that he did *not* use. Much of Agassiz's *Essay* features a litany of claims in this negative mode: as only two alternatives exist, any argument against production of form by physical laws (a mode of origin that would induce a functional correlation of morphology and environment on the broadest scale) must provide support for the organization of relationships as categories of divine will and thought. After an introductory chapter, for example, the first two sections of Agassiz's *Essay* present a contrast with a common intent. How can physical laws simply produce the "best" solution for each particular circumstance if (1) identical environments house creatures of all four great body plans, and if (2) each of the body plans manages to inhabit all major environments? Agassiz summarizes: "The simultaneous exis-

---

*Agassiz stuck resolutely to this system until his dying day, despite collapsing evidence, particularly for the union of Coelenterata and Echinodermata within the Radiata (a subject of much personal research by Agassiz). His last, posthumous article, "Evolution and Permanence of Type" (1874), mounts an attack on Darwin from the perspective of this fourfold taxonomy. A good deal of filial piety must lie behind Agassiz's loyalty, for Cuvier, at the end of his life, had befriended the young and inexperienced Swiss naturalist, and even passed on to Agassiz the project that would assure his later fame—a monograph of all fossil fishes. This union also includes a powerful irony, for Cuvier was the most prominent of all functionalist thinkers, and Agassiz uses their shared taxonomic framework as an ultimate defense of formalism. But the richness of great and expansive systems (like Cuvier's) allows such multiplicity of use and interpretation—and sets the curious phenomenon of strange bedfellows, both in politics and intellectual life. Cuvier fancied himself as a committed empiricist. He took the quadripartite system (so amenable to formalist interpretation) as a "given" not particularly subject to analysis at all, and he then lavished his functionalist interpretations on the myriad modifications for adaptive purpose *within* each plan.

tence of the most diversified types under identical circumstances exhibits thought, the ability to adapt a great variety of structures to the most uniform conditions. The repetition of similar types, under the most diversified circumstances, shows an immaterial connection between them; it exhibits thought, proving directly how completely the Creative Mind is independent of the influence of a material world" (p. 132).

Agassiz claimed even stronger support from the geological record. Environmental change exhibits no directional pattern through time, but life's history features progressive change (via successive creations, not by evolution) within each of the four immutable types. How could unaltered physical laws and nondirectional physical change fashion a progressive history of life?

> Who could, in the presence of such facts, assume any causal connection between two series of phenomena, the one of which is ever obeying the same laws, while the other presents at every successive period new relations, an ever changing gradation of new combinations, leading to a final climax with the appearance of Man? Who does not see, on the contrary, that this identity of the products of physical agents in all ages, totally disproves any influence on their part in the production of these ever-changing beings, which constitute the organic world, and which exhibit, as a whole, such striking evidence of connected thoughts! (p. 101).

I do not claim that the refutation of Paleyan natural theology motivated this line of argument. As his major aim, Agassiz tried to debunk his caricatured version of "materialism" by showing that organisms cannot be directly constructed by physical laws. Agassiz advances his argument primarily by invoking numerous variations on the same theme: organisms do not "match" the physical world in the way that ice forms as the predictable and appropriate state for water at certain temperatures and pressures; thus, we see "how completely the Creative Mind is independent of the influence of a material world" (p. 132 as quoted above).

Agassiz begins his explicit attack on functionalism by acknowledging Paley's style of natural theology as the more common argument for God's existence and benevolence (Agassiz cites the Bridgewater Treatises, the primary Paleyan documents of his generation), but then holding that adaptation cannot represent God's primary mark upon natural history for two reasons: (1) Good correlation of function to environment would not illustrate God's care in any case, for such a relation may only record the production of form by physical causes. (2) Adaptationism fails as a generality because too many constraints, imposed by unity of type, limit any organic approach to optimality:

> The argument for the existence of an Intelligent Creator is generally drawn from the adaptation of means to ends, upon which the Bridgewater Treatises, for example, have been based. But . . . beyond certain limits, it is not even true. We find organs without functions, as, for instance, the teeth of the whale, which never cut through the gum, the

breast in all males of the class of Mammalia; these and similar organs are preserved in obedience to a certain uniformity of fundamental structure, true to the original formula of that division of animal life, even when not essential to its mode of existence. The organ remains, not for the performance of a function, but with reference to a plan (pp. 9–10).

Adaptation exists, of course, but only as a superficial and secondary overlay upon unity of type—the deeper and true reflection of God's majestic order: "When naturalists have investigated the influence of physical causes upon living beings, they constantly overlooked the fact that the features which are thus modified are only of secondary importance in the life of animals and plants, and that neither the plan of their structure, nor the various complications of that structure, are ever affected by such influences" (p. 17).

Most importantly, this deeper unity of type not only represents a natural principle in dichotomous opposition to adaptation, but also proves that creative thought, not mere mapping upon physical conditions, establishes organic order.

In all these animals and plants, there is one side of their organization which has an immediate reference to the elements in which they live, and another which has no such connection, and yet it is precisely this part of the structure of animals and plants, which has no direct bearing upon the conditions in which they are placed in nature, which constitutes their essential, their typical character. This proves beyond the possibility of an objection, that the elements in which animals and plants live . . . cannot in any way be considered as the cause of their existence (p. 33).

Having cleared away the notion that something so trivial as adaptation might represent God's signature in nature, Agassiz can now complete his ultimate defense of taxonomy as the custodian of God's presence in nature, as manifested in the broad relationships sanctioned by unity of type. Consider how much we may know of God's nature—a veritable volley of adjectives— once we locate his correct signature at the appropriate pole of nature's great dichotomy:

The products of what are commonly called physical agents are everywhere the same, (that is, upon the whole surface of the globe) and have always been the same (that is, during all geological periods); while organized beings are everywhere different and have differed in all ages. Between two such theories of phenomena there can be no causal or genetic connection. The combination of space and time of all these thoughtful conceptions exhibits not only thought, it shows also premeditation, power, wisdom, greatness, prescience, omniscience, providence. In one word, all these facts in their natural connection proclaim aloud the One God, whom man may know, adore, and love; and Natural History must, in good time, become the analysis of the thoughts of the Creator of the Universe, as manifested in the animal and vegetable kingdoms (p. 135).

Moreover, in understanding taxonomy as an incarnation of divine thought, we also sense our own importance in the cosmos. For if our taxonomy can mirror God's order so well, then our minds must also resemble His in principle, however infinitely poorer in capacity:

> Do we not find in this adaptability of the human intellect to the fact of creation, by which we become instinctively, and, as I have said, unconsciously, the translators of the thoughts of God, the most conclusive proof of our affinity with the Divine Mind? And is not this intellectual and spiritual connection with the Almighty worthy of our deep consideration? If there is any truth in the belief that man is made in the image of God, it is surely not amiss for the philosopher to endeavor, by the study of his own mental operation, to approximate the workings of the Divine Reason, learning, from the nature of his own mind, better to understand the Infinite Intellect from which it is derived. Such a suggestion may, at first sight, appear irreverent. But, which is the truly humble? He who, penetrating into the secrets of creation, arranges them under a formula which he proudly calls his scientific system? Or he who, in the same pursuit, recognizes his glorious affinity with the Creator, and, in deepest gratitude for so sublime a birth right, strives to be the faithful interpreter of that Divine Intellect with whom he is permitted, nay, with whom he is intended, according to the laws of his being, to enter into communion? (p. 8).

With so much at stake, from the basis of natural order to confidence in our mental affinity with God himself, the primacy of broad taxonomic formalism over local adaptationism (however exquisite) becomes an issue of highest moment and passion. Darwin's added dimension of history would derail Agassiz's grand design just three months after the *Essay* received its definitive printing, but we should remember Agassiz's effort, and grasp his argument, as perhaps the noblest brief ever presented for the centrality of systematics among the sciences.

### AN EPILOG ON THE DICHOTOMY

While acknowledging some historical interest in the contrast, modern evolutionists might question, on two grounds of supposed irrelevance to current issues, the time I have taken to contrast Paley with Agassiz: (1) Paley and Agassiz struggled to find the proper signature of God in nature, and such an effort no longer counts as part of science; (2) Darwin added a third, historical dimension, thereby fracturing the old dichotomy of form and function, and rendering its terms obsolete.

I would argue, in response, that Darwin's addition, though surely the most important and revolutionary event in the history of biology, scarcely rendered the old dichotomy irrelevant (see pp. 251–260 for a fuller development of this point). As Figure 4-3 shows, any morphology attributed to Darwin's historical dimension must still, by recursion, be judged by the dichotomy at its time of origin—that is, we must still know whether an ancestral form arose by adaptation or constraint (or by what mixture of the two poles). Thus, we may say that Darwin's new dimension expanded the scope of the dichotomy by compelling its application to two domains—past and present—when we analyze the basis of any trait in a living organism.

Evolution does not establish an ultimate divide for all transitions in the history of biology. Several themes pass right through this great revision, only altering their terms and explanations. Formalism vs. functionalism may be the most prominent and persistent of issues too grand even for evolution to undo (or fully resolve). Paley and Agassiz once fought this battle in grand style; Dawkins and Goodwin cannot cast so broad a conceptual net, or muster the same stylistic panache today, but they pursue the same conflict. Paley vs. Agassiz remains relevant to modern evolutionists by the primary criterion of genealogical continuity.

If Paley and Agassiz represent the yin and yang of totality for the analysis of form, then Darwin, though a pluralist who understood both poles, did ultimately cast his lot with the Paleyan yin, in filial piety with a British tradition that has spanned centuries, and still continues today. This imbalance, and the struggle for redress that now commands so much discussion in contemporary evolutionary biology, defines one of the three major issues that led me to write this book. The formalist alternative, as embodied in the subject now generally called "constraint," provides a counterweight to stabilize the second leg in Darwin's essential tripod of support—the primacy of adaptation in asserting the creativity of natural selection at overwhelming relative frequency among the causes of evolutionary change.

The past holds sufficient interest and capacity for illumination all by itself, and no justification in terms of present enlightenment need ever be given. Still, as a practicing scientist, I do favor the use of history as a current guide—while I struggle not to wrench the meaning and motivation of arguments from the primary matrix of their own time. I don't know how else to proceed when tides of history overwhelm a worthy subject for little reason beyond the vagaries of fashion and contingency. Scientists too often become convinced that inexorable logic or irrefutable data have closed a subject forever. Even worse, given our propensity for historical ignorance, we often collectively forget that an alternative ever existed at all. In such cases, I know no better tactic for reopening an important subject than the record of history—the proof that brilliant scientists (so worthy of our admiration that we cannot belittle their concerns) devoted their concentrated attention to an issue that never achieved true settlement, but only veered towards transient "resolution" by sociological complexities of shifting preferences, rather than logic of proof or exigencies of data. I believe that structuralist and formalist approaches to anatomy fell out of favor for such invalid reasons of fashion, and that the full range of this primary dichotomy must now be reestablished. And I unabashedly call upon the great formalists of history to state their case; while I ask modern evolutionists to make the proper translation to modern terms.

To reassert the importance of both poles in this dichotomy, I again cite my primary candidate for the unenviable title of most worthy "invisible man"—an important and influential thinker and educator in his day, but now entirely forgotten. I tried to resurrect the Reverend James McCosh, president of Princeton University, in Chapter 2 (pp. 116–118), and I now again want to call upon his fine book, published in 1869 in collaboration with George Dickie: *Typical Forms and Special Ends in Creation*. The Greek inscription on the title page—*typos kai telos* (type and purpose)—epitomizes the argument. The two poles of the dichotomy inhere in all natural objects, and full explanation demands attention to both:

> In taking an enlarged view of the constitution of the material universe, so far as it falls under our notice, it may be discovered that attention, at once extensive and minute, is paid to two great principles or methods of procedure. The one is the Principle of Order, or a general plan, pattern, or type, to which every given object is made to conform with more or less precision. The other is the Principle of Special Adaptation, or particular end, by which each object, while constructed after a general model, is, at the same time, accommodated to the situation which it has to occupy, and a purpose which it is intended to serve. These two principles . . . meet in the structure of every plant and every animal (McCosh and Dickie, 1869, p. 1).

McCosh also recognized the contingent and socially embedded nature of national preferences. He notes that English tradition—from Robert Boyle and John Ray through Paley to the Bridgewater Treatises—has favored the adaptationist theme. Thus, he argues, recent discoveries in formalist morphology have been viewed as threatening by some biologists (McCosh cites the French and German schools of ideal or transcendental morphology, especially in their English translation through the work of Richard Owen, whom I treat later in this chapter): "The arguments and illustrations adduced by British writers for the last age or two in behalf of divine existence, have been taken almost exclusively from the indications in nature of special adaptation of parts. Hence, when traces were discovered in the last age of a general pattern, which had no reference to the comfort of the animal or the functions of the particular plant, the discovery was represented by some as overturning the whole doctrine of final cause; not a few viewed the new doctrine with suspicion or alarm" (McCosh and Dickie, 1869, pp. 6–7).

But McCosh regards this perceived threat as false, and urges that formalist insights be welcomed—for full explanation demands attention to both poles. McCosh expresses the two key ideas in religious terms as natural illustrations of "lofty wisdom" (formalism) and "providential care" (functionalism). We call the same themes constraint and adaptation, but the image of exquisite balance remains every bit as valid today:

> We do not know whether to admire most the all-pervading order which runs through the whole of nature, through all the parts of the plant and

animal, and through the hundreds of thousands of different species of plants and animals, or the skillful accommodation of every part, and of every organ, in every species, to the purpose which it is meant to serve. The one leads us to discover the lofty wisdom which planned all things from the beginning, and the enlarged beneficence reaching over all without respect of persons; whereas the other impresses us more with the providential care and special beneficence which, in attending to the whole, has not overlooked any part, but has made provision for every individual member of the myriads of animated beings (p. 439).

Though mercilessly savaged for intellectual mediocrity by W. S. Gilbert and other satirists and activists, the British peerage did turn out an occasional scholar or two. The Duke of Argyll might have won his title fair and square if Gilbert's ultimate recommendation had ever been instituted. (The Fairy Queen in *Iolanthe,* royally pissed off at a group of nobles, threatens: "peers shall teem in Christendom, and a Duke's exalted station be attainable by competitive examination!") In a presidential address to the British Association, the good Duke, as a prominent critic of evolution and author of several books still worth reading today, argued that relations between both necessary poles of the dichotomy still persisted as a key issue in Darwin's new biology: "What is the meaning of that great law of adherence to type and pattern, standing behind, as it were, and in reserve, of that other law by which organic structures are specially adapted to special modes of life? What is the relation between these two laws; and can any light be cast upon it derived from the history of extinct forms; or from the conditions to which we find that existing forms are subjected?" (quoted in McCosh and Dickie, 1869, p. 68).

Since then, countless events, from meanderings of history to permanencies of empirical discovery, have rocked this subject back and forth. But an equilibrium at a center of dynamic tension, not of complacent rest, may foster our best biological understanding, and the Duke's question could not be more current, more *à propos.*

## Unity of Plan as the Strongest Version of Formalism: The Pre-Darwinian Debate

### *MEHR LICHT* ON GOETHE'S LEAF

A prevalent myth of our time proclaims that broad and interdisciplinary visions, though held in disrepute today, were once valued in a more ecumenical age that celebrated the "Renaissance man." But the motto that "a cobbler should stick to his last"* dates from the 4th century BC, and people who wan-

---

*The original version comes from Pliny, quite a "Renaissance man" himself, who cited Apelles from ca. 325 BC. A last, by the way, is a shoemaker's model foot, not an abstract statement about stubbornness. The original—*ne supra crepidam sutor iudicaret*—literally states that a cobbler should not judge above his last, and therefore includes some social bias amidst its narrowness.

der outside their primary field have always attracted suspicion or ridicule. In 1831, near the end of a long life, a poet who had ventured into science deplored his failure to obtain a fair hearing, but defended his forays as internally necessary for a broad and searching intellect:

> The public was taken aback, for inasmuch as it wishes to be served well and uniformly, it demands that every man remain in his own field. This demand is well grounded, for a man who wishes to achieve excellence, which is infinite in its scope, ought not to venture on the very paths that God and nature do. For this reason it is expected that a person who has distinguished himself in one field, whose manner and style are generally recognized and esteemed, will not leave his field, much less venture into one entirely unrelated. Should an individual attempt this, no gratitude is shown him; indeed, even when he does his task well, he is given no special praise. But a man of lively intellect feels that he exists not for the public's sake, but for his own. He does not care to tire himself out and wear himself down by doing the same thing over and over again. Moreover, every energetic man of talent has something universal in him, causing him to cast about here and there and to select his field of activity according to his own desire (1831 essay, in Mueller and Engard, 1952, p. 169).

We might ignore this statement, if its author stood among the many hopefuls whom history fails to memorialize in either their chosen or their adopted professions. But the writer cited above, J. W. von Goethe, wrote a thing or two of enduring merit! Moreover, and in retrospect, his ventures into science far transcended the brief forays of an amateur dabbler.

In any case, Goethe did not suffer complete neglect from scientists during his lifetime. In 1831, the great anatomist Etienne Geoffroy Saint-Hilaire, praised Goethe's science as the work of "a poet trying to sing the grandeur of the universe in another form" ("un poète s'essayant de chanter sous une autre forme les grandeurs de l'univers"—1831, p. 189). Geoffroy continued (1831, p. 193): "If Goethe had not already amassed enough titles to be proclaimed the greatest genius of his century, he would have added, to his crown of great poet and profound moralist, the fame of a wise naturalist—due to him for the profundity of his views, and for the philosophical force of his opinions on the subject of botanical analogies."*

But Geoffroy's praise (see Fig. 4-8) cannot be reckoned as entirely disinterested, for Goethe had just favored his side in the greatest brouhaha of early 19th century zoology—the celebrated 1830 debate with Cuvier before the

---

*Geoffroy wrote this work before Owen's codification of the terms "homology" and "analogy" in their modern meaning. Unfortunately—for the resulting situation could not be more confusing—Geoffroy used the word analogy (*théorie des analogues*) for similarities of common generating type that we now call homologies. In this chapter, I will use the modern terminology, and only retain Geoffroy's name in direct quotations (but always with a reminder that he speaks of homology by our reckoning).

*Académie des sciences.* Geoffroy needed all the help he could get (and he would later recruit other literary figures, including the novelists Balzac and George Sand, to his cause as well). Cuvier, after all, was no ordinary opponent, and the subject of their argument—the age-old dispute of formalism and functionalism—could not have been more central to natural history.

Geoffroy, with good cause, viewed Goethe as the doyen and spiritual leader of formalist morphology. Not only had Goethe coined the word "morphology," but he had, long before, defended for plants the central proposition that Geoffroy championed for animals as the starting point for his anatomical views—the reduction of form to a single generating archetype (the leaf for Goethe, the vertebra for Geoffroy). While the young Geoffroy worried about establishing a career and surviving a revolution, Goethe was travelling in Italy and developing the theory of his 1790 work, *Versuch die Metamorphose der Pflanzen zu erklären* (Fig. 4-9). (This work, little more than a pamphlet, consists of 123 numbered and almost aphoristic paragraphs. I shall quote by number from the standard translation of Mueller and Engard (1952). But I have read and own a copy of the original, which I highly recommend to anyone who appreciates the fusion of great writing and fascinating science.)

Goethe had been strongly interested in morphology throughout his life, and his preferences had always tended towards formalism, particularly towards the strongest version of the argument (and subject of this section)—the vision of a single, generating archetypal form, setting both the bounds and the possibilities of realized morphology. His two most famous forays into animal anatomy both rested upon a formalist foundation: (1) his early support for the vertebral theory of the skull, a conviction that he traced to 1791 when he examined "a battered sheep skull from the sand of the dunelike Jewish

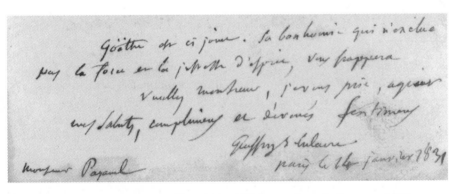

4-8. A letter from Etienne Geoffroy Saint-Hilaire, written in January 1831 to a Mr. Payaud, with Goethe featured in the penultimate paragraph. (Geoffroy presumably enclosed some writings of Goethe along with this letter to Payaud.) The text of the paragraph reads as follows: *"Goethe est ci-joint. Sa bonhomie qui n'exclue pas la force et la justesse d'esprit, vous frappera."* (Goethe is enclosed. His good nature, which does not preclude strength and fairness of spirit, will strike you.) (Author's collection.)

Versuch

die

## Metamorphose der Pflanzen

zu erklären

von

J.  W.  von  Göthe.

Gotha.
Ettingersche Buchhandlung.

1790.

4-9. Title page of Goethe's 1790 pamphlet on the growth and archetypal expressions of plants. (Author's collection.)

cemetery in Venice" (1823 essay, in Mueller and Engard, 1952, p. 237); and (2) his discovery of the human premaxillary bone in 1784, based on its presence in other mammals and his convictions about unity of type. (Goethe called this bone the intermaxillary; others referred to it as "Goethe's bone." In an essay written in 1832, the year of his death, Goethe recalled this discovery as "the first battle and the first triumph of my youth" (Goethe, 1832, p. 573).)

But Goethe chose botany for his most extensive study in formalism, and probably his finest contribution to science. In this important work, Goethe applied to plants the same vision that Geoffroy and Owen would later advance in trying to reduce the great complexity and diversity of animal (or at least vertebrate) form to the single generating pattern of an archetypal vertebra (see Geoffroy, 1831, for a homage to Goethe). For Goethe, the leaf represented an archetypal form for all plant parts growing from the central stem*—from cotyledons, to stem leaves, to sepals, petals, pistils and stamens, and fruit.

The common epitome of Goethe's system—all is leaf—should not be taken literally as the actual reduction of all serial diversity to the actual form of a stem leaf. Such a reading would contravene the Platonic character of archetypes in formalist theory. The "leaf" represents an abstract generating principle, from which stem leaves depart least in actual expression. Goethe writes: "We ought to have a general term with which to designate this diversely metamorphosed organ and with which to compare all manifestations of its form. . . . We might equally well say that a stamen is a contracted petal, as that a petal is a stamen in a state of expansion; or that a sepal is a contracted stem leaf approaching a certain stage of refinement, as that a stem leaf is a sepal expanded by the influx of cruder saps" (1790, No. 120).

Goethe expressed the epitome of his system in measured tone within his essay (1790, No. 119): "The organs of the vegetating and flowering plant, though seemingly dissimilar, all originate from a single organ, namely, the leaf." In private, he became more effusive: "[I have traced] the manifold specific phenomena in the magnificent garden of the universe back to one simple general principle" (1831 essay, in Mueller and Engard, 1952, p. 168). To friends, as to the philosopher J. G. Herder, he could become positively effusive (dare I say florid): "The archetypal plant as I see it will be the most wonderful creation in the whole world, and nature herself will envy me for it. With this model and the key to it, one will be able to invent plants . . . which, even if they do not actually exist, nevertheless might exist and which are not merely picturesque or poetic visions and illusions, but have inner truth and logic. The same law will permit itself to be applied to everything that is living" (letter of 1787, quoted in Mueller and Engard, 1952, p. 14).

Goethe dissects and compares, trying to find the leaf-like basis of apparently diversified and disparate structures. The anastomosed sepals, forming the calyx at the base of a flower, must be leaves that fail to separate when a cutoff of nutriment stops expansion of the stem: "If the flowering were retarded by the infiltration of superfluous nutriment, the leaves would be separated and would assume their original shape. Thus, in the calyx, nature forms

---

*Goethe's theory encompasses only the lateral and terminal organs, not the supporting roots and stems. Goethe was often castigated for this omission, unfairly I think. (A theory for all appended parts cannot be dismissed as a mean thing, even if the framework remains unaddressed.) Goethe became quite sensitive to such criticism, and defended his failure to consider roots by heaping disdain upon such lowly parts: "My critics have taken me to task for not considering the root in my treatment of plant metamorphosis . . . I was not concerned with it at all, for what had I to do with an organ which takes the form of strings, ropes, bulbs, and knots, and manifests itself in such unsatisfying alternation, an organ where endless varieties make their appearance and where none advance. And it is advance solely that could attract me, hold me, and sweep me along on my course. Let everyone to his own way. Let him, if he can, look back upon forty years of accomplishment, such as the Good Genius has granted me" (from unedited paragraph written in 1824, quoted in Mueller and Engard, 1952).

no new organ but merely combines and modifies organs already known to us" (1790, no. 38).

When parts become too distinct to show connection and reduction to the leaf archetype in one species, Goethe uses the comparative approach to find transitional forms in other taxa. The seedpod and sexual organs are manifestly unleaflike in many plants, but Goethe establishes transitional series to species with, for example, leaflike seedpods, or fertile stem leaves (as in ferns). Consider his exposition of the comparative method for "difficult" seedpods: "Nature obscures the similarity to the leaf most when she makes the seed containers soft and juicy or firm or woody; however, the similarity will not escape our attention if we contrive to follow it in all its transitional stages" (1790, No. 79). Or for the even more divergent cotyledons that eventually grow into tolerably leaflike form:

> 12. They are often misshapen, crammed, as it were, with crude matter, and as much expanded in thickness as in breadth; their vessels are unrecognizable and scarcely distinguishable from the mass as a whole. They bear almost no resemblance to a leaf, and we might be misled into regarding them as special organs.
>
> 13. Yet in many plants the cotyledons approach leaf form: they flatten out; exposed to light and air, they assume a deeper shade of green; their vessels become distinct and begin to resemble veins.
>
> 14. Finally they appear before us as true leaves: their vessels are capable of the finest development; their similarity to the subsequent leaves will not permit us to consider them separate organs; and we recognize them instead as the first leaves of the stem (1790, Nos. 12–14).

If Goethe's system really advocated, as often misportrayed, a simple and exclusive concept of the archetypal leaf, his theory could stake no claim for interesting completeness—for this central principle cannot explain systematic variation in form up the stem, and therefore could not operate as a full explanation for both similarities *and* characteristic differences in the parts of plants. But, in his most fascinating intellectual move, Goethe proposes a complete account by grafting two additional principles onto the underlying notion of archetype: the progressive refinement of sap, and cycles of expansion and contraction. We may regard these principles as *ad hoc* or incorrect today, but the power of their conjunction with the archetypal idea can still be appreciated with much profit.

These two additional principles embody both necessary sides of the primary Western metaphor for intelligibility in any growing, or historically advancing, system—arrows of direction and cycles of repeatability (I called these conjoined principles "time's arrow" and "time's cycle" in my book on the discovery of geological time—Gould, 1987b). We must, in any temporal process, be able to identify both sources of story and order: vectors of change (lest time have no history, defined as distinctness of moments), and underlying constant or cyclical principles (lest the temporal sequence proceed only as one uniqueness after another, leaving nothing general to identify at all). Goethe, faced with observations of both directionality and repeatability up the stem, recognized the need for both poles of this dichotomy.

REFINEMENT OF SAP AS A DIRECTIONAL PRINCIPLE. Up and down, heaven and hell, brain and psyche vs. bowels and excrement, tuberculosis as a noble disease of airy lungs vs. cancer as the unspeakable malady of nether parts (see Susan Sonntag's *Illness as Metaphor* for a brilliant analysis of these conventional images). Almost irresistibly, we apply this major metaphorical apparatus of Western culture to plants as well—with gnarly roots and tubers as lowly objects of the ground, and fragrant, noble flowers as topmost parts, straining towards heaven. Goethe, by no means immune to such thinking in an age of *Naturphilosophie*, viewed the growth of a plant as progressing towards refinement from cotyledon to flower. He explained this directionality by postulating that, moving up the stem, each successive leaf modification progressively filters an initially crude sap. Inflorescence cannot occur until these impurities have been removed. The cotyledons begin both with minimal organization and refinement, and with maximal crudity of sap: "We have found that the cotyledons, which are produced in the enclosed seed coat and are filled to the brim, as it were, with a very crude sap, are scarcely organized and developed at all, or at best roughly so" (1790, No. 24).

The plant then grows towards a floral apotheosis, but too much nutriment delays the process of filtering sap—as material rushes in and more stem leaves must be produced for drainage. A decline in nutriment finally allows filtering to attain the upper hand, and the sap becomes sufficiently pure for inflorescence: "As long as cruder saps remain in the plant, all possible plant organs are compelled to become instruments for draining them off. If excessive nutriment forces its way in, the draining operation must be repeated again and again, rendering inflorescence almost impossible. If the plant is deprived of nourishment, this operation of nature is facilitated" (1790, No. 30). Finally, the plant achieves its topmost goal: "While the cruder fluids are in this manner continually drained off and replaced by pure ones, the plant, step by step, achieves the status prescribed by nature. We see the leaves finally reach their fullest expansion and elaboration, and soon thereafter we become aware of a new aspect, apprising us that the epoch we have been studying has drawn to a close and that a second is approaching—the epoch of the flower" (1790, No. 28).

CYCLES OF EXPANSION AND CONTRACTION. If the directional force worked alone, then a plant's morphology would only express this smooth continuum of progressive refinement up the stem. Since, manifestly, plants do not display such a pattern, some other force must be operating.* Goethe de-

---

*Note the isomorphism between Goethe on plant morphology and Lamarck on the mechanics of evolution (see Chapter 3). Both argued that a progressive force dominated the entire system—yielding, if unopposed, a march up the chain of being for Lamarck and a progressive refinement of organs up the stem for Goethe. But both scientists bowed to the empirical data of greater complexity and messiness in nature. Both therefore needed to postulate orthogonal or deflecting forces to produce discontinuities and lateral morphologies—local adaptation for Lamarck and cycles of expansion and contraction for Goethe.

scribes this second force as cyclical, in opposition to the directional principle of refining sap. He envisages three full cycles of contraction and expansion during ontogeny. The interplay of these progressive and cyclical forces produces the full pattern of a general refinement up the stem, but impacted by discontinuities and transitions that express no directional pattern ("contraction" of stem-leaves to sepals by bunching together in a circlet, for example). The cotyledons begin in a retracted state. The main leaves, and their substantial spacing on the stem, represent the first expansion. The bunching of leaves to form the sepals at the base of the flower marks the second contraction, and the subsequent elaboration of petals the second expansion. The reduction of archetypal leaf size to form pistils and stamens marks the third contraction, and the formation of fruit the last and most exuberant expansion. The contracted seed within the fruit then starts the cycle again in the next generation. Put these three formative principles together—the archetypal leaf, progressive refinement of sap up the stem, and three expansion-contraction cycles of vegetation, blossoming, and bearing fruit—and the vast botanical diversity of our planet falls under the chief vision of formalism: production of realized variety from interaction of a few abstract, general, and internally based (not externally imposed and adaptationally driven) morphological laws: "Whether the plant vegetates, blossoms, or bears fruit, it nevertheless is always the same organs with varying functions and with frequent changes in form, that fulfill the dictates of nature. The same organ which expanded on the stem as a leaf and assumed a highly diverse form, will contract in the calyx, expand again in the petal, contract in the reproductive organs, and expand for the last time as fruit" (1790, No. 115).

This formalist commitment implies an aversion to primary explanation by adaptation, function or final cause. In accord with all the great formalists, Goethe often expressed his dislike of explanations based upon the externality of fit between form and function (though he delighted in the evident fact of such fit, as formalists also generally do, for such an admission poses no threat to the chief formalist argument for *primacy* of morphological order—see Chapter 11 on exaptation).

Goethe's statements on final cause often attack the larger idea of manufacture for explicitly human ends—not the chief complaint of formalist morphology, but worth recording, if only for the power of Goethe's prose:

> For several centuries down to the present, we have been retarded in our philosophic views of natural phenomena by the idea that living organisms are created and shaped to certain ends by a teleological life force. ... Why should he not call a plant a weed, when from his point of view it really ought not to exist: He will much more readily attribute the existence of thistles hampering his work in the field to the curse of an enraged benevolent spirit, or to the malice of a sinister one, than simply regard them as children of the universal nature, cherished as much by her as the wheat he carefully cultivates and values so highly (from essay of 1790, in Mueller and Engard, 1952).

But Goethe also attacked adaptationist primacy in the more focused realm of explaining morphology: "It is not a question of whether the concept of final causes is convenient, or even indispensable, to some people, or whether it may not have good and useful results when applied to the moral realm; rather, it is a question of whether it is an aid or a deterrent to physiologists in their study of organized bodies. I make bold to assert that it does deter them, therefore avoided it myself and considered it my duty to warn others against it" (2nd essay on plant metamorphosis, written in 1790, in Mueller and Engard, 1952, p. 80).

Citing a perennial complaint, then and now, against adaptationist explanations—that such efforts tell good stories in the speculationist mode, but do not explain morphology—Goethe compares final causes with Linnaeus' fanciful descriptions of sexual anatomy in plants: "For example, Linné calls flower petals 'curtains of the nuptial bed,' a parable that would do honor to a poet. But after all! the discovery of the true physiological nature of such parts is completely blocked in this way, just as it is by the convenient and false espousal of the theory of final causes" (2nd essay on plant metamorphosis, written in 1790, in Mueller and Engard, 1952, pp. 79–80).

Proper morphological explanations, Goethe asserts, must be sought on internal and formalist principles; external fit, though of great importance, can only be regarded as secondary: "In my opinion, the chief concept underlying all observation of life—one from which we must not deviate—is that a creature is self-sufficient, that its parts are inevitably interrelated, and that nothing mechanical, as it were, is built up or produced from without, although it is true that the parts affect their environment and are in turn affected by it" (2nd essay on plant metamorphosis, written in 1790, in Mueller and Engard, 1952, p. 80).

In a remarkable passage, that could serve as a credo for modern formalism as well, Goethe asserts his central claim for internalist primacy, while also specifying the vital, but secondary, role of adaptation. Internal formation acts as a primary source that "must find external conditions." Adaptation may then shape a range of diversity from an underlying form, but the archetypal pattern cannot be explained by these secondary modifications, and the adaptations themselves can only express a superficial restructuring of inherent order:

Man, in considering all things with reference to himself, is obliged to assume that external forms are determined from within, and this assumption is all the easier for him in that no single living thing is conceivable without complete organization. Internally, this complete organization is clearly defined; thus it must find external conditions that are just as clear and definite, for its external existence is possible only under certain conditions and in certain situations. . . . An animal possesses external usefulness precisely because it has been shaped from without as well as from within, and—more important and quite natural—because the external element can more readily adapt the external form to its own purposes than it can reshape the internal form. We can best see this in a species of seal whose exterior has taken on a great deal of the fish character while its skeleton still represents the perfect quadruped (2nd essay on plant metamorphosis, written 1790, in Mueller and Engard, 1952, p. 83).

Goethe's views therefore provide a "test case" for a primary thesis of this book. We should, I believe, recognize the space of our intellectual world as inherently structured, by some combination of our evolved mental quirks and the dictates of logic, into a discontinuous array of possible, coherent positions—hence the *double entendre* in the title of this book. These mental positions express "morphologies," just as organisms do. The chief components of these "morphologies" must reside together and interact to build the "essence" of any powerful intellectual system. The components of a theory's essence should be recognized as both deep and minimal; with other less important and potentially dispensable principles allied to them in secondary webs subject to "restructuring" by "adaptation." (Thus I advocate a minimal set of three principles for defining the essence of Darwinism, while regarding other components of the usual Darwinian nexus as conjoined more loosely and less central intellectually.) These essential and minimal components remain correlated, although arising independently and in reiterated fashion, across languages, centuries and cultural traditions. Such firm linkages define the structure of these few nucleating positions in the intellectual landscape.

In formalist or structuralist theories, the strongest correlation unites a commitment to generative laws of form with an aversion to adaptationist explanation as the primary goal of morphology. The two commitments need not conjoin in logic or empirical necessity; indeed, Darwin found a brilliant argument to drive them asunder by identifying most (though not all) generating principles as past adaptations, and relegating remaining laws of form to a peripheral or secondary status (see section 1 of this chapter). But almost every formalist theory of morphology also views adaptation as secondary tinkering rather than primary structuring.

I regard Goethe as an exemplar of this approach to major scientific theories because he was, in an important sense, an outsider to the swirling debates of formalism vs. functionalism in his time. He understood, of course, his affinity with the formalism of German *Naturphilosophie*. But he did not attend the debates, publish in the journals, or use the lingo of developing scientific professionalism. He viewed himself as apart and neglected. In fact, he didn't even regard the debate between Cuvier and Geoffroy, which fascinated him so keenly at the end of his life (see pages 310–312), primarily as a struggle between formalism and functionalism, but rather as a contest between the empiricism of Cuvier and the intuitionism of transcendental morphology—and his explicit preference for Geoffroy invoked his poet's concern with the primacy of abstract ideas as much as his morphologist's attention to the primacy of form.

In this context of Goethe's separation from the core of scientific controversy in 1830, we should not treat his own formalism as derivative, imitative, or simply imbibed from the stated mores of a recognized intellectual brother-hood. If his views also feature—as they do—a linkage of interest in laws of form with an antipathy to adaptationist explanation, then we may interpret the correlation as independently generated, at least in part, and therefore as good evidence for a link based upon intrinsic intellectual entailment in the "morphology" of formalism as a key "nucleating" idea in biology.

Indeed, Goethe showed a strong appreciation for the morphology (and, in this case, the utility) of dichotomy in intellectual life. In discussing his under-standing of the division between Cuvier and Geoffroy, Goethe noted that each man defended not a single idea or a unitary position, but rather a nexus or complex of mutually entailed notions, causing a precipitation at one of two foci—with these two aggregations then opposing each other like the poles of a magnet. For Goethe, the systems of Cuvier and Geoffroy formed "two different doctrines, which are so ordinarily and so necessarily separated that little chance exists for finding them together in a single person. On the contrary, it is of their essence that they not be well allied" (Goethe, 1831, p. 181). For Darwin, discontinuity originates by historical contingency (fol-lowing extinction of intermediate forms) in a fully accessible and isotropic morphospace. *Natura non facit saltum.* But the universe of formalism—in ideas and in morphology—views discontinuity as inherent in the structure of inhabitable space.

## GEOFFROY AND CUVIER

### *Cuvier and conditions of existence*

The struggle of Cuvier and Geoffroy continues to rivet our attention (from Russell, 1916, to the modern book of Appel, 1987) because this conflict fea-tures the two central elements of intellectual drama: a clash of two superior minds within the primal tale of professional ontogeny: two scholars begin as warm friends fired with the idealism of youth, and end as wily, cynical, politi-cally astute opponents. (The conventional view interprets Cuvier as a clear winner by virtue of such astuteness and Geoffroy as loser by naiveté and woolliness. I shall defend a different version of interesting, disparately styled, equality.)

When the French revolutionary government established the *Muséum d'histoire naturelle* as the world's finest in 1793, Etienne Geoffroy Saint-Hilaire, installed as the first curator for vertebrates, played a primary role in bringing Georges Cuvier to Paris, thus launching his scientific career. The two men enjoyed close friendship, sharing living quarters and making idealistic plans for the reform and flowering of natural history. In 1798, as Geoffroy embarked with Napoleon upon a long expedition to Egypt, he wrote to Cuvier: "Goodbye my friend, love me always. Do not cease to consider me as a brother" (in Appel, 1987, p. 73).

But their differences in temperament, intellect and style eventually and in-evitably drove them apart. Cuvier became one of the most powerful, politi-cally conservative figures ever to operate in Western science. The oft-quoted statement of the awestruck Charles Lyell, visiting Cuvier at the height of his influence, provides insight into the nature of his power:

> I got into Cuvier's sanctum sanctorum yesterday, and it is truly characteristic of the man. In every part it displays that extraordinary power of methodizing which is the grand secret of the prodigious feats which he performs annually without appearing to give himself the least trouble . . . There is first the museum of natural history opposite his house, and admirably arranged by himself, then the anatomy museum connected with his dwelling. In the latter is a library disposed in a suite of rooms, each containing works on one subject. There is one where there are all the works on ornithology, in another room all on ichthyology, in another osteology, in another *law* books! etc. . . . The ordinary studio contains no bookshelves. It is a longish room comfortably furnished, lighted from above, with eleven desks to stand to, and two low tables, like a public office for so many clerks. But all is for the one man, who multiplies himself as author, and admitting no one into this room, moves as he finds necessary, or as fancy inclines him, from one occupation to another. Each desk is furnished with a complete establishment of inkstand, pens, etc. . . . There is a separate bell to several desks. The low tables are to sit to when he is tired. The collaborators are not numerous, but always chosen well. They save him every mechanical labour, find references, etc., are rarely admitted to the study, receive orders and speak not (in Adams, 1938, p. 267).

Appel notes an interesting source of Cuvier's accumulated influence: "Cuvier was able to remain on the Council [of State] through the Empire, three kings, and several ministries because he held no extreme opinions and was willing to support whatever regime was in power" (1987, p. 53). Yet, lest we view this chameleonic shifting merely as cynical and self-serving, much like the Vicar of Bray in the old song about maintaining office through all the vicissitudes of 17th century British politics, Appel points out the underlying consistency of a true political and biological conservative: after a bloody and traumatic revolution, any hierarchical order, proceeding from any source holding promise for stability, must be preferred over potential anarchy and populism.

Appel designates three broad domains of difference between Cuvier and Geoffroy: Cuvier's conservative connection to substantial political power, his insistence (largely for rhetorical purposes, since science cannot operate in such a manner) that the profession restrict itself to reporting positive facts and shunning speculation, and his commitment to one of the purest forms of functionalism ever maintained in the study of morphology. Appel notes the evident connection between political élitism and the call for a descriptive, factually-based science of experts:

> In a politically volatile country which had recently experienced traumatic revolution, Cuvier justly feared that speculative theories, most of

which had a materialist tinge, would be exploited in the name of science and undermine religion and promote social unrest. If science could be limited to experts and restricted to accumulating "positive facts" then it might achieve a measure of autonomy, while at the same time the questioning that might lead to heretical theories would be eliminated. As Cuvier became increasingly concerned about the danger posed by certain biological theories, he became increasingly insistent on the restraints imposed by proper scientific method (Appel, 1987, pp. 52–53).

The third theme of morphological explanation, though supported by other roots, also melds into the Cuvierian totality of politics, method, and theory—for Cuvier's functionalism views organisms as discrete, untransformable entities, designed for specific conditions of life *and no other.* By contrast, Geoffroy held opposite attitudes on all three accounts—as an outsider in politics, both academic and national; a dreamer and visionary in methodology, a man who explicitly exalted the power of ideas to guide and even to channel factual inquiry; and a resolute formalist in morphology, with a theory of robust generation and transformation along lines set by overarching laws of structure and archetypal form.

To grasp the purity of Cuvierian functionalism, we must break through a century's commitment to genealogical models of relationship. We are now so wedded (properly of course) to the homological basis of deep similarity by descent, that we can scarcely imagine any other theory of *Bauplan.* After all, what could the sequence of humerus, to radius and ulna, to carpals, metacarpals and phalanges denote except inheritance by common descent when expressed over so broad a functional range as dolphin, dog and bat. Even the most rabid panselectionist would not identify phylum-level homologies (broad symplesiomorphies) as indications of *current* function. At most, following Darwin (see pp. 253–260), they would view such features as originating by adaptation in distant ancestors. Current function will then be expressed in particular modifications of homologies *within* each line.

Yet Cuvier actually believed that common features of current *Baupläne* recorded such immediate functional rules of correlation. Cuvier acknowledged that science does not yet understand organic physics well enough to know the logical basis of these rules, and must therefore work empirically from comparative anatomy, but the regularities must be rooted in function and will, one day, be resolved analytically. Start with a carnivore's claw (or canine tooth, or any other tool of its trade), and all other items of anatomy follow by mechanical necessity. One part implies the next, and eventually the entire skeleton, according to correlations set by functional rules alone. Type records broad function; specific adaptation denotes local function. No part exists "in vain" or merely to indicate conformity to plan (vestigial organs, developmental sequelae). Evolution becomes literally inconceivable because change in one part requires corresponding change in every other intimate detail—and no one can imagine a mechanism for such globally coordinated alteration. (Nor can one, even today, gainsay this excellent argument. If evolution were not mosaic, transmutation would be inconceivable, and would not occur precisely for the reasons stated by Cuvier.)

In Cuvier's remarkable output of publications, three works stand out as powerful, comprehensive documents that established professions and set a good part of the course of 19th century biology—the 5 volume *Leçons d'anatomie comparée* of 1800–1805, the 4 volume *Recherches sur les ossemens fossiles* of 1812, and *Le règne animal* of 1817. The pivotal role of these three works has always been acknowledged, but their common philosophical grounding in Cuvier's overarching functionalism has not been adequately recorded.

The *Leçons* of 1800–1805 arranges natural history in functional terms by shunning the usual taxonomic order and proceeding instead by organ systems considered in operational rather than morphological terms. Volume one treats locomotion, functionally focused and defined *("les organes du mouvement"),* while subsequent volumes proceed through sensation, digestion, circulation, respiration, voice, generation, and excretion.

The very first lesson, functionally organized as *"considérations sur l'économie animale,"* presents the heart of Cuvier's approach. His theory of function cannot be characterized as a crude, "democratic" adaptationism, part by part with each item separately optimized, but rather as a more subtle, hierarchical system that renders both structural regularities and correlations in functional terms. Primary functions, common to all organisms, lie at the base—origin by generation, growth by nutrition, and termination by death (see Russell, 1916, p. 31). Secondary functions—feeling and moving—build a layer above and set the morphology of organs for their manner of operation. These secondary, or "animal," functions, with their neuromuscular expression, determine a yet higher level of "vital functions"—digestive, circulatory, and respiratory, in that order. Feeling and movement require a set of organs to hold and process food; digestion then implies a system of distribution (circulation). Higher levels may then feed back "in a type of circle" (Cuvier, 1805, p. 47) to influence the logically prior foundation. Power of movement affects mode of generation and *"fluide nerveux"* of secondary status flows through channels of tertiary circulation. Above all, function holds priority and determines structure; coordination and correlation among structures records the hierarchical ordering of interrelated functions (see particularly Cuvier, 1805, pp. 45–60).

Cuvier states the functional foundation of his morphology in bold terms (1805, p. 47): "The laws that determine the relationships of organs are founded upon this mutual dependence of functions, and upon the aid that they lend to each other. These laws have a necessity equal to laws of metaphysics and mathematics. For it is evident that a proper harmony among organs that act upon each other is a necessary condition of existence* for the

---

*Conditions d'existence* became Cuvier's motto for his functionalist credo. By this phrase, as evident in this quotation, he did not only designate adaptation to external environment, but also coordination of parts by and for the pursuit of proper function. Note that Darwin used Cuvier's phrase in identifying the functionalist pole of the dichotomy (1859, p. 206, and Section 1 of this chapter).

creature to which they belong. If one of these functions were modified in a manner incompatible with modifications of other organs, this creature could not exist."

This statement of analytically necessary functional laws, and ineluctable correlation of parts, echoes the philosophy better known from the justly celebrated *Discours préliminaire* of Cuvier's 1812 *Recherches*, the document that founded modern paleontology by establishing the fact of extinction and organic succession through time. The laws of organic form have a purely functional basis. One anatomical part implies all others, for proper function (not abstract laws of structure) demands such interdependence.* Animals therefore cannot undergo substantial change by evolution because such a complex and precisely coordinated transformation of all parts could not occur—especially under functionalist theories of the independent and adaptational origin of each part (rather than the coordinated change of all parts of an archetypal form along preestablished lines of possibility, thus making evolution far easier to conceive under the formalist philosophies that Cuvier rejected). Therefore, when geological conditions change drastically, many species die and can never reappear or continue in any way. The sequence of extinctions through time gives the earth a history by establishing a vector of directional change. Geology, now furnished with an alphabet, can finally become a science. Cuvier expresses the functional basis of correlation:

> Every organized individual forms an entire system of its own, all the parts of which mutually correspond, and concur to produce a certain definite purpose, by reciprocal reaction, or by combining towards the same end. Hence none of these separate parts can change their forms without a corresponding change in the other parts of the same animal, and consequently each of these parts, taken separately, indicates all the other parts to which it has belonged (from the standard Jameson translation, 1818, p. 99).

> In short, the shape and structure of the teeth regulate the forms of the condyle, of the shoulder-blade, and of the claws, in the same manner as the equation of a curve regulates all its other properties (1818, p. 102) ... Anyone who observes merely the print of a cloven hoof, may conclude that it has been left by a ruminant animal, and regard the conclusion as equally certain with any other in physics or in morals (p. 105).

The relationship of the third great work—*Le règne animal* of 1817—to this functionalist nexus seems more obscure at first. Here Cuvier codifies the system of animal taxonomy that he first published in 1812—the abandonment

---

*This doctrine, called "correlation of parts," spawned the legend, much abetted by Cuvier's overenthusiasm in the *Discours préliminaire*, that paleontologists can reconstruct entire skeletons from single bits of bone. (We may do so *inductively* by knowing that a tooth of distinctive form only grows in a rhino's jaw, but we cannot—as Cuvier hoped and hyped—make such reconstructions analytically. Cuvier, in fact, admitted as much by stating that current practice, in the light of our analytical ignorance, must be empirical—and by bragging that he could outdo any competitor by virtue of his superb collection of skeletons at the *Muséum!*—see Gould, 1991b.)

of the old bipartite division of vertebrate and invertebrate (and the equation of vertebrate classes with invertebrate phyla), for a system of four equal *embranchements* based on necessarily separate and untransformable anatomical plans: Radiata, Articulata, Mollusca, and Vertebrata. This appeal to limited and untransformable anatomical designs as a basis for taxonomic order smacks of structuralism, but Cuvier, true to his guiding philosophy, presents a purely functional interpretation. Appel (1987, p. 45) explains: "The unity within an *embranchement* came not from a comprehensive unity of plan, but from a common arrangement of the nervous system, functionally the most important system of the animal. The forms of the other major systems remain constant throughout an *embranchement* because the other systems—respiration, circulation, etc.—were functionally subordinate to the nervous system and determined by the requirement of the nervous system. Animals within an *embranchement* could vary almost arbitrarily in their accessory parts, precisely because accessory parts were not necessitated by the choice of the nervous system." Both unity and diversity therefore achieve a functional interpretation—unity by operational design, diversity by local adaptation. Conditions of existence set both major aspects of taxonomy.

I emphasize a primary intellectual correlation throughout this chapter—formalism with commitment to internal constraint (in the positive sense of channeling change, not only the negative definition of restriction). To render this connection meaningful, the converse must also hold: functionalism must correlate with denial of constraint. Cuvier's arguments test and affirm this implication.

In an overly broad (and therefore operationally meaningless) construction of constraint, all biologists acknowledge some restriction on organic form, if only because all conceivable shapes and sizes have not been realized. But we usually do not apply this term to nature's avoidance of obviously unworkable creatures (flying elephants or large dinosaurs with pencil-thin legs in Galileo's world of laws regulating the ratio of surface to volume), for no one disputes the underlying physical basis for their nonexistence. (For historically contingent reasons of modern professional life within a Darwinian functionalist paradigm, we currently apply the term "constraint" primarily to internal channels and limitations not set by adaptation—see my full argument for this usage in Chapter 10, pp. 1027–1037. That is, we apply the concept of "constraint" to sources of influence *outside* a favored explanation—see Gould, 1989a.)

Thus, Cuvier cheerfully acknowledged limits set by function, but did not view such boundaries as constraining because aborted, unworkable creatures offend the very notion of a rational creating force. Instead, and thereby affirming the link of functionalism to a denial of constraint, Cuvier clearly cherished his general theory as a principle for maximizing God's liberty to create (translated as "adaptation to alter" in the modern evolutionary version of functionalism). Cuvier wrote in an 1825 essay on "Nature": "If we look back to the Author of all things, what other law could actuate Him but the necessity of providing to each being whose existence is to be continued the means of assuring that existence? And why could He not vary His materials and His instruments? Certain laws of coexistence of organs were therefore necessary, but that was all. For to establish others there must have been a want of freedom in the action of the organizing principle, which we have shown to be only a chimera" (in Appel, 1987, p. 138).

Cuvier, a severe rationalist (see Fig. 4-10 for an interesting and previously unpublished illustration of Cuvier's rationalism and hostility to florid meta-

4-10. A remarkable note, written by Cuvier in his own hand, and indicating how much this rationalist thinker rejected and ridiculed silly metaphorical uses of poetic imagery as a substitute for rigor, or for saying anything of real substance. Here Cuvier jotted down two such fatuous and metaphorical uses of "sphere"—obviously stored away for later use in satire or ridicule. (Author's collection.)

Definition of life by M. Virey. Life is a circular movement, sustained and measured by time; time, that infinite sphere, of which God alone is the center, and where living beings are placed on the circumference, describing in their rapid orbit, the circle of their destiny.

Definition of poetry by Mme. de Staël. Poetry is the winged mediator, which moves distant nations and ancient times in a sublime sphere where admiration takes the place of sympathy.

phor), rarely waxed poetic about nature's abundances, but he surely rejoiced that organic form knew no limits beyond good design.

> While always remaining within the boundaries prescribed by necessary conditions of existence, nature abandons herself to all fecundity not limited by these conditions; and without ever departing from the small number of possible combinations for modification of important organs, she seems in all accessory parts, to be limitlessly endowed . . . Thus we find that as we move away from the principal organs, and approach those that are less important, varieties are multiplied; and when we arrive at the surface, where the nature of things ordains that the least important parts be placed, and where any damage is least dangerous, the number of varieties becomes so great that all the work of naturalists has not succeeded in giving us any idea of its magnitude (1805, p. 58).

### Geoffroy's formalist vision

Since the modes and practices of science inevitably reflect a surrounding social environment, we should scarcely be surprised that the early to mid 19th century world of revolution in politics, and romanticism in art, literature, and music, also inspired a series of biological movements called *Naturphilosophie* in Germany and romantic, idealistic, transcendental, or philosophical anatomy elsewhere. A scientific movement may begin under strong social influence and little compulsion by data, but its empirical adequacy may ultimately rank high nonetheless. (Evolutionists, above all other professionals, should be optimally preprogrammed to appreciate the difference between reasons for origin, and assessment of eventual value—see pp. 1214–1218 particularly for Nietzsche's analysis of this vital issue in historiography.) Geoffroy, as the most important of the transcendental morphologists, heard the songs of his time, but he also composed a flawed symphony that plays better today than to the previous generation that built the Modern Synthetic theory of evolution, and that improves even more when we recover and refurbish the original instruments of its initial performance.

The story has been told many times and in many contexts (think of Don Quixote), but romantic dreamers often temporize and lose ground while practical schemers reap the benefits of accumulated diligence. Cuvier, three years younger than Geoffroy, began his *Muséum* career in a clearly subordinate professional status. But while Geoffroy followed his bliss in Egypt, Cuvier built his career in Paris. Cuvier soon overtook his former protector, and Geoffroy brooded. (Cuvier, for example, entered the *Académie des sciences*, the forthcoming stage for the great 1830 debate, in 1795, while Geoffroy did not win membership until 1807.) By 1805, Cuvier had already published his *Leçons d'anatomie comparée* in five volumes, while Geoffroy had produced no major counterweight. Geoffroy, strong in ambition whatever his shortcomings in political acumen, knew that he needed a distinctive approach or discovery to secure his renown, and he found a guiding light in formalism, the "philosophie anatomique" of the book (1818) that would secure his reputation.

Geoffroy began by applying the chief formalist notion of unity of type to the vertebrate skeleton. Reptiles, birds, and mammals presented minimal difficulty, but fishes posed the key challenge to such a comprehensive view. Compared with terrestrial vertebrates, fishes seemed so different in their anatomy of skull, fins, and shoulder girdle, and so disparate in mode of respiration, that any notion of a common plan must be deemed untenable if not fatuous *prima facie*. Cuvier had argued on functional grounds that the uniqueness of several skeletal elements in fishes testified to their fitness for swimming and breathing in water.

Geoffroy published a group of memoirs on the anatomy of fishes in 1807, the first successes of his research program. Working primarily with bones of the shoulder girdle, he found a putative homologue of the furcula (wishbone) in birds. The functionalist credo that such a bone must exist "for" flight must therefore be false. Rather, the furcula in birds, and its homolog in fishes (operating as an additional rib in some species, and as an aid to opening the gills in others), must be specialized representatives of an abstract element in the archetype of all vertebrates. The form of the archetype holds priority, whereas diversified functional utility only represents a set of secondary modifications, superimposed by conditions of existence upon the primacy of underlying form. Thus, in his first foray into formalism, Geoffroy codified the key idea of structural constraint: form exerts both logical and temporal priority upon function; good designs exist in abundance because the archetype includes this potential for secondary modification; function does not create form, rather form finds function: "Without a direct object in swimming animals, without a utility determined in advance, and thrown, so to say, by chance into the field of organization, the furcula enters into connection with the organs near it; and according to the manner in which this association is formed, it takes on uses which are in some sense prescribed by them" (Geoffroy, 1807, quoted in Appel, 1987, p. 87).

The boldest version of the formalist argument for vertebrates, strongly upheld and extended by Geoffroy, hypothesizes a comprehensive unity of type across the entire phylum—with all elements present in all species (if only in embryos, or fused in adults), and with no new elements originating for specific functions. This strict account embodies both meanings of constraint in their strongest versions—the negative sense of limitation in restriction of elements to pieces of the archetypal jigsaw puzzle; and the positive sense of directed channels providing numerous, though ordered, possibilities for modified shapes (including forms as yet unrealized on our planet, but predictable from the channels, and implied by observed developmental pathways).

Geoffroy wrote in 1807 (quoted in Appel, 1987, p. 89): "It is known that nature works constantly with the same materials. She is ingenious to vary only the forms . . . One sees her tend always to cause the same elements to reappear, in the same number, in the same circumstances, and with the same connections."

Talk is cheap, and romantic notions of abstract, overarching unity can easily be verbalized. Lorenz Oken, the leader of German *Naturphilosophie*, wrote wondrous aphorisms (1809–1811, English translation, 1847), and produced solid empirical work early in his career (1806), but never established a methodological program or built a factual foundation for his formalist philosophy. Meckel, Carus, and other *Naturphilosophen* extended the empirical side, but we rightly honor Geoffroy as the legitimate focus of this movement by our primary scientific criterion of fruitful utility. Geoffroy won his fame as a formalist because he managed to "cash out" the common ideas of transcendentalism in a workable program of research. His program included the two elements demanded of any good theory in natural history: a method for identifying the central phenomenon, and a reasonable explanation for exceptions.

The paradox and pitfall of unity of type as a working research program lies in the vast range of modifications that the archetype experiences under the widely varied adaptive regimes of our planet. Elements of the archetype should, in principle, be named and identified by their *form*, but the idealized archetype may be modified into incompatibility and unrecognizability along the copious adaptive pathways of concrete earthly biology—and we therefore face the dilemma that archetypal elements cannot always be identified by their shapes, or even by their discreteness (for elements fuse, or appear in embryos and then drop out during ontogeny). Some other criterion must be developed.

Geoffroy's major productive insight (still a favored basis for recognizing anatomical homologies—see Riedl, 1978) lay in his "principle of connections"—the claim that homology must be identified by the relative positions and spatial interrelationships of elements, rather than primarily by form. Parts may expand and contract according to utility, but topology remains unaltered, and the archetype can be traced by unvarying spatial order.

Yet, as so often happens amidst the exuberant diversity of natural history, the criterion must be nuanced as "traced by unvarying spatial order . . . except when you can't." Just as Haeckel bolstered recapitulation by bounding and taxonomizing exceptions (heterochronies and heterotopies in his terminology—see Gould, 1977b), and as Darwin specified forces other than, but clearly subsidiary to, natural selection (1859, p. 6), Geoffroy recognized a key class of exceptions to the principle of connections in his concept of metastasis (we use the word in a different, medical sense today, but the general meaning of movement to anomalous places has not altered). Connections can break and blocks of elements can move (though topology within blocks does not alter). For example, the shoulder girdle attaches to the rear of the head in fishes. But, in tetrapods, this connection breaks and several vertebrae may be interposed between skull and forelimbs (see p. 320 for the central role of this metastasis in Owen's interpretation of the vertebral archetype).

In addition, Geoffroy tried to codify rules for secondary adaptive modification of archetypal form. Why do elements vary so much in size, and why can they fuse or even disappear? Geoffroy relied primarily upon a *loi de balancement*, or principle of compensation. Only so much general material can be commandeered to construct the archetypal elements. If one part becomes hypertrophied by utility, others must atrophy to secure the constancy of the common fund. Geoffroy wrote succinctly in 1829 (quoted in Russell, 1916, p. 72): "There is only a single animal modified by the inverse reciprocal variation of all or some of its parts."

Most systems of thought achieve their exemplification in a canonical document; how would natural selection be defined without the *Origin?* Geoffroy's formalism received its codification in an 1818 book with a majestic title—*Philosophie anatomique* (or Anatomical Philosophy, explicitly not the less grandiose *Anatomie philosophique,* or Philosophical Anatomy). Geoffroy's subtitle brought the subject down to earth and bone with the best test case that vertebrates can offer—*Pièces osseuses des organes respiratoires* (the bony elements of the respiratory organs). Geoffroy began this work with an interesting example that highlights the contrast of formalism and functionalism, and that sowed the seeds for his later public debate with Cuvier.

Cuvier had named and described four bones in the opercular series of teleost fishes—operculum, preoperculum, suboperculum, and interoperculum. And he, given his functionalist perspective, had treated these bones as unique to fishes and necessarily present for their utility in respiration by gills. Geoffroy developed a contrary interpretation based on his commitment to unity of the vertebral type, and to a primary implication that these bones must be homologues of elements with different functions in other vertebrates—for all vertebrates possess the same archetypal pieces, and none can be gained or lost. Geoffroy worked on the opercular bones from 1809 to 1812, without resolution. In 1812, a good year for wars and overtures, Cuvier argued that he had located, in the skulls of fishes, the homologs for all bones in the mammalian head—leaving no mammalian structures to serve as potential homologues of opercular bones in fishes. Henri de Blainville, Geoffroy's chief formalist supporter, then argued for homology between the opercular elements of fishes and bones of the tetrapod lower jaw. But, in 1817, Cuvier showed Geoffroy a preparation of a pike and convinced him that all bones of the lower jaw could be matched with jaw elements of tetrapods, again leaving no tetrapod bones to interpret as transformations of the opercular series. Geoffroy then returned to this crucial problem and realized, in a flash as he later stated, that he would have to investigate the only remaining elements in the tetrapod head for a solution: the opercular bones must be homologs of the mammalian middle ear bones! Geoffroy reminisced in 1830 (quoted in Appel, 1987, p. 97): "I regained courage and recommenced my studies, never to abandon them again."

Respiratory bones set the crucial experiment for vertebrate unity of type because such deciding tests must provoke maximal dangers of disconfirmation and grapple with the most difficult issues of validation. The respiratory bones from fish to tetrapods pose *prima facie* challenges to unity of type—for they present no apparent homology from sea to land, and they also exhibit maximal difference of function within vertebrates. If these bones could be won for formalism, then the rest of the skeleton would fall into order.

The *Philosophie anatomique* includes five monographs on homologies between fishes and tetrapods for bones of (and around) the respiratory elements: first on putative homology of opercular and hearing bones (incorrect in retrospect of course), second on the sternum, third on the hyoid, fourth on the branchial arches and their derivates (including the true homologs of the hearing bones, as biologists later learned), and fifth on the shoulder girdle. Cuvier, not yet perceiving (or at least not publicly admitting) his colleague's work as a comprehensive threat to his functionalist system, proclaimed Geoffroy's publication as bold, challenging, and worthy of respect—though almost surely wrong. He strongly doubted Geoffroy's focal homology of opercular and hearing bones, for how could such large, central, and functionally necessary bones of fishes represent the same elements as tiny, almost superfluous nubbins ensconced within organs of differing utility in tetrapods?

Geoffroy, who certainly equalled Cuvier in lack of modesty, proclaimed in the introduction to the *Philosophie anatomique* that his work marked "a new epoch, to which the publication of this book fixes the date" (1818, p. xxxi). He also admitted, tweaking Cuvier's allegiance to the primacy of positive facts, that formalist commitments come first. With such a proper conceptual key, the bones of the skeleton fall into place: "I do not hide it; my direction has been given to me by an *a priori* principle" (1818, p. 11)—although, he hastens to add, unity of type has worked so well that the principle could now be inferred inductively from the skeleton itself!

Geoffroy proceeds immediately to battle in defending formalism by explicit contrast with false assumptions in Cuvier's functionalist alternative. He tells us on page 3 that fishes, by virtue of their functional differences from other vertebrates, seem to possess an anatomy of irreducible uniqueness. "It might appear to the observer . . . that fishes, in order to exist, must call upon the intervention of new organs, and could only be complete in their construction by means of elements destined for them alone, bones created uniquely for their profit" (1818).

But Geoffroy counterposes formalist unity of type to the functionalist alternative of special organs for novel uses: "I can and will satisfy you by showing you that all the elements used in the composition of fishes are exactly and entirely the same as those that enter into the formation of man, mammals, birds, and reptiles" (1818, p. 9).

Geoffroy then takes up the two key challenges previously mentioned. He must first explicate the undeniable fact that the girdle, forelimbs and trunk organs have shifted back in tetrapods relative to their anterior position in fishes—as a result of the interposition of vertebral elements, and in apparent contradiction to the law of connections. Geoffroy invokes his exceptional principle of metastasis and argues that "the trunk exists under the milieu [*sous le milieu*] of the vertebral column" as a whole (1818, p. 9), but not under any particular element in the series. Second, he acknowledges that respiratory organs—with their maximally varying forms and functions in vertebrates—do pose the chief challenge for his system: "In this case, it would be entirely natural to assume in advance that the action of external conditions would impose requirements capable of placing the respiratory apparatus outside the condition of the other organs. From the two modes so rigidly or-

dained [*impérieusement exigés*] for respiration, one might and could conclude [*on a dû et pu conclure*, in more euphonious French] that two different
organic systems must exist . . . In this situation, respiration sets the most important question to treat under our views" (1818, pp. 12–13).

After presenting his homological solutions, Geoffroy buttresses his formalist philosophy by explicit defense of the key claim that internal laws and
constraints establish a primary and controlling pattern, with adaptive modification as secondary, consequential and limited: "This influence of the exterior world, if ever called upon to become a perturbing cause of organization, must be bound necessarily within very straight limits: animals must
oppose to them [exterior forces] several attributes inherent to their nature . . .
This struggle cannot fail to end to the advantage of interior organization,
which has laws [*droits*] against which nothing can prevail" (1818, pp. 208–
209).

Finally, Geoffroy cuts to the heart of the deepest philosophical issue in the
debate by noting that the realized use of archetypal elements for two such different purposes as breathing in air and water requires that the elements themselves be fashioned with great redundancy in potential utility: "double means
have been prepared for a single function" (1818, p. 448). The archetype, by
maintaining this potential for a full range of eventual expression, cannot be
optimal for any particular role. Since archetypes exert logical and temporal
priority over any particular expression, pure form endowed with redundancy
of functional expression must hold sway over utility and adaptation: "Nature
has conceived her plan for construction of a vertebrate animal under a double
point of view: she had to choose a form of composition, so that the ideal animal could accommodate itself equally to the two environments that envelop
our globe. Above all, it was necessary . . . that these two domains of the external world, which so rigorously impose two such different modes of respiration, must call upon the single and only basis of [morphological] organization" (1818, p. 448).

Perhaps Geoffroy had anticipated the forthcoming struggle when he made
his florid appeal, in the closing words of his *Discours préliminaire* (1818,
p. xxix), to the attention and approbation of the next generation: "Oh might
I learn that [my conclusions] have been useful for the youth of our schools.
What group in our beautiful France is more worthy of interest? What devotion, what application, what ardor for study! Oh admirable youth, so occupied with noble productions of the mind, you seem absorbed in a single
thought, in the thought that led Virgil to say: *Felix qui potuit rerum cogno-
scere causas* [Happy is he who can know the causes of things]."

Throughout the 1820's, as events moved to their eventual climax in the
1830 debate, Geoffroy continued to codify and fortify his formalist philosophy—while Cuvier, who had once viewed Geoffroy's work with mild interest
from a different perspective, moved (or felt pushed) to overt opposition.*

---

*Since their debate has so often been misunderstood as an argument about evolution,
something should be said about this anachronistic error. Cuvier's philosophy did foreclose
any possibility of evolution. Geoffroy did accept a limited form of transmutationism, and

Geoffroy emphasized, again and again, the two themes most calculated to rouse Cuvier's ire, and most central to the formalist conviction that adaptation for function can only record a secondary matching (following production by laws of form), not a primary construction.

1. Function follows form. Consider Geoffroy's favorite motto: "such is the organ, such will be its function." Or two more specific statements from 1829 (quoted in Russell, 1916, p. 77): "Animals have no habits but those that result from the structure of their organs . . . A vegetarian regime is imposed upon the Quadrumana by their possession of a somewhat ample stomach, and intestines of moderate length."

2. Final causes cannot serve as explanatory principles. Consider Geoffroy's second favorite motto: *"Je me garde de prêter à Dieu aucune intention"* (I take care not to ascribe any intentions to God).

Could Cuvier, who thought he knew God's ways (or at least the extent of His freedoms), remain silent against the taunts of such a turbulent priest?

### The debate of 1830: foreplay and aftermath

Geoffroy, who loved aphorisms amidst his intense wordiness, often stated: "there is, philosophically speaking, only a single animal." When he applied this radical notion of archetype only within the Vertebrata (as he did in his seminal work of 1818), Cuvier remained at relative peace, albeit in opposition. But, two years later, Geoffroy took the fatal step that first elicited overt opposition from his former friend. Upholding formalism in explicit contrast to the functionalist credo had riled Cuvier to a considerable extent, but so long as this apostasy did not invade the schema of four unbridgeable *embranchements*, the foundation of Cuvier's taxonomy (1817), Geoffroy could be tolerated.

In 1820, Geoffroy made a crucial move that filled him with the joy of unification, and struck Cuvier as an act of unbridled imperialism: he extended the vertebral archetype to encompass arthropods as well, thus bringing two of the four *embranchements* under a common generating form—the vertebra itself. In his key article of 1822, Geoffroy described this "discovery" as "one

---

he did write on the subject (particularly in his monographs on fossil crocodiles). We might also allow that his brand of formalism did encourage an acceptance of evolution as a possibility (as Owen's approbation also testifies)—for the generation of great and continuous diversity (within channels) from an underlying archetype does establish a friendly climate for transmutation (at least within *Baupläne*), while Cuvier's optimized functionalism discourages any thought of evolutionary intermediacy. Evolution played a very minor role in the 1830 debate. Still, we will never understand the great antithesis of functionalism and formalism—a subject that has pervaded the history of biology—if we misread this dichotomy in the later light of evolutionary theory. The debate between Cuvier and Geoffroy centered upon the primacy of form or function in morphology, hardly at all on evolution. Russell (1916, p. 66) has neatly summarized Geoffroy's views on evolution: "That he did believe in evolution to a limited extent is certain; that his theory of evolution was, as it were, a by-product of his lifework, is also certain. Geoffroy was primarily a morphologist and a seeker after the unity hidden under the diversity of organic form. His theory of evolution had as good as no influence on his morphology, for he did not to any extent interpret unity of plan as being due to community of descent."

of the greatest joys that I ever felt in my life" (1822, p. 99), for he had realized that "insects formed another class of vertebrated animals, and that they were, consequently, brought under the common law of uniformity of organization" (1822, p. 99).

An attempt to homologize the *Baupläne* of insects and vertebrates implies some wondrous correspondences, and Geoffroy did not shrink from the logically necessary, but inherently curious implications. He argued that both phyla are fundamentally metameric, with the idealized vertebra itself acting as the archetype of each segment—and therefore, in repetition and regional specialization, of the entire animal. If Goethe's leaf could generate all the organs of a plant, then Geoffroy's vertebra would prefigure the entire skeleton of animals.

Since the arthropod skeleton overlies the internal organs, whereas vertebrate bones lie below flesh and blood, their homology implies a remarkable conclusion, endorsed in yet another motto devised by Geoffroy: insects must live *within* their own vertebrae! Geoffroy wrote in 1822: "Every animal lives outside or inside its vertebral column" *(Tout animal habite en dehors ou en dedans de sa colonne vertébrale).* Continuing the string of surprising implications, all explicitly endorsed by Geoffroy, if exoskeletal body rings must be treated as homologs of vertebrae in the spinal column, then arthropod appendages must be equated with vertebrate ribs—and insects walk on their ribs!

Geoffroy's attempted comparisons posed two major challenges to his own "law of connections" for establishing homology: how can inside and outside become reversed if topology be inviolate; and how can arthropods, with their ventral nerve tracts, be brought into structural harmony with the dorsal nerve cord of vertebrates? Geoffroy proposed an ingenious explanation for the metastasis of inside and outside: He argued that hard parts (and other organs) develop as deposits or exudates on the outside of vessels. In vertebrates, the dorsal nerve cord secretes the vertebrae around itself, while other organs emerge as exudates around vessels of the circulatory system. But insects lack a heart, making their circulatory system too weak to build organs. Therefore, only the nervous system can carry material for the deposition of hard parts— and all other organs must form within the resulting outer set of rings.

The inversion of orientation did not require such an elaborate rationale, but only a repositioning of viewpoint. Geoffroy regarded "top" and "bottom" as subjective terms of a secondary and derivative functionalism. (Geoffroy never intended his homology of vertebrates and inverted arthropods as an evolutionary claim for the origin of vertebrates from an arthropod that literally turned over onto its back. For Geoffroy, the two orientations represented ecological alternatives for a common design.) The primary topology of formalism puts little store by the derivative ecology and function that leads the same side of an invariant organization to turn towards the sun in some groups, and towards the ground in others.

This solution is not, by any means, problem-free, for such changes of topology must also be rationalized. In particular, and as a major stumbling-block (especially for later, evolutionary versions in the "worm that turned" theory of vertebrate origins), the mouth lies below the brain and spinal cord in vertebrates, but above the nerve cords in arthropods, with the esophagus piercing through the cords. Turn an insect over, and the mouth should lie on top, above the brain. Geoffroy and later supporters of this homology generally argued that the old arthropod mouth and nerve-piercing esophagus simply closed, while the digestive tube formed an entirely new ventral "mouth" (therefore not homologous with the arthropod orifice). In any case, and with an almost wondrous and partly humorous irony as I shall show in Chapter 10, Geoffroy's fundamental homology has been validated (in modern guise) after more than a century of calumny. The genetic determinants of dorso-ventral patterning may well be homologous but reversed in expression in arthropods and vertebrates (see pp. 1117–1123).

If Geoffroy belittled Cuvier's functionalism in his argument about orientation, he also attacked the deeper postulate of adaptational primacy by arguing, once again, that the archetypal vertebra comes first, with any use of the resulting structures developing only later as a consequence. Why, he asks, do arthropods use their "ribs" for locomotion? And he answers with the old cliché about mountain climbing—because they are there. Geoffroy wrote in 1820 (quoted in Russell, 1916, p. 77): "From the circumstance that the vertebra is external, it results that the ribs must be so too; and, as it is impossible that organs of such a size can remain passive and absolutely functionless, these great arms, hanging there continually at the disposition of the animal, are pressed into the service of progression, and become its efficient instruments."

Cuvier may have been offended by the arthropod connection, but he was too smart a rhetorician, and too much a figure of the establishment, to be drawn into the limelight of public debate, thus giving even more publicity to Geoffroy's apostasy. Geoffroy goaded him throughout the 1820's, but with no public response. Finally, the dam burst in 1830. Meyranx and Laurencet, two young provincial naturalists with an eye on advancement, presented a monograph, "Some considerations on the organization of mollusks," to the *Académie des sciences*—the standard path for career building at the time. They suggested that the anatomy of a squid might be homologized with a vertebrate bent back upon itself at the middle of the vertebral column, so that the buttocks and base of the spine lined up with the nape of the neck.

Their original paper has been lost, and we do not know how far they meant to carry the comparison, or how much they had intended to inject themselves into the formalist-functionalist controversy. But we do know that Geoffroy, appointed as one of two commissioners to prepare a public report for the *Académie*, rejoiced at this entering wedge for a second imperialistic raid upon Cuvier's *embranchements*. The inclusion of arthropods had once seemed radical enough, but if mollusks could also be brought under the vertebral archetype, then three of four phyla would be reduced to common design, and a final absorption of the Radiata could not lag far behind. The dream of total unification now seemed within Geoffroy's grasp.

Therefore, at the weekly, Monday afternoon meeting of February 15, 1830, Geoffroy presented an enthusiastic endorsement of Meyranx and Laurencet's work, perhaps extending their conclusions far beyond their own intents and desires. And Cuvier finally reached his breaking point.

On February 22, Cuvier appeared, colored charts in hand, to demolish the proposed homology of mollusk and vertebrate. He showed, with devastating effect, that although some organs may look similar and bear the same name, they occupy entirely different topological positions in the two phyla and therefore cannot be homologized by Geoffroy's own criterion of connection. Moreover the anatomy of cephalopods features several organs not found in vertebrates at all. Strongly attacking Geoffroy and his pretensions, Cuvier stated (in Geoffroy, 1830, p. 243):

> One of our learned colleagues, Monsieur Geoffroy Saint-Hilaire avidly seized upon this new view and announced that it completely refuted all that I had said on the distance that separates mollusks and verte-brates. Going much further than the authors of the memoir [Meyranx and Laurencet], he concluded that, up to now, zoology had had no solid base, that it had been an edifice constructed upon sand, and that the only true basis, henceforth indestructible, shall be a certain principle that he calls unity of composition [*unité de composition*], and which, he assures us, will have a universal application.

Following this flourish of controlled contempt, Cuvier presented his specific rebuttal (in Geoffroy, 1830, p. 257): "Cephalopods have several organs in common with vertebrates, and fulfilling similar functions; but the organs are differently arranged in mollusks, often constructed in a different manner, and accompanied by several other organs that vertebrates do not possess."

Poor Meyranx and Laurencet. They became the ultimate victims of numerous clichés, hackneyed by virtue of their fundamental truth—bit off more than they could chew, sacrificial lambs, caught in the middle, between Scylla and Charybdis, a rock and a hard place. The two young men disappeared, forthwith and permanently, both from immediate view, and from later history. Of poor Laurencet, we do not even know his first name (official reports, in those days, spoke only of M.—for Monsieur—so-and-so). Of the equally wretched Meyranx, we know only his attempt to rend his garments before the powerful Cuvier. He wrote, in an abject letter to Cuvier (quoted in Appel, 1987, p. 147): "I cannot find words to express how devastated I am that our Memoir has given rise to disputes. We could scarcely believe that anyone could draw such exaggerated consequences from a single, simple consideration on the organization of mollusks." He then added that the memoir contained nothing "which contradicts the admirable work that you have written and that we regard as the best guide in this matter."

Cuvier, having demolished a specific argument about mollusks, and grasping the deeper issue with his usual clarity, set the groundwork for expanding the debate by defending his functionalist view against the true subject of Geoffroy's primary concern—the defense and hegemony of formalist morphology. Cuvier stated (in Geoffroy, 1830, pp. 248–249), belittling the idea of unity of type: "It [unity of type] is only a principle subordinate to another that is much more important and much more fecund—that of the conditions of existence, of the fitting of parts and their coordination for the role that the animal must play in nature. This is the true philosophical principle, from which flows the possibility of certain resemblances, and the impossibility of others; this is the rational principle from which one may deduce that of analogies [homologies in modern usage] of plan and composition."

With Meyranx, Laurencet and mollusks forgotten (to Cuvier's satisfaction and Geoffroy's relief), the debate moved on to greater generalities of formalism vs. functionalism. Geoffroy replied briefly on February 22, and then, on March 1, presented a general defense of his *théorie des analogues* (homologies), illustrated primarily with his old favorite example of hyoid bones in fishes and tetrapods. On March 8, Geoffroy had fallen ill and Cuvier refused to deliver his rebuttal in his colleague's absence. The large crowd, lured by the promised fireworks more than the putative content, dispersed in disappointment. Cuvier replied in kind by failing to attend the following week; tit for tat. The debate finally resumed on March 22, with Cuvier's rebuttal of Geoffroy's claims for hyoid homologies. Geoffroy defended himself on March 29, but he had clearly tired of the affair, for he stated, with more than a whiff of disingenuous disengagement, that "a meeting of the disciples of the Portico" had regretfully turned into theater—"a pit applauding the outrageous comedies of Aristophanes" (quoted in Appel, 1987, pp. 154–155). Cuvier replied one last time on April 5, but Geoffroy ended the public debate by announcing that he would not respond. Cuvier surely enjoyed advantages as a brilliant debater and consummate politician, but we must not consider Geoffroy either devoid of wiliness, or willing to surrender. He merely shifted ground to the more comfortable medium of print. By April 15, in a fit of zeal and celerity as impressive as anything achieved with our current technology of instant books, Geoffroy had sent to the printers the text of his *Principes de philosophie zoologique*, containing all papers and commentary presented by Cuvier and himself during the public debate. (Several years ago, I had the good fortune to purchase Cuvier's own copy of this work. The book bears Cuvier's library stamp (Fig. 4-11), but I find no sign, in marginalia or any other indication of use, that the great man ever consulted the volume!)

Intellectual debates of such grand scale and diverse content can never be won or lost in the unambiguous fashion of more worldly events, as when Joe Louis knocked out Max Schmeling. Most biological texts (Russell, 1916, for example) proclaim Cuvier the victor—surely a fair judgment for the narrow, initiating topic of molluscan homologies. But the debate quickly moved from this immediate instigation to the broadest question of formalism vs. functionalism—an issue that cannot be resolved as total victory or defeat. Moreover, the debate embodied a hundred subtexts in sociology, philosophy, and politics—open vs. closed meetings of the *Académie*, facts vs. theory in science, élites vs. populism in research—and all these swirling, largely orthogonal themes could not fall into a single pattern of victory for one side.

Furthermore, and in a curious sense, the debate didn't seem so tumultuous, fierce, or epochal at the moment of its actual unfolding—six meetings, two misses, and an abrupt, unfinished ending. This incident became central to the later history of biology largely through machinations and unplanned consequences of its aftermath, and primarily because a mixture of good luck and a special kind of insight allowed the dreamer Geoffroy to recoup everything he had lost in immediate debate to the magisterial Cuvier, and to attain some sort of victory in retrospect, or at least a "draw" with great advantages.

On the ledger of luck, Geoffroy gained history's greatest and most conventional form of advantage when Cuvier died in 1832, thus awarding Geoffroy 12 additional years to reconstruct the incident, unopposed and in his fashion (a kind of poetic justice in this case, since Cuvier had so adroitly used the same power with such effect—as in his infamous *éloge* of Lamarck, see pp. 170–173). Secondly, Geoffroy obtained the finest free publicity conceivable when Europe's greatest literary figure, the aged Goethe, expressed such

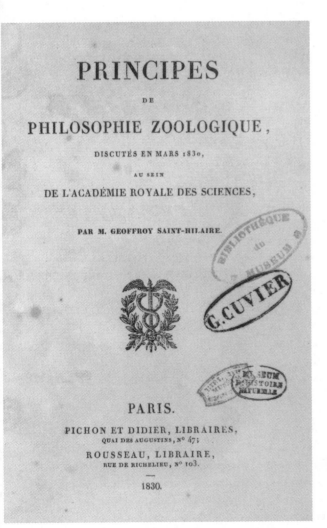

4-11. A remarkable artifact in the history of evolutionary biology! This is the title page of Cuvier's personal copy—a "gift" from Geoffroy—of the volume that Geoffroy so quickly edited and published after their debate of 1830 at the Academy of Sciences. (Author's collection.) I can find no evidence that Cuvier, after placing his library stamp upon this volume, ever read a word of the text. Following Cuvier's death, this book became the property of the Library of the *Muséum d'Histoire naturelle*, which later sold it as a duplicate.)

lively interest in the debate and wrote two articles on the subject, including the very last piece before his death (Goethe, 1832). Although Goethe declared no victor, his basic sympathy lay with Geoffroy—a kindred soul who defended poetic insight against pure empiricism, and who had, in a real sense that inspired their deepest intellectual bond, completed for animals (with the archetypal vertebra) the program that Goethe had begun so brilliantly for plants (and their archetypal leaf).

The debate of Geoffroy and Cuvier unfolded during one of the most important and tumultuous events of 19th century French history—the revolution of 1830. This coincidence prompted the most famous anecdote of the entire episode—a tale that documents the extent of Goethe's involvement. Goethe's friend Frédéric Soret recalled:

> The news of the Revolution . . . reached Weimar today, and set everyone in a commotion. I went in the course of the afternoon to Goethe's. "Now," he exclaimed as I entered, "what do you think of this great event? The volcano has come to an eruption; everything is in flames, and we no longer have a transaction behind closed doors!" "A frightful story," I replied. "But what else could be expected under such notorious circumstances, and with such a ministry, than that matters would end with the expulsion of the royal family?" "We do not appear to understand each other, my good friend," replied Goethe. "I am not speaking of those people at all, but of something entirely different. I am speaking of the contest, of the highest importance for science, between Cuvier and Geoffroy Saint-Hilaire, which has come to an open rupture in the Academy" (quoted in Appel, 1987, p. 1).

In his first article (1831, p. 179), Goethe conjured up the scene of debate: "In this sanctuary of science . . . where all is order and decorum, where one encounters people of high culture, where one responds with moderation . . . lively debate has broken out, debate which appears to lead only to personal dissention, but which, viewed from a higher perspective, has more value and more future worth." He then epitomized the differences between the protagonists in both method and theory (1831, p. 180): "Cuvier presents himself as having an indefatigable zeal for distinction and description . . . Geoffroy devotes himself to the hidden affinities of creatures . . . The totality is always present in an interior sense, from which follows the conviction that the particular can arise from the totality."

Goethe also recognized the link between a commitment to formalist constraints and channels, and a reluctance to explain morphology by utility and adaptation—for he had promoted the same correlation in his own work on plants. Presenting the most essential aspect of Geoffroy's methodology, Goethe writes (1832, p. 62): "It is necessary to cite, as most important, his having shown the uselessness of explanations in terms of final causes."

Geoffroy surely enjoyed good fortune in Cuvier's death and Goethe's interest, but he also actively campaigned in an unconventional yet strikingly effective manner, for elevating the importance of the debate and rewriting its story to his advantage. As stated above, his book of documents went to the printers just 10 days after the debate ended. Although he did not importune Goethe directly, he surely took full advantage of this fortuitous involvement (see Fig. 4-8 for some private evidence). Speaking before the *Académie* later in 1830, Geoffroy noted Goethe's favorable commentary on his "instant" book, referring to the great poet as "the first authority of Germany . . . the celebrated Goethe . . . who has just accorded my work the greatest honor that a French book can receive" (quoted in Appel, 1987, pp. 166–167).

Taking a cue from Goethe's clout, Geoffroy actively recruited support from literary leaders in France. Balzac dedicated *Père Goriot,* perhaps his most famous novel, "to the great and illustrious Geoffroy Saint-Hilaire as a tribute of admiration for his labors and his genius." The *avant-propos* to Balzac's *La comédie humaine* (1842) contains the following description of Geoffroy's system:

> There is only one animal. The Creator has used only a single pattern for all organized beings. The animal is a principle which takes its external form, or, to be more exact, the differences in its form, from the milieus in which it is obliged to develop. Zoological species are the result of these differences. The proclamation and defense of this system, which is, moreover, in harmony with our ideas of divine power, will be the eternal glory of Geoffroy Saint-Hilaire, the victor over Cuvier in this point of higher science, and whose triumph has been hailed by the last article written by the great Goethe (in Appel, 1987, p. 192).

Balzac then used Geoffroy's centerpiece, in a timely and fascinating way, in this and other novels—arguing that all people partake of a single human essence, with individual variation best explained by environmental differences.

Geoffroy also courted the friendship and publicity of George Sand. Of a meeting with Geoffroy at the Jardin des Plantes in 1836, she wrote: "The old Geoffroy is for his part a rather curious beast, as ugly as the orangutan, as talkative as a magpie, but for all that full of genius" (in Appel, 1987, p. 189). Geoffroy sent Sand several of his publications; she declared herself unable to do them technical justice, but still proclaimed them "broad and magnificent," throwing Cuvier "to the ground . . . for anyone who detests meanness in the arts" (in Appel, 1987, p. 189).

So many questions about historical influence find their best resolution in our understanding of time frames and time scales. Cuvier may have won at least a rhetorical victory in a scientific debate of great intellectual import but limited duration and public impact during two months in 1830. But this event in clock time then yielded to a literary tradition of retelling, orchestrated in large part by the sole surviving protagonist. The original technical issues evoked little interest or understanding among the chief literary retellers who, in Appel's words (1987, p. 175), "came to see Geoffroy as a heroic figure, Cuvier as a paltry fact collector, and the debate as a major event in French intellectual history."

What version, then, should we embrace if we must address the largely meaningless, but endlessly fascinating, issue of "who won?" Something happened in 1830; a factual basis of word and gesture once unfolded. But we cannot recover the original scene, and the actual debate would be subject to endless interpretation in any case. If later judgments and interpolations loom so large in the mythological versions learned and accepted by students for nearly two centuries since the actual events (and I use "mythological" in the primal and powerful, not the pejorative, sense)—Geoffroy's triumph among the literati, Cuvier's in most scientific accounts—then these constructions replace the unattainable original and become an important reality in their own right.

Finally (as I shall stress throughout this book), in the deepest sense, and by a plethora of disparate criteria, neither Geoffroy nor Cuvier could have "won" because neither man held tools of triumph in principle. Formalism and functionalism represent poles of a timeless dichotomy, each expressing a valid way of representing reality. Both poles can only be regarded as deeply right, and each needs the other because the full axis of the dichotomy operates as a lance thrown through, and then anchoring, the empirical world. If one pole "wins" for contingent reasons of a transient historical moment, then the advantage can only be temporary and intellectually limited. Such an ephemeral victory did occur in the recent history of evolutionary theory—the exaltation of functionalism in the hardening adaptationism of the Modern Synthesis, codified in the late 1950's and early 1960's, and marked by celebrations of the Darwinian centennial in 1959 (see Chapter 7). In fact, this very historical context led me to emphasize structuralism and formalism in this book (because its insights have been neglected in modern evolutionary biology, not because formalist approaches could ever be labeled as "more true"), and to revivify the great formalist thinkers, from Geoffroy to Owen to Galton, Bateson, and Goldschmidt—not as an antiquarian indulgence, but for the *current* utility of their ideas. (Other ages have needed to rescue functionalism from equally limited formalist domination.)

But I would not spend so much time on this endeavor for reasons of selfish and personal interest alone. Formalism *resartus* has been externally motivated by great advances in genetics and development (see Chapters 10 and 11)—a system of knowledge that requires a structure of explanation based as much on how organisms *can* be built, as on how they *do* work. We should give the last word to Goethe who, in choosing the debate between Geoffroy and Cuvier for his swansong to the world, eloquently defended the claims of both men. He wrote primarily of ideas and facts, but also of archetype and adaptation, arguing that "the more vitally these two functions of the mind are related, like inhaling and exhaling, the better will be the outlook for the sciences and their friends."

## RICHARD OWEN AND ENGLISH FORMALISM: THE ARCHETYPE OF VERTEBRATES

### No formalism please, we're British

I own a letter (see Fig. 4-12) written in October 1879 by Richard Owen and addressed simply: "J. Pearson Langshaw, Esq. (in or near) Lancaster." The stamp cost only a penny, and Her Majesty's post managed to make the delivery with such minimal information. Owen announced that he had a "crowlet to pluck" with his friend for not visiting on a recent excursion near his whereabouts (Owen suspected a fear of further humiliation at the checkerboard as the reason for Langshaw's avoidance). Owen then spoke of a visit to Ireland, describing a new locality for *Megaceros* (the "Irish Elk"), and noting that Britain had formed part of a mainland during Pleistocene low sea levels. He ended with a note of chauvinism at the height of British imperial and industrial expansion: "I very much enjoyed a fortnight with the Tory Member for the County of Wicklow; visited a new locality of *Megaceros*, confirmatory of its antiquity, and coevality with the Elephants and Rhinoceroses which roamed over the continent represented now by certain Islands that set the rest of the world to rights."

Yet, for all his political and institutional allegiance to his native land, Britain's greatest vertebrate anatomist cast his intellectual lot with the continent that lay abreast of those "certain islands" and championed the strongest version of formalism—the theory of single generating archetypes, at least for all vertebrates—in the land so well adapted for the functionalism of Paley and the Bridgewater Treatises.* Owen sensed his incongruity and recognized that his formalist message would be better heard in France or Germany than in his own country. On the very first page of his greatest formalist monograph, *On the Nature of Limbs* (1849), Owen wrote: "I became fully conscious how foreign to our English philosophy were those ideas or trains of thought concerned in the discovery of the anatomical truths, one of which I propose to explain on the present occasion in reference to the limbs or locomotive ex-

---

*Owen conceived his mission as marrying the schools of "morphology and teleology," or formalism and functionalism; but, as we shall see, he forged this union with a clearly dominant formalist partner, and therefore wins primary allegiance to this school by the proper criteria of relative frequency and primacy of cause. This casting may seem ironic or contradictory in the light of Owen's common designation as the "British Cuvier," and of his youthful visit to Cuvier in 1831. But this common appellation primarily honored Owen's professional skill and his domination of the discipline, not his ideology (the intended comparison being to Cuvier's power, not to his ideas). Records of the Paris visit paradoxically affirm Cuvier's lack of influence; for, although Owen frequently visited Cuvier, his notes and diaries feature a strange lack of commentary on Cuvier's science or thought. Owen's grandson wrote in the standard life and letters of his grandfather (1894, vol. 1, p. 50): "His rough diary, which he kept during his stay in Paris, seldom mentions the fossil vertebrate collection, and shows that his interviews with Baron Cuvier were for the most part of a purely social character. It notes, for example, that he attended pretty regularly Cuvier's soirees held on Saturday evenings, and that he enjoyed the music. With the diary agree his letters. Both devote page after page to the sights and amusement of Paris. Owen, in fact, seems to have regarded his stay in Paris as an exceedingly pleasant and entertaining holiday."

Huxley's remarkably fair *éloge* of Owen makes the same point: "It was not uncommon to hear our countryman called 'the British Cuvier.' . . . But when we consider Owen's contribution to "philosophical anatomy," I think the epithet ceases to be appropriate. For there can be no question tht he was deeply influenced by, and inclined towards, those speculations of Oken and Geoffroy St. Hilaire, of which Cuvier was the declared antagonist and often the bitter critic" (in Owen, 1894, vol. 2, p. 312).

tremities. A German anatomist, addressing an audience of his countrymen, would feel none of the difficulty which I experienced" (1849, p. 1). And when Owen attempted to secure a German translation of this work, he wrote to Rudolf Wagner of Göttingen (quoted in Desmond, 1982, p. 48): "The subject is better adapted for the character of mind and thought of a German audience than for our matter of fact English."

As Ospovat (1981) and other scholars have shown, Owen was not the only British scientist who caught (and rode) the wave of formalist excitement ema-

4-12. For one penny and a very approximate address, the British postal service managed to deliver Richard Owen's letter to his friend Langshaw. (Author's collection.) Note his playful, but quite chauvinistic comments about England versus the continent on the last page of the letter.

nating from the continent; in such a strongly social profession as science, even the most profoundly idiosyncratic thinker lies embedded in his contemporary world (see Chapter 11 on the leading 20th century British formalist, D'Arcy Thompson). Yet, as I have emphasized throughout this chapter—for the theme must play a central role in any proper understanding of Darwin and the essence of Darwinian theory—formalism remained a minority position in Britain, poorly adapted and fundamentally alien to a culture freighted with several centuries of functionalist preference.

As a fitting illustration of this functionalist milieu, consider Britain's pre-eminent philosopher of science in the generation of Owen and Darwin: William Whewell. (As author of a Bridgewater Treatise, Whewell cannot be considered a neutral commentator.) The first edition (1837) of his most important and comprehensive work, *History of the Inductive Sciences*, presents a functionalist perspective, almost as rigid and exclusive as Paley's. By the third edition (1869), however, Whewell claimed a change of heart, stating that Owen had provoked the alteration. Yet Whewell's "revised" attitude remains quintessentially functionalist, if anything even more so because he now recognizes and understands the formalist alternative, but relegates this "newcomer" to marginality. Whewell begins by providing a fair and concise contrast of the two schools as portrayed in the debate between Geoffroy and Cuvier—though he scarcely hides his preferences in judging functionalism as "truths which are irresistibly apparent and which may therefore be safely taken as the bases of our reasonings":

> According to this theory [Geoffroy's], the structure and functions of animals are to be studied by the guidance of their analogy only [homology in modern parlance]; our attention is to be turned, not to the fitness of the organization for any end of life or action, but to its resemblance of the other organizations by which it is gradually derived from the original type . . . On the other hand, the plan of the animal, the purpose of its organization in the support of its life, the necessity of the functions to its existence, are truths which are irresistibly apparent, and which may therefore be safely taken as the bases of our reasonings. This view has been put forward as the doctrine of the conditions of existence: it may also be described as the principle of a purpose in organization; the structure being considered as having the function for its end (1869, p. 483).

Whewell then states the chief claim for functionalist primacy: body parts exist primarily "for" their useful action: "That the parts of the body of animals are made in order to discharge their respective offices, is a conviction which we cannot believe to be otherwise than an irremovable principle of the philosophy of organization, when we see the manner in which it has constantly forced itself upon the minds of zoologists and anatomists in all ages" (1869, p. 489). "In the organized world," Whewell adds (1869, p. 491), "we may and must adopt the belief, that organization exists for its purpose, and that the apprehension of the purpose may guide us in seeing the meaning of the organization."

Whewell ends with a striking musical analogy, arguing that formalism conveys a certain pleasure and appreciation, but that full delight and instruction require apprehension in terms of purpose:

> To us this doctrine [of final causes or functionalism] appears like the natural cadence of the tones to which we have so long been listening: and without such a final strain our ears would have been left craving and unsatisfied. We have been lingering long amid the harmonies of law and symmetry, constancy and development; and these notes, though their music was sweet and deep, must too often have sounded to the ear of our moral nature, as vague and unmeaning melodies, floating in the air around us, but conveying no definite thought, molded into no intelligible announcement (1869, p. 495).

### The vertebrate archetype: constraint and nonadaptation

All biologists know that Richard Owen defined the terms analogy and homology in their modern sense, and that he made a tripartite division of the second category into general, special and serial (thus demonstrating the generative and developmental, rather than the evolutionary, basis of his underlying concept—see Chapter 10, pp. 1070–1076, for an extensive analysis of Owen's categories in the light of modern developmental biology). With this framework, constructed specifically in the light of the formalist-functionalist debate, Owen could engage the problem that Darwin would later designate as paramount in morphology—the special homology of similar parts with divergent functions. "What can be more curious," Darwin would write (1859, p. 434 and p. 112 of this book), "than that the hand of a man, formed for grasping, that of a mole for digging, the leg of the horse, the paddle of the porpoise, and the wing of the bat, should all be constructed on the same pattern."

Surely, Owen reasoned, these underlying structural similarities could not be explained by common utility—the category that he had designated as "analogy." And thus, the "British Cuvier" explicitly contradicted the central belief of his eponym by denying a functional explanation for homology: "The attempt to explain, by the Cuverian principles, the facts of special homology on the hypothesis of the subserviency of the parts so determined to similar ends in different animals—to say that the same or answerable bones occur in them because they have to perform similar functions—involve [sic] many difficulties, and are opposed by numerous phenomena" (Owen, 1848, p. 73).

Owen clearly accepts the common conceptual taxonomy of his generation, for he argues that functionalism and formalism represent the only intelligible interpretations of morphology. We rightly reject functionalism for special homology, but if we deny formalism as well, then we retain, for explanation, nothing but a stochastic "slough of despond": "With regard to the structural correspondences manifested in the locomotive members; if the principle of special adaptation fails to explain them, and we reject the idea of these correspondences as manifestations of some archetypal exemplar on which it has pleased the Creator to frame certain of his living creatures, there remains only the alternative that the organic atoms have concurred fortuitously to produce such harmony. But from this Epicurean slough of despond every healthy mind naturally recoils" (1849, p. 40).

Owen chooses the formalist exit from Bunyan's swamp, and calls upon the guidance of Plato to pull him out (an even classier assist than Dante's employment of Virgil). Special homology can only be resolved by recognizing the common generating pattern for all specific manifestations—the Platonic archetype (or general homology) behind the variety of worldly incarnations. The archetype does not denote an object or an ancestor, but an abstract generating formula, a blueprint, a formal cause. Owen engraved his version of the vertebrate archetype upon a seal and wrote to his sister Maria in 1852, trying to explain this arcane concept in layperson's terms: "It represents the archetype, or primal pattern—what Plato would have called the 'divine idea' on which the osseous frame of all vertebrate animals—i.e. all animals that have bones—has been constructed. The motto is 'the one in the manifold,' expressive of the unity of plan which may be traced through all the modifications of the pattern, by which it is adapted to the very habits and modes of life of fishes, reptiles, birds, beasts, and human kind" (in Owen, 1894, vol. 1, p. 388).

In 1849, Owen published his treatise, *On the Nature of Limbs*, originally delivered as a lecture on February 9 at the Royal Institution. I regard this book as the best expression of Owen's archetypal theory, the most interesting document ever written in English to defend this strongest version of formalist theory in biology.

Despite the title (aptly chosen and cleverly constructed as we shall see), Owen's treatise attempts to reduce the *entire* vertebrate skeleton, in all its manifold variety, to a single archetypal element, multiply repeated and specialized. Owen writes: "General anatomical science reveals the unity which pervades the diversity, and demonstrates the whole skeleton of man to be the harmonized sum of a series of essentially similar segments, although each segment differs from the other, and all vary from their archetype" (1849, p. 119).

For the naming and essence of this archetypal element, Owen agrees with Geoffroy in designating the vertebra. We must conceptualize Owen's "vertebra" not only as a spinal disc, but as a set of highly generalized elements (a central disc surrounded by various bars and rods) ripe for modification along myriad pathways. Owen's archetypal unit (Fig. 4-13) operates as an abstract blueprint of bursting potential. (For example, in the "vertebra" that makes the shoulder girdle, the pleurapophysis lateral to the centrum becomes the scapula, while the haemapophysis below forms the coracoid, and the lowermost haemal spine makes the front of the sternum.) Owen writes: "I have satisfactorily demonstrated that a vertebra is a natural group of bones, that it may be recognized as a primary division or segment of the endoskeleton, and that the parts of that group are definable and recognizable under all their teleological modifications, their essential relations and characters appearing through every adaptive mask." (Note the strong claim—the key and continuing relevance of formalism in critiquing Darwinian traditions—that any specialization for utility, or "teleological modification," imposes an "adaptive mask" upon the generating archetype. In such phrases, we grasp the essential difference between formalism and functionalism. Adaptive modification, the architect of morphology in Darwinian functionalism, becomes, in formalist thought, a secondary, superficial and confusing overprint upon the underlying essence.)

For anyone wishing to explain the human skeleton by genesis from a vertebral archetype, three great groups of bones must be resolved in different ways and with varying degrees of difficulty: the vertebral column itself, the skull, and the limbs with their associated girdles. The archetypal model obviously works for the vertebral column, the empirical source of the theory in the first place. The skull and the limbs therefore become crucial experiments for testing the model of archetypal genesis.

The attempt to depict the skull as a profound modification of a few vertebrae substantially predates Owen (see p. 283 on Goethe's allegiance, dating from observations on a sheep's skull made in 1790). The subject had been much aired and debated, with the number of proposed vertebrae ranging from one (Duméril in 1808) to seven (Geoffroy). The most common resolution proposed four vertebrae, a number popularized by Oken and accepted by Owen. Oken had named the four elements from back to front—occipital, parietal, frontal, and nasal—and he had associated each with a primary sense: auditory, lingual, ocular, and olfactory. Owen accepted these four names. He argued that lateral and ventral elements of the occipital vertebra

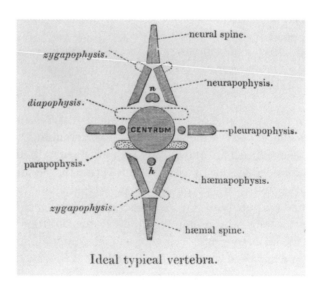

Ideal typical vertebra.

4-13. Owen's picture of the ideal or archetypal vertebra, interpreted by him as the ground plan for all parts of the vertebrate skeleton. From Owen, 1849. (Author's collection.)

formed the pectoral girdle. He kept Oken's names—parietal, frontal, and nasal—for neural halves of the three anterior vertebrae, and designated the haemal (ventral) halves by their associated structures: hyoid, mandibular, and maxillary.

Having thus resolved one of the two problematic bone groups by traditional arguments, Owen turned to the single remaining issue for completion of the archetypal research program—the explanation of limbs and associated girdles as modified vertebral parts. In this sense, *The Nature of Limbs* should not be read as a specialized treatise on one part of the body, but as an attempt to complete the most radical version of formalism by bringing the last outpost of the vertebrate skeleton under the vertebral archetype.

Owen's argument for limbs might strike us today as contrived and peculiar. His effort does not represent the high water mark of formalist logic even in its own terms and times, but may still win our grudging respect for ingeniousness and pure chutzpah. The apparent problem, after all, can only be described as daunting. After using all the lateral and ventral parts of a vertebra to build the girdle, what remains for constructing the prominent complexity of humerus, radius and ulna, carpals and metacarpals, right down to the most distal phalanx of the digits? Surely these bones can only be "novel" structures unrelated to the archetypal vertebra, and the integrative program of archetypal reduction and genesis must fail. We might choose to downplay the supernumerary status of a stapes or hyoid bone; but we can scarcely disregard the need to encompass limbs within any general theory of the vertebrate skeleton—for an archetype that omits such a major structure can only provide a partial and paltry explanation indeed.

Owen's improbable solution homologizes the vertebrate limb, in all its complexity, with a simple, unbranched projection from the haemapophysis (see Fig. 4-13), called a diverging ray (note the ray on each vertebra of the archetype in Fig. 4-14).

But how could Owen justify a comparison of so many articulated bones with a hypothetical single rod? Owen used the time-honored comparative method by attempting to trace back the complexity of vertebrate limbs in a structural series of simplification, leading to the lungfish *Lepidosiren* and its minimal pectoral ray. Lest this series be rejected as a concatenation of heterogeneous objects, Owen presented a tripartite argument: (1) the structural series denotes a descent by simplification; (2) simplification occurs by "arrest of development," bringing the reduced form closer to an embryonic state; (3) the embryo, following von Baer's principles, reveals the generating archetype in a way that the complexly modified adult cannot. Sensing that opponents might view the proposition as a "transcendental dream," Owen defended his structural series (Fig. 4-15) as a voyage to the archetype: "It is no mere transcendental dream, but true knowledge and legitimate fruit of inductive research, that clear insight into the essential nature of the organ, which is acquired by tracing it step by step from the unbranched pectoral ray of the lepidosiren to the equally small and slender but bifid pectoral ray of the amphiuma, thence to the similar but trifid ray of the proteus and through the

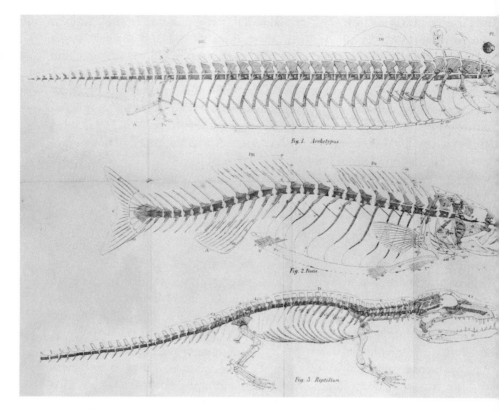

4-14. The key plate from Owen's 1849 monograph on the nature of limbs. The archetype, built of a series of vertebrae, shown at the top, with skeletons of a fish and a reptile below. In his boldest move, Owen attempts to derive the entire limb of later vertebrates from the single diverging ray of each vertebral element in the archetypal form. The diverging ray is the little spike projecting upward at about 10 o'clock from the junction of two vertical elements below each vertebral centrum of the archetype. (Author's collection.)

progressively superadded structures and perfections in higher reptiles and in mammals" (p. 70).

But these strained homologies then incur a second, equally serious problem in requiring a pronounced shift of position among vertebrate classes—an interpretation inconsistent with the formalist principle that topology and connection serve as the primary criteria of homology. Geoffroy had developed his concept of metastasis (see p. 300) to explain exceptions in the same troubling example, and Owen followed this continental solution. The pectoral girdle of fishes attaches to the rear of the skull. In fact, Owen regarded the bones of the girdle as the haemal portions of the fourth, or occipital, skull vertebra. (The arm and hand arise from the diverging ray of this vertebra and also become parts of the head by homology.) Owen recognizes the counterintuitive oddity of such a claim, but must follow the formalist logic: "However strange and paradoxical the proposition may sound, the scapular arch and its appendages, down the last phalanx of the little finger, are truly and essentially bones of the skull" (p. 112).

But all tetrapods separate the shoulder girdle and front appendages from the skull by a sequence of intercalated vertebrae. Owen argues, as did Geoffroy, that the haemal portion of the fourth skull vertebra has migrated back in the terrestrial classes. And why should this degree of transposition be

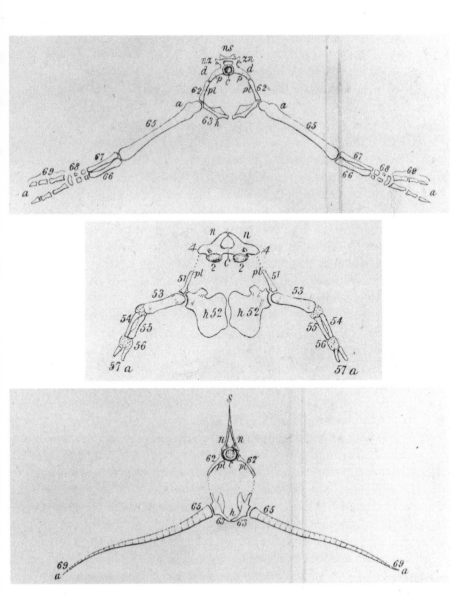

4-15. Owen justifies his bold claim that the entire limb might be derived from a diverging ray by showing a series of maximally reduced limbs in living vertebrates from *Proteus* with several elements but only two digits, to *Amphiuma,* also with two digits but fewer elements, and finally to *Lepidosiren* with a single element maximally similar to the hypothetical diverging ray. (Author's collection.)

regarded as improbable? After all, the pelvic fin of many living teleost fishes, the undeniable serial homolog of the pectoral, moves sufficiently far forward to lie in front of the pectoral. If this more radical movement occurs in modern species, why balk at a less profound metastasis to separate forelimb from skull? "But it may be objected that the ordinary costal or haemal arch has been detached from its centrum for the purpose of this comparison. True! And the scapular arch in mammals, birds and reptiles, is a haemal arch so dislocated,—a statement which I do not hesitate to make under a pledge to demonstrate the proper centrum and the rest of the segment or vertebra to which it belongs" (p. 50).

Having thus brought the entire skeleton, with all its complexity and adaptive variety, into homology with the archetypal vertebra, Owen could proclaim both the glory and the generality of formalist morphology. On the largely rhetorical subject of glory, Owen joined Agassiz in refuting the Paleyan link of functionalism to God's beneficence. Shall we not regard a generalized archetype, a sublime and abstract pattern for all manifest variety, as a loftier testimony to a truly omnipotent God than the mean material fitting, however exact, of some unique and particular object to an immediate environment? "The satisfaction felt by the rightly constituted mind must ever be great in recognizing the fitness of parts for their appropriate functions; but when this fitness is gained, as in the great toe of the foot of man or the ostrich, by a structure which at the same time betokens harmonious concord with a common type, the prescient operation of the One Cause of all organization becomes strikingly manifested to our limited intelligence" (p. 38).

Archetypal thinking also exalts our own status, for if God ordained the archetype, he certainly recognized all potential modifications in advance, and the concept of human existence therefore long predated our actual appearance: "The recognition of an ideal Exemplar for the vertebrated animals proves that the knowledge of such a being as man must have existed before man appeared. For the Divine mind which planned the Archetype also foreknew all its modifications" (pp. 85–86).

In fact, the entire geological history of vertebrates may be interpreted as a movement towards humanity, guided by natural forces ordained by God as secondary causes. Owen's oft-quoted last paragraph provides a genuine expression of evolutionary views in this limited sense (transformations within an archetypal framework under unknown, but natural, laws established by God to implement His plans of progress):

> To what natural laws or secondary causes the orderly succession and progression of such organic phenomena may have been committed we as yet are ignorant. But if, without derogation of the Divine power, we may conceive the existence of such ministers, and personify them by the term "Nature," we learn from the past history of our globe that she has advanced with slow and stately steps, guided by the archetypal light, amidst the wreck of worlds, from the first embodiment of the Vertebrate

idea under its old Ichthyic vestment, until it became arrayed in the glorious garb of the Human form (p. 86).

Owen drew from his archetype all the standard implications that set the research program of formalism—the correlations that define the essence (and continuing relevance) of this pole on the dichotomy: importance of constraint and mistrust of adaptationism (accompanied by demotion to secondary status).

CONSTRAINT. This term includes two distinct meanings, both in vernacular English usage (see Gould, 1989a, and Chapter 10, pp. 1025–1061) and in biological jargon—the negative concept of restriction, and the positive sense of channeling. Those who belittle the evolutionary importance of the subject do not deny the phenomenon itself, but rather limit their concept to the negative meaning.

Owen properly depicted constraint as both limitation and channeling. In the former meaning, for example, he notes that the first digit of the generalized mammalian hand or foot develops only two phalanges, while the others grow three. These numbers do not change, even when utility would dictate otherwise—as in elephants where the first and fifth toes do not differ in length, and all digits are enclosed in a large pad (1849, p. 37); or in humans, where the first toe becomes massive and weight-bearing (but cannot gain an additional phalanx), and the little toe almost vestigial (while still retaining its full complement of three phalanges).

On the positive theme of channels, Owen regards an archetype as a blueprint of myriad possibilities (made all the more intelligible by limiting their range to products of common elements in unvarying topological order). All realized examples on earth therefore include only a small subset of possible forms. Owen even felt free to speculate about the anatomy of life on other worlds, provided that the vertebral archetype can lay claim to universal status: "Our thoughts are free to soar as far as any legitimate analogy may seem to guide them rightly in the boundless ocean of unknown truth. And if censure be merited for here indulging, even for a moment, in pure speculation, it may, perhaps, be disarmed by the reflection that the discovery of the vertebrate archetype could not fail to suggest to the anatomist many possible modifications of it beyond those that we know to have been realized in this little orb of ours" (1849, p. 83).

For example, no earthly vertebrate grows more than two pairs of limbs, but the archetype bears diverging rays (the source of limbs by general homology), on each vertebra, and additional pairs therefore become possible:

We have been accustomed to regard the vertebrate animals as being characterized by the limitation of their limbs to two pairs, and it is true that no more diverging appendages are developed for station, locomotion and manipulation. But the rudiments of many more pairs are present in many species. Although they may never be developed as such in this planet, it is quite conceivable that certain of them may be so developed, if the Vertebrate type should be that on which any of the inhabitants of other planets of our system are organized. The conceivable modifications of the vertebrate archetype are very far from being exhausted by any of the forms that now inhabit the earth, or that are known to have existed here at any period (1849, p. 83).

ADAPTATION. Neither Owen nor any prominent formalist has ever denied interest or importance to the manifestly obvious phenomenon of adaptation. Formalists do not question the high frequency of adaptation, but only dispute the relative ranking of utility as a causal argument. In the view of functionalists, from creationists like Paley to evolutionists like Darwin, adaptation embodies the source and cause of morphological order and change. For formalists, adaptation becomes a secondary phenomenon, imposed upon primary and underlying laws of form to fit a particular organism to an immediate environment. Adaptation remains vital; for without such specific utility, the organic world would feature only abstract models, but no real creatures in their stunning variety. Yet, adaptation still works in a sequential and secondary fashion to place an overlay upon the archetype. Thus, while Owen continually speaks of morphology *and* teleology, we must not view him as a mushy pluralist, advocating equality of the two poles. His mode of blending ranks the poles, with adaptation distinctly subservient to archetype in the classical mixture of formalist thought.

Owen argued that two great laws build actual organisms from the archetypal form. The first, called irrelative (or vegetative) repetition, iterates the archetypal element into a series of similar parts. The second, adaptive or teleological force, then modifies the various segments in different ways demanded by their mode of life.

Since the adaptive force imposes secondary modifications upon an initial string of identical archetypal elements, we must penetrate behind this imposed veil of specific utility and specialization if we wish to apprehend the archetype itself. Various formalist principles lead us to fruitful strategies for peering behind the adaptive mask: embryos as more archetypal than adults; early and simple forms as closer than later and more complex creatures, following "the law that the Archetype is progressively departed from as the organization is more and more modified in adaptation to higher and more varied powers and actions" (p. 49).

This secondary and derivative character of adaptation leads to a linguistic convention of structuralism, where functional fit becomes an impediment to research upon laws of form. The movement of the tetrapod forelimb away from its initial position within the last vertebra of the skull, for example, shows "the antagonizing power of adaptive modification by the removal of that arch from its proper segment" (p. 59). We focus on embryos and simple anatomies in our study of the archetype because, in these forms, "the archetype is least obscured by purposive adaptations" (p. 55).

The derivative nature of adaptation also debars this important phenomenon from any role as a primary organizing principle of morphology. Owen begins his treatise with an incisive argument, cutting to the heart of Paley's error in ignoring relationships among organisms and speaking only of particular designs and their individual excellences. Using Paley's own device of analogy to machines, Owen undermines functionalism from within. Manufactured structures may be individually optimized for their utility; therefore, such contrivances of human technology will not be strongly constrained by homological elements of common design. If organisms had been similarly built on mechanical principles of optimality in adaptation, they would show more structural variation, and not be morphologically clustered as varied manifestations of archetypes. The archetypes themselves, therefore, cannot represent principles of merely functional design:

> To break its ocean-bounds, the islander fabricates his craft, and glides over the water by means of the oar, the sail, or the paddle-wheel. To quit the dull earth man inflates the balloon, and soars aloft, and, perhaps, endeavors to steer or guide his course by the action of broad expanded sheets, like wings. With the arched shield and the spade or pick he bores the tunnel: and his modes of accelerating his speed in moving over the surface of the ground are many and various. But by whatever means or instruments man aids, or supersedes, his natural locomotive organs, such instruments are adapted expressly and immediately to the end proposed. He does not fetter himself by the trammels of any common type of locomotive instrument, and increase his pains by having to adjust the parts and compensate their proportions, so as best to perform the end required without deviating from the pattern previously laid down for all. . . . Nor should we anticipate, if animated in our researches by the quest of final causes in the belief that they were the sole governing principle of organization, a much greater amount of conformity in the construction of the natural instruments by means of which these different elements are traversed by different animals. The teleologist would rather expect to find the same direct and purposive adaptation of the limb to its office as in the machine (1849, pp. 9–10).

Moreover, to stress the key methodological point, immediate utility does not imply design for a current end. Complex shapes and anatomies, developed under formalist rules of structural transformation, may find utility *after* arising for nonadaptive reasons. The delayed fusion of mammalian skull bones may now serve as a prerequisite for parturition through a small birth canal, but birds and reptiles show a similar delay, and this "adaptive" feature did not arise "for" its current and indispensable use in mammals:

> I think it will be obvious that the principle of final adaptation fails to satisfy all the conditions of the problem. That every segment in almost every bone which is present in the human hand and arm should exist in the fin of the whale, solely because it is assumed that they were required in such number and collocation for the support and movements of that undivided and inflexible paddle, squares as little with our idea of the simplest

mode of effecting the purpose, as the reason which might be assigned for the great number of bones in the cranium of the chick, viz. to allow the safe compression of the brain case during the act of exclusion, squares with the requirements of that act. Such a final purpose is indeed readily perceived and admitted in regard of the multiplied points of ossification of the skull of the human fetus, and their relation to safe parturition, but when we find that the same ossific centers are established, and in similar order, in the skull of the embryo kangaroo, which is born when an inch in length, and in that of the callow bird that breaks the brittle egg, we feel the truth of Bacon's comparison of "final causes" to the Vestal Virgins, and perceive that they would be barren and unproductive of the fruits we are laboring to attain, and would yield us no clue to the comprehension of that law of conformity of which we are in quest (1849, pp. 39–40).

### Owen and Darwin

The changing and uncertain relationship between Darwin and Owen presents an intriguing story in Victorian scientific sociology. Darwin's statement in his autobiography has been frequently quoted: "I often saw Owen, whilst living in London, and admired him greatly, but was never able to understand his character and never became intimate with him. After the publication of the *Origin of Species* he became my bitter enemy, not owing to any quarrel between us, but as far as I could judge out of jealousy at its success."

Owen and Darwin first met for professional reasons after Darwin's return on the *Beagle:* Darwin had gathered the material (important bones of South American fossil mammals), and Owen possessed the anatomical skills. Lyell wrote to Owen on October 26, 1836, inviting him to dinner: "Among others you will meet Mr. Charles Darwin, whom I believe you have seen, just returned from South America, where he has labored for zoologists as well as for hammer-bearers" (in Owen, 1894, vol. 1, p. 102). The two men met and liked each other well enough. Darwin entrusted his *Beagle* material to his anatomical colleague, and Owen became the taxonomic author of *Toxodon* and Darwin's other spectacular finds.

Their later antagonism arose for several reasons, some obvious, others less clear. Owen could surely be devious, arrogant and unpleasant. Darwin had struck a blow to the heart of Owen's system by substituting a flesh and blood ancestor, a concrete beastly thing, for the lovely, abstract, Platonic archetype. But something deeper and more intellectually honorable than simple jealousy lay at the core of their growing antipathy.

Owen often enters the false dichotomies of standard histories as a virulent anti-evolutionist, the man who whispered into Wilberforce's ear before the famous debate with T. H. Huxley. If true, simple jealousy might provide an adequate motive: he who overturns my world, and (implicitly at least) makes me a fool in a profession I once dominated, can scarcely remain my companion. Darwin contributed to this impression of Owen as a special creationist by so identifying him in the *Origin* and other writings—and Darwin cannot be entirely blamed for this mischaracterization. For Owen, never the clearest of writers, and ever the diplomat in an aristocratic world where he hob-nobbed with skill as a social climber and a seeker of support for his Museum, could be infuriatingly opaque in his stated commitments. In fact, Darwin, al-though characteristically genial and conciliatory to a fault in his writings, permitted himself a rare burst of trenchant irony in expressing his frustration at Owen's slippery attitude toward evolution. In the historical sketch added to later editions of the *Origin,* Darwin wrote:

> When the first edition of this work was published, I was so completely deceived, as were many others, by such expressions as "the continuous operation of creative power," that I included Professor Owen with other paleontologists as being firmly convinced of the immutability of species; but it appears that this was on my part a preposterous error. In the last edition of this work, I inferred, and the inference still seems to me per-fectly just, . . . that Professor Owen admitted that natural selection may have done something in the formation of a new species; but this it ap-pears is inaccurate and without evidence. I also gave some extracts from a correspondence between Professor Owen and the editor of the "Lon-don Review" from which it appeared manifest to the editor as well as to myself, that Professor Owen claimed to have promulgated the theory of natural selection before I had done so; and I expressed my surprise and satisfaction at this announcement; but as far as it is possible to un-derstand certain recently published passages I have either partially or wholly again fallen into error. It is consolatory to me that others find Professor Owen's controversial writings as difficult to understand and to reconcile with each other, as I do (Darwin, 1872b, pp. xvii–xviii).

One can certainly appreciate Darwin's frustrations. Owen did tailor his statements to circumstances and audiences, appearing cautious or critical as the case warranted, and always taking as much credit as possible. For exam-ple, Owen wrote a particularly nasty notice of the *Origin* in the April 1860 issue of the *Edinburgh Review* (published anonymously, following the tradi-tion of several leading journals at the time. Guessing the identity of review-ers—not at all difficult in this case—became a favorite Victorian intellectual pastime).

In this commentary, Owen did proclaim the origin of species as the greatest of biological problems: "The origin of species is the question of questions in zoology; the supreme problem which the most untiring of our original la-borers, the clearest zoological thinkers, and the most successful generalizers have never lost sight of, whilst they have approached it with due reverence" (Owen, 1860, in Hull, 1973, p. 77).

Writing anonymously, Owen praised himself as Darwin's unacknowledged predecessor in accepting the fact of evolution, but more cautious, and there-fore more worthy and philosophical, on the question of mechanisms:

The great names to which the steady inductive advance of zoology has been due during these periods, have kept aloof from any hypothesis on the origin of species. One only, in connection with his paleontological discoveries, with his development of the law of irrelative repetition and of homologies, including the relation of the latter to an archetype, has pronounced in favor of the view of the origin of species by continuously operative creational law; but he, at the same time, has set forth some of the strongest objections or exceptions to the hypothesis of the nature of that law as a progressively and gradually transmutational one (Owen, 1860, in Hull, 1973, p. 184).

Lest the term "creational law" still seem ambiguous, and lest anyone put the wrong name to the description, Owen later became more explicit. (In the jargon of Owen's day, a secondary cause operated under natural law, thus representing the subject matter of science. God, as "first cause," may have established natural laws, and therefore secondary causes, at the beginning of time, but nature then unfolds under these invariable laws, thus defining the domain of science): "Owen has long since stated his belief that some preordained law or secondary cause is operative in bringing about the change" (Owen, 1860, in Hull, 1973, p. 210).

I acknowledge Owen's opacity and shiftiness, but I also think that we should take him at face value here, for his claim follows his earlier writings, and also accords with the standard view of most formalist thinkers, especially Geoffroy. I believe that Owen had, for more than a decade before the *Origin* appeared, accepted a limited form of evolution—within archetypes, and along channels preordained by archetypal constraints. He never accepted global transmutation, for his brand of limited evolution could not generate the archetypes themselves (which stand as primitive terms, or "givens" in his system), but could only produce variety within their permitted channels. I don't know how to read the famous last lines of the *Nature of Limbs* (see p. 322), except as a genuine statement of the usual formalist commitment to evolution in this admittedly restricted but entirely legitimate sense: "To what natural laws or secondary causes the orderly succession and progression of such organic phenomena may have been committed we are as yet ignorant. . . . But if, without derogation of the Divine power, we may conceive the existence of such ministers, and personify them by the term 'Nature' . . ." (1849, p. 86).

If we thus accept Owen as at least a halfhearted evolutionist long before the *Origin,* then the basis of his unhappiness with Darwin becomes easier to grasp. Darwin mocked Owen's caution, consumed his precious archetype, made evolution global, and then proposed a central mechanism of change (natural selection) in the functionalist mode, diametrically opposed to Owen's formalist inclinations. Owen despised the extent and character of Darwin's evolutionism, but not the idea of evolution itself. All ideologists know that the enemy within provokes more intellectual danger and emotional distress than the enemy without.

Consider Darwin's two principal apostasies from Owen's point of view. First, Darwin reconfigured the abstract archetype as a material ancestor, thus converting Platonism to materialism. Darwin wrote his challenge right in the margin of his personal copy of Owen's *Nature of Limbs* (quoted in Ospovat, 1981, p. 146): "I look at Owen's archetypes as more than ideal, as a real representation as far as the most consummate skill and loftiest generalization can represent the parent form of the Vertebrata."

Second, and even more disturbingly, Darwin inverted Owen's system, and the entire formalist program, by explaining the archetype in functional terms as a congeries of past adaptations materially inherited by descendants. Darwin practically mocked Owen's formalism in the crucial paragraph on p. 206 that I cited to introduce this chapter (pp. 251–257). For Darwin used the jargon of the formalist-functionalist dichotomy—unity of type vs. conditions of existence—and then sank the formalist pole (along with Owen's most precious concept of the archetype) into the alien functionalist sea of natural selection. In an important sense, Owen *vs.* Darwin on evolution replays Geoffroy vs. Cuvier on morphology.

Others appreciated this reincarnation of formalism vs. functionalism, and cheered from the sidelines or within the fray. St. George Mivart, regarded by Darwin as his most cogent critic (see pp. 1218–1224), construed Owen's formalism as a proper evolutionary exit from Darwin's bankrupt natural selection:

> Owen . . . spread abroad in England the perception that a deep significance underlies the structure of animals—a significance for which no stress or strain and no influence of heredity, and certainly no mere practical utility, can account. The temporary overclouding of this perception through the retrograde influence of Darwin's hypothesis of "natural selection" is now slowly but surely beginning to pass away . . . Homologies for which neither heredity nor utility will account reveal themselves in the limbs of chelonians, birds, beasts, and most notably in those of man (from an 1893 statement by Mivart quoted in Owen, 1894, vol. 2, pp. 94–95).

From the other side, Asa Gray understood Darwin's central contribution as the proper reintroduction of purpose, or functionalism, into biology. In 1874, Gray wrote to *Nature* (quoted in Ospovat, 1981, p. 148) that Darwin had done great service for biology by "bringing back to it teleology; so that, instead of morphology vs. teleology, we shall have morphology wedded to teleology"—in other words, functionalist hegemony by proper criteria of primacy and relative frequency. Darwin certainly appreciated the argument, for he wrote back to Gray: "What you say about teleology pleases me especially."

In his usual perceptive manner, Russell (1916, p. 78) wrote: "The problem as Geoffroy and Cuvier understood it was not an evolutionary one. But the problem exists unchanged for the evolutionist, and evolution theory is essentially an attempt to solve it in one direction or the other." So the problem appeared to Owen and Darwin; and so it remains for us today.

# Darwin's Strong but Limited Interest in Structural Constraint

## DARWIN'S DEBT TO BOTH POLES OF THE DICHOTOMY

Darwin, who understood so well that all organisms must be shaped by their history, could scarcely build a great structure in the realm of ideas without subjecting his mental edifice to the same causal influences. Darwin's theory grew in the context of the formalist-functionalist struggle, and he knew and appreciated the issues and terminology (see Section I of this chapter).

A cardinal premise of this book holds that Darwin must be ranked in the functionalist line—for the causal mechanics of his theory grant such clear primacy to adaptation, however subtly the argument develops.* But any evolutionary theory, in adding a historical dimension to reshape the simpler world of the formalist-functionalist dichotomy, would necessarily draw upon both axes of the old system to build the new, orthogonal dimension of temporal change. In two senses, formalist thought included great potential to influence Darwin's evolutionary views.

First, and most obvious in its possibilities (though largely unrealized by Darwin, I shall argue), classical formalism, with its key concept of transformational channels within the bounds of archetypal design, followed a logic intrinsically favorable to a limited form of evolution, while the optimalist functionalism of Paley, or of Cuvier, rooted the impossibility of transmutation in the core of their central argument (Cuvier on correlation of parts, for example). Many of the leading pre-Darwinian formalists supported evolution in this restricted sense—Geoffroy in France, Meckel in Germany, Owen in England. Yet, although Darwin could not have been isolated from this influence, I see no strong evidence that his decision to embrace evolutionism derived from this source (Gruber and Barrett, 1974; Schweber, 1977; Sulloway, 1982). Moreover, the most prominent and fully elaborated evolutionary theory available to Darwin (however strongly he rejected the formulation—see pp. 192–197) resided in the opposite camp of Lamarck's *functionalism*. Finally, the content of Darwin's theory, from his earliest codification in 1838, stood clearly outside formalist thinking—both in replacing archetypes with flesh and blood ancestors built by adaptation, and in advocating the functionalism of natural selection itself.

Secondly, Ospovat (1981) presented the important thesis that Darwin's chief intellectual change within his theory of natural selection between codi-

---

*The subtleties, centering upon two arguments, express great intellectual force, and constitute the power of natural selection in Darwin's version. First, since selection makes nothing, and can only choose among variations supplied by other causes, Darwin developed a theory of variation (copious, small in extent and isotropic—see pp.141–146) that reduced this evolutionary factor to a status as raw material only, and granted all power of *change* to the functional force of natural selection. Second, Darwin brilliantly used the classical relative frequency arguments of "nooks and crannies" and "sequelae" to place formalist influence upon evolutionary change at a periphery of unimportance relative to selection (see pp. 255–256).

fication in 1838 and publication in 1859, lay in abandoning an original belief in perfect adaptation, and in accepting the crucial concepts of relative adaptation ("locally better than") and imperfection imposed by constraints of phyletic history—the argument that completed Darwin's rejection of Paleyan optimality, first in accepting evolution rather than God as the architect of morphology, and only later in recognizing that history implies imperfection in current design. I accept Ospovat's claim that this change must be interpreted as fundamental to the structure of Darwin's theory—see my own extended treatment of Darwin's crucial use of imperfection as primary evidence for evolution itself (pp. 111–116). Finally, Ospovat demonstrates that Owen's disparagement of teleology, and the formalist notion of constraint imparted by archetypes and rules for their transformation, played an important role in Darwin's shift of attitude.

If Owen's formalism influenced Darwin in this manner, then why should Darwin be placed so firmly in the functionalist line? The first and most definitive answer must cite the obvious statement that Darwin explicitly so identified himself—in using the defining terms of the formalist-functionalist debate (unity of type vs. conditions of existence), and in declaring his allegiance to the functionalist proposition as a "higher law" subsuming unity of type as past adaptation (see pp. 251–260).

The broader reason lies in a decision to concentrate on the causal mechanics of explanatory theories for evolutionary change. Evolution opens its umbrella over a vast subject, with many concerns and meanings. A taxonomy of attitudes might designate several alternative criteria for subdivisions. I believe that the *logic* of the *inner workings* of the *primary causal theory*—the *structure* of evolutionary theory, as I named this book—should hold primacy in definitions.* Darwin developed strong ideas about history, attitudes towards analogy, convictions about geology, and philosophical grounding in much that the Victorian age held dear. But his mechanism of evolutionary change—the theory of natural selection—rests upon a central logic, a mode of work-

---

*Pardon a personal footnote, but this criterion—mechanics of the causal theory as a basis for a taxonomy of ideas—fills me with *déjà vu*. I pursued the same argument in my first book, *Ontogeny and Phylogeny* (1977b), when I made a primary separation between von Baer's laws of embryonic repetition and later divergence, and Haeckel's doctrine of recapitulation. Since these two principles often yield the same data for phylogenetic inferences (greater similarity of embryonic stages to ancestral morphologies), many biologists had tended to lump the two accounts together. Yet their causal mechanics are not only different, but strictly opposed—von Baer's as a theory of embryonic retention by unaltered inheritance, Haeckel's as a theory of active evolutionary change by acceleration of previously adult morphologies into early stages of descendant's ontogenies. I was able to show: (1) historically, that major actors in the great late 19th century debate on this topic supported my division by viewing von Baer and Haeckel as opposed; and (2) intrinsically, that the logic of my argument fell into a greatly clarified, and much more useful, structure under such a primary division by causal mechanics. I believe that the general acceptance of my division, and the emergence of heterochrony (the active Haeckelian theme) as a useful and well defined concept in evolutionary theory (Alberch *et al.*, 1979; McKinney, 1999; Gould, 2000e) have vindicated my approach.

ing, that should define the basic attributes of Darwinism. The theory of natural selection is functionalist, by Darwin's own recognition and definition, and by routes (and roots) of causality inherent in its proposed mechanics. Intrinsic factors supply copious, small-scale, isotropic variation—raw material only, and no direct cause or impetus of change. Evolutionary change occurs by natural selection, as organisms adapt to modified local environments. The mechanism of evolutionary change therefore remains functionalist in Darwin's theory; selection powers change, and organisms adapt as a primary result. Darwin also defined his major problem squarely within the functionalist theme of adaptation, as he wrote, in the oft-quoted statement in the Introduction to the *Origin* (1859, p. 3), that much evidence, from many sources, could validate the factuality of evolution itself, but that "nevertheless, such a conclusion, even if well founded, would be unsatisfactory until it could be shown how the innumerable species inhabiting this world have been modified, so as to acquire that perfection of structure and coadaptation which most justly excites our admiration." The argument from imperfection (pp. 111–116) demonstrates the factuality of evolution, but we need to explain adaptation if we wish to understand the *mechanism*.

### DARWIN ON CORRELATION OF PARTS

Many evolutionists can cite the specific example given by Darwin in Chapter 6 of the *Origin* ("Difficulties on Theory") in allowing that not all useful structures arise by natural selection for their current role, however essential that function may be to the life of the organism: "The sutures in the skulls of young mammals have been advanced as a beautiful adaptation for aiding parturition, and no doubt they facilitate, or may be indispensable for this act; but as sutures occur in the skulls of young birds and reptiles, which have only to escape from a broken egg, we may infer that this structure has arisen from the laws of growth, and has been taken advantage of in the parturition of the higher animals" (1859, p. 197).

But I suspect that few of our colleagues know that Darwin took this example directly from the much longer and more detailed treatment in Richard Owen's famous essay (1849) *On the Nature of Limbs* (see quote from Owen on p. 325). This example not only provides direct evidence for an influence of the leading British structuralist thinker upon Darwin (a link often denied because we misread Owen as a creationist, and a leader of the rearguard), but also illustrates a fascinating change of emphasis between Owen's invocation and Darwin's borrowing. For Owen, the example lies at the heart of nature's primary causal structure, and serves as a point of entry to his most trenchant critique of functionalism in general—particularly his claim that explanations of organic form in terms of utility match the barrenness of the Vestal Virgins in Bacon's famous aphorism. For Darwin, on the other hand, the example illustrates an exception to natural selection included within a chapter entitled "Difficulties on Theory."

But Darwin's serious concern with structural constraint cannot be denied (and his borrowed example from Owen surely indicates respect and attention). Darwin linked the efficacy of natural selection to a set of assumptions about the nature of variation (see pp. 141–146). But he could not be satisfied with such an abstraction, and he recognized that any complete theory required an understanding of mechanisms of variation. Darwin presented his major discussion of constraint in Chapter 5 on "Laws of Variation"—for any exception to his trio of necessary properties for variation (copious, small in extent, and undirected) would compromise the exclusive power of natural selection by granting a role to "internal" principles of variation in the direction of evolutionary change. Any exception, in short, would represent a "law of variation" acting not only as a source of raw material, but also as a subsidiary to natural selection among causes of change. (In both the *Origin* (1859) and in his extended treatise on *The Variation of Animals and Plants Under Domestication* (1868), Darwin employed the phrase "laws of variation" to specify properties that could produce evolutionary change independent of natural selection. He proposed a threefold taxonomy of "use and disuse," "direct action of the external environment"—the bases of evolutionary theories often attributed to Lamarck and Geoffroy respectively—and "correlation of growth," or structural constraint as discussed in this chapter. Properties of variation that merely supplied raw material for natural selection—the indispensable trio of copious, small, and undirected—apparently did not count among the "laws," for Darwin viewed these properties as auxiliaries or handmaidens of selection.)

Darwin begins his discussion by admitting contemporary ignorance about causes of variation: "I have hitherto sometimes spoken as if the variations—so common and multiform in organic beings under domestication, and in a lesser degree in those in the state of nature—had been due to chance. This, of course, is a wholly incorrect expression, but it serves to acknowledge plainly our ignorance of the cause of each particular variation" (1859, p. 131).

When we do not know the underlying causal bases of important and related phenomena, taxonomies based upon overt expressions become our best practical procedure for specification and understanding. Darwin therefore tries to gather the phenomena of variation into categories. Most categories either enhance adaptation by routes other than natural selection (use and disuse, and direct effects of the environment—Lamarckism and Geoffroyism in later parlance), or, at most, serve to enhance or slow down selection by affecting the amount of available raw material in variation. Only one category truly challenges the functionalist credo by embracing the primary structuralist theme of internal *constraint* upon adaptation, with consequential nonfunctionality for certain features. Darwin names this category "correlations of growth," and offers a definition: "I mean by this expression that the whole organization is so tied together during its growth and development, that when slight variations in any one part occur, and are accumulated through natural selection, other parts become modified. This is a very important subject, most imperfectly understood" (1859, p. 143). Note, even here, how Darwin defends the primacy of selection (by the "sequelae" argument of the relative frequency tradition—see pp. 255–256). Natural selection builds a feature, and others follow by correlational linkage to this generating cause.

Darwin clearly defines correlation of growth as a category contrary to natural selection, for he explicitly excludes the common case of taxonomically correlated structures that arise by separate selection on each feature, with later joint propagation by simple inheritance (wings and beaks in birds). He defines correlation of growth, on the other hand, as *structurally forced* association independent of immediate selection: "We may often falsely attribute to correlation of growth, structures which are common to whole groups of species, and which in truth are simply due to inheritance; for an ancient progenitor may have acquired through natural selection some one modification in structure, and, after thousands of generations, some other and independent modification; and these two modifications, having been transmitted to a whole group of descendants with diverse habits, would necessarily be thought to be correlated in some necessary manner" (1859, p. 146).

Darwin's genuine interest in correlations of growth arose from several sources, including the mystery surrounding the subject. "The nature of the bond of correlation is very frequently quite obscure . . . What can be more singular than the relation between blue eyes and deafness in cats, and the tortoise-shell color with the female sex?" (1859, p. 144). Again, as for the larger issue of variation in general, Darwin follows a fruitful operational strategy: when you can't specify causes, at least gather overt phenomena into reasonable categories. The short epitome of the *Origin* (where correlation of growth receives seven pages of discussion) lists four major categories:

1. Adaptive modifications of early ontogeny that propagate effects into later growth: "The most obvious case is, that modifications accumulated solely for the good of the young or larvae, will, it may safely be concluded, affect the structure of the adult; in the same manner as any malconformation affecting the early embryo, seriously affects the whole organization of the adult" (1859, p. 143).

2. Correlated variation in serially homologous and symmetrical structures of the body. (Note how Darwin, following the popular structuralist theory of the vertebral archetype, viewed correlated variation of jaws and limbs as potentially homologous): "The several parts of the body which are homologous, and which, at an early embryonic period, are alike, seem liable to vary in an allied manner: we see this in the right and left sides of the body varying in the same manner; in the front and hind legs, and even in the jaws and limbs, varying together, for the lower jaw is believed to be homologous with the limbs" (1859, p. 143).

Although he cited the argument of ontogenetic correlation as first and most obvious, Darwin devoted more interest and attention to this second category of homologous variation. He presents (especially in the longer version of 1868—see pp. 336–341) a variety of wide-ranging and intriguing cases (not all correct, of course). For example, in regarding teeth and hair as homologous, Darwin conjectures: "I think it can hardly be accidental, that if we pick out the two orders of Mammalia which are most abnormal in their dermal covering, viz. *Cetacea* (whales) and *Edentata* (armadilloes, scaly anteaters, etc.), that these are likewise the most abnormal in their teeth" (1859, p. 144).

Nonetheless, since Darwin remains eager to assert the primacy of natural selection as the centerpiece of his worldview, he reminds us that correlation can only be subsidiary in impact—ever present to be sure, but always subject to cancellation if selection favors dissociation: "These tendencies, I do not doubt, may be mastered more or less completely by natural selection: thus a family of stags once existed with an antler only on one side; and if this had been of any great use to the breed it might probably have been rendered permanent by natural selection" (1859, p. 143).

3. Homologous parts not only vary together, but also tend to join or fuse. "Homologous parts, as has been remarked by some authors, tend to cohere . . . nothing is more common than the union of homologous parts in normal structures, as the union of the petals of the corolla into a tube" (1859, pp. 143–144).

4. One part (usually hard upon soft) may impress its form upon another: "Hard parts seem to affect the form of adjoining soft parts; it is believed by some authors that the diversity in the shape of the pelvis in birds causes the remarkable diversity in the shape of their kidneys. Others believe that the shape of the pelvis in the human mother influences by pressure the shape of the head of the child" (1859, p. 144).

Darwin considered one further category, strongly emphasized by Goethe and, later, by Geoffroy as the *"Loi de balancement"* or compensation: "If nourishment flows to one part or organ in excess, it rarely flows, at least in excess, to another part; thus it is difficult to get a cow to give much milk and to fatten readily. The same varieties of the cabbage do not yield abundant and nutritious foliage and a copious supply of oil-bearing seeds" (p. 147). But Darwin, while acknowledging the importance and intellectual pedigree of this principle, wisely chose to exclude compensation from his discussion of structural correlation because he could state no clear criterion (and the problem remains just as vexatious today) for separating negative interaction due to selection from forced and nonadaptive correlation due to limited resources: "For I hardly see any way of distinguishing between the effects, on the one hand, of a part being largely developed through natural selection and another and adjoining part being reduced by this same process or by disuse, and, on the other hand, the actual withdrawal of nutriment from one part owing to the excess of growth in another and adjoining part" (p. 147).

Although limited space and numerous hedges clearly indicate the subordinate status of constraint to adaptation in Darwin's evolutionary views, he evidently did take serious interest in correlations of growth, and he did identify the theme as contrary to, or at least independent of, natural selection—as in this statement: "I know of no case better adapted to show the importance of the laws of correlation in modifying important structures, independently of utility, and therefore, of natural selection, than that of the difference between the outer and inner flowers in some Compositous and Umbelliferous plants"

(1859, p. 144). (Note Darwin's interesting linguistic choice in designating this example as so well "adapted" to illustrate constraint.)

Darwin also allowed that taxonomically important, and not just trivial, characters could be shaped by correlations of growth: "Hence we see that modifications of structure, viewed by systematists as of high value, may be wholly due to unknown laws of correlated growth, and without being, as far as we can see, of the slightest service to the species" (1859, p. 146).

Darwin, as all professional evolutionists know, had been writing a much fuller version of his evolutionary views (a book that would have been about as long as Lyell's three-volume *Principles of Geology*) when Wallace's note from Ternate arrived in 1858. The hurried *Origin of Species* (1859) is an epitome (of 490 pages!) without formal references. Darwin intended to complete and publish the longer version, but never realized this project. Instead, he took most of the material designated for the early part of *Natural Selection* (Darwin's putative title for the full treatment), expanded his coverage, and published his longest book in 1868, the two-volume *Variation of Animals and Plants Under Domestication*. (Incidentally, the fourteen-page introductory chapter to volume 1 presents Darwin's clearest and most cogent general summary of his evolutionary views and methodological principles. This largely unread essay should be assigned to all students of evolution.)

Volume one provides a chapter by chapter treatment of various domesticated species (with Chapter 5 on "domestic pigeons," unsurprisingly given Darwin's interest, as the most illuminating). This material largely represents an expansion of Chapter 1 in the *Origin*, "Variation under domestication." Volume two then presents Darwin's general ideas on variation and inheritance (an expansion of Chapters 4 and 5 of the *Origin*, with much added material). Darwin actually published more about inheritance than about natural selection, thus refuting the common statement that he neglected the subject of heredity—a myth engendered by the retrospective fallacy, given his success with selection and his limited impact in resolving the principles of heredity. Volume two includes sets of chapters on four subjects: inheritance, crossing, selection, and causes of variation, all capped with the chapter that would become (along with his geological error on the "parallel roads" of Glen Roy) his nemesis—Chapter 27 on "the provisional hypothesis of pangenesis" (Darwin's Lamarckian conjecture about the nature of inheritance). Darwin included his material on structural constraint within this volume, providing an expanded version of his views in Chapters 25 and 26 on "correlated variability," a term that he now regards as preferable to "the somewhat vague expression of correlation of growth" (1868, vol. 2, p. 319), used previously.

However, despite extensive elaboration and addition of examples, Darwin's treatment scarcely differs from the short version in the *Origin*. He presents the same taxonomy for modes of "correlation of growth." Again, early modification with propagating effects through ontogeny wins first place, although Darwin now adds some interesting examples: "with short-muzzled races of the dog certain histological changes in the basal elements of the bones arrest their development and shorten them, and this affects the position of the subsequently developed molar teeth" (1868, vol. 2, p. 321).

Darwin then inserts a new category, strangely overlooked in 1859 (a testimony, perhaps, to his subsidiary concern for the general subject of structural constraint)—allometric effects of change in size, either of the whole body, or of parts:

> Another simple case of correlation is that with the increased or decreased dimensions of the whole body, or of any particular part, certain organs are increased or diminished in number, or are otherwise modified. Thus pigeon-fanciers have gone on selecting pouters for length of body, and we have seen that their vertebrae are generally increased in number, and their ribs in breadth. . . . In Germany it has been observed that the period of gestation is longer in large-sized than in small-sized breeds of cattle. With our highly improved animals of all kinds the period of maturity has advanced . . . and, in correspondence with this, the teeth are now developed earlier than formerly, so that, to the surprise of agriculturists, the ancient rules for judging the age of an animal by the state of its teeth are no longer trustworthy (1868, vol. 2, p. 321).

As in 1859, Darwin devotes most attention to correlated variability in homologous structures. In part, this preference can claim a methodological basis, for Darwin finds less mystery in the bonding or joint variation of homologs than in most other cases of structural constraint:

> In many cases of slight deviations of structure as well as of grave monstrosities, we cannot even conjecture what is the nature of the bond of connection. But between homologous parts—between the fore and hind limbs—between the hair, hooves, horns, and teeth—we can see that parts which are closely similar during their early development, and which are exposed to similar conditions, would be liable to be modified in the same manner. Homologous parts, from having the same nature, are apt to blend together, and, when many exist, to vary in number (1868, vol. 2, pp. 419–420).

Amidst a plethora of interesting examples, Darwin cites pigeons that develop feathers and incipient wing-like membranes on portions of the foot corresponding to the position of wings on the forelimbs: "In feather-footed pigeons, not only does the exterior surface support a row of long feathers, like wing feathers, but the very same digits which in the wing are completely united by skin become partially united by skin in the feet; and thus by the law of the correlated variation of homologous parts we can understand the curious connection of feathered legs and membrane between the two outer toes" (1868, vol. 2, p. 323).

Citing the theory of vertebral archetypes again, Darwin tries to correlate variation in head and limb bones, while acknowledging that not all biologists accept the idea: "If those naturalists are correct who maintain that the jawbones are homologous with the limb bones, then we can understand why the head and limbs tend to vary together in shape and even in color; but several highly competent judges dispute the correctness of this view" (1868, vol. 2, p. 324). He also argues that joint slimness in head and limbs of greyhounds, and similar thickness of both structures in draft horses, might record structural correlation between putative homologs.

Proceeding further, Darwin considers hair, feathers, hooves, horns and teeth as "homologous over the whole body" (p. 326). He argues that sheep with a tendency to grow multiple horns also have "great length and coarseness of fleece." He also reports that sheep with more curly wool bear more spirally twisted horns, and that many breeds of hairless dogs grow deficient teeth.

Moving to humans, Darwin tries to relate some forms of inherited baldness to weakness of dentition, and even claims that rare cases of restored hair in old age may be accompanied by renewal of teeth. In his most intriguing example, Darwin discusses Julia Pastrana, the famous "bearded lady" of circus sideshows. Proclaiming her, with true Victorian sensibility, "a remarkably fine woman" (p. 328), Darwin continues: ". . . but she had a thick masculine beard and a hairy forehead; she was photographed, and her stuffed skin was exhibited as a show; but what concerns us is, that she had in both the upper and lower jaw an irregular double set of teeth, one row being placed within the other . . . From the redundancy of the teeth her mouth projected, and her face had a gorilla-like appearance" (p. 328).

Darwin even invokes putative homology—this time between organs of sight and hearing—to explain the case that he had always found personally most bothersome: deafness correlated with blue eyes in cats:

> The organs of sight and hearing are generally admitted to be homologous, both with each other and with the various dermal appendages; hence these parts are liable to be abnormally affected in conjunction . . . Here is a more curious case: white cats if they have blue eyes, are almost always deaf . . . This case of correlation in cats has struck many persons as marvelous . . . [But,] we have already seen that the organs of sight and hearing are often simultaneously affected. In the present instance, the cause probably lies in a slight arrest of development in the nervous system in connection with the sense organs. Kittens during the first nine days, whilst their eyes are closed, appear to be completely deaf; I have made a great clanging noise with a poker and shovel close to their heads, both when they were asleep and awake, without producing any effect. . . . Now, as long as the eyes continue closed, the iris is no doubt blue, for in all the kittens which I have seen this color remains for some time after the eyelids open. Hence if we suppose the development of the organs of sight and hearing to be arrested at the stage of the closed eyelids, the eyes would remain permanently blue and the ears would be incapable of perceiving sound; and we should thus understand this curious case (pp. 328–329).

And yet, though Darwin now presents more examples of correlated variability than he had provided in 1859, the main thrust of the 1868 volumes, as we shall see in the next section, only accentuates the dominance of selection over these exceptions to its domination.

## THE "QUITE SUBORDINATE POSITION" OF CONSTRAINT TO SELECTION

Paradoxically perhaps, an absence of strong concern for a subject may be better expressed by inadequacy of treatment than by total evasion. Had Darwin simply omitted the subject of correlated variability altogether, we would know little about his attitude toward constraint (however much we might infer from his indifference). But the weakness of his limited discussion reveals far more. As E. S. Russell rightly remarks (1916, p. 240): "Darwin's conception of correlation was singularly incomplete. As examples of correlation he advanced such trivial cases as the relation between albinism, deafness and blue eyes in cats, or between the tortoise-shell color and the female sex. He used the word only in connection with what he called 'correlated variation.' . . . He took it for granted that the 'correlated variations' would be adapted to the original variation which was acted upon by natural selection." (I find Russell a bit harsh in this particular claim. Correlated variability may be "adapted" to the primary target of selection in the sense that organic coherence in growth must be maintained, for the correlation marks an inherited pathway. But Darwin clearly designates the realized feature as potentially independent of utility, though admittedly not harmful, for selection would then work to eliminate it.)

Darwin's discussion of variation in the *Origin* clearly illustrates his conviction about the primacy of selection over any internal drive in supply of variation or constraint. Variation (with its three key properties) must be available as raw material, but all shaping and change arise by natural selection. In those rare cases when change can be traced directly to variability itself—that is, when a phenomenon caused by variability becomes visible as a property of a population, rather than the oddity of an individual—we must seek an explanation in the relaxation of selection's usual vigilance and control. Rudimentary parts, for example, become free from selective control, and may therefore manifest the influence of subsidiary factors, including constraint, that selection would ordinarily mask: "Rudimentary parts, it has been stated by some authors, and I believe with truth, are apt to be highly variable . . . their variability seems to be owing to their uselessness, and therefore to natural selection having no power to check deviations in their structure. Thus rudimentary parts are left to the free play of the various laws of growth, to the effects of long continued disuse, and to the tendency to reversion" (1859, pp. 149–150).

The concluding statement in Chapter 5 on "Laws of Variation" clearly expresses the domination of selection: "Whatever the cause may be of each slight difference in the offspring from their parents, and a cause for each must exist, it is the steady accumulation, through natural selection, of such differences, when beneficial to the individual, that gives rise to all the more important modifications of structure, by which the innumerable beings on the face of this earth are enabled to struggle with each other, and the best adapted to survive" (1859, p. 170).

But the fuller treatment of the 1868 work—Darwin's longest book, and primarily written as a treatise on variation, not selection—asserts this fundamental claim even more forcefully. As in the *Origin*, Darwin does allow (1868, vol. 2, p. 320 and p. 355, for example) that correlations of growth produce features (including taxonomically important markers) independent of utility. But the domination of selection, by arguments of relative frequency and importance, now becomes even more explicit, as in this statement from the introductory chapter: "I shall in this volume treat, as fully as my materials permit, the whole subject of variation under domestication. We may thus hope to obtain some light, little though it be, on the causes of variability. . . . During this investigation we shall see that the principle of Selection is all important. Although man does not cause variability and cannot even prevent it, he can select, preserve, and accumulate the variations given to him by the hand of nature in any way which he chooses; and thus he can certainly produce a great result" (1868, vol. 1, p. 3).

Darwin, as I argued previously, may not be a consistently brilliant writer in the tradition of Huxley or Lyell. But he did exceed his more stylish colleagues in a literary gift for inventing metaphors that capture the essence of complex ideas. Most of Darwin's enduring lines fall into this category—the face of nature bright with gladness, struggle for existence, survival of the fittest, wedging as an image for competition. The *Origin* introduces strikingly appropriate and beautifully crafted metaphors in crucial places—the tree of life in the summary of natural selection at the end of Chapter 4 (pp. 129–130), and the entangled bank of the book's final flourish (p. 489).

Darwin also developed a remarkable metaphor to summarize his conviction about the relative importance of selection and variation. He introduced this long passage at the end of his chapter on selection in his 1868 work—the "interloper" chapter, if you will, where the dominating force surveys an entire volume devoted to subservients. We might label this image as the metaphor of building stones for the house of morphology. No other passage in all Darwin's writing so strongly illustrates the domination of selection over raw material.

All components for the primacy of selection, and for the inconsequentiality of constraint (and other internal factors), flow together in this striking image: Selection may depend upon variation, but the character of variation hardly matters (so long as appropriate amounts and styles be present), given the power of selection. Variation cannot be truly random, and we should interest ourselves in its particular forms and biases (the shapes of stones used by the mason). But, in the deepest sense, these preferred forms exert no influence upon the final building when selection (the architect) takes charge. For laws of variation (shapes of stones) "bear no relation to the living structure which is slowly built up" (the form of the building). An architect, armed with a blueprint and enough stones, can build the desired structure, whatever the shapes of pieces available for construction. Thus, "variability sinks to a quite subordinate position in importance in comparison with selection."

> Throughout this chapter and elsewhere I have spoken of selection as the paramount power, yet its action absolutely depends on what we in our ignorance call spontaneous or accidental variability. Let an architect be compelled to build an edifice with uncut stones, fallen from a precipice. The shape of each fragment may be called accidental; yet the shape of each has been determined by the force of gravity, the nature of the rock, and the slope of the precipice—events and circumstances, all of which depend on natural laws; but there is no relation between these laws and the purpose for which each fragment is used by the builder. In the same manner the variations of each creature are determined by fixed and immutable laws; but these bear no relation to the living structure which is slowly built up through the power of selection, whether this be natural or artificial selection.
>
> If our architect succeeded in rearing a noble edifice, using the rough wedge-shaped fragments for the arches, the longer stones for the lintels and so forth, we should admire his skill even in a higher degree than if he had used stones shaped for the purpose. So it is with selection, whether applied by man or by nature; for though variability is indispensably necessary, yet, when we look at some highly complex and excellently adapted organism, variability sinks to a quite subordinate position in importance in comparison with selection, in the same manner as the shape of each fragment used by our supposed architect is unimportant in comparison with his skill (1868, vol. 2, pp. 248–249).

I suggest, as a major theme of this book, that Darwinian evolutionists, ever since, have placed too much confidence in this edifice. Darwin's metaphorical structure, fully shaped by the architect of natural selection, cannot be dismissed as a house of cards, but the walls have developed some cracks and may even be ripe for a breach.

# The Fruitful Facets of Galton's Polyhedron: Channels and Saltations in Post-Darwinian Formalism

## Galton's Polyhedron

Charles Darwin often remarked, as in the *Descent of Man* (1871, vol. 1, pp. 152–153), that he had pursued two different goals as his life's work: "I had two distinct objects in view; firstly, to show that species had not been separately created, and secondly, that natural selection had been the chief agent of change." Darwin spoke wisely (and practically) in these lines. He lies in Westminster Abbey for his unbounded success in the first endeavor; whereas, unbeknownst to many evolutionists who have experienced only the age of natural selection's triumph since the 1930's, Darwin's theory, his second endeavor, never enjoyed much success in his lifetime, and never attracted more than a modest number of adherents. (The titles of Peter Bowler's excellent historical treatises on late 19th century evolutionary thought capture this paradox well—*The Eclipse of Darwinism*, 1983; and *The Non-Darwinian Revolution*, 1988.)

As discussed in Chapter 2 (see pp. 137–141), Darwin's evolutionary critics encountered their greatest stumbling block in their inability to envision natural selection as a creative force. Natural selection could surely serve as an executioner or headsman—the eliminator of the unfit. But such a negative role must occupy a distinctly secondary rank in the panoply of evolutionary forces. The central question of evolutionary theory remains: what creates the fit? The difficulty of this question, and the supposed inadequacy of natural selection as a solution, inspired a vast literature, including two famous Darwinian title parodies by two leaders of his opposition—*The Origin of the Fittest* (1887) by the American Neo-Lamarckian E. D. Cope, and *The Genesis of Species* (1871) by the British structuralist and saltationist St. George Mivart. (Darwin regarded Mivart's criticism as especially serious; the only chapter that he ever added to later editions of the *Origin*—Chapter 7 on "Miscellaneous objections"—largely presents a rebuttal of Mivart's critique—see Chapter 11, pp. 1218–1224, for a full analysis.)

I cannot present a complete taxonomy of alternative proposals in late 19th century thought (lest this section become a multivolumed book in itself). I will, instead, commit the primary historical "sin" of self-serving retrospection, and focus on those critiques of creativity that stressed formalist or structuralist themes now relevant in modern reformulations of evolutionary theory. Thus, I ignore several of the most important currents in late 19th century debate, particularly the strong Lamarckism of many thinkers, and the various threads of theistic and other forms of finalistic directionality.

Since Darwin's essential trio of assumptions about variability—copious, small, and undirected (see pp. 141–146)—does permit natural selection to act as the creative force of change, non-Darwinian alternatives, by logical necessity, deny one or more of these assertions. The diverse formalist theories of this chapter gain conceptual unity in granting *directional* power to internal factors, and not only to the interaction of environment with isotropic raw material. Darwinian claims for the small size and nondirectional character of variations become the obvious candidates for confutation—for formalist alternatives to these Darwinian bastions grant directional power to internal causes (whereas a denial of the third claim of copiousness only places limits upon natural selection without supplying any substitute as a cause of change). Thus, in this late 19th century heyday of alternatives to Darwinism, formalist and structuralist thought centered upon claims for the evolutionary importance of *saltational* and *directional* variation.

The most striking model and epitome for this formalist opposition derives from a source that will strike many evolutionists as anomalous or paradoxical—Darwin's brilliant and eccentric cousin Francis Galton. (The two men shared Erasmus Darwin as a grandfather, surely the most eminent member of the family before Charles himself.) Galton did study continuous variation extensively, and he therefore gained a reputation as guiding spirit for the leading biometricians, Pearson and Weldon. Moreover, his longterm trumpeting of eugenic improvement also promoted the assumption that he favored insensibly gradual and continuous change in evolution.

But Galton, a pluralist in his views on evolutionary causality, viewed discontinuous variation as even more efficacious. Echoing Huxley's frequent plea to Darwin for a larger permissible size in useful variants (advice that Darwin explicitly rejected because he understood so well that the creativity of natural selection would be seriously compromised thereby), Galton wrote (1889, p. 32) that evolutionary theory "might dispense with a restriction for which it is difficult to see either the need or the justification, namely, that the course of evolution always proceeds by steps that are severally minute, and that become effective only through accumulation. That the steps *may* be small and that they *must* be small are very different views; it is only to the latter that I object."

We all recognize Galton's main contribution to the study of continuous variation in his recognition and elucidation of the crucial concept of regression toward the mean. But Galton did not interpret regression in a modern genetic light. For him, regression guaranteed that continuous variation could not yield progressive evolutionary change, because all favorable extremes will, in subsequent generations, regress towards the mean, and no permanent or directional modification can therefore accrue. Substantial change to new "types" and taxa must occur by the occasional production of true-breeding "sports"—discontinuous variants that do not meld to intermediacy in hybrid offspring, and are therefore not subject to regression. In thus emphasizing "sports" for evolutionarily efficacious variation, and stability for taxa at other times (as regression to the mean holds continuous variants in check), Galton also became a hero of the early Mendelians, Bateson and de Vries, particularly for his role in formulating a general rationale for their non-Darwinian concept of saltational origin for new species by macromutation.

Galton subsumed both non-Darwinian formalist themes of discontinuity in effective variation and internally-generated, preferred channels of change (constraints) in a brilliant metaphor that I have called "Galton's polyhedron" (see Gould and Lewontin, 1979). This image had been forgotten by 20th century biologists, but many of Galton's contemporaries discussed the model and its implications. Mivart (1871) invoked the polyhedron as a centerpiece of the critique that most attracted Darwin's attention and response (see Mivart, 1871, pp. 97, 113, and 228); W. K. Brooks (1883, p. 296) cited this image in the most important American treatise on variation, a book that strongly influenced Brooks's visiting student, William Bateson. Bateson (1894, p. 42) then described "the metaphor which Galton has used so well—and which may prove hereafter to be more than a metaphor." Kellogg (1907), speaking of Galton's "familiar analogy" (p. 332), considered the polyhedron as an ideal illustration for the key non-Darwinian challenge of heterogenesis (saltational evolution). And de Vries (1909, p. 53) stated that Galton's polyhedron expressed his own view of variation "in a very beautiful way."

Galton introduced the metaphor of the polyhedron in his eugenic manifesto and most influential book, *Hereditary Genius* (1869). In discussing "stability of types" in the closing chapter on "general considerations," Galton presented his model in an overtly material, and petrological, form:

> The mechanical conception would be that of a rough stone, having, in consequence of its roughness, a vast number of natural facets, on any one of which it might rest in "stable" equilibrium. That is to say, when pushed it would somewhat yield, when pushed much harder it would again yield, but in a less degree; in either case, on the pressure being withdrawn, it would fall back into its first position. But, if by a powerful effort the stone is compelled to overpass the limits of the facet on which it has hitherto found rest, it will tumble over into a new position of stability, whence just the same proceedings must be gone through as before, before it can be dislodged and rolled another step onwards. The various positions of stable equilibrium may be looked upon as so many typical attitudes of the stone, the type being more durable as the limits of its stability are wider. We also see clearly that there is no violation of the law of continuity in the movements of the stone, though it can only repose in certain widely separated places (1884 edition, p. 369).

Twenty years later, in *Natural Inheritance* (1889), the metaphor moved from an afterthought in the back of the book to the focal argument of an early chapter on "organic stability" (pp. 18–34). Galton now granted the image an abstract and formal geometry—as a polyhedron (based on a model that he actually built). He also supplied an illustration (reproduced as Fig. 5-1).

It is a polygonal slab that can be made to stand on any one of its edges when set upon a level table . . . The model and the organic structure have the cardinal fact in common, that if either is disturbed without transgressing the range of its stability, it will tend to re-establish itself, but if the range is overpassed it will topple over into a new position . . . Though a long established race habitually breeds true to its kind, subject to small unstable deviations, yet every now and then the offspring of these deviations do not tend to revert, but possess some small stability of their own. They therefore have the character of sub-types, always, however, with a reserved tendency under strained conditions, to revert to the earlier type. The model further illustrates the fact that sometimes a sport may occur of such marked peculiarity and stability as to rank as a new type, capable of becoming the origin of a new race with very little assistance on the part of natural selection . . .

When the slab rests . . . on the edge AB . . . it stands in its most stable position . . . So long as it is merely tilted it will fall back on being left alone, and its position when merely tilted corresponds to a simple deviation. But when it is pushed with sufficient force, it will tumble on to the next edge, BC, into a new position of stability. It will rest there, but less securely than in its first position; moreover its range of stability will no longer be disposed symmetrically. A comparatively slight push from the front will suffice to make it tumble back, a comparatively heavy push from behind is needed to make it tumble forward. . . . If, however, the slab is at length brought to rest on the edge CD . . . the next onward push, which may be very slight, will suffice to topple it over into an en-

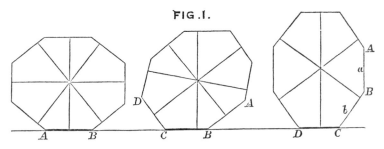

FIG. I.

5-1. Galton's own illustration of his model of the polyhedron. Note how the themes of saltation, or facet flipping, and constraint in strictly limited pathways available for change arise from a similar geometric basis in this mode of depiction.

tirely new system of stability; in other words, a "sport" comes suddenly into existence.

In Galton's own conception, the polyhedron embodies four major implications that assert and codify the power of formalist constraint as an evolutionary agent of *change* (not just an impediment), while controverting the essential Darwinian claim that natural selection alone builds new forms in a creative and accumulative fashion.

1. Occasional large variations (sports) are more important for evolution than omnipresent, normally distributed small variation. This substitution of big for small forces a major compromise, and may even represent a fatal weakness, in Darwin's theory of natural selection.

Galton first introduced his polyhedron to question Darwin's key claim for insensible gradation in evolutionary continuity. (Galton, of course, did not deny continuity, but he wished to substitute a series of jumps—facet flipping, if you will—for Darwin's smoothness): "It is shown by Mr. Darwin, in his great theory of 'The Origin of Species,' that all forms of organic life are in some sense convertible into one another, for all have, according to his views, sprung from common ancestry. . . . Yet the changes are not by insensible gradations; there are many, but not an infinite number of intermediate links; how is the law of continuity to be satisfied by a series of changes in jerks?" (1884 edition, p. 369).

In a later article on "Discontinuity in evolution," Galton posed the fundamental question of change in Darwin's favored style: "By what steps did A change into B? Was it necessarily through the accumulation of a long succession of alterations, individually so small as to be almost imperceptible, though large and conspicuous in the aggregate, or could there ever have been abrupt changes?" (1894, p. 363). Acknowledging the criterion of relative frequency for resolving debates in natural history, Galton correctly notes that Darwin did catalog exceptions, but only to log their peripheral character and to assert the domination of gradualistic accretion by natural selection leading to adaptation:

> Notwithstanding a multitude of striking cases of the above description collected by Darwin, the most marked impression left on the mind by the sum of all his investigations was the paramount effect of the accumulation of a succession of petty differences through the influence of natural selection. This is certainly the prevalent idea among his successors at the present day, with the corollary that the Evolution of races and species has always been an enormously protracted process. I have myself written many times during the last few years in an opposite sense to this.

Galton then strongly asserts that most evolutionary novelty, in opposition to Darwin, probably arises *per saltum:* "Many, if not most breeds, have had their origin in sports" (p. 365). Galton bases his rationale on the argument that continuous, small-scale Darwinian variability, though omnipresent, cannot be effective because regression toward the mean precludes accumulation in favored directions. Galton introduces the term "transilience" (literally "going between," and recently revived by Templeton, 1982, for a different mechanism in the same spirit) to describe his favored concept of non-Darwinian saltatory variation, or facet-flipping:

> No variation can establish itself unless it be of the character of a sport, that is, by a leap from one position of organic stability to another, or as we may phrase it, through "transilient" variation. If there be no such leap the variation is, so to speak, a mere blend or divergence from the parent form, towards which the offspring in the next generation will tend to regress; it may therefore be called a "divergent" variation. . . . I am unable to conceive the possibility of evolutionary progress except by transiliences, for, if they were merely divergences, each subsequent generation would tend to regress backwards towards the typical center (p. 368).

2. Internal factors establish a hierarchy of stabilities, discontinuous in origin, and explaining differing degrees of divergence among typical forms.

The nonhomogeneity of morphospace seems so "obviously" intrinsic to nature (lions close to tigers, with a big jump separating all cats from dogs and wolves), that we rarely consider the puzzles thereby raised. Once evolution itself becomes paradigmatic, simple inheritance and descent become the obvious, first-level reason for ordering the resemblances portrayed in our taxonomic hierarchies. But simple descent does not solve all problems of "clumping" in phenotypic space; we still want to know why certain forms "attract" such big clumps of diversity, and why such large empty spaces exist in conceivable, and not obviously malfunctional, regions of potential morphospace. The functionalist and adaptationist perspective (see Dobzhansky, 1958, quoted herein on p. 527, for a particularly striking metaphor of this view) ties this clumping to available environments, and to shaping by natural selection. Structuralists and formalists wonder if some clumping might not record broader principles, at least partly separate from a simple history of descent with adaptation—principles of genetics, of development, or of physical laws transcending biological organization.

Galton proposed his polyhedron to explain a hierarchy of stabilities as internally generated, not externally shaped by gradual natural selection. As the long quotation (p. 345) and figure indicate, Galton purposely shaped his model to encompass both small islands of stability within species ("subtypes" in his terms, and easily subject to reversion by tipping back to the primary type—as in facet BC of Fig. 5-1, which easily falls back to AB, but can move onward to a new type only by a much bigger push), and also to cover the origin of new taxa that cannot revert (the push from facet CD to any position across the long axis and its major facet AB). Of course, Galton's polyhedron does include a role for external forces like natural selection: something must push, or the polyhedron can't move at all. But selection, in this model, provides only an impetus: both discontinuity and directionality of change follow internal rules (completely for discontinuity, and partly for directionality by establishing channels—trajectories between facets—that must translate the impetus into actual change).

In an interesting discussion, Galton distinguishes internally enjoined positions of stability from clumping by simple descent, and he rejects the Darwinian argument that taxonomic structure records differential shaping by natural selection: "In the first place each race has a solidarity due to common ancestors and frequent interbreeding. Secondly, it may be thought by some, though not by myself, to have been pruned into permanent shape by the long-continued action of natural selection. But, in addition to these, I have for some years past maintained that a third cause exists more potent than the other two, and sufficient by itself to mold a race, namely that of definite positions of organic stability" (1894, p. 364).

Obviously struggling with a difficult concept that eludes precise formulation, Galton seizes upon a variety of metaphors from "governments, crowds, landscapes, and even from cookery and . . . from mechanical inventions" (1889, p. 22) to argue that workable solutions may be viewed as isolated and stable islands in a sea of largely empty space (impossible combinations). These "nucleating points" mark physically possible places, predetermined by the structure of matter and space, not *a posteriori* results of natural selection working as a contingent force in local environments. (Today, of course, we would formulate this idea in Darwinian terms as "multiple adaptive peaks," but Galton struggles with an alternate view, worthy of our respect as an interesting option, of discontinuous solutions, internally set. In Galton's mechanism, natural selection would still operate—but only in the negative role of policing, and the facilitating task of providing a push into a preset channel.) Only a few forms of government can lead to internal stability; only some landscapes cohere; only a few combinations make flavorful dishes, despite a wide range of ingredients. In a memorable defense of formalist preference for the timelessness of distinct and stable configurations, and for downplaying the role of historical specificity, Galton writes of crowds and public rituals: "Every variety of crowd has its own characteristic features. At a national pageant, an evening party, a race-course, a marriage, or a funeral, the groupings in each case recur so habitually that it sometimes appears to me as if time had no existence, and that the ceremony in which I am taking part is identical with others at which I had been present one year, ten years, twenty years, or any other time ago" (1889, p. 23). Misidentifying (as King Solomon) the Preacher of Ecclesiastes, Galton regrets the sameness that such structural ahistoricity imposes. But he cannot disparage the way of the world: "It is the triteness of these experiences that makes the most varied life monotonous after a time, and many old men as well as Solomon have frequent occasion to lament that there is nothing new under the sun" (1889, p. 24).

Morphotypes represent islands of stability, rare combinations of coherence among available parts (for history only becomes relevant by providing an inventory of availability). Natural selection may preserve these morphotypes once they form by facet-flipping, but their origin must be regarded as discontinuous and internally mediated. The public carriages of London take discrete forms as omnibuses, hansoms, and four-wheelers. All these forms can be improved, but the boundaries of the types seem inviolable, for "the old familiar patterns cannot, as it thus far appears, be changed with advantage, taking the circumstances of London as they are." (Galton, obviously, did not anticipate the macromutation of the internal combustion engine!) The three "islands" arose as discontinuous inventions and cannot be transformed one into the other, for intermediary steps would be structurally inviable: "A useful blend between a four-wheeler and a hansom would be impossible; it would have to run on three wheels and the half-way position for the driver would be upon its roof" (1889, p. 31). (In the old hansom, or two-wheeled cab, the driver sat on an elevated platform behind the passenger cabin, and reins to the horses ran over the roof. Drivers of four-wheeled cabs sat in front.)

3. The positions of stability, for both subtypes (discontinuities within species) and types, are not honed by natural selection, but internally preset as rare configurations of coherence among parts. The causal basis of stable form must therefore be explained by internal integrity, not by adaptation.

In his first formulation (1869, 1884 edition, p. 370), Galton proposed an internal, correlational basis for morphological stability: "It is easy to form a general idea of the conditions of stable equilibrium in the organic world where one element is so correlated with another that there must be an enormous number of unstable combinations for each that is capable of maintaining itself unchanged, generation after generation."

As his most enduring practical legacy, Galton (1892) pioneered the use of fingerprints in identification and criminology. He also found in these "papillary ridges" an ideal example of discontinuous stable "islands" (subtypes in this case) that could not be attributed to direct natural selection. Galton extols the value of these patterns as stable and traceable throughout ontogeny: "We know nothing by observation about the persistence of any internal character, because it is not feasible to dissect a man in his boyhood, and a second or third time in his after life, whereas finger prints can be taken as often as desired" (1894, p. 366).

Fingerprints serve as ideal examples of internally generated structural islands for "they fall into three definite and widely different classes . . . transitional forms between them being rare and the typical forms being frequent, while the frequency of deviations from the several typical centers . . . correspond approximately with the normal law of frequency" (1894, pp. 366–367). Individual patterns remain stable throughout life "and are consequently not unimportant in spite of their minute character" (p. 366).

Nonetheless, the stabilization of these three typical forms cannot be attributed to natural selection:

Notwithstanding the early appearance of the patterns in fetal life and their apparent importance, they are totally independent of any quality upon which either natural selection or marriage selection can be conceived to depend. For example, I find the same general run of patterns in

English, Welsh, Jews, Basques, Hindoos, Negroes, men of culture, farm laborers, criminals, and idiots. I have failed to observe the slightest correlation between the patterns and any single personal quality whether physical or mental. They are therefore to be looked upon as purely local peculiarities, with a slight tendency towards transmission of inheritance (p. 366).

Galton concludes his treatment of this case with a formalist flourish: "I therefore insisted that the continual appearance of these well-marked and very distinct patterns proved the reality of the alleged positions of organic stability, and that the latter were competent to mold races without any help whatever from the process of selection, whether natural or sexual."

4. Galton's polyhedron also highlights the theme of internally-based directionality, not only of discontinuous change.

Galton often stated that his model did not deny continuity in change (1869, 1884 edition, p. 369), but only confuted the insensible character of transitions—for the polyhedron tumbles in a jerky fashion by facet flipping, and does not roll smoothly towards "better" positions. Galton's conception of change does grant a role to natural selection; some force has to push the polyhedron.

But a status as provider of an impetus scarcely fulfills Darwinian requirements for selection's power. In a metaphor for illustrating pure Darwinism, organisms may be represented as billiard balls, with natural selection as the pool cue. A perfectly round ball denotes Darwinian isotropic variation; the organism only supplies raw material, and cannot set its own direction of change. The ball's trajectory depends upon the pool cue of natural selection and the form of the surface (local environment). (The surface of this old table may be channeled and pitted, representing directions favored by external environments.) The pool cue supplies propulsion, and the ball rolls with no internal control over its own direction of motion.

But Galton's polyhedron pushes back. Absent an impetus, the polyhedron cannot tumble at all, but the pusher (the "pool cue" in Galton's model) doesn't set the direction of motion (or at least can only push effectively in a strictly limited number of trajectories set by the configuration of facets on the morphologically complex "billiard ball"). The direction of tumbling will therefore be determined as much by the internal structure of the polyhedron as by the coordinates and strength of the impetus. Only certain, internally established channels of change can be realized, even if natural selection must always initiate the tumbling of the polyhedron—a very different image from setting the smooth billiard ball in motion! In this sense, Galton's polyhedron weds the theme of directionality with the idea of discontinuity.

Galton emphasizes this dual concern in his initial presentation. Just after describing the first version of his polyhedron (the stone with many facets), he introduces another metaphor to reinforce the theme of directionality combined with facet flipping—an image that may not be so apt or fruitful as the polyhedron itself, however expressive of Galton's intent:

Now for another metaphor, taken from a more complex system of forces. We have all known what it is to be jammed in the midst of a great crowd, struggling and pushing and swerving to and fro, in its endeavor to make a way through some narrow passage. There is a dead lock; each member of the crowd is pushing, the mass is agitated, but there is no progress . . . At length, by some accidental unison of effort, the dead lock yields, a forward movement is made, the elements of the crowd fall into slightly varied combinations, but in a few seconds there is another dead lock, which is relieved, after a while, through just the same process as before. Each of these formations of the crowd, in which they have found themselves in a dead lock, is a position of stable equilibrium and represents a typical attitude (1869, 1884 edition, pp. 369–370).

Formalism, as the preceding chapter documents, boasts a long and distinguished pedigree, well antedating both Darwin and any explicit discussion of evolution. Darwinism rendered many formalist concerns irrelevant, but key features of the structuralist agenda could not be encompassed, or even well addressed, by natural selection and its functionalist mechanics. Galton's polyhedron provides a strikingly apt metaphor for the two great themes of formalism that continue to demand attention within evolutionary theory, and that the Darwinism of his day could not adequately comprehend—discontinuous evolution, and internally generated pathways: in other words, *saltation* and *channels*. Both themes express the more general conception that internal properties of organisms "push back" against external selection, thereby rendering evolution as a dialectic of inside and outside—that organisms, in other words, must be conceived as polyhedrons, not billiard balls.

I shall show, in the rest of this chapter, that formalist, or structuralist, evolutionary thought, from the immediate post-Darwinian years to the codification of the Modern Synthesis, continued to emphasize the twinned concepts of saltation and channels—for these notions represent two sides of the same conviction that internal structure can set and constrain the pathways of change. The modern plea to put history and organic integrity back into evolutionary theory echoes the same call. If "constraint" has become a buzz-word of contemporary evolutionary theory, then I must assume that the shades of Galton and Geoffroy are smiling, for the structure of their thought has withstood the formalist's ultimate test of timelessness.

# Orthogenesis as a Theory of Channels and One-Way Streets: The Marginalization of Darwinism

## MISCONCEPTIONS AND RELATIVE FREQUENCIES

The German zoologist Wilhelm Haacke devised the word "orthogenesis" in 1893 (see Kellogg, 1907; Bowler, 1983); but the concept implicitly motivated the entire formalist tradition that sandwiched Darwin (in a chronological sense) between Goethe and Geoffroy on one side, and searchers for the mechanical rules of embryological development on the other. The word literally means "straight (line) generation," but the term never bore a merely descriptive meaning, and all evolutionists understood the wider import. Orthogenesis denotes the claim that evolution proceeds along defined and restricted pathways because internal factors limit and bias variation into specified channels. In this key sense, orthogenesis must be regarded as a formalist theory, standing against the central Darwinian principle that natural selection imparts direction by shaping isotropic variation (and doesn't only act in a negative and subsidiary way to eliminate the unfit while some other process creates the fit). (Evolutionists recognized, of course, that natural selection could also produce a directional anagenesis—first called "orthoselection" by Ludwig Plate—and that claims for orthogenesis must therefore demonstrate a causal basis for internal channeling beyond the power of natural selection to shape, and not only record the simple pattern of monotonic change itself).

Later on, after the Modern Synthesis congealed, and a latter day Darwinian consensus needed to recruit some whipping boys from the past, orthogenesis became a convenient foil for illustrating the bad old days of failure to grasp selection's power. Ever since, most textbook one-liners have dismissed orthogenesis as a theistic remnant operating as a mild pollutant within science, an almost mystical theory of arcane and inexorable direction. I shall present the arguments of three prominent supporters—G. H. T. Eimer, A. Hyatt, and C. O. Whitman—to explicate orthogenesis as a viable and well-wrought formalist alternative (or supplement) to Darwinism at a time when natural selection could muster no compelling defense. But in hopes of encouraging a more sympathetic hearing, I begin with three points raised to allay our conventional misconceptions in a more general way.

1. Some prominent non-Darwinians may justly be designated as "theistic evolutionists"—St. George Mivart and Pierre Teilhard de Chardin, for example. But orthogenesis does not fall into this category. Rather, and entirely to the contrary, all leading orthogeneticists insisted vociferously that their arguments for internal directionality included no teleological or theistic component. Most leading orthogeneticists held strictly mechanistic views in the mainstream of the highly deterministic late 19th century scientific consensus. They argued that internal channels arose as products of conventional, physical causes, based upon properties of hereditary and developmental systems. (These properties may have been unknown, hence "mysterious" in the vernacular sense, but certainly not spiritual or teleological.) In stating claims for predictability of phyletic directions, and for parallelism of numerous independent lineages constrained by the same internal mechanics, most orthogeneticists considered themselves in better tune with the physical and deterministic spirit of the age, whereas Darwinians fell into disharmony by committing themselves to undirected variation and unpredictable contingency of change. Moreover, the orthogeneticists argued, how could a charge of theistic progressionism be leveled when orthogenetic channels drove lineages to extinction as often as to complexity?

2. If the concept of internally constrained channels only represented a

vague theoretical notion invented to oppose Darwin's principle of undirected variation, then the usual charge of vacuousness might apply. But all leading orthogeneticists, as my three case studies will demonstrate, identified a primary channel fully consistent with a late 19th century biological consensus. Haeckel's biogenetic law had become widely accepted as the preeminent principle for tracing phylogeny (Gould, 1977b). This law of recapitulation required that new evolutionary features be added to the ends of previous ontogenies, and that early stages of development be continually speeded up (law of acceleration) to provide room for these additions; adult stages of ancestors therefore migrated "backwards" in ontogeny to become juvenile features of descendants. This principle validated a primary ontogenetic channel as the major determinant of highly constrained evolutionary variability. Features consistent with established ontogenetic trajectories—as additions (by hypermorphosis) or regressions (by secondary slowing of developmental rate)—might easily arise as new evolutionary variants. Other configurations, even if potentially useful to organisms, might never arise in such a constrained system. In short, ontogeny itself served as the primary channel of constrained variation in orthogenetic theory.

3. The "hard version" of orthogenesis (held by some proponents, including Hyatt in my case studies) of inexorable one-way streets leading straight to extinction by degeneration of form compels little attention today (and enjoyed little support even in its own time). But the "softer" version of two-way channels—furrows of constrained variation that provided biased material to natural selection, but could not drive a trend to extinction all by itself and against Darwinian forces—expresses a primary and enduring theme of a formalist and structuralist biology that should still engage our close attention.

To explicate orthogenesis, I turn once again to Vernon Kellogg (1907), author of the finest book on varieties of evolutionary theories and their distinctions (in a time of maximum disaffection with Darwinism and general agnosticism about alternatives). As discussed on pages 163–169, my framework for this book owes much to Kellogg's argument that Darwinism should be defined by a meaningful essence of minimal commitments—and that other notions can therefore be classified as basically helpful ("auxiliary" in his terminology) or contrary ("alternative").

Kellogg identified three major "alternatives" to natural selection—one as functionalist as Darwinism, but offering a different explanation of adaptation (Lamarckism), and the other two as structuralist denials that adaptation must guide the origin of new species (orthogenesis, and heterogenesis or saltationism in de Vries's macromutational style). But Kellogg showed particular sensitivity to the nuances, shadings and subtleties that arguments about relative frequency always impart to natural history. He clearly stated that the logic of all three alternatives stands squarely against natural selection if we argue for prevailing strength of effect and relative frequency: "All of these theories offer distinctly substitutional methods of species forming" (1907, p. 262).

However, Kellogg also recognized that "milder" versions might be seen as auxiliary—not consonant to be sure (for the nonselectionist logic cannot be contravened)—but supplementary rather than substitutional. Thus, for example, if acquired characters are inherited only rarely and weakly, then Lamarckism might aid natural selection in developing adaptation more quickly (by secondary reinforcement from a different source)—a position advocated by Darwin himself throughout the *Origin* (1859, pp. 134–139, for example). But if acquired characters are inherited faithfully all the time, then natural selection will be overwhelmed and Lamarckism becomes a refutation of Darwinism. Relative frequency determines the distinction. Kellogg writes, in a statement just before the quotation of the last paragraph (1907, p. 262): "Few biologists would hold any of these theories to be exclusively alternative with natural selection; de Vries himself would restrict natural selection but little in its large and effective control or determination of the general course of descent."

A similar situation prevails for orthogenesis. In Hyatt's "hard" (and truly anti-selectionist) version, the internal pathway dominates all lineages, literally pushing the impotent force of natural selection aside, and forcing lineages to extinction by phyletic senility. Natural selection exists as a "true" force, but can only operate as a peripheral factor that can, at most, delay the inevitable. In milder versions, the relative frequencies equalize (and the orthogenetic pathways need not lead so clearly to inadaptive forms). In soft versions, still defendable today (though we have ceased, for good reasons, to use the term "orthogenesis"), such internal drives become auxiliaries to selection in providing an initial boost of directed variation for the "incipient stages" of useful structures that posed so many problems for early Darwinians (see Mivart, 1871)—and then letting the ultimately more powerful force of natural selection prevail in the larger realm of evident utility. Kellogg writes of such potentially "friendly" versions for Darwinism:

> In true orthogenesis the variation, and hence the lines of modification, are predetermined. It seems obvious, however, to any believer in natural selection that sooner or later the fate of these lines of development will come into the hands of selection. And most orthogeneticists do indeed admit this. But it is precisely in the making of a start in modification that orthogenesis fills a long-felt want, and if capable of proof, should be gladly received by Darwinians as an important *auxiliary* theory in the explanation of modification, species-forming, and descent [my italics, and an interesting choice of words since Kellogg classifies orthogenesis as an alternative but recognizes here that a sufficiently mild version would fall into his other category of auxiliaries to Darwinism] (1907, p. 276).

None of the three versions discussed below presents quite so mild a view (though C. O. Whitman approaches such a formulation). All three conceive orthogenesis as a competitor to Darwinism and a more powerful force than natural selection. But these versions do illustrate a spectrum from the "centrist" Eimer (the primary popularizer of the name and idea), who viewed natural selection as deprived of power but not contravened; to the "hard liner" Hyatt on one side, who interpreted the orthogenetic drive as contrary to selection; to the conciliatory Whitman on the other, who hoped to find appropriate and mutually reinforcing status for all viable contenders in a pluralistic evolutionary theory.

I present orthogenesis as a spectrum grounded in relative strength and frequency because I believe that a potential role for modern versions of such structuralist theories should be judged in the same manner. A mild formalism of constraint, akin to some ideas within the unfairly reviled theory of orthogenesis, may now enrich our Darwinian world (see Chapters 10 and 11)—and the potential fusion, in its richest form, would not be designated as strict selectionism with a little bit added, but would be recognized as a potentially integrated theory of a new kind (with a persistent Darwinian core). In this light, an understanding of the original formulations of orthogenesis and their varying relationships with Darwinism may enlighten us in our current struggle to integrate structuralist and functionalist approaches to evolutionary causality.

## THEODOR EIMER AND THE *OHNMACHT* OF SELECTION

Theodor Eimer's evolving views followed a channel every bit as directional (though inadaptive by current, and perhaps transient, standards) as the constraining orthogenetic pathways that he ascribed to organisms. His empirical work of the 1870's, on coloration of lizards from Capri, presented a predominantly functionalist argument, with a boost from internal channels to foster movement through incipient stages, and to reinforce the process along the way. In the first of his two volumes on orthogenesis—published in German in 1888 and translated into English in 1890 as "Organic evolution as the result of the inheritance of acquired characters according to the laws of organic growth"—Eimer stressed internal channels, relegated Darwin to a periphery, but still sought a genuine fusion (as the title proclaims) of formalist and functionalist perspectives. The second and last volume (for Eimer died soon thereafter)—published in 1897 as *Orthogenesis der Schmetterlinge: ein Beweis bestimmt gerichteter Entwickelung und Ohnmacht der Natürlichen Zuchtwahl bei der Artbildung*—presents an anti-selectionist polemic (directed more at his sparring partner Weismann than at Darwin) and a defense of internal direction as preeminent. Its title proclaims Eimer's change of emphasis from fusion to exclusivity: *Orthogenesis of butterflies: a proof of definitely directed development and the weakness of natural selection in the origin of species.*

Gustav Heinrich Theodor Eimer (1843–1898) was born near Zürich and eventually became professor of zoology at Tübingen, Germany. He imbibed the late 19th century mechanistic tradition that so permeated German science (in such movements as *Entwicklungsmechanik*)—an attitude strongly opposed to speculative phylogenizing, the main thrust of the "Darwinian" (read Haeckelian) tradition in Germany (see Gould, 1977b, chapter 6). But Eimer also expressed sympathy for "our great philosopher Oken" (1890, p. 433), the early 19th century formalist oracle of *Naturphilosophie,* who had proceeded him at Tübingen, and had taught Agassiz, among others. Thus, Eimer doubted Darwinian functionalism from both sides of 19th century German biology—the romanticism of early 19th century formalist morphology, and the mechanism of late 19th century experimentalism.

All sources agree that Eimer's treatise (1890 translation) became the major English language source for the theory of orthogenesis. Contemporaries either set their own discussion in its light (see Whitman, 1919—a posthumous publication of work done before 1910), or recommended its primary study to those unfamiliar with, or hostile to, orthogenesis (Kellogg, 1907, p. 322). Modern historians of sciences (Bowler, 1979, 1983) continue to view Eimer as "the major popularizer" of orthogenesis (Bowler, 1983, p. 141, in his book on non-Darwinian evolutionists of the late 19th century). I shall therefore treat Eimer's views first.

Eimer's mechanist side led him to reject any vitalist or "teleological" tinge to orthogenesis. "I repudiate any special internal force of evolution. According to my view, everything in evolution is due to perfectly natural processes, to material, physical causes" (1890, p. 64). In fact, Eimer's philosophical defense of orthogenesis relies largely on its putative superiority over Darwinism as an evolutionary mechanics in the determinist tradition; for a discovery of lawlike order and direction in the key domain that Darwin had surrendered to chance—the origin of variation—would represent a notable triumph for a physicalist worldview. Eimer's opening words (1890, p. 1) set his entire argument in this context: "It seemed to me long ago of the greatest importance to undertake an investigation of the question whether the modification (variation) of the species of animals is not governed by definite laws." Eimer, of course, concluded in the affirmative (1890, p. 1): "If the principles of Darwinism are true because they can be shown to follow from natural laws, then it was to be expected that obedience to laws would also be discovered in that province which Darwin had surrendered to chance. But if variation were shown to follow certain laws, the same demonstration would apply to the origins of species."

If the directions of variation are strongly channeled and lawlike, then evolutionary history may one day achieve the predictability of physical science (in its late 19th century deterministic version): "The evolution—the growth—of species one from another proceeds onwards as though following a plan drawn out beforehand" (1890, p. 29). This leaning towards predictability flows from the particular theory of channeling adopted by all leading orthogeneticists—phyletic cooptation of the ontogenetic pathway. By virtually synonymizing "evolution" and "growth" in the statement cited above, Eimer expressed the common Haeckelian belief that if ontogeny and phylogeny cannot be exactly equated, both processes proceed under a common nexus of causes ("phylogeny is the mechanical cause of ontogeny," to cite a familiar Haeckelian maxim—see Gould, 1977b).

Since the predictable character of ontogeny cannot be denied, this comparison establishes a *prima facie* case for orthogenetically channeled evolutionary change. Eimer (1890, p. 379) emphasizes the distinction without a difference: "We have to distinguish from one another (a) individual (personal) growth, (b) the growth of the race (the species), or phyletic growth. The latter is, however, merely the sum of the modifications due to growth which the individuals of a line of descent have undergone in course of time." Then, in a stronger statement of unity, he adds:

> The individual growth of every plant, every animal is a brief and rapid repetition, under the continued influence of similar stimulation, of the series of effects produced by external stimuli in the course of vast periods of time on the tissues of its ancestors. The character of the individual growth of every living being therefore depends essentially on phyletic growth, the individual growth includes phyletic growth in itself. Since the individual growth of every living being is thus a stage of phyletic growth, since the latter . . . presents a sum of individual growths, both are traced back to one and the same process—fundamentally they cannot be separated (pp. 381–382).

Eimer used the ontogenetic channel as a device to elucidate a range of evolutionary phenomena beyond simple directionality of change. Why, for example, do some populations of a species "move on" to more advanced stages of phylogeny (and to formal status as a new species), while others languish in stability? Some contemporaries had argued that populations must first become isolated and then may diverge as selection dictates (while parental forms remain stable in their unchanged environment). But Eimer denied allopatry as a precondition for speciation: some groups within a species simply show more phyletic "activity" in varying beyond the ancestral ontogenetic trajectory. As these groups advance in form, they proceed further than their stable neighbors, eventually to a distance beyond the range of interbreedability with parental forms. Eimer thus argued that orthogenesis could explain both directional change and diversification.* Speciation marks the fractioning of a phylogenetic sequence into separate segments representing persistent and altered populations. "Varieties and species are therefore in reality nothing but groups of forms standing at different stages of evolution, that is, at different stages of phyletic growth, whether it be that they outstripped their fellows or their fellows them in the process of evolution, so that connection by intermediate forms was lost. . . . The essential cause of the separation of species is seen to be the persistence of a number of individuals of a definite lower grade of this evolution, while the rest advance farther in modification" (pp. 30–31).

Partly as a rhetorical device to be sure, but largely from deep conviction about the essence (and attractiveness) of his system, Eimer presented his style of orthogenesis as a reasonable and happy intermediate, an Aristotelian

---

*Darwin faced the same issue when he realized that natural selection, as an agent of anagenesis, could not fully encompass evolution without a separate explanation for multiplication of species—a gap that Darwin attempted to fill with his "principle of divergence" (see pp. 224–236).

golden mean, between the two extreme views of purely internal and exclusively external forces as causes of evolutionary change. At one counterproductive extreme of strict internalism, his colleague Nägeli acknowledged external forces of environmental shaping and adaptation only as hindrances to the expansive and progressive character of inherently driven evolution. Nägeli granted but two roles to externality: the accountant's job of crafting clearer taxonomic groups by eliminating intermediates; and the grim reaper's task of pruning a bush that would otherwise have grown even more luxuriant. All evolutionary advance and diversification arises from intrinsic properties of life, particularly from a *Vervollkommungskraft*, or perfecting force.

Nägeli expressed this most internalist of all post-Darwinian evolutionary theories with a striking metaphor of gardeners and growing bushes—an iconography well worth remembering as a guide to this style of theorizing (note the obviously intended comparison of Darwinians with children):

> Still better may we compare the vegetable kingdom to a great tree branching from its base upwards, of which the ends of the twigs represent the plant forms living at one time. This tree has an enormous power of sprouting, and it would, if it could develop without hindrance, form an inextricable bush-like confusion of innumerable branchings. Extermination in the struggle for existence, like a gardener, prunes the tree continually, takes twigs and branches away, and produces an orderly arrangement with clearly distinguishable parts. Children who see the gardener daily at his task may well suppose that he is the cause of the formation of the branches and twigs. Yet the tree, without the constant pruning of the gardener, would have been much greater, not in height, but in extent, and in the richness and complexity of its branching. In the perfecting process (progression) and adaptation lie the mechanical impulses which lead to the abundance of forms; in competition and extermination, or in Darwinism proper, only the mechanical cause of the formation of gaps in the two organic kingdoms (quoted in Eimer, 1890, p. 19).

Incidentally, and to show the power of mechanistic thinking in scientific culture at the time, Nägeli (though usually cited as a leading vitalist today) insisted as vociferously as Eimer that his orthogenetic forces, though unresolved, must be entirely natural consequences of the physiochemical construction of organisms.

At the other extreme of overextended externalism, strict Darwinians hold that organisms contribute only isotropic raw material (hence no direction from internal forces), and that natural selection produces all evolutionary change as adaptation. Eimer did not engage Darwin himself as a chief opponent—for Darwin had held more pluralistic views and was, moreover, no longer available for polemical battle. Instead, Eimer focused his anti-selectionist arguments against August Weismann, chief disciple of the exclusivistic version that adopted the label of "neo-Darwinism" (see Romanes, 1900, on the difference, non-continuity, and "non-homology" of this "brand" of Neo-Darwinism—Weismann and Wallace's hyperadaptationism—with the same term as adopted by the Modern Synthesis of the 1940's, and continuing today). Weismann, as discussed extensively in Chapter 3, strongly advocated the *Allmacht* ("all-power," or omnipotence) of natural selection. In a rhetorical dig, Eimer countered with the *Ohnmacht* ("without-power," or weakness) of Darwin's primary force.*

Eimer attempted to land squarely in the middle of this debate (1890, p. 63): "Neither Nägeli's view, which ascribes the principle of utility an almost infinitesimal effect, nor Weismann's which regards adaptation as all powerful, can be unreservedly accepted. The truth lies between them." Eimer joins inside and outside with a model of internal orthogenesis as the architect of possible pathways for change (with ontogenetic trajectories as the most important channels) and environmental forces as potentiators of the channels into actual expression as evolutionary alteration. As a simple, but instructive example, channels feature two directions. Orthogenesis builds the channel—thus constraining change to two paths in a potential infinity—but doesn't specify the direction. Environment supplies the required push, thus assuring that the prefigured change will be either adaptive or at least neutral, for selection operates as an efficient executioner of the ill-designed. (Or, taking an even larger role, environment might choose the channel, if several stand open for possible entry.)

---

*In an obvious foray against Weismann, Eimer (1897) used this word in the subtitle of his second and final volume on orthogenesis—*Ein Beweis bestimmt gerichteter Entwickelung und Ohnmacht der natürlichen Zuchtwahl bei der Artbildung* ("a proof of definitely directed development and the weakness of natural selection in the origin of species"). I focus here on the more balanced view of external and internal forces presented in Eimer's first volume of 1888, his only major work translated into English, and therefore the main source of his influence among anglophone evolutionists. Natural selection gets short enough shrift in this balanced view (for Eimer, as we shall see, grants most external power to Lamarckian forces and little to Darwinian selection). But in the 1897 work, and largely (I suspect) as a result of long and bitter polemics with Weismann, Eimer became ever more dismissive about natural selection (while still giving lip service to the importance of environment, though largely through use and disuse). Eimer continually contrasts the *Ohnmacht der Selection* with the *Herrschaft* (domination) *der Orthogenese*. He considers natural selection (1897, p. i) "*von der geringsten Bedeutung*" (of the smallest significance) as a factor of change. He cites the standard argument—that selection creates nothing but can only choose among variants presented by another process—as the *Fundamental Einwurf* (objection) to Darwin's system: "Selection can create nothing new, but can only work with characters that already exist and are useful in and of themselves" (1897, p. iii). Eimer waxes polemical about "the exaggerated, blinkered presentation of the principle of selection, right up to the proclamation of *Allmacht*" (1897, p. iv), and he particularly lambastes assumptions made by strict selectionists about adaptation: "They satisfy themselves either in simply stating that this or that is 'adaptive,' thereby bringing investigation to an end; or they begin a round of groundless speculation, which surely has nothing to do with exact science" (1897, p. iv). Further comparison of Eimer's earlier "balanced" version of 1888 with his anti-Weismann polemic of 1897 would make an interesting historical project. The Weismann connection, as well as the polemical tone, may be sampled in Eimer's motto: "orthogenesis is the mortal enemy *[Todfeind]* not only of the omnipotence *[Allmacht]* of natural selection, but also of the hypothesis of the germ plasm that is based upon it." (1897, pp. xiv–xv).

Morphological "degeneration" provides the best example of environmental impetus within Eimer's concept of balance between internal and external forces. The ontogenetic channel supplies the internal component, since simplification can be achieved most easily by backing down the ontogenetic pathway. But environment must provide the downward push—for change usually proceeds in the other (and upward) direction, as the biogenetic law asserts. Environment, in this case, might act in a positive sense (when simplification leads to local adaptation, as in parasitism), or in a more negative way (when unfavorable climates prevent the full passage of ontogeny, and a resulting juvenilization then becomes inherited in Lamarckian fashion). Eimer writes: "The abundance of the species which have been formed by degeneration, by retrogression, is known to every zoologist. It is self-evident that their origin is to be traced to the action of external conditions" (1890, p. 53).

Eimer found his best example of balance, or environmentally triggered orthogenesis, in a favorite case of all early evolutionists—the Mexican axolotl, or sexually mature tadpole of the genus *Ambystoma*. (Axolotl formerly bore the separate generic name *Siredon*, while the parental form often received a misplaced "l" as *Amblystoma*—as in the quote below.) Ontogeny set the orthogenetic channel and change could only proceed in the prefigured direction, up or down. But the actual path taken depends upon the environment—down for warm and permanent ponds, up in more terrestrial climates:

> *Siredon* is for the problem of evolution one of the most important living animals, in that it brings so beautifully before our eyes the transition of lower sexually mature into a higher sexually mature form, and at the same time shows so clearly the causes of the transformation. We discern these causes simply in the reaction of the organism under external conditions, the increased exertion of an organ already in process of formation (the lung), and the disappearance of another (gills) in consequence of definite demands of the environment, and changes connected with these by correlation. *Amblystoma* appears where the axolotl has too little water to live in, where it is compelled to live on dry land. That this is the case is proved by the possibility of artificially rearing Amblystoma from the axolotl by gradually withdrawing water (1890, p. 46).

At this point, Darwinian forces might have played a major role in Eimer's system. He needed external potentiators to trigger prefigured orthogenetic pathways. Had he chosen selection as a preferred external force, his resulting amalgam would still have departed from a Darwinian rubric—for the idea of internal orthogenetic channels runs so contrary to the key Darwinian postulate of undirected variation. But selection would still have played an important role as an instigator of trends.

However, Eimer chose Lamarckism as the preferred external potentiator for his "brand" of functionalism. He opted for use and disuse with inheritance of acquired characters as the standard mode for transferring environmental information to organisms—and thus as the primary mechanism for adaptive change. Selection, to Eimer, therefore became a negative force—an eliminator of the unfit, once other factors had channeled a trend.

Eimer's 1890 subtitle neatly expressed his balanced view of external and internal forces—". . . the result of the inheritance of acquired characters according to the laws of organic growth." The laws of organic growth act as internal (structural) channels of orthogenesis; the catch-phrase of the Lamarckian system then identifies the primary external pusher. Natural selection therefore becomes marginalized to the periphery of half a theory.

Eimer's low opinion of selection represents a common viewpoint among evolutionists of his generation, as expressed in the most familiar of all late 19th century critiques (see pp. 137–141)—the denial of "creativity" to natural selection (viewed entirely as a negative force, while new and favorable features must arise by some other "creative" process). Eimer places himself squarely in this majoritarian tradition when he writes (1890, pp. 383–384): "Natural selection can, as I have repeatedly remarked, create nothing new. It only so far contributes to the growth of the organic world that it selects the forms which are most fitted for life, and preserves them for the future action of new stimuli and of crossing . . . Thus the power of selection lies chiefly in the promotion and diversification of organic growth. It is . . . only an indirect cause of the evolution of living beings."

Eimer's original contribution lay in his characterization and defense of the creative force that could displace selection to such an insignificant periphery. Eimer understood the crucial role of undirected variability in the logic of Darwinian argument. He knew that his orthogenetic channels would derail the Darwinian system by vesting creativity for change in the process of variation itself; selection could then only speed up or hinder what internally generated directionality had previously supplied. Eimer contrasts orthogenesis against selectionism, while also reemphasizing the mechanistic, and non-vitalistic, character of channeled variation:

> This conclusion is in a certain sense opposed to Darwin's, since it recognizes a perfectly definite direction in the evolution and continuous modification of organisms, which even down to the smallest detail is prescribed by the material composition (constitution) of the body. According to this conclusion, the real Darwinian principle, that of selection depending on utility, is only effective within the limits which are prescribed by the material composition of the body, that is, by the fixed directions of evolution. Accordingly there is nothing fortuitous, but everything in evolution to the smallest detail is governed by laws (1890, p. 431).

If the creative force of evolution resides in the process of directed variation itself, then the nature of internal channeling assumes crucial importance. In the absence of a documented mechanistic theory for the nature of inheritance, Eimer and all leading orthogeneticists followed the empirical tradition of inducing supposed regularities of channeled variation from common features of case studies. Eimer presented his list, with varying numbers and orderings of categories, in all his major publications (1890, pp. 28–30; 1897, pp. 18–21).

Each principal "law" of channeled variation identifies a putative ontogenetic pathway, and therefore indicates the status of individual growth as the primary determinant of evolutionary direction. Eimer includes in his list:

1. As an overarching principle, the biogenetic law of recapitulation, which specifies that the sequence of terminally added features in phylogeny shall, through acceleration of development, become the ontogenetic channel of future changes. But what regularities specify the character of these terminal additions?

2. The invariant series of changes in color markings from longitudinal stripes, to spots, to transverse stripes, and finally (in most cases, when intensification occurs) to darker and more uniform coloring. Today, we may be surprised and puzzled (as, by the way, were many scientists at the time, for example Whitman, 1919) by Eimer's decision to present so apparently particular and contingent a sequence as not only effectively universal, but also as the major specific rule of channeling stressed in all his publications. Eimer first enunciated this principle in his early work on lizards. The doctrine then became the foundation for his most important study on coloration of butterfly wings (the focus of the final volume (1897) of his treatise on orthogenesis and evolutionary theory). Eimer's specific claims for butterflies did not gain a wide following, as many naturalists recognized the circular character of his argument. (He assumed the law of longitudinal striping→spots→ transverse striping *a priori* and then used this principle to establish "phyletic" sequences of living species with no other criterion of cladistic order).

3. "The law of wave-like evolution, or law of undulation" (1890, p. 29). New characters appear at particular parts of the body—almost always the posterior end—and then pass forward during growth. A series of progressive waves may sweep over the body during a single ontogeny. This law, apparently so arbitrary and riddled with exceptions, also met with little favor, even among fellow orthogeneticists. Whitman (1919), for example, accepted the notion of spatial "waves" in ontogeny as channels of variation, but argued that progressive color variation passed from front to back in pigeons, in direct opposition to Eimer's pathway. (A "rational" basis for spatial waves could be sought in the biogenetic law, as older parts of the body should act as the source for new phyletic features added terminally. But the posterior end of an animal may be either old or young depending on the body's mode of growth.)

4. "The law of male preponderance." In Eimer's words (1890, p. 28): "that where new characters appear, the males, and indeed the vigorous old males, acquire them first, that the females on the contrary remain always at a more juvenile lower stage, and that the males transmit these new characters to the species." This principle could claim a rationale beyond simple sexism (though this social context should not be disregarded as a source either), for male preponderance followed from the general theme behind all Eimer's channels— the biogenetic law. If, as most biologists believed in Eimer's time, males tend to move beyond terminal female stages in a common ontogenetic channel,

then new features will first appear in males, since novelty must be added terminally under Haeckel's law.

The specification of such channels as constraints on adaptation does not, of itself, push natural selection to a periphery of unimportance; for, as Darwin argued in another context (see pp. 251–260), existing channels must have evolved for some adaptive reason in ancestors—and if selection constructs adaptation, then Darwinian forces can reclaim their creative power, whatever limits these inherited channels may place upon current adaptation. In fact, Eimer did toy with the idea that selection might have built the primary channels—doing so in a burst of fanciful speculation, much in the mode that he himself would later castigate as the primary weakness of adaptationist thinking.

For the primary law of longitudinal stripes to spots to transverse stripes in coloration, for example, Eimer suggested that the initial state might represent an adaptation of ancestors (under the false assumption that monocots precede dicots in the geological record of angiosperms): "The fact of the original prevalence of longitudinal striping might be connected with the original predominance of the monocotyledonous plants, whose linear organs and linear shadows would have corresponded with the linear stripes of the animals" (1890, p. 57). Eimer then extended this speculation by guessing that conversion to spots "might be connected with the development of a vegetation which cast spotted shadows"; and the final transition to transverse stripes "with the shadows, for example, of the branches of woody plants—thus the marking of the wild cat escapes notice among the branches of trees" (p. 57).

From this phyletic fancy, Eimer moved to more conventional Darwinian reasons for other channels. Male preponderance "might possibly be explained by the fact that the males fight the battle of existence more than the females, and therefore must always be first to respond to new demands" (p. 58). And ornamental waves moving from posterior to anterior might also gain a selectionist basis "by the fact that the part of the body farthest from the head is most in need of mimicry, because it is least protected in other ways by the sense organs, and because it is at a special disadvantage; that it is the last part to be withdrawn from the pursuit of an enemy" (p. 58).

But Eimer could not, ultimately, grant even this much power to selection and adaptation for two major reasons. First, he decided that several of his channels expressed no evident utility and, even if adaptive in final expression, could not have possessed any selective value in incipient states (1890, p. 59):

> But all this does not explain the first occurrence of the new characters, nor the undeviating course of the evolution in a particular direction. For when a number of varying individuals are compared it is seen that the variations of all tend to a definite end, and that the majority of the intermediate forms show stages in the development of the characters which are absolutely without use to them. This cannot be explained except by

natural growth, whose operations are changed, intensified, or diminished to a certain extent by the stress of adaptation, and may also at times be entirely restrained.

(Linguistic choices can be highly illuminating. Note how Eimer refers to adaptation as a "stress" rather than a determining cause—that is, a push from the outside that can, at most, speed up, slow down, or change in minor ways the primary pathway of internal direction.)

Second, although adaptationist guesses can be formulated for the observed channels, formalist alternatives also exist, and seem eminently more reasonable. The law of undulation from back to front, for example, only reflects the metameric growth of many animals. The back segments form last, and late stages, under the biogenetic law, develop progressive characters (1890, pp. 60, 69).

In sum, orthogenesis derails Darwinian functionalism by denying the crucial requirement for undirected variability. "The variation of species takes place not in all kinds of directions irregularly, but always in definite directions, and indeed in each species in a given time in only a few directions" (1890, p. 20). Ultimately, Eimer rejected Darwinism for the most common of all 19th century reasons—the critique of "creativity"—for he clearly understood the deathblow that such powerfully channeled and limited variation would deal to Darwinian hopes for awarding a dominant relative frequency to natural selection in creating evolutionary change: "Natural selection becomes weak [ohnmächtig] thereby. It cannot be depicted as an active major cause [Hauptmittel] in the transformation of forms; at most, it can be an auxiliary cause [Nebenmittel] of such transformation, as it can only perfect what orthogenesis ordains" (1897, pp. 14–15).

The metaphors and images chosen by scientists to illustrate their complex views often provide our best insight into their relative weighting of interacting forces and foci. Eimer developed what he and others considered a "liberal" or a "compromise" version of orthogenesis, where internal forces of directed variation do not determine the entire shape and rate of phylogeny (as in Hyatt's more extreme views), but rather work in balance with environmental determinants—a model of external triggers and choosers for a strictly limited number of internally set pathways. This balanced view, as noted above, might have awarded important space to Darwinian functionalism as an external force—except for Eimer's particular commitment to Lamarckian causes as primary determinants of the external component, with selection acting only as an auxiliary force, largely confined to the negative role of eliminating the unfit.

Eimer summarized these weightings in a metaphor that compared evolution with the migration of a population. The direction of movement represents the orthogenetic pathway, the dominant cause of the entire process. External forces of environment act primarily in finetuning this predetermined direction. Adaptation may play an important role, but the Lamarckian inheritance of acquired characters trumps Darwinian factors in this realm of functional influence. Selection only sharpens the edge of success by eliminating the failures.

> Thus we may compare the whole process of the modification of forms to the results of the migration of the people over an extensive foreign territory. Some tribes, not having the strength to follow, soon, others later, remain behind, others again reach a distant goal. Some retain their characters in their new home or strengthen them, even modify them by correlation, others change under the influence of external conditions and adapt themselves to the environment—all that is not sufficiently capable of endurance is left lying by the way and perishes, and if the struggle for existence is at all severe only the toughest of all survive (1890, p. 55).

## ALPHEUS HYATT: AN ORTHOGENETIC HARD LINE FROM THE WORLD OF MOLLUSKS

Standard examples of orthogenesis all took the same form: paleontologically based, and therefore temporally extensive, monotonic trends towards clearly inadaptive features, leading inevitably to the extinction of the afflicted lineage. Supposed "best cases" included the enlarging antlers of the "Irish Elk" (but see Gould, 1974), the extended canines of saber-toothed cats, and the self-strangulation of the oyster *Gryphaea,* as overcoiling of one valve clamped the other shut, immuring the animal in its own shell (but see Gould, 1972).

These lurid stories, in their textbook versions, are the caricatures of a serious non-Darwinian theory once quite popular among paleontologists from the late 19th century through the 1930's: the idea that trends, though adaptively initiated, might break from environmental control and run, inexorably along the same path, eventually to extinction—if the proximal cause that originally responded to the adaptive pressure became so entrenched that selection could no longer halt or reverse the trend. W. D. Lang (1923), for example, proposed that an originally adaptive increase in rates of shell secretion might become unreversible, leading to *Gryphaea*'s deathly dilemma: "These trends, even if at first encouraged by the environment because they are of use to the organism, are soon out of the environment's control; they are lapses which may overtake *Ostrea* [the supposed ancestor of *Gryphaea*] at any moment of its evolution—trends which having once started continue inevitably to the point when their exaggeration puts the organism so much out of harmony with its environment as to cause its extinction." A. E. Trueman, who had developed the empirical case for *Gryphaea* (1922), also asserted the anti-Darwinian character of internally driven trends: "Excessive development implies that the evolution was out of the control of the environment and it may be presumed that some internal factor was responsible" (1940, p. 93). Trueman proposed no mystical or vitalist explanation (as the standard caricature maintains), but sought a cause consistent with modern genetics. (He suggested that such inadaptive orthogenesis might arise by mutation pressure too high for natural selection to overcome.)

An even stronger anti-Darwinian version of the theory—a maximal departure from functionalist explanation—held that all stages of a trend might follow a prefigured, internally programmed path, and that environmental selection need not be invoked at all, even for an initiating push. These versions usually anchored their argument in a more than metaphorical reading of phylogeny as akin to ontogeny and imbued with similar, inevitable stages of youth, maturity and old age.

To illustrate this hardest-line account of orthogenesis (and to set up a contrast with the opposite and most accommodating version of C. O. Whitman, discussed in the next section), I turn to an influential monograph on Miocene snails (see, for example, the lengthy popular and well illustrated account in Le Conte, 1888, pp. 236–239), published in 1880 by the American paleontologist Alpheus Hyatt. The case becomes particularly relevant for this historical chapter because Darwin and Hyatt had engaged in a long and frustrating correspondence, full of misunderstanding, about the theory of ontogenetically programmed phylogeny (see F. Darwin, 1903, pp. 338–348). Hyatt later sent Darwin a copy of his 1880 monograph, and Darwin replied, less than a year before his death, with an uncharacteristically ungracious acknowledgment (reprinted in F. Darwin, 1903, p. 393—letter of May 8, 1881): "I am much obliged for your kind gift . . . which I shall be glad to read, as the case has always seemed to me a very curious one. It is all the kinder in you to send me this book, as I am aware that you think that I have done nothing to advance the good cause of the Descent-theory."

Hyatt, deeply stung, donned his hair shirt, and quickly penned a response on May 23: "I tell you that your strongest supporters can hardly give you greater esteem and honor. I have striven to get a just idea of your theory, but no doubt have failed to convey this in my publications as it ought to be done." Francis Darwin, Charles' son and editor of this volume, then adds, after quoting Hyatt: "We find other equally strong and genuine expressions of respect in Prof. Hyatt's letters." But, genuine respect notwithstanding—and I don't doubt Hyatt's *bonae fides* for a moment—no version of orthogenesis could be more contrary to Darwinism than the theory of internally programmed phylogeny.

From a parochial American standpoint at least, the evolutionary theories devised and promoted by paleontologists E. D. Cope and Alpheus Hyatt occupy an important historical position. For a nation still coming of age as a scientific power, and still bearing a reputation, at least in natural history, as supplier of data to the theory-mills of a more sophisticated Europe, the rise of an American movement, centered in a novel theoretical perspective, and gaining both attention and respect in Europe, marked an important gain in maturity. Cope and Hyatt led this co-called "American school," often identified as "Neo-Lamarckism."

Cope and Hyatt did accept the inheritance of acquired characters, but a new view on the mechanism of recapitulation, and a distinctive argument about ontogeny in general, built the truly central and characteristic argument of their theory (see Gould, 1977b, Chapter 4). These two scientists proposed a new mechanics for the biogenetic law, and this "principle of acceleration" forms the core of their theory and the basis of its orthogenetic implications. Ontogeny will recapitulate phylogeny provided that two necessary principles can be validated. All supporters of the biogenetic law promoted some version of these principles. First, new characters must arise in phylogeny as additions to the end of previous ontogenies. This principle of "terminal addition" (Gould, 1977b) would, of itself, engender recapitulation, as descendants pass through the adult stages of ancestors before accreting their own novelties in ontogeny—except for a logical problem that required the second principle for a full and coherent theory. The phylogeny of a lineage unfolds through thousands of steps in geological immensity; new stages cannot be added indefinitely to the unaltered ends of previous ontogenies, lest growth to adulthood take untold years to reach completion. Some process—some law of heredity—must produce a general speeding-up of development, so that ancestral ontogenies can unfold more rapidly, leaving time at the end for addition of novel features.

All recapitulationists necessarily defended some form of speeding up for ancestral ontogenies through phyletic time. Haeckel himself, who thought a great deal about genealogies but precious little about mechanisms, advocated a differential dropping out of stages, with compression of the remaining steps to a shortened ancestral ontogeny (Gould, 1977b). Cope and Hyatt, who both devised a theory of recapitulation in 1866 (independently of each other and of Haeckel), first proposed an ultimately more popular and plausible version of ontogenetic quickening—the "principle of acceleration," or general increase in rate of development (with no necessary excision of stages). This law of acceleration (as a foundation for recapitulation) became the most important theoretical contribution of the American school (Fig. 5-2).

The law of acceleration held that ancestral ontogenies unroll more and more rapidly in successive descendants, thus making room for new stages in phylogeny. (The gill slits of a human embryo can therefore represent the compressed and accelerated adult form of our piscine ancestry.) With ontogeny thus depicted as a quickening treadmill through time, attention could shift to the nature of stages that accumulated in successive terminal additions—for the accreted stages form a series that defines the lineage's phylogeny. If the sequence of accreted stages represents the unfolding of an internal "program," then phylogeny can be justly called orthogenetic—for the law of acceleration makes room on the treadmill, and the next stage of a predictable sequence then struts its hour upon the adult stage.

Later versions of "Neo-Lamarckism" in this American style cannot be called orthogenetic because they did not propose an internally programmed series of new stages. These later accounts embody a view of causality every bit as functionalist as Darwin's, though relying upon Lamarckian mechanisms rather than natural selection. New features arise adaptively as organisms actively respond to needs imposed by altered environments. The sequence of novel stages accretes as a contingent series, mapping the functional requirements of a changing external world.

5-2. Mechanisms proposed by recapitulationists to compress late stages of an ancestral ontogeny into earlier stages of the embryology of descendants. In Cope and Hyatt's solution, shown bottom left at B, stages are compressed by acceleration. In Haeckel's solution, bottom right at C, some stages are simply omitted by deletion. From Gould, 1977b.

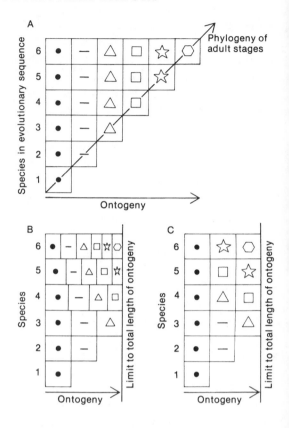

But earlier versions of Cope's views (and Hyatt's opinion throughout his life) posit an evolutionary mechanism diametrically opposed to this later functionalism—an internal dynamic yielding an orthogenetic phylogeny of predetermined stages, with the source of predictable novelty inherent in ontogeny itself. This history of lineages unfolds along a "grand potential ontogeny" much longer than the realized portion of early species in a phyletic series. The adult stage of the initiating species does not reach beyond an early phase of the potential sequence. (Suppose, for example, that the full series includes 100 stages, ending in predictable extinction. The ancestral species may only progress from stages 1 to 10 in its own ontogeny, leaving 90 available steps for successive terminal additions in the phylogeny of subsequent species.) This concept of an extended potential ontogeny as the source of predictable phyletic additions became the most powerful version within a class of non-Darwinian theories generally regarded today as purely fanciful or falsely analogic, and without conceivable mechanism—the idea of racial ontogenies and life cycles. But the concept of a genealogical ontogeny, however indefensible by modern standards, once possessed an interesting rationale in this recapitulatory context.*

---

*As a small footnote in the logic of evolutionary theory and the history of Darwinian arguments, this notion of "phyletic life cycles" provides the best historiographic refutation of

In Cope's early view, lineages arose with a latent phyletic life cycle extending far beyond an initially realized ontogeny. So long as acceleration prevailed in phylogeny, old ontogenies unfolded more rapidly and new stages of the grand potential ontogeny accreted at the end. A much less common form of regressive evolution could foster a slowing down of ontogeny ("retardation" in Cope's terminology) and a retention of previously juvenile stages as adult forms of descendants (the older "degenerative" interpretation of a phenomenon now generally viewed more positively as "neoteny"). Since acceleration occurs far more commonly than retardation in evolution (as progress generally prevails over regress), the general vector of genealogy proceeds as an unrolling of phyletic life cycles. The stages of phylogeny are, in any case, internally programmed and predictable. Cope wrote in 1869 (quoted in Cope, 1887, p. 123): "Genera have been produced by a system of retardation or acceleration in the development of individuals: the former on pre-established, the latter on preconceived lines of direction."* Cope recognized the formalist implication of this view, as expressed in the key postulate that the origin of a structure precedes its use—in opposition to the cardinal principle of any functionalist theory (including both Darwinism and Lamarckism) that functional

---

the old canard that natural selection is a tautology and therefore empty of content (see Bethel, 1976, and refutation in Gould, 1977c). This hoary claim, still a favored gambit of creationists, brands selection as a useless concept because its watchword—"survival of the fittest"—becomes meaningless when fitness is defined in terms of survival. The argument can be refuted in several ways (including the value of tautology in many scientific contexts—see Sober, 1993), but Darwin's own rebuttal seems most compelling to me. Darwin did not define fitness retrospectively by observed survival. He insisted, in principle at least, that fitter organisms could be identified before any environmental test by features of presumed biomechanical or ecological advantage. (The speediest deer can be specified beforehand, and their differential survival in a world of wolves can then be tested empirically.) Some critics dismiss Darwin's claim by arguing that no one would be foolish enough to predict differential survival of the less adapted—and that such a *Gedanken* experiment therefore becomes meaningless. But the theory of racial life cycles proves empirically that several leading evolutionary thinkers once made predictions of exactly this type as central propositions of influential theories. As an essential postulate, Hyatt's theory held that less fit forms (by Darwin's *a priori* definition) would prevail over better adapted individuals during periods of phyletic old age and racial senescence. "Survival of the fittest" cannot be dismissed as an empty statement if alternative empirical claims not only can be formulated in principle, but also actually build the core of historically important theories.

*Cope recognized, of course, that many small scale adaptations (changes in color and proportion, for example) could not be rendered as stages in a phyletic series of acceleration and terminal addition. These functional contingencies became the basis for recognizing differences at the species level—whereas new steps in the programmed sequence (or retreats down the staircase by retardation) set the chief criterion for establishing new genera. This attempt to designate taxonomic rank by the theoretical status of new features, while fundamentally misguided by current views, gives us insight into older concepts about progress and predictability in evolution. In this feature, and ironically, Cope's early system is more truly Lamarckian than his later and more explicit "Neo-Lamarckism"—as this distinction between criteria of central and superficial change mirrors Lamarck's central concept of different causes for upward and tangential evolution (see Chapter 3, pp. 186–189). Cope's view also entailed a logical paradox that helped to sink the theory—for Cope could not deny that two genera might reside in the same species if a small heterochronic change, and no other, occurred in one population of a lineage.

needs impel adaptive change. (Cope's subsequent shift to a functionalist position prompted the development of his later Neo-Lamarckism.) Cope wrote in 1870 (quoted in Cope, 1887, pp. 145–146): "We look upon progress as the result of the expenditure of some force forearranged for that end. It may become, then, a question whether in characters of high grade the habit or use is not rather the result of the acquisition of the structure than the structure the result of the encouragement it offered to its assumed beginnings by its use."

Hyatt differed from Cope in an essential manner that made his theory the most uncompromisingly recapitulatory of all 19th century views, the most committed of all proposed evolutionary mechanics to "programmed" racial life cycles, and the most orthogenetic. Cope had provided the obvious interpretation for a standard 19th century perception that, whereas most lineages progressed by complexification, others regressed to greater simplicity. Cope argued that progressive lineages undergo acceleration and therefore "gain room" to add new stages to the end of old ontogenies; regressive lineages, on the other hand, experience ontogenetic retardation and never surpass the juvenile stages of their ancestry. Hyatt, in a paradox resolved by an ingenious argument, tried to render both progressive and regressive evolution as results of acceleration alone. For Hyatt, the law of acceleration reigned virtually without exception; no more extensive or uncompromising version of universal recapitulation has ever been offered.

Hyatt resolved the apparent paradox with an argument that he affectionately called his "old age theory." The programmed steps of a potential phylogeny proceed through a sequence more than merely analogous to the phases of ontogeny. Adult stages of early species in a lineage exhibit traits of phyletic youth; adults in a lineage's geological midlife display the features of phyletic maturity; while adults of species near the extinction of a lineage finally develop unmistakable signs of phyletic senility. Hyatt's hardest-line, internally programmed version of orthogenesis rests upon this notion of a phyletic life cycle. The stages of a phylogeny become as predictable and predetermined as the phases of an ontogenetic sequence. Environment must be sufficiently favorable to permit the unfolding (just as a fetus will not grow without adequate nutrition), but the sequence of stages is internally ordained, not functionally entrained by interaction with a surrounding environment. Hyatt wrote (1897, pp. 91–92): "There is a rise of the individual through progressive stages of development to the adult and a decline through old age to extinction. In the evolution of the stock to which the individual belongs there is a similar law, a rise through progressive stages of evolution to an acme and a decline through retrogressive stages to extinction . . . The type, like the individual, has only a limited store of vitality, and both must progress and retrogress, complete a cycle and finally die out, in obedience to the same law."

In this scheme, even the simplified ontogenies of regressive evolution can arise by acceleration (see Fig. 5-3). The adult stages of phyletic old age resemble, by analogy to the "second childhood" of our own senility, the simple features of youth (although these recurring traits signify decline and extinction, rather than exuberance, as they now appear in an exhausted stock). By this time, acceleration has become so intense that most middle stages in ontogeny begin to drop out entirely. The earliest embryonic stages, however, remain stubbornly persistent. As acceleration intensifies, newly-introduced senile features push back the older progressive traits of phyletic midlife, until these middle stages encounter the persistent juvenile features. Pushed at one end by senile features, and pressed against the impenetrable wall of persistent embryonic traits at the other end, these progressive middle stages finally tumble off the phyletic conveyor belt. Characters of phyletic old age now merge with juvenile features (see Fig. 5-3). Ontogeny becomes shorter (by excision of intermediate stages), and simpler (because the remaining juvenile and old-age stages resemble each other in external appearance). In this way, the simplified ontogeny of regressive lineages does not represent a retardation or truncation of development, but rather an acceleration so intense that all intermediate complexity (once intercalated between true youth and "second childhood") disappears by compression. Hyatt writes (1889, p. x): "Acceleration produces first, the earlier development of some of the progressive characteristics

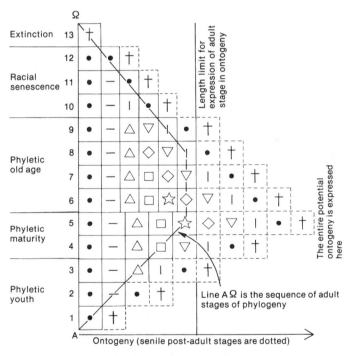

5-3. Hyatt combined his concept of a predetermined phyletic life cycle with his principle of universal acceleration to explain how even the simplified ontogenies of regressive evolution can originate by acceleration. In Hyatt's "old age theory" (his designation) as extinction nears, senile stages of phyletic youth and maturity become the adult stages of a waning stock in racial senescence. Ontogeny becomes so shortened by acceleration and deletion that senile stages merge with persistent juvenile stages to produce a greatly simplified and senile course of life. From Gould, 1977b.

combined with geratologous characteristics; secondly, the earlier development of geratologous characteristics and their fusion with larval characteristics, which occasions the complete replacement of progressive characters, and occurs only in the extreme forms of retrogressive series."

Hyatt applied his old-age theory with abandon to all spheres of life and culture. I find no usage more curious than his invocation of phylogerontism to argue against voting for women (1897b). Hyatt claims that "in the early history of mankind the women and men led lives more nearly alike and were consequently more alike physically and mentally, than they have become subsequently in the history of highly civilized peoples. This divergence of the sexes is a marked characteristic of progression among highly civilized races." Ontogenetic old age tends to blur sexual differences as men become less hirsute and develop larger breasts, while sexual activity declines equally (or so Hyatt claimed) in both sexes. Since phyletic sequences mirror ontogeny (and since the human race has become dangerously phylogerontic already), we must beware any culturally enhanced blurring of distinctions between the sexes—for androgyny of any form (physical, cultural or conceptual) denotes racial senescence. Giving the vote to women will enhance this dangerous tendency towards equalization of roles:

> Such changes [women's suffrage] . . . might lead to what we might now consider as intellectual advance, [but] this would not in any way alter the facts that women would be tending to become virified and men to become effeminized, and both would have, therefore, entered upon the retrogressive period of their evolution . . . The danger to women cannot be exaggerated, nor too carefully considered, in view of the fact that advanced women have adopted the standard of men, and have not tried as yet to originate feminine ideals to guide them in their new careers and thus maintain the divergence of the sexes (1897, p. 91).

This notion of an internal program for phylogeny (including the time bomb of inherent racial senescence and extinction) ran so contrary to Darwin's convictions about functionalism and contingency that he couldn't grasp Hyatt's conception at all (more, I think, through disbelief at the content, than inability to comprehend the argument). In 1872, as Darwin grappled with views of the American school in preparing the 6th and last edition of the *Origin*, he engaged Hyatt in a long correspondence about acceleration and racial life cycles (in F. Darwin, 1903, pp. 338–348). Darwin expressed his perplexity in the first letter: "I confess that I have never been able to grasp fully what you [that is, Cope and Hyatt] wish to show, and I presume that this must be owing to some dulness on my part" (in F. Darwin, 1903, p. 339).

After several exchanges of letters and diagrams (with some gain in clarification), Darwin remained puzzled by the most anti-selectionist and non-functionalist theme in Hyatt's system: the explanation of simplified ontogenies in phyletic old age by intensified acceleration, with senile adult features interpreted as nonadaptive preludes to extinction. Darwin conjectured in response: why not propose the far simpler interpretation that these shortened ontogenies and "degraded" adult forms represent adaptations to conditions of life that also characterized early stages of the lineage. This stark contrast and mutual incomprehension illustrate, in a striking manner, the difference between Hyatt's formalist orthogenesis as an ultimate drive to phyletic death, and Darwin's functionalism, with extinction as failure to adapt. Darwin wrote to Hyatt (in F. Darwin, 1903, pp. 343–344): "With respect to degradation of species towards the close of a series, I have nothing to say, except that before I arrived at the end of your letter, it occurred to me that the earlier and simpler ammonites must have been well adapted to their conditions, and that when the species were verging towards extinction (owing probably to the presence of some more successful competitors) they would naturally become readapted to simpler conditions."

Later in the same letter, Darwin pens the most famous line of this correspondence—a lovely contrast between the contingency of environmentally entrained adaptation and the predictability of "hardline" formalism as a theory of internal necessity: "After long reflection I cannot avoid the conviction that no innate tendency to progressive development exists" (in F. Darwin, 1903, volume 1, p. 344).

Through the density of theoretical discussion in these letters, another theme circulates. Hyatt expresses his plans to restudy one of the most famous paleontological series of presumed stratigraphic continuity in an isolated setting: the Miocene fresh-water planorbid pulmonates of the Steinheim lake in Germany (then interpreted as a volcanic caldera, but now recognized as a meteor crater—see Reif, 1976). The German paleontologist Hilgendorf had published an already classical account in 1866, including one of the first genealogical diagrams to reflect Darwin's new world order. Hyatt proposed a restudy and Darwin opined: "I earnestly hope that you may visit Hilgendorf's famous deposit . . . I most sincerely wish you success in your valuable and difficult researches" (in F. Darwin, 1903, p. 344). Hyatt proceeded, and eventually provoked Darwin's last and bitter response to his orthogenetic ideas by sending Darwin a copy of his monograph (1880) on the Steinheim planorbids.

Hyatt's proposed phylogeny could hardly differ more from Hilgendorf's original interpretation (Fig. 5-4). Where Hilgendorf drew a conventional branching tree with a monophyletic root, Hyatt presented four lineages, separate at the base and evolving in strict parallel. Such a striking difference should, in our conventional view of scientific change, record Hyatt's improved observations at the site. Hyatt did make some empirical changes (although we would have to view his effort, in retrospect, as a continuity in steady state rather than an improvement over Hilgendorf, for he corrected some errors but introduced just as many others). But Hyatt's alterations primarily record the application of a different theory to the same data. Iconography often provides a powerful guide to conceptual frameworks because pictures frequently make explicit what our psyches fail to acknowledge in the verbal mode (Rudwick, 1992; Gould, 1989c, 1993d, 1996a). Hyatt's phylogeny of the Steinheim planorbids (Fig. 5-5) epitomizes hardline orthogenesis under the guidance of phyletic life cycles.

Hyatt depicts four lineages within the lake beds, each beginning from an ancestral *Planorbis levis* stock (not shown). This putative phylogeny rests upon two principles derived from his orthogenetic "old age" theory (and showing that Hyatt's convictions directed his observations, rather than the conventionally touted vice versa).

1. Hyatt distinguished the four lineages on the basis of supposedly progressive and retrogressive characters. The rationale for these designations probably owed more to vague and general cultural conventions (largely the folklore of more and less as better and worse) than to any explicitly biological argument. Progressive characters include increase in size, shell thickness, strength of ornamentation, and change of shape from planispiral (flat like an ammonite) to trochiform (domed like a conventional snail). In other words, small, thin, smooth, and flat specifies a primitive state; whereas large, thick,

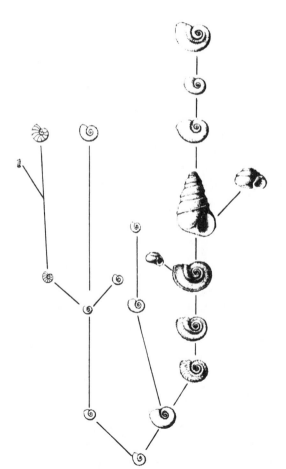

5-4. Hilgendorf's original interpretation of 1866 for the evolution of the Steinheim planorbids. Note how Hilgendorf depicts his phylogeny in a conventionally tree-like form.

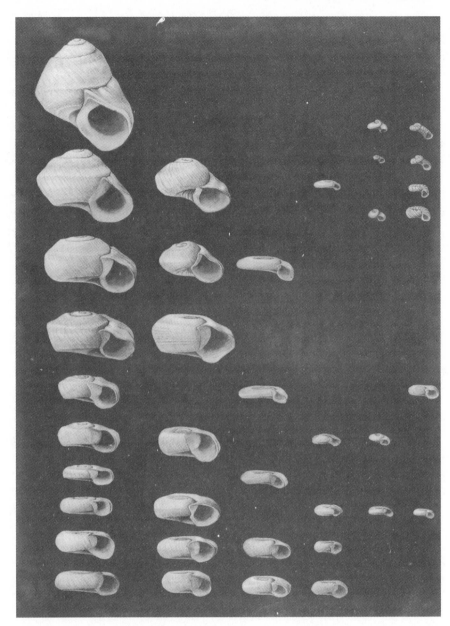

5-5. Hyatt's phylogeny for the Steinheim planorbids could not differ more from Hilgendorf's. Hyatt envisages several parallel lineages each in different stages of the same phyletic life cycle, with some lineages becoming stronger (as expressed in their larger and thicker shells) and others becoming evolutionarily senile in their thinner, smaller, and irregularly coiled shells. This interesting figure comes from a glass plate that Hyatt prepared for his 1880 monograph but did not publish. The printed version is much cruder and less informative. I here publish Hyatt's original for the first time. (I now occupy his office and I found this plate in the drawer that still contains his Steinheim specimens.) The three rightmost lineages represent three sublines of a single degenerating stock, hence giving four lineages *in toto* (as the text states).

bumpy and blocky equals advanced. By contrast, regressive characters include decrease in size, loss of ornamentation, thinning of the shell and, above all, a tendency for irregular growth by uncoiling (loss of order as a sign of both ontogenetic and phyletic senility).

Hyatt identifies three of his four lineages as progressive, distinguishing them by different combinations of the key characters. The most purely progressive *steinheimensis-trochiformis* lineage advances by all criteria to greater size, thickness, quadrate shape with keels and carinae (rather than a smooth whorl profile) and, as the formal name states, a domed outline. Shells of the *oxystomus-supremus* lineage become larger and more ornamented; the spire does not increase in height, but the shell still grows taller because the underside of the whorl profile becomes more inflated. The *parvus-crescens* lineage shows less advance, as the shell remains flat and smooth, but increase in size establishes the primarily progressive character.

Retrogressive tendencies appear in the three sublineages of the remaining branch, all derived from *P. minutus*. The *turbinatus* sublineage shows a mixture of progressive and retrogressive characters (Hyatt, 1880, p. 17, refers to this melding as "the battle of the tendencies"), with modest size increase and some strengthening of ornament offsetting a basic decline. The middle, or *denudatus*, sublineage is purely retrogressive, as shells become smaller, smoother and irregular in growth by increasingly erratic coiling. Finally, the *distortus* lineage also mixes phyletic strength and weakness. Ornament remains strong and size increases in portions of the lineage; but, as the name implies, coiling becomes irregular as the stock declines.

2. Although later convention (and emerging practice in his own time) would lead us to read Hyatt's chart as a stratigraphic sequence, his phylogeny employs an unconventional iconography. Vertical position does not represent time or stratigraphy, but rather stage in an orthogenetic sequence. Snails drawn at the same level did not necessarily live at the same time, but show common "attainment" in a phyletic series. Thus, for example, *P. trochiformis*, the ultimate stage of the most progressive series, lived near the bottom of the stratigraphic sequence—implying that this lineage ran its full course with geological rapidity at the base of the section. Conversely, *P. oxystomus*, the initial form of the second progressive lineage, makes an initial appearance high in the sequence.

This unconventional iconography illustrates the power of theory to channel perception—orthogenesis as an organizing principle, in this specific case. Consider the immense confidence that a scientist must be willing to invest in the validity of a chosen surrogate to substitute any other criterion for the eminently available (and obviously meaningful) stratigraphic order of time as the measuring rod for vertical position in phyletic charts. (Cladists have created quite a fuss in our day by using inferred branching order in preference to time of observed paleontological appearance, if they include fossils in their phylogenies at all—see Schaeffer, Hecht, and Eldredge, 1972. In the heyday of overweening confidence in recapitulation, several paleontologists reversed the conventional geological procedure and inferred stratigraphic order from presumed phyletic stage based on ontogenetic repetition of ancestral adult stages—see Smith, 1898, on ammonite phylogeny, for example. Thus, Hyatt's procedure, while interestingly unconventional, scarcely lacks precedent—or consequent.)

In sum, Hyatt presents a picture of multiple lineages, evolving in parallel but at different times (though in the same lake), through a preset sequence of stages—with some lineages displaying an upward march to progressive characters, and others a downward slide to regressive states of the same features. Hyatt justifies all these claims under his old-age theory of orthogenetic unfolding: phylogenies proceed inexorably from periods of phyletic youth and vigor, through maturity to racial senescence and extinction. Consider a series of questions, all resolved by the orthogenetic interpretation (and all refuting Darwinism, or any other functional account):

1. Why do the separate lineages go through similar stages? The causes cannot reside in functional entrainment by common environmental pressures (either by Darwinian selection or Lamarckian response to perceived needs) because the same stages occur at different times in various lineages (same response in different environments), while different lineages (progressive vs. regressive) often evolve disparate forms at the same time (different response in the same environment). Hyatt argues that the cause of parallelism must therefore be sought in an internal shove, not an external (environmental) push:

> While the perpetuation and survival of the differential characteristics can be thus accounted for [by natural selection], we must look to other causes for the production of the parallel forms and the regularity of succession of these forms, as shown in the arrangement in the different series, and in the development of the individual. This cause lies in some law of growth and heredity which reacts against the tendency of the physical environment to produce variations and differences, and produces parallelism in the development of different individuals of the same species, of different species in the same series, and in the succession of forms in the different series, and also limits the tendency to variation within definite boundaries in the species (1880, p. 26).

2. Why does the invariant series of stages follow this characteristic sequence, with either a march to progressive features in shell size, thickness, shape and coiling, or a fall to increasingly degenerate states of the same characters? Again, an answer cannot be provided by functional adaptation in either the Darwinian or Lamarckian mode, for extended regressive sequences, by their inadaptive nature, could not then occur. Instead, this sequence of up and down marks the full scope of the "grand potential ontogeny," defining the orthogenetic phylogeny of the entire fauna: "Thus, we can readily understand that each of these series, whether progressive or retrogressive, can so far as its collective life is concerned, be compared in the closest manner with the life of an individual, and similar correspondence be traced in both, and also that the tendencies exhibited are of two kinds in each, one towards

building up of the organization and the other directly opposed to this" (1880, pp. 17–18).

Hyatt also tries to provide direct evidence for the ontogenetic construction of phylogeny. He notes, for example, that ordinary adults of regressive lineages reach stages found only in the most degenerate individuals of progressive populations—those that have grown far past their normal adult form to a marked senility. The ruinous dotage of a progressive individual therefore corresponds with the ordinary adult form of a phylogerontic race (1880, p. 17).

3. Why, in the same lake and during the same general period, do some lineages progress while other closely related lines regress? Again, no functional or adaptationist answer can suffice, for the same times and environments should not engender opposite responses in such closely allied lineages. The solution must reside in internal orthogenesis. Lineages progress or regress according to their internal state—particularly, their status in the unrolling of the grand potential ontogeny. Progressive lineages, in their phyletic youth, can resist a harsh environment; but regressive sequences, in their phyletic dotage, must succumb. Hyatt's four lineages achieve their distinctions by occupying different positions in the grand potential ontogeny. His three progressive lineages evolve in the vigor of their phyletic youth or maturity; meanwhile, the regressive series decline in their phylogerontic senility.

Environment does not, as in functionalist theories, operate as an aid or entrainer, but rather as a clear detriment and degrading force. Lineages in their phyletic youth can prevail by innate virility against the incessant storm:

> How shall we account for the progression of the progressive series? How then could this environment act upon such closely allied shells, in such an opposite way as to cause the decrease of some races and be entirely healthy for others? We habitually refer such questions among animals, and in man, to the innate strength or pliability of the constitution of the race or the individual, and account for the survival, growth, and development of races and individuals by this reference to their supposed ability either to resist change in their surroundings, or to become modified in accordance therewith . . . Precisely the same environment, therefore, may produce results diametrically opposed to each other, even upon different individuals of the same species or closely allied forms, provided there is anything in the constitution either directly acquired or inherited, which enables the organization of one to resist or fit itself to conditions which the other cannot healthfully endure (1880, p. 16).

But lineages in phyletic senility cannot prosper; their decreases in size denote waning viability, and their irregularities of growth signify the last gasp of faltering strength. A wounded or senile individual in a progressive race may uncoil in injury or dotage, whereas the same fate awaits all ordinary adults in regressive stocks. "Retrogressive characteristics . . . could be compared with the pathological conditions, normal or abnormal, of occasional diseased and senile individuals of the progressive series. They [regressive features] are . . .

inherited with ever increasing effect in successive species, occasioning distortions and retrograde metamorphoses, and finally leading to the extinction of the race" (1880, p. 14).

Early in his monograph, Hyatt advances the conventional claim of scientific methodology—that his phylogeny should be judged and accepted on the criterion of objectivity in research: "These series, having been the result of no preconceived plan of arrangement as far as the author could judge, were considered to be approximately natural" (1880, p. 8). Yet Hyatt clearly falls victim to his own admonition. He did not establish his four-lineage scheme by any principle of ordering specimens in a manner that could be called "objective" or even bound by rules independent of his phyletic preferences. Hyatt's scheme of multiple parallel lineages represents a theoretical construction, dictated by his orthogenetic conviction about racial life cycles, not a proclamation of nature.

First of all, Hyatt could not separate his lineages with accuracy or confidence. He uses a method of "eyeballing" and, following the limits of his time, presents no statistical arguments. Such a failure to quantify need not derail a study in principle; we need no measuring rods to sort a mixed pile of sparrows and elephants into two groups. But the Steinheim planorbids interfinger and intergrade in the most complex manner, both spatially and temporally. The shells occupy a grand clump of morphospace, with subclumps here and there to be sure, but with no clear or persistent piles. No researcher since Hyatt has been able to specify four distinct lineages, each maintaining integrity through time.

Second and most important, Hyatt admitted that he could not use the standard method for establishing lineages—temporal succession in the fossil record. For he could not identify any clear stratigraphic sequences at all! Following his words about natural series and lack of preconceptions, Hyatt wrote: "These series . . . were assumed to be a reliable basis for working hypotheses, in spite of the fact, that no certain data with regard to succession in time were obtainable" (1880, p. 8). How then can phyletic series be established?

Hyatt then admits that, in the absence of stratigraphy, phyletic sequences must be identified by the expectations of a biological principle—dare we label it a preconception?—namely, the biogenetic law itself.* Hyatt even permitted himself to construct phyletic sequences from specimens found on the same bedding plane if successive stages of a recapitulatory series could be identified: "This assumption rests largely upon well known laws of heredity, such

---

*Hyatt's apparent illogic can be comprehended when we recognize that most scientists of his day regarded the biogenetic law as sufficiently well validated to represent a fact of nature. Many paleontologists maintained this conviction to quite recent times. My late senior colleague Bernie Kummel told me of an argument he once had with the most powerful American paleontologist in the generation just before his own—R. C. Moore. Kummel, as a young Turk, had disputed the universal validity of recapitulation over cocktails one evening. Moore slammed down his glass and bellowed: "Bernie, do you deny the Law of Gravity!"

as these, that an animal found to repeat the stages of another animal of a closely allied species in the young, with the addition of new characteristics in the adult, may be considered to be either a lineal descendant of that species, or of some form common to both; that in such cases as these, whether the form or species occur mixed on the same level, or on different levels, there is but one natural arrangement" (1880, p. 8).

If recapitulation entered the argument as an *a priori* assumption, then so did the cognate notion (in Hyatt's mind at least) of racial life cycles. Hyatt assumed the very phenomenon he hoped to prove when he built his phyletic sequences in the absence of stratigraphic resolution, often from specimens on the same bedding plane. He resolved all violations of stratigraphic order by dictates of the "old age" theory. For example, both Hyatt and Hilgendorf regarded *P. oxystomus* as a derived branch of the main stock. But Hyatt interpreted this species as the base of an extensive progressive series, built upon variation within a single bedding plane because he could establish no empirical continuity to still higher stratigraphic levels. Hyatt's preferred order helped his argument immensely—for the idea of parallel yet non-synchronous lineages strongly supported his claim for internal necessity, not environmental entrainment, in the unfolding of phyletic stages. But he presented no evidence beyond the ordering power of the preconceived theory itself!

In a second example that engendered bitter debate with the followers of Hilgendorf, Hyatt claimed additional evidence for non-synchronous parallelism by finding the most advanced specimens of *P. trochiformis* (the acme of the main progressive lineage) in the *lowest* stratigraphic levels—leading to the assertion that this entire lineage unfolded with great rapidity (and providing more evidence for internal programming since other lineages would then be evolving much more slowly, and Hyatt could therefore identify differential phyletic vigor, rather than variation in environmental pressure, as the cause for disparities in evolutionary rate). But other researchers could only find *P. trochiformis* at high stratigraphic levels (Hyatt's lower specimens may have eroded from upper levels and washed down)—and therefore interpreted this lineage as evolving much more slowly.

If Hyatt maintained such overweening faith in the validity of his orthogenetic theory of ontogenetic programming, we can scarcely be surprised that he also read the supposed empirics of the Steinheim planorbids as a disproof of Darwinism—hence Darwin's negative response in receiving the monograph (see p. 372). Hyatt attempted to specify both the potential and the limits of natural selection for the Steinheim planorbids—and the restrictions overwhelmed the possibilities. Following the usual formalist critique of selection's creativity, Hyatt allowed that selection might explain why four lineages, rather than fewer or many more, became established, and why they continued to propagate. The origin of four lineages in the Steinheim lake, Hyatt tells us:

> . . . appears therefore to be perfectly well accounted for by Darwin's theory of natural selection. In no other way can we possibly account for the

selection of but four out of the varieties of *Pl. levis,* and the continuous propagation and increasing intensity of the differences which they exhibit. An examination . . . will show anyone how many variations are lost in each form or species of the series, and how few are continued. This can only be accounted for upon the supposition that those which survived possessed in some way advantages indicated by their peculiar variations, which enable them to propagate those variations, and suppress their less fortunate neighbours (1880, p. 26).

But how shall we explain the far more important issue of causes for actual directions of modification in the lineages—that is, the evolutionary changes themselves? Hyatt asks whether selection could be effective here: "Are these parallelisms adaptations, and can they possibly be attributed to the direct action of the uniform external environment upon the forms of the different series?" (1880, p. 19). Hyatt denies any formative power to the Darwinian mechanism. Selection can do the negative work of weeding and separating, but cannot perform the positive action—the essence of evolution—of changing and progressing.

Hyatt offers two major critiques of selection in the light of ontogenetically programmed orthogenesis. First, how can the functional premise of adaptation be supported when so much change occurs in the regressive mode following phyletic maturity? "Nothing can exceed the confidence with which the strict Darwinist assumes, without any appeal to observation, that all characteristics which are inherited are necessarily advantageous. Exactly the reverse is very often true" (1880, p. 101). Second, he recognizes that Darwin's system entails a crucial assumption of isotropic, undirected variability. Since the conveyor belt of the grand potential ontogeny introduces new characters with a decided bias, creativity resides in the internal directionality of variation. Natural selection can only work as a subsidiary force to the primary agent of directed variation. "Natural selection, in fact, is simply one of the transient conditions of the physical surroundings, having no value as a cause of origin of characteristics" (1880, p. 102). The ontogenetic conveyor belt feeds new characters and creates evolutionary novelty; selection can only eliminate, separate and impose a little cosmetic shaping upon the internally generated trend:

> Thus it may be said that the struggle for existence, and the survival of the fittest, is a secondary law grafted upon laws of growth, and governed by them in all its manifestations. The law of natural selection, as generally understood, assumes in the first place the existence of an animal type, of its descendants, and of a tendency to variation (indefinite and unlimited) in every one and all of these descendants, from which (an indefinite and unlimited) selection may take place during the struggle for existence between competing forms, destroying the weak and permitting only the strongest and fittest of these variations to survive. The truth is, as far as my studies have gone, that there is no such thing as indefinite or unlimited variations in any species. . . . This obvious proposition, if admitted,

leads at once to the question, what are the limits within which a species may vary? . . . The limits of variation in the species have been found to correspond to the growth changes in an individual (1880, p. 20).

Hyatt viewed his orthogenetic theory as a contribution to the larger vision of 19th century mechanistic science—a hope that when we finally learn the laws of heredity and discover the principles of biased variability, then evolution shall become as predictable as ontogeny: "In all cases the individual and its series must change by growth along certain lines of modification, which it is but reasonable to suppose we shall someday be able to map out beforehand for a series of forms with the same precision that we can now forecast the metamorphoses of any individual in a given species" (1880, p. 18). Charles Darwin, who understood the contingent character of history, could not have disagreed more forcefully.

Hyatt's hardline version of orthogenesis offered no quarter for fruitful interaction with Darwinism. If all structuralist and formalist thinking, and all theories of channeled variation, existed only (and in principle) in this adversarial mode, then the important 19th century debate on orthogenesis could teach us little today. But my account of Eimer's orthogenetic theory has already illustrated the potential for useful interaction between internal orthogenesis and external adaptation—even though Eimer chose Lamarck rather than Darwin for the basis of his functionalist component. I shall next present the even more accommodating version of C. O. Whitman, to illustrate a maximal contrast with Hyatt, and to emphasize the possibility of Darwinian insight from orthogenesis.

Kurt Vonnegut introduced the useful word "karass" to describe groups of people who may not explicitly interact, or even know each other, but whose lives seem tied together by action and circumstance. Hyatt and Whitman (who did, in fact, know each other well) must have belonged to the same karass. Both studied under Louis Agassiz. (Hyatt spent his career in the Boston area, worked primarily on ammonites, and occupied the office that I now inhabit at the Museum of Comparative Zoology.) Both participated in the early days of summer courses in New England natural history and marine biology (Whitman studied with Agassiz on Penikese Island; Hyatt ran a teacher's school of natural history in Annisquam). Both men rank as the two key figures in the early days of the Marine Biological Laboratory at Woods Hole, Hyatt as first president of the board of trustees, Whitman as first director. Both devoted their major research efforts to formulating theories of orthogenesis, Hyatt with ammonites and snails, Whitman with pigeons. But their orthogenetic theories could not have differed more profoundly, particularly in their divergent attitudes toward complementarity with Darwinism. In understanding why the adversarial Hyatt reached a dead end, and in grasping the insight offered by the accommodating Whitman (who, for unfortunate historical reasons, gained very little historical impact for his orthogenetic views), we may better appreciate both the blind alleys and the vital themes of orthogenesis (and of formalism in general) in our struggle to develop a more adequate modern theory of evolution.

## C. O. WHITMAN: AN ORTHOGENETIC DOVE IN DARWIN'S WORLD OF PIGEONS

The old benediction—"may you live in interesting times"—has been regarded as either a blessing or a curse. Charles Otis Whitman certainly merited such an epithet, for his professional life spanned the greatest range of opposites through the grandest transition in ideas that biology has ever experienced. He began by studying with Louis Agassiz, last of the great and legitimate creationists, and ended as the chief American promoter of mechanistic embryology in the German tradition. He made his primary reputation in "cell lineage" studies of the fates and products of the earliest blastomeres. But, unlike most experimentalists of the time, he also pursued other research as a gifted natural historian and evolutionary theorist. His major work in later years, while he served as professor at the University of Chicago, treated a subject that could not have been more canonical for Darwinian evolutionary biology in the naturalistic tradition—heredity, variation, and evolution in Darwin's own chosen organism, the domestic pigeon. Whitman, however, used Darwin's pigeons to support orthogenesis, and to deny selection a primary or formative role in evolution.

C. O. Whitman died in 1910, of pneumonia contracted after working furiously on the first cold day of winter to provide shelter for his birds. (F. R. Lillie, once his assistant and later his successor at Woods Hole, eulogized his old boss: "In his zeal for his pigeons, he forgot himself.") Whitman had never published an extensive defense of his orthogenetic theories. His diverse and voluminous writings were finally collated and published posthumously as a large three-volume monograph by the Carnegie Institute of Washington in 1919. The evolutionary debates of the early 20th century had been fierce, and finally won by the followers of Darwinian theory. I have often wondered how this history might have differed if this paramount biologist (Kellogg, 1907, p. 288, called Whitman "the Nestor of American zoologists") had lived to publish what might have been the best empirical defense of orthogenesis. In any case, the posthumous and much delayed 1919 monograph was too disjointed, too incomplete and above all, too late, to win any potential influence.

Whitman accepted Kellogg's classification of evolutionary theories as auxiliary or alternative to Darwinism. He also agreed with Kellogg that three major alternatives fueled the great debate as the century turned: Lamarckism, orthogenesis, and macromutationism. Since Whitman rejected the inheritance of acquired characters with all Weismann's zeal, his own list of viable alternatives included only two theories. Since he also believed in the strict continuity of Darwinian gradualism, de Vries's mutationism held no appeal for him either (though he regarded the theory as a viable contender, while treating Lamarckism as a dead issue). These rejections left only orthogenesis as a potentially valid challenge to Darwinism. The first and most important volume of Whitman's monograph bears the title: "Orthogenetic Evolution in Pigeons."

Whitman provides our best example for disproving the false equation of orthogenesis with some form of theistic teleology—the main source for current derision, and for our failure to grasp the strengths and serious recommendations of this approach. The link to teleology can be dismissed as not only wrong, but entirely backwards. No case could be clearer than Whitman's, for he spent a maximally distinguished career as one of the great mechanists of American experimental embryology. He did not conceive orthogenetic trends as mystical impulses from outside, but as mechanistic drives from within, based upon admittedly unknown laws of genetics and embryology. Consider the last words of his 1919 monograph (p. 194):

> If we are to draw the line sharply between science and all transcendental and telistic mysticism, we must regard the germ-organism as wholly mundane in origin and nature. If the germ is a thing of evolution from purely physical foundations—and any contrary assumption is a denial of the evolution principle, then we may say that it is a self-building within the limits of physical conditions, and just as truly autonomic in its form and behavior as is the crystal. In the formation of a crystal self-determination is ever present, and so it must be in the case of the organism.

Orthogenesis therefore emerges as a favored *a priori* prediction of deterministic science. Whitman's opening words strike the same theme with a note of triumphal optimism:

> Progress in science is better indicated by the viewpoints we attain than by massive accumulation of facts. Darwin's perspective made him a prodigy in the assimilation of facts and an easy victor in the greatest conflict science has thus far had to meet. His triumph has won for us a common height from which we see the whole world of living beings as well as all inorganic nature; phenomena of every order we now regard as expressions of natural causes. The supernatural has no longer a standing in science; it has vanished like a dream, and the halls consecrated to its thraldom of the intellect are becoming radiant with a more cheerful faith (1919, p. 3).

Moreover, the particular character and personal history of Whitman's mechanistic outlook suggested the specific form of his orthogenetic argument. His work on cell lineages had mapped the fate of the earliest blastomeres, and had indicated that the source of eventual organs could be specified even in minute and formless clumps of initial cells. If embryos grew so predictably, why should evolutionary change be devoid of similar order? Ontogeny, in other words, should serve both as a model and a source for evolution—a joint vision of directional change from within. Whitman, in fact, argued that ontogeny and phylogeny represent the same essential process: "Development is the one word that seems to me to best circumscribe the

more general problems of biology. It is also the one word that best emphasizes the essential unity of ontogeny and phylogeny. These two terms have
been used as if they stood for two distinct series of phenomena, when in reality they apply to one and the same series" (1919, p. 177). Phylogeny therefore
becomes as determinate as growth itself: "Not only is the direction of the
change hitherto discoverable, but its future course is predictable" (1919,
p. 38).

In his most telling statement, and in response to de Vries's oft-repeated but
invalid argument (see p. 445) that orthogenesis revitalizes teleology, Whitman invokes the ontogenetic comparison to defend orthogenesis as the position most consistent with a mechanistic worldview:

> I take exception here only to the implication that a definite variation ten-
> dency must be considered to be teleological because it is not "orderless."
> I venture to assert that variation is sometimes orderly and at other times
> rather disorderly, and that the one is just as free from teleology as the
> other. In our aversion to the old teleology, so effectually banished from
> science by Darwin, we should not forget that the world is full of order,
> the organic no less than the inorganic. Indeed, what is the whole devel-
> opment of an organism if not strictly and marvelously orderly? Is not ev-
> ery stage, from the primordial germ onward, and the whole sequence of
> stages, rigidly orthogenetic? . . . If a developmental process may run on
> throughout life, . . . what wonder if we find a whole species gravitating
> slowly in one or a few directions? . . . If a designer sets limits to variation
> in order to reach a definite end, the direction of events is teleological; but
> if organization and the laws of development exclude some lines of varia-
> tion and favor others, there is certainly nothing supernatural in this
> (1919, p. 11).

Darwin had begun the *Origin of Species* in a most honorable way that affirmed the necessary, and heretofore largely lacking, empirical foundation of
evolutionary argument. Darwin's first chapter did not announce to the world
his sweeping reform of all life and thought; instead, he wrote about pigeons
(1859, p. 20): "Believing that it is always best to study some special group, I
have, after deliberation, taken up domestic pigeons. I have kept every breed
which I could purchase, or obtain, and have been most kindly favored with
skins from several quarters of the world . . . I have associated with several eminent fanciers, and have been permitted to join two of the London pigeon
clubs."

Darwin used pigeons to advance the two primary and distinct arguments of
his book: (1) the factual claim that evolution had occurred, and represented
the source of organic relationships, and (2) the theoretical assertion that natural selection operated as the primary cause of evolutionary change. He supported the first contention by proving that the full range of extensive diversity
in modern domesticated breeds had descended from a common wild source,
the rock-pigeon *Columba livia*. (Darwin then added the crucial analogical argument that such intraspecific change could, by extension, serve as a model

for evolution at all scales and times). He buttressed natural selection by noting that breeders had produced this extensive range of results by propagating favored forms from a cornucopia of essentially isotropic and undirected variation.

Whitman, of course, accepted the first contention, but refuted the second by challenging one of Darwin's smaller claims. Darwin had observed two major patterns of coloration within *Columba livia*—(1) "two-barred," with two black bands on the front edges of the wings and uniform gray color elsewhere (Fig. 5-6); and (2) "checkered" (spelled "chequered" by both Darwin and Whitman), with black splotches on some or all wing feathers (Fig. 5-7), but also retaining the two bars (usually in more indistinct form). Darwin regarded the two-barred state as ancestral, and the checkered pattern as derived. Whitman reversed this sequence, writing:

> The wild rock pigeons, universally regarded as the ancestral stock of all our domestic pigeons, exhibit two very distinct color patterns, one consisting of black chequers uniformly distributed to the feathers of the wing and the back, the other consisting of two black wing bars on a slate-gray ground. The latter was regarded by Darwin as the typical wing

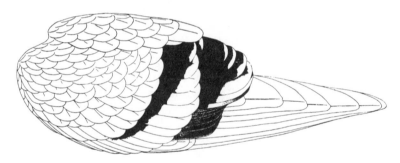

5-6. Whitman's figure of the two-barred wing pattern, which Darwin regarded as ancestral and Whitman interpreted as an advanced stage in his orthogenetic sequence. From Whitman, 1919.

5-7. The checkered pattern, viewed by Darwin as derived and by Whitman as the primitive state in the evolution of pigeon wing colors. From Whitman, 1919.

pattern for *Columba livia*; the former was supposed to be a variation arising therefrom, a frequent occurrence but of no importance. Just the contrary is true; the chequered pigeon represents the more ancient type, from which the two-barred type has been derived. . . . The direction of evolution in pattern in the rock pigeons has been from a condition of relative uniformity to one of regional differentiation (1919, p. 49).

Whitman's inversion of Darwin's sequence lay embedded within a theory of evolutionary change that Darwin would also have rejected. Whitman based his reversal on a more general concept of directional change in coloration from an initial homogeneity (checkers on all wing feathers) towards regional differentiation (elimination of checkers over most of the wing, with strengthening and coalescence to bars at the distal edge. The bars form by enlargement and alignment of checkers on adjacent feathers; a bar, in other words, arises from a row of checkers that "flow together in a single band"— Whitman, 1919, p. 99).

Whitman then expanded his sequence of reduction plus regional differentiation beyond the patterns of domesticated pigeons to identify an ineluctable, orthogenetic tendency in the entire family Columbidae (with the portion displayed by domestic pigeons as just a small part of a much more extensive trend). He identified a prototype for the entire series in the "turtle-dove" pattern, a homogeneous field of feathers, each with a dark spot in the center (see Fig. 5-8): "This ancestral mark is a dark spot filling the whole central part of the feather, leaving only a narrow distal edge of a lighter color. This mark is still well preserved in some of the old world turtle-doves—best in the Oriental turtle-dove of China and Japan. The chequer of *Columba livia* differs from the dark center of *Turtur orientalis* only in form and in having a lateral position" (1919, p. 23).

From this beginning, the trend moved towards an inexorable end, guided

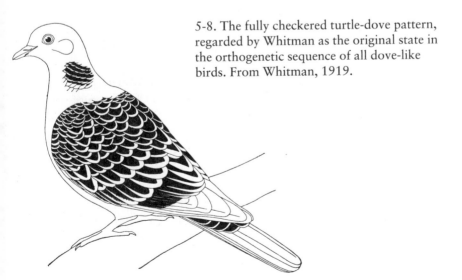

5-8. The fully checkered turtle-dove pattern, regarded by Whitman as the original state in the orthogenetic sequence of all dove-like birds. From Whitman, 1919.

by two criteria: general reduction in coloration, and concentration of remaining color into regionally differentiated bands or swaths. Thus, in the first stages, spots disappear from much of the plumage, while local areas may develop strong concentrations. Subsequently, these local intensities diminish in number and area (bars, for example, may become both narrower and fewer in number). Finally, the regional concentrations become effaced as well, and the bird turns light and monochromatic.

In a bold move towards complete generality, Whitman then tried to extend this idea of an archetype, followed by an orthogenetic trend, beyond pigeons and doves to the entire field of avian plumage. He postulated that the uniform turtle-dove pattern should be regarded as ancestral for all coloration in birds. With differing degrees of heterochrony in the inexorable process of reduction, and varying places and styles of regional concentrations, all observed plumages might then be rendered as extensive variations upon a single orthogenetic trend. Even the ocelli of peacocks, for example, can be interpreted as altered spots of the turtle-dove pattern, while their restriction to limited areas of the plumage (however showy and conspicuous the result) denotes one form of participation in the universal trend: "With this [turtle-dove] pattern as an archetype it is possible to get an orientation of the whole field of avian patterns and to thread our way through what before seemed an impenetrable maze of multifarious variations, with no discoverable beginning or end of order" (1919, p. 58).

The mere claim for a trend, even such a pervasive and inevitable series, does not of itself complete an argument for orthogenesis, or internally directed variation. After all, both the archetypal turtle-dove pattern and all subsequent stages of reduction might be adaptations, externally selected from isotropic Darwinian variation. But Whitman well understood the ingredients required to distinguish true orthogenesis from orthoselection or some other functionalist explanation of trends—namely, (1) evidence that the trend proceeds independently from (or even despite) adaptive pressures from local environments (selection may alter rates or add details, but cannot derail the basic route); and (2) data supporting an internally based directionality of variation available for shaping into evolutionary change (ontogenetic channeling in Whitman's view, as we shall see).

On the first criterion, Whitman upheld the inexorable character of a trend towards local differentiation and final effacement, independent of what environment might favor in functional terms.

> The process of evolution in color patterns has been a sweeping one, involving the whole surface and taking the same general direction. The stages reached are various, ranging all the way from the full chequered to the wholly unchequered state; from chequers and bars combined in different proportions to bars alone; from many bars to three, two, one, a remnant, or none; and in all shades of brown, black, gray, red, to pure white. Nowhere in this field of variations do we find any indications that

chequers originated in the form of bars at the posterior end of the wing and then spread from behind forward (1919, p. 55).

The trend, Whitman argues, is pervasive and entirely general. He presented (1919) a remarkable vision of inexorable movement through the entire family of pigeons, from a uniformly spotted archetype to some idealized, albinized version of the Holy Ghost, depicted as a pure white dove in many medieval paintings: "When we see all these stages multiplied and varied through some 400 to 500 wild species and 100 to 200 domestic breeds, and in general tending to the same goal, we begin to realize that they are . . . slowly passing phases in the progress of an orthogenetic process of evolution, which seems to have no fixed goal this side of an immaculate monochrome—possibly none short of complete albinism." Can one conceive a more unpigeonlike state (in both appearance and deed)—at least in our metaphors—than "immaculate"? To the primary spotting agent of cities throughout the world, Whitman thus gave a higher aspiration and the promise of a purer form. As the Psalmist wrote, "Behold, I was shapen in iniquity . . . Wash me, and I shall be whiter than snow" (Psalm 51).

The hypothesized putative inexorability of this trend allowed Whitman to explain or interpret many otherwise puzzling phenomena. He seemed most pleased about the unification thereby provided for the two main patterns of domestic pigeons—the empirical source of the entire study. If Darwin had been right, Whitman argued, then the origin of Darwin's postulated ancestral pattern (two-barred) would remain a mystery, and we would also lack any explanation for the subsequent evolution of checkers *de novo*. But if the trend begins with the turtle-dove pattern as ancestral, then the checkered state may be easily derived therefrom, and the two-barred condition becomes just a further step in reduction and concentration of pigment along an orthogenetic series. (This argument, of course, leaves the origin of the turtle-dove pattern itself as an unexplained "primitive term," but methods of phyletic analysis can at least establish its ancestral status.) "We could not explain how two bars could arise *de novo* in a clear gray wing surface; but we can see how a sweeping reduction process, antero-posterior in direction, would leave two or more rows of chequers cut to dimensions that would coalesce in transverse bars at the posterior end of the wing" (1919, p. 61).

To cite another example of variational puzzles resolved by the orthogenetic trend, Whitman notes that the stock dove, *Columba aenas*, develops weaker bars than the domestic pigeon and never exhibits any checkers. The trend, having so far surpassed the stage reached by domestic pigeons, no longer permits the development of any checkers, even as an occasional variant in highly colored individuals: "We can readily understand why the stock dove, which has, at least in many cases, a vestigial third bar, quite like that in domestic pigeons, never appears in chequered dress. It is moving in the other direction, and no reversal of course is now open to it" (1919, pp. 60–61).

On the second criterion of channeled variability, Whitman cites three lines of evidence for viewing his orthogenetic sequence as an extension of the ontogenetic pathway. First, in comparison with their own juvenile plumages, adult birds generally develop patterns of coloration that may be designated as "further along" the orthogenetic pathway. In a world of recapitulation, this palpable change in the course of weeks becomes a surrogate for the invisible alterations of past millennia. Whitman uses this ontogenetic evidence to assert orthogenesis against his two chief rivals, Darwin and de Vries:

> Moreover—and this is as close as we can hope to get to actual seeing— we find that progress of just the kind we are looking for is certainly made in passing from the juvenal [alternate spelling of juvenile, generally archaic in English, but still occasionally used in ornithology] to the adult plumage. This is an ontogenetic change of a few weeks, which we can easily demonstrate by experiment to be progressive and continuous (p. 33). . . . Even in the widest departures, when every spot has vanished in the adult plumage, the young bird frequently exhibits more or less perfect traces of the old marking and sometimes requires several molts to reach its mature condition (p. 58). . . . Juvenile phases of color patterns become luminous as recapitulations in the sense of the biogenetic law and do not stand as isolated prodigies of natural selection or as meaningless exhibitions of mutations (p. 65).

Second, as Eimer also maintained (and for complex reasons rooted both in cultural biases about sexual differences, and in the facts of embryology), Whitman viewed sexual distinctions as products of the same ontogenetic trajectory, with males "further along" than females. He then argued that plumages of adult male pigeons display more advanced stages of the orthogenetic series than females of the same population. Third, Whitman argued that, within the adult plumage of individual birds, last-formed feathers developed more advanced characters—as the biogenetic law required, based on its key principle of terminal addition for evolutionary novelty. Thus, three criteria, all interpreted as manifestations of ontogeny—differences among molts of a single bird, between sexes of adults, and within the adult plumage by order of formation—indicated a pervasive channel of variation, virtually compelling evolution into an extended ontogenetic pathway directed towards reduced and concentrated coloration.

To these categories of ontogenetic evidence, Whitman then added two additional sources of data to buttress his orthogenetic series: (1) From comparative anatomy, he asserted that phyletic series, established by criteria independent of coloration, illustrated the orthogenetic sequence in many parallel lineages. (I doubt the claimed independence in many cases, and Whitman may therefore have advanced a largely circular argument by basing phyletic inferences on the supposed trends in color themselves.) (2) From breeding, Whitman found that selection along the orthogenetic trajectory could only move "forward" from checkers to bars, and never in Darwin's proposed order from bars to checkers. Selection must push, but the phyletic sequence can only proceed in one direction—down the channel of orthogenetic variation:

The conclusion supported by comparative study admits of experimental confirmation. We may take pigeons of the two-barred type, and try to advance from this condition to that of the chequered type, by selecting in each generation birds with the widest bars, and especially any that may have a trace of a third bar. This I have tried continuously for 6 years and with several different stocks. I have not been able to establish a third bar, or to extend chequers in front of the vestigial third bar, which is often found. With pure bred birds, not allowed to mingle with chequered birds, I believe it is impossible to advance from bars to the chequered bird state. With chequered pigeons, on the other hand, it is fairly easy to advance in the opposite direction, gradually clearing the field and leaving two bars. The process has been carried to the point of completely eliminating the bars (1919, p. 60).

These accumulated sources of evidence led Whitman to strong assertions about the primacy of orthogenesis among rival theories: "The orthogenetic process is the primary and fundamental one" (1919, p. 35). In his boldest statement (1919, p. 191), Whitman advocated a model of inevitable evolutionary flow, and explicitly limited the role of natural selection to tinkering with the style and rate of a determined sequence:

The steps are seriated in a causal, genetic order—an order that admits of no transpositions, no reversals, no mutation-skips, no unpredictable chance intrusions. This series may conceivably be lengthened or shortened, strengthened or weakened; indeed, we may multiply the number of steps at will; that is, we may provoke one or more steps to arise between any two normal steps; but in that case the new steps will be measured true to the time and place of introduction, and their direction will invariably coincide with that of the series as a whole, so that if the time and place of origin are noted, the nature and extent of the strides may be approximately predicted.

Whitman wrote most of his work on pigeons and orthogenesis between 1900 and 1910, the period of greatest agnosticism and debate about evolutionary mechanisms (see Kellogg, 1907). He therefore upheld orthogenesis as an explicit preference among competing theories. Rejecting Lamarckism, Whitman faced the macromutationism of de Vries and the selectionism of Darwin as chief rivals. And with Darwin's own pigeons reinterpreted as the bulwark of orthogenesis, we can hardly be surprised that Whitman singled out natural selection for special criticism: "To attempt to explain all this as the work of natural selection would lead into an endless tangle of conjecture that would leave even the simplest facts as unapproachable mysteries. Natural selection has probably had most to do with the end stages in the evolution of characters, but little or no direct influence in originating them. The two-barred condition has been reached in the simplest possible way, not by accidental variation or chance mutation, but by progressive modification of a chequered condition previously established" (1919, p. 61).

Whitman's attitude towards natural selection bears closer scrutiny as an aid (still useful today) for clarifying the borderline between two intergrading yet contradictory strategies: (1), using the structuralist and formalist concept of channels in pluralistic reinforcement with natural selection to forge helpful revisions of basic Darwinian theory (the position advocated in this book); or (2) viewing channels as so deep, so unidirectional, and so limiting that such constraints impel evolutionary change from within, leaving selection only to tinker with minor details (a truly anti-Darwinian theory that led the Modern Synthesis to reject orthogenesis completely). I cannot place Whitman on either end of this continuum—for he argued both sides and usually rested with ambivalence at some middle position. But his writings provide our best illustration of this important concept in the logic and historiography of theories.

Whitman's usual account of natural selection grants a distinctly subsidiary role to Darwin's process. By adaptation's dumb luck, the inexorable process of reduction in color may occasionally generate a form with utility. At this point, natural selection may intervene to tinker, rearrange, and even strengthen the valued colors. But selection cannot long prevail, for even a useful concentration of color must eventually move towards orthogenetic effacement: "Even in cases where natural selection has probably played a conspicuous part in modifying and beautifying these marks . . . we find that the reducing process has not been brought to a standstill" (1919, p. 62).

Whitman asks us to consider examples from both major components of color—bars and checkers: The bars may be useful as marks of recognition, but they arise by orthogenetic reduction, not natural selection, and Darwinian forces cannot maintain them against the stronger internal push to effacement. (Note the interesting admission at the end of this statement that we have yet to fathom the mechanism of a process powerful enough to overcome selection.)

> Standing alone on a pale gray ground, these bars would gain immensely in conspicuity [sic] and utility as ornamental recognition marks. The advantage of all this to the species, whatever it be, would be merely an accident of the situation presented at this particular point in a progressive series of modifications. It is conceivable that the utility of the bars might be great enough to give natural selection a chance to step in and bar [pun intended?] the way to further reduction. But the process of obliteration has certainly gone much farther in many other species. There may be stages in the process which suggest utility; but when we consider the whole series of stages and note that the process runs on, sweeping away the stages which we imagine to be most useful, we are left with the conviction that some general principle underlying the course of events has not yet been fathomed (1919, p. 61).

Whitman stresses the same point, in even stronger form, for the few, conspicuous and apparently adaptive checkers of mourning doves. These remaining marks of color are "hanging tough," stubbornly resisting the orthogenetic washout—but natural selection encountered and preserved this pattern by mere good fortune and must eventually let go:

> It is here that we may with some reason suspect the intervention of natural selection. It would, in this case, come in, not as a primary factor to originate a new character, but adventitiously, by invitation, as it were, of favoring predeterminations and environmental conditions. . . . The ornamental value of these few chequers and their utility as recognition marks would obviously be enhanced by their isolation in a plain ground, just as a few trees, concealed in a large forest, become conspicuous when left standing alone. These chequers, being on the larger feathers, would have the advantage of size, and so their preeminence, attained without the aid of natural selection, would be an open door through which it might enter and contribute to their improvement. The part possibly taken, however, could at most be but a late and inconsiderable share of the total achievement summed up in these spots; and the course of events in at least one of the allied forms . . . indicates that these marks are destined to be washed out (1919, p. 56).

(Note, once again, the literary theme that authors often reveal their basic commitments, probably quite unconsciously, in their choice of words. In this statement, Whitman refers to natural selection as an "intervention"—an externality imposed upon the essential process of orthogenesis.)

Nonetheless, Whitman does acknowledge exceptions. In one case (but only here), he does allow that selection may have reversed, albeit in a minor way, the orthogenetic sequence. He notes that iridescence heightens the value of color in adaptive display. Iridescent spots become unusually conspicuous and potentially useful—so much so, that selection may actually strengthen them against the orthogenetic tide. Thus, when the independent trait of iridescence becomes conjoined with pigmentation, the orthogenetic sequence can be meaningfully impacted by selection. (Whitman properly uses his own criteria, as previously discussed, to gauge the importance of this exception. Juvenile plumages, in this case, develop less conspicuous spotting than adult feathers—so the ontogenetic path belies the orthogenetic sequence): "Iridescence thus appears to be a phenomenon tending to elevate the spots and bring them within the sphere of utility. It seems not only to put a check upon the reduction of pigment, but also to actually turn the tide in the opposite direction, for the reduction in this region is not carried so far in the old as in the young male and female, as we shall presently see. As the acquisition of metallic brilliancy is accompanied by an exceptional love of display in the male, the chief directing factor in its development may well be natural selection" (1919, p. 43).

These statements might lead a modern evolutionist to view Whitman's orthogenesis as irrelevant to current debates (if not risible in any context). But if Whitman did not come to praise Darwin, he did not write to bury the founding father either. Within his chosen context of primacy for the orthogenetic pathway, Whitman sought a maximal and fruitful interaction with

Darwinism. I know no other orthogeneticist who remained so open to the prospect of a pluralistic consensus (for, in contrast with Whitman, most scientists of this school entered the fray with strong anti-Darwinian inclinations). As one indication of his more conciliatory stance, Whitman followed the usual attitude of naturalists in accepting adaptation as a central phenomenon. He speaks of "the most remarkable phenomenon of the organic world, namely adaptation" (1919, p. 40). He also recognized that orthogenesis cannot be construed as inherently adaptive, whereas Darwin's force actively creates utility. He admits the conundrum that orthogenesis, while true by observation, does not explain the progressive and adaptive character of life—and he realizes that evolutionary biology needs an account, as yet unavailable, for a probable bias towards adaptation in the stages of orthogenetic channels: "But how comes it to pass that these advances are, on the whole, adaptive and progressively so? Recapitulation can only conserve what is given. If it moves on within a progressive way, there must be some way of limiting germinal variations to lines of accumulative improvement. Here we find ourselves confronted with the difficulty which has long led investigation and theory, and the solution is yet a long way ahead" (1919, p. 180).

Beyond this central acknowledgment, two features of Whitman's thinking open his particular version of orthogenesis to a broad synthesis with Darwinism.

1. Like Eimer, Whitman developed an interpretation of orthogenesis that could fuse external pushes with internal channels. Eimer also sought a fusion with functionalist views (see pp. 360–365), but he opted for a Lamarckian push as his external source, and explicitly relegated Darwin to an insignificant periphery among sources of adaptation. But Whitman rejected Lamarckism and located his external push in natural selection.

Whitman directly criticized Eimer for his negative view of Darwinism, and for subjecting the entire theory of orthogenesis to undeserved derision thereby:

> Among the rival theories of natural selection two are especially noteworthy. One of these is now generally known as orthogenesis. Theodor Eimer was one of the early champions of this theory . . . Eimer's intemperate ferocity toward the views of Darwin and Weismann, coupled with an equally intemperate advocacy of the notion that organic evolution depends upon the inheritance of acquired characters, was enough to prejudice the whole case of orthogenesis. Moreover, the controversial setting given to the idea of definitely directed variation, without the aid of utility and natural selection, made it difficult to escape the conclusion that orthogenesis was only a new form of the old teleology, from the paralyzing domination of which Darwin and Lyell and their followers had rescued science. Thus, handicapped, the theory of orthogenesis has found little favor (1919, p. 9).

2. Eimer was a polemicist by dint of personality. Whitman, as a great administrator, displayed an opposite temperament in his inclination to seek compromise among *good* ideas. Whitman believed that major systems, as

devised and supported by such brilliant men as Darwin and de Vries, must inevitably hold at least partial value—and he sought a fruitful union of these systems with his own favored theory of orthogenesis. "Natural selection, orthogenesis, and mutation appear to present fundamental contradictions, but I believe that each stands for truth, and that reconciliation is not distant" (1919, p. 10).

We all know that the theories of de Vries and Darwin eventually reached peace through a recognition that micromutations could act as the source of isotropic Darwinian variation. We regard this fusion as the basis for the Modern Synthesis. A similar and vital task has only begun in our time, but we now live in an age struggling for further union—to join the success of this Modern Synthesis with neglected structuralist and formalist themes of developmental constraint and channeled variation (see Chapters 10 and 11 for my effort in this direction). Whitman surely erred in interpreting a channel of variation— a pathway of potential evolution in either direction—as a one-way street of inevitable change. (The reinterpretation of orthogenetic "one way streets" as "channels" of preferred variability establishes a key "translation" for updating this older and valuable literature into relevance for our modern debates. I also strongly suspect, in opposition to both Darwin and Whitman, that ancestral pigeons were neither two-barred nor checkered, but both. After all, ancestors exist as populations, not archetypes. Both states persist in continuous gradation within many modern populations of pigeons—and this entire channel may well have been expressed among variable adults in ancestral populations.)

However, Whitman's notion that selection does not encounter a full range of isotropic variation, but must work instead with material strongly biased by internal constraint, may supply a key theme for an even higher synthesis of external and internal forces—a theory that will preserve a Darwinian core, but finally and properly incorporate the formalist themes, advocated as central to evolutionary understanding by many of the finest biologists from the very beginning of our profession. Whitman succinctly stated the basis of this synthesis, but we are only now beginning to learn enough about genetics and development to vindicate his hunches: "Natural selection waits for opportunities to be supplied, not by multifarious variation or orderless mutation, but by continuous evolutional processes advancing in definite directions" (1919, p. 13).*

---

*We should give the last word, if only in a footnote, to the ever-perceptive T. H. Huxley, Darwin's stoutest supporter, but an incisive critic for several aspects of natural selection in its strict form. Whitman cited this passage from Huxley *(Darwiniana)* as an epigraphic quotation to one of his articles on orthogenesis (1919, p. 64): "But the causes and conditions of variation have yet to be thoroughly explored, and the importance of natural selection will not be impaired even if further inquiries should prove that variability is definite and is determined in certain directions rather than in others, by conditions inherent in that which varies. It is quite conceivable that every species tends to produce varieties of a limited number and kind and that the effect of natural selection is to favor the development of some of these, while it opposes the development of others along their predetermined lines of modification."

## Saltation as a Theory of Internal Impetus: A Second Formalist Strategy for Pushing Darwinism to a Causal Periphery

### WILLIAM BATESON: THE DOCUMENTATION OF INHERENT DISCONTINUITY

Darwin, as noted earlier, viewed his own accomplishment as dual and distinguishable: establishing the fact of evolution by copious data, and devising a theory, natural selection, to explain the mechanism of change. Darwin also stated that the first achievement must be ranked as more fundamental, for the deepest and most disturbing implications flow from the simple fact of genealogical continuity itself, whatever the philosophically radical character of natural selection as a cause of change.

William Bateson, speaking at the major Darwinian centennial celebration of 1909, made the same point—and the same assessment of the two achievements:

> Darwin's work has the property of greatness in that it may be admired for more aspects than one. For some the perception of the principle of natural selection stands out as his most wonderful achievement to which all the rest is subordinate. Others, among whom I would range myself, look up to him rather as the first who plainly distinguished, collected, and comprehensively studied that new class of evidence from which hereafter a true understanding of the process of evolution may be developed. We each prefer our own standpoint of admiration; but I think that it will be in their wider aspect that his labors will most command the veneration of posterity . . . We shall honor most in him not the rounded merit of finite accomplishment, but the creative power by which he inaugurated a line of discovery endless in variety and extension (Bateson, 1909, p. 85).

Bateson's and Darwin's motives, however, could scarcely have been more different. Darwin, while ranking his joys, took great pride in both achievements. But Bateson viewed natural selection as an insignificant force and a methodological disaster. In downpeddling natural selection, Bateson presented his argument as an attempt to save Darwin's wider viewpoint from its own worst error, thus preserving the centennial season as a time of triumph.

William Bateson (1861–1926), son of a classical scholar who served as master of St. John's College, Cambridge, shared with Charles Darwin both the enormous advantages of birth and the potential impediment of a slow educational start. Darwin's father reproached him in 1825: "You care for nothing but shooting, dogs, and rat-catching, and you will be a disgrace to yourself and all your family." Bateson, at a similar stage in his education, was branded as "a vague and aimless boy" by his headmaster at Rugby. Yet Bateson finally focussed his interests on zoology and morphology, studying with Sedgwick and Weldon at Cambridge, and from 1883 to 1884 (in an ineresting reversal at a time when most aspiring American scholars traveled to Europe for postgraduate work) with W. K. Brooks, the finest American zoologist of his time, at Johns Hopkins.

I generally shun psychological or intellectual biography in this book (both for limitations of space and authorial competence), but I have long been fascinated by the structural principle that groups of ideas seem to cohere just as morphological parts often correlate—with possession of one trait implying a set of logically or mechanically appended consequences. The rationale for this book depends, in large part, on such a structural isomorphism between nucleating centers of mutually implicating ideas and the integrity of organic *Baupläne,* for my notion of a Darwinian essence, construed as a minimal but distinctive set of interpenetrating and almost necessarily correlated concepts, builds the organizational framework of this book.

Whatever the validity of this general framework, I think we will all admit that ideas do coagulate in implicating sets, and that fascination with one—and we get hooked for the damnedest of impenetrable reasons—attracts us to the others as well. The Darwinian set implies a basic intrigue with functional and adaptational arguments and includes preferences for gradualism of change, separability of parts, and efficiency of competition. An opposing set—an aggregation that exerted a far lesser, but still identifiable, pull upon Darwin himself (see pp. 330–341), and that motivates the formalist "nucleating center" of this chapter—includes fascination with structurally based correlation, evolution by internally generated sources of variation, and suspicion of adaptational scenarios as primary explanations for basic organic design.

For whatever reasons, and from his earliest days in zoology, Bateson felt drawn to the structuralist set* (and to consequent disfavor for Darwinian mechanisms). He quoted and admired the literature on distrust of functional and teleological arguments, from Bacon to Voltaire. Bateson's wife remarked in her memoir (1928, p. 13): "I think he never travelled without a copy of *Candide* in his pocket." In 1888, at the beginning of his career, he wrote to his sister: "My brain boils with evolution." But note the main theme that emerged from this cauldron—the necessary breadth and extent of the network of correlations enjoined by any primary change, with inevitable swamping of the primary trigger by the sequelae (a keen foreshadowing of

---

*I also feel the strong tug of this theme, and for a personal reason. I was trained as a strict adaptationist, and I accepted and vigorously promoted this worldview in my early papers. These works now embarrass me, with such statements as: "I acknowledge a nearly complete bias for seeking causes framed in terms of adaptation" (1966, p. 588—at least I labelled the preference as a "bias"); and ". . . the fundamental problem of evolutionary paleontology—the explanation of form in terms of adaptation" (1967, p. 385). Yet I also felt the pull and fascination of the opposite set—though I had no inkling of the coordinated force behind the varied concepts, no explicit idea of the coherence (or even the terminology), and certainly no sense of the challenge thus posed to my juvenile certainties. I do not know why I felt the tug so strongly. But the ideas must cohere intrinsically if a young scholar can be so pulled by all of them, yet so unaware of their aggregation or their import. I discuss these personal aspects further in Chapter 1.

my personal favorite among modern structuralist themes, as embodied in the concept of exaptation—see Chapter 11):

> My brain boils with evolution. It is becoming a perfect nightmare to me. I believe now that it is an axiomatic truth that no variation, however small, can occur in any part without other variation occurring in correlation to it in all other parts; or, rather, that no system, in which a variation of one part had occurred without such correlated variation in all other parts, could continue to be a system. This follows from what one knows of the nature of an "individual," whatever that may be . . . Further, any variation must always consist chiefly of the secondary correlated variations and to an infinitely small degree of an original primary variation (in Bateson, 1928, p. 39).

Bateson put his formalist and non-Darwinian thoughts together in one of the most interesting biological works of the late 19th century—*Materials for the Study of Variation* (1894). This work has been read primarily as a defense of saltational variation and change, a brief for structuralism using the "facet flipping" theme of Galton's polyhedron (see pp. 342–351). I will not challenge this primacy, but I do wish to demonstrate that Bateson's book integrates a broader set of formalist themes (including distrust of adaptation and suspicion of historical contingency) under a primary concern for saltation and discontinuity as a counter to Darwinism.

Bateson's *Materials* may be a famous book in retrospect (primarily, I suspect, because scholars want to grasp how the man who later invented the word genetics looked upon variation in the last pre-Mendelian decade). But the volume failed in its own time as a long compendium by an unknown author, and a financial disaster. Bateson's wife remembered (1928, pp. 57–58): "The book was not a success—the professors and lecturers of the day did not introduce their students to it. Perhaps, that they should not was to be expected. For a few years the annual arrival of the publisher's account was a dismal event, and the book was put "in remainder" and dropped out. The second volume as such was never written."

As a rhetorical strategy, many comprehensive works begin with a small logical puzzle or anomalous observation. For example (see Gould, 1987b, for details), Burnet and Hutton presented their grand geological systems as solutions, in Burnet's case, to the problem of sources for water in Noah's flood; or, for Hutton, as a logical dilemma in final cause (why would a benevolent God, attuned to human needs, allow soil to be made by a process that must eventually erode the earth away). Bateson's *Materials* also poses its central argument as the solution to a particular puzzle: how can evolution produce a world of taxonomic discontinuity when environmental gradients, as potential impetuses for change, are generally continuous: "The differences between species . . . are differences of kind, forming a discontinuous series, while the diversities of environment to which they are subject are on the whole differences of degree, and form a continuous series" (1894, p. 16).

Bateson recognized that both Lamarckism and Darwinism, as functionalist mechanisms, posited a flow of information from environment to organism as a basis of adaptive transformation. How then could a continuous environment yield our world of thinly populated morphospace, with vast gaps between realized designs? "According to both theories [Lamarckism and Darwinism], specific diversity of form is consequent upon diversity of environment, and diversity of environment is thus the ultimate measure of diversity of specific form. Here then we meet the difficulty that diverse environments often shade into each other insensibly and form a continuous series, whereas the specific forms of life which are subject to them on the whole form a discontinuous series. The immense significance of this difficulty will be made more apparent in the course of this work" (1894, p. 5).

For Bateson, a general solution could be derived from logical implications of the argument, prior to any search for causes: nature's discontinuity must arise from the internal workings of organisms:* "Such discontinuity is not in the environment; may it not, then be in the living thing itself?" (1894, p. 17).

Bateson, as a young Turk inspired by ideals of German mechanism and American experimentalism, but working in a more traditional world of descriptive natural history, knew what he didn't like about the inferential procedures of most Darwinians in his generation: the conjoined tactic of speculation based on embryology for phylogenetic reconstruction, and guesswork about utility for inferences about adaptation by natural selection. Bateson lists the two pitfalls of such sterile work: "The first of these is the embryological method, and the second may be spoken of as the study of adaptation. The pursuit of these two methods was the direct outcome of Darwin's work" (1894, p. 7).

Bateson longed to apply the mechanistic style of experimental science to the causes of evolution. If guesswork about externalities had served the field so poorly, why not look to the intrinsic characters of organisms, features that might be resolved by manipulation and by understanding the mechanics of heredity. Variation itself must be taken as a primary phenomenon. Why, at least as an initial strategy, look beyond this palpable and measurable property of populations? Perhaps the causes of evolutionary change lie in variation itself and not in a superimposed external sorting, as the more complex Darwinian mechanism proposed: "Variation, in fact, is evolution. The readiest way, then of solving the problem of evolution is to study the facts of variation" (1894, p. 6).

---

*Central though this question may be to Bateson's inquiry (and sympathetic as I am to his book), I confess that I have never understood why Bateson regards this point as so telling and decisive. I think that Darwin presented a simple and perfectly satisfactory solution to this dilemma (which he clearly recognized and discussed at length in early chapters of the *Origin*)—namely, that forms once filling the gaps between modern discontinuities have now become extinct. (Most intermediates, after all, are not contemporary creatures, but graded series on two lineages of ancestors running back to a common branching point, often deep in the geological distance.)

Bateson notes the fascination of his colleagues for the causes of variation as expressed in the nature of heredity, but holds that the absence of any hard evidence has mired the subject in fanciful and speculative hypotheses, from Darwin's pangenesis to Haeckel's perigenesis. The quest should be postponed for now (though Bateson wouldn't have long to wait, as the Mendelian revival lay just around the corner of the coming century). Meanwhile, an empirical approach might yield great benefits, at least by providing an inductive entry to the difficult subject of causes. Why not, in short, simply gather the facts of variation: "It is especially strange that while few take much heed of the modes of variation or of the visible facts of descent, everyone is interested in the causes of variation and the nature of 'heredity,' a subject of extreme and peculiar difficulty. In the absence of special knowledge, these things are discussed with enthusiasm, even by the public at large. But if we are to make way with this problem, special knowledge is the first need. We must know what special evidence each group of animals and plants can give, and this specialists alone can tell us" (1894, p. ix).

Bateson chose to express this strategy of empirical compilation in the title of his book: "To collect and codify the facts of variation is, I submit, the first duty of the naturalist" (1894, p. vi). Brave words, to be sure, but Bateson recognized that such a complex and multifarious subject could not be resolved simply by toting the relative frequencies of an empirical list. He also understood that the very idea of a totally unbiased listing could only operate as a self-serving fiction to bolster a myth of perfect scientific objectivity. Bateson recognized a pervasive bias in traditional accounts of variation—strong preferences for continuity and gradualism, as expressed in the old Leibnizian and Linnaean aphorism, *natura non facit saltum:* "First there is in the minds of some persons an inherent conviction that all natural processes are continuous . . . Secondly, variation has been supposed to be always continuous and to proceed by minute steps because changes of this kind are so common in variation" (1894, p. 16). Bateson's list, therefore, would be a purposive account of a particular sort of variation: "If facts of the old kind will not help, let us seek facts of a new kind" (1894, p. vi).

Since Bateson sought the causes of evolution in variation itself, and since he viewed discontinuity as the primary fact of natural history, discontinuous variation among organisms within populations became his favored source of evolutionary change. *Materials for the Study of Variation* is not an unbiased compendium of all organic mutability, but rather an explicit attempt to catalog discontinuous variation as a source of insight into internally driven causes of evolution. The subtitle of the book explicitly refutes any claim to balance or comprehensiveness among styles: *"Treated with Especial Regard to Discontinuity in the Origin of Species."*

Bateson divided variation into meristic (for serially and symmetrically repeated, countable and discontinuous structures) and substantive (for ordinary continuous variability). As an obvious ploy to promote his preference for locating the causes of evolution in discontinuity of variation, the meristic ategory commanded his primary attention. Bateson devotes the entire text of *Materials* to compiling examples of meristic variation. He planned, but never wrote or even seriously began, a second volume on substantive variation.

(Bateson left a substantial legacy of biological terminology. He is best known, of course, as the inventor of the word *genetics,* but *Materials* includes two new terms of later importance—*meristic* for this general style of variation, and *homeosis* for a subset that has since become central in modern evolutionary and developmental biology; see Chapter 10.)

The evolutionary import of Bateson's book may be summarized in three characteristic features of his argument for saltational variation and change, with a subsequent fourth theme then centered upon his discussion of implications for Darwinism.

NATURE OF THE EXAMPLES. Materials is, above all else, a compendium of examples of discontinuous variation in meristic characters. Bateson begins with a long sequence of chapters (pp. 87–422) on linear series, starting with arthropod segments, moving through vertebrae and ribs, where Bateson presents the "type" cases of homeosis, and proceeding to branchial openings, mammae, teeth and digits. The second sequence of chapters (pp. 423–566) treats symmetrically repeated structures under three headings: radial series, bilateral series, and secondary symmetry and duplication. Although Bateson adopted a convention of presenting facts in small type and interpretation in a larger font, he remained true to his own version of the Kantian dictum that percepts without concepts are blind—for even the factual small-type listings contain implicit interpretations for his worldview, and against Darwin's.

As his primary theme, Bateson emphasizes a basic implication of meristic variation. Segments must be conceptualized as discrete anatomical forms, and supernumeraries (or deletions) are therefore usually complete (or entirely suppressed). Half an added segment usually denotes a structural and functional absurdity, in principle. Merism, by its very nature, implies discontinuity in construction and change. Bateson writes, for example, about 12-jointed antennae within a group of normally 11-jointed beetles:

> Would it be expected that the longicorn Prionidae, most of which have the unusual number of 12 antennary joints, did, as they separated from the other longicorns which have 11 joints, gradually first acquire a new joint as a rudiment which in successive generations increased? . . . If anyone will try to apply such a view to hundreds of like examples in arthropods, of difference in number of joints and appendages of near allies . . . he will find that by this supposition of continuity in variation he is led into endless absurdity. Surely it must be clear that in many such cases to suppose that the limb came through a phase in which one of its divisions was half-made or one of its joints half-grown, is to suppose that in the comparatively near past it was an instrument of totally different character from that which it has in either of the two perfect forms. But no such supposition is called for. With evidence that transitions of this na-

ture may be discontinuously effected the difficulty is removed (1894, pp. 410–411).

Whereas this basic argument contravenes gradualism, a second implication then disputes natural selection. Supposed anomalies of discontinuous variation often seem no less structurally "perfect" than normal forms. Why, then, do we so confidently ascribe the "normal" forms to long honing by natural selection working upon continuous variation? Should we not rather conclude that, since normal and anomalous variants may be equally well constructed, both categories arise by internal regulation? Of roaches with four-jointed tarsi (instead of the normal five), Bateson writes:*

> The four-jointed tarsus occurring thus sporadically, as a variety, is not less definitely constituted than the five-jointed type, and the proportion of its several joints are not less constant. It is scarcely necessary to point out that these facts give no support to the view that the exactness or perfection with which the proportions of the normal form are approached is a consequence of selection. It appears rather, that there are two possible conditions, the one of five joints and the other with four, each being a position of organic stability. Into either of these the tarsus may fall; and though it is still conceivable that the final choice between these two may have been made by selection, yet it cannot be supposed that the accuracy and completeness with which either condition is assumed is the work of selection, for the "sport" is as definite as the normal (1894, pp. 64–65).

In a further move to complete the basic argument within a supposed compendium of objectively listed facts, Bateson emphasizes homology between the teratology of individuals in one population and the normal morphology of a related species—with an obvious implication of transformation by saltational variation. Fusion of bilaterally symmetrical organs in the mid line (or fission of singletons into pairs on the antimeres) establishes a major class of cases:

> A normally unpaired organ standing in the middle line of a bilateral symmetry may divide into two so as to form a pair of organs; and conversely, a pair of organs normally placed apart from each other on either side of a middle line may be compounded together so as to form a single organ in the middle line. In animals and plants nothing is more common than for different forms to be distinguished from each other by the fact that an organ standing in the middle line of one is in another represented by

---

*Note Bateson's use of the term "position of organic stability." Bateson here cites Galton's phrase, borrowed from his passages on the polyhedron metaphor. Bateson uses the phrase throughout his text, often with quotation marks to indicate its source. Thus, Bateson clearly embraced, as did many of his contemporaries, Galton's doubled-edged metaphor for formalism against pure Darwinian externalism—though Bateson emphasized the facet-flipping (saltational) rather than the inherently directional (orthogenetic) theme of the metaphor.

two organs, one on either side. The facility therefore with which each of these two conditions may arise from the other by discontinuous variation is of considerable importance (1894, p. 448).

THE PHYSICAL BASIS OF DISCONTINUITY. Early in his studies, Bateson developed an insight, a sort of epiphany in his own assessment, that would shape his views (haunt might be a better word) throughout his career. His emphasis on discontinuity, his dislike of Darwinism, his inability to come to terms with chromosomal theory, all reflect this central vision of his thinking. Bateson decided that discontinuously repeated organic structures bore isomorphic, and therefore common causal, similarity to physical phenomena produced by waves and vibrations. He therefore sought a physical cause for heredity in some wave-like form of energy—the "vibratory theory" in his own words—and he could therefore never fully accept a particulate basis for genetics. In 1891, he wrote with great excitement to his sister, stating that he dared not even hope to have an idea of such import ever again:

> Did I tell you anything about my new vibratory theory of repetition of parts in animals and plants? I have been turning it over again lately, and feel sure there is something in it. It is the best idea I ever had or am likely to have—do you see what I mean?—divisions between segments, petals, etc. are internodal lines like those in sand figures made by sound, i.e. lines of maximum vibratory strain, while the mid-segmental lines and the petals, etc. are the nodal lines, or places of minimum movement. Hence all the patterns and recurrence of patterns in animals and plants—hence the perfection of symmetry—hence bilaterally symmetrical variation, and the completeness of repetition whether of a part repeated in a radial or linear series etc. etc. I am, as you see, in a great fluster (in Bateson, 1928, p. 42).

In his next letter, he added: "You'll see—it will be a commonplace of education, like the multiplication table or Shakespeare, before long!"

*Materials* does not discuss the vibratory theme at length, if only because Bateson chose to organize the book as a compendium of data—and he could present no factual support for his suspicions about the production and inheritance of discontinuous variation. Still, he advanced several conjectures about the construction of phenotypic discontinuity from underlying continuity by a simple physical or chemical impetus. He attempts, for example, to analogize the discrete concentric rings of eye-spots on lepidopteran wings with pond ripples (metaphorically) and, in greater hope of causal isomorphism, with chemical reactions:

> A whole eye-spot may come, or it may go . . . leaving the field of the cell plain and without a speck. The suggestion is strong that the whole series of rings may have been formed by some one central disturbance, somewhat as a series of concentric waves may be formed by the splash of a stone thrown into a pool. It is especially interesting to remember that the

formation even of a number of concentric rings of different colors from an animal pigment by the even diffusion of one reagent from a center occurs actually in Gmelin's test for bile-pigments. Bile is spread on a white plate and a drop of nitric acid yellow with nitrous acid is dropped on it. As the acid diffuses itself distinct rings of yellow, red, violet, blue, and green are formed concentrically around it by the progressive oxidation of the bile-pigment . . . This example is merely given as an illustration of the possibility that a series of discontinuous chemical effects may be produced in concentric zones by a single central disturbance (1894, p. 292).

Rarely missing an opportunity for a dig at Darwinism, Bateson then adds that, with the chemical analogy, we at least hope to find a cause, whereas the standard adaptational speculation can make no such claim: "As to the function of the ocellar markings nothing is known, and I am not aware that any suggestion has been made which calls for serious notice" (1894, p. 294).

Writing more generally on the same theme, Bateson tries to attribute discrete color classes directly to the chemical stability of pigments—and to dismiss the alternative functional explanation of adaptational guesswork about the utility of discontinuous difference. (This passage occurs in Bateson's only short discussion of substantive rather than meristic variation—the category intrinsically less favorable to his preference for discontinuity. Thus, he strives for explanatory generality across all classes of variation): "It would, I think, be simpler to regard the constancy of the tints of the several species and the rarity of the intermediate varieties as a direct manifestation of the chemical stability or instability of the coloring matters, rather than as the consequences of environmental selection for some special fitness as to whose nature we can make no guess. For we do know the phenomenon of chemical discontinuity, whatever may be its ultimate causes, but of these hypothetical fitnesses we know nothing, not even whether they exist or no" (1894, p. 48).

In his summation, Bateson reiterates his conviction that meristic discontinuity may represent a necessary phenotypic expression of an underlying mechanical regularity, and not a set of adaptations gradually crafted by natural selection: "To sum up: There is a possibility that meristic division may be a strictly mechanical phenomenon, and that the perfection and symmetry of the process, whether in type or in variety, may be an expression of the fact that the forms of the type or of the variety represent positions in which the forces of division are in a condition of mechanical stability" (1894, p. 71). In addition, Bateson could not resist a final and explicit anti-Darwinian dig in stating that such symmetrical forms would "owe their perfection to mechanical conditions and not to selection or to any other gradual process" (1894, p. 70).

SUSPICION OF HISTORY (AS WELL AS ADAPTATION) AS A CAUSE OF MORPHOLOGY. The pure formalist or structuralist not only rejects functional accounts based on slow building for utility, but also tries to avoid any appeal to deep history in explaining the origin of morphology. The pure structuralist prefers ahistorical accounts (see Chapter 11, Section 1, for modrn versions), and seeks to explain form in terms of chemical and mechanical forces now operating through development. (Structuralists do not deny, of course, that history sets the presence of one mechanics, rather than another, in any particular lineage. But the analysis of immediate causes for current anatomies must still invoke present and intrinsic workings.)

Bateson best illustrates his full allegiance to the structuralist program by interpreting a range of putatively historical phenomena in terms of contemporary mechanics. He admits, for example, that highly variable parts, like mammalian third molars, often have little functional utility—and may therefore be "permitted" to range widely by a history of failing function. But Bateson still seeks a primary explanation in terms of current construction: "The oft-repeated statement that 'useless' parts are specially variable, finds little support in the facts of variation, except in as far as it is a misrepresentation of another principle. The examples taken to support this statement are commonly organs standing at the end of a meristic series of parts, in which there is a progression or increase of size and degree of development, starting from a small terminal member" (1894, pp. 78–79).

Similarly, the features that we call atavisms and attribute to past echoes should be viewed as alternate mechanical pathways. In this case, a Darwinian might invoke both styles of explanation, for even a historical vestige must be built along a current developmental route. But Bateson did not seem to grasp this necessary duality, and he often used the second aspect (mechanical pathway of building) to castigate the first (adaptational basis)—thus illustrating, by his error, the near exclusivity of his structuralist interpretations. Bateson uses simple mechanical analogs to make his point:

> But all that we know is that now and then it shoots wide and hits another mark, and we assume from this that it would not have hit if it had not aimed at it in a bygone age. To apply this to any other matter would be absurd. We might as well say that a bubble would not be round if the air in it had not learned the trick of roundness by having been in a bubble before: that if in a bag after pulling out a lot of white balls I find a totally red one, this proves that the bag must have once been full of red balls, or that the white ones must all have been red in the past (1894, p. 78).

EVOLUTIONARY CONCLUSIONS AND IMPLICATIONS FOR DARWINISM. The central thesis of *Materials* can be stated positively and succinctly: much variation (or at least the evolutionarily significant fraction) is discontinuous, mechanically and chemically built through heredity, and often well formed (and therefore potentially useful) by intrinsic construction. The primary cause of evolution, a process that also tends to be discontinuous, must therefore be located directly in the rules, patterns and directions of variation: "Is it not then possible that the discontinuity of species may be a consequence and expression of the discontinuity of variation?" (1894, p. 69).

But Bateson makes few positive claims in this mode. Rather he presents most arguments in a context of refutation—with Darwinian natural selection as the prime target, particularly the themes of insensibly gradual change and selective pressure guided by utility. In so proceeding, Bateson expresses no hostility for Darwin himself. Moreover, his strategy in separating Darwin's factual and theoretical achievements in order to render high personal praise also makes sense in this context. Bateson's style of refutation extends far beyond biology into broader and contemporary themes of science. Bateson regarded himself as an experimental modernist, upholding ideals of tractable science against a sterile speculative tradition that had taken hold in two areas of natural history—the guesswork of phyletic reconstruction and the hypothetical assignment of adaptive utility.

Bateson's attitude towards natural selection and adaptation provides a good indication of his procedures and prejudices. As stated above, Bateson uses several facts of discontinuous variation to downplay selection as a creative force. Consider just two arguments:

1. If rare and discontinuous variants may originate as well formed and potentially useful at their sudden appearance, why assume that normal forms must be gradually crafted to perfection by natural selection: "The existence of sudden and discontinuous variation, the existence, that is to say, of new forms having from their first beginning more or less of the kind of perfection that we associate with normality, is a fact that disposes, once and for all, of the attempt to interpret all perfection and definiteness of form as the work of selection. The study of variation leads us into the presence of whole classes of phenomena that are plainly incapable of such interpretation" (1894, p. 568).

2. If variation is inherently discontinuous and often large in effect, then selection can only choose among alternatives presented by internal causes, and therefore cannot operate as a creative force in evolutionary change. (Here, of course, Bateson merely recounts the standard argument on "creativity" advanced by nearly all non-Darwinian theorists.) Bateson, for example, writes about butterfly wingtips that exhibit either red or purple, but nothing in between: "It is easier to suppose that the change from red to purple was from the first complete, and that the choice offered to selection was between red and purple" (1894, p. 73).

But Bateson devotes his main thrust of argument to a methodological theme—to designating the tradition of adaptationist "story telling" as a poor substitute for experiment and proof. Some of the most powerful statements against this conventional, and still all too common, form of evolutionary conjecturing may be found in Bateson's 1894 book and later writings.

Bateson acknowledges the allure and fascination of adaptationist conjecture: "This study of adaptation and of the utility of structures exercises an extraordinary fascination over the minds of some . . . The amount of evidence collected with this object is now enormous, and most astonishing ingenuity has been evoked in the interpretation of it" (1894, p. 10).

Yet this so-called evidence, Bateson then asserts, represents little more than a set of conjectures about possible benefits, not a proof of actual (and gradual) construction for utility:

In these discussions we are continually stopped by such phrases as, "if such and such a variation then took place and was favorable," or, "we may easily suppose circumstances in which such and such a variation if it occurred might be beneficial," and the like. The whole argument is based on such assumptions as these—assumptions which, were they found in the arguments of Paley or of Butler, we could not too scornfully ridicule. "If," say we with much circumlocution "the course of nature followed the lines we have suggested, then, in short, it did." That is the sum of our argument (1894, p. v).

We might recognize the bankruptcy of such an approach, Bateson claims, if we looked inside ourselves and acknowledged that we undoubtedly could, and almost surely would, concoct an adaptationist scenario for any case better explained in another way (as in the establishment, through random processes, of a rare discontinuous variant as the norm of a small island population):

In any case of variation there are a hundred ways in which it may be beneficial or detrimental. For instance, if the "hairy" variety of the moorhen became established on an island, as many strange varieties have been, I do not doubt that ingenious persons would invite us to see how the hairiness fitted the bird in some special way for life in that island in particular. Their contention would be hard to deny, for on this class of speculation the only limitations are those of the ingenuity of the author (1894, p. 79).

This lamentable practice, Bateson argues, giving natural history such a low reputation among the sciences, will only end when naturalists accept an alternate structuralist biology—for the key concepts of discontinuity and correlation must dismantle the strict adaptationist's necessary (but often unstated) view of organisms as malleable aggregations of independently improvable parts: "For the crude belief that living beings are plastic conglomerates of miscellaneous attributes, and that order of form and symmetry have been impressed upon this medley by selection only; and that by variation any of these attributes may be subtracted or any other attribute added in indefinite proportion, is a fancy which the study of variation does not support" (1894, p. 80).

Though Bateson, throughout the book, uses the rhetorical device of opposing discontinuity in variation to Darwinian gradualism, he also stresses the positive theme that internally generated saltations may represent, in themselves, the long-sought creative component of evolutionary change (with selection then operating as a subsidiary device to spread these novel features through populations): "If the evidence went no further than this the result would be of use, though its use would be rather to destroy than to build up. But besides this negative result there is a positive result too, and the same discontinuity which in the old structure had no place, may be made the framework round which a new structure may be built" (1894, p. 568). As a final contrast, and with a positive finish, Bateson writes: "A presumption is created that the discontinuity of which species is an expression has its origin not in the environment, nor in any phenomenon of adaptation, but in the intrinsic nature of organisms themselves, manifested in the original discontinuity of variation" (1894, p. 567).

Bateson ends his *Materials* with a striking plea (akin to an equally passionate statement in the preface to Simpson, 1944) for an end to the dichotomy of valuation that breeds discord and miscommunication between experimentalists and field naturalists. Bateson himself favored the experimental approach, and wrote his book to compel an appreciation of this "unfamiliar" methodology by his fellow naturalists. But he also understood that laboratory work cannot solve the problems of evolution without detailed knowledge of natural history derived from the field—and his book presents a magnificent compendium of empirical examples, spanning nearly 1000 pages, and mostly drawn from traditional descriptive literature. The integration so devoutly to be wished, Bateson argues, will arise from the study of variation—for naturalists can record the variety and understand its sway and distribution, while experimentalists can manipulate the results and hope to learn causes. Above all, variation must become the focus of union because the causes of evolution, including the origin of species, must lie within these *Materials,* properly ordered, manipulated, and explained. Bateson's plea deserves citation *in extenso:*

> These things attract men of two classes, in tastes and temperament distinct, each having little sympathy or even acquaintance with the work of the other. Those of the one class have felt the attraction of the problem. It is the challenge of nature that calls them to work. But disgusted with the superficiality of "naturalists" they sit down in the laboratory to the solution of the problem, hoping that the closer they look, the more truly will they see. For the living things out-of-doors, they care little. . . . With the other class it is the living thing that attracts, not the problem. To them the methods of the first school are frigid and narrow. . . . With senses quickened by the range and fresh air of their own work, they feel keenly how crude and inadequate are these poor generalities, and for what a small and conventional world they are devised. Disappointed with the results, they condemn the methods of the others, knowing nothing of their real strength. . . . Beginning as naturalists they end as collectors, despairing of the problem, turning for relief to the tangible business of classification, accounting themselves happy if they can keep their species apart, . . . Thus each class misses that which in the other is good. But when once it is seen that, whatever be the truth as to the modes of evolution, it is by the study of variation alone that the problem can be attacked, and that to this study both classes of observation must equally contribute, there is once more a place for both crafts side by side: for though many things spoken of in the course of this work are matters of

doubt or of controversy, of this one thing there is no doubt, that if the problem of species is to be solved at all it must be by the study of variation (1894, pp. 574–575).

For biologists, 1900 marks far more than the arbitrary turning of a century (in one mode of reckoning) because, in that year, the barrier that all evolutionists recognized as the chief impediment to further insight—ignorance about the causal basis of heredity—began to crumble with the rediscovery of Mendel's principles. Bateson himself well understood the strict limits that necessarily impeded further progress until the basis of heredity and variation could be established. He wrote *Materials* as an empirical list, largely because he could propose no causal guide to variation, and therefore hoped that a compendium might suggest some hints, or at least prove useful *faute de mieux*. He expressed frustration about this missing key at the end of *Materials*, and proposed that the basis of heredity be sought in breeding experiments: "But beyond a general impression, in this, the most fascinating part of the whole problem, there is still no guide. The only way in which we may hope to get at the truth is by the organization of systematic experiments in breeding, a class of research that calls perhaps for more patience and more resources than any other form of biological inquiry. Sooner or later, such investigations will be undertaken and then we shall begin to know" (1894, p. 574).

Bateson invented the word "genetics" (in 1905). He then fought for discrete inheritance against the biometrical school of Pearson and Weldon, made many important Mendelian discoveries during the first decade of the new science (application to animals as well as plants, elucidation of the phenomena of epistasis and linkage), founded the *Journal of Genetics,* and served as an effective spokesman for the new world order. In a late address of 1924 (published in 1928), he contrasted the Mendelian before and after, with special reference to *Materials:* "Only those who remember the utter darkness before the Mendelian dawn can appreciate what has happened. Stories which then seemed mere fantasies, are now common sense. When I was collecting examples of variation in 1890, I remember well reading the fanciers' tales about dun tumbler pigeons being almost always hens, and about the 'curious effects of crossing' with cinnamon canaries, but I would never have dared to repeat them" (in Bateson, 1928, pp. 405–406).

In this light, the continuing saga of Bateson and evolutionary theory should tell a tale of pleasure and progress. He should, like de Vries (see pp. 425–439), posit an identity of his favored discontinuous variants with major Mendelian mutations, argue that the riddle of evolution has been solved in his terms, and proceed forward to greater discovery and satisfaction (at least until Fisher and others inaugurate the Modern Synthesis by upholding the efficacy, and Mendelian character, of small-scale continuous variation as well—a recognition that did not dawn widely until after Bateson's death).

In fact, Bateson's later career did not follow this happy scenario at all. Instead, he walked the all too common ontogenetic trajectory from young Turk to old fogey. He did rejoice in the potentially Mendelian character of discontinuity, but he altered none of his earlier views and became, in the eyes of most younger contemporaries, an increasingly dyspeptic and conservative force (Arthur Koestler's characterization of Bateson, as a vitriolic opponent in the sad case of the Neo-Lamarckian Paul Kammerer, paints a colorful, though not entirely fair, portrait of Bateson's later career—see Koestler, 1971). Bateson continued his hostility to Darwinian and all other forms of functionalist explanation (hence his brutal opposition to Kammerer and other Lamarckians). Above all, the new genetics eventually ran away from him, primarily because he would not budge from his old, controlling belief in physical causation of phenotypic discontinuity by underlying wave-like or vibratory motions. Bateson took this idea so literally that he could never accept the "materialistic" chromosomal account of inheritance. Thus, he continued to insist, following his beloved "vibratory" theory, that transmission of hereditary information, while obeying the Mendelian rules of course, must be promulgated by force and motion, rather than by discrete particles.

Bateson delivered the Silliman lectures at Yale University in 1907 and, after considerable delay, published the text as *Problems of Genetics* in 1913, his major post-Mendelian statement on heredity and evolution. His views had changed very little from the themes and claims discussed in *Materials* in 1894. He rejoices in the Mendelian discovery, and gives a good account of early work, while focusing on the limits for evolutionary theory—particularly on his old problem of explaining meristic discontinuity, and his hope for a mechanical explanation based on waves and energy. "In Mendelian analysis we have now, it is true, something comparable with the clue of chemistry, but there is still little prospect of penetrating the obscurity which envelops the mechanical aspect of our phenomena" (1913, p. 32).

But if we cannot yet fathom meristic variation, at least we may infer that inheritance must be vibratory, not particulate. Ironically, then, Bateson commits his greatest error in thinking about his favorite phenomenon—for he never suspected the integrating theme that particles might code for substances controlling rates of processes:

> When however we pass from the substantive to the meristic characters, the conception that the character depends on the possession by the germ of a particle of a specific material becomes even less plausible. Hardly by any effort of imagination can we see any way by which the division of the vertebral column into x segments or into y segments, or of a Medusa into four segments or into six, can be determined by the possession or by the want of a material particle. The distinction must surely be of a different order. If we are to look for a physical analogy at all we should rather be led to suppose that these differences in segmental numbers correspond with changes in the amplitude or number of dividing waves than with any change in the substance or material divided (1913, p. 35).

In a statement reminiscent of D'Arcy Thompson (see p. 1179), Bateson expresses his hopes for mathematical analysis in morphology:

> It is in the geometrical phenomena of life that the most hopeful field for the introduction of mathematics will be found. If anyone will compare one of our animal patterns, say that of a zebra's hide, with patterns known to be of a purely mechanical production, he will need no argument to convince him that there must be an essential similarity between the processes by which the two kinds of patterns were made . . . Patterns mechanically produced are of many and very diverse kinds. One of the most familiar examples, and one presenting some especially striking analogies to organic patterns, is that provided by the ripples of a mackerel sky, or those made in a flat sandy beach by the wind or the ebbing tide. With a little research we can find among the ripple marks, and in other patterns produced by simple physical means, the closest parallels to all the phenomena of striping as we see them in our animals . . . We cannot tell what in the zebra corresponds to the wind or the flow of the current, but we can perceive that in the distribution of the pigments . . . a rhythmical disturbance has been set up which has produced the pattern we see; and I think we are entitled to the inference that in the formation of patterns in animals and plants mechanical forces are operating which ought to be, and will prove to be, capable of mathematical analysis (1913, p. 36).

Though Bateson never found his underlying vibrations, his faith in their existence as bearers of heredity fueled his primary anti-Darwinian argument for discontinuous variation, and consequent evolution by internally-generated saltation:

> When the essential analogy between these various classes of phenomena is perceived, no one will be astonished at, or reluctant to admit, the reality of discontinuity in variation, and if we are as far as ever from knowing the actual causation of pattern we ought not to feel surprised that it may arise suddenly or be suddenly modified in descent. Biologists have felt it easier to conceive the evolution of a striped animal like a zebra from a self-colored type like a horse . . . as a process involving many intergradational steps; but so far as the pattern is concerned, the change may have been decided by a single event, just as the multitudinous and ordered rippling of a beach may be created or obliterated at one tide (1913, pp. 36–37).

Bateson remained obstinate, and no closer to a solution, as the chromosomal theory became a foundation of modern biology. In 1924, he wrote to the great mathematician G. H. Hardy: "We have had some absurd attempts— mostly from biometricians—to apply mathematics to biology, but as I said my hope is still that I may live to see mathematics applied to biology properly. The most promising place for a beginning, I believe, is the mechanism of pattern."

Bateson became more despondent about evolutionary theory in his later years, while remaining as stubborn as ever about his personal certainties. He visited Canada in 1922, and delivered a famous address entitled "Evolutionary faith and modern doubts" to the annual meeting of the American Association for the Advancement of Science. These were not happy times of consensus for evolutionary theory in general, but Bateson promulgated a particularly bleak vision, clearly colored by the failure of his personal hopes for a vibratory theory of heredity. He lamented that his closing plea of 1894, for integration of lab and field, had been thwarted thus far: "I had expected that genetics would provide at once common ground for the systematist and the laboratory worker. This hope has been disappointed. Each still keeps apart. Systematic literature grows precisely as if the genetical discoveries had never been made and the geneticists more and more withdraw each into his special 'claim'—a most lamentable result. Both are to blame . . . The separation between the laboratory men and the systematists already imperils the work, I might almost say the sanity, of both" (1922, in 1928, p. 397).

Bateson then issued his famous pronouncement—one of the most widely repeated lines in the history of evolutionary writing: "Less and less was heard about evolution in genetical circles, and now the topic is dropped. When students of other sciences ask us what is now currently believed about the origin of species we have no clear answer to give. Faith has given place to agnosticism" (1922, in 1928, p. 391).

Bateson did not fully understand the political and distinctively American context in which he had uttered these lines—the early days of agitation by William Jennings Bryan and the creationist movement for the first wave of anti-evolution laws that culminated in the Scopes Trial of 1925. Creationists seized upon Bateson's words, with their favored and unvarying tactic (still continuing today!) of willful distortion for rhetorical effect. What! A world's leading expert, British no less, from Darwin's own land, claiming to be (dare the word be uttered) agnostic about evolution! Bateson, appalled and angered, spent much time writing letters and articles to stress the point that we must still emphasize today against the rhetoric and similar distortions of modern creationists: theoretical doubt and debate do not alter the factual status of a subject; the fact of evolution and the theory of natural selection do not build the indivisible Eng and Chang of natural history, but rather specify claims of a different order.

But if Bateson became suffused with doubt about evolutionary mechanisms, he never wavered in his conviction that functionalist accounts in general, and Darwinian gradualism in particular, must rank as subsidiary and peripheral to a more valid formalism. Characteristically (for he never shunned controversy), Bateson chose the occasion of Darwin's most important centenary celebration—at Cambridge University in 1909—to present his strongest critique of adaptation from a formalist perspective. He begins—using a favored physical analogy—with the venerable and standard critique of creativity: natural selection, as a negative force, can make nothing, but can only choose among variants produced by another process:

To begin with, we must relegate selection to its proper place. Selection permits the viable to continue and decides that the nonviable shall perish; just as the temperature of our atmosphere decides that no liquid carbon shall be found on the face of the earth: but we do not suppose that the form of the diamond has been gradually achieved by a process of selection. So again, as the course of descent branches in the successive generations, selection determines along which branch evolution shall proceed, but it does not decide what novelties that branch shall bring forth (1909, p. 96).

In a briefer epitome, Bateson had previously written (1904, in 1928, p. 238): "Selection is a true phenomenon; but its function is to select, not to create."

Bateson then launches a two-pronged attack. In a first methodological critique, bordering on meanness (despite his cogent point), Bateson inverts Darwin's intent in proposing small-scale, continuous, isotropic variability as the source of evolutionary change. Darwin used this claim as a brilliant ploy for tractability in a context of ignorance about the nature of variation (see pp. 141–146)—for the assumption of isotropy allowed variation to play the role of supplying "raw material" only, thus permitting the search for mechanisms of evolutionary *change* to proceed notwithstanding. (An insistence that knowledge of the mechanisms of variation must underlie any explanation of phyletic *change* would have stymied the development of evolutionary theory and practice, for Darwin's world knew effectively nothing about the causes of variation, and possessed no techniques for obtaining the requisite information.) But for Bateson, the Darwinian claim of isotropy could only impede the development of a proper theory—for Bateson believed that the causes of change lay in variation, and an appeal to look elsewhere must therefore foreclose progress. Moreover, the Darwinian's favored "elsewhere" too often encouraged sterile exercises in adaptational guesswork, rather than a rigorous approach to assessing utility. Making an analogy to his favorite work of Voltaire (see p. 397), Bateson wrote:

While it could be said that species arise by an insensible and imperceptible process of variation, there was clearly no use in tiring ourselves by trying to perceive that process. This labor saving counsel found great favor. All that had to be done to develop evolution theory was to discover the good in everything, a task which, in the complete absence of any control or test whereby to check the truth of the discovery, is not very onerous. The doctrine "que tout est au mieux" [that all is for the best—the Leibnizian line that Voltaire places in the mouth of Dr. Pangloss] was therefore preached with fresh vigor, and examples of that illuminating principle were discovered with a facility that Pangloss himself might have envied, till at last even the spectators wearied of such dazzling performances (1909, pp. 99–100).

In a second substantive critique, Bateson sought to limit the domain of natural selection—a standard tactic based upon an argument of relative frequency. He had already extirpated the heart of Darwinian importance by labeling selection as a negative force. He now sought a further restriction by shrinking the frequency of selection's application even further. Many creatures may not be so well adapted as tradition dictates; natural selection need not always be working ("daily and hourly scrutinizing" in Darwin's words, 1859, p. 84), or, if working, not necessarily operating with substantial power: "May not our present ideas of the universality and precision of adaptation be greatly exaggerated? The fit of organism to its environment is not after all so very close—a proposition unwelcome perhaps, but one which could be illustrated by very copious evidence. Natural selection is stern, but she has her tolerant moods" (1909, p. 100).

Moreover, many structures usually regarded as direct adaptations may originate as sequelae or side-consequences of other changes ("spandrels" in my terminology—Chapter 11, and Gould and Lewontin, 1979). Organic integration, indissoluble by selection, may represent a more important morphological phenomenon than selective scrutiny part by part: "I feel quite sure that we shall be rightly interpreting the facts of nature if we cease to expect to find purposefulness wherever we meet with definite structures or patterns. Such things are, as often as not, I suspect rather of the nature of toolmarks, mere incidents of manufacture, benefiting their possessor not more than the wire-marks in a sheet of paper, or the ribbing on the bottom of an oriental plate renders those objects more attractive in our eyes" (1909, pp. 100–101).

I have presented this exegesis of Bateson in such detail because he so explicitly presented the formalist viewpoint as a direct alternative to Darwinism. His own style emphasized the facet-flipping (or saltational) theme of Galton's polyhedron, but he understood the place of directional variation in the general argument, and he expressed support for the second theme of orthogenesis with a conventional formalist emphasis on predictability and internally generated order, writing for example (1924, in 1928, p. 407): "What we have learned of variation, especially of the incidents of parallel variations, has taught us that many varietal forms owe their origin to a process of unpacking a definite pre-existing complex, with the consequence that, given the series of varieties to which one species is liable, successful predictions may sometimes be made as to the terms which will be found in allied series . . . These symptoms of order and variation have prepared our minds, and there may well be a sense in which orthogenesis will be found to denote a valid principle."

Bateson therefore defended the purest example I know, among major 20th century thinkers, of a conscious and fully developed formalist philosophy harnessed to an explicitly anti-Darwinian theory. His formulation demonstrates that the dichotomy between structuralist and functionalist thought, the conceptual basis and primary theme of this chapter, cannot be regarded as an idiosyncratic or artificial device of rhetoric or textual organization, but rather denotes a widely perceived antithesis between two coherent worldviews about nature.

In *Problems of Genetics*, Bateson lays out the dichotomy most clearly, even using the terms "external" and "innate" to contrast the Darwinian functionalism that he rejects with his own favored structuralism. This remarkable passage encapsulates the tradition of argument by relative frequency in natural history. Bateson acknowledges that natural selection occurs, of course, but relegates Darwin's force to a periphery of unimportance as an arbiter among novelties generated internally. Bateson manages, in this single passage, to attribute *both* stability and variation to internal causes, and to brand selection as a secondary tinkerer upon patterns established thereby:

> We may ascribe the difference either to causes external to the organisms, primarily, that is to say, to a difference in the exigencies of adaptation under natural selection; or on the other hand, we may conceive the difference as due to innate distinctions in the chemical and physiological constitutions of the fixed and the variable respectively. There is truth undoubtedly in both conceptions. If the mole were physiologically incapable of producing an albino, that variety would not have come into being, and if the albino were totally incapable of getting its living it would not be able to hold its own . . . I incline to the view that the variability of polymorphic forms should be regarded rather as a thing tolerated than as an element contributing directly to their chances of life; and on the other hand, that the fixity of the monomorphic forms should be looked upon not so much as a proof that natural selection controls them with a greater stringency, but rather as evidence of a natural and intrinsic stability of chemical constitution (1913, p. 28).

Bateson presents an even more striking contrast in later passages of the same book, when he develops an image for a great, if undoable, thought experiment—the perfectly controlled account of evolution under uniform conditions, unbuffeted by any of the Darwinian externalities that make real results so untidy and unpredictable: "No one disputes that the adaptation of organisms to their surroundings is one of the great problems of nature, but it is not the primary problem of descent. Moreover, until the normal and undisturbed course of descent under uniform conditions is ascertained with some exactness, it is useless to attempt a survey of the consequences of external interference" (1913, p. 187).

I am somehow stunned by this structuralist audacity in branding the functionalist panoply as mere "external interference"—and of imagining a formalist internal order so set and predictable that pathways of evolution might become as regular and predictable as planetary orbits, if only we could remove all these pesky environmental influences. The impetus and *sine qua non* of change for Darwin becomes, for Bateson, a mere disturbance that sullies an otherwise lovely experiment.

## HUGO DE VRIES: A MOST RELUCTANT NON-DARWINIAN

### Dousing the great party of 1909

It must have been a grand show. Wallace and Hooker still lived, and happily attended to present their memories and current views. Darwin's son Francis helped with arrangements; while Sir George, his most academically accomplished offspring (and Plumian Professor of Astronomy at Cambridge), contributed an article on "The Genesis of Double Stars." Charles Darwin (just plain Charles, for Victoria never did grant him a knighthood) had the good sense to publish his greatest work at age 50—and the centenary of his birth therefore coincided with the 50th anniversary of the *Origin*. A grand occasion for a double celebration.

Cambridge University Press, at Darwin's *alma mater*, published the proceedings of his centennial party without delay in the right year of 1909, under the editorship of botanist A. C. Seward. The choice of participants had been ecumenical, ranging in profession from the great anthropologist J. G. Frazer (of *The Golden Bough*) to the equally celebrated historian J. B. Bury, and in attitude from such Darwinian stalwarts as Hooker and Weismann to such active opponents as William Bateson. All participants had their say and delivered both their praises and their criticisms. In this medley of maximal diversity, however, only one statement seemed so egregious to the editor that he felt compelled to make a public statement.

In his short preface, editor Seward acknowledged the pluralism of his volume in expressing "the divergence of views among biologists in regard to the origin of species" (1909, p. vii). Then, in a single sour note, he cried "foul" about one passage: "In regard to the interpretation of a passage in the *Origin of Species* quoted on page 71, it seemed advisable to add an editorial footnote; but, with this exception, I have not felt it necessary to record any opinion on views stated in these essays" (Seward, 1909, p. v).

Turning to page 71, we find ourselves in the midst of an article on "Variation" by the Dutch botanist Hugo de Vries (following Weismann and preceding Bateson in a fascinating bridge between opposites). De Vries had won widespread fame for his "Mutation Theory" on the origin of species (2-volume German edition in 1901 and 1903; English translation in 1909). Most biologists viewed this saltational proposal (correctly so, I shall argue) as anti-Darwinian in mechanism. Yet de Vries persisted in trying to cover himself with the mantle of Darwin's presumed (though posthumous) approval. In the offending passage, de Vries twisted both logic and literary interpretation to argue that Darwin had really meant to identify saltational variation as the source of evolutionary change—whereas plain sense and everyone else's reading indicated that Darwin had favored insensible variation and rejected sports. De Vries wrote:

> Returning to the variations which afford the material for . . . natural selection, we may distinguish two main kinds. . . . Certain variations constantly occur, especially such as are connected with size, weight, color, etc. They are usually too small for natural selection to act upon, having hardly any influence in the struggle for life: others are more rare, occurring only from time to time, perhaps once or twice in a century, perhaps even only once in a thousand years. Moreover, these are of another type, not simply affecting size, number or weight, but bringing about something new, which may be useful or not. . . . In his criticism of miscella-

neous objections brought forward against the theory of natural selection after the publication of the first edition of the *Origin of Species,* Darwin stated his view on this point very clearly:—"The doctrine of natural selection or the survival of the fittest, which implies when variations or individual differences of a beneficial nature happen to arise, these will be preserved." In this sense the words "happen to arise" appear to me of prominent significance. . . . A distinction is indicated between ordinary fluctuations which are always present, and such variations as "happen to arise" from time to time. The latter afford the material for natural selection to act upon on the broad lines of organic development, but the first do not. Fortuitous variations are the species producing kind, which the theory requires; continuous fluctuations constitute, in this respect, a useless type. . . . Darwin's variations, which from time to time happen to arise, are mutations, the opposite type being commonly designed fluctuations (de Vries, 1909b, pp. 70–72).

Seward responded in his unique footnote, and with annoyance barely concealed:

I think it right to point out that the interpretation of this passage from the *Origin* by Prof. De Vries is not accepted as correct either by Mr. Francis Darwin or by myself. We do not believe that Darwin intended to draw any distinction between two types of variation; the words 'when variations or individual differences of a beneficial nature happen to arise' are not in our opinion meant to imply a distinction between ordinary fluctuations and variations which 'happen to arise.' . . . The statement in this passage that 'Darwin was well aware that ordinary variability has nothing to do with evolution, but that other kinds of variation were necessary' is contradicted by many passages in the *Origin* (Seward, 1909, p. 71).

Why did de Vries so covet a linkage with Darwin that he would torture and distort his hero's words to forge the supposed bond? And why did Seward single out this passage among others more overtly hostile to the source of this centennial celebration? To resolve this small puzzle, we must explore the wider context of de Vries' background and purposes. In particular, we must rescue de Vries from his conventional "sound bite" status as "Mendel's rediscoverer," and recognize this near accident in his career as distinctly secondary to a much deeper, older, and direct inspiration from Darwin. We shall see that the profundity of de Vries' intellectual break with Darwin, combined with his psychological inability to sever overt homage, set the deeper source of Seward's legitimate annoyance about a single passage. An understanding of de Vries' reluctant apostasy provides our best biographically based insight into the nature of Darwinian logic—and of the persistent power and attraction of formalist alternatives (stressing the facet-flipping mode of Galton's polyhedron in this case).

### *The (not so contradictory) sources of the mutation theory*

Hugo de Vries became the world's most celebrated evolutionist during the early 20th century. His *Mutationstheorie* received wide approbation as the most important proposal about evolutionary mechanisms since the *Origin of Species* and the theory of natural selection. He made three triumphant visits to the United States (in 1904, 1906, and 1912) and published two books in English (de Vries, 1905, 1907a) based on summer lectures at the University of California, Berkeley. His views therefore became especially well known to American and other English-speaking audiences. All professionals continue to recognize his name today, but his ideas have suffered a nearly total eclipse for two major reasons, one legitimate and one unfair.

For the legitimate source, de Vries chose an unfortunate research strategy— a botanical equivalent of putting all his eggs in one basket. His theory—based largely on results from a single species, the evening primrose *Oenothera Lamarckiana*—requires that his chosen exemplar represent a biological norm in order to grant the required generality to his proposed mechanism. But the "species forming" saltations of *Oenothera* were soon revealed as oddities of an unusual chromosomal system, for *Oenothera Lamarckiana* is a permanent heterozygote, a hybrid with chromosomes of each component linked in rings, and thus segregating together in meiosis. (Only half the seeds are viable, because both homozygotes are lethal.)

For the unfair reason, de Vries has been so identified with an almost accidental moment of enduring fame that we have lost the main thrust and rationale of his life's work. Such moments often inspire catechistic one-liners that persist as the instant legacy of great thinkers. De Vries has suffered even more than most scholars caught in such a predicament, for he became the subject of two knee-jerk catechisms:

1. Mendel's forgotten work of 1865 was independently rediscovered in 1900 by Hugo de Vries, Carl Correns, and Erich Tschermak-Seysenegg.

2. Ironically, the first application of Mendelism to evolutionary theory did not help to affirm Darwinism, but to assert yet another alternative mechanism of change—the saltatory origin of new species by macromutation. The Modern Synthesis, the true fusion of Darwin and Mendel, began two decades later when scientists finally realized that small-scale (Darwinian) variation could also claim a particulate basis, and that macromutations played no important role in evolution. We may praise de Vries as a rediscoverer of Mendel, but his own interpretation of particulate inheritance delayed a proper resolution.

Neither of these conventional accounts can be dismissed as false, but neither properly expresses the reasons for de Vries' interests and discoveries. His link with Mendel represents a relatively minor encounter *en passant* in a career dedicated to other concerns. De Vries did discover the Mendelian segregation laws during the 1890's, and he did demonstrate their occurrence in some 20 species. As he prepared to publish this work in early 1900, his colleague, Professor Beyerinck at Delft, sent him an old paper with the following note: "I know that you are studying hybrids, so perhaps the enclosed reprint of the year 1865 by a certain Mendel, which I happen to possess, is still of some interest to you" (quoted in Stomps, 1954—Stomps succeeded de Vries as professor of botany in Amsterdam). De Vries therefore reported Mendel's forgotten priority as he began to publish his results in 1900.

The link to Mendel, while surely true, must be labeled as unfair when cited as an exclusive epitome of de Vries' career. Discovery of the segregation laws did excite him, but this work represented only a sidelight to his major interest. Most early studies based on the Mendelian rediscovery did oppose Darwinism at first, but the deep irony of de Vries's contribution lies in the fact that he had taken up the study of heredity as a direct consequence of his concern—indeed, in his own words, his "love"—for Darwin as a man and a scholar. Darwin's theory of pangenesis (Darwin, 1868) served as de Vries' inspiration—and de Vries' first major book (his best in the judgment of many distinguished biologists, both then and now) presented a brilliant reformulation of Darwin's insight (*Intracellular Pangenesis,* de Vries, 1889).

De Vries turned to the study of heredity in order to probe the mechanisms of evolution. But he never considered the Mendelian segregation laws as particularly relevant to this goal—for these principles only regulated the distribution of characters by hybridization among differing phenotypes, whereas evolution required the origin of new variation. De Vries did not continue his work on Mendel's principles, and his two great books on evolution (de Vries, 1905 and 1909a) cite Mendel only rarely, and only in contexts peripheral to his main arguments about saltation and evolutionary novelty. De Vries's biographer van der Pas rightly comments (1970, p. 99): "After the rediscovery of Mendel's laws, many investigators took up the subject. De Vries was not among them, however. He believed that hybridization only causes redistribution of existing characters and for that reason cannot explain the appearance of new species. Therefore, he concentrated on the phenomenon of mutation, which he believed explained the origin of new species and therefore gave necessary support to the theory of evolution."

If Mendel only represented a sidelight in de Vries's career, two main sources stand out as inspirations for his interests and strategies—his teacher Julius Sachs and his mentor and guru Charles Darwin. De Vries' long life (1848–1935) spanned the years from Darwin to Dobzhansky. Unhappy with the quality of his initial botanical education in the Netherlands, de Vries decided, in the early 1870's, to continue his studies in Germany. Beginning in 1871, while teaching in an Amsterdam secondary school, de Vries began spending his long summer vacations in the laboratory of the leading plant physiologist, Julius Sachs. De Vries wrote a series of monographs and performed elegant experiments on osmosis in plant cells, the basis for his later work on the role of cellular turgor in the motions of growing plants. Sachs considered de Vries as his best student and helped to secure for him the first instructorship in plant physiology in the Netherlands, when the Amsterdam Athenaeum became a full university in 1877. De Vries later served as professor of botany and taught at Amsterdam until his retirement in 1918, though he remained professionally active until his death, working from an experimental garden and laboratory that he built in the remote village of Lunteren.

Through the 1870's, de Vries worked exclusively on problems in mechanical and chemical physiology. But, in the early 1880's, inspired directly by Darwin, he began to shift his interests to evolution and heredity. From 1885 to 1887, he published a series of 19 articles on "improving races of our cultivated plants" for a Dutch agricultural journal. He first found mutations of *Oenothera* in 1886, and worked steadily on his evolutionary views until his major work, *Die Mutationstheorie* appeared in two volumes in 1901 and 1903 (the Mendelian rediscovery only occurred as the book neared completion and could not have inspired much of de Vries' conceptual apparatus). De Vries credited his *Intracellular Pangenesis* of 1889 as the source of his theoretical views on evolution. By 1890, he had abandoned work in physiology, and he then spent the rest of his career as a student of evolution and heredity.

From Sachs and his colleagues, de Vries absorbed the leading philosophical tenets of late 19th century German science, then the envy and model of the Western world—experimentalism and the mechanical world view. Throughout his later career in evolutionary biology, de Vries insisted that his success derived from his attempt to substitute an active, experimental and quantitative methodology for the older comparative and descriptive approaches of natural history.

In the frontispiece of his first American book (1905), de Vries shunned humility and ranked himself, by virtue of his experimentalism, at the pinnacle of progress in the history of evolutionary studies. (*The Mutation Theory* did not appear in English translation until 1909, and these published Berkeley lectures therefore represent the first extensive account of de Vries' views in English. *Intracellular Pangenesis* first appeared in English translation in 1910.) De Vries wrote:

> The origin of species is a natural phenomenon—Lamarck.
> The origin of species is an object of inquiry—Darwin.
> The origin of species is an object of experimental investigation—de Vries (1905, frontispiece).

De Vries' rhetorical expression of this theme in his major work (1909a translation of 1901 German edition) follows an interesting course—self-serving to be sure, but revealing. Failure to progress in evolutionary studies, he argues, may be attributed to an antiquated methodology: "We have a doctrine of descent resting on a morphological foundation. The time has come to erect one on an experimental basis" (1909a, volume 1, p. 207). (De Vries maintained a generous view of the experimental domain—for he usually applied the term to the rigorous recording of well-tracked pedigrees in garden plots, rather than to more classical manipulation in sterilized buildings under controlled conditions. He wrote (1905, p. 463): "The exact methods of the laboratory must be used, and in this case the garden is the laboratory.")

De Vries then extended his argument in two directions from this central methodological premise. First, to gain broadest generality in aligning evolutionary studies with physical sciences of higher status, de Vries opens *The Mutation Theory* with a claim that theories based on discrete, atomized particles suggest better experiments than hypotheses about continua. Moreover, such theories can also ally biology with more prestigious fields like chemistry: "By the mutation theory I mean the proposition that the attributes of organisms consist of distinct, separate and independent units. These units can be associated in groups and we find, in allied species, the same units and groups of units. Transitions, such as we so frequently meet with in the external form, both of animals and plants, are as completely absent between these units as they are between the molecules of the chemist" (1909a, volume 1, p. 3). He then expressed the same argument more strongly in a popular article (1907b, p. 17): "This principle of mutations is conducive to the assumption of distinct units in the characters of plants and animals. Even as chemistry has reached its present high development chiefly through the assumption of atoms and molecules as definite units, the qualities of which would be measurable and could be expressed in figures, in the same way systematic botany and the allied comparative studies are in need of a basis for measurement and calculations."

Second, and in an odd conflation of proper methodology and empirical truth value, de Vries argues in his *Preface* that Darwinian gradualism should be rejected (or at least strongly disfavored *a priori*) because insensible change over millennia cannot easily become the subject of experiment!

> The origin of species has so far been the object of comparative studies only. It is generally believed that this highly important phenomenon does not lend itself to direct observation, and, much less, to experimental investigation. This belief has its root in the prevalent form of the conception of species and in the opinion that the species of animals and plants have originated by imperceptible gradations. These changes are indeed believed to be so slow that the life of a man is not long enough to enable him to witness the origin of a new form. The object of the present book is to show that species arise by saltations and that the individual saltations are occurrences which can be observed like any other physiological process . . . In this way we may hope to realize the possibility of elucidating, by experiment, the laws to which the origin of new species conform (1909a, volume 1, p. vii).

With such negativity towards the methodology and worldview of natural selection and gradualism, how could Charles Darwin serve de Vries as chief intellectual guru, even surpassing the influence of Sachs and experimentalism? We cannot grasp de Vries' convictions and contradictions until we understand the powerful extent and threefold nature of Darwin's largely psychological hold upon him.

First of all, we often forget the extent of Darwin's work on plant physiology—largely published during the 1870's as de Vries began his career, and primarily in the same areas, particularly the proximate causes of movement, that most interested de Vries. Darwin wrote books on *Insectivorous Plants* (1875a), *The Movement and Habits of Climbing Plants* (1875b), and *The Power of Movement in Plants* (1880a). In these technical studies, the two men could not have stood closer as intellectual colleagues—even though their enduring fame would arise elsewhere, from their disparate studies of evolution.

Second, de Vries directly courted and won Darwin's admiration and friendship. The two men exchanged extensive correspondence (reprinted in Van der Pas, 1970). Darwin extended much effort to help de Vries. He sent complimentary copies of his books to de Vries, and he wrote to Asa Gray for seeds, so that de Vries could pursue some experiments on movement in tendrils. De Vries, for his part, repeated and extended many of Darwin's experiments on the physiological basis of movement and insectivory in several species. For example, de Vries wrote to Darwin on December 8, 1880:

> I am very much obliged to you for your great kindness of sending me your work on the Power of Movement in Plants . . . I was especially interested by your experiments on the movements and the curious sensitiveness of the roots and plumules of young seedling plants, which I hope to repeat as soon as I shall have an occasion . . . I always remember the great pleasure I had in repeating the experiments, described in your work on insectivorous plants. . . . In your work, you often speak of my papers, . . . and I am much indebted to you for your kind judgment of them, which will be a stimulus to me in endeavoring to contribute my part to the advancement of science.

In the summer of 1878, just before promotion to his Amsterdam professorship, de Vries visited England and fulfilled his fondest hope of meeting Darwin. He first called on Hooker and Thistleton-Dyer in London, but found them cool, however kind and correct. By contrast, de Vries greatly enjoyed a memorable, if short, visit with Darwin at Dorking, the home of Darwin's brother-in-law, Sir Thomas H. Farrer. De Vries described this visit in a charming letter to his fiancée on August 14, 1878:

> Today I have visited Darwin; I am happy that it happened and I must say that Darwin was so very cordial and friendly . . . The conversation was quite easy; they all spoke very slowly and clearly and they gave me the time to speak up; thus I did better in speaking English than I expected . . . In the garden there were hothouses with peaches and grapes. Darwin told me a long story about the peaches and immediately offered me one of them; it was delicious.
>
> During our scientific conversation, there was the same laughing mania as you have seen with Sachs; Sachs laughs all the time, Darwin somewhat less but as merrily. He was very interested in what I have done lately . . .
>
> He puts a footstool on a chair before he sits down on it, for he gets headaches if he sits low—the poor soul! Mr. Farrer told me that today

Darwin felt exceptionally well and happy and that I was lucky. Mrs. Darwin takes good care of him and will never allow him to become too tired; she simply sends him to bed!

He has deepset eyes and in addition very protruding eyebrows, much more than one would say from his portrait. He is tall and thin and has thin hands, he walks slowly and uses a cane and has to stop from time to time. . . . His speech is very lively, merry and cordial, not too quick and very clear.

It is remarkable how soon one feels at home with people who are friendly and cordial. What a difference with Hooker and Dyer; they were cold and I did not care about them. But I enjoyed my visit with Darwin and I feel much more happy these last days. It is such a pleasure to find that somebody is really interested in you and that he cares about what you have discovered.

Third, and most importantly, Darwin also directly inspired de Vries' shift from physiological to evolutionary and genetic studies. With fond memories in old age, de Vries told an interviewer in 1925 (quoted in Van der Pas, 1970, p. 192): "I was led to the study of heredity by my love for Darwin."

The source of de Vries' inspiration did not lie in the *Origin of Species* or the *Descent of Man*, but rather in Darwin's speculation on heredity, the "provisional hypothesis of pangenesis" as Darwin characterized his own proposal, published as the last chapter of his two-volume 1868 treatise on *Variation in Plants and Animals Under Domestication*. In his last letter to Darwin (who died six months later), de Vries wrote, up to date as ever in commenting on Darwin's last book *The Formation of Vegetable Mold Through the Action of Worms*, but also mentioning a stronger interest in Darwin's older views on pangenesis: "After reading the first chapter of your book, I have been attending to the habits of worms, and had the good fortune of repeating some of your interesting observations . . . For some time I have been studying the causes of the variations of animals and plants, as described in your treatise . . . I have always been especially interested in your hypothesis of Pangenesis and have collected a series of facts in favor of it" (de Vries to Darwin, October 15, 1881).

Darwin's hypothesis of pangenesis served as a speculation that could validate Lamarckian inheritance, a mode of transmission that Darwin deemphasized but did not contest. According to pangenesis, the basis of hereditary characters resides within tiny cellular particles called gemmules. All cells produce gemmules during growth and later life. Gemmules then migrate from somatic to germ cells, where they collect to pass inherited characters to the next generation. The germ cells therefore store "actual physical representatives of all the cells which have existed during the whole life of the parent body" (Kellogg, 1907, p. 218). Since gemmules become modified in somatic cells by conditions of life and the actions of organisms, acquired characters can be inherited.

De Vries suggested a fundamental (and correct) modification that turned

Darwin's theory into something quite different. He abandoned Darwin's key notion of the migration of gemmules across cell boundaries—thus removing the rationale for Lamarckian inheritance. In de Vries' revised concept, the hypothesized hereditary particles behaved so differently that they required a different name; thus, still honoring Darwin, de Vries rechristened the gemmules as "pangenes."

In de Vries' concept of *Intracellular Pangenesis* (the title of his 1889 book), the nucleus of each cell contains all particles (pangenes) needed to construct an organism. But only some pangenes are expressed in each cell, thus explaining the differential morphogenesis of parts. Expressed pangenes migrate out of the nucleus into the cytoplasm, where they orchestrate the appropriate embryology. In no case can pangenes move between cells. De Vries wrote: "The hypothesis that all living protoplasm is built up of pangenes, I call intracellular pangenesis. In the nucleus every kind of pangene of the given individual is represented; the remaining protoplasm in every cell contains chiefly only those that ought to become active in it" (1889, in 1910 translation, p. 215).

This remarkably prescient theory comes as close to the secret of heredity as anyone had managed in the speculative tradition before the elucidation of genes and chromosomes. Whiggish historians nearly always regard *Intracellular Pangenesis* as de Vries' greatest book. In abstract concept, his nuclear pangenes differ little from the particles of heredity that would soon be recognized and named as genes, especially since de Vries viewed his pangenes as a minimal set of basic instructions, not naively as a collection of items for specifying each overt phenotypic part. His notion of active and latent pangenes recalls dominant and recessive alleles—and one might justly argue that de Vries had therefore been "preadapted" to appreciate Mendel. Fortune, as Pasteur famously said, favors the prepared mind. Moreover, the notion that all instructions reside in the nucleus (with passage to the cytoplasm, at appropriate times and places, for transmission of local messages) bears remarkable isomorphism with our modern mechanism of DNA, RNA, and the differentiation of cells.

Two further aspects of *Intracellular Pangenesis* play important roles in this story. First, de Vries' theory became the source of our modern term "gene"—for Johannsen explicitly derived the shortened name directly from de Vries' "pangene." Moreover, since de Vries' "pangene" honored Darwin's name for his speculative particle of heredity, Darwin himself becomes the ultimate source (via de Vries) for this basic biological term. Few evolutionary biologists recognize this curious terminological odyssey, making Darwin himself the ultimate, if indirect, source of our modern term "gene."

Second, we note in de Vries' treatment of Darwin a microcosm of the strange and almost painfully ambivalent fealty that tied him emotionally (and verbally) to Darwin even while he devised a contradictory theory—the source of Seward's anger, as described in my opening remarks. De Vries' theory, despite its personal source in Darwin's pangenesis, became a fundamentally different intellectual entity. In Darwin's pangenesis, gemmules move from somatic to germ cells and provide a mechanism for Lamarckian inheritance. In de Vries' intracellular pangenesis, pangenes move from the nucleus to the cytoplasm of the same cell and specify a theory of cellular differentiation.

Yet de Vries insisted on downplaying this difference as a minor variation. For rhetorical purposes, he asserted that his denial of intercellular movement for gemmules constituted only a minor reform in Darwin's ideas. His new theory could therefore remain entirely in Darwin's spirit. Throughout his life, de Vries could not break verbal fealty with the primary hero and inspiration of his youth. In a late work of 1922, de Vries wrote:

> Freed from the assumption of a transportation of gemmules through the organism, the conception of Pangenesis is the clear basis of the present manifold theories of heredity. An organic being is a microcosm, says Darwin, a little universe, formed of a host of self-propagating organisms, inconceivably minute, and numerous as the stars of heaven. In honor of Darwin, I have proposed to call these minute organisms Pangenes, and this name has now been generally accepted under the shortened form of genes. They are assumed to be the material bearers of the unit characters of species and varieties (1922, p. 222).

### The mutation theory: origin and central tenets

Evolutionists usually assume that, since de Vries ranks within the trio of Mendel's resurrectionists, his "mutation theory"—with its genetic title and deserved status as the greatest challenge to Darwinism from the early 20th century—must be traceable to a Mendelian inspiration. But de Vries always insisted that his theory, and almost everything else he valued, could boast a Darwinian source. In particular, he asserted later in his career that the root of the Mutation Theory lay in an insight about two distinctly different kinds of variation that he had obtained from Darwin's theory of pangenesis, and then developed within his own *Intracellular Pangenesis* of 1889. I am not confident that this link can be defended, for considerable (and rather tortured) exegesis must be applied to so interpret the actual text of de Vries' 1889 book, whatever his later memories. But de Vries' debt and psychological fealty to Darwin can only be called pervasive, while the timing of de Vries' interpretation can also be defended (for the Mendelian discovery postdated the genesis of the mutation theory).

In the English version (1910) of *Intracellular Pangenesis*, de Vries wrote a note to his translator, pointing to a passage that he identified as the source of the Mutation Theory: "An altered numerical relation of the pangenes already present, and the formation of new kinds of pangenes must form the two main factors of variability" (1910, p. 74). De Vries interpreted this passage as presaging the key claim of his later mutation theory—that new species arise suddenly by a distinct and special kind of saltational variation (called mutation), while ordinary, imperceptible, omnipresent, Darwinian variability cannot forge evolutionary novelties. Late in his career, de Vries wrote (1922,

p. 222): "Species and varieties have originated by mutation, but are, at present, not known to have originated in any other way. Originally this conception has been derived from the hypothesis of unit characters as deduced from Darwin's Pangenesis, which led to the expectation of two different kinds of variability, one slow and one sudden."

Whatever the intellectual roots, de Vries eventually centered his illustration and defense of the mutation theory upon a single source, the evening primrose, *Oenothera Lamarckiana*—so we can identify a particular empirical basis for the defense of his views. (This strategy, as mentioned previously, ultimately backfired when researchers explained the unusual mutability in O. *Lamarckiana* as a peculiarity of the plant's hybrid nature and chromosomal system, and not as the generality that de Vries required.) In 1886, de Vries found odd and distinct mutational variations growing among a wild field of evening primroses at Hilversum, near Amsterdam. He later described his mixture of good fortune and conscious preparation (1905, p. 27):

> Cultivated plants of course, had only a small chance to exhibit new qualities, as they have been so strictly controlled during so many years. Moreover, their purity of origin is in many cases doubtful. . . . For this reason I have limited myself to the trial of wild plants of Holland, and have had the good fortune to find among them at least one species in a state of mutability. It was not really a native plant, but one probably introduced from America or at least belonging to an American genus. It was the great evening primrose or the primrose of Lamarck. A strain of this beautiful species is growing on an abandoned field in the vicinity of Hilversum, at a short distance from Amsterdam. Here it has escaped from a park and multiplied. In doing so it has produced, and is still producing quite a number of new types . . . This interesting plant has afforded me the means of observing directly how new species originate, and of studying the laws of these changes.

De Vries' method for finding and propagating *Oenothera* mutants included a mixture of experimental care and hard work in the Burbankian mode. To find new mutants, he sowed prodigious numbers of seeds. For example, in his 1888 sowing, he tested 15,000 seedlings and found 10 mutations. To propagate and breed his new forms, and to test for their purity in inheritance, de Vries stringently followed the obvious rules for tracing pedigrees: fertilize each plant with known pollen, prevent insect pollination, save and sow all seeds separately.

De Vries, in crisp summary, presented the essence of his theory as exhibited by the *Oenothera* mutants—sudden, fully constituted, nonadaptive, observable, experimentally ascertainable origin of new species: "They came into existence at once, fully equipped, without preparation or intermediate steps. No series of generations, no selection, no struggle for existence was needed. It was a sudden leap into another type, a sport in the best acceptation of the word. It fulfilled my hopes, and at once gave proof of the possibility of the direct observation of the origin of species, and of the experimental control thereof" (1905, p. 550).

De Vries reported different numbers and names for new *Oenothera* species in various publications, but the seven cited in his 1904 Berkeley lectures provide a good feel for his categories and criteria (see Fig. 5-9). Above all, the mutants appeared suddenly and bred true (other mutants reverted, hybridized, or segregated in only a proportion of progeny). De Vries placed his seven new species into three categories. First, he designated two "true elementary species"—the red-veined *O. rubrinervis* and the giant *O. gigas* (a tetraploid by later discovery). De Vries described these new forms as "robust and stark species which seem to be equal in vigor to the parent plant, while diverging from it in striking characters" (1905, p. 533). In a second category of "retrograde varieties," de Vries identified the smooth-leaved *O. laevifolia*, the short-styled *O. brevistylis*, and the dwarf *O. nanella*, "a most attractive little plant" (1905, p. 531). These new forms also arose with abrupt distinctness, and bred as true as the two "elementary species," but these three mutants differed from the parent *O. Lamarckiana* by loss or diminution of an ancestral character, not by addition of novelty—hence their separate, and less admirable, category. (I am intrigued by our cultural bias that designates increased bulk as a novelty *(O. gigas)*, while ranking shortness as mere subtraction *(O. nanella)*.) De Vries then placed two forms into a third (and inconsequential) class of "weak species" that "have no manifest chance of self-maintenance in the wild state" (1905, p. 537). The whitish *O. albida* "grows too slowly and is overgrown," while the oblong-leaved *O. oblongata* "bear small fruits and few seeds."

All three central tenets of the mutation theory work in direct opposition to Darwinian gradualism and sequential shaping by natural selection in the origin of species—hence the widespread and proper interpretation of de Vries' theory as a powerful refutation of Darwinism (Kellogg, 1907, who lists the views of de Vries as an alternative, not an auxiliary, to Darwinism).

1. Above all, the mutation theory embodies the most unabashedly saltational notion ever seriously regarded as an evolutionary mechanism. New species arise in a single step by a sudden, discontinuous, fully formed and true breeding leap in phenotype. De Vries could not have been clearer in his introductory comment to the *Mutation Theory* (1909a, p. 3): "Species have arisen from one another by a discontinuous, as opposed to a continuous process. Each new unit, forming a fresh step in this process, sharply and completely separates the new form as an independent species from that from which it sprang. The new species appears all at once; it originates from the parent species without any visible preparation and without any obvious series of transitional forms."

De Vries explicitly contrasts his new view with Darwinian gradualism: "A current belief assumes that species are slowly changed into new types. In contradiction to this conception the theory of mutation assumes that new species and varieties are produced from existing forms by sudden leaps" (1905, p. vii). Such sudden inception precludes any role for natural selection in the origin of species. Taking the traditional anti-Darwinian line, and employing a striking metaphor, de Vries claims that selection cannot construct anything new, but can only operate as a sieve to preserve favorable forms produced by some other process: "Natural selection is a sieve. It creates nothing, as is so often assumed; it only sifts. It retains only what variability puts into the sieve. Whence the material comes that is put into it, should be kept separate from the theory of its selection. How the struggle for existence sifts is one question; how that which is sifted arose is another" (1909a, volume 2, p. 609).

2. If new species originate in single leaps, then their origin must be non-adaptive, however much their future survival may require a fortuitous match with local environments: "It explains in a very simple way the existence of the vast number of specific characters which are quite useless or as to the use of which we have no idea at all . . . According to the commonly accepted theory of selection only characters advantageous to their possessors should arise; ac-

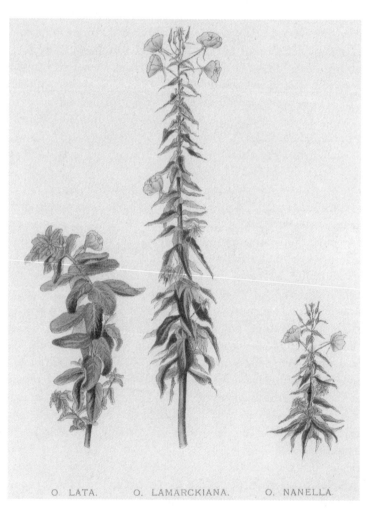

5-9A. *Oenothera Lamarckiana* contrasted with two of DeVries' mutations. From Volume 2 of DeVries' *Mutation Theory.*

cording to the theory of mutation on the other hand useless or even disadvantageous ones may also appear" (1909a, volume 1, pp. 209–210)

3. Slow Darwinian changes cannot be observed on human and experimental time scales, but the mutation theory brings evolution into the domain of observational science; "the origin of species may be seen as easily as any other phenomenon" (1905, p. 26). Emphasizing a direct contrast between the virtues of his operational theory and the fatal intractability of Darwinian gradualism (here attributed to Wallace, for de Vries could not bear to saddle his hero with such a negative assessment), de Vries brands gradualism as obstructionist, and compliments his own view as liberating: "I shall try to prove that sudden mutation is the normal way in which nature produces new species and new varieties. These mutations are more readily accessible to observation and experiment than the slow and gradual changes surmised by Wallace and his followers, which are entirely beyond our present and future experience.

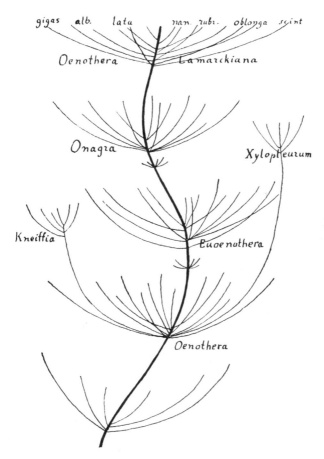

5-9B. At the apex of this phylogeny, DeVries shows the 7 mutant forms derived from *Oenothera Lamarckiana*. The rest of the diagram illustrates DeVries' general view of evolution, with most lineages stable nearly all the time, but entering short mutational episodes when several new species may arise virtually at the same time. From Volume 2 of DeVries' *Mutation Theory*.

The theory of mutations is a starting point for direct investigation while the general belief in slow changes has held back science from such investigations during half a century" (1905, p. 30).

As an experimental reductionist, committed to finding the mechanism for large-scale evolution of phenotypes in the smallest cellular parts, de Vries sought the causal basis for his mutation theory in the character of variation and its putative causes. As a foundation for all his theorizing, de Vries proposed a strict separation between two distinct types of variation: fluctuating and mutational. (This division, of course, establishes the same false dichotomy that prompted the famous "biometrician" vs. "Mendelian" debate— a struggle that de Vries' context did much to promote). De Vries stated that, in the early 1870's, he had read Quetelet's work on normal curves and Galtonian regression to the mean—and had determined thereby that the omnipresent, small scale or, in his favored term, "fluctuating" variation could not be parlayed into directional evolutionary change, as Darwin's theory required (Fig. 5-10). Evolution must therefore require a conceptually and causally distinct mode of sudden, larger-scale, true breeding and non-regressing variation—a necessary source eventually found in the "mutations" that yielded distinct new phenotypes in *Oenothera*.

De Vries acknowledged that selection of fluctuating variation could produce new agricultural races and stocks of domesticated animals. But this Darwinian alteration can only yield a minor change from the mean of a parental stock: "It is responsible only for the smallest lateral branches of the pedigree, but has nothing in common with the evolution of the main stems. It is of very subordinate importance" (1905, p. 801). These new races, if not constantly superintended, will rapidly revert towards parental characters by regression to the mean (1909a, volume 1, pp. 88–89). De Vries, who understood the logic of Darwinism so keenly (see pp. 446–451), knew that the most promis-

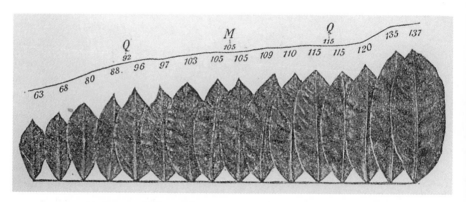

5-10. DeVries' illustration of continuous "fluctuating" variation, which he regarded as ineffective as a source for evolutionary change. From volume 1 of DeVries' *Mutation Theory*.

ing Darwinian escape from this paradox lay in a claim that fluctuating variation would be reconstituted symmetrically about a newly-established mean—
and that continual, directional selection to great cumulative effect could
therefore be achieved. De Vries, following the classic argument about a "rigid
sphere" of variation, as famously formulated by Fleeming Jenkin (1867), simply denied that such reconstitution could occur. Again, using Wallace as Darwin's surrogate, de Vries stated (1909a, volume 1, p. 42): "I admit that with
this assumption [of limitless and reconstitutable variation] it would be very
easy and simple to account for the phenomenon of adaptation . . . If
[Wallace's] assumption is once granted everything follows. But it is, as a matter of fact, fallacious."

De Vries summarized his views on the inefficacy of fluctuating variation
(which, he suspected, did not have a particulate Mendelian basis and arose by
influences that we would now call ecophenotypic). How could substantial
and permanent evolutionary change originate from a style of variation that
(1) always regressed toward the mean; (2) arose in strictly limited extent,
with preponderance near the mean, and only rarely at a useful phenotypic
difference (the normal curve); (3) enjoined a strictly linear set of effects, only
producing more or less of a feature, while "creative" evolution required the
development of true novelty: "Individual variability, when tested by sowing,
reverts to its original mean, the forms of its variants are connected together,
are coherent and not discontinuous. It is centripetal in as much as the variations are grouped most densely around a mean. Finally—and this is very important—it is linear; because the deviations occur in only two directions—
less or more."

By default therefore, but not at all as a negative argument, evolutionary
novelty must arise by a phenomenologically and causally distinct style of
variation that de Vries called "mutational"—i.e., sudden, fortuitous (and
therefore nonadaptive), true breeding and nonreverting saltations. De Vries
called these saltational variants "species," but we must understand (as he did)
that such units cannot be equated with traditional Linnaean taxa of the same
name. With his mutation theory, de Vries entered (and largely shaped, though
he did not originate) a major debate in systematics.

Obviously, a de Vriesian saltation does not, *in se,* make a new species in our
usual sense of the term—for the single mutant plant is only an individual with
a discontinuous phenotype, however true breeding in self-fertilization. In
what sense, then, could de Vries insist that he had discovered the mechanism
for the origin of new species?

In large part, de Vries based his claim upon an attempted redefinition.
He argued that the traditional Linnaean species encompasses an imprecise,
compound aggregation including varying numbers of phenotypically distinct,
true-breeding entities (and a fair amount of continuous variability as well,
based on the fluctuating style). The true-breeding subtypes represent nature's
genuine units and should be so designated. They arise by discontinuous saltation, without intermediates, and should be called "elementary species." As a

practical point, de Vries did not propose that all traditional taxonomy be restructured. He would allow the Linnaean names to persist as "species," with his smaller, "real" units termed "elementary species." (De Vries did not follow his own recommendation consistently, for he gave new species names, in traditional Linnaean form, to his mutational variants. For example, the large tetraploid became *Oenothera gigas* derived from *O. Lamarckiana*). In this argument, de Vries supported a movement, then current in systematics, to designate the traditional Linnaean units as linneons and the true-breeding subtypes as jordanons (to honor the botanist Alexis Jordan)—recognizing both as species of different sorts and by different criteria (with the jordanon as "more" biological and the linneon as tolerated by practical necessity). De Vries wrote:

> We may conclude that systematic species, as they are accepted nowadays, are as a rule compound groups. Sometimes they consist of two or three, or a few, elementary types, but in other cases they comprise 20, or 50, or even hundreds of constant and well-differentiated forms (1905, p. 38).
>
> The systematic species are the practical units of the systematists and florists, and all friends of wild nature should do their utmost to preserve them as Linnaeus has proposed them. These units, however, are not really existing entities; they have as little claim to be regarded as such as the genera and families have. The real units are the elementary species . . . Pedigree culture is the method required and any form which remains constant and distinct from its allies in the garden is to be considered as an elementary species (1905, p. 12).

De Vries' historical argument for changing emphasis from the linneon (ordinary species) to the jordanon (de Vriesian elementary species) provides an interesting insight into his worldview and rhetorical style. Before Linnaeus, he claims, genera stood as the "natural" units of common discourse: "The old vulgar names of plants, such as roses and clover, poplars and oaks, nearly all refer to genera" (1905, p. 33). Linnaeus, also searching for the natural unit, failed to extend his argument far enough. He began with genera and then moved "down" to species. He knew that he might proceed to still smaller units, but chose to go no further:

> Afterwards Linnaeus changed his opinion on this important point, and adopted species as the units of the system. He declared them to be the created forms, and by this decree at once reduced the genera to the rank of artificial groups. Linnaeus was well aware that this conception was wholly arbitrary, and that even the species are not real indivisible entities. But he simply forbade the study of lesser subdivisions. At this time, he was quite justified in doing so, because the first task of the systematic botanists was the clearing up of the chaos of forms and the bringing of them in connection with their real allies (1905, p. 34).

Just as the establishment of Linnaean species had made genera artificial, so too does the recognition of de Vriesian elementary species relegate the conventional Linnaean species to a congeries with no natural status. De Vries argues that this theoretical progress from larger to successively smaller units of natural "reality" illustrates the general advance of science as a reductionistic enterprise. A "stepping down" from the Linnaean species to the de Vriesian elementary species can claim both the sanction of history and the virtue of utility: "What is to guide us in the choice of the material? The answer may only be expected from a consideration of elementary species. For it is obvious that they only can be observed to originate, and that the systematic species, because they are only artificial groups of lower unities, can never become the subject of successful experimental inquiry" (1905, p. 517).

This redefinition of species as discontinuous saltations inevitably raised the issue of whether de Vries' new units ("elementary species," or "jordanons" of other terminologies) always originated from single monophyletic sources, or represented discrete phenotypes that could arise more than once—thus divorcing this supposedly "most real" taxon from the usual genealogical criterion of monophyly for a basic unit in a phylogenetic system. De Vries, following both the logic of his argument and his observation that elementary species of *Oenothera* arose again and again, accepted the implication (so strange to modern "population thinking" and genealogically based taxonomy) that the same species could, and usually did, arise many times. In fact, such a propensity for multiple origin established a major criterion for potential success. De Vries noted as central to his concept (1909a, p. 208) "the assumption that the new form or species does not arise merely once from the parent species but . . . a great many times and with some degree of regularity."

De Vries devised an interesting set of subtypes for his saltations—thus revealing another aspect of the philosophical complexity of his ideas (not always expressed with consistency, but often replete with interesting psychological and sociological influences). If "stepping down" from the linneon to the jordanon revealed a reductionist bias usually interpreted as "modernist" in his time, then de Vries's classification of mutations reveals an allegiance to notions of progress and regress that might be deemed archaic in its implied fascination for the *scala naturae*.

De Vries recognized some of his mutations as starkly different (in a qualitative sense) from the parental form. But others could be linked in a series, with the parent as prototype, either by loss of ancestral characters or by simple quantitative alteration. All categories included equally "good" species in the causal or genetic sense—that is, equally discontinuous entities, formed suddenly without intermediates, and true breeding under self-fertilization. But only the first category established genuine novelty in evolution; thus, only these truly different species contributed to the progress of life's history. All other categories comprised variations on parental forms (usually based on loss or diminution of characters), and could only constitute a series around the parental prototype. Therefore, in the oldest taxonomic ploy of evolutionary thought (dating to Lamarck's distinction of progressive increments from lateral branches—see Chapter 3), de Vries subdivided, by their presumed phylogenetic effect, these taxa of similar genetic status. He labelled mutations yielding phyletic novelties as "elementary species," while phenotypic departures still linked to parental morphologies became "varieties." De Vries then made a further subdivision among varieties, distinguishing taxa formed by loss of a character ("retrograde varieties") from those that may seem more advanced than the parent but really display nothing new (atavistic reappearance of characters present in closely related species, for example, or simple enhancement of a character already present). De Vries wrote (1905, pp. 246–247):

> There is a real difference between *elementary species* and *varieties*. The first are of equal rank, and together constitute the collective or systematic species. The latter are usually derived from real and still existing types. Elementary species are in a sense independent of each other, while varieties are of a derivative nature . . . We have assumed that the first came into existence by the production of something new, by the acquirement of a character hitherto unnoticed in the line of their ancestors. On the contrary, varieties, in most cases, evidently owe their origin to the loss of an already existing character, or in other less frequent cases, to the reassumption of a quality formerly lost. Some may originate in a negative way, others in a positive manner, but in both cases nothing really new is acquired.

In his most forthright statement about the differing phyletic roles of progressive and regressive mutations, de Vries then stated (1905, p. 15):

> Many instances could be given to prove that progression and retrogression are the two main principles of evolution at large. Hence the conclusion that our analysis must dissect the complicated phenomena of evolution so far as to show the separate functions of these two contrasting principles. Hundreds of steps were needed to evolve the family of the orchids, but the experimenter must take the single steps for the object of his inquiry. He finds that some are progressive and others retrogressive, and so his investigation falls under two heads, the origin of progressive characters, and the subsequent loss of the same. Progressive steps are the marks of elementary species, while retrograde varieties are distinguished by apparent losses.

The logic of de Vries' system may be sound, but he faced—as he well understood—a major empirical dilemma. He had found consistent mutations in only one lineage, the genus *Oenothera*, and with high frequency only in the species *O. Lamarckiana*. (He noted an isolated example or two in other lineages, particularly in a plant with the intriguing name of "peloric toad-flax," but found no consistently mutable form besides *Oenothera*. He also tested other species of *Oenothera*, but found most immutable or, in the case of *O. biennis* for example, much less subject to alteration than *O. Lamarckiana*. In an interesting later paper (de Vries, 1915), he tried to calculate the "coefficient of mutation in *Oenothera biennis*" vs. *O. Lamarckiana*, concluding that the latter species showed a 6 to 10 fold increase in mutability. De Vries attributed this augmentation to a transition of one or more pangenes from stable to labile positions—a pure speculation, but again consistent with his system and logic.)

Why only one? Did such rarity mean that de Vries had only discovered an oddball, with no general message for evolution? De Vries recognized that such an inference would destroy his system, and he therefore argued that all (or at least many) species maintain potential for entering a "mutable period," but that very few actually exist in such a state at any moment. (We know, after all, that most species are stable in both current and paleontological perspective. If all lineages were as mutable as *O. Lamarckiana*, we would never be able to designate Linnaean taxa, for nature would then present a constantly changing and unbreakable continuum, rather than a set of discrete and recognizable populations.) De Vries considered himself fortunate that he had located even one species in such a state—for if "mutable periods" constitute an almost incalculably tiny fraction of a species' lifespan, the probability of finding any given species in such a state at any moment becomes effectively zero. In trying to turn the tables on his adversaries, de Vries argued that the discovery of even one case presupposes a generality for extremely rare "mutable periods"—for if such mutability could be dismissed as simply freakish and unique, he could not have expected to encounter even a single example!

> The view that it might be an isolated case, lying outside of the usual procedure of nature, can hardly be sustained. On such a supposition it would be far too rare to be disclosed by the investigation of a small number of plants from a limited area. Its appearance within the limited field of inquiry of a single man would have been almost a miracle. . . . The mutable condition . . . must be a universal phenomenon, although affecting a small proportion of the inhabitants of any region at one time: perhaps not more than one in a hundred species, or perhaps not more than one in a thousand, or even fewer may be expected to exhibit it (1905, p. 687).

But why should a species enter a rare mutable period, and why should most species be stable nearly all the time? What triggers a transition into this exceedingly uncommon state of evolutionary promise? On this crucial point of his entire system, de Vries fell almost eerily silent, for he could offer nothing precise. He supposed (1909a, volume 1, pp. 206–207) that some external trigger of environmental change—isolation by colonization of new areas, for example—must initiate phyletic lability, but amounts and directions of mutation must be attributed to internal states of pangenes (1905, p. 691). He offered a few general words about pangenes becoming mutable, or moving to a position of high changeability, or arising *de novo* with such a propensity. But he still couldn't cite anything physical, or propose anything testable. De Vries' distress and unease about this crucial subject even inspired a rare burst of almost religious romanticism, an odd rhetorical strategy (and smokescreen) for such a severe rationalist: "The view of permanency represents life as being surrounded with unavoidable death, the principle of periodicity follows in the same way the idea of resurrection, granting the possibility of future progression for all living beings. At the same time it yields a more hopeful prospect for experimental inquiry" (1905, p. 693).

The complexities, interrelationships (in some cases amounting to near contradictions) and comprehensive character of ideas at the core of de Vries' mutation theory may best be illustrated by his attempt to epitomize his system as a set of seven laws (1905, pp. 558–571, 578). Note particularly the tug of war (both logical and psychological) between his understanding that the theory refutes Darwinian principles on one hand, and his desire, on the other hand, to retain fealty with Darwin as a personal hero.

1. "New elementary species appear suddenly, without intermediate steps" (p. 558). De Vries's first paragraph of description boldly expresses the contradiction between this statement and Darwinian principles. (But note how he declines to attach Darwin's name to the orthodoxy he opposes—speaking instead only of "current scientific belief," or "the ordinary conception"):

> This is a striking point, and the one that is in the most immediate contradiction to current scientific belief. The ordinary conception assumes very slow changes, in fact, so slow that centuries are supposed to be required to make the differences appreciable. If this were true, all chance of ever seeing a new species arise would be hopelessly small. Fortunately, the evening primroses exhibit contrary tendencies . . . The mutants, that constitute the first representatives of their race, exhibit all the attributes of the new type in full display at once. No series of generations, no selection, no struggle for existence are needed to reach this end (1905, p. 558).

2. "New forms spring laterally from the main stem" (p. 560). De Vries presents a cogent defense of cladogenetic vs. anagenetic modes for conceptualizing evolutionary change, including the full set of implications that continue to evoke debate today:

> The current conception concerning the origin of species assumes that species are slowly converted into others. The conversion is assumed to affect all the individuals in the same direction and in the same degree. The whole group changes its character, acquiring new attributes. . . . The birth of the new species necessarily seemed to involve the death of the old one . . . The general belief is not supported by the evidence of the evening primroses. There is neither a slow nor sudden change of all the individuals. On the contrary, the vast majority remain unchanged; thousands are seen exactly repeating the original prototype yearly, both in the native field and in my garden. There is no danger that Lamarckiana might die out from the act of mutating, nor that the mutating strain itself would be exposed to ultimate destruction from this cause (pp. 560–561).

3. "New elementary species attain their full constancy at once" (p. 562). Again, de Vries states his first words of explanation in a forthrightly and explicitly anti-Darwinian manner, in this case confuting gradualism and the adaptationist perspective. "Constancy is not the result of selection or of improvement. It is a quality of its own. It can neither be constrained by selection if it is absent from the beginning nor does it need any natural or artificial aid if it is present" (pp. 562–563).

4. "Some of the new strains are evidently elementary species, while others are to be considered as varieties" (p. 564). De Vries regarded his taxonomy of relative merit and evolutionary potency of mutations (discussed on pp. 431–434) as sufficiently important to rank as one of the seven cardinal statements.

5. "The same new species are produced in a large number of individuals" (p. 566). De Vries also recognized the importance of his distinctive, non-genealogical principle (see p. 433) that mutations forming new "elementary species" may arise several times (thus imparting a greater chance of success to the novel taxon).

6. "The relation between mutability and fluctuating variability" (p. 568)—so stated as a phrase rather than a declarative sentence. De Vries recognized this causal claim for a fundamental distinction between two modes of variation as the focus of his theory (see pp. 430–432). No other point received so much discussion in his texts. I do not know why he placed this fundamental statement of his reductionist program in 6th position among 7 statements.

7. "The mutations take place in nearly all directions" (p. 570—I shall present a more extensive discussion of this claim on pp. 446–451). De Vries emphasized this statement as his major tactic for maintaining fealty with Darwin at macroevolutionary scales, while destroying his mentor's theory for the origin of species. If the phenotypic ranges of new species form an isotropic distribution about the parental type, then the manifest directionality of evolution at geological scales must record the action of a higher selection process upon these species-level variations. Can a form of Darwinian argument therefore prevail among species (to produce trends), even while the sudden origin of new species precludes selectionism in Darwin's own favored realm?

From this conceptual foundation, de Vries reached further to promote his mutation theory as the basis for an overarching worldview. To illustrate the range of implications explicitly developed by de Vries, consider just two issues of widely differing import. On the first, and practical, question of benefits to agriculture, the mutation theory suggested that conventional selection (on fluctuating variation) could only yield limited and easily reversed improvements. But new mutations might secure large and permanent benefits. Yet, as a practical dilemma, new mutations are rare and cannot be induced by our efforts. What benefit can emerge from scientific horticulture if this discipline must wait patiently for good fortune, and can then only apply the journeyman's procedure of preservation and propagation: "the practice of the horticulturist in producing new varieties is limited to isolation, whenever chance affords them" (1905, p. 606). As a legitimate escape from this disabling consequence of his theory, de Vries proposed that future knowledge of the causes of mutation might place evolutionary alteration under our control, and give us power so far beyond the scope of selection that we might truly become the masters of nature: "We may search for mutable plants in nature, or we may hope to induce species to become mutable by artificial methods. The first promises to yield results most quickly, but the scope of the second is much greater and it may yield results of far more importance. Indeed, if it should once become possible to bring plants to mutate at our will and perhaps even in arbitrarily chosen directions, there is no limit to the power we may finally hope to gain over nature" (1905, p. 688).

On the second, and theoretical, issue of insights from evolutionary theory for human cultural and racial differences, de Vries stated, with principled consistency, that his views on the origin of species suggested no implications whatever: "Our knowledge of the origin of species in nature has no bearing on social questions" (1909a, volume 1, p. 156). De Vries regarded human racial distinctions as arising entirely from selection (or drift) upon fluctuating variability. *Homo sapiens* resides, with the vast majority of species, in a long-standing phase of stability; not, like the evening primrose, in a rare state of mutability: "Since the beginning of the diluvial period, man has not given rise to any new races or types. He is, in fact, immutable, albeit highly variable" (1909a, volume 1, p. 156). This fluctuating variation provides a source for all racial differences which, however "profound" in phenotypic appearance, must therefore remain as limited and changeable as any alteration fashioned in this weak Darwinian mode:

> Many mistakes may in the future be avoided if a clear distinction be drawn between mutability and variability in the ordinary sense. The variability exhibited by man is of the fluctuating kind: whereas species arise by mutation. The two phenomena are fundamentally different. The assumption that human variability bears any relation to the variation which has or is supposed to have caused the origin of species is to my mind absolutely unjustified. Man is a permanent type, like the vast majority of species of animals and plants . . . As we have seen, it is characteristic of these types to exhibit a certain amount of fluctuating variability. Man is no exception to this rule. Therefore all that we can apply to the treatment of social questions is our knowledge of ordinary variability. The facts of specific differentiation are interesting but not relevant (1909a, volume 1, pp. 154–155).

In sum, de Vries' Mutation Theory became the most important set of concepts in evolutionary biology during the early 20th century. The theory attained this central status by (1) its radically different and non-Darwinian view of the origin of species; (2) the breadth of its concerns, ranging from variation at the smallest scale to modes of geological pattern at the largest; (3) the range of its implications, as illustrated above, thus expanding the doctrine from a scientific theory to a comprehensive worldview.

## Darwinism and the mutation theory

CONFUSING RHETORIC, AND THE PERSONAL FACTOR. The Mutation Theory—in its logic, on its face (and clearly in the eyes of de Vries' contemporaries)—seems so evidently contrary to the central tenets of Darwinism. Kellogg classified de Vries' theory as one of the three major alternatives to natural selection (with Lamarckism and orthogenesis as the other candidates). De Vries himself, and with relish, explained his theory in the light of Galton's polyhedron (pp. 342–351), the primary anti-Darwinian metaphor of his day:

> Little shocks make it totter; it oscillates round its position of equilibrium and finally returns to it. A slightly stronger push however can make it go so far that it comes to lie on a new side. The oscillations round a position of equilibrium are the fluctuations, the transitions from one position of equilibrium to another correspond to the mutations. The track left behind by the rolling polyhedron can be regarded as the line of descent of the species; each subdivision of this track, corresponding to a side of the polyhedron, representing a particular elementary species; each transitional movement to a new position of mutation (1909a, volume 1, p. 55).

Yet I began this section on de Vries with a strange story about the uniquely sour note that he introduced into Darwin's biggest centenary party by torturing a Darwinian quotation to gain his master's supposed approval for a manifestly un-Darwinian view about the nature of variability (pp. 415–417). I then discussed the powerful psychological and intellectual hold that Darwin exerted upon de Vries through his status as personal hero (pp. 421–423). I claim no insight into the subtleties of psychology, but de Vries' relationship with Darwin surely ranks as the most complex, enigmatic, and contradictory personal interaction discussed in this book. Other paired opponents—Cuvier and Geoffroy, Weismann and Spencer, for example—battled in public and provide the usual stuff of controversy. But de Vries met Darwin only once, and their struggle unfolded later, and largely within de Vries' own head.

De Vries managed (and apparently needed) to support several contradictory propositions, to play several roles at the same time: a loyal disciple, who would neither propagate nor tolerate any diminution of his master's fame and rightness; a shrewd compromiser, who would bring a glorious past into harmony with later discoveries; a novel revolutionary, who could sweep aside the old and establish a startlingly different theory as a source of personal fame. In documenting the range of de Vries' rhetorical strategies, one can only experience the frustration of any careful and attentive reader in trying to locate a coherent center among the welter of contradictory claims.* Consider de Vries' several positions:

---

*For this reason, I emphasize the logic of argument, rather than the psychologic of presentation, throughout this book. If a minimal Darwinian "essence" resides within the logic of my three key statements about levels of selection, creativity in selection, and extrapolationism, then the Darwinian commitments of other scientists can be judged by degree of

1. The mutation theory is fully Darwinian. (Darwin might have flirted with a false idea about the efficacy of fluctuating variation, but he ultimately recognized the primacy, if not the exclusivity, of mutational variation.)

> It is in fullest harmony with the great principle laid down by Darwin. In order to be acted upon by that complex of environmental forces, which Darwin has called natural selection, the changes must obviously first be there. The manner in which they are produced is of secondary importance and has hardly any bearing on the theory of descent with modification. A critical survey of all the facts of variability of plants in nature as well as under cultivation had led me to the conviction, that Darwin was right in stating that those rare beneficial variations, which from time to time happen to arise—the now so-called mutations—are the real source of progress in the whole realm of the organic world (1909a, volume 1, p. 74).

This comment advances the stubborn position that so frustrated A. C. Seward, and inspired his justified rejoinder at Darwin's birthday party. De Vries' argument must be designated as a remarkable example of "stonewalling," a word just entering the English language in de Vries' time (according to the *OED*) from a combination of British cricket slang, Australian political terminology, and the memory of Confederate General Stonewall Jackson. In the light of Darwin's firm and consistent emphasis on imperceptibly gradual change, and his equally clear denial of the efficacy of "sports" (a term that de Vries acknowledged as synonymous with his mutations), how could de Vries count Darwin as a saltationist? De Vries admits that Darwin didn't really say the proper words, but he then claims that Darwin must have intended to do so: "Darwin's view, although he never definitely formulated it, was that it was these occasional single variations which brought about the continual differentiation of living forms" (1909a, volume 1, pp. 86–87). De Vries then attributed the truly Darwinian belief in gradualism to Wallace and other less worthy epigones: "Wallace's view is that the material for species forming selection is furnished by fluctuating variability; and that these infinitesimal differences are gradually heaped up in the same direction until ultimately they attain the dimensions of specific differences" (ibid., p. 87)—a good definition of Darwin's actual view!

2. The mutation theory is Darwinian, while also incorporating a few minor glosses and corrections of Darwin's own views.

> The mutation theory is intended to be a support and corollary to the selection theory of Darwin. There can be no doubt that Darwin correctly set forth the essential steps in the evolutionary process and that changes in his views mostly relate to those minor points, for which, at this time,

---

consonance with this necessary foundation. By this proper criterion, de Vries' saltational mechanism can only be labeled as anti-Darwinian (as the most astute of all commentators, Vernon Kellogg, recognized and illustrated)—whatever de Vries' psychological need for fealty to a personal hero.

the material of facts was not adequate to a correct decision. The mutation theory claims to remove many of the difficulties, inherent to the Darwinian doctrine, as e.g. the general occurrence of useless characters and the impossibility of explaining the first beginning of a selection on the ground of its usefulness (1922, p. 223).

3. The mutation theory is Darwinian. We must admit Darwin's errors on some issues, even for important points. But we can't blame him, given the limitations of knowledge in his time: "My work claims to be in full accord with the principles laid down by Darwin and to give a thorough and sharp analysis of some of the ideas of variability, inheritance, selection, and mutation, which were necessarily vague at his time. It is only just to state, that Darwin established so broad a basis for scientific research upon these subjects, that after half a century many problems of major interest remained to be taken up" (1905, p. ix).

4. Darwin recognized both fluctuating and mutational variation, but he never formulated a judgment about their relative importance (a direct contradiction of claim 1 on the dominance or exclusivity of mutational variation): "Darwin almost always speaks of these two types in his discussion on selection but never separates them, and is always in doubt as to their relative importance in the origin of species" (1909a, volume 1, p. 31).

In a variation upon this position, de Vries sometimes claimed that Darwin's agnosticism about the relative importance of these two types of variation holds no significance for evolution, and indicates no weakness in Darwin's logic, because the issue of how variation arises becomes subordinate to the role of natural selection once we feel confident that organisms will generate sufficient variation in any case: "Darwin has left the decision on this difficult and obviously subordinate point to his followers" (1909b, p. 84).

5. Darwin recognized both fluctuating and mutational variation and regarded both modes as important. Wallace later restricted "Darwinism" to the fluctuating mode alone (de Vries, 1905, p. 8). True Darwinians, who continue to recognize both modes, tend to be favorable towards the Mutation Theory (though they must revise their views on the relative significance of fluctuating variation):

Unlike the prevailing form of the theory of selection, the doctrine of mutation lays stress on sudden or discontinuous changes, and regards only these as active in the formation of species. The Darwinian form of the theory of selection regards both these and fluctuating variations as operative in the origin of new forms, whilst Wallace favors the other extreme, according to which all formation of species goes by a slow and gradual process of change. The two schools of thought naturally adopt different attitudes towards the doctrine of mutation. It is at once rejected by Wallace's adherents, whilst those who incline to Darwin's own form of the theory are less unreservedly inimical; many of them have even greeted it with open arms (1909a, volume 2, p. 599).

6. Darwin recognized both kinds of variation. Early in his career he correctly emphasized the mutational mode. Unfortunately, his critics later badgered him into a more extreme and less generous commitment to the primacy of fluctuating variation (note the direct contradiction to claim 4 and, in turn, the contradiction of 4 and 1—quite an intellectual odyssey: from exclusivity of one form, to pluralism, to exclusivity of the other. Note also that de Vries often stated all these positions in the same publication; thus, I am not only recording a consistent and legitimate ontogenetic change in opinion): "To sum up, we see that Darwin always distinguished between individual differences and single variations and that he ascribed to the latter at least a very considerable role in the origin of species. It was only by the pressure of criticism that he finally gave up this view and gave the place of honor to the ever-present individual variations" (1909a, volume 1, p. 39).

7. Darwin erred in advocating natural selection for the origin of species. The Mutation Theory has corrected this basic mistake, and therefore represents a novel direction for evolutionary thought. Darwin deserves our highest praise for his historical role, but he has now been superseded:

> The theory of descent aims at the scientific explanation of systematic relationship. It is Darwin's immortal service to have obtained general recognition for this generalization. By doing this he revolutionized the whole of biological, systematic, embryological and paleontological science. Tapping inexhaustible sources for new investigation and discovering everywhere mines where new facts were to be had for the picking up. The several propositions and hypotheses which Darwin employed as supports for this theory should be regarded now only as such, since their interest is mainly historical. They have served their purpose and are thereby fully justified. . . . This is especially true of the theory of selection, which now has served its time as an argument for the theory of descent; happily this theory no longer stands in need of such support (1909a, volume 1, p. 28).

8. The theory of natural selection is erroneous. The theory of mutation is correct. "The mutation theory is opposed to that conception of the theory of selection which is now prevalent. According to the latter view the material for the origin of new species is afforded by ordinary or so-called individual variation. According to the mutation theory individual variation has nothing to do with the origin of species. This form of variation, as I hope to show, cannot even by the most rigid and sustained selection lead to a genuine overstepping of the limits of the species and still less to the origin of new and constant characters" (1909a, volume 1, p. 4).

One might conjecture that this full range of viewpoints represents a transition from heart (in the early entries of the sequence) to mind—that is, from de Vries' need to express fealty with his personal hero to his recognition of the oppositional logic within his own system. (One might also be more cynical and interpret the early entries as diplomatic attempts to court favor with Darwinians—but I don't think that this interpretation can be fairly defended, for

Darwinism did not rank as a dominant philosophy at the time, and issues of heart hold such evident prominence for de Vries.) The Mutation Theory, with its explicitly saltational and nonadaptational origin of species, must be read as a confutation of Darwinism on the central question (the title of Darwin's book, after all) of "The Origin of Species." And yet, by an interesting argument (developed in the next section), de Vries did provide a genuine and ample field for Darwinian logic in another realm, even while he tried to extirpate natural selection without compromise on its own original turf.

THE LOGIC OF DARWINISM AND ITS DIFFERENT PLACE IN DE VRIES' SYSTEM. I have documented the psychological vacillation in de Vries's assessment of Darwin, but a stark contrast must be drawn between this frustrating emotional indefiniteness and de Vries' clear and subtle understanding of selectionist logic. I think that only two other early evolutionists—Weismann and Darwin himself—ever grasped so fully, both in basic logic and expanding implications, the rich meanings of selectionism. De Vries' personal dilemma lay in his unwillingness to tar his personal hero with the brush of selectionist errors (in his judgment), not in any softness or vacillation in understanding selection itself. Thus, he tried to distance Darwin from Darwin's own beliefs, grasping at straws (often of de Vries' own construction) in tortuous exegetical efforts to remake Darwin as a closet saltationist.

One can hardly deny that de Vries' Mutation Theory represents, in principle, about as anti-Darwinian a mechanism as anyone could construct at the crucial level of Darwin's own concerns (and chosen book title)—the origin of species. Neither selection nor adaptation can play a creative role in evolutionary change if new species arise in single, fortuitous leaps.

Yet de Vries insisted that his theory followed Darwinian principles at the larger scale of full unrolling of life's history—and here he presented a sound and fascinating argument that his contemporaries never understood (in their failure to grasp the generality of selectionist logic) and that later history therefore, and unfortunately, forgot. I argued in Chapter 2 that the operation of a selectionist mechanism makes three crucial demands upon the nature of internal "raw material": that variation be copious, small in scope (relative to the unit of incremental change at the scale under consideration), and undirected (isotropic). At the level of speciation, de Vries' Mutation Theory becomes decidedly anti-Darwinian by failing the second test—for single mutations generate new species in one step, and no creative role can be assigned to selection or adaptation at all.

But suppose that we "promote" our gaze and consider evolutionary trends through geological time as the relevant scale of change. Then a species-forming mutational step might be considered sufficiently small (relative to the full trend) to fit into Darwinian *logic*—though not into Darwin's own *theory*, which explicitly requires, as its central tenet, that a process of organismal selection must govern the origin of species.

But if we can regard speciational steps as small increments in macroevolution, then the applicability of Darwinian logic to trends would depend upon the validity of the remaining two criteria at this higher level: copiousness and isotropy. The criterion of copiousness will surely be fulfilled in lineages undergoing a de Vriesian period of mutability. The full operation of Darwinian logic at this larger scale will then stand or fall upon the remaining criterion of isotropy. De Vries analyzed the issue in these appropriate terms and, by stating strong support for isotropy. He therefore identified a *proper* fealty with Darwinian logic (in a domain different from Darwin's own application, after denying the efficacy of the same general logic in Darwin's own favored realm).

De Vries often states the principle of isotropy as one of his central conclusions: all characters may mutate in all directions. "Single variations [mutations] seem to be presented by all characters, to proceed in every direction and to be apparently without limit" (1909a, volume 1, p. 33). Moreover, de Vries expresses this view not only as an empirical conclusion but as a logical consequence of his theory: "The mutation theory demands that organisms should exhibit mutability in almost all directions" (1909a, volume 1, p. 204).

Even more significantly, de Vries recognizes that isotropy must be asserted to validate Darwinian selection at the higher level of evolutionary trends. Tying isotropy directly to the efficacy of selection, de Vries writes (1905, p. 574): "Nearly all qualities vary in opposite directions and our group of mutants affords wide material for the sifting process of natural selection." Selection can only operate as a sieve, but if variations are copious, isotropic and small (a good description for the status of de Vriesian mutations relative to the full extent of a geological trend), then these species-forming mutations can forge no large-scale cladal trend by themselves—and even a sieve, by extended directional accumulation, becomes a creative mechanism. (This metaphor beautifully restates Darwin's basic argument about the creativity of selection—see Chapter 2. But de Vries denies selection at Darwin's own favored level of organisms in populations, and grants power to Darwin's mechanism only at the higher level of sustained trends among species in clades.)

> According to Darwin, changes occur in all directions, quite independently of the prevailing circumstances. Some may be favorable, others detrimental, many of them without significance, neither useful nor injurious. Some of them will sooner or later be destroyed, while others will survive, and which of them will survive is obviously incumbent on the question, whether their particular changes agree with the existing conditions or not. This is what Darwin has called the struggle for life. It is a large sieve, and it only acts as such. Some fall through and are annihilated, others remain above and are selected, as the phrase goes. Many are selected, but more are destroyed: daily observation does not leave any doubt upon this point. How the differences originate is quite another question. It has nothing to do with the theory of natural selection, nor with the struggle for life (1905, p. 571).

De Vries strongly rejected any notion of directed variability (nonisotropy) in the production of trends—and therefore maintained no sympathy at all for the orthogenetic school. He even slipped into the philosophic fallacy that Whitman criticized (see p. 385), by equating nonisotropy with teleology in the strong sense of inherently unscientific assertion: "According to the Darwinian principle, species forming variability—mutability—does not take place in definite directions. According to that theory, deviations take place in almost every direction without preference for any particular one, and especially without preference for that direction along which differentiation happens to be proceeding. Every hypothesis which differs from Darwin's in this respect must be rejected as teleological and unscientific" (1909a, volume 1, p. 198).

In a later, remarkable passage, de Vries gathered all these elements together, identifying nonisotropy as both a central Darwinian claim and a necessary bulwark in the struggle against supernaturalism. He also acknowledges that Darwin applied the resulting selectionist mechanism at the organismal level, whereas he favors the species level, thus devising a fundamentally different theory:

> We are strongly opposed to the concept of a definite "tendency to vary" which would bring about useful changes, or at least favor their appearance. The great service which Darwin did was that he demonstrated the possibility of accounting for the evolution of the whole animal and vegetable kingdom without invoking the aid of supernatural agencies. According to him, species forming variability exists without any reference to the fitness of the forms to which it gives rise. It simply provides material for natural selection to operate on. And whether this selection takes place between individuals, as Darwin and Wallace thought, or whether it decides between the existence of whole species, as I think; it is the possibility of existence under given external conditions which determines whether a new form shall survive or not. . . . The mutability of *Oenothera Lamarckiana* satisfies all these theoretical conditions perfectly. Nearly all organs and all characters mutate, and in almost every conceivable direction and combination (1909a, volume 1, p. 257).

I have presented this extended treatment of de Vries for a reason embedded in the plan of this chapter, and crucial to the logic of this book. I argue that "internalism" poses two separate challenges to pure Darwinian functionalism: saltational change arising from internal forces of mutability, and inherent directionality of variation (corresponding to facet-flipping and channeling on Galton's polyhedron). Most internalists ("structuralists," "formalists," "laws of form" theorists in other terminologies) emphasize the second theme of channels and preferred directionality of variation (now most often expressed in the popularity of "constraints" as a subject in modern evolutionary literature—see Chapters 10 and 11). This style of internalism represents the primary theme of Goethe, of Geoffroy, of Owen, and of the orthogeneticists. Fewer internalists emphasize the saltational theme—and those who do, like Bateson, tend to support channeling as well as facet-flipping (for the two themes fuse well into a coherent anti-Darwinian philosophy, as Bateson recognized and articulated).

When such basic themes commingle, our experimental traditions lead us to search for "pure" end-member cases—examples of one item without the other, so that we may assess the unadulterated contribution of each theme treated separately. We can identify several pure channeling theorists—biologists who extolled directional variation, but supported gradualism and rejected saltation (several orthogeneticists fall into this category). But we rarely encounter a "pure" saltationist who accepts isotropy for large-scale changes and rejects all notions of preferred directionality. (Since saltation also implies an internal control upon change, the allied theme of preferred directions usually gains assent as well—as in Bateson's arguments on homeotic variation.) But de Vries represents our instructive "pure" case of one without the other—a saltationist who accepted the isotropy of these immediate and substantial changes. Moreover, de Vries explicated the logic of his unusual commitment to recognize the interesting role that such a "nonstandard combination" must imply for Darwinism. (Saltationism precluded any role for selection in the origin of species; but mutational isotropy resurrected the Darwinian apparatus at the higher level of evolutionary trends.) As a "test case," therefore, de Vries—all by himself—balances the rest of this chapter, with its primary emphasis on internally channeled variation. De Vries's brilliance, particularly his clear attention to Darwinian logic, makes him a particularly attractive and instructive figure in our quest to understand the components of internalist thinking.

### De Vries on macroevolution

De Vries never placed primary emphasis upon macroevolutionary themes in his books, but these issues do receive more than passing attention. By his own intent and reckoning, de Vries' main contribution to macroevolution lay in his resolution of Kelvin's paradox—an earth too young to permit evolution by Darwinian gradualism (see Chapter 6). Obviously, if new species arise *per saltum* in a single generation, limits on the earth's age would not preclude the evolutionary work actually accomplished. "The demands of the biologists and the results of the physicists are harmonized on the ground of the theory of mutation" (1905, p. 712). "One of the greatest objections to the Darwinian theory of descent arose from the length of time it would require if all evolution was to be explained on the ground of slow and nearly invisible changes. This difficulty is at once met, and fully surmounted by the hypothesis of periodical but sudden and quite noticeable steps. This assumption requires only a limited number of mutative periods, which might occur within the time allowed by physicists and geologists for the existence of animal and vegetable life on the earth" (1905, p. 29).

De Vries even made a semi-quantitative assessment, based again on *Oenothera*. Given Kelvin's 24 million year estimate for a habitable earth, even one mutation (adding a character) every 4,000 years would provide 6,000 new features for any extant lineage. Since neither *Oenothera*, nor any other creature, bears so many distinct features ("a number far higher than comparative and systematic science can by any means accumulate in its descriptions"—1909a, volume 2, p. 665), geological time becomes positively bounteous for requirements of the mutation theory. De Vries epitomized his view in an "equation" (not really a mathematical formula, but a set of concepts epitomized by letters): $M \times L = BT$, or number of mutations times length of intervals between mutations equals "biological time." So long as biological time doesn't exceed available geological time, Kelvin's paradox may be resolved.

But de Vries' main contribution to macroevolution does not lie in this explicit geological aid (soon made irrelevant by radioactive extension of the earth's age in any case). Rather, de Vries formulated a fully articulated macroevolutionary theory based on the application of Darwinian logic to the higher level of species and trends—a true theory of "species selection" (even so named by de Vries!). I do not think that de Vries ever recognized the full import of what he had done—for he never grants the theme real prominence, and pieces of the argument lie scattered throughout his writing. But de Vries developed all the parts, and they do cohere. His acute understanding of Darwinian logic (a grasp that also led him to reject selection where he felt that such a mechanism couldn't apply) drove him on and informed all the disparate elements of his thinking.

De Vries developed his macroevolutionary concepts as a set of logical implications from the Mutation Theory. He recognized, first of all, that an origin of new species by recurrent and effectively identical saltation provides no variation for the working of any Darwinian process of selection: "[Perhaps] the mutants of one type . . . would not be pure . . . but would exhibit different degrees of deviation from the parent. The best would then have to be chosen in order to get the new type in its pure condition. Nothing of the kind, however, was observed. All the *oblonga* mutants were pure *oblongas*. The pedigree shows hundreds of them in the succeeding years, but no difference was seen and no material for selection was afforded" (1905, pp. 559–560).

All effective evolutionary variation exists only *among the "elementary species"* formed by mutation from a parental stock, *not among individual organisms* within these species. In principle, therefore, if selection works as an evolutionary force at all, Darwin's process can only sort elementary species, not organisms within populations. De Vries clearly recognizes the difference between the conventional Darwinian form of organismal selection and his proposal for selection among elementary species: "The struggle for existence, that is to say the competition for the means of subsistence, may refer to two entirely different things. On the one hand, the struggle takes place between the individuals of one and the same elementary species, on the other between the various species themselves. The former is a struggle between fluctuations, the latter between mutations" (1909a, volume 1, pp. 211–212).

Horticulturists, de Vries noted, recognize the two modes with different names (1905, pp. 604–605)—"race-breeding" for the limited and relatively ineffective Darwinian sorting based on fluctuating variation among organisms in a population; and "variety-testing" for sorting among discrete elementary species produced by mutation. In a remarkable passage, de Vries then names this second mode "species selection,"* and also denies that this higher-level process can be equated with natural selection:

> The word selection has come to have more than one meaning. Facts have accumulated enormously since the time of Darwin, a more thorough knowledge has brought about distinction, and divisions at a rapidly increasing rate, with which terminology has not kept pace. Selection includes all kinds of choice. . . . Selection must, in the first place, make a choice between the elementary species of the same systematic form. *This selection of species or species selection* [my italics] is now in general use in practice where it has received the name of variety testing. This clear and unequivocal term however, can hardly be included under the head of natural selection. The poetic terminology of selection by nature, has already brought about many difficulties that should be avoided in the future. On the other hand, the designation of the process as a natural selection of species complies as closely as possible with existing terminology, and does not seem liable to any misunderstanding. It is a selection *between species.* Opposed to it is the selection *within the species* (1905, pp. 742–744).

De Vries then develops this concept of species selection as a set of guidelines for a general theory of macroevolution. He argues that sustained evolutionary trends must arise by species selection for two reasons: (1) Variation among species represents the only available "fuel" for an effective process of selection. (2) Trends are clearly adaptive and accumulative, but the mutational origin of elementary species is both nonadaptive and discretely sudden. Mutations, therefore, cannot produce trends by themselves; a "higher-order" selection upon discrete mutational phenotypes must occur:

> The differentiating characteristics of elementary species are only very small. How widely distant they are from the beautiful adaptive organizations of orchids, of insectivorous plants and of so many others! Here the difference lies in the accumulation of numerous elementary characters, which all contribute to the same end. Chance must have produced them, and this would seem absolutely improbable, even impossible, were it not

---

*De Vries wrote this book in English, so "species selection" represents his own chosen term and not the product of a translation. I note this fact with personal chagrin. "Species selection" has been a central component in the debates about punctuated equilibrium, and paleontologists have been discussing this idea intensely since the mid 1970's. We have all attributed the term to Stanley (1975), who brilliantly articulated the concept and its implications. No one recognized that de Vries—in a book written in English by the world's leading evolutionist at the time—not only developed the idea, but designated the concept by the same name (and realized the full set of implications). De Vries did not emphasize "species selection" or discuss the concept at length, but I can offer no excuse beyond my own inadequate research for my ignorance of this point. See Gould (1993b) for my attempt at amends to this great scientist.

for Darwin's ingenious theory. Chance there is, but no more than any-where else. It is not by mere chance that the variations move in the re-quired direction. They do go, according to Darwin's view, in all direc-tions, or at least in many. If these include the usual ones, and if this is repeated a number of times, cumulation is possible; if not, there is simply no progression, and the type remains stable through the ages. Natural se-lection is continually acting as a sieve, throwing out the useless changes and retaining the real improvements. Hence the accumulation in appar-ently predisposed directions, and hence the increasing adaptations to the more specialized conditions of life (1905, p. 572).

De Vries also recognizes that species selection must include two compo-nents, corresponding to birth and death biases in conventional organismic se-lection (1909a, volume 1, p. 200, and volume 2, p. 660): Species selection will favor those lineages that (1) produce more elementary species by muta-tion (birth bias), and (2) generate phenotypes fortuitously adapted to chang-ing local conditions (persistence bias).

The need for strong birth biases, combined with the central claim (see p. 435) that periods of mutability affect only a few species (and for a very short time relative to their geological longevity), led de Vries to embrace the importance and near universality of long-term stasis within species—an argu-ment strikingly isomorphic with the apparatus that we introduced much later in developing punctuated equilibrium (Eldredge and Gould, 1972; Gould and Eldredge, 1977, 1993—and, again, much to my chagrin for not knowing about de Vries' earlier version. The two accounts invoke entirely different principles of saltational vs. allopatric speciation, but the two arguments still employ an isomorphic logic).

De Vries cites several supports for the empirics of stasis: the ability of sys-tematists to define most taxa unambiguously; the persistence of identical phenotypes for centuries in populations that have become widely isolated (1909a, volume 1, p. 206); geological persistence through such epochs of ex-tensive climatic change as ice ages (1905, p. 696); documented longevity of many species through several geological periods (1905, pp. 698–699). De Vries then chides Darwinians for asserting imperceptible transmutation in the face of manifest, documented constancy. Darwinians have been driven to this inconsistency, de Vries asserts, by the gradualistic implications of their theory, but a more accurate view of evolutionary mechanisms affirms stasis as an ex-pectation, not an embarrassment to be ignored, or explained away by appeals to an imperfect fossil record:

Many facts plead in favor of the constancy of species. This principle has always been recognized by systematists. Temporarily the current form of the theory of natural selection has assumed species to be inconstant, ever changing and continuously improved and adapted to the requirements of the life conditions. The followers of the theory of descent believed that this conclusion was unavoidable and were induced to deny the manifest fact that species are constant entities. The mutation theory gives a clue to the final combination of the two contending ideas. Reducing the change-ability of the species to distinct and probably short periods, it at once explains how the stability of species perfectly agrees with the principle of descent through modification (1905, p. 694).

In his earlier writings (1905, 1909a), de Vries strongly supported adaptation as the primary result of trends forged by species selection. But he altered this conviction in later articles, and thereby came to espouse the full range of internalist critiques against Darwin (in questioning both gradualism and functionalism). In 1922, de Vries contributed a short, but highly revealing chapter to J. C. Willis' famous critique of adaptation based on the correlation of geographic distribution and geological longevity: *Age and Area* (Willis, 1922).

Of course, de Vries had always opposed adaptationism for the origin of species (1922, p. 224) because selection cannot craft good design if species arise in single steps: "Specific characters have evolved without any relation to their possible significance in the struggle for life. The facts are contrary to the main principle of the selection theory of Darwin" (1922, p. 226). De Vries interpreted Willis' argument as a final proof for this linchpin of his theory—and he expressed delight: "The general belief in adaptation as one of the chief causes of the evolution of specific characters is best directly contradicted by the statistical studies of Willis . . . This result must be considered as the one great proof, which the mutation theory still wanted for its acceptance in the field of systematic zoology and botany" (1922, p. 227).

But differentia of higher taxa arise by cumulation during species selection, and may therefore be adaptive. "It is a curious fact that most of the striking instances of beautiful adaptation to special forms of life are characters of genera and sub-genera, or even of whole families, but not of single species. Climbing plants and tendrils, insectivorous plants, desert types—submerged water plants, and numerous other instances could be adduced" (1922, p. 22). (De Vries regarded the distinction between nonadaptation in most defining traits of species, and adaptation for the differentia of higher taxa, as virtual proof for the nonselectionist origin of species by saltation and the functional origin of higher taxa by species selection.)

Still later, however, and further inspired by Willis, de Vries reassessed even this restricted role for adaptation—as he recognized that an *a posteriori* functional correlation of form and environment (especially for broad characters of higher taxa) need not indicate adaptive fashioning under current circumstances. Using an argument of exaptation (Gould and Vrba, 1982, and Chapter 11), de Vries recognized and embraced the alternative view that such characters arise for an immediate reason (often nonadaptive), then radiate out randomly by Willis' argument, and finally survive in environments that, by good fortune for the species, favor a set of characters originally evolved for reasons unrelated to current function:

> Everywhere in nature, in geological periods as well as at present, the
> morphological characters of newly originated types have no special sig-

nificance in the struggle for life. They are not known to aid them in their initial dispersal. They may afterwards prove to be useful or useless, but this has no influence upon their evolution. Obvious instances of usefulness occur, as a rule, only at much later periods during the wandering of the new forms, when unexpectedly they arrive in environments specially fitted for them. The usual phrase, that species are adapted to their environment, should therefore be read inversely, stating that most species are now found to live under conditions fit for them. The adaptation is not on the side of the species, but on that of the environment. In a popular way we could say that in the long run species choose their best environment (1922, pp. 226–227).

I ran into Ernst Mayr as I was completing this chapter and asked him if he had ever met de Vries. "No," he said, "botanists and zoologists didn't talk to each other very much in those days, and, anyway, I was a Lamarckian then." The reasons for their failure to meet may have been non-ideological (largely generational and disciplinary), but I treasure, nonetheless, the image of the world's greatest Darwinian at the close of the 20th century, then about the same age as de Vries at this most non-Darwinian endpoint of his career, speaking about their non-interaction. De Vries did come to inhabit a different world. Whatever his love and fealty for Darwin, de Vries expunged the guiding concept of natural selection from his hero's own realm of the origin of species. But de Vries then reinserted selection into the higher domain of macroevolution—at least until he eventually dropped the functional theme from this world as well. De Vries developed cogent critiques, though his alternative mechanism can no longer be defended. His bannings and separations must now be judged as too stark. Instead, we need to expand and modify Darwin's world to a hierarchical view of selection operating differently and simultaneously at several levels of nature's individuality—and not segregate natural selection to exclusive operation in a single domain, whether organismal (for Darwin) or speciational (for de Vries).

## RICHARD GOLDSCHMIDT'S APPROPRIATE ROLE AS A FORMALIST EMBODIMENT OF ALL THAT PURE DARWINISM MUST OPPOSE

"It was nearly eleven hundred, and in the Records Department, where Winston worked, they were dragging the chairs out of the cubicles and grouping them in the center of the hall, opposite the big telescreen, in preparation for the Two Minutes Hate . . . A hideous, grinding screech . . . burst from the telescreen . . . The Hate had started. As usual, the face of Emmanuel Goldstein, the Enemy of the People, had flashed onto the screen" (George Orwell, *1984*). In my own factual version of this fictional archetype, we snickered over a deluded man rather than screaming at a potentially dangerous enemy—but the expectation of group reaction, based on little more than our ignorance combined with the prompting of our leaders, still evokes an eerie and uncomfortable feeling of similarity in my memory.

I began my graduate work at Columbia University in 1963 (after undergraduate study at small, iconoclastic Antioch College), just a few years after the codification of the hardest versions of the Modern Synthesis in and around the Darwin centennial celebrations of 1959 (see Chapter 7). I had never heard of Richard Goldschmidt. Yet his name surfaced in almost every course—never with any explication of his views, but only in a fleeting and derisive reference to something called a "hopeful monster." Students then responded with a derisive sign of recognition—as our professors seemed to expect as a badge of membership in some inner circle. I found the oft-repeated exercise—one might almost have called it a ritual—offensive and demeaning, both to Goldschmidt and to any notion of my potentially independent intelligence.

My memories cannot be deemed either exaggerated or idiosyncratic. Frazzetta (1975, p. 85) recalled similar experiences from another university: "No one stopped to consider whether in all of Goldschmidt's assailable propositions, there existed anything worth thinking about. There was no time for such consideration as long as there was so much merry mayhem to be carried out. In my university classes, the name 'Goldschmidt' was always introduced as a kind of biological 'in joke,' and all we students laughed and snickered dutifully to prove that we were not guilty of either ignorance or heresy." Guy Bush (1982) corroborates our memories: "When his name did come up it was inevitably in the context of 'hopeful monsters' and to the accompaniment of subdued snickers and knowing nods. It didn't take long to learn that Richard B. Goldschmidt was not to be taken seriously as an evolutionary biologist."

A few years later, when I unearthed Goldschmidt's *Material Basis of Evolution* from our library (and found much of value amidst some admitted nonsense), a senior colleague and former professor decided to check his own copy to see if he had formerly dismissed the book too harshly. He could not find the volume on his shelf, and only then remembered that he had discarded the book several years earlier as containing nothing of value!

Every orthodoxy needs a whipping boy, but why Goldschmidt? A question of personality, perhaps? Students and colleagues tend to remember Goldschmidt (1878–1958) as kind and even courtly, but also as arrogant and imperious, thus fulfilling anyone's stereotypical image of Herr Doktor, the German Professor. He did indeed hold such an official and topmost status, as first director of genetics at the Kaiser Wilhelm Institute for Biology in Berlin—until his Jewish ancestry forced a relocation to Berkeley, and the start of a second career, in the late 1930's. Viktor Hamburger told me that fellow students called Goldschmidt "the Pope," in reference (not deference) to his imperiousness. (This apparently anomalous title may not be so peculiar for a prominent, established German-speaking Jew, a group that often surpassed the average Prussian in loyalty and patriotism. I can still hear the acid words of my Yiddish-speaking Hungarian grandmother, recalling the snubs of well-bred Viennese girls, after her father sent her to a Jewish school in the capital of the

Austro-Hungarian empire.) Goldschmidt certainly practiced his penchant for pious proclamations *ex cathedra:* "I am certain that in the end I shall turn out to have been right" (1960, p. 307).

These factors of personality may have heightened his candidacy, and exacerbated the depth of collegial reaction, but Goldschmidt surely became a whipping boy primarily, and properly, for ideological reasons. Goldschmidt's 1940 book, *The Material Basis of Evolution,* based on the Silliman Lectures given at Yale in 1939, became the standard text for his apostasy. We may specify several rationales, based on the major claims of this volume, for Goldschmidt's anathematized status among the synthesists.

1. Above all, and in his characteristically uncompromising manner, Goldschmidt held that new species arose saltationally, by a mode of genetic change different in kind from the alterations that yield adaptive modification within species. (The controversial nature of this difference in "kind" identifies the key issue for a proper assessment of Goldschmidt, as we shall see.)

2. In Goldschmidt's view, Darwin had correctly described change within species as gradual, adaptive, and diversifying—but this mode of evolution leads only to the establishment of a *Rassenkreis* (a polytypic species), never to the formation of a new species. True species must be separated by "bridgeless gaps." Goldschmidt organized *The Material Basis of Evolution* in two sequential sets of chapters, entitled Microevolution and Macroevolution. In a scheme of argument that could not have been "better" designed to rouse ascendant neo-Darwinians to anger, the first part extols Darwinian processes in their strictly limited domain, while the second emphasizes their impotence in producing new species (while proposing workable alternatives in the saltationist mode). Goldschmidt links the two sections with the following paragraph—an anti-Darwinian clarion call that he printed entirely in italics: "*Subspecies are actually, therefore, neither incipient species nor models for the origin of species. They are more or less diversified blind alleys within the species. The decisive step in evolution, the first step toward macroevolution, the step from one species to another, requires another evolutionary method than that of sheer accumulation of micromutations*" (1940, p. 183).

3. Apostates may generate maximal anger, but not every opponent can gain such an anathematized status. A fool by nature, or a scholar who displays ignorance in the field of his chosen iconoclasm, cannot qualify, and will attract more pity than rage. Apostates must be smart, skilled, potentially effective (and therefore feared)—and also former adherents to the orthodoxy they now reject. Goldschmidt could not be dismissed as an ignorant "lab man," unacquainted with the source of strongest Darwinian arguments—field data of natural history. He had undertaken one of the most thorough studies ever attempted on the empirics of geographic variation in a single species, the gypsy moth *Lymantria dispar.* He states that he had expected to affirm the Darwinian apparatus at all scales: "As a convinced Darwinian I believed geographic races to be incipient species. I hoped to prove by such an analysis the correctness of this idea. I was completely acquainted with what twenty years

later was rediscovered as 'the new systematics,' and my convictions, as expressed in 1920 and 1923, were practically the same as those of present-day
Neo-Darwinians" (1960, p. 318).

If Goldschmidt had simply rejected Darwinism outright, he would have angered the synthesists quite sufficiently. But Goldschmidt proceeded further to
an argument almost guaranteed to arouse far deeper frustration. He proclaimed the selectionist mechanism as completely sufficient to account for *all*
differentiation within species—and then announced that this basic style of
microevolution bore no causal relevance to the production of new species. In
other words, he denied the cardinal extrapolationist premise that evolution in
the small—the only mode routinely subject to direct observation—could, by
extension in time, produce the entire panoply of life's history. To many synthesists, Goldschmidt's ideas ranked as an ultimate council of despair. How
can science proceed at all without such a uniformitarian and operational
premise?

Goldschmidt, as an *enfant terrible,* clearly enjoyed the fuss that he had engendered: "I certainly had struck a hornet's nest. The Neo-Darwinians reacted savagely. This time I was not only crazy but almost a criminal" (1960,
p. 324). He also provoked a vigorous and extended reaction. Most evolutionists know, for example, that Ernst Mayr wrote one of the great classics of the
Synthesis, *Systematics and the Origin of Species* (1942), as a direct response
to Goldschmidt's *Material Basis.* Mayr recalled (1980, p. 420): "Even though
personally I got along very well with Goldschmidt, I was thoroughly furious
at his book, and much of the first draft of *Systematics and the Origin of Species* was written in angry reaction to Goldschmidt's total neglect of such overwhelming and convincing evidence."

We may best illustrate the depth of feeling (and the perceived extent of
Goldschmidt's apostasy) by examining the review of *Material Basis* written
for *Science* by Th. Dobzhansky (1940), whose own *Genetics and the Origin
of Species* had codified the developing Synthesis three years earlier. The rhetorical strategy of this review embodies the general reaction of the emerging
Neo-Darwinian consensus. Dobzhansky grants warmest praise to Goldschmidt's persona and to the sweep of his effort. He begins by stating: "This
book contains the only basically new theory of organic transformation propounded during the current century" (1940, p. 356)—a peculiar statement,
given the former popularity of de Vriesian saltationism (although Dobzhansky may have viewed de Vries's Mutation Theory, not formally printed until
1901, as a late 19th century formulation, especially since de Vries had published his major empirical work on *Oenothera* in the 1880's and 1890's). For
Dobzhansky, only three serious theories of evolutionary mechanics precede
Goldschmidt's book—Lamarckism, which "has become obsolete owing to its
basic assumption having fallen short of experimental verification" (p. 356);
autogenesis (orthogenesis), dismissed as "in conflict with the principle of
causality in vogue in the materialistically-minded modern science" *(loc. cit.);*
and Darwinism, which Dobzhansky accepts, and which "underwent great
changes because of the forward strides of genetics, but the unbroken continu-

ity of ideas between the 'neo-Darwinism' and Darwin's original theory is evident." Goldschmidt's book now "connotes an at least temporary end of the undivided reign of neo-Darwinian theories" *(loc. cit.).*

Dobzhansky covered his review in a patina of judiciousness, even of approbation. He also presents a fair and clear epitome of Goldschmidt's major points. Dobzhansky refers to Goldschmidt's theory as "brilliantly developed and masterfully presented" (p. 357), and he adds: "Goldschmidt's keenly critical analysis has emphasized the weaknesses and deficiencies of the neo-Darwinian conception of evolution, which are numerous, as even partisans ought to have the courage to admit" (p. 358).

But, in the deeper theme and purpose of his review, Dobzhansky's rejection could not have been more total or dismissive (much as he advocates a close reading and study of Goldschmidt's book). First of all, he does not count Goldschmidt's ideas as an *evolutionary* theory at all, as expressed in the title of Dobzhansky's review: "Catastrophism versus evolutionism." The first sentence, as quoted above, presents Goldschmidt's theory as the first new view of "organic transformation," not of evolution—words that Dobzhansky chose very carefully and purposefully. He then explains: "Goldschmidt not only relegates natural selection to a place of relative unimportance, but in effect rejects evolution beyond the narrow confines in which it has been admitted to exist by Linnaeus and many creationists. His theory belongs to the realm of catastrophism, not to that of evolutionism."

Recalling the stereotypical cry of the stadium vendor—"you can't tell the players without a scorecard"—later scholars often need a historical primer of definitions to identify certain claims properly. Dobzhansky refers here to Lyell's rhetorical strategy for specifying the requirements of a scientific geology. A truly scientific theory based on *verae causae* (true causes), Lyell tells us, must embrace the uniformitarian postulate that small-scale changes, observable on our current earth, can produce, by gradual accumulation through geological time, all the grand events of our planet's history. Evolution, for Dobzhansky, defines all theories of biological change set within this proper uniformitarian mode. The catastrophic alternative—that occasional paroxysms sweep the earth to produce most important change, whereas the daily accumulation of tiny, observable alterations can lead to nothing substantial—represents a retreat to the bad old days of useless speculation and theological influence. Goldschmidt's saltational theory of the "hopeful monster" falls into this basically unscientific mode. As Lyell wrote in his magisterial prose (1833, p. 6): "Never was there a dogma more calculated to foster indolence, and to blunt the keen edge of curiosity . . . We see the ancient spirit of speculation revived, and a desire manifested to cut, rather than patiently to untie, the Gordian knot." Therefore, by placing Goldschmidt's book within the catastrophist tradition, Dobzhansky almost denies any scientific status to the theory at all.

Dobzhansky then reinforces his dismissal by declining even to present any counterarguments: "It is impossible to attempt here a critique of Goldschmidt's theory, for this would require a book approximately the same size as his own" (p. 358—indeed, Mayr would soon write such a rebuttal). But Dobzhansky's demurral does not prevent him from trying to annihilate Goldschmidt with the unkindest cut of all—an explicit removal of scientific status: "But in the reviewer's opinion the simplicity of Goldschmidt's theory is that of a belief in miracles" (p. 358).

No one can deny that Goldschmidt's theory merits historical attention for its palpable and extensive influence, at least upon the psyches of his major opponents. But we also need to assess whether anything in Goldschmidt's theory merits our respect and study today. We must therefore clarify a primary issue that Goldschmidt himself unfortunately plunged into deep confusion: how shall we characterize the genetic source of saltations that make new species? Beginning in the 1930's, and extending with increasing scope, unconventionality, and self-assurance to his death, Goldschmidt developed an idiosyncratic concept of genetics that eventually sought to refute the "particulate" or "corpuscular" gene entirely (see the culmination of this development in Goldschmidt's last and least cogent book—*Theoretical Genetics,* 1955. A comparison between Goldschmidt's "holistic" view and Bateson's unwillingness to abandon his "vibrational" theory of heredity would provide an interesting subject of research. I suspect that more than mere coincidence must inhere in the observation that Bateson and Goldschmidt—the most thoroughly non-Darwinian thinkers among important 20th century evolutionists, particularly as expressed in their full and coordinated support for the channeling *and* the facet-flipping themes of Galton's polyhedron—both insisted upon a holistic concept of genetic material).

In short, Goldschmidt finally concluded that the underlying basis for all mutational change must be sought in alterations of chromosomal patterns. If inversions, translocations, and other chromosomal changes can exert such a marked effect upon phenotypes in the absence of alteration within supposed genes, why should genes exist at all as discrete and bounded entities? Perhaps all genetic changes arise as alterations in pattern, with mappable, so-called micromutations as modifications of minimal spatial extent and phenotypic effect. (Goldschmidt, of course, did not deny the methodology of locating and mapping "genes" on chromosomes. He merely considered these loci as operationally definable spots on an indivisible chromosome. Order must be conserved for normal development. The "mutations" of conventional terminology must represent disruptions of this standard order, not material changes within discrete entities.) Eventually, Goldschmidt even regarded individual chromosomes as mere segments of a more comprehensive, holistically acting, system. As he touted this concept with increasing vigor and assurance, even after Watson and Crick's resolution of the structure of DNA in 1953, Goldschmidt became more and more marginalized within his field.

This idiosyncratic view of genetics bears an obvious relationship to Goldschmidt's saltational concept. If all genetic change can be rendered as alteration of pattern within a single integrated system, then some changes must be great enough in scope to reorient the entire program of development (while others, with only local effect, correspond to micromutations of standard interpretations). Goldschmidt called these pervasive changes "systemic muta-tions," and he identified them as the underlying source of saltational events that produce new species by transcending the ineffective Darwinian diversi-fication of races:

> For a long time I have been convinced that macroevolution must proceed by a different genetic method . . . A pattern change in the chromosomes, completely independent of gene mutations, nay, even of the concept of the gene, will furnish this new method of macroevolution . . . So-called gene mutation and recombination within an interbreeding population may lead to a kaleidoscopic diversification within the species, which may find expression in the production of subspecific categories . . . The change from species to species is not a change involving more and more additional atomistic changes, but a complete change of the primary pat-tern or reaction system into a new one, which afterwards may again pro-duce intraspecific variation by micromutation. One might call this differ-ent type of genetic change a *systemic mutation* . . . Whatever genes or gene mutations might be, they do not enter this picture at all. Only the arrangement of the serial chemical constituents of the chromosome into a new, spatially different order, i.e., a new chromosomal pattern, is in-volved (1940, pp. 205–206).

This bold statement highlights the key issue surrounding Goldschmidt's role in current reformulations of evolutionary theory. Goldschmidt clearly ties his phenotypic concept of the "hopeful monster" to his genetic hypothe-sis of "systemic mutation" as a cause. If these two notions are indissolubly linked, and if the hopeful monster can only be conceived as a phenotypic manifestation of this deeply fallacious genetic theory, then we may dismiss this colorful term as a historical curiosity. I place Goldschmidt's denial of cor-puscular genes, and his attempt to construct a holistic genetics based upon position effects in a fully integrated interchromosomal system, into the inter-esting category of major ideas that we may honor as "gloriously wrong." Goldschmidt made a grand, not a paltry, error—for his system proposes an entirely different way of knowing, with intellectual and scientific ramifica-tions at broadest scale. But this generous breadth of vision doesn't make Goldschmidt's genetic system any less wrong, and the obvious argument re-mains: If hopeful monsters and systemic mutations only represent two aspects of the same phenomenon, then we must place the unitary concept aside, how-ever gently and with sympathetic interest.

But even a cursory investigation of Goldschmidt's career, and a first-pass analysis of his writings, reveals a separate, older and more important theme behind the concept of the hopeful monster. Goldschmidt sets most of his macroevolutionary discussions in the context of developmental systems and their ontogenies, not of idiosyncratic genetics and their operation. The confu-sions and conflations within *The Material Basis of Evolution* remain both palpable and frustrating—and must be regarded as Goldschmidt's own doing (and undoing). In this book, he sometimes speaks of systemic mutations as causes of macroevolution—alterations, by chromosomal repatterning, of entire genetic systems. But, more often, he discusses macroevolutionary change as a consequence of alterations in developmental ontogeny. His language then becomes dramatically different. Goldschmidt now refers to the genetic basis of large, species-forming phenotypic changes as "mutations"—and he now speaks of conventional alterations at specific sites, not of holistic repatternings. He often describes these mutational changes as "small," and he argues for far-reaching consequences because genes affect rates of development, and small changes occurring early in growth can trigger cascading results throughout ontogeny. "A single mutational step affecting the right process at the right moment can accomplish everything provided that it is able to set in motion the ever-present potentialities of embryonic regulation" (1940, p. 297). These developmental themes, of course, would be regarded as interesting and acceptable to orthodox synthesists (however underemphasized within the traditions of this theory). "The physiological balanced system of development is such that in many cases a single upset leads automatically to a whole series of consecutive changes of development in which the ability for embryonic regulation, as well as purely mechanical and topographical moments, come into play; there is in addition the shift in proper timing of integrating processes. If the result is not, as it frequently is, a monstrosity incapable of completing development or surviving, a completely new anatomical construction may emerge in one step from such a change" (1940, p. 486).

How then shall the hopeful monster be defined: the product of an illusory systemic mutation (and therefore a chimaera to be set aside), or as the result of a small genetic change that, by working early in ontogeny, produces a substantial final effect (and therefore an acceptable idea to stretch the Neo-Darwinian envelope)?

This confusion epitomizes the key issue for evaluating Goldschmidt's book, since the importance of his macroevolutionary ideas depends upon a resolution. Many readers have noted and commented upon this frustration, and Goldschmidt himself remarked (1955, quoted in Frazzetta, 1975, p. 116): "I have been reproached for not having made clear in my book *The Material Basis of Evolution* whether I was speaking of systemic mutation (scrambling of the chromosomal pattern) or of ordinary mutations of a macroevolutionary type, and of being confused myself on what I meant."

In one sense, of course, any resolution based on Goldschmidt's own intellectual ontogeny must admit a genuine incoherence, even a contradiction, between the different parts of his book. After all, systemic mutations differ markedly from small genetic changes that cascade to large effect by acting early in ontogeny—and Goldschmidt clearly grants each phenomenon, in different passages, *the* dominant role in macroevolutionary change! But I think that several persuasive arguments can be made, including the existence of a genuine literary "smoking gun," for regarding the developmental theme as more important, both in Goldschmidt's career and in his 1940 book (even though systemic mutation, as a much more radical concept—albeit ultimately false and detrimental to his influence—became the chief obsession of Goldschmidt's old age).

1. *Material Basis* grants the developmental theme a clear prominence in both place and space. In his introductory pages, Goldschmidt complains that evolutionary thought suffers by disregarding a vital subject: "There is, finally, another field which has been neglected almost completely in evolutionary discussions; namely, experimental embryology. The material of evolution consists of hereditary changes of the organism. Any such change, however, means a definite change in the development of the organism" (pp. 5–6). "A change in the hereditary type can occur only within the possibilities and limitations set by the normal process of control of development" (p. 1).

The macroevolutionary half of the book—the search for the "other evolutionary method" behind the origin of species—cobbles two strikingly different discussions together: the first (pp. 184–250) on systemic mutation and the non-existence of the "corpuscular gene"; and the second (pp. 250–396) on constraints and opportunities of developmental systems, and the potential macroevolutionary consequences of mutations affecting early development. Goldschmidt strongly emphasized the second discussion (at more than twice the length devoted to the first theme). Moreover, he situated his entire treatment of hopeful monsters (pp. 390–393) within the developmental section. In the opening paragraph on hopeful monsters, Goldschmidt lists his favorite potential examples—concrescence of tail vertebrae to produce a fanlike arrangement of feathers in ancestral birds, the passage of both eyes to one side of the head in flatfishes, and an achondroplastic bow-legged dog that ranks as a mere monster until humans need to extract badgers from dens and therefore breed dachshunds. Goldschmidt interprets these examples entirely in terms of small mutations affecting early development. He states, following the citation of dachshunds (p. 391): "Here, then, we have another example of evolution in single large steps on the basis of shifts in embryonic processes produced by one mutation . . . This basis is furnished by the existence of mutants producing monstrosities of the required type and the knowledge of embryonic determination, which permits a small rate change in early embryonic processes to produce a large effect embodying considerable parts of the organism."

2. The developmental theme pervaded Goldschmidt's career, in both duration of work and centrality of focus. I mentioned above (p. 453) that Goldschmidt centered his empirical studies upon the gypsy moth, *Lymantria dispar.* His voluminous research on the genetics of geographic variation, published under the collective title *Untersuchungen zur Genetik der geographischen Variation,* convinced him that diversification within species, though Darwinian and adaptive, did not lead to the origin of new species. But Goldschmidt pursued a second line of work on *Lymantria* for more than 20 years and through equally voluminous publication—studies on sexual determination and intersexuality (culminating in a long series of papers collectively titled, *Untersuchungen über Intersexualität*).

In these developmental studies, Goldschmidt recognized normal sexuality as a quantitative phenomenon produced by a balance of male and female sex determiners. He produced a series of graded intersexes by altering these balances experimentally. At a time (between 1910 and 1920) when the great majority of geneticists focused their work upon basic principles of transmission, Goldschmidt had already begun to study gene function and development—in order to establish a profession that he called "physiological genetics." He recognized that genes work by controlling the rates of chemical processes. Normal development requires a proper balance and definite timing of substances; evolutionary change occurs by alteration in the timing of development. Goldschmidt initiated and extended the concept of "rate genes," and the germ of the hopeful monster clearly lies within this crucial concept of his career.

This work also led Goldschmidt to what many scientists regard as his most enduring contribution—the naming and characterization of "phenocopies." If genes affect timing, then experimental manipulations of temperature and chemical environment might induce changes identical with those found in mutants, thus confirming the rate hypothesis. Goldschmidt produced such "mutant" phenotypes without mutations and christened them "phenocopies." He maintained great fondness for this subject and for his discoveries in this area. Indeed, the very last words of his posthumously published autobiography (1960, p. 326) do not proffer cosmic advice, but merely state: "It is my greatest intellectual happiness that I can still work in my laboratory and even make interesting discoveries in the field of chemically induced phenocopies."

Goldschmidt's later apostasy on macroevolution may be traced to this personal source in his early work on development, particularly to his early recognition that small genetic changes, operating early enough in ontogeny, may engender cascading effects towards a large phenotypic jump in a single genetic step.

3. The very term "hopeful monster," and the form of evidence adduced by Goldschmidt in support, establish the developmental theme as primary. Why did Goldschmidt use such an odd term at all, an apparently flippant phrase (and, therefore, a poor rhetorical strategy for pushing heterodox views) almost guaranteed to generate rebuke from upholders of orthodox (and dull) academic prose? In part, of course, the term began in whimsy, and then flowed too far on winds of circumstance. Goldschmidt recalled in his autobiography (1960, p. 318): "I spoke half jokingly of the hopeful monster in my first publication on the subject." But, in another sense, the term could not have been more apt or appropriate once one recognizes that ontogenetic development—*not* systemic mutation—undergirds the concept.

What makes a monster hopeful? Goldschmidt identifies two necessary and sufficient conditions:

(1) The mutant must, by good fortune, be well fitted for a particular, previously unexploited environment in its vicinity—the Darwinian, or functional theme. A mutant rat with fused tail vertebrae is just a monster; a proto-bird, with feathers better positioned for flight as a fortuitous consequence of a similar fusion, becomes a hopeful monster. An ordinary nektonic teleost fish with

both eyes on one side of the head is only a monster; a benthic flatfish with both eyes on the head's upper surface, with better scanning of surroundings as a lucky result, becomes a hopeful monster. A short and bow-legged dog is merely a monster; a frankfurter that can drag badgers from their holes is a hopeful monster.

(2) More importantly, hopeful monsters must pass a developmental criterion. The vast majority of mutations with large phenotypic effects are lethal—that is, just monsters. However, certain rare mutations will produce extensive, but viable, phenotypic changes *because they operate within the confines of a well regulated developmental system.* Such changes yield workable organisms (which may thrive if they become lucky enough to find a welcoming environment), rather than inviable hodge-podges of unintegrated systems in varying phases of ontogeny. The fecund macroevolutionary monster becomes potentially "hopeful" when all phenotypic effects unfold in a coordinated manner within a regulated developmental system. In his autobiographical statement, just before admitting that he had originated the term "hopeful monster" in partial jest, Goldschmidt linked his concept firmly to the developmental theme: "What addition to Darwinism was needed in order to account for the macroevolutionary processes? The solution was the existence of macromutations, which, in rare cases, could affect early embryonic processes so that through the features of embryonic regulation and integration at once a major step in evolution could be accomplished and fixed under certain conditions" (1960, p. 318).

Invoking the classical formalist theme of constraints and channels, Goldschmidt argues that a knowledge of developmental systems and their norms of reaction can specify the range of perturbations that might yield hopeful monsters—a clear invocation of "developmental constraint" in its positive mode of enabling (see Chapter 10): "Within a constant genotype the potentialities of individual development may include a range of variation of the same phenotypic order of magnitude which otherwise characterizes large evolutionary steps based upon changes in the genotype. The norm of reaction thus shows that paths are available for changes in the genotype (mutations in the broadest sense) without upsetting normal developmental processes" (p. 260).

Goldschmidt designates this creative constraining force in the last phrase— "without upsetting normal developmental processes." If shifts to alternate pathways discombobulate development, then any resulting monster must be hopeless. Many intricately complex systems simply fall apart or change in injurious ways under the impact of major perturbations. But the regulation of organic systems has evolved to accommodate impacts and to integrate changes into canalized and viable pathways. Goldschmidt's famous phrase transcends whimsy or nonsense—once we grasp the intended developmental theme. Goldschmidt granted hope to his monsters because regulation can integrate certain large alterations of phenotype into viable systems of development.

4. The origin and subsequent ontogeny of hopeful monsters (both the term and the concept) reveal a "smoking gun" for centrality of the developmen-

tal theme. Evolutionary biologists should honor world's fairs, despite their hoopla and crass commercialism—for Goldschmidt's work provides a second example of their spur to scientific progress. C. O. Whitman presented his most cogent defense of orthogenesis in pigeons (see pp. 383–394) in an address delivered at a meeting held in conjunction with the St. Louis fair of 1904 (also the source of the ice cream cone, several Scott Joplin rags, and the song "Meet me in St. Louis, Louis"). Richard Goldschmidt christened the term "hopeful monster" in an address at the AAAS meeting of 1933, held in conjunction with Chicago's World's Fair to celebrate a "century of progress." Goldschmidt, representing the "Kaiser Wilhelm Institute for Biology, Berlin-Dahlem," spoke on "some aspects of evolution." In his closing paragraph (1933, p. 547), he coined the fateful term in summarizing his entire paper:

> I chose . . . first, an aspect where I had to express skepticism in regard to well-established beliefs. I tried to show on the basis of large experimental evidence that the formation of subspecies or geographic races is not a step towards the formation of species but only a method to allow the spreading of a species to different environments by forming preadaptational mutations and combinations of such, which, however, always remain within the confines of the species. The second aspect which I discussed was one where I felt again optimistic. I tried to emphasize the importance of the methods of normal embryonic development for an understanding of possible evolutionary changes. I tried to show that a directed orthogenetic evolution is a necessary consequence of the embryonic system which allows only certain avenues for transformation. I further emphasized the importance of rare but extremely consequential mutations affecting rates of decisive embryonic processes which might give rise to what one might term hopeful monsters, monsters which could start a new evolutionary line if fitting into some empty environmental niche.

Two features of this citation (and of the whole article) clinch my argument. First, the article's structure provides an epitome for the book that Goldschmidt would publish seven years later, and that would seal and symbolize his apostasy. *The Material Basis of Evolution* must have been written as an expansion of this outline—a two-part structure, with the first half (Goldschmidt's self-styled "skepticism") on the Darwinian character, but macroevolutionary inefficacy, of adaptive differentiation within species; and the second half (Goldschmidt's proclaimed "optimism") on a different style for macroevolution based on occasional saltation in a rare but viable mode, as embodied in the slightly whimsical phrase "hopeful monster." Goldschmidt wrote the following statement in 1933, but he could not have composed better jacket copy for his 1940 book:

> At the beginning of this lecture I said that my mind, like that of many geneticists, is oscillating between skepticism and optimism with regard to the views on the means of evolution as derived from genetical work. I

have now presented to you examples of both states of mind: First, a bit of skepticism with regard to the role which the formation of geographic races or subspecies may have played in evolution; and then a bit of optimism in trying to show that the physiological system underlying orderly development, on the basis of the genetic constitution, allows some of the larger steps in evolution to be understood as sudden changes by single mutations concerning the rate of certain embryological processes (1933, p. 546).

This quotation would work as full jacket copy for Goldschmidt's later book—except for one omission. The quotation contains no statement at all, about systemic mutations or the attempt to construct a revolutionary, holistic genetics by denying the corpuscular gene. In other words, Goldschmidt developed the full intellectual framework of his argument for the strict separation of micro- and macroevolution, and for the saltational basis of macroevolution, by invoking the developmental theme alone—that is, *before* he initiated his campaign for a revolutionary genetics (beginning in the late 1930's, and then continuing and intensifying to his death). The developmental theme enjoys both *temporal priority* and *complete sufficiency*. Goldschmidt devised the hopeful monster (both the term and concept) before he ever formulated his radical genetics. Moreover, the developmental theme can carry the argument for saltational macroevolution all by itself. This conclusion, I think, resolves the puzzle of textual confusion in *The Material Basis of Evolution*. Goldschmidt had constructed his outline by 1933, based on the developmental theme alone. He began to formulate his radical genetics later, and then interpolated this material into a structure already established. These interpolations often seem hasty or haphazard, and Goldschmidt's chapters on systemic mutation do not always cohere with the earlier material. Ironically, the passages on systemic mutation in *The Material Basis* work much like an ordinary "hopeless monster" in the organic world. They do not mesh with the coherent outline or developmental program of a book planned and coordinated long before!

In introducing the developmental theme to carry his ideas on macroevolution, Goldschmidt (1933, p. 543) states that biologists have long recognized the need to understand the genetic basis and selective advantage of major evolutionary changes—but that a crucial third component has been missing: "But there is a third point, often neglected, which lies, I think, at the basis of the whole problem, namely, the nature of the developmental system of the organism which is to undergo evolutionary change." Goldschmidt then argues that his macroevolutionary ideas arose "as a logical consequence of my views on gene-controlled development" (p. 544), with a key in the concept of alterations in rate: "The most probable mutational change with a chance to lead to a normal organism is a change in the typical rate of certain developmental processes" (p. 544). He then praises D'Arcy Thompson for locating the phyletic meaning of these ideas in small mutational changes in rates, operating early in development to yield a saltational origin of new adult phenotypes: "Translated into phylogenetic language, this would mean that immense evolutionary effects could be brought about by changing the differential growth rates of the whole body or organ at an early point in development, with all the necessary secondary effects of such a change" (p. 545). In rare cases, such ontogenetic cascades will produce a viable organism (by working within developmental channels) lucky enough to find a favorable environment—in other words, a hopeful monster: "We certainly know of many cases of mutational shifts of the rate of certain developmental processes leading to non-viable results, for example, caterpillars with pupal antennae, larvae of beetles with wings . . . But I cannot see any objection to the belief that occasionally, though extremely rarely, such a mutation may act on one of the few open avenues of differentiation and actually start a new evolutionary line" (p. 544).

My pleasure in locating this resolution (in a 1933 article) for the textual difficulties in Goldschmidt's 1940 book then became enhanced when I noted another theme, by no means absent from the later book, but stressed in 1933 to a far greater extent, and with clearer purpose. I have repeatedly emphasized, as the central notion of this chapter, that the full formalist (or internalist) critique of Darwinian functionalism embraces two themes, both illustrated by Galton's incisive metaphor of the herky-jerky polyhedron—facet flipping (saltationism) and channeling (constraint in the positive sense of preferred directions for change).

One might expect that the chief apostate and whipping boy of an orthodoxy would embrace the full range of a coherent opposing philosophy. We usually view Goldschmidt as a pure saltationist, and the vehemence of orthodox reaction to only half a loaf might seem puzzling. But in the 1933 article, Goldschmidt gives equal weight to both internalist arguments—as he repeatedly, and explicitly, ties the theme of channeling to its strongest version of orthogenesis. Saltation and channeling march in tandem throughout his argument, and the entirety builds a satisfying version of the full formalist critique.

I have already cited one Goldschmidtian invocation of orthogenesis linked to saltationism, from the concluding paragraph: "I tried to show that a directed orthogenetic evolution is a necessary consequence of the embryonic system which allows only certain avenues for transformation" (p. 547). But the two themes remain indissolubly connected throughout the article, and channeling receives as much attention as saltation—whereas Goldschmidt did emphasize saltation and downplay channeling in his later writings. In fact, in the 1933 article, Goldschmidt invokes channeling at the very beginning of his macroevolutionary discussion, just after citing the importance of development and even before he introduces the argument for saltation:

A considerable number of developmental processes between egg and adult have to be changed, in order to lead to a different organization. Development, however, within a species is, we know, considerably one-tracked. The individual developmental processes are so carefully interwoven and arranged so orderly in time and space that the typical result is only possible if the whole process of development is in any single case set in motion and carried out upon the same material basis.* Changes in this developmental system leading to new stable forms are only possible as far as they do not destroy or interfere with the orderly progress of developmental processes (p. 543).

The explicit invocation of orthogenesis then follows (p. 544): "If there are only a few avenues free for the action of mutational changes without knocking out of order the whole properly balanced system of reactions, the probability is exceedingly high that repeated mutations will go in the same direction, will be orthogenetic . . . We have pointed out a long time ago and still hold that orthogenesis is not the result of the action of selection or of a mystical trend, but a necessary consequence of the way in which the genes control orderly development—a way which makes only a few directions available to mutational changes."

Only now, after explicating the theme of channeling, does Goldschmidt introduce the rationale for saltation in its context: "But how about the possibility of occasional successful mutational changes acting upon earlier developmental processes? Would such a change, if possible at all without breaking up the whole system of the orderly sequence of development, not at once have the consequence of changing the whole organization and bridging with one step the gap between taxonomically widely different forms?" (p. 544).

Thereafter, as in the summary statement cited previously, Goldschmidt combines the two themes. He conjectures, for example, how saltation and orthogenesis might jointly explain phyletic sequences of limb rudimentation:

> Let us assume a mutational change in rate of differentiation of the limb-bud of a vertebrate . . . The consequent rudimentation of the organ would probably not interfere with orderly development of the organism. Here, then, an avenue would be open to considerable evolutionary change with a single basic step, provided that the new form could stand the test of selection, and that a proper environmental niche could be found to which the newly formed monstrosity would be preadapted and where, once occupied, other mutations might improve the new type. And in addition, the possibility for an orthogenetic line of limb-rudimentation would be a further consequence.

In the extensive reading required to compose a chapter like this, one acquires great respect for rare scientists with the mental power, and basic thoughtfulness, to explore and integrate the full set of implications and ramifications within great themes—and formalist vs. functionalist thinking must rank among the greatest of all biological themes (if only because this contrast expresses an attitude towards nature so deep and basic that the most important watershed in the history of biology—the development and acceptance of

---

*Note how Goldschmidt here uses the words that would become his title in 1940—but only for the developmental theme in this passage.

evolutionary theory—did not disrupt the discussion, but only reclothed an old antithesis in new language and causality).

In the post-Darwinian debates, I feel that only four evolutionists in my study fully plumbed the depth of this dichotomy (Galton specified all the themes, and even developed the canonical metaphor of the polyhedron, but he was only dipping): C. O. Whitman, Hugo de Vries, William Bateson, and Richard Goldschmidt. The views of these four men embody important differences, and Goldschmidt emerges as the best standard bearer for the full version of formalism. Two of the four embraced one aspect of Galton's polyhedron, but rejected the other for interesting reasons: Whitman as an orthogeneticist and gradualist; and de Vries as a saltationist who accepted the isotropy of species-level variation (and therefore constructed a higher-level Darwinism for trends among species). Bateson understood the connection and brought the themes together, but his generation hadn't gained enough knowledge about potential mechanisms to suggest more than an abstract and speculative synthesis. (Interestingly, Goldschmidt begins his 1933 article, his best presentation of the full critique, with a reference to Bateson's famous 1914 address to the British Association.)

Richard Goldschmidt understood all the connections and, however flawed the result, developed a coherent theory for a full internalist alternative to gradualist and Darwinian functionalism, a view that integrated both themes—facet-flipping and channeling—of Galton's polyhedron. Goldschmidt became the chief focus for vocal opposition by the synthesists, a symbol for all the bad old ways of outdated, typological thinking. I do not write to defend his specific ideas. The particulars of his genetic theory were deeply wrong, and disproved even in his lifetime, though he would not change his commitments. But I do maintain that his fully articulated critique remains as powerful as ever, and must be integrated with Darwinian orthodoxy to form a true and higher synthesis. In choosing Goldschmidt as the focus of their derision, the synthesists selected the right person for the best reason of all: Goldschmidt developed and fully understood all pieces of the critique, and he knew how the arguments cohered. Does the best fit always survive?

We need iconoclasts, if only to keep us thinking and probing. At the end of the Two Minutes Hate in *1984*, Emmanuel Goldstein's "hostile figure melted into the face of Big Brother, . . . full of power and mysterious calm, and so vast that it almost filled up the screen. Nobody heard what Big Brother was saying. It was merely a few words of encouragement, the sort of words that are uttered in the din of a battle, not distinguishable individually but restoring confidence by the fact of being spoken." Ignorance is not strength.

# CHAPTER SIX

# Pattern and Progress on the Geological Stage

## Darwin and the Fruits of Biotic Competition

### A GEOLOGICAL LICENSE FOR PROGRESS

A plethora of mottoes reminds us that political revolutions are never tidy—not a gentleman says one, no good omelette without cracking eggshells says another. Intellectual revolutions may avoid trails of blood (or they may not, at least metaphorically), but transitions in ideas can become as messy and complex as overthrows of temporal government. One world cannot be substituted for another without leaving some loose ends and some substantial pieces of an uncompleted puzzle.

Darwin got a great deal right, and he organized even more material into an internally coherent logic of argument. But he failed to achieve resolution on several important issues (especially when cultural convention clashed with implications of his theory), including some questions of great salience for him. Following his customary frankness, Darwin made no false claims for consistency, and ambiguities remain in his writing. In so doing, he followed the prescription for greatness in two famous statements by celebrated Americans: "A foolish consistency is the hobgoblin of little minds" (R. W. Emerson, 1841, *Self-Reliance*).

> Do I contradict myself?
> Very well then, I contradict myself,
> (I am large, I contain multitudes).
> (Walt Whitman, *Song of Myself*, from *Leaves of Grass*)

Darwin's greatest failure of resolution centered on an issue that assumed cardinal importance in Victorian culture—progress (both its definition, and its empirical and theoretical justification). Our current world of nuclear weaponry and global pollution does not rank this issue so centrally, but we have never escaped its allure. Several key figures of the Modern Synthesis devoted books to the subject (Huxley, 1953; Simpson, 1947; Dobzhansky, 1967; Stebbins, 1969), while symposia and volumes still appear with great regularity (Nitecki, 1988; Ruse, 1996; and Gould, 1996a, for a contrary view).

Darwin's dilemma can be stated easily: The bare-bones mechanics of the theory of natural selection provides no rationale for progress because the theory speaks only of adaptation to changing local environments. (The morphological degeneration of a parasite may enhance local adaptation as surely as any intricate biomechanical improvement in a bird's wing.) Moreover, Darwin regarded the banishment of inherent progress as perhaps his greatest conceptual advance over previous evolutionary theories—and he said so, often and forcefully, as in this epistolary comment, previously cited on page 373, to the American progressionist paleontologist Alpheus Hyatt on December 4, 1872: "After long reflection I cannot avoid the conviction that no innate tendency to progressive development exists" (in F. Darwin, 1903, vol. 1, p. 344).

On the other hand, Darwin was not prepared to abandon his culture's central concern with progress, if only to respect a central metaphor that appealed so irresistibly to most of his contemporaries—that if the history of life embodied predictable advance, then imperial expansion and industrial growth might be validated, at least by analogy, as the inherent consummation of Victorian desire and destiny, and not merely as an odd and ephemeral bump on the surface of history. And so Darwin penned other statements with equal assurance, as in this famous comment at the close of the *Origin:* "As natural selection works solely by and for the good of each being, all corporeal and mental endowments will tend to progress towards perfection" (1859, p. 489). Both opinions appear prominently and often in Darwin's writing, and they do not jibe.

This ambivalence on the specific question of progress highlights a broader issue at the center of Darwinism. Amidst the various meanings of Lyell's "uniformitarianism," one concept has been judged as paramount by many scholars (notably Rudwick, 1969): "non-progressionism" or uniformity of state—the proposition that the earth remains in a dynamic steady-state of constant, pulsating, cyclical change without direction: a strange kind of ahistoricism at the heart of ceaseless motion. Darwin owed a profound intellectual debt to Lyell, including far more than the expropriation of a geological stage to support the play of natural selection (see both Chapter 2 and later sections of this chapter). By transfer and analogy, Lyellian uniformity also provided a methodology for the general formulation and application of natural selection itself. Lyell's view of change gave Darwin a framework not only for the obvious features of gradualism, incrementalism, and extrapolationism (as often noted), but also for the less recognized ahistoricism of evolutionary mechanics. The bare bones of natural selection supply no vector for the pathway of life: environments change in their non-directional manner, and organisms respond in a continuous dance of local adjustment.

But the history of life includes some manifestly directional properties—and we have never been satisfied with evolutionary theories that do not take this feature of life into account (see Gould, Gilinsky and German, 1988). (Indeed, the stubbornly vectorial properties of paleontological change eventually led Lyell to surrender this key aspect of uniformity in later editions of the *Principles of Geology*—the most significant alteration of his intellectual ontogeny; see Gould, 1987b.) Darwin felt that natural selection could not be accepted as a thoroughly sufficient theory of evolution unless this mechanism could also explain evidences of pattern and vector in life's history. But how could Darwin meet such a requirement if natural selection—as a central attribute of its radical character and not a peripheral aspect easily withdrawn or compromised—had been devised as a biological analog for Lyell's uniformity of state, or non-directionalism?

I have just epitomized Darwin's dilemma in its most abstract form. In the immediate practice of his century, one prominent example consumed nearly all discussion of the general subject of vector and pattern—the concept of progress. If Darwin could validate progress by natural selection, then he would solve his dilemma of how to extract directional pattern from an apparently ahistorical theory.

This context of validating a concept of progress in macroevolution establishes an unconventional locus for a discussion—now to follow—on the key Darwinian subject of "struggle" and the nature of competition in general, but I am convinced that this topic finds its best fit at this point within the basic logic of Darwinian argument, and that a failure to recognize this appropriate place has led many evolutionists to underappreciate the theoretical significance of much that geology and paleontology have provided of late towards the reformulation of our subject—particularly the significance of mass extinction as an agent of change, and the central role of vectorial patterns as a subject in itself. (To readers who wondered why I treated struggle so cursorily in Chapter 2 on the essentials of Darwinian argument, I apologize for any puzzlement, while asserting that the subject—meriting all its traditional importance—belongs here.)

Evolutionary biologists should never lose sight of a cardinal principle linking history and function—that historical origin and immediate utility represent independent subjects with no necessary connection (see Chapter 11 for an extended discussion of this principle). Struggle and competition entered the ontogeny of Darwin's thought for a variety of reasons related to Malthus, the necessary hecatomb for powering natural selection, views on the plenitude of nature, etc. Struggle also serves many functions in the logic of Darwin's completed theory. But I believe that one role may be designated as paramount. Darwin used his distinctive views on struggle to validate the concept of progress as a cardinal vector in the history of life. He invoked his own interpretation of struggle—in particular, his conviction about the predominance of biotic competition—as an "added" principle to guarantee a pattern of progress that could not be derived, without such an auxiliary, from natural selection in its most abstract and generalized form.

But logics of argument form webs, and no benefit accrues without a price, or at least a set of implications. The dominance of biotic competition could validate progress—and thus, in a vital sense, "complete" the Darwinian system. But the adoption of such an argument required that a premise be imported from a field external to the biological logic of selection—and such increases in the logical complexity of theories also court danger. In this particular case, the domination of biotic competition as a patterning agent requires that the earth's geological history proceed in a particular way: for the stage of environmental change must permit the Darwinian play to operate (and dominate) in our real world. Even the most logical and brilliant theory can do no explanatory work if surrounding conditions never permit its results to emerge. (The expansion of $H_2O$ upon freezing may be both true and abstractly important, but irrelevant on a hot planet that has never experienced a temperature approaching $0°$ C.)

Darwin himself may not have felt the press or worry of this added commitment to a Lyellian earth, for his belief in such a world had deep roots, well antedating his formulation of natural selection (see his first three geological books on coral reefs, volcanic islands, and the geology of South America—1842, 1844, and 1846). Still, the conceptual constraint of requiring an external license for an internally consistent mechanism has operated as a distinctive and problematical claim throughout the history of Darwinism. I shall, in this chapter, first explicate Darwin's argument about biotic competition and progress, then discuss the required geological license more directly, summarize the strengths and character of the unfairly maligned catastrophist alternative, and suggest how an alteration of the geological stage might modify or expand the tenets of Darwinism.

## THE PREDOMINANCE OF BIOTIC COMPETITION AND ITS SEQUELAE

We all know that the most vulgar misinterpretation of Darwin, often willfully made for martial ends, holds that "survival of the fittest" mandates the subjugation and extermination of people and nations considered inferior. We also know the conventional and proper response to this harmful distortion: Darwin conceived "struggle" as a metaphorical concept defined in terms of reproductive success, not bloody battle. We can all cite the famous and standard quotation: "I should premise that I use the term Struggle for Existence in a large and metaphorical sense, including dependence of one being on another, and including (which is more important) not only the life of the individual, but success in leaving progeny. Two canine animals in a time of dearth, may be truly said to struggle with each other which shall get food and live. But a plant on the edge of a desert is said to struggle for life against the drought, though more properly it should be said to be dependent upon the moisture" (1859, p. 62).

Still, the link of struggle with overt battle does play a crucial role in Darwin's thought. He did include both biotic competition (the domain of overt battle) and prevalence in difficult environments (the plant at the edge of a desert) within his larger concept of struggle. And he did regard all forms of biotic competition, including symbiosis and symbolic posturing for success in mating—not only combat leading to death or injury—as modes of struggle. Nonetheless, by strongly emphasizing biotic over abiotic competition, and by stressing examples leading to the death of losers, Darwin did favor the close analogs of battle. Thus, his friend and supporter T. H. Huxley frequently referred to natural selection as the "gladiatorial theory" of existence and, in his famous essay on ethics and evolution (1893), urged human beings, as a primary ethical precept, to determine nature's ways and then act in an opposite manner. (Gladiators, by the way, and to make Huxley's etymological point, are not happy or grateful people, but warriors who fight to the death with a *gladius,* or sword in Latin.)

Darwin's decided choice in advocating a predominant relative frequency for biotic competition as a mode of struggle forms the crucial link in a chain of argument that stretches back to basic beliefs about the fullness of nature and points forward to a rationale for progress and the need for a uniformitarian geological stage. Consider a sequence of five consecutive, but interrelated subjects:

### The rule of biotic competition

Prince Peter Kropotkin, the charming Russian anarchist who spent 30 years in English exile, has generally been viewed as idiosyncratic and politically motivated in his famous attack on Darwinian competition, and his advocacy of cooperation as the norm of nature—*Mutual Aid* (1902). In fact, Kropotkin, who was well trained in biology, spoke for a Russian consensus in arguing that density-independent regulation by occasional, but severe, environmental stress will tend to encourage intraspecific cooperation as a mode of natural selection (Todes, 1988; Gould, 1991b). The harsh environments of the vast Russian steppes and tundras often elicited such a generalized belief; Kropotkin and colleagues had observed well in a local context, but had erred in overgeneralization. But Darwin and Wallace, schooled in the more stable and diversely populated tropics, may have made an equally parochial error in advocating such a dominant role for biotic struggle over limited resources in crowded space (Todes, 1988).

The shaping of diversity and the powering of natural selection by biotic competition—and not primarily by simple selective response to changing physical conditions—forms a central and recurring argument in the *Origin.* Three primary themes record its sway:

THE NECESSARY PREREQUISITE OF PLENITUDE. If populations generally stand at their carrying capacity, with numbers not fluctuating greatly, then biotic competition must dominate, for no group can increase except at the expense of others (while Kropotkin's underinhabited world can support more of any population if, by mutual aid, their members can counteract environmental stress). Darwin strongly subscribed to this version of the ancient principle of plenitude (see p. 229), arguing from his favored Malthusian base that a population's geometric capacity for growth guarantees the geologically instantaneous achievement of optimal numbers: "From the high geometrical powers of increase of all organic beings, each area is already fully stocked with inhabitants" (1859, p. 109).

METAPHORS OF COMPETITION. Since Darwin used metaphor so effectively, we can often infer his primary commitments from his choice of images. Darwin, as noted before, does cast a broad net in spreading "struggle" across biotic and abiotic realms—but war and conquest, combat and death, provide the principal examples of competition throughout the *Origin*. Why else did Tennyson's earlier line from *In Memoriam* (1850)—"nature red in tooth and claw"—become the canonical characterization of Darwin's world (see Gould, 1992a)? We may not know the particular reasons for success, but victory and battle set the appropriate context: "Probably in no one case could we precisely say why one species has been victorious over another in the great battle of life" (Darwin, 1859, p. 76). Two species, previously isolated and meeting for the first time, "are maiden knights who have not fought with each other the great battle for life or death. But, whenever . . . they meet, and come into competition, if one has the slightest advantage over the other, that other will decrease in numbers or be quite swept away" (*Natural Selection*, 1856–1858, 1975 edition, edited by Stauffer, p. 227). Calmness and cooperation may seem to hold sway, but lift the veil and observe the struggles to death in this vale of tears: "We behold the face of nature bright with gladness, we often see superabundance of food; we do not see, or we forget, that the birds which are idly singing round us mostly live on insects or seeds, and are thus constantly destroying life; or we forget how largely these songsters, or their eggs, or their nestlings, are destroyed by birds and beasts of prey" (1859, p. 62).

EXPLICIT STATEMENTS OF RELATIVE FREQUENCY. Darwin often contrasts the relative strengths of relationships among organisms vs. response to physical conditions. In each case, he stresses the greater importance of biotic competition. Migration, for example, will affect species more by forcing them into competition with other creatures than by exposing them to new physical environments: "These principles come into play only by bringing organisms into new relations with each other, and in a lesser degree with the surrounding physical conditions" (p. 351). Speaking of movement to oceanic islands, Darwin notes the "deeply-seated error of considering the physical conditions of a country as the most important for its inhabitants; whereas it cannot, I think, be disputed that the nature of the other inhabitants, with which each has to compete, is at least as important, and generally a far more important element of success" (p. 400). And, in his baldest statement, Darwin asserts (p. 477) that "the relation of organism to organism is the most important of all relations."

Darwin's caveats, whenever he presents a *prima facie* case for abiotic control, are even more revealing. Reports of biotic competition elicit only simple approbation, while putative examples of response to physical circumstances often provoke reminders that we may not be viewing the matter correctly, and that biotic competition may still be exerting a hidden sway.

> The structure of every organic being is related, in the most essential yet often hidden manner, to that of all other organic beings, with which it comes into competition for food or residence, or from which it has to escape, or on which it preys. This is obvious in the structure of the teeth and talons of the tiger; and in that of the legs and claws of the parasite which clings to the hair on the tiger's body. But in the beautifully plumed

seed of the dandelion, and in the flattened and fringed legs of the water-beetle, the relation seems at first confined to the elements of air and water. Yet the advantage of plumed seeds no doubt stands in the closest relation to the land being already thickly clothed by other plants; so that the seeds may be widely distributed and fall on unoccupied ground. In the water-beetle, the structure of its legs, so well adapted for diving, allows it to compete with other aquatic insects, to hunt for its own prey, and to escape serving as prey to other animals (p. 77).

And epidemic extirpations, not generally attributed to microorganisms in Darwin's day and therefore an apparently clear case of regulation by non-competitive forces, may be caused by "parasitic worms"—an organism large enough to engender thoughts about overt and visible competition between parasite and host: "When a species, owing to highly favorable circumstances, increases inordinately in numbers in a small tract, epidemics—at least, this seems generally to occur with our game animals—often ensue: and here we have a limiting check independent of the struggle for life. But even some of these so-called epidemics appear to be due to parasitic worms, which have from some cause, possibly in part through facility of diffusion amongst the crowded animals, been disproportionably favored: and here comes in a sort of struggle between the parasite and its prey" (p. 70).

### Wedging and the causes of extinction

Darwin's most striking metaphor for biotic competition, invoked from his very first jotting about natural selection (after his Malthusian insight of October 1838) to the *Origin of Species,* imagines a surface packed tightly with wedges, representing nature chock full to its carrying capacity. Such a maximally crowded world provides only one path for entry—by forcing ("wedging") another creature out. Biotic competition rules with a vengeance: "The face of Nature may be compared to a yielding surface, with ten thousand sharp wedges packed close together and driven inwards by incessant blows, sometimes one wedge being struck, and then another with greater force" (1859, p. 67).

The longer version of *Natural Selection,* Darwin's original manuscript, presents an even more revealing characterization, replete with almost frantic images of crowding, an explicit focus on species (in other passages, Darwin construes the wedges as individual organisms), and the relegation of physical limitation to an underlying layer, usually not penetrated, while the real work of nature proceeds by biotic struggle in the visible region above (*Natural Selection,* 1856–1858, 1975 edition, edited by Stauffer, p. 208):

Nature may be compared to a surface covered with ten-thousand sharp wedges, many of the same shape and many of different shapes representing different species, all packed closely together and all driven in by incessant blows: the blows being far severer at one time than at another; sometimes a wedge of one form and sometimes another being struck; and one driven deeply in forcing out others; with the jar and shock often transmitted very far to other wedges in many lines of direction: beneath the surface we may suppose that there lies a hard layer, fluctuating in its level, and which may represent the minimum amount of food required by each living being, and which layer will be impenetrable by the sharpest wedge.

In this exigent world of intense and ubiquitous competition, severity of struggle will be directly proportional to degree of relationship—most intense among members of the same species, strong between individuals of closely related species, and generally tapering with genealogical distance (and ecological dissimilarity). As a result, new species tend to eliminate their ancestors and closest relatives: "Each new variety or species, during the progress of its formation, will generally press hardest on its nearest kindred, and tend to exterminate them" (1859, p. 110).

Extinction therefore becomes a consequence of failure in biotic struggle, for ecosystems generally stand chock full, and new wedges must be poised to make their move whenever a chink appears. All species become enmeshed in a perpetual upward spiral, running continuously just to keep pace with their fellows—the Red Queen hypothesis (Van Valen, 1973): "For as all organic beings are striving, it may be said, to seize on each place in the economy of Nature, if any one species does not become modified and improved in a corresponding degree with its competitors, it will soon be exterminated" (1859, p. 102).

### The geological extension of wedging

If wedging rules the moment in a crowded world, then the extension of wedging through time should build patterns of origination and extinction in the fossil record. Following the dictates of the wedge, Darwin presents extinction as gradual and natural—not as rapid elimination in the wake of environmental catastrophe, but as slow diminution in the face of competition from "superior" forms, usually of close genealogical relationship (see Chapter 12, pp. 1296–1303, for a further development of this argument). Darwin chides us for ever regarding extinction as unusual, and draws an analogy to the inevitability of death, usually following a gradual course of prolonged weakening: "I may repeat what I published in 1845, namely, that to admit that species generally become rare before they become extinct—to feel no surprise at the rarity of a species, and yet to marvel greatly when it ceases to exist, is much the same as to admit that sickness in the individual is the forerunner of death—to feel no surprise at sickness, but when the sick man dies, to wonder and to suspect that he died by some unknown deed of violence" (p. 320).

Darwin counterposes this view of extinction as gradual failure in biotic competition to the alternative that both he and Lyell so strongly rejected—catastrophic global paroxysm and resulting mass extirpation: "On the theory of natural selection the extinction of old forms and the production of new and improved forms are intimately connected together. The old notion of all the inhabitants of the earth having been swept away at successive periods by catastrophes, is very generally given up" (p. 317).

Darwin centers his second geological chapter ("On the succession of organic beings," Chapter 10) upon an argument, framed in his usual mode of probing behind the literal appearance of an imperfect record, for the prevalence of a pattern that would validate gradual, biotically-driven extinction as a norm for the history of clades. Darwin denies that much extinction occurs by simultaneous or coordinated removal of unrelated forms. On the contrary, he argues, groups wane slowly and individually as superior competitors wax, producing a distinctive pattern of "megawedging" through geological time. "The extinction of old forms is the almost inevitable consequence of the production of new forms" (p. 343). "As new species in the course of time are formed through natural selection, others will become rarer and rarer, and finally extinct. The forms which stand in closest competition with those undergoing modification and improvement, will naturally suffer most" (p. 110).

### The validation of progress

For two reasons, Darwin could not find a rationale for progress in abiotic, physically-driven extinction and adaptation: first, a non-directional vector of environmental change can only elicit a set of meandering responses in the adaptive adjustments of organisms; second, the more serious challenge of catastrophe and mass extinction raises the specter of randomness and death for reasons unrelated to the adaptive struggles of normal times—the wheel of fortune vs. the wedge of progress (Gould, 1989d).

But victory over other creatures in an intense and unrelenting struggle for limited resources does permit an inference about progress. Now species triumph because, in some sense admittedly difficult to define, winners are "better" than the forms they vanquish. And the more uniformitarian the larger picture—the more that macroevolutionary pattern arises as a simple summation of immediate struggles—so do we gain increasing confidence that replacement and extinction must record the differential success of globally improved species. Thus, progress becomes an ecological concept for Darwin—not a deduction from the inevitable mechanics of natural selection, but a mode of operation for natural selection in a *particular kind* of ecological world. If crowded habitats, where creatures must struggle to the death for limited resources, represent an ecological norm on earth, and if geological change usually proceeds at a sufficiently stately and unobtrusive pace to permit the fruits of biotic competition to accumulate into patterns of origination and extinction through time, then we may understand why "organization on the whole has progressed" (p. 345). Darwin links all his statements about progress firmly to his ecological theory of plenitude and to the prevalence of biotic competition.

Consider Darwin's language and imagery ("inferior" forms "beaten" by "victorious" relatives) as he presents his key argument for linking the gradual geological decline of groups to the success of closely related competitors (a claim now strongly compromised by accumulating data on mass extinction—see Chapter 12): "The forms which are beaten and which yield their places to the new and victorious forms, will generally be allied in groups, from inheriting some inferiority in common; and therefore as new and improved groups spread throughout the world, old groups will disappear from the world; and the succession of forms in both ways will everywhere tend to correspond" (p. 327). Moreover, note Darwin's continual emphasis on advantage and competition in crowded ecosystems: "As natural selection acts solely by the preservation of profitable modifications, each new form will tend in a fully-stocked country to take the place of, and finally to exterminate, its own less improved parent or other less-favored forms with which it comes into competition" (p. 172).

The link of progress to biotic competition in a crowded world had permeated Darwin's thought from his first formulation of natural selection, as this passage from the E Notebook (January 18, 1839) indicates: "The enormous number of animals in the world depends on their varied structure and complexity.—Hence as the forms became complicated, they opened fresh means of adding to their complexity.—But yet there is no necessary tendency in the simple animals to become complicated although all perhaps will have done so from the new relations caused by the advancing complexity of others."

In the *Origin of Species*, all explicit statements about progress invoke a rationale of biotic competition, and employ a metaphor of battle. I find Darwin's conviction especially revealing in the light of his frank admission that he can neither formulate a way to test his proposal, nor specify a criterion by which progress might be measured:

> But in one particular sense the more recent forms must, on my theory, be higher than the more ancient; for each new species is formed by having had some advantage in the struggle for life over other and preceding forms. If under a nearly similar climate, the eocene inhabitants of one quarter of the world were put into competition with the existing inhabitants of the same or some other quarter, the eocene fauna or flora would certainly be beaten and exterminated; as would a secondary fauna by an eocene, and a paleozoic fauna by a secondary fauna. I do not doubt that this process of improvement has affected in a marked and sensible manner the organization of the more recent and victorious forms of life, in comparison with the ancient and beaten forms; but I can see no way of testing this sort of progress (pp. 336–337).

Darwin's most widely-quoted statement about progress appears in the summary to his two geological chapters. This famous passage also includes an odd mixture of firm conviction based on metaphors of competition ("the race for life" in this case), combined with some discomfort about the absence of a crisp definition: "The inhabitants of each successive period in the world's history have beaten their predecessors in the race for life, and are, in so far, higher in the scale of nature; and this may account for that vague yet ill-defined sentiment, felt by many paleontologists, that organization on the whole has progressed" (p. 345).

### Sequelae

The central importance to Darwin of a link between progress and biotic competition seems especially clear in the various ramifications that branch so richly from his basic proposition. All these sequelae point to a certain "smoothness," a form of predictability, an accumulation through time of the reasonable and little into the sensible and big. Nature is not capricious; superior forms prevail for cause; their triumph breeds further success and wider expansion; change proceeds in an orderly fashion—not in a clocklike manner to be sure, but at least decorously.

Widespread and speciose genera usually include the ancestral stocks of later successes, for extended geographic ranges and large populations indicate triumph in competition, and good mettle for future progress: "The great and flourishing genera both of plants and animals, which now play so important a part in nature, thus viewed become doubly interesting, for they include the ancestors of future conquering races. In the great scheme of nature, to that which has much, much will be given" (*Natural Selection*, 1856–1858, 1975 edition, edited by Stauffer, p. 248).

If brought into competition after previous isolation, big clades from large regions will prevail over less speciose groups from smaller areas because their members have been tested in hotter fires of competition: "For in the larger country there will have existed more individuals, and more diversified forms, and the competition will have been severer, and thus the standard of perfection will have been rendered higher" (p. 206). Thus the success of North American mammals in South America following the rise of the Isthmus of Panama "is due to the greater extent of land in the north, and to the northern forms having existed in their own homes in greater numbers, and having consequently been advanced through natural selection and competition to a higher stage of perfection or dominating power, than the southern forms" (p. 379). In a revealing metaphor, Darwin then praises "the larger areas and more efficient workshops of the north" (p. 380).

Looking at the complementary theme of failure, aberrant genera include few species because such creatures have been beaten by superior forms in competition (and not for a variety of other potential reasons including limited speciation, or specialization to rare and unusual environments): "Such richness in species, as I find after some investigation, does not commonly fall to the lot of aberrant genera. We can, I think, account for this fact only by looking at aberrant forms as failing groups conquered by more successful competitors, with a few members preserved by some unusual coincidence of favorable circumstances" (p. 429).

Since competition will be ubiquitous, efficient, and unrelenting in a crowded world, steady change should represent a norm, while stasis must record the unusual circumstance of reduced competition—as in the "living fossils" explicitly dubbed "anomalous" by Darwin: "These anomalous forms may almost be called living fossils; they have endured to the present day from having inhabited a confined area, and from having thus been exposed to less severe competition" (p. 107). In explaining why so few pairs of living species consist of one highly modified descendant and one unchanged surviving ancestor, Darwin invokes the high probability of substantial change, due to biotic competition, in *both* lineages stemming from a common root:

> It is just possible by my theory, that one of two living forms might have descended from the other; for instance, a horse from a tapir; and in this case direct intermediate links will have existed between them. But such a case would imply that one form had remained for a very long period unaltered, whilst its descendants had undergone a vast amount of change; and the principle of competition between organism and organism, between child and parent, will render this a very rare event; for in all cases the new and improved forms of life will tend to supplant the old and unimproved forms (p. 281).

Darwin's thought lies best revealed in a remarkable paragraph from the *Origin*'s final summary. All themes of this section now flow together—the denial of mass extinction (as Darwin borrows Lyell's favorite rhetorical trick of conflating this concept with nonscientific views of creation), the linkage of improvement in some groups to the extermination of competitors, and the strongest statement in the entire *Origin* about the predominant relative frequency of biotic competition vs. response to altered physical conditions. For Darwin now makes the boldest possible claim of all—an assertion that the ubiquity, continuity, and gradualism of biotic competition might actually permit us to use morphological change as a rough measure of elapsed time!*

> As species are produced and exterminated by slowly acting and still existing causes, and not by miraculous acts of creation and by catastrophes; and as the most important of all causes for organic change is one which is almost independent of altered and perhaps suddenly altered physical conditions, namely, the mutual relation of organism to organism—the improvement of one being entailing the improvement or the extermination of others; it follows, that the amount of organic change in the fossils of consecutive formations probably serves as a fair measure of the lapse of actual time (pp. 487–488).

In summary, Darwin's link of progress to biotic competition completes his argument against evolutionary systems (like Lamarck's) that propose separate forces for progress and adaptation, and that, as an unintended result, fall

---

*Every time Darwin makes such an overextended statement, his own honesty and subtlety draw him back immediately. The very next line presents the obvious caveat: "A number of species, however, keeping in a body might remain for a long period unchanged, whilst within the same period, several of these species, by migrating into new countries and coming into competition with foreign associates, might become modified; so that we must not overrate the accuracy of organic change as a measure of time" (p. 488).

into the disabling paradox analyzed in Chapter 2: *palpable phenomena are unimportant; while important phenomena remain intractable.* Lamarck distinguished a lateral process of local adaptation from a linear force of progress. Adaptation, as a local event of potentially rapid occurrence, could be observed, but this diversionary change provided no insight into the orthogonal and more important vector of progress through substantial time.

Darwin wanders on the fringes of the same dilemma. He identifies natural selection as a force of local adaptation. He wishes to escape the Lamarckian paradox of orthogonal causes by arguing for strict uniformity and extrapolation. The palpable and local force of adaptation therefore becomes, by smooth extension, the source of *all* evolutionary change at *all* levels. But how then could Darwin render progress—an idea that we might dismiss today as a cultural bias (Gould, 1996a), but that Darwin, as an eminent Victorian, did not wish to abandon (see Richards, 1992; Ruse, 1996)? Natural selection cannot provide the answer all by itself and without auxiliary principles, for this force must work in Lyell's world of non-directional uniformity. Natural selection, at the "bare bones" of its mechanism, only builds adaptation to changing local environments; the principle includes no statement about inherent directionality of any kind, not to mention progress.

Darwin resolved this tug of war between the logic of his theory and the needs of his century by invoking a particular ecological context as the normal stage for natural selection. If most ecosystems are chock full of life, and if selection usually operates in a regime of biotic competition, then the constant removal of inferior by superior forms will impart a progressive direction to evolutionary change in the long run. In opposition to most of his evolutionary predecessors (Lamarck in particular), who postulated a higher (and impalpable) realm of causality to encompass progress, Darwin stuck with his single level of immediate and testable natural selection—and ensured progress by adding a boundary condition about the state of ecology, rather than by devising an additional and untestable causal apparatus. By this ingenious strategy, Darwin managed to have his cake of unified theory at a single, accessible level, and also to satisfy his culture's hunger for rationalizing progress.

## Uniformity on the Geological Stage

### LYELL'S VICTORY IN FACT AND RHETORIC

I spoke in Chapter 2 of a "Goldilocks problem" in Darwin's views on the nature of environment and geological change. Since Darwin uses "trial and error" (with the organism proposing and environment disposing) as the chief metaphor in his predominantly externalist theory of change, the outer environment (biotic and abiotic) assumes a more important role in the theory of natural selection than in most other evolutionary accounts of the 19th century. For Darwin, environmental change must be neither too little, lest the external prod fail, nor too great, lest the prod become a determinant in itself, thus demoting the role of the organism. In practice, too little change only emerged as a serious option in a metaphorical sense, as embodied in Kelvin's argument for the restriction of geological time (see pp. 492–502). But too much change characterized the core of a geological system—catastrophism—that, if generally valid, would severely compromise Darwinism by the fundamental criterion of relative frequency.

Darwin's need for a "golden mean" of geological change flows from his extrapolationist premise that observable and small-scale natural selection can provide, by extension, the causal basis for life's history at grand scales of morphological transformation through geological time. Darwin rooted his defense of this premise in the validity of uniformitarianism, as preached by his guru, Charles Lyell. The uniformitarian defense of extrapolationism therefore undergirds the third leg of my proposed tripod for an "essence" of Darwinian theory. This Lyellian assumption buttresses the ordinary operation of natural selection in the immediacy of any ecological moment, but the theme of the first section of this chapter raises the ante by including Darwin's treatment of pattern on a geological stage. The raw mechanism of natural selection provides no direction for organic change, and yields no predictable order for life's history through time. However, by adding a set of distinctive ecological arguments to the bare-bones mechanics—notably the domination of overt biotic competition as a primary mode of struggle within perpetually crowded communities—Darwin could validate the central belief of his surrounding culture, the concept of progress, as a primary signal of life's history.

Thus, the "golden mean" of geological change became doubly important to Darwin, because both the general operation of natural selection, and his particular rationale for progress in macroevolution, require a Lyellian geological world. The specter of catastrophism also became much more potent in the light of Darwin's stipulation that biotic competition acts as the chief agent of direction in life's history. For if mass extinction (and other phenomena of "too much" environmental change) establish patterns in the history of life at too high a relative frequency, then biotic competition will be demoted, if not replaced, by an ordering force of opposite meaning—for mass extinctions introduce a powerfully confusing and potentially confuting new actor: the tumbling, whimsical wheel of fortune rather than the slow and steady wedge of progress.

The norms of science dictate that major works be presented as objective explorations of data, with general conclusions derived from empirical evidence and devised late in the process of discovery. But most seminal books in the history of science can be read as briefs for passionately-held, elegantly-articulated, brilliantly-advocated (and, to be sure, well-defended) views of nature. As a premier example, Charles Lyell, a lawyer by profession, may have presented his epochal *Principles of Geology* in the conventional style of humble factual documentation. But this great work must be understood as perhaps the most explicit and most able brief ever presented in the guise of a major scientific treatise.

The sources of Lyell's success in promoting his uniformitarian view—which later emerged as such a fitting solution to Darwin's Goldilocks problem by describing an earth with a "just right," intermediate and dependable level of geological change—have been extensively explored by scholars in the past thirty years, and a general consensus has emerged (Hooykaas, 1963; Simpson, 1963; Porter, 1976; and especially Rudwick, 1969; my own first publication developed some of the same ideas independently and in the midst of some embarrassing juvenilia—see Gould, 1965). Lyell presented a plethora of compelling and well-presented evidence in his favor (gradualism does, after all, maintain at least a respectable relative frequency among patterns of geological change); but he triumphed as much by force of rhetoric, as by strength in documentation. Two features of his rhetoric stand out for effectiveness.

1. He invented a persuasive dichotomy, pitting uniformity and rectitude on one side, against catastrophism and reaction on the other. Catastrophism, Lyell argued, represented everything that had stifled the development of geology in a dismal past—not only for the falsity of claims for worldwide paroxysmal change, but also (and especially) for the sterility of a method that sought to explain the past by causes that do not operate today on our slowly changing earth. In attacking his cardboard version of catastrophism, Lyell penned some of the finest polemical lines ever written by a scientist: "Never was there a dogma more calculated to foster indolence, and to blunt the keen edge of curiosity, than this assumption of the discordance between the former and existing causes of change." Catastrophist geology became "a boundless field for speculation" that could "never rise to the rank of an exact science." Lyell ended this volley with his most famous metaphor: "We see the ancient spirit of speculation revived, and a desire manifested to cut, rather than patiently to untie, the Gordian knot" (1833, volume 3, p. 6).

2. He took advantage of a "creative confusion" by extending the umbrella of his single term "uniformity" over a variety of concepts with differing status—thereby attempting to win assent for claims of dubious merit by giving them the same name as other arguments that all scientists accept as valid. In particular, Lyell stoutly defended—and defined as "uniformity"—a set of methodological assumptions included within any full and proper definition of science (and embraced with equal vigor by all serious catastrophists as well; see Gould, 1987b): especially the spatiotemporal invariance of natural law and the actualistic principle that hypothetical causes should not be postulated so long as observable modern processes can generate the phenomenon in question, at least in principle. But Lyell also extended the term "uniformity" to a set of empirical claims about the natural world—testable statements that might be true or false, but emphatically cannot be treated as methodological assumptions, necessarily embraced *a priori* as a license to practice science at all. Two of these "substantive uniformities" influenced Darwin greatly, and have echoed loudly through the 20th century as well: gradualism, or uniformity of rate (especially the production of large-scale phenomena by accumulation of ordinary, daily effects through immense stretches of time); and non-directionalism, or uniformity of state (the empirical pattern of ceaseless, often cycling modifications, without vectors of directional change). Lyell eventually abandoned uniformity of state, when he finally became convinced,

in the 1850's, that the fossil record of vertebrates exhibited vectorial change, and when Darwin then argued so persuasively that evolution could serve Lyell as a strategy of minimal retreat, permitting him to retain all other major components of his world view, while moving to acknowledge life's directional history. Gradualism then became—and has remained ever since—the sole surviving cardinal claim of uniformity in the substantive mode.

In his most clever, and devastatingly effective, trope of rhetoric, Lyell argued that the substantive claims of "uniformity" must be valid because the basic practice of science requires that we accept a set of methodological assumptions bearing the same name despite their truly different status ("uniformity" of law and process). In so doing, Lyell managed to elevate a testable claim about gradualism to the status of a received *a priori* doctrine vital to the successful practice of science itself. This subtle conflation has exerted a profound, and largely negative, influence upon geology ever since, often serving to limit and stifle hypotheses about rates of processes, and to bring derision upon those who advocated even local catastrophes. (Consider the now standard story of J Harlen Bretz and his long reviled, but later vindicated, catastrophic explanation for the channelled scablands of Washington by sudden flooding—Gould, 1980d; Baker and Nummedal, 1978.) In the obvious contemporary example (see Chapter 12 for details), no one can comprehend the emotional vigor of the debate engendered by Alvarez's proposal for catastrophic mass extinction by extraterrestrial impact (Alvarez *et al.*, 1980, and the oral history of Glen, 1994) without understanding the historical legacy of Lyell's successful and tricky rhetorical argument against catastrophism.

Ernst Mach and many others have truly (and famously) noted that, for "big" issues, scientific reform proceeds largely by persuading the next generation. Mach's claim has usually been cited in the somewhat cynical mode: one must wait for the old generation to die because nothing can change their minds. But the same transgenerational theme applies, in an oddly backwards manner, to false characterizations that win assent by force of rhetoric. Such misattributions don't persuade contemporaries who understand the subtleties of the real issues by direct experience. But, since historical memory tends to occupy only a narrow range from nonexistent to short among scientists, false versions begin to prevail as soon as the actual practitioners die, and cardboard can quickly replace flesh. Thus, anyone who knew Cuvier, Elie de Beaumont, or d'Orbigny, recognized their mental power, their scientific integrity, and the considerable empirical support enjoyed by their systems. But when these men died, Lyell's characterization persisted, and "catastrophism" became equated with anti-science and dogmatic theological reaction. The label stuck, and Lyell's rhetorical triumph placed catastrophism beyond the pale of scientific respectability.

The arms of misreason extend across generations. When primary documents disappear from sight,* textbook pap can clone itself, and resulting leg-

---

*Not nefariously, in this case, for the great works of Cuvier and other catastrophists have always remained on library shelves, and have been much valued by historians and col-

ends then beget further fantasy with little hope for correction within an established system of belief. Thus Cuvier, one of the greatest intellects of 19th century science, a child of the Enlightenment and a champion of rationality, became a miracle-mongering apologist for ecclesiastical reactionaries who had thrust their fingers into the crumbling dike of superstition in a vain effort to stem the inexorable advance of Lyellian science. Consider just two characterizations of Cuvier from leading geological textbooks of the last generation: Gilluly, Waters and Woodford (1959) on catastrophes: "These, he [Cuvier] believed, destroyed all existing life, and following each a whole new fauna was created: this doctrine, called Catastrophism, was unquestionably inspired by the Biblical story of the Deluge." Or Stokes (1973, p. 37) on the progress of science: "Cuvier believed that Noah's flood was universal and had prepared the earth for its present inhabitants. The Church was happy to have the support of such an eminent scientist, and there is no doubt that Cuvier's great reputation delayed the acceptance of the more accurate views that ultimately prevailed."

I don't raise this example in the abstract interest of intellectual justice. The acknowledgment of catastrophism as a viable alternative to Darwin's geological requirements establishes an important theme of this book, and a potent reforming force within modern evolutionary theory. We might pursue this issue only by assessing the validity of modern arguments in the catastrophic mode, thus continuing to ignore earlier history (usually because we accept the cardboard characterization of Cuvier and colleagues, and therefore regard contemporary claims as viable for the first time). I will discuss some modern rationales in Chapter 12; but, in this historical part of the book, I need to demonstrate that catastrophism contained important elements of validity from the start—elements that rebut Darwin's crucial claim for gradual accumulation of changes induced by biotic competition as the predominant vector of life's history. I therefore present the basic argument in the most important of all catastrophist texts, the *Discours préliminaire* of Georges Cuvier (1812, but in its canonical English translation by Jameson, 1818). I do not, of

---

lectors. But never doubt the power of false characterization to ban effective consideration of the readily available. A scientist beyond the pale becomes an object of ridicule without being read—and the force of silence should never be underestimated. To cite just one personal anecdote about Cuvier and his *Discours préliminaire:* The stereotyped Cuvier stands accused in most textbooks for arguing that catastrophes wipe all life off the face of the earth, and that God then creates new biotas from scratch. But Cuvier never advanced such a claim. No doubt, when pressed, he would have accepted some new creation to replenish a depleted world. But he attributed much local faunal change across stratigraphic boundaries to migration from previously isolated areas following geographic alterations that accompany episodes of rapid geological change (citing, as a potential example, the migration of Asian mammals to Australia should a land bridge ever connect these continents). Cuvier didn't hide this argument; he presents his viewpoint prominently in Section 30 of the canonical Jameson translation (1818, pp. 128–129). Yet, at least a half dozen times in my professional life, colleagues ranging from graduate students to senior professors have approached me with excitement, thinking that they had just made an important and original discovery: "Hey, look at this. Cuvier didn't believe in complete replacement by new creation . . ." "Yes," I reply, "page 128; the passage has always been there."

course, claim that the demonization of catastrophism arose as a Darwinian plot, for Lyell's effort long predates the acceptance of evolutionary theory. But I will argue that Darwin needed Lyellian geology to grant natural selection (and biotic competition) a dominant role in setting macroevolutionary pattern. Moreover, I do not claim that Cuvier was right and Lyell wrong—thereby "correcting" Lyell's persisting unfairness with a modern version that would be equally false and one-dimensional in the other direction. Cuvier should not be resurrected as more right than Lyell, but his views must be reassessed as sufficiently valid to revoke the license that Darwin recognized as so crucial for granting a dominant relative frequency to gradual geological change.

### CATASTROPHISM AS GOOD SCIENCE: CUVIER'S ESSAY

A central irony pervades the story of Lyell's rhetorical victory over catastrophism. Textbook pap, extending the exaggeration even beyond Lyell's mischaracterization, has leveled two major charges against the catastrophists: first, that they downplayed or distorted geological facts to defend their *a priori* beliefs; second, that they invented theories primarily to support a religious traditionalism linked to a restricted time scale and the defense of Noah's flood and other Biblical stories.

I call this description ironic because all leading catastrophists embraced a general conception of science entirely contrary to this mischaracterization. The catastrophist synthesis, as a working theory, rested upon two pillars, one substantive and the other methodological. Substantively, as the name implies, the theory regarded major geological change as concentrated in infrequent bursts of global paroxysm.* But, in a coordinating theme of equal importance, this sequence of catastrophes imparted a directional history to the earth and life.

Most catastrophists viewed the series of paroxysms as diminishing in intensity through time. They also postulated a geological dynamics to explain the link of directionality with occasional paroxysm. The theory of the French geologist Elie de Beaumont summarized the postulated mechanics of catastrophism. The earth, as a result of "hot" formation under the nebular hypothesis of Kant and Laplace, has cooled continuously through time, thus establishing a primary directional vector of change. This secular cooling engenders a catastrophist dynamics, for the outer crust solidifies into a rigid shell, while the inner matter, still molten, contracts in cooling. The "pulling away" of this inner core from the rigid outer crust creates an instability, resolved not by grad-

---

*The old canard about advocating a short, even a Mosaic, time scale arises from an illogical extension of this claim. A short time scale does require paroxysm to encompass events of the geological record within such a limited span. But the converse of this argument—the claim falsely attributed to catastrophists—does not follow: for a dynamics of paroxysm does not require or even imply a doctrine of limited time. The earth may be millions or billions of years old, as the catastrophists believed, and still concentrate its major changes in brief bursts.

ual change, but by rare, global paroxysms, when the crust fractures and collapses upon the shrunken core. Life's vector of progress records an increasing adaptation to harsher climates of a cooling earth.

Since a coordinating vector of temporal cooling generates the entire system, and since many scientists and historians regard the theme of directionalism as even more central than the dynamics of paroxysm, several scholars have urged, in recent years, that the entire movement be redesignated as the "directionalist synthesis," rather than "catastrophism." The major "catastrophists" never defined themselves as a school opposed to a dichotomous Lyellian alternative, and therefore never gave their movement a name. The construction of such a dichotomy, with moral values attached to each side, set a major aspect of Lyell's rhetorical strategy.

Methodologically, all leading catastrophists adopted a distinctive attitude towards the geological record. They preached a radical empirical literalism: interpret what you see as a true and accurate record of actual events, and interpolate nothing. If horizontal strata overlie a sequence of broken and tilted beds, then a catastrophe must have terminated one world and initiated another, as the geological discontinuity implies. If one fauna disappeared at such a boundary, and younger beds contain fossils of different creatures, then a mass extinction must have eradicated the older fauna. The catastrophists advocated directionalism as a primary theme for the earth's history, and empirical literalism as a fundamental approach to science.

How ironic, then, that modern textbook cardboard should misidentify Lyell as an empiricist who, by laborious fieldwork and close attention to objective information, drove the dogmatists of catastrophism out of science. To the contrary, the catastrophists were the empirical literalists of their time! Lyell and Darwin opposed catastrophism by probing "behind appearance" to interpret, rather than simply to record, the data of geology. For Lyell and Darwin, the geological record must be treated as imperfect to an extreme degree—in the standard metaphor developed by Lyell and propagated by Darwin, like a book with few pages preserved and only a few letters surviving on each of these pages. Moreover, Lyell argued, the geological record has also become distorted in a systematic way that would foster a false concept of change if we attempted a literal reading. Geological unconformities and local extinctions look paroxysmal, but only because slow, daily changes rarely leave any evidentiary trace at all. We therefore can observe only the infrequently preserved waystations of a true continuity, and we misinterpret the massive lacunae as evidence for rapid change. If, to cite Lyell's example, Vesuvius erupted again and buried a modern Italian town directly atop Pompeii, would we interpret history by the literal evidence of a Latin culture suddenly extirpated in a (potentially global) episode of volcanism, then followed by the saltational origin of a distinct, but clearly allied, Italian civilization, accompanied by such new cultural artifacts as beer cans and electric bulbs?

Proper procedure in geology, Lyell asserted, requires that we interpolate into a systematically impoverished record the unpreserved events implied by our best theoretical understanding. Lyell and Darwin worked by interpretation and interpolation; the catastrophists preached empirical literalism! (I do not raise this issue to denigrate Lyell and Darwin, for I support their procedure as a general statement about scientific methodology. Slavish literalism should be shunned in general, and not only (as in this geological case) when we have reason to regard a preserved record as systematically imperfect. Still, I know no greater irony in the history of science than the inverted posthumous reputations awarded to Lyell and the catastrophists for their supposed positions on "objectivism" in science.)

In paleontology, catastrophism reached an apogee in Georges Cuvier's *Discours préliminaire*, originally written as a preface to his great four-volume compendium on fossil vertebrates (*Recherches sur les ossemens fossiles*, 1812), but published and republished separately as an "Essay on the theory of the earth." Cuvier did not present his *Essay* as a textbook of catastrophism, but as a statement about the roles that paleontology and geology should play in unravelling the history of the earth. Nonetheless, Cuvier's *Essay* exposes all characteristic features of catastrophism as a science, and illustrates the incompatibility of this geological approach with Darwin's prerequisites for natural selection as a chief agent of macroevolutionary pattern.

On the substantive side of catastrophism, Cuvier devoted most attention to demonstrating life's temporal directionality, and to illustrating the value of such a vector for inferring geological history and stratigraphic order. As his greatest contribution, Cuvier proved that species could become extinct (a phenomenon still widely doubted at the inception of the 19th century). In his major source of evidence, Cuvier demonstrated that the anatomy of some fossil quadrupeds lay outside the boundaries of variation within modern species. He also traced a stratigraphic sequence of increasing similarity to modern faunas in successively younger beds, thus documenting a directional pattern within sequences of extinction, and providing the earth with a meaningful history. Cuvier begins the *Essay* by castigating his predecessors for combining their grandiose speculative theorizing with an inattention to fossils and their stratigraphic positions. He then presents his concept of proper procedure in the form of a list of questions, mostly centered upon historical pattern and direction in stratigraphy. "Are there certain animals and plants peculiar to certain strata, and not found in others? What are the species that appear first in order, and those which succeed? Do these two kinds of species ever accompany one another? Are there alterations* in their appearances; or, in other words, does the first species appear a second time, and does the second species then disappear?" (1818, p. 65).

Cuvier's answer, leading to the birth of modern paleontology, affirms directionality in two senses: fossils from successively older strata become increas-

---

*Cuvier's original text reads: "Y a-t-il des alternatives dans leur retour," so "alterations," in Jameson's standard translation, should probably read "alternations," thus rendering Cuvier's inquiry as a question about directionality. He wants to know whether fossil species mark unique episodes of time, a proposition that would be disproved if faunas appeared and reappeared in alternation.

ingly less like modern forms, and thus ever more "primitive" by conventional definitions of progress:

> It is, in the first place, clearly ascertained, that the oviparous quadrupeds are found considerably earlier, or in more ancient strata, than those of the viviparous class . . . The most celebrated of the unknown species belonging to known genera, or to genera nearly allied to those that are known, as the fossil elephant, rhinoceros, hippopotamus, and mastodon, are never found along with the more ancient genera; but are only contained in alluvial formations . . . Lastly, the bones of species which are apparently the same with those that still exist alive, are never found except in the very latest alluvial depositions (1818, pp. 112–115).

Cuvier expresses a similar interest in the directionality of physical history. He argues (following the Wernerian system) for systematically changing mineralogy through time, and for a pattern of increasing restriction of effect, as an original and universal ocean shrinks, thus decreasing the intensity of catastrophes as well: "The sea has not always deposited stony substances of the same kind. It has observed a regular succession as to the nature of its deposits; the more ancient the strata are, so much the more uniform and extensive are they; and the more recent they are, the more limited are they, and the more variation is observed in them at small distances" (1818, p. 34).

Cuvier treats directionality as his principal theme, but the validation of catastrophe does not rank far behind, and the two subjects mesh into a distinctive and comprehensive view. Cuvier opens the *Essay* with an exposition of catastrophist dynamics. Interestingly, he begins, as Lyell did from the other side (and as good advocates so frequently do), with a potent rhetorical device: we see the world in an inherently biased way from our limited and daily perspective, but deeper investigation reveals that opposite forces prevail in the fullness of time. Lyell began by questioning our undue focus on civil catastrophes of death, famine and war, and by arguing that we overemphasize such tragedies as a consequence of their personal impact. We therefore fail to appreciate the far greater power of ordinary events to render history by accumulation through time. Cuvier, in reversed perspective, claims that we grant too much power to the calm of daily life because we live within its immediate, surrounding pervasiveness. We therefore fail to realize that rare and unusual events set the basic pattern of history. After a preliminary discussion about the data and power of natural history as a science, Cuvier begins his *Essay* with a striking image devised to equate catastrophism with a broad and generous vision of reality:

> When the traveller passes through those fertile plains where gently-flowing streams nourish in their course an abundant vegetation, and where the soil, inhabited by a numerous population, adorned with flourishing villages, opulent cities, and superb monuments, is never disturbed except by the ravages of war and the oppression of tyrants, he is not led to suspect that nature also has had her intestine wars, and that the surface of

the globe has been much convulsed by successive revolutions and various catastrophes. But his ideas change as soon as he digs into that soil which presents such a peaceful aspect, or ascends the hills which border the plain; they are expanded, if I may use the expression, in proportion to the expansion of his view; and they begin to embrace the full extent and grandeur of those ancient events (pp. 29–30).

The next several sections of the *Essay* present, in sequence, a framework for regarding catastrophe as the primary agent of geological change—"proofs of revolutions"; "proofs that revolutions have been numerous"; "proofs that revolutions have been sudden"; "proofs of the occurrence of revolutions before the existence of living beings." Cuvier then examines, as Lyell did but to reach an opposite conclusion, the efficacy of modern causes, declaring them insufficient to render the events of catastrophic episodes. Cuvier's words—in the most famous passage of his entire *oeuvre*—have usually been cited out of context, to equate catastrophism with despair and even with hostility to scientific explanation. But, clearly, Cuvier harbored no such intent. He does express some regret at the discordance between catastrophic and daily causes—for the task of science would become easier if present forces sufficed. But the significance of the geological record lies in its potential for documenting the catastrophic causes:

> It has long been considered possible to explain the more ancient revolutions on its surface by means of these still existing causes; in the same manner as it is found easy to explain past events in political history, by an acquaintance with the passions and intrigues of the present day. But we shall presently see that unfortunately this is not the case in physical history; the thread of operation is here broken, the march of nature is changed, and none of the agents that she now employs were sufficient for the production of her ancient works [or, to cite Cuvier's most famous line in its French original—*le fil des opérations est rompu; la marche de la nature est changée; et aucun des agens qu'elle emploie aujourd'hui ne lui auroit suffi pour produire ses anciens ouvrages*] (1818, p. 44; 1812, p. 17).

Note Cuvier's careful choice of words. He does not appeal to mystery by stating that current causes didn't work in an uninterpretable past; rather, he deems modern causes insufficient to explain the evidence for historical catastrophes. We must therefore study the geological record directly if we wish to resolve the causes of catastrophes. I find nothing objectionable, or contrary to good scientific methodology, in this argument.

Cuvier summarizes the substantive part of his *Essay* in a paragraph that unites catastrophe with directionality, and the physical record with the biological history of the earth:

> Life, therefore, has been often disturbed on this earth by terrible events—calamities which, at their commencement, have perhaps moved and overturned to a great depth the entire outer crust of the globe, but

which, since these first commotions, have uniformly acted at a less depth [sic] and less generally. Numberless living beings have been the victims of these catastrophes; some have been destroyed by sudden inundations, others have been laid dry in consequence of the bottom of the seas being instantaneously elevated. Their races even have become extinct, and have left no memorial of them except some small fragment which the naturalist can scarcely recognize (1818, p. 38).

The last line of this quotation helps to explain the longest and most brilliant (though ultimately incorrect) final section of the *Essay*—the source of so much misunderstanding about Cuvier, and about catastrophism. Given the fragmentary nature of geological evidence, and the tendency for such evidence to become more and more inadequate as we penetrate deeper into time, our best empirical hope for understanding catastrophes lies in a detailed study of the most recent event. By coordinating two sources of evidence— natural history for estimating the effect of ordinary causes since the last paroxysm, and civil history (because the last catastrophe occurred within human memory)—we might characterize at least one event well enough to build a model for the generality. Cuvier therefore scans the oldest records of all cultures, rejecting some as fabulous, adjusting and coordinating others, and finally reaching the conclusion that "the crust of our globe has been subjected to a great and sudden revolution, the epoch of which cannot be dated much farther back than five or six thousand years ago" (1818, p. 166). Since Western culture recorded this event as Noah's flood, and since Cuvier used the Bible as one source of information about this episode, posterity has interpreted this section of the *Essay* as a tortured exercise in Christian apologetics—thus affirming the usual interpretation of catastrophism as theological reaction.

But Cuvier proceeds with a precisely opposite intent, and his *Essay* therefore becomes a seminal work of Enlightenment humanism. Cuvier does not marshall geological and civil history to support the biblical account of Noah's flood; rather, he uses the Old Testament as one source among many in a broad effort to unite the traditions of disparate fields towards a common intellectual goal. Of course Cuvier cites the Bible—as one legitimate source for making historical inferences, but with no favored status. Cuvier pays as much attention to the traditions of the Assyrians, the Parsis and the Hindus, and he grants even more credence to the records of ancient China. His evidence for the last catastrophe does not rest upon scriptural assertion, but on a supposed confluence of differing empirical sources, as the lengthy title for this chapter of the *Essay* proclaims: "The concurrence of historical and traditionary testimonies, respecting a comparatively recent renewal of the human race, and their agreement with the proofs that are furnished by the operations of nature."

Cuvier's *Essay* also stresses the methodology of catastrophism, particularly the empirical literalism of its favored approach to the geological record. Cuvier, even more strongly than Lyell, dismisses the speculative tradition pursued by previous generations, with their grandiloquent claims for comprehensive grasp, but so little attention paid to facts of the stratigraphic record. (Cuvier even includes Lyell's hero, James Hutton, among the speculative system builders, though praising him faintly for proceeding "with more caution.") True scientific progress now demands a methodological revolt—a scaling down of explanatory focus from fuzzy and grand theorizing to immediate and palpable observation, a move from the armchair to the field and museum. Progress also demands the specific coordination of two empirical sources—stratigraphic succession as documented by fieldwork, and taxonomic knowledge of organic diversity as revealed in large museum collections. We may designate the knowledge of causes as our final goal, but we must proceed by voluminous and coordinated study of the empirical record:

> If, from the want of sufficient evidence, these questions cannot be satisfactorily answered, how shall we be able to explain the causes of the presently existing state of our globe . . . Naturalists seem to have scarcely any idea of the propriety of investigating facts before they construct their systems. The cause of this strange procedure may be discovered by considering that all geologists hitherto have either been mere cabinet naturalists, who had themselves hardly paid any attention to the structure of mountains, or mere mineralogists, who had not studied in sufficient detail the innumerable diversity of animals, and the almost infinite complexity of their various parts and organs. The former of these have only constructed systems; while the latter have made excellent collections of observations, and have laid the foundations of true geological science, but have been unable to raise and complete the edifice (1818, pp. 66–67).

Cuvier's unwillingness to proceed much beyond immediate data deprives the *Essay* of any "grand" conclusion—perhaps for the better. Cuvier never specifies how the two substantive themes of directionalism and successive catastrophes might unite to forge a general theory of the earth's behavior. He presents no proposal, like Elie de Beaumont's of later years, for a general theory of planetary dynamics. Cuvier's final section includes no general summary, no stirring plea for ultimate solutions, but only presents some practical suggestions for fruitful empirical work, accompanied by a list of potential examples. We should now focus, Cuvier argues, not on the most recent strata (which have been intensely studied already), and not on the earth's beginnings (which remain too distant and too different for adequate resolution), but on fossiliferous rocks of intermediate age—on the gypsum quarries of Aix, the sand-hills of the Apennines, and the "stinkstone slate of Oeningen." "It appears to me," Cuvier concludes, "that a consecutive history of such singular deposits would be infinitely more valuable than so many contradictory conjectures respecting the first origin of the world and other planets and respecting phenomena which have confessedly no resemblance whatever to those of the present physical state of the world" (1818, p. 173).

Lyell would not have disagreed with these sentiments; homilies about the primacy of observation, after all, top the list of clichés in scientific prose. As their primary query, both Cuvier and Lyell asked not whether, but rather how, science should treat the empirics of the geological record—and Cuvier's chief difference with Lyell and Darwin centered upon his empirical literalism versus their commitment to probing behind the appearances of a systematically imperfect record. A dramatic example of this distinction occurs at a key point in Cuvier's *Essay,* and serves to illustrate the real and continuing contrast between legitimate themes in catastrophism versus Darwin's need for a uniformitarian geology based upon the accumulation of small effects. In Section 30 on "proofs that the extinct species of quadrupeds are not varieties of the presently existing species," Cuvier considers a potential argument against the reality of mass extinctions produced by geological catastrophes. No one can deny that many fossil quadrupeds represent species no longer living—for Cuvier had proved this point beyond a doubt. Only one logical alternative therefore remained to challenge Cuvier's own conclusion that these ancient species had perished—namely, evolution. Perhaps these forms never died, but gradually changed into different species now extant. Lamarck, Cuvier's closest colleague, had been advocating this idea for more than a decade, and straining their former friendship thereby (see Chapter 3). But Cuvier replied with an empirical rejoinder that could scarcely be gainsaid so long as the fossil record could be treated as literally accurate. No intermediary forms have been found as fossils, while new species occur in strata directly atop the doomed faunas:

> This objection may appear strong to those who believe in the indefinite possibility of change of forms in organized bodies, and think that during a succession of ages, and by alterations of habitudes, all the species may change into each other, or one of them give birth to all the rest. Yet to these persons the following answer may be given from their own system: If the species have changed by degrees, as they assume, we ought to find traces of this gradual modification. Thus, between the *Palaeotherium* and the species of our own days, we should be able to discover some intermediate forms; and yet no such discovery has ever been made. Since the bowels of the earth have not preserved monuments of this strange genealogy, we have a right to conclude, That the ancient and now extinct species were as permanent in their forms and characters as those which exist at present; or at least, that the catastrophe which destroyed them did not leave sufficient time for the production of the changes that are alleged to have taken place (1818, p. 119).

Darwin, as we all know, did not challenge Cuvier's literal description, but argued that a woefully imperfect record had failed to preserve insensibly graded intermediates in almost all cases. And Darwin, with his characteristic honesty, also admitted that his entire system depended upon the validity of this approach to the fossil record: "He who rejects these views on the nature of the geological record, will rightly reject my whole theory" (1859, p. 342).

In another confessional paragraph in the same mode, Darwin wrote of his debt to Lyell: "He who can read Sir Charles Lyell's grand work on the Principles of Geology, which the future historian will recognize as having produced a revolution in natural science, yet does not admit how incomprehensibly vast have been the past periods of time, may at once close this volume" (1859, p. 282). Of course, we now acknowledge the immensity of time and therefore continue to open the *Origin*. We also dismiss Cuvier's argument that lack of recorded intermediacy precludes evolution.

But Cuvier's claim retains force in a more restricted sense: the earth's measured history does not provide enough time to dissolve the appearance of mass extinction into gradual change masked by an imperfect record. Geology therefore does run "too fast" to explain these crucial episodes by extended gradualism through an imperfect record—and if catastrophic mass extinctions are, as many paleontologists now argue (see Chapter 12), more frequent, profound, rapid, and different in their effects than we had previously admitted, then natural selection in Darwin's accumulative mode, with biotic competition as a primary source of order, may become seriously demoted in relative frequency among the causes of macroevolutionary pattern. In this sense, the pace of geological change remains a vital subject for evolutionary theory, and Cuvier's last line still sounds a valid warning. For evolutionists committed to biotic competition by wedging as the primary source of macroevolutionary pattern, new data on mass extinction may "not leave sufficient time for the production of the changes that are alleged to have taken place."

Fortunately, we may proceed beyond conjecture in trying to discern Darwin's personal response to geological systems that threatened the efficacy of natural selection as an adequate source for the larger pattern of life's history. Cuvier died while Darwin worked on the *Beagle* and Lyell thrust his metaphorical sword (at least to Darwin's satisfaction) through the heart of traditional catastrophism. But Darwin could not rest easy, in full confidence that geological change would always plod along at optimal slowness for natural selection. For a formidable challenge, similar in broad concept to Cuvier's but very different in overt claims, greatly worried Darwin during the last fifteen years of his life—the "odious spectre" of Lord Kelvin. Darwin's widely-misunderstood response proves, once again, how clearly he had pondered and assimilated the logic of his theory—and how much he required the slowness and uniformity of geological change.

### DARWIN'S GEOLOGICAL NEED AND KELVIN'S ODIOUS SPECTRE

The familiar story of Lord Kelvin's incursion into geology has usually been recounted as a morality play. In basic outline, arrogant physics invades, but beleaguered natural history holds the line and triumphs, ultimately in a twist of delicious irony. I won't dispute this basic outline, but an attention to detail does compromise the traditional moral message, while also providing a striking example of Darwin's geological commitments.

In 1866, William Thomson, the future Lord Kelvin, published one of the most arrogant documents in the history of science—a one-paragraph paper (with an appended calculation) boldly entitled "The 'Doctrine of Uniformity' in Geology Briefly Refuted." This manifesto became the stalking horse in Kelvin's 40-year campaign to refute the substantive uniformities of rate and state (see pp. 479–484) by arguing that the earth's limited age did not provide enough time for explanations based solely on the accumulation of small effects produced by causes acting at current rates. Over the years, Kelvin developed a set of arguments, yielding broad ranges rather than precise ages, and based on limits to the age of both the sun (derived from estimates of meteoric influx), and of the earth (derived from outflow of heat and rotational slowing by tidal friction). Kelvin originally favored a date of some 100 million years (with an upper bound at 400 million), but he refined his estimates downward as the years passed, and finally settled upon a limited span of only 10–30 million years (at least for the duration of a solidified outer crust).

In Kelvin's most famous argument (and sole subject of the 1866 note), measurements of interior heat restrict the earth's age on the assumption that outflow represents continued cooling from an initially molten state. By measuring rates of outflow, we should be able to set an outer limit of maximal age for the origin of life by specifying the initial time of formation for a solid planetary surface. (In practice, such a calculation must remain highly uncertain due to inhomogeneities of planetary composition and our own ignorance about the earth's interior.) More importantly, the entire argument rests upon an assumption that no sources of novel heat exist, and that all current efflux must therefore represent a residual flow from the original fireball of an initially molten planet. When the discovery of radioactivity revealed an engine of new heat, Kelvin's argument collapsed. In the delicious irony mentioned above, the same force that dethroned Kelvin's limited duration soon provided a clock to measure the earth's actual age—and the billions favored by many geologists triumphed over Kelvin's long campaign for restriction. (Kelvin lived into the age of radioactivity, but never publicly acknowledged his defeat. Lord Rutherford tells an interesting story of an early lecture that he delivered in 1904 on determining the age of the earth by radioactive decay—see Gould, 1985c. Rutherford spotted the aged Kelvin in his audience and realized that he was "in for trouble." "To my relief," Rutherford writes, "Kelvin fell fast asleep, but as I came to the important point, I saw the old bird sit up, open an eye and cock a baleful glance at me. Then a sudden inspiration came, and I said Lord Kelvin had limited the age of the earth, provided no new source of heat was discovered. That prophetic utterance refers to what we are now considering tonight, radium!")

At this point in the conventional morality play, the story becomes a homily in the Manichean mode: The elegant mathematics of an arrogant physicist expires on the Achilles' heel of a false assumption. The humble and patient observers of nature, who always knew, in the bones of their rich empirical experience, that Kelvin must be wrong, but who dared not oppose such a powerful foe, triumph in the end. Empiricism wins the day: "speak to the earth, and it shall teach thee" (Job 12:8). In this canonical version, Darwin stands with his fellow geologists and biologists, forging a common front of natural historians against an intruder with no feel for the empirics of history. But this version requires revision to the point of near reversal—for the battle lines have been drawn falsely for two prominent reasons that have largely disappeared from historical memory.

First, Kelvin's efforts did not inspire fear and ridicule among most natural historians. The great majority of biologists and geologists welcomed his attempt to replace a vague feeling about immensity with actual limits. Moreover, most naturalists regarded Kelvin's figure of 100 million years as quite generous and fully sufficient to render geological history at any rate suggested by the empirical record. (Only later, in the 1890's, as Kelvin revised his estimate drastically downward, did significant numbers of geologists demur, arguing that their record could not fit into the time now allotted. Darwin's persistent but idiosyncratic opposition, documented below as the chief claim of this section, has been mistaken as a general consensus of geologists, thus leading to the main error of the canonical version.) In fact, many natural scientists breathed a sigh of relief at the amplitude of Kelvin's early allotments. To cite one particularly astute observer from general culture, and to illustrate the diffusion of a common impression of sufficiency, Mark Twain stated in his famous essay, "The damned human race" (largely written as a satirical response to A. R. Wallace's argument for intrinsic human meaning in the cosmos, as justified by an early version of the anthropic principle): "According to these [Kelvin's] figures, it took 99,968,000 years to prepare the world for man, impatient as the Creator doubtless was to see him and admire him. But a large enterprise like this has to be conducted warily, painstakingly, logically."

Second, Kelvin declared no general warfare against a hidebound science of geology; the implications that he drew from his own estimates of the earth's age contained little to offend most earth scientists. Kelvin did haughtily dismiss one style of argument carelessly pursued by many geologists, particularly the most committed of uniformitarians: the treatment of time as so vast that no practical limit could be placed upon any process—a kind of heuristic eternity, if you will. Most geologists accepted Kelvin's chastisement on this point, and happily altered their language because they felt unthreatened by Kelvin's estimate. One hundred million years seemed quite sufficient to accomplish any observed or inferred geological work.

Kelvin saw right through the shaky basis of Lyell's conceptual edifice: the conflation of substantive and methodological meanings of uniformity. Spatiotemporal invariance (uniformity) of natural law must be assumed as a basis for scientific inquiry into the past, but uniformity of state (non-directionalism of earth's history) cannot be inferred as a consequence, and can, moreover and to the contrary, be directly refuted by nature's own invariable laws. On this particular issue, Lyell could not have encountered a tougher opponent than the man who had formulated the second law of thermodynamics! The very constancy of the second law, and the attendant vector of time that follows as a result, smashed any hope for a long-term, steady-state in the earth's physical appearance. In other words, uniformity of law disproved uniformity of state. (Eddington later gave elegant expression to this argument by describing Kelvin's second law of thermodynamics as "time's arrow.")

Diplomatically perhaps, Kelvin attacked Hutton's earlier version of the conflation (via Playfair's famous exegesis of 1802), rather than Lyell's contemporary presentation. Kelvin quotes Playfair's (1802) most famous lines: "The Author of nature has not given laws to the universe, which, like the institutions of men, carry in themselves the elements of their own destruction. He has not permitted in His works any symptoms of infancy, or of old age, or any sign by which we may estimate either their future or their past duration." Kelvin responds by demolishing this false rationale for a non-directional earth (Thompson, 1868, p. 2): "Nothing could possibly be further from the truth than this statement. It is pervaded by a confusion between 'present order,' or 'present system,' and 'laws now existing'—between destruction of the earth as a place habitable to beings such as now live on it, and a decline or failure of law and order in the universe."

Later, in discussing the secular slowing of the earth's rotation due to tidal friction imposed by the moon (a trend now supported empirically by evidence from daily and yearly growth lines of fossil organisms), Kelvin strongly asserts that even a small directional effect, measured as seconds of slowing per century, strongly compromises the entire Lyellian system: "It is quite certain that a great mistake has been made—that British popular geology at the present time is in direct opposition to the principles of natural philosophy . . . There cannot be uniformity. The earth is filled with evidences that it has not been going on forever in this present state, and that there is progress of events towards a state infinitely different from the present" (1868, p. 16).

Kelvin criticized Lyell more directly for his claim that heat lost by radiation into space can always be reconstituted from other sources (chemical and electrical)—for such an article of non-directional faith violates the second law of thermodynamics: "These statements are directly opposed to the general principle of the dissipation of energy: and the hypothesis which they suggest is very inconsistent with our special knowledge of the conduction and radiation of heat, of thermoelectric currents, of chemical action, and of physical astronomy" (1868, p. 231).

Kelvin then invokes his second law to identify the main vector of physical change through time: as entropy and disorder increase, the energy of most causes must diminish. Therefore, most geological processes must have acted with substantially more vigor on the early earth, leading to an expectation for more rapid biological change at this time, if changes in the physical world potentiate biological evolution.

I earnestly beg Professor Huxley, and those in whose name he speaks, to reconsider their opinion, that the secular cooling of the earth and of the sun "had made no practical difference to the earth during the period of which a record is preserved in stratified deposits." There is, surely, good ground for Sir Roderick Murchison's opinion that metamorphic causes have been more active in ancient times than at present, because of more rapid augmentation of temperature downwards below the earth's surface; and it cannot be reasonably urged that a hotter sun is not a probable explanation of the supposed warmer climate of the Paleozoic ages (1868, p. 230).

Kelvin's argument brings us to the nub of Darwin's objections. Darwin's strong opposition to Kelvin, a reaction that could almost be described as fear and loathing, has often been recorded, but rarely understood. To appreciate Darwin's intense reaction, we must focus upon the geological prerequisite for his account of evolution—a steady and intermediate rate of change, enough to prod, but not too much to overwhelm, natural selection. Kelvin's directional geology did not invoke the paroxysmal specter of traditional catastrophism. But he had raised, in many ways, an even greater threat—for his challenge operated as a double-edged sword to attack natural selection from both sides of Darwin's geological needs.

### A question of time (too little geology)

Kelvin's estimate of time seemed generous to most naturalists, even to supporters of natural selection like Wallace, and to Darwin's self-appointed spokesman, T. H. Huxley. But Darwin envisioned natural selection as working so slowly, especially in its progressive mode of extended biotic competition, that any talk about limitation made him intensely nervous—for too little time could be equated with insufficient geological impetus for evolutionary change. Darwin had shown his hand in calculating a greatly exaggerated 300 million years for the denudation of the Weald (see p. 153), a figure that he expunged with embarrassment from later editions of the *Origin* (see Burchfield, 1975, pp. 70–72). Darwin had also urged readers to close his volume if they could not accept Lyell's views on the "incomprehensibly vast" time available for natural selection.

Kelvin's 100 million years sounded sufficiently long, but Darwin harbored deep doubts, especially in the light of Kelvin's related argument for a vector of diminishing rates of physical change through time. Kelvin's directional vector clashed with the most important item in Darwin's apologetics for the imperfection of the geological record. Darwin had been troubled, for example, by the abrupt appearance of so many complicated anatomies in the Cambrian explosion: "The case at present must remain inexplicable; and may be truly urged as a valid argument against the views here entertained" (1859, p. 308). Darwin concluded that Precambrian seas had "swarmed with living creatures" not yet found as fossils. But the anatomical distance from the first living molecule to a trilobite certainly exceeded all later change from Cambrian forms to modern organisms. Following the gradualistic premise that amounts of change provide a rough measure of time, Darwin concluded that most of the earth's history had passed before the Cambrian explosion: "If my theory be true, it is indisputable that before the lowest Silurian stratum was deposited [Cambrian of modern terminology], long periods elapsed, as long as, or probably far longer than, the whole interval from the Silurian age to the present day" (1859, p. 307).

Thus, Kelvin's 100 million years since the initial consolidation of the earth's crust implied substantially less than 50 million years for the entire fossil record since the Cambrian explosion. The situation only worsened as Darwin pondered Kelvin's views. If early changes had been so rapid and intense, then the earth must have spent most of its history just "calming down" enough to enter a realm where natural selection might work. Even more of Kelvin's limited time must therefore be allotted to a pre-Darwinian earth, and even less to a severely restricted later world amenable to natural selection as the cause of life's pattern. In January 1869, Darwin wrote to James Croll: "Notwithstanding your excellent remarks on the work which can be effected within a million years, I am greatly troubled at the short duration of the world according to Sir W. Thomson, for I require for my theoretical views a very long period before the Cambrian formation" (in F. Darwin and Seward, 1903, volume 2, p. 163). And, more graphically, Darwin wrote to A. R. Wallace in 1871: "I can say nothing more about missing links than I have said. I should rely much on pre-Silurian times; but then comes Sir W. Thomson like an odious spectre" (in Marchant, 1916, Letters of Wallace, volume 1, p. 268).

Lest we construe Darwin's anxiety as exaggerated, consider the corroboration that he received for his distress right from the horse's mouth. Kelvin himself addressed Darwin's greatest fear in his own direct and succinct way. The great physicist did not dispute evolution *per se*, but Kelvin argued that his own limitations upon time had effectively debarred natural selection as an important mechanism. Kelvin, following a cultural tradition that Darwin had transcended, believed that some spiritual force must be guiding the progress of evolution, if only because geology provided insufficient time for producing the observed order by mechanical processes. As Burchfield observes (1975, p. 73): "Kelvin was convinced that the complexity of life bore witness to the work of a Creative Intelligence. He was equally convinced that whereas natural selection would require almost endless time, divine guidance would enable evolution to produce the diversity of life in a relatively short period. Thus as far as evolution was concerned, his arguments for limiting the earth's age were also proofs of design in nature." Kelvin wrote: "A correction of this kind [on the duration of geological time] cannot be said to be unimportant in reference to biological speculation. The limitation of geological periods, imposed by physical science, cannot, of course, disprove the hypothesis of transmutation of species; but it does seem sufficient to disprove the doctrine that transmutation has taken place through 'descent with modification by natural selection'" (Thompson, 1868, p. 222).

### *A question of direction (too much geology)*
Thus, the vector imposed by Kelvin's Second Law—intense energy of early causes, diminishing continuously through time—troubled Darwin by compressing the fossil record into a restricted realm of adequate calmness. But the

same vector also challenged Darwin by stressing the importance of early rapidity itself. Restricted time implied too little geology when considered in terms of Darwin's need for an intermediate amount of environmental change. But the rapid change on Kelvin's early earth implied too much geology on the flip side of impermissibility. Intense heat flows and high rates of volcanism and erosion on an early earth raised the specter of traditional catastrophism, with its paroxysms of mass extinction. This argument about rapidity opposed Darwin at both essential poles of practice and theory (see Chapter 2). Could a process like natural selection really operate effectively on a planet engulfed in such perpetual turmoil? Must other causes of biological change be postulated for the early earth—a particularly distressing prospect for Darwin, who yearned, above all, to establish a temporally invariant and fully general account of evolution. Moreover, the prospect of such a prominent vector clashed with Darwin's Lyellian vision of an earth operating with sufficient constancy of change that the study of modern causes would suffice for explaining the past as well. Kelvin, astute as ever, had explicitly proclaimed the victory of "catastrophism" more for its explanation of a vector of diminishing intensity than for its theory of paroxysms (1868, pp. 231–232). And Darwin, who had so assiduously explored the ramifications of all major ideas in natural history, rejected the directionalist aspect of catastrophism as firmly as he dismissed the paroxysmal claim.

Yet Darwin could not escape the directionalist implication, and he eventually compromised on this single point alone. Putting his best face on adversity, Darwin added a passage to later editions of the *Origin*. He now acknowledged the vector of diminishing intensity in change, admitted higher rates of evolution on an early earth, and made peace with Kelvin's temporal restrictions by awarding evolution an ontogeny with a speedy childhood: "It is, however, probable, as Sir William Thompson [sic] insists, that the world at a very early period was subjected to more rapid and violent changes in its physical conditions than those now occurring; and such changes would have tended to induce changes at a corresponding rate in the organisms which then existed" (*Origin*, 1872, 6th edition).

In a reciprocal move, Darwin also quietly dropped the following passage of his first edition, with its opposing claim for slower initial rates of evolution— no doubt with regret, for this original argument had embodied his favored theme of biotic control: "During early periods of the earth's history, when the forms of life were probably fewer and simpler, the rate of change was probably slower; and at the first dawn of life, when very few forms of the simplest structure existed, the rate of change may have been slow in an extreme degree" (1859, p. 488).

If all evolutionists had reacted to Kelvin in the same way, we might have learned something about conflicts across scientific disciplines, but little about Darwin's distinctive view. However, in a fascinating and little known aspect of this familiar story, Darwin stood virtually alone in loathing Kelvin's restriction—and his reaction therefore reveals some important implications of his own strict selectionist logic applied with firm gradualist commitments.

Darwin, as I have emphasized throughout, was dogged and relentless, fiercely honest and logical in his thinking. He wrestled with every major difficulty, working and reworking, fretting and fretting again, until he achieved closure or at least understood why a solution eluded him. He often became obsessed with problems (levels of selection, for example) that his supporters either didn't grasp at all, or didn't understand as sources of interest or trouble. Sometimes, as with Kelvin, he probably worried an issue too far, but when we grasp the source of his exaggerated concern, we understand the logic of his theory in a more complete way.

Darwin's opposition to Kelvin has been well recorded, but the missing piece in this historical puzzle, and the key for revising the false but canonical version of this story as a morality play, lies in Darwin's loneliness. I have cited the approbation of most geologists for Kelvin's efforts, but when we note the similar acquiescence of Darwin's two major English supporters—Huxley and Wallace—then the story becomes even more interesting (and the falsity of the conventional version even more apparent).

Huxley devoted his 1868 presidential address for the *Geological Society of London* to defending this profession against Kelvin's charges. Yet by Huxleyan standards—as the greatest literary polemicist (equalled, perhaps, by Buffon) in the history of biology—this particular address packs little punch. Huxley does not assert a distinctive geological way of thought against Kelvin's unwelcome intrusions. Instead, he simply accepts Kelvin's claims, and defends geology only against Kelvin's characterizations. No one, he argues, not even Lyell (who had, by then, abandoned uniformity of state by admitting vectors in the history of life), maintains so strict and comprehensive a view of uniformity. In fact, the old dichotomy of uniformity vs. catastrophe has largely been swept aside, with both views yielding to an "evolutionism" based on slow, continuous and directional change on an ancient earth.

Huxley argues that neither of Kelvin's two major claims contradicts this new, evolutionist synthesis. Geologists can accept Kelvin's directionalism because uniformitarianism has abandoned any former flirtation with the doctrine of an earth in steady state. Geologists would be distressed by a truly serious limitation upon time, but 100 million years provides more than sufficient amplitude for any legitimate geological purpose.

Coming to his key point, Huxley allows that some people (he does not mention his friend Darwin by name) feel a tug between Kelvin's dates and a greater age supposedly implied by the extreme slowness of evolutionary change. But this feeling, Huxley assures us, cannot be defended. We can only assess the speed of evolution by calibration against elapsed geological time. Previous assertions of extreme slowness flowed from geological convictions about immensity—as no purely biological data exist for a truly independent calibration. If Kelvin has now demonstrated that time must be shorter, we can only conclude that evolution has generally been faster.

But, it may be said that it is biology, and not geology, which asks for so much time—that the succession of life demands great intervals; but this appears to me to be reasoning in a circle. Biology takes her time from geology. The only reason we have for believing in the slow rate of the change in living forms is the fact that they persist through a series of deposits which, geology informs us, have taken a long while to make. If the geological clock is wrong, all the naturalist will have to do is to modify his notions of the rapidity of change accordingly. And I venture to point out that, when we are told that the limitation of the period during which living beings inhabited this planet to one, two, or three hundred million years requires a complete revolution in geological speculation, the *onus probandi* rests on the maker of the assertion, who brings forward not a shadow of evidence in its support (1869, year of publication of 1868 address, in Huxley, 1894, pp. 328–329).

And so Huxley, bowing to the physicists, concludes that geology has developed no legitimate reason for discomfort with Kelvin's dates, while biology has similarly failed to offer valid objection. Therefore, a little terminological misunderstanding, and a minor battle over professional turf, can be resolved into the sweetness and light of intellectual agreement.*

Huxley's acquiescence may not surprise us. After all, given his attraction to saltationist ideas, Huxley had never shared Darwin's commitment to gradualism and the attendant need for such ample geological time. But when we learn that A. R. Wallace, a stouter defender of natural selection than Darwin himself, also readily accepted Kelvin's dates, then we can better sense the idiosyncrasy of Darwin's concern. Wallace ventured even further than Huxley. He published several letters in *Nature* on the age of the earth and the measurement of geological time (1870, 1892, 1893, 1895a and b). Wallace willingly accepted Kelvin's dates, and he presented an ingenious argument to explain why some biologists had been fooled into believing that evolution proceeded so slowly.

Working from a theory by James Croll linking ice ages to changes in the earth's orbit, Wallace claimed that the last 60,000 years had experienced extraordinary climatic stability due to unusually low orbital eccentricity. Before then, and probably for most of geological time, orbital eccentricity had been more pronounced, prompting climatic fluctuations in local areas, attendant deaths and migrations of faunas, and a greatly accelerated rate of evolution. As an unfortunate consequence of our generally valid uniformitarian method of measuring current rates and extrapolating backwards, we developed the

---

*Huxley could never leave a conflict with such calm and no vitriol. Thus, as a sample of his prose and personality, consider (from the same address) this incisive passage on what a later generation would call GIGO. I would also read this statement as an implied criticism of arrogance among physical scientists, an exercise in wound licking after Kelvin's effective salvos: "Mathematics may be compared to a mill of exquisite workmanship, which grinds you stuff of any degree of fineness; but, nevertheless, what you get out depends upon what you put in it; and as the grandest mill in the world will not extract wheatflour from peascods, so pages of formulae will not get a definite result out of loose data" (1869, in 1894, p. 333).

false impression from our recent (and unusual) climatic stability that evolution must always proceed with imperceptible slowness. But the general pace of natural selection can be much faster, matching the usual rate of climatic fluctuation. When we recognize this "slower change of species since the glacial epoch than at any former period" (1870, p. 455), we can correct our false allegiance to evolution's perpetual sloth, "thus allowing us to suppose change of form in the organic world to go on more rapidly than we had before thought possible" (1870, p. 455).

Wallace then proposes a correction to Lyell's estimates of time based on turnover in molluscan species. He obtains an age of 24 million years for the base of the Cambrian, proclaims himself in agreement with Kelvin's 100 million years since the solidification of the earth's surface (leaving some 75 million years for the Precambrian interval that Darwin had proclaimed longer than all subsequent time), and urges Darwin to accept this framework with good grace and confidence in natural selection:

> These figures will seem very small to some geologists who have been accustomed to speak of "millions" as small matters; but I hope I have shown that, so far as we have any means at present of measuring geological time, they may be amply sufficient. Taking Sir William Thomson's allowance of a hundred million years for the time during which the earth can have been fit for life, it yet allows Mr. Darwin, for the process of development from the primordial germ, three times as many years anterior to the Cambrian epoch as have elapsed since that date, an amount of time which, I believe, will fully satisfy him (1870, pp. 454–455).

I find an ironic similarity between this tale of Darwin and his intellectual brethren, and the story of Solomon and the baby claimed by two women. The true parent loves her child so fiercely that she would rather give her offspring to an imposter than to see the infant dismembered in mock compromise. The false mother feels no such protective love, thus permitting Solomon's wise identification. I do not wish to push the analogy too far. Huxley and Wallace cannot be called false parents, but the child of natural selection had not originated in their womb (or, in Wallace's case, at least not for so long a gestation, or so slow and prideful a growth). They had not nurtured this idea through so many years of thought, through such struggles of upbringing, to build such an extensive and coherent edifice of logic and implication. They could therefore compromise, sacrifice a bit here, trim a bit there, to preserve the entity, however tarnished. But, for Darwin, any departure from full integrity became unthinkable. Darwin also understood, however, that intransigence compromises the spirit of science, and that all good thinkers must make and admit mistakes. For this reason, Darwin separated his core commitments (the "essences" of my treatment) from postulates more easily compromised. To abandon, or even seriously to mitigate, any of the core commitments might be equated with the unacceptable solution of dividing the child. A conviction about the generally slow and steady character of geological change, a prerequisite for the third leg on my tripod of Darwinian essentials, lay among the core commitments—for natural selection could not shape macroevolutionary pattern without this auxiliary support from another profession. We can only understand the intensity, and the loneliness, of Darwin's reaction against Kelvin when we properly read his response as a cry from the heart of his entire system.

# CHAPTER SEVEN

# The Modern Synthesis as a Limited Consensus

## Why Synthesis?

A rose may retain its fragrance under all vicissitudes of human taxonomy, but never doubt the power of a name to shape and direct our thoughts. The evolutionary consensus that became such a bulwark of orthodoxy by the time of the Darwinian centennial celebrations of 1959 featured no recognized name in its early days. Fisher offered no general designation for his genetically revivified Darwinism in 1930, nor did Haldane in 1932. Dobzhansky, beginning the second wave of integration in 1937, proposed no label, either for the theoretical center or for the general movement.

The accepted name emerged later, and without conscious intent. The "synthetic theory," or the "modern synthetic theory"—in many ways an oddly uninformative and overly broad name—derives from the title of a book written by the grandson of Darwin's most effective defender: *Evolution, The Modern Synthesis,* published by Julian Huxley in 1942. (Historian of science B. Smocovitis (1996) points out that, as a general goal among scholars, synthesis enjoyed a great vogue during these years, especially as a central theme for measuring intellectual maturity, as expressed in the "unity of science" movement expounded by positivist philosophers of the Vienna Circle and supported by biological pundits like J. H. Woodger. Ernst Mayr, for example, strongly supported the unity of science movement early in his career, but changed his mind when he began to fear that misplaced claims for grander synthesis would bury natural history in a reductionist scheme to uphold the primacy of physics and chemistry.)

Huxley obviously felt that the morphology of evolutionary consensus could best be described as a synthesis—that is, a gathering together of previously disparate elements around a central core. Following Smocovitis's argument on the favorable *Zeitgeist* for "synthesis" provided by the "unity of science" movement, note how Huxley places his chosen name within this wider context by extolling the general virtues of synthesis.

Biology in the last twenty years, after a period in which new disciplines were taken up in turn and worked out in comparative isolation, has become a more unified science. It has embarked upon a period of synthesis, until today it no longer presents the spectacle of a number of semi-independent and largely contradictory sub-sciences, but is coming to rival the unity of older sciences like physics, in which advance in any one branch leads almost at once to advance in all other fields, and theory and experiment march hand-in-hand. As one chief result, there has been a rebirth of Darwinism (1942, p. 26).

Thus we evolutionists have, ever since, called our central theory by a name fashioned for a largely superseded set of concerns. In another sense, however, Huxley chose his name to emphasize a particular scenario for evolutionary science—and a characterization of this history as a "synthesis" seems both reasonable and accurate. Huxley viewed the synthesis, in an interpretation that became common in his time and has remained conventional ever since, as a two-stage process of integration around a renewed Darwinian core. He emphasized this Darwinian center at the close of the passage cited above, and then offered praise, in purpler prose, to "this reborn Darwinism, this mutated phoenix risen from the ashes of the pyre kindled by men so unlike as Bateson and Bergson" (1942, p. 28).

The *first phase*, reasonably called a *synthesis of Mendel and Darwin*, required the extinction of Mendelism's first major episode of evolutionary employment in the non-Darwinian saltational theory of de Vries (see Chapter 5). Huxley cites three major steps (1942, p. 25): the recognition that Mendelian principles operate in all organisms, unicellular and multicellular, plant and animal; the key insight that small scale, continuous Darwinian variability also maintains a Mendelian basis; and the mathematical demonstration that small selection pressures acting on minor genetic differences can render evolutionary change. This work culminated in the origin of theoretical population genetics (see Provine, 1971), and led to our almost catechistic invocation of a trinity—Fisher, Haldane, and Wright—as heroes of this first phase. (I do not challenge the attribution, but have long been amused by the almost formulaic citation; even the order of names has become invariant—as in another more famous Trinity!)

The *second phase*, also a "synthesis" in the vernacular sense of the word, began with Dobzhansky's masterful book (1937) and proceeded as *a linking of traditional subdisciplines in biology to the core theory forged during the first phase.* This activity, already in full swing when Huxley wrote his book, eventually included such classics as Mayr (1942) for systematics, Simpson (1944) for paleontology, White (1945) for cytology, Rensch (1947) for morphology, and Stebbins (1950) for botany. Huxley made a good prediction, and hoped that his own effort would fuel the grand integration: "The time is ripe for a rapid advance in our understanding of evolution. Genetics, developmental physiology, ecology, systematics, paleontology, cytology, mathematical analysis, have all provided new facts or new tools of research: the need today is for concerted attack and synthesis. If this book contributes to such a synthetic point of view, I shall be well content" (1942, p. 8).

I accept this traditional account of what the Synthesis synthesized, but

I prefer to view the history of the Synthesis under a different rubric and terminology developed by historian of science Will Provine and by myself—
namely, (1) *restriction* followed by (2) *hardening*, with the first process
viewed as largely admirable, the second as mostly dubious.

Provine has argued (1986) that the first phase of synthesizing, the integration of Mendel and Darwin to a core discipline of population genetics, should
be viewed as a welcome restriction (a "constriction" in his favored term)—
for biologists could now shed the several competing and truly contradictory
theories (primarily orthogenetic and saltational) that had made the domain of
evolutionary studies seem so anarchic at the Darwinian celebrations of 1909.

But the first phase also included a vigorously pluralistic range of permissible mechanisms within the primary restriction. The original synthesists
wanted to render all of evolution by known genetic mechanisms; but they
tended to agnosticism about relative frequencies among the legitimate phenomena, notably on the issue of drift (and other random phenomena) vs.
selection.

The second phase began with this pluralism intact, as the first wave of
books from the late 1930's and 1940's clearly illustrates. But this broad net
then tightened, as the leading synthesists promoted natural selection, first to a
commanding frequency and then to virtual exclusivity as an agent of evolutionary change. This consensus hardened to an orthodoxy, often accompanied by strong and largely rhetorical dismissal of dissenting views—a position
that reached its acme in the Darwinian centennial celebrations of 1959. The
pluralism of "consistent with genetics" eventually narrowed to a restrictive
faith in what Weismann had called the "all-sufficiency" of natural selection,
with the accompanying requirement that phenotypes be analyzed as problems in adaptation. I almost feel as if the arbiters of fashion and theory had
distributed sandpaper to all evolutionists—and that ordinary professional activity shifted to a vigorous smoothing out of all remaining bumps, facets, and
channels of Galton's polyhedron (see pp. 342–351), as organisms became unfettered billiard balls, rolling wherever the pool cue of natural selection dictated upon the crowded table of ecology.

This hardening extended beyond overconfidence in adaptation to a more
general, and sometimes rather smug, feeling that truth had now been discovered, and that a full account of evolution only required some mopping up and
adumbration of details. The paeans of self-satisfaction penned for the 1959
centennial of the *Origin* (see pp. 569–576) almost invite a parody of Hamlet's
insightful comment about his mother's rhetorical overstatements: the writers
praised too much, one thinks.

## Synthesis as Restriction

### THE INITIAL GOAL OF REJECTING OLD ALTERNATIVES

William Bateson, to be sure, had his own particular axe to grind when he so
dramatically expressed his pessimism about evolutionary theory in his famous address to the American Association for the Advancement of Science in

1922: "faith has given place to agnosticism" (see p. 412 for the full quote). The first rumblings of synthesis had already occurred. As the most obvious and important example, Fisher's pivotal 1918 paper on "The Correlation Between Relatives on the Supposition of Mendelian Inheritance" had demonstrated the potential Mendelian (polygenic) basis for the small-scale, continuous variation that Darwinians had always identified as the primary source of evolutionary change—and that the early Mendelians had either ignored as supposedly non-particulate, or denigrated as evolutionarily insignificant (see pp. 429–431 on de Vries's attitude). Fisher's demonstration established a basis for resolving the increasingly fruitless debate between "biometricians" (Darwinian supporters of continuous variation) and "Mendelians" (upholders of saltational change, in the first evolutionary use of principles that would later blend with Darwinism)—a seminal, if ultimately barren, dichotomy in early 20th century biology. By using the principles of one side (Mendelian particulate inheritance) to explain the supposedly contradictory phenomena of the other (continuous and isotropic variation in phenotypes)—and then by banning the opposite (saltational) phenomena originally attached to the principles—Fisher staged a brilliant coup that surely deserves the label of "synthesis."

If Bateson had fallen a bit behind the times by 1922, he did accurately record the near anarchy that had prevailed in the study of evolutionary mechanisms just a few years earlier. Using Kellogg's (1907) framework, as I have done throughout this book (see pp. 163–169 for an account and rationale), we may classify attitudes outside the central core of Darwinism either as helpful expansions or as contradictory substitutions—auxiliaries and alternatives in Kellogg's terminology. Kellogg identified three major alternatives to Darwinism—Lamarckism, saltationism and orthogenesis. At the 1909 celebrations for Darwin's centennial, all three alternatives enjoyed substantial support, probably equal in extent, and in the reputation of leading supporters, to the popularity of Darwinism itself. As a supreme and well known irony, the original impact of the Mendelian rediscovery—the ultimate source of restriction and synthesis—had only increased the range and intensity of evolutionary debate by revivifying the saltational alternative and thus augmenting the roster of major challenges to Kellogg's triple threat.

Clearly, the state of evolutionary theory required restriction for further advance—either a settling upon one of the four contenders (Darwinism plus the three challenges), or a new formulation. The first phase of the synthesis accomplished this goal in three major moves: (1) by choosing the Darwinian central core as a proper and fundamental theory; (2) by reading Mendelism in a different way to validate, rather than to confute, this central core; and (3) by utilizing this fusion to ban the three alternatives of Lamarckism, saltation, and orthogenesis.

Darwinism is a functionalist theory with an operational core that must place primary weight upon building adaptation by the mechanism of natural selection (see Chapters 2 and 4). In logic and principle, therefore, the theory could be confuted in either of two ways: by affirming an alternative mechanism for adaptation, or by denying the primacy of functionalism, and proposing a dominant role for internal, structuralist shaping of evolutionary change. Provine (1983) has designated the first phase of synthesis as a restriction because the fusion of Darwin with Mendel validated the rejection of all three Kelloggian alternatives, and in both of their more general categories.

ALTERNATIVE THEORIES OF FUNCTIONALISM. Most importantly, this first-stage fusion finally gained enough knowledge and proof to depose the longest and most severe of all challenges to Darwinism—the doctrine that preceded natural selection, and that Darwin himself had both accepted as a subsidiary player and granted an ever-increasing role through subsequent revisions of *The Origin of Species*. This so-called, if misnamed, "Lamarckian" theory of soft inheritance and direct production of adaptation by inheritance of acquired characters, had stubbornly persisted through all vicissitudes of evolutionary debate to remain a favorite in certain circles (at least among field naturalists, if not among experimentalists) well into the 1920's.

Mendelians had always rejected Lamarckism as inconsistent with the mechanisms of inheritance, but Darwinians before the synthesis had generally downplayed, ignored or actively rejected Mendelism. The first-stage fusion finally gave Darwinian functionalists a tight rationale for outright rejection of this alternative Lamarckian route to adaptation. This clear solution struck most Darwinians as far more satisfactory than the uneasy pluralism that functionalists (including Darwin himself) had been forced to espouse before—for Darwinism and Lamarckism, as pathways to the same result of adaptation, had coexisted no more peacefully than most competitors for the same prize. An old motto proclaims that the enemy without should always be preferred to the enemy (or apparent helpmate) who saps your strength from within.*

DEMOTING INTERNALISM. The dismissal of Lamarckism left two strong alternatives in Kellogg's triad: saltationism and orthogenesis (see Chapters 4 and 5). These two challenges to Darwinian functionalism have always been linked in their common focus on internal sources of change and direction—as jointly embodied in the metaphor of Galton's polyhedron (see Chapter 5. The reformulation of Mendelism as a source for pervasive, small-scale, continuous variability (and not only for mutations of substantial extent) provoked a strong distaste for saltationism, as early population geneti-

---

*In writing only these two paragraphs on this dismissal of Lamarckism, I imply no judgment about relative historical importance. Scarcely any event in the history of evolutionary theory could be more vital or central than the formulation of a rationale for expunging from orthodoxy (and rendering virtually inconceivable in theory) the most venerable of all evolutionary mechanisms. I downplay this subject here only because this book treats the history of *valid* auxiliaries and alternatives to strict Darwinian *functionalism*—and Lamarckism, as an *invalid functionalism*, therefore becomes tangential to my concerns on both grounds (while remaining central to the larger, general history of evolutionary theory). Similarly, this chapter gives relatively little space to the formulation of the Modern Synthesis for the same reason: this history, so well covered in other sources (Provine, 1971; Mayr and Provine, 1980), specifies the formulation of modern orthodoxy, while this book treats the persistent sources of heterodoxy.

cists denied evolutionary importance to mutations of large effect. At the same time, the Mendelian explanation of this copious and small-scale variability left little scope for orthogenesis. Darwinism surely welcomed this further Mendelian aid, for all functionalist theories must try to smooth Galton's polyhedron.

Yet, while internalist theories endured a pronounced demotion during this first phase of restriction, these alternatives to Darwinism did not suffer the Lamarckian fate of complete dismissal in theory. Saltations could not be banished or denied, but only declared unimportant in evolution. Orthogenesis could not be overthrown in principle, for mutation pressure could conceivably boost the frequency of alleles from within. Moreover, and more importantly as an enjoined consequence of its own premises, population genetics had to acknowledge another potentially substantial source for non-adaptive evolutionary change: the effects of sampling errors, primarily in small populations.

Thus, the first phase of restriction invoked a fusion of Mendel and Darwin to dismiss or downplay the traditional roster of alternatives to Darwinian functionalism. But the resulting theory remained open and pluralistic in welcoming all notions consistent with the new formulation of Mendelism. I shall devote the rest of this section to three illustrations of the two key properties in this first phase of the synthesis: (i) the revivified Darwinian core, and (ii) toleration of a broad phenomenology, including substantial nonadaptation, so long as results could be rendered by known genetic mechanisms. I will discuss the two most important books of early population genetics: Fisher (1930) and Haldane (1932). Huxley's compendium (1942) should be read as a transitional document, and may belong more properly to the beginning of the second phase of hardening, but I include Huxley's book here as a summation of early ideas in the Synthesis, if only for its symbolic role in supplying the developing theory with a name.

### R. A. FISHER AND THE DARWINIAN CORE

The history of the Modern Synthesis holds special, one might almost say inspirational, interest in the good fortune of its construction by such a fascinating group of scientists—so different in personality, so diverse in philosophical attitude amidst their defining agreement, and so brilliant. I experienced the great intellectual privilege of knowing most of the second-phase founders—and I have tried to understand the personal components of scientific greatness by, for example, contrasting Dobzhansky's infectious enthusiasm with Mayr's fierce commitment and encyclopedism. I have also pondered the intra-individual variety in trying to square the warm and expansive humanism of Simpson's writing with the irascibility of his personality.

The leaders of the first phase were equally fascinating and diverse. Their differences have been discussed extensively, perhaps most cogently by one of the actors himself (Wright, 1978). But R. A. Fisher holds a special status as author of the movement's first major book (1930), and as the most thoroughgoing and uncompromising strict Darwinian among early synthesists. As a great logician and statistician, Fisher explored the Darwinian consequences of small changes in large populations with a consistency and completeness never achieved before. We may choose, in retrospect, to regard the views of other evolutionists as more subtle, or more attuned to nature's diversity and ambiguity—but no one can match Fisher as an advocate of the main line in its pure form. We must therefore treat him as inaugural, in both senses of priority and centrality.

Fisher begins *The Genetical Theory of Natural Selection* by explaining how the logic of Darwinism requires a particulate theory of inheritance in order to achieve self-sustainability as a mechanics of change without special pleading and additional forces. He states in his introduction (1930, pp. vii–viii): "That an independent study of Natural Selection is now possible is principally due to the great advance which our generation has seen in the science of genetics."

As an opening gambit, Fisher develops the intriguing argument that Darwin's forced allegiance to blending inheritance entailed far more than mere inconvenience and surmountable difficulty, thus confuting the usual claim, advanced by scientists who have traditionally viewed the *Origin* as a complete and fully workable theory of natural selection. Rather, Fisher maintains, blending inheritance represented a central impediment, debarring natural selection as the chief agent of evolution. By constantly degrading favorable genotypes, blending requires an enormous input of new mutation to fuel any process of change. But mutation rates this high could easily overwhelm any reasonable force of natural selection. Evolution would then be propelled largely from within, and natural selection, as a theory of "trial and error externalism," could not operate as the major cause of change. But a particulate theory provides raw material for favorable change without degradation—and a vastly lowered rate of new input now suffices. Evolution need not be driven from within, and may now be pulled by the external force of natural selection.

Fisher conceived his argument as a proof by elimination. Once we dispense with a need for such lavish internal stirring, no credible force but natural selection remains. Mendelism therefore validates Darwinism: "The whole group of theories which ascribe to hypothetical physiological mechanisms, controlling the occurrence of mutations, a power of directing the course of evolution, must be set aside, once the blending theory of inheritance is abandoned. The sole surviving theory is that of Natural Selection" (1930, p. 20).

Fisher illustrates the restrictive character of this first synthetic phase by invoking the fusion of Darwin and Mendel to discard each alternative of Kellogg's triad. Lamarckism (as noted before) goes quickly and gently into that good night—for soft inheritance cannot exist under the newly validated mechanics of Mendelism. Saltationism surrenders when small-scale continuous variability gains a Mendelian basis and wins evolutionary effectiveness thereby. This small-scale Mendelian component provides superior evolutionary raw material for two reasons: (1) overwhelming predominance in sheer amount (saltations occur only rarely, but minor variants are ubiquitous); and (2) greater potential for utility (small changes can often be advantageous, but large excursions will almost always be disruptive—see Fig. 7-1, from Fisher). When DeVries and other biologists had disparaged small-scale variation as ineffective and different in kind from Mendelian effects, saltations had prevailed *faute de mieux*. But once we can assert a continuum in effect and a uniformity in genetic mechanism for variation at all scales, then we must prefer an omnipresent and potentially advantageous mode to an exceedingly rare and almost always deleterious extreme in the spectrum. Fisher writes: "The chance of improvement, for very small displacements, tends to the limiting value one-half, while it falls off rapidly for increasing displacements (p. 39). In any highly adapted organism the probability of advantage through any considerable evolutionary step (saltation) rapidly becomes infinitesimal as the step is increased in magnitude" (p. 114).

Orthogenesis, the third and last of Kellogg's alternatives to Darwinism, expires with Fisher's initial argument on the virtues of particulate inheritance. The rates of mutation required by Darwinism become much lower under particulate models (and observation confirms that such a workably modest rate of mutation exists in nature). Since orthogenesis can only operate when mutation pressure becomes high enough to act as an agent of evolutionary change, empirical data on low mutation rates sound the death-knell of internalism. Fisher writes: "For mutations to dominate the trend of evolution it is thus necessary to postulate mutation rates immensely greater than those which are known to occur, and of an order of magnitude which, in general, would be incompatible with particulate inheritance" (1930, p. 20).

Fisher, with his mathematical training, reigned as the master of abstract

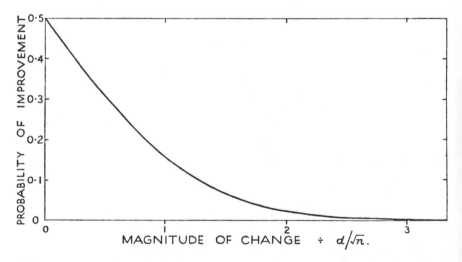

7-1. R. A. Fisher's classic illustration of the negative relationship between magnitude of change and probability of evolutionary utility. The vast majority of large mutations are deleterious; small mutations are both far more frequent and more likely to be useful. From Fisher, 1930.

generality among the early synthesists. He favored infinitesimal changes in large panmictic populations, as illustrated by physical metaphors and analogies; and he maintained little interest in the historical quirks, vagaries and complex structuring of actual populations. When subject to such constant and unconstrained natural selection in large panmictic populations, living creatures become exquisitely fitted to their environments: "Organisms in general are, in fact, marvelously and intricately adapted, both in their internal mechanisms, and in their relations to external nature" (p. 41).

In Fisher's world of panadaptationism and pure Darwinian generality, neutralism can only maintain an insignificant relative frequency, and natural selection must reign. Within such an unrelenting context of slow changes and large populations, Darwinian functionalism must triumph: "The very small range of selective intensity in which a factor may be regarded as effectively neutral suggests that such a condition must in general be extremely transient. The slow changes which must always be in progress, altering the genetic constitution and environmental conditions of each species, must also alter the selective advantage of each gene contrast" (1930, p. 95).

Many examples could be cited from Fisher's *Genetical Theory*, but two features strike me as particularly revealing in illustrating Fisher's maximally exclusive and general Darwinism.

THE ANALOGY OF FISHER'S "FUNDAMENTAL THEOREM" WITH THE SECOND LAW OF THERMODYNAMICS. After explicating and justifying his "fundamental theorem" of natural selection—*"the rate of increase in fitness of any organism at any time is equal to its genetic variance in fitness at that time"* (p. 35, Fisher's italics)—Fisher, enamored as ever with physical analogies, compares this central principle of his own construction with the second law of thermodynamics:

> It will be noticed that the fundamental theorem proved above bears some remarkable resemblances to the second law of thermodynamics. Both are properties of populations, or aggregates, true irrespective of the nature of the units which compose them; both are statistical laws; each requires the constant increase of a measurable quantity, in the one case the entropy of a physical system and in the other the fitness . . . of a biological population . . . Professor Eddington has recently remarked that "The law that entropy always increases—the second law of thermodynamics—holds, I think, the supreme position among the laws of nature." It is not a little instructive that so similar a law should hold the supreme position among the biological sciences (1930, p. 36).

In a curious and even ironic sense, the most striking feature of this analogy lies in its imprecision and inaptness (especially as devised and presented by such an exacting thinker)—as Fisher himself admits directly after the statement quoted above, when he lists, as exceptions, several "profound differences" (p. 37). (The claim for such generality in physical terms must have struck Fisher as vitally important if he chose to make the comparison, declare its status as fundamental, and then immediately proceed to hedge or retract every important similarity!) Consider the first three exceptions listed by Fisher (1930, p. 37):

1. The systems considered in thermodynamics are permanent; species on the contrary are liable to extinction, although biological improvement must be expected to occur up to the end of their existence.

2. Fitness, although measured by a uniform method, is qualitatively different for every different organism, whereas entropy, like temperature, is taken to have the same meaning for all physical systems.

3. Fitness may be increased or decreased by changes in the environment, without reacting quantitatively upon that environment.

But what do these exceptions express in ordinary biological parlance? Contingency, individuality, and interaction, for the three points respectively. Could anyone have presented a better list of the peculiarly biological properties that make organisms and their history so intrinsically *unlike* simpler physical systems that operate by timeless and general laws? Do these differences between physical thermodynamics and Darwinian biology then rank as exceptions or invalidations?

THE EUGENICAL CHAPTERS. Fisher's *Genetical Theory* has generally been acknowledged, and properly so, as the keystone of 20th century evolutionary theory. Yet few contemporary biologists have actually read the book *in extenso*, and one feature of this common neglect seems especially puzzling. The last five chapters, nearly 40 percent of the entire volume, present a single coherent (if fatally flawed) argument in eugenics—a claim that modern industrial society (particularly the British version) has entered a potentially fatal decline as a result of "social promotion of the relatively infertile." In essence, Fisher argues that people who rise socially, by dint of moral or intellectual superiority, also tend to express ineluctable genetic propensities (not reversible, environmentally induced preferences) for infertility. This superior upper stratum will therefore be swamped by greater reproduction of less worthy social classes. Throughout human history, most great civilizations have declined for this reason. Society should fight this decay by rewarding gifted *and highly fecund* members of the lower classes, thereby helping them to rise and rejuvenate the reproduction of higher social strata.

A tradition of discreet silence has enveloped these chapters. Provine (1971), for example, relegates this material to a single sentence in his important book (pp. 153–154): "In the concluding five chapters he extended his genetical ideas to human populations." This discretion, I suppose, reflects our embarrassment that such a paragon of our profession should have ended his canonical book with such a long argument for a politically discredited movement (see Gould, 1991c, for an analysis and critique of Fisher's eugenical arguments). This professional silence surely cannot reflect a belief that these chapters bear no connection to the rest of the book, and that Fisher merely appended this material to grind his political axe—for Fisher states that he could have dispersed this 40 percent among the other chapters, and then adds

(p. x): "The deductions respecting Man are strictly inseparable from the more general chapters."

I regard the conspiracy of silence about these chapters as both unscholarly and overly fastidious. First of all, how can we justify silence about integral parts of an important thinker's work because we now recoil at his beliefs? (Wagner's anti-semitism retains intimate linkage with his musical productions, but we cannot ban such glorious operas.) Second, even if we wish to defend such posthumous cleansing, Fisher's eugenics can only be judged as "garden variety" material for his time, and not as especially benighted or vengeful. His visions of proper social stratification may surely be judged élitist (scarcely a rare attitude for an Oxbridge don in class-conscious Britain), but anachronistic exponents of modern political correctness will appreciate other facets of his argument. (Fisher, for example, cautiously advocates racial mixing for its role in increasing genetic variance, thereby supplying more material at the right tail of the human distribution, even though admixture with a "lower race"—Fisher did not espouse egalitarian beliefs!—might depress the mean.)

But the central relevance of these final chapters lies in the consonance of Fisher's eugenical argument with his commitment to a general and statistical Darwinism. Fisher's eugenics provides our most interesting and incisive affirmation of his evolutionary philosophy. Darwinian triumph must be measured as differential reproductive success, statistically defined in large populations—not as particularistic victory for nifty bits of morphology (or mentality) in Tennyson's world of "nature red in tooth and claw." Moreover, Fisher maintains that our current pattern of degeneration arises from differentials in birth rate, not from selective superiority in resisting death—so Darwinian "success" can only be viewed as statistical leverage in components of reproductive advantage, not as improvement in any social or vernacular sense: "Even the highest death-rate in this period, that in the first year of life, must be quite unimportant compared with slight differences in reproduction; for the infantile-death rate has been reduced in our country to about seven percent of the births, and even a doubling of this rate would make only about a third as much difference to survival as an increase in the family from three children to four" (1930, p. 194).

Finally, this eugenical example illuminates the central Darwinian claim for the power of slight statistical advantage. A truly effective, and truly Darwinian, eugenics, Fisher argues, will focus on apparently tiny reproductive differentials, and not on the elimination of rare and overt "saltations"—sterilization of the genetically diseased or mentally defective, as in the programs favored by most eugenicists who did not grasp the Darwinian imperative. We might regard small differences in birth rates as trifling, and unlikely to exert much effect upon the rapid time scales of human history. But anything that can be measured at all over the minimal span of a generation or two translates to an enormous effect in evolutionary time. Thus, the social promotion of relative infertility, however "invisible" in comparison to the devastation of war or the progress of technology, will yield an evolutionary degeneration far in excess of almost any other Darwinian change in nature. In evolutionary time, Fisher laments, our social structures disintegrate rapidly; we had better pay heed: "Civilized man, in fact, judging by the fertility statistics of our own time, is apparently subjected to a selective process of an intensity approaching a hundredfold the intensities we can expect to find among wild animals, with the possible exception of groups which have suffered a recent and profound change in their environment" (1930, p. 199).

### J. B. S. HALDANE AND THE INITIAL PLURALISM OF THE SYNTHESIS

Haldane purposely included a plural in the title of his book—*The Causes of Evolution* (1932)—for he believed that nothing so encompassing could be so unifactorial. But Haldane wrote his book in the tradition of restriction, primarily to debunk Kellogg's triad of alternatives by showing the power of natural selection. He states (p. v) that his book began as a series of lectures entitled "A Re-examination of Darwinism," and he then announces his primary aim in the preface (p. vi): "To prove that mutation, Lamarckian transformation, and so on, cannot prevail against natural selection of even moderate intensity." (Haldane treats the same subject more formally in the book's lengthy mathematical appendix, thus uniting both the front and back matter for a single purpose.)

Haldane presents a conventional account of the revivification of Darwinism and the rejection of alternatives. Darwinism had fallen on bad times before the synthesis: "Criticism of Darwinism has been so thoroughgoing that a few biologists and many laymen regard it as more or less exploded" (p. 32). The Darwinian resurrection followed from the recognition that continuous, small-scale variation could also claim a Mendelian basis (p. 71) and, especially, that tiny selection pressures, working in a cumulative manner on such minor variations, could effectively explain all evolution: "But however small may be the selective advantage the new character will spread, provided it is present in enough individuals of the population to prevent its disappearance by mere random extinction. . . . An average advantage of one in a million will be quite effective in most species" (1932, p. 100).

The development of mathematical population genetics establishes the centerpiece of Darwinian revival. Haldane even begins the tradition of a founding trinity in stating, however immodestly (p. 33): "I can write of natural selection with authority because I am one of the three people who know most about its mathematical theory."

However, in contrast to Fisher's quest for pervasive and abstract generality, Haldane felt compelled to bring the smaller and more particular puzzles of natural history under his theoretical umbrella. Here he allows a substantial range of exceptions to Darwinism, albeit at subsidiary frequency—thus illustrating the predominant pluralism of the early synthesis. Haldane rejects Lamarckism outright, as contrary in principle to the known workings of inheritance. But, in a remarkable passage, he finds some space, in chinks and corners of the new world of fusion between Darwin and Mendel, for the two internalist theories in Kellogg's triad of alternatives—saltation and orthogenesis. (In fact, Haldane even repeats the "standard" anti-Darwinian claim for selection's merely subsidiary and negative role in enhancing and stabilizing a saltational change arising by other means—even though Haldane regards this alternative mechanism as rare in nature.) Galton's polyhedron cannot be fully rounded by the emerging Darwinian consensus:

> But if we come to the conclusion that natural selection is probably the main cause of change in a population, we certainly need not go back completely to Darwin's point of view. In the first place, we have every reason to believe that new species may arise quite suddenly, sometimes by hybridization, sometimes perhaps by other means. Such species do not arise, as Darwin thought, by natural selection. When they have arisen they must justify their existence before the tribunal of natural selection, but that is a different matter. . . . Secondly, natural selection can only act on the variations available, and these are not, as Darwin thought, in every direction. In the first place, most mutations lead to a loss of complexity (*e.g.* substitution of leaves for tendrils in the pea and sweet pea) or reduction in size of some organ (*e.g.* wings in *Drosophila*). . . . Mutations only seem to occur along certain lines (1932, pp. 138–139).

Two modes of non-Darwinian change especially intrigued Haldane. First, though he tried to reinterpret as many cases as possible in a Darwinian manner, Haldane accepted some paleontological claims for supposedly orthogenetic trends, and he admitted that the developing Darwinian synthesis could find no place for such phenomena: "Many such cases—for example the development of large size or large horns—can, I think, be put down to the ill effects of competition between members of the same species. Others, such as the exaggerated coiling of *Gryphaea* cannot at present be explained with any strong degree of likelihood" (1932, p. 141). (This example seems especially ironic in retrospect, because *Gryphaea*'s supposed overcoiling to necessary extinction never occurred, and the claim rested upon misreported and misinterpreted data—see Chapter 10, pp. 1040–1045 and Gould, 1972.)

As his favorite general argument for awarding a small space to orthogenesis, Haldane cited the putatively higher frequency of degenerational over progressive evolution, arguing that such a tendency probably required an internalist explanation rooted in a bias for deletional mutations: "Degeneration is a far commoner phenomenon than progress. It is less striking because a progressive type, such as the first bird, has left many different species as progeny, while degeneration often leads to extinction, and rarely to a widespread production of new forms . . . But if we consider any given evolutionary level we generally find one or two lines leading up to it, and dozens leading down" (1932, pp. 152–153).

Second, Haldane accepted the common wisdom of taxonomists in his generation that most differentia of species expressed no adaptive significance. He also acknowledged this factual substrate as a primary source of legitimate doubts, then common among taxonomists, about Darwinism: "But when we have pushed our analysis as far as possible, there is no doubt that innumerable characters show no sign of possessing selective value, and, moreover, these are exactly the characters which enable a taxonomist to distinguish one species from another. This had led many able zoologists and botanists to give up Darwinism" (1932, pp. 113–114).

Haldane even presents the interesting argument that we have been fooled into accepting a dominant frequency for adaptation by a pronounced bias in the fossil record—the differential preservation of species with persistently large populations subject to control by small Fisherian differentials in natural selection. Perhaps most species exist as much smaller populations, and therefore become subject to Wrightian dynamics of genetic drift—even if such species rarely enter the fossil record and therefore fail to leave evidence for their dominant relative frequency. Haldane even cites the highest of all authorities to buttress this idea:

> But Wright's theory certainly supports the view taken in this book that the evolution in large random-mating populations, which is recorded by paleontology, is not representative of evolution in general, and perhaps gives a false impression of the events occurring in less numerous species. It is a striking fact that none of the extinct species, which, from the abundance of their fossil remains, are well known to us, appear to have been in our own ancestral line. Our ancestors were mostly rather rare creatures. "Blessed are the meek: for they shall inherit the earth" (1932, pp. 213–214).

### J. S. HUXLEY: PLURALISM OF THE TYPE

As with Haldane, Huxley also credited a well received lecture that he had presented on Darwinism as the stimulus for writing his much longer book—a 1936 presidential address to the British Association on "Natural selection and evolutionary progress." Huxley maintained the focus of this lecture in presenting a thoughtful, but partisan, defense of Darwinism throughout *Evolution, The Modern Synthesis,* beginning with a wry comment on the extensive pessimism so common before the movement he christened: "The death of Darwinism has been proclaimed not only from the pulpit, but from the biological laboratory; but, as in the case of Mark Twain, the reports seem to have been greatly exaggerated, since today Darwinism is very much alive" (1942, p. 22).

Huxley encapsulates the central logic of Darwinism in much the same way, and with the same intent, that I advocate in this book. He recognizes the three main characteristics of variation as central (pp. 22–24)—copiousness (though not pervasive enough for mutation pressure to overwhelm selection), smallness of phenotypic effect, and nondirectionality—and he credits Mendelism with providing the physical explanation for what Darwin could only deduce from first principles of natural selection, while hoping for later confirmation from discoveries about the basis of heredity.

In an interesting discussion on the nature of theories and their central logic, Huxley disputes Lancelot Hogben's claim that the Mendelian fusion had so altered Darwin's own notion of mechanics, that the reformulation of Fisher, Haldane, and Wright should neither bear Darwin's name nor even retain the term "natural selection" for its central mechanism. Huxley replies that all theories must change by growth, but that the proper standard for maintenance of a name must be defined by continuity in key precepts in a central logic:

> Hogben is perfectly right in stressing the fact of the important differences in content and implication between the Darwinism of Darwin or Weismann and that of Fisher or Haldane. We may, however, reflect that the term *atom* is still in current use and the atomic theory not yet rejected by physicists, in spite of the supposedly indivisible units having been divided. This is because modern physicists still find that the particles called atoms by their predecessors do play an important role, even if they are compound and do occasionally lose or gain particles and even change their nature. If this is so, biologists may with a good heart continue to be Darwinians and to employ the term Natural Selection, even if Darwin knew nothing of mendelizing mutation (1942, p. 28).

Huxley also follows the English tradition (see pp. 116–119) for central emphasis upon adaptation in the definition of evolutionary mechanisms. He speaks of "a functionally-guided course of evolution" (p. 39), and almost claims an *a priori* status for panadaptationism: "Our enumeration will also serve as a reminder of the omnipresence of adaptation. Adaptation cannot but be universal among organisms, and every organism cannot be other than a bundle of adaptations, more or less detailed and efficient, coordinated in greater or lesser degree" (1942, p. 420).

But as further evidence for pluralism in the early synthesis, and despite this emphasis upon the ubiquity of adaptation, Huxley then speaks favorably of the same challenges and exceptions that intrigued Haldane—orthogenesis and nonadaptation. Whereas he does claim (correctly) that most cases of supposed orthogenesis only represent instances of phyletic constraint, he also provides an interesting taxonomy of genuine examples. Mirroring our modern distinction between positive and negative meanings of constraint (see Chapter 10, pp. 1025–1061, and Gould, 1989a), Huxley speaks of dominant and subsidiary orthogenetic restriction:

> True orthogenetic restriction depends on a restriction of the type and quantity of genetic variation. When dominant it prescribes the direction of evolution: when subsidiary it merely limits its possibilities. . . . Dominant orthogenetic restriction [is] very rare, if indeed it exists at all. . . .

Subsidiary orthogenetic restriction is probably frequent, but we are not yet able to be sure in most cases whether a limitation of variation as actually found in a group is due to a limitation in the supply of mutations or to selection, or to other causes. It is, however, certain that some mutational effects recur regularly in some allied species, and probable that this phenomenon is widespread (1942, p. 524).

Huxley also cites overcoiling in *Gryphaea* (Trueman, 1922)—the classic case of his time, though since invalidated (Gould, 1972)—as a primary puzzlement and most promising example for "dominant orthogenesis":

We must provisionally face an explanation in terms of orthogenesis—i.e. of evolution predetermined to proceed within certain narrow limits, irrespective of selective disadvantage except where this leads to total extinction. It should be noted that, even if the existence of orthogenesis in this cause [sic, for case] be confirmed, it appears to be a rare and exceptional phenomenon, and that we have no inkling of any mechanism by which it may be brought about. It is a description, not an explanation. Indeed its existence runs counter to fundamental selectionist principles (1942, p. 509).

Despite his general commitment to adaptation, Huxley also granted some importance (beyond mere existence) to Wright's genetic drift in the formation of species with small population sizes (p. 58). He even extended the power of this non-adaptational force to the origin of generic differences, though not beyond: "It may be presumed, on somewhat indirect evidence, that 'useless' non-adaptive differences due to isolation of small groups may be enlarged by the addition of further differences of the same sort to give generic distinction, though it seems probable that differences of family or higher rank are always or almost always essentially adaptive in nature" (1942, p. 44).

Thus, the early synthesis, in the view of both its founders and its namegiver, reinstated Darwinism as the centerpiece of evolutionary theory by rejecting any substantial role for the full spate of previously popular alternatives. (I should say "instated," for Darwinism had never before attained majority appeal as a mechanism, even during Darwin's lifetime.) But the early synthesists, with Fisher's exception, also left a few facets intact on Galton's polyhedron. Their interest lay in showing that our increasing knowledge of the Mendelian world could establish natural selection as the primary cause of evolutionary change, not in staking a claim for Darwinian exclusivity.

## Synthesis as Hardening

### THE LATER GOAL OF EXALTING SELECTION'S POWER

Evolutionists have generally depicted the second phase of the Synthesis as a gathering of traditional subdisciplines under an umbrella constructed during the first phase by fusing Mendel with Darwin. I learned something fundamental about this second phase as a participant at the conference, entitled *Workshop on the Evolutionary Synthesis,* that Ernst Mayr convened in Boston in 1974. This conference—an amazing experience for a young evolutionist at the beginning of a career—included every major living participant in the Synthesis except Bernhard Rensch, who was ill; G. G. Simpson, who was angry; and Sewall Wright, whom Mayr simply would not invite, despite pleas from yours truly and several others. I don't think I ever experienced a greater moment of pure "academic awe" than my first impression, when I looked across from "our" side of the table (where Mayr had placed the "young" historians and evolutionists) and saw Dobzhansky, Mayr, Stebbins, Ford, and Darlington all together on the other side.

This marvelous conference was marred (in terms of its stated purpose) only by a severe difficulty in keeping these men to the intended subject of their reminiscences about past accomplishments. They all remained so passionately involved in modern research that, whenever the planned reminiscences began, someone would make a reference to the latest paper revising some view or another—and they would immediately begin a learned discussion about current events, fueled by delight at new findings that forced revisions of their old certainties! (A difficulty for the conference's stated aim perhaps, but personally one of the most memorable events that I have ever witnessed. If the best practitioners can maintain such openness and involvement to the end of their lives, then scholarship need not fear ossification. Such traits do not, however and alas, represent the norm in science—so I did come to understand the special excellence of these extraordinary men, and I did achieve some visceral grasp of why they, and not others, made the Synthesis.)

I had always viewed the books of the second phase as coequal. But the conference discussions emphasized a major point previously unclear to me: the preeminence of Dobzhansky's 1937 book, *Genetics and the Origin of Species.* This volume did not merely happen to enjoy the luck of first publication in a series—a temporal *primus inter pares,* so to speak. Dobzhansky's volume provided a direct and primary inspiration for the books that followed. Speaker after speaker rose to state that his own contribution had been prodded by reading Dobzhansky's account first.

And now the irony—and the key point about disjunction between the two phases of the Synthesis. If we wish to argue that the first phase of synthesis featured the construction of population genetics by Fisher, Haldane and Wright, while the second phase brought traditional subdisciplines into this framework, we should expect the primary translator to be fluent in the language of transfer. In one sense, Dobzhansky did possess the requisite fluency—uniquely (at least for English-speaking scientists), and for an interesting reason of national traditions. As I also learned at the 1974 conference, only in Russia had Mendelian experimental work been merged, extensively and successfully, with traditional taxonomy and natural history. Dobzhansky, after all, had developed expertise as both a skilled *Drosophila* experimentalist and a specialist on the taxonomy of coccinellid beetles (ladybirds). In America and Western Europe, experimentalism and field biology occupied two different and largely hostile worlds. Could the second phase of the synthesis have emerged from a Western *Drosophila* lab like T. H. Morgan's (where field biology held low status and enjoyed no practice—see p. 532), or a museum program in comprehensive systematics (with virtually no experimental facilities)? Dobzhansky exported a fusion that Western science, in ignorance of the Russian language and in hostility to communist politics, had failed to recognize—even though H. J. Muller had brought the first *Drosophila* stocks to Russia, thus fueling Dobzhansky's optimal training with a Western trigger.

But if Dobzhansky could integrate the Mendelian experimental world with natural history, what about the supposed centerpiece of mathematical population genetics? Here, by his own repeated, almost gleeful, admission, Dobzhansky remained a near dunce. He did not study, nor could he even understand, the details of this literature. Of his long and fruitful collaboration with Sewall Wright, Dobzhansky simply said that he had followed the principle of "father knows best"—that is, he bypassed Wright's mathematical manipulations and accepted his English explanations on faith. In fact, of all the great second-phase synthesists only G. G. Simpson possessed sufficient mathematical background to read and understand these papers.

Dobzhansky's willingness to accept an incomprehensible literature, and the later acquiescence of so many leaders from other subdisciplines (largely via Dobzhansky's "translation"), testify to a powerful shared culture among evolutionists—a set of assumptions accepted without fundamental questioning or perceived need to grasp the underlying mechanics. Such a sense of community can lead to exhilarating, active science (but largely in the accumulative mode, as examples cascade to illustrate accepted principles). As a downside, however, remaining difficulties, puzzles, anomalies, unresolved corners, and bits of illogic may retreat to the sidelines—rarely disputed and largely forgotten (or, by the next generation, never learned). This situation may sow seeds of an orthodoxy that can then become sufficiently set and unchallenged to verge on dogma—as happened in many circles, at least among large numbers of epigones, at the acme of the Synthesis in the late 1950's and 1960's.

In this section, I shall try to illustrate one example *in extenso*—the central and defining case, I believe—of the narrowing suffered by a synthesis that became augmented in power but downgraded in the art and tactic of questioning. I call this increasing confidence, bordering on smugness, the "hardening" of the Synthesis. Thus I contrast the positive restriction of the first phase—the elaboration of a generous and comprehensive theory, and the invalidation of false and fruitless alternatives—with the negative tightening that occurred during the ontogeny of the second phase. This hardening—still our legacy today—must serve as a starting point for any current attempt to introduce more amplitude into evolutionary theory. The hard version of the Synthesis provides a standard for judging (by contrast) the interest and importance of modern revisions—from neutralism,* to punctuated equilibrium, to a common feeling that the theme of developmental constraints not only gives substance to an old truth, but also confutes the hardened version's commitment to Darwin's (I should really say Fisher's) billiard ball against Galton's polyhedron.

My example shall trace the transformation of adaptation from an option to be ascertained (albeit favored and granted a dominant relative frequency) to an *a priori* assumption of near ubiquity (save in trivial or derivative situations without evolutionary importance)—in other words the burnishing of Galton's polyhedron to the billiard ball of pure functionalism (allowing no significant pushing back from internal structure upon the direction of evolutionary change). This hardening buttressed (or rather, in my view, overly rigidified and sclerotised) one leg on the essential Darwinian tripod of support—the second theme of functionalism against internalist and structuralist forces (see Chapters 2, 4, and 5).

But hardening pervaded all major themes of Darwinian central logic, and the other two legs of the tripod also experienced their own form of petrifaction (treated in less detail in Section 4 of this chapter). Pluralistic (and, admittedly, often loose) thinking about levels of selection yielded to an explicit promulgation of organismic selection as the only acceptable mode—as a virtual campaign to root out group selection accompanied the battle of Williams (1966) against Wynne-Edwards (1962). Thirdly, a willingness to grant some independence, or at least some puzzlement, to patterns in macroevolution (see Haldane and Huxley's respectful view of orthogenesis, as discussed in the last section), ceded to the hard view that all phenomena measured in millions of years must be explained by smooth extrapolation from palpable causes on generational scales in modern populations—and that the paleontological record can therefore only present a pageant of products generated by known causes, and not provide an independent theory or even a set of additional causal principles.

I have used a particular method to demonstrate the hardening of the Synthesis—textual comparison of early and later works by key authors. Ontogeny can be an unconscious trickster. In trying to forge sense and continuity in

---

*If the Synthesis had retained the pluralism of its early years, Kimura's neutral theory would have been welcomed from the first, under the criterion that any result legitimated by the mechanics and mathematics of known genetic processes thereby secured a rightful place (Wrightful in this particular case)—though Kimura's claim would have been viewed as surprising in the light of adaptationist preferences. But when the Synthesis hardened, and adaptationism itself became the primary criterion for acceptability, Kimura's theory seemed beyond the pale to many evolutionists. I shall never forget a decisive moment in my own early career, when I began to understand the difference between theoretical power and potentially dangerous overconfidence: Ernst Mayr rising (at the annual meeting of the Evolution Society in New York) to confute the claim for neutralism in synonymous third position substitutions. Such changes could not, *a priori* and in principle he stated, be neutral. Alterations in the third position must impart some difference, perhaps energetic, that selection can "see" even if the coded amino acid does not alter. This must be so, he stated, because we now know that all substantial change is adaptive.

our own lives, we often forget or "reconstruct" the actualities of our early years—thus subtly recasting our former selves as miniatures of our current beliefs. Therefore, direct interview can be a notoriously unreliable technique (while representing, ostensibly, the most direct and empirical of all scholarly sources)—for an older person may become a very unreliable chronicler of his own past. But written records stand as frozen testimonies, unaltered fossils of a time that may not be personally recoverable with high accuracy.

I received my first insight into the hardening of the Synthesis by a proper (if gentle) pedagogical correction. During my graduate student years, I presented a report on paleontology in the Modern Synthesis to a seminar at the American Museum of Natural History. In the characteristically naive manner of a young and awestruck protoscholar, I explicated the views of Simpson and others as jewels of reforming consistency, *lux in tenebris* and complete from the first. Bobb Schaeffer, a wonderful teacher, stopped me as I was explaining Simpson's complex idea of "quantum evolution" (see p. 530). I had done well, he said, for the concept as presented in Simpson's 1953 book, but had I ever studied the original version in *Tempo and Mode in Evolution* (1944)? I replied that I had not read this initial formulation, for I had assumed that the first account could only represent a less developed, and therefore pale and trifling, version of later subtlety. Schaeffer said that the two discussions differed fundamentally, but that Simpson had minimized the appearance of change by retaining the same terms while profoundly altering their meaning. (Schaeffer also told me that he had argued the issue with Simpson for years, and that the essence of Simpson's change, for which Schaeffer took some credit—a shift from nonadaptionist to selectionist interpretation of intermediate forms in major phyletic transitions—had only strengthened the general argument, even though Simpson had covered up his changes.) I did not believe that most of the profession could have missed such a major shift, but I checked. Schaeffer was entirely right.

My personal failure piqued my interest and I began to wonder whether Simpson's change had been idiosyncratic or part of an unrecognized pattern. I began to check early and late works of other key figures, particularly Dobzhansky and Mayr. All had moved from pluralism to strict adaptationism—and along a remarkably similar path. I began to view this transition as the major ontogenetic event of the Synthesis during its second phase. I christened this change as the "hardening" of the Synthesis and wrote four papers on the subject (Gould, 1980e, 1982d, 1983b and c). The rest of this section documents my three favorite cases—Dobzhansky through the three editions (1937, 1941, and 1951) of his seminal book, Mayr (1942 vs. 1963), and Simpson (1944 vs. 1953)—and reproduces a good deal of material from my earlier articles.

Several historians have tested my hypothesis by application to other key figures, and have affirmed the adaptational hardening as general (e.g., Beatty, 1988, and Smocovitis, 1996, 2000). Sewall Wright, subject of Provine's massive biography (1986; see also Provine, 1971), provides the most interesting and revealing case. Wright's name, of course, immediately evokes the phenomenon of genetic drift, generally called the "Sewall Wright effect" in articles of the early Synthesis. One would therefore regard Wright as the man most likely to speak for the importance of nonadaptation, and against any functionalist hardening.

In fact, when interviewed late in life, as both Provine and I can attest, Wright complained bitterly that his views on the evolutionary role of genetic drift had been consistently misinterpreted (Wright died in 1988 at age 98, sharp as ever to the very end). Since genetic drift describes stochastic change in gene frequencies by sampling error, one might assume that Wright had advocated a radically non-Darwinian approach to evolutionary change by demoting selection and adaptation, and boosting the importance of accident. But Wright strongly denied such an interpretation of his views. He argued, with evident justice apparent to anyone who reads the works of his last thirty years, that his theory of "shifting balance," while providing an important role for genetic drift, remains strongly adaptationist—though adaptation generally arises at a level higher than the traditional Darwinian focus on organisms.

In brief (see p. 555 for a fuller account), Wright asserted that he had invoked genetic drift primarily as a generator of raw material to fuel an adaptationist process of interdemic selection. If the founding deme of a new species occupies one adaptive peak on a complex landscape (to use standard Wrightian imagery), movement to additional peaks requires genetic drift—for this stochastic process permits small demes to descend slopes and enter valleys, where selection can then draw a deme up to another peak. When demes within a single species populate several peaks, interdemic selection can operate as a powerful mechanism of adaptation.

Wright therefore (and accurately) depicted his later shifting balance theory as adaptationist, and as invoking drift only for a source of variation among demes. But Wright, though estranged in many ways from the developing synthesis (see Section 4), followed its trend toward increasingly exclusive emphasis upon adaptation in evolutionary change. The version of shifting balance that Wright advocated during the last 30 years of his life did not originate by sudden creation, complete in this final form. Shifting balance emphasized different themes and arguments in Wright's earlier work, and these articles, written during the pluralistic phase of the synthesis, granted a much greater role to randomness and nonadaptation in evolutionary *change*. In fact, Wright often, and explicitly, invoked drift as a non-Darwinian agent of change in articles written during the early pluralistic phase of the synthesis.

Wright presents a striking example of the principle that later recollections may be inferior, as historical sources, to written testimony from the time in question. Provine (1986) has catalogued Wright's ambiguities and multiple intents during the crucial period of 1929–1932. The later selectionist view already stands in the wings, but most passages of these early articles advocate the nonadaptationist role for drift that Wright would later reject (and deny he ever held). Wright wrote in 1931 (p. 158), for example, that shifting balance "originates new species differing for the most part in nonadaptive respects." In the following year, he stated (1932): "That evolution involves nonadaptive differentiation to a large extent at the subspecies and even the species level is indicated by the kinds of differences by which such groups are actually distinguished by systematists. It is only at the subfamily and family levels that clearcut adaptive differences become the rule. The principal evolutionary mechanism in the origin of species must then be an essentially nonadaptive one" (pp. 363–364). Provine (1986) concludes: "The careful reader in 1932 would almost certainly conclude that Wright believed nonadaptive random drift was a primary mechanism in the origin of races, subspecies, species, and perhaps genera. Wright's more recent view that the shifting balance theory should lead to adaptive responses at least by the subspecies level is found nowhere in the 1931 and 1932 paper."

## INCREASING EMPHASIS ON SELECTION AND ADAPTATION BETWEEN THE FIRST (1937) AND LAST (1951) EDITION OF DOBZHANSKY'S *GENETICS AND THE ORIGIN OF SPECIES*

Dobzhansky's original probe (1937) toward synthesis operated more as a methodological claim for the sufficiency of genetics than a strong substantive advocacy of any particular causal argument—although he clearly states his general Darwinian preferences in this first edition. Dobzhansky held, contrary to his own Russian mentor Filipchenko, that the methods of experimental genetics can provide enough principles to encompass evolution at all levels. But Dobzhansky did not play favorites among the admitted set of legitimate principles. He did not, in particular, proclaim the pervasive power of natural selection leading to adaptation as a predominant style and outcome of evolutionary change.

Some inkling of the chaotic and depressed state of evolutionary theory before the Synthesis can be glimpsed in a simple list of previously popular arguments that Dobzhansky regarded as sufficiently important to refute—claims that denied his hope for synthesis by treating Mendelian processes observed in the laboratory as different from the genetic modes for regulating "important" evolutionary change in nature. Dobzhansky rebuts the following arguments explicitly: Continuous variation in nature is non-Mendelian and different in kind from discrete mutational variation in laboratory stocks (p. 57); Mendelian variation can only generate differences between taxa of low rank (races to genera), while higher taxa owe their distinctions to another (and unknown) genetic process (p. 68); chromosomal changes are always destructive and can only lead to degeneration of stocks (p. 83); differences between taxa of low rank are directly induced by the environment and have no genetic or evolutionary basis (p. 146); Johannsen's experiments on pure lines prove the ineffectiveness of natural selection as a mechanism of evolutionary change (p. 150); selection is too slow in large populations to render evolution, even in geological time (p. 178); genetic principles cannot account for the origin of reproductive isolation (p. 255).

Dobzhansky's fifth chapter, on "variation in natural populations," stresses the pluralism of the early synthesis. Observable genetic phenomena provide a source for *all* evolution; we can trace full continuity from studies in the laboratory, to variation within natural populations, to formation of races and species:

> It is now clear that gene mutations and structural and numerical chromosome changes are the principal sources of variation. Studies of these phenomena have been of necessity confined mainly to the laboratory and to organisms that are satisfactory as laboratory objects. Nevertheless, there can be no reasonable doubt that the same agencies have supplied the materials for the actual historical process of evolution. This is attested by the fact that the organic diversity existing in nature, the differences between individuals, races, and species, are experimentally resolvable into genic and chromosomal elements which resemble in all respects the mutations and the chromosomal changes that arise in the laboratory (1937, p. 118).

But what forces mold and preserve this variation in nature? Dobzhansky stresses natural selection (p. 120), but he does not grant this process the dominant role that later "hard" versions of the synthesis would confer. He emphasizes genetic drift (which he calls "scattering of the variability") as a fundamental mode of evolutionary change in nature, not as an odd phenomenon occurring in populations too small to leave any historical legacy. He argues that local races can form without influence from natural selection, and he supports Crampton's (1916, 1932) interpretation of the nonadaptive and indeterminate character of substantial racial differentiation in the Pacific land snail *Partula*. He emphasizes that evolutionary dynamics depend, in large measure, upon the size of populations *because* selection does not always control the outcome (and we therefore need information about numbers of individuals and their mobility in order to assess the effects of drift, migration, and isolation). He coins the term "microgeographic race" and argues that most group distinctions at this level may be both nonadaptive and genetically based, contrary to the opinions of many naturalists who then regarded such races as adaptive and nongenetic.

The sixth chapter then treats natural selection explicitly. Dobzhansky begins by clearing away some early Mendelian misconceptions about the impotence of natural selection (logical errors in interpreting Johannsen's experiments on pure lines, for example). He then poses a central question: Darwin devised the theory of natural selection to explain adaptation; admitting Darwin's success in this area, may we then extrapolate and argue that selection controls the direction of all evolutionary change (p. 150)? Dobzhansky answers that we cannot defend such an extension of selection's power. He then criticizes the strict selectionism of Fisher (p. 151), and praises a book that would later be castigated by all leading synthesists as a remnant of older and unproductive ways of thought—Robson and Richards (1936), with their defense of a nonadaptive origin for most subspecific and even interspecific differences in closely related forms.

A long concluding section (pp. 185–191) supports Wright's "island model" of selection among semi-isolated demes occupying different peaks of an adaptive landscape. Dobzhansky pleads for more study of "the physiology of populations" since Wright's model proclaims three factors as important in different ways, while not granting inherent predominance to any: genetic drift, migration, and natural selection: "Since evolution as a biogenic process obviously involves an interaction of all of the above agents, the problem of the relative importance of the different agents unavoidably presents itself. For years this problem has been the subject of discussion. The results of this discussion so far are notoriously inconclusive; the 'theories of evolution' arrived at by different investigators seem to depend upon the personal predilections of the theorist" (p. 186). Dobzhansky does, however, suggest that Wright's model may validate the common conviction of naturalists that the morphological differentia of races and species must often be nonadaptive.

*Genetics and the Origin of Species* went through three editions (1937, 1941, and 1951). As in the successive versions of Darwin's *Origin*, the differences among these editions cannot be dismissed as trivial or cosmetic, for they convey a major change in emphasis—an alteration that set the research program for most evolutionary biologists until the past few years. As the Synthesis developed, the adaptationist program grew in influence and prestige, and other modes of evolutionary change fell into disrepute, or became redefined as locally operative but unimportant in the overall picture.

Dobzhansky's third edition (1951) clearly reflects this hardening. He still insists, of course, that not all change can be called adaptive. He attributes the frequency of some traits to equilibrium between opposed mutation rates (p. 156) and doubts the adaptive nature of racial variation in blood types. He asserts the importance of genetic drift (pp. 165, 176) and does not accept as proof of panselectionism one of the centerpieces of the adaptationist program—A. J. Cain's work on frequencies of banding morphs in the British land snail *Cepaea* (p. 170).

But inserted passages and shifting coverage convey, as their common focus, Dobzhansky's increasing faith in the scope and power of natural selection, and in the adaptive nature of most evolutionary change. He deletes the two chapters that contained most material on nonadaptive or nonselected phenomena (polyploidy and chromosomal changes, though he includes their material, in much reduced form, within other chapters). He adds a new chapter on "adaptive polymorphism" (pp. 108–134). Moreover, he now argues that anagenesis, or "progressive" evolution, works only through the optimizing, winnowing agency of selection based on competitive deaths; species adapting by increased fecundity in unpredictably fluctuating environments do not contribute to anagenesis (p. 283).

But the most remarkable addition occurs right at the beginning. I label these passages remarkable because I doubt that Dobzhansky really believed what he literally said; I feel confident that he would have modified his words had anyone pointed out how his increasing fascination for adaptationism had led him to downgrade the deepest and oldest of evolutionary themes to effective invisibility (see Chapter 10, pp. 1175–1178 for the modern relevance and refutation of this striking image).

Dobzhansky poses the key question of organic form and taxonomy: why do organisms form discrete and clearly nonrandom "clumps" in populating morphological space? Why does the domain of mammalian carnivores contain a large cluster of cats, another of dogs, a third of bears, leaving so much unoccupied morphological space between? Dobzhansky begins by "promoting" Wright's model of the "adaptive landscape" to an inappropriate level. In so doing, Dobzhansky subtly shifts the model's meaning from an explanation for nonoptimality (with important aspects of nonadaptation) to an adaptationist argument about best solutions. Wright devised his model to explain differentiation among demes *within* a species. He proposed the metaphorical landscape to justify a fundamentally nonadaptationist claim: If a "best solution" exists for the phenotype of a species (the highest peak in the landscape), why don't all demes reside there? But if we "upgrade" the model to encompass differences *between* species within a clade, then metaphorical landscapes mutate into a framework for strict adaptationism. Each peak now becomes the optimal form for a single species (not the nonoptimal form for some demes within a species). And related peaks represent a set of best solutions as the various adaptations of separate evolutionary entities within a clade.

Dobzhansky then attempts to solve the problem of clumping with an adaptationist argument based upon the organization of ecological space into preexisting optimal "places" where good design may find a successful home. Evolution has produced a cluster of cats because an "adaptive range," studded with adjacent peaks, exists in the economy of nature, waiting, if you will, for creatures to move in. In other words, discontinuity in taxonomic space maps discontinuity in optimal form for available environments, with adaptation as the agent for mapping.

> The enormous diversity of organisms may be envisaged as correlated with the immense variety of environments and of ecological niches which exist on earth. But the variety of ecological niches is not only immense, it is also discontinuous . . . The adaptive peaks and valleys are not interspersed at random. "Adjacent" adaptive peaks are arranged in groups, which may be likened to mountain ranges in which the separate pinnacles are divided by relatively shallow notches [*sic*, Dobzhansky does indeed mean "notches" in this passage, not "niches" (as later in the quotation)]. Thus, the ecological niche occupied by the species "lion" is relatively much closer to those occupied by tiger, puma, and leopard than to those occupied by wolf, coyote, and jackal. The feline adaptive peaks form a group different from the group of the canine "peaks." But the feline, canine, ursine, musteline, and certain other groups of peaks form together the adaptive "range" of carnivores, which is separated by deep adaptive valleys from the "ranges" of rodents, bats, ungulates, primates, and others. In turn, these "ranges" are again members of the

adaptive system of mammals, which are ecologically and biologically segregated, as a group, from the adaptive systems of birds, reptiles, etc. The hierarchic nature of the biological classification reflects the objectively ascertainable discontinuity of adaptive niches, in other words the discontinuity of ways and means by which organisms that inhabit the world derive their livelihood from the environment (pp. 9–10).

Thus, Dobzhansky renders the hierarchical structure of taxonomy as a fitting of clades into preexisting ecological spaces. Discontinuity emerges not so much as a function of history, but as a reflection of adaptive topography. But this interpretation cannot hold; surely, the cluster of cats exists primarily as a consequence of homology and historical constraint. All felines share a basic morphology because they arose from the common ancestor of this clade alone. We doubt neither the excellent adaptation of this common ancestor nor the claim that all descendants may fit equally well into their current environments. But the feline group and the gaps that separate this cluster from other families of carnivores reflect history above all, not the current organization of ecological topography. All feline species have inherited the unique *Bauplan* of cats, and cannot deviate far from this commonality as they adapt, each in its own particular way. Genealogy, not current adaptation, provides the primary source for clumped distribution in morphological space.

### THE SHIFT IN G. G. SIMPSON'S EXPLANATION OF "QUANTUM EVOLUTION" FROM DRIFT AND NONADAPTATION (1944) TO THE EMBODIMENT OF STRICT ADAPTATION (1953)

Although Simpson, probably more than Dobzhansky, personally favored selectionist arguments in the initial version of his seminal work (1944), he also adopted a pluralistic stance at first. In fact, at the crux of his book, Simpson proposed an explicitly nonadaptationist theory to resolve the greatest anomaly in the fossil record; he also considered this theory of "quantum evolution" as the crowning achievement of his book.

Like Dobzhansky in his first edition (1937), Simpson (1944) espoused consistency of all evolutionary change with principles of modern genetics as his primary assertion for a general and synthetic theory. The major challenge to unity and consistency arose from the infamous "gaps" or discontinuities of the fossil record—particularly at the largest scale of appearances for new *Baupläne* without fossil intermediates. Simpson wrote:

> The most important difference of opinion, at present, is between those who believe that discontinuity arises by intensification or combination of the differentiating processes already effective within a potentially or really continuous population and those who maintain that some essentially different factors are involved. This is related to the old but still vital problem of micro-evolution as opposed to macro-evolution . . . If the two proved to be basically different, the innumerable studies of micro-

evolution would become relatively unimportant and would have minor value to the study of evolution as a whole (1944, p. 97).

To explain these discontinuities, Simpson relied, in part, upon the classical argument of an imperfect fossil record. But he also conceded that such a prominent pattern could not be interpreted as entirely artificial—and he recognized that his favored process of gradualistic Darwinian selection in the phyletic mode would not provide a full explanation. He therefore proposed his book's only major departure from explanations based upon selection leading directly to adaptation—and thus, in his most striking and original contribution, framed the hypothesis of quantum evolution.

Simpson clearly took great pride in this novel theory, for he ended his book with a twelve-page defense of quantum evolution, identified as "perhaps the most important outcome of this investigation, but also the most controversial and hypothetical" (p. 206). Faced with the prospect of abandoning strict selection in the gradual, phyletic mode, he framed a hypothesis that adhered rigidly to his more important goal—the proviso that macroevolution must be rendered by genetical models and mechanisms operating within species, and amenable to study in living populations. Thus, he focused upon the only major phenomenon in the literature of population genetics that permitted a mechanism other than selection to serve as a basis for directional change—Sewall Wright's genetic drift.

He envisaged major transitions as occurring within small populations (where drift might be effective and preservation in the fossil record virtually inconceivable). He chose the phrase "quantum evolution" because he conceived the process as an "all-or-none reaction" (p. 199) propelling a small population across an "inadaptive phase"—explicitly so named—from one stable adaptive peak to another. Since selection could not initiate this departure from an ancestral peak, he called upon drift to carry the population into an unstable intermediary position, where it must either die, retreat, or be drawn rapidly by selection to a new stable position. Simpson felt that, with quantum evolution, he had carried his consistency argument to completion by showing that genetical models could encompass the most resistant and mysterious of all evolutionary events—the rapid origin of novel phenotypes at high taxonomic levels. Quantum evolution, he wrote, is "believed to be the dominant and most essential process in the origin of taxonomic units of relatively high rank, such as families, orders, and classes. It is believed to include circumstances that explain the mystery that hovers over the origins of such major groups" (p. 206). Simpson could, therefore, conclude: "The materials for evolution and the factors inducing and directing it are also believed to be the same at all levels and to differ in mega-evolution only in combination and in intensity" (p. 124).

Simpson's emphasis on quantum evolution underscores a central feature of his explanatory preferences in 1944—his pluralistic view of evolutionary mechanisms. He wished to render all of macroevolution as the potential consequence of microevolutionary processes, not to rely dogmatically upon any single process. Although he favored selection leading to adaptation as a primary theme, he explicitly denied that all evolution could be adaptive and under selective control. He concluded: "The aspects of tempo and mode that have now been discussed give little support to the extreme dictum that all evolution is primarily adaptive. Selection is a truly creative force and not solely negative in action. It is one of the crucial determinants of evolution, although under special circumstances it may be ineffective, and the rise of characters indifferent or even opposed to selection is explicable and does not contradict this usually decisive influence" (1944, p. 180).

When pressured for a new edition of *Tempo and Mode*, Simpson realized that evolutionary theory had developed too much in the intervening ten years to permit a reissue or even a simple revision. The field that he pioneered had stabilized and flourished: "It was [in the late 1930s] to me a new and exciting idea to try to apply population genetics to interpretation of the fossil record and conversely to check the broader validity of genetical theory and to extend its field by means of the fossil record. That idea is now a commonplace" (1953, p. ix). Thus, Simpson followed the outline of *Tempo and Mode*, but wrote a new book more than double the length of its ancestor—*The Major Features of Evolution*, published in 1953.

The two books differ in many ways (see p. 522 for my personal and professional introduction to the distinctions), most notably in Simpson's increasing confidence that selection within phyletic lineages must represent the only important cause of substantial change. Consider the following addition to the 1953 book, a speculative comment on trends in titanothere horns, with its prompt dismissal—tinged with impatience, if not incipient dogmatism—of the venerable argument that no evident function can be ascribed to the incipient stages of useful structures: "This long seemed an extremely forceful argument, but now it can be dismissed with little serious discussion. If a trend is advantageous at any point, even its earliest stages have *some* advantage. Thus if an animal butts others with its head, as titanotheres surely did, the slightest thickening as presage of later horns already reduced danger of fractures by however small an amount" (p. 270).

But the most dramatic difference between the two books lies in Simpson's demotion to insignificance of the concept that had formerly been, by his own reckoning and explicit announcement, his delight and greatest pride—quantum evolution. This hypothesis embodied the pluralism of his original approach—a reliance on a *range* of genetical models. For he had advocated genetic drift to propel small populations off adaptive peaks into an ultimately untenable inadaptive phase. And he had explicitly christened quantum evolution as a mode different in *kind*, not only in rate, from phyletic transformation within lineages. But now, as the adaptationist program of the Synthesis hardened, Simpson decided that genetic drift could not trigger any major evolutionary event: "Genetic drift is certainly not involved in all or in most origins of higher categories, even of very high categories such as classes or phyla" (p. 355).

In an "intermediate stage" of his personal ontogeny—his presentation to the Princeton conference on genetics, paleontology and evolution—Simpson (1949, p. 224) had emphasized the dominance of selection in quantum evolution, while not denying other factors. But by 1953, he had completed his personal transition. Quantum evolution now merits only four pages in an enlarged final chapter on modes of evolution. More importantly, this concept has now mutated to a meaning that Simpson had explicitly denied before: merely a name for phyletic evolution when the process operates at a maximal rate—an evolutionary tempo differing only in degree from the leisurely, gradual transformation of populations in ordinary geological time. Quantum evolution, he now writes, "is not a different sort of evolution from phyletic evolution, or even a distinctly different element of the total phylogenetic pattern. It is a special, more or less extreme and limiting case of phyletic evolution" (p. 389). He lists quantum evolution as one category among the four styles of phyletic evolution (p. 385)—with all four characterized by "the continuous maintenance of adaptation." The bold hypothesis (1944) of an *absolutely* inadaptive phase has been replaced by the semantic notion of a *relatively* inadaptive phase (an intermediary stage inferior in design to either the ancestral or the descendant *Bauplan*). But *relative* inadaptation poses no threat to the adaptationist paradigm. Even the strictest Darwinian will feel no *Angst* if the fit of phenotype to environment decreases for an intermediate form in a new habitat, relative to the ancestor in a *different* original place; (the two forms, after all, cannot directly compete). Even less *Angst* will then accompany an acknowledgment that this intermediate form may be less well designed than its own future descendant (for selection should engender increasing adaptation through time, especially as a population adjusts to a strikingly new environment). In short, such relatively inadaptive populations can only be regarded as adequately adapted to their own environments at their own time (unsubjected, as they must be, to competition with better adapted ancestors in a different habitat, or with improved future descendants in this new world). Quantum evolution, by linguistic redefinition, therefore moves comfortably under the umbrella of the adaptationist program. Simpson now even suggests that quantum evolution may be *more* rigidly controlled by selection than any other mode of evolution (though he still invokes inadaptation for the initial trigger): "Indeed the relatively rapid change in such a shift is more rigidly adaptive than are slower phases of phyletic change, for the direction and the rate of change result from strong selection pressure once the threshold is crossed" (p. 391).

## MAYR AT THE INCEPTION (1942) AND CODIFICATION (1963): SHIFTING FROM THE "GENETIC CONSISTENCY" TO THE "ADAPTATIONIST" PARADIGM

If we consider the synthesis as a fusion of three equally robust disciplines—experimental genetics, population genetics, and studies of natural history expressed primarily by systematics (and not as an imposition of the first two, as modernisms, upon a hidebound, or even moribund, third mode of study)—

then the role played by Mayr and other field naturalists in building the synthesis becomes fully constitutive and not only derivative. Mayr (1980), wearing his historian's hat, has strongly defended such an account of the Synthesis against the reductionist tradition that regards genetics as paramount, and the second phase of the Synthesis largely as a whipping of older disciplines into line. I do not deny Mayr's partisan motives in advancing this interpretation, but I also concur with his judgment.

Dobzhansky, as argued above, became the beacon of the second phase because he represented the only tradition, Russian genetics, that tried to fuse experimental Mendelism with systematics and natural history, rather than imposing the first upon the second (or ignoring the second entirely). At Mayr's 1974 conference, Dobzhansky vividly recalled the impediments to synthesis within American traditions. He had originally left Russia to work with Thomas Hunt Morgan, America's premier experimental geneticist. Dobzhansky recalled Morgan's attitude to natural history:

> "Naturalist" was a word almost of contempt with him, the antonym of "scientist." Yet Morgan himself was an excellent naturalist, not only knowing animals and plants but aesthetically enjoying them . . . Morgan was profoundly skeptical about species as biological and evolutionary realities. The species problem simply did not interest him. . . . Biology had to be strictly reductionistic. Biological phenomena had to be explained in terms of chemistry and physics. Morgan himself knew little chemistry, but the less he knew the more he was fascinated by the powers he believed chemistry to possess. There was no surer way to impress him than to talk about biological phenomena in ostensibly chemical terms (1980, p. 446).

Morgan, Dobzhansky also remembered, "liked to say that genetics can be studied without any reference to evolution." Could the Synthesis have taken root in such soil?

Dobzhansky brilliantly set a different task for evolutionary theory—an enterprise embodied in Darwin's title (but not treated as a major theme in his book), and emerging from traditions of systematics and natural history (while scarcely conceivable for someone with Morgan's, and to a large extent Darwin's, views on the unreality of species): how can a theory originally constructed to describe continuous change in natural populations also explain the discontinuous structure of nature's taxonomic diversity? The central problem of evolution, Dobzhansky asserted, is the origin of discontinuity among species.

This statement sounds commonplace today, but only because Dobzhansky and the Synthesis moved the question to center stage. Morgan and virtually all experimentalists had argued that the origin and nature of variation, and its manner of spread through populations, defined the key issues in evolutionary theory. Morgan disavowed the species problem as, at best, a hang-up of dull taxonomists and, at worst, a bogus issue because species have no reality in the flow of nature. (We name species, under this view, only because our poor minds can't handle continuity.)

Dobzhansky didn't deny the importance of Morgan's questions. But he argued that evolution operates on a series of levels, and that the primary gaze of natural history must not be focused upon these lower levels, but upon the broader phenomenon of the origin of species itself (Darwin's title, after all). Diversity represents the primary fact of nature (and the first topic of chapter 1 in Dobzhansky's book). Diversity arises by the splitting of lineages—that is by speciation. Speciation produces discontinuity in nature. How can a continuous process of genetic change yield such bounded separations? The origin of discontinuities between species must therefore be recast as the key problem in evolutionary theory. Only a naturalist (better yet, a trained systematist) could have reset the stage for synthesis in such a fruitful way.

> The origin of hereditary variations is, however, only a part of the mechanism of evolution. . . . These variations may be compared with building materials, but the presence of an unlimited supply of materials does not in itself give assurance that a building is going to be constructed . . . Mutations and chromosomal changes are constantly arising at a finite rate, presumably in all organisms. But in nature we do not find a single greatly variable population of living beings which becomes more and more variable as time goes on; instead, the organic world is segregated into more than a million separate species, each of which possesses its own limited supply of variability which it does not share with the others. . . . The origin of species . . . constitutes a problem which is logically distinct from that of the origin of hereditary variation (Dobzhansky, 1937, p. 119).

Mayr (1942) then furthered Dobzhansky's program by dedicating an entire book to modes of speciation, and to realigning taxonomic practice with insights of the developing Synthesis. He even formulated his title in conscious parallel to Dobzhansky's (while both, of course, also claim and honor Darwin)—and as a manifesto for the centrality of his field: *Systematics and the Origin of Species*. Mayr's first paragraph (1942, p. 3) sets his theme and tone:

> The rise of genetics during the first thirty years of this century had a rather unfortunate effect on the prestige of systematics. The spectacular success of experimental work in unraveling the principles of inheritance and the obvious applicability of these results in explaining evolution have tended to push systematics into the background. There was a tendency among laboratory workers to think rather contemptuously of the museum man, who spent his time counting hairs or drawing bristles, and whose final aim seemed to be merely the correct naming of his specimens. A welcome improvement in the mutual understanding between geneticists and systematists has occurred in recent years.

Mayr (1942) follows the characteristic pluralism of the early synthesis in listing all valid evolutionary principles that can explain the data of systematics. His major aim therefore follows the program of "healthy restriction"—

the focus of the first phase of the Synthesis (see pp. 503–508). Thus, Mayr explicitly rejects such fallacies as Larmackian inheritance, and the idea that higher taxa arise by different and mysterious routes—thereby invoking an argument by elimination to make evolutionary change at all levels fully consistent with principles of genetics at work in modern populations and subject to experiment in the laboratory or observation in the field. Whereas Mayr's major themes remain Darwinian, he still invokes the full panoply of legitimate genetic forces. Note in particular that selection (leading to adaptation), while listed first, represents only one force in an array collectively responsible for the formation of species. Adaptation holds no exclusivity, or even any particular pride of place: "First, there is available in nature an almost unlimited supply of various kinds of mutations. Second, the variability within the smallest taxonomic units has the same genetic basis as the differences between the subspecies, species, and higher categories. And third, selection, random gene loss, and similar factors, together with isolation, make it possible to explain species formation on the basis of mutability, without any recourse to Lamarckian forces" (1942, p. 70).

Mayr reemphasizes this pluralistic theme at the end of his book in asserting the essential integrative claim that all phenomena of macroevolution can also be subsumed by the Synthesis. Inclusion within the Synthesis implies explanation by principles of modern genetics, not a commitment to any particular mode of genetic change: "It is feasible to interpret the findings and generalizations of the macroevolutionists on the basis of the known genetic facts (random mutation) without recourse to any other intrinsic factors" (1942, p. 292). Mayr then lists the eight key principles of modern genetics that he regards as necessary for accomplishing the integration. Only one, number seven on the list, mentions selection and adaptation (p. 293).

As a more positive argument against adaptationist exclusivity, Mayr's own taxonomy of "factors involved in speciation" (p. 216) grants explicit and equal weight to adaptation and nonadaptation as the two primary categories of divergence. He writes (p. 216): "We may classify these factors as (1) those that either produce or eliminate discontinuities and (2) those that promote or impede divergence. The latter may be subdivided further into adaptive (selection) and non-adaptive factors."

Within this important category of nonadaptation, Mayr includes many prominent phenomena that he would later ascribe to selection.

1. Nearly all polymorphism within species:

> There is, however, considerable indirect evidence that most of the characters that are involved in polymorphism are completely neutral, as far as survival value is concerned. There is, for example, no reason to believe that the presence or absence of a band on a snail shell would be a noticeable selective advantage or disadvantage. Among the many species of birds which occur in several clear-cut color phases, there is, with one or two exceptions, no evidence for selective mating or any other advantage of any of the phases (p. 75).

2. Most geographic variation in clines:

> It is difficult to see why the gradual decrease from the north to the south in the number of the bridled individuals *(ringvia)* in populations of the Atlantic murre *(Uria aalge)* should have an adaptational significance . . . The convergent development in several species of *Draco* also seems to belong to the category of non-adaptive clines (p. 96).

3. Much geographic variation in general:

> It should not be assumed that all the differences between populations and species are purely adaptational and that they owe their existence to their superior selective qualities. . . . Many combinations of color patterns, spots, and bands, as well as extra bristles and wing veins, are probably largely accidental. This is particularly true in regions with many stationary, small, and well-isolated populations, such as we find commonly in tropical and insular species. . . . We must stress the point that not all geographic variation is adaptive (p. 86).

Mayr's later book (1963) expanded to more than twice the number of pages, and became even more weighty in its assurances. This work shaped my own evolutionary thinking more than any other book—and I am confident that most naturalists of my generation would offer the same testimony. As I reread *Animal Species and Evolution* in preparing to write this chapter—and examined my old marginalia, penciled in preparation for the deciding oral exam of my Ph.D. program—I came to appreciate even more (now that I know the genre's difficulty through personal experience) the enormous labor and creative thought involved in bringing so much material together. And I finally understood the defining word that once puzzled me in Julian Huxley's review of Mayr's book—"magisterial." (The etymological source does not reside in "magnificent" or "majestic," though Mayr's book surely merits either of these accolades, but in *magister,* the Latin word for teacher. A great magister is not a schoolroom pedant, but a wise preceptor who holds mastery within his teaching authority, or *magisterium.* Magisterial, above all else, means authoritative. And to what greater virtue, after all, may an author aspire?)

Although Mayr's 1963 book covers the same general material, and in similar order, as the 1942 version, the works differ profoundly, and Mayr chose a new title (just as Simpson had done in noting the changes between his 1944 and 1953 volumes). I would specify two thematic changes as most important.

1. The primary role of geographic isolation as a *sine qua non,* and the consequent near universality of allopatric speciation, has consistently formed the centerpiece of Mayr's worldview. But, in 1942, pure continuationism reigned. Populations split into roughly equal divisions and each subgroup then functioned as a microcosm of the ancestral mass—as in the model now called "dumb-bell allopatry" and considered (by Mayr at least) both rare and relatively ineffective in producing new species. In other words, Mayr (1942) originally identified no distinctive properties promoting speciation in certain kinds of isolated populations vs. others. Isolation itself, and the severing of

gene flow, rendered any population ripe for speciation: "The big gaps which we find between species are preceded by the little gaps which we find between subspecies and by the still lesser gaps which we find between populations. Of course, if these populations are distributed as a complete *continuum,* there are no gaps. But with the least isolation, the first minor gaps will appear" (1942, p. 159).

But by 1963, Mayr had developed the full apparatus of the distinctive theory that he later called "peripatric speciation" to emphasize a sharp separation from his original, continuationist version of allopatry. For the peripatric model promotes the role of *small* populations, isolated at the *periphery* of parental ranges, and subject to a special maelstrom of influences including greatly enhanced selection and random effects of the founder principle—all leading to potential achievement of specific status with relative speed by a "genetic revolution." Mayr says (personal communication) that he introduced this new apparatus in a paper (1954) that achieved no impact, but nonetheless represents his most important idea and best work. (*Nihil sub sole novum.* He published this paper in a symposium volume—the greatest repository of unread literature, both then and now.)

2. Mayr's 1942 book included little explicit material about adaptation, since this volume emphasized the origin and development of discontinuity between species, and said little about anagenetic change within populations. This context of minimal consideration reflects Mayr's pluralism and lack of commitment to strict adaptationism at this time. (This claim may sound paradoxical, but should not be so read. Views expressed in passing—by their simple acknowledgment of an unchallenged belief—tend to record a professional consensus more clearly than material explicitly touted as central and distinctive.) But, in 1963, Mayr added a full consideration of variation and change within populations—the main reason for a much longer book. Here the hardened, panadaptationist position of the later Synthesis reigns supreme, perhaps more strongly than in any other book of comparable influence.

In the mid 1990's, Mayr himself (*in litt.* and personal communications—see end of this section), while continuing to explicate and defend his favored themes of 1963, denies any substantial change between the volumes of 1942 and 1963 on questions of adaptation. This difference between current memory and textual record, previously discussed as a general principle (see p. 521), provides a fascinating illustration of how scholars can slowly and unconsciously imbibe a shifting professional consensus, thus imposing a subjective and personal impression of stability upon a virtual transmogrification. I find this unconscious alteration all the more ironic in Mayr's case because his first category of major change in ideas about speciation—his intellectual move from the dumbbell to the peripatric model—so strongly encourages a widened space for nonadaptationist themes (for many evolutionists have interpreted his notions of genetic revolutions and founder effects in small peripheral isolates as a powerful antidote to the classical panadaptationist model of Fisherian panmixia in large populations). Yet Mayr never translated the implications of these changes in his own ideas about speciation into doubts about adaptation in his chapters on variation and change within populations.

No good naturalist, living in our complex universe of relative frequencies, could ever become an uncompromising dogmatist on the subject of adaptation. Mayr therefore mentions occasional inadaptive features (1963, p. 156), or acknowledges the importance of developmental constraint (p. 608). But these statements function more as footnotes or placeholders in the logic of an argument; for Mayr does not treat alternatives to adaptation as operational imperatives in the ordinary analysis of cases. Moreover, Mayr laces his pluralistic admissions with hedges and caveats. Note, for example, how Mayr frames his main admission of potential nonadaptation only as an argument against optimality, not as a denial of selection—and how his closing hedge anticipates a movement of even these least promising cases into the adaptationist camp:

> Each local population is the product of a continuing selection process. By definition, then, the genotype of each local population has been selected for the production of a well-adapted phenotype. It does not follow from this conclusion, however, that every detail of the phenotype is maximally adaptive. If a given subspecies of ladybird beetles has more spots on the elytra than another subspecies, it does not necessarily mean that the extra spots are essential for survival in the range of that subspecies. It merely means that the genotype that has evolved in this area as the result of selection develops additional spots on the elytra . . . Yet close analysis often reveals unsuspected adaptive qualities even in minute details of the phenotype (1963, p. 311).

Selection holds primacy of place as the ruling force of evolution: "Every species is the product of a long history of selection and is thus well adapted to the environment in which it lives. There is no doubt that the phenotype as a whole, including its physiological properties, is adaptive and is produced by a genotype that is the result of natural selection. This is not contradicted by the fact that an occasional component of the phenotype is adaptively irrelevant" (1963, p. 60).

Above all else, Mayr regards one conclusion as especially well confirmed by observation: adaptation rules in "every local population" as selection to "exacting requirements" of local environments produces an "optimal phenotype." One could hardly state the adaptationist position more boldly: "One conclusion emerges from these observations more strongly than any other: every local population is very precisely adjusted in its phenotype to the exacting requirements of the local environment. This adjustment is the result of a selection of genes producing an optimal phenotype" (1963, p. 318).

Mayr's treatment of potential alternatives illustrates his adherence to the rule of adaptation, both as a methodological preference and an empirical claim. Geographic trends that he formerly attributed to incidental allometries have now become active adaptations: "A particularly impressive result of studies of ecogeographical rules is the discovery of the extreme sensitivity of body proportions to natural selection. The former belief that proportions are determined by 'built-in' allometry factors and change automatically with changes in body size is not supported by these findings" (1963, p. 324).

Neutral genes become improbable, almost nonsensical in principle, once we recognize the pervasive monitoring of nature by selection:

> Entirely neutral genes are improbable for physiological reasons. Every gene elaborates a "gene product," a chemical that enters the developmental stream. It seems unrealistic to me to assume that the nature of the particular chemical (enzyme or other product) should be without any effect whatsoever on the fitness of the ultimate phenotype. A gene may be selectively neutral when placed on a particular genetic background in a particular temporary physical and biotic environment. However, genetic background as well as environment change continually in natural populations and I consider it therefore exceedingly unlikely that any gene will remain selectively neutral for any length of time (1963, p. 207).

Consequently, even the most apparently trivial features probably originated by direct selection. "One can never assert with confidence that a given structure does not have selective significance. The peculiar tarsal combs of the males in certain species of *Drosophila* turned out to have an important function during copulation; the color patterns of *Cepaea* snails have cryptic significance, mitigating predator pressure" (1963, p. 190).

In 1963, Mayr repudiated all three major classes of nonadaptation that he had defended in 1942: polymorphisms, clines, and much geographic variation in general. Explicitly refuting his own former view, Mayr now argues (1963, p. 162) that the ubiquity of selection must imply an adaptive basis for polymorphisms (see also pp. 158 and 167): "Such neutral polymorphism, it was claimed, was maintained by 'accident.' Now that the cryptic physiological effects of 'neutral' genes have been discovered, it is evident that such genes are anything but selectively neutral. It is altogether unlikely that two genes would have identical selective values under all the conditions in which they may coexist in a population."

In a remarkable statement, he then urges that polymorphisms and clines be viewed as evidence for adaptation *a priori*: "Selective neutrality can be excluded almost automatically wherever polymorphism or character clines are found in natural populations . . . Virtually every case quoted in the past as caused by genetic drift due to errors of sampling has more recently been reinterpreted in terms of selection pressures" (1963, pp. 207–208).

As for geographic variation, what else could such a phenomenon represent but adaptation to an altered environment, with selection as an efficient and omnipresent watchdog: "The geographic variation of species is the inevitable consequence of the geographic variation of the environment. A species must adapt itself in different parts of its range to the demands of the local environment. Every local population is under continuous selection pressure for maximal fitness in the particular area where it occurs. . . . Each local environment exerts a continuous selection pressure on the localized demes of every species and models them thereby into adaptedness" (1963, pp. 311–312).

Throughout Mayr's 1963 book—with a cadence that sounds, at times, almost like a morality play—phenomenon after phenomenon falls to the explanatory unity of adaptation, as the light of nature's truth expands into previous darkness: non-genetic variation (p. 139), homeostasis (pp. 57, 61), prevention of hybridization (p. 109). Former standard bearers of the opposition fall into disarray, finally succumbing to defeat almost by definition: "It is now evident that the term 'drift' was ill-chosen and that all or virtually all of the cases listed in the literature as 'evolutionary change due to genetic drift' are to be interpreted in terms of selection" (p. 214). All particular Goliaths have been slain (although later genetic studies would revivify this particular old warrior): "The human blood-group genes have in the past been held up as an exemplary case of 'neutral genes,' that is, genes of no selective significance. This assumption has now been thoroughly disproved" (p. 161).

However, Mayr's most interesting expression of movement towards a hardened adaptationism occurs not so much in these explicit claims for near ubiquity, but even more forcefully in the subtle redefinition of all evolutionary problems as issues in adaptation. The very meaning of terms, questions, groupings and weights of phenomena, now enter evolutionary discourse under adaptationist presumptions. Not only have alternatives to adaptation been routed on an objective playing field, Mayr claims in 1963, but the conceptual space of evolutionary inquiry has also become so reconfigured that hardly any room (or even language) remains for considering, or even formulating, a potential way to consider answers outside an adaptationist framework.

Major subjects, the origin of evolutionary novelty for example, now reside exclusively within an adaptationist framework by purely functional definition: "We may begin by defining evolutionary novelty as any newly acquired structure or property that permits the performance of a new function, which, in turn, will open a new adaptive zone" (p. 602). In a world of rapid and precise adaptation, morphological similarity between distantly related groups can only arise through convergence imposed by similar adaptive regimes upon fundamentally different genetic material. The older, internalist view (constraint-based and potentially nonadaptationist)—the claim that we might attribute such similarities to parallelism produced by homologous genes—is dismissed as both old-fashioned and wrong-headed. (In modern hindsight, this claim provides a particularly compelling example of how hardened adaptationism can suppress interesting questions—for such homologues have now been found in abundance. Their discovery ranks as one of the most important events in modern evolutionary science—see Chapter 10, p. 1092, where we will revisit this particular Mayrian claim): "In the early days of Mendelism there was much search for homologous genes that would account for such similarities. Much that has been learned about gene physiology makes it evident that the search for homologous genes is quite futile except in very close relatives" (1963, p. 609).

Subjects that might have seemed challenging or exceptional now achieve a place within the adaptationist framework by expanded definition. Nonfunctional pleiotropic consequences, for example, become an aspect of orthodoxy because they now enter a hardened discourse in the redefined guise of features subsidiary to a main effect of adaptive significance. (I do not challenge the particular assertion in this case, but I do feel that such an important subject deserves consideration from a structuralist perspective as well): "Pleiotropic gene action is the key to the solution of many other puzzling phenomena . . . Color, pattern, or some structural detail may be merely an incidental by-product of a gene maintained in the gene pool for other physiological properties. The curious evolutionary success of seemingly insignificant characters now appears in a new light" (1963, p. 162).

All potential anomalies yield to a more complex selectionist scenario, often presented as a "just-so-story." Why did the crown height of molars increase so slowly, if hypsodonty became so advantageous once horses shifted to vegetational regimes of newly-evolved grasses with high silica content? Mayr devises a story—sensible, though empirically wrong in this case—and regards such a hypothetical claim for plausibility as an adequate reason to affirm a selectionist cause. (The average increase may have been as small as the figure cited by Mayr, but horses did not change in anagenetic continuity at constant rates. Horses probably evolved predominantly by punctuated equilibrium—see Prothero and Shubin, 1989, and Chapter 9. The average of a millimeter per million years represents a meaningless amalgam of geological moments of rapid change during speciation mixed with long periods of stasis): "An increase in tooth length (hypsodonty) was of selective advantage to primitive horses shifting from browsing to grazing in an increasingly arid environment. However, such a change in feeding habits required a larger jaw and stronger jaw muscles, hence a bigger and heavier skull supported by heavier neck muscles, as well as shifts in the intestinal tract. Too rapid an increase in tooth length was consequently opposed by selection, and indeed the increase averaged only about 1 millimeter per million years" (1963, p. 238).

In 1991, I asked Ernst Mayr about changes between his 1942 and 1963 books. He acknowledged the structural alterations, of course—particularly his addition of several chapters emphasizing adaptational themes. But he strongly denied any personal augmentation of adaptationist preferences through the intervening years, citing the interesting argument that, as a Lamarckian in his evolutionary youth (well before both books), he had always favored adaptationism. He even wrote me a fascinating letter the day after our lunchtime conversation:

Dear Steve,

I gave considerable thought to your question how my 1963 book differed from the 1942 one, and why adaptation was so much more featured in the later volume. I think I now have the answer.

Remember that I consider evolution by and large to consist of two processes: 1) the maintenance and improvement of adaptedness, and 2) the origin and development of diversity.

Since (2) was so almost totally ignored by the pre-Synthesis geneticists, I focussed in 1942 on (2). By the 1950s the study of diversity had been fully admitted to evolutionary biology, owing to the efforts of Dobzhansky, myself, Rensch and Stebbins, and in my 1963 book I could devote a good deal of attention to (1). This was rather easy because, as you know, I used to be a Lamarckian. And Lamarckians are adaptationists. Hence, it is not that from 1942 to 1963 I had become an adaptationist, rather I reconciled in 1963 my adaptationist inclination with the Darwinian mechanism (Letter of December 20, 1991).

(Mayr then added a handwritten footnote, demoting to insignificance the one subject for which he did acknowledge a reversal of opinion between the two books: "Neutral polymorphism is an infinitesimal percentage of all evolutionary phenomena. Don't make a mountain out of this little mole-hill.")

I do not deny Mayr's stable adaptationist preferences (through his ontogenetic change in explanatory preferences from Lamarck to Darwin). This personal stability provides an even better reason for regarding as important, and therefore generally indicative, the textual evidence of transition from pluralism in 1942 to adaptationist hardening in 1963 (for Mayr's 1942 text may therefore, by implications of his own testimony, be reporting the conventional pluralistic wisdom of the time despite Mayr's own personal preference for adaptationism). On the subject of adaptation—not the major concern of either book (for both treat speciation and the production of diversity as their primary topics)—Mayr recorded a professional consensus both times, largely passively I suspect (hence his personal inattention to the alteration). Scientists must struggle to identify and understand these influences of "shared culture," for such a "background" consensus fuels the sources of unconscious bias for each of us at every moment of our careers.

## WHY HARDENING?

I have documented the adaptationist hardening of the Modern Synthesis in some detail, but I have not addressed an obvious and pressing question: why did this conceptual trend occur? Several aspects of an answer seem clear, but I can offer no full or satisfying resolution.

The culture of science trains us to believe that such major shifts of emphasis record improvements in knowledge won by empirical research and discovery. I do not deny that observation did play a significant role, at least in illustrating, with some elegant examples, the power of adaptation. Consider, for example, the "ecological genetics" of E. B. Ford and his panselectionist school in England. Their commitment to adaptationist explanations of effectively all variation among populations, and their documentation of strong selection coefficients in nature, buoyed the strict Darwinian faith. Dobzhansky's own empirical work increased his belief in the power of selection. In 1937, he tended to attribute chromosomal inversion frequencies in natural populations of *Drosophila* to genetic drift, but he then discovered that these frequencies fluctuate in a regular and repeatable way from season to season, and he therefore decided (with evident justice) that such systematic and iterated change must be adaptive.

But empirical discovery cannot supply the entire (or even, I think, the major) reason for adaptationist hardening, for each favorable case can be matched by a failure (often hedged or unacknowledged), and no adequate assessment of an overall relative frequency has ever been achieved—to this day. Thus, any judgment, in either direction, must represent the fashionable imposition of a few well-documented cases upon an unstudied plethora. For example, A. J. Cain and colleagues did win a major victory for adaptation by showing that banding-morph frequencies in the land snail *Cepaea*, a former mainstay for claims about genetic drift, reflected selection based upon visual predation by birds, and upon climatic factors (Cain and Sheppard, 1950, 1952, 1954).

But Cain and his colleagues then recognized and named the outstanding pattern of "area effects" (Cain and Currey, 1963a and b)—abrupt geographic changes in banding-morph frequencies occurring with no perceptible alteration in any environmental factor that might impose a selection pressure. In what can only be labeled an article of faith, Cain attributed area effects to selection based upon "cryptic [meaning *truly* unmeasured and unperceived by any investigator, not merely subtle] environmental differences"—a remarkable affirmation of an *a priori* preference based upon *not finding* the necessary empirical confirmation. Good evidence has since been presented for a non-adaptive explanation of area effects as historical remnants of previous patterns in land use, and not as an outcome of current regimes in selection (Cameron *et al.*, 1980; see review of the entire case in Gould and Woodruff, 1990). (Area effects rank as anomalies under selectionist presuppositions—hence Cain's need to supply an orthodox adaptationist explanation, even in the absence of required evidence. Under a "legacy of history" explanation, such discordance of morphology with present geography presents no anomaly and need not even receive a special name.)

If adaptationist hardening cannot be explained as simply and empirically driven, we might turn to historical and sociological themes. Smocovitis (1996), as previously mentioned (see p. 503), presents the intriguing thesis that renewed optimism following the wreckage of World War II (including the hope inspired by the newly constituted United Nations) launched a strong push for scientific defenses of potential human improvement and evolutionary progress—an impetus that became a semi-official movement spurred by positivistic theories of knowledge proffered as antidotes for older irrationalisms. Smocovitis writes:

> If selection had enough agency (and at the same time were a mechanical principle) then all the more rapid and possible the "improvement" of hu-

mans. . . . More strongly selectionist models would be favored by biologists who modelled themselves after physicists at the same time they pointed the way to the "improvement" of humanity and painted a progressive and optimistic picture of the world . . . Evolutionary models favoring random genetic drift, which enforced a stochastic view of evolution—and culture—would not be favored in a post-war frame of mind seeking to "improve" the world. So powerful would be the need for a progressive and selectionist framework in the 1940's that even Dobzhansky and Wright were to adopt more strongly selectionist models.*

Some complex mixture of empirical and sociological themes may explain the adaptationist hardening of the synthesis, but we must not neglect the additional impetus of a cultural analog to drift and founder effects in small populations. The community of evolutionary biologists is sufficiently small, and sufficiently stratified—a few lead and many follow, as in most human activities—that we need not necessarily invoke some deep and general scientific or societal trend to explain a change in opinion by a substantial community of evolutionists in different nations. A reassessment by a few key people, bound in close contact and mutual influence, might trigger a general response. The three leading exponents of hardening in America—Dobzhansky, Simpson, and Mayr—worked together as colleagues in a "New York Mafia" centered at Columbia University and the American Museum of Natural History. Add another seemingly eternal principle of human affairs—that founders tend to be brilliant and subtle, and to keep all major difficulties constantly in mind, while epigones generally promulgate the faith and disregard, or never learn, the problems, exceptions, and nuances—and we may then wish to view the adaptationist hardening as ultimately inadaptive for the broadest goal of understanding evolution aright. Bandwagons might well be construed as cultural analogs of internalist drives in nonfunctional orthogenesis. Theories can grow tired. Theories can also harden and lose their bearings when complacency occupies the driver's seat.

## Hardening on the Other Two Legs of the Darwinian Tripod

To illustrate the hardening of the Modern Synthesis, I have documented its most significant ontogenetic trend *in extenso*—increasingly exclusive emphasis on adaptation as the sign of natural selection's pervasive power. But if we epitomize the Synthesis as Darwinism reclothed in Mendelian understanding,

---

*I am not generally drawn to sociological proposals in this mode, and I reacted negatively at first to Smocovitis's suggestion. But I have since read widely in the just post-World War II literature, and I only now understand the fervor and hope of "never again," following all the devastation, and the heartrending impact (and inspired shame) as knowledge of the Holocaust surfaced. I was too young, when the war ended, to experience viscerally both this horror and hope, but I do grasp the character of this unusual time with a pervasive theme and agenda—and I accept the idea that humanistically inclined scientists must have hoped fervently that their own field might contribute to the reconstruction.

then, following this book's focal argument that the minimal commitments of Darwinian logic encompass three central themes, the other two legs of this essential tripod should experience corresponding changes as the Synthesis hardened. I will not provide so extensive a discussion of these other legs—levels of selection and extrapolation into geological time—but I do wish to record that the literature of these subjects also experienced the same ontogeny of solidification (and unjustified neglect of reasonable alternatives).

## LEVELS OF SELECTION

Darwin, as we have seen (pp. 125–137), viewed organisms as nearly exclusive agents of selection—for deep reasons situated at the core of both the logic (the invisible hand of Adam Smith transferred to nature) and the psychology (the inversion of Paley's world) of his theory. But few Darwinians grasped the rationale or centrality of this principle, and a tradition of vagueness and loose thinking about levels of selection developed. Some, like R. A. Fisher, rode Darwin's wave and wrote explicitly and cogently about reasons for choosing individual organisms as the proper locus, and for disregarding, as effectively impotent, other levels that must be deemed conceivable in theory (1958, on species selection—see my critique of Fisher on pp. 644–652). But others, dating back to A. R. Wallace himself (see pp. 131–132), never understood the full logic and implications of this issue, and ranged indiscriminately up and down potential levels, without grasping the theoretical problems entailed by such excursions.

Thus, a fluid situation prevailed on this issue at the time of the Darwinian centennial celebrations of 1959—my point of reference for the triumphal height of the Modern Synthesis in its strongly adaptationist version. Adaptation had become all the rage, but vagueness shrouded the key issue of selection's focus and level—and for two reasons.

First, and less important because the position attracted few supporters, a few evolutionists explicitly advocated a multi-level view of both selection and adaptation. A group of Chicago ecologists, authors of an important textbook known by its acronym of AEPPS (Allee, Emerson, Park, Park and Schmidt, *Principles of Animal Ecology,* 1949), generated and led this small movement. Emerson spoke at the Chicago centennial symposium, and presented his multi-level view in both content and title: "The evolution of adaptation in population systems."

Emerson begins by acknowledging the conventional Darwinian preference for individual organisms (and reminding us that he will not neglect this usual argument). But he then stakes his higher claim: "It is my intention in this essay to emphasize the evolution of adaptation in population systems without, however, negating the data or the major interpretations of the roles of individuals in evolutionary history or processes" (1960, p. 307).

I find Emerson's article frustrating, for his arguments are so reasonable in some places, and so very wrong, to the point of illogic, in others. On the one hand, he presents a defendable and properly philosophical criterion for higher-level selection based on features of populations that cannot be explicated as additive results of organismal properties—in other words, "emergent" characters. He correctly defines a population ripe for selection at its own level as "an inclusive entity with emergent characteristics that transcend the summation of the attributes of the component individuals" (1960, p. 307).

But, having legitimately defined the problem, he then launches into an almost rhapsodic, and simply illogical, claim that almost anything with definable boundaries can be recognized as a unit of natural selection: "Natural selection operates at each level of integration from the gene and complex polygenic characters within the individual, to the whole individual, and to various levels of intraspecific population systems and interspecific interadapted community systems and ecosystems" (1960, p. 340).

This argument can be defended in theory so long as the higher unit operates as an interactor with surrounding environments and remains in a genealogical nexus engaged in differential reproduction (see Chapter 8). But how can Darwinian selection possibly operate directly on an ecosystem? However we may choose to define such an entity, ecosystems do not mate and do not produce children (see Chapter 8, pp. 597–613 on criteria of Darwinian individuality). No argument can be made about their differential reproductive success, and no Darwinian calculus can therefore be applied to their history through time.

Emerson doesn't seem to grasp that selection works by differential reproductive success, not by design for immediate, self-serving utility: "It would be extremely difficult," he writes (1960, p. 319), "to explain the evolution of the uterus and mammary glands in mammals or the nest-building instincts of birds as the result of natural selection of the fittest individual." But if milk-rich mammary glands promote the survival of offspring, then the mother acts in her own Darwinian interest. In short, Emerson's paper gives us an unintended insight into the confusing lack of definition that natural selection has always suffered, even at the moment of its greatest explicit influence.

Second—and more important in its virtual ubiquity—leading evolutionists, though well aware that orthodoxy identified individual organisms as the focus of selection, did not grasp the logical necessity or centrality of such a claim in Darwinian theory, and therefore often indulged in vague, perhaps unconscious, and often fuzzy, statements about the efficacy of higher levels. (I say "fuzzy" because most of these claims about populations and groups only invoked the non-emergent effects of organismic characters—and therefore do not necessarily qualify as valid statements about higher-level selection. I don't think that many evolutionists had properly formulated this crucial issue at the time.)

Dobzhansky, for example (1957, p. 392), states that selection operates on organisms, but then proposes that such phenomena as heterozygote advantage might record some populational "extra" in exposing the reduced fitness of homozygotes as a kind of organismic sacrifice "for" the group: "Natural selection operates through differential survival and differential fertility of individuals, and yet at some times brings about such forms of integration of the gene pool of the population which lead to the sacrifice of some of the individual members of the population. The phenomenon of balanced polymorphism, with highly fit heterozygotes contrasting with less fit homozygotes, is one of such forms of genetic integration of Mendelian populations."

Mayr's most authoritative book (1963) provides an excellent illustration of organismic orthodoxy amidst statements, lacking clear definition, about selection at higher levels. Mayr surely recognizes the usual form of proper Darwinian argument—that apparent benefits to populations should be explained, whenever possible, as effects of selection upon organisms: "The solution usually proposed for the difficulty raised by the conflict between a benefit for the individual and one for the population is to make the population rather than the individual the unit of selection. . . . It would seem preferable to search for solutions based on the selective advantage of individual genotypes, such as Fisher's explanation of an even sex ratio" (1963, pp. 198–199).

In rereading Mayr's 1963 book with the hindsight of thirty years, however, I was struck by the number of passages and arguments that either speak loosely about explicit advantages to groups and populations (rather than fortuitous beneficial effects arising as side consequences of selection on organisms), or seem to state an explicit claim for selection at the population level. Most of these statements focus on the virtues of genetic variability. Mayr asks (1963, p. 158): "Why are not all individuals of a population identical in appearance? Is it because diversity is of selective advantage to the population?" He then argues (1963, p. 308) that the *primary* function of chromosomal variation lies in the flexibility thus accorded to populations: "They appear to have, as primary function, either the increase of adaptability and adaptedness of these populations through balanced polymorphism of entire chromosome sections or the regulation of the amount of recombination." In fact, Mayr's main justification for regarding polymorphism as adaptive—a major shift in his own adaptationist hardening from the examples used to support nonselectionist claims in his 1942 book—focuses on advantages to populations (1963, p. 251).

> Polymorphism is based on and produced by definite genetic mechanisms, such as genes for differential niche selection and the heterosis of heterozygotes. A population that has not responded to selection for such mechanisms and therefore lacks polymorphic diversity is more narrowly adapted, more specialized, and therefore more vulnerable to extermination. The widespread occurrence of genetic mechanisms that produce and maintain polymorphism is directly due to selection and is in itself a component of adaptiveness. It seems appropriate, therefore, to speak of "adaptive polymorphism."

A rally around the flag of organismic selection, and an explicit (and vociferous) denial of higher levels, became a major movement in evolutionary theory during the 1960's. The hardening of adaptationism had occurred largely during the 1940's and 1950's (in time for the Darwinian centennial of 1959); but the refinement of adaptationist arguments to nearly exclusive operation at the organismic level followed later. This reform* emerged largely within the field of animal behavior, where the ethological tradition, particularly in the work of Konrad Lorenz, had long promulgated a loose and largely unconsidered approach to multilevel selection.

The primary impetus to explicit debate appeared with the publication, in 1962, of V. C. Wynne-Edwards's *Animal Dispersion in Relation to Social Behavior.* Since most evolutionists now regard Wynne-Edwards's primary argument as wrong (and I do not dispute this consensus), we greatly undervalue his work and misconstrue its importance. Most evolutionists today, many of whom have never read Wynne-Edwards, know his book only by reputation as a dumb argument for group selection that George Williams and others thoroughly demolished. I regard this assessment as entirely unfair. Wynne-Edwards's claim for group selection may be wrong, but I can cite few other theories, presented within evolutionary biology during my career, that could be deemed so challenging in implication, so comprehensive in claims, so fascinating in extension, and so thought-provoking.

First of all, and essential for grasping the book's sweep, "Animal Dispersion" presents a theory about organic self-regulation of population numbers, not primarily an argument for group selection in general (although group selection serves as a fundamental feature of the proposed mechanism). Wynne-Edwards begins, as so many others have done (including Darwin), with an analogy to human institutions. When predators show no restraint in the midst of plenty, ecosystems may crumble as both predators and prey succumb. He speaks with feeling about the collapse of Arctic whaling (1962, p. 5): "The stocks of the two right whales have never recovered, and the population of Greenland-whaling men and of those who ministered to them has become effectively extinct."

Wynne-Edwards then generalizes from carnivory to food-based limitation in any kind of eating, and to the transcendent need for regulating population sizes of consumers in any ecosystem. I am fascinated that Wynne-Edwards, in this statement, invokes the same, largely metaphysical, argument that Darwin proposed in specifying a *summum bonum* for the construction of nature (a situation established by very different mechanisms in Wynne-Edwards's and Darwin's systems): the old principle of plenitude, or maximization of the kinds and numbers of organisms in any given segment of earthly real estate (see p. 229 for Darwin's version):

---

*Although I strongly advocate a hierarchical model of multilevel selection (see Chapters 8 and 9), I regard this restriction to organismic selection as an important and positive reform. Earlier claims for group and higher level selection had been formulated so vaguely and falsely that they impeded our understanding of both this important concept and of the theory of selection in general. This salutary reform tore down erroneous standards and insisted that no further claims be made until the logical edifice could be properly rebuilt—no examples without a proper substructure; no paintings without a strong frame.

The need for restraint in the midst of plenty, as it turns out, must apply to all animals whose numbers are ultimately limited by food whether they are predators in the ordinary sense of the word or not. . . . Where we can still find nature undisturbed by human interference, whether under some stable climax of vegetation or proceeding through a natural succession, there is generally no indication whatever that the habitat is run down or destructively overtaxed. On the contrary the whole trend of ecological evolution seems to be in the very opposite direction, leading towards the highest state of productivity that can possibly be built up within the limitations set by the inorganic environment. Judging by appearances, chronic over-exploitation and mass poverty intrude themselves on a mutually-balanced and thriving natural world only as a kind of adventitious disease, almost certain to be swiftly suppressed by natural selection. It is easy to appreciate that if each species maintains an optimum population-density on its own account, not only will it be providing the most favorable conditions for its own survival, but it will automatically offer the best possible living to species higher up the chain that depend on it in turn for food. Such *prima facie* argument leads to the conclusion that it must be highly advantageous to survival, and thus strongly favored by selection, for animal species (1) to control their own population-densities, and (2) to keep them as near as possible to the optimum level for each habitat they occupy (1962, pp. 8–9).

In Darwinism, this regulation proceeds by a fundamentally Malthusian method—imposed from outside by a hecatomb on populations that outstrip resources. Wynne-Edwards holds that such an indirect and inefficient mode of external imposition wreaks havoc upon ecosystems, and that existing stabilities therefore imply the operation of an entirely different system for regulating populations—internally, by complex sets of behaviors that limit reproduction and match population sizes to appropriate resources. Since the Darwinian imperative leads organisms to maximize their own reproductive success, such internal limitation can only be achieved by mechanisms of group selection powerful enough to counteract the personal gains of individual organisms from conventional Darwinian selection.

The ingenuity of Wynne-Edwards's theory lies largely in the range of behavioral phenomena that he interprets as devices evolved by group selection for limitation of population size. In fact, Wynne-Edwards ascribes the origin of social organization itself to this need for limitation upon the size of populations. Note that by "conventional competition" he does not mean the vernacular "orthodox" or "ordinary" (which would then become Darwinian, or the opposite of his personal intention), but rather apparent competition by bluff, ritual and display—convention in this sense—rather than actual (and potentially destructive) fighting:

Undisguised contest for food inevitably leads in the end to overexploitation, so that a conventional goal for competition has to be evolved in its stead; and it is precisely in this—surprising though it may appear at first sight—that social organization and the primitive seeds of all social behavior have their origin . . . Putting the situation the other way around, a society can be defined for our purposes as an organization capable of providing conventional competition: this, at least, appears to be its original, most primitive function, which indeed survives more or less thinly veiled even in the civilized societies of man (1962, p. 14).

Almost all the rich repertoire of putative Darwinian behaviors become, for Wynne-Edwards, devices evolved by groups of organisms to limit their population size. Dominance hierarchies and pecking orders become group-selected controls, exercised by denying reproductive rights to large numbers of potential breeders. The chorusing of frogs, insects and birds become censusing devices, whereby populations may judge their numbers and trigger appropriate behaviors of regulation.

Such homeostatic adaptations exist in astonishing profusion and diversity, above all in the two great phyla of arthropods and vertebrates. There we shall find machinery for regulating the reproductive output or recruitment rate of the population in a dozen different ways—by varying the quota of breeders, the number of eggs, the resorption of embryos, survival of the newborn and so on; for accelerating or retarding growth-rate and maturity; for limiting the density of colonization or settlement of the habitat; for ejecting surplus members of the population, and even for encompassing their deaths in some cases in order to retrieve the correct balance between population-density and resources (1962, p. 9).

Such massive suppression of the Darwinian game could only be achieved by group selection—that is, by the differential success of groups with emergent social behaviors that debar reproduction for many members, thus limiting population size from within, and winning temporal persistence by avoiding collapse through overexploitation: "We have met already with a number of situations—and shall later meet many more—in which the interests of the individual are actually submerged or subordinated to those of the community as a whole" (1962, p. 18).

Wynne-Edwards surely understood the stringent requirements for such a mechanism. He recognized, for example, that demes or social groups must be persistent and genealogically exclusive in order to act as higher-level "individuals" in a selective process—as in this epitome of his views on group selection (1962, p. 144).

To understand group-selection we ought first to recognize that normally local populations are largely of common descent, self-perpetuating and potentially immortal. They are the smallest subdivisions of the species of which this is true, and can be adapted to safeguard their own future. What is actually passed from parent to offspring is the mechanism for responding correctly in the interests of the group in a wide range of circumstances. What is at stake is whether the group itself can survive or will become extinct. If its social adaptations prove inadequate, the stock will decline or disappear and its ground be colonized by neighboring stocks with more successful systems: it must be by this process that group-characters slowly evolve.

The abstract logic of this argument cannot be faulted, but we must ask if the required conditions are encountered frequently enough in nature. Do social groups remain sufficiently exclusive; is group selection strong enough to overcome Darwinian organismic selection; do social behaviors originate by organismal or group selection? Nearly all evolutionists would now agree that groups rarely maintain the required cohesion, and that group selection (in Wynne-Edwards's mode) will usually be far too weak a force to prevail over the conventional Darwinian mechanism of organismic selection.

George Williams's brilliant book, *Adaptation and Natural Selection* (1966), provided the historical focus for general rejection of Wynne-Edwards's theory. Williams wrote with other sources and targets in mind as well. (He told me, for example, that he had originally been most strongly motivated by the false arguments of Allee, Emerson and the Chicago school.) But Wynne-Edwards stood out as the main group-selectionist game in town when Williams wrote his book.

Williams's book won deserved influence by its incisiveness of logic and argument, and for its persuasive style of composition. (I know no better example of a work that prevailed primarily by the entirely honorable sense of the unfairly maligned word "rhetoric," properly defined as "the effective use of language.") Williams begins by characterizing adaptation as an "onerous" concept that should be invoked only when all simpler explanation fails. We should then become all the more impressed when we find that we need to invoke adaptation so often! Having established one "tough" criterion by permitting the invocation of adaptation only when all else fails, Williams then proposes another—this time governing the advocacy of levels higher than the conventional Darwinian focus upon organisms. In short, Williams states, don't make a claim about higher levels unless both logic and empirics permit no other alternative. Adaptation is onerous enough in any case; if we must call upon such a mechanism, we should do so at the lowest possible level of the genealogical hierarchy. This appeal to some form of parsimony or reduction leads Williams to reject all claims for group selection, so long as Darwinian organismic selection can render the same phenomenon in principle. Williams presents his argument largely as a theoretical proposition, and only rarely as an empirical claim. If the phenomenology of a situation can be rendered by an organismic interpretation, he asserts, one should then advocate this lower level of causality—even if a group selectionist scenario violates no tenet of logic or plausibility.

In his introductory pages, Williams tells us that claims for group selection have inspired his attempt at cleansing and simplification:

> Even among those who have expressed the opinion that selection is the sole creative force in evolution, there are some inconsistent uses of the concept. With some minor qualifications to be discussed later, it can be said that there is no escape from the conclusion that natural selection . . .

can only produce adaptations for the genetic survival of individuals. Many biologists have recognized adaptations of a higher than individual level of organization. A few workers . . . postulate that selection at the level of alternative populations must also be an important source of adaptation, and that such selection must be recognized to account for adaptations that work for the benefit of groups instead of individuals. I will argue . . . that the recognition of mechanisms for group benefit is based on misinterpretation, and that the higher levels of selection are impotent and not an appreciable factor in the production and maintenance of adaptation (1966, pp. 7–8).

This statement includes several interesting features, suggesting useful definitions and frameworks that I will follow throughout this book (while often disagreeing with Williams's own conclusions): (1) The term natural selection shall refer only to Darwin's process at the organismic level; selection at higher levels requires a different name. (2) The logic of group (and higher level) selection cannot be denied; we may only reject the process as impotent relative to natural selection, not as inconceivable. (3) The criterion advanced by group selectionists—the existence of properties that work for the benefit of groups, but at the expense of individual organisms—may be sound in theory but inapplicable in fact, for virtually all proposed cases either have been misinterpreted or remain subject to recasting in terms of advantages for organisms alone.

Williams then states his "doctrine," frankly so designated: "The ground rule—or perhaps *doctrine* would be a better term—is that adaptation is a special and onerous concept that should be used only where it is really necessary. When it must be recognized, it should be attributed to no higher a level of organization than is demanded by the evidence. In explaining adaptation, one should assume the adequacy of the simplest form of natural selection, that of alternative alleles in Mendelian populations, unless the evidence clearly shows that this theory does not suffice" (1966, pp. 4–5).

Williams's doctrine then serves as a hammer against group selection. This higher level process poses no problem in theory, for "there can be no sane doubt about the reality of the process. Rational criticism must center on the importance of the process and on its adequacy in explaining the phenomena attributed to it" (1966, p. 109). But group adaptation is both methodologically onerous (more so than Darwinian adaptation, which is onerous enough already), and theoretically impotent (though potentially operative).

If there are many adaptations of obvious group benefit which cannot be explained on the basis of genic selection, it must be conceded that group selection has been operative and important. If there are no such adaptations, we must conclude that group selection has not been important, and that only genic selection—natural selection in its most austere form—need be recognized as the creative force in evolution. We must always bear in mind that group selection and biotic adaptation are more onerous principles than genic selection and organic adaptation. They should only be invoked when the simpler explanation is clearly inadequate (1966, pp. 123–124).

(Note that Williams here introduces the ultimate causal reduction to genes as units of selection. He speaks of adaptation at the organismic level—but only as the consequence of genic selection. Thus Williams's book also becomes the manifesto for the ultimate—and, I think, erroneous—Darwinian reductionism still popular today as "selfish gene" thinking in such fields as sociobiology and evolutionary psychology. See Chapter 8 for a critique.)

In closing, Williams waxes messianic in his pointed comparison of natural selection with the teachings of Jesus (see John 8:12 and 14:6): "Perhaps today's theory of natural selection, which is essentially that provided more than 30 years ago by Fisher, Haldane, and Wright, is somewhat like Dalton's atomic theory. It may not, in any absolute or permanent sense, represent the truth, but I am convinced that it is the light and the way" (1966, p. 273).

I love Williams's book; his austere and incisive argument shaped my thinking and that of all evolutionists in my generation. But Williams's central thesis includes a disabling problem in logic, one that produced unfortunate effects in evolutionary practice. Parsimony, or Occam's razor, embodies an important logical principle when properly applied. William of Occam, a 14th century English philosopher and divine (a Franciscan), strongly espoused nominalism against the Platonic concept of ideal types as entities in a realm higher than material existence. (For nominalists, our designations of general categories only have standing as names [*nomina*] based on abstraction from objects in the material world, not as ideal and "excess" archetypes in a non-material realm.) Occam devised his famous motto, *"non sunt multiplicanda entia praeter necessitatem"* (entities are not to be multiplied beyond necessity), as a weapon in this philosophical battle—an argument against the existence of an ideal Platonic realm (for nominalists regard names of categories only as mental abstractions from material objects, and not as descriptions of higher realities, requiring an additional set of unobserved ideal entities, or essences).

Occam's razor, in its legitimate application, therefore operates as a logical principle about the complexity of argument, not as an empirical claim that nature must be maximally simple. Williams's key invocation of parsimony—to reject group selection when an explanation based on organismic selection can be devised for the same results—fails as a general argument, and does not use Occam's razor in a valid manner, on two grounds:

1. Whereas Occam's razor holds that we should not impose complexities upon nature from non-empirical sources of human argument, the factual phenomena of nature need not be maximally simple—and the Razor does not address this completely different issue at all. The Lamarckian one-step route to adaptation, for example, operates more simply and directly than the Darwinian two-step process of variation and selection. But nature happens to follow Darwin's path. Similarly, the simultaneous operation of several hierarchical levels in selection may represent a more complex system than the idea that se-

lection works only on organisms. But nature may (and does) work in this hierachical manner.

2. We should recognize Williams's claim as a statement about reductionism, not (as he thought) as an invocation of parsimony. Organismic selection is not intrinsically "simpler" than group or species selection. (One could only call organismic selection "simpler" in the obviously invalid psychological sense of affirming our habits and legacies as Darwin's intellectual children.) Consider Williams's argument: "Various levels of adaptive organization, from the subcellular to the biospheric, might conceivably be recognized, but the principle of parsimony demands that we recognize adaptation at the level necessitated by the facts and no higher. It is my position that adaptation need almost never be recognized at any level above that of a pair of parents and associated offspring" (1966, p. 19).

Lower levels in a hierarchy cannot be deemed inherently simpler, either to conceive or to operationalize, than higher levels. If we had been brought up in an intellectual world that emphasized populations, rather than organisms, as primary entities, we would probably regard interdemic selection as maximally simple, and organismic selection as an unwelcome complication. *A priori* preference for lower levels represents a claim for reductionism, not parsimony. I do not say that such a preference therefore becomes invalid; I simply ask evolutionists to recognize the proper status of Williams's claim as an argument about reductionism—and also to acknowledge that reductionism, as a cultural prejudice, may be far harder to defend than true parsimony, when properly invoked as a logical principle (though aspects of our preferences for parsimony may rank as cultural prejudice as well).*

In Western science, which developed with such strong traditions for explanation by analytic division into constituent parts, claims for reduction have often been mistakenly advanced in the name of parsimony—most notably in biophilosopher C. Lloyd Morgan's early 20th century dictum that no human activity should be explained by a higher psychological faculty when a lower faculty suffices.

This inappropriate invocation of parsimony did not disable Williams's argument because he usually proceeded beyond this theoretical point. That is, Williams extended his argument further by presenting direct evidence favoring the organismic mode in particular cases. He wrote: "This conclusion seldom has to rest on appeals to parsimony alone, but is usually supported by specific evidence" (1966, p. 19). But subsequent developments force us to

---

*Footnote added in proof stage: Just as I submitted this completed book to the publisher, the press conference on Darwin's birthday (Feb. 12, 2001), announcing the very low number of genes in the human genome, struck the deepest blow of our lifetimes against the conventions of reductionism, and for the irreducibility of proteomic (and full phenotypic) explanation to simple properties of codes at lower levels. Combinations, replete with emergent properties, and the specifics of contingent phyletic histories, must become a key partner, if not a primary locus, for biological explanation (see Gould, 2001).

consider one of the most troubling phenomena in the sociology of science—the principle of epigones and bandwagons.

Williams himself did not abuse, rigidify, misconstrue, or unduly simplify his criteria—but his followers did, both early and often (to cite the classical principle for voting in Boston local elections), as Williams's "doctrine" became a dogma among his epigones. Few aspects of academic life can be more distressing and ironic than the common observation that a fine scholar often becomes a victim of his own success in this manner—but subtle positions can be trivialized to sound bites in science as well as in political culture.

"Genic or organismic selection only" became the bandwagon slogan of the late 60's and 70's. Combined with a strong preference, already established as the Synthesis developed, for hardline adaptationism in general (see previous section), this restriction set a predisposition strong and exclusive enough to be labeled as a dogma: interpret all substantial phenotypic characters as adaptations built by natural selection in the organismic mode (or lower). This dictum did not always function as a cleansing wind in a former stable, but all too often as a narrow and misdirected tunnel that carried a necessary reform too far. Moreover, many epigones used the dogma as a kind of linguistic game rather than a guide to research: "Can I tell a clever story to render this or that puzzling phenomenon as an organismic, rather than a group, adaptation?" For some evolutionists, the ability to spin such a tale, and to answer such a challenge as a theoretical affirmation, became the goal of a supposedly scientific effort. I have never witnessed a more distressing bandwagon in science, or seen any idea of such salutary origin pushed so far in the direction of thoughtless orthodoxy.

(Pardon a personal incident, but I remember raising a question, early in my career, at a session of the first ICSEB meeting in 1973. I asked a speaker, following his formal presentation, if the dwarfed size of Pleistocene mammals on Mediterranean islands might have been favored by resistance to extinction afforded by the correlated effect of larger population sizes (than full-bodied hippos and elephants could have maintained in such small places). I hadn't thought the issue through, and I may well have been making a dumb suggestion, but the speaker's response floored me (and stunned me into silence at this ontogenetic stage of early diffidence). He said this and only this—and his words, with their intended dripping irony, still cut through me—"are you really satisfied with a group selectionist argument like that?" He made no attempt to rebut my suggestion with any content whatever; the stigma of group selection sufficed for refutation.)

As a final illustration of how reform, once established, can turn into the opposite phenomenon of rigidification, I interviewed Sewall Wright several times during the last decade of his life. He felt hurt by what he interpreted as his exclusion from the Modern Synthesis (beyond the ritualistic invocation of his name within the founding trinity of population genetics). "I was out of it," he told me. He explained this passage into obscurity as the failure of a new generation of evolutionists to understand either his intended role for genetic drift, or his proposed mode for the operation of drift within his "shifting balance" theory of evolution.

Most evolutionists of the 1960's viewed genetic drift only as a random force of evolutionary change—a prime anomaly under adaptationist hardening (or at least a factor relegated to the marginal role of efficacy only in tiny populations at the brink of extinction). Since genetic drift bore such a prominent association with Wright's name (extending to its original designation as the "Sewall Wright effect" in early days of the Synthesis), such demotion to marginality relegated the author to a similar fate under hardline adaptationism.

This situation is surely unfair enough in itself, but—and now the irony—Wright had also participated in the adaptationist hardening of the Synthesis (see pp. 522–524), and his later interpretation of genetic drift invoked this concept primarily as an aid to an enlarged style of adaptationism, and not as a contrary force in evolutionary change (as he had originally argued). So if Wright had tried to be helpful in the service of orthodoxy, why did he become so misunderstood and relegated to the sidelines?

Many reasons might explain Wright's fall into limbo, including the opaque character of his highly mathematical writing and the fact that he *had* invoked genetic drift as a nonadaptive force for change in his earlier work (see p. 523). But the major factor almost surely resides in a failure of evolutionists to understand the multi-leveled character of his theory—the aspect that allows drift to serve as an input to an adaptive process. Wright told me (in an interview in 1981) that he originally intended to call his shifting-balance process the "two-level theory," for the full process relies on essential components of both organismic and interdemic sorting. (Wright also told me that he regarded "exclusive focus on individual selection" as the major error of the Synthesis.)

In Wright's later formulation of "shifting balance," drift enters into a creative and selective process in the following manner: The founding population of a new species moves by selection to an adjacent adaptive peak in a larger landscape. The chief problem for adaptationism then intrudes: this initial peak will probably not represent the most favored spot in the landscape (for other unoccupied peaks probably stand higher), but how can daughter demes ever move to these better places? Valleys of lesser fitness surround this local peak. If evolution always proceeds towards adaptation, then the initial population must remain stuck on the first peak forever. But drift allows small groups to enter valleys, and to cross troughs into areas where selection may then draw populations up to a higher peak. When expanding populations, by this process, occupy several peaks in the landscape, a process of interdemic (interpeak) sorting can occur, eventually leading to a mean increase in adaptation for the species as a whole. Thus, genetic drift does not operate as a random force against adaptation in Wright's mature theory, but as a source of variability for fueling a higher level process of interdemic sorting. In other words, drift operates as part of a process that enhances adaptation through higher-level sorting.

This argument seems clear enough in logic (its validity, or relative frequency, in the world of real populations raises a different issue that can only be resolved empirically). Why was Wright's theory misunderstood or, even worse, simply ignored? I suggest the following (and unfortunate) primary reason: How could Wright's argument be grasped by an intellectual community now committed to the exclusivity of organismic selection? The conceptual tools no longer exist under such a stricture: adaptation arises by "struggle" for reproductive success among genes or organisms; drift causes a population to depart from a place attained by such struggle. How then could drift possibly act as a helpmate to adaptation? To grasp Wright's view, one must allow for a higher level of interdemic sorting, and one must understand the logic of hierarchical models, with sorting operating at several nested levels.

The intellectual space for viewing drift as an aid to interdemic selection doesn't exist in a context of exclusive commitment to selection at genic or organismic levels. Wright's idea then gets demoted to a status even lower than "merely wrong" in such a world; "shifting balance," in Wright's own sense, becomes inconceivable, and therefore intellectually inaccessible. Erroneous ideas can at least be expressed and made available to others with potentially different opinions. But the definitions of orthodoxy simply erased Wright's multiple-level theory—in much the same way that evolutionary stasis could not be recognized as interesting, or even grasped as a phenomenon at all, within a community committed to gradualism. When we think an idea through, and then reject the notion, we have at least made an intellectual decision (perhaps wrong, perhaps overly rigid). But when we maintain an unarticulated and unexamined commitment, and then use such a premise, albeit unconsciously, to render interesting ideas inconceivable, then we have fallen under the spell of dogma.

Sewall Wright—unlike Schubert, Wegener and a host of historical figures deemed tragic—lived long enough to witness his vindication (and to participate mightily in his renewed respect by writing a four-volume mathematical treatise, largely during his eighties—see Wright, 1978). But his period of unjustified eclipse should warn us all about the dangers of bandwagons and unexamined commitments.

### EXTRAPOLATION INTO GEOLOGICAL TIME

A good flavor of the confidence, even the dogmatism, of the hardened synthesis, as presented at the Darwinian centennial celebrations of 1959 (see pp. 569–576), shines forth in Mayr's introductory proclamation from his 1963 book. Mayr pronounces the "complete unanimity" of competent professional opinion, the "colossal ignorance" of the "few dissenters," and the consequent "waste of time" involved in any refutation of the intellectual stragglers:

> When we reread the volumes published in 1909, on the occasion of the
> 50th anniversary of the *Origin of Species,* we realize how little agree-

ment there was at that time among the evolutionists. The change since then has been startling. Symposia and conferences were held all over the world in 1959 in honor of the Darwin centennial, and were attended by all the leading students of evolution. If we read the volumes resulting from these meetings . . . we are almost startled at the complete unanimity in the interpretation of evolution presented by the participants. Nothing could show more clearly how internally consistent and firmly established the synthetic theory is. The few dissenters, the few who still operate with Lamarckian and finalistic concepts display such colossal ignorance of the principles of genetics and of the entire modern literature that it would be a waste of time to refute them. The essentials of the modern theory are to such an extent consistent with the facts of genetics, systematics, and paleontology that one can hardly question their correctness (1963, p. 8).

Later on, Mayr proposes a succinct definition of the Synthesis, emphasizing all three legs (or branches) of the essential tripod (or tree) of Darwinian logic "The proponents of the synthetic theory maintain that all evolution is due to the accumulation of small genetic changes, guided by natural selection, and that transpecific evolution is nothing but an extrapolation and magnification of the events that take place within populations and species" (1963, p. 586).

I regard Mayr's balance of emphases as particularly revealing. His definition includes two phrases. The first statement buttresses the two legs of Darwin's tripod that receive most attention in this work—control of direction by external selection rather than internal constraint (with attendant gradualism of change), and operation of the process through differential reproductive success of organisms (implicit in the term "natural," as opposed to some other form or level of selection). But the second, and longer, statement affirms Mayr's appreciation for the importance of the third leg—the complete sufficiency of microevolutionary theory to explain the entire history of life—so long as the earth's geological behavior sets a proper stage, and does not derail a full extrapolation of microevolutionary mechanics into all geological time and to the entire extent of phylogenetic change. For how can we celebrate the power and generality of a beautiful, sufficient and completely validated mechanism of change, established for the immediacy of an ecological here and now, if the same processes cannot render the broad pattern of life's history as well?

This theme of extrapolation becomes, in many ways, the most comprehensive issue of all. In Chapter 2, I distinguished two separate aspects of Darwin's radicalism in proposing the theory of natural selection—a methodological pole to grant operational status to the study of evolution by asserting that observable events, however apparently trivial and inconsequential, can explain change at all scales by extension; and an ideological pole to present a radical mechanism of evolutionary change for rendering all "higher" attributes of good design and organic harmony as side-consequences of a process working only by a struggle among organisms for personal reproductive success. Strict Darwinians must defend extrapolation as crucial at both poles—for observable events at small scale cannot generate the full panoply of phylogeny without such a principle; while daily happenings cannot accumulate into totalities if the background setting does not "behave" properly—if, for example, the ecological stage explodes every once in a geological while, fortuitously dumping most of the improved and accumulated inhabitants into a vat of extinction.

For these two reasons—to validate the entire methodological pole and to uphold the third leg of the Darwinian tripod supporting the ideological pole—the principle of extrapolation represents the key to the validity of the Synthesis as a fully general theory of evolution. And extrapolation—the essential Lyellian postulate that Darwin imbibed from his most important mentor (see p. 94)—embodies two central aspects of what Lyell and his school called "uniformitarianism": (i) the complete theoretical sufficiency of currently acting small-scale changes to produce, by successive and imperceptible increments, the entire panoply of large-scale phenomena; and (ii) the proper "behavior" of the earth, with geological change sufficiently slow and steady that trends produced by gradualistic, accumulative natural selection will not be derailed often enough to yield a history of life patterned more by these geological upsets than by biological accumulations.

For this last leg of the tripod, the Synthesis did not so much harden as become emboldened during its ontogeny. We have seen (pp. 514–518) how Haldane and Huxley, in the early days of the Synthesis, still respected (even feared) apparent paleontological exceptions to natural selection as the cause of trends. But the balance of power had shifted by the 1960's. Simpson (1944, 1953) and others had forged their "consistency argument"—the principle that all known phenomena of the fossil record can, in principle, be explained by modern mechanisms of genetics and selection (even though no direct proof of sufficiency can be derived from paleontological evidence). Paleontology had been tamed, taken in by the synthesis, and told to behave. (And I do intend "taken in" in the metaphorical sense as well, as I shall argue more explicitly in Chapter 9.) Paleontology could retain the archives of actual phenomenology as its particular bailiwick in exchange for giving up the conceit of believing that the fossil record could say anything distinctive about the causes of evolutionary change.

The distinguished panel on "the evolution of life" at the Chicago centennial celebration of 1959 included Julian Huxley as chairman, Th. Dobzhansky, E. B. Ford, Ernst Mayr, Ledyard Stebbins, and Sewall Wright. After an orthodox discussion of mechanisms, Huxley shifted the topic to "the course, the process, of evolution as shown in the fossils" (Tax and Callender, 1960, volume 3, p. 127)—and paleobotanist D. I. Axelrod rose to present a summary in advance, characterizing the history of life as a stately process of unfolding to more and better: "I think most of us are in full agreement about the gradual change in time: increasing diversification; then, gradual transformation, so new categories gradually arise, first at smaller and then at higher levels" (in Tax and Callender, 1960, volume 3, p. 127).

In the panel's stated agenda of 16 points, only one even hints at non-adaptive phenomena, and only as an adjunct to selectionist orthodoxy. Huxley addressed this topic midway through the discussion: "Natural selection may lead to side effects, which at the same time are of no adaptive value but may later provide the basis for adaptive changes" (Tax and Callender, 1960, volume 3, p. 125). Vertebrate paleontologist E. C. Olson, the symposium's lone (and very gentle) doubter (see pp. 574–576), ventured a perceptive comment on this point. But his words, as Moses said of Pharaoh's chariots, promptly "sank as lead in the mighty waters." The entire discussion of this topic occupies less than a page. Olson said: "This is the general area in which we can include events that are random with respect to the adaptive value of the genotype of populations. I refer to the simple matter of accident—for example, the effects of a forest fire on a population ... This sort of side effect, the impact of accidents and other factors producing non-adaptive shifts, may cause very rapid changes and give completely new shape to the course of evolution. I think this is an extremely important evolutionary factor" (in Tax and Callender, 1960, volume 3, p. 125).

If the Synthesis viewed the entire history of life, the full tree itself, as growing by an adaptive and stately unfolding, then the history of single branches —trends in lineages, the primary topic of macroevolutionary study—also received a thoroughly adaptationist reading in the extrapolationist mode. At the same Chicago conference, Simpson defended the adaptationist postulate for all geometries, parallel as well as diverging lineages, and even (in principle and without direct evidence) for "erratic" features where selective control "is not apparent."

> The selectionist theory is that a trend is adaptive for the lineage involved, that it continues only as long as it is adaptive, that it stops when adaptation is as complete as selection can make it in given circumstances, and that it changes or the group becomes extinct if a different direction of evolution becomes adaptive. Often the adaptive nature of a trend seems apparent. Often it is not apparent, but the postulate seems required to account for otherwise erratic features of trends. In instances of parallel evolution the selectionist theory is that changes actually occurring in parallel are adaptive over the whole ecological range occupied by the group, while those divergent (radiating) within the group are adaptations to special niches within that range (Simpson, 1960, pp. 170–171).

Under these precepts, a procedure of building scenarios in the strictly adaptationist mode, based on assumption and conjecture, often passed for adequate explanation. The second half of the Chicago panel on the "evolution of life," supposedly dedicated to the actual record of evolutionary change, included almost no discussion about paleontology, and relied on theoretical inferences about the past based on knowledge of modern organisms. Ernst

Mayr, however, did offer the following conjecture—wrong in many details (as we now know), yet firm in its confident adaptationist scenario—for the evolution of lungs. (Devonian fishes already possessed lungs, for the trait is symplesiomorphic in tetrapods and their aquatic ancestors, with the swim bladder of later fishes as its derived homolog. But note Mayr's confidence in his erroneous conjecture for the easy construction of such a novelty—from scratch, gradually, and in pure adaptive continuity with unchanging function):

> I think the development of lungs is now pretty well understood. Certain fishes during the Devonian period lived in stagnant, fresh water swamps, where oxygen was so scant that respiration through the skin and the gills no longer provided the necessary oxygen. Apparently they came to the surface and gulped air, from which the membranes of the digestive tract took up oxygen. When that stage was reached, there was a tremendous selection pressure for developing diverticles and enlarging this respiratory surface of the digestive tract. As soon as the necessary gene combination providing such diverticles appeared, selection pressure could push this tendency further and further, and this led quite naturally to the development of lungs (in Tax, 1960, volume 3, p. 136).

As documented in Chapter 6, the putative domination of biotic over abiotic competition provided Darwin with a rationale for defending general progress in the history of life. The synthesists upheld this orthodoxy as well, thereby imparting broad predictability to the stately unfolding of life. Huxley offered a clear assessment of relative frequencies in the founding document (1942, p. 495): "Sometimes the inorganic environment changes markedly, as when there is a climatic revolution, such as occurred at the end of the Cretaceous; but in general it is the organic environment which shows the more rapid and important alterations."

In his concluding address to the entire Chicago symposium, Huxley then remarked, with the expanded scope and surer resolve of nearly two decades in hardening: "Improved organization gives biological advantage. Accordingly, the new type becomes a successful or dominant group. It spreads and multiplies and differentiates into a multiplicity of branches. This new biological success is usually achieved at the biological expense of the older dominant group from which it sprang or whose place it had usurped. Thus, the rise of the placental mammals was correlated with the decline of the terrestrial reptiles, and the birds replaced the pterosaurs as dominant in the air" (1960, p. 250).

In citing the canonical example of dinosaurs and mammals, Huxley exposes the heart of the extrapolationist error—the assumption that large-scale pattern can be inferred by extending, through immense time, the small effects of observable processes (in this case the supposed general and overall "superiority" of mammals over reptiles in most cases of immediate competition). In explaining trends, the greatest threat to this orthodoxy lies in occasional but profound environmental catastrophe that disrupts and resets the pattern accumulating during "normal" times. The theory and factuality of catastrophic mass extinction has now broken this orthodoxy (see Chapter 12), but simple knowledge of mass extinction had always posed a threat, especially in Cuvier's original paroxysmal interpretation (see pp. 484–492). The synthesists therefore treated this apparent phenomenon in the conventional and congenial way, either by dismissing a catastrophic cause, or by "spreading out" the time of extinction so that all deaths could be encompassed by traditional competitive mechanisms, perhaps enhanced in intensity by rapid environmental changes, and therefore propelling adaptive evolution even more rigorously. Huxley (1942, p. 446), for example, held that tough physical conditions only accelerated the competitive takeover by superior groups: "The worsening of the climate at the end of the Mesozoic reduced the general adaptiveness of the dinosaurs, pterosaurs, and other reptilian groups, while increasing that of the early mammals and birds."

Ernst Mayr, in his characteristically forthright way, then linked the denial of catastrophic extinction, via uniformitarianism, to the crucial second statement in his definition of the Synthesis (see p. 557), the requirement for extrapolation into geological vastness: "Yet it has become clear that there is nothing in the past history of the earth that cannot be interpreted in terms of the processes that are known to occur in the Recent fauna. There is no need to invoke unknown vital forces, mutational avalanches, or cosmic catastrophes. Geographic speciation, adaptation to the available niches (guided by selection), and competition are largely responsible for the observed phenomena"(1963, p. 617).

In short, by viewing trends as adaptive and anagenetic phenomena, propelled by competition and building, by a lengthy process of stepwise summation, the principal pattern of life's history, the Synthesis encompassed the most salient phenomenon of paleontology within its favored framework of extrapolation. All causality could reside in the accessible here and now. How then, we must ask, did the Synthesis treat the two phenomena—speciation and extinction—now viewed as crucial in breaking the extrapolationist orthodoxy (for if trends must be expressed as the differential success of some kinds of species vs. others, with most species formed in geological moments, then the adaptive struggles of populations don't extrapolate smoothly to changes of mean and modal phenotypes within clades)?

The developing orthodoxy generally acknowledged speciation and then demoted its importance and distinctiveness. According to Huxley, for example, all radiations should be treated as adaptive and each event of speciation therefore represents an independent, gradualistic expression of an anagenetic trend (1942, p. 487): "The adaptive radiation is seen to be the result of a number of gradual evolutionary trends, each tending to greater specialization—in other words to greater adaptive efficiency in various mechanisms subservient to some particular mode of life. . . . Each single adaptive trend also shows the phenomenon of successional speciation."

In a statement that I find charming, however wrongheaded, Nicholson (1960, p. 518, at the Chicago centennial symposium) extolled speciation as a device to provide more opportunities for adaptation to work its "untrammeled" magic: "The splitting of organisms into the genetically isolated groups we call 'species' has played a very important part in evolution, for it has permitted selection to proceed untrammeled within each group, so permitting adaptations of innumerable kinds in the different groups. Had organisms not divided into genetically isolated groups, the numerous and beautiful adaptations so characteristic of living things could not have evolved, nor could organisms have used the resources of the world in the efficient way they do" (Nicholson, 1960, p. 518).

More commonly, however, speciation received short shrift rather than glory. Evolution required such a process of multiplication, of course, lest favorable trends disappear through the extinction of single species bearing their fruits. Speciation therefore became a hedge against death by parcelling out, into several iterated lines, a set of adaptations built anagenetically—so that the extinction of one species could not abort the general benefit. The trend itself remains anagenetic (see Fig. 7-2), for speciation does not contribute to the directionality of evolutionary change. (Under later views, including punctuated equilibrium, differential speciation *constructs* the trend, and anagenetic main trunks do not even exist.)

Simpson held strongly to this view, and even ventured a quantitative defense, in his repeated assertions that speciation represents only a minor mode in evolution because 90 percent of important changes arise anagenetically in the phyletic mode (1944, 1953; Simpson recognized three major modes of change: speciation, phyletic evolution, and quantum evolution). Huxley, in a grand prose flourish, then branded speciation as a pretty little epiphenomenon, a luxurious patina upon the grand pattern of evolution—never realiz-

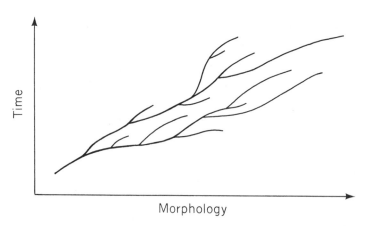

7-2. Standard view of the role of speciation in evolutionary trends under the Modern Synthesis. Speciation certainly plays an important part in iterating favorable variations produced by anagenesis within species. (If this iteration did not occur, lineages would quickly become extinct because individual species must eventually die.) But the trend in morphology arises almost entirely by anagenetic directionalism within the geological duration of individual species.

ing that the pattern itself might be built by higher-level sorting, operating through the differential success of certain kinds of species!

> The formation of many geographically isolated and most genetically isolated species is thus without any bearing upon the main processes of evolution. . . . Much of the minor systematic diversity to be observed in nature is irrelevant to the main course of evolution, a mere thrill of variety superimposed upon its broad pattern. We may thus say that, while it is inevitable that life should be divided up into species, and that the broad processes of evolution should operate with species as units of organization, the number thus necessitated is far less than the number which actually exist. Species-formation constitutes one aspect of evolution; but a large fraction of it is in a sense an accident, a biological luxury, without bearing upon the major and continuing trends of the evolutionary process (Huxley, 1942, p. 389).

Amidst this attempt to relegate the origin of the primary unit of macroevolution to irrelevancy at larger scales, one prominent voice within the Synthesis spoke up for the centrality of speciation in constructing large-scale pattern. In a cautious, but prophetic statement, Ernst Mayr (1963, p. 587) wrote: "To state the problems of macroevolution in terms of species and populations as 'units of evolution' reveals previously neglected problems and sometimes leads to an emphasis on different aspects." (Much of macroevolutionary theory, as developed later, begins with this proposition, and Mayr therefore becomes an inspiration—ironically in a sense, for several key concepts in this developing body of thought have challenged other aspects of the Synthesis that Mayr so strongly championed. For example, the theory of punctuated equilibrium rests upon a proper translation into geological time of Mayr's peripatric theory of speciation—see Eldredge and Gould, 1972, and Chapter 9.

Directly refuting Huxley's charge that speciation only ranks as a frill and luxury in the overall pattern of evolutionary change, Mayr wrote:

> I feel that it is the very process of creating so many species which leads to evolutionary progress. Species, in the sense of evolution, are quite comparable to mutations. They also are a necessity for evolutionary progress, even though only one out of many mutations leads to a significant improvement of the genotype. Since each coadapted gene complex has different properties and since these properties are, so to speak, not predictable, it requires the creation of a large number of such gene complexes before one is achieved that will lead to real evolutionary advance. Seen in this light, it appears then that a prodigious multiplication of species is a prerequisite for evolutionary progress. . . . Without speciation, there would be no diversification of the organic world, no adaptive radiation, and very little evolutionary progress. The species, then, is the keystone of evolution (1963, p. 621).

A world of difference separates the negative view held by most synthesists—that speciation merely iterates (and therefore buffers) adaptations produced by a different, anagenetic process—from Mayr's recognition that adaptations may be pieced together through accumulated events of speciation, each chancy in itself and not directed towards the eventual novel phenotype. In this sense, Mayr's view becomes the root for those branches of modern macroevolutionary theory that treat speciation as a higher-order analog of organismic birth—leading to a concept of trends as the product of differential sorting within the multitude of units thereby produced, and not as the extrapolated result of organismic selection within anagenetic lineages.

If most of the synthesists viewed speciation as trivial, they didn't grant even this modicum of concern to the counterpart process of extinction. Although they acknowledged the death of species (for a process affecting 99 percent of all species that ever lived can't be entirely ignored), they viewed extinction entirely in a negative light—as a loss of adaptation, and therefore as a failure in evolution, something to be recognized but not extensively discussed in polite company. Even Ernst Mayr, who understood so clearly how speciation could enter a higher-level process of sorting, didn't grasp the logical corollary—that any selective process must pair survivals with eliminations, and that "defeats" can therefore teach us as much as "victories." Instead, Mayr professed puzzlement as to why such a profoundly negative phenomenon should be so common:

> We find so many cases of extreme sensitivity of natural selection, doing the most incredible and impossible things; and yet the whole pathway of evolution is strewn left and right with the bodies of extinct types. The frequency of extinction is a great puzzle to me (in Tax, 1960, volume 3, p. 141).
>
> Natural selection comes up with the right answer so often that one is sometimes tempted to forget its failures. Yet the history of the earth is a history of extinction, and every extinction is in part a defeat for natural selection. . . . Natural selection does *not* always produce the needed improvements (1960, pp. 375–376).

The Synthetic approach to macroevolution can be encapsulated in a few dicta: view life as stately unfolding under adaptive control; depict trends as accumulative and anagenetic within lineages according to the extrapolationist model; downplay or ignore the macroevolutionary calculus of birth and death of species. These propositions leave little role for the actual archives of life's history—the fossil record—beyond the *documentation* of change. The *causes* of change must be ascertained elsewhere, and entirely by neontologists (my profession's term for the folks who study modern organisms). Thus the Synthesis held paleontology at arm's length. (I suppose we deserved this denigration in retaliation for the plethora of poorly-conceived, anti-Darwinian assertions and speculations that so many earlier paleontologists had falsely based upon the fossil record—see Chapter 4. In this sense, our later demotion, however unfairly extended, became part of the salutary cleansing accomplished by the early Synthesis in its first phase of restriction—see pp. 503–508.) Huxley (1942, p. 41) spoke of "the illegitimacy of using data on the course of evolution to make assertions as to its mechanism."* He continued:

> As admitted by various paleontologists . . . a study of the course of evolution cannot be decisive in regard to the method of evolution. All that paleontology can do in this latter field is to assert that, as regards the type of organisms which it studies, the evolutionary methods suggested by the geneticists and evolutionists shall not contradict its data. For instance, in face of the gradualness of transformation revealed by paleontology in sea-urchins or horses it is no good suggesting that large mutations of the sort envisaged by de Vries shall have played a major part in providing the material for evolutionary change (1942, p. 38).

Even so iconoclastic a morphologist as D. Dwight Davis, who would later tweak strict adaptationism so effectively in discussing formal and historical constraints in his classic monograph on the giant panda (Davis, 1964), wrote for the Princeton meeting on genetics, paleontology, and evolution (1949, p. 77): "Paleontology supplies factual data on the actual rates of change in the skeleton and the patterns of phyletic change in the skeleton. Because of the inherent limitations of paleontological data, however, it cannot perceive the factors producing such changes. Attempts to do so merely represent a superimposition of neobiological concepts on paleontological data."

I admit, of course, that paleontologists have no access to mechanisms requiring *direct* observation of ontogeny and ecological interaction. But to say, as Davis does, that we cannot ever derive concepts of evolutionary mechanisms from paleontological data—and must therefore gain all our causal understanding from "neobiology"—seems excessively pessimistic, and consigns paleontology to impotence. If paleontologists cannot gain insights about mechanisms, then historical *science* of any kind becomes impossible, for all scientific study of the past must make causal inferences from results of processes that cannot be directly observed.

Moreover, if historical data hold such limited promise, then the consequences become even more serious for science in general. For if we acknowledge that extrapolationism can't suffice in principle because much of macroevolution proceeds by patterns of differential birth and death among species, and if we cannot generate any theory about such higher-level sorting because we cannot observe the constituent events directly, then much of evolution becomes unknowable in principle. Fortunately, such pessimism may be firmly

---

*But Huxley's own later contentions belie this strong claim. For example, he argues against uniform internal drives in parallelism, and for control by external selection, by noting that characters do not always correlate in the same manner within parallel trends observed in different lineages of fossils: "In all cases where fossils are abundantly preserved over a considerable period, we find the same phenomena. The change of form is very gradual. It is often along similar lines in related types. And in general it appears that different characters vary independently" (1942, p. 32). But doesn't this statement qualify as an example of "using data on the course of evolution to make assertions as to its mechanism"?

rejected (see Gould, 1986, on Darwin's use of historical science, and 1989c, on applications to the history of life at greatest scale). Much of the best science—inevitably and properly—relies upon inference and insight, not always upon direct sight.

Young scientists can easily succumb to the thrall of such proclamations by leaders. The stupidest passage I ever wrote occurs in the heart of a contribution to independent macroevolutionary theory—our original piece on punctuated equilibrium, where I stated (this excerpt comes from my part of a joint text with Niles Eldredge):

> First, we must emphasize that mechanisms of speciation can be studied directly only with experimental and field techniques applied to living organisms. No theory of evolutionary mechanisms can be generated directly from paleontological data. Instead, theories developed by students of the modern biota generate predictions about the course of evolution in time. . . . We can apply and test, but we cannot generate new mechanisms. If discrepancies are found between paleontological data and the expected patterns, we may be able to identify those aspects of a general theory that need improvement. But we cannot formulate these improvements ourselves (Eldredge and Gould, 1972, pp. 93–94).

Stanley (1975) then properly rebuked us for such unwarranted subservience. Our current partnership with "neobiology," based on the "bonded independence" of macroevolutionary theory—the recognition that we can generate and test novel concepts but cannot come close to a fully adequate account of macroevolution without the vital input of microevolutionary theory—produces a better balance of subdisciplines. This mutually sustaining interaction must benefit paleontology, but such an enlarged view will also aid anyone, in any evolutionary subdiscipline, who wishes to comprehend the "grandeur in this view of life."

## From Overstressed Doubt to Overextended Certainty

### A TALE OF TWO CENTENNIALS

Darwin did all Americans a mnemonic favor by entering the world on the same day as Abraham Lincoln—February 12, 1809. He also made life simpler for conference organizers by publishing the *Origin of Species* in 1859, at age 50—thus intensifying the force of commemorations and cutting their required number in half. We have indeed celebrated mightily at the requisite times, with the usual array of resulting *Festschriften*. As others have noted, and as I have stated throughout this chapter, the two celebrations of the 20th century occurred at maximally disparate moments in the history of evolutionary theory: in 1909 at the heyday of doubt about natural selection as a potent mechanism, and in 1959 at the apotheosis of certainty about the nearly exclusive power of selection as an agent of evolutionary change. A comparison of the two centennials therefore provides a striking example and epitome of the success (and rigidification) of the Modern Synthesis.

Consider two of the leading symposia in 1909: the "official" celebration held in Cambridge (Seward, 1909), and the major American vernacular *Festschrift*, published as a special issue of *Popular Science Monthly* in 1909. The cardinal message reeks with ambiguity (for a celebration of Darwin's accomplishments): complete confidence in the fact of evolution; lavish praise for Darwin as midwife of the factual confirmation; admission that no consensus has been reached on mechanisms of evolutionary change; and a general feeling that natural selection plays, at most, a minor role.

A few strong selectionists restated their claims, most notably the two surviving members of Darwin's inner circle: Joseph Hooker and Alfred Russel Wallace. But even Wallace, the most ardent of selectionists, could no longer muster the confidence and enthusiasm of former years. The qualifiers in his "triumphalist" statement could not be more revealing—for he can now only assert that selection has been adopted as a "satisfying" solution by "a large number" of qualified experts: "And this brings me to the very interesting question: Why did so many of the greatest intellects fail, while Darwin and myself hit upon the solution of this problem—a solution which this celebration proves to have been (and still to be) a satisfying one to a large number of those best able to form a judgment on its merits" (Wallace, 1909, p. 398).

The range of opinions expressed at the Cambridge symposium illustrates the turmoil in evolutionary theory at Darwin's 100th birthday. Participants spanned the full spectrum from August Weismann's defense of selection's *Allmacht* (all-sufficiency) to Bateson's claims for selection's impotence (accompanied by lavish praise for Darwin's other achievement in establishing the fact of evolution—see p. 396). More commonly, authors tried to assimilate Darwin to their own disparate views, thereby turning the profession's hero into a chameleon. For de Vries, Darwin became a closet saltationist (see p. 416 on editor Seward's annoyance at de Vries' false and self-serving reinterpretation of Darwin). For Haeckel, Darwin ranked as a pluralist, a true kin to the speaker who had dedicated volume 2 of his *Generelle Morphologie* collectively to Darwin, Lamarck, and Goethe! (Haeckel, 1866). Haeckel wrote (1909, pp. 140–141), trying to distance Darwin from Weismann's position (called "neoDarwinism" at the time), and to reinvent the symposium's hero as a man in the middle between selectionism and Lamarckism: "It seems to me quite improper to describe this [Weismann's] hypothetical structure as 'NeoDarwinism.' Darwin was just as convinced as Lamarck of the transmission of acquired characters and its great importance in the scheme of evolution . . . Natural selection does not of itself give the solution of all our evolutionary problems. It has to be taken in conjunction with the transformism of Lamarck, with which it is in complete harmony."

The strategy of Henry Fairfield Osborn (in heaping praise on Darwin while denying any substantial power to natural selection) well illustrates the most consistent theme of both 1909 symposia. "There is no denying," Osborn writes (1909, p. 332), "that there is today a wide reaction against the central feature of Darwin's thought and this leads us to consider the merit of this re-action." Osborn then invoked a time-honored diplomatic tactic by defining Darwin's achievement as threefold: establishment of the "law of evolution" itself, documentation of the fact of evolution, and development of the theory of natural selection. Since the first two propositions cannot be gainsaid, why fuss over the third, even if Darwin overemphasized the role of natural selection? Osborn notes (1909, p. 332): "There is some lack of perspective, some egotism, much onesidedness in modern criticism. The very announcement, 'Darwin deposed,' attracts such attention as would the notice 'Mt. Blanc removed.'"

Osborn correctly identifies two claims at the core of Darwinian theory: (1) selection operates on undirected variability to cause evolutionary change (legs one and two of my tripod); and (2) gradualism rules in geological time (leg three and the methodological pole): "In the operations of this intimate circle of minute variations within organisms, he was inclined to believe two things: first that the fit or adaptive always arises out of the accidental, or that out of large and minute variations without direction selection brings direction and fitness; second, as a consistent pupil of Lyell, he was inclined to believe that the chief changes in evolution are slow and continuous" (Osborn, 1909).

But Osborn then gently chides Darwin for putting too much faith in the power of selection: "There can be no question, however, that Darwin did love his selection theory, and sometimes overestimated its importance" (1909, p. 336). Then, as a consummate politician and administrator, Osborn put a positive spin on his criticism—granting ultimate praise with only a faint damn. He emphasized the most common of all anti-Darwinian arguments— that selection can only operate as a negative force ("judicial" rather than "creative"). But he then converted Darwin's weakness to centennial strength with a remarkable diplomatic move: Darwin's problems arose from his ignorance of heredity, but he set a great task for us thereby, and we must persevere:

> Selection is not a creative principle, it is a judicial principle. It is one of Darwin's many triumphs that he positively demonstrated that this judicial principle is one of the great factors of evolution. Then he clearly set our task before us in pointing out that the unknown lies in the laws of variation and a stupendous task it is. At the same time he left us a legacy in his inductive and experimental methods by which we may blaze our trail. Therefore, in this anniversary year, we do not see any decline in the force of Darwinism but rather a renewed stimulus to progressive search.

This diplomatic theme—that Darwin did not discover an adequate mechanism of evolution, but we celebrate his centennial because he opened up a new world of research—became a virtual litany for symposiasts. For example, William Morton Wheeler virtually threw selection in the ashcan as he praised Darwin: "And even if we go so far as to say that natural selection may eventually prove to be an unimportant factor in evolution, to be consigned to the limbo of defunct hypotheses, together with Darwin's views on Pangenesis, sexual selection and the origin of species from fluctuating variations, we must, I believe, still admit that the great English naturalist opened up before us a vast new world of thought and endeavor" (Wheeler, 1909, p. 385).

T. H. Morgan, who would later become a strong supporter of natural selection, began his centennial contribution by expressing the standard argument that Darwin's importance transcends the limitations of natural selection: "The loyalty that every man of science feels towards Darwin is something greater than any special theory. I shall call it the spirit of Darwinism, the point of view, the method, the procedure of Darwin" (Morgan, 1909, p. 367). Morgan then ended his article, entitled "For Darwin," by heaping tangential scorn on natural selection while praising the liberating generality of evolution itself: "We stand today on the foundations laid 50 years ago. Darwin's method is our method, the way he pointed out we follow, not as the advocates of a dogma, not as the disciples of any particular creed, but the avowed adherents of a method of investigation whose inauguration we owe chiefly to Charles Darwin. For it is this spirit of Darwinism, not its formulae, that we proclaim as our best heritage" (1909, p. 380).

William Bateson, the least Darwinian of the symposiasts, began his article on the same theme, and then stated his own view right up front: "Darwin's work has the property of greatness in that it may be admired from more aspects than one. For some the perception of the principle of Natural Selection stands out as his most wonderful achievement to which all the rest is subordinate. Others, among whom I would range myself, look up to him rather as the first who plainly distinguished, collected, and comprehensively studied that new class of evidence from which hereafter a true understanding of the process of Evolution may be developed" (Bateson, 1909, p. 85). Bateson then added (see p. 596 for more on this quotation and Bateson's general views), in a statement that strikes me as the most appropriately generous, genuine and cogent expression of an argument that could be advanced with equal validity today: "We shall honor most in him not the rounded merit of finite accomplishment, but the creative power by which he inaugurated a line of discovery endless in variety and extension" (1909, p. 85).

Finally, consider the dilemma of A. C. Seward, general editor of the "official" Cambridge celebration, who followed a British tradition of fairness in inviting all sides, but then struggled to find some coherence amidst the Babel of papers he received: "The divergence of views among biologists in regard to the origin of species and as to the most promising directions in which to seek for truth is illustrated by the different opinions of contributors. Whether Darwin's views on the *modus operandi* of evolutionary forces receive further confirmation in the future, or whether they are materially modified, in no way affects the truth of the statement that, by employing his life 'in adding a little to Natural Science,' he revolutionized the world of thought" (Seward, 1909, p. vii).

I can imagine no contrast more stark, no reversal so complete, as the comparison of these doubts in 1909 with the confidence and near unanimity expressed fifty years later at the *Origin*'s centennial in 1959. The success of the Modern Synthesis established the difference. Beginning as a pluralistic marriage of Darwin and Mendel in the 1930's, the Synthesis had hardened by 1959 into a set of core commitments that, at least among epigones and acolytes, had become formulaic and almost catechistic, if not outright dogmatic.

Again, I will consider two leading *Festschriften* of this later centennial, the two major American celebrations in this case: the American Philosophical Society's annual general meeting in Philadelphia, and the elaborate festival held in Chicago in 1959 (published as a three volume compendium, edited by Sol Tax, in 1960). Major speakers at both meetings attributed the remarkable uniformity of opinion on all major issues to the success of the Synthesis, particularly to a consensus on the paramount, virtually exclusive, role of natural selection as the cause of evolutionary change. Ledyard Stebbins, appropriately for the City of Brotherly Love, spoke in Philadelphia about the unifying power of natural selection: "The last quarter of the century which has elapsed since the publication of *The Origin of Species* has seen the gradual spread and an almost universal acceptance by biologists actively working with problems of evolution of some form of the neodarwinian concept of evolutionary dynamics. This concept may be broadly defined as one which, like Darwin's original concept, maintains that the direction and rate of evolution have been largely determined by natural selection" (Stebbins, 1959, p. 231). Meanwhile, in Chicago, Julian Huxley gave a capsule history of Darwinism, ascribing the same binding role to natural selection: "The emergence of Darwinism, I would say, covered the fourteen-year period from 1859 to 1872; and it was in full flower until the 1890's, when Bateson initiated the anti-Darwinian reaction. This in turn lasted for about a quarter of a century, to be succeeded by the present phase of Neo-Darwinism, in which the central Darwinian concept of natural selection has been successfully related to the facts and principles of modern genetics, ecology, and paleontology" (Huxley, 1960, p. 10).

Michael Lerner's development of the argument (in the Philadelphia symposium) may be viewed as typical for this confident time. He begins with the venerable (if cryptic) motto of the Greek poet Archilochus: "The fox knows many things, but the hedgehog knows one big thing." As a profession, Lerner states, we marched along the path from Darwin to the Modern Synthesis as urchins, following the "one big thing" of natural selection, and ultimately rejecting the major alternatives as "sins against Occam's razor." Lerner wrote:

> Their one big thing, natural selection, set at rest the doctrine of special creation. In combination with our knowledge of Mendelian inheritance acquired since Darwin's day, it rendered obsolete such alternative theories of evolution as were based on extra-mechanical agencies, or on direct adaptation of organisms to their immediate environment (that is, on inheritance of acquired characters), and exposed them as sins against Occam's razor. Natural selection furnished the binding principle for a general or unified theory of historical change in the living world (Lerner, 1959, p. 173).

Lerner felt so confident that he proclaimed natural selection necessarily dominant *a priori,* not merely validated by evidence: "There is no longer any doubt that natural selection is more than a theoretical possibility—it is unquestionably a logically imperative necessity in any accounting for evolution" (1959, p. 174). He acknowledges, of course, that selection cannot manufacture, but can only shape, the physical material of organisms, but he compares selection's role to Michelangelo's claim that a great sculptor works to liberate beautiful forms from the blocks of stone that begin as their raw material—a lovely, poetic rendition of the standard argument for selection's creativity (see Chapter 2). In so doing, Lerner trivializes the role of potential constraints, even suggesting that a sow's ear might not represent an impossible starting point for a silk purse:

> In the same way, natural selection does not originate its own building blocks in the form of mutations of genes. But from them it does create complexes; it solves in a diversity of ways the great variety of problems that successful individuals and populations face; it builds step by step, even if by trial and error, entities of infinite complexity, ingenuity, and be one inclined to say so, beauty. Granted that it needs appropriate raw materials, that it may not necessarily be able to make a silk purse out of a sow's ear; yet, interacting with other evolutionary mechanisms, it has created the human species out of stuff which in its primordial stage may have looked no more promising (1959, p. 179).

Lerner's conclusion bears more than a whiff of similarity to the apostolic creed, suitable for multiple repetition by the faithful: "Evolution is the most fundamental biological law yet discovered. Natural selection is the basic mechanism implementing it. The principle of descent with modification, creatively, albeit opportunistically, husbanded by natural selection, is as firmly established as any concept in biology" (1959, pp. 181–182). (I don't disagree with the content; I just don't feel fully at ease with the triumphalist presentation.)

If Lerner verges on the overconfident, some centennial expressions treated any conceivable alternative with disdain. I have already cited Mayr's assertion of "complete unanimity" in competent professional opinion and of the "colossal ignorance" of remaining doubters. In Chicago, Mayr even resorted to theological language in citing "the opposing evils of Lamarckism and saltationism" (Mayr, 1960, p. 350). Others noted, but with some sense of unfairness, the vilification of Lamarck. C. H. Waddington regretted that "Lamarck is the only major figure in the history of biology whose name has become, to all intents and purposes, a term of abuse" (1960, p. 383); while Marston Bates noted that "Lamarck remains some kind of horrible example of wrong thinking in the introductory textbooks" (1960, p. 548).

But the prevailing tenor of these symposia does not display pugnaciousness towards opponents (which would imply an existing and meaningful conflict of uncertain resolution), but smugness in the confidence that a total victory has, at last, been achieved after a long battle (cigars and a drink around the fireplace at night, to cite an androphilic metaphor of past and privileged generations). Evolutionary theory is now essentially complete; we know how the process works and now only need to supply some details. G. G. Simpson had written in a 1950 essay (reprinted in Simpson, 1964, p. 14): "This general theory is now supported by an imposing array of paleontologists, geneticists, and other biological specialists. Differences of opinion on relatively minor points naturally persist and many details remain to be filled in, but the essentials of the explanation of the history of life have probably now been achieved."

The Chicago symposiasts continually asserted their agreement with this confident consensus. Tinbergen spoke of "the all-pervading power of selection" (1960, p. 609). Huxley (in Tax and Callender, 1960, volume 3, p. 45) defined the future task of evolutionary biology as filling in the blanks: "We are no longer having to bother about establishing the fact of evolution, and we know that natural selection is the major factor causing evolutionary change. Our problems now concern working out in detail how natural selection operates, defining what we mean by 'increase of organization,' tracing the general trends that appear in the course of evolution, and so on." He then described the range of phenomena that selection can fashion—in short, everything that might happen in evolution! "It produces branching; it produces increasing adaptation, improvement, progress, or whatever you like to call it; and it produces horizontal persistence of branches, or stabilization" (in Tax and Callender, 1960, volume 3, p. 139).

I argued in the last section that "hardening" of the Synthesis gains clearest expression in an increasing faith that adaptation must be both the impetus and result of nearly any evolutionary change. The 1959 symposia continually stress this theme. Panadaptationism became a premise for Chicago's major panel discussion on "the evolution of life," not an issue to be adjudicated by participants. Panelists received a list of assumptions, including the statement that "transformation always leads to adaptive or, better, teleonomic results" (in Tax and Callender, 1960, volume 3, p. 109).

Confidence in adaptation grew so great that many symposiasts presented their arguments in a "can't fail" manner, by delimiting a set of supposedly inclusive outcomes, each validating adaptation for any conceivable result.* Mayr, for example, argued that the general ecological rules of Bergmann and others enjoy good adaptive explanations, but that the numerous exceptions also affirm adaptation because local (and opposite) factors can override the

---

*But what scientific good can derive from a theory that includes no possibility of refutation from within? (A Mormon friend once told me that archaeologists of his church would either one day find direct evidence that the people of Mormon and Moroni had migrated from the Near East and lived in the New World until the 4th century AD, which would support the testimony of the Book of Mormon, or they would not find such evidence, which would also support the doctrines of his church by illustrating God's challenge to his people to keep faith in the absence of empirical support.) In such cases, one can only suggest alternative theories from without, and try to persuade people of good will that these alternatives provide better explanations for the purely empirical evidence.

general trend: "In recent years the analysis of these rules has shown that, as we stressed earlier, all phenotypes are compromises among a variety of conflicting selection pressures. As a result, there are many so-called exceptions to such rules, where a new selection pressure takes over and adjusts an organism or a local population in a different way" (in Tax and Callender, 1960, volume 3, p. 138).

In another example of victory by virtual definition, Tinbergen acknowledged that randomness might provide a theoretical alternative to adaptation in the evolution of a behavior. But since he construed randomness only as an absence of evidence for selection, and since he regarded the variety of conceivable adaptationist explanations as effectively inexhaustible, how would one ever validate randomness in any particular case? "This task [of explaining the results of evolution] really amounts to an assessment of the relative importance of the contribution made by random variation, on the one hand, and by adaptation directed by selection, on the other. Since randomness is, per definition, detectable only by elimination of every conceivable directedness, it is natural that this approach should lead to a quest for directedness" (Tinbergen, 1960, p. 602).

Adaptation pervades Tinbergen's discourse and world of thought. He even proposes a turnabout from Darwin's own, eminently sensible, view that nonadaptive features of conservative inheritance (deep homologies) provide optimal characters for taxonomic definition, since more recent adaptations tend to be homoplastic (as easily convergeable with similar features in independent lineages) and nondistinctive. Tinbergen, on the other hand, states that his paper will focus upon "the extent to which taxonomic characters must be assumed to be due to natural selection" (1960, p. 595). He then carries his adaptationist paean even further by arguing that evolutionists (as opposed to other scientists who might need to classify for different reasons) must divide organisms and designate their characters in terms of adaptive complexes, thus assuring his preferred interpretation by predefining the structure of observation itself:

> The conclusion that adapted features are systems of functionally related components forces us to reconsider once more the question What is a taxonomic character? The answer is, of course, that it depends on the aims which the scientist has in mind. The classifier is fully entitled to use, e.g., the tameness of the kittiwakes, their nest-building behavior, the black neck band of their young, and their nidicolous habits as four separate characters. But the evolutionist is not entitled to treat them as four independent characters. To him, the correct description of the characteristics of the species would be in terms of adapted systems, such as (1) cliff breeding; (2) pelagic feeding; (3) orange inside of the mouth and related characteristics of posturing (Tinbergen, 1960, p. 609).

These symposia not only featured adaptation as the centerpiece of the biological world, but also extended the concept to all other fields included within their program of lectures. Robert McC. Adams, then a young anthropology professor at Chicago, but later the Secretary of the Smithsonian Institution, confessed an initial skepticism about the symposium, based on "uncertainty about finding anything in common to talk about with representatives of other disciplines" (in Tax and Callender, 1960, volume 3, p. 268). But he discovered relevance in learning to view human societies as "adaptive mechanisms," and in using this idea to grant an "evolutionary role" to culture, thus equating adaptation with the entire realm of potential evolutionary insight:

> As man evolves, he superadds culture to his genetic equipment, and by this new addition he is enabled to adapt in a whole series of much more effective and complex ways—to spread himself over the entire globe, to construct very complex societies, and, in fact, frequently to direct the evolution of species all around him. Human societies are adaptive mechanisms; they have to be understood as having an evolutionary role rather than as uniquely human creations that are not to be compared with the evolutionary development of other organisms (Adams in Tax and Callender, 1960, volume 3, p. 268).

Only one "interloper," historian Ilza Veith, dared to suggest that nonadaptive phenomena might be important in evolution, but Julian Huxley firmly dismissed these worries:

> *Veith:* In my field, perhaps the most rewarding line would be to find those moments or those evolutionary processes that will present weaknesses, where maladaptation will occur, and where the mind will not continue to function in its normal manner.
> *Huxley:* I am sorry you wish to concentrate on maladaptation. I should think it would be much better to concentrate on adaptation from the positive angle (Tax and Callender, 1960, volume. 3, p. 269).

Sweetness surely triumphed in Chicago, but perhaps at the expense of light. The panel discussions ended in a virtual orgy of agreement, with Darwin as hero and adaptation as king. Even Sewall Wright, who had approached selection with ambiguity for years but had finally made his peace with the hardened consensus (though in his own idiosyncratic way—see pp. 522–524), ended his paper by writing: "From a more general standpoint, all of this is merely an elaboration in terms of modern genetics of the conception of evolution by natural selection advanced by Darwin in the *Origin of Species* a hundred years ago" (Wright, 1960, p. 471). Wright became even more accommodating in his role on the "evolution of life" panel. As the discussion wound down, Wright presented a simple comment as a last word before Julian Huxley's summary: "I agree with everybody."

Yet a bit of rain, as our mottoes proclaim, must fall on any long parade. One skeptic and whistle blower did speak out in Chicago, unsurprisingly a paleontologist who doubted the sufficiency of synthetic adaptationism as a complete explanation for the fullness of events in geological time. The American vertebrate paleontologist E. C. Olson had become disturbed by the increasingly dogmatic, peremptory, and exclusivist tone that many synthesists had adopted in this period of hardening. He spoke, with some irony, of the consensus that "has come to be known as the 'synthetic theory of evolution' but has also been variously termed 'selection theory,' 'neo-Mendelian theory,' and 'neo-Darwinian theory.' It is unfortunate that occasionally it is called '*the* theory of evolution,' as if no other could exist" (Olson, 1960, p. 524).

Olson then identified three aspects of the logic and sociology of the synthetic theory that, in veering towards dogmatism, made him uncomfortable. First, the theory had become flexible enough to encompass all possible results almost *a priori,* thus setting itself no challenges for potential refutation:

> The feeling of a slight sense of frustration in the elasticity involved in developing a universal explanation is hard to avoid . . . There is little or nothing that cannot be explained under the selection theory, and, at present, this theory appears to be unique in this respect (1960, p. 530) . . . This possible danger is amply revealed in some studies of the last decade which seem more concerned with fitting results into the current theory than with evaluation of results in terms of a broader outlook. Further, of course, much research is conceived and carried out within the framework of the theory, and, no matter what its excellence, is not likely to break out of this framework (1960, p. 536).

Second, the synthesists themselves often haughtily dismiss those who disagree as misguided, if not obtuse: "The statement is made, in effect, that those who do *not* agree with the synthetic theory do *not* understand evolution and are incapable of so doing, in most cases because they think typologically . . . Some avid proponents of the synthetic theory would appear to . . . eliminate as competent students of evolution, because of their inability to understand *the* theory, those who may disagree" (1960, pp. 526–527; Olson's italics).

Third, the success of consensus and consequent derision has silenced most doubters, but their numbers may be large and their questions cogent: "There exists, as well, a generally silent group of students engaged in biological pursuits who tend to disagree with much of the current thought but say and write little . . . It is, of course, difficult to judge the size and composition of this silent segment, but there is no doubt that the numbers are not inconsiderable. Wrong or right as such opinion may be, its existence is important and cannot be ignored or eliminated as a force in the study of evolution" (1960, pp. 523–524). As a paleontologist, Olson expressed most unhappiness with the "consistency argument" that awarded the synthetic theory hegemony over all scales of macroevolution—a misplaced confidence achieved by extrapolating, by fiat more than by evidence, a process that undoubtedly works in the ecological here and now to a sufficient explanation for all major changes occurring over hundreds of millions of years (1960, pp. 531 and 533).

Yet, however cogent Olson's doubts, his attempt to inject more pluralism and skepticism into evolutionary theory ultimately failed—and for a valid reason from the orthodox point of view. A successful whistle blower must proceed beyond the exposure of faults in his boss's domain; he must also suggest a path towards greater accuracy and fuller explanation. And, on this constructive side, Olson could offer nothing. He ventured a few comments about cytoplasmic inheritance as a possible mechanism that might not follow all synthetic rules, but such a limited and inadequate speculation could not fuel such a comprehensive set of doubts! Revisionists would gain no hearing until they could propose an extensive and positive set of extensions or alternatives—and I write this book because I believe that such an affirmative program has now been formulated. Olson's critique achieved no currency, and the hardened version descended from the empyrean academy into the vernacular world of textbooks, the ultimate test of establishment by social imposition as well as by professional consensus.

### ALL QUIET ON THE TEXTBOOK FRONT

Professional writing tends to be nuanced and judicious. Even the strongest partisan finesses his commitment and adds at least a footnote or tangential comment, so that any charge of oversimplification or dogmatism may be countered by stating: "but look on page 381 (in the small print); you see, I raised that caveat myself."

To learn the unvarnished commitments of an age, one must turn to the textbooks that provide "straight stuff" for introductory students. Yes, textbooks truly oversimplify their subjects, but textbooks also present the central tenets of a field without subtlety or apology—and we can grasp thereby what each generation of neophytes first imbibes as the essence of a field. Moreover, many textbooks boast authorship by the same professionals who fill their technical writings with exceptions, caveats, and complexities.

I have long felt that surveys of textbooks offer our best guide to the central convictions of any era. What single line could be more revealing, more attuned to the core commitment of a profession that bathed in the blessings of Victorian progressivism, and aspired to scientific status in Darwin's century, than the epigram that Alfred Marshall placed on the title page to innumerable editions of his canonical textbook, *Principles of Economics: "natura non facit saltum."*

The changing foci of 20th century textbooks provide direct insight into the history of evolutionary thought and the eventual triumph of Darwinism. In particular, if the Synthesis truly hardened, as I have argued, then texts following the 1959 centennial celebrations—the apogee of strict selectionism— should describe evolution in unambiguously panadaptationist language, and should extol the sufficiency of natural selection to craft the entire range of evolutionary phenomena at all scales, ecological to geological.

This section does not present a systematic survey of texts, though I have consulted everything I could find, including nearly all major American books for introductory college biology (and several high school textbooks as well). A more complete search, extended back in time to cover the early days of the Synthesis, and the pre-synthetic period as well, would provide a fascinating topic for a dissertation in the history of science or education. This field of vernacular expression has been neglected by scholars, though the subject would yield great insight (for such material obviously represents the only formal contact that most students ever receive with any given discipline).

I apologize for my almost anecdotal approach, but I think that I have identified a robust pattern supporting the hypothesis of hardening. I will focus on the two topics that authors of texts found most congenial in their efforts to explain synthetic evolutionism to introductory audiences: the centrality of adaptation, and the sufficiency of synthetic microevolution to explain events at all scales. (I consider here only the evolution chapters of comprehensive biology texts for introductory courses, not entire textbooks on evolution. These short, unvarnished and straight-line accounts of adaptation and extrapolation appear in the context of such epitomes. Full texts on evolution, which cannot be called "introductory" or "elementary" (for such courses have always been taught at intermediate or advanced levels in American universities), do treat the subject more comprehensively, with a proper listing, often called "textbooky" in our jargon, of divergent views.)

### Adaptation and natural selection

In this age of sound bites, even short chapters include final summaries to tell students the pith of what they must remember. Consider the following from Nelson, Robinson, and Boolootian (1967, p. 249), written to summarize a chapter entitled "Evolution, Evidences and Theories." I cite the entire statement, not an excerpt:

Principles

1. Charles Darwin proposed a theory of evolution based on variation, competition, and consequent natural selection.

2. The basic mechanism of evolution is now known to be changes in gene frequencies of populations through time, guided by natural selection.

On the subject of exclusivity, Darling and Darling (1961, p. 199) tell us that "any organism is a bundle of interacting adaptations. Most all the features of all living things are adaptations." Howells (1959, p. 24), a great evolutionary anthropologist publishing his popular text in Darwin's centennial year, discussed natural selection with his usual panache and good humor, but also in the all-encompassing celebratory mode: "So much for natural selection, the external force, that finger beckoning to the otherwise unguided heredity of an animal type. All other principles and facts of evolution may be satisfactorily related to it or explained by it, and the century following 1859 has seen Darwin triumphant."

Simpson, Pittendrigh and Tiffany (1957, p. 405), an excellent text that dominated the market for years (and featured a leading architect of the Synthesis as first author), also stated that any nonrandom evolution must be adaptive: "The evolutionary changes that result from nonrandom reproduction are clearly adaptive: the changes are always, necessarily, of such a kind as to improve the average ability of the population to survive and reproduce in the environments that they inhabit."

Many paeans to adaptation proceeded beyond mere claims about omnipresence to assert optimalized excellence, or near organic perfection, as well. Convictions about the exclusive power (as well as the range) of natural selection emerge most clearly from such statements, as by Telford and Kennedy (1965, p. 3) (Kennedy later became the president of Stanford University and editor of *Science* magazine):

> It is of profound importance for the nature of the organism that, due to natural selection, the evolutionary changes in organisms have either moved relentlessly in the direction of efficiency or have kept them attuned to a changing environment. . . . Evolutionary adaptation thus suggests an extremely fine attunement between organism and environment. The organism doesn't merely get along; its whole life mode has been tempered and refined by the successful competition of generations of its ancestors with a multitude of differing genotypes. Thus even in the finest details of their organization, organisms are constructed and operate in a manner which makes sense in terms of the way they make their living.

From this assertion of omnipresence for adaptation in morphologies, physiologies and behaviors of the moment, these texts then proceed to ascribe the second great phenomenon of evolution—the production of diversity—to natural selection as well. Simpson et al. (1957, p. 405) extend selection's scope to all phenomena at all scales by writing: "The evolutionary process, viewed in broad perspective, is characterized by two major features: it produces diversity among living things, and it gives rise to their adaptation, their fitness to survive and reproduce efficiently in the environments they inhabit. These two features are interdependent: life's diversity is largely a diversity in adaptation."

Speciation, although replete with nonadaptive elements in Mayr's canonical formulation, usually receives a textbook description as an even stronger affirmation of natural selection (because the process now operates in *two* separated lines, working its differential effects to produce just the right adaptations in both distinct and varying environments). Nelson *et al.* entitling their section "Speciation: The Results of Adaptation," write in summary (1967, p. 235): "Natural selection operating on the variability present in the genotypes of populations can cause better adaptation of organisms to their environment. Coupled with reproductive isolation, these adaptations bring about speciation."

Jones and Gaudin (1977, p. 548) introduce their discussion of speciation with a scenario of pure adaptation and extrapolation. (Their full text discusses other mechanisms, including polyploidy—but note the pride of place awarded to adaptation, and the argument that so separate and important a phenomenon as geographic isolation only provides an impetus by setting new selection pressures in a different environment):

> The accumulation of adaptations can lead to the production of new species, a process called speciation. . . . Suppose a population of gophers liv-

ing in a valley is divided in two by a river that cuts a channel through their valley. The two segments of the population are now effectively separated from one another, and any environmental differences that exist between the two regions of the valley will result in adaptations restricted to one side or the other of the river. . . . Different selective pressures now will be operating on opposite sides of the river. Given sufficient time, the two gopher populations may diverge quite extensively.

With speciation thus explained as an extended consequence of adaptation under certain environmental circumstances, the same argument can then be smoothly extended to life's full pattern in geological time. Alexander (1962, p. 826) tells students that all phylogeny flows from "the fact of adaptation." I can hardly imagine a more gradualistic and meliorist account of evolution, with all death for improved existence, and all life in continual motion towards more and better:

> We need only accept the fact of adaptation—the idea that organisms are fitted for the particular environments in which they live—to see the necessity for a process of organic evolution. The environment in which organisms live has not been constant . . . Organisms, of course, do not exist under conditions for which they are not adapted. They have, therefore, met these various conditions at different times and places; in order to persist under a changing environment they themselves have had to change. We may think of organic evolution, therefore, as the progressive change of plants and animals in harmony with the changes in their environments. The unadapted die out and disappear. Those organisms whose descendants can fit into the new conditions survive, expand in numbers and kinds, and take over the changing habitat.

### Reduction and trivialization of macroevolution

The hypothesis of selection's *Allmacht,* and adaptation's ubiquity, rests upon the validity of extrapolation to the full range of geological time, for what power (or generality) can a well-formulated theory of local adaptation assert if the same process, by uniformitarian extension, cannot explain the origin of multicellularity, the rise of mammals, and the eventual emergence of human intelligence? Paradoxically perhaps, this extrapolationist assertion becomes, at the same time, the most vulnerable and the most essential of all synthetic propositions—vulnerable in necessary reliance upon a "consistency argument" in the absence of empirical proof, and essential because the theory becomes such a paltry and limited device if its explanatory range cannot extend beyond the compass of its directly observable effects.

No evolutionary assertion has been more commonly advanced in textbooks, or more superficially (and almost nonchalantly) proclaimed by fiat, than the claim that adaptation by natural selection must be fully sufficient to render life's entire history. In the last section, I documented the "promotion" of arguments about pervasiveness of adaptation in local circumstances, to speciation, to the entire tree of life. Capping this sequence, Howells (1959,

p. 28) writes that "all this exploring, stopping, and rushing, in the pursuit of profitable adaptation, has resulted in the great family tree of the animals."

Nelson *et al.* (1967, p. 239) briefly extol the full sequence—from the rule of selection in local populations, through speciation, to the origin and diversification of phyla: "Evolution in its simplest and broadest sense means changes in gene frequency over a period of time. Natural selection guides these changes . . . Over long periods the accumulation of changes may be sufficient to separate once similar populations into distinct groups. In the course of evolutionary history this divergence has apparently led to different classes (mammals, birds, fish, etc.), different phyla (insects and corals, for example), and even different kingdoms (plants and animals)."

The dominant high school text of the 1960's and 70's depicts the standard equine example of macroevolution as anagenetic gradualism guided by natural selection, thus making any definition of chronospecies arbitrary: "The fossil record shows that all these differences are the result of a series of many gradual changes. Each change that became established through natural selection must have been very slight; only when many such changes accumulated did they result in detectable differences. How can this long sequence of horses be divided into species?" (Biological Sciences Curriculum Study, Green Version, 1973, p. 621). The accompanying figure of the phylogeny of horses depicts the actual (and copiously branching) bush as a smooth ladder of progress (see Fig. 7-3).

Bonner (1962, pp. 52–53), another leading evolutionist who also wrote a popular text, argued that paleontologists can't study the mechanics of evolution directly, but professed complete confidence in the efficacy of microevolutionary selection:

> Paleontologists as well as ecologists have been for some years studying the evolutionary factors we have discussed, and have continuously attempted to see how the fossil record, on the one hand, or the present-day distribution of animals and plants, on the other, fit in with this scheme. There seem to be no major discrepancies, and a general feeling that the mechanism of evolution is understood prevails, particularly in regard to the importance of selection and the method of formation of new species. . . . Some groups such as the mollusks have been exceedingly slow in their progress while others, such as the mammals, have been very rapid. Again this can be totally understood in terms of selection in particular environments. No other hypothetical mechanisms seem to be necessary to account for the facts as we know them.

Such confidence in microevolutionary sufficiency can only lead to a downgrading of paleontology—either to theoretical irrelevance, or to a status as a mere repository for results of processes that can only be elucidated by studying modern organisms (and may then be smoothly extrapolated across a million millennia). I do not think that this derogatory judgment originated by the conscious intent of most textbook authors. Rather, the marginalization of paleontology flows directly from the logic of pure extrapolation. The basic

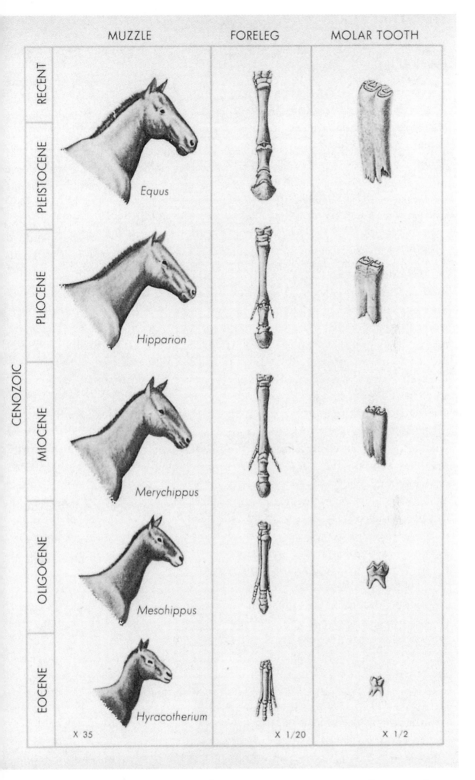

| MUZZLE | FORELEG | MOLAR TOOTH |

**CENOZOIC**

RECENT

PLEISTOCENE — *Equus*

PLIOCENE — *Hipparion*

MIOCENE — *Merychippus*

OLIGOCENE — *Mesohippus*

EOCENE — *Hyracotherium*

X 35     X 1/20     X 1/2

7-3. Standard textbook misdepiction of a copiously branching evolutionary lineage as a ladder of progress. This canonical view of the evolution of horses appeared in the 1973 *Green Version* of the most popular and most widely respected high school textbook produced by the *Biological Sciences Curriculum Study*.

argument takes two forms. Some authors explicitly exclude paleontology from the theoretical game:

> Evolution can be studied on the population level only with living organisms. The fossil record provides too few data to allow such treatment; it merely allows paleontologists to reconstruct the history of animal and plant groups. The population approach makes it possible to ask such questions as: what is the rate of evolution in a given species? What factors influence the course or rate of evolution? What conditions are necessary for evolution to begin or cease? (Baker and Allen, 1968, p. 524). [I do not see why paleontologists cannot address all three of these questions with data from the morphology of fossils and their temporal distribution.]

But I must confess that a stronger and more focused form of this argument has long evoked my deeper distress, and has served, in substantial measure, as the impetus for personal career choices in research, and for my eventual decision to write this book. I refer to the claim, repeated almost as a catechism, and obviously copied from textbook to textbook, that macroevolution poses no problem not resolvable by a further understanding of allelic substitutions directed by natural selection in contemporary populations. We may move smoothly from one gene to an entire *Bauplan,* and extrapolate upwards from a few generations to a geological era. No additional problems arise in temporal vastness. Macroevolution becomes little more than industrial melanism writ large. But can we even imagine, in a world dominated by effects of scale, that such a maximal extension of form and time will engage not a single force or principle beyond the factors fully in evidence at the lowest level? Can the smallest scales really provide an entirely sufficient model for the largest? Can a uniformitarianism this rigid truly be sustained? If so, then paleontology only represents a playground for the full display of microevolutionary muscle—and textbooks need not consider the fossil record as more than an archive of the pathways carved by this power.

Most standard textbooks make this confident assertion based on little beyond hope and tradition—thus making macroevolution a nonsubject. Bonner (1962, p. 48), for example, writes: "There is no reason to believe that these large changes are not the result of the very same mechanics of the small changes of industrial melanism. One involves a small step over a few years; the other involves many many thousands of steps over millions of years."

Curtis (1962, p. 712), in a best-selling text of the 60's, begins her short section on macroevolution by stating: "Can the same processes that slowly shape the seed of mustard weed or change the color of the peppered moth create the differences between elephants and daisies or between butterflies and redwood trees? Darwin believed so—all he felt that was needed was time, millions of years of slow change. Today, almost all evolutionists are, in principle, in general agreement with Darwin's conclusions."

Several texts even present this canonical argument as their *only* statement about macroevolution. I end this chapter by quoting two striking examples of this trivialization and marginalization of macroevolution, each from the most important source in its respective genre. As mentioned above, BSCS textbooks (written by a semiofficial consortium of private and governmental sources, The Biological Sciences Curriculum Study) virtually cornered high-school markets during post-Sputnik years of the 1960's and 1970's. The 1968 version of *Biological Science: An Inquiry Into Life* includes a heading on "The Origin of Genera and Larger Groups." But the text contains only two paragraphs, fully reproduced below:

> The final question which we must ask about the forces of evolution is this: can mutation, recombination, selection, and barriers to cross-breeding explain the major trends of evolution, such as the divergence of cat-like from doglike animals and the evolution of the horse from its small primitive ancestors?
>
> The mechanisms that govern these major trends of evolution cannot be studied directly: they took place many thousands or millions of years ago. Nevertheless, a study of populations today, and of fossils, provides strong evidence that the same evolutionary forces in operation today have guided evolution in the past. One species evolves into two (or more). All the new species continue to evolve, becoming more different from one another until eventually we would classify them as different genera (1968, p. 203).

*Life on Earth* (1973) surely ranks as the most distinguished textbook of introductory college biology published during the 1970's. Written by a team of eight authors, and headed by two of the world's leading evolutionists (E. O. Wilson and T. Eisner), this book staked an explicit claim for groundbreaking novelty by linking appropriate expertise at the highest level with accessibility in style, and excellence in design and illustration. Chapter 28 on "The Process of Evolution" ends with the heading "Macroevolution." The quotation below may seem limited in content, particularly for a college text, but I do not cite an excerpt. I have reproduced the book's *entire* section on macroevolution!

In this passage, the history of life becomes a simple extension of the story of the raspberry eye-color gene. (For the second edition, the authors switched to the standard case of industrial melanism, but did not alter the general argument at all.) Paleontologists may be burdened with an incomplete record, the authors assert, but as they look more carefully, the gap between the raspberry gene and the Cambrian explosion closes continually. I can only express my astonishment at such a limited, but definitive, assertion by applying Ethel Barrymore's famous closing line to this dismissal of macroevolution as a subject: "That's all there is, there isn't any more."

> Each of the examples of microevolution examined, involving shifts in the frequencies of small numbers of genes, could be multiplied a hundred-fold from reports in the scientific literature. Biologists have been privileged to witness the beginnings of evolutionary change in many kinds

of plants and animals and under a variety of situations, and they have used this opportunity to test the assumptions of population genetics that form the foundations of modern evolutionary theory. The question that should be asked before we proceed to new ideas is whether more extensive evolutionary change, macroevolution, can be explained as an outcome of these microevolutionary shifts. Did birds really arise from reptiles by an accumulation of gene substitutions of the kind illustrated by the raspberry eye-color gene?

The answer is that it is entirely plausible, and no one has come up with a better explanation consistent with the known biological facts. One must keep in mind the enormous difference in time scale between the observed cases of microevolution and macroevolution. Under natural conditions the nearly complete substitution of the melanic gene of the peppered moth took 50 years. Evolution of the magnitude of the origin of the birds usually, perhaps invariably, takes many millions of years. As paleontologists explore the fossil record with increasing care, transitions are being documented between increasing numbers of species, genera, and higher taxonomic groups. The reading from these fossil archives suggests that macroevolution is indeed gradual, paced at a rate that leads to the conclusion that it is based upon hundreds or thousands of gene substitutions no different in kind from the ones examined in our case histories (1973, p. 792).

But, *pace* Ms. Barrymore, there is *so* much more—as research in the vibrant field of macroevolution, filling the pages of numerous journals (all founded after these dismissive comments), attests; as the development of a tight and powerful theory of hierarchical selection embodies (see Chapters 8 and 9); as the union of developmental with evolutionary biology displays (see Chapters 10 and 11); as our advancing understanding of genomic complexity asserts. Can we not feel the frustration of E. C. Olson as he queried the titans of the Modern Synthesis in Chicago? Can we not understand why a few iconoclasts never made their peace with such a comfortable and limiting orthodoxy? Can we not gain a visceral (and not only an intellectual) sense of C. H. Waddington's isolation and irritation when he made his famous comment on the limitations of population genetics (Waddington, 1967), and won admiration for his panache but no consideration for his content: "The whole real guts of evolution—which is, how do you come to have horses and tigers, and things—is outside the mathematical theory."

# Segue to Part Two

Contemporary challenges to all three central commitments of Darwinism (the legs of the tripod in my chosen metaphor, or the "essence" of the theory in the legitimate use of a word generally shunned by evolutionary biologists) prompted me to write this book. Such forms of debate set the mainsail of scholarly life, and cynics may be excused for suspecting the academic equivalent of glitz and grandstanding when their colleagues proclaim major unhappiness with received wisdom. This cynicism merits special attention when Charles Darwin serves as a target—for the demise of Darwinism has been trumpeted more often than the guard changes at Buckingham Palace, notwithstanding the evident fact that both seem to stand firm as venerable British institutions. (I state nothing new here: Kellogg (1907) began his wonderful book, the basis for my style of exposition, by refuting a German proclamation, then current, about the *Sterbelager*, or death-bed, of Darwinism—see Dennert, 1904, for an English translation of the book that inspired Kellogg's long and celebrated rejoinder.)

If continuity breeds respect—and what other criterion could an evolutionist propose in this volatile vale of tears?—then the most persuasive rejoinder to a charge of superficial and ephemeral grandstanding must lie in the documentation of long persistence and serious attention for a given critique. Persistence, of course, need not imply cogency; Lord only knows the lengthy pedigree of stupidity. But an analog to natural selection also operates in the world of ideas, and truly silly notions do get weeded out at certain levels of intellectual competence. Moreover, only a small subset of our forebears rank as brilliant thinkers. When we can designate a critique as both longstanding in general and seriously supported by scholars in this subset, then such arguments should command our respect and attention. (Brilliance, of course, only implies cogency, not correctness. Cultural biases and simple lack of information can lead even the most gifted minds to firm convictions that seem risible today. But I do assert that brilliant scholars, while often as wrong as anyone else, devise their positions for interesting and instructive reasons. We may now reject Lyell's strict views about substantive uniformity, and Paley will find few modern devotees for his natural theology. But we must not write these men off—and we will learn much by studying the reasons for their distinctive attitudes.)

In this context, I ask readers to consider two points about the tripod of necessary support for Darwinism:

*First,* even during the period of current orthodoxy, beginning with the coalescence and spread of the Modern Synthesis; the three supports have never been particularly firm, or adequately defended. The first support—restriction of selection to the organismal level—received little explicit defense, but rather prevailed in a fuzzy sort of way by convention in practice. Sloppy statements implying group selection abounded in the literature (as documented in Chapter 7, pp. 544–556), and some disciplines, notably the classical ethology of Lorenz and his disciples, frequently cited arguments about supraorganismal selection without understanding their consequences for Darwinian theory. This situation changed dramatically when Wynne-Edwards (1962) advanced his explicit argument for group selection at a predominant relative frequency—and Williams (1966) wrote his spirited defense of the Darwinian straight and narrow by setting out the centrality of organismal selection so forcefully.

The second support—the validation of selection as a nearly exclusive mechanism of evolutionary change, as embodied in the adaptationist program—received strong verbal approbation, and elegant illustration in a few cases, but won orthodox status largely as a bandwagon effect prompted by the urgings of a few central figures, notably Mayr and Dobzhansky, and the subsequent acquiescence of most professionals to the assertion of such leading figures, and not to the data of convincing demonstrations (see Chapter 7 for a detailed defense of this claim, as embodied in my hypothesis on the "hardening" of the Modern Synthesis). In particular, taxonomic orthodoxy just before the Synthesis (Kinsey, 1936; Robson and Richards, 1936) regarded most geographic variation within species as nonadaptive. The opposite opinion triumphed as the Synthesis reached a height of prestige and orthodoxy, but few actual cases had been overturned by data. Rather, a shifting theoretical preference led to assertions of dominant relative frequency based on documentation inadequate to affirm either view. (I do not regard earlier arguments for nonadaptation as inherently more cogent. On the contrary, I am convinced that we still have no good idea about the relative frequencies of adaptive and nonadaptive effects in geographic variation. I only claim that the shift to adaptationist preferences resulted more from a bandwagon effect than from direct evidence—and that this second leg of the tripod therefore never enjoyed adequate buttressing.)

The third support—extrapolation, explicitly discussed here in terms of the surrogate proposition of geological uniformity, and so necessary to provide a stage that would nurture, or at least not disrupt, Darwin's hope for explaining the entire history of life by "pure" extension of principles derived from the small and palpable—prevailed more by assumption than by active validation, with Kelvin's defeat and Rutherford's proof of the earth's great age read as adequate defense (a logically insufficient argument, by the way, because time may be long, but change still concentrated in rare catastrophic episodes). By the largely arbitrary and contingent sociology of disciplines, paleontology belonged to the geologists, and students of fossils therefore received virtually no training in evolution. With few exceptions, notably the work of G. G. Simpson, paleontology (at least during the first half of the 20th century) played little role in the development of a theory to account for its own subject matter.

*Second,* and more important in summarizing the first half of this book: not only can we identify basic logical weaknesses in standard defenses for the three Darwinian supports, but cogent critiques have also been persistent, indeed omnipresent though changing in form, throughout the history of evolutionary thought. These histories may not be widely known to current practitioners, but the best minds of our profession have struggled continuously with themes of the essential tripod—and their arguments deserve our attention and respect. (Without this knowledge, we tend to imbue orthodoxies with false permanence or, even worse, to lose sight of basic principles in the surrounding silence, thereby converting dubious but central postulates into hidden assumptions. History can and should be liberating.)

For the first leg of the tripod of essential support, hierarchy theory and multiple levels of selection do not represent only a modern gloss upon Darwinism. Rather, contemporary versions of these concepts have reinvigorated the oldest issue of our profession. In Chapter 3, I showed that the first two evolutionary systems well known to English-speaking naturalists—Lamarck's and Chambers's—relied on a causal hierarchy that contrasted progress with diversification. Darwin explicitly combatted these ideas with a single-level theory based on extrapolating the small and observable results of natural selection, operating on organisms in local populations, to render all evolutionary phenomena at all scales of time and effect. Weismann, and Darwin himself as he struggled to explain diversity, then considered hierarchical models of selection formulated very much in the spirit of modern versions. Weismann, after much internal debate, leading to eventual rejection of his previous commitment to the strict Darwinism of single-level selection on organisms alone, eventually advocated a full hierarchy of levels, explicitly citing this concept as the most distinctive innovation and centerpiece of his mature evolutionary views.

The internalist critique of adaptationism (the main subject of modern criticism on the second, or "creativity," leg of the tripod) boasts an even more venerable pedigree. I showed, in Chapters 4 and 5, how this critique defined the major difference between British (Paleyan) and continental versions of natural theology in preevolutionary days. I then demonstrated that the same division, transformed as the structuralist-functionalist dichotomy, served as a focus for evolutionary debate—pitting the functionalism (adaptationism) of such disparate theories as Darwinian selectionism and Lamarckian soft inheritance against the great continental schools of structuralism, as advocated by such scientists as Goethe (and most of the German *Naturphilosophen*), Geoffroy Saint-Hilaire (and the French transcendental morphologists), and, in a rare move across the Channel, in major themes of the complex and much misunderstood evolutionary views of Richard Owen. Finally, I traced the two

major lines of post-Darwinian internalist thought (orthogenesis and saltationism), and documented the continuity of this pedigree, after the Mendelian
rediscovery, in the macromutationism of Hugo de Vries and (combining both
strands of constraint and saltation) in the apostasy of Richard Goldschmidt.
Modern critiques of adaptationism rest upon an ancient legacy.

For the third leg of extrapolationism, as illustrated here by the surrogate
theme of Darwin's geological needs (for the other major aspects of extrapolationism fall into the theoretical domains of the first two essential postulates),
I showed, in Chapter 6, that classical catastrophism operated as good science,
and also represented the literal empiricism of its time. I then argued that
Lyell's uniformitarian victory arose largely as a triumph of skilled but dubious rhetoric. The aspects of catastrophism that posed the strongest challenges
to Darwin's ideas on the origin of macroevolutionary pattern never received a
convincing critique, and have now experienced a legitimate rebirth in modern
views on mass extinction.

Nonetheless, although each leg of the Darwinian tripod faces a venerable
indictment from the fullness of history, the path of modern reform surely does
not lie with these classical critiques, for each embodies fatal flaws in each of
two debilitating ways:

ANCHORS IN CULTURAL BIASES. The attempt to validate human superiority by the doctrine of progress identifies the heaviest burden imposed by
Western culture upon evolutionary views of all stripes. In their nineteenth
century versions, all three critiques of the essential postulates of Darwinism
sunk their major root (and fallacy) in the concept of progress. For the first leg,
the original hierarchical models of Lamarck and Chambers construed their
higher level of large-scale change as a force of progress orthogonal to a palpable cause of local adaptation. For both Lamarck and Chambers, the two
forces of general progress and local adaptation are not only geometrically orthogonal (upwards vs. sidewards), but also conceptually opposed, as the lateral force "pulls" lineages from their upward course into dead ends of local
specialization. For the second leg, most structuralist visions postulated an inherent increase of complexity and progress mediated by laws of form and internal principles of living matter. These internalist theories proved attractive
because, in contrast with the Darwinian contingency of shifting local adaptation, they offered more promise as validations for the great psychic balm of
progress. On the third leg, catastrophism might seem inherently opposed to
ideas of regulated and predictable increase in life's complexity, but the classical versions advocated an intimate connection between progressionism and
paroxysmal change—for pre-Darwinian catastrophists generally postulated
renewed faunas of increasing excellence after each episode of extinction. This
theme became such an intrinsic component of classical catastrophism that
many scholars now designate this movement as the "directionalist synthesis"
or as "progressionism," and not by the paroxysmal dynamics of catastrophic
change.

FAILURES OF LOGIC. The three critiques, in their nineteenth century
versions, are explicitly anti-Darwinian. That is, they propose alternative

causes of evolution that either deny natural selection entirely, or demote Darwin's preferred cause to insignificance. These alternative forces are, in any case, opposed to—clearly not synergistic with—Darwin's principle of natural selection at the organismic level.

The explicit, often vociferous, invocation of these critiques against Darwinism set the primary agenda for scientific debate from the very beginning of modern evolutionary theory. On the first leg, the Lamarck-Chambers tradition of a primary force of progress (effectively inaccessible to empirical study), opposed to a palpable cause of immediate adaptation (eminently operational for research, but decidedly secondary in significance), acted as a major spur to Darwin's development of a fully operational theory with causes working at a single and accessible level. On the second leg, internalist notions of orthogenesis and saltation denied creativity to natural selection and defined the major versions of late 19th century anti-Darwinism. On the third leg, classical catastrophism became Darwin's personal *bête noire*, an obstacle that he surmounted by allegiance to Lyellian uniformity. Later in Darwin's career, the appearance of a similar threat in a different guise—the claim for an earth too young to render the results of evolution by natural selection in a gradualistic mode—led Darwin to characterize Lord Kelvin as an "odious specter."

I believe that the historical tradition for using these critiques as supposed confutations of Darwinism has engendered a great deal of unnecessary and unproductive wrangling in our own time, as markedly different versions of the same critiques needlessly evoke old fears. I also believe that we can find the way to a better (and healing) taxonomy by following the lead of Kellogg's fine presentation (1907), already much praised in this book. Kellogg, as previously discussed (pp. 163–169), divided critical commentary about Darwinism into arguments "auxiliary to" and "alternative to" natural selection—enlargements and confutations, if you will. In the past, critiques of the Darwinian tripod have usually been advocated in Kellogg's alternative, or destructive, mode—and a tradition for quick (often ill-considered) and defensive reaction by Darwinians has developed whenever the critical buzz-words rise again: rapid change, group selection, mass extinction, directed mutation, for example. But all these critiques can also generate powerful versions in Kellogg's auxiliary, or helpful and expansive, mode—as Kellogg himself recognized when he classified Weismann's theory of hierarchy as one of the two most significant *auxiliary* propositions of his time.

The older, alternative mode of these critiques did lead to a series of dead ends, rightly rejected by the resurgent Darwinism of the Modern Synthesis. Older versions of hierarchy (the first leg) foundered in the mysticism of superorganisms and harmonious ecologies; constraint and laws of form (the second leg) became mired in invalid macromutationism or lingering orthogenesis; catastrophic geology (representing the third leg) languished in the failure of all proposed mechanisms for global paroxysm. The old versions, freighted by the cultural bias of progress, and rooted in false arguments for the demise of Darwinism, richly deserved the rejection they received.

But modern forms of these critiques are now being advanced in different and helpful versions within Kellogg's auxiliary mode—that is, as ideas to expand, while substantially changing, the Darwinian core. For the first leg, and most importantly, the hierarchical theory of multi-level selection retains Darwin's emphasis on the centrality of selection as a mechanism, but rejects the notion that the organismal level must hold nearly exclusive sway as a causal locus of change (while wondering if this conventional Darwinian level can even claim dominant status—Chapters 8 and 9). On the second leg, modern ideas of constraint and channeling deny the crucial isotropy of variation, so necessary to the logic of selection as the primary directional force in evolution, and therefore envision important roles for structural and internal causes as patterning agents of evolutionary change (Chapters 10 and 11). These internal channels work with selection as conduits for its impetus—that is, as auxiliary (not alternative) forces to natural selection. For the third leg, current notions of mass extinction do not challenge the Darwinian mechanism of selection *per se*, but suggest that any full explanation of macroevolutionary pattern must integrate the accumulated Darwinian effects of normal times with the profound restructurings of diversity that occur in environmental episodes too rapid or too intense for adaptive response by many species and clades (Chapter 12).

Therefore, in modern versions of the three critiques, classical Darwinism either becomes expanded (in the theory of hierarchical selection), or dynamically counterposed with other causal forces working in concert with selection to produce the patterns of life's history, either at a conventional microevolutionary scale (internal channels as conduits for selection) or as interacting regimes through geological time (mass extinctions and selective replacements). But we cannot fairly portray these expanded views as pure sweetness and light for orthodox Darwinism. Much that has been enormously comfortable must be sacrificed to accept this enlarged theory with a retained Darwinian core—particularly the neat and clean, the simple and unifocal, notion that natural selection on organisms represents *the* cause of evolutionary change, and (by extrapolation) the *only* important agent of macroevolutionary pattern.

On the first leg, the theory of hierarchical selection differs substantially from classical Darwinism in basic logic and concept—for explanations of both stability and change must now be framed as compound results of a balanced interaction of levels, working in all possible ways (in concert, in conflict, or orthogonally), and not as shifting optimalities built at a single level. On the second leg, an emphasis on constraints and channels implies a new set of operational concerns, and a revamping of the evolutionary research program. Internally imposed biases upon directions of change become a major subject for study—and the role of developmental patterns must again become prominent in evolutionary theory. (I must confess great personal pleasure in observing the rapid progress of this integration, as the wall between these two subjects seemed so frustratingly impenetrable when I published my first book, *Ontogeny and Phylogeny*, not so long ago in 1977.) On the third leg, a renewed appreciation for the shaping power of mass extinction must reinstate paleontology as a source of theory, and not merely a repository for the historical unfolding of processes fully illuminated by microevolutionary studies. Thus, the new theory produced by the confluence of these critiques and their integration with classical Darwinism will not emerge from a simple act of generosity or *noblesse oblige* by previous orthodoxy. The new theory may remain Darwinian in spirit (a "higher Darwinism," see Gould, 1982b), but its development requires a wrenching from several key assumptions of classical Darwinism—not simply a smooth evolution from conventional precepts—as embodied in both the tripod of essential theoretical support, and the methodology of uniformitarian extrapolation (the theoretical and methodological poles of Darwinism, as discussed in Chapter 2).

Substantial change in any domain usually follows such a scenario, and cannot unfold in smooth and untroubled gradualistic continuity. The venerable Hegelian triad of thesis→antithesis→synthesis may not adequately describe all examples of important change (see p. 23 for more on this general concept), but this classic philosophical model of tension and (often episodic) resolution seems more in tune with nature—or, to use the terminology of this book, higher in relative frequency among patterns of change. Human thought, unlike the evolution of life, does include the prospect of meaningful progress as a predictable outcome, especially in science where increasingly better understanding of an external reality can impose a fundamental organizing vector upon a historical process otherwise awash in quirks of individual personalities, and changing fashions of cultural preferences. Surely our views on the nature of taxonomic order have progressed (in the sense of better consonance with the true causes of diversity) from the eclecticism of Aldrovandi, to the coherent creationism of Linnaeus, to Cuvier's addition of a temporal dimension, to Darwin's evolutionary synthesis of space, form, and time.

The Hegelian triad proceeds by confrontation between old and new systems (thesis and antithesis), and by their melding into a novel theory preserving worthy aspects of both—synthesis. But the continuing interplay of confrontation and reconstitution does not spin in an endless circle, proceeding nowhere. Useful synthesis builds a transformed structure, and does not merely shuffle an unaltered deck (or raise an unstable house of cards precariously built from unaltered parts). Darwin constructed a powerful antithesis to older evolutionary views rooted in predictable progress and internal drives. Modern versions of the three critiques now present a worthy antithesis to the limitations of strict Darwinism. The second part of this book presents this Hegelian antithesis, written in the hope and expectation of synthesis and improvement. The synthesis that must eventually emerge will build a distinct theoretical architecture, offering renewed pride in Darwin's vision and in the power of persistent critiques—a reconstitution and an improvement, waiting for the next antithesis that must lead us onward to the next of many future syntheses in the wondrous, eternal play of mind and nature.

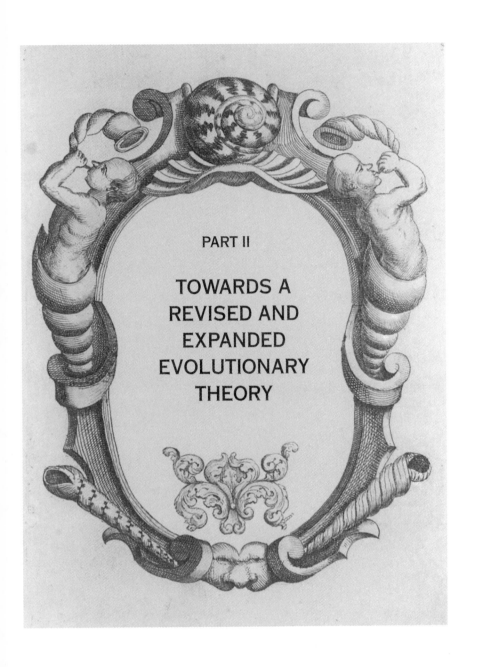

PART II

TOWARDS A
REVISED AND
EXPANDED
EVOLUTIONARY
THEORY

# CHAPTER EIGHT

# Species as Individuals in the Hierarchical Theory of Selection

## The Evolutionary Definition of Individuality

### AN INDIVIDUALISTIC PROLEGOMENON

The perceived excesses of the French Revolution may have sapped English enthusiasm for the tenets of Enlightenment Rationalism—the faith of Darwin's grandfather Erasmus. The subsequent romantic movement stressed opposite themes of emotion vs. logic, and national variety vs. universal reason. Charles Darwin, who revered his grandfather but also loved Wordsworth's poetry, received a firm grounding in both great philosophical and aesthetic traditions. He also—and perhaps as a direct result—maintained strong fascination for a central theme common to both movements, but for different reasons: the role of individuals as agents of change in larger systems. (The Enlightenment focussed on individuals as effective intellectual agents and inherent bearers of rights—"unalienable" in Jefferson's memorable phrase—and therefore as primary causal and moral agents in themselves, not as expendable items of a larger collectivity. The Romantics exalted individual effort as the motive force of social change through the actions of occasional heroes of higher sensibility.)

In any case, and whatever the deeper source, we do know that, as Darwin stitched together his theory of natural selection in 1838, he centered his major intellectual struggle in the few weeks before his "Malthusian" insight (Schweber, 1977) upon the role of individuals as primary causal agents of evolutionary pattern, even at largest scales (see full discussion in Chapter 2). He first studied the economic theory of Adam Smith through the major secondary source then available—Dugald Stewart's *On the Life and Writing of Adam Smith*. He expressed special fascination for Smith's distinctive notion that the overall optimality of an economy might best (and paradoxically) be fostered by allowing individuals to maximize personal profit without restraint (the doctrine known ever since as *laissez faire*, or "let do"—more roughly, "leave 'em alone"). He then read an extensive analysis of the work of the great Belgian statistician Adolphe Quetelet—particularly his central notion of *l'homme moyen* (average man), based on the aggregation of individual attributes into collectivities.

In the context of this conscious and directed search, we should not be surprised that Darwin's theory of natural selection rests upon the same central paradox that fueled Adam Smith's system: postulate a cause based on individuals ruthlessly pursuing their own benefits; an ordered polity will then arise as an incidental side consequence. No dismissal of Paley's omniscient God as the direct creator of general order could possibly have been more incisive, or more radical. We can therefore understand why Darwin insisted so strongly upon a single-level theory of natural selection—with struggle among individual bodies as a virtually exclusive locus of causality (see Chapters 2 and 3 for extended analysis). The downward shift of agency, from a purposively benevolent deity to the amoral self-interest of organisms, embodies the most distinctive and radical aspect of Darwinism.

Given Darwin's intense and conscious desire to restrict causality to competing organisms, I have been particularly struck, in researching and writing this book, by the inability of all the most diligent, and most thoughtful, early selectionists to make such a system work fully and consistently—despite intense and clearly focussed efforts to "cash out" Darwin's vision. As discussed in Chapters 3 and 5, only two early evolutionists fully grasped the meaning of natural selection, and the logic behind Darwin's restriction to the level of organisms. The first, August Weismann, defended Darwin's system with utmost zeal, as he spoke with pride about the *Allmacht* (all-might, or omnipotence) of natural selection. He began by advocating rigid adherence to Darwin's level of organisms. But, in fighting the resurgent Lamarckism of late 19th century thought, Weismann had to descend a notch to postulate important "germinal selection" at the level of hereditary units. Late in his career, he recognized the logical generality enjoined by his admission of a second locus—that selection can work on objects with requisite properties at any level of the genealogical hierarchy. Weismann therefore articulated a fully hierarchical model of selection operating at several levels both below and above individual organisms. Moreover, he developed this full theory not in retreat or as a hedge, but as a compelling extension of selection's central logic—and as further testimony to an *Allmacht* even more inclusive.

The second, Hugo de Vries, postulated a macromutational mechanism that logically precluded the production of new species by gradual selection of intrapopulational variation. In so doing, he committed intellectual parricide against his personal hero, Charles Darwin—and this mental act caused him great psychological distress. He assuaged his feelings of guilt, and showed his understanding of the abstract logic of selection, by insisting that he remained loyal to Darwinism—but at a higher level of selection among species, rather than among organisms (see pp. 446–451).

We must also not neglect the man who had invested the most effort in holding the line at organismal selection, and who had the most to lose if such a restriction could not work—Charles Darwin himself. Darwin struggled mightily to bring all evolutionary phenomena, including a host of apparently exceptional items from hymenopteran colonies to prevention of interspecific hybridization, under the umbrella of organismic selection, often with truly ingenious formulations—and he ultimately failed, as all others had. The logic of his argument led Darwin to postulate higher level selection at two crucial points (see pp. 127–137)—to explain the evolution of altruism in human societies by interdemic selection, and to encompass multiplication of species under his essential "principle of divergence" by a partial appeal to species selection. If none of the most rigorous and savvy early Darwinians could render evolution without some appeal to selection at levels higher than individual organisms, shall we not tentatively conclude that both the logic of the theory and nature's empirical record compel such an expansion, and the attendant notion of hierarchy?

## THE MEANING OF INDIVIDUALITY AND THE EXPANSION OF THE DARWINIAN RESEARCH PROGRAM

We may agree with the strictest formulation of agency in Darwinian theory: natural selection works by a struggle (actual or metaphorical) among individuals for personal reproductive success. In other words, selection occurs when properties of a relevant individual interact with the environment in a causal way to influence the relative representation of whatever the individual contributes to the hereditary make-up of future generations. If we place the theory's causal focus so squarely upon individuals as agents, then we might suppose that Darwin's unitary perspective must apply: all results at all evolutionary scales must cascade from the causal process of selection among individuals, defined in the conventional vernacular manner as the bodies of organisms.

But, as Hamlet said of the fears that prevent suicide (an adaptation, some would no doubt argue, for keeping humans viable as Darwinian agents), "ay, there's the rub." What is an individual? Are vernacular bodies the only objects in nature that merit such a designation—especially when discrete "bodiness" doesn't always define an unambiguous individual at the focal level of Darwin's intent (not to mention the difficulties encountered in trying to characterize entities at levels above and below bodies in the genealogical hierarchy of nature)?

For example, biologists spent more than a fruitless century trying to decide whether the parts of siphonophores are "persons" in a colony or organs of an organism—only to recognize that the question cannot be answered because both solutions can justly claim crucial and partial merit (see Gould, 1984d). Are grass blades or bamboo stalks bodies in their own right (as some aspects of functional organization suggest), or parts (called ramets) of a larger evolutionary individual (called a genet)? Do our feelings about definition shift when ramets become spatially discrete and therefore look just like conventional bodies—as in the parthenogenetic offspring of an aphid stem-mother (designated, in their totality, as a single EI, or evolutionary individual, by Janzen, 1977)? And what shall we do with discrete bodies that maintain some genetic variation among themselves (and cannot, therefore, form a set of identical ramets), but operate together as differentiated items (analogs of organs) in a larger "totality" like a beehive or ant colony with a single queen? Wilson and Sober (1989) have urged a revival for the old concept of "super-organism" in such circumstances.

As so much uncertainty surrounds the issue of how we define an "individual" at the supposedly unambiguous level of Darwin's own intent, we should not be surprised that attempts to restrict the concept to organic bodies have yielded more confusion than resolution. Perhaps we should try a different and more general approach. Perhaps we should attempt to specify a set of minimal properties required to designate an organic entity as an "individual"—and then ask whether any objects at levels above or below traditional bodies possess these properties, and therefore qualify for inclusion under an expanded concept of individuality. If so, we might obtain a useful definition divorced from the happenstances of scale, and therefore sufficiently general to provide a deeper (and clearer) understanding for this central concept in Darwinism.

This subject has generated an enormous and often confusing literature throughout the history of Darwinian thought—and more so than ever before during the past twenty years. Some colleagues may wish to throw up their hands and brand the entire enterprise with labels usually invoked pejoratively by scientists—merely "semantic" or "philosophical." Indeed, several of the finest contemporary philosophers of science have devoted considerable attention to this issue—see, for example and in alphabetical order, Brandon (1982), Hull (1980), Lloyd (1988), Sober (1984), and Wimsatt (1981). But I believe that both the volume and the confusion arise for two reasons that compel primary attention to the subject: the issue is both exceedingly difficult and enormously important.* The best scholars tend to gravitate to the most fascinating and portentous questions—and the confluence of extensive consideration by the most prominent philosophers of science (as mentioned above) and the most thoughtful evolutionary biologists from early days (Darwin, Weismann, de Vries, as discussed above) to current times cannot be accidental or wrong-headed. As a testimony to this current concern, and again in alphabetical order, I cite as a small sample: Arnold and Fristrup

---

*I have struggled with this issue all my professional life, and have often wondered why the questions raised seem so much more recalcitrant, and so much more cascading in implications, than for any other major problem in Darwinian theory. I don't think that mere personal stupidity underlies my puzzlement—or rather, if so, the mental limitations must be largely collective, because other participants share the same struggle and express the same frustrations. I don't mean to sound either grandiloquent or exculpatory, but I seriously wonder if some of the difficulties might not arise largely from limitations in the common mental machinery of *Homo sapiens*. Levi-Strauss and the French structuralists may well be correct in holding that human brains work best as dichotomizing machines at single levels. We make our fundamental divisions by two (nature and culture or "the raw and the cooked" in Levi-Strauss's terms, night and day, male and female), and we therefore experience great mental difficulty with continua, and with any system other than a two-valued logic (hence Aristotle's law of the excluded middle, and other similar guides). We are especially ill-equipped to think hierarchically, and to juggle simultaneous influences from several nested levels upon the foci of our interest. The hierarchical theory of natural selection rests upon all these intrinsically difficult modes of reasoning.

(1982), Dawkins (1976, 1982), Eldredge (1985a), Fisher (1958), Ghiselin (1974a and b), Leigh (1977), Lewontin (1970), Maynard Smith (1976), Stanley (1975, 1979), Vrba (1980; Gould and Vrba, 1982; Vrba and Gould, 1986), Williams (1966, 1994), D. S. Wilson (1983), and Wright (1980). Collaborations between philosophers and biologists have also added to the interest (for example, Wilson and Sober, 1994; Sober and Wilson, 1998; Lloyd and Gould, 1993; Gould and Lloyd, 1999).

Discussion of this most difficult and most important subject may be organized in a hundred different ways. I have chosen a point of entry that may seem peculiar or indulgent as an abstract philosophical question tenuously related to the "real" biology of organic objects: are species individuals or classes? As a twofold justification for this strategy, I found, first and personally, that I could best organize this material and place all subjects into logical sequence, if I started here and worked systematically outward through a particular net of implications. (Others, no doubt, would choose different beginnings and construct just as sensible and comprehensive a sequence.) Second and collectively, this particular philosophical question has been widely and passionately discussed in the biological literature, and has struck several scientists (e.g. Eldredge, 1995) as a potential centerpiece unwisely relegated to a peculiar periphery by many scholars.

In 1974, Michael T. Ghiselin published an article in *Systematic Zoology* under a title that I found insufferably self-indulgent at the time (especially since his manuscript directly followed my own densely empirical article on local geographic variation in the land snail *Cerion bendalli* on the Bahamian island of Abaco), but have since come to view as adequately justified: "A radical solution to the species problem." In short, Ghiselin argued that many classical problems about species (not primarily or especially related to this chapter's topic of levels in selection) could be instantly resolved if we—in the Pauline manner of "scales falling from the eyes"—reversed our customary definition of species as classes (or universal categories that can "house" objects) and reconceptualized them instead as individuals (or particular things). A species then becomes a singular item—an *evolutionary* entity defined by both a unique historical genesis and a current particular cohesion.

I will not trace the large and complex trail that Ghiselin's proposal generated in the scientific and philosophical literature (see, for example, Ghiselin, 1987). In my reading and understanding, I do not think that any clean resolution can be stated, or even any consensus described. Perhaps we might best acknowledge, with Mayr (1982a and b), that the term "species," as conventionally used and understood, includes statements about both classes and individuals. In this sense, the extensive discussion of Ghiselin's proposal sharpened our thinking, but provided no closure.

In another sense, however, and following a common (largely sociological) pathway in science, the explicit airing of such an interesting theme launched, or at least impacted in major ways, a substantial set of theoretical issues, including two of central importance to this book: the nature of evolution as a historical discipline, and the definition of individuality as crucial to the "units of selection" problem. In his initial article, Ghiselin (1974b, p. 543) dimly perceived the key implication for hierarchical selection if species be construed as individuals rather than classes, and selection (by Darwin's definition) works on individuals—namely, that selection must also operate among species-individuals (and, by extension, potentially at several levels in a hierarchy of units, each properly construed as an "individual"). But Ghiselin did not complete his argument and grant full *evolutionary* individuality to species. "Species are units, and they have evolutionary importance, but the same may be said of organisms. Doubtless both organisms and species specialize. And probably organisms become adapted but species do not, except in so far as they consist of adapted organisms" (Ghiselin, 1974b, p. 543).

David Hull (1976), in the first major philosophical extension of Ghiselin's proposal, firmly linked the concept of species as individuals to the older issue of units (or levels) of selection, thus properly tying the rationale for a causal theory of hierarchical selection to the generalization of Darwin's key insight that selection can only operate by the differential reproductive success of "individuals": "Entities at various levels of organization can function as units of selection if they possess the sort of organization most clearly exhibited by organisms: and such units of selection are individuals, not classes" (Hull, 1976, p. 182). In his important later article—the *locus classicus* of the pivotal distinction between "replicators" and "interactors" (see next section of this chapter)—Hull then added (1980, p. 315): "Individuality wanders from level to level, and as it does, so too does the level at which selection can occur."

If the rationale for a hierarchical theory of selection resides in the expansion of "individuality" to several levels of biological organization (see Gould and Lloyd, 1999), then we must specify a set of criteria that any material configuration must meet to merit designation as an "individual." We may, I think (see Gould, 1994), most usefully divide these criteria into two categories: (1) requirements in ordinary language for granting individuality to any configuration (vernacular criteria); and (2) requirements in Darwinian theory for regarding any entity as an evolutionary individual, or potential agent of selection (evolutionary criteria). (I trust that, despite a traditional ethos contrary to such an admission, all thoughtful and self-introspective scientists no longer feel threatened or disloyal in acknowledging that all definitions must be theory laden—see Kuhn, 1962, for the classic statement.)

We must also resolve one other terminological confusion before listing the criteria of individuality. What word shall we use as a general term for the discrete "thing" that can serve as a unit of selection at various hierarchical levels? In an important article, a manifesto for reviving interest in the power of group selection and the validity of the hierarchical model in general, D. S. Wilson and E. Sober (1994) suggest that we use the term "organism" for the generality (and therefore speak of "species organisms," "gene organisms," and so forth), while restricting the word "individual" to organic bodies (you, me, the oak tree, and the barnacle) at the conventional level of Darwinian concern. They choose this definition because they emphasize—I would say overemphasize (see Gould and Lloyd, 1999)—functional cohesiveness among their general criteria of "thingness," a property better captured by "organism" than by "individual" in vernacular English.

I strongly urge the opposite and more conventional solution. This issue, I fully recognize, only concerns words, not the empirical world. But we get so muddled, and waste so much time, when we fail to be clear about words and definitions—especially when various scholars use the same word in different, or even opposite, ways—as in the classic confusion generated when molecular biologists began to use "homology" for the percent of similarity in genetic sequence between two organisms, rather than for the well-established and entirely different concept of joint possession due to common ancestry. Fortunately, in this case, we evolutionists have apparently managed to persuade our molecular colleagues to respect conventional usage, and to call their important concept "sequence similarity," or some such.

No one would create such a muddle on purpose, but this particular confusion already exists—and some common ground must therefore be established if we wish to address this growing and important literature without a perennial need to stop, translate, and bear linguistic idiosyncrasies continually in mind whenever we read a paper. At the moment, most authors use "organism" for the Darwinian body (me and thee), and "individual" for the generalized unit of selection at any hierarchical level—while others (like Wilson and Sober) employ reversed definitions.

I strongly urge the former course—organisms as conventional vernacular bodies, and individual for the generalized term—for two reasons. First, this decision represents more common usage, both in vernacular English and among biologists. (Several academic departments include the phrase "organismic biology" in their title to defend a continuing focus on entire bodies against molecular claims for hegemony. But if genes are organisms as well, this ploy will not work!) Second, the technical definition of an "individual" in academic philosophy, and the spread of this terminology in the large literature inspired by Ghiselin's arguments for species as individuals (summarized above), grants priority to "individual" as the general term, and "organism" as the restricted body. Ghiselin (1974b, p. 536) clearly defended this usage in his original definition: "In logic, 'individual' is not a synonym for 'organism.' Rather, it means a particular thing." And Hull (1976, p. 175) explicitly labeled the application of "organism" to higher-level objects as misleading because vernacular language so strongly equates "organism" with discrete bodies. He then urged "individual" as the general term, as advocated here: "From the point of view of human perception, organisms are paradigm individuals. In fact, biologists tend to use the terms 'organism' and 'individual' interchangeably. Thus, biologists who wish to indicate the individualistic character of species are reduced to terming them 'superorganisms.' The same claim can be expressed less misleadingly by stating that both organisms and species are individuals."

In discussing criteria of individuality, I will focus on species as paradigms for higher-level evolutionary entities for two reasons: (1) because I believe that a proper theory of macroevolution, the central concern of this book, rests upon such a proposition; and (2) because species seem so maximally unlike discrete "things" to many biologists, thus making the correction of this false impression a prerequisite for accepting the full hierarchical model of selection. But species can claim no favored status in the hierarchical model, and I use them here only as an example—so that the argument may then proceed to a full set of levels, each characterized by a valid kind of individual acting in a distinctive way.

### Criteria for vernacular individuality

When we apply the term "individual" in ordinary English, we envisage a set of properties centered upon uniqueness, discreteness, functionality, and cohesion considered both spatially and temporally. To be a unique "thing," and not just a part of a continuum, a named object must clearly begin and end— and must remain its definable self throughout a continuous existence. We may, I think, best summarize this intuition in three criteria. To be called an individual, a material entity must have:

- a discrete and definable beginning, or birth;
- an equally discrete and definable ending, or death; and
- sufficient stability (defined as coherence of substance and constancy of form) during its lifetime to merit continuous recognition as the same "thing."

I realize, of course, that the third criterion amalgamates several crucial notions into a single statement. We might specify at least four properties involved in our ordinary concept of "sufficient stability" for individuality:

CHANGE. An individual may undergo some, even substantial, change during its lifetime, but not so much either to become unrecognizable or to encourage redefinition as a different thing—and, particularly, for temporal sequences of individuals, not so much alteration that late stages come to resemble the next-named individual in a sequence more than the early stages of the same individual.

DISCRETENESS AND COHESION. An individual must maintain clear and coherent boundaries during its lifetime. Parts should not "ooze out" into other individuals, while components of other individuals should not enter and become incorporated.

CONTINUITY. An individual cannot fade in and out of existence during its lifetime, but must maintain material continuity throughout. Members of classes, on the other hand, are not so constrained, for classes are defined by common properties, not by historical continuity. As Hull (1980) argues, the class of gold atoms does not require continuity or filiation. If all gold disappeared, its position on the periodic table would remain—and an element later reconstituted with the right atomic particles and requisite properties would still, and legitimately, be called gold. But if all peacocks die, the species-individual *Pavo cristatus* disappears forever. Even if some human engineer retained an electronic record of the entire *Pavo cristatus* genome, and future technology permitted chemical reconstitution from nucleotides, we couldn't call the resurrected creature a member of *Pavo cristatus*, even if the reconstituted object looked and acted like an extinct peacock of old.

FUNCTIONALITY, OR ORGANIZATION. We expect that, at least in some crucial ways, the parts of an individual will work together so that the individual functions in a distinctive and cohesive way. This criterion, though crucial as we will see to the second set of evolutionary criteria, may be the least important (perhaps even dispensable) for vernacular definitions. If a bounded object maintained all the other listed properties, but failed to do anything as an entity (and acted, instead, largely as a repository of separate parts), we would still call the object an individual, however inert and uninteresting.

Conventional organisms certainly possess all these properties—as well they must, for the bodies of complex animals established our vernacular Western paradigm for the general concept of individuality. Yet note that, even here, at the point of maximal clarity, some fuzziness and indefiniteness plague every criterion. Consider human bodies, the inevitable exemplar of the paradigm. Our lives have reasonably discrete beginnings—but if a true moment could be defined without ambiguity, then our social and political debates about abortion would require new terms and engage different issues. Death might seem even more definable and momentary—but, again, fuzziness and ambiguity plague our definitions, leading to complex, and often heart wrenching, medical and legal wrangles. Perhaps our bodies pass the criterion of "sufficient stability" with more clarity. We don't fuse with others, or rise again from the dead (at least in a material world that science can adjudicate). We are certainly designed to operate discretely, even if our actions be dysfunctional. All our chemicals, and most of our cells, undergo periodic replacement—but I remain myself and continue to look sufficiently like my baby pictures (though not much like my early embryonic form with tail and gill slits!).

So organisms surely pass muster as individuals. But we encounter problems, including several classic issues subject to endless discussion in the literature, when we try to assign individuality at other (particularly higher) levels of the organic hierarchy. For example, the standard objection to interdemic selection (see pp. 648–652) holds that too many demes fail the criterion of "sufficient stability"—for they may not persist long enough to matter in evolution, and their borders may be too "leaky" as organisms move in and out in the absence of reproductive isolation between the parts (organisms) of different demes. All too frequently, demes may operate, in Dawkin's apt metaphor (1976, p. 36), "like clouds in the sky or dust storms in the desert . . . temporary aggregations or federations."

Defenders of classical "group" (i.e. interdemic) selection recognize these problems of course, and all workable models have been purposely constructed to overcome such objections by specifying conditions that will permit demes to fulfill the defining criteria listed above. In fact, the classical empirical issue of our literature on group selection asks whether demes can "hold together" long and discretely enough so that the differential proliferation and survival of some demes vs. others can propel the general increase of an "altruistic" allele (promoting demic success), even while the allele's frequency declines within groups as "selfish" alternatives prevail in conventional natural selection among bodies. If demes can "hold together" by this operational criterion of evolutionary outcomes, then they possess "sufficient stability" to be regarded as individuals on functional grounds within selectionist theory.

Traditionally, biologists have not been willing to imbue species with these requisite criteria for individuality. Species, in an argument dating to both Lamarck and Darwin, have often been construed as mere names of convenience attached to segments of evolving continua without clear borders. Under this gradualistic and anagenetic view (see Fig. 8-1), a species near the end of its arbitrary existence must be phenotypically more similar to a forthcoming descendant than to the initial ancestor. (Indeed, under strict gradualism, we even face the definitional absurdity that the last generation of an ancestor should be reproductively isolated from its own offspring—that is, the first generation of a new descendant. Some creatures may eschew such incest on moral or adaptive grounds, but no one would gainsay the biological possibility.) Thus, on this traditional view, species cannot maintain sufficient temporal stability to be called individuals. In addition, species do not have discrete birth points if they branch from their ancestors at rates no different from characteristic tempos of transformation during their subsequent anagenetic lifetimes (see Fig. 8-1). At most, some species display clear termination in extinction (but others evolve gradually to descendants.) Thus, species do not function as good vernacular individuals if gradualism and anagenesis pervade the history of life.

Even so—or as long as most species arise by splitting of lineages rather

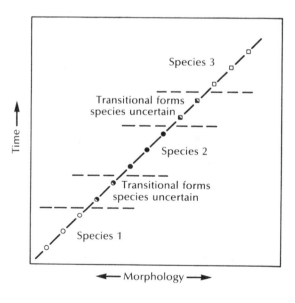

8-1. The traditional view (as depicted, but not defended, in Lipps, 1993) of why species cannot be construed as proper biological individuals but only as arbitrary segments of a smooth and unbreakable continuum.

than by wholesale transformation, no matter how gradual the tempo of branching—the individuality of species may be maintained in some technical sense, though only by violation of our vernacular intuitions. After all (see Fig. 8-2), so long as branching points (or fuzzy intervals) can be temporally located at all, then species do have definable intervals of existence, and can be individuated on this basis, even if their life courses violate our usual notions about sufficient morphological stability.

Many evolutionary biologists have failed to recognize that the so-called cladistic revolution in systematics rests largely upon this insistence that species (and all taxa) be defined as discrete historical individuals by branching (leading to the rule of strict monophyly)—and not as classes with "essential" properties by appearance (leading to the acceptance of paraphyletic groups). Many biologists reject (and regard as nonsense) the cladistic principle that no species name can survive the branching off of a descendant—and that both branches must receive new names after such an event, even if the ancestral line remains phenotypically unchanged. But this counterintuitive rule makes sense within cladistic logic—for cladists define new entities only as products of branching (the word *clade* derives from a Greek term for *branch*). A transforming species that does not branch cannot receive a new name even if the final form bears no phenotypic resemblance or functional similarity to the original ancestor. Thus, if such extensive transformation occurs in unbranched lineages, a cladist, by failing to designate a truly different anatomy with a distinctive name, retains the technical individuality of species at the price of a severe assault against legitimate intuition.

Can we find any solution to this dilemma? Must we either deny that species can be viewed as individuals, or else accept a logically "pure" definition based on branching, but strongly in violation of vernacular usage? I suggest that this issue can be resolved empirically, and need not persist as a definitional or philosophical conundrum. If gradualism and anagenesis prevail in

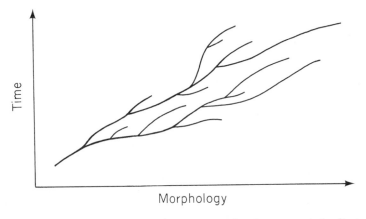

8-2. A repeat of Figure 7-2 to show that, even under the most gradualistic and anagenetic models, species can still be individuated under a conception of evolution as a branching process at the species level.

nature, then all the aforementioned problems cannot be avoided and apply in force. But suppose, as Eldredge and I have long argued in our theory of punctuated equilibrium (see Chapter 9), that gradualistic anagenesis occurs only rarely in nature, and that the great majority of species remain essentially stable throughout their geological lifetimes. (Our concept of stasis recognizes that species fluctuate mildly throughout time, to an extent no different from ordinary geographic variation among demes of a species at any one moment, but we hold that mean values of phenotypes generally do not change in a cumulative or directional manner.) Suppose also that species, on geological scales, branch in unresolvable "moments." (In nearly all geological circumstances, single bedding planes amalgamate the events and accumulated results of several thousand years.) If species tend to originate in thousands to tens of thousands of years—that is, with glacial slowness by the inappropriate criterion of a human lifetime in potential observation—and then to persist in stasis for millions of years, their origin becomes instantaneous in geological time, and species arise as discrete individuals at this proper macroevolutionary scale. Of course, some fuzziness must attend the origin of a species, for we acknowledge that macromutational beginnings in leaps of a single generation rarely, if ever, occur. But when "fuzziness" occupies only a thousand years in a million—that is one tenth of one percent of later existence in stasis—then geological indefiniteness surely does not exceed even the relative duration of the fuzziness (9 months in some 80 years) attending the embryological beginning of human personhood!

Under punctuated equilibrium, the remaining criterion of discrete death achieves even clearer definition—for nearly all species disappear by extinction ("living on" only through their progeny of daughter species with new names and individualities), and not by gradual bodily transformation into something else. Species deaths, at geological scales, are surely more discrete and "momentary" than human deaths scaled against the lengths of our lifetimes.

In summary, then, species that originate by branching can be individuated even under the assumption that gradualistic anagenesis prevails during the history of most species lifetimes (but only by violating our vernacular conception of "personhood" or individuality). However, if punctuated equilibrium prevails as an empirical proposition (see Chapter 9 for defense of this contention), then species are individuals—in some cases much "better" individuals than conventional bodies of organisms—by all vernacular criteria. Under punctuated equilibrium, species originate at points of birth with initial fuzziness confined to an insignificant (usually unmeasurable) moment properly scaled against later existence in stasis. They experience even clearer moments of death, for nearly all species terminate by true extinction and not by transformational passage into a descendant that vernacular (non-cladistic) usage will wish to recognize with a different name (a phenomenon called "pseudo-extinction" by paleontologists). And species surely maintain "sufficient stability" during their geological lifetimes by all criteria outlined on page 602. They remain discrete by reproductive isolation (conventionally cited, ever since Buffon, as the chief criterion of "specieshood"). They function as a unit and persist continuously. Above all, they do not change substantially in phenotype—the crucial concept of stasis. Surely, the average species in stasis undergoes less temporal change (with less directionality) than human bodies experience in our passage from babyhood through adult vigor and into senescence. If humans retain discrete personhood through all these slings and arrows of ontogeny, then species (under punctuated equilibrium) function as equally good or even better individuals by the same criteria of vernacular definition.

In describing exceptions and fuzzinesses in the application of these vernacular criteria to organisms, and acknowledging that species face the same difficulties of definition, Hull (1976, p. 177) wrote: "However, exactly the same questions arise for both. If organisms can count as individuals in the face of such difficulties, then so can species." But Hull assumed that these common problems plague species far more intensely than they threaten organisms. I would suggest that the opposite situation may prevail in nature: species may be even better individuated than organisms when punctuated equilibrium applies (and we consider species at their appropriate scales of geological time). This issue unites these two chapters in a crucial link between the theory of punctuated equilibrium (Chapter 9) and the classical debate about "units" or "levels" of selection (Chapter 8)—a conjunction that underlies my views on the importance and validation of macroevolutionary theory.

Interestingly, albeit through a glass darkly, Hull (1976) grasped the logical link between the phenomenology of punctuated equilibrium and the definition of species as individuals in his first important paper on this subject—even though he had not, by this time, encountered our empirical and theoretical arguments for such a pattern (Eldredge, 1971; Gould and Eldredge, 1971; Eldredge and Gould, 1972). (In his more inclusive review of 1980, Hull then explicitly joined our particular claims to the defense of species as individuals.) Hull begins by stating the problem (1976, p. 185): "Earlier I described individuals as reasonably discrete, spatiotemporally continuous and unitary entities individuated on the basis of spatiotemporal location rather than similarity of some kind. But one might object that species lack these characteristics. For example, in most cases new species arise gradually."

Hull then recognized that some neontological models of speciation accelerate the rate of branching relative to the supposedly standard rate of anagenesis within species—and that such an acceleration will sharpen the definability of species by the criterion of discrete birth: "But there are processes in nature which serve to narrow the boundaries between ancestral and descendant species . . . The end result is that the number of organisms intermediate between the ancestral and descendant species is reduced considerably" (1976, p. 185). Finally, Hull stresses the important point that all individuation, at any appropriate scale, entails some fuzziness at the boundaries—and that species therefore need not be construed as "worse" individuals than bodies (1976, p. 185). "If processes similar to those just described are common in nature, then the boundaries between ancestral and descendant species can be narrowed considerably, though not to a one-dimensional Euclidean line. But, of course, the replication of organisms does not happen instantaneously either. If absolutely discrete boundaries are required for individuals, then there are no individuals in nature. It is only our relative size and duration which make the boundaries between organisms look so much sharper than those between species."

But, to continue the Euclidean metaphor, and using an appropriate ruler with (say) a minimally noticeable geological increment equal to 10,000 years, the boundaries of many species do become momentary under punctuated equilibrium. Stasis persists for a long run of increments. At a commonly observed duration of 5 to 10 million years for marine invertebrate species in the fossil record (Raup, 1985; Stanley, 1985), one thousand increments of stasis would represent the geometry of a species lifetime, while even a million for the much shorter average duration of terrestrial mammalian species yields 100 increments. By comparison, many (probably most) events of speciation unroll *within* a single increment—leading to abrupt and momentary origin at geological scales, and the right-angle convention that has become standard for plotting the emergence of species under punctuated equilibrium (see Fig. 8-3).

### Criteria for evolutionary individuality

The vernacular criteria discussed above provide necessary, but obviously insufficient, conditions for identifying an entity as an evolutionary "individual" with the capacity to act as a causal agent in a process of Darwinian selection. Most unambiguous vernacular individuals cannot operate as Darwinian actors. The earth, for example, surely merits designation as a well-defined individual—with a specifiable birth (perhaps attended by some initial fuzziness as a primordial fireball), sufficient stability over billions of years (including enough climatic homeostasis to provide a stage for the history of life), and a forthcoming rapid death (presumably by absorption after the sun burns out some five billion years from now, and expands in diameter at least to the orbit of Jupiter). But the earth remains "infertile" in the crucial Darwinian sense of reproductive potentiality. Planets do not have children, and therefore cannot function as Darwinian individuals.

I do not cite this example to win an argument by ridicule, but rather to emphasize, once again, that all definitions must be embedded within theories. Mere vernacular individuality does not suffice for identification as a causal actor in Darwinian theory. Evolutionary individuality (or, more strictly, Darwinian individuality, for different theories of biological change may entail other criteria) requires an additional set of attributes rooted in two features of Darwin's world: the genealogical basis of evolution as a branching tree, and the causal efficacy of selection as the leading process of evolutionary change.

REPRODUCTION. Darwinian individuals must be able to bear children. Biological evolution is defined as a genealogical process. Darwinian evolution operates by the differential increase of your progeny (or whatever you pass into future generations) relative to the progeny of other individuals within the larger entity of your membership.

INHERITANCE. Your children must, on average, be more like you than like other parents of your generation—so that evolution may proceed by the differential increase of your own heritable attributes (a requirement of Darwinian systems, not of all conceivable evolutionary mechanisms). In other words, a principle of inheritance must prevail to permit the tracing of genealogical patterns—so that the relative reproductive success of ancestors may be assessed.

VARIATION. This criterion lies so deeply, and so fundamentally, within the constitution of Darwinism as a revolutionary ontology (and not just as a theory of evolution), that we should, perhaps, not even list variation as a separate criterion, but merely state that this conception underlies all Darwinian thinking. We can hardly imagine a more radical restructuring of the material

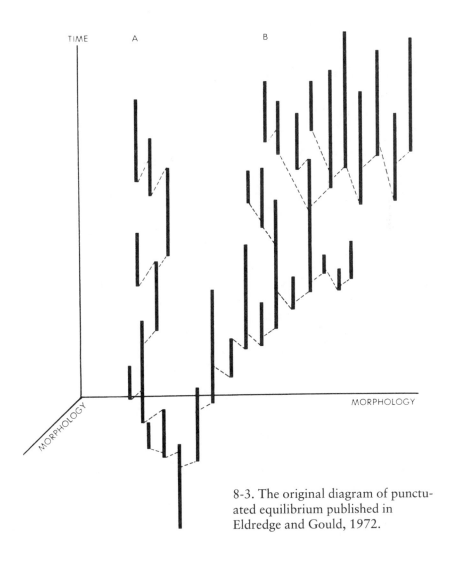

8-3. The original diagram of punctuated equilibrium published in Eldredge and Gould, 1972.

world than the Darwinian shift to variation among members of a population as an ultimate and irreducible reality (see Mayr, 1982b; Gould, 1996a)—a reversal of the old Platonic notion that essences (approximated empirically by measuring mean values, or by trying to construct an abstract ideal form and then searching for a closest actual embodiment) define the nature of things, and that variation among actual individuals (organisms in populations, in our most relevant example) can only be construed as "accidental," and judged by relative departure from a materially unattainable ideal.

Heredity and reproduction work in concert with variation to empower Darwinian selection in genealogically recognizable lineages. The failure of any criterion debars Darwinian evolution as a genealogical process. An absence of reproduction, for example, enforces an oddly limited form of "evolution" restricted to rules (or vagaries) of change within one or a number of individuals, all separately constructed at the outset. Vernacular usage, in fact, does apply the term "evolution" to some nongenealogical systems of this sort—as in the "evolution" of stars along the H-R sequence. But the causes of such systematic temporal changes, unfolding predictably under laws of nature (and not by the contingencies of variational history), differ so profoundly from Darwinian evolution that we really should insist upon different words for these maximally disparate modes of history (Gould, 2000a). (A great burden of misunderstanding, in both popular and professional cultures, must be ascribed to our confusing use of common terminology for such different causes. Many interested laypeople feel that biological evolution must unfold by internal necessity just as stars follow their predictable sequences and as galaxies expand following the big bang. And many professional evolutionists, suffering from the common affliction of physics envy, and immured in the reductionistic biases of Western scientific culture, have tried to find progressive patterns directly imposed by natural law, where Darwinian contingency actually reigns.)

An absence of variation also stymies Darwinian change by eliminating the raw material or substrate for any selective mechanism. Evolution in nonvarying populations might be treelike and genealogical, but such a process could not be Darwinian. One would have to imagine some very unearthlike way to generate change and diversity—for example, random dispersal of initially identical creatures to varying environments, followed by a Lamarckian or directly inductive process of heritable environmental stamping upon all members of a population.

Variation without heredity (that is, an absence of correlation between properties of offspring and parents) also stymies Darwinian causality. Selection could occur in a single generation. That is, the biggest or the ugliest might outreproduce all others, or even ruthlessly murder all small and beautiful conspecifics—but to what evolutionary avail, if the offspring of survivors then reconstitute all the original variation in original proportions? If variation occurred without correlation to parental constitution, but with inherent bias in a given direction—so that even random mortality produced a trend—

then evolution would occur. But we have always labeled such styles of internally-directed change as non-Darwinian, with Lamarckism as a primary and historically most influential example.

INTERACTION. At each level, the varying individuals of an evolving population (organisms of a deme, demes of species, species of a clade) must interact with the environment in such a way that some individuals achieve relatively greater reproductive success as a causal result of heritable properties manifested by these fitter, and not manifested (or not as effectively expressed) by less fit individuals. This causal claim embodies the key feature of natural selection as an active process. In other words, we must be able to devise a testable causal scenario about why the differential possession of certain heritable properties yields increased reproductive success.

These statements inevitably engage the crucial issue of whether we should define selection by this causal interaction of individuals and environments, or by the product actually transmitted to future generations (see next section). The logic of Darwinism dictates that the form of heredity's product—however fascinating in variety across nature's scales—cannot specify agency of selection. Interaction with environment defines agency (Lloyd and Gould, 1993; Gould and Lloyd, 1999)—and agents must be individuals (by both vernacular and evolutionary criteria). Some interacting individuals (like genes) usually pass faithful copies to the next generation. Others (like species) pass inevitably modified copies that are still more like themselves than like any other individual at their level. Still others (like sexual organisms) disaggregate their personhood and pass hereditary pieces and particles.

All these different strategies for hereditary passage permit us to recognize interacting individuals as causal agents of Darwinian selection. The special and unusual tactic of sexual organisms may seem curiously indirect (and we all know the enormous and confusing literature devoted to this subject), but disaggregation works as well as relatively faithful passage, so long as the essential Darwinian imperative remains in force: that is, so long as selectively successful individuals manage to bias the next generation with relatively more of their own hereditary material—however that material be passed or packaged. The "goal" of natural selection cannot be defined by faithful replication, but rather by relative "plurifaction," or "more-making."* The individual that plurifies by increasing the percentage of its contribution to the heredity of the next generation (however the units or items of heredity be constituted) gains in the evolutionary game. And we call the game Darwinian if plurifaction occurs by a causal interaction between properties of the successful individual and its environment.

---

*In the late 1970's and early 1980's, I engaged in long and vociferous arguments with my graduate students Tony Arnold and Kurt Fristrup about the criteria of species selection. (As discussed on pp. 656–670, I now believe that they were right, and I was wrong.) In the course of these discussions, we developed this idea and name of "more-making" or *plurifaction*. (If manufacture means, literally, making it by hand, and petrifaction means turning it into stone; then plurifaction simply means making more of it.) I do not now remember who first devised the word, or who contributed most to the concept's codification.

As for the vernacular criteria previously discussed (see pp. 602–603), these specifically evolutionary criteria teach us that organisms are not the only individuals capable of acting as units of Darwinian selection. In particular, and continuing to use species as a "type" example of individuality at higher levels, all evolutionary criteria apply to the species as a basic unit of macroevolution. Species have children by branching (in our professional jargon, we even engender these offspring as "daughter species"). Speciation surely obeys principles of hereditary, for daughters, by strong constraints of homology, originate with phenotypes and genotypes closer to those of their parent than to any other species of a collateral lineage. Species certainly vary, for the defining property of reproductive isolation demands genetic differentiation from parents and collateral relatives. Finally, species interact with the environment in a causal way that can influence rates of birth (speciation) and death (extinction).

As a further benefit for thus codifying the criteria of evolutionary individuality, we can immediately cut through the foolishness surrounding several distressingly common, but artlessly and rather thoughtlessly contrived, claims (or, rather, loose metaphors) about the Darwinian character of large items in nature—an attractive idea for many people, particularly for romantics and "new-agers" who yearn for meaningful agency at the highest levels. We can dismiss these claims because the object hypothesized as an agent of selection fails several crucial requirements for designation as an evolutionary individual. As an obvious example, many proponents of the so-called Gaia hypothesis wax poetic about the earth and atmosphere as a homeostatic system robustly balanced by interaction with life to secure and stabilize the conditions required by organisms for diversification and geological persistence. Supporters often assume that such functional coherence must make the earth sufficiently like an organism to merit designation as a living entity. Some have even stated that the earth must therefore be recognized as the largest and most inclusive product of Darwinian selection—or even that the earth should, in fact, be viewed as a true Darwinian individual. This woolly notion confuses a gut feeling about functionality or adaptive "optimality" (for support of life) with the requirements of Darwinian agency. The earth does not generate children, and did not arise by competitive prowess as the sole survivor among defeated brethren (who must have died or been expelled, I suppose, from the solar system long ago). Therefore, among a plethora of other reasons, the earth cannot be construed as a Darwinian agent or unit of selection.

More plausibly, and more interestingly, communities and ecosystems have sometimes been designated as potential units of selection. In this instance, at least, a case could be conceived—for communities do maintain some functional coherence, some boundaries (however loose), and some potential for splitting off "daughter" communities with sufficient resemblance to a parent. But I can hardly imagine a set of circumstances that would allow such ecological units to express enough criteria of individuality to qualify for Darwinian agency. Communities are not (for the most part) genealogically constructed or filiated. They can rarely maintain sufficient coherence or persistence, for constituent species move in and out in relative independence. Williams (1992,

p. 55) writes, for example: "The reason must be that communities lack the necessary high rates of reproduction and replacement and especially the high level of heritability required for effective selection. They change their make-up so rapidly that selection among communities must be overwhelmed by endogenous change."

But these principled exclusions leave us with a rich hierarchy of legitimate biological individuals, all related by the fascinating property of nested inclusion within evolution's genealogical system. In appropriate circumstances, broad enough for vital agency in the evolution of life on earth, individuals at many levels—including genes, cell lineages, organisms, demes, species, and clades—can act as units of Darwinian selection. I doubt that we can defend any longer—or as any more than a convenient and parochial preference based on the happenstances of size and duration for a human body—the central Darwinian conviction that organisms represent the fundamental level of Darwinian individuality, with all other levels either nonexistent, impotent, entirely subservient, or operating only in odd and restricted circumstances.

## The Evolutionary Definition of Selective Agency and the Fallacy of the Selfish Gene

### A FRUITFUL ERROR OF LOGIC

Science thrives upon the continuous correction of error. Most errors arise from inadequate knowledge of the empirical world, or (if grounded in a theoretical prejudice) at least persist because we have no means (conceptual or technological) to secure their empirical refutation. For example, we once lacked the technology to prove that buried organic matter might petrify, and that wood made of stone might therefore represent the remains of ancient plants, and not the power of rocks to mimic organic design by a process analogous to crystallization.

Only rarely, however, do professions get sidetracked by pursuing an extensive and longlasting program of research initiated by an error in reasoning rather than an inadequacy of empirical knowledge. Yet I think that the gene-centered approach to natural selection—based on the central contention that genes, as persistent and faithful replicators, must be fundamental (or even exclusive) units of selection—represents a purely conceptual error of this unusual kind. In beginning with Williams's manifesto (1966)—based on a mode of thinking rooted in the brilliantly consistent, if limited, worldview of R. A. Fisher (1930), but immediately inspired by the remarkable work of W. D. Hamilton (1964)—and proceeding through the codification of Dawkins (1976), to numerous works both popular (especially Cronin, 1991) and technical (Dennett, 1995), this gene-based approach to selective agency has inspired both fervent following of a quasi-religious nature (see R. Wright, 1994), and strong opposition from many evolutionists, who tend to regard the uncompromising version as a form of Darwinian fundamentalism resurgent (see Gould, 1997d), variously designated as ultradarwinism (Eldredge, 1995) or hyperdarwinism.

I shall show in this section that, while genes may be appropriately designated as fundamental replicators (under a defendable but nonexclusive strategy of research), replicators simply aren't units of selection or, for that matter, causal agents at all under our usual notions of mechanism in science. The misidentification of replicators as causal agents of selection—the foundation of the gene-centered approach—rests upon a logical error best characterized as a confusion of bookkeeping with causality.

We fall into another serious fault of reasoning when we accept the common conceptual taxonomy that relegates error itself to a purely negative category of unfortunate blunder. Some errors do lead only to blind alleys and wasted time. But others, as thoughtful scientists have always recognized, serve as essential prods and directors of progress through correction. Darwin's famous words, distinguishing harmful from salutary error, have frequently been cited in this context: "False facts are highly injurious to the progress of science, for they often endure long; but false views, if supported by some evidence, do little harm, for every one takes a salutary pleasure in proving their falseness" (from the *Descent of Man*). I prefer the stronger statement of the great Italian economist, Vilfredo Pareto: "Give me a fruitful error any time, full of seeds, bursting with its own corrections. You can keep your sterile truth for yourself."

During my career in evolutionary science, no error has proven more fruitful in Pareto's sense than the gene-centered approach to selection. The central claim, clearly expressed, forced us to reconceptualize the entire domain of evolutionary causality. The outrageous character of such an ultimate reduction compelled us to rethink our subject by explicitly rejecting the oldest, most traditional and entirely commonsensical notions about our own bodies as agents. (Yet the reductionistic cast of the theory fit so well with conventional ideas about the goals of science that many biologists "caught the spirit" and "followed the program" despite its assault upon ordinary intuition.) Nevertheless, the theory could not work. However stubborn and heroic the attempt, explanation inevitably faltered upon the central logical error—especially when selection had clearly worked upon emergent properties of higher-level individuals, and no verbal legerdemain could recast the story in terms of genes as causal agents. If "Pareto errors" contain the seeds that burst their own boundaries, then such uncommon errors of fallacious reason (rather than absent fact) qualify best for this status. Empirical correction usually requires a period of waiting for new technologies or new discoveries (as the sources of resolution do not lie within the argument), but logical errors always carry the seeds of correction within the fruit of their own structure.

## HIERARCHICAL VS. GENIC SELECTION

The fallacy of gene selectionism, and the consequent validity of the alternative (and opposite) hierarchical model of selection, can best be expressed in a series of seven arguments and vignettes—of different length, but all connected in a logical order, and all developed for the same import and purpose:

### The distinction of replicators and interactors
### as a framework for discussion

Both leading founders of modern gene selectionism as a general view of evolution (Williams, 1966; Dawkins, 1976) drew a crucial distinction between reproductive units of heredity, and entities that interact with the environment to bias the transmission of reproductive units into the next generation. Williams viewed nearly all evolution as proceeding *via* genes as reproductive units, with adaptation of organisms (the interacting entities) construed as a result—a duality that he usually labeled (1966, p. 124 for example) as "genic selection and organic adaptation." Dawkins (1976) agreed entirely, and drew a more colorful and explicit distinction between "replicators," considered as units of selection and identified as genes—and "vehicles," considered as merely passive repositories built by replicators for their own purposes, and identified as bodies of organisms. In other words, both Williams and Dawkins invoked a criterion of replication to identify genes as the active and fundamental agents of natural selection.

In his 1980 review on "Individuality and selection," David Hull formalized this distinction in a manner that has—quite usefully and properly in my view—organized the professional discussion on units of selection ever since. Hull (1980, p. 318) defined a replicator as "an entity that passes on its structure directly in replication"; and an interactor as "an entity that directly interacts as a cohesive whole with its environment in such a way that replication is differential." Hull then defined selection with reference to both attributes: "a process in which the differential extinction and proliferation of interactors cause the differential perpetuation of the replicators that produced them."

Hull insisted that a causal account of selection must include both concepts (1980, pp. 319–320): "Evolution of sorts could result from replication alone, but evolution through natural selection requires an interplay between replication and interaction. Both processes are necessary. Neither process by itself is sufficient. Omitting reference to replication leaves out the mechanism by which structure is passed from one generation to the next. Omitting reference to the causal mechanisms that bias the distribution of replicators reduces the evolutionary process to the 'gavotte of the chromosomes,' to use Hamilton's propitious phrase." Later, Hull (1994, pp. 627–628) continued to espouse this view: "According to the terminology I prefer, there are no units of selection because selection is composed of two subprocesses—replication and interaction. Selection results from the interplay of these two subprocesses. Genes are certainly the primary (possibly sole) units of replication, whereas interaction can occur at a variety of levels from genes and cells through organisms to colonies, demes, and possibly entire species."

I shall argue in this section that the causality of selection resides in interaction, not in replication, and that the hierarchical model almost automatically prevails once we accept this analysis of causality. Moreover, Hull's intuitions ran in this direction from the start, for even while he insisted upon the "relevancy" of both replication and interaction, Hull always acknowledged that the classical argument for multiple levels of selection only invokes inter-actors. He wrote in his original paper (1980, p. 325): "In most cases when biologists argue that entities more inclusive than single genes function in the evolutionary process, they have interaction in mind." And Hull (1994, p. 628) directly followed his defense of duality (quoted just above) with this sentence: "The units-of-selection controversy concerns levels of interaction, not levels of replication." I shall defend and develop Hull's intuition in the rest of this section. Only interactors can be deemed *causal* agents in any cus-tomary or reasonable use of this central term. Replicators are important in evolution, but in a different role as items for bookkeeping. Replicators are not causal agents. If causality resides in interactors, and interactors at several levels rank as legitimate evolutionary individuals, then the hierarchical theory of selection becomes unassailable as a coherent logical structure, subject to the ultimate scientific test of empirical verification (or invalidation) in nature.

### Faithful replication as the central criterion for the gene-centered view of evolution

As noted above, both Williams and Dawkins chose to define units of selection as replicators rather than interactors. I shall explain under argument three why I am confident that they made the wrong choice—thus committing the fruitful "Pareto error" discussed at the outset of this section. Having thus de-cided, and correctly understanding that selection can only work on "individ-uals" as previously defined, what replicating individuals would Williams and Dawkins then designate as units?

We all know that they chose genes as fundamental—and effectively ex-clusive—replicators, and therefore as *the* unit of selection in Darwinian the-ory (in maximal contrast with the hierarchical theory of multiple, simul-taneously-acting levels, as defended in this book). I will discuss the stated reasons for their choice, but I cannot know the deeper motivations of their philosophical and psychological preferences. I strongly suspect that they, and all defenders of strict gene selectionism, feel drawn to the traditional reduc-tionism of science. They understood that Darwin himself went as far as he could in this direction, by breaking down the Paleyan edifice of highest-level intentionality (God himself) to the lowest level then practical—organisms struggling for reproductive success (see Chapter 2). They also recognized that this breakdown had produced revolutionary consequences for Western thought, particularly in reconceptualizing all perceived natural "benevo-lence" (especially the good design of organisms and the harmony of ecosys-tems) as a side-consequence of struggle for personal success among lowest-level individuals, rather than as an explicit intention of a loving and omnipo-tent deity. I imagine that the more thoughtful gene selectionists then worked by analogy, reasoning that if they could break causality down even further, below the level of the organism, similarly interesting, and perhaps revolution-ary, consequences might follow. I can't gainsay either the intuition or the am-bition—but I can fault the resulting argument for an erroneous choice of both category and of level.

If a search for ultimate reduction below the Darwinian body set the deeper motivation for choosing genes as units of selection, what particular rationales did proponents of this theory offer? Both Williams and Dawkins began by arguing that the conventional unit of Darwinian theory—bodies of organisms—cannot properly occupy this role because organisms lack a key feature that genes possess. The bodies of sexual organisms disaggregate in reproduction, making only half an appearance (so to speak) in the genetic constitution of offspring. How can something so ephemeral be a unit of selection? But genes pass faithful copies of themselves into future generations, and therefore maintain the integrity required of an agent of natural selection in their definition.

Both Williams and Dawkins advance the same argument in three steps: (1) the unit of selection must be a replicator; (2) replicators must transmit faithful, or minimally altered, copies of themselves across generations; (3) sexual organisms disaggregate across generations and therefore cannot be units of selection, but genes qualify by faithful replication. Williams developed this argument in his first book (1966), and continues his verbal defense to this day, despite remarkable movement, as we shall see, towards the position advocated in this volume. But Williams still employs the language of gene-selectionism, particularly in the identification of genes as "units of selection" by virtue of faithful replication (so different from Hull's pluralistic view that the definition of a unit must include both replication and interaction): "These complications are best handled by regarding individual [i.e. organismic] selection, not as a level of selection in addition to that of the gene, but as the primary mechanism of selection at the genic level. Because genotypes do not replicate themselves in sexual reproduction (cannot be modeled by dendrograms), they cannot be units of selection" (Williams, 1992, p. 16).

Dawkins (1978) advances the same argument, with the same designation of genes as units of selection: "However complex and intricate the organism may be, however much we may agree that the organism is a unit of *function*, I still think it misleading to call it a unit of *selection*. Genes may interact, even 'blend' in their effects on embryonic development, as much as you please. But they do not blend when it comes to being passed on to future generations."

In a later book (1982, p. 91), Dawkins affirms the terminology of genes as units of selection, by making a strong link to his favorite subject of adaptation: "The whole purpose of our search for a 'unit of selection' is to discover a suitable actor to play the leading role in our metaphors of purpose. We look at an adaptation and want to say, 'It is for the good of . . .' Our quest . . . is for the right way to complete that sentence . . . I am suggesting here that, since we must speak of adaptations as being for the good of something, the correct something is the active, germ-like (*sic,* but clearly a misprint for the intended 'germ-line') replicator."

Dawkins's extended defense of genes as *the* unit of selection invokes a set of related criteria bearing unmistakable concordance with primal virtues of our culture, another extrascientific reason for the argument's appeal—namely, faithfulness, (near) immortality, and ancestral priority. Dawkins enlarges the basic argument about faithfulness—sexual organisms disaggregate across generations but genes transmit accurate copies—into a paean about genetic immortality compared with the tragic transiency of our personal lives:

> It does not grow senile, it is no more likely to die when it is a million years old than when it is only a hundred. It leaps from body to body down the generations, manipulating body after body in its own way and for its own ends, abandoning a succession of mortal bodies before they sink in senility and death. The genes are the immortals, or rather, they are defined as genetic entities which come close to deserving the title. We, the individual survival machines in the world, can expect to live a few more decades. But the genes in the world have an expectation of life which must be measured not in decades but in thousands and millions of years. In sexually reproducing species, the individual is too large and temporary a genetic unit to qualify as a significant unit of natural selection. The group of individuals is an even larger unit. Genetically speaking, individuals and groups are like clouds in the sky or dust storms in the desert. They are temporary aggregations or federations. They are not stable through evolutionary time (1976, p. 36).

Dawkins then commits one of the classical errors in historical reasoning by arguing that because genes preceded organisms in time, and then aggregated to form cells and organisms, genes must therefore control organisms—a confusion of historical priority with current domination (see Chapter 11, and Gould and Vrba, 1982, for a full discussion of this common fallacy). But Dawkins's argument collapses for many reasons, most notably the issue of emergence. A higher unit may form historically by aggregation of lower units. But so long as the higher unit develops emergent properties by nonadditive interaction among parts (lower units), the higher unit becomes, by definition, an independent agent in its own right, and not the passive "slave" of controlling constituents. In advancing this false argument, Dawkins closes with a statement that can only compete with some choice Haeckelian effusions for the title of purplest prose passage in the history of evolutionary writing:

> Replicators began not merely to exist, but to construct for themselves containers, vehicles for their continued existence. The replicators which survived were the ones which built survival machines for themselves to live in . . . Survival machines got bigger and more elaborate, and the process was cumulative and progressive . . . Four thousand million years on, what was to be the fate of the ancient replicators? They did not die out, for they are past masters of the survival arts. But do not look for them floating loose in the sea; they gave up that cavalier freedom long ago. Now they swarm in huge colonies, safe inside gigantic lumbering robots, sealed off from the outside world, communicating with it by tortuous indirect routes, manipulating it by remote control. They are in you and me; they created us, body and mind; and their preservation is the ultimate rationale for our existence. They have come a long way, those replicators.

Now they go by the name of genes, and we are their survival machines (1976, p. 21).

One might dismiss this rhetorical flourish as harmless enthusiasm. But we must also recognize that, however extended the metaphors, Dawkins's images do accurately express his false theory of selective agency—for if genes can be depicted as exclusive units of selection, then they become *the* causal agents of evolution; and if bodies are Darwinian ciphers both for their transiency and by their lethargy relative to the "lean and mean" genes living within, then bodies might as well be described as inert and manipulated repositories ("lumbering robots").

Dawkins writes in his introduction (1976, p. ix): "We are survival machines—robot vehicles blindly programmed to preserve the selfish molecules known as genes. This is a truth which still fills me with astonishment. Though I have known it for years, I never seem to get fully used to it." I can only regard this honest admission as a striking example of the triumph of false consistency over legitimate intuition.

### Sieves, plurifiers, and the nature of selection: the rejection of replication as a criterion of agency

The linkage of selective agency to faithful replication has been urged with such force and frequency that the argument now functions as a virtual mantra for many evolutionary biologists. But when we consider the character of natural selection as a causal process, we can only wonder why so many people confused a need for measuring the results of natural selection by counting the differential increase of some hereditary attribute (bookkeeping) with the mechanism that produces relative reproductive success (causality). Replicators cannot be equated with causal agents (unless they also happen to be interactors, for only interactors can be agents). Units of selection must be actors within the guts of the mechanism, not items in a calculus of results.

Genes struck many people as promising units for a twofold reason that does record something of vital evolutionary importance, but bears little relationship to the issue of selective agency. Persistence and replication do lie among the necessary (but not sufficient) criteria for calling any biological entity an evolutionary individual. Since evolution requires hereditary passage, and since genes transmit faithful copies of themselves, and also represent the smallest functional unit of physical continuity between generations of sexual organisms (the kind of individuals we know best for obvious parochial reasons), many biologists assumed that genes must therefore act as the basic (or even the only) unit of selection.

This interesting error arises from two common fallacies in human reasoning:

THE CONFUSION OF NECESSARY WITH SUFFICIENT CONDITIONS. We all agree that units of selection must be evolutionary individuals in Darwinian theory—and that status as an evolutionary individual depends upon a set of criteria discussed on pages 602–613. These criteria do include hereditary passage and sufficient persistence—the properties most strikingly exhibited by genes. But evolutionary individuals, to act as units of selection, must also display other properties that genes do not generally possess. In particular, a unit of selection must interact "directly . . . as a cohesive whole with its environment in such a way that replication is differential"—to quote Hull's definition once again (1980, p. 318).

But in sexual organisms, and in other higher-level individuals, genes do not usually interact *directly* with the environment. Rather, they operate via the organisms that function as true agents in the "struggle for existence." Organisms live, die, compete and reproduce; as a result, genes move differentially to the next generation.

Of course genes influence organisms; one might even say, metaphorically to be sure, that genes act as blueprints to build organisms. But such statements do not substantiate the critically necessary claim that, therefore, genes interact directly with the environment when organisms struggle for existence. The issue before us—the venerable problem of "emergence"—is largely philosophical and logical, and only partly empirical. Genes would interact directly only if organisms developed no emergent properties—that is, if genes built organisms in an entirely additive fashion, with no nonlinear interaction among genes at all. In such a situation, organisms would be passive repositories, and genes could be construed as units of selection—for anything done by organisms could then be causally reduced to the properties of individual genes.

This aspect of the question must be decided empirically. But the issue is also quite settled (and was never really controversial): organisms are stuffed full of emergent properties; our sense of organismic functionality and intentionality largely arises from our appreciation of these emergent features. Thus, since genes interact with the environment only indirectly through selection upon organisms, and since selection on organisms operates largely upon emergent characters, genes cannot be units of selection when they function in their customary manner as faithful and differential replicators in the process of ordinary natural selection among organisms. Dawkins's metaphors of selfish genes and manipulated organisms may be colorful, but such images are also fatefully misleading because Dawkins has reversed nature's causality: organisms are active units of selection; genes, while lending a helping hand as architects, remain stuck within these genuine units.

THE THEORY-BOUND NATURE OF CONCEPTS AND DEFINITIONS. We are drawn to the faithfulness of gene replication, especially when compared with the contrasting transiency of sexual organisms, who must disaggregate to reach the next generation. We might therefore assume that genes become primary candidates for units of selection as a consequence of their potential immortality, while organisms fall from further consideration by the brevity of their coherent lives.

"Sufficient stability" surely ranks as an important criterion for the "evolutionary individuality" required of a "unit of selection." But, in Darwinian theory and the search for units of selection, "sufficient" stability can only be defined as enough coherence to participate as an unchanged individual in the causal process of struggle for differential reproductive success. To be causal

units under this criterion, organisms need only persist for the single generation of their lifetimes—as they do. This endurance may not strike us as a long
time in some intuitively appealing psychological sense, or relative to the persistence of faithful gene replicates, or considered in comparison with geological scales—but these temporal frameworks are irrelevant to the question and
theory at hand. Organisms last long enough to act as units of selection in a
Darwinian process; they therefore possess the "sufficient stability" required
of evolutionary individuals.

Of course, evolutionary individuals must all be able to pass—differentially
and in a heritable manner—their favorable properties into future generations.
But no aspect of this requirement implies or requires that units of selection
must pass copies of *themselves,* bodily and in their entirety, into the next generation. The criterion of heredity only demands that units of selection be able
to bias the genetic makeup of the next generation towards features that secured the differential reproductive success of parental individuals. Units of selection only need to plurify their own representation in the next generation;
they need not copy themselves. Sexual organisms happen to plurify by disaggregation and subsequent differential passage of genes and chromosomes.
Other kinds of individuals, including genes, asexual organisms and species,
plurify more coherently. This common confusion of plurifaction with faithful
replication has erected a serious stumbling block to proper understanding of
the hierarchical theory of selection.

We can best clarify this crucial issue of the relationship between selective
agency and criteria of faithful replication vs. plurifaction if we drop, for a
moment, the conventional framework of replication vs. interaction, and return instead to a different metaphor commonly invoked during 19th century
debates about the nature of Darwinism and natural selection—namely sieves.

We may use the classical metaphor of sieving to illustrate the inappropriateness of faithful replication as a criterion for defining units of selection. The
"goal" of a unit of selection is not unitary persistence (faithful replication)—
and I can't quite figure out why so many late 20th century Darwinians ever
tried to formulate the concept in this manner. The "goal" of a unit of selection is concentration by plurifaction—that is, the differential passage of
"youness" into the next generation, an increase in relative representation of
your heritable attributes (whether you pass yourself on as a whole, or in
disaggregated form, into the future of your lineage).

In the favored metaphor of Darwin's day, selection works like a sieve laden
with all the individuals of one generation. Surrounding environments shake
the sieve, and particles of a certain size become concentrated, while others
pass through the webbing (lost by selection). Sieving represents the causal act
of selection—the interaction of the environment (shaking the sieve) with
varying individuals of a population (particles on the sieve). As a result of this
interaction, some individuals live (remain on the sieve), while others die (pass
through the sieve)—and survival depends causally upon variation in emergent properties of the particles (in this simplest case, large particles remain,
and small particles pass through to oblivion).

The surviving particles need to reproduce in genealogical systems of evolutionary individuals. They may do so by fissioning (faithful passage) or by disaggregation and reconstitution of new individuals as mixtures of hereditary parts of previous individuals. The individuals of the old generation eventually die and evaporate. The individuals of the new generation now live on the sieve, waiting for the next shake.

But this specification of the varied modes for constituting new individuals does not represent what we mean by selection. An entity must be able to reproduce to be defined as an evolutionary individual, but this entity need not replicate faithful copies of itself. Rather, it needs to be able to plurify—that is, to increase, relative to other individuals, the representation of its hereditary contribution to the next generation. Integral "you" may be disaggregated in the process, but so long as the next generation contains a relative increase in your contributions, and so long as you operated as an active causal agent of the Darwinian struggle while you lived, then you qualify as a unit of selection (and a winning unit in this case).

An interesting episode in the history of Darwinism clarifies this concept in a striking manner. We all know that Darwin accepted the idea of "blending inheritance," or the averaging of parental characteristics in the offspring of sexual reproduction. Now blending inheritance marks an ultimate denial of faithful replication—for the hereditary basis of any selected character becomes degraded by half in breeding with an average individual. A paradox therefore arises. If units of selection must be faithful replicators, and if Darwin both understood natural selection and believed in blending inheritance, then why did he ever imagine that selection could work as a mechanism?

We can only resolve this conundrum by recognizing that faithful replication is not—and never was—the defining characteristic (or even a necessary property) of a unit of selection. Darwin, even given his belief in blending inheritance, could view sexual organisms as primary units of selection because he understood agency in a different way that remains valid today: units of selection are evolutionary individuals that interact with the environment and plurify as a causal result. We may return to the metaphor of sieving. Natural selection can work under blending inheritance because shaking the sieve favors the possessors of advantageous traits in each generation—for any individual with a phenotype biased in the favored direction gains a better chance of remaining on the sieve. The offspring of the most favored individuals will blend substantially back to the mean, but this style of inheritance only slows the process of selection—for, as a result of differential survival and reproduction in each generation, the mean itself still gradually moves in the favored direction.

### Interaction as the proper criterion for identifying units of selection

The aforementioned arguments about sieves, plurifaction, and the inappropriateness of faithful replication for designating units of selection lead to a simple conclusion: we can only understand the causal nature of selection when we recognize that units of selection must be defined as *interactors, not as replicators*. Hull's distinction has great merit, but he fell into an overgenerous pluralism in arguing that identification of causal agency must include statements about both the faithfulness of replicators and the potency of interactors. Individuals need not replicate themselves faithfully to be units of selection. Rather, they must contribute to the next generation by hereditary passage, and they must plurify their contributions relative to those of other individuals. But the contributions themselves can be wholes or parts, faithful replicates or disaggregated bits of functional heredity. Selection demands plurifaction, not faithful replication.

The simple observation of plurifaction—the relative increase of an individual's representation in the heredity of subsequent generations—does not suffice to identify the operation of natural selection, for plurifaction can occur by nonselective means, and phenotypes can increase in frequency but then be unable to plurify. Consider the primary example of each phenomenon. First, individuals may plurify by accidents of genetic drift. Suppose that individuals fall through the sieve of selection at random, but survivors show increased frequency of certain heritable traits by accident. These surviving individuals will plurify, but they have not operated as active units of selection. Second, individuals may increase in frequency for phenotypic reasons unrelated to heredity. Suppose that large individuals remain differentially on the sieve, but that individuals grow larger than average for purely ecophenotypic reasons uncorrelated with any aspect of heredity that can pass to subsequent generations. Large phenotypes have increased in frequency for causal reasons—but they will not be able to plurify because they cannot bias the heredity of subsequent generations.

So selection demands plurifaction because evolutionary individuals must maintain lineages by hereditary passage, and selection occurs by increase in relative representation. But plurifaction can only represent a necessary condition, not a cause. We define selection as occurring when plurifaction results *from a causal interaction between traits of an evolutionary individual* (a unit of selection) *and the environment* in a manner that enhances the differential reproductive success of the individual. Thus, and finally, *units of selection must, above all, be interactors.* Selection is a causal process, not a calculus of results—and the causality of selection resides in interaction between evolutionary individuals and surrounding environments. The study and documentation of group and higher-level selection has been stymied and thrown into disfavor by our confusion over these issues—and especially by the blind alley of a logically false argument that identified replicators rather than interactors as units of selection, and then constructed a fallacious, reductionistic theory, precisely opposite in structure to the hierarchical model, by specifying genes (because they replicate faithfully) as ultimate or exclusive units of selection. In this context, I note with delight that group selection has risen from the ashes to receive a vigorous rehearing (Sober and Wilson, 1998, for a full treatment; Lewin, 1996, for a popular account under the title "Evolution's new heretics"; and Gould and Lloyd, 1999, for resolution of a final logical problem). This potent revival rests upon two proposals that, as centerpieces of this book, could not gain my stronger assent: the identification of evolutionary individuals as interactors, causal agents, and units of selection; and the validation of a hierarchical theory of natural selection based upon a principled understanding that evolutionary individuals exist at several levels of organization—including genes, cell lineages, organisms, demes, species, and clades.

D. S. Wilson has most vigorously championed this revival (Wilson, 1980, 1983), while his collaboration with philosopher E. Sober has produced a particularly important paper and a subsequent book on the subject (Wilson and Sober, 1994, with 33 accompanying commentaries and the authors' response; Sober and Wilson, 1998). Wilson and Sober anchor their argument by insisting that units of selection must be defined as interactors, not replicators.

I must raise only one mild quarrel with Wilson and Sober. I agree entirely that units of selection must be defined as interactors, but I prefer a "looser" or "broader" concept of interaction that fosters the proper identification of highest-level individuals in species and clade selection. Wilson and Sober stress the "organism-like" properties of interactors, and therefore make the confusing and regrettable linguistic decision to use "individual" for conventional bodies, and "organism" as the general name for a unit of selection at any hierarchical level; whereas I and most biologists (see Gould and Lloyd, 1999) advocate a reversed terminology. In characterizing the evolutionary principle of interaction, I would stress the potential for a rich panoply of emergent fitnesses, and for the consequent capacity of plurifaction.

Their chosen stress on "organism-like" properties leads Wilson and Sober to emphasize direct modes of interaction based on actual contact of sympatric individuals—the old vision of two gladiators duking it out to the finish. But interaction does not require physical contact. Interaction occurs between individuals and environments, not necessarily between individual and individual. The interaction must be able to yield plurifaction for causal reasons based on properties that enhance differential reproductive success—but, again, competing individuals need not interact directly with each other. Rather, to speak of selection, competing individuals only need to plurify at different relative rates based on similar causal interactions with environments. But the environments may be spatially separate and broadly defined. This issue does not often arise at the traditional level of Darwin's chosen evolutionary individuals—that is, organisms. But higher-level individuals, particularly species and clades, do often compete without contact—and our notion of units of selection must include this important mode of interaction.

Several thoughtful biologists have stressed this point, and I have compiled a small file of such statements. I shall present here only the forceful argument of Williams (1992, p. 25), who has changed his view substantially since formulating the theory of gene selectionism in 1966:

> There are many further questions on the meaning and limits of clade selection. One issue is whether the populations that bear the gene pools need be in ecological competition with each other. I believe that this is

not required, any more than individuals within a population need interact ecologically to be subject to individual selection. The reproductive
success or failure of a soil arthropod, with an expected lifetime dispersal
of a few meters, will hardly influence prospects for a conspecific a hundred meters away. But the descendants of these two individuals might
compete, and genes passed on by one may ultimately prevail over those
passed on by the other. Selective elimination of one and survival of the
other a hundred meters away is individual selection as long as the two
arthropods can be assigned to the same population and their genes to the
same gene pool. . . . In the same way, two gene pools in allopatry can be
subject to natural selection if, as must always be true, their descendants
might be alternatives for representation in the biota . . . The ultimate
prize for which all clades are in competition is representation in the
biota.

### The internal incoherence of gene selectionism

I regard the heyday of gene selectionism as an unusual episode in the history
of science—for I am convinced that the theory's central argument is logically
incoherent, whatever the attraction (and partial validity) of several tenets,
and despite the value of a mental exercise that tries to reconceptualize all nature from a gene's point of view. Close textual analysis* of this theory's leading documents reveals persistent internal problems, explicitly recognized by
authors and invariably met by arguments so flawed in construction that even
the defenders seem embarrassed, or at least well aware of the glaring insufficiency.

I am not alone in noting this peculiar situation, and in calling for some serious consideration by historians. Wilson and Sober (1994, p. 590) write: "The
situation is so extraordinary that historians of science should study it in detail: a giant edifice is built on the foundation of genes as replicators, and
therefore as the 'fundamental' unit of selection, which seems to obliterate the
concept of groups as organisms. In truth, however, the replicator concept
cannot even account for the organismic properties of individuals. Almost as

---

*This kind of textual exegesis, a standard mode of scholarly work in the humanities,
should be pursued more often in scientific discussion as well. Scientists tend to reject such
an approach, I suppose, because we believe that forms of argument and rhetorical styles
only lend a superficial patina to the "real" substance of logic and evidence, and therefore
can teach us nothing of interest. I think that we have thereby missed a major source of insight about the operation of science—a source that would not only deepen our understanding of history and procedure, but would also help us to judge and analyze such contemporary issues as the logic of selectionist theory. If we locate consistent slips, foibles, jagged
edges, strains, or near apologies—as presented verbally—then we can often pinpoint weaknesses in logic or failure of empirical support. I show, in this section, that all major supporters of gene selectionism fall into such verbal patterns at the theory's main loci of inconsistency. In previous books, I have tried to use this mode of analysis to explicate such issues as
the nature of geological time (Gould, 1987b), the logic of biological determinism (1981a),
and the concept of evolutionary progress (1996a) and predictability (1989c).

an afterthought, the vehicle concept is tacked onto the edifice to reflect the harmonious organization of individuals, but it is not extended to the level of groups."

The central problem lies as deep as our definition of the key concept of "cause" in science. Aristotle proposed a broad concept of causality divided into four aspects, which he called material, efficient, formal and final (or, roughly, stuff, action, plan and purpose—that is, the bricks, the mason, the blueprint and the function, in the standard "parable of the house," used for more than two millennia to explicate Aristotle's concept). As many historians have noted, modern science may virtually be defined by a revision of this broad view, and a restriction of "cause," as a concept and definition, to the aspect that Aristotle called "efficient." (The word "efficient" derives from the Latin *facere*, to make or to do. Efficient causes are actual movers and shakers, the agents that apply the forces. Aristotle's term does not engage the modern English meaning of doing something well, as opposed to doing something at all.)

The Cartesian or Newtonian world view, the basis of modern science, banned final cause for physical objects (while retaining the concept of purpose for biological adaptation, so long as mechanical causes, rather than conscious external agencies, could be identified—a problem solved by natural selection in the 19th century). As for Aristotle's material and formal causes, these notions retained their relevance, but lost their status as "causes" under a mechanical world view that restricted causal status to active agents. The material and formal causes of a house continue to matter: brick or sticks fashion different kinds of buildings, while the bricks just remain in a pile, absent a plan for construction. But we no longer refer to these aspects of building as "causes." Material and formal attributes have become background conditions or operational constraints in the logic and terminology of modern science.

I present this apparent digression because the chief error of gene selectionism lies in a failed attempt to depict genes as efficient causes in ordinary natural selection—and the chief "textual mark" of failure can be located in tortuous and clearly discomforting (even to the authors!) arguments advanced by all leading gene selectionists in a valiant struggle to "get through" this impediment. For no matter how an author might choose to honor genes as basic units, as carriers of heredity to the next generation, as faithful replicators, or whatever, one cannot deny a fundamental fact of nature: in ordinary, garden-variety natural selection—Darwin's observational basis and legacy—organisms, and not genes, operate as the "things out there" that live and die, reproduce or fail to propagate, in the interaction with environments that we call "natural selection." Organisms act as Aristotle's efficient causes—the actors and doers—in the standard form of Darwin's great and universal game.

Gene selectionists know this, of course—so they must then struggle to construct an argument for saying that, even though organisms do the explicit work, genes may somehow still be construed as primary "units of selection," or causal agents in the Darwinian process. This misguided search arises from a legitimate intuition—that genes are vitally important in evolution, and clearly central to the process of natural selection—followed by the false inference that genes should therefore be designated as primary causes. Needless to say, no biologist wishes to deny the centrality or importance of genes, just as this intuition holds. But genes simply cannot operate as *efficient* causes in Darwin's process of organismic selection. Genes, as carriers of continuity to the next generation, may be designated as material causes in Aristotle's abandoned terminology. But we no longer refer to the material aspects of natural processes as "causes." Organisms "struggle" as agents or efficient causes; their "reward" may be measured by greater representation of their genes, or material legacies, in future generations. Genes represent the product, not the agent—the stuff of continuity, not the cause of throughput.

The standard gambit of gene selectionists, in the light of this recognized problem, invokes two arguments, both indefensible.

ATTEMPTS TO ASSIGN AGENCY TO GENES BY DENYING EMERGENT PROPERTIES TO ORGANISMS. Once one admits, as all gene selectionists must and do, that genes propagate via selection on organisms as interactors, how then can one possibly ascribe direct causal agency to genes rather than to bodies? Only one logical exit from this conundrum exists: the assertion that each gene stands as an optimal product in its own place, and that bodies impose no consequences upon individual genes beyond providing a home for joint action. If such a view could be defended, then bodies would become passive aggregates of genes—mere packaging—and selection on a body could then be read as a convenient shorthand summary for selection on all resident genes, considered individually.

But such a reductionistic view can only apply if genes build bodies without nonlinear or nonadditive interactions in developmental architecture. Any nonlinearity precludes the causal decomposition of a body into genes considered individually—for bodies then become, in the old adage, "more than the sum of their parts." In technical parlance, nonlinearity leads to "emergent" properties and fitness at the organismic level—and when selection works upon such emergent features, then causal reduction to individual genes and their independent summations becomes logically impossible. I trust that the empirical resolution of this issue will not strike anyone as controversial, for we all understand that organisms are stuffed full of emergent features—an old intuition stunningly affirmed by the first fruits of mapping the human genome (see the full issues of *Science* and *Nature* in February 2001 and my own initial reaction for general audiences in Gould, 2001). What else is developmental biology but the attempt to elucidate such nonlinearities? The error of gene selectionists does not lie in their stubborn assertion of pure additivity in the face of such knowledge, but rather in their conceptual failure to recognize that this noncontroversial nonlinearity destroys their theory.

Dawkins admits the apparent problem (1976, p. 40): "But now we seem to have a paradox. If building a baby is such an intricate venture, and if every gene needs several thousands of fellow genes to complete its task, how can we reconcile this with my picture of indivisible genes, springing like immortal chamois from body to body down the ages: the free, untrammeled, and self-seekings agents of life?"

Dawkins attempts a lame resolution by invoking the quintessentially Oxbridge metaphor of rowing, with the nine men (eight oarsmen and a cox) as genes, and the boat as a body. Of innumerable candidate rowers, we put together the best boat "by random shuffling of the candidates for each position"—and then running large numbers of trials until the finest combination emerges. Of course the rowers must cooperate in a joint task, but we generate no nonlinearities because localized optimality prevails, and the winning boat ends up with the best possible oarsman in each place. Dawkins then transfers this image back to biology and asserts his view of selection as piecemeal optimization—so that each locus (each seat in the boat) eventually houses the best candidate: "Many a good gene gets into bad company, and finds itself sharing a body with a lethal gene, which kills the body off in childhood. Then the good gene is destroyed along with the rest. But this is only one body, and replicas of the same good gene live on in other bodies which lack the lethal gene . . . Many [good genes] perish through other forms of ill luck, say when their body is struck by lightning. But by definition luck, good and bad, strikes at random, and a gene which is *consistently* on the losing side is not unlucky; it is a bad gene" (1976, p. 41).

Such a notion of individualized genetic optimality must be rejected as empirically false; but even if true, this concept still wouldn't support the required claim for nonexistence of emergent organismic features. Even Dawkins admits (in the quotation just above) that selection can only optimize "phenotypic consequences" (1982, p. 237)—and if phenotypes arise (as they do) by complex nonadditivity among genetic effects, then the genes in your body cannot maintain the essential property of independence represented by Dawkins's metaphor of optimal goats, hopping happily and separately across the generations.

In any case, this false view of organisms as additive consequences of individually optimized genes underlies the familiar metaphorical language developed by Dawkins over the years: "I am treating a mother as a machine programmed to do everything in its power to propagate copies of the genes which reside inside it" (1976, p. 132). Or "A monkey is a machine which preserves genes up trees; a fish is a machine which preserves genes in the water; there is even a small worm which preserves genes in German beer mats. DNA works in mysterious ways" (1976, p. 22). These colorful images misstate actual pathways of causality. *Organisms* work in *wondrous* ways, and they operate via emergent properties that invalidate Dawkins's concept of genes as primary agents.

THE *CETERIS PARIBUS* DODGE. When the logic of an argument requires that the empirical world operate in a certain manner, and nature then refuses to cooperate, unwavering supporters often try to maintain their advocacy by employing the tactic of conjectural "as if." That is, you admit the failure of complex nature to meet your theoretical needs, but then claim that you will simplify the actual circumstances "as if" the system under study operated in the required way. The classical "as if" argument goes by its Latin title of *ceteris paribus*, or "all other things being equal." *Ceteris paribus* imposes additivity upon a system truly made of complexly interacting parts. You isolate one factor and state that you will analyze its independent effects by holding all other factors constant.

*Ceteris paribus* ranks among the oldest of scholarly devices, an indispensable tactic for any student of complex systems. I am certainly not trying to mount a general assault upon this venerable and valuable mode of exemplification. Two common circumstances define the legitimate domain of *ceteris paribus*: (1) as a heuristic or exploratory device for approaching systems of such complexity that you don't even know how to think about influences of particular parts, unless you can hypothetically assign all others to a theoretical background of constancy; and (2) as a powerful experimental tool when you can actually hold other factors constant and perturb the system by varying your studied factor alone.

But *ceteris paribus* becomes an illegitimate dodge, an invalid prop to make a potentially false argument unbeatable by definition, in systems dominated by nonadditivity—that is, where the very act of holding all other factors constant may make your favored factor work in a manner contrary to its actual operation in a real world of interaction. If A conquers B only when the two entities share a field alone, but usually loses to B when C also dwells on the field, and if real fields invariably include C, then we cannot crown A as absolutely superior to B on the basis of a single and artificial *ceteris paribus* trial that excluded C from action and consideration.

The use of *ceteris paribus* to support gene selectionism constitutes a similar denial of a known reality. This tactic represents a fallback position after acknowledging the impossibility of asserting a genuine claim for nonadditivity in the translation of genes to organisms. In other words, you admit that massive nonlinearity actually exists, but then state that, for purposes of discussion, you will counterfactually impose *ceteris paribus* so that genes can be equated with linear effects. For example, Dawkins explicitly invokes the key phrase (in English rather than Latin) in defending his requisite (but fallacious) notion that genes may be identified as operating "for" particular parts of phenotypes, thus creating the impression that organisms may be treated as additive aggregates rather than entities defined by nonlinear interaction.

> For purposes of argument it will be necessary to speculate about genes "for" doing all sorts of improbable things. If I speak, for example, of a hypothetical gene "for saving companions from drowning," and you find such a concept incredible, . . . recall that we are not talking about the gene as the sole antecedent cause of all the complex muscular contractions, sensory integrations, and even conscious decisions, which are involved in saving somebody from drowning. We are saying nothing

about the question of whether learning, experience, or environmental influences enter into the development of the behavior. All you have to concede is that it is possible for a single gene, other things being equal and lots of other essential genes and environmental factors being present, to make a body more likely to save somebody from drowning than its allele would (1976, p. 66).

In another passage (1976, p. 39), Dawkins unwittingly surrenders this necessary tactic by admitting that we dare not discuss the interactive web of embryonic development, lest we be unable to speak of genes "for" particular aspects of organismal phenotypes:

> Everybody knows that wheat plants grow bigger in the presence of nitrate than in its absence. But nobody would be so foolish as to claim that, on its own, nitrate can make a wheat plant. Seed, soil, sun, water, and various minerals are obviously all necessary as well. But if all these other factors are held constant, and even if they are allowed to vary within limits, addition of nitrate will make the wheat plants grow bigger. So it is with single genes in the development of an embryo. Embryonic development is controlled by an interlocking web of relationships so complex that we had best not contemplate it.

As a striking demonstration that *ceteris paribus* cannot rescue gene selectionism from logical paradoxes and violations of ordinary linguistic usage, Dawkins (1982, p. 164) addresses the problem of how to treat a genetic deletion favored by natural selection at the organismic level, if genes represent the fundamental units of selection, and if we must be able to treat each genetic item as a Darwinian individual with a distinct and independent history. If "gene language" must prevail, and if we need to specify the selective value of such a deletion, what can we call the loss but "a replicating absence"! Should we not, at this point, admit instead that organisms are the relevant causal agents in this case, and that organisms have achieved a selective benefit by the alternate but orthodox genetic route of deletion rather than substitution? Some humans have done well with "plenty of nothing," but I don't think we should root our ontology in taxonomies for various kinds and forms of faithfully propagating absences.

> Any organism that happened to experience a random deletion of part of its selfish DNA would, by definition, be a mutant organism. The deletion itself would be a mutation, and it would be favored by natural selection to the extent that organisms possessing it benefited from it, presumably because they did not suffer the economic wastage of space, materials, and time that selfish DNA brings. Mutant organisms would, other things being equal [*ceteris paribus* again!], reproduce at a higher rate than the loaded down "wild type" individuals, and the deletion would consequently become more common in the gene pool. Here we are recognizing that the deletion itself, the absence of the selfish DNA, is itself a replicating entity (a replicating absence!), which can be favored by selection.

All major proponents of gene selectionism have unintentionally illustrated the theory's incoherence by trying to "cash out" their system, and failing at the crucial point of assigning causal agency in natural selection. For, however these proponents may talk about genes as primary agents or units of selection, they cannot deny that nature's Darwinian action generally unfolds between discrete organisms and their environments. These authors therefore acknowledge this basic fact and then tend to lapse into verbal obfuscation on the gene's behalf. I have already noted a prime example in Williams's claim, quoted previously in another context, that organismic selection should be regarded not "as a level of selection in addition to that of the gene, but as the primary mechanism of selection at the genic level." But what does this statement mean? Williams recognizes organismic selection as the "primary mechanism" by which genes pass differentially from one generation to the next. But primary mechanisms are efficient causes in any standard analysis of the logic of science. Williams (1992, p. 38) presents an accurate epitome of selection in the following passage: he states that selection must always operate on interactors (and he knows, as the previous quotation shows, that organisms usually constitute the relevant interactors in cases that he wishes to describe as genic selection); he also recognizes that information must pass to future generations by faithful heredity, and he seems to acknowledge that such biased passage defines the result, not the cause, of selection. Yet he fails to take the final step of acknowledging that these statements debar gene selectionism as the mechanism of evolution. "Natural selection must always act on physical entities (interactors) that vary in aptitude for reproduction, either because they differ in the machinery of reproduction or in that of survival and resource capture on which reproduction depends. It is also necessary that there be what Darwin called 'the strong principle of inheritance,' so that events in the material domain can influence the codical record. Offspring must tend to resemble their own parents more than those of other offspring. Whenever these conditions are found there will be natural selection."

Over the years, Dawkins has developed a litany of similar admissions. Of course organisms must be regarded as the foci of selection, but since biased gene passage occurs as a result of this process, we may identify genes as agents of selection. (But results are not causes, although foci of action surely are): "Just as whole boats win or lose races, it is indeed individuals [organisms] who live or die, and the immediate manifestation of natural selection is nearly always the individual level. But the long-term consequences of nonrandom individual death and reproductive success are manifested in the form of changing gene frequencies in the gene pool" (1976, p. 48).

Dawkins then apologizes for framing his descriptions in terms of organisms as causal actors, excusing himself for succumbing to temptations of convenience. (But perhaps we find this mode "convenient" because we achieve the best description of a causal reality thereby—while the genic mode remains tortuous and uncomfortable because we sense the central error in such formulation): "In practice it is usually convenient, as an approximation, to regard the individual body as an agent 'trying' to increase the number of all its genes in future generations. I shall use the language of convenience. Unless otherwise stated, 'altruistic behavior' and 'selfish behavior' will mean behavior directed by one animal body towards another [p. 50] . . . We shall continue to treat the individual as a selfish machine, programmed to do whatever is best for his genes as a whole. This is the language of convenience [p. 71]."

In a later book (1982, p. 4) Dawkins admits that gene replicators must be selected by proxy—that is, via organisms as causal actors: "The most important kind of replicator is the 'gene' or small genetic fragment. Replicators are not, of course, selected directly, but by proxy; they are judged by their phenotypic effects."

This argument, I think, has truly become an inadaptive meme, destined for eventual extinction, but propagated wherever gene selectionism survives, whether in technical literature or popular presentation. A major popular book on this topic holds (Cronin, 1991, p. 289): "If organisms are not replicators, what are they? The answer is that they are vehicles of replicators . . . Groups, too, are vehicles, but far less distinct, less unified . . . In this weak sense, then, 'group selection' could occur . . . [but it] would in no way undermine the status of genes as the only units of replicator selection. This does not mean that higher level entities are unimportant in evolution. They are important, but in a different way: as vehicles."

### Bookkeeping and causality: the fundamental error of gene selectionism

The error and the incoherence of gene selectionism, as documented above, can be summarized in a single statement illustrating the fruitful, "Pareto-like" character of the central fallacy: proponents of gene selectionism have *confused bookkeeping with causality*. This error achieves its Pareto status of substantial utility because changes recorded at the genetic level do play a fundamental part in characterizing evolution, and records of these changes (bookkeeping) do maintain an important role in evolutionary theory. But the error remains: bookkeeping* is not causality; natural selection is a causal process, and units or agents of selection must be defined as overt actors in the mechanism, not merely as preferred items for tabulating results.

No one has ever stated the issue more accurately or succinctly than George Williams himself (1992, p. 13), thus increasing my puzzlement at his failure to recognize how his own formulation invalidates the gene selectionism that still wins his lip service: "For natural selection to occur and be a factor in evolution, replicators must manifest themselves in interactors, the concrete realities that confront a biologist. The truth and usefulness of a biological theory

---

*Working through the logic and problems of this vexatious issue has been pursued as a collective enterprise among many biologists and philosophers for more than 20 years. I have used the terminology of bookkeeping and causality for some time (Gould, 1994), and have developed or sharpened some of the arguments. But I do not think that I devised the labels. I believe that I first picked up the terminology of bookkeeping from arguments presented by the University of Chicago philosopher Bill Wimsatt. Many authors have used this fruitful distinction for some time.

must be evaluated on the basis of its success in explaining and predicting material phenomena. It is equally true that replicators (codices) are a concept of great interest and usefulness and must be considered with great care for any formal theory of evolution, either cultural or biological." Williams's statement agrees completely with the position that I have advanced in this book—an attitude that, by general consensus, leads logically and directly to the hierarchical model of selection, and the invalidity of single-level, gene-based views. Williams allows that interactors represent the "concrete realities" confronting biologists (and chapter 4 of his 1992 book eloquently defends the concept of legitimate interactors at several hierarchical levels of increasing genealogical inclusion). He admits that both the "truth and usefulness" of a biological theory, natural selection in this case, depends upon the explanation of material phenomena—that is, interactors operating as agents. He does not include replicators—the basis of gene selectionism—in this category, for his last sentence grants them a separate but equal status in evolutionary theory: "It is equally true that replicators (codices) are a concept of great interest" needed "for any formal theory of evolution." Now, if replicators are not causal agents, but are vital for any full account of evolution—then what are they? I suggest that we view gene-level replicators as basic units for keeping the books of evolutionary change—as "atoms" in the tables of recorded results.

Williams did not slip or misspeak in the quotation cited above. He repeats this separation of a causal agent from a basis of hereditary transmission—with interactors as agents and replicators as transmitters—in several other passages, including (1992, p. 38) "Natural selection must always act on physical entities (interactors) . . . It is also necessary that there be what Darwin called 'the strong principle of inheritance' . . ."

Whereas Williams makes valid separations and defines proper roles, but then seems unwilling to own the theoretical consequences, Dawkins, on the other hand, seems merely confused. In discussing group selection (1982, p. 115), for example, Dawkins writes: "The end result of the selection discussed is a change in gene frequencies, for example, an increase of 'altruistic genes' at the expense of 'selfish genes.' It is still the genes that are regarded as the replicators which actually survive (or fail to survive) as a consequence of the (vehicle) selection process."

By putting the word "vehicle" in parentheses, as a reminder of selection's intrinsic nature rather than a mere modifying adjective, Dawkins admits that interactors (vehicles in his terminology), not replicators, operate as agents of selection. He describes the differential survival of replicators as a consequential result of this causal process—therefore as units for bookkeeping rather than agents of causality—but he then fails to disentangle these two different aspects of evolution, while continuing to grant favored status to genes.

We may indeed, and legitimately as a practical measure, decide to keep track of an organism's success in selection by counting the relative representation of its genes in future generations. (In large part, we count at the genic level for the reason always emphasized by Williams and Dawkins—because sexual organisms do not replicate faithfully and therefore cannot be traced as discrete entities across generations.) But this practical decision for counting does not deprive the organism of status as a causal agent, nor does such a procedure grant causality to the objects counted. The listing of accounts is bookkeeping—a vitally important subject in evolutionary biology, but not a form of causality.

If, as I have argued (see also Wilson and Sober, 1994), the incoherence of gene selectionism denotes a rare case in science of an influential theory felled by a logical error—in this case the confusion of bookkeeping with causality—rather than a fallacious proposal about the empirical world, then we must ask why so many people fell into this error so readily, and why the fallacy did not become more quickly apparent. I suspect that three major reasons underlie not only the error of gene selectionism, but also the strong willingness, even the fervor, expressed by so many evolutionists in embracing the concept. The first two reasons may claim a social basis in traditions of scientific inquiry. But the third reason, and surely the most intriguing from both a scientific and philosophical perspective, emanates directly from the logical structure of hierarchies, the conceptual framework that must replace gene selectionism.

The two reasons rooted in traditions of scientific procedure include the most general of statements and a preference peculiar to traditions of Anglophonic evolutionary biology. For the generality, I state nothing profound or original in pointing out that a decision to privilege the level of genes plays into the strongest of all preferences in Western science: our traditions of reductionism, or the desire to explain larger-scale phenomena by properties of the smallest constituent particles.

The allure of reductionism encourages the following kind of error, or sloppy thinking: we correctly note that genes play a fundamental role in evolution (as preferred items for a calculus of change—the bookkeeping function); we also recognize that genes lie at the base of a causal cascade in the development of organisms; finally, and most generally, we view genes as the closest biological approach to an "atom" of basic structure, and therefore as the cardinal entity of a reductionistic research program. From these statements, we easily slip into the unwarranted inference that genes must also be fundamental units or agents in natural selection, the primary causal theory of our profession—all the while forgetting the criteria of individuality and interaction that define units or agents within the logic of the theory itself.

The second, and more particular, reason flows from explicit traditions of the Modern Synthesis, especially from the approach favored by Fisherian population genetics (see Chapter 7). The heuristics of this field prospered greatly with models that kept track of gene frequencies without worrying much about the locus of selective action. A common fallacy in science then conflates a practical basis for success with the causal structure of nature. Jim Crow (1994, p. 616), one of the world's most thoughtful geneticists, expressed this point particularly well, but then also failed to distinguish bookkeeping from causality. Writing "In praise of replicators"—and well should we praise them, but, I would argue, as excellent agents for accounting!—

Crow explained why our traditions have favored the genetic level of analysis (1994, p. 616):

> The reason, I think, is that these pioneers [Fisher and other founders of the Synthesis] and their intellectual heirs have been concerned, not with selection as an end in itself, but with selection as a way of changing gene frequencies. Selection acts in many ways: it can be stabilizing; it can be diversifying; it can be directional; it can be between organelles; it can be between individuals; it can be between groups . . . But the bottom line has always been how much selection changes allele frequencies and through these, how much it changes phenotypes. This suggests that we should judge the effectiveness of selection at different levels by its effects on gene frequencies.

I could not ask for a better statement of (unconscious) support for the position here maintained. Again, as I noted in several other quotations from gene selectionists, Crow allows that selection, as a causal force ("selection as an end in itself," in his words), operates on interactors at several hierarchical levels of individuality, including groups. He also admits that changes in gene frequencies arise as a result of such selection. He then states—and again I don't object—that these alterations in allelic frequencies should be read as a "bottom line" in judgments about selection's effect. Nicely said, but a bottom line for what? Crow then gives his crucial answer—for keeping the books of evolutionary change: "we should judge the effectiveness of selection at different levels by its effects on gene frequencies." Altered gene frequencies are therefore results (for bookkeeping), while selection (the cause of the changes) operates upon interactors "at different levels" of individuality.

This notion of a "bottom line" also provides our best entrée into the third and most important reason for choosing genes as units of bookkeeping: the intrinsically asymmetrical nature of causal flow in hierarchies of inclusion. I particularly appreciate the doubly amplified utility of hierarchy theory in this example—for the hierarchical view, as I shall show, both serves as a replacement for gene selectionism, but also (in a situation not devoid of irony)* provides the rationale for why many biologists chose, albeit for fallacious reasons, to focus on genes in the first place!

We do need to keep the books of evolutionary change, and bookkeeping does require a basic unit of accounting. Candidates for this status must obey the primary criterion always stressed by gene selectionists: faithful replication. But genes do not exhaust the range of faithful replicators. Asexual organisms and species also rank as sufficiently faithful. Reductionistic preferences in general, and claims for relatively greater faithfulness of genic vs. higher-level replication, might set a preference for genes—but another crucial argument, usually unrecognized or unmentioned, seals the case.

---

*The logic of this case recalls the celebrated example (see pp. 492–502) of Rutherford's use of radioactivity both to impugn the theoretical basis for Kelvin's young age for the earth and then to provide the empirical basis for measuring a revised and much older age.

Because bookkeeping is not the same enterprise as causality, and because we are not, in simply counting, trying to establish the causes of differential success, we want to make sure, above all, that we choose a unit better suited than any other to record *all* evolutionary changes, whatever their causal basis. No single unit of bookkeeping can monitor every conceivable change, but the gene becomes our unit of choice because the nature of hierarchies dictates that genes inevitably provide the most comprehensive record of changes at all levels. (Even so, gene records will miss certain kinds of changes that we generally call evolutionary. For example, as Wilson and Sober (1994) point out, assortative mating of organisms within a population may greatly increase the ratio of homozygotes to heterozygotes at many loci, but need not change gene frequencies in the population.)

Hierarchies are allometric, not fractal (see Gould and Lloyd, 1999), and various levels translate a common set of causes to strikingly different results and frequencies. Moreover, hierarchies are directional, and therefore not indifferent to the nature of the flow of influence. As the most important of all such asymmetries, change at a low level may or may not produce an effect at higher levels—"upward causation" in the standard terminology (see Campbell, 1974; Vrba, 1989). But change at a higher level must always sort the included units of all lower levels—by the analogous process of "downward causation."

If a gene increases in copy number within a genome by duplication and lateral spread (gene selection in the genuine sense), phenotypes of organisms may or may not be affected. But selection on higher-level individuals *always* sorts the lower-level individuals included within. If ugly organisms outcompete beautiful conspecifics, then genes for ugliness increase in the population. If stenotypic species prevail over eurytopes in species selection, then genes associated with stenotypy increase within the lineage. If species of polychaetes eliminate species of priapulids in competition over geological time, then polychaete genes increase in the marine biota.

Given this intrinsic asymmetry, what single unit would a good bookkeeper choose? Obviously not the organism, or any high-level individual, because we would then miss many changes at lower levels—and a good bookkeeper wants, as the chief desideratum of his profession, to record all changes. As noted above, low-level selection need not impose any effect upon higher levels at all. Equally obviously, our optimal bookkeeper will choose genes—not because genes are intrinsically more basic (the reductionist fallacy); not because genes are primary causal agents, or causal agents at all (the gene selectionist fallacy); and not because genes replicate faithfully (for other kinds of individuals do so as well); but, rather, because genes, as the lowest-level individuals in a hierarchy, manifest the unique property of recording all changes. Thus, the intrinsic nature of hierarchies sets our preference for genes as units of bookkeeping—for only genes act as nearly ubiquitous recorders of all evolutionary alterations, whatever their level or cause of occurrence.

Finally, we must note one other property that, while strongly favoring genes as units of bookkeeping, shows even more clearly why genes cannot be exclusive units of selection, or causal agents. Bookkeepers must, above all, be objectivists, not sleuths or storytellers. A good bookkeeper wants an unimpeachable record, not a causal hypothesis (that can always be wrong). Books kept in terms of gene frequencies become the best objective records of "descent with modification" because they do *not* make causal attributions, but only count changes ("just the facts, ma'am," to cite a famous detective from the early days of television). The hierarchical nature of evolutionary mechanics, and the simultaneous action of selection on individuals at several levels, implies our inability to know the causal basis of change from records of altered gene frequencies alone.

Genetic change cannot, of itself, specify the causal level of sorting because selection at any higher level sorts individuals at all included lower levels as an automatic effect, and not necessarily for direct causal reasons at all. Two basic considerations bar inferences of cause from the genetic account books alone. First, an observation of genetic sorting doesn't specify the relevant causal level. A gene associated with strong jaws may increase in frequency within the class Polychaeta because polychaete organisms with strong jaws outcompete weaker-jawed conspecifics in organismic selection; because polychaete species with strong jaws also develop emergent populational characters that defeat weak-jawed species in species selection; or because strong-jawed polychaetes do especially well in driving out jawless priapulids by clade selection.

Second, even when we can identify the level of causality for an incident of genetic sorting, we cannot know (from the increase in frequency alone) whether the gene sorted positively has prevailed by a selected effect upon the phenotype, or for a set of possibly nonadaptive reasons. Does the gene associated with strong jaws actually promote the construction of this phenotypic basis for organismic selection, or has this gene hitchhiked to greater frequency by close linkage with another gene that does build the selected phenotype? Nonadaptive possibilities only increase for selection at higher levels. If polychaetes have increased by clade selection over priapulids, does the plurified polychaete gene big-A build part of the relevant priapulid-beating phenotype, or does big-A just count as one of the myriad polychaete genes that happen to specify, by homology and a few hundred million years of evolutionary separation, the historical uniqueness of the clade?

The nature of hierarchies dictates a choice of genes as optimal units of bookkeeping. The nature of hierarchies also creates a possibility—then realized in nature for fascinating reasons largely unknown, and mostly beyond the scope of this book (but see Buss, 1987)—for structuring the world of biology as a hierarchy of individuals, each encompassing the ones below as new levels accrete in evolution, and each capable of acting as a unit of selection, a causal agent of Darwin's expanded theory.

### Gambits of reform and retreat by gene selectionists

As I have emphasized throughout this section, gene selectionism can't be made to work as a general philosophy. The logic of the theory does not cohere, and the system cannot attain consistent completion. Yet the allure of the gene remains powerful, largely for reasons of general preference in our culture, rather than for any observed power or intrinsic biological status possessed by evolutionary individuals of this lowest level. When an incoherent argument remains intriguing, and supporters cannot bear the wrench of total abandonment, a favored theory must be festooned with compromises and "howevers," or so changed in form that only lip service remains to cover a truly altered substance. Often, given human tendencies to paint a bright face on adversity, gene selectionists have made their necessary retreats, but presented them as refinements or elaborations of the original theory. In this closing section, I shall show that the two most prominent "revisions" of gene selectionism—Dawkins's extended phenotype (1982) and Williams's codical selection (1992)—represent defeats rather than improvements as advertised.

DAWKINS ON THE "EXTENDED PHENOTYPE." I always admired the chutzpah of Senator Aikens' brilliant solution to the morass of our involvement in the Vietnamese War. At the height of our reverses and misfortunes, he advised that we should simply declare victory and get out. Richard Dawkins got in with his 1976 book, *The Selfish Gene*. He declared victory with *The Extended Phenotype* in 1982—although he had really, at least with respect to the needs and logic of his original argument, gotten out.

With admirable clarity, and no ambivalence, Dawkins proclaimed the doctrine of exclusive gene selectionism in 1976: "I must argue for my belief that the best way to look at evolution is in terms of selection occurring at the lowest level of all . . . I shall argue that the fundamental unit of selection, and therefore of self-interest, is not the species, nor the group, nor even, strictly, the individual. It is the gene, the unit of heredity (1976, p. 12). So selection occurs at only one lowest level—the gene, labelled as 'the fundamental unit of selection.' Nothing more inclusive, not even an organism, can be called a unit of selection."

Dawkins presented his later work, *The Extended Phenotype*, as an extension and elaboration of gene selectionism: "This book," he wrote, "is in some ways the sequel to my previous book, *The Selfish Gene*" (1982, p. v). Dawkins had admitted, in 1976, that genes work through phenotypes of the "lumbering robots" (organisms) serving as their passive homes. But if genes are nature's real actors, and phenotypes only their means of expression, then why limit phenotypes to bodies? Any consequence of a gene should be equally capable of carrying the gene's interest in a process of selection. Dawkins admitted of course that most aspects of this extended phenotype—the footprint of a shorebird in the sand, for example—will be too ephemeral, or too by-the-by, to be effective in the gene's interest. But other parts of the extended phenotype (with the beaver's dam as Dawkins's favorite example) do contribute to the success of beaver genes, and should be included within "the extended phenotype" that the gene—the ultimate and only unit of selection (at least in 1976)—can manipulate in its full range of machinations for replicative success.

Dawkins (1982, pp. iv–vii) therefore insisted that the viewpoint of *The*

*Extended Phenotype* evolved gradually and progressively from *The Selfish Gene.* "The present book," he tells us, "goes further," presumably in the same direction:

> This belief—that if adaptations are to be treated as "for the good of" something, that something is the gene—was the fundamental assumption of my previous book. The present book goes further. To dramatize it a bit, it attempts to free the selfish gene from the individual organism which has been its conceptional prison. The phenotypic effects of a gene are the tools by which it levers itself into the next generation, and these tools may "extend" far outside the body in which the gene sits, even reaching deep into the nervous system of other organisms. Since it is not a factual position I am advocating, but a way of seeing facts, I wanted to warn the reader not to expect "evidence" in the normal sense of the word.

So genes have become even more fundamental, and bodies even more inconsequential: "Fundamentally, what it going on is that replicating molecules ensure their survival by means of phenotypic effects on the world. It is only incidentally true that these phenotypic effects happen to be packaged up into units called individual organisms" (Dawkins, 1982, pp. 4–5).*

But now the argument begins to unravel. Just when the gene seems poised to swallow the organism entirely as just one incidental aspect of the gene's armamentarium (the fully extended phenotype), Dawkins turns around, and tells us that we may treat organisms as focal entitites after all, and describe evolution from the organism's point of view just as well: "I am not saying that the selfish organism view is necessarily wrong, but my argument, in its strong form, is that it is looking at the matter the wrong way up . . . I am pretty confident that to look at life in terms of genetic replicators preserving themselves by means of their extended phenotypes is at least as satisfactory as to look at it in terms of selfish organisms maximizing their inclusive fitness" (1982, pp. 6–7).

Shall we then favor the gene or the organism? Dawkins claims to prefer genes and to find greater insight in this formulation. But he allows that you or I might prefer organisms—and it really doesn't matter. In a telling analogy, Dawkins compares genes and organism to the two possible versions (different

---

*But it is not "only incidentally true" that genes generally come packaged into organisms on our planet—and that, in full extension, organic matter coagulates into evolutionary individuals at several levels of an inclusive hierarchy: genes, cell lineages, organisms, demes, species, and clades. This process of coagulation has occurred for active and interesting structural reasons only dimly understood (Buss, 1987). But how could we regard this most fundamental feature of the organic world, constituting the basis of evolutionary causality in units of selection, as "only incidentally true"? This structure may well be contingently true—in the sense that we can imagine an alternative world composed only of naked genes—but our planet's biological reality surely cannot be designated as incidental in the usual sense of unimportant or not fundamental. Indeed, the origin of such hierarchical structure may not even be contingently true, but broadly predictable (see Kauffman, 1993; Maynard Smith and Szathmáry, 1995).

cerebral resolutions of the same visual reality) in the famous optical illusion known as the Necker Cube:

> After a few more seconds the mental image flips back and it continues to alternate as long as we look at the picture. The point is that neither of the two perceptions of the cube is the correct or "true" one. They are equally correct. Similarly the vision of life that I advocate, and label with the name of the extended phenotype, is not probably more correct than the orthodox view. It is a different view and I suspect that, at least in some respects, it provides a deeper understanding. But I doubt that there is any experiment that could be done to prove my claim (1982, p. 1).

Moreover, we really needn't quarrel over our choices because the issue can achieve no empirical resolution in any case. I'll push my preference (and hope to persuade you of its greater capacity for mind stretching, its superior literary charm, or its greater tickling of the fancy); and you can then advocate your opposite, and equally valid, version. Dawkins begins his book: "This is a work of unabashed advocacy. I want to argue in favor of a particular way of looking at animals and plants, and a particular way of wondering why they do the things that they do. What I am advocating is not a new theory, not a hypothesis which can be verified or falsified, not a model which can be judged by its predictions, . . . I am not trying to convince anyone of the truth of any factual proposition" (1982, p. 1). This argument about equally valid, but quite inverse, perspectives on a common reality pervades the entire book, as in this late passage (1982, p. 232): "The whole story could have been told in . . . the language of individual manipulation. The language of extended genetics is not demonstrably more correct. It is a different way of saying the same thing. The Necker Cube has flipped. Readers must decide for themselves whether they like the new view better than the old."

Among professional philosophers, such Necker-Cube thinking goes by the name of *conventionalism*, an argument that frameworks of explanation cannot be judged as true or false, or even more or less empirically adequate—but only as equally correct, and only as more or less preferable by such nonfactual criteria as depth of insight provided or satisfaction gained in understanding. Conventionalism may offer an interesting and fruitful approach, especially for some scientific debates that seem especially refractory to empirical resolution—and also (more generally) for teaching people that ideas and attitudes influence science; and that "naive realism," with its assumption that improved theories arise only from observation, represents a silly and bankrupt approach to the natural world.

But conventionalism cannot apply to this case because an empirical resolution exists, and the apparent Necker-cube duality of gene or organism does not denote, as Dawkins mistakenly argues, two equally valid perspectives on the same issue, but rather expresses a correct *vs.* a false view of the nature of causality in Darwinian theory. Dawkins has misconstrued his categories in judging gene-based and organism-based viewpoints as alternative versions of the same kind of explanation. The gene-based view works best for bookkeeping, while the organism-based view represents one legitimate level of causality—the one regarded as effectively ubiquitous and exclusive by Darwin himself. In this sense, both views are valid; but they are not comparable—and genes *vs.* organisms do not represent alternatives on an identical playing field of common explanatory intent.

Moreover, Dawkins's shift from the selfish gene to the extended phenotype does not reflect a simple extension or elaboration of a consistent and developing viewpoint. He tries to save face with such a portrayal, but his strategy fails. The conventionalism of *The Extended Phenotype* negates and denies the explicit defense of gene selectionism as an empirical reality, as presented in *The Selfish Gene*. Dawkins's first book says, in no uncertain terms (see quotation on page 618), that genes are exclusive units of selection (or causal agents), and that bodies, as passive lumbering robots, cannot play such a role. The second book says that we can view evolution equally well from either the gene's or the organism's point of view, that Dawkins still prefers genes, but that others remain free to favor bodies with just as much claim to empirical adequacy. The disparate logic of these two formulations precludes their interpretation as developing versions of the same view of life, and one theory is not a subtler extension of the other. These two positions connote logically contrasting, and mutually exclusive, accounts of causality in evolution. I do not happen to regard either as correct, but I think we can all agree that Dawkins's later view of the extended phenotype derails and controverts his earlier defense of gene selectionism as nature's true way.

I do not know why Dawkins altered his view so radically. But may I suggest that he simply could not—because no one can after a proper analysis of the basic logic of the case—maintain full allegiance to the fallacious argument of strict gene selectionism. Dawkins tried hard in 1976, but ultimately needed to make so many statements from the organism's point of view that he must have begun to wonder whether he could really continue to regard such organismal language as a mere convenience, while touting the genic formulation as a unique reality. Perhaps he finally decided that if organism-based language seemed so stubbornly ineluctable, then organism-based causality might be equally inevitable, at least as a legitimate option. With such an admission, the selfish gene becomes an impotent meme.

**WILLIAMS'S CODICAL HIERARCHY.** Williams's epochal book of 1966 set the intellectual basis for gene selectionism, and may justly be called the founding document for this ultimate version of Darwinian reductionism. But by 1992, Williams had realized that interactors, and not replicators, constitute units of selection, or causal agents in the usual sense of the term—and that hierarchy must hold because no level of interaction can be deemed exclusive, or even primary. Williams, however, did not wish to abandon his old apparatus for viewing genes as fundamental and preferred units of selection. But *que faire?* Genes are replicators in their only universal role (they can also be interactors in the much more restricted status of one legitimate level in an extensive hierarchy, as discussed on pp. 689–695)—and interactors, not replicators, are units of selection in the causal sense.

Williams therefore tried an interesting gambit. He admitted that interactors form a hierarchy of evolutionary individuals at several levels, and that these interactors are units of selection in our usual sense of material entities participating in a causal process. These interactors build a material hierarchy—and gene selectionism cannot apply to this legitimate domain. Williams therefore established a different and parallel hierarchy* for abstract units of information (as opposed to material entities)—and he construed genes as basic "units of selection" in this alternative and parallel domain, which he called codical (the adjectival form of codices, the plural of codex, his term for a single unit

---

*This interesting idea of parallel hierarchies to separate the replicative and interactive criteria of evolutionary individuality originated with Eldredge (1989; see also Vrba and Eldredge, 1984), who spoke of genealogical and economic hierarchies. The scheme continues here with Williams's similar distinction of material and codical systems. I find the idea of dual hierarchies both interesting and challenging, but ultimately flawed and counterproductive in the introduction of unnecessary complexity. (My rejection of this scheme defines my only major difference with my closest colleague Niles Eldredge, who has worked with me for 25 years on problems of macroevolutionary theory.)

Eldredge's "economic" and Williams's "material" hierarchies include the interactors defined as proper units of selection in this book—and also in Wilson and Sober (1994), and (by unintended verbal admission, though not explicitly) by such gene selectionists as Dawkins and Williams, as I have shown throughout this section. (Eldredge calls this hierarchy "economic" to stress the doing and dying of such entities in nature's ecosystems.) Eldredge's "genealogical" and Williams's "codical" hierarchies express the concept of replication (as nonmaterial units of information for Williams, but as an alternative hierarchy of replicating material entities for Eldredge).

I find the framework of dual sequences unnecessarily complex and divisive because a single theme unites our search to define units of selection, and a single hierarchy expresses this theme in the best and clearest way. Units of selection must be evolutionary individuals by the criteria outlined on pages 608–613. Above all, such individuals must be interactors in order to function as units of selection in a causal process. They must also possess a mechanism of plurifaction—that is, interactors must be able to bias the heredity of subsequent generations towards more of their own contribution, however these contributions be packaged. This need for plurifaction underlies our sense that replication plays a vital role in evolutionary individuality—a role sufficiently important to be mistaken as causal and primary by gene selectionists, or at least to warrant a separate hierarchy (by Eldredge). But I raise two points to obviate the need for a separate hierarchy of replicators: (1) replication (or some other form of hereditary passage) constitutes only one of several necessary criteria for defining evolutionary individuality; and (2) this criterion of hereditary passage only demands that interactors possess a means of plurifaction; faithful replication represents one style of hereditary passage, but not a necessary mode for attribution of evolutionary individuality or designation as a unit of selection. Sexual organisms plurify by disaggregation and differential passage of genes; other kinds of evolutionary individuals plurify by faithful passage.

We should formulate a single hierarchy—call it material, genealogical, or perhaps simply evolutionary—composed of interactors with adequate modes of plurifaction. These evolutionary individuals build a hierarchy of inclusion, with each higher level encompassing the individuals beneath as parts. Most units in Eldredge's parallel hierarchies appear in both his economic and genealogical arrays—and therefore represent the evolutionary individuals we seek for a single hierarchy—for these are the entities that possess both the interactive (economic) and hereditary (genealogical) properties required of any evolutionary individual.

of information). If genes can't claim exclusivity (or even causal status at all) as units of selection in the usual domain of material objects, then Williams would establish a new and separate hierarchy for nonmaterial units of information—and here the gene could continue to reign.

Williams therefore proposed a fundamental distinction between entities and information, speaking of "two mutually exclusive domains of selection, one that deals with material entities and another that deals with information and might be termed the codical domain" (1992, p. 10). But I do not think that the codical domain can claim either meaning or existence as a locus for causal units of selection, for two reasons:

ODD MAPPING UPON LEGITIMATE INTUITIONS. Williams continues his allegiance to the nemesis of gene selectionism, the false criterion that has always doomed the theory to incoherence: faithful replication as the defining property for a "unit of selection"—now reformatted as a unit that only exists in the newly formulated codical domain, for Williams has now admitted that replicators are not causal agents in the usual realm of material entities. Williams promotes his old standard—faithful replication—as the primary criterion for "unithood" in his codical domain, thus leading to the following peculiar position: *genes* are units of selection (as the replicating consequence in the codical domain of selection upon organisms in the material domain); *gene pools* are also units of selection (as replicating consequences of higher-level selection upon groups to clades); but *genotypes*, in an intermediate category, are not units of selection (except in asexual organisms, where replication is faithful). Thus the codical domain skips a space in the hierarchy, and contains no organismic level of selection (except for asexual creatures) because the corresponding codex is impersistent.

THE OLD ERROR OF CONFUSING BOOKKEEPING WITH CAUSALITY. Williams's complex move in devising a separate hierarchy for nonmaterial units of information (and then juxtaposing this new sequence against the conventional hierarchy of evidently material and admittedly causal units), amounts to little beyond an elaborate and superfluous effort to rescue the unsalvageable theory of gene selectionism by granting both primacy and causal status (but only linguistically) to genes as replicators. But nothing new has been added beyond some terminology. The old error remains in full force—if anything, even heightened by the counterintuitive complexities and mental manipulations required to operationalize the scheme of dual hierarchies. A parallel hierarchy for nonmaterial entities of information? What can such a claim mean? Take the idea apart; pull the codical clothing off this new emperor, and whom do we find naked underneath? our old friend, the bookkeeper. Why must he continually try to play on the field of material objects engaged in nature's grand game of causality? Why should he be ashamed of his vital but different role? Bookkeeping is also a necessary, and entirely honorable, activity. The results of causal processes must be tabulated, and we rightly treasure the lists. We continue to stand in awe before "60" in Babe Ruth's home run column for 1927, and "70" in Mark McGwire's for 1998. But 70 is a record, not a cause—a summary of a great achievement, not the bat itself, or the muscles in a pair of strong arms. As nonmaterial objects suited for recording, codices are units of bookkeeping.

The history of gene selectionism has provided a grand intellectual adventure for evolutionary theory—from inception as a manifesto (Williams, 1966), through numerous excursions into pop culture, to valiant (though doomed) attempts to work through the logical barriers and to develop a consistent and workable theory (Dawkins, 1982; Williams, 1992). "Pareto-errors" always inspire a good race. No one really loses—though false theories like gene selectionism must eventually yield—because the resulting clarifications can only strengthen a field, and interestingly fallacious ideas often yield important insights. Without this debate, evolutionary biologists might never have properly clarified the differing roles of replicators and interactors, items for bookkeeping and units of selection. And we might not have developed a consistent theory of hierarchical selection without the stimulus of an opposite claim that genes could function as exclusive causal agents.

Some evolutionists, largely perhaps in fealty to their own pasts, continue to use the language of gene selectionism, even while their revised accounts elucidate and unconsciously promote the hierarchical view (see, in particular, Williams's excellent fourth chapter, in his 1992 book, on selection upon multiple interactors at several levels). Williams, to use a locution of our times, may still be talking the talk of gene selectionism, but he is no longer walking the walk.

Nearly all major participants in this discussion met at Ohio State University in the summer of 1988. There I witnessed a wonderful little vignette that may serve as an epitome for this section. George Williams presented his new views (the substance of his 1992 book), and surprised many people with his conceptual move towards hierarchy (within his unaltered terminology). I could not imagine two more different personalities in the brief and telling interchange that followed. Marjorie Grene—the great student of Aristotle, *grande dame* of philosophy, one of the feistiest and toughest people I have ever known, and a supporter of the hierarchical view—looked at Williams and simply said: "You've changed a lot." George Williams, one of the calmest and most laconic of men, replied: "It's been a long time."

# Logical and Empirical Foundations for the Theory of Hierarchical Selection

## LOGICAL VALIDATION AND EMPIRICAL CHALLENGES

### R. A. *Fisher and the compelling logic of species selection*

R. A. Fisher added a short section entitled "the benefit of species" to the second edition (1958) of his founding document for the Modern Synthesis: *The Genetical Theory of Natural Selection* (first published in 1930). I do not know why he did so, but the result could not be more favorable for fruitful debate—for Fisher, in these few additional paragraphs, grants to the concept of species selection the two requisite properties for any healthy and controversial theory. In presenting his argument, Fisher proclaims the logic of species selection unassailable, and then denies that this genuine phenomenon could have any substantial importance in the empirical record of evolution on our planet. No situation can be more propitious for useful debate about a scientific theory than validation in logic accompanied by controversy about actual evidence! (Obviously, I do not share Fisher's pessimism about empirical importance, and shall devote this section to explaining why.)

Fisher begins this interpolated passage by stating that Natural Selection (in his upper-case letters), in its conventional Darwinian mode of action among organisms, cannot explicitly build any features for "the benefit of the species" (though organismic adaptation may engender such effects as side consequences). Speaking of instinctual behaviors, Fisher writes (1958, p. 50): "Natural Selection can only explain these instincts in so far as they are individually beneficial, and leaves entirely open the question as to whether in the aggregate they are a benefit or an injury to the species." But Fisher then recognizes that, in principle, selection among species could occur, and could lead to higher-level adaptations directly beneficial to species. However, lest this logical imperative derail his strict Darwinian commitments to the primacy of organismic selection, Fisher then adds that species selection—though clearly valid in logic and therefore subject to realization in nature—must be far too weak (relative to organismic selection) to have any practical effect upon evolution. I regard the following lines (Fisher, 1958, p. 50) as one of the "great quotations" in the history of evolutionary thought:

> There would, however, be some warrant on historical grounds for saying that the term Natural Selection should include not only the selective survival of individuals of the same species, but of mutually competing species of the same genus or family. The relative unimportance of this as an evolutionary factor would seem to follow decisively from the small *number* of closely related species which in fact do come into competition, as compared to the number of individuals in the same species; and from the vastly greater *duration* of the species compared to the individual.

Fisher's theoretical validation of the logic behind species selection has never been effectively challenged. Even the most ardent gene selectionists have granted Fisher's point, and have then dismissed species selection from extensive consideration (as did Fisher) only for its presumed weakness relative to their favored genic level, and not because they doubt the theoretical validity, or even the empirical reality, of selection at this higher level. Dawkins (1982, pp. 106–107) has emphasized Fisher's argument about impotence by noting that, at most, species selection might accentuate some relatively "uninteresting" linear trends (like size increase among species in a lineage), but could not possibly "put together complex [organismal] adaptations such as eyes and brains." Dawkins continues:

> When we consider the species . . . the replacement cycle time is the interval from speciation event to speciation event, and may be measured in

thousands of years, tens of thousands, hundreds of thousands. In any given period of geological time, the number of selective species extinctions that can have taken place is many orders of magnitude less than the number of selective allele replacements that can have taken place . . . We shall have to make a quantitative judgment taking into account the vastly greater cycle time between replicator deaths in the species selection case than in the gene selection case.

I strongly support Dawkins's last statement, but will argue (see pages 703–712) that, when we factor punctuated equilibria into the equation, species selection emerges as a powerful force in macroevolution (though not as an architect of complex organismic adaptations).

Williams has also supported Fisher's argument about the logic of higher level selection—even in his gene selectionist manifesto of 1966, where he defends the possibility, but then denies the actuality: "If a group of adequately stable populations is available, group selection can theoretically produce biotic adaptations, for the same reason that genic selection can produce organic adaptations" (1966, p. 110). In his later book, however, Williams becomes much more positive about the importance and reality of selection at several hierarchical levels: "To Darwin and most of his immediate and later followers, the physical entities of interest for the theory of natural selection were discrete individual organisms. This restricted range of attention has never been logically defensible" (1992, p. 38).

The developing literature has added three "classical" arguments against higher-level selection to supplement Fisher's point that cycle times for species must be incomparably long relative to the lives of organisms. All these arguments share the favorable property of accepting a common logic but challenging the empirical importance of legitimate phenomena—a good substrate for productive debate in science, in contrast with the confusion about concepts and definitions that so often reigns. In the rest of this section, I shall summarize the four classical arguments (Fisher's original plus the three additions); note that they can all be effectively challenged at the level for which they were devised ("group," or interdemic, selection); and then demonstrate that none has any strong force, in principle, against the empirical importance of the still higher level of species selection.

### The classical arguments against efficacy of higher-level selection

The usual arguments against higher-level selection admit that such phenomena must be possible in principle, but deny any meaningful efficacy on grounds of rarity and weakness relative to ordinary natural selection upon organisms.

WEAKNESS (BASED ON CYCLE TIME). R. A. Fisher's classical argument: How could species selection exert any measurable effect upon evolution? Rate and effect depend upon numbers and timings of births and deaths—to provide a sufficient population of items for differential sorting. But species endure for thousands or millions of years, and clades count their "populations" of component species in tens, or at most hundreds, and not as the millions or billions of organisms in many populations. How could species selection yield any measurable effect at all (relative to ordinary organismic selection) when, on average, billions of organismic births and deaths occur for each species origin or extinction, and when populations of organisms contain orders of magnitude more members than populations of related species in a clade?

WEAKNESS (BASED ON VARIABILITY). Hamilton (1971, 1987), in devising arguments against interdemic selection, pointed out that variation among demic mean values for genetically relevant and selected aspects of organismic phenotypes will generally be lower than variation among organisms within a population for the same features. Group selection cannot become a strong force if the mean phenotypes of such higher-level individuals express such limited variation to serve as raw material for selective change.

INSTABILITY, AS IN DAWKINS'S METAPHORS OF DUSTSTORMS IN THE DESERT AND CLOUDS IN THE SKY. This argument has also been most frequently advanced against interdemic selection. Demes, by definition, maintain porous borders because organisms in the same species can interbreed, and members of one deme can therefore, in principle, invade and join another in a reproductive role. If such invasions become frequent and numerous, the deme ceases to function as a discrete entity, and cannot be called an evolutionary individual. Moreover, many demes lack cohesion on their own account, and not only by susceptibility to incursion. Demes may arise as entirely temporary and adventitious aggregates of organisms, devoid of any inherent mechanism for cohesion, and defined only by the transient and clumpy nature of appropriate habitats that may not even persist for a requisite generation—as in the deme of all mice in a haystack, or all cockroaches in an urban kitchen.

INVASIBILITY FROM OTHER MORE POTENT LEVELS, USUALLY FROM BELOW. This standard argument, related both to Fisher's first point about cycle time and to the third point about invasibility discussed just above —and classically used to question the potential evolution of altruism by interdemic selection—asks how higher-level selection could possibly become effective if its operation inherently creates a situation where more powerful, lower-level invaders can cancel any result by working in the opposite direction. Suppose that interdemic selection, cranking along at its characteristic pace, increases the overall frequency of altruistic alleles in the entire species because demes with altruists enjoy differential success in competition against demes without altruists. This "leisurely" pace works well enough, but as soon as a selfish mutant arises in any deme with altruists, the advantage of this mutant in organismic selection against the altruistic allele should be so great that the frequency of altruistic genes must plummet *within* the deme, even while the deme profits in group selection from the presence of altruistic organisms. By Fisher's argument of cycle time, organismic selection of the self-serving should trump interdemic selection for altruism.

### Overcoming these classical arguments, in practice for interdemic selection, but in principle for species selection

Since the bulk of modern debate about higher-level selection has addressed interdemic (or so-called group) selection, the classical arguments have been framed mainly at the level just above our conventional focus upon organisms (though I predict that emphasis will shift to higher levels, particularly to species selection, as macroevolutionary theory develops). All four arguments have force, and do spell impotence for interdemic selection in many circumstances. But, as full generalities, these arguments have failed either to disprove interdemic selection as a meaningful process worthy of consideration at all, or to deny the efficacy of interdemic selection in several important circumstances.

I shall not review this enormous literature here (as my primary concern rests at still higher levels of selection), but I wish to note that two classes of argument grant interdemic selection sufficient strength and presence to count as a potentially major force in evolution (see Wade, 1978; Sober and Wilson, 1998). First, much mathematical modelling (and some experimental work) have adequately shown that, under reasonable conditions of potentially frequent occurrence in nature, group selection can assert its sway against the legitimate power of the four classical objections. In the cardinal example, under several plausible models, the frequency of altruistic alleles can increase within a species, so long as the rate of differential survival and propagation of demes with altruistic members (by group selection) overcome the admitted decline in frequency of altruists within successful demes by organismic selection. The overall frequency may rise within the species even while the frequency within each surviving deme declines.

Second, some well-documented patterns in nature seem hard to explain without a strong component of interdemic selection. Female-biased sex ratios, as discussed by Wilson and Sober (1994, pp. 640–641), provide the classic example because two adjacent levels make opposite and easily tested predictions: conventional organismic selection should favor a 1:1 ratio by Fisher's famous argument (1930); while interdemic selection should promote strongly female-biased ratios to enhance the productivity of groups. Williams (1966) accepted this framework, which he proposed as a kind of acid test for the existence of group selection. He allowed that female-biased ratios would point to group selection, but denied that any had, in fact, been documented, thus validating empirically the theoretical arguments he had developed for the impotence of group selection. Williams concluded (1966, p. 151): "Close conformity with the theory is certainly the rule, and there is no convincing evidence that sex ratios ever behave as a biotic adaptation." But empirical examples of female-biased ratios were soon discovered aplenty (see Colwell, 1981, and numerous references in Wilson and Sober, 1994, p. 592). Some authors (Maynard Smith, 1987, for example) tried to interpret this evidence without invoking group selection, but I think that all major participants in the discussion now admit a strong component of interdemic selection in such results—and reported cases now number in the hundreds, so this phenomenon cannot be dismissed as an odd anomaly in a tiny corner of nature. Williams now accepts this interpretation (1992, p. 49), writing "that selection in female-biased Mendelian populations favors males, and that it is only the selection among such groups that can favor the female bias."

The primary appeal of this admirably documented example lies in the usual finding of only moderate female biases—more than organismic selection could allow (obviously, since any bias at all would establish the point), but less than models of purely interdemic selection predict. Thus, the empirical evidence suggests a balance between adjacent and opposing levels of selection—with alleles for female-biased sex ratios reduced in frequency by organismic selection within demes, but boosted above the Fisherian balance (across species as a whole) because they increase the productivity of demes in which they reside, however transiently, at high frequency.

When we move to the level of species selection, the most important for macroevolutionary theory, we encounter an even more favorable situation. For interdemic selection, the classical contrary arguments had legitimate force, but could be overcome under conditions broad enough to grant the phenomenon considerable importance. For species selection, on the other hand, three of the classical arguments don't even apply in principle—while the fourth (weakness due to cycle time) becomes irrelevant if punctuated equilibrium prevails at a dominant relative frequency.

Proceeding through the classical objections in reverse order, the fourth argument about invasibility from below has strength only in particular contexts—when, in principle, a favored direction of higher-level selection will usually be opposed by stronger selection at the level immediately below. (In the classic case, selfish organismal "cheaters" derail group selection for altruism. Nonetheless, while the argument of invasibility may hold for this particular case—and while, for contingent reasons in the history of science, this example became the paradigm for discussion of interdemic selection—I see no reason in principle for thinking that organismal selection must always, or even usually, oppose interdemic selection. The two levels may operate simultaneously and in the same direction, or at least orthogonally—see Wade's (1978) classic work on this subject.)

In any case, no general reason has been advanced for thinking that organismic or interdemic selection should characteristically oppose species selection—and the argument of invasibility therefore collapses. Of course, organismic selection *may* operate contrary to the direction of species selection—and must frequently do so, particularly in the phenomenon that older textbooks called "overspecialization," or the development of narrowly focussed and complex adaptations (the peacock's tail as a classic example) that enhance the reproductive success of individual organisms, but virtually guarantee a decreased geological life span for the species. Other equally common modes of organismic selection, however, either tend to increase geological longevity (improvements in general biomechanical design, for example) or to operate orthogonally, and therefore "beneath the notice" of species selection. Since our best examples of species selection work through differential rates of speciation rather than varying propensities for extinction, and since most organismal adaptations probably don't strongly influence a population's rate of speciation (or at least don't manifest any bias for decreasing the rate), essential orthogonality of the two levels will often prevail in evolution.

The third argument of instability, while potent for demes, clearly does not apply to species. Sexual species are as well bounded as organisms. Just as genes and cell lineages generally do not wander from organism to organism (whereas organisms often move readily from deme to deme), neither can organisms or demes wander from species to species. The reasons for such tightness of bounding differ between organism and species, but these two evolutionary individuals probably exceed all others in the strength of this key criterion. Species maintain and "police" their borders just as well as organisms do.

The tight bounding of an organism arises from functional integration among constituent parts, including an impermeable outer covering in most cases, and often an internal immune system to keep out invaders. The tight bounding of a species (as classically defined for sexually reproducing eukaryotes) arises from reproductive interaction among parts (organisms), with firm exclusion of parts from any other species. Moreover, this exclusion is actively maintained, not merely passively propagated, by traits that became a favorite subject of study among founders of the Modern Synthesis, especially Dobzhansky and Mayr—so-called "isolating mechanisms." Species may lack a literal skin, but they remain just as well-bounded as organisms in the sense required by the theory of natural selection.

This discussion on the validity and centrality of species as units of selection highlights my only major unhappiness with Wilson and Sober's (1994) superb defense of hierarchical selection, otherwise followed closely in this book. They insist upon functional integration as the main criterion for identifying units of selection (vehicles in their terminology, interactors or evolutionary individuals for others). They insist that the following question "is and always was at the heart of the group selection controversy—can groups be like individuals in the harmony and coordination of their parts" (1994, p. 591).

I do not object to the invocation of functionality itself, but rather to their narrow definition, too parochially based upon the kind of functionality that organisms display. The cohesion (or "functionality") of species does not lie in the style of interaction and homeostasis that unites organisms by the integration of their tissues and organs. Rather, the cohesion of species lies in their active maintenance of distinctive properties, achieved by joining their parts (organisms) through sexual reproduction, while excluding the parts of other species by evolution of isolating mechanisms.

I much prefer and support Wilson and Sober's more general definition (1994, p. 599): "Groups are real to the extent that they become functionally organized by natural selection at the group level." Species meet this criterion by evolving species-level properties that maintain their cohesion as evolutionary individuals. The key to a broader concept of "functionality" (that is, the ability to operate discretely as a unit of selection) lies in the evolution of active devices for cohesion, not in any particular style of accomplishment—either the reproductive barriers that maintain species, or the homeostatic mechanisms that maintain organisms.

The second argument of weakness based on lack of sufficient variability among group mean values also doesn't apply to species. Demes of mice from separated but adjacent haystacks may differ so little in group properties that the survival of only one deme, with replenishment of all haystacks by migrants from this successful group, might scarcely alter either allelic frequencies across the entire species, or even average differences among demes. But new species must differ, by definition, from all others—at least to an extent that prevents the reproductive merging of members. Thus, the differential success of some species in a clade must alter—usually substantially—the average properties of the clade (whereas, one level down, the differential success of some demes need not change the average properties of the species very much, if at all).

The first argument about weakness due to long cycle time and small populations therefore remains as the only classical objection with potential force against species selection. And, at first glance, Fisher's argument would seem both potent and decisive. The basic observation cannot be faulted: billions of organism births usually occur for each species birth; and populations of organisms within a species almost always vastly exceed populations of species in a clade. How then could species selection, despite its impeccable logic, maintain any measurable importance when conventional organismal selection holds the tools for such greater strength?

The logic of Fisher's argument cannot be denied, but we must also consult the empirical world. Organismic selection must overwhelm species selection when both processes operate steadily and towards the same adaptive "goal"—for if both levels work in the same direction, then species selection can only add the merest increment to the vastly greater power of organismic selection; whereas, if the two levels work in opposite directions, organismic selection must overwhelm and cancel the effect of species selection.

But the empirical record of the great majority of well-documented fossil species affirms stasis throughout the geological range (see next chapter). The causes for observed nondirectionality within species have not been fully resolved, and the phenomenon remains compatible with the continuous operation of strong organismic selection—for two common explanations of stasis as a central component of punctuated equilibrium include general prevalence of stabilizing selection, and fluctuating directional selection with no overall linear component due to effectively random changes of relevant environments through time. In any case, however, the observation of general stasis seems well established at high relative frequency (I would say dominant, but I also must confess my partisanship).

In this factual circumstance, since change does not generally accumulate through time within a species, organismic selection in the conventional gradualistic and anagenetic mode cannot contribute much to the direction of trends within a clade. Change must therefore be concentrated in events of branching speciation, and trends must arise by the differential sorting of species with favored attributes. If new species generally arise in geological moments, as the theory of punctuated equilibrium holds, then trends owe their explanation even more clearly to higher-level sorting among species-individuals acting as discrete entities with momentary births and stable durations in geological time.

Organismic selection may trump species selection in principle when both processes operate at maximal efficiency, but if change associated with speciation operates as "the only game in town," then a weak force prevails while a potentially stronger force lies dormant. Nuclear bombs certainly make conventional firearms look risible as instruments of war, but if we choose not to employ the nukes, then bullets can be devastatingly effective. The empirical pattern of punctuated equilibrium therefore becomes the factual "weapon" that overcomes Fisher's strong theoretical objection to the efficacy of species selection.

(This argument provides a second example for the importance of punctuated equilibrium in validating the independence of macroevolutionary theory by failure of pure extrapolationism from microevolutionary dynamics. We saw previously (pp. 604–608) that punctuated equilibrium strongly fosters the argument for species as evolutionary individuals capable of operating as units of selection. We now note that punctuated equilibrium also affirms the potential strength of species selection against a cogent theoretical claim for its impotence.)

In summary, three of four classical arguments against higher-level selection do not apply to species, while the fourth loses its force in a world dominated by punctuated equilibrium. I see no barrier to the cardinal importance of species selection in the history of life.

## EMERGENCE AND THE PROPER CRITERION FOR SPECIES SELECTION

### Differential proliferation or downward effect?

This subject and its literature, as I have noted throughout the chapter, have been plagued to an unusual degree by conceptual confusions and disputes about basic definitions and terminology. As an important example, and as many participants have noted (see especially Damuth and Heisler, 1988; and Brandon, 1988, 1990), two quite different criteria for the definition of higher-level selection have circulated through the literature. (In most cases, they yield the same conclusion, so this situation has not produced anarchy; but in a few cases, some crucial, they may lead to different assertions, so the situation has fostered confusion.)

In the first approach, one chooses a focal level of analysis (conventionally one of the two lower levels of organism or gene), and then considers the effect of membership in a higher-level group upon fitness values of the chosen lower-level unit. If, for an identical organism, life in one kind of deme yields a fitness different from life in another kind of deme, then selection includes a group effect from the deme level. (We invoke this formulation, for example, if we argue for group selection by showing that *organisms* in a deme with altruists do better than identical organisms in a group lacking altruists.)

In the second approach, strongly favored here, we hold firm to the classical bare-bones Darwinian definition, but recognize that selection can work on evolutionary individuals at many hierarchical levels. Selection has traditionally been defined as the differential reproductive success of evolutionary individuals based on the fitnesses of their traits in interaction with the environment. Thus, we recognize higher-level selection by the differential proliferation of some higher-level individuals (demes, species, clades) over others—just as we define conventional natural selection by the differential reproductive success of some organisms based on phenotypic traits that confer fitness.

These two approaches often yield concordant results for the obvious reason that differential proliferation of higher-level units (the second criterion) often defines the group effect that influences the fitness of lower-level individuals chosen as a focus (the first criterion). But the two criteria need not correspond, leading to situations where we would identify group selection by one criterion, but deny the same process by the other. For example, under the first criterion of group effects on lower-level fitness, some higher-level properties of groups can influence lower units without causing any differential reproduction of the groups themselves. On this criterion, for example, some biologists have held that frequency dependent selection must be viewed, *ipso facto,* as an example of group selection—a claim simply incomprehensible under the alternative criterion of differential group proliferation. (The unresolved, and perhaps largely semantic, issue of whether kin selection should be interpreted as a form of group selection, or only an extension of conventional lower-level selection, also presupposes this criterion of group effect upon lower-level fitness—see Wilson and Sober, 1994.)

A predominantly sociological issue has often set preferences between these criteria. Paleontologists, and other students of species selection, myself included, have strongly advocated the criterion of differential reproduction for higher-level individuals as a strict and obvious analog of ordinary natural selection as conventionally understood. Neontologists and students of group selection have generally (though not always) preferred the criterion of "group effect on gene or organismal fitness," both from fealty to Darwinian traditions for using organisms as a primary focus, and because certain contentious issues, especially the evolution of altruism, have generally been posed in organismal terms—"why can saintly Joe be so nice if he loses reproductive success thereby?"

Three major reasons lead to my strong preference for the criterion of differential proliferation correlated with properties of relevant evolutionary individuals that confer fitness in interaction with their environment. First, we thereby follow standard definitions of selection, which have always been based on causal plurifaction, not on mere effect. Second, why would we ever prefer an elaborate and indirect definition—in terms of effects on something else at a scale far removed from the causal interaction—over a simpler account rooted in the direct result of the causal process itself? Considered in these terms, the criterion of "group effect on organismal fitness" seems downright peculiar. We only entertain such a standard for contingent reasons of history and philosophical preference—the Darwinian tradition for focusing on organisms, and our larger scientific allegiance to reductionism. Third, we can too easily lose the force and location of causality when we study a phenomenon through indirect effects expressed elsewhere, rather than by immediate operation. True, we are supposed to assess the separate effects upon lower-level focal units—from deme membership, or species membership, for example. But since several higher levels may simultaneously affect a lower focal unit, we may not be able to untangle the differences, and we may end up with an account of consequences, rather than causes.

As an obvious example of these pitfalls, I point out that gene-selectionism, with all its fallacies, arises from an erroneous inversion in the criterion of "group effect on lower-level fitness." One begins with the basic statement that membership in higher-level units affects the fitness of genes. So far, so good. But if one then makes the error of assuming that replicators, rather than interactors, should be units of selection—and then chooses genes as fundamental replicators both by general reductionistic preference, and by allegiance to faithfulness in replication as a necessary criterion—then one becomes tempted to misidentify effects as causes. The gene selectionist then slides down the following slippery slope: why should I talk about higher-level interactors affecting gene fitness? why don't I just consider higher-level interactors as one aspect of the gene's environment? in that case, why should I talk about higher-level interactors as entities at all? environment is environment, however constituted, and whether clumped into interactors housing the genes or not? in fact, why even try to identify the environment's forms of clumpiness? why not, instead, simply average the gene's fitness over all aspects of environment to achieve a single measure of the gene's evolutionary prowess?

This line of argument, as its least attractive feature, relentlessly dissolves causality. We begin with the causal agents of selection—interactors at various hierarchical levels. (Even the most ardent gene selectionists, as I show on pages 631–632, cannot avoid discussing the causal process of selection in terms of these interactors.) We then represent interactors by their effects on genes. Next, we decide to consider interactors only as environments of genes. Then we lose interest in their nature and action because "environmental clumping" (the "expression" of interactors in this view) does not appear to represent an important issue. Finally, we dissolve the interactors entirely by deciding to average the fitness of genes across all aspects of the environment. And, before we notice what we have really done, causality has disappeared.

In a vigorous defense of gene selection against the hierarchical view of Wilson and Sober (1994), Dawkins (tongue-in-cheek to be sure) pretends to be "baffled" by "the sheer, wanton, head-in-bag perversity of the position that they champion" (commentary in Wilson and Sober, 1994, p. 617). Such a sense of strong psychological frustration must arise when you and your opponents seem to be saying the same thing, but in such utterly different ways, and to such radically different effect. Thus, Dawkins presents his gene-selectionist reformulation of Wilson and Sober's *Weltanschauung* (mine as well, by the way):

> Selection chooses only replicators . . . Replicators are judged by their phenotypic effect. Phenotypic effects may happen to be bundled, together with the phenotypic effects of other replicators, in vehicles. Those vehicles often turn out to be the objects that we recognize as organisms, but this did not have to be so. . . . There did not have to be any vehicles at all. . . . The environment of a replicator includes the outside world, but it also includes, most importantly, other replicators, other genes in the same organism and in different organisms, and their phenotypic products.

(Note that I did not exaggerate or caricature in my previous summary; gene selectionists do regard "clumping" into vehicles as beside the point, and they do dissolve these vehicles—the true units of selection—into "environment" considered as the sum of contexts for any gene.)

Wilson and Sober (1994, p. 641) responded to Dawkins with their own frustration:

> Dawkins remains so near, yet so far . . . We could not ask for a better summary of the gene-centered view. The question is, are vehicles of selection absent from this account or have they merely been reconceptualized as environments of the genes. The answer to this question is obvious at the individual [organism] level, because Dawkins acknowledged long ago that individuals [organisms] can be vehicles of selection . . . despite the fact that they are also environments of the genes. The answer is just as obvious at the group level . . . [Dawkins's] passage does not refute the existence of vehicles, but merely assumes that the vehicle concept can be dispensed with and that natural selection can be studied entirely in terms of average genic effects.

Is this brouhaha much ado about nothing? Are the two views—selection on a hierarchy of interactors, and representation of all selective forces in terms of gene fitnesses, with interactors treated as environments of genes—truly equivalent, and our decision just a matter of preference, or a question of psychological judgment about superior sources of insight? Is this twofold choice just another manifestation of Dawkins's old Necker Cube (see p. 640)—a flipping between two equivalent facets of reality, an example of conventionalism in philosophy?

The answer, I think, must be a clear and resounding "no." The two alternatives represent strikingly different views about the *nature of reality and causality*. We all agree that we need to know causes—and natural selection is a causal process. Gene selection confuses bookkeeping (properly done at the genic level) with causality (a question of evolutionary individuals plurifying differentially, based on interaction of their phenotypes with the environment). If we dissolve interactors into an overall "environment" of the genes, and then average a gene's fitness across all environments—the procedure of gene selectionism—then we lose causality.

Wilson and Sober (1994, p. 642) also reject the purely pluralist, or Necker Cube view: "There is no room for pluralism on these substantive empirical issues . . . Group-level adaptations can be represented at the individual [organism] and gene level by averaging the fitness of lower level units across higher level units. Gene- and individual-level adaptations cannot be interpreted as group adaptations without committing the errors of naive group selection, but the gene's-eye view and the individual's-eye view cannot deny the existence of group-level adaptations (when groups are vehicles of selection) without being just plain wrong."

Arnold and Fristrup (1982, p. 115) make the same point for the intrinsic reality—and not just preferential status *vs.* other equivalent representations—of species selection: "The characters that increase individual [organismic] fitness do not necessarily cause speciation or prevent extinction. Thus, it is misleading to adopt the convention of expressing all higher level trends in terms of individual [organism] level fitness."

For all these reasons, I strongly advocate that we define higher-level selection as the differential proliferation of relevant evolutionary individuals based on causal interaction of their properties with surrounding environments—rather than by representing the effect of higher-level membership on the fitness of a designated lower-level individual. Only in this way will we avoid a set of confusions, and two pitfalls that easily follow, one after the other, with the first as a kindly delusion, and the second as an outright error: first, a falsely pluralistic belief in the equivalency of alternative representations at different levels; and, second, the siren song of gene selection as defining the only legitimate level of causal analysis in evolution. Only in this way will we achieve a clear and unified view that treats each level in the same manner, and approaches each evolutionary individual with the same set of questions. With this apparatus of analysis, we can attain a coherent and comprehensive theory of hierarchical selection—the most potentially fruitful, promising, and proper expansion of the Darwinian research program now before us.

### Shall emergent characters or emergent fitnesses define the operation of species selection?

Once we agree to define higher-level selection by differential proliferation of relevant units based on interaction between their traits and the environment, then we must (above all) develop clear criteria for the definition and recognition of traits in the unfamiliar world of higher-level individuals. Since we encounter enough trouble in trying to define and parse traits for the kind of individuals we know best—integral, complex, and continuous organisms like ourselves—we should not be surprised that this issue becomes particularly refractory at higher levels, and thus acts as a considerable impediment to the development of a rigorous theory of hierarchical selection. In particular, what should count, for purposes of defining evolutionary interaction with the environment, as a trait of a species?

The developing literature on this subject has featured a rich and interesting debate between two quite different approaches that, nonetheless, can be united in a coherent way to form the basis of a unified macroevolutionary theory of selection: the "emergent character" approach, as particularly championed by Elizabeth Vrba (1983, 1984b, 1989; Vrba and Eldredge, 1984; Vrba and Gould, 1986); and the "emergent fitness" approach inherent in the classic paper of Lewontin (1970), developed and defended in the important work of Arnold and Fristrup (1982), given further mathematical form in Damuth (1985), and Damuth and Heisler (1988), and most fully codified and expressed by Lloyd (1988—see also Lloyd and Gould, 1993; and Gould and Lloyd, 1999).

Grantham (1995), in an excellent review of hierarchical theories of macroevolution, has christened this discussion "The Lloyd-Vrba Debate," so the issue has now even acquired a proper name. The codification makes me feel a bit strange, since I have written papers on the subject with both protagonists (Gould and Vrba, 1982; Vrba and Gould, 1986; Lloyd and Gould, 1993; Gould and Lloyd, 1999), and do not view the issue as dichotomous; though the two viewpoints are surely distinct, and I have changed my mind—as a former supporter of Vrba's "strict construction," who became convinced that Lloyd's more inclusive formulation forges a better match with conventional definitions of selection, and provides more promise for constructing an operational theory. But Lloyd does not disprove Vrba; rather, Vrba's exclusive domain becomes a subset of "best cases" in Lloyd's formulation. In this crucial sense, the theories sensibly intermesh.

Vrba's "emergent character" approach requires that a trait functioning in species selection be emergent at the species level—basically defined as origin by non-additive interaction among lower-level constituents. Since all science works within particular sociological and historical circumstances, we must understand that the greatest appeal of this strict criterion lies in its ability to "fend off" the conventional objection to species selection in a Darwinian and reductionistic world—namely, that the trait in question, although describable as characterizing a species, "really" belongs to the constituent, lower-level parts—and that the causal process therefore reduces to ordinary Darwinian natural selection on organisms or genes. For, when Vrba's criterion of emergence holds, one can't, in principle, ascribe the trait in question to lower levels. The trait, after all, does not exist at these lower levels. It makes a "first appearance" at the species level, for the trait arises through non-additive interaction of component lower-level parts or influences. If one species proliferates differentially within a clade by higher rates of speciation based upon such populational traits as geographic range, or density of packing among organisms, then we cannot ascribe selection to the organismic level—for organisms, by the logic of definition, cannot possess a population density, while the geographic range of a species need not correlate at all, or in any simple way, with the size of an organism's personal territory during its lifetime.

The strength of the "emergent character" criterion lies in its ability to identify a set of hard-line, unambiguous cases for species selection. For we must speak of selection among species if the relevant trait not only doesn't exist at any lower level, but can't even be represented as a linear combination of lower-level parts—for the nonadditive interactions that build the populational trait only arise within the population, and make no sense outside such an aggregation.

But we soon begin to worry that such a criterion may be too restrictive in eliminating a wide variety of traits that we intuitively view as features of populations, but that do not arise by nonlinear interaction of subparts, and do not therefore qualify as emergent by Vrba's criterion (which also matches the standard definition of the important concept of emergence in philosophy). Species and other higher-level individuals also develop features that seem to "belong" to them as an entity, but that arise additively as "aggregate" or "sum-of-the-parts" characters. Consider the mean value of a trait? This figure belongs to no individual and becomes, in this legitimate sense, a character of the population. But a mean value doesn't "emerge" as a functional "organ" of the population by nonlinear interactions among organisms. A mean value represents an aggregate character, calculated by simple summation, followed by division.

And how shall we treat variability—an even more "intuitive" candidate for a species-level character that may be important in survival and proliferation of species? An individual organism doesn't possess a variability, so the property belongs to the species. But variability also represents an aggregate character—another average of a sum-of-the-parts. Do we not want to talk about species selection when species B dies because constituent organisms show no variation for a trait that has become strongly inadaptive in the face of environmental change—while species A lives and later multiplies because the same trait varies widely, and includes some states that can prosper in the new circumstances? Yes, species B dies because each of its parts (organisms) expires. In this sense, we can *represent* extinction as a summation of deaths for organismal reasons. But don't we also want to say that A survived by virtue of greater variability—a trait that does not exist at the organismal level, but that surely interacted with the new environment to preserve the species?

Vrba's solution, which I greatly respect but now regard as less useful than the alternative formulation, requires that we not designate differential proliferation of species based on aggregate characters of populations as species selection—but rather that we interpret such cases as upward causation from the traditional organismal level. Vrba (1980 *et seq.*) has coined and developed the term "effect hypothesis" for such situations—since the differential proliferation of species A vs. species B arises as an effect of organismal properties (of the individuals in species A that vary in the "right" direction), resulting in the survival of species A.

Vrba, and (I think) all other major workers in this area, have always regarded the effect hypothesis as a macroevolutionary theory because, in a heuristic and descriptive sense, one must apply the notion to species considered as items of evolutionary history. But events under the effect hypothesis are causally reducible to the traditional organismic level. (This kind of situation represents the minimal claim for an independent macroevolutionary theory—the need for descriptive engagement at the level of species, even if no distinct causality emerges at this higher level. This book defends the stronger claim for important causal uniqueness at the species level and above. Vrba, of course, also advocates this stronger version because she argues that some cases of differential species proliferation arise by the effect hypothesis, while others occur by true species selection based on emergent characters. I advocate a much larger role for causal uniqueness by defending the emergent fitness approach, a criterion that greatly expands the frequency and importance of species selection.)

To facilitate this distinction, Vrba and I developed a terminology to resolve a common confusion in evolutionary theory between the simple, and purely descriptive, observation of differential reproductive success—which we named "sorting"—and the causal claim—always and properly called "selection"—that observed success arises from interaction between properties of the relevant evolutionary individual and its environment (see Vrba and Gould, 1986). Evolutionary biology needs this distinction because students of the field have often—with misplaced confidence in selection's ubiquity and exclusivity—made a case for selection based on nothing more than an observation of differential reproductive success (sorting), without any attempt to elucidate the cause of such sorting. A leading textbook, for example, proclaimed that "selection . . . is differential survival and reproduction—and no more" (Futuyma, 1979, p. 292).

Under Vrba's criterion of emergent characters, differential species proliferation by the effect hypothesis counts only as sorting at the species level—since the characters responsible for selection belong to organisms, but transfer an effect to the species level by upward causation. On the other hand, differential species proliferation based on emergent species characters does count as selection at the species level. However, under the broader criterion of emergent fitness, any species-level trait that imparts an irreducible fitness to species in their interaction with the environment defines a true process of selection at the species level, whether the trait itself be aggregate or emergent.

In the "emergent fitness" approach, we do not inquire into the history of species-level traits that interact with the environment to secure differential proliferation. We do not ask where the traits originated in a structural or temporal sense—that is, whether such traits arose by emergence at the species level, or as aggregate features by summation of properties in component organisms or demes. We only require that these traits characterize the species and influence its differential rate of proliferation in interaction with the environment. In other words, we only demand that aspects of the *fitness* of the species be emergent and irreducible to the fitnesses of component organisms. For cases where species function as interactors, or potential units of selection, Lloyd and Gould write (1993, pp. 595–596):

> Interactors, and hence selection processes themselves, are individuated by the contributions of their traits to fitness values in evolutionary models; the trait itself can be an emergent group property or a simple summation of organismic properties. This definition of an entity undergoing selection is much more inclusive than in the emergent character approach, since an entity might have either aggregate or emergent characters (or both) . . . The emergent fitness approach requires only that a trait have a specified relation to fitness in order to support the claim that a selection process is occurring at that level. . . . In other words, the interactor's fitness covaries with the trait in question.

In a classic example, much discussed in the literature (Arnold and Fristrup, 1982; Gould, 1982c; Lloyd and Gould, 1993; Grantham, 1995), several clades of Tertiary gastropods show trends to substantial decrease in relative frequency of species with planktotrophic larvae vs. species that brood their young. In one common explanation (by no means universally accepted), this reduction occurs by species sorting based on the lower speciation rate of planktotrophic species—an hypothesized consequence of the lower probability for formation of isolates in species with such widespread and promiscuous larval dispersal. The sorting clearly occurs by selection, since low speciation rate arises as a consequence of interaction between traits of interactors and their environment. But at what level does selection occur?

Under the emergent character approach, the case becomes frustrating and ambiguous. Does the crucial property of "low speciation rate" in planktotrophs result from an emergent species character? In one sense, we are tempted to answer "yes." Organisms, after all, don't speciate; only populations do—so mustn't the trait be emergent at the population level? But, in another sense, low speciation rate arises as a consequence of population structures induced by planktotrophy, a presumed adaptation at the organismal level—so perhaps the key character can be reduced to simple properties of organisms after all.

I have gone round and round this example for twenty years, often feeling confident that I have finally found a clear resolution, only to recognize that a different (and equally reasonable) formulation yields the opposite interpretation. All other participants in this debate seem equally afflicted by frustration, so perhaps, the fault lies in the concepts, and not in ourselves that we seem to be underlings, unable to achieve closure.

However, if we invoke the broader criterion of emergent fitness, the problem gains a clear resolution in favor of species selection. A structural feature of populations, leading to a low frequency of isolation for new demes, must be treated as a character of populations in any conventional usage of language. As stated above, individual organisms don't speciate; only populations do—so the character belongs to the species. However, the character may represent an aggregate rather than an emergent feature—thus debarring species selection under the emergent character approach. But, under the emergent fitness approach, so long as the character (whether aggregate or emergent) belongs to the species, and so long as the fitness of the species covaries with the character—and no one denies the covariation in this case—we have detected an instance of species selection.

Arnold and Fristrup (1982, p. 114) present this argument in a clear and forceful way:

> The critical characters—larval strategies—may well have arisen for reasons that can be seen as adaptive in a traditional Darwinian sense. However, regardless of the mechanism by which they became fixed, these strategies behave as properties of species in that they result in distributions of rates of speciation and extinction within this group . . . It might be tempting to assume that there are fewer planktotrophic species because the individuals in these species were somehow less fit than the individuals in non-planktotrophic species. However, it is obvious that the same result could obtain even if planktotrophic and non-planktotrophic individuals [organisms] have equal fitnesses, by virtue of the population structures that are concomitants of these larval strategies. Thus, the observed distribution of species types within these gastropods is not predicted from individuals level fitness alone, underscoring the necessity of the higher level of analysis.

In other words, the relative frequency of planktotrophic species falls not because planktotrophic organisms must be less fit (they may, in fact, be more fit on average across the clade), but because a character fixed by organismic selection yields the effect of lowering the speciation rate at a higher level. The population structure produced by planktotrophy may not rank as an emergent character, but does confer an emergent fitness at the species level—a fitness irrelevant to individual organisms, which, to emphasize the obvious point one more time, do not speciate.

Finally, we may seal the case by citing Grantham's important argument (1995, p. 301) that "species selection does not require emergent traits because higher-level selection acting on aggregate traits can oppose lower-level selection." Vrba herself has argued (1989, p. 80) that "the acid test of a higher level selection process is whether it can in principle oppose selection at the next lower level." Surely such an opposition can arise "in principle" (and probably in actuality) in this case—for planktotrophy could be positively selected at the organismic level, but may, through its strong effect on population structure, and the resulting consequences for rates of speciation, enjoin negative selection at the species level.

To summarize, we all agree that an independent theory of macroevolution must identify higher-level causal processes that are not reducible to (or simple effects of) causes operating at conventional lower levels of gene and organism. This premise defines the theoretical salience of the debate about species selection—for if such a process exists, and can also be validated as both common in evolution and irreducible in principle, then macroevolutionary theory has been achieved. For this reason, evolutionary biologists, who usually eschew academic philosophy (as the mildly philistinistic culture of science generally dictates), have joined in such classical philosophical debates as the meaning of reduction and emergence.

Vrba's criterion of emergent characters establishes an obvious case for irreducibility because the trait that causes species selection can claim neither existence nor representation at the conventional organismic level. Grantham writes (1995, p. 308): "When a component of species-level fitness is correlated with an emergent trait, this correlation cannot be reduced because the trait cannot be represented at the lower level." But Lloyd's broader criterion of emergent fitness also establishes irreducibility, even if the trait involved in the correlation between trait and fitness is reducible under the effect hypothesis. In Lloyd's case, the *fitness* is irreducible (as shown practically in the previous example of gastropod lineages, where higher-level fitness based on speciation rate opposes lower-level fitness based on the same trait of larval adaptation). The technical point may be summarized in the following way: selection is defined by the *correlation* between a species-level trait and species-level fitness; therefore, the irreducibility of *either* component of the correlation establishes irreducibility for the selection process. Grantham notes (1995, p. 308): "Emergent traits are not, however, necessary for species selection. If an aggregate trait affects a component of species-level fitness (e.g. rate of speciation) and this component of fitness is irreducible, then the trait-fitness correlation will be irreducible."

Vrba's emergent character approach embodies one great strength, but two disarming weaknesses. This criterion does identify the most irrefutable, and in many ways the most interesting, subset of cases for species selection—examples based on genuine species *adaptations* (for an emergent character that evolved as a consequence of its value in fitness is, *ipso facto*, an adaptation); whereas nonemergent characters that contribute to species fitness via the effect hypothesis are exaptations (Gould and Vrba, 1982; Gould and Lloyd, 1999), at the species level (and adaptations at the lower level of their origin).

But the emergent character criterion suffers from two problems that would render the theory of species selection, if framed exclusively in its light, eternally contentious and, perhaps, relatively unimportant as well. First, by including only the "hardest-line" cases within the concept, we may be unduly limiting species selection to an unfairly small compass. (For example, and as an analogy, we wouldn't want to restrict the concept of "adaptation" only to the small subset of true biomechanical optima—for most adaptations only hold the status of "better than," not *ne plus ultra*). Second, emergence can often be extremely difficult to document for characters—so, in practice, the concept may be untestable in most circumstances. To differentiate between a truly emergent species character and an effect of a lower-level character, one often needs a great density of narrative information about the actual history of the lineage in question—information only rarely available in the fossil record, not to mention our spotty archives for living species.

By contrast, the emergent fitness approach enjoys the great virtue of fully general applicability. For, when one only has to consider current circumstances (the trait-fitness correlation), and need not reconstruct prior history (as the designation of emergence for a species-level character so often requires), then we can study any present reality that offers enough information for a resolution. We certainly use this most broadly applicable, nonhistorical approach in traditional studies of natural selection at the organismic level— that is, we identify current selective value whether the feature conferring differential reproductive success arose as an adaptation for its current contribution to fitness, or got coopted for its present role from some other origin or utility. (In other words, both preadaptations and spandrels—features that arose as adaptations for something else, or for no adaptive purpose at all— can function just as well in a regime of current selection as true adaptations forged by the current regime.) The historical origin of characters, and their later shifts in utility, constitute a central and fascinating question in evolutionary theory—and provide a main theme for Chapter 11 of this book. But we define the *process* of selection ahistorically—as differential reproductive success based on current interaction between traits of evolutionary individuals and their environments—that is, the concept of selection remains agnostic with respect to the historical origin of the traits involved.

The emergent fitness approach presents four favorable features that establish species selection as a central, fully operational, and vitally important subject in evolutionary biology—thereby validating both the necessity and the distinctness of macroevolutionary theory.

1. Rather than depending upon a documentation of prior history in the narrative mode (often untestable for lack of information), we move to a fully general mathematical model that can, in principle, identify components of higher-level selection in any case where we can obtain sufficient data on the current operation of a selection process. Arnold and Fristrup (1982) expanded Price's (1970, 1972) covariance formulae to encompass a set of nested levels, and devised an approach closely allied to analysis of covariance, considering selection at one level as a "treatment effect" upon selection at an adjacent level. Damuth and Heisler (1988) developed a similar method, also based on covariances (or regression of fitness values on characters); this procedure has been expanded by Lloyd (1988; Lloyd and Gould, 1993). As Lloyd and Gould (1993, p. 596) describe the method: "This is done by describing interactors at the lower level first. If a higher-level interactor exists, the higher-level correlation of fitness and trait will appear as a residual fitness contribution at the lower level; we must then go to the higher level in order to represent the correlation between higher-level trait and higher-level fitness."

Lest this method seem to fall into the very reductionistic trap that species selection strives to overcome—because we begin at the lowest level and only move higher if we find a residual fitness—I point out that we use this procedure only as a convenient and operational research method, and decidedly not with the reductionistic hope that no residuals will appear, and that the lowest level will therefore suffice for a full explanation. We may be stuck with the technical term "residual" as a common statistical usage in such circumstances—but there is nothing conceptually residual about higher-level selection. Selection at lower levels cannot be designated as more true or basic, with higher levels then superadded if necessary. The statistical "residual" of our procedure exists as a separate but equal natural reality in our fascinating world of hierarchical selection.

2. The emergent fitness approach establishes a large and general realm for the operation of species selection. Any evolutionary trend that must be described, at least in part, as a result of species sorting automatically becomes subject to the analytical apparatus here proposed, and therefore a candidate for identification of species selection. (And I can hardly imagine that any important trend unfolds without a major—I would say almost always predominant (see Chapter 9)—component of species sorting, for extensive anagenesis rarely occurs in single lineages, and none can persist very long without branching in any case.)

3. The emergent fitness approach allows us to use a single, familiar, and minimalist definition of selection in the same manner at each level—differential proliferation of evolutionary individuals based on interactions of their traits with the environment. We therefore achieve a unified theory of selection at all scales of nature. The availability of a fully operational analytical apparatus, connected with this definition, greatly enhances the scientific utility of emergent fitness as a definition of species selection.

4. As an admittedly more subjective and personal point, the emergent fitness approach allows us to encompass under the rubric of species selection several attributes of populations that many participants in this debate have intuitively wished to include within the causal compass of species acting as evolutionary individuals, but which the more restrictive emergent character approach rules out. Many of us have felt that two distinct kinds of species properties should figure in species selection because, for different reasons, such features cannot function at the lower and traditional level of organismic selection. In the first category, emergent characters of species obviously can't operate at the organismic level because they don't exist for organisms. These features clearly serve as criteria of species selection in either the emergent character or the emergent fitness approach.

In a second category, some important aggregate characters of species can't function in selection at the organismic level, not because they have no expression at this lower level (for they clearly exist as organismic properties, at least in the form of traits that aggregate additively to a different expression at the species level), but because such properties do not *vary* among organisms, and therefore supply no raw material for selection's necessary fuel. I speak here of a common phenomenon recognized by different jargons in various subdisciplines of our field—autapomorphies for cladists, or invariant *Bauplan* characters for structuralists. Suppose that each species in a clade has evolved a unique state of a homologous character—and that, within each species, all organisms develop the same state of the character, without meaningful variation. In this situation, all variation for the homologous character occurs among species, and none at all within species. If a trend now develops within the clade when some species live and proliferate because they possess their unique state of the character, while others die because their equally distinct and unvarying state has become maladaptive in a changed environment, should we call such a result species selection—for each species manifests a single attribute, and all variation occurs among species? Interestingly, de Vries originally coined the term species selection (see pages 448–451) for precisely this situation, where no relevant variation exists within species, and all variation occurs among species.

To summarize: in the first situation, the character *doesn't exist* at the organismal level, and each species develops only one state of the (emergent) character because the character belongs to the species as a whole. Therefore, selection for this character can only occur among species. In the second situation, the character *doesn't vary* at the organismal level, and each species in a clade has evolved a unique and different state of the character. Again, selection can only occur among species. In either situation, each species manifests one different and unvarying state of a feature that cannot operate in organismic selection—so selection for this feature can only occur among species.

The emergent status of the character leads us to designate the first situation as species selection without any ambiguity or alternative. But we balk at designating the second situation as species selection because the relevant species-level character (lack of variation) represents an aggregate, not an emergent, feature. The emergent fitness criterion rescues us from this dilemma, and forges an intuitive union between the two situations by designating both as species selection. Lack of variation—the aggregate species character—interacts with the environment to influence differential rates of proliferation among species. This character imparts an emergent fitness to the species, and therefore becomes an agent of species selection. (After all, the species doesn't die because organism A, or B, or C, possesses a trait that has become maladaptive; the species dies because none of its parts (organisms) can develop any other form of the trait—and this lack of variation characterizes the species, not any of its individual organisms.)

I believe that such "species selection on variability"—the title that Lloyd and I gave to our 1993 paper—will prove to be a potent style of selection at this level. (When I was struggling with the issue of whether such an aggregate character as variability could count as a property of species, I asked Egbert Leigh, a brilliant evolutionist and the leading late 20th century disciple of R. A. Fisher, whether he thought that variability could operate as a character in species selection—and he replied: "if variability isn't clearly a character of a species, then I don't know what is.")

To cite just one hypothetical example that I have often used to illustrate this issue and to argue for species selection on variability: Suppose that a wondrously optimal fish, a marvel of hydrodynamic perfection, lives in a pond. This species has been honed by millennia of conventional Darwinian selection, based on fierce competition, to this optimal organismic state. The gills work in an exemplary fashion, but do not vary among individual organisms for any option other than breathing in well-aerated, flowing water. Another species of fish—the middling species—ekes out a marginal existence in the same pond. The gills don't work as well, but their structure varies greatly among organisms. In particular, a few members of the species can breathe in quite stagnant and muddy waters.

Organismic selection favors the optimal fish, a proud creature who has lorded it over all brethren, especially the middling fish, for ages untold. But now the pond dries up, and only a few shallow, muddy pools remain. The optimal fish becomes extinct. The middling species persists because a few of its members can survive in the muddy residua. (Next decade, the deep, well-aerated waters may return, but the optimal fish no longer exists to reestablish its domination.)

Can we explain the persistence of the middling species, and the death of the optimal form, only by organismic selection? I don't think so. The middling species survives, in large part, as a result of the greater variability that allowed some members to hunker down in the muddy pools. (We may even argue that the optimal fish always prevailed against most members of the middling species, even at the worst time, so that most middlings died quickly when the pond dried, while the optimals hung on longer, but eventually succumbed.) The middling species survived *qua* species because the gills varied among its parts (organisms), not because all its members gained advantage when the environment changed. (For most middling organisms continued to fare worse than the optimal fishes.) We may *represent* this story at the organismal level by discussing the gills of the few middling fishes that carried the species through the crisis. But the middling species prevailed by species selection on variability—for this greater variability imparted an emergent fitness to the interaction of the species with the changed environment.

Species selection on variability also possesses the salutary property of uniting the two major themes of this book, the concepts that I regard as the most important revisions now needed to mend and strengthen the two main legs of the essential Darwinian tripod: the hierarchical theory of natural selection as a vibrant expansion of Darwin's focus on the organismal level, and the centrality of constraint as a channeler of evolutionary direction in concert with natural selection (which can no longer maintain the exclusivity that strict Darwinians wished to impart). An important component for explaining the patterning of life's history lies in limitations and channels imposed and retained by developmental architecture—and these constraints do much of their work at higher levels, in large part by influencing "species selection on variability."

I close this discussion with three points that validate the status of species selection as an irreducible macroevolutionary force, and place the proposed criteria of emergent characters and emergent fitnesses under a common rubric.

**THE FALLACY OF "NECKER CUBING"** The philosophical doctrine of conventionalism, as expressed by Dawkins (1982) in his Necker Cube metaphor (see pages 640–641), presents an important challenge to claims for an independent macroevolutionary theory based on higher-level selection. For if all cases of higher-level selection, however cogently defended, represent only one legitimate way to describe a process that can always be causally expressed in terms of selection at conventional lower levels as well, then why bother (except for the fun of it, or for the psychological insight thus provided) with the alternative higher level, when the traditional Darwinian locus invariably works just as well?

I do not doubt that some evolutionary events can be alternatively expressed (and I shall mention one category under my second point below), but Necker cubing will not apply to genuine cases of irreducible species selection because the nature of the world (not the conventions of our language) regulates the locus of causality. Two reasons debar the Necker cube for true cases of species selection. First, for Vrba's "hardest" category of species selection based on emergent characters, no expression at conventional lower levels can be formulated because the relevant species character does not exist at the usual Darwinian locus of organisms. Second, for Lloyd's broader category of species selection based on the emergent fitness associated with aggregate species characters, the "Necker cubers" commit a basic error in logic. They correctly note that the aggregate *character* can be represented at the organismic level—so they invoke the conventionalism of alternative and equally valid expression. But, as discussed on page 659, the species-level *fitness* imparted by the aggregate character, not the character itself, denotes the irreducible feature that defines species selection on this criterion.

In other words, Necker cubers commit the same error in this case that Dawkins made in his original use of the metaphor to claim that all organismal selection can also be expressed in terms of gene selection. The metaphor of the Necker cube only applies when the *same thing* attains equal and alternative representation, not when the Cube's two versions represent genuinely different aspects of a common phenomenon. In Dawkins's original error, *something* can always be represented at the gene level—but that something counts as bookkeeping, not as the causality of selection, which remains organismal in his standard cases. Similarly, for aggregate species-level characters involved in selection, *something* can always be represented at the organismic level—but that something, in this case, only involves the *composition* of the character, not the *causal process* of selection, which occurs irreducibly at the species level as identified by emergent species-level fitnesses.

**A UNIFIED PICTURE OF SPECIES SELECTION** In advocating such an expanded role for species selection, we must guard against the ultimate fallacy of claiming too much—for if we turn all forms of species sorting into species selection by verbal legerdemain, then we falsely "win" by definition, but actually lose by an overly imperialistic extension that permits no distinctions and therefore sacrifices all utility as an empirical proposition in science. Fortunately, we can unite both criteria of emergent fitnesses and characters into a unified scheme that establishes two realms of species selection, one more inclusive than the other, but that also identifies a domain of species sorting leading us to reject causation by species selection.

Grantham (1995) has presented such a scheme, reproduced here as Figure 8-4. (I had independently developed the same system, almost with the same picture, in preparing to write this chapter. I mention this not to compromise Grantham's originality or priority in any way—for priority is chronology, and his cannot be gainsaid!—but to express the firm and almost eerie satisfaction that such a "multiple" formulation brings (see Merton, 1965), and to offer this example as proof that the inherent logic of a complex argument often drives independent researchers to a definite and almost ineluctable result— validating in this case the coherence of this "take" on species selection.)

Grantham's diagram circumscribes two realms of species selection, labeled as "hierarchical explanations." The A realm contains Vrba's firmest examples based on emergent characters, while the B realm adds Lloyd's cases based on the emergent fitnesses associated with aggregate species-level characters. (Vrba, of course, would restrict species selection to the A realm, and ascribe the B realm to the "effect hypothesis"—but everyone seems to agree on the structure and relationships of the realms.) The A realm seems firmer because emergent characters count as adaptations of species, and maintain no expres-

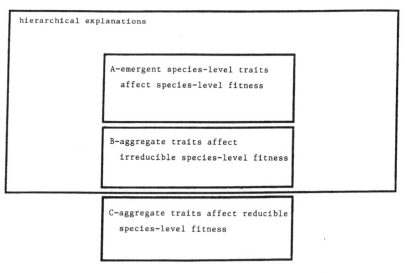

8-4. Grantham's 1995 epitome of criteria for invoking species selection in hierarchical models. The A domain includes rare best cases of species selection based on emergent species-level traits. The B domain adds aggregate traits that affect irreducible species-level fitness, and therefore also participate in species selection under the interactor approach. The aggregate and reducible traits of the C domain belong only to organisms and cannot figure in arguments for species selection.

sion at lower levels. The B realm seems "looser" because these aggregate species characters can be represented at the organismic level, even though they may also rise by upward causation to become exaptations of species (Gould and Vrba, 1982; Vrba and Gould, 1986; Gould and Lloyd, 1999). But, in any case, the resulting species-level fitnesses are irreducible—so the B realm also represents species selection by standard definitions of selection as a causal process.

The C realm includes cases of species sorting based on aggregate species-level characters that impart only a reducible fitness at the species level—and therefore do not count as species selection. One might add a D realm at the base for cases describable as species sorting, but not associated with any higher-level character, either aggregate or emergent, and therefore not qualifying for consideration as species selection on any definition of species as evolutionary individuals and interactors. The D realm, which may be quite large, includes several categories, most obviously species sorting based on the higher-level analog of drift—or random differentials in survival and death of species within a clade (see my summary chart, pp. 718–720).

As for any scientific theory, we want, most of all, to be able to make clear and testable distinctions at the crucial boundary between cases that affirm and cases that fall outside the hypothesis under consideration—in this case, between the B and C realms separating irreducible species selection from species sorting reducible to organismic selection. In these formative days for the theory of species selection, we have not yet developed a full set of firm criteria for making these crucial allocations. But let me suggest one guidepost at the outset. Ever since this literature began, astute workers have developed a strong intuition that species sorting based on events of differential birth (speciation rates) will usually represent true species selection, while species sorting based upon differential death (extinction) will often be reducible to organismic level (see Gilinsky, 1981; Arnold and Fristrup, 1982; Vrba and Eldredge, 1984; Grantham, 1995; Gould and Eldredge, 1977; Gould, 1983c).

The source of this intuition—which may turn out to be both wrong, and superficially based—arises from a sense that the extinction of a species may often be adequately explained simply as the summed deaths of all organisms, each for entirely organismal reasons and with no significant contribution from any species-level property. When the last reproductive organism dies, the species becomes extinct. But how could a new species *originate* without some involvement of population-level features? After all, individual organisms do not speciate; only populations do. But individual organisms die, and the extinction of a species might, at least in principle, represent no more than the summation of these deaths. Grantham expresses this common intuition particularly well (1995, pp. 309–310):

The concept of "speciation rate" cannot be expressed at the organismic level because there is no simple set of organismic traits that determine speciation rate. Rather, a diverse set of organismic and population-level traits (including dispersal ability, population structure, and behavioral compatibility between members of distant populations) affect gene flow and therefore affect speciation rates. Because of the large variety of factors affecting speciation rate . . . the higher level property of "speciation rate" is, at best, extraordinarily difficult to express in organismic terms. The speciation rate of a taxon is irreducible . . . A species goes extinct if and only if every individual dies. Whereas differences in speciation rates cannot be expressed in organismic terms, differences in extinction rates will often be reducible (unless population-level traits such as variation matter).

Thus, I suspect that the A and B realms will be heavily populated with cases based on differential speciation, whereas the C realm will feature cases based on differential extinction.

A PERSONAL ODYSSEY Many historians of science, particularly feminists like Donna Harraway (1989, 1991), have forcefully argued that scholars can strike their most effective blows against the myth of pure objectivism by being candid about the interaction of their own autobiographies with their current claims—thus exposing the inevitable (and basically welcome) cultural and psychological embeddedness of science, while opening an author's prejudice both to his own scrutiny, and to the examination of his readers. To do so obsessively or promiscuously in a book of this sort would only clutter a text that would then become even more insufferably egocentric or idiosyncratic—so I have usually desisted (except for some parts of Chapter 1, and the dubious indulgence of my appendix to Chapter 9). But I will follow Harraway's recommendation in this particular case, because no other subject in evolutionary theory has so engaged and confused me, throughout my career, as the definition and elucidation of species selection. For no other problem have I made so many published mistakes, and undergone so many changes of viewpoint before settling on what I now consider a satisfactory framework. Moreover, my basic reason for current satisfaction rests upon an interesting correction from within my own body of work—and, though I remain heartily embarrassed for not grasping both the inconsistency and the necessary resolution many years earlier, I do take some pleasure in my eventual arrival—and I do think that the story may help to illustrate the intellectual coherence of the framework now proposed in this book.

I made two sequential errors of opposite import. When Niles Eldredge and I first formulated punctuated equilibrium, I was most excited by the insight that trends would need to be reconceptualized as differential success of species, rather than anagenesis within lineages (a theme only dimly grasped in Eldredge and Gould, 1972, but fully developed in Gould and Eldredge, 1977, after much help from Stanley, 1975, and later from Vrba, 1980). I then committed the common fallacy of extending personal excitement too far—and I made the error (as we all did in these early days of "species selection" under punctuated equilibrium) of labelling as species selection any pattern that needed to be *described* in terms of differential success for species treated (under punctuated equilibrium) as stable entities. In other words, we failed to distinguish selection from sorting, and used the mere existence of sorting at the species level as a criterion for identifying species selection. This definition of species selection must be rejected as clearly wrong—particularly for the invalid "promotion" of several cases properly viewed as effects of causes fully reducible to conventional organismic selection.

In reaction to this previous excess, I then retreated too far in the other direction, by restricting species selection too severely—i.e., only to cases based on characters emergent at the species level (Gould, 1983c; Vrba and Gould, 1986). My later work with Elizabeth Lloyd (Lloyd and Gould, 1993; Gould and Lloyd, 1999) convinced me that emergent characters, while properly identifying species selection, only identified a subset of genuine cases, and that emergent fitness, as defended in this section, provided a conceptually broader, and empirically more testable criterion.

In preparing this chapter, I finally realized why I had originally erred in restricting species selection to emergent characters. The source for amending Vrba and Gould (1986) lay in an earlier paper that I had written with Elisabeth Vrba (Gould and Vrba, 1982), particularly in the codification of adaptation (or the origin of a character directly for its current utility) and exaptation (or the cooptation of a preexisting character for an altered current utility) as subsets of the more inclusive phenomenon of aptation (any form of current utility, whatever the historical origin).

We developed this terminology, which has now been widely accepted (see extensive discussion in Chapter 11), in order to make a crucial, but often disregarded, distinction between "reasons for historical origin" and "basis of current utility." The common conflation of these entirely separate notions has engendered enormous confusion in evolutionary theory—a situation that we documented and tried to correct in our paper (Vrba and Gould, 1986). Hardly any principle in general historical reasoning (not only in evolutionary theory) can be more important than clear separation between the historical basis of a phenomenon and its current operation. For example, crucial components of current utility often arose nonadaptively as spandrels, or side-consequences, of other features actively constructed or evolved (Gould and Lewontin, 1979).

I felt so enlightened by this distinction, and so committed (as a paleontologist and historian) to the special role of historical origin, that I longed to apply this notion to the important concept of species selection. I therefore concluded that we should not speak of species selection unless the character that imparted the relevant fitness could be identified as a true adaptation at the species level—that is, as a feature belonging to the species as a higher-level Darwinian individual, and evolved directly for current utility in promoting the differential success of the species. Emergent species characters qualify as adaptations—and I therefore felt drawn to this narrow criterion for identifying species selection.

In so doing, I committed a basic logical error about the nature of selection.

However much I may love history, selection cannot be, and has never been, defined as a historical relationship of character and result. Selection must be defined by *present operation,* as identified by an observable differential in reproductive success based on the current interaction of a trait of a Darwinian individual with its environment. This definition includes no reference to the historical origin of any relevant trait, which may be either an adaptation or an exaptation. Damuth and Heisler (1988) emphasize this crucial point, with an apt literary flourish at the end to note the irrelevancy of a trait's "aristocracy" (depth of historical origin, or "blue-blooded" continuity) to the hierarchy of selection:

> The historical origin of a character is irrelevant to the way that it functions in a selection process. Thus, the issue of whether a character is a group or individual "adaptation," and whether it has been shaped for its present role by any particular process, is of no importance in the study of the selection mechanism. There may certainly be historical significance in such observations about the origin of characters. Nevertheless, selection evaluates characters in terms of their current relationship to fitness only, not in terms of their history. There is hierarchy in the world of natural selection, but no aristocracy.

Once I recognized the irrelevancy of historical origin to the identification of selection—my only previous rationale for insisting that characters for species selection must be species-level adaptations, and therefore emergent at the species level—I understood that the "emergent character" criterion must be rejected as too restrictive (while correctly identifying the firmest subset of cases for species selection), whereas the "emergent fitness" criterion must be preferred, as not only legitimately broader in scope, but also properly formulated in terms of conventional definitions of selection. In my own preferred nomenclature, species-level characters that are exaptations rather than adaptations can function perfectly well in species selection. Aggregate species-level characters originate as exaptations of species because they arise at the organismal level and pass upwards as effects to the species level. When I mistakenly thought that characters for species selection had to be species-level adaptations, I had excluded aggregate characters (as species-level exaptations), and therefore falsely rejected the emergent fitness approach (see Gould and Lloyd, 1999, for an elaboration of this argument).

In the early 1980's, my own students Tony Arnold and Kurt Fristrup had strongly urged the criterion of emergent fitness upon me, and I well remember my bitter disappointment that I could not convince them to use the restrictive criterion of emergent characters! (I had not yet developed the nomenclature of adaptation and exaptation, and therefore did not yet possess the personal tools for a conceptual resolution.) Thus, my error reflected an active commitment (not a passive consequence of inattention), maintained in the face of an available correction that I now regard as one of the finest papers ever published on the subject (Arnold and Fristrup, 1982). I did not grasp, for another decade, how the terminology developed by Vrba and me also derailed the criterion that we both preferred. To sum up: selection operates on current utilities, and remains agnostic about historical origins in utilizing both adaptations and exaptations with equal facility. Emergent species-level characters will generally count as adaptations, thus clearly available for species selection. But all aggregate species-level characters represent potential exaptations, and therefore become equally available for species selection under the proper criterion of emergent fitness.

I would, however, salvage a lesson from this odyssey of errors. Vrba and I were not wrong in identifying emergent characters as especially interesting (we only erred in deeming them necessary for species selection). Emergent characters belong exclusively to the species. As adaptations, they become part of the defining cohesion that permits a species to function as an evolutionary individual. Emergent characters thus stand out in designating the style of individuality maintained by species. Aggregate characters, on the other hand, do not clearly define a species as a functional entity (variability, for example, represents an attribute, not an "organ," of a species)—for aggregate characters belong as much to the component organisms, as to the entire species. Thus, emergent characters are special and fascinating (though not essential to the definition and recognition of species as legitimate Darwinian individuals—see Gould and Lloyd, 1999). Emergent characters do deserve primary consideration in discussions about the structural basis of species both as natural entities in general, and as the basic individuals of macroevolution in particular. But we do not require emergent characters to identify a process of selection.

As a final note, and as one contribution to recognizing the crucial and characteristic differences among Darwinian individuals at the six primary levels of the evolutionary hierarchy, we should suspect that species selection will emphasize exaptations, whereas organismal selection employs a higher relative frequency of adaptations—for species, as more loosely organized in functional terms than organisms, probably possess far fewer emergent characters than organisms. But species "make up" for their relative paucity of adaptations by developing a higher frequency of exaptations. Most of these exaptations derive their raw material from adaptations at the organismal level that cascade upwards to effects at the species level. By joining fewer adaptations (emergent characters) with more exaptations (usually based on aggregate characters), species may become just as rich as organisms in features that can serve as a basis for selection. Species selection may therefore become just as strong and decisive as conventional Darwinian selection at the level of organisms—a process whose power we do not doubt, and whose range we once falsely extended to encompass all of nature.

## HIERARCHY AND THE SIXFOLD WAY

### *A literary prologue for the two major properties of hierarchies*

Our vernacular language recognizes a triad of terms for the structural description of any phenomenon that we wish to designate as a unitary item or "thing." The thing itself becomes our focus, and we call it an object, an entity, an individual, an organism, or any one of a hundred similar terms, depending on the substance and circumstance. The subunits that build the individual are then called "parts" (or units, or components, or organs, etc., depending upon the nature of the focal item); while any recognized grouping of similar individuals becomes a "collectivity" (or aggregation, society, organization, etc.). In other words, and in epitome, individuals are made of parts and aggregate into collectivities.

The hierarchical theory of selection recognizes many kinds of evolutionary individuals, banded together in a rising series of increasingly greater inclusion, one within the next—genes in cells, cells in organisms, organisms in demes, demes in species, species in clades. The focal unit of each level is an *individual,* and we may choose to direct our evolutionary attention to any of the levels. Once we designate a focal level as primary for a particular study, then the unit of that level—the gene, or the organism, or the species, etc.—becomes our relevant or focal individual, and its constituent units become parts, while the next higher unit becomes its collectivity. Thus, if I place my focus at the conventional organismic level, genes and cells become parts, while demes and species become collectivities. But if my study enjoins a focus on species as individuals, then organisms become parts, and clades become collectivities. In other words, the triad of part—individual—collectivity will shift, as a threefold entirety, up and down the hierarchy, depending upon the chosen subjects and objects of any particular study.

This dry linguistic point becomes important for a fundamental reason of psychological habit. We humans are hidebound creatures of convention, particularly tied to the spatial and temporal scales most palpably familiar in our personal lives. Among nature's vastly different realms of time, from the femtoseconds of some atomic phenomena to the aeons of stellar and geological time, we really grasp, in a visceral sense, only a small span from the seconds of our incidents to the decades of our lives. We can formulate other scales in mathematical terms; we can document their existence and the processes that unfold in their domains. But we experience enormous difficulty in trying to bring these alien scales into the guts of or our understanding—largely for the parochial reason of personal inexperience.

We make frequent and legendary errors because we tend to extrapolate the styles and modes of our own scale into the different realms of the incomprehensibly immense or tiny in size, vast or fleeting in time. Geologists, for example, well appreciate the enormous difficulties that most people encounter (including our professional selves, despite so many years of training and experience) in trying to visualize or understand the meaning of any ordinary statement in "deep" or "earth time"—that a landscape took millions of years to develop, or that a lineage exhibits a trend to increasing size throughout the Cretaceous period. All of us—professionals and laypeople alike—continue to make the damnedest mistakes. I have, for example, struggled for thirty years against the conventional misreading of punctuated equilibrium as a saltational theory in the generational terms usually applied to such a concept in evolutionary studies. The theory's punctuations are only saltational on geological scales—in the sense that most species arise during an unmeasurable geological moment (meaning, in operational terms, that all the evidence appears on a single bedding plane). But geological moments usually include thousands of human years—more than enough time for a continuous process that we would regard as glacially slow by the measure of our lives (see Goodfriend and Gould, 1996, for an example). Thus, punctuated equilibrium represents the proper geological scaling of speciation events that may span several thousand years, not a slavish promotion of "instantaneity," as conventionally measured in a human time frame, to the origin of species.

As we misunderstand scales of time, we fail just as badly with viscerally unfamiliar realms of size. Our bodies lie in the middle of a continuum ranging from the angstroms of atoms to the light years of galaxies. Individuality exists in all these domains, but when we try to understand the phenomenon of "thingness" at any distant scale, we easily fall under the thrall of the greatest of all parochialisms. We know one kind of individual so intimately and with such familiarity—our own bodies—that we tend to impose the characteristic properties of this level upon the very different styles of being that other scales generate. This inevitable human foible provokes endless trouble, if only because organismal bodies represent a very peculiar kind of individual, serving as a very poor model for the comparable phenomenon at most other scales.

The "feel" of individuality at other scales becomes so elusive that most of the best exploration has been accomplished by literary figures, not by scientists. The tradition extends at least as far back as Lemuel Gulliver, whose "alien" contacts did not depart greatly from our kind of body and our norm of size. This theme has best been promoted, in our generation, within the genre of science fiction. I particularly recommend two "cult" films, *Fantastic Voyage* and *Inner Space*, both about humans reduced to cellular size and injected into the body of another unaltered conspecific. This ordinary body becomes the environment of the shrunken protagonists, a "collectivity" rather than a discrete entity—while the "parts" of this body become individuals to the shrunken guests. When Raquel Welch fights a bevy of antibodies to the death in *Fantastic Voyage*, we understand how location along the triadic continuum of part—individual—collectivity depends upon circumstance and concern. A tiny, if crucial, part to me at about two meters tall becomes an entire and ultimately dangerous individual to Ms. Welch at a tiny fraction of a millimeter.

The parochiality of time has served us badly enough, but the parochiality of bodily size has, for two reasons, placed even more imposing barriers in our path to an improved and generalized evolutionary theory—a formulation well within our grasp if we can learn how to expand the Darwinian perspective to all levels of nature's hierarchy. First, we know almost viscerally what our bodies do best as Darwinian agents—and we then grant universal importance to these properties both by denying interest to the different "best" properties of individuals at other levels, and by assuming that our "bests" must, by extension, power Darwinian systems wherever they work. Our bodies are best at developing adaptations in the complex and coordinated form that we call "organic." Many evolutionists therefore argue, in the worst parochialism of all, that only adaptations matter as an explanatory goal of Darwinism, and that such adaptations must therefore drive evolution at all levels. I don't even think that such a perspective works well for organisms—surely the locus of most promising application (Gould and Lewontin, 1979)—but this attitude will surely stymie any understanding of individuality at other levels, where complex adaptations do not figure so prominently. Evolutionists will not be able to appreciate the different individuality of species, where exaptive effects hold at least equal sway with adaptations, if they continue to regard spandrels, sequelae, and side consequences as phenomena generated by "the boring by-product theory" (Dawkins, 1982, p. 215).

Second, we just don't comprehend the scale-bound realities in other domains of size, and we err by imposing our own perceptions when we try to think about the world of a gene, or of a species. In one of the most famous statements of 20th century biology, D'Arcy Thompson (1942, p. 77) ended his chapter "On Magnitude" (in his classic work, *On Growth and Form*—see the first section of Chapter 11 for a general analysis of his work) by noting how badly we misread the world of smaller organisms because our large size places us in gravity's domain (a result of low surface/volume ratios in larger creatures, but not a significant feature in other realms of size). If we encounter so much trouble for extremes within our own level of organismic individuality, how will we grasp the even more distant worlds of other kinds of evolutionary individuals? D'Arcy Thompson wrote (1942, p. 771):

> Life has a range of magnitude narrow indeed compared to that with which physical science deals; but it is wide enough to include three such discrepant conditions as those in which a man, an insect, and a bacillus have their being and play their several roles. Man is ruled by gravitation, and rests on mother earth. A water-beetle finds the surface of a pool a matter of life and death, a perilous entanglement or an indispensable support. In a third world, where the bacillus lives, gravitation is forgotten, and the viscosity of the liquid, the resistance defined by Stokes's law, the molecular shocks of the Brownian movement, doubtless also the electric charges of the ionized medium, make up the physical environment and have their potent and immediate influence on the organism. The predominant factors are no longer those of our scale; we have come to the edge of a world of which we have no experience, and where all our preconceptions must be recast.

As one example, consider the difficulty we experience, despite our preferences for reductionism in science, when we try to visualize the world of our genes, where nucleotides function as active and substitutable evolutionary parts—and where chromosomes build a first encasement, followed by nuclei and cells, with our body now serving as an entire universe, whose death will also destroy any gene still resident within. Think of the initial resistance that most of us felt towards Kimura's neutralist theory—largely because we falsely "downloaded" our adaptationist views about organisms into this different domain, where high frequencies of neutral substitution become so reasonable once we grasp the weirdly (to us) divergent nature of life at such infinitude. And if we fare so badly for the small and immediate, supposedly so valued by our reductionist preferences, how can we comprehend an opposite extension into the longer life, the larger size, and the markedly different character of species-individuals—a world that we have usually viewed exclusively as a collectivity, an aggregation of our bodies, and not as a different kind of individual in any sense at all?

I like to play a game of "science fiction" by imagining myself as an individual of another scale (not just as a human being shrunken or enlarged for a visit to such a *terra incognita*). But I do not know how far I can succeed. As organisms, we have eyes to see the world of selection and adaptation as expressed in the good design of wings, legs, and brains. But randomness may predominate in the world of genes—and we might interpret the universe very differently if our primary vantage point resided at this lower level. We might then note a world of largely independent items, drifting in and out by the luck of the draw—but with little islands dotted about here and there, where selection slows down the ordinary tempo and embryology ties things together. How, then, shall we comprehend the still different order of a world much larger than ourselves? If we missed the strange world of genic neutrality because we are too big, then what passes above our gaze because we are too small? Perhaps we become stymied, like genes trying to grasp the much larger world of change in bodies, when we, as bodies, try to contemplate the domain of evolution among species in the vastness of geological time? What are we missing in trying to read this world by the inappropriate scale of our small bodies and minuscule lifetimes?

Once we have become mentally prepared to seek and appreciate (and not to ignore or devalue) the structural and causal differences among nature's richly various scales, we can formulate more fruitfully the two cardinal properties of hierarchies that make the theory of hierarchical selection both so interesting and so different from the conventional single-level Darwinism of organismal selection. The key to both properties lies in "interdependence with difference"—for the hierarchical levels of causality, while bonded in interaction, are also (for some attributes) fairly independent in modality. Moreover, these levels invariably diverge, one from the other, despite unifying principles, like selection, applicable to all levels. Allometry, not pure fractality, rules among the scales of nature.

1. Selection at one level may enhance, counteract, or just be orthogonal to selection at any adjacent level. All modes of interaction prevail among levels and make prominent imprints in nature.

I emphasize this crucial point because many students of the subject have focussed so strongly on negative interaction between levels—for a sensible and practical reason—that they verge on the serious error of equating an operational advantage with a theoretical restriction, and almost seem to deny the other modes of positive (synergistic) and orthogonal (independent) interaction. Negative interaction wins primary heuristic attention because this mode provides our most cogent evidence, not merely for simultaneous action of two levels, but especially for the operation of a controversial or unsuspected level. If two levels work in synergism, then we easily miss the one we do not expect to see, and attribute the full effect to an unsuspected strength for the level we know. But if the controversial level yields an unexpected effect contrary to the known direction of selection at a familiar level, then we may be able to specify and measure the disputed phenomenon.

In the example cited previously, individual selection favors a balanced sex ratio, while interdemic selection leads to female bias in many circumstances. Our best evidence for the reality of interdemic selection emerges from the discovery of such biases—not so strong as purely interdemic selection would produce (for organismic selection operates simultaneously in the other direction), but firm enough to demonstrate the existence of a controversial phenomenon. But if interdemic selection also worked towards a 1:1 ratio, we could attribute such an empirical finding exclusively to the conventional operation of organismic selection.

Negative interaction, however, does yield one distinguishing consequence to highlight this mode as especially important in the revisions to evolutionary theory that the hierarchical model will engender. In conventional one-level Darwinism, stabilities generally receive interpretation as adaptive peaks or optima, thus enhancing the functionalist bias inherent in the theory. The major structuralist intrusion into this theme ordinarily occurs when we have been willing to allow that natural selection can't surmount a constraint—elephants too heavy to fly even if genetic variability for wings existed; insects confined to small sizes by the inherited *Bauplan* of an exoskeleton that must be molted, and a respiratory system of skeletal invaginations that would become too extensive at the surface/volume ratio of large organisms. But the constraints in these cases act as passive walls, not active agents.

The hierarchical theory of selection suggests a theoretically quite different and dynamic reason for many of nature's stabilities: an achieved balance, at an intermediary point optimal for neither, between two levels of selection working in opposite directions. Several important phenomena may be so explained: weak female bias as the negative interaction of organismal and interdemic selection (see above); restriction of multiple copy number in "selfish DNA" as a balance between positive selection at the gene level, suppressed by negative selection (based, perhaps, on energetic costs of producing so many copies irrelevant to the phenotype) at the organismic level. I also suspect that stable and distinctive features of species and clades must represent balances between positive organismic selection that would drive a feature to further elaboration, and negative species selection to limit the geological longevity of such "overspecialized" forms. In any case, a world of conceptual difference exists between stabilities read as optima of a single process, and stabilities interpreted as compromises between active and opposed forces.

As an example of overemphasis upon negative interaction, Wilson and Sober (1994, p. 592) ask: "Why aren't examples of within-individual [organism] selection more common?" They mention the most familiar case of meiotic drive, and then discuss the conventional argument for rarity of such phenomena: the integrity of complex organisms implies strong balance and homeostasis among parts; therefore, any part that begins to proliferate independently will threaten this stability, and must therefore be disfavored by organismic selection, a force generally strong enough to eliminate such a threat from below.

If selection within bodies generally opposes the organismic level, as this discussion implies, then we properly expect a low frequency for the phenomenon, since evolution has endowed the organismic level with a plethora of devices for resisting such dysfunctional invasion from within. Although I accept this argument for a low frequency of selection *contrary* to the interests of enclosing organisms, selection within bodies may not be so rare when we include the other modalities of *synergistic* and *orthogonal* directions. The most interesting hypothesis for extensive selection at the gene level, the notion originally dubbed "selfish DNA" (Orgel and Crick, 1980; Doolittle and Sapienza, 1980), attributes the observed copy number of much middle-repetitive DNA to orthogonal gene-level selection initially "unnoticed" by the organism, though eventually suppressed by negative selection from above when copies reach sufficient numbers to exact an energetic drain upon construction of the phenotype (see fuller discussion on pp. 693–695). In fact, I suspect that organismic complexity could never have evolved without extensive gene-level selection in this orthogonal (or synergistic) mode. For if we accept the common argument that freedom to evolve new phenotypic complexity requires genetic duplication to "liberate" copies for modification in novel directions, then how could such redundancy arise if organismic selection worked with such watchdog efficiency that even a single "extra" copy, initially unneeded by the organismic phenotype, induced strong negative selection from above, and immediately got flushed out—thus, in an odd sense, making the organism a delayed Kamikaze, killing its "invader" now and, by summation of such consequences, itself later?

Leo Buss (1987), in a fascinating book on the role of hierarchical selection in the phylogenetic history of development (see pp. 696–700 for further discussion), offers a compelling case for the vital importance of *both* synergistic and negative selection between levels in the history of life, which he views largely as a tale of sequential addition in hierarchical levels—so that nature's current hierarchy becomes a problem for historical explanation, not an inherent structure fully present throughout time. Buss argues that synergism must fuel the first steps in adding a new level atop a preexisting hierarchy (for initial negativity against the previous highest level would preclude the origin of a new level). But, having once achieved a tentative foothold, the new level stabilizes best by imposing negative selection against differential proliferation of individuals at the level just below—for these individuals have now become parts of the new level's integrity, and selection at the new level will tend to check any dysfunctional imbalance caused by differential proliferation from below.

2. Each hierarchical level differs from all others in substantial and interesting ways, both in the style and frequency of patterns in change and causal modes. Nature's hierarchy, for all the commonality of its unifying principles (selection, for example, acting at each level), does not display fractal structure with self-similarity across levels.

As the theory of hierarchical selection develops, I predict that no subject within its aegis will prove more fascinating than the varying strengths and modalities among levels. Just as the study of allometry has recorded characteristic and predictable scale-dependent differences in structure and function of organisms at strongly contrasting sizes—a prominent subject in biology ever since Galileo formulated the principle of surfaces and volumes in 1638, and so elegantly codified in D'Arcy Thompson's masterpiece of both prose and concept, *On Growth and Form* (1917, second edition, 1942)—so too does individuality as a tiny gene imply substantially different properties for a unit of selection than "personhood" as a large species or an even larger clade. Allometric effects across hierarchical levels should greatly exceed the familiar (and extensive) differences between tiny and gigantic organisms for two unsurprising reasons (see Gould and Lloyd, 1999, for a detailed development of this argument). First, the size ranges among levels are far greater still. Second, organisms share many common properties simply by occupying a common level of evolutionary individuality despite an immense range of size; but the levels themselves differ strongly in basic modes of individuality, and therefore develop far greater disparity.

But this promise also implies a corresponding danger. In some famous lines composed for a quite different, but interestingly related purpose, Alexander Pope explored the paradox of man's intermediary status between two such disparate extremes, both so desperately needed to know and to understand (the bestial and the godly in Pope's concern)—but both so inscrutable as so far from our own being:

> Placed on this isthmus of a middle state,
> A being darkly wise and rudely great . . .
> He hangs between; in doubt to act or rest;
> In doubt to deem himself a god, or beast . . .
> Created half to rise, and half to fall;
> Great lord of all things, yet a prey to all;
> Sole judge of truth, in endless error hurl'd;
> The glory, jest, and riddle of the world!

I appreciate this image of an "isthmus of a middle state"—a narrow standing place linking two larger worlds of smaller and greater. Pope's dilemma may pack more emotional punch in its moral meaning (since his greater and lesser worlds define questions of value rather than geometry), but our problem features greater intellectual depth—for, surely, a larger conceptual chasm separates the gene from the clade in modes of evolutionary mechanics, than the bestial from the virtuous in styles of human behavior.

The problem can be summarized with another, and much older, classical quotation. "Man is," as Protagoras wrote in his wonderfully ambiguous epigram, "the measure of all things"—ambiguous, that is, in embodying both positive and negative meanings: positive for humanistic reasons of ubiquitous self-valuing that might lead to some form of universal brotherhood and compassion; but negative because our own "measure" can be so parochially limiting, and therefore so conducive to misunderstanding other scales if we must assess these various domains by the allometric properties of our limited estate.

This issue becomes especially serious for the hierarchical theory of selection. Humans hold status as both evolutionary individuals and organisms— yet all other "separate but equal" evolutionary individuals at other hierarchical levels are *not* organisms. Unfortunately, organisms constitute a very special and distinctly odd kind of evolutionary individual, imbued with unique properties absent from (or much weaker in) other individuals (at other levels) that are equally potent as evolutionary agents. But if we mistakenly regard our own unique properties as indispensable traits for *any* kind of evolutionary individual—the classic error of parochialism—then we will devalue, or even fail to identify, other individuals defined by different properties and resident at other levels.

I shall explore some of these crucial differences in the next two sections (disparate properties of the six major levels; and extensive comparison of organisms and species as evolutionary individuals). In this introductory comment, I only wish to emphasize that the uniqueness of the organism as a unit of selection lies in securing individuality by maximal homeostatic interaction among parts, an integration that ties each subpart to the fate of all, and therefore strongly discourages any "breakout" or differential proliferation (by suborganismic selection) from within. To be sure, such integration represents a powerful strategy for individuation, but this strategy does not specify the only legitimate path, and other potent evolutionary individuals use other mechanisms. For this reason, I regret that Wilson and Sober (1994) so emphasize these "organic" properties of individuality in their general definition, meant to apply to all levels. This parochial focus leads them to downplay the individuality of units of selection at other levels, where different definitional criteria predominate—in species, for example, where the maintenance of boundaries by reproductive isolating mechanisms, and the mixture of subparts in replenishment (sexual reproduction), maintain cohesion and stability just as well as organisms do by the different strategies of homeostasis and functional interaction of subparts.

### Redressing the tyranny of the organism: comments on characteristic features and differences among six primary levels

I have little tolerance for numerical mysticism. I feel no special affinity for threes (as trinities), fours (Jung's primal archetype), fives (for fingers or echinoderms), sevens (for notes of the musical scale, planets in the Ptolemaic system, and so much else), or nines (the trinity of trinities). Similarly, in recognizing six hierarchical levels for this discussion—genes, cells, organisms, demes, species, and clades—I only utilize a device of convenience, and do not make any assertion about a fixed number of units in the expanded hierarchy of Darwinian action.

Any such claim of definity could only rank as both foolish and incoherent for at least two reasons. First, the hierarchy has not been set by structural or logical principles, but historically evolved in a contingent manner. Thus, before the inventions of sexual reproduction and multicellular organisms, neither species nor organisms (as a level distinct from cells) existed, and a quadripartite hierarchy held sway (and still does today in the dominant world of asexual unicells)—gene, cell, clone, and clade. Second, several of the levels discussed here coagulate numerous phenomena because they lie between two clear boundaries. As Buss (1987) points out, for example, we might, in certain contexts, recognize several items that encase genes but serve as parts of multicellular organisms: chromosomes, organelles, cells, organs, etc. Before the multicellular organism evolved, and began to act with such effectiveness as a suppressor of intraorganismic selection, we might have construed this domain of "proto-individuality" quite differently, and with finer resolution.

As a second argument against granting necessary or inherent status to these six levels, I have followed nearly all students of this field in preferring a fully nested hierarchy of increasing inclusion, to other legitimate interactors that function only occasionally, transiently, or in special circumstances. This fully nested hierarchy operates with Linnaean logic in requiring that lower units amalgamate completely, and under strict genealogical constraint—so that no lower unit can belong to more than one higher unit, while no higher unit can "forage" outside its hereditary line to incorporate the lower units of other distinct evolutionary branches at the same level. Just as a genus can't belong to two families, a species of flies cannot incorporate some onychophores and a few myriapods to construct a more versatile species-individual.

We logically require this property of nesting to correlate the nonhistorical process of selection with a set of quintessentially historical phenomena in evolutionary biology, including phylogenetics and the study of adaptation. Without such a fully inclusive hierarchy, for example, we could not use one level as a surrogate or convenient descriptor for events at other levels in the same nest—as when we choose the gene level for keeping the general books of evolution (see pp. 632–637 on the error of gene selectionism).

Nature, of course, does not always obey this logical stricture, though we may appeal to the empirical success of this formulation as an indicator that nature does comply at a preponderant relative frequency. If life did not generally work within a hierarchy of inclusion, the biotic world would present such a different appearance that our conventional ordering devices would not operate usefully, and would never have been proposed or accepted. (I am not a naive realist, and I have argued throughout this book that we impose our social preferences upon nature in constructing our theories. But nature does provide a strong input, and does impose a powerful constraint upon our formulations.) No one would ever have suggested a nested system like Linnaeus's, if common experience proclaimed that novel taxa generally arise by distant amalgamation—if, for example, each new mammal arose by a principle of "disparate thirds," say with equal mixtures of dugong, aardvark, and howler monkey. (We all know, of course, though we rarely discuss the subject in polite company, that the Linnaean logic, which presupposes a topology of branching without amalgamation, cannot apply to groups that do show massive mixture, as in some families of plants with extensive hybridization, or especially in prokaryotes evolving with frequent lateral transfer—a phenomenon that, on accumulating evidence, may be common enough to truly discombobulate the Linnaean version for the pre-multicellular majority of life's tree (see Doolittle, 1999), with practical and theoretical consequences as broad as any revolutionary discoveries in the recent history of evolutionary biology.) Similarly, we all appreciate the conceptual difficulties imposed by some prominent cases in evolution, mostly at the genic or cellular level, that do violate the hierarchy of inclusion—most notably, the origin of some organelles as symbiotic prokaryotes.

Since units of selection operate as interactors with the environment, and since entities obeying the criteria of "personhood" (see pages 602–613) do occasionally cohere by distant genealogical amalgamation, nature does present some exceptions to the principle of a fully nested hierarchy for evolutionary individuals. But these exceptions truly function as the "rule provers" of our mottoes (in the sense of probing, or testing, our generalities), and not as falsifiers. The most widespread cases, including the origin of cellular organelles by endosymbiosis, represent "frozen" phenomena of history, not active amalgamations presently building evolutionary individuals by junction of disparate genealogical lines. (However, genic exceptions, as noted above, may be rife if lateral transport occurs as frequently as current theory and data are now beginning to suggest, especially for prokaryotes.) The most common, active cases involve symbiotic and coevolutionary unions tight enough to obey the Biblical rule of Naomi and Ruth: "whither thou goest, I will go." Wilson and Sober (1989), for example, present a fascinating discussion of "phoretic associations," or obligate carriage, by wingless insects as they move among resource patches, of various mites, nematodes, fungi, and microorganisms. In some cases, the load of these "hangers on" can disable or even kill the insect, and conventional Darwinism will then work in its usual, competitive, and organismic mode. But the phoretic associates may be limited to densities that do not affect the insect, and may also provide resources indispensable for successful colonization of new patches—in which case, the entire association may be evolving as a "superorganism."

With these caveats in mind—the somewhat arbitrary division of the evolutionary hierarchy into six levels, and the acknowledgment of interesting exceptions to full nesting among nature's various individuals—I shall try to specify some distinctive "allometric" properties of the levels and their interactions:

THE GENE-INDIVIDUAL As we enter this first unfamiliar world of such great, and literally basic, importance to evolution, we encounter an initial rung of strong difference from the organism-individuals that, if only for psychological reasons, must stand as prototypes for our parochial concept of how a proper Darwinian unit must function in natural selection. If we could ever truly grasp the gene's world, with full sympathy and appreciation for relative frequencies, hard-line selectionism would yield to a fascinating enlargement that would actually strengthen selectionist theory by synergism with other (non-contradictory) forces—so this subject should therefore not intimidate strict Darwinians. For the most part, however, the necessary acknowledgment of different gene-level processes has unfolded within the traditional perspective of organismic selection—with three basic categories of interpretation as "good" for organisms, and acceptable on this basis; "bad" for organisms, and a destabilizing danger that must be conquered; or irrelevant to organisms and therefore unimportant.* The implications for a hierarchical reconstruction of evolutionary theory have therefore been missed or downplayed. Consider the two major themes of recent literature:

MOTOO KIMURA AND THE "NEUTRAL THEORY OF MOLECULAR EVOLUTION." Although I have called this book "*the* structure of evolutionary theory," I have propagated my own lamentable parochialism under a pretense of generality. For this book, despite its exuberant length, largely restricts itself to the Darwinian tradition of conventional causal explanations based on selection as a central mechanism. I do, to be sure, treat the major critiques of unbridled selectionism (constraints as channels, failure of pure extrapolationism into geological time), but I conduct this discussion within a Darwinian world, and do not adequately consider truly alternative mechanisms of change and their domains of operation. Since selection is a causal theory of change based on distinctive traits of definable individuals within specified environments (quite apart from any stochastic sources for the variation that provides raw materials of change), the obvious first-line alternative to selection must lie in *random reasons for change* itself.

As a basic statement in the logic of an argument, this point can hardly be denied, and therefore enjoys a long history of recognition in evolutionary thought. But recognition scarcely implies acceptance. The Victorian age, basking in the triumph of an industrial and military might rooted in technology and mechanical engineering, granted little conceptual space to random events, so the issue barely arose in Darwin's own time. (Darwin got into enough trouble by invoking randomness for sources of raw material; he wasn't about to propose stochastic causes for *change* as well! To this day, a distressingly familiar vernacular misunderstanding of Darwinism rests upon confusing these two components (sources of raw material and causes of change)—as in the common charge that Darwinism must be wrong because human complexity couldn't arise by purely random processes. Nineteenth

---

*Anyone, like me, who grew up in America with fiercely traditional immigrant grandparents from "the old country" will appreciate the humor of such limited and inappropriate reference points. My grandmother's only concern for any cultural or historical event (all of which she followed with great interest and intensity) stood out in her single, invariant question: "Is it good for the Jews?"

century theories of probability also eschewed ontological randomness in favor of causal production by interaction of so many fundamentally orthogonal mechanisms that stochastic formulations would best fit the observed results—the philosophical solution traditionally adopted by the scientific determinists who invented probability theory, most notably by Laplace himself.)

For these primarily societal reasons, theories of random change enjoyed little currency before our own century, when for both external reasons of a new cultural context (spawned by such events as the breakup of colonial empires, the devastation of World Wars, and the consequent questioning of predictable progress as time's direction), and internal prods from the mathematical apparatus of population genetics, random models of change became a major and controversial subject in evolutionary theory. I shall not review this well-known story, centering on the life and work of Sewall Wright (see Wright's own magnificent four-volume summing up, and Provine's fine biography). I only need to remind readers that genetic drift (often called "the Sewall Wright effect" in early literature), while unimpeachable in theory, and therefore surely operative in nature, received very short shrift, especially as the Modern Synthesis hardened around its adaptationist core (see Chapter 7). The Synthesis did not and could not deny genetic drift; instead, supporters resorted to the classical argument for dismissal in natural history—relegation to insignificant relative frequency. I learned the argument as a near mantra in all my graduate classes during the mid 1960's: fixation by genetic drift can only occur in populations so tiny that most will already be on the brink of extinction.

Genetic drift at the traditional organismic level enjoys far more respect and currency today, but the basic argument of the Synthesis does have merit at this hierarchical level. Sexually-reproducing, multicellular organisms generally share two properties that greatly limit the efficacy of genetic drift: they live in populations far too large for random fixation in the face of nearly any measurable selection pressure; moreover, the style of individuality manifested by organisms, based on well-balanced functional integration among subparts, renders the traits of these interactors particularly subject to scrutiny by natural selection.

Do these good reasons for demoting random change at the organismic level doom this alternative style of evolution to weakness or impotency throughout the hierarchy? Clearly not, as the recent history of our profession proves; moreover, we may even invert the standard hope for extrapolation from the level we know best, and assert instead that the organismic level discourages random change as a peculiarity of individuality in this realm—and that analogs of genetic drift at other levels should expect healthy, if not dominant, relative frequencies.

All evolutionists also know that ideas of random change have enjoyed greatest success, based on inherent plausibility, at the genic level, where the so-called "neutral theory of molecular evolution," most strongly associated with the great Japanese geneticist Motoo Kimura (1968, 1983, 1985, 1991a and b), but initiated and developed by others as well (Jukes, 1991), has often been hailed as the most interesting revision of evolutionary theory since Darwin.

When we consider the two properties of organisms that depress the frequency of fixation by drift at this level, we easily spot the difference that makes randomness so important at the lower genic level. Population size, also characteristically large for gene-individuals, cannot supply the reason. But the workings of DNA establish a strong supposition for absence of selective pressure from the organismal level at a high percentage of nucleotide sites, where alternative states do not influence the phenotypes of organisms—hence the designation of drift at this level as the *neutral* theory of molecular evolution.

Kimura's classical categories of evidence all depend upon the observation that maximal rates of nucleotide change occur at sites that do not influence the organismal phenotype—on the reasonable assumption that organismal selection usually acts in the stabilizing mode to preserve favorable sequences, and that sites under selective influence must therefore change at less than the maximal rate. The threefold confirmation of this prediction provides powerful evidence for the neutral theory—(1) for synonymous substitutions of the third nucleotide in a triplet; (2) for much higher rates of change in untranslated introns than in surrounding exons; and (3) for entirely untranslated pseudogenes, where rates at all three positions of triplets match the rapid third-position rate for translated DNA.

The move from mere plausibility to the important claim for high, or even dominant, relative frequency arises both by implication from the basic theory, and from observation. The three phenomena described above, after all, include a large percentage of all nucleotide changes—so neutralism must maintain a high relative frequency at this level if we have interpreted the rates of change correctly. At the broadest scale of geological time, the (admittedly approximate) ticking of the molecular clock in so many phylogenetic studies achieves its most plausible reading as a consequence of generally comparable rates for the high percentage of neutral substitutions. (The alternate explanation of averaging out for fluctuating selective control over sufficiently long periods of time cannot be dismissed *a priori*, but smacks of special pleading—whereas neutralism expects this result as the consequence of a central proposition.)

Kimura has always stressed the *high frequency* of neutral substitutions as his main challenge to Darwinian traditions. He writes, for example (1991a, p. 367), that "in sharp contrast to the Darwinian theory of evolution by natural selection, the neutral theory claims that the overwhelming majority of evolutionary changes at the molecular level are caused by random fixation (due to random sampling drift in finite populations) of selectively neutral (i.e., selectively equivalent) mutants under continued inputs of mutations." At the same time, Kimura also consistently insisted—and not, I think, merely for diplomacy's sake, or for any lack of resolve, but rather with genuine conviction (I discussed the matter several times with Kimura in person, so I will also stand as witness)—that the neutral theory did not contradict or dethrone Darwinism, but should rather be integrated with natural selection into a more complete and more generous account of evolution. Most neutral changes, after all, occur "below" the level of visibility to conventional Darwinian processes acting at the organismic level. Moreover, although most nucleotide changes may be neutral at their origin, the variability thus provided may then become indispensable for adaptive evolution of phenotypes if environmental change promotes formerly neutral substitutions to organismic visibility—an important style of cross-level exaptation (Vrba and Gould, 1986; Gould and Lloyd, 1999) that may serve as a chief prerequisite to the evolution of substantial phenotypic novelty. Kimura writes, for example (1985, p. 43): "Of course, Darwinian change is necessary to explain change at the phenotypic level—fish becoming man—but in terms of molecules, the vast majority of them are not like that. My view is that in every species, there is an enormous amount of molecular change. Eventually, some changes become phenotypically important; if the environment changes, some of the neutral molecules may be selected and this of course follows the Darwinian scheme." Thus, Kimura's statement exemplifies the central principle that the various levels of evolution's hierarchy work in characteristically different ways—and that levels can interact fruitfully in these disparate modes.

The chronological reaction of Darwinian hardliners to the neutral theory can be epitomized in a famous, if sardonic, observation about the fate of controversial theories. Tradition attributes this rueful observation to T. H. Huxley, but some form of the statement may well date to antiquity, the usual situation for such "universal" maxims. In any case, the earliest reference I know comes from the great embryologist von Baer, who attributed the line to Agassiz (von Baer, 1866, p. 63, my translation): "Agassiz says that when a new doctrine is presented, it must go through three stages. First, people say that it isn't true, then that it is against religion, and, in the third stage, that it has long been known."

The first two stages unfolded in their conventional manner, with quizzical denial followed by principled refutation in theory (see p. 521 on Mayr's argument that neutralism cannot be true because we now know the ubiquity of selection). However, the third stage—still stubbornly occupied by some strict Darwinians—arose with an interesting twist, providing a cardinal illustration for this section's major theme: the dangers of parochialism, particularly the tendency to interpret all evolution from an organismal vantage point. Instead of simply stating that neutralism has long been known (so what's the big deal?), detractors now tend to say: "well, yes, it's true, and let's be generous and give Kimura and company due credit. But, after all, neutral substitutions only occur at sites without consequence for organismic phenotypes. So why focus upon such changes? Without any organismal effect, they can't be important in evolution. And no one can blame Darwin or Darwinian tradition for ignoring an invisible phenomenon."

This exculpation of Darwin cannot be faulted in logic, but the rest of the argument reflects a narrow and discouraging attitude. Isn't the claim of unimportance absurd *prima facie*? How can anyone advance an argument for downgrading, as marginal, a process potentially responsible for more than half of all nucleotide substitutions—the supposed basis of evolution within a scientific ethos centered on reductionist preferences? Only a lingering prejudice for viewing organisms as a unique and intrinsic focal level could possibly generate such a claim.

Yes, an organism might view the world of its own compatriots as stable islands rising above an invisible sea—and choose to disregard random change within this swirling ocean of underlying, constant activity. But (if I may pursue this strained metaphor for a moment), any dynamic particle in the ocean could just as well, and perhaps with more merit, view the islands as rare and insignificant pedestals intruding into the truly fundamental substrate. May I just note the sterility of such a subjective argument, and state that any process with so strong an impact on change at any level cannot be unimportant in a world judged by relative frequencies.

As an illustration of the importance (and separability) of hierarchical levels, we may invoke balances produced by negative interaction among levels as a measure for the indispensability of molecular neutrality in full explanations of evolutionary phenomena. Just as a stable balance may arise by opposite forces of *selection* at adjacent levels, different processes—in this case neutrality at one level vs. selection at another—can also produce an intermediary result testifying to the importance of *both* styles of change. In such cases, moreover, neutrality enjoys a special heuristic advantage because random models yield general, quantitative predictions, while selectionist explanations usually require knowledge of particular circumstances that are much harder to decipher, and often impossible to quantify (for lack of requisite historical information).

For example, *Spalax ehrenbergi*, a blind Near Eastern mole rat, develops a rudimentary eye with an irregular lens that cannot focus an image. The eye is covered by thick skin and hair, and the animal shows no neurological response to powerful flashes of light (see p. 1282 for fuller discussion of this case in a different context). As expected under the neutral theory, the major lens protein, $\alpha$A-crystallin, evolves much faster in *S. ehrenbergi* than in other murine rodents with normal vision (Hendricks et al., 1987)—nine amino acid replacements in a sequence of 173, over 40 million years of evolutionary separation, whereas the other nine rodents of this study show identical amino acid sequences, with no alterations at all from the ancestral state. But this rate of change for *Spalax* represents only 20 percent of the average for true pseudogenes, our best standard for a maximal and purely neutral pace of evolution. At a rate of alteration too fast for stabilizing selection, but too slow for pure neutrality, the results imply a dynamic balance between molecular drift and weakened selective control at the organismic level. (Suggestions for continued utility of a non-seeing lens include possible function in adjusting physiology to seasonal cues from changing day lengths, though we know no mechanism for perceiving such fluctuations without vision, see Haim et al., 1983; and developmental constraint based on formation of eyes as a necessary inducer of some later and fully functional feature in embryology.)

I close this woefully insufficient commentary by reemphasizing the point that our discomfort or disinterest in random change largely reflects the peculiarity of the individual and level that we know best—organisms—and does not record any rarity or impotence for stochastic forces as agents of phyletic change in evolution. Processes of drift probably exert least influence upon the organismic level, for the two reasons cited earlier: large population sizes, and a style of individuality that forges coherence by strict functional coordination of subparts, and therefore makes nearly every trait of the organism subject to selection strong enough to overwhelm drift.

But the organism is a unique and peculiar kind of individual—and these strictures upon drift do not apply so strongly at any other level. We have seen, in this section, how structural features of DNA impose neutrality or near-neutrality upon selection at a large percentage of sites, perhaps a majority. For this reason (and not by limitation of population size), randomness becomes a fundamental process of evolutionary change at the genic level, however weak such a force may be (or, indeed, may *not* be!) at the organismic level. We shall see that, at the highest levels of species and clades, randomness again attains a high relative frequency—but this time mostly as a result of low N for species in clades. If such different causes grant randomness a high relative frequency at several important levels of evolution's hierarchy—and if we can only assert low relative frequency at one level, and for reasons rooted in the peculiar character of individuality in this realm alone—then have we not committed a great conceptual error, and seriously narrowed our general view of evolution and the history of life, by giving short shrift to this most obvious of all alternatives to selection as a cause of change?

TRUE GENIC SELECTION. When future historians chronicle the interesting failure of exclusive gene selectionism (based largely on the confusion of bookkeeping with causality), and the growing acceptance of an opposite hierarchical model, I predict that they will identify a central irony in the embrace by gene selectionists of a special class of data, mistakenly read as crucial support, but actually providing strong evidence of their central error. Gene selectionists have always welcomed genuine cases of a phenomenon that they then falsely generalize to all evolution—that is, differential proliferation of genes within genomes for reasons acting at the genic level, and independent of effects introduced by downward causation from selection at any higher level.

Gene selectionists have naively embraced these examples as apparent confirmations of their belief that effectively all selection operates at this lowest level. If genes can work their magic even without a boost from the vehicles they usually employ as lumbering robots subject to their will, then our appreciation for their omnipotence can only increase. But such superficial admiration obscures a true distinction that actually illustrates the bankruptcy of exclusive gene selectionism. These examples do not showcase the maximal power of a ubiquitous phenomenon; rather, and quite to the contrary, they represent the only class of instances where pure and untrammeled gene selection can operate at all!

As argued previously in this chapter (pp. 613–644), when gene selectionists speak of genes using organisms as their vehicles, they commit a deep error by inverting causality and ascribing to genes (which only *record* the causal result, and therefore serve as good units of bookkeeping) the agency in natural selection that really belongs to the organism—for vehicles (or interactors) operate as units of selection, or causal agents of Darwinian evolution. But when genes do not use organisms as vehicles and engage in differential proliferation on their own accord, then the genes themselves do act as vehicles—and, consequently, can become units of selection. Gene selection *only* exists when genes can operate as vehicles (interactors); thus, these cases illustrate the restricted range of a process that gene selectionists naively regard as optimal illustrations of a ubiquitous phenomenon. The resulting irony deserves emphasis. Supposed best cases become only cases, and therefore disproofs of a generality when properly interpreted. Wilson and Sober (1994, p. 592) put the point well: "These examples have been received with great fanfare by gene-centered theorists as some sort of confirmation of their theory. However, they do not confirm the thesis that genes are replicators—all genes are replicators by definition and no documentation is needed. These examples are remarkable because they show that genes can sometimes be vehicles. They seem bizarre and disorienting because they violate our deeply rooted notion that individuals are organisms."

Devotees of the genic level may eventually accept the defeat of their theory of exclusivity with good grace—for the supplanting hierarchical model provides more than enough room for true (and fascinating) examples of genuine genic selection, perhaps at quite high relative frequency once we acknowledge and learn to recognize the synergistic and orthogonal modes, as well as the better-documented examples of genic selection that harms organisms.

Moreover, when we recognize that many kinds and aggregations of genetic units can function in selection, the scope of this level becomes even wider. Selection may operate at the lowest unit of the nucleotide itself, if preferential substitution arises, for example, by differential production and consequently greater availability of one nucleotide *vs.* alternatives (the analog of natural selection by birth biasing). Selection among entire genes and other DNA segments of comparable length may also hold great significance in evolution—as in Dover's important hypothesis of "molecular drive" (Dover, 1982).

In fact, we may be impeding a proper recognition of the substantial frequency of selection within genomes by naming the phenomenon for only one mode among many—"gene selection." In the early days of Watson and Crick, biologists tended to conceptualize genomes as linear arrays of functional units (tightly strung beads with no spaces between in the usual metaphor). But we now know that most genes of eukaryotes, with their structure of exons separated by introns, do not maintain strict spatial continuity. Moreover, the functional genes of most complex metazoans represent, in any case, just a few percent of the full genome. All other kinds of genomic elements, forming an overwhelming majority of sites, can also evolve by processes of drift and selection.

For this reason, Brosius and Gould (1992) suggested that we use a more general term—"nuon" for nucleic acid sequence (DNA or RNA)—to recognize any stretch of nucleic acid, functional or not in organismic terms, that can evolve by differential origin or replication:

> Genomes do not consist only of genes. Sequences located between and also within gene boundaries, accounting for a large portion of the genomes of higher Eucarya, are not being addressed in a similar manner, partly due to the widespread opinion that these sequences are without function . . . We propose to name all identifiable structures represented by a nucleic acid sequence (DNA or RNA) as "nuons." A nuon can be a gene, intergenic region, exon, intron, promotor, enhancer, terminator, pseudogene, short or long interspersed element . . . or any other retroelement, transposon, or telomer—in short, any unit from a few nucleotides to thousands of base pairs in length.

Proceeding upwards, aggregates of genes can also function as units of selection—including, as prominent agents in evolution, chromosomes (Nei, 1987), and organelles and bacterial plasmids within cells (Eberhard, 1980, 1990).

Organismic selection generally works with great effectiveness in suppressing "revolts" to organismic integrity by differential proliferation of elements from within (see Buss, 1987; Leigh, 1991). Most of the characteristic properties of genomic organization and embryological development—from Hamilton's "gavotte of the chromosomes" in meiosis, to such phenomena as germ line sequestration and maternal determination in embryogenesis—may have evolved largely to suppress suborganismic selection, thereby assuring the integrity of multicellular organisms. Meiosis itself presumably evolved to place one copy of each gene "in the same [gametic] boat," thus converting organisms, rather than genes, into a primary unit of selection by the Musketeer's criterion of "all for one." But, once achieved, meiosis must be actively guarded by organismal selection against destabilizing drivers and distorters—all to preserve what Leigh (1991, p. 258) calls "the genome's common interest in honest meiosis."

Nonetheless, the evolutionary literature abounds with cases, both "classic" and new, of meiotic drivers, chromosomal segregation distorters, and other phenomena that favor the plurifaction of individual genes or sequences (including entire chromosomes) within the genome or population of genomes—usually with negative consequences for organismal selection above. Perhaps such cases must be relatively rare in nature, and only prominent in our literature for their intriguing oddity and exceptional status in the light of organismal selection's usual power to suppress such "outlaws."

Driving genes and chromosomes use a variety of devices to increase their relative representation by suborganismal selection. Some, including the classic *t*-allele of house mice (Lewontin, 1970), cause dysfunction in sperm carrying the nondriving homologue; others, like the supernumerary chromosomes of rye, segregate preferentially into functional gametes. Werren (1991, p. 393) attributes the interest generated by these cases to implications for the hierarchical model of selection: "Driving chromosomes are of general interest in population genetics as examples of 'selfish' or 'parasitic' genetic elements. Such elements challenge the concept of the individual genome as a 'cooperative' unit because they gain a transmission advantage relative to the rest of an individual's genome but are often detrimental to individual organisms."

Werren (1991; Werren, Nur, and Wu, 1988; Werren and Beukeboom, 1993) has also discovered and developed one of the most elaborate and interesting cases of suborganismal selection, a testimony to the complexity of interaction among levels of selection as well. In the parasitoid wasp *Nasonia vitripennis*, a supernumerary chromosome called PSR (paternal sex ratio) has evolved "an extreme and unusual form of transmission drive" (Werren, 1991, p. 392). This chromosome, carried in sperm, induces supercondensation of all male chromosomes (except itself) into a chromatin mass before the fertilized egg's first mitotic division. These chromosomes are then eliminated, while PSR survives. Since wasps are haplodiploid, this elimination converts an egg that would have become a diploid female into a haploid male (with PSR). This procedure obviously gives PSR a selective advantage in transmission drive because this unpaired chromosome will always be transmitted by males produced from fertilized eggs.

Just as obviously, organismic selection must oppose PSR, lest the entire population become both male and extinct. Werren (1991) had modeled conditions of maximal opposition from organismal selection. Subdivision of populations will be most effective in producing increased competition among PSR males, with reduced availability of females. But the story becomes even more complicated because suborganismal competition against PSR has also evolved by at least two devices that bias the sex ratio in a female direction (Werren and Beukeboom, 1993): (1) a maternally transmitted bacterium, called son-killer, that prevents the development of unfertilized (male) eggs; and (2) a cytoplasmically inherited agent of unknown structure and origin, that induces female wasps to produce nearly 100 percent daughters (called MSR, for maternal sex ratio). The possibilities introduced by haplodiploidy surely influence this variety and complexity in competing selection among suborganismic units—so stories this elaborate may not be common in nature. Still, as students of teratology in anatomy have always argued, we test and illustrate general rules by studying such cases at the limits.

But the main weight of gene selection in nature—the category that establishes a high relative frequency for the phenomenon—probably resides in cases that are synergistic with, or orthogonal to, organismic selection, and therefore not opposed by this powerful, conventional mode. Any genetic element that can propagate itself within the genome, either by iteration in tandem or by duplication and transposition to other chromosomes, works thereby as a vehicle of its own relative increase—and therefore as an agent of positive Darwinian selection at the genic level. If this propagation encounters no resistance at some other level (particularly by the watchful organism)—

either because bodies don't "notice" the increase (at least while the number of genic copies remains "within bounds"), or because higher-level selection also benefits from such differential genic proliferation (if, for a hypothetical example, an X-driving chromosome helped to generate the female bias that interdemic selection also favored)—then genic selection can be quite rapid and powerful. This general phenomenon, perhaps of great importance in evolution, has acquired the unfortunate name of "selfish DNA," as designated in two seminal papers, representing independent and simultaneous discovery, and published back to back in *Nature* in 1980 (Orgel and Crick, 1980; Doolittle and Sapienza, 1980). These authors proposed that such genic selection, orthogonal (at first) to organismal selection, might account for most of the middle-repetitive DNA—some 15 to 30 percent of the genome in humans and *Drosophila,* and usually existing as tens to a few hundred copies per sequence, with copies often widely dispersed among several chromosomes.

Other hypotheses might explain this phenomenon, particularly as a potential organismic need for enhanced levels of any products ultimately made by any gene. (In a purely organismal view, all genes may be able to proliferate, but not to fix their multiple copies unless organismic selection favors the increase. However, the two levels might also act synergistically, with genic drive evolving only in some genes, and for Darwinian benefit at this basal level, but with proliferation then enhanced by positive organismic selection upon bodies carrying more copies).

The "selfish DNA" hypothesis includes an attractive feature, rooted in the hierarchical theory of selection, for explaining stabilization of copy number at tens to hundreds, rather than an ultimately suicidal proliferation to inevitable death of the organism and all gene-individuals contained therein. Genic selection may begin in the orthogonal mode, as initial increases impose no consequences upon the phenotype. But organisms must eventually take notice, if only for the energetic drain, and presumed slowing of ontogenetic development, imposed by replication of so many unneeded copies with every cellular division. Original orthogonality must therefore eventually yield to a situation of genic selection contrary to organismic interest. At this point, negative selection at the organismic level should stabilize and limit further increase—the presumed explanation, within the theory, for the intermediary copy number of middle-repetitive DNA.

Although I regard the hypothesis of "selfish DNA" as powerful, probably correct in many cases, and therefore as our best argument for substantially important selection at the genic level, two features in its initial promotion distress me because they embody (without conscious intent, I assume) the persistent parochialism of organismic bias, even among those who explicitly promote the hierarchical alternative. First, consider the unfortunate choices of names. Proliferating genic elements have generally been called "outlaws," "renegades," or "parasites"; and the general phenomenon entered our literature as the hypothesis of "*selfish* DNA." Orgel and Crick imposed a double whammy of opprobrium in the title of their original article: "Selfish DNA: the Ultimate Parasite." The only reason that I can imagine for such derogatory terms resides in the unstated (and probably unconscious) notion that benefits for organisms define the ultimate goal and purpose of evolution as a general phenomenon. Thus, anything that can evolve, but either hurts the organism actively, or even just manages to sneak past organismal scrutiny, must be designated as selfishness, nastiness, or even usurpation—as promoted by some reprobate object that would place its own propagation above the general good of evolution.

Surely, we must reject such parochial thinking and terminology. Propagating genic elements should not be described as parasites or renegades; nor can they be defined as "selfish" in any meaningful or general sense. Rather, propagating genes follow the Darwinian imperative at their own level, and therefore act as any good Darwinian agent "should"—that is, to increase their own representation within their own environment, the genome in this case. As Darwinians, we should honor their pluck in such a difficult endeavor (for organisms do tend to be watchful and suppressive), rather than heaping derogatory terms upon them. Such genes could only be deemed "selfish," "parasitic," etc., from a false and limited perspective that values the organism alone as an agent of evolutionary success. After all, we don't call a peacock selfish for evolving such a beautiful tail, and thus limiting the geological longevity of the species.

To fully embrace the hierarchical model, a concept that marks a fundamental shift in theory, not just an interesting new wrinkle upon an unaltered concept of nature's basic construction, we must reconceptualize all of evolution, and revise both our worldview, and our language, accordingly.

Second, even in terms of our conventional focus on organisms, genic selection may provide crucial and indispensable flexibility for evolution of any substantial organismic novelty, including features conventionally placed in our most vaunted category of "increasing complexity." The general argument has become traditional in evolutionary theory (since the pioneering book of Ohno, 1970), and represents a solution to the following, otherwise disabling, paradox: Organismal selection on the earth's original prokaryotic biota might have constructed an optimal cell, "mean and lean" as could be, with a single copy of each gene to make, in the best possible way, one product indispensable for cellular success and propagation. But how could such an inflexible organism ever change beyond minor adjustment to altered environmental circumstances? As Ohno wrote (1970): "from a bacterium only numerous forms of bacteria would have emerged." But duplicated copies can provide requisite redundancy, permitting one copy to manufacture the needed product, while others become free to change—and to add new functions, thus providing a potential route to increasing complexity.

But if selection only works at the organismic level, and our "mean and lean" bacterial prototype has attained an optimal configuration, what process provides evolution with the multiple copies needed for flexible addition of functions? We gain nothing from noting that duplications provide later blessings, since evolution cannot operate for the benefit of unknown and unpredictable futures, unless our basic view of scientific causality needs fundamental revision, and the future can determine the present.

Hierarchical selection provides the most promising exit from this substantial paradox: multiple copies cannot originate for future organismic benefit, but they can evolve by present genic selection! (Later exaptive utilization in the generation of organismal complexity illustrates the important historical principle that reasons for origin must be sharply separated from current utility—see Chapter 11 for extensive discussion. Evolution continually recycles, in different and creative ways, many structures built for radically different initial reasons.) In 1970, Ohno wrote with great prescience: "The creation of a new gene from a redundant copy of an old gene is the most important role that gene duplication played in evolution."

Thus, if duplication requires genic selection in many or most cases, then the first level of evolution's hierarchy not only operates with respectable relative frequency, but even provides an indispensable boost for generating the *summum bonum* of our deepest prejudices—the complex organism, with eventual evolution of a single strange mammalian species endowed with a unique capacity for self-reflection, but occupying an isthmus of a middle state, a good vantage point for looking down with thanks to duplicating genes, and up with awe to a tree of life that could generate such an interesting and accidental little twig.

THE CELL-INDIVIDUAL I speak here not of free-living unicells (where cell and organism represent the same unit of evolutionary individuality), but of cells that generally house full genomes and form the environment of genes at the level below, while also serving as parts and building blocks of multicellular organisms at the level above. From our limited viewpoint as highly complex metazoans built by intricate and integrated programs of embryological development, we tend to neglect this intermediary level of differential cellular proliferation (not just to build bigger organs in the somatic environment, for such a process yields no evolutionary reward in competition with other cells for representation in future generations, but rather to gain preferential access to the germ line, and thus to achieve evolutionary success by positive selection at the cell level). We neglect this subject because positive selection now so rarely occurs at this level in complex metazoans—and for a reason continually emphasized in this chapter: the effectiveness of multicellular organisms in suppressing the differential propagation of subparts as a necessary strategy for maintaining functional integrity, the definitive property of individuality at the organismal level.

This suppression has been so effective, while the consequences of failure remain so devastating, that human organisms have coined a word for the cell lineage's major category of escape from this constraint, a name with power to terrify stable human organisms beyond any other threat to integrity and persistence—cancer. I suspect that we would learn much more about this large class of diseases (mistakenly viewed by most of the public as a single entity) if we treated the subject in evolutionary terms as a historical result of the cell's initial capacity, retained from its phylogenetic past as an entire organism, for differential proliferation over other cells (formerly competitors as separate organisms, not compatriots as components of other organs). Of course, modern human cells that escape this constraint do themselves no ultimate good, for they have no access to the germ line, and their unrestrained growth eventually eliminates both their own lineage and the entire surrounding organism. To this extent, the organism's general strategies do eventually prevail, following an initially successful assault by a cell lineage. But what a pyrrhic victory! Nonetheless, the double effectiveness of a virulent cancerous cell lineage—crowding out in place and distant metastasis to other locations in the body—recalls the more "benign" strategies of other successful evolutionary plurifiers within a constrained space (genic proliferation by tandem duplication and transposition; budding off of new demes and "capture" of existing demes by immigration and transformation).

If selection at the level of cell lineages now plays only a minor role in most groups of multicellular organisms, we should not view this hierarchical level as intrinsically impotent, but rather as historically suppressed in "the evolution of [multicellular] individuality," to cite the title of Leo Buss's seminal book on this intriguing subject (Buss, 1987). In Buss's terminology selection upon cells must now unfold in the "somatic environment," where suppression reigns in the service of organismic integrity, whereas such selection once occurred in the "external environment," where unicellular organisms could experience the full independence and competitive range of Darwin's world. (In fact, since most organisms on earth remain unicellular—see Gould, 1996a, on the persistence of the bacterial mode throughout the history of life—this transition has never occurred for the vast majority of organisms on earth.)

This cellular level therefore provides our best demonstration that the current evolutionary hierarchy in styles of individuality arose both historically and contingently, and not with necessity as a timeless, predictable, invariant consequence of natural law. Levels have surely been added sequentially through time, as Buss has emphasized. If life began with naked replicators at the genic or subgenic level, then these earliest times for life may have featured, uniquely for this initial interval, the property that strict Darwinians have tried so hard to impose upon our richer world of modern life—selection at one level only. The evolution of cells led to a tripartite hierarchy that characterized most of life's 3.5 billion year history, and still regulates the majority of earthly organisms: genes, cells, and clones. The evolution of sexual reproduction added species, while the complex processes that constructed the multicellular individual then added the organism (the body that encloses cells and cell lineages).

Suppression of cell lineage selection by the multicellular organism has greatly restricted a once vibrant and multifarious level. I must confess to my own parochialism in recognizing just one unit, the cell, as a surrogate for all entities that enclose genomes and form parts of organisms. Certainly organelles (or at least the mitochondria and chloroplasts that began their evolutionary history as symbiotic prokaryotes), and sometimes tissues and modules of embryological development, can also act (in principle) as suborganismal units of selection. I make the amalgamation because this level has been largely suppressed, and therefore doesn't often come to our attention in studies of modern multicellular organisms. In so doing, I feel some allegiance to the folk taxonomist who (as so often recorded for indigenous cultures) assiduously names each species (much as a trained Linnaean systematist would do) for creatures important to his life, but then lumps into large categories (weeds, butterflies, bugs) the organisms of no great moment in his world.

As the central premise of his fascinating and seminal book, Buss (1987) argues that the multicellular individual arose by "the interplay between selection at the level of the individual and selection at the level of the cell lineage" (p. 29). More specifically, he attributes the distinctive features of metazoan development to an initial competition among cell lineages, eventually tamed and regulated by organismic selection in the interests of bodily integrity. Buss writes: "The thesis developed here is that the complex interdependent processes which we refer to as development are reflections of ancient interactions between cell lineages in their quest for increased replication. Those variants which had a synergistic effect and those variants which acted to limit subsequent conflicts are seen today as patterns in metazoan cleavage, gastrulation, mosaicism, and epigenesis" (p. 29).

Clearly, such a concept becomes intelligible only under the aegis of a hierarchical model of selection, as defended in this book's central thesis. Buss recognizes this conceptual link, of course, and his work becomes a strong confirmation of both the efficacy and necessity of this basic reconstruction in evolutionary theory. In terms similar to the views expressed here, Buss writes (pp. 5–6): "The logical structure of Darwin's argument allows any unit to evolve if it replicates with high fidelity, and if selection distinguishes between the variants. Species, populations, and lineages of individuals, cells, organelles, and gene sequences can all potentially evolve. Yet we have been largely content to attribute the whole of biological diversity to selection upon individuals [organisms]. The once comfortable cloak of the Modern Synthesis has become restrictive." (I am also grateful to Buss for recognizing the role of my profession, particularly in the work of Eldredge, Jablonski, Stanley, Vrba, and myself, in developing the hierarchical theory of selection. He writes (p. ix): "Indeed, hierarchical perspectives on evolution are undergoing a rebirth among paleontologists at the moment.")

In Buss's model of historical and sequential construction for nature's hierarchy, new levels arise to enclose the individuals of older levels by a two-step process. The initial features of the nascent level must originate in synergism, or positive interaction, with selection at the level just below, which formerly stood topmost, but will now be superseded (in the literal sense of "sat upon") by the newly-emerging style of organization. New levels must begin with such a helpful boost, for the initial tentative and unformed steps cannot yet possess enough power to suppress or regulate a well-established level beneath. But stabilization of the new level, implying a power to suppress at least some forms of harmful proliferation from within, then requires negative interaction, once the new and higher level achieves enough coherence to act in its own right.

Since we have no direct data for key transitions that occurred so long ago and left no fossil evidence (so far as we know), Buss constructs some hypothetical examples of how such a process could work. (Such entirely speculative scenarios must be understood within their acknowledged limits—that is, as hypothetical stories, "cartoons" in Buss's words, invented to illuminate a potential mode, and not as claims about any historical actuality.) For example, if the first tentative multicellular organisms evolved as little more than spherical colonies of identical protists floating in the ocean, how might essential organismic properties like cellular differentiation emerge? Suppose that a variant cell lineage arose in such a loosely-knit, hollow sphere of cells, causing members of the new line to enter the sphere's center, where proliferation could continue. In this way, a new cell lineage (and the beginning of cellular differentiation for the organism) could originate and proliferate by selection at the cell level. Buss then supposes that such an event might also be beneficial for the organism, and he draws an analogy to the ontogeny of some modern sponges:

> The origin of a variant cell line which entered the center of such a sphere to continue cell division . . . may have produced a structure which was sufficiently negatively buoyant to fall to the sea floor. Many modern sponges . . . do just this. A flagellated sphere populated by amoeboid cells simply drops to the ocean bottom . . . The pelago-benthic life cycle of sponges may have arisen as a consequence of variants which, in pursuing their own replication, fortuitously presented the individual with a benthic existence and all the attendant opportunities inherent in the invasion of a new adaptive zone.

This move toward a more complex and better integrated organism begins with an initial synergism between cellular and organismic selection (origin of a new cell lineage by invasion and proliferation in the organism's hollow center, leading to organismal advantages through an imposed change of habitat). But later stabilization of this innovation requires the suppression of cell lineage selection by the organismic level—for if the two cell lineages (at the sphere's periphery and center) engage in an anarchic battle for ever greater representation in cellular percentages, either the organism will lose coherence and die, or one lineage will win and the organism will return to its previous state of minimal differentiation.

Moving away from speculation and towards an explanation of metazoan development, Buss interprets several defining features of many (but not all) metazoan phyla as records of successful suppression of cell-lineage selection by organismal selection from above. In particular, he views early germ-line sequestration (Weismann's crucial criterion in his defense of Darwin against resurgent late 19th century Lamarckism, see Chapter 3), and maternal predestination, as organismal devices evolved to set and stabilize the course of development as early in ontogeny as possible, thus greatly reducing the potential for new forms of differential cellular proliferation either to arise at all in later ontogeny, or to reach the germ line and act in cell-lineage selection even if they do manage to originate. Buss sums up his thesis:

> Selection at the level of the individual has opposed selection at the level of the cell lineage by acting to set the timing of terminal somatic differentiation as far back in ontogeny as possible—whenever possible into the maternal cytoplasm itself. (p. 5). . . . The release of the totipotent germinative lineage from the task of producing somatic tissues meant that the number of divisions made by the totipotent lineage could be reduced and, consequently, the opportunity for variants to arise to become severely restricted (p. 100) . . . Metazoans, by the twin devices of maternal predestination and germ-line sequestration, have effectively closed their ontogenies to heritable intrusion arising in the course of that ontogeny. A novel epigenetic program can only arise if a mutation of extraordinarily improbable precision and autonomy occurs in the germ cells themselves (p. 102).

But nothing can be won without a price in our complex world of interacting levels, either in evolution or in human society. In stabilizing the organismic level with such effective devices to suppress cellular and other forms of suborganismic selection, organisms have greatly reduced their flexibility for future evolutionary change of more than a superficial nature. For these mechanisms of development do not suppress only the forms of cell-lineage selection that would harm the organism; rather, they impede *any* effective cellular selection at all, whether beneficial or harmful. These policing devices of the organism therefore close off an avenue once open for substantial change in basic designs, thus restricting maximal potency to the iteration of essentially similar species (as in such famous examples as the cichlids of African lakes, or the Galapagos finches), now representing evolution in its most vigorous contemporary mode. *Ou sont les neiges d'antan?* "The clear implication is that evolution of cellular differentiation fueled the evolution of controls over variants which fail to behave altruistically. The mechanisms which metazoans employ to limit the heritability of variants which fail to contribute to somatic functions are blind to the traits which a variant might express. *Potentially beneficial variants are as limited as are potentially detrimental ones*" (p. 103, Buss's italics).

This perspective implies a striking limitation upon the strictly Darwinian style of extrapolative and gradualistic selection that the Modern Synthesis promulgated as an adequate explanation for evolution at all scales of time and effect (see quotation from Wilson et al. on p. 583). If Buss's views are valid, then conventional Neo-Darwinian evolution must work within strictures of essentially established ontogenies that can surely generate exuberant adaptive variations upon set themes, but may be effectively unable to construct major innovations that establish the outlines of macroevolution. Once again, we grasp the need for independent macroevolutionary theory—and Buss has supplied an important piece of the general argument with his concept of a correlation between such major innovations and the origin of new hierarchical levels, a theme that obviously requires the hierarchical model and cannot be encompassed within the strict Darwinism of the Modern Synthesis. Buss concludes (p. 188, his italics): "Synergisms between the units drove the elaboration of a higher unit and conflicts arising between units were minimized by adaptations limiting further variation. This conclusion has the fascinating and crucial corollary that *the major features of evolution were shaped during periods of transition between units of selection.*"

THE ORGANISM-INDIVIDUAL  As virtually the entire history of Darwinian thought has unfolded under the assumption that organisms act as nearly exclusive agents of selection (or at least that our interest in evolution centers upon the alterations and fates of organisms), I shall not dwell upon this canonical individual here. I want only to reemphasize the unique and decidedly peculiar features of our kind of entity (in contrast to the characteristic properties of individuals at other levels): maximal cohesion based on functional integration, including relatively inflexible spatial orientations of subparts (spatiotemporal if we include embryogenesis). This style of integrity enables the organism to be particularly effective in suppressing selection against its interests by potential evolutionary individuals dwelling within and forming its parts. As noted above, the virtual "extinction" of effective cell lineage selection in complex metazoan phyla occurred as a historical result of the evolutionary "invention" of the intricate organism—perhaps the only example of an "endangered level" in the entire history of evolution!

As another portentous implication of individuality in this mode, organisms become chock full of adaptations as a consequence, under natural selection, of building coherence by functional integration. This local phenomenon at one level of Darwinian individuality has generated an understandable and commanding concern with adaptation, leading to doctrines of exclusivism in extreme cases (all too common, given our psychological preferences for simple and unifying worldviews—a need traditionally met theologically, but sometimes, particularly in our increasingly secular age, scientistically). If, as some strict Darwinians believe, "organized adaptive complexity" represents both the primary result of evolution and the cause of all other patterns in the history of life, then we will fail to understand nature for two cardinal reasons: (1) because we have adopted a criterion too strict even for its organismal level of most promising application (see Chapters 10 and 11); and (2) because the criterion of "organized adaptive complexity" does not strongly characterize the nature or definition of individuality at most other levels of the hierarchy.

Nature's hierarchy is not fractal; each level, to express the point metaphorically, does some things well, and other things poorly or not at all—and the evolutionary pattern of nature features many essential things. In our mother's house—the Earth—are many mansions. Gene selection is "good" at iterating elements—an important input of raw material for generating "organized adaptive complexity" at a higher level. Organisms are good at building complex adaptations. Species are good at forging temporal trends of geological duration, and their efforts largely regulate the relative diversity among phyla (why so many beetles, and so few pogonophorans). To say (as Dawkins, Williams, and other detractors often do) that species selection must be unimportant because such a process can't build organismal complexity reminds me of the cook who didn't like opera because singing couldn't boil water.

THE DEME-INDIVIDUAL This kind of individual has borne the brunt of the general argument about higher-level selection ever since Darwin awarded the idea a strictly limited amount of conceptual space in trying to puzzle out the origins of human altruism (see pp. 133–137). The subject has been extensively reviewed and controverted (Wynne-Edwards, 1962, vs. Williams, 1966, for an early and generally unacceptable version; Wade, 1978, 1985; D. S. Wilson, 1980, 1983, 1989; Wilson and Sober, 1994; Sober and Wilson, 1998, for reviews). I shall therefore provide only an idiosyncratic sketch here, for the terms and concepts of this discussion permeate the chapter, while my own interest as a paleontologist flows to the still higher levels that have not been extensively studied.

In a curious way, the development and acceptance of hierarchy theory has been impeded because the classical treatment of this subject has been focussed so strongly, indeed almost exclusively, on this level—and demes are the hardest of all individuals to validate and justify within the evolutionary hierarchy. All other individuals build better boundaries (to retain their own subparts, or lower-level individuals, and to exclude the subparts of other individuals at their level), and experience less difficulty in remaining sufficiently stable for the requisite time until reproduction. But demes are especially vulnerable to the classic objection (see p. 647) that, lacking strong internal mechanisms for coherence, their individuality may be too fleeting and subject to change by loss or invasion—as in Dawkins's well-formulated and memorable image of dust storms in the desert or clouds in the sky. Indeed, as I argued previously (p. 648), the classic defense of interdemic selection depends upon the identification of plausible conditions that would allow such adventitious groups to remain stable long enough to act as units of selection. The centering of the general argument for higher-level selection upon demes has, by false and unfortunate implication, led to the widespread impression that *any* kind of supraorganismal selection must face the same difficulties—perhaps with problems growing ever more intense as individuals become more inclusive. But this argument, based on illogical assumptions about linear extrapolation, does not hold because demes (in most circumstances) are uniquely unstable in the evolutionary hierarchy. Species, for example, usually attain as much stability and coherence as organisms, though by different mechanisms (see pp. 703–705).

Group selection has traditionally been invoked under our organismic biases as an explanation for bodily behaviors—with altruism as a paradigm, ever since Darwin himself (see Chapter 2, pp. 133–136)—that seem, *prima facie,* difficult to explain as beneficial to organisms, but can easily be construed as valuable for groups. But we should recognize such restricted invocation (only for cases that trouble organismic traditionalists) as yet another parochial limitation, and we should acknowledge a potentially general role for interdemic selection within any species of appropriate population structure. (Under such a criterion of judgment by relative frequency, we must ask a different, and quite unanswered, fundamental question: how many higher taxa generally maintain population structures that promote interdemic selection; in what environments; and with what correlations to such factors as phylogenetic status, body size, behavioral complexity, etc.)

If various arguments for the rarity of extensive evolution within large panmictic populations hold merit, and if Sewall Wright's shifting balance theory applies to a high percentage of populations, then interdemic selection may become a major mechanism for evolution within species through time. However, if punctuated equilibrium generally holds (see Chapter 9 for a defense of this view), then anagenesis within species will be rare in any case (whether by transformation via organismic selection under panmixia, or by shifting balance via interdemic selection in appropriately subdivided populations). Or perhaps, as an intermediate position, panmictic transformation is rare, but shifting balance frequent, in species that meet the criteria for appropriate population structure. The high relative frequency of punctuated equilibrium would then measure the relative rarity of such population structures, and the few groups that show extensive gradualism within species may generally subdivide their populations according to Wrightian criteria. This conjecture has not been tested, but could be, and with an interesting mixture of paleontological data on the history of species and neontological information on population structures within modern representatives of the same groups.

In any case, even if Wright's criteria don't hold often enough within the central range of species during the heart of their geological life, Mayr's peripatric model of speciation suggests that the origin of most species may occur by a process close to interdemic selection, and operating near a blurred borderline with species selection. If many species spawn large numbers of peripherally isolated demes, but only a few of these demes become species; and if the small class of successful speciators possess traits at the population level that encourage full speciation in interaction with the environment; then species will arise by selection and differential preservation on a just a few "winners" within a set of populations that begin as demes of an ancestral species (as best illustrated by the probable main reason for failure of others to speciate— reincorporation of a peripherally isolated deme into the larger parental population).

For all these reasons, I suspect that selection among deme-individuals holds an importance as yet unrealized (and perhaps occurring in modes as yet unconceptualized) within our general picture of evolution. I have, in my career, witnessed three examples of widespread dismissal by ridicule as part of a professional ethos: the rejection of continental drift as physically inconceivable; the shunning of Goldschmidt's macroevolutionary ideas as dangerous to the Darwinian consensus; and the dismissal of group selection as addlepated nonsense (see pp. 553–556). Nothing in my intellectual life has made me feel more uncomfortable.

I take great pleasure in the comeuppance of the smug ridiculers in all three cases. Plate tectonics has validated continental drift to become a new paradigm for geology. Goldschmidt's particular genetics win no general plaudits, but his views on a conceptual break between micro- and macroevolution now enjoy substantial support. The vindication of group selection has been slower, but now moves on apace (see Sober and Wilson, 1998)—with a vigorous professional discussion finally occurring, and with general attention now accorded, both in the popular press (Lewin, 1996), and in the commentary sections of general professional journals (Morrell, 1996). *Sic semper tyrannis.*

THE SPECIES-INDIVIDUAL I propose, as the central proposition of macroevolution, that species play the same role of fundamental individual that organisms assume in microevolution. Species represent the basic units in theories and mechanisms of macroevolutionary change. In this formulation, the origins and extinctions of species become strictly analogous to the births and deaths of organisms—and just as natural selection works through differential proliferation based on schedules of organismal births and deaths, so too does species selection operate upon the frequencies and timetables of origins and extinctions. The next section of this chapter—entitled "the grand analogy"—shall complete this argument by attempting to cash out this comparison in detail, with all the intriguing differences that arise when disparate individuals at two such different levels work by the same abstract mechanism.

I will therefore confine this preliminary discussion to the three major objections that have been raised against the foundational idea that species can act as important evolutionary individuals. These objections treat, in reverse order, the three words in the key phrase, "important evolutionary individuals." The first objection holds that species cannot be construed as proper individuals; the second admits that species are individuals, but argues that they cannot operate as interactors (as required for units of selection); while the third allows that species may be recognized as both individuals and interactors, but insists that they must remain effectively impotent in both roles.

SPECIES AS INDIVIDUALS. The classic argument of evolutionary gradualism denies real existence to species because they can only be defined as arbitrarily delineated segments of a lineage in continual anagenesis. Both Lamarck and Darwin, despite their maximally different views about proposed evolutionary mechanisms, strongly supported the nominalistic claim that only organisms exist as natural units, and that species must therefore represent abstractions, formally designated only for human convenience. (As many historians have remarked, Darwin chose an odd title for his revolutionary book—for he focusses upon the explanation of substantial change by anagenesis, and says little about speciation by branching of lineages.)

I presented the case for treating species as individuals in an earlier section of this chapter (pp. 603–608), noting that punctuated equilibrium greatly aids such a delineation, but then extending and generalizing the argument by holding that species can be individuated under any scheme that depicts their origin as an event of branching, rather than anagenetic transformation. Critics of this view, particularly Williams (who does not dispute the truly necessary claim for origin by branching), continue to raise standard objections, especially "an absence of a decisive beginning for a species" (Williams, 1992, p. 121). But Williams, in advancing this argument, commits the classic error of failure to appreciate proper scales. His claim for fatal fuzziness in origins views the question from a generational perspective at the scale of human lifetimes. The great majority of species, however, arise in geological moments (thousands of years, and thus overly long only at the inappropriate scale of our personal lives)—a shorter period of ambiguity (relative to later duration as a clearly separate entity) than we note for most asexual organisms that reproduce by budding (the proper organismic analog for the origin of a new species by branching)!

Most other published objections to species as individuals also express little beyond our psychological difficulty in making a transition to different criteria at unfamiliar scales. Some critics have argued, for example, that species can't develop the requisite property of heritability, because no mechanism can be analogized with the well-known Mendelian basis for this phenomenon at the organismic level. But heritability measures the correlation between parents and offspring based on direct transmission of formative properties—and daughter species surely inherit parental characteristics by this standard route. The required correlation arises by transmission of autapomorphic characters through retained homology—the appropriate mechanism of heritability at this higher scale, and in no way "worse" than Mendelian criteria for the construction of organisms. Moreover, species heritability can be measured in the same general way, and with the same potential accuracy, as standard organismic heritability—as Jablonski (1987) has done in our best-recorded case of species selection for the evolution of marine mollusks in normal times and episodes of mass extinction.

Other critics charge that species are too spatially diffuse, or too lacking in mechanisms of internal coherence, to count as individuals. But, again, these arguments only arise from failure to conceptualize this different scale in an appropriate manner—a mental foible rooted in our parochial allegiance to the particular (and poorly-scaling) criteria of individuality for organisms. Species don't build a physical skin, but reproductive isolating mechanisms maintain their borders just as sharply. Species don't evolve immune systems and other forms of "policing" against outside invaders, but the constant admixture among their parts via sexual reproduction maintains coherence with more than adequate force.

SPECIES AS INTERACTORS. This more interesting and challenging argument has unfolded among supporters of macroevolutionary theory as an "in-house" debate. Most discussants, including Brandon, Gould, Jablonski, Lloyd, Stanley, and Vrba, strongly support the concept of species as units of selection; while Damuth and Eldredge grant species a role as replicators, but not as interactors, and therefore not as agents of selection. Grantham (1995) has tried to mediate these positions with a compromise that will, I suspect, satisfy neither side.

Critics allow that species may be "fundamental units" of macroevolution in some sense—but, they say, only as the replicators that serve as "atoms" of cladistic phylogeny, and not as interacting units that forge macroevolutionary change by active competition in natural environments. (Eldredge, for example, includes species in the genealogical column of his two-hierarchy scheme —see page 642 for a critique—but not in his economic column of interactors.) A species, the critics continue, may live in too broad a range of environments, and over too wide a geographic range—often discontinuous to boot—to serve as an interactor, or unit of selection. Moreover, although individual populations of two species may compete sympatrically over a well-delineated geographic range, entire species rarely maintain sufficient overlap to interact with each other as complete units.

To resolve this apparent dilemma, Damuth (1985) proposes that we define a new interactor corresponding most closely to the hierarchical level where species serve as replicators. Using a criterion of direct competition in sympatry, Damuth proposes the term "avatar" for such interactors, defined as sympatric populations in ecological competition, and therefore interpretable as alternatives subject to selection. Grantham's (1995) "compromise" position maintains allegiance to Damuth's insistence upon potential interaction in sympatry. Grantham defends species selection, and regards species as potential interactors—but he would restrict any particular study of species selection to members of clades living in the same broad region. He writes (1995, p. 311): "I suggest that paleontologists focus on geographically constrained portions of monophyletic clades."

I would raise three arguments against this proliferation of terms and categories—and for the status of species as adequate interactors.

1. A standard mode of construing competition among organisms has beguiled us into thinking that interaction requires sympatry. As argued in Chapter 6 (pp. 470–477), Darwin strongly asserted the predominance of biotic over abiotic competition as the only promising path for a defense of progress in evolution. This preference has passed through the Victorian fascination with overt battle as a defining mode of competition, right into our present times, with continuing Tennysonian metaphors about "nature red in tooth and claw" (see Gould, 1992a), and newspaper stories about firms engaged in Darwinian struggles to the death as they vie directly for the allegiance of a limited population of consumers. (As I revised this chapter in the summer of 2000, a new magazine for "business evolving in the information age" made its debut under the name *Darwin*—also available on line at www.darwin-mag.com.)

But this focus on the biotic mode has always been indefensible as a claim for exclusivity, or even dominant relative frequency. In Darwin's own time, Huxley ridiculed this notion as "the gladiatorial theory of existence," while Kropotkin (1902) and others constructed alternatives based on cooperation in sympatry and the prevalence of abiotic competition in most environments (see Todes, 1988; Gould, 1991b). Darwin himself clearly favored an expansive concept of interaction with environments in natural selection—as when he insisted, in a famous passage (1859, p. 62), that "a plant on the edge of a desert" struggles for existence against the drought and other features of the physical environment just as surely as "two canine animals in a time of dearth" struggle more overtly for a limited supply of meat.

This point becomes important when we try to translate this debate about organisms to a definition of higher-level interactors. Biotic competition does require sympatry for direct and literal struggle, while abiotic competition imposes no such conditions, and must often occur among organisms that never encounter each other, even while living in sympatry. If we use biotic competition as our (often unconscious) paradigm for the entire, and far broader, concept of interaction, then we too easily become unduly committed to the false restriction that interactors must be able to duke it out directly. In upward translation, this bias leads to the idea that species-individuals can't be interactors unless they live in the same place, and thus maintain a potential for engaging in some analog of overt battle.

But interaction at the canonical level of organisms doesn't demand direct contact, or even life in the same place—and no one has denied that organisms operate as quintessential interactors, and units of selection. If abiotic competition dominates the history of life—as many distinguished researchers insist (see references in Allmon and Ross, 1990), at least for many groups in many circumstances—then potential for direct contact cannot be invoked as a primary criterion for defining interactors.

Williams (1992) has strongly asserted the non-necessity of sympatry (and resulting potential for direct "struggle") in defining higher-level interactors—and he uses the same analogy here advanced for asserting a similar non-necessity at the organismal level. I presented the full quote before, but repeat the operative line here (1992, p. 25): "One issue is whether the populations that bear the gene pools need be in ecological competition with each other. I believe that this is not required, any more than individuals within a population need interact ecologically to be subject to individual selection." Later, Williams specifically criticizes Damuth's definition of avatars on this basis. Speaking of populations not in direct competition, but subject to similar stresses (a common predator in their separate environments in this hypothetical case), Williams writes (1992, p. 52): "I am inclined to recognize that clade selection is operating even here, unlike Damuth, who maintains that only sympatric avatars, populations in ecological competition, can be alternatives subject to selection. Allopatric forms may not be ecological competitors, for the inattention of a predator or anything else, but they compete for representation in the biota, the ultimate prize in clade selection."

2. Although I recognize that some notion of a common environment must be invoked when we wish to define allopatric species as competing interactors, I do not view such a requirement as either rarely met or particularly difficult to specify. (I only mean, by the last phrase, "any more difficult to specify than for sympatric interactors." We cannot know, in fully adequate detail, how individuals at any level react to all nuances of the environment, in all their horrendously complex and nonlinear interactions. Who can say whether two sympatric organisms, given their inevitable differences, perceive the same local change of environment—even such a linear effect as a falling temperature—in the same way? I am only arguing that we face the same difficulty for sympatric, as for allopatric, interactors in this respect.)

At least two strong arguments support the notion of adequate environmental similarity in allopatry:

(i) Environments cannot be conceptualized (or even operationalized) as objective places or circumstances in a world fully external to the organisms involved. First of all, environments include all interactions with other organisms, both conspecific and belonging to different taxa, and not just the climates, substrates, and other more measurable properties of a surrounding physical world. Second, and more important, as Lewontin has emphasized so forcefully (1978, 2000), environments are intrinsically referential, and actively constructed by the organisms in question. Environments, in short, are made, not found. Thus, important properties of the environment must be sufficiently comparable in a set of closely related and partly allopatric species engaged in a process of species selection. These species share key traits as autapomorphies of their clade—and since these traits help to construct the relevant environment, sufficient similarity becomes, in part, an active construction of related organisms, not only a happenstance of common externalities.

(ii) Organisms needn't occupy the same turf in order to be impacted in similar ways by the kinds of broad environmental changes that seem so important at geological scales. To choose an extreme example, when, 65 million years ago, a large bolide struck the earth in the region now occupied by the Yucatán peninsula, I suspect that *Tyrannosaurus rex* in the western United States, and its recently discovered sister taxon in Africa, experienced consequences sufficiently common and negative to influence their extinction (while some small-bodied mammals, living there and elsewhere, survived as a consequence of organismal or higher-level characters that also do not require sympatry with dinosaurs for meaningful comparison). Again, Williams (1992, p. 25) explains the issue succinctly and at more immediate scales: "Suppose a climatic change causes the brown trout of the upper Rhine to die out but lets the brown trout of the upper Danube survive. Suppose further that the difference in fate is attributable to some difference in gene frequency that causes a difference in vulnerability to the change. That is surely clade selection. The ultimate prize for which all clades are in competition is representation in the biota."

3. In many cases of species selection, the success of one species over an-

other cannot be explained by competition between their sympatric populations, but depends upon a species-level trait of the species's full range—in other words, species selection of the whole, not of avatars or sympatric subsections. I present in Figure 8-5 a hypothetical case developed by Robert N. Brandon (personal communication, 1988, Ohio State meeting).

The three species of a clade live on four adjacent volcanic islands. Species 2 can move readily across small oceanic gaps and inhabits all four islands. Species 1 and 3 have limited mobility and live on only one island each. (Species 2 gains no necessary advantage of the moment thereby.) The population of Species 1 on Island A, and of Species 3 on Island C, may each exceed the total number of organisms in Species 2 on all four islands. In fact, on any individual island, either Species 1 or Species 3 may always fare better than Species 2. Each island maintains an active central volcano; when the volcano erupts, all life on the island dies, but the adjacent islands remain unaffected. One fine day, the volcanoes of Islands A and C erupt. As a consequence, Species 1 and

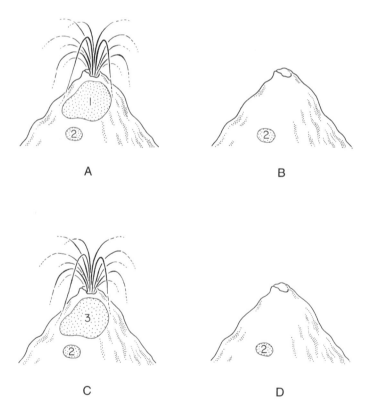

8-5. A hypothetical example of species selection based on traits that belong to entire species—in this case the full geographic range—and not to avatars or subpopulations thereof. See text for details of this verbal case developed by R. N. Brandon. Species 2 survives by virtue of its ability to spread among islands, even though any other species dominates over species 2 on any island of joint occurrence.

Species 3 become extinct, but Species 2 survives thanks to populations on Islands B and D—that is, only by virtue of populations allopatric with Species 1 and 3.

Clearly, Species 2 has survived as a result of greater geographic range, caused by whatever organismal, deme, or species traits permitted the colonization of all islands. Geographic range may be either an emergent or aggregate trait of successful Species 2; but, in any case, this trait exists at the species level and confers an irreducible fitness based on superior range (obviously a property of the species, and not of any individual organism, deme, or avatar). This hypothetical case presents a potential and plausible example of species selection based on a trait of the entire species and its complete range—and explicitly not on any sympatric avatar, or any other subsection of the full entity.

SPECIES SELECTION AS POTENT. Two separate arguments, one empirical and the other theoretical, have been raised against the efficacy of species selection. The first, which I regard as unfair, claims that a paucity of currently recorded empirical examples must indicate the rarity of the phenomenon. I would respond, first of all, that a few excellent (and elegant) cases have been well documented, so this process cannot rank as a distant plausibility waiting for an improbable verification, as some critics have charged. Jablonski (1987), for example, performed a pioneering study on species selection in Cretaceous mollusks during the long background interval preceding the mass extinction at the period's end. He found that species with planktotrophic larvae (defined as floating and feeding, and therefore remaining aloft for substantial time) generally have larger geographic ranges and longer geological durations than species with nonplanktotrophic larvae (defined as either never planktonic, or floating without feeding, and therefore aloft for only a short period).

Jablonski supplies good inferential evidence for the two key claims that a hypothesis of species selection requires. First, he presents a strong case that geographic range not only correlates with longevity, but helps to cause the extended duration. Species tend to reach their maximal range soon after their origin, and to maintain this breadth thereafter as a potent hedge against extinction. Second, he calculated a strong heritability for geographic range by assessing the parent-offspring regression for this character. Geographic range surely constitutes a character of the species, not (obviously) of individual organisms. This trait confers an emergent fitness on species that gain increasing longevity thereby. All necessary attributes for an interpretation based on species selection have therefore been identified.

(The case also includes interesting complexities. As mentioned previously for similar examples in Tertiary mollusks, nonplanktotrophic species generally experience shorter longevity and maintain smaller populations in their more restricted ranges; but they also speciate more frequently, a presumed consequence of greater ease in forming isolated populations—for their evolution of larvae without extensive periods of flotation restricts gene flow among demes. Thus, the greater longevity of planktotrophic species need not imply increasing dominance of such species within the clade, for this positive trait can be counterbalanced by the higher speciation rates of shorter-lived nonplanktotrophic species. Moreover, Jablonski also showed that selective forces can change radically during episodes of mass extinction. In the great dying at the end of the Cretaceous period, geographic range of species shows no correlation with survivorship through the event. But, interestingly, geographic range of entire molluscan clades (though not of their component species) does correlate positively with persistence through the mass extinction— a potential example of clade selection.)

I freely admit that well-documented cases of species selection do not permeate the literature. But I regard this infrequency as a great opportunity, rather than a restrictive limitation or an indication that the phenomenon scarcely exists. We have barely begun to acknowledge (much less to define or operationalize) this process, and we have still not entirely agreed upon criteria for recognition. We face the tradition of a full century spent *not* considering causes at this level (indeed, actively denying the existence of such levels at all). We are just learning how to look—or, to state the issue more incisively, we have just begun to recognize *that* we should be looking at all! We face all the promise of a rich but unploughed field—and (to commit two literary barbarisms of mixed metaphors and parodied quotations at the same time), we should summon up the courage of John Paul Jones and recognize that we have not yet begun to think.

I regard the second, or theoretical, objection as even more unfair in its purely traditionalist grounding in the parochialism of viewing organisms as exclusive agents of evolutionary interest or importance—more an aesthetic defense about comfort or preference than an intellectual argument about mechanisms. Several Darwinian strict constructionists, Richard Dawkins and Daniel Dennett in particular, hold that almost everything of interest in evolutionary biology either inheres in, or flows from, natural selection's power to craft the intricate and excellent design of organisms—"organized adaptive complexity," in Dawkins's favorite phrase. "Biology is engineering," Dennett tells us again and again in his narrowly focussed book (Dennett, 1995).

I do not deny either the wonder, or the powerful importance, of organized adaptive complexity. I recognize that we know no mechanism for the origin of such organismal features other than conventional natural selection at the organismic level—for the sheer intricacy and elaboration of good biomechanical design surely preclude either random production, or incidental origin as a side consequence of active processes at other levels. But I decry the parochialism of basking so strongly in the wonder of organismic complexity that nothing else in evolution seems to matter. Yet many Darwinian adaptationists adopt this narrow and celebratory stance in holding, for example, that neutrality may reign at the nucleotide level, but still be "insignificant" for evolution because such changes impose no immediate effects upon organismal phenotypes; or that species selection can regulate longstanding and extensive trends in single characters, but still maintains no "importance" in evolution because such a process can't construct an intricate organismal phenotype of numerous, developmentally correlated traits.

Dawkins (1982, pp. 106–108), for example, damns species selection with faint praise in these terms:

> I shall argue that a belief in the power of species selection to shape simple major trends is not the same as a belief in its power to put together complex adaptations such as eyes and brains . . . The species selectionist may retreat and invoke ordinary low level natural selection to weed out ill-coadapted combinations of change, so that speciation events only serve up already tried and proved combinations to the sieve of species selection. But this "species selectionist" . . . has conceded that all the interesting evolutionary change results from inter-allele selection and not from interspecies selection, albeit it may be concentrated in brief bursts punctuating stasis . . . The theory of species selection . . . is a stimulating idea which may well explain some single dimensions of quantitative change in macroevolution. I would be very surprised if it could be used to explain the sort of complex multi-dimensional adaptation that I find so interesting.

This statement commits the classic intentional fallacy of the prosecutor: attributing beliefs not held to adversaries, and then castigating them for apostasy (or praising them for good sense in recantation)—as illustrated by the paradigm for an opening thrust in a line of inquiry: "when did you stop beating your wife?" Dawkins finds the adaptive complexity of organisms uniquely interesting. I also regard the subject as fascinating, and I would never attribute this quintessential property of organisms to selection at some other level. I fully acknowledge, as do all species selectionists, that the adaptive complexity of organisms arises primarily by causal processes operating at the organismic level.

But this pluralistic principle applies equally well to other levels. If adaptive complexity marks "what organisms do," and must therefore be explained at the organismic level—then "what species do" implies a consideration of causation at the species level. Species "do" two primary things in macroevolution: they carry trends within clades across long geological stretches of time, and they stand as basic units (geological "atoms" if you will) for counting the waxing and waning of differential diversity through time (why does our current biota feature 500,000 named species of beetles, but fewer than 50 of priapulids?). As a paleontologist, I regard these two phenomena as surpassingly important, while I remain happy to grant Dawkins's commanding interest in the adaptive complexity of organisms. But just as I try not to impose my causes (for other scales and levels) upon his material *a priori*, I ask him to acknowledge the importance of my favored themes within a comprehensive evolutionary theory (even if they do not engage his personal concern), and therefore to recognize the efficacy of different appropriate causes at this paleontological level. In short, Dawkins and others commit a classic psychological fallacy in denying status to species selection by confusing personal interest with general importance.

Only one line of defense remains open to those who still wish to deny the importance of species-level processes after correcting this psychological fallacy, and admitting that trends and changing patterns in diversity rank as vital subjects in a complete evolutionary theory, and also represent "what species do." Such a Darwinian stalwart must argue that all (or nearly all) phenomena at the species level find their causes in upward translation from ordinary natural selection on organisms. Thus, if current biotas feature half a million species of beetles, this plethora can only imply that beetle organisms maintain a particularly favorable adaptive design. And if geological trends privilege increasing body size, larger brains, more complex ammonite sutures, more symmetrical crinoid cups, fewer horse toes, and a thousand other documented patterns, these features must triumph by their adaptive value to organisms. I shall make no further arguments against such a narrow perspective here (to save my rebuttal for Chapter 9, pp. 886–893), and will only quote a great American character, Sportin' Life in *Porgy and Bess*, to remind us that received wisdom does not always prevail:

> The things that you're liable to read in the Bible
> It ain't necessarily so.

THE CLADE-INDIVIDUAL    Although a logical space must exist in our structure of explanation for this highest level of the evolutionary hierarchy, I am not sure that clade selection plays a major role in evolution. Most clades contain so few parts (species) that their waxing and waning must often occur by processes that either operate as random inputs to the clade level, or result from selection among subparts (species selection, or lower-level selection), and therefore appear as drives at the clade level (and not as selection among entire clades treated as individuals). Secondly, while I have advocated a plurality of mechanisms for coherence of individuals at various levels in the hierarchy, I do have trouble in conceptualizing an adequate "glue" for clades, especially since their parts (species) may live in such complete independence, and in such different ecologies, on distant continents. Finally, clades maintain the peculiar property (perhaps only an odd "allometric" consequence of necessary structure at this highest level, and not any compromise in efficacy) of necessarily originating as a single subpart—the founding species, and gaining definition (as a full level) only retrospectively, after adding new parts (more species) sequentially.

How then, given all these difficulties, could clades compete, *qua* clades as discrete and integral evolutionary items, even under the broad definition (see p. 706) that does not require direct contact or even life in sympatry? Is a clade, uniquely among evolutionary individuals of the hierarchy, more a "holding firm" for subparts than a coherent entity frequently subject to selection at its own level?

One route to claiming a potential importance for clade selection remains open, but I am not confident that the argument can prevail (though Williams, 1992, despite his past as an ardent gene selectionist, has become a strong advocate of this view). What do we mean, for example, when we say that dinosaurs died and mammals survived, or that brachiopods dwindled to a remnant while clams continually expanded? Do these descriptive statements imply clade selection? A general argument would have to be framed in the following way: any distinct clade maintains defining autapomorphic characters expressed by all subparts (species). If a clade survives, while another living in roughly comparable habitats, dies—and if survival can be tied to autapomorphic characters held by the persisting clade (and absent in the extinct clade)—may we not speak of clade selection based on a range of variability that includes the key characters in the surviving case, but precludes their expression in the extinct clade?

For example, if mammals survived in part by virtue of small body sizes, and dinosaurs died for a set of consequences related to invariably (and substantially) larger body size, couldn't we say that mammals, as a clade, possessed genetic determinants (shared by homology in all subparts, with homology as the "glue" of cladal coherence) that all dinosaurs lacked as a result of their own evolved cladal distinctions? If such a scenario can count as clade selection (rather than just clade sorting, as an obviously valid description), then selection at this highest level becomes common in nature—for many clades yield in geological time to phylogenetically distant clades that share sufficient similarity in habitat and function to rank as genuine "replacements."

I am not comfortable with this general argument, for no one has yet articulated firm and operational criteria for distinguishing true clade selection (based on irreducible fitness conferred by a clade-level property) from descriptive clade sorting (or differential survival as an effect of lower level properties belonging to species or organisms, but translating upwards to success or failure of a clade as a geologically persistent entity). Some examples probably do represent genuine clade selection—as in Jablonski's (1987) case of clade survival (through mass extinction), correlated with geographic range of the entire clade, but *not* with ranges of component species. Most other examples, however, may not invoke any genuine clade-level character (either aggregate or emergent), but only represent the death of each species, item by item (part by part in cladal terms, for this highest-level individual also maintains the peculiar property of relative immunity, especially in clades with large numbers of widely distributed subparts, to the fate of individual subparts). We may frame our best descriptions for such cases in terms of clade sorting, but do they also qualify as cases of clade selection?

At a minimum, however, such arguments illustrate a need for macroevolutionary accounts at all levels, even when causality arises from lower levels and merely affects the fate of higher-level individuals. Thus, the explicit study of macroevolution would remain vital even if traditionalists had been correct in ascribing all causality to organismic selection. But we needn't take refuge in this "minimalist" defense. Causal processes—and not only selection, as I shall demonstrate in the next section—do operate at substantial (often controlling) relative frequency at all levels (with the possible exception of some dubiety about the importance of clade selection as expressed above, and some recognition that organismic selection has effectively squashed most cell-lineage selection in many phyla of multicellular organisms).

I therefore end this section with two statements from George Williams (1992), who once rejected higher-level selection with such verve and skill (1966), but who (while properly reasserting his excellent arguments against the old form of so-called "naive group selection," or interdemic selection in the Wynne-Edwards modality) now strongly defends both the importance of selection at the species level ("clade selection" of lowest rank in his terminology, because he rejects species as units), and our lamentable failure to consider this vital process in our previous theorizing. Echoing my methodological point that a rarity of recorded examples does not imply any actual weakness in nature, Williams writes (1992, p. 35): "Only the barest beginnings have been made in searching the fossil record for evidence of clade selection. The record can be searched for statistically significant trends in diversity and abundance of particular clades . . . It can also be searched for consistent selection of certain characters."

In an expansive and forceful plea for pluralism—representing the finest form of support that a paleontologist could obtain from colleagues engaged in the study of microevolution—Williams (1992, p. 31) then states that allelic change in populations cannot account for evolution because gene-pools function in nature through their entrapment within higher-level individuals operating and interacting as coherent and distinct entities in macroevolution.

> The natural selection of alternative alleles, acting largely independently at each locus, is the only force tending to maintain or improve adaptations shown by the ephemeral organisms formed by the ephemeral genotypes. If one could look back to the evolution of our own or any other sexually reproducing species, back to well before the Cambrian, no other fitness enhancing process of any importance would be found. Having taken that position, I must take another. The microevolutionary process that adequately describes evolution in a population is an utterly inadequate account of the evolution of the earth's biota. It is inadequate because the evolution of the biota is more than the mutational origin and subsequent survival or extinction of genes in gene pools. Biotic evolution is also the cladogenetic origin and subsequent survival and extinction of gene pools in the biota.

# The Grand Analogy: A Speciational Basis for Macroevolution

### PRESENTATION OF THE CHART FOR MACROEVOLUTIONARY DISTINCTIVENESS

When Niles Eldredge and I first formulated the theory of punctuated equilibrium in the early 1970's (Eldredge, 1971; Gould and Eldredge, 1971; Eldredge and Gould, 1972; Gould and Eldredge, 1977), we had only the germ of an insight that its tenets could lend support to a generalized theory of macroevolution, then entirely undeveloped. We did, however, dimly grasp the key notion that punctuated equilibrium might help to grant species a sufficient stability and coherence for status as what we would now call an evolutionary individual, or unit of selection. We developed this insight by groping towards an analogy that, when generalized and fully fleshed out (with apologies for another parochial organismic metaphor of common language!), sets a foundation for macroevolutionary theory. We dimly recognized, in short, that if species act as stable units of geological scales, then evolutionary trends—the fundamental phenomenon of macroevolution—could be conceptualized as results of a "higher order" selection upon a pool of speciational events that might occur at random with respect to the direction of a trend. In such a case, the role of species in a trend would become directly comparable with the classical status of organisms as units of change within a population under natural selection. We wrote (1972, p. 112):

> A reconciliation of allopatric speciation with long-term trends can be formulated . . . We envision multiple . . . invasions, on a stochastic basis, of new environments by peripheral isolates. There is nothing inherently directional about these invasions. However, a subset of these new environments might . . . lead to new and improved efficiency . . . The overall effect would then be one of net, apparently directional change: but, as with the case of selection upon mutations, the initial variations [species] would be stochastic with respect to the change [trend].

Several paleontologists groped towards a generalization during the next few years, but Stanley (1975, 1979) made the greatest headway in appreciating the full generality of such an analogistic procedure for macroevolutionary theory: "In this higher-level process species become analogous to individuals, and speciation replaces reproduction. The random aspects of speciation take the place of mutation. Whereas, natural selection operates upon individuals within populations, a process that can be termed species selection operates upon species within higher taxa, determining statistical trends" (Stanley, 1975, p. 648).

Stanley preceded this statement with a claim that I regard as fully justified and prescient, but that became a lightning rod for unfair criticism: "Macroevolution is decoupled from microevolution, and we must envision the process governing its course as being analogous to natural selection but operating at a higher level of organization" (1975, p. 648). Largely on the basis of this claim about "decoupling," Stanley, Eldredge and I, and others, were often accused of trying to scuttle Darwinism, and to invent an entirely new (and fatuously speculative) causal apparatus for evolutionary change (meaning, and explicitly so stated in this reductionistic critique, a new genetics).

We made no such claim, and the words quoted above speak for themselves. We were trying to explore the different workings of selection on individuals at levels of the evolutionary hierarchy higher than the conventional Darwinian focus upon organisms. Not only do I continue to regard this procedure as fruitful and fully justified, but I would also defend such an effort as the basis for an independent macroevolutionary theory that can harmoniously expand our conventional and exclusive focus on organisms to yield a more satisfactory general account of life's workings and history.

I also continue to regard the *individuality of species* as the *central proposition* of such an expanded theory. If organisms are the traditional units of selection in classical Darwinian microevolution within populations, then species operate in the same manner as basic units of macroevolutionary change. This perspective establishes an irreducible hierarchical structure in nature, precluding the smooth upward extrapolation of microevolutionary change within populations to explain evolution at all scales, particularly phenotypic trends and patterns of diversity displayed in geological time—the proposition that true devotees of microevolutionary exclusivism rightly feared. If species, as stable units and genuine evolutionary individuals, interpose themselves between populational anagenesis and trends within clades, then the lower-level process cannot smoothly encompass the higher-level phenomenon. For this fundamental (and excellent) reason—and not because any "new" genetics or anti-Darwinian forces reign in a threatening world of macroevolution—Stanley introduced his key notion of "decoupling."

The levels become decoupled because macroevolution must employ species as "atoms," or stable and basic units of change. Decoupling then becomes intensified because higher levels exhibit allometric properties that distinguish their phenomenology from the workings of lower levels. Thus, macroevolution with species as individuals must differ, in deep and interesting ways, from microevolution with organisms as individuals. These differences, and not any fatuous claims about "new genetics," express the uniqueness of macroevolution, and the validity of our argument for decoupling.

An extensive analogy—"the grand analogy," if you will (see Gould and Eldredge, 1977, p. 142)—between organismal microevolution and speciational macroevolution provides a good tool for assessing the differences imposed by scaling among the levels. Stanley (1975, p. 649) and Gould and Eldredge (1977, pp. 142–145) proposed some partial and preliminary schemes, and several others have added components along the way (Stanley, 1979; Vrba, 1980; Grantham, 1995, for example). I present this grand analogy below, largely in the form of a chart contrasting the key features of organic structure and evolution in their organismal and speciational manifestations. For each major category, I list the most important differences between the levels. A fuller explication of all items on the chart follows.

## THE PARTICULARS OF MACROEVOLUTIONARY EXPLANATION

### The structural basis

The first category of structural differences seems straightforward enough. In order to construct the analogy, we ratchet the focal level of individuality up from the organism to the species, thus redefining both lower components and higher contexts in the structural triad of part-individual-collectivity (see page

| Feature | Organismal Level | Species Level |
|---|---|---|
| **The Triad of Structure** | | |
| Individual | Organism | Species |
| Part | Gene, cell | Organism, deme |
| Collectivity | Deme, species | Clade |
| Usual result of proliferation of one part to crowd out others | Cancer | Immediately adaptive anagenesis |
| **The Criteria of Individuality** | | |
| Production of new individuals | Birth | Speciation |
| Elimination of individuals | Death | Extinction |
| Sources of cohesion | | |
| Stability of individual | Physiological homeostasis in ontogeny | Sources of stasis in punctuated equilibrium |
| Boundaries against invasion | Skin to delineate; immune system to police | Reproductive isolating mechanisms |
| "Glue" of subparts | Functional integration & division of labor | Social structure & behavioral interaction among parts (organisms); recombination in sexual reproduction to mix parts in their replication |
| Inheritance | Asexual by budding from one individual, or sexual by mixture of two individuals | "Asexual" by budding from one individual |
| Source of new variation in newborn individuals | Mutation | Geographic (or some other form of) isolation (a precondition); drift & selection (mechanisms), causing differences that break reproductive integrity |
| Spread of new variation to other individuals in the collectivity | Recombination in sexual reproduction | Generally absent except for hybridization between species in some clades |
| Frequency of new variation in replicated individuals | Very rare for any single trait | Inherent in birth process and always present |
| **Modes of Change in the Collectivity** | | |

*Drives,* or Directional Variation Within or Between Individuals

| | | |
|---|---|---|
| Heritable ontogenetic change within the individual = ontogenetic drive | Lamarckism—powerful if it occurred, but precluded by nature of heredity | Anagenesis (gradualism within species); rare by punctuated equilibrium |

Table 8-1 (continued)

| Feature | Organismal Level | Species Level |
|---|---|---|
| 2. Biased production of new individuals=reproductive drive | Mutation pressure | Directional speciation |
| 2a. Frequency of biased production | Very low (if harmful to organism) because organismal selection effectively suppresses lower levels | Potentially common f( two reasons: 1) spe( processes don't stro suppress lower-level lection; 2) new indi uals must originate with change from p ent |

B. *Selection,* or Differential Proliferation Due to Traits of Interactors

| Feature | Organismal Level | Species Level |
|---|---|---|
| 1. Name of process | Natural (organismal) selection | Species selection |
| 2. Basis in birth | Differential birth | Differential speciation |
| 3. Basis in death | Differential death | Differential extinction |
| 2a. Reason for non-directionality of variation as precondition of selection's power | Inherent in nature of mutation as unrelated to needs of organism | No necessary reason; benefits of organism species frequently co cide. No relation if mediate adaptive co texts of new species uncorrelated with di tion of trend. Testab as Wright's Rule |
| 2b. Distinctive feature of birth bias | Usually internal to organism; need not lead to adaptation to environment | Usually irreducible as based on traits of pc ulations, not organis |
| 3a. Distinctive feature of death bias | Usually yields adaptation to local environment | Often reducible as sim summation of organ deaths |

C. *Drift,* or Random Differential Proliferation

| Feature | Organismal Level | Species Level |
|---|---|---|
| 1. Within the collectivity | Genetic drift | Species drift |
| 2. In founding of new collectivities | Founder effect | Founder drift |
| 1a. Frequency | Rare except in special circumstances of small populations, or neutrality of many genic sites | Common because most clades have low $N$; intensified by reducti of $N$ in mass extinct |
| 2a. Frequency | Common; depends on $N$ of founding population | Very common for two reasons: 1) necessary (and often large) diff ence from ancestor a each founding; 2) greatly different pote tials in allopatric regions |

| Feature | Organismal Level | Species Level |
|---|---|---|
| **External and Internal Environments** | | |
| **Competition and the External Environment** | | |
| In direct contact | Most often biotic | More likely to produce an effect by differential elimination |
| Not in direct contact; often allopatric | More often abiotic | More likely to produce an effect by differential birth |
| Main feature | Major source of occasional biomechanical or general progress | Often reducible to organismal level |
| Main feature | Adaptation to local circumstances; no general vector | Usually irreducible |
| **Constraint and the Internal Environment** | | |
| Limits on runaway change by directional evolution of parts | Lamarckian inheritance doesn't occur | Punctuated equilibrium suppresses anagenesis by stasis |
| Structural brakes upon change | Design limits of Bauplan | Positive correlation of frequency of speciation and extinction apparently unbreakable |
| Variational brakes | Rarity of new mutation allayed by recombination in sexuals; serious in asexuals (allayed by short generations in many unicells) | Sufficient change per new individual, but low $N$ of species in clades |
| Developmental brakes | Von Baer's laws of complex ontogenesis | Hold of homology |
| Positive channeling by structure | Heterochrony and preferred ontogenetic extensions | Differential ease & permissibility of Bauplan modifications |
| Positive channeling by variation | Not important, given rarity of directional variation | Frequent correlation of directional speciation with differential proliferation |
| Size of exaptive pool | High in such crucial circumstances as genetic redundancy usable in evolution of complexity | Generally high because lower levels not suppressed and frequently correlative |

673). But this basic ratcheting already reveals some pivotal differences between the evolution of organism-individuals and species-individuals. In Table 8-1, line I2a, for example, notes the profoundly different outcome that usually ensues when particular parts of the individual proliferate differentially and crowd out other parts. Such a process usually spells disaster for a complex multicellular organism—and we call the result cancer—because parts lack independent viability (and therefore harm both themselves and their collectivity, the organism, by unchecked proliferation), and because organisms build coherence (an important criterion of individuality) by functional integration and division of labor among parts. But species achieve equal coherence by other routes. The parts of a species—that is, its component organisms—do have independent viability; moreover, their interests in proliferation often coincide with the health of the enclosing species. Thus, in a species-individual, differential proliferation of some parts at the expense of other parts does not lead to death of the full entity, but usually to adaptation by anagenesis.

### Criteria for individuality

Moving to the second category of criteria for individuality (see pp. 602–613 of this chapter), we may regard the species-level analogs of organismal birth and death (lines II1–2)—speciation and extinction—as both evident and well recognized. But the different causes of cohesion (line II3) are both fascinating and portentous throughout the chart. I only remind readers that the mechanisms used by species, while not clamping down so hard on lower levels, and therefore providing substantial "play" for interaction between organismal and species selection, provide species with as much coherence and stability as the "standard" devices of morphological boundaries, internal policing and functional integration among parts, do for organisms.

Important differences arise in the mode of production for novel variation in newborn individuals. Mutation supplies this attribute at the organismal level. (Following conventional usage, I consider recombination in sexual organisms as a device for spreading variation among individuals, although I recognize, of course, that novel combinations also arise thereby. In asexual organisms, a better analog for species in any case, mutation alone supplies new variation.) Speciation itself is not the proper analog of mutation at the species level (an error previously made both by me, in Gould and Eldredge, 1977, and by Stanley, 1975). Speciation, the production of a new species-individual by budding, is the analog of organismal birth, particularly the birth of asexual organisms. We made this error by inadequately interpreting one of the most interesting differences between organisms and species as evolutionary individuals. The birth of a new organism, particularly in asexuals, may or may not engender any substantial difference from parental form or genetics. But the birth of a new species necessarily includes the generation of enough difference from ancestors to preclude reproductive amalgamation between the parts (organisms) of the two species. We therefore mistook a forced correlate of birth at the species level (change at speciation) with the process of birth itself (speciation)—and equated the correlate at one level with the phenomenon at the other. The proper analog of mutation, a source of variation for new individuals, is the change that insures reproductive isolation between species (with geographic isolation as a usual precondition, and drift and selection as mechanisms)—see line II5.

This difference underlies the two important disparities listed as lines II4a and II5a—one favoring the evolutionary capacity of organisms, the other of species. Sexual organisms can spread favorable variation to other individuals in the collectivity by recombination. But the favorable features of new species remain stuck in the species and its lineal descendants, and cannot be spread to other species in the clade—except in the infrequent circumstance of hybridization among species in a clade of multicellular forms. (By contrast, lateral transfer seems to be common in the evolution of prokaryotic lineages.) This preclusion of lateral spread puts a strong damper upon evolution within clades. The same limitation, of course, affects asexual vs. sexual organisms—and represents a standard argument for the great advantage of sex, and its evolutionary prevalence, in complex metazoans (Williams, 1975; Maynard Smith, 1978).

Species do gain advantages, on the other hand, in the necessary association of birth with change (sometimes small in extent because reproductive isolation can develop with minimal genetic change, but usually quite substantial). This input helps to offset the disadvantages of small population sizes (species in clades) for species selection. The asexual budding of a new species always yields novelty; the asexual budding of a new organism usually yields clonal identity, and only produces novelty if mutation intervenes.

### Contrasting modalities of change: the basic categories

The greatest interest in this analogy lies in the third category of contrasting modalities. Since individuals vary and collectivities evolve (by cumulative changes in their contained individuals) in the standard formulation, I shall first define the three major styles of change within collectivities (populations for the organismal level, and clades for the species level).

DRIVE. This term has often been used for particular cases—meiotic drive, or molecular drive, for example—but deserves to be formalized and generalized. A driving process transforms a collectivity by directionally changing its contained individuals from within. Drives should be construed as opposite or at least orthogonal to selection. Drives produce change by directional transformation of relevant individuals, not by differential proliferation of some kinds of individuals over others. Thus, in pure cases of drive, change occurs without any differential proliferation. In the paradigm cases of drive, either an individual alters in the course of ontogeny and passes these modifications to offspring, while all individuals produce the same number of offspring with the same reproductive capacities (so no selection can take place); or an individual produces offspring endowed with directional differences from its own constitution—but again, no differential reproduction occurs, and no selection can take place. (As one complexity—an ineluctable consequence of the hierarchical perspective, but a blessing in richness rather than a nuisance in confusion—drives at one level can result from selection at a lower level. In the obvious case, anagenesis within a species—a drive at the species level—traditionally arises from selection among organisms within the species.)

SORTING (SELECTION AND DRIFT). This descriptive term generalizes our usual notion of evolutionary change in a collectivity by differential proliferation of some kinds of individuals vs. others. Sorting, as previously defined (p. 659), is a causally neutral and purely descriptive term for any evolution by differential proliferation, whatever the mechanism involved (see original formulation in Vrba and Gould, 1986). Of the two major modes of sorting, *selection,* based on causal interaction of traits with environments, ranks as the canonical style of evolution, the essence of Darwin's insight, and the foundation of modern theorizing. But sorting can also proceed randomly, a process termed *drift.* In the hierarchical model, both selection and drift can occur at all levels, under appropriate conditions. I discussed previously, for example, how selectively-based sorting of species can occur either by upward causation from selection at the organismic level ("the effect hypothesis" of Vrba, 1980, also called "effect macroevolution"), or by selection based on irreducible fitness of species-level traits in their interaction with environments (true species selection)—see pp. 652–670.

### Ontogenetic drive: the analogy of Lamarckism and anagenesis

The two categories of drive present some of the most consequential and counterintuitive pairings in the entire table (at least they stimulated my own thoughts substantially). In a first category, line IIIA, we must acknowledge as an instance of "drive" any consistently directional change that occurs during the ontogeny of an individual, and then passes by inheritance to offspring.

We do not usually include such a process in our standard account of evolution for an interesting reason based on the history of evolutionary thought and the nature of Mendelian genetics: We generally focus our causal accounts exclusively on organisms in the Darwinian tradition; at the organismic level, a drive of this character would validate the most anathematized and fallacious of alternatives to Darwinism—namely Lamarckism, with "soft" inheritance of acquired characters (see Chapter 3 on Weismann's use of hierarchical thinking to counteract Lamarckism). Thus, ontogenetic drives based on phenotypic changes that are generated by organic activity and then passed to offspring, probably don't exist at the organismal level due to the nature of DNA and the mechanics of heredity. The defeat of Lamarckism—ontogenetic drive in this context—marks one of the great episodes in the history of evolutionary thought. If evolution did proceed in the Lamarckian mode, the geological history of life would assume an entirely different appearance, primarily by enormously accelerated rates of change, and suppleness of adaptive modification. I doubt, for example, that we would find any stable higher-level entities like species in a Lamarckian world. (Human cultural change compares so poorly with Darwinian evolution primarily because our customs and technologies do evolve in this vastly more rapid and flexible Lamarckian mode. Whatever we invent in one generation, we pass directly to the next by emulation and instruction.)

The proper species-level analog for ontogenetic drive, or Lamarckian evolution, sounds a bit bizarre at first—but probably only for the irrelevant psychological reason that we have so firmly rejected the organismic example, while promoting the species-level version as a standard mode of change. With species as individuals and organisms as parts, the gradual transformation (without branching) of an entire species by organismal selection—the standard, canonical description of "evolution" itself—becomes the legitimate analog, at the species level, of heritable ontogenetic alteration, or Lamarckian change, at the organismal level! If, as tradition used to hold, such ontogenetic drive dominates macroevolution, then we must record this striking difference in pattern between levels.

I would argue, however (and under my admittedly partisan commitment to punctuated equilibrium), that this standard impression is fictional, and that ontogenetic drive occurs only rarely at the species level. Differences in frequency will, of course, persist—for mechanisms of inheritance preclude ontogenetic drive in theory at the organismic level, while the analogous process remains possible in principle, though rare in fact, given the nature of populations and their modes of change, at the species level. Thus, small importance remains a common theme at both levels. Most species originate in a geological moment, and persist in stasis thereafter (with, at most, mild fluctuation about an unvarying mean, but no directional change, as the concept of drive requires—see Chapter 9).

I would also venture an analogy to the organismal level in support of inherent reasonableness for the rarity of anagenesis (ontogenetic drive) at the species level. As argued above, the Lamarckian mode works with extraordinary rapidity and efficiency: if organisms changed in this way, we could not fail to notice, because evolution would then operate so differently. I would suggest that we approach macroevolution at the species level in the same way. If species changed gradually most of the time, the pageant of life's history, as shown by the fossil record, would present an entirely different appearance. The most extensive transformations would occur in a few million years at most. (Many hypothetical calculations have been made to illustrate this point—for example, that a small, four-footed, terrestrial mammal can evolve into the largest whale in a fraction of Tertiary time, so long as a single population in transformation maintains the smallest effectively measurable selection coefficient, unabated and without change in direction.)* Stable

---

*In my favorite more specific example, Williams (1992, p. 129)—who is, to say the least, no general champion of punctuated equilibrium or detractor of anagenesis—points out that mean morphological changes in some North American populations of English sparrows during their century of residence on our side of the Atlantic reach a maximum of about 5% increase for the lengths of long bones of the wings and legs. This anagenetic increment, so small that "no birdwatchers will notice in their old age that the bird looks any different from what they remember from childhood" (Williams, 1992, p. 129), would, nonetheless, if maintained only for the geologically trivial interval of one million years, be

clades could not dominate the history of life, as they manifestly do, particularly in marine invertebrates (clams, snails, horseshoe crabs, brachiopods, all from the Paleozoic to the present); nor, among more rapidly changing terrestrial clades, could dinosaurs (not to mention the more stable insects) persist and rule for so long in a world where most species evolved continually by the analog of ontogenetic drive in the Lamarckian mode.

Of course, we could posit other reasons for braking the rapidity and efficiency of change by ontogenetic drive in macroevolution—disruption of trends by mass extinction; high frequency of trends that benefit organisms but harm species (peacock's tails), for example. But I suspect that the simplest of all reasons will explain the evident pattern: the species-level analog of ontogenetic drive—gradual transformation within a species—just doesn't occur very often.

Finally, I note that R. A. Fisher's classic argument for the impotence of species selection rests on the standard assumption that this mode of driving does prevail in evolution. For, if most species, most of the time, changed gradually from within (see pp. 644–646), then selection among species would be, as Fisher rightly noted, an operative but impotent process, capable of generating only an insignificant amount of change relative to the dominant and ubiquitous drives of anagenesis. But if anagenesis rarely occurs, Fisher's argument collapses. I wonder if Fisher ever explicitly realized that anagenesis would trump species-selection because anagensis is Lamarckian at the species level, while species selection is Darwinian at the same level—for Lamarckian processes can always overwhelm the much weaker force of Darwinian change if both operate generally and in an unimpeded manner at the same level.

### Reproductive drive: directional speciation as an important and irreducible macroevolutionary mode separate from species selection

Thus, the first category of ontogenetic drive illustrated interesting differences in style between levels, but little variation in effect—for I conclude that this mode has scant impact upon evolution at either level. But when we consider the second category of reproductive drive (biased production of offspring that vary in a given direction from parents), we encounter one of the chart's most striking disparities—a crucial, yet almost entirely unrecognized and unexplored difference in basic pattern between micro- and macroevolution. To choose a hypothetical example of simplified form but maximal clarity: Suppose that each collectivity (a population for the organismal level, or a clade for the species level) contains ten individuals (organisms or species, for the two levels). Each individual gives birth to a single offspring, and all offspring have identical life spans and reproductive capacities. Thus, *no selection at all*

---

"capable of turning sparrow-size into ostrich-size bones, and back again, about 54 times." Clearly, anagenesis at virtually any rate high enough to stand out above measurement error over a human lifetime cannot be sustained in a unidirectional manner for meaningful intervals in geological time.

can take place. Now suppose that a strong bias exists in production of off-spring, so that 80 percent arise with smaller bodies than their parent (lower weight for the offspring organisms, lower average body weight for the off-spring species). Suppose also that this pattern continues from generation to generation. This driving process would generate a strong trend to smaller bodies in the collectivities at both levels—a gradual trend to decreased body size in the population at the organismal level; and to species with smaller average body sizes within the clade at the species level.

As discussed previously (p. 691), reproductive drives of this kind can occur at the organismal level, and a variety of names for such processes exist, including mutation pressure and meiotic drive. But the Darwinian tradition has always regarded such phenomena as insignificant as a consequence of their rarity. Indeed, these processes must be rare in a fully Darwinian world, because reproductive drives violate the necessary precondition of undirected variability for natural selection (see pp. 144–146). Darwinians did not win this debate by simple logic or evident factuality, but only by a great intellectual struggle marking a crucial episode in the history of evolutionary thought. The classical debate about orthogenesis, for example (see Chapter 5), centered upon the Darwinian denial of such reproductive drives, which, as the competing orthogeneticists all realized, would overwhelm selection by higher efficacy—if they existed. Perhaps such reproductive drives rarely occur at this level in nature because, having no known basis for inherent adaptivity, they have been actively suppressed by organismal selection—another potential example of the most distinctive feature of organismic individuality: the power evolved by functional integrity to suppress lower-level selection from within.

However, when we move to the species level, the analogous driving phenomenon of *directional speciation* suffers no constraint or suppression—and may represent one of the most common modes of macroevolution. Two major reasons underlie the high potential frequency for directional speciation (as opposed to the rarity of its analog at the organismal level—see line III2a on the chart). First, as noted in several other contexts, the species-individual does not maintain integrity (as the organism does) by suppressing differential proliferation of some parts over others. Since drives at an upper level arise by differential proliferation of lower-level units, this absence of suppression leaves a large open field for driving processes to operate at the birth of new species. Second, since new species-individuals must arise with sufficient heritable novelty to win reproductive isolation from their parent (whereas children of asexual organisms may be clonally identical with parents), all species births include genetic change as an automatic consequence. Any statistical directionality in such changes among species in a clade will produce a trend by drive.*

---

*At the risk of an unwarranted metaphorical excursion into anthropomorphic imagery, one might contrast limited change at organismal birth with necessary change at species birth in the following manner: New metazoan organisms arise by a process of complex development, which must discourage change for reasons recognized ever since von Baer formulated his laws of embryology (1828). At the organismal level, the new individual sepa-

We may postulate any number of plausible circumstances that would generate directional biases in the origin of new species—thus producing a cladal trend without any contribution from species selection. Moreover, potential causes for directional bias exist at all levels—organismic, demic, or species—thus greatly expanding the scope of the phenomenon. As a central theoretical point, directional speciation, when based on irreducible species-level properties, represents a style of independent and causal macroevolution *not* based on species selection. Thus, the claim for an independent body of macroevolutionary theory does not depend upon the validity and high relative frequency of the Darwinian analog most often discussed as a paradigm case, namely species selection. Directional speciation, when based on irreducible species-level traits or processes, designates another category of intrinsically macroevolutionary change.

To continue in the hypothetical mode with the example cited previously, one can easily imagine how a cladal trend, attributable entirely to reproductive drive (and not at all to selection), and leading to decreasing average (organismal) body size, might be caused at either the organismal, demic, or species level. At the conventional organismic level, a pervasive environmental change over the entire region of a clade's occupancy might favor natural selection for smaller bodies. (Perhaps, to choose a somewhat cardboard example, a temperate region has become tropical, and smaller organisms now gain advantages within each species of a clade by the adaptive correlates of Bergmann's Rule.) Each species produces a single daughter species and then dies out—so no selection can occur at the species level. But if most species, in the new climatic regime, originate at smaller average body size because natural selection favors this trait among the organisms in each species, then the cladal trend arises by directional speciation with a cause based on selection at the organismal level—the classic case of a drive at a higher level produced by directional selection among contained parts at a lower level.

For a hypothetical case based on interdemic selection, suppose that each species in a clade develops ten small and isolated demes at the periphery of the parental range. Suppose that average body size in these peripheral isolates varies randomly around the parental mean. Suppose further that, for each species, only one of the ten peripheral demes survives, intensifies its differences, and eventually becomes a new species—while the parent and the other nine peripheral isolates all die. Again, no species selection can take place, for

rates intrinsically from the parent; how then, may this offspring be kept sufficiently like the parent to preserve the collectivity of the population? An opposite problem attends the birth of species. At the species level, new individuals are born by speciation, which enhances change. But species do not separate intrinsically from their parents. They are born in fuzzy continuity. Their separation may be difficult. They must be cast out, or they will reintegrate. Necessary change at speciation enhances this defining process of casting out from the parent. The newly born species faces a structural problem opposite from the neonatal organism's dilemma: how may the new species-individual become sufficiently unlike the parent to be cast out, thus enhancing the collectivity of the clade by adding another part? In short, the new metazoan organism forms outside the parent: how can it be kept close? The new species separates with difficulty from the parent: how can it be cast out?

each parent spawns one and only one daughter. (I realize, of course, that these strictures sound absurd if construed as actual and coordinated occurrences in nature; I am only following the time-honored heuristic method in science of constructing "pure" end-member hypotheticals to help clarify our thoughts.)

Finally, let us say that, in general, the surviving (and ultimately speciating) deme lies at or near the smaller-bodied end of the random distribution (around the parental mean) of average body size. Let us also posit that the smaller average body size of new species arises as a consequence of a deme-level property conveying differential success in interdemic selection among the ten peripheral isolates initially spun off from each species. An obvious (and not implausible) reason might be found in a strong correlation between small bodies and larger $N$ in any population (the "more ants than elephants" principle, albeit in a more restricted range). The surviving deme might owe its success to generally larger population size in a tough peripheral environment.

The cladal trend to smaller body size among species would then arise by a drive of directional speciation (new species biased to originate with smaller-bodied organisms than those of their ancestors, in a situation where no species selection can occur). The cause of the trend, in this hypothetical case, will be interdemic selection—for the ten peripheral isolates arise as demes of the parental species. Selection among these demes favors those with smaller average body size, based on correlation with the causally controlling deme-level property of larger population size. This deme-level property confers an irreducible fitness upon demes in their interaction with the environment. (In an extreme, albeit improbable, case, interdemic selection based on larger population size could even outweigh negative organismic selection against small bodies.) Again, a drive at the species level arises by selection among lower-level parts, in this case demes rather than organisms.

Finally, a drive at the species level may be caused by an irreducible species-level character. Suppose that each species spins off only one peripherally isolated population and that, invariably, the parental population dies while the peripheral isolate becomes a new species. Suppose that the single peripheral deme, generated by each species, generally features organisms with a smaller average body size than the organisms of the parental population. Suppose that this directional bias arises as a result of a species-level trait in the parental population. Perhaps, for example, the social structuring or territorial system of the parental population preferentially excludes smaller organisms of both sexes, and that these smaller organisms therefore tend to migrate to the species border, where they aggregate to form the isolated population that will generate a new species. (Again, the case merely requires conceivability for purposes of illustration, not plausibility.) In this situation, no selection can occur at the species level because each parental species produces one daughter species and then dies. The cladal trend to species with smaller average body size arises by the driving process of directional speciation—and the cause lies in a species-level trait of the parental population. As stated above, we here encounter a case of irreducible macroevolution not based on species selection. Examples of this kind illustrate that the domain of independent macroevolutionary theory extends well beyond the phenomenology of Darwinian analogs in species selection.

As an additional argument for the importance of directional speciation as a driving force in evolution—and as an example of interesting complexity engendered by the hierarchical model (and of differences in the character of explanation between this hierarchical reformulation, and the traditional one-level world of Darwinian evolution)—note what often happens when causes at one level correlate with emergent properties involved in causes at a higher level; for we then encounter the fascinating situation of disparate theoretical meanings for inexorably linked phenomena at two levels. (We have already discussed one common example in the causation of higher-level drives by lower-level selection.) Another important example, potentially encompassing one of the dominant phenomena of macroevolution, translates the results of ordinary selection at the organismal level into strong constraints acting as causes of directional speciation at the species level. In this sense, when considered at the appropriate higher level, macroevolutionary pattern results much more from immediate constraint, and less from the traditional selectionist mode, than we have generally been willing to allow—thus suggesting another potentially important reform and expansion of Darwinian thinking (see Chapter 10 for a fuller discussion).

Consider two cases of cladal trends produced by the driving cause of directional speciation. Figure 8-6 depicts the common pattern in both examples. At a starting point, the clade contains two kinds of species in equal numbers—those bearing trait A, and those bearing trait B. Every reproducing species generates two daughters and no variation exists for differences in species birth rates among those species that have offspring—so no species selection can occur. Evolution proceeds rapidly by directional speciation because A-Species can only produce A-Daughters, while B-Species produce 50 percent

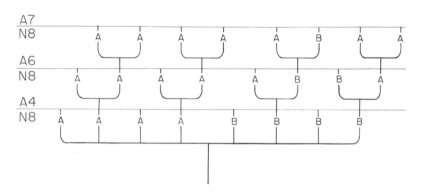

8-6. A cladal trend produced entirely by directional speciation with no species selection. *A* species can only produce *A* daughters, while *B* species produce 50% *A* daughters and 50% *B* daughters. Under these conditions of strongly directional speciation, a powerful trend towards *A* leading to quick disappearance of *B* from the clade, will arise, even under a regime of random mortality among species.

A-Daughters and 50 percent B-Daughters. (If we posit random mortality of a given percentage of species before they split into their daughters, as in Fig. 8-6, then B-Species will eventually be entirely eliminated, and A-Species will become fixed in the clade.)

A first kind of example would not disturb the tranquility of any committed adaptationist, for a functionalist theme translates well across the levels. Suppose that Cope's Rule were true in the classical sense—it is not, by the way (Stanley, 1973; Jablonski, 1987, 1997; McShea, 1994; Gould, 1988b, 1997b, and pp. 902–905 of this book)—and that organismal selection always favored size increase because big organisms prevail in competition. A-Species are large and B-Species are small; A's only give rise to other A's, while B's give rise either to A's (given the pervasive advantage of increasing size), or to B's at equal frequency (for small size may still be favored in some habitats of the clade). The strong Cope's-Rule trend in the clade occurs by directional speciation. The adaptationist theme prevails at both levels. Average organisms in the clade become larger because bigger is better; and species increase in average body size because their parts (organisms) do better at larger size. (No species-level trait exists in regulating this trend, and the entire phenomenon arises by conventional organismal selection based on advantages of increased body size.)

But a second kind of example—undoubtedly quite common in evolution—would perturb a strict adaptationist by translating selection at the organismic level to regulation of the cladal trend by constraint. Suppose now—and such an explanation has been urged as an alternative to species selection for the increase of nonplanktotrophic species within Tertiary clades of gastropods (Strathmann, 1978, 1988)—that a molluscan clade begins with an equal number of species of nonplanktotrophs (A-Species) and plankotrophs (B-Species). Planktotrophic larvae stay aloft through the motion of complex ciliary bands that beat in concert. Selection pressures for nonplanktotrophy lead to loss of these bands, and consequent benthic development of a maternally-protected brood. Plankotrophs can always, in principle, convert to nonplanktotrophy because the bands can be lost; but the transition cannot proceed in the other direction because ciliary bands can't be reconstituted once they have disappeared in evolution (see Gould, 1970b, on the proper meaning of Dollo's Law of irreversibility in evolution).

The origin of each species may be governed entirely by the conventional route of adaptation based on natural selection of organisms. But a structural limitation in possible directions of change produces the cladal trend by directional speciation towards increasing frequency of nonplanktotrophic species—for a planktotrophic parent species can generate either planktotrophic or nonplanktotrophic daughters, while a nonplanktotrophic parent can only produce nonplanktotrophic daughters. The numerical situation corresponds exactly with Figure 8-6 and the previous example based on Cope's Rule (with A-Species now read as nonplanktotrophs, and B-Species as planktotrophs), but the explanation at the cladal level differs crucially—for the trend arises by structural constraint upon possible directions of change, not from any general or global advantage for nonplanktotrophic organisms. (In fact, planktotrophic species might hold a small advantage in species selection for longevity, and the trend to nonplanktotrophy might still arise by directional speciation under this potent constraint.)

I strongly suspect that trends driven by structural constraints within large systems, and not by adaptational advantages to organisms, pervade evolution, but have been missed because we focus on means or extremes in a distribution and not on the full range of variation as a more telling "reality" (see Gould, 1996a, for an entire book on this subject, written for popular readers; and Gould, 1988b, for a technical account). The vaunted trend to increasing complexity in the history of life, for example, only records the small and extending tail of an increasingly right-skewed distribution through time—but with a strong and persistent bacterial mode that has never altered during life's entire 3.5 billion year history, leaving this planet now, as always, in the Age of Bacteria (see pp. 897–901 for a further development of this example). This extending right tail may record little more than the constraint of life's origin right next to the lower bound of preservable complexity in the fossil record. Only one direction—towards greater complexity—remained open to "invasion," and a small number of species dribble in that direction through time, thus extending the right tail of the skewed distribution.* But no evidence now exists to support an argument that higher complexity should be construed as a "good thing in general" (in adaptive terms, or otherwise), either at the organismal or species level. In fact, the few studies based on patterns of speciation in clades where founding members lie far from any upper or lower structural boundary, and therefore impose no constraint upon either decreasing or increasing complexity, show no trend at all towards increasing complexity. Approximately equal numbers of species arise with less complex and with more complex phenotypes than their ancestor (see McShea, 1993,

---

*Examples of this sort illustrate the important point that drives of directional speciation do not necessarily require a differential number of speciation events along the route of the trend. A directional bias may also arise if numbers of speciation events occur with equal frequency in either direction, but the average phenotypic magnitude of the trending half exceeds the amount of change in the half oriented away from the trend. Such cases may be common when a founding lineage lies near a boundary, and amounts of change become severely constrained in one direction. Thus, for the bacterial mode of life, for example, we may easily imagine (data for an adequate test do not exist, so far as I know) that as many speciation events yield a less complex as a more complex descendant. But so little room exists between the mode and the lower limit that changes to reduced complexity cannot depart far from the ancestral state, while an open range to the right of the mode permits a far greater magnitude of change in the direction of greater complexity. For an actual example, Wagner (1996) documented a general trend to increasing spire height in Paleozoic gastropods, but found an equal frequency of speciation events towards lower-spired and higher-spired daughters. The trend, however, records a bias in amounts of change. For some reason, gastropods that become high spired also experience a marked reduction in the amount of change per speciation event, even though they continue to produce equal numbers of daughters in both directions—whereas low-spired ancestors generate much higher average change per speciation event. The mean spire height of the entire clade therefore increases.

on mammalian vertebral columns; McShea, Hall, Grimsson and Gingerich, 1995, on mammalian teeth; and Boyajian and Lutz, 1992, on ammonite sutures).

### Species selection, Wright's Rule, and the power of interaction with directional speciation

I have long regarded species selection as the most challenging and interesting of macroevolutionary phenomena, and the most promising centerpiece for macroevolutionary theory. While I continue to espouse this view, my rethinking for this chapter has led me to appreciate the significant power of two other species-level processes: drives of directional speciation as just discussed (see also Gould, 1982c), and species drift, the higher-level analog of genetic drift. I would now argue that the interaction of these three processes sets the distinctive character of macroevolution.

As for natural selection at the organismic level, the two major modes of species selection operate by differential rates of generating daughter species (the analog of birth biases in natural selection) and differential geological longevity before extinction (the analog of death biases in natural selection). At the species level, however, the difference between these two modes does not rest upon the same basis that distinguishes their analogs at the organismic level.

At the organismal level, natural selection by birth bias works mainly upon such "internal" traits as reproductive rate and brood size, and often doesn't increase adaptation in the conventional sense of phenotypic molding to better biomechanical design for local environments. For example, an organism gains a large selective advantage merely by breeding a bit earlier, though nothing else about the phenotype need alter (Gould and Lewontin, 1979, referred to this mode as "selection without adaptation"). But natural selection by death bias among organisms usually yields phenotypic adaptation for better fit to the ambient environment.

At the species level, however, our main concern moves to an interesting difference in causal locus. Most cases of selection by differential speciation operate by the interaction of an irreducible species-level character—some feature of population structure—with the environment, and therefore represent genuine species selection. After all, and as stated before, organisms don't speciate; only populations do. But for selection by differential extinction, a higher frequency of cases can probably be explained as the simple summation of organismal deaths, and may therefore be causally rendered at this conventional lower level—for both organisms and species die. Thus, students of species selection have rightly focussed on differential speciation as their most promising category (see Gilinsky, 1981, for both theoretical arguments and empirical examples).

However, the most interesting of all differences between organismal and species selection may lie not in the phenomena themselves, but rather in the character of their interaction with the two other primary modes of evolutionary change: drives, and drift (I shall discuss drift in the next section). Our sense of the commanding potency of organismal selection rests upon the conformity of Mendelian genetics to one of the cardinal prerequisites of Darwinian systems (see Chapter 2): that the variation serving as raw material for natural selection be "random" (with an operational meaning of "undirected towards adaptive states," not "equally likely in all directions")—so that selection, rather than biases inherent in variation, can become the "creative" force in evolutionary change (see p. 144 for further discussion in a related context). This crucial condition can be validated at the organismic level—not because mutations (and other sources of genetic variation) are truly random in the mathematical sense, but because mutation represents a process so different from natural selection, and operating on material (the structure of DNA) so disparate from the bodies of organisms (integrated tissues and organs), that we cannot postulate a reason why favored directions of mutation should correspond in any way to the needs of organisms.

But no comparable argument exists for any *a priori* expectation that the analogous variation (among species within a clade) made available for species selection should also be random with respect to the direction of a trend. Species do not discourage drives among their parts (organisms), while organisms usually do suppress directional variation at lower levels (because the proliferative "interests" of individual genes and cell lineages generally run counter to the adaptive needs of organisms). Moreover, the adaptive features of organisms often confer benefits upon their species as well—as when species live longer because their well-designed organisms prevail in competition. Therefore, we cannot defend an *a priori* basis for asserting randomness in the variation that serves as raw material for species selection.

This situation creates both a problem and a challenge for the analog of Darwinism at the species level—for maximal efficiency of species selection does demand undirected variability, and by the same classical argument originally devised for the organismic level. The randomness of species-level variation with respect to the direction of a trend therefore becomes a matter for empirical testing, rather than a claim predictably flowing from the nature of materials and processes. Such a test should also receive high priority for anyone interested in discovering the frequency and strength of species selection in the explanation of evolutionary trends.

For these reasons, Gould and Eldredge (1977) formulated such a test under the name of "Wright's Rule." We took our cue from a prescient statement by Sewall Wright (1967) that the direction of speciation might be random with respect to the origin of higher taxa, just as we consider mutation to be random relative to the direction of natural selection. Wright's Rule, in our formulation, therefore asserts either that drives of directional speciation do not exist at all in a given situation (the strong version), or at least that any existing directional bias not occur along the vector of an established trend (a weaker version, but fully adequate for assertions of species selection). If Wright's Rule holds, then trends must be attributed to differential proliferation of certain kinds of species (by selection or drift), and not to any drives from within based on directional variation arising from lower-level processes.

Wright's Rule represents a strong test for putative species selection, but I now realize that its failure does not eliminate species selection from consideration. When Wright's Rule holds, a trend must be attributed to species sorting, for no directional component exists at the lower level of variation among units of sorting. But if Wright's Rule fails in any particular case, then species selection cannot forge the trend *exclusively*—although species selection may still operate as one contribution in a hierarchical system. A speciational drive may act synergistically with species selection to intensify a trend. (Since drives tend to be more potent than selection, a powerful drive, with strong violation of Wright's Rule, will probably relegate species selection to an insignificant role. But small departures from Wright's Rule permit a substantial intensification of the trend by species selection. In any case, and in situations of unusually complete paleontological data, we should be able to measure the relative strengths of drive and sorting when the two modes act synergistically.)

Wright's Rule has been tested in some cases, but not often enough—and the subject remains ripe for future research, including several Ph.D. theses! Gould and Eldredge (1977) found Wright's Rule validated for Gingerich's data on early Tertiary *Hyopsodus*. MacFadden (1986) failed to confirm Wright's Rule in the evolution of horses, where a directional bias exists for descendant species to arise at larger body sizes than their ancestors. Arnold, Kelly, and Parker (1995) validated Wright's Rule for a remarkably complete data set of 342 ancestral-descendant pairs in Cenozoic planktonic foraminifera. An equal number of species arose at larger and at smaller sizes than ancestors; see Figure 8-7. In a pioneering study, notable for completeness and density of data (and a consequent capacity to distinguish among *all* the various modes of evolutionary change), Wagner (1996) documented three general and speciational trends in the evolution of gastropods during the lower Paleozoic (Cambrian through Silurian): towards higher spires, more inclined apertures and narrower sinuses. For 276 ancestor-descendant pairs over the entire clade, Wagner confirmed Wright's Rule for spire height and inclination, where as many species differed from ancestors in a direction away from the general trend, as along the ultimately favored route. But data for sinus width, where a statistically significant bias exists for speciation in the direction of

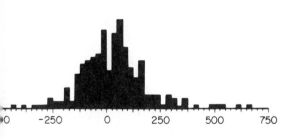

CENOZOIC MEAN = 29.7 MICROMETERS

-250        0        250        500        750

MAGNITUDE OF SIZE CHANGE AT SPECIATION

8-7. Validation of Wright's Rule in a study by Arnold et al. (1995) of 342 ancestral-descendant pairs of Cenozoic planktonic foraminifera. Descendant species originate, with equal frequency, at larger and smaller sizes than their ancestors.

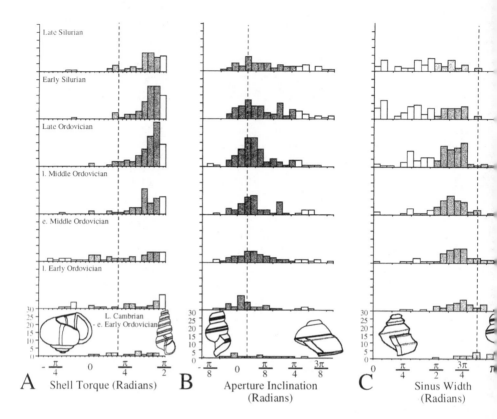

8-8. Top: Clear speciational trends in Lower Paleozoic gastropods towards higher spires (A—measured as shell torque), more inclined apertures (B) and narrower sinuses (C). The bottom diagram demonstrates that the first two trends obey Wright's Rule in showing no bias in species origins in the direction of the trend. However, the trend for sinus width does show a bias for new species to originate in the direction of narrower sinuses—thus yielding a complex trend, partly produced by directional speciation, and not entirely by species selection. From Wagner, 1996.

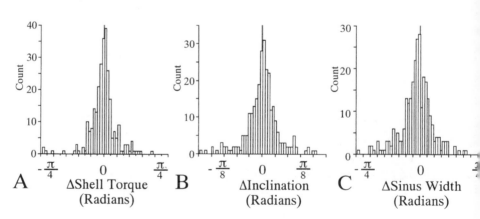

narrower sinuses, falsified Wright's Rule (see Fig. 8-8) and documented a drive of directional speciation. Wagner further demonstrated (1996, p. 1000) that "this bias is distributed throughout the entire clade," for three major subclades all display the drive. Nonetheless—and showing the power of such data to identify and tease apart the different components of a trend into their relative quantitative strengths—Wagner also documented a component of species sorting in the overall trend to narrower sinuses, for "species with wide sinuses were significantly less likely to survive the end-Ordovician mass extinction" (1996, p. 990).

I would go further and suggest that synergisms of drive and sorting (as Wagner has documented for the trend to narrower sinuses in Paleozoic gastropods) should be common in the history of many clades, and probably mark a powerful mode of macroevolution distinct from conventional microevolution, where such synergism must be rare. Good *a priori* reasons exist for supposing that features biasing the directionality of speciation might also favor sorting towards the same end. Such synergism should be most evident when the causes of both bias and sorting work at the same (usually organismic) level—as when, for example, a trait under strongly positive organismic selection (like large body size) arises preferentially in speciation events, and then promotes the greater longevity of species so originating. But such synergisms may also be common when causes differ in level—as when, for example, a drive occurs by organismal selection, and species-selection then causes sorting in the same direction. For, unlike the situation at the next lower pairing of levels (where genic and cell lineage selection so often run counter to the interest of organismal selection, and consequently become suppressed), selection at the organismal level does not conflict in principle with selection at the species level. Selection at these two levels should, therefore, be synergistic as often as opposed. Such synergisms should therefore be frequent and powerful in macroevolution.

### Species-level drifts as more powerful than the analogous phenomena in microevolution

At the organismal level, the second major mode of sorting—drift by random processes—operates in two ways that should be distinguished both for potentially different roles and frequencies at this level, and because the species-level analogs diverge even more clearly. We may distinguish random shift within the collectivity—called genetic drift at the conventional organismal level—from random effects introduced at the founding of new demes or species by small numbers of organisms. Mayr (1942) introduced the term "founder effect" to distinguish this second category (though the basic mechanism does not differ from ordinary genetic drift), and to emphasize that the differences initiating a new species need not arise entirely by natural selection, but may be significantly enhanced by random effects at the outset, because a small number of founders will, for stochastic reasons, surely not begin a new population with the same gene frequencies as the ancestral population, while some alleles (even if favorable) will be lost by random non-inclusion in all founding organisms.

Although both genetic drift and founder effects obviously occur at the organismal level, our traditions have tended to downplay the role of random processes vs. selection as sources of sorting—so the phenomena generally receive short shrift. Some conventional arguments for genuine rarity at the organismal level may be valid, particularly given the requirement for either small populations or effective neutrality of drifting sites. (The initiating criterion of low $N$ may, however, be quite generally met if Mayr's theory of peripatric speciation holds, hence his emphasis on the "founder principle." Similarly, if bottlenecking to very small numbers typically occurs during the history of many species, then genetic drift also becomes important in anagenesis. The argument for effective neutrality, as discussed previously (pp. 684–689), works best at the genic level, where drift may predominate by Kimura's neutral theory of molecular evolution.)

However, at the species level, these traditional objections to high frequency for drift become invalid, and we should anticipate a major role for this second cause of sorting. Low population size (number of species in a clade) provides the enabling criterion for important drift in both categories at the species level. The analog of genetic drift—which I shall call "species drift"—must act both frequently and powerfully in macroevolution. Most clades do not contain large numbers of species. Therefore, trends may often originate for effectively random reasons. Consider a trend produced by random deaths (a comparable argument can easily be made for random birth differentials), based on Raup's "field of bullets" model (1991 and Chapter 12). Suppose, for example, that each of the ten species of a clade lives in a small area, with each species allopatric to all others. Over a certain period of time, a bolide (or some gentler environmental change with power to drive a local species to extinction) strikes half the areas at random and eliminates the resident species of the clade while each of the species in the five safe areas branches off a daughter, thus restoring the cladal population of 10 species. At an $N$ this low, some trends (and perhaps a substantial number) will inevitably arise by this mode of random removal. Perhaps, for example, four of the five species with mean body size below the cladal average will happen to die. A substantial random trend to increased body size then occurs within the clade.

When we move from the homogenous "field of bullets" model to a scaling of effects in the real world, and consider the consequences of infrequent, but severe, mass extinction on a global scale, the potential role of random trends by elimination only increases—for random effects based on small numbers will be greatly intensified. (The reduction of species number in mass extinction may be conventionally causal, but the final death of the clade, after reduction to less than a handful of species, may then be effectively random. For example, so few trilobite species still lived when the great Permian extinction occurred that I'm not sure we need to seek a "trilobite specific" cause for the final elimination of this previously dominant group.)

When we move to the second category of random results achieved by sorting in the colonization of new places—the analog of the founder effect—then comparison with the organismal version becomes less straightforward, although we may be confident that the species-level version holds potential for great importance in evolution. The species-level analog, which I will call "founder drift" (see lines IIIC2 and IIIC2a), does not work through a simple phenotypic difference between a colonizing species and the parental stay-at-home—for all species differ by definition, and disparities arise by the usual combination of selective and random effects, usually expressed at the organismic level. The stochastic analog to Mayr's "founder effect" at the organismal level lies in random aspects of the differential capacity for proliferation of new species in allopatric regions of a clade's full range.

A hypothetical example will illustrate this unfamiliar concept. Suppose that a clade contains only two species, living in adjacent islands with similar environments. The islands, however, lie on different oceanic plates, and movements of plate tectonics cause the coalescence of one island with a large neighboring continent, while leaving the other island in the midst of the ocean. The species on the continent proliferates into a large subclade of new species, while the species on the island, lacking any room for expansion, remains as the only species of its subclade. Because the process of speciation yields phenotypic disparity intrinsically, the founding continental species will differ from its insular sister species. Therefore, the clade will show a strong trend in the direction of autapomorphic traits possessed by the continental founder. But such a trend will often be entirely random with respect to the plurified traits of the continental founder. That is, these spreading traits may be completely neutral in the crucial sense that if the other (insular) species had colonized the island that coalesced with the continent instead, its autapomorphies would have proliferated, and the cladal trend would have proceeded with the same force, but in the opposite direction. Only the luck of residence on one island rather than the other (and not any preferential interaction of some traits *vs.* others with the environment) leads to the differential proliferation of one species's traits over those of the sister species.

Situations of this sort must be common, if not virtually canonical, in evolution. Almost any two geographic regions must maintain differential capacity to house species of a given clade. If both regions are colonized by founding species, and, many million years later, one region holds substantially more species than the other, the random component of spatial and ecological opportunity must often play a greater role in differential speciation than the selective force of greater capacity for differential proliferation in one subclade *vs.* the other based on interactions of traits with environments. I use the term random in a special, but surely legitimate, sense. Suppose that a large and ecologically diverse Region 1 can accommodate 50 species of a subclade, while smaller and more homogeneous Region 2 can only maintain 10 (I realize that species create their own environments, and that regions don't maintain fixed numbers of available addresses, but I invoke this simplification for the sake of argument). Subclade A invades Region 1, while Subclade B begins in Region 2. The resulting strong cladal trend toward the autapomorphic characters of Subclade A cannot be called accidental in the global sense—for Region 1 does predictably accommodate more species. But the trend may be accidental in the sense that Subclade A, rather than Subclade B, happened to invade the more prosperous region—and that if Subclade B had been the colonizer, its progeny would have done equally well, and would have dominated the cladal trend with the same force actually shown by Subclade A. In this case, we call the trend random because A's success does not arise from any superiority of an interacting trait (*vs.* B's phenotype), but only from the accident of colonizing a more propitious place (see Eble, 1999, and Chapter 11 of this book for a discussion of this evolutionary meaning of "random").

As with the relationship between directional speciation and species selection, these two forms of species-level drift must often interact with the other main cause of sorting—i.e., selection—to produce a trend (as when Subclade A, in the example just above, increases *both* by the good fortune of greater opportunity, *and* by selective benefits conferred by its traits). The organismic level may experience a higher relative frequency of domination by selective forces, but the world of species evolves by complex interactions among the processes of drive, selection, and drift.

### The scaling of external and internal environments

I have not tried to develop an exhaustive comparison between levels for influences of external and internal environments upon the modes of change discussed in previous sections. But I offer a few sketchy comments to encourage further work in this area.

For environmental factors that induce competition among individuals and therefore establish selection pressures (line IVA of the chart), I contrast modes that involve direct contact among individuals with those that can proceed in allopatry. At the organismic level, this contrast exposes a strong correlation between prevalence of biotic factors in direct contact and abiotic factors in allopatry. At the species level, a different correlation may dominate: the association of selection by differential elimination with direct contact, and selection by differential birth with allopatry (lines IVA1 and IVA2).

This contrast also leads to different implications at the two levels. At the organismic level, as Darwin himself argued in his primary justification for progress in the history of life (see Chapter 6), the biotic mode correlates more often with adaptation by general biomechanical improvement, and the abiotic mode with adaptation to local circumstances of the physical environment, with no vectorial component as environments fluctuate randomly through time. At the species level, we may expect to find a strong correlation of selection by differential elimination with potential reduction to the organismal level, while selection by differential birth represents the most promising domain for true and irreducible species selection.

For constraints of internal environments (line IVB), I make a distinction between negative factors that limit amounts and directions of change, and positive properties that channel change in certain directions, or provide particular opportunities for evolutionary novelties and breakthroughs. (I also base Chapter 10, this book's major discussion of constraint, on the same distinction.) The operation of these constraints often differs in interesting ways at the two levels.

For some of the important limits, line IVB1 specifies a major shaping force of life's structure, a factor not often explicitly acknowledged. Why does the world contain stable individuals at all, and at any level? Why doesn't evolution work as continuous flux at all scales, rather than primarily by selection upon individuals stable enough to persist, at least through one round of differential sorting? Comparable reasons can be stated at both the organismic and species levels, thus giving evolution its primary shape or structure: Lamarckian inheritance does not occur at the organismal level, thus stabilizing the ontogeny of heritable variability. At the species level, punctuated equilibrium suppresses anagenesis by maintaining species-individuals in stasis.

When we explore the structural brakes that limit amounts of change in most trends (line IVB2), several factors could be mentioned, but I just list, as an example, the single property that I consider most important. For organisms, those paragons of individuality by the criterion of structural and functional integrity, design limits of the *Bauplan* (both internally by structural constraint, and externally by adaptive possibilities) place strong brakes upon almost any evolutionary trend. Contrary to the themes of several popular films, elephants will never fly, and insects will not reach elephantine proportions and engulf our cities as a plague of megalocusts.

At the species level, Stanley (1979) made an important observation that has not been sufficiently appreciated for its defining force in limiting the possibilities of species selection. If we consider the two major modes of positive species selection—enhancing the rate of production for new species, and extending the geological longevity of existing species—why shouldn't some lineages be able to maximize both properties simultaneously, thus becoming gigantic megaclades, dominating the earth's biota? (Perhaps, of course, a few clades have been able to approach this ideal—thus explaining the great success of beetles and nematodes.) In other words, why don't clades ratchet themselves towards this pinnacle by species selection—by working both ends of the game, and evolving species of extraordinary durability and fantastic rates of branching, superspecies that live for several geological periods and spawn large numbers of daughters all along the way?

Stanley (1979) argues, with extensive data in support, that the nature of speciation as a process, and the general rules of ecology, engender a strong, and effectively unbreakable, negative correlation between speciation and extinction rates. Unfortunately for ambitious species with dreams of megacladal domination (but happily for any ideal of a richly varied biota), the major factors that boost speciation rates also raise the probability of extinction;

while features that enhance longevity also suppress the rate of speciation. For example, small populations in stressful environments are especially prone to *both* speciating *and* dying; while large, global populations of marked stability and great mobility (like *Homo sapiens* and *Rattus rattus*) are remarkably resistant to extinction (unless, like one of the above, they evolve an odd capacity for potential self-destruction), but ill-equipped to form the isolated populations that can generate new species.

For a third limiting constraint of brakes on the amount of available variation for selectional processes (line IVB3), infrequency of new mutation may play an important role at the organismal level (not so often in sexual forms, where recombination greatly boosts the amount of variability among individuals, but usually a defining limit in asexuals, and perhaps the major reason for the rarity and marginal status of asexuals among complex Metazoa, but not in unicells with short generations). At the species level, variation per individual may be more than adequate (given the forced correlation of birth with change), but many clades contain too few individuals, giving birth too rarely, for very efficient selection (Fisher's argument—see page 645).

For a final factor among limiting constraints (in this abbreviated list), brakes on development act strongly at both levels (line IVB4). Ever since the inception of modern embryology, von Baer's (1828) laws have defined the hold placed by ontogenetic intricacy upon potentials for change in complex Metazoa. At the species level, the hold of homology (as expressed in all the factors, genetic and otherwise, that limit the amount of change per speciation event) functions as a developmental constraint in the same basic manner—that is, by limiting the difference that can separate a parent and its immediate offspring.

All these sources of limitation also contribute to the more important positive aspects of constraint as channeling or enhancing preferred directions for change. In the category of positive channeling by structure (line IVB5), ontogenetic pathways already established in the lives of organisms provide by extension, or by relative shuffling of rates among components (see Gould, 1977b), the classic mode of constrained and substantial change in organismic evolution—thus explaining the importance of heterochrony as a morphogenetic phenomenon (Jones and Gould, 1999; McNamara and McKinney, 1991). At the upper level of speciational trends within clades, structural rules and differential ease of modifiability among parts and correlations of *Baupläne* play the same role of directing and accelerating change along certain preferred pathways. Liem (1973), for example, showed how a set of small and accessible changes in a jaw muscle, the fourth *levator externi*, could greatly alter the adaptive feeding devices of cichlid fishes (but not of other related groups), thus helping to explain the rapidly evolved species flocks of this clade in several African lakes.

In a second category of positive channeling by directed variability from levels below, the organismic level experiences no important effect because such drives will generally be suppressed by organismic selection. But the driving force of directional speciation can greatly enhance and channel cladal trends by working synergistically with such species-level modes of change as species selection.

Perhaps the most important positive constraint, acting similarly at both levels, lies in the large size of "exaptive pools" (see full discussion in Chapter 11), or nonrandom variation made available through evolutionary processes acting on other features or at other levels, but later exploitable by organisms or species for their own exaptive benefits. The redundancy supplied by genetic duplication for organismal flexibility serves as the classic illustration of this phenomenon at the traditional level of natural selection. The exaptive pool of the species level may become even larger because species do not suppress lower levels of change, while these genic and organismal directionalities frequently act in synergism with advantages at the species level (see Gould and Lloyd, 1999, for a detailed development of this argument).

### Summary comments on the strengths of species selection and its interaction with other macroevolutionary causes of change

Species selection, the Darwinian analog at this higher level, but by no means the only irreducible force of macroevolutionary change, differs from conventional natural selection at the organismic level both in character and in general strength. The major aspect of character—as I have emphasized throughout this chapter in stressing the non-fractality of hierarchical levels—lies in the potency of species selection for governing "what species do." Species selection does not, and cannot, build the complex adaptive phenotypes of organisms, but this common statement only recognizes the general nature of hierarchical organization and does not represent a fair criticism of the efficacy of species selection, despite the claims of Dawkins (1982) and others (see p. 711 for a discussion of this point).

The primary force of species selection lies in its power to promote trends within clades, and to regulate the waxing and waning of differential species diversity within and among clades through time. The influence of species selection upon trends will also be enhanced because this process not only builds trends in species-level characters directly, but also establishes correlated trends in any character of the organismal phenotype that either helps to determine the species-level property, or merely hitchhikes upon the trend by linkages of homology within the phylogenetic structure of evolutionary trees—a very common phenomenon, as Raup and Gould, 1974, showed in theory and practice. This insight about trends, which I shall explore more thoroughly in the next chapter (pp. 886–893), may provide a key for explaining one of the most puzzling phenomena in paleontology—persistent and pervasive cladal trends (such as decreasing stipe number in graptolites, or increasing symmetry of crinoidal cups) that have defied all attempts at explanation in traditional terms of biomechanical advantages to organisms.

As for general strength, species selection (in primary comparison with the traditional level of Darwinian natural selection on organisms) includes certain features that diminish its influence, and others that enhance its power. Among factors that weaken the potential of species selection, we may mention:

(1) The generally low population size of species in clades, and the generally long life of species-individuals, both factors limiting the amount of variation usually available for a process of selection.

(2) Unlike the organism, the species-individual does not actively suppress selection at lower levels within itself. Since the individuals of lower levels, by their shorter cycle times, present much more variation for selection (per given unit of time), this unsuppressed lower-level selection may overwhelm the operation of species selection.

(3) Species selection, as the analog of *asexual* reproduction at the organismal level, becomes subject to the same important limit that favorable traits arising in one individual cannot be transferred laterally (for mixing and matching) to other individuals, but only vertically to direct descendants.

(4) Species selection is limited by particular structural constraints, encountered only at this higher level, most notably the apparently unbreakable correlation between origination and extinction rates, thus tying together by negative interaction the very two phenomena that, if *positively* associated—that is, high speciation with *low* extinction—could so powerfully accelerate any trend produced by species selection.

But these negative forces and limits will be counteracted by several features that grant potential strength to species selection:

(1) Species selection may be theoretically weak relative to the power of transformation by continuous selection of lower-level individuals (organisms in this case) within species. But, in fact, such transformation by anagenesis rarely occurs in nature, as the great majority of species exhibit stasis during their geological lifetimes. With general anagenesis usually weak or inoperative, and with effective organismal selection concentrated at the origin of new species and their differentia (and thus also limited to the cycle time of species themselves), species selection can become a predominant process.

(2) The population size of species in clades may be low, but each event of speciation must produce difference from parental traits (at least enough to yield reproductive isolation)—whereas events of organismal birth need not add any new variation to the population. The amount of change per speciational event may be large, even providing a potential macroevolutionary analog to macromutation. (At such a point, however, we must also allow some possible weakening of selection's power as well, for macromutation, by producing a completed form of change in one step, deprives selection of its creative role in building adaptation gradually—see Chapter 2.)

(3) At the species level, not only does each birth of a new individual include novel variation that may be substantial, but the variation also arises in an adaptive context (whereas mutation, the source of variation at the organismic level, will usually be detrimental to the organism). Of course, the adaptive component in the production of a new species-individual need only exist at the level of its own causal origin—often the organismal level, rather than the

species level itself. But the new variation will often be adaptive at the species level for two reasons: first, because species-level rather than organismic processes often underlie the genesis of the variation; and second, because variation caused at the organismic level will often be synergistic with species advantages, whereas mutational variation rarely enjoys synergism with the benefits of organisms.

(4) The common synergism of organismal with species advantages produces a powerful acceleration of macroevolution (Gould and Lloyd, 1999). Drives of directional speciation (often based on organismal adaptation) frequently foster species selection along the same pathway by accelerating the speciation rate or, perhaps more commonly, by enhancing the longevity of species arising in the direction of the drive. On the other hand, when organismal selection runs counter to the interests of species, negative species selection may provide the only effective higher-level force that can act as a governor to slow or stymie the trend—probably a common feature in phylogeny, and previously given (in textbooks of my student days, but now rarely used) the unfelicitous and unfortunate name of "overspecialization."

As a final point and guide to understanding the essential role of the species-individual in macroevolution, we must remind ourselves of the highly unusual character of the individual conventionally (and usually unthinkingly) taken as a paradigm for all evolutionary causality—the organism. If we view evolutionary change as tripartite in causal nature—with drive, selection, and drift as the three major modes—then we may say that the organism allows selection to reign nearly supreme by "clearing out" the surrounding space of the other two processes. Drives do not seem important at the organismal level, because drives emerge from below, and organisms, as repeatedly emphasized in this chapter, work so effectively as suppressors of lower-level selection. At the same time, drift produces limited impact at the organismal level because population sizes tend to be too large in most circumstances, and because the high degree of functional integration within organisms grants a selective significance to nearly every part, thus lowering the relative frequency of substantial neutrality in potential sites for drift. Therefore, selection based on organismal properties reigns at this canonical level—thus engendering the two great parochial prejudices of the strict Darwinian world view: the adaptationist program as a guide to nearly all evolutionary phenomena, and the virtual restriction of causality to natural selection working at the single level of organisms (two of the three legs of Darwin's tripod in the terms of this book).

But when we turn to the species level, we find an interesting partnership among the three causal forces of drive, selection and drift. Selection at the species level does not "clear out" these surrounding forces. Drives from below exert great influence in the phenomenon of directional speciation. Drift maintains similar impact in both its major manifestations: as species drift for the transformation of collectivities (clades); and as founder drift in differential proliferation or reduction of subclades by accidents of propitious or limiting colonization. This absence of "clearing out" denotes no failure or weakness of selection at the species level, but should rather be viewed as a different "strategy" for the distinct and effective world of macroevolution. Higher-level selection does not bestride this larger world like the colossus of its analog at the organismal level. But higher-level selection gains a different kind of strength and interest in its fascinating and fruitful synergism (and opposition) with drives from below and with drift in the collegiality of its own domain.

# CHAPTER NINE

# Punctuated Equilibrium and the Validation of Macroevolutionary Theory

## What Every Paleontologist Knows

### AN INTRODUCTORY EXAMPLE

If Hugh Falconer (1808–1865) had not died before writing his major and synthetic works, he might be remembered today as perhaps the greatest vertebrate paleontologist of the late 19th century. Falconer went to India in 1830 as a surgeon for the East India Company, but spent most of his time as a naturalist in two very different realms. In 1832, he became superintendent of the botanical garden at Saharanpur, at the base of the Siwaliks, a "foothill" range of the Himalayas. There he played a major role in fostering the cultivation of Indian tea, but he also collected and described one of the most famous and important of all fossil faunas, the Tertiary mammalian remains of the Siwalik Hills. Broken health forced a return to England in 1842, where he worked for several years on the collection of Indian fossils at the British Museum. He then returned to India, this time as professor of botany at Calcutta Medical College, but declining health forced his permanent repatriation to England in 1855. During the last decade of his life, Falconer studied the late Tertiary and Quaternary mammals of Europe and North America, particularly the history of fossil elephants.

Colleagues revered Falconer for his prodigious memory, his gargantuan capacity for work, and his inexhaustible attention to the minutest details. Darwin, as discussed in Chapter 1 (pp. 1–6), held immense respect for Falconer, and invested much hope and trepidation in the prospect that such a master of detail might be persuaded about the probable truth of evolution.

Among all his observations and general conclusions, Falconer took greatest interest in the stability he observed in species of fossil vertebrates, often through long geological periods, and across such maximal changes of environment as the recent glacial ages. Falconer, of course, began with the usual assumption that such stability implied creation and permanence of species. Darwin included him among the great paleontologists who supported such a view. Noting the strength of this opposition to evolution, Darwin wrote

745

(1859, p. 310): "We see this in the plainest manner by the fact that all the most eminent paleontologists, namely Cuvier, Agassiz, Barrande, Falconer, E. Forbes, *etc.* . . . have unanimously, often vehemently, maintained the immutability of species."

Darwin sent Falconer a copy of the first edition of the *Origin of Species,* preceded by the following note (letter of November 11, 1859): "Lord, how savage you will be, if you read it, and how you will long to crucify me alive! I fear it will produce no other effect on you; but if it should stagger you in ever so slight a degree, in this case, I am fully convinced that you will become, year after year, less fixed in your belief in the immutability of species. With this audacious and presumptuous conviction, I remain, my dear Falconer, Yours most truly, Charles Darwin." (Several years before, Darwin had chosen Falconer as one of the very few scientists to whom he confided his beliefs about evolution. Falconer had not, to say the least, reacted positively. In a letter to Hooker on October 13, 1858, Darwin had written of Falconer's jocular, but entirely serious, response: ". . . dear old Falconer, who some few years ago once told me that I should do more harm than any ten other naturalists would do good, [and] that I had half-spoiled you already!")

Falconer wrote to Darwin on June 23, 1861, expressing his great respect (and that of so many others) for the *Origin,* though not his agreement: "My dear Darwin, I have been rambling through the north of Italy, and Germany lately. Everywhere have I heard your views and your admirable essay canvassed—the views of course often dissented from, according to the special bias of the speaker—but the work, its honesty of purpose, grandeur of conception, felicity of illustration, and courageous exposition, always referred to in terms of the highest admiration. And among your warmest friends no one rejoiced more heartily in the just appreciation of Charles Darwin than did, Yours very truly, H. Falconer." Darwin, greatly relieved, replied the next day: "I shall keep your note amongst a very few precious letters. Your kindness has quite touched me."

Hugh Falconer did reassess his worldview, and did accept the principle of evolution (though not causality by natural selection)—but only within the context of the one overarching phenomenon that so strongly governed the nature of the fossil record according to his extensive and meticulous observations: the longterm stability of fossil species, even through major environmental changes. Falconer published his reassessment in an 1863 monograph entitled: "On the American fossil elephant of the regions bordering the Gulf of Mexico (*E. columbi,* Falc.); with general observations on the living and extinct species." But he first sent a copy of the manuscript to Darwin (on September 24, 1862), in eager anticipation of Darwin's reaction to his new views. In the first paragraph of his letter, Falconer reemphasized the stability of species through great climatic changes, arguing that any evolutionary account must deal with this primary fact of paleontology:

Do not be frightened at the enclosure. I wish to set myself right by you before I go to press. I am bringing out a heavy memoir on elephants—an *omnium gatherum* affair, with observations on the fossil and recent species. One section is devoted to the persistence in time of the specific characters of the mammoth. I trace him from before the Glacial period, through it and after it, unchangeable and unchanged as far as the organs of digestion (teeth) and locomotion are concerned. Now, the Glacial period was no joke: it would have made ducks and drakes of your dear pigeons and doves.

Darwin, of course, was delighted. He wrote to Lyell on October 1, 1862: "I found here a short and very kind note of Falconer, with some pages of his 'Elephant Memoir,' which will be published, in which he treats admirably on long persistence of type. I thought he was going to make a good and crushing attack on me, but, to my great satisfaction, he ends by pointing out a loophole, and adds, '. . . The most rational view seems to be that they [Mammoths] are the modified descendants of earlier progenitors, etc.' This is capital. There will not be soon one good paleontologist who believes in immutability."

If we turn to the key section of Falconer's 1863 monograph, entitled "persistence in time of the distinctive characters of the European fossil elephants," we can trace the development of an important evolutionary argument (I am quoting from the posthumous two-volume 1868 collection of Falconer's complete works). Falconer begins with his basic claim about the constancy of species: "If there is one fact, which is impressed on the conviction of the observer with more force than any other, it is the persistence and uniformity of the characters of the molar teeth in the earliest known Mammoth, and his most modern successor" (p. 252). Falconer then extends his observations from this single species to the entire clade of European fossil elephants: "Taking the group of four European fossil species . . . do they show any signs, in the successive deposits of a transition from the one form into the other? Here again, the result of my observation, in so far as it has extended over the European area, is, that the specific characters of the molars are constant in each, within a moderate range of variation, and that we nowhere meet with intermediate forms" (p. 253).

Falconer finds this constancy all the more significant, given the extreme climatic variation of the glacial ages: "If we cast a glance back on the long vista of physical changes which our planet has undergone since the Neozoic Epoch, we can nowhere detect signs of a revolution more sudden and pronounced, or more important in its results, than the intercalation and subsequent disappearance of the Glacial period. Yet the 'dicyclotherian' Mammoth lived before it, and passed through the ordeal of all the hard extremities which it involved, bearing his organs of locomotion and digestion all but unchanged" (pp. 252–253).

But Falconer then declines to use these observations of stability and sudden geological appearance without intermediates as evidence for special creation. He proclaims himself satisfied with Darwin's basic evolutionary premise, and draws the obvious inference that new species of elephants did not evolve by transformation of older European species, but must have emerged from other stocks:

The inferences which I draw from these facts are not opposed to one of the leading propositions of Darwin's theory. With him I have no faith in the opinion that the Mammoth and other extinct Elephants made their appearance suddenly, after the type in which their fossil remains are presented to us. The most rational view seems to be, that they are in some shape the modified descendants of earlier progenitors. But if the asserted facts be correct, they seem clearly to indicate that the older Elephants of Europe . . . were not the stocks from which the later species . . . sprung, and that we must look elsewhere for their origin (pp. 253–254).

Falconer thus anticipates a primary inference of punctuated equilibrium—that a local pattern of abrupt replacement does not signify macromutational transformation *in situ,* but an origin of the later species from an ancestral population living elsewhere, followed by migration into the local region. Falconer suggests that the ancestry of later European species may be sought among Miocene species in India: "The nearest affinity, and that a very close one . . . is with the Miocene . . . of India" (p. 254).

Falconer then summarizes the puzzles that such stability—of such long-lasting, widespread forms in such variable environments—raises for evolutionary theory: "The whole range of the Mammalia, fossil and recent, cannot furnish a species which has had a wider geographical distribution, and at the same time passed through a longer term of time, and through more extreme changes of climatal (*sic*) conditions, than the Mammoth. If species are so unstable, and so susceptible of mutation through such influences, why does that extinct form stand out so signally, a monument of stability?" (p. 254).

Darwin's reaction to these famous pages in the history of paleontology make fascinating reading, especially in the light of persistence (or reemergence) of all major issues in our modern debate about punctuated equilibrium. First, with his usual insight into the mechanics of his own theory, Darwin expresses special surprise that teeth should be so stable within species—for the same features vary so greatly among species. As many modern evolutionists have remarked—though Darwin did not use the same terminology—natural selection works by converting variation within populations to differences among populations: a primary expression of the extrapolationist principle in Darwinian logic. But the stasis of species challenges such continuationism. (Darwin included his remarks in a long letter to Falconer, written on October 1, 1862, as a response to the manuscript on elephants that Falconer had sent Darwin, and that would become the 1863 publication quoted above): "Your case seems the most striking one which I have met with of the persistence of specific characters. It is very much the more striking as it relates to the molar teeth, which differ so much in the species of the genus, and in which consequently I should have expected variation."

Darwin then searches for ways to mitigate the surprise of such stasis in the face of environmental changes that should have altered selective pressures. He suggests, first, that the global fluctuations of ice-age climates might not have seemed so extensive to elephants. Perhaps they migrated with a favored climatic belt, therefore experiencing little fluctuation, and perhaps no major selective pressures for change: "You speak of these animals as having being exposed to a vast range of climatal changes from before to after the Glacial period. I should have thought, from analogy of sea-shells, that by migration (or local extinction when migration is not possible) these animals might and would have kept under nearly the same climate."

Searching for another way to explain the absence of anticipated (and gradual) change, Darwin then argued that altering climates may generally imply evolutionary modification, but that groups in serious decline, including elephants, often become stalled in their capacity to vary, and especially to form new taxa: "A rather more important consideration, as it seems to me, is that the whole proboscidean group may, I presume, be looked at as verging towards extinction . . . Numerous considerations and facts have led me in the *Origin* to conclude that it is the flourishing or dominant members of each order which generally give rise to new races, sub-species, and species; and under this point of view I am not at all surprised at the constancy of your species." But if Darwin had not been surprised, or at least disturbed, why did he try so hard to reconcile this unexpected phenomenon with his general theory? Falconer, in any case, replied that elephants remained in vigor, and could not be considered as a group on the verge of elimination.

I recount this story at some length, as an introduction to punctuated equilibrium, both because Falconer and Darwin presage in such a striking manner, the main positions of supporters and opponents (respectively) of punctuated equilibrium in our generation, and because the tale itself illustrates the central fact of the fossil record so well—geologically abrupt origin and subsequent extended stasis of most species. Falconer, especially, illustrates the transition from too easy a false resolution under creationist premises, to recognizing a puzzle (and proposing some interesting solutions) within the new world of evolutionary explanation. Most importantly, this tale exemplifies what may be called the cardinal and dominant fact of the fossil record, something that professional paleontologists learned as soon as they developed tools for an adequate stratigraphic tracing of fossils through time: the great majority of species appear with geological abruptness in the fossil record and then persist in stasis until their extinction. Anatomy may fluctuate through time, but the last remnants of a species usually look pretty much like the first representatives. In proposing punctuated equilibrium, Eldredge and I did not discover, or even rediscover, this fundamental fact of the fossil record. Paleontologists have always recognized the longterm stability of most species, but we had become more than a bit ashamed by this strong and literal signal, for the dominant theory of our scientific culture told us to look for the opposite result of gradualism as the primary empirical expression of every biologist's favorite subject—evolution itself.

## TESTIMONIALS TO COMMON KNOWLEDGE

The common knowledge of a profession often goes unrecorded in technical literature for two reasons: one need not preach commonplaces to the initiated; and one should not attempt to inform the uninitiated in publications they do not read. The longterm stasis, following a geologically abrupt origin, of most fossil morphospecies, has always been recognized by professional paleontologists, as the previous story of Hugh Falconer testifies. This fact, as discussed on the next page, established a basis for bistratigraphic practice, the primary professional role for paleontology during most of its history.

But another reason, beyond tacitly shared knowledge, soon arose to drive stasis more actively into textual silence. Darwinian evolution became the great intellectual novelty of the later 19th century, and paleontology held the archives of life's history. Darwin proclaimed insensibly gradual transition as the canonical expectation for evolution's expression in the fossil record. He knew, of course, that the detailed histories of species rarely show such a pattern, so he explained the literal appearance of stasis and abrupt replacement as an artifact of a woefully imperfect fossil record. Thus, paleontologists could be good Darwinians and still acknowledge the primary fact of their profession—but only at the price of sheepishness or embarrassment. No one can take great comfort when the primary observation of their discipline becomes an artifact of limited evidence rather than an expression of nature's ways. Thus, once gradualism emerged as the expected pattern for documenting evolution—with an evident implication that the fossil record's dominant signal of stasis and abrupt replacement can only be a sign of evidentiary poverty—paleontologists became cowed or puzzled, and even less likely to showcase their primary datum.

But this puzzlement did sometimes break through to overt statement. For example, in 1903, H. F. Cleland, a paleontologist's paleontologist—that is, a respected expert on local minutiae, but not a general theorist—wrote of the famous Devonian Hamilton section in New York State (which has since become the "type" for an important potential extension of punctuated equilibrium to the integrated behavior of entire faunas, the hypothesis of "coordinated stasis"—see pp. 916–922):

> In a section such as that of the Hamilton formation at Cayuga Lake . . . if the statement *natura non facit saltum* is granted, one should, with some confidence, expect to find many—at least some—evidences of evolution. A careful examination of the fossils of all the zones, from the lowest to the highest, failed to reveal any evolutionary changes, with the possible exception of *Ambocoelia praeumbona* [a brachiopod]. The species are as distinct or as variable in one portion of the section as in another. Species varied in shape, in size, and in surface markings, but these changes were not progressive. The conclusion must be that . . . the evolution of brachiopods, gastropods, and pelecypods either does not take place at all or takes place very seldom, and that it makes little difference how much time elapses so long as the conditions of environment remain unchanged (quoted in Brett, Ivany, and Schopf, 1996, p. 2).

But far better than such explicit testimonies—and following various gastronomical metaphors about the primacy of practice (knowing by fruits, proofs of the pudding, etc.)—the most persuasive testimony about dominant stasis and abrupt appearance inheres, without conscious intent or formulation, in methods developed by the people who use fossils in their daily, practical work. Evolutionary theory may be a wonderful intellectual frill, but workaday paleontology, until very recently, used fossils primarily in the immensely useful activity (in mining, mapping, finding oil, etc.) of dating rocks and determining their stratigraphic sequence. These practical paleontologists dared not be wrong in setting their criteria for designating ages and environments. They had to develop the most precise system that empirical recognition could supply for specifying the age of a stratum; they could not let theory dictate a fancy expectation unsupported by observation. Whom would you hire if you wanted to build a bridge across your local stream—the mason with a hundred spans to his credit, or the abstract geometer who has never left his ivory tower? When in doubt, trust the practitioner.

If most fossil species changed gradually during their geological lifetimes, biostratigraphers would have codified "stage of evolution" as the primary criterion for dating by fossils. In a world dominated by gradualism, maximal resolution would be obtained by specifying a precise stratigraphic position within a continuum of steady change, and much information would be lost by listing only the general name of a species rather than its immediate state within a smooth transition. But, in fact, biostratigraphers treat species as stable entities throughout their documented ranges—because the vast majority so appear in the empirical record. Finer resolution can then be obtained by two major strategies: first, by identifying species with unusually short durations, but wide geographic spread (so-called "index fossils"); and, second, by documenting the differing ranges of many species within a fauna and then using the principle of "overlapping range zones" to designate geological moments of joint occurrence for several taxa (see Fig. 9-1).

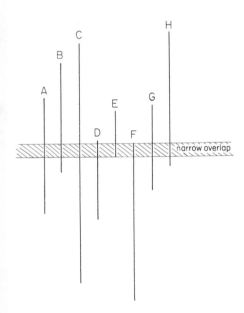

9-1. Knowledge by working biostratigraphers of stasis observed in the vast majority of fossil species led these field scientists, for practical and not for theoretical reasons, to use the criterion of "overlapping range zones" for maximal precision in stratigraphic correlation. If most species changed gradualistically within their geological lifetime, stage of evolution within individual species would provide a better criterion for correlation.

This peculiar situation of discordance between the knowledge of practical experts and the expectation of theorists impressed Eldredge and me deeply when we formulated punctuated equilibrium. We therefore made the following remarks in closing our first paper on the application of our model to biostratigraphy (Eldredge and Gould, 1977):

> [We] wondered why evolutionary paleontologists have continued to seek, for over a century and almost always in vain, the "insensibly graded series" that Darwin told us to find. Biostratigraphers have known for years that morphological stability, particularly in characters that allow us to recognize species-level taxa, is the rule, not the exception. It is time for evolutionary theory to catch up with empirical paleontology, to confront the phenomenon of evolutionary non-change, and to incorporate it into our theory, rather than simply explain it away . . . We believe that, unconsciously, biostratigraphic methodology has been evolutionarily based all along, since biostratigraphers have always treated their data as if species do not change much during their [residence in any local section], are tolerably distinguishable from their nearest relatives, and do not grade insensibly into their close relatives in adjacent stratigraphic horizons . . . Biostratigraphers, thankfully, have ignored theories of speciation, since the only one traditionally available to them has not made much sense. To date, evolutionary theory owes more to biostratigraphy than vice versa. Perhaps in the future evolutionary theory can begin to repay its debt.

Finally, the witness of experts engaged in a lifelong study of particular groups and times provides especially persuasive testimony because, as I have emphasized throughout this book, natural history is a science of relative frequencies, not of unique cases, however well documented. We* have never doubted that examples of both gradualism and punctuation can be found in the history of almost any group. The debate about punctuated equilibrium rests upon our claim for a dominant relative frequency, not for mere occurrence. The summed experiences of long and distinguished careers therefore provide a good basis for proper assessment.

The paleontological literature, particularly in the "summing up" articles of dedicated specialists, abounds in testimony for predominant stasis, often viewed as surprising, anomalous, or even a bit embarrassing, because such experts had been trained to expect gradualism, particularly as the reward of diligent study. To choose some examples in just three prominent fossil groups representing the full span of conventional "complexity" in the invertebrate record, most microorganisms seem to show predominant stasis—despite the excellent documentation of a few "best cases" of gradualism in Cenozoic planktonic Foraminifera (see pp. 803–810). For example, MacGillavry

---

*I may be an arrogant man, but I would never be so pompous as to use the 'royal' we. I cannot separate my views on punctuated equilibrium from those of my colleague and partner in this venture from the start, Niles Eldredge. When I write 'we' in this section, I mean 'Eldredge and Gould.'

(1968, p. 70) wrote from long practical experience: "During my work as an oil paleontologist, I had the opportunity to study sections meeting these rigid requirements [of continuous sedimentation and sufficient span of time]. As an ardent student of evolution, moreover, I was continually on the watch for evidence of evolutionary change . . . The great majority of species do not show any appreciable evolutionary change at all. These species appear in the section (first occurrence) without obvious ancestors in underlying beds, are stable once established, and disappear higher up without leaving any descendants."

Echoing the hopes and disappointments of many paleontologists (including both Eldredge and me), who trained themselves in statistical methods primarily to find the "subtle" cases of gradualism that had eluded traditional, subjective observation, Reyment (1975, p. 665) wrote: "The occurrences of long sequences within species are common in boreholes and it is possible to exploit the statistical properties of such sequences in detailed biostratigraphy. It is noteworthy that gradual, directed transitions from one species to another do not seem to exist in borehole samples of microorganisms."

Moving to a metazoan group generally regarded as relatively "simple" in form, and especially prominent in the fossil record, particularly in Paleozoic strata, Roberts (1981, p. 123) concluded from many years of studying Australian Carboniferous brachiopods: "There is no evidence of 'gradualistic' evolutionary processes affecting brachiopod species either within or between zones, and the succession of faunas can be regarded as 'punctuated.'"

Johnson (1975), inspired by Ziegler's (1966) documentation of one putatively gradualistic sequence in the brachiopod *Eocoelia*, decided to search for others—and found only examples of punctuation and stasis throughout the Paleozoic record. He wrote (1975, p. 657):

> After completion of Ziegler's paper we talked a number of times about the possibilities of duplicating his efforts with other fossils and in other times. It was a heady prospect . . . In subsequent years many workers have attempted to seek out and define lineages of brachiopod species and other megafossils in the lower and middle Paleozoic with little success. My conclusion, subjective in many ways, is that speciation of brachiopods in the mid-Paleozoic via a phyletic mode has been rare. Rather, it is probable that most new brachiopod species of this age originated by allopatric speciation.

Derek Ager, the world leader in studying later Mesozoic brachiopods, summed up his lifelong effort in several papers towards the end of his career. He wrote (1973, p. 20): "In twenty years work on the Mesozoic Brachiopods, I have found plenty of relationships, but few if any evolving lineages . . . What it seems to mean is that evolution did not normally proceed by a process of gradual change of one species into another over long periods of time." Ten years later (1983, p. 563), Ager reiterated: "The general picture seems to fit in with the Gouldian doctrine of 'hardly ever' [that is, documentation of gradualism only very rarely]. Certainly there is no evidence in the group as a whole of phyletic gradualism happening throughout a species at any one moment in time. Species A never changes into species B everywhere simultaneously and gradually."

When we consider trilobites, the exemplars of Paleozoic invertebrate "complexity," Robison (1975, p. 220) concluded from extensive study of Middle Cambrian agnostid trilobites in Western North America: "I have found a conspicuous lack of intergradation in species-specific characters, and I have also found little or no change in these characters throughout the observed stratigraphic ranges of most species."

Fortey (1985) spent many years studying a particularly favorable sequence for fine-scale temporal resolution from the early Ordovician of Spitzbergen. He examined 111 trilobite and 56 graptoloid species, finding a predominance of punctuated equilibrium in both groups—with gradualism in "less than 10 percent of the total" for trilobites, and, for graptoloid species, with punctuational origins "at least four times as important as gradualistic ones" (1985, p. 27). Fortey's case becomes particularly convincing because he could calibrate punctuational sequences against rarer cases of gradualism in the same strata—and therefore be confident that punctuations do not merely represent the missing strata of conventional gradualistic rates. In a later paper, Fortey, who is, by the way, no partisan of punctuated equilibrium, reaches the following general conclusion, and also affirms our point about respect for the age-old knowledge of biostratigraphic practitioners: "Many invertebrate paleontologists would agree that the fossil record of species of their groups is dominated by lack of change—by stasis—and that where phylogenies have been worked out then the evidence direct from the rocks shows punctuated lineages in a majority of cases. For reasons I have explained, it is likely that stratigraphic paleontologists would *always* have maintained such a view, but the difference is that now this would be accepted by paleobiologists as well" (1988, p. 13).

Moving to a different arthropod group from another time, Coope's famous studies of Late Cenozoic fossil beetles (summarized in Coope, 1979) provide one of our best cases for dominance of the punctuational mode. Unusually good preservation greatly increases the power of this example. Coope discusses his best case (for beetles extracted from the carcasses of woolly rhinos in the western Ukraine), but then extends his argument to most examples:

> Here the complete beetles were preserved down to the tarsal and antennal joints; when the elytra were raised, the wings could be unfolded and mounted; and parasitic mites, both larvae and adults, were found underneath the wings. Although this was quite exceptional preservation, it is common to find intact abdomens from which the genitalia can be dissected; the frequently transparent integument often reveals detailed structures of the internal sclerites. Preservation is frequently adequate to enable details of the microstructure of the surface of the hairs and scales to be examined with scanning electron microscopy (1979, p. 248).

Coope concluded that most species showed extensive stasis, even with such detail available for observation: "The early Pleistocene fossils, probably dating from over a million years ago, are referable to living species and some existing species extend well back into the late Tertiary" (1979, p. 250).

In what I regard as the most fascinating and revealing comment of all, George Gaylord Simpson, the greatest and most biologically astute paleontologist of the 20th century (and a strong opponent of punctuated equilibrium in his later years), acknowledged the literal appearance of stasis and geologically abrupt origin as *the* outstanding general fact of the fossil record, and as a pattern that would "pose one of the most important theoretical problems in the whole history of life" if Darwin's argument for artifactual status failed. Simpson stated at the 1959 Chicago centennial celebration for the *Origin of Species* (in Tax, 1960, p. 149):

> It is a feature of the known fossil record that most taxa appear abruptly. They are not, as a rule, led up to by a sequence of almost imperceptibly changing forerunners such as Darwin believed should be usual in evolution. A great many sequences of two or a few temporally intergrading species are known, but even at this level most species appear without known intermediate ancestors, and really, perfectly complete sequences of numerous species are exceedingly rare. . . . These peculiarities of the record pose one of the most important theoretical problems in the whole history of life: is the sudden appearance . . . a phenomenon of evolution or of the record only, due to sampling bias and other inadequacies?

Such a discordance between theoretical expectation and actual observation surely falls within the category of troubling "anomalies" that, in Kuhn's celebrated view of scientific change (1962), often spur a major reformulation.

## DARWINIAN SOLUTIONS AND PARADOXES

Only one chapter of the *Origin of Species* bears an apologetic title—ironically, for the subject that should have provided the crown of direct evidence for evolution in the large: the archive of life's actual history as displayed in the fossil record. Yet Darwin entitled Chapter 9 "On the Imperfection of the Geological Record."

In Chapter 2 (pp. 146–155), I discussed Darwin's convictions about gradualism, and the crucial link between his defense of natural selection and one of the three major and disparate claims subsumed within this complex concept: the insensibility of intermediacy. The theory of punctuated equilibrium does not engage this important meaning for two reasons: first, our theory does not question the operation of natural selection at its conventional organismic level; second, as a theory about the deployment of speciation events in macroevolutionary time, punctuated equilibrium explains how the insensible intermediacy of human timescales can yield a punctuational pattern in geological perspective—thus requiring the treatment of species as evolutionary individuals, and precluding the explanation of trends and other macroevolutionary patterns as extrapolations of anagenesis within populations.

Rather, punctuated equilibrium refutes the third and most general meaning of Darwinian gradualism, designated in Chapter 2 (see pp. 152–155) as "slowness and smoothness (but not constancy) of rate." Natural selection does not require or imply this degree of geological sloth and smoothness, though Darwin frequently, and falsely, linked the two concepts—as Huxley tried so forcefully to advise him, though in vain, with his famous warning: "you have loaded yourself with an unnecessary difficulty in adopting *Natura non facit saltum* so unreservedly." The crucial error of Dawkins (1986) and several other critics lies in their failure to recognize the theoretical importance of this third meaning, the domain that punctuated equilibrium does challenge. Dawkins correctly notes that we do not question the second meaning of insensible intermediacy. But since his extrapolationist view leads him to regard only this second meaning as vital to the rule of natural selection, he dismisses the third meaning—which we do confute—as trivial. Since Dawkins rejects the hierarchical model of selection, he does not grant himself the conceptual space for weighing the claim that punctuated equilibrium's critique of the third meaning undermines the crucial Darwinian strategy for rendering all scales of evolution by smooth extrapolation from the organismic level. For this refutation of extrapolation by punctuated equilibrium validates the treatment of species as evolutionary individuals, and establishes the level of species selection as a potentially important contributor to macroevolutionary pattern.

This broadest third meaning of gradualism may not be required for natural selection at the organismic level, but gradualism as slowness and smoothness of rate (not just as insensible intermediacy between endpoints of a transition) forms the centerpiece of Darwin's larger worldview, indeed of his entire ontology—as illustrated (again, see Chapter 2) in the crucial role played by this style of gradualism throughout the corpus of his works—from his first book on the origin of coral atolls (1842) to his last on the formation of topsoil by the action of worms (1881).

Lest anyone doubt that Darwin strongly advocated this most inclusive form of gradualism as slowness and smoothness (in addition to the less comprehensive claim for insensible intermediacy of transitions), I shall cite a few examples from the full documentation of Chapter 2—cases where Darwin clearly meant "slow and steady over geological scales," not just "insensibly intermediate at whatever rate."

For example Darwin argues that species may arise so slowly that the process generally takes longer than the *entire* duration of a geological formation (usually several million years)—thus explaining apparent stasis *within* a formation as gradual evolution over insufficient time to record visible change! Darwin writes (1859, p. 293): "Although each formation may mark a very long lapse of years, each perhaps is short compared with the period requisite to change one species into another." Darwin even argued that the pace of evolutionary change might be sufficiently steady to serve as a rough geological clock: "The amount of organic change in the fossils of consecutive formations probably serves as a fair measure of the lapse of actual time" (1859, p. 488). I also show in Chapter 2 that Darwin's conviction about extreme slowness and steadiness of change can be grasped, perhaps best of all, as the common source of his major errors—particularly his fivefold overestimate for the denudation of the Weald, and his conjecture that complex metazoan life of modern form must have undergone an unrecorded Precambrian history as long, or longer, than its known Phanerozoic duration.

Despite this strong belief in geological gradualism, Darwin knew perfectly well—as all paleontologists always have—that stasis and abrupt appearance represent a norm for the *observed* history of most species. I needn't rehearse Darwin's solution to this dilemma, for his familiar argument represents more than a twice-told tale. Following the lead of his mentor, Charles Lyell, Darwin attributed this striking discordance between theoretical expectation and actual observation to the extreme imperfection of the fossil record.

(As discussed more fully on pages 479–484, this argument served as the centerpiece for Lyell's system, and for the entire uniformitarian school. But then, what alternative could they embrace? The literal appearance of the geological record so often suggested catastrophe, or at least "moments" of substantial change, especially in faunal turnover. To assert a gradualism of geological rate against this sensory evidence, one had to declare the evidence illusory by advancing the general claim—quite legitimate as a philosophical proposition—that science must often work by probing "behind appearance" to impose the expectations of a valid theory upon an empirical record that, for one reason or another, cannot directly express the actual mechanisms of nature. Moreover, the "argument from imperfection" holds substantial merit and cannot be dismissed as "special pleading." Like most chronicles of history, and far more so than many others, the geological record is extremely spotty. To cite Lyell's famous metaphor once again, if Vesuvius erupted again and buried a modern Italian city atop Pompeii, later stratigraphers might find a sequence of Roman ruins capped by layers of volcanic ash and followed by the debris of modern Italy. Taken literally, this sequence would suggest a catastrophic end to Rome followed by a saltation, linguistically and technologically, to the industrial age—an artifact of nearly 2000 years of missing data that would have recorded the evolution of Italian from Latin and a gradual passage from walled cities to traffic jams.)

To quote the two most famous statements on this subject from the *Origin of Species*, Darwin summarizes his entire argument by closing Chapter 9 with Lyell's metaphor of the book (1859, pp. 310–311):

> For my part, following out Lyell's metaphor, I look at the natural geological record, as a history of the world imperfectly kept, and written in a changing dialect; of this history we possess the last volume alone, relating only to two or three countries. Of this volume, only here and there a short chapter has been preserved; and of each page, only here and there a few lines. Each word of the slowly-changing language, in which the his-

tory is supposed to be written, being more or less different in the interrupted succession of chapters, may represent the apparently abruptly changed forms of life, entombed in our consecutive, but widely separated, formations.

In epitomizing both geological chapters, Darwin begins with a long list of reasons for such an imperfect record, and then concludes with his characteristic honesty (1859, p. 342): "All these causes taken conjointly, must have tended to make the geological record extremely imperfect, and will to a large extent explain why we do not find interminable varieties, connecting together all the extinct and existing forms of life by the finest graduated steps. He who rejects these views on the nature of the geological record, will rightly reject my whole theory." (Huxley must have been thinking of this line when he issued his warning that Darwin's unswerving support of *natura non facit saltum* represented "an unnecessary difficulty." Darwin's "whole theory"— the mechanism of natural selection—does not require, as Huxley pointed out, this geological style of gradualism in rate.)

The paradoxes set by Darwin's solution for the current practice of paleontology and macroevolutionary theory receive their clearest expression in another remarkable statement from the *Origin of Species* (1859, p. 302), a testimony to Darwin's sophisticated understanding that nature's "facts" do not stand before us in pristine objectivity, but must be embedded within theories to make any sense, or even to be "seen" at all. Darwin acknowledges that he only understood the extreme imperfection of the geological record when paleontological evidence of stasis and abrupt appearance threatened to confute the gradualism that he "knew" to be true: "But I do not pretend that I should ever have suspected how poor a record of the mutations of life, the best preserved geological section presented, had not the difficulty of our not discovering innumerable transitional links between the species which appeared at the commencement and close of each formation, pressed so hardly on my theory."

### The paradox of insulation from disproof

The "argument from imperfection" (with its preposition purposefully chosen by analogy to the "argument from design") works adequately as a device to save gradualism in the face of an empirical signal of quite stunning contrariness when read at face value. But if we adopt openness to empirical falsification as a criterion for strong and active theories in science, consider the empty protection awarded to gradualism by Darwin's strategy. For the data that should, *prima facie*, rank as the most basic empirical counterweight to gradualism—namely the catalog of cases, and the resulting relative frequency, for observed stasis and geologically abrupt appearances of fossil morphospecies—receive *a priori* interpretation as signs of an inadequate empirical record. How then could gradualism be refuted from within?

The situation became even more insidious in subtle practice than a bald statement of the dilemma might suggest. Abrupt appearance (the punctuations of punctuated equilibrium) might well be attributed to the admittedly gross imperfection of our geological archives. The argument makes logical sense, must certainly be true in many instances, and can be tested in a variety of ways on a case by case basis (particularly when we can obtain independent evidence about rates of sedimentation).

But how can imperfection possibly explain away stasis (the equilibrium of punctuated equilibrium)? Abrupt appearance may record an absence of information, but *stasis is data*. Eldredge and I became so frustrated by the failure of many colleagues to grasp this evident point—though a quarter century of subsequent debate has finally propelled our claim to general acceptance (while much else about punctuated equilibrium remains controversial)—that we urged the incorporation of this little phrase as a mantra or motto. Say it ten times before breakfast every day for a week, and the argument will surely seep in by osmosis: "stasis is data; stasis is data . . ."

The fossil record may, after all, be 99 percent imperfect, but if you can, nonetheless, sample a species at a large number of horizons well spread over several million years, and if these samples record no net change, with beginning and end points substantially the same, and with only mild and errant fluctuation among the numerous collections in between, then a conclusion of stasis rests on the *presence* of data, not on absence! In such cases, we must limit our lament about imperfection to a wry observation that nature, rather than human design, has established a sampling scheme by providing only occasional snapshots over a full interval. We might have preferred a more even temporal spacing of these snapshots, but so long as our samples span the temporal range of a species, with reasonable representation throughout, why grouse at nature's failure to match optimal experimental design—when she has, in fact, been very kind to us in supplying abundant information. *Stasis is data.*

So if stasis could not be explained away as missing information, how could gradualism face this most prominent signal from the fossil record? The most negative of all strategies—a quite unconscious conspiracy of silence—dictated the canonical response of paleontologists to their observations of stasis. Again, a "culprit" may be identified in the ineluctable embedding of observation within theory. Facts have no independent existence in science, or in any human endeavor; theories grant differing weights, values, and descriptions, even to the most empirical and undeniable of observations. Darwin's expectations defined evolution as gradual change. Generations of paleontologists learned to equate the potential documentation of evolution with the discovery of insensible intermediacy in a sequence of fossils. In this context, stasis can only record sorrow and disappointment.

Paleontologists therefore came to view stasis as just another failure to document evolution. Stasis existed in overwhelming abundance, as every paleontologist always knew. But this primary signal of the fossil record, defined as an absence of data for evolution, only highlighted our frustration—and certainly did not represent anything worth publishing. Paleontology therefore fell into a literally absurd vicious circle. No one ventured to document or quantify—indeed, hardly anyone even bothered to mention or publish at all—the most common pattern in the fossil record: the stasis of most morphospecies throughout their geological duration.

All paleontologists recognized the phenomenon, but few scientists write papers about failure to document a desired result. As a consequence, most nonpaleontologists never learned about the predominance of stasis, and simply assumed that gradualism must prevail, as illustrated by the exceedingly few cases that became textbook "classics": the coiling of *Gryphaea*, the increasing body size of horses, etc. (Interestingly, nearly all these "classics" have since been disproved, thus providing another testimony for the temporary triumph of hope and expectation over evidence—see Gould, 1972.) Thus, when punctuated equilibrium finally granted theoretical space and importance to stasis, and this fundamental phenomenon finally emerged from the closet, nonpaleontologists were often astounded and incredulous. Mayr (1992, p. 32) wrote, for example: "Of all the claims made in the punctuationalist theory of Eldredge and Gould, the one that encountered the greatest opposition was the observation of 'pronounced stasis as the usual fate of most species,' after having completed the phase of origination . . . I agree with Gould that the frequency of stasis in fossil species revealed by the recent analysis was unexpected by most evolutionary biologists."

(To cite a personal incident that engraved this paradox upon my consciousness early in my career, John Imbrie served as one of my Ph.D. advisors at Columbia University. This distinguished paleoclimatologist began his career as an evolutionary paleontologist. He accepted the canonical equation of evolution with gradualism, but conjectured that our documentary failures had arisen from the subtlety of gradual change, and the consequent need for statistical analysis in a field still dominated by an "old-fashioned" style of verbal description. He schooled himself in quantitative methods and applied this apparatus, then so exciting and novel, to the classic sequence of Devonian brachiopods from the Michigan Basin—where rates of sedimentation had been sufficiently slow and continuous to record any hypothetical gradualism. He studied more than 30 species in this novel and rigorous way—and found that all but one had remained stable throughout the interval, while the single exception exhibited an ambiguous pattern. But Imbrie did not publish a triumphant paper documenting the important phenomenon of stasis. Instead, he just become disappointed at such "negative" results after so much effort. He buried his data in a technical taxonomic monograph that no working biologist would ever encounter (and that made no evolutionary claims at all)—and eventually left the profession for something more "productive.")

Paradoxes of this sort can only be resolved by input from outside—for gradualism, having defined contrary data either as marks of imperfection or documents of disappointment, could not be refuted from within. Reassessment required a different theory that respected stasis as a potentially fascinating phenomenon worthy of rigorous documentation, not merely as a failure to find "evolution." Eldredge and I proposed punctuated equilibrium in this explicit context—as a framework and different theory that, if true, could validate the primary signal of the fossil record as valuable information rather than frustrating failure. We therefore began our original article (Eldredge and Gould, 1972) with a philosophical discussion, based on work of Kuhn (1962) and Hanson (1961), on the necessary interbedding of fact and theory. We ended this introductory section by writing (1972, p. 86):

> The inductivist view forces us into a vicious circle. A theory often compels us to see the world in its light and support. Yet we think we see objectively and therefore interpret each new datum as an independent confirmation of our theory. Although our theory may be wrong, we cannot confute it. To extract ourselves from this dilemma, we must bring in a more adequate theory; it will not arise from facts collected in the old way . . . Science progresses more by the introduction of new world-views or "pictures" than by the steady accumulation of information . . . We believe that an inadequate picture has been guiding our thoughts on speciation for 100 years. We hold that its influence has been all the more tenacious because paleontologists, in claiming that they see objectively, have not recognized its guiding sway. We contend that a notion developed elsewhere, the theory of allopatric speciation, supplies a more satisfactory picture for the ordering of paleontological data.

### The paradox of stymied practice

This second paradox cascades from the first. If a theory—geologically insensible gradualism as the anticipated expression of evolution in the fossil record, in this case—can insulate itself against disproof from within by defining contrary data as artifactual, then proper assessments of relative frequencies can never be achieved—for how many scientists will devote a large chunk of a limited career to documenting a phenomenon that they view as a cardinal restriction recording a poverty of available information?

Paleontological studies of evolution therefore became warped in a lamentable way that precluded any proper use of the fossil record, but seemed entirely honorable at the time. We practitioners of historical sciences, as emphasized throughout this book, work in fields that decide key issues by assessment of relative frequencies among numerous possible outcomes, and only rarely by the more "classical" technique of "crucial experiments" to validate universal phenomena. Therefore, any method that grossly distorts a relative frequency by excluding a common and genuine pattern from consideration must seriously stymie our work. When traditional paleontologists eliminated examples of abrupt appearance and stasis from the documentation of evolution, they only followed a conventional precept—for they believed that both patterns recorded an artifact of imperfect data, thus debarring such cases from consideration. The relative distributions of evolutionary rates would therefore emerge only from cases of gradualism—the sole examples judged as sufficiently data-rich to record the process of evolution in adequate empirical detail.

But this project could not even succeed in its own terms, for gradualism occurs too rarely to generate enough cases for calculating a distribution of rates. Instead, paleontologists worked by the false method of exemplification: validation by a "textbook case" or two, provided that the chosen instances be sufficiently persuasive. And even here, at this utterly minimal level of documentation, the method failed. A few examples did enter the literature (see Fig. 9-2 for comparison of an original claim with a secondary textbook version)— where they replicated by endless republication in the time-honored fashion of textbook copying (see Gould, 1988a). But, in a final irony, almost all these famous exemplars turned out to be false on rigorous restudy—see Hallam, 1968, and Gould, 1972, for stasis rather than gradual increase in coiling in the Liassic oyster *Gryphaea*; Prothero and Shubin, 1989, on stasis within all documented species of fossil horses, and with frequent overlap between ancestors and descendants, indicating branching by punctuational speciation rather than anagenetic gradualism; and Gould, 1974, on complete absence of data for the common impression that the enormous antlers of *Megaloceros*

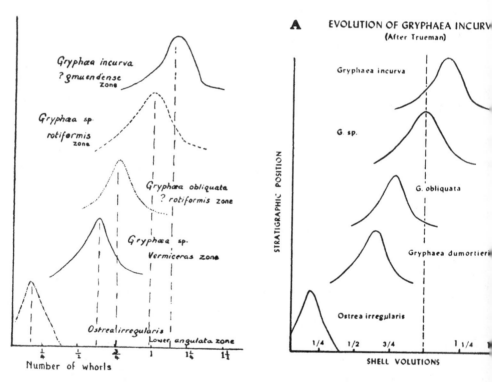

9-2. Trueman's original claim for phyletic gradualism in the increased coiling of *Gryphaea* in Lower Jurassic rocks of England (left). To the right, a textbook smoothing and simplification of the same figure. Trueman's claim has been invalidated for two reasons: first, *Gryphaea* did not evolve from *Ostrea*; and, second, subsequent studies have not validated any increase of coiling within *Gryphaea*, despite Trueman's graphs. Nonetheless, once such figures become ensconced in textbooks, they tend to persist even when their empirical justification has long been refuted in professional literature.

(the "Irish Elk") increased gradually in phylogeny, with positive allometry as body size enlarged.

Traditional paleontology therefore placed itself into a straightjacket that made the practice of science effectively impossible: only a tiny percentage of cases passed muster for study at all, while the stories generated for this minuscule minority rested so precariously upon hope for finding a rare phenomenon—and received such limited definition by the primitive statistical methods then available (or, more commonly, remained unidentified by any statistical practice at all)—that even these textbook exemplars collapsed upon restudy with proper quantitative procedures. But consider what might have occurred, if only paleontologists had recognized that stasis is data (I will grant some validity to the standard rationale for regarding the second phenomenon of punctuation as an artifact of an imperfect record). As Hallam said to me many years ago, after he had disproved the classical story of gradualism in *Gryphaea*: more than 100 other species of mollusks, many with records as rich as *Gryphaea*'s, occur in the same Liassic rocks, yet no one ever documented the stratigraphic history of even a single one in any study of evolution, for all demonstrate stasis. Scientists picked out the only species that seemed to illustrate gradualism, and even this case failed.

Despite the widespread use of proper quantitative methods today, and despite increasing attention to the validity of stasis as an evolutionary phenomenon, this bias still persists. I do not doubt that several species of Cenozoic planktonic Foraminifera display gradual transitions (see pp. 803–810), but I know that these examples have been extracted for study from a much larger potential sample of species never documented in detail because their apparent stasis seems "boring" to students of evolution. An eager young statistician goes to a lifelong expert and says: I want to devote my doctoral thesis to a statistical study of evolution in a species of foram (the most promising of major taxa, thanks to a hyperabundance of specimens and excellent stratigraphic data in oceanic cores); which species shall I choose? And the expert advises: why not study *Graduloconoides gradualississima*; I know that this species shows interesting changes during the upper Miocene in cores A through Z. Meanwhile, poor old boring *Stasigerina punctiphora*, just as abundant in the same cores, and just as worthy of study, gets bypassed in silence.

I find this situation particularly frustrating as paleontology's primary example of an insidious phenomenon in science that simply has not been recognized for the serious and distorting results perpetrated under its aegis. Most scientists do not even recognize the problem—though some do, particularly in the medical and social sciences, where the error has been named "publication bias," and has inspired a small but important literature (Begg and Berlin, 1988). In publication bias, prejudices arising from hope, cultural expectation, or the definitions of a particular scientific theory dictate that only certain kinds of data will be viewed as worthy of publication, or even of documentation at all. Publication bias bears no relationship whatever with the simply immoral practice of fraud; but, paradoxically, publication bias may exert a far more serious effect (largely because the phenomenon must be so much more common than fraud)—for scientists affected by publication bias do not recognize their errors (and their bias may be widely shared among colleagues), while a perpetrator of fraud operates with conscious intent, and the wrath of a colleague will be tremendous upon any discovery.

Begg and Berlin (1988) cite several documented cases of publication bias. We can hardly doubt, for example, that a correlation exists between socioeconomic status and academic achievement, but the strength and nature of this association can provide important information, for both political practice and social theory. White (1982, cited in Begg and Berlin) found a progressively increasing intensity of correlation with prestige and permanence of published source. Studies published in books reported an average correlation coefficient of 0.51 between academic achievement and socioeconomic status; articles in journals gave an average of 0.34, while unpublished studies yielded a value of 0.24. Similarly, Coursol and Wagner (1986, cited in Begg and Berlin) found publication bias both in the decision to submit an article at all, and in the probability for acceptance. In a survey of outcomes in psychotherapy, they noted that 82 percent of studies with positive outcomes led to submission of papers to a journal, while only 43 percent of negative outcomes provoked an attempt at publication. Of papers submitted, 80 percent that report positive outcomes were accepted for publication, but the figure fell to 50 percent for papers claiming negative results.

In my favorite study of publication bias, Fausto-Sterling (1985) tabulated claims in the literature for consistent differences in cognitive and emotional styles between men and women. She does not deny that genuine differences often exist, and in the direction conventionally reported. But she then, so to speak, surveys her colleagues' file drawers for studies not published, or for negative results published and then ignored, and often finds that a great majority report either a smaller and insignificant disparity between sexes, or no differences at all. When she collated all studies, rather than only those published, the much-vaunted differences often dissolved into statistical insignificance or triviality.

For example, a recent favorite theme of pop psychology attributed different cognitive styles in men and women to the less lateralized brains of women. Some studies have indeed reported a small effect of greater male lateralization; none has found more lateralized brains in women. But most experiments, as Fausto-Sterling shows, detected no measurable differences in lateralization at all and this dominant relative frequency (even in published literature) should be prominently reported in the press and in popular books, but tends to be ignored as "no story."

Paleontology's primary example of publication bias—the nonreporting of stasis under the false belief that such stability represents "no data" for evolution—illustrates a particularly potent form of the general phenomenon, a category that I have called "Cordelia's dilemma" (Gould, 1995) to memorialize the plight of King Lear's honest but rejected daughter. When asked by Lear for a fulsome protestation of love in order to secure her inheritance, Cordelia, disgusted by the false and exaggerated speeches of her sisters Goneril and

Regan, chose to say nothing, for she knew that "my love's more ponderous than my tongue." But Lear mistook her silence for hatred or indifference, and cut her off entirely (with tragic consequences later manifested in his own madness, blindness, and death), in proclaiming that "nothing will come of nothing."

Cordelia's dilemma arises in science when an important (and often predominant) signal from nature isn't seen or reported at all because scientists read the pattern as "no data," literally as nothing at all. This odd status of "hidden in plain sight" had been the fate of stasis in fossil morphospecies until punctuated equilibrium gave this primary signal some theoretical space for existence. Apparent silence—the overt nothing that actually records the strongest something—can embody the deepest and most vital meaning of all. What, in western history, has been more eloquent than the silence of Jesus before Pilate, or Saint Thomas More's date with the headsman because he acknowledged that fealty forbade criticism of Henry VIII's marriage to Anne Boleyn, but maintained, literally to the death, his right to remain silent, and not to approve?

In summary, the potentially reformative role of punctuated equilibrium resides in an unusual property among scientific innovations. Most new theories in science arise from fresh information that cannot be accommodated under an old explanatory rubric. But punctuated equilibrium merely honored the firmest and oldest of all paleontological observations—the documentable stasis of most fossil morphospecies—by promoting this pattern to central recognition as an expected result of evolution's proper expression at the scale of geological time. This reformulation cast a bright light upon stasis, a preeminent fact that had formerly been mired in Cordelia's dilemma as a grand disappointment, and therefore as "no data" at all, a pattern fit only for silence in a profession that accepted Darwin's argument for gradualism as the canonical expression of evolution in the fossil record.

## The Primary Claims of Punctuated Equilibrium

### DATA AND DEFINITIONS

First of all, the theory of punctuated equilibrium treats a particular level of structural analysis tied to a particular temporal frame. G. K. Chesterton (1874–1936), the famous English author and essayist, wrote that all art is limitation, for the essence of any painting lies in its frame. The same principle operates in science, where claiming too much, or too broad a scope of application, often condemns a good idea to mushy indefiniteness and consequent vacuity.

Punctuated equilibrium is not a theory about all forms of rapidity, at any scale or level, in biology. Punctuated equilibrium addresses the origin and deployment of species in geological time. Punctuational styles of change characterize other phenomena at other scales as well (see Section V of this chapter)—catastrophic mass extinction triggered by bolide impacts, for example—and proponents of punctuated equilibrium would become dull specialists if they did not take an interest in the different mechanisms responsible for similarities in the general features of stability and change across nature's varied domains, for science has always sought unity in this form of abstraction. But punctuated equilibrium—a particular punctuational theory of change and stability for one central phenomenon of evolution—does not directly address the potentially coordinated history of faunas, or the limits of viable mutational change between a parental organism and its offspring in the next generation.

The theory of punctuated equilibrium attempts to explain the macroevolutionary role of species and speciation as expressed in geological time. Its statements about rapidity and stability describe the history of individual species; and its claims about rates and styles of change treat the mapping of these individual histories into the unfamiliar domain of "deep" or geological time—where the span of a human life passes beneath all possible notice, and the entire history of human civilization stands to the duration of primate phylogeny as an eyeblink to a human lifetime. The claims of punctuated equilibrium presuppose the proper scaling of microevolutionary processes into this geological immensity—the central point that Darwin missed when he falsely assumed that "slowness" of modification in domesticated animals or crop plants, as measured in ordinary human time (where all of our history, and so many human generations, have witnessed substantial change within populations, but no origin of new species), would translate into geological time as the continuity and slowness of phyletic gradualism.

Once we recognize that definitions for the two key concepts of stasis and punctuation describe the *history of individual species scaled into geological time,* we can establish sensible and operational criteria. As a central proposition, punctuated equilibrium holds that the great majority of species, as evidenced by their anatomical and geographical histories in the fossil record, originate in geological moments (punctuations) and then persist in stasis throughout their long durations (Sepkoski, 1997, gives a low estimate of 4 million years for the average duration of fossil species; mean values vary widely across groups and times, with terrestrial vertebrates at lesser durations and most marine invertebrates in the higher ranges; in any case, geological longevity achieves its primary measure in millions of years, not thousands). As the primary macroevolutionary implication of this pattern, species meet all definitional criteria for operating as Darwinian individuals (see pp. 602–613) in the domain of macroevolution.

This central proposition embodies three concepts requiring definite operational meanings: stasis, punctuation, and dominant relative frequency. (I am not forgetting the thorny problems associated with the definition of species from fossil data, where anatomy prevails as a major criterion and reproductive isolation can almost never be assessed directly—and also with the putative correspondence of morphological "packages" that paleontologists designate as species with the concept as understood and practiced by students of modern populations of sexually reproducing organisms. I shall treat these issues on pages 784–796.)

Stasis does not mean "rock stability" or utter invariance of average values for all traits through time. In the macroevolutionary context of punctuated equilibrium, we need to know, above all, whether or not morphological change tends to accumulate through the geological lifetime of a species and, if so, what part of the average difference between an ancestral and descendant species can be attributed to incremental change of the ancestor during its anagenetic history. Punctuated equilibrium makes the strong claim that, in most cases, effectively no change accumulates at all. A species, at its last appearance before extinction, does not differ systematically from the anatomy of its initial entry into the fossil record, usually several million years before.

Of course we recognize that mean values will fluctuate through time. After all, measured means would vary even if true population values remained utterly constant—which they do not. And, with enough samples in a vertical sequence, some must include mean values (for some characters) outside conventional bounds of statistically insignificant difference from means for the oldest sample. Such fluctuation also implies that the final population will not be identical with the initial sample.

In operational terms, therefore, we need to set criteria for permissible fluctuation in average values through time. Two issues must be resolved: the amount of allowable difference between beginning and ending samples of a species, and the range of permissible fluctuation through time. Since we wish to test a hypothesis that little or no change accumulates by anagenesis during the history of most species, and since we have no statistical right to expect that (under this hypothesis) the last samples will be identical with the first, we should predict either that (i) the final samples shall not differ statistically, by some conventionally chosen criterion, from the initial forms; and at very least (ii) that the final samples shall not generally lie outside the range of fluctuation observed during the history of the species. (If final samples tend to lie outside the envelope of fluctuation for most of the species's history, then anagenesis has occurred.)

For the permissible range of fluctuation, we should, ideally, look to the extent of geographic variation among contemporary populations within the species, or its closest living relative. If the temporal range of variation stays within the spatial range for any one time, then the species has remained in stasis. Obviously, we cannot apply this optimal criterion for groups long extinct, but a variety of proxies should be available, including comparison of a full temporal range with the known geographic variation of a well-documented and widespread nearest living relative. Studies of stasis in Neogene species can often use the optimal criterion because the actual species, or at least some very close relatives, are often still extant. In the most elegant documentation of stasis for an entire fauna of molluscan species, Stanley and Yang (1987) used this best criterion to find that temporal fluctuation remained within the range of modern geographic variation for the same species. They could therefore affirm stasis in the most biologically convincing manner.

Since stasis is data, but punctuation generally records an unresolvable transition when assessed by the usual expression of fossil data in geological time, we need to formulate an appropriate definition of rapidity. (Punctuated equilibrium makes no claim about the possibility of substantial change at rates that would be called rapid by measuring rods of a human lifetime. Therefore, and especially, punctuated equilibrium provides no insight into the old and contentious issue of saltational or macromutational speciation.) As a first approach, the duration of a bedding plane represents the practical limit of geological resolution. Any event of speciation that occurs within the span of time recorded by most bedding planes will rarely be resolvable because evidence for the entire transition will be compressed onto a single stratigraphic layer, or "geological moment."

However, the limits of stratigraphic resolution vary widely, with bedding planes representing years or seasons in rare and optimal cases of varved sediments, but several thousand years in most circumstances. We therefore cannot formulate a definition equating punctuation with "bedding plane simultaneity." (After all, such a definition would, almost perversely, preclude the "dissection" of a punctuation in admittedly rare, but precious, cases of sedimentation so complete and so rapid that an event of speciation will not be compressed, as usual, onto a single bedding plane, but will "spread out" over a sufficient stratigraphic interval to permit the documentation of its rapid history.)

Punctuations must, instead, be defined relative to the subsequent duration of the derived species in stasis—for punctuated equilibrium, as a theory of relative timing, holds that species develop their distinctive features effectively "at birth," and then retain them in stasis for geologically long lifetimes. (These timings play an important role in the recognition of species as Darwinian individuals—see discussion on "vernacular" criteria of definable birth, death, and sufficient stability for individuation—Chapter 8, pp. 602–608).

I know no rigorous way to transcend the arbitrary in trying to define the permissible interval for punctuational origin. Since definitions must be theory-bound, and since the possibility of recognizing species as Darwinian individuals in macroevolution marks the major theoretical interest of punctuated equilibrium, an analogy between speciation and gestation of an organism may not be ill conceived. As the gestation time of a human being represents 1–2 percent of an ordinary lifetime, perhaps we should permit the same general range for punctuational speciation relative to later duration in stasis. At an average species lifetime of 4 million years, a 1-percent criterion allows 40,000 years for speciation. When we recognize that such a span of time would be viewed as gradualistic—and extremely slow paced at that—by any conventional microevolutionary scaling in human time; and when we also acknowledge that the same span represents the resolvable moment of a single bedding plane in a great majority of geological circumstances; then we can understand why the punctuations of punctuated equilibrium do not represent de Vriesian saltations, but rather denote the proper scaling of ordinary speciation into geological time.

Punctuation does suffer the disadvantage of frequently compressed recording on a single bedding plane (so that the temporal pattern of the full event cannot be dissected); moreover, an observed punctuation often represents the even less desirable circumstance of missing record (Darwin's classic argument from imperfection), or only partial pattern (as when a punctuation in a single geological section marks the first influx by migration of a species that originated earlier and elsewhere). Since stasis, on the other hand, provides an active (and often excellent) record of stability, empirical defenses of punctuated equilibrium have understandably focussed on the more easily documentable claims for equilibrium, and less frequently on more elusive predictions about punctuation. But we must not conclude, as some authors have suggested, that punctuation therefore becomes untestable or even impervious to documentation—and that the thesis of punctuated equilibrium must therefore depend for its empirical support only upon the partial data of stasis. The documentation of punctuation may be both more difficult and less frequently possible, but many good cases have been affirmed and several methods of rigorous testing have been developed.

In the first of two general methods, one may document the reality of a punctuation (as opposed to interpretation as a Darwinian artifact based on gaps in sedimentation) by finding cases of gradualism within a stratigraphic sequence (which must then be sufficiently complete to record such an anagenetic transition), and then documenting punctuational origins for other species in the same strata. Using this technique for Ordovician trilobites from Spitzbergen, Fortey (1985) found a ratio of about 10:1 for cases of punctuation compared with gradualism.

In a second, and more frequently employed, method, one searches explicitly for rare stratigraphic situations, where sedimentation has been sufficiently rapid and continuous to spread the usual results of a single bedding plane into a vertical sequence of strata. Williamson (1981), for example, published a famous series of studies on speciation of freshwater mollusks in African Pleistocene lakes. (These articles provoked considerable debate (Fryer, Greenwood and Peake, 1983), and Williamson died young before he could complete his work. However, in my admittedly partisan judgment, Williamson more than adequately rebutted his critics (1985, 1987).)

These African lakes form in rift valleys, where sedimentation rates are unusually high because the rift-block foundations of the lake sink continuously, and sediments can therefore accumulate above, without interruption. Thus, the thousand-year duration of a speciation event may span several layers of foundering sediment. With this unusual degree of resolution, Williamson was even able to demonstrate a remarkable phenomenon in change of variability within a speciating population—a pattern that appeared over and over again in several events of speciation, and may therefore be viewed as potentially general (see Fig. 9-3): Williamson found limited variation around parental mean values in the oldest samples; intermediacy of mean values within speciating samples, but accompanied by a greatly expanded range of variation (though still normal in distribution); and subsequent "settling down" of variation to the reduced level of the ancestral population, but now distributed around the altered mean value of the derived species.

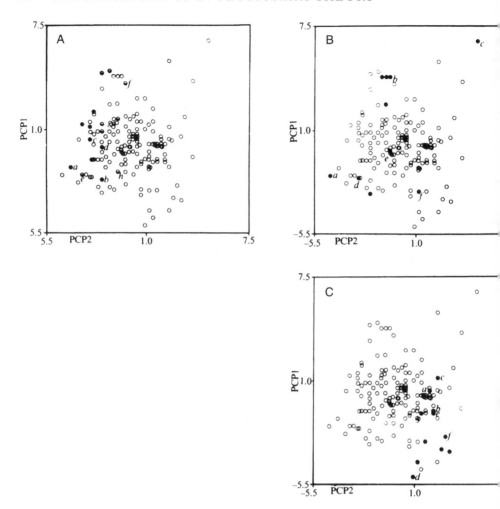

9-3A. The dissection of a punctuation made possible by unusually high sedimentation rates. Williamson's analysis of variation and central tendency during a punctuation in the *B. unicolor* lineage of Pleistocene fresh water pulmonate snails from the African rift valley. Each diagram shows all the specimens from the entire sequence, with only those specimens for the relevant interval depicted in black. A. Parental form before the punctuation with multivariate modal morphology concentrated to the left of the range. B. Expanded variation throughout the range during the time of the punctuation itself. C. Restricted variation again, but settling down upon the morphology of a new taxon following the punctuation, as seen in the reduction of variation with change in modal position towards the right side of the array. From Williamson, 1981.

If this kind of unusual circumstance spreads a punctuational event of speciation through a sufficient stratigraphic interval for resolution, another strategy of research will sometimes permit the dissection of a punctuation in conventional cases of full representation on a single bedding plane. Goodfriend and Gould (1996) documented such a case because they could estab-

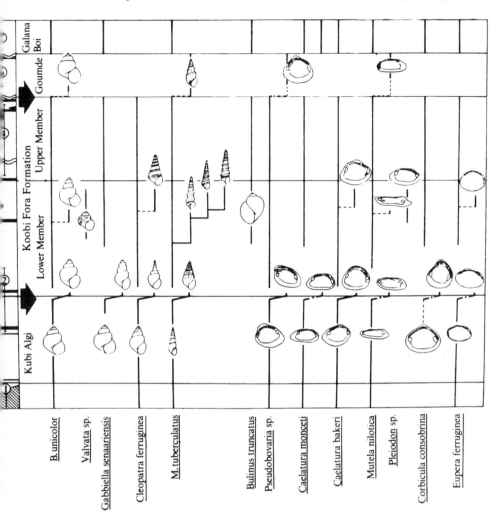

9-3B. Relative timings of punctuational events throughout Williamson's entire series. From Williamson, 1981.

lish absolute dates for the individual shells on a single bedding plane. (Admittedly, this technique cannot be generally applied—especially to sediments of appreciable age, where errors of measurement for any method of dating must greatly exceed the full span of the bedding plane. But this method can be used for late Pleistocene and Holocene samples.)

On a single mud flat (a modern "bedding plane," if you will) on the island of Great Inagua, we found a complete morphological transition between the extinct fossil pulmonate species *Cerion excelsior* and the modern species *Cerion rubicundum*. Many lines of evidence indicate that this transition occurred by hybridization, as *C. rubicundum* migrated to an island previously inhabited only by *C. excelsior* among large species of *Cerion*. Ordinarily, we would find such a complete morphological transition on a single bedding plane, but be unable to perform any fine scale analysis in the absence of methods for dating individual shells. That is, we would be unable to discover whether the unusual morphological range represented a temporal transition or a standing population with enhanced variation. But Goodfriend and I could date the individual shells by amino acid racemization for all specimens, keyed to radiocarbon dates for a smaller set of marker shells. We found an excellent correlation between measured age and multivariate morphometric position on the continuum between ancestral *C. excelsior* and descendant *C. rubicundum* (see Fig. 9-4). The transition lasted between 15,000 and 20,000 years—a good average value for a punctuational event, and a fact that we could ascertain only because the individual specimens of a single bedding plane could be chemically dated independently of their morphology.

We can therefore define stasis and punctuation in operational terms, with stasis available for test in almost any species with a good fossil record, but punctuation requiring an unusual density of information, and therefore not routinely testable, but requiring a search for appropriate cases (not an unusual situation in sciences of natural history, where nature sets the experiments, and scientists must therefore seek cases with adequate data). The third key issue of relative frequency may be easier to operationalize—as one need only tabulate cases pro and con within well-documented faunas—but remains harder to define.

As the most important ground rule, the theory of punctuated equilibrium makes a claim about dominating pattern, or relative frequency, not just an as-

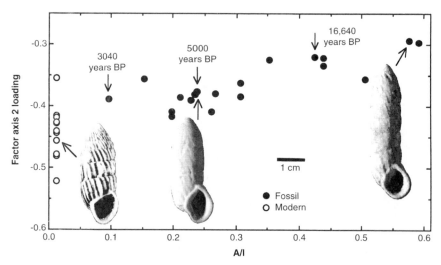

9-4. Another way to dissect a punctuation by obtaining absolute age dates for all specimens on a bedding plane, and thus obtaining temporal distinctions within the compression. The ancestral and high spired *Cerion excelsior,* over no more than 15,000 to 20,000 years (well within the range of punctuational dynamics), hybridizes with invading *Cerion rubicundum,* with gradual fading out of all morphometric influence from the unusually shaped ancestor.

sertion for the existence of a phenomenon. Such issues cannot be resolved by anecdote, or the documentation, however elegant, of individual cases. If anyone ever doubted that punctuated equilibrium exists as a phenomenon, then this issue, at least, has been put to rest by two decades of study following the presentation of our theory, and by clear and copious documentation of many cases (see pp. 822–874). Nonetheless, as pleased as Eldredge and I have been both by the extent of this research and the frequency of its success, the "ideal case study" method cannot validate our theory.

Punctuated equilibrium does not merely assert the existence of a phenomenon, but ventures a stronger claim for a dominant role as a macroevolutionary pattern in geological time. But how can this vernacular notion of "dominant" be translated into a quantitative prediction for testing? At this point in the argument, we encounter the difficult (and pervasive) methodological issue of assessing relative frequency in sciences of natural history. If species were like identical beans in the beanbag of classical thought experiments in probability, then we could devise a sampling scheme based on enumerative induction. Enough randomly selected cases could establish a pattern at a desired level of statistical resolution. But species are irreducibly unique, and the set of all species does not exhibit a distribution consistent with requirement of standard statistical procedures. It matters crucially whether we study a clam or a mammal, a Cambrian or a Tertiary taxon, a species in the stable tropics, or at volatile high latitudes. Moreover—and especially—the "ideal case study" method has often failed, and led to parochialisms and false generalities, precisely because we tend to select unusual cases and ignore, often quite unconsciously, a dominant pattern. Indeed, proclamations for the supposed "truth" of gradualism—asserted against every working paleontologist's knowledge of its rarity—emerged largely from such a restriction of attention to exceedingly rare cases under the false belief that they alone provided a record of evolution at all! The falsification of most "textbook classics" upon restudy only accentuates the fallacy of the "case study" method, and its root in prior expectation rather than objective reading of the fossil record.

Punctuated equilibrium must therefore be tested by relative frequencies among all taxa (or in a truly randomized subset) in a particular fauna, a particular clade, a particular place and time, etc. If we can say, as Ager did (see p. 753) that all but one Mesozoic brachiopod species displays stasis, or as Imbrie did (see p. 760) that all but one Devonian species from the Michigan Basin shows no change, then we have specified a dominant pattern, at least within a particular, well-defined and evolutionarily meaningful package. I cannot give a firm percentage for what constitutes a "dominant" relative frequency—for, again, we encounter a theory-bound claim, where "dominant" specifies a weight, beyond which the morphological history of a clade must be explicated primarily by the differential success of species treated as stable entities, or Darwinian individuals in macroevolution—and not by anagenetic change within species. More research must be done, largely in the testing of mathematical models under realistic circumstances, to learn the relative frequencies and rates required to impart such dominance to species-individuals in the course of macroevolution. For now, and for empirically minded paleontologists, the study of relative frequencies in entire faunas, rather than the extraction of apparently idealized cases, should be pursued as a primary strategy of research.

Critics have sometimes stated that punctuated equilibrium rests upon declaration rather than documentation. (Maynard Smith once compared the theory to "Aunt Jobisca's" maxim about ancient verities "known" by folk wisdom *a priori*.) We do indeed assert that working paleontologists know the fact of dominant stasis in their bones—but this claim represents a fair consensus about the history of a field, and does underscore a paradox of nonconcordance between deep practical knowledge and imposed theoretical expectation. We have never tried to argue that such a "professional feeling" constitutes documentation for punctuated equilibrium. As with all scientific theories, punctuated equilibrium will live or die by concrete and quantifiable evidence. As with any good hypothesis, punctuated equilibrium becomes operational when workable definitions can be provided for key claims and expectations—in this case, for *stasis, punctuation,* and *relative frequency.* Contrary to the impression of some critics who have not followed the primary literature of paleobiology during the last 25 years, punctuated equilibrium has proven its fruitfulness and operational worth by being tested—and usually confirmed, but sometimes confuted—in a voluminous literature of richly documented cases (see pp. 822–874).

### Microevolutionary links

Eldredge and I coined the term punctuated equilibrium in a paper first presented (Gould and Eldredge, 1971) at a symposium entitled "Models in Paleobiology" at the 1971 Annual Meeting of the Geological Society of America. T. J. M. Schopf, the organizer of the symposium, conceived the enterprise as a tutorial in modern evolutionary theory for professional invertebrate paleontologists. By accidents of history, invertebrate paleontologists generally receive their advanced academic degrees from geology departments, not from biology. Fossils became primary tools for stratigraphic correlation long before the development of evolutionary theory, and even before all scientists had accepted them as remains of ancient organisms! Given traditions of narrowness in postgraduate education—particularly in Europe, where students often attend no formal courses at all, and certainly no courses for credit, outside the department that will grant their degree—most paleontologists, before the present generation, did not receive any explicit training in evolutionary biology, and could not articulate the basic concepts of population genetics or theories of speciation. In paleontological usage, "evolution" designated little more than the inferred pathway of phylogeny. This "little learning" often became the "dangerous thing" of Alexander Pope's classic couplet, as paleontologists derived their understanding of evolution from memories of old textbooks, or from shared impressions amounting to little more than the blind leading the blind. This situation has now changed dramatically—and Eldredge and I do take pride in the role played by punctuated equilibrium in encouraging this shift of interest—as a profession of paleobiology, supported by several new journals dedicated to the subject (*Paleobiology, Historical Biology, Lethaea, Palaios,* and *Palaeogeography Palaeoclimatology Palaeoecology* (or *P-cubed* to aficionados), for example), has arisen to accommodate burgeoning research in the application of evolutionary theory to the fossil record, and in enlarging and revising the theory in the light of novel macroevolutionary data.

In any case, Schopf's symposium featured a series of presentations, each suggesting how one aspect of paleontological work might be enlightened by modern microevolutionary theory, particularly as expressed in the application of models, preferably quantitative in nature. Eldredge and I drew the topic of species and speciation—and our original article on punctuated equilibrium (Eldredge and Gould, 1972) emerged as a result. (As I have often stated, the basic idea had been presented in Eldredge, 1971. We had been graduate students together at the American Museum of Natural History, under the tutelage of Norman D. Newell. We had discussed these issues often and intensely throughout our graduate years. We had been particularly frustrated—for we had both struggled to master statistical and other quantitative methods—with the difficulty of locating gradualistic sequences for applying these techniques, and therefore for documenting "evolution" as paleontological tradition then defined the term and activity. When I received Schopf's invitation to talk on models of speciation, I felt that Eldredge's 1971 publication had presented the only new and interesting ideas on paleontological implications of the subject—so I asked Schopf if we could present the paper jointly. I wrote most of our 1972 paper, and I did coin the term punctuated equilibrium—but the basic structure of the theory belongs to Eldredge, with priority established in his 1971 paper.)

I mention this background to clarify the original context and continuing focus of the theory of punctuated equilibrium—a notion rooted in the explicit goal that Eldredge and I set for ourselves: to apply microevolutionary ideas about *speciation* to the data of the fossil record and the scale of geological time. Before we proposed the theory of punctuated equilibrium, most paleontologists assumed that the bulk of evolutionary change proceeded in the anagenetic mode—that is, by continuous transformation of a unitary population through time (see Fig. 9-5). In this context, most paleontological discussion about species centered itself upon a contentious issue that constantly circulated throughout our literature (see Imbrie, 1957; Weller, 1961; McAlester, 1962; Shaw, 1969) and even generated entire symposia dedicated to potential solutions (see Sylvester-Bradley, 1956): the so-called species problem in paleontology.

This supposed problem—more philosophical and definitional than empirical (once one accepts the underlying assumptions about anagenesis as a dominant factual reality)—arises because a true continuum cannot be unambiguously divided into segments with discrete names. If population A changes so extensively by anagenesis that we feel impelled to provide the resulting popu-

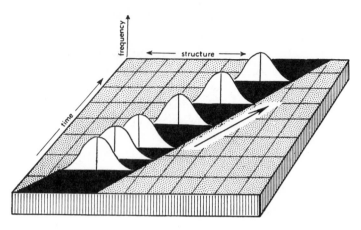

9-5. Typical textbook illustration of evolution by continuous anagenetic transformation of an unbranched population through time. This textbook labels the figure explicitly and exclusively as its icon of "evolution" itself, not of gradualism or any other subcategory of evolutionary change. From the standard paleontological textbook of my student generation, Moore, Lalicker, and Fischer, 1953.

lation with a new Linnaean name (as species B), then where should we place the breakpoint between A and B? Any boundary must be arbitrary—if only by the illogic of the unavoidable implication that the last parental generation of species A could not, in principle, breed with its own immediate offspring in species B. (We may abhor human incest for social reasons, but we can scarcely deny the biological possibility—hence the perceived societal need for a taboo.) This problem generated a large, tedious, and fruitless literature, primarily because the issue always remained available, unresolved and therefore ripe for yet another go-round whenever a paleontologist needed to deliver a general address and couldn't think of anything else to say.

Punctuated equilibrium took a radically different approach by admitting unresolvability under the stated assumptions, but then denying the focal empirical premise that new species usually (or even often) arise by gradualistic anagenesis. Instead, Eldredge and I argued that the vast majority of species originate by splitting, and that the standard tempo of speciation, when expressed in geological time, features origin in a geological moment followed by long persistence in stasis. Thus, the classic and endlessly-fretted "species problem in paleontology" disappears because species act as well-defined Darwinian individuals, not as arbitrary subdivisions of a continuum. Species then gain definability because they almost always arise by speciation (that is, by splitting, or geographic isolation of a daughter population followed by genetic differentiation from the parental population), not by anagenesis (or transformation of the entire mass of an ancestral species). To be sure, a new species must pass through a short period of ambiguity during its initial differentiation from an ancestral population, but, in the proper scaling of macroevolutionary time, this period passes so quickly (almost always in the unresolvable geological moment of a single bedding plane), that operational definability encounters no threat.

Of course, gradualists did not deny that speciation often occurs by branching. They just didn't grant this process of splitting any formative role in the accumulation of macroevolutionary change for three reasons. First, they conceived speciation only as an engine for generating diversity, not as an agent for changing average form within a clade (that is, for the key macroevolutionary phenomenon of trends—see quotes of Huxley and Ayala, and Mayr's response, on p. 563). Trends arose by anagenesis (see Fig. 9-6), and speciation only served the subsidiary (if essential) function of iterating a favorable feature, initially evolved by anagenesis, into more than one taxon—thus providing a hedge against extinction.

Second, they granted little quantitative weight to the role of speciation (splitting as opposed to anagenesis) in the totality of evolutionary change. In a famous estimate that became canonical, Simpson (1944) stated that about 10 percent of evolutionary change occurred by speciation, and 90 percent by anagenesis.

Third, when gradualists portrayed speciation at all (see Fig. 9-7), they depicted the process as two events of anagenesis proceeding at characteristically slow rates. Thus, they identified nothing distinctively different about change by speciation. Some contingency of history, they argued, splits a population into two separate units, and each proceeds along its ordinary anagenetic way. Punctuated equilibrium, on the other hand, proposes that the geological tempo of speciation differs radically from gradualistic anagenesis. (We also argue, of course, that such anagenesis rarely occurs at all!)

The theory of punctuated equilibrium therefore began as a faithful response to Schopf's original charge to Eldredge and me: to show how standard

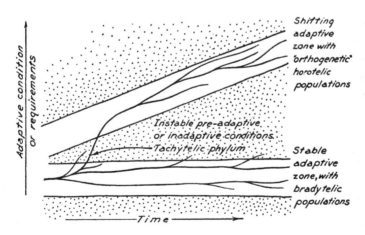

9-6. A standard illustration from Simpson (1944), showing that all trends, and all stability for that matter, originate primarily in the anagenetic mode—that is, by change during the lifetime of individual species, with branching serving primarily to diversify and iterate the favorable designs originated by anagenesis, and thus to prevent extinction of the lineage.

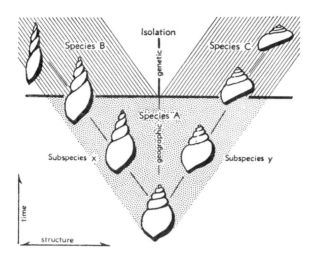

9-7. Another illustration from the standard student paleontological textbook of the 1950's, with speciation depicted merely as two events of gradualistic change, following a separation of lineages. From Moore, Lalicker, and Fischer, 1953.

microevolutionary views about speciation, then unfamiliar to the great majority of working paleontologists, might help our profession to interpret the history of life more adequately. (As a best testimony to this unfamiliarity, I note that most paleontologists didn't even recognize the conceptual and terminological distinction between "speciation" defined as a process of splitting, and the accumulation of enough change by anagenesis to provoke the coining of a new Linnaean name for an unbranched single population.)

In this crucial sense, the theory of punctuated equilibrium adopts a very conservative position. The theory asserts no novel claim about modes or mechanisms of speciation; punctuated equilibrium merely takes a standard microevolutionary model and elucidates its expected expression when properly scaled into geological time. This scaling, however, did provoke a radical reinterpretation of paleontological data—for we argued that the literal appearance of the fossil record, though conventionally dismissed as an artifact of imperfect evidence, may actually be recording the workings of evolution as understood by neontologists.* This empowering switch enabled paleontologists to cherish their basic data as adequate and revealing, rather than pitifully fragmentary and inevitably obfuscating. Paleontology could emerge from the intellectual sloth of debarment from theoretical insight imposed by poor data—a self-generated torpor that had confined the field to a descriptive role in documenting the actual pathways of life's history. Paleontology could now take a deserved and active place among the evolutionary sciences.

The major and persisting misunderstanding of punctuated equilibrium among neontologists—a great frustration for us, and one that we have tried

---

*All professions maintain their parochialisms, and I trust that nonpaleontological readers will forgive our major manifestation. We are paleontologists, so we need a name to contrast ourselves with all you folks who study modern organisms in human or ecological time. You therefore become neontologists. We do recognize the unbalanced and parochial nature of this dichotomous division—much like my grandmother's parsing of *Homo sapiens* into the two categories of 'Jews' and 'non-Jews.'

to explicate and resolve again and again (Gould and Eldredge, 1977, 1993; Gould, 1982c, 1989e), though without conspicuous success—involves the false assumption that if we are really saying something radical, we must be staking a claim for a novel mechanism of speciation, or for a different (read non-Darwinian) style of genetic change. When our critics then join this false assumption to our terminology of "unresolvable geological moments" or "punctuations," they begin to fear that the dreaded specter of saltationism must be lurking just around the corner, trying yet again to raise its ugly head after such a well-deserved burial. Vituperation then trumps logic, angry assumption precludes careful reading, and punctuated equilibrium becomes a loathed doctrine of ignorant and grandstanding paleontologists who ought to stay in their own limited bailiwick, and get on with the job of documenting large scale patterns generated by mechanisms that can be recognized and comprehended only by neontologists.

But punctuated equilibrium makes no iconoclastic claim about speciation at all. The radicalism of punctuated equilibrium lies in the extensive consequences of its key implication that conventional mechanisms of speciation scale into geological time as the observed punctuations and stasis of most species, and not as the elusive gradualism that a century of largely fruitless paleontological effort had sought as the only true expression of evolution in the fossil record. The central intellectual strategy of our original 1972 paper rests upon this premise. We took Mayr's allopatric theory (as expressed in his classic treatise of 1963, deemed "magisterial" by Huxley), and tried to elucidate its implied expression when scaled into geological time. We did not select this theory to fit a paleontological pattern that we wished to validate. We choose Mayr's formulation because his allopatric theory represented the most orthodox and conventional view of speciation then available in neontological literature—and we had been given the task of applying *standard* evolutionary views to the fossil record. I recognize, with 30 years of hindsight, that our original assessment both of Mayr's theory and of professional consensus may have been both naive and overly dichotomous, but we could not have stated our intent more clearly—the reform of paleontological practice by the paradoxical route of applying a *fully conventional* apparatus of neontological theory. We wrote (1972, p. 94): "During the past thirty years, the allopatric theory has grown in popularity to become, for the vast majority of biologists, *the* theory of speciation. Its only serious challenger is the sympatric theory. Here we discuss only the implications of the allopatric theory for interpreting the fossil record of sexually-reproducing metazoans. We do this simply because it is the allopatric, rather than the sympatric, theory that is preferred by biologists."

Mayr's version of allopatry fit the paleontological pattern of punctuation and stasis particularly well. If most new species arise from small populations peripherally isolated at the edges of a parental range, then we cannot expect to document a gradual transition by analyzing the stratigraphic sequence of samples for a common species. For we will usually be collecting from the population's central range during its period of stability. Daughter species originate in three circumstances that virtually guarantee a punctuational expression in the fossil record: (1) they arise rapidly (usually instantaneously) in geological time, and they originate both (2) in a small geographic region (the peripheral isolate), and (3) elsewhere (beyond the borders of the parental range that provides the exclusive source for standard paleontological collections). The "sudden" entrance of a daughter species into strata previously occupied by parents usually represents the inward migration of a peripheral isolate, now "promoted" by reproductive isolation to full separation, not the origin of a new species *in situ.*

Eldredge and I have often been asked what we think of sympatric speciation, or of various models, like polyploidy, for rapid origin even in human time. We do not mean to be evasive or obscure in our assertions of agnosticism. (I am intensely interested in the literature on speciation, and I would love to know the relative frequencies of these other models vs. classical Mayrian peripatry. But this important issue does not strongly impact punctuated equilibrium, and surely cannot be resolved by paleontological data.) Punctuated equilibrium simply requires that any asserted mechanism of speciation, whatever its mode or style, be sufficiently rapid and localized to appear as a punctuation *when scaled into geological time.* If I understand them correctly, most alternative models to peripatry generally operate even more rapidly than the conventional Mayrian mode that we invoked to anchor our theory—as obviously true for polyploidy, and also for most versions of sympatric speciation (if only because the constant threat of dilution by gene flow from surrounding parentals can best be overcome by rapid achievement of reproductive isolation in ecological time). Therefore, punctuated equilibrium can only gain strength if these alternative mechanisms become validated at meaningful relative frequencies. (The faster the better, one might say.) But punctuated equilibrium does not require this boost—and we therefore remain agnostic—because the most conventional form of Mayrian peripatry already yields the full set of phenomena predicted by punctuated equilibrium when properly scaled into the immensity of geological time. (Punctuated equilibrium, on the other hand, does not maintain a similar agnosticism towards any putative mechanism of speciation that conceives the process of splitting as no more rapid than imagined rates for the gradual anagenesis of large central populations. Some models of so-called "dumbbell allopatry"—or the splitting of a parental population into two effectively equal moieties, with subsequent anagenesis in each—do construe speciation as consequently slow in geological expression, and therefore do threaten punctuated equilibrium. But I do not think that such models enjoy much support among biologists, especially for operation at a high relative frequency.)

Geological time can be both a wonder and a snare because we grasp the idea in our heads (all scientists know how many zeroes follow the one in expressing millions or billions), but we face a primal, and fundamentally psychological, difficulty in trying to incorporate this central concept into the guts of our intuition. We can lose information in upward scaling when glacial slowness in human history becomes a passing and unresolvable geological moment. But we can also gain when operational invisibility at our scale (inability to distinguish a small effect from measurement error) becomes palpable and prominent in the large, or when the almost inconceivable rarity of an event that averages one expression in ten thousand years achieves guaranteed repetition across millions.

## MACROEVOLUTIONARY IMPLICATIONS

If punctuated equilibrium has broader utility beyond the reform of paleontological practice, then we must look to potential implications for macroevolutionary theory, and for consequent enrichment in our general understanding of mechanisms that regulate the history of life. I have linked my treatments of punctuated equilibrium and the hierarchical theory of natural selection to form the longest section of this book (presented as two chapters, 8 and 9) because I believe that punctuated equilibrium supplies the central argument for viewing species as effective Darwinian individuals at a relative frequency high enough to be regarded as general—thereby validating the level of species as a domain of evolutionary causality, and establishing the effectiveness and independence of macroevolution by two of the three criteria featured throughout this book as indispensable foundations of Darwinism.

*First,* punctuated equilibrium secures the hierarchical expansion of selectionist theory to the level of species, thus moving beyond Darwin's preference for restricting causality effectively to the organismic realm alone (leg one on the essential tripod). *Second,* by defining species as the basic units or atoms of macroevolution—as stable "things" (Darwinian individuals) rather than as arbitrary segments of continua—punctuated equilibrium precludes the explanation of all evolutionary patterns by extrapolation from mechanisms operating on local populations, at human timescales, and at organismic and lower levels (leg three on the tripod of Darwinian essentials). Thus, as emphasized in the last section, punctuated equilibrium presents no radical proposal in the domain of microevolutionary mechanics—in particular (and as so often misunderstood), the theory advances no defenses for saltational models of speciation, and no claims for novel genetic processes. Moreover, punctuated equilibrium does not attempt to specify or criticize the conventional mechanisms of microevolution at all (for punctuated equilibrium emerges as the anticipated expression, by proper scaling, of microevolutionary theories about speciation into the radically different domain of "deep," or geological time). But punctuated equilibrium does maintain, as the kernel of its potential novelty for biological theory, that these unrevised microevolutionary mechanisms do not hold exclusive sway in evolutionary explanation, and that their domain of action must be restricted (or at least shared) at the level of macroevolutionary pattern over geological scales— for punctuated equilibrium ratifies an effective realm of macroevolutionary mechanics based on recognizing species as Darwinian individuals. In other words, *punctuated equilibrium makes its major contribution to evolutionary theory, not by revising microevolutionary mechanics, but by individuating species (and thereby establishing the basis for an independent theoretical domain of macroevolution).*

As discussed in Chapter 8 (see pp. 648–652), punctuated equilibrium wins this role by refuting Fisher's otherwise decisive argument for the impotence (despite the undeniable existence) of species selection. So long as most new species arise by branching (speciation) rather than by transformation (anagenesis), species can be individuated by their uniquely personal duration, bounded by birth in branching and death by extinction. But if anagenesis, fueled by Darwinian organismic selection, operates to substantial effect during the lifetimes of most species, then, by Fisher's argument, such microevolutionary transformation must overwhelm species selection in building the overall pattern of macroevolutionary change—for the number of organism-births must exceed species-births by several orders of magnitude, and if every event of birthing, at each level, supplies effective variation for evolutionary transformation, then the level of species can contribute virtually nothing to the totality of change. But if stasis rules and anagenesis rarely occurs, then speciation becomes the more effective level of evolutionary variation. And if speciation unfolds in geological moments, then species in geological time match organisms on our ordinary yearly scales in both distinctness and discreteness. Thus, the pattern of punctuated equilibrium establishes species as effective individuals and potential Darwinian agents in the mechanisms of macroevolution.

In summary, G. G. Simpson gave a singularly appropriate title to his epochal 1944 book that defined the potential of paleontology to devise insights about evolutionary mechanisms: *Tempo and Mode in Evolution.* If we accept Simpson's focus on tempo and mode as primary subjects, then punctuated equilibrium has provoked substantial revisions of macroevolutionary theory and practice in both domains.

### Tempo and the significance of stasis

For tempo, punctuated equilibrium reverses our basic perspective. We must abandon our concept of constant change operating within a sensible, stately range of rates as the normal condition of an evolving entity. We must then reformulate evolutionary change as a set of rare episodes, short in duration relative to periods of stasis between. Stability becomes the normal state of a lineage, with change recast as an infrequent and concentrated event that, nonetheless, renders phylogeny as a set of summed episodes through time. The implications of this fundamental shift resonate afar by impacting a set of issues ranging from the most immediately practical to the most broadly philosophical (including, in the latter category, an interesting consonance with the atomism and quantization invoked to define the general intellectual movement known as "modernism"—as expressed in disparate disciplines from Seurat's pointillism in art, to Schönberg's serial style in music; and as opposed to the smooth continuationism favored by earlier mechanistic views of causality). In a theme more immediately relevant to biology, the same shift ineluctably places much greater emphasis upon chance and contingency, rather than predictability by extrapolation—for the ordinary condition of stasis provides little insight into when and how the next punctuation will occur, whereas the fractal character of gradualism suggests that causes of change at any moment will, by extrapolation, predict and explain the larger effects accumulated through longer times.

On the practical side, punctuated equilibrium's formulation of tempo has validated the study of stasis—paleontology's prevalent pattern within species—as a source of insight about evolution, rather than a cause of chagrin best bypassed and ignored as a testimony to an embarrassing poverty of evidence. Punctuated equilibrium has broken "Cordelia's Dilemma" of silence about the supposed "nothing" of stasis, and has established a burgeoning subfield of research in the documentation of stability at several levels. In pursuing and valuing this documentation, scientists then feel compelled to postulate explanations for the puzzling frequency of this previously "invisible" phenomenon—and theoretical inquiry about the "why" of stasis has also flourished following the prod from punctuated equilibrium (see pp. 877–885 for fuller discussion).

### Mode and the speciational foundation of macroevolution

For mode, as discussed throughout this chapter, punctuated equilibrium has established a speciational basis for macroevolution. By supplying crucial data and arguments for defining species as effective Darwinian individuals—that is, as basic units for describing macroevolution in Darwinian terms as an outcome of patterns in differential birth and death of species treated as stable individuals, just as microevolution works by the same process applied to births and deaths of organisms—punctuated equilibrium validates the hierarchical theory of selection. This hierarchical theory (explicated in Chapter 8) establishes the independence of macroevolution as a theoretical subject (not just as a domain of description for accumulated microevolutionary mechanics), thereby precluding the full explanation of evolution by extrapolation of microevolutionary processes to all scales and times.

In practical terms, the implications of punctuated equilibrium for evolutionary mode have strongly impacted two prominent subjects, heretofore almost always rendered by extrapolation as consequences of adaptation within populations writ large: evolutionary trends within clades, and relative waxing and waning of diversity within supposedly competing clades through time. Punctuated equilibrium suggests novel, and irreducibly macroevolutionary, explanations for both phenomena (see pp. 885–916).

Finally, the role of punctuated equilibrium in establishing an independent field of macroevolution includes both a weak and a strong version. The first, undoubtedly valid as a generality, "uncouples" macro from microevolution as a descriptive necessity, while not establishing independent causal principles of macroevolution. The second clearly regulates many cases, but has not yet been validated as commanding a high relative frequency; this second, or strong, version establishes irreducible causal principles of macroevolution.

The weak version, based on "species sorting" rather than "species selection," holds that evolution must be described as differential success in birth and death of stable species, but allows that the causality behind reasons for differential success might emerge from the conventional Darwinian level of struggling organisms within successful populations—the effect hypothesis of Vrba (see p. 658). In this version, we need a descriptive, but not a causal, account of macroevolution based on species as individuals.

However, in the strong version, based on true species selection, the differential success of species arises from irreducible fitness defined by the interaction of species-individuals with their environments. Chapter 8 presents an extensive argument for the efficacy of true species selection at high relative frequency. Validation of this argument would establish a genuinely causal and irreducible theory of macroevolution. This difficult issue stands far from resolution, but represents the most exciting potential for punctuated equilibrium as an impetus in formulating a revised structure for evolutionary theory.

## The Scientific Debate on Punctuated Equilibrium: Critiques and Responses

### CRITIQUES BASED ON THE DEFINABILITY OF PALEONTOLOGICAL SPECIES

#### Empirical affirmation

The issue of whether true biospecies (or entities operationally close enough to biospecies) can be recognized in fossils has prompted long and intense debate in paleontology (see Sylvester-Bradley, 1956, and other references previously cited), and does not represent a new or special difficulty raised by punctuated equilibrium. But given the reliance of punctuated equilibrium on speciation as the mechanism behind the pattern, this old problem does legitimately assume a central place in debates about our theory (as emphasized in all negative commentary, particularly clearly by Turner, 1986, and in the book-length critiques of Levinton, 1988, and Hoffman, 1989).

At least we may begin by exposing the canonical issue of the older literature as a *Scheinproblem* (literally an "appearance problem" with no real content): the logical impossibility of defining a species boundary within a gradualistic continuum (see my previous discussion on p. 775). I think we may now accept that the punctuational pattern exists at high relative frequency, and that few gradualistic and anagenetic continua have been documented between fossil species. Turner's (1986) sharp critique, for example (and I do accept his formulation, though not his resolution), depicts the chief claims of punctuated equilibrium as a three-pronged fork. He accepts the first tine—the existence of the punctuational pattern itself—as sufficiently demonstrated by enough empirical cases in the fossil record. He regards the third tine—macroevolutionary invocation of the theory to explain trends by species sorting—as "an important extension of evolutionary theory into a hitherto little explored territory" (1986, p. 206). But he then rejects the second tine as both unlikely and too difficult to test in any case—explanation of the punctuational pattern as a consequence of speciation scaled into geological time.

If we accept that temporal sequences of fossils generally don't appear in the geological record as unbreakable continua, but usually as morphological "packages" with reasonably defined boundaries and sufficient stability within an extended duration, how can we assert that these packages represent biospecies, or at least that they approximate these neontologically defined units with sufficient closeness to bear comparison? After all, we cannot apply conventional tests of observed ecological interaction or interbreeding to fossils—and, whereas biospecies may be *recognized* by morphological differentia in everyday practice, they are not supposed to be so *defined*. Can the temporally extended "morphospecies" of paleontology really be equated with the "nondimensional species concept" (Mayr's words) of neontology?

I certainly accept the centrality and difficulty of these issues, but I do not regard them as insuperable, and I do not view the species concept as untestable with fossils. After all, the overwhelming majority of modern species in our literature and museum drawers have also been phenotypically, not ecologically, defined. Once we accept that no special paleontological riddles arise from the *Scheinproblem* of temporal continua, then most paleospecies have been no worse characterized than the majority of neospecies. Still, I will not advance this excuse as exculpatory for the fossil record, for a neontologist could reply, with impeccable logic, that neospecies so defined should also be regarded as uncertain, if not vacuous, and that no paleontological defense can be mounted by arguing that ordinary practice with fossils follows the worst habits (majoritarian though they may be) of neontological taxonomy.

But a best defense of phenotypically defined neospecies would follow from demonstrations that taxa so established usually do match true biospecies upon proper behavioral and ecological study—a line of research often pursued with success (see references in Jackson and Cheetham, 1994, and in Jablonski, 1999). Similarly, my main source for confidence about paleospecies arises from proven correspondences with true biospecies in favorable cases providing sufficient information for such a test (particularly for extant species with lengthy fossil records). I do not, of course, argue that all named paleospecies are true biospecies, or that I can even estimate the percentage properly so defined (any more than we know the relative frequency of modern taxa that represent true biospecies). But I do not see why the probability that well-defined paleospecies, based on good collections from many times and places, might represent proper biospecies should be any lower than the corresponding figure for equally well documented, but entirely morphologically defined, modern taxa. (In fact, one might argue that well-documented paleospecies probably maintain a higher probability for representing biospecies, because we know their phenotypes, and have measured their stability, across long periods of time and wide ranges of environment—whereas modern "morphospecies" may arise as ecophenotypic expressions of a single time and place, therefore ranking only as local populations, rather than true species.)

When well-defined paleospecies have been tested for their correspondence with modern biospecies, such status has often been persuasively affirmed. Two recent studies seem particularly convincing. Michaux (1989) studied four living species of the marine gastropod genus *Amalda* from New Zealand. Fossils of this genus date to the upper Eocene of this region, while all four species extend at least to the Miocene-Pliocene boundary. The four taxa represent good biospecies, based on absence of hybrids in sympatry, and on extensive electrophoretic study (Michaux, 1987) showing distinct separation among species and "no detectable cryptic groupings" (Michaux, 1989, p. 241) within any species. Michaux then used canonical discriminant analysis to achieve clear morphometric distinction among the species based on 10 shell measurements for each of 671 live specimens.

He then made the same measurements on 662 fossil specimens from three of the species (the fourth did not yield enough shells for adequate characterization). Mean values, in multivariate expression based on all 10 variables, fluctuated mildly through time (see Fig. 9-8), but never departed from the range of variation within extant populations—an excellent demonstration of stasis as dynamic maintenance within well-defined biospecies through several million years. Michaux concluded (1989, pp. 246–248): "Fossil members of three biologically distinct species fall within the range of variation that is exhibited by extant members of these species. The phenotypic trajectory of each species is shown to oscillate around the modern mean through the time period under consideration. This pattern demonstrates oscillatory change in phenotype within prescribed limits, that is, phenotypic stasis."

Jackson and Cheetham's (1990, 1994) extensive studies of cheilostome bryozoan species provide even more gratifying affirmation, especially since these "simple" sessile and colonial forms potentially express all the attributes of extensive ecophenotypic variation (especially in molding of colonies to substrates, and in effects of crowding) and morphological simplicity (lack of enough complex skeletal characters for good definition of taxa) generally regarded as rendering the identification of biospecies hazardous, if not effectively impossible, in fossils. Moreover, Cheetham had begun his paleontological studies (see discussion on pp. 867–870) under the assumption that careful work would reveal predominant gradualism and refute the "new" hypothesis of punctuated equilibrium—so the conclusions eventually reached were not favored by any *a priori* preference!

In a first study—devoted to determining whether biospecies could be recognized from skeletal characters (of the sort used to define fossil taxa) in several species within three genera of extant Caribbean cheilostomes—Jackson and Cheetham (1990) examined heritability for skeletal characters in seven species. In a "common garden" experiment (under effectively identical conditions at a single experimental site), they grew $F_1$ and $F_2$ generations from embryos derived from known maternal colonies collected in disparate environments and places. Multivariate discriminant analysis assigned all but 9 of 507 offspring into the same morphospecies as their maternal parent. The authors then used electrophoretic methods to study enzyme variation in 402 colonies representing 8 species in the three genera. They found clear and complete correspondence between genetic and morphometric clusterings, and also determined (p. 581) that "genetic distances between morphospecies are consistently much higher than between populations of the same morphospecies"; moreover, they found no evidence for any cryptic division (potential "sibling species") within skeletally defined morphospecies.

In a concluding and gratifying observation—indicating that paleontolo-

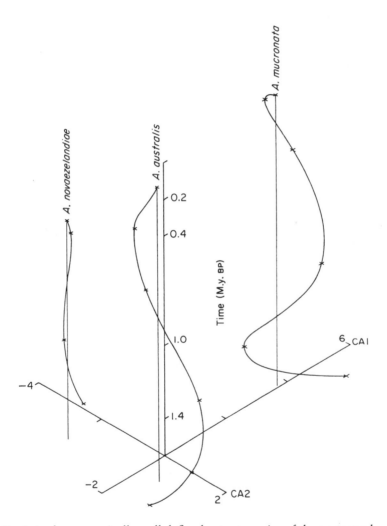

9-8. Stasis in three genetically well defined extant species of the gastropod *Amalda* from New Zealand, based on 662 fossil specimens. Mean values in multivariate expression based on all ten variables fluctuate mildly through time, but never depart from the range of variation within extant populations. From Michaud, 1989.

gists need not always humble themselves before the power of neontological genetic analysis of biospecies—Jackson and Cheetham (1990) p. 582) make an empirical observation about the capacity of morphometric data (of the sort generated from fossils):

> The identity of quantitatively defined morphospecies of cheilostome bryozoans is both heritable and unambiguously distinct genetically. The importance of rigorous quantitative analysis was underlined by our discovery of three species of *Stylopoma* previously classified as one, a separation subsequently confirmed genetically. The widely supposed lack of correspondence between morphospecies and biospecies may result as much out of uncritical acceptance of outdated, subjectively defined taxa as from any fundamental biologic differences between the two kinds of species.

Jackson and Cheetham (1994) then followed this study with a more extensive documentaiton, this time using large numbers of fossil species as well as living forms, of phylogenetic patterns in two Caribbean cheilostome genera, *Stylopoma* (included in the first study as well), and *Metrarabdotos* (the subject of Cheetham's earlier and elegant affirmations of punctuated equilibrium from morphometric data alone—Cheetham, 1986 and 1987, and extensively discussed on pp. 867–870). Again, and for both genera, they found strict correspondence between genetically defined clusters and taxa established by skeletal characters accessible from fossils.

With increased confidence that the taxa of his classical studies on punctuated equilibrium in *Metrarabdotos* represent true biospecies, Cheetham (now writing with Jackson) could affirm his earlier work (Jackson and Cheetham, 1994, p. 420): "Morphological stasis over millions of years punctuated by relatively sudden appearances of new morphospecies was demonstrated previously for *Metrarabdotos*. Our updated results strengthen confidence in that pattern, with 11 morphospecies persisting unchanged for 2–6 m.y., all at p > 0.99, and no evidence that intraspecific rates of morphological change can account for differences between species."

For *Stylopoma*, where fossil evidence had not previously been analyzed morphometrically, results also affirmed punctuated equilibrium throughout (1994, p. 420): "The excellent agreement between morphologically and genetically defined species used in this taxonomy suggests that morphological stasis reflects genuine species survival over millions of years, rather than a series of morphologically cryptic species. Morover, eleven of the 19 species originate fully formed at p > 0.9, with no evidence of morphologically intermediate forms, and all ancestral species but one survived unchanged all with their descendants."

In a concluding paragraph about both genera, Jackson and Cheetham wrote (p. 407): "Stratigraphically rooted trees suggest that most well-sampled *Metrarabdotos* and *Stylopoma* species originated fully differentiated morphologically and persisted unchanged for > 1 to > 16 m.y., typically alongside their putative ancestors. Moreover, the tight correlation between phenetic, cladistic, and genetic distances among living *Stylopoma* species suggests that changes in all three variables occurred together during speciation. All of these observations support the punctuated equilibrium model of speciation."

Despite the encouragement provided by these and other cases, problems continue to surround the definition of paleontological species—a subject of central importance to punctuated equilibrium, given our invocation of speciation as the quantum of change for life's macroevolutionary history, and the source of raw material for higher-level selection and sorting. These problems center upon three main issues (in both the inherent logic of the case and by recorded debate in the literature): the first untroubling, the second potentially serious, and the third largely resolved in empirical terms. All three issues raise the possibility that paleospecies systematically misrepresent the nature and number of actual biospecies. (If paleospecies don't correspond with biospecies in all cases—an undeniable proposition of course—but if these discrepancies show no pattern and produce no systematic bias, then we need not be troubled unless the relative frequency of noncorrespondence becomes overwhelmingly high, an unlikely situation given the excellent alignments found in the few studies explicitly done to investigate this problem, as discussed just above.) The following three subsections treat these three remaining issues *seriatim*.

### Reasons for a potential systematic underestimation of biospecies by paleospecies

Might we be missing a high percentage of actual speciation events because paleontologists can only recognize a cladogenetic branch with clear phenotypic consequences (for characters preserved as fossils), whereas many new species arise without substantial morphological divergence from their ancestors? In the clearest case, paleontologists (obviously) cannot detect sibling species, a common phenomenon in evolution (see Mayr, 1963, for the classic statement). Moreover, we may also miss subtle changes in phenotype, or substantial alterations (of color, for example) in features that are often important in recognizing species, but do not achieve expression in the fossil record.

Our harshest critics have urged this point as particularly telling against punctuated equilibrium. Levinton, for example (1988, p. 182), holds that "the vast majority of speciation events probably beget no significant change." He then views the consequences as effectively fatal for punctuated equilibrium (1988, p. 211): "The punctuated equilibrium model argues that morphological change is associated with speciation and that species are static during their history due to some internal stabilizing mechanism. There is no evidence coming from living species to support this. If anything, recent research has demonstrated that speciation occurs typically with little or no morphological change; hence the large-scale occurrence of sibling species."

Hoffman (1989, p. 115) invokes this argument to assert the untestability, hence the nonscientific status, of punctuated equilibrium:

> Long-term evolutionary stasis of species, however, simply cannot be tested in the fossil record. Paleontological data consist solely of a small sample of phenotypic traits—little more than morphology of the skeletal parts—which does not allow us to make any inference about changes in a species's genetic pool or even about changes of the frequency distribution of phenotypes in a phyletic lineage. The non-preserved portion of the phenotype of each fossil species is so extensive that it may always undergo considerable evolutionary changes that remain undetectable by the paleontologist. What appears then to the paleontologist as a species in complete evolutionary stasis may in fact represent a succession of fossil species or perhaps a whole cluster of species, a phylogenetic tree with a sizable number of branching points, or speciation events.

While I freely admit all these arguments for underrepresentation of true species in fossil data, I do not comprehend how punctuated equilibrium could be thus rendered untestable, or even seriously compromised (see further arguments in Gould, 1982c and 1989e; and Gould and Eldredge, 1993). I base my argument on two logical and methodological principles, not on the probable empirical record (where I largely agree with our critics).

THE PROPER STUDY OF MACROEVOLUTION. By consensus, and accepting a criterion of testability, science does not include, within its compass of inquiry, fascinating questions that cannot be answered (even if they address potentially empirical subjects). For example, and for the moment at least, we know no way to ask a scientific quesiton about what happened before the big bang, for compression of universal matter to a single point of origin wipes out all traces of any previous history. (Perhaps we will eventually devise a way to obtain such data, or perhaps the big bang theory will be discarded. The question might then become scientifically tractable.) Similarly, we know that many kinds of evolutionary events leave no empirical record—and that we therefore cannot formulate scientific questions about them. (For example, I doubt that we will be able to resolve the origins of human language, unless written expression occurred far earlier than current belief and evidence now indicate.)

The nature of the fossil record leads us to define macroevolution as the study of phenotypic change (and any inferable correlates or sequelae) in lineages and clades throughout geological time. Punctuated equilibrium proposes that such changes generally occur in discrete units or quanta in geological time, and that these quanta represent events of branching speciation. Thus, we do identify speciation as the source of raw material for macroevolutionary change in lineages. But we do not, and cannot, argue (or attempt to adjudicate at all) the quite different proposition that all speciation events produce measurable quanta of macroevolutionary change. The statement—our proposition—that nearly all macroevolutionary change occurs in increments of speciation carries no implications for the unrelated claim, often imputed to punctuated equilibrium by our critics (but largely irrelevant to our theory), that nearly all events of speciation produce an increment of macroevolutionary change. This conclusion flows from elementary logic, not from empirical science. The argument that all B comes from A does not imply that all A leads to B. All human births (at least before modern interventions of medical technology) derived from acts of sexual intercourse, but all acts of intercourse don't lead to births.

To draw a more relevant analogy: in the strict version of Mayr's peripatric theory of speciation, nearly all new species arise from small populations isolated at the periphery of the parental range. But the vast majority of peripheral isolates never form new species; for they either die out or reamalgamate with the parental population. Similarly, most new species may never be recorded in the fossil record; but, if the theory of punctuated equilibrium holds, when changes do appear in lineages of fossils, speciation provides the source of input in a great majority of cases. Thus, most speciation could be cryptic (and unknowable from fossil evidence), while effectively all macroevolutionary change still arises from the minority of speciation events with phenotypic consequences. Just as peripheral isolates might represent "the only game in town" for forming new species (though few isolates ever speciate), cladogenetic speciation may be "the only game in town" for inputting phenotypic change into macroevolution (though few new species exhibit such change).

THE TREATMENT OF INELUCTABLE NATURAL BIAS IN SCIENCE. In an ideal world—the one we try to construct in controlled laboratory experiments—no systematic bias distorts the relative frequency of potential results. But the real world of nature meets us on her own terms, and we must accept any distortions of actual frequencies that directional biases of recording or preservation inflict upon the archives of our evidence. At best, we may be able to correct such biases if we can make a quantitative estimate of their strength. (This general procedure, for example, has been widely followed to correct the systematic undermeasurement of geological ranges imposed by the evident fact that observed first and last occurrences of a fossil species can only provide a minimal estimate for actual origins and extinctions, for the observed geological range of a species must be shorter (and at least cannot be longer) than the actual duration. Studies of "waiting times" between sequential samples within the observed range, combined with mathematical models for constructing error bars around first and last occurrences, have been widely used to treat this important problem—see Sadler, 1981; Schindel, 1982; Marshall, 1994.)

Often, however, we can specify the direction of a bias, but do not know how to make a quantitative correction. In such cases, the sciences of natural history must follow a cardinal rule: if the direction of bias coincides with the predicted effect of the theory under test, then researchers face a serious, perhaps insurmountable, problem; but if a systematic bias works against a theory, then researchers encounter an acceptable impediment—for if the theory can still be affirmed in the face of unmeasurable biases working *against* a favored explanation, then the case for the theory gains strength.

For proponents of punctuated equilibrium, speciation represents the primary source for morphological changes that, by summation of increments, build trends in the history of lineages. If a systematic bias in the nature of paleontological evidence leads us to underestimate the number of speciation events, and if we can still explain trends by this observed number (necessarily less than the actual frequency), then the case for punctuated equilibrium becomes stronger by affirmation in the face of a bias working against full expression of the theory's effect. Thus, although we regret the existence of any bias that we cannot correct, a systematic underrepresentation of speciation events does not subvert punctuated equilibrium because such a natural skewing of evidence makes the hypothesis even more difficult to affirm—and support for punctuated equilibrium therefore emerges in a context even more challenging than the unbiassed world of controlled experimentation.

Moreover, one might even stress the bright side and recognize that such biases may exist for interesting reasons in themselves—reasons that might even enhance the importance of punctuated equilibrium and its implications. I doubt that Levinton (1988, p. 379) intended the following passage in such a positive light, but I would suggest such a reading: "One cannot rule out the possibility that speciation is rampant, but morphological evolution only occurs occasionally when a population is forced into a marginal environment and subjected to rapid directional selection. What then becomes interesting is why the character complexes evolved in the daughter species remain constant. This is, again, the issue of stasis, which I believe to be the legitimate problem spawned by the punctuated equilibrium model."

Finally, I am not sure that fossil species do strongly and generally underestimate the frequency of true biospeciation—although I do accept that a bias, if present at all, probably operates in this direction. The most rigorous empirical studies on correspondence between well-defined paleospecies and true biospecies—the works of Michaux and of Jackson and Cheetham discussed above—affirm a one-to-one link between paleontological morphospecies and extant, genetically-defined biospecies.

### Reasons for a potential systematic overestimation of biospecies by paleospecies

If a bias did exist in this opposite direction, the consequences for punctuated equilibrium would be troubling (as implied in the previous section on acceptable and unacceptable forms of unavoidable natural biasing). For if we systematically name too many species by paleontological criteria, then we might be affirming punctuated equilibrium by skewing data in the direction of our favored theory, rather than by genuine evidence from the fossil record. However, I doubt that such a problem exists for punctuated equilibrium, especially since all experts—both strong advocates and fierce critics alike (as the preceding discussion documented)—seem to agree that if any systematic bias exists, the probable direction lies in the acceptable opposite claim for underestimation of biospecies by paleospecies.

I don't doubt, of course, that past taxonomic practice, often favoring the erection of a species name for every recognizable morphological variant (even for odd individuals rather than populations), has greatly inflated the roster of legitimate names in many cases, particularly for fossil groups last monographed several generations ago. (Our literature even recognizes the half-facetious term "monographic burst" for peaks of diversity thus artificially created. But this problem of past oversplitting cannot be construed as either uniquely or even especially paleontological, for neontological systematics then followed the same practices as well.) The grossly uneven, and often greatly oversplit, construction of species-level taxonomy in paleontology has acted as a strong impediment for the entire research program of the prominent school of "taxon-counting" (Raup, 1975, 1985). For this reason, the genus has traditionally been regarded as the lowest unit of rough comparability in paleontological data (see Newell, 1949). Sepkoski (1982) therefore compiled his two great compendia—the basis for so much research in the history of life's fluctuating diversity—at the family, and then at the genus, level (but explicitly not at the species level in recognition of frequent oversplitting and extreme imbalance in practice of research among specialists on various groups).

Although this problem has proved far more serious for taxon-counters than for proponents of punctuated equilibrium, a potential bias towards overrepresentation also poses a threat for our theory, as Levinton (1988, p. 364) rightly recognizes: "The problem is not very new. Meyer (1878) claimed that the ability to recognize gradual evolutionary change in *Micraster* [a famous sequence of Cretaceous echinoids] was obscured by the rampant naming of separate species by previous taxonomists."

This issue would cause me serious concern—for the claim of overestimation does, after all, fall into the worrisome category of biases favoring a preferred hypothesis under test—if two arguments and realities did not obviate the danger. First, if supporters of punctuated equilibrium did try to affirm their hypothesis by using names recorded in the literature as primary data for judging the strength and effect of speciation upon evolutionary trends, then we would face a serious difficulty. But I cannot think of any study that utilized this invalid approach—for paleontologists recognize and generally avoid the dangers of this well-known directional bias. Punctuated equilibrium, to my knowledge, has never been defended by taxon counting at the species level. All confirmatory studies employ measured morphometric patterns, not the geological ranges of names recorded in literature.

Second, as stated above, all students of this subject seem to agree that if a systematic bias exists in relative numbers of paleospecies and biospecies, fossil data should be skewed in the opposite direction of recognizing *fewer* paleospecies than biospecies—an acceptable bias operating against the confirmation of punctuated equilibrium.

### Reasons why an observed punctuational pattern might not represent speciation

Suppose that we have empirical evidence for a punctuational event separating two distinct morphological packages regarded as both different enough to be designated as separate paleospecies by any standard criterion, and also genealogically close enough to support a hypothesis of direct ancestry and descent. What more do we need? Does this situation not affirm punctuated equilibrium *ipso facto?*

But critics charge (and I must agree) that such evidence cannot be persuasive by itself, because punctuated equilibrium explicitly links punctuational patterns to events of branching speciation. Therefore, recorded punctuations produced for other reasons do not affirm punctuated equilibrium—and may even challenge the theory if their frequency be high and, especially, if they cannot be distinguished in principle (or frequently enough in practice) from events of cladogenetic branching.

Punctuational patterns often originate (at all scales in evolutionary hierarchies of levels and times) for reasons other than geologically instantaneous speciation—and I welcome such evidence as an affirmation of pervasive importance (see p. 922 *et seq.*) for a general style of nongradualistic change, with punctuated equilibrium as its usual mode of expression at the speciational scale under consideration in this chapter. But testable, and generally applicable, criteria have been formulated for distinguishing punctuated equilibrium from other reasons for punctuational patterns—and available evidence amply confirms the importance and high relative frequency of punctuated equilibrium.

Of the two major reasons for punctuational patterns not due to speciation, Darwin's own classic argument of imperfection—geological gradualism that appears punctuational because most steps of a continuum have not been preserved in the fossil record—retains pride of place by venerable ancestry. I have already presented my reasons for regarding this argument as inconsequential (see pp. 765–774). I do not, of course, deny that many (or most) breaks in geological sequences only reflect missing evidence. But proponents of punctuated equilibrium do not base their claims on such inadequate examples that cannot be decided in either direction. The test cases of our best literature—whether their outcomes be punctuational or gradualistic—have been generated from stratigraphic situations where temporal resolution and density of sampling can make appropriate distinctions by recorded evidence, not conjectures about missing data.

The second reason has been highlighted by some critics, but unfairly I think, because punctuated equilibrium has always recognized the argument and has, moreover, enunciated and explicitly tested proper criteria for making the necessary distinctions. To state the supposed problem: what can we conclude when we document a truly punctuational sequence that cannot be attributed to imperfections of the fossil record? How do we know that such a pattern records an event of branching speciation, as the theory of punctuated equilibrium requires? When ancestral Species A abruptly yields to descendant Species B in a vertical sequence of strata, we may only be witnessing an anagenetic transformation through a population bottleneck, or perhaps an event of migration, where Species B, having evolved gradualistically from Species A in another region, invades the geographic range, and abruptly wipes out its ancestor.

But an appropriate and non-arbitrary criterion exists—and has been fully enunciated, featured as crucial, and subjected to frequent test, from the early days of punctuated equilibrium. We can distinguish the punctuations of rapid anagenesis from those of branching speciation by invoking the eminently testable criterion of ancestral survival following the origin of a descendant species. If the ancestor survives, then the new species has arisen by branching. If the ancestor does not survive, then we must count the case either as indecisive, or as good evidence for rapid anagenesis—but, in any instance, certainly not as evidence for punctuated equilibrium.

Moreover, by using this criterion, we obey the methodological requirement that existing biases must work against a theory under test. When ancestors do not survive following the first appearance of descendants, the pattern may still be recording an event of branching speciation—hence affirmation for punctuated equilibrium. But we cannot count such cases in our favor, for the plausible alternative of rapid anagenesis cannot be disproven. By restricting affirmations to cases where ancestors demonstrably survive, we accept only a subset of events actually caused by speciation. Thus, we underestimate the frequency of punctuated equilibrium—as we must do in the face of an unresolvable bias affecting a hypothesis under test.

In our first papers, we did not recognize or articulate the importance of tabulating cases of ancestral survival following punctuational origin of a descendant as a criterion for distinguishing punctuated equilibrium from other forms of punctuational change. (Both of our original examples in Eldredge and Gould, 1972, did feature—and prominently discuss—ancestral survival as an important aspect of the total pattern. We had a proper "gut feeling" about best cases, but we did not formalize the criterion.) But, beginning in 1982, and continuing thereafter, we have stressed the centrality of this criterion in claims for speciation as the mechanism of punctuated equilibrium. Contrasting the difference in paleontological expression between Wright's shifting balance and punctuated equilibrium by speciation, for example, I wrote (Gould, 1982c, p. 100): "Since punctuational events can occur in the phyletic mode under shifting balance, but by branching speciation under punctuated equilibrium, the persistence of ancestors following the abrupt appearance of a descendant is the surest sign of punctuated equilibrium."

This criterion has been actively applied, in an increasingly routine manner (as researchers recognize its importance), in the expanding literature on empirical study of evolutionary tempos and modes in well-documented fossil sequences. Cases of probable anagenetic transformation have been documented (no ancestral survival when good stratigraphic resolution should have recorded such persistence, had it occurred), especially in planktic marine Foraminifera, where long oceanic cores often provide unusually complete evidence (Banner and Lowry, 1985; Malmgren and Kennett, 1981, who coined the appropriate term "punctuated anagenesis" for this phenomenon).

However, abundant cases of ancestral survival, and consequent punctuational origin of descendant taxa by branching speciation, have also been affirmed as illustrations of punctuated equilibrium. These examples span the gamut of taxonomies and ecologies, ranging from marine microfossils (Cronin, 1985, on ostracodes); to "standard" macroscopic marine invertebrates (with Cheetham's famous studies of bryozoans, 1986 and 1987, as classic and multiply documented examples), to freshwater invertebrates (Williamson's 1981 work on multiple events of speciation in African lake mollusks, where ancestral species reinvade upon coalescence of lakes following periods of isolation that provided conditions for speciation); to terrestrial vertebrates (Flynn, 1986, on rodents; Prothero and Shubin, 1989, on horses). I shall discuss this important issue in more detail within the forthcoming section on evidence for punctuated equilibrium (see pp. 822–874), but I have been particularly (if parochially) gratified by the increasing application of punctuated equilibrium to the resolution of hominid phylogeny. The criterion of ancestral survival has been prominently featured in this literature, as by McHenry (1994), who notes that "ancestral species overlap in time with descendants in most cases in hominid evolution, which is not what would be expected from gradual transformations by anagenesis."

In any case, punctuated equilibrium can be adequately and generally recognized by firm evidence linking observed punctuational patterns to branching speciation as a cause. The theory of punctuated equilibrium is eminently testable and has, indeed, passed such trials in cases now so numerous that a high relative frequency for this important evolutionary phenomenon can no longer be denied (see Gould and Eldredge, 1993).

### CRITIQUES BASED ON DENYING EVENTS OF SPECIATION AS THE PRIMARY LOCUS OF CHANGE

Once we overcome the problem of definability for species in the fossil record, punctuated equilibrium still faces a major issue rooted in the crucial subject of speciation. Punctuated equilibrium affirms, as a primary statement, that ordinary biological speciation, when properly scaled into geological time, produces the characteristic punctuational pattern of our fossil record. We must therefore be able to defend the central implication that morphological change should be preferentially associated with events of branching speciation. Our critics have strongly argued that such a proposition cannot be justified by our best understanding of evolutionary processes and mechanisms.

I believe that our critics have been correct in this argument, and that Eldredge and I made a major error by advocating, in the original formulation of our theory, a direct acceleration of evolutionary rate by the processes of speciation. This claim, I now think, represents one of the two most important errors that we committed in advocating punctuated equilibrium during the past 25 years. (The other error, as discussed and corrected on pages 670–673, lay first in our failure to recognize the phenomenon of species selection as distinct (by hierarchical reasoning) from classical Darwinian organismic selection, and then (see Gould and Eldredge, 1977) in our decision to advocate an overly broad and purely descriptive definition rather than a properly limited meaning based on emergent characters or fitnesses—see pages 656–670.)

We did not urge this correlation between speciation events and morphological change in a self-serving and circular manner—*i.e.*, only because the pattern of punctuated equilibrium could be best defended thereby. We did, of course, recognize the logical link, as in the following statement from Gould, 1982c, p. 87 (see also Gould and Eldredge, 1977, p. 137): "Reproductive isolation and the morphological gaps that define species for paleontologists are not equivalent. Punctuated equilibrium requires either that most morphological change arise in coincidence with speciation itself, or that the morphological adaptations made possible by reproductive isolation arise rapidly thereafter." But we based our defense of this proposition upon a large, and then quite standard, literature advocating a strong negative correlation between capacity for rapid evolutionary change and population size. Small populations, under these models, maintained maximal prospects for rapid transformation based on several factors, including potentially rapid fixation of favorable variants, and enhancement of differences from ancestral populations by interaction of intense selection with stochastic reasons for change (particularly the founder effect) that can only occur with such effective speed in small populations. Large and stable populations, by the converse of these arguments, should be sluggish and resistant to change.

This literature culminated in Mayr's spirited defense for "genetic revolution" as a common component of speciation (first proposed in a famous 1954 article, and then defended *in extenso* in the 1963 book that served as the closest analog to a "bible" for graduate students of my generation). Since Mayr (who coined the name "founder effect" in this context) also linked his concept of "genetic revolution" to the small, peripherally isolated populations that served as "incipient species" in his influential theory of peripatric speciation—and since we had invoked this theory in our original formulation of punctuated equilibrium (Eldredge and Gould, 1972)—our defense of a link between speciation and concentrated episodes of genetic (and phenotypic) change flowed logically from the evolutionary views we had embraced. Thus, we correlated punctuations with the extensive changes that often occurred during events of speciation in small, peripherally isolated populations; and we linked stasis with the expected stability of large and successful populations following their more volatile and punctuational origins as small isolates.

I can claim no expertise in this aspect of neontological evolutionary theory, but I certainly acknowledge, and must therefore provisionally accept, the revised consensus of the past twenty years that has challenged this body of thought, and rejected any general rationale for equating the bulk of evolutionary change with events of speciation in small populations, or with small populations in any sense. As I read the current literature, most evolutionists now view large populations as equally prone to evolutionary transformation, and also find no reason to equate times of speciation—the attainment of reproductive isolation—with acceleration in general rates of genetic or phenotypic change (see, for example, Ridley, 1993; and Williams, 1992). (I do, however, continue to wonder whether the Mayrian viewpoint might still hold some validity, and might now be subject to overly curt and confident dismissal.)

This situation creates a paradox for our theory. The pattern of punctuated equilibrium has been well documented and shown to predominate in many situations (see pp. 822–874), but its most obvious theoretical rationale has now fallen under strong skepticism. So either punctuated equilibrium is wrong—a proposition that this partisan views as unlikely (although obviously possible), especially in the face of such strong documentation—or we must identify another reason for the prominence of punctuated equilibrium as a pattern in the history of life. In our article on the "majority" (21st birthday!) of punctuated equilibrium, Eldredge and I expressed this dilemma in the following manner (Gould and Eldredge, 1993, p. 226): "The pattern of punctuated equilibrium exists (at predominant relative frequency, we would argue) and is robust. *Eppur non si muove;* but why then? For the association of morphological change with speciation remains as a major pattern in the fossil record." (Our Italian parody, missed by many readers of the original article, alters Galileo's famous, but almost surely legendary, rebuke to the Inquisition, delivered secretly and *sotto voce* after he had been forced to recant his Copernican views in public: *Eppur si muove*—nevertheless it does move. Our parody says "nevertheless it does *not* move"—a reference to the overwhelming evidence for predominant stasis in the history of species, even if our original evolutionary rationale, based on population size, must be reassessed.)

This paradox permits several approaches, including the following two that I would not favor. One might simply argue that the pattern of punctuated equilibrium demonstrably exists, so the task falls to evolutionary theorists to find a proper explanation. The current absence of a satisfactory account does not threaten the empirical record, but rather directs inquiry by posing a problem. Or one might doubt that any single explanation can render the phenomenon, and suspect that many rationales will yield the observed pattern (including Mayrian genetic revolutions, even if we now regard their relative frequency as low). Thus, we need to identify a set of enabling criteria from evolutionary theory, and then argue that their combination may render the observed phenomena of the fossil record.

Most researchers would regard a third approach as preferable in science: an alternate general explanation of different form from the previous, but now rejected, leading candidate. I believe that such a resolution has been provided by Douglas Futuyma (1986, 1988a and b, but especially 1987),* although his

---

*Futuyma remains quite skeptical of punctuated equilibrium in general, and I would place him more among our critics than our supporters. But he does accept the empirical pattern, and he is an expert on speciation. Thus, when he developed an original way to resolve the paradox of why punctuations might correlate with events of speciation, even if processes of speciation don't accelerate the rate of evolution, he published his ideas as a constructive contribution to the general debate. Even though Futuyma disagrees with our claims for the general importance of punctuated equilibrium (while he, obviously, does not deny the phenomenon), he has granted us serious attention and has acknowledged the intellectual interest of the debate we provoked—and no one could ask for more from a good critic. Futuyma wrote (1988, p. 225), in stressing the need to integrate "synchronic" ap-

simple, yet profound, argument has not infused the consciousness of evolutionists because the implied and required hierarchical style of thinking remains so unfamiliar and elusive to most of us. (In fact, and with some shame, I am chagrined that I never recognized this evident and elegant resolution myself. After all, I am supposedly steeped in this alternative hierarchical mode of thinking—and I certainly have a strong stake in the problems of punctuated equilibrium.)

In short, Futuyma argues that we have been running on the wrong track, and thinking at the wrong level, in trying to locate the reason for a correlation between paleontological punctuations and events of speciation in a direct mechanism of accelerated change promoted by the process of speciation itself. Yet Futuyma does agree that a strong correlation exists (and has been demonstrated, in large part by research and literature generated by debate about punctuated equilibrium). Since we all understand (but do not always put into practice!) the important logical principle that correlation does not imply causality (the *post hoc* fallacy), an acknowledgement of the genuine link doesn't commit us to any particular causal scheme—especially, in this case, to the apparently false claim that mechanisms of speciation inherently enhance evolutionary rates.

Futuyma begins by arguing that morphological change may accumulate anywhere along the temporal trajectory of a species, and not exclusively (or even preferentially) during the geological moment of its origin. What then could produce such a strong correlation between events of branching speciation and morphological change from an ancestral phenotype to the subsequent stasis of an altered descendant? Futuyma—and I am somewhat rephrasing and extending his argument here—draws an insightful and original analogy between macroevolution and the conventional Darwinism of natural selection in populations.

The operation of natural selection requires that Darwinian individuals interact with environments in such a manner that distinct features of these individuals bias their reproductive success relative to others in the population. As a defining criterion of Darwinian individuality, entities that interact with the environment must show "sufficient stability" (see discussion on pp. 611–613)—defined in terms of the theory and mechanism under discussion as enough coherence to perform as an interactor in the process of natural selection.

Darwin recognized that organisms operate as fundamental interactors for

---

proaches as pursued by neontologists interested in evolutionary mechanisms with the "historical" themes favored by systematists and paleontologists—all (to borrow a line from elsewhere) "in order to form a more perfect union."

> We need to identify and to define rigorously questions to which both synchronic and historical evolution can make truly indispensable contributions. Some such questions have already been posed, so we now find systematists and population geneticists converging on the analysis of macromolecular sequences, geneticists publishing in *Paleobiology* (thanks to the healthy stimulus of punctuated equilibrium), systematists and students of adaptation finding a *rapprochement* in the use of phylogenetic information to test hypotheses of behavioral, physiological, and other adaptations.

microevolution within populations. (Gene selectionists make a crucial error in arguing that sexual organisms are not stable enough to be regarded as units of selection because they must disaggregate in forming the next generation. But units of selection are interactors, and the "sufficient stability" required by the theory only demands persistence through one episode (generational at this level) of selective interaction to bias reproductive success—as organisms do in the classical Darwinian "struggle for existence," see full discussion on pages 619–625.) Organisms achieve this stability through ordinary mechanisms of bodily coherence (a protective skin, functional integration of parts, a regulated developmental program, etc.).

What, then, produces a corresponding stability for units of macroevolution? Species-individuals are constructed as complex units, composed of numerous local populations, each potentially separate (at any moment) due to limited gene flow, and each capable of adaptation to unique and immediate environments. Thus, in principle, substantial evolution can occur in any local population at any time during the geological trajectory of a species. A large and developing literature, much beloved by popular sources (media and textbooks) for illustrating the efficacy of evolution in the flesh of immediacy (that is, within a time frame viscerally understood by human beings), has documented these rapid and adaptive changes in isolated local populations—substantial evolution of body size in guppies (Reznick et al., 1997), or of leg length in anolid lizards (Losos et al., 1997), for example (see Gould, 1997f).

But these changes in local populations cannot gain any sustained macroevolutionary expression unless they become "locked up" in a Darwinian individual with sufficient stability to act as a unit of selection in geological time. Local populations—as a primary feature of their definition—do not maintain such coherence. They can in principle—and do, in the fullness of geological time, almost invariably in practice—interbreed with other local populations of their species. The distinctively evolved adaptations of local populations must therefore be ephemeral in geological terms, unless these features can be stabilized by individuation—that is, by protection against amalgamation with other Darwinian individuals. Speciation—as the core of its macroevolutionary meaning—provides such individuation by "locking up" evolved changes in reproductively isolated populations that can, thereafter, no longer amalgamate with others. The Darwinian individuation of organisms occurs by bodily coherence for structural and functional reasons. The Darwinian individuation of species occurs by reproductive coherence among parts (organisms), and by prevention of intermingling between these parts and the parts of other macroevolutionary individuals (that is, organisms of other species).

Rapid evolution in local population of guppies and anoles illustrates a fascinating phenomenon that teaches us many important lessons about the general process of evolution. But such changes can only be ephemeral unless they then become stabilized in coherent higher-level Darwinian individuals with sufficient stability to participate in macroevolutionary selection. These local populations usually strut and fret their short hour on the geological stage, and then disappear by death or amalgamation. They produce the ubiquitous and geologically momentary fluctuations that characterize and embellish the long-term stasis of species. They are, to use Mandelbrot's famous metaphor for fractals, the squiggles and jiggles on the coastline of Maine depicted at a scale that measures the distance around every boulder on every beach along the shore, and not at the resolution properly enjoined when the entire state appears on a single page in an atlas. Macroevolution represents the page of the atlas. The distance around each boulder (marking substantial but ephemeral changes in local populations of guppies and lizards)—however important in the immediacy of an ecological moment—becomes invisible and irrelevant (as the transient fluctuations of stasis) in the domain of sustained macroevolutionary change (Fig. 9-9).

In other words, morphological change correlates so strongly with speciation not because cladogenesis accelerates evolutionary rates, but rather because such changes, which can occur at any time in the life of a local population, cannot be retained (and sufficiently stabilized to participate in selection) without the protection provided by individuation—and speciation, via reproductive isolation, represents nature's preeminent mechanism for generating macroevolutionary individuals. Speciation does not necessarily promote evolutionary change; rather, speciation "gathers in" and guards evolutionary change by locking and stabilization for sufficient geological time within a Darwinian individual of the appropriate scale. If a change in a local population does not gain such protection, it becomes—to borrow Dawkins's metaphor at a macroevolutionary scale—a transient duststorm in the desert of time, a passing cloud without borders, integrity, or even the capacity to act as a unit of selection, in the panorama of life's phylogeny.

To cite Futuyma's summary of his powerful idea (1987, p. 465): "I propose that because the spatial locations of habitats shift in time, extinction of and interbreeding among local populations makes much of the geographic differentiation of populations ephemeral, whereas reproductive isolation confers sufficient permanence on morphological changes for them to be discerned in

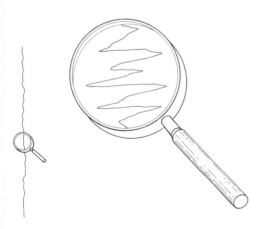

9-9. Stasis does not imply absolute stability, but rather directionless fluctuation that generally does not stray beyond the boundaries of geographic variation within similar species and, particularly, does not trend in any given direction, especially towards the modal morphology of descendant forms. This figure shows that, when a small segment in geological stasis becomes magnified so that change may be visualized on a generational scale, the natural fluctuations within local populations become more visible—but still do not, at the proper geological focus, exceed the bounds of stasis within the species.

the fossil record." Futuyma directly follows this statement with the key implication of punctuated equilibrium for the explanation of evolutionary trends: "Long-term anagenetic change in some characters is then the consequence of a succession of speciation events."

Later in his article, Futuyma (p. 467) explicitly links speciation with sufficient stability (individuation) for macroevolutionary expression: "In the absence of reproductive isolation, differentiation is broken down by recombination. Given reproductive isolation, however, a species can retain its distinctive complex of characters as its spatial distribution changes along with that of its habitat or niche . . . Although speciation does not accelerate evolution within populations, it provides morphological changes with enough permanence to be registered in the fossil record. Thus, it is plausible to expect many evolutionary changes in the fossil record to be associated with speciation." And, at the end of his article, Futuyma (p. 470) notes the crucial link between punctuated equilibrium and the possibility of sustained evolutionary trends: "Each step has had a more than ephemeral existence only because reproductive isolation prevented the slippage consequent on interbreeding with other populations . . . Speciation may facilitate anagenesis by retaining, stepwise, the advances made in any one direction . . . Successive speciation events are the pitons affixed to the slopes of an adaptive peak."

I hope that Futuyma's simple yet profound insight may help to heal the remaining rifts, thereby promoting the integration of punctuated equilibrium into an evolutionary theory hierarchically enriched in its light.

### CRITIQUES BASED UPON SUPPOSED FAILURES OF EMPIRICAL RESULTS TO AFFIRM PREDICTIONS OF PUNCTUATED EQUILIBRIUM

I shall treat the specifics of this topic primarily in the next section on "the data of punctuated equilibrium." But the logic of this chapter's development also requires that I state the major arguments and my responses in this account of principal critiques directed at the theory—for the totality of attempted rebuttals has not only posited theoretical objections in an effort to undermine the theory's logic or testability (as discussed in the first two parts of this section), but has also proceeded by accepting the theory's program of research as valid, and then arguing that the bulk of data thus accumulated refutes punctuated equilibrium empirically. I shall summarize discussion on the two major strategies pursued under this rubric: refutation by accumulation of important cases, and rejection by failure of actual data to fit models for predicted phylogenetic patterns.

#### *Claims for empirical refutation by cases*
PHENOTYPES. Despite some early misunderstandings, long since resolved by all parties to the discussion, we recognize that no individual case for or against punctuated equilibrium, however elegantly documented, can serve as a "crucial experiment" for questions in natural history that must be decided

by relative frequencies. No exquisite case of punctuated equilibrium—and many have been documented—can "prove" our theory; while no beautiful example of gradualism—and such have been discovered as well—can refute us. The key question has never been "whether," but rather "how often," "with what range of variation in what circumstances of time, taxon, and environment," and especially, "to what degree of control over patterns in phylogeny?" A single good case can only validate the reality of the phenomenon—and the simple claim for existence has not, surely, been an issue for more than 20 years. Similarly, an opposite case of gradualism can only prove that punctuated equilibrium lacks universal validity, and neither we nor anyone else ever made such a foolish and vainglorious claim in the first place. The empirical debate about punctuated equilibrium has always, and properly, focussed upon issues of relative frequency.

I shall present the empirical arguments for asserting dominant relative frequency, rather than mere occurrence, for punctuated equilibrium on pages 854–874. If we ask, by contrast, whether strong evidence for predominant gradualism has been asserted for any major taxon, time or environment, one case stands out as a potentially general refutation of punctuated equilibrium in one important domain at least: the claim for anagenetic gradualism as a primary phylogenetic pattern in the evolution of Cenozoic planktonic Foraminifera.

This case gains potential power and generality from the unusually favorable stratigraphic context, and the consequent nature of sampling, in such studies. The data come from deep oceanic cores, with stratigraphic records presumably unmatched in general completeness, for these environments receive a continuous supply of sediment (including foraminiferal tests) from the water column above. Moreover, these microscopic organisms can usually be extracted in large and closely spaced samples (sieved from disaggregated sediments), even from the restricted volume of a single oceanic core. Thus, forams in oceanic cores should provide our most consistently satisfactory information—in terms of large samples with good stratigraphic resolution—for the study of phylogenetic pattern. If gradualistic anagenesis prevails in such situations of maximal information—even if punctuated equilibrium predominates in the conventional fossil record of marine invertebrates from shallow water sediments—shouldn't we then conclude that Darwin's old argument must be valid after all; that punctuational patterns represent an artifact of missing data; and that more complete information will affirm genuine gradualism as the characteristic signal of phylogeny?

I acknowledge the highest relative frequency of recorded gradualism for foraminiferal data of this type, and I also admire the procedural rigor and informational richness in several of these studies. But I do not regard this case as a general argument against punctuated equilibrium—and neither, I think, do most of my paleontological colleagues, whatever their overall opinion about our theory, for the following reasons based upon well-known features of the fossil record in general, and the biology of forams in particular.

1. As emphasized in my previous discussion of publication bias (see p. 763), I remain unconvinced that a predominant relative frequency for gradualism—as opposed to genuine documentation of several convincing cases—has been established, even for this maximally promising taxon. No one has ever compiled an adequately random, or even an adequately numerous, sample of planktonic species drawn from the entire clade. Gradualistic lineages have been highlighted for study as a consequence of their greater "interest" under conventional views, while putatively stable lineages have tended to remain in unexamined limbo as supposedly uninformative, or even dull. Thus, the fact that gradualism prevails in a high percentage of published studies tells us little about the relative frequency of gradualism in the clade as a whole.

A telling analogy may be drawn with a crucial episode in the history of genetics. With classical techniques based on the Mendelian analysis of pedigrees, only variable genes could be identified. (If every *Drosophila* individual had red eyes, earlier researchers could legitimately assume some genetic bases for the invariance, but no genes could be specified because traits could not be traced through pedigrees. But once a white-eyed mutant fly appeared in the population, geneticists gained a necessary tool for identifying relevant genes by crossbreeding the two forms and tracing the alternate phenotypes through successive generations. In other words, genes had to vary before they could be specified at all.)

Therefore, under these methodological constraints (which prevailed during most of the 20th century history of genetics), a dominant *measured* frequency for variable genes taught us nothing about the *actual* frequency of variable genes across an entire genome—for we knew no way to generate a random or unbiased sample by selecting genes for study *prior* to any knowledge about whether or not they varied. The fact of variation in all *known* genes only recorded a methodological limitation that precluded the identification of nonvariable genes.

I don't, of course, claim that methodological strictures on paleontological lineages have ever been so strong—that is, we could always have selected stable lineages for study, had we chosen to do so. But, in practice, I'm not sure that the actual procedural bias has operated with much less force in paleontology than in genetics, so long as researchers confined their attention to lineages that appeared (by initial qualitative impression) to evolve by gradual anagenesis. Just as all known genes might be variable (while variable genes actually represent only a few percent of the total complement, because the remaining 95 percent of invariant genes could not be recognized at all), most studied species might illustrate gradual trends (while gradualistic species represent a small minority of all lineages because no one chooses to study stable species).

Genetics resolved this problem by inventing techniques—with electrophoresis as the first and historically most important—for identifying genes prior to any knowledge about whether or not they varied. This methodological advance permitted the resolution of several old and troubling questions, most notably the calculation of average genetic differences among human races.

This central problem of early Mendelian genetics could never be addressed—even to counter the worst abuses of biological determinism and social Darwinism—because biologists could not generate random samples of genes, and could therefore only overestimate average distances by ignoring the unknowable invariant genes among races, while studying the (potentially small) fraction capable of recording differences among groups. With electrophoretic techniques, and the attendant generation of a random sample with respect to potential variability, geneticists soon calculated the average genetic differences among races as remarkably small and insignificant—a conclusion of no mean practical importance in a xenophobic world. Similarly, a truly random sample (with respect to the distribution of anagenetic rates) might show a predominance for stasis, even if previous studies (with their strong bias for preselection of variable species) had generally affirmed gradualism.

I am encouraged to accept the probable validity of this argument by the important study of Wagner and Erwin (1995), who used the different and comprehensive technique of compiling full cladograms for two prominent Neogene families of planktonic forams: Globigerinidae and Globorotaliidae. In applying a set of methods for inferring probable evolutionary mode from cladistic topology (see full discussion and details on pp. 820–822), they found that, in both families, branching speciation in the mode favored by punctuated equilibrium (divergence of descendants with survival of ancestors in stasis) vastly predominated over the origin of new species by anagenetic transformation. Thus, the literature's apparent preference for anagenesis in tabulated studies of individual lineages may only record an artifact of biased selection in material for research.

2. Even if gradualism truly does prevail in planktonic forams, we could not infer that the observed predominance of punctuated equilibrium in marine Metazoa must therefore reflect the artifact of an imperfect geological record. The difference might record a characteristic disparity between the taxa, not a general distinction in quality of geological evidence between deep oceanic cores and conventional continental sequences—a proposition defended in the third argument, just below. The deep oceanic record may *usually* be more complete, but the subset of best cases from conventional sequences surely matches the foram data in quality—and convincing studies of punctuated equilibrium and gradualism generally use these best records. Thus, the subset of most adequate metazoan examples should match, in quality of evidence, the usual records of forams from oceanic cores.

3. A third argument completes the trio of logical possibilities (all partially valid, I suspect, though I would grant most weight to this third point) for denying that a currently recorded maximal frequency of gradualism for planktonic foraminiferal lineages casts doubt on the general importance of punctuated equilibrium in evolution. The first argument attributes an apparent high frequency to biased sampling in the preselection, for rigorous study, of lineages already highlighted by taxonomic experts for suspected gradual change. The second and third arguments, on the other hand, hold that if high frequency truly characterizes this group, no general rebuttal of punctu-

ated equilibrium follows thereby. The second argument denies the common assumption that high-frequency records uniquely complete geological evidence—and that gradualism will therefore prevail whenever the fossil record becomes good enough to preserve its true domination (with a high frequency for punctuated equilibrium then construed, by Darwin's original argument, as the artifact of a gappy record). This second argument maintains that, while foram data may be more complete on average, the best metazoan examples of punctuated equilibrium have been validated with excellent samples from admittedly rarer but equally complete geological sequences, thus precluding the explanation of punctuated equilibrium as artifactual.

The third argument also grants the reality of higher-relative frequency for gradualism in forams, but argues against extrapolation to larger multicellular organisms on grounds of genuine difference in evolutionary mode, based on important biological distinctions between these single-celled creatures of the oceanic plankton and sexually reproducing metazoan species that, however parochially, have served as the basis for most of our evolutionary theory and, in any case, form the bulk of the known fossil record.

This third argument should not be viewed as special pleading by partisans, but as a positive opportunity for developing hypotheses about the importance (or insignificance) of punctuated equilibrium based on the correlation between differences in frequency and distinctive biological properties of various taxonomic groups—particularly in features related to speciation, the presumed evolutionary basis of punctuated equilibrium. Planktonic forams, with their asexuality, their small size and rapid turnover of generations, their unicellularity, their vast populations, and their geographic links to water masses, display maximal difference from most metazoans, and may therefore be especially suited for helping us to understand, by contrast, the prevailing mechanisms of evolution in multicellular and sexually reproducing organisms. The general nature of these differences does indeed point to a set of factors tied to the definition and division of populations, therefore granting plausibility to the claim that so-called "species" of planktonic forams should show more gradualism than metazoan taxa, while punctuated equilibrium may prevail in sexually reproducing multicellular species. The subject deserves much more attention and rigor, but to sketch a few suggested factors:

(1) Population characteristics. We conventionally name Linnaean species of asexual protistans, but even if adequately stable "packages" of form or genetic distinctness exist in sufficiently extended domains of space and time to merit a vernacular designation as "populations," what comparison do such entities bear with species of sexually reproducing multicellular organisms? (Needless to say, I raise no new issue here, but only recycle the perennial question of "the species problem" in asexual organisms.) Punctuated equilibrium posits a link of observed evolutionary rates to properties of branching speciation in populations. I don't even know how to think about such issues in planktonic forams, where vast populations may be coextensive with entire oceanic water masses, and where numbers must run into untold billions

of organisms for every tiny subsection of a geographic range. How do new populations become isolated? How do favorable (or, for that matter, neutral) traits ever spread through populations so extensive in both space and number?

(2) Morphology and definition. If metazoan stasis can be attributed, at least in part, to developmental buffering, what (if any) corresponding phenomenon can keep the phenotypes of simple unicells stable? Perhaps foraminiferal phenotypes manifest substantial plasticity for shaping by forces of temperature, salinity, etc., in surrounding water masses (see Greiner, 1974)— as D'Arcy Thompson (1917, 1942) proposed for most of nature in his wonderfully iconoclastic classic, *On Growth and Form*—see pp. 1179–1208. (Thompson's claim that physical forces shape organisms directly holds limited validity for complex and internally buffered multicellular forms, but his views may not be so implausible for several features of simpler unicells.) Could many examples of foraminiferal gradualism (compared with metazoan stasis in similar circumstances) reflect the plasticity of these protists in the face of gradual changes in the physical properties of enveloping oceanic water masses through time? If so, such gradual trends would not be recording evolutionary change in the usual genetic sense.

(3) Most interestingly (as a potential illustration of the main theoretical concern of this book), we must consider the potential for strongly allometric scaling of effects from a defined locus of change to other levels of an evolutionary hierarchy. To reiterate a claim that runs, almost like a mantra, throughout this text: punctuated equilibrium is a particular theory about a definite level of organization at a specified scale of time: the origin and deployment of species in geological perspective. The punctuational character of such change does not imply—and may even, in certain extrapolations to other scales, explicitly deny—a pervasive punctuational style for all change at any level or scale. In particular, punctuated equilibrium posits that tolerably gradual trends in the overall history of phenotypes within major lineages and clades (including such traditional tales as augmenting body size in hominids, increasing sutural complexity in ammonoids, or symmetry of the cup in crinoids) should reveal a punctuational fine texture when placed "under the microscope" of dissection to visualize the individual (speciational) "building blocks" of the totality—what we have long called the "climbing up a staircase" rather than the "rolling a ball up an inclined plane" model of fine structure for trends.

Similarly, in asking about evolutionary causality under selective models (see Chapter 8), we need to identify the primary locus of Darwinian individuality for the causal agents of any particular process—for only properly defined Darwinian individuals can operate as "interactors" in a selective process: that is, can interact with environments in such a way that their own genetic material becomes plurified in future generations because certain distinctive properties confer emergent fitness upon the individual in its "struggle for existence" (see pp. 656–667). Punctuated equilibrium maintains that species, as well-defined Darwinian individuals, hold this causal status as irreducible components, or "atoms," of evolutionary trends in clades. The apparatus of punctuated equilibrium then explains why trends, when necessarily described as speciational, display a punctuated pattern at geological scales (as expressed in the theory's basic components of stasis and geologically abrupt appearance). In a larger sense, punctuational accounts of trends propose a similar allometric model for any relevant scale—that is, any microscope placed over higher-level smoothness may reveal an underlying "stair-step" pattern among constituent causal individuals acting as Darwinian agents of the trend.

In sexually reproducing metazoa, species clearly play this role as causal individuals (see Chapter 8). The theoretical validity of punctuated equilibrium depends upon such a claim and model. But when we turn to such asexually reproducing unicells as planktonic forams, designated "species" cannot be construed as proper Darwinian individuals, and therefore cannot be primary causal agents (or interactors) in evolutionary trends. To locate the proper agent, the legitimate analog of the metazoan species, we must move "down" a level to the clone—to what Janzen (1977), in a seminal paper, called the EI, or "evolutionary individual."

When we execute this conceptual downshift in levels to locate the focal evolutionary individual in asexual and unicellular lineages, we recognize that the foram "species" acts as an analog to the metazoan lineage or clade, not to the metazoan species. The foram "species" represents a temporal collectivity whose evolutionary pattern arises as a summed history of the Darwinian individuals—clones in this case—acting as primary causal agents.

We can now shift the entire causal apparatus one level down to posit a different locus of punctuational change in planktonic forams. Just as the punctuational history of species generates smooth trends in the collectivity of a lineage or clade in Metazoa, so too might the punctuational history of clones yield gradualism in the collectivities so dubiously designated as "species" in asexual unicells. In other words, foram "species" may exhibit gradualism because these supposed entities are really results or collectivities, not proper Darwinian individuals or causal agents. Eldredge and I first presented this argument in our initial commentary on the debate about punctuated equilibrium (Gould and Eldredge, 1977, p. 142, and Fig. 9-10):

> We predict more gradualism in asexual forms on biological grounds. Their history should be, in terms of their own unit, as punctuational as the history of sexual Metazoa. But their unit is a clone, not a species. Their evolutionary mode is probably intermediate between natural selection in populations, and species selection in clades: variability arises via new clones produced rapidly (in this case, truly suddenly) by mutation. The phenotypic distribution of these new clones may be random with respect to selection within an asexual lineage (usually termed a "species," but not truly analogous with sexual species composed of interacting individuals). Evolution proceeds by selecting subsets within the group of competing clones. If we could enter the protists' world, we would view this process of "clone selection" as punctuational. But we study

their evolution from our own biased perspective of species, and see their gradualism as truly phyletic—while it is really the clonal analog of a gradual evolutionary trend produced by punctuated equilibria and species selection.

Lenski's remarkable studies on controlled evolution of bacteria under laboratory conditions of replication provide striking evidence for this claim (see full discussion of this work on pp. 931–936). Lenski and colleagues (Lenski and Travisano, 1994; Papadopoulos *et al.*, 1999) monitored average cell size for 10,000 generations in 12 lineages of *E. coli*. Cell size increased asymptotically in each lineage, steadily for the first 3000 generations or so, but remaining relatively stable thereafter. The fine structure of increase, however, proceeded in a punctuational manner in each lineage—a step-like pattern of stability in average cell size, followed by rapid ratcheting of the full population up to larger dimensions. This punctuational pattern presumably occurred because clones act as primary Darwinian individuals in this system. The full lineage must "wait" for sudden introduction of favorable variation in the form of occasional mutations, initiating novel clones that can then sweep through the entire lineage to yield a punctuational step in the overall phylogeny (at a scale of 10,000 generations in phenotypic history). Predictable, replicable size increase occurs by punctuational clone selection in each case (see Fig. 9-11). Lenski's powerful result does not illustrate a case of punctuated equilibrium, *sensu stricto*, but he does provide a challenging and in-

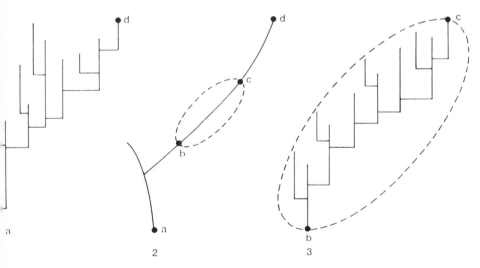

9-10. The supposed gradualism noted in many foram species may represent a view, from too high a level, of an overall trend within a phyletic sequence properly analyzed in terms of punctuational events at the level of clone selection—the appropriate mode in such asexual forms. 1 shows a conventional metazoan lineage in punctuated equilibrium. 2 shows the apparent gradualism in a foram lineage. 3 shows a gradualistic segment between B and C magnified so that the appropriate process of punctuational clone selection becomes visible. From Gould and Eldredge, 1977.

structive argument for considering the validity of punctuational change at all levels.

Just as the careful watchdog at any scientific meeting will unhesitatingly call out "what's the scale" when a colleague fails to include a measurement bar on a slide of any important object, we must always ask "what's the level" when we analyze the causal basis of any evolutionary pattern. Punctuational clone selection can yield gradualism within collectivities conventionally (if dubiously) called "species," just as punctuated equilibrium, acting on species as Darwinian individuals, can produce gradual trends in the overall history of lineages and clades.

GENOTYPES. Punctuated equilibrium is a theory about the evolution of phenotypes (both in concept and in operational testability for paleontological hypotheses), and correlations with genotypic patterns provide neither a crucial test nor even any necessary prediction. For example, critics of punctuated equilibrium have often argued that the apparently cumulative character of overall genetic distances among members of an evolving clade, expressed as a high correlation between measured disparities and independently derived times since divergence from a common ancestor—the kind of information that, in idealized (but rarely encountered) situations, yields a rough "molecular clock"—should argue strongly against punctuational styles of evolution, while affirming anagenetic gradualism.

But, leaving aside the highly questionable empirical status of these claims, the hypothesis of punctuated equilibrium would not be affected by positive outcomes, even at much higher relative frequency than the known history of life apparently validates. In supposing that "molecular clocks" tick against the requirements of punctuated equilibrium, we fall into two bad habits of thinking that impede macroevolutionary theory in general, and therefore rank as important conceptual barriers against the theses of this book. First, reductionistic biases often lead us to seek an "underlying" genetic basis for any overt phenomenon at any scale, and then to view data at this level as a fundamental locus for proper evolutionary explanation. (But consider only two among many rebuttals of such a position: (1) a genetic pattern may be

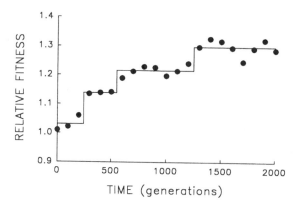

9-11. From Lenski and Travi[s] 1994. At too broad a scale, t[he] increases within Lenski's bac[terial] lineages seem gradual. But a [finer] scale analysis shows a stepwi[se] punctuational pattern of clo[nal] selection with stasis recordin[g] waiting time between favora[ble] mutations, and punctuation [caused] by rapid sweep of these rarel[y] favorable mutants through th[e] population.

non-causally correlated with coincident evolutionary expressions at other scales; and (2) in principle, genetic expressions of a common causal structure do not rank as intrinsically more "deep," "real," "fundamental," or "basic" than other manifestations in different forms and at different levels; causal relevance depends upon the questions we ask and the processes that organisms undergo.)

Second, the "allometric" effects of scaling either render the same process in a very different manner at various scales, or (perhaps more frequently, and primarily in this case at least) generate the distinctive patterns of different scales by independent processes, acting simultaneously, but with each process primarily responsible for results at its own appropriate level.

If I could affirm, as may well often be the case, that punctuated equilibrium regulated the phenotypic pattern of evolution in a given clade, while genotypic distances conformed closely to a "molecular clock," I would not conclude that punctuated equilibrium had therefore been downgraded, or exposed as incorrect, superficial, or illusory—with genetic continuity as a physically underlying (and conceptually overarching) reality. Rather, I would regard each result as true and appropriate for its own scale and realm—with the full pattern of legitimate difference standing as an intriguing example of resolvable complexity in evolutionary scaling and causality. Moreover, this particular pattern might easily result from a highly plausible scenario of complex and multileveled causation—namely, that neutral substitutions at the nucleotide level impart a signal sufficiently like a genomic metronome to dominate the molecular results, while ordinary speciation both regulates the phenotypic history of populations, and works by the expected pattern of punctuated equilibrium. The genomic results, *in principle*, need *not* extrapolate to encompass the pattern of speciational (macroevolutionary) change. After all, we do understand that gene trees do not entirely match organism trees in phylogeny!

In this way—as in the foregoing example of predictable differences between asexual unicells and sexually reproducing metazoa—punctuated equilibrium proves its value primarily by hypothesizing sensible distinctions: that is, by operating at scales and biological conditions where cladogenetic speciation plausibly sets evolutionary pattern. Punctuated equilibrium should not prevail where species cannot exist as Darwinian individuals, or where continuously occurring, and largely nonadaptive, substitution of nucleotides probably regulates the bulk of genomic change. In this crucial sense, punctuated equilibrium becomes a valuable hypothesis by delineating such testable distinctions, rather than allowing evolution to be conceptualized monistically as a single style of alteration, or a single kind of process either flowing from, or applicable to, all scales of change.

The question of consistency between observed genetic patterns in living species, and the relative frequency of punctuated equilibrium in their phylogeny, shall be treated in the next section on the correspondence of punctuated equilibrium with predictions of evolutionary modeling. But one genetic issue has been widely discussed in the literature, and should be included in this section on empirical results. Several researchers have noted that punctuated equilibrium implies a primary prediction about patterns of genetic differences among species: if most change accumulates at ruptures of stasis during events of speciation, and not continuously along the anagenetic history of a population, then overall genetic differences between pairs of species should correlate more closely with the estimated number of speciation events separating them, than with chronological time since divergence from common ancestry. (This prediction might be clouded by several factors, including the foregoing discussion on attributing the bulk of genomic change to continuity at a lower level, and a number of potential reasons for discordance between phenotypic effect and extent of responsible genetic change. But I certainly will not quibble, and I do allow that punctuated equilibrium suggests the broad generality of such a result.)

In the early days of debate about punctuated equilibrium, Avise (1977) performed an interesting and widely discussed test. In comparing genetic and morphological differences among species in two fish clades of apparently equal age but markedly different frequencies of speciation, Avise found a higher correlation of distances with age than with frequency of branching, and therefore favored gradualism over punctuated equilibrium as an explanation of his results. But Mayden (1986) then showed that Avise's test did not apply well to his chosen case (primarily because we cannot be sure of roughly equal antiquity for the two clades). He then argued, as several supporters of cladistic methodology had urged, that such tests should be applied only to well-confirmed cladistic sister groups—for, in such cases, even if paleontological data permit no certainty about the actual time of joint origin from common ancestry, at least we can be confident that the two clades are equally old! Mindel et al. (1989) then performed such a properly constituted test on the reptilian genus *Sceloporus,* and more loosely on allozymic data in general, and found a positive correlation between evolutionary distance and frequency of speciation—thus validating the primary prediction of punctuated equilibrium.

### Empirical tests of conformity with models

Limitations of the fossil record restrict prospects for testing punctuated equilibrium by inductive enumeration of individual species and lineages. Cases with sufficient resolution may not be common enough to establish a robust relative frequency; or systematic biases based on imperfections in the fossil record may lead to artifactual preferences for punctuated equilibrium—thus making the data unusable as a fair test for a minimal frequency. (I do not regard these problems as particularly serious, and I will provide several examples of adequate resolution in the next section of this chapter. But we should, in the light of these difficulties, also be exploring other ways of testing punctuated equilibrium, as considered below.)

In another strategy that has been pursued by some researchers, but could (and should) be exploited to a much wider and more varied extent, we might characterize, in quantitative fashion, broader patterns in the deployment of

diversity through time and space in major taxonomic groups—and then devise tests to distinguish among contrasting causes: anagenetic vs. cladogenetic; gradual vs. punctuational. If certain well defined patterns can only be
generated, say, by branching speciation rather than by anagenetic transformation (or vice versa, of course), then we can use the fit of broad results with
distinctive models, rather than minute documentation on a case-by-case basis, to establish the relative frequency of punctuated equilibrium.

In an important study, for example, Lemen and Freeman (1989) investigated "the properties of cladistic data sets from small monophyletic groups
6–12 species) . . . using computer simulations of macroevolution" (p. 1538).
They contrasted the differing outcomes of data generated under anagenetic
gradualism vs. punctuated equilibrium, and then examined cladograms of extant monophyletic groups for consistency with these "abstract, end-member"
alternatives. They claimed better support for gradualism, but several flaws in
their logic and data render their conclusions moot, much as their pioneering
approach may be applauded and recommended for further study.

Lemen and Freeman tested actual data against three modelled differences
between cladograms generated by gradualism *vs.* punctuated equilibrium.

1. Punctuated equilibrium should produce a strongly positive correlation
between number of branching points and apomorphies of species—because
change occurs at speciation and does not further correlate with passage of
time *per se*. Lemen and Freeman's models did affirm this expected result, and
real data, somewhat ironically, revealed "higher correlations of apomorphies
and branch points than could be explained by either mode of macroevolution" (p. 1549). But the authors rejected an interpretation of this information as favorable to punctuated equilibrium because anagenesis, under
certain conditions, also yields positive correlations, and because high correlations can also arise artificially by errors in establishing cladograms: "a
consistent error in polarity can profoundly affect the correlation of total
apomorphies and branch points" (p. 1551). Fair enough, but Lemen and
Freeman are not nearly so circumspect when equally flawed data seem to favor their preferred alternative of gradualism.

2. The modal (but not the mean or median) number of autapomorphies
will always be zero under strict punctuated equilibrium. This odd-sounding
situation arises because, in the cladograms, an event of branching produces a
daughter species with some autapomorphies and a persisting parental species
remaining in stasis with none. With no change except at branching points, the
value of zero autapomorphies must remain most common across all species
on the cladogram. Under gradualism, autapomorphies simply accumulate
through time, whatever the pattern of branching, so zero should not mark a
preferred or particularly common value.

Lemen and Freeman never found a mode of zero autapomorphies in real
data, and therefore rejected punctuated equilibrium as a predominant style of
evolution. But had they pursued explanations based on artifacts (as they did
so assiduously when the data seemed to favor punctuated equilibrium), they
would have realized that taxonomic practice precludes the definition (or even

the recognition) of species without autapomorphies. Such species arise frequently in the modelled system as a necessary consequence of the chosen rules of generation and the general logic of cladistic analysis. But, in neontological practices of naming, a species without autapomorphies represents an oxymoronic concept, and such taxa could never be designated at all. Lemen and Freeman recognize this point in writing about their various forms of gradualistic modelling (p. 1551): "When distinctness of species is demanded the lack of autapomorphies may not be the most expected condition."

3. Under punctuated equilibrium, "as the number of characters used in the analysis increases, the distribution of the number of autapomorphies per species becomes bimodal. Under gradualism, the distribution of autapomorphies remains unimodal under all conditions" (1538). This situation, a spinoff from their second criterion, arises because each branch, in an event of punctuated equilibrium, produces one changed descendant and one persisting ancestor—and the more characters you measure, the more you pick up the differences between stasis on one branch and change on the other. Under gradualism, total change correlates only with elapsed time, so accumulating autapomorphies should form a unimodal distribution so long as species duration remains unimodal as well.

Lemen and Freeman found no bimodal distributions in real data, and therefore concluded again in favor of gradualism. But, once more, the differences between idealized modeling and data from real organisms scuttles this conclusion. In the models, we know for sure that long arms without branching are truly so constituted, for we have perfect information of all simulated events. These unbranched arms, under punctuated equilibrium, should accumulate no autapomorphies—and the low mode of the bimodal distribution arises thereby. But, in real data of cladograms based on living organisms, long unbranched arms usually (I would say, virtually always) record our ignorance of numerous and transient speciational branchings that quickly became extinct and left no fossil record. (Moreover, since Lemen and Freeman's cladograms only include living organisms, even if successful and well-represented fossil species existed, they would not be included.) When we note a long arm without branches on a modern cladogram, and then assume (as Lemen and Freeman did) that accumulated autapomorphies between node and terminus must have arisen gradually and anagenetically, we commit a major blunder. We have no idea how many unrecorded speciation events separate node and terminus, and we cannot assert that recorded autapomorphies did not occur at these (probably frequent) branchings. In other words, Lemen and Freeman's bimodality test assumes that unbranched arms of their cladograms truly feature no speciation events along their routes, whereas numerous transient and extinct species must populate effectively all of these pathways.

Other applications of this method—modeling of alternative outcomes and testing of contrasting predictions against patterns of real data—have yielded results favorable to punctuated equilibrium. In a pathbreaking paper, Stanley (1975—see elaboration in Stanley, 1979 and 1982) first proposed this style of testing and developed four putative criteria, all affirming punctuated equilibrium. (Stanley's tests may be reduced to three, as his second "test of the Pontian cockles" represents a particular instance of his first "test of adaptive radiations." Stanley argued:

TEST OF ADAPTIVE RADIATION. After calculating average species durations from the fossil record, one can affirm that pure anagenetic gradualism (or temporal stacking of species end-to-end) cannot account for the magnitude of recorded adaptive radiations in the time available—so rapid cladogenesis must be invoked.

TEST OF LIVING FOSSILS. Punctuated equilibrium associates realized amounts of change primarily with frequency of speciation, anagenetic gradualism primarily with elapsed time. If so-called "living fossils"—ancient groups with little recorded change—also show unvarying low diversity through time, then we can affirm the primarily prediction of punctuated equilibrium, and refute the corresponding expectation of gradualism (for these groups are ancient). Stanley then documented such a correlation between clades identified as "living fossils" and persistently low diversity in these clades.

TEST OF GENERATION TIME. Under gradualism, amounts of realized evolution should correlate strongly with generation time—for the time that should mark accumulated evolutionary change does not tick by an abstract Newtonian clock, but by number of elapsed generations, representing the number of opportunities for natural selection to operate. But, under punctuated equilibrium, amount of change correlates primarily with frequency of speciation—a property with no known relationship to generation time. Stanley then cited the well-documented lack of correlation between evolutionary rate and generation time as evidence for the prevalence of punctuated equilibrium (fast-evolving elephants *vs.* stable invertebrates with short generations).

Much as I regard Stanley's arguments as suggestive, I cannot accept them as conclusive for two basic reasons. First, other plausible explanations exist for the patterns noted. For example, many reasons other than the prevalence of punctuated equilibrium might explain a lack of correlation between realized evolution and generation time, even in a world of anagenetic gradualism. The correlation might simply be weak or too easily overwhelmed (and therefore rendered invisible) by such other systematic factors as variation in the intensity of selection. (Maybe elephants, on average, experience selection pressures higher by an order of magnitude than those affecting short-lived invertebrates; maybe population size overwhelms the factor of generation time.)

Second, most of Stanley's tests (particularly his key claim about adaptive radiation) don't really oppose punctuated equilibrium to gradualism, but rather contrast a more general claim about the speciational basis of change (whatever the mode of speciation) with anagenesis. Moreover, the tests employ a somewhat unfairly caricatured concept of gradualism. I doubt that the most committed gradualist ever tried to encompass the maximal change between ancestor and any descendant in an adaptive radiation by stacking species end to end, and then calculating whether the full effect could arise in the allotted time. A committed gradualist might fairly say of an adaptive radiation: "of course the magnitude of change in both form and diversity corre-lates with number of branching events (what else could a 'radiation' mean). But adaptive radiations only accelerate the frequency of branching in re-sponse to ecological opportunity ('open' environments just invaded or just cleared out by extinction); they do not affect the modality of change. I will al-low that, in adaptive radiations, most new species arise in less time than usual, but still gradualistically. If full speciation takes half the average time (one million rather than a modal two million years, for example), but still oc-curs imperceptibly and still occupies a large percentage of an average species's lifetime, then gradualism encounters no threat in adaptive radiation."

However, in another crucial sense, at least one of Stanley's tests does illus-trate the most salutary potential role for punctuated equilibrium: its capacity to act as a prod for expansive thought and new hypotheses, whatever the out-come of the empirical debate about relative frequency. Paleontologists had been truly stymied in their thinking about the important and contentious topic of "living fossils." Neither of the two conventional explanations could claim any real plausibility. Every textbook that I ever consulted as a student dutifully repeated the old saw that living fossils had probably achieved opti-mal adaptation to their environment. Therefore, no alternative construction could selectively replace an ideal form achieved so long ago. But no one ever presented any even vaguely plausible evidence for such a confident assertion. Why should horseshoe crabs lie closer to optimality than any other arthro-pod? What works so well in the design of lingulid *vs.* other brachiopods? What superiority can a lungfish assert over a marlin or tuna? In fact, since liv-ing fossils also (by traditional depiction) present such a "primitive" or "ar-chaic" look, the claim for optimality seemed specially puzzling.

The other obvious explanation, in a gradualistic and anagenetic world ruled by conventional selection, held that living fossils had stagnated because they lacked genetic variation, and therefore presented insufficient fuel for Darwinian change. This more plausible idea seemed sufficiently intriguing that Selander et al. (1970), in the early days of electrophoresis as a novel method for measuring overall genetic variation, immediately applied the tech-nique to *Limulus*, the horseshoe crab—and found no lowering of genetic variability relative to known levels for other arthropods. This negative pat-tern has held, and no standard lineage of living fossils exhibits depauperate levels of genetic variability.

But punctuated equilibrium suggests another, remarkably simple, explana-tion once you begin to think in this alternative mode—an insight that ranks in the exhilarating, yet frustrating, category of obvious "scales falling from eyes" propositions, once one grasps the new phrasing of a basic question. If evolutionary rate correlates primarily with frequency of speciation—the car-dinal prediction of punctuated equilibrium—then living fossils may simply represent those groups at the left tail of the distribution for numbers of speciation events through time. In other words, living fossils may be groups that have persisted through geological time at consistently and unvaryingly low species diversity. (Average species longevity need not be particularly high, for low species numbers, if consistently maintained without geological bursts of radiation, will yield the full effect.) Such groups cannot be common—for consistently low diversity makes a taxon maximally subject to extinction in our contingent world of unpredictable fortune, where spread and number represent the best hedges against disappearance, especially in episodes of mass extinction—but every bell curve has a left tail.

This explanation holds remarkably well, and probably provides a basic explanation of "living fossils." Such groups are neither mysteriously optimal, nor unfortunately devoid of variability. They simply represent the few higher taxa of life's history that have persisted for a long time at consistently low species number—and have therefore never experienced substantial opportunity for extensive change in modal morphology because species provide the raw material for change at this level, and these groups have never contained many species.

Westoll (1949), for example, published a classic study, summarized again and again in treatises and textbooks (Fig. 9-12), showing that lungfishes evolved very rapidly during their early history, but have stagnated ever since. The literature abounds in hypothesized explanations based on adaptation and ecological opportunity in an anagenetic world. The obvious alternative stares us in the face, but rises to consciousness only when theories like punctuated equilibrium encourage us to reconceptualize macroevolution in speciational terms: in their early period of rapid evolution, lungfishes maintained high species diversity, and could therefore change quickly in modal morphology. Their epoch of later stagnation correlates perfectly with a sharp reduction of diversity to very low levels (only three genera living today, for example) with little temporal fluctuation in numbers—thus depriving macroevolution of fuel for selection (at the species level), and relegating lungfishes to the category of living fossils.

A breakthrough in the application of quantitative modelling to cladistic patterns of evolution directly recorded in the fossil record has been achieved by Wagner (1995 and 1999) and Wagner and Erwin (1995). These authors show, first of all, the pitfalls of working only with cladistic information from living organisms, and they illustrate the benefits of incorporating stratophenetic data from the fossil record into any complete analysis (see Wagner and Erwin, 1995, pp. 96–98, in a section entitled "why cladistic topology is insufficient for discerning patterns of speciation"). They then build models based on three alternative modes of evolution, and characterize the differences in cladistic pattern expected from each: anagenetic gradualism, speciation by "bifurcation" (where, after branching, the two descendant species both accumulate differences from an ancestor then recorded as extinct), and speciation by "cladogenesis" (where one daughter species arises with autapomorphic differences, but the ancestral species persists in stasis). Cladogenesis is usually defined—both in this book and in the evolutionary literature in general—as any style of evolution by branching of lineages rather than by transformation of a single lineage (anagenesis). Wagner and Erwin restrict the term "cladogenesis" to the mode of speciation predicted by punctuated equi-

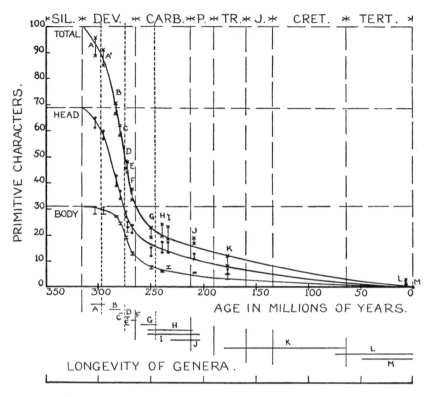

9-12A. The famous figure from Westoll (1949) showing rapid morphological change early in the history of lungfishes, followed by prolonged stagnation thereafter.

librium—branching off of a descendant, leaving a persisting and unaltered ancestor. They contrast this mode with bifurcation—the style of speciation predicted by gradualism: splitting of an ancestral population into two descendant species, both diverging steadily from the ancestor (which becomes extinct). I follow Wagner and Erwin's restricted use of "cladogenesis" only in discussing their work, and use the broader definition throughout the rest of this book.

The last two modes of bifurcation and cladogenesis both depict branching speciation in the definitional sense that two species emerge, where only one existed before. But note the crucial difference: bifurcation represents the operation of speciation in a gradualistic world, where an event of branching may be considered equivalent to two cases of gradualism following a separation of populations, and where the separation itself need not correlate with any acceleration in rate of evolutionary transformation. Cladogenesis, on the other hand, represents the predictions and expectations of punctuated equilibrium. Therefore, if we can model the differences between bifurcation and cladogenesis, and test these distinctive expectations against real patterns in nature, we may achieve our best and fairest potential evaluation for the relative frequency of punctuated equilibrium—for punctuated equilibrium can-

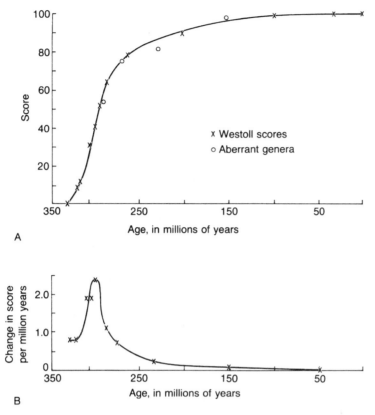

9-12B. Redrawing and simplification of these data in the excellent paleontological textbook of Raup and Stanley (1971). The bottom icon, showing an early
mode and a right skew, has become canonical in textbooks. The data are firm
and fascinating, but the interpretation has general been faulty as a result of
gradualistic and anagenetic assumptions. Lineages did not stagnate in any
anagenetic sense; rather, species diversity became so dramatically lowered (and
has always stayed so—only three genera of lungfishes remain extant today) that
speciational processes have never again had enough fuel to power further extensive phyletic change.

not be affirmed merely by showing that realized evolutionary patterns must
record speciation and cannot be rendered by anagenetic, end-to-end stacking.
Even the most committed anagenetic gradualists never denied the importance
and prevalence of speciation. They hold, rather, that speciation generally occurs in the gradualistic mode—as two cases of divergence at characteristic
rates for unbranching lineages—and not, as supporters of punctuated equilibrium maintain, as geologically momentary bursts representing the budding of
descendant populations from unchanged, and usually persisting, ancestral
species in stasis. Thus, the best possible test for punctuated equilibrium must
distinguish between the expectations of bifurcating vs. cladogenetic models
of speciation.

I am embarrassed to say that neither I nor my colleagues working on the
validation of punctuated equilibrium ever conceptualized the simple and ob-

vious best test for distinguishing the bifurcating model of speciational gradualism from the cladogenetic model of punctuated equilibrium. In this case, the impediment may be clear, but I can offer no legitimate excuse for my opacity—and I congratulate Wagner and Erwin on their formulation.

The solution lies in the distribution and frequency of "hard" polytomies in cladogenetic topologies. I failed to appreciate the following point: under punctuated equilibrium, new species branch off from unchanged and persisting ancestors. The successful ancestor remains in stasis and may live for a long time. Therefore, these "stem" species may generate numerous descendants during their geological tenure, while remaining unchanged themselves. Now what cladistic pattern must emerge from such a situation? A group of species branching at different times from an unchanged ancestor must yield a cladistic polytomy. Cladograms cannot distinguish different times of origin from an unaltered ancestor, and can therefore only record the phenetic constancy of the common and unchanging ancestor as a polytomy, for all branches emerge from an invariant source. Bifurcation, on the other hand, can produce a range of cladistic topologies (Wagner and Erwin, 1995, p. 92), but not domination of the overall pattern by polytomies. Thus, gradualistic *vs.* punctuational models of speciation should be distinguishable by distributions of polytomies in the resulting cladogram.

I suspect that many of us never recognized this point because we have been trained to view polytomies negatively as an expression of insufficient data to resolve a true set of ordered dichotomies. (Shades of our profession's former failure to conceptualize punctuated equilibrium because we had been trained to view geologically rapid appearances as artifacts of an imperfect fossil record!) Thus, we never recognized that polytomies might also be denoting a positive and resolvable pattern—multiple branching through time of several species from an unaltered ancestral source. Of course—and, again, just as with punctuated equilibrium itself—polytomy can also result from imperfection, and we need criteria to separate "real" polytomies representing a signal from the history of life from polytomies that only record artifacts of an imperfect record. Wagner and Erwin (1995) develop such a criterion by distinguishing between "hard" polytomies that include the persisting ancestor and "soft" polytomies that arise from an inability to resolve true sets of ordered dichotomies.

Wagner and Erwin's modelling demonstrates the translation of punctuational speciation to a cladistic pattern of predominant polytomies. (Wagner and Erwin used my own model of punctuational phylogenies, done with D. M. Raup in the 1970's (Raup and Gould, 1974), to show this mapping of punctuational phylogeny to a polytomous cladogram—see Figure 9-13—but I had never made the connection myself.)

Wagner and Erwin then applied their modelled differences to cladograms for two well-resolved, but maximally different (in taxon and time) species-level phylogenies in the fossil record: two Neogene clades of planktonic foraminifers (Globigerinidae and Globorotaliidae), and Ordovician representatives of the gastropod family Lophospiridae. In both cases, the cladograms indicated an overwhelming predominance of speciation by cladogenesis as a cause of phylogenetic patterning—thus affirming the predictions of punctuated equilibrium. For globigerinids, the cladistic topology revealed 40 speciation events, 5 probably anagenetic, 35 cladogenetic, and none bifurcating. Wagner and Erwin did not present full tabulations for the globorotaliids, but stated (p. 105): "The results are not presented here, but they were similar to those found for globigerinids: cladogenesis is significantly more common than anagenesis, a positive association exists between having long temporal ranges and leaving cladogenetic descendants, and no such association exists for anagenetic ancestors."

For the lophospirid gastropods, they write (p. 106): "Our preferred cladogram for lophospirids is rife with polytomies. Of the eleven polytomies, only

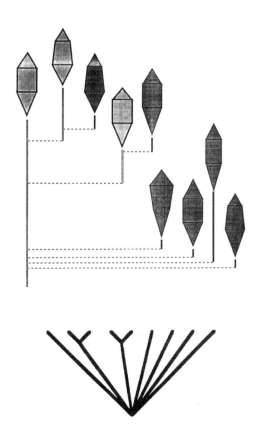

9-13. To my embarrassment, Wagner and Erwin (1995)—for I had not seen the obvious implication that would have enormously helped my argument—showed how phylogenies based upon iteration of several species from an unchanged parent stock (as Raup and Gould, 1974, had generated, and Wagner and Erwin reproduced, at the top of this figure) must yield, in cladistic representation, a polytomy. Thus, polytomies may provide evidence for punctuated equilibrium and do not necessarily represent the "signature" of missing data needed to resolve the system into dichotomies. If the ancestral form doesn't change throughout its geological range, all descendants must in principle arise at a polytomous junction of a cladogram.

two do not include plesiomorphic species. Thus, nine may represent hard polytomies." Of 42 implied speciation events, a maximum of six may have been anagenetic, while only one may represent a bifurcation. Again, cladogenetic speciation, the expectation of punctuated equilibrium, dominates the phylogenetic pattern.

Wagner and Erwin's overall conclusion accords fully with patterns expected in phylogenies built primarily—one might say overwhelmingly—by punctuated equilibrium (Wagner and Erwin, p. 110):

- Cladogenesis is significantly more common than anagenesis.
- Species with longer temporal and geographic ranges are more likely to leave descendants via cladogenesis or the factors contributing to wider temporal and geographic ranges also contribute to the likelihood of cladogenetic evolution.
- If anagenesis occurs, it only applies to species with restricted temporal and geographic ranges.
- Bifurcation accounts for a negligible amount of speciation.

We cannot often obtain well-resolved species level phylogenies from paleontological data, and inferences from higher taxa will probably remain too murky and insecure to permit general use of such models for testing hypotheses explicitly based on the evolutionary behavior of species. Still, other data sets do exist in fair absolute abundance (while representing a low percentage of the total number of potential lineages in life's history). Studies like Wagner and Erwin's can be replicated and extended for many taxa—and such a strategy can provide powerful tests for the relative importance of punctuated equilibrium in the history of life and the generation of phylogenetic patterns. The first tests have been highly favorable, but we have scarcely any idea what an extended effort might teach us about the basic modalities of macroevolution.

## Sources of Data for Testing Punctuated Equilibrium

### PREAMBLE

Punctuated equilibrium has generated a fruitful and far ranging, if sometimes acrimonious, debate within evolutionary theory (see appendix to this chapter). While we feel much pride (mixed with occasional frustration) for the role that punctuated equilibrium has played in instigating such extensive rethinking about the definitions and causes of macroevolution, we take even more pleasure in the volume of empirical study provoked by the theory of punctuated equilibrium, and pursued by paleontologists throughout the world. These carefully documented case studies (both pro and con) build a framework of proof for the value of punctuated equilibrium, as illustrated by the most important of all scientific criteria—operational utility. Such cases have been featured in numerous symposia and books dedicated to the empirical basis of punctuated equilibrium. This literature includes: the 1982 symposium in Dijon, France, entitled *Modalités, rythmes, méchanismes de l'évolution biologique: gradualisme phyletique ou équilibres ponctués* and published as Chaline, 1983; the 1983 Swansea symposium of the Palaeontological Association (United Kingdom) on "Evolutionary case histories from the fossil record" and published as Cope and Skelton (1985); the book *The Dynamics of Evolution: The Punctuated Equilibrium Debate in the Natural and Social Sciences* (Somit and Peterson, 1992) that began as a symposium for the annual meeting of the American Association for the Advancement of Science, and then appeared as a special issue (1989) of the *Journal of Biological and Social Structures;* the 1992 symposium of the Geological Society of America on "Speciation in the Fossil Record," held to celebrate the 20th anniversary of punctuated equilibrium, and published in book form as Erwin and Anstey (1995); and the 1994 Geological Society of America symposium on coordinated stasis, published in a special issue of the journal *Palaeogeography, Palaeoclimatology, Palaeoecology* in 1996 (volume 120, with Ivany and Schopf as editors). Several other unpublished symposia, including the notorious Chicago macroevolution meeting of 1980 (see pages 981–984), focused upon the topic of punctuated equilibrium. Finally, several books have treated punctuated equilibrium as an exclusive or major topic, including the favorable accounts of Stanley (1979), Eldredge (1985, 1995), and Vrba (1985a), and the strongly negative reactions of Dawkins (1986), Dennett (1995), Hoffman (1989), and Levinton (1988).

As emphasized throughout this book, most general hypotheses in natural history, with punctuated equilibrium as a typical example, cannot be tested with any single "crucial experiment" (that is, by saying "yea" or "nay" to a generality after resolving a case with impeccable documentation), but must stand or fall by an assessment of relative frequency. Moreover, we can't establish a decisive relative frequency by simple enumerative induction (as in classical "beans in a bag" tests of probability)—for individual species cannot be treated as random samples drawn from a totality with a normal (or any other kind of simple) distribution, but represent unique items built by long, complex and contingent histories. Time, taxon, environment and many other factors strongly "matter," and no global evaluation can be made by counting all cases equally. We may, however, be able to reach robust solutions for full populations within each factor—for planktonic forams, terrestrial mammals, Devonian brachiopods, or species of the Cambrian explosion, for example. Part C of this section reports several such studies, nearly all finding a predominant relative frequency for punctuated equilibrium.

Nonetheless, hundreds of individual cases have been documented since we proposed punctuated equilibrium in 1972. I do not think that most authors pursue such work under any illusion that they might thus resolve the general debate, but rather for the usual, and excellent reasons of ordinary scientific practice. Researchers pursue such studies in order to apply promising general concepts to cases of special interest that draw upon their unique skills and expertise. Such studies are pursued, in other words, to resolve patterns within *Australopithecus afarensis,* or among species in the genus *Miohippus,* not to adjudicate general issues in evolutionary theory.

Nonetheless, compendia of such studies do provide a "feel" for generalities of data in admittedly non-randomized samples, and they do establish archives of intriguing and well-documented cases both for pedagogical illustration, and simply for the general delight that all naturalists take in cases well treated and conclusively resolved. I shall therefore discuss this mode of documentation as practiced for two categories central to punctuated equilibrium: patterns of gradualism or stasis within unbranched taxa (part B of this section), and tempos and modes of branching events in the fossil record (part C). Part D will then treat the more decisive theme of relative frequencies.

### THE EQUILIBRIUM IN PUNCTUATED EQUILIBRIUM: QUANTITATIVELY DOCUMENTED PATTERNS OF STASIS IN UNBRANCHED SEGMENTS OF LINEAGES

As previously discussed (see pp. 758–765), the main contribution of punctuated equilibrium to this topic lies in constructing the theoretical space that made such research a valid and recognized subject at all. When paleontologists equated evolution with gradual change, the well-known stasis of most lineages only flaunted a supposed absence of desired information, and could not be conceptualized as a positive topic for test and study. By representing stasis as an active, interesting, and predictable feature of most lineages most of the time, punctuated equilibrium converted an unconceptualized negative to an intriguing, and highly-charged positive, thereby forging a field of study.

Nonetheless, we cannot argue that a proven predominance of stasis within lineages can establish the theory of punctuated equilibrium by itself. Punctuated equilibrium implies and requires such stasis, but remains, primarily, a theory about characteristic tempos and modes of branching events, and the primary patterning of phyletic change by differential birth and death of species.

Stasis has emerged from the closet of disappointment and consequent non-recording. At the very least, paleontologists now write, and editors of journals now accept, papers dedicated to the rigorous documentation of stasis in particular cases—so skeptics, and scientists unfamiliar with the fossil record, need not accept on faith the assurances of experienced paleontologists about predominant stasis in fossil morphospecies (see pp. 752–755). Moreover, stasis has also become a subject of substantial theoretical interest (see pp. 874–885), if only as a formerly unexpected result now documented at far too high a frequency for resolution as an anticipated outcome within random systems (Paul, 1985); stasis must therefore be actively maintained. In any case, paleontologists are now free to publish papers with such titles as: "*Cosomys primus:* a case for stasis" (Lich, 1990), and "Apparent prolonged evolutionary stasis in the middle Eocene hoofed mammal *Hyopsodus*" (West, 1979).

The study of McKinney and Jones (1983) may be taken as a standard and symbol for hundreds of similar cases representing a characteristic mixture of satisfaction and frustration. These authors documented a sequence of three successional species of oligopygoid echinoids from the Upper Eocene Ocala Limestone of Florida. The two stratigraphic transitions are abrupt, and therefore literally punctuational. But available evidence cannot distinguish among the mutually contradictory explanations for such passages: gradualism, with transitions representing stratigraphic gaps; rapid anagenesis for a variety of plausible reasons including population bottlenecks or substantial environmental change; punctuated equilibrium based on allopatric speciation elsewhere (or unresolvably *in situ,* given coarse stratigraphic preservation), and migration of new species to the ancestral range. Hence, frustration. (Moreover, as this pattern represents the most frequent situation in most ordinary sequences of fossils, we can readily understand why the testing of punctuational claims within the theory of punctuated equilibrium requires selection of cases—fortunately numerous enough *in toto,* however modest in relative frequency—with unusual richness in both spatial and temporal resolution.)

At the same time, however, we gain satisfaction in eminent testability for the set of claims representing the second key concept of stasis. Any species, if well represented throughout a considerable vertical span marking the hundreds of thousands to millions of years for an average duration, can be reliably assessed for stasis *vs.* anagenetic gradualism by criteria outlined previously (pp. 765–774). McKinney and Jones (1983) compiled excellent evidence for stasis in each of their three species—the basis, after all, for using these taxa in establishing biozones for this section. (As argued on pp. 751–752, biostratigraphers have always used criteria of stasis and overlapping range zones in their practical work on the relative dating of strata.) McKinney and Jones conclude (1983, p. 21): "These observations suggest there is little chance of species misidentification due to ontogenetic or phylogenetic effects when using this lineage for biostratigraphic purposes."

Smith and Paul (1985) studied vertical variation of the irregular echinoid *Discoides subucula* in a remarkably complete and well-resolved sequence of Upper Cretaceous sands. The species occurred throughout 8.6 m of section, apparently representing continuous sedimentation within one ammonite zone spanning less than 2 million years. The authors were able to sample meter by meter through a section with an interesting inferred environmental history: "The sediment that was then being deposited changed from clean, well-washed sand to a very muddy sand, and so one might expect to find evidence of phyletic gradualism in response to these changes" (1985, p. 36).

Smith and Paul did measure a steady change in shape towards a more conical form, a common response of irregular echinoids to muddy environments. But such an alteration can be ecophenotypically induced during ontogeny, and the authors see no reason to attribute this single modification to genetically based evolution (while not, of course, disproving the possibility of such genuine gradualism). Otherwise, stasis prevails throughout the section: "In other, more important characters, *D. subucula* remains morphologically static and shows no evidence of phyletic gradualism" (1985, p. 29).

This case becomes particularly interesting, and merits consideration here, as a demonstration of how far reliable inference can extend, even when the tempo and mode of origin for a descendant species cannot be directly resolved (the usual situation in paleontology). The potential descendant, *D. favrina*, enters the section near the top and overlaps in range with *D. subucula*, thus implying cladogenetic origin rather than anagenesis. The descendant's larger size and hypermorphic morphology suggest a simple heterochronic mechanism for the production of all major differences, hence increasing our confidence in (although clearly not proving) a hypothesis of direct evolutionary filiation. Finally, the fact that no morphological differentia of the species undergo any phyletic transformation within the lifetime of the putative ancestor further underscores the punctuational character of the transition, whatever the mode followed. The one character that does change during the tenure of *D. subucula* (perhaps only ecophenotypically, as discussed above) does not move towards the morphology of descendant *D. favrina*. The authors conclude (1985, pp. 36–37):

> Clearly the sedimentary record is complete enough and represents a sufficiently long period of time to be able to detect phyletic gradualism. Yet throughout this period *D. subucula* remains otherwise morphologically static. Characters that have been modified in closely related species show no evidence of undergoing gradual transformation within the duration of the species . . . The overlapping ranges of the two species and the total absence of phyletic gradualism in the characters that serve to distinguish the species suggests that punctuated equilibrium is a better model for speciation in this particular case.

In a later section (pp. 854–874), I shall discuss the generality of stasis within taxa or times under the more appropriate heading of empirical work on relative frequencies. But I shall also note this broader argument here, and in passing, if only to underscore the strong psychological bias that still pervades the field, thereby conveying a widespread impression that gradualism maintains a roughly equal relative frequency with punctuated equilibrium, whereas I would argue that, in most faunas, only a small minority of cases (surely a good deal less than 10 percent in my judgment) show evidence of gradualism. Under this largely unconscious bias, most researchers still single out rare cases of apparent gradualism for explicit study, while bypassing apparently static lineages as less interesting.

Johnson (1985), for example, studied 34 European Jurassic scallop species, and concluded (p. 91): "One case . . . was discovered where . . . the sudden appearance of a descendant form could fairly be ascribed to rapid evolution (within no more than one million years). Inconclusive evidence of gradual change over some 25 million years was discovered in one of the other lineages studied . . . but in the remaining 32 lineages morphology appears often to have been static." Yet Johnson virtually confines his biometrical study to the two cases of putative change, presenting only a single figure for just one of the

32 species in stasis. Johnson's title for his excellent article also records this bias in degrees of relative interest—for he sets the unmentioned but over-whelmingly predominant theme of stasis in opposition to his label for the entire work: "The rate of evolutionary change in European Jurassic scallops."

The most brilliantly persuasive, and most meticulously documented, example ever presented for predominant (in this case, exclusive) punctuated equilibrium in a full lineage—Cheetham's work on the bryozoan *Metrarabdotos*, more fully treated on pp. 843–845 and 868–870—began as an attempt to illustrate apparent gradualism. Cheetham wrote (*in litt.* to Ken McKinney, and quoted with permission from both colleagues): "The chronocline I thought was represented . . . is perhaps the most conspicuous casualty of the restudy, which shows that the supposed cline members largely overlap each other in time. Eldredge and Gould were certainly right about the danger of stringing a series of chronologically isolated populations together with a gradualist's expectations." Cheetham's biometry led him to the opposite conclusion of exclusive stasis: "In nine comparisons of ancestor-descendant species pairs, all show within-species rates of morphological change that do not vary significantly from zero, hence accounting for none of the across-species difference" (Cheetham, 1986, p. 190).

The establishment of stasis as an operational and quantifiable subject behooves us to develop methods and standards of depiction and characterization. Several studies have simply presented mean values for single characters in a vertical succession, but such minimalism scarcely seems adequate. At the very least, variances should be calculated (and included in published diagrams in the form of error bars, histograms, etc.)—if only to permit statistical assessment of significances for mean differences between levels, and for correlations of mean values with time.

Smith and Paul (1985), for example, presented both ontogenetic regressions and histograms for samples from each meter of sediment to illustrate stasis in relative size of the peristome in *Discoides subucula* (Fig. 9-14). Cronin (1985) also used both central tendency and variation to illustrate stasis throughout 200,000 years of intense climatic fluctuation (during Pleistocene ice cycles) for the ostracode *Puriana mesacostalis*. Cronin (Fig. 9-15) encircled all specimens of the species at three expanding levels of time in a multivariate plot of the first two canonical axes (encompassing 92 percent of total information): (1) variation in a single sample spanning 100 to 1000 years; (2) in one formation encompassing 20,000 to 50,000 years; and (3) across two formations, representing 100,000 to 200,000 years. Two features of this pattern provide insight into the anatomy of stasis: first, relatively small increase in the full range of variation over such marked extensions in lengths of time; second, the concentric nature of the enlarging ellipses, indicating no preferred direction in added variation, but merely the regular expansion anticipated in any random system with increasing sample size. As Cronin notes, this lack of directionality seems all the more surprising when we recognize that this lineage persisted in stasis through several ice-age cycles. Stasis must be construed as a genuine phenomenon, actively maintained—and not as an absence of anything. Cronin writes (1985, pp. 60–61): "Total within-sample variability representing $10^2$ to $10^3$ years is only slightly less than variability over $10^5$ to $2 \times 10^5$ years. *Puriana mesacostalis* shows no secular trends in its morphology over this time interval that might be evident from a lack of concentricity of the ovals—stasis is directionless. Yet high-amplitude environmental fluctuations occurred during this time that could have catalyzed speciation or caused extinction."

Once we construe stasis as an interesting evolutionary phenomenon, actively promoted within species, we then become eager to know more about its fine-scale anatomy and potential causes. A remarkable series of studies by Michael A. Bell on the Miocene stickleback fish *Gasterosteus doryssus* (Bell and Haglund, 1982; Bell, Baumgartner and Olson, 1985; Bell and Legendre, 1987) provide evidence at a maximal level of paleontological resolution, for these fossils occur in abundance in varved sediments with yearly bands— surely a *summum bonum* for attainable temporal precision! Bell and col-

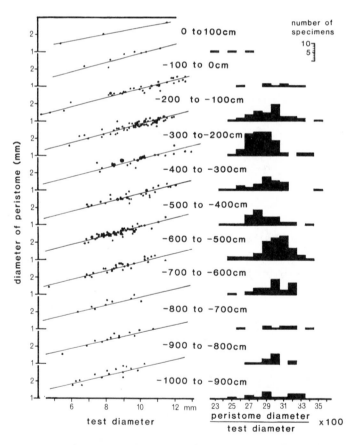

9-14. An impressive demonstration of stasis in the peristome size of the echinoid *Discoides subucula*. From Smith and Paul (1985). All specimens are shown for each narrow collecting interval, spaced one meter apart.

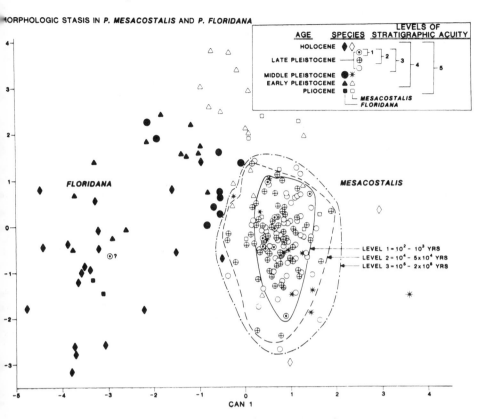

MORPHOLOGIC STASIS IN *P. MESACOSTALIS* AND *P. FLORIDANA*

9-15. Two hundred thousand years of stasis, during intense climatic fluctuation of Pleistocene ice cycles, in the ostracod *Puriana mesacostalis*. From Cronin (1985). The three circles around the specimens of this species show increased variation with expanding amounts of time: a single sample representing 100 to 1000 years, one formation encompassing 20,000 to 50,000 years, and two formations representing 100,000 to 200,000 years. The range of variation expands but the modal values do not change at all—as a hypothesis of stasis would predict.

leagues have documented extensive and complex temporal variability, both for single characters and correlated complexes, over tens of thousands of years (the 1985 study, for example, included several sampling pits covering about 1/3 of the total sedimentary record in a full sequence of approximately 110,000 years). They found some gradual trends in parts of the sequence, a great deal of fluctuation, and a few levels of abrupt alteration, often completely reversing a gradual change built through most of preceding time. In sum, this extensive and multifarious variation includes no sustained or accomplished directionality, and means for most single characters end up about where they began, whatever the internal wanderings between endpoints. Bell et al. (1985, p. 264) conclude: "Despite the temporal trends and heterogeneity of all characters through time, the end members [oldest and youngest sample] of only two of the time series (i.e. dorsal spine and dorsal fin ray numbers) are significantly different from each other; most characters return to their original states."

This complex anatomy of stasis again illustrates an active process of maintenance. Perhaps Futuyma's insight (see pp. 796–802) about linkage between speciation and achieved, stable morphological change can also help in this case. I suspect that much of the fluctuation, especially the occasional abrupt changes, represents a complex mosaic of shifting geographic borders for transient local populations. Vertical sequences probably record a mixture of temporary change within local populations and successive migrations of distinct local populations in and out over the same geographic spot (sample pit in this case). But both of these sources—short-term changes within a population and the mosaic of differences between demes—will be transient and fluctuating unless a set of differentia can be "locked up" by reproductive isolation within a newly-formed species. Since Bell's sequence includes no events of speciation, sustained changes do not accrue. The unusual extent of directionless fluctuation then records the especially high degree of temporal resolution.

Bell and colleagues (1985, p. 258) conclude correctly that "the irregular patterns and great magnitude of phenotypic changes that are observed indicate that conventional paleontological samples may miss important evolutionary phenomena and are not comparable to shorter-term evolution in extant populations." Fair enough; I asserted the same argument earlier (see p. 801) by metaphorical comparison to our current cliché for illustrating different scales of fractal self-similarity: one cannot measure around every headland of every sea-cove (transient changes over years in local populations) when calculating the coastline of Maine (macroevolutionary trends over millions of years) at the scale of a single page in an atlas. But one must firmly reject the tempting implication that either scale can be judged "better" or more complete. Yes, the paleontological scale misses "important evolutionary phenomena" of transient fluctuation in local populations. But measurement of details at this local scale cannot be extrapolated to encompass or explain a macroevolutionary trend either—and such local details therefore miss "important evolutionary phenomena" as well. Rather than accusing *any* level of insufficiency for its inevitable inability to resolve events at other unrecorded scales, we should simply acknowledge that any full understanding of evolution requires direct study and integration of the fascinating uniquenesses (as well as the common features) of all hierarchical levels in time and structure.

I have emphasized that one cannot achieve a proper "feel" for the relative frequency of punctuated equilibrium merely by tabulating published cases for individual lineages (whereas the study of relative frequency in a well-bounded, taxon, time, or environment—provided that researchers do not preselect their circumstances based on well-documented subjective appearance in favor of one side or the other—has yielded valuable data, as discussed on pp. 854–874). Claims for gradualism attain their highest frequencies at "opposite" ends of the conventional chain of being—that is, for foraminifers and for mammals. In my partisan way, I suspect that the former case may be valid, but attributable to biological differences that predict gradualism within asexual protist "species" as the expected consequence of punctuational clone selection, and therefore a proper analog of gradual trends in metazoan lineages that arise from cumulated punctuational speciation (see pp. 807–810). For mammals, I suspect that published reports follow traditions of work and expectation more closely than they record actual relative frequencies of nature.

I certainly accept the numerous cases of well-documented gradualism in foraminiferal lineages, and I acknowledge MacLeod's argument (1991) that abrupt transitions without branching in some sequences (the "punctuated anagenesis" of Malmgren et al., 1983) may arise as artifacts of condensed intervals of sedimentation within truly gradualistic trends. But punctuated equilibrium has also been demonstrated with data of equal abundance and completeness (the elegant case of Wei and Kennett, 1988, stands out for thorough documentation of both the geography of origin *and* subsequent history of nontrending in *Globorotalia (Globoconella) pliozea)*, and we have little idea, and no firm data, about overall relative frequencies for tempo and mode of evolution in this group.

Stasis has been demonstrated in other microfossil "species" with equally dense documentation—e.g., Nichols (1982) on lower Tertiary pollen, Wiggins (1986) on upper Miocene dinoflagellates, and Sorhannus (1990) on the Pliocene diatom *Rhizosolenia praebergonii*. Ross (1990) suggested that putative differences in relative frequencies might be tested by comparing foraminiferal lineages with microfossils of sexual metazoans preserved in comparable abundance in the same sediments—and that forams *vs.* ostracodes might provide a good test. Indeed, published cases for ostracodes seem to speak strongly for stasis and punctuation as a predominant pattern (in contrast with foraminiferal data). I have already discussed Cronin's (1985) work on Cenozoic ostracodes (see p. 827), and now cite his general conclusion (p. 60):

> Morphologic and paleozoogeographic analysis of Cenozoic marine Ostracoda from the Atlantic, Caribbean, and Pacific indicates that climatic change modulates evolution by disrupting long-term stasis and catalyzing speciation during sustained, unidirectional climatic transitions and, conversely, by maintaining morphologic stasis during rapid, high-frequency climatic oscillations. In the middle Pliocene, 4 to 3 million years ago, at least six new species of *Puriana* suddenly appeared as the Isthmus of Panama closed, changing oceanographic circulation and global climate. Since then morphologic stasis has characterized ancestral and descendant species during many glacial-interglacial cycles.

The origin of new species by branching in response to geographic opportunity (rise of the Panamanian isthmus), rather than by anagenetic gradualism as a selective consequence of changing environments, matches the predictions of punctuated equilibrium. Of the contribution made by stasis to this conclusion, Cronin writes (p. 61): "Morphologic stasis characterizes most shallow marine ostracodes from the western Atlantic that were subjected to these climatic changes, suggesting a pattern predicted by the model of punctuated equilibrium."

Whatley's (1985) study of the common and speciose ostracode genera *Poseidonamicus* and *Bradleya* in Tertiary and Quaternary sediments of the southwest Pacific also match the expectations of punctuated equilibrium throughout. Whatley found some gradualism in size changes (a common pattern), but only stasis for all defining features of shape and ornament. Whatley concludes for *Poseidonamicus* (p. 108): "Although over some 55 million years, the ornament of the genus underwent considerable change, several of its species remained morphologically very stable over long periods of time: 10 to 15 million years being not uncommon . . . This would seem to be evidence of virtual stasis between speciation events with respect to the evolution of the ornament of the various species." In an interesting comment, relating stasis to the major prediction of punctuated equilibrium for evolutionary trends—the stairstep rather than the ball-up-the-inclined-plane model—Whatley writes (p. 109):

> Although the morphological change from the ornate *P. rudis* to the smooth *P. nudus* [I do love the rhyme as well] took place over a time span of more than 50 million years and, therefore, from a generic standpoint represents a very gradual change, it must be emphasized that the individual species within this evolutionary series are effectively invariable with respect to their ornament. Morphological change was abrupt and coincided with speciation and further speciation was required to bring about yet further ornamental change. Ornamental change is clearly saltatory, very abrupt, and punctuated.

Gingerich's (1974, 1976) cases of gradualism in tooth size for several lower Eocene mammalian lineages in the Big Horn Basin of Wyoming instituted the empirical debate about punctuated equilibrium (see our response and critique in Gould and Eldredge, 1977). The tracing of gradualistic sequences for densely sampled series of small mammals (also based on dental evidence) then became an important research program for French paleontologists (see Godinot, 1985; and Chaline and Laurin, 1986, for sources more accessible to anglophonic readers). Large mammals have also furnished evidence for gradual anagenesis within species, as in Lister's study (1993a and b) of mammoths and moose—though he acknowledges that small sample sizes preclude a rigorous distinction of this interpretation from an alternative reading of several cladogenetic events, each perhaps punctuated, and all leading in the same direction of change (Lister, 1993a, p. 77).

But numerous examples of stasis in equally well sampled strata have also been documented for mammals (see pp. 854–870 for commentary on relative frequencies). The rodent sequences that form the empirical basis for most gradualistic studies of the French school have also yielded several examples of stasis (Lich, 1990; Flynn, 1986). Summarizing his work on rhizomyid rodents from the Miocene Siwalik deposits in Pakistan, Flynn (1986, p. 273) wrote: "Most early rhizomyid species survive on the order of millions of years, with at least two spanning about five million years, and display apparent stasis in most characters."

Several analogs of Gingerich's classic studies on gradualism have provided strong evidence for stasis, thus proving diversity of modes, even where gradualism had been most strongly asserted as an exclusive pattern. Gingerich had studied the small condylarth *Hyopsodus* in early Eocene rocks from the Big Horn Basin of northwestern Wyoming. West (1979), however, found only stasis for the same genus from slightly younger Middle Eocene rocks from the Bridger Formation of southwestern Wyoming. West concluded (1979, p. 252): "Bridger Formation *Hyopsodus* data seems to show little size change through approximately one million years. This stasis or equilibrium condition . . . is the only well developed pattern in Bridger *Hyopsodus*." Schankler (1981) then analyzed another genus, the condylarth *Phenacodus*, from the Big Horn Basin strata used by Gingerich to document gradualism in different taxa, and found only stasis within species (with abrupt transitions between species—a pattern that Schankler interpreted, correctly in my view, as a probable result of migration into a local area, rather than punctuational speciation *in situ*). He concluded (1981, p. 137): "The long-term stasis in morphology and size shown by the four species of *Phenacodus* conforms to the pattern expected in a model of evolution by punctuated equilibria."

As for the mammal we all love best (see pp. 908–916 for a more complete analysis), gradualism had long reigned as an unquestioned (and often quite unconscious) assumption in hominid evolution. An extensive, historically sanctioned set of dogmata, from ideas about "missing links" to the "single species hypothesis," presupposed gradualism as a philosophical foundation. An early study by Cronin et al. (1980)—which would not be defended by several of its coauthors today—made the classic error of regarding a monotonically changing set of mean values as virtual proof for anagenetic gradualism. (Such data cannot distinguish the stair steps of punctuated equilibrium from the same empirical pattern produced by gradualism in highly incomplete sections.)

The spotty data of hominids offer little opportunity for adequate testing of such ideas (and we wouldn't even think of applying an apparatus of this kind to such a poor example if we didn't care so much about the particular case). Nonetheless, I am gratified by some strong hints of substantial stasis in several hominid species, especially for increasingly persuasive data on the importance of apparently punctuational speciation in this small clade during a crucial million year African interval (ca. 2–3 my B.P.) that featured the putative origin of at least half a dozen hominid species. Rightmire's early claims (1981, 1986) for stasis in *Homo erectus* have been strongly challenged (Wolpoff, 1984), though the jury has surely not yet come in (despite a tentative vote from this juror, despite his general biases in the other direction, for at least some fairly persuasive gradualism within this species).

But two apparently sound cases of stasis have attracted substantial attention while we should also not neglect, if only for its radical meaning in the light of previous assumptions, the short-term stasis of *Homo sapiens*, at least from the earliest Cro-Magnon records in Europe (about 40,000 years B.P. to our present circumstances). When we realize that the cave painters of Chauvet, Lascaux, and Altamira do not differ from us in any phenotypic features, their stunning achievement seems less mysterious. For the two more substantial cases, the 0.9 to 1.0 million years of stasis in the first well documented hominid species, *Australopithecus afarensis* (aka "Lucy"), has been presented with much data and commentary (Kimbel, Johanson and Rak, 1994; see discussion of popular misapprehensions in Gould, 1995). Grine (1993) has also recorded 0.8 million years of stasis in *Australopithecus robustus* from Swartkrans cave in South Africa.

I am, in any case, gratified to note the changing presuppositions of this small, contentious and vital field of paleoanthropology. In early years of this debate, after refuting the Cronin et al. (1980) hypothesis, Jacobs and Godfrey (1982, p. 85) wrote: "The Hominidae can no longer be blissfully assumed to be safely above the punctuationist challenge to the gradualist orthodoxy." Just twelve years later, McHenry could assert in the closing line of his review (1994): "It is interesting, however, how little change occurs within most hominid species through time."

This elevation of stasis to visibility, respectability and even to expectation, has generated subtle and interesting repercussions for gradualism. When gradualism enjoyed high status as a virtually definitional consequence of evolution itself, few researchers thought to question such an anticipated result (but simply rejoiced in any rare instance of affirmation). However, once stasis emerges as an alternative norm, with gradualism designated as uncommon by the same analysis, then gradualism itself must fall under scrutiny for the first time.

With this shift of perspective, a paradox that should have been obvious from the start finally emerged into clear view: gradualism, *prima facie*, represents a "weird" result, not an anticipated and automatic macroevolutionary expression of natural selection—thus, perhaps, accounting for its rarity. Geological gradualism operates far too slowly to yield any workable effect at all when properly scaled down and translated to the immediacy of natural selection in local populations! (See Jablonski, 1999, for a forceful assertion of this paradox.)

Again, we encounter the major dilemma that I call (Gould, 1997f) "the paradox of the visibly irrelevant"—that is, phenomena prominent enough to be detectable and measurable at all in local populations during ordinary human time must cascade to instantaneous completion when scaled into geological time, whereas truly gradual effects in geological time must be effectively invisible at scales of human observation in ecological time. Consequently, what we see in our world can't be the direct stuff, by simple extrapolation, of sustained macroevolutionary change—while what we view as slow and steady in the geological record can't be visible at all (in the same form) by the measuring rod of our own life's duration.

Eldredge and I first raised this point explicitly in 1977 (Gould and Eldredge, 1977), for we had missed this implication in our original formulation of 1972. Here, on this issue, we finally caught the attention of many neontological colleagues who, before then, had been unmoved by punctuated equilibrium. How can geological gradualism be the extrapolated expression of natural selection within populations? Surely, if a doubling of tooth size (say) requires 2 million years to reach completion, then the process must be providing so small an increment of potential advantage in each generation that natural selection couldn't possibly "see" the effect in terms of reliably enhanced reproductive success on a generational basis. Can a tooth elongated by a tiny fraction of a single millimeter possibly confer any evolutionary advantage in a selective episode during one generation of a population's history? Conversely, if bigger teeth provide such sustained advantages, why stretch the process over millions of years? Neontological studies have amply confirmed that natural selection can be a powerful force—the lesson, after all, of our entire, and burgeoning, literature of measurable change in Darwin's finches, anolid lizards, peppered moths, etc. So why shouldn't such a doubling of tooth length be achieved over the palpable span of a few human generations? Of course we all recognize a host of standard arguments for reining in the speed of selective response: negative consequences through discoordination with other parts of the body, slowing by networks of correlated effects upon other anatomical features. But I doubt that even the summation of all such effects could generate sufficient restraining power to spread the blessings of a moment over two million years of plodding achievement. (See, however, p. 540 for Mayr's confident assertion, *a priori* and without evidence, of this evolutionary style and rate as canonical).

In other words, gradualism should be viewed as a *problem and a potential anomaly,* not as an expectation. In an important early recognition of this principle, Lande (1976), who (to say the least) is no friend of punctuated equilibrium, calculated that Gingerich's measured trends confer such a small effect upon the immediacy of ecological moments that, for one case, Lande calculated an advantage corresponding to elimination of individuals four or more standard deviations from the mean in regimes of truncation selection! However unrealistic one might deem such a model, no one should miss the "bottom line": most populations don't include *any* viable individuals four standard deviations from the mean—and one can hardly imagine that the removal of such occasional misfits or anomalies could slowly move the mean value of a population to new adaptive heights over a million sustained years.

I do not mean to say that this paradox cannot be resolved to make gradualism intelligible once again, but I do hold that any revalidation demands a substantial reconceptualization for this venerable phenomenon. The obvious solution lies embedded in results such as Bell and Haglund (1982) on the fine-scale structure of stasis. Selection in the immediacy of ecological moments cannot be measured as either the net nontrending of stasis or the steady accumulations of changing means in anagenetic gradualism. Any local population constantly jiggles to and fro in selective accommodation to changing local environment (as when mean coloration for peppered moths becomes darker for a few centuries, but then lighter again, and back to previous values, when lichens return to trees after abatement of industrial pollution). The extent of selection in an anagenetic sequence must be cumulated through each and every one of these jiggles, not measured by calculating the coefficients needed simply to change one endpoint into another. (Such a tactic would lead to the evidently false conclusion that little or no selection had ever occurred in peppered moths.) In other words, perhaps we must construe gradualism itself as a "higher level" phenomenon of net accumulation through the jiggles, not as an expression of ordinary directional selection summed through the ages.

But such a conclusion then raises a different (and broader) question: what, then, *is* ordinary geological gradualism after all? How can such a minuscule directional effect persist through all the swings and jiggles? And what does such a phenomenon represent? Must we interpret such slow net change as caused by drift, as Lande's models made conceivable? Such a conclusion would seem unlikely given the common impression that certain features, size increase in particular, occur preferentially and nonrandomly in gradualistic sequences (but see Jablonski, 1997, and Gould, 1997b, on the apparent falsity, and status as a psychological artifact, of this venerable claim known as "Cope's Rule"). Can we even argue for natural selection as the primary cause of classical gradualism at all? I am confident that selection remains a good candidate, but of what sort, and at what level? The selective basis of gradualism surely cannot be ascribed to the extrapolated advantage at every given moment of traits so enhanced over the long run. Rather, the selective edge must lie in some form of more general benefit not consistently visible in ecological moments, but somehow skewed to a higher probability of immediate occurrence that can then cumulate to a consistent trend in macroevolutionary time.

One might be tempted to equate this skewing agent with some form of general biomechanical improvement that might hold cumulative sway above the jiggling of momentary advantage in any direction. But then the kinds of features that seem to prevail in gradual anagenesis do not stand out for potential membership in this category. Perhaps we need to consider selection on supraorganismal units, or perhaps we should entertain nonselectionist alternatives, especially in the light of Lande's modelling for drift. (Such hypotheses of random change would require a far better knowledge of relative frequencies, both for characters within a taxon and among taxa themselves, than we now possess or even know how to generate.) In any case, I do not think we have even begun to explore the range of potential explanations for the puzzling phenomenon of anagenetic gradualism. I, at least, find the subject very confusing and challenging.

Finally, once we recognize gradualism as an interesting puzzle rather than a dull expectation, we may be led to "dissect" the phylogenetic "anatomy" of such trends more carefully, thus adding an operational benefit to the renewed theoretical interest. In a striking example, Kucera and Malmgren (1998) published an elegant study of morphological change in the late Cretaceous *Contusotruncana* lineage of planktonic forams. After several million years of stasis, the defining feature of "mean shell conicity" increased in a gradualistic manner (see Fig. 9-16) for 3.5 million years, beginning 68.5 million years ago, in this anagenetic lineage.

The mean values of Figure 9-16 record a conventional gradualistic sequence, but the greater detail of Figure 9-17, illustrating the morphology of all specimens, not only the means for each level, reveals fascinating details that suggest novel interpretations. In short, the range of variation, after remaining stable during the preceding period of morphological stasis, increased rapidly during the half million year interval from 68.5 to 68.0 million years ago. The subsequent gradual trend then developed *within* the envelope of this expanded range—a spread in variation that had already reached its full extent at the onset of the gradualistic interval. In other words, variation increased rapidly, and the gradual trend then unfurled into the enlarged morphospace of this new range. In fact, as Kucera and Malmgren point out, the upper endpoint of variation never expands after the initial surge, and the trend in mean values records a loss of variation by removal of flattened shells at the lower end.

I do not mention these details as a punctuational partisan trying to downgrade this example of gradualism, or to reinterpret the trend as a "mere" consequence of a punctuationally expanded range of variation. The gradual trend is both genuine and well documented—but the mapping of variation into its space gives us new insight into potential mechanisms of gradualism (while also imparting an important lesson about the significance of variation, the perils of not recording such data, and the potential for misreading patterns in expansion and contraction of variation as conventionally directed trends in mean values—see Gould, 1996a). The gradual trend to greater conicity in *Contusotruncana* probably warrants a conventional selectionist ex-

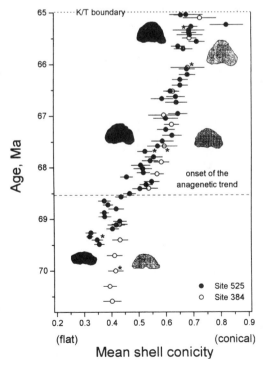

9-16. From Kucera and Malmgren (1998). A good example of gradualism for mean shell conicity in the planktonic foram *Contusotruncana*. After several million years of stasis, this trait increases in a gradualistic manner for 3.5 million years beginning 68.5 million years ago.

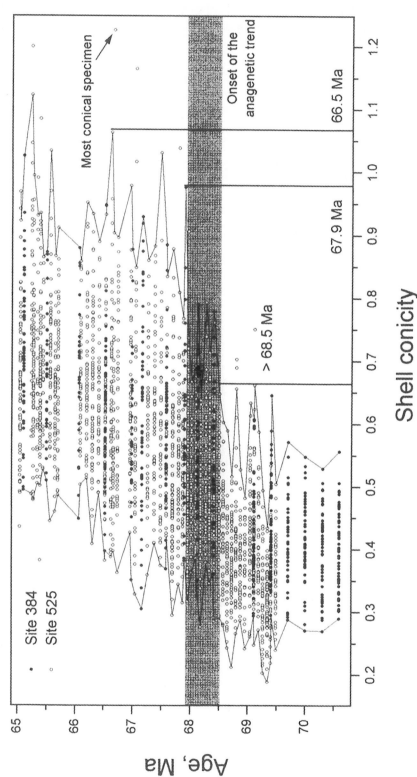

9-17. When we plot the variation for all specimens abstracted as mean values in Figure 9-16, a fascinating pattern emerges (from Kucera and Malmgren, 1998), the variation remained stable during stasis in the ancestral lineage. Variation then increased rapidly during the half million year interval from 68.5 to 68.0 million years ago. The subsequent gradual trend then developed *within* the envelope of this explanation, in part—shifting means resulting from selective removal of disadvantaged flatter specimens. But we also need to understand the potentiating condition established by an initial (and geologically rapid) expansion in the range of variation. What mechanisms underlie such change in a variational spectrum? Evolution can't anticipate future needs for altered means, so the enlarged range can only be exaptive for the subsequent trend. What, then, lies behind such rarely documented (but eminently testable) expansions and contractions of variational ranges?

## THE PUNCTUATIONS OF PUNCTUATED EQUILIBRIUM: TEMPO AND MODE IN THE ORIGIN OF PALEOSPECIES

Stasis is data, and potentially documentable in any well-sampled series of persistently abundant fossils spanning a requisite range of time—i.e., most of the duration of an average species in a given taxon, ranging roughly from a million years or so for such rapidly evolving forms (or perhaps just more closely scrutinized, or richer in visibly complex characters) as ammonites and mammals to an average of 5 to 10 million years for "conventional" marine invertebrates. But punctuation may only record an absence of intermediary data. Thus, as noted several times before, the second word of our theory stands more open to general test than the first, and this operational constraint inevitably skews the relative abundances of published information.

If punctuational claims were truly untestable, or subject to empirical documentation only in the rarest of special circumstances, then the entire theory would be severely compromised. Fortunately, many cases at the upper end of a spectrum in richness of data (both in abundance of specimens and fine-scale temporal resolution) offer adequate materials for distinguishing the causes and modalities of punctuation. The testable cases do not nearly approach a majority of available species, but neither do they stand out as preciously unusual. Thus, we face a situation no different from most experimental testing in science, especially when we cannot construct ideal conditions in a laboratory, and must use nature's own "experiments" instead. That is, we must pick and choose cases with adequate information for resolution, and without inherent biases that falsely presuppose one solution over others. But experimentalists in "hard science" seek the same unusual resolvability when they "improve" upon the ordinary situations of nature by establishing fixed, simplified, and measurable circumstances in a laboratory. Natural historians proceed no differently, and with no more artificiality or rarity of acceptable conditions for testing—except that we must ask nature to set the controls (and must therefore live by her whims rather than our manipulations. In this sense, the naturalist's tactic of "choosing spots" selectively corresponds with the experimentalist's strategy of establishing controls in laboratories).

The testing of punctuations has generated two primary themes of research: the establishment of criteria for distinguishing among the potential causes of literal punctuations in the fossil record; and the specification of fine-scale "anatomy" (data of timing, mode, morphology, geography, etc.) for punctuational events.

### The inference of cladogenesis by the criterion of ancestral survival

I doubt that any professional paleontologist would dispute the statement that a great majority of paleospecies make a geologically abrupt first appearance in the fossil record. But this statement about an observed, literal pattern carries almost no interpretive weight because the phenomenon so described can be explained by such a wide variety of putative causes, including the following distinctly different proposals:

THE TRADITIONAL GRADUALIST VIEW. The species arose by geologically gradual transformation of an ancestral population, but our woefully imperfect fossil record did not preserve the intermediary stages.

THE CLADOGENETIC PROPOSAL OF PUNCTUATED EQUILIBRIUM. The literal record represents the expected geological scaling of biological processes responsible for the origin of species. New species arise by isolation and branching of a segment of the ancestral population. The branch evolves to a new species by continuous transformation, but at a rate that, however "slow" by the inappropriate standard of a human lifetime, runs to completion during the "geological moment" of a bedding plane in most cases.

PUNCTUATED ANAGENESIS. The new species arises by continuous transformation, *in toto* and without branching, of an ancestral species, but at a rate too rapid for geological resolution of intermediary stages.

SUDDEN APPEARANCE BY MIGRATION INTO A LOCAL SECTION. The new species arose in another region at a rate and mode that cannot be determined from local evidence. A punctuational first appearance in any particular geographic region records a process of migration from another area of earlier origin.

These four proposals have strikingly different implications for the validation of punctuated equilibrium. The first opposes punctuated equilibrium unambiguously and would disprove the theory, or at least consign it to irrelevancy as a cause of pattern in the history of life, if this mode of classical gradualism could be affirmed at dominant relative frequency for the origin of new species.

Punctuated equilibrium predicts that the second explanation must hold as the primary generator of the dominant empirical signal of punctuational origin for paleospecies. If most species did not arise by rapid cladogenesis at appropriate geological scales, then punctuated equilibrium would be disproven as a major cause of evolutionary pattern (and would be relegated to a status of marginality and insignificance in the history of life).

The third explanation may fall within the "spirit" of punctuated equilibrium, by identifying a genuine geological punctuation, rather than a false appearance based on missing data in gradualistic sequences, as the source for an empirical observation of abrupt origin. But if punctuated anagenesis could be validated at high relative frequency for the origin of paleospecies, then punctuated equilibrium—a theory about cladogenesis—would be demoted or negated, while the important ancillary concept of explaining trends, within a hierarchical model (see Chapter 8), as differential success of species within clades, would also become marginalized.

The fourth, or migrational, alternative may resolve a local issue in a given section, but can only indicate, for questions about tempo and mode of speciation, a need for additional information of wider geographic scope. For we must still learn whether the new species, arriving as a punctuational migrant, arose by anagenesis or cladogenesis, and at either a gradual or punctuational tempo, in its natal area. But if the migrant invades the territory of a surviving ancestor—a common pattern in recorded literature—then, at least, we have documented the cladogenetic origin that punctuated equilibrium requires.

Against the charge that our theory cannot be adequately tested, participants in the empirical debate about punctuated equilibrium have long recognized, and generally utilized, an excellent criterion possessing the two cardinal virtues of a probing agent for scientific hypotheses: ready (and unambiguous) application in most cases, and an inherent bias against punctuated equilibrium by underrepresentation of actual cases. I presented this tool—ancestral survival following punctuational origin—earlier in the chapter (see pages 793–796), while leaving the primary documentation for this section.

The criterion of ancestral survival invokes paleontological data of the most conventional and easily acquired kind—specimens in local sections, forming samples of sufficient size for basic taxonomic identifications—and not distant inferences from models or from fossil data to unobservable correlates in behavior or physiology. Moreover, the criterion is properly biased against punctuated equilibrium in recognizing only the subset of legitimate cases with documented ancestral survival (therefore leaving in limbo all genuine cases where ancestors may have survived in other regions, or may not yet have been found). Any tabulation based on the criterion of ancestral survival must therefore underestimate the true relative frequency of punctuated equilibrium.

One potential biasing factor, however, might lead to an overestimate for punctuated equilibrium under this criterion, and must therefore be scrutinized and avoided. Under the fourth explanation presented above for literal observations of punctuation between a descendant and a surviving ancestor in a local section, migration of the descendant from a different region (where it might have originated gradually), rather than punctuational evolution *in situ*, could produce an artificial boost in frequency if falsely counted as a proven case of punctuated equilibrium. (I suspect that most cases in this mode do represented punctuated equilibrium, based on general arguments that most speciation events in unobserved regions of the geographic range will themselves be punctuational, but we obviously cannot count these examples favorably, because our entire case would then become circular by assuming the premise supposedly under test.)

The proper solution to such unresolvable cases lies in proper scrutiny, and in declining to count them as proven support for punctuated equilibrium. Most paleontologists recognize and follow this recommended practice. For example, to cite three titles from opposite ends of the conventional taxonomic spectrum, Sorhannus (1990) could not determine whether the punctuational origin of the diatom *Rhizosolenia praebergonii* from ancestral *R. bergonii* 2.9 million years ago in the Indian Ocean occurred *in situ* or by migration from the central Pacific. He entitled his article: "Punctuational morphological change in a Neogene diatom lineage: 'local' evolution or migration?" Schankler (1981), as previously reported, attributed punctuational patterns of Eocene condylarth *Phenacodus* to probable migration, and called his paper: "Local extinction and ecological re-entry of early Eocene mammals." And Flynn (1986) documented an excellent case of ancestral survival in Miocene rodents from Pakistan (in a group frequently cited for high relative frequencies of gradualism), but couldn't distinguish evolution *in situ* from migration as the cause of observed cladogenesis. He therefore only cited the literal pattern itself in his title: "Species longevity, stasis, and stairsteps in rhizomyid rodents."

Among affirmations of punctuated equilibrium by the criterion of ancestral survival, and ordering my discussion along a conventional taxonomic spectrum (for no reason beyond antiquated custom), Wei and Kennett's classic study (1988) illustrates how geographic data can be integrated with vertical sequences to resolve evolutionary modes not deducible from data of single sections. These authors showed that the upper Miocene planktonic foram *Globorotalia (Globoconella) conomiozea terminalis* evolved gradually into *G. (G.) sphericomiozea* during a 0.2 million year interval in central parts of its geographic range.

At the same time, intensification of the Tasman Front (Subtropical Divergence) separated peripheral populations of the warm subtropics from the central stock. The isolated population then branched rapidly into a new species, *G. (G.) pliozea*, in less than 0.01 million years, or 5 percent of the time taken for anagenetic transformation of the ancestral stock at the center of its range. The anagenetic trend proceeded in a direction (loss of keel and development of a more conical test) opposite to the morphological innovations (flattened test and more pronounced keel) of the allopatrically speciating peripheral form. The new species, following its punctuational origin, persisted in stasis for more than a million years. About halfway through this interval, a descendant of the central stock migrated into the warm subtropical region of *G. (G.) pliozea*. The two species then coexisted for half a million years without apparent intermixing, and with no interruption of stasis.

The rich data of microfossils from oceanic cores, often providing good resolution for both geographic and temporal variation, have also documented punctuational speciation (usually allopatric) with ancestral survival in several other cases. Cronin (1985) correlated the punctuational origin of six species in the ostracode *Puriana* with changes in oceanographic circulation engendered by the Pliocene rise of the Isthmus of Panama. Cronin comments

(p. 60) that "since speciation occurred, ancestral species and their descendants have coexisted, in some cases sympatrically." In a study of Miocene deep sea ostracodes from the southwest Pacific, Whatley (1985, p. 109) documented two cases of allopatric and punctuational origin for new species followed by migration back to the parental range and subsequent coexistence with the ancestral species.

Alan Cheetham's work on American Cenozoic clades of the cheilostome bryozoan genera *Metrarabdotos* and *Stylopoma* (Cheetham, 1986, 1987; Jackson and Cheetham, 1994; Cheetham and Jackson, 1995) merits citation at several points in this chapter for its unparalleled documentation of all major tenets of punctuated equilibrium—both in clarity of conclusions and richness of empirical evidence. I present a general summary in the section on relative frequency (p. 868), but Cheetham's fruitful use of ancestral persistence should be noted here. Jackson and Cheetham (1995, p. 204) cite three primary empirical sources for documenting punctuated equilibrium from paleontological data: "The geologically abrupt appearance of species in the record, the static morphologies of species for millions of years, and the extensive temporal overlap between apparent ancestor-descendant species pairs."

Their summary of overwhelming support for punctuated equilibrium from the last source (Jackson and Cheetham, 1994, p. 407) states that "most well-sampled *Metrarabdotos* and *Stylopoma* species originated fully differentiated morphologically and persisted unchanged for > 1 to > 16 million years, typically alongside their putative ancestors."

On Cheetham's celebrated and frequently reprinted diagram of evolution and cladogenesis in the *Metrarabdotos* clade (Fig. 9-18, and redundant in citing "evolution and cladogenesis" because all phyletic change occurs by cladogenesis in this lineage), ancestors persist after the origin of descendants in 7 of the 9 cases where Cheetham felt confident enough to assert a phylogenetic claim for direct filiation. (Marshall's important challenge (1995) to assessments of stratigraphic range in several cases does not counter Cheetham's hypotheses about filiation, and certainly does not challenge the assertion of overlap, a claim based on direct observation of joint occurrence, not on inference.) The two cases where ancestral persistence has not been directly observed (see Fig. 9-18), but may well have occurred (the derivation of *M. tenue* from *sp. 10*, and of *M. unguiculatum* from *M. lacrymosum*), both fall "outside the interval of dense sampling" (Cheetham, 1986, p. 201), where Cheetham achieved a stratigraphic resolution by Sadler's (1981) criteria of 0.63. For *Stylopoma*, "eleven of the nineteen species originate fully formed at $p > 0.9$, with no evidence of morphologically intermediate forms, and all ancestral species but one survived unchanged along with their descendants" (Jackson and Cheetham, 1994, p. 420).

By dense sampling in both vertical sequence and geographic spread, Nehm and Geary (1994) demonstrated the punctuational origin of the gastropod *Prunum christineladdae* from its ancestor *P. coniforme* in a small part of its

Caribbean range, during a short interval (0.6 to 2.5 percent) of ancestral persistence in stasis. Following the descendant's origin, ancestors continued to thrive in central areas of the range.

In a common pattern found in many taxa, punctuated equilibrium can be confirmed, even in local sections, and even when ancestors do not occur in the same strata as their descendants. Frequently, a population from an ancestral species of known and widespread geographic range branches punctuationally to a descendant that maintains exclusive occupancy of the range for a time, but then becomes extinct. The ancestor subsequently reinvades the range, thus establishing earlier coexistence during the descendant's geological tenure. For example, Bergstrom and Levi-Setti (1978) documented the threefold reappearance of the Middle Cambrian trilobite *Paradoxides davidis*

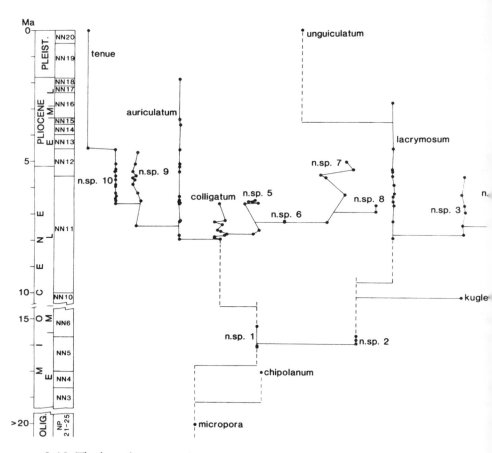

9-18. The best documented, indeed already canonical, example of punctuated equilibrium as an invariant pattern for an entire clade across its full geographic range—the research of Cheetham on Tertiary and Quaternary Caribbean species of *Metrarabdotos*. Each point depicts a multivariate centroid based on all characters, not just a single feature. All species express stasis, several for extended periods and a large number of samples. Ancestors persist after the origin of descendants in 7 of 9 cases where Cheetham felt confident enough to assert a claim for direct filiation.

*davidis* following local and allopatric origins of derived taxa that then become extinct at diastems, with the ancestor reappearing in strata just above.

Similarly, Ager (1983) traced the allopatric origin of late Pliensbachian brachiopod species from the central stock of *Homoeorhynchia acuta,* and the later Toarcian migration of the descendant *H. meridionalis* into the ancestral region. Williamson's (1981) celebrated and controverted study (see pp. 769–771) of punctuational origin for several pulmonate snail species in African Pleistocene lakes invokes the same kind of evidence—as the ancestral species migrates back (in several separate episodes, moreover) after a coalescence of lakes and the extinction of descendant species that had originated in previous times of isolation.

When all evidence derives from a restricted region, the separation of punctuation *in situ* from migrational incursion (with origin elsewhere at an unspecified tempo) becomes more difficult, but some criteria of admittedly uncertain inference may still be useful. For example, Smith and Paul (1985) argue that the sudden appearance of the descendant echinoid *Discoides favrina* in strata still holding ancestral *D. subucula* may represent an event of punctuational speciation on morphological grounds—for the descendant species, though visually distinct in many features, can be easily derived, given allometric patterns shared by both forms, through a simple heterochronic process of hypermorphosis.

In graptolites, the pattern of ancestral survival after cladogenetic origin of a descendant taxon has been noted frequently enough to inspire its own terminology as the concept of "dithyrial populations" (Finney, 1986), or samples from the same stratum containing two directly filiated and noninter-grading species.

The widespread geographic distribution of many late Tertiary and Quaternary mammalian lineages provides several examples of geographically resolvable allopatric origin followed by later survival with the ancestral species. For example, *Mammuthus trogontherii,* the presumed ancestor of the woolly mammoth *M. primigenius,* first appears in northeastern Siberia while its presumed ancestor, *M. meridionalis,* continued to survive in Europe (Lister, 1993a, p. 209).

Other forms of evidence can lead to strong inferences from data of ancestral survival to origin of descendants by punctuated equilibrium, even in the absence of such firm geographic data. I previously mentioned the growing evidence for rapid cladogenesis as the primary pattern in hominid evolution (see p. 833), based on several criteria, including the high relative frequency of observed overlap, the limited time available for cladogenetic origin (even when place and geological moment have not been clearly specified), and our confidence that all events (at least preceding the origin of *Homo erectus*) occurred in Africa. In his review, McHenry (1994, p. 6785) stated that "ancestral species overlap in time with descendants in most cases in hominid evolution, which is not what would be expected from gradual transformations by anagenesis." McHenry's summary diagram (reproduced here as Fig. 9-19) shows a clear pattern of dominant relative frequency for rapid cladogenesis—

a weight that has only increased in the light of discoveries since then (see Leakey et al., 2001), particularly for a vigorous phase of cladogenesis 2–3 million years ago, leading to at least half a dozen hominid species (see Johanson and Edgar, 1996).

On the same subject of punctuational and cladogenetic reformulations for classic evolutionary trends previously framed (and widely celebrated in both textbook and story) as exemplars of anagenetic gradualism, the phylogeny of horses has been rewritten as a copious cladogenetic bush replete with ancestral survival in the very parts of the sequence once most firmly read as a tale of linear progress. For example, Prothero and Shubin (1989) have shown that the Oligocene transition from *Mesohippus* to *Miohippus* conforms to punctuated equilibrium, with stasis in all species of both lines, transition by rapid branching rather than phyletic transformation, and stratigraphic overlap of both genera (one set of beds in Wyoming has yielded three species of *Meso-*

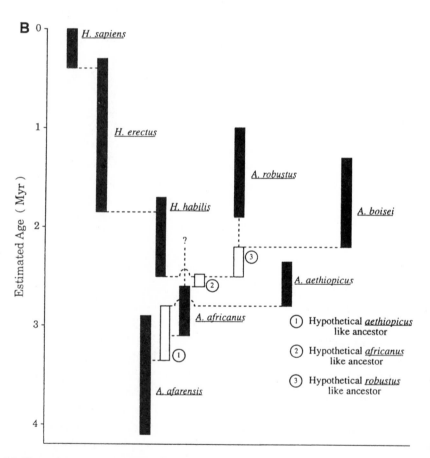

9-19. From McHenry (1994). The hominid record is spotty, but the basic pattern of substantial stasis within several species—particularly *A. afarensis*—and numerous branching points with persistence of putative ancestors lends support to the model of punctuated equilibrium.

*hippus* and two of *Miohippus*, all contemporaries). Prothero and Shubin conclude: "This is contrary to the widely-held myth about horse species as gradualistically-varying parts of a continuum, with no real distinctions between species. Throughout the history of horses, the species are well-marked and static over millions of years. At high resolution, the gradualistic picture of horse evolution becomes a complex bush of overlapping, closely related species."

To end this section with a particularly instructive example, punctuated equilibrium has frequently been saddled with the charge that inherent limitations of paleontological data yield biased results, artificially and superficially favorable to the theory—with Darwin's classic argument against a literal reading of punctuations as the conventional antidote. However, an opposite bias may also be significant, and may lead to serious underestimation of punctuated equilibrium in a circumstance likely to be quite common: when a descendant, fully distinct at its origin but initially rare, enters the ancestral area, and then increases steadily in relative abundance as the ancestor declines to extinction. The true evolutionary pattern will be fully punctuational, with stasis in both ancestor and descendant throughout, and with abrupt geological origin of the descendant. But if we misread the event as a tale of anagenetic transformation, and if the two species overlap extensively in ranges of variation, then we will misinterpret the full pattern as transformation by anagenesis, rather than replacement with steadily increasing relative abundance of the descendant species.

The important distinction between these interpretations can be made with appropriate statistical tools applied to samples of sufficient size—but the punctuational alternative must be conceptually available to suggest such a test. In this subtle sense, among so many other more overt reasons explored in this book, expectations of gradualism seriously restrict our range of potential explanations for evolutionary modes and tempos—and punctuated equilibrium therefore becomes both suggestive and expansive, whether or not the hypothesis holds in any particular case.

In an elegant demonstration of this principle, Heaton (1993 and 1996 for data *in extenso*) showed that a classic case of supposedly gradualistic anagenesis in Oligocene rodents from the western United States really represents a case of replacement. Heaton writes (1993, p. 297): "Statistical investigation of large samples suggests instead that two closely related species coexisted, and the shift in mean size that was thought to represent anagenesis actually represents replacement."

Heaton demonstrated the distinct character of the two taxa both by bimodality in their joint occurrences (Fig. 9-20), and by showing that the two species maintained distinctly different geographic ranges (with overlap in Nebraska and eastern Wyoming, but only the descendant taxon living at the same time in the Dakotas—see Figs. 9-21 and 9-22). The small species, *Ischyromys parvidens*, predominates in the early Orellan, although the larger *I. typus* already occurs low frequency in the same strata. *I. typus* then in-

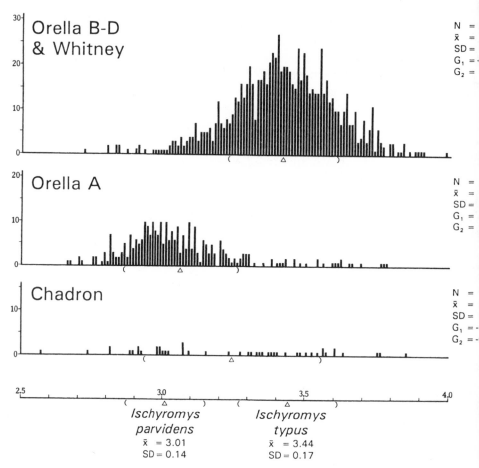

9-20. From Heaton (1993). Data that had, in the past, been interpreted as a gradualistic evolution of increasing size within a single species actually represent a change in relative abundance of two species, each stable throughout its interval—with the species of larger body size gradually becoming more common in the local section.

creases, as ancestral *I. parvidens* declines, throughout the remainder of Orellan times.* Interestingly, *I. typus* does undergo a small anagenetic increase following the extinction of *I. parvidens,* "but this change is minor and not deserving of chronospecies recognition" (Heaton, 1993, p. 297), and the species, in any case, becomes extinct soon thereafter—a common pattern,

---

*As an example of the conceptual stranglehold that gradualism once imposed upon such data, the major study done before punctuated equilibrium on the evolution of these rodents presupposed anagenetic gradualism at a constant rate: "This treatment assumes that a regular increase in size continued at approximately the same rate throughout Orellan time" (Howe, 1956, p. 74). When Howe then detected accelerated change at two paleosols marking boundaries of substages within the Orellan, he assumed (without any direct evidence) that the paleosols must mark diastems, or time gaps, compressing a true gradualism of change into the literal appearance of a small hiccup. But Heaton found no evidence for any temporal hiatus at these boundaries.

Heaton's results included both punctuated equilibrium (two stable species changing only

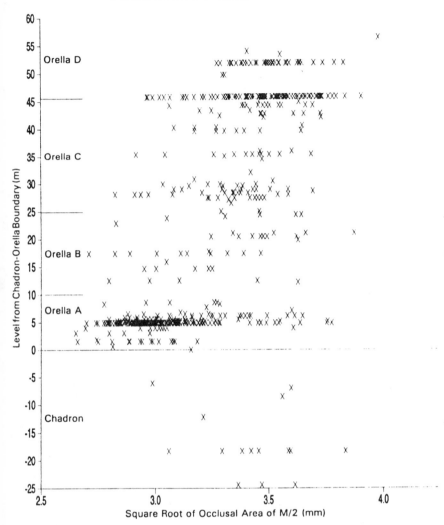

## Toadstool Park, Nebraska

9-21. Note that both species of *Ischyromys* live sympatrically and remain in stasis in some parts of their range, particularly in Nebraska and Wyoming.

in relative abundance) and gradualism (minor size increase within the larger species at the end of its range) in a total pattern, but he concluded (correctly, I think, in my own biased way) that punctuated equilibrium had shown greater utility in challenging previous assumptions that had stymied proper conclusions. He closed his paper by writing (1993, p. 307): "So, *Ischyromys* displays features of both 'punctuated equilibria' and 'phyletic gradualism' as defined by Eldredge and Gould (1972). But the primary revelation of this study is that what was thought to be a single gradually evolving lineage must now be seen as the replacement of one stable species by another." (In the fairness of full disclosure, Heaton did his graduate work under my direction. But I really do encourage independence and contrariness, and some of my students have documented gradualism, even in their Ph.D. dissertations, when truly (and for the only time) beholden to my "official" approval—e.g. Arnold, 1982.)

also strongly implicating punctuated equilibrium as the major generator of larger trends, if only because the fruits of anagenesis get plucked so quickly unless they can be "locked up" in cladogenetic iterations.

### The "dissection" of punctuations to infer both existence and modality

Once a literal punctuation has been noted, and a cladogenetic origin inferred by such criteria as the documentation of ancestral survival, further testing of punctuated equilibrium as the mode of origin for the new species may be achieved by several standards that might be characterized (somewhat meta-

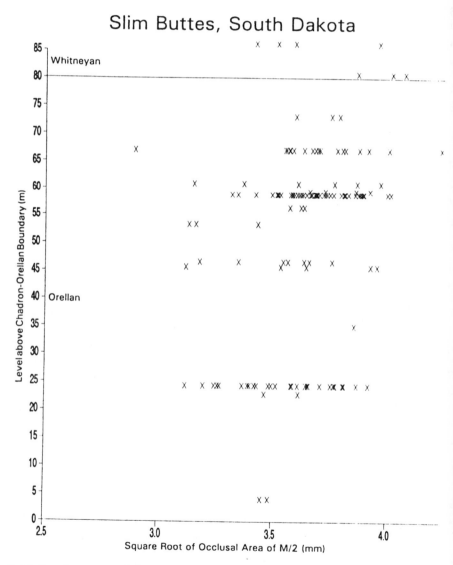

9-22. In other areas of the range, as here in South Dakota, only the descendant species lives during the entire interval.

phorically) as devices for "dissecting" the punctuation by revealing an internal "fine structure" with probative value for inferences about evolutionary causes. Three major modes of dissection have been featured in the existing literature (although the theme has not been organized in this manner before), each explicitly invoked as a tool for the potential validation or refutation of punctuated equilibrium.

TIME. I discussed the operational definition of punctuation as scaled to periods of stasis (see pp. 765–774). The obvious barrier to testing this primary requirement of the theory lies in our inability to specify requisite information about time in "standard" paleontological situations, where the duration of speciation lies beneath the resolving power of our basic operational "moment"—the bedding plane. Therefore, to achieve a proper dissection, we must search for unusual situations that permit an adequate resolution of time in one of two manners dictated by the logic of the problem: either by finding a way to date individual specimens compressed on a single bedding plane, or by locating situations of unusually rapid sedimentation, where a sequence of events usually collapsed onto a single bedding plane can be expressed in true temporal order through a vertical sequence.

The first tactic can be applied only in highly unusual circumstances effectively limited to nearly modern bedding planes with specimens that can be dated individually, for the error bars associated with most radiometric techniques exceed the entire duration of most bedding planes (except for isotopes with very short half lives, which can then only be applied to Pleistocene or Holocene specimens). However, in a recent example (discussed more fully on p. 771 and Fig. 9-4), Goodfriend and Gould (1996) traced a species transition by hybridization in the land snail *Cerion* on the Bahamian island of Great Inagua. We found all specimens jumbled together on a modern mudflat (a bedding-plane-to-be, if you will), and we then used a combination of radiocarbon dating and amino acid racemization to determine that the smooth and complete species transition occupied 15,000 to 20,000 years—a reasonable figure for a punctuational event (here compressed, as usual, into a single geological "moment," which we, thanks to the rare combination of recent occurrence and availability of dating techniques, were able to disaggregate and resolve).

In the far more common situation of sedimentation rates high enough to spread the usual compressions of single bedding planes into resolvable vertical sequences, assessments have been made in both relative and absolute terms. I previously cited Fortey's (1985) conclusion in the relative mode (see p. 769), based on calibrating punctuational origins against gradual transitions observed for other taxa in the same strata (thus obviating the usual claim that literal punctuations probably represent a geologically slow gradualism that extremely spotty sedimentation cannot record). With this technique, Fortey's found about a 10:1 ratio for punctuated *vs.* gradual origins of species in Ordovician trilobites from Spitzbergen.

The best examples in the more satisfactory absolute mode do not arise from direct paleontological records, but as firm inferences based on species flocks in lakes or on islands of known and recent origin (with African cichlid fishes as the classic case of modern evolutionary biology). These evolutionary "explosions" often produce several hundred species in just a few thousand years, and must be ranked as punctuational with a luxurious vengeance! But such circumstances do not represent a norm for most speciation in most clades, and such an unusual phenomenon, however stunning and however well documented, cannot suffice to validate a proposed generality.

The punctuational origin of many species can be accurately timed with direct paleontological data. Lister (1996) calculated a maximum of only 5000 years for the Quaternary evolution of dwarfed woolly mammoths on Wrangel Island, and 6000 years for the dwarfed red deer of Jersey. The punctuational origin of the marginellid gastropod *Prunum christineladdae*, based on the study of Nehm and Geary (1994), took 73,000 to 275,000 years, and spanned 0.6 to 2.5 percent of the full duration of the ancestral species. Reyment (1982) calculated outside limits of 100,000 to 200,000 years (perhaps a good deal less) for origin of the Cretaceous ostracode *Oertiella chouberti* from its ancestor, *O. tarfayaensis*.

GEOGRAPHY. I have already discussed this important tool for validating punctuated equilibrium by gathering data at a more inclusive and finer scale than the local documentation of a literal punctuation. On pp. 840–845, I described cases where geographic data affirmed an allopatric and punctuational event of cladogenesis, thus demonstrating that the abrupt appearance of a descendant species truly represents punctuated equilibrium, and not just a migrational incursion of a species that originated by an uncertain mode in an unknown place (Wei and Kennett, 1988, for protistans; Williamson, 1981, for mollusks; Lister, 1996, and Heaton, 1993, for mammals, among many others. Albanesi and Barnes (2000) present a particularly well documented case both for allopatric and punctuational origin of new taxa and survival of ancestors in their original regions for a lineage of Ordovician conodonts).

MORPHOMETRIC MODE. By quantitative study of patterns in morphological transition between ancestral and descendant species, several criteria of inference can increase our confidence in the identification of punctuated equilibrium, both by establishing a case for direct filiation rather than simple replacement by a taxon evolved elsewhere, and by indicating a punctuational mode for the cladogenetic event. As illustrations of this approach, consider three effective morphometric arguments:

1. Visually extensive change (supposedly requiring many independent inputs expressed over substantial time) can arise as coordinated consequences of one, or few, generating factors, and can therefore readily be accomplished at a punctuational tempo. This "standard" argument has a long pedigree, and serves many purposes, in evolutionary theory (see Chapter 10, for example, on "positive" constraints). In the context of punctuated equilibrium,

this theme establishes plausibility for temporal compression of visually substantial change into a single cladogenetic event at a punctuational tempo. The argument also proceeds in several modes, including inductive approaches based on "covariance sets" (Gould, 1984b) of correlated characters transformed together along a single multivariate axis. For example, the transition documented by Goodfriend and Gould (1996) involved several measures coordinated by a single change in direction and rate of growth along the axis of coiling—a coherently correlated pattern running orthogonal to, and therefore independent of, the standard covariance set representing shell size alone. Since the cladogenetic event altered shape (as expressed along the axis of coiling), but not size, this multivariate separation established the source of morphometric change and also revealed its unitary nature in modified growth.

Among examples of the opposite deductive approach, based on fitting an apparent complexity of observed changes to a simple model of underlying generation, Smith and Paul (1985) recognized a suite of alterations through a punctuational event in Cretaceous echinoids as coordinated consequences of a single heterochronic change; and Benson (1983) explained a "punctuational event" (1983, p. 398) in the ostracode *Poseidonamicus* as a set of secondary, and mechanically automatic, accommodations of the carapace to a primary change in shape.

2. Patterns of changing variation through a "dissected" punctuation may reveal a cladogenetic mode of direct filiation. Empirical patterns of variation may permit distinction between punctuational incursion from elsewhere (a migrational event of uncertain interpretation) and cladogenesis *in situ*. To contrast two studies previously reported, Heaton (1993) found consistent and unaltered bimodality in a vertical sequence of Oligocene rodents, with an appearance of overall gradualism arising as an artifact of directionally changing relative abundances. Such a pattern—while confirming stasis and showing the utility of punctuated equilibrium as a generator of alternative hypotheses—cannot resolve the descendant's mode of origin, for the new species enters the fossil record in full and complete distinction. In Williamson's study of African lake mollusks (1981), however, a distinctive pattern of variation (see previous discussion on page 769) implicates cladogenesis *in situ*—for Williamson's fine-scale vertical resolution allowed him to discern an initial period of expanded intrapopulational variation followed by a reduction back to ancestral levels, but now centered about the altered mean of the new species.

3. Comparison of within- and between-species variation as a test of extrapolationism. If new species arise by gradualistic anagenesis, then the direction of selection within populations and the pattern of temporal variation during the life of a species should mirror the morphological changes between species in a geological trend. But Shapiro (1978) estimated natural selection in the Miocene scallop *Chesapecten coccynelus* by comparing specimens that died as juveniles to those that survived to adulthood. He measured the direction of implied change as not only different from, but actually orthogonal to, the distinction between this species and its descendant, *Chesapecten nefrens*.

Shapiro could not resolve the tempo or mode of the cladogenetic event itself, but he showed that its direction cannot be extrapolated from an inferred pattern of change by natural selection within the ancestral species.

Kelley (1983, 1984) then conducted a more extensive study of all molluscan species with sufficiently rich vertical records in these classical and well-studied Miocene beds of Maryland. She found trending within species for only 16 of 90 cases (17.8%) as defined by well-determined rank correlation coefficients between a shell measurement and stratigraphic position. But for the majority of positive coefficients, she found that the trend within a species is "oriented opposite to the direction of the succeeding species's morphology, indicating a decoupling of macroevolution from microevolution in those cases" (1983, p. 581).

## PROPER AND ADEQUATE TESTS OF RELATIVE FREQUENCIES: THE STRONG EMPIRICAL VALIDATION OF PUNCTUATED EQUILIBRIUM

### *The indispensability of data on relative frequencies*

As stated before (p. 823), proponents of punctuated equilibrium have always recognized that the theory cannot be proven, and can win only two minimal validations—proof of plausibility and promise of testability—from documentations, however rigorous and complete, of individual cases (as presented in the last two sections). As its primary claim, therefore, punctuated equilibrium must assert a dominant role for stasis within species and rapid cladogenesis between species in the construction of macroevolutionary patterns at the appropriate scale of speciation and trends across species within clades. This assertion requires that punctuated equilibrium maintain a *dominant relative frequency* in the origin of new paleospecies. Tests of the theory must therefore focus upon percentages of occurrence in exhaustive, or at least statistically definitive, surveys of particular taxa, faunas, and times.

Species cannot be conceptualized as indistinguishable beans in the conventional bag of our standard metaphor for problems in probability—for the nature of history grants uniqueness to times and taxa, and therefore precludes any simple tabulation by global enumerative induction. We may, however, assess relative frequencies for well-bounded situations restricted to taxa of a given fauna, species within a monophyletic clade, or representatives of a particular time or geological formation. Several such studies have been carried out, and effectively all have found the clear signal of a dominant relative frequency for stasis and punctuation, as predicted by the theory of punctuated equilibrium. I regard these data as our most convincing indication of the validity and importance of punctuated equilibrium as a primary generator of pattern in the history of life. I am also surprised that this clear signal has not been more widely appreciated as the most decisive result in a quarter century of research and debate about punctuated equilibrium.

For reasons previously discussed under the heading of "publication bias" or "Cordelia's dilemma" (see pp. 763–765), proper tests of relative frequencies cannot be made by a "catch as catch can" style of simple enumeration based on previously published studies done for other reasons. Until quite recently in paleontology, strong and pervasive biases equated evolution with gradual change, and regarded stasis as "no data," and therefore not worth recording. Tabulations of older literature will inevitably favor gradualism both because no other style of evolution attracted study, and (even more problematically) because paleontologists, expecting only gradualism, tended to misread other patterns in this conventional light. Proper (and noncircular) testing—as in any statistical study—requires that the items chosen for sampling display no bias (imposed by human choice or preference) away from their relative frequencies in nature. When this ideal cannot be realized in natural experiments, which necessarily lack the rigor of laboratory controls, we should at least insist that unavoidable biases be directed against the hypothesis under test.

Thus, one cannot achieve a reliable relative frequency for punctuated equilibrium by tabulating cases from an existing literature, where strong biases in favor of gradualism may reasonably be suspected (or, to put the issue more accurately, virtually guaranteed). May I simply restate Tony Hallam's comment to me on why evolutionary studies of mollusks in English Liassic beds have concentrated with near exclusivity on *Gryphaea* (which, ironically, does not, after all, display the kind or direction of gradualism that initiated this literature in Trueman's famous (1922) paper—see Hallam, 1968; Gould, 1972; Jones and Gould, 1999): "Why hasn't anyone ever examined any of the 100 or so *other* molluscan species, many with equally good records, in the same strata?" Hallam then answered his own rhetorical question: "Because they seem to show stasis, and were therefore regarded as uninteresting"(see Johnson's (1985) affirmation of this stasis).

As an example of major differences between adequate and biased modes of sampling, two contrasting studies were presented at the North American Paleontological Convention, Boulder, Colorado, 1986. Barnovsky calculated the relative frequency of punctuated equilibrium *vs.* anagenetic transformation for Pleistocene mammals based exclusively on previously published reports in the literature. The two modes were supported at close to equal frequency.* Prothero then reported his field study for *all* mammalian lineages in Oligocene rocks of the Big Badlands of South Dakota. (See pages 861–865 for a full discussion of Prothero's refined and extended results—an even more impressive validation of punctuation equilibrium by well-established relative frequencies.) Nearly all lineages remained in stasis, and all new forms entered the record with geological abruptness. Prothero found very few cases of gradual anagenesis. Of course the differences might be real; perhaps the Pleisto-

---

*Interestingly, Barnosky's (1987) published version of his oral presentation refined his conclusion and tabulated a strong majority for punctuated equilibrium, even when compiled from an existing literature biased by previous traditions for ignoring stasis as non-data, and favoring apparent cases of gradualism. In his compendium for Quaternary mammals, Barnosky (1987) found punctuated equilibrium "supported twice as often as phyletic gradualism . . . The majority of species considered exhibit most of their morphological change near a speciation event, and most species seem to be discrete entities."

cene did witness a much higher frequency of gradualism. But I suspect that Barnovsky's result records a bias in older literature, when paleontologists tended to publish only when they found "interesting" lineages in the midst of change. But Prothero studied *all* lineages for a time and place, without preconception about modes or tempos—and his relative frequencies matched the predictions of punctuated equilibrium.

Proper empirical tests of relative frequencies impose two crucial requirements: *first*, that cases be sampled without any preselection in favor of one outcome or the other; and *second*, that cases be sufficiently numerous to establish a statistically significant relative frequency for a totality. The "totalities," "universes," or "populations" that inspire studies of relative frequencies for testing punctuated equilibrium constitute the "usual suspects" of evolutionary research: all species in a monophyletic taxon (genealogical criterion), or all species (perhaps of restricted taxonomic scope) in a given biota over a specified time and area (temporal and geographic criteria).

### Relative frequencies for higher taxa in entire biotas

I previously cited the admittedly subjective testimony of many leading experts about the overwhelming predominance of punctuated equilibrium among all lineages in the group of their lifelong expertise and specialization (not just those featured in published studies)—see pages 752–755. Some paleontologists have tried to provide a rough quantification for this "feel." For example, Fortey (1985) states that, for graptolites and trilobiles, "the gradualistic mode does occur especially in pelagic or planktonic forms, but accounts for 10% or less of observations of phyletic change, and is relatively slow." J. Jackson (cited in Kerr, 1995, p. 1422) attempted to separate out only the most persuasive cases of unbiased sampling in faunal studies of relative frequencies. Of this subset, he remarked: "I'm imposing pretty strict criteria, but in the few cases I know [that meet these criteria], it's perhaps 10 to 1 punctuated." Later, and after a more rigorous attempt to compile best documented cases for the time and general environment best suited for supplying the requisite density of data—Neogene benthonic species of macroinvertebrates—Jackson concluded (in Jackson and Cheetham, 1999, p. 75): "Overall, 29 out of 31 species of Neogene benthos for which phylogenetic data are available exhibited punctuated morphological change at cladogenesis that is consistent with the theory of punctuated equilibria. Cases of punctuation more than double if we include extended morphological stasis . . . Thus, most but not all cases of speciation in the sea are punctuational."

The most persuasive studies have applied morphometric methods to large numbers of species in exhaustive (or at least statistically well validated) tabulations for the full diversity of higher taxa within particular faunas or spans of time. Hallam (1978), for example, tabulated data for all adequately defined European Jurassic bivalve species, forming a compendium of 329 taxa. He found "overwhelming support" (p. 17) for punctuated equilibrium, with the single exception that 15 to 20 percent of his species showed phyletic size increase—but no changes in shape—during their geological tenure. Only

>ne lineage, the famous oyster *Gryphaea,* showed a corresponding gradual :hange in shape as well—a consequence of heterochronic linkage to phyletic variation in size (see affirmation of Jones and Gould, 1999). Hallam concluded, in persuasive support of punctuated equilibrium by the proper criterion of relative frequency, and with explicit attention to important and potentially confounding issues of geographic variation and missing data due to gaps in the geological record (1978, p. 17):

> The results of my analysis of 329 European Jurassic species provide, with an important exception, overwhelming support for the punctuated equilibria model. Species whose morphology appears to persist unchanged for long periods are abruptly terminated usually with one or more species of the same genus succeeding the older species with marked morphological discontinuity. The species ranges are long compared with the ammonites that allow fine stratigraphic subdivision and can be used to eliminate the possibility of significant stratigraphic gaps in the rock succession. Geographic variation within Europe is negligible, and more cursory examination of data from other continents provides no encouragement for the view that gradualistic events linking the "punctuated equilibria" in time took place outside Europe.

I trust, however, that Tony Hallam, one of my best friends in science, will not think me fractious or ungrateful if I point out that he then devoted the empirical content of his paper to documenting phyletic size increase in several species and, especially, to tracing gradual evolutionary changes within *Gryphaea*—in other words to the 1 lineage among 329 that illustrated phyletic gradualism. He presented no morphometric data for the overwhelming majority of species that remained in stasis throughout their existence. He wrote (1978, p. 17): "The succeeding sections of this paper are devoted primarily to this aspect of phyletic gradualism [size increase] and its implications in the broader context of environmental control of speciation, starting with the detailed analysis of *Gryphaea.*"

The unconsciously imbibed power of gradualism thus remained so strong during these early years of the punctuated equilibrium debate, that Hallam could declare "overwhelming support for the punctuated equilibria model" as his primary conclusion and the focus of his study—and then follow conventional practice in applying morphometric methods only to rare examples of gradualism within his sample, even though the predominant signal of stasis could be validated just as rigorously by the same methods. For all the theoretical uncertainties that still animate the punctuated equilibrium debate, at least we have made substantial headway on this operational issue since the 1970's. Any similar study, done now, would almost surely include the documentation of stasis.

Kelley (1983, 1984) studied all molluscan lineages with adequate samples over sufficiently long ranges in one of the most famous and widely studied of all fossil faunas: the Miocene deposits of the Chesapeake Group in Maryland (Shattuck, 1904, for the classic statement; Schoonover, 1941, for the standard stratigraphic study). In an initial study of rank correlations between stratigraphic position and values of unit characters at a standard shell length estimated from bivariate regressions, she found no directional change for 82 percent of characters within species of 8 lineages. Of the 18 percent showing significant rank correlations with time, most directional changes either become reversed later in the same sequence, or run in a direction opposite to the net transformation between the measured species and its descendant. In other words, such changes, however genuine, should be read either as mild fluctuations within a pattern of stasis, or as intraspecific temporal variation unrelated to the trend of the larger lineage. For example, shells of the bivalve *Lucina anodonta* become gradually less inflated from the Calvert into the overlying Choptank Formation. But the same species then regains its ancestral degree of inflation in the succeeding St. Mary's Formation (p. 587). Kelley (1983, p. 596) concluded with both substantive and methodological comments: "Within these middle Miocene mollusc species, then, changes are more commonly oscillatory than unidirectional . . . Most variables follow a pattern of fluctuation within a narrow range of values through time . . . In order to approach the goal of unbiased assessment of entire faunas, I examined all taxa of the mollusc faunas which were abundant enough for statistical analysis. Because no other bias controls the taxa chosen for study, these data provide strong evidence for punctuated equilibria."

Kelley's subsequent study (1984) affirms these patterns from a multivariate perspective based on discriminant analysis of 10 characters through 14 to 20 stratigraphic levels. Figure 9-23 (from Kelley, 1984, p. 1247) shows the stratigraphic distribution of centroids for each lineage at each level, as projected upon the first discriminant axis. Stasis prevails within most species (shown as unbroken vertical plots), while the four lineages composed of two or more successional species through the sequence generally show a stairstep pattern across transitions, and stasis within the bounds of species. In a very few cases, notably the transition from the lower to the middle species of *Anadara*, a trend within an ancestral species does move gradually towards the descendant's mean value. But even in this case, the third and uppermost species of the lineage then reverses the trend and moves back towards the beginning value.

Kelley (1984) also used patterns of misclassification for individual specimens to illustrate the character of predominant stasis. In the three successive species of *Astarte,* for example, 96.7 percent of specimens fall nearest the centroid of their own species—thus indicating sharp and clear division between successive species. But variation within species showed the opposite pattern. Only 42.1 percent of specimens fell nearest the centroid for their own stratigraphic level. Most remarkably, only 36.7 percent of misidentified specimens fall closest to centroids for samples of either the same or an immediately adjacent stratigraphic level. In other words, nearly $\frac{2}{3}$ of misidentified specimens stood closer to the centroids of stratigraphically distant populations than to the centroids for samples adjacent to their own time. This pattern of nonlirectional distribution throughout the full vertical range of species—compared with sharp divisions between species—illustrates the strength and character of stasis in these well-known fossil lineages.

Perhaps the most impressive and definitive study of pervasive stasis in molluscan faunas has been presented by Stanley and Yang (1987) for Neogene bivalves from the Western Atlantic region. They studied 24 variables (normalized for shell size) in 19 lineages, for a total of more than 43,000 measurements. Stanley and Yang followed a comprehensive sampling method, unbiased with respect to likelihood of punctuation and stasis, and including all species within four bivalve taxa (Lucinidae, Tellinacea, Veneridae and Arcticacea) with shells sufficiently large and geometrically tractable (flat to only weakly convex) for their measurement protocol, and with adequate numbers of well-preserved specimens (almost always more than 20 per sample, with a minimum of 16) over a sufficient range of time (at least 4 million years from early Pliocene to Recent).

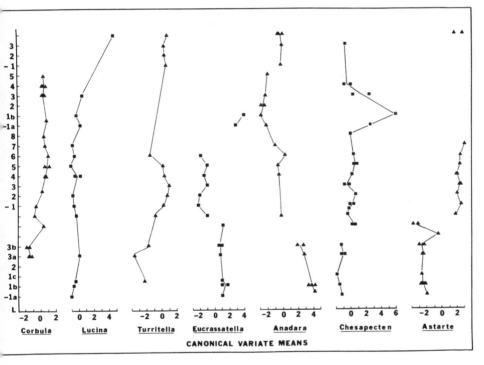

9-23. Multivariate changes based on discriminant analysis of 10 characters throughout 14 to 20 temporal units in the evolution of seven molluscan genera in Miocene strata of Maryland. From Kelly, 1984. Stasis prevails within a large majority of species. For most lineages where a descendant replaces an ancestor, a stair-step punctuation characterizes the transition. In a particularly interesting case of three successional species, ancestral *Anadara* does seem to move anagenetically towards the morphology of its descendant, which then remains quite stable. But the third and uppermost species, arising punctuationally, returns virtually to the morphology of the initial form.

Two additional features enhance the methodological value of this study: first, these species belong to the best known and most intensely studied of all molluscan faunas; secondly, all species either still exist (12 of 19 cases) or can be compared with a close living relative (almost surely the immediate descendant in 4 cases, and perhaps directly filiated in the other 3). Thus, in the most important innovation of this study, temporal variation can be directly scaled against current geographic variation of the same species, or of a close relative. In testing whether temporal fluctuations exceed the limits of stasis, comparison with the range of geographic variation among current populations of the same species should serve as our best anchor and standard.

Using eigenshape analysis for multivariate representation of shell form, Stanley and Yang first compared variation among modern populations for each species with differences between these modern populations and early Pliocene (circa 4 million years old) samples of the same species. In a convincing demonstration of stasis properly scaled to realized intraspecific variation, Figure 9-24 (from Stanley and Yang, 1987, p. 124) shows histograms for overlap of eigenshape areas in comparing modern geographic variation with 4 million year distances between early Pliocene and modern samples. The temporal mode slightly exceeds the geographic value, but the ranges overlap completely, and the difference in central tendency is very small. The authors conclude (p. 113) that "with minor exceptions, the distribution of morphologic distances between 4 million year old and Recent populations resembled the distribution of distances between conspecific Recent populations." "Approximate morphological stasis has been the rule for the taxa considered" (p. 124).

Stanley and Yang then extended their study (for species with available data) back to Miocene samples up to 17 million years old. Even for this extended duration, they found the same pattern of mild fluctuation, rarely extending outside the range of modern geographic variation, and with no

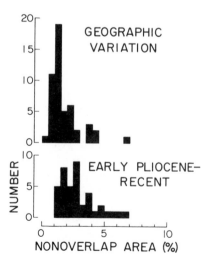

9-24. Non-overlap percentages based on comparisons of eigenshapes—from Stanley and Yang, 1987. The top histogram shows all conspecific pairs for the geographic variation of Recent populations. The bottom histogram shows differences between Recent populations and their presumed early Pleistocene ancestors. Note that the spread for the full geological range barely exceeds the spread for differences among geographic variants of living populations.

ccumulative directional effect. For example, Figure 9-25 (from Stanley and ʾang, p. 132) shows the temporal distribution of mean values for each of ne 24 characters over 17 million years in the venerid bivalve *Macrocallista maculata*. For most characters, the full temporal range lies within the variaⱽonal scope of living populations (noted by the "forks" for separate geoɽaphic samples at the top of the trajectory). They conclude (p. 113): "We ɑalculated net rates of evolution separating pairs of populations that belong ɔ single lineages. For all intervals of time, the distribution of differences beǂween population means for individual variables is remarkably similar to a ɔomparable distribution representing the comparison of pairs of conspecific ʀecent populations from separate geographic regions . . . Evolution has folɔwed a weak zigzag course, yielding only trivial net trends."

A particularly impressive study by Prothero and Heaton (1996) documents ɦe overwhelming dominance of punctuated equilibrium in a full tabulaɟion of one of the most prominent fossil faunas—a study that also gives ᵫs good insight into how biased reporting in general, and Cordelia's diᶒemma in particular (p. 763), can so strongly skew tabulated results to apɞearances of equal frequency or only mild domination by punctuated equiᶔbrium. These authors studied one of the world's richest and best known ɱammalian sequences—the upper Eocene and Oligocene White River Group ɔf the American High Plains, particularly as exposed in the Big Badlands of ɔouth Dakota—"one of the densest and most complete records of mammaᶨan evolution anywhere in the world . . . The spectacularly stark and beautiᵫul outcrops . . . have been a Mecca for fossil collectors ever since the first fosᵫils were described in 1846 . . . Enormous collections have accumulated, and White River fossils are found in nearly every rock shop and mineral show ᵫcross the country" (p. 259). This large mammalian assemblage seems to posᵫess sufficient long-term coherence (from Duchesnean strata of the late midᶨle Eocene into Arikareean strata of late Oligocene times) for designation as ɦe White River Chronofauna (Emry, 1981).

The authors spent more than a decade conducting "an unbiased survey of

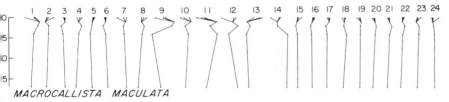

MACROCALLISTA MACULATA

ϱ-25. From Stanley and Yang, 1987. A history of change during 17 million years ϝor each of 24 measured characters in the bivalve *Macrocallista maculata*. For ɽhe great majority of characters, the entire temporal spread lies within the scope ɔf variation in the geographic range of living populations (represented by the "forks" for separate samples at the top of the trajectory). This form of compariɔon provides an excellent documentation of stasis by the criterion of scaling to ɽhe full range of geographic variation at a single time within the same taxon.

all fossil mammal lineages . . . which have large enough sample sizes and recent systematic revision" (p. 258) for the seven million year period (37–30 million years ago) across the Eocene-Oligocene transition (Chadronian to Whitneyan North American land mammal "ages"). The protocol also includes the two other factors—good geographic spread and temporal resolution—most essential for proper studies of punctuated equilibrium, but all too rarely realized: "This study considers geographic variation over a wide area (from western Montana and North Dakota in the north and west, to Colorado in the south), with very fine-scale chronostratigraphic control from magnetic stratigraphy and $^{40}Ar/^{39}Ar$ dating" (p. 258). Finally, and fortunately, this interval includes a major global climatic change (with no disruption of continuity in sedimentation), thus permitting researchers to study how such an external input influences rates of speciation and styles of phyletic change.

Prothero and Heaton found near exclusivity for punctuated equilibrium in the 177 well-documented mammalian species of this fauna. "Most species are static for 2–4 million years on average, and some persist much longer" (p. 257). "Only three examples of gradualism can be documented in the entire fauna, and these are mostly size changes" (p. 257). The details of these three cases also illustrate the exceptional status of gradualism, even at the smaller scale of their own taxonomic context:

1. The lagomorph *Palaeolagus* undergoes reduction in size of upper molars, accompanied by loss of their roots, during the early Orellan, but maintains stasis for much longer invervals both above and below: "Chadronian *Palaeolagus* shows about 2 m.y. of stasis, followed by gradual reduction in size and development of rootless upper molars during the early Orellan. From Orella B onward, several species of *Palaeolagus* are present, and except for slight changes, they are static for several million years" (p. 273).

2. The artiodactyl *Leptomeryx* experiences "subtle, gradual change in a number of characters" (p. 263) in the transition from *L. speciosus* to *L. evansi,* but both species show stasis throughout most of their substantial history—so we do not here witness the "classic" continuous anagenesis that supposedly makes the definition of species so arbitrary in temporal sequences: "While the transition from *L. speciosus* to *L. evansi* is not stratigraphically instantaneous, it occurs in a relatively short time compared to the long durations of both species."

3. The oreodont *Merycoidodon* does seem to undergo extensive and gradual dwarfing (30 percent size reduction) over a one million year interval in the early Orellan. I accept this case as a good example of extended gradualism (see Fig. 9-26 taken from Prothero and Heaton, 1996, p. 262), but also note that the trend occurs within a common genus, including several species otherwise showing predominant stasis—and that the dwarfing trend only involved the labile character of size, without concomitant changes of shape, a common finding among exceptional cases of gradualism in faunas dominated by punctuated equilibrium (see previous discussion of Jurassic bivalves on page 855).

Such exhaustive and unprejudiced tabulations can give us insight into the limited value—that is, for establishing proper relative frequencies, not for resolutions of particular cases—of trying to infer the quantitative distribution of rates and modes for all taxa from previously published research carried out within the "best case" tradition, usually with strong (and unacknowledged) preference for defining evolution only as geologically gradual change. Before

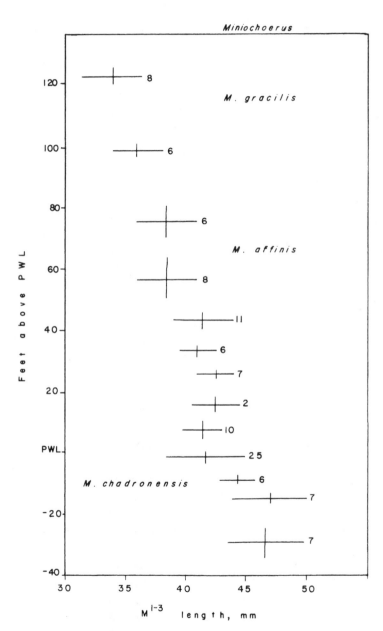

9-26. A rare case of gradualism amongst the overwhelming domination in relative frequency for stasis in 177 well-documented mammalian species of the Big Badlands fauna of South Dakota. This species of *Miniochoerus* (previously known as *Merycoidodon*) does undergo gradual dwarfing to 30% size reduction over a one million year interval. From Prothero and Heaton, 1996.

the punctuated equilibrium debate began, how would an evolutionary pale-ontologist have treated the White River Fauna? Almost surely, any expert on these strata would have selected the cases of apparent gradualism for study and publication, while ignoring the others as negative instances of no evolu-tion, worth only a side comment at best, if noted at all, and suited for explicit mention (but without any quantitative analysis) only in formal taxonomic treatises. Thus, readers with no personal knowledge of the entire fauna—es-pecially non-paleontological readers unaware of strong signals for stasis and punctuation in virtually all faunas—would almost surely assume that the three reported studies characterized the usual situation for the history of fos-sil species, rather than representing the only examples of a rare phenomenon.

Prothero and Heaton (1996, p. 258) raise the important point that the ex-amples of gradualism most widely featured, and most frequently cited to urge the general case against punctuated equilibrium, derive from such faunas—where they stand as unusual examples against an unstudied (or simply non-discussed) but overwhelming prevalence for stasis and punctuation among all species. They remind us, for example, that Gingerich's most famous half dozen or so cases from lower Eocene beds of the northern Bighorn Basin "are just part of a fauna of over a hundred genera. Detailed monographs by Bown (1979), Schankler (1980), and Gingerich (1989) [in his very own taxonomic work] have shown that stasis is prevalent among most of the taxa not fea-tured by Gingerich (1976, 1980, 1987)" (p. 258). Of another famous claim for gradualism (one that I do not challenge as a single case, while asking that relative frequencies also be acknowledged), Prothero and Heaton (1996, p. 258) write: "Krishtalka and Stucky (1985, 1986) reported a gradualistic transformation in the early Eocene artiodactyl *Diacodexis*. However, this is a single lineage from the same faunas described by Schankler, Gingerich, and Bown, so these studies do not address the overall prevalence of gradualism vs. stasis."

Finally, although this issue belongs more to the forthcoming discussion of faunal stasis as an extension of punctuated equilibrium (see pp. 916–922), Prothero and Heaton's (natural) experimental design in choosing the White River Chronofauna for such intensive study included the existence, in the midst of the fauna's duration, of one of the most profound and rapid climatic changes in Tertiary North America—"the earliest Oligocene climatic crash" (p. 257) at 33.2 million years ago, where "vegetation changed from dense forests to open forested grassland, mean annual temperatures dropped 13°C, and conditions got much drier and more seasonal" (p. 257). The nondis-turbance of stasis—indeed, the virtual "ignorance" of this event by most spe-cies, at least by observable changes in diversity or skeletal anatomy—also illustrates the strength of stasis and the apparently active (rather than merely passive) sources of its maintenance. Prothero and Heaton write (1996, p. 257): "Only a few mammalian lineages speciated, a few more went extinct, and the vast majority (62 out of 70) persisted through this climatic event with no observable response whatsoever." The authors then end their paper by throwing down the gauntlet to supporters of traditional evolutionary views about response to environmental change: "Evolutionary theory has come a long way since Eldredge and Gould (1972) first pointed out that stasis is the norm in the fossil record, and the data cannot be simply dismissed or explained away . . . In fact, stasis and resistance to change is so ingrained that species can actually pass through the most significant climatic change of the last 65 million years as if nothing happened."

In recent years, studies of stasis and punctuation in entire faunas have blossomed, especially with the introduction and testing of two partly complementary, but partly dissonant, explanations for apparently concerted stability of entire faunas over substantial intervals—the *turnover-pulse hypothesis* of Vrba (1985), and the theory of *coordinated stasis*, developed by Brett and Baird (1995 and several other works), and extensively (often contentiously) treated in the symposium of Ivany and Schopf, 1996. I shall treat the theory itself in the last section of this chapter (pages 916–922), but will record here the convincing documentation of extensive faunal stasis that established the evidentiary base for these ideas.

The famous Middle Devonian (Givetian) Hamilton fauna of the Appalachian Basin has provided a "type" case for coordinated stasis. The Hamilton fauna includes more than 330 species of mostly typical Paleozoic invertebrates, ranging through about 9 million years of strata in a series of about twenty identifiable "communities" or biofacies. About 80 percent of these species persist throughout the entire Hamilton, while fewer than 20 percent carry over from the fauna just below. Ever since Cleland's original and wistful comment in 1903 (quoted on p. 750), students of the Hamilton fauna have recognized the overwhelming signal of stasis presented by almost every species of the assemblage—although this original wistfulness has now ceded to considerable positive interest! Brett and Baird write (1995, p. 301): "Individual lineages within particular biofacies of the Hamilton biotas appear to display very little morphological change, and that which is observed is neither progressive nor directional."

Several taxa of this fauna have now been analyzed in great morphometric detail, beginning with Eldredge's classic study of the trilobite *Phacops rana*, one of the "founding" examples of punctuated equilibrium (see Eldredge, 1971; Eldredge and Gould, 1972). Eldredge found stasis in more than 50 characters, and directional, but punctuational, change only in one—reduction in rows of eye facets in two punctuational steps during a 5–6 million year period otherwise marked by stasis for this feature as well. Other quantitative studies of stasis in Hamilton species include Pandolfi and Burke (1989) for tabulate corals, Lieberman (1995) on trilobites, and Lieberman, Brett, and Eldredge (1994) on brachiopods. The last study considered 8 characters in two species using principal components and canonical discriminant analysis. The authors found fluctuating variation, correlated neither with age nor facies, throughout the interval. However, and ironically given past expectations of gradualism, the uppermost samples plotted closer to the lowermost than to any intervening population. For the entire fauna, Brett and Baird conclude (1995, p. 303): "Taken together, these studies indicate that a majority of Hamilton lineages display virtual stasis from oldest to youngest samples. Slight nondirectional change is observed in some cases. Such variation seemingly records very minor evolutionary fluctuation . . . However, it clearly does not lead to development of major new grades of morphological development . . . Several of the species appear abruptly in the Appalachian Basin near the beginning of the Hamilton fauna or become locally extinct at its end."

### Relative frequencies for entire clades

We add another component to studies of relative frequencies when, in addition to the thoroughness provided by assessing all lineages within a given time or region, we add the phylogenetic component of complete coverage for clades (preferably monophyletic of course, but sometimes paraphyletic in the existing literature). Obviously, we feel most secure about such phylogenetic assertions when truly cladistic, or at least stratophenetic, criteria have been used for definitions, but many studies in this mode employ a standard that, albeit and admittedly less preferable, probably provides as much confidence in practical utility: investigations of distinctive taxa known on reliable biogeographic grounds to be restricted to a region exhaustively studied. Clades confined to isolated islands, lakes, or other such distinct and coherent places and environments constitute our best cases under this criterion.

I have discussed nearly all the best examples in this mode under other headings of this chapter, and will only make brief reference here. Several "classic" mammalian lineages fall into this category of excellent cladistic definition and overwhelming domination by the punctuational pattern of stasis within species and geologically abrupt transitions between—all despite (or rather, in a punctuational reformulation, because of!) such celebrated evidence for sustained and important trends. I include here the excellent evidence for horses (see p. 905) and the spottier but still persuasive data on hominid evolution (see pp. 908–916)—in each case, for clades well delimited both by morphology and geography.

Among such geographically confined clades, Vrba's classic studies (1984a and b) of African antelopes stand out for detailed data on one of the most successful and speciose of vertebrate higher taxa. In the maximally diverse tribe Alcelaphini (including blesbucks, hartebeests, and wildebeests), the Quaternary record includes 25 species, all with a geologically sudden origin in recorded data, and with cladogenesis as a reliably inferred mode of origin for at least 18 species. Several species lived for 2 million years or longer in stasis, and no ancestors with incrementally transitional morphologies have been found for any of these forms.

I continue to be amazed by the skewed interpretation often imposed by gradualistic expectations upon data for clades that seem, at least in my partisan judgement, clearly dominated by punctuated equilibrium in overall relative frequency. For example, in a well-known work, White and Harris (1977) used the Plio-Pleistocene record of African pigs for supposed validation of gradualism as a primary guide in biostratigraphic resolution (particularly of some important hominid-bearing strata). They did document one or two cases of gradual change, notably an increase in third molar length for *Meso-choerus limnetes*. But the clade includes 16 species during this short period of no more than 4 million years, 8 of which arise by punctuational cladogenesis even in White and Harris's own diagram (1977, p. 14). The authors' comments, unwittingly I suspect, frequently point to the domination of evolutionary history in this clade by cladogenetic events and their consequences. They write (1977, p. 14), for example, that "*Metridochoerus* underwent a substantial adaptive radiation during the early Pleistocene, and at one point four distinct metridochoere species existed contemporaneously."

Many of the best invertebrate examples fall into the same category of unique and endemic taxa confined to isolated places, and therefore forming, by strong inference, a complete and coherent phylogenetic unit—as in Williamson's study (1981), cited several times previously, of speciation in pulmonate snails from separated African lakes. In another example from a famous sequence of much greater temporal extent, Geary (1990, 1995) studied the evolution of melanopsid gastropods in the Middle to Late Miocene beds (spanning 5 to 10 million years) of the Pannonian Basin in eastern Europe. In one case of gradualism, following a much longer interval of at least 7 million years in stasis for the ancestral form, *Melanopsis impressa* transformed to *M. fossilis* by directional increases in shell size and shouldering over a two million year interval. However, within the same Pannonian Stage, at least six new melanopsid species arose by punctuation: "their first appearances are abrupt, and preceded by no intermediate forms" (Geary, 1995, p. 68). Geary (p. 69) regards the stratigraphic resolution as "not particularly good," but still fixes the origin of these species to within "tens of thousands of years"—a clear punctuation by the criterion of scaling against average species duration in stasis within the clade. Figure 9-27 (from Geary, 1995, p. 68) depicts Geary's results for this geographically isolated evolutionary radiation.

As mentioned many times in several contexts within this chapter, Alan Cheetham's studies of the bryozoan *Metrarabdotos* (1986, 1987), now supplemented with the work of Jackson and Cheetham (1994, and Cheetham and Jackson, 1995) on *Stylopoma*, have set a standard of excellence and confidence for empirical studies of relative frequency. All major desiderata for such research have been realized in these genera—a group with a well-resolved phylogeny, in a clade restricted to a geographic region, and exhaustively sampled in strata of unusually complete resolution over a long period. Moreover, Cheetham's multivariate morphometrics permit us to assess stasis and punctuation as a morphological totality, not only as a potentially biased impression based on a few preselected characters. Finally, studies with Jackson on the ecology and genetics of extant species demonstrate (see page 786) that the morphology of paleospecies almost surely provides a good surrogate and identifier for true biospecies in this clade. (As a personal note, I am also gratified that Cheetham began these studies with the intention of proving his suspicions for gradualism in the context of the developing debate about punctuated equilibrium—and ended up with the finest data-driven evidence ever gathered for the domination of a total evolutionary pattern by punctuated equilibrium.)

To recapitulate the major conclusions of these studies (see also Figure 9-18 for the phylogeny of *Metrarabdotos,* with morphology expressed as multivariate Euclidian distances between samples based on all canonical scores of a discriminant analysis, and connecting nearest morphological neighbors in stratigraphic sequence), Cheetham measured 46 characters in 17 species of *Metrarabdotos* over a duration of 15 million years, with intense sampling for a 4.5 million year interval of Upper Miocene to Lower Pliocene sediments (3.5 to 8.0 million years ago) in the Dominican Republic. Cheetham (1986, p. 195) specified the favorable features of *Metrarabdotos* on both geographic and phylogenetic grounds: "The ascophoran genus *Metrarabdotos* is a favorable subject for detailed analysis of evolutionary pattern because of its diversity and wide distribution during much of Miocene and Pliocene time. Caribbean species . . . form an apparently monophyletic subset within which phylogenetic relationships can be inferred independently of evolutionary events in eastern Atlantic-Mediterranean congeneric species groups."

Moreover, the unusual resolution for the detailed sampling interval permits "a fine-scale comparison of successive populations similar to those made with oceanic planktonic groups in deep-sea cores" (1986, p. 195), thus dispelling the common argument of gradualists that stasis in metazoans from conventional sediments must arise as an artifact of coarseness of resolution, while

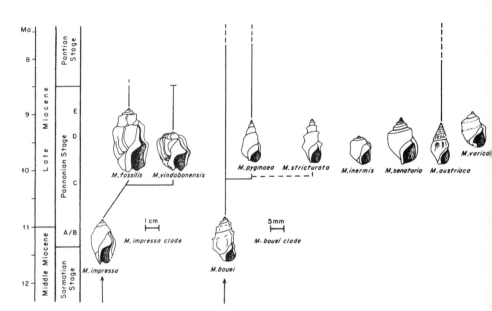

9-27. From Geary, 1995. In the radiation of melanopsid gastropods in Middle to Late Miocene beds (spanning 5–10 million years) of the isolated Pannonian Basin, Geary found one case of gradualism where, after at least 7 million years of stasis, ancestral *M. impressa* transformed to *M. fossilis* by directional increase in shell size and shouldering over a million year interval. However, during the same time, at least six new melanopsid species arose by punctuation, as also shown.

gradualism of microfossils in deep-sea cores must record a general pattern that would be seen wherever stratigraphic sampling could attain such completeness. Cheetham (1986) calculated 160,000 years for average spacing between successively sampled populations in the intensely collected interval—for a stratigraphic completeness of 0.63 by Sadler's (1981) criteria.

In *Metrarabdotos* (again see Fig. 9-18), 11 of the 17 species persist in stasis for 2–6 million years, and all originate punctuationally within the limit of resolution (at least in the intensely sampled interval) of 160,000 years—undoubtedly in far less time for many branching events, since 160,000 years represents a maximum figure based on the available unit of measurement. Again for the intensely sampled interval, Cheetham writes (1986, p. 190) that "nine comparisons of ancestor-descendant species pairs all show within-species rates of morphologic change that do not vary significantly from zero, hence accounting for none of the across-species difference. In all cases, the ratio of within-species fluctuation to across-species difference is low enough to allow the punctuated pattern to be distinguished with virtual certainty. In at least seven of the cases, ancestor species persisted after giving rise to descendants, in conformity with the punctuated equilibrium mode of evolution."

The morphometric details can only increase confidence in "the remarkably clear-cut evidence for a punctuated evolutionary pattern in these *Metrarabdotos* species" (1986, p. 201). The reported central tendencies of samples integrate data from 46 measurements, providing a good assessment of general anatomical distance (based on characters considered important in the taxonomy and functional morphology of these organisms), and not on selected single characters (see Cheetham, 1987, for affirmation of punctuated equilibrium from analyses of temporal trends in individual characters as well). In supplementary affirmation, Cheetham studied the fine-scale pattern of temporal variation by computing autocorrelations between mean scores of stratigraphically successive pairs of populations: "In all cases, the autocorrelations of mean scores of successive populations are nonsignificant and near zero, and the autocorrelations of rate deviations are negative and (except in one case) nonsignificant. These autocorrelations clearly indicate that changes within species are fluctuations around a near-zero, otherwise unchanging rate" (1986, p. 201).

Finally, some authors (see Marshall, 1995) have challenged Cheetham's phylogeny for its stratophenetic basis. But a purely cladistic analysis, as now preferred by many researchers, not only changes the previous scheme in only minor ways (Cheetham and Jackson, 1995, p. 192), but also—and the point becomes almost amusingly obvious once one grasps the different criteria used by the two methods—leads to an even stronger pattern of punctuated equilibrium, for the stratophenetic phylogeny minimizes the mean morphologic distances between putative ancestor-descendant pairs, while the cladistic phylogeny makes no such assumption and must therefore yield a larger mean difference between species. Since the documented stasis *within* species is not affected in either case, the cladistic scheme must increase the average magnitude of punctuational events, thus only decreasing the likelihood that between-species differences could be extrapolated from temporal variation within species (see Cheetham and Jackson, 1995, p. 192, for an elaboration of this argument with appropriate data).

The corroborative study of a second bryozoan genus, *Stylopoma,* from the same beds yields an identical conclusion of overwhelming predominance—indeed, exclusivity—for punctuated equilibrium. Jackson and Cheetham (1994) used 12 morphological features to identify 19 species in this rarer genus. Despite their more limited information, Cheetham and Jackson (1995, p. 195) found that "temporal overlap between putative ancestor-descendant species pairs is even greater than for *Metrarabdotos,* with 10 species surviving beyond the detailed sampling interval more than 6 million years to the Holocene." Moreover, "no evidence of morphologically intermediate forms" (Jackson and Cheetham, 1994, p. 420) has been found for any transition; all species origins are fully punctuational at the scale of detailed sampling.

Finally, since *Stylopoma* provided Jackson and Cheetham's principal data for the correspondence of genetically defined biospecies with morphologically designated paleospecies (modern specimens of *Metrarabdotos* are much less common and not so well suited for genetic work), this second study provides strong additional support for punctuated equilibrium by coordinating several potentially independent indicators of evolutionary change with rapid events of branching speciation: "Moreover, the tight correlation between phenetic, cladistic, and genetic distances among living *Stylopoma* species suggests that changes in all three variables occurred together during speciation. All of these observations support the punctuated equilibrium model of speciation."

I regard these empirical studies of relative frequencies as the strongest evidence now available for the most important and revisionary claim made by the theory of punctuated equilibrium: the overwhelming domination of evolutionary patterns in geological time by events at the species level (or higher), and the consequent need to explain macroevolution by patterns of sorting among species rather than by extrapolated trends of anagenetic transformation within continuous lineages.

### Causal clues from differential patterns of relative frequencies

Once we set our focus of inquiry on determining the relative frequencies of punctuated equilibrium in different times, places, environments, and taxa, we can ask the classic question of natural history, a subject rooted in the concept of variation: do we note characteristic differences in relative frequencies based on any of these factors and, if so, can we draw any causal inferences (useful to evolutionary theory) from these patterns. I have already raised this question in a number of preceding contexts in this chapter, most extensively for the observed higher frequency of gradualism in predominantly asexual oceanic protistan lineages, where I argued that this unusual result may not record the greater completeness of strata in oceanic cores (as traditional views have assumed), but probably arises from interesting biological differences that have led us to look for a truly underlying punctuational pattern at the wrong scale in this case (see pp. 803–810).

Most discussion on the linkage of differences in frequencies to distinctive structural and functional characteristics of organisms (rather than to types of environments *per se*) has focussed on claims for variation among broad taxonomic groups (high frequency for gradualism in rodents vs. rarity in bovids, for example), but I strongly suspect that this genealogical emphasis reflects traditions of specialization in research more than any inherent preference for taxonomic parsing in such a search. We should also consider more general features of organisms that cut across taxonomic lines, and we should therefore examine broader differentia potentially related to chosen environments or tendencies to speciate. Schoch (1984), for example, suggested a link between high frequencies for punctuational speciation and intense social competition, arguing that selection on such features tends to proceed so rapidly, even in ecological time, that speciation would almost surely occur in a geological instant. Breton (1996) linked punctuational modes to evolution of "pioneer" structures (evolutionary novelties tied to morphological reorganizations), and gradualism to "stabilization" and "settlement" structures (refinements and improvements in local adaptations). I suspect that such arguments may apply better to the different issue of average amounts of change per speciation event, than to questions about the relative frequency of punctuational events (at whatever degree of alteration) *per se*.

Most arguments about patterns of differences in relative frequencies have invoked "externalist" claims about characteristic environments, rather than "internalist" correlations with structural features of organisms (although the two subjects may, of course, be correlated and need not stand in antithesis). In a first attempt, Johnson (1975, 1982), working with Devonian brachiopods and conodonts but generalizing more widely in an important set of papers, linked higher frequencies of gradualism to pelagic environments and greater prevalence of punctuated equilibrium to benthic habitats. He then justified the ecological correlations by linking characteristic evolutionary modes with relative stability of environments: "Among marine invertebrates, pelagic organisms are the most likely to have inhabited extensive, gradually changing environments and are therefore the most likely to have evolved by a rate and pattern that can be described as phyletic gradualism . . . Post-larval, attached and stationary benthic organisms are the most likely to have inhabited environments that are subject to relatively abrupt changes and are therefore the most likely to have evolved by a rate and pattern that could be described as punctuated equilibria."

In a series of papers, Parsons (well summarized in 1993) suggested a similar linkage, while proposing a different, but generally concordant, explanation based on a putative correlation between environmental "stress" and patterns of genetic variation and available "metabolic energy." (I put Parsons's last factor in quotation marks because I have trouble grasping both the definition and operationality of such a concept.) Parsons writes (1993, p. 328): "In moderately stressed and narrowly fluctuating environments, sufficient genetic variability and metabolic energy should be available to permit adaptation. In these environments, phyletic gradualism is expected. In highly stressed and widely fluctuating environments, a punctuated evolutionary pattern is expected whereby stasis occurs most of the time."

In one of the most important recent papers on punctuated equilibrium, Sheldon (1996) has generalized a superficially paradoxical link of morphological stasis to highly fluctuating environments, and gradualism to more stable and narrowly fluctuating environments, as the *"plus ça change"* model—citing the sardonic French motto that "the more things change, the more everything's the same," a reference to the proposed link of morphological stasis with highly variable environments. Sheldon (1996, p. 772) explained the basis and resolution of the paradox: "One might expect a changing environment to lead to changing morphology, and a stable environment to stable morphology. But over long intervals the opposite may often occur . . . Perhaps gradual phyletic evolution can only be sustained by organisms living in or able to track narrowly fluctuating, slowly changing environments, whereas stasis, almost paradoxically, seems to prevail in more widely-fluctuating, rapidly changing environments." Sheldon's figure (reproduced here as 9-28) will make his argument clear. Species in highly variable habitats must adapt to pervasive and rapid fluctuations, and generally do so by evolving a stable and generalized morphology suited to the full environmental range. But when external fluctuation exceeds a certain limit of internal toleration, rapid speciation may be the only viable response. On the other hand, mildly fluctuating environments may enhance selection for more precisely tuned adaptations capable of tracking long-term climatic trends by gradual adjustment.

Sheldon (1987) began his work by publishing one of the most widely discussed empirical defenses of gradualism, based on several lineages of Ordovician trilobites from the Builth Inlier, an environment interpreted as generally stable and only narrowly fluctuating. (I appreciate the richness of Sheldon's data, but regard his interpretations as ambiguous, for most of his published trajectories seem to me—from my partisan standpoint (as I keep repeating to

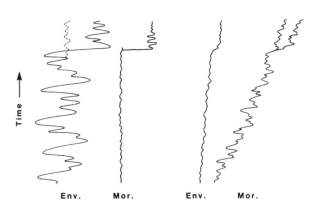

9-28. An epitome of Sheldon's argument (1996) that, paradoxically, highly fluctuating environments may induce stasis and punctuation, with gradualism more commonly found in environments undergoing slower but more steady change.

remind readers to be especially critical)—more consistent with expectations of fluctuating stasis with little or no net change, see Eldredge and Gould, 1988.) I am particularly grateful that Sheldon, even while developing one of the most famous data sets against the theory, has always accepted the importance of establishing relative frequencies for different groups and situations, and has consistently regarded punctuated equilibrium as a valuable theory (to which he has made major contributions), with important implications for our understanding of macroevolution.

Such broad arguments about environmental correlations have been notoriously difficult to document because, even when the effect can be validated as both real and pervasive, so many other factors will be operating in any particular case (including immediate and local influences able to overwhelm the smaller impact of the generality under test) that the signal may be lost in surrounding noise. But I am strongly attracted to Sheldon's *plus ça change* hypothesis for two primary reasons. First, the concept makes good sense of patterns that have often been noted empirically, but regarded as confusing in interpretation—particularly the common finding of pronounced stasis through major climatic fluctuations, including Pleistocene ice age cycles (Cronin, 1985, on ostracodes; Coope, 1980, 1994 on beetles), and the largest climatic crash in Tertiary North America (Prothero and Heaton, 1996). The presentation of a hypothesis like Sheldon's prompts researchers to focus studies on interesting issues, and to seek wider implications. For example, Wei (1994) used Sheldon's hypothesis to explain the link of stasis to intensification of ice-age climatic fluctuations in the planktonic foram *Globoconella inflata*. Wei (1994, p. 81) suggests that stasis may represent "a compromise for the species as an attempt to meet with both glacial and interglacial extremes."

Second, Sheldon's hypothesis predicts a large suite of definite correlations subject to empirical test. *Plus ça change* predicts linkages of different relative frequencies for punctuated equilibrium and gradualism to geographic gradients (with more punctuated equilibrium expected in temperate areas, and more gradualism in the topics), environmental distinctions (with more punctuated equilibrium in nearshore shallow-water strata and more gradualism in offshore regions, as Johnson had earlier predicted), and evolved responses of organisms and populations (with, *ceteris paribus*, more punctuated equilibrium in eurytopes and *r*-strategists, and more gradualism for stenotopes and *K*-strategists). Needless to say, *ceteris paribus* does not always hold—but with so many expected consequences, the probability of finding patterns (if they exist) does rise substantially.

Finally, as for any good hypothesis in science, Sheldon's *plus ça change* suggests several interesting extensions. For example, Sheldon raises an intriguing argument for linking these putative correlations with patterns of genuine selection at the species level or above:

Perhaps the most important (and perhaps the most controversial) mechanism I am suggesting here is a type of lineage selection with two stages:

(1) if an established or an incipient species experiences a widely fluctuating environment on geological timescales, the evolutionary response (morphological change) tends to become damped with time, and (2) those species that are least sensitive to environmental change (the most "generalized" in a long-term sense) are the ones that tend to persist, remaining in morphological stasis until a threshold is reached (Sheldon, 1996, p. 218).

In a second extension, Sheldon makes an almost quizzical, but oddly compelling, argument based on another important source of potential correlations, previously unaddressed here but perhaps quite important: time itself, expressed either as the absolute time of particular intervals in the earth's history, or as the relative time of distinctive segments in the general "ontogeny" of a species's duration. Many biologists have noted the apparent paradox that so little sustained and directional evolution (as opposed to abundant evidence for rapid and adaptive fluctuations in such characters as bill form in Darwin's finches or wing colors in peppered moths) has been noted for species in historic, and recent prehistoric, times during the tenure of modern humans (who have also remained in stasis) on earth. I would, of course, attribute this phenomenon mostly to a general prediction for stasis in the vast majority of lineages at any time (while charging our puzzlement only to the false equation of evolution with gradual change). But Sheldon raises the interesting ancillary argument that this general expectation may now be enhanced by special advantages for stasis in the regimes of strong and rapid worldwide climatic fluctuation that our earth has been experiencing in these geologically unusual times: "Given the Quaternary climate upheavals, relatively little evolution may be occurring worldwide at present (except for evolution induced by humans)" (Sheldon, 1996, p. 209). I can only hope that the more punctuated equilibrium induces change in our evolutionary views, the more things will *not* be the same in our interpretations of the history of life.

## The Broader Implications of Punctuated Equilibrium for Evolutionary Theory and General Notions of Change

### WHAT CHANGES MAY PUNCTUATED EQUILIBRIUM INSTIGATE IN OUR VIEWS ABOUT EVOLUTIONARY MECHANISMS AND THE HISTORY OF LIFE?

#### The explanation and broader meaning of stasis

As emphasized throughout this chapter, the stress placed by punctuated equilibrium upon the phenomenon of stasis may emerge as the theory's most important contribution to evolutionary science. The material world does not impact our senses as naturally and objectively parsed categories. We can make accurate observations and measures of particular "things," but the ordering of "things" into categories must be construed largely as a mental operation based on our theories and attitudes towards "reality." Moreover, we must also apply mental screening to select "things" meriting our attention within nature's potential infinity, and even to recognize a configuration of matter as a "thing" in the first place. Therefore, phenomena without names, and without theories marking them as worthy of notice, will probably not be recognized at all.

The phenomenon always existed "out there" in nature, of course, but punctuated equilibrium largely "created" the category of stasis as an important item in evolutionary theory through a four-step process of (1) defining stasis as a positive "thing" with properties and boundaries, a phenomenon rather than an unnamed and unrecorded absence of evolution; (2) bringing stasis to visibility as the expectation of a particular theory of evolutionary modalities; (3) suggesting methods for the active and rigorous study of stasis, so that the concept could be operationalized as a subject for empirical research; and (4) granting interest and importance to stasis as a controversial topic with broad implications for revising traditional modes of thought in evolutionary biology.

Before Eldredge and I published our first paper in 1972, most paleontologists treated stasis as an embarrassment, imposed by the poverty of the fossil record upon hopes for recording evolution (defined as gradualistic anagenesis), and therefore as not meriting active study, or even explicit recognition as a discrete phenomenon. Just a decade later, the situation had changed so dramatically that Wake, Roth and Wake (1983) could write, "perhaps no phenomenon is as challenging to evolutionary biologists as what has been termed 'stasis'" (p. 212), defined by them as "the maintenance of a standard morphology over vast periods of time during which much environmental change has taken place" (p. 211). Illustrating my claim that a phenomenon becomes interesting only in the light of defining theories, Wake *et al.* (1983, p. 212) then stated: "With natural selection operating in a changing environment as an agent of adaptation, we expect to see changes at the organismal, ultimately physiological and morphological, level. How, though, can we explain the paradoxical situation in which environments change, even dramatically, but organisms do not?"

As I now survey the subject, a quarter century after our initial presentation and definition, stasis has become an even more general and important issue in evolutionary theory for three principal reasons:

FREQUENCY. Once the phenomenon had been named, and criteria established for recognition and study, researchers documented stasis at far too high a relative frequency to represent anything other than an evolutionary norm and expectation. Such predominance also implicates stasis as a property *actively* maintained by species—thus leading to a substantial literature (discussed at the end of this subsection) on the causes of non-change. Several authors, notably Paul (1985) and Jackson and Cheetham (1994, also Cheetham and Jackson, 1995), developed models and data sets to prove that stasis occurs too frequently for explanation under random models (including pure neutralism with no natural selection), and therefore must be caused by active forces promoting such a result, either directly, or as a consequence of some important linked property of organisms or populations.

This growth in emphasis has been so vigorous since 1972 that geological gradualism, once the unquestioned expectation of evolution itself, is now generally regarded as an infrequent, if not anomalous, phenomenon requiring a special explanation in the light of anticipated stasis. Geary (1990, p. 507) after documenting a case of gradualism within a clade showing a much higher frequency of stasis, wryly noted: "Given that past studies were assumed complete only if gradual change was apparent, it seems somewhat ironic that unseen mechanisms or events, however realistic, must now be invoked in order to explain an instance of gradual change!" Gradualism, in short, has become both a rarity and a puzzle. (Much as I take a rather wickedly and secret personal pleasure in this sea change, I'm not sure that I can, in good scientific conscience, regard such *a priori* mental downgrading of gradualism as a "good thing." *A posteriori* downgrading based on documented rarity represents nature's chief signal in my view, but I do think that any study should begin with equal potential welcome for either result!)

GENERALITY. The interest in stasis, originally generated by punctuated equilibrium for inquiries at the appropriate level of species durations through time, has since expanded to other domains of size and time, and to more comprehensive questions about the nature of change itself. Causes operating at punctuated equilibrium's proper scale will not explain other forms of stasis, but the generalized definition and inquiry did arise by expansion from our theory (at least as a sociological phenomenon), while we may also anticipate the identification of some common causes or constraints (see further discussion on conceptual "homology," pp. 928–931)—that is, in evolutionary parlance, causal parallelisms, based on structural homologies, rather than convergences or mere analogies of appearance—behind the deeper generality (with different immediate forces producing similar and partly homologous results at various levels). I shall discuss some of these other scales in Part B of this subsection. These extensions include: punctuational anagenesis for directional changes in lineages of asexual organisms by clonal sorting (in a domain below punctuated equilibrium, which, *sensu stricto,* only operates at the level of speciation to explain trends in multicellular sexual lineages by species sorting); longterm morphological stability for basic anatomical features of larger clades (at a level above punctuated equilibrium and within monophyletic lineages—see Chapter 10); putative "lock-step" stasis for the great majority of defining species within larger faunas through significant geological intervals (at a still higher level above punctuated equilibrium and across genealogical lineages to a consideration of faunal dynamics—see pp. 916–922). Interest has also extended beyond evolutionary systems to the meaning and causes of stasis in stairstep patterns of ontogenetic growth, stubbornly persistent plateaus followed by thresholds of rapid change in response to continuous input in human learning, and active stasis followed by punctuational breakdown in the history of human ideas and social organization (see pp. 952–967).

CAUSALITY. Fruitful debate about the causes of stasis must first specify the level manifesting the common phenomenology. (Obviously, causes of learning plateaus in piano playing cannot be strict homologs of ecological reasons for joint stability of species in coordinated stasis, even though the graphed pattern of change may manifest the same geometrical form.) In this section, I confine my discussion to punctuated equilibrium *sensu stricto*, and not to the general pattern of punctuational change at any level—that is, to proposed reasons for the observed high relative frequency of stasis during the full geological range of metazoan species as preserved in the fossil record.

But first, and as an example of how discussion can proceed at cross purposes when proper scales have not been specified, I must note that most of the literature proclaiming punctuated equilibrium as "old hat" (Lewin, 1986) or something long known and merely hyped by ill-informed paleontologists, has only analyzed ecologically rapid anagenesis in populations rather than the relevant phenomenon of cladogenesis by speciation scaled against subsequent geological duration in the stasis of species so generated. Most notably (in terms of subsequent commentary), two papers of the mid-80's (Newman, Cohen and Kipnis, 1985; and Lande, 1986) developed mathematical models to show that single populations could move rapidly (in the "ecological time" of a human career) from one adaptive peak to another in the absence of environmental change. (The major previous stumbling block had been set by problems in envisaging how a population could move *down* an adaptive peak, against any force of selection, to inhabit a valley, and therefore become subject to selection up an adjacent peak. The basic solution—that the descent must be rapid—allows sufficient impetus against selective forces, and also links the models to themes of speedy anagenesis.)

Lewin (1986) used these studies to write a news and views feature for *Science* entitled "Punctuated equilibrium is now old hat," while also recognizing that ecologically rapid anagenesis does not address the scale or level (not to mention the reality of *changing* environments in our actual world) of punctuated equilibrium's central concern. We welcome such plausible models of ecologically punctuated anagenesis as a contribution to understanding the panoply of causes that yield punctuational change at other levels. But this smaller-scale phenomenon, however fascinating and important, bears little relevance to the causes of stasis within species during geological time (or to the cladogenetic sources of geological punctuation as a slow branching event in ecological time).

We may order the major propositions for explaining stasis at the scale of punctuated equilibrium as an array running from conventional resolutions based on Darwinian organismic selection to more iconoclastic proposals invoking either higher levels of causation or less control by selection and adaptation. (Much of the genuine interest in the otherwise tedious and tendentious debate on the theoretical novelty of punctuated equilibrium lies in the legitimate weights that will eventually be assigned to the various proposals of this array.)

STABILIZING SELECTION. For most evolutionists who chose to see nothing new in punctuated equilibrium, the previously unacknowledged frequency of stasis (admitted, albeit sometimes begrudgingly, as an unexpected finding) could only indicate a stronger role than previously envisaged for the conventional mechanism of stabilizing selection. Although this putative explanation of stasis within paleospecies achieved an almost canonical status among evolutionists who tried to forge complete compatibility between punctuated equilibrium and the Modern Synthesis, and although we all acknowledge stabilizing selection as too important and pervasive a phenomenon to hold no relevance for this issue, a complete explanation of stasis in these conventional terms seems implausible both on empirical grounds, and also by the basic logic of proper scaling.

As often emphasized in this chapter, if stasis merely reflects excellent adaptation to environment, then why do we frequently observe such profound stasis during major climatic shifts like ice-age cycles (Cronin, 1985), or through the largest environmental change in a major interval of time (Prothero and Heaton, 1996)? More importantly, conventional arguments about stabilizing selection have been framed for discrete populations on adaptive peaks, not for the *totality* of a species—the proper scale of punctuated equilibrium—so often composed of numerous, and at least semi-independent, subpopulations. A form of stabilizing selection acting among rather than within subpopulations may offer more promise—as Williams (1992) has proposed (see discussion under point 6)—but such forms of supraorganismal selection fall into a domain of heterodoxies, not into this category of conventional explanations that would leave the Modern Synthesis entirely unaffected by the recognition of stasis as a paleontological norm.

DEVELOPMENTAL AND ECOLOGICAL PLASTICITY. If stabilizing selection holds that species don't change because they have achieved such excellence in current adaptation, this second proposal (of Wake, Roth and Wake, 1983) proposes that species don't change (in an evolutionary and genetic sense) because they can usually accommodate to environmental alteration by exploiting the plasticity (behavioral and developmental) permitted within their existing genetic and ontogenetic system—thus calling upon the physiologist's entirely different meaning of the term "adaptation" (improvement in functionality by exploiting possibilities within a norm of reaction, as in the enlarged lungs of people who inhabit the high Andes), rather than the usual evolutionary meaning in our profession.

(Although I have roughly ordered this list of proposed explanations for stasis from Darwinian conventionality towards more challenging proposals, I don't regard any item as excluding any other—indeed, I would be surprised if all cannot claim at least some measure of validity, for once again we deal with an issue of relative frequencies and differential circumstances—and I don't regard any pair as establishing a contradiction. In particular, these first two proposals, although different in implications about styles and reasons for limited change in species, remain primarily complementary in their common attribution of stasis to reasons based on satisfactory current status—the first on immediate optimality of overt features, the second on inherent plasticity within a current, and presumably adaptive, norm of reaction.)

Wake et al. (1983), for example, document how salamanders, artificially raised to encounter only fixed potential prey, will learn to eat immobile objects, thus contradicting "the widespread assumption that amphibians feed only on moving prey" (p. 216), and also permitting substantial "adaptation" (physiologist's sense again) to feeding regimes without disturbing the stasis of evolved form. In a thought-provoking conclusion, Wake et al. (1983, p. 219) site (and cite) stasis as one component of a more general attitude towards stability of systems and preference for non-change, with evolution conceptualized as a "default option" in the history of life—in contrast with the usual view of active and normative change embodied in the first explanation of stabilizing selection:

> Stasis is but the most rigid form of the stability that pervades living systems. Thus organisms have evolved as systems resistant to change, even genetic change. While changing environmental conditions may ultimately necessitate change in the system, until some critical point the system remains stable and compensating. The living system is sometimes envisioned metaphorically as a kind of puppet, with enormous numbers of strings, each controlled genetically, or as a blob of putty that can flow in any direction given sufficient force (selection). Our metaphor is the living system as a balloon, with the environment impinging as countless blunt probes. The system compensates environmental and genetic changes, and persists by evolving minimally.

DEVELOPMENTAL CONSTRAINT. This proposal veers more towards heterodoxy in ascribing stasis to an internally specified inability to change (thereby implying frequent suboptimality of adaptation), rather than to lack of adaptive impetus for change due to current optimality (explanation one) or flexibility within a current constitution (explanation two). (This notion of inability stands forth most clearly in the strict definition—too strict in my view (see Chapter 10)—of constraint as absence of genetic variation for a particular and potentially useful alteration, as in the consensus concept of Maynard Smith et al., 1985.)

In our original paper on punctuated equilibrium, Eldredge and I (1972), basing our arguments partly on Mayr's (1954, 1963) concept of genetic revolutions in speciation of peripherally isolated populations, but more on Lerner's notions (1954) of ontogenetic or developmental, but especially of genetic, "homeostasis," proposed such constraint as the primary reason for stasis. We wrote (1972, pp. 114–115):

> If we view a species as a set of subpopulations, all ready and able to differentiate but held in check only by the rein of gene flow, then the stabil-

ity of species is a tenuous thing indeed. But if that stability is an inherent property both of individual development and the genetic structure of populations, then its power is immeasurably enhanced, for the basic property of homeostatic systems, or steady states, is that they resist change by self-regulation. That local populations do not differentiate into species, even though no external bar prevents it, stands as strong testimony to the inherent stability of species in time.

This proposal became one of the most widely controverted aspects of punctuated equilibrium, especially in linkage with other, largely independent concepts like the prevalence of neutral change (Kimura, 1968), and the exaptation of originally nonadaptive spandrels (Gould and Lewontin, 1979), also viewed as challenges to the more strictly adaptationist concept of Darwinian evolution then prevalent. I now believe that these criticisms, with respect to the issue of stasis in paleospecies through geological time, were largely justified—and that the theme of constraint, while not irrelevant to the causes of stasis in punctuated equilibrium, does not play the strong role that I initially advocated. (However—and perverse as this may seem to some detractors—my conviction about the general importance of constraint *vs.* adaptationism at other more appropriate scales has only intensified, particularly in the context of revolutionary findings in developmental genetics—see Chapter 10.)

I have changed my initial view for two primary reasons. First, the arguments of Mayr and Lerner, the intellectual underpinnings of our initial proposals about constraint, have not held up well under further scrutiny, particularly in the privileging of small populations as especially, if not uniquely, endowed with properties that permit the breaking of stasis. Further modelling has led most evolutionists to deny that any major impediment for such change can be ascribed to large populations. Second, I now realize that my arguments for the channeling of potential direction and limitation of change apply primarily to levels above species—to aspects of the developmental *Baupläne* of anatomical designs that usually transcend species boundaries, rather than to resistance of populations against incorporating enough genetic change to yield reproductive isolation from sister populations.

THE ECOLOGY OF HABITAT TRACKING. This explanation for stasis, long favored by my colleague Niles Eldredge (1995, 1999), offers a first alternative (in this list) based on the structuring of species-individuals as ecological entities, rather than on adaptations or capacities of component organisms—thus taking explanation to a higher descriptive level of the evolutionary hierarchy. Otherwise, however, habitat tracking ranks as a conventional Darwinian explanation in calling upon stabilizing selection to confer stasis upon populations that react to environmental change in their geographic locale not by evolutionary alteration to new conditions, but rather by moving with their favored habitat to remain in an unchanged relationship with their environment of adaptation. Eldredge writes (1999, p. 142): "Paradoxically (and contrary to at least superficial Darwinian expectations) . . . stabilizing natural selection will be the norm even as environmental conditions change—

so long, that is, as species are free to relocate and 'track' the familiar habitats to which they are already adapted. Rather than remaining in a single place and adapting to changing conditions, species move. And so they tend to remain more or less the same even if the environment keeps on changing."

I place this otherwise conventional explanation towards the heterodoxical end of my list because habitat tracking embodies the remarkably simple and obvious (in one sense), yet profound and unconventional view (in another sense) that evolutionary change represents a last resort, and not a norm for most times, in the response of populations to their environments. (The second explanation of plasticity also invokes this theme, but from the organism's, rather than the population's, perspective.) Habitat tracking also emphasizes the cohesion, and evolutionary reality, of supraorganismic individuals—an essential theme in the hierarchical reconstruction of Darwinian theory (see Chapter 8). This subtly unconventional notion of change as a last resort or default option puts one's mind in a much more receptive state towards the reality of stasis as a genuine and fundamental phenomenon in evolutionary theory.

THE NATURE OF SUBDIVIDED POPULATIONS. With this fifth category, we finally enter the realm of truly—that is, causally—macroevolutionary explanations based on the reality of supraorganismal individuals as Darwinian agents in processes of selection. In a brilliant paper that may well become a breakthrough document on this perplexing subject, Lieberman and Dudgeon (1996) have explained stasis as an expected response to the action of natural selection upon species subdivided (as most probably are) into at least transiently semi-autonomous populations, each adapted (or randomly drifted) to a particular relationship with a habitat in a subsection of the entire species's geographic range.

Lieberman and Dudgeon derived their ideas (see also McKinney and Allmon, 1995, for interesting support) in the context of Lieberman's extensive multivariate morphometric analysis of two brachiopod species from the famous Devonian Hamilton fauna of New York State (see pp. 916–922). Lieberman noted profound stasis (with much morphological "jiggling" to and fro but no net change) over 6 million years (Lieberman, Brett, and Eldredge, 1994, 1995); but he also studied samples of each species from each of several paleoenvironments through time. Paradoxically (at least at first glance), Lieberman documented several cases of measurable change in single discrete and continuous paleoenvironments through the section—but not for the entire species integrated over all paleoenvironments (an argument against habitat tracking, explanation 4 above, as a primary explanation for stasis). "It was found," Lieberman and Dudgeon write (1996, p. 231), "that more change occurred through time within a single paleoenvironment than across all paleoenvironments."

Interestingly, such a conclusion also builds a strong argument against the standard explanation of stabilizing selection (number one of this list) for stasis in paleospecies—because demes tracking single and stable environments through time should show no, or at least less, change than the species as a whole, not more. Lieberman and Dudgeon write (p. 231): "If stabilizing selection played a prominent role in maintaining stasis one would expect to find relatively little morphological change through time within a single environment." Williams (1992) has made a similar argument, at a lower scale, against stabilizing selection by emphasizing that the copiously, and lovingly, documented efficacy of natural selection in short-term situations of human observation—from beaks of Darwin's finches to industrial melanism in *Biston betularia*—makes stabilizing selection doubtful as a general explanation for such a pervasive phenomenon as stasis within paleospecies.

But when we consider this finding in supraorganismal terms, with demes as Darwinian individuals, an evident and sensible interpretation immediately emerges. A temporally coherent population may adapt gradually and continually while tracking one of several paleoenvironments inhabited by a species. But how can these anagenetic changes spread adaptively through an entire species composed of several other subpopulations, each adapted to (and tracking) its own paleoenvironment through time? No single morphology can represent a functional optimum for all habitats. In this common, and probably canonical, situation for species in nature, stability emerges as a form of "compromise" in most circumstances, a norm among "competing" minor changes that are, themselves, probably distributed more or less at random around a standard configuration, with each particular solution generally incapable, in any case, of spreading through all other demes of the species in the face of better locally adaptive configurations in most of these demes.

Of course, one can think of several obvious alternative structures where gradual change might be noted—lack of metapopulational division, with the entire species acting as a single deme, or some accessible and general biomechanical advantage that might be adaptive in all demes. But such circumstances may be uncommon—however important by cumulation in the overall history of life—in any general sample of species within a clade at any given time, thus accounting for the predominant relative frequency of stasis among all species, and for the relative rarity of anagenetic change within species as well.

Lieberman and Dudgeon summarize their proposed explanation by writing (1996, p. 231)

> Stasis may emerge from the way in which species are organized into reproductive groups occurring in separate environments. . . . The morphology of organisms within each of these demes may change through time due to local adaptation or drift, but the net sum of these independent changes will often cancel out, leading to overall net stasis . . . Only if all morphological changes across all environments were in the same direction in morphospace, or if morphological changes in a few environments were very dramatic and in the same direction, would there be significant net change in species morphology over time. . . . Thus, as long as a species occurs in several different environments one would predict on average it should be resistant to change.

The theoretical modelling of Allen, Schaffer and Rosko (1993) offers intriguing support in an implication not discussed by the authors. Allen et al. argue that the demic structure required for Lieberman's explanation of stasis strongly buffers species against extinction in chaotic ecological regimes. As an evident corollary, species selection must favor this architecture for species if such chaotic circumstances often prevail (or even just occur sporadically enough to impact a species' fate) over the geographic and temporal ranges of most species in nature. Thus, stasis would attain a predominant relative frequency among paleospecies because higher-level selection so strongly favors the persistence of species composed of multiple, semi-independent demes— the architecture that, as a consequence, engenders stasis by Lieberman's argument. Allen et al. (1993, p. 229) write:

> Even when chaos is associated with frequent rarity, its consequences to survival are necessarily deleterious only in the case of species composed of a single population. Of course, the majority of real world species . . . consist of multiple populations weakly coupled by migration, and in this circumstance chaos can actually reduce the probability of extinction . . . Although low densities lead to more frequent extinction at the local level, the decorrelating effect of chaotic oscillations reduces the degree of synchrony among populations and thus the likelihood that all are simultaneously extinguished.

NORMALIZING CLADE SELECTION. I cite Williams's (1992, p. 132) term for what most evolutionists would identify as a form of interdemic selection within species. (Williams uses "clade selection" as a general descriptor for all forms of selection among gene pools rather than among genes or gene combinations in organisms.) Williams also notes, as did Lieberman in a different context, the paradox of such strong empirical evidence for predominant stasis in the light of abundant data on substantial change within populations during the geological eyeblink of human careers in observation and experiment.

Williams therefore proposes, using Bell's work on stickleback fishes as a paradigm, that the environments of many demes within most species tend to be highly transient in geological terms, whereas one primary environment (often the original context of adaptation for the species) often tends to be highly persistent. (This phenomenon, however well recorded in sticklebacks, need not extend to a generality for species in nature, as Williams would readily admit in citing sticklebacks as a paradigm, not a claim for nature's normality. Sticklebacks exhibit this pattern because they generate successful, but also transient, freshwater demes from a persisting saltwater stock of lower population density.) Williams (1992, p. 134) therefore argues: "Clade selection acts against freshwater populations either because they cannot compete in mature freshwater faunas or because their habitats and ecological niches are ephemeral. The freshwater forms come and go in rapid succession, but the species complex endures in much the same form for long periods of time . . . [based on] the implied rapid extinction and intense clade selection against all but the conservative marine form . . . The appearance of stasis in the fossil record would result from an enormous variability in the persistence of ecological niches."

I am more attracted to Lieberman's suggestions, based on averaging among demes with no net change among persistent demes adapted to differing habitats, than to Williams's hypothesis, based on differential survival of one stable deme in a persistent habitat—if only because Lieberman has generated empirical evidence for longterm survival in several habitats within his two brachiopod species, whereas Williams's stickleback example may represent an unusual situation in the drastically different habitats (fresh *vs.* marine waters) of his transient vs. persistent demes. Still, I applaud these two suggestions for stasis based on the structuring of species-individuals as collections of deme-individuals, with differential selection acting upon demes in an irreducibly macroevolutionary mode. These proposals therefore occupy the heterodox end in a spectrum of proposed explanations for stasis—for they challenge the Darwinian orthodoxy of primacy or exclusivity for organismal selection. I especially appreciate Williams's openness towards explanations in this form, given his previous and highly influential preferences for formulating all evolutionary explanation, except when absolutely unavoidable, at the level of genic selection (in his famous book, Williams, 1966, as discussed on pp. 550–554).

In summary, then, the assertion of predominant stasis in the geological history of most paleospecies—one of the two primary claims of punctuated equilibrium—has provoked an interesting debate in evolutionary theory, with implications for some of the most basic concepts and perspectives in our science. First, and if only as a comment about the contemporary sociology of science, the recognition of stasis as a norm of controlling relative frequency at the level of punctuated equilibrium (at least for conventional sexual species of Metazoa), has spurred general interest in phenomena of stability and non-change throughout other levels of evolutionary inquiry (see, for example, Maynard Smith, 1983). We do not yet know (see fuller discussion on pp. 928–931) whether or rather how much, stasis across all scales might be attributed to structural similarity in nature's materials and processes—thus rendering this common pattern as an interesting parallelism (to use our evolutionary jargon) with genuinely homologous causal elements across scales, rather than a fortuitous convergence of similar overt patterns for disparate and merely analogous reasons. But at least we stand at the threshold of such an inquiry.

Second, and even more generally, the validation of predominant stasis as a norm would impel us to recast the basic problematic of evolution itself. If, following our conventional assumptions from Darwin to now, change represents the norm for a population through time, then our task, as evolutionary biologists, lies in specifying how this expected and universal phenomenon operates. But if, as punctuated equilibrium suggests, stasis represents the norm for most populations at most times; and if, moreover, stasis emerges as an active norm, not merely a passive consequence (as the modelling of Jackson and

Cheetham, 1995, strongly suggests in documenting stasis at too high a relative frequency for models based on neutralism, directional selection, or any set of assumptions that do *not* include some active force promoting stasis directly)—then evolutionary change itself must be reconceptualized as the infrequent breaking of a conventional and expected state, rather than as an inherent and continually operating property of biological materials, ecologies and populations.

A phenomenon marking the disruption of normality holds a very different philosophical status than a phenomenon representing the ordinary architecture of biological space and time. Evolutionary change, regarded as an occasional disrupter of stasis, requires a different set of explanatory concepts and mechanisms—a different view of life, really—from evolutionary change, defined as an anagenetic expectation intrinsically operating in most populations most of the time. Punctuated equilibrium proposes that the macroevolutionary key to this new formulation lies in speciation, or the birth of new higher-level individuals at discrete geological moments (corresponding to long intervals at the scale of a human lifetime). Macroevolution, in this view, becomes an inquiry into modes and mechanisms for breaking the stasis of existing species, and generating new species, conceived and defined as discrete higher-level Darwinian individuals—and not a question about how species-individuals gradually change their parts and constitutions through time (as in conventional Darwinism). But even if this particular formulation at geological scales eventually yields more limited impact or utility than proponents of punctuated equilibrium suspect, the more general redefinition of evolution as a set of rare incidents in the breaking of stasis, rather than the pervasive movement of an expected and canonical flow, still poses an interesting challenge for rethinking a fundamental proposition about the nature and history of life.

### Punctuation, the origin of new macroevolutionary individuals, and resulting implications for evolutionary theory

I have argued throughout this chapter that sets of related implications for expanding and reformulating the structure of Darwinian theory, particularly in applications to macroevolution, flow from each of the two major components of punctuated equilibrium—stasis as a norm for the duration of paleo-species, and punctuation (on geological scales) for their cladogenetic origin. The punctuational origin of species by cladogenesis provides our strongest rationale for regarding species as true evolutionary individuals in Darwin's causal world—rather than as arbitrarily delineated segments of transforming continua, and therefore not as genuine entities at all (a position maintained by both Darwin and Lamarck in some of their most forceful passages). If, following what I called the "grand analogy," species represent "items" or "atoms" of macroevolution in the same sense that organisms operate as fundamental interactors for natural selection in microevolution (see pp. 714–744), then many features of the mechanics and patterning of macroevolution must be reformulated. For macroevolution then becomes a process irreducibly fueled by the differential birth and death of species (just as microevolution, under natural selection, is powered by the differential reproductive success of organisms)—and not, as Darwin and his successors have long held, a phenomenology ultimately built by, and extending causally from, the accumulating consequences of continuous organismic adaptation in transforming populations.

In this sense, punctuated equilibrium—by crowning the case for stable species as atoms of macroevolution—challenges all three legs of the essential Darwinian tripod: the first leg of organismal focus most directly, by establishing the higher-level species-individual as a potent causal agent of evolution as well; the second leg of functionalism more indirectly by affirming, as generators of macroevolutionary patterns, several modes of explanation that do not flow from organismal adaptation, or even rest upon an adaptational base at all; and, most comprehensively, the third leg of extrapolationism by validating a hierarchical view of pattern and causality, and by denying that the mechanisms of macroevolutionary change all flow from our uniformitarian understanding of how natural selection, working in the organismal mode, can alter populations on the scale of human observation in historical time.*

To illustrate the expansive and reformative potential of the species-organism as a causal agent in macroevolution, I will discuss the three major topics that punctuated equilibrium has helped to redefine during the past two decades:

TRENDS. In Chapter 8, I proposed that trends among species in clades may differ substantially from trends among organisms in populations as an "allometric" result of varying weights attached to the three major causal processes at disparate scales of organism-individuals and species-individuals—drives, and the two sources of sorting, drift, and selection (see pp. 714–744 for full development of an argument only summarized here). At the conventional organismic level, drives from below assume little importance because the

---

*I also wish to reemphasize that I assert no exclusivistic claim in this formulation. Supporters of the hierarchical theory must not repeat the parochial error of their forebears by arguing that their newly-specific, higher-level mechanisms can explain everything by reaching down, just as Darwinian traditionalists tried to develop a complete causal theory by extrapolating up. Thus, we do not challenge either the efficacy or the cardinal importance of organismal selection. As previously discussed, I fully agree with Dawkins (1986) and others that one cannot invoke a higher-level force like species selection to explain "things that organisms do"—in particular, the stunning panoply of organismic adaptations that has always motivated our sense of wonder about the natural world, and that Darwin described, in one of his most famous lines (1859, p. 3) as "that perfection of structure and coadaptation which most justly excites our admiration." But should we not regard as equally foolish, and equally vain (in both senses of the word), any proposal that insists upon explaining all "things that species and clades do" as extrapolated consequences of organismic adaptation? I would not invoke species selection to explain the marvelous mechanics of beetle elytra, but the same theme of appropriate scale also leads me to equal confidence that the excellent adaptive design of beetle organisms cannot fully explain why this order so vastly predominates in species diversity, even among the most speciose of all metazoan classes—to the point of inspiring Haldane's canonical quip about God's "inordinate fondness" for these creatures (see Gould, 1993a, for an exegesis of this famous anecdote).

organism-individual so effectively suppresses the selective proliferation of lower-level individuals within its own body. In most circumstances, the sorting process of drift also contributes little to sustained trends because population sizes (of organisms in demes) usually exceed the small numbers required for maximal efficacy of such a stochastic force. Thus, of the three potential mechanisms, trends at the organismal level usually arise by selection.

But this understandable, and theoretically defendable, domination of selection at the focal level favored by traditional Darwinism does not extrapolate well to the higher causal level of species (as Darwinian individuals) in clades (as populations). When we shift our focus to this upper level, all three processes can claim significant potential weight in theory. (We cannot yet estimate the actual empirical weights due to paucity of research on a topic so recently defined—but see Wagner, 1996, for a breakthrough study based on quantitative and statistical discrimination of all three modes for various trends in the evolution of Paleozoic gastropods—see pp. 733–735 for a summary of his particular conclusions.) Since the species-individual does not preferentially suppress its own transformation by directional alteration of subparts (organisms), macroevolutionary trends may often be propelled by drives from below. Such drives may arise either by the orthodox route of anagenetic transformation in populations *via* organismic natural selection ("ontogenetic drive" in my terminology of Table 8-1), or by the unorthodox process of directional speciation ("reproductive drive" in my terminology).

When Wright's Rule holds (see pp. 731–735), and species arise at random with respect to the direction of a sustained trend in a clade, then we must invoke sorting processes among species. Sorting by drift can be highly effective at the species level because $N$ tends to be small in relevant populations (species within clades), in contrast with the traditional Darwinian level (organisms within demes), where the magnitude of $N$ usually precludes effective drift for major traits of organismal phenotypes.

A traditional Darwinian perspective might therefore lead us to denigrate the efficacy of the species level as a locus of causation for trends. If species do not marshall sufficient "strength" to stifle their own transformation by drives from below, or sufficient numbers to "prevent" the propagation of a cladal trend by random sorting, then species must pale as evolutionary agents before the strength of organisms (which manifest enough functional integrity to resist any differential proliferation of subparts, and also maintain sufficient population size to forestall random, and potentially nonadaptive, transformation of their collectivities).

I would, however, suggest that such an attitude stymies evolutionary theory as a restricting bias in the category that Francis Bacon called *idola theatri,* or idols of the theater, in his brilliant early 17th century analysis of mental impediments to understanding the empirical world. Bacon defined idols of the theater as constraining mental habits bred by allegiance to conventionalized systems of thought. In the present case, we fall into the bad habit of reading susceptibility to drive and drift as signs of weakness in an evolutionary individual because the Darwinian agent that we understand best, and that we have previously deemed exclusive—the organism—happens to resist these modes of change as an active consequence of its inherent structure. But nature builds her scales with strong allometry, and not in a fractal manner with every higher level formed as an isometrically enlarged version of each lower level enfolded within (Gould and Lloyd, 1999).

I suggest that we challenge this idol of our traditional Darwinian conceptualizations, and at least open ourselves to the opposite view that the species-individual's capacity for change by drive and drift, as well as by selection, defines a potential source of strength for this hierarchical level as an exploiter of the full panoply of available causes for trends. Perhaps we should pity the poor organism for its self-imposed restrictions. Or perhaps, rather, we should praise the organism for managing to achieve so much with such a limited range of mechanisms! (Pardon my metaphorical lapses. I am, of course, suggesting that we view the different interplay of potential forces at various hierarchical levels as sources of distinctiveness and strength for each. We will gain a better understanding of evolutionary mechanics when we try to identify the particular capabilities of each level rather than attempting to establish a single "gatekeeper's" criterion for ranking levels in linear order by their quantity of a single enabling power analogous to such fictions as IQ.)

In any case, this allometric expansion of potential mechanisms for trending at the species level offers significant promise for fracturing by redefinition (rather than solving in conventional terms) one of the great conundrums of paleontology—an issue much fretted over, and bruited about, but usually (and finally) cast aside with vague statements of hopeful confidence that traditional explanations will suffice once we finally record enough details in any given case. At least in terms of dedicated pages in our professional literature,* trends represent the cardinal subject of macroevolution (with differential waxing and waning of diversity within and among clades, especially as influenced by episodes of mass extinction, as the second great theme of evolutionary discussion in paleontology).

Paleontology has long been trapped in the dilemma of recognizing only one conventional model for the explanation of trends, and then finding little cred-

---

*One might argue that this focus only records another of Bacon's idols rather than an evident empirical reality. Bacon's *idola tribus*, or idols of the tribe, refer to mental biases deeply rooted in inherent modes of mental functioning, or human nature itself. Humans are pattern-seeking and story-telling creatures—and we prefer to tell our stories in certain modes that may reflect particular cultural traditions as well as universal preferences of thought. We shun randomness and non-directionality in favor of stories about movement in particular directions for definable reasons subject to moral judgment. We compiled the entire Bible as a grand and extended narrative in this mode, and then granted just one uncomfortable chapter to Ecclesiastes as the loyal opposition, where "time and chance happeneth to all" and "there is no new thing under the sun." Thus, our chosen focus upon trends in the paleontological record may only record their salience in piquing our interests and preferences—and not a genuinely high relative frequency among all clades in nature. This subject deserves a great deal of thought and extended study. A remarkable article by Budd and Coates, 1992, on the predominance of non-trending in the evolution of montastraeid corals may point the way to substantial reform—see p. 937.

ible evidence for the model's adequacy. By the expectations of all three central precepts in Darwinian logic, and by our habits of restricting explanations of sustained organismic trends to selectionist causes (given valid arguments for rejecting the alternatives of drift and drive at the organismic level, as discussed above), increasing adaptation of organisms must also propel macroevolutionary trends under extrapolationist premises.

(All Darwinians understand, of course, that natural selection only yields adaptation to immediate environments—a notion not conducive to sustained directional trends through geological time, given the effectively random fluctuation of most environmental configurations through substantial geological intervals. Consequently, most sustained trends have been interpreted as generalized biomechanical improvements conferring advantages across most or all experienced environments, and arising from Darwin's own preference for domination of biotic over nonbiotic competition in the history of life. See the discussion of Darwin's rationale for this defense of "progress" in evolution, Chapter 6, pp. 467–479.)

Discourse about trends dominates the traditional literature of evolutionary paleontology, both at the most general level of universal phenomenology (Cope's law of increase of size, Dollo's law of irreversibility, Williston's law of reduction and specialization of modular segments, etc.), and as a dominating theme for the history of almost any individual clade. We all know the particular tales for textbook groups—increasing brain size in hominids; larger body size, fewer toes, and higher crowned cheek teeth in horses; increasing symmetry of the cup in Paleozoic crinoids, with eventual expulsion of the anal ray to the top of the calyx; complexification in ammonite suture lines; reduction in number of stipes in graptolite colonies. These summary themes for clades, all based on the concept of general biomechanical improvement through time, distill the essence of traditional paleontology.

I do not deny that generalized organismic advantage may explain some of these classic trends. I do not, for example, challenge the traditional notion that increasing perfection of radial symmetry may confer adaptive benefits in feeding upon sessile organisms like stalked crinoids (see Moore and Landon, 1943, for the classic statement). But I also note that other classic trends, apparently ripe for explanation in biomechanical terms, have stubbornly resisted any reasonable hypothesis ever proposed—most notably the complexifying ammonite suture, which does not clearly confer greater resistance to shell crushing, and does not evidently aid the growing animal by increasing surface area of tissues covering the septa.

But other trends, despite their prominence, have never generated even a plausible hypothesis of biomechanical advantage. Why should fewer-stiped graptolite colonies "do better" in any usual sense of organismic (or, in this case, astogenetic) advantage? In such circumstances, we need to expand our explanatory net by considering alternative causal resolutions based on differential success of species as Darwinian individuals engaged in processes of sorting. Instead of focussing upon the putative biomechanical virtues of fewer stipes, we should be asking how and why such a character might correlate with the propensities of species for branching or for resistance to extinction—the "birth" and "death" processes that regulate sorting at this higher level.

I am not even confident that we should preferentially attribute traits with more plausible organismic advantages—including the enlarging brain of hominids as an obvious example—to conventional microevolutionary explanations, without seriously considering unorthodox possibilities based on causal correlations of such traits with propensities for speciation, or on the sheer good fortune of nonadaptive hitchhiking due to fortuitous presence in the subclade growing to domination for other reasons. We should be paying more attention to interesting and plausible proposals like Sacher's (1966) on lifespan and developmental timing as the primary target of selection in hominid evolution, with large brains on a facilitating causal pathway to advantageous retardations of development (Gould, 1977b).

In any case, and most generally, the need to describe trends—when punctuated equilibrium dominates the geometry of evolutionary change within a clade—as differential success of stable species, rather than as extrapolated anagenesis of populations, requires, in itself and as a "one liner" of extensive reformatory power, a radical reformulation (in the literal sense of reconstruction from the very *radices,* or roots, of the subject on up) of the primary topic in our macroevolutionary literature. Jackson and Cheetham (1999, p. 76) conclude: "Granted the prevalence of punctuated equilibria, macroevolutionary trends must arise through differential rates of origination and extinction, and not by adaptive evolution within single species. All of this is compatible with traditional neodarwinian evolutionary biology, but was unexpected before the theory of punctuated equilibria."

In summary, the efficacy of drifts and drives, in addition to selection, for generating trends at the hierarchical level of species as Darwinian individuals, suggests a rich, and virtually unexploited, domain of alternative explanations that might break through the disabling paradox of our current inability to resolve such a salient phenomenon in our preferred mode of adaptive advantages to organisms. Species-level explanations of trends in organismal phenotypes add at least two categories of potential resolution to our usual search for organismic benefit.

First, the trending character may be causally significant not for its phenotypic consequences to the organism, but for its role in influencing rates and directions of speciation in populations of organisms bearing the trait. If fossorial features of burrowing rodents (Gilinsky, 1986), or nonplanktonic lifestyles of marine molluscan larvae (Hansen, 1978, 1980), help to generate populational traits that enhance speciation rates, then a trend spreading such organismic features through a clade may arise by positive sorting of species rather than by general adaptive advantage of the phenotype itself. The organismal phenotypes may enjoy no general advantage at all, and may only produce adaptation to relatively ephemeral habitats within the clade's potential range. In fact, both fossorial rodent species (for reasons of small population size) and nonplanktonic molluscan species (for limited geographic ranges) may experience a reduced average geological longevity relative to surface dwelling (for rodents) and planktonic (for mollusks) clade members. But these impediments may be overbalanced by enhancement of speciation rates, thus driving the trend.

Second, both the low $N$ for species in clades (relative to organisms in most populations), and the remarkably (and, for most people, counterintuitively) high frequency of fortuitous but significant correlations between pairs of traits in systems built by genealogical branching (Raup and Gould, 1974) virtually guarantee that trending by drift will be much more common in sorting of species-individuals than in conventional sorting of organism-individuals. After all, branching evolution imparts a set of autapomorphic traits (through a unique common ancestor) to any subclade of species—and we can scarcely believe that each of these traits establishes the basis of selective existence and success for each species in the entire monophyletic group. Therefore, any process that favors the relative proliferation of any subclade for any reason will automatically engender a positive trend for any included autapomorphic character, whatever the causal basis of the general trend.

When we combine this spur to drift by hitchhiking with the observation that many successful clades go through severe bottlenecks (often as single surviving species) during their geological existence, we obtain even more compelling reasons for considering drift as a major source of macroevolutionary trends, however much we may reject analogous processes as substantial generators of trends in phenotypic characters controlled by organismal selection. For example, ammonites endured two severe bottlenecks at two major mass extinction events—suffering reduction, perhaps to two surviving lineages in the Permo-Triassic debacle, and to a single lineage in the closing Triassic event. In this light, why should we regard explanations based on general biomechanical advantages for organisms as preferable to the obvious blessing of good fortune upon any trait belonging to the phenotype of single lineages that manage to squeak through such profound bottlenecks? Few other evolutionary processes can promote traits from partial representation to exclusivity within a population (of species-individuals within a clade in this case) so quickly and so decisively.

The paleontological literature has just begun to reconceptualize trends in speciational terms. Initial results offer much encouragement, both for revising traditional explanations of particular temporal sequences, and for posing new questions requiring tests by different kinds of data. New insights often emerge just by framing the subject in terms of numbers and longevities of taxa rather than gradual fluxes of form. In their study of trends in Mississippian crinoids, for example, Kammer, Baumiller, and Ausich (1997) reach a conclusion that surprised them only because such a reasonable idea had not previously been formulated in operational terms (p. 221): "Results of this study indicate that among Mississippian crinoids niche generalists had greater species longevity than niche specialists. Although logical, few data have previously been developed to test this relationship."

The complexity of the subject then becomes apparent when the authors discuss why the obvious implication—that a trend towards generalists should sweep through the clade—fails because *ceteris paribus* (all other things equal) does not hold at several levels and strengths of correlation. First of all, "niche generalists tend to have fewer species per clade than niche specialists" (Kammer et al., 1997, p. 221), thus illustrating the most pervasive and powerful macroevolutionary constraint recognized so far at the speciational level (see general discussion on pages 739–741): the forced and intrinsic negative correlation between speciation rate and longevity. Secondly, these patterns hold "only during times of background extinction when Darwinian natural selection prevails" (p. 221). Mass extinctions may then impact species at random, without preferential regard for their ecological status or prospects for longevity in normal times.

Similar questions may be asked at all scales, including general patterns for life itself. For example, many robust paleontological data sets show a general tendency for increased longevity in marine invertebrate species through geological time. Even a famously iconoclastic thinker like David Raup, who devoted so much of his career to exploring the power of random systems to render observed patterns of the fossil record, interpreted this result as our best case for a meaningful concept of "progress," defined as increasing adaptive excellence of organisms (and leading to greater resistance to extinction). But several authors (Valentine, 1990; Gilinsky, 1994; Jablonski, Lidgard and Taylor, 1997) have reread this result in terms of species sorting as a tendency "for high-turnover taxa to be replaced over geologic time by low-turnover taxa" (Jablonski, et al., 1997, p. 515).

As Gilinsky (1994) notes, such a pattern, thus reformulated, may bear little or no relationship to general adaptive excellence at the organismal level. If speciation and extinction rates generally operate in balance (as they do), then some clades may be designated more "volatile" (in Gilinsky's terminology) as marked by their high rates of speciation and extinction, and others as more stable for lower rates of both these defining processes at the species level. In an abstract and general sense, both "strategies" may be regarded as equal in yielding the same result of steady cladal persistence, but with volatile clades showing more variation around a stable mean. However, in our real world of fluctuating environments and, especially, mass extinctions, volatility may doom clades in the long run because any reduction to zero (however "temporary" and reversible in abstract modelling) extinguishes all futures in our actual world of material entities. Since volatile clades, on average, must cross the zero line more frequently than stable clades, a general trend to increasing species longevity may only arise as an indirect consequence of the higher vulnerability of volatile clades and the consequent accumulation of stable clades through geological time.

Several recent articles on bryozoan evolution illustrate the utility of a speciational approach to trends—if only as a method for setting base lines and making distinctions, all the better to document patterns *not* attributable to differential speciation. For example, Jablonski, Lidgard and Taylor (1997) found that "the generation of low-level novelties are effectively driven by speciation rates" (p. 514), whereas the origin of major apomorphies of larger groups do not correlate so clearly with numbers of speciation events, but rather with their magnitude (with such rare events favored in certain times and environments).

In documenting a speciational basis as a "null hypothesis" of sorts for identifying evolutionary patterns arising by other routes, McKinney et al. (1998) compared the differential successes of cheilostome and cyclostome bryozoan clades through time. In a fascinating discovery, they noted that, in times of joint decline for both clades (late Cretaceous and post-Paleocene after a Danian spike in diversification), "the relative skeletal mass of cyclostomes declined much more precipitously than did relative species richness" (p. 808). The authors could therefore identify an important trend by standardizing species numbers: "There is a long-term trend for the average cheilostome species to generate a progressively greater skeletal mass than the average cyclostome species. This could result from a gradual trend toward relatively larger colony sizes within cheilostomes, a greater number of colonies per cheilostome species, or both" (pp. 808–809).

In a similar vein, but at the larger scale of the phylum's initial Ordovician radiation, Anstey and Pachut (1995) found no relationship between number of speciation events and the establishment of defining apomorphies among major subgroups at the base of the clade. They write (1995, pp. 262–263): "The morphologies recognized as higher taxa of bryozoans were not built up through a gradual accumulation of species differences but appear to have diverged very rapidly in the initial radiation of the phylum . . . The processes producing the major branching events and familial apomorphies, therefore, apparently were not driven by speciation, and likewise could not have resulted from species selection or species sorting."

THE SPECIATIONAL REFORMULATION OF MACROEVOLUTION. Beyond the immediate expansion of explanations for the signature phenomenon of trends, the recasting of macroevolution as a discourse about the differential fates of stable species (treated as Darwinian individuals) carries extensive implications for rethinking both the pageant of life's history and the causes of stability and change in geological time.

I am particularly grateful that Ernst Mayr (1992, p. 48)—the doyen of late 20th century evolutionary biologists, and the inspirer of punctuated equilibrium through his views on peripatric speciation (though scarcely an avid supporter of our developed theory)—has identified the required speciational reformulation of macroevolution as the principal component of "what had not been recognized [in evolutionary theory] before" Eldredge and I codified punctuated equilibrium. Mayr continues, stressing the species as the macroevolutionary analog of the organism considered as the "atom" of microevolution:

> It was generally recognized that regular variational evolution in the Darwinian sense takes place at the level of the individual and population,

but that a similar variational evolution occurs at the level of species was generally ignored. Transformational evolution of species (phyletic gradualism) is not nearly as important in evolution as the production of a rich diversity of species and the establishment of evolutionary advance by selection among these species. In other words, speciational evolution is Darwinian evolution at a higher hierarchical level. The importance of this insight can hardly be exaggerated.

As a most general statement, and extending Mayr's views from his specific words (cited just above) to his most characteristic philosophical observation, Darwinism's major impact upon western thinking transcends the replacement of a fixed and created universe by an evolutionary flux. As thoughtful evolutionists have always noted, and as Mayr has particularly stressed in our times by contrasting "essentialist" and "populational" ways of thinking, a fundamental revision in our concept of the essence of reality—from the Platonic archetype to the variable population—may represent Darwin's most pervasive and enduring contribution to human understanding. For what could be more profound or portentous than our switch from "fixed essences" to "sensibly united groups of varying items," as our explanation for the reality behind our names for categories in our parsings of the natural world?

Yet however successful we have been in executing this great philosophical shift at the level of microevolution—where we understand that no archetype for a seahorse, a sequoia or a human being exists; where an enterprise named "population genetics" stands at the core of an explanatory system; and where we have all been explicitly taught to view change as the conversion of intrapopulational variation into interpopulational differences—we have scarcely begun to execute an equally important reconceptualization for our descriptions and explanations of macroevolution. We still encapsulate the pageant of life's history largely as a set of stories about the trajectories of abstracted designs through time. Beetles and angiosperms flourish; trilobites and ammonites disappear. Horses build bigger bodies and fewer toes; humans evolve larger brains and smaller teeth. Meanwhile, in grand epitome, life itself experiences a general rise in mean complexity as a primary definition (in popular culture at least) of evolution itself.

We know, of course, that flourishing implies more species, while extinction marks an ultimate reduction to zero. But we still tend to visualize these patterns of changing diversity as consequences of the status of designs, with the Darwinian optimum standing for the old Platonic archetype as an ideal and guarantor. (And if we manage to construe such quintessentially populational phenomena as the waxing and waning of groups in this Platonic mode, then we will surely not be tempted to reformulate anatomical trends in average form within groups—a phenomenon far more congenial to our essentialist mythologies—in the Darwinian language of changing frequencies in variable populations.) In this way, the two major phenomena of macroevolution—phenotypic trends within groups and changes in relative diversities among groups—have stubbornly resisted the reformative power of Darwin's deepest insight. Moreover, this resistance arises for the most ironical of recursive reasons—namely, that other biases of the Darwinian tradition, particularly the reductionist and extrapolationist premises discussed throughout this book as the essential Darwinian tripod, have forestalled an application of Darwin's deepest insight to nature's grandest scale!

We have conceptualized macroevolution as temporal fluxes of adaptive form, with either some notion of an average phenotype for a clade (corresponding to vernacular ideas of the "ordinary" or "normal"), or some extreme value (representing our view of the "promise" or "potential" of a collectivity) standing as a surrogate or summary for all variation, in both form and number of species, within the clade under consideration. This attitude has led us to embrace a set of patent absurdities that rank, nonetheless, as received wisdom about the history of life, and that we continue to support in a passive way because the fallacies can only become apparent when we reconceive macroevolution as an inquiry centered upon the changing sizes and variabilities of clades based on the fates of their component Darwinian individuals—their species, construed as "atoms" of macroevolution (at least for sexually reproducing organisms).

Thus, we usually summarize the history of life as a drama about generally increasing complexity of form (with bacteria, beetles and fishes left successively behind, despite their evident prosperity) when, at the very most, such a theme can only apply to a small group of species on the right tail of life's general distribution—and when, by any fair criterion generally employed by evolutionists, bacteria have always dominated the history of life from their origins in exclusivity more than 3.5 billion years ago to their current mastery of a much more diverse world. And we have chosen horses as our textbook and museum hall example of progressive evolution triumphant, when modern equids represent a pitiable remnant of past diversity, a small clade entirely extinguished in its original, and formerly speciose, New World home, and now surviving as only a half dozen or so species of horses, asses and zebras at the more hospitable termini of past migrations. Horses, moreover, represent failures within a failure—for the once proud order of Perissodactyla now persists as only three small clades of threatened species (tapirs, rhinoceroses and equids, with the last receiving an artificial boost for human purposes), while the once minor order of Artiodactyla now dominates the guild of large, hoofed herbivorous mammals as one of evolution's great success stories. But we will not grasp these evident patterns, truly generated by changing diversity of species, if we continue to dwell in a conceptual prison that frames the history of life as a flux from monad to man, and the phylogeny of horses as a stately race from little splay-footed eohippus to the one-toed nobility of Man O' War.

The key to a more expansive formulation—also a more accurate depiction in the language of probable causes based on genuine evolutionary agents—lies in recasting this discourse about fluxes of means or extreme values within clades as a history of the differential origins and fates of species, as organized by nature's genealogical system (that is, evolution itself) into the monophyletic groups of life's tree. If we can accomplish this speciational reformulation of macroevolution, we will understand that many classical "trends" emerge as passive consequences of temporal variation in numbers of species within a clade often enhanced when structural constraints channel the potential directions of such variation, and not as selectively driven vectors in the biomechanical form of "average" organisms within the clade. We should be asking questions about numbers of actual species, rather than rates of flux for anatomical archetypes or abstractions. We should be studying the dynamics of differential species success as the causal basis of macroevolutionary pattern, not placing our hopes for explanation upon undefinable optima for competitive triumph of organic designs.

We will finally recognize that causes for the evolution of form and the evolution of diversity do not interact in the conceptual opposition that defined Lamarck's original formulation of evolution (see Chapter 3), and that persists today in common statements (see G. G. Simpson on p. 562, J. S. Huxley on p. 563, or F. Ayala, 1982) that speciation (or cladogenesis) builds the luxury of iterated variation, while a different, and altogether more important, process of transformational anagenesis fashions the trends of form that culminate in such glories as the human brain and the dance language of bees. We will then grasp that many—though not all—phenomena in the evolution of form arise as noncausally correlated consequences of patterns in the changing diversity of species.

If heaven exists (and the management let Darwin in), he must be greeting this prospect with the same thought that the founders of America emblazoned on the new nation's Great Seal: *annuit coeptis* (he smiles on our beginnings). For Darwin tried (see Chapters 2 and 3) to disassemble Lamarck's sterile dichotomy between fundamental, but probably unresolvable, causes of progressive evolution in form, and secondary, albeit testable, causes of lateral diversification—and to reformulate all evolution in terms of the previously trivialized tangent, while branding the supposed main line as illusory. I am proposing an analogous reform at the level of macroevolution—with the "diversity machine" of speciation, previously labelled as secondary and merely luxurious, also recognized as the generator of what we perceive as trends in form within clades. Again, the humble and testable factor, once relegated to a playground of triviality, becomes the cause of a supposedly higher process formerly judged orthogonal, if not oppositional. But this time, both the atom of agency and the cause of change reside at the higher level of macroevolution, and must therefore be accessed in the unfamiliar framework of deep time, rather than directly observed in human time. Our intellectual resources are not unequal to such a task, and we could not ask for a better leader than Darwin.

To illustrate how such a speciational reformulation might proceed, let us consider, at three levels of inquiry, the consequences of documenting macroevolution as expansion, contraction and changing form in the distribution of all species within clades through time. I regard this unfamiliar categorization as an empowering substitute for the usual tactic of summarizing the history of a clade by some measure of central tendency, or some salient extreme that catches our fancy, and then plotting the trajectory of this archetypal value through time.

LIFE ITSELF. In popular descriptions of evolution, from media to museum halls, but also in most technical sources, from textbook pedagogy to monographic research, we have presented the history of life as a sequence of increasing complexity, with an initial chapter on unicellular organisms and a final chapter on the evolution of hominids. I do not deny that such a device captures something about evolution. That is, the sequence of bacterium, jellyfish, trilobite, eurypterid, fish, dinosaur, mammoth, and human does, I suppose, express "the temporal history of the most complex creature" in a rough and rather anthropomorphic perspective. (Needless to say, such a sequence doesn't even come close to representing a system of direct phylogenetic filiation.)

But can such a sequence represent the history of life, or even stand as a surrogate either for the fundamental feature of that history, or for the central causal processes of evolutionary change? When we shift our focus to the full range of life's diversity, rather than the upper terminus alone, we immediately grasp the treacherous limitations imposed by misreading the history of an extreme value as the epitome of an entire system. The sequence of increasingly right-skewed distributions with a constant modal value firmly centered on bacteria throughout the history of life (Fig. 9-29) represents only a cartoon or

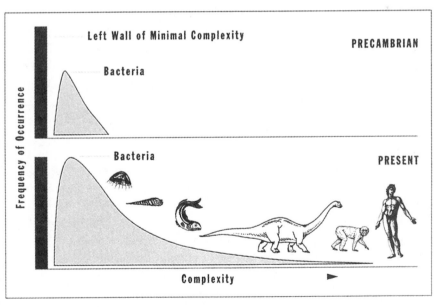

9-29. This cartoon of changing form and range in the histogram of complexity through life's phylogeny illustrates how we fall into error when we treat extreme values as surrogates or epitomes of entire systems. A view that emphasizes speciation and diversity might recognize the constancy of the bacterial mode as the outstanding feature of life's history. From Gould, 1996a.

an icon for an argument, not a quantification. The full vernacular understanding of complexity cannot be represented as a linear scale, although meaningful and operational surrogates for certain isolated aspects of the vernacular concept have been successfully designated for particular cases—see McShea, 1994.

I do not see how anyone could mistake the extreme value of a small tail in an increasingly skewed distribution through time for the evident essence, or even the most important feature, of the entire system. The error of construing this conventional trend of extremes as the essential feature of life's history becomes more apparent when we switch to the more adequate iconography of entire ranges of diversity through time. Consider just three implications of the full view, but rendered invisible when the sequential featuring of extremes falsely fronts for the history of the whole.

1. The salience of the bacterial mode. Although any designation of most salient features must reflect the interests of the observer, I challenge anyone with professional training in evolutionary theory to defend the extending tip of the right tail as more definitive or more portentous than the persistence in place, and constant growth in height, of the bacterial mode. The recorded history of life began with bacteria 3.5 billion years ago, continued as a tale of prokaryotic unicells alone for probably more than a billion years, and has never experienced a shift in the modal position of complexity. We do not live in what older books called "the age of man" (1 species), or "the age of mammals" (4000 species among more than a million for the animal kingdom alone), or even in "the age of arthropods" (a proper designation if we restrict our focus to the Metazoa, but surely not appropriate if we include all life on earth). We live, if we must designate an exemplar at all, in a persisting "age of bacteria"—the organisms that were in the beginning, are now, and probably ever shall be (until the sun runs out of fuel) the dominant creatures on earth by any standard evolutionary criterion of biochemical diversity, range of habitats, resistance to extinction, and perhaps, if the "deep hot biosphere" (Gold, 1999) of bacteria within subsurface rocks matches the upper estimates for spread and abundance, even in biomass (see Gould, 1996a, for a full development of this argument). I will only remind colleagues of Woese's "three domain" model for life's full genealogy (see Fig. 9-30), a previously surprising but now fully accepted, and genetically documented, scheme displaying the phylogenetic triviality of all multicellular existence (a different issue, I fully admit, from ecological importance). Life's tree is, effectively, a bacterial bush. Two of the three domains belong to prokaryotes alone, while the three kingdoms of multicellular eukaryotes (plants, animals, and fungi) appear as three twigs at the terminus of the third domain.

2. The cause of the bacterial mode. "Bacteria," as a general term for the grade of prokaryotic unicells lacking a complex internal architecture of organelles, represent an almost ineluctable starting point for a recognizable fossil record of preservable anatomy. As a consequence of the basic physics of self-organizing systems and the chemistry of living matter—and under any

popular model for life's origins, from the old primordial soup of Haldane and Oparin, to Cairns-Smith's clay templates (1971), to preferences for deep-sea vents as a primary locale—life can hardly begin in any other morphological status than just adjacent to what I have called (see Fig. 9-29) the "left wall" of minimal conceivable preservable complexity, that is, effectively, as bacteria (at least in terms of entities that might be preserved as fossils). I can hardly imagine a scenario that could begin with the precipitation of a hippopotamus from the primordial soup.

Once life originates, by physico-chemical necessity, in a location adjacent to this left wall (see Kauffman, 1993), the subsequent history of right-skewed expansion arises predictably as a fundamental geometric constraint of this initial condition combined with the principles of Darwinian evolution—that is, so long as the most genuine trend of life's history then prevails: "success" measured variationally, in true Darwinian fashion, as expansion in diversity and range through time.

If life continues to add taxa and habitats, then structural constraints of the system virtually guarantee that a right tail of complexity will develop and increase in skew through time as a geometric inevitability, and not necessarily for any overall advantage conferred by complexity. As noted above, life must begin, for physico-chemical reasons, next to the left wall of minimal complexity. Little or no "space" therefore exists between the initial bacterial mode and this natural lower limit; variation can expand only into the "open" domain of greater complexity. The vaunted trend to life's increasing complexity must be reconceived, therefore, as a drift of a small percentage of species from the constant mode of life's central tendency towards the only open di-

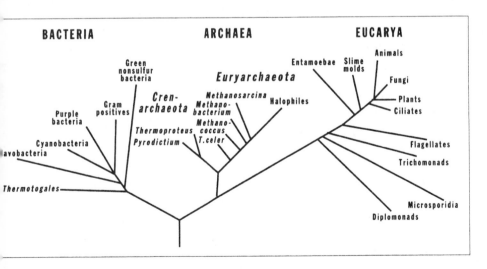

9-30. In life's full genealogy, all three multicellular kingdoms grow as twigs at the terminus of just one branch among the three great domains of life's history. The other two domains are entirely prokaryotic. From work of Woese and colleagues, as presented in Gould, 1996a.

rection for expansion. To be able to formulate this alternative view at all, we must reconceive the history of life as expansion and contraction of a full range of taxa under constraints of systems and environments, rather than as a flux of central tendencies, valued extremes, or salient features.

3. The right tail as predictable, but passively generated. A critic might respond that he accepts the reformulation but still wishes to assert a vector of progress as life's central feature in the following, admittedly downgraded, way: yes, the vector of progress must be construed as the expanding right tail of a distribution with a constant mode, not as a general thrust of the whole. But this expanding tail still arises as a predictable feature of the system, even if we must interpret its origin and intensification as the drift of a minority away from a constraining wall, rather than the active trending of a totality. The right tail had to expand so long as life grew in variety. This tail therefore originated and extended for a reason; and humans now reside at its present terminus. Such a formulation may not capture the full glory of Psalm 8 ("Thou hast made him a little lower than the angels"), but a dedicated anthropocentrist could still live with this version of human excellence and domination.

But the variational reformulation of life's system suggests a further implication that may not sit well with this expression of human vanity. Yes, the right tail arises predictably, but random systems generate predictable consequences for passive reasons—so the necessity of the right tail does not imply active construction based on overt Darwinian virtues of complexity. Of course the right tail might be driven by adaptive evolution, but the same configuration will also arise in a fully random system with a constraining boundary. The issue of proper explanations must be resolved empirically.

By "random" in this context, I only mean to assert the hypothesis of no overall preference for increasing complexity among items added to the distribution—that is, a system in which each speciation event has an equal probability of leading either to greater or to lesser complexity from the ancestral design. I do not deny, of course, that individual lineages in such systems may develop increasing complexity for conventional adaptive reasons, from the benefits of sharp claws to the virtues of human cognition. I only hold that the entire system (all of life, that is) need not display any overall bias—for just as many individual lineages may become less complex for equally adaptive reasons. In a world where so many parasitic species usually exhibit less complexity than their freeliving ancestors, and where no obvious argument exists for a contrary trend in any equally large guild, why should we target increasing complexity as a favored hypothesis for a general pattern in the history of life?

The location of an initial mode next to a constraining wall guarantees a temporal drift away from the wall in random systems of this kind. This situation corresponds to the standard paradigm of the "drunkard's walk" (Fig. 9-31), used by generations of statistics teachers to illustrate the canonical random process of coin tossing. A drunkard exits from a bar and staggers, entirely at random, along a line extending from the bar wall to the gutter (where he passes out and ends the "experiment"). He winds up in the gutter on every iteration of this sequence (and with a predictable distribution of arrival times) simply because he cannot penetrate the bar wall and must eventually "reflect off" whenever he hits this boundary. (Of course, he will also end up in the gutter even if he moves preferentially towards the bar wall; in this case, the average time of arrival will be longer, but the result just as inevitable.)

The issue of active drive (a small bias in relative frequency fueled by the general Darwinian advantages of complexity) vs. passive drift (predictable movement in a random system based on the model of the drunkard's walk) for the expansion of the right tail must be resolved empirically. But the macroevolutionary reformulation of life's history in variational terms establishes a conceptual framework that permits this question to be asked, or even conceived at all, for the first time. Initial studies on mammalian vertebrae and teeth, foraminiferal sizes, and ammonite sutures have been summarized in Gould, 1996a, based on pioneering studies of McShea, Boyajian, Arnold, and Gingerich. This initial research has found no departure from the random model, and no overall preference for increase in complexity in studies that tabulate *all* events of speciation.

GENERAL RULES. The older literature of paleobiology focused on the recognition and explanation of supposedly general "rules" or "laws" regulating the overt phenomenology of life's macroevolutionary pattern. As the modern synthesis developed its core of Darwinian explanation, several leading theorists (see especially Haldane, 1932, and Rensch, 1947, 1960) tried to render these laws as large-scale expressions of evolution's control by adaptive anagenesis in populations under Darwinian natural selection.

This subject fell out of favor for several reasons, but in large part because non-adaptationist explanations, deemed less interesting (and certainly less coordinating) than accounts based on natural selection, provided an adequate compass for most of these "laws." Thus, for example, Dollo's law of ir-

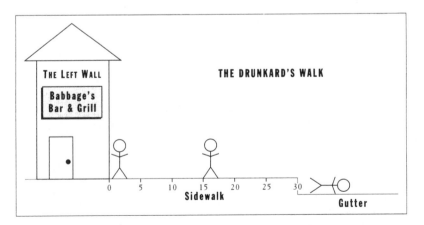

9-31. The standard statistical model of the "drunkard's walk" shows that even the expanding right tail of life's right-skewed histogram of complexity may arise within a random system with equal probabilities for the movement of any descendant towards either greater or lesser complexity. From Gould, 1966a.

reversibility (see Gould, 1970b) only restates the general principles of mathematical probability for the specific case of temporal changes based on large numbers of relatively independent components. And Williston's law of reduction and specialization in modular segments may only record a structural constraint in random systems, thus following the same principles as my previous argument about the expanding right tail of complexity for life's totality. Suppose that, in overall frequency within the arthropod clade, modular species (with large numbers of similar segments) and tagmatized species (with fewer fused and specialized groupings of former segments) always enjoy equal status in the sense that 50 percent of habitats favor one design, and 50 percent the other. (I am, of course, only presenting an abstract "thought experiment," not an operational possibility for research. Niches don't exist independent of species.) But suppose also that, for structural reasons, modular designs can evolve toward tagmatization, but tagmatized species cannot revert to their original modularity—an entirely reasonable assumption under Dollo's law (founded upon the basic statements of probability theory) and generalities of biological development. Then, even though tagmatization enjoys no general selective advantage over modularity, a powerful trend to tagmatization must pervade the clade's history, ultimately running to completion when the last modular species dies or transforms.

However, one of these older general rules has retained its hold upon evolutionary theory, probably for its putative resolvability in more conventional Darwinian terms of general organismic advantage: Cope's Law, or the claim that a substantial majority of lineages undergo phyletic size increase, thus imparting a strong bias of relative frequency to the genealogy of most clades—a vector of directionality that might establish an arrow of time for the history of life.

A century of literature on this subject had been dominated by proposed explanations in the conventional mode of organismic adaptation fueled by natural selection. Why, commentators asked almost exclusively, should larger size enjoy enough general advantage to prevail in a majority of lineages? Proposed explanations cited, for example, the putative benefits enjoyed by larger organisms in predatory ability, mating success, or capacity to resist extreme environmental fluctuations (Hallam, 1990; Brown and Maurer, 1986).

The speciational reformulation of macroevolution has impacted this subject perhaps more than any other, not because the theme exudes any special propensity for such rethinking—for I suspect that almost any conventional "truth" of macroevolution holds promise for substantial revision in this light—but because its salience as a "flagship," but annoyingly unresolved, issue inspired overt attention. Moreover, the conventional explanations in terms of organismal advantage had never seemed fully satisfactory to most paleontologists.

The rethinking has proceeded in two interesting stages. First, Stanley (1973), in a landmark paper, proposed that Cope's Law emerges as a passive consequence of Cope's other famous, and previously unrelated, "Law of the Unspecialized"—the claim that most lineages spring from founding species with generalized anatomies, under the additional, and quite reasonable, as-

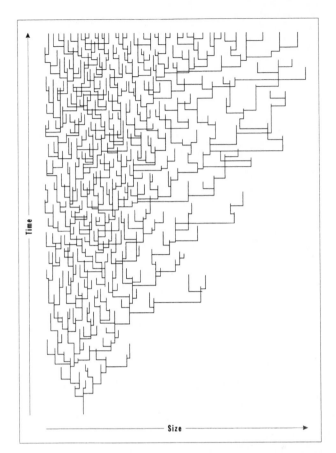

9-32. Cope's Law shown, under a speciational perspective, as a differential movement of speciation events towards larger size from a constraining boundary imparted by a small founding member of the lineage. Adapted from Stanley, 1973.

sumption that the majority of generalized species also tend to be relatively small in body size within their clades.

These statements still suggest nothing new so long as we continue to frame Cope's Rule as anagenetic flux in an average value through time—that is, as a conventional "trend" under lingering Platonic approaches to macroevolution. But when we reformulate the problem in speciational terms—with the history of a Cope's Law clade depicted as the distribution of all its species at all times, and with novelty introduced by punctuational events of speciation rather than anagenetic flux—then a strikingly different hypothesis leaps forth, for we now can recognize a situation precisely analogous (at one fractal level down) to the previous construction of life's entire history: an evolving population of species (treated as stable individuals), in a system with a left wall of minimal size (for the given *Bauplan*), and a tendency for founding members to originate near this left wall (Fig. 9-32).

Therefore, just as for all of life in my previous example, if the clade prospers with an increasing number of species, and even if new species show no directional tendency for increasing size (with as many species arising smaller than, as larger than, their ancestors), then the *mean* size among species in the clade must drift to the right, even though the *mode* may not move from initial smallness, just because the space of possible change includes substantial room in the domain of larger size, and little or no space between the founding lineage and the left wall. Thus, as Stanley (1973) stated so incisively, Cope's Law receives a reversed interpretation as the structurally constrained and passive evolution (of an abstracted central tendency, I might add) *from small size*, rather than as active evolution *towards large size* based on the organismic advantages of greater bodily bulk under natural selection.

But we must then carry the revision one step further and ask an even more iconoclastic question: does Cope's Law hold at all? Could our impressions about its validity arise as a psychological artifact of our preferential focus upon lineages that grow larger, while we ignore those that remain in stasis or get smaller—just as we focus on fishes, then dinosaurs, then mammoths, then humans, all the while ignoring the bacteria that have always dominated the diversity of life from the pinnacle of their unchanging mode throughout geological time?

Again, we cannot even ask this question until we reformulate the entire issue in speciational terms. If we view a temporal vector of a single number as adequate support for Cope's Law, we will not be tempted to study *all* species in a monophyletic clade that includes signature lineages showing the documented increase in size. But when we know, via Stanley's argument, that Cope's Law can be generated as a summary statement about passively drifting central tendencies in random systems with constraining boundaries, then we must formulate our tests in terms of the fates of all species in monophyletic groups. Jablonski (1997) has published such a study for late Cretaceous mollusks of the Gulf and Atlantic coasts (a rich and well-studied fauna of 1086 species in Jablonski's tabulation) and has, indeed, determined that, for this prominent group at least, prior assertions of Cope's Law only represent an artifact of biased attention (see commentary of Gould, 1997b). Jablonski found that 27–30 percent of genera do increase in mean size through the sequence of strata. But the same percentage of genera (26–27 percent) also decrease in mean size—although no one, heretofore, had sought them out for equal examination and tabulation

Moreover, and more notably for its capacity to lead us astray when we operate within a conceptual box defined by anagenetic flux rather than variation in numbers of taxa, an additional 25–28 percent of genera fall into a third category of generally and symmetrically increasing variation through the sequence—that is, the final range for all species within the genus includes species both smaller and larger than the extremes of the ancestral spread. I strongly suspect that a previous inclusion of these genera as affirmations of Cope's Law engendered the false result of dominant relative frequency for phyletic size increase. Older treatments of the topic usually considered extreme values only, and affirmed Cope's Law if *any* later species exceeded the common ancestor in size—thus repeating, in miniature, the same error generally committed for life's totality by ignoring the continuing domination of bacteria, and using the motley sequence of trilobite to dinosaur to human as evidence for a central and defining thrust. Obviously, from a variational and speciational perspective, successful genera with substantially increasing numbers of species through time will probably expand their range at both extremes of size, thus undergoing a speciational trend in variation, not an anagenetic march to larger sizes!

PARTICULAR CASES. This lowest macroevolutionary level of individual monophyletic clades has defined the soul of paleontological discourse through the centuries. Only the histories of particular groups can capture the details that all vivid story-telling requires; the "why" of horses and humans certainly elicits more passion than the explanation (or denial) of Cope's Law, or the pattern of increasing mean species longevity in marine invertebrates through time. Yet, even here on such familiar ground, our explanations remain so near and yet so far—for these "closer" stories of particular histories must also be reexamined in a speciational light. Consider just two "classics" and their potential revisions.

1. HORSES AS THE EXEMPLAR OF "LIFE'S LITTLE JOKE." As noted above (Fig. 7-3, and 580–581), the line of horses, proceeding via three major trends of size, toes and teeth from dog-sized, many-toed, "eohippus" with low-crowned molars to one-toed *Equus* with high-crowned molars (see Fig. 9-33 for W. D. Matthew's classic icon, linearly ordered by stratigraphy) still marches through our textbooks and museums as the standard-bearer for adaptive trending towards bigger and better.

I do not deny that, even in a refomulated speciational context, several aspects of the traditional story continue to hold. MacFadden (1986), for example, has documented a clear cladal bias towards the punctuational origin of new species at larger sizes than their immediate ancestors, so both the iconic transition from ladders to bushes, and the recognition that several speciational events lead to smaller sizes, even to dwarfed species, throughout the range of the lineage, does not threaten (but rather reinterprets in interesting ways) the conventional conclusion that horses have generally increased in size through the Cenozoic. Moreover, I do not doubt the usual adaptational scenario that a transition from browsing in soft-turfed woodlands to grazing on newly-evolved grassy plains (grasses did not evolve until mid-Tertiary times) largely explains the adaptive context for both the general reduction in numbers of toes from splayed feet on soft ground to hoofs on harder substrates, and the increasing height of cheek teeth to prevent premature wear from eating grasses of high silica content.

Nonetheless, a speciational reformulation in terms of changing diversity as well as anatomical trending tells a strikingly different, and mostly opposite, story for the clade as a whole. Modern perissodactyls represent but a shade of their former glory. This clade once dominated the guild of large-bodied mammalian herbivores, with speciose and successful groups, especially the titanotheres, that soon became extinct, and with diversity in existing groups far exceeding modern levels. (The rhinoceros clade once included agile running forms, the hyrachiids, wallowing hippo-like species, and the indricotheres, the largest land mammals that ever lived.) Modern perissodactyls exist as three small clades of threatened species: horses, rhinos, and tapirs.

Horses have declined precipitously from their maximal mid-Tertiary abun-

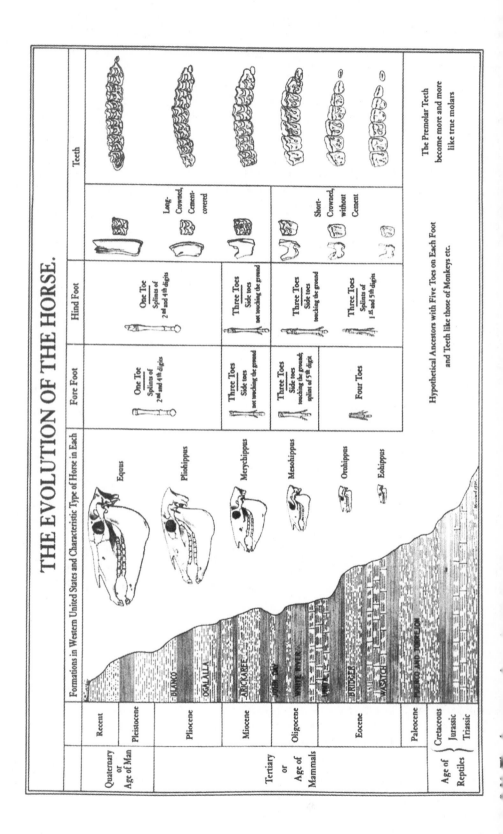

dance to modern marginality in both place and number. As O. C. Marsh proved to T. H. Huxley in a famous incident in the history of paleontology (see Gould, 1996a, pp. 57–58), horses evolved in America only to peter out and disappear completely in their native land, surviving only as a few lines of Old World migrants. In North America alone, from 8 to 15 million years ago, an average of 16 species lived contemporaneously. As recently as 5 million years ago, at least six sympatric species of horses lived in Florida alone (MacFadden et al., 1999). Only 6–8 species of horses inhabit the entire earth today.

Moreover, if one wished to argue that a particular affection for *Equus* permits a restricted focus on this pathway through the bush alone, with the pruning of all other pathways regarded as irrelevant; and if one then claimed that this sole surviving pathway illustrates the predictable excellence of increasing adaptation by expressing a pervasive cladal trend (so that *any* surviving lineage would still tell the same adaptive tale for its own particular sequence); then the speciational view must debunk this last potential version of the old triumphalist attitude as well. If all paths through the equid bush led to the same Rome of modern *Equus,* or even if all major and prosperous paths—as measured by species range, diversity, or any conventional attribute of phyletic success—moved in the same general direction, then one might separate issues of species numbers, where the decline of equids through time cannot be denied, from the question of cladal direction, where the classic trend could still be asserted as predictable and progressive, even while the clade declined in diversity.

But reasonable, and very different, alternative scenarios for cladal direction remained viable until the recent restriction of the clade to the single genus *Equus.* MacFadden's (1988) study of all identifiable ancestral-descendant pairs showed 5 of 24, or more than 20 percent, leading towards decreased size. The general bias remains clear, but alternative scenarios also remain numerous and entirely reasonable (albeit, obviously, unrealized in the unpredictable history of life).

For example, the dwarfed genus *Nannippus* lived as a highly successful equid subclade for eight million years (far longer than *Equus* so far) at substantial species diversity (at least four named taxa) in North America, becoming extinct only 2 million years ago. Suppose that, in the reduction of the clade to one genus, *Nannippus* had survived and the forebears of *Equus* died? What, with this small and plausible alteration by "counterfactual history," would we then make of the vaunted and canonical trend in the evolution of horses? *Nannippus* did not exceed ancestral eohippus in size, and still grew three toes on each foot. *Nannippus* molars were as relatively high-crowned as those of modern *Equus,* but I doubt that any biologist would challenge my noncynical scenario based on psychological salience: If *Equus* had died, and *Nannippus* survived as the only genus of modern horses, this "insignificant" clade would now be passing beneath our pedagogical notice as just one more "unsuccessful" mammalian line—like aardvarks, pangolins, hyraxes, dugongs, and several others—equally reduced to a shadow of former success.

The situation becomes even more paradoxical, and also entirely general, when we take an honest look at our iconographic prejudices in the light of speciational reformulations for macroevolution. Consider the true "success stories" of mammalian evolution—the luxurious clades of rodents, chiropterans (bats), and, among large-bodied forms, the antelopes of the artiodactyl clade that expanded so vigorously as perissodactyls constantly declined within the guild of large-bodied hoofed herbivores. Has anyone ever seen, either in a textbook or a museum hall, a chart or a picture of these truly dominating clades among modern mammals? We don't depict these stories because we don't know how to draw them under our restrictive anagenetic conventions.

If we define evolution as anagenetic trending to a "better" place, how can we depict a successful group with copious modern branches extending in all directions within the cladal morphospace? Instead, and entirely unconsciously of course, for we would laugh at ourselves if we recognized the fallacy, our conventions lead us to search out the histories of highly unsuccessful clades—those now reduced to a *single* surviving lineage—as exemplars of triumphant evolution. We take this only extant and labyrinthine path through the phyletic bush, use the steamroller of our preconceptions to linearize such a tortuous route as a main highway, and then depict this straggling last gasp as the progressive thrust of a pervasive trend. I refer to this pattern of "life's *little* joke" (see Gould, 1996a).

2. RETHINKING HUMAN EVOLUTION. Ever since Protagoras proclaimed that "man [meaning all of us] is the measure of all things," Western intellectual traditions, bolstering our often unconscious emotional needs, have invariably applied the general biases of our analytic procedures, with special energy and focussed intensity, to our own particular history. As a cardinal example, concepts of human evolution long labored under the restrictive purview (now known to be empirically false) of the so-called "single species hypothesis" (see Brace, 1977)—the explicit claim that a maximal niche breadth, implied by the origin of hominid consciousness, made the coexistence of more than one species impossible, since no two species can share the same exact habitat under the "competitive exclusion" principle, and since consciousness must have expanded the effective habitat of hominid species into the full ecological range of conceivable living space. Human evolution could therefore be viewed as a single progressive series gradually trending towards our current pinnacle.

Needless to say, such a scheme precluded any speciational account of human history because only one taxon could exist at any one time, and trends had to record the anagenetic transformation of the only existing entity. (I do not think that such a relentlessly limiting scenario has ever been presented *a priori*, or so stoutly defended in principle even by a special name of its own, for any other lineage in the history of life.) This theoretical stricture enforced several episodes of special pleading for apparently contrary data. For example, early evidence for two distinct and contemporaneous australopithecine lineages (the so-called gracile and robust forms, now universally regarded as separate taxa) inspired dubious proposals about sexual dimorphism. The same theoretical constraint led many researchers to regard European neanderthals as necessarily transitional between *Homo erectus* and modern humans, even though the empirical record indicated a punctuational replacement in Europe about 40,000 years ago, with no evidence for any anatomical intermediacy.

I freely confess my partisan attitude, but I do think that reforms of the past two decades have centered upon a rethinking of this phylogeny in speciational terms, with punctuated equilibrium acting both as a spur for reformulation and a hypothesis with growing empirical support. Claims for stasis have been advanced, and much debated, for the only two species with appreciable longevity of a million years or more—*Australopithecus afarensis*, aka "Lucy," and *Homo erectus* (Rightmire, 1981, 1986, for support; Wolpoff, 1984, for denial). Lucy's earlier case seems well founded, while claims for our immediate ancestor, *Homo erectus*, have inspired more controversy, with arguments for gradual trending towards *Homo sapiens*, at least for Asian populations, generating substantial support within the profession. (But if the reported date—see Swisher et al., 1996—of 30,000 to 50,000 years for the Solo specimens holds, then the geologically youngest Asian *H. erectus* does not stand at the apex of a supposed trend.)

More importantly, hominid "bushiness" has sprouted on all major rungs of the previous ladder implied by the single-species hypothesis. The hominid tree may grow fewer branches than the comparable bush of horses, but multiple events of speciation now seem to operate as the primary drivers of human phylogeny (see Leakey et al., 2001, for a striking extension to the base of the known hominid bush in the fossil record), while humans also share with horses the interesting feature of present restriction to a single surviving lineage, albeit temporarily successful, of a once more copious array. (I would also suggest, on the theme of "life's little joke," that the contingent happenstance of this current restriction has skewed our thinking about human phylogeny away from more productive scenarios based on differential speciation.)

Speciation has replaced linearity as the dominant theme for all three major phases of hominid evolution (see Tattersall and Schwartz, 2000; and Johanson and Edgar, 1996, for booklength retellings of hominid history centered upon this revisionary theme of bushiness *vs.* linearity). First, before the origin of the genus *Homo*, the australopithecine clade differentiated into several, often contemporaneous, species, including, at a minimum, three taxa of "robust" appearance (*A. boisei, A. robustus,* and *A. ethiopicus*), and at least two of more "gracile" form (*A. afarensis* and *A. africanus*). Some of these species, including at least two of "robust" form, survived as contemporaries of *Homo*. Historically speaking, the death of the single species hypothesis may be traced to Richard Leakey's discovery, in the mid 1970's, of two undeniably different species in the same strata: the most "hyper-robust" australopithecine and the most "advanced" *Homo* of the time (the African form of *Homo erectus*, often given separate status on cladistic criteria as *Homo ergaster*).

(Needless to say, no true consensus exists in this most contentious of all scientific professions—an almost inevitable situation, given the high stakes of scientific importance and several well known propensities of human nature, in a field that features more minds at work than bones to study. Nonetheless, despite endless bickering about details, I don't think that any leading expert would now deny the theme of extensive hominid speciation as a central phenomenon of our phylogeny—see Johanson and Edgar, 1996.)

Second, the crucial period of 2–3 million years ago, spanning the origin of initial diversification of the genus *Homo*, also represents the time of maximal bushiness for the hominid clade, then living exclusively in Africa. The correspondence of a time of maximal speciation with anatomical change of greatest pith and moment in our eyes—that is, the origin of our own genus *Homo*, with an extensive expansion of cranial capacity—probably records a causal process of central importance in our evolution, not just an accident of coincidental correlation. Vrba's turnover-pulse hypothesis, an extension of punctuated equilibrium (see pp. 918–922), represents just one causal proposal for this linkage. As many as six hominid species may have coexisted in Africa during this interval, including three members of the genus *Homo*.

Third, the central theme of bushiness persisted far longer than previous conceptions of human evolution had ever allowed, right to the dawn of historical consciousness. Under earlier anagenetic views, European neanderthals marked a transitional stage in a global passage. But under speciational reformulations, and acknowledging extensive anatomical distance between neanderthals and moderns, *Homo neandertalensis* must be construed as a European offshoot of local *Homo erectus* populations, with *Homo sapiens* evolving in a separate episode of speciation, probably in Africa on strong genetic and more tenuous paleontological grounds, and then replacing neanderthals by migration. Similarly, Asian *Homo erectus* populations may not have passed anagenetically into *Homo sapiens*, but may also have been replaced by migrating stocks of *Homo sapiens*, originally from Africa. If the redating of the Solo *Homo erectus* specimens can be confirmed (as mentioned above), then this Asian replacement occurred at about the same time as the death of neanderthals in Europe.

This information implies the astonishing conclusion, at least with respect to previous certainties, that three human species still inhabited the globe as recently as 40,000 years ago—*Homo neanderthalensis* as the descendant of *Homo erectus* in Europe, persisting *Homo erectus* in Asia, and modern *Homo sapiens*, continuing its relentless spread across the habitable world. This contemporaneity of three species does not match the richness of an entirely African bush with some half a dozen species about 2 million years ago, but such recent coexistence of three human species does require a major reassessment of conventional thinking. The current status of our clade as a single species represents an oddity, not a generality. Only one human species now inhabits this planet, but most of hominid history featured a multiplicity, not a unity—and such multiplicities constitute the raw material of macroevolution.

This recasting of human evolution in speciational terms documents the extent of proffered revision, but the scope of reform gains even greater clarity when we recognize the pervasive nature of the speciational theme as a guide for resolving paradoxes, understanding puzzles of popular misconception, and offering new formulations to break impasses in almost every nook and cranny of discussion about the evolutionary history of our own lineage. When common claims seem askew or confused, I would venture to suggest that the first and best strategy for breakthrough will usually lie in a speciational reformulation for any puzzling issue. Consider just three bugbears of popular confusion in serious newspapers and magazines, and in books on general science for lay audiences, all finding a potentially simple and elegant resolution in speciational terms.

1. The ordinariness of "out of Africa." During the 1990's, popular articles on human evolution, at least in American media, focussed upon one issue above all others: the supposed dichotomy between two models for the origin of *Homo sapiens*. The press labelled these alternatives as, first, the "multiregional" hypothesis (also dubbed the "candelabra" or the "menorah" scheme to honor our religious pluralism in metaphorical choice) representing the claim that *Homo erectus* migrated from Africa about 1.5 to 2.0 million years ago and established populations on the three continents of Africa, Europe, and Asia. All three subunits then evolved in parallel (with enough gene flow to maintain cohesion) towards more gracile and substantially bigger-brained *Homo sapiens*. The second position, usually called "out of Africa" or "Noah's ark," holds that *Homo sapiens* emerged in one coherent place (presumably Africa from genetic evidence of relative similarities among modern humans) as a small, speciating population geographically isolated from the ancestral *Homo erectus* stock. This new species, probably arising less than 200,000 years ago, then migrated out of Africa to spread throughout the habitable world, displacing, or perhaps partly amalgamating with, any surviving stocks of *Homo erectus* encountered along the way.

The dichotomous division, as presented by the media, may have been a bit stark and unsubtly formulated. But I can raise no major objection either to the basic categorization or to most press reports about details of explanation and evidence. Still, I became more and more puzzled, and eventually amused in a quizzical or sardonic way, by a remarkable fallacy in basic interpretation that pervaded virtually every article on the subject. Almost invariably, popular presentations labelled the multiregional view as the conventional expectation of evolutionary theory, and out-of-Africa as astonishingly iconoclastic, if not revolutionary.

But all professionals must recognize that the exactly opposite situation prevails. Out-of-Africa presents a particular account, "customized" for human evolution, of the most ordinary of all macroevolutionary events—the origin of a new species from an isolated, geographically restricted population branching off from an ancestral range, and then, if successful, spreading to other suitable and accessible regions of the globe. By contrast, multiregionalism should be labelled as iconoclastic, if not a bit bizarre. How could a new species evolve in lockstep parallelism from three ancestral populations spread over more than half the globe? Three groups, each moving in the same direction, and all still able to interbreed and constitute a single species after more than a million years of change? (I know that multiregionalists posit limited gene flow to circumvent this problem, but can such a claim represent more than necessary special pleading in the face of a disabling theoretical difficulty?) Do we advocate such a scenario for the evolution of any other global species? Do we ever suspect that rats evolved on several continents, with each subgroup moving in the same manner towards greater ratitude? Do pigeons trend globally towards increased pigeonosity? When we restate the thesis in terms of non-human species, the absurdity becomes apparent. Why, then, did our media not grasp the singular oddity of multiregionalism and recognize out-of-Africa—especially given the cascade of supporting evidence in its favor—as the most ordinary of evolutionary propositions?

I can only conclude that popular views of evolution conceptualize the process, in general to be sure but particularly as applied to humans, as a linear and gradual transformation of single entities. Most consumers would be willing to entertain the speciational alternative for lineages (like rats and pigeons) that impose no emotional weight upon our psyches. But when we ask the great Biblical question—"what is man that thou art mindful of him?" (Psalm 8)—we particularly yearn for explanations based on anticipated global progress, rather than contingent origins of small and isolated populations in limited local regions. We want to regard our origin as the necessary, or at least predictable, crest of a planetary flux, not as the chancy outcome of a single event unfolding in a unique time and place.

This example, I believe, best illustrates the deep-seated nature of prejudices that must be overcome if we wish to grasp the truly Darwinian character of macroevolution as change wrought by differential success of favored individuals (species) within variable populations (clades)—thus finally breaking the Platonic chain of defining evolution as improvement of an archetypal form. Eldredge (1979) has advocated this transition by contrasting "taxic" approaches to evolution with the older "transformational" view. In the particular context of human evolution (Gould, 1998b), I have labelled multiregionalism as a "tendency theory," and out-of-Africa as an "entity theory." However much we may yearn to regard ourselves as the apotheosis of an inherent tendency in the unfolding of evolution, we must someday come to terms with our actual status as a discrete and singular item in the contingent and unpredictable flow of history. If we could bring ourselves to view this prospect as exhilarating rather than frightening, we might attain the psychological prerequisite for intellectual reform.

2. The unsurprising stability of *Homo sapiens* over tens of thousands of years. In a second example of a subject generally reported by the press with factual accuracy accompanied by ludicrously backwards explanation, the supposedly astonishing stability of human bodily form from Cro-Magnon cave painters to now has provided notable fodder for science journalists during the past decade. Consider, for example, the lead article in the prestigious "Science Times" section of the *New York Times* for March 14, 1995, entitled:

"Evolution of Humans May at Last Be Faltering." The opening sentence reads: "Natural evolutionary forces are losing much of their power to shape the human species, scientists say, and the realization is raising tantalizing questions about where humanity will go from here." (Professionals should always become suspicious when we read the universal and anonymous justification, "scientists say.")

These stories begin from the same foundational fallacy and then proceed in an identically erroneous way. They start with the most dangerous of mental traps: a hidden assumption, depicted as self-evident, if recognized at all— namely, a basic definition of evolution as continuous flux. Under this premise, the correct observation that *Homo sapiens* has experienced no directional trending for at least 40,000 years seems outstandingly anomalous. Reporters therefore assume that this unique feature of human evolutionary history requires a special explanation rooted in mental features that we share with no other species. They then generally assume, to complete the cycle of false argument, that human culture, by permitting the survival of marginal people who would perish in the unforgiving world of raw Darwinian competition, has so relaxed the power of natural selection that evolutionary change (popularly defined as "improvement") can no longer occur in *Homo sapiens*. This situation supposedly raises a forest of ethical questions about double-edged swords in the cure of diseases arising from genetic predisposition, the spread of genes for poor vision in a world of cheap eyeglasses, *et cetera ad infinitum*. (Pardon my cynicism based on some knowledge of the history of such arguments, but the neo-eugenical implications of these claims, however unintended in modern versions, cannot be ignored or regarded as just benignly foolish.)

This entire line of fallacious reasoning, with all its burgeoning implications, immediately collapses under a speciational reformulation. Once people understand *Homo sapiens* as a biological species, not a transitory point of passage in the continuous evolutionary progression of nature's finest achievement, the apparent paradox disappears by conceptual transformation into an expectation of conventional theory. Most species—especially those with large, successful, highly mobile, globally spread, environmentally diverse, and effective panmictic populations—remain stable throughout their history, at least following their origin and initial spread, and especially under the model of punctuated equilibrium that seems to apply to most hominid taxa. Change occurs by punctuational speciation of isolated subgroups, not by geologically slow anagenetic transformation of an entirety.

So if speciation usually requires isolated populations, how (barring science fiction scenarios about small groups of people spending generations in space ships hurtling towards distant stars) can a global species like *Homo sapiens*, endowed with both maximal mobility and an apparently unbreakable propensity for interbreeding wherever its members travel, ever expect to generate substantial and directional biological change in its current state? Most species should evoke predictions of stability, but *Homo sapiens* must lie at an extreme end of confidence for such an expectation. Thus, there is no solution to the supposed paradox of human stability because there is no problem. *Homo sapiens* has been stable for tens of thousands of years, and any proper understanding of macroevolution as a speciational process must yield this very expectation.

The same resolution applies to the extensive, and almost preciously silly, literature on human biological futures. (I do not speak of the real issues surrounding genetic engineering as an interaction of culture and nature, but of the fallacious and conjectural scenarios that treat presumptive human futures under a continuing regime of natural selection.) We have all seen reconstructions of improved future humans with bigger brains and disappearing little toes (perhaps balanced by the calloused butts of perennial couch potatoes). We also note the same features in reconstructions of advanced extraterrestrial aliens like ET, and in conjectural restorations of the hyper-brainy bipedal dinosaurs that might now rule the world if a bolide hadn't struck the earth 65 million years ago.

Again, this entire theme is moonshine. The only sensible biological prediction about human futures envisions continued stability into any time close enough to warrant any meaningful speculation. In any case, cultural change, in its explosive Lamarckian mode, has now so trumped biological evolution, that any directional trend in any allelic frequency can only rank as risibly insignificant in the general scheme of things. For example, during the past ten thousand years, any distinctive alleles in the population of native Australians must have declined sharply in global frequency as relative numbers of this subgroup continue to shrink within the human population as a whole. At the same time, cultural change has brought most of us through hunting and gathering, past the explosive new world triggered by agriculture, and into the age of atomic weaponry, air transportation and the electronic revolution, not to mention our prospects for genetic engineering and our capacity for environmental destruction on a global scale.

We have done all this, for better or for worse, with a brain of unaltered structure and capacity—the same brain that enabled some of us to paint the caves of Chauvet and the ceiling of the Sistine Chapel. What could purely biological and Darwinian change accomplish, even at a maximal rate (a mere thought experiment in any case, since we can only predict future stability for the short times that can justify any reasonable claims for insight), in the face of this explosive cultural transformation that our unchanging brains have unleashed and accomplished?

3. The conventional rate (and unconventional mode) of supposedly rapid trends traditionally cited as testaments to our uniqueness. When we recognize that human evolution occurred largely by differential success and replacement among species within a phyletic bush—and not by anagenetic transformation in measures of central tendency for a single, coherent entity in constant flux—then almost every standard claim about the tempo and mode of this process must be reformulated, and often substantially revised. My first two examples treated broad issues of maximal public attention. I now close this section with a smaller, but stubbornly persistent, error in order to clarify

and exemplify the fractal reach of this speciational reformulation into the smallest nooks and crannies of conventional wisdom.

Throughout my professional life, I have read, over and over again, the almost catechistic claim that the increase in cranial capacity from *Homo erectus* to *Homo sapiens* (variously specified as a 50 percent growth from about 1000 cc to 1500 cc as a lower estimate, to a doubling from 750 to 1500 as an upper bound) represents a stunning example of evolution at maximal rate, something so unusual and unprecedented that we must seek a cause in the particular adaptive value, and potential for feedback, of human consciousness. (Again, I suspect that our mental predisposition to commit such errors arises from an overwhelming desire to find something unique not only in the *result*—a defensible proposition—but also in the biological *mechanism* of human consciousness.) This claim for a maximal rate presumes anagenetic transformation, with the high value then calculated by spreading the total increment over the short amount of available time (from 100,000 years as a lower bound to a million years as an upper limit).

Two major errors, one obvious and almost ludicrous, the other more subtle and speciational, promote this "urban legend" that would disappear immediately from the professional literature if people only stopped to think before they copied canonical lines into their textbook manuscripts. First, the claim isn't even true within its own assumption of anagenetic transformation. Such a rate should not be designated as rapid at all when we recognize the proper scaling between our estimates of selection's strength in ecological time and its effect in geological time. Williams (1992, p. 132), for example, cites a standard claim and then presents some calculations:

> Even some widely recognized examples of rapid evolution are really extremely slow. Data on Pleistocene human evolution are interpretable in various ways, but it is possible that the cerebrum doubled in size in as little as 100,000 years, or perhaps 3000 generations (Rightmire, 1985). This, according to Whiten and Byrne (1988) is "a unique and staggering acceleration in brain size." How rapid a change was it really? Even with conservative assumptions on coefficient of variation (e.g. 10%) and heritability (30%) in this character, it would take only rather weak selection ($s = 0.03$) to give a 1% change in a generation. This would permit a doubling in 70 generations. An early hominid brain could have increased to the modern size, and back again, about 21 times while the actual evolution took place. Indeed, it is plausible that a random walk of 1% increases and decreases could double a quantitative character in less than 3000 generations.

If this first error of scaling has been identified before, the second and deeper fallacy of false assumptions about mode has generally escaped the notice of critics. The anagenetic assumption that trends represent the flux of a central tendency in a species's global transition to a better form must be replaced, in most cases, with a speciational account of trends as differential success of certain species within a clade. When we add the additional observation that punctuated equilibrium describes the history of most species, the absurdity of the conventional claim becomes immediately apparent.

The change from *Homo erectus* to *Homo sapiens* did not occur in a gradual and global flux throughout the range of *Homo erectus*, but as an event of speciation, geologically rapid under punctuated equilibrium and local in geography (probably in Africa). By Williams's argument above, the incorporation of the entire increase into the geological moment of speciation represents no surprise whatsoever. Why, then, do we misconstrue the rate as maximal over a much longer anagenetic transformation? We fall into this trap as a simple artifact of the happenstance that this particular event of speciation occurred very recently—100,000 to 250,000 years ago by most current estimates. The full transition probably occurred during the geological moment of speciation, but we erroneously interpret the change as progressing in gradualist fashion, and at constant rate, from this time of origin until the present day. Since the moment of origin happened to occur only a short time ago, the false anagenetic rate becomes very high (for we integrate the full change over a small interval). But if the identical punctuational event had occurred, say, 2 million years ago—an entirely plausible, but actually unrealized, situation—then the same episode of speciation would be read as anagenetically slow because the identical amount of change would now be falsely spread over a much longer interval. In other words, the claim for a maximal anagenetic rate in a trait marking the apotheosis of human success only records a false interpretation of global anagenesis for the recent cladogenetic origin of this trait in a punctuational event of speciation.

ECOLOGICAL AND HIGHER-LEVEL EXTENSIONS. The basic logic and formulation of punctuated equilibrium does not proceed beyond the level of species treated as independent individuals, or "atoms" of macroevolution. All biologists recognize, of course, that extensive ecological interactions bind each "atom" to others in complex ways. The "bare bones" structure of punctuated equilibrium does not include specific claims about ecological aggregations made of species as component parts. In this sense, punctuated equilibrium operates as a "null hypothesis" of sorts, by regarding each species as making its own way through geological time, with interactions treated as important components within the set of background conditions needed to explain the particulars of any history. In this sense, punctuated equilibrium treats time homogeneously, and species as independent agents; the theory therefore includes no inherent, or logically enjoined, predictions about the nature of temporal clumping in the ecological interactions among species.

But we know, as a basic fact and preeminent source of perennial controversy about the fossil record, that clumping occurs as a major pattern in the history of life—from minimal (and obvious) expression in joint disappearances during mass extinctions, to various theories that stress more pervasive and tighter clumping at several lower levels, all coordinated by the broad notion that ecosystems (or "communities," or other terminological alternatives, each with different theoretical implications) operate coherently during normal times as well, keeping species together in "organic" ways that falsify the null hypothesis of independent status.

As I write this chapter at the end of the 1990's, the implication of punctuated equilibrium for higher-level theories about the pulsing and clumping of species in putatively stable communities through considerable stretches of geological time has become the most controversial and widely discussed "outreach" from the basic theory that Eldredge and I formulated in 1972. Whatever my own opinions on the major alternatives now under debate— and I confess to a somewhat cautious, if not downright conservative, stance retaining a maximal role for the null hypothesis of species as independent Darwinian individuals making their own way through geological time—I take pride in the role that punctuated equilibrium has played in building an intellectual context for making such a debate possible in the first place.

One can't even pose meaningful questions about higher-level aggregations of species unless species themselves can be construed as stable and effective ecological or evolutionary agents—a status best conferred by punctuated equilibrium's recognition of species as true Darwinian individuals (see pp. 604–608). Under Darwin's personal view of species as largely arbitrary names for transitory segments of lineages in continuous anagenetic flux, such questions make no sense, and the entire potential subject remains undefined. (Some tradition exists for paleontological study of clumping in the distribution of named species through time, but this literature has never achieved prominence because researchers could not shake an apologetic feeling that they had based their studies on chimerical abstractions. For reasons well beyond accident or non-causal correlation, the beginning of serious and extensive research on this subject has coincided with the development and acceptance of punctuated equilibrium, a theory that recognizes these species as genuine evolutionary individuals.)

Precedents for studies of coordination above the species level can be found in such formulations as Boucot's (1983) twelve EEU's (ecological evolutionary units) for the entire Phanerozoic, or Sepkoski's (1988, 1991) three successive EF's (evolutionary faunas) for the same interval. But these works address the different subject of how major environmental shifts, including (but not restricted to) the substrates of mass extinctions, impact biotas defined at the family level and above. The subject of temporal interactions among species as basic macroevolutionary units raises a different set of questions about the nature of the "glue" that binds such sets of Darwinian individuals together at time scales matching their average durations (as contrasted with global geological changes that may coordinate—but probably not actively "bind"— biotas for intervals greatly exceeding the average lifespan of species). Some pioneering studies—most notably Olson's (1952) remarkable work on "chronofaunas" of late Paleozoic terrestrial vertebrates—have offered intriguing suggestions about the potential "glue" that may bind species into evolutionary ecosystems. However, as stated above, the subject could not be readily conceptualized until punctuated equilibrium provided a theoretical rationale for viewing species as legitimate Darwinian individuals.

The intense discussion of punctuational patterns at the level of species aggregations or ecosystems (extensive stability of species composition in regional faunas, followed by geologically rapid overturns and replacements of large percentages of these species) has centered upon two theoretical schemes and their proposed exemplars in the fossil record (see extensive symposium of 18 articles, edited by Ivany and Schopf, 1996, and entitled "New perspectives on faunal stability in the fossil record"). Working with the famous and maximally documented Hamilton faunas (Devonian) of New York State, and then extending their work up and down the stratigraphic record for a 70 million year interval of Paleozoic time in the Appalachian Basin, Brett and Baird (1995) documented 13 successive faunas, each including 50 to 335 invertebrate species, and each showing considerable stability both for the history of any species throughout its range (the predicted stasis of punctuated equilibrium, see Lieberman, Brett and Eldredge, 1995, for quantitative evidence), and, more importantly in this context, in the virtually constant composition of species throughout the fauna's range.

Each fauna persists for 5 to 7 million years until replaced, with geological rapidity, by another strikingly different fauna including only 20 percent or fewer carryovers from the preceding unit. As a defining attribute, 70 to 85 percent of species in the fauna persist from the earliest strata to the very end, remaining in apparently stable ecological associations (with characteristic numerical dominances of taxa) to forge a pattern that Brett and Baird call "coordinated stasis." Eldredge (1999, p. 159) writes of coordinated stasis: "It is a true, repeated pattern, the most compelling and at the same time underappreciated pattern in the annals of biological evolutionary history."

Vrba (1985) found a similar pattern in the maximally different ecosystem of vertebrates in Pliocene terrestrial environments of southern and eastern Africa. The geologically rapid faunal replacement, following an extensive period of previous stability, occurred in conjunction with a 10–15 degree drop in global temperatures that lasted some 200,000 to 300,000 years, and began about 2.7 to 2.8 million years ago. In generalizing this pattern as the "turnover-pulse hypothesis," Vrba emphasizes the role of environmental disruption in prompting the transition and, especially, the coordinated effects of both extinction and speciation as consequences of disruption—extinction by rapid change and removal of habitats favored by species of the foregoing fauna, and origination by fragmentation of habitats and resulting opportunities for speciation by geographic isolation of allopatric populations. As an example of the role granted to increased propensities for speciation, as promoted by the same environmental events that decimated the previous faunas, Vrba links the origin and initially rapid speciation (at least three taxa) of the genus *Homo* to this Pliocene turnover-pulse, a proposition that has generated substantial interest and debate.

The two propositions—Brett and Baird's coordinated stasis, and Vrba's turnover-pulse—identify similar patterns in prolonged stasis and punctuational replacement for linked groups of species. However, the two formulations differ in proposed explanations for this common pattern. Brett and Baird identify rapid environmental turnover as the trigger for collapse of incumbent faunas, but tend to view the prolonged stability of each fauna as a consequence, at least in large part, of internal ecological dynamics. New faunas come together largely by migration of separate elements from other regions (rather than originating primarily by local speciation *in situ,* as in Vrba's model), but then maintain stability by ecological interactions. Vrba, on the other hand, tends to attribute both fundamental aspects of the pattern— prolonged stasis and abrupt replacement—to vicissitudes and stabilities of the physical environment. As noted above, she also attributes the construction of new faunas to local speciation following fragmentation of habitats induced by environmental change, whereas Brett and Baird stress migration for the aggregation of new faunal associations. In Vrba's view, rapid physical changes induce the turnover by (at least local) extinction, and then also engender the subsequent stability as a propagated effect—because interspecific interactions play little role in regulating faunal stability, which must then arise as a basic expectation from punctuated equilibrium about the independent behavior of individual species. Ivany (1996, p. 4) accurately describes this aspect of Vrba's model: "stasis intervals between are in essence side consequences requiring no additional explanation beyond that required to explain stasis in individual lineages."

The developing debate in the paleontological literature has focussed upon two issues of markedly different status. First, does the pattern actually exist— with sufficiently crisp and operational definition in any single case, and with sufficiently frequent occurrence among all cases—to warrant an assertion of evolutionary generality (see numerous examples and discussion in the Ivany and Schopf symposium, 1996, cited above)? I remain entirely optimistic on this point, if only because the "type" example of the Hamilton Fauna seems well and extensively documented.

However, and to confess a personal bias, my feelings of caution about unmitigated endorsement also arise from a substantial worry under this heading. If a capacity for individuation establishes the basis, or even just ranks as an important criterion, for status as an evolutionary agent in a Darwinian world, then our logical inability to render the faunas of coordinated stasis or turnover-pulse as coherent individuals does cause me concern. I do recognize that higher units of ecological hierarchies (see Eldredge, 1989) generally lack the coherence of individuals defining most levels of the standard genealogical hierarchy—because ecological associations cannot "hold" their component members as tightly as genealogical individuals enfold their subparts. Species, at a high level of the genealogical hierarchy, function as excellent Darwinian individuals because their subparts (organisms) remain tightly bounded by potential for interbreeding within, and prevention without. But ecological units like "faunas" must be constructed in a more "leaky" manner, for I cannot imagine a force that could hold taxonomically disparate forms together by ecological interaction with anything like the strength that species can muster to "glue" component organisms into a higher individual. However, and in response to my own doubt, the demic level of the genealogical hierarchy manifests a similar intrinsic leakiness because no strong "glue" exists, in principle, to prevent the passage of component parts (organisms) from one deme to another. Still, and after much debate, the efficacy of interdemic selection now seems well established, at least in certain important evolutionary settings (see pp. 648–652 and Sober and Wilson, 1998).

Second—and more importantly in raising a theoretical issue at the heart of evolutionary studies—if the pattern of coordinated stasis and turnover-pulse does exist with sufficient clarity and frequency, then what forces hold faunas together at such intensity and for such long intervals, especially in the light of intrinsic capacity for "leakiness," as mentioned above? (Theoretical debate on this issue has rightly centered upon the putative causes of coordination in faunal stability, not on the rapidity of overturn. All formulations agree in ascribing quick transitions between faunas to direct effects of environmental perturbation.) Roughly speaking, two proposals of strikingly different import have dominated this debate. Some authors—in what we may call the "conservative" view, not for any intrinsic stodginess, but for envisioning no new or unconventional explanatory principles—hold that faunal stasis requires no active coordinating force at all, but arises as a side consequence of the environmentally triggered overturns themselves. (Vrba's formulation, as noted above, tends to this interpretation.) All active control then falls to the extrinsic causes of rapid overturn, with the coordination in between merely recording the predicted behavior, under punctuated equilibrium, of species acting as independent entities. In other words, we see temporal "packages" of coordinated stasis because external forces impose coincident endings and beginnings.

But other authors (see Morris, 1996; Morris et al., 1995) advocate active causal mechanisms, at the level of interaction among species, for holding the components of ecosystems together during periods of stasis—a notion generally called "ecological locking," and envisaging an explicit and intrinsic "glue" to build and then to hold the coordination of coordinated stasis. Morris, for example, cites the work of O'Neill et al. (1986) on mathematical theories of ecological hierarchies, in advancing a "claim that ecosystems in frequently disturbed settings become hierarchically organized such that the effects of large, low-frequency disturbances do not propagate through the system and cause disruption" (Ivany, 1996, p. 7). Other proposals for "intrinsic" mechanisms of coordination have invoked the general concept of "incumbency," and tried to designate theoretical reasons why established associations of species, even if non-optimal and only contingently or adventitiously built, may resist displacement by active mechanisms rooted in the behavior and construction of such aggregations.

These admittedly somewhat fuzzy and operationally ill-defined proposals address, nonetheless, the core of a vitally important issue within the developing hierarchical extension of Darwinian theory: how far "up" a hierarchy of levels do active causal forces of evolutionary change and stability extend? Do such causes generally weaken, or become restricted to peripheral impacts, at these higher levels? If so, can we attribute such diminution to increasingly limited opportunities for devising "glues" that might bind components into coherent individuals at these higher levels? Can "glues" for higher units in ecological hierarchies be strong enough, even in theory, to achieve the sufficient bounding (and bonding) that higher levels of the genealogical hierarchy (like species) can and do attain?

I do not know how this debate will develop, and how, or even whether, these questions can be operationally defined and activated. We remain stymied, at the moment, because so little thought, and so little empirical work, has been devoted to operational criteria for distinguishing alternatives—particularly for defining the different expectations of coordination as a passive consequence of joint endings vs. an active result of ecological locking during intervals of stasis. Perhaps such distinctions can be defined and recognized in the statistics of faunal associations (varying strengths and numbers of paired correlations, similarities in joint ranges and relative abundances of groups of species: what numbers of taxa, and what intensities of coordination, imply active locking beyond the power of passive response among independent items to accomplish?). Given the notorious imperfections of the geological record, and the daunting problems of consensus in taxonomic definition, I recognize the extreme difficulty of such questions. But the issues raised are neither untestable nor non-operational, and the concepts involved could not be more central to evolutionary theory. Whatever the future direction of this debate, punctuated equilibrium has proven its mettle in prompting important extensions beyond its original purview, and in proposing a fruitful strategy of research, based on a new way of viewing the fossil record, that broke some longstanding impasses in paleontological practice. At the very least, punctuated equilibrium has raised some interesting and testable questions that could not be framed under previous assumptions about evolutionary mechanisms and the patterns of life's history.

As a final note and postscript, either extreme alternative for the explanation of faunal stasis—passive consequence or active ecological locking—bears an interesting implication for the significance of punctuated equilibrium. (Of course, I would be shocked if either extreme eventually prevailed, or if a future consensus simply melded aspects of both proposals into harmony. I suspect that the reasons behind coordinated stasis are complex, multifarious, and informed by other modes and styles of explanation as yet unconceived.) If coordination arises as a passive consequence, then our original version of punctuated equilibrium, proposed to explain the pattern of individual species, also suffices to render this analogous pattern at the higher level of faunas as well—thus increasing the range and strength of our mechanism. But if coordination must be forged by higher-level mechanisms of active ecological locking, then punctuated equilibrium provided the basis, both logically and historically, for regarding species as evolutionary individuals, the conceptual prerequisite for Darwinian theories of causation at the level of aggregations of species.

## PUNCTUATION ALL THE WAY UP AND DOWN? THE GENERALIZATION AND BROADER UTILITY OF PUNCTUATED EQUILIBRIUM (IN MORE THAN A METAPHORICAL SENSE) AT OTHER LEVELS OF EVOLUTION, AND FOR OTHER DISCIPLINES IN AND OUTSIDE THE NATURAL SCIENCES

### General models for punctuated equilibrium

If the distinctive style of change described by punctuated equilibrium at the level of speciation—concentration in discrete periods of extremely short duration relative to prolonged stasis as the normal and actively maintained state of systems—can be identified in a meaningful way at other levels (that is, with sufficient similarity in form to merit the same description, and with enough common causality to warrant the application in more than a metaphorical manner), then general mathematical models for change in systems with the same fundamental properties as species might also be expected to generate a pattern of punctuated equilibrium under assumptions and conditions broad enough to include nature's own. In this case, we might learn something important about the general status and range of application of such a pattern—thus proceeding beyond the particular constraints and idiosyncrasies of any biological system known to generate this result at high relative frequency.

Many scholarly sources in the humanities and social sciences, with Thomas Kuhn's theory of scientific revolutions as the most overt and influential, have combined with many realities of late 20th century life (from the juggernaut of the internet's spread to the surprising, almost sudden collapse of communism in the Soviet Union, largely from within) to raise the general critique of gradualism, and the comprehensive acceptability of punctuational change, to a high level of awareness, if not quite to orthodoxy. But the greatest spur to converting this former heresy into a commonplace, at least within science, has surely arisen from a series of mathematical approaches, some leading to little utility despite an initial flurry of interest, but others of apparently enduring worth and broad applicability. These efforts share a common intent to formalize the pattern of small and continuous inputs, long resisted or accommodated by minimal alteration, but eventually engendering rapid breaks, flips, splits or excursions in systems under study: in other words, a punctuational style of change. These proposals have included René Thom's catastrophe theory, Ilya Prigogine's bifurcations, several aspects of Benoit Mandelbrot's fractal geometry, and the chief themes behind a suite of fruitful ideas united under such notions as chaos theory, non-linear dynamics, and complexity theory.

Empirical science has also contributed to this developing general movement by providing models and factual confirmations at several levels of analysis and for several kinds of systems, with catastrophic mass extinction theory first seriously proposed in 1980 and then strongly promoted by increasingly firm evidence for bolide triggering of the late Cretaceous event, as an obvious input from paleontology. I take pride in the role that punctuated equilibrium has played as a spur for this larger intellectual transformation—for our 1972 proposal, formulated at one level of biological change, provided some general guidelines, definitions and terminology, and also provoked a good deal of interest in the application of this general style of change to other fields of study and other levels of causality. This extension has proceeded so far that some scientists and scholars from other disciplines (see Gersick, 1991; Mokyr, 1990; Den Tex, 1990; Rubinstein, 1995, for example) now use punctuated equilibrium as the general term for this style of change (while we would prefer that punctuated equilibrium retain its more specific meaning for the level of speciation, with punctuational change or punctuationalism used for the generality).

Some recent mathematical work has explicitly tried to model punctuational change at the level and phenomenology of our original theory. Rand and Wilson (1993, p. 137), for example, following Rand et al.'s (1993) "general mathematical" model for "Darwinian evolution in ecosystems," applied their basic apparatus to the problem of speciation in individual taxa within ecosystems, primarily to test whether or not the pattern of punctuational equilibrium would emerge. "We do not mean," they write (1993, p. 137), "a large multispecies extinction event but rather the sudden disappearance of an evolutionary stable state causing a species to undergo very rapid evolution to a different state."

Under basic trade-off "rules" of bioenergetics and ecological interaction, which they call "constraints" (correlations, for example, between a prey's increase in population size and the exposure of individuals to a predator's attention), punctuated equilibrium emerges as a general pattern. Gradual change may prevail in systems without such constraints, but as the authors state, and to say the very least about nature's evident complexity(!), "the absence of such constraints is biologically unrealistic" (1993, p. 138). Moreover, they argue, reasonable features of the model's internal operation, and recognized properties of natural ecosystems, suggest a general status for punctuational change at all levels: "In this note," Rand and Wilson write (1993, p. 137), "we wish to address the important issue of gradualism against punctualism in evolutionary theory. We discuss this in terms of a simple illustrative example, but emphasize that . . . our results apply quite generally and are ubiquitous and wide ranging."

Explicit modelling of other levels has also yielded punctuational change as an expectation and generality under realistic assumptions. Elsewhere, I discuss models for punctuational anagenesis within populations in ecological time (see p. 877), erroneously interpreted by some critics as a demonstration that punctuated equilibrium emerges from ordinary microevolutionary dynamics and therefore embodies nothing original—although such studies should be interpreted as illustrating the potential generality of punctuational change by rendering the same pattern as an anticipated result of different processes (transformation of a deme rather than origin of a new species by branching) at a lower hierarchical level (intra rather than interspecific change).

Punctuational change has been modelled more frequently at the most evident level above punctuated equilibrium—coordinated and rapid change in several species (or the analogs of separate taxa in modelled systems) within communities or faunas. Per Bak's "sandpile" model of self-organized criticality (see Bak and Sneppen, 1993; Sneppen et al., 1994; and commentary of Maddux, 1994) have generated both particular interest and legitimate criticism. Maddux (1994, p. 197), noting the "minor avalanche of articles on the theme," began his commentary by writing: "That physicists are itching to take over biology is now well attested . . . But surely only a brave physicist would take on Darwin on his home ground, the theory of evolution, let alone Gould and Eldredge on punctuated equilibrium."

Bak's models operate by analogy to metastable sandpiles, where grains may accumulate for long periods without forcing major readjustment (the analog of community stability), whereas, at a critical point, just one or a few added grains will trigger an avalanche, forcing the entire pile to a new and more stable configuration (the analog of mass extinction and establishment of new faunas, not to mention the straw that broke the camel's back). In his basic model, Bak assigns random fitnesses, chooses the "species" with the smallest fitness, and then reassigns another random number both to this item and to the two neighboring species of its line (to stimulate interactions among taxa in communities). He also randomly selects a certain number of other points for similar reassignment (to acknowledge that interconnections among taxa need not link only the most obviously related or adjacent forms).

This procedure often generates waves of rapidly cascading readjustments, propagated when some species receive small numbers in the reassignment of fitnesses, and then must change, taking their neighbors and also some distant forms with them, by the rules of this particular game—a play that admittedly cannot mimic nature closely (if only because the model includes no analogs of extinction or branching), but that may give us insight into expected rates and patterns of change within simple systems of partly, and largely stochastically, linked entities. In any case, Bak and his colleagues have formalized the general notion that small inputs (random reassignment of fitness to just one entity of lowest value) to simple systems of limited connectivity among parts (changes induced in a few other entities by this initial input) can lead to punctuational reformation of the entire system.

In a similar spirit, the substantial research program known as Artificial Life (AL to aficionados) takes an empirical, if only virtual, approach to such questions by generating and tracking evolving systems operating under simple rules in cyberspace. I regard such work as of great potential value, but often philosophically confused because researchers have not always been clear about which of two fundamentally different intentions they espouse: (1) to build systems that mimic life with enough fidelity to state something useful about actual evolution on earth; or (2) to construct alternative virtual worlds so explicitly unlike actual life in their minimality that we can ferret out some abstract properties, applicable to any genealogical system, by using models that permit perfect tracking of results and also operate with sufficient simplicity to identify the role of any single component.

Ray (1992, a pioneer in these studies of "evolution in a bottle" or "synthetic life in a computer" (1992, p. 372), started his "Tierra" system by designing a block of RAM memory as "a 'soup' which can be inoculated with creatures" (1992, p. 374), and then beginning with a "prototype creature [that] consists of 80 machine instructions," with "the 'genome' of the creatures consisting of the sequence of machine instructions that make up the creature's self-replicating algorithm."

"When the simulator is run over long periods of time, hundreds of millions or billions of instructions, various patterns emerge" (1992, p. 387). Obviously, the results depend crucially on the human mental protoplasm that sets the particular rules and idiosyncrasies of the virtual system. Ray found, for example (and unsurprisingly), that "under selection for small sizes, there is a proliferation of small parasites and a rather interesting ecology." Similarly, "selection for large creatures has usually led to continuous incrementally increasing sizes . . . This evolutionary pattern might be described as phyletic gradualism" (p. 387).

But under the much more "reality mimicking" condition of no consistent directional selection for size, Ray found "a pattern which could be described as periods of stasis punctuated by periods of rapid evolutionary change, which appears to parallel the pattern of punctuated equilibrium described by Eldredge and Gould" (p. 387). Ray then describes his frequently replicated and longest running results in more detail (pp. 387–390):

> Initially these communities are dominated by creatures with genome sizes in the 80's. This represents a period of relative stasis, which has lasted from 178 million to 1.44 billion instructions . . . The system then very abruptly (in a span of 1 or 2 million instructions) evolves into communities dominated by sizes ranging from about 400 to about 800. These communities have not yet been seen to evolve into communities dominated by either smaller or substantially larger size ranges. The communities of creatures in the 400 to 800 size range also show a long-term pattern of punctuated equilibrium. These communities regularly come to be dominated by one or two size classes, and remain in that condition for long periods of time. However, they inevitably break out of that stasis and enter a period where no size class dominates . . . Eventually the system will settle down to another period of stasis dominated by one or a few size classes which breed true.

All models previously discussed have generated punctuational patterns at explicit and particular levels of evolutionary change (anagenetically within demes, for the origin of species by branching, and in coordinated behavior of groups of species within entire faunas and ecosystems). This range of success suggests that the apparent ubiquity of punctuational patterns at substantial, if not dominant, relative frequencies may be telling us something about general properties of change itself, and about the nature of systems built of interacting components that propagate themselves through history. Some preliminary work has attempted to formalize these regularities, or even just to identify them through a glass darkly (see, for example, Chau, 1994, on Bak's models).

Bak has tried to specify two "signatures of punctuated equilibrium" in very general properties of systems: "a power-law distribution of event sizes where there is no characteristic size for events, but the number of events of a certain size is inversely proportional to some power of that size"; and a property that Bak calls $1/f$ noise, "where events are distributed over all time-scales, but the power or size of events is inversely proportional to some power of their frequency" (Shalizi, 1998, p. 9). Since we can document such inverse relationships between magnitude and frequency in many natural systems—indeed, R. A. Fisher (1930) began his classic defense of Darwinism with a denial of efficacy for macromutations based on their extreme rarity under such a regularity—punctuational change may emerge as predictably general across all scales if Bak's conditions hold.

The intellectual movement dedicated to the study of complex dynamical systems and their putative tendencies to generate spontaneous order from initial randomness—a prominent fad of the 1990's, centered at the Santa Fe Institute and replete, as all fashions must be, with cascades of nonsense, but also imbued with vital, perhaps revolutionary, insights—has identified punctuated equilibrium as a central subject of inquiry. A defining workshop, held in Santa Fe in 1990, specified three primary illustrations or consequences of this discipline's central principle, "the tendency of complex dynamical systems to fall into an ordered state without any selection pressure whatsoever": the origin of life, the "self-regulation of the genome to produce well defined cell types"; and "the postulated sudden waves of evolutionary change known as 'punctuated equilibrium.'"

Stuart Kauffman, the leading biological theorist and mathematical modeler of this movement (see Chapter 11, pp. 1208–1214 for a discussion of his work on structuralist approaches to adaptive systems), stressed the generality of punctuational change by beginning with simple models of coevolution and then obtaining punctuational change at all levels as a consequence. *Science* magazine's report of this 1990 meeting linked Kauffman's multilevel work to the ubiquitous emergence of punctuated equilibrium from models of highly disparate systems and processes—all suggesting a generality and an intrinsic character transcending any particular scale or phenomenology: "This pattern of change and stasis itself evolves," says Kauffman. "In the subtly shifting network of competition and cooperation, predator and prey, a fast-evolving species might suddenly freeze and cease to evolve for a time, while a formerly stable species might suddenly be forced to transform itself into something new. The fossil record of the latter process would then resemble 'punctuated equilibrium': a pattern of stasis interrupted by sudden change, which some paleontologists now believe to be the norm in real evolution. . . . This same pattern of stasis punctuated by sudden change also showed up in a number of other ecosystem models presented at the workshop, even when those models seemed superficially quite different. Does this mean some more general mechanism is at work, some theory that could account for the behavior of these models—and perhaps real life—no matter how they are structured?"

A false and counterproductive argument has enveloped this work during the past few years. Bak, in particular, has noted that punctuations at the highest level, corresponding to simultaneous extinction of a high percentage of components in a system, can be generated from internal dynamics alone, and require no environmental trigger of corresponding (or even of any) magnitude. He and others then draw the overextended inference that because such large scale punctuations can arise endogenously, the actual mass extinctions of the fossil record therefore need no exogenous trigger of environmental catastrophe, or any other external prod. This claim, emanating from a theoretical physicist with little knowledge of the empirical archives of geology and paleontology, and emerging just as persuasive evidence seems to have sealed the case for bolide impact as a trigger of at least one actual mass extinction (the end Cretaceous event 65 million years ago), could hardly fail to raise the hackles of observationally minded scientists who, for reasons both understandable and lamentable, already bear considerable animus towards any pure theoretician's claim that success in modelling logically entails reification in nature.

The obvious solution—if human emotions matched human logic in clarity, or the empirical world in complexity—would welcome the mathematical validation of potential endogenous triggers (often of small initial extent) for punctuational change as a partner with well-documented exogenous triggers (of great extent in one well documented case, but perhaps also of potentially small magnitude as well). Instead of waging a false battle for preference or exclusivity of one alternative between two plausible arguments, we should recognize instead the complementary and general theme behind both proposals—their common role as sources for punctuational change (which then achieves higher status as a truly general pattern in nature), and in their mutual reinforcement for revising and expanding the Darwinian paradigm on all three supporting legs of its essential tripod. For the punctuational style of change—disfavored by Darwin, who recognized the necessary status of gradualism within the logic of his world view—now emerges as a primary consequence of repairs and reinforcements upon all legs of the tripod: the expansion beyond small uniformitarian inputs for the external triggers and causes of leg three (thus granting environment an even greater role than Darwin himself, who so brilliantly introduced the concept to defeat previous internalist theories of change, had envisioned); and the recognition that constraints of systems (leg two)—not only overt natural selection—acting at all levels of a causal and genealogical hierarchy (leg one) can also generate punctuations from within.

### Punctuational change at other levels and scales of evolution

A PRELIMINARY NOTE ON HOMOLOGY AND ANALOGY IN THE CONCEPTUAL REALM. The simple documentation of punctuational patterns at scales other than the speciational status of punctuated equilibrium (and, therefore, presumably attributable to different causes as well) gives us little insight into the key question of whether or not punctuated equilibrium, in either its observed phenomenology or its proposed mechanics, can lay claim to meaningful generality in evolutionary studies. Rather, the overt similarity in pattern must be promoted to importance through an additional claim, akin in the world of ideas to the weight that an assertion of homology would carry in assessing the value of taxonomic characters. What, then, would make an example of punctuational change from another scale (where the immediate speciational cause of punctuated equilibrium could not apply) effectively "homologous" to punctuated equilibrium—that is, sufficiently similar by reason of phenomenological "kinship" that the similar pattern across disparate scales may be read as revealing the shared components of a common explanation?

We rank some similarities across scales as capricious enough to be deemed accidental, and therefore devoid of causal meaning. The appearance of a "face" on a large mesa on the surface of Mars—an actual case by the way, often invoked by fringe enthusiasts of extraterrestrial intelligence—bears no such conceptual homology to faces of animals on earth. We label the similarity in pattern as accidentally analogous—even though the perceived likeness can teach us something about innate preferences in our neural wiring for reading all simple patterns in this configuration (a line below two adjacent circles) as faces. (An actual face and the accidental set of holes on the mesa top may stimulate the same pathway in our brain, but the two patterns cannot be deemed causally similar in their own generation—that is, as faces.)

Identity of specific cause will rarely be available to provide a basis for asserting meaningful homology, rather than misleading analogy, between common patterns at disparate scales. Punctuated equilibrium, for example, gains power and testability in proposing a particular scale-bound reason for an observed phenomenon—the expression of ordinary speciation in geological time, in this case. Since most theories win strength by such specificity, conceptual homologies across scales must seek other definitions and rationales. A punctuational pattern below the scale of punctuated equilibrium (change within a single deme for example), or above (temporal clumping in the origin or extinction of many species within a fauna, could not, in principle, be explained by the specific causes of punctuated equilibrium itself.

Therefore, meaningful "homology" in this conceptual sense must generally be sought in properties that are genuinely held in common across systems and scales, and that operate to channel the different causes of these various scales into the same recognizable and distinctive pattern. Moreover, such homology becomes all the more interesting if the particular efficient causes at different scales—the actual pushers and movers of immediate change in each case—remain evidently disparate, thus implying, by elimination, that the observed commonality of pattern may arise from constraining channels of similar structural properties across scales. If all roads lead to Rome, then the eternal city ranks as a dominant and ineluctable attractor!

In the present case of punctuational patterns across markedly different scales of time and component entities, claims for conceptual homology rest upon an overarching hypothesis that punctuation records something quite general about *the* nature of change itself, and that the differing causes of punctuational change at each level—the waiting time between favorable mutations in bacterial anagenesis (see next section), the scaling of ordinary speciation as a geological moment in punctuated equilibrium, or the simultaneity of species deaths in mass extinction—must run in a common structural channel that sets and constrains the episodic nature of alteration.

If punctuated equilibrium gains this generality by conceptual homology, then both components of its name should achieve such transfer across nature's numerous scales of size and time. (The general mathematical models discussed in the last section presume such meaningful transfer as a primary rationale for their relevance.) The equilibrial component wins potential generality if active resistance to change can be validated as an important structural property of systems discrete and stable enough to be named and recognized as entities at any scale of nature (whatever the causes of stability, whether internal to self-integration or imposed from without upon an intrinsically less coherent structure—a fascinating question that should become an object of research, not the subject of prior definition). This property of *active* maintenance underlies our primary claim about stasis in punctuated equilibrium, and our insistence that stasis must be conceptualized and defined as a positive phenomenon, not as a disappointing or uninteresting absence of anticipated change. (Throughout this chapter, I have provided evidence, primarily in observed relative frequencies—far too high to originate either passively or randomly in a world of natural selection and genetic drift—for interpreting stasis as an actively generated property of systems, embodied in species at the scale of punctuated equilibrium, but necessarily recognized in structures of different status at scales both below and above species.)

The punctuational component, operationally measured by its short duration relative to periods of stasis within definitive structures of the same scale, would then achieve homological generality as the obverse to proposed reasons for stasis: the reinterpretation of change—at least in its usual, if not canonical, expression—as a rare and rapid event experienced by systems only when their previous stabilities have been stretched beyond any capacity for equilibrial return, and when they must therefore undertake a rapid excursion to a new position of stability under changed conditions.

Obviously, these "brave" statements about conceptual homology across disparate scales and immediate causalities must remain empty and meaningless without operational criteria for distinguishing—if I may again use the conventional evolutionary jargon in this wider context—meaningful similarity of genesis (homology) from misleading superficiality of appearance (analogy). As a first rule and guideline, we might look to the same basic precept of probability that regulates our general procedures in the study of overt similarity among separate phenomena: co-occurrence of substantial numbers of potentially independent parts as a sign of meaningful genetic (and conceptual) connection *vs.* resemblance based upon single or simple features, however visually striking, as far more likely to be unconnected, separately built, and perhaps not even meaningfully alike in any causal or functional sense (the complex and identical topology of arm bones in a bat's wing and a horse's foreleg as meaningfully homological *vs.* my face and the same disposition of holes on the Martian mesa as meaninglessly analogical). Thus, and in a practical sense, I focus much of the following discussion upon a search for what I will call "conjoints," or sets of independent features whose joint occurrence predisposes us to consider meaningful conceptual homology in punctuational patterns of change produced by different immediate causes at disparate scales of size and time. (I have used the same form of argument frequently throughout this book—as in emphasizing the usual conjunction of openness to saltational change, belief in the importance of internal channeling, and suspicion about adaptationist explanation in defining the biological worldview of structuralist thinkers—see Chapters 4 and 5).

When such broad "homologies" of common structural constraint have been established across several realms of size and time, then we can most fruitfully ask some second-order questions about systematic, or "allometric," differences (see Gould and Lloyd, 1999) in the expression of common patterns along continua ordered by increasing magnitude among scales under consideration. For example, do internal forces of cohesion among subparts set the primary basis for active stasis, or does the "fit" of a structure into a balanced and well-buffered environment, made of numerous interacting entities, prevent change in a system otherwise fully capable of continuous alteration in the absence of such externalities? Does the balance between these internalist and externalist explanations change as we mount through scales of magnitude? Is the change systematic (and therefore "allometric" in the usual sense), or capricious with respect to scale itself, having no correlation with magnitude?

The important principle that meaningful similarity may reside in homology of structural constraint across scales, while particular causes that "push" phenomena through these constrained channels may vary greatly, has rarely been stated or exemplified with proper care, and has therefore usually been ignored by commentators on the role of events at one scale in the interpretation of others. Most regretfully, a frequent misunderstanding has then led to dismissal of meaningful commonality in pattern because a critic notes a strong difference in immediate causes for a pattern at two scales and then rejects, on this erroneous basis, any notion of an informative or integrative status for the similarity. Or, even worse, the critic may become intrigued by a cause just elucidated at one scale and then assume that the significance of such a discovery can only lie in extrapolating this particular and strongly scale-bound cause to debunk a different mechanism previously proposed to explain the same pattern at another level—rather than exploring the more fruitful and integrating hypothesis that a genuine basis for meaningful similarity in pattern might reside in homologous structural constraints that channel different causes to similar outcomes at the two distinct scales.

For example, an excellent science reporter for the *New York Times* erroneously argued that punctuations caused by long waiting times between rapidly-sweeping, favorable mutations in bacterial anagenesis on a scale of months should lead us to reinterpret the speciational breaks of punctuated equilibrium (at geological scales) as similarly caused by quick and simple genetic changes! "The finding that all it takes is a few mutations and a little natural selection to generate punctuated evolution comes as a surprise. Researchers say numerous theories that are considerably more complex have been put forth to explain what might produce the punctuation seen in the fossil record. If bacteria are any indication, the rapid evolution documented in the fossil record might be the product of a very few simple, if quick, genetic changes" (Yoon, 1996).

But R. E. Lenski, the chief scientist in the bacterial study (Elena, Cooper, and Lenski, 1996), properly sought commonality with punctuated equilibrium in the domain of homologous reasons for punctuational patterns. Recognizing the disparities in scale, and the different causes thus implied, they rightly declined to apply the term punctuated equilibrium to their findings. Instead, they invoked the general term for the pattern itself as the title for their paper (Elena, Cooper, and Lenski, 1996): "Punctuated evolution caused by selection of rare beneficial mutations."

**PUNCTUATION BELOW THE SPECIES LEVEL.** I have, at several points in this and the preceding chapter, discussed various empirical and theoretical studies that validate the pattern of substantial stability followed by rapid peak shifts in the anagenetic transformation of single populations during the microevolutionary time of potential human observation (see p. 877). I have also urged (to reiterate the theme of the preceding section) that such an important conclusion should not be read as an argument that punctuated equilibrium holds no interest for evolutionary theory because ordinary population genetics can produce patterns of stasis and punctuation—a common but erroneous claim rooted in the misinterpretation of punctuated equilibrium as a saltational theory in ecological time. Rather, this small-scale anagenetic conclusion for another domain of size and time should be read as welcome confirmation—based on causes different from the generators of punctuated equilibrium at a larger scale—for the broader claim that punctuational patterns may be common and robust across several spatial and temporal realms in nature.

But the most impressive affirmations of punctuational patterns at scales below punctuated equilibrium have emerged, in recent years, from a domain unparalleled (and unmatchable) for richness of empirical data on evolution over a sufficient number of generations to claim potential linkage with scales of substantial evolutionary change in nature: well-controlled experimental studies of bacterial lineages.

This field is now developing so rapidly that any particular study, as discussed here, will, no doubt, seem quite rudimentary by the time this book reaches the presses. But, as I write in 1999, an impressive case may be taken as indicative of possibilities and directions. By using strains of *E. coli* that pass through six generations in a single day, Elena, Cooper, and Lenski (1996; see also Lenski and Travisano, 1994) were able to study evolution in cell size for 10,000 continuous generations. By imposing constancy of environment (to limits of experimental perceptibility of course), and using a strain lacking any mechanism for genetic exchange (Elena et al., 1996), mutation becomes the sole, and experimentally well isolated, source of genetic variations.

The experimenters have followed 12 replicate populations, each founded by a single cell from an asexual clone, and each grown under the same regimen (of daily serial transfer, with growth for 24 hours in 10 ml of a glucose-limited minimal salts medium that can support *ca.* $5 \times 10^7$ cells per ml). At an average of 6.6 bacterial generations per cycle, the population undergoes a daily transition from lag phase following transfer, to sustained increase, to depletion of limiting glucose and subsequent starvation. At each serial transfer, a 1:100 dilution begins the next daily cycle with a minimal bacterial population of *ca.* $5 \times 10^6$ cells. Samples of the common ancestral population, and of selected stages in the history of each population, were stored at $-80°$ C, and can be revived for competition experiments with the continually evolving populations—a situation that can only fill a paleontologist with envy, and with thoughts of beautiful and utterly undoable experiments from life's multicellular history (neanderthals or australopithecines released in New York City; tyrannosaurs revived to compete against lions in a field of zebras, etc.).

In each of the 12 populations, both fitness and cell volume increased in a punctuational manner through the 10,000 generations of the experiment. (The experimenters measured cell volume by displacement (Lenski and Travisano, 1994, p. 6809), and mean fitness of populations by the Malthusian parameter of realized rate of increase in competition against resuscitated populations of the common ancestor.) The general path of increase followed the same trajectory in all populations, but with fascinating differences of both form and genetics in each case—a remarkable commentary, at such a small and well-controlled scale, of the roles of detailed contingency and broad predictability in evolution (see the explicit discussion of Lenski and Travisano, 1994, on this point).*

---

*Soon after I wrote this section, *Science* published a special issue on evolution (25 June 1999), featuring the work of Lenski's lab in a news article entitled, "Test tube evolution catches time in a bottle." The twelve populations have now been evolving for 24,000 generations. Although all have shown similar increases in cell size and fitness, the genetic bases of change have been highly varied and unpredictable. Moreover, alteration in environmental and adaptive regimes yields no common response. When, after 2000 generations of growth on glucose (with similar evolutionary responses), the 12 populations were switched

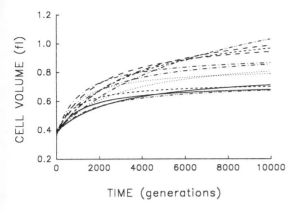

9-34. Punctuation in clonal evolution below the species level—from Lenski and Travisano, 1994. In 12 replicate populations of *E. coli* during ten thousand generations, mean cell size increases rapidly during the first two thousand generations, and then very little during subsequent evolution.

Punctuational patterns occur at the two different scales of overall trajectory and detailed dynamics, even within the limited scope of this study. In the 1994 paper, Lenski and Travisano sampled each of the 12 populations once every 500 generations. They noted a rapid increase in mean cell size, well fit by a hyperbolic model (see Fig. 9-34), for the first 2000 generations in each population, followed by several thousand subsequent generations of little or no further increase—a pattern that they described as punctuational in one of the major conclusions of their paper (1994, p. 6809): "For *ca.* 2000 generations after its introduction into the experimental environment, cell size increased quite rapidly. But after the environment was unchanged for several thousand generations more, any further evolution in cell size was imperceptible . . . These data therefore indicate a rapid bout of morphological evolution after the population was placed in the experimental environment, followed by evolutionary stasis (or near stasis)."

But, as reported in a later paper (Elena, Cooper and Lenski, 1996), they then sampled each population at a much finer scale—every 100 generations for the first 3000 generations of the experiment. Now they found clear evidence of a punctuational "step pattern" (see Fig. 9-11, discussed previously) within the initial phase of rapid increase that they had previously fit with a simple hyperbolic model. The authors noted: "When mean fitness was measured every 100 generations over the period of faster change . . . a step function model, in which periods of stasis were interrupted by episodes of rapid change, gave a better fit to the data than did the hyperbolic model. Evidently,

---

to a different sugar (maltose), some populations flourished, but others grew poorly. After 1000 generations on maltose, all twelve populations did improve in fitness, but not nearly so much (and, more importantly, not nearly so consistently) as on glucose. The starting genotype for the 12 populations had been identical for the first experiment with glucose, but different (after 2000 generations of evolution on glucose for each population) for the initiation of the subsequent maltose experiment. Apparently, any departure from simple and controlled experimental conditions towards the genetic and environmental variation invariably encountered in the natural world greatly decreases the predictability, while emphasizing the contingency, of outcomes.

it was necessary to make measurements at sufficiently high frequency to resolve the punctuated dynamics."

This gain of insight by finer sampling raises the important methodological theme that proper choice for a scale of inquiry depends crucially upon the resolution needed to identify and characterize the underlying causal process of the observed pattern—in particular, to specify the *natural unit* or entity of change in the given system. (Such studies often face the paradox that, whereas the recognition of this principle requires no act of genius, empirical adequacy often founders upon a conceptual dilemma: We can specify a proper scale if we know the causal basis beforehand. But, more often than not, we undertake such research in order to discover an unresolved causal basis—thus bringing upon ourselves the classical problem of a single equation with two unknowns: the causal basis and the scale required for its identification, to complete the analogy.)

At the macroevolutionary scale of punctuated equilibrium *sensu stricto*, events of speciation represent the natural unit, and geological resolution must be sufficient to identify the occurrence and timing (relative to stasis, or any other pattern, in the species's subsequent geological history) of origination for these macroevolutionary "atoms." Several published studies have been fatally marred by the basic flaw of using a scale so coarse that a trend generated by multiple events of staircase speciation could not be distinguished in principle from the same result achieved by smooth anagenesis in an unbranched lineage. In the most widely discussed fallacy thus engendered, Cronin et al. (1980) claimed gradualism (explicitly against punctuated equilibrium) for major trends in hominid evolution because a temporal sequence of mean values moved in the same direction. But the successive points were so widely separated in time and morphology that the authors could not determine whether they had measured mean values of successive species during their periods of stasis, or had sampled points in an anagenetic continuum. (Punctuated equilibrium, after all, was proposed as an alternative explanation for phyletic trends of this kind, not as a denial of their existence!) The scale of measurement used by Cronin et al. may be compared with Lenski's first procedure of sampling every 500 generations. Both schemes are too coarse to "catch" the underlying causal unit of change—speciation, if punctuated equilibrium holds, for the macroevolutionary case of hominids; the infrequent origin and sweep of favorable mutations in bacterial anagenesis.

In a fascinating study, extending (to an utterly different realm of inquiry) the generality of this important point about appropriate scale of measurement for the recognition of punctuations, Lampl et al. (1992) note that human growth in body length has generally, and for centuries of study, been regarded as smooth (albeit highly variable in rate at different states of ontogeny) because "individuals have been traditionally measured at quarterly intervals during infancy, and annually or biannually during childhood and adolescence. Physiological data are mathematically smoothed and growth is represented as a continuous curve" (1992, p. 801). But by measuring a sample of 31 "clinically normal" Caucasian children at intervals ranging from daily to weekly between the ages of 3 days and 21 months, Lampl et al., using the language and concepts of punctuated equilibrium, found that "90 to 95 percent of normal development during infancy is growth-free and length accretion is a distinctly saltatory process of incremental bursts punctuating background stasis" (p. 801). In fact, Lampl et al. did not detect the pattern of quickness in change and prevalence of stasis until they measured subjects at their finest daily scales (for even the semi-weekly and weekly measurements smoothed out punctuations over intervals of stasis). They conclude (Lampl et al., 1992, p. 802): "Human length growth during the first two years occurs during short (less than 24 hours) intervals that punctuate a background of stasis. Contrary to the previous assumption that the absence of growth in developing organisms is necessarily pathological, we postulate that stasis may be part of the normal temporal structure of growth and development."

Lenski's bacterial populations generate large numbers of mutations (some $10^6$ every day in each population, by the estimate of Elena et al., 1996). But the step-dynamics revealed by the finer scheme of sampling—a pattern "predicted . . . by a simple model in which successive beneficial mutations sweep through an evolving population by natural selection" (Elena et al., 1996)— presumably occur for two reasons: first, the well-known exponential principle, however intuitively paradoxical for most people untrained in science, that "many generations are required for the beneficial allele to reach a frequency at which it has an appreciable effect on mean fitness, but then relatively few generations are required for that allele to become numerically dominant" (op. cit.); second, and probably more important, the extreme rarity of favorable variants amidst the daily plethora of mutations, leading to "a substantial waiting period before a beneficial mutation even occurs" (op. cit.). Thus, at the proper scale (for resolving the causal mechanism) of sampling every 100 generations, the plateaux of stasis mark the waiting times between favorable sweeps, while punctuations express the sweep itself.

Finally—and to place some substantial empirical weight upon the keystone of my argument for the potential generality of punctuational change—Lenski and his colleagues have greatly increased our understanding of evolution by developing an artificial (in the best and fully positive sense of the word) experimental system purposely reduced to an absolute "bare bones" of Darwinian causal minimalism. With an identical genetic starting point for each replicate, an asexual clonal system that permits no genetic exchange among cells, and an unchanged environment (the regimen of daily transfer by controlled dilutions into a constant growth medium), this experiment leaves only two factors free to work and vary as potential agents of evolutionary change: the paired and essential Darwinian components of new mutations as raw material, and shaping by natural selection among cells that vary as a consequence of these mutations. The fact that punctuational dynamics prevail in this first truly adequate experiment in pure Darwinian minimalism must at least evoke a suspicion—even among biologists who, by custom and *faute de mieux*, have never questioned gradualism—that this episodic mode might be expressing something important about the general nature of change itself across the varying scales of nature's evolutionary construction.

PUNCTUATION ABOVE THE SPECIES LEVEL. Punctuated equilibrium stands on an "isthmus of a middle state" (to quote Alexander Pope out of specific context, but in proper structural analogy, see page 680)—a speciational bridge linking the microevolutionary history of discrete populations with the macroevolutionary waxing and waning of clades through geological time. I believe that the prevalence of punctuational change on the bridge itself (punctuated equilibrium *sensu stricto*), combined with a strong case for punctuational dynamics in Darwinian processes stripped to a lean and clean minimality in microevolution (see preceding section), behooves us to consider a generalization across all scales, by suggesting an examination of larger realms beyond the bridge of speciation. I shall therefore discuss potentially instructive examples (not comprehensive statistical generalities, a worthy goal not nearly in current reach) at three expanding levels: (1) consequences of accumulated events of ordinary speciation within the history of individual clades; (2) the origin of phenotypically complex and extensive evolutionary novelties; and (3) the history of biotas in ecological and evolutionary time.

STASIS ANALOGS: TRENDING AND NON-TRENDING IN THE GEOLOGICAL HISTORY OF CLADES. Do we find cladal patterns, generated by different causal mechanisms, that might be sufficiently "homologous" (see pp. 928–931) with punctuated equilibrium to warrant comparison based on a common deep structure? (I am, of course, not considering or reiterating here (see full discussion, pp. 886–893) the most important and direct impact of punctuated equilibrium upon cladal histories—its ability to explain trends as the differential success of species rather than the extrapolated result of adaptive anagenesis. This section treats *other* scales and causes of change with meaningful structural parallels to punctuated equilibrium at the species level.) Possible parallels for the punctuational aspect will be treated in subsequent chapters—rapid origin of extensive evolutionary novelties for cladal beginnings (Chapter 10), and patterns of mass extinction for endings (Chapter 12)—but we should also consider the almost entirely neglected analogs of stasis at the cladal level.

An apt comparison for clades may be made to philosophical and sociological reasons for the previous failure of evolutionary biology to study, or even to acknowledge, the phenomenon of stasis as the predominant feature of phyletic history in a large majority of species. Stasis, construed as absence of evolution, once designated a negative result unworthy of a category, or even a name. In a similar way (and I cast no stones from a sinless state, for I have followed this tradition myself throughout my career), the study of trends has consumed nearly all research on the history of clades. And why not? Trends tell stories, and evolution is a narrative science. Western tradition, if not universal human nature itself, has always favored directional tales of conquest and valor (with Darwinian analogs of competition and adaptation), while experiencing great discomfort with the aimless cyclicity of Ecclesiastes, however much we may admire the literary power: "The thing that hath been, it is that which shall be; and that which is done is that which shall be done; and there is no new thing under the sun" (Ecclesiastes 1:9).

But the undeniable salience of trends—a psychological comment about our focus of attention—bears no necessary relationship to the relative frequency or causal weight of this phenomenon in the natural history of clades. How many monophyletic clades feature sustained and substantial trends in major characters of functional importance—and what percentage of characters participates in trends that do exist in such clades? Indeed, we have no idea whatever, for no neutral compilations exist. No one has ever tabulated the number or percentage of non-trending clades within larger monophyletic groups. The concept of a non-trending clade—the higher level analog of a species in stasis—has never been explicitly formulated at all. If only one percent of clades exhibited sustained trends, we would still focus our attention upon this tiny minority in telling our favored version of the story of life's history.

(Ironically, stable lineages become salient enough to catch our attention only at the extreme that we call "living fossils"—species or lineages supposedly unchanged during such long stretches of geological time that their stability becomes a paradox in a world of Darwinian evolutionary flux and continuity. As a double irony—see pages 817–820 for a full discussion in the light of punctuated equilibrium's different reading—we have also thoroughly misinterpreted this phenomenon under the same gradualistic bias that inspired our notice in the first place! The classical "living fossils" (the inarticulate brachiopod *Lingula,* the horseshoe crab *Limulus polyphemus,* the extant coelacanth) are not long-lived as species (*Limulus polyphemus,* for example, has no fossil record at all), but rather belong to clades with such a low speciation rate that little raw material for cladal trending has been generated over the ages.)

I suspect that most clades, while waxing and waning in species diversity through time, show no outstanding overall directionality. But we do not know because the literature has never recognized, or attempted to tabulate, the frequency of such "Ecclesiastical" clades that change all the time, but "go" nowhere in particular during their evolutionary peregrinations. Paleontologists achieved no sense of the strength of punctuated equilibrium, even though Eldredge and I had formulated the theoretical apparatus, until researchers studied the relative frequencies of stasis and punctuation in entire faunas, or entire clades, by full sampling and with no predisposing bias to favor one kind of result—see discussion on pages 854–874 for this extensive and growing literature. Similarly, we will not know the general fate of clades until we ratchet this methodology one notch higher, and sample sets of clades not identified by our prior sense of their evolutionary "interest." Stasis is data at the species level. Non-trending is data at the clade level.

Budd and Coates (1992) broke conceptual ground in devoting an entire paper to such Ecclesiastical non-directionality in the actively evolving and speciating clade of montastraeid corals during 80 million years of Cretaceous time. As a rationale for their study, the authors state an analogy to the lower level phenomenon of punctuated equilibrium: "Just as the study of stasis within species has facilitated understanding of morphologic changes associated with speciation, we show that study of nonprogressive evolution offers valuable insight into how the causes of trends interact and thereby produce complex evolutionary patterns within clades, regardless of their overall direction."

The central theme of non-trending, identified by Budd and Coates for this large clade of massive, reef-building corals, stands as an empirical pattern in any case, but the (admittedly somewhat speculative) explanation proposed by the authors also builds an interesting framework for regarding such a signal as predictable and unsurprising, rather than anomalous. Their proposal also integrates the two principal Darwinian critiques of this book by attributing a causal pattern generated at the species level (the hierarchical expansion of my first theme) to the effects of architectural or developmental constraint (the structuralist or internalist perspective of my second theme) in channeling the possibilities and directions of natural selection.

Budd and Coates (1992) propose that monstastraeid species vary within a range set by minimal and maximal size of individual corallites on these large colonies. Such a notion does not debar classical trending, for the clade could originate in one small portion of the permitted range, and then strongly trend towards the other domain. But Budd and Coates argue that Cretaceous montestraeids already inhabited the full range, and that each end represented an adaptive configuration continually available and exploited throughout the clade's duration. Therefore, phenotypic evolution fluctuated between the two realized potentials of a fully populated domain of workable solutions.

The authors argue that "large-corallite" species (3.5 to 8.0 mm in diameter) maximize efficiency in removal of sediment, and tend to dominate in turbid waters; while "small-corallite" species (2.0 to 3.5 mm in diameter) prevail in clearer waters of the reef crest. Moreover, large-corallite species derive most nutrition by direct carnivory, whereas small-corallite species tend to feed upon their own symbiotic zooxanthellae. Budd and Coates then advance the claim—the more speculative aspect of their scenario—that montastraeid species remain constrained with this range by limitations at either end: an inability of still smaller corallites to develop and function adequately, and a restriction in septal number and strength that would not grant sufficient biomechanical support to still larger corallites.

These two arguments may validate and explain the basis of active non-trending in a persistently vigorous and successful clade. For if such constraints limit the range of corallite size, and if each end enjoys advantages in different environments continuously available in major parts of the habitat, then evolution might oscillate back and forth, with no persistent directional component, throughout cladal history.

Budd and Coates document such a directionless oscillation within the clade's developmental and adaptive boundaries during four successive divisions of Cretaceous time. The transition from interval 1 to interval 2 featured a differential production of small-corallite species from large-corallite ancestors, as well as a southward expansion of the clade's geographic range. Limited and directionless speciation, accompanied by predominant stasis within established species prevailed during intervals 2 and 3. Between intervals 3 and 4, large-corallite species radiated from small-corallite ancestors, and the geographic range of the clade became more restricted. In other words, the general pattern at the end of interval 4 differed little from the initial spread of morphology and geographic range at the outset of interval one, the inception of the study itself.

But the montastraeids remained a vigorous, successful, and evolutionarily active clade throughout Cretaceous times. Who are we to proclaim their pattern "boring" or unworthy of study because the evolutionary history of these corals does not resonate well with human preferences about "interesting" or "instructive" stories? Perhaps we should force ourselves to learn that patterns traditionally shunned for such quirky reasons of human appetite may hold unusual interest and capacity to teach—precisely because we have never sought messages that might challenge our complacencies. The predominant pattern of life's history cannot fail to be instructive—and such non-trending may well mark a norm of this magnitude, even if heretofore hidden in plain sight because we also see with our minds, and conventional concepts can be more blinding than mere ocular failure.

PUNCTUATIONAL ANALOGS IN LINEAGES: THE PACE OF MORPHO-LOGICAL INNOVATION. I do not wish to resuscitate one of the oldest canards, and least fruitful themes of evolutionary debate: the claim for truly saltational, or macromutational—that is, effectively one generational, or ecologically "overnight"—origin of new species or morphotypes. This ultimate extreme in punctuational change has never been supported by punctuated equilibrium, or by any sensible modern account of punctuational change in any form. Even if older evolutionists did advocate this mode of change (see my discussions of de Vries on pp. 415–451 and of Goldschmidt on pp. 451–466), they granted no exclusivity to its operation, and they also defended more continuationist, and more structurally plausible, accounts of rapid origin for morphological novelties—as in the developmentally rooted and theoretically sensible concept, based on mutations in "rate genes," embodied in Goldschmidt's unwisely named "hopeful monster," in contrast with the speculative scenario that he built upon his later concept of "systemic mutations"—see Schwartz (1999) for an interesting modern retelling and defense of this notion.

Just as punctuated equilibrium scales the geologically abrupt (but ecologically slow and continuous) process of speciation against the long duration of most species in subsequent stasis, punctuational hypotheses at higher levels regard the pacing of substantial phenotypic change in the origin of novel morphotypes as similarly episodic, with origins concentrated in very short episodes relative to periods of stability in basic design during the normal waxing and waning of clades—and perhaps with the ecologies, or the structural and developmental bases, of such episodes belonging to a distinctly different class of circumstances from those that regulate the ordinary pace of flux and speciation during the long-term geological history of most clades and morphotypes. In other words, a timekeeper with a metronome beating at an appropriate frequency for discerning the units and causes of evolution at each scale of nature's hierarchy might recognize an episodically (and rarely) pulsating, rather than an equably flowing, tempo as the dominant signal of change in all realms.

For example, Erdtmann (1986, p. 139) proposed that the active cladogenesis of early Ordovician planktic graptolites (an extinct subphylum of colonial organisms close to the chordate lineage) "operated on two levels: gradualistic change involving species-level and intergeneric clades, and punctualistic (anagenetic) changes operating on supergeneric levels." He linked the rapid and extensive morphogenetic innovations of the punctuational mode, involving such basic features of colonial form as loss of bithecae and reduction in number of stipes, to major environmental changes marked by rapid eustatic shifts in sealevel. Moreover, these punctuational innovations arise by an astogenetic mode (a term for the ontogeny of colonies) different from the developmental basis of most smaller changes that mark the flux of speciation during "normal" geological intervals. The punctuational innovations that produce new developmental patterns begin at the proximal end of the colony—that is, they affect the early ontogeny of the initial organisms of the developing aggregate, thereby pervading the life cycles of both the organism and the colony, and strongly affecting the global phenotype of the entire structure. The lower-level changes (smaller variations within existing developmental themes), on the other hand, tend to begin at the colony's distal end—that is, they arise late in the ontogeny of older organisms in the colony (for new organisms of the colony arise proximally, pushing older organisms to progressively more distal positions), and then proceed to earlier phylogenetic expression in both the colony's astogeny and the individual organism's ontogenies. These lower-level changes therefore affect only a small, and astogenetically late, portion of the colony's form—hence their much more limited capacity for yielding major morphological change.

This case provides an interesting astogenetic analog to the common claim that heterochronic changes in the early ontogeny of organisms gain a distinctive status among evolutionary mechanisms in their potential for rendering substantial phenotypic change (at a punctuational tempo) with minimal genetic alteration. In fact, the lability inherent in early ontogenetic changes of rate and regulation undergirds most theorizing about qualitatively different categories of evolutionary outcomes based on similar underlying magnitudes of raw genetic alteration—the most promising basis for a dominant punctuational tempo in the history of morphological innovation in evolution.

By linking constraints of preferred developmental channels with a punctuational tempo that precludes accumulative incremental selection as the sole cause of extensive evolutionary change, this familiar argument unites the two central themes of this book—hierarchical models of evolutionary mechanics with structuralist accounts of evolutionary stasis and directionality. Several recent volumes have explored the growing power and prestige of this argument, as provided by breakthroughs in unraveling the genetics of development (see Chapter 10), combined with classical data of allometry and heterochrony (Raff, 1996; Schwartz, 1999; McNamara, 1997; McKinney and McNamara, 1991; McKinney, 1988).

Wray (1995) has recently summarized an emerging generality that integrates all components of the argument across a wide variety of organisms. His chosen title—"Punctuated evolution of embryos"—underscores the putative generality of change in this mode, with punctuated equilibrium as its major expression at the level of ordinary speciation, and his proposed linkage of development and ecology as its hypothesized primary source for the rarer, but highly consequential, phenomenon of the origin of morphotypic novelty.

Nearly two centuries of tradition proclaim the conservatism of early larval and embryonic phases of the life cycles—from von Baer's enunciation of his celebrated laws (1828, see discussion in Gould, 1977b) to the standard evolutionary rationale that formative stages of early ontogeny become virtually impervious to change because cascading consequences, even of apparently minor alterations, would discombobulate the subtle complexities of development. Recent discoveries of "deep" genetic homologies and developmental pathways among animal phyla separated for more than 500 million years (see Chapter 10) have tended to highlight this conventional view.

But Wray (1995) summarizes several recent studies of broad taxonomic scope—with best examples from the sea urchin *Heliocidaris,* the frogs *Eleutherodactylus* and *Gastrotheca,* and the tunicate *Mogula*—all showing that "similar species have . . . modifications in a variety of crucial developmental processes . . . that have traditionally been viewed as invariant within particular classes or phyla" (Wray, 1995, p. 1115). These substantial changes in the development of closely related forms exhibit three common properties: (1) They usually affect traits of timing and regulation in early development, including specification of cell fates and movement of cells during gastrulation. (2) They yield substantial changes in larval forms and modes of life, but often leave the adult phenotype largely unaltered. (3) They are associated with major changes in the ecology and life history strategies of larval or early developmental forms, and involve such major changes as loss of larval feeding ability (the echinoderm and frog examples) or capacity to disperse (tunicates).

Wray presents evidence that alterations in larval ecology "drive changes in development," not vice versa. Moreover, and most importantly, comparison of molecular and phenotypic modification shows that these "functionally profound changes in developmental mechanics can evolve quite rapidly" (ibid., p. 1116). For species of *Heliocidaris* with strikingly different developmental mechanisms, fewer than 10 million years have elapsed since divergence from common ancestry. Wray draws a general and punctuational conclusion from this evidence for ecologically driven change in mechanisms of early development—events that can occur very rapidly and do not compromise the conserved life styles of later development due to greater dissociation than traditional views would allow between successive phases of ontogeny: "Long periods of little net change, with functionally minor modifications in developmental mechanisms and larvae, seem to be the normal mode of evolution. This near stasis is interrupted on occasion by rapid, extensive, and mechanically significant changes that coincide with switches in life history strategy . . . Rapid modifications can arise in developmental mechanisms that have been conserved for hundreds of millions of years."

Note the striking similarity of language (with analogs of stasis and punctuation)—and of evolutionary style in tempo and mode—between punctuated equilibrium and Wray's description of phenotypic and ecological shift at these much larger scales of morphological change and temporal extent (with the analog of stasis persisting for hundreds of millions of years). These similarities in style and import seem to mark a genuine conceptual "homology" based on similar structural principles regulating the nature of change in complex systems.

As a final example of the fruitfulness and detailed testability of punctuational models for the origin of morphogenetic novelty, Blackburn's (1995) remarkable study of "saltationist and punctuated equilibrium models for the evolution of viviparity and placentation" deserves special notice for the author's clarity in specification of hypotheses, and for his richness and rigor in attendant documentation. Blackburn treats the multiple evolution of viviparity in squamate reptiles (lizards and snakes)—a much better case for studying this phenomenon than the group that most of us emphasize for parochial reasons (the Mammalia), for extant squamate species include many examples in all stages of the process. This taxonomic richness permits clear distinction of gradualistic vs. punctuational alternatives, and also provides sufficient data to distinguish between punctuated equilibrium and saltation as the dominant punctuational style. Moreover, and largely because the subject has been embraced as a workable surrogate for unresolvable questions about mammalian evolution, the origin of viviparity in squamate reptiles has become a classical case, and has therefore generated an extensive literature to illustrate (however unintentionally) some major biases of evolutionary argumentation. The power of Blackburn's study resides in three interrelated themes:

1. Gradualistic scenarios have dominated the classical literature in ways that authors rarely defend, or even recognize. In particular, previous workers have assumed that three apparent stages in the "perfection" of live bearing must represent an actual historical sequence gradually and incrementally evolved by all lineages that reach the "last stage"—viviparity or live birth, placentation for gas exchange and water intake, and placentotrophy for embryonic nutrition. Blackburn writes (1995, pp. 199–201): "Viviparity and placentation in squamates have stood for over half of a century as examples of gradual evolution . . . Even recent reviewers have not considered the applicability of alternative evolutionary models."

The supposed evidence for such gradualism consists largely of inferences drawn from structural series of extant forms, without affirmation, or even consideration, of an explicit phylogenetic hypothesis that the successive stages represent a cladistic sequence. (Even worse, a phylogenetic inference has often been based only upon the series itself—a flagrantly circular argument that validates the conclusion by the hypothesis supposedly under test.) For example, extant oviparous species do vary substantially in the stage of development at which the eggs are laid—and researchers have generally assumed that a linear ordering of such a series must represent an evolutionary continuum "on the way" to viviparity: "The inferred continuum of developmental stages at oviposition among squamates commonly is interpreted as evidence for a gradual increase in the proportion of development occurring in the female reproductive tract" (ibid., p. 201).

2. Blackburn marshalls an impressive array of data from a broad range of fields—taxonomy, development and geology, in particular—to affirm an alternative punctuational scenario for the evolution of live birth, with simple viviparity, placentation and placentotrophy as three distinct modes, not three way stations in a progressive sequence. (I suspect that our gradualistic biases have been particularly intrusive in this case because we unconsciously read the squamate story in a mammalian perspective that makes placentotrophy the "obvious" goal of any trend to live bearing.)

*In taxonomy,* viviparity has originated more than 100 times among squamate reptiles (ibid., p. 202). But cladistic data have provided not a single case of correspondence between branching order and the four structural stages of the hypothetical trend: ovipary, vivipary (live birth without placentation of embryos), placentation, and placentotrophy. Blackburn writes (1995, pp. 201–202): "Clines of phenotypic variation that can be invoked to support gradualistic evolution of viviparity and placentotrophy tend to be composites of unrelated species representing multiple lineages . . . Despite the documentation of over 100 evolutionary origins of viviparity in squamates . . . available evidence has not yet permitted construction of a single, complete phenocline of parity modes and embryonic nutritional patterns out of representatives of a single clade."

*In structure and development,* Blackburn coordinates several lines of evidence to argue that intermediary forms between any two stages in the hypothetical trend either cannot be found, or exist only rarely and in a tenuous state (because such transitional phenotypes would experience either architectural problems in construction or adaptive insufficiencies in function). For example, if viviparity evolved by progressive delay of oviposition, then we might expect, among extant species, "a full continuum of developmental stages . . . representing steps in the parallel evolutionary transformations that have occurred independently (and perhaps to different degrees) in various lineages" (p. 202). Instead (see Fig. 9-35), the distribution of developmental stages at oviposition shows marked bimodality, with species either depositing eggs containing embryos in the pharygula/limb bud stages (with near normality or minor left skewing for this lower mode, and no right skew in the direction of the putative trend) or else retaining the eggs to term and then giving birth to live young.

Moreover, the supposed development of placentation, and then of placentotrophy, only after the origin of live birth also derives no support from documented intermediary stages. In the traditional view, the shell membrane between fetal and maternal tissues must thin gradually, permitting an initial function of placental organs in uptake of water and exchange of gases. Placentotrophy then evolves later "as the placental supply of nutrients first supplements and then supplants provision by the yolk" (p. 208). But evidence from at least 19 independent clades of viviparous squamates indicates that all "have anatomically recognizable placentae derived from both the chorioallantois and the yolk sac" (p. 208). Thus Blackburn concludes, "the existence of a truly non-placental viviparous squamate has not been documented in over a century of investigation . . . The universal occurrence of placentae in viviparous squamates is most consistent with the view that placental organs that accomplish gas exchange and water uptake evolve simultaneously with viviparity" (p. 209).

Similarly, no purely lecithotrophic (yolk feeding) placental squamates have been discovered, and all viviparous forms derive at least some nutrition through the placental organs. Thus, "available data are most consistent with the hypothesis that incipient placentotrophy is a necessary correlate of viviparity." The three "logical" steps of the hypothetically gradual trend become telescoped into a single structural transition, with an evident implication of punctuational origin.

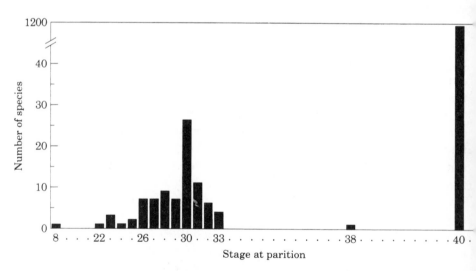

9-35. Punctuational change in the morphological evolution of lineages in squamate reptiles, Ovoviviparity does not evolve by progressive and gradualistic delay of oviposition, but rather shows marked bimodality with females either depositing eggs with embryos in their early limb bud stages or else retaining the eggs within their body to term, and then giving birth to live young (the right mode). The existence of a few intermediary species shows that the full sequence proceeds by punctuational steps and not by full saltation.

Blackburn supplements this empirical evidence for a punctuational divide with both structural and functional rationales for the inviability of putative intermediary stages. Viviparity without placentation may be structurally unattainable because live birth requires that the eggshell become sufficiently thin "to permit gas exchange in the hypoxic uterine environment" (p. 211)—while such reduction may entail gas and water exchange (that is, incipient placentation) as a virtually automatic consequence. Thus, Blackburn argues (p. 210), "placentation is best viewed as a necessary correlate of viviparity, not as a 'reproductive strategy' *per se.*"

Intermediacy may be equally unlikely in functional and adaptationist terms as well. Both endpoints entail costs as well as putative benefits—oviparity in dangers and energetic requirements of nesting behavior, and in maternal loss of calcium in making eggshells (p. 211); viviparity in decreased maternal mobility, fecundity, or clutch size. A hypothetical intermediate that incurs both sets of costs—for example, the calcium drain from internal shells of hypothetical stage one (still too thick for placentation), combined with a heightened susceptibility to predators caused by compromised mobility—without winning greater compensation in combined benefits, could not compete against either end member of the supposed trend, and probably would not survive even if patterns of development permitted evolutionary access to this putatively transitional design.

Finally, *in geology,* the recent origin of most viviparous lines strongly supports a punctuational inference. A few origins of squamate viviparity may date to late Mesozoic or early Cenozoic times (p. 207), but most represent Pliocene or Pleistocene events. Moreover, taxonomic distribution fully supports the rapidity of full transition to placentotrophy. More than 60 percent of origins for viviparity "have occurred at subgeneric levels, and virtually all have arisen at subfamilial levels." Several origins can be traced to populations of a single species (with other populations remaining oviparous)—for example, a Pleistocene event within *Lacerta vivipara,* and an origin within the past 11 to 25 thousand years within the *Sceloporus aeneus* complex (p. 207). The extent of structural differences between oviparous and viviparous populations of these minimally distant forms (both temporally and phylogenetically) fully matches the phenotypic separation noted for the same features in cladistically distant lineages.

3. As an indication that data of natural history can provide combined criteria to permit fine and testable distinctions, Blackburn has been able to reject saltation and defend punctuated equilibrium as the probable cause and temporal basis of this well-documented punctuational pattern. Blackburn notes that "under the punctuated equilibrium model, typical oviparity and viviparity could represent regions of stasis, with prolonged oviparous egg-retention being a transitory, intermediate stage between them" (p. 206). The structurally well integrated and functionally well adapted end members "would contrast with the instability of the evolutionary intermediate, and prolonged oviparous egg-retention would be a relatively short-lived (and hence scarce) pattern" (p. 206). By contrast, of course, saltational models predict the struc-

tural unattainability and adaptive inviability of intermediates, and therefore envisage a one-step transition (with substantial opportunity, perhaps, for later adaptive fine-tuning).

In favor of punctuational origins (rapid transition between domains, based on structural properties of endpoints as well coordinated states that actively resist change, and with intermediary forms as unstable, and "driven" towards one of the endpoints) versus saltational events (truly sudden transition, scaled to the magnitude of the unit of change and enforced by the absolute structural inaccessibility of intermediary states), Blackburn cites several features of squamate viviparity. The putative intermediary stage of "prolonged oviparous egg retention" (p. 206), while empirically rare, structurally unstable, and adaptively compromised (as discussed above), does exist as the characteristic form of a few species in nature, not just as a facultative or transient state of a population in transformation. As the bimodality of Figure 9-35 shows, prolonged oviparous egg retention does represent an attainable intermediary state between two endpoints. But few forms occupy this uncertain ground between advantageous configurations.

In a second indication of punctuational rather than saltational change, Blackburn notes that a viable first step to intermediacy does exist in nature as a strategy that can be activated under certain environmental conditions: "facultative egg-retention with continued intra-oviductal development" (p. 206). Several squamate species exhibit this phenotypic flexibility in development. "Such facultative retention could provide raw material for natural selection, making more likely the evolution of a pattern in which prolonged egg-retention was obligative."

As a third confirmed prediction, favoring punctuation over saltation when joined with the two previous arguments (but unable to make the distinction otherwise, while confuting gradualism in any case), the phenotypic similarity of oviparous and viviparous congeners affirms the relative ease and accessibility of transition: "As an evolutionary unstable pattern, prolonged egg-retention might lead to viviparity, or might revert to typical oviparity; thus the less genotypic change involved, the more probable the origin of viviparity would be" (p. 206).

Blackburn's final paragraph (p. 212) serves as an apt reminder about the restrictive nature of gradualistic bias, and of the power and inherent probability of punctuational alternatives in a world that may favor this mode of change as a general structural property of material organization at all scales: "For over 60 years, research on amniotes has assumed that gradualistic change is the sole mechanism by which viviparity, placentation, and placentotrophy could have evolved. Future empirical and theoretical analyses of reproduction in squamates and other vertebrates should not overlook the potential of non-gradualistic models as explanations for evolutionary change and the biological patterns it has produced."

PUNCTUATIONAL ANALOGS IN FAUNAS AND ECOSYSTEMS. If stable locations, reached by rapid movement though "perilous" intermediary terrain, sets the structural basis of punctuational change for the "internalities" of evolution in complex organic phenotypes, then a similarly episodic, rather than evenly flowing, mode of change might characterize the "externalities" of environments that regulate any coordinated evolutionary tempo among components of biotas and ecosystems. I have already considered the scale of environmental punctuation most immediately relevant to punctuated equilibrium—the geological history of regional biotas (see pp. 916–922 on coordinated stasis, the turnover-pulse hypothesis, and other notions of faunal transition by coincidence of punctuational extinction and origination in a high percentage of species within a previously stable biota). But the generality of such punctuational tempos in external controls might also extend to levels both below and above the direct mechanics of speciation itself.

In a general argument strikingly similar to Blackburn's for the evolution of squamate viviparity, Smith (1994) holds that gradualistic assumptions have stymied our understanding of evolutionary processes at the small scale of ecological immediacy in deep-sea faunas. No other environment has been so conventionally associated with plodding, incremental change through substantial periods of time. Smith begins his article by noting the "the deep-sea floor is traditionally perceived as a habitat where low food flux and sluggish bottom currents force life to proceed at slow, steady rates. In this view, benthic community structure is controlled by equilibrium processes, such as extreme levels of habitat partitioning, made possible by remarkable ecosystem stability" (Smith, 1994, p. 3).

As indicated by the title of his article—"Tempo and mode in deep-sea benthic ecology: punctuated equilibrium revisited"—Smith holds that we must revise this traditional view, and reconceive the deep-sea as a punctuational domain where "endogenous disturbances may be relatively frequent, and where pulses of food reach the seafloor from the upper ocean" (p. 3). In what he labels as a "parallel argument" to our punctuated equilibrium from a much lower scale of analysis—in other words, as a claim for conceptual homology of constraining structural principles (in the language of this section)—Smith discusses three examples of "pulsed events that 'punctuate' the apparent 'equilibrium' of the deep-sea floor" (p. 3), and that "may substantially influence processes of modern and past ecological significance including (1) maintenance of macrofaunal diversity and population structure, (2) deposit-feeder–microbe interactions and associated trace production, and (3) dispersal and biography of chemosynthetic communities at the deep-sea floor" (p. 3).

First, Smith documents the importance of "pulsed physical disturbance" in benthic faunas of the Nova Scotian Rise (4750–4950 m)—particularly of erosional "storms" that scour and redeposit sediments "to depths of millimeters-centimeters over areas encompassing at least tens of square kilometers" (p. 7), and that strongly influence both the composition and successional stage of local faunas.

In a second microbiotal example, Smith documents the importance of "phytodetrital pulses" in nutrition for the deep-sea macrofauna. The "slow and steady 'drizzle'" usually regarded as the gradual (and meager) planktonic contribution to sustaining deep-sea life "can be punctuated by downpours of 'phylodetritus' (i.e., detrital material composed primarily of relatively fresh phytoplanktonic remains), during which the flux of labile particulate organic carbon to the seafloor temporarily exceeds biological demand, yielding a carpet of 'food'" (p. 7).

Finally, and to add a third punctuational source of maximally different character from the physical and microfloral cases discussed above, Smith argues (p. 10) that "whale falls" produce occasional and (obviously) "huge local pulses" of organic matter that may decay to produce distinctive "chemosynthetic habitats" supporting faunal associations much like those documented at deep-sea vents. For example, in 1987, his team discovered a 21-meter whale skeleton at a depth of 1240 m: "The bones were covered with mats of sulfur bacteria and clusters of small mussels and limpets; nearby sediments harbored large vesicomyid clams" (p. 10)—for a total of 42 macrofaunal species, only nine of which also inhabited surrounding sediments. Smith concludes that "sunken whales may provide dispersal stepping stones for at least some of the species dependent on sulfide-based chemosynthesis."

Strong circumstantial evidence indicates considerable temporal and spatial influence for this source that most of us would surely have regarded as dubious, if not risible, at apparent face value of relative importance. A fossilized chemosynthetic community has been reported from a 35 million year old whale fall on the Northeast Pacific ocean floor (p. 10). "Whale skeletons," Smith concludes (p. 10), "may be the dominant source of chemosynthetic habitats over the vast sediment plains constituting most of the ocean floor."

At the opposite end of a hierarchy in spatial and temporal scales, punctuational models continue to gain in strength and acceptability for events that impact entire biotas at regional or even planetary scales—with catastrophic mass extinction as a "flagship" notion, spurred by nearly conclusive evidence for bolide impact as the trigger of the K-T global dying 65 million years ago (see Chapter 12 for full treatment). An expansion of research away from the extinctions themselves, and towards the subsequent recovery phases as well, has strongly accentuated the episodic and punctuational character of this most comprehensive signal in the history of life.

Even after the Alvarez's impact hypothesis forced paleontologists to acknowledge the potentially catastrophic nature of at least some mass extinctions, students of fossils usually assumed that subsequent recoveries of global faunas must have been tolerably gradual. This expectation has not been fulfilled, and episodes of recovery from maximum decimation at the extinction to full reestablishment of previous levels of diversity occur more quickly, and in a much shorter percentage of the "normal" time (until the next mass extinction), than previously suspected. (Of course, no one expects that recoveries which require successive events of branching can be nearly as rapid as truly catastrophic extinctions, which can feature truly simultaneous killings—so the complete record of an extinction-recovery cycle will surely remain asymmetric. But the recoveries now seem to occur rapidly enough, in most cases, to invoke the central concept of punctuational change: origin in a tiny fraction of later existence in stasis.)

For example, Kerr (1994) begins his report on Peter Sheehan's work (in a commentary entitled "Between extinctions, evolutionary stasis") by writing (Kerr, 1994, p. 29): "More and more, paleontologists are learning that the full measure of a mass extinction can't be found in its immediate toll. Just as important is the wholesale reorganization of living communities that takes place afterward. And those brief recovery periods, lasting just a few million years, are all the more important because during the tens of or hundreds of millions of years that follow, until the next mass extinction, not much may happen."

Sheehan divides the last 640 million years into six major faunal packages that he calls EEU's, or Ecologic Evolutionary Units. Each lasts for 35 to 147 million years, and each ends at a mass extinction. The subsequent recovery periods for the new units occupy only 3 to 8 million years.

This recent affirmation of a strongly punctuational character for change (primarily extinction) at the highest level has led to a tendency, probably overextended—and I blame myself, in part, for propagating the theme, see Gould, 1985a—for ascribing a dualistic character to the pulse of evolution, with punctuations of mass extinction alternating with a more stately flow in "normal" times between these macropulses. But this view may prove to be overly simplistic, although not wrong. When we assess each level of change by its own appropriate measuring rod (scaled to emphasize the relevant unit or units), all may be punctuational. We must dismiss as irrelevant and misleading the fact that punctuations at a small scale may "smooth out" to more gradual and continuous trends when inappropriately measured at too large a scale to reveal the causal mechanics, or even to identify the relevant unit, of change—a theme that I have emphasized throughout this chapter, in such examples as punctuated bacterial anagenesis, viewed as gradual when sampled too infrequently to note the steps of mutational sweeps; and cladal trends, viewed as anagenetic when sampled too broadly to discern the speciational jumps of punctuated equilibrium.

In a provocative work, Raup (1992) played devil's advocate by asking if all extinctions at all levels, from single local populations to global faunas, might be catastrophic—for he could not reject the "null hypothesis" of his "field of bullets" model (random and catastrophic removal, triggered by "bolides" of various sizes randomly shot towards the earth at frequencies inversely proportional to their size and effect) in favor of the traditional Darwinian model of gradual declines mediated by competitive inferiority in biotic interactions. I do not believe that such extreme punctuationalism could rule so completely (see full discussion of this argument in Chapter 12, pp. 1323–1326). But finer analysis of the most famous cases of supposedly gradual, and biotically controlled, events may well require such a punctuational reinterpretation. Most outstandingly, perhaps the two most widely discussed and most generally accepted examples of geologically slow global diversification—the Ordovician spread of the great Paleozoic marine invertebrate fauna, and the Mesozoic "modernization" of invertebrate predators and prey (the classic example of a supposed biotic and gradualistic "arms race")—now seem to occur far more abruptly in each separate geographic region, with the previous impression of gradual construction based on a blurring of the different times of transition in each region (Jablonski, 1999, p. 2114).

In an important paper, Miller (1998) has generalized this claim by summarizing the increasing evidence for punctuational tempos in faunal change (both locally and regionally, and for both extinctions and the necessarily slower rediversifications)—with our conventional notions of gradual flux, particularly for build-ups, arising as an artifact of summation over displaced timings for rapid pulses in several regions. Miller first states the general observation and emerging principles (1998, pp. 1158–1159): "In recent years, local and regional studies of marine faunal patterns have converged on a similar theme—that biotic turnover occurred episodically through investigated stratigraphic intervals. There were comparatively broad intervals with little net turnover, punctuated by narrower intervals in which many taxa either emigrated or became extinct and were replaced by a roster of taxa that either originated in the area or immigrated into it. . . . Episodicity appears to be a general feature of regional stratigraphic packages." He then uses this finding to correct what may be a substantial error in traditional views (1998, p. 1159): "Thus, major faunal transitions in global-scale compilations, which seem to have transpired over protracted intervals of geological time, took place far more rapidly and episodically when evaluated regionally or locally. The transitions only appear gradual on a global scale because of variations in their timing from venue to venue."

Finally, Miller asserts a general "fractal" conclusion about punctuational change (ibid., p. 1159): "The processes that produced major mass extinctions simply represented the largest and most globally extensive of a spectrum of perturbations that produced episodic biotic transitions."

As a closing note in this context, Miller also offers a similar punctuational reinterpretation for the putatively best documented and most widely accepted case of global, geologically gradual, and broadly progressive change in life's history—the pattern that Vermeij (1987) has called "escalation" (largely, and with good reason, to avoid false implications and arguments in the traditional notion of "progress"), based on relayed "arms race" between predators and their prey, and on other kinds of similarly reciprocal biotic interaction through extensive time. This entirely sensible concept of escalation seemed to provide the best available argument for two deeply rooted and strongly held themes of traditional Darwinian extrapolationism: the predominant power of biotic interactions to shape patterns in the history of life, and trends towards the slow accumulation of biomechanically improved designs in major lineages.

The general argument sounds so reasonable, but when we rethink macroevolution as a process based upon geologically rapid production of higher-level individuals by punctuational speciation as the primary units of change, then the mechanics of this usual interpretation of escalation become elusive. The pattern certainly exists—especially for Vermeij's (1977) classic case of increasing strength and efficiency in crab claws matched by growing intricacy and sophistication of adaptive defenses in molluscan shells. But how can such an arms race operate if the full trend proceeds by stepping stones of punctuational speciation for any increment, and not in the style of tit-for-tat anagenetic escalation, based on immediate organismal competition and more familiar to us through human models of "anything you can do, I can do better"—a point that Vermeij himself recognizes and finds puzzling (1987)?

Miller, on the other hand, affirms the gradual trend to escalation in biomechanical improvement—and I don't think that any party to this debate denies the reality of the pattern (for we have been arguing about mechanisms)—but finds the same unconventional (and punctuational), finer-scale theme upon "dissecting" the full result into component causal units. Again, each step in escalation seems episodic in each region, with the full trend thus rendered as a summation of punctuational events. Miller writes (p. 1159): "Although the case for these kinds of transitions over the sweep of the Phanerozoic is difficult to deny, the manner in which they transpired over shorter intervals is less certain. There is little evidence of gradual escalation through stratigraphic intervals at local or regional levels. The introduction of escalated forms appears to have occurred episodically, in concert with the broader class of changes in taxonomic composition discussed earlier, which suggests a role for physical mediation."

Two general points provide a fitting close for this section:

1. The probable generality of punctuation and stasis as a powerful—if not predominant—style of change across all scales must lead us to reassess our previous convictions about "important" and "interesting" phenomena in evolutionary theory and the history of life. Kerr admits the potential generality, in reporting punctuation at lower levels to complement Sheehan's similar claim for the broadest scale. But he closes his report by writing (1994, p. 29): "Sheehan sees these intervals as analogous to his longer, global EEU's reaffirming that stability—as boring as it may be—is the evolutionary norm." But, to restate my mantra, and to emphasize its implications for understanding the history of earthly life and the psychology of human discovery, stasis is data—and data of such high generality, such unanticipated occurrence, and such theoretical interest simply cannot be boring.

2. The ubiquity—and the possibly canonical character—of punctuational change at all scales, from the shortest trends of bacterial anagenesis in single clonal lineages over weeks to months, to the broadest patterns of global waxing and waning of biotas through the history of life in deep time, can only recall the familiar tale, by now a cliché, of the Eastern sage who revealed the nature of the cosmos to his disciple: the globe of the earth rests on the back of an elephant who stands, in turn, on the back of a turtle. When asked by the disciple what one might find under the turtle, from its feet to the ultimate source of being, the sage simply replies: "it's turtles all the way down." I suspect that it is also punctuational change all the way down, from Permian

extinctions to mutational sweeps through little laboratory populations of *E. coli.*

### Punctuational models in other disciplines: towards a general theory of change

PRINCIPLES FOR A CHOICE OF EXAMPLES. In their symposium for the American Association for the Advancement of Science, and in their subsequent book, Somit and Peterson (1992) explored the wider role of punctuated equilibrium in suggesting similar modes of change in other disciplines. (Their edited book bears the title: *The Dynamics of Evolution: The Punctuated Equilibrium Debate in the Natural and Social Sciences.*) In discussing the "manner in which punctuated equilibrium theory renders its greatest contribution to the behavioral sciences" (1992, p. 12), they suggested (loc. cit.): "By providing a different metaphor for explaining social phenomena, the theory may assist us in better understanding human behavior in all of its manifestations."

I don't question either the widespread invocation or the extensive utility of the metaphorical linkage, and I list elsewhere (pp. 976–979) a range of such invocations across disciplines from economics to cartooning to guidelines for the self-help movement. But in discussing the application of punctuated equilibrium to other disciplines, I am more interested in exploring ways in which the theory might supply truly causal insights about other scales and styles of change, based on conceptual and structural "homologies" (as defined and discussed on pp. 928–931), rather than broader metaphors that can surely nudge the mind into productive channels, but that make no explicit claim for causal continuity or unification. Thus, in discussing the influence of punctuated equilibrium upon other disciplines, I will focus upon two kinds of potentially homological proposals.

First, where authors proceed beyond simple claims for broad similarities in jerky tempos of change to identify additional and explicit overlaps in the set of collateral principles that I called "conjoints" (see p. 930) in defining conceptual homology vs. analogy—including, for example, (1) claims that link punctuations to the origin of discretely individuated units arising by branching (a conceptual homolog of speciation), (2) discussions of the difference between punctuational and saltational modes, and (3) proposals about active causes for the maintenance of stasis. And second, where authors use the similarities between punctuated equilibrium and punctuational tempos in their own discipline to advance more than vaguely metaphorical suggestions for general theories about the nature of change in systems that may be said to "evolve," and to display historical continuity.

EXAMPLES FROM THE HISTORY OF HUMAN ARTIFACTS AND CULTURES. I presented arguments for punctuational models of human biological evolution in a previous section (pp. 908–916). But I have also been struck by the frequency of punctuational explanations advanced for patterns in the development of human artifacts and cultural history, processes that must "evolve" under causes and mechanisms quite different from the genetic variation and natural selection that regulate our Darwinian biology. Moreover, the Lamarckian character of human cultural change—the inheritance by teaching of useful innovations acquired during the life of an inventor—provides an entirely plausible mechanism for a more accumulative, progressive and gradual style of change in this realm than the Darwinian character of physical evolution (and the explicit denial of Lamarckian effects) should permit for the history of our anatomical changes. Thus, the discovery of punctuational patterns in cultural change might be viewed as even more surprising than the application of punctuated equilibrium to our morphological evolution.

For example, although more gradual and accumulative change may prevail in the history of tools following the origin of *Homo sapiens* (and, one presumes, a markedly increased capacity for cultural transmission), many scholars have noted, usually with surprise, a remarkable lack of change in the *Homo erectus* tool kit during more than a million years. For example, Mazur (1992, p. 229; see also Johanson and Edey, 1981, and Roe, 1980) states: "the early tool cultures were remarkably stable over long periods of time. The constancy of the Acheulean hand-ax tradition has been especially noted, for hand axes have been found at sites widely separated in distance and across a million years of *Homo erectus's* existence look very similar to one another, their uniformity more striking than regional differences." Such collateral data support a view of *Homo erectus* as an individuated biological species, an entity rather than an arbitrarily defined segment of a continuity in anagenetic advance.

The history of scholarly research on European Paleolithic cave art provides an especially interesting example of how belief in progressivistic and gradual anagenesis can operate as a limiting preconception, and how punctuated equilibrium can play a salutary role as a potential corrective, or at least as a source of novel hypotheses for consideration. No aspect of the prehistory of artifacts has stunned or moved modern humans more than the parietal (wall) art of the great caves of Lascaux, Altamira, and many others, with their subtle and beautiful animal paintings that establish an immediate visceral link of aesthetic equality between the anonymous prehistoric artists and a Leonardo or Picasso. At least by standards of human history, these caves span a considerable range of time, from Chauvet at greater than 30,000 years BP (by radiocarbon evidence) to several at about 10,000 years of age.

Unsurprisingly, all great scholars of cave art have wanted to learn if any "evolution" can be discerned in the temporal sequence of these images. Two preeminent scientists built sequential schools of thought that virtually define the intellectual history of this subject in the 20th century. These men, the Abbé Henri Breuil and André Leroi-Gourhan, shared a firm belief that gradualistic evolution through a series of progressive stages provides a primary organizing theme for the history of parietal art (see Gould, 1998b, for an epitome of their beliefs), even though, in other philosophical respects, their worldviews could not have been more different. (In fairness, no techniques of absolute dating were available to these scientists, so they used the traditional methods of art history in attempting to establish chronology by a series of developmental stages. But for human art in historical times, we can back up such aesthetic theories with known dates of composition, so the argument does not become intrinsically circular.)

The Abbé Breuil viewed the paintings functionally as a form of hunting magic (if you draw them properly, they will come). He accepted the linearly progressivist view of evolution that his late 19th century education had inculcated, and that his religious convictions about human perfectability also favored. He therefore conceived the chronology of cave art as a series in styles of improvement, leading to rigidification and a final "senile" decline. In an early article of 1906, he wrote: "Paleolithic art, after an almost infantile beginning, rapidly developed a lively way of depicting animal forms, but didn't perfect its painting techniques until an advanced stage."

Leroi-Gourhan, a devoted follower of Lévi-Strauss and his functionalist school, embraced the opposite concept that cave art embodies timeless and integrative themes of human consciousness, based on dichotomous divisions that define our innate mental style of ordering the complex world around us. Thus, we separate nature from culture (the raw *vs.* the cooked in Lévi-Strauss's famous metaphor), light from darkness, and, above all, male from female. Leroi-Gourhan therefore read the caves as sanctuaries where the numbers and positions of animals (with, for example, horses as male symbols, and bisons as female) reflected our unchanging sense of natural order based on a primary sexual dichotomy—with an appended set of symbolic, and similarly dichotomized, attributes, including the conventional and sexist active *vs.* passive, and rational *vs.* emotional.

Given his view of cave art as representing the unchanging structure of human mentality, Leroi-Gourhan might have emphasized an implication of stasis for the duration of this form of expression. In fact, matching Breuil in commitment to the notion of gradualistic progress, Leroi-Gourhan contrasted a stability in conceptual intent with continual improvement in fidelity of artistic rendering for images of unchanged significance—that is, a gradual progression in overt phenotypes (the only aspect of change that an "evolutionist" might note and measure) contrasted with a constancy in symbolic meaning. Leroi-Gourhan wrote in 1967: "The theory . . . is logical and rational: art apparently began with simple outlines, then developed more elaborate forms to achieve modeling, and then developed a polychrome or bichromate painting before it eventually fell into decadence."

This scenario sounds eminently reasonable until one subjects the argument to further scrutiny, with an explicit effort to identify and question gradualistic biases. After all, we are not examining a lineage of enormous geological extent spanning a range of phenotypic complexity from amoeba to mammal, or even from one species to another. We are tracing about 20,000 years in the history of a single species, *Homo sapiens,* that remained anatomically stable throughout this time. Of course, cultural achievements can accumulate progressively while Darwinian biology remains unaltered. And, of course again, we assume that the first person who ever took ochre to wall could not render a mammoth with all the subtlety developed by later artists; some substantial learning and development must have occurred. But then, the earliest known cave paintings do not record these initial steps, for our oldest data probably represent a tradition already in full flower—so that we observe, in the total range now available to us, something analogous to the history of Western art from Phidias to Picasso (with much change in style, but not directional progress), not the full sweep from the first hominid who ever pierced a hole in a bear tooth and then strung the object around his neck, to the Desmoiselles of Avignon. Why then should we ever have anticipated a linear sequence of change in the known history of Paleolithic parietal art?

Indeed, and to shorten a longer story, the discovery and dating of Chauvet cave in 1994, abetted by improvements in radiocarbon methods that provide accurate results from tiny samples, have now disproven the controlling gradualist and progressivist assumption in an entire tradition of research. The paintings at Chauvet, dated as the oldest of all known sites (30,000 to 34,000 years BP), include all features previously regarded as identifying the highest, and latest, stage of achievement in a sequence of increased artistry (as found in the much younger caves of Lascaux and Altamira). In other words, the full range of styles extended throughout the entire interval of dated caves, with the most sophisticated forms fully present at the oldest site now known.

Bahn and Vertut (1988) invoked punctuated equilibrium in their prescient anticipation of the disproof that would soon follow. They also made an astute argument—based on punctuated equilibrium's concept of species as discrete individuals with considerable capacity for spatial variation at any one time *vs.* the tendency of anagenetic thinking to regard the phenotype of any moment as a uniform stage in a temporal continuum—that geographic variation, in itself, should have precluded any expectation of a simple chronological sequence, even if a general trend did pervade the entire series. After all, why should areas as distant as southern Spain, northeastern France and southeastern Italy go through a series of progressive stages in lockstep over 20 thousand years? Regional and individual variation can swamp general trends, even today in our globally connected world of airplanes and televisions. Why did we ever think that evolution should imply a pervasive signal of uniform advance? Bahn and Vertut (1988) write:

> The development of Paleolithic art was probably akin to evolution itself: not a straight line or ladder, but a much more circuitous path—a complex growth like a bush, with parallel shoots and a mass of offshoots; not a slow, gradual change, but a "punctuated equilibrium," with occasional flashes of brilliance . . . Each period of the Upper Paleolithic almost certainly saw the coexistence and fluctuating importance of a number of styles and techniques, . . . as well as a wide range of talent and

ability . . . Consequently, not every apparently "primitive" or "archaic" figure is necessarily old . . . and some of the earliest art will probably look quite sophisticated.

A similar reconceptualization and corrective, for a more restricted region at a smaller scale of centuries rather than millennia, has been offered, citing punctuated equilibrium as a source of ideas, in Berry's (1982) treatise on the history of the Anasazi people of western North America. Berry treats the Anasazi as a geographically variable cultural entity, in many ways akin to a biological species under punctuated equilibrium, and not, as in most previous writing, as a group in continuous flux, with nearly all variation expressed temporally. Eldredge and Grene (1992, p. 118) write of Berry's work:

> Rather than interpreting the pattern as a linear history, in which change sometimes occurred rapidly and at other times at a more leisurely pace, Berry argues that the patterns of stasis interrupted by spurts of rather profound cultural change do not represent linear evolution, but rather a sequence of habitation and replacement. The Anasazi are a historical whole, as regionally diverse and as temporally modified as they were. They were replaced by another cultural system, not as a smooth evolutionary outgrowth but because the Anasazi were eventually (and rather abruptly) no longer able to occupy their territory.

Several social scientists have used the model of punctuated equilibrium as a guide to reconstructing patterns in social and technological development as punctuationally disrupted and then reformulated, rather than gradually altering—as in Weiss and Bradley (2001) on climatic forcing as a prod to rapid societal collapse in early civilizations throughout the world, over several millennia of time and types of organization. Adams (2000) has generalized this argument about "accelerated technological change in archaeology and ancient history." He explicitly cites the Lyellian tradition as a former impediment to recognizing and resolving such social punctuations (2000, pp. 95–96): "Built into traditional Darwinian 'descent with modification' was an acceptance of the standpoint of Lyell's geological gradualism. In its time, his assumption of uniformitarian change in the earth's geological history carried the day against competing doctrines of catastrophism. Today, however, there is increasing recognition of great diversity in rates of evolutionary change . . . Accelerated phases of change, often referred to in evolutionary biology as punctuations, invite closer study by students of human as well as natural history."

Finally, as a generality for the key transition to agriculture that marks (through the accumulation of wealth leading to social stratification, and the initiation of fixed-placed dwelling leading to towns and cities) the multiple inception of what, for better or worse, we generally call "civilization"— Boulding (1992, p. 181) cites active stasis and rapid punctuation as the predominant pattern, in opposition to a uniformitarian tradition most famously promulgated in Alfred Marshall's *Principles of Economics,* one of the most nfluential textbooks ever written, and a volume that, through decades and numerous editions, bore on its title page the Leibnizian, Linnaean, and Darwinian maxim: *Natura non facit saltum*. Boulding writes: "In the economy we certainly find periods of relative stability, in which society is getting neither much richer nor much poorer, but these periods of stability do seem to be punctuated by periods of very rapid economic development. The transition from hunting-gathering societies to agriculture at any particular locality seems to set off a period of rapid economic growth. This transition was usually rather rapid and, it would seem, irreversible."

EXAMPLES FROM HUMAN INSTITUTIONS AND THEORIES ABOUT THE NATURAL WORLD. If relatively prolonged periods of actively maintained stability, followed by episodic transition to new positions of repose, mark the most characteristic style of change across nature's scales, and if we have generally tried to impose a gradualistic and progressivistic model of change upon this different reality, then we must often face anomalies that engender confusion and frustration in our personal efforts to improve our lives or to master some skill. To cite two mundane examples from my own experience, I spent several, ultimately rather fruitless years learning to play the piano. Whenever I tried to master a piece, I would become intensely frustrated at my minimal progress for long periods, and then exhilarated when everything "came together" so quickly, and I could finally play the piece. I also liked to memorize long passages of poetry and great literature, primarily Shakespeare and the Bible, an activity then practiced and honored in the public primary and secondary schools of America. I would get nowhere forever, or so it seemed—and then, one fine day, I would simply know the entire passage.

Only years later—and perhaps serving as a spur to my later interest in punctuated equilibrium—did I conceptualize the possibility that plateaus of stagnation and bursts of achievement might express a standard pattern for human learning, and that my previous frustration (at the long plateaus), and my exhilaration (at the quick and rather mysterious bursts), might only have reflected a false expectation that I had carried so long inside my head—the idea that every day, in every way, I should be getting just a little bit better and better.

I don't know that explicit instruction in the higher probability of punctuational change, and the consequent appeasement of frustration combined with a better understanding of exhilaration, would improve the quality of our lives. (For all I know, the frustration and exhilaration yield important psychological benefits that outweigh their inadequate mapping of nature.) But I do suspect that a general recognition of the principles of punctuational change—leading us to understand that learning generally proceeds through plateaus of breakthroughs, and that important changes in our lives occur more often by rapid transition than by gradual accretion—might provide some distinct service in our struggles to fulfill the ancient and honorable Socratic injunction: know thyself.

I also think that an explicit application of punctuational models to many aspects of change in human institutions and technologies might improve our grasp and handling of the social and political systems that surround and include us. For example, in a stimulating paper emanating from her own research on "project groups" formed to study and initiate organizational change, Gersick (1991) explored the commonalities of punctuational change at six distinct levels of increasing scope—in the lives of individuals, the structures of groups (her own work), the history of human organizations, the history of ideas, biological evolution (our work on punctuated equilibrium), and general theory in physical science (Prigogine on bifurcation theory). Her paper, published in the *Academy of Management Review,* bears the title, "Revolutionary change theories: a multilevel exploration of the punctuated equilibrium paradigm," and begins by stating: "Research on how organizational systems develop and change is shaped, at every level of analysis, by traditional assumptions about how change works. New theories in several fields are challenging some of the most pervasive of these assumptions, by conceptualizing change as a punctuated equilibrium: an alternation between long periods when stable infrastructures permit only incremental adaptations, and brief periods of revolutionary upheaval." (See also Wollin, 1996, on the utility of punctuated equilibrium for resolving the dynamics of growth of complex human and social systems in general: "A hierarchy based approach to punctuated equilibrium: an alternative to thermodynamic self-organization in explaining complexity.")

In her own work on "task groups," Gersick (1988) reached a surprising conclusion. These associations did not proceed incrementally towards their assigned goals, but rather tended to hem, haw and dither until they reached a particular, and temporally definable, point of quick transition towards a solution. "Project groups," she writes (1991, p. 24), "with life spans ranging from one hour to several months reliably initiated major transitions in their work precisely halfway between their start-ups and expected deadlines. Transitions were triggered by participants' (sometimes unconscious) use of the midpoint as a milestone, signifying 'time to move.'"

These particular results inspired Gersick's more general consideration of punctuational models. In so doing, she explicitly follows both approaches previously advocated (p. 928) as strategies for transcending metaphor and discovering causally meaningful connections among punctuational phenotypes of change across levels and disciplines: the identification of "conjoints," or properties correlated with the basic punctuational pattern (the basic strategy of documenting a complexity in number and interaction of parts too high to attribute to causally accidental resemblance, and therefore necessarily based on homology); and the proposal of a general rationale, transcending the particular of any scale of analysis or class of objects, for the punctuational character of change.

For example, Gersick emphasizes the need for active resistance towards change as a validation of stasis, and she makes the same link that Eldredge and I have stressed for biological evolution between punctuational change in hierarchical systems and structural constraint viewed as partly contrary to an adaptationist paradigm. She writes (1991, p. 12): "Gradualist paradigms imply that systems can 'accept' virtually any change, any time, as long as it is small enough; big changes result from the insensible accumulation of small ones. In contrast, punctuated equilibrium suggests that, for most of systems' histories, there are limits beyond which change is actively prevented, rather than always potential but merely suppressed because no adaptive advantage would accrue."

Gersick's lists of commonalities among her six levels, both for periods of stasis and for episodes of punctuation, satisfy the strategy of conjoints, while her ranked list of scales, and especially her linkages of particular categories to the two most overarching theories of punctuational change—Kuhn's for human thought and Prigogine's for the natural world—meet the criterion of generalization. For example, her chart of comparison among the six levels for "equilibrium periods" cites both commonalities and conjoints, with an interestingly different emphasis (from our concerns with biological systems) upon the potential for strong limitation placed upon incremental pathways within a plateau—an important theme for the Lamarckian character of human cultural change. She lists as "commonalities" (p. 17): "During equilibrium periods, systems maintain and carry out the choices of their deep structure. Systems make adjustments that preserve the deep structure against internal and external perturbations, and move incrementally along paths built into the deep structure. Pursuit of stable deep structure choices may result in behavior that is turbulent on the surface."

Similarly, her chart for punctuational episodes stresses the unpredictability and potential nonprogressionism of outcomes, a surprising theme for human systems based on supposed and explicit goals, but a notion that we did not emphasize in formulating punctuated equilibrium because biological evolution proceeds in a highly contingent manner for so many other reasons (see Gould, 1989c), some recognized and emphasized by Darwin himself. Thus, this important theme, while equally central within the structure of our theory, did not have similar salience for us—and we thank Gersick for her insight and generalization. Gersick writes in her heading (p. 20): "Revolutions are relatively brief periods when a system's deep structure comes apart, leaving it in disarray until the period ends, with the 'choices' around which a new deep structure forms. Revolutionary outcomes, based on interactions of systems' historical resources with current events, are not predictable; they may or may not leave a system better off. Revolutions vary in magnitude."

As another example of fruitful borrowing across disciplines, Mokyr (1990, p. 350) begins his study of technological change by noting that Alfred Marshall's advocacy of gradualism in economics played a similar role in the human sciences to Darwin's impact upon the natural sciences: "Charles Darwin and Alfred Marshall were both extremely influential men . . . Darwin and Marshall both believed that nature does not make leaps. Both were influenced by a long and venerable tradition that harked back to Leibniz rooted in the Aristotelian notion of the continuity of space and time." (On the same subject, see also Loch, 1999, on "A Punctuated equilibrium model of technology diffusion.")

The development of improved systems for human communication in the past century must represent one of the most goal-directed and clearly progressionist sequences identifiable in either the human or natural sciences—hence the apparent consonance with gradualist models, and the unpromising character, at first glance, of punctuationalist alternatives. But Mokyr defends a punctuational reformulation by centering his descriptions and explanations on a comparison of stages in this technological trend with the discrete origin of biological species as genuine entities, and then citing punctuated equilibrium for emphasizing this theme as a reform within Darwinian biology (Mokyr, 1990, p. 351).

While not denying the clearly goal-oriented and progressive nature of the trend—and while also (along with Gersick) noting the Lamarckian capacity of human culture to change directionally and incrementally within plateaus, but also stressing the qualitative differences between this limited gradualism and the much larger and quicker transitions of goal-directed punctuations—Mokyr (p. 354) describes the basic outline of this history as four stages separated by punctuational breaks. He also, again as with Gersick, stresses the nonpredictability of outcomes, and then notes, continuing the analogy with speciation, that punctuational models enjoin the study of particular conditions favorable to leaps of change, an issue that does not arise in explanations based upon gradualistic anagenesis:

> Long-distance communications thus illustrate the abruptness of technological change. There was no natural transition from semaphore to the electrical telegraph, nor a gradual movement from the telegraph to the first radio transmission by Marconi in 1894, nor a smooth natural development from the long-wave radio used in the first thirty years to the shortwave systems of the later 1920s. Each of the three systems was subsequently perfected by a long sequence of microinventions, but these would not have occurred without the initial breakthroughs. Their concept was novel, they made things possible that were previously impossible, and they were pregnant of more to come. Therein lies the essence of a macroinvention. . . . Many macroinventions, just like the emergence of species, were the result of chance discoveries, luck, and inspiration. Biologists agree that certain environments are more conducive to speciation than others.

Just as for technological change, we tend to view the history of scientific ideas on particular subjects as, in principle, the most incrementally progressionist of all human activities by the empiricist paradigm of ever-closer approximation to natural truth through objective accumulation of data under unchanging principles of "the scientific method"—an idea famously challenged by Kuhn (1962) in the most influential punctuationalist theory in 20th century scholarship—see p. 967. Thus, punctuationalist reformulations in this domain tend to strike people as especially surprising. The distinguished

Dutch petrologist E. Den Tex (1990), in summing up his career of study on the nature of granites, credited punctuated equilibrium with reorganizing his lifelong attempt to make sense of a complicated history, stretching back to the late 18th century (the neptunist-plutonist debate between the schools of Werner and Hutton), and featuring fluctuations between poles of two dichotomies (sedimentary *vs.* igneous formation, and recruitment from deep magmas *vs.* transformation from existing crustal rocks), complicated by shifting allegiances and amalgamations, followed by breakages, of separable aspects of all end-member theories.

In his fascinating and highly personal paper, entitled "Punctuated equilibria between rival concepts of granite genesis," Den Tex (1990) notes that he had first tried to apply other models of noncontinuous and progressive change, especially the celebrated Hegelian notion of successive syntheses reached by opposition between a thesis and its antithesis. He then found a better general description, with new and fruitful hints for explanation, in our model of punctuated equilibrium, particularly in the parsing of history as discrete steps, analogous to individuated species with definite sets of properties—a process "in dynamic equilibrium . . . punctuated from time to time by allopatric speciation, *i.e.* by rapid, random, discrete steps taking place in locations isolated from the main stem" (1990, p. 216).

Finally, since punctuated equilibrium arose as a theory about change in the natural world (not in the history of human understanding thereof), Eldredge and I have been gratified by the utility of our theory in suggesting structurally homologous modes of change in other branches of natural science. I have been especially pleased by geological examples distant from our own paleontological concerns, because no other field can match anglophonic geology—resulting largely from the legacy of Lyellian uniformitarianism (see Chapter 6)—in explicit, and often exclusive, fealty to strictly gradualistic models.

Lawless (1998, and a good name for iconoclasts), in an article entitled "Punctuated equilibrium and paleohydrology," notes the hold that gradualistic models have imposed on the history of ideas about hydrothermal ore deposits, particularly of gold. He begins by expressing a paradox: if ores accumulate gradually in such systems, and given the average amount of gold carried in most percolating waters, minable deposits should be much more common—indeed almost ubiquitous in hydrothermal systems that persist for at least 25,000 years. But the much rarer distribution of such deposits suggests to Lawless that periods of accumulation must be limited to brief episodes "which cause vigorous boiling through a restricted volume of the reservoir" (p. 165). Lawless views the general history of hydrothermal systems—including the development of ore deposits as just one feature among many—as punctuational, and caused by rapid, intense, rarely-acting forces: "Such disturbances caused by tectonic activity, magmatism, volcanic activity, erosion, climatic changes or other processes may occur at long intervals, but be responsible for producing some of the most significant characteristics of the system, including economic mineral deposits."

As a matter of potential practical importance, Lawless recognizes that, just as punctuated equilibrium must not be construed as an argument against predictable trends, but rather as a different mechanism for the episodic production of such trends, so too might the punctuational origin of hydrothermal ore deposits be reconceptualized as directional but episodic. He writes (p. 168): "Recognition of the quasi-cyclic and episodic nature of these events within the lifetime of a hydrothermal system could be described as leading to more 'catastrophic' models. There are similarities to the concepts of punctuated equilibrium recently proposed in paleontology and biological evolution . . . these concepts emphasize the importance of specific events which are of random occurrence on a short time scale but statistically predictable on a longer scale."

**TWO CONCLUDING EXAMPLES, A GENERAL STATEMENT, AND A CODA.** As final examples in this chapter, two recent authors have used punctuated equilibrium as the central organizing principle for books on subjects of different scale, but of great importance in human life and history—Kilgour (1998) on *The Evolution of the Book,* and Thurow (1996) on *The Future of Capitalism.* Moreover, each author uses punctuated equilibrium not as a vague metaphor, but as a specific model of episodic change offering casual insights through the identification of structural homologies as defined in this chapter.

Kilgour notes that a theme of greater efficiency marks the history of bookmaking (and might be misread as evidence for anagenetic gradualism, just as trends in the evolutionary history of clades have often been similarly misconstrued when a punctuational model of successive plateaus defined by discrete events of branching actually applies). He writes (p. 4): "Form aside, the major change throughout the entire history of the book has been in the continuous increase in speed of production: from the days required to handwrite a single copy, to the minutes to machine-print thousands of copies, to the seconds to compose and display text on an electronic screen."

But, as Kilgour knows, and adopts as the major theme of his book, form cannot be put "aside." When one probes through these progressive improvements in function to underlying bases in form, the history of the book becomes strongly punctuational. In a pictorial summary for his central thesis (Fig. 9-36), Kilgour views the evolution of the book—defined (p. 3) as "a storehouse of human knowledge intended for dissemination in the form of an artifact that is portable, or at least transportable, and that contains arrangements of signs that convey information"—as a sequence of four great punctuations: the clay tablet, the papyrus roll, the codex (modern book), and the electronic "book" (with no canonical form as yet since we are now enjoying, or fretting our way through, the rare privilege of living within a punctuation), with three "subspeciational" punctuations within the long domination of the codex (Gutenberg's invention of printing with movable type in the mid 15th century, and the enormous additional increases in production made possible first by the introduction of steam power at the beginning of the 19th century, and then by the development of offset printing in the mid 20th century).

As I have emphasized throughout this discussion of human cultural and technological change, the Lamarckian nature of inheritance for these processes permits more directional accumulation within periods of overall stasis in basic design than analogous chronologies of biological evolution can probably exhibit. Thus, while the codex (that is, the familiar "bound book") enjoyed its millennium and a half of domination in fundamentally unchanged form, several innovations both in design (the introduction of pagination and indexes) and in human practice and collateral discovery (the invention of eyeglasses at the end of the 13th century, and the spread of the "newfangled" practice of silent reading in the 15th century) greatly expanded the utility of a product that remained stable in form. (While books remained scarce, people read aloud and, apparently, did not even imagine a possibility that seems obvious to us—namely, that one might read without speaking the words. By reading aloud, one copy could be shared with many, but silent reading demands a copy for each participant.)

But the four great designs (or at least the three we know, for the fourth has not yet stabilized) have experienced histories strikingly akin, in more than vaguely metaphorical ways, to the origin and persistence of biological species treated as discrete individuals. In the first of three major similarities, each principal form persisted in effectively unchanged design for long periods of time by standards of human technological innovation. Moreover, each transition introduced a great improvement by solving an inherent structural problem in the previous design. Therefore, the extended persistence of each flawed

| | | |
|---|---|---|
| Clay Tablet | First Punctuation | 2500 BC |
| Papyrus Roll | Second Punctuation | 2000 BC |
| Codex | Third Punctuation | AD 150 |
| Printing | Fourth Punctuation | 1450 |
| Steam Power | Fifth Punctuation | 1800 |
| Offset Printing | Sixth Punctuation | 1970 |
| Electronic Book | Seventh Punctuation | 2000 |

9-36. Kilgour, in his 1998 work on the evolution of the book, presents a punctuational model of clay tablet, to papyrus roll, to codex, to printing, and onward to the undetermined future of the electronic book. All major features of the biological model apply here in adequate isomorphism for causal insight, including the survival of ancestors after the branching of descendants.

design illustrates an important reason, more "environmental" than structural, for the existence of stasis in natural systems: the advantages of incumbency. "The extinction of clay tablets," Kilgour writes (pp. 4–5) "was ensured by the difficulty of inscribing curvilinear alphabet-like symbols on clay"; while "the need to find information more rapidly than is possible in a papyrus-roll-form book initiated the development of the Greco-Roman codex in the second century A.D." Of this predominant stasis, Kilgour writes (p. 4): "Extremely long periods of stability characterize the first three shapes of the book; clay tablets and papyrus-roll books existed for twenty-five hundred years, and the codex for nearly two thousand years. An Egyptian of the twentieth century B.C. would immediately have recognized, could he have seen it, a Greek or Roman papyrus-roll book of the time of Christ; similarly, a Greek or Roman living in the second century A.D. who had become familiar with the then new handwritten codex would have no trouble recognizing our machine-printed book of the twentieth century."

Secondly, the successive stages do not specify segments of an anagenetic flow (whatever the punctuational character of each introduction), but rather arise as discrete forms in particular areas—thus following the pattern of branching speciation so vital to the validation of punctuated equilibrium, and also meeting the chief operational criterion for distinguishing punctuated equilibrium from punctuated anagenesis: the survival of ancestral forms after the origin of new species. Kilgour notes (p. 158) that "clay tablets and papyrus-roll books coexisted for two thousand years, much as two biological species may live together in the same environment." He also notes, both wryly and a bit ruefully, "that books on paper and books on electronic screens, will, like clay tablets and papyrus books, coexist for some time, but for decades rather than centuries" (p. 159).

Third, each form—at least before improved communication of the past two centuries made such localization virtually inconceivable—originated in a particular time and place, and in consonance with features of the immediately surrounding environment, a meaningful analog to the locally adaptive origin of biological species. Kilgour writes (p. 4): "the Sumerians invented writing toward the end of the fourth millennium B.C. and from their ubiquitous clay developed the tablet on which to inscribe it. The Egyptians soon afterward learned of writing from the Mesopotamians and used the papyrus plant, which existed only in Egypt, to develop the papyrus roll on which to write."

I know Lester C. Thurow as a colleague from another institution in Cambridge, MA, but I had never discussed punctuated equilibrium with him, and was surprised when he used our theory as one of two defining metaphors in his book, *The Future of Capitalism*. In distinction to most examples in this chapter, Thurow does invoke punctuated equilibrium in a frankly metaphorical and imagistic manner, but he also shows a keen appreciation for the conjoints of punctuated equilibrium applied to his subject of macroeconomics. Thurow writes in the context of the collapse of communism in the Soviet Union, and in the survival of capitalism as a distinctive, effectively universal, and perhaps uniquely workable system of human economic organization within a highly technological matrix. Yet capitalism now faces a crisis of reorganization in a world where major change (both natural and social) occurs by punctuated equilibrium and not by slow incrementation. Thurow therefore presents two metaphors from the natural sciences to ground his argument: "To understand the dynamics of this new economic world, it is useful to borrow two concepts from the physical sciences—plate tectonics from geology and punctuated equilibrium from biology" (p. 6).

In an obviously imagistic metaphor with no causal meaning, Thurow identifies the five "economic tectonic plates" that will incite the next punctuation by crunching and grinding as their rigid borders crash: "the end of communism," "the technological shift to an era dominated by man-made brainpower industries," "a demography never before seen" (increasing average age, greater movement of populations to cities, etc.), "a global economy," and "an era where there is no dominant economic, political or military power" (pp. 8–9).

But his invocation of punctuated equilibrium shows more structural and causal connection to the "parent" phenomenon from the natural sciences. Thurow (p. 7) notes both the long plateaus and the tendency for rapid historical shifts in transitions between macroeconomic systems that organize entire societies:

Periods of punctuated equilibrium are equally visible in human history. Although they came almost two thousand years later, Napoleon's armies could move no faster than those of Julius Caesar—both depended upon horses and carts. But seventy years after Napoleon's death, steam trains could reach speeds of over 112 miles per hour. The industrial revolution was well under way and the economic era of agriculture, thousands of years old, was in less than a century replaced by the industrial age. A survival-of-the-fittest social system, feudalism, that had lasted for hundred of years was quickly replaced by capitalism.

More notably, and marking Thurow's fruitful use, he stresses important conjoints of punctuated equilibrium as the most relevant—and practical—themes for our current and dangerous situation. First, the common phenotype of punctuation leads him to recognize that general structural rules must underlie both the maintenance of stasis and, through their fracturing, the episodes of punctuation as well. The rules will differ in social and natural systems, but the general principle applies across domains. Thurow argues that stasis requires a meshing of technology and ideology, while their radical divergence initiates punctuation, a situation that we face today. (Marx, in an entirely different context, held a similar view about both the character of the rule and the punctuational outcome.)

Technology and ideology are shaking the foundations of twenty-first-century capitalism. Technology is making skills and knowledge the only sources of sustainable strategic advantage. Abetted by the electronic media, ideology is moving toward a radical form of short-run individual consumption maximization at precisely a time when economic success will depend upon the willingness and ability to make long-run social investments in skills, education, knowledge, and infrastructure. When technology and ideology start moving apart, the only question is when will the "big one" (the earthquake that rocks the system) occur. Paradoxically, at precisely the time when capitalism finds itself with no social competitors—its former competitors, socialism or communism, having died—it will have to undergo a profound metamorphosis (p. 326).

Second, Thurow continually stresses the contingency implied by the scale change between causes of alteration in "normal" times (with greater potential and range in social than in natural systems due to the Lamarckian character of cultural inheritance) and the different modes and mechanisms of punctuational episodes. Thus, the rules we may establish from our experience of ordinary times cannot predict the nature or direction of any forthcoming punctuation.

In a period of punctuated equilibrium no one knows that new social behavior patterns will allow humans to prosper and survive. But since old patterns don't seem to be working, experiments with different new ones have to be tried (p. 236) . . . How is capitalism to function when the important types of capital cannot be owned? Who is going to make the necessary long-run investments in skills, infrastructure, and research and development? How do the skilled teams that are necessary for success get formed? In periods of punctuated equilibrium there are questions without obvious answers that have to be answered (p. 309).

Thurow therefore stresses the virtues of flexibility, ending his book (pp. 326–327) with an appropriate image for "the period of punctuated equilibrium" that "the tectonic forces altering the economic surface of the earth have created":

Columbus knew that the world was round, but he . . . thought that the diameter of the world was only three quarters as big as it really is. He also overestimated the eastward land distance to Asia and therefore by subtraction grossly underestimated the westward water distance to Asia . . . Given the amount of water put on board, without the Americas Columbus and all his men would have died of thirst and been unknown in our history books. Columbus goes down in history as the world's greatest explorer . . . because he found the completely unexpected, the Americas, and they happened to be full of gold. One moral of the story is that it is important to be smart, but that it is even more important to be lucky. But ultimately Columbus did not succeed because he was lucky. He succeeded because he made the effort to set sail in a direction never before taken despite a lot of resistance from those around him. Without that enormous effort he could not have been in the position to have a colossal piece of good luck.

To end on a more personal note, if I were to cite any one factor as probably most important among the numerous influences that predisposed my own mind toward joining Niles Eldredge in the formulation of punctuated equilibrium, I would mention my reading, as a first year graduate student in 1963, of one of the 20th century's most influential works at the interface of philosophy, sociology and the history of ideas: Thomas S. Kuhn's *The Structure of Scientific Revolutions* (1962). (My friend Mike Ross, then studying with the eminent sociologist of science R. K. Merton in the building next to Columbia's geology department, ran up to me one day in excitement, saying "you just have to read this book right away." I usually ignore such breathless admonitions, but I respected Mike's judgment, and I'm surely glad that I followed his advice. In fact, I went right to the bookstore and bought a copy of Kuhn's slim volume.)

Of course, Kuhn's notion of the history of change in scientific concepts advances a punctuational theory for the history of ideas—going from stable "paradigms" of "normal science" in the "puzzle solving" mode, through accumulating anomalies that build anxiety but do not yet force the basic structure to change, through rapid transitions to new paradigms so different from the old that even "conserved" technical terms change their meaning to a sufficient extent that the two successive theories become "incommensurable." The book has also served, I suspect, as the single most important scholarly impetus towards punctuational thinking in other disciplines.

Since the appearance of our initial paper on punctuated equilibrium in 1972, several colleagues have pointed out to me that Kuhn himself, in a single passage, used the word "punctuated" to epitomize the style of change described by his theory. These colleagues have wondered if I borrowed the term, either consciously or unconsciously, from this foundational source. But I could not have done so. (I do not say this in an exculpatory way—for if I had so borrowed, I would be honored to say so, given my enormous respect and personal affection for Kuhn, and the pleasure I take in being part of his intellectual lineage. We did, but in an entirely different discussion about the definition of paradigms, cite Kuhn's book in our 1972 paper.) Kuhn used the word "punctuated" in the 1969 "postscript" that he added to the second edition of his book (see quotation below). In 1972, I had only read the first edition.

I mention this final point not as pure self-indulgence, but largely because Kuhn's single use of the word "punctuated," located in the closing paragraphs of the seventh and last section (entitled "the nature of science") of his postscript, expresses a surprising opinion that seems eminently exaptable as an appropriate finale to this chapter. Like all scholars whose works become widely known through constantly degraded repetition that strays further and further from unread original sources, Kuhn could become quite prickly about fallacious interpretations, and even more perturbed by bastardized and simplistic readings that caricatured his original richness.

He therefore ended his postscript by discussing two "recurrent reactions" (1969, p. 207) to his original text. He regarded the first reaction (irrelevant to this chapter) as simply unwarranted, so he just tried to correct his critics. But the second reaction bothered him more—as "favorable" (p. 207) to his work but, in his view, "not quite right" in granting him too much credit. He states that, whereas he had certainly hoped to catch the attention of scientists, he cannot understand why scholars in nonscientific disciplines should have found his work enlightening. After all, he claims (p. 208), he had only tried to explain the classical punctuational views of the arts and humanities to practitioners of another field—science—where social and philosophical tradition had long clouded this evident point. (The word "punctuated" appears in this passage):

> To one last reaction to this book, my answer must be of a different sort. A number of those who have taken pleasure from it have done so less because it illuminates science than because they read its main theses as applicable to many other fields as well. I see what they mean and would not like to discourage their attempts to extend the position, but their reaction has nevertheless puzzled me. To the extent that the book portrays scientific development as a succession of tradition-bound periods punctuated by noncumulative breaks, its theses are undoubtedly of wide applicability. But they should be, for they are borrowed from other fields. Historians of literature, of music, of the arts, of political development, and of many other human activities have long described their subjects in the same way. Periodization in terms of revolutionary breaks in style, taste, and institutional structure have been among their standard tools. If I have been original with respect to concepts like these, it has mainly been by applying them to the sciences, fields which had been widely thought to develop in a different way.

To this passage, I can only respond that I see what *he* means, but I also think that Kuhn undercuts the range of his own originality and influence by misreading the very social context that inspired his work. Indeed, the ethos of science—the conviction that our history may be read as an ever closer approach to an objective natural reality obtained by making better and better factual observations under the unvarying guidance of a timeless and rational procedure called "the scientific method"—does establish the most receptive context imaginable for mistaken notions about gradualistic and linear progressionism in the history of human thought. Indeed also, the traditions of disciplines that practice or study human artistic creativity—with their concepts of discrete styles and revolutionary breaks triggered by "genius" innovators—should have established punctuational models as preferable, if not canonical.

But, as Kuhn acknowledges, his book enjoyed a substantial vogue among artists and humanists, who also felt surprised and enlightened by his punctuational theory for the history of ideas. I think that Kuhn underestimated the "back influence" of science upon preexisting fields in humanistic study. Eiseley (1958) labeled the formative era of evolutionary theory as "Darwin's century," and we must never underestimate the influence of the Darwinian revolution, and of other 19th century notions, particularly the uniformitarian geology of Lyell, upon reconceptualizations of modes of discovery and forms of content in other disciplines, no matter how distant. Progressivistic gradualism—so central to most late 19th century versions of biological and geological history, and so strongly abetted by the appearance of "progress" (at least to Westerners) in the industrial and colonial expansion of Western nations, at least before the senseless destruction of the First World War shattered all such illusions forever—became a paradigm for all disciplines, not just for the sciences. Kuhn may have called upon some classical notions from the arts and humanities to construct a great reform for science; but his corrective also, and legitimately, worked back upon a source that had strayed from a crucial root idea to become beguiled by a contrary notion about change that seemed more "modern" and "prestigious."*

In addition, and finally, I think that Kuhn underestimated the potential role of scientific ideas in resolving old puzzles that have long stymied humanistic understanding of artistic creativity, and that remain seriously burdened by the hold of theories as ancient as the Platonic notion of essences and universals. In a lovely passage, directly following the "punctuational" quotation just cited, Kuhn acknowledges that the Darwinian concept of species as varying

---

*I dare not even begin to enter the deepest and most difficult of all issues raised by differences between scientific and humanistic practice: why does the history of scientific ideas, even when proceeding in a punctuational mode, marked by quirky, unpredictable and revolutionary shifts, undeniably move to better understanding (at least as measured operationally by our technological successes)—that is, and not to mince words, to progress in knowledge—whereas no similar vector can be discerned in the history of the arts, at least in the sense that Picasso doesn't (either by any objective measure or by simple subjective consensus) trump Leonardo, and Stravinsky doesn't surpass Bach (although later ages may add new methods and styles to the arsenals of previous achievement). The naive answer—that science searches for a knowable, objective, external reality that may justly be called "true," whereas art's comparable standard of beauty must, to cite the cliché, lie in the eye of the beholder—is probably basically sound, and probably explains a great deal more of this apparent dilemma than most academic sophisticates would care to admit. (In this belief, I remain an old-fashioned, unreconstructed scientific realist—but then we all must take oaths of fealty to our chosen profession.)

But I also acknowledge that the question remains far more complicated, and far more enigmatic, than this fluffy claim of such charming naievete would indicate. After all, we only "see" through our minds (not to mention our social organizations and their pervasive biases). And our minds are freighted with a massive cargo of all the inherent structural baggage that Kant called the *synthetic a priori,* and that modern biologists would translate as structures inherited from ancestral brains that built no adaptations *for* what we designate as "consciousness." In this light, why should we be "good" at knowing external reality? After all, our vaunted consensuses—and on this point, Kant remains as modern as the latest computer chip—may record as much about how our quirkily constructed brains must parse this "reality," as about how external nature truly "works." But enough of unanswerable questions! I only note that Kuhn himself raises this great issue in his closing thoughts on the special character of science: "It is not only the scientific community that must be special. The world of which that community is a part must also possess quite special characteristics, and we are no closer than we were at the start to knowing what these must be. That problem—What must the world be like in order that man may know it?—was not, however, created by this essay. On the contrary, it is as old as science itself, and it remains unanswered."

populations without essences (and even without abstractions, like mean values, to act as preferred or defining states) might break through a powerful and constraining prejudice, ultimately rooted in Platonic essentialism, that leads us to search for chimaerical idealizations as ultimate standards of comparison in the definition and evaluation of artistic "style." Kuhn writes, now viewing styles as paradigms and paradigms as higher level individuals like biological species (pp. 208–209): "Conceivably the notion of a paradigm as a concrete achievement, an exemplar, is a second contribution [of my book]. I suspect, for example, that some of the notorious difficulties surrounding the notion of style in the arts may vanish if paintings can be seen to be modeled on one another rather than produced in conformity to some abstracted canons of style."

Punctuated equilibrium represents just one localized contribution, from one level of one discipline, to a much broader punctuational paradigm about the nature of change—a worldview that may, among scholars of the new millennium, be judged as a distinctive and important movement within the intellectual history of the later 20th century. I am pleased that our particular formulation did gain a hearing and did, for that reason, encourage other scholars over a wide range of scientific and nonscientific disciplines (as illustrated in this chapter) to consider the larger implications of the more general punctuational model for change. I am especially gratified that many of these scholars did not just borrow punctuated equilibrium as a vague metaphor, however useful, but also understood, and found fruitful, some of the more specific "conjoints" distinctive to the level and phenomenon of punctuated equilibrium, but also applicable elsewhere. For the punctuational paradigm encompasses much more than a loose and purely descriptive claim about phenotypes of pulsed change, but also embodies a set of convictions about how the structures and processes of nature must be organized across all scales and causes to yield this commonality of observed results. Only in this sense—punctuated equilibrium as a distinctive contribution to a much larger and ongoing effort—can I understand Ruse's gracious reappraisal of his initial negativity toward punctuated equilibrium: "Grant then that there is indeed something going on that looks like a paradigm (or paradigm difference) in action. People (like my former self) who dismissed the idea were wrong—and missing something rather interesting to boot" (Ruse, 1992, p. 162).

From the more restricted perspective of the aims of this particular book, I can at least assert that punctuated equilibrium unites the three definitive themes of this volume—the three legs of my tripod of support for an expansion of Darwinian theory, thereby leading me to conclude that an empirically legitimate and logically sound structure does encompass and unite these three arguments into a coherent and general reformulation and extension of the Darwinian paradigm: the hierarchical theory of selection on leg one, the structuralist critique of Darwinian functionalism and adaptationism on leg two, and the paleontologist's conviction (leg three) that general macroevolutionary processes and mechanisms cannot be fully elucidated by uniformitarian extrapolation from the smallest scale of our experiments and personal observations. Punctuated equilibrium has proven its mettle in:

1. Elucidating and epitomizing what may be the primary process of a distinctive level in the evolutionary hierarchy: the role of species as Darwinian individuals, and the speciational reformulation of macroevolution—for leg one.

2. Defining (and, in part, thereby creating) the issue of stasis as a subject for study, and in helping to explicate the structural rules that hold entities in active stasis at various levels, but then permit rapid transition to qualitatively different states—for leg two.

3. Stressing that level-bound punctuational breaks preclude the prediction or full understanding of extensive temporal change from principles of anagenetic transformation at the lowest level (a mode of evolution, moreover, that punctuated equilibrium regards as rare in any case), thus emphasizing contingency and denying extrapolationist premises and methodologies—for leg three.

In developing this set of implications, I do hold, in my obviously biased way, that punctuated equilibrium has performed some worthy intellectual service. The relative frequency of its truth value, of course, must be regarded as another matter entirely, and an issue that only time can fully resolve. But I would maintain that, in the quarter century following its original formulation, punctuated equilibrium has at least prevailed, against an initial skepticism of active and general force and frequency, in three central empirical claims (quite independent of any theoretical weight that evolutionary biology may ultimately wish to assign): (1) documentation of the basic mechanism in cases now too numerous and too minutely affirmed to deny status as an important phenomenon in macroevolutionary pattern; (2) validation of stasis as a genuine, pervasive, and active phenomenon in the geological history of most species; and (3) establishment of predominant relative frequency in enough comprehensive and well-bounded domains to assure the control of punctuated equilibrium over substantial aspects of the phyletic geometry of macroevolution. A fourth, and ultimately more important, issue for evolutionary theory remains unresolved: the implications of these empirical findings for the role of genuine selection among species-individuals (rather than merely descriptive species-sorting as an upwardly cascading expression of conventional Darwinian selection acting at the organismic level) as the causal foundation of macroevolutionary pattern.

Let me therefore end this chapter by restating the last paragraph of the review article for *Nature* that Eldredge and I wrote (Gould and Eldredge, 1993, p. 227) to celebrate the true majority, or coming of age—that is, the 21st birthday—of punctuated equilibrium. We wrote this paragraph to assess the role of punctuated equilibrium within a larger and far more general intellectual (and cultural) movement that, obviously, punctuated equilibrium did not create or even instigate, but that our theory didn't simply or slavishly follow either. We did, I think, contribute some terms and concepts to the larger enterprise, and we did encourage scholars in distant fields to apply a mode of thinking, and a model of change, that had formerly been as unconventional (or even denigrated) in their fields as in ours. But we do find ourselves in the paradoxical, and at least mildly uncomfortable, position—as we tried to express in these closing words of our earlier article—of having developed a theory with empirical power, and at least some theoretical interest, in its own evolutionary realm, but that must largely depend, for any ultimate historical assessment, upon the fate and efficacy of more general intellectual currents (including both dangerous winds of fashion and solid strata of documentation) well beyond our control of competence.

> In summarizing the impact of recent theories upon human concepts of nature's order, we cannot yet know whether we have witnessed a mighty gain in insight about the natural world (against anthropocentric hopes and biases that always hold us down), or just another transient blip in the history of correspondence between misperceptions of nature and prevailing social realities of war and uncertainty. Nonetheless, contemporary science has massively substituted notions of indeterminacy, historical contingency, chaos and punctuation for previous convictions about gradual, progressive, predictable determinism. These transitions have occurred in field after field. Punctuated equilibrium, in this light, is only paleontology's contribution to a *Zeitgeist,* and *Zeitgeists,* as (literally) transient ghosts of time, should never be trusted. Thus, in developing punctuated equilibrium, we have either been toadies and panderers to fashion, and therefore destined for history's ashheap, or we had a spark of insight about nature's constitution. Only the punctuational and unpredictable future can tell.

## Appendix: A Largely Sociological (and Fully Partisan) History of the Impact and Critique of Punctuated Equilibrium

### THE ENTRANCE OF PUNCTUATED EQUILIBRIUM INTO COMMON LANGUAGE AND GENERAL CULTURE

As a personal indulgence, after nearly 20 years' work on this book, I wish to present an unabashedly subjective, but in no sense either consciously inaccurate or even incomplete, account of the extra-scientific impact and criticism of punctuated equilibrium during its first quarter century. As extra-scientific, I include both the spread and influence of punctuated equilibrium into non-biological fields and into general culture, and also the subset of opinions voiced by biological colleagues, that, in my judgment, are not based on logical or empirical argument, but rather on personal feelings spanning the gamut from appreciation to bitter jealousy and anger. I realize that such an effort, which I do regard as self-indulgent, may be viewed as unseemly by some colleagues. I would only reply, first—speaking personally—that I have, perhaps, earned the right after so much deprecation (matched or exceeded, to be sure, by a great deal of support); and second—speaking generally—that such an effort may have value for people interested in metacommentary upon science (historians, sociologists, scientific colleagues with an introspective bent), if only because few scientific theories garner so wide a spate of reactions, both popular and professional (and for reasons both worthy and lamentable). Moreover, the comments, while necessarily and admittedly partisan, of an originator of the theory (who has also kept a chronological file, as complete as he could manage, on the developing discussion) might have some worth as primary source material (in contrast with more objective, but secondary, analysis and interpretation). Therefore, I do not, in this section, include any overt discussion of the rich and numerous scientific critiques, issues, extensions, and arguments inspired by punctuated equilibrium. These subjects have already been treated in the main body of this chapter.

Needless to say, density and intensity of discussion bear no necessary correlation with the worth or validity of a subject; after all, many theological phenomena that have provoked wars, filled libraries, and consumed the lives of countless scholars, may not exist at all. Still, if only for naïve reasons, I take a generally hopeful view about human intelligence and discernment—at least to the extent of believing that when large numbers of thoughtful people choose to devote substantial segments of careers to the consideration of a new idea, this expenditure probably records the idea's genuine value and interest, and does not represent a pure snare or delusion.

Thus, above all else, I take pleasure in the perceived and expressed utility of punctuated equilibrium in altering a field that had largely languished in doldrums of little to do (as gradualism had defined the domain of recognized empirics for fine-scale evolution, and very few cases of paleontological gradualism could ever be documented)—and in providing an operational base for fruitful study by showing that the primary empirical signals of stasis and punctuation represented meaningful data on the tempo and mode of evolution, and not just a mocking signal from nature about the discouraging imperfection of the fossil record. The greatest success of punctuated equilibrium lies not in any torrent of words provoked by the theory, but in the volume of empirical study pursued under its aegis by paleontologists throughout the world (see Section IV of this chapter, for an account of this literature).

While this professional debate unfolded in full force, the name and concept of punctuated equilibrium also moved from the scientific literature into general culture, at least on the intellectual edges, but often into more popular consciousness as well. Consider five categories recording this spread:

1. ESTABLISHMENT OF THE NAME IN DICTIONARIES AND LITERATURE. Punctuated equilibrium has won entry to latest editions of standard general dictionaries of the English language, including the Addenda to *Webster's Third New International Dictionary* (1986), where it shares a page with such other neologisms as *psychedelic, psychobabble, pump iron, putz, quark, rabbit ears* (which, as a name for those old indoor TV antennas, will no doubt pass quickly into oblivion along with the item itself), and *race walking*. I con-

gratulate *Webster's* on a better and more accurate definition than many professional colleagues have misdevised for their denigrations. *Webster's* suggests: "a lineage of evolutionary descent characterized by long periods of stability in characteristics of the organism and short periods of rapid change during which new forms appear esp. from small subpopulations of the ancestral form in restricted parts of its geographic range." Punctuated equilibrium also occurs in most alphabetical compendia of scientific terms and concepts, including the *World Information Systems Almanac of Science and Technology* (Golob and Brus, 1990), *The Penguin Dictionary of Biology* (Thain and Hickman, 1990), and the *Oxford Dictionary of Natural History* (Allaby, 1985).

As a further mark of general recognition, several novelists have made casual references to punctuated equilibrium (in works for mass audiences, not arcana for the literati). Stephen King mentions punctuated equilibrium in chapter 30, "Thayer gets weird," of *The Talisman*. The celebrated English novelist John Fowles included the following passage in his novel, *A Maggot*. (Fowles, a distinguished amateur paleontologist himself, also created fiction's outstanding paleontologist, the hero of his novel, *The French Lieutenant's Woman*): "This particular last day of April falls in a year very nearly equidistant from 1689, the culmination of the English Revolution, and 1789, the start of the French; in a sort of dozing solstitial standstill, a stasis of the kind predicted by those today who see all evolution as a punctuated equilibrium." In his 1982 crime novel, *The Man at the Wheel*, Michael Kenyon doesn't cite the name, and does veer towards the common saltational confusion, but presumably wrote this passage in the light of public discussion about punctuated equilibrium. "Now there're biologists saying Darwin got it wrong, or at any rate not wholly right, because evolution isn't slow, continuous change, it's sudden bursts of change after millions of years of nothing, so if the polar bear happened suddenly, why not the world?"

2. LISTING THE THEORY AS AN EVENT IN CHRONOLOGICAL ACCOUNTS OF THE GROWTH OF 20TH CENTURY KNOWLEDGE. Isaac Asimov cited punctuated equilibrium among the seven events of world science chosen to characterize 1972 in his book, *Asimov's Chronology of Science and Technology* (1989). Rensberger (1986) included punctuated equilibrium in his alphabetical compendium *How the World Works: A Guide to Science's Greatest Discoveries*. *In Our Times* (Glennon, 1995), a lavishly illustrated "coffee table" book on the cultural history of the 20th century, used the Darwinian centennial of 1982 for discussing the strength of evolution (contra creationism) and the status of the field. Among three "takes" to mark the year (evolution shared space with acid rain and Madonna's first single recording, while a sidebar list of "new in 1982" includes liposuction and Halcion sleeping pills), the column on evolution bears the title "Darwin refined," and concentrates exclusively on punctuated equilibrium. In a refreshing departure from common journalistic accounts, the epitome is both incisive and generally accurate: ". . . The theory of punctuated equilibria reconciled Darwinism with

paleontological reality . . . The debate could be traced back to Darwin, who'd candidly admitted that gradual evolution did not square with the fossil record. Gould emphasized that Darwinism was 'incomplete, not incorrect.' The theory of punctuated equilibria, however, proved a crucial refinement of Darwinian thought, as well as a useful model for other disciplines from anthropology to political science."

In a long article for *The New Yorker* on the history of the American Museum of Natural History, Traub (1995) emphasized the frustration of staff scientists when public supporters only recognize exhibits and have no inkling that the Museum also operates as a distinguished institute of scientific research. In listing accomplishments from Boaz to Mead in anthropology, and from Osborn to Simpson and Mayr for paleontology and evolution, Traub mentioned only punctuated equilibrium to mark the continuation of this tradition into recent years: "And in the early seventies, Stephen Jay Gould and Niles Eldredge accounted for anomalies in the fossil record by arguing that evolution proceeded not steadily but sporadically—a theory known as 'punctuated equilibria.'"

I acknowledge, of course, the blatant unfairness of this selectivity, especially the legitimate grievances prompted thereby among those who built cladistic theory at the Museum during these same years. I cite this example (with some mixture of embarrassment and, to be honest, a tinge of pride as well) to point out how punctuated equilibrium became, in popular culture, a synecdoche for professional discussion about evolution. I also recognize that this journalistic ploy rightly angers and inspires jealousy among colleagues; I merely point out that Eldredge and I cannot fairly be blamed for this cultural phenomenon. Neither of us ever organized a symposium, or even called a reporter, to discuss punctuated equilibrium in public (and neither of us was interviewed by the author of this *New Yorker* article).

3. INTERNATIONAL SPREAD. Punctuated equilibrium has been prominently discussed in newspaper and magazine articles of nearly all major Western nations—in France as *équilibres intermittents* (Blanc, 1982) or *équilibres ponctués* (the leading newspaper *Le Monde* of May 26, 1982; Devillers and Chaline, 1989; Courtillot, 1995); in Spain and Latin America as *equilibrio interrumpido* (Sequieros, 1981) or *equilibrio punctuado* (Valdecasas and Herreros, 1982; Franco, 1985); in Italy as *equilibri punteggiati* (Salvatori, 1984); and in Germany as *Unterbrochenen Gleichgewichts* (Glaubrecht, 1995). (All these citations come from popular articles about the theory, not from biopic pieces about the authors, or from technical literature.)

Punctuated equilibrium has also been featured in Western but non Indo-European, accounts in Hungary, Finland and Turkey, and in several articles of the non-Western press, notably in India, Japan, Korea and China. The theory even penetrated the strongest of political iron curtains to emerge, on March 21, 1983, as a feature article in Maoist Mainland China's major newspaper *Ren Min Ri Bao (The People's Daily)*. They wrote, in a commentary not notable for accuracy: "Theories against Darwin have taken the opportunity to make their appearances. The most typical of all this is the theory of 'punctuated equilibrium' . . . This theory holds that organic evolution proceeds by leaps and bounds and not through continuous change."

4. TEXTBOOKS. This criterion may be viewed as the most unenviable of all, but when a new idea enters textbooks as a "standard," almost obligatory item (remember that no other written genre ranks as more conservative or more cloned through endless copying and regimentation by publishers' requirements), then we may affirm that the notion has flowed into a cultural mainstream. As I shall document on pages 994–999, punctuated equilibrium has become a standard entry in textbooks at both the college and high school levels in America.

5. AN ITEM IN GENERAL CULTURE. When the National Center for Science Education, America's leading anti-creationist organization, put out two bumper stickers as sardonic comments upon the favored evangelical "Honk if you love Jesus," they chose "Honk if you love Darwin" and "Honk if you understand punctuated equilibrium." (Niles Eldredge tells me that, in his car one day, he became frightened by a persistent honker; when he ventured a sheepish glance, fearing an encounter with a gun, or at least an upraised third finger, he noted only a smile on the other driver's face, and a finger pointing downward to the bumper sticker.) My colleagues may be satirizing punctuated equilibrium as terminological mumbo jumbo, but at least they thought they could raise some money (and some laughs) with this item!

Although not always understood or properly employed (but often, to my surprise and gratification, excellently epitomized and tactfully used), punctuated equilibrium has become a recognized term and concept both in scholarship of widely disparate fields, and in popular culture. I first noted this spread in 1978, a few years before punctuated equilibrium splashed into public recognition, when nationally syndicated columnist Ellen Goodman featured punctuated equilibrium in an op-ed piece entitled "Crisis is a way of life bringing sudden change." I feel that Goodman, then unknown to me but now a respected colleague and friend, captured the essence of punctuated equilibrium's suggestions about the general nature of change, and did so with clarity and insight—a good beginning, not always followed in later commentary. Goodman wrote (in part):

> I am not normally the sort of person who curls up in front of the fire with a good science book. The last time I found Charles Darwin interesting was in "Inherit the Wind." But I was still intrigued by Stephen Ray [sic] Gould's thoughts about evolution . . . [for he] has written about natural change in a way that makes sense out of our current lives and not just out of fossils. Gould thinks Darwin's view of evolution . . . was actually "a philosophy of change, not an indication from nature." he says that "gradualism" was part of the 19th century prejudice in favor of orderliness . . . In that sense, I suppose we are all still Darwinians. How many of us harbor the hope that the change in our lives will be gradual, rather

like being promoted from the seventh grade to the eighth. We would like our lives to be an accumulation of skills and wisdom . . .

[But] people may go through the greatest changes in their lives in the shortest chunks of time. I have known someone who, after years of stagnation, raced through a decade of personal growth in the first year of a new career. I have known others who experienced a generation's worth of change in six post-divorce months . . .

We often underestimate the suddenness, even the randomness, of the change itself. I suppose that our observations are no more colored than Darwin's. We see gradual change, in part, because we go looking for it. We find it because we need it. Our research into the past reflects our fear of the future . . . Natural history is, as [Gould] puts it, "a series of plateaus punctuated by rare and seminal events that shift systems from one level to another." In that way, I suspect, people have a lot in common with rocks.

Punctuated equilibrium has often been the explicit focus of scholarship in distant fields (see pp. 952–967 for a technical discussion with extended examples; I only mention the range of invocations here)—including a lead article by Gans (1987) on "punctuated equilibria and political science" with five commentaries by other scholars and a response by Gans in the journal *Politics and the Life Sciences*; an analysis of my rhetorical style by Lyne and Howe (1986) in the *Quarterly Journal of Speech*; and an exchange between Thompson (1983) and Stidd (1985) in the journal *Philosophy of Science*. However, the general spread of punctuated equilibrium into vernacular culture will not be best illustrated by such explicit treatments (which may be read as didactic efforts to instruct), but rather by more casual comments implying a shared context of presumed understanding before the fact. I therefore present a partial and eclectic list, united only by the chancy criterion that someone called the items to my attention:

1. In economics, the distinguished columnist David Warsh used punctuated equilibrium to illuminate episodic change and long plateaus in the history of markets and prices ("What goes up sometimes levels off," *Boston Globe*, 1990, and a good epitome for stasis *vs.* progressivism), and also to support the general concept of punctuational change at all levels, in a defense of capitalism with the ironic title "Redeeming Karl Marx" (*Boston Globe*, May 3, 1992). In a recent bestselling book, *The Future of Capitalism*, MIT economist Lester Thurow centered his argument upon two concepts borrowed from the evolutionary and geological sciences—punctuated equilibrium and plate tectonics (see further comment on pp. 964–966).

2. In political theory, the scholarly book of Carmines and Stimson (1989) argues for an episodic model of change based on case studies of the New Deal and race relations in America. "Dynamic evolutions," they write (1989, p. 157), "thus represent the political equivalent of biology's punctuated equilibrium."

3. In sociology, Savage and Lombard's (1983) model "of the process of change in the structure of work groups" cites a key prod from punctuated equilibrium:

Social scientists, at least in small-group studies, generally follow the uniformitarians' view. In recent years studies in several fields have led to revisions in arguments about these classic views. Paleobiology . . . continues to provide some of the most specific and convincing of the newer studies. Even though the field is far removed from the study of changes in work groups in South America, it is informative to examine some of them. Writing in 1972 about the fossil record of mollusks, Eldredge and Gould concluded that in the development of a new species "the alternative picture [to gradual and continuous change is] of stasis punctuated by episodic events."

4. In history, Levine (1991) used our term and concept to center his argument about the history of working-class families in an article entitled: "Punctuated Equilibrium: The modernization of the proletarian family in the age of ascendant capitalism."

5. In literary criticism, Moretti (1996) cited punctuated equilibrium to epitomize the history of the epic as a literary genre, the principal subject of his book: "It is an undulating curve; a discontinuous history that soars, then gets stuck. Overall, it is the conception illustrated by Gould and Eldredge with the theory of 'punctuated equilibria'" (Moretti, 1996, p. 75).

6. In art history, Bahn and Vertut (1988) used punctuated equilibrium to refute the standard gradualist and progressivist views of the greatest scholars of Paleolithic cave painting, the Abbé H. Breuil and André Leroi-Gourhan (see further comments on pp. 953–956).

7. In the dubious, but popular, literature of "self help" Connie Gersick's fascinating and thought-provoking work (see pp. 958–959) links individual and organizational growth to patterns of punctuated equilibrium. But her subtlety was badly sandbagged in the news bulletin of the University of California (where she teaches at UCLA) for February 21, 1989: "Gersick likened this transition to a midlife crisis, which, she said, is part of a phenomenon known as 'punctuated equilibrium' . . . For organizations which rely on the results of creative efforts, Gersick notes that understanding the transitions within the creative process can help groups to work more effectively. 'Managers may be able to build more punctuation points into the process.'"

8. In humor (and to restore equilibrium after the last quotation), Weller properly situates punctuated equilibrium between gradualism and true saltationism in his book *Science Made Stupid* (for another example, see Fig. 9-37).

These citations obviously vary greatly in cogency and utility, but they do indicate that punctuated equilibrium has struck a chord of consonance with themes in contemporary culture that many analysts view as central and troubling. Some usages amount to mere misguided metaphorical fluff, but others may direct and focus major critiques. In any case, since people are not stupid (at least not consistently so over such a broad range of disciplines), I must conclude that punctuated equilibrium has something general, perhaps even important, to say.

## AN EPISODIC HISTORY OF PUNCTUATED EQUILIBRIUM

### *Early stages and future contexts*

I have never enjoyed a reputation for modesty, so I believe that the following introductory comments represent a genuine memory, not a bias following a bent of personality. I was proud of our 1972 paper, and of my initiating oral presentation at the 1971 meeting of the Geological Society of America in Washington, D.C. I hoped that punctuated equilibrium would influence the practice of paleontology by showing that the fossil record, read literally, might depict the process of evolution as understood by neontologists, and not only reflect an absence of evidence pervasive and discouraging enough to make the empirical study of macroevolution virtually impossible at fine scale. Among the ordinary run of papers, this goal cannot be called modest, so I maintained some hope for punctuated equilibrium from the start. But I had no premonition about the hubbub that punctuated equilibrium would generate—for two reasons internal to the theory and to professional life, and also for our general inability to know the contingencies of external history.

For the internal reasons, I simply did not grasp, at first, the broader implications of punctuated equilibrium for evolutionary theory, as embodied in our proposals about stasis and the necessary explanation of macroevolutionary pattern by species sorting. I do clearly remember—and this recollection

Poorly punctuated equilibruim.

9-37. A humorous perspective on punctuated equilibrium.

continues to strike me as viscerally eerie—that I felt something significant lurking in the short section on trends (1972, pp. 111–112) that Eldredge had written. I somehow knew that this section included the most important claim in the paper, but I just couldn't articulate why. Second, I never imagined that the paper would generate any readership beyond the small profession of paleontology.

Two incidents reinforce my memory of modest expectations. My father, a brilliant and self-taught man who never had much opportunity for formal education and therefore grasped the logic of arguments far better than the sociological realities of the academy, got excited when he read the 1972 paper in manuscript, and said to me: "this is terrific; this will really make a splash; this will change things." I replied with mild cynicism, and with the distinctive haughtiness of an "overwise" youngster who views his parents as naive, that such a hope could not be fulfilled because so few scientists ever bother to read papers carefully, or to mull over the implications of an argument not rooted exclusively in graphs and tables. As a second example, my former thesis advisor John Imbrie stopped to congratulate me on my "well argued non-Neodarwinian argument about paleontology and evolution" after my original oral presentation in 1971. I appreciated the praise, but remained mystified by why he thought that an argument for operational paleobiology, based on proper scaling of allopatric speciation, could be viewed as theoretically iconoclastic as well.

The early history of punctuated equilibrium unfolded in a fairly conventional manner for ideas that "catch on" within a field. The debate remained pretty much restricted to paleontology (and largely pursued in the new journal founded by the Paleontological Society to publish research in the growing field of evolutionary studies—*Paleobiology*). Theoretical implications received an airing, but most discussion, to our pride and delight, arose from empirical and quantitative studies done explicitly to test the rival claims of gradualism vs. punctuation and stasis in data-rich fossil sequences. Most important were the critical studies of Gingerich (1974, 1976) on putative gradualism in Tertiary mammalian sequences from the western United States. In any case, our hopes for a fruitful unleashing of empirical studies based on new respect for the power and adequacy of the fossil record were surely fulfilled.

Enough data, argument, and misconception as well had accumulated by the summer of 1976 that Eldredge and I decided to write a retrospective and follow-up—a longer article dedicated mostly to the detailed analysis of published data, and appearing in *Paleobiology* under the title: "Punctuated equilibria: the tempo and mode of evolution reconsidered" (Gould and Eldredge, 1977). Meanwhile, we couldn't fail to note that the arguments of punctuated equilibrium, substantially aided by the support and extension of our colleague S. M. Stanley in a widely discussed *PNAS* article of 1975 that introduced the term "species selection" in a modern context (and developed the implications that I had been unable to articulate from our original section on evolutionary trends), were now beginning to attract attention in the larger field of neontological evolutionary studies. Stanley then followed with an important book on macroevolution (1979).

From an isolated South Africa, Elisabeth Vrba published an astonishing paper (1980) that gave an even more cogent and comprehensive voice to the macroevolutionary implications of punctuated equilibrium. (Following British custom from a former colony, she published as E. S. Vrba; Eldredge and I had never heard of her work and didn't even know her gender. The paper burst upon us as a most wonderful surprise.) In 1980, to fulfill an invitation from the editors to celebrate the 5th anniversary of our new journal *Paleobiology,* I then published a general article on the potential reform of evolutionary theory, a pretty modest proposal I thought, but, oh my, did neo-Darwinian hackles rise (see pp. 1002–1004).

At this point, the story becomes more like ordinary history in the crucial sense that predictable components, driven by the internal logic of a system, interact with peculiar contingencies to yield a result that no one could have anticipated. Punctuated equilibrium did begin to receive general commentary in professional journals (with Ridley's 1980 *News and Views* piece for *Nature* as a first example), but I am sure that our theory would never have become such a public spectacle if this interest had not coincided with two other events (or rather one event and a surrounding political context).

In October 1980, Chicago's Field Museum of Natural History held a large international conference on Macroevolution. This meeting, inspired in good part (but by no means entirely, or even mainly) by the developing debate over punctuated equilibrium, would have been a major event in our profession in any case. But the Chicago meeting escalated to become something of a cultural *cause célèbre* because, and quite coincidentally, the symposium occurred at the height of renewed political influence for the creationist movement in America.

This fundamentalist movement, dedicated (as a major political goal) to suppressing the teaching of evolution in America's public schools, had flourished in the early 1920's under the leadership of William Jennings Bryan, had culminated in the famous Scopes trial in Tennessee in 1925, but had then petered out and become relatively inactive, especially following the 1968 Supreme Court decision, *Epperson vs. Arkansas,* that finally overturned the anti-evolution laws of the Scopes era on First Amendment grounds.

But creationism surged again in the 1970's, largely in response to an increasingly conservative political climate, and to the growing political savvy and organizational skills of the evangelical right. Creationists enjoyed a second round of success in the late 1970's, culminating in the passage of "equal time" laws for creationism and evolution in the states of Arkansas and Louisiana. We would eventually win this battle, first by overturning the Arkansas law in early 1982 (see pp. 986–990 for the role of punctuated equilibrium in this trial), and then by securing a resounding Supreme Court victory in 1987 in *Edwards vs. Aguillard.* But, in 1980 as the Chicago meeting unfolded, creationists were enjoying the height of their renewed political influence, and evolutionists were both justly furious and rightly worried.

Even with this temporal conjunction, the Chicago meeting wouldn't have attracted public attention if the press had not been alerted by accidental circumstances (neither the participants nor the organizers invited general journalists to the meeting). At most, reports would have appeared in the *News and Views* sections of *Nature* and *Science,* and professional history might have been tweaked or even altered a bit.

But the general press caught on and grossly misread the forthcoming meeting as a sign of deep trouble in the evolutionary sciences (rather than the fruitful product of a time of unusual interest and theoretical reassessment for a factual basis that no one doubted), and therefore as an indication that creationism might actually represent a genuine alternative, or at least a position that stood to benefit from any perceived confusion among evolutionists.

No single source can be blamed for thus alerting and misinforming the press, but an unfortunate article by James Gorman appeared in the popular magazine *Discover* just a month before the meeting ("The tortoise or the hare," October, 1980), leading with the following confused and unfortunate paragraph:

> Charles Darwin's brilliant theory of evolution, published in 1859, had a stunning impact on scientific and religious thought and forever changed man's perception of himself. Now that hallowed theory is not only under attack by fundamentalist Christians, but is also being questioned by reputable scientists. Among paleontologists, scientists who study the fossil record, there is growing dissent from the prevailing view of Darwinism. Partly as a result of the disagreement among scientists, the fundamentalists are successfully reintroducing creationism into textbooks and schoolrooms across the U.S. In October, a hundred or so scientists from half a dozen different disciplines will gather at Chicago's Field Museum . . .

This misconstruction yielded two unfortunate consequences—first, in inspiring a substantial contingent of the general press to attend the Chicago meeting under the false assumption that these technical proceedings would yield newsworthy stories about the success and status of creationism; and, second, by creating a blatantly false taxonomy that dichotomized natural historians into two categories: true-blue Darwinians *vs.* anyone with any desire to revise anything about pure Darwinism (including the strangest bedfellows of evolutionary revisionists and creationist ignoramuses). We must never doubt the potency of such false taxonomies, especially when promulgated by a general press that grasps the true issues poorly, and also plays to an audience too prone to read any dispute as a dichotomous pairing of good and evil. (Consider, for example, the harm done when scientific fraud, the worst of conscious betrayals for all we hold dear as a profession, gets linked with scientific error, a correctable and unavoidable consequence of any boldness in inquiry, because both lead to false conclusions. The pairing of punctuated equilibrium and creationism because both deny pure Darwinian gradualism, falls into the same category.)

The Chicago meeting also produced many good and responsible commentaries in the general press (Rensberger in the *New York Times*, November 4, 1980, for example) and in professional journals (Lewin in *Science* for November 21, 1980, for example). But some very bad accounts also appeared, especially unfortunate in their linkage of success for punctuated equilibrium with the spread of creationism. For example, a lead article in *Newsweek* (November 2, 1980), perhaps the most widely read of all reports, did properly brand the link as a confusion, and also stated that punctuated equilibrium represents a revision, not a refutation, of evolution, but such passing "subtleties" can easily be missed when subjects become so tightly juxtaposed, as in the *Newsweek* story: "At a conference in mid-October at Chicago's Field Museum of Natural History, the majority of 160 of the world's top paleontologists, anatomists, evolutionary geneticists and developmental biologists supported some form of this theory of 'punctuated equilibria.' While the scientists have been refining the theory of evolution in the past decade, some nonscientists have been spreading anew the gospel of creationism."

This kind of reporting kindled the understandable wrath of orthodox Darwinians and champions of the Modern Synthesis. They became justifiably infuriated by two outrageous claims, both falsely linked to punctuated equilibrium by some press reports. First, some absurdly hyped popular accounts simply proclaimed the death of Darwinism (with punctuated equilibrium as the primary assassin), rather than reporting the more accurate but less arresting news about extensions and partial revisions. For example, the same *Newsweek* article stated that "some scientists are still fighting a rear-guard action on behalf of Darwinism," and "it is no wonder that scientists part reluctantly with Darwin." Moreover, even the best and most balanced articles often carry exaggerated and distorted headlines (most scientists, I suspect, don't know that reporters are not generally permitted to write their own headlines). Boyce Rensberger's *New York Times* story on the Chicago conference could not have been more fair or accurate, but the hyped headline proclaimed: "Recent studies spark revolution in interpretation of evolution." Since the article focussed on punctuated equilibrium, some colleagues then blamed Eldredge and me for an exaggeration promulgated neither by ourselves nor by the reporter.

Second, since punctuated equilibrium had served as the most general and accessible topic among the many questions debated at the Chicago Macroevolution meeting, our theory became the public symbol and stalking horse for all debate within evolutionary theory. Moreover, since popular impression now falsely linked the supposed "trouble" within evolutionary theory to the rise of creationism, some intemperate colleagues began to blame Eldredge and me for the growing strength of creationism! Thus, we stood falsely accused by some colleagues both for dishonestly exaggerating our theory to proclaim the death of Darwin (presumably in our own cynical quest for fame), and for unwittingly fostering the scourge of creationism as well. I believe that the strong feelings generated by punctuated equilibrium ever since cannot be divorced from this unfortunate historical context. (I also believe, of course, that the intense interest—as opposed to those intense feelings—arises largely from the challenging intellectual content of the theory itself.)

I don't want to advance the exculpating argument that all unfortunate parts of this debate can be traced to purely external and unpreventable press hyping, and that we and our colleagues, in arguing both for and against punctuated equilibrium, have always walked Simon pure on the intellectual high road. I will discuss on pages 1010–1012 the extent to which our own actions may have contributed to the unseemly side of the discussion. But I do maintain that this truly uncontrollable external context set the primary reason for extended and unwarranted emotionality over the subjects of punctuated equilibrium and macroevolutionary challenges to conventional Darwinism.

Meanwhile, in a simultaneous unfolding of the tragedy and the farce (in contrast with the famous epigram that historical tragedies generally experience *later* replay as farces), a truly risible episode of intense public discussion about punctuated equilibrium erupted in England. American creationism may not rank as a full tragedy, although any suppression of a cardinal subject in public schools surely qualifies as an academic equivalent of murder. By contrast, the great British Museum debate can only be viewed as comical. In an epitome that risks caricature—although the full story veered as close to pure absurdity, and therefore to unalloyed comedy, as anything I have ever witnessed in the sociology of science—the British Museum (Natural History) opened a new exhibit on dinosaurs, based almost exclusively on the rigid cladism espoused by the Museum curators. Beverly Halstead—a man who might be judged as utterly infuriating and even cruelly meddling had he not been so charming and so personally warm and generous—hated these exhibits with all his heart, for Beverly was an unabashed Simpsonian and a devotee of adaptationist biology. So Beverly, following a uniquely British tradition for generating tempests in teapots by inflationary prose fashioned of pure bombast—just where do you think that Blake's famous lines about seeing the world "in a grain of sand" and heaven "in a wild flower" came from?—decided to float the following blessed absurdity, a guarantee of public attention rather than instant burial, in the letters column of the *Times*. He accused—and I swear that I do not exaggerate—the British Museum of foisting Marxism upon an unwitting public in this new exhibit, because cladism can be equated with punctuated equilibrium, and everyone knows that punctuated equilibrium, by advocating the orthodoxy of revolutionary change, represents a Marxist plot.

Well, the press bit, and a glorious volley of ever more orotund letters appeared, both in the general press and in the professional pages of *Science, Nature,* and the *New Scientist.* Since I don't wish to prolong discussion of this peculiar byway (I doubt that any of the Museum curators had any abiding interest in politics beyond the academy, or personally stood one inch to the left of Harold Wilson), and since this chapter represents my own partisan account, let me simply reproduce my own letter to *Nature*—part of the final volley, just before the editors wisely and forcefully cut off all further fulmination forthwith:

Sir—I have been following the "great museum debate" in your pages with a profound sense of detached amusement. But as matters are quickly reaching a level of absurdity that may inspire me to write the 15th Gilbert and Sullivan opera, and as I am, in a sense, the focal point for Halstead's glorious uproarious misunderstanding, I suppose I should have my say.

Halstead began all this by charging that the venerable Natural History Museum is now purveying Marxist ideology by presenting cladism in its exhibition halls. The charge is based on two contentions: (1) a supposed link between the theory of punctuated equilibrium, proposed by Niles Eldredge and myself, and cladistic philosophies of classification; and (2) an argument, simply silly beyond words, that punctuated equilibrium, because it advocates rapid changes in evolution, is a Marxist plot. For the first, there is no necessary link unless I am an inconsistent fool; for I, the co-author of punctuated equilibrium, am not a cladist (and Eldredge, by the way, is not a Marxist, whatever that label means, as if it mattered). Under cladism, branching events may proceed as slowly as the imperceptible phyletic transitions advocated by the old school. Punctuated equilibrium does accept branching as the primary mode of evolution, but it is, fundamentally, a theory about the characteristic rate of such branching—an issue which cladism does not address.

For Halstead's second charge, I did not develop the theory of punctuated equilibrium as part of a sinister plot to foment world revolution, but rather as an attempt to resolve the oldest empirical dilemma impeding an integration of paleontology into modern evolutionary thought: the phenomena of stasis within successful fossil species, and abrupt replacement by descendants. I did briefly discuss the congeniality of punctuational change and Marxist thought (*Paleobiology*, 1977) but only to illustrate that all science, as historians know so well and scientists hate to admit, is socially embedded. I couldn't very well charge that gradualists reflected the politics of their time and then claim that I had discovered unsullied truth . . .

I saw the cladistic exhibits last December. I did not care for them. I found them one-sided and simplistic, but surely not evil or nefarious. I also felt, as a Victorian aficionado who pays homage to St. Pancras [a wonderously ornate late nineteenth century railroad station] on every visit to London, that most of the newer exhibits are working against, rather than with, the magnificent interior that houses them. But I would not envelop these complaints in ideological hyperbole; Halstead has said enough.

We can best explore the consequences of this historically contingent context by examining the use of punctuated equilibrium in two domains that, in their contrast, span the range of public influence from the ridiculous to the potentially sublime: the political propaganda of creationist literature, and the developing treatment of punctuated equilibrium in journalism, and high school and college textbooks of biology.

### Creationist misappropriation of punctuated equilibrium

Since modern creationists, particularly the "young earth" dogmatists who must cram an entire geological record into the few thousand years of a literal Biblical chronology, can advance no conceivable argument in the domain of proper logic or accurate empirics, they have always relied, as a primary strategy, upon the misquotation of scientific sources. They have shamelessly distorted all major evolutionists in their behalf, including the most committed gradualists of the Modern Synthesis (their appropriations of Dobzhansky and Simpson make particularly amusing reading). Since punctuated equilibrium provides an even easier target for this form of intellectual dishonesty (or crass stupidity if a charge of dishonesty grants them too much acumen), no one should be surprised that our views have become grist for their mills and skills of distortion. I have been told that Duane Gish, their leading propagandist, refers to his compendium of partial and distorted quotations from my work as his "Goulden file."

Standard creationist literature on punctuated equilibrium rarely goes beyond the continuous recycling of two false characterizations: the conflation of punctuated equilibrium with the true saltationism of Goldschmidt's hopeful monsters, and the misscaling of punctuated equilibrium's genuine breaks between species to the claim that no intermediates exist for the largest morphological transitions between classes and phyla. I regard the latter distortion as particularly egregious because we formulated punctuated equilibrium as a positive theory about the nature of intermediacy in such large-scale structural trends—the "stairstep" rather than the "ball-up-the-inclined-plane" model, if you will. Moreover, I have written numerous essays in my popular series, spanning ten printed volumes, on the documentation of this style of intermediacy in a variety of lineages, including the transition to terrestriality in vertebrates, the origin of birds, and the evolution of mammals, whales and humans—the very cases that the usual creationist literature has proclaimed impossible.

To choose a standard example by the movement's "heavies" (Bliss, Parker and Gish, 1980, p. 60), the following text embodies the first standard error, while their accompanying illustration (Fig. 9-38) records the second error by equating punctuated equilibrium with the saltational origin of each vertebrate class (if anyone has any lingering doubt about the pseudoscientific character of this movement, try to make any sense at all of this figure, a supposed expression of their proper practice of the graphical and quantitative approach to science): "Gould and Eldredge state that fossils, like living forms, vary only mildly around the average or 'equilibrium' for each kind. But, they say, the appearance of a 'hopeful monster' can interrupt or 'punctuate' this equilibrium. According to the new concept of 'punctuated equilibrium,' fossils are not supposed to show in-between forms. The new forms appeared suddenly, in large steps."

We may at least label creationist Everett Williams as timely in adding the insult of misnaming to the injury of the same distortion in a 1980 newspaper column: "The latest version of the process is called 'punctual [sic] evolution.' In this version, evolution is seen as moving in giant surges and then becoming stagnant for eons."

A broadsheet from Hillsborough, North Carolina, entitled "Harvard scientists agree: Evolution is a hoax!!!" goes whole hog in assimilating us to its own version of the rock of (small) ages: "The facts of 'punctuated equilibrium' which Gould, Eldredge, Stanley and other top biologists are forcing the Darwinists to swallow fit the picture that Bryan insisted on, and which God has revealed to us in the Bible. Every species of organism was separately created during the six 'days' of creation . . . This is the doctrine taught by Scripture and by Cuvier (the father of paleontology) alike, and modern biology is forcing the Darwinists to accept it."

In *The Genesis Connection*, J. L. Wiester commits the same error of scaling (in maximal degree this time), and then also cites us as hidden supporters of his one true way: "The theory of punctuated equilibrium holds that life did not evolve in the slow uniform method that Darwin envisioned but rather in rapid evolutionary bursts of major change called adaptive radiations. The Cambrian explosion of marine life was such an adaptive radiation . . . The new theory of punctuated equilibrium brings the thinking of science remarkably closer to the biblical view. It is notable that the more evidence scientists discover (or fail to discover), the closer scientific theory moves toward the unchanging biblical pattern."

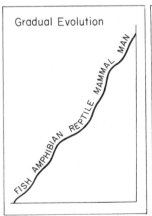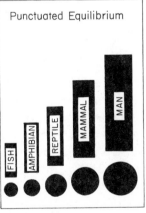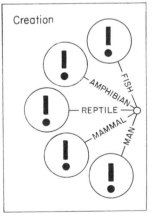

9-38. Creationist distortion of punctuated equilibrium modified from original in "Fossils: key to the present" by Bliss, Parker, and Gish, 1980. They misdepict punctuated equilibrium as a saltationist theory, with all vertebrate classes arising in single steps, all at the same time.

If observers ever hoped for more accuracy or fairness from "official" publications of the two largest creationist organizations in America (in contrast with the "independents" previously cited), they shall be disappointed. The Jehovah's Witnesses journal *Awake!* reported on the Chicago macroevolution meeting in its issue for September 22, 1981, rooting the story in error number one, the Goldschmidt equation: "This revised view of evolution is called 'punctuated equilibrium,' meaning one species remains for millions of years in the fossil record, suddenly disappears and a new species just as suddenly appears in the record. This, however, is not really a new proposal. Richard Goldschmidt advanced it in the 1930's, called it the 'hopeful monsters' hypothesis, and was much maligned for it. 'Punctuated equilibrium' is a much more impressive designation."

Writing in the September 1982 issue of *Signs,* the leading journal of the Seventh Day Adventists, H. W. Clark discussed the Arkansas creationism trial in terms of error number two, false upward scaling to denial of transitional forms between classes and phyla. Clark equated our punctuations with faunal breaks between geological periods. A sidebar then misnames the author as well as distorting the theory: "Thank you, Dr. Jay Gould: Dr. Jay Gould is the distinguished Harvard Paleontologist who has raised a storm in evolutionary circles with his new theory of 'punctuated equilibrium.' Without intending to do so, he has told the scientific world that Darwin was wrong and the creationists are right. Not that he planned to, of course! Darwin recognized that the theory of evolution needs an unbroken line of gradually changing fossils. Now along comes Dr. Jay Gould and agrees with the creationists: the missing links aren't there and never were. Thank you, Dr. Gould!" You're quite welcome.

On the same theme of shoddiness in supposed creationist "scholarship," I was quite struck by a photograph, supposedly of me, that appeared in M. Bailey's creationist book for children (Greenhaven Press, 1990), *Evolution: Opposing Viewpoints.* The gentleman depicted sports a flowing beard and bald pate—while my head hair has a precisely opposite distribution. He is also considerably older (and, I fancy, a good deal uglier) than I. I finally realized that he is the 19th century robber-baron Jay Gould (no relation, by the way).

While America deals almost exclusively with creationists of the Protestant fundamentalist line (at least among the movement's chief political activists), other religions have their own similar crosses to bear. I was sent a 1983 Hindu version by one Satyaraja dasa entitled "Puncturing the jerk theory." An article by Barbara Sofer in a recent issue of *Hadassah Magazine* reports on the rare phenomenon of Israeli creationism, and cites one adherent: "Schroeder points out that the newer theories of punctuated evolution come closer to the biblical description."

Such inane and basically harmless perorations may boil the blood, but creationist attempts to use punctuated equilibrium in their campaigns for suppressing the teaching of evolution raise genuine worries. Nonetheless, here we can fight back directly—and we have always won. Elijah, after all, taught us how to fight fire (or rather the inability of reprobates to kindle any real flame) with fire—and the splendid man cited by creationists as their own primary hero did promise that truth would make us free.

In the Texas textbook hearings of 1984, for example, Mel and Norma Gabler, the infamous professional propagandists for forcing a right-wing evangelical agenda into textbooks, lobbied for four imposed changes in any evolutionary passages within biology texts. One of the items rested upon the second standard creationist mischaracterization of punctuated equilibrium: "There are systematic gaps in the fossil record, showing absences of inter-mediate links. Punctuated equilibrium was devised to explain these gaps. Therefore presentation of evolutionary lineages, such as from amphibians to reptiles to birds and mammals, cannot be supported with evidence. The text-books should be revised to reflect this understanding."

But our side holds a strong weapon in such public cases, for we can also testify, and therefore expose. We have never failed in these circumstances. The Gablers' proposal lost and the State of Texas endorsed good biology books.

Given the power of the First Amendment, and the fairness and intellectual stature of federal judge William Overton, our success in overturning the Ar-kansas equal time law, in a trial held in Little Rock in late 1981, seemed as-sured. I can only regard my own role, as one of six expert witnesses for sci-ence and religion, as both minor and probably irrelevant to the inevitable decision. But I was able to speak for paleontology and to add our unique tem-poral perspective to the documentation.

Clarence Darrow's scientific witnesses were not permitted to testify in the Scopes Trial of 1925. The Louisiana creationism law, the only other statute passed in modern times by a state legislature, and a virtual copy of the Arkan-sas law, was dismissed by summary judgment following our success in Arkan-sas, and was therefore never tried in court. The State of Louisiana appealed this case to the Supreme Court, where oral argument consumes but one hour, and only the principal lawyers may testify, with no witnesses called. There-fore, for the first and only time in American history, the Arkansas trial per-mitted full scale testimony about creationism in a court of law. I feel honored that I had the opportunity to help present the case for evolution as natural knowledge, and for creationism as pseudoscience, in the only legal venue ever provided to experts in the relevant professions throughout this long and im-portant episode in 20th century American history.

My testimony and cross examination at the Arkansas trial lasted for the better part of a day, and focussed upon two subjects: the absurdity of attrib-uting the entire fossil record to the single incident of Noah's flood (a favored creationist ploy for cramming the entire geological history of the earth into a mere 6000 years or so), and creationist pseudoscientific practice as illustrated by their clearly willful distortions of the theory of punctuated equilibrium. (We did not, in this trial, try to "prove" evolution—a subject scarcely in need of such treatment, and not for a court of law to adjudicate in any case—but only to expose creationism as a narrow form of dogmatic religion, masquerading as science in an attempt to subvert First Amendment guarantees against the establishment of religion in public institutions.)

Creationists continue to distort punctuated equilibrium, but we continue to win by exposing them in fair forums. For example, in 1997, Rep. Russell Capps of the North Carolina General Assembly used a "standard" misquotation from one of my essays about punctuated equilibrium in arguing before the legislature for a law that would ban the teaching of evolution as a fact (although teachers could still present the subject as a hypothesis). I suspect that Capps simply lifted the quote from Duane Gish's *Evolution: The Fossils Say No!* (I do love that title!) and never read my essay, because his version used exactly the same deletions as Gish's. Rep. Bob Hensley, an opponent of the bill, asked for my aid, and I wrote a letter, which he read to the assembly, detailing this dishonest treatment of my writing. I stated, in part (letter of April 4, 1997):

> [My] article is not an attack on evolution at all, but an attempt to explain how evolution, properly interpreted, yields the results that we actually see in the fossil record. The first part of the quotation is accurate, but about rates of change, not whether or not evolution occurs. The second part of the quotation after the three dots—"it was never seen in the rocks"—seems then to deny that evolution occurs. But if you read my full text and look for the material left out, it is obvious that my word "it" refers to gradualism as a style of evolutionary change, and not to evolution itself. If one reads the rest of the essay, the intent is abundantly clear. For example, I state on page 182: "The modern theory of evolution does not require gradual change. In fact, the operation of Darwinian processes should yield exactly what we see in the fossil record. It is gradualism that we must reject, not Darwinism." . . . Thus you can see that my essay actually says exactly the opposite of the false quotation cited by your colleague. This is so typical of the intellectual level of most creationist literature. Do we really want our students to be taught by this form of dishonest argument?

The counterattack succeeded. Rep. Hensley wrote to me on April 21, 1997: "Because of your efforts, the Bill has now been withdrawn from consideration in the House Education Committee." Shabby and dishonest argument can win a fragile and transient advantage, but so long as we fight back, we will win. God (who, as a self-respecting deity, must honor and embrace empirical truth) really is on our side.

### Punctuated equilibrium in journalism and textbooks

All scientists have read egregiously bad, hyped and distorted press commentaries about the more subtle and nuanced work of their field. I too get annoyed at such stories, but I have also learned to appreciate that most journalists take their job seriously, follow the ethics of the field, and tend to turn out good stories, on balance. When hype occurs, the fault lies just as often with scientists who simplify and overpromote their work, as with reporters who accept what they hear too uncritically. (Journalists should check, of course, and must therefore bear part of the blame, but scientists should begin any general critique of the press with an acknowledgment of our own *in camera* foibles.)

The extensive press coverage of punctuated equilibrium has generally maintained adequate to high quality. Ironically, though, the most common errors—which like the old soldiers, cats and bad pennies of our mottoes, never seem to fade but turn up, however sparsely, again and again with no diminution in frequency—match the mistakes cited by creationists for utterly different purposes. If willful misuse and unintentional, albeit careless, error repeat the same false arguments, then what serves as a common source amidst such different motives (and different frequencies of occurrence, of course—pervasive for creationists, rare for reputable journalists)? Deep constraints on human mentality (common difficulties with concepts of scaling and probability, for example)? Persistent historical and cultural prejudices (about progress and gradualism, for example)? The malfeasance and hyped misleading of original authors (as our severest critics like to claim)? In any case, I am fascinated by the entire issue of commonality in errors across such a maximal range of motives, and I believe that something deep about the nature of mentality and the sociology of knowledge lies exposed therein.

Schemes of oversimplification must rank as the *bête noire* of journalism, at least in the eyes of scientists and other scholars. Since dichotomization stands as our primary mode of taxonomic oversimplification, probably imposed by the deep structure of the human mind, we should not be surprised that journalists have tended to treat the punctuated equilibrium debate as a dichotomous struggle between gradualists and punctuationalists, superimposed upon another false dichotomy (with supposedly perfect mapping between the two) of Darwinians (read gradualists) against anti-Darwinians (read punctuationalists). This struggle then occurs within a political dichotomy—a genuine division this time—of evolution *vs.* creationism. (The misappropriation of punctuated equilibrium by creationists, as documented in the last section, violates this last dichotomy and can thus be easily grasped as unfair by nearly everyone.)

The error of dichotomy appears most starkly in the minimal length and maximal hype of advertising copy for books. Pergamon's come-on for Nield and Tucker's *Paleontology*, for example, promises that "the approach in the evolutionary discussions is fully in line with the most recent understandings of the punctuated equilibrium/phyletic gradualism debate." The blurb for Oliver Mayo's *Natural Selection and Its Constraints* proclaims: "Among other topical matters, he touches upon the controversial question of 'punctuated equilibrium' or 'phyletic gradualism' as a mechanism for major evolutionary change."

A prominent cultural legend (with "The Emperor's New Clothes" as a prototype) celebrates the young and honest naif as exposer of an evident truth that hidebound adults will not or cannot admit. True to this scenario, the Summer 1993 publicity blurb sheet of Mount Holyoke College reports the happy story of Heather Winklemann, a senior who had just won a prestigious Marshall Fellowship for graduate study in England. The article focuses on her intimidating but successful oral interview, held in San Francisco. As a prospective paleontologist, the committee asked her: "Does evolution work by punctuated equilibrium? Answer yes or no?" Ms. Winklemann replied cogently by exposing the dichotomy as false—and she got her fellowship. The article ends: "'That question took me by surprise,' Winkelmann recalls, 'because if you know anything about the topic you know it can't be answered with a yes or a no. Were they trying to catch me on the question? I told the committee that I couldn't give a yes-or-no answer and why.' Heather Winklemann's answer evidently was what the committee was hoping to hear."

Beyond dichotomy, a failure to recognize the theory's proper scale stands as the most common journalistic error about punctuated equilibrium, in accounts both positive and negative. Many reporters continue to regard yearly or generational changes in populations as a crucial test for punctuated equilibrium. Thus, Keith Hindley reported the fascinating work of Peter Grant and colleagues on changes in population means for species of Darwin's Galápagos finches following widespread mortality due to extreme climatic stresses. Hindley placed the entire story in the irrelevant light of punctuated equilibrium (which cannot even "see" such transient fluctuations in population means from year to year): "Striking new evidence has refuelled the heated scientific debate about the process of evolution . . . The followers of Stephen Gould of Harvard claim that such rapid changes or 'jumps,' caused by environmental pressures, are the key to the emergence of new species . . . This episode has provided Gould's supporters with some of the ammunition their theory has so far lacked: good examples of sudden evolution among species alive today."

Negative accounts of punctuated equilibrium often make the same error. In reviewing a book by Ernst Mayr in the *New York Times,* Princeton biologist J. L. Gould (no relation) discusses the link of punctuated equilibrium to Mayr's views on allopatric speciation. But he then attacks punctuated equilibrium because "its authors seem to believe that species-level changes can occur in one generation, presumably by the production of what the embryologist Richard Goldschmidt called 'hopeful monsters.'"

Among human foibles, our tendency to excoriate a bad job in public, but merely to smile in private at good work, imposes a marked asymmetry upon the overt reporting of relative frequencies in human conduct and intellect. In truth, although I have singled out some "howlers" for quotation in this section, most press reports of punctuated equilibrium have been accurate, while a few have been outstanding. Consequently, I close this section on punctuated equilibrium and the press with two extensive quotations from two leading science writers, one British and one American—with thanks for confirming my faith in the coherence and accessibility of the ideas and implications of punctuated equilibrium. In *The Listener* (magazine of the BBC) for July 19, 1986, Colin Tudge beautifully explains the key concepts and general reforms proposed by punctuated equilibrium, while also giving the critics their due:

A third modification of the neo-Darwinian orthodoxy is embraced in the hypothesis of punctuated equilibrium, proposed in the early 1970's by the American biologists Niles Eldredge and Stephen Jay Gould. The idea of punctuated equilibrium is not intended to dispute Darwin's central notion that the evolutionary destinies of plants and animals are shaped largely or mostly by natural selection. But it does take issue with two of his subsidiary notions: the idea that evolutionary change brought about by natural selection is necessarily gradual; and the idea that natural selection can operate only at the level of the individual.

No idea in biology has caused more contention and indeed rancor over the past 15 years. Some opponents of Gould and Eldredge argue that their observation is just plain wrong—that evolution *is* gradual. Some argue that even if it were true it would be trivial. And some suggest that even if it were true and not trivial, then it is in any case untestable, and therefore not worth considering.

In truth, the paleontological record sometimes seems to show that one form of animal may gradually turn into another, in Darwinian fashion, but often it seems to show precisely the pattern that Gould and Eldredge propose. . . .

It's at this point that some biologists say "So what?" Who ever doubted that evolution can at times proceed more quickly than at others? Even if true (in some cases), the observation is trivial. This, however, is a severe misrepresentation of Gould and Eldredge's idea, for they are not simply making the banal observation that evolution is sometimes fast and sometimes slow. They are suggesting that the "jumps" that can be observed in the fossil record represent the emergence of new species—that is, of groups of organisms that reproduce sexually with each other but not with other groups. . . .

Indeed, Gould and Eldredge go further than this. They suggest that when a species divides to form several new species, this is analogous to the birth of new individuals; and just as natural selection tends to weed out weak individuals in favor of the strong, so it serves to weed out new experimental species. Thus, they suggest, natural selection can operate at the level of the species ("species selection") and not simply at the level of the individual, as Darwin proposed. This is not a trivial observation. . . .

The attacks on punctuated equilibrium seem powerful. But Gould gives as good as he gets, and my own betting is that the theory of punctuated equilibrium, with a bit more buffering from biologists at large, will take its place as an important modification of Darwin's basic ideas.

In an article on Peter Sheldon's claims for extensive gradualism in trilobites, and therefore generally critical of punctuated equilibrium, James Gleick states that our theory has provoked "the most passionate debate in evolutionary theory over the last decade," and then provides a fine summary of our key ideas, and of the intellectual depth of the resulting debate (*New York Times*, December 22, 1987):

Steady flow or fits and starts—the division between these conceptions of evolution has dominated the debate over evolutionary theory. The punctuated equilibrium model has stimulated much research and drawn many adherents. Some of its central notions have taken firm hold.

Even the most traditional Darwinians, for example, acknowledge that punctuated equilibrium has become an important part of the picture of evolution. Some species do little of evolutionary interest for millions of years at a time. . . .

But the debate continues to rage, because it concerns far more than speed itself. At stake are the fundamental questions of evolution: when and why does a creature change from one form to another? Is most evolution the slow, unceasing accumulation of the small changes a geneticist sees in laboratory fruit flies, or does it occur in episodes, when a small population, perhaps isolated geographically, suddenly changes enough to give rise to a new species?

Suddenly, in paleontological terms, can mean hundreds of thousands of years. . . . Proponents of punctuated equilibrium take pains to stress that such events rely mainly on the Darwinian principles of natural selection among individuals varying randomly from one another. Even so, to some biologists, punctuated equilibrium seems like a resort to some process apart from the usual rules—"mutations that appear to be magic," Dr. Maynard Smith said.

"They have argued that their results mean that evolution as seen on the large scale is not just the summing up of small events," he said, "but a series of quite special things that people like me"—population geneticists—"don't see. We don't want to be written out of the script."

The movement of scientific ideas into textbooks may provide our best insight into social forces that direct the passage from maximal professional independence into the most conservative of print genres. To be successful, textbooks must sell large numbers of copies to audiences highly constrained by set curricula, teachers who hesitate to revise courses and lessons substantially, and conservative communities that shun scholastic novelty. These external reasons reinforce the internal propensities of publishers who are happy to jazz up or dumb down, but not to innovate, and authors who experience great pressure to follow the conventions of textbook cloning, and not to depart from the standard takes, examples, illustrations, and sequences. Did you ever see a high school biology textbook that doesn't start the evolution chapter with Lamarck's errors, Darwin's truths, and giraffes' necks in that order?

In this context, I delight in the rapid passage of punctuated equilibrium from professional debate to nearly obligatory treatment in the evolution chapter of biology textbooks. I could put a cynical spin on this phenomenon, but prefer an interpretation, in my admittedly partisan manner, based on the successful ontogeny of punctuated equilibrium from a controversial idea to a firm item of natural knowledge, however undecided the issues of relative frequency and importance remain.

But I am also not surprised that textbooks encourage promulgation of standard errors—a tendency arising from pressures to simplify ideas, downplay controversy, favor bland consensus, and generate a fairly uniform treatment from text to text. We often encounter, for example, the same oversimplification by dichotomy that compromises so many press reports. Villee and collaborators (1989) state, for example: "Some scientists believe that evolution is a gradual process, while others think evolution occurs in a series of rapid changes." The headings of entire sections often bear this burden, as in Tamarin's (1986) title for his pages on evolutionary rates: "Phyletic Gradualism Versus Punctuated Equilibrium."

However, the bland consensus favored by textbooks (and euphemistically called "balance") often imposes a peculiar resolution foreign to most journalistic accounts, where controversy tends to be exaggerated rather than defanged to a weak and toothless smile of agreement at a meaningless center. Textbooks therefore tend to present the dichotomy and then to state that "I am right and you are right and everything is quite correct," to quote Pish-Tush in *The Mikado*—as average reality rests upon the blandest version of a meaningless golden mean. The 1996 edition of J. L. Gould and W. T. Keeton proclaims (p. 511) that "the usual tempo of speciation probably lies somewhere between the gradual-change and the punctuated equilibrium models." (But such a various phenomenon as speciation has no "usual tempo," or any single meaningful measure of central tendency at all. Blandness, in this case, reduces to incoherence.)

In another example, Levin (1991, p. 112) concludes with pure textbook boilerplate that could be glued over almost any scientific controversy: "The final chapter on the question of punctuated evolution versus phyletic gradualism has not been written. At present, the proponents of punctuated evolution appear to be more numerous than those of phyletic gradualism. Like most controversies in science, however, the answer need not lie totally in one camp, and it is evident that instances of phyletic gradualism can also be recognized in the fossil record of certain groups of plants and animals."

If we consider dichotomy as a general mental error of oversimplified organizational logic, then the most common scientific fallacy in textbook accounts of punctuated equilibrium resides, once again, in false scaling by application of the theory to levels either below or above the appropriate subject of speciation in geological time. As before, the conflation of punctuated equilibrium (speciation in geological moments) with true saltation (speciation in a single generation, or moment of human perception) persists as the greatest of all scaling errors. I am discouraged by this error for three basic reasons: (1) It has been exposed and explained so many times, both by the authors of punctuated equilibrium and by many others; so continued propagation can only record carelessness. (2) Saltation at any appreciable relative frequency surely represents a false theory, so punctuated equilibrium becomes tied to a patently erroneous idea; whereas misapplication of punctuated equilibrium to higher levels may at least misassociate the name with a true phenomenon (like catastrophic mass extinction). (3) This particular error of scaling embodies our worst mental habit of interpreting other ranges of size, or other domains of time, in our own limited terms.

For example, Mettler, Gregg, and Schaffer's textbook on *Population Genetics and Evolution* (1988, p. 304) states: "The punctuated equilibrium theory, on the other hand, holds that sudden appearance is due to rapid selection, rather than rapid spread, and that stasis results because evolutionary change occurs in large discrete jumps rather than by a series of gene substitutions. There really are no gradual changes or intermediate stages." In their volume on *Sexual Selection* for the prestigious *Scientific American* series (1989, p. 83), Gould and Gould (no relation) write: "The proven ability of selection to operate quickly in at least some cases, has led to the widely publicized theory of punctuated evolution. According to the original version, no intermediate forms are preserved simply because there are no halfway creatures in the first place: new species come into being in single steps."

Turning to misscaling in the other direction, Wessells and Hopson (1988, pp. 1073–1074) equate punctuated equilibrium with the origin of new *Baupläne* and faunal turnovers in mass extinction: "The central tenet of punctuated equilibrium is that a lineage of organisms arises by some dramatic changes—say, the rapid acquisition of body segmentation in annelids—after which there is a lengthy period with far fewer radical changes taking place." They then write of two great evolutionary bursts in the history of sea urchins (following the late Cambrian and Late Triassic mass extinctions). "One might interpret this record to reflect two 'punctuations' in the Ordovician and early Jurassic periods. And the 'equilibrium' times would be from the Ordovician through the Triassic and, perhaps, from the Jurassic to today. This record may be consistent with the punctuated equilibrium hypothesis."

Chaisson's ambitious textbook on nearly everything—*Universe: An Evolutionary Approach to Astronomy*—equates punctuated equilibrium with faunal turnovers in mass extinction. His section entitled punctuated equilibrium (1988, p. 481) begins by stating: "The fossil record of the history of life on Earth clearly documents many periods of mass extinction." He then adds (p. 483): "Punctuated equilibrium merely emphasizes that the rate of evolutionary change is not gradual. Instead, the 'motor of evolution' occasionally speeds up during periods of dramatic environmental change—such as cometary impacts, reversals of Earth's magnetism, and the like. We might say that evolution is imperceptibly gradual most of the time and shockingly sudden some of the time." But Chaisson's "imperceptibly gradual" times—the intervals of so-called "normal" evolution between episodes of mass extinction—build their incremental trends by stair steps based on the true rhythm of punctuated equilibrium in rapid origin and subsequent stasis of individual species.

However, even in this maximally constrained and conservative world of textbooks, some reform has emerged from punctuated equilibrium. Above all, the debate on punctuated equilibrium prodded the authors of nearly all major textbooks to include (often as entirely new sections) substantial and explicit material on macroevolution—in contrast with the appalling absence or shortest shrift awarded to the topic in standard textbooks of the 1950's and 1960's (as documented in Chapter 7, pages 579–584).

At the level of details and content, many textbooks provide gratifyingly accurate (if often critical) definitions and appraisals of punctuated equilibrium. Unsurprisingly, textbooks written by paleontologists have generally provided the clearest treatments. Nield and Tucker (1985, p. 162), for example, stress the role of punctuated equilibrium in rendering the fossil record operational for evolutionary studies: "We usually witness sudden appearance of new species, followed by long static periods and ultimate extinction. Formerly it was supposed that this fact reflected the incompleteness of the fossil record, but the belief now is that it represents something very important about the evolutionary process." Similarly, Dott and Prothero (1994, p. 61) end their section on "The fossil record and evolution" by stating:

> To some paleontologists, species are more than just populations and genes. They are real entities that seem to have some kind of internal stabilizing mechanism preventing much phenotypic change, even when selection forces change. Clearly, the fossil record produces some unexpected results that are not yet consistent with everything we know about living animals and laboratory experiments. This is good news. If the fossil record taught us nothing that we didn't know already by biology, there wouldn't be much point to evolutionary paleontology.

Finally, Dodson and Dodson (1990, p. 520) provide an excellent summary on the implications of punctuated equilibrium for evolutionary theory:

> Most evolutionary biologists are prepared to acknowledge that punctuated equilibrium is an important phenomenon, even if somewhat less so than its more enthusiastic advocates claim. And population geneticists, who have labored mainly to clarify the genetic basis of evolutionary change, may now have to give greater attention to the problem of evolutionary stasis . . . Thus, the question is not whether punctuated equilibria occur, but how general they are and whether they can be absorbed into the modern evolutionary synthesis.

Among the best treatments of punctuated equilibrium in textbooks, I would cite Kraus's book for high school biology (1983), the continuing efforts of Alters and McComas (1994) to design a high school curriculum based on punctuated equilibrium, and the college textbooks by Avers (1989) and Price (1996). Much of the graphical material has also been highly useful—as in Price's ingenious inclusion of both spatial and temporal dimensions to show how allopatric speciation yields both stasis and punctuation in the fossil record—see Figure 9-39.

As a model of excellence, and of clear, accurate, stylish writing as well, the treatment of punctuated equilibrium in the most popular textbook of the 1980's embodies the reasons for this volume's well-deserved status. Helena Curtis was a thoughtful writer, not a professional biologist, but she mastered the material and could write circles around her competition. (I also know, from personal conversation, that she initially felt quite skeptical about the importance of punctuated equilibrium—so her generous treatment records the judgment of a critical observer, not a partisan.) I reproduce below most of Curtis and Barnes's (1985, pp. 556–557) section on "Punctuated Equilibria," the closing topic in their chapter on "evolution." If these authors could be so fair and accurate, then textbooks can achieve excellence as a genre, and punctuated equilibrium lies safely within the domain of the understandable, the informative, and the interesting:

> Although the fossil record documents many important stages in evolutionary history, there are numerous gaps . . . Many more fossils have, of course, been discovered in the 100 years since Darwin's death. Nevertheless, fewer examples of gradual change within forms have been found than might have been expected. Until recently, the discrepancy between the model of slow phyletic change and the poor documentation of such change in much of the fossil record has been ascribed to the imperfection of the fossil record itself.
>
> About a decade ago, two young scientists, Niles Eldredge of the American Museum of Natural History and Stephen Jay Gould of Harvard University, ventured the radical proposal that perhaps the fossil record is not so imperfect after all. Both Eldredge and Gould have backgrounds in geology and invertebrate paleontology, and both were impressed with the fact that there was very little evidence of phyletic change in the fossil species they studied. Typically, a species would appear abruptly in the fossil strata, last 5 million to 10 million years, and disappear, apparently

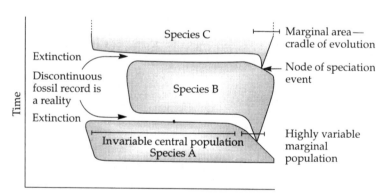

Distribution of species

9-39. An excellent textbook figure of punctuated equilibrium from Price, 1996. He includes, in a way that had never occurred to me or Eldredge, both spatial and temporal dimensions to show how allopatric speciation yields both stasis and punctuation in the fossil record. His own caption reads: "A general scenario for the punctuated equilibrium concept of evolutionary change. Visualize a species change pattern that appears when time is measured vertically through a stratigraphic section of rock, and ponder how many such rock sections would be needed to reveal at least the distribution of species in the central population."

not much different than when it first appeared. Another species, related but distinctly different—"fully formed"—would take its place, persist with little change, and disappear equally abruptly. Suppose, Eldredge and Gould argued, that these long periods of no change ("stasis" is the word they use) punctuated by gaps are not flaws in the record but *are* the record, the evidence of what really happens.

How could it be that a new species would make such a sudden appearance? They found their answer in the model of allopatric speciation. If new species formed principally in small populations on the geographic periphery of the range of the species, if speciation occurred rapidly (by rapidly, paleontologists mean in thousands rather than millions of years), and if the new species then outcompeted the old one, taking over its geographic range, the resulting fossil pattern would be the one observed . . .

As the new model has become more fully developed, particularly by Steven M. Stanley of Johns Hopkins (also a paleontologist), it has become more radical. Its proponents now argue that not only is cladogenesis the principal mode of evolutionary change (as Mayr stated some 30 years ago) but that natural selection occurs among species as well as among individuals. . . . In this new formulation, species take the place of individuals, and speciation and extinction substitute for birth and death. In short, there are two mechanisms of evolution, according to this proposal: in one, natural selection acts on the individual, and in the other, it acts on the species.

Will the punctuated equilibrium model be assimilated into the synthetic theory? Or will some radical new concept of evolutionary mechanisms spread through the scientific strata, outcompeting the old ideas? At this writing, it is too early to tell. All that is clear is that this proposal has stimulated a vigorous debate, a reexamination of evolutionary mechanisms as currently understood, and a reappraisal of the evidence. All of this indicates that evolutionary biology is alive and well and that scientists are doing what they are supposed to be doing—asking questions. Darwin, we think, would have been delighted.

## THE PERSONAL ASPECT OF PROFESSIONAL REACTION

Among false dichotomies, the strict division of a professional's reaction into scientific conclusions based on legitimate judgment and personal reasons rooted in emotional feelings represents a particularly naïve and misleading parsing of human motivation. Our analytic schemes do require some heuristic divisions, but the notion that good reason stands in primal antithesis to bad feelings surely caricatures the depth and complexity of human reactions. All scientific critiques arise in concert with a complex and often unconscious range of emotional responses (not to mention a social and cultural context, which scientists, trained to absorb the myth of objectivity, are particularly disinclined to recognize). The fact that we can analyze the pure logic of an argument *a posteriori* tells us little about the ineluctably nonlogical motives and feelings behind any decision to frame such an argument at a given time, and in the hope of a particular outcome.

Nonetheless, because punctuated equilibrium has provoked so much commentary of a personal nature from scientific colleagues, often expressed with unusual intensity both pro and con (for statements published in professional literature), I don't know how else to parse the content in this case. I discussed the rich and numerous intellectual critiques in the main body of this chapter, but what can be done with the large residuum of unusual personal commentary? I cannot simply ignore it, both because the discussion would then be so selectively incomplete, and also (for personal reasons of course) because I find so much of the most negative commentary so false and unfair—and I do wish to exercise what Roberts Rules calls a "point of personal privilege" as expressed in a basic right of reply. Thus, I have tried to separate personal commentary (in this section) from the critical discourse of ideas, while acknowledging the small psychological sense of such a division. The heuristic advantages of thus splitting each side's clutter from the other's content may justify this procedure.

### The case ad hominem against punctuated equilibrium

I should state up front that I regard this discourse as rooted in little more than complex fallout from professional jealousy, often unrecognized and therefore especially potent. I shall, in the next subsection (pp. 1010–1012), own what I regard as the share of responsibility that Eldredge and I bear for standard misconceptions about punctuated equilibrium, but I believe that the *ad hominem* literature on this subject primarily records inchoate and unanalyzed feelings and habits of thought among our most negatively inclined colleagues.

The common denominator to all these expressions lies in a charge—the basis of most claims on the low road of accusation *ad hominem*—that punctuated equilibrium is false, empty, or trivial, and that the volume of discussion, both in professional literature and general culture, can only record our trickery, our bombast, our dishonesty, our quest for personal fame, or (in the kindest version) our massive confusion. (But what then must these detractors conclude about the intellectual acumen of so many of their peers who support punctuated equilibrium, or at least find the discussion interesting?) I read the case *ad hominem* as a brief composed of two charges, culminating in what has almost become an "urban legend" equivalent in veracity to those alligators in the sewers of New York City, indefensible in fact or logic, but propagated by confident repetition within the club of true believers. I will respond to each point by analyzing the passages from my writing that have become virtually canonical as supposed confirmation.

1. In the kindest version, we are depicted as merely confused and overly hopeful. We develop a good little modest idea that might help the benighted community of paleontologists, but we then begin to suffer delusions of grandeur, and to believe that we might have something to say about evolution in general. (We really don't of course, for punctuated equilibrium only confirms

all the beliefs and predictions of the Modern Synthesis.) We now make the crucial error of deciding that our punctuations must require a new evolutionary mechanism unsuspected by Darwinian gradualism—probably a new style of genetic change directing the process of speciation. But we have only made a fundamental mistake in scaling, for our punctuations are slow enough in microevolutionary time to record the ordinary workings of natural selection.

I cringe when I read characterizations like this because such statements only indicate that the perpetrators haven't read our papers, and must either be expressing their fears or some undocumented gossip that passes for wisdom along academic grapevines. As quotations throughout this chapter amply demonstrate, we have always taken a position contrary to these charges. We didn't err in failing to recognize that a paleontologist's punctuation equals a microevolutionist's continuity. Rather, *we based our theory upon this very idea from the start,* by demonstrating that the conventional allopatric model of speciation scales as a punctuation, not as gradual change through a long sequence of strata, in geological time. Clearly, we could not have located anything theoretically radical in the punctuations of our theory—since we built our model by equating these punctuations with ordinary microevolutionary events of peripatric speciation!

It is true that we staked no unconventional claims for evolutionary theory in our original paper (Eldredge and Gould, 1972)—while urging substantial reform of paleontological practice—but only because we hadn't yet recognized the implications of punctuated equilibrium in this domain. It is also true that we began to urge theoretical reform in subsequent papers (beginning in Gould and Eldredge, 1977, and continuing in Gould, 1982c, 1989e, and Gould and Eldredge, 1993), but we have never based these proposals on the speed or nature of punctuations. Again, as demonstrated by citations throughout this chapter, we locate any revisionary status for punctuated equilibrium in its suggestions about the nature of stasis, and particularly its implications for attributing macroevolutionary phenomena to causes operating on the differential success of species treated as Darwinian individuals. Ordinary speciation remains fully adequate to explain the causes and phenomenology of punctuation.

2. If (as argument one holds) punctuated equilibrium includes no theoretical novelty, and if the theory has enjoyed such intense discussion in both popular and professional literature, then we must have created this anomaly by using rhetorical skills to flog our empty notions in a quest for personal fame. So we hyped, and the media followed like sheep. Dawkins (1986) writes, for example: "Punctuationism is widely thought to be revolutionary and antithetical to neo-Darwinism for the simple reason that its chief advocates have said that it is: said so, moreover, in loud and eloquent voices, making frequent and skillful use of the mass media. The theory, in short, stands out from other glosses on the neo-Darwinian synthesis in one respect only: it has enjoyed brilliant public relations and stage management." (Do I detect a whiff of jealousy in this expostulation?)

I reject this argument about mass media on two grounds: first, for its condescending assumption of such pervasive and universal incompetence within the fourth estate (my section on the press, pages 990–994, includes several examples of highly accurate and critical coverage); and second, for its false and unflattering conjectures about our procedures and integrity. Neither I nor Eldredge has ever engaged in "skillful use" or "stage management" of media.

I have no personal objection to active courting of journalists by scientists, so long as fairness and integrity do not become compromised, especially by caricature, oversimplification or dumbing down. The public is not stupid and can handle scientific material at full conceptual complexity. (Necessary simplification of terminology, and avoidance of jargon, need not imply any sacrifice of intellectual content.) But, as a matter of personal preference, I have never approached the media in this manner. I have never arranged a press conference or meeting, or even placed a phone call to a reporter. I try to be responsive when approached, but I have been entirely reactive in my contact with media on the subject of punctuated equilibrium. Moreover, although I occupied the most "bully pulpit" in America for popular writing about evolution—my monthly column in *Natural History Magazine,* published from January 1974 to January 2001—I never used this forum to push punctuated equilibrium. Of 300 successive essays, I devoted only two to this subject. No ethical or intellectual barrier stood against more extensive treatment, but I preferred to use the great privilege of this forum to learn about new evolutionary byways that I would otherwise not have had time to study, rather than to advocate what I had already treated in the greater depth of professional journals.

So if this second *ad hominem* argument won't even wash for a presumably naive press, how can colleagues regard the attention of a far more sophisticated professional community as nothing but a spinoff from our hype and rhetoric?

When this charge has been laid against me, cited evidence almost always rests upon two supposed claims (and their canonical quotations) expressed in my putatively most radical paper of 1980, entitled: "Is a new and general theory of evolution emerging?" I wrote this paper for the 5th anniversary of *Paleobiology,* as a companion piece to a longer analysis of biological research in our profession: "The promise of paleobiology as a nomothetic, evolutionary discipline" (see Gould, 1980a and b).

The received legend about this paper—I really do wonder how many colleagues have ever based their comments on reading this article with any care, or even at all—holds that I wrote a propagandistic screed featuring two outrageously exaggerated claims: first, the impending death of the Modern Synthesis; and second, the identification of punctuated equilibrium as the exterminating angel (or devil). I do not, in fact and in retrospect (but not in understatement), regard this 1980 paper as among the strongest, in the sense of most cogent or successful, that I have ever written—but neither do I reread it with any shame today. Some of my predictions have fared poorly, and I would now reject them—scarcely surprising for a paper that tried to summarize all major theoretical revisions then under discussion among evolutionists.

For example, I then read the literature on speciation as beginning to favor sympatric alternatives to allopatric orthodoxies at substantial relative frequency, and I predicted that views on this subject would change substantially, particularly towards favoring mechanisms that would be regarded as rapid even in microevolutionary time. I now believe that I was wrong in this prediction.

But the relatively short section devoted to punctuated equilibrium (Gould, 1980b, pp. 125–126) presents this subject in a standard and unsurprising manner, and I would not change any major statement in this part of the paper. (My reassessment away from high relative frequency for rapid speciation in microevolutionary time, and back to the peripatric orthodoxy of our original views, represents a rethinking of another section of this 1980 paper, and does not speak to the validity of punctuated equilibrium. As I have emphasized throughout this chapter, punctuated equilibrium was formulated as the expected macroevolutionary expression of conventional allopatric speciation— so a return to this conventional model can scarcely threaten the theory's validity!)

THE SUPPOSED GENERAL DEATH OF THE SYNTHESIS. Given the furor provoked, I would probably tone down—but not change in content—the quotation that has come to haunt me in continual miscitation and misunderstanding by critics: "I have been reluctant to admit it—since beguiling is often forever—but if Mayr's characterization of the synthetic theory is accurate, then that theory, as a general proposition, is effectively dead, despite its persistence as textbook orthodoxy" (Gould, 1980b, p. 120). (I guess I should have written the blander and more conventional "due for a major reassessment" or "now subject to critical scrutiny and revision," rather than "effectively dead." But, as the great Persian poet said, "the moving finger writes, and having writ . . ." and neither my evident piety nor obvious wit can call back the line—nor would tears serve as a good emulsifier for washing out anything I ever wrote!)

Yes, the rhetoric was too strong (if only because I should have anticipated the emotional reaction that would then preclude careful reading of what I actually said). But I will defend the content of the quotation as just and accurate. First of all, I do not claim that the synthetic theory of evolution is wrong, or headed for complete oblivion on the ashheap of history; rather, I contend that the synthesis can no longer assert full sufficiency to explain evolution at all scales (remember that my paper was published in a paleobiological journal dedicated to studies of macroevolution). Two statements in the quotation should make this limitation clear. First of all, I advanced this opinion only with respect to a particular, but (I thought) quite authoritative, definition of the synthesis: "if Mayr's characterization of the synthetic theory is accurate." Moreover, I had quoted Mayr's definition just two paragraphs earlier. The definition begins Mayr's chapter on "species and transspecific evolution" from his 1963 classic—the definition that paleobiologists would accept as most applicable to their concerns. Mayr wrote (as I explicitly quoted): "The proponents of the synthetic theory maintain that all evolution is due to the accumulation of small genetic changes, guided by natural selection, and that transspecific evolution is nothing but an extrapolation and magnification of the events that take place within populations and species."

Second, I talked about the theory being dead "as a general proposition," not dead period. In the full context of my commentary on Mayr's definition, and my qualification about death *as a full generality,* what is wrong with my statement? I did not proclaim the death of Darwinism, or even of the strictest form of the Modern Synthesis. I stated, for an audience interested in macroevolutionary theory, that Mayr's definition (not the extreme statement of a marginal figure, but an explicit characterization by the world's greatest expert in his most famous book)—with its two restrictive claims for (1) "*all* evolution" due to natural selection of small genetic changes, and (2) transspecific evolution as "nothing but" the extrapolation of microevolutionary events—must be firmly rejected if macroevolutionary theory merits any independent status, or features any phenomenology requiring causal explanation in its own domain. If we embrace Mayr's definition, then the synthesis is "effectively dead" "as a general proposition"—that is, as a theory capable of providing a full and exclusive explanation of macroevolutionary phenomena. Wouldn't most evolutionary biologists agree with my statement today?

Nonetheless, I was reviled in many quarters, and in prose far more intemperate and personal than anything I ever wrote, for proclaiming the death of Darwinism, and the forthcoming enshrinement of my own theory as a replacement (see, for example, A. Huxley, 1982; Thompson, 1983; Cain, 1988; Vogel, 1983; Ayala, 1982; Stebbins and Ayala, 1981a and b; Mayr, 1982a; and Grant, 1983, under the title: "The synthetic theory strikes back").

Many reasons underlie this error, and I do accept some responsibility for my flavorful prose (but not for any lack of clarity in intended meaning, or for any statement stronger than Mayr's dismissive words about my own profession of macroevolution). One common reason, perhaps the most prominent of all, arises from careless scholarship and cannot be laid at my doorstep. I provided the full quotation that offended so many colleagues, along with Mayr's accompanying words, so necessary to grasp the definition that I used. But my statement is usually quoted in deceptively abridged form, leading to a false reading clearly opposite to what I intended. I usually find my words cited in the following abridgment: "The synthetic theory . . . is effectively dead, despite its persistence as textbook orthodoxy." Much commentary has been based upon this truncated and distorted version, not on my actual words. Fill in those three dots before you fire.

HOMO UNIUS LIBRI. An old and anonymous Latin proverb states: *cave ab homine unius libri*—beware the man of one book. I do appreciate the attention that punctuated equilibrium has received, and, as a fallible mortal, I am not adverse to the recognition that this debate has brought me. But as a curious and general consequence of extensive publicity for a single achievement, the totality of one's work then tends to be read as a long and unitary commentary upon this singular idea or accomplishment. The Latin motto should therefore be read from both ends: we should be wary of a person who has only one good idea, but we should also not automatically assimilate an entire life by synecdoche to the single aspect we know best. Leonardo's war machines bear little relationship to the *Mona Lisa;* Newton's chronology of ancient kingdoms never mentions gravity or the inverse square law; and Mickey Mantle was also the best drag bunter and fastest runner in baseball.

Perhaps I should be flattered by the implied importance thus accorded to punctuated equilibrium, but I do maintain interests, some just as consuming, and some (I hope) just as replete with implications for evolutionary theory. Critics generally complete their misunderstanding of my 1980 paper by first imagining that I proclaimed the total overthrow of Darwinism, and then supposing that I intended punctuated equilibrium as both the agent of destruction and the replacement. But punctuated equilibrium does not occupy a major, or even a prominent, place in my 1980 paper.

This article tried to present a general account of propositions within the Modern Synthesis that, in my judgment, might require extensive revision or enlargement, especially from the domain of macroevolution. I did speak extensively—often quite critically—about the reviled work of Richard Goldschmidt, particularly about aspects of his thought that might merit a rehearing. This material has often been confused with punctuated equilibrium by people who miss the crucial issue of scaling, and therefore regard all statements about rapidity at any level as necessarily unitary, and necessarily flowing from punctuated equilibrium. In fact, as the long treatment in Chapter 5 of this book should make clear, my interest in Goldschmidt resides in issues bearing little relationship with punctuated equilibrium, but invested instead in developmental questions that prompted my first book, *Ontogeny and Phylogeny* (Gould, 1977b). The two subjects, after all, are quite separate, and rooted in different scales of rapidity—hopeful monsters in genuine saltation, and punctuated equilibrium in macroevolutionary punctuation (produced by ordinary allopatric speciation). I do strive to avoid the label of *homo unius libri*. I have even written a book about baseball, and another about calendrics and the new millennium.

The section on punctuated equilibrium in my 1980 paper is both short in extent, and little different in content from my treatment of the subject elsewhere. I began with the usual definition: "Our model of 'punctuated equilibria' holds that evolution is concentrated in events of speciation and that successful speciation is an infrequent event punctuating the stasis of large populations that do not alter in fundamental ways during the millions of years that they endure" (p. 125). I then made my usual linkage to ordinary allopatric speciation, not to any novel or controversial mechanism of microevolution. Moreover, I emphasized the scaling error that so often leads people to confuse punctuated equilibrium with saltationism:

Speciation, the basis of macroevolution, is a process of branching. And this branching, under any current model of speciation—conventional allopatry to chromosomal saltation—is so rapid in geological translation (thousands of years at most compared with millions for the duration of most fossil species) that its results should generally lie on a bedding plane, not through the thick sedimentary sequence of a long hillslope . . . It [gradualism] represents, first of all, an incorrect translation of conventional allopatry. Allopatric speciation seems so slow and gradual in ecological time that most paleontologists never recognized it as a challenge to the style of gradualism—steady change over millions of years—promulgated by custom as a model for the history of life (p. 125).

Finally, I stressed that the radical implications of punctuated equilibrium lay in proposed explanations for such macroevolutionary phenomena as cladal trends, not in any proposal for altered mechanisms of microevolution: "Evolutionary trends therefore represent a third level superposed upon speciation and change within demes . . . Since trends 'use' species as their raw material, they represent a process at a higher level than speciation itself. They reflect a sorting out of speciation events . . . What we call 'anagenesis,' and often attempt to delineate as a separate phyletic process leading to 'progress,' is just accumulated cladogenesis filtered through the directing force of species selection."

IN TRES PARTES DIVISA EST: THE 'URBAN LEGEND' OF PUNCTUATED EQUILIBRIUM'S THREEFOLD HISTORY. The opponents of punctuated equilibrium have constructed a fictional history of the theory, primarily (I suppose) as a largely unconscious expression of their hope for its minor importance, and their jealousy towards its authors. This history even features a definite sequence of stages, constructed to match a classic theme of Western sagas: the growth, exposure and mortification of hubris (try Macbeth as a prototype, but he dies before reaching the final stage of penance; so try Faust instead, who lusts for the world and ends up finding satisfaction in draining a swamp). This supposed threefold history of punctuated equilibrium also ranks about as close to pure fiction as any recent commentary by scientists has ever generated.

In stage one, the story goes, we were properly modest, obedient to the theoretical hegemony of the Modern Synthesis, and merely trying to bring paleontology into the fold. But the prospect of worldly fame beguiled us, so we broke our ties of fealty and tried, in stage two, to usurp power by painting punctuated equilibrium as a revolutionary doctrine that would dethrone the Synthesis, resurrect the memory of the exiled martyr (Richard Goldschmidt), and reign over a reconstructed realm of theory. But we were too big for our breeches, and the old guard still retained some life. They fought back mightily and effectively, exposing our bombast and emptiness. We began to hedge, retreat, and apologize, and have been doing so ever since in an effort to regain grace and, chastened in stage three, to sit again, in heaven or Valhalla, with the evolutionary elite.

Such farfetched fiction suffers most of all from an internal construction that precludes exposure and falsification among true believers, whatever the evidence. Purveyors of this myth even name the three stages, thus solidifying the false taxonomy. Dawkins (1986), for example, speaks of the "grandiloquent era . . . of middle-period punctuationism [which] gave abundant aid and comfort to creationists and other enemies of scientific truth." In the other major strategy of insulation from refutation, supporters of this "urban legend" about the modest origin, bombastic rise, and spectacular fall of punctuated equilibrium forge a tale that allows them to read any potential disconfirmation as an event within the fiction itself. (Old style gradualism pursued exactly the same strategy in reading contrary data as marks of imperfect evidence within the accepted theory—and thus could not be refuted from within. I am struck by the eerie similarity between the structure of the old theory and the historical gloss invented by opponents of a proposed replacement.)

In particular, and most offensive to me, the urban legend rests on the false belief that radical, "middle-period" punctuated equilibrium became a saltational theory wedded to Goldschmidt's hopeful monsters as a mechanism. I have labored to refute this nonsensical charge from the day I first heard it. But my efforts are doomed within the self-affirming structure of the urban legend. We all know, for so the legend proclaims, that I once took the Goldschmidtian plunge. So if I ever deny the link, I can only be retreating from an embarrassing error. And if I continue to deny the link with force and gusto, well, then I am only backtracking even harder (into stage 3) and apologizing (or obfuscating) all the more. How about the obvious (and accurate) alternative: that we never made the Goldschmidtian link; that this common error embodies a false construction; and that our efforts at correction have always represented an honorable attempt to relieve the confusion of others.

But the urban legend remains too simplistically neat, and too resonant with a favorite theme of Western sagas, to permit refutation by mere evidence. So Dennett (1995, pp. 283–284) writes: "There was no mention in the first paper of any radical theory of speciation or mutation. But later, about 1980, Gould decided that punctuated equilibrium was a revolutionary idea after all . . . [But] it was too revolutionary, and it was hooted down with the same sort of ferocity the establishment reserves for heretics like Elaine Morgan. Gould backpedaled hard, offering repeated denials that he has ever meant anything so outrageous." And Halstead (1985, p. 318) wrote of me (with equal poverty in both logic and grammar): "He seems to be setting up a face-saving formula to enable him to retreat from his earlier aggressive saltationism, having had a bit of a thrashing, his current tack is to suggest that perhaps we should keep the door open in case he can find some evidence to support his pet theories so let us be 'pluralist.'"

I do not, of course, claim that our views about punctuated equilibrium have never changed through the years of debate (only a dull and uninteresting theory could remain so static in the face of such wide discussion). Nor do I maintain a position that would be even sillier—namely, that we made no important errors requiring corrections to the theory. Of course we made mistakes, and of course we have tried to amend them. But I look upon the history of punctuated equilibrium (from my partisan vantage point of course) as a fairly standard development for successful theories in science. We did, indeed, begin modestly and expand outward thereafter. (In this sense, punctuated equilibrium has grown in theoretical scope, primarily as macroevolutionary theory developed and became better integrated with the rest of evolutionary thought—and largely through articulation of the hierarchical model, as discussed in the previous chapter).

We started small as a consequence of our ignorance and lack of perspective, not from modesty of basic temperament. As stated before, we simply didn't recognize, at first, the interesting implications of punctuated equilibrium for macroevolutionary theory—primarily gained in treating species as Darwinian individuals for the explanation of trends, and in exploring the extent and causes of stasis. With the help of S. M. Stanley, E. S. Vrba and other colleagues, we developed these implications over the years, and the theory grew accordingly. But we never proposed a radical theory for punctuations (ordinary speciation scaled into geological time), and we never linked punctuations to microevolutionary saltationism.

Of course we made mistakes—serious ones in at least two cases—and the theory has changed and improved by correcting these errors. In particular, and as documented extensively in Chapter 8, we were terribly muddled for several years about the proper way to treat, and even to define, selection at the level of species—the most important of all theoretical spinoffs from punctuated equilibrium. We confused sorting with selection (see Vrba and Gould, 1986, for a resolution). We also did not properly formulate the concept of emergence at first; and we remained confused for a long time about emergence of characters vs. emergence of fitness as criteria for species selection (Lloyd and Gould, 1993; Gould and Lloyd, 1999). In retrospect, I am chagrined by the long duration of our confusion, and its expression in many of our papers. But I think that we have now resolved these difficult issues.

Secondly, as discussed on pages 796–798, I think that we originally proposed an incorrect reason for the association of rapid change with speciation. But I believe that we portrayed the phenomenology correctly, and that we have, with the help of Futuyma's (1987) suggestions, now developed a proper explanation. Thus, the theory of punctuated equilibrium has altered substantially to correct these two errors. Interestingly (and ironically), however, these important changes do not figure at all in the deprecating claims of the urban legend about our supposed retreats and chameleon-like redefinitions—for our detractors hardly recognize the existence of punctuated equilibrium's truly radical claim for evolutionary theory: its implications for selection above the species level, and for the explanation of trends.

Punctuated equilibrium, in short, has enjoyed true Darwinian success through the years: it has struggled, survived, changed and expanded. But the theoretical evolution of punctuated equilibrium belongs to the sphere of cultural change with its Lamarckian mode of transmission by direct passage of acquired improvements. Thus, the theory need not remain in Darwinian stasis, but may grow—as it has—in (gulp!) a gradualist and progressive manner.

The saltationist canard has persisted as our incubus. The charge could never be supported by proper documentation, for we never made the link or claim. All attempts collapse upon close examination. Dennett, for example, who insists (1997, p. 64) that "for a while he [Gould] had presented punctuated equilibrium as a revolutionary 'saltationist' alternative to standard neo-Darwinism," documents his supposed best case by assuring readers (1995, p. 285) that "for a while, Gould was proposing that the first step in the establishment of any new species was a doozy—a non-Darwinian saltation." Dennett directly follows this claim with his putative proof, yet another quotation from my 1980 paper, which he renders as follows: "Speciation is not always an extension of gradual, adaptive allelic substitution to greater effect, but may represent, as Goldschmidt argued, a different style of genetic change—rapid reorganization of the genome, perhaps non-adaptive" (Gould, 1980b, p. 119).

I regard Dennett's case as pitiful, but the urban legend can offer no better. First of all, this quotation doesn't even refer to punctuated equilibrium, but comes from a section of my 1980 paper on the microevolutionary mechanics of speciation. Secondly, Dennett obviously misreads my statement in a backwards manner. I am trying to carve out a *small* theoretical space for a style of microevolutionary rapidity at low relative frequency—as clearly stated in my phrase "not always an extension of gradual . . ." But Dennett states that I am proposing this mechanism as a general replacement for gradual microevolutionary change in *all* cases of speciation—"the first step in the establishment of *any* new species" in *his* words. But *my* chosen phrase—"not always"—clearly means "most of the time," and cannot be read as "never." In short, I made a plea for pluralism, and Dennett charges me with usurpation. Then, when I try to explain, I am accused of beating a retreat to save face. When placed in such a double bind, one can only smile and remember Schiller's famous dictum: *Mit Dummheit kämpfen die Götter selbst vergebens.*

Finally, the claim that we equated punctuated equilibrium with saltation makes no sense within the logical structure of our theory—so, unless we are fools, how could we ever have asserted such a proposition? Our theory holds, as a defining statement, that ordinary allopatric speciation, unfolding gradually at microevolutionary scales, translates to punctuation in geological time. Microevolutionary saltation also scales as a punctuation—so the distinction between saltation and standard allopatry becomes irrelevant for punctuated equilibrium, since both yield the same favored result!

Moreover, the chronology of debate proves that we did not issue disclaimers on this subject only to cover our asses as we retreated from exaggerations of our supposed second phase, because we have been asserting this clarification from the very beginning—that is, from the first paper we ever wrote to comment upon published reactions to punctuated equilibrium. Our first response appeared in 1977, long before we issued the supposed clarion call of our false revolution in 1980. We wrote (Gould and Eldredge, 1977, p. 121), under the heading "Invalid claims of gradualism made at the wrong scale": "The model of punctuated equilibria does not maintain that nothing occurs gradually at any level of evolution. It is a theory about speciation and its deployment in the fossil record. It claims that an important pattern, continuous at higher levels—the 'classic' macroevolutionary trend—is a consequence of punctuation in the evolution of species. It does not deny that allopatric speciation occurs gradually in ecological time (though it might not—see Carson, 1975), but only asserts that this scale is a geological microsecond."

We have never changed this conviction, and we have always tried to correct any confusion of scaling between saltation and punctuation, even in papers written during the supposed apogee of our revolutionary ardor, during illusory stage 2 of the urban legend. For example, under the heading of "The relationship of punctuated equilibrium to macromutation," I wrote in 1982c (p. 88): "Punctuated equilibrium is not a theory of macromutation . . . it is not a theory of any genetic process . . . It is a theory about larger-scale patterns—the geometry of speciation in geological time. As with ecologically rapid modes of speciation, punctuated equilibrium welcomes macromutation as a source for the initiation of species: the faster the better. But punctuated equilibrium clearly does not require or imply macromutation, since it was formulated as the expected geological consequence of Mayrian allopatry."

### An interlude on sources of error

With such limited skills in sociology and psychology, and from too close a personal and partisan standpoint, I cannot claim much insight into the general sources of persistent nonscientific errors among professional colleagues. But I wish to offer a few thoughts, at least to separate what Eldredge and I must own from the truly unfair, and often intemperate, charges so often made against us.

Any complex situation arises from multiple causes, with inevitable shortcomings on both sides of any basically dichotomous issue. But when I list our own faults and failures, I find nothing of great depth, and no indication of any sustained stupidity, carelessness, lack of clarity, or malfeasance. Thus, I continue to feel far more aggrieved than intemperate—although I wouldn't give up this lifetime's intellectual adventure for any alternative construction of a scientific career.

For our part, I think that critics can identify three sources of potential confusion that might legitimately be laid at our doorstep, and might have been prevented had our crystal ball been clearer.

1. In our original paper (Eldredge and Gould, 1972), but not subsequently, we failed to explain, in a sufficiently didactic and explicit manner, that when paleontologists use such terms as "rapid," "sudden," or "instantaneous," they refer to expressions of events at geological scales, and not to rates of change in microevolutionary time. But we cannot be blamed for anything more than a failure to anticipate the range of interest that our paper would generate. After all, we wrote this paper for paleontologists, and never expected a wider audience. We used the standard terminology of our profession, well known and understood by all members of the clan. Indeed, few nonpaleontologists ever read this original article, published in an obscure symposium volume with a small press run. From 1977 on, in all papers widely read

by neontologists, and serving as a basis for enlarged discussion, we clearly explained the differences in scaling between micro- and macroevolutionary rates.

2. As acknowledged on pages 1002–1004, I did use some prose flourishes that, in a context of considerable suspicion and growing jealousy, probably fanned the flames of confusion. Although I never stated anything unclearly, and committed no logical errors that could legitimately have inspired a resulting misreading, I should have toned down my style in a few crucial places.

3. We may have sown some confusion by using partially overlapping terminology for a specific theory (punctuated equilibrium), and for the larger generality (punctuational styles of change) in which that theory lies embedded. But this taxonomic usage does stress a legitimate commonality that we wished to emphasize. We also chose and used our terms with explicit consistency and clear definitions—so careful reading should have precluded any misunderstanding.

The testing and development of punctuated equilibrium—a well-defined and circumscribed theory about the origin and deployment of speciation events in geological time—has always been our major concern. But as students of evolution, we have also been interested in the range of applicability for the geometric generalization represented by this theory—the unfolding of change as occasional punctuation within prevailing stasis, rather than as gradualistic continuity—to other scales of space and time, and for other causes and phenomena of life's history. We have called this more general and abstract style of change "punctuational," and have referred to the hypothesis favoring its generality as "punctuationalism."

We have always been careful and clear about the differences between our specific theory of punctuated equilibrium and the general proposition of punctuational change. (In fact, we strove to be explicit, even didactic, about this distinction because we recognized the confusion that might arise otherwise.) But perhaps the words are too close to expect general understanding of the distinction, particularly from hostile critics who have invested their emotional ire in the legend that we have been pursuing an imperialistic, grandstanding quest to enshrine punctuated equilibrium as a new paradigm for all the evolutionary sciences.

Still, as a statement of a basic intellectual principle, why should we allow ourselves to be forced into suboptimal decisions by the least thoughtful and most emotionally driven forms of misunderstanding among critics? Punctuationalism *is* the right and best word for the general style of change expressed by punctuated equilibrium as a specific example at a circumscribed level and phenomenology. As long as we take special care to be clear and explicit about the distinction, why should we sacrifice this most appropriate form of naming? I believe that we have been scrupulous in characterizing and highlighting this point, right from our first introduction in 1977, when we began a section entitled "Towards a general philosophy of change" with these words: "Punctuated equilibria is a model for discontinuous tempos of change at one biological level only: the process of speciation and the deployment of species in geological time. Nonetheless, we believe that a general theory of punctuational change is broadly, though by no means exclusively, valid throughout biology" (Gould and Eldredge, 1977, p. 145). In 1982, in the midst of illusory stage 2, when I was supposedly touting macromutation as the cause of punctuated equilibrium in order to dethrone Darwinism, I explicitly drew the same distinction in order to separate the phenomena, while noting an interesting similarity in abstract geometric style of change across scales (Gould, 1982c, p. 90):

> These legitimate styles of macromutation are related to punctuated equilibrium only insofar as both represent different and unconnected examples of a general style of thinking that I have called punctuational (as opposed to gradualist or continuationist thought). I take it that no one would deny the constraining impact of gradualistic biases upon evolutionary theorizing. Punctuational thinking focusses upon the stability of structure, the difficulty of its transformation, and the idea of change as a transition between stable states. Evolutionists are now discussing punctuational theories at several levels: for morphological shifts (legitimate macromutation), speciation (various theories for rapid attainment of reproductive isolation), and general morphological pattern in geological time (punctuated equilibrium). These are not logically interrelated, but manifestations of a style of thought that I regard as promising and, at least, expansive in its challenge to conventional ideas. Any manifestation may be true or false, or of high or low relative frequency, without affecting the prospects of any other. I do commend the general style of thought (now becoming popular in other disciplines as well) as a fruitful source for hypotheses.

However, when I turn to factors that must be laid at our critics' doorstep, I can compile a longer and more serious list, including attitudes and practices that do compromise the ideals of scholarship. (Remember that I deal, in this section, only with personal and nonscientific critiques of punctuated equilibrium. We have also been properly subjected to very sharp, entirely appropriate, and fully welcomed criticism of a technical and scientific nature—and the theory of punctuated equilibrium has only been altered and improved thereby. I have discussed these legitimate criticisms in Section IV of this chapter.)

1. THE EMOTIONAL SOURCE OF JEALOUSY. Given the vehemence of many deprecations, combined with a weakness or absence of logical or scientific content, I must conclude that the primary motivating factor lies in simple jealousy—that most distressing, yet most quintessentially human, of all destructive emotions ("as cruel as the grave" according to the *Song of Songs*; "the jaundice of the soul," in Dryden's metaphor; and, in the most memorable definition of all, Shakespeare's words of warning to Othello: "It is the green-eyed monster which doth mock the meat it feeds on").

Punctuated equilibrium has generated a large and public volume of commentary. I am confident that genuine interest and content has generated the bulk of this publicity, but I understand the all too human tendency to view achievements of perceived rivals as imposture rooted in base motivation. Moreover, jealousy gains a particularly potent expression in science for the ironic reason that professional norms do not permit us to acknowledge such feelings or motivations, even to ourselves. Our negativities are supposed to arise from perceived fallacies in the logic or empirical content of hypotheses we dislike, not from personal expressions of envy. Thus, if our emotions exude distress and anger, but we cannot admit, or even recognize, jealousy as a source, then we must impute our genuine envy to the supposed intellectual malfeasance of our opponents—and our internal feeling becomes falsely objectified as their failing. This form of transference leads to larger problems in the sociology of science than we have generally been willing to admit.

2. THE PHILOSOPHICALLY INTERESTING ISSUE OF LIMITED CONCEPTUAL SPACE. I have long faced a paradox in trying to understand why many intelligent critics seem unable to understand or acknowledge our reiterated insistence that the radical claim of punctuated equilibrium lies not in any proposal for revised microevolutionary mechanisms (especially not in any novel explanations for punctuations), but rather at the level of macroevolution, in claims for efficacy of higher-level selection based on the status of species, under punctuated equilibrium, as genuine Darwinian individuals.

When smart people don't "get it," one must conclude that the argument lies outside whatever "conceptual space" they maintain for assessing novel ideas in a given area. Many evolutionists, particularly those committed to the strict Darwinism of unifocal causation at Darwin's own organismic level, or below at the genic level, have never considered the hierarchical model, and apparently maintain no conceptual space for the notion of effective selection at higher levels. These scientists then face the following situation: (1) they note correctly that punctuated equilibrium stakes some claim for novelty within evolutionary theory; (2) their concept of "evolutionary theory" does not extend to causation above the organismic level, so they do not grasp the actual content of our claim; (3) they correctly understand that punctuated equilibrium offers no radical statement about microevolutionary mechanics; (4) Q.E.D., the authors of punctuated equilibrium must be grandstanding by asserting a radical claim without content. But the limit lies within the conceptual space of our critics, not in the character of our rhetoric.

3. THE PARTICULAR PREJUDICE THAT FANS THE FLAMES. Certain words embody unusual power, for reasons both practical and emotional—"fire" in a crowded theater, or "communist" at right-wing pep rallies of old. For reasons of impeccable historical pedigree, thoroughly explored in Chapters 2–6 of this book, and rooted largely in Darwin's own philosophical preferences, the most incendiary words for dedicated Darwinians (once we get past Lamarckism, creationism, and a few others) must be the various synonyms of "sudden"—"rapid," "instantaneous," "quick," "discontinuous," and the like. Proponents of punctuated equilibrium do use these words—but at an appropriate scale of geological time, to express microevolutionary continuities that translate to punctuations in this larger temporal realm. Nonetheless,

some orthodox Darwinians react with knee-jerk negativity towards any claim at all about rapidity. Any invocation of "rapid" must conjure up saltation and Goldschmidt, and must be met by counterattack. How else can we explain such a persistent confusion based on a false construction, then elevated to an urban legend, that the originators of punctuated equilibrium have always tried to identify and dispel?

### The wages of jealousy

THE DESCENT TO NASTINESS. I treated the general *ad hominem* case against punctuated equilibrium in the last section. But some specific charges against punctuated equilibrium have bordered on the inane, or even the potentially actionable in our litigious world. To mention a few highlights along this low road:

THE CHARGE OF DISHONESTY. The following event unfolds with lamentable predictability in our imperfect world: when a controversy becomes impassioned, someone will eventually try to land the lowest academic blow of all by launching a charge of plagiarism or dishonest quotation. The debate about punctuated equilibrium reached this nadir when Penny (1983) accused us of cooking a quote from the *Origin of Species* by omitting passages without noting the deletion, and thereby changing Darwin's meaning to suit our purposes. Penny quoted from the 6th edition of the *Origin* to back up his claim. We, however, had used the first edition—and had rendered Darwin's words accurately (Gould and Eldredge, 1983). Enough said.

THE CHARGE OF RIP-OFF. A more conventional strategy for those who wish to deny a colleague's originality consists in claiming that a putative novelty really has an old pedigree—a twice told tale, said long before, preferably by a leading scientific light, and not in an obscure source (so that those under question cannot claim forgivable ignorance of minutiae). I suppose, therefore, that when we began to arouse substantial jealousy, someone was bound to argue that Darwin himself had said it all before.

The litany of this claim may hold some sociological interest for the time and energy invested by several commentators (Penny, 1983, 1985; Gingerich, 1984a and b, 1985; Scudo, 1985). These authors did point out some legitimate similarities between certain Darwinian statements and the tenets of punctuated equilibrium—including a significant one-sentence addition to later editions of the *Origin* (which we had indeed missed), acknowledging the occurrence of punctuational tempos, and apparently inspired by Falconer's objections, as highlighted in the introductory section of this chapter.

I regard this case as fundamentally misguided for general historiographic reasons, outlined in Gould and Eldredge (1983, p. 444):

> One simply cannot do history by searching for footnotes and incidental statements, particularly in later editions that compromise original statements. As with the Bible, most anything can be found somewhere in Darwin. General tenor, not occasional commentary, must be the criterion for judging a scientist's basic conceptions. If Darwin historians agree on a

single point (for example, see Gruber [1974] and Mayr [1982b]), it is the importance and pervasiveness of Darwin's gradualism—a commitment far stronger than his allegiance to natural selection as an evolutionary mechanism.

Fortunately, one needn't take my partisan word in refutation. Frank Rhodes, then the president of Cornell University, but a distinguished paleontologist by training and first career, became interested in punctuated equilibrium and its links to the history of evolutionary thought. He therefore spent a sabbatical term researching the relationship of Darwin's thinking to the claims and tenets of punctuated equilibrium. He did find many genuine Darwinian resonances, while affirming our originality and concluding that "the hypothesis of punctuated equilibrium is of major importance for paleontological theory and practice" (Rhodes, 1983, p. 272).

When Gingerich (1984, p. 116) wrote a commentary on Rhodes's article, dedicated to denying our originality and asserting once again that Darwin had said it all before, Rhodes replied with generosity and firmness (under Gingerich, 1985, p. 116):

> I do argue that punctuated equilibrium—whether true or false—is a "hypothesis of major importance" and that it has had a beneficial impact on the quality of recent paleontological studies. Gingerich asks, "Which nuances [of punctuated equilibrium] were unanticipated by Darwin?" From a long list, I suggest the following: its relationship to the genetics of stasis and the punctuation, morphological stasis and developmental constraints, evolutionary models in relation to paleoecology, stratigraphical correlation, species selection, mathematical models of evolutionary rates, selection of RNA molecules, phylogenetic divergence, and the evolution of communities. These topics, and many more studied from the viewpoint of punctuated equilibrium, have been the subject of recent papers . . . To suggest that there was no nuance of punctuated equilibrium which was "unanticipated by Darwin" is to make an icon of Darwin and to adopt an extravagantly Whiggish view of the history of Darwin's particular contribution—great as that was.

Some critics then followed a substitutional strategy: if one denigration fails, try another in the same form. If "Darwin said it all" fails as an optimal dismissal, then try the best available paleontological version: "G. G. Simpson said it all." Again, the search for reinterpretations and footnotes began, as this new version of denigration began to coagulate among our most committed detractors: Simpson (1944) devoted his seminal book to documenting the great variation in evolutionary rates, and punctuated equilibrium therefore has nothing new to say.

But, punctuated equilibrium was never formulated as a hypothesis about great variability in anagenetic rates (which, indeed, everyone has long acknowledged). Punctuated equilibrium presents a specific hypothesis about the location of most evolutionary change in punctuational cladogenesis, followed by pronounced stasis. Yet Simpson, as documented in Chapter 7, pages 562–563, denied major importance to cladogenesis at all, and held that 90 percent of evolutionary change occurred in the anagenetic mode. Moreover (see pages 528–531), Simpson's important hypothesis of "quantum evolution"—the idea that our detractors usually try to depict as equivalent to punctuated equilibrium—treats a vitally important, but entirely different phenomenon of different mode at a different scale: the anagenetic origin of major structural innovations, not the pacing of ordinary speciation.

Several authors, in their desire to name Simpson as the true author of punctuated equilibrium, completely misunderstood his work. Andrew Huxley, for example, misinterpreted the well-known paleontological concept of a *Stufenreihe*. Huxley quoted from Simpson's 1944 book (Huxley, 1982, p. 145): "He says (pp. 194–195): 'The pattern of step-like evolution, an appearance of successive structural steps, rather than direct sequential phyletic transitions, is a peculiarity of paleontological data more nearly universal than true rectilinearity and often mistaken for the latter,' and quotes the name *Stufenreihe* given to this mode of evolution by Abel in 1929. This is exactly equivalent to 'punctuated equilibria.'" But a *Stufenreihe* is a stratigraphic sequence displaying an evolutionary trend constructed of collateral relatives rather than direct ancestors (called, by contrast, an *Ahnenreihe*). For example, *Australopithecus robustus*, *Homo neanderthalensis* and *Homo sapiens* form a *Stufenreihe*, while *A. afarensis*, *Homo ergaster*, and *Homo sapiens* build a putative *Ahnenreihe*. *Stufenreihen* are necessarily discontinuous because they pile a cousin on top of an uncle on top of a grandfather, while true *Ahnenreihen* record genuine genealogical descent without breaks. In any case, the contrast bears no relationship to the concept of punctuated equilibrium (which is a hypothesis about the geometry of *Ahnenreihen*).

Mettler, Gregg and Schaffer (1988, p. 288) even grant Simpson authorship of our name! "Finally, there is the punctuated equilibrium view of Eldredge and Gould (1972), and Vrba (1983). Even though the term was coined by Simpson, these authors have given it new emphasis."

At least pathos can be balanced by bathos in our wondrously varied world. The irrepressible Beverly Halstead, labelling me with my all-time favorite epithet of "petty obnoxious infauna," while depicting Simpson as a deity watching over his loyal epigones from on high, reviewed Simpson's last book with a panegyric that left even his earlier excoriation of the British Museum in the rhetorical dust (Halstead, 1984, p. 40):

> Indeed, the original presentation of punctuated equilibrium was in anti-neo-Darwinian language but the substance was nonetheless easily accommodated within the framework given long ago by Simpson . . . It has been Simpson's overwhelming reasonableness and commonsense, as exemplified in this book, that has done so much to entrench the Modern Synthesis in the consciousness of most paleontologists, the literary pyrotechnics of Steve Gould notwithstanding. Simpson's humility before his

fossils, a special kind of innocence, is perhaps one of his most endearing traits. Because he is kind and tolerant, he finds it nigh impossible to believe that some of the supporting framework of our discipline is infested with some petty obnoxious infauna. My only criticism of Simpson in his book is his apparent unwillingness to contemplate the existence of real nasties emerging from the woodwork . . . Let him [Simpson] look down from the commanding heights knowing that the citadel of neo-Darwinism still has its staunch defenders in this more combative age. We will do our best not to let him down.

THE CHARGE OF ULTERIOR MOTIVATION. When charges of dishonesty or lack of originality fail, a committed detractor can still label his opponents as unconcerned with scientific truth, but motivated by some ulterior (and nefarious) goal. Speculations about our "real" reasons have varied widely in content, but little in their shared mean spirit (see, for example, Turner, 1984; Konner, 1986; and Dennett, 1995). I will discuss only one of these peculiar speculations—the charge that punctuated equilibrium originated from my political commitments rather than from any honorable feeling about the empirical world—because, once again, the claim rests upon a canonical misquotation and exposes the apparent unwillingness or inability of our unscientific critics to read a clear text with care.

I have already discussed Halstead's version of the political charge in the great and farcical British-Museum-cum-cladism-cum-Marxism debate (see pages 984–985). The supposed justification for this construction lies in another quotation from my writing, second in false invocation only to the "death of the Synthesis" statement discussed earlier (p. 1003).

I do not see how any careful reader could have missed the narrowly focused intent of the last section in our 1977 paper, a discussion of the central and unexceptionable principle, embraced by all professional historians of science, that theories must reflect a surrounding social and cultural context. We began the section by trying to identify the cultural roots of gradualism in larger beliefs of Victorian society. We wrote (Gould and Eldredge, 1977, p. 145): "The general preference that so many of us hold for gradualism is a metaphysical stance embedded in the modern history of Western cultures: it is not a high-order empirical observation, induced from the objective study of nature . . . We mention this not to discredit Darwin in any way, but merely to point out that even the greatest scientific achievements are rooted in their cultural contexts—and to argue that gradualism was part of the cultural context, not of nature."

We couldn't then assert, with any pretense to fairness or openness to self-scrutiny, that gradualism represents cultural context, while our punctuational preferences only record unvarnished empirical truth. If all general theories embody a complex mixture of contingent context with factual adequacy, then we had to consider the cultural embeddedness of preferences for punctuational change as well. We therefore began by writing (p. 145) that "alternative conceptions of change have respectable pedigrees in philosophy." We then discussed the most obvious candidate in the history of Western thought: the Hegelian dialectic and its redefinition by Marx and Engels as a theory of revolutionary social change in human history. We cited a silly, propagandistic defense of punctuational change from the official Soviet handbook of Marxism-Leninism, in order to stress our point about the potential political employment of all general theories of change. We concluded (p. 146): "It is easy to see the explicit ideology lurking behind this general statement about the nature of change. May we not also discern the implicit ideology in our Western preference for gradualism?"

But the argument required one further step for full disclosure. We needed to say something about why we, rather than other paleontologists at other times, had developed the concept of punctuated equilibrium. We raised this point as sociological commentary about the *origin* of ideas, not as a scientific argument for the *validity* of the same ideas. An identification of cultural or ontogenetic sources says nothing about truth value, an issue that can only be settled by standard scientific procedures of observation, experiment and empirical test. So I mentioned a personal factor that probably predisposed me to openness towards, or at least an explicit awareness of, a punctuational alternative to conventional gradualistic models of change: "It may also not be irrelevant to our personal preferences that one of us learned his Marxism, literally at his daddy's knee."

I have often seen this statement quoted, always completely out of context, as supposed proof that I advanced punctuated equilibrium in order to foster a personal political agenda. I resent this absurd misreading. I spoke only about a fact of my intellectual ontogeny; I said nothing about my political beliefs (very different from my father's, by the way, and a private matter that I do not choose to discuss in this forum). I included this line within a discussion of personal and cultural reasons that might predispose certain scientists towards consideration of punctuational models—just as I had identified similar contexts behind more conventional preferences for gradualism. In the next paragraph, I stated my own personal conclusions about the general validity of punctuational change—but critics never quote these words, and only cite my father's postcranial anatomy out of context instead:

> We emphatically do not assert the "truth" of this alternate metaphysic of punctuational change. Any attempt to support the exclusive validity of such a monistic, *a priori*, grandiose notion would verge on the nonsensical. We believe that gradual change characterizes some hierarchical levels, even though we may attribute it to punctuation at a lower level—the macroevolutionary trend produced by species selection, for example. We make a simple plea for pluralism in guiding philosophies—and for the basic recognition that such philosophies, however hidden and inarticulated, do constrain all our thought. Nonetheless, we do believe that the punctuational metaphysic may prove to map tempos of change in our world better and more often than any of its competitors—if only because

systems in steady state are not only common but also so highly resistant to change.

**THE MOST UNKINDEST CUT OF ALL.** If none of the foregoing charges can bear scrutiny, strategists of personal denigration still hold an old and conventional tactic in reserve: they can proclaim a despised theory both trivial and devoid of content. This charge is so distasteful to any intellectual that one might wonder why detractors don't try such a tactic more often, and right up front at the outset. But I think we can identify a solution: the "triviality caper" tends to backfire and to hoist a critic with his own petard—for if the idea you hate is so trivial, then why bother to refute it with such intensity? Leave the idea strictly alone and it will surely go away all by itself. Why fulminate against tongue piercing, goldfish swallowing, skateboarding, or any other transient fad with no possible staying power?

Nonetheless, perhaps from desperation, or from severe frustration that something regarded as personally odious doesn't seem to be fading away, this charge of triviality has been advanced against punctuated equilibrium, apparently to small effect. To cite a classic example of backfiring, Gingerich (1984a, 1984b) tried to dismiss punctuated equilibrium as meaningless and untestable by definition—and to validate gradualism *a priori* as "commitment to empiricism and dedication to the principal [sic] of testability in science" (1984a, p. 338), with stasis redefined, oxymoronically in my judgment, as "gradualism at zero rate" (1984a, p. 338). Gingerich then concludes (1984b, p. 116): "Punctuated equilibrium is unscaled, and by nature untestable. It hardly deserves recognition as a conjecture of 'major importance for paleontological theory and practice.' . . . Hypotheses that cannot be tested are of little value in science."

But how can Gingerich square this attempted dismissal with his own dedication of a decade in his career to testing punctuated equilibrium by fine-scale quantitative analysis of Tertiary mammals from the western United States (Gingerich, 1974, 1976)? These studies, which advanced a strong claim for gradualism, represent the most important empirical research published in the early phase of the punctuated equilibrium debate. Gingerich then recognized punctuated equilibrium as an interesting and testable hypothesis, for he spent enormous time and effort testing and rejecting our ideas for particular mammalian phylogenies. He then argued explicitly (1978, p. 454): "Their [Eldredge and Gould's] view of speciation differs considerably from the traditional paleontological view of dynamic species with gradual evolutionary transitions, but it *can* be tested by study of the fossil record."

Among Darwinian fundamentalists (see my terminology in Gould, 1997d), charges of triviality have been advanced most prominently and insistently by Dawkins (1986, p. 251) who evaluates punctuated equilibrium metaphorically as "an interesting but minor wrinkle on the surface of neo-Darwinian theory"; and by Dennett (1995, p. 290) who calls punctuated equilibrium "a false-alarm revolution that was largely if not entirely in the eyes of the beholders."

But a close analysis of Dawkins's and Dennett's arguments exposes the parochiality of their judgment. They regard punctuated equilibrium as trivial because our theory doesn't speak to the restricted subset of evolutionary questions that, for them, defines an exclusive domain of interest for the entire subject. These men virtually equate evolution with the origin of intricately adaptive organic design—"organized adaptive complexity," or O.A.C. in Dawkins's terminology. They then dismiss punctuated equilibrium on the narrow criterion: "if it doesn't explain the focus of my interests, then it must be trivial." Dawkins (1984, p. 684), for example, properly notes the implications of punctuated equilibrium for validation of higher-level selection, but then writes: "Species-level selection can't explain the evolution of adaptations: eyes, ears, knee joints, spider webs, behavior patterns, everything, in short, that many of us want a theory of evolution to explain. Species selection may happen, but it doesn't seem to *do* anything much." "Everything"? Does nothing else but adaptive organismal design excite Dawkins's fancy in the entire and maximally various realm of evolutionary biology and the history of life—the "endless forms most beautiful and most wonderful" of Darwin's closing words (1859, p. 490).

But the truly curious aspect of both Dawkins's and Dennett's charge lies in their subsequent recognition, and fair discussion, of the important theoretical implication of punctuated equilibrium: the establishment of species as Darwinian individuals, and the consequent validation of species sorting and selection as a prominent process in a hierarchical theory of Darwinian evolution. In 1984, Dawkins acknowledged that this aspect of punctuated equilibrium "does, in a sense, move outside the neo-Darwinian synthesis, narrowly interpreted. This is about whether a form of natural selection operates at the level of entire lineages, as well as at the level of individual reproduction stressed by Darwin and neo-Darwinism." In his 1986 book, Dawkins then devotes a substantial part of the chapter following his rejection of punctuated equilibrium to an evaluation of species selection. But he finishes his exploration by reimmersion in the same parochial trap of denying importance because the phenomenon doesn't explain his exclusive interest in adaptive organismal design: "To conclude the discussion of species selection, it could account for the pattern of species existing in the world at any particular time. It follows that it could also account for changing patterns of species as geological ages give way to later ages, that is, for changing patterns in the fossil record. But it is not a significant force in the evolution of the complex machinery of life . . . As I have put it before, species selection may occur but it doesn't seem to do anything much!" (Dawkins, 1986, pp. 268–269). But doesn't "the pattern of species existing in the world at any particular time" and "changing patterns in the fossil record" represent something of evolutionary importance?

At the end of his long riff against punctuated equilibrium, Dennett also pauses for breath and catches a glimmer of the concept that seems important and theoretically intriguing to many students of macroevolution (Dennett, 1995, pp. 297–298):

The right level at which to look for evolutionary trends, he [Gould] could then claim [indeed I do], is not the level of the gene, or the organism, but the whole species or clade. Instead of looking at the loss of particular genes from gene pools, or the differential death of particular genotypes within a population, look at the differential extinction rate of whole species and the differential "birth" rate of species—the rate at which a lineage can speciate into daughter species. This is an interesting idea . . . It may be true that the best way of seeing the long-term macroevolutionary pattern is to look for differences in "lineage fecundity" instead of looking at the transformations in the individual lineages. This is a powerful proposal worth taking seriously.

I am puzzled by the discordance and inconsistency, but gratified by the outcome. Dawkins and Dennett, smart men both, seem unable to look past the parochial boundaries of their personal interest in evolution, or their feelings of jealousy towards whatever effectiveness my public questioning of their sacred cow of Darwinian fundamentalism may have enjoyed (see Gould, 1997d)—so they must brand punctuated equilibrium as trivial. But they cannot deny the logic of Darwinian argument, and they do manage to work their way to the genuine theoretical interest of punctuated equilibrium's major implication, the source of our primary excitement about the idea from the start.

**THE WISDOM OF AGASSIZ'S AND VON BAER'S THREEFOLD HISTORY OF SCIENTIFIC IDEAS.** When I was writing *Ontogeny and Phylogeny*, I came across a wonderful, if playfully cynical, statement by the great embryologist Karl Ernst von Baer (1866, p. 63) about Louis Agassiz's view on the ontogeny of scientific theories (also quoted on p. 687): "Deswegen sagt Agassiz, dass wann eine neue Lehre vorgebracht würde, sie drei Stadien durchzumachen habe; zuerst sage man, sie sei nicht wahr, dann, sie sei gegen die Religion, and im dritten Stadium, sie sei längst bekannt gewesen." [Therefore, Agassiz says that when a new doctrine is presented, it must go through three stages. First, people say that it isn't true, then that it is against religion, and, in the third stage, that it has long been known.]

I won't vouch for the generality of this scenario, but Agassiz's rule certainly applies to the history of nonscientific debate about punctuated equilibrium, particularly to the aspect governed by jealousy of critics—as Eldredge and I recognized in a previous publication entitled: "Punctuated equilibrium at the third stage" (Gould and Eldredge, 1986). The first stage of empirical denial, extending roughly from our original publication in 1972 to the Chicago macroevolution meeting of 1980, featured studies of fossil sequences by paleontologists (notably Gingerich, 1974 and 1976), many of whom tried to deny that punctuated equilibrium occurred very frequently, if at all, by documenting cases of gradualism.

During the second phase, spanning the first half of the 1980's, the primary subject of this section, punctuated equilibrium, was vociferously dismissed as contrary to religion—that is, as apostate anti-Darwinian nonsense. Our theory, falsely read as a saltationist doctrine proclaiming the overthrow of the Modern Synthesis, if not of Darwinism itself, received a hefty dose of anathematization in tones usually reserved for demonizing religious heterodoxies.

The third phase then began in the mid-1980's, as documented in Sections III–V of this chapter, and has continued ever since. The evidence became too great, and we withstood all ideological attacks without sustaining appreciable damage. Punctuated equilibrium now seemed both coherent in argument, and supported by a sufficient number of empirical studies to become a recognized evolutionary phenomenon—though at a relative frequency as yet undetermined. Such a situation must cause critics to remember the old cliché: if you can't beat 'em, join 'em (but don't grant 'em too much credit for innovation or originality). Move instead to phase three of Agassiz's continuum—"sure it's true, but we always knew this; punctuated equilibrium amounts to no more than a little wrinkle on the skin of neo-Darwinism."

As an initiating episode of the third phase, Darwinian biologists began to construct models that rendered punctuational patterns (though not always cladogenetic events of true punctuated equilibrium) by standard formulae of population genetics under certain reasonable assumptions and conditions. We have always welcomed these formulations, for we never sought the radical content of punctuated equilibrium in novel microevolutionary processes, as I have emphasized throughout this chapter—and any demonstrated mechanism for punctuational patterns evokes both our interest and satisfaction. The first two studies in this genre appeared in 1985—Newman, Cohen and Kipnis (1985) and Lande (1985). In 1986, Roger Lewin wrote a "news and views" commentary for *Science* entitled: "Punctuated equilibrium is now old hat." He ended with a gratifying comment by Joel Cohen: "In terms of the tenor of the debate, which at times has been strident, the new results will bring the various parties closer together. Cohen readily concedes that population geneticists very probably would not have applied their mathematical tools to the issue in this way had there not been such a big fuss stirred up by the paleontologists' claims. 'They deserve credit for that,' he says."

So I guess we won by Agassiz's scenario, even if personal motivations of an ungenerous nature lead our severest critics to belittle our achievement as true after all, but trivial from the outset. But why then did they ever make such a fuss?

**A CODA ON THE KINDNESS AND GENEROSITY OF MOST COLLEAGUES.** This section, devoted to unscientific critiques by professional colleagues, centers on unhappy themes of jealousy, pettiness and meanness of spirit. But I do not wish to leave the impression that these unpleasantnesses have dominated the totality of discussion about punctuated equilibrium. Quite to the contrary, in fact—and I have already discussed the numerous tough, spirited, helpful and scientific critiques of punctuated equilibrium in sections III–V of this chapter. Intense and useful debate has predominated throughout the history of punctuated equilibrium. Most of our colleagues have unstintingly followed the norms and ideals of scientific discussion, and we have primarily wrestled with good argument and content, not primarily with deprecation and personal attack.

I have mentioned and cited several of these generous reactions, these fair and accurate characterizations, throughout this section—the journalistic accounts of Tudge and Gleick (pp. 993, 994); the excellent textbook epitome of Curtis and Barnes (p. 998); the generous assessment of punctuated equilibrium's scientific importance by Rhodes (1983); and the acknowledgement of punctuated equilibrium's prod to further exploration of formulae in population genetics by Cohen (p. 1022). I also wish to emphasize that most professional colleagues have always given us generous credit, and have applauded both the debate and the interest generated by punctuated equilibrium.

I have particularly appreciated the fairness of severe critics who generally oppose punctuated equilibrium, but who freely acknowledge its legitimacy as a potentially important proposition with interesting implications, and as a testable notion that must be adjudicated in its own macroevolutionary realm. Ayala (1982) has been especially clear and gracious on this point:

> If macroevolutionary theory were deducible from microevolutionary principles, it would be possible to decide between competing macroevolutionary models simply by examining the logical implications of microevolutionary theory. But the theory of population genetics is compatible with both punctualism and gradualism; and, hence, logically it entails neither. Whether the tempo and mode of evolution occur predominantly according to the model of punctuated equilibria or according to the model of phyletic gradualism is an issue to be decided by studying macroevolutionary patterns, not by inference from microevolutionary processes. In other words, macroevolutionary theories are not reducible (at least at the present state of knowledge) to microevolution. Hence, macroevolution and microevolution are decoupled in the sense (which is epistemologically most important) that macroevolution is an autonomous field of study that must develop and test its own theories.

Such statements stand in welcome contrast to the frequent grousing of strict Darwinians who often say something like: "but we know all this, and I said so right here in the footnote to page 582 of my 1967 paper; you have stated nothing new, nothing that can alter the practice of the field." I will never forget the climactic moment of the Chicago macroevolution meeting, when John Maynard Smith rose to make such an ungenerous statement about punctuated equilibrium and macroevolutionary theory in general—and George Oster responded to him: "Yes, John, you may have had the bicycle, but you didn't ride it." In the same vein, I appreciate the comment of Marjorie Grene (quoted in Stidd, 1985), a famous philosopher who has greatly aided the clarification of evolutionary theory:

> Yet on both these counts—gradualism and neutralism—evolutionists other than paleontologists now appear unmoved. Just what we've said all along, they cry, even though, I swear, I've heard, and seen, them

emphatically assert just the opposite, time and time again. No sudden changes, no non-adaptive changes, they used to exclaim, while now they ask cheerfully: why not stasis, sudden change, and neutral mutations all over the place except for an adaptive innovation here and there, now and then? We always knew it was like that. Nothing really new, no revolution here.

Finally, I am heartened by the many top-ranking biologists who have found fruitful ideas and new wrinkles in the concept of punctuated equilibrium and its macroevolutionary implications—for utility in practice remains the ultimate criterion of judgment in science. I appreciate Dan Janzen's affirmation in his article "on ecological fitting" (Janzen, 1985, p. 308): "I suddenly realize that I have blundered through the front door of the turmoil over punctuated equilibria. We don't have to dig at the fossil record; punctuated equilibria are right here in front of us, represented by most of the species that you and I have anything to do with."

I welcome the generous assessment of Kenneth Korey (1984) in the preface to his compendium of Darwin's best writing, *The Essential Darwin*:

> Unquestionably no single challenge to the synthesis has provoked more attention than the theory of punctuated equilibrium advanced by Niles Eldredge and Stephen Gould. . . . It is true that punctuated equilibrium was not a prediction of the synthesis; on the contrary, Simpson emphasized continuous, phyletic evolution as the most pervasive feature of evolution at this level . . . On the macroevolutionary front . . . punctuated equilibrium as an empirical proposition is not, perforce, in conflict with the synthesis, although if its wide province becomes established, then a more complete theoretical explanation for stasis will certainly be wanted. Species selection, in its present form, would seem to require the most profound reworking of evolutionary theory.

And I thank Paul Ehrlich (1986) for recognizing the genuine novelty and utility of punctuated equilibrium in his book, *The Machinery of Nature*:

> The jury is not in on the punctuated-equilibria controversy. That the "snapshot" of differentiation we see today seems to reveal all stages of differentiation does not necessarily signal a win for the gradualists . . . And it is not fair to swallow the punctuationist view within the gradualist orthodoxy simply because the possibility of rapid speciation has always been part of that orthodoxy. The punctuationist view is about dominant patterns, not about what is possible—and it represents a genuine challenge to one widely held tenet within evolutionary theory.

It has been a wonderful ride on Oster's bicycle, and we still have such a long road to travel.

# The Integration and Adaptation (Structure and Function) in Ontogeny and Phylogeny: Historical Constraints and the Evolution of Development

## Constraint As a Positive Concept

### TWO KINDS OF POSITIVITY

#### An etymological introduction

After Job has endured, and countered with remarkable success in his straightened circumstance, three cycles of argument from each of his three supposed friends and comforters, a fourth participant in this moral and intellectual debate for the ages—the problem of theodicy, or why should the righteous suffer as much bodily torment and material deprivation as the unjust?—steps forward to make his pitch. Elihu, the son of Barachel the Buzite, offers an only slightly more comforting argument in admitting that Job's suffering may well be undeserved, but urging that Job view his distress as a salutary discipline leading to reconciliation with God (no physical relief, to be sure, but a damned sight more encouraging than the previous insistence of Zophar, Eliphaz and Bildad that Job must have sinned if God had so punished him).

Elihu states that he had hesitated to intervene previously because his youth demanded forbearance. (Modern Biblical scholars, on the other hand, regard Elihu's argument as so inconsistent with the rest of the book, in both style and content, that these late chapters probably represent a subsequent interpolation, thus explaining the curious fact that Elihu's name appears nowhere else in the entire book. Elihu probably "waited" his turn until the end because he didn't exist in the original story.) But he will now speak because he must. An internal force demands that he remain silent no longer: "I also will shew mine opinion. For I am full of matter; the spirit within me constraineth me. Behold my belly is as wine which hath no vent; it is ready to burst" (Job 32:17–19).

I choose this unconventional mode of beginning a scientific discussion with Biblical exegesis because this chapter (and a central theme in the logic of this entire book) rests upon a particular definition and construction of the concept of constraint—a meaning easily defended both terminologically and factually, but often so buried in a confusing and contentious literature that the centrality of the argument becomes lost in collegial frustration.

In short, I emphasize two major premises at the outset: (1) The concept of constraint must be sharpened and restricted in meaning to a coherent set of causal factors that can promote evolutionary change from a structuralist perspective different from—in the helpful sense of "in addition to" or "in conjunction with, and yielding interesting nonlinear conclusions in the amalgamation," rather than "in opposition to"—the functionalist logic of Darwinian natural selection. (2) The concept of constraint must include theoretically legitimate and factually important *positive* meanings—*i.e.*, constraints as *directing causes* of particular *evolutionary changes*—rather than only the negative connotations of structural limitations that prevent natural selection from crafting an alteration that would otherwise be favored and achieved.

The passage from Job, needless to say, only provides an etymological justification for this crucial positivity of meaning. The case for actual existence, and important relative frequency, of these positive aspects then becomes the organizing theme of this chapter. But etymology provides a good beginning, because we must first establish the coherence of a case in language and logic before we can ask, with appropriate clarity, whether nature assents to such a reasonable and testable hypothesis.

The meanings and derivations of "constraint" are varied and complex. The Latin root *stringere* means both to compress or to draw tight (the negative connotations), but also to move, affect or touch (the positive aspects). The prefix *con*, meaning "with" or "together," brings several items into the field of change or compression. Thus, constraints can surely be negative—as when we toss a group of miscreants into a jail cell in order to keep them close and restrict their movements. But constraints can also be positive, as when we force a group of items into closer conjunction so that their combined power and speed can grow and also become more focused in a particular direction towards a definite goal—as in the increased speed of fluids in narrowed pipes, according to Bernoulli's principle.

I do not deny that modern English usage favors the negative connotations—hence my rationale for this introduction. But the positive meanings remain current, and certainly sanctioned both historically and linguistically. I began with Elihu's statement because, when I first studied the Bible as a teenager, this passage confused me. I did not yet know the positive meaning of constraint, and therefore couldn't figure out why Elihu, although practically bursting from his need to speak, felt so constrained that he dared not do so (even though he did so in the very next verse!). Of course, Elihu meant "constraineth" in the opposite and positive sense that his need would force, or constrain, the desired result—and that his words would pour forth in a definite and channeled direction, not as a random spewing.* This active sense

---

*Interestingly, the *King James Bible* (a 17th century document) uses the word constraint ten times, nine in the positive sense of directing or forcing an action in a particular way. The most popular of 20th century "updates," the *Revised Standard Version,* keeps the word "constrains" in some passages, but often changes the King James entry to "urge," "compel," or "make," to emphasize the obviously intended positive meaning.

of "constraint" as promoting change in particular directions marks the positive meaning that motivates this chapter and raises an important issue for Darwinian theory in providing a best general case and opportunity for positive interaction with the functionalist precepts of natural selection.

### The first (empirical) positive meaning of channeling

Orthodox Darwinian functionalists have often reacted by arguing "why the big fuss, we know (and both admit and use) the concept of constraint already" to evolutionists of a structuralist bent who claim that a properly formulated version of "constraint" should evoke great interest and provoke substantial reform. I would accept this jaded reaction if the version of "constraint" proffered by structuralist critics stayed within the negative meanings described above. But positive meanings of constraint—and I will outline two different constructions of positivity in this section—can lead to important extensions of evolutionary theory by questioning and reformulating what I have called the second branch, or the second tripod leg, of Darwinism's essential triad of indispensable arguments: the functionalist attribution of effectively all substantial evolutionary change to natural selection.

I agree that negative constructions of constraint do not seriously challenge this major precept (while supplying some interesting subtleties and wrinkles that orthodox functionalists can use and appreciate) for the following set of interconnected reasons: Mainline Darwinism is a functionalist theory of "trial-and-error externalism" (in R. C. Lewontin's phrase). The organism "proposes" by generating variation, ultimately by mutation (and subsequently by distribution in sexual recombination for organisms traditionally deemed "higher"), among members of populations. This variation acts as raw material—the "chance" in Monod's famous metaphor of "chance and necessity"; the "error" in Lewontin's "trial-and-error"—for a causal process of natural selection (the "necessity" in Monod's pairing; the "trial" of Lewontin's joining). That is, the organism proposes, and the environment (interacting with the organism) disposes.

The organism's generation of variation provides the internal component of evolution; the environment's process of selection marks the external contribution. These internal and external factors play strikingly different roles in Darwinian theory—a contrast well epitomized (as noted above) in Monod's phrase "chance and necessity." The internal component can only supply raw material and does not establish the rates or vectors of change. This claim—that variation provides potential, but not direction—sets a fundamental postulate of Darwinian mechanics and philosophy. Natural selection, the external component, carries full responsibility for the direction—and also, ultimately, for the modes and rates—of evolutionary change.

As discussed at length in Chapter 2, Darwin's central insight that variation must be "isotropic"—particularly, that it be copious in amount, small in extent, and undirected towards adaptive configurations—underlies his brilliant grasp of what selection requires from variation to permit a functionalist theory to operate in principle, and also to dominate the causes of evolutionary change. If variation is truly isotropic in Darwin's hypothesized sense, then selection gains free rein (and reign) as the cause of change—in sharp contrast to nearly all competing 19th century evolutionary theories, with their stress on internally generated directionality.

To epitomize the theoretical importance of constraint in a single sentence, the concept of non-isotropy in variation may be roughly synonymized with notions of "constraint"—that is, with claims that internal factors restrict the freedom of natural selection to establish and control the direction of evolutionary change.

The crucial importance of constraint in evolutionary theory therefore centers upon this potential challenge, particularly upon the nature and scope of the restrictions. In Chapters 4 and 5—the longest section of this book's first half, because the subject consistently commanded the principal attention of Darwin's most cogent critics—I considered both pre and post-Darwinian theories of evolutionary internalism, as rooted in the common claim that any significant non-isotropy of variation would reduce, or even cancel, the creative role of natural selection by identifying potent internal forces of evolutionary change in either *directed* or *saltatory* variation. These two persistent bugbears of "quick" and "channeled" received their most memorable (and influential) joint expression in the challenging metaphor of Galton's polyhedron (see pages 342–351 of Chapter 5).

Of course, no sophisticated Darwinian ever denied some limited domain of validity to the concept of internally generated "constraints." (As I have emphasized throughout this book, validation in natural history rarely follows the criterion of "never in principle for this would violate nature's laws," as favored in some constructions of the so-called exact sciences, but rather the standard of "conceivable in principle, but not occurring often enough to matter," as followed in historical sciences that formulate most basic judgments by analysis of relative frequencies.) Rather, Darwinian functionalists have tended to admit certain kinds of constraints, and have then tried to limit their modes of occurrence and domains of action in such a manner that the central principle of Darwinian theory—the control of evolutionary *change* by natural selection—will not be threatened.

In short, and to summarize these few pages of argument in a paragraph, orthodox Darwinians have not balked at *negative* constructions of constraint as *limits* and *impediments* to the power of natural selection in certain definable situations. But they have been far less willing to embrace *positive* meanings of constraint as *promoters, suppliers,* and *causes* of evolutionary *direction* and *change*. This distinction follows logically from the basic premises of Darwinian functionalism, because the admission of a potent and positive version of constraint would compromise the fundamental principle that variation (the structuralist and internalist component of evolution) only proposes, while selection (the functionalist and externalist force) disposes as the only effective cause of change.

In considering how structural constraints might limit the power of natural selection to adapt each feature of an organism to each local environment, we recognize that some modes will rank as "benign" for Darwinian functionalists; others as less benign but not subversive to orthodox theory; whereas still others—particularly the positive modes that promote, and do not just limit, evolutionary change—do pose a deeper theoretical challenge, and have therefore set the major battleground of a subject that has, of late, become both highly confusing and maximally contentious in the literature of evolutionary biology (Gould, 1980c; Alberch, 1982; Maynard Smith et al., 1985; Stearns, 1986; Antonovics and van Tienderen, 1991; Schwenk, 1995; Duboule and Wilkins, 1998; Eble, 1999).

The most benign category does not restrict the organism's potential for reaching a best adapted overall phenotype for a particular environmental background—and therefore only counts as "constraint" with respect to un-realizable and idealized abstractions. To cite two examples, so called "trade offs" preclude *separate* optimization of each part because natural selection works upon the entire organism as a totality. The best-adapted whole cannot evolve as a simple summation of separately optimized parts because, in an integrated structure that must function as a single coherent entity, the "perfection" of some parts can only be achieved at the expense of others. Therefore, to cite an old conjecture for illustrating the obvious, the optimal size for a human brain at birth may be too big to allow the passage of a neonate through the birth canal. But such structural constraints, imposed by a selected whole upon individual parts not subject to independent optimization in any case, do not challenge, but rather affirm, the central Darwinian postulate that selection works on organisms.

Secondly, mechanical limits (also structural or formal in character) obviously preclude certain solutions that might offer abstract advantages in adaptation. Zebras could avoid feline predators by flying away, but even if genetic variation existed (as it almost surely does not!) for constructing a supernumerary pair of limbs in winglike form, zebras clearly exceed permissible weight limits under the venerable Galilean principle of declining surface to volume ratios in large creatures.

I cite both these examples tongue-in-cheek because no one would view such obvious, and evidently "benign," classes of structural constraint as challenges to Darwinian adaptationism (or even as particularly interesting in any intellectual sense). Darwinian functionalism works by local adaptation of integral organisms to immediate environments. Neither biomechanical optimization part by part (prevented by "trade-offs" or integral constraints), nor putatively advantageous configurations outside the limits of mechanical possibility (physical or formal constraints), poses any challenge to the tenets of Darwinian functionalism.

In a less benign, theoretically relevant (although ultimately not debilitating), and widely discussed category, limitations based upon absence of sufficient variability to provide raw material for natural selection (and usually called genetic or developmental constraints) do operate widely in nature. (Since natural selection "makes nothing" by itself, but can only operate upon raw material supplied by an independent process of variation—a statement familiar enough to rank as a "mantra" among Darwinian evolutionists—a shortage of building stones can slow, or even derail, the construction of a well designed house.)

In the "consensus paper" of Maynard Smith and eight prominent colleagues associated with a wide range of views from Darwinian orthodoxy (Lande and Maynard Smith himself) to serious structuralist heterodoxy (Raup, Goodwin and Kauffman), "limitations on phenotypic variability" (1985, p. 269) became the nucleating point of agreement in their remarkable exercise in intellectual diplomacy. Therefore, "developmental constraint" in this sense of limitation in necessary raw material to fuel the workings of natural selection, has become (in a minimalist interpretation that I do not challenge) the canonical "base-line" or "common ground" definition for this important structural component in Darwinian theory. (In fairness, Maynard Smith et al., do acknowledge positive meanings of constraint as both legitimate and more interesting (see p. 1037), while advocating this "negative" definition as a minimal standard that all evolutionists can embrace, and that no Darwinian need regard as dangerously debilitating.)

Thus, to return to my previous and facetious example, zebra wings would not work for the reason cited above, but natural selection will presumably never encounter an opportunity even to attempt their construction because sufficient variability for a supernumerary pair of limbs presumably does not exist in the genetic and developmental systems of tetrapods. In a more meaningful category—representing a frustrating and unresolved issue that has permeated Darwinian discussion ever since the eponym himself—the limited range of realized phenotypes in some clades (with domestic breeds of cats vs. dogs as the classic example) may reflect a structural limit in variation, rather than a lack of selective opportunity or advantage.

An important functionalist principle of natural selection, frequently (and quite explicitly) emphasized by Darwin, holds that we may, in operational terms as the "null hypothesis" of our initial assumptions for empirical testing, treat populations as though they always possess sufficient variation to permit natural selection an unimpeded range of action. As a practical expression of this basic Darwinian belief, rates of evolutionary change fall under the control of natural selection, not of limitations (or superfluities) in raw material. Clades that either change slowly, or fail to generate many species, should be regarded as subject to little selective pressure, not as limited by intrapopulational variation. If cats have developed far fewer varieties than dogs, then the differential selective efforts of human breeders, rather than any disparity in the ranges of available raw material, should explain the striking difference.*

To cite just one among his many explicit statements of this crucial claim,

---

*To make a personal statement, I was a dyed-in-the-wool adaptationist during my undergraduate and graduate years, but this particular claim—advanced by all my teachers as an article of faith—always bothered me. I saw no reason beyond an overweening faith in the strength and ubiquity of selection (Weismann's *Allmacht* all over again) to assume that variation in supply of raw material should exert so little effect on *rates* of evolutionary

Darwin wrote to Lyell in 1862 (quoted in F. Darwin, 1903, vol. 2, p. 338—see pp. 330–341 for further discussion, and especially p. 341 for Darwin's comprehensive architectural metaphor for dismissing constraint as a theoretical challenge to natural selection):

> Mere variability, though the necessary foundation of all modifications, I believe to be almost always present, enough to allow of any amount of selected change; so that it does not seem to me at all incompatible that a group which at any one period (or during all successive periods) varies less, should in the long course of time have undergone more modification than a group which is generally more variable.
>
> Placental animals, e.g. might be at each period less variable than Marsupials, and nevertheless have undergone more *differentiation* and development than marsupials, owing to some advantage, probably brain development.

I label this negative category of constraint based on lack of sufficient variability as "less benign" because its operation does place a genuine damper, both actual and theoretical, upon the exclusivity of natural selection as the cause of evolutionary change. Darwin himself certainly read the issue in this light, as the above quotation indicates—for if his argument fails, and constraint often trumps selection as a regulator of evolutionary rate, then his resulting disappointment, in needing to recalibrate and downgrade the relative importance of natural selection, will evidently be severe.

But this far even the most devoted selectionist must proceed—for the logic of this "less benign" category cannot be gainsaid. The basic formulation of the theory of natural selection does require structural input of raw material by variation to fuel the functional outcome of evolutionary change by selection. (And since natural selection cannot, in principle, manufacture this necessary fuel for its own operation, lack of input can stymie output—just as the niftiest motor car can't move if you run out of gas in the middle of the Sahara, hopelessly far from the nearest petrol station in Timbuctu.) Therefore, one cannot brand limitation in raw material as an incoherent concept, or even an empirical rarity, *a priori*.

And yet, if selectionists can hold the line—as they generally attempt to do—at this negative definition of constraint, their theory, while deniably impacted, suffers no serious setback to its truly essential postulate that natural selection controls the *direction* of evolutionary change. At most, the negative forces of constraint may slow down, or even prevent, modifications. But so long as these structural factors do not operate in a positive sense—either to determine important variation in rates and extents of change or, more threateningly, to impact or set the actual direction of change—then the fundamental Darwinian rule still prevails: variation proposes, but only natural

---

change—even though I remained quite willing to believe that natural selection always set the *direction* of change.

selection can dispose. Internal forces supply the possibilities; but natural selection builds the pattern (because the possibilities nearly always exist in sufficient abundance to fuel the changes that natural selection might favor). Constraints impede, but do not direct.

Thus, a standard Darwinian "truce" accepts the notion, even the potential importance, of constraint as a negative force that can impede rates and amounts of change (and therefore cannot be dismissed as irrelevant to the generation of evolutionary pattern in this limited sense). But strict Darwinian functionalists generally try to hold the line by denying importance, or even legitimacy, to positive definitions of constraint as causes of directionality in evolutionary change.

In one of the best examples I have ever encountered of the vital (almost morally enjoined) principle that the forebears of our current struggles demand our continuing study and respect—and that we often gain, as recompense for this fealty, substantial practical benefit in getting our own thoughts straight, and channeled in useful directions—the rich history of debate about Darwinian theory has brought the theme of positive constraints into sharp focus (see Chapters 4–5). This clarity emerges from the common emphasis placed by all major structuralist critics upon (1) the difference between positive and negative meanings of constraint (accompanied by specifications that only the positive meanings could pose serious difficulties for Darwinism), and (2) the parsing of positivity into two essential themes of speed, or enhancement of rates beyond the power of natural selection to instigate, and channeling, or the preferential (perhaps even requisite) flow of change in particular directions set by internal possibilities, even if natural selection must supply an initial impetus. We should also note the ironic sense in which this argument inverts the canonical roles of the two central components in Darwinian theory. In natural selection, an internal source of variation provides the impetus, whereas selection determines direction. In channeled change by constraint, natural selection supplies the impetus by "getting the ball rolling" (to use Galton's metaphor of the pool table), but the directionality of evolutionary change, or "where the ball rolls," emerges from internal channels that, so to speak, "use" natural selection as their convenient source of power. In short, variation as raw material and selection as the shaper of change in Darwinism; *vs.* selection as raw power, and channeled variation for shaping in theories of positive constraint. Such an epitome, needless to say, remains far too simple to resolve nature's ways—but this formulation does embody a clear and useful conceptual dichotomy for clarifying our thoughts.

### *The second (definitional) positive meaning of causes outside accepted mechanisms*

A second, and conceptually quite distinct, sense of positivity for the concept of constraint also arises from a vernacular meaning of the word, but embodies a philosophical position about the general nature of theories and arguments in science, rather than a specific empirical claim about the nature of evolution.

Consider ordinary linguistic usage for the following scenario: a favored or orthodox theory undergirds the basic research program of a discipline—the usual situation that Kuhn (1962) calls "normal science" as practiced under the influence of a reigning paradigm. All scientists know, of course, that rare episodes of transition between explanatory systems, often occurring with sufficient speed and upset—both structural (to theories) and emotional (to practitioners)—to be deemed revolutionary, mark our most interesting times. Moreover, nearly all scientists, if not utterly devoid of ambition or intellectual verve, regard the development of a new explanatory system as the highest form of achievement in their profession. Nonetheless, the full working careers of most scientists proceed in the usual mode of research within a basic paradigm—a "good life" full of interest and intellectual excitement, as any rich paradigm features forests of unsolved puzzles, and byways (or even substantial roads) of expansion and originality.

Within such a ruling theory, a set of accepted causes and mechanisms operates to yield a range of outcomes specified as permissible. (When too many inexplicable results become well documented outside this permissible range, ruling theories become strained, and an interesting time of theoretical transition may soon be at hand.) Now, as a purely linguistic point, what should we call a set of anomalous results that would not have occurred if our reigning theory held the dominant or exclusive sway usually granted to its precepts? What, for example, would we say about our inability to turn mercury into gold if our causal theory proclaimed the possibility of so doing, or (to choose a case of expansion rather than restriction) what would we call our newfound ability to generate living insects from decaying flesh if our theory dictated that only plants, but not animals, could originate by spontaneous generation?

We might, of course, eventually abandon our old theory for a novel system of explanation. But what if we do not choose to do so, at least not yet, and especially if we know that our theory really does work well, and as specified, for a large range of well documented cases? We would have to acknowledge that the old theory does not enjoy so wide or exclusive a domain of application as we had previously asserted. What would we then call the classes of exceptions—particularly the results of unorthodox causes that forced us to accept limitations upon the old beliefs? We generally label such exceptions as "constraints" because they restrict the range and power of our orthodox explanations.

I regard this conceptual meaning of constraint—the imposition of limits upon the range of orthodox theories by documentation of exceptions and demonstration of unorthodox causes—as undeniably "positive" in the important intellectual and psychological sense that any scientist worth his salt must cherish such upsetting discoveries for the conceptual challenges thus unleashed. Thus, if the Darwinian functionalism of natural selection acts as a reigning theory, then any documented constraint from internal channeling of variation—whether positive or negative in the empirical sense discussed in the last section—must be viewed as intellectually positive for questioning our orthodoxy and documenting something new and interesting that shouldn't have happened under our usual views. Thus, this second, conceptually positive meaning of "constraint" embodies a relative concept that might be feared by those who enjoy the comforts of power, but should provoke the delight of all scientists.

To cite two recent examples of this relative meaning from recent evolutionary literature, Weiss (1990) wrote an iconoclastic paper arguing that geometric laws of serial repetition, combined with limited structural paths of alteration, dominate directions of phyletic change. (By the way, I disagree with his conclusion and am only discussing his terminology.) In this context, the usual orthodoxy of ordinary natural selection, working towards optimality in local adaptation, will be judged in an opposite manner as an annoying trifle that might falsify the grand pattern and temporarily hide its effects—in other words, as a constraint upon the regularity of geometrically predictable transformation. Weiss (1990, p. 21) acknowledged that natural selection occurs, but he dismissed the process as local distortion: "The pervasiveness of metameric 'duplication with variation' shows that it is a central principle of evolution. . . . Despite . . . the pattern-distorting effects of selection and drift, this evolutionary strategy is essentially unidirectional."

Weiss's taxonomy of concepts, so peculiar to those of us with Darwinian training, makes sense in his system. We would never think uniting selection and drift into the same category, for we view them as opposite processes with respect to our primary interest in adaptation. But, within a theory of predictable linear change enjoined by geometric principles, both drift and selection operate as local oddities that distort a broader and fundamental pattern. In any case, when we note how selection becomes a constraint upon a structuralist theory of geometrically rule-bound transformation, we can understand more easily why internal channels of preferred variation would be labeled as constraints upon a theory that ascribes all evolutionary direction to natural selection. (When a rebel labels one's own central belief as a limiting constraint, the generality of the usage becomes startlingly clear!)

In another example, Jackson and Cheetham (1999) cite punctuated equilibrium as constraining because phylogenetic patterns generated by this theory preclude several classes of results predicted by orthodox selectionist models of gradualistic anagenesis in populations. They write (1999, p. 72): "The realities of punctuation and stasis need to be better incorporated into evolutionary studies. Punctuated speciation does not contradict conventional neodarwinian mechanisms, but it does constrain the range of probable evolutionary scenarios for speciation, evolution of life histories and macroevolutionary trends." "Macroevolutionary trends," they add in explanation (p. 76), "must arise through differential rates of origination and extinction, and not by adaptive evolution within single species."

Several participants in debates about the evolutionary meaning of constraint (see Gould, 1989a) have explicitly embraced this relative definition. Stearns (1986), for example, properly rejected a usage so overly broad that the term would then lose all meaning—namely, the designation of all cause as "constraint" because any active force must direct change in one way rather than down other conceivable paths. "The meaning of the word would [then] vanish," Stearns notes (1986, p. 35). He therefore recommends: "We can preserve it in a relative sense if we recognize that it only has meaning in a local context where one concentrates on the possibilities latent in certain processes and views the limitations on those possibilities as arising from outside that context." Therefore, in considering revisions and expansions of Darwinian theory, ordinary natural selection becomes the context and any force (like internally channeled variation) limiting its exclusive sway in directing evolutionary change, becomes a constraint.

Antonovics and van Tienderen (1991), in an influential article that cleared away much of the accumulating nonsense in definitional debates about constraint in evolution, also favored this relative concept as a solution. They agreed with my argument (Gould, 1989a) that "it is those factors that influence the process but are external to the favored theory that should be termed constraints" (Antonovics and van Tienderen, 1991, p. 167). But, choosing a terminology that struck them as more consistent with the ethos of scientific neutrality, they preferred the term "null model" to my "favored theory." (I would reply that we do not usually refer to strong theories—like natural selection—based on particular and well-articulated causes, as "null models." I would also argue that nothing negative attends the admission that disciplines operate under favored theories—a "good thing" for science, so long as we retain flexibility for change and do not equate "favored" with "established"; and, especially, so long as we treat the status of "favored" as an impetus for challenge rather than passive acquiescence, as we manifestly do when we invoke constraints to rebut overly strict versions of natural selection.)

In any case, Antonovics and van Tienderen survey the literature and find, in support of the argument developed here, that "the overall null model used by most authors was one of evolution by natural selection (irrespective of the level of selection)" (p. 167), and that nearly all explicit claims for "constraints" upon change within populations and lineages "dealt with evolutionary constraints to adaptation by natural selection" (p. 166). They also noted the "odd" feature of relative definitions that strikes many scientists as paradoxical, but would not be so regarded if we accepted the honorable and inevitable principle, so familiar to philosophers of science and language, but still faced with discomfort by many scientific professionals, that all terminology must be "theory bound"—specifically, in this case, that orthodox results of one theory become constraints in other theories. They write (p. 167): "Given evolution by random drift as a null model, natural selection now becomes a constraint!" Yes, and appropriately so—with no exclamation point needed to register surprise.

Although I disagree with his particular recommendation, Eble (1999) published a thoughtful and conceptually innovative paper rooted in this important principle of the inevitability and appropriateness of theory-bound terminology. Eble notes, and brilliantly analyzes, two entirely distinct, but all too frequently conflated, meanings of "chance" and "randomness" in evolutionary theory. His article, entitled "On the dual nature of chance in evolutionary biology and paleontology," distinguishes the conventional statistical meaning from a particular and distinctive sense frequently employed in Darwinian literature—namely, "chance" defined as events occurring for reasons unrelated to the canonical mechanism of natural selection. Eble writes (1999, p. 77): "The gist of the evolutionary notion of chance is that events are independent of an organism's need and of the direction provided by natural selection in the process of adaptation."

Eble discusses examples ranging widely across all scales of evolution, but we all know (and we all make excuses for the resulting confusion in our lectures to elementary courses) the most troubling and common case—the claim that mutational variation in populations, the fuel of natural selection, is "random." Of course, we know perfectly well that such usage does not invoke the usual mathematical concept of randomness, and that we only mean "unrelated to the direction of natural selection"—a point emphasized in Chapter 2 in my discussion of Darwin's need for isotropy in variation (see pp. 144–146). Eble (1999, p. 78) cites the acknowledgment of many biologists, and the analyses of such leading philosophers as Popper and Sober, of this almost "studied" confusion, including a quotation of my own statement (Gould, 1982b, p. 386): "By 'random' in this context, evolutionists mean only that variation is not inherently directed towards adaptation, not that all mutational changes are equally likely."

(Eble recommends that we retain the words "chance" and "random" for both meanings, and then enforce the separation with the restricting adjectives "statistical" *vs.* "evolutionary" chance. He argues (p. 75) that "evolutionary studies . . . can benefit from the simultaneous application of statistical and evolutionary notions of chance"—defining the second concept as "independence from adaptation and the directionality imposed by natural selection," a definition as clearly and explicitly "theory-bound" as any I have ever read. I agree entirely with Eble's analysis. I dissent only from his terminological decision to retain the word "chance" for both concepts, and to rely upon moderating adjectives to enforce the distinction. I would prefer the codification of a different name for the evolutionary meaning both because I don't trust the power of subsidiary adjectives to clarify the vital distinction, and because the statistical meaning represents such an important concept, in both science and practical human life, that exclusivity of usage might aid our uphill battle to educate people about the basic meaning of probability. But my terminological disagreement with Eble does not detract from my admiration for his clear characterization of the distinction, and his rich discussion of the largely unrecognized confusions thus generated.)

In any case, Eble's characterization of evolutionary "chance," and his documentation of such extensive usage in a sense so contrary to the basic mathematical meaning of a fundamental term in science, only underscores the enormous range and influence of natural selection as our canonical theory. If organismal selection, and its key consequence of adaptation, have become so prototypical in defining how evolution works, and what evolution does, that we usually designate any other result as a "chance" phenomenon—even though the outcome may have been generated in a deterministic manner by a process that would be called "causal" in any standard scientific usage—then we achieve a better understanding of how subtle, and how extensive, the clutches of convention can become, even among people committed to innovation and the value of novelty.

When language unconsciously promotes orthodox mechanisms, setting barriers against our examination of alternative modes of causality, then we should vigorously analyze our terminological usages to seek a clarity that might open new possibilities. When we understand the relative meaning of constraint as a theory-bound term, expressing the orthodoxy of selection and designating all other causes of change as limitations upon an expectation*— and when we come to view this relative sense of "constraint" as a positive definition that urges us to explore alternatives to standard explanations— then we can stand a terminological bias on its head, for potential use against the same conceptual lock that engendered such a peculiar terminology in the first place.

## HETEROCHRONY AND ALLOMETRY AS THE *LOCUS CLASSICUS* OF THE FIRST POSITIVE (EMPIRICAL) MEANING: CHANNELED DIRECTIONALITY BY CONSTRAINT

I advocate nothing original in asking evolutionists to focus upon the empirically positive concept of constraint as channels for change, rather than (as in the negative meaning) limits to natural selection imposed by insufficient raw material in variation. The "consensus paper" of Maynard Smith et al. (1985), while stressing a minimalist definition of absent variability for change in certain directions (as a strategy for achieving a "least common denominator" of agreement among authors of very disparate opinions), emphasized both the legitimacy and greater interest of the positive meaning: "Does development merely prevent evolution from following particular paths or does it also serve as a directing force, accounting in part for oriented features of various trends and patterns?" (Maynard Smith et al., 1985, p. 281). Alberch (1982, p. 313) also accentuated the positive by stressing the two great themes—saltations and channels, or speed and directionality—that have always anchored the formalist or structuralist critique of Darwinian functionalism (see Chapters 4–5): "Development does not only define the apportionment of phenotypic

---

*As one more example of how "constraint" terminology can be biased by assumptions that equate "ordinary" evolution with selection and adaptation, Schwenk (1995, p. 251) argued that constraints "can have either negative or positive evolutionary effects at the lineage level (*i.e.*, hamper or promote organismal adaptation)." I doubt that Schwenk truly believes what he literally says—that any "evolutionary effect" hampering organismal adaptation must be labeled as "negative"—but his statement illustrates the common conception that evolution "ought" to build better adapted organisms, and that any other result must be regarded as disappointing or somehow wrongheaded. But all manner of highly interesting phenomena beyond (and sometimes opposed to) organismal adaptation pervade such a richly varied, universal, many-leveled, highly complex process as organic evolution. Do we really want to label this lifeblood of fascination for our favorite subject as negative?

variation upon which selection operates, but it can result in discontinuities and directionality in morphological transformations." For further discussion of the positive meaning of constraint, see Riedl (1978), Gould (1980c, 1989a), and Wagner (1988). Note also that both statements, cited just above, define the positive meaning in explicit contrast with the more usual negative reading, while emphasizing the far greater evolutionary interest of the positive sense.

The familiar and conceptually conjoined realms of allometry and heterochrony define a *locus classicus* for positive constraints in providing a sensible link between the two central themes of speed (for ease) and channeling (for direction). If we wish to argue that biased channels of internally-set variation can aid natural selection or any other functional theme in evolution, where could we find a better example than ontogeny itself, especially when the course of life features substantial allometry across a broad range of size, and often of environment as well (especially for organisms with distinct phases of larva and adult, for example). After all, this fundamental channel already generates a series of well-adapted stages each time an organism grows to maturity, for all parts of the life cycle must "work" in the Darwinian world of environmental interaction, or else the organism would not exist. (See Chapter 5 for my historical discussion of orthogenesis, as advocated by Eimer, Hyatt, and Whitman, for longstanding recognition of ontogenetic allometry as the primary source of positively channeled constraints.)

If any of these phenotypes would benefit the organism at a different size or stage of life, or if any different combination of characters (reachable by retuning the rates of development among relevant features), might yield increased adaptation, then the existing channel of ordinary ontogeny already holds the raw material in a particularly effective state for evolutionary change. And the more pronounced the allometry, the greater the potential extent of such realizable change.

If allometric ontogenies establish channels of positive constraint, then heterochrony supplies a convenient and effective mechanism for evolutionary utilization. By selective acceleration or retardation of single traits, small to large complexes of correlated characters, or even entire phenotypic stages, heterochrony can differentially extend or compress features across ontogenetic trajectories, and can also "mix and match" the characteristics of several stages into a transformed phenotype. (Contrary to a popular impression, for example, the evolutionary power of progenesis does not lie in full "promotion" of a functioning larva to sexual maturity, but rather in the almost invariable, and sometimes adventitiously beneficial, combination of characters that progenesis yields—with some features "left behind" in the early ontogenetic stages appropriate to the truncated age of sexual maturation, and others accelerated to appear in a phenotypically more adult form through correlation with the early achievement of sexual maturity—see Gould, 1977b.)

For these reasons, heterochrony has long been a favored concept among evolutionists searching for mechanisms to accelerate evolutionary rates in complexes of characters—for simple changes in "rate genes" (to use Goldschmidt's old phrase) may yield extensive consequences for entire organismal phenotypes, as suites of correlated characters change in concert with altered rates of development. Thus, with strongly allometric ontogenies as favored channels, and with heterochrony available as a mechanism to move sets of characters quickly along these channels, organisms often meet both classical criteria—channeling and speed—for utilizing constraint as a positive accelerator of evolutionary change.

For these and other reasons, the subject of heterochrony has generated a long, memorable and voluminous literature (see De Beer, 1930 for the classic 20th century statement; Gould, 1977b, for a historical and then-current summary; and McKinney, 1988, McKinney and McNamara, 1991, and McNamara, 1997, for subsequent views). In a later summary, McKinney (1999) notes that three major themes have marked the fruitful use of heterochrony within macroevolutionary studies in recent years: heterochronoclines (or trends caused by temporal displacements of developmental rates), heterochronic biases within clades, and the origin of novelties.

In summarizing the extensive literature on heterochronoclines, McKinney emphasizes the same point stressed here—heterochrony and allometry as convenient and available mechanisms, whereby selection can accelerate and intensify adaptive change (making positive constraint a "partner," not an "antagonist," of selection in many cases). He writes (1999, p. 150): "Heterochronic variation is a very rapid, easy way to produce coadapted suites of traits. It makes sense that simple extrapolations (or truncations) of major environmental parameters (such as water depth, sediment size and temperature) could select for relatively simple extrapolations (or truncations) along the ontogenetic trajectory of a population (= cladogenesis) or species (= anagenesis)."

The frequency of neoteny in salamanders, potentiated by unusual ease in dissociation of sexual maturation from somatic development, represents the classic case of heterochronic biases. In another sensible correlation of positive constraints in heterochrony with adaptive utility, Whiteman (1994) demonstrates that amphibian paedomorphs generally arise when the aquatic habitat of larvae becomes more productive, or more stable, than the terrestrial environment of adults. At a larger scale, and in an intriguing macroevolutionary speculation, McNamara (1997) surveys the known examples of heterochrony among trilobites, and finds that paedomorphosis predominates in Cambrian lineages, while the opposite processes of peramorphosis seems to gain the higher relative frequency in later Paleozoic lines. McNamara wonders if this pattern might not reflect changes in the organization and activity of homeotic genes in times of evolutionary turmoil in and after the Cambrian explosion *vs.* the relative "calm" of the later Paleozoic evolutionary world.

For the third theme of evolutionary novelties, the classic literature has stressed the role of global paedomorphosis, usually in the progenetic mode (as noted in Gould, 1977b), in shedding the "excess baggage" of adult complexity and reverting to the more labile phenotypes that often characterize juvenile forms—an "escape from specialization" in the classic description. Such examples of "wiping the slate clean" may explain the origin of some major groups (see Mooi, 1990, on sand dollars among echinoids), but McKinney (1999) rightly points out that an even more powerful, and almost surely more frequent, heterochronic boost to the origin of novelties may lie in the potential for what I previously called the "mix and match" of characters produced by varying modes of heterochrony in different features and complexes within the same organism.

As a poor and parochial surrogate for adequate review of an immense literature, I limit myself to two examples from my own research that have enlightened me about the evolutionary implications of positive constraint in this allometric and heterochronic form.

### *The two structural themes of internally set channels and ease of transformation as potentially synergistic with functional causality by natural selection: increasing shell stability in the* Gryphaea *heterochronocline*

In quantitative studies of fossil invertebrates, no case has commanded nearly so much attention as the evolution of coiled Jurassic oysters of the genus *Gryphaea* in the British Isles (see Trueman, 1922 for the classic statement). I collected only the major papers of this debate into a full book (Gould, 1980f), while a volume of equal extent has been published since then, leading to what I regard, with obvious self-serving bias, as a genuine solution in Jones and Gould (1999). I will not discuss earlier errors and struggles (see Gould, 1972, for a compendium), and will simply note a consensus reached by the 1970's— that the complete lower Jurassic sequence from *Gryphaea incurva* to *Gryphaea gigantea* features a basic trend in a set of phyletically correlated characters, including substantial increase in body size, decrease in coiling, and increasing relative width of the valves.

These trends, at least in a descriptive sense, certainly seem to embody the heterochronic result of paedomorphosis, as all sustained changes in form led to progressive juvenilization in adult phenotypes of later phylogenetic stages (Fig. 10-1). The strong allometry of increased coiling through ontogeny permits an easy identification of this trend, as larval shells cement briefly to a hard object, with the young organism then breaking free and coiling throughout life on a muddy substrate. Juvenile lower valves (after breaking their initial cementation) therefore begin growth as relatively flat, and then coil progressively throughout life. The phyletic trend to flattening strongly resembles a progressive excursion to earlier and less coiled stages of ontogeny.

But this descriptive consensus remained stymied by a common technical problem in heterochronic studies within paleontology. The causal distinctions within heterochrony can only be specified with reference to the chronological age of specimens, and few fossils record the months and years of their growth in a recoverable manner (see discussion of this dilemma in McKinney and McNamara, 1991; and Jones and Gould, 1999). Without information about the age of specimens, we could not tell whether increasing body size simply represented a chronological extension of the life cycle (with juvenilization of form then unresolvable in mode of origin), or an increase in rates of growth over an unchanged length of life (with the paedomorphic result then attributable to the heterochronic process of neoteny based on prolongation of rapid juvenile growth rates and attendant retention of characteristic morphologies associated with these rates).

I had the privilege of working with Douglas Jones, who developed the first reliable procedures for inferring ages from growth banding (by matching isotopic cycles, interpreted as seasonal, with morphological banding, Fig. 10-2—because simple counts of banding, the standard procedure of past sclerochronological study, had never yielded firm results). We were able to break this conceptual logjam by determining the ages of shells throughout the trend (Jones and Gould, 1999), and resolving the problem, thanks to unusual cooperation from nature (who rarely provides clear answers at one extreme of a potential continuum). The larger adult shells of later phylogenetic stages showed no increase at all in length of life, but died at the same age as adult shells in the earliest stages in the trend (Fig. 10-3). Thus, we could identify the correlated phenotypic trend in size and shape as a genuine case of neoteny

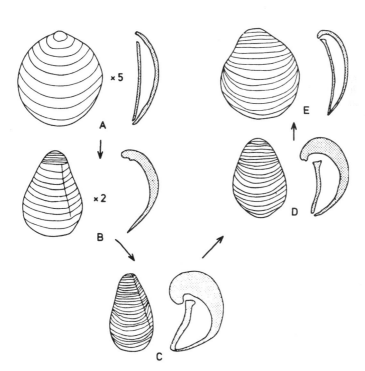

10-1. Paedomorphosis in lower Jurassic *Gryphaea*. The left sequence (top to bottom) shows ontogenetic stages of the ancestral species drawn at the same size as adults of the phylogenetic series (the right sequence, bottom to top). From Gould, 2000e; adapted from Hallam.

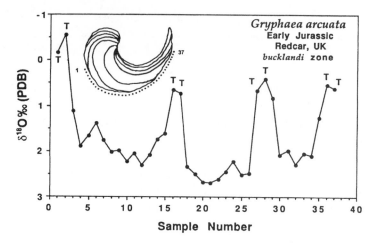

10-2. Determination of age of individual *Gryphaea* shells by oxygen isotope profiles across annual growth increments. Solving this problem of sclerochronology allowed us to determine, for the first time, the actual mode of heterochrony in *Gryphaea*. From Jones and Gould, 1999.

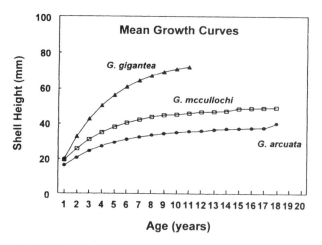

10-3. Increase in shell size (measured as shell height) in the phyletic sequence *G. arcuata* to *G. gigantea*. Although the shells augment markedly in size, this increase does not reflect longer periods of growth as descendants are larger at each comparable age, and the average adult descendant dies before reaching the final age of the average adult ancestor. From Gould, 2000e.

(Fig. 10-4), using the allometric channel of *Gryphaea*'s ontogeny to evolve a broader and less coiled adult shell in later stages of the sequence.

When we combine this structural analysis of the evolutionary trend with a well documented scenario for its adaptive basis, the positive aspect of constraint as an adjunct to selection stands forth in an unusually clear manner. The environmental correlation of flat and cemented *Ostrea* with clear waters and hard substrates, and of coiled and free-living *Gryphaea* with muddy sub-

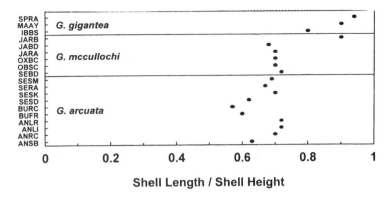

10-4. Evidence of neoteny for the same sequence of *G. arcuata* to *G. gigantea* as measured by increasing juvenilization of form in the increasing length to height ratio—implying that descendant adults become less coiled and therefore more like the flatter juvenile shells. Our analysis of absolute ages for each shell allowed us to specify this case of paedomorphosis as neotenic. From Gould, 2000c.

strates, establishes a long-enduring, iteratively-evolved pattern demanding functional explanation. The adaptive value of coiling in *Gryphaea* has long been ascribed with much evidence in support (see Hallam, 1968, p. 119) and little dissent among experts, to the animal's need to keep the shell commisure above the muddy substrate, lest the shell become entombed or clogged, leading to the animal's death.

Coiling represents an excellent morphological means—presumably the best available given the limitations of bivalved molluscan design—for continually raising the commisure above the substrate as the shell grows. But coiling, particularly if intensified in a relatively narrow shell, also entails the negative consequence of increasing instability, for a narrow object, shaped like the rocker of a hobby horse, can easily be tipped over from its presumed upright life position (see Fig. 10-5, with the plane of bilateral symmetry orthogonal to the substrate). In fact, in highly coiled *Gryphaea*, a shell tipped over onto its side (Fig. 10-6) lies in a position of greater stability than a shell in this presumed, and only viable, life orientation. Some early German paleobiologists, after discovering this fact from hydrodynamic experiments, actually postulated that *Gryphaea* might have lived in such a side-down position. But Hallam (1968) and others argue convincingly that a shell on its side would soon become clogged with mud, rendering the animal unable to feed. Moreover, once the heavy shell is tipped, the animal cannot right itself—so quick death would seem to follow as an inevitable consequence of such displacement from the bilateral living position.

We therefore assume that stabilization of a shell that must coil to rise above a muddy substrate represents a fundamental functional problem for gryphaeate oysters. (Indeed, the most strongly coiled *Gryphaea incurva*, the ancestral state of the Jurassic sequence, developed an especially thickened lower valve, presumably to gain stability by ballasting such a non-optimal form.)

Hallam's persuasive flow tank experiments (1968) identified the morphological changes that could provide greater shell stability in an evolutionary lineage beginning with the problematical *G. incurva:* namely, larger size, broader shells, and decreased coiling. The evident conclusion now simply follows: if natural selection favored all these traits as conjoined enhancers of essential stability, wouldn't its action be greatly aided by any internal mechanism that happened to bias variation in these directions, or to forge correlations among these jointly beneficial characters?

Fortunately, all the morphological features already exist within the ontogenetic (and strongly allometric) channel of the founding member, *G. incurva*—for young shells of this ancestral species are relatively broader and less coiled than the adults that will develop from them. If these features can be brought forward by paedomorphosis into later ontogeny, greater stability can be achieved.

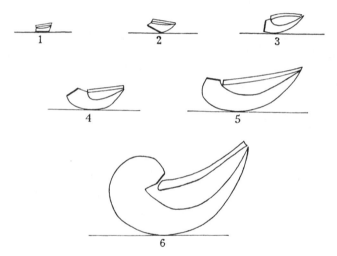

10-5. The life position of *Gryphaea,* with the implied adaptive advantage of coiling in keeping the aperture of the shell above the muddy surface. From flow channel experiments of Hallam, 1968.

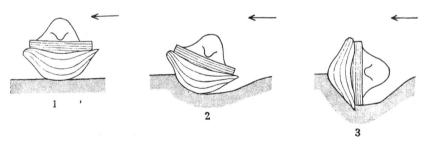

10-6. A *Gryphaea* shell tipped onto its side—an inviable position—is actually more stable than a shell in its life position of Figure 10-5. From Hallam, 1968.

Heterochrony now seals the case and intensifies the joint benefit. Neoteny often operates through a correlation of juvenile form with rapid growth rates of the young organism. If these rapid rates can be prolonged into later ontogeny, then juvenile form can also be "promoted" to the adult stage—and the adult shell will also increase substantially in size (at any age in common with ancestors). Thus, as an automatic consequence of heterochronic correlations, working within a pre-set allometric channel of ancestral ontogeny, all three adaptive desiderata for greater shell stability can evolve in tandem as consequences of a single focus of selection—that is, for prolongation of rapid juvenile growth rates. By thus linking all the valued characters, and evoking their common expression by one basic developmental change, positive constraints work synergistically with natural selection to produce an apparently complex set of adaptive changes with relative ease and speed.

### *Ontogenetically channeled allometric constraint as a primary basis of expressed evolutionary variation: the full geographic and morphological range of* Cerion uva

Since snail shells preserve a complete record of ontogeny in a unitary and rigid structure that generally cannot be modified after initial formation (at least in exterior expression, whereas the shell interior can often be altered through secondary resorption and deposition by appressing soft tissues), this taxon presents unusual opportunities for the study of developmental constraints. Evolution in any character, whether caused by selection or not, must automatically elicit a suite of correlated responses throughout such an integral and integrated structure.

However, the isometric growth model of the logarithmic spiral, so often assumed to apply to the actual growth of most mollusks will greatly limit the expression of such constraints based on "correlations of growth" (Darwin's own phrase), because such a logarithmic shell does not change its shape through ontogeny (D'Arcy Thompson, 1917), and heterochrony therefore loses its power to alter the form of descendants by general retardation or acceleration—for the juvenile shell looks just like a scaled-down adult, and global paedomorphosis, for example, would therefore exert no effect upon form. But when shells grow with pronounced allometry, then positive channels of constraint attain great potential for influencing or directing the evolution of phenotypes (as expressed in a complex, rigid structure, preserving a complete and unaltered record of ontogeny, where any change must elicit a cascade of correlated consequences, and where strong allometries establish a rich playing field for effective heterochrony).

In fact, molluscan shells rarely grow as idealized logarithmic spirals, and nearly all forms, even such prototypes of supposed isometry as the gastropod genus *Turritella,* actually grow with measurable allometry (Andrews, 1971). Moreover, Vermeij (1980) and Kemp and Bertness (1984) have presented strong theoretical arguments for regarding an allometry of doming (relative increase of height to width) as a predictable consequence of general modes of growth in gastropods. In any case, the West Indian land snail *Cerion*, the primary subject of my own research, and perhaps the most phenotypically diverse genus of land snails, grows with strong and invariant allometry in three recognizable stages (Fig. 10-7): an early button-shaped or triangular phase with width increasing far more rapidly than height; an intermediary "barrel" phase, where width increases slowly or not at all, and height grows rapidly; and a final twisting of the aperture (phase 3 of Fig. 10-7), before deposition of the definitive adult lip. In fact, *Cerion* owes its name—from the Greek word for wax, in reference to the characteristic shape of beehives—to the form produced by the first two allometric phases, particularly in species with a relatively sharp transition between the upper button and the middle barrel.

I have used inductive multivariate biometry to identify, and then to judge the extent of influence for groups of ontogenetically correlated characters ("covariance sets" in my terminology) that are both mechanically enforced by allometric growth, and that also exert substantial, often controlling, impact upon patterns of temporal and geographic variation in the phenotypic history of species. These covariance sets usually dominate several major axes of orthogonal variation detected in such techniques as factor and discriminant analysis (see Gould and Woodruff, 1986, for a detailed application; Gould, 1989a, for a general statement; and Gould, 1992b, for an analysis of the infamous "square snails" in the *Cerion dimidiatum* complex of Cuba, a phenotype that Maynard Smith had declared "impossible" at the 1980 Chicago macroevolution meeting in order to acknowledge that even he, as a strict Darwinian, did not deny a role for constraint in precluding access to certain regions of morphospace. His principle cannot be gainsaid, but his application failed because he assumed a logarithmic spiral model, thus forbidding the square shape that can be attained only by intense allometry; but *Cerion*'s allometry leads precisely to such squareness at the extreme of contrast and sudden transition between allometric phases one and two).

To cite an example from one species (as a prototype for demonstrating the dominating relative influence that such constraints can exert in particular situations), one of the major covariance sets of *Cerion*'s allometry—the "jigsaw

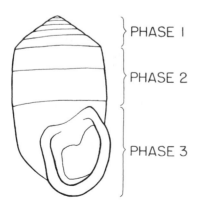

PHASE 1

PHASE 2

PHASE 3

10-7. The three standard phases in the allometric growth of *Cerion*, responsible for major changes in adult form induced by small alterations early in development. From Gould, 1989a.

constraint" of my terminology (Gould, 1989a)—may seem almost trivial in its obvious nature, but still exerts great influence in setting patterns of variation within *Cerion* at all levels, from intrapopulational variation, to geographic variation within a species, to chronoclines, to regional patterns of differentiation in species complexes (see Gould, 1989a, for details). If different shells reach virtually the same final size—and *Cerion*, as one of its major biometric advantages does, unlike most invertebrates, reach a final size marked by the secretion of a thickened lip in the third allometric phase—then shells with larger whorls must end their growth with fewer whorls. (In two jigsaw puzzles with frames of the same size, the one with smaller pieces must use more pieces to fill the common space—hence my name for the covariance set.)

The basic principle might be regarded as both obvious and entirely unprofound. Its operation would also impose scant effect upon any molluscan shell that grew in close conformity with the idealized logarithmic spiral—for two shells of the same size, one with few and the other with many whorls, would then display the same shape, and no substantial differences (beyond the number of whorls) would be apparent. But *Cerion*'s extensive and distinctive allometry triggers a large and visually striking set of correlated changes, necessarily leading to obvious differences in form between few and many whorled shells of the same size. (Such distinctions can be readily characterized, and judged in relative strength, on factor or discriminant axes of multivariate biometric studies.) For example, large-whorled specimens grow fewer whorls and therefore undergo a later transition to the second allometric phase (which invariably occurs between the 5th and 6th whorl), thereby yielding a more triangular adult shell, as relatively less of the total growth occurs during the "barrel" of the second allometric phase.

This single constraint, with its complex sequelae, explains virtually all the interregional geographic variation in one of the most interesting, and certainly the most intensely studied, species of *Cerion*—the geographic and morphological outlier (also the holotype of the genus, and a species named by Linnaeus himself), *Cerion uva* from Aruba, Bonaire and Curaçao. Moreover, recognition of the jigsaw constraint allowed me to resolve, in a manner congenial to all parties, the most substantial and longstanding debate in the history of *Cerion* studies.

In a large monograph, published in 1924, H. B. Baker, a great descriptive malacologist, claimed that he could distinguish four geographic domains of variation by subtle but entirely characteristic differences in shell form: Aruba, Bonaire, Eastern Curaçao and Western Curaçao. (The island of Curaçao, shaped like a dumbbell with eastern and western portions joined by a much narrower neck of land, may be sensibly so divided; the two halves were probably separated by higher sealevels of former interglacial epochs.) Baker used the classical and subjective criterion of a taxonomist's "good eye," and could therefore not defend his impressions in the face of extensive biometrical studies by Hummelinck (1940), then extended and confirmed by De Vries (1974). These Dutch researchers, unable to identify covariance sets with their univariate and bivariate techniques, claimed that intrasample variation in mean shell size swamped all other factors. Moreover, they could locate no clear evidence at all for Baker's interregional distinctions.

By using multivariate methods to study the influence of covariance sets, particularly the jigsaw constraint, upon geographic variation, I was able to resolve this question (Gould, 1984b) in a way that honored (as partial) the findings of all these excellent researchers. The Dutch scientists correctly noted the strong influence of variation in mean shell size among samples. But I was able to show: (1) that this variation can be isolated on a single factor axis; (2) that size ranges among samples are effectively equal, and influence the shells in essentially the same way in each of the four regions; and (3) that these intraregional differences in size almost surely arise for ecophenotypic reasons (see argument and documentation in Gould, 1984b), based on more vigorous and continuous growth of shells in moist and well vegetated microhabitats.

But I also discovered that each of Baker's four regions could be clearly identified by the evolution of differences that may be small in a genetic sense (a common situation for geographic variation within species), but that produce substantial effects upon the adult phenotype by altering several key characters in tandem through constraints of ontogenetic channels identified by covariance sets. For example, shells from Bonaire (see Fig. 10-8) grow a distinctively jutting apertural lip, a consequence of conjoined modification in characters building the third allometric phase.

Effectively all other geographic variation could be ascribed to the jigsaw constraint. For reasons that I could not resolve, *Cerion* develops virtually no variation in average adult shell size (measured as height plus width) within local populations in each of Baker's four regions—with a range from 29.79 mm in Eastern Curaçao to 30.69 mm for Aruba, giving a maximum interregional difference of only 1.6 percent. This contingently evolved (and not, obviously, geometrically necessary) invariance of size triggers a maximal effect for the jigsaw constraint—that is, so long as substantial variation exists in the sizes of whorls.

*Cerion uva* does, in fact, exhibit extensive and geographically distinctive variation in whorl sizes, with regional means spanning almost a full whorl, and ranging from 8.56 whorls in Western Curaçao to 9.35 in Aruba. The maximal "play" thus accorded to the jigsaw constraint then establishes the interregional distinctions that Baker had correctly noted but could not adequately characterize. Figure 10-9 shows minimum convex polygons drawn around the multivariate means for samples in each region (in a study based on 135 samples of 19 measures on each of 20 snails). The corners of the triangular diagram represent the first three axes of a factor analysis for mean vectors of samples. The three axes hold nearly equal explanatory power (30.5, 34.2, and 32.6 percent respectively, for a total of 97.3 percent of all information in the 19 measures among samples).

The second axis absorbs the intersample differences in size that led Hummelinck and de Vries to miss the regional distinctions. The extensive variation on this dimension does not differentiate the four regions, as indicated by the similar orientation of each polygon along this axis. (In an interesting exception, a few ancient samples (2000 to 3000 years old by radiocarbon) from Indian middens on Curaçao (see Gould, 1971a), shown in the small polygon marked I on Fig. 10-9, include far larger shells with sample means well outside the range of modern variation—as shown by their localized and maximized values on this second axis.)

The first and third axes express different aspects of the jigsaw constraint. By these methods, we can isolate this interregional component (on truly independent, mathematically orthogonal axes) from the substantial intraregional variation in size that obscured the broader geographic pattern in the studies of Hummelinck and de Vries. We can also assess the relative strengths of these two sources in compositing the total amount of difference among sample means. Baker's interregional differences explain about ⅔ of the total variation among sample means. And, with the exception of Bonaire's distinction by a

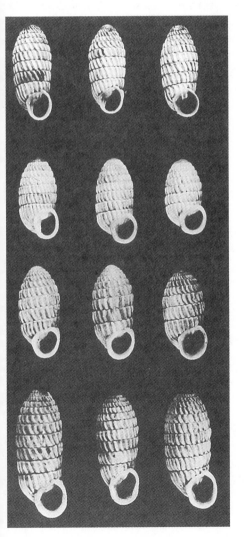

10-8. Geographic variation in *Cerion uva*. Top row, Aruba; second row, Bonaire; third row, western Curaçao; bottom row, Indian middens (3,000 years old) on Curaçao. The Bonaire shells grow a jutting apertural lip arising from changes in the third allometric phase. All other major variation can be ascribed to the jigsaw constraint. The Aruba snails, with their small whorls, reach the same final size as modern Curaçao specimens—with all major aspects of their markedly different shape arising as consequences of different lengths of residence, based on whorl sizes, in the first two allometric phases.

different covariance set, expressed on the fourth axis of this analysis (as discussed previously), effectively all interregional difference arises from the operation of the jigsaw constraint—as generated by minor differences in whorl sizes promoted to substantial overall phenotypic effect through the allometric consequences of *Cerion*'s ontogenetic channel, so long as average adult shells, as they do in this case, reach the same final size.

Note, in Figure 10-9, the separation of polygons for Aruba, Eastern Curaçao and Western Curaçao by the first and third axes that express the jigsaw constraint, while each polygon shows a similar extension along the second axis, representing the different and separable component of intraregional (and ecophenotypic) variation in mean shell size. Figure 10-8 shows characteristic shells for the regions. Note the jutting apertures from Bonaire (second row) and, especially, the contrast, built by the jigsaw constraint, between many-whorled specimens from Aruba in the top row (longer relative residence in the second allometric phrase which, distinctively in *Cerion uva*, induces an absolute narrowing in later whorls, leading to a "barrel" shape for the entire shell, fat in the middle and narrowing at both ends)—and the fewer whorled, but same sized, specimens from Western Curaçao in the third row (which pass less of their ontogeny in the second "barrel" phase and therefore do not become constricted towards the end of growth, as in the Aruba speci-

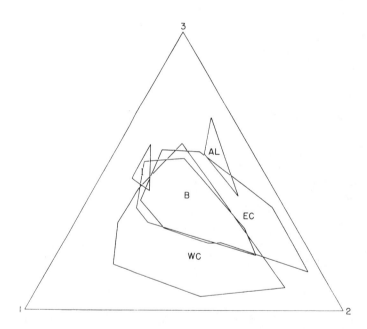

10-9. Minimum convex polygons drawn around the multivariate means for samples in each region. The ecophenotypic factor of size makes no distinctions as each polygon becomes elongated along this second axis, and as the truly larger fossil shells from Aruba occupy a separate position at the high size end of this spectrum. But the first and third axes express the jigsaw constraint, and the defining regional geographic variation within this type species of the genus achieves clear expression in the separation of polygons on these axes.

mens). If a constraint engendered by an allometric channel in ontogeny can so control the regional pattern of geographic variation in an important species of such a well-studied genus, then we cannot deny a major role to this positive mode of evolutionary change by developmental constraint.

## THE APTIVE TRIANGLE AND THE SECOND POSITIVE MEANING: CONSTRAINT AS A THEORY-BOUND TERM FOR PATTERNS AND DIRECTIONS NOT BUILT EXCLUSIVELY (OR SOMETIMES EVEN AT ALL) BY NATURAL SELECTION

### *The model of the aptive triangle*

In a classic line of contemporary American literature, W. P. Kinsella writes of a midwestern farmer so beguiled by the legend of the great baseball hero Shoeless Joe Jackson that he constructs a stadium in his wheatfield because he heard a voice saying to him: "if you build it, he will come." I often feel that many modern evolutionary biologists unconsciously obey a similar mantra in their approach to the phenotypic features of organisms: "if it works well, then natural selection made it."

In two historical discussion of this book's first part—my analysis of Darwin's fateful words at the end of chapter 6 of the *Origin of Species* (pp. 251–260), and my presentation of "Galton's Polyhedron" as the most effective formalist or structuralist metaphor for illustrating missing alternatives in schemes of evolutionary causality that consider natural selection as the only mechanism of change (pp. 342–351)—I presented triangular models of causal poles for the origin of phenotypic features: a representation well suited for portraying alternatives and complements to natural selection as the causal basis of organic form.

Let me now propose a slightly different triangular model with the same three poles, but now representing only organismal features that "work well" both in the classical sense of good biomechanical design, and the technical meaning of conferring fitness upon organisms in their interaction with environments—in other words, to the features that biological terminology, and ordinary vernacular usage, call "adaptations," but that I would rather designate as "aptations" (see Gould and Vrba, 1982), a more general term that acknowledges their current utility while remaining agnostic about their source of origin. I will therefore designate this model (modified from Seilacher, 1970, and ultimately traceable to Galton's Polyhydron) as "the aptive triangle" (though I will submit to standard "loose" (or *sensu lato*) usage, and usually refer to the features plotted upon this diagram by the only term that current language recognizes—namely, adaptations).

The basic diagram (Fig. 10-10, presented before, in part, as Fig. 4-6) designates three vertices as idealized end members and also recognizes, of course, that almost any actual feature will plot either along an edge (influenced by two vertices), or, more frequently, in the triangle's interior (where all three end members contribute). This mode of ternary plotting has been used most frequently by petrologists for depicting the composition of actual rocks as amalgamations of three idealized end-members, in full expectation that few, if any, real rocks will include only one end-member and plot right at one of the triangle's vertices.

In keeping with my previous discussion, and with Seilacher's original conception, we may call these idealized end-members "functional," "historical," and "structural." In other words, any phenotypic feature now "working well" for an organism may have been constructed by a process that directly crafted the feature for its current function (the first corner), inherited from an ancestral form (the second corner), or built by some structural mechanism or process not directly related to, or engendered by, the functional needs of the organism.

As discussed in my previous analysis of Darwin's brilliant argument in Chapter 6 of the *Origin* (pp. 251–260), the argument for natural selection as the dominant cause of evolutionary change must be made in the following way under the aegis of this model (as Darwin did, but without constructing any formal picture like Fig. 10-10): At the functional vertex, natural selection stands alone as the only known and effective cause in this mode. If the Lamarckian mechanism operated in nature, then inheritance of acquired and adaptive characters would provide another functionalist option for explaining the origin of working design. But inheritance does not so operate, on this planet at least. (Darwin took the more generous view that Lamarckian inheritance might exist, but at a relative frequency distinctly subsidiary to natural selection.)

At the historical vertex, working features passively inherited from ancestors did not originate to meet *current* functional needs. But so long as these

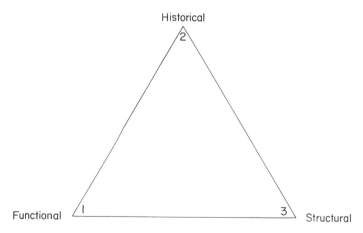

10-10. Standard triangular diagram for depicting basic causes of form as functional (immediate adaptation to current circumstances), historical (inherited by homology, whatever the basis of ancestral origin), and structural, or arising either as physical consequence of other features or directly from the nature of physical forces acting on biological materials. All vertices may yield aptive traits of great utility to the organism.

features arose initially by natural selection in the ancestral line, then their ultimate origin remains functional—and natural selection, as previously noted, represents the only known (for Darwin, the only effective) cause of functional change. Finally, at the structural vertex, Darwin allowed that features not arising for functional reasons, but only coopted for current utility, must be admitted as genuine exceptions to the principle that adaptive features can only originate for functional reasons (with natural selection as the only known and sufficiently powerful functional mechanism). But he then demoted this class of real exceptions by the standard argument in studies of natural history: he claimed, invoking the classical justifications of "sequelae" and "nooks and crannies" (see p. 1249), that currently adaptive features with nonadaptive structural origins must, by their rarity, reach only an insignificant relative frequency among evolved traits of organisms.

The impeccable logic of this formulation can help critics by clarifying how any potential argument against this hegemony of natural selection must proceed. At the functional vertex, one would have to identify other important mechanisms in addition to natural selection—and none have been proposed, at least to the satisfaction of this author (although the argument for "a little bit of bacterial Lamarckism"—as I like to characterize the controversial claims of Cairns et al. (1988)—may have some merit in a limited domain).

At the historical vertex, one would have to reject the contention that constraining homologies of inheritance, and the resulting heterogeneous clumping of species in organic morphospace, record the consequences of natural selection in constructing the novel traits of ancestral forms, followed by the continuing control of selection upon subsequent patterns of phyletic change in descendant lineages—an argument that I will advance in Section II of this chapter. Finally, at the structural vertex, one would have to counter Darwin's argument by asserting that he greatly underestimated the relative frequencies of these admitted exceptions to natural selection for the origin of currently functional features—a claim that I will advance in Chapter 11. Thus, the form of Chapters 10 and 11, and my argument for the importance of structural constraints at high relative frequency in the origin of currently adaptive organismal characters, will center upon recent arguments for a reconceptualization of the historical vertex, and for a reevaluation of relative frequency at the structural vertex of the aptive triangle.

### Distinguishing and sharpening the two great questions

THE STRUCTURAL VERTEX. I shall begin with the simple and more direct question posed from the structural vertex: does the existence of current adaptations necessarily imply their functional origin at all (either as a direct response to current environments, or by inheritance of traits with a functionalist origin at their ancestral inception)? How, in other words, can good Darwinian design (aptations in my favored restriction, adaptations in the vernacular) arise by processes that do not involve functional adaptation?

Since this category is defined negatively—to designate causes of functional characters *not* evolved by functionalist mechanisms—the structural vertex becomes something of a miscellaneous repository, as Seilacher (1970) recognized for the same end member (that he called *bautechnischer,* or architectural). In particular, the structural vertex includes two strikingly different subcategories, united in their common appeal to physical consequences rather than functional crafting, but otherwise strikingly disparate, even diametrically opposite, in their implications for a key question, admittedly subsidiary to our present inquiry about structural vs. functional origin, but of central importance to evolutionary study in general: namely, the role of history and contingency in the interpretation of evolutionary lineages.

I shall consider these two structural categories *seriatim,* and in depth, in Chapter 11, and will therefore only present the basic conceptual framework here. In the first category, some adaptive features of organisms may be directly molded by, or may originate as immediate and deterministic consequences of, the physical properties of matter and the dynamical nature of forces—in other words, not by an accumulative process of functional honing through selection, and not (for that matter) by any uniquely biological process at all. When Williams (1966) famously, if a bit facetiously, remarked that we shouldn't consider a flying fish's capacity to fall back into the water as an adaptation because their descent represents a necessary consequence of physical mass—even though this capacity may be vital to the continued life of the fish, and therefore strongly aptive—he invoked a direct physical property of matter (not subject to alteration by selection at all in this case!).

D'Arcy Thompson's (1917, 1942) theory—that physical forces directly impose an optimal biomechanical form upon plastic organic material—marks the admittedly idiosyncratic *locus classicus* for this general attitude (see pp. 1179–1208). Stuart Kauffman's (1993) interesting concept of "order for free" (good design automatically generated by nature's laws, with no need for laborious construction by a particular biological process like natural selection) provides the most fruitful current context for this approach to aptive organic design.

In the second category that Darwin designated as "correlations of growth," and Gould and Lewontin (1979) called spandrels, features arise nonadaptively as physically necessary consequences of other changes that may (and, in all probability, usually do) have an adaptive basis, or as inevitable and unselected sequelae of general organic designs (that, again, generally arise for conventional functional reasons).

Nonselected origin for structural reasons defines the common ground of the two categories—directly generated by physical forces in the first, indirectly developed as correlated consequences in the second. But the philosophical implications of these two bases could not be more different in one crucial respect—hence the oppositional stance often adopted between champions of the two modes, despite their acknowledged common ground at the vertex of nonselected origin by physical necessity.

Pure D'Arcy Thompsonians maintain little interest in history and phylogeny, and may even become overtly hostile to the commanding influence of these concepts in evolutionary biology. After all, if a trait arises by physical necessity, why should we care about the specific contingencies that brought this or that lineage into the domain of the particular physical law under study. At any time, any lineage located in this domain must behave in the same way. This version of structuralism embraces the classical spatiotemporal invariance of natural law, and cares little (if at all) about historical pathways that happen to potentiate the law's operation in any particular case. Most evolutionists (including the author of this book) are historians at heart, and must view such derisory dismissal of phylogeny as anathema, however fascinating they find (as I do) the partial validity of this theme, and however much they may admire (and bravo again from this observer) the inimitable power of D'Arcy Thompson's prose style.

Spandrelists, in strong contrast, generally share the evolutionary biologist's traditional fascination for contingent details of history in individual lineages under study. Spandrels do express general and predictable properties, but they originate as necessary consequences of particular triggers that can only be understood in a historical and phylogenetic context. If Julia Pastrana grew two rows of teeth as a correlated consequence of her abnormal hairiness (see p. 338 for a discussion of this example from Darwin's writing), then the forced correlation, set (in Darwin's view) by the constraining homology of hair and teeth, records and reflects the phyletic uniqueness of mammalian development (not the operation of invariant, universal laws), even if the extra teeth grew by enforced physical necessity. And even though the spandrels of San Marco must be built once the architects decide to mount hemisphaerical domes on four adjacently orthogonal rounded arches, we can only understand the basic blueprint that necessarily engendered the spandrels by studying the particular history of ecclesiastical architecture.

THE HISTORICAL VERTEX. The structural vertex poses a direct question about the origin of currently adaptive features themselves: what percentage of items in this category did *not* originate by a process of adaptation, but were coopted for present utility from non-adaptive beginnings? If we can determine a high relative frequency in general, or even if we could only specify a subset of crucial evolutionary situations for such nonadaptive origins, then an exclusively adaptationist theory for the genesis of aptive structures will no longer suffice, and evolutionary theory will require enrichment from structuralist alternatives promoted to a more than marginal or peripheral status.

The historical vertex, on the other hand, poses a more indirect challenge that might better be designated as a metaquestion: Given a functional origin for presently adaptive features (either by immediate construction for a current role, or by adaptive origin in an ancestor, with subsequent maintenance by homology in descent), may we also regard the markedly inhomogeneous distributions of organisms across the potential morphospace of good organic design as a best set of solutions to functional problems, or do we need to invoke internal constraints and channels to explain substantial aspects of this decidedly "clumped," and decisively non-random, occupation of a theoretical "design space"?

In other words—and I label the inquiry as a "metaquestion" for this reason—in what ways does the skewed and partial occupancy of the attainable morphospace of adaptive design record the operation of internal constraints (both negative limitations and positive channels), and not only the simple failure of a limited number of unconstrained lineages to reach all possible positions in the allotted time? (Geological history may be long, and the number of evolutionary lineages immense, but even these substantial quantities must be risibly small compared with the number of spatiotemporal positions in potentially "colonizable" morphospace.)

In attempting to explain such non-random clumping in adaptive morphospace, Darwinians have traditionally emphasized the contingency of limited time and numbers—rather than any failure to populate accessible regions as a consequence of active constraint—because their functionalist theory presupposes the power of natural selection to break such constraints (whose existence, needless to say, they cannot and do not wish to deny), and the consequent accessibility (at different levels of effort and probability, to be sure) of all physically possible adaptive designs.

If the influence of historical constraints must be integrated with the conventional mechanism of unfettered adaptive exploration (limited by accidents of historical opportunity) to explain the markedly non-random clumping of actual organisms in the potential morphospace of adaptive and theoretically accessible organic form, then this metaquestion about nonfunctional causes for the *distribution* of adaptive features poses a different kind of challenge to our usual views about the power and range of natural selection in the explanation of functional design.

In my presentation thus far, I have epitomized this crucial issue in an abstract way, but I shall, in the final section of this chapter, present the empirical results—primarily from the burgeoning study of genetic bases for the major developmental patterns of organic *Baupläne* ("evo-devo," or evolution of development to its devotees)—that have so surprised the biological sciences in recent years, bringing this new study of ancient themes to the forefront of our science.

I shall argue that two prominent discoveries have magnified the importance of historical constraints *vs.* the free operation of natural selection to a point where this historical aspect of constraint can no longer be denied prominence in designating the causes of evolutionary *change:* First, "*deep homology*" or the discovery that major phyla, separated by more than 500 million years of independent evolutionary history, still share substantial (if not predominant) channels of development based on levels of genetic retention that proponents of the Modern Synthesis had specifically declared inconceivable, given the presumed power of natural selection to modify any independent line in its own uniquely adaptive direction. Second, the importance of *parallelism* (a concept rooted in internal constraint) for explaining independently evolved features of distant phyla, traits long touted as textbook examples of convergence (a concept rooted in externally conditioned adaptation).

I shall also argue that deep homology often embodies the "negative" empirical theme of constraints as limitation, while parallelism features the "positive" empirical theme of constraints as enabling channels. Both themes, however, have forced evolutionary biologists to reassess the importance of constraints at the historical vertex for explaining the actual distribution of adaptive form within potential organic morphospace. In this sense, both themes count as "positive" in my second (or conceptual) sense that any powerful argument challenging a stale and limiting consensus must be treasured in science.

No good or experienced naturalist could ever fully espouse the reductionistic belief that all problems of organic form might be answered by dissolving organisms into separate features, each with a specified function, and each optimized independently by natural selection. But theories do drive, or at least nudge, adherents towards their extreme formulations—and even such sophisticated versions of Darwinism as the Modern Synthesis (see Chapter 7) biased the perspectives of biologists in this direction by advocating natural selection as, effectively, the sole cause of evolutionary change. Various pleas, heard with increasing frequency during the past generation (Goodwin, 1994, for example), to "put the organism back into evolution," or to "reestablish a meaningful science of morphology," should be understood as expressions of a growing conviction that theories of part-by-part functionalism cannot explain the major patterns of life's history and current morphological distribution.

We do need to reformulate, in modern and operational ways, the old notions of organic integrity, and structural determination from the "inside" of genetics and development, thus balancing our former functionalist faith in the full efficacy of adaptationism with positive concepts of internal and structural constraint. Only in this way can we forge a unified science of form to integrate the architecture and history of organisms with their daily struggles to survive, prosper, and propagate in a complex ecological surround—a world that Western culture once perceived as "the face of nature bright with gladness" (Darwin, 1859, p. 62), but that we now recognize as the material domain of natural selection, a process carrying no moral implications for human life (thus permitting us to throw aside the crutch of comforting imagery that Darwin so rightly rejected), but operating with relentless (though not exclusive) force throughout living nature.

### An epitome for the theory-bound nature of constraint terminology

We may use the model of the aptive triangle to illustrate how my second, or conceptually "positive," meaning of constraint (including both the "positive" and "negative" empirical modes of channels and limits) rests upon the theory-bound nature of all scientific terms and definitions. As I argued previously (see pp. 1032–1037), if we designate a set of causes as canonical within an orthodox theory, then vernacular usage designates other causes lying outside the theory, but nonetheless influencing phenomena that should fall under the aegis of orthodoxy, as "constraints" upon the power or validity of the standard theory. (My designation of such constraints as "positive" then follows the usual ethos of science in placing special value on new ideas that challenge complacent conventionalities.) In this crucial sense, the identification of certain causes as constraints depends upon the claims and nature of a standard theory.

The historical and structural vertices of the aptive triangle become sources of constraint when this standard theory emanates from the third, or functional, vertex—as classical Darwinism does, and as Darwin himself so clearly specified in his own analysis of the three vertices (without explicitly identifying such a model) in closing Chapter 6, entitled "Difficulties on Theory," in the *Origin of Species*—again, see pp. 251–260 of this book for my analysis of this key Darwinian argument). In making this "pure end-member" analysis of canonical causes *vs.* constraints, I am consciously presenting extreme, or cardboard, versions of central theories in order to clarify a logical (and terminological) point about the naming of expectations and exceptions. (Just as I argued previously that no actual empirical case would fall precisely at a vertex of total determination by one factor alone, I also acknowledge that no subtle thinker's theory will fall right on a vertex either. Nonetheless, discussion in terms of pure end-members may be defended as a conceptual device for clarifying the central content and primary commitment of more complex theories.)

Figure 10-11 illustrates the changing terminology of constraint and convention under three pure end-member theories of causation at vertices of the aptive triangle. For the pure adaptationist, committed to natural selection as the controlling and functionalist mechanism of evolutionary change, all causes of currently adaptive form that *cannot* be attributed to direct selection for immediate utility must count as constraints. (Fig. 10-11a depicts this version, with the canonical cause placed at the functional vertex, and with other vertices making contributions that then be called constraints upon the full and free operation of current natural selection to forge immediate utility.)

As another virtue of these simplified representations, we can also grasp how any pure end-member theorist must treat the exceptions ("constraints") that cannot be denied as causal contributors to currently adaptive form. In this case (Fig. 10-11a), I have already noted Darwin's own excellent strategy (see pp. 251–260): admit the historical inputs, but attribute their cause to natural selection in the past; then admit the structural inputs as genuine exceptions, but relegate them to a low and insignificant relative frequency. Thus, all "constraints" either record the operation of the canonical mechanism in the past, or stand as genuine exceptions rendered impotent by their rarity.

But if I were committed to a view that the direct action of physical forces (as expressed in the spatiotemporal invariance of natural law) builds the adaptive forms of organisms directly, and without any appeal to functionalist or distinctively biological principles like natural selection—D'Arcy Thompson, in fact, advocated this general view in as pure a form as any 20th century biologist dared to espouse (see pp. 1179–1208)—then the structural

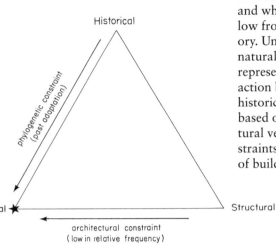

10-11a. What we label as primary causes and what we consider as constraints follow from the logic of our preferred theory. Under strictly functional views of natural selection, the functional vertex represents the orthodox cause of current action by natural selection, whereas the historical vertex imposes constraints based on past adaptations and the structural vertex imposes architectural constraints based on limitation in the nature of building materials.

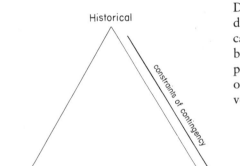

10-11b. For a structuralist thinker like D'Arcy Thompson, optimal forms built directly by physical laws become canonical, and influences from other vertices become constraints—either of ecological particulars from the functional vertex, or of contingency from the historical vertex.

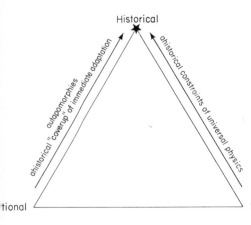

10-11c. For a cladist interested in historical reconstruction of branching sequences, the historical vertex becomes primary, and influences from the other two vertices represent constraints—ahistorical constraints of universal physics from the structural vertex and autapomorphies of particular and immediate adaptation from the functional vertex.

vertex becomes the locus of canonical causes, while contributions from the other two vertices become constraints upon full determination by general laws of nature.

Under this theory (Fig. 10-11b), adaptive form arises from the operation of general laws upon biological materials. But if, to understand any current adaptation, we need to invoke strictures based either upon: (1) passive inheritance within a specifically designated genealogical system (a constraint from the historical vertex, imposed by a unique and contingent biological particular, thus detracting from a claim for full causation by general laws); or (2) upon the immediate construction of a particular adaptation by a biological process tied to specifics of adaptive pressures in one environment at one time (a constraint from the functional vertex of current natural selection)—then the full power of the purely physical model becomes compromised. Thus, as I show in Figure 10-11b, a purely physicalist theory for the direct generation of adaptive form by spatiotemporally invariant laws of nature, places its canonical mechanism at the structural vertex, and regards inputs from both the historical vertex (strictures from past particulars) and the functional vertex (strictures imposed by the specifics of current biological situations) as constraints.

Finally (Fig. 10-11c), a pure (and caricatured) cladist, who believes that the reconstruction of genealogical pattern (without reference to modes of causation) defines the goal and purpose of evolutionary biology, would locate his canonical mechanisms at the historical vertex, and view contributions from the other two vertices as constraints upon his ability to detect a pure genealogical signal in the currently adaptive traits of organisms. Influences from the structural vertex must be counted as constraints because their timeless generality covers or distorts the desired signal of particular history with an unwanted contribution from causes with no specific genealogical content. And influences from the functional vertex impose a confusing immediate particular—an autapomorphy offering no help at all in the reconstruction of lineages, and therefore conventionally omitted in cladistic analysis—degrading a phyletic signal that might otherwise map the organism's position in the genealogical system of a more general lineage. (This insight about particular and immediately adaptive features—autapomorphies in cladistic terminology—has long been regarded as a truism in taxonomic practice. Darwin himself frequently emphasized (1859, chapter 13) that new and unique adaptations can only confound taxonomic relationships, and that systematists must privilege characters with broad homological residence in the taxa of larger genealogical groupings.)

Thus, the cardboard Darwinian functionalist, the cardboard physical structuralist, and the cardboard genealogical cladist each chooses a different vertex for canonical causation, and must then define influences from the other two vertices as constraints upon the efficacy of his orthodox mechanism. My examples are purposefully cartoonish, but the principle thus illustrated represents an important, and insufficiently appreciated, generality in science—the theory-bound nature of terms, and particularly, in this case, the designation as "constraint" of all evolutionary causes lying outside the range of orthodox mechanisms, thereby compromising their power and generality. We should regard this terminological notion of constraint as positive in its capacity to question accepted ways of thinking—the theme that shall now structure the remainder of this chapter.

## Deep Homology and Pervasive Parallelism: Historical Constraint as the Primary Gatekeeper and Guardian of Morphospace

### A HISTORICAL AND CONCEPTUAL ANALYSIS OF THE UNDERAPPRECIATED IMPORTANCE OF PARALLELISM FOR EVOLUTIONARY THEORY

#### A *context for excitement*

The last chapter of my first book, *Ontogeny and Phylogeny*, published in 1977, amounted to little more than a terminal exercise in frustration. I had written 500 pages on the history and evolutionary meaning of heterochrony, and had then been stymied, at the point of potential synthesis, by an inability to relate the well-documented (and reasonably well understood) subject of macroevolutionary changes wrought by shifts in developmental timing to any viable analysis (or even description) of the underlying genetic and embryological mechanisms.

I could only wave my hands and write a few vague paragraphs about the putative importance of "regulatory" genes—then an almost purely abstract concept (at least for eukaryotic development), based on no direct documentation of any worth, and supported only by three inferential forms of argument: analogies to rudimentary knowledge about the different systems of prokaryotic regulation (primarily the work of Jacob and Monod); general models suggesting the necessity of a regulatory hierarchy, with some genes operating as primary controls on rates and placements of structural genes and their products (Britten and Davidson, 1971); and such conclusions by negative inference as King and Wilson's (1975) famous calculation of more than 99 percent identity between human and chimp polypeptides, implying that the considerable phenotypic differences between the two species must therefore reside in the action of a small class of unknown regulatory genes.

Of course, this frustration only recorded a technological inability to specify these regulators, not any failure to grasp the centrality of the subject. I wrote (1977b, p. 406): "The most important event in evolutionary biology during the past decade has been the development of electrophoretic techniques for the routine measurement of genetic variation in natural populations. Yet this imposing edifice of new data and interpretation rests upon the shaky foundation of its concentration on structural genes alone (*faute de mieux*, to be sure;

it is notoriously difficult to measure differences in genes that vary only in the timing and amount of their products in ontogeny, while genes that code for stable proteins are easily assessed)."

Barely 20 years later, this statement reads like a quaint conceptual fossil from an "ancient" time of crossbows and arquebuses, when we could only reconstruct the anatomy of genes from their protein products (and could not recognize regulatory genes that did not deposit such results in explicit flesh and blood). I therefore succumbed to the necessity of technical limits and ended a long book with the weakest of conclusions—a future hope, however heartfelt and (in retrospect) accurately surmised: "I believe that an understanding of regulation must lie at the center of any rapprochement between molecular and evolutionary biology; for a synthesis of the two biologies will surely take place, if it occurs at all, on the common field of development" (Gould, 1977b, p. 408).

Now, as I begin this chapter in the summer of 1999, I can only express both my joy and astonishment at a subsequent speed of resolution and discovery that has sustained my predictions, but also made my earlier book effectively obsolete, not only within my own lifetime, but during my active mid-career. The field of evolutionary developmental biology (known as "evo-devo" to practitioners), while still in its infancy, has invented the tools—and already cashed out a host of stunning and unexpected examples—for decoding the basic genetic structure of regulation, and for tracing the locations and timings of regulatory networks in the early development of complex multicellular creatures.

But this very pace of growth and excitement presents a problem for a book like this, with a "lead time" measured in months to years, rather than the professional journal's weeks to months or the popular press's days. The discoveries of deep homology and pervasive parallelism among phyla separated for more than 500 million years continue to accumulate at an accelerating pace, based on methodological refinements and extensions, in both speed and accuracy, that could hardly have been conceptualized even a decade ago.

This situation places me in a quandary (although I could hardly imagine a happier form of puzzlement). The data of evo-devo constitute the largest and most exciting body of novel empirics to support this book's general thesis. Since I have tried to provide thorough overviews of empirical documentation for other central elements of my overall theory, I should now be tabulating and evaluating these cases of deep genetic homology *in extenso*. But I am hoist by my own petard of emphasis on appropriate scales. The data of evo-devo accumulate and improve at such a pace that any thought of a "review article" written more than two years before anticipated publication can only be regarded as absurd. In other words, this book's timescale of production must be labeled as geological compared with a pace of discovery that can only be measured in ecological time.

I will therefore adopt the following strategy as appropriate to the circumstance. I will exemplify the best and most informative of current empirical cases, but I ask readers to heed the following label of warning: "I wrote this section in the closing months of the last millennium. The cases discussed herein represented a 'state of the art' at this historical moment. This 'state' will be obsolete and superseded by the book's publication, but I am confident that the general themes and directions will hold and grow. Please consider the empirical discussion as exemplification, not as fulfillment."

However, the timescale of this book also permits a luxury not afforded to authors of journal articles. For I can balance this guaranteed empirical super-annuation against a discussion of general significance that, if properly situated within this book's broader subject of the history and structure of macro-evolutionary theory, may succeed in exemplifying the signal importance of evo-devo in changing and expanding our basic conception of evolutionary causality (even while I must fail to capture what the favored cliché of the moment calls the "cutting edge" of actual discovery). I will therefore focus my treatment of evo-devo upon some crucial issues in the structure of evolutionary theory—all rooted in the concept of "constraint" in relationship to natural selection—that have frequently been overlooked, bypassed, or short-changed in the midst of immediate excitement generated by the novel data of this burgeoning field.

What features generally lead scientists to strongly shared feelings about the unusual importance of a set of discoveries? We might nominate sheer novelty as an initial, base-level property—especially when enhanced by a conquest over nature's previous taunt to scientists: you know where to look in theory, but you haven't developed the proper tools for perception. (The canonical example of such rare triumphs, Galileo's *Sidereus nuncius* of 1610, comes inevitably to mind—a mere "pamphlet" that packed more oomph per paragraph than any other document in the history of printing. After all, the first telescopic look at a previously invisible cosmos necessarily "skimmed off" a set of magnificent and unexpected novelties, including the composition of the Milky Way as a sea of stars, the satellites of Jupiter, the phases of Venus, and the topography of the Moon.)

Much of our fascination with the data of evo-devo arises from the sheer novelty of discovery in biological domains that had been previously and totally inaccessible. These empirical gems also illustrate, even in these early days, the integrating power of scientific conclusions to translate a previous descriptive chaos into explanatory sensibility. As an example, consider the name given to the truly elegant theory of floral genesis, as developed by students of *Arabidopsis,* the *"Drosophila"* of angiosperm biology—the ABC Model (Coen and Meyerowitz, 1991; Weigel and Meyerowitz, 1994; Jürgens, 1997; Busch, Bomblies, and Weigel, 1999; Wagner, Sablowski, and Meyerowitz, 1999).

In this elegantly simple model (see Fig. 10-12), based on genes with homeotic effects upon serially repeated structures arranged in systematic order (with repetition in concentric whorls rather than linearly along a body axis), *A* genes operating alone determine the form of the outermost whorl of leaf-like sepals; *A* plus *B* genes regulate petals in the next whorl within; *B* plus *C* genes mark the male stamens, while *C* genes working alone determine the most interior female carpels. Moreover, *leafy*, a "higher control" gene previously recognized as an initiator or suppressor of floral growth and placement in general (Weigel and Nilsson, 1995), apparently also regulates the more specific operation of the ABC series. (Busch et al., 1999, demonstrate that a protein produced by *leafy* bonds directly to a particular DNA segment of a C gene responsible for the generation of carpels.)

This model enjoys obvious significance for the full gamut of evolutionary issues, ranging from the most theoretical (in "updating" Goethe's formalist theory (see pp. 281–291) that all parts regulated by the ABC series conform to a generalized "leaf" archetype), to the most practical (hopes of florists to enhance AB interactions and grow flowers with larger and more numerous petals). But in the present context, I merely wish to highlight a linguistic point: the selected terminology of ABC surely encapsulates the accurate impressions (and the excitement) of researchers who recognize their role as pioneers engaged in the construction of a basic alphabet for a new understanding of nature.

The pure discoveries of evo-devo may fit the heroic image of science as conqueror of previous ignorance (the *tabula rasa* model of light upon previous darkness or, literally, the first writing on a blank slate). But the most stunning of scientific novelties surely gain their status by virtue of their unexpected or surprising character—that is, their failure to match, or even their power to mock, the anticipated constitution of a part of the natural world previously

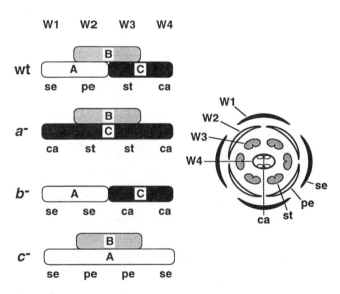

10-12. The elegant simplicity of the *ABC* model, from Weigel and Meyerowitz, 1994. In this model, based upon the genetics and development of *Arabidopsis*, the four circlets of sepals, petals, stamens, and carpels achieve their distinctive forms under the following influences, as shown in the diagram. *A* genes determine the development of sepals; *A* and *B* petals; *B* and *C* stamens; and *C* alone, carpels.

inaccessible to investigation. (The most stunning property of Jupiter's four large moons, when first seen by Galileo, lay not in their mere existence, but in the recognition that their revolution about the planet would fracture the crystalline sphere that, in the "certain" knowledge of previous views, marked Jupiter's domain in a geocentric universe—and that such a sphere, therefore, could not exist.) In the same manner, the central significance of our dawning understanding of the genetics of development lies not in the simple discovery of something utterly unknown (the ABC floral model, or the specification of antero-posterior differentiation by arthropod Hox genes), but in the explicitly unexpected character of these findings, and in the revisions and extensions thus required of evolutionary theory.

The discovery that has so discombobulated the confident expectations of orthodox theory can be stated briefly and baldly: the extensive "deep homology" now documented in both the genetic structure and developmental architecture of phyla separated at least since the Cambrian explosion (*ca.* 530 million years ago) should not, and cannot, exist under conventional concepts of natural selection as the dominant cause of evolutionary change. Natural selection must therefore operate in a context of far greater constraint (in both the "negative" sense of limits upon freedom to craft particular adaptive solutions, and in the "positive" sense of synergism in the specification of preexisting or preferred internal channels) than the usual functionalist characterizations of Darwinian theory envisage.

I am not trying to construct straw men or cardboard images for easy demolition. Of course, no good Darwinian naturalist ever conceptualized organic matter as pure putty molded by natural selection to local optimality. The hold of phenotypic homology has always fascinated evolutionary biologists and served as the basis for classification and phylogenetic reconstruction. Even the most orthodox Darwinian systematists have always recognized that "putty-like" characters—maximally labile and malleable by natural selection in an unconstrained way—must be shunned in phyletic reconstruction (as sources of autapomorphic traits and manifestors of convergence), while taxonomists must base their hierarchical orderings on nested levels of homological retention among related taxa.

But two classical views about homology have traditionally served to integrate this cardinal principle of historical constraint with a functionalist theory of evolutionary mechanisms. First, as previously discussed in more detail (pp. 251 and 1058), Darwinian biology attributes the origin of shared homologous characters to ordinary adaptation by natural selection in a common ancestor. Moreover, homologous characters not only continue to express their adaptive origin, but also remain fully subject to further adaptive change—even to the point of losing their ready identity as homologies— if they become inadaptive in the environment of any descendant lineage. Homological similarity in related taxa living in different environments therefore indicates a lack of selective pressure for alteration, not a limitation upon the power of selection to generate such changes. (At the Chicago Macroevolution meeting of 1980, for example, Maynard Smith acknowledged the allometric basis of many homologies, but stated that the attribution of such similarity to "developmental constraint" would represent what he proposed to christen as the "Gould-Lewontin fallacy"—for natural selection can unlock any inherited developmental correlation if adaptation to immediate environments favors such an alteration.)

Second, homological holds must be limited in taxonomic and structural extent to close relatives of similar *Bauplan* and functional design. The basic architectural building blocks of life—the DNA code, or the biomolecular structure of fundamental organic compounds, for example—may be widely shared by homology among phyla. But the particular blueprints of actual designs and the pathways of their construction—the form of the Gothic cathedral rather than the chemical formula of calcite in the facing stone (see pp. 1134–1142 for extensive treatment of this point)—must be limited to clades of closer relationship. The identical topology of bones in mammalian forearms of markedly different utility (the whale's flipper, the horse's leg, the bat's wing, and my typing, or literally manipulating, digits) can be homologous, but we expect no comparable hold of history upon the more generally similar segmentation of arthropod metameres and vertebrate somites (not to mention the non-homology of bones in mammalian forearms and teleost forefins).

Any wider hold of homology would have to inspire suspicions that the central tenet of orthodox Darwinism can no longer be sustained: the control of rates and directions of evolutionary change by the functional force of natural selection. In a particularly revealing quote within the greatest summary document of the Modern Synthesis, for example, Mayr (1963, p. 609) formulated the issue in a forthright manner (see p. 539 for previous discussion of this statement). After all, he argued, more than 500 million years of independent evolution must erase any extensive genetic homology among phyla if natural selection holds such power to generate favorable change. Adaptive evolution, over these long intervals, must have crafted and recrafted every genetic locus, indeed every nucleotide position, time and time again to meet the constantly changing selective requirements of continually varying environments. At this degree of cladistic separation, any independently evolved phenotypic similarity in basic adaptive architecture must represent the selective power of separate shaping by convergence, and cannot record the conserved influence of retained genetic sequences, or common generation by parallelism: "In the early days of Mendelism there was much search for homologous genes that would account for such similarities. Much that has been learned about gene physiology makes it evident that the search for homologous genes is quite futile except in very close relatives."

But we now know that extensive genetic homology for fundamental features of development does hold across the most disparate animal phyla. For an orthodox Darwinian functionalist, only one fallback position remains viable in this new and undeniable light (and Ernst Mayr, vigorous as ever at age 95 as I write these words, would be the first to welcome this illumination wholeheartedly, and to laugh at his old cloudy crystal ball on this single issue). One can admit the high frequency and great importance of such genetic constraints (and also designate their discovery as stunningly unexpected), while continuing to claim that natural selection holds exclusive sway over evolutionary *change* because deep homology only imposes limits upon styles and ranges of developmental pathways, but cannot power any particular phyletic alteration. Natural selection can still reign supreme as the pool cue of actual evolutionary motion.

But a formalist defender of positive constraint will reply that such unanticipated deep homology also channels change in positive ways—and that the key to this central argument resides in an old distinction that, unfortunately, cannot be matched for both conceptual and terminological confusion, and for consequent failure of most evolutionists to engage the issue seriously: namely, the differences in causal meaning (not just in geometric pattern) between parallelism and convergence. The next section shall treat the history and logic of this issue in detail, but I shall first present the following basic formulation in relevant terms of balances between constraint and selection:

Even the most committed adaptationist would not deny that the independent evolution of similar phenotypic features (in both form and function) in two closely related lineages may be facilitated by the presence, in both ancestors, of the same genes and developmental pathways inherited from a recent common ancestor. (The independently evolved features of these two lineages cannot be called homologous on basic definitional grounds, but the features may still be built by homologous genes and along homologous developmental pathways.)

For example, no adaptationist would be fazed by the suggestion that relative increase in antler size within two separate cervid lineages undergoing phyletic increase in body size occurred because both lineages retained an ancestral allometry that may well be homologically pervasive within the Cervidae (Huxley, 1932, for this classic case of positive allometry). Inherited constraint may set a preferred channel, but selection must still guide any lineage into such an internally biased path. So a functionalist may view such undeniably positive constraints as, at most, helpmeets or facilitators of natural selection, while continuing to regard selection as a necessary instigator, and therefore as the primary cause of change.

But—and now we come to the nub of the issue, and to the central role of positive developmental constraint as a major challenge to selectionist orthodoxy—the attribution of similar evolutionary changes in independent lineages to internal constraint of homologous genes and developmental pathways, and not only to an external impetus of common selective pressures, must be limited to very close relatives still capable of maintaining substantial genetic identity as a consequence of recent common ancestry. Mayr's characterization of selectionist orthodoxy comes again to mind: distantly related lineages cannot be subject to such internal limitation or channeling because the pervasive scrutiny and ruthless efficiency of natural selection, operating on every feature over countless generations in geological immensity, must have fractured any homological hold by underlying genes and developmental pathways over the freedom of phenotypes to follow wherever selection leads.

Darwin's famous words, so often quoted, haunt the background of this discussion (1859, p. 84): "It may be said that natural selection is daily and hourly scrutinising, throughout the world, every variation, even the slightest; rejecting that which is bad, preserving and adding up all that is good; silently and insensibly working, whenever and wherever opportunity offers, at the improvement of each organic being in relation to its organic and inorganic conditions of life."

Therefore, any uncannily detailed phenotypic similarity evolved between distantly related groups must arise by convergence from substrates of non-homologous genotypes—thus affirming our usual view of selection's over-arching power, especially if common function for the two similar forms can validate the hypothesis of generation within a comparable adaptational matrix. (Note the logical danger of circularity that intrudes upon the argument at this point, for this extent of detailed similarity—the very datum that, in an unbiased approach, would lead one to entertain parallelism based upon common internal constraint as a viable alternative to convergence based on similar adaptive needs—now becomes an *a priori* affirmation of selection's power, the hypothesis supposedly under test.)

For this reason, such detailed functional and structural similarities, evolved independently in distantly related lineages, have become "poster boy" examples of convergence—itself the "poster boy" phenomenon and general concept for showcasing selection's dominant sway—precisely because similarities evolved in this mode cannot, by Mayr's argument, be ascribed to parallelism based on positive constraint imposed by homologous genetic and developmental pathways. With internal channeling thus theoretically barred as a potential source of impressive similarity, convergence becomes the favored explanation by default. The argument, surely "tight" in logic and principle, seems incontrovertible.

Since I do not wish to dwell on the previous errors that we all committed on this issue, let me simply illustrate the older view, and the magnitude of current reversal, with one of my own mistakes—from a 1976 paper extolling convergence as evolutionary biology's closest natural analog to replication in the experimental sciences (Gould, 1976, p. 177):

> The convergent evolution of similar structures fulfills, at least imperfectly, the criterion of independent replication that any experiment requires. An adequate theory of functional morphology must explain adaptive design by studying how different organisms react to the same selective regime. If we want to know whether plate tectonics is a true, universal physics of large bodies or only a descriptive account of this planet's history, other planets must be studied. If we want to know whether the biochemical unities of all life on earth have general import as optimal designs (given the nature of universal chemistry) or merely reflect the monophyletic origin of life on earth (and the homologous status of ATP and left-handed amino acids), then we shall have to hope for life on Mars. If, to retreat to something more immediate, one wishes to as-

sess the functional limits or mechanical constraint upon the human eye, one would do well—as J. Z. Young and others have done—to study the octopus.

But, as I shall discuss on pp. 1123–1132, one of the major discoveries of evo-devo has revealed a deep genetic homology underlying and promoting the separate evolution of lens eyes in cephalopods and vertebrates. The overt phenotypes do record substantial convergence (for different body tissues build corresponding structures in the two groups), but both phyla share key underlying genes and developmental pathways as homologies, and the example has lost its former status as the principal textbook case of natural selection's power to craft stunning similarities from utterly disparate raw materials. Eyes of such strikingly similar design owe their independent origin as much to genetic and developmental *parallelism,* based on internal constraints of homologous genes and developmental pathways, as to selection's capacity for iterating nearly identical adaptations from scratch by *convergence.*

With this "one liner" of maximal force—evo-devo has reinterpreted several textbook examples of convergence as consequences of substantial parallelism—we can encapsulate the depth of theoretical disturbance introduced by this subject into the heart of Darwinian theory. Our former best examples of full efficacy for the functional force of natural selection only exist because internal constraints of homologous genes and developmental pathways have kept fruitful channels of change open and parallel, even in the most disparate and most genealogically distant bilaterian phyla. The homological hold of historical constraint channels change at all levels, even for the broadest patterning of morphospace, and not only for details of parallel evolution in very closely related groups.

### A terminological excursis on the meaning of parallelism
THE NINE FATEFUL LITTLE WORDS OF E. RAY LANKESTER. The transforming power of this discovery upon evolutionary theory would stand out more clearly if the key terms and concepts had not become so muddled in our literature, and therefore so widely misunderstood or disregarded by modern researchers. (This situation cannot validate the graybeard's perennial lament: "them young fellers just don't keep up with the views of the older guys, like we did when we wuz gettin' started." The concepts and terminology surrounding the origin and status of similar structures in different lineages have inspired particular difficulty and unclear thinking ever since Darwin, and even before. In their classic paper on the subject, still the best treatment ever published, Haas and Simpson (1946) devoted the bulk of their long text to the history of confusion over differences between parallelism and convergence—with the two authors finally agreeing to disagree about the most fruitful definitions, even as they resolved the conceptual confusions.)

We should begin by recalling a central distinction that we all know, and probably all regard as refreshingly free from conceptual ambiguity: the difference between *homology* and *homoplasy.* Homologous structures are similar by inheritance from a common ancestor. Homoplastic structures are similar by independent evolution, for we can infer that the common ancestor did not possess the structure. In other words, the dichotomy of these two terms captures the essential difference between common ancestry and independent origin. At a first level of interpretation (but here we immediately plunge a toe into troubled waters, as we shall soon see), the dichotomy also marks a conceptual distinction between the hold of history and the power of adaptation.

So far so good—and I will not challenge the accepted and codified current definition of these two terms for describing an important logical distinction in evolutionary biology. But we often gain better understanding—and do not merely indulge an antiquarian passion for trivial and superannuated detail—when we explore the historical origin of a word, and then discover a marked discrepancy between initial and current usage. I intend no criticism of current usage in making such an observation. Words change their meanings, just as organisms evolve. We would impose an enormous burden upon our economy if we insisted upon payment in cattle every time we identified a bonus as a pecuniary advantage (from the Latin *pecus*, or cattle, a verbal fossil from a former commercial reality).

E. Ray Lankester, T. H. Huxley's protégé and the finest evolutionary morphologist in the generation just after Darwin, proposed the concept of homoplasy in 1870 (see Lester and Bowler, 1995, and Gould, 1999a, on Lankester's life and general views). I suspect that most evolutionary biologists could cite Lankester as a source, but I will wager a substantial sum that very few colleagues could identify (to their pecuniary benefit) a supreme irony in Lankester's original paper, entitled "On the use of the term homology in modern zoology, and the distinction between homogenetic and homoplastic agreements"—namely, that he defined homoplasy as a subcategory of homology, in apparent defiance of current usage (which, I repeat, I do not challenge) of homology and homoplasy as dichotomous opposites. The reasons for his distinctions, and for subsequent changes and refinements of meaning, tell an interesting story that can unlock the essential distinction between parallelism and convergence, and also explain the significance of evo-devo in unleashing the capacity of parallelism to rebalance formalist and functionalist causes within evolutionary theory.

Richard Owen enjoyed the height of his influence when Lankester wrote his paper, and the younger morphologist properly went to the source of all later usage (Owen, 1848 and 1849) in defining his terms. Lankester, as a Darwinian acolyte, also correctly noted the philosophical difficulty facing anyone who sought to translate such vital terms as homology into the new evolutionary context. Owen, as Lankester notes, defined homology within a Platonic theory of archetypal form (see discussion of Owen's concepts on pp. 312–329). How can the term be carried over into Darwin's world? Lankester (1870, p. 34) began his paper by stating this problem:

> Whilst the adoption of the theory of evolution has broken down the notions at one time held by zoologists and botanists as to the existence of

more or less symmetrical classes and groups in the organic world, established by some inherent law of Nature which limited her productive powers to arbitrary special plans or types of structure, and has taught us to see, in the variously isolated and variously connected kinds of animals and plants, simply the parts of one great genealogical tree, which have become detached and separated from one another in a thousand different degrees, through the operation of the great destroyer Time, yet certain terms and ideas are still in use which belonged to the old Platonic school, and have not been defined afresh in accordance with the doctrine of descent.

In particular, Owen had specified three categories of homology: special, general, and serial. (His classical and definitive 1848 treatise, "On the archetype and homologies of the vertebrate skeleton," comprises three chapters, titled "special homology," "general homology," and "serial homology" respectively.)

Owen's famously vague and broad definition of "homologue" as "the same organ in different animals under every variety of form and function" (1848, p. 7, repeated from 1843, p. 374) invokes a Platonic notion of sameness as "proceeding from a common archetype." Lankester had the good sense and vision to recognize (and we continue to assent today) that this concept did enjoy philosophical coherence, and could be translated into evolutionary terms—but that the Darwinian version implied different distinctions, requiring a subdivision of meanings and significances within a general notion that remained usefully unitary.

Owen's three categories share tighter bonding in the idea that parts can be called homologous so long as they can be construed as expressions or embodiments of the same idealized archetype (a key pre-evolutionary notion of formalist, as opposed to functionalist, thinking, and therefore particularly difficult to translate into a functionalist theory of evolution like natural selection). Obviously, as a first pass for evolutionary translation, we should redefine the Owenian archetype as the Darwinian common ancestor—thus substituting the real flesh and blood of physical continuity for a Platonic notion of formal identity. We can then proceed, as Lankester notes, with evolutionary versions of Owen's three categories.

For Owen (1848, p. 7), special homology refers to "the correspondence of a part or organ, determined by its relative position and connections, with a part or organ in a different animal." In evolutionary terms, we regard these two parts (in two different organisms) as homologous because they descend from the same feature in a common ancestor. Lankester (1870, p. 36, first paragraph) recognized this criterion of common ancestry as paramount—the definition that "without doubt the majority of evolutionists" would assign to the concept of homology. Lankester proposed—although his name never took hold—that this aspect of Owen's broader concept ("special homology") be called *homogeny* (or homology *sensu stricto*.) What then becomes of Owen's other two categories?

Owen defined general homology as "a higher relation . . . in which a part or series of parts stands to the fundamental or general type, and its enunciation involves and implies a knowledge of the type on which a natural group of animals, the vertebrate for example, is constructed" (1848, p. 7). This idea that we can assert a form of homology between two parts (in two organisms) because both express the same general archetype, rather than because one part can be designated as the "same" as the other part (as in special homology), does not translate by a similar criterion of descent from common ancestry because, as Owen noted, general homology records "a higher relation." For example, Owen regarded arms, legs, and heads for that matter, as derivations from a common vertebral archetype. Thus, the forearm of an aardvark is a general homolog of my leg (not to mention the shrew's head and the whale's flipper).

Obviously, these pairings do not represent homology by direct descent from common ancestry. And yet, we would not deny that some legitimate evolutionary commonality links my leg and aardvark's forearm (forget the shrew's head, although I would not be shocked if the old vertebral theory for the origin of the skull reemerges some day in a renewed form of validity). Moreover, we can be quite confident that the similarity marks a genuine *historical hold*, not a fortuity, or a convergence separately evolved from different archetypal bases. Still, the hold cannot be equated with true common ancestry, and must arise instead as a constraint based on common genesis from a source that imposes limitations or sets preferred channels of change from within. In his Platonic perspective, Owen called this common source an archetype. We would identify such a generating source as a developmental constraint from the historical vertex of the aptive triangle—perhaps arising from homologous genes or homologous developmental pathways in the two separated lineages.

Finally, Owen defined serial homology as the iteration of an archetypal form *within* the same organism in a set of repeated parts, perhaps each specialized for a particular function, but still bearing signs of the common architectural plan—as in the biramous appendages of arthropods, whether specialized as antennae, mouth parts, walking legs or genital claspers; and in the arms and legs of tetrapods.

When asked how he could square serial homology with his basic definition of "the same organ in different animals . . ." Owen would reply that the criterion of sameness trumped the requirement for different organisms, and that different places within the same organism would suffice. And we do not judge this response today as flippant or invalid because we share Owen's feeling that some common principle—although not common ancestry—validates a legitimate comparison of my leg and the aardvark's forearm (general homology) and the aardvark's forearm with the aardvark's own leg (serial homology). We would also identify this common principle as developmental constraint based on homologous genes and embryological pathways (whether expressed as arms and legs in the same animal, or as similar structures in two animals—although we would call such structures nonhomologous because they do not descend from a common ancestor, even though they owe their structural similarity, in large part, to construction by homologous genes and developmental pathways. After all, if no common genetics or development influenced the ontogeny of the aardvark's arm and leg, we would view the two limbs as purely convergent and unaffected by internal constraint, while Owen, in such circumstances, would not have defined them as serially homologous).

We may now return to Lankester's problem: Owen's special homology translates easily into evolutionary language as descent from common ancestry. This phenomenon presents no conceptual problem, and Lankester therefore chose to separate this subcategory as the unambiguous "best case" of *homogeny*. But how shall general and special homology be represented in evolutionary language? On the one hand, Owen applied these terms to separate structures that do not descend from the same structure in a common ancestor. They should therefore be distinguished from special homology (Lankester's homogeny) because we must be able to identify true and unbroken continuity in physical descent as a basis for phylogenetic reconstruction.

On the other hand, general and serial homology do record a hold of history over descent within clades. Some common property, present in a clade as a consequence of phylogenetic history, does generate the similarities in these two other Owenian categories. But this property can only be identified as a common generating pattern, a common constraint, a common pathway of development, or a common set of hereditary tendencies—and explicitly *not* as an overt common ancestral structure retained by descent in subsequent branches of the clade.

What then shall we call these products of common phylogenetic patterns in organic architecture—these separately evolved results of common developmental constraints, we would say today—but *not* of overt and expressed common ancestral phenotypes? Lankester proposed that we call them *homoplasts* in contrast with the *homogens* of common structural origin, and that we designate the process of their production as *homoplasy*, as distinguished from the *homogeny* of strict descent from common ancestral structures. But he considered both processes as subdivisions of a larger and coherent concept of homology—homogeny for Owen's special homology, and homoplasy for Owen's general and serial homology.

In defending his placement of homoplasy within a broad but coherent concept of homology, Lankester asks (1870, p. 38): "What is the other quantity covered by the term homology over and above homogeny?" Lankester answers that many similarities not due to inheritance of common ancestral structures nonetheless arise as consequences of the inheritance of unique, phylogenetically constrained building patterns—and therefore deserve inclusion within a broader category of similarity based upon descent (as opposed to similarity derived purely by independent adaptation, with no contribution by constraint from an organism's past history). These independently evolved, but historically constrained, similarities—we would now call them parallelisms—define Lankester's original concept of homoplasy. Lankester does acknowledge that homoplastic similarities must be evoked by similar environmental pressures (the pool cue of natural selection, in Darwinian terms), but he stresses the internal basis of inherited common building patterns and materials (1870, p. 42): "Under the term 'homology,' belonging to another philosophy, evolutionists have described and do describe two kinds of agreement—the one, now proposed to be called 'homogeny,' depending simply on the inheritance of a common part, the other, proposed to be called 'homoplasy,' depending on a common action of evoking causes or moulding environment on such homogenous parts, or on parts which for other reasons offer a likeness of material to begin with."

The foregoing exegesis raises an obvious question: if Lankester restricted homoplasy to independent origin of similar features based on common and phyletically distinctive internal constraints (though not common ancestral structures) in two or more lineages—thus drawing the phenomenon close enough to the essential and defining theme of homology (the "hold of history") to rank, in Lankester's system, as a subcategory of homology (broadly defined)—then how did the term migrate to the opposite meaning now universally and unambiguously understood today? In other words, how did homoplasy move from a subcategory of homology to become the diametric opposite of homology, with the domain of homology then shrinking to encompass only Lankester's narrower category of homogeny, and the domain of homoplasy expanding to include all similarities evolved independently and not directly inherited from a common ancestral structure?

What looks like an enormous difference—the expulsion of homoplasy as a subcategory of homology *(sensu lato),* and its establishment as a phenomenon directly contrary to homology *(sensu stricto)*—actually rests upon a small point: the migration of convergence *into* the category of homoplasy as now defined. If we decide that the crucial distinction between homology and homoplasy should rest upon common ancestry *vs.* independent origin, then one important phenomenon, necessarily included within homoplasy by the defining criterion of independent origin for similar structures, shares too much conceptual overlap with homology to permit a clear and comfortable *theoretical* separation (however firm the *descriptive* division): independent origin channeled by common internal constraints of homologous genes or developmental pathways—in other words, the phenomenon known as *parallelism.*

But Lankester originally defined homoplasy exclusively on the basis of phenomena that we would attribute to parallelism (Owen's general and serial homologies). Therefore, for him, homoplasy could legitimately count as a subcategory of homology *(sensu lato),* even though he recognized that he had to separate homoplasy from homogeny (homology *sensu stricto*) by the genealogical criterion of common ancestry vs. independent origin. But, if the scope of homoplasy ever expanded to embrace convergences as well—a defendable move because convergences also record an independent origin of similarities—then the combination of parallelisms plus convergences into one category would destroy the conceptual linkage of homoplasy with homology. With the addition of convergence (based on explicit denial of common internal constraints, and exclusive focus on a common external context of adaptation), homoplasy loses its former common ground with homology (in positing historical hold—whether of structures in homogeny, or of genes, pathways and potential in homoplasy—as the shared causal basis of similar structures in two lineages). Moreover, this expanded category of homoplasy now includes only one universal feature to define its own coherence: independent origin, in both parallelism and convergence. This feature also places homoplasy into antithesis with homology (common ancestry vs. independent origin).

Convergence did join parallelism to build an expanded category of homoplasy, thus setting the opposition that continues to define these terms as an exhaustive and dichotomous division today. Ironically, as a final point, Lankester himself—in a logical inconsistency within his own paper— spawned this dramatic shift in his own intended parsings by adding those "nine fateful little words" (of my title to this subsection) to the end of one statement in his original article. On page 41, as he tries to distinguish his newly formulated concept of homoplasy from the older notion of analogy, he presents (in this single passage) such a broad definition of homoplasy (in the midst of a "generous" attempt to show that analogy must be construed as broader still, and therefore not synonymous with homoplasy no matter how far we extend the concept) that he actually includes independent evolution by convergence—not a subcategory of homology by any stretch of the imagination!—in those nine words at the end (presented in italics below, but not in Lankester's original): "Homoplasy includes all cases of close resemblances of form which are not traceable to homogeny, all *details* of agreement not homogenous, in structures which are broadly homogenous, *as well as in structures having no genetic affinity.*"

Once pure convergence had been added to homoplasy via these nine fateful little words, the linkage of homoplasy to homology could no longer be defended—and the two concepts moved from their initial union to their current antithesis.

I recount this story at some length because I know no better way to illustrate the central tension and conceptual confusion within the concept of homoplasy. Parallelism and convergence do share the common descriptive feature of defining an independent origin for similar structures in two lineages. But in causal terms, particularly for assessing the relative weights of formal *vs.* functional factors in evolutionary change, the conceptual difference could not be more important—for parallelism marks the formal influence of internal constraint, while convergence reflects the functional operation of natural selection upon two substrates different enough to exclude internal factors as influences upon the resulting similarity.

This recognition of internal channeling as the root cause of *parallelism*— the principal basis for ascribing evolutionary change, and not only limitation, to historical constraint—lies at the heart of evo-devo's theoretical novelty and importance to the Darwinian worldview. This context behooves us to formulate, and to clarify, the causal distinction between parallelism and convergence—and not just to lump these two principles together by their single common property of specifying an independent origin for similar features in separate lineages. I believe that the history and logic of debate about the meaning of parallelism provide our best path to understanding this important revision in evolutionary theory.

**THE TERMINOLOGICAL ORIGIN AND DEBATE ABOUT THE MEANING AND UTILITY OF PARALLELISM.** After struggling through dense paragraphs and conceptual thickets, G. G. Simpson (in Haas and Simpson, 1946, p. 325) finally conceded that traditional confusion about the evolutionary meaning of similarity rested upon a logical dilemma, not an absence of empirical data for resolution of a factual issue. Why, to state the dilemma succinctly, does parallelism resist easy fitting into a coherent conceptual structure for the terminology of evolutionary similarity? In particular, why, when parallelism comfortably joins convergence to establish a coherent larger category (called homoplasy) for similar structures evolved independently, have so many good biologists, from the first formulation of the concept until today, continued to "feel in their bones" that something about parallelism veers off towards the supposedly opposite category of homology? Why, to use a vernacular expression often invoked in this discussion (as by Patterson, 1988), does parallelism seem to occupy a "gray zone" between the clear homology of evident retention by common descent, and the clear homoplasy of convergence by selective production of strikingly similar structures (in both form and function) from entirely different points of origin (the "cup coral" shape of rudistid bivalves, prorichthofeniid brachiopods and rugosan corals, for example)?

I have used Simpson's insight to construct the enlarged chart presented as Table 10-1. In an incisive footnote, explicating his differences with coauthor Otto Haas, Simpson makes the logical point that, although homology and homoplasy do cohere as dichotomous opposites encompassing all cases, they bear to each other the odd relationship of a positive claim (A) contrasted with an absence thereof (not-A). Nothing in logic forbids such a taxonomy, but scientists, maintaining a deeply engrained (if unconscious) preference for classification by causes, feel discomfited, in a way that they may not even be able to articulate, about a scheme that contrasts a positive assertion (homology as descent from common ancestry) with its descriptive absence (homoplasy as similarity *not* by descent from the same structure in a common ancestor). Simpson wrote (in Haas and Simpson, 1946, p. 325):

> Homology, as we agree, is best defined as similarity interpreted as due to common ancestry. Homoplasy, as we also agree, is best defined as similarity (or as including any process leading to similarity) that is not explicitly interpreted as due to common ancestry. Both terms rather than being purely descriptive . . . express an opinion, one positive and one negative. Homology expresses an opinion as to how the similarity arose. Homo-

plasy expresses an opinion as to how the similarity *did not* arise, i.e., that it did not arise by homology, but it does not express an opinion as to how the similarity *did* arise. I do not . . . see these as alternatives at the same categorical level. The set is not positive, "a" and "b" as mutually exclusive categories, but is a dichotomy of "a" and "not-a." Under "not-a" it is still possible to have a sequence of alternatives, such as "b," "c," etc., that are positive categories on the same level as "a."

The standard literature does include a venerable term—analogy—that might establish a contrast with homology in the causal sense that wins our almost visceral assent as more satisfactory, with homology as positive A (similarity due to common descent, with no need to invoke direct selective molding), and analogy as oppositely positive B (similarity due to common pressures of natural selection upon backgrounds of no common descent).

We now encounter the logical dilemma that underlies nearly all our extensive and lamentable confusion on this issue. Homoplasy and analogy might strike us, at first, as fully synonymous, for both invoke natural selection as the source of separate evolution for similar structures in two lineages. This synonymy certainly applies for convergence. But homoplasy comes in two flavors: parallelism and convergence—with parallelism as the historical root (in Lankester's original definition of homoplasy), but only convergence carrying the full flavor of synonymy. That is, convergence stands opposite to homology by both criteria—the negative not-A of origin *not* by common descent, and the positive B of origin by natural selection working in a similar way upon two unrelated substrates.

Unfortunately, a common error of human thinking leads us to define broad and variable categories by their clearest extreme cases. Thus, many scientists have assumed that all homoplasy, whether by parallelism or by convergence, must originate entirely for functional reasons, and not at all by constraint (the B category of exclusively dichotomous logic); whereas, the "not-A" of independent origin identifies the only property truly required for inclusion within the broad definition of homoplasy. Simpson continues (in Simpson and Haas, 1946, ibid.):

> Moreover, the implication is usually present to some degree and it has sometimes been explicitly stated that the structural similarity here in question is not due to homology but is correlated with community of function as opposed to community of ancestry. It is in this sense that analogy is a true alternative (but not the only alternative) to homology as a positive category on the same level, a "b" category rather than a "not-a" category or something on a different level altogether. That is, analogy, when used in this way, expresses a positive opinion, or theory, that a structural resemblance is correlated with function, just as homology expresses the view that it is correlated with common ancestry. Unlike homoplasy, analogy offers an alternative theory as to the basis of the resemblance in question.

The flavor of parallelism, however, lies in the gray zone of the A-B type classification that our traditions favor (see Table 10-1). In a descriptive sense, parallelism surely ranks within not-A by definition. But when we assess parallelism's relationship to the B category of our alternative scheme, we run right into the aforementioned logical conundrum. Parallelism partakes of B in its invocation of similar selective regimes to produce homoplastic *structures* from two separate starting points lacking the structure. But parallelism also includes too much A-ness (that is, claims for genuine homology of some sort) to rank as pure B.

The logical solution, had the issue been properly formulated, is not, and never has been, particularly arcane or difficult. Parallelism lies in a gray zone as a consequence of its different status within two conceptual schemes that do not parse nature in exactly the same way, but that we tend to conflate (because both capture important properties of phyletic change) when we consider the meaning of similarity in evolution. Parallelism includes aspects of *both constraint and independent selection*—not as a wishy-washy mixture in one grand pluralistic glop of all-things-for-all-people, but in rigorously different parsings for different levels of consideration (again, as evident in Table 10–1. Several authors have stressed the dependency of these terms on the hi-

Table 10-1. Evolutionary Similarities in Different Lineages Classified by Two Logical Types (A vs. B or A vs. Not-A) and by Two Criteria (in Realized Structur or in Underlying Generators of the Structures

| | | | |
|---|---|---|---|
| A) *Logic* | | | |
| Type 1 | Zone A | Gray Zone | Zone B |
| Type 2 | Zone A | Zone Not-A | |
| B) *Name of Phenomenon* | | | |
| In Type 1 | Homology | Parallelism | Converger |
| In Type 2 | Homology | Homoplasy | Homoplas |
| C) *Basis for Similarity* | | | |
| In Type 1 | Negative constraint | Positive constraint plus independent selection | Independe selectior |
| In Type 2 | Conserved descent | Independent origin | Independe origin |
| D) *Origin of Similarity* for: | | | |
| Realized structures | Conserved descent | Independent selection | Independe selectior |
| Underlying generators of the structures | Conserved descent | Conserved descent | Independe selectior |
| E) *Cause of Similarity for Realized Structure* | | | |
| Inheritance | Yes | No | No |
| Current selection | No | Yes | Yes |
| F) *Cause of Similarity for Underlying Generator* | | | |
| Inheritance | Yes | Yes | No |
| Current selection | No | No | Yes |

erarchical level under consideration—see Roth, 1991; Wagner, 1989; Bolker and Raff, 1996. Wing of bats and birds are, after all, convergent as wings, but homologous as forearms).

At the level of an overt phenotypic structure under explicit consideration, parallelism denies homology and asserts independent origin. But, at the level of the generators for the overt feature—the genes regulating its architecture, and the developmental pathways defining its construction—parallelism affirms homology as the concept's fundamental meaning and *raison d'être*, and the basis for its dichotomous contrast with convergence as alternatives within the more inclusive category of homoplasy. Thus, parallelism does require independent regimes of similar selection, but the resulting phenotypic likenesses must also be channeled from within by homologous generators.

(In an odd sense, one might view this old issue of differences between parallelism and convergence as a grand foreshadowing for an important debate that evolutionary biologists have only recently clarified in their minds—but that might have achieved earlier resolution had we all remembered this older discussion: the recognition that cladistic gene-trees do not correspond entirely with organism-trees. The capacity for parallelism rests upon organismal branching before gene branching. Continuing the argument, one might also view the first steps in the opposite mode of gene branching before organism branching as a molecular representation of Owen's old concept of serial homology. Paralogs within one organism are serial homologs; different paralogs in two organisms are general homologs; only orthologs in two organisms are special homologs, the heart of the modern concept of pure homology, or Lankester's homogeny—see p. 1071.)

Framed this way, the maddening complexities and counterclaims of the literature gain immediate clarification. One must then ask why the distinction between parallelism and convergence has bred so much conceptual trouble in the past. In particular, the two terms have often been purposefully combined (and demoted) to merely descriptive names for stages in a continuum. The terms will then only designate the trivial geometric difference between features evolved independently in two lines that remain at about the same distance in overall phenotype (parallelism) and lineages that become more similar as a consequence of their separate evolution of such functionally comparable features (convergence). One can only wonder, then, why biologists ever bothered to devise explicit terms for mere geometric waystations in a continuum with no interesting causal distinctions. Yet Haas, for example, defended this descriptive and geometric meaning, while his co-author G. G. Simpson demurred (in Haas and Simpson, 1946). And Willey (1911), in a first book entirely dedicated to the subject (and title) of "Convergence in Evolution," also denied a meaningful distinction in choosing his single term to encompass the entire subject of separately evolved similarities. Willey wrote (1911, p. xi): "I have used the word convergence in a wide sense . . . The [traditional] definitions leave us in the dark as to what degrees of relationship would entitle a given case to be classed as one of parallelism or of convergence."

In my judgment, Wake (1991, pp. 543–544) has correctly explained why many biologists blurred the theoretical difference between parallelism and convergence, and then relegated the terms to descriptive waystations in a continuum of results for a single causal process. When the subject of internal constraint faded to a periphery of interest (or even of active denial) within the functionalist orthodoxy of selection's overarching power and adaptation's empirical preeminence at the height of the Modern Synthesis (see Chapter 7 on hardline versions of the Synthesis that peaked in the late 1950's and 1960's), the conceptual distinction of parallelism as a manifestation of internal channeling became uninteresting to most evolutionists (or, in the worst effects of biasing by restrictive theories, even unperceivable). With the defining feature of parallelism thus banished to a limbo of theoretical irrelevance, biologists limited their concern to the support provided for adaptationist preferences by the common feature of all homoplasies: the guiding power of independent selective regimes, whether aided by homologous internal channels (parallelism) or not (convergence), to fashion the same functional result in separate lineages. Wake wrote (1991, pp. 543–544):

> My central theme is the phenomenon of nondivergent evolutionary change among lineages, including convergent morphological evolution, parallelism, and some kinds of reversal—in other words, what phylogeneticists term homoplasy . . . Convergence and parallelism often are considered to constitute strong evidence of the functioning of natural selection. Patterson stated "The general explanation for convergence is functional adaptation to similar environments" (1988, pp. 616–17), but I argue that alternatives must always be considered. In recent years increasing attention has been given to the possibility that parallelism is a manifestation of internal design constraints, and so both functionalist and structuralist constructs predict its occurrence.

As Wake's statement implies, two reasons—one "good" and the other "bad" in the conventional, if simplistic, terms, usually applied to such assessments in science—underlie this movement of parallelism to a periphery of limited interest, or to conflation with convergence, a phenomenon of opposite theoretical import in judging the differential weights of constraint and adaptation in the origin of homoplastic similarities. Wake correctly identifies the "bad" reason as an overemphasis on functionalist themes that limited the scope of evolutionary theory during the mid-century's height of enthusiasm for a "hardened" version of the "Modern Synthesis." Phenomena like parallelism, defined by components of internal constraint, did not elicit the attention of many evolutionists during this period.

But, as Wake recognizes in the last sentence of his statement, parallelism also received limited attention for the eminently "good" reason that, however well defined in a conceptual sense, the crucial distinction between parallelism and convergence could not be cashed out in operational terms until recently—for biologists could not identify the "homologies of underlying generators" (the shared genetic and developmental bases of independently evolved structures) needed to distinguish parallelism from the purely adaptational phenomenon of convergence. But evo-devo has become an active field, while the subject of parallelism has been catapulted from a periphery of forced inattention (as a clearly defined but non-operational concept) into the center of evolutionary studies, largely because biologists have now developed criteria for distinguishing the internal constraints of parallelism from the purely selective basis of convergence.

In short, more than a century after recognizing the important conceptual distinction, we can finally resolve actual cases by assessing the different contributions to homoplastic similarity made by constraining channels based on homologous generators and directing pathways based on common regimes of selection. I shall present the evidence of best cases in the next section, but will first close this section on conceptual and terminological analysis by citing five chronological episodes in the history of evolutionary debate about parallelism. These linked episodes all exemplify a crucial argument about the importance to general evolutionary theory of current research on the genetics of development: Despite all subsequent confusion and denigration, the concept of parallelism arose as a causal claim about channels of constraint *vs.* purely functionalist explanations rooted in natural selection (or some other adaptationist mechanism, as NeoLamarckism remained popular in the early years of this debate) for the evolution of homoplastic resemblance.

The interesting literature on parallelism (as opposed to some of the meaningless wrangling over terminology) never lost this theoretical context throughout a century of research and commentary. The delay in resolution, and the prolongation of theoretical discussion, did not reflect any lack of clarity on the part of evolutionists, especially as explicated by G. G. Simpson, who understood and promoted the concept of parallelism and its potentially radical implications for Darwinian theory. Rather, the persisting frustration about parallelism primarily recorded the inability of geneticists and developmental biologists to identify the generators posited as the basis of "latent" or "underlying" homology in the evolution of homoplastic structures deemed parallel rather than convergent. This bolted door of stymied practice has now been unlocked, and we have crossed a threshold into a period of amazingly fruitful research on parallelism in particular, and on the role of developmental constraint based on deep homology in general, for establishing the markedly nonrandom clumping of actual organisms within life's potential morphospace.

THE ORIGIN OF THE TERM "PARALLELISM." Interestingly, this term first entered evolutionary theory with an entirely different meaning—but for another concept, indeed a far stronger version, of internal channels as major determinants of trends in the history of life: the theory of recapitulation in embryology. In pre-evolutionary versions, Agassiz had spoken of a "threefold parallel" of embryological, taxonomic, and paleontological series within larger types. The American paleontologist and evolutionary theorist E. D. Cope then formalized an evolutionary version of the "law of parallelism" within recapitulatory theory (see Gould, 1977b, for relevant sources and quotations).

"The relation of genera," Cope writes (1887, p. 45), "which are simply steps in one and the same line of development, may be called exact parallelism." In other words, different genera belonging to the same parallel series will run, during their full ontogenies, down varying lengths of a common developmental (and phyletic) trackway. In this sense, the adult of one genus may be virtually identical (exactly parallel in Cope's terms) with the juvenile form of another genus that runs further along the common track during its own ontogeny. Obviously, these common trackways, regulating both the ontogeny and phylogeny of entire series of related genera, invoke a concept of internal constraint with a vengeance. Cope, in this early version of his developing ideas, placed far more stress on internal channeling to explain taxonomic relationships than his later attraction for the functionalist theory of Neo-Lamarckism would allow (see Gould, 1977b and 1981b, for an analysis of Cope's changing views on the relative importance of constraint and function).

The first use of parallelism in its modern meaning, including its dichotomous pairing with convergence, can also be traced to two of the greatest American vertebrate paleontologists of the late 19th century: W. B. Scott and H. F. Osborn. If the concept can claim a "founding" quotation at all, Scott (1891, p. 362), in a long and famous article on the osteology of early perissodactyls and artiodactyls, invoked degree of taxonomic relationship to distinguish parallelism from convergence, while emphasizing their common attributes as homoplastic confounders of phylogeny: "But if the various species of the ancestral genus may acquire the new character independently of each other (parallelism), or if the species of widely different genera may gradually assume a common likeness (convergence), then it is plain that such a genus is an artificial assemblage of forms of polyphyletic origin."

In his 1895 summer lecture at the Marine Biological Laboratory in Woods Hole, Scott (1896, p. 56) provided a more formal definition: "By parallelism is meant the independent acquisition of similar structure in forms which are themselves nearly related, and by convergence such acquisition in forms which are not closely related, and thus in one or more respects come to be more nearly alike than were their ancestors."

More importantly, Scott then explicitly argued that he needed to distinguish these two categories of homoplasy because parallelism, based on constraints of inherited channels for preferred change, will generally confound phylogeny less than convergences that arise by similar functional impact upon truly different starting points (1896, p. 58): "It seems the most obvious of commonplaces to say that numerous and close resemblances of structure are *prima facie* evidences of relationship. Yet the statement is true, even though the resemblances have been independently acquired, because parallelism is a more frequently observed phenomenon than convergence, and because the more nearly related any two organisms are, the more likely are they to undergo similar modifications."

Osborn, the patrician "kingmaker" of American paleontology (and quite a potentate in American science in general), cited Scott's definitions in several papers, paying special attention—in the context of his own pluralistic views on the importance of both formal and functional factors as evolutionary causes—to the role of parallelism in combining the push of selection (or some other functionalist cause) with the internal channeling of constraint as the architect of preferred pathways for any agent of "pushing." For example, in his 1902 paper on "Homoplasy as a law of latent or potential homology," Osborn had already identified parallelism as falling into a gray zone between the pure analogy of convergence and the pure homology of unaltered inheritance. With parallelism's notion of "predeterminate variation" (1902, p. 270), Osborn argues, "I think we have to deal with homology or, more strictly, with a principle intermediate between homology and analogy."

In a 1905 article on "The ideas and terms of modern philosophical anatomy," Osborn then presented a first chart (reproduced here as Fig. 10-13) of relations among these terms, including parallelism and convergence as subcategories of analogous resemblance (in contrast with homologous resemblance here restricted to Lankester's notion of homogeny). His chart depicts the geometrical distinction between parallelism and convergence, but his definitions follow Scott in relying not on the descriptive difference between parallel and converging lines, but on "similar characters arising independently in similar or related animals or organs" for parallelism, *vs.* "similar adaptations arising independently in dissimilar or unrelated animals or organs" for convergence.

These foundational statements indicate both the conceptual clarity and the

I. HOMOLOGOUS, *i. e.*, *Homogenous.*

II. ANALOGOUS.
*Parallel.* Analogous adaptations, *i. e.*, similar characters arising independently in *similar or related animals or organs*, causing a similar evolution, and resulting in parallelisms.

*Convergent.* Similar adaptations arising independently in *dissimilar or unrelated animals or organs*, causing a secondary similarity or approximation of type, resulting in *convergence.*

III. NON-ANALOGOUS.
*Divergent.* Increasing specialization and differentiation resulting in 'divergence' or 'adaptive radiation.'

10-13. In his 1905 article, H. F. Osborn treats parallelism and convergence as subcategories of analogous resemblance. But note how he follows Scott in defining parallelism by common possession of underlying generating factors, and not by the mere geometry of results.

nonoperational nature—about the most frustrating situation one can face in science—of the distinction originally made in defining parallelism and convergence. Both Scott and Osborn grasped the importance of separating homoplasy due to underlying homology of generators ("latent or potential homology" in Osborn's apt phrase in the title of his 1902 paper) from homoplasy rooted exclusively in a similar external context of adaptation. But the biology of their time provided no way to specify or identify these generators. Scott and Osborn therefore had to invoke the entirely unsatisfactory, indirect and vague surrogate of "degree" of taxonomic resemblance—arguing (quite properly of course, however nonoperationally) that the closer the relationship between two separated lineages, the more likely that any homoplastic characters will arise by parallelism. Scott expressed his frustration at this unsurmountable situation on the page following his initial definitions (1891, p. 363): "The distinction between the two classes of phenomena [parallelism and convergence] is obviously one of degree rather than of kind, and it will therefore be convenient to consider them together."

THE GREATER SALIENCE OF PARALLELISM FOR NON-DARWINIAN FORMALISTS, AND FOR ANTI-DARWINIAN THEORISTS OF VARIOUS STRIPES, IN LATE 19TH AND EARLY 20TH CENTURY EVOLUTIONARY DEBATES. We understand why parallelism faded from general consideration when the strict adaptationism of later and hardened versions of the Modern Synthesis pushed the general subject of internal constraint to a periphery of intellectual concern and presumed relevance. Similarly, we should easily comprehend why the same phenomenon—and the importance of distinguishing its component of constraint from the purely adaptational basis of convergence—would have generated more interest and greater clarity of definition during the period of its initial formulation (1890's to 1920's), when non-Darwinian formalist, and more overtly anti-Darwinian orthogenetic and saltational, theories enjoyed considerable vogue as adjuncts or alternatives to natural selection.

Two linguistically and geographically defined traditions of argument reinforce my contention that parallelism has always been understood and debated as a theory of constraint based on homologous generators for the origin of homoplastic similarities. First, the American paleontologists who initially codified the concept of parallelism, did so in the context of pluralistic support for non-Darwinian internal mechanisms of evolutionary change (working in conjunction, or potentially in opposition, to natural selection, which they also accepted as a valid mechanism). We find parallelism sufficiently interesting today as an indicator of preferred internal channels that selection can exploit in coordinated evolutionary change. Imagine the even greater theoretical interest of parallelism for evolutionists who hoped to discover, in its workings, new principles and mechanisms of change that might fundamentally enrich or alter the basis of evolutionary theory.

In the article that first defined parallelism, for example, Scott (1891, pp. 370–371) argued that the orthogenetic linearity of parallel series implied a primary non-selectionist cause for phylogenetic transformation, since lineages under the control of natural selection should exhibit more temporal fluctuation: "So far as the series of fossil mammals which we have been considering are concerned, the developmental history appears to be very direct, and subject to comparatively little fluctuation, advancing steadily in a definite direction, though with slight deviations."

In his 1902 article, Osborn invoked parallelism more explicitly as a central argument for internal control of phylogenetic directionality, and against natural selection as a primary cause of change. In fact, following a standard tradition of continental non-Darwinian argument, Osborn demoted natural selection to a mere "exciting cause" ("exciting," that is, in the literal sense of "initiating," not in the modern meaning of "thrilling") that can arouse the inherent channels of necessary change, and provoke homoplastic evolution along parallel paths. In his typically regal way, Osborn begins his paper by quoting his own prophetic words of 1897: "My study of teeth in a great many phyla of Mammalia in past times has convinced me that there are fundamental predispositions to vary in certain directions; that the evolution of teeth is marked out beforehand by hereditary influences which extend back hundreds of thousands of years. These predispositions are aroused under certain exciting causes [note his verbal demotion of natural selection] and the progress of tooth development takes a certain form converting into actuality what has hitherto been potentiality."

Osborn then ends his paper (1902, p. 270) by explicitly citing the "latent or potential" homology of parallelism as an alternative to natural selection among causes of evolutionary change:

> These homoplastic cusps [of teeth in independent lineages of mammalian evolution] do not arise from selection out of fortuitous variations, because they develop directly and are not picked from a number of alternates . . . We are forced to the conclusion that in the original tritubercular constitution of the teeth there is some principle which unifies the subsequent variation and evolution up to a certain point. Herein lies the appropriateness of Lankester's phrase, "a likeness of material to begin with." Philosophically, predeterminate variation and evolution brings us upon dangerous ground. If all that is involved in the Tertiary molar tooth is included in a latent or potential form in the Cretaceous molar tooth we are nearing the *emboitement* hypothesis of Bonnet or the archetype of Oken and Owen.

Second, continental European theorists in the formalist tradition (see Chapters 4 and 5) had always emphasized constraint channeled by laws of form as a primary alternative to functionalist theories like natural selection. These scientists should therefore have taken a particular interest in parallelism, especially in its distinction from convergence for the origin of homoplastic similarity—for convergence exalts natural selection, while parallelism stresses internal channeling and supports the standard continental view of selection as a mere potentiator, or at most a minor diverter, of predictable and lawlike changes that must follow internally specified rules of morphogenetic transformation.

Haas and Simpson (1946) cite all the major evolutionary theorists among continental paleontologists of the early to mid 20th century—particularly Abel, Dacqué and Schindewolf—in support of these weights and definitions. In 1921, for example, Dacqué compared parallelism with Eimer's anti-Darwinian concept of orthogenesis (see pp. 355–365 for full discussion on Eimer's views), while stressing the distinction of parallelism and convergence by the predominant causality of constraint *vs.* adaptation (Haas and Simpson, 1946, p. 335).

G. G. SIMPSON AND THE CAUSAL *VS.* GEOMETRIC DEFINITION OF PARALLELISM. With parallelism thus falsely depicted as somehow contrary to selection, one can hardly blame the resurgent Darwinians of the Modern Synthesis for their diminished attention to a phenomenon that had been unfairly cited against the cause of change that they now wished to reassert as primary, if not virtually exclusive. (This history provides another concrete illustration of a general argument about older *vs.* modern versions of constraint that I advance throughout this book. The older versions interpreted constraint as contrary to selection, thus earning the indifference or enmity of Darwinian theorists when they regained ascendancy during the 1930's and afterwards. This unfortunate historical situation clouded the utility of constraint *within* Darwinian theory as an adjunct, a potentiator, or (at most distinction) an orthogonal source of evolutionary change. Modern versions of constraint can overcome this unfortunate division and reunite these two vital sources, formalist and functionalist, into an expanded and more general theory of Darwinian evolution.)

But the most perceptive of Darwinian theorists would not let such a contingent historical happenstance extinguish an important concept and distinction within the scope of evolutionary causality. In particular, G. G. Simpson—indisputably the most brilliant and biologically sophisticated of 20th century evolutionary paleontologists—continually emphasized the significance of a causal concept of parallelism based upon constraint, and the importance of distinguishing this mode of homoplasy from the opposite style of convergence based entirely upon shared adaptive contexts rather than shared homologous generators.

In his epochal 1945 treatise on principles of taxonomy and classification of mammals, Simpson drew a sharply dichotomous distinction between homology and convergence (1945, p. 9): "Animals may resemble one another because they have inherited like characters, homology, or because they have independently acquired like characters, convergence." Simpson then spoke of parallelism as "a third sort of process [that] also produces similarities" (p. 9)—for he recognized the "hybrid" nature of a concept that required independent episodes of similar selection, but nonetheless constructed homoplastic likenesses from homologous generators in two separate lines. With his usual insight, Simpson made the proper theoretical separation, but then ran right into the old wall of stymied practice—for the biology of his day knew no methods for identifying the homologous generators that could mark a homoplastic similarity as parallel rather than convergent. Unable to cash out his theoretical clarity in actual practice, Simpson threw in the towel and admitted operational defeat (1945, p. 9):

> It is a complication that a third sort of process also produces similarities: parallelism. The term is descriptive rather than explanatory and refers to the fact that distinct groups of common origin frequently evolve in much the same direction after the discontinuity between them has arisen, so that at a later stage the phyla may have characters in common that were not visible in the common ancestry but that tend, nevertheless, to be more or less in proportion to the nearness of that ancestry. This proportional tendency distinguishes parallelism from convergence, but the distinction is far from absolute. The two phenomena intergrade continuously and are often indistinguishable in practice.

Simpson (1945, p. 10) also stressed the intermediate nature of parallelism in phylogenetic inference, recognizing that even homoplastic characters usually record reasonably close genealogical affinity (in their common origin from homologous generators) in cases of parallelism, but must be regarded as confounders of affinity in cases of convergence: "Homology is always valid evidence of affinity. Parallelism is less direct and reliable, but it is also valid evidence within somewhat broader limits. It may lead to overestimates of degree of affinity, but it is not likely to induce belief in wholly false affinity. Convergence, however, may be wholly misleading, and a principal problem of morphological classification on a phylogenetic basis is the selection of characters that are homologous or parallel and not convergent."

In his 1961 book on *Principles of Animal Taxonomy*, Simpson continued to express his frustration at the conceptual need, but operational impossibility, of distinguishing parallelism from convergence. "The distinction of parallelism from convergence is vital," he writes (1961b, p. 106). Fifteen years after his joint paper with Haas, and their disagreement over geometrical versus causal definitions of the terms, Simpson stated in frustration (1961b, p. 103): "Parallelism is the independent occurrence of similar changes in groups from a common ancestry and *because* they had a common ancestry. Some students (for example, Haas in Haas and Simpson, 1946) have preferred a more purely descriptive definition, especially by the geometrical model of parallel lines, symbolizing two lineages both changing but not becoming significantly either more or less similar. . . . Most taxonomists do, however, consider that the term parallelism should be used only when community of ancestry is pertinent to the phenomenon."

Simpson concludes his discussion (1961b, p. 106) with the clearest statement I have ever read for citing homology of underlying generators as the basis of parallelism, and on the joint operation of both overt selection and underlying homology in the evolution of homoplastic structures by parallelism:

Parallelism has several theoretical bases that help one to understand and also to recognize it. The structure of an ancestral group inevitably restricts the lines of possible evolutionary change. That simple fact greatly increases the probability that among the number of descendant lineages several or all will follow one line. That probability will be further reinforced by natural selection in a geographically expanding and actively speciating group if the ecologies of diverse lineages remain similar in respect to the adaptations involved in the parallelism. The degree of dependence on similar ecology resembles that of convergence, but the retention of homologous characters from the relatively near common ancestry usually distinguishes parallelism. The parallel lineages (unlike those only convergent) furthermore start out with closely similar coadapted genetic systems, and similar changes are more likely to keep the system adequately coadapted.

PARALLELISM AS A "GRAY ZONE" BETWEEN HOMOLOGY AND CONVERGENCE. Despite Simpson's careful separations, and his stress on their theoretical importance, many biologists ignored the important theoretical differences between these two subcategories of homoplasy. If they recognized parallelism and convergence as distinctive terms at all, they often could not state any rationale for the terminology beyond the triviality of an abstract and formal geometric difference between parallel and converging lines.

But thoughtful evolutionists continued to struggle with the "hybrid" character of parallelism. Michener (1949), for example, in the finest technical application of the concept, honored the causal (rather than geometric) distinction: "The potentiality for similar changes, resulting in parallel characters, no doubt results from the fact that related animals have homologous chromosomes and genes" (1949, p. 140).

The cladistic revolution in taxonomic practice also forced renewed attention to the distinction, and to the "intermediate" status of parallelism in producing homoplastic structures based on homologous generators—leading, for example, to Saether's (1983) concept of "underlying synapomorphies," defined as "the capacity to develop synapomorphy" or "close parallelism as a result of inherited factors within a monophyletic group" (Saether, 1983, p. 343).

The acknowledgment of homologous generators actually led some taxonomists, including such leaders as Mayr (1974), to include parallelism *within* a broader definition of homology, while most researchers continued to rank parallelism as an uncomfortable subcategory of homoplasy (Patterson, 1988), or as a "hybrid" notion based on homoplastic origin from homologous generators (as in Saether, 1983). Perhaps Patterson (1988, p. 619) put the matter best by writing: "In morphology, the 'gray zone' between homology and nonhomology concerns congruence—or inferred common ancestry—and whether parallelism (which does invoke common ancestry) should be included or excluded from homology."

THE OPERATIONAL RESCUE OF PARALLELISM BY EVO-DEVO. The

ulmination of more than a century of conceptual and terminological struggle
may now be epitomized in a triumphalist tone usually shunned in science, but
clearly justified in this rare case: the development of genetic and developmen-
al techniques that established the field of evo-devo have finally allowed biologists to identify the homologous generators that always specified the con-
ept of parallelism in theoretical terms. Parallelism has now, and finally after
. century of terminological recognition, become an operational subject for
volutionary research. Moreover, the first flood of results has revealed a
lepth and extent of parallelism among distant phyla that strict Darwinians
ad explicitly deemed inconceivable, and that even the most enthusiastic
vell-wishers and partisans of constraint did not dare to imagine in their fond-
st dreams (unless their capacity for imagination greatly exceeded the scope
of this particular rooter—see Gould, 1977b).

## A SYMPHONY IN FOUR MOVEMENTS ON THE ROLE OF HISTORICAL CONSTRAINT IN EVOLUTION: TOWARDS THE HARMONIOUS REBALANCING OF FORM AND FUNCTION IN EVOLUTIONARY THEORY

As a literary device, metaphor spans a particularly broad band of relative
merit—from treacherous comparisons virtually guaranteed to confuse or mis-
tate a causal analysis, to illuminating analogies intended to explicate the un-
amiliar or to impose a useful and sensible order upon an otherwise inchoate
mass of ideas and information. By invoking the following risky comparison
of the major ideas and putative theoretical reforms of evo-devo to the four
movements of a classical symphony, I mean to highlight some aspects of the
comparison, while abjuring others. I do not, in the most obviously nonadaptive feature of the metaphor, claim any chronological basis, or any numerical ordering of importance, for the four sequential themes.

Rather, I rest my case for the utility of this organizing device upon an admittedly peculiar isomorphism between these disparate realms. I believe that
he burgeoning literature on the genetics of development can be explicated
most usefully (in terms of a probable enduring influence upon evolutionary
heory) as a set of four subjects—and that these subjects, presented in their
most sensible and logical order, invite a close comparison with the "standard" sequence and thematic progression of the four movements in a classical
symphony: statement, development, scherzo, and generalization—or, for the
iterature of evo-devo, deep homology, pervasive parallelism (for features
once deemed convergent), saltational musings, and reasons for the markedly
nhomogeneous occupation of morphospace among animal phyla.

### Movement one, Statement: deep homology across phyla: Mayr's functional certainty and Geoffroy's structural vindication.

DEEP HOMOLOGY, ARCHETYPAL THEORIES, AND HISTORICAL CON-
STRAINT. In the most important general book on evo-devo written in the
last decade of a millennium, Raff (1996, p. 428) astutely epitomized the

importance of constraint for an enriched and revised version of Darwinian
theory.

> A long-standing and important theoretical conception of the relation-
> ship between development and evolution is that of developmental con-
> straints. The idea that developmental rules can direct or constrain the
> course of evolution has two origins. A number of evolutionists, particu-
> larly in the generation following Darwin, took antiselectionist positions,
> and posited that internal forces direct evolution and produce long-term
> trends independent of the external environment. That is not a tenable
> position, but neither is extreme selectionism. Internal genetic and devel-
> opmental constraints of various kinds must exist, but . . . they are diverse
> and poorly understood. Yet if internal factors constrain evolution, they
> are hardly a minor issue. The acceptance of internal constraints does not
> mean that Darwinian selection is unimportant, but it does mean that the
> variation presented to selection is not random.

Two aspects of this statement capture both the optimism and the theoretical importance of this emerging field. By defining the subject of constraint as
collateral and helpful to selectionism (rather than oppositional, if not sub
stitutional, as in most 19th century versions of internalism, as Raff mentions
above and as I document extensively in Chapters 4 and 5), Raff depicts the
growth of evo-devo as interactive building in a different architectural style,
rather than as demolition. Secondly, by summarizing the main import of con
straint for Darwinian theory in the claim "that the variation presented to
selection is not random," Raff correctly identifies the locus of greatest importance for evolutionary theory—for the logic of pure selectionism does presuppose nondirectional variability (see pp. 144–146), and the existence of
strongly preferred channels, based on the architecture and history of development, does require an important restructuring (not just a minor nuancing) of
Darwinian logic.

I argued in the last section that development establishes preferred channels
of variation in two primary modes, both "positive" in their salutary contribution to a more accurate and sophisticated evolutionary theory. But by mechanistic criteria of channels as limitations or impetuses, we might deem the first
mode—based on the surprising discovery of "deep homology" in the genetic
basis of conserved developmental pathways among distantly related animal
phyla—as "negative," in highlighting the limitations thus imposed upon directions of change. (Nonetheless, combinatorial possibilities remain as broad
as realized bilaterian diversity, so these limits may direct, but surely do not
seem to throttle life—see Kirschner and Gerhart, 1998 and references therein,
on flexibility and evolvability.) The second mode—based on the equally surprising discovery of common genetic pathways underlying several textbook
cases of supposed convergence, thus recasting these homoplasies as parallelisms potentiated by common developmental architecture—then achieves its
best explication as a set of positive impetuses for channeling adaptive change
into accessible pathways.

However, in Raff's larger sense, both modes express the common, cardinal feature of nonisotropic, or channeled, variation—thus imposing a preferred structure, from the "inside" of organic development, upon the raw material that external forces of Darwinian selection must utilize. Both modes also delighted (or disturbed) evolutionary biologists with the greatest surprises in the last generation of our science—based on results that were actively unexpected in theory, not merely unsuspected for lack of imagination. I shall, in this first section, discuss deep homology as the more general and fundamental of the two modes. My second section shall then extend Raff's theme of directed variation as the focus of constraint within evolutionary theory, this time through the positive channel of unanticipated parallelisms.

In a famous line from the prologue of *Faust,* Goethe wrote: *Es irrt der Mensch, so lang er strebt*—we err, so long as our struggle lasts. Goethe probably intended this celebrated statement as a romantic effusion about human striving in general, but we may apply his words to the nearly universal attitude of fellow biologists, at least since Darwin's watershed of 1859, towards Goethe's own brainchild in developmental biology, and towards the general approach to morphology—a word of Goethe's own invention—embodied within such theories.

As discussed *in extenso* in Chapter 4, Goethe's theory of the leaf as a botanical archetype for all lateral structures off the angiosperm stem (including cotyledons and all flower parts) presented the most famous botanical proposal among a set of archetypal theories that would soon sweep the world of animal morphology as well, culminating in the vertebral archetype advocated by Owen for all major parts of the vertebrate skeleton (including the skull and limbs) and, most extensively, by Etienne Geoffroy Saint-Hilaire for the generative basis of all animal form (first for all vertebrates, then adding arthropods, then mollusks, and no doubt proceeding further had he not then encountered the wrath and active opposition of Cuvier and his functionalist theory of adaptive form—see Chapter 4, pp. 291–312).

The argument that structural and morphological archetypes underlie, and actively generate, a basic and common architecture in taxonomically distant groups defines—both as a fact of our profession's actual history and as a dictate of the logic of our explanatory theories—the strongest kind of claim for developmental constraint as a major factor in patterns of evolutionary change and the occupation of morphospace. I suspect that the depth of this challenge has always been recognized, but the empirical case for such constraining archetypes has remained so weak, since the heyday of Geoffroy and Owen some 150 years ago, that the issue simply didn't generate much serious concern—and rightly so.

The concept of interphylum archetypes, deemed too bizarre to warrant active refutation, experienced the curt and derisive dismissal reserved for crackpot ideas in science. (Goldschmidt's saltational apostasy, on the other hand, inspired voluminous and impassioned denial because his ideas did seem sufficiently and dangerously plausible to the Modern Synthesis—see pp. 451–466.) Indeed, the notion of interphylum archetypes struck most biologists as so inconceivable in theory that empirical counterclaims hardly seemed necessary. After all, the notion required extensive genetic homology among phyla, and the power of natural selection, working on different paths for a minimum of 530 million years since the origin of distinct phyla in the Cambrian explosion, seemed to guarantee such thorough change at effectively every nucleotide position that the requisite common foundation could not possibly have been maintained (see Mayr, 1963, p. 609, as previously discussed on pp. 539 and 1066).

When, in the mid-1980's, initial studies began to discern deep homology between arthropod and vertebrate *Hox* genes, I well remember saying to myself (amidst my astonishment about a result so consonant with the theoretical framework that I had espoused in 1977 in my first book, *Ontogeny and Phylogeny,* but had not dared to view as subject to empirical validation in my lifetime): yes, perhaps for some functional commonality in the broadest construction of basic body axes (A-P in particular), but surely not for the more detailed structural homology—particularly between arthropod metameres and vertebrate somites—demanded by the old archetypal theories. But, only 15 years later, central nuggets of validity had been affirmed for nearly all the classical archetypal theories, even the most farfetched. Needless to say, the archetypes do not function as their inventors claimed. The differences between leaves and floral parts do not arise by progressive refinement of sap up the stem; and the abstract vertebra does not function as a generator for all major features of the axial skeleton (including ribs and appendages) in vertebrates and arthropods.

Moreover, at least two prominent claims for the vertebral archetype probably hold little, if any, validity. The distinctive features of the vertebrate skull and forebrain seem to arise, in large part, under the formative influence of the distinctive neural crest (see the classical statement of Gans and Northcutt, 1983), and not as a complex fusion (much like an arthropod tagma) of a definable number of rostral vertebrae (from 3 to 8 in various formulations). And although some broad homologies may set the basic axes of limbs in both arthropods and vertebrates (see pp. 1138–1142), the structures cannot be regarded as basically homologous, even in underlying developmental pathways; nor can they be derived from any particular component of a generalized vertebra. Nonetheless, all three major archetypal theories of Goethe and Geoffroy—the classical sources of ridicule for the general concept—have now been confirmed in aspects that cannot be dismissed as superficial or secondary.

**MEHR LICHT (MORE LIGHT) ON GOETHE'S ANGIOSPERM ARCHETYPE.** Students of the mustard *Arabidopsis* have discovered unexpected validity in central features of Goethe's founding theory of the archetypal leaf (see Pelaz et al., 2000). Starting at the bottom, Meinke (1992) studied the *lec* (leafy cotyledon) mutant that partially transforms cotyledons into leaves. He argued that the wild-type allele (*LEC)* activates "a wide range of embryo-specific

pathways in higher plants" (p. 1647), and that suppression by the *lec* mutation therefore causes reversion to a ground state—which, as Goethe proposed
so long ago, most closely resembles a stem leaf in basic form. (Biologists with
a zoomorphic bias, including the author of this book, may be confused by a
claim that embryonic features might thus be conceived as departures from a
ground state. The directionality of bilaterian ontogeny, with embryonic features as transient and formative, leads us to equate embryonic forms with any
sensible concept of a "ground state." But plants maintain embryonic tissues
throughout life as restricted and persistently specialized regions on differentiated foundations that animal biologists might tend to regard as "adult."
Therefore, a botanical rationale for viewing these foundations as a ground
state, with embryonic tissues as a specialization, can easily be defended.)

Meinke (1992, p. 1649) concludes: "The phenotype of *leafy cotyledon* suggests that the difference between leaves and cotyledons in *Arabidopsis* is controlled by a single regulatory gene *(LEC)* expressed only during embryogenesis." Then, in a statement strikingly evocative of Goethe's archetypal
theory, he portrays (1992, p. 1649) the ordinary stem leaf as a ground state,
with all its serial homologs (to apply this zoomorphic term to cotyledons
and, putatively, to flower parts) as specializations thereupon: "The preferred
model is that *LEC* functions to activate a wide range of embryo-specific pathways in plants. Loss of gene function disrupts embryonic maturation and returns mutant cotyledons to a basal developmental state. The leafy appearance
of mutant cotyledons was unexpected because there was no evidence that cotyledons defective in maturation should be transformed into foliage leaves.
However, this observation is consistent with the origin of cotyledons as specialized leaves during plant evolution and the homology of embryonic cotyledons and vegetative leaves."

For the more complex organs of inflorescence at the other end, Weigel and
Meyerowitz, in their classic review (1994) of the ABC model (see pp. 1063–
1065) for floral development in *Arabidopsis* (and many other angiosperms,
though perhaps not all, see Kramer and Irish, 1999), posed a first key extension beyond the model's basic elucidation: "The ABC model left one complication, though: what happens in the absence of all organ identity activity"
(p. 203). Weigel and Meyerowitz then turned to Goethe for the classic prediction based on notions of the archetypal leaf: "Goethe (1790) had proposed
that floral organs represent modified leaves, suggesting that a vegetative leaf
is the ground state of floral organs."

Weigel and Meyerowitz presented striking evidence to confirm this Goethian prediction that suppression of all ABC activity should cause presumptive floral parts to approach the ground state of stem leaves. The sequential
action of ABC genes permits a simple formulation of tests for this hypothesis.
AC double mutants, for example, should knock out determinants for the outermost sepals of whorl 1 (triggered by A genes alone) and the innermost carpels of whorl 4 (C genes alone), but impose less effect upon the petals and stamens of whorls 2 and 3, which also require the influence of B genes (see

p. 1063 and Fig. 10-12). Experiments then confirmed this precise, and rather odd, prediction: "Indeed, organs in these two whorls are very much like vegetative leaves—they develop with stipules, are green and covered with branched hairs, and senesce slowly, all characteristics of leaves but not of floral organs" (Weigel and Meyerowitz, 1994, p. 203). By the same logic, triple mutants should grow all floral parts in leaf-like form—as they do: "In triple mutants that lack A, B, and C activities, all floral organs resemble leaves" (pp. 203–204—and Fig. 10-14), thus supporting (Pelaz et al., 2000, p. 202) "the theory that flower organs are simply modified leaves." Theissen and Saedler (2001, p. 470) add, with specific homage to Goethe: "combined loss-of-function of class A, B, and C genes results in a transformation of all floral organs to leaves, corroborating Goethe's view that leaves are a developmental ground state."

Moreover, gain of function mutations also confirm the model by imposing inner floral expression upon outer parts, thus resembling the action, for a different symmetry of radial whorls, of classical homeotic mutations of *Drosophila*, expressed in a linear, antero-posterior array. Overexpression of C genes, for example (1994, p. 206), represses A functions in whorls 1 and 2, "with carpels where sepals are usually found, and stamens in the places ordinarily occupied by petals" (p. 206).

Later work has revealed some of the upstream regulators of this system. For example, Pelaz et al. (2000) identified three genes (named *SEP1/2/3*) required for the action of B and C genus that regulate the inner three whorls of petals, stamens and carpels. In triple mutant *Aribadopsis* plants that suppress the action of *SEP1/2/3*, all floral whorls develop as sepals (which are regulated by A genes). (See Honma and Goto, 2001, for later data on the even

10-14. Mutations that delete activity of all *ABC* genes cause all floral organs to develop as leaves. Ordinary flower at *A*; triple mutant with all flower parts replaced by leaves at *B*. From Weigel and Meyerowitz, 1994.

broader role of SEP genes in "providing flower-specific activity" (p. 528) in combination with genes of the ABC series.)

Other studies provide additional confirmation (in modern genetic form) for Goethe's original formalist notion of leaves as a ground state. A "meristem identity factor" *LEAFY* (*LFY*) potentiates *APETALA1* (*AP1*), which, in turn, activates the ABC floral genes. Wagner et al. (1999, p. 582) demonstrate that this sequence of *LEAFY* to *AP1* is "necessary and sufficient for this transition" (p. 582). Standard techniques for documenting the effects of both loss and gain-of-function mutants confirm this cascade. In the *lfy-6* mutant, suppressing the action of *LFY*, "most flowers are replaced by leaves and second-order shoots"; while overexpression of either *LFY* or *AP1* "results in formation of flowers or leaves and flowers in positions normally occupied by leaves" (Wagner et al., 1999, p. 582. See further confirmations in Busch et al., 1999).

Extending the model to other angiosperm clades, Hofer et al. (1997) studied *PEAFLO,* the pea homolog of *LFY.* They performed several experiments to extend Goethe's formalist concept of morphological serial homology, now abetted by new data on genetic and developmental homology, between leaves and flower parts. They state: "A striking comparison can be made between the similar developmental units of compound leaves and flowers: both arise laterally from primordia derived from the shoot apical meristem; both produce lateral, leaf-like organs; and both are determinate." Hofer et al. (1997) then affirmed and extended the evidence for developmental homology by (1) identifying pleiotropic mutants that affect both leaf and floral development in similar ways, and (2) by studying homeotic mutations that "result in the conversion of floral organs to leaf-like structures" (p. 581). Their concluding remark, reinforced by a later observation of Theissen and Saedler (2001, p. 469), might have caused Faust to lose his bet with Mephistopheles—by inducing such delight that this restless, archetypal romantic might finally have savored a present moment with sufficient gusto to blurt out the fateful phrase that would seal his doom: *"verweile doch, du bist so schön"* (stay awhile, thou art so beautiful). Hofer et al. write (1997, p. 586): "Compound leaves and flowers can thus be considered to be derivatives of the same ancestral structure." Theissen and Saedler simply conclude: "Goethe was right when he proposed that flowers are modified leaves."

## HOXOLOGY AND GEOFFROY'S FIRST ARCHETYPAL THEORY OF SEGMENTAL HOMOLOGY

AN EPITOME AND CAPSULE HISTORY OF HOXOLOGY. These Goethian confirmations extend, at least for now, little beyond the serial homology of apparently disparate parts on the same plant. But archetypal claims for homology across distantly related phyla raise far more serious theoretical problems. No shockwaves attended the discovery of common genetic and developmental pathways for the serial array of arthropod appendages, despite their functional differentiation as antennae, mouth parts, legs, genital claspers, etc. But the discovery that homological pathways also persist among animal phyla that have evolved independently since the Cambrian explosion has reversed previous certainties and brought Geoffroy's despised archetypal theories into renewed respectability.

The roots of this great discovery extend back (at least terminologically) to another key figure of this book, the English geneticist William Bateson (see Chapter 5, pp. 396–415). Bateson became fascinated by a class of mutations with the peculiar, and often large, effect of causing the characteristic form of one member in a serial array to develop in a different location usually occupied by another member of the same array. Bateson called such mutations "homeotic," and their peculiar forms, almost humorous in some cases, gave them a special salience among geneticists. Unsurprisingly—for arthropods are serial organisms *par excellence,* while this particular insect became the lynchpin of genetics—the homeotic mutations of *Drosophila* became classics of the genre, famous for their oddness as well as their utility (for geneticists, not for the afflicted flies!).

We all remember our undergraduate textbook pictures—and the attendant, inevitable thoughts of Hollywood monster movies—of flies with such mutations as *antennapedia* (legs where antennae "ought" to be), *bithorax* (with another pair of wings rather than halteres on the third thoracic segment, thus seeming to "revert" the fly—a false interpretation as we shall see—to the ancestral four winged condition), and *bithoraxoid* (with a supernumerary pair of legs on the first abdominal segment, thus giving eight legs *in toto* and seeming to mock the very definition of the class Hexapoda). In my favorite example, a homeotic mutation in mosquitoes actually replaces the biting stylets with a pair of legs, thus rendering the creature "ouchless." I entertained various fantasies about breeding these lovely mutants, introducing them into natural populations, and destroying this scourge of humanity from within. But, alas and unsurprisingly, the scheme would never work, and I couldn't interest a single venture capitalist—for the mutation is effectively lethal; a mosquito that cannot bite to draw blood cannot feed at all.

E. B. Lewis used such homeotic mutations to develop his model for the evolution and operation of the *bithorax* complex in *Drosophila,* the breakthrough that effectively began the modern study of evo-devo and that won a most deserved Nobel prize for its pioneer. (The Nobel awards include no category for evolutionary studies. Only twice has a prize been given for work in evolutionary biology, each time by nuancing the definition of medicine to include work with legitimate consequences for health, but scarcely in the mainstream of medical research—first to Lorenz, Tinbergen, and von Frisch, for foundational studies in ethology, and second to my dear colleagues Ed Lewis, Christiane Nüsslein-Volhard, and Eric Wieschaus for unlocking the genetic basis of fundamental architectures in animal development.)

In a simple and brilliant model, Lewis (1978) inferred that the *bithorax* complex evolved by gene duplication, with all members (up to eight) remaining aligned in a tandem array on the third chromosome. Since these *BX-C* genes regulated developmental positions in the posterior part of the thorax and throughout the abdomen (see Fig. 10-15), Lewis assumed that the functional basis of duplication lay in the need for more genes to achieve evolutionary differentiation from the ancestral homonomy of repeated and similar (if not identical) appendages on each body segment—in this case, and for dipterans in general, to suppress the development of legs on abdominal segments and to convert the second pair of wings (on the third thoracic segment) into the small pair of balancing halteres.

The model then implied an elegant mechanism for gene functioning in morphological differentiation. Lewis argued that the first gene in the array turned on in the second thoracic segment and in all posterior segments, with each subsequent gene having its anterior boundary of expression one or two segments further back, but then turning on from there to the posterior end of the fly. Clearly, such a system would build a simple and linear gradient with least gene product at the anterior end of expression for the entire array (where only the first gene turns on), and most product at the posterior end of the animal (where all genes are active).

The further beauty of this model then lies in the simple testability of the im-

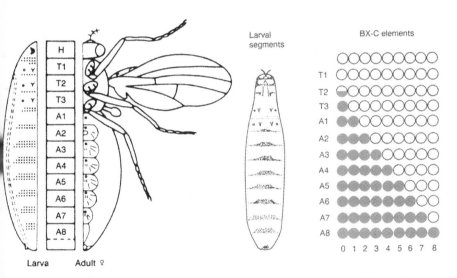

10-15. E. B. Lewis's original, brilliant, but not entirely correct model for developmental action of genes in the *bithorax* complex of *Drosophila*. Lewis assumed that differentiation of complexity from original homonomy, particularly the conversion of the second pair of wings to halteres, and the suppression of legs on the abdominal segments, required a duplication of further genes in the set. He proposed a tandem array of up to eight genes, each turning on in sequence, but with expression beginning in successively more posterior parts of the developing larva—thus establishing a gradient with ever more gene product accumulating towards the rear of the animal. Therefore—and this part of the model remains basically correct—loss-of-function mutations should weaken the gradient and cause anterior structures to develop in a more posterior position; while gain-of-function mutations should intensify the gradient and cause posterior structures to develop in more anterior sites.

plied mechanism for initiating appropriate structures in each segment: the more gene product, the more posterior the appearance (given a linear gradient with greatest concentration at the rear end). Thus, any loss-of-function mutation, leading to a weakening of the gradient, should cause anterior structures to develop in a more posterior position. In a corresponding manner, gain-of-function mutations, or ectopically induced overexpressions, should intensify the gradient and cause posterior structures to grow in more anterior positions. Shifts in both these directions would produce homeotic effects under Bateson's original definition—and the *BX-C* complex had originally been recognized by a set of arresting homeotic mutations.

Lewis's model neatly explained the most famous and puzzling homeotic transformations, both based on loss-of-function mutations. *Bithorax*, the celebrated four-winged fly, does not represent an atavistic reversion to the ancestral state, but arose by a weakening of the gradient that caused the third thoracic segment (usually bearing the much reduced second set of wings in the derived form of balancing halteres) to develop instead as a supernumerary second thoracic. Since second thoracics bear ordinary wings, a fly with two second thoracics will grow two pairs of wings. Similarly, the equally peculiar eight-legged, or *bithoraxoid,* fly developed by another loss-of-function mutation under the same rules of Lewis's gradient. The gradient became sufficiently weakened in the first abdominal segment to cause this normally legless module to develop instead as a supernumerary third thoracic. Since each thoracic segment bears a pair of legs (giving insects their defining six for the animal's three thoracic segments), a fly with (effectively) four thoracic segments would grow eight legs.

As a virtually definitional consequence of truly great theories developed in a previous *terra incognita,* several aspects of an original formulation invariably turn out to be wrong, while central concepts persist in greatly improved form. The most interesting development since the classical formulation (Lewis, 1978), has reversed Lewis's argument that the duplications arose to provide positional cues needed to potentiate the evolution of the distinctive insect body plan (in particular, to suppress legs on the abdomen and convert wings to halteres on the last thoracic segment). In formulating his original hypothesis, Lewis (1978) made the conventional assumption of both Darwinian and ordinary vernacular reasoning: that greater specialization of the phenotype would correlate with increase in the number of generating units. But the idea that morphological novelties must "await" the provision of new genetic material by duplication (or some other process) has been disproven by the fascinating discovery—with central implications for my general argument about constraint, to be developed in the concluding fourth "movement" of this "symphony" (pp. 1147–1178)—that all major arthropod *Hox* genes had already appeared before the separation of arthropod classes and, for that matter, of protostome phyla as well.

Homologs for all 8 insect *Hox* genes have been found in other arthropod classes, including the maximally homonomous (identically segmented)

Myriapoda and, for that matter, in the equally homonomous sister phylum of onychophorans (Grenier et al., 1997). De Rosa et al. (1999) conclude that the full complement must be even more ancient, as phyletic analysis indicates a minimum of 7 *Hox* genes for the bilaterian ancestor, and at least 8 for the common ancestor of protostome phyla.

Thus, the differentiation of distinctive bilaterian body plans has occurred not by the duplication or recruitment of additional *Hox* genes, but by changes in their regulation and their downstream targets. Presumably, *Hox* genes "read" positional information to set the location of differentiating structures, thereby triggering the cascade of downsteam architects, but not building the varied structures themselves. As Warren et al. (1994, p. 461) write: *Hox* genes "provided a pre-existing groundplan upon which insect segmental diversity evolved." Carroll (1995, p. 483) therefore restated the Lewis hypothesis as follows: "What has evolved in the course of insect and fly evolution are not new genes but new regulatory interactions between *BX-C* proteins and genes involved in limb formation and wing morphogenesis."

The discovery of the homeobox—a 180 base pair unit coding for a 60 amino acid homeodomain with important regulatory action as a DNA binding protein—as a common constituent of *Hox* genes (and others as well) opened the floodgates of this amazingly fruitful research in the early 1980's. By probing for homeoboxes, *Hox* genes could quickly be located and characterized, and (even more crucially for evolutionary analysis) their homology to genes of other organisms (even in other phyla) established. The two homeotic complexes of *Drosophila*—*Antennapedia (ANT-C)* and *Bithorax (BX-C)*—were quickly revealed as separated subunits, controlling the positioning of anterior and posterior structures along the A-P axis respectively, of a single *Hox* cluster that maintains its integrity in the beetle *Tribolium*, and in other non-dipteran insects. Powers et al. (2000) show that the mosquito *Anopheles gambiae* also retains a single and undivided *Hox* cluster, so the *Drosophila* subdivision does not characterize Diptera in general.

The established rules of "hoxology"* vindicated the central principles of morphogenesis in Lewis's model, though under an interestingly different genetic regime. (The *BX-C* component of the *Drosophila Hox* sequence contains only three genes, and if they arose, one from the other, by tandem duplication, these events probably preceded the separation of protostome and deuterostome phyla.) But Lewis could not have been more prescient in recognizing the essential sequence and form of *Hox* action, and in specifying the implied consequences and tests. Lewis's principle established the basis for discovering homologous genes (and homologous actions) in distant groups, thus potentiating evo-devo's greatest and most surprising discovery of "deep homology" among animal phyla—the key to the reevaluation of

---

*The verbal pun upon "doxology"—the short, formulaic, and unvarying, prayers of Christian liturgy—inevitably comes to mind, although we trust that the norms of science will prevail to impose substantial and interesting improvements upon the current hoxology.

historical constraint as an essential component of evolutionary theory and pattern.

Manak and Scott's (1994, p. 63) epitome of "hoxology" illustrates the centrality of Lewis's original conceptions in a different guise:

Several rules governing homeotic gene function have been fairly well conserved. (1) Genes are ordered along the chromosome in the same order as their expression and function along the anterior-posterior axis of the animal. (2) More genes are usually expressed in more posterior regions. (3) Loss of gene function leads to loss of structures or to development of anterior structures where more posterior structures should have formed. (4) Activation of genes where they should be off, i.e. gain-of-function mutations, leads to posterior structures developing where more anterior structures would normally be found. To these generalizations we may add some molecular data. (5) Each homeotic gene contains a single homeobox and encodes a sequence-specific DNA-binding protein which acts as a transcription factor. (6) Most of the homeotic genes are transcribed in the same direction, with the 5' ends of transcription units oriented toward the posterior end of the *Hox* cluster.

The perfect colinearity of spatial order along the chromosome with the sequence of morphological differentiation along the developing animal's antero-posterior axis summarizes the most stunning conclusion of this research, and also generates most other hoxological regularities. This central property of colinearity supplies a rationale for Lewis's original concept of a gradient generated by tandem duplicates turning on in spatial order along the chromosome. (The spatial sequence usually reflects a temporal order as well, as morphologically anterior and genetically 3' units generally operate first in ontogeny, with differentiation then proceeding temporally towards the posterior. Some models of *Hox* evolution regard the temporal factor as primary (see Duboule, 1992; Dollé et al., 1993; Deutsch and Le Guyader, 1998), and I shall discuss this issue further in the last part of this section.)

The other morphological rules also follow from this central precept of colinearity (Lewis could not have known about items 5 and 6 in the above list when he devised his model). The rules for loss and gain-of-function mutations express this key property in a particularly convincing manner. I have already discussed the classic cases of four-winged and eight-legged flies as anteriorizations of posterior segments caused by loss-of-function mutations. The ultimate loss, a fly developing with no *Hox* gene function at all, leads to lethality, with the dead embryo as a grim and fascinating manifestation of expected rules: a misfit bearing antennae on each of its segments (Shubin et al., 1997, p. 644). (The antennae, or most anterior appendages, normally develop with no *Hox* activity at all.) The famous *Antennapedia* mutant introduces *Hox* activity into this anterior region and thus grows a leg in the antennal position. Other gain-of-function mutations also cause posterior structures to move forward, as expected. The first discovered gain mutation in the *Hox* genes, *Contrabithorax (Cbx),* causes the second thoracic segment to dif-

erentiate as another third thoracic—and the fly therefore grows two pairs of
halteres and no wings (Lewis, 1992, p. 1530)!

Raff (1996, p. 307) has expressed the surprise of colinearity, and its evolutionary implications for constraint, in the opening words of his section on
"frozen controls?":

> Constraint in gene organization is a clouded topic at best, but disturbing
> observations loom up like logging trucks on a foggy mountain road. The
> Hox genes have presented the most puzzling instance of deeply con-
> served gene order. In all phyla so far examined (arthropods, nematodes,
> and vertebrates), the *Antennapedia* and *Bithorax* homeotic gene homo-
> logues are clustered, they have the same transcriptional orientation and
> order of activation, and their transcription is colinear with the body axis.
> The conservation of a set of clustered genes over half a billion years is
> difficult enough to accept, but colinearity with body axis defies credibil-
> ity. Yet it's true.

VERTEBRATE HOMOLOGS IN STRUCTURE AND ACTION. So far, the
formalist or archetypal content of this discussion has been largely limited to
the Goethian theme of common bases for the generation of differentiated serial homologs in a *single* organism—in other words, to internal constraints
and channels in the evolutionary history of particular forms and lineages. But
the more radical archetypal theories—including both of Geoffroy's derided
arguments about vertebral foundations and dorso-ventral inversions—postulate the maintenance of such constraints in phyla of distant taxonomic separation and immensely long periods of independent evolution. Such theories of
constraining homologies among groups focus our attention upon the quite
different and larger issue of inhomogeneities in the morphospace of animal
designs. Does the markedly nonrandom clumping of organisms within this
morphospace record historical constraint (where organisms have been, and
where, in consequence, they then cannot go), and not only the power of selection (where organisms do best, with all workable positions accessible)?

The discovery of homeoboxes, and the development of simple probes for
their identification, provoked a grand "fishing expedition" (or "gold rush"
for a more positive metaphor) throughout the taxonomic pool of organisms. When such procedures become easy, efficient and inexpensive enough,
scientists will be tempted to try experiments that would otherwise be deemed
foolish.

As an obvious candidate for crazy experiments, especially in the persistently dim light of Geoffroy's archetypal hypothesis for arthropods and vertebrates, a search for vertebrate homologs of arthropod *Hox* genes could
hardly have remained unthought or undone, although I doubt that anyone dared to anticipate success (again, see Mayr's canonical quotation on
p. 1066). As we all now know and utilize the stunning successes of these experiments, a reminder of the initial astonishment, and of the tentative nature
of first conclusions, dramatically illustrates how far this research has pro-

ceeded in 15 years (and will no doubt extend, thereby rendering these pages obsolete, in just a few additional years).

Not only do *Hox* genes exist in vertebrates, but homologs for all *Drosophila Hox* genes have been found, arranged in the same linear order on chromosomes, and acting with the same colinearity in development along the A-P axis of the vertebrate body. Moreover, vertebrate *Hox* genes have undergone fourfold replication and exist as four paralogous sequences on four different chromosomes. (The vertebrate sister taxon, amphioxus, has but a single *Hox* cluster, so we can make good inferences about the timing of amplification in our lineage. The agnathan lamprey probably has only three *Hox* sequences. Interestingly, and uniquely among deuterostomes, or any other animal, the single *Hox* cluster of amphioxus has an "extra" or 14th *Hox* gene at the 5' end—see Ferrier et al., 2000.) The vertebrate *Hox* genes can be arranged into 13 paralogy groups. (No vertebrate genome includes all 13 genes in any single cluster. The mouse, for example, has 39 of the 52 possible genes—Ferrier et al., 2000. The single sequence of amphioxus, however, does include a copy of each *Hox* gene. The increase in potential number within each group occurred largely by duplications of the posteriormost (5') homologs of *Drosophila Hox* genes.)

Lewis (1992, p. 1529) captured the excitement of this work in a single opening adverb: "Astonishingly, mice and humans not only have cognates of the *BX-C* and *ANT-C* genes in a single *HOM-C*, but the complexes occur in four sets, each in a different chromosome." Slack et al. (1997, p. 867) echoes a consensus in designating this discovery of deep homology as "the most spectacular achievement of molecular developmental biology." Yet initial expectations certainly did not forecast emerging realities. In a 1990 review, De Robertis described the decision to undertake an experiment leading to the discovery of the first vertebrate homeobox gene in *Xenopus laevis* (Carrasco, McGinnis, Gehring, and De Robertis, 1984—a good Orwellian year). I was a bit saddened (but mostly amused) by the closing observation on the counterintuitively negative correlation that often emerges (or gets imposed by the realities of laboratory culture) between youth and willingness to think the unthinkable. To any graduate student reading this book, I can only say: *Verbum sapientiae . . .* "We decided to try what seemed, at the time, a crazy experiment: to isolate a gene similar to *Antennapedia* from frog DNA with McGinnis and Gehring's fruit fly homeobox probes. There was little reason to believe that the frog DNA contained such a gene or that the genes of such unrelated species would be significantly similar. Still, we felt it was worth the attempt. Some of our colleagues were skeptical that such an experiment could ever work, and two of our students declined to help on those grounds."

The initial discovery of homology in genetic structure for arthropod and vertebrate *Hox* did not seal the case for evolutionary meaning, since no one yet knew how vertebrate *Hox* genes operated. Carrasco et al. (1984, p. 409) wrote of their original discovery: "If the frog gene cloned here eventually turns out to have functions similar to those of the fruit fly genes, it would represent the first development-controlling gene identified in vertebrates." Evidence for similarity of action soon followed, thus securing the argument for meaningful morphogenetic conservation across at least 530 million years, and almost maximal bilaterian separation.

The vertebrate *Hox* genes also exhibit the crucial colinearity between sequential order on the chromosome and site of action along the body's A-P axis. Moreover, and most impressively, several early studies confirmed that the familiar arthropod rules for loss-of-function (anterior structures move back) and gain-of-function (posterior structures more forward) generally apply to vertebrate development as well (although unique and non-homeotic effects have also been demonstrated, as in Pollock et al., 1995). For example, in loss-of-function experiments, Le Mouellic et al. (1992) deactivated the mouse *Hoxc-8* gene (previously, as in this 1992 paper, called *Hox-3.1*) and noted anteriorization of vertebral form throughout a substantial region of the body axis extending from the 14th to the 21st vertebra (T7 to L1). In the most striking effect, a supernumerary pair of ribs (characteristic of thoracic vertebrae) grew on the first lumbar vertebra. In general, "vertebrae and ribs displayed more or less pronounced transformations, turning them into structures resembling those characteristic of the adjacent anterior segment" (1992, p. 251).

Rancourt et al. (1995) also observed anteriorization towards the adjacent segment in mice with disrupted expression of *Hoxb-5* and *Hoxb-6*. The first thoracic segments often lost their rib heads and grew altered lateral processes "making them indistinguishable from C7" (1995, p. 112). Since, with the rarest exceptions of 6 to 9 in sloths and 6 in manatees, all mammals possess 7 cervical vertebrae (yes, including giraffes, who grow very long cervicals but don't augment their number!), this homeotic transformation of the first thoracic to the form of a supernumerary (or eighth) cervical seems as curiously in violation of basic taxonomic signatures as the more famous four-winged and eight-legged *Drosophila*.

In an interesting temporal analog, illustrating the common coincidence of spatial and temporal ordering in the expression of *Hox* sequences, Dollé et al. (1993) disrupted the most 5′ (and therefore last acting) *Hoxd-13* gene in mice, and noted a variety of effects upon the limbs, all interpretable as neotenic changes expressing developmental delays evoked by deactivating the last stages of a normal temporal sequence in ontogeny. (I particularly appreciate Dollé et al.'s conscious linkage of these genetic results to the classical data on heterochrony (see Gould, 1977b) as a morphological approach to questions about the regulation of development.) Dollé et al. (1993, p. 438) note an interesting relationship between these genetic results and common pathways of evolutionary change in heterochronic phenotypes, thus invoking this chapter's central theme of positive constraints based on internal channels:

In such evolutionary modifications, the first skeletal elements to be lost are usually those that are formed last during the establishment of the chondrogenic pattern. In *Hoxd-13* mutants, the missing skeletal elements are precisely those that appear last during the development of the autopods. There is therefore a correlation between the extreme 5′ loca-tion of the *Hoxd-13* gene within its complex, its last position in the tem-poral sequence of activation and its involvement in the patterning of the last-appearing structures. The *Hoxd-13* phenotype may thus be consid-ered as resulting from a block in a developmental sequence. This arrest occurs at the end of the process and corresponds to the time at which this gene is supposed to become active. Consequently, only those structures appearing at the end of the process, or parts of those structures still de-veloping at this stage, will be altered.

In a corresponding manner, gain-of-function mutations often yield the ex-pected effects of posteriorization. Kessel et al. (1990) induced overexpression of the mouse *Hoxa-7* gene (previously called *Hox-1.1*) by inserting a pro-moter sequence of chicken DNA. Two results indicate a forward movement of posterior structures: (i) the first two vertebrae, the atlas and axis, became simplified, assuming a "structure characteristic of more posterior vertebrae" (1990, p. 302); (ii) the last cervical vertebra of one animal developed a pair of ribs and assumed the form of the next posterior series of thoracic vertebrae.

Kessel and Gruss (1991) then induced overexpression by application of retinoic acid. "Posterior transformations occurred along the complete body axis after RA administration on day 7 of gestation and were accompanied by anterior shifts of *Hox* gene expression domains in embryos" (1991, p. 89). In a particularly interesting result, Lufkin et al. (1992) ectopically expressed *Hoxd-4* (previously *Hox-4.2*) in regions of the developing head anterior to its usual boundary of expression in somites of the cervical vertebrae. "This ectopic expression results in a homeotic transformation of the occipital bones towards a more posterior phenotype into structures that resemble cervical vertebrae" (p. 835). Phyletic inference is treacherous, and absurd claims have been made in misanalogies between phyletic history and developmental anomaly. But a transformation of skull bones towards the identity of verte-brae does induce thoughts of a presumably more homonomous ancestral ver-tebrate.

Interestingly, the A-P axis of the vertebrate limb also seems to follow the same rules of colinearity. Morgan and Tabin (1994) demonstrated the impor-tance of the *Hoxd* series in differentiation of the chick limb bud. They ob-served expression of successive 5′ genes in progressively more posterior re-gions. Overexpression of *Hoxd-11* in regions anterior to its normal domain led to the growth of an additional phalanx in digit 1 (which normally has one, while subsequent digits have 2, 3, and 4 respectively, excluding the ter-minal claw)—"leading to a morphology similar to that of digit 2" (p. 183), a posteriorization anticipated in gain-of-function regimes. Ectopic expression of *Hoxd-11* in anterior regions of the chick wing that normally grow no skel-etal elements at all induced the growth of a supernumerary digit (resembling digit 2 in morphology) at the wing's anterior edge.

Tickle (1992) noted the similarity of *Hoxd* expression in the chick wing to

Lewis's gradient model for establishing domains of differentiation. Of the genes at the 5′ end of the complex, she wrote (1992, p. 188): "Cells in the posterior part of the bud that will give rise to posterior structures such as a 'little finger' express all the genes, whereas anterior cells that will give rise to the anterior 'thumb' express only *Hox-4.4*" (*Hoxd-9* in modern terminology). These rules, apparently pervasive (at least in bilaterally symmetrical Bilateria with an A-P axis), also explain several well known empirical regularities in the classical literature on experimental embryology. Citing the correlation of spatial order and temporal sequence, Tickle (1992, p. 188) notes: "Because activation can proceed in only one direction along the complex, this explained why manipulations can convert anterior structures into posterior ones, but never posterior into anterior."

The high degree of sequence similarity often found between homologous arthropod and vertebrate *Hox* genes (amounting to near identity of homeodomains in some cases) leads to the remarkable, but (by now) scarcely surprising, interphylum substitutability revealed by so many experiments (and further discussed as evidence for parallelism in the evolution of eyes on pages 1123–1132). Fly *Hox* genes, expressed in vertebrates, usually broker the same developmental sequences as their vertebrate homologs—and vice versa. Needless to say, such experiments yield the "correct" morphologies for each phylum, thus reinforcing the well-established conclusion that *Hox* genes specify proper positions and regulate downstream cascades, but do not build anatomical structures themselves. If *Hox* genes worked as architects as well as specifiers, then the frights of Hollywood horror movies might become realities, and the fly with a human head might really scream "please help me" from the despair of his spider-web prison.

As one example among so many, the *Drosophila Hox* gene *Antennapedia* promotes leg identity, presumably by repressing previously unknown antennal genes. Casares and Mann (1998) have now identified two antennal determiners, including *homothorax (hth)*. As one line of evidence, they cloned *Meisl*, the mouse homolog of *Drosophila hth*, and expressed it ectopically in the fly's anal primodium, which normally develops without expressing any *Hox* genes. The anal plates of these flies then grew as antennae. (Most *Hox* genes suppress antennae, so ectopic expression of *Meisl* in *Hox* domains does not generate antennae in odd places, but induces other malformations, including markedly truncated legs on the thoracic segments.)

As a person with literary pretensions, I am always fascinated by the sure signal of scientific progress conveyed by the evolution of a rationalized and simplified terminology. The original *Hox* terminologies were eclectic and specific. Students of *Drosophila* first identified two clusters of homeotic genes, but could not recognize them as separated parts of a single ancestral sequence. So they awarded different names: *Antennapedia* complex *(ANT-C)* for genes regulating anterior structures, and *Bithorax* complex *(BX-C)* for genes operating in the fly's rear half. When homologs of both were detected as a single sequence in beetles, terminology began to coalesce, and the entire series assumed the name of *HOM-C*. When researchers discovered vertebrate homologs, they did not want to use the same names at first, for they had not yet affirmed the corresponding similarities of colinearity and action (and were probably still reeling from the basic shock of the discovery itself). So the vertebrate homologs became *Hox* genes. This potentially anarchic situation deteriorated further when, after finding four *Hox* complexes in vertebrates, researchers started naming the genes in each complex by their order of discovery, and not by their invariant spatial positions along the chromosome. (Perhaps they did not yet believe that colinearity could prevail here as well.) Thus, *Hox-1.1* denoted the first discovered, not the most 3', gene of the first *Hox* series.

Happily, these discrepancies and illogicalities have now been sorted out and—like the standardization of railroad gauges, or the choice of an internal combustion engine for all cars (thus abandoning a host of other early and workable devices)—a common and integrated terminology has developed, not by the official fiat of any particular meeting or official commission, but by obvious advantages in daily use. The four vertebrate complexes have been renamed *Hoxa* to *Hoxd* and the genes within each have been numbered from 1 to 13 in their proper A-P, or 3' to 5', order. Meanwhile, acknowledging the proven homologies of gene structure, position and action, the fly folks have dropped their different name for the complex, and now also denote their sequence as *Hox*, rather than *HOM-C*. This congelation of a simple and unified taxonomy, replacing the previous promiscuity of different and uncoordinated names for each gene, marks the coherence and maturation of an important field from an initiating chaos of uncoordinated empirical promise.

SEGMENTAL HOMOLOGIES OF ARTHROPODS AND VERTEBRATES: GEOFFROY'S VINDICATION. The discovery of these deep homologies in genetic structure and action among phyla (particularly between vertebrates and arthropods) brings us back to Geoffroy's daring theory of the vertebral archetype. Researchers have documented homology in key regulatory genes of development, and have also shown the conservation of basic developmental patterns between the two phyla, particularly in differentiation of structures along the A-P axis under the influence of homologous *Hox* genes and their principles of colinearity. But Geoffroy's formalist theory rests upon an additional and crucial premise—one that continued to strike most researchers as unlikely, even after the first discovery of these broad commonalities in development. For Geoffroy postulated that the segment (the vertebra in Geoffroy's terminology) represents a fundamental—and truly homological—unit of construction in both phyla. Therefore, to validate the basic premise of Geoffroy's theory, the vertebrate somite must also be homologous with the insect metamere (similar patterns of differentiation along the A-P axis cannot suffice), and such a close comparison seemed exceedingly unlikely, if not anathematic, to most biologists. In the classic pre evo-devo book on the origin of the coelom and segmentation, Clark (1964) described the independent origin of arthropod and vertebrate segments as "universally accepted." And

Moore and Willmer (1997, p. 34) although writing after most of the genetic discoveries discussed in this section, affirmed the independent evolution of segmentation as virtually beyond dispute, and therefore an exemplar and "type case" for good pedagogy in phyletic inference: "As an object lesson to begin with, it is evident . . . that the character we score as 'segmentation' has to have arisen at least twice, since it occurs in the protostome annelid/arthropod grouping and again in the very distant deuterostome chordates, but not in any of their possible common ancestors." (Their confidence, presumably, would only be increased by the subsequent discovery of a fundamental split among the protostome phyla, with arthropods on one branch and annelids on the other—thus implying a third independent origin of segmentation.)

But now, at a dawning millennium in human calendrics, two sequential sets of discoveries have provoked a rethinking even of this most "settled" issue, and some genuine segmental homology between arthropods and vertebrates now seems almost inescapable. No simple one-for-one correspondence of somite with metamere can be specified down the A-P axes of these phyla, and no archetypal form like Geoffroy's "vertebra" can be reconstructed as an ancestral prototype for all segments. Moreover, vertebrate somites do not seem to be constructed by the arthropod cascade of gap, pair rule, segment polarity genes, etc.—see p. 1110 for more detail on these differences. But anatomical homologies between these two segmented phyla on maximally divergent boughs of the bilaterian tree extend well beyond mere positioning and pattern of A-P differentiation, and also include important aspects of segmentation as well. If the common ancestor of arthropods and vertebrates did not already possess a segmented body, this "urbilaterian" (in the terminology of De Robertis, 1997) had probably established the fundamental genetic pathways behind segmentation and the differentiation and specialization of segments— a system maintained ever since in both phyla, and based in large part on the *Hox* sequences and their colinearity.

1. REDISCOVERING THE VERTEBRATE RHOMBOMERES. Initial data on the mode of action of vertebrate *Hox* genes seemed, at first, to support the traditional conclusion that no segmental homology existed between the two phyla. The primary sites of *Hox* action generally correlate with the anterior expression boundary of each gene—and these boundaries extended past the developing vertebral column into anterior regions of the embryo. Some enterprising geneticists then rediscovered an important fact, established in the 19th century by the great German school of descriptive anatomists, and then forgotten by several subsequent generations who dismissed such work as the dullest form of cataloguing done at the least causally relevant scale by the most hidebound methodology of holistic observation. For these 19th century anatomists had discovered that the vertebrate hindbrain eventually develops into a unitary structure, but begins as a linear series of 7 or 8 segments called rhombomeres. Moreover, specific rhombomeres seem to control (or at least correlate with) the development of important aspects of anterior anatomy, including the deployment of the cranial nerves. Finally, as the spur

to renewed respect for such "trivial" data of gross anatomy, the anterior expression boundaries of several *Hox* genes map consistently to specific rhombomeres.*

The striking similarity between the action of vertebrate *Hox* in rhombomeres and insect *Hox* in metameres generates strong suspicions of homology. For example, some vertebrate *Hox* sequences follow the common insect pattern that the anterior expression boundary of each successive 5' gene "skips" a segment, appearing two segments towards the animal's posterior. In mice, *Hoxb-2* turns on in the third rhombomere, *Hoxb-3* in the fifth, and *Hoxb-4* in the seventh. Moreover, cell populations of the rhombomeres seem to follow the same "compartment" rules of insect parasegments—i.e., cells originating before the formation of rhombomere boundaries may place progeny in several rhombomeres, but the clones of all cells formed after the development of a rhombomere boundary do not transgress into adjacent rhombomeres.

These observations may lead a skeptic to admit that some segmental homology exists, but only between the bulk of an arthropod's body and a relatively insignificant portion of a vertebrate's anterior end (and not even to the crucial face or forebrain). At this point, however, a key paleontological fact should convert skepticism into strong interest. The rhombomeres of the embryonic hindbrain correlate directly with the pharyngeal arches developing just alongside (Fig. 10-16). In fact, each pharyngeal arch corresponds with two rhombomeres (Raff, 1996, p. 343). As we should remember from our elementary courses, all early vertebrate embryos develop pharyngeal arches, or gill slits. Tetrapods lose these structures in later embryology, but their positions determine important aspects of embryological topology (including migratory paths of neural crest cells and the subsequent locations of cranial nerves, as mentioned above), while some of their parts transform into important organs of gnathostome vertebrates. (Most famously, the jaw arises from the first gill arch, while an element of the second arch becomes, in jawed fishes, the hyomandibula (suspending the upper jaw to the braincase) and later, in tetrapods, the stapes, or hearing bone.)

But, more importantly for acknowledging a meaningful segmental homology between arthropods and vertebrates, the rhombomeres and their underlying *Hox* codes do not only generate some important features of later tetra-

---

*"Whole animal biologists," including the author of this book, can only experience enormous hope and gratification when colleagues trained in molecular and experimental traditions recognize the utility of data so often ignored and disparaged as antiquarian or superannuated. In fact, such a pattern has often been repeated in the history of science, as when the initial recognition of Mendelian mutations led early geneticists to reexhume old data long dismissed as mere description of phenomenological oddity—for example, the literature (dating to the earliest days of scientific publishing) on developmental anomalies. When molecular biologists value such classical data more highly (and utilize them more fruitfully) than practitioners in the classical fields manage to do themselves, then we may truly hope for an integrated biology based on the prospect, so often expressed but so little realized until recently, that molecular and organismic biology might finally consummate a union on the common field of development.

pod anatomy. They also constitute, in the earliest agnathan vertebrates, the major functional aspect and structural extent of the organism's segmental anatomy—and not just a small portion of the anterior end. The region of the agnathan gill slits occupied more than half the body's length in many early forms. Moreover, the pharyngeal clefts functioned not only in breathing, but also, as the branchial basket, in gathering and filtering food. In fact, these earliest vertebrates may have fed in the manner of many arthropods, by passing food along a series of segments and their appendages, from posterior to anterior towards the jawless mouth (rather than in the reverse direction that we know so well from our own experience!). For many of the earliest agnathan vertebrates, and without gross exaggeration, one might be tempted to regard the posterior vertebral column (behind the branchial basket) as an add-on and afterthought. In this historical sense, if insect metameres are homologs of rhombomeres in the developing hindbrain of vertebrates, then segmental homology between the two phyla governs the major primordial system of vertebrate segmentation, even if most later gnathostome clades deemphasized this anterior system and strengthened the somites of the subsequent and posterior vertebral column.

2. MORE EXTENSIVE HOMOLOGIES THROUGHOUT THE DEVELOPING SOMITES. If homologies based on the *Hox* code place vertebrate rhombomeres into phylogenetic union with arthropod metameres, must we conclude that the far more prominent somites of the gnathostome vertebral column bear no relationship of homology with arthropod segments? Such a conclusion need not follow, for the obvious reason that development and specification of arthropod segments requires the operation of several genetic systems prior to and beyond the activation of *Hox* genes. The *Hox* genes, after all, do not regulate the formation, number and timing of segments, and

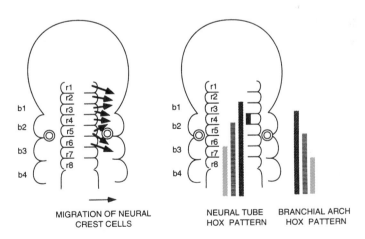

10-16. Schematic diagram from Raff, 1996, showing that each rhombomere in the developing embryonic hindbrain of vertebrates correlates directly with the pharyngeal arches developing just alongside, with each arch corresponding to two rhombomeres.

they do not code for the actual structures built within each segment. The *Hox* genes turn on *after* the segments have been generated by other systems. They then act to regulate the appropriate (and different) downstream cascades that actually build the specialized structures of each segment. Thus, we may also search for homologies between vertebrates and arthropods in the prior systems that specify numbers and positions of segments before *Hox* genes begin their work in regulating specific fates.

Although the long germ-band style of segmentation in *Drosophila* (all segments forming simultaneously as divisions of an embryo with a fully established A-P axis) represents a highly derived condition with respect to the plesiomorphic state of most insects (short germ-band development, with new segments added in a temporal sequence, one by one at the posterior end), we almost inevitably turn to *Drosophila* as an arthropod model of segmentation because our knowledge of this fly so exceeds our understanding of any other arthropod's development. The identities and differentiation of *Drosophila*'s segments occur in a programmed cascade of linked and ever-finer specifications that always draws my mind to the basic model of Genesis I (by which I intend no statement about creation, needless to say, but refer only to the geometric style of building complexity by successive division and differentiation out of primal homogeneity, rather than by addition). In this primal tale of Western culture, the cosmos begins "without form and void," and its products then originate by compartmentalization and increasing specification of units: light from darkness on day one; earthly from heavenly waters on day two; earthly land from earthly water on day three; and division of heavenly light into sun and moon on day four.

*Drosophila*'s first specification even begins in a prior generation, for protein products of maternal genes like *bicoid* and *nanos* appear in the egg cytoplasm to designate the anterior and posterior embryonic poles. These maternal genes activate gap genes like *hunchback* that specify broad regions along the A-P axis. Gap genes then regulate the expression of pair-rule genes, whose bands of activity establish the embryo's parasegment boundaries. These pair-rule genes express themselves in every other segment, but also regulate the next level of differentiation in the genetic cascade: segment-polarity genes like *engrailed* and *wingless*. The action of segment-polarity genes finally establishes the anterior and posterior domains of each segment. Now that segment boundaries have been set, and the spatial domains of each segment determined, *Hox* genes can finally establish segment identities by regulating downstream cascades of appropriate architects.

Interestingly, although evidence remains limited to a few taxa and effects as I write this section in January 2000, some apparent vertebrate homologs of these segmental cascades have been detected with reasonable confidence.

(1) Pair-rule genes and somite formation in zebrafish and chicks. Müller et al. (1996) studied the expression of *her1*, a homolog of the *Drosophila* pair-rule gene *hairy*, in the zebrafish *Danio rerio*. Expression of *her1* occurs in transient stipes within the presomitic mesoderm. Although more than 10

bands eventually form, no more than three are expressed at any one time, because older (anterior) bands fade as new bands appear at the tail bud (see description in Kimmel, 1996). Since the *her1* bands form and fade before the appearance of somites, Müller et al. (1996) traced the fate of cells from the *her1* bands in later embryos. In a particularly gratifying result, cells of the first *her1* band formed somite 5, while cells of the second band generated somite 7, thus confirming homology of action for pair-rule genes (expression in every other somite) as well as homology of genetic sequence.

Pennisi (1997) then described the work of Pourquié and colleagues on *chairy* (for chick hairy), the chick homolog of the same *Drosophila* pair-rule gene *hairy*. This study added important data on the timing of gene action, again linking the spatial order along a major body axis to a temporal sequence that can easily implicate heterochrony, the classical rubric for elucidating relationships between ontogeny and phylogeny, as a pathway (and preferred channel) for evolutionary change. The early chick embryo grows an elongated region where about 50 somites will originate, one at a time, starting at the anterior end, and taking about 90 minutes for each to form. Pourquié and colleagues found that *chairy* first becomes active in the rear 70% or so of the entire elongated region. The band of expression then narrows and shifts forward towards the head, finally becoming concentrated in a thin stripe at the rear edge of the next somite to form. After this stripe appears, the gene turns on again over the same broad region, beginning the cycle anew and ending in a sharp stripe at the next posterior segment in the developing array.

(2) A segment polarity gene in amphioxus. Although vertebrate homologs of arthropod segment polarity genes do not seem to function in establishing segmentation (for their expression begins only after somitic boundaries have formed), *AmphiEn*, the only amphioxus homolog of the *Drosophila* segment polarity gene *engrailed*, appears in stripes at the posterior border of the first eight somites to develop (Holland et al., 1997). (*Drosophila engrailed* appears in a similar position at the anterior borders of developing parasegments, which become the posterior borders of adult segments, since each final segment forms from the junction of the posterior half of one parasegment with the anterior half of the next parasegment in the A-P array.) Holland et al. (1997, p. 1723) draw a strong inference about segmental homology: "The segmental expression of *AmphiEn* in forming somites suggests that the functions of *engrailed* homologs in establishing and maintaining a metameric body plan may have arisen only once during animal evolution. If so, the protostomes and deuterostomes probably shared a common segmented ancestor."

(3) Does resegmentation occur in developing vertebrae, and could such a process be homologous with the conversion of embryonic insect parasegments to adult segments? As De Robertis (1997) reminds us, anatomical data known for more than a century indicate that a subset of cells in each somite (called the sclerotome) forms a vertebra. But each adult vertebra arises by

"resegmentation" as the posterior half of one sclerotome fuses to the anterior half of the next sclerotome along the A-P axis. "The end result is a phase shift of the vertebra with respect to the muscle, so that the segmental muscles can span, and move, adjoining vertebrae" (De Robertis, 1997).

These anatomical data, never satisfactorily verified, have now been confirmed by cell lineage studies in birds. De Robertis argues that such vertebral resegmentation may be homologous, and not merely analogous, with the similar construction of insect segments from conjoined halves of adjacent parasegments. De Robertis concludes (1997): "It seems improbable that such a complicated way of making individual metameres would have arisen independently twice in evolution."

3. SOME CAVEATS AND TENTATIVE CONCLUSIONS. I need hardly remind my fellow evolutionary biologists that these results, no matter how fascinating and surprising, show only limited and partial homology, in the strict sense needed to affirm Geoffroy's archetypal notions, between arthropod metameres and vertebrate somites. To cite the two most important caveats: First, even the most impressive finding, the mapping of *Hox* activity to rhombomeres of the developing vertebrate hindbrain, does not establish full homology between particular arthropod and vertebrate segments. We may, I think, legitimately speak of homology in the basic function, and in the spatiotemporal operation of the *Hox* genes themselves, and therefore in the fundamental patterning of the A-P axis. But the segments along this axis have already been established by this point in development, and the action of *Hox* genes (as discussed previously on p. 1107) does not build the segments, but rather turns on downstream cascades that differentiate the "right" structures in the appropriate places.

At this point, we have no evidence for, and some substantial (albeit negative) evidence against, the building of rhombomeres along genetic pathways homologous with those that determine arthropod segments. No data suggest that gap genes, pair-rule genes, and segment polarity genes—the temporal cascades responsible for the development of arthropod segments—also build vertebrate rhombomeres. Thus, in the overall case for homology between vertebrate and arthropod segments, the rhombomeres can only claim an architectural status as "preformed" compartments in which a homologous set of genes then operates to regulate the further differentiation of appropriate structures within each segment. But we cannot claim homology in the pathways of genetic construction for the compartments themselves.

Second, although some impressive homologies may now be asserted for structures along the main A-P axis of arthropods and vertebrates (despite their major differences in adult appearance and function), two important comparisons in Geoffroy's hypothesis cannot, for different reasons, be defended as support for strongly constraining homology: The relationship of insect appendages with vertebrate limbs, and the interpretation of the vertebrate head as an amalgam of several vertebrae (which might then be viewed as potentially homologous with the arthropod head, construed as a tagma of several segments).

As for limbs, I will argue in the next section (pp. 1134–1142) that the putative homology of some genetic pathways resides in such generalized rules of morphological organization (for the initiation of any "outpouching" orthogonal to a major axis, for example) that little support for particular historical constraints can be drawn from the claim for genetic retention. (After all, properties so pervasive and general as the structure of DNA, or so broad as the necessary physical geometry of elongation and outpouching, do not manifest the specificity required to identify limitations or channels arising from definite historical positions on life's phyletic tree.)

As for the vertebrate head, current knowledge favors a status even less congenial to claims for homology across phyla, and to strong historical constraint: interpretation of this definitive vertebrate structure as a true novelty and neomorph, and not as a highly modified organ constructed from parts homologous to units of the arthropod *Bauplan*. The foundation of this argument rests upon distinctive features of the vertebrate neural crest and its astonishing range of developmental derivatives and influences (Gans and Northcutt, 1983). Thus, despite important homologies in products of the developing hindbrain and its rhombomeres, the vertebrate mid and forebrain seems to represent a largely "suradded" structure, unique to the vertebrate (or at least to the chordate) lineage.

I do not challenge this general argument, but some aspects of the vertebrate fore and midbrain may exhibit developmental homology with anterior segmentation in protostome phyla. In particular (see Simeone et al., 1992; Holland et al., 1992; and Raff, 1996, pp. 199–200), the *Drosophila* gap genes orthodenticle *(otd)* and empty spiracles *(ems)* operate in the establishment of head segmentation at the fly's front end, anterior to the domain of expression for *Hox* genes. Two homologs of each of these homeobox genes (*Otx1* and 2, and *Emx1* and 2) have now been identified in mice, and their domain of action also maps to the forebrain and midbrain, anterior to the expression of *Hox* genes in the rhombomeres of the hindbrain (see Fig. 10-17, taken from Holland et al., 1992, p. 627). But we do not yet know if these genes encode common modes of action (in addition to their similarity in genetic structure and locus of operation).

On this chapter's central subject of degrees of constraint, Holland et al. (1992) offer the interesting suggestion that these gap gene homologies might enforce less channeling upon patterns of development than the *Hox* genes impose, and that the greater independence, flexibility and subsequent novelty and variety of the vertebrate head might flow, in part, from the absence of more constraining *Hox* action in the mid and forebrain regions. (In particular, as Figure 10-17 illustrates, "the four *Otx* and *Emx* genes show a nested series of posterior expression boundaries, in contrast to clearly nested anterior expression boundaries in the *Hox* genes" (Holland et al., 1992, p. 627.) Moreover, whereas the anatomical expression of *Hox* genes strictly parallels their spatial order on the chromosome, *Otx* and *Emx* show no evidence for similar clustering in the genome). Holland et al. (1992, p. 628) therefore hypothesize:

If roles for *Hox, Otx* and *Emx* genes in body regionalization evolved early in metazoan radiation, the fundamental molecular dichotomy within the vertebrate neural tube is a legacy from events preceding the evolution of the vertebrate head; nonetheless, there are likely to be adaptive consequences. If the different homeobox gene families are under different modes of regulation (for example if the tight clustering of *Hox* genes restricts mutational change) then subsequent variation and adaptive radiation will have been constrained to different extents anterior and posterior of the midbrain/hindbrain boundary. We suggest this could be a molecular basis for the comparative evolutionary plasticity of the vertebrate forebrain and midbrain, but conservation of hindbrain morphology, during vertebrate evolution.

Lumsden and Krumlauf (1996) discuss another prospect for potential homology in genetic action at the anterior end of arthropods and vertebrates (although such examples, as for the previous case of *Otx* and *Emx,* bear limited application to Geoffroy's particular theory about the segmental basis of

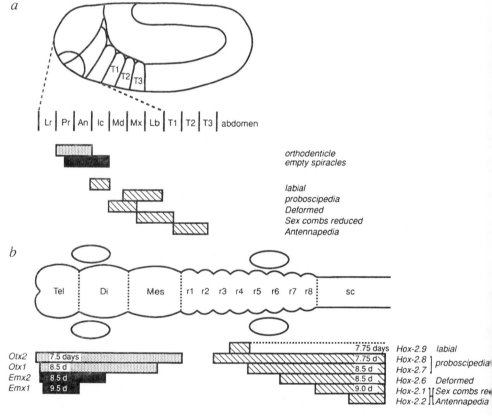

10-17. Note, in the rhombomeres of the developing mouse brain (part B of the figure), the nested anterior expression boundaries of *Hox* genes, as opposed to the posterior nesting of expression boundaries in *Otx* and *Emx.* From Holland et al., 1992.

anatomical homology, because any similar action occurs within segments at an arthropod's frontal end, but operates within the apparently unsegmented mid and forebrain of developing vertebrates). In the chick midbrain, rostral to the anterior limit of *Hox* action, a long-range signaling region, located at the isthmic constriction between the posterior end of the midbrain and the rhombomeres behind, regulates AP patterning within the unsegmented field of the developing midbrain. Signals from this isthmus regulate the action of *En-1* and *En-2*, two engrailed genes homologous with the prominent segment-polarity engrailed gene of *Drosophila*. In chicks, the engrailed gradient (see Fig. 10-18) spreads from the isthmus in both directions, decreasing anteriorly through the mesencephalic vesicle and also posteriorly through the first rhombomere (Lumsden and Krumlauf, 1996, p. 1112). Moreover, these authors add (p. 1112), "*En* expression is the earliest known marker for mesencephalic polarity."

Finally, although this argument only applies to the relationship of vertebrates with other chordates, and not to any protostome group, the forebrain, and even the neural crest, may not be so confined to true vertebrates as previous views generally assumed. In overt appearance, the anterior end of amphioxus does not include any organs comparable with the vertebrate mid or forebrain. But Holland and Holland (1998) report that amphioxus homologs of two genes with important action in the vertebrate fore and midbrain also operate in generating the anteriormost cerebral structures of amphioxus. (*AmphiOtx*, the homolog of the vertebrate *Otx* that operates in both fore and midbrain, is expressed at the anterior end, and in the ventral and lateral walls, of the cerebral vesicle in amphioxus. *AmphiDll*, the homolog of vertebrate *Dlx* that operates in the forebrain, is expressed at the extreme anterior end of the cerebral vesicle, and also in the dorsal wall.)

Holland and Holland conclude (1998, p. 651) that "the expression patterns of these amphioxus genes suggest that the cerebral vesicle is largely homologous to the vertebrate forebrain, but cannot rule out a midbrain homolog." Moreover, the expression pattern of *AmphiDll* during neurulation implies "that the epidermal cells bordering the neural plate may represent a

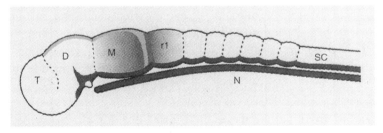

10-18. Possible homology in genetic action at the anterior end of both arthropods and vertebrates. In chicks, a gradient of *Engrailed* expression spreads from the isthmus of the developing brain in both directions, decreasing anteriorly through the mesencephalic vesicle, and also decreasing posteriorly through the first rhombomere. From Lumsden and Krumlauf, 1996.

phylogenetic precursor of the vertebrate neural crest" (p. 648). Nonetheless, emphasizing the novelty of vertebrate usage, whatever the homology of development with amphioxus, "the migrating cells of amphioxus do not differentiate into the wide variety of cell types known to originate from the vertebrate neural crest, but eventually remain part of the epidermis" (Holland and Holland, 1998, p. 654).

Despite these caveats, we can only conclude that the first fruits of evo-devo have revealed some remarkable, and extensive, homolgy in both genetic structure and action among animal phyla (particularly between arthropods and chordates, the former prototypes of separation in our traditions and literature), and that these data have confirmed some important aspects of the most ridiculed formalist theory of constraint in the history of morphological and evolutionary thought: Geoffroy's claim for homology between vertebrate and arthropod segments, with the idealized segment itself regarded as the basic unit of generation.

The history of our dawning realization—as expressed in an acknowledgment that this heresy of homology between phyla must be reclothed in modern genetic language as a partial reality—also followed an interesting pathway of strong reluctance gradually yielding to bemused acceptance of ever widening scope. From the initial Mayrian stance of near theoretical impossibility for recognizable genetic homology, we first admitted (with the discovery of vertebrate *Hox* genes) that footprints of common ancestry could be preserved during more than 500 million years and such substantial anatomical divergence. But we still doubted that such genetic resemblances could continue to encode phenotypic homologies in these disparate phyla. Then, in a second step, we acknowledged the potential homology of *Hox* action in general spatial organization along the A-P axis, but still declined to accept any common basis for segmentation in arthropods and vertebrates (the key to Geoffroy's hypothesis).

In a third stage, the similarity of *Hox* action in patterning vertebrate rhombomeres and arthropod segments demonstrated at least some homology between modes of differentiation in arthropod segments and in the compartmentalized organization of the vertebrate hindbrain and its extensive derivatives. In a fourth stage, researchers then discovered some partial and limited homologies in earlier determinants of segmentation itself (pair-rule and segment-polarity genes) between arthropods and some aspects of segmental development in the main vertebrate body axis posterior to the rhombomeres.

In short, and in a story to be, no doubt, extensively continued (expanded, contracted, changed, reinterpreted, etc.), Geoffroy's theory of complete and overarching homology based on a common vertebral archetype surely will not prevail in anything like its original form. But this most ridiculed of all heresies, so contrary in principle to strict Darwinian expectations of the Modern Synthesis, and so widely dismissed as a romantic delusion until just a few years ago, has now resurfaced, in appropriately revised terms, as a primary, and initially surprising example of the unanticipated durability of ancient genetic pathways, and of their continuing power to constrain the subsequent phylogeny of life along broad and fruitful (but still limited) routes of wonderfully diverse, but historically rule-bound, adaptive designs.

GEOFFROY'S SECOND ARCHETYPAL THEORY OF DORSO-VENTRAL IN-VERSION OF THE COMMON BILATERIAN GROUNDPLAN. The ridicule heaped upon Geoffroy's second archetypal theory for homologizing arthropods and vertebrates did not descend entirely from his genuine and original argument, but from an explicitly phyletic version championed by later evolutionists, but never intended by Geoffroy himself. (Geoffroy maintained a generally supportive attitude towards an evolutionary world view, and even gave his name to a theory of causation that he did advocate in passing, but never really developed *in extenso*—the idea that soft inheritance could operate by immediate impress of external conditions upon parts of organisms, yielding inheritable changes directly, rather than by the more indirect route of Lamarckian organic response to "felt needs." In fact, late 19th century discussions of evolutionary mechanisms often listed three primary contenders: Darwinism for the theory of natural selection, Lamarckism for soft inheritance by organic response, and so-called "Geoffroyism" for soft inheritance following direct imposition. In this light of Geoffroy's positive attitude to evolution, his failure to cite, as support for transmutation, his archetypal theory of dorso-ventral inversion between insects and vertebrates provides strong evidence that he did not intend this anatomical comparison to be read in a phylogenetic context.)

After resolving (to his satisfaction) the common structure of vertebrate and arthropod segmentation, and dealing with the inconvenient fact, for a hypothesis of homology, that vertebrates grow internal hard parts and arthropods an exoskeleton (see pp. 304–306 for Geoffroy's ingenious, and gloriously wrong, resolution), Geoffroy moved to a second archetypal theory for the outstanding remaining difference in basic anatomy between the phyla: their apparently reversed dorso-ventral orientations—for the two main nerve cords of arthropods run along the ventral surface below the central gut, while the single nerve tube of vertebrates runs along the dorsal surface, above the gut. As we all learned in Biology 1 (but usually not with proper respect or understanding for Geoffroy's interesting conjecture), Geoffroy resolved this striking difference by suggesting that the same groundplan underlay the development of both phyla, but that this common design appeared in reversed orientation, with arthropods interpreted as, essentially, vertebrates turned on their backs (see Fig. 10-19).

We also learned, quite correctly—for Geoffroy's supporters never resolved this issue with any plausibility—that such a reversal scarcely brings the two phyla into complete correspondence. In the knottiest remaining difficulty, the front end of the arthropod gut passes between the nerve cords and emerges on the lower surface as a ventral mouth. In an overturned position, the vertebrate mouth should therefore pass by the dorsal nerve cord and emerge on the upper surface. But vertebrate mouths are also ventral. So Geoffroy weakly argued that the mouths of the two phyla are not homologous—and that the original mouth, which would have opened dorsally in vertebrates, simply closed up, while a separate ventral mouth originated as a neomorph.

This history, discussed extensively in Chapter 4, becomes crucially relevant to the modern validation of Geoffroy's own centerpiece for his theory of dorso-ventral inversion. I must also address this question in the largely non-historical second half of this book because later exegetes have seriously misrepresented Geoffroy's intent by substituting a later phyletic version that modern research in evo-devo rightly rejects—and that, if conflated with Geoffroy's actual theory, will lead to continued and unjustified dismissal of his second, and remarkably ingenious, archetypal formulation.

In short, Geoffroy never advanced (and, I suspect, never even conceptualized) a historical argument about direct evolutionary transformation: the claim that vertebrates evolved from arthropods when an ancestral trilobite or merostome literally flipped upside down during its phyletic ascent. However, the American morphologist William Patten did popularize such an evolutionary account in arguing that the first prominent group of putative fossil vertebrates, the jawless "ostracoderms," did not belong to a completed vertebrate line, but represented an intermediate stage between arthropods and fishes along the "great highway of organic evolution" (his words, see Patten, 1912, 1920), with the transition literally achieved by anatomical inversion, as an arthropod that swam on its back settled to the bottom, thus converting its original dorsal side to a new belly. (The ostracoderms, with their external plates,

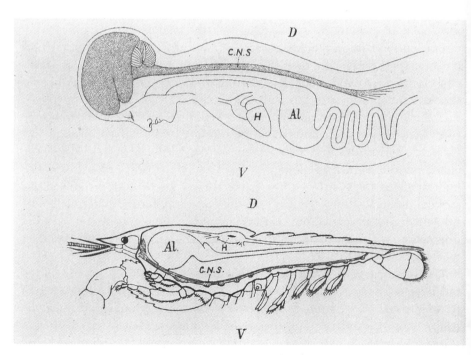

10-19. An unintentionally amusing illustration from Gaskell, 1908, showing the inverted topology of vertebrates and arthropods—with major nerve cord above the gut in vertebrates and below in arthropods.

do resemble, but by convergence, the eurypterid arthropods in several features of external form and function.)

For Geoffroy, however, the inversion of axes between vertebrates and arthropods does not denote an evolutionary transition in either direction, but represents instead (as the archetypal mode of thinking would imply) two opposite specializations upon a shared abstract groundplan that generated both great phyla along predictable pathways of internally specified laws of form and their permissible transformations.

As a structuralist thinker, committed to a formal, rather than a functional, approach to the explanation of organic design and variation, Geoffroy argued that the apparently fundamental difference in disposition of organs between arthropods and vertebrates should be reconceptualized as both secondary and superficial—a consequence of opposite ecological orientations for the same archetypal structure. The shared and constraining pattern specifies a central gut and a peripheral major channel for the nervous system, both oriented parallel to the body's A-P axis. Vertebrates, so to speak (and befitting their higher status and dignity), have oriented their main nerve tract upwards toward the sun and surface, while the humbler arthropods have directed the same peripheral aspect of archetypal form downward towards the earth and ocean bottom.

A functional theory, like Darwinian natural selection, would tend to interpret this ecological correlation as a primary impetus for the later evolutionary fixation of these two opposite arrangements. But to a structuralist thinker like Geoffroy, the same ecological situation becomes both derivative and temporally consequential (not to mention ideologically inconsequential as well). The established differences represent a realized subset of possible transformations for an archetypal form under structurally determined rules of geometric constraint and possibility. The happenstance of opposite ecological orientation for a common archetypal design only records a later adaptive overlay—a diversity of form arising for structural reasons and then finding both an appropriate suite of functions and the right environments for their realization. Thus, in Geoffroy's view, the inversion of dorso-ventral axes in arthropods *vs.* vertebrates does not validate a direct flip of evolutionary transformation, but rather represents two separately developed expressions of a common archetypal structure, one oriented up towards the sun, the other down towards the earth, in a secondary ecological specialization that can only obscure an underlying, and truly ruling, unity of constrained design.

The modern version of Geoffroy's vision—so different in genetic and evolutionary (as opposed to formal and archetypal) evidence, yet so eerily similar in philosophical style (as a structuralist account based on internal channels of transformational constraint)—originated in the mid 1990's based on unanticipated discoveries of genetic homology in genes that operate in patterning dorsal and ventral surfaces and structures in *Drosophila* and *Xenopus*. (See Sasai, et al., 1994; Holley et al., 1995; and De Robertis and Sasai, 1996, for the pioneering work of De Robertis's lab at UCLA, and François et al., 1994; and François and Bier, 1995, for studies of similar import from Bier's lab at the University of California, San Diego. See also the general commentary of Hogan, 1995; De Robertis, 1997; and Gould, 1997c.)

The chordin *(chd)* gene of *Xenopus* codes for a protein that operates in patterning the dorsal side of the developing embryo, and also plays an important role in formation of the dorsal nerve cord. But *sog,* the homolog of *chd* in *Drosophila*, is expressed on the *ventral* side of the developing larva, where it acts to induce the formation of ventral nerve cords. Thus, the same gene by evolutionary ancestry acts in the development of both the dorsal nerve tube in vertebrates and the ventral nerve cords in *Drosophila*—in conformity with Geoffroy's old claim that the two phyla can be brought into structural correspondence by inversion.

Two further discoveries then promoted this intriguing hint into a strong case. First, a major gene acting in development and specification of the dorsal surface in flies (decapentaplegic, or *dpp*), has a vertebrate homolog (*Bmp-4*) that patterns the ventral side of *Xenopus*. Moreover, the entire system seems to operate in a similar manner—but inverted—in the two phyla. That is, *dpp*, diffusing from the top to the bottom, can antagonize *sog* and suppress the formation of the ventral nerve cords in *Drosophila*—while *Bmp-4* (the homolog of *dpp*) diffusing from the bottom to the top, can antagonize chordin (the homologue of *sog*) and suppress the formation of the dorsal nerve cord in vertebrates (see Fig. 10-20).

Second, the fly genes work in vertebrates, and vice versa. Vertebrate *chordin* can induce the formation of central nerve tissue in flies, while fly *sog* can induce dorsal nerve tissue in vertebrates. These three discoveries, taken

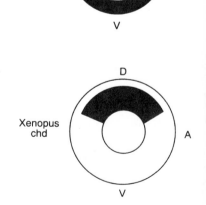

10-20. A highly idealized and schematic illustration of the common developmental pattern in the inverted topologies for gut and nerve cords in vertebrates and arthropods. *Chd* in *Xenopus* is the homolog of *Sog* in *Drosophila*—and both help to regulate the development of the nerve cords, dorsally in vertebrates and ventrally in arthropods.

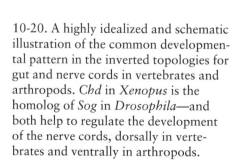

together, offer strong support for a modern recasting of Geoffroy's old theory of inversion.

These studies have aroused considerable excitement and controversy, and a substantial set of alternative interpretations has been proposed. But, in my judgment, some of these objections attack the wrong target (Patten's hypothesis of direct evolutionary transition, not Geoffroy's argument for common structural design), while others raise legitimate questions of an interesting and fundamental nature, although not yet resolvable by information now in hand. To cite a most cogent example in each category:

Jacobs et al. (1998) compare the neural organization of arthropods and vertebrates to a platyhelminth outgroup bearing a potentially plesiomorphic design: "The flatworm nervous system is often conceived of as having an anterior nerve ring with four major nerves emanating posteriorly from it" (1998, p. 348). They then point out, citing Bier (1997), that "the default condition of the ectoderm is neurectoderm" (p. 349), and that the development of non-neural ectoderm therefore requires additional and apomorphic downregulation, now largely accomplished, in vertebrates and arthropods, by the *chd/sog* and *dpp/BMP-4* systems described above. Jacobs et al. (1998) therefore interpret the relatively inverted systems of neural development in arthropods and vertebrates as two different specializations from the plesiomorphic (flatworm) condition of four major nerves extending posteriorly in radial symmetry around the anterior ring. Thus, arthropods retain the two ventral cords (and suppress dorsal neurectoderm by *dpp* action described above), whereas vertebrates keep the plesiomorphic state in a dorsal position and suppress ventral neurectoderm by the action of *BMP-4*. Jacobs et al. (1998, pp. 349–350) conclude: "The bilaterian central nervous system would be the product of concentrating the nervous organization in part of the ectoderm, by eliminating it from other regions . . . If this were the case, then the ventral nervous system in protostomes could derive from the ventral pair of nerves in the orthogon [the fourfold system of flatworms] and the dorsal system in vertebrates from the dorsal pair."

So far so good, and so reasonable. But Jacobs et al. (1998, p. 350) then make a false inference about Geoffroy's views: "The above scenario explains the available data without invoking an instantaneous dorsoventral inversion as envisioned in the transcendental scheme of Geoffroy." But Jacobs et al., while correctly criticizing Patten's theory of flipover in direct phyletic transition, misattribute this view to Geoffroy. In fact, the scenario of Jacobs et al. fits splendidly with Geoffroy's *actual* hypothesis of separate and different transformation, constrained by structural rules of growth, from a common archetype—in this particular case, suppression of either the two dorsal, or the two ventral, nerve cords in an originally radially symmetrical circlet of four. (Gerhart, 2000, questions the inversion hypothesis with an alternative strikingly similar to the proposal of Jacobs et al., but equally, and truly, consonant with Geoffroy's actual claim.)

In a different potential criticism, Bang et al. (2000) accept the description of homologous systems of neural generators and suppressors operating at opposite poles of the dorsoventral axis in arthropods and vertebrates. But they question the status of this system as an ancient and conserved primary marker and definition of the body axis throughout the history of bilaterian animals, from the time of the ancestral "urbilaterian" (De Robertis and Sasai, 1996) through the differentiation of arthropods, vertebrates, and all other phyla deriving from this common node.

Perhaps, they argue, the *dpp/sog* and *BMP-4/chd* interaction expresses a much more general (and perhaps more ancient) signaling pathway "that has been conserved in evolution but coopted for patterning very different aspects of the body" (Bang et al., 2000, p. 23). In potential support, they note (see Yu et al., 1996) that, in *Drosophila*, "*dpp* is expressed in vein precursor cells in the pupa, whereas *sog* is expressed in the intervein-cells and suppresses the formation of veins." The separate cooption, in arthropods and vertebrates, but in reversed orientation, of such a general signaling pathway would represent a parallelism based on so broad and abstract a homology of underlying genetic routes of development that an evolutionary interpretation in terms of constraint would become uninformative because the "hold of history" would then become so loose and unspecific. (I shall devote the entire second part of the next section—pp. 1134–1142—to this central issue, by elucidating the contrast between the genuine but uninteresting homology of Pharaonic bricks and the important historical mark and constraint of Corinthian columns. I will therefore let this example stand as a prelude to this forthcoming discussion, while also adding an incisive comment from Wray and Lowe (2000, p. 48): "The existence of developmental modules that are reapplied in functionally similar contexts in nonhomologous structures poses a very real problem for testing hypotheses of homology among morphological structures.")

For now—and so much more shall be discovered in the first years of our new millennium—we may recapitulate the stunning novelty of this first theme by contrasting Mayr's conventional 1963 statement that genetic homology between phyla may be dismissed *a priori* and in principle, based on our general understanding of the power of natural selection, with a 1996 statement by Kimmel (p. 329), not at all intended as a "gotcha" or an ironic commentary on Mayr's misplaced confidence, but certainly appropriate as an opening sentence for a 1996 article on a new view of life: "We have come to find it more remarkable to learn that a homolog of our favorite regulatory gene in a mouse is not, in fact, present in *Drosophila* than if it is, given the large degree of evolutionary conservation in developmentally acting genes."

### *Movement two, Elaboration: parallelism of underlying generators. Deep homology builds positive channels of constraint*

PARALLELISM ALL THE WAY DOWN: SHINING A LIGHT AND FEEDING THE WALK. The deep homologies discussed in the last section operate as shared starting points and subsequent conduits for historically constrained change. But, in an even more positive role for the historical shaping of evolution from within, homologous developmental pathways can also be employed (and deployed) as active facilitators of homoplastic adaptations that might otherwise be very difficult, if not impossible, to construct in such a strikingly similar form from such different starting points across such immense phyletic gaps. In short, this fascinating evolutionary phenomenon, long discussed under the rubric of convergence in our literature, now stands ripe for reinterpretation, in several key cases, as the positively constrained outcome of remarkable homologies in underlying pathways of genetic and developmental construction.

This general shift in viewpoint—from a preference for atomistic adaptationism (favoring the explanation of each part as an independent and relatively unconstrained event of crafting by natural selection for current utility) to a recognition that homologous developmental pathways (retained from a deep and different past, whatever the original adaptive context) strongly shape current possibilities "from the inside"—has permeated phylogenetic studies at all levels, from similarities among the most disparate phyla to diversity among species within small monophyletic segments of life's tree. No case has received more attention, generated more surprise, rested upon firmer data, or so altered previous "certainties," than the discovery of an important and clearly homologous developmental pathway underlying the ubiquitous and venerable paradigm of convergence in our textbooks: the independent evolution of image-forming lens eyes in several phyla, with the stunning anatomical similarities of single-lens eyes in cephalopods and vertebrates as the most salient illustration. As Tomarev et al. (1997, p. 2421) write: "The complex eyes of cephalopod molluscs and vertebrates have been considered a classical example of convergent evolution." (Assertions of anatomical convergence remain valid in the restricted domain of final products, whereas a phenomenon of opposite theoretical import now holds sway for the pathway of construction itself.)

PARALLELISM IN THE LARGE: *PAX-6* AND THE HOMOLOGY OF DEVELOPMENTAL PATHWAYS IN HOMOPLASTIC EYES OF SEVERAL PHYLA

DATA AND DISCOVERY. Salvini-Plawen and Mayr (1977), in a classical article nearly always cited in this context, argued that photoreceptors of some form have evolved independently some 40 to 60 times among animals, with six phyla developing complex image-forming eyes, ranging from cubomedusoids among the Cnidaria, through annelids, onychophores, arthropods and mollusks to vertebrates along the conventional chain of life. In the early 1990's, using *Drosophila* probes, researchers cloned a family of mammalian *Pax* genes, most notably *Pax-6*, which includes both a paired box and a homeobox (Walther and Gruss, 1991). Soon thereafter, several recognized mutations in the form and function of eyes were traced to alterations in *Pax-6*. For example, mice heterozygous for Small eye *(Sey)* have reduced eyes, whereas lethal homozygotes, before their death, develop neither eyes nor nose. Similarly, human Aniridia *(An)* causes reduced eyes, sometimes lacking the iris, in heterozygote form, while the lethal homozygotes also develop no eyes at all. Further studies then demonstrated expression of *Pax-6* in the spinal cord, several parts of the brain, and especially in the morphogenesis of vertebrate eyes, "first in the optic sulcus, then in the optic vesicle, the pigmented and the neural retina, the iris, in the lens and finally in the cornea" (Gehring, 1996, p. 12).

Although the existence of a *Drosophila* homolog could certainly have been anticipated—the *Pax* genes, after all, were found with fly probes—few researchers expected that a *Drosophila* version would also function in the same basic way. But the *Drosophila Pax-6* homolog mapped to the eyeless *(ey)* locus (Quiring, et al., 1994), named for a mutation discovered early in the 20th century, and producing, in homozygous state, flies with strongly reduced eyes, or lacking eyes entirely. Moreover, the conservation between mammalian and insect *Pax-6* sequences is impressively high, with 94 percent amino acid identity in the paired box and 90 percent in the homeobox (Gehring, 1996).

The similar function of these *Pax-6* homologs in different phyla was then dramatically affirmed by expressing the mouse gene in *Drosophila* (Halder et al., 1995), and finding that the mammalian version could still induce the formation of normal fly eyes. Noting that *Pax-6* acts as an upstream regulator of a large set of more specific genes, Gehring (1996, p. 14) makes the obvious, but important, point: "Of course, the eyes that are induced by the mouse gene are *Drosophila* compound eyes, since the mouse gene is only the switch gene and another 2500 genes from *Drosophila* are required to assemble an eye."

In the boldest of all experiments, leading to results that attracted substantial and well-deserved public attention, Gehring and colleagues then found that ectopic expression of either the murine or *Drosophila* version of *Pax-6* could induce supernumerary eyes on the antennae, legs and wings of flies (Fig. 10-21), thus supporting Gehring's designation of *Pax-6* as a "master control gene" for the development of eyes. Gehring (1996, p. 13) wrote that these "ectopic eyes are morphologically normal with normal photoreceptors,

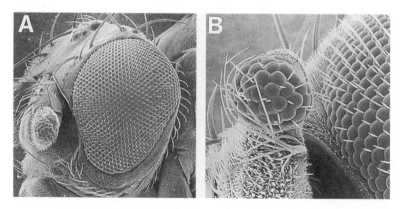

10-21. From Gehring, 1966. One of the most remarkable discoveries from the early days of evo-devo. An ectopic eye (smaller and to the left of the normal eye in A; enlarged in B) can be induced in *Drosophila* by targeted expression of the mouse homolog of *Pax-6* in flies.

lens, cone and pigment cells and an electroretinogram as it is typical for photoreceptor cells can be recorded, when the ectopic eyes are exposed to light." Nonetheless, as these ectopic eyes are not neurologically "wired up," the fly presumably cannot use them for vision. (To give some sense of the excitement and weirdness of these results upon their initial discovery—for we have, a mere five years later, already become accustomed to such findings—I include as Fig. 10-22 the "Post-it" note that Gehring penciled when he sent me his first reprint announcing this achievement.)

But this conserved developmental pathway for insect and vertebrate eyes, however surprising in the light of previous assumptions about the impossibility of such genetic homology between phyla, did not yet directly address the theoretical issue of convergence in evolution. After all, the single-lens eye of vertebrates bears little anatomical similarity to the multifaceted fly eye, and no claims for convergence had been staked upon this case. But the discovery

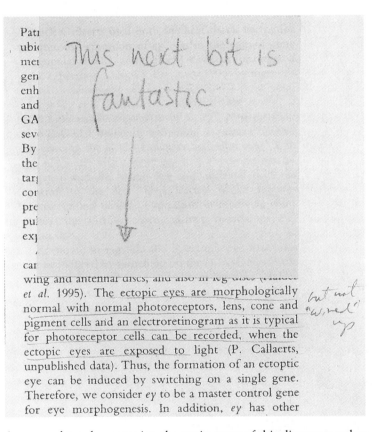

10-22. A personal touch expressing the excitement of this discovery: when Gehring sent me his 1996 reprint on inducing ectopic fly eyes with mouse genes, he inserted this Post-it just above his finding that the ectopic eyes are morphologically normal and react to light in the normal way. I added the marginal notation—not to pour water on a great discovery—that these eyes are not wired to the brain and therefore will not function.

of a homologous developmental pathway for the disparate eyes of two such different phyla raised an obvious question about the generality of *Pax-6* as a "master control gene" (using Gehring's phrase again) for all complex eyes, including the "paradigm for convergent evolution" (Gehring, 1996, p. 14): the remarkable similarities in function and structure between the single-lens eyes of vertebrates and cephalopods.

This case has persisted as a classic, ever since the formulation of convergence as a concept, because the two eyes look so much alike, and work so similarly, despite their separate origins from different tissues: the vertebrate eye as an evagination of the brain, and the cephalopod eye by invagination of the epidermis. The squid eye forms from a monolayer of epidermis that becomes thickened, multilayered and internalized on the dorsal side of the head lobe. The outer ectodermal layer forms the iris and the outer lens portion, while the inner half of the lens arises from the inner ectodermal layer. Thus, the adult lens contains two parts, divided by a septum. Meanwhile, the cornea, also of ectodermal origin, derives from a quite different source on the edge of the arms, as they grow forward. In vertebrates, by contrast, the optic vesicle arises as an evagination of the diencephalon, whereas the lens then develops from overlying ectoderm. As the most interesting consequence of these differences—well known, perhaps, because the vertebrate eye seems more "jury-rigged" than the eye of the conventionally "inferior" squid on this basis—the polarity of photoreceptors becomes inverted in vertebrates, but remains everted (an apparently superior design) in cephalopods.

In a keenly anticipated result, Tomarev et al. (1997) found a homolog of vertebrate and arthropod *Pax-6* in the squid *Loligo opalescens*. This gene is expressed in the development of the embryonic eyes, olfactory organs, brain and arms. (This common expression in visual and olfactory systems bears further study, especially given the common ectodermal origin of both organs in vertebrates and their embryonic interaction with adjacent regions of the

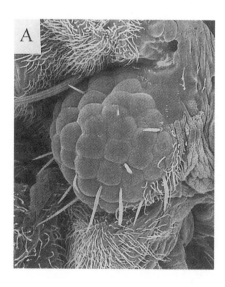

10-23. The squid version of *Pax-6* also induces the development of ectopic eyes in *Drosophila*. From Tomarev et al., 1997.

developing brain.) In the most satisfying result (see Fig. 10-23), ectopic expression of squid *Pax-6* also induced supernumerary eyes in *Drosophila*. Tomarev et al. write (1997, p. 2424): "Squid *Pax-6* is able to induce ectopic *Drosophila* eyes on wings, antennae, and legs, as was previously demonstrated for *Drosophila eyeless* and mouse *Pax-6*. All *Drosophila* eye-specific structures including cornea, pigment cells, cone cells and photoreceptors with rhabdomeres were formed in the ectopic eyes induced by squid *Pax-6* DNA."

THEORETICAL ISSUES. Two related questions have dominated the emerging discussion of these *Pax-6* homologies: the putative status of this gene as a "master control" for eyes, and the impact of *Pax-6* upon the claim for independent evolution of eyes by convergence. Gehring (1996, 1998) bases his terminology of "master control" upon three properties of *Pax-6*: its status as upstream regulator of a substantial cascade of more specific eye-forming gene products; its interchangeability among phyla, while always acting as a trigger to the downstream production of the "right" eyes for any given animal; and its general ability to trigger the formation of supernumerary eyes in odd places.

We may legitimately quibble, as Jacobs et al. (1998) and many others have done, that upstream position in a cascade should not be equated with either causal or temporal primacy, for novel regulatory elements can be introduced by evolution into any position of a developing sequence. Nonetheless, I would not begrudge a researcher the right to bestow an incisive name upon such an important discovery. *Pax-6* may be no more important than hundreds of other genes in the sense that usable eyes will not form, absent its normal operation. But the designation, as "master control," of such early and such general action (including the key property of interchangeability among phyla) does no violence to ordinary linguistic usage.

However, this very generality raises the crucial issue (see forthcoming pages for a fuller discussion) of whether the action of *Pax-6* must then be regarded as too broad and too universal to sustain any argument for meaningful constraint upon the evolution of eyes in disparate phyla. After all, if activating *Pax-6* represents little more than flicking on a master switch at the power plant (with animal development than analogized to the operation of any electrical device thereafter), then its admittedly necessary action fails to identify any channel of development specific enough to warrant designation as a dedicated impetus for the evolution of one adaptive solution over other attainable possibilities. However, in this case (but not in others, as I shall argue on pp. 1034–1042), the actions of *Pax-6* are sufficiently specific and precise to set a definite channel among conceivable alternatives, and not just to open a floodgate through which subsequent cascades might flow in any direction (see discussion of this point and listing of criteria for specificity by Jacobs et al., 1998, p. 334, who conclude that this "documentation of eye homology was quite a coup").

This acknowledgment of sufficient specificity for *Pax-6* then raises a final question about the extent of revision thus required in evolutionary concepts of convergence *vs.* constraint. Some enthusiasts have claimed that genetic homology in such a crucial and early-acting developmental pathway requires a wholesale reinterpretation of this classic convergence as a pure case of parallel evolution based upon underlying constraint. For example, Tomarev et al. (1997, p. 2426) end their important paper on squid *Pax-6* by stating: "Our data support the idea that morphologically distinct eyes of different species have arisen through elaboration of a common conserved *Pax-6*-dependent mechanism that is operative at early stages of eye development and that the anatomical differences among eyes arose later in evolution. Consequently, we believe that eyes in cephalopods and vertebrates have a common evolutionary origin and are products of parallel rather than convergent evolution."

This question would be unresolvable, and would become a source of endless terminological wrangling if a single and exclusive answer—either independent adaptations by convergence or similar solutions by constraints of parallelism—had to emerge as the explanation for a unitary phenomenon. (Several participants in the developing debate have operated upon just such a contentious assumption, hence the need for explicit treatment of this eminently resolvable question.) But the issue of how evolution can generate such similar and highly complex eyes in disparate phyla requires an invocation of both phenomena at different levels of analysis. The conventional view of convergence cannot be denied for the final products of adult anatomy, as documented in my previous discussion of the fascinating differences in form and developmental origin for the strikingly similar eyes of cephalopods and vertebrates. But the traditional claim for exclusive convergence at all scales implies a purely functional explanation, positing an independent evolution of eyes along entirely separate and internally unconstrained sequences of natural selection, with no aid from any common starting point or channel of development.

However, the *Pax-6* story has now furnished an important homological basis in underlying developmental pathways for generating complex eyes in cephalopods and vertebrates. Thus, a channel of inherited internal constraint has strongly facilitated the resulting, nearly identical solution in two phyla, and evolutionists can no longer argue that such similar eyes originated along entirely separate routes, directed only by natural selection, and without benefit of any common channel of shared developmental architecture. But just as the advocates of pure convergence erred in claiming exclusive rights of explanation, the discovery of *Pax-6* homologies does not permit a complete flip to exclusive explanation by constraint.

As so often happens in our world of biological hierarchy, convergence prevails at one level, and constraint at another. The similarities in adult anatomy are primarily convergent, but *Pax-6* establishes an important homology in underlying pathways of generation. We thus encounter a case of homoplasy in final results based upon significant homology in underlying developmental architecture. As discussed extensively on pages 1061–1089, and as presented in tabular form on page 1078, this common circumstance, however muddled by a century of confusion in our literature, nevertheless enjoys a clear and simple solution in proper formulation of the concept of parallelism, or homoplasy of results based on homology of underlying generators. This recasting of the paradigm case for pure convergence as an outcome of substantial parallelism in a key developmental channel has now highlighted the neglected role of constraint as a strongly positive force in organismic adaptation. For we must now grant strong probability to the proposition that, absent an "internal" direction supplied by the preexisting *Pax-6* developmental channel, natural selection could not have crafted such exquisitely similar, and beautifully adapted, final products from scratch, and purely "from the outside."

Moreover, two studies published after my initial composition of this section strongly reinforce the increasing emphasis on constraint and parallelism, rather than independent adaptation and convergence, in the evolution of complex eyes in widely separated phyla of animals. First, Pineda et al. (2000) report homologs of both *Pax-6* and *sine oculis* in the planarian *Girardia tigrina*. These genes operate in the same cascade, with *Pax-6* directly regulating *sine oculis*, as in phyla with complex lens eyes. But the much simpler visual system of *Girardia* includes no lens. Pineda et al. (2000, p. 4525) write: "The eye spots of planarians are one of the most ancestral and simple types of visual systems, close to the prototypic eye proposed by Charles Darwin. The planarian eye spots consist of two cell types: a bipolar nerve cell with a rhabdomere as a photoreceptive structure and a cup-shaped structure composed of pigment cells."

Thus, the basic genetic cascade had already originated, and already regulated visual systems, before the evolution of complex lens eyes, indicating the preexistence of the developmental pathway as a positive constraint of parallelism. Pineda et al. show that repression of the *sine oculis* homolog completely suppresses the development of eyes in regenerating planarians, thus demonstrating commonality of function as well as structure in the developmental genetics of some of the simplest and most complex eyes among disparate animal phyla.

Second, and from the other end of the logic of the general argument, further aspects of underlying developmental homology have been found in the general construction of anatomically divergent lens eyes of arthropods and vertebrates—so the evidentiary basis of parallelism now extends well beyond the *Pax-6* system itself. Neumann and Nüsslein-Volhard (2000) show that the retinas of both *Drosophila* and zebrafish are patterned by a morphogenetic wave of strikingly similar form and timing—driven by *Hedgehog* in *Drosophila* and by its homolog *Sonic Hedgehog* in zebrafish—both inducing a cascade of neurogenesis across the retina. The strikingly unexpected finding of this additional homology in patterning for such anatomically different products led the authors to conclude (2000, p. 2139):

Analysis of the *Pax6/Eyeless* gene has indicated that the mechanism of eye induction may be conserved across the animal kingdom. However, the dramatic variation of the eye structure not only between vertebrates and invertebrates, but also within the vertebrate lineage, has suggested that events downstream of eye induction may have evolved independently. Our results show that the role played by *Hh* signalling in retinal differentiation is conserved between flies and fish. This suggests that *Hh* was already used to pattern a primordial eye structure before vertebrate and invertebrate eye lineages diverged, and thus supports a common evolutionary origin of the animal eye.

A QUESTION OF PRIORITY. This emerging story of *Pax-6* homologies directly engages one of the classic conundrums of macroevolutionary theory, an issue that troubled Darwin himself, and that elicited a famous treatment—based, in a fascinating but not really surprising coincidence, upon the evolution of eyes!—in one of the *Origin*'s most brilliant passages (in Chapter 6, entitled "Difficulties on Theory"): how can "organs of extreme perfection" ever arise if crucial components of the final product could not have functioned in their current manner in any conceivable ancestral form of simpler design? Darwin's general answer established the important evolutionary principle of cooptation: the component in question must have originally functioned in another, perhaps related, manner, and then been coopted for its current role (see pp. 1218–1224 for full treatment).

But this general solution then engendered a second problem of even broader import: how can a trend towards a highly complex organ ever get started at all, if the initial stages can bear so little structural or functional similarity to the final product? In this case, how could eyes ever form if the simplest incipient state in the founding member of a trend couldn't function for anything even roughly analogous to vision? (How, in other words, can evolution ever take the first step to a simple light-sensing organ, not to mention the much later development of image-forming devices?) How can evolution "know" where to start when faced with millions of potentially alterable molecules and processes, none manifesting even the first selected step of a forthcoming trend? (Natural selection may power the trend after step one has been reached, but how can this initial entrance be effected?)

To resolve this deeper problem, Darwin advanced the brilliant hypothesis—in the sense of a wonderfully simple idea once formulated, but quite nonobvious beforehand—that first steps must rely upon purely fortuitous variation, or fortuitous cooptability, in the favorable direction. Writing of Batesian mimicry in butterflies, for example, Darwin notes that the adaptive value of a tasty mimic to a noxious model cannot be gainsaid, but what, he then asks, can get the process started? Why, in particular, did the ancestor of the mimic choose this particular model among scores of other noxious species in the same fauna? Darwin answers that the first step must rely upon a slight fortuitous resemblance to one particular model—thus setting an initial (and accidental) tiny advantage that natural selection can "notice" and thenceforth enhance.

In a famous passage, Darwin uses this argument to defend the evolution of complex lens-eyes by natural selection:

> To suppose that the eye, with all its inimitable contrivances . . . could have been formed by natural selection, seems, I freely confess, absurd in

the highest possible degree. Yet reason tells me, that if numerous grada-tions from a perfect and complex eye to one very imperfect and simple, each grade being useful to its possessor, can be shown to exist . . . then the difficulty of believing that a perfect and complex eye could be formed by natural selection, though insuperable by our imagination, can hardly be considered real (1859, pp. 186–187).

Expanding his discussion in later editions, Darwin specifies a potential starting point of maximal simplicity in structure and function: "The simplest organ which can be called an eye consists of an optic nerve, surrounded by pigment-cells and covered by translucent skin, but without any lens or other refractive body. We may, however, . . . descend even a step lower and find ag-gregates of pigment-cells, apparently serving as organs of vision, without any nerves, and resting merely on sarcodic tissue. Eyes of the above simple nature are not capable of distinct vision, and serve only to distinguish light from darkness" (1872b, p. 135).

Therefore, following this theme, if we wish to develop a complete evolu-tionary explanation for the role of *Pax-6* homologies in regulating the forma-tion of complex structures in the lens-eyes of several phyla, we also need to understand its ancestral role in species with much simpler organs of vision, or without eyes at all. Why, in short, did *Pax-6*, rather than some other mole-cule, become the homologous "master control gene" of such complex struc-tures, especially if the common ancestor of modern phyla with lens-eyes had only evolved eyes of much simpler form and function?

Fortunately, even at our current embryonic stage of research, some intrigu-ing hints exist for a resolution, thus completing the intellectual structure of an evolutionary argument for important parallelism in the evolution of eyes, as regulated by the positive channel of *Pax-6* homologies. *Pax-6* homologs have been cloned from three cnidarian genera—from a jellyfish and a hydra (Sun et al., 1997, though questioned by Catmull et al., 1998). Catmull et al. specu-late (1998, p. 355) that "the capture of a homeobox by an ancestral *Pax* gene probably permitted a transition from functions in cell-fate specification to roles in anterior patterning," and later to still more specialized roles in the development of the central nervous system and finally in the specification of eyes. Because *Acropora* lacks eyes, despite showing sensitivity to light, Catmull et al. suspect that the *Pax-6* homolog of this cnidarian may regulate anterior (and distal) patterning of the nervous system. (They also conjecture, on the same grounds, that the *Pax-6* homolog in the blind nematode *C. elegans* may operate as a plesiomorphic regulator of the head region, rather than as a sign of heritage from an eyed ancestor.)

But Sheng et al. (1997), in an intriguing discovery that might link *Pax-6* to an ancestral function tied more closely to vision, found that *Drosophila Pax-6* directly regulates the expression of the visual pigment rhodopsin in photoreceptor cells. Sheng et al. (1997, p. 1122) therefore propose that "the evolutionarily ancient role of *Pax-6* was to regulate structural genes (e.g. rho-dopsin) in primitive photoreceptors, and only later did it expand its function to regulate the morphogenesis of divergent and complex eye structures." "*Pax-6*," they continue (p. 1129), "is locked in the regulatory pathway of eye development because of its more ancient function in the direct regulation of terminal photoreception genes like *rh*. Later in evolution, genes specific for each type of eye may have been added to this regulatory pathway to specify divergent and complex eye structures."

This appealing hypothesis, if validated, would address both issues in Darwin's dilemma, as described above, for the origin of organs of extreme complexity. First, *Pax-6* would manifest a plesiomorphic function related to vision in much simpler ancestral structures that can detect light or motion, but do not form images—for rhodopsin operates as a major visual pigment in such organs. Second, and proceeding phylogenetically further back to a potential utility even before the origin of vision (analogous to the initial choice of a model in the evolution of mimicry), rhodopsin operates in sensitivity towards light in all three multicellular kingdoms—suggesting a symplesiomorphy of great phyletic depth!—even when the physiological basis of response cannot be meaningfully compared with vision in animals. Rhodopsin, for example, acts in phototaxis to guide the swimming of green algae towards or away from light. Moreover, Saranak and Foster (1997) show that rhodopsin also guides the zoospores of the fungus *Allomyces reticulatus* towards light—suggesting (Saranak and Foster, 1997, p. 465) that "the origin of vision might have been the phototaxis of their unicellular ancestors."

PARALLELISM IN THE SMALL: THE ORIGIN OF CRUSTACEAN FEEDING ORGANS. Although interphylum parallelisms, based on homologies of developmental pathways, may provide greater *éclat* for their status as both utterly unanticipated in traditional Darwinian theory, and also a bit "weird" to boot, the greater importance and transformative power of this principle for the ordinary practice of evolutionary research will surely reside in the far more numerous and precisely defined cases of parallel evolution within much smaller monophyletic clades. In these instances, a parallel rather than convergent basis for similar adaptations does not provoke the same sort of surprise (for this alternative had always been plausible in theory for taxa of shared *Bauplan* and relatively recent common ancestry), but the value of parallelism becomes greatly increased by the operational basis thus granted to firm and testable explanations—by moving away from adaptationist scenarios in the largely speculative mode, and towards morphogenetic rules with specifiable, even predictable, realizations.

Ultimately, I suspect that the major reformatory significance in such accumulating examples of parallelism "in the small" will lie primarily in their capacity to resuscitate, and place upon center stage, the once derided formalist concept that taxonomic order largely represents the realized manifestations of more general developmental rules and pathways ("laws of form" in the archaic, but not entirely invalid, terminology of Geoffroy's biology), rather than the adaptive nuclei where environmental advantage reins in a much more promiscuous range of possibilities. (In this formalist or structuralist view, adaptation by natural selection surely sets the actual points of occupation along potential pathways of realizable form, but basic taxonomic order reflects the limits and preferred channels of internal potential as much, or more, than the happenstances of immediate selective advantage.)

To cite just one impressive case of extensive parallelism in the taxonomic order of a substantial clade within a phylum, Averof and Patel (1997) have studied the action of the *Hox* genes *Ubx* and *abdA* in specifying the form of gnathal and thoracic appendages in Crustacea. In most arthropods, the gnathal appendages (maxillae) are specialized for feeding, and the larger thoracic appendages for locomotion. But in numerous and phyletically varied crustacean taxa, appendages of the anterior thoracic segments have been reduced in size and specialized for primary utility in feeding. These thoracic feeding limbs are called maxillipeds, and phyletic analysis clearly illustrates their multiple independent evolution within the Crustacea.

As a general, and presumably ancestral, pattern in Crustacea, the transition between gnathal and thoracic segments marks the anterior expression boundary for *Ubx* and *abdA,* and these genes do not operate in the gnathal region, where smaller feeding appendages (maxillae) develop. The branchiopods, for example, follow this basic scheme: *Ubx* and *abdA* are expressed throughout the thorax, and no maxillipeds form (Averof and Akam, 1995). These genes do not operate in the anterior gnathal segments, which develop the smaller maxillae. In other groups, the generation of maxillipeds on anterior thoracic segments correlates precisely with the suppression of *Ubx* and *abdA* in these segments alone, and these specialized appendages then grow to resemble the smaller maxillae of the adjacent anterior (gnathal) region of the AP axis, where *Ubx* and *abdA* are not expressed in normal development.

The precision of this correlation is impressive, and presumably causal. Among the malacostracans, for example, the leptostracans also develop no maxillipeds, and *Ubx* and *abdA* are expressed throughout the thorax. But in peracarids, the first, and sometimes the second, of eight thoracic appendages develop as maxillipeds. Averof and Patel (1997) document the suppression of both *Ubx* and *abdA* in these anterior thoracic segments with maxillipeds. In *Mysidium colombiae,* for example, T1 generates a maxilliped, whereas the appendage of T2 remains primarily a swimming organ, but develops gnathal features at its distal end. Averof and Patel found that *Ubx* and *abdA* are entirely repressed in T1, but expressed in the proximal portion of the T2 endopod, while being excluded from the distal portion that acquires the gnathal features of a maxilliped.

The familiar decapods (lobsters, crabs, shrimp) generally bear eight thoracic segments, the anterior three with maxillipeds and the posterior five with walking legs (hence the name of the group, meaning 10-footed). This situation correlates perfectly with the suppression of *Ubx* and *abdA* in the first three thoracic segments by backward shifting of their joint anterior expression boundary. The finer scale variations within the clade also follow the same developmental rule. For example, although adult lobsters of the most familiar (and edible) *Homarus americanus* bear the usual five pairs of large thoracic limbs and three pairs of anterior maxillipeds, only T1 and T2 show limb reduction at hatching, while the appendage of T3, at this early stage, continues to resemble a walking leg in size and form. Averof and Patel found that, at this intermediate point in development, *Ubx* and *abdA* are repressed only in T1 and T2. (The repression presumably extends to T3 during later molts, but Averof and Patel do not present data for these later stages.)

Maxillipeds also develop in several other crustacean groups, widely dispersed within the taxonomic space of the clade. In a non-exhaustive compilation, Averof and Patel found no exceptions to the rule that maxillipeds develop instead of walking legs when the anterior expression boundary of *Ubx* and *abdA* shifts back, thus suppressing the action of these *Hox* genes in a specified number of anterior thoracic segments. For example, Averof and Patel (1997) studied two copepod species with maxillipeds only on T1. They found (unsurprisingly by now) that the anterior expression boundary of *Ubx* and *abdA* had shifted back only one segment, with activity beginning in T2.

The important evolutionary message of these findings follows from the clear implication, based on cladistic analysis, that maxillipeds have arisen several times, and independently, in crustacean phylogeny—but always, as Averof and Patel's data illustrate so impressively, under control of the same homologous developmental rule, presumably a plesiomorphic trait of the clade. Thus, this striking example of clearly adaptive, multiply repeated, and effectively identical, transformations of anterior thoracic walking legs to feeding appendages represents a striking case of parallel evolution based on frequent evocation of a homologous developmental pathway, and *not* a demonstration of convergent evolution rooted in similar pressures of natural selection acting upon unconstrained and "random" variation in each case.

Averof and Patel (1997, p. 686) affirm this interpretation, but unfortunately introduce some terminological confusion (albeit minor, and easily correctable) in summarizing their splendid study: "Our findings indicate that such convergent changes may have been achieved by similar developmental changes (involving similar posterior shifts in the expression boundary of *Ubx-abdA*) on several independent occasions. This suggests that, given a particular developmental system, there may be limited ways for achieving a particular morphological result."

But these changes are fully parallel, and not convergent, in both developmental pathway *and* phenotypic result because maxillipeds arise by independent recruitment and expression of the same, homologically retained developmental rule among the taxa that independently evolve appendages of the same basic form and anatomical structure along a strongly positive and clearly adaptive internal channel of constraint. (For lens eyes of squid and vertebrates, on the other hand, homologous generators build similar structures from different tissues—thus making the eyes largely convergent as adult phenotypes and largely parallel in developmental architecture.)

**PHARAONIC BRICKS AND CORINTHIAN COLUMNS.** In both gastronomy and the academy, too much of a good thing can quickly pall. A concept, to be useful and interesting, must make distinctions and define categories of exclusion. A rubric for all possible cases explains nothing—as Gilbert and Sullivan's Grand Inquisitor Don Alhambra explained to the naively egalitarian Gondolier Kings of Barataria: "When everyone is somebody, then no one's anybody."

It is certainly understandable, and probably psychologically inevitable, that exciting discoveries tend to become overextended in the first flush of reformatory application. Our recently acquired ability to identify genetic homologies in the developmental pathways of homoplastic final structures has sometimes engendered a misplaced enthusiasm for reinterpreting similarities previously ascribed to pure convergence as examples of parallel evolution (defined as homoplastic results based on homology of underlying generators).

But, as I have argued throughout this book, concepts only become interesting in contexts set by the logic and the history of theoretical issues thus addressed. The primary significance in recasting convergence as parallelism lies in the very different implications of the two processes both within Darwinian theory and for the larger question of the relative weights that should be assigned to internal structural constraint and functional adaptation in populating the morphospace of life's history. Pure convergence stands at a Darwinian functional extreme, where uncanny similarities of phyletically distant taxa arise from entirely different starting points, propelled by selective pressures alone, without any boost from internal channeling. Parallelism, by contrast and with reversed evolutionary meaning, attributes the identical result, at least in large part, to a homologous generating channel that guides two independent sequences of selection down the same path from within.

Parallelism will only define an interesting antithesis to convergence if the underlying homology prescribes a highly distinctive, detailed, and strongly determinative channel of constraint—for only then will the homoplastic result owe its primary form to the structure of the internal channel, and not to the functional processes of adaptation acting from outside. But if, on the other hand, the underlying homology only generates a simple immediate product, leading to a broad and non-specific range of potential outcomes, the homology establishes no meaningful channel of internal *constraint,* and makes no contribution to the revisionary power of this theme within evolutionary theory—that is, to move away from an endpoint of pure Darwinian functionalism towards a more comprehensive theory, enriched by contrasting perspectives based on structural principles of internal channeling.

In the light of our burgeoning knowledge of genetic sequences and their actions, homology of some sort or level will always be found in underlying generators of similar end products—if only (however much the example becomes a *reductio ad absurdum*) because all organisms share the same genetic code by common ancestry. But no one would argue that we should redescribe a classical case of convergence as parallelism simply because the markedly different developmental pathways of the two adaptations both rest upon the action of genes made of DNA!

The analogy in the title of this subsection may help to clarify the central issue of this important discussion. We need to develop criteria for ordering and evaluating our highly varied and ever-growing compendium of homoplastic results generated along homologous developmental pathways—for these cases fall along a continuum from narrow and controlling channels of constraint to insignificant sharing of nonspecific building blocks. When Pharaoh "made the children of Israel serve with rigor" (Exodus 1:13), they fabricated bricks to use in a full range of buildings: "And they built for Pharaoh treasure cities, Pithom and Raamses" (Exodus 1:11). Now if these bricks built every structure in the city, from great pyramids to public toilets, we might identify a homologous generator of all final products (bricks of the same composition made by the same people in the same way over a continuous stretch of time). But we could scarcely argue that these homologous generators exerted any important constraint over the differing forms of Pharaoh's final products—if only because all realized architectural diversity shared the same building blocks.

But if I note a majestic portico of Corinthian columns in front of a building in modern Manhattan, I recognize a strong internal constraint imposed by an architectural module of very different status. The Corinthian column, last and most ornate of the classical orders, consists of a slender fluted shaft (with 24 flutes in "standard" examples), capped by a striking, distinctive, and elaborate capital (the defining "species" character in a taxonomic analogy) adorned with stylized acanthus leaves. Few Greek examples survive, but the Romans then used Corinthian columns extensively and for several centuries. Vetruvius, who wrote the only surviving work on classical architecture, described such columns in detail in the 1st century BC, and later builders of the Italian Renaissance replicated the design in all aspects of form and proportion, whence, ever since, buildings in classical style have often used Corinthian columns on their facades, porticos and lobbies.

Like Pharaonic bricks, Corinthian columns hold clear status as homologous underlying generators for their continuous phyletic history and stable form. But whereas Pharaonic bricks did little to constrain a resulting building by their form or structural character, and would not therefore sustain an interesting interpretation of parallelism for two similar buildings that happened to employ them in construction (if only because many other, very unsimilar, buildings in town also use the same bricks), Corinthian columns do exert a strong structural constraint from an inherited past (a homology) that can help us to identify and distinguish buildings, even 2500 years after the invention of this unchanged form.

First of all, we can identify the lineage of the element just by looking (for the acanthus leaves mark the species), so I know the source of constraint before proceeding any further, whereas a brick may be difficult to peg as Pharaonic, or as a product of independent invention for a simple, obvious and utilitarian form (a Chinese version, for example). Second, tradition and intrinsic form dictate that this large and elaborate column only be used in limited ways (whereas my Pharaonic brick can build almost anything)—so the choice of the column constrains the form and function of the building.

For example, in commercial areas of Manhattan, we can be pretty sure that a set of Corinthian columns will be fronting a bank, a government building or a church—so the generator constrains the function. In residential areas, we can be confident that a similar set of columns will mark the domicile of wealthy folks, not the entrance to a public housing project—so the generator also constrains the location and social setting.

In short, and emphasizing the evolutionary analogy, two similar buildings made of identical bricks in the cities of Pithom or Raamses are not constrained to be alike by the structural properties of their admittedly homologous building blocks. The bricks represent a lowest level, non-specific, non-constraining homologous generator, and the similarity between the two buildings is convergent, a result of architectural decisions about good form for an intended purpose—that is, a product of external selection based on required function, not of internal channeling imposed by component parts. But the similar form of a town hall in modern America and a market hall in ancient Rome, as highlighted by their nearly identical facades of Corinthian columns, must be attributed, in large part, to the complex, highly specific, phyletically stable design of these chosen architectural modules, which therefore do constrain the form and function of the buildings in important ways. We may therefore ascribe much of the similarity to parallelism, based on the common choice of a homologous building element that establishes a channel of expectation, and has done so for millennia (and also includes too much complexity and too little flexibility to be used in many ways beyond the traditional employment).*

Thus, examples of homologous underlying generators form a continuum from Pharaonic bricks, which are too simple, general, and multipurpose to constrain a final result in important ways, to Corinthian columns, which are sufficiently complex, structurally limited in potential utility, and restricted by a long and stable history of traditional employment, to channel any building into just a few recognized forms and functions. When underlying homologous generators operate like Corinthian columns, they entail interpretations of parallelism, rather than pure convergence, for structural and functional similarities in resulting adult anatomies. But when homologous generators operate like Pharaonic bricks, they usually do not strongly constrain similarities between two independent structures built with their aid, and we would

---

*Such metaphors from distant professions always pay the price of their utility in suggestive analogy by their occasional capacity to be confusing in their invocation of different systems with different causal bases. At this point, an architect who loves the strict Darwinian model would say: "but why is he calling a Corinthian column an internal constraint rather than an adaptation? After all, I was commissioned to build a bank so I chose this element as a good fit with my project. The column is therefore an optimal adaptation based on my skilled selection." But I then reply, "Yes, for you and your building in a non-Darwinian system. But if an organism carries the genes for Corinthian columns as a deeply intrinsic aspect of its developmental system, enmeshed in both upstream and downstream cascades of regulation as both result and promoter, then the organism is stuck with this eminently serviceable device, and can only construct itself in certain ways under the constraints imposed by this inherited, internal element."

not automatically, on this basis, ascribe such similarities to parallel evolution. I do not know how to ordain hard and fast rules for breaking this smooth continuum into sharp domains of bricks that permit interpretations of convergence *vs.* columns that imply parallelism—but I trust that the analogy will clarify the issues involved, which must then be adjudicated on a case-by-case basis.

Heretofore, as the argument of this chapter demanded, I have been presenting cases of biological equivalents to Corinthian columns, leading to reassignments of convergence to parallelism (although I did raise the "brick" issue in wondering whether the signaling system behind dorso-ventral inversion of arthropods and vertebrates might be too broad to bear Geoffroy's interpretation—because such a general system may regulate many other distinctions as well, and may therefore become prone to independent cooptation (in different form) by two separate groups. Therefore, the facts of DV inversion do not yet guarantee an explanation as two different specializations of a homologous and archetypal ancestral state—see p. 1122).

But many examples of homologous generators acting more like bricks than columns have also been accumulating in the literature. Such examples imply interpretations more favorable to adaptation (and convergence) than to constraint (and parallelism) for two distinct reasons: first, because bricks are too general and non-specific in their operation to exert much constraint upon the complex form of a final product; and, second, because bricks are sufficiently simple and multifarious in their range of potential developmental utility that each of two lineages now using the same brick in the same way, may have cooopted this architectural module independently, and from a different ancestral use in each case. In this second circumstance, the functional similarity of bricks in the two lineages would not even be homologous, given their independent cooptation from different sources, although the bricks remain homologous in genetic structure (by attribution of the requisite similarity in nucleotide sequences to a more distant common ancestor).

To cite just two examples of bricks (that is, very general and effectively non-constraining homologies of genetic and developmental architecture) from the recently published genome of the nematode, *C. elegans:* First, for homologies going "down" life's traditional ladder, Chalfie (1998, p. 620) noted the brick-like nature of worm genes with homologs in yeast: "Most orthologues [to yeast] in the worm are needed for . . . core functions, such as intermediary metabolism, DNA-, RNA- and protein-metabolism, transport and secretion, and cytoskeletal structure. In contrast, yeast has no orthologues for many of the proteins involved in intercellular signaling and gene regulation in *C. elegans.*" Second, for brick-like homologies going "up" the same fallacious ladder, Böhm et al. (1997) found that the *par-1* gene of *C. elegans,* which codes for a protein that activates the markedly asymmetrical division of cells in the first embryonic cleavage, has a mammalian homolog that regulates the polarization of epithelial cells.

This central issue of Pharaonic bricks and Corinthian columns has become most salient in the fascinating and rapidly developing literature on the extent and meaning of similarities in development between the appendages of arthropods and vertebrates (Tickle, 1992; Tabin, Carroll and Panganiban, 1999; Panganiban et al., 1997; Shubin, Tabin and Carroll, 1997; Arthur, Jewett and Panchen, 1999; Minelli, 2000; for example). I will state my own tentative reading of these preliminary data up front: most documented homologies are too brick-like to impart a sufficiently strong and specific constraint for validating either the actual homolgy of limbs themselves, or even a claim for predominant parallelism in the evolution of homoplastic appendages. This situation may be contrasted with the highly and specifically channeled developmental homologies underlying the establishment and differentiation of major body axes, several aspects of segmentation itself, and the evolution of important homoplastic organs at several levels, including eyes among phyla and maxillipeds among crustacean taxa. These homologies are more than sufficiently column-like to validate channels of internal constraint as primary determinants of specific final products. However, some attributes of homoplastic features in arthropod and vertebrate appendages do offer intriguing hints that, even here, developmental homologies may be sufficiently column-like in some cases to implicate constraints of internal generating channels as major causes of similarity in adult structures.

As a prime example of a brick-like developmental homology, now regarded as too broad and loosely constraining to specify important details of final products as outcomes shaped by internal channels, but often regarded as more column-like in the first excitement of discovery, the *Drosophila* distal-less gene *(Dll)* is expressed at the distal tip of developing appendages and seems important in regulating their outgrowth from the body axis (Cohen et al., 1989). In the mid 1990's, researchers found a mammalian homolog (called *Dlx*) that seems to operate in virtually the same way, with expression along the distal edge of the chick wing bud (Carroll et al., 1994; Panganiban et al., 1995).

But as studies proceeded, an *embarras de richesses* soon became apparent, as distal-less homologs were found at the terminal regions of almost any structure that grows out from a central mass or body axis in all three great groups of bilaterians—including annelid parapodia, onychophoran lobopodia, tunicate ampullae and echinoderm tube feet (Panganiban et al., 1997). Lee and Jacobs (1999) then pointed out that not only does distal-less seem to regulate the proximo-distal axis of any outgrowth, but it also tends to show preferred action in early embryos (including maternal transcripts in several cases), in animal poles and anterior regions of developing embryos, and in ectodermal germ layers. Thus, distal-less may not only display the broad function of regulating outgrowths at their distal tips; it may also operate in the service of even more basic distinctions that can only be designated as early, anterior and top. Distal-less, in this sense, must be regarded as a quintessential Pharaonic brick of protean character, or just about as non-specifically unconstraining as an internal developmental element can be. If anyone wanted to argue that insect and vertebrate appendages should be deemed homologous because both are regulated by distal-less homologs, then these structures are only homologous as outgrowths in all Bilateria, and the claim becomes almost as meaninglessly broad as saying that I am homologous to each of my *E. coli* residents because we are both made of DNA inherited from a common ancestor. Moreover, with such a broad range of functions, and such ubiquity of occurrence, distal-less genes might have been independently coopted from different copies with different utilities, rather than commonly employed from the same ancestral source, in the arthropod and vertebrate forebears that first used them to regulate the outgrowth of appendages.

Panganiban et al. (1997, p. 5165) state the case for such a broad and relatively unconstraining homology: "The most straightforward explanation for these observations is that the last common ancestor of the protostomes and deuterostomes had some primitive type of body wall outgrowths, *e.g.*, a sensory or perhaps a simple locomotory appendage, and that the genetic circuitry governing the outgrowth of this structure was deployed at new sites many times during evolution." Shubin et al. (1997, p. 647) then add a reasonable, but admittedly indecisive, argument for favoring common ancestry over independent cooptation: "The expression of *Dll*-related genes could represent convergent utilization of the gene. However, the fact that out of the hundreds of transcription factors that potentially could have been used, *Dll* is expressed in the distal portions of appendages in six coelomate phyla makes it more likely that *Dll* was already involved in regulating body wall outgrowths in a common ancestor of these taxa."

On the other hand, when homologies of underlying generators (for homoplastic structures between phyla) begin to involve several genes and their complex interactions—rather than just one product expressed at the distal tip of any outgrowth—then the homology attains sufficient definition and specificity to act as a constraining Corinthian column of positive evolutionary channeling, rather than as an all-purpose Pharaonic brick for building nearly any kind of structure that natural selection might favor. For example, no one would argue that the chick forearm and fly wing are homologous as flight appendages, if only for the obvious and compelling reason that basal chordates—not only as inferred from living surrogates, but also as reasonably well represented in the fossil record of the Cambrian explosion and its sequelae—lack paired appendages entirely. But accumulating evidence now indicates that all three major axes (antero-posterior, dorso-ventral, and proximo-distal) may be established (or at least strongly regulated) by homologous, and respectably complex, genes and their interactions—a strong case for meaningful column-like constraint in this important anatomical system as well.

In the wing or leg imaginal disc of *Drosophila*, *hedgehog* acts to establish the AP axis by initial expression in the posterior compartment of the disc. In response to *Hedgehog*, a thin layer of cells along the border of both anterior and posterior compartments produces another protein encoded by the *dpp* gene. "*dpp*, in turn, is a long-range signal providing positional information, and hence differential AP fates, to cells in both compartments" (Shubin et al.,

1997, p. 645). The vertebrate limb develops in a similar way under homologous influences. *Sonic Hedgehog (Shh)*, a homolog of *Drosophila hedgehog*, is also located in the posterior part of the limb bud, and also induces a homolog of *dpp*, called *Bmp-2* in its vertebrate version, which then acts in differentiating the AP axis. Interestingly, and in a striking confirmation of important homology in function as well as structure, misexpression of either *Shh* or *hedgehog* on the anterior rather than the posterior edge of the developing appendage causes the same striking malformation: mirror image duplications at the anterior border.

In organizing the proximo-distal structure of the *Drosophila* wing, a specialized set of cells, called the wing margin, runs along the DV border of the imaginal wing disc. The *fringe* gene establishes the edge of the developing wing at the interface between cells that do and do not express *fringe*. The vertebrate apical ectodermal ridge (AER), like the *Drosophila* wing margin, also runs along the DV border of the developing limb bud. *Radical fringe*, a vertebrate homolog of *fringe*, is also expressed in the dorsal region limb ectoderm before the AER forms. Moreover, at the border of cells that do and do not express *radical fringe*, *Ser-2*, a homologue of *Drosophila serrate* (an important element in the downstream cascade regulated by *fringe*) becomes expressed, and the AER then forms.

Data for the DV axis are less well developed, but "genes specifying DV polarity in both groups have been identified" (Shubin et al., 1997, p. 646). For example, in *Drosophila*, the expression of *apterous* helps to define the dorsal compartment of the wing disc and also specifies dorsal cell fates. A related, but not clearly homologous vertebrate gene, *Lmx-1*, defines a dorsal compartment of the vertebrate limb, and also conveys dorsal cell fate.

Shubin *et al.* (1997, p. 646) summarize the evolutionary meaning of these complex, column-like developmental constraints: "The simplest phylogenetic implication to draw from these comparisons is that individual genes that are expressed in the three orthogonal axes are more ancient than either insect or vertebrate limbs . . . either similar genetic circuits were convergently recruited to make the limbs of different taxa or a set of these signaling and regulatory systems are ancient and patterned a structure in the common ancestor of protostomes and deuterostomes."

Their stated preference for the second alternative (Shubin et al., 1997, p. 647) illustrates the intermediary character of this example, as recording too broad a constraint to permit the identification of any particular vertebrate limb as "the same" structure as any given arthropod appendage, but also as based upon a sufficiently detailed and complex set of genetic homologies to set a common developmental channel with more specificity than a Pharaonic brick, albeit not at the level of constraint of a full Corinthian column: "This ancestral structure need not have been homologous to arthropod or vertebrate limbs; the regulatory system could have originally patterned any one of a number of outgrowths of the body wall in a primitive bilaterian for example . . . The key step in animal limb evolution was the establishment of an integrated genetic system to promote and pattern the development of certain outgrowths. Once established, this system provided the genetic and developmental foundation for the evolution of structures as diverse as wings, fins, antennae and lobopodia."

### Movement three, Scherzo: Does evolutionary change often proceed by saltation down channels of historical constraint?

As I documented in Chapters 4 and 5, internally channeled evolution (orthogenesis) has been intimately linked with discontinuous change (saltationism) in the history of structuralist thought (with the model of "Galton's polyhedron" serving as the classical image for the connection). The linkage isn't physically necessary or logically impelled, for some orthogeneticists have favored gradualism (C. O. Whitman, pp. 383–395), whereas some saltationists have rejected internal directionality (Hugo de Vries, pp. 415–451). But in expanding the causes of evolutionary change beyond the incremental gradualism of externally directed Darwinian selection, and in regarding internal channels of developmental constraint as important mediators of phyletic trending, most advocates of formalist or structuralist explanation (Bateson, D'Arcy Thompson, and Goldschmidt, for example) have supported some linkage of channeled directionality with at least the possibility of saltational movement down the channels—if only because the potential phyletic analog of such ontogenetic phenomena as metamorphosis seems intriguing and worth exploration.

Thus, with the reintroduction of internal channeling by historical constraint (based on genetic homology) into our explanatory schemes, we must ask whether saltational themes (that had been even more firmly rejected by the Darwinian Synthesis) can also advance a strong case for a rehearing. My own conclusions are primarily negative (hence my parsing of this theme as a scherzo, and as the shortest movement of my analysis), but the subject clearly merits some airing (and undoubtedly holds limited validity), if only as a sign of respect for the intuition of so many fine evolutionists, throughout the history of our subject, that structural channeling—now clearly affirmed as a theme of central importance—implies a serious consideration of saltational mechanics.

As discussed extensively in Chapter 9, in the context of debate over punctuated equilibrium, notions of "rapidity" depend strongly upon the time scales of their context. Invocations of suddenness raise quite different evolutionary issues at each level of consideration. In this section, I shall discuss true saltation (discontinuous changes, potentially across a single generation, and usually mediated by small genetic alterations with major developmental effects), and not punctuational patterns at larger scales of time (continuous changes that would be regarded as slow and gradual across human lifetimes, but that appear instantaneous when scaled against the millions of years in stasis for a resulting species or developmental *Bauplan*).

Nonetheless, I note in passing the relevance of developmental themes to punctuational patterns at these larger, and very different, scales of explanation. For example, several authors have argued that our emerging concepts of deep homology might help to elucidate such macroevolutionary "classics" of large-scale rapidity as the Cambrian explosion. Under Lewis's (1978) original model of evolution from ancestral homonomy (multiple, identical segments) by accretion of duplicated *Hox* genes to achieve differentiation of specialized parts along the body axis, the *Baupläne* of the major animal phyla must originate separately and gradually, as each added developmental component permits further differentiation. How, then, could so many basic designs make such a coordinated first appearance in five to ten million years, unless some genetic glitch or unknown environmental trigger initiated a rampant episode of duplication in many lineages simultaneously, or unless the pattern only represents an artifact of preservation, rather than an actual macroevolutionary event?

But the first fruits of evo-devo (see next movement, pp. 1147–1178 for full discussion) have reversed this scenario by documenting a full complement of *Hox* genes in the most homonomously segmented invertebrate bilaterian phyla, thus suggesting the opposite process of loss and divergence for the differentiation of numerous complex and specialized patterns from initial homonomy (De Rosa et al., 1999). The punctuational character of the Cambrian explosion seems far easier to understand if the basic regulatory structure already existed in ancestral homonomous taxa, and the subsequent diversification of *Baupläne* therefore marks the specialization and regionalization of potentials already present, rather than a dedicated and individualized addition for each major novelty of each new *Bauplan*. The Cambrian explosion still requires a trigger (see Knoll and Carroll, 1999, for a discussion of possible environmental mediators, including the classical idea of an achieved threshold in atmospheric oxygen), but our understanding of the geological rapidity of this most puzzling and portentous event in the evolution of animals will certainly be facilitated if the developmental prerequisites already existed in an ancestral taxon.

Knoll and Carroll (1999, p. 2134) stress this point in a section of their article entitled "Cambrian diversification: So many arthropods, so little time." They add (loc. cit., see also Grenier et al., 1997, p. 551):

> The entire onychophoran-arthropod clade possesses essentially the same set of *Hox* genes that pattern the main body axis. Thus, Cambrian and recent diversity evolved around an ancient and conserved set of *Hox* genes . . . Increase in segment diversity is correlated with changes in the relative domains of *Hox* gene expression along the main body axis . . . Most body plan evolution arose in the context of very similar sets of *Hox* genes, and thus was not driven by *Hox* gene duplication . . . Bilaterian body plan diversification has occurred primarily through changes in developmental regulatory networks rather than the genes themselves, which evolved much earlier.

But returning to the very different, central subject of this section—the possibility and meaning of evolutionary saltation at the organismic level of discontinuity across generations—we may at least assert a case for plausibility, so that, at the very least, this perennially contentious subject will not be dismissed *a priori*. First of all, we cannot deny either the existence of such large and discontinuous phenotypic shifts in mutant organisms, or the conventional basis assigned to them: small genetic alterations with major developmental consequences. For example, a single base substitution in *bicoid*, the maternal gene product that sets the AP axis by supplying positional information within the *Drosophila* larva, can reverse the axes of symmetry (Frohnhöfer and Nüsslein-Volhard, 1986; Struhl et al., 1989). Of this and other cases, Akam et al., in the introduction to their 1994 book on *The Evolution of Developmental Mechanics* write (1994, p. ii): "It is a commonplace of developmental genetics that minimal genetic change can lead to the most dramatic morphological effect."

Second, we can also posit believable mechanisms that avoid the classical problems of specifying how genes with such disruptive effects could ever be integrated into the intricate and finely tuned development of a complex metazoan ontogeny and, even if viable, how such saltations could spread through populations following their mutational origin in single individuals, given the almost inevitable fitness depression that must accompany any interbreeding with modal individuals. Schwartz (1999), for example, presents a modern version of the old argument for origin in a nonlethal recessive state, followed by accumulation without expression in heterozygotes until the achievement of a critical frequency permits an effectively simultaneous overt appearance of the phenotype in numerous homozygotes. Addressing the second aspect of this problem, and basing their case on a viable homeotic mutant in a homonomous species of centipedes, Kettle et al. (1999, p. 393) argue that most workable mutations of such large effect may arise in homonomous ancestors of more specialized groups: "Perhaps the severe fitness depression accompanying homeotic transformation would have been less pronounced, even absent, in a primitive arthropod with many similar segments."

But mere plausibility doesn't imply likelihood, and two strong arguments would seem to indicate a minimal role for the evolutionary efficacy of such developmental saltations: first, a negative statement based on the fallacy of usual sources of inference; and, second, a positive argument about more plausible alternatives for the same sources of inference.

For the negative statement, both common arguments for inferring saltational origins from modern developmental patterns falter upon the general fallacy (discussed in detail in Chapter 11) of invoking current circumstances to make unwarranted inferences about historical origins:

1. The fact that major phenotypic effects accompany the repression or alteration of key developmental switches in modern organisms does not imply a saltational origin either for the switch itself, or for any extensive consequences of its mutational variations. For example, the fact that *Hox* genes now repress the expression of *Dll* in the insect abdomen, thus suppressing the development of appendages—and that a single mutation (albeit ultimately lethal) in one *Hox* gene can reverse this effect and emplace leg rudiments on each larval segment (Lewis, 1978)—does not permit the inference that insects lost their abdominal legs in one phylogenetic saltation.

2. Reasonable inferences about saltational losses cannot be theoretically in-

verted into hypotheses about sudden origins for the developmental cascades thus repressed. (We note here a developmental version of the error so commonly committed as a thoughtless consequence of naming the normal function of genes for the results of their discombobulation. If the mutational silencing of a gene precludes the development of a child's ability to read, we have not thereby identified a "reading gene," although such taxonomies and inferences remain all too frequent in our literature, particularly on human cognitive abilities.) For example, the fact (Swalla and Jeffery, 1996) that a loss-of-function mutation in the *Manx* gene of the tunicate *Mogula* can repress chordate features and lead to a tailless (anural) larva—and that this function can be restored in interspecific hybrids with other *Mogula* species that develop with a tailed (urodele) larval form—does not imply that the tunicate larval tail and notocord (once a popular theme in theories about the origin of vertebrates, as in the classic paper of Garstang, 1928) arose by saltational introduction of *Manx* activity. (Swalla and Jeffery make no such inference, of course, and I hate to see their fascinating discovery so misused, as in many press reports.)

As a general structural principle, applicable across a full range of natural phenomena, from cosmology to human social organization, complex systems can usually collapse catastrophically, whereas the construction of such functional intricacy can only occur by sequential accumulation—a pattern that I have called "the great asymmetry" (Gould, 1998a).

For the positive argument, more plausible continuationist scenarios can explain the modern phenomena that most often tempt us to invoke hypothetical saltation to resolve their origin. In an important article, for example, Akam (1998) discusses the property of *Hox* action that has often led to saltational inferences: "The *Hox* genes might justifiably be considered master control genes (Gehring, 1996) for segment identity. For most segments of the insect trunk, they provide the only conduit for channeling axial information from the early embryo to cells at the later stages of development." Akam then exposes the same fallacy that I discussed above as negative point 2: "It is tempting to shift this process into reverse, and to assume that segment diversification has been achieved by a series of overt homeotic mutations generating novel complexity."

Akam then develops a much more plausible evolutionary model for the incremental origin of developmental patterns mediated by so-called "selector genes," which have generally been viewed "as stable binary switches that direct lineages of cells to adopt alternative developmental fates" (Akam, 1998, p. 445). Akam proposes an alternative concept "for the regulation of *Hox* genes within compartments" by "enhancer modules," conceptualized as "local signals, hormone receptors or any of the other stimuli that commonly mediate gene regulation. In this regard, it makes the *Hox* genes like any other genes. It predicts that small changes, particularly in the structure of their promotor modules, will change the phenotype of segments" (p. 448). "By accepting a role for the regulation of *Hox* genes within compartments," Akam adds (p. 448), "we demote them from their privileged status as stable binary switches."

Akam envisages the gradual evolution of different enhancers (or different levels and mixtures of the same enhancers) in various arthropod segments, leading to a phyletically diverging regulation of the same *Hox* gene down the body axis, but with continued expression in a segment-specific manner. "*Ubx* for example is regulated by the 'abx' enhancers in parasegment 5, which integrate patterning information in one way, but by the 'bxd' enhancers in parasegment 6, which specify a different within-segment pattern." As these alterations in expression evolved gradually within different segments, "the change would not necessarily be recognized as a 'homeotic mutation'" (p. 448). These and other models reinforce the important principle that extensive and discontinuous phenotypic effects in the development of modern organisms do not imply the saltational origin of these features in phylogeny.

Other models, however, permit more space for an important frequency of saltational shifts in evolution. Duboule and Wilkins (1998), for example, tie an increasing propensity for saltational change to functional recruitment of genes for multiple tasks: "Transitions from gradual to discontinuous rates of evolutionary change are an inevitable consequence of the multiple use of genes through evolutionary tinkering, given appropriate selective pressures" (1998, p. 54). Interestingly, in the context of this chapter and the history of this subject, their model explicitly links this increasing frequency of saltation to the "hardening" of internal constraints that arise as a consequence of incorporating key genes into multiple networks of regulation. They write (p. 58): "The greater the number of networks that a gene product is involved in, the smaller the scope for new variations to be offered to natural selection. The idea of internal constraints leading to restrictions in the production of evolutionary novelties is not new. However, we would like to argue that internal constraints result, indirectly but inevitably, from the increasing work load imposed by successive recruitment of genes to new functions." In networks of such complexity, they conclude, any "novel equilibrium will have to be established as a one-step event and not through the accumulation, in time and space, of many mutations of small effect, or gradualistic change" (p. 58).

In any case, and however important such saltational changes may be in establishing fundamental evolutionary novelties (my own betting money goes on a minor and infrequent role), phyletic discontinuity at lower taxonomic levels, based on small genetic changes with large regulatory effects, has been documented in several cases. In a fascinating example, Ford and Gottlieb (1992) found that about 20–30 percent of several hundred *Clarkia concinna* plants growing in a single locality at Point Reyes, California displayed the *bicalyx* mutant, a homeotic variation that replaces the usual circlet of four bright pink petals with a second circlet of sepals.

By Mendelian analysis of ratios in cross breeding between normal and *bicalyx* plants, Ford and Gottlieb established that a single point mutation produces the *bicalyx* phenotype. Moreover, and in contrast to many homeotic mutations, the *bicalyx* plants show no developmental abnormalities and no apparent fitness depressions; insect pollinators continue to visit *bicalyx* flowers with no apparent reduction in frequency. Ford and Gottlieb note (1992, p. 673) that "homeotic mutants have been found and propagated in gardens, but have almost never been reported as natural populations"—thus giving special interest to this case of viability in nature. (They also mention the intriguing historical footnote that Linnaeus himself found a peloric *Linaria* (with normal upper petals replaced by spurred lower petals), apparently growing in abundance within its habitat, but probably seed sterile, and therefore arising as a vegetatively propagated clone.)

Ford and Gottlieb outline a reasonable scenario for the promotion of this fecund, and naturally growing, homeotic form to specific status (1992, p. 673):

> The absence of deleterious pleiotropy or fitness-reducing epistatic interactions in *bicalyx* suggests that mutations with extensive morphological consequences can be successfully accommodated by plant developmental systems. If such mutants were to become associated with chromosomal rearrangements reducing the fertility of hybrids between them and their progenitors, a process that has occurred repeatedly in *Clarkia,* the new population would probably be accorded species status (p. 673). *Bicalyx* demonstrates that a large morphological difference governed by a simple genetic change can become established in a natural plant population (p. 671).

Since *bicalyx* presumably establishes its large and homeotic effect through the developmental channel of the ABC system or some homolog (see previous discussion on pp. 1063–1065), this case also illustrates the common linkage between internal channels as positive constraints and potential speed of phyletic movement down the channel—as Galton first proposed in his model of the polyhedron (see pp. 342–351). In a zoological example of the same linkage, Fitch (1997, p. 166) documents a "topological constraint" limiting the number and positions, and channeling the potential directions of evolutionary change, for rays in the male tail of the nematode *C. elegans.* Not only are the directions of potential change both limited and most plausibly attained by saltation, but the transitions can also be generated by single point mutations. Of this interesting correlation between constraint and saltation, Fitch concludes (1997, pp. 166–167): "Because single genetic changes can be postulated for some of the evolutionary change in the male tail, I predict that many evolutionary changes in morphology will have resulted mainly from changes at single loci . . . Because the power of selection is limited by variation, such developmental constraints could cause significant bias in the evolution of form."

*Movement four, Recapitulation and Summary: Early rules and the inhomogeneous population of morphospace: Dobzhansky's landscape as primarily structural and historical, not functional and immediate.*

BILATERIAN HISTORY AS TOP-DOWN BY TINKERING OF AN INITIAL SET OF RULES, NOT BOTTOM-UP BY ADDING INCREMENTS OF COMPLEXITY. As a common pattern in the history of science, great and unexpected theoretical discoveries often elicit fairly conservative theoretical interpretations at first—if only because most of us can only imagine so much novelty at once. As discussed before in different contexts, when Ed Lewis (1978) recognized the *Hox* genes as linearly ordered in both space and time, and inferred their origin by tandem duplication from a precursor, he interpreted the evolutionary significance of these amplifications in an "obvious" and conventional manner (the proper "first pass" procedure in science, and therefore recalled here without critical intent, while noting that our later explanatory reversal only underscores the importance of Lewis's discovery). Lewis proposed that the addition of duplicated *Hox* genes could be directly and causally correlated with the specialization and differentiation of appendages along the arthropod AP axis, as an originally homonomous ancestor evolved into the diverse *Baupläne* of major arthropod taxa. In particular, Lewis argued that an addition of *Hox* genes allowed evolving members of the insect line to suppress the growth of legs on the abdominal segments, and (in Diptera) to convert the wings of the second thoracic segment to halteres. Lewis's original scenario matches our conventional view of evolution, and of complex systems in general—particularly in the assumption that a history of increasing elaboration in overt products (the phenotypes of complex bilaterian phyla) should be underlain by a growth in the number and intricacy of generating factors (genes regulating developmental outcomes).

In one of the most important early discoveries of evo-devo, this entirely reasonable scenario has been overturned and, in large measure, reversed. Two strong sources of evidence now indicate that a full complement of *Hox* genes had already evolved in the presumably homonomous common ancestor, not only of the protostome phyla, but of the entire bilaterian line (thus further exemplifying the homologies of arthropods and vertebrates). The multiple and independent evolution from homonomy towards complexly specialized and differentiated *Baupläne* in several phyla did not entail any increase in the number of *Hox* genes, but rather a regionalization and decrease in the range of action of individual genes and, especially, changes in both the regulation and content of the downstream cascades differently engaged by the various *Hox* genes.

1. Modern homonomous organisms share the full complement of *Hox* genes with closest relatives among classically differentiated invertebrates. The genome of myriapods, the homonomous sister group of insects for example, includes a full set of insect *Hox* homologs (Raff, 1996). At the higher level of a sister group to the entire arthropod phylum, the undoubted "standard" for highly differentiated body plans along the AP axis, the genome of the homonomous Onychophora also includes all insect *Hox* genes, as well as an ortholog of the pair-rule gene *fushi tarazu* (Grenier et al., 1997). (Modern onychophores include only a few Gondwana species, restricted to moist terrestrial habitats. The generic name of the most famous modern form, *Peripatus,* honors the homonomy of the numerous lobopods, the pair of leglike structures on each segment. But the Onychophora included a prominent and diverse group of marine representatives in the earliest faunas of the Cambrian period.) Grenier et al. (1997, p. 549) conclude that "the segmental

diversity of arthropods evolved without an increase in *Hox* gene number. The evolution of arthropod segmental diversity must therefore have involved regulatory changes in *Hox* genes and/or their targets."

2. Phylogenetic reconstruction affirms a full *Hox* complement for the bilaterian common ancestor, with restriction and occasional unemployment far more prominent than addition in subsequent evolution. The analysis of De Rosa et al. (1999) indicates at least 8 *Hox* genes for the protostome common ancestor, and at least 7 for the bilaterian progenitor (see Fig. 10-24). Moreover, comparisons at greater detail support the growing consensus (see Halanych et al., 1995; and Aguinaldo et al., 1997, based on 18S ribosomal RNA) that the protostome phyla split into two great genealogical groups, the ecdysozoans or molting phyla (including arthropods, nematodes and priapulids), and the lophotrochozoans (including annelids, mollusks, brachiopods, platyhelminths, and nemerteans). The ecdysozoan genome includes the posterior *Hox* gene *Abd-B*, whereas lophotochozoans have two "Abd-B-

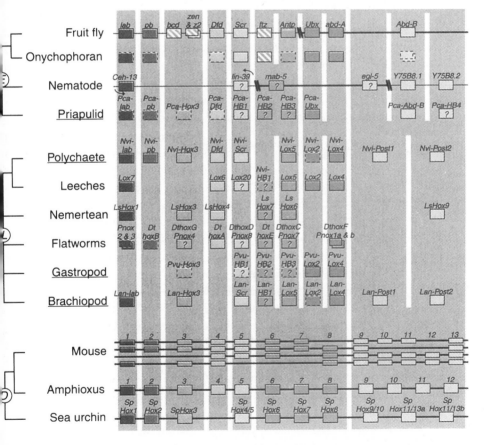

10-24. By this analysis and cladogram, the protostome common ancestor must have possessed at least eight *Hox* genes, whereas the bilaterian progenitor must have had seven. From De Rosa et al., 1999.

like" genes named *Post1* and *Post2*. In addition, ecdysozoan genomes include *Ubx* in the central cluster, whereas lophotrochozonas share the different, but closely related, *Lox2*.

Although De Rosa et al. (1999) cite 7 as a minimum for the common bilaterian ancestor (*lab/Hox1, pb/Hox2, Hox3, Dfd/Hox4, Scr/Hox5*, one additional central gene and one posterior gene), Figure 10-24 also indicates at least 10 shared *Hox* genes for the stem lophotrochozoan and 8 for the stem ecdysozoan. Using a more generous estimate based on a hypothesis that "most or all of the *Hox* genes that are present in extant bilaterians may have been present in the common ancestor, but that some orthology relationships have become obscured" (De Rosa et al., 1999, p. 775), the protostome common ancestor might have possessed ten *Hox* genes (the 7 listed above plus 2 central and 1 posterior), or even more if the deuterostome situation of multiple posterior *Hox* genes is primitive rather than derived. In any case, either a minimum of 7 or a maximum of 10 or more provides ample support for the key conclusion that a full *Hox* complex had already evolved before the establishment of distinctive features of the major bilaterian *Baupläne*. De Rosa et al. (1999, p. 775) conclude their article by stating: "The subsequent bilaterian history of *Hox* genes would have been primarily one of functional divergence and gene loss, rather than gene duplication. Regardless of the exact number of *Hox* genes in the bilaterian ancestor, the major period of progressive expansion of the *Hox* cluster due to tandem duplication events predated the radiation that generated the bilaterian crown phyla, concurrent with radical evolutionary changes in body architecture and development."

As a fascinating footnote to the rich phyletic information contained in the conservation of *Hox* genes, the Mesozoa have long presented a deep puzzle in the study of animal phylogeny. These creatures lack body cavities and effectively all the characteristic organs of animals, including a gut or a nervous system. Their maximally simplified development even proceeds without gastrulation or the differentiation of germ layers. Many zoologists have therefore considered their organization as primitive, and have even regarded the Mesozoa as a surviving key to the phyletic transition between unicellularity and the evolution of truly multicellular organization with differentiation of tissues and organs—hence their name, Mesozoa (from the Greek *meso*, meaning "middle"), as a potential intermediate between the protistans, formerly called Protozoa, and the true Metazoa. But the mesozoans are parasites of metazoans, and parasites often become extremely simplified in phenotype. Thus, the opposite interpretation of descent from an ordinary and complex metazoan ancestor has remained entirely plausible. Unfortunately, the highly simplified and autapomorphic anatomy of mesozoans has provided no clues about ancestry, despite more than a century of extensive study.

But Kobayashi et al. (1999) have isolated a *Hox* gene, *DoxC*, from a dicyemid mesozoan (parasites of cephalopod renal sacs). PCR analysis shows that *DoxC* is an ortholog of the "middle group" *Hox* series. The middle group *Hox* genes have only been found in triploblasts, and do not exist in Cnidaria. Hence, these data would seem to validate the hypothesis that mesozoans are secondarily simplified bilaterians, and not the sole survivors of the vaunted intermediary group between protists and metazoans. Moreover, further study of *DoxC* affirms its orthology with the lophotrochozoan *Lox5*, rather than with a middle *Hox* gene of the ecdysozoan clade. Therefore, not only are the mesozoans revealed as simplified metazoan parasites, but we may also place the ancestry of these formerly enigmatic forms after the major split in protostome phylogeny, and into one of the two great groups as a relative of platyhelminths, brachiopods, and annelids, rather than arthropods or onychophores.

These revolutionary discoveries have inspired a growing literature on the hypothetical phenotype, or at least the shared developmental architecture, of a stem bilaterian, or even a stem animal. Slack et al. (1993) tried to define a "zootype" as the "defining character, or synapomorphy" of the kingdom Animalia (p. 491), with maximal expression in ontogeny at a "phylotopic stage . . . at which all major body parts are represented in their final positions as undifferentiated cell condensations . . . or the stage at which all members of the phylum show the maximum degree of similarity" (loc. cit.). They base this concept on common possession of "a system of gene expression patterns, comprising the *Hox* cluster type genes and some others [encoding] relative position in all animals" (loc. cit.). As discussed just below, such a concept may apply to all triploblasts, but probably not to diploblasts (whose *Hox* homologs show some common properties of individual action, but not the integrated spatial and temporal colinearity found in most triploblasts studied so far).

The less ambitious attempt to define the phenotype and organization of a common bilaterian ancestor (named Urbilateria by de Robertis and Sasai, 1996; see also Kimmel, 1996; de Robertis, 1997; and Pennisi and Roush, 1997) may be more tractable, but specific arguments about whether this common ancestor had already developed recognizably modern versions of segments (Kimmel, 1996), antennae, photoreceptors, or a heart (see Fig. 10-25 for a cartoon of alternative possibilities from Pennisi and Roush, 1997)— with obvious implications for views about the homology of adult phenotypes, beyond the already established homology of underlying generators—are, in my opinion, premature.

For example, in a challenging proposal, Arendt et al., 2001, propose a homologous origin for tube-shaped guts of primary larvae in both protostomes and deuterostromes—structures long granted an independent origin in conventional evolutionary thinking. They base their argument upon "the shared, and very specific, expression of *brachyury* in ventral developing foreguts of the starfish bipinnaria, echinoid pluteus, enteropneust tornaria and polychaete trochophore larva," suggesting "common ancestry (homology) of larval foreguts in Protostomia and Deuterostomia, despite the different developmental origin of these structures" (p. 84). The authors then claim additional support from equally specific actions of two other developmental genes—a common expression of *goosecoid* in the foreguts of various bilaterians, and of *otx* along pre– and postoral ciliary bands. As an obvious point of contention, these putative homologies, if expressive of common ancestry, require that primary larvae be regarded as the basal anatomical form of bilaterian animals—a hypothesis supported by some (Peterson and Davidson, 2000, for example), but vigorously rejected by others (Valentine and Collins, 2000, for example).

We cannot yet determine—if genetic and developmental data of modern organisms could allow us, in principle, to resolve such questions at all— whether a hypothetical urbilaterian already possessed highly developed phenotypic structures (in either larvae or adults) acting like the Corinthian columns of my metaphor on pages 1134–1142, or whether this common ancestor had only expressed its extensive genetic and developmental pathways, preserved forever after as homologies in all bilaterian phyla, as phenotypic Pharaonic bricks of limited specification and extensive flexibility. The solution to such puzzles requires paleontological data (not yet available, but eminently attainable in principle). In any case, I regard this issue as a largely speculative sidelight that does not affect—and must not lead us to forget or put aside—the striking reformulation of evolutionary theory implied by the well-documented genetic and developmental homologies alone. De Robertis expresses this key argument in the final line of his 1997 article on the ancestry of segmentation: "The realization that all Bilateria are derived from a complex ancestor represents a major change in evolutionary thinking, suggesting that the constraints imposed by the previous history of species played a greater role in the outcome of animal evolution than anyone would have predicted until recently."

The *Hox* genes of diploblasts apparently do not show the intergenic organization of colinearity that defines the key developmental homology of bilaterians, but cnidarians do possess *Hox* genes with some commonality of action to their bilaterian homologs. Martinez et al. (1998) found four *Hox*

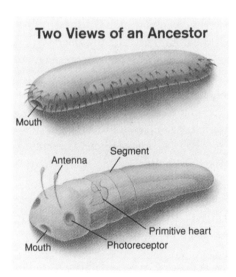

10-25. A cartoon from Pennisi and Roush, 1997, on our minimal and maximal homologies with an ancestral bilaterian. The maximal version is segmented, and has evolved a prototypical heart, eyes, a mouth, and antennae.

genes among various Cnidaria, corresponding to one medial group precursor, the anterior *Drosophila* genes *labial* and *proboscipedia,* and the posterior *Abdominal-B* (see also Miller and Miles, 1993). But even though these cnidarian genes follow the same 3′ to 5′ order as their bilaterian homologs, no evidence exists for colinearity of action in the development of any body structure (particularly the oral-aboral axis). Moreover, cnidarians apparently lack several key bilaterian *Hox* elements. Martinez et al. (1998, p. 748) write that "the genes in the middle of the [bilaterian] *Hox* clusters form a monophyletic group that includes no cnidarian genes. This is most readily explained by derivation of these genes through duplication of a single precursor after the origin of Cnidaria."

A fascinating study by Cartwright et al. (1999), however, does affirm some general similarity of *Hox* action in cnidarians and bilaterians by demonstrating a formative role for cnidarian *Hox* in specification of the oral-aboral axis in two distantly related hydrozoans, *Hydra* itself and *Hydractinia symbiolongicarpus.* (This single cnidarian *Hox* gene, *Cnox-2,* specifies full differentiation along the oral-aboral axis of polyps, whereas the sequential colinear activation of the full bilaterian *Hox* suite specifies differentiation along the bilaterian AP axis—thus illustrating once again that the primary novelty of bilaterian origins resides in the spatial sequence of *Hox* genes and the evolution of their coordinated action. We have, in any case, no reason to view the cnidarian oral-aboral axis as homologous to the AP axis of bilaterians.)

The study of Cartwright et al. gains strength from the multiple possibilities for natural and laboratory experiments inherent in the fourfold polymorphism of *Hydractinia* polyps, and in the ease of experimental transformation of one type into another. The "normal" feeding polyp of *Hydractinia,* the gastrozooid, corresponds to that of *Hydra,* and shows full oral-aboral differentiation from the distal mouth and hypostome to the body column and foot at the proximal end. In both *Hydra* and *Hydractinia, Cnox-2* is expressed at high levels in the foot and body column and at successively lower levels up the axis towards the head, which shows very weak *Cnox-2* expression.

But *Hydractinia symbiolongicarpus* also develops three polymorphic variants, clearly interpretable as intensifications of either the oral or aboral ends of the main polypary axis (see Fig. 10-26). Gonozooids and dactylozooids are specialized, respectively, for sexual reproduction and for capturing eggs of the colony's hermit crab host. Both lack a hypostome and tentacles and seem to represent "an expansion of the body column to the exclusion of oral regions" (Cartwright et al., 1999, p. 2183). The authors found "no detectable difference in *Cnox-2* expression along the aboral-oral axis in either the gonozooid or dactylozooid" (p. 2185), and general levels of expression equaled those found at the base of the gastrozooid—thus affirming the anatomical inference that both polymorphs develop by extending the specialized aboral end of the axis to the full length of the polyp, and suppressing the head region entirely.

In satisfying contrast, a fourth polymorph, the tentaculozooid that plays a role in defending the colony, resembles a single gastrozooid tentacle, and therefore appears to represent "all head"—an expanded part of the oral end only, with the aboral end suppressed. Cartwright et al. (1999) found very weak expression of *Cnox-2* throughout the full length of tentaculozooids "at approximately the same level seen in the tentacles of gastrozooids" (p. 2185).

In an additional affirmation by experimental manipulation, dactylozooids can be removed from the colony and induced thereby to transform into gastrozooids. *Cnox-2* expression initially decreased in the developing hypostome of the transforming polyp, and then in the tentacle region, but not at the aboral end, "until ultimately, dactylozooids that fully transformed into gastrozooids displayed aboral-oral *Cnox-2* expression patterns indistinguishable from that of normal gastrozooids" (p. 2185).

I believe that this crucial discovery about early emplacement of key developmental patterns—at least in bilaterian ancestry (and to a lesser extent in all animals)—combined with a central fact of timing in phylogeny, establishes a framework for understanding the primary importance of historical constraint, and of formalist (or internalist) perspectives in general, for explicating both the subsequent pathways of animal evolution and the resulting, markedly inhomogeneous habitation of potential morphospace in the history of life. Three logically connected and sequential arguments (presented as the last three sections of this final movement) combine to reset the balance of structure and function, or constraint and selection, in evolutionary theory—

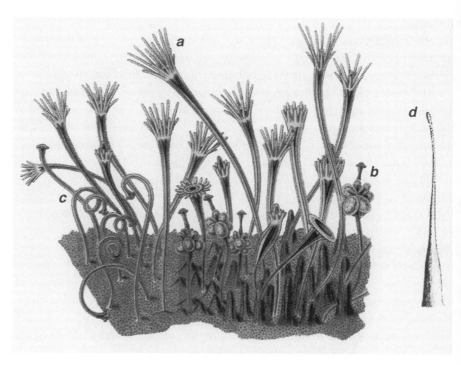

10-26. The four major polymorphic types of polyps in *Hydractinia*. A, gastrozooid; B, gonozooid; C, dactylozooid; and D, tentaculozooid. From Cartwright et al., 1999.

so that structure and constraint, the formerly disfavored and neglected first terms of each pairing, can achieve the same attention and respect that we properly accord to the proven potency of Darwinian forces represented by the second term in each pairing.

SETTING OF HISTORICAL CONSTRAINTS IN THE CAMBRIAN EXPLO-SION. Hughes (2000, p. 65) has expressed this cardinal discovery of evo-devo in phyletic and paleontological terms: "It is hard to escape the suspicion that what we witness in the Cambrian is mainly tinkering with developmental systems already firmly established by the time these Cambrian beasts showed up." As a reminder for non-paleontologists, all major bilaterian phyla with conspicuously fossilizable hard parts make their first appearance in the fossil record within the remarkably short interval (5–10 million years, but probably near or below the lower value) of the so-called Cambrian explosion (535–525 million years ago). (The single exception, the Bryozoa, first appear in the subsequent Ordovician Period.)

Unfortunately, however, as the data of molecular phylogeny accumulate, a conceptual error has begun to permeate the field, and to stymie the integration of this new source with direct information from the fossil record of early animal life, a field that has also enjoyed a renaissance in both methodology and discovery during the past twenty years (Gould, 1989c; Conway Morris, 1998; Knoll and Carroll, 1999). Although some molecular estimates for the divergence times of animal phyla correspond closely with the Cambrian explosion itself—Ayala et al. (1998), for example, cite 670 million years for the chordate vs. echinoderm division within deuterostomes—the majority of sources posit a much earlier set of divisions, deeply within Precambrian times. Wray et al. (1996) give 1.2 billion for protostomes vs. deuterostomes, and 1.0 billion for echinoderms vs. chordates; while Bromham et al. (1998) calculate confidence intervals broadly consistent with Wray et al.'s earlier dates. The 680 million year upper bound of their intervals (with much older means, of course) still suggests a minimal splitting age at least 150 million years before the explosion itself.

I do not possess the requisite skills to evaluate these different estimates, and the current literature seems too labile for a confident conclusion in any case. But I can assert that proponents of the older dates have muddied conceptual waters by supposing that their deeply Precambrian splitting times somehow either invalidate, or at least strongly compromise, the reality of the Cambrian explosion. For example, Wray et al. (1996) write: "Our results cast doubt on the prevailing notion that the animal phyla diverged explosively during the Cambrian or late Vendian, and instead suggest that there was an extended period of divergence . . . commencing about a billion years ago."

Bromham et al. (1998) codify this fallacy by inventing a straw man called "the Cambrian explosion hypothesis," defined as a claim "that the phyla and even classes of the animal kingdom originated in a rapid evolutionary radiation at the base of the Cambrian" (p. 12386). They then present their early splitting dates as a refutation of this conjecture: "We can use our results to confidently reject the Cambrian explosion hypothesis, which rests on a literal interpretation of the fossil record" (p. 12388). Of this paleontological record, they conclude (p. 12388): "It seems probable that metazoan diversity is recorded for the first time in the Cambrian because of a combination of ideal fossilization conditions and the advent of hard parts, or larger bodies, or both, that make many animal lineages 'visible' in the fossil record."

But I don't know a single paleontologist who would ever have formulated such a "Cambrian explosion hypothesis"—if only because the claim makes no logical sense, and can be confuted, in any case, by well-known paleontological data. Paleontologists have never regarded the Cambrian explosion as a genealogical event—that is, as the actual time of initial splitting for bilaterian phyla from a single common ancestor that, so to speak, crawled across the Precambrian-Cambrian boundary all by its lonesome. The Cambrian explosion, as paleontologists propose and understand the concept, marks an anatomical transition in the overt phenotypes of bilaterian organisms—that is, a geologically abrupt origin of the major *Baupläne* of bilaterian phyla and classes—not a claim about times of initial phyletic branching. The facts of the Cambrian explosion remain quite agnostic with respect to the two views about branching times now contending in the literature—Ayala *et al.*'s (1998) claim for divisions quite near the anatomical explosion, and Wray *et al.*'s (1996) and Bromham *et al.*'s (1998) argument for earlier splittings more than a billion years ago. After all, genealogical splitting and anatomical divergence of basic design represent quite different (albeit related) phenomena with no necessarily strict correlation, as exemplified in the following analogy:

If a group of Martian paleontologists had visited the earth during the Eocene epoch, they would have encountered two coexisting, and scarcely distinguishable, species of the genus *Hyracotherium*. If they had then followed the subsequent history of the lineages, they would have watched one species differentiate into the clade of rhinoceroses and the other into the clade of horses. But if a modern commentator then concluded that horses and rhinos had existed as distinct designs in their modern form (lithe runners *vs.* horned behemoths) since the Eocene, we would laugh at such a silly confusion, and point out that splitting times cannot be equated with completed anatomical divergence—especially under conventional views of Darwinian gradualism! After all, the Eocene visitors had only observed two effectively identical cousins, and could not have known that each would serve as progenitor for a highly distinct clade.

Similarly, the facts of the Cambrian explosion cannot distinguish whether —to continue with my earlier image—one tiny worm, or ten tiny worms, crawled across the Cambrian boundary as bilaterian precursors. The Cambrian explosion, as an anatomical argument, merely holds that if ten Precambrian worms formed the pool of Cambrian ancestors, they probably looked as alike as those two *Hyracotherium* species that engendered horses and rhinos.

I do not claim that the issue of one *vs.* ten tiny worms holds no relevance for other aspects of evolutionary theory, but only that the factuality of the

Cambrian explosion as an anatomical episode in the differentiation of *Baupläne* remains equally comfortable with either genealogical alternative. The question of one *vs.* ten does, however, bear strongly upon the important question of internal *vs.* external triggers for the explosion. If only one lineage generated all Cambrian diversity, then an internal trigger based upon some genetic or developmental "invention" becomes plausible. But unless lateral transfer can be validated at this multicellular level, or unless inventions of this magnitude can be so massively and coincidentally convergent, then the transformation of 10 tiny worms into the larger and well differentiated *Baupläne* of Cambrian phyla suggests an external trigger—the hypothesis traditionally favored by paleontologists in any case. (The venerable oxygen hypothesis maintains pride of ancestry, but the recent claim for melting of a "snowball earth" sometime before the Cambrian transition may well represent an even more plausible environmental trigger—see Hoffman et al., 1998; Hyde et al., 2000.)

In any case, and *pace* Bromham et al., the argument for the reality of the Cambrian explosion as an anatomical event does not depend upon the purely negative evidence of unfound fossil complexity in earlier strata, but includes several strongly positive paleontological assertions. In Darwin's time, and for nearly 100 years thereafter until the 1950's, the Precambrian fossil record stood entirely and embarrassingly blank. But paleontologists have not kept their subsequent discoveries hidden as a trade secret, and the richness of our current Precambrian record, particularly for the 100 million years preceding the Cambrian explosion, has been widely reported (Conway Morris, 1998; Gould, 1989c; McMenamin and McMenamin, 1990, for just a few among several entire books, written for general audiences on the subject). Thus, the absence of complex bilaterians before the Cambrian explosion rests upon extensive examination of appropriate sediments replete with other kinds of fossils, and located on all continents.

For example, the earth's first prominent assemblage of animals, named the Ediacara fauna for the Australian locality of its first discovery but now known from all continents, lived from about 600 million years ago right up to the explosion, with perhaps a few forms surviving beyond. These large creatures (up to a meter in length in one case, though most specimens occupy the range of centimeters to decimeters) tend to be highly flattened in form, composed of numerous sections that seem to be "quilted" together (certainly not segmented in any metameric way), and appear to possess no body openings. Although some researchers have sought the origin of a few bilaterian phyla within this fauna (Fedonkin and Waggoner, 1997), the comparisons seem farfetched and many paleontologists regard the Ediacaran animals as an early expression of pre-bilaterian possibilities of diploblast design (with modern cindarians and a few other groups surviving as a remnant of this fuller diversity), while other experts have regarded them as an entirely separate (and failed) experiment in multicellular life (Seilacher, 1989) or even as a group of marine lichen (Retallak, 1993)!

In any case, these Ediacaran fossils are soft bodied, and their preservation on all continents surely suggests that any coeval bilaterians with hard parts (or even with soft anatomy to match the Ediacarans) should be easily collectable. We do, in fact, have strong evidence for bilaterian presence in late Precambrian times, but not in a form that would lead us to postulate the anatomical complexity and specificity that first appears in the Cambrian explosion itself. Xiao et al. (1998) reported the discovery of embryos representing the blastomeres of the first few cleavage stages of apparent bilaterians (from rocks about 570 million years old, in early Ediacaran times), and from strata with a style of phosphatic replacement that can only preserve such tiny organisms. (See Chen et al., 2000, for expansion and corroboration of this interpretation.)

More importantly, paleontologists have documented a fairly rich record of benthic tracks and trails (but no body fossils) that could not have been made by the sessile or planktonic Ediacaran organisms and have, by consensus of all experts, been regarded as bilaterian in origin. But—and here's the rub—these trackways are very small, measuring 5 mm in diameter at a maximum, with most only 1 mm or so in width (see Valentine and Collins, 2000). Moreover, these tracks and trails do not extend deeply into Precambrian time. Hughes (2000, p. 64) states: "Traces made by bilaterians extend back to about 550 million years at least, but earlier sediments are famous for their undisturbed sedimentary lamination. The rise of animals able to mine organic resources in sediments in complex ways officially defines the base of the Cambrian."

Thus, positive evidence indicates only a late Precambrian origin for bilaterians of any kind. The same data imply that all Precambrian bilaterians ranged in size from the microscopic to the barely visible, and that the Cambrian boundary marks a real and geologically sudden appearance of both large complex bilaterian body fossils, and a major change in the size and complexity of their tracks and trails (Knoll and Carroll, 1999). We must then ask whether, in our highly non-fractal and allometric world, the anatomical complexity underlying and potentiating the scope of the Cambrian explosion could have originated in such tiny animals. (The fact that substantial complexity can be retained in some miniaturized offshoots of large bilaterians does not permit the reverse inference of initial invention at such small sizes.)

Most experts have argued that the complexity and diversity of bilaterian anatomy, as achieved in the Cambrian explosion, could not have evolved in creatures limited to a few mm at most in their major body axis. (Moreover, the simplicity of Precambrian tracks and trails also suggests limited styles of motion and feeding strategies in the tiny creatures that made the trace fossils.) The most popular and interesting conjecture for a biological trigger to a non-artifactual Cambrian explosion (Davidson et al., 1995; Peterson et al., 1997; Peterson and Davidson, 2000) calls upon markedly increased body size potentiated by the evolution of set-aside cells—a mechanism that permitted the tiny and anatomically simple ancestral bilaterians to circumvent ancient constraints of size and to enter a domain of magnitude where modern anatomical complexity could evolve. (But see Valentine and Collins, 2000, who challenge Davidson et al.'s key assumption that the tiny larvae of indirectly-developing modern bilaterians represent plesiomorphic models for ancestral adults before the evolution of set-aside cells.)

Thus, given that the Cambrian explosion was a real event, and that the basic homologies and developmental rules of bilaterian design (particularly as manifested in the spatial and temporal colinearity of hoxology) had already been established in the ancestors of the explosion (those one to ten tiny worms, if you will), then we may infer that bilaterian diversity unfolded along the channels of developmental patterns held in common from the beginning of this holophyletic clade. Bilaterian diversity, in other words, represents an extensive set of modifications and tinkerings upon a basic pattern set by history at the outset, and then adumbrated in one geologically brief episode to establish all fundamental building plans. Forever after, for more than half a billion years, the subsequent evolution of complex animals—that is, all bilaterian history since the Cambrian explosion—has been restricted to much more limited permutation within the confines of these early, congealed designs (however glorious and richly varied the range of ecological results).

Once we accept these premises, one broad question, rather more philosophical in nature and famously contentious given the assumptions of our cultural histories and our anthropophilic propensities, must be aired (see Gould, 1989c, and Conway Morris, 1998, for the alternative positions, and also our explicit debate in Conway Morris and Gould, 1998): If the basic developmental patterns of bilaterians arose quickly, and have remained fixed in basic form since then, do these historical invariants represent a set of mechanically limited and excellent, perhaps even optimal, designs that natural selection would have established in much the same way at any time and under any ecological or geological regime? Or do they represent just one possible solution among numerous entirely plausible alternatives of strikingly different form, each yielding a subsequent history of life entirely different from the outcome actually experienced on earth? In the second alternative, life's history unfolds with much of the unpredictability and contingency so famously displayed, for example, in the history of human cultural diversity—and the accident of a common developmental starting point for subsequent bilaterian diversity then assumes even more importance as a golden happenstance directly responsible for the particulars of the world we know.

Historical constraint based on developmental homology assumes great importance in either case, but if the particular constraints that actually set the channels of bilaterian diversity could only have arisen within a narrow range of basically similar and workable states, then much of life's pageant unfolds by predictable regularities of natural selection. If, however, the developmental plans actually established in the Cambrian explosion—albeit eminently workable, and therefore exploited by natural selection to build the particulars of life's later successes and failures—represent only one contingently-achieved set among a broad realm of alternatives (each "equally pleasing" to natural selection), then life's actual pageant on earth becomes highly unpredictable, and the happenstance of a realized beginning (the historical constraints of bilaterian developmental homology) assumes a far more prominent role in shaping the subsequent history of life.

My own arguments for contingency have been well aired (Gould, 1989c; see subsequent debate on the key technical issue in Gould, 1991a; Briggs et al., 1992, with response by Foote and Gould, 1992), and this debate only addresses the role of historical constraints in setting the actual pathways of life's singular history on earth, and not the existence of the constraints themselves (the subject of this section). Thus, I will not discuss the important question of predictability vs. contingency much further, except to clarify the problem by noting that questions of contingency enter our understanding of evolutionary pattern at two levels of inquiry about the Cambrian explosion and its consequences.

*First,* we must ask if the basic bilaterian homologies themselves, particularly the *Hox* rules, represent an optimal solution that natural selection would have constructed in any case, or a workable happenstance among many alternatives. The very fact that some homonomous bilaterian phyla possess a complete complement of *Hox* genes, and that the original function of these genes therefore cannot match their present role in controlling the various downstream cascades that specialize and differentiate the sequence of structures along the AP axis, speaks strongly for contingency—because current utilities must therefore represent cooptations from different original functions, rather than primary adaptations. Such cooptation, expressing the principle of "quirky functional shift" (see Chapter 11, pp. 1218–1229 for full discussion), inevitably suggests (but admittedly does not prove) a high degree of fortuity, as implied by the required capacity of features built for one function to act in another way that could not have influenced or regulated their original construction by any functional evolutionary mechanism like natural selection.

In this particular case, for example, Deutsch and Le Guyader (1998) have suggested a historically prior function for *Hox* (and other zootype) genes in designing "an appropriate neuronal network in bilaterian animals" (p. 713). Recognizing the relevance of this idea to the issue of contingency and the Cambrian explosion (1998, p. 716), they write: "Hence, the presence, before the Cambrian explosion, of a large number of *Hox* genes, whose domains of activity extend from the post-oral head to the abdomen, cannot be accounted for by a function in driving morphological diversity. Another role has to be assumed for the ancestral function of the *Hox* genes. We postulate that the zootype genes primitively specified neural identity."

*Second,* we must ask if the realized variants that congealed so quickly as specialized and differentiated body plans (the major bilaterian *Baupläne*)— permitting no further origin of novel anatomies sufficiently distinct to warrant taxonomic recognition as phyla—represent a predictable set of "best solutions" within the broad possibilities of historical constraint permitted by shared developmental rules? Or do they constitute a subset of workable, but basically fortuitous, survivals among a much larger set that could have functioned just as well, but either never arose, or lost their opportunities, by historical happenstance? I admit my partisanship for the latter position (Gould, 1989c) and freely acknowledge that my judgments have won some support, but no consensus to say the least (Conway Morris, 1998). I would only point out that even the strongest opponents of contingency admit that arthropod disparity (the measured range of anatomical designs, not the number of species) had reached a fully modern range in the Burgess Shale faunas (Middle Cambrian, about 10 million years after the explosion)—and that more than 500 million years of additional arthropod evolution has not expanded the scope of anatomical disparity at all (Briggs et al., 1992; Foote and Gould, 1992, present evidence for the counter view that Cambrian disparity exceeded modern levels, despite much lower species diversity). I would also urge my colleagues to spend more time studying Cambrian "oddballs" that do not easily fit into recognized higher taxa, including *Xidazoon* among "orphan" taxa (Shu et al., 1999), or *Fuxianhuia* among arthropods that do not belong to any recognized class (Chen et al., 1995), and not to focus so strongly, as most studies have done in recent years, upon cladistic attempts to place all Cambrian forms at least into the stem regions of major phyla, if shared derived markers of crown groupings bar their entry—a strategy that leads researchers to ignore the autapomorphies of these peculiar taxa, and to coax other features into plesiomorphy with modern taxa.

**CHANNELING THE SUBSEQUENT DIRECTIONS OF BILATERIAN HISTORY FROM THE INSIDE.** If the bilaterian ancestor possessed a full complement of *Hox* genes, and if all major variants upon this initial system had already congealed by the end of the Cambrian explosion, then subsequent bilaterian evolution must unfold within the secondary strictures of these realized specializations upon an underlying plan already channeled by primary constraints of the common ancestral pattern. But lest we begin to suspect that rigid limitation must represent the major evolutionary implication of such constraint, I must reemphasize the positive aspect of constraint as fruitful channeling along lines of favorable variation that can accelerate or enhance the work of natural selection. Moreover, the evolutionary flexibility of developmental channels achieves its most impressive range—as Chapter 11 will discuss as its primary subject—through the crucial principle of cooptation, or the extensive and inherent capacity of genes evolved for one particular function to operate, through evolutionary redeployment, in strikingly different adaptive ways.

Among "higher" triploblast phyla of markedly divergent design, echinoderms represent the obvious test case for studying the flexibility of homologous developmental genes. With their remarkable autapomorphies of radial symmetry, calcitic endoskeleton, and a water vascular system for circulation, how could these creatures evolve within the confines of a genetic regulatory system that builds bilaterial, axially specialized organisms with blood vascular systems in both their immediate sister phylum (the vertebrates) and in plesiomorphic taxa of more distant common ancestry (the protostome phyla on both major branches). Did echinoderms delete their ancestral determinants to evolve such an aberrant morphology or did they acquire entirely new regulatory genes and developmental rules?

Few data now exist to address this important issue, but preliminary results suggest that echinoderms have retained their genetic homologies with other bilaterian phyla, while coopting several of these genes (with stable function in other phyla) for different roles in their own unique development. In a pioneering study, Lowe and Wray (1997) documented the expression in echinoderms of orthologs of three important regulatory genes that encode transcription factors with a homeodomain, and that generally function in the same broad way in both vertebrates and arthropods (and must therefore be plesiomorphic to any derived condition in echinoderms): *distal-less* for proximodistal patterning in outgrowth of limbs, *engrailed* for neurogenesis along the axis of the CNS, and *orthodenticle* for the differentiation of anterior structures.

Lowe and Wray documented a full spectrum of results, ranging from retention to cooption for markedly different echninoderm roles. At an extreme of retention, the brittle star *Amphipholis squamata* expresses *engrailed* in neuronal cell bodies along the five radial nerves. Lowe and Wray note (1997, pp. 719–720): "This expression is superficially similar to that in bilaterial animals, in which *engrailed* is expressed within a serially repeated subset of ganglionic neurons along the antero-posterior axis. It is possible that a neurogenic role for *engrailed* is widely conserved among triploblastic animals."

In the intermediary state of a retained general role transferred to novel organs, sea urchins express *distal-less* at the distal ends of the five primary podia (tube feet) soon after their formation—thus preserving the standard function of regulating outgrowths from a body axis by expression at their distal tips, but now applied to an autapomorphic outgrowth with no homolog in any other bilaterian phylum! Finally, at the extreme of full cooptation (for new functions in new organs), brittle stars express *orthodenticle* in ectoderm overlying the terminal ossicles at the ends of the arms—a position with only tenuous and hypothetical connection to the anterior end of the AP axis in bilaterian phyla. Moreover, at least for *engrailed* and *orthodenticle* (the copy-number of *distal-less* remains undetermined in echinoderms), only one ortholog exists in any echinoderm studied so far—so new functions cannot be ascribed to the cooptation of duplicated copies.

Lowe and Wray's final statement (1997, p. 721) emphasizes the important conclusion that genetic and developmental homologies of triploblast animals still permit enormous flexibility in evolutionary diversification—primarily by the principle of cooptation: "The highly derived body architecture of echinoderms evolved at least in part through extensive modifications in the roles and expression domains of regulatory genes inherited from their bilaterial ancestors. Even the limited number of genes and species we examined demonstrates a remarkable evolutionary flexibility in genes that have previously been considered interesting mainly for their conserved roles in arthropods and chordates."

If we now turn our attention to these "conserved roles in arthropods and chordates," at least three sources of evidence underscore the central conclusion of this section: that the evolution of differentiated and specialized *Baupläne* from a presumably homonomous common ancestor proceeds—paradoxically, and contrary to the scenario of Lewis's (1978) original hypothesis about *Hox* genes—by reduction and restriction, rather than by addition of genes or expansion of their domains of activity.

1. Specialization of anatomy sometimes correlates with deletion or unemployment of *Hox* genes, or their redeployment to other functions. For example, *Zen,* the insect ortholog (in both structure and position) to vertebrate *Hox3,* exhibits no *Hox* function and is not expressed along the AP axis of the developing larva, but plays some role instead in the formation of extraembryonic membranes. In a paper that wins, by acclamation, the Steinbeckian prize for title parodies ("Of mites and Zen"), Telford and Thomas (1998) cloned the homolog of *Drosophila Zen* in the orbatid mite *Archegozetes.* They found expression of this chelicerate homolog "in a discrete anteroposterior region of the body with an anterior boundary coinciding with that of the chelicerate homolog of the *Drosophila Hox* gene *proboscipedia*" (p. 591). This fascinating result suggests that *zen* may have lost its *Hox* function in *Drosophila* as a consequence of functional redundancy due to overlap with another *Hox* gene.

Taking the argument further, Telford and Thomas present evidence that the *Drosophila* pair-rule gene *fushi tarazu* may also be "a divergent *Hox* gene that has adopted a new role" (p. 594; see also Dawes et al., 1994, on a locust homolog of *fushi tarazu* that shows no pair-rule function). These observations on the original roles of *Drosophila zen* and *fushi tarazu* suggest "that the original complement of arthropod *Hox* genes must be revised from eight to ten" (Telford and Thomas, 1998, p. 594), thus emphasizing the role of gene loss in the specialization of body plans.

2. Stasis or slow change in *Hox* genes indicates their conserved role in evolution. Akam et al. (1994), in a section of their paper entitled "*Hox* genes that got away," contrast the conservation of *Hox* genes in insects (as documented by high levels of sequence similarity among taxa) with much higher rates of divergence in homeobox genes that do not now function within the *Drosophila Hox* series, but may have belonged to the *Hox* cluster of an arthropod common ancestor: the maternally expressed *bicoid* (encoding the morphogen that produces a crucial AP gradient in the early syncytial embryo), the pair-rule gene *fushi tarazu,* and the two *zen* genes. The putative orthologs of *fushi tarazu* in other insects "are almost unrecognizable outside of their homeodomains, and have accumulated approximately 10 times as many changes in their homeodomains as have homeotic [i.e., *Hox*] genes in the same comparisons" (Akam et al., 1994, p. 209). The authors then generalize about these non-*Hox* homeobox genes (p. 214): "We think that these genes may be derived . . . from *Hox* genes which, in the lineage leading to *Drosophila,* have escaped from the conservative selection that characterizes homeotic genes."

3. Overexpression, in both position and amount, of vertebrate *Hox* genes has generated atavisms in several experiments, thus suggesting that derived specializations evolve by tighter regionalization and restriction of expression in individual *Hox* genes. Pollock et al. (1995) studied the influence of overexpression for *Hoxb-8* and *Hoxc-8* upon the skeletal development of mice, concluding that "many of the morphological consequences of expanding the mesodermal domain and magnitude of expression of either gene were atavistic" (p. 4492). For example, the earliest Paleozoic vertebrates grew "free ribs" (independent from and articulating with the vertebrae) along the entire body axis, from the base of the skull to the tail. Many subsequent tetrapod lineages, particularly among mammals, reduced the number of free ribs dramatically. But vestiges of the ancestral free ribs sometimes remain as small units fused with the vertebrae. In particular, the lumbar pleurapophyses of posterior mammalian vertebrae "most likely represent an ancestral rib that has fused with the lateral portion of the vertebrae and now serves as a point of attachment for muscle groups of the back" (p. 4495). Pollock et al. (1995) documented "the reappearance of free ribs at the expense of lumbar pleurapophyses" in *Hoxb-8* transgenic mice—"a clear example of atavism" (p. 4495). In another experiment, mice developing with overexpression of both *Hoxb-8* and *Hoxc-8* grew costal tubercles on their lower thoracic ribs. Costal tubercles represent a vestige of the second head of the articulating boss in free ribs. Normal mice develop no costal tubercles on these ribs at all.

In a similar experiment, Lufkin et al. (1992) ectopically expressed a *Hoxd* gene "more rostrally than its normal mesoderm anterior boundary of expression" (p. 835) at the level of the first cervical somites. This anomalous anterior expression generated "a homeotic transformation of the occipital bones towards a more posterior phenotype into structures that resemble cervical vertebrae" (p. 835). One should not read too much evolutionary meaning into one experimental manipulation, but since the same ectopic expression also induced other changes of a potentially atavistic nature (particularly "the presence of clearly segmented neural arches arising from the most anterior somites," p. 840), and since the vertebrate skull and forebrain probably arose as novel features at the anterior end of a more homonomous ancestor, any transformation of skull parts towards the phenotype of more homonomous posterior vertebrae can hardly fail to elicit thoughts about the phylogeny of the vertebrate skull—especially when these potential atavisms arise by reversing the presumed phyletic restriction and posterior localization of *Hox* action. Lufkin et al. (1992), in what I can only regard as an expression of chutzpah (but still worth pondering), even ask: "Would ectopic expression of additional *Hox* genes be required to convert fully the neurocranium into vertebrae?" (p. 840).

Pollock et al. (1995, pp. 4495–4496) also reach a bold conclusion that may go too far, but that merits careful consideration:

> The observation that expansion of the functional domain of a *Hox* gene
> can result in the transformation of a modern costal structure to a more

ancient form suggests that regional repression of *Hox* gene expression
could have played a role in the evolution of the vertebral column . . .
We propose that, in antecedent vertebrates, *Hox* genes involved in patterning the axial skeleton were expressed in relatively broad regions of
paraxial mesoderm. The resulting less complex *Hox* code would have established the similar vertebral identities observed in broad regions of
early vertebrate skeletons. Regionalization of the vertebral column subsequently evolved in concert with the evolution of restricted patterns of
*Hox* gene expression in paraxial mesoderm.

When we turn to specific examples in the evolution of differentiated and
complexified axial structures in arthropods and vertebrates, we commonly
find correlations between morphological inventions and the restriction and
regionalization of *Hox* gene expression. I have already discussed several cases
in previous sections of this chapter—particularly the complex and elegantly
documented story of multiple parallel evolution of crustacean maxillipeds
from ancestral limbs that were more homonomous with the rest of the posterior thoracic series. This complexification originates by suppression of *Ubx*
and *abdA* in just those anterior thoracic segments that develop maxillipeds in
each case.

Enlarging upon this example, Averof and Akam (1995) found that three
middle *Hox* genes—*Antp, Ubx* and *abdA*—"are expressed in largely overlapping domains in the uniform thoracic region" of branchiopod crustaceans
(1995, p. 420, and see Fig. 10-27). In the more highly differentiated insects,
on the other hand, the same three genes show more restricted expression in
discrete domains, where "they specify distinct segment types within the thorax and abdomen" (p. 420). In particular, *Antp* turns on in thoracic segments
(that develop legs and wings) but not in abdominal segments. Similarly, both
*Ubx* and *abdA* are expressed in all abdominal segments, where they repress
*distal-less* and therefore presumably regulate the repression of legs on all abdominal segments.

In an elegant affirmation that has already become a classic of the evo-devo
literature, Warren et al. (1994) demonstrated that lepidopteran larvae develop prolegs on their abdominal segments by localized deletion of *Ubx* and
*abdA* expression, followed by subsequent derepression of *distal-less* in the
small bilateral patches on each abdominal segment, from which the prolegs
then grow. They conclude (1994, p. 458) that "abdominal limb formation in
butterflies has been made possible by the evolution of a regulatory mechanism for shutting off these two *BX-C* [*Hox* in modern terminology] genes in
selected cell populations, which then permits *Dll* and *Antp* to be expressed."

Warren et al. then raise an obvious question that exemplifies a related principle in the evolution of differentiated complexity from homonomous ancestry: Why didn't butterflies evolve their abdominal prolegs by the "easier"
route of fixing a *Dll* mutation to release repression by *Ubx* and *abdA*? Why
follow the more complex scenario of first repressing the two *Hox* genes in local patches, and then permitting the ordinary action of *Dll*? Warren et al. of-

fer the reasonable explanation that derepression of *Dll* will not be sufficient, by itself, to build a full proleg. A much more extensive cascade of downstream genes generates the proleg (presumably with a crucial boost from *Dll*)—and upstream *Hox* repression must be released in order to potentiate the full downstream cascade. Thus, differentiation from homonomy may proceed either by regionalizing and restricting the *Hox* genes themselves, or by altering and specializing the downstream cascades regulated by *Hox* genes.

As a further example of this second process, Warren et al. (1994) also documented the absence of *Ubx* from *Drosophila* wing imaginal discs, whereas *Ubx* occurs at high levels in the section of the dorsal metathoracic disc that generates the halteres (markedly reduced wings that function as balancing organs) on the third thoracic segment. This observation led to the conjecture that, just as *Hox* genes regulate the appearance or repression of legs on particular segments, *Hox* genes might also determine the presence or absence of wings in a similarly direct manner. Thus, if *Ubx* prevents full wing development in *Drosophila* T3, perhaps a suppression of *Ubx* permits the generation of a large and complete second pair of wings on the homologous T3 segment of Lepidoptera.

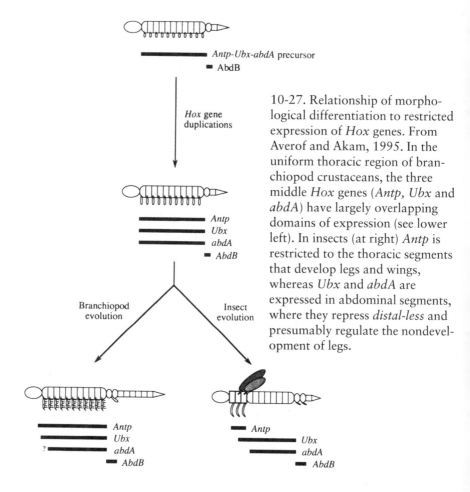

10-27. Relationship of morphological differentiation to restricted expression of *Hox* genes. From Averof and Akam, 1995. In the uniform thoracic region of branchiopod crustaceans, the three middle *Hox* genes (*Antp, Ubx* and *abdA*) have largely overlapping domains of expression (see lower left). In insects (at right) *Antp* is restricted to the thoracic segments that develop legs and wings, whereas *Ubx* and *abdA* are expressed in abdominal segments, where they repress *distal-less* and presumably regulate the nondevelopment of legs.

But this reasonable conjecture was then falsified because, just as in dipterans, *Ubx* does not turn on in the lepidopteran forewing imaginal disc, but achieves high levels of expression in the hindwing disc. Therefore, the growth of lepidopteran hindwings in the presence of *Ubx* must depend upon differences in the downstream T3 cascade of flies *vs.* butterflies. Warren et al. (1994, p. 461) conclude: "The most logical explanation is that the sets of downstream wing-patterning genes regulated by *Ubx* in these orders have diverged. In this view, *Ubx* operates in butterflies upon pattern regulating genes to differentiate hindwings from forewings, and in flies upon a different set of genes to distinguish halteres from wings."

In further confirmation from detailed patterns at lower taxonomic levels, Weatherbee et al. (1999) then studied *Hindsight,* a homeotic mutation in butterflies that transforms parts of the hindwing into forewing identity. They found that these hindwing transformations in color and scale morphology occur in regions of the forewing where *Ubx* expression has been lost, thus sensibly explaining, under the general rule for *Hox* expression in butterfly wings, the apparent forewing identities of these altered regions. Reemphasizing the important principle previously illustrated for echinoderms *vs.* other triploblast phyla, but at this lower taxonomic level—that channels of internal homology also promote flexibility, not just limitation, through such mechanisms as cooptation and diversification of downstream cascades—Weatherbee et al. (1999, p. 113) write: "The diversity of insect hindwing patterns illustrates the broad range of possible morphologies that can evolve in homologous structures that are regulated by the same *Hox* gene."

I turn, finally, to the two canonical and most anatomically extensive examples of evolution from homonomy to regional specialization and complexity in the evolution of insects and other arthropods—evolution from the plesiomorphic state of walking legs on all post-oral segments and, for pterygotes, from the ancestral condition (as revealed in the fossil record and preserved in modern mayfly larvae) of wings on all thoracic and abdominal segments. Data from evo-devo have effectively resolved the old debate about whether insect wings evolved as novel structures from hypothesized rigid extensions of the body wall in terrestrial forebears (the paranotal theory), or from dorsal branches of polyramous appendages of ancestral forms (the limb-exite theory).

Genetic data support the exite theory and provide a fascinating example of cooptation in evolution (the general subject of the subsequent Chapter 11). In the exite theory, insect wings and legs are, in some sense, serially homologous as specializations of different parts of an ancestral polyramous appendage— the wings from the dorsalmost branch (the exite) and the leg from the ventralmost walking branch. In their major topological difference, wings develop as sheets, and legs as tubes. In *Drosophila*, the wing grows under the crucial influence of *apterous*, which is expressed only in dorsal cells and therefore maintains clear distinction between dorsal and ventral surfaces, thus abetting the growth of a sheet-like structure (Shubin et al., 1997). But the *apterous* gene does not function in the growth of tubular legs in *Drosophila*. However,

Averof and Cohen (1997) found *apterous* expression in the sheet-like dorsal branch of respiratory epipodites in a branchiopod crustacean, thus supporting the exite theory for the origin of wings. Shubin et al. (1997, p. 645) draw a reasonable phyletic conclusion consistent with this section's theme of evolution from homonomy to specialization: "This suggests that Recent wings evolved from the respiratory lobe of an ancestral polyramous limb, probably first appearing in the immature aquatic stages as gill-like structures, such as those found on all trunk segments of extinct Paleodictyoptera or extant mayfly larvae."

On the related issue of evolutionary suppression of wings on most segments and their restriction to two pairs in most insects, and to one in dipterans (see Fig. 10-28), Carroll et al. (1995, p. 58) demonstrate that "wing formation is not promoted by any homeotic gene, but is repressed in different segments by different homeotic genes." Against the older view (consistent with Lewis's original additive model of phenotypic complexification) that *Antp* positively regulates the formation of wings and halteres on T2 and T3 of *Drosophila*, Carroll et al. (1995) present evidence that other *Hox* genes repress wing primordia on the remaining body segments. For example, *Scr* is expressed in both labial and T1 segments; in mutant embryos lacking *Scr* expression, "flight appendage primordia arise in the T1 segment" (p. 58). As for suppression of more posterior wings, the oldest information about homeotic mutations in *Drosophila* documented the development of complete wings on T3, where the vestigial halteres usually form. We now know that this *bithorax* phenotype (which gave its name to the previous designation of the posterior *Drosophila Hox* series as the "bithorax complex" or *BX-C*) results from a mutation that represses the *Ubx* gene in T3. Another mutation of *Ubx* leads to the growth of wing primordia on A1 as well.

Carroll et al. (1995) propose that when wings existed on all post-oral body segments of a homonomous ancestor, "there was no homeotic gene input into their number or design" (p. 59). Carroll et al. then hypothesize that elimination of wings from most segments occurred as the *Hox* genes became

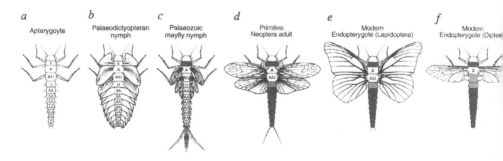

10-28. Differentiation of flight in the evolution of insects as a consequence of repression of wings on posterior segments by various *Hox* genes. The fossil nymph in B possessed wings on all segments. These are reduced but still present on all segments in the fossil mayfly nymph at C. From Carroll et al., 1995.

regionalized and individualized, leading to suppression by different elements in various parts of the body: "The evolution of *Scr*-responsive elements led to the modification or elimination of prothoracic wings and the evolution of *abdA* and *Ubx*-responsive elements led to the elimination of abdominal wings and, in the Diptera, to the reduction of metathoracic wings."

For the most general question of specialization of appendages for a wide variety of forms and functions from their uniform state on all segments posterior to the head of a homonomous ancestor, several lines of evidence identify walking appendages (either uniramous, or biramous with an upper gill branch) as the ancestral state for a homonomous ancestor, and as a continuing "ground state" for modern more differentiated forms as well. First, the most homonomous modern groups—the Myriapoda among the arthropods, and the Onychophora as a sister group to the entire arthropod phylum—bear leg-like structures on each segment. Second, numerous Cambrian arthropods that cannot be placed into modern groups share the common property of nearly identical biramous appendages on all postoral segments, and only a pair or two of antennae on any preoral segments—as in *Marrella*, the most common fossil in the Burgess Shale (see Fig. 10-29 and Gould, 1989c). Third, as discussed previously (pp. 1132–1134), the extensive suite of thoracic segments that bear identical leg-like appendages in many modern Crustacea also show extensive and complete overlap of expression for several *Hox* genes.

Proceeding down the AP axis of complexified arthropods, we first note that antennae develop in the most anterior segments where no *Hox* expression occurs. This situation probably marks a retention of the ancestral condition. Even the most homonomous forms, including myriapods and onychophores, exhibit some specialization in the head segments at the anterior end, and only grow identical appendages on subsequent postoral segments. Thus, the original *Hox* complex probably never regulated development at the extreme anterior end around the mouth, and antennae probably represent the plesiomorphic condition for segments with no *Hox* action. Interestingly, and in confirmation, the suppression of all *Hox* activity in *Tribolium* yields the le-

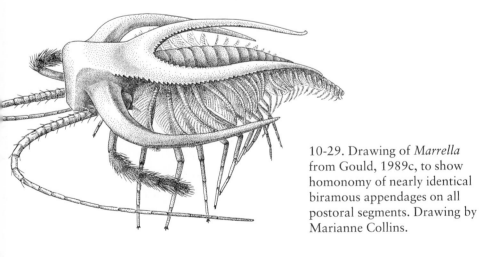

10-29. Drawing of *Marrella* from Gould, 1989c, to show homonomy of nearly identical biramous appendages on all postoral segments. Drawing by Marianne Collins.

thal consequence of a dead larva with antennae on all segments (Stuart et al., 1991; Shubin et al., 1997, p. 664; see also Cassares and Mann, 1998, on antennal-determining genes repressed by *Hox* action in *Drosophila*).

All other specializations down the AP axis are apparently derived and dependent upon differentiation and regionalization, or elimination in some cases, of expression in various *Hox* genes. For example, in gnathal segments just posterior to segments bearing antennae in many groups, the most homonomous modern forms develop mouth parts of essentially leg-like form (as in myriapods). These leg-like appendages express *Distal-less* at their distal tips, the typical situation for ordinary arthropod legs. But *Distal-less* is not expressed at the distal ends of more specialized (and non leg-like) feeding appendages of insects and crustaceans. "These data," Shubin et al. write (1997, p. 644), "agree with fossil evidence suggesting that crustacean and insect mandibles were reduced from the primitive whole-limb mandible by truncation of the mandibular proximodistal axis."

I have already discussed, in previous parts of this section, the role of *Hox* restrictions and repressions in the evolution of all other outstanding phenotypic specializations in more posterior regions of arthropod bodies, including the differentiation of maxillipeds from legs on the previously homonomous crustacean thorax (pp. 1132–1134), the restriction of wings to just one or two thoracic segments in insects (p. 1165), and the complete suppression of legs on the insect abdomen, with localized *Hox* repression to permit the growth of prolegs on the abdominal segments of lepidopteran larvae (p. 1165).

When we turn to the history of vertebrates, we first encounter an apparent exception to the generality that phenotypic specialization correlates with reduction in number of *Hox* genes and regionalization of their action. Amphioxus, the modern cephalochordate surrogate for an ancestral form, has only one *Hox* cluster, while gnathostome vertebrates have four—so duplication, occurring at least twice, clearly marks a major feature of vertebrate evolution, with obvious implications for correlating the complexity of our phylum with this marked increase in the total number of *Hox* genes, and in apparent contradiction to the opposite relation of phenotypic elaboration with genetic restriction, as discussed throughout this section.

But the single cluster of amphioxus contains homologs of the first 10 paralogy groups of vertebrate *Hox* genes, arranged in the usual colinear array. Moreover, the amphioxus genome includes at least two *AbdB*-like genes, indicating that tandem duplication of these posterior *Hox* elements was already underway in the cephalochordates, even though true vertebrates have carried the process further (Carroll, 1995; Coates and Cohn, 1998). Therefore, essentially the full *Hox* complement had already been established when the genome of an immediate vertebrate ancestor included only one set of *Hox* genes. Moreover, the full fourfold amplification had already been completed by the origin of jaws in early fishes because all modern gnathostomes—that is, all living species of vertebrates except for the two small lineages of agnathan fishes, the lampreys, with three *Hox* sets, and the hagfishes—have four sets.

Thus, the common ancestor of all 40,000 or so modern gnathostome species already had four *Hox* sets, and only the handful of agnathan species has fewer sets among modern vertebrates. Thus, our large clade of 40,000 species evolved under the general rule featured throughout this section: phenotypic specialization correlated with *Hox* deletions and restrictions of expression. Or, to put the matter somewhat facetiously, you start with all you will ever get, and work "down" from there—an optimal formula for the evolutionary importance of historical constraint.

As Coates and Cohn (1998, p. 375) write (see also Fig. 10-30): "During the period since these gene duplication episodes, jawed vertebrate *Hox* cluster evolution seems to have been characterized by gene deletions." Moreover, as Figure 10-30 also shows, teleost fishes, which did not originate until Mesozoic times, evolve different patterns of deletion from those found in mammals, a group with a Paleozoic ancestry from a very different vertebrate lineage—thus "indicating quite separate patterns of gene loss in tetrapod and teleost lineages" (Coates and Cohn, 1998, p. 375).

The relatively homonomous architecture of the postcranial skeleton of many early fishes (and many early tetrapods as well) has evolved in the conventionally "higher" tetrapods, primarily in mammals, into a more complex, specialized and regionalized axial skeleton with clear and often quite sharp distinctions, in both form and function, from cervical to thoracic to lumbar, sacral and caudal regions of the vertebral column. Burke et al. (1995) have demonstrated an interesting basis for much of this phenotypic complexity and regionalization in the establishment of definite boundaries of action for particular paralogy groups of the *Hox* clusters, thus repeating the general arthropod correlation of *Hox* regionalization with phenotypic specialization along the AP axis.

For example, different groups of vertebrates vary greatly in the number of vertebrae per region, but the boundary between regions may still remain

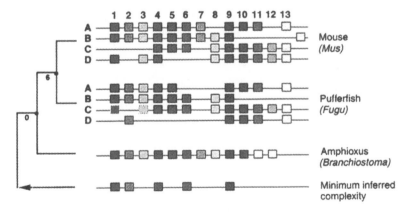

10-30. Following the evolution of four *Hox* clusters in vertebrates, the major pattern of change has not resulted in further addition, but rather in elimination —in different patterns in various groups. From Coates and Cohn, 1998, p. 375.

sharp (Burke et al. refer to the shifts of these boundaries to emplace more or fewer vertebrae within any given region, as "transpositions"). In their most intriguing conclusion, Burke et al. (1995) found that some of these phenotypic transitions correlate precisely with anterior expression boundaries of particular *Hox* genes. For example, *Hoxc-6* marks the transition between cervical and thoracic vertebrae, despite the highly variable number of cervical vertebrae, ranging from 3 or 4 in frogs, to 7 in mice (as in virtually all mammals, including giraffes), to 17 in geese. The thoracic-lumbar transition generally correlates with the expression of *Hoxa-9, Hoxb-9,* and *Hoxc-9,* whereas the *Hoxd-9* boundary tends to be shifted backwards to the lumbo-sacral transition. Carroll (1995, p. 483) comments on these differences in the ninth paralogy group: "This may be significant because the thoracic-lumbar distinction is not general among tetrapods. It may be that shifts within the *Hox-9* group were important in the evolution of this transition from a more uniform trunk, perhaps even in the evolution of the tetrapods from fish."

These regularities of *Hox* regionalization may help us to understand both the limitations and flexibilities of vertebrate anatomy in terms of historical constraint. In an early article, for example, Tabin (1992) suggested that tetrapod limbs may now be constrained to five digits per limb (despite the presence of up to 8 digits in the earliest tetrapods of the Late Devonian Period—see Coates and Clack, 1990; and Gould, 1993e) because the *Hoxd* series that plays such a major role in patterning limbs may now only generate five "addresses" for the development of distinct digits. Many polydactylous mutants (and experimental manipulations) exist in vertebrates, but the supernumeraries are always phenotypic replicates of one of the five distinct digits, so the general hypothesis holds (see also Shubin et al., 1997, pp. 642–643).

A related classical question asks why tetrapods, honoring their name, never grow more than four limbs, whereas the other major terrestrial group of arthropods usually evolves phenotypes with more appendages (even though we might imagine, on functional grounds, that an increased number of supports would be even more valuable in large vertebrates with much lower area to volume ratios—for supporting strength of bone scales as cross-sectional area). Coates and Cohn (1998, p. 379) note that "the nearest approach to a third pair of lateral appendages may be the lateral caudal keels of certain fishes, such as tuna and various sharks." But a true anatomical third pair has never evolved in any tetrapod or extant fish. (The extinct acanthodian fishes evolved the only vertebrate departure from the principle of two primary limb pairs.)

But as *Hox* rules constrain, they can also be tweaked to win interesting flexibility. Cohn and Tickle (1999), for example, studied *Hox* expression in the axial skeleton of pythons, which can grow more than 300 essentially identical vertebrae, and which retain hindlimb rudiments but express no forelimb development at all. Except for the atlas, every vertebra anterior to the rudimentary hindlimb develops ribs (a thoracic feature) as well as ventral hypopophyses (a cervical feature), suggesting to Cohn and Tickle (1999, p. 474) that "information encoding thoracic identity may have extended into the cervical region and partially transformed these segments. Thus the entire trunk resembles an elongated thorax."

Cohn and Tickle studied the expression of *Hoxb-5, Hoxc-6,* and *Hoxc-8* in the ontogeny of pythons. In both teleosts and tetrapods, the anterior expression boundaries of all three genes in lateral plate mesoderm occurs "at the forelimb/pectoral fin level, where they are involved in specifying forelimb position and shoulder development" (p. 475). But pythons develop no phenotypic expression of the forelimb at all, thus suggesting that a suppression of this positional boundary, and a forward expansion of expression in these genes, might be causally related to the vast increase in number, and identity in thoracic form, of snake vertebrae—and potentially helping to explain one of the most striking functional novelties ever evolved in vertebrates.

Indeed, Cohn and Tickle (1999, p. 475) found *Hoxc-8* and *Hoxb-5* expression "throughout the python lateral plate mesoderm, with expression terminating at the very anterior limit of the trunk. Thus, the entire vertebral column anterior to the cloaca exhibits patterns of *Hox* gene expression consistent with thoracic identity, and we were unable to detect restricted *Hox* expression patterns in the lateral plate mesoderm associated with forelimb position in other tetrapods." In interesting and confirming contrast, they detected a sharp posterior boundary of *Hoxc-8* expression right at the level of the hindlimb rudiments, "which coincides with the last thoracic vertebra in older animals." Their phyletic hypothesis underscores the power of constraining rules as positive channels that can be tweaked in rare and interesting ways to yield remarkable phenotypic and functional excursions into novelty—but always under the rubric of *Hox* rules and their potentiating, but also directional, flexibilities: "Expansion of these *Hox* gene expression domains in both paraxial and lateral plate mesoderm may be the mechanism which transformed the entire snake trunk towards a thoracic/flank identity and led directly to the absence of forelimb development during snake evolution" (1999, p. 475).

AN EPILOG ON DOBZHANSKY'S LANDSCAPE AND THE DOMINANT ROLE OF HISTORICAL CONSTRAINT IN THE CLUMPED POPULATION OF MOR-PHOSPACE. As I emphasized in setting out the varied categories and evolutionary significances of constraint (see pp. 1151–1161 and Figs. 10-10 and 10-11), the historical vertex treated in this Chapter 10 does not refute the functional or adaptational premises of traditional Darwinism by asserting a nonadaptive origin for the constraints thus generated—for this more direct challenge arises from the structural vertex that will be treated in Chapter 11. That is, I do not doubt that most, or nearly all, constraints from the historical vertex originate as direct adaptations in the ancestral taxon of their initial appearance. But, having thus originated, these adaptations may then "congeal" to limit directions of potential alteration in descendant taxa (the negative meanings), or to channel future change in preferred directions that often accelerate or grant easier access to adaptive solutions (the positive meanings). In terms of the classical model of Galton's polyhedron, the pool cue of natural selection may always do the actual pushing, but if internal channels—set by history, and grafted into the genetic and developmental architecture of current organisms—designate a limited set of possible pathways as conduits for selection's pushing, then these internal constraints can surely claim equal weight with natural selection in any full account of the causes of any particular evolutionary change.

But if the challenge posed by historical constraint to traditional Darwinian functionalism does not lie in an argument about nonadaptive origins, then how can this category of constraint rectify and expand evolutionary theory beyond the narrowness imposed by overly adaptationist versions of Darwinism favored during the heyday of the Modern Synthesis (see Chapter 7)?

In describing my basic framework of argument, I asserted (see pp. 1055–1057) that the challenge of historical constraint resides in a "metaquestion" about the role of adaptation in establishing the clumpiness of occupied morphospace, not in a direct inquiry about the adaptive status of each evolutionary novelty considered one-by-one. In short, I argued that the markedly inhomogeneous occupation of morphospace—surely one of the cardinal, most theoretically important, and most viscerally fascinating aspects of life's history on earth—must be explained largely by the limits and channels of historical constraint, and not by the traditional mapping of organisms upon the clumped and nonrandom distribution of adaptive peaks in our current ecological landscapes. In other words, the inhomogeneous occupation of morphospace largely records the influence of structural rules and regularities emerging "from the inside" of inherited genetic and developmental systems of organisms, and does not only (or even primarily) reflect the action of functional principles realized by the mechanism of natural selection imposed "from the outside."

In a recent article, Arthur and Farrow (1999, p. 183) pose the key issue in much the same terms: "Why do animal take the forms they do, and not others? Why . . . are all land vertebrates 'tetrapods'—except for cases of secondary loss, for example snakes—while none have six, eight, or many legs? Why is the situation precisely reversed for land arthropods? In general, why are certain areas of multicellular morphospace densely populated with many representative species, while other areas, apparently characterizing viable designs, are unoccupied by any extant or extinct animals?" Then, although I would label their distinctions as overly dichotomized and too mutually exclusive (for I seek a fusion of structural and functional influences), Arthur and Farrow also pose the alternatives (1999, p. 183) in much the same manner followed here:

> There are two very different answers to these questions, representing two opposing schools of thought on the relative importance of natural selection and developmental constraint in determining the actual distribution of morphologies that we observe . . . One is the "pan-selectionist" view that variation is potentially available in all directions from any given phyletic starting-point, and that selection determines which subset

of variants prevails. The alternative is the "developmental constraint" view that many of the gaps we observe between different morphologies do not arise from the non-adaptiveness of the absent forms but rather from the difficulty of making them through an ontogenetic process.

I began this "symphony" of evo-devo with a quotation from one of the great architects of the Modern Synthesis—Mayr's statement, based on adaptationist premises then both reasonable and conventional, that any search for genetic homology between distantly-related animal phyla would be doomed *a priori* and in theory by selection's controlling power, a mechanism that would surely recycle every nucleotide position (often several times) during so long a period of independent evolution between two lines. The new data of evo-devo have falsified this claim and revised our basic theory to admit a great, and often controlling, power for historical constraints based on conserved developmental patterns coded by the very genetic homologies that Mayr had deemed impossible.

For the sake of both symmetry and logic, it seems fitting to end this section by recalling another quotation by another great architect of the Synthesis, based on the same panadaptationist assumptions about natural selection's controlling power—and also falsified, since then, by new information on historical constraints, impelling renewed respect for formalist themes in revising and expanding our theories of evolutionary mechanisms. But Dobzhansky's closing statement (1951) differs from Mayr's opener (1963) in one crucial way: Mayr's denial of genetic homology represented a sensible consensus for his time; whereas Dobzhansky's assertion of purely adaptational mapping upon ecological places to explain the clumpy population of morphospace made little sense, even at the height of enthusiasm for natural selection's exclusive power—and I can only conclude (as discussed more fully in Chapter 7, pp. 526–528) that Dobzhansky, in his enthusiasm for strict Darwinian theory, had temporarily undervalued a cardinal fact of natural history that his initial training as a systematist had certainly infused into the marrow of his understanding.

In a brilliant opening move, Dobzhansky began the third (1951) edition of his founding document for the Synthesis, *Genetics and the Origin of Species*, by recognizing the diversity of modern organisms, and the striking discontinuities within this plethora of form, as the central problem of evolutionary biology—at a time when most colleagues would surely have cited modes of continuous transformation, or mechanisms for changes in gene frequencies, within single populations instead. (Despite this unconventionality in subject and level of focus, Dobzhansky opted for a traditional selectionist explanation by titling the first subsection of his book: "diversity and adaptedness.")

As a wise and wonderful human being, and as a humanist at heart, Dobzhansky began his book with a generous perspective on the meaning and importance of organic diversity. The opening paragraph (1951, p. 3) reads: "Man has always been fascinated by the great diversity of organisms which live in the world around him. Many attempts have been made to understand the meaning of this diversity and the causes that bring it about. To many minds this problem possesses an irresistible aesthetic appeal. Inasmuch as scientific inquiry is a form of aesthetic endeavor, biology owes its existence in part to this appeal."

After stating the key issue, Dobzhansky then cites the discontinuities within this diversity as the crucial phenomenon demanding explanation. But he begins this second subsection, entitled "discontinuity," by unconsciously showing his Darwinian commitments in citing organisms as the "prime reality" of biology (whereas, in a hierarchical reformulation of Darwinian theory, several evolutionary levels feature other biological individuals just as interesting, and just as well constituted—with Dobzhansky's beloved species, the quanta of his concern for diversity, as a primary example of individuality at a higher level). Dobzhansky writes (1951, p. 4): "Although individuals [*i.e.*, organisms] limited in existence to only a short interval of time, are the prime reality with which a biologist is confronted, a more intimate acquaintance with the living world discloses a fact almost as striking as the diversity itself. This is the discontinuity of the variation among organisms."

Dobzhansky then commits his conceptual error in proposing a purely selectionist explanation—externalist at an extreme in its appeal to environmental topography as the sole mapping function for discontinuities in organic diversity—for the crucial fact of clumpiness in the habitation of morphospace. I discussed this passage extensively in Chapter 7 (pp. 526–528), and will only present a summary here. Dobzhansky begins by changing the level of application for Sewall Wright's nonadaptationist model, originally devised to explain why the varied demes of single species may reside upon several discontinuous peaks of an adaptive genetic landscape. By promoting this model to the species level (see Fig. 10-31), and regarding the inhabitants of each peak as a species instead of a deme (and then reconfiguring the peaks as adaptive optima in an ecological terrain, rather than sets of workable genetic combinations among demes, with only the highest peak representing an optimum position for the species), Dobzhansky converted the theoretical meaning of Wright's model from an explanation for why so many demes have *not* obtained a best possible configuration into a paean for the adaptive optimality of each element in a fauna.

In describing this promoted adaptive landscape, as presented in Figure 10-31, Dobzhansky commits his panadaptationist fallacy by attempting to render the inhomogeneous occupation of morphospace as a simple one-to-one "mapping" of discontinuity upon the external "terrain" that set the selective pressures responsible for crafting all aspects of organic diversity.*

> The enormous diversity of organisms may be envisaged as correlated with the immense variety of environments and of ecological niches which exist on earth. But the variety of ecological niches is not only im-

---

*I apologize for this second citation of a long quote in an overly ample book—see p. 527; but its uncanny appropriateness in these two different contexts leads me to beg your indulgence for this redundancy.

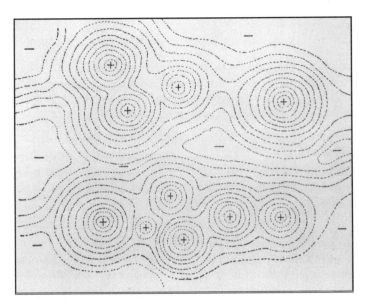

10-31. Sewall Wright's model of the adaptive landscape "promoted" by Dobzhansky to adaptive peaks for optimal residence of species in an environmental landscape. From the third edition of Dobzhansky's *Genetics and the Origin of Species.*

mense, it is also discontinuous. One species of insect may feed on, for example, oak leaves, and another species on pine needles; an insect that would require food intermediate between oak and pine would probably starve to death. Hence, the living world is not a formless mass of randomly combining genes and traits, but a great array of families of related gene combinations, which are clustered on a large but finite number of adaptive peaks. Each living species may be thought of as occupying one of the available peaks in the field of gene combinations. The adaptive valleys are deserted and empty.

Furthermore, the adaptive peaks and valleys are not interspersed at random. "Adjacent" adaptive peaks are arranged in groups, which may be likened to mountain ranges in which the separate pinnacles are divided by relatively shallow notches. Thus, the ecological niche occupied by the species "lion" is relatively much closer to those occupied by tiger, puma, and leopard than to those occupied by wolf, coyote, and jackal. The feline adaptive peaks form a group different from the group of the canine "peaks." But the feline, canine, ursine, musteline, and certain other groups of peaks form together the adaptive "range" of carnivores, which is separated by deep adaptive valleys from the "ranges" of rodents, bats, ungulates, primates, and others. In turn, these "ranges" are again members of the adaptive system of mammals, which are ecologically and biologically segregated, as a group, from the adaptive systems of birds, reptiles, etc. The hierarchic nature of the biological classification reflects the objectively ascertainable discontinuity of adaptive niches, in other words the discontinuity of ways and means by which organisms that inhabit the world derive their livelihood from the environment.

But the striking discontinuities in morphospace, and their ordering into taxonomic hierarchies, surely don't, at least primarily, "reflect the objectively ascertainable discontinuity of adaptive niches"—and Dobzhansky certainly understood the unstated major reason for such inhomogeneity, even though the strict adaptationism so favored at this time had momentarily clouded his excellent judgment, thus explaining his curious omission. Cats, lions and tigers work admirably well, with each species displaying excellent adaptation to its immediate environment. But the set of all feline species does not clump closely together in morphospace because the summits of an underlying topography now happen to lie at such close mutual proximity in an external world "out there." Felines form a tight cluster because they share, by historical constraints of ordinary genealogy, a large set of distinctive traits, unique to them alone by virtue of "propinquity of descent," to cite Darwin's own description of the phenomenon—although each of these traits probably arose for good and conventional adaptationist reasons in a common ancestor. And the larger gap between felines and canines also records, as its primary *raison d'être,* a greater separation in *history,* not the architecture of spacing between two groups of peaks in the *current* mountain range of worldly ecology.

Sometimes we need the press of new data to inspire the recollection of old truths. I do not doubt that many discontinuities in morphospace represent the colonization by optimal phenotypes of widely dispersed peaks in maximal biomechanical efficiency. But I am equally confident that more of nature's evidently nonrandom, and oddly dispersed, clusters in morphospace, bearing such enormously different weights ranging from single "outliers" to millions of species, primarily record the historical constraints imposed by workable solutions with adaptive origins—developmental designs that then congealed, enforcing reiteration and change within their internally directed channels forever after. Five, for all I know, may be optimal for the radial symmetry of echinoderms, and therefore predictable for any phylum in their domain. But can we argue that the sixfold way of a much, much larger clump marks an optimal and inevitable number for walking, and that elytra represent the only possible design for joint excellence in flight and protection? God, as one of our most celebrated colleagues famously exclaimed, must have an inordinate fondness for these particular creatures if he allowed one design among so many conceivable alternatives to congeal so hard, and then to iterate so often, in nature's wondrous interplay of constraint and adaptation.

# The Integration of Constraint and Adaptation (Structure and Function) in Ontogeny and Phylogeny: Structural Constraints, Spandrels, and the Centrality of Exaptation in Macroevolution

## The Timeless Physics of Evolved Function

### STRUCTURALISM'S ODD MAN OUTSIDE

In a famous passage from the Introduction to the *Origin of Species*, Darwin identified the intricately adaptive character of most anatomical features as the primary phenomenon that any theory of evolutionary mechanisms must explain. Many other sources of information, he states, can easily prove evolution's factuality, but we will not understand the causes of change until we can explain "how the innumerable species inhabiting this world have been modified so as to acquire that perfection of structure and coadaptation which most justly excites our admiration" (1859, p. 3).

Darwin felt that this striking and pervasive functionality of organic design required an explicit functional theory of evolutionary causes rooted in the proposition that adaptive structures originate "for" their utility. As functionalist theories, both Lamarckian soft inheritance and Darwinian natural selection share a defining premise that environmental information about adaptive design somehow passes to organisms, and that organisms then respond by fashioning traits to enhance their competitive ability within these environments. (Above all, functionalist theories require explicit interaction of organism and environment in the service of improving local adaptation. The pure imposition of one side upon the yielding properties of the other side does not qualify.)

The strikingly different mechanisms of the two major functionalist theories—organic response to felt needs for Lamarck, natural selection upon isotropic variation for Darwin—should not obscure their agreement on the key functional principle that adaptation drives evolution as organisms change to

*1179*

secure better fit to their environments. We prefer Darwin and reject Lamarck because nature's mechanisms of heredity and variation validate the efficacy of natural selection and disprove the existence of soft inheritance, not because we can specify any basic difference in their shared commitment to a functionalist account of evolutionary mechanics.

Structuralist or formalist theories, on the other hand, generally seek to explain the origin of adaptive design in terms of such internal forces as constraint and directed variability. In the strictest versions of these theories, external causes can only act as editors to distill the most workable phenotypes from the full range of potential shapes that structural rules engender. Function may therefore determine what lives and what dies, but not what can (and does) originate.

Chapters 4 and 5 discussed the two major structuralist theories that we now reject for their operation as strict alternatives and denials of Darwinian functionalism: (1) orthogenesis, with its central claim that evolutionary trends follow internal drives in variation, and that selection can only accelerate or retard these inherent and inevitable pathways; and (2) saltation, with its premise that occasional fortuitous discontinuities in variation create new species all at once, and that selection can only intensify the process by preserving lucky sports and eliminating old, superseded designs.

Chapters 4 and 5, and much of Chapter 10 as well, also discuss more acceptable forms of structuralism that do not attempt to replace natural selection, but rather work in concert with known Darwinian mechanisms to channel possible directions of evolutionary change "from the inside" along pathways of variation that record constraints of history or principles of physical construction. I shall return to this theme of internal constraints in the last two sections of this chapter, where I discuss the important structural principle of non-adaptive origin followed by cooptation for a descendant's utility.

This initial section of Chapter 11, however, represents an interlude between historical constraints (Chapter 10) and structural constraints based on mechanically forced (or, at the opposite end of this spectrum, simply inherited) correlations with actively selected features (Sections II and III of this chapter). This interlude also discusses a form of structural constraint—but of a markedly different nature: direct molding by physical laws and forces acting upon the developing organism. This "maverick" theme has played only a small role in the history of evolutionary thought (a fact that should elicit no judgment about actual importance, for we all recognize that today's ignored or ridiculed theme can become the centerpiece of tomorrow's revolutionary theory).

If adaptive phenotypes originate directly and immediately from the imposition of physical forces upon the "yielding putty" (if you will) of organic material, then we need no functionalist account of "that perfection of structure and coadaptation"—for good form emerges automatically from the nature of physical reality (by external forces imposed upon the organism, or internal forces exerted from within as the organism grows). We get, in Kauffman's memorable phrase (1993) "order for free," and need not posit any explicit organismal mechanisms, as functionalist theories propose (Darwinian selection or Lamarckian inheritance), for "reading" the selective requirements of local environments and responding by evolutionary change.

Put another way, this structuralist theory of direct imposition asks nothing of organic matter beyond its malleability for passive shaping by physical forces. (One might argue that the malleability itself arose by a functionalist mechanism like natural selection, but the discussion of this section focuses upon current modes of change, not the origin of preconditions that make such changes possible.)

This theory of adaptive design by direct imposition differs from most formalist accounts of evolutionary change in two major ways (while agreeing on the central premise—the basis for my taxonomy of theories in the first place—that structural rules, rather than functional responses, generate organismal phenotypes):

1. Most structuralist theories identify the sources of adaptive order as residing largely "inside" the organism in the form of constraining genetic and developmental homologies, or the allometric and consequential rules that Darwin called "correlations of growth." (For this reason, I have used "formalist," "structuralist," and "internalist" as virtually synonymous terms throughout this book.) But the structuralist theory of direct imposition locates the causes of adaptive order in physical laws of nature lying "outside" (and prior to) the specific architectural blueprints of each particular *Bauplan* (even though these physical laws may impose their shaping powers "from the inside" during growth).

2. In an even more iconoclastic claim (discussed previously on pp. 1053–1055 as the defining peculiarity of this way of thought), proponents of adaptive design by direct imposition tend to ignore, and often to devalue quite explicitly, the role of phylogeny, or any kind of historical analysis, in setting the *Baupläne* or developmental rules that channel and constrain patterns of evolution in any particular group. If physical forces shape organisms directly, then their prior histories don't matter, and we need only consider the immediate impress of current circumstances upon malleable organic materials. After all, we don't invoke any aspect of history or genealogical connection to explain why Cambrian quartz from Asia exhibits the same crystal structure as Recent quartz from America. So why should we not attribute the logarithmic spirals of Paleozoic and modern gastropods to the same spatiotemporal invariance of physical laws?

One might say, in epitome, that the first argument opposes this theory to all other, and more conventional, forms of structuralist thought; whereas the second statement, far more radical in scope, opposes this theory to the central concept of evolutionary biology itself (in both structuralist and functionalist accounts): the role of history, and the importance of phylogeny in understanding both present forms and future prospects.

I doubt that this theory of adaptive design by direct physical imposition could ever stand as a complete, or even a dominant, explanation of evolution. (We shall see that even the most celebrated exponent of this view, D'Arcy Wentworth Thompson in *On Growth and Form*, ultimately ceded the major turf of explanation, at least for complex organisms, to phylogeny and heredity rather than to immediate physical imposition.) But I suspect that direct physical assembly or imposition may supply an important, perhaps even a controlling, theme in two major and rather different arenas of evolutionary theory: (1) The origin of life and the initial assembly of basic and universal components of cellular organization and genetic structure. Do we not sense that much of life's initial history falls into the domain of "universal chemistry" and the general physics of self-organizing systems, whereas the actual, divergent pathways of metazoan phyla then fall under the control of historical contingency? (2) Broad predictabilities of life's pattern through time, transcending the contingent particularities of any individual lineage. The structure of ecological pyramids must display some physical predictability, whereas the occupation of the top carnivore apex by lions or tigers or bears (or phorusrhachids or borhyaenids) demands knowledge of historical particulars. The increasingly right skewed distribution of life's complexity, with stability of the bacterial mode throughout (Gould, 1996a), speaks more to the general physics of reflecting boundaries (of minimal complexity in this case), the physically necessary origin of life at this minimal complexity, and the stochasticity of random walks, than to any historical detail of uniquely earthly existence.

I am a historian at heart, and although the theme of immediate physical assembly intrigues me—and no one with literary pretensions could remain unmoved by the coincidence that D'Arcy Thompson's *Growth and Form,* the most stylish book in the history of anglophonic biology, also happens to be the "Bible" of this particular view of life—I don't think that the hypothesis of direct physical construction will play a large part in the expansion of Darwinian theory advocated as my central argument in this book. I therefore view this section as a "place holder" in the logic of my case, and not at all as a complete or even adequate account of an important subject. I will occupy this particular place in an idiosyncratic manner by analyzing D'Arcy Thompson's great work (1917, 1942) and then discussing, much more briefly, the most important modern expressions of this view of life in the works of Goodwin (1994) and Kauffman (1993). But method does lie in the sanity of this choice, for one could not ask for a better vehicle than D'Arcy Thompson's brilliant argument and stunning prose. His magisterial (if idiosyncratic and, at times, even cranky) book embodies an entire world view within its ample scope. His specific examples may be wrong or dated, but no one has ever presented a more complete and coherent version of this approach to the explanation of evolution, including explicit discussion of all major implications for general theory. In this sense, an exegesis of D'Arcy Thompson may well represent the most modern and relevant way to discuss this important corner of evolutionary thought.

### D'ARCY THOMPSON'S SCIENCE OF FORM

#### The structure of an argument

In 1945, the Public Orator of Oxford lauded D'Arcy Thompson as *unicum disciplinae liberalioris exemplar* (the outstanding example of a man of liberal education—at the ceremony for his receipt of an honorary degree as Doctor in Civil Law). In 1969, the Whole Earth Catalog, the commercial bible of the "green" movement in America, called his major work "a paradigm classic." Few people can list such diverse distinctions in their compendium of honors. But then, few people have displayed so wide a range of talent. D'Arcy Wentworth Thompson (1860–1948), Professor of Natural History at the Scottish universities of Dundee and St. Andrews, translated Aristotle's *Historia animalium,* wrote glossaries of Greek birds and fishes, compiled statistics for the Fishery Board of Scotland and contributed the article on pycnogonids to the Cambridge Natural History.

But D'Arcy Thompson's current reputation rests almost entirely upon a book of a thousand pages, revered by artists and architects as well as by engineers and biologists—the "paradigm classic," *On Growth and Form* (1917, 2nd edition, 1942). P. B. Medawar (1967, p. 232) lauded this volume as "beyond comparison the finest work of literature in all the annals of science that have been recorded in the English tongue." G. Evelyn Hutchinson (1948, p. 579) regarded *Growth and Form* as "one of the very few books on a scientific matter written in this century which will, one may be confident, last as long as our too fragile culture."

Although I have studied D'Arcy Thompson's wonderful book throughout my career (see Gould, 1971b, for my first, and in retrospect embarrassingly puerile, publication in a journal of the humanities), I originally made a major error in siting him within the history of biology. All intellectuals love a courageous loner, and I had been beguiled by D'Arcy Thompson's seemingly anachronistic peculiarities—his flowery, sometimes overblown, but often soaring and powerful, Victorian prose; his expertise at fully professional levels in Latin and Greek; even his lifelong residence in an outlying region that, in my false mental geography, might well have been located above the Arctic Circle. In retrospect, I had unthinkingly conflated my sense of his intellectual distance from conventional thought with an assumption about physical isolation as well. When I finally visited the University of St. Andrews (for the humbling experience of receiving an honorary degree in D'Arcy Thompson's own bailiwick), I recognized its proximity to Edinburgh, and its easy access by rail. (As a further confirmation of St. Andrews's central location within the contemporary world, I began to write this section on the very day that Tiger Woods won the British Open on the world's original, and still most famous, golf course of St. Andrews.)

I had therefore viewed D'Arcy Thompson as the ultimate man out of time—a Greek geometer and classical scholar, a Victorian prose stylist at the dawn of modernism's lean and cynical attitude (for the first edition of *Growth and Form* appeared in 1917 in the midst of Word War I, while another and even more destructive war greeted the second edition of 1942; many historians have noted that, in a meaningfully ideological, rather than an arbitrarily calendrical, reckoning, the 20th century really begins with World War I and the end of illusions about progress and the benevolent hegemony of European control).

I shall not try to rob D'Arcy Thompson of his genuine singularities, but when we place his biological views into the context of evolutionary debate in his own time, we find, underneath his quirky stylistic uniqueness, a standard critique of Darwinism based upon the common argument that natural selection cannot fashion novel features, but can only eliminate the unfit, bolstered by a claim for saltation as a common mode in the origin of highly distinct taxa and anatomical groundplans. D'Arcy Thompson did link this standard critique to an uncommon solution—his central claim that physical forces shape adaptive form directly—but we should regard his theory as an unusual solution to the standard conundrums of his time, and not as an anachronistic importation from Pythagorean Greece, clothed in the prose of Dickens or Thackeray.

*On Growth and Form* is a weighty tome (793 pages in the original edition of 1917, enlarged to 1116 pages in 1942), but D'Arcy Thompson presents his central thesis as a tight argument, expressed in clear logical order, with proper attention to inherent difficulties—and, above all, artfully developed throughout. (My analysis here, with two labeled exceptions, follows the first edition.)

In a common conceit (in the non-pejorative sense of a fanciful device), scientists often clothe a truly radical idea in the falsely modest garb of merely useful technicality. Thus, D'Arcy Thompson asserts that he wrote *Growth and Form* only to make biologists a bit more comfortable with the mathematical description of morphology. He states in his epilog (1917, p. 778):

> The fact that I set little store by certain postulates (often deemed to be fundamental) of our present-day biology the reader will have discovered and I have not endeavored to conceal. But it is not for the sake of polemical argument that I have written, and the doctrines which I do not subscribe to I have only spoken of by the way. My task is finished if I have been able to show that a certain mathematical aspect of morphology, to which as yet the morphologist gives little heed, is interwoven with his problems, complementary to his descriptive task, and helpful, nay essential, to his proper study and comprehension of Form. *Hic artem remumque repono.* *

Beginning his assault upon biological traditions of explanation, D'Arcy Thompson reminds us that we feel no discomfort in ascribing the elegant and well-fitting forms of inorganic objects to physical forces that can mold them directly, and that also embody the advantage (for our comprehension) of simple mathematical description. Why, then, when organic forms display equally elegant and simple geometries, and when these biological shapes also match the expected impress of physical forces, do we shy from invoking the same explanation of direct production that we apply without hesitation to identical forms in nonorganic nature? (1917, pp. 7–8):

---

*D'Arcy Thompson couldn't resist a frequent Latin (or Greek, or Italian, or French, or German, or whatever) quotation, always untranslated. This little paragraph-ending phrase continues his theme of false modesty by proclaiming: At this point I close my composition and put back my oar.

The physicist proclaims aloud that the physical phenomena which meet us by the way have their manifestations of form, not less beautiful and scarce less varied than those which move us to admiration among living things. The waves of the sea, the little ripples on the shore, the sweeping curve of the sandy bay between its headlands, the outline of the hills, the shape of the clouds, all these are so many riddles of form, so many problems of morphology, and all of them the physicist can more or less easily read and adequately solve: solving them by reference to their antecedent phenomena, in the material system of mechanical forces to which they belong, and to which we interpret them as being due . . .

Nor is it otherwise with the material forms of living things. Cell and tissues, shell and bone, leaf and flower, are so many portions of matter, and it is in obedience to the laws of physics that their particles have been moved, molded and conformed. . . . Their problems of form are in the first instance mathematical problems, and their problems of growth are essentially physical problems, and the morphologist is, *ipso facto,* a student of physical science.

Our reluctance, D'Arcy Thompson claims, arises largely from conventional beliefs about the "special" character of life, based on a traditional assumption that organic shapes embody purposes and therefore demand teleological explanation, whereas inorganic forms exert no action of their own, and can only be explained as passive records of physical forces. We assert organic uniqueness by invoking both an active and passive argument. The passive argument sets living things apart, without specifying any uniquely biological causes or processes (p. 2):

The reasons for this difference lie deep, and in part are rooted in old traditions. The zoologist has scarce begun to dream of defining, in mathematical language, even the simpler organic forms. When he finds a simple geometrical construction, for instance in the honey-comb, he would fain refer it to psychical instinct or design rather than to the operation of physical forces; when he sees in snail, or nautilus, or tiny foraminiferal or radiolarian shell, a close approach to the perfect sphere or spiral, he is prone, of old habit, to believe that it is after all something more than a spiral or a sphere, and that in this "something more" there lies what neither physics nor mathematics can explain. In short he is deeply reluctant to compare the living with the dead, or to explain by geometry or by dynamics the things which have their part in the mystery of life.

Biologists then advance the active argument to posit a set of distinctively organic causes that, in their outcomes, mimic the same forms that physical forces, left to their own devices, would impose upon any plastic material. At this point, D'Arcy Thompson introduces his critique of Darwinism and of functionalist evolutionary thought in general. In the paragraph following the last citation, D'Arcy Thompson identifies the two main culprits in our erroneous convictions about special biological forces behind good organic design: phyletic solutions (or any kind of historical explanation), and adaptationist speculation leading to false assumptions about the need for such functionalist mechanisms as natural selection (pp. 2–3):

> He [the morphologist] has the help of many fascinating theories within the bounds of his own science, which, though a little lacking in precision, serve the purpose of ordering his thoughts and of suggesting new objects of enquiry. His art of classification becomes a ceaseless and an endless search after the blood-relationships of things living, and the pedigrees of things dead and gone. The facts of embryology become . . . a record not only of the life-history of the individual but of the annals of its race. . . . Every nesting bird, every ant-hill or spider's web displays its psychological problems of instinct or intelligence. Above all, in things both great and small, the naturalist is rightfully impressed, and finally engrossed, by the peculiar beauty which is manifested in apparent fitness or "adaptation,"—the flower for the bee, the berry for the bird.

For all its Victorian amplitude, D'Arcy Thompson usually imposes adequate restraint upon his literary talents. But he does occasionally soar "over the top," and nothing incites this tendency more than his aversion to Darwinian speculation in the adaptationist mode—as in the following example (pp. 671–672):

> Some dangerous and malignant animals are said (in sober earnest) to wear a perpetual war-paint. The wasp and the hornet, in gallant black and gold, are terrible as an army with banners; and the Gila Monster (the poison-lizard of the Arizona desert) is splashed with scarlet—its dread and black complexion stained with heraldry more dismal. But the wasp-like livery of the noisy, idle hover-flies and drone-flies is but stage armour, and in their tinsel suits the little counterfeit cowardly knaves mimic the fighting crew.
>
> The jewelled splendour of the peacock and the humming-bird, and the less effulgent glory of the lyre-bird and the Argus pheasant, are ascribed to the unquestioned prevalence of vanity in the one sex and wantonness in the other.
>
> The zebra is striped that it may graze unnoticed on the plain, the tiger that it may lurk undiscovered in the jungle; the banded Chaetodont and Pomacentrid fishes are further bedizened to the hues of the coral-reefs in which they dwell. The tawny lion is yellow as the desert sand; but the leopard wears its dappled hide to blend, as it crouches on the branch, with the sun-flecks peeping through the leaves. . . .
>
> To buttress the action of natural selection the same instances of "adaptation" (and many more) are used, which in an earlier but not distant age testified to the wisdom of the creator and revealed to simple piety the high purpose of God.

At this turning point in his argument, D'Arcy Thompson calls upon his expertise in classics to invoke Aristotle's exegesis of causality in his favor. We may acknowledge that biological forms embody "purposes" expressed as adaptive utility in the Darwinian struggle for life ("final causes" in Aristotle's terminology, with "final" referring to utilitarian, not temporal, ends). The Darwinian functionalist, D'Arcy Thompson then claims, makes his key error in assuming that the identification of utility (final cause) automatically specifies the process by which such utility originates—a false inference from purpose to mechanism. But, as Aristotle pointed out, a full explanation for natural objects and phenomena requires the identification of several distinct kinds of causes. In particular, the *final cause* (utility) of an object does not specify the *efficient cause,* or mechanism, that actually (and actively) constructed the object ("efficient," that is, in the technical sense of making or "effecting," rather than the more restricted vernacular sense of doing something well).

When we identify "not sinking into the mud" as the adaptive value (final cause) of webbing on the feet of shore birds, we have not proven thereby that the efficient cause of webbing must be functionalist in nature, and explicitly tied to the purpose (final cause) now served by this feature. After all, webbing might have arisen by any one of numerous, and entirely plausible, nonfunctionalist mechanisms (or by functionalist mechanisms unrelated to current utility for standing on mud)—and then been happily and fortuitously available for cooptation to its current purpose. D'Arcy Thompson preferred efficient causes of direct physical imposition (an improbable alternative in this particular case), but his general point cannot be gainsaid. The correct description of a final cause does not, by itself, identify the mechanism by which this utility originated (p. 5):

> The use of the teleological principle is but one way, not the whole or the only way, by which we may seek to learn how things came to be, and to take their places in the harmonious complexity of the world. To seek not for ends but for "antecedents" is the way of the physicist, who finds "causes" in what he has learned to recognize as fundamental properties, or inseparable concomitants, or unchanging laws, of matter and of energy. In Aristotle's parable, the house is there that men may live in it; but it is also there because the builders have laid one stone upon another: and it is as a *mechanism,* or a mechanical construction, that the physicist looks upon the world. Like warp and woof, mechanism and teleology are interwoven together, and we must not cleave to the one and despise the other.

Moving from Aristotle to his own nation's greatest philosopher at the dawn of modern science, D'Arcy Thompson then cites Bacon's famous disparagement of final causes (as vestal virgins with empty downturned cups) falsely cited to explain mechanisms of production (pp. 5–6)—and he blames a knee-jerk style of Darwinian adaptationism for this common conflation in evolutionary science (that is, for erroneous assumptions that functional utilities automatically identify structural or mechanical origins by natural selection):

Nevertheless, when philosophy bids us hearken and obey the lessons both of mechanical and of teleological interpretation, the precept is hard to follow: so that oftentimes it has come to pass, just as in Bacon's day, that a leaning to the side of the final cause "hath intercepted the severe and diligent inquiry of all real and physical causes," and has brought it about that "the search of the physical cause hath been neglected and passed in silence." So long and so far as "fortuitous variation" and the "survival of the fittest" remain engrained as fundamental and satisfactory hypotheses in the philosophy of biology, so long will these "satisfactory and specious causes" tend to stay "severe and diligent inquiry," "to the great arrest and prejudice of future discovery."

D'Arcy Thompson's citation of Bacon's famous critique does not imply any personal distaste for the subject of excellent adaptation or final causation in general. Quite to the contrary, D'Arcy Thompson's focus on geometric beauty and mechanical optimality led him to emphasize the loveliest and most stunningly efficient of organic designs. Thus, his complaint did not lie with the existence of adaptation, but with the too-facile Darwinian assumption that such final causes imply a mode of construction explicitly powered by the value of the developing adaptation itself—in other words, a functionalist mechanism like natural selection. The nub of D'Arcy Thompson's system, and his reason for emphasizing the different statuses of efficient and final causation, resides in his conviction that efficient causes of physical construction craft final causes as automatic consequences—thus obviating the need for a special category of mechanisms (again like natural selection) to explain biological adaptation. D'Arcy Thompson expresses his admiration and feeling for final causes in one of his loveliest prose flourishes (p. 3):

Time out of mind, it has been by way of the "final cause," by the teleological concept of "end," of "purpose," or of "design," in one or another of its many forms (for its moods are many), that men have been chiefly wont to explain the phenomena of the living world; and it will be so while men have eyes to see and ears to hear withal. With Galen, as with Aristotle, it was the physician's way; with John Ray, as with Aristotle, it was the naturalist's way; with Kant as with Aristotle, it was the philosopher's way. It was the old Hebrew way, and has its splendid setting in the story that God made "every plant of the field before it was in the earth, and every herb of the field before it grew." It is a common way, and a great way; for it brings with it a glimpse of a great vision, and it lies deep as the love of nature in the hearts of men.

As the last step in his general argument, D'Arcy Thompson then asks us to consider how far the simplest and most direct style of efficient causation might carry us in explaining adaptive organic form. Perhaps many features owe their geometric optimality—leading to maximization of utility, or final cause, as well—to the simplest mechanism of direct shaping by the physical forces most relevant to the behaviors of the organism in its daily struggles for life. By invoking an analogy to the limits of science in aesthetic and moral arguments, D'Arcy Thompson allows that his favored theme of physical imposition may not carry us as far as we (or at least he) would like to go. But he makes a strong argument for this kind of minimalism (final causes generated by physical imposition, thus obviating the need for special mechanisms to secure adaptation) as an appropriate first approach (pp. 8–9):

> How far, even then, mathematics will suffice to describe, and physics to explain, the fabric of the body no man can foresee. It may be that all the laws of energy, and all the properties of matter, and all the chemistry of all the colloids are as powerless to explain the body as they are impotent to comprehend the soul. For my part, I think it is not so. Of how it is that the soul informs the body, physical science teaches me nothing: consciousness is not explained to my comprehension by all the nerve-paths and "neurons" of the physiologist; nor do I ask of physics how goodness shines in one man's face, and evil betrays itself in another. But of the construction and growth and working of the body, as of all that is of the earth earthy, physical science is, in my humble opinion, our only teacher and guide.

### The tactic and application of an argument

D'Arcy Thompson followed a definite strategy in attempting to carve out the largest possible empirical role for his "minimalist" structural theory on the genesis of good design and adaptive form in organisms. He would begin with his best "shot"—the outward shapes of simple unicellular organisms—and then sally forth from this plausible beginning. Again, he initiates the search in his overtly modest mode (p. 10): "My sole purpose is to correlate with mathematical statement and physical law certain of the simpler outward phenomena of organic growth and structure or form: while all the while regarding, *ex hypothesi,* for the purposes of this correlation, the fabric of the organism as a material and mechanical configuration."

The empirical chapters of *Growth and Form* embody this plan by first elucidating a most promising principle (surface/volume ratios), applying it to a best potential case (protistan form), and then moving from this position of initial strength into ever less likely realms of application, always trying to capture the largest possible domain for explaining final causes (adaptive forms) as automatic consequences of the direct action of physical forces (efficient causes) upon yielding organic material.

After a short introductory statement, presenting the basic argument as summarized in my preceding pages, D'Arcy Thompson composes two lengthy chapters to set a context for the empirical cases to follow. The first, entitled "on magnitude" and devoted to an elegant explication, still read in many undergraduate courses, of Galileo's principle of necessarily declining surface/volume ratios as geometrically similar objects increase in size, holds a central place in the logic of D'Arcy Thompson's general theory. (Ironically, many of the most fervent admirers of this chapter, especially those who encounter it out of context in a book of course readings, have no inkling of its anchoring purpose in a much broader theory that they would, no doubt, heartily reject.) If physical forces shape organisms directly, then our best test resides in the preeminence of the S/V principle, and the linear scaling of this ratio with increasing organismal size. Tiny animals must dwell in a world dominated by forces acting upon their surfaces, while large animals will be ruled by gravitational forces operating upon volumes. We can therefore test the efficacy of physical forces by noting whether organisms show the "right" conformations for direct molding by the appropriate relative strengths of these forces at their size.

The following chapter, entitled "the rate of growth," then develops the dynamic argument that physical forces will be exerted upon vectors of growth during an organism's ontogeny, not merely upon a realized final form. The subsequent 15 chapters then follow a sequence, beginning with single cells, where growth plays a minimal role and forms may be construed as simple responses to a small number of constraining conditions and imposing forces, as in D'Arcy Thompson's most famous comparison (Figure 11-1) of protistan cells to Plateau's surfaces of revolution—a set of shapes exhibiting minimal areas in designs that are radially symmetrical about a single axis.

D'Arcy Thompson then moves on to simple aggregations of cells or units, but proceeding no "further" (up the traditional chain of complexity) than fairly uniform tissues of a single organ, minimally differentiated metazoans like sponges, and colonial organisms made of similar units crowded together. He presents a wide taxonomic range of putative cases for direct mechanical construction, but with strong emphasis upon the most plausible circumstance of geometric forms automatically engendered by closest packing of malleable units of the same basic size and composition (the "soap-bubble" paradigm, if you will)—including an ingenious analysis of sponge and holothurian spicules as mineralized maps of the junctions between units, and not as

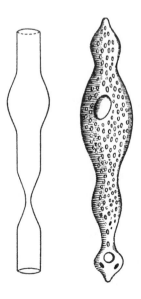

11-1. D'Arcy Thompson's famous comparison of protistan cells to Plateau's surfaces of revolution—shapes of minimal area with radial symmetry around a single axis. From D'Arcy Thompson, 1917.

phyletically unique and distinctive forms (Figure 11-2); an explanation of honeycomb cells as optimal balances of strength and holding capacity; and a convincing claim that the hexagonal closest packing of corallites in Paleozoic colonial rugosans cannot be read as distinctive phyletic adaptations of particular lineages, because corallites assume circular cross sections when they grow in less confined spaces with no contact between individuals (Figure 11-3).

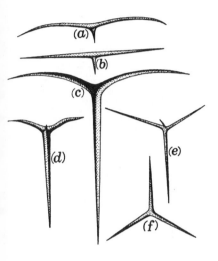

11-2. Sponge spicules shown as "maps" of junctions among cells in "soap bubble" models. From D'Arcy Thompson, 1942.

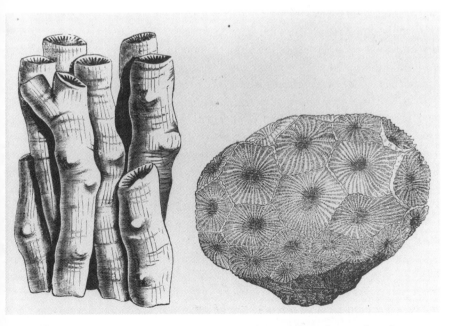

11-3. Ecophenotypic shaping of corallites by physical crowding rather than genetic coding. Cross sections are circular when the corallites do not touch, and hexagonal when they grow in a tightly packed configuration. From D'Arcy Thompson, 1917.

In a third category, D'Arcy Thompson then considers geometrically regular growth patterns of more complex creatures in cases where an observed order might record the operation of a simple building principle plausibly regulated by direct mechanical production—as in his most famous chapter on the logarithmic spiral (mainly in molluscan shells, but also for unicellular forams and ruminant horns) as the paradigm curve that increases in size without changing its shape; and his largely derivative discussion of phyllotaxis, with obedience to the Fibonacci series explained not as a Pythagorean mystery, but as an automatic consequence of initiating each new spiral in a radiating series by setting its founding element into the largest available space at the generating center.

But when, in a final set of cases, D'Arcy Thompson must discuss the complex features of "higher" metazoan phyla that cannot be reduced to consequences of single principles in growth—in other words, the difficult problems of morphology that have always been regarded as paramount to the enterprise—he makes much less headway, and largely confines his attention to "peripheral" questions, including the ordering of differences among forms as expressions of relatively simple transformation gradients (but leaving the core form as an unexplained "primitive term" or "given" in the analysis), and the correlation of obviously ecophenotypic or epigenetic modifications (the healing of broken bones, for example) with forces acting upon the object during this secondary modification. (We shall see (pp. 1196–1200) how his inability to treat the shared properties of complex taxonomic *Baupläne* as more than unexplained inputs into his theory of transformation scuttles any hope that his system might enjoy controlling, or even general, application as a theory of biological form.)

To illustrate how D'Arcy Thompson applies his central argument across this empirical range, we should consider his own favored principle of surface/volume ratios as an exemplar because the fundamental property of size itself establishes a basic prediction for testing the efficacy of physical forces. Allometric "corrections" and accommodations can only proceed so far, and small creatures should therefore be predominantly shaped by forces acting on surfaces, and large animals by forces acting on volumes. Creatures of intermediate size might record a "tug of war," displaying the work of both sets. I therefore consider three famous examples of sensible correlations with increasing size.

1. For tiny creatures living fully in the realm of surficial forces, D'Arcy Thompson documents the conformity of many organisms (across a wide taxonomic spectrum) to the shape of "such unduloids as develop themselves when we suspend an oil-globule between two unequal rings, or blow a soap-bubble between two unequal pipes" (p. 247). (Obviously, D'Arcy Thompson must identify, in each case, the specific organic constraint corresponding to the two terminal rings of unequal size in his physical models. In one case, for example, he writes (p. 247) that "the surface of our Vorticella bell finds its terminal supports, on the one hand in its attachment to its narrow stalk, and on the other in the thickened ring from which spring its circumoral cilia.")

I choose this case (Figure 11-4) because D'Arcy Thompson invoked the conformity of so many tiny organisms with the well-known and easily-produced unduloid to draw an explicit contrast between his explanatory preferences and the conventional Darwinian account of adaptive design (pp. 248–249) (note, especially, how he highlighted these differences for resolving both the adaptive status of the basic form itself, and the genesis of a rich set of taxonomically designated variants upon the basic form):

> Here we have an excellent illustration of the contrast between the different ways in which such a structure may be regarded and interpreted. The teleological explanation is that it is developed for the sake of protection, as a domicile and shelter for the little organism within. The mechanical explanation of the physicist (seeking only after the "efficient," and not the "final" cause), is that it is present, and has its actual conformation, by reason of certain chemico-physical conditions: that it was inevitable, under the given conditions, that certain constituent substances actually present in the protoplasm should be aggregated by molecular forces in its surface layer; that under this adsorptive process, the conditions continuing favorable, the particles should accumulate and concentrate till they formed a membrane, thicker or thinner as the case might be; that this

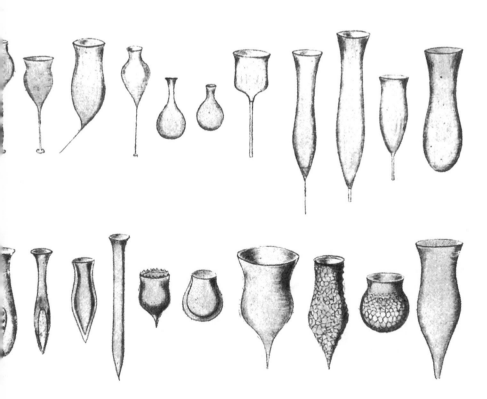

11-4. Single celled protists assuming the form of unduloids—and taken by D'Arcy Thompson as proof of immediate physical construction rather than genetic encoding. See text for details. From D'Arcy Thompson, 1917.

membrane was inevitably bound, by molecular forces, to become a surface of the least possible area which the circumstances permitted; that in the present case, the symmetry and "freedom" of the system permitted, and *ipso facto* caused, this surface to be a surface of revolution; and that of the few surfaces of revolution which, as being also surfaces *minimae areae,* were available, the unduloid was manifestly the one permitted, and *ipso facto* caused, by the dimensions of the organisms. We also see that the actual outline of this or that particular unduloid is also a very subordinate matter, such as physico-chemical variants of a minute kind would suffice to bring about; for between the various unduloids which the various species of Vorticella represent, there is no more real difference than that difference of ratio or degree which exists between two circles of different diameter, or two lines of unequal length.

In cases like this, strict selectionists conventionally assert that no genuine problem exists, but only a conceptual or terminological confusion. After all, any devotee of natural selection knows that adaptive shapes must be explained both in terms of survival value and immediate mode of construction. Therefore, the selectionist would continue, "I am happy to suppose that natural selection built the adaptive unduloid by fostering the differential reproductive success of growth variants that could attain the advantageous form along a single dimension of selection, rather than by having to construct each property in a piecemeal fashion, character by character."

In fact, selectionists can even cite a terminology to bolster their understanding that any adaptation requires, for its full explication, an account of both survival value and mode of construction (which, in truth, only reflects Aristotle's old distinction of final and efficient causes, and might as well bear these names rather than their currently favored neologisms). Selectionists generally refer to these two complementary modalities as "ultimate" and "proximate" causes—often supposing that they have, by this terminology, won some preciously new insight to clear away the conceptual fog of centuries. However, as stated above, the distinction only codifies a particular expression of Aristotle's argument on the multiple meanings and aspects of causality.

Nonetheless, even if not new, this argument about the complementarity and non-oppositional nature of ultimate and proximate causation cannot be gainsaid—and Darwinians advance this point with complete justice. That is, when selectionists cite the adaptive advantage of a form, they surely do not deny the need for a different statement about the immediate mode of genetic and developmental origin in any individual as well. However, we also need to recognize that this legitimate defense of adaptationist language does not apply to D'Arcy Thompson's point of genuine contention (logically genuine that is, not necessarily empirically correct in any given instance).

D'Arcy Thompson does not merely argue that he has found the mode by which natural selection worked to build adaptive unduloids. Rather, he advances the radically different, and truly oppositional, argument that natural selection need not be invoked at all, and as any kind of cause in this case. For he holds that physical forces shaped the unduloid directly, without *any* selection of favored forms from a range of variants. In other words, he believes that the efficient cause of mechanical imposition constructs the final cause automatically, thus obviating the need for any separate and explicitly biological or functional explanation to fashion the adaptive shape of the unduloid. (In selectionist jargon, D'Arcy Thompson argues that the proximate cause fashions the ultimate cause all by itself, thus explaining two properties for the price of one mechanism. I also happen to think that D'Arcy Thompson was probably wrong in this case, and that the traditional Darwinian scheme, with different forms of explanation needed for ultimate and proximate causes, probably applies to this case. But, the logic of D'Arcy Thompson's argument remains sound.)

2. At intermediary sizes, the automatically realized forms of inorganic objects often map the "conflicting" expressions of surficial forces holding things up and volumetric forces pulling them down. In a fascinating section added to the 2nd edition of 1942, D'Arcy Thompson studied drops of more viscid material falling through water. He compares the resulting forms (balancing surface tensions that retard descent and spread out the drops, with gravitational forces that pull the dense drops towards the bottom of the vessel) to the strikingly similar (and often quite complex) radially symmetrical shapes of jellyfishes (Figure 11-5). D'Arcy Thompson wrote (1942, pp. 397–398): "Not only do we recognize in a vorticoid drop a 'schema' or analog of medusoid form, but we seem able to discover various actual phases of the splash or drop in the all but innumerable living types of jellyfish . . . It is hard indeed to say how much or little all these analogies imply. But they indicate, at the very least, how certain simple organic forms might be naturally assumed by one fluid mass within another, when gravity, surface tension and fluid friction play their part."

3. At still larger sizes, surface tension becomes so negligible that rigid hard parts become necessary to maintain shape, lest gravity create a world of pan-

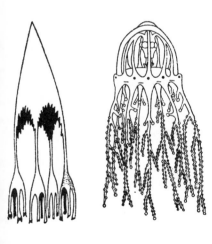

11-5. The jellyfish as a map of physical forces for a creature of intermediary size, and therefore subject both to forces that act on its surface and on its volume. The dense protoplasm is pulled down by gravity, but sufficiently retarded in its fall to spread out under forces of surface tension. From D'Arcy Thompson, 1942.

cakes. In what may be his most famous example, D'Arcy Thompson proved a case of direct construction in response to immediate gravitational forces by showing that internal trabeculae in the head of the human femur strengthen the bone along the precise lines of its greatest need for buttressing against compressive forces—for when bones break and heal improperly, the trabeculae are absorbed and then reform along lines of stress dictated by the limping walk of suboptimal reknitting. No one, in this case, could make the usual claim for phyletic determination by natural selection (at least not for these particular trabeculae in these unfavorable circumstances, although one might identify selection as the basis for this underlying lability in trabecular formation). D'Arcy Thompson writes (p. 687):

> Our bone is not only a living, but a highly plastic structure; the little trabeculae are constantly being formed and deformed, demolished and formed anew. Here, for once, it is safe to say that "heredity" need not and cannot be invoked to account for the configuration and arrangement of the trabeculae: for we can see them, at any time of life, in the making, under the direct action and control of the forces to which the system is exposed . . . Herein then lies, so far as we can discern it, a great part at least of the physical causation of what at first sight strikes us as a purely functional adaptation.

### *The admitted limitation and ultimate failure of an argument*

As a common theme in the tragedies of human literature and history, entities of all sorts (from bodies, to cities, to structures of ideas) often unravel at the height of their apparent triumph, for the surface of success may fail to anchor any roots in the general substrate below. Embodied within the very undeniability of D'Arcy Thompson's explanation for the direct mechanical shaping of optimally positioned trabeculae in the human femur, we can also locate the source of a strictly limited applicability that D'Arcy Thompson himself eventually had to own.

After all, the trabeculae can be explained as direct consequences of immediate mechanical forces *because* they cannot be construed as inherited aspects of a phenotype that might be subject to natural selection, or to any process of truly evolutionary modification for that matter. They represent labile responses of the ontogenetic moment, and will therefore be subject to specification by immediate forces (and, consequentially and crucially, not candidates for hereditary transmission in our non-Lamarckian earthly biology). But when we study stable and inherited features with equal claim to adaptive optimality—the main kinds of characters that theories of adaptive evolution try to explain—how can we make an equally strong case (or even a case of any plausibility at all) for their immediate construction by physical forces acting upon the organism during growth? Immediate physical forces may build my trabeculae, but how can they shape a set of stable and inherited traits that made a first appearance when I was a tiny embryo in utero, long before the gravitational forces to which these features are adapted could play any role in building my body by direct imposition?

D'Arcy Thompson had to face this critical problem head on—and he did so in a forthright manner by putting the best possible face upon adversity via two arguments that fatally compromised his dream of winning generality for his idiosyncratic theory of structuralist evolution from the outside. First and foremost, he simply admitted that his principle of direct imposition couldn't explain the complex forms of the multicellular phyla that, for however parochial a reason, have always defined the central subject and puzzlement of morphology. D'Arcy Thompson continued to maintain—and he may well have been right in some cases—that good matches between simple organic conformations (primarily the outward forms of unicellular creatures) and geometric shapes of well known mathematical definition and easily accomplished mechanical construction probably illustrate his favored principle of direct imposition by physical forces. But he had to admit that he could not apply this line of reasoning to the basic form of a horse or a tuna.

Interestingly, he "came clean" on this point right after his strong argument about the production of unicellular unduloids by forces of surface tension (cited on p. 1193). He begins by quoting a conventional defense of phylogenetic reasoning by E. Ray Lankester, then refutes the argument for his unduloids, but cannot deny its application to large and complex multicellular forms (pp. 251–252):

> "The fact that we are able to classify organisms at all in accordance with the structural characteristics which they present, is due to the fact of their being related by descent." But this great generalisation is apt in my opinion, to carry us too far. It may be safe and sure and helpful and illuminating when we apply it to such complex entities,—such thousand-fold resultants of the combination and permutation of many variable characters,—as a horse, a lion or an eagle; but (to my mind) it has a very different look, and a far less firm foundation, when we attempt to extend it to minute organisms whose specific characters are few and simple, whose simplicity becomes much more manifest when we regard it from the point of view of physical and mathematical description and analysis, and whose form is referable, or (to say the least of it) is very largely referable, to the direct and immediate action of a particular physical force.

But D'Arcy Thompson truly throws in the towel during the most poignant and appropriate round—right at the end of his last empirical chapter, as he reaches the apex of complexity in his analysis of vertebrate skeletons (chapter 16 "on form and mechanical efficiency"), and just before he recoups relevance in his brilliant final chapter on the theory of transformed coordinates. He begins by admitting that he cannot describe the skeleton as "a resultant of immediate and direct physical or mechanical conditions" precisely because the very biological principle that he has tried to deny (or at least to underplay) throughout the book—the phyletic inheritance, rather than immediate construction, of the underlying *Bauplan*—cannot be gainsaid at this level of complexity. (With refreshing candor, D'Arcy Thompson admits that he had tried his best "to circumscribe the employment of the latter [that is, of heredity] as a working hypothesis in morphology.") D'Arcy Thompson writes, in his key statement (p. 715):

> It would, I dare say, be a gross exaggeration to see in every bone nothing more than a resultant of immediate and direct physical or mechanical conditions; for to do so would be to deny the existence, in this connection, of a principle of heredity. And though I have tried throughout this book to lay emphasis on the direct action of causes other than heredity, in short to circumscribe the employment of the latter as a working hypothesis in morphology, there can still be no question whatsoever but that heredity is a vastly important as well as a mysterious thing; it is *one* of the great factors in biology. . . . But I maintain that it is no less an exaggeration if we tend to neglect these direct physical and mechanical modes of causation altogether, and to see in the characters of a bone merely the results of variation and of heredity, and to trust, in consequence, to those characters as a sure and certain and unquestioned guide to affinity and phylogeny.

This admission then leads to a recovery of relevance via the second argument, presented in his last and, in the judgment of most biologists, his most important chapter "on the theory of transformations, or the comparison of related forms." All professional evolutionists know D'Arcy Thompson's famous diagrams of related organisms compared by imposing a Cartesian grid upon one form, treated as a reference, and then rendering other forms as results of simple distortions and transformations of the grid lines (see Figure 11-6 for an example). But I think that most of us have not understood the logical and theoretical reasons behind D'Arcy Thompson's invention, largely because (as for the chapter "on magnitude") we read this section out of context, and do not grasp its intimate relation (as an apotheosis, given the limitations he had to admit) to his general and idiosyncratic theory of form.

That is, we tend to interpret these Thompsonian transformed coordinates as a crude, and ultimately failed, attempt to operationalize (by pictorialization) a good intuition about the multivariate nature of evolutionary change before the development of appropriate statistical techniques, and the invention of computers, permitted us to apply genuine multivariate mathematics to problems of form. Most of us, I think, envisage the deformed coordinate grid as a mere residuum of a qualitative analysis focused on the transformed bodies themselves—just a set of guidelines needed to make a crude map of the organisms under consideration.

In so doing, we misunderstand D'Arcy Thompson's intention in a precisely backwards manner. His interest lay primarily in the lines of the stretched and deformed grids, for he had remained true to his theory that physical forces shape organisms directly. He had made a painful and necessary surrender of that theory—by bowing to conventional evolutionary resolutions in terms of heredity and phylogeny—for explaining the complex and intricately multi-variate *Baupläne* of complex metazoan animals. In other words, he accepted that the basic designs must be admitted as "primitive terms" or "background conditions" within his theory—as "givens" to be acknowledged (and attributed to other kinds of causes), and as basic inputs before any further analysis could be conducted in his favored terms.

In making such an admission D'Arcy Thompson swallowed a bitter pill. He had to accept the existence and contrary construction of "hipponess" or "eagleness" at the outset, and then to determine what field might be left to his favored causes of direct mapping by physical forces. The theory of transformed coordinates presents his positive approach to this dilemma, his attempt to keep his theory maximally relevant in the light of his enforced concession to historicism in general, and to Darwinism in particular. He could not lay claim to the basic forms themselves, but he would still make a play for the taxonomic variety produced by their transformations.

If the differences among related species could be expressed as simple distor-

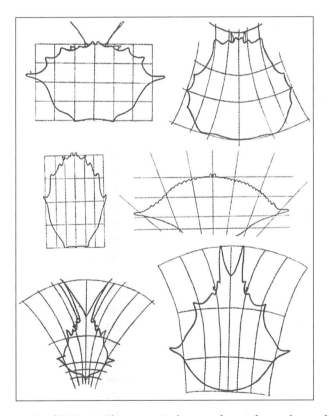

11-6. An example of D'Arcy Thompson's theory of transformed coordinates. To understand his view, we must recognize that these figures are meant to feature the transformation grid lines themselves—as indications of physical forces that directly impose phyletic changes upon the organisms. From D'Arcy Thompson, 1917.

tions of a grid, then the transformed grid itself would become a picture of the lines of forces responsible for the evolutionary deformation. Thus, D'Arcy Thompson valued the lines of the transformed grids above the altered organisms themselves, for he hoped that his pictures of simple and coordinated transformation would revivify his theory in a more limited domain. The lines of transformation would map the forces that converted the initial form into its descendants or relatives—and by D'Arcy Thompson's theory of direct imposition, those lines would then identify the geometric operation of the *actual forces* that caused the evolutionary changes by direct imposition. He would not win hipponess (or full happiness) for his theory, but he might encompass the set of realized variations upon hipponess. Put another way, perhaps D'Arcy Thompson could "have it all" for the simple forms of some unicells; but he would have to settle for the variations (leaving the fundamental configurations to history and heredity) when he treated the complexities of vertebrate organization.

Sweet indeed are the uses of adversity, as D'Arcy Thompson put his best possible spin upon a positive ending to a most unwelcome admission (p. 723):

> But in a very large part of morphology, our essential task lies in the comparison of related forms rather than in the precise definition of each; and the *deformation* of a complicated figure may be a phenomenon easy of comprehension, though the figure itself have to be left unanalysed and undefined. This process of comparison, of recognising in one form a definite permutation or *deformation* of another, apart altogether from a precise and adequate understanding of the original "type" or standard of comparison, lies within the immediate province of mathematics, and finds its solution in the elementary use of a certain method of the mathematician. This method is the Method of Co-ordinates, on which is based the Theory of Transformations.

### Odd Man In (D'Arcy Thompson's structuralist critique of Darwinism) and Odd Man Out (his disparagement of historicism)

I premised this discussion by arguing that D'Arcy Thompson's showy singularity in both style and substance—the quirks and anachronisms that seem to place him "out of time"—must not blind us to the fact that these uniquenesses overlie a rather standard early 20th century structuralist critique of Darwinian functionalism. D'Arcy Thompson's residence within his own time therefore becomes as informative as his idiosyncrasies.

In particular, I have emphasized, throughout this book, the logical and almost ineluctable linkage in structuralist thought between defenses of internally channeled directionality and saltationist mechanics (as particularly exemplified in the definitive model of Galton's Polyhedron—see pp. 342–351). I have also stressed the equally tight relationship of these views to a critique of adaptations—not of their preeminent existence (which D'Arcy Thompson embraces and celebrates), but of their necessary construction "for" utility by a functionalist mechanism like natural selection (see my previous discussion of D'Arcy Thompson's contemporary, William Bateson, pp. 396–415).

On the theme of constraint, D'Arcy Thompson frequently argues that once one accepts a more than analogical comparison between the good designs, based on idealized geometry, of organic and inorganic objects, one must also defend (not apologetically, but for the utility of the conclusion in understanding anatomical order and taxonomic interrelationships) the limitation of organic morphology to channels of transformation that the physical causes of morphogenesis must follow (p. 137): "The world of things living, like the world of things inanimate, grows of itself, and pursues its ceaseless course of creative evolution. It has room, wide but not unbounded, for variety of living form and structure, as these tend towards their seemingly endless, but yet strictly limited, possibilities of permutation and degree."

Once these directing channels of evolution have been established by physical forces governing the changes, we have no reason to restrict our concept of movement down these channels to the slow and continuous pacing of Darwinian gradualism. Relevant physical laws will regulate the path, "be the journey taken fast or slow." In the following passage (p. 155), D'Arcy Thompson begins to link his model of channeled discontinuity to critiques of specific principles within evolutionary theory—in this case to the prevalent idea that similarities between ontogeny and phylogeny must arise for phyletic or historical reasons. Why introduce this "extra" and superfluous hypothesis, D'Arcy Thompson asks, if the same physical law, "caring little" whether it works during the growth of individuals or the evolution of lineages, mandates the same changes in any separate case:

> The differences of form, and changes of form, which are brought about by varying rates (or "laws") of growth, are essentially the same phenomenon whether they be, so to speak, episodes in the life-history of the individual, or manifest themselves as the normal and distinctive characteristics of what we call separate species of the race. From one form, or ratio of magnitude, to another there is but one straight and direct road of transformation, be the journey taken fast or slow; and if the transformation take place at all, it will in all likelihood proceed in the self-same way, whether it occur within the life-time of an individual or during the long ancestral history of a race.

But for sheer iconoclasm, nothing in all of D'Arcy Thompson's prose matches the section that he added as "conclusion" to the chapter on transformed coordinates in the second (1942) edition of *Growth and Form*—a "great quotation" well worth citing *in extenso*. Here he makes his most incisive and forthright linkage between the two great structuralist themes of channeling and saltational discontinuity. Taxonomic groups of organisms may be compared (truly "homologized" in the conceptual sense of being regulated by the same causal principle) with families of mathematical curves generated by definite parameters of construction. Complete continuity may reign within a family, but the gaps between can only be crossed *per saltum*. D'Arcy

Thompson emphasizes that "there is no argument against the theory of evolutionary descent" (1942, p. 1094) in such a concept, but he also maintains that this view of life's conformity to ordinary physical principles suggests a wide and radical range of non-Darwinian implications, including the rejection of Darwin's views on imperfection of the geological record, a defense of saltational transition between *Baupläne* (contrasted with potential continuity of taxonomic order within), and a solution to the problem of life's inhomogeneous occupation of potential morphospace as an organic incarnation of real mathematical discontinuities in the geometry of nature.

> There is one last lesson which coordinate geometry helps us to learn. . . .
> In the study of evolution, and in all attempts to trace the descent of the animal kingdom, fourscore years' study of the *Origin of Species* has had an unlooked-for and disappointing result. . . .
> This failure to solve the cardinal problem of evolutionary biology is a very curious thing . . . We used to be told, and were content to believe, that the old record was of necessity imperfect—we could not expect it to be otherwise; the story was hard to read because every here and there a page had been lost or torn away . . . But there is a deeper reason. When we begin to draw comparisons between our algebraic curves and attempt to transform one into another, we find ourselves limited by the very nature of the case . . .
> An algebraic curve has its fundamental formula, which defines the family to which it belongs . . . With some extension of the meaning of parameters, we may say the same of the families, or genera, or other classificatory groups of plants and animals . . . We never think of "transforming" a helicoid into an ellipsoid, or a circle into a frequency-curve. So it is with the forms of animals. We *cannot* transform an invertebrate into a vertebrate, nor a coelenterate into a worm, by any simple and legitimate deformation. . . .
> A "principle of discontinuity," then, is inherent in all our classifications, whether mathematical, physical or biological; and the infinitude of possible forms, always limited, may be further reduced and discontinuity further revealed . . . The lines of the spectrum, the six families of crystals, Dalton's atomic law, the chemical elements themselves, all illustrate this principle of discontinuity. In short, nature proceeds *from one type to another* among organic as well as inorganic forms; and these types vary according to their own parameters, and are defined by physico-mathematical conditions of possibility. In natural history Cuvier's "types" may not be perfectly chosen nor numerous enough, but *types* they are; and to seek for stepping-stones across the gaps between is to seek in vain, forever. . . .
> Our geometrical analogies weigh heavily against Darwin's conception of endless small continuous variations; they help to show that discontinuous variations are a natural thing, that "mutations"—or sudden changes, greater or less—are bound to have taken place, and new

"types" to have arisen, now and then. Our argument indicates, if it does not prove, that such mutations [occur] on comparatively few definite lines, or plain alternatives, of physico-mathematical possibility.

If D'Arcy Thompson ended the final chapter on transformed coordinates with his most trenchant critique of Darwinian functionalism, he also lets us know, within the body of the chapter, that he built his entire theory as an alternative to the major implication of Darwinism for the daily practice of evolutionary biology and our conventional manner of conceptualizing organisms—for Darwinism implies separability of traits, and the subsequent potential for their independent optimization by natural selection, whereas the very possibility of relating creatures by such simple transformation grids points to structural channeling by overarching laws of growth (p. 727):

> This independent variability of parts and organs . . . would appear to be implicit in our ordinary accepted notions regarding variation; and, unless I am greatly mistaken, it is precisely on such a conception of the easy, frequent, and normal independent variability of parts that our conception of the process of natural selection is fundamentally based . . . But if, on the other hand, diverse and dissimilar fishes can be referred as a whole to identical functions of very different coordinate systems, this fact will of itself constitute a proof that variation has proceeded on definite and orderly lines, that a comprehensive "law of growth" has pervaded the whole structure in its integrity, and that some more or less simple and recognisable system of forces has been at work.

In discussing morphological variety among Radiolaria, where he suspects that realized taxonomic diversity most closely approaches the filling of all slots permitted by generating laws of form, D'Arcy Thompson extends his critique to the ultimate step of even doubting whether many configurations, as occupants of geometrically attainable positions in a predictable series, even need to be interpreted as adaptive at all (p. 607):

> In few other groups do we seem to possess so nearly complete a picture of all possible transitions between form and form, and of the whole branching system of the evolutionary tree: as though little or nothing of it had ever perished, and the whole web of life, past and present, were as complete as ever. It leads one to imagine that these shells have grown according to laws so simple, so much in harmony with their material, with their environment, and with all the forces internal and external to which they are exposed, that none is better than another and none fitter or less fit to survive. It invites one also to contemplate the possibility of the lines of possible variation being here so narrow and determinate that identical forms may have come independently into being again and again.

D'Arcy Thompson bases these critiques of natural selection directly upon his own idiosyncratic theory of form. But, in his more conventional participation in the general debate of his time, he also presents the standard anti-

Darwinian arguments that prevailed within evolutionary biology before the Modern Synthesis reinstated natural selection at the center of the subject. In a brief commentary on the extinction of dinosaurs, for example, D'Arcy Thompson reiterates the common claim that natural selection, although a genuine force in evolution, can only play the minor and negative role of eliminating the unfit, and not the central part of creating the fit. "We begin to see," he states (p. 137), "that it is in order to account, not for the appearance, but for the disappearance of such forms as these that natural selection must be invoked." He then embellishes the argument with one of his characteristic prose flourishes (pp. 137–138):

> But there comes a time when "variation," in form, dimensions, or other qualities of the organism, goes farther than is compatible with all the means at hand of health and welfare for the individual and the stock; when, under the active and creative stimulus of forces from within and from without, the active and creative energies of growth pass the bounds of physical and physiological equilibrium: Then, at last, we are entitled to use the customary metaphor, and to see in natural selection an inexorable force, whose function is not to create but to destroy,—to weed, to prune, to cut down, and to cast into the fire.

But if this general and multifaceted critique of Darwinian functionalism—arising from his contrary structuralist account of the direct production of adaptive form by physical forces—harmonized well with major trends of thought in the evolutionary biology of his generation, the second prominent implication that he drew from his idiosyncratic theory of form could not have stood in more oppositional relationship to an even deeper, and even more general, assumption of the evolutionary sciences, both in D'Arcy Thompson's day and in our own. For D'Arcy Thompson used his theory to disparage, indeed virtually to abolish, phylogenetic reasoning and historical explanations in general. He did not, of course, deny that phylogeny happened, and that a tree of life existed; and he did not challenge the fact that each species stands atop a historical series of ancestral forms. But he did argue that these common features of biological reality possessed virtually no explanatory value for understanding either the morphology of individual species, or the anatomical relationships among species, as depicted in his own diagrams of transformed coordinates.

For if immediate physical forces shape the adaptive configurations of each modern species directly, then what relevance can be found in extinct ancestral shapes that responded to different physical forces of prior times, and that do not constrain the current forms of their descendants? The sensible, rational, and constrained order of related living forms in taxonomic groups represents a set of realizable positions within the mathematical boundaries of laws that currently shape the relevant organisms, not a set of phyletic constraints inherited from the past and now resident within organisms in the form of inherited genetic and developmental patterns that limit and channel the taxonomic structure of the living world.

For these reasons, I have labeled D'Arcy Thompson's "physicalist" and "externalist" form of structuralism as deeply out of harmony with the internalist and historically based theories of structural constraint that are now enjoying such a revival within evolutionary theory, and that constitute the subject of the rest of this chapter. I have also not hidden my general approbation for these more popular internalist and historicist versions. Paleontologists are, after all, historians at heart and by profession, and I regard historical causation as the most powerful and distinctive mode of reasoning—indeed the *raison d'être*—of the evolutionary sciences. But I must also confess my lifelong attraction to the prose, and the chutzpah, behind D'Arcy Thompson's iconoclasm. His work has much to teach us, although I do not think that his general theory of form can be validated as more than a peripheral aid and secondary shaper within a primary nexus of historicism.

In any case, D'Arcy Thompson wrote many of his stylistically strongest and philosophically most interesting disquisitions in a form that can only be labeled as "riffs" for different taxonomic groups on the common subject of the irrelevance of phylogeny in general, and the uselessness of Darwinian mechanisms in particular, for explaining either their anatomical ground plan or the ordered array of forms defining their taxonomic structure. He presents variants on the same riff for unduloid protists, radiolarians, the hard parts of foraminifers, sponge spicules, bird eggs, the multifarious variations in shape among species of Mesozoic ammonoids, and the spiral patterns of stems in climbing plants.

For example, in the final paragraph of his chapter on logarithmic spirals, D'Arcy Thompson notes that the same set of varieties upon this universal curve populate the seas throughout Phanerozoic times. Therefore, he can't help wondering whether molluscan shells of this form can be meaningfully parsed into historical series of descent, or even accorded particular adaptive values (p. 586):

> Again, we find the same forms, or forms which (save for external ornament) are mathematically identical, repeating themselves in all periods of the world's geological history; and, irrespective of climate or local conditions, we see them mixed up, one with another, in the depths and on the shores of every sea. It is hard indeed (to my mind) to see where Natural Selection necessarily enters in, or to admit that it has had any share whatsoever in the production of these varied conformations. Unless indeed we use the term Natural Selection in a sense so wide as to deprive it of any purely biological significance; and so recognise as a sort of natural selection whatsoever nexus of causes suffices to differentiate between the likely and the unlikely, the scarce and the frequent, the easy and the hard: and leads accordingly, under the peculiar conditions, limitations and restraints which we call "ordinary circumstances," one type of crystal, one form of cloud, one chemical compound, to be of frequent occurrence and another to be rare.

In my opinion, D'Arcy Thompson's most powerful statement (also most likely to include important elements of validity) occupies several pages at the end of his next chapter 12 "on the spiral shells of the foraminifera." In this "conclusion" (pp. 607–611), he decries the standard tendency of the time (now happily abandoned, but mostly for reasons other than D'Arcy Thompson's critique) to establish speculative phylogenies not so much on stratigraphic successions, but on idealized series moving from simple forms like the spherical *Orbulina* to ever more complex designs, and with attributions of increasing adaptive value (usually for greater test strength) to these more elaborate forms. Instead, D'Arcy Thompson views the diversity of foram shells as so many incarnations of permitted designs under the shaping rules that actually direct the building of tests. Since these rules do not change through time, the potential forms remain immanent in nature, and become occupied again and again, throughout the history of the entire lineage, by various clades that fall under the relevant set of forces. These forms are therefore no more arrangeable from worse to better, and no more subject to specific accounts of historical filiation, than are the varied shapes of snowflakes or quartz crystals. (I particularly like this "riff" because D'Arcy Thompson admits that phylogenies for complex metazoans may be better founded, and because he may well be right that, for these simplest creatures of the fossil record, repeated shapes in widely separated times and places may represent massive homoplasy based on the ease of reevolving certain basic designs. After all, a hippo is only a hippo and will evolve just once, but the form of a microscopic calcareous sphere, floating in the ocean, may be attainable along many routes—pp. 610–611):

> The theorem of Organic Evolution is one thing; the problem of deciphering the lines of evolution, the order of phylogeny, the degrees of relationships and consanguinity, is quite another. Among the higher organisms we arrive at conclusions regarding these things by weighing much circumstantial evidence, by dealing with the resultant of many variations, and by considering the probability or improbability of many coincidences of cause and effect. . . .
>
> But in so far as forms can be shown to depend on the play of physical forces, and the variations of form to be directly due to simple quantitative variations in these, just so far are we thrown back on our guard before the biological conception of consanguinity, and compelled to revise the vague canons which connect classification with phylogeny.
>
> The physicist explains in terms of the properties of matter, and classifies according to a mathematical analysis, all the drops and forms of drops and associations of drops, all the kinds of froth and foam, which he may discover among inanimate things; and his task ends there. But when such forms, such conformations and configurations, occur among *living* things, then at once the biologist introduces his concepts of heredity, of historical evolution, of succession in time . . . if fitness for a function, of adaptation to an environment, of higher and lower, of "better" and "worse." This is the fundamental difference between the "explanations" of the physicist and those of the biologist.
>
> In the order of physical and mathematical complexity there is no ques-

tion of the sequence of historic time. The forces that bring about the sphere, the cylinder or the ellipsoid are the same yesterday and to-morrow. A snow-crystal is the same to-day as when the first snows fell. The physical forces which mould the forms of Orbulina, of Astrorhiza, of Lagena or of Nodosaria to-day were still the same, and for aught we have reason to believe the physical conditions under which they worked were not appreciably different, in that yesterday which we call the Creta-ceous epoch; or, for aught we know, throughout all that duration of time which is marked, but not measured, by the geological record.

### An epilog to an argument

We can, today, easily identify D'Arcy Thompson's primary error as an expres-sion of the venerable *post hoc* fallacy ("after this, therefore because of this"). He had correctly noted a strong correlation, throughout organic nature, be-tween the forms of organisms and the shapes that inorganic objects assume under direct molding by physical causes acting upon them. He therefore ad-vocated the simplest hypothesis that these physical causes had directly fash-ioned the organic forms (just as we would unhesitatingly assert for inorganic objects). Here he made an empirical rather than a logical error. That is, we cannot accuse D'Arcy Thompson of not recognizing the potential fallacy of drawing a causal inference (direct physical production) from the observation of a correlation (between realized organic forms and the idealized optima constructed by these physical forces in the inorganic realm). He understood perfectly well that biologists preferred a different and more complex expla-nation for the same generality—that the inorganic objects may be directly crafted, but that organisms generally achieve the same result by operation of a different kind of biological force, natural selection working by differential reproductive success and survival of the fittest. In the following passage, for example, D'Arcy Thompson separates the two arguments: first, the false inference of direct organic production, followed by the correct observation that organic forms obey physical laws (p. 10): "We want to see how, in some cases at least, the forms of living things, and of the parts of living things, can be explained by physical considerations, and to realise that, in general, no organic forms exist save such as are in conformity with ordinary physical laws."

In other words, and using D'Arcy Thompson's favored Aristotelian termi-nology, he tried to depict the physical laws to which organic shapes conform so well as the actual efficient causes of these shapes. But, in general, Darwin-ians were right all along. These physical laws are formal causes, or blueprints of optimal adaptive designs for given circumstances of size, materials and ecology. The laws give us insight into the adaptive values, or final causes, of organic designs. But the efficient cause of good organic design is usually natu-ral selection.

Ironically, this great student of Aristotle (D'Arcy Thompson wrote stan-dard translations, still in print, for two of Aristotle's biological treatises) guessed wrong about the category of causes embodied in the correlation of organic shapes with the optimal forms directly produced by physical forces. He regarded the correlation as a map of the actual efficient cause. This aspect of good organic design does express a final cause in adaptation, but any evolutionary changes must still must be crafted by an efficient cause—and Darwinian natural selection generally acts as the efficient cause we seek for our explanations.

But before we dismiss *Growth and Form* as a brilliant and wonderfully written disquisition rooted in a central error, we should pause to reflect upon the partial validity of D'Arcy Thompson's theory of direct impress, or "order for free" in current parlance (Kauffman, 1993). D'Arcy Thompson admitted that he could not apply his theory to explain the groundforms of complex creatures, which he then accepted as "givens" in his analysis of transformed coordinates. This admission scuttled more of his hopes for generality than he was ever willing to acknowledge. But D'Arcy Thompson's theory cannot be rejected as entirely, or even generally, wrong. Surely his arguments for the hexagonal forms of crowded corallites, and the conformity of the ends of hive cells to the Maraldi angle, are correct: organisms don't have genes "for" hexagonality *per se*. Developmental genetics may regulate the types of materials, and their rates and places of production. But hexagonal shapes probably arise, just as for inorganic materials in similar conditions, by direct shaping under laws of closest packing.

I am confident that biologists can trace the lineage of hippos deep into an artiodactyl past in the early Cenozoic—as a genuine historical particular of unique form, requiring a phylogenetic explanation. But when someone tells me that a particular form of bacterium has not changed for 3.5 billion years because the oldest of all fossils displays the same shape as some modern species, then I doubt that this correct observation teaches me anything about filiation in a particular and continuous lineage; whereas I may learn something about basic forms, homoplastically attained again and again during the history of life, and therefore bearing no particular phyletic message. And I may use D'Arcy Thompson's procedures to establish the probable reasons behind these shapes, whether or not my explanations lie in direct shaping by these physical forces (a strong possibility for the bacteria, but not for the hippos), or in the adaptive values of designs actually built by the efficient cause of natural selection.

### ORDER FOR FREE AND REALMS OF RELEVANCE FOR THOMPSONIAN STRUCTURALISM

In the most important modern work in the D'Arcy Thompsonian tradition, Kauffman (1993, p. 443) "invites our attention to the central theme of this [that is, his] book: Order in organisms may largely reflect spontaneous order in complex systems." Kauffman retains strong feality to D'Arcy Thompson's central principle that the adaptive order of biological systems arises by direct imposition of physical forces—thus advocating an "externalist" form of structuralist thought that has not played a large role in the history of evolutionary biology. But Kauffman's specific insights and foci of concern differ widely from D'Arcy Thompson's emphasis upon morphometric geometry under laws of classical Newtonian dynamics. In the two major disparities, Kauffman first calls upon a different set of physical principles that has recently inspired both professional activity and public interest. In a second, and welcome, difference, Kauffman wishes to abet selection by supplying "order for free" from the inherent nature of the physical world, whereas D'Arcy Thompson tried to develop a largely substitutional theory of adaptive form that would relegate natural selection of effective insignificance.

In stating his fealty and staking out his differences, Kauffman (1993, p. 643) pays homage to D'Arcy Thompson and acknowledges the "small trickling of intellectual tradition" that this "outlier" species of structuralism has engendered, although Kauffman would surely wish to enlarge the flow to Mississippian proportions:

> D'Arcy Thompson's famous and elegant book *On Growth and Form* stands as one of the best efforts to find aspects of organismic order which can be understood as aspects which we might, on good grounds, expect. His enquiry, which led him to consider minimal energy surfaces, transformations of coordinate systems as a function of differential growth, and a whole beautiful panoply of phenomena, has stood as a persistent spring for a small trickling of intellectual tradition down through contemporary biology. Thompson applied classical physics to biology. It has been said that a weakness of some biologists is persistent physics-envy: the seeking of a deep structure to biology.

Kauffman then extols the virtues of physics-envy, while recommending that biologists redirect their jealousy away from the Newtonian mechanics that D'Arcy Thompson revered (p. 644): "There is a new physics aborning, and it is time to again fall open victim to physics-envy. For want of a better name, the area which is emerging is something like a theory of complex systems . . . This book is an effort to continue in Thompson's tradition with the spirit now animating parts of physics. It seeks origins of order in the generic properties of complex systems."

By pluralizing his title, and by being even more explicit in his subtitle, Kauffman emphasizes his different aim of arranging a marriage between selection and inherent order, with the latter as the older and more experienced partner who encourages a younger spouse to invigorate and direct the united effort emerging from a preexisting substrate: *The Origins of Order. Self-Organization and Selection in Evolution.* "I have made bold to suggest that much of the order seen in organisms is precisely the spontaneous order in the systems of which we are composed. Such order has beauty and elegance, casting an image of permanence and underlying law over biology. Evolution is not just 'chance caught on the wing.' It is not just a tinkering of the ad hoc, of bricolage, of contraption. It is emergent order honored and honed by selection" (p. 644).

"My own aim," Kauffman adds (p. 26), "is not so much to challenge as to broaden the neo-Darwinian tradition. For, despite its resilience, that tradition has surely grown without seriously attempting to integrate the ways in which simple and complex systems may spontaneously exhibit order." "My aim throughout is to attempt to characterize . . . those aspects which may reflect the self-organized properties of the . . . system and those which reflect selection—and to determine a way of recognizing the marriage between the two" (p. 407).

Kauffman continually invokes two key phrases to epitomize his understanding of direct physical molding in the evolution of adaptive form in anatomy, ontogeny and interacting biological systems in general. First, he seeks to explicate the spontaneous "order for free," to which systems naturally conform, and which provides natural selection with a rich substrate for fine tuning and more specific molding. In stating his intentions for ontogeny, and recalling his similar conclusions for ecosystems, Kauffman writes (p. 409). (I like his phrase "gratuitously present" as a description of order for free):

> Highly constrained, poised cell types and ordered patterns of gene activity, each able to change to only a few others, are gratuitously present in a vast class of genomic regulatory systems . . . The phase transition from one regime to another is governed by simple parameters of the system, such as richness of coupling among the variables. The order seen in ontogeny, I shall suggest, is just that which arises spontaneously in the powerfully ordered regime found in parallel-processing networks. Selection, I shall further suggest, by achieving genomic systems in the ordered regime near the boundary of chaos, is likely to have optimized the capacity of such systems to perform complex gene-coordination tasks and evolve effectively.

In the other important book from the 1990's on this view of life, Goodwin (1994, p. 186) emphasizes the "generic" nature of order for free: "Much (and perhaps most) of the order that we see in living nature is an expression of properties intrinsic to complex dynamic systems organized by simple rules of interaction among large numbers of elements. This order is generic, and what we see in evolution may be primarily an emergence of states generic to the dynamics of living systems."

With his second phrase, Kauffman emphasizes evolvability rather than form or organization *per se* in arguing that biological systems naturally evolve to "adaptation at the edge of chaos." He holds (p. 645) that "the capacity to evolve is itself subject to evolution and may have its own lawful properties. The construction principles permitting adaptation, too, may emerge as universals. Adaptation to the edge of chaos is just such a candidate construction principle." Kauffman continually stresses the abstract, general and timeless nature of those aspects of biological order that he would ascribe to "the nature of things" rather than to any distinctively organic mechanism like natural selection (which can then act upon the inherent and generic properties to construct more specific utilities in particular environments).

For these reasons of non-competition at their different scales (generic order *vs.* specific design) and the sequential nature of their interaction (generic order as given, selective order as superimposed refinement), the structural order generated by physical necessity meshes well with the functional order built by natural selection (p. xv): "Selection achieves and maintains complex systems poised on the boundary, or edge, between order and chaos. These systems are best able to coordinate complex tasks and evolve in a complex environment. The typical, or generic, properties of such poised systems emerge as potential ahistorical universals in biology."

I have emphasized throughout this section that the unusual structuralist theme of spontaneous order externally imposed by physical law (and therefore so different from structuralism's conventional focus upon internal channeling set by phyletic history, and then encoded in genetic and developmental programs) enjoys its greatest potential strength in two areas where other styles of explanation either don't apply in principle, or have simply (and so far) failed to yield robust results: the origin of life and its early history up to the construction of a prokaryotic cell; and the explanation of broad, recurrent, and potentially ahistorical, or at least not phylogenetically constrained, patterns (the form of ecological pyramids, rather than the particular top carnivore resident thereupon, the right skewed distribution of life's complexity rather than the occupation of its realized neurological extremity by *Homo sapiens*). The structure of Kauffman's book affirms these foci (and the resulting promise of this approach through noninterference with the large and legitimate domain of necessarily historical explanation).

Life's origin and precellular history occupies one of three major sections in *The Origins of Order*. Kauffman stresses his message of physical necessity in designating this set of chapters as "The Crystallization of Life." He structures his argument by attacking a straw man that, at least in the understanding of most paleontologists, fell from popularity more than 20 years ago when the fossil record yielded cells of bacterial form in the most ancient sediments that could preserve organic structures (3.5 to 3.6 billion years old). But this old and superseded claim of exceedingly low probability for life's origin does serve as a convenient foil for Kauffman's (and virtually the entire profession's) search for different answers rooted in the predictable and generic nature of organic chemistry and the physics of self-organizing systems (p. 285): "The second part of this book . . . explores a heretical possibility. The origin of life, rather than having been vastly improbable, is instead an expected collective property of complex systems of catalytic polymers and the molecules on which they act. Life, in a deep sense, crystallized as a collective self-reproducing metabolism in a space of possible organic reactions. If this is true, then the routes to life are many and its origin is profound, yet simple."

Kauffman then devotes the other two sections of his book to the second theme of broad and timeless structural generalities behind the specific adaptive solutions crafted by the functionalist mechanism of natural selection: Part 1 on models for interaction on rugged fitness landscapes, and Part 3 on the order of ontogeny. As an example of structural generality behind functional specificity, Kauffman's model predicts that the waiting time for successful long-step jumping to higher, but distant, adaptive peaks (contrasted with different rules for simple expansion to adjacent peaks) doubles after each successful transition, thus forecasting, for example, that if a first success emerges from two tries on average, the tenth will require a mean wait equal to the time needed for more than 1000 attempts. In this manner, and very sensibly in my opinion (see Gould, 1989c), Kauffman tries to account for the origin of all fossilizable metazoan phyla in the Cambrian explosion, followed by more than 500 million years featuring no further origin of body plans of such distinctly different design. But this level of generality offers no insight into the particular historical questions of taxon and time that have always defined the guts and soul of biology: why arthropods, and why then rather than a billion years before or after (with the latter, contingently plausible, scenario precluding my writing and your reading this book, among other differences between our actual world and innumerable sensible, but unrealized, alternatives).

In explicating this feature of broad generality for the legitimate realm of physical imposition and predictability, Kauffman (pp. 13–14) draws an apt analogy that cannot be denied, but that also identifies the limits of his favored approach:

> There is no doubt that our awareness of historical contingency is proper. The question we must address is whether there might be statistical order within such historical processes. A loose analogy makes this point. Imagine a set of identical roundtopped hills, each subjected to rain. Each hill will develop a particular pattern of rivulets which branch and converge to drain the hill. Thus the particular branching pattern will be unique to each hill, a consequence of particular contingencies in rock placement, wind direction, and other factors. The particular history of the evolving pattern of rivulets will be unique to each hill. But viewed from above, the statistical features of the branching patterns may be very similar. Therefore, we might hope to develop a theory of the statistical features of such branching patterns, if not of the particular pattern on one hill.

An ironic solution to controversy between this form of timeless structuralism and historical particularism could emerge from a treaty that rigidly relegated each domain to its proper space within the analogy. But such a clearly defined and well-patrolled truce would also deprive biology of much interest and legitimate skirmishing, for our deepest puzzles and most fascinating inquiries often fall into a no-man's land not clearly commanded by either party—while we must also admit (and treasure) our human inclination to expand the domains of personally favored explanations ("pushing the envelope" in a favored cliché of our times). Historicists do claim that much of what we have often interpreted as timeless and predictable generality (with evolutionary "progress" towards some form of consciousness as a prominent example) truly falls into the domain of contingent, but still fully explainable, good fortune in the particular history of this specific planet (whatever the percentage of inhabited worlds that eventually evolve consciousness in some form).

Similarly—for the subject would otherwise evoke no interesting debate—D'Arcy Thompson, Kauffman, Goodwin, and all biologists attracted to this view of life have extended their putative realm of predictable generality based on universal physical structure into examples coveted by historicists as central features of their domain. For example, I have no quarrel with Goodwin's (1994, p. 132) attribution of patterns in phyllotaxis, particularly the transition from distichy to spiral ordering, to simple constraining geometries of necessary spatial filling based on the size and rate of origin for new units at the generating center—if only because the promiscuous phyletic scattering of these transitions, and their consequent correlation to immediate rates and sizes rather than to historical context, points to physical automaticity rather than to genealogical constraint. "The frequency of the different phyllotactic patterns in nature," Goodwin writes, "may simply reflect the relative probabilities of the morphogenetic trajectories of the various forms and have little to do with natural selection."

But I balk when Goodwin then wishes to extend this claim for physical generality to such a phyletically localized, complex, and historically particular structure as the tetrapod limb (whereas I acknowledge, of course, the fascination and utility of recent data from evo-devo—see Chapter 10—on general rules that this historical particular then uses to craft its uniquenesses). I don't deny Goodwin's following statement about generation from rules (1994, p. 155), but I would maintain that these particular rules originated as consequences of contingent events in vertebrate history (now expressed as regularities in the development of limb buds), and not as simple properties of the universal order of geometry: "Tetrapod limbs are defined as the set of possible forms generated by the rules of focal condensation, branching bifurcation, and segmentation in the morphogenetic field of the limb bud. All forms are equivalent under transformations that use only these generative processes. With this we arrive at a logical definition of tetrapod limbs that is independent of history. The idea of a common ancestral form as a special structure occupying a unique branch point on the tree of life ceases to have taxonomic significance."

Several colleagues have complained that phrases like "adaptation to the edge of chaos," while incorporating some currently fashionable imagery and terminology, lack clear scientific definition and operational utility. I regard this judgment as overly harsh and would argue to the contrary, that Kauffman and his colleagues at the Santa Fe Institute for the study of complex systems are groping towards something important. If we have been unable, thus far, to achieve a rigorous formulation, we should at least recognize that science itself has been so tuned to other, largely reductionist, modes of thought, that the basic conceptual tools have never been developed. I welcome this exploration in terra largely incognita and would only like to point out, in ending this section, that the implications for evolutionary theory may extend even further than the major protagonists have recognized.

In particular, and for its engagement with a dominant theme of this book (the hierarchical reformulation of selectionist theory as the first leg on a tripod of Darwinian central logic), if species gravitate to a position of best possible balance between optimization for the moment and flexibility for future change, then Darwinian organismic selection cannot directly fashion such adaptations with advantages only measurable in terms of capacity for success in the face of future environmental changes, a central component in Kauffman's concept of benefits provided by residence at the edge of chaos (p. 409): "Selection, I shall further suggest, by achieving genomic systems in the ordered regime near the boundary of chaos, is likely to have optimized the capacity of such systems to perform complex gene-coordination tasks and evolve effectively." A Darwinian can argue that flexibility linked to future capacity for change arises exaptively as a lucky consequence of features actively evolved for immediate organismic advantage. But such capacities can also evolve by direct selection, at a higher level, for species-individuals who win differential reproductive success by their propensity for living through external crises that consign closely related species-individuals to extinction.

## Exapting the Rich and Inevitable Spandrels of History

### NIETZSCHE'S MOST IMPORTANT PROPOSITION OF HISTORICAL METHOD

A. N. Whitehead famously remarked (see previous citation on p. 57) that all later philosophy might well be described as a footnote to Plato. How often, indeed, must any decent scholar invent a formulation with pride in systematic analysis, and with hope for originality—only to discover that one of history's truly great thinkers had established the same principle, recognized its importance, and even specified its full range of application. I described a case of this ultimately humbling experience (see p. 51), when I discovered that, for all the struggles of several macroevolutionists, none more intense than my own, to define a workable concept of species selection in the 1970's, Hugo de Vries had formulated the idea, and even applied the same name, in his 1905 book, written in English and therefore scarcely qualifying as light hidden under a bushel for anglophonic readers.

I knew that a claim for originality could never be asserted for my various writings on the key structural and historical principle of inherent differences between current utility and causes of origin, and on the consequent impossibility of inferring reasons for evolutionary construction only from current adaptive roles (Darwin, after all and as we shall see (pp. 1218–1224), invoked this principle to disperse a theoretical objection that he regarded as the most potent challenge to the central logic of natural selection). I have written about this subject for more than 30 years, and with a growing attempt at systematization, moving from my naïve, albeit accurate, distinction of "immediate" and "retrospective" significance in *Ontogeny and Phylogeny* (1977b), to the spandrels of San Marco (Gould and Lewontin, 1979), to the codification of exaptation as a missing term in the science of form (Gould and Vrba, 1982). Only my work on punctuated equilibrium has attracted more citations, and I ventured to hope that I had, at least, presented a richer and more systematic analysis to demonstrate the centrality of this underappreciated principle, which operates so effectively as a bulwark for structuralist perspectives in evolutionary theory—thus setting the location of the topic within this book. (I wrote in 1982, epitomizing the meaning of this theme for the general subject of constraint as a structural channeler of adaptation (Gould and Vrba, 1982, p. 13): "Exaptive possibilities define the 'internal' contribution that organisms make to their own evolutionary future.")

Then, in 1998 and thanks to the broader vision of my (then) graduate student Margaret Yacobucci, I discovered that Friedrich Nietzsche had brilliantly elucidated this principle, with its full spate of implications, in one of his most celebrated works, *The Genealogy of Morals*, first published in 1887.

Throughout his career, Nietzsche (1844–1900) struggled to identify and define the root motives behind our conventional beliefs about morality, philosophy and religion in Western traditions. He viewed these beliefs as secondary and functional expressions of a primary, generating source: "the essence of life, its will to power" (1967 edition, p. 56). And he recognized that we would never understand the nature and character of this primary source if we only analyzed the current utility of its secondary manifestations.

Nietzsche has received a bum rap from history, and for reasons clearly beyond his intention or control. In identifying traditional beliefs as secondary expressions of a will to power, he did not wish to deny their potency or their value, but only to make a proper logical separation so that their sources of origin, which we must also understand if we wish to achieve a full appreciation of their history and status, might be disentangled from their current utility. The later fascist misreading of Nietzsche did try to validate the worth, and to promote the pure expression, of a will to power on no basis beyond its mere existence—the very illogical step that Nietzsche analyzed so clearly in the work discussed in this section.

Nietzsche became mentally incapacitated in 1889, and lived the last years of his life under his sister's care. She later became an ardent Nazi, and used her control over Nietzsche's literary estate to further her own purposes, including the publication of notes that Nietzsche had discarded, and even some minor forgeries of her own. We owe Nietzsche far more respect and admiration than he receives from those who know him only for a common misunderstanding of his concept of the "*Übermensch*" or "superman" (not a Hitlerian defense of domination by the more powerful, but Nietzsche's ascetic description of a person who could accept complete repetition ("eternal recurrence" in his terms) of life, with all its horrors, rather than wishing for an edited version); and from those who may feel ambivalent towards Richard Strauss' tone poem on *Also Sprach Zarathustra* (Nietzsche's treatise on the *Übermensch*), the source for the stunning opening theme of Stanley Kubrick's film *2001*, and a bit scary in its apparent glorying of a transcendence that might not always be kind to the majority left behind.

In the *Genealogy of Morals*, Nietzsche addresses the differences between historical origin and current utility in section 12 on the nature and meaning of punishment. He begins by outlining the various opinions of moral philosophers on the function or purpose of punishment in current society—"for example, revenge or deterrence" (p. 55), or more specifically (p. 57) "as a means of rendering harmless, of preventing further harm . . . as payment of a debt to the creditor in any form (even one of emotional compensation) . . . as a means of isolating disturbance of balance."

Nietzsche does not deny the force of these current utilities (and may well approve them as a matter of personal or public morality). Rather, he wants to resolve the different issue of the historical origin of punishment in human evolution (a quest highlighted in his book's title, *The Genealogy of Morals*). He recognizes that confusion between his question of historical origin and the unchallenged documentation of current utility poses the greatest barrier towards resolution. The opening sentences of this section outline the problem (p. 54): "Now another word on the origin and purpose of punishment—two problems which are separate, or ought to be: unfortunately people usually throw them together. How have the moral genealogists reacted so far in this matter? Naively, as is their wont: they highlight some 'purpose' in punishment, for example revenge or deterrence, then innocently place the purpose at the start, as *causa fiendi* [cause of making] of punishment—and have finished. But 'purpose of law' is the last thing we should apply to the history of the emergence of law."

I would not, in this book, so highlight this crisp dissection of a key problem in evolutionary biology as well—the distinction between historical origin and current utility—if Nietzsche had not generalized the issue as central to all historical study, and if he had not so clearly explicated both the biological meanings, and the implications for adaptationist analysis as well.

Nietzsche labels the need to distinguish historical origin from current utility as "the major point of historical method" (p. 57). "There is no more important proposition for all kinds of historical research" (p. 55), he adds, just before presenting his clearest statement of the general issue: "Namely, that the origin or the emergence of a thing and its ultimate usefulness, its practical application and incorporation into a system of ends, are *toto coelo* [entirely, or literally 'to the highest heavens'] separate; that anything in existence, having somehow come about, is continually interpreted anew, requisitioned anew, transformed and directed to a new purpose."

To resolve his particular issue, Nietzsche needs to make this separation because he wishes to locate the *origin* of punishment in the almost inevitable manifestation of a primal will to power. But if we make the mistake of equating an admitted and efficacious modern utility (in deterrence or resolution of debt, for example) with the ground of origin as well, we will never understand the *genealogy* of morals. Again, and contrary to the common misunderstanding, Nietzsche does not wish to advocate historical origin as a source of validation. Quite to the contrary, he argues that we need to understand the reasons for origin in order to analyze the source and strength of the underlying motivation (whatever the current utility), thus giving us better insight into our actions and natures.

In a fascinating passage, Nietzsche then uses the biological example of eye and hand to assert his specific point about punishment, and to introduce a relative ranking, with the adaptation of current utility regarded as a secondary imprint upon a more fundamental original source:

> No matter how perfectly you have understood the usefulness of any physiological organ (or legal institution, social custom, political usage, art form or religious rite) you have not yet thereby grasped how it emerged: uncomfortable and unpleasant as this may sound . . . for people down the ages have believed that the obvious purpose of a thing, its utility, form and shape are its reason for existence: the eye is made to see, the hand to grasp. So people think punishment has evolved for the purpose of punishing. But every purpose and use is just a sign that the will to power has achieved mastery over something less powerful.

Two other aspects of Nietzsche's extraordinary analysis show how completely he had grasped this key principle of historical explanation with all its far-reaching implications, each of equal importance in evolutionary biology as well. First, he recognizes (as Darwin did) that the disengagement of current utility from historical origin establishes the ground of contingency and unpredictability in history—for if any organ, during its history, undergoes a series of quirky shifts in function, then we can neither predict the next use from a current value, nor can we easily work backwards to elucidate the reasons behind the origin of the trait. Note, in the following passage, how Nietzsche refers to the chain of secondary utilities as "adaptations"; how he specifies that the steps in the sequence of utilities follow each other "at random" (in Eble's (1999) sense of unrelated to, and unpredictable from, previous states and not in the strict mathematical sense); and how he clearly recognizes the significance of this principle for dispersing any hope that a phyletic history might be interpreted as a "progressus towards a goal," another almost eerie similarity with Darwin's understanding of the meaning of contingency in evolution:

> The whole history of a "thing", an organ, a tradition can to this extent be a continuous chain of signs, continually revealing new interpretations and adaptations, the causes of which need not be connected even amongst themselves, but rather sometimes just follow and replace one another at random. The "development" of a thing, a tradition, an organ is therefore certainly not its *progressus* towards a goal, still less is it a logical *progressus*, taking the shortest route with least expenditure of energy and cost,—instead it is a succession of more or less profound, more or less mutually independent processes of subjugation exacted on the thing.

Second, Nietzsche promulgates an ordering of importance, with reasons for origin as primary in more than a merely temporal sense, and current utilities as sets of secondary "adaptations" (his description) with only transient status and less influence (than the persisting force behind the primal origin) upon any future state. I would not defend this ranking for an application to evolutionary theory, if only because Nietzsche's formative "will to power" identifies a persisting force that must influence any subsequent adaptation as well, whereas the original context of an evolved phenotypic feature need not exert such a continuing hold upon later history. But I do appreciate Nietzsche's point, which can be translated into evolutionary terms as the source of constraint. The original reason does continue to exert a hold upon history through the structural constraints that channel later usages. Once feathers originate for thermoregulation, the form of any later utility for flight will be influenced by features built for the original context.

Nietzsche therefore criticizes those "genealogists" who mistake a current utility for a source of origin—for this erroneous argument "forces 'adaptation' into the foreground, which is a second-rate activity, just a reactivity . . . This is to misunderstand the essence of life, its will to power. We overlook the prime importance which the spontaneous, aggressive, expansive, reinterpreting, redirecting and formative powers have, which 'adaptation' only follows when they have had their effect."

Finally, Nietzsche reasserts the biological analog of the hand to reinforce his rankings and to reemphasize the importance of understanding historical origin, and of establishing criteria for separating origins from later utilities (in the face of difficulties outlined at the end of the quotation) in any truly historical study:

> The procedure itself will be something older, predating its use as punishment, that the latter was only *inserted* and interpreted into the procedure (which had existed for a long time though it was thought of in a different way), in short, that the matter is *not* to be understood in the way our naïve moral and legal genealogists assumed up till now, who all thought the procedure had been *invented* for the purpose of punishment, just as people used to think that the hand had been invented for the purpose of grasping. With regard to the other element in punishment, the fluid one, its "meaning," the concept "punishment" presents, at a very late stage of culture (for example, in Europe today), not just one meaning but a whole synthesis of "meanings": the history of punishment up to now in general, the history of its use for a variety of purposes, finally crystallizes in a kind of unity which is difficult to dissolve back into its elements.

### EXAPTATION AND THE PRINCIPLE OF QUIRKY FUNCTIONAL SHIFT: THE RESTRICTED DARWINIAN VERSION AS THE GROUND OF CONTINGENCY

#### How Darwin resolved Mivart's challenge of incipient stages

Darwin treated this issue of discordance between historical origin and current utility in his catch-all Chapter 6 entitled *Difficulties on Theory*. Although this chapter amalgamates a potpourri of objections, and we may therefore conclude that Darwin regarded none of them as sufficiently central for separation as a full topic in its own right, he does grant special prominence to functional shift as a solution to a range of issues, including the overtly opposite pairing of "organs of extreme perfection" and "organs of small importance." He even dignifies the principle with a rarity in his own prose conventions, an adjective of intensification: "In considering transitions of organs, it is *so* important to bear in mind the probability of conversion from one function to another" (1859, p. 191, my italics).

But Darwin only came to appreciate the centrality of this principle when the book that he considered most cogent as a general critique of natural selection—St. George Mivart's *On the Genesis of Species* (1871)—led him to compose, for the 6th and final edition of the *Origin of Species* (1872), the only chapter ever added to his book, largely as a point by point refutation of Mivart's claims: the interpolated Chapter 7 entitled "Miscellaneous objections to the theory of natural selection." As a further observation on stylistic questions, I'm sure that Mivart's decision to name his own book with a parody on Darwin's title must have caught Darwin's special attention, and perhaps his ire. In calling his work *On the Genesis of Species* (rather than the *Origin*), Mivart needled Darwin with the common taunt of the times (see pp. 139–140): that natural selection could play a minor and negative role in eliminating the unfit, but that some other "positive" force must generate the fit. (I suspect that most of us would prefer to have our ideas rejected as dangerously wrong, but at least interesting and worthy of anathematization, rather than dismissed as correct, but trivial.)

Mivart expresses the thoroughness of his condemnation (speaking of himself in the third person) with a common rhetorical strategy in Victorian science: claiming the ultimate fairness of an initially favorable impression, only dispelled and reversed by careful and objective consideration of a catalogue of empirical evidence. In this passage, Mivart asserts the "secondary and subordinate" role of natural selection, while claiming that another "positive" mechanism, able to generate the fit (of his book's title), must be sought (1871, p. 225):

> He was not originally disposed to reject Mr. Darwin's fascinating theory. Reiterate endeavours to solve its difficulties have, however, had the effect of convincing him that that theory as the one or as the leading explanation of the successive evolution and manifestation of specific forms, is untenable. At the same time he admits fully that "Natural Selection" acts and must act, and that it plays in the organic world a certain though a secondary and subordinate part.
>
> The one *modus operandi* yet suggested having been found insufficient, the question arises, Can another be substituted in its place? If not, can anything that is positive, and if anything, what, be said as to the question of specific origination?

St. George Mivart (1817–1900) became something of a tragic figure in Victorian biology. He devoted much of his career to reconciling biology and religion in terms of his unconventional attitudes in each discipline—only to meet ultimate rejection by both camps. At age seventeen, he abandoned his Anglican upbringing, became a Roman Catholic, and consequently (in a less tolerant age of state religion) lost his opportunity for training in natural history at Oxford or Cambridge. He became a lawyer but managed to carve out a distinguished career as an anatomist nonetheless. He embraced evolution and won firm support from the powerful T. H. Huxley, but his strongly expressed and idiosyncratic anti-Darwinian views incurred the wrath of Britain's biological establishment. He tried to unite his biology with his religion in a series of books and essays, and ended up excommunicated for his trouble six weeks before his death.

The ever-perceptive Darwin had chosen well for his primary source of worry. In this book, Mivart presents a thorough and logically inclusive account of structuralist evolutionary thought as a substitute for natural selection (complete with the usual linkage of channeling and saltation into a coherent primary critique). Mivart centered his attack upon an argument embodied in a phrase that still persists as virtually his only commonly recognized legacy to the history of evolutionary thought. He introduces this phrase as the title to his first substantive chapter (after some opening pages entitled "introduction"): "The incompetency of 'natural selection' to account for the incipient stages of useful structures." (In my popular writing, I have referred to this critique as the "5 percent of a wing principle," as expressed in the common layman's objection: "I can understand how wings work for flight once they originate, but how can evolution ever make a wing in Darwin's gradualist and adaptationist mode if five percent of a wing can't possibly provide any benefit for flight?")

Mivart constructs this chapter as a compendium of examples where, in his judgment, no putative value could be assigned to early incipient stages (lepidopteran mimicry, flight, the placement of both eyes on the upper side of flatfishes, etc.). He then concludes (1871, p. 61): "That minute, fortuitous, and indefinite variations could have brought about such special forms and modifications as have been enumerated in this chapter, seems to contradict not imagination, but reason."

Mivart then attempts to resolve this problem in the most obvious manner—by the saltationist claim that intermediary stages never existed, and that novel adaptations may arise in single steps. Interestingly, and as mentioned in Chapter 5 (p. 344), Mivart invokes the compelling structuralist model and image of Galton's Polyhedron (see pp. 342–351) to illustrate this centerpiece of his system (pp. 97–98):

> Arguments may yet be advanced in favor of the view that new species have from time to time manifested themselves with suddenness, and by modifications appearing at once (as great in degree as are those which separate *Hipparion* from *Equus*), the species remaining stable in the intervals of such modifications: by stable being meant that their variations only extend for a certain degree in various directions, like oscillations in a stable equilibrium. This is the conception of Mr. Galton, who com-

pares the development of species with a many faceted spheroid tumbling over from one facet, or stable equilibrium, to another. The existence of internal conditions in animals corresponding with such facets is denied by pure Darwinians.

Mivart recognizes that any saltationist claim must resolve the two cardinal objections that lead many scientists to embrace the gradualism of Darwin and Lyell: (1) How, short of invoking something miraculous, can so many parts be altered together, and all at once, to produce a harmonious and adaptive result in a highly modified descendant; and (2) how can any theory of one-step transformation explain the parallel or convergent modification of sets of co-ordinated features, as found in many independent lineages?

Mivart faces these difficulties along the standard structuralist route, by calling upon the other concept so often twinned with claims for saltation: internally channeled change. If alterations, like isotropic Darwinian mutations, could modify an organism in any direction, then the difficulties stated above would become insuperable. But if jumps can only occur along certain limited routes, set by the internal structure of organisms and predisposed towards harmonious alteration of coordinated parts, then saltations become both limited in directional expression, and biased towards workability. Again, Galton's Polyhedron suggests such a linkage of saltation (facet flipping) with internally limited directionality (restriction of routes of change to positions underlain by adjacent facets). Mivart states (1871, p. 143): "All these difficulties are avoided if we admit that new forms of animal life of all degrees of complexity appear from time to time with comparative suddenness, being evolved according to laws in part depending on surrounding conditions, in part internal—similar to the way in which crystals (and, perhaps from recent researches, the lowest forms of life) build themselves up according to the internal laws of their component substance, and in harmony and correspondence with all environing influences and conditions."

But what operational good can emanate from an invocation of such internal forces (even granting the logical soundness of the argument), if the nature of these forces remains unknown and mysterious, thus reducing their status to special pleading? Mivart even cites the classical literary spoof of such foolish arguments, as presented by the phony doctors of Moliere's hypochondriac: "But it may be again objected that to say that species arise by the help of an innate power possessed by organisms is no explanation, but is a reproduction of the absurdity, *l'opium endormit parcequ'il a une vertu soporifique* (p. 230)" (opium puts you to sleep because it possesses a soporific virtue).

In reply, Mivart points out that we also know nothing about the physical nature of Newtonian gravity, but still find the concept useful because such mathematical regularities as the inverse square law have explanatory, predictive and integrative power. Similarly, we may not know the actual workings of heredity, but we can, by empirical cataloguing and experimentation, determine sets of observed regularities in the variations of modern species. Following the traditions of 19th century structuralism, Mivart recommends the study of natural "sports," both the occasional large variants that can survive in nature, and the teratological malformations that may not be viable, but that illustrate the potential pathways of internally coordinated variation, following recognizable channels of ontogeny, sexual variation, etc. "It is probable therefore that new species may arise from some constitutional affection of parental forms—an affection mainly, if not exclusively, of their generative system" (p. 233).

In the added chapter to his 6th and final edition, Darwin refers to Mivart as "a distinguished zoologist," and admits that he has presented all viable objections to natural selection "with admirable art and force" (1872b, p. 164). He then summarizes Mivart's structuralist alternative, describing first the claim for channeling, and then the argument for saltation. He rejects both, primarily because they lack either a known mechanism or verified cases, Darwin then reasserts his belief in the efficacy of gradualistic natural selection, working upon isotropic and undirected variation (pp. 187–188):

> At the present day almost all naturalists admit evolution under some form. Mr. Mivart believes that species change through "an internal force or tendency," about which it is not pretended that anything is known. That species have a capacity for change will be admitted by all evolutionists; but there is no need, as it seems to me, to invoke any internal force beyond the tendency to ordinary variability, which through the aid of selection by man has given rise to many well-adapted domestic races, and which through the aid of natural selection would equally well give rise by graduated steps to natural races or species . . .
>
> Mr. Mivart is rather inclined to believe, and some naturalists agree with him, that new species manifest themselves "with suddenness and by modifications appearing at once." For instance, he supposed that the differences between the extinct three-toed Hipparion and the horse arose suddenly. He thinks it difficult to believe that the wing of a bird "was developed in any other way than by a comparatively sudden modification of a marked and important kind"; and apparently he would extend the same view to the wings of bats and pterodactyls. This conclusion, which implies great breaks or discontinuity in the series, appears to me improbable in the highest degree.

Darwin acknowledges that Mivart's argument about incipient stages had been particularly troubling (p. 165): "The one new point which appears to have struck many readers is, 'natural selection is incompetent to account for the incipient stages of useful structures.' This subject is intimately connected with that of the gradation of characters, often accompanied by a change of function—for instance, the conversion of a swim-bladder into lungs." Darwin notes that he had dealt with this issue in his original Chapter 6 on "difficulties," admits that he had not paid the subject sufficient heed, and praises Mivart for an opportunity to correct his previous slighting (p. 185): "A good opportunity has thus been afforded for enlarging a little on gradations of structure, often associated with changed functions,—an important subject, which was not treated at sufficient length in the former editions of this work."

I will not describe Darwin's solutions to the problem of "incipient stages" at length, because I have already done so in Chapter 2, my exegesis of the *Origin of Species*. But I do need to emphasize the relevant point for this chapter—that both of his effective arguments against Mivart's supposed trump card invoke structural principles of constraint rooted in Nietzsche's "major point of historical method": the discordance between historical origin and current function.

In his first argument, Darwin readily admits the "5 percent of a wing" problem, and then presents the incisive solution that becomes enshrined in evolutionary theory (by no fault of Darwin, who never used the term) under the most unfortunate name of "preadaptation." Yes, five percent of a wing offers no conceivable aerodynamic benefit, and could not therefore either be formed, or converted into a full wing, under a smooth regime of natural selection *for flight*. But sequences forged by selection only presuppose continuity in differential reproductive success, not continuity in a single function. Thus, the incipient stages may have performed a different function, for which their 5 percent of a wing imparted benefits. Eventually, the enlarging proto-wing entered the domain of aerodynamic benefit, and the original function changed to the primary utility now exploited by most birds. Current function cannot be equated with reasons for historical origin. Mivart's cardinal objection disappears, thus explaining why "it is *so* important to bear in mind the probability of conversion from one function to another."

Darwin roots his second argument in the related, but even more generalized, structural principle of redundancy—the inherent capacity (based on intrinsic structure rather than current function) of most organs to work in more than one way (either at the same time, providing dual benefits, or with one utility overtly exploited by natural selection, and the other latent, providing unselected flexibility for future change). Darwin presents this argument in a fascinating manner by coupling two apparently opposite facts about redundancy: that a single function can be performed by more than one organ, and that a single organ can perform more than one function. Thus, an organ need not invent an entirely new function in some mysterious manner, but may evolve by intensifying a previously minor use, or even by recruiting an inherent but unexpressed potential. Meanwhile, the modified organ can abandon its previous major function because other organs can continue (or intensify) their former operation in the service of the same necessary task.

Thus, reptilian jaw bones can become mammalian ear bones because they already played some role in sound transmission while they functioned primarily to articulate the jaw of therapsid forebears (the principle of two functions for one structure). They then become free to move into the middle ear because the transitional forms (as demonstrated empirically by such fossils as *Diarthrognathus*, and not only as a reasonable conjecture) possessed a double jaw joint (the reciprocal principle of two structures for one function)—and the bones of the old quadrate-articular joint could then become the malleus and incus of the ear because the new dentary-squamosal joint was already "up and running," thus avoiding the specter of an inconceivable intermediate with an unhinged jaw.

Interestingly, Darwin's own favorite example for coupling these two principles of redundancy, repeated many times in the *Origin of Species* and still resident in most biology textbooks today, is not only wrong, but backwards. He invoked this coupling to explain the supposed conversion of piscine air bladders into lungs. But the same argument works just as well, in reverse order, for the actual transformation of lungs in plesiomorphic fishes to swim bladders in highly derived teleosts that did not originate until the Triassic (Liem, 1988). Darwin wrote (1859, pp. 204–205): "A swim-bladder has apparently been converted into an air-breathing lung. The same organ having performed simultaneously very different functions, and then having been specialized for one function; and two very distinct organs having performed at the same time the same function, the one having been perfected whilst aided by the other, must often have largely facilitated transitions." (Ancestral fish lungs can indeed also function for buoyancy, whereas, and more obviously, gills work as well as lungs for breathing. Modern lungfishes retain both systems, as their formal name, Dipnoi (or "two breathing") testifies.)

### The two great historical and structural implications of quirky functional shift

This principle of functional shift deserves far more prominence, and explicit recognition, than it has ever received among evolutionary theorists. I have tried to emphasize its vital role in establishing the contingency and unpredictability of evolutionary change by an adjectival strategy of designation as "quirky functional shift." In operational terms, we should acknowledge, most of all, the property of major functional alteration based upon far more limited (in extreme cases, virtually absent) structural change—another way of expressing the structuralist concept of inherent flexibility in natural forms and designs (to different degrees that should be subject to specification on a case by case basis). In any event, however textually underemphasized, this principle has always played two important roles in standard Darwinian theory:

AS THE GROUND OF CONTINGENCY FOR LIFE'S HISTORY. Thoughtful Darwinians, no matter how confidently they have identified natural selection as the exclusive cause of evolutionary change, have always recognized that their theory necessarily underpredicts the actual pathways of life's history—and that explanations for the byways of individual lineages (and major aspects of the highways as well) can only be located in the factual record of particulars. This concept potentiated the tacit truce that, until recent years, held between paleontologists and Darwinian theorists under the Modern Synthesis. Under accepted terms, the theorists said to the paleontologists: "give up your old claims about special macroevolutionary mechanisms, and admit our contention that microevolutionary population genetics and natural selection hold full theoretical sufficiency. We will then grant you control over the actual pageant of life's history by allowing that no nomothetic theory (and ours is 'as good as it gets') can specify actual pathways without factoring in the historical particulars that only your record preserves." (I have scarcely hid my conviction, either in this book or elsewhere, that this truce always operated as a "lousy deal" for the science of paleontology.)

In a minimal sense, Darwinian theory must grant this space to contingency—if only because even the most basic and least sophisticated form of the theory holds that organisms adapt to changing local environments (and do not follow preestablished routes towards "progress" or any other goal). Since we all admit that local environments change on an erratic and contingent vector through time, life's overall pathway must be dominated by contingent factors, even if every immediate event of natural selection could, in principle, achieve a deterministic explanation in local environmental terms. (After all, this feature of Darwinism as emphasized in Chapter 2, and as long appreciated by intellectual historians, established the most radical aspect of natural selection from the start—as contrary to all earlier evolutionary speculations, with their assumptions about lawlike directionality, usually regulated by divine intent.)

But if contingency resided only in this basic aspect of environmental scaling, then the principle, though sound enough, would not run so deep in Darwinian traditions. Rather, contingency gains its greatest force through the principle of quirky functional shift: the discordance between historical origin and current utility, and the consequent fallacy of direct inference from modern status to initial meaning. Nietzsche emphasized the primary role of this discordance in the study of history by writing (as quoted more fully on p. 1217) that "the development of a thing, a tradition, an organ is therefore certainly not its *progressus* towards a goal," and that the inevitability of functional shift makes any important historical sequence "instead . . . a succession of more or less profound, more or less mutually independent processes of subjugation exacted on the thing."

Even a unidirectional sequence of changing form with basically unaltered function would require explicit knowledge of contingent environmental histories for anything close to full or satisfactory explanation. The addition of quirky functional shift, usually in several episodes for each organ in any complex phylogeny, guarantees a cardinal role for historical explanation in any major lineage (again, as Nietzsche recognized). In a personally favorite example, for combining the canonical general case with a particular ending twist, the African black heron, *Egretta ardesiaca,* uses its wings largely to shade the shallow water of its habitat, thus providing a clear view of available prey (my thanks to E. Vrba for this example, as discussed in Gould and Vrba, 1982).

Any intelligent person with a sense of history's length and meaning could identify the author of the following statement as a modern Parsifal, or perfect fool: "Aha, now I know why herons evolved wings—in order to eat, for they would starve if they couldn't shade the water and see their food." Our savvy interlocutor would offer the obvious refutation that most birds use their wings to fly (as does the egret, albeit as a "demoted" secondary utility in the species' current habitat). Any knowledgeable biologist would then add—hence my fondness for this example of *double* quirky functional shift—that feathers were initially coopted for flight in a much older functional shift, a defining transition in avian phylogeny from a different initial role, perhaps in thermoregulation. This old hypothetical argument has now been fortified by two sources of evidence: experimental data on the thermodynamic, with no accompanying aerodynamic, advantages of tiny protowings (Kingsolver and Koehl, 1985); and historical data on the probable origin of birds from the smallest-bodied lineage of running dinosaurs, a group that might have experienced the greatest functional need for supplementary thermoregulation, given their highest levels of activity combined with highest surface/volume ratios (as a consequence of minimal body size), of any lineage within the dinosaurian clade. Indeed, this example has served as the canonical illustration, ever since Mivart's defining book of 1871, for "the problem of incipient stages of useful structures," or quirky functional shift (and also as the eponym for my designation, in nontechnical writing, as "the 5 percent of a wing problem").

Faced with this argument, our Parsifal might continue to reject contingency, and embrace predictability, on the false assumption that natural knowledge, as "scientific," must be so constituted. Even after we mock his previous conviction that he knows why egrets developed wings when he understands their present use, we discover that he has not generalized the message, for he now argues from the opposite temporal end: "But if I, as a great scientist with full knowledge of evolutionary theory, had visited the Earth in early Jurassic times and observed the avian ancestor as a small running dinosaur using feathers on protowings for supplementary thermoregulation, I surely could have concluded that this animal would evolve larger wings and eventually enter a realm of cooptable utility for flight. I would also know that, for 150 million years, the ancestors of African herons would use those wings for flight, and then, on a continent to be called Africa [for our seer also knows the future history of plate tectonics on Earth] this one little avian lineage would redeploy those old thermoregulatory organs for yet another novel task of shading water." At this point, and continuing the literary analogy, we could only hope that our Parsifal finds the Holy Grail of quirky functional shift, and abandons his foolish ways for the path of wisdom!

Needless to say, the actual history of any key organ in any major lineage far surpasses this avian cartoon both in complexity and in number of episodes of functional shifting. Just consider, for starters, the passage from a cartilaginous rod-like element, functioning to support the agnathan gill, to the hyomandibula of gnathostome fishes (used primarily to suspend the upper jaw from the cranium), to the stapes of tetrapods (following fusion of the upper jaw to the skull, and responding to a functional need for a different mode of sound perception in air *vs.* water).

This quirky historical character of major evolutionary change in particular lineages—thoroughly explainable after the fact, however unpredictable in principle beforehand—constitutes the greatest fascination of the subject for many practitioners, myself included. Yet, this same inherent historicity has saddened scientists of other temperaments and predilections. For people who find greatest satisfaction in those aspects of nature that achieve full meaning and explanation under invariant and timeless laws, but who cannot resist the fascination of evolutionary biology as a career, the irreducibly contingent aspect of their chosen subject defines its least congenial attribute. Such scientists have therefore tended to underplay (or even, in extreme cases, largely to deny) contingency, or to focus on those broader aspects of the subject, far from the fascination of the to-ings and fro-ings of real history in concrete lineages, that do fall into the more conventional realm of predictability under natural law. Indeed, the previous section of this chapter treated this species of structuralist thought—and though I did not hide my own lack of affinity for this approach, I trust that I did grant the subject my genuine respect and acknowledgment of partial validity (while also expressing my abiding admiration for the sheer iconoclasm and beautiful prose of D'Arcy Thompson). I may be a historian at heart, but I do understand Kauffman's frustration, and his point, when he recognizes the intellectual linkage of natural selection to contingency, and then writes (1993, p. 26):

> We have come to think of selection as essentially the only source of order in the biological world. . . . It follows that, in our current view, organisms are largely ad hoc solutions to design problems cobbled together by selection. It follows that most properties which are widespread in organisms are widespread by virtue of common descent from a tinkered-together ancestor, with selective maintenance of the useful tinkerings. It follows that we see organisms as overwhelmingly contingent historical accidents, abetted by design . . .
>
> My own aim is not so much to challenge as to broaden the neo-Darwinian tradition. For, despite its resilience, that tradition has surely grown without seriously attempting to integrate the ways in which simple and complex systems may spontaneously exhibit order.

AS ONE OF THE TWO MAJOR SOURCES OF STRUCTURALIST INPUT INTO THE PRIMARILY FUNCTIONALIST BASIS OF DARWINIAN THEORY. I treated Darwin's primary acknowledgment of a subsidiary role for structuralist, and at least partly non-adaptationist, thinking within the theory of natural selection—his treatment of "correlations of growth," or non-adaptative side-consequences of adaptive change—in Chapter 4, pages 330–341. Darwin's discussion of quirky functional shift, and his recognition of this principle's indispensability for including the evolution of major novelties within the compass of natural selection by gradual change, marks his second substantial foray into subsidiary themes of a primarily formalist or structuralist character—in modern terms, his acknowledgment of an important role for internal constraint (as a precondition and helpmeet for natural selection) in directing the history of evolutionary lineages.

The role played by historical constraint in quirky functional shift lies implicit within the previous discussion of contingency, and therefore needs little additional elaboration. Suffice to say that if a capacity for utilization in markedly different ways did not lie within the inherent or formal structure of most primary adaptations, then evolution would never be able to reach a novel "there" from its present "here"—and life's history would stagnate in transient perfection (and then expire when surrounding environments underwent their occasional substantial alterations).

After all, natural selection cannot act as a magic wand for the immediate construction of any urgent need. The adaptability—or, in the more general term now finally receiving substantial and deserved attention from organismal biologists (see p. 1270), the "evolvability"—of any phenotype must depend, in large part, on a flexibility for future change that simply cannot arise, if we understand the nature of causality itself aright, by direct natural selection at the usual Darwinian level of organismal phenotypes. Therefore, a large component of evolvability must be attributed to inherent structural properties of features that originated by natural selection for one reason, but also manifest a capacity for subsequent recruitment (with minimal change) to substantially different and novel functions. The study and systematization of these formal and structural reasons for evolvability sets an important agenda, now largely unfulfilled but attracting considerable interest, for evolutionary biology.

To return to my previous example, the agnathan ancestor that built a series of v-shaped, backward-pointing gill arches, each made of several rod-like elements, for pumping water to breathe and feed, evolved these features for its own immediate needs, and not (obviously) with any forethought about modifiability into jaws that might one day surround its unsupported mouth. But if the elements of the foremost arch had not inherently possessed the form, the positioning, the coordination, and the developmental potential to move to a more anterior position surrounding the mouth, the gnathostome lineage would never have emerged, the agnathans might have remained a relatively minor component of marine faunas (or become extinct entirely), and terrestrial environments, to this day, might have remained the domain of plants and insects—perfectly competent and "happy" ecosystems, building a lovely earth teeming with life, but evolving nothing conscious to proclaim its aesthetic, extol its virtues, or to record, perhaps even to seal, its doom. We must thank both this contingent good fortune, and the latent structural possibilities of gill arches, for this shot at our own particular brand of record keeping (even of "immortality" in some operationally meaningful sense of the term).

The story, of course, continues from there (and for each lineage), with a constant twinning of contingency and structural potentiality. If one marginal group of fishes had not evolved a peculiar fin, with a branching central element orthogonal to the body's antero-posterior axis (rather than parallel to the axis, as in most self-respecting members of the clade and guild), no support firm enough to build the centerpiece of a limb for terrestrial life might ever have emerged within the lineage of vertebrates. And if these resulting tetrapods had never evolved their forelimb for terrestrial locomotion, the celebrated convergence of aerodynamic form in the wings of bats, birds and pterosaurs—the supposed disproof of contingency's dominant role in evolution!—would have died aborning for want of a common and contingent substrate on which to hang these adaptive marvels of similarly excellent design.

In wondering why this principle of quirky functional shift, or the discordance between reasons for origin and current utility, has received such short shrift in Darwinian traditions, despite its pivotal importance for these two central aspects of natural selection, I can only conclude that its status has, heretofore, rested only upon its acknowledged capacity for auxiliary aid, and not upon a claim for conceptual novelty thus supplied to evolutionary theory. Such an evaluation flows easily from an understanding that quirky functional shift, when confined to this Darwinian formulation, remains entirely within the ordinary functionalist and adaptationist framework of the general theory. That is, Darwin's version of the principle, overly restrictive as I shall show in the next subsection, remains fully adaptational in confining its compass to functional shift *from one utility to another.* The feature in question initially evolves as a conventional adaptation for one function, and then becomes coopted for a different role. This shift may validate the centrality of contingency for a Darwinian explanation of history, but the process remains under adaptational direction at all times—and the fundamental mechanism of Darwinian evolutionary change never cedes any control. In other words, features that undergo this Darwinian style of quirky shift retain full functionality throughout, and their changes remain under the government of natural selection at all times. Thus, the analysis of history may be enriched, but the mechanisms of evolution do not alter or augment.

Nonetheless, although I may recognize why Darwinian tradition has underemphasized quirky functional shift, I still believe that this inattention has created substantial problems in our understanding of the logic of evolutionary change. (As a psychological inference, I also suspect that this neglect flows from the status of quirky functional shift as a slightly uncomfortable "odd man out," exuding a structuralist odor within an apparatus deemed powerful and intellectually intriguing for its functionalist basis and mechanics—just as Darwin's other structuralist principle of "correlations of growth" has received similarly little regard in the history of our field, at least until evo-devo made constraint an operational concept, thus inspiring both our interest and attention.)

### How exaptation completes and rationalizes the terminology of evolutionary change by functional shifting

Following my idiosyncratic interest in the use of language as an underappreciated guide to the history and relative ranking of concepts, I wish to cite a gaping hole in the logical terminology of Darwinian evolution as primary evidence for this long, and regrettable, undervaluation of quirky functional shift. (This section follows the argument of Gould and Vrba, 1982.)

As a staple of anglophonic biology, long predating Darwin's explanatory spin, and extending back to Paley at the dawn of the 19th century, and to Boyle in the heart of the scientific revolution of the late 17th century, the process of crafting (usually construed as God's creating) a feature for a particular utility has been called "adaptation"—following the etymology of fashioning for *(ad)* a use *(aptus)*. No problem so far; only an apt choice of terminology. But if "adaptation" denotes the process of crafting or creating for a use, what shall we call the resulting structure so used. We usually call the structure an "adaptation" as well—again no intrinsic problem, for we often use the same noun for a process and its results ("construction," "building," etc. to cite some analogs in the same architectural domain).

But problems may arise in historical systems if the current utility of an adaptation (the noun used for the result) did not arise by the process (also called adaptation) that built the result at its initial appearance—for, under the Nietzsche-Darwin principle of quirky functional shift, the form of the current adaptation (feature) may have arisen by adaptation (process) for a very different role. (Similarly, the *building* on my corner now serves as a shelter and soup kitchen for homeless people—a utility not directly related to the purpose of its initial *building* as a church. We may not become confused in this case because we know the short history of this site, and the current use does not stray far from the stated ideals of the broader institution that originally raised the structure. But we could make some serious errors if we maintained a strong interest in long histories with spotty records featuring multiple episodes of functional shifting, and then assumed that the use of a current building automatically revealed the intention of its original building. When we recognized and generalized the error in such reasoning, we might even want to make a terminological distinction between our name for the current object and our name for the process of its original construction.)

In fact, I am not inventing an abstract or overfine distinction here. This very problem has been directly, even urgently, addressed in some of the most widely read and respected writings in evolutionary biology. When Williams (1966) composed his classic defense and explication of adaptation, he wisely identified adaptation as an "onerous" concept, to be invoked only when truly necessary, and restricted to a clear domain of unambiguous definition and use. He recommended, in particular, that the term be applied to a current feature only when we can "attribute the origin and perfection of this design to a long period of selection for effectiveness in this particular role" (1966, p. 6). He even advocated a terminological distinction between the use of such a genuine adaptation (its "function") and the use, potentially just as crucial to an organism's survival, of a feature *not* crafted by selection for its current role (and therefore not an adaptation in Williams's restricted terminology). Williams suggested that we call this second form of fortuitous utility an "effect"—giving as an incisive, if somewhat facetious, example the propensity of flying fishes to fall back into the water as an effect (not the function) of the organism's mass. In other words, Williams invoked the term "effect" to designate the operation of a useful character not built by selection for its current role.

Although Darwin never formalized the issue, he clearly intended to restrict "adaptation" to Williams's sense of structures built by selection for their current utility—for he explicitly denies that an "indispensable" feature of mammalian development should be called an adaptation (1859, p. 197): "The sutures in the skulls of young mammals have been advanced as a beautiful adaptation for aiding parturition, and no doubt they facilitate, or may be indispensable for this act; but as sutures occur in the skulls of young birds and reptiles, which have only to escape from a broken egg, we may infer that this structure has arisen from the laws of growth, and has been taken advantage of in the parturition of the higher animals."

But if we follow, as I believe we should, this restrictive clarification advocated by Darwin and Williams, what shall we call a feature that initially arose for a reason different from the selective basis of its current operation—even though this present utility may be as crucial to the organism's adaptive success as the function of any organ built directly by selection for its current role? Indeed, and curiously, the lexicon of evolutionary biology, until recently, included no name for a feature that now contributes to an organism's fitness in natural selection, but that arose for a different reason—the very kind of outcome whose explication had been recognized by Nietzsche as "the major point of historical method." In other words, we cannot claim that the previous absence of such a term merely recorded the irrelevancy or peripheral status of the concept.

Evolutionary biology has long recognized a name for a related aspect of this phenomenon—but I cannot think of a more infelicitous term in our entire lexicon, explicitly so lamented and identified by scores of biologists. Because we have acknowledged the principle of quirky functional shift, if only for Darwin's own need in refuting Mivart's critique, we have felt some pressure to recognize a term for the potential utilities inherent in original uses. What should we call a feather's potential for flight while it still resides on the forearm of a small running dinosaur, functioning only for thermoregulation? Evolutionary biology has generally referred to such latent potentials as "preadaptations."*

For two reasons, "preadaptation" cannot fulfill our need for a term to designate features that arose for reasons different from their current utility.

---

*During the past 20 years, this term has been fading from use, in part (I believe) as a consequence of such critiques as this present section (and Gould and Vrba, 1982). Current graduate students may now encounter this term only rarely. But when I was a graduate student in the mid 1960's—admittedly a while ago, but not exactly Mesozoic either—preadaptation was a standard term in constant and continual use (Bock, 1959, for example). On the same theme of shifting terminology, reflecting a declining faith in the exclusivity of adaptation by natural selection as the basis of all evolutionary results, another common usage of my graduate years has completely disappeared. (I find it hard, even a bit embarrassing, to recall that we ever spoke so uncritically.) But, in these years (as all evolutionists of my generation will affirm), we used the term "adaptation" as a simple descriptive synonym—indeed the preferred name in professional circles—for *any* feature of a phenotype, with no intended implication about the origin or utility of the item. Merely to exist was to be an adaptation. We would, talking only descriptively about morphological features, the forelimbs of theropod dinosaurs, for example, say: "This adaptation was larger in *Allosaurus* than in *Tyrannosaurus*."

1. Preadaptation can only describe a potential future utility of a feature operating in a different manner in an ancestor. The thermoregulating feather may be called a preadaptation for flight. But when birds then coopt feathers as essential components of an airborne wing, we surely cannot continue to call them preadaptations for their *current* utility! I simply refuse to be called a "wannabe scientist" (or even a "promising scientist") because I once had a dream, and even (in retrospect) some inherent capacity for its realization, as a kid on the streets of New York.

2. What term in all our lexicon has ever come to us so inherently "prepackaged" for inevitable trouble and misunderstanding? The motivation behind the name may be clear and fair enough—the desire to recognize a different potential in a current actuality. But in our real world, where we so often allow our hopes for intrinsic meaning to obscure the realities of a natural order—random and senseless in human terms, and replete with "bad things happening to good people"—we guarantee ourselves nothing but trouble when we invent a word with a "plain meaning" of foreordination as a description and definition of our best examples to illustrate the precisely opposite concepts of fortuity and contingency. The resulting, entirely predictable, confusions became legion in biology classrooms, and professors developed a tradition for explicating and apologizing in advance whenever they mentioned "preadaptation." Terms that automatically evoke such embarrassment must be fatally flawed and fit only for the favored anathematization of my childhood years: "good riddance to bad garbage."

I could present a catalog of such textbook apologies, but will cite only Frazzetta's lament (1975, p. 212) to prove that my fulminations at least cannot be called idiosyncratic: "The association between the word 'preadaptation' and dubious teleology still lingers, and I can often produce a wave of nausea in some evolutionary biologists when I use the word unless I am quick to say what I mean by it."

To rectify this odd situation of a missing term at the center of a key subject in evolutionary biology, Vrba and I proposed that features coopted for a current utility following an origin for a different function (or for no function at all) be called *exaptations*—that is, useful (or *aptus*) as a consequence of *(ex)* their form—in contrast with adaptations, or features directly crafted for their current utility. Adaptations have functions, and exaptations, following Williams's recommendation, have effects. We summarize our recommendations in Table 11-1 (from Gould and Vrba, 1982).

This coinage completes a logical structure that has been recognized ever since Darwin (and made explicit ever since Nietzsche) but that never included a term for one of the central rooms in the edifice. (My reasoning may be both simplistic and self-serving, but I can imagine only one explanation for such a curious situation: following Darwinian traditions, and especially under the orthodoxy of the "hardened" version of the Modern Synthesis, biologists became so accustomed to regarding all evolutionary change as adaptation directed by natural selection that they lost sight of the importance, or even the existence, of an undeniable corollary—that many (indeed most) features, as a

Table 11-1. A Taxonomy of Fitness.

| Process | Character | Usage |
| --- | --- | --- |
| Natural selection shapes the character for a current use—adaptation | Adaptation | Function |
| A character, previously shaped by natural selection for a particular function (an adaptation), is co-opted for a new use—cooptation | Exaptation | Effect |
| A character whose origin cannot be ascribed to the direct action of natural selection (a nonaptation), is coopted for a current use—cooptation | Exaptation | Effect |

consequence of quirky functional shift, do not reveal their original evolutionary context in their current utility. If all phenotypic traits are adaptive, and built by adaptation, why bother to make a formal distinction between features crafted, and features merely coopted, for their current utility?

But our renewed respect and attention to structuralist themes now makes such a formal distinction essential. Thus, Vrba and I recommended that features crafted for current use continue to be called *adaptations* (adopting the restriction advocated by Darwin and Williams), and that features coopted for current use, following an origin for some other reason, be called *exaptations*. We would also prefer that biologists embrace "aptation" rather than "adaptation" as the general descriptive term for a character now contributing to fitness, with exaptation and adaptation defined as the two subcategories of aptation, thus designated to recognize the crucial distinction between cooptation and direct shaping in the historical construction of characters.

This simple terminological strategy addresses the fair criticism that we can often only know the current basis of fitness—when we do not have enough evidence to determine whether a character developed as an exaptation or adaptation. In such cases, under our scheme, we refer to the character as an "aptation" and leave the further specification of its origin unaddressed. (Our current terminological conventions operate in this manner after all, for I may call a character an adaptation whether I accept Paley's belief in divine creation or Darwin's mechanism of origin by natural selection. Both authors did, indeed, call useful structures "adaptations.")

In an ideal world (and if I held the powers of a czar, which would, of course, then make such a world unideal *ipso facto*) I would fight for the full scheme, and campaign to replace "adaptation" with "aptation" as a base-level description (with no implication about mode of origin) for features now contributing to fitness. But I know the odds against unseating centuries of usage for a word that not only serves as a staple of vernacular speech, but also enjoys unparalleled professional salience as a standard-bearer for our preferred evolutionary theory. I don't play the lottery, and I don't understand the recreational appeal of skydiving or bungee jumping. So I will mortify my desires and learn to live with the traditionally broad use, thus facing with stoicism all the attendant confusion between "adaptation" as a general "state term" for useful features (whatever their mode of origin), and "adaptation" as a more restricted "process term" for the subset of aptive features that arose in the context of their current utility. However, a few people do win lotteries and survive horrendous falls, so I will not surrender entirely. *Dum spiro, spero.*

On the other hand, and to paraphrase Mr. Huxley in a famous context, I am prepared to go to the stake for exaptation—for this new term stands in important contrast with adaptation, defining a distinction at the heart of evolutionary theory, and also plugging an embarrassing hole in our previous lexicon for basic processes in the history of life.

### *Key criteria and examples of exaptation*

I cannot present a "review article" of empirical cases of exaptation, for the defining notion of quirky functional shift might almost be equated with evolutionary change itself, or at least with the broad and venerable subject of, in textbook parlance, "the origin of evolutionary novelties." I will therefore focus on the fate and utility of "exaptation" as a term for describing the evolutionary result of functional cooptation from a different source of origin. Our term (first defined in Gould and Vrba, 1982, p. 4) has not swept the field, as I might have hoped in my arrogant or naïve mode, but "exaptation" has certainly attracted a good share of attention and fruitful use—and may therefore be designated as adaptive for its original "intent."

Above all, biologists have subjected the term to intense criticism and scrutiny (see, for example, Coddington, 1988, and Buss et al., 1998), from which "exaptation" has emerged with strength and proven utility. In my opinion (partisan, of course), Arnold (1994) has presented the best single illustration of exaptation's importance as a concept and its operationality as a tool of research. He begins by recognizing the need to distinguish exaptation from adaptation as subcategories of the more general phenomenon of "aptation" (for he accepts and utilizes our suggested name for the encompassing concept as well). He also emphasizes the crucial methodological point, as previously discussed for the comparable case of the invisibility of stasis under conventional definitions of evolution (Chapter 9), that exaptation must be explicitly defined within a revised theoretical framework, and cannot simply be "discovered" by researchers working within the paradigm of the hardened Modern Synthesis—because anything that "works" will be called an "adaptation" in the conventional theory, and will therefore be scrutinized no further for its potentially exaptive status (Arnold, 1994, p. 128): "One of the main reasons for trying to recognize exaptations is precisely because they are so easily mistaken for adaptations. If the two kinds of aptation are not differentiated, we risk the possibility of exaggerating the undoubted importance of adaptation in fitting organisms to their environments and of ignoring a phenomenon which, like advantageous mutation, is one of the main sources of beneficial accident in the evolutionary process."

Arnold then turns to the operational utility of exaptation, first refuting the arguments of those who found the concept intelligible, even interesting, but doubted that sufficient information could generally be obtained about the history of evidently functional features to make a proper distinction based upon inferences about prior states. Arnold argues that several advances, particularly the codification of cladistic techniques for phyletic ordering, have made the distinction operational in a sufficiently high percentage of cases: the formerly broad definition of "adaptation for all useful traits whatever their origin was reasonable at a time when it was difficult to find out how advantageous traits had arisen, unless this had been observed in recent populations. However, phylogeny reconstruction now allows many individual exaptations to be recognized with some certainty, and makes distinction of exaptive and adaptive origin of performance advantage appropriate" (p. 126).

The relative timings for the origin of a form and for the inception of its current function—as inferred either from the branching points of a cladistic analysis, or from direct knowledge of historical sequences—provide the main criteria for distinction of exaptation from adaptation. "For adaptation," Arnold writes (p. 132), "a hypothesis is refuted if the new trait develops before the relevant selective regime. If the test is passed, it is possible to check whether the new trait really confers an advantage in the new regime that its plesiomorphic state does not."

Arnold then asks how we should interpret the opposite phenomenon "in which a derived trait and a regime in which it gives a performance advantage first appear concurrently on the same node on a lineage" (p. 133). This situation of coincidence in cladogeny between form and function would seem to point to adaptation, but Arnold notes (p. 133) that an imperfect record must fail to provide evidence for several (probably most) events of speciation, and that traits may arise before selective regimes at these missing nodes, and then be compressed into coincidence with a current selective basis at the first recorded node of their joint occurrence. I accept this point as evidently valid, but favorable for tests of the importance of exaptation. As emphasized in other contexts within this book, ineradicable biases in testing a hypothesis present little problem, and may even constitute a blessing in disguise, when their direction works against the hypothesis under test—because the hypothesis gains stronger affirmation by success in the face of such unfavorable odds. Since missing nodes must, by Arnold's argument, induce an underestimate for the frequency of exaptations by redirecting some genuine cases into the opposite category of adaptations *(sensu stricto)*, this bias does not pose problems for tests of exaptation. For example, and using these twinned criteria, Arnold found 70 percent of 61 apomorphies in the lacertid lizard genus *Meroles* to be "concurrent with occupation of the environmental situation" (p. 133) of their present function—therefore requiring that they be ranked as adaptations.

Cases of multiple utility for a single feature offer special promise for resolution by these criteria of temporal or cladogenetic sequencing. Arnold cites the case of "the aberrant arboreal tropical African lacertid lizard, *Holaspis guentheri*," whose extremely flattened head "allows it to hunt and hide in narrow crevices beneath bark, and also constitutes an aerofoil which enables it to glide from tree to tree. Phylogenetic analysis shows that the flattening first developed in the context of crevice use and was only later coopted to gliding" (p. 139).

Moreover, such primary adaptations as head flattening for penetration of crevices usually work in synergy with other coopted features that operate as exaptations in the complex and multifaceted "fit" of the organism to its new environment. For example, when lizard heads become flattened, "the eyes do not usually become correspondingly smaller and, in normal activity, bulge upwards above the skull surface" (p. 139). Arnold then continues to describe the remarkable exaptation of a mouthful of eye: "However, when a lizard flees into a narrow crevice the eyes must be accommodated within the depth of the flattened head. They are most usually pushed downwards by the ceiling of the crevice as the lizard moves deeper into it, so their upper margins are flush with the skull roof and their lower sections bulge through the palate into the buccal cavity (p. 139)." In scincids and lacertids, the eye bulges vertically downward into the suborbital foramen. This aptation depends upon the preexistence of this opening (obviously evolved for other reasons), as indicated by its general distribution on the cladogram of lizards. Therefore, "as the occurrence of the foramen on the phylogeny of the forms concerned precedes occupation of crevices, its use for accommodating the eye within the reduced depth of the skull is an exaptation" (p. 139).

A common, but unfounded, objection to exaptation enters the logical structure of argument at this point. Several colleagues (Coddington, 1988, for example) have claimed that since almost any exapted structure will undergo secondary modification for its new role, and since these subsequent changes must count as adaptations, the concept of exaptation becomes either useless or confusing because any primarily exapted structure must then accrete secondary adaptations to be fully "fit" for its new role. I raise this issue here because, as Arnold points out, the suborbital "foramen is initially small and triangular, allowing only limited projection of the eye into the buccal cavity" (p. 139). This hole then undergoes a secondary adaptive enlargement to accommodate the eye more completely.

I am confident that this common objection cannot be sustained, because hierarchical sequences of processes, each with a different name and status, practically define the nature of complex historical change, and pose no conceptual problems (but rather help us to understand and sort out these sequences), provided that we can specify the order of temporal precedence and hierarchical nesting. Exactly the same issue arises for homology and convergence, and for plesiomorphy and apomorphy. The front appendages of bats and birds are homologous as forearms and convergent as wings; live birth is plesiomorphic for the clade of marsupial and placental mammals, and apomorphic for the same clade within the Tetrapoda. Similarly, the suborbital foramen of lizards is exaptive as a preexisting receptacle for the pushed-down eyes of lizards with flat heads, whereas the subsequent enlargement of the hole may be adaptive for better accommodation of the eyes. The two aspects can easily be separated, and their named distinction helps us to understand the probable sequence of evolutionary events. As Arnold states (p. 139): "there is subsequent, presumably adaptive, modification improving the initial exaptation, the foramen becoming larger and more rounded."

Interestingly, and as further confirmation of the primarily exaptive nature of these vacuities, crevice-dwelling cordylid lizards, representing another separate evolutionary entrance into this habitat, shift their eyes medially into the interpterygoid vacuity, rather than downward into the suborbital foramen. Arnold argues (see Figure 11-7 on these dual routes to exaptation in crevice-dwelling lizards) that cordylids may have utilized this alternative strategy because this group happens to possess a large interpterygoid space (as lacertids and scincids do not) that can accommodate the eye without the secondary modification required to "house" eyes in the suborbital space. Of this sidewards exaptation of cordylids, Arnold writes (p. 140): "This again turns out to be an exaptation, for examination of the phylogeny of cordylids shows that expansion of the interpterygoid vacuity evolves before crevice use, although after the origin of the suborbital foramen. The vacuity may have been utilized instead of the foramen because, being large, it provided immediate housing for a large portion of the eye, whereas the foramen would only have been able to provide this after some modification, as in lacertids and scincids."

In his richest and most extensive analysis, Arnold then discusses six separate evolutionary innovations of "sand-diving" among lizards (quick entry into aeolian dunes to escape predators). He finds all six to be equivalent, both in efficacy and as solutions to the same functional requirement. "In no instance," he writes (p. 156), "is there evidence that the different methods employed reflect different mechanical problems." In his combination of functional and cladistic analysis, he interprets the mechanisms used for sand-

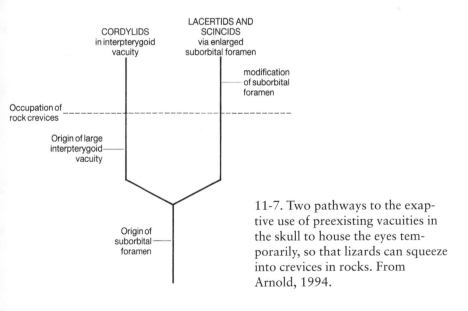

11-7. Two pathways to the exaptive use of preexisting vacuities in the skull to house the eyes temporarily, so that lizards can squeeze into crevices in rocks. From Arnold, 1994.

diving as directly adaptive (that is, first evolved in conjunction with, and presumably for, the behavior) in two cases, but as exaptive in the other four sequences. Interestingly, the exaptive lineages coopted their sand-diving movements from two different functional sources in ancestral lineages: from "part of the drilling mechanism used in firm substrates" (p. 161) by ancestral lineages that hide themselves in harder grounds, and "from the burial pattern adopted before periods of inactivity" (p. 161) in other ancestral lineages (that is, from movements carried out much more slowly, in a clandestine fashion, and for purposes of dormancy rather than escape from predators).

Two additional criteria might be cited as evidence for both the use and usefulness of exaptation as a concept in evolutionary biology, and in other forms of historical study as well.

1. Utility in fields distant from evolutionary biology. Markey (1997) invoked our concept of exaptation to explain the peculiar history of the letter *perth* in the runic alphabet of futhark. (The word "futhark" is an acronym for the first six letters of the runic alphabet, just as "alphabet" itself combines the first two letters of the Greek sequence, alpha and beta.) *Perth* must have had ordinary phonemic value in a still earlier system, but the letter is never glossed in futhark texts and has left no descendants in any Germanic language. (Actually, the letter occurs only one time in all runic literature—in the *English Rune Poem*.)

Markey (1997) argues that, having lost its original phonemic use, "the *p*-rune appears to have been a redundant luxury" (p. 10). Some versions of later futhark alphabets simply eliminated the symbol, but others retained the *p*-rune, apparently for an interesting structural reason with excellent literary analogy to quirky functional shift in biological features. The *p*-rune happened to stand right at the middle of the futhark alphabet, where its phonemic suppression encouraged a different use as a place marker or mnemonic guide, at the halfway point of a long sequence more easily recalled in two divided halves. (Markey shows that several versions of futhark added letters, but always kept the *p*-rune right in the middle of the sequence.)

Markey argues that the *p*-rune, phonemically extinct in Germanic tongues, then lost its exapted function as a place marker when these languages replaced the runic alphabet with our current Latin system. At this point, the *p*-rune again resisted extinction by another exaptation, this time for spelling Latin loan words with an initial *p* sound before a vowel (as in *papa* for "pope," or *pater* for "priest"), a phonemic combination not found in Germanic words. (Runic-*p* served the same function for some English loan words of non Indo-European origin, as in "pebble.") In any case, Markey (p. 11) found our biological concept of exaptation useful in describing this complex double quirky functional shift from ordinary phonemic value in a hypothetical ancestor, to place marking in futhark, to renewed phonemic value for loan words when the Latin alphabet replaced futhark: "Exaptation is manifested by functional bifurcation. The primary function of feathers was warmth, the secondary function flight. The primary function of runic-*p* appears to have been that of a boundary marker, while its secondary function was loan-word spelling. Runic-$p$ has every appearance of having been coopted. Non-essential in the runic system, it must have been essential in some system, presumably the parent of Older Futhark."

In a fascinating exegesis of Emil Durkheim's seminal, late 19th century sociological studies of the division of labor, Catton (1998) invoked our concept of exaptation to explain a central error that Durkheim might not have committed if he had recognized the principle of functional shift, either from Darwin (whom he studied intensely), or from the more nearly contemporary Nietzsche. Durkheim recognized (correctly) that division of labor, and the attendant specialization of tasks in society, can greatly reduce competition and lead to "organic solidarity" (Catton, 1998, p. 89). But he then erred in assuming that this current utility also permits the inference that division of labor arose, in explicit analogy with speciation, as a direct adaptation for its current function of reducing competition and stabilizing both social and economic systems. "To Durkheim," Catton explains (p. 117), "it seemed abatement of competition by means of differentiation was the necessary removal of an otherwise insurmountable barrier to mutualistic interdependence. That was why division of labor was supposed to result in organic solidarity."

But Catton then exposes the dilemma and logical error entailed by Durkheim's commitment to an evolutionary analogy with speciation. For biologists argue, and have demonstrated in many cases, that mutualistic interactions often evolve from initial antagonisms and exploitations: "Evolutionary ecologists now know that mutualism can evolve from antagonism . . . by some modification of structure or behavior that changes the outcome of an interaction from which the parties cannot withdraw."

The secondarily-evolved cooperation may remain "good" for both parties, while so altering the initial state of the system that origins cannot be inferred from this current utility. Catton (1998, p. 118) found our discussion of exaptation useful in explaining this important concept to his colleagues. (He also shows his appreciation of the corollary that secondary adaptation for a new role does not impeach the exaptive origin of the coopted utility): "An adaptation has a function. An exaptation has an effect. Once that effect becomes important in the life of an organic type (in its new environment), natural selection may 'improve' the exapted trait, eventually making it an adaptation, and converting the effect into a true function."

2. Passage of the term from explicitly cited novelty to general and unreferenced usage in evolutionary literature. The sequence may be bittersweet for originators, but only the most narcissistic or insecure scientist could fail to take pleasure when a concept of his invention, or an experiment of his doing, loses explicit connection to his authorship by "evolution" into an ordinary term of art within the profession. This form of acquired anonymity crowns the diffusion to general success of a suggestion or innovation with a "point source" of origin, now happily forgotten and relegated to the domain of antiquarian or historical studies.

Exaptation has already passed through the three major stages of this sequence. In a first stage, exaptation, as a novel term, became an explicit focus for studies to test or illustrate its utility—as in Arnold's title (to his 1994 paper, discussed above): "Investigating the origins of performance advantage: adaptation, exaptation and lineage effects"; or in Almada and Santos (1995): "Parental care in the rocky intertidal: a case study of adaptation and exaptation in Mediterranean and Atlantic blennies."

In a second stage, the concept serves, *inter alia*, as part of an ordinary analysis, and not as an explicit focus of study—but the term still requires a citation to its source of origin, or at least must be defined and presented within quotation marks. For example, Chatterjee (1997), advocating the currently less popular "trees down" arboreal (rather than the "ground up" terrestrial) theory for the origin of birds, argued that many climbing adaptations of tree-dwelling ancestors "were exapted for gliding" in the transitional stages towards full flight. The language of Chatterjee's full sentence records an interesting, and undoubtedly unconscious, intermediary stage in the acceptance of functional shift as a principle in evolutionary analysis: "Surprisingly, many of these arboreal innovations were exapted for gliding" (p. 311). But such cooptations and functional shifts can only be deemed "surprising" when contrasted with expectations of continuous improvement within a single "adaptive zone" (to use Simpson's classical terminology of 1944). Once we recognize functional shift and cooptation as important components in almost any extensive evolutionary sequence, we will no longer label exaptations as surprising.

In an example from the most fecund realm of exaptation in molecular evolution, Weiner and Maizels (1999) explained to their biochemical colleagues who may not be *au courant* with the literature of evolutionary theory: "Those with an evolutionary bent sometimes use the word 'exaptation' to describe the appropriation of a molecule with one job for a completely different purpose. Exaptation contrasts with 'adaptation,' a seemingly natural extension of preexisting functions" (1999, p. 64). Their article, entitled "a deadly double life," documents the fascinating "remarkable discovery" (1999, p. 63) that the carboxyl-terminal domain of human tyrosyl-transfer RNA synthetase (the enzyme that catalyzes the attachment of the amino acid tyrosine to the appropriate tRNA molecule prior to protein synthesis) shows clear homology (49 percent sequence similarity) with a cytokine performing the quite different—one might say conceptually opposite—function of attracting phagocytic cells to sites of apoptosis, suggesting in a broader sense that "secretion of tyrosyl-tRNA synthetase may help to shut down residual protein synthesis in the dying cell" (p. 63).

In this case, the synthetase activity seems primary (for a set of reasons elaborated in Weiner and Maizels, 1999), and the "opposite" role in cell death secondary, following gene duplication, and leading to the molecule's "deadly double life." Weiner and Maizels (1999) argue that the utility in apoptosis originates as an exaptation recruited from an "accidental" effect of the gene's primary activity: "The recruitment of tryrosyl-tRNA synthetase as an extracellular death messenger" (p. 64) follows from its primary role and allows these molecules to "serve as harbingers of impending cell death when released from their normal cellular compartments": "Release of proteins from their normal locations in the cell may have originally been a symptom of cell death, rather than a cause of it. Evolution may then have exploited the *accidental* [their italics] release of these proteins (and possibly others) to build, amplify, and eventually fine-tune the death circuitry."

Finally, in a third phase, the term enters the literature as a standard item of professional lingo, requiring no further citation of original sources (probably unknown to the authors in any case) than any other word of professional jargon. Thus, Jablonski and Chaplin (1999, p. 836) view manual dexterity and eventual tool use as human exaptations of a bipedal posture that originally arose as part of a common threat display in ancestral apes, and Roy (1996) analyzes exaptations for defense in the fossil history of stromboid gastropod shells.

I take greatest pleasure, however, in the spread of exaptation as a term of art in the most pervasively expressed domain of molecular evolution. For example, in a review article on "interspersed repeats and other mementos of transposable elements in mammalian genomes," Smit (1999) consistently uses exaptation to describe coopted utilities of multiply repeated and dispersed transposable elements (the classic molecular items that inspired the concept of "selfish DNA," see pp. 693–695 for further discussion).

In a section of a paper on "domestication of individual transposable elements" (p. 661, and I do appreciate his witty and apposite metaphors from the vernacular), Smit writes: "Throughout time, host genomes have rummaged through the novel sequences accumulated by transposition and have recruited numerous elements . . . Far from merely expanding genomes with interspersed repeats, their legacy ranges from spliceosomal introns and antigen-specific immunity to many recent recruits in highly specialized functions" (1999, pp. 661–662). Although Smit notes an apparently reduced exaptive role for such transposons in humans vs. mice, "leading to speculations on host defense mechanisms" (p. 657) in our species, he also lists an impressive array of potential human examples, "some with household names" (p. 661). For example, Smit cites a fascinating exaptation of all higher primates, probably essential to the existence of this book and any reader's kind and current attention thereto: "BC200, the only known fully recruited SINE in humans, is a brain-specific RNA that is part of a ribonucleoprotein complex preferentially located in the dendrites of all higher primates. It is presumably derived about 50 million years ago from a monomeric Alu and has since been selectively conserved in all studied descendants."

Note Smit's proper conceptual separation of initial cooptation from later retention by natural selection. Following both Darwin and Nietzsche, the secondary adaptive enhancement by natural selection (in this case, perhaps, merely a selective retention without further structural change)—that is, the promotion of the *current utility*—does not permit a conclusion about the different cause of *historical origin* (in this case, presumably as an exaptation from repeated copies of a transposon, replicated and amplified for quite different reasons, and probably initially nonadaptive at the primary Darwinian level of the organismal phenotype).

The concepts of functional shifting, the structuralist implications thereof, the classical examples (especially the exaptation of feathers for flight), and the terminology of cooptation and exaptation, were all worked out on the conventional "playing field" of anatomies and behaviors of complex multicellular organisms. But this rubric of theory and argument will surely enjoy its greatest application in the domain of molecular evolution, where the functional redundancy of multiple usage for most gene products and the structural redundancy of duplications and repeated elements enlarge the scope of functional shifting away from mere striking illustrations, and towards ubiquity. (Note how this pairing, at the molecular level, of several functions for one gene with several genes for one function precisely matches the logic of Darwin's anatomical argument for the structural prerequisites of quirky functional shift—see p. 107 for my discussion in terms of Darwin's favorite example of lungs and swim bladders in the evolution of fishes.) Classical cases are already beginning to emerge from this level of analysis, perhaps none more complex and fascinating, or more widely cited, than the work of J. Piatigorsky, G. Wistow and many others on the eye-lens crystallins of both vertebrates and invertebrates.

The crystallins are structural elements that constitute about 90 percent of the total soluble protein of eye lenses in most vertebrates. Most crystallins are found in lens fibers, which lose their nuclei (and other organelles) and must therefore, since they cannot replenish by division, remain stable through the organism's life (Piatigorsky, 1992). In beginning an invited review to a society of ophthalmologists, Piatigorsky (1993a) explained (to a group of professional biologists who generally lack specialized training in evolutionary theory) how functionalist and adaptationist biases had led to longstanding assumptions that must now be discarded (1993a, p. 283):

> As scientists and physicians we are accustomed to seeking order and purpose in the world. It is commonplace for us to think that specialized tasks require custom made instrumentation, the more sophisticated the mission, the more honed the result . . . And so for approximately 100 years vision scientists have considered the crystallins as a very limited set of highly specialized proteins especially chosen and designed for their ability to confer the required refractive properties onto the transparent lens. We have grown up with the idea that crystallins are as specialized as the eye itself.

Molecular and genetic studies of the 1980's and 1990's have dramatically reversed this view by identifying crystallins as a diverse set of exapted enzymes and proteins with strikingly different original functions, often still maintained. Piatigorsky continued (1993a, p. 283): "Recent studies, however, have changed this restricted view and have shown that crystallins are essentially borrowed proteins of diverse origins. These lens structural proteins not only play a refractive role in the lens, but they have important non-refractive functions within and outside of the eye."

Vertebrate crystallins have been divided into two groups (Wistow, 1993; Lee et al., 1994): the structural stress-proteins of the alpha and beta/gamma crystallin group found in most vertebrate lenses; and the highly diverse, so-called "taxon specific" crystallins, generally found in more restricted lineages, and exapted from enzymes that continue to operate in their earlier manner elsewhere in the body (and often in the lens as well).

The structural proteins of the first group also represent exaptations, rather than direct adaptations, for vision. This cascade of reinterpretation began in the early 1980's with the discovery that alpha crystallins are homologs of the small heat shock proteins of *Drosophila*. One of the two alpha crystallin genes continues to produce a heat shock protein (Piatigorsky et al., 1994), while the other has become more specialized for lens functions, although both also continue to act as molecular chaperones. The beta/gamma crystallins (Piatigorsky and Wistow, 1991) are more distantly related to microbial dormancy proteins, also inducible by osmotic shock and other stresses.

But the second group of "taxon specific" crystallins show far more diversity in their multiple routes of exaptation from previous functions (often still retained) as enzymes. For example, delta crystallin of chickens is argininosuccinate lyase; epsilon crystallin of ducks is identical with the metabolic enzyme lactate dehydrogenase; tau crystallin of turtles is alpha-enolase; and mu crystallin, found in many marsupials, is ornithine cyclodeaminase (Piatigorsky, 1993b). In a proof of multiple recruitability in independent events across great phyletic distances, the eta crystallin that constitutes more than 25 percent of soluble proteins in the lens of elephant shrews is the enzyme cytoplasmic aldehyde dehydrogenase (Piatigorsky and Wistow, 1991), whereas the omega crystallin of octopuses has also been exapted from aldehyde dehydrogenase in a separate cephalopod event (Piatigorsky et al., 1994). The theme of lens proteins as exapted enzymes then extends to further phyletic diversity, for the major lens component of squid S crystallin, is related to the detoxification enzyme glutathione S-transferase.

Piatigorsky (1993a, p. 284) summarizes the dominant role of exaptation for the origin and status of lens crystallins: "A number of the crystallins have been shown to be expressed outside of the lens and to possess its original nonrefractive activity. Indeed, a hallmark of an enzyme-crystallin is that it is expressed at high concentration in the refractive lens and at a much lower concentration in other tissues, where it has at least one other non-refractive role."

Since examples of exaptation always raise the structuralist theme of preconditions for recruitment, we must ask what common properties of proteins and enzymes in this large array of highly disparate sources prompts or facilitates cooptation as lens crystallins. Some evident requirements—with transparency as the most obvious property—probably represent merely incidental and nonadaptive consequences of molecular structures evolved for other reasons, in the same evident sense that natural selection did not make our blood red or our bones white for any directly adaptive reason rooted in the colors themselves. But although transparency surely stands as a primary prerequisite, many other enzymes and proteins share this necessary property, but have never been recruited as lens crystallins—so more specific preconditions must be sought (Wistow, 1993; Piatigorsky, 1993a and b). Piatigorsky (1993a, p. 285) lists "high solubility in water to achieve the high concentrations necessary to attain the appropriate refractive index and thermodynamic stability, since loss of cell nuclei in the fiber cells prevents turnover in this region of the lens."

Even more specifically, Wistow (1993, pp. 303–304) notes that all cellular lenses require unusually elongated cells as building blocks—a common property or potential, as Wistow argues, of the original utilities from which lens crystallins have been exapted: "As the lens evolved, the necessary refractive power must have been achieved by recruiting genes that are active under the prevailing conditions of cell elongation and whose protein products fit the broad requirements of their new role. Osmotic stress proteins, cytoskeleton chaperones and easily inducible detoxification enzymes would have been good candidates. Such an origin could have engendered underlying similarities in gene expression for groups of crystallins."

But the case of crystallins owes its emerging status as a "classic" of exaptation largely to the strong evidence gathered for a range of structural prerequisites and preconditions that can facilitate such functional shifts. Arnold (1994) has proposed a set of subcategories for sources and styles of exaptation (see also Gould and Vrba, 1982), and I shall devote the final section of this chapter (pp. 1277–1294) to the further development of such a taxonomy and to exploring its implications for macroevolutionary patterns and possibilities. The subject has assumed some urgency in studies of molecular evolution because the crucially important mechanism of gene duplication has frequently been overextended and interpreted as virtually the only possible basis for exaptation—when a gene with an important function duplicates (Ohno, 1970, for the classic statement), thereby "freeing" one copy for cooptation to a different utility. Exaptation does occur by duplication in the evolution of some lens crystallins, but other exapted crystallins are products of a single gene that continues to make the critical enzyme of its presumably original function—a process that Piatigorsky and Wistow (1989) called "gene sharing," and that Darwin explicitly recognized in citing organs with two distinct functions as good candidates for quirky functional shift (see p. 1223).

For example, the duck genome includes only a single gene to code for both the exapted lens crystallin and the original enzyme in at least two cases: epsilon crystallin (lactate dehydrogenase B) and tau crystallin (alpha-enolase). In some cases, the lens crystallins even retain their enzymatic activity within the eye. The zeta crystallin of several hystricomorph rodents is quinone oxidoreductase, and may protect the eye against oxidation "or even filter UV radiation" (Wistow, 1993, p. 301). The amount of episilon crystallin in many birds, produced by the same single-copy gene that codes for the enzyme lactate dehydrogenase B, also correlates well with exposure to light, and may provide enzymatic protection as well as visual refraction (Wistow, 1993, p. 301).

As an example of exaptation associated with duplication, two genes produce delta crystallin in chickens and ducks, with the delta1 gene specialized for lens expression, and the delta2 gene producing the same enzyme, argininosuccinate lyase, in non-lens tissue, but also generating some lens crystallin as well. Interestingly, both genes are equally active in the duck lens, which thus includes ASL enzyme activity (through the enhanced action of the delta2 gene) at a 1500 fold higher level than in chicken lenses (for no understood function as yet).

A presumably much older duplication occurred in the alpha crystallins present in most vertebrates, with the alphaA and alphaB genes now residing on different chromosomes. The alphaA gene has specialized for production of its lens crystallin, but maintains some activity in other organs of some species. However, the alphaB gene has retained more of its original function in generating a heat shock protein, while also coding for lens crystallin.

Interestingly, and beginning to unite crucial themes of the last two chapters, the alphaA crystallin gene of chickens is regulated by at least 5 control sites (Cvekl et al., 1994). Sites C and E bind *Pax-6* (the famously homologous "master regulator" of eye development in squids, arthropods, and vertebrates—see pp. 1123–1132) in the lens to stimulate alphaA crystallin promotor activity, thus controlling high expression of this gene in the lens and repression in fibroblasts (Cvekl et al., 1994, p. 7363). These authors also report that *Pax-6* binds to the lens-specific regulatory enhancer of the delta1 crystallin gene of chickens, and to the lens-specific regulatory sequence of the zeta crystallin gene of guinea pigs.

Finally, some lens crystallin genes undergo more extensive duplication, usually followed by a further specialization of some copies for lens functions, as expected. "For the beta and gamma crystallins, multiple gene duplications have led to gene families with six or more members that seem to be specialized for lens" (Piatigorsky and Wistow, 1991, p. 1079). In the most extensive example of duplication, the squid genome includes at least 10 S-crystallin genes, all derived from the gene that produces the glutathione S-transferase (GST) enzyme. These S-crystallin genes are expressed only in the lens and cornea and now lack enzymatic function, with one exception of "very slight GST activity" in a single S-crystallin (Piatigorsky et al., 1994, p. 243): "The S-crystallin genes encoding the inactive enzyme derivatives have acquired an additional exon which probably contributes to the loss of enzyme activity of the crystallin."

To close this long section, and these details of a developing classic, with a lovely corroborative tale in the venerable tradition of natural history, the squid *Euprymna scolopes* collects phosphorescent bacterial symbionts in a "light organ" located in the center of its mantle cavity. "The squid uses the light emitted by the symbiotic bacteria in its behavior, presumably in anti-predatory displays and/or intraspecific communication" (Montgomery and McFall-Ngai, 1992, p. 21000). The light organ also includes a lens formed as a thick pad of transparent tissue, and built by the squid as a derivative of muscle from its hindgut. "The tissue functions as a convex lens to refract the light from the localized bacterial source over the ventral surface of the squid" (Montgomery and McFall-Ngai, 1992, p. 21000). (Incidentally, I shall never forget the kindness of these authors, or the eerie fascination of this system in these remarkable animals, when I had the privilege of visiting their lab in the early 1990's.)

The ocular lens of squid is epidermally derived and used for sight. This second and entirely different kind of lens, both in development and function, is built from muscle tissue and operates to enhance and refract the light generated by symbiotic bacteria! Yet the lens crystallin of the muscle-derived light enhancer, called L-crystallin by Montgomery and McFall-Ngai, is apparently exapted from an ALDH-like enzyme, as is the eta crystallin of the ocular lens of elephant shrews and the omega crystallin of the ocular lens of octopuses, both discussed previously. Montgomery and McFall-Ngai (1992, p. 21003) argue that enzymatic activity of ALDH may be preserved for protection against peroxidative damage. They end their paper with both an observation and a challenge: "Possibly, ALDH was first recruited for such a purpose, and then secondarily converted in some species to serve a largely structural role. However, as is the case with all other enzyme/crystallins discovered, why ALDH was selected [I would say exapted] as a structural protein is unknown."

## THE COMPLETE VERSION, REPLETE WITH SPANDRELS: EXAPTATION AND THE TERMINOLOGY OF NONADAPTIVE ORIGIN

### *The more radical category of exapted features with truly nonadaptive origins as structural constraints*

Throughout the previous section, I emphasized how the theme of quirky functional shift, and the resulting discordance between reasons for historical origin and the adaptive basis of current utility, introduced an important structuralist component into the otherwise functionalist logic of Darwinian theory. In particular, the developmental prerequisites and structural potentials of any ancestral state—and not only the adaptive pressures emanating from present environments—must be factored in as both limits and facilitators for evolutionary change, thereby acting as constraints (in both positive and negative senses) upon phylogenetic pathways.

Nonetheless, as also emphasized throughout (and in the subsection's title of "the restricted Darwinian version"), the basic concept of exaptation remains consistent with orthodox Darwinism (while expanding its purview and adding some structural clarification and sophistication) for an obvious reason: the principle of quirky *functional* shift does not challenge the control of evolution by natural selection as an adaptational process. Unpredictable shift of function may establish the ground of contingency, and may imply a role for structural constraints upon phyletic pathways. But this principle does not undermine the functionalist basis of evolutionary change because features so affected remain adaptive throughout: they originate for one function (presumably by natural selection), and then undergo quirky shift to a different utility.

However, the principle of functional shift, combined with Nietzsche's argument about the invalidity of inferring historical origin from current utility, implies a disarmingly simple and logical extension that does challenge the rule of Darwinian mechanics and functionalist control over evolutionary change. Ironically, the very simplicity of the argument has often led to its dismissal as too obvious to hold any theoretical importance—a "feeling" that I shall try to refute in this section, and whose disproof represents an important step in the central logic of this book.

The deeper challenge posed to orthodox Darwinism by the principle of functional shifting flows from the implication that, if current *utility* does not reveal reasons for historical origin, then these initial reasons *need not be adaptational or functional at all*—for features with current adaptive status may have originated for nonadaptive reasons in an ancestral form. In other words, and in the terminology of Table 11-1, when current aptations rank as exaptations rather than adaptations, their coopted source will be identifiable as an ancestral structure with either adaptive origins (for a different function) or nonadaptive origins (for no function at all). (I do accept the standard view that strongly *in*adaptive features hold little prospect for an evolutionary legacy because natural selection must soon eliminate them. But *non*adaptive—that is, effectively or nearly neutral—features may persist for several reasons, including the "invisibility" of true neutrality to pressures of selection, and the status of many nonaptations as automatic architectural byproducts, as in Darwin's "correlations of growth" or Gould and Lewontin's "spandrels.")

The logical validity and evident application of this simple argument cannot be gainsaid. Indeed, several examples, mentioned *inter alia* in the preceding section on exaptation, fall into this category of features with current utility exapted from a nonadaptive ancestral status. The optical property of transparency, shared by all the diverse proteins and enzymes that have been exapted as lens crystallins, may represent a trivial and automatic consequence of physical and chemical structures evolved for other reasons. But this purely derivative and nonadaptive feature still stands as a gatekeeper and prerequisite for exaptation to vision. We may regard Darwin's example of non-fusion of skull sutures in mammalian neonates as a far richer and less obvious case. For we need, in this example, to unravel enough specific history of mammalian descent to know that this property arose in ancestors born from eggs, and therefore cannot be a direct adaptation, initially evolved to compress the head and permit passage through the narrow mammalian birth canal. (We should also remember that Darwin explicitly declined to call non-fusion an "adaptation" for this reason, even while he acknowledged the functional necessity for such a property in the evolution of mammalian live birth.)

The general conclusion may be stated in a simple manner, but I believe that the resulting implications for evolutionary theory are both profound and curiously underappreciated: If many features that operate as adaptations under present regimes of natural selection were exapted from ancestral features with nonadaptive origins—and were not built as direct adaptations for their current use (or exapted from ancestral features with adaptive origins for different functions)—then we cannot explain all pathways of evolutionary change under functionalist mechanics of the theory of natural selection. Instead, we must allow that many important (and currently adaptive) traits originated for nonadaptive reasons that cannot be attributed to the direct action of natural selection at all and, moreover, cannot be inferred from the exaptive utility of the trait in living species. Because the subject of evolutionary biology must engage many crucial questions about the *origins* of features, and cannot be confined to the study of current utilities and selective regimes, nonadaptationist themes therefore assume an important role in any full account of life's history and the mechanisms of evolutionary change.

The key to this expansion of evolutionary theory therefore lies in the category of currently useful traits with nonadaptive origins—my rationale for prefacing this topic with two sections to develop the prerequisites of the argument: one on Nietzsche's principle of general discordance between bases of current utility and reasons for historical origin, and another on the terminology of exaptation as a framework for describing and appreciating the importance of currently useful structures coopted from a different ancestral status, rather than directly evolved for their present function. By introducing, in this final section, the theme of exaptations based upon features with nonadaptive origins, I complete the chart of Table 11-1 (see p. 1233) by recognizing two subcategories of exaptation: (1) Cooptations of features that originated for different adaptive reasons—the principle of "quirky functional shift" that enriches (with structuralist "flavor"), but does not challenge, the functionalist control of evolution by natural selection, and that also establishes the ground of contingency within the Darwinian world view. (2) Cooptations of features with nonadaptive origins—the theoretically radical category that precludes any complete explanation of evolution in adaptationist terms, and that provides a nonadaptationist alternative for evolutionary inquiry about the origins of currently adaptive biological features.

But if this argument is so simple to state, so airtight in logic, and at least interesting (I would say profound, but not with everyone's initial approbation!) in its implications for evolutionary theory, then why hasn't the category of currently useful features with nonadaptive origins been perceived as more troublesome by orthodox Darwinians? Why, to sharpen the paradox, did the concept not even receive a name in conventional theory? This complex question embraces many dimensions, including psychological and historical influences lying well beyond my professional competence and the scope of this book. But, for the relevant dimension of the structure of Darwinian theory, two strong reasons—both invalid in my view, with their refutation as the primary aim of this section—have permitted most defenders of strict natural selection to acknowledge the existence of nonadaptive features, but then to relegate them to a periphery of rarity and impotence where they can exert no effective role in setting pathways for the history of life, or in specifying principles of evolutionary theory. Both arguments flow from the assumption (which I do not challenge) that most nonadaptive traits of organisms arise as structural consequences of selection upon other features, or upon other aspects of the same trait. (I speak here of traits with enough stability and complexity to become exapted for meaningful utility in a descendant lineage. No one doubts, of course, that many truly trivial features of organisms have no bearing upon fitness, and lie well beneath the visibility of natural selection):

NOOKS AND CRANNIES. Darwinians have generally argued that most structural consequences of natural selection on other features can survive as true nonadaptations (and not be eliminated as inadaptive) only if they occupy little space or require minimal metabolic input. The mold marks on old bottles (made in two-piece molds) surely testify to mechanisms of manufacture, but serve no purpose. If we regarded them as ugly, or if they disrupted a bottle's utility, we could easily remove them. But we allow them to persist in their inconsequential triviality.

SECONDARY AND CONSEQUENTIAL STATUS. This common argument commits the historical error that inspired Nietzsche's warning. Many people have assumed that nonadaptive origin as a consequence of selection upon another feature relegates a trait to permanent insignificance because it arose in a passive and sequential manner—a clear conflation of reasons for historical origin with potentials for subsequent utility. Perhaps, in an ancestral pelycosaur, the skull sutures remained unfused (at birth from an egg) as a trivial consequence of some developmental necessity in a forthcoming and free-living ontogeny. But this nonadaptive property may later become the prerequisite to the success of mammalian descendants, when live birth became the autapomorphic key to a new and highly successful mode of life.

### *Defining and defending spandrels: a revisit to San Marco*

When I chose the notorious spandrels of San Marco to illustrate these fallacies with an architectural analogy that might not automatically raise the hackles of orthodox Darwinians (or thicken the scales over their eyes, to cite the other common biological metaphor for resistance to unfamiliar ideas), and also to introduce a term for the most common category of nonadaptive features with high potential for subsequent exaptive utility, I did not make a capricious selection for idiosyncratic reasons of personal interest (Gould and Lewontin, 1979). Rather, I chose this Venetian example because the historical record and current status of these particular four pendentives includes a sufficient richness and certainty of documentation to refute all the common objections raised against similar claims for the evolutionary insignificance of nonadaptive features in biological systems.*

---

*In retrospect, I suppose that I made an adaptive choice, at least by the usual standard of subsequent discussion and brouhaha in the scientific literature—a legacy that has surely helped to clarify this important issue, and to focus attention upon alternatives to adaptationist explanation in evolutionary biology. But I must also confess that Lewontin and I never anticipated so many exaptive spinoffs from this introductory image—including

When a hemispherical dome stands on a set of four rounded arches meeting at right angles to form a square—a very common design in ecclesiastical architecture and many other buildings—four tapering triangular spaces must appear directly under the dome, each formed by the space left over between the dome itself (above) and the pair of adjacent arches (at the sides) meeting at right angles (see Figure 11-8). These four spaces, called *pendentives,* must form as a structurally necessary side consequence of the architect's basic decision to mount a dome on four rounded arches, so arrayed. Lewontin and I never claimed that these spaces did nothing useful (for obvious and trivial starters, their roofing keeps out the birds and the rain). But we did argue that their ineluctable size and shape—their number and triangular form—arose as side consequences of a previous architectural decision, and could not be viewed as adaptations in themselves. (The four pendentives, in other words and by analogy with my previous example, hold the same status as mold marks on an old bottle—necessary side consequences of an architectural decision, not functional features in themselves.)

The general architectural term for such "spaces left over" is *spandrel*—a lovely name derived from the primordial human tool of measurement, the span of our own hand (or of the corresponding feature in an anthropomorphized divine architect—as in Isaiah's God (40:11–12) "who hath measured the waters in the hollow of his hand, and meted out heaven with the span"). Classical spandrels are two-dimensional spaces left over (Figure 11-9) including the vertical boards between steps of a staircase, the triangular spaces between arches arrayed in a linear row, and the flat horizontal stretches (called "spandrel courses") on large office buildings, located between the tops of windows on the floor below and the bottoms of windows in the next story just above. Apparently—see documentation in Gould, 1997e—some architects restrict the term "spandrel" to two-dimensional spaces left over, whereas others, particularly in European usage, extend the term to any space that arises as a side-consequence of a prior decision, and not as an explicitly designed feature in itself, thus including the three-dimensional pendentives of San Marco and thousands of other buildings. In any case, I consciously decided to apply this remarkably appropriate term to San Marco's pendentives because their shared property with classical two-dimensional spandrels—their status as architectural byproducts (at least for their forms, numbers, and placements)—cannot be denied. I also felt that biology needed a term for such architectural sequelae of "adaptive" decisions, and that this well known term from a related discipline could serve admirably.

---

an entire book by linguistic scholars on our (mostly unconscious) literary tactics (Selzer, 1993); a wise commentary by a noted scholar of medieval buildings (Mark, 1996), and, wonder of wonders in our faintly philistine (and avowedly secular) professional community, a burgeoning interest in at least two humanistic subjects generally shunned by scientists for reasons of passive ignorance, or even active distaste: church architecture (Dennett, 1995; Houston, 1997) and literary parody of the puerile, "ain't-I-clever," sort embodied in two titles, "The Scandals of San Marco" and "The Spaniels of St. Marx." Ouch! (Borgia, 1994; Queller, 1995).

11-8. Four spandrels with brilliant mosaics of the evangelists under a circular dome of the Cathedral of San Marco in Venice. From Demus, 1984.

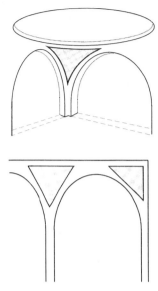

11-9. A three-dimensional spandrel, or pendentive, contrasted with a more conventional two dimensional spandrel between arches in a linear row (below). From Gould, 1997e.

In any case, although spandrels must originate as necessary side-consequences of an architectural decision, and not as forms explicitly chosen to serve a purpose, they still exist in undeniable abundance, and can therefore be secondarily used in important and interesting ways by clever architects, artists and patrons of buildings. (I grew up in New York City, and have always admired the lovely ornamentations on the spandrel courses of many of our finest art-deco skyscrapers, particularly the zoological motifs on the Chanin Building just opposite Grand Central Station.) In my "holotype" of the central dome of San Marco's cathedral in Venice, where the entire interior space has been covered in glorious mosaics, the four pendentives of the central dome have been ornamented in a complex way, stunningly well "fit" both for the space occupied, and for the symbolic meanings portrayed by the mosaic designs. The four evangelists (including St. Mark, the cathedral's patron and most celebrated resident interred therein) occupy the widest top sections of the spandrels, under a motto in Latin doggerel: *Sic actus Christi, describuunt quatuor isti*—thus did these four describe the acts of Christ. Beneath the evangelists, in the narrowing triangular spaces, personifications of Eden's four rivers (see Genesis 2) hold amphoras (Greek pitchers) over their shoulders, each pouring water onto a single flower wedged into the narrow and elongated space at the base of each pendentive.

The design—beautiful, complex, and particularly appropriate both to Christian symbolic representation and to the surrounding space—exudes utility. But no one would make the mistake of arguing that the spandrels exist to house the evangelists. The spandrels originated as a nonadaptive side-consequence of a prior architectural decision. These originally nonadaptive spaces were then coopted (several centuries later, in this case) as "canvasses" for wonderfully appropriate designs. In biological terms, the mosaic designs are secondary adaptations, and the spandrels themselves then become exaptations for the residency of these designs. Again, I chose this architectural analog because I felt that a similarly unambiguous case in organisms might become conceptually muddled—for biologists have been trained to regard anything that "works well" as an adaptation, and might therefore not "see" the originally nonadaptive nature of the spaces. But I felt—rightly, in retrospect—that this quite precise architectural analog would not generate enough emotional salience to act as a barrier against understanding the intended point.

A desire for clarity in illustration served as my primary motive, but our original article (Gould and Lewontin, 1979) does not sufficiently emphasize my other major reason for choosing this example as a "holotype" to illustrate the important category of nonadaptive features originating as architectural side consequences. I chose the San Marco spandrels because they so evidently refute, in terms of the architectural analogy, the two standard arguments raised against a similar importance for nonadaptive structures in biological morphology (as outlined above on p. 1249):

NOOKS AND CRANNIES.   One cannot brand the spandrels as trivial in occupied space or peripheral position—as one might legitimately hold for the mold marks on a bottle. The four spandrels under any dome occupy a substantial area, surely equal to a large percentage (if not the totality) of the area of the dome above. As we shall see, this generous size and central placement also refutes a major aspect of the second dismissal based on consequentiality.

CONSEQUENTIALITY.   The false inference of evolutionary insignificance from secondary original status as a side-consequence of a primary adaptation includes two arguments of different logical standing, but each equally invalid:

1. THE EMPIRICAL CLAIM.   For biological examples, many scientists have assumed that the temporally sequential status of any exapted utility (as imposed upon a primary nonadaptation) must relegate any subsequent use to marginal importance. Thus, the evangelists and rivers are adaptive in their purposeful and lovely fit within a preexisting (and initially nonadaptive) space, and in their important message as conveyed within the larger aim of the building's role as a Christian church. But these mosaic designs are just as surely secondary and sequential—as later adaptations restricted by prior constraints upon the number and form of an initially nonadaptive housing. The mosaicists made an adaptive choice, but preexisting constraints strongly limited their options. The four spandrels could not house, in any easy or adaptive way, the three children saved from the fiery furnace or the five books of Moses (not to mention the inelegance of setting 2.5 Commandments in each spandrel). Biologists often conflate a genuine limitation upon options with a false inference that constrained solutions, however adaptive, cannot generate structures of major importance either to the current working of organisms, or to their future evolutionary potential.

But, as Nietzsche argued, a secondary and constrained origin implies nothing about potential for either present or future importance—and the designs in San Marco's spandrels clearly expose this fallacy. Extensive feedback from the spandrels to the mosaics of the dome proves that secondary features can exert pervasive influence upon the basic design of a totality. The domes of San Marco are radially symmetrical and therefore provide, *in se* and considered alone, no reason for favoring a quadripartite mosaic design. Yet all but one of San Marco's five domes contain mosaics arranged in four-part symmetry— clearly, in each case, to harmonize with the iconography in the four triangular spandrels below. For example, in the mosaic design of our "holotype" central dome, three circles of figures radiate out from a central image of Christ: angels, disciples, and virtues. Each circle is divided into quadrants, even though the dome itself is radially symmetrical in structure. Each quadrant meets one of the four spandrels in the arches below the dome.

Another dome contains angels in the spandrels and the twelve apostles in the dome, arranged in four groups of three, with each group centered on one of the four spandrels below. Yet another dome presents four male saints in the dome and four female saints in the spandrels, with each male perfectly centered between two of the females. Thus, an ineluctable architectural byproduct can, nonetheless, determine the fundamental design of a totality that ordained its consequential origin. The natural world abounds in recursions and feedbacks of this sort. Mustn't the ever cascading spandrels of the human brain be more weighty than the putative primary adaptations of ancient African hunter-gatherer ancestors in setting the outlines of what we now call "human nature"?

2. THE METHODOLOGICAL CLAIM BASED ON OPERATIONALITY. As discussed throughout this section, biologists have often been reluctant to base terminological distinctions on differing historical pathways to a similar current result—for the good reason that a poverty of historical records often denies us the data needed to reach a firm conclusion, whereas current situations can always be directly observed or experimentally manipulated. Thus, a biologist might argue that the distinction between exaptation and adaptation, although logically sound and conceptually interesting, cannot be "cashed out" in a sufficiently high percentage of cases because we so often lack enough historical data to determine whether a currently useful structure originated by natural selection for its present function (adaptation) or got coopted to its current role from an initial status either as an adaptation for some other function, or as a nonadaptive spandrel (exaptation).

I accept this point, of course, and have given my response in a previous section (p. 1233): when we cannot resolve the historical antecedents of a currently useful feature, we need not apply the terminology of adaptation *vs.* exaptation at all. The feature remains an aptation (in its current status), and may be so named, whatever its unresolved historical origin. But I also admit that if the distinction between exaptation and adaptation can be drawn in only a small percentage of cases, and only under unusually favorable circumstances, then the concepts enjoy little practical or operational use, and the terms might as well be abandoned in the actual practice of science.

However, I am confident that the distinction can be rigorously made in a high percentage of cases, probably a firm majority. If direct knowledge of historical sequences from paleontological data established the only path to resolution, then imperfections of the fossil record would preclude resolution at sufficient relative frequency. But evolutionary biologists can also reach firm conclusions about historical sequences from cladistic reconstructions of phyletic topography based upon the distribution of traits among living organisms. (I did not participate in the cladistic revolution within systematics, and I always maintained a cautiously critical, if basically supportive, attitude towards this scientific reform. But, and now in retrospect, I must credit cladistics with the signal achievement of devising a workable methodology for inferring historical and genealogical pathways from the distribution of features among modern organisms—thereby making the reconstruction of biological history operational as a generality, and not only in special cases of adequate evidence from the fossil record. Such an accomplishment marks a fundamental advance for a historical science like evolutionary biology—and cladistics must therefore be celebrated, if for no other reason, as one of the great achievements in the history of evolutionary thought.) In any case, my previous discussion of Arnold's work on sand-diving and crevice-dwelling lizards (pp. 1234–1238) illustrated the use of both criteria (direct historical sequencing and cladistic reconstruction) in distinguishing adaptation from exaptation in biological data.

The spandrels of San Marco provide an even clearer case—relatively free from biasing preferences imposed by our engrained assumptions about biological structures—for the testability, and hence the operationality, of distinctions between adaptation and exaptation, using both major criteria of direct data and inferences from taxonomic structure. In fact, my initial choice of this example stemmed explicitly from the availability of definitive data in the rarely available category of preserved historical records of actual genealogical sequences.

To state the conceptual problem: One might strongly favor the hypothesis, based on structural arguments alone, that the spandrels originated as nonadaptive side consequences, and only later achieved utility in housing the evangelists—thus identifying this derived function as exaptive rather than adaptive. But the static evidence of architectural form cannot decide the historical issue, for the alternative interpretation remains logically unimpeachable, however unlikely—and resolution therefore requires evidence about actual historical sequences. That is, one might reverse the flow of causality and argue: why must I regard the spandrels as primary nonadaptations constraining a later choice of aptive ornamentation? Perhaps the four evangelists represent a primary impetus rather than a secondary accommodation. Perhaps the architect chose to build his church with domes mounted on sets of four rounded arches because he had such a terrific idea for festooning the resulting spandrels of the central dome with mosaics of the four evangelists and the four rivers of Eden. In this case, the spandrels would exist to house the evangelists, and the mosaic designs would become primary adaptations.

I chose the San Marco spandrels because, in this case, we have firm evidence in both categories to reject this alternative possibility, to affirm the origin of spandrels as nonadaptive side consequences of a larger architectural decision, and to understand the mosaic designs as secondary adaptations within a space exapted for their utility.

(1) Direct historical data. We know that the spandrels were not built to house the evangelists because San Marco stood and operated in its appointed role as a church for at least three centuries before the mosaicists applied their astonishing work to a series of constrained and previously unornamented spaces (see Demus's classic four volume monograph (1984) on the history, architecture and iconography of San Marco).

(2) Inferences from taxonomic structure. Human buildings cannot be ordered as branching genealogical systems following both Linnaean and Darwinian logic—so we cannot base our inferences upon true cladograms in this case. But a taxonomy of ecclesiastical, and other large public, buildings does permit us to invoke a series of standard arguments, long utilized in the comparative anatomy of organisms as well, and all indicating both a nonadaptive original status for the spandrels, and a secondarily adaptive role for the evangelists as good designs in exapted spaces:

1. Ubiquity *vs.* occasionality as evidence for both priority and necessity. Thousands of Western buildings feature domes atop rounded arches—and every single one of them generates tapering triangular spaces at the intersections. These spandrels are ornamented in a wide variety of ways, each appropriate to the local circumstance. Many carry no ornaments at all (indicating that spandrels must be generated but need not bear "adaptive" designs).

2. Constancy of form *vs.* variety in usage. How could such a diversity of employment always generate the same housing? This particular distribution of anatomical "features" indicates that a constant form preceded the various ornaments thereof (both historically in phyletic time in some cases, and structurally in ontogenetic and architectural sequence in all cases). I have, for example, noted various religious foursomes in the spandrels of other churches —the four major Old Testament prophets, Isaiah, Jeremiah, Ezekiel, and Daniel in many cases; or, in San Ignazio in Rome (and "politically correct" by current standards of gender equality), four Old Testament heroes and their weapons: David with his sling, Judith with her sword (to behead Holofernes), Samson with his jawbone, and Jael with her tentpost (to transfix Sisera through the head). I also have noted secular themes in civic or scientific building—the four continents of Africa, Europe, Asia, and America under the main dome of the Victor Emanuel arcade in Milan; four classical lawmakers (Justinian, Pericles, Solon, and Cicero) under the glass dome in the Victorian courtroom of the Landmark Center, St. Paul, Minnesota; four mainstays of civilization (peace, justice, industry, and agriculture) in the County Arcade of Leeds, England, built in 1900; or the four Greek elements in the pendentives under the main dome at the headquarters of the National Academy of Sciences in Washington, DC!

3. Suboptimal or ill-fitting designs as evidence of historical sequencing. As emphasized in Chapter 2, in my analysis of Darwin's rich and subtle methodology for historical inference, perfection covers the tracks of history, whereas oddities and imperfections often reveal both the direction and the stages of temporal sequences. When four evangelists establish an optimal design in four spandrels, we cannot determine the sequence of events from the structure alone: either the spaces came first, and the evangelists fit in later, or plans to depict the evangelists came first, and architects then fashioned the spaces as appropriate housing. But a peculiar, ill-fitting, or suboptimal design might suggest an order of historical precedence. If three spandrels housed elegant mosaics of Genesis, Exodus, and Leviticus, while similar designs for Numbers and Deuteronomy appeared all scrunched together in an ugly and overcrowded fourth spandrel, we might assume (at least as a preferred hypothesis for further test) that the four spandrels originated for a different and prior reason, and that a later mosaicist miscalculated badly in formulating plans for placing symbols of the Torah into these preexisting spaces.

Similarly, the foursomes in several sets of spandrels in European churches seem rather forced or even ill-fitting, thus indicating that a fixed number of spaces (and their form) preceded any decision about embellishment. In the 16th century church of San Fedele in Milan, for example, four concepts, personified as women, decorate the spandrels under the central dome—the famous biblical trio of faith, hope, and charity (1 Corinthians, 13), with the remaining fourth spandrel occupied by religion. Three spandrels might have carried the intended design better, but architectural constraints dictated a quartet, so the designers had to draft a fourth participant, however unsanctioned by a very famous quotation. By etymology, "religion" may mean "tying together," but this particular woman seems more out of place than integrative at San Fedele. Interestingly, the design of the great Romanesque pulpit in the Duomo of Pisa (the building adjacent to the structurally inadaptive but touristically highly exaptive *Torre Pendente*, or Leaning Tower) imposes no architectural necessity for a quartet of spandrels. Its lectern rests upon a tripartite column that expands to three ornamental spaces at the top. The three spaces carry heads representing faith, hope, and charity—all by their proper selves this time, with no fourth interloper to complete the occupation of a preexisting architectural constraint.

4. Invariable correlation of a specific form under discussion with broader structural features of the totality. As an unprofound point in this case, the number of spandrels always correlates perfectly with the number of arches supporting an overlying dome—thus identifying the spandrels as automatic side consequences of a broader architectural decision. We might question this argument if all buildings mounted their domes on four arches, and therefore always generated pendentives in sets of four. But, although four remains by far the most common number, the comparative anatomy of large public buildings includes some variety. The central dome of St. Paul's Cathedral in London rests upon eight arches, and the resulting eight spandrels feature the

four evangelists at the eastern end (where the sun rises upon novelty), contrasted with the four great Old Testament prophets to the west (where the sun sets on ancient ways).

### Three major reasons for the centrality of spandrels, and therefore of nonadaptation, in evolutionary theory

I therefore find the concept of spandrels, or features of nonadaptive origin as structural byproducts or side consequences of other architectural decisions, to be both coherently definable and eminently testable. The importance of spandrels in evolutionary biology must then rest on two further attributes: (1) their engagement with conventional theory in a challenging way that suggests potentially important changes or expansions in our general understanding of evolution; and (2), their establishment as sufficiently common to constitute a high percentage of biologically and evolutionarily relevant traits of organisms and other biological individuals.

I think that spandrels pass the first test in a robust manner, for their existence at high relative frequency (the claim of the second test) would challenge a key procedure of the adaptationist program that has long served as the day-to-day working methodology of Darwinian biologists engaged in the explanation of particulars. At the most basic level, we simply cannot gain an adequate evolutionary explanation for a trait by elucidating, however elegantly, however experimentally, and however quantitatively, its contribution to the fitness of the organisms or populations in which it now resides. Purely adaptationist analysis therefore cannot resolve history for two major reasons:

1. Through the principle of quirky functional shift, and Nietzsche's discordance between reasons for current utility and sources of historical origin, our understanding of how a current trait works cannot elucidate its mode of origin—an ineluctably and logically central task of evolutionary explanation, and one of the most interesting questions that any historical science can pose.

2. Adding insult to injury, even the most sophisticated documentation of adaptive value in a current feature gives us no right to assert similar adaptational control over its past states—even admitting the principle of quirky functional shift, and the possibility of strikingly different past usages, with current functions emerging as exaptations. Rather, the principle of spandrels suggests that a high percentage of traits now contributing in important ways to fitness arose for no adaptive reason at all, but rather as automatic side consequences of other forces (usually selection on other aspects of the organism to be sure, but with no direct selection on the trait under study). The adaptationist program cannot provide a full accounting of evolutionary change if a high percentage of traits originated as nonadaptive spandrels.

We must then pose the second question about the relative frequency of spandrels. If rare, they remain conceptually interesting, but minor in actual importance for the evolutionary understanding of particular lineages—the bread and butter of daily practice in our science. My broader case for the high frequency, indeed for the near ubiquity, of spandrels occupies the last section of this chapter, but I raise three points here to set a framework of plausibility for frequencies far too high to ignore.

**RELATIONSHIPS TO GEOMETRY AND ARCHITECTURE.** As Geoffroy and other formalist thinkers recognized from the inception of evolutionary studies in biology, organisms are integrated entities, not hodge-podges of independent attributes each dedicated to a separate function. For two major reasons, this evident and venerable notion implies a great importance and high relative frequency for spandrels. *First,* any change in one part of the body must propagate correlated alterations to other parts. Selection may generate the original change for adaptive reasons, but many automatic consequences will probably be spandrels. *Second,* any adaptive feature of one organ will also express inherent and ineluctable attributes that must rank as spandrels. Most of these sequelae, although surely more numerous than adaptive aspects of the same feature, will probably remain forever irrelevant to evolutionary success of the lineage. (Bones are made of calcite and apatite for adaptive reasons, but bones are also white because the chemistry of these compounds so dictates. Invisible during life, this spandrel property of whiteness will probably never influence the evolutionary history of the surrounding organism. But evolution can also generate surprises in the same domain. Prior to the evolution of eyes, who would have predicted that the optical transparency of several enzymes and proteins might one day become relevant to their suitability for cooptation as lens crystallins?)

Thus, even the simplest and universal geometries of filling space must generate a host of spandrels to accompany any basically adaptive style of growth or biomechanical form. Moreover, by using criteria of direct historical records (in infrequent but best cases) or inference of genealogical order from cladistic reconstructions based on living species (a strong, if indirect, mode of argument almost always potentially available), we should be able to divide the useful features of organisms into direct adaptations, coopted adaptations with different original uses, and coopted spandrels—with the last category embodying a crucial challenge to strict adaptationist thinking: currently useful features with nonadaptive origins. Consider a simple example where the geometric nature of the spandrel can easily be defined, and where we may infer nonadaptive origin from the evidence of cladograms. (Examples of this kind could probably be multiplied indefinitely, and for any organism, were biologists more inclined to grant the subject more attention and explicit study):

All snails that grow by coiling a tube around an axis must generate a cylindrical space, called an umbilicus, along the axis. The umbilicus may be narrow and entirely filled with calcite (then called a columella). But the space often remains open, especially in land snails. A few species use the open umbilicus as a brooding chamber to protect their eggs (Lindberg and Dobberten, 1981).

Is the umbilical brooding chamber a coopted spandrel—a space that arose as a nonadaptive, geometric byproduct of winding a tube around an axis? Or did snails initially evolve their spiral coiling as part of an actively selected de-sign centered upon the direct advantages of protected eggs in a cigar-shaped central space? We cannot use the first method of data from actual historical sequences to resolve this question because we do not know whether the first coiled snails brooded their eggs in an umbilical chamber. But the second method of cladistics and comparative anatomy seems decisive in this case, however inferential: the cladogram of gastropods includes thousands of spe-cies, all with umbilical spaces (often filled as a solid columella and there-fore unavailable for brooding) but only a very few with umbilical brooding. Moreover, the umbilical brooders occupy only a few tips on distinct and late-arising twigs of the cladogram, not a central position near the root of the tree. We must therefore conclude—both from geometric logic (ineluctable produc-tion of the umbilicus, given coiling of the shell) and from the distribution of umbilical brooding on the cladogram—that the umbilical space arose as a spandrel and then became coopted for later utility in a few lines of brooders.

In an equally evident example of an automatic side consequence generated as a geometrical necessity, *Megaloceros giganteus,* the so-called "Irish Elk," elongated the neural spines of its shoulder vertebrae for an immediately adap-tive reason that seems well documented in fossil evidence, and rigorously val-idated by biomechanical analysis (Lister, 1994). These deer grew the largest antlers in the history of the Cervidae (up to 35 kilograms in weight on a 2 kg skull). Many big herbivorous mammals with heavy heads extend their neural spines to provide an increased area of attachment for enlarged *ligamentum nuchae* muscles that hold up the head and neck—a problem that probably af-fected *Megaloceros* more than any other deer.

The elongated spines are clearly adaptive (for internal insertion of larger muscles), but the outward expression of these enlarged bones—a raised area at the shoulders, covered with hair—probably existed as a nonadaptive span-drel at its initial phyletic appearance, an inevitable consequence of the geome-try of physical space. And so this feature would have remained, as the vast majority of spandrels do, until the species's demise—except for a coopted utility that converted the original spandrel into an exaptation in this case. This raised area—literally a patch of skin spanning a space enlarged for inter-nal reasons—became enhanced, altered in shape to a more prominent and lo-calized hump, and festooned with distinctive stripes and colors, all (presum-ably) for coopted utility in mating display. I freely admit that the exaptive potential of such simple bumps and spaces must be limited, and may never exceed the primary adaptation (which originally engendered the feature in question as a spandrel) in evolutionary importance. (I also confess that I love this example largely for humanistic reasons (see Gould, 1996b). Fatty humps and coat colors do not fossilize, and *Megaloceros* became extinct many thou-sands of years ago. We only know about the hump and its colors because our Cro-Magnon ancestors painted a few of these animals on French and Spanish cave walls—see Figure 11-10.)

When we move from simple tubes and sheets (aspects of universal geome-try, even though evolved modes of growth in particular lineages must elicit the forms) to more complex developmental architectures that record the contingencies of particular lineages (rather than the general geometry of Euclidian space), both the range and the number of potential spandrels, and their capacity for future exaptive utility, must broaden enormously. I have already mentioned Darwin's (and Owen's) example of non-fusion in neonatal mammalian skull sutures (p. 326).

In complex sexual animals, a particularly interesting class of spandrels originates as consequences of a developmental constraint leading both sexes along an initially shared embryological pathway that later branches to differentiate a set of homologous structures into the two major facies of a species's sexual dimorphism. In the most widely discussed (literally since Aristotle), and still very much unresolved, case of so-called "male mimicking" genitalia in female spotted hyenas, the peniform clitoris, and the *labia majora* that fold and fuse in the midline to resemble a scrotal sac, may have originated as spandrels of high testosterone levels that, in female development, correlate with the attainment of a larger adult size than males, and with behavioral domination over males. (We should all remember a lesson from our introductory Biology 1 course: the penis and clitoris, and the scrotal sac and *labia majora*, are homologous pairs of organs, specialized along divergent paths by disparate hormonal titers in the development of the two sexes.)

This fascinating case remains controversial (see Frank, 1997), and some recent information indicates that high levels of androgens may not induce "male-like" features in female development (R. Wrangham, personal communication). But the explicit formulation of such a nonadaptationist hypothesis, based upon spandrels, after more than 2000 years of unchallenged adaptationist speculation, has certainly sparked the debate and inspired a vast outpouring of research that will eventually resolve the issue. At least we may

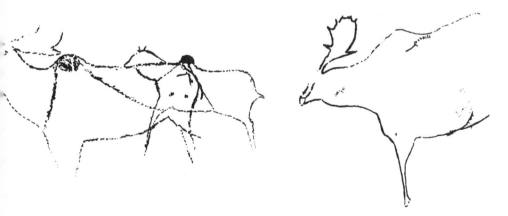

11-10. Exaptation of raised area at the shoulders as a hump (with secondary adaptation of coloring), presumably significant in sexual selection in *Megaloceros*. We only were able to learn about this unfossilizable feature of this extinct deer because our Cro-Magnon ancestors painted these animals on cave walls, as here from Cougnac.

be confident that no future researcher of such high status and such empirical expertise will again commit the bald version of the Nietzschian fallacy of confusing current utility with historical origin—as did Kruuk (1972), the most eminent student of spotted hyenas, when he argued that he had resolved the adaptive origin of "male-mimicking" genitalia by documenting the undoubted utility of these features in facilitating recognition of previously solitary clan members in the "meeting ceremony."

I wanted to call my popular article (Gould, 1987a) on my favorite human example "tits and clits," but desisted because readers would have assumed a sexist bias when I really meant to designate *male* tits, not to denigrate women. (I eventually settled upon the rhythmically catchy, but not quite so pithy, "male nipples and clitoral ripples.") The male question may be largely facetious, although the issue has evoked substantial and explicit discussion ever since Buffon: why do men grow apparently non-functional nipples? The puzzlement of so many people, including several accomplished scientists, flowed from adaptationist biases that demanded an explanation in utilitarian terms: perhaps males can suckle babies in certain circumstances; or perhaps they once did, and male nipples persist as a vestige? But the probable resolution, based on a quite different (albeit simple) perspective, requires the codification of a concept of nonadaptive spandrels for recognition and understanding: males probably grow nipples because females need them for an evident purpose, and many aspects of development follow a single pathway. So females grow nipples as adaptations for suckling, and males grow smaller and unused nipples as a spandrel based upon the value of single developmental channels.

The female counterpart, however, has evoked much argument, and imposed substantial grief upon millions of women during the 20th century: why do most female orgasms emanate from a clitoral, rather than a vaginal (or some other), site? The male biologists who fretted over this question simply assumed that a deeply vaginal site, nearer the region of fertilization, would offer greater selective benefit to the Darwinian *summum bonum* of enhanced reproductive success—hence the supposed puzzle. In the tragedy heaped upon this error, Sigmund Freud defined the almost anatomically impossible transition from clitoral to vaginal orgasm as the defining criterion of sexual maturity in women. He even regarded, and so labelled, the persistence of perfectly satisfactory and exciting clitoral orgasm as a form of frigidity. Under his influence, millions of women with normal sexual responses struggled to meet this impossible criterion of "true" realization—with consequences ranging from the sad to the tragic.

But a more probable resolution—if any mystery remains once we shed our adaptationist biases—might identify the clitoral site as a spandrel by the same argument applied to male nipples. The female clitoris is the developmental homolog of the male penis, and the adaptive value of male orgasm seems no more problematical than the biological function of the female breast. The clitoral site of female orgasm need not hold any special adaptive value *per se,* but may arise as a developmental consequence of selection upon the same organ in males.

This argument about the nonadaptive nature of the clitoral site (as advanced by Symons, 1979; and Gould, 1987a) has been widely misunderstood as a denial of either the adaptive value of female orgasm in general, or even as a claim that female orgasms lack significance in some broader sense. Only a grossly overextended commitment to exclusivity for the simplest form of adaptationist argument could ever have led to such foolish misunderstanding. I cannot speak from direct experience of course, but I accept the clear testimony that clitoral orgasm plays a pleasurable and central role in female sexuality and its joys. All these favorable attributes, however, emerge just as clearly and just as easily, whether the clitoral site of orgasm arose as a spandrel or an adaptation. (As a spandrel, the clitoral site would represent the different expression of a male adaptation, just as male nipples may be the spandrels of a female adaptation.) After all, we defined the concept of spandrels largely in terms of their rich potential for exaptive *utility* (Gould and Vrba, 1982). To state Nietzsche's principle in another way: origin as a spandrel implies no diminution of potential for crucial and joyous exaptive use later on.

As to the important question of a potentially adaptive nature for female orgasm in general, I have no firm opinion, and certainly feel no hostility towards functional hypotheses in conventional Darwinian terms of enhanced reproductive success. (I have only questioned the adaptive interpretation of the specifically clitoral site, since basic developmental anatomy would seem to dictate such a placement for other, and prior, reasons. Alfred Kinsey (1953), a very fine evolutionary entomologist who achieved far greater fame in his "second career" as a sociologist of sexual behavior, upbraided Freud, who had begun his own professional life as a comparative anatomist (with a doctoral thesis on the neurology of amphioxus), for failing to draw the obvious inference from a developmental homology that he knew perfectly well, and from easily available information about the rich innervation of the clitoris, and the virtual anesthesia of the vaginal canal.

Some common hypotheses for the adaptive nature of female orgasm have not fared well, including the standard argument that muscular contractions during orgasm help to draw sperm down the vaginal canal towards the site of fertilization (see Buss et al., 1998). Other arguments of a more psychological nature, especially the claim that orgasm may have positive value in stimulating pair bonding, make more sense in a too obvious way that always manages to incite my skepticism. In any case, the establishment of an important and potentially exaptive value of female orgasm does not challenge the hypothesis that the clitoral site originated as a nonadaptive spandrel (in Darwinian terms for enhancing reproductive success, not in human terms of female pleasure) of selection upon the homologous organ in males, the penis.

RELATIONSHIP TO COMPLEXITY. As a primary correlation regulating the distribution and importance of spandrels *vs.* primary adaptations, increasing complexity of an organ must imply a rising relative frequency of nonadaptive side consequences with potential future utility. With greater complexity in number and form of components, cooptable side consequences must rise to exceed, or even to overwhelm, primary adaptations. The chief example in biology may be a unique feature of only one species, but we obviously (and properly) care for legitimate reasons of parochial concern. The human brain may have reached its current size by ordinary adaptive processes keyed to specific benefits of more complex mentalities for our hunter-gatherer ancestors on African savannahs. But the implicit spandrels in an organ of such complexity must exceed the overt functional reasons for its origin. (Just consider the obvious analogy to much less powerful computers. I may buy my home computer only for word processing and keeping the family spread sheet, but the machine, by virtue of its inherent internal complexity, can also perform computational tasks exceeding by orders of magnitude the items of my original intentions—the primary adaptations, if you will—in purchasing the device.)

A failure to appreciate the central role of spandrels, and the general importance of nonadaptation in the origin of evolutionary novelties, has often operated as the principal impediment in efforts to construct a proper evolutionary theory for the biological basis of universal traits in *Homo sapiens*—or what our vernacular calls "human nature."

I welcome the acknowledgment of self-proclaimed "evolutionary psychologists" (compared with the greater stress placed by the "sociobiology" of the 1970's on a search for current adaptive value) that many universal traits of human behavior and cognition need not be viewed as current adaptations, but may rather be judged as misfits, or even maladaptations, to the current complexities of human culture. But most evolutionary psychologists have coupled this acknowledgment with a belief that the origins of such features must be sought in their adaptive value to our hunter-gatherer African ancestors. (Much of the daily practice of current "evolutionary psychology" focuses upon efforts to identify and characterize the EEA (their term), or "environment of evolutionary adaptation," for the origin of cognitive universals as direct adaptations in the common ancestral population of all modern humans.)

I applaud this use and recognition of the Nietzsche-Darwin principle of discordance between reasons for historical origin and bases of current utility (or disutility). But I also believe that "evolutionary psychology" will remain limited and stymied in its worthy and vital goal—to understand the human mind in evolutionary terms—so long as its practitioners place such unwarranted and effectively exclusive weight upon conventional adaptationist explanations for the *origin* of universal cognitive traits, and fail to recognize the central role (I would say dominant, but the issue obviously remains open) of constraints and nonadaptations in the initial construction of the cognitive and emotional modules and attributes that we collectively designate as "human nature."

A central principle about constraint from each of my two chapters (10 and 11) on the subject would broaden the range of hypotheses and lead to a richer and ultimately more accurate "evolutionary" psychology, both in immediate empirical terms of understanding the human mind, and in conformity with the true depth and range of modern evolutionary theory, rather than invoking an almost caricatured version of adaptationism as the only ground of evolutionary explanation for the origin of traits.

1. At a sufficient depth and distance, original adaptations now act primarily as historical constraints, and must be so characterized and analyzed (the central theme of Chapter 10). When we recognize a cognitive universal of human mentality as ill-fit to the complexities of modern social life, we do not then achieve an explanation of its *human* origin in adaptationist terms simply because we can state a good case for its initial phyletic appearance as an adaptation. We need to specify the evolutionary distance and the environmental context of initial appearance before we can render any judgment. In general, I would accept the statement that if we can locate the feature's adaptational origin in the last common ancestor of *Homo sapiens*, or even as far back as the common ancestor of the hominid line (after splitting from the lineage of great apes), then we may legitimately argue that this initial adaptive context establishes the "evolutionary meaning" of the feature in our quest to understand its appearance in human phyletic history.

But suppose that the feature had a far more ancient, but still fully adaptational, origin in a distant ancestor of very different form and neurological function, and also living in a very different environment—say, in the basal gnathostome fish of early Paleozoic times. Suppose also that this mental attribute has persisted ever since as a plesiomorphic aspect of the basic operation of the vertebrate brain. When we then try to explain the evolutionary significance of this mental mode in contemporary human life—especially when we try to identify its role in quirky and clearly suboptimal characteristics of human reasoning in the modern world—would we wish to claim that an adaptational analysis (in recognition of the feature's Darwinian origin in such a distant ancestor) will provide our best understanding? Clearly, we do not so proceed in most evolutionary analyses—and for good reasons discussed at length in Chapter 10 on the evolution of development. Rather, we treat such features predominantly as historical constraints because, as invariant and plesiomorphic traits of our entire clade (not only of all hominids and primates, but also of all mammals and tetrapods), they operate as unchanging constraints upon any subsequent evolution of mental modes, despite their adaptational origin in such a distant ancestor of such different form and environment.

I suspect that many puzzling features of human mentality would be better resolved if we conceptualized them as historical constraints derived from distant adaptational origins. To cite a hypothetical example (that would attract my substantial and favorable wager were I a betting man): I agree with a major theme of structuralist philosophy and research, as developed most cogently in our times by Claude Levi-Strauss and his followers, that identifies our tendencies to parse natural variety into pairs of opposed and dichotomous categories as an inherent property of human mental functioning—with male and female, night and day, and culture *vs.* nature as primary examples. I think that most people would also identify this strong preference as a constraint with highly unfortunate consequences for human life—not only because we so often construct invalid dichotomous taxonomies in our real world of complex continua, but primarily because we so often impose another conceptual module for moral judgment upon our pairings (the Manichean good *vs.* bad), and then proceed to identify one side of the dichotomy (including ourselves and our preferences) as righteous, and the other side (including "foreigners" and competitors) as worthy of anathematization or even ripe for burning. (I need hardly add that yet another aspect of human mentality, our capacity to devise grisly means of death and torture, and our technological ability to apply such means to large numbers of people in short periods of time, makes our innate preferences for dichotomization particularly dangerous.)

Now I am perfectly willing to believe that our brain's preference for dichotomization arose as a highly adaptive attribute in a very distant and ancient small-brained ancestor that, to enhance its prospects for survival, needed to make limited, quick, and twofold decisions that exhausted the maximal capacity of its judgment in any case: mate or wait, eat or sleep, fight or flee. But, whatever the adaptational basis of origin, dichotomization then persisted throughout the subsequent phylogeny of vertebrates as a historical constraint that became more and more quirky, and more and more limiting, as the brain enlarged into the much more sophisticated instrument of a lineage that eventually generated our exalted, but curiously freighted, selves.

2. At the level of immediate reasons for persistence and flourishing of the hunter-gatherer common ancestor of *Homo sapiens* in Africa, many distinctive mental attributes of our species, including major features of "human nature" that define our evolutionary success, must have arisen as nonadaptive spandrels (later exapted, in several cases, as vital bases of our current domination), and not as primary adaptations (the central theme of Chapter 11). This conclusion necessarily follows from the previous argument that, at the level of maximal natural complexity represented by the human brain, consequential spandrels must, at least in number, overwhelm the primary adaptations that generate them. Therefore, in terms of exaptive potential for evolutionary futures, the brain includes more cooptable spandrels than primary adaptations. Any "evolutionary psychology" that neglects the nonadaptational origin of many features now useful (or at least used, however dubiously), and that limits the domain of evolutionary inquiry to arguments (often speculative) about initial adaptive causes and benefits, will become more misleading than enlightening in restricting investigation to such a narrow scope of inquiry. We must abandon the largely unconscious bias of an overly strict Darwinian approach that equates all "evolutionary" explanation with adaptationist analysis.

**THE UBIQUITY OF SPANDRELS UNDER A HIERARCHICAL CONCEPTION OF EVOLUTIONARY LEVELS AND CAUSALITY.** If Darwin's own view of natural selection as a single-level process operating on organisms had prevailed, spandrels would still be pervasive in nature and important in evolutionary theory. But the scope of nonadaptive side consequences would then be limited to overt structures, physiologies and behaviors of bodies. That is, spandrels would be altered bits and pieces (often substantial) of organic "stuff," molded as the propagated effects of primarily adaptational changes wrought by selection upon other parts of the body.

But under the revised and expanded hierarchical theory (Chapter 8), where selection works simultaneously on a nested hierarchy of biological individuals (genes, cell lineages, organisms, demes, species, clades), the domain of spandrels becomes much larger, and their importance to evolutionary theory expands accordingly, and for an interesting reason that has not been adequately addressed in the literature (see Gould and Lloyd, 1999), but will occupy most of the next and last section of this chapter. The expansion of spandrels under a hierarchical theory of selection establishes the most interesting and intricate union between the two central themes of this book—the defense of hierarchical selection (as an extension and alteration of Darwin's single-level organismal theory) on the first leg of the tripod of essential components in Darwinian logic; and the centrality of structural constraint (with nonadaptively originating spandrels as a primary constituent) for a rebalancing of relevant themes, and as a correction to the overly functionalist mechanics of selection on the second leg of the tripod (or branch of the tree—see Figure 1-4).

To epitomize the central argument: under a hierarchical theory of selection, any novelty introduced for any reason (usually adaptational) at any level, must propagate a series of effects to biological individuals at other levels of the hierarchy. Duplication of genetic elements by direct selection at the gene level, for example, propagates redundancies to the organismic level; any organismal adaptation at Darwin's level propagates changes to the encompassing species-individual, as expressed in such species traits as population size, geographic range, and coherence among subparts (organisms). These propagated effects must be defined and treated as spandrels. As "injections" from another level (where the initiating change probably had an adaptational, or physically automatic, basis), these propagated effects cannot be viewed as adaptations at the level under consideration. Moreover, because these effects exist as true properties at the level under consideration—that is, as actual "stuff" rather than unused potentials of features now operating in different ways—they must be treated as initially nonadaptive "things" or spandrels, rather than as mere potentialities of some hypothetical future utility. Therefore, under a hierarchical model, spandrels include both the architectural side consequences of adaptational changes at the level of their origin, and the large realm of effects propagated to other levels as nonadaptive consequences of changes wrought for directly causal reasons.

This concept of cross-level spandrels neatly explicates a variety of phenomena that have long been recognized as both true and essential, but that have remained puzzling or anomalous under the conventional Darwinian rubric of a functionalist, single-leveled theory of selection. For example, our canonical, almost mantra-like, statement about the deleterious nature of most mutations achieves such an evident explanation that the resulting "aha" seems almost humorous in its suddenly obvious character. Each mutation arises for a perfectly good reason (usually chemical rather than adaptational in this case) at the gene level. But the effects then imposed upon organismal phenotypes must be designated as spandrels—that is, as nonadaptive side consequences expressed at another level. These effects will usually be deleterious, because the organism, as a highly complex, well integrated, and biochemically efficient object, will more often be hindered than helped by a change that arises as an "injection" from elsewhere, established by causes directly operating only in this elsewhere, and not subject to initial scrutiny at the level of injection. For the same reason, and in another mantra, we designate mutations as "random"—not in the mathematical sense of equally likely in all directions, but in the special evolutionary sense (see Eble, 1999) that such mutations originate without reference to the adaptive needs of the organism. When we recognize the phenotypic expression of mutations as cross-level spandrels, this property of "randomness" becomes entirely sensible, and no longer puzzling as a supposed sign of organic inefficiency.

Gene duplication, and other modes of origin for the repeated elements that constitute such a high percentage of the genomes of complex organisms (and that have been so puzzling under Darwinian assumptions about organisms as "lean and mean" machines honed to optimality by the relentless power of natural selection), represent the most important "playing field" yet identified for the evolutionary importance of cross-level spandrels. (I thank my colleague Jürgen Brosius for helping me to understand and work through the implications of this concept—see Brosius and Gould, 1992; Brosius, 1999; and Brosius and Tiedge, 1996.)

In some cases, of course, gene amplication originates as an immediate adaptation at the organismal level, especially when the availability of more gene product provides a selective advantage to the organism. But, more commonly, amplification occurs for causal reasons at the genic level itself, often by the conventional Darwinian mechanism of increased reproductive success, in this case by generating more copies of oneself, and inserting them into various places in one's surrounding totality—that is, in the genome itself. (Such an argument about direct Darwinian selection at the gene level provides the rationale, as previously discussed (p. 693), for the important hypothesis of "selfish DNA"—see Orgel and Crick, 1980; and Doolittle and Sapienza, 1980 for the original publications.) Yet evolutionists have also recognized (see Ohno, 1970 for the classic statement) that these extra copies may strongly impact the evolutionary future of organisms by supplying flexibility for change through their redundancy. But this otherwise sensible argument also seems to raise a central dilemma in causality itself—since flexibility for future change cannot cause the current origin or maintenance of any feature! We can resolve this problem by recognizing augmented copies as nonadaptive (and cross-level) spandrels at the time of their initial expression at the organismic level. Later recruitment and utilization of spandrels represents a perfectly sensible, indeed inevitable, concept under notions of constraint and hierarchical selection. I have, of course, and throughout this chapter, referred to such later utilization as exaptation—in this case by the cooptation of initially nonadaptive spandrels.

I have, in the past, objected to the usual terminology of such amplified elements as "junk DNA," feeling that such a dismissive term could only record an adaptational bias towards viewing such currently "superfluous" stuff as an insult to Darwinian optimality. I wrote (in Brosius and Gould, 1992, p. 10706): "Genes duplicated or amplified by the tens to the thousands . . . have been named in an ambiguous or even derogatory manner (*e.g.*, pseudogene or 'junk DNA'). Such names do not reflect the significance of retroposed sequences as large valuable assets for the future evolvability of species; and, as a result, it is more difficult to contemplate their significance, impact, and function."

But I have changed my mind after reading an insightful commentary by Sydney Brenner (1999) on my 1997 paper about the meaning and significance of spandrels in evolution. Brenner begins by acknowledging the role of adaptational biases in our misunderstanding of the meaning and significance of amplified DNA:

> There is a strong and widely held belief that all organisms are perfect and that everything within them is there for a function. Believers ascribe to the Darwinian natural selection process a fastidious prescience that it cannot possibly have and some go so far as to think that patently useless features of existing organisms are there as investments for the future . . .
>
> Even today, long after the discovery of repetitive sequences and introns, pointing out that 25% of our genome consists of millions of copies of one boring sequence, fails to move audiences. They are all convinced by the argument that if this DNA were totally useless, natural selection would already have removed it. Consequently, it must have a function that still remains to be discovered. Some think that it could even be there for evolution in the future—that is, to allow the creation of new genes. As this was done in the past, they argue, why not in the future?

But Brenner then defends the traditional terminology of junk DNA with an argument (based on the contrast of junk and garbage in vernacular English) that I had not considered, and that now strikes me as wise and useful:

> Some years ago I noticed that there are two kinds or rubbish in the world and that most languages have different words to distinguish them. There is the rubbish we keep, which is junk, and the rubbish we throw away, which is garbage. The excess DNA in our genomes is junk, and it is there because it is harmless, as well as being useless, and because the molecular processes generating extra DNA outpace those getting rid of it. Were the extra DNA to become disadvantageous, it would become subject to selection, just as junk that takes up too much space, or is beginning to smell, is instantly converted to garbage.

Brenner then ribs my literary and terminological pretensions (and I accept his criticism). But he also finds a resolution to the conceptual puzzles surrounding junk DNA in recognizing that such amplified sequences, when they arise causally at the gene level and then get propagated as effects to the organismal level, are nonadaptive spandrels with great potential for later exaptation to utility. Therefore, their designation as junk—that is, as currently useless, but harmless (as opposed to garbage), and replete with potential future value—seems entirely appropriate, and I belatedly embrace this term as a proper implication flowing from the definition and meaning of spandrels:

> The paper [Gould, 1997e] has an important message and I strongly urge my readers at least to look at it, even if all the words in it can't be understood. I offer this brief summary as a guide.
>
> The term spandrel originates in architecture and is used to describe spaces left over as a consequence of some other design decision, such as the triangles that remain behind when a rectangular wall is pierced by an arched opening. No self-respecting architect would simply leave such spaces, especially in a grand cathedral with a rich patron. Instead they would be decorated, as is the case of the four pendentives under the dome of San Marco in Venice, which are decorated with the four evangelists. This example is a good one, because the historical sequence of events is known. The spandrels are the consequence of a structural design decision, a by-product of placing a dome on rounded arches; three centuries later, mosaicists decorated these spaces. Thus spandrels are not primary adaptations but, because they can have later uses, they become in Gould's terminology, exaptations.

## The Exaptive Pool: The Proper Conceptual Formula and Ground of Evolvability

### RESOLVING THE PARADOX OF EVOLVABILITY AND DEFINING THE EXAPTIVE POOL

Conventional Darwinian organismal selection adapts creatures to their immediate local environments—a process of specialization, usually operating to produce particular contrivances that reduce organismal flexibility for future evolution to radically altered conditions, especially when adaptation leads to simplification and loss of structures, as in extreme, but common, cases of internal parasitism. In a subset of situations—especially emphasized by Darwin as the potential ground of general "progress" in the history of life (see p. 467)—local adaptation may be achieved by a generalized improvement in biomechanical design that might be construed as promoting future prospects and flexibilities, rather than restricting phyletic options by specialization. But few evolutionists would doubt that organismal selection leads far more often to diminution of future prospects by specializations and losses than to enlargement by general biomechanical improvement. Natural selection in the organismal mode can only construct local adaptations in the here and now. We can all conjure up the conventional image of highly specialized and gorgeously adapted forms revelling in their successes of the moment, but then dying in the fullness of geological time, as marginal generalists parlay their staying power into phyletic persistence. (Natural selection, of course, may also favor such generalists in certain momentary environments, but not for their future prospects.)

And yet, flexibility for future change manifestly exists in differential degrees among organisms. This flexibility contributes mightily to the longterm macroevolutionary success of lineages, but cannot be directly built or maintained by ordinary natural selection in the organismal mode. We designate this differential capacity for success and extent of future change by the vague and loosely-defined name of "evolvability"—a concept that, until recently, remained unpopular among Darwinian biologists by evoking feelings of discomfort and confusion. The reasons for this usual distaste flow from the failure of conventional theory to provide a context that could make such a concept intelligible rather than paradoxical. After all, if "evolvability" seems contrary to the general workings of natural selection, and if natural selection represents the fundamental mechanism of evolutionary change in populations and lineages, then how could "evolvability" be defined or characterized as anything other than an accident or a passive residuum? Phenomena without direct mechanisms generally do not win much interest or approbation from working scientists.

For example, in their important 1998 article entitled "Evolvability," Kirschner and Gerhart express both apparent paradoxes attending this crucial concept within a strict Darwinian context: the seemingly logical need to impute selective advantages to supraorganismic levels (within a theory committed to the primacy, or even exclusivity, of organismal selection), and the almost unavoidable "feeling" that benefits of evolvability can only accrue to future states (which, in any standard account of causality itself, cannot be influencing the present evolution of beneficial features). Kirschner and Gerhart write (1998, p. 8426): "The proposal that evolvability has been selected in metazoan evolution raises difficulties because it seems to be a trait of lineages or clades rather than individuals. Clade selection is often considered an 'explanation of last resort.' Also, evolvability seems to confer future rather than present benefit to the individual."

But if we follow the expanded Darwinian logic of this book, the paradoxes become only apparent because the theoretical revisions developed herein validate both apparent peculiarities of "evolvability," thus bringing this crucial concept within the rubric of a revised evolutionary theory. First, selection at higher levels *is* an important force in evolution—and evolvability can therefore originate directly at the level of its evident advantage. Second, the structuralist validation of exaptation establishes, as a central aspect of evolutionary theory, the future cooptation of features initially evolved for other reasons. Thus, hierarchy and positive constraint—the two primary revisions to the first two components of Darwinian central logic—show their mettle in providing theoretical resolutions for each of the superficially paradoxical aspects of "evolvability," a concept that evolutionary biologists have lately recognized as vital, but treated so gingerly, or even apologetically, in the absence of a proper theoretical framework for admitting something so evidently important among accepted modes of causality and explanation.

Indeed, a remarkable change has been brewing for the last decade or so in evolutionary studies. "Evolvability" has suddenly become a hot topic, even among the most orthodox of modern Darwinians (Dawkins, 1996, for example). This change has occurred for at least three good reasons listed below, each reflecting one of the major topics of these two chapters on the biology of evolutionary constraint. But I feel, as stated just above, that the subject still languishes for want of a proper theoretical context in revisions and expansions of a Darwinian world view that had become too narrow in its focus on organismal adaptation and the sufficiency of known microevolutionary mechanisms to explain all scales of evolutionary change. In this final section, I therefore try to provide a context for evolvability by combining the two central theoretical reformulations of this book: (1) hierarchical models of selection; and (2) the importance of structuralist approaches to biological form and function, as expressed in concepts of constraint and, especially for elucidating this particular topic, in the importance of spandrels as nonadaptively originating side consequences, then available for later cooptation to utility as exaptations.

1. From studies of evo-devo, the discovery of extensive genetic and developmental homology among distantly related phyla, especially the common presence, spatial orientation, and mode of action of *Hox* genes in bilaterian phyla, has focussed attention upon the flexibility inherent in the great range of interesting, workable, and often realized permutations that can be generated from developmental rules shared by all complex animal phyla. In particular, and as discussed previously, the disproof and subsequent reversal of Lewis's original "bottoms up" hypothesis of sequential addition and differentiation of *Hox* genes in causal concert with the complexification and differentiation of arthropod phenotypes, has emphasized the enormous flexibility inherent in broad rules emplaced at the outset, and plesiomorphic among all bilaterian phyla—for the common ancestor of protostomes and deuterostomes already possessed a full complement of *Hox* genes, as do the most homonomous of living groups, the Onychophora and the Myriapoda (see pp. 1147–1150). Thus, the realized diversity of bilaterians evolved in a "top down" fashion (at least for features regulated by the *Hox* series) from a common ancestor with a full set of basic components and their rules of action already in place.

The constraints of these rules have provided more flexibility in their fecund channels than limitations through their "forbidden places"—a theme rightly emphasized in the finest book yet written on the relationship of homologously shared and rule-bound developmental architecture to flexibility and evolvability in the phyletic richness of subsequent life (Gerhart and Kirschner,

1997). But the importance of constraint and preexisting opportunity in channeling the pathways of change should not be underestimated—for the message of evo-devo does not proclaim that "all is possible under such flexible rules." One might, for example, denigrate the importance of constraint in noting that a rod-like element of an ancestral agnathan gill arch exhibited sufficient malleability in form and function to become the tiny, disparately shaped and divergently functioning malleus of the mammalian middle ear. The gill-arch elements may therefore work as general building blocks of unconstraining Pharaonic bricks in the metaphor of my treatment of this subject (pp. 1134–1142). But we must also remember that, absent *some* skeletal element of appropriate form and position (whatever its capacity for future modification to almost any other shape or function), vertebrates would probably never have evolved jaw bones that could transmute to ear bones—and our lineage (if it had survived at all) might have remained an insignificant component of bottom-dwelling mud-sucking marine faunas, thus precluding, at least for this planet at this time, the evolution of any species with enough cognitive capacity to fret about such issues.

2. From studies of genetic structures and sequences, the astonishingly high relative frequency of multiply repeated elements (with respect to previous assumptions about the nature of genomes), and the multiplicity of ways, from gene duplication to retrotransposition, for generating them, have documented enormous redundancy and combinatorial flexibility within genomes —even if we have designated the objects of this discovery with the disrespectful name of "junk DNA" (but see p. 1269 for my favorable reevaluation of this term). Retrotransposons, for example, constitute about 40 percent of mammalian genomes (Kazazian, 2000). The human genome includes about 500,000 truncated versions and some 3000 to 5000 full copies of the LINE-1 long terminal repeat. Chromosome 2 of *Arabidopsis* includes 239 tandem duplications involving 593 genes. A larger duplication of almost 2.5 million bases appears on two chromosomes in four large blocks. A long stretch of chromosome 4, including 37 genes, has been duplicated on chromosome 5. Chromosome 2 also contains a region with 75 percent of the mitochondrial genome, reflecting a recent transfer of a substantial block of DNA from an organelle to the nucleus (Meyerowitz, 1999)—quite a "gift" of "play" from the genic to the organismal level!

3. From studies of the properties of populations, communities and interactions among evolving entities by mathematical modeling and computer simulation. Several researchers have used these methods in attempts to identify the abstract and general conditions that might confer flexibility, persistence, and capacity for change upon an evolving population or group of entities. Most notably, as discussed previously (p. 1210), Kauffman's (1993) claim that successful systems move towards "adaptation at the edge of chaos" rests upon attention to evolvability as a key ingredient of longterm success. Such systems must be adaptive, but too much (and too precise) a local fitting may freeze a system in transient optimality with insufficient capacity for future change. Too much chaos may prove fatal by excessive and unpredictable fluctuation,

both in external environments and internal states. But a capacity to adjust to chaotic situations also confers evolvability. Adaptation at the edge of chaos balances both desiderata of current functionality and potential for future change, or evolvability.

Yet, however much biologists can document and articulate these components of evolvability, the general concept itself has remained uncomfortable, even paradoxical, because the clear existence of "flexibility for future benefit," and the equally obvious importance of evolvability to the long-term persistence and success of lineages, cannot be rendered as a directly causal and explicit outcome of the Darwinian mechanisms that we have viewed as fully sufficient for understanding the causes of evolutionary change. Organismal selection for traits that confer differential reproductive success in the ecological moment simply cannot generate, in any active or direct manner, a set of features that achieves evolutionary significance only by imparting flexibility for change in distant futures. We cannot deny that these features of evolvability deeply "matter" in the history of lineages; but how can benefits for futures arise by any causal process in the here and now? Moreover, and adding insult to anomaly, the major components of evolvability apparently reside in "superfluous" genetic elements, and in supernumerary items of anatomy, that almost seem to mock our usual concepts of the stark efficiency of selection as a natural arbiter between the immediately useful and the discardable junk. As an example of this discomfort and puzzlement, just consider the language of Kazazian's excellent review of mammalian retrotransposons (2000, pp. 1152–1153), as he struggles to grasp and communicate the apparent discordance between current irrelevancy (at the level of necessary generation) and future benefits: "Although retrotransposons have been viewed as selfish DNAs that provide no benefit to their host cell, we now know that over evolutionary time they have increased the diversity of the genome through a variety of mechanisms providing it with considerable 'added value.' . . . It is clear that L1 elements are the master mammalian retrotransposons. Although L1s may be selfish, they are clearly not junk, for they have played a major part in our evolution and the evolution of our genomes." But the genuine junk of today can be exapted for the triumphs of tomorrow. The spores of penicillin didn't do us much good (and even imposed substantial harm in spoiling our foods) until Dr. Fleming made his fortuitous discovery of a previously unrecognized, and now eminently lifesaving, property.

As its central theme and purpose, this book proposes a set of expansions and revisions to Darwinian theory that, among other salutary features, can resolve the paradox of evolvability by exposing the issue as a *Scheinproblem*, or problem of appearance—that is, a spurious appearance within an overly restricted theory presently lacking the language and concepts for granting causal intelligibility to this evidently vital theme in evolutionary studies. The key revisions proposed on each leg of the logical structure that I have called the tripod of essential components in Darwinian theory provide, in their ensemble, an explanation of evolvability in hierarchical and structuralist terms.

1. The expansion of selection to a hierarchical theory of simultaneous op-

eration on Darwinian individuals at several nested levels: Future flexibilities cannot be targets of conventional organismal selection; nor can such an attribute enter any calculation of the differential organismal fitness that fuels Darwin's mechanism. But this conclusion does not imply that attributes of evolvability must remain uninvolved as agents in any form of selection—for the most commonly cited components of evolvability can act as positive traits to enhance the fitness of species-individuals in selection at the species level. Just consider the two most widely recognized boosts to evolvability at the species level: the propensity of some species to generate relatively more daughter species, and the resistance to extinction conferred upon some species by such organismal features as generalized anatomy and wide ecological tolerance. These two properties represent the primary analogs, at the species level, of the two cardinal attributes—birth rates and death rates—that virtually define the fitness of organisms in the calculus of Darwinian benefits for traits that must ultimately express their selective advantages by correlation with enhanced birth or retarded death of organisms. Organismal selection cannot craft propensities for speciation or resistances to extinction in any direct way, but these properties act as primary features in the higher-level process of species selection, and therefore figure prominently and directly in the calculus of selective advantage under hierarchical models.

But a potential problem still remains. Species selection can certainly *utilize* and *maintain* these important traits of species-individuals, but species selection cannot always *fashion* such traits, especially when they emerge into this higher level as spandrels of selection upon organisms—as must occur when emergent fitness at the species level resides in the higher-level expression of organismal traits that can only originate by conventional organismal selection (see Lloyd and Gould, 1993; Gould and Lloyd, 1999 for an analysis of the logic of this issue). For example, in the case cited above, the enhanced resistance of a species to extinction emerges as a consequence of such organismal traits as generalized anatomy and broad physiological tolerance.

But when we probe further, and draw the appropriate analogy to the level we understand best, we recognize that ability to utilize, combined with inability to fashion, represents a norm within Darwinian theory, not a distinctive anomaly provoked by evolvability at the species level. Ordinary natural selection doesn't manufacture its variational raw material either. Darwinians have always understood that their theory's most quirky, most original, and most brilliant intellectual "move" explains how a process that creates nothing directly can, nonetheless, operate on raw material of different origin to become a "creative" force in the construction of novel and useful features. Indeed, I devoted much of Chapter 2 to illustrating how Darwin's contemporaries rarely grasped this intensely paradoxical, but defining, property of natural selection's genuine creativity, based only upon its power to enhance or eliminate, but not to craft variation in a direct and dedicated manner. And I pointed out that the isotropy (or undirectedness) of variational raw material acts as a prerequisite for granting natural selection this potentially creative role.

Thus, Darwinians have always argued that mutational raw material must be generated by a process other than organismal selection, and must be "random" (in the crucial sense of undirected towards adaptive states) with respect to realized pathways of evolutionary change. Traits that confer evolvability upon species-individuals, but arise by selection upon organisms, provide a precise analog at the species level to the classical role of mutation at the organismal level. Because these traits of species evolvability arise by a different process (organismal selection), unrelated to the selective needs of species, they may emerge at the species level as "random" raw material, potentially utilizable as traits for species selection.

The phenotypic effects of mutations are, in exactly the same manner, spandrels at the organismal level—that is, nonadaptive and automatic manifestations at a higher level of different kinds of causes acting directly at a lower level. The exaptation of a small and beneficial subset of these spandrels virtually defines the process of natural selection. Why else do we so commonly refer to the theory of natural selection as an interplay of "chance" (for the spandrels of raw material in mutational variation) and "necessity" (for the locally predictable directions of selection towards adaptation). Similarly, species selection operates by exapting emergent spandrels from causal processes acting upon organisms.

2. The acknowledgment of structural components as joint causes and specifiers, along with natural selection, for the directions of evolutionary change: (In the most radical version, these structural inputs operate as positive constraints generated as consequences of features with nonadaptive origins, thus precluding a purely functionalist or adaptationist account for both the origins, and the channels of subsequent change, of organismal traits.) To summarize this major claim of the preceding section: if important components of evolvability must emerge as spandrels of natural selection on other features (for they cannot be fashioned as direct products of a process that cannot explicitly make "things" for potential future benefits), and if these spandrels serve as exaptable raw material for higher-level processes of change, then we will need to describe macroevolution at least partly in the language of channeling by historical and structural constraint (often based upon features with nonadaptive origins), and not entirely in conventional functionalist terms of selective modelling to current environments, potentiated by an effectively unfettered capacity of mutational raw material to provide the wherewithal for evolutionary movement in any immediately adaptive direction.

3. The conventional mechanisms of microevolution cannot, in their extrapolation through geological immensity, fully explain the causes and patterns of evolutionary change at larger scales: I have argued throughout this book that the biological components of nonextrapolation lie within critiques of the first two legs of the tripod—and I have therefore invoked the external "disobedience" of the geological stage to represent this theme as a surrogate from the domain of another relevant subject (particularly for paleontologists like myself). Similarly, for this case of evolvability, species level selection on the first leg, and nonadaptive spandrels on the second leg, identify the major barriers

to explaining evolvability by extrapolated microevolutionary Darwinian processes.

Above all, and to move forward towards a clear and operational definition of concepts, the crucial subject of evolvability requires a taxonomy for its numerous modes, and their strengths and distributions. The key to an inclusive accounting lies in the general notion of *usable* features (for promoting long-term diversification and success) now *unused*. What kinds of attributes thus contribute to evolvability? What encourages their production or augments their number (especially since natural selection in the organismal mode cannot craft such features directly)? Who has more of them and why?

Clearly, organisms and populations maintain what we might call a "fund" or "pool" of potential utilities now doing something else, or at least doing no harm. I propose that we designate this ground of evolvability as "The Exaptive Pool,"* and that we try to establish a logical, interesting, and empirically workable taxonomy for the various attributes in this fund of exaptable potential. The exaptive pool represents the structural basis of evolvability, the potential vouchsafed to future episodes of selection (at all levels) in a world of strongly polyhedral objects always and ineluctably built with interesting corners and facets that facilitate and channel the directions of evolutionary movement.

## THE TAXONOMY OF THE EXAPTIVE POOL

### *Franklins and Miltons, or inherent potentials vs. available things*

The American dime, the smallest and thinnest (but not the least valuable) of our coins, can't purchase much these days, but still functions as legal tender, the primary cause of its manufacture and useful persistence. American dimes have also, starting a few years after his death near the end of World War II, borne a representation of Franklin D. Roosevelt on their recto ("heads") side. As an unintended consequence of their thinness, American dimes (and no other indigenous coins) also happen to fit snugly into the operative groove on the head of a standard screw—and dimes therefore work very well as adventitious screwdrivers. (I strongly suspect that virtually every adult American has used a dime in this well-known supplementary way, just as many of us know

---

*Because exaptations are coopted utilities, and because the attributes in this pool of potential remain as yet uncoopted, some people might object to my designation of this ground of evolvability as "the *exaptive* pool." Shouldn't this fund be called the "preadaptive pool" (not that ugly and misleading word again!), or the "potaptive pool" (but why invent a neologism if a current term will suffice)? However, in a purely linguistic and etymological sense, "exaptive pool" represents proper usage, and does not convey any confusing implication of names applied before actions that justify the names. The suffix *"ive"* refers to "something that performs or tends towards or serves to accomplish an action" (per *Webster's*)—a fine description of attributes that bring organisms towards evolvability and serve to help them accomplish this happy state. Direct*ives* are meant to be followed, but perhaps not just yet. Act*ive* sockets will work when you plug in an appliance. Sedat*ive* pills will calm you down after you ingest them. Exapt*ive* pools will supply their bearers with a leg up in the longterm sweepstakes of evolution.

how to jimmy a doorlatch with a credit card.) American dimes are therefore adaptations as money, and exaptations as screwdrivers.

This potential for supplementary use as a screwdriver represents an inherent capacity of the dime's size and shape, not a separate and unused entity arising as a side consequence of some other change. The inherent potentials of any object (for uses other than their intended purpose of manufacture) establish a large and important category of attributes in the exaptive pool of any biological individual. I propose that we refer to these inherent (but currently, or at least usually, unexploited) potential functions as "franklins" to honor the most famous exaptation of the American dime. (Indeed, if our currency inflates much further, dimes may become virtually useless as money, and their exaptive role as screwdrivers may achieve a primary status in immediate function. When Russia's currency hyperinflated after the collapse of the Soviet Union, I witnessed a remarkable example of this phenomenon. Five kopeck coins had become truly worthless as currency, but only objects of their size, shape and weight could operate public telephones. Sharp entrepreneurs therefore stood at telephone booths, offering to sell old five kopeck coins for 100 rubles each—2000 times their official monetary value, but when ya gotta call, ya gotta call.)

Franklins are not actual but unemployed "things out there." Franklins are alternative potential functions of objects now being used in another way. In the words of my subtitle, franklins are inherent potentials, not available things. This distinction may seem trivial, inordinately fine as an exercise in logic chopping, or even parodic to hard-nosed scientists committed to the professional ethos of "just give me the facts, and leave linguistic and theoretical niceties to effete humanists who lack the luxury of an objectively empirical subject matter." But I will attempt to show, in the next two sections, that the distinction between inherent potentials and available things, as the two fundamental categories of the exaptive pool, provides the conceptual key to understanding the importance of spandrels, and for recognizing the strong weight that must be applied to structuralist, and particularly to nonadaptationist, elements in the exaptive pool—thus defining the revisionary power of this concept in evolutionary theory, and exposing the depth of different explanations required to understand evolvability *vs.* ordinary adaptation by natural selection.

I will just mention, for now, that franklins constitute the theoretically untroubling category embodied in the Darwinian notion of quirky functional shift (as discussed on pp. 1218–1229 of this chapter). Franklins capture the important concept so poorly expressed in the old term "preadaptation"— that is, suitability for another function not presently exploited because the feature has been adapted by natural selection for a different utility. When feathers function for enhanced thermoregulation on the arm of a small running dinosaur, their potential aerodynamic benefits are franklins, or inherent but unused potentials. When Michelin's rubber works as an automobile tire, its suitability for manufacturing cheap and durable sandals for poor children in developing countries is a franklin. In short, franklins represent future potentials within structures now adapted to a different utility. When evolution then coopts the structure for actual service in this formerly potential role, the franklin becomes an exaptation. The entire process remains under the control of Darwinian selection and adaptation throughout. Franklins do not vie with current utilities, and they cannot be construed as "things" (whether adaptive or nonadaptive) waiting for a potential place in the adaptive sun. Franklins are the inherent potentials that permit Darwinian pathways, under the control of natural selection leading to adaptation at all times, to undergo quirky and unpredictable shifts from one function to a qualitatively different utility.

The attributes in the second major category of the exaptive pool are, by important contrast, actual entities, pieces of stuff, material things that have become parts of biological individuals for a variety of reasons (to be exemplified in the next section), but that have no current use (and also cause no substantial harm, thereby avoiding elimination by selection). Items in this second category of "available things" can originate in several ways—as nonadaptive spandrels (the most important subcategory, I shall argue), as previously useful structures that have become vestigial, or as neutral features fortuitously introduced "beneath" the notice of selection.

I propose that we refer to these available but currently unused material organs and attributes as "miltons" to honor one of the most famous lines in the history of English poetry. John Milton ended his famous sonnet *On His Blindness,* written in 1652, by contrasting two styles of service to God: the frenetic activity of evangelists and conquerors and the internal righteousness of people with more limited access to worldly action:

> . . . thousands at his bidding speed,
> And post o'er land and ocean without rest;
> They also serve who only stand and wait.

Miltons, in short, are actual things, presently without function, but holding within their inherent "goodness" the rich seeds of potential future utility. Now, they only stand and wait; tomorrow, they may be exapted as key innovations of great evolutionary lineages.

Miltons constitute the radical counterpart to the conventionality of franklins within the exaptive pool. Miltons break the exclusivity of the adaptationist program by basing a large component of evolvability not upon the potential of already functional (and adaptive) features to perform in other ways, but rather upon the existence of a substantial array of truly nonadaptive features—unused things in themselves rather than alternative potentials of features now functioning in other ways (and regulated by natural selection at all times). If features with truly nonadaptive origins occupy a substantial area in the full domain of evolvability, then we must grant this structuralist (and nonadaptationist) theme a generous and extensive space within the logic and mechanics of evolutionary theory. For this reason, I have argued that spandrels—already the most important category of miltons, but made far more significant by the inclusion, under their rubric, of cross-level effects of features originating at other levels (see pp. 1286–1294 of this section)—win their central status in revising and expanding evolutionary theory because they represent the primary input of an overtly structuralist and nonadaptationist concept into the central logic of evolutionary causation.

### Choosing a fundamentum divisionis for a taxonomy: an apparently arcane and linguistic matter that actually embodies a central scientific decision

Table 11-2 presents my sketchy and preliminary proposal for a taxonomy of subcategories in the exaptive pool. I shall retain franklins, or inherent potentials, as an integrity for now, not because I doubt that they could be usefully divided into subcategories, but simply because I wish to focus upon miltons, or available things, as the component of the exaptive pool that holds most reformatory promise within evolutionary theory.

I divide miltons into two major categories according to their different modalities of generation: *structurally,* as automatic and nonadaptive sequelae or side consequences of changes in other features, or at other levels; and *historically,* as nonadaptive features not linked by structural or mechanical necessity to another feature of a biological individual, but rather introduced sequentially in time by processes that can generate and tolerate such nonadaptive entities.

I then divide each of these two subcategories into two further groups. Structural miltons, as necessarily and automatically consequential, are all, and collectively, spandrels. But spandrels come in two different "flavors," with the second group of cross-level effects (newly so categorized here) representing, in my judgment, the key addition that elevates spandrels to a position of central importance in evolutionary theory. (I will present my full argument for considering cross-level effects as spandrels in the next section, pp. 1286–1294.) In any case, the first structural group of at-level, or architectural, spandrels includes the mechanical and automatic side consequences, deployed throughout the rest of the individual, of any primary change (usually adaptive) evolved by other features of the same individual. When I originally defined the biological concept of spandrels (Gould and Lewontin,

Table 11-2. A Taxonomy for the Exaptive Pool

| | |
|---|---|
| A. Inherent potentials | |
| *Franklins* | |
| B. Available things | |
| *Miltons* | |
| 1. As architectural consequences | Structurally |
| *Spandrels* | |
| i) At-level spandrels by geometry | |
| ii) Cross-level spandrels by injection | |
| 2. As historical unemployments | Historically |
| *Manumissions* | |
| 3. As invisible introductions | Historically |
| *Insinuations* | |

1979), with the pendentives of San Marco as my "holotype," I described only this group—not for any good or principled reason, but simply because I had not recognized the status of cross-level effects in a hierarchical theory of selection. Obviously, I already regarded this restricted set of at-level architectural byproducts as potentially significant in evolution. But the inclusion of cross-level effects as a second category moves the concept of spandrels from an edge of interest to a center of potential importance in evolutionary theory.

The second grouping of cross-level, or propagated, spandrels includes the expressed effects upon biological individuals of changes introduced for a definite reason (whether adaptational or not) at a different initiating level. I include such cross-level effects under the rubric of miltonic spandrels, rather than franklinian potentialities (as they have usually, if unconsciously, been regarded, when conceptualized at all), because, like the architectural sequelae of my at-level group, they are actual, initially unused, nonadaptive things that also arise as side consequences—even though the side consequence in this second group of cross-level spandrels are propagated effects to other levels (often with no expression at the level of origin for the primary change that generated them), whereas the side consequences in the first group of at-level spandrels are immediate mechanical correlates of a primary change in the same individual, and therefore easier to spot and define.

I then divide the second subcategory of historical miltons into two groupings of markedly different status:

1. Features that lose an original utility without gaining a new function. In a first group of "unemployments" or "manumissions," previously utilized features become liberated from functional or selective control, and gain freedom to become exapted for other uses. Currently nonadaptive as a historical result of their altered status, they fall out of selective control and into the exaptive pool as actual items that must now "stand and wait," but might serve again in an altered evolutionary future.

We have generally granted little evolutionary potential to such vestigial unemployments because relatively quick reduction and loss—as in the standard example of eyeless cave fishes—seems to follow as an inevitable injury of degeneration added to the initial insult of unemployment. But such manumitted miltons may be quite common, particularly at the gene level (where the process has achieved greater recognition). Many multiply repeated evolutionary phenomena—the deletion of larval forms in the evolution of direct development, or the shedding of adult stages in progenetic lineages, for example—must leave a substantial number of genes in such an "unemployed" state. Yet, as we also recognize, full unemployment may occur only rarely because most genes function in more than one way. (This fact, however, should be regarded as salutary for future exaptive potentials in keeping partially "unemployed" genes in an active state of resistance to true operational discombobulation by accumulation of neutral mutations.)

In the most fascinating confirmation of our literature (see p. 688 for more details in a different context), Hendriks et al. (1987) sequenced the alphaA crystallin gene in the blind mole rat *Spalax ehrenbergi* (which still grows a lens in its vestigial eye, although the lens's irregular shape cannot focus an image, and the eye remains buried under skin and hair in any case). They found that the blind mole rat has accumulated mutational changes in alphaA crystallin at a rate far higher than that observed in nine other rodents—but still (to emphasize the intended point) at only 20 percent of the characteristic rate for truly neutral pseudogenes. The authors conclude that although the alphaA crystallin gene emplaces substantial gene product into a nonoperational lens (at least for vision), the gene must still serve some function, supported by stabilizing selection, to resist the full neutral rate of change.

In another context on p. 1245, I noted that, although the alphaA crystallin gene is more specialized for generating lens protein than its paralog alphaB crystallin, its product still appears in other organs in some mammalian species. Moreover, the nonseeing eye of blind mole rats may continue to function in other ways. Haim et al. (1983) show that blind mole rats still perceive light (and not by the obvious nonvisual route of correlates to temperature) in regulating their circadian clocks to photoperiod. Hendriks et al. (1987) argue that melatonin, secreted by the equally nonvisual retina, may act as a prime circadian regulator—and they therefore suggest that the developmental pathway leading to the eye and its lens, including the action of alphaA crystallin, may be adaptively preserved because the eye also performs essential nonvisual functions.

2. *Features introduced beneath selection's scrutiny.* In the second group of "insinuations," nonadaptive features may enter the exaptive pool by neutral drift. Just as we have unfairly discounted the role of manumitted miltons, we have probably underestimated the relative frequency of insinuated miltons, although for a different reason. We have regarded neutral changes as both peripheral and rare (largely restricted to tiny populations on the verge of elimination in any case) because we have granted too much power to selective control—thus permitting our orthodoxy to become self-fulfilling by circular argument. But the frequency of insinuated miltons may actually be quite high, especially given the inevitability of their substantial introduction via founder effects.

In a fascinating recent example (Tsutsui et al., 2000; Queller, 2000), the Argentine ant *Linepithema humile*, "a superb invader of non-native habitats" (Queller, 2000, p. 519), has become firmly established in Mediterranean environments of California, much to the detriment of several native species and to the distress of humans. Queller notes *(op. cit.)*: "We rarely understand why invading species succeed, although a common advantage is that they leave their predators, parasites and pathogens behind. Although this explanation could apply to Argentine ants, it seems that the most serious enemies left behind were the warring clans of its own species."

In Argentina, ants of different colonies fight, and these antagonisms, and the resulting elimination of many colonies, keep the entire species at relatively low population densities, thus leading to an ecosystem that includes many other equally successful ant species. But, in California, *Linepithema humile* does not fragment into mutually antagonistic colonies, but lives as a single aggregate. Queller states *(op. cit.):* "They form a vast supercolony within which there is little aggression and extensive interchange of both workers and queens. Such ants are called unicolonial because a whole population effectively becomes one colony." The Californian success of *Linepithema humile* does not arise from their prowess as fighting individuals (for these relatively small ants do not seem especially gifted as gladiators in combat by the myrmecine equivalent of *mano a mano*), but from the sheer power of their numbers—"because of their ability to rapidly recruit legions of troops from their network of nests. So peace with flanking nests generates advantages in competition with other species" (Queller, 2000, p. 519).

Fighting among different colonies in native Argentina depends upon the ants's ability to recognize degrees of genetic relationship as assessed by differences at seven microsatellite loci of nuclear DNA. (Interestingly, in another form of exaptation—a cross-level spandrel in this instance—the usefulness of these markers presumably arises from the rapid and substantial variability thus imparted among populations, and flowing upwards as an effect from the neutral status of most mutations in these microsatellites.) Thus, in Argentina, ants tolerate conspecifics of closely related colonies and fight with genealogically more distant conspecifics in colonies of greater genetic disparity.

However, and unsurprisingly, the ancestors of the California invaders passed through a genetic bottleneck in their initially small population, probably derived from a single native colony. The California ants lost about two-thirds of the genetic variability at microsatellite loci. "So all of the ants are genetically alike, and those applying the old similarity-tolerance rule could be fooled into accepting everyone as kin" (Queller, 2000, p. 519). Thus, as an example of an insinuated milton (a feature introduced nonadaptatively by neutral drift) that has been exapted for marked, if transient, utility, the success of *Linepithema humile* in California seems to rest upon a loss of genetic variability that induced the ants to form a single supercolony operating as a military phalanx (to continue the dubious tradition of anthropomorphic description for behaviors of social insects) of remarkable power and efficiency. Queller notes (p. 5190): "Paradoxically, the ecological success of the introduced populations stems not from adaptation but from the loss of an adaptation—colony recognition—due to genetic drift." But, then, if we recognized the hierarchical nature of selection, and repressed our adaptationist biases, this situation might not appear so paradoxical, and we might even search explicitly, encouraged by a conceptual reform and expectation, for phenomena that may be very common in nature, but that elude our notice because we find them anomalous, and do not recognize their existence until they stare us in the face.

Finally, this case also suggests an interesting flip-side of potential longterm disadvantages for such transient success—a near necessity in our hierarchical world of potentially conflicting levels, lest victory at one position propagate throughout the system to full and permanent triumph. (The scientifically astute Tennyson may have experienced a more universal insight than we usually acknowledge in penning his famous line in *Morte d'Arthur* (1842): "The old order changeth, yielding place to new . . . lest one good custom should corrupt the world.") If reduced conflict builds supercolonies of such competitive success, why does the same species live so differently, and so much more modestly, in its native Argentinian abode?

Queller (2000, p. 520) presents the interesting and cogent explanation that, by losing their capacity to identify degrees of kinship, the ants "cannot improve or even sustain their cooperative behavior. Without relatedness, adaptive modifications of cooperative worker behavior cannot be favored and maladaptive ones cannot be disfavored." Queller *(op. cit.)* therefore concludes that the randomly established feature (the miltonic insinuation) behind transient triumph must also spell eventual doom: "Random drift will become important again, this time because of the absence of any opposing force rather than a small population size. The ants are like a casino gambler who is lucky once but cannot quit: chance got them their stake, but over the long run it can only lead to ruin." Thus, either the California ants will eventually restore their genetic variability, split into fighting colonies, and become restricted to smaller overall populations with greater staying power; or they will suffer "the lingering death of decay by drift" (p. 520). Tsutsui et al. (2000) propose what, to most people, would sound like an absurd and counterintuitive strategy for control, although the suggestion clearly makes theoretical (but perhaps not practical) sense to anyone schooled in the intricacies of evolutionary theory: introduce *more* ants with substantial variation at those microsatellite loci, thus encouraging self-regulation by the reinitiation of conflict!

Thus, in summarizing the categories in my proposed taxonomy of the exaptive pool, the currently unused but eminently usable features that build the ground of differential evolvability fall into two groups of inherent potentials (franklins) and available things (miltons). Miltons, in turn, include three distinct groupings ordered by different sources of origin. The exaptive pool therefore contains:

- Potentials (franklins)
- Consequences (miltons arising as spandrels)
- Manumissions (miltons arising by unemployment)
- Insinuations (miltons arising by random drift)

All taxonomies—thus embodying the richness of fascination of systematics as a scientific subject (Gould, 2000c)—mix aspects of nature's objective order with human preferences for utility or intelligibility. Even if we allow that these four categories exist "out there" in nature—and even I, although I developed this scheme, would not go so far in trying to craft a naturalistic defense, for I recognize that the objective items of the exaptive pool could be parsed in other ways—our decisions about their ranking and secondary ordering require a choice among several logically legitimate alternatives. All taxonomies base such choices on the designation of a *fundamentum divisionis,* or basis of primary ordering. The differences among alternative *fundamenta* reflect the theories we favor as most useful in understanding and explaining the phenomena encompassed by the taxonomy. (Thus, as Ernst Mayr has so forcefully argued throughout his career, taxonomies are theories about the basis of order, not boring and neutral hatracks, pigeonholes or stamp albums for accommodating the objective and uniquely arranged items of nature.)

For the exaptive pool, I can imagine three fairly obvious candidates for *fundamenta* (and I am sure that other possibilities have eluded me). My choice of franklins *vs.* miltons reflects my convictions and arguments about the most appropriate theoretical context for understanding the evolutionary meaning and importance of the exaptive pool.

First, suppose I were working within a more orthodox context that assumed effective control over evolution by natural selection. Suppose further that, within this context, I still appreciated the importance of specifying the currently unused attributes of organisms that might contribute to future success. From such a standpoint, I would probably be tempted to choose a *fundamentum* that drew a primary distinction between the ineluctable unused attributes that arise even when natural selection works in its "leanest and meanest" mode of optimization, and a second groups of unused attributes that arise either from the cessation or the weakness (or at least the non-exclusivity) of natural selection. In such a taxonomic system, I would place all franklins together with all miltonic spandrels in my first category of ineluctable attributes of pure selection (inherent potentials of features optimized by selection for other functions, and inevitable side consequences of similarly optimizing selection). I would then devote my second category to the other two groupings of miltons—manumissions and insinuations—arising as historical results of selection's relinquished control: features falling out of its purview for the first group, and features insinuated beneath its notice in the second group.

In a second *fundamentum* that might also follow from stronger selectionist commitments, I might decide to make a primary division between attributes that originated either as direct adaptations or as inherencies of direct adaptations for the first group, and attributes with truly nonadaptive origins for a second group. In this case, I would place miltonic manumissions and franklins in the first category (as adaptations that become unemployed, and as additional potentials of features that arose as direct adaptations). I would then unite miltonic spandrels and insinuations into a second category of attributes originating as nonadaptations—spandrels as features that are nonadaptive in their own isolated selves (whatever the adaptive status of other features that generated them as side consequences), and insinuations as nonadaptive features that originated beneath the notice and malleability of selection.

I have chosen a quite different third alternative—inherent potentials (franklins) *vs.* available things (miltons)—because I wish to emphasize the structuralist and nonadaptionist components of evolutionary theory that have, in Darwinian traditions, been downplayed or ignored. This *fundamentum* stresses the primary difference between consequences of purely adaptationist mechanisms applied to directly adaptive features (the franklins, or inherent potentials, that, ever since Darwin's excellent arguments against Mivart, have been recognized as the basis of quirky functional shift in unpredictable evolutionary sequences remaining entirely under selectionist control); and the extensive and influential set of material features that actually "stand and wait" by themselves as nonadaptive attributes (not as additional but now only potential uses of present adaptations). In other words, I choose this *fundamentum* because miltons (and not franklins) pose a genuine challenge to the exclusivity of adaptationist mechanisms. Franklins enlarge the scope and sophistication of selectionist argument, adding a genuine flavor of formalist limitation and potentiation to an otherwise naively functionalist theory based only upon organic accommodation to selective pressures of an external environment. But miltons emplace a genuinely nonadaptionist component into the heart of evolutionary explanation—for if many features originate as nonadaptations, and if nonadaptations, as material items of miltonic "stuff," stand and wait while occupying a substantial percentage of the exaptive pool, then evolutionary explanations for both the origin of novelties, and for the differential capacity of lineages to enjoy future phyletic expansion and success, will require a revised and expanded version of Darwinism, enriched by nonselectionist themes of a formalist and structuralist research program. I therefore choose my *fundamentum* as the best taxonomic device for exploring the role of nonadaptation and structural constraint in the exaptive pool of evolvability.

### Cross-level effects as Miltonic spandrels, not Franklinian potentials: the nub of integration and radical importance

As one of its most interesting and potentially reformatory implications, the hierarchical expansion of selectionist theory introduces an extensive array of features into the exaptive pool (and upon the consciousness of evolutionary biologists) as effects propagated to other levels by features that arise for directly causal reasons at a focal level. (One need not challenge the conventional view that most direct reasons at focal levels will be selectionist and adaptational—for the propagated effects may still assume a different and nonadaptive status.) What shall we call such effects that only become manifest at other levels and may be truly invisible at the focal level of their generation as consequences? Shall we interpret them as franklins, or inherent and presently unexploited utilities of features originating for other functions—with their only difference from more conventional franklins (flight capacities of thermoregulatory feathers) resting upon the fact that they happen to reside in biological individuals at levels other than the level of the feature that generated them?

I do not think that such cross-level effects can be interpreted as franklins, or mere potentials. Rather, cross-level effects are available *things*, albeit available only to biological individuals at other levels. Cross-level effects are therefore miltons, not franklins. Franklins are potentials, not things. As such, franklins can only be recruited sequentially in time from the primary adaptations in which they inhere as alternative utilities (feathers for flight *following* feathers for thermoregulation to cite the canonical example once again). (If feathers performed both functions from the start—a perfectly plausible scenario, of course—then we would never have designated their aerodynamic role as a franklin, for this function would have been part of their adaptive expression *ab initio*.) But cross-level effects, as miltonic things standing and waiting in the exaptive pool, become available as separately cooptable attributes right from their origin. They can therefore be utilized (by exaptation at their different level) *simultaneously* with the continuing primary adaptive function of the generating feature at its focal level of origin.

For example, the duplicated gene that arose by gene level selection, achieving an adaptive advantage thereby at the genic level (by definition, and through its plurifaction), may be simultaneously exapted at the organismal level by undergoing a mutational change (only now of potential benefit to the organism as a consequence of the gene's redundancy), and contributing thereby to a new organismal function. Similarly, the organismal form that adapted to its immediate environment by evolving a lecithotrophic bottom-dwelling larva from a planktotrophic ancestor, may simultaneously impart an exaptive effect to its species by enhancing the speciation rate *via* the altered demic structure of isolated subpopulations that no longer experience the gene flow previously potentiated by floating planktonic larvae. I am not, by the way, inventing these cases as personal speculations. Each represents the most widely discussed potential example of cross-level exaptation for its pair of levels.

A recent example (Podos, 2001; Ryan, 2001) illustrates the range and probable ubiquity of simultaneous emplacement of spandrels to other levels as consequences of primary adaptations at a focal level. In the best known case of Darwin's finches, Podos (2001) shows that ordinary adaptation of bill sizes and shapes in response to climatic changes and competition with other species (as so superbly documented in the continuing work of the Grants and others—1986 for an early summary, for example) imposes automatic consequences upon the form, style, and range of the resulting song, because "two functional systems—that used for feeding and that used for singing—share a common morphology, the beak" (Podos, 2001, p. 186). Sharper and narrower beaks permit wider ranges and precision of song, whereas heavier and blunter beaks impose greater "constraint" (Podos's term) upon potentials and specificities of resulting vocal production. Since songs function as powerful premating isolating mechanisms, the automatic divergence of song, arising as a side-consequence of ordinary adaptation of bills in feeding, and the different degrees of distinctiveness attached to specific forms of the bill, may have profound consequences in a resulting (and ultimately highly exaptive) differential capacity for speciation among different subclades of this classic group (based upon varying capacities of the resulting song to act as an effective signal for mate recognition). Ryan's commentary acknowledges Podos's inference about exaptive effects on speciation rate as necessarily conjectural for now, but as the most interesting larger implication of this important study.

This property of simultaneous utilization carries two important implications for the status of cross-level effects. First, cross-level effects are thereby identified as things rather than potentials—that is, as miltons rather than franklins. (Differential variation in range of song among subclades is a "thing" that may impose emergent fitness at the species level at the same time as its generating adaptations in the form of the bill operate in their ordinary Darwinian manner at the organismal level.) Second, cross-level effects, as things, generate a potentially radical challenge to conventional Darwinian concepts. Unused potentials, as argued several times previously, remain fully within the adaptationist program as possible future uses of features that arose as adaptations and will always be adaptations (albeit for a different function in the future). But unused things begin as nonadaptations right at their origin (whatever their future importance as exaptations)—and the demonstration of their high relative frequency in the exaptive pool, and of their importance to the evolutionary history of many lineages, would introduce a significant non-adaptationist element into evolutionary theory.

So if we agree that cross-level effects are miltonic things and not frank-linian potentials, into what category of miltons (see Table 11-2) do they fall? If the previous components of this argument prove acceptable, then this final question enjoys a simple and unambiguous resolution: they are spandrels. The key property of spandrels lies in their automatically consequential character as things necessarily enjoined by other changes. Cross-level effects fit this definition in all ways, for they arise in concert with the primary change, and as a necessary consequence thereof. But they do manifest an interestingly different property from such conventional at-level spandrels as the penden-tives of San Marco or the cylindrical space that can be exapted as a brooding chamber at the center of a gastropod shell: they express themselves in a biological individual at a different level from the individual bearing the feature that generated the effect.

As stated before, and as the heart of my argument, I believe that the designation of cross-level effects as a second class of spandrels greatly increases the range and importance of this concept—for cross-level spandrels are probably far more common, and of far more frequent importance in potentiating the future direction of evolutionary lineages, than the at-level spandrels that provoked the initial formulation of this concept. As an almost naively evident defense of this claim, I would point out that more cross-levels exist than focal levels (obviously, as focal levels must be singular!), so cross-level spandrels have more "places to go" than at-level spandrels. Moreover, any cross-level attribute holds greater potential, *prima facie,* for manifestation as a quirky and oddly nonadaptive feature—for anything arising at one level and injected into another must enter its new domain adventitiously and without reference to the norms and needs of its adopted "home"; whereas an at-level spandrel can only be tolerated if it meshes reasonably well with a design already established for its kind of entity.

These attributes of cross-level spandrels embody their importance in revising our usual understanding of evolution, particularly at the macroevolutionary level of the species-individual. As argued above, at-level spandrels must be nearly neutral (or at least not too burdensome) because they "come with the territory" and must immediately integrate themselves as parts of the larger structure built at their level—for at-level spandrels share the same kind of form, and are made of the same sorts of materials, as the primary adaptations themselves. If pendentives made buildings collapse, or doubled their cost without compromising their mechanical function, then architects would not choose the designs that generate them as necessary consequences of otherwise favorable properties.

But cross-level spandrels, especially when injected upwards into more slowly cycling biological individuals at higher levels, can establish themselves more easily, and beyond any screening power of the higher-level individual, by pressure of numerous introductions within single generations of the slowly-cycling higher-level individual. I have been emphasizing the future exaptive potential of nonadaptive spandrels, but nonadaptive features can also work in an opposite manner, becoming detrimental (truly inadaptive) to their bearers—a fate that probably befalls at-level spandrels only rarely (because inadaptive effects will generally preclude their introduction in the first place), but that may represent a common outcome of cross-level spandrels injected into higher-level individuals, and not readily suppressed, at least initially, because they can become rooted before any episode of generational cycling reveals their disutility to the higher-level individual.

The claims of the last paragraph may seem arcane and distant (in my abstract formulation) from empirical reality. But this phenomenon has long been recognized at the species level, even though evolutionary biology previously lacked the conceptual apparatus to offer a general explanation. We all acknowledge that many organismal adaptations impose strongly negative consequences upon the geological longevity of their lineage. Any highly complex, metabolically expensive, and intensely specialized adaptation (the peacock's tail, or virtually any elaborate contrivance of runaway sexual selection); any alteration, especially involving the loss and simplification of complex ancestral structures, that adapts an organism to a transient and highly specialized environment (the "degenerate" parasite utterly dependent on a unique and unusual host); must strongly compromise the geological potential of a subclade bearing its autapomorphy, relative to a sister subclade retaining an ancestral and generalized morphology and ecology. In fact, any feature—and they must be legion—that provides adaptive benefits at the organismal level, but that simultaneously injects such "negative" spandrels into the encompassing species-individual (either suppressing its rate of speciation or decreasing its geological longevity thereby), will be inadaptive at the higher level, but unpreventable by insertion before the species-individual can "notice" and reject the feature.

This pairing of organismal adaptation with injected spandrels that prove inadaptive to the encompassing species-individual sets the proper conceptual context, under the hierarchical theory of selection, for what our literature has long called, in an ambiguous and merely descriptive way, the "opportunism" of evolution. "Opportunism," like "preadaptation," should be recognized as an odd and inappropriately anthropocentric term, designating one subtype in the large and general category of features established for good causal reasons at one level that then impose effects at other times, or upon other levels, either positively (as in "preadaptation") or negatively (as in "opportunism" for immediate advantages, leading to extinction in the long run). Moreover, although the term has been fading from use of late, all textbooks of evolution used to include an explicit section on a phenomenon called "overspecialization"—another nonsense phrase devised to treat, when theory lacked the proper concepts, the important observation that many organismal adaptations impose inadaptive effects as spandrels upon the encompassing species-individual. What in this pairing could possibly be called "overspecialization" with any justification? The organism becomes adaptively specialized to its own immediate benefit, and the species suffers as a nonadaptive (ultimately inadaptive) side consequence of spandrels representing the adaptation's expression at the species level.*

But I don't want to leave the impression that upwardly-injected cross-level spandrels always spell dissolution, or even doom, at the higher level—and that the species-level, in particular, suffers from this phenomenon, even to the point of becoming a weakened or ineffective locus of evolution thereby (for such a conclusion would greatly compromise the hierarchical theory itself by an argument akin to Fisher's rejected claim (see pp. 644–652) for the logical ineluctability, but practical insignificance, of species selection). Two related arguments reinforce the evolutionary importance of cross-level spandrels, while also reaffirming the power of the species-individual as a biological agent in evolution.

1. We must not view upwardly-cascading effects as uniquely detrimental for species, but generally neutral or even positive at other levels. I suspect, rather, that a majority of upwardly-cascading effects will be negative at any higher level of expression. Indeed, as argued previously, we have long recognized this phenomenon as a fundamental property of Darwinism (though, again, we have lacked the conceptual apparatus for explaining the results in these appropriate terms). The phenotypic expressions of mutations are spandrels at the organismal level, and we have long recognized the vast majority as deleterious for the organism. But we do not regard this inevitable property (of anything injected adventitiously into a different level) as globally detrimental to organisms. Populations of organisms are large enough, and the generational cycling time of organisms short enough, to tolerate a substantial load of general disadvantage in exchange for the occasional opportunity

---

*In some way, and through a very dark glass, I had some inkling of this problem as an undergraduate. In an initial embarrassing episode of juvenilia, my first undergraduate term paper in an evolutionary course, I treated this very subject, but could proceed no further than suggesting the more appropriate description of "ultraspecialization," having at least recognized that the process commits no active "wrong." In some psychological sense that I feel strongly but cannot clearly define, I view the genesis of this entire book as personal expiation for the puerility of this initial effort!

thereby achieved in encountering a favorable spandrel that can then be exapted in the canonical organismal process of natural selection. (This phenomenon gains its extreme expression in bacterial evolution, where the rarity of favorable mutants hardly curtails rapid evolution because populations are so large, and generations so short, that highly infrequent exaptable effects occur often enough to drive substantial evolutionary change by sheer brevity of the waiting time between favorable injections, even in the absence of any mechanism of recombination to spread the benefits among individuals.)

Thus, the organismal level can usually well afford this carnage of generally deleterious mutational effects in order to win its fuel of positive variants for natural selection. Indeed, Darwin's world offers the organism no other choice —for natural selection makes nothing directly and can only operate creatively in the generation of evolutionary change if some other process supplies enough undirected variation to power its odd mode of negative construction by elimination. One might even "applaud" this inevitable property of cross-level spandrels as "just the thing" that natural selection needs to become nature's potent driver of evolution despite its weakness, its dependencies, and its peculiar style of operation. Three interrelated facts establish and undergird natural selection's capacity to power organismal evolution in the face of these limitations: (1) natural selection requires "random" fuel undirected towards adaptive states (lest such an internal force for automatic organic good overwhelm the weak power of selection to produce similar results in a characteristically slow, gradual and roundabout way); (2) the fuel supplied by phenotypic effects of mutation expresses itself as cross-level spandrels injected from the gene level into organismal phenotypes; and (3) cross-level spandrels manifest the required property of noncorrelation with benefits at their new level of injection. Would nature be Darwinian at all, absent these interesting properties of cross-level spandrels that must supply the fuel of natural selection—thus establishing the category of "chance" in the duality of "chance" (effects of cross-level spandrels manifested as mutational phenotypes) and "necessity" (direct action of natural selection for adaptation to local environments) that we generally cite as an epitome of the Darwinian mechanism and worldview? If we just remember that the phenotypic expressions of mutations are cross-level spandrels, we will hold an important key for unlocking the curiosities of Darwinism.

2. I don't wish to imply that species-individuals can only be weakened, given their generally small population sizes within their clades, and their slow cycling times, by the nonadaptive (and perhaps usually inadaptive) character of most effects that supply a component of emergent fitness to species selection (Gould and Lloyd, 1999), and that arise as cross-level spandrels injected upwardly from the organismal level below. On the contrary, although species may often suffer (in terms of their ability to overcome inadaptive spandrels) by their small population numbers and slow cycling times—whereas organisms vanquish this impediment by large populations and quick cycling—species also, and in "trade-off," obtain a tremendous "leg up" over the organismal level by expressing an inherent "allometric" prop-

erty (see Gould and Lloyd, 1999 for this extension of allometry) possessed by virtue of their sizes and cyclings, but absent in organisms for equally inherent reasons of their own constitution. As argued at length in Chapter 8, organisms maintain their biological individuality as discrete entities by strategies of intricate and precise functional interrelationships among constituent parts, and by maintenance of internal and external boundaries (immune systems and skins) to exclude the subparts of other individuals from their geographic space. Thus, and most characteristically, if not uniquely, the organism maintains its integrity by rigorous policing, and by active suppression of differential proliferation among its subparts, a result that could otherwise follow from positive selection upon the individuals of lower levels within the organism itself (especially upon cell lineages, with the failure of such organismal policing leading to the result that we call cancer).

Thus, organisms sacrifice the benefits of including more upwardly cascading effects as components of their exaptive pool because they work so hard to preserve their distinctive style of individuality by suppressing the churning of lower levels of selection within their bodies. But, by important contrast, species construct their equally powerful integrity and discreteness by different means that do not require such suppression of upwardly cascading effects. Species maintain their boundaries primarily by reproductive isolation of their subparts (the organisms that constitute their populations) from subparts of other species-individuals. By permitting their subparts to reproduce only with each other, and not with subparts of other species-individuals, a species maintains its integrity, and constructs its boundary, with just as much clarity and efficiency as organisms can muster by different strategies of functional integration among organs and active exclusion of invaders. This different, but equally efficient, modality of species does not include the suppression of lower level selection as a consequence. Therefore, species remain open, as organisms do not, to experiencing a full and rich range of cross-level spandrels, injected as consequences of selection acting on lower-level individuals (primarily upon organisms) within their boundaries.

Now—and here's the rub—we have generally viewed this openness of species as a negative sign of impotence at this higher level of biological organization. For, reasoning from our parochial perspective as organisms—and falsely supposing that organismal ways must be universally best ways—we have viewed the species's non-suppression of upwardly-cascading spandrels as a mark of its failings as a potential unit of selection. But perhaps we should reverse this perspective and learn to respect the distinct allometric properties of species individuality as potential strengths for a different mode of selection—with the higher frequency of cross-level spandrels viewed as a source of rich potential denied to organisms, rather than as a mark of inefficiency imposed upon species.

The species-individual, as a Darwinian interactor in selection at its own level, operates largely with cross-level exaptations arising from unsuppressed evolution of subparts (primarily organisms) at lower levels within itself. Such nonsuppression acts as a source of evolutionary potential by permitting species to draw upon a wider pool of features than organisms can access.

By *not* suppressing this evolutionary churning from within, the species-individual gains enormous flexibility in remaining open to help from below, expressed as exaptive effects that confer emergent fitness. Rather than viewing this nonsuppression of aid from other levels, with the accompanying failure to build many active adaptations at its own level, as a sign of wimpy weakness for the species—construed as a "poor organism" in the implication of most traditional thought—we should rather interpret these allometrically driven properties as cardinal strengths, and recognize the species as a "rich-but-different" Darwinian individual. The species, in this view, acts as a shelter or arbor that holds itself fast by active utilization of the properties that build its well-defined individuality. By fostering internal change, and thereby gaining a large supply of upwardly cascading exaptive effects, species use the features of all contained lower-level individuals through the manifestation of their effects on the shelter itself. The species, through its own distinctive features of individuality, and requiring neither indulgence nor apologia from human understanding, will continue to operate as a powerful agent in Darwin's world whether or not we parochial organisms, limited by our visceral feelings and traditions of language, choose to expand our view and recognize the sources of evolutionary potency at distant scales of nature's hierarchy.

In conclusion, and to reiterate my rationale for placing so much emphasis on cross-level spandrels in evolution, this single theme, more than any other in this book, unites and exemplifies the weaknesses on all three legs of the tripod of essential postulates in conventional Darwinian logic, while also pointing the way towards revisions that will expand and strengthen these three supports to produce an improved, and more comprehensive, evolutionary theory by retaining its Darwinian basis in the expanded form of a fully hierarchical theory, while adding, to its preserved selectionist core, several aids and flavors from alternative traditions of structuralist thought.

Thus, the hierarchical theory of selection (for strengthening and expansion of the first leg) greatly augments the role of spandrels by adding the cross-level category as more potent in numbers and more various in potential results than the traditional at-level category. For at-level spandrels, in the usual Darwinian account of selection as a uniquely organismal process, must remain confined to structural byproducts and space fillers within a context of integral adaptation of the body. But if selection works simultaneously at several hierarchically ordered levels of biological individuality, then the domain of spandrels expands to include any enjoined expression (upon Darwinian individuals at other levels) of changes causally introduced at a focal level—for these injected and adventitious expressions must originate nonadaptively (and "randomly" in our usual loose parlance) relative to their causal reasons of origin at the focal level of their construction.

By the same token, spandrels become structural constraints rather than direct adaptations (or even alternative potentials of direct adaptations)—that is, they enter the exaptive pool as miltonic things rather than franklinian potentials. Moreover, as constraints of this type, they add a structuralist and nonadaptationist component to the workings of evolution—thus strengthening the tripod on leg two by including aspects of this formerly rejected mode of causation among the totality of devices that generate creative change in evolution. On a more specific note, if cross-level spandrels maintain an important relative frequency among the components of evolutionary change, then these automatic expressions at other levels—introduced separately from, and simultaneously with, the primary changes that generate them at a different focal level—may largely control the possibilities and directions of evolution from a structural "inside," rather than only from the functional "outside" of natural selection.

Thus, if the positive structural constraints of spandrels—particularly in their cross-level mode as effects propagated to various levels of the evolutionary hierarchy—can help to explain the phenomenon of evolvability and the parsing of categories in the exaptive pool, then reforms on legs one and two of the essential Darwinian tripod will also illustrate why the extrapolationist premise of the third leg cannot suffice to explain evolution either. For macroevolution cannot simply be scaled-up from microevolutionary mechanics if the phenomenology of this larger scale depends as much upon the potentials of evolvability as upon the impositions of selection, and if the exaptive pool promotes evolvability largely by the later utility at the same level (or simultaneous exaptive use at other levels) of spandrels that originate for nonadaptive reasons. The explanation of macroevolution requires structuralist and hierarchical inputs from various scales, and cannot be fully rendered as an extension of organismal adaptation, smoothly scaled up through the immensity of geological time.

### A CLOSING COMMENT TO RESOLVE THE MACROEVOLUTIONARY PARADOX THAT CONSTRAINT ENSURES FLEXIBILITY WHEREAS SELECTION CRAFTS RESTRICTION

In closing this section by reiterating the opening argument (p. 1270) in another context, I should extend my previous statement on the bounded independence of macroevolution to stress the positive theme of interesting differences, and not only a negative claim for the necessary limitation of any explanation based upon pure extrapolation from microevolutionary mechanics. For the failure of microevolutionary extrapolation resides in something far deeper than mere insufficiency. Rather, and thus operating to intensify the explanatory gap, a cardinal feature of microevolution works directly against the potentials for macroevolution defined by the exaptive pool—thereby requiring that macroevolution proceed by actively overcoming this microevolutionary limitation, and not only by "adding value" to its mere insufficiency.

Darwinian evolutionists have known this all along in their heart of hearts, and have tended to escape the resulting paradox by a leap of faith into the enabling power of geological time to accomplish anything by accumulation of small inputs. Most events of microevolutionary adaptation—that is, of ordinary Darwinian natural selection in the organismal mode—work against evolvability by locking organisms into transient specializations and reducing the flexibility of the exaptive pool. This fact engenders the central paradox noted above: that immediate organismal processes tend to derail prospects for longterm evolutionary success at the species and clade level. Darwinian traditions have tried to surmount this stumbling block by arguing that, however the process of specialization might restrict future prospects, natural selection still makes "better" organisms by rewarding success in direct competition against conspecifics. And at least one component of this "betterness," albeit a minority component—the occasional achievement of local adaptation by general biomechanical improvement, rather than by limiting specialization—must provide the major source of increments for macroevolutionary patterning by extrapolation. But this argument is bankrupt, and I have, throughout this book, chronicled a host of reasons for its failure.

Therefore, the macroevolutionary success of species and clades must arise, in large part, by active utilization of selective processes at their own higher levels, and in opposition to the generally restricting implications and sequelae of microevolution. Moreover, in fueling these macroevolutionary successes, species must exapt the rich potentials supplied by structural and historical constraints of spandrels and other miltonic "things" emplaced into the exaptive pool against (or orthogonally to) these restricting tendencies of natural selection—in other words, by exploiting the components of a phenomenon that we have loosely called "evolvability" and vaguely recognized as something apart from natural selection. And thus, an expansion of the first leg by hierarchical selection, and a strengthening of the second leg by structural constraint, really does build a "higher Darwinism" of greater sophistication and explanatory power—an indispensable basis in our struggles to understand "this view of life," the evolutionary process that made us, and imbued us with all the spandrels of body and soul that force us to ask such difficult questions about the meaning of our own existence and of nature's ways. These spandrels of historical ancestry and structural inevitability may impede our search for solutions by imposing such quirky modalities upon our mental operations, but they also grant us more than sufficient power to overcome and prevail. Sweet, and adaptive, are the uses of adversity. Shakespeare, after all, in the words that follow this famous statement, parodied just above, promised us salvation, or at least succor, in natural history, where we would find "tongues in trees, books in the running brooks, sermons in stones, and good in every thing."

# Tiers of Time and Trials of Extrapolationism, With an Epilog on the Interaction of General Theory and Contingent History

## Failure of Extrapolationism in the Non-Isotropy of Time and Geology

### THE SPECTER OF CATASTROPHIC MASS EXTINCTION: DARWIN TO CHICXULUB

Greatness shines brightest in adversity. The logic of the *Origin*'s 9th Chapter (1859, pp. 279–311)—"On the Imperfection of the Geological Record"— shows Darwin's reasoning at its very best and most systematic, all in the service of resolving his worst problem. For, in this chapter, he must explain why the subject that should, at first glance, have provided the strongest and most direct confirmation of evolution, seems to mock, in its opposite message (at least if read in an empirically literal manner), the gradual and incremental style of change touted throughout his book both as a validation of natural selection, and as the primary empirical basis for confidence in the factuality of evolution itself.

Darwin's logic proceeds in a systematic way within the norms of scientific discourse, moving linearly and relentlessly through the chapter from the problem with the easiest resolution (why do we not find living intermediary forms between modern species) to the most difficult appearance of all (episodes of mass origination and extinction in the fossil record). The opening issue, representing a misconception and not a true threat, achieves an easy solution: evolution is a process of branching, not of linear transformation, as the brilliant "tree of life" metaphor, closing the operative chapter four on the mechanics of natural selection, had so well exemplified (1859, pp. 129–130). Few, if any, living species are the unaltered forebears of another modern form; rather, any two sister species have branched and diverged from an ancient common ancestor. Therefore, we do not expect to encounter living intermediates between extant species, for, although transitional forms surely existed, they died long ago and should only be found in the fossil record.

Having disposed of this simple error in thought, Darwin must now face the real dilemma that intermediates rarely occur in the fossil record either, where they should exist in abundance: "Why then is not every geological formation and every stratum full of such intermediate links? Geology assuredly does not reveal any such finely graduated organic chain; and this, perhaps, is the most obvious and gravest objection which can be urged against my theory" (p. 280).

Darwin's general answer stands out in the title to his chapter, and he invokes the crucial claim to solve each of the three issues that, in their increasing difficulty, define the flow and logic of his treatment. The very next sentence, following the quotation cited just above, states this comprehensive solution: "The explanation lies, as I believe, in the extreme imperfection of the geological record" (p. 280). This "argument from imperfection" became so indispensable to Darwin's dispersal of his major challenge from the fossil record that, later in the chapter, he ventured one of the most honest and revealing statements in all our scientific literature: a striking admission that the needs of theory had provoked an understanding of paleontological data that he otherwise might never even have considered. Such feedbacks always occur in science, but the empirical ethos of our profession leads us to underplay, or never to recognize in our own mental processing at all, this reverse flow from the expectations of theory to the perception and interpretation of factuality. Darwin's admission strikes me as wonderfully honest and self-scrutinizing, but also as potentially triggering a trap of circular reasoning, as the dictate of theory, mandating an expectation of imperfection, biases the reading of a fossil record that might actually be displaying more genuine signal than the "noise" of absence: "But I do not pretend that I should ever have suspected how poor a record of the mutations of life, the best preserved geological section presented, had not the difficulty of our not discovering innumerable transitional links between the species which appeared at the commencement and close of each formation, pressed so hardly on my theory" (p. 302).

First, Darwin presents his easier defense of gross imperfection at global scales, as he argues convincingly that we lack an adequate sample of life's full richness because both natural limitations (non-deposition of strata during most intervals) and insufficient human effort in a very young science (poverty of existing collections relative to fossils that could be gathered, exploration of only a small and geographically restricted percentage of the earth's fossil bearing strata) preclude any adequate sampling of life's full richness.

But Darwin must then admit that such general reasons do not resolve the second, and local, issue of why we do not find a "finely graduated organic chain" within single formations that do seem to preserve a continuous record of strata: "From the foregoing considerations it cannot be doubted that the geological record, viewed as a whole, is extremely imperfect; but if we confine our attention to any one formation, it becomes more difficult to understand, why we do not therein find closely graduated varieties between the allied species which lived at its commencement and at its close" (pp. 292–293).

Again, for this second issue, Darwin stresses the record's imperfection, now relying on discontinuity of deposition and environment in single regions through geological intervals. Strata may seem continuous, he correctly argues, but most times contributed no sediment because accumulation only occurs during gentle subsidence of basins, and such conditions do not generally prevail. By the same token, the local environment occupied by a species in anagenetic transformation will not generally persist long enough in any single place, and organisms will track their moving habitats. For example, the shallow marine habitat of so many invertebrate species can only continue (under conditions that also accumulate sediments with fossils) when rates of deposition for strata evenly match rates of subsidence for substrates—and how often, and for how long, can such a fine balance be maintained in any one place?

Having ascribed, to his satisfaction, both global and local absences of transitional forms to the general imperfection of geological records, Darwin must finally, in the closing sections of Chapter 9 (and spilling over into substantial parts of Chapter 10), rebut a third geological challenge to evolution, and especially to gradualistic explanations framed in terms of natural selection. To overcome this last obstacle, Darwin must tackle the harder problem of an apparently positive signal against his expectations, rather than (as in the first two cases) a negative result of failure to locate an anticipated confirmation. To complete his argument, Darwin must now explain away the evidence for global episodes of apparently sudden mass extinction or origin of entire faunas.

We should first pause to ask why Darwin even considered this signal from the fossil record as such a problem, especially for episodes of mass extinction. Why did he view the prospect of simultaneous extirpation as an issue at all, either for evolution or for natural selection? Natural selection does not guarantee the power of adaptation in all circumstances—and if environments change rapidly and profoundly enough, these alterations may exceed the power of adaptation by natural selection, with extinction of most forms as the expected result, even in the most strictly Darwinian of circumstances.

As a general answer—and as the primary reason for treating this subject within a chapter on modern critiques of the third leg, or extrapolationist premise, of Darwinian central logic—Darwin's hostility to catastrophic mass extinction does not arise primarily from threats posed to the mechanism of natural selection itself, but more from the challenges raised by the prospect of sudden global change to the key uniformitarian, or extrapolationist, assumption that observable processes at work in modern populations can, given the amplitude of geological time, render the full panoply of macroevolutionary results by prolonged accretion and accumulation.

The problem of mass extinction became acute for Darwin because geological paroxysm threatened something quite particular, vitally important, and therefore of much greater immediate pith and moment than his general methodological preference for locating all causality in the palpable observation of microevolution (see Chapter 2). Global catastrophe could undermine the ecological argument that Darwin had so carefully devised (see Chapter 6, pp. 467–479) to validate something more particular but no less important: his culture's central belief in progress, especially when Darwin had so increased the difficulty of the problem by constructing a theory (natural selection itself) that could not render this consummately desired result through its bare-bones mechanics. (For this reason, I discuss mass extinction in this relatively short Chapter 12, conceived as a counterpart to the equally brief Chapter 6 of this book's historical half—for these chapters feature the surrogate geological defenses of extrapolation, rather than the arguments from biological theory inherent in the first two legs of Darwin's logical tripod, and treated *in extenso* in Chapters 3–5 of the historical half, and 8–11 of this second half on contemporary debates.)

To explain the general pattern of life's history, Darwin sought to extrapolate the results of competition ordained by the immediacies of natural selection in ecological moments. In particular (as discussed and documented in Chapter 6, pp. 467–479), he used his "metaphor of the wedge" to argue that most competition, in a world chock full of species, unfolds in the biotic mode of direct battle for limited resources, *mano a mano* so to speak, and not in the abiotic mode of struggle to survive in difficult physical conditions. If struggle by overt battle (which favors mental and biomechanical improvement) trumps struggle against inclement environment (which often favors cooperation rather than battle and usually leads, in any case, to specialized local adaptation rather than to general improvement), then a broad vector of progress should pervade the history of life.

These two geological chapters (9 and 10) include nearly all of Darwin's passages and notable arguments for linking general progress to the extrapolation of momentary biotic competition through geological time. "The theory of natural selection is grounded on the belief that each new variety, and ultimately each new species, is produced and maintained by having some advantage over those with which it comes into competition; and the consequent extinction of less-favored forms almost inevitably follows" (p. 320). Or consider this passage, with multiple metaphors of victory and defeat: "In one particular sense the more recent forms must, on my theory, be higher than the more ancient; for each new species is formed by having had some advantage in the struggle for life over other and preceding forms . . . I do not doubt that this process of improvement has affected in a marked and sensible manner the organisation of the more recent and victorious forms of life, in comparison with the ancient and beaten forms" (pp. 336–337). Most famously, Darwin writes in the summary of both chapters (p. 345): "The inhabitants of each successive period in the world's history have beaten their predecessors in the race for life, and are, in so far, higher in the scale of nature; and this may account for that vague yet ill-defined sentiment, felt by many paleontologists, that organisation on the whole has progressed."

To bring the literal appearance of mass extinction back under the rubric of extrapolation, Darwin realizes that he does not have to deny episodes of markedly increased extinction entirely. Rather, he need only "spread out" the appearance of true simultaneity into a period long enough to permit explanation by biotic competition, perhaps intensified by tough physical times that make organismal battles even more stringent than usual, while remaining within the ordinary range and mode. After all, and in the anachronism of modern slang, "when the going gets tough, the tough get going."

Thus, high rates of change in physical environments, so long as they stay within permissible uniformitarian limits, will enhance extinction by "turning up the gain" on the dial of input from the geological stage. But, to emphasize the key point of Darwin's efforts, a world of conceptual difference separates a false appearance of catastrophe that can, by invoking the imperfection of geological records, be spread over sufficient time to remain within the uniformitarian range (leading to intensification of evolutionary rates by ordinary modes), and a true catastrophe that must impose its burden of extinction by direct environmental impress under rules different from those regulating the primarily biotic competition of normal times and ordinary ecology (and generating a macroevolutionary vector of progress as a result). Darwin would therefore turn to his old standby of an imperfect geological record to disperse this third and greatest challenge to his extrapolationist vision.

Darwin presents two basic arguments—the first more theoretical and biological, and the second far more practical, crucial, operational and geological—to buttress his claim that a threatening appearance of simultaneity in mass extinction and origination should be "spread out" to occupy enough time for explanation on uniformitarian premises by the ordinary operation of natural selection. First, theory dictates that old species generally become extinct by biotic competition with new and improved forms, not by direct extirpation through marked changes in the physical environment. "The extinction of old forms is the almost inevitable consequence of the production of new forms" (p. 343). Darwin even denies a possible "escape route" for the less fit by asserting that mean global diversity has remained fairly constant through time—so the poorly adapted must go to the wall in clearing limited space for improved forms, and cannot hang on at the peripheries of a general expansion that welcomes the new without necessarily destroying the old: "Thus the appearance of new forms and the disappearance of old forms, both natural and artificial, are bound together. . . . We know that the number of species has not gone on indefinitely increasing, at least during the later geological periods, so that looking to later times we may believe that the production of new forms has caused the extinction of about the same number of old forms" (p. 320). Moreover, to enhance the implausibility of truly catastrophic mass dying, Darwin holds that "the complete extinction of the species of a group is generally a slower process than their production" (p. 318).

In a long discussion on pages 325–327, Darwin collates all aspects of his biological argument that ordinary competition will explain the literal appearance of simultaneous global extinction and origination. The final paragraph summarizes his extrapolationist convictions (p. 327):

Thus, as it seems to me, the parallel, and, taken in a large sense, simultaneous, succession of the same forms of life throughout the world, accords well with the principle of new species having been formed by dominant species spreading widely and varying; the new species thus produced being themselves dominant owing to inheritance, and to having already had some advantage over their parents or over other species; these again spreading, varying, and producing new species. The forms which are beaten and which yield their places to the new and victorious forms, will generally be allied in groups, from inheriting some inferiority in common; and therefore as new and improved groups spread throughout the world, old groups will disappear from the world; and the succession of forms in both ways will everywhere tend to correspond.

But Darwin's success hinges upon the second and more important geological argument—for his biological rationale only presents a theoretical defense, whereas he must overturn a strong signal from a literal reading of the fossil record: the appearance of true global simultaneity in mass extinction of entire groups and faunas, at a rate far too fast for any biological mechanism based on ordinary competition. At this crux, Darwin calls upon his standard argument from imperfection to "spread out" this apparent moment into sufficient time for uniformitarian explanation.

Darwin admits the literal signal (p. 322): "Scarcely any paleontological discovery is more striking than the fact, that the forms of life change almost simultaneously throughout the world." But this impression must be an artifact produced by the markedly incomplete preservation of more gradual and continuous change in a woefully imperfect geological record (pp. 317–318): "The old notion of all the inhabitants of the earth having been swept away at successive periods by catastrophes is very generally given up, even by those geologists . . . whose general views would naturally lead them to this conclusion. On the contrary, we have every reason to believe, from the study of the tertiary formations, that species and groups of species gradually disappear, one after the other, first from one spot, then from another, and finally from the world."

Among the many relevant aspects of imperfection, Darwin stresses two systematic factors that can compress a gradual transformation into a false appearance of simultaneity. First, sediments do not accumulate continuously, even in stratigraphic successions that look complete and uninterrupted. Strata pile up in continuity only when their basin of deposition slowly subsides, and this geological situation can occupy only a small percentage of total time. Thus, most intervals will generate no sediments at all, and a group slowly petering out to extinction may seem to disappear all at once because sedimentation ceased when the group still included several declining species. If strata didn't begin to accumulate again until much later, all these species may have slowly dribbled out of existence during the intervening period of nondeposition: "We do not make due allowance for the enormous intervals of time, which have probably elapsed between our consecutive formations—

longer perhaps in some cases than the time required for the accumulation of each formation. These intervals will have given time for the multiplication of species from some one or some few parent-forms; and in the succeeding formation such species will appear as if suddenly created" (pp. 302–303).

The second, and more sophisticated, argument follows from this principle of non-deposition during most intervals. We fall, Darwin argues, into circular reasoning in claiming that similar events in widely separated regions must have occurred simultaneously—for we make our judgment of temporal coincidence from the geological similarity alone, and not from any independent measure of time. For example, if we note the disappearance of several brachiopods in one stratum and the first appearance of several clams in the stratum just above, and we find the same pattern in a distant region on the other side of the earth, we might be tempted to proclaim a truly momentary wipeout followed by effectively simultaneous origin of functionally similar creatures. But this transition might actually occur very slowly in any single place, and leave no record of its true pace because a long interval of nondeposition followed the last preserved stratum of brachiopods. Moreover, this truly slow transition, prompted by ordinary biological competition of superior clams against inferior brachiopods, one species at a time, might have unfolded at quite different times in separated regions of the globe—for the process can only begin when the clam fauna migrates to a new area, and these migrations may span a considerable range of time (falsely compressed to simultaneity by our error in viewing the first stratum with clams as coeval throughout the world). Darwin summarizes this complex argument (pp. 327–329):

> Therefore as new and improved groups spread throughout the world, old groups will disappear from the world; and the succession of forms in both ways will everywhere tend to correspond . . . If the several formations in these regions have not been deposited during the same exact period,—a formation in one region often corresponding with a blank interval in the other . . . in this case, the several formations in the two regions could be arranged in the same order, in accordance with the general succession of the form of life, and the order would appear to be strictly parallel.

In a striking example, summarizing both the biological argument for gradual replacement by competition and the geological claim for false appearance of simultaneity by imperfection of preserved records, Darwin makes his plausible case for extrapolation and uniformitarian explanation, even for the two most famous cases of mass extinction for formerly prominent groups (see pp. 1314–1316 for a modern perspective on the demise of these taxa): trilobites at the Permo-Triassic event, and ammonites at the Cretaceous-Tertiary mass dying (pp. 321–322):

> With respect to the apparently sudden extermination of whole families or orders, as of Trilobites at the close of the paleozoic period and of Ammonites at the close of the secondary period, we must remember

what has been already said on the probable wide intervals of time between our consecutive formations; and in these intervals there may have been much slow extermination. Moreover, when by sudden immigration or by unusually rapid development, many species of a new group have taken possession of a new area, they will have exterminated in a correspondingly rapid manner many of the old inhabitants; and the forms which thus yield their places will commonly be allied, for they will partake of some inferiority in common. Thus, as it seems to me, the manner in which single species and whole groups of species become extinct, accords well with the theory of natural selection.

I have discussed Darwin's defense of uniformitarian extrapolation in detail because his argument, in this case, proved so successful in directing more than a century of research away from any consideration of truly catastrophic mass extinction, and towards a virtually unchallenged effort to spread the deaths over sufficient time to warrant an ordinary gradualistic explanation in conventional Darwinian terms, with any environmentally triggered acceleration of rate only serving to intensify the effects of ordinary competition, species by species. I can't think of any other prominent subject in paleontology where uniformitarian presuppositions clamped such a tight and efficient lid upon any consideration of empirically legitimate and conceptually plausible catastrophic scenarios. Merely to suggest such a thing (as even so prominent a scientist as Schindewolf, 1963, discovered) was to commit an almost risible apostasy.

In particular, these uniformitarian assumptions about the extended duration of apparent mass extinctions led geologists and paleontologists to favor earth-based rather than cosmic physical inputs (for most plausible extraterrestrial causes work with greater speed and intensity), and to focus upon telluric influences (like changing climates and sea levels) that could most easily be rendered as gradualistic in style . So strongly entrenched did this prejudice remain, even spilling over into popular culture as well, that a few years after Alvarez et al. (1980) published their plausible, and by then increasingly well affirmed, scenario of extraterrestrial impact as a catastrophic trigger for the Cretaceous-Tertiary event, the *New York Times* even ridiculed the idea in their editorial pages, proclaiming (April 2, 1985) that "terrestrial events, like volcanic activity or changes in climate or sea level, are the most immediate possible causes of mass extinctions. Astronomers should leave to astrologers the task of seeking the cause of earthly events in the stars."*

*I'm usually quite unshockable, and nothing from the fourth estate (even so high a denizen thereof as the *New York Times*) ever surprises me. But I was amazed that America's most distinguished newspaper would editorialize against a theory so clearly subject to empirical test and so eminently interesting as well. I ended up by sounding off in a popular commentary for *Discover* Magazine (October, 1985) with a title parody on the *Times*'s venerable motto: "All the News That's Fit to Print and Some Opinions that Aren't." I commented (Gould, 1985b) that the absurdity of their overreach might best be grasped by comparison with a hypothetical editorial that might have appeared in the *Osservatore Romano* (the official Vatican newspaper) for June 22, 1633: "Now that Signor Galileo, albeit under

Thus, for example, when Gilluly (1949) published one of the most famous and influential geological papers of the mid-twentieth century, arguing that one popular physical theory for mass extinction—episodes of orogeny, or mountain building—could not be construed as either global in effect or simultaneous in occurrence on all continents, but should rather be interpreted as sequences of more limited local events liberally spread out in time, he ended his paper with a stirring manifesto: "Long live Charles Lyell and his doctrine of uniformitarianism!" And, if I may cite an embarrassing incident from my own graduate career, when my mentor Norman Newell decided to invest considerable effort in compiling data from faunal lists in the world's paleontological literature to see if the maligned and effectively abandoned theme of mass extinction held any validity, I thought that the old man had taken leave of his good scientific sense in wasting so much time on a truly settled issue. For didn't we all know that the extinctions really spanned considerable intervals of time, and that any blip detected from faunal lists could only be recording an artifact of longer periods artificially compressed into simultaneity by imperfections of the fossil record?

Only with this understanding of the historical impact and persistence of Darwin's uniformitarian and extrapolationist view of extinction in the fossil record can we grasp the conceptual reforming power (and not merely the phenomenological fascination) of the improving case—from a wild idea rejected out of hand by nearly all paleontologists in 1980, to a firmly documented virtual fact of nature by 2000—for the triggering of at least one mass extinction, the Cretaceous-Tertiary event, by impact of a large extraterrestrial object (see Alvarez et al., 1980, for the original proposal, and Glen, 1994, for history of science in progress in a book entitled: *The Mass Extinction Debates: How Science Works in a Crisis*).

Two comments on the K-T (Cretaceous-Tertiary) transition, one new and one old, may be taken as emblematic of the magnitude of both theoretical and practical reformulation. First, M. J. S. Rudwick, a prominent systematist of fossil brachiopods early in his career and the world's leading historian of geology in later years, commented to Glen with the professional skills and "feel" of both segments of his ontogeny: "It never crossed my mind that . . . the brachiopod groups I worked with expired suddenly by modern K/T boundary standards. I thought that the brachs went out suddenly, but that 'suddenly' . . . in 1967 meant a few million years, which was considered geologically sudden" (quoted in Glen, 1994, p. 41).

Second, an argument prominently advanced by Charles Lyell himself dramatically illustrates the difference between strictly uniformitarian expectations and the implications of truly catastrophic triggers for mass extinction.

---

slight inducement, has renounced his heretical belief in the earth's motion, perhaps students of physics will return to the practical problems of armaments and navigation, and leave the solution of cosmological problems to those learned in the infallible sacred texts." I also suggested that, as a *quid pro quo* of ultimate fairness, the *Times* might award to the Paleontological Society the right to determine the date and amount of their next price increase.

Lyell, as well known and recorded (see Gould, 1987b, for example), named the epochs of the Tertiary Era by a statistical method based on the percentage of molluscan species still extant—from Eocene (or "dawn of the recent" for the lowest percentage) to Pliocene (or "more of the recent" for the much higher percentages of later strata). He then noted that the uppermost Cretaceous (Maastrichtian) and lowermost Tertiary beds held no species in common at all. By his argument of statistical gradualism, this complete non-overlap could only be explained by a vast gap of missing time—a period long enough to remove all Cretaceous species, one by one, at the same rate as the Tertiary demise. But since the entire Tertiary did not suffice to overturn the molluscan fauna completely (as a few Eocene species still survive), Lyell reasoned that a globally unrecorded interval of time, longer than the full Tertiary span (so well represented by a voluminous paleontological record throughout the world), probably intervened between the latest Cretaceous and earliest Tertiary strata. As Lyell's hypothetical missing interval of more than 65 million years actually spans only a geological moment under the impact scenario, I would nominate Lyell's following statement (1833, p. 328) as the worst forecast ever made under the uniformitarian method of extrapolating from a range of observed rates!

> There appears, then, to be a greater chasm between the organic remains of the Eocene and Maastricht beds, than between the Eocene and Recent strata; for there are some living shells in the Eocene formations, while there are no Eocene fossils in the newest secondary [that is, Maastrichtian or uppermost Cretaceous] group. It is not improbable that a greater interval of time may be indicated by this greater dissimilarity in fossil remains . . . We may, perhaps, hereafter detect an equal, or even greater series, intermediate between the Maastricht beds and the Eocene strata.

Despite the uniformitarian consensus from Darwin's time until the late 20th century, occasional scholars of high reputation continued to float catastrophic proposals for the unresolved puzzle of mass extinctions that, despite the orthodox conviction about "spreading out" into missing geological time, never meshed well with gradualist presuppositions and continued, like the proverbial sore thumb, to stick out above a comfortable background. But, in fairness, we cannot blame geologists and paleontologists for rejecting these proposals because, to cite a familiar motto, extraordinary claims require extraordinary evidence—and these attempts to resuscitate catastrophism remained entirely speculative (or at least undocumented by anything beyond the basic data for mass extinction itself, an evidentiary source that had already, to the satisfaction of an entire profession, been rendered consistent with uniformitarian presuppositions).

To cite the two most notable examples from the generation before Alvarez, Schindewolf (1963), in an article entitled "Neokatastrophismus," proposed bursts of cosmic radiation as the paroxysmal mechanism of mass extinction—with direct nuclear death (for the exterminations) and vast increases in mutation rates among survivors (for subsequent replacements by highly altered forms). But, to show the frustration (and scientific nonoperationality) of such proposals, Schindewolf actually stated—thus providing a favorite case that I have used for decades to illustrate the difference between science and speculation—that he had postulated cosmic radiation explicitly because such a cause would leave *no* empirical sign (then known to geologists) in the record of strata and fossils. (For Schindewolf had to admit that the empirical record revealed no direct evidence at all for a catastrophic mechanism of mass extinction, and he therefore had to seek a potential cause that would leave no testable sign of its operation! Can one possibly imagine an unhappier situation for science?—to face the prospect of a plausible explanation that does not, in principle, leave evidence for its validation.)

In a second example, well remembered by paleontologists of my generation, Digby McLaren used his presidential address to the Paleontological Society in 1970 to hypothesize that a bolide impact had triggered the Devonian mass extinction. In the light of Alvarez's later triumph with a similar explanation for the K-T event, one might be tempted to view this address as prophetic. But, much as I admire both Digby himself and iconoclasts in general, I'm sure McLaren would admit that he simply "lucked out" in this case. For, like Schindewolf, McLaren could present no evidence at all for his bolide, and simply slipped this proposal into the end of his talk in an almost apologetic manner, after documenting the style and extent of the extinction itself (the main focus of his paper). (Moreover, to this day and despite excellent evidence for bolide triggering of the K-T event, we have no satisfactory explanation of the Devonian extinction, and no credible data for its causation by impact.)

By contrast, the genesis of the Alvarez's hypothesis for the K-T mass extinction could not have been more different, or more exemplary for science. For the K-T bolide proposal began with an unanticipated empirical discovery—generated, ironically, during a test for an opposite hypothesis, and therefore surely not gathered under the aegis of any iconoclastic theoretical thoughts. Geologist Walter Alvarez, trying to test an idea about latest Cretaceous sedimentation rates under traditional gradualist views of the extinction, asked his father, the Nobel laureate in physics Luis Alvarez, whether any isotopic signature might provide evidence for the following conjecture: Walter wondered whether a false appearance of rapidity in extinction might arise from an unusual slowdown in sedimentation rates, thereby compressing a "standard" amount of extinction into an unusually short stratigraphic interval.

Luis proposed a measurement or iridium, an element virtually absent from the earth's indigenous surficial rocks. (Presumably, the earth formed with iridium at standard cosmic abundance, but this indigenous iridium, as such a heavy and unreactive element, quickly sank well below the surface, especially since the earth's crust was effectively molten early in the planet's history.) Luis therefore supposed that throughout Phanerozoic time, iridium has entered the earth's surface only through cosmic influx, and at the effectively constant rate of uniformitarian assumptions—the "gentle cosmic rain from heaven" in radiation and tiny particles, as standard views and terminology then held. Luis reasoned that Walter's hypothesis could therefore be tested by measuring iridium concentrations in latest Cretaceous sediments, arguing that a small positive excursion would validate Walter's supposition, as the constant cosmic influx became diluted by less than the usual amount of iridium-free terrestrial sediment.

But when the Alvarez team measured iridium in boundary-layer sediments, they found a value so high that they had to invert their initial assumption in the most radical manner. Terrestrial sedimentation would have to cease for longer than the earth's entire history to produce such a high spike from a low and constant cosmic influx. Rather, they now reasoned, a true and sudden influx of iridium must have occurred right at the K-T boundary itself—with the obvious "culprit" as a large extraterrestrial body striking the earth and implacing an enormous and momentary dose of cosmic iridium. Thus, the revival of catastrophic theories for mass extinction began with an empirical surprise generated during a test for a conventional gradualistic hypothesis—the exact opposite (in both form and utility) of previous exercises in evidence-free and catastrophically-driven speculation.

This thoroughly different character of the Alvarez hypothesis—as an evidence-driven claim bursting with seeds of testability, rather than a sterile speculation—should have caught the attention and intrigue of all scientists from the start. But the anti-catastrophic biases of Lyellian and Darwinian traditions ran so deep, and the knee-jerk fear and disdain of paleontologists therefore stood so high, that even this welcome novelty of operationality did not allay rejection and outright disdain from nearly all established professional students of the fossil record (whereas other relevant subdisciplines with other traditions, planetary scientists and students of the physics and engineering of impacts, for example, reacted in markedly more mixed or positive ways—see Glen, 1994). I will never forget a 1979 phone conversation (as the preprints of Alvarez et al., 1980, circulated) with David Raup, perhaps the only other invertebrate paleontologist of my generation who reacted with initial warmth to the impact hypothesis:* It was one of those laconic affairs,

*Since I have far more often been wrong than right in my intuitions about iconoclastic ideas, perhaps I may be excused for taking some pride, in retrospect, in calling this affair, even for the right reasons. I can take no credit whatsoever for the theory's success, since I work in other areas of paleontology and never (unlike Raup and others) did any primary research on the subject. At most, I did write many private letters to participants, and several popular essays, trying to explain both to my colleagues and to general readers, why they should take impact seriously, and why the Alvarez hypothesis differed so dramatically from previous speculations. In a climate of such negativity, these efforts may have done some good—and Glen (1994, p. 49) was kind enough to write: "By contrast, the paleontologist Stephen Jay Gould, perhaps earth's science's most widely read spokesman, was intellectually predisposed to welcome the impact hypothesis straightaway, in part because his unorthodox theory of punctuated equilibrium . . . articulated well with impact theory. Gould thus provided welcome and vital encouragement through sustained communication with the Alvarez group early on, when only the iridium evidence was at hand and paleontologic backlash against the theory was strong." Moreover, I cannot assert any general claim for anything inellectually admirable in myself or the very few initial paleontological supporters of impact—as we all had purely personal reasons for favorable predispositions

where no more need be said: Raup: "This time it's different you know." Me, in reply: "Yes, of course, the iridium."

In the retrospect of a mere twenty years between initial proposal and such substantial success (although by no means total, in either its own hopes or terms), we may identify several reasons to honor the conventional criteria used by scientists to judge the strength and importance of hypotheses—criteria based on empirical affirmation, fruitful extension, and widening intellectual scope, rather than on such nonoperational notions as progress towards absolute truth. Science is, as P. B. Medawar stated in the title to his finest book, the *Art of the Soluble:*

1. At their initial decision to publish, the Alvarezes had detected an iridium spike only at two nearby localities in Denmark and Italy, and couldn't even be confident in their theory's crucial prediction of a worldwide enhancement. (Fortunately for them, evidence for a third and virtually antipodal spike from New Zealand arrived in time for inclusion in the original publication.) But, within a decade, affirmation had accumulated from so many collateral sources—all independent of, and unpredictable from, the iridium spike itself—that the case for impact had effectively been sealed. These additions ranged from shocked quartz in the K-T boundary layer throughout the world (with silica tetrahedra arranged in an unusual manner only associated, so far as we know, with high pressures of impact, including the initial discovery of such forms in nuclear bomb craters), to the "smoking gun" of a gigantic crater of exactly appropriate age—the Chicxulub structure off the Yucatan peninsula in Mexico. As this evidence accumulated, the alternative volcanist scenario (with iridium recruited from the earth's interior in extrusive events of a magnitude never witnessed in historical times), which had provided such good material for fruitful debate, yielded the floor (although volcanic action initiated by impact may have played an important role in the full scenario of explanation for mass extinctions). Needless to say, other exciting hypotheses generated by the impact debate—particularly Raup and Sepkoski's claim for a 26 million year periodicity in extinction, with a subsequent set of wondrous astronomical hypotheses as potential explanations, including the actions of a previously unknown dwarf companion star to the sun—have not fared so well (although a few jurors, with a few good arguments, are still holding out).

2. Enhanced and surprising interdisciplinary communication offers no guarantee of scientific rectitude or success, but we can only celebrate the veritable orgy of exciting, and at least intellectually fruitful, discussion and collaboration inspired by the impact hypothesis among scientists in subdisciplines that had never read each other's work, hardly even knew the names of the most reputable leaders in the disparate domains, and could barely speak

---

(me for the punctuational themes that Glen notes above, Raup for his work on random processes, vertebrate paleontologist Dale Russell for his prior, if speculative, writing on a potential role for impacts). Always look to reasons of personal interest, rather than general wisdom, in such cases—and remember the wise words of W. S. Gilbert's most honorable, if unlikeable, man: King Gama of *Princess Ida:* "A charitable action I can skillfully dissect; And interested motives I'm delighted to detect."

the same scientific language (most notably of Linnaean Latin *vs.* Newtonian triple integral signs). Paleontologists met, and even eventually published papers with, colleagues who had formerly received little more than a grunted "hello" in 20 years of hallway passing, or who had only been seen as strangers across a crowded room during some unenchanted evening at a faculty party. The several Snowbird conferences in Utah will never be forgotten by anyone who enjoyed the privilege of attendance, and who participated in the warm fellowship, and sometimes antic debate, with (among many others) nuclear physicists, taxonomic paleontologists, and historians of science.

3. Most importantly, and diagnostically for scientific practice, the impact hypothesis proved its mettle (at least for me) in the explicit suggestions and prods that it provided for particular (and ultimately highly fruitful and exciting) paleontological research that would never even have been conceptualized without its nudge and encouragement. I have argued throughout this book that the broad world-views of scientists (with gradualism, uniformitarianism, and strict Darwinian adaptationism as the major examples in this context) do not merely act as passive summaries of general beliefs, but serve as active definers of permissible subjects for study, and modes for their examination. At best, a potent context continually provokes more fruitful work. But at worst, and (unfortunately) ever so often in the history of science, such world-views direct and constrain research by actively defining out of existence, or simply placing outside the realm of conceptualization, a large set of interesting subjects and approaches, often including the very classes of data best suited to act as potential refutations of the world-view. Such self-referential affirmations are not promoted cynically, or (for the most part) even consciously, but they do, nonetheless, operate as strong impediments to scientific change.

As argued throughout Chapter 9, my greatest pride in punctuated equilibrium lies in the theoretical space it created for active study of subjects that could win neither definition nor existence under gradualistic presuppositions: particularly stasis (previously viewed as an embarrassing failure to detect evolution, and therefore as a non-subject), now generally seen as an important and surprising result at several levels in the history of life; and the punctuational explanation of trends by differential success of species treated as discrete Darwinian individuals (an alternative with explanatory options that simply didn't exist under older models of trends defined exclusively as anagenetic transformation). In several similar ways—I will cite just two here—the catastrophic impact hypothesis of mass extinction created an enlarged intellectual space that forced paleontologists to reevaluate data once viewed as comfortably consistent with gradualist assumptions, but clearly subject to extension and better definition as tests for gradualism *vs.* catastrophe. In this vital way, self-fulfilling claims for convention became sources for discrimination among rival hypotheses about some of the most important questions in the history of life.

For example, in a justly influential paper, Signor and Lipps (1982) recognized that the well-documented literal signal of taxa slowly "petering out" in the stratigraphic record before a mass extinction boundary might actually be consistent with truly catastrophic removal—a "counterintuitive," but actually rather obvious, option that paleontologists had never conceptualized because the literal signal matched their expectations, and they therefore never questioned the meaning. Signor and Lipps argued that if all taxa under consideration had truly died at once, their recorded disappearance in the stratigraphic record would still be sequential, as based on probability of fossilization. After all, a top, large-bodied carnivore like *Tyrannosaurus*, with a local geographic range and a relatively small N, might only yield a fossil once every several meters during the stratigraphic range of its actual existence; whereas a foraminifer that lived as billions of individuals at every moment of a continuous oceanic core, should provide abundant specimens in every mm of its stratigraphic existence. Thus, even if the dinosaur and the foram species died at the same instant in a worldwide catastrophe, the last dinosaur fossil might still appear several meters below the extinction boundary, while foram fossils should persist to the last stratum.

This artifactual pattern, now appropriately named the Signor-Lipps effect, can be distinguished from a genuine petering out that truly gradual extinction would produce—most obviously by testing for correlations between time of disappearance before the boundary and expectations of preservability in the fossil record (as measured by "waiting time" between fossils in normal strata between extinction events, not by working with abundances near the boundary itself, where circular reasoning may so easily intervene). Regardless of the outcome in any particular application—and some studies have yielded consistency with Signor-Lipps, whereas others seem to show genuine petering out—paleontologists had never formulated, or even conceptualized, this important methodology until the catastrophic impact scenario forced their attention to such questions.

As a different approach to the same basic situation, one might decide instead to take a group long interpreted as showing a clear literal signal of petering out—and then go to the most promising stratigraphic record in the world, pulling apart every single bedding plane to see if surpassingly rare specimens might occur nearer, if not right up to, the extinction boundary (for you only need one to disprove complete extinction). This "needle (or dinosaur) in the haystack" strategy represents the "flip-side" of the Signor-Lipps approach to the general problem of sampling in science—either use a global and statistical method to extract a clear signal from broad data, or sample with such intensity in a more limited area that you can effectively survey the available "universe," and no longer even require the art of statistical inference. But why would one even think of sampling with such intensity, absent the prod of the impact hypothesis and its prediction of putative success. After all, if your world-view enjoins petering out, and your data (read literally) clearly display just such a pattern, why would you petition the National Science Foundation for cash, and then spend several summers sweating in a desert, pulling apart every bedding plane in a single place virtually guaranteed to yield nothing. Such behavior could only point to an unsound mind—

unless, of course, you had a really good reason to believe that a diamond lay hidden somewhere in that particular stratigraphic haystack.

This method of hyperintense sampling in optimal places has been applied, with great success in validating substantial, even fully maintained, abundance right up to the K-T boundary itself for two groups whose ostensible petering out had provided a mainstay of empirically based opposition to catastrophic mass extinction—ammonites as affirmed by Peter Ward's work in France and Spain (Ward, 1992), and dinosaurs as indicated by collections of Sheehan et al. (1991) in Montana and North Dakota. Seek and (perhaps) ye shall find, but in a real world of such limited time and opportunity in scientific careers, one does need a theoretical license, as well as a landowner's specific permission, to seek.

Despite my personal excitement at the theoretical and practical import of the impact hypothesis, I must confess my initial surprise at a statement that historian of science Bill Glen made to me in the early 1990's. For he asserted that the reforming power of the impact theory would surpass even that of Plate Tectonics in the history of geology. (And one cannot accuse Glen either of sour grapes or parochialism, for he not only wrote the book on "living" history of science for the impact debate (Glen, 1994), but had previously written an even more highly acclaimed history in progress for plate tectonics as well (Glen, 1982). Now I still don't fully agree with Glen, if only for two primary reasons:

First, despite all its successes in 20 years, the Alvarez scenario still applies, with proven power, only to the single event of the K-T mass extinction. None of the other four great mass dyings show clear iridium spikes or other evidence for triggering by impact (although some of the smaller extinction events have been more plausibly linked to possible impact). Thus, the Alvarez scenario remains a historical explanation, however elegantly affirmed, for a single event, and not a general theory of mass extinction. (And, as I came to know and understand Luis Alvarez late in his life, I rather suspect that this situation would have frustrated him intensely, for, as a theoretical physicist by trade, he remained committed to the view that science can only attain its true goal by establishing general explanations rooted in the spatio-temporal invariance of natural law. To learn that he had become godfather to the contingent explanation of a great event, and not to the formulation of a general theory of mass extinction, would have left him unamused.)

Second, even though we now have confidence in the factuality of impact as a trigger for the K-T extinction, we still cannot specify a satisfactory "killing scenario" to explain the timing and differential susceptibility to dying among life's various taxa—a scarcely surprising circumstance, given the complexity of the event and the potential number of dire consequences that impact might unleash, but still a damper upon any feeling of full satisfaction. Indeed, I remain amused by how the competing (and, to be sure, partly complementary) ideas follow the canonical scenarios for disaster in Western culture—the ten plagues of Moses. And whenever our scientific preferences so clearly match our culturally inherited stories, we should begin to question the complex intellectual and psychological bases of the hypotheses we have chosen to test.

Moses tried them all (and eventually triumphed, of course). For the most popular theme of "nuclear winter" and the cutting off of light (and photosynthesis) by a persisting worldwide dust cloud raised by the impact, "he sent a thick darkness over all the land, even darkness which might be felt." For acid rain and worldwide fires, "he gave them hailstones for rain; fire mingled with the hail ran along upon the ground." For poisoning of the oceans, "he turned their water into blood." For killing plagues induced by the few successful survivors, "their land brought forth frogs, yea even unto their King's chambers." And for selective diseases unleashed in this novel environment, we have the analog of Moses' ultimate weapon: "He smote all the firstborn of Egypt, the chief of all their strength."

Nonetheless, I do appreciate the undeniably valid part of Glen's contention. For many scientists and historians have noted that, whereas plate tectonics changed our view of the earth's structure more profoundly than any previous theory had ever done—and may even be described as supplying our first adequate account for the physics of earth-sized bodies in general—plate tectonics also, albeit ironically, promulgated a conservative reform in not challenging in the slightest way, but rather supplying a mechanism to validate, the deepest of all geological presuppositions: uniformitarianism itself. For if the earth's largest mountains can rise as a result of two plates crunching together at their characteristic moving rate of millimeters per year, and if the great faultlines, earthquakes and volcanic provinces also mark the edges of such slowly drifting plates, then the central Lyellian dictum of explaining all grand, and all apparently quick, events by the accumulation of slow movements, utterly imperceptible at the scale of our daily lives, gains a kind of planetary and mathematical validity that can only be deemed awesome in its phenomenological range and intellectual reach.

On the other hand, Glen continued, the impact theory—even if we never succeed in establishing this mechanism as a general theory, and even if such catastrophes remain confined to explanations of particular events—directly fractured Lyellian uniformity, therefore penetrating far deeper in its iconoclasm than the admittedly more comprehensive theory of plate tectonics could ever bore. In any case, and in the terms and concerns of this book, the validation of a truly catastrophic triggering mechanism for at least some events of mass extinction dramatically fractured the support that Darwin needed from the kind of geological stage necessarily set for playing out his preferred game of life. The vital extrapolationist premise of the third leg on the tripod of essential Darwinian logic must fail if global paroxysm can undo, redirect, or even substantially impact a pattern of life's history that, in a fully Darwinian scheme of explanation, must scale up in full continuity from the microevolutionary realities of competition in observable ecological time.

I have, in my own writings, tried to summarize the theoretical importance of readmitting truly catastrophic scenarios of mass extinction back into scientific respectability (after 150 years of successful Lyellian anathematization) by stating an emerging consensus about four crucial and general features of such events, each strongly negative (and, in their ensemble, probably fatal) for the key extrapolationist premise needed to maintain a claim of exclusivity for a strictly Darwinian theory of evolutionary process (with descriptive aspects of life's pageant still left for paleontological documentation, because a general theory must underpredict an actual outcome in a historically contingent world): mass extinctions are *more frequent, more rapid, more intense, and more different in their effects* than paleontologists had suspected, and that Lyellian geology and Darwinian biology could permit.

In terms of the broad categories of pattern in life's history that will now require at least partial explanation from catastrophic theories of mass extinction, and will probably not be rendered on Darwinian assumptions of extrapolation from microevolutionary theory,* I would summarize the most impor-

---

*May I, for one last time, repeat the rationale, from the internal logic of this book, that leads me to the otherwise odd strategy of treating the refutation of the third, or extrapolationist, leg of central Darwinian logic at reduced length (in Chapter 6 in the historical half and the present Chapter 12 in the modern half) relative to the other two legs of hierarchy and levels of selection (Chapter 3 in the first half, and 8–9 in the second half) and structuralist thinking and constraint *vs.* strict selectionism (Chapters 4–5 in the first half and 10–11 in the second half). My reasons rest upon four arguments, the first three intertwined and intellectual, the fourth separate and personal. *First,* this book treats the *biological* structure of evolutionary, particularly Darwinian, theory. Of the three central components designated as legs on a tripod of conceptual definition for Darwiniasm—(1) levels of selection (Darwinian organismal *vs.* modern notions of a full hierarchy); (2) functional *vs.* structural approaches (externalist selection as virtually the sole creative force in evolutionary change *vs.* the importance of internal constraint from several sources, not all selectionist); and (3) uniformitarian extrapolation of microevolutionary styles of natural selection to explain the full panoply of life's changes through geological time—the *biological* aspects of the critique for the third theme of extrapolationism lie mainly within expansions and revisions of Darwinism on the first two legs. *Second,* I therefore decided to confine my explicit discussion of the third leg (since the biological critiques had already been treated under the first two legs) to the surrogate theme of geological requirements to potentiate the biological mechanism (the kind of stage needed to display the proposed play). The surrogate theme of a different profession does not require such detailed treatment as the central biological critiques. *Third,* and almost as a historical footnote (but a truly intriguing point missed in most discussions of Darwinism and the 19th century debates about evolution), Darwin himself became uncomfortable with his need to call upon such surrogate themes from other disciplines, and he tried (in later years) to avoid such ancillaries, and to devise a theory that would render his conclusions about life's history entirely by biological principles. Most importantly, he abandoned his early satisfaction with allopatric modes of speciation (that called upon geographic surrogacy to ensure a world with sufficient opportunity for isolation of populations), and moved towards the sympatric theory embodied in his "principle of divergence" (see Chapter 4 of the *Origin,* including the book's sole figure), and considered by him (in the famous "eureka" statement of his carriage ride in 1854, see p. 224) as second in importance only to his formulation of natural selection itself in devising his full theory. Because this sympatric theory, relying upon the generality of selective benefits for extreme forms in any environment, proved ultimately quite unsatisfactory, and even illogical, we can grasp, with even greater clarity, the importance of this issue in Darwin's mind—as he so rarely allowed himself to place such weight upon a patently dubious argument (and I do mean patent, for he fretted overtly about the inherent problems, see

tant changes under two rubrics: the "random" and the "different rules" models for alternating regimes of faunal turnover in mass extinction *vs.* ordinary, sequential (and, in Darwin's own preferences, overtly competitive) replacement in "normal" times between these infrequent but intense episodes.

In the purely random model, which I do not consider of great importance in life's history, a group might die in a mass extinction for no reason of particular sensitivity towards the catastrophic agents of extermination, but for reasons little beyond the luck of the draw—the "bad luck" *vs.* the traditional "bad genes" in Raup's amusing characterization (Raup, 1991a). (I consider this random model as less important than deterministic reactions to the "different rules" of these parlous times because, even in these extreme episodes, the species number of major taxa generally remains large enough to preclude full removal by random inclusion of all members of one kind within a percentage of totality. All ten red beans in a bag of 100 may disappear often enough in a random destruction of 75 beans. But we would not expect all 10,000 dinosaur species in a fauna of 100,000 land taxa to die in a truly random reduction to 25,000 taxa.)

Nonetheless, a fact of the fossil record, not widely known or appreciated outside the community of professional paleontologists, may grant the random model an important role in a few crucial circumstances during life's history (and such moments make all the difference in contingent sequences, as Jimmy Stewart discovered when his guardian angel showed him the alternate history of his town, absent his existence, in *It's A Wonderful Life*): some groups, formerly dominant in their habitats and therefore viewed as persistently "major" in our conceptions, fluctuated enormously in diversity throughout geological time, and just happened to be surviving at very low N (a situation from which they had always rebounded before, and in normal times) when a mass extinction intervened. And when an event of latest Permian magnitude occurs—the largest of all mass dyings, with estimates of species loss ranging up to 96 percent (see Raup, 1992)—and your group contributes only a lineage or two to the global fauna, you can easily disappear, entirely and forever, for little reason beyond bad luck (those two red beans, both included among the 96 percent of dead benthic taxa, but entirely equivalent in ordinary Darwinian prowess to the 4 percent that survived).

Among several plausible cases in this mode, the two specifically cited by

---

pp. 236–248), unless its conclusion played a central role in his full system, and he could devise no other path toward a desired outcome. Hence, and in any case, in thus deemphasizing the themes of geological surrogacy, I am only paying homage to Darwin's own preferences in the structuring of his theory. *Fourth,* disconnectedly, finally, and entirely personally, I have done no technical paleontological research on these geological themes (beyond some early, and largely philosophical, analysis of the concept of uniformitarianism, including my first published paper of Gould, 1965), whereas I have devoted my career to paleontological studies of the first two biological legs of hierarchy and constraint. Thus, I lack the competence and the "feel" of personal expertise (or the folly of personal engulfment) to treat this theme at comparable length, for which all readers should rejoice (as even this footnote has hypertrophied quite enough already)!

Darwin (see p. 1302) to express his confidence in gradualistic and uniformitarian models—trilobites and ammonites as the "signatures" of the Permian and Cretaceous extinctions respectively—may, ironically, fit this potential scenario best. Trilobites abounded in most Paleozoic faunas (at least in the affections of fossil collectors), but they had been reduced to only two lineages of low diversity by latest Permian times. The death of these two, and the consequent termination of one among only four great arthropod classes, may reveal no insufficiency of trilobite anatomy, ecology or development, and may only record the failure of any among a very few coins to come up heads at a weird moment that diverted 96 percent of all flips into a tailspin towards oblivion.

The case of ammonites is both more complex and more instructive. They suffered greatly during the last three of five great Phanerozoic dyings—latest Permian, latest Triassic, and latest Cretaceous. They barely survived the first two, with only one or two lineages persisting in each case, and then died entirely in the K-T event (although their less speciose relatives, the chambered nautiloids, survived to this day to become favored items in aquariums and shell collections, and to become, as the "ship of pearl," Oliver Wendell Holmes's celebrated metaphor of eternity: "Build thee more stately mansions, O my soul . . .").

After the first two restrictions, the ammonites reradiated mightily and became major components of Triassic, and then of Jurassic-Cretaceous, faunas. Perhaps we do need an "ammonite-specific" reason for the final Cretaceous death, as Ward has shown (1992) that their latest Maastrichtian diversity remained respectable. But I raise a different point here: Can we specify any "real" mathematical or biological difference among reduction to 1, 2 or 0 lineages? That is, can we say that the Cretaceous extinction was causally worse for ammonites than the preceding Permian or Triassic events? Surely not from the numbers themselves (and we have no other basis for such an assertion, at least at this moment of research and understanding). We can state no causal or statistical difference among 0, 1 or 2 survivors. These results are effectively equivalent. We might as well put slips with the three numbers in a bag and let nature choose one at random for each of the last three mass extinctions. And yet, in a real world of genealogical history, zero *vs.* anything else makes all the difference in the universe: absolute termination absolutely forever, *vs.* the possibility of redemption and reassertion. In this sense as well, and again in a sequence so influenced by contingency, effectively random removal based on small numbers in the face of catastrophe can impact the history of life in truly fateful and permanent ways. Principles regulate ranges of events, but particular and unpredictable events make history.

Nonetheless, I am confident, and I think nearly all paleontologists would agree, that the "different rules" model—a more conventionally causal view of history—plays a far more important role in expressing the power of catastrophic mass extinction to fracture the crucial extrapolationist premise of Darwinian central logic. I devoted the first part of this section (pp. 1296–

1303) to Darwin's own emphasis on the importance of "spreading out" the timing of apparent "mass extinction" into sufficient spans for explanation by ordinary Darwinian competition, with sequential and individual deaths of species occurring in conventional realms of natural selection, at most against a background of unusual environmental perturbation that may "turn up the gain" of intensity for standard causes, but will not change the usual rules or reasons. I also showed, in particular, how Darwin needed this extrapolationist argument to validate a concept of progress through life's history that his cultural context demanded (and to which he personally assented), but that he recognized (and, in his philosophical radicalism, greatly appreciated) as underivable from the "pure" operation of natural selection, and therefore recoverable only by an additional ecological postulate about the predominance of biotic competition in a perpetually crowded biota (the "metaphor of the wedge").

The key phenomenon for this entire discussion—whether explanation be offered by Darwinian extrapolation, or by the random or different rules alternatives—has always resided in the observed selectivity of mass extinctions: why do some taxa flourish and others die, especially since the observed patterns of mass extinctions do not simply intensify the tendencies of normal times (that is, mass dyings do not preferentially remove those groups already on the wane by competition with superior forms during "background" times. Mammals were not expanding, with dinosaur retreating, during the long span of Cretacous life).

Under the "different rules" model, extirpated groups die for definable reasons of conventional anatomy, physiology, behavior, or population structure. But their death follows from the unpredictable, and suddenly instituted, "different rules" of catastrophically altered environments in episodes of mass extinction, and does not occur because these taxa had evolved properties that would have doomed them in the same manner (albeit more slowly and sequentially) for ordinary reasons of Darwinian competition during normal times. In fact, and even worse for conventional arguments about progress, the traits that spell doom in catastrophically altered circumstances may just as well have originated as the adaptive features that secured success, and competitive superiority in the normal Darwinian times just preceding. In this important sense, if the previous Darwinian "best" often die for unpredictable but deterministic reasons in the suddenly altered worlds of catastrophic mass extinction, then Darwin's crucial argument for progress (always weak and suspect because it could not flow from the abstract logic of natural selection itself, and required an additional ecological belief in plenitude and biotic struggle) collapses through the disruption, or even the reversal of its vector, as imposed by these dramatic episodes with their different rules for who flourishes and who goes to the wall. And if these episodes are sufficiently numerous, profound, rapid and different (my four criteria of p. 1313), then their accumulated impact may balance, or even reverse, the Darwinian accumulation during much longer stretches of normal times, thereby imbuing the full pattern of life's history with a very different form (perhaps entirely devoid of any meaningful vector of progress) than the extrapolationist premise of Darwinian central logic had assumed.

Thus, the influence of the "different rules" model in helping to explain the waxing and waning of taxa in macroevolution represents the most interesting and far-reaching modification of Darwinian expectations unleashed by catastrophism's renewed respectability, and by the resulting inadequacy of uniformitarian extrapolation from Darwinian microevolution to supply a full explanation for the causes of pattern in life's history. After all, and in admitted caricature, if the size of dinosaurs had marked their success over mammals in habitats of large terrestrial vertebrates for more than 130 million years; and if this established basis for Darwinian success then became an important causal factor in their differential death, with small size as an equally vital reason for mammalian survival; then the tables truly turned with the institution of these particular different rules of a K-T moment—as marks of failure (or at least of limitation) in the long background of Darwinian competition became fortuitous substrates for survival, while the primary features of former and such prolonged domination became unalterable portents of doom (whether directly or, more probably, via such correlated consequences as generally smaller population sizes and generally greater ecological specialization). "How are the mighty fallen" exclaimed David in his famous lament (2 Samuel:1). But Goliath had died by David's superior wit (and good aim), whereas Saul expired as a consequence of his own madness. Dinosaurs, on the other hand, may have fallen by their might, but surely not by their fault.

Thus, for these good reasons, the paleontological literature on the biological implications of mass extinction has, for the past 20 years, rightly focused upon the documentation of "different rules" in such potentially catastrophic episodes, and on their impact upon pattern in the history of life. I cannot, in the context of this chapter, present a compendium of these ever increasing, and ever more sophisticated, studies. I shall therefore cite just a few of the early results that have already become classic in a burgeoning field. I should also at least acknowledge—to open up a real can of worms that I shall not even attempt to close—that, at least partly as a "back formation" from doubts inspired by the debate on mass extinction, the entire subject of the efficacy of competition, in normal times as well as in geological perspective, has fallen under increasing question (see Simberloff, 1983, 1984 for some explicitly microevolutionary doubts; Gould and Calloway, 1980, for a paleontological example; Benton, 1996, for a general argument from the fossil record; but also Sepkoski, 1996, for a strong and more nuanced defense of competition in paleontologically normal times, and Sepkoski et al., 2000, for a beautifully documented and fascinating example, and Gould, 2000b, for commentary thereupon).

Jablonski published the most important early work on this subject, establishing a methodology and terminology in his classic paper (Jablonski, 1986b) on "Background and mass extinctions: the alternation of macroevolutionary regimes." Some of the specifically K-T patterns will strike no one as surprising, given the context; but the readjustment of diversity under such a pronounced moment of different rules still imposes a major signal upon the overall pattern of life's history. For example, land plants reproducing by seeds, rhizomes, or any other mode of propagation by bodies that can lay dormant under soil, tended to survive at higher frequency (Wolfe, 1990). And Sheehan and Hansen (1986) found higher death rates among animals directly linked in their feeding to a supply of living plants, whereas feeders on dead plant material, scavengers and detritivores tended to fare better.

Jablonski's studies break more general conceptual ground in their broader scope across several extinction events and in their search for commonalities across mass dyings rather than reactions to a specific impact scenario of the K-T event. He argues (in Jablonski and Bottjer, 1983, for example) that species-rich clades tend to increase in numbers of taxa during background times (largely by species selection within subclades whose species-individuals generate relatively more daughters, and therefore already a macroevolutionary claim that cannot be extrapolated from Darwinian organismal selection). However, these species-rich clades then tend to fail differentially in mass extinction because the same properties that enhance the capacity for speciation in background times—stenotopy and limited capacity for dispersal, for example—make taxa more susceptible to removal in catastrophic episodes.

In later extensions of the same theme, Jablonski (1986a, 1987) established the best documented case of genuine clade selection in evolution (working contrary, or at least orthogonally, to species selection, and therefore atop the hierarchy of potential levels at an apex that, at least in my judgment, probably operates only rarely in nature—see pp. 712–714 for further discussion). He found no consistent relationship between the properties of species within molluscan clades and the full clade's propensity for survival through mass extinctions. However, the clade's entire geographic range (but not the individual ranges of its component species) correlated strongly and positively with survivorship through mass extinctions.

Jablonski (1996) then documented a further disconnect between microevolutionary expectations in extrapolation, and macroevolutionary realities. In his rich data base of late Cretaceous mollusks, Jablonski could find no evidence for the most venerable of all supposed generalities in trending—Cope's rule, or the tendency of lineages to increase in body size (see pp. 902–905 for further discussion). Nonetheless, good microevolutionary reasons—and data—can be cited for claiming a general selective benefit for increased body size that should *(ceteris paribus)*, in a Darwinian world of extrapolation, yield Cope's rule in macroevolutionary extension. The reasons for this extrapolationist failure may be formulated at several levels, including forces operating in background times as well. But one important factor also intervenes in the different rules of mass extinction, where susceptibility of clades "does not appear to have been size selective" (Jablonski, 1996, p. 279). Jablonski concludes, also citing his earlier work on geographic ranges (1996, p. 279): "Survivorship [in mass extinction] appears to have hinged on other factors such as broad geographic range, which shows little or no correlation with body size in marine invertebrates. The analyses presented here reinforce the view that macroevolutionary patterns need not be simple extensions of those seen at the level of individual organisms over microevolutionary time."

I would also make the further observation that these discordances between background and mass rules (the "alternation" of regimes in Jablonski's terminology) must play a major role in the most general patterning of life's history, absent which the tree of earthly life would have grown in a markedly different shape. (In this most crucial sense, of course, I reassert the primary theme of this book that macroevolutionary theory matters profoundly.) For if different rules did not impose their signals at levels above microevolutionary extrapolation, the powerful themes of Darwin's world would push through to completion in life's phylogeny. For example, the species-rich clades of background times, with their dual advantages in organismic and species-level selection, would eventually eliminate the species-poor clades entirely if a still higher-level component of advantage for at least some species-poor clades did not "kick in" during episodes of different rules in mass extinction. And if the general, albeit slight, statistical edge of larger body size scaled straight up from local populations to the global biota at geological scales, what would guarantee a world enriched with all the little shrews and hummingbirds of our delight, not to mention the continued existence of short people, including the author of this book.

These generalities enter the corpus of macroevolutionary theory, but the different rules of mass extinction must still work through the specificities of various causes that provoke the rare, but potent, catastrophes of planetary history (with impact at the K-T boundary as the only firmly established case so far). Thus, some of the most interesting, if hypothetical, invocations have been proposed as explanations for specific and otherwise puzzling results of the K-T event. I have long been intrigued, for example, by the striking pattern of differential extinction in the oceanic plankton—with 73 percent of coccolithophorid genera, 85 percent of radiolarians, and 92 percent of forams failing to survive, while diatoms suffered only a 23 percent loss of genera.

Kitchell, Clark and Gombos (1986) made the interesting argument, later supported by direct data of Griffis and Chapman (1988) on survivorship of phytoplankton in conditions of prolonged darkness, that the differential success of diatoms probably bears no relationship to any notion of cosmic "betterness" or general "progress," but may only record the fortuity in exaptive use (under the different rules of K-T darkness) of adaptations evolved for ordinary, short-term microevolutionary advantages in background times (and not, obviously, in anticipation of any additional edge in forthcoming catastrophes!) Kitchell et al. (1986) argue that most diatom species have evolved mechanisms of dormancy (formation of resting spores, for example), permitting these photosynthetic organisms to survive extended periods of darkness, including several polar months per year for species living at high latitudes. Moreover, since diatoms build their skeletons of silica, which they can extract most readily in oceanic zones of upwelling that can be uncertain in placement and impersistent in timing, the capacity to "shut down" for periods of dormancy between such favorable moments gains a second important adaptive value.

But these two reasons for dormancy—to survive seasonal periods of darkness at high latitudes and to "wait out" silica-poor times between episodes of upwelling—operate entirely in the microevolutionary world of ordinary Darwinian selection. In this lucky case, these particular advantages did scale up to good fortune, given the factor favored by many researchers for the primary agent of the K-T killing scenario—extended darkness from a global dust cloud generated by the impacting bolide and its excavated and elevated earthly products. Thus, in this hypothesis, the relative prosperity of diatoms *vs.* the relative destruction of forams and other planktonic groups arose by the fortuity of how key adaptations for background conditions happened to "play" in an unpredictable and utterly different world of catastrophic impact. These "different rules" happened to suit diatoms and spell disaster for forams.

And if diatoms prevailed by such good fortune, let us not forget a key reason behind the possibility of this most immediate interaction between me as writer and you as reader. Dinosaurs and mammals had shared the earth for more than 130 million years, fully double the subsequent period of mammalian success that led to the possibility of *Homo sapiens* among some 4000 other living species in our mammalian clade. If the data of Sheehan et al. (1991) hold, and dinosaurs did persist in respectable abundance right to the moment of impact, then we may reasonably conjecture that, absent this ultimate random bolt from the blue, dinosaurs would still dominate the habitats of large terrestrial vertebrates, and mammals would still be rat-size creatures living in the ecological interstices of their world. In this most vitally personal of all cases, we really should thank our lucky stars that, at least in one cogent interpretation, certain marks of our ancestral incompetence—persistently small size in a dinosaurian world, for example—suddenly turned into a crucial and fortuitous advantage under the different rules of K-T impact, while the former source of triumph for dinosaurs may have spelled their doom under these same newly imposed rules. To be sure, this speculative scenario only references a particular event, and its much later impact upon the possibility of origin for one odd species. Yes, of course, we seek general theory as the goal of science, not the explanation of such odd particulars. But this tale, above all, happens to be *our* particular, and the most precious source of our possibility. Enough said.

## THE PARADOX OF THE FIRST TIER: TOWARDS A GENERAL THEORY OF TIERS OF TIME

Although I have tried to present a critical exegesis of both the sources and the logic of Darwin's argument for general progress as a broad statistical and accumulating consequence of biotic competition in a crowded world, I must also confess that Darwin's rationale and development seem basically sound to me. And yet, I do not think that the actual history of life maps the expectations of his argument, thus leaving us with a central evolutionary paradox. Nearly all vernacular understanding of evolution, and much professional interpretation as well, would deny my last statement and affirm a broad signal of such progress. Although I do appreciate the appeal of an argument based on "ratcheting" for an accretion of levels in complexity through time—for certain kinds of more elaborate aggregation and integration cannot viably disassemble once conjoined, although any level can be lost by extinction (my interpretation of the claims presented by Maynard Smith and Szathmary, 1995, for "the major transitions in evolution")—I fail to find any rationale beyond anthropocentric hope and social tradition for viewing such a sequence as a fundamental signal, or an expression of the main weights and tendencies in life's history. After all, two of the three great boughs on life's phyletic tree remain prokaryotic, while all three multicellular kingdoms extend as twigs from the terminus of the third bough. If we regard intellectual skepticism against anthropocentrism as a worthy cause, I don't see how we can deny that the persisting domination, and continued rosy prospects, of prokaryotes epitomize the primary aspect of life's history (see Gould, 1996a, and pp. 897–901 of this book for an elaboration of this argument). And if we must honor animals in our parochialism, arthropods surely hold an enormous edge over vertebrates.

I have referred (Gould, 1985a) to this failure of Darwin's sensible argument to impress itself upon the actual history of life as "the paradox of the first tier"—thus also giving away my preference between the two major possibilities for resolution (see forthcoming discussion, and my defense of nonfractal "tiers" of time with different predominating causes and patterns, with Darwin's good argument operating only at the first tier, and unable to "push through" to impose a pervasive vector upon the history of life). If we accept my characterization of this situation as a paradox, then we must ask why a valid argument for progress, based upon the uniformitarian extrapolation through geological time of the microevolutionary mechanics of natural selection, fails to make its anticipated mark upon earthly phylogeny. As an abstract issue in logic, two "pure" end-member solutions can be specified because the basic proposition includes two assertions, with the falsification of either being sufficient to destroy the full argument even if the other assertion remains entirely valid. (Needless to say, the actual resolution of the paradox in our messy "real world" will, no doubt, combine aspects of both with numerous other factors as well.)

The first assertion holds that natural selection, operating primarily by biotic competition under ecological plenitude, will indeed generate a bias towards "progress" by granting a statistical edge to general mental and biomechanical improvement. The second assertion then holds that this microevolutionary edge should accumulate smoothly, through time and up levels, to yield the general vector of life's progress that Darwin described in his geological chapters (see quotes on pp. 467–479). Thus, refutation at the first end-member accepts full fractality and extrapolation from microevolutionary generality to macroevolutionary pattern, but denies that micro-evolution works in Darwin's required manner. The opposite refutation of the second end-member fully accepts Darwin's argument for the operation of microevolution, but denies the extrapolationist premise of scaling in continuity to impart the same theme of progress to higher levels, and thus to life's broadest history.

If only for the obvious reason that Darwinian selection has been so over-whelmingly validated, both empirically and theoretically, as a dominant mechanism of evolutionary change in populations at generational time scales, the first end-member refutation has not garnered much consideration or support, although Raup's principled exploration (1991b, 1992, 1996), to be discussed just below, deserves considerable respect and attention, if only to remind us that apparently absurd propositions may hold far more plausibility than our knee-jerk reactions allow. But the second end-member, based on the opposite premise of non-extrapolatability for a Darwinian mechanism fully valid at its primary, smallest scale, tier of time, has been—correctly in my judgment—the standard, if usually inchoate or inarticulated, view of paleontologists uncomfortable with the full sufficiency of microevolutionary principles to explain the entire history of life. The scaling of time's tiers, in this second position (and to cite a pair of metaphors), is neither fractal nor isometric.

Before defending this second position as the key to solving the paradox of the first tier, I should say that, despite some initial enthusiasm as our profession first embraced the renewed respectability of catastrophism, I doubt that any paleontologist would now defend a dichotomous division of time into two tiers—an ordinary or "background" world, granted entirely to Darwinian mechanisms, between catastrophic episodes; and a few, but markedly effective, momentary disruptions of this generality in episodes of mass extinction. This model of an alternation between background and wipeout regimes presents far too simple a picture, while admittedly capturing the central principle of higher and rarer modes that must be titrated with ordinary Darwinism to generate the full pattern. As with the cognate theme of hierarchical levels of selection, explored and defended throughout this book, we must try to render time as a series of rising tiers, each featuring distinctive modes of evolution, and each functioning as a gatekeeper to bar full passage, a ringmaster to add new acts to the mix, and a facilitator to alter, in interesting ways, the expression of conventional Darwinism in its domain.

Time's higher tiers, in other words, introduce causes and phenomena to expand the modalities of evolution, not to restrict or refute the powerful Darwinian forces that rise from the organismal level in ecological time, but do not maintain their pervasive sway in these broader realms. Again, as with rising hierarchical levels of structure, our increasing understanding of evolution at time's upper tiers establishes the architecture of a larger, sturdier, and interestingly different Darwinian edifice, and does not operate as a demolition team for razing old domiciles and then building some hurried heap of superficial appeal, thrown up without a foundation, and therefore destined to topple as soon as the inevitable winds of fashion change their capricious course.

### *Fractal iconoclasm scaled down*

In evolutionary theory, the canonical observation of inverse relationship between frequency of occurrence and intensity of effect has generally been used to deny the force, or even the existence, of the largest pulses as too rare to matter, even in the amplitude of geological time. (Fisher's citation, invoked to privilege small events by their overwhelming frequency, opens the general argument in his *Genetical Theory of Natural Selection* (see pp. 508–514), and remains the classical example in Darwinian literature.) Interestingly, the application of this inverse relationship to assess the range of catastrophic models had proceeded in precisely the opposite manner—from accepting the magnitude of a proven single event (the K-T extinction) at maximal scale, and then extrapolating *down* to ask whether much more common small events of the same type could yield enough oomph and frequency to provide an alternative account for our Darwinian preferences at the ecological scale of natural selection's supposed and undoubted domination.

My dearest paleontological colleague David M. Raup has delighted me throughout my career, and kept an entire profession on its collective toes, by acting as Peck's Bad Boy to express outrageous and unthinkable ideas in the form of testable hypotheses. I confess that I have never quite figured out whether Dave believes in, or would place more than a minuscule probability upon, the hypotheses behind several of his tests; or whether he just loves to play the role that the Church wished to assign to Galileo—that is, to present the almost surely untrue as a hypothetical claim in mathematical form, thereby to sharpen our empirical and logical skills in finding best arguments for truthful propositions. Only once did I ever win an argument against one of his null hypotheses, or "beans in a bag" models, for random worlds composed of identical objects. Raup held for several years (before the Alvarez data convinced him of the reality of a K-T event) that mass extinctions might be entirely artifactual, representing only an occasional extreme in sampling from an actual set of equally-sized extinction pulses. "But Dave," I would say in frustration, "perhaps the Permo-Triassic grandaddy of all extinctions *can* be rendered statistically as no more than an extreme sample from a uniform pool. Still, you can't deny that, on Earth, the Triassic organisms that actually reappear are so different from their Permian forebears. Something 'real' must have happened then." I think that he finally acquiesced to this point!

Raup developed his extreme model as a thought experiment because mass extinction by bolide impact might, at least in principle, be regarded as random in both of Eble's (1999) senses previously called, in this particular context, the "random" and "different rules" models (see p. 1314)—that is, either truly so in the formal statistical sense, or only so in the vernacular sense that reasons for differential success in such catastrophically altered moments must be exaptive with respect to Darwinian bases for evolving the relevant features in the first place, and in background times. Raup therefore posed the following sly question: if this broad sense of randomness applies to the largest event, and if the famous inverse curve of frequency *vs.* effect implies a continuity in causality as well, then maybe we should extrapolate *down* and reconceptualize the smaller and much more common extinctions as equally random in their *raison d'être* (in opposition to the kneejerk view that local Darwinian determinism rules for ecological moments in competition by wedging, whereas randomness can only enter at higher levels, where the speed and intensity of an input can catch a *Bauplan* unawares).

Such a model of fractal continuity in extinction, triggered by sudden impact at all scales and levels, might be conceptualized as a "field of bullets" (Raup, 1991a)—with agents of destruction raining from the sky and death as a random consequence of residence in the wrong place at the wrong time (when each member of the population expresses exactly the same properties as any other, and with each independent of all others). One might conceptualize the agents of catastrophic destruction (the field of bullets) in either of two ways:

First, the random shooter in the sky may, for each episode of the game, release a varying number of simultaneous bullets of identical form, with continually decreasing probability of a larger number (following the inverse curve of frequency and magnitude). Thus, as a lazy or compassionate character, he hurls only one bullet most often, but must occasionally release such a dense load that few inhabitants can escape annihilation. Thus, the vast majority of moments feature none, one, or just a few extinctions, easily equated with our usual idea of a "background," but with causes just as random in their "selection" of targets, and just as sudden in their effects, as in the largest event of mass extinction. Once in a great while, following the dictates of the same distribution and its implied continuity in causality, bullets reach the extirpating density of the nearly continuous sheet of arrows launched by the English longbowmen at Agincourt in 1415, where the French suffered some 6000 deaths to an English handful. Second, the random shooter might always release the same number of projectiles, this time following the inverse curve by using smaller bullets (covering a tiny percentage of territory) most of the time, and large bombs (flattening most of life's field) only rarely, at the much lower frequency of mass extinctions.

In practice, Raup (1991b, 1992, 1996) derived a "kill curve" (his chosen term) from the empirical compendium of generic level extinctions per geological stage developed by J. J. Sepkoski, and widely used by the entire "taxon counting" school of modern paleobiology (see Figure 12-1). The frequency distribution, based on Sepkoski's data, assumes the expected inverse form, monotonic and strongly right skewed, with about half the 106 geological units (with their average duration of 6 million years) plotting in the leftmost interval, and showing less than 10 percent extinction of genera.

Raup's kill curve (Figure 12-2) follows the familiar form (the inverse relationship of frequency and magnitude again) that generates such vernacular concepts as the "100 year flood"—so often, and so tragically, misunderstood by so many people who, for lack of education to undo one of the most stubborn of our inherent mental foibles, do not grasp the basic meaning of probability and assume, for example, that they may safely build their house on the floodplain because the 100 year deluge swept through the region five years ago and therefore cannot recur for almost another century. Raup freely admits that the orderly and monotonic form of the kill curve does not specify any style of causality by itself—and, most relevantly and especially, does not permit an inference about random effects at lowest levels merely because we can advance a powerful case for this style in one event at the highest level.

Nonetheless, the numerical specifics of Earth's particular curve does im-

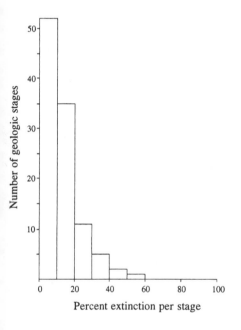

12-1. From Raup, 1996. Typical negative correlation of intensity of extinction and frequency of occurrence, with small extinctions common and large events rare.

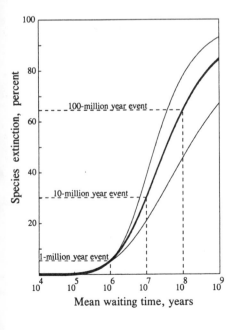

12-2. From Raup, 1996. The "kill curve" derived from Figure 12-1, showing expected waiting time between extinction events of various magnitudes—with longer waiting times for larger events.

pose some restriction of possibilities, and does suggest some causal inferences. Raup notes (1996), in particular, that if each species, following Darwin's explicit claims (cited at the beginning of this chapter), pursued its own
independent history of origin, expansion, reduction and death in terms of its
own special competitive prowess *vs.* other unique species in singular faunas—
thus implying that no general cause can impact all species at once (or at least
that such general causes only nudge, and do not set or determine, the overall
pattern)—then the kill curve, while continuing to obey its predictable monotonic form, would never reach such high percentages of death in its rarest upper episodes. Such concentration in mass extinction does imply a coordinated
cause of some sort. Raup writes (1996, p. 422):

> If each genus died out independently of the others, the probability of
> producing a range of from near zero to 52 percent extinction in 106 six
> million year stages would be negligible. This means that pulses of extinc-
> tions of genera must be connected in some way . . . because of common
> factors, such as ecological interdependence or shared physical stress. We
> thus see a picture of episodic extinction wherein the more intense an ex-
> tinction episode, the rarer it is. To describe extinction only as back-
> ground or mass extinction, as is commonly done, is to hide much of the
> structure of the extinction phenomenon.

### The nonfractal tiering of time

Strict Darwinism implies continuity in the style and causal structure of
change from successive generations in populations to the waxing and waning
of faunas across geological eras. An alternate construction of time as a series
of discrete tiers, or at least of rising "regions of coagulation" that pull phenomena away from boundaries and towards more central nucleating places
—with each tier then featuring different weights and styles, or even truly distinct modes, of causality—would seriously challenge the crucial extrapolationist premise of Darwinian logic (the third leg on my tripod of support). Although I know the quibbles and inconsistencies, and I recognize that many of
my neontological colleagues regard such problems as central flaws worth a
lifetime's research, I have no personal quarrel with Darwin's argument that a
vector of general progress would pervade the history of life if all scales of time
record the competitive styles of natural selection supposedly prevailing at the
first tier of anagenetic change within the history of single populations. We
should therefore take firm notice, and regard as highly paradoxical, the failure of life's history to feature such a vector as an obvious organizing principle
and predominant signal of phylogeny.

I reject, for the most part, the Raupian solution of fractality in time, with
Darwinian inefficacy throughout. Instead, I favor the alternative view that
Darwinism basically works as advertised in its own realm at the first tier of
time, but cannot "push through" to impose its characteristic signal upon processes and phenomena of higher tiers. Such a proposal implies a very different
attitude towards time and change—an attitude inspired and encouraged by

two controversial topics of the last quarter of the 20th century: punctuated equilibrium and catastrophic mass extinction, our two most notorious hypotheses about the non-homogeneity of time's principal modes.

In proposing that we conceptualize time as a rising set of tiers, I do not argue (thus hoping to forestall, by this explicit statement, the same misunderstanding kindled by the debate about hierarchical levels of selection) that entirely new, and truly anti-Darwinian, forces emerge at each higher tier. I am quite content to allow that no fundamental laws of nature, and no entirely novel causes or phenomena, make their first appearance in larger slices of time. But, at these broader scales and intervals, the known principles of genetics, and the documented mechanisms of selection, may operate by distinct and emergent rules that, as a consequence of time's tiering, cannot be fully predicted from the operation of the same kinds of causes at lower levels. The logic of this critique flows from potential fallacies in Darwinian assumptions about extrapolation across time's putative smoothness and causality's supposed invariance. (Selection on species-individuals, for example, follows all the abstract and general principles required by Darwinism for this central mechanism of the general theory, but the modes and regularities of selection at this higher level differ strongly, and cannot be predicted, from our canonical understanding of Darwinian organismal selection within populations—see Chapter 8.)

The dilemma, and eventual insufficiency of the Modern Synthesis for paleontology lay in this third crucial Darwinian claim that all theory could be extrapolated from the first tier, thus converting macroevolution from a source of theory into a pure phenomenology—a body of information to document and to render consistent with a theoretical edifice derived elsewhere. But if the tiers of time, and the hierarchy of life's structure, create pattern by emergent rules not predictable from processes and activities at lower tiers, then paleontology will add insights, and augment theory, without contradicting the principles of lower tiers.

We need to distinguish between the stability and continuity of causal principles (the spatiotemporal invariance of natural law in our usual jargon) and the potentially discontinuous (and disparate) expression of these principles across a spectrum of time that may be strongly stepped (as nucleating points attract surrounding events, and as nature's inherent structure clears out space at the unstable positions between), rather than fully and smoothly continuous. I began my original article (Gould, 1985a, p. 2) on the tiering of time with an admittedly humble analogy that may still help to clarify the important principle of noncontradiction between structural tiering and a unitary order of underlying entities and causes:

In the glory days of Victoria's reign, when a pound bought more than five American dollars, the English economy operated on two distinct tiers. The working man, paid weekly for his labor and without benefit of banking or hope of accumulation, might pass his entire life without ever seeing a pound note, for he received his wage in shillings and pence and its total never reached a full pound. Meanwhile, bankers in the City of London transacted the world's imperial business in pounds. India today operates on two similar and largely noninteracting tiers—the 100-rupee notes (about ten dollars) of the hotel shops and the bustling economy of the bazaars, where 10 rupees buy at least one of anything and no one ever sees (or could cash) a 100-rupee note.

Our world of times and amounts is not always continuous. Its metrics usually extend smoothly from one end to the other (shillings did grade to pounds and rupees are rupees), but its activities are often sharply concentrated in definite regions of a potential spectrum, with large open spaces between. Systems often drive in opposite directions away from break points; location on one or another side of a threshold inevitably pushes toward an equilibrium far above or below.

I do not know if we should make a formal attempt to specify a definite number of tiers in time—as the principle seems sound, whereas definite criteria for precise designation seem elusive. (I remain far more comfortable and confident with the concept of hierarchy in levels of selection than with the increasing scale of tiers in time—for the genealogical hierarchy can be elucidated with fair precision by criteria of Darwinian individuality (see Chapter 8), whereas time's tiers lack such a unitary concept for coordination. For the same reason, I do not follow my closest colleague Niles Eldredge's attempt to identify dual or parallel hierarchies of genealogy and ecology, for the genealogical units enjoy clear definition, whereas ecological levels, like time's tiers, lack a coherent *fundamentum divisionis* for unambiguous specification.)

But if I may make an analogy to the geological time scale for the Phanerozoic eon, we may at least be confident about the few broadest tiers (the three geological eras in my analogy), even while we argue about some boundaries and specifications for the smaller units within these broad domains (geological periods in my analogy). Darwinian organismal selection, with an overall statistical edge granted to biotic competition in crowded ecosystems, dominates the first tier of anagenesis within populations during ecological time. If the fractal principle of Darwin's central belief in extrapolation held, then the norms of this process at the first tier would accumulate in a linear fashion through time to yield a history of life with the same basic form and causal structure, but scaled up in smooth continuity to generate all patterns of phylogeny. I have designated life's failure to display this pattern, particularly its disinclination to feature a clear signal of overall progress, as the "paradox of the first tier."

As stated above, I would resolve this paradox by accepting Darwin's basic view of pattern and causality at the first tier, but then asserting that distinct modes of change and balances of causes, operating at higher tiers, introduce enough systematic difference to cancel out the first tier's vector. If the first tier governs anagenesis within populations, then the second tier features trends within monophyletic clades. Darwinian tradition holds (as discussed throughout this book, but primarily in Chapters 2 and 9) that such phyletic trends be explained, by simple extension and extrapolation, as adaptive anagenesis carried further through time's geological fullness. I shall not here rehearse the lengthy arguments and documentation of Chapter 9, but I believe that punctuated equilibrium, as the dominant pattern and process of the second tier during millions of "background" years between pulses of mass extinction, undoes the first tier's vector of progress by supplying a radically different general explanation for phyletic trends: the differential success of certain species within clades, with each species treated as a stable individual during the several million years of its average geological existence. Since the reasons for differential success of species extend so far beyond (while also including) the traditional citation of adaptive biomechanical advantages for constituent organisms; and since several of these reasons tend to run orthogonal, or even counter, to general expectations of organismal progress; the crafting of trends by punctuated equilibrium derails extrapolation from the first tier, thus providing the second tier with a different set of explanations for its central phenomenon of cladal trending.

This barrier at the second tier would be sufficient, by itself, to resolve the paradox of the first tier, but even the second tier's rules for trending do not accumulate through broader realms of time to explain, by smooth extrapolation, the patterns of waxing and waning for major taxonomic groups. That is, the causes of cladal trends through long geological intervals of "normal" time do not accumulate to patterns of success and failure through the full Phanerozoic range. For, as the preceding section of this chapter documented, the "random" and "different rules" models of differential success in episodes of catastrophic extinction then derail the phyletic trends of the second tier for the same basic reason that these cladal trends previously undid the anagenetic accumulations of the first tier: that is, by introducing new patterns and rules at these rare moments of maximal impact. Thus, in summary, adaptive anagenesis of a single lineage at the first tier cannot be extrapolated to cladal trends within a monophyletic group of species at the second tier; and these cladal trends of the second tier then cannot be extrapolated through episodes of mass extinction to explain patterns of differential success for life's higher taxa throughout Phanerozoic time. Punctuated equilibrium undoes anagenesis, and catastrophic mass extinction derails punctuated equilibrium.

As with the workings of hierarchical levels in selection, the effects of adjacent tiers of time may interact in all possible ways. Higher tiers do not automatically counteract lower tiers. Adjacent tiers may also reinforce each other to intensify a signal in life's history—as I illustrated, for example, in reporting Jablonski's claim (1987) that species selection on species-rich subclades at the second tier often reinforces the adaptive advantages gained by the organisms of these species in Darwinian selection at the first tier. Similarly, if only for exaptive merits at the higher tier, diatoms flourished by virtue of the same features at both the first and third tiers (if my argument on page 1319 holds)—as the evolution of mechanisms for dormancy proved adaptive at the first tier in aiding survival during high-latitude months of darkness and in times of low silica between episodes of upwelling; whereas the same feature may have favored the persistence of diatom species during a prolonged period of darkness imposed by a global dust cloud excavated in bolide impact, the key ingredient advocated by many researchers for the killing scenario of the K-T event.

But adjacent tiers may also act in an orthogonal manner, with invocation of the "random" model at the third tier as an obvious general case—for truly random differential survival must run orthogonal to deterministic Darwinian reasons for evolving the clade's distinctive traits at the first tier, or equally deterministic (but different) reasons for establishing the defining features of cladal trends by punctuated equilibrium at the second tier. Finally, the evident possibility of opposing reasons at adjacent tiers has sparked our interest in catastrophic mass extinction from the start—with dinosaurian superiority over mammals maintained at the first and second tiers throughout Mesozoic times, and mammalian success then achieved by the "different rules" model at the third tier of differential passage through mass extinction, putatively based upon the very features that marked the first and second tier failures of mammals for 130 million previous years.

Although limited space and personal competence prevent my proceeding beyond this sketchy and cartoonish model of three tiers, I suspect that future work will identify several inhomogeneities and subtiers, particularly between the history of independent and individual clades at the second tier and the coordinated impact of environments upon entire biotas at the third tier. Several intermediate modes and processes, affecting groups of species during times of unusual externalities in particular geographic regions, but not global biotas at catastrophic moments, must "intervene" between the pure influence of punctuated equilibrium upon a single clade and the full impact of worldwide catastrophe upon a global biota. For example, the model of coordinated stasis (discussed on pp. 916–922) argues that cladal trends do not always maintain the implied freedom of punctuated equilibrium to proceed independently, and in an unconstrained manner by differential success in the generation of new species at the second tier, but will often be subject to a form of community stasis that must first be broken by disruptions smaller than mass extinction, and resident within its own tier. Similarly, between the first and second tier, ordinary anagenesis often cannot "push through," even to the point of disruption by punctuated equilibrium at the second tier, because the ecological communities that set the anagenetic regime for a single species become disrupted on a scale of hundreds of thousands of years by the climatic fluctuations of Milankovitch cycles that break up communities and quickly disperse their elements into new arrangements in different places.

To conclude this section with a historical example of the profound distinction between traditional untiered views of selective domination smoothly scaled up to all times and magnitudes, and the alternative nonfractal concept of a much broader range of potential outcomes engendered by interaction among the characteristic modes and processes at different tiers, consider the famous conclusion from Hatcher, Marsh and Lull's 1907 monograph on the anatomy, taxonomy, evolution and extinction of ceratopsian dinosaurs. In ending their text with a section on "probable causes of extinction," these three great paleontologists only allowed themselves to consider the two standard hypotheses of the conventional uniformitarian school of pure extrapolation, complete with its Darwinian assumption that vectors of general progress must pervade such systems (in this instance by the replacement of dinosaurs with mammals, for reasons of warm-blooded superiority in anatomical design and mentality). In either case, they argue, the extinction must spread through an extended time in several episodes, step by step. The two reasons themselves—biotic competition from a superior group, and adaptive failure in the face of changes in the physical environment imposed by ordinary terrestrial forces (acting at greater than their usual intensity, but still in their characteristic mode)—cause inferior groups to peter out as life gets better and better through each geological day:

> Several theories have been advanced as to the probable causes of extinction of the Ceratopsia . . . It seems that animals of another race, or hordes of creatures which emigrated from another region, would be more likely to exterminate their predecessors. The mammals fulfill the requirements of a new foe, and the development of the frill in the Ceratopsia has been considered as meeting the necessity for a better protection of the neck blood vessels from the weasel-like attack of small but bloodthirsty quadrupeds. Another notion . . . was that the Cretaceous mammals sought out the eggs of dinosaurs and destroyed them—Cope even going so far as to suggest the Multituberculata, with their long, sharp anterior teeth, as the probable offenders . . .
>
> By far the most reasonable cause . . . seems to be changing climatic conditions and a contracting and draining of the swamp and delta regions caused by the orographic upheavals which occurred toward the close of the Cretaceous. The Ceratopsidae and their nearest allies, the Trachodontidae, both highly specialized plant feeders, were unable to adapt themselves to a profoundly changed environment because of this very specialization, and, as a consequence, perished.
>
> That the Ceratopsia made a gallant struggle for survival seems evident, for they lived through the first series of upheavals at the close of the Laramie and also the second series at the close of the Arapahoe, which were accompanied by great volcanic outbursts in the Colorado region; but the changes accompanying the final upheaval which formed most of the great western mountain chains and closed the Mesozoic era gave the death blow to this remarkable race (Hatcher, Marsh and Lull, 1907, p. 195).

But new perspectives from two higher tiers have reversed this conventionality, particularly for a vigorous group like the ceratopsians. At the second tier, these particular dinosaurs, among all other subclades, remained in maximal flourish of expansion and speciation right to the close of Cretaceous times, and surely cannot be marked as doomed, or even in decline with respect to mammals, during such a period of maintenance and expansion by copious speciation, or introduction of new Darwinian species-individuals at this macroevolutionary level. And then, with catastrophism reintroduced at the third tier as a hypothesis of renewed respectability, the ceratopsians died, in concert with all other dinosaurs (leaving the anatomically divergent birds as sole survivors of their monophyletic clade), when an unpredictable paroxysmal change radically altered earthly environments and drove several groups to extinction through no adaptive failure of their own, while imparting fortuitous exaptive success to creatures that had lived throughout the long reign of dinosaurs, and never made any headway towards displacement, or even towards shared domination with one of the most successful vertebrate groups in the history of life.

## An Epilog on Theory and History in Creating the Grandeur of This View of Life

This comfortable view of ceratopsian (and all dinosaurian) demise engendered smug feelings among evolutionists and paleontologists of previous generations for two reasons, both lamentable. First, the implied pattern of a lawlike and predictable vector of progress, culminating in mammalian victory over dinosaurs and crowned by the eventual evolution of a single conscious scribe within the triumphant clade, validated the oldest social traditions and deepest psychological hopes of Western cultures—the strongest possible reason for turning our brightest beacon of skepticism upon so congenial a conclusion defended by so little beyond emotional satisfaction. Second, the supposed underpinning (and virtual guarantee) of this happy result by a putative general law of nature, enhanced the meaning and centrality of the particular outcome as a dictate of universal science, not merely a fortuitous circumstance, or even a special dictate of an arcane controlling power whose comprehensive reasons can never be entirely known (and whose future actions can therefore never be fully anticipated).

If, however, as the central thesis of this book maintains and the *Zeitgeist* of our dawning millennium no longer rejects, we cannot validate the actuality of mammalian success by general principles, but only as a happy (albeit entirely sensible) contingency of a historical process with innumerable alternatives that didn't happen to attain expression (despite their equal plausibility before the fact), then we must face the philosophical question of whether we have surrendered too much in developing a more complex and nuanced view of causality in the history of life.

What is science, after all, if not the attempt to understand the natural world by explaining its phenomenology as causal consequences of spatio-temporally invariant laws? We may need to know the particularities of a given set of initial conditions in order to infer the details of later states reached by the operation of these laws, but we do not regard the resolution of such details as essential or causal components of the explanation itself. (I confess that, after 30 years of teaching at a major university, I remain surprised by the unquestioned acceptance of this view of science—which, by the way, I strongly reject for reasons exemplified just below—both among students headed for a life in this profession, and among intellectually inclined people in general. If, as a teacher, I suggest to students that they might wish to construe probability and contingency as ontological properties of nature, they often become confused, or even angry, and almost invariably respond with some version of the old Laplacean claim. In short, they insist that our use of probabilistic inference can only, and in principle, be an epistemological consequence of our mental limitations, and simply cannot represent an irreducible property of nature, which must, if science works at all, be truly deterministic.)

Natural historians have too often been apologetic—but most emphatically should not be—in supporting a plurality of legitimately scientific modes, including a narrative or historical style that explicitly links the explanation of outcomes not only to spatiotemporally invariant laws of nature, but also, if not primarily, to the specific contingencies of antecedent states, which, if constituted differently, could not have generated the observed result. As these antecedent states are, themselves, particulars of history rather than necessary expressions of law, and as subsequent configurations can cascade in innumerable directions, each crucially dependent upon tiny differences in the antecedent states, we regard these subsequent outcomes as unpredictable in principle (as an ontological property of nature's probabilistic constitution, not as a limitation of our minds, or as a sign of the inferior status of historical science), however fully explainable they will become, at least in principle, after their occurrence as the single actualized result among innumerable unrealized possibilities.

In order to gain entry into the hallowed halls of science (often defined, far too parochially, in terms of quantified predictability as a *summum bonum*), natural historians have often been too willing to accept an inferior status, based on the principled unpredictability of their largely contingent phenomenology, in order to gain recognition as practitioners of science at all. (For in this Laplacean construction, the frequency of probabilistic inference correlates directly with the weakness of scientific apparatus—for we live, under this fallacy, in a genuinely deterministic world, and the extent of our recourse to probability therefore maps our relative inability to define the true determinisms of any particular process.)

Wise natural historians, with Darwin himself as a most articulate champion, have always rejected this disparagement, and its attendant relegation to inferior status—and have defended historical explanation, with its claims for contingency and the ontological status of probabilistic structure, as a fascinating, even inspirational, property of complex nature. Such contingency, moreover, in no way compromises the power of legitimate explanation, for our inability to predict before the fact only records the true character of this complexity, whereas our subsequent capacity to explain after the fact can reach the same level of confidence as any physical resolution under invariant law, provided that we can obtain enough factual detail about antecedent states to resolve their causal relation to the observed outcome. In fact, and on this very subject, Darwin made a striking exception to his astonishingly calm and genial temperament, and permitted himself a rare excursion into satire and sharp criticism. Moreover, he expressed these partisan thoughts in the most prominent of all possible places—the very last line of the *Origin of Species*, where he rejected the traditional claims of quantitative physical science to represent the apotheosis of sophistication, and awarded higher honor to his own discipline of natural history and evolutionary biology, as embodied in the gnarly and meandering icon of the luxuriantly, but contingently, branching tree of life.

But Darwin, ever so sly in his Victorian propriety, enshrouded this terminal line in such a flourish of benign prose that most readers, for more than a century, have construed his famous closing sentence as a poetic metaphor, intended only to ornament a revolution with a coda of ecumenical kindness. In fact, I am convinced that Darwin conceived this finale primarily as a mordent critique of the haughtiness and narrowness of physical scientists in debasing natural history, and as a defense of the greater interest and relevance of his own chosen profession. (I need only recall Darwin's extreme discomfort at Lord Kelvin's arrogant dismissal both of natural selection in particular, and Lyellian geology in general—see Chapter 6 for details. This famous incident should remind readers that Darwin may well have harbored angry feelings about the pretensions of mathematical physics and celestial mechanics to superior status over natural history among the sciences.)

Note how Darwin contrasts the dull repetitiveness of planetary cycling (despite the elegance and simplicity of its quantitative expression) with the gutsy glory of rich diversity on life's ever rising and expanding tree. Darwin even gives his metaphor a geometric flavor, as he contrasts the horizontal solar system, its planets cycling around a central sun to nowhere, with the vertical tree of life, starting in utmost simplicity at the bottom, and rising right through the horizontality of this repetitive physical setting towards the heavenly heights of magnificent and ever expanding diversity, into a contingent and unpredictable future of still greater possibility: "There is grandeur in this view of life . . . [and] whilst this planet has gone cycling on according to the fixed law of gravity, from so simple a beginning endless forms most beautiful and most wonderful have been, and are being, evolved" (1859, p. 490).

Throughout the *Origin of Species*, Darwin stresses the beauty, and especially the simplifying power, of historical explanation in evolutionary science as a cardinal feature of his view of life (as opposed to other versions of evolution). In one of the most striking examples of "less is more" in the history of science (eloquently and elegantly more in this case), Darwin continually emphasizes that the age-old perception of a "natural system" among organisms had always presumed a basis of order that must be complex, arcane and abstract; intricately numerical and geometrically lawlike; or divinely ordained, and therefore of literally highest and deepest significance. Louis Agassiz, Darwin's near contemporary and the last truly sophisticated scientific creationist in biological theory, had even argued (see Chapter 3 for an exegesis of his view) that since each species represents a single divine idea incarnated on earth, the "natural system" of taxonomic order among species must literally record the character of God's mind, for taxonomy discovers the principles of higher structuring among God's own unitary items of thought.

But Darwin's profound, and wonderfully simple, alternative cuts through centuries of assumptions about the unresolvable depth and complexity of natural order with a breathtakingly direct and concrete resolution: the "natural system," or taxonomic order among species, just records the history of an unbroken genealogical sequence of historical descent, the arborescent topology of the tree of life. The height of arcane mystery becomes a record of simple history: "As all the organic beings, extinct and recent, which have ever lived on this earth have to be classed together, and as all have been connected by the finest gradations, the best, or indeed, if our collections were nearly perfect, the only possible arrangement, would be genealogical. Descent being on my view the hidden bond of connection which naturalists have been seeking under the term of the natural system" (1859, pp. 448–449). Moreover, this conclusion has important operational consequences, not just philosophical implications. If, for example, life's order records the connected pathways of a contingent and "messy" history, then a variety of formerly popular numerological schemes (like the "quinarian system" based on organizing taxa into rigid and invariable groups of five for each higher level) cease to make scientific sense.

Over and over again, throughout the *Origin*, Darwin stresses that, for a large class of problems about species and interacting groups, answers must be sought in the particular and contingent prior histories of individual lineages, and not in general laws of nature that must affect all taxa in a coordinated and identical way (1859, p. 314):

> I believe in no fixed law of development, causing all the inhabitants of a country to change abruptly, or simultaneously, or to an equal degree . . . The variability of each species is quite independent of that of all others. Whether such variability be taken advantage of by natural selection, and whether the variations be accumulated to a greater or lesser amount, thus causing a greater or lesser amount of modification in the varying species, depends on many complex contingencies—on the variability being of a beneficial nature, on the power of intercrossing, on the rate of breeding, on the slowly changing physical conditions of the country, and more especially on the nature of the other inhabitants with which the varying species comes into competition.

Interestingly, one of the strongest modern critics of historicism in evolutionary science (Kauffman, 1993, as extensively discussed in Chapter 11), has explicitly identified the contingent status of the branching tree of life as his primary source of discomfort with Darwin's system (Kauffman, 1993, pp. 5–6 in a section entitled "evolutionism, branching phylogenies, and Darwin"). Kauffman, of course, does not deny that the icon of branching correctly expresses the topology of life's history (at least for eukaryotic organisms). But he does argue, in the tradition of his intellectual mentor D'Arcy Thompson (see Chapter 11, pp. 1182–1208), that our Darwinian tradition places too much emphasis upon the particular history of a lineage to explain various evolutionary features that should, in his judgment, be encompassed under timeless and general laws as expressions of universal physics, and not explained as contingencies of unpredictable and individual pathways. Thus, although Darwin's own commitment to contingency has been underemphasized, or even unrecognized, by his later followers (largely in their own attempt to win more prestige for evolution under the misconception that science, in its "highest" form, explains by general laws and not by particular narrations), I am scarcely alone in identifying this central (and, in my judgment, entirely laudable) aspect of Darwin's view of life.

Kauffman, on the other hand, makes the same identification as a sharp critic. His single page of discussion, devoted to doubts about particularism rooted in the tree of life, cites a form of the word "branching" no less than twenty times, a sure mark of Kauffman's discomfort with this model, and his good insight about an appropriate target for criticism. Kauffman writes (1993, p. 5), for example:

> The onset of evolutionism brought with it the concept of branching phylogenies. The branching image, so clear and succinct, has come to underlie all our thinking about organisms and evolution . . . With the onset of fullblown evolutionism and Darwin's outlook based on branching phylogenies, the very notion that biology might harbor ahistorical universal laws other than "chance and necessity" has become simple nonsense. Darwin's ascension marks a transition to a view of organisms as ultimately accidental and historically contingent. Our purposes have become analysis of branching evolutionary paths and their causes on one hand, and reductionistic unraveling of the details of organismic machinery accumulated on the long evolutionary march on the other.

It is important to recognize, and I'm sure that Kauffman and other critics would concur, that this debate between immanent vs. narrative styles of explanation contrasts different modes of factual knowing, and that *both* alternatives stand firmly opposed to trendy and nonsensical claims about the relativity of empirical "truth" in the light of social embeddedness for any transiently privileged intellectual procedure. When a champion of contingency (for the large chunk of nature properly falling under the aegis of narrative explanation) argues that he can explain with rigor after the fact what he could not have predicted in principle before the fact, he presents his best judgment about the empirical structure of historical complexity. Moreover, he does not confess thereby either any limitation imposed by an inferior form of science, or any irreducible subjectivity engendered by the admittedly ineluctable interaction of human perception and mentality with external "reality" in all efforts to understand nature's ways.

I would rather, and in the opposite direction, contend that our increasing willingness to take narrative explanations seriously has sparked a great potential gain, through admitting a pluralism of relevant and appropriate styles of explanation, in our accurate understanding of nature's wondrous amalgam of rulebound generalities and fascinating particulars. If I may return once more to Hatcher et al. (1907) on the extinction of ceratopsian dinosaurs (see p. 1331), the authors's inarticulated assumption that explanation must flow from general principles of evolutionary biology and uniformitarian geology allowed no intellectual space beyond the most conventional proposals about vectors of organic progress generated by the extrapolation of natural selection in microevolutionary time, and on climatic change wrought by (at most) some intensification of ordinary geological processes. These presuppositions, in our current judgment, led Hatcher and his colleagues to factually incorrect conclusions based on false premises about inherent dinosaurian inferiority. In this case, I would argue that the introduction of narrative perspectives—particularly the idea that the K-T event should be explained as a singularity triggered by a bolide impact, and imposing its major effects fortuitously and exaptively upon particular features evolved in other contexts and "for" different reasons—has enlarged our armamentarium of potential explanations, and has surely led to a gain in factual understanding through an increased range of permissible scientific approaches.

As a first, and overly simplified, conclusion, one might then say that more adequate explanation in the evolutionary sciences demands that we titrate these two essential metacomponents of general theory and narrative particulars, or invariant predictability and contingent singularity, to achieve any satisfactory understanding of our primary subject matter—broad phenomena that embody sufficient regularity to exemplify the basic principles of theory, but that also engage, in their explicit reference to particular times, places and taxa, enough of the fascinating detail of historical events to ensure that even the most committed generalist will learn to appreciate, perhaps even to cherish, the antecedent details that ultimately fashion the empirical objects and events through which those basic principles become manifest.

I would not argue that all *conceivable* evolutionary questions must invoke enough historical particulars to require a large contingent component in their full explanation. After all, a paleontologist could claim that he only cares about mass extinction in general, and remains entirely indifferent to the question of why trilobites died in the Permian and ammonites in the Cretaceous. But what a heartless, gutless and uncurious soul he would then become. Indeed, James Hutton came pretty close to such total unconcern with the particular histories of geological sections in his "Theory of the Earth"— see Gould, 1987b. But then, Hutton's imperviousness to the fascination of history struck his friends and contemporaries as downright peculiar and mysterious; and the longstanding impression about his opacity and unreadability stems as much from this peculiarly desiccated focus, as from any supposed inadequacy in his prose style. Even in his own time, Hutton's friends felt that he could never prevail by his own wits, and that they would have to write "ponies" to make his ideas accessible. The most famous of these guides (Playfair, 1802), one of the great works in the history of geology in itself, succeeded largely by applying Hutton's theoretical ideas to explain puzzling particulars that historically minded scholars had long found anomalous.

In any case, and as a purely factual observation about the likes and habits of practicing scientists, hardly a natural historian, dead or alive, has ever failed to locate his chief delight in the lovely puzzles, the enchanting beauty, and the excruciating complexity and intractability of actual organisms in real places. We become natural historians because we loved those dinosaurs in museums, scrambled after those beetles in our backyard, or smelled the flowers of a hundred particular delights. Thus, we yearn to know, and cannot be satisfied until we do, both the general principles of how mass extinction helps to craft the patterns of life's history, and the particular reason why Pete the *Protoceratops* perished that day in the sands of the Gobi.

This perspective on mixing immanent and historical styles of scientific explanation in the evolutionary sciences, places me, in concluding this book, into an oddly paradoxical situation, exemplifiable in four statements. *First,* I have championed the cause, and equal claim, of contingency (particularly in Gould, 1989c and 1996a) to the point of my ready identification as a proponent of this position (and with no complaint on my part, and no feeling that my critics have been unfair in any oversimplification). *Second,* the standard strategy for invoking contingency in natural history employs a device of argument legitimately deemed restrictive in its negative criterion, and surely slated for abandonment as students of contingency develop their armamentarium of positive methods and preferential means of identification—but now accepted *faute de mieux* and in acknowledgment of current practice. That is, we tend to begin with a preference for explanation by predictability and subsumption under spatiotemporally invariant laws of nature, and to move towards contingency only when we fail. Contingency therefore becomes a residual domain for details left unexplained by general laws.

(Even so sophisticated a historian as McPherson (1988), studying so richly documented an episode as the American Civil War, grants the crucial Northern victory at Gettysburg to contingency largely because all classically proposed general reasons, either for the Union's triumph in the entire war, or for success in this key battle in particular, have conspicuously failed. This being said, the host of fascinating details then evinced to explain Northern success at Gettysburg—each apparently trivial, each unpredictable, and each eminently changeable before its occurrence by the tiniest of different circumstances—seems particularly impressive and conclusive as an example of contingent explanation, even for the most important events in history. Nonetheless—for this key point remains especially troubling, and should serve as a sharp spur to both thought and action—however satisfactory the final interpretation, we might never have gotten to contingency at all unless the alternative mode of explanation, so strongly privileged *a priori,* had failed. And I need hardly remind evolutionary biologists that such approaches, based upon prejudicially ordered preferences, remain dangerous because the strengths of our (frequently unconscious) assumptions, and the "flexibilities" of nature in seeming to bow to our biases (because we push too hard, and often unawares), may preclude any access to alternatives at all, as in our failure to consider fruitful and operational hypotheses that do not ascribe organismal traits to adaptation (Gould and Lewontin, 1979).)

*Third,* Darwin himself followed this strategy in the *Origin,* opening up an admittedly considerable space for contingency when he could not devise a testable generality, or when he felt that he had reached a level of uniqueness in detail that required a similar uniqueness in antecedent generating conditions. *Fourth,* and finally, I therefore find myself in what most of my friends and colleagues—but not my own assessment of my deeper interests and concerns—might construe as the anomalous position of trying to "win back" for general theory a substantial realm of macroevolutionary phenomenology that, in its failure to emerge predictably from microevolutionary principles of strict Darwinism, would be granted (under point two) to the very realm of contingency that I have tried so strenuously to promote and enlarge.

But I embrace this apparent paradox with delight. I have championed contingency, and will continue to do so, because its large realm and legitimate claims have been so poorly attended by evolutionary scientists who cannot discern the beat of this different drummer while their brains and ears remain tuned only to the sounds of general theory. But this book—entitled *The Structure of Evolutionary Theory*—does not address the realm of contingency as a central subject, and does fire my very best shot in the service of my lifelong fascination for the fierce beauty and sheer intellectual satisfaction of timeless and general theory. I am a child of the streets of New York City; and although I reveled in a million details of molding on the spandrel panels of Manhattan skyscrapers, and while I marveled at the inch of difference between a forgotten foul ball and an immortal home run, I guess I always thrilled more to the power of coordination than to the delight of a strange moment—or I would not have devoted 20 years and the longest project of my life to macroevolutionary theory rather than paleontological pageant.

So yes, guilty as charged, and immensely proud of it! The most adequate one-sentence description of my intent in writing this volume flows best as a refutation to the claim of paradox just above: This book attempts to expand and alter the premises of Darwinism, in order to build an enlarged and distinctive evolutionary theory that, while remaining within the tradition, and under the logic, of Darwinian argument, can also explain a wide range of macroevolutionary phenomena lying outside the explanatory power of extrapolated modes and mechanisms of microevolution, and that would therefore be assigned to contingent explanation if these microevolutionary principles necessarily build the complete corpus of general theory in principle. To restate just the two most obvious examples at the higher tiers of time exemplified in this chapter: (1) punctuated equilibrium establishes, at the second tier, a general speciational theory of cladal trending, capable of explaining a cardinal macroevolutionary phenomenon that has remained stubbornly resistant to conventional resolution in terms of adaptive advantages to organisms, generated by natural selection and extrapolated through geological time; (2) catastrophic mass extinction at the third tier suggests a general theory of faunal coordination far in excess (see Raup's quantitative argument on p. 1326) of what Darwinian microevolutionary assumptions about the independent history of lineages under competitive models of natural selection could possibly generate.

In most general terms, and in order to form a more perfect union among evolution's hierarchy of structural levels and tiers of time, this revised theory rests upon an expansion and substantial reformation of all three central principles that build the tripod of support for Darwinian logic: (1) the expansion of Darwin's reliance upon organismal selection into a hierarchical model of simultaneous selection at several levels of Darwinian individuality (gene, cell lineage, organism, deme, species and clade); (2) the construction of an interactive model to explain the sources of creative evolutionary change by fusing the positive constraints of structural and historical pathways internal to the anatomy and development of organisms (the formalist approach) with the external guidance of natural selection (the functionalist approach); and (3) the generation of theories appropriate to the characteristic rates and modalities of time's higher tiers to explain the extensive range of macroevolutionary phenomena (particularly the restructuring of global biotas in episodes of mass extinction) that cannot be rendered as simple extrapolated consequences of microevolutionary principles.

And yet, as an epilog to this epilog and, honest to God, a true end to this interminable book, I risk a final statement about contingency, both to explicate the appeal of this subject, and to permit a recursion to my starting point in the most remarkable person and career of Charles Robert Darwin. Although contingency has been consistently underrated (or even unacknowledged) in stereotypical descriptions of scientific practice, the same subject remains a perennial favorite among literary folk, from the most snootily arcane to the most vigorously vernacular—and it behooves us to ask why.

Our greatest novelists have reveled in this theme, as Tolstoy devoted both prefaces of *War and Peace* to explaining why Napoleon's defeat in Moscow in 1812 rested upon a thicket of apparently inconsequential and independent details, and not upon any broad and abstract claim about the souls of nations or the predictable efficacy of Russia's two greatest generals, November and December. And *Wuthering Heights* would have lost both its story line and existence if poor Heathcliffe had not overheard, and utterly misunderstood, a conversation not intended for his ears in any case. And where would our occasionally philosophical movies find a subject if they couldn't mine the contingent fascinations of alternative and unrealized histories, either of little towns *(It's a Wonderful Life)* or of otherwise inconsequential people (the *Back to the Future* trilogy). And how could satire flourish if contingency movies couldn't generate an opposing parody *(Groundhog Day)*, based upon a day that, in its repetition, cannot be changed at all, even by the most portentous act of murder or suicide that its utterly frustrated protagonist can devise to extract himself from this nightmare of no novelty—until, of course, he finally understands the wisdom behind the only consistent definition that a philosophical determinist can possibly devise for liberty: Spinoza's conception of freedom as "the recognition of necessity."

If we then ask why literary, but not scientific, people have taken such a shine to contingency, I doubt that we need probe much beyond the most obvious of all reasons, the framework for the conventional stereotype of each discipline, and the putative difference between them as well. Science supposedly rests upon the objective generality of nature's laws and the utter insignificance of a practitioner's personality, or even his identity (beyond our vulgar and personal need to count coup, and also to count the prospects of future funding, prizes, privileges and parking places). Why else have we been trained to write our professional papers in the unstylish passive voice, as if "I" didn't exist at all, and every datum "was discovered" in some disembodied manner? After all, although some particular somebody has to do it, the "it" is out there, and objectively knowable. Thus, it will be found, and within a narrow range of predictable time, largely dependent upon the development of technologies that initially make the discovery possible.

The equally silly and simplistic stereotype of the "other" side holds that literary people view the world as completely inchoate and unstructured (beyond the ideologically uninteresting, if practically portentous, compendium of observed regularities, suggesting, for example, that we will splatter if we fall off the roof of a 20-story building, or crunch if we happen to insert ourselves between a speeding vehicle and a concrete wall). Therefore, the argument continues, we make our own way in a subjective and unconstraining world. We alone are the architects and responsible agents of both our personal and our collective destinies.

As exaggerated as these characterizations may be, they do reflect some genuine cultural, and even partly justifiable, differences between two important, even noble, enterprises in their uncaricatured state. And, in this case, science could learn an important lesson from the literati—who love contingency for the same basic reason that scientists tend to regard the theme with suspicion. Because, in contingency lies the power of each person, no matter how apparently insignificant he may seem, to make a difference in an unconstrained world bristling with possibilities, and nudgeable by the smallest of unpredictable inputs into markedly different channels spelling either vast improvement or potential disaster.

And so, if Joshua Lawrence Chamberlain, former professor at Bowdoin College, and now commander of the gallant 20th Regiment of the State of Maine had not led one of the last successful bayonet charges in the history of warfare (because he had run out of ammunition and could only hope to prevail by a bluff of this sort), thus preventing the outflanking of the Union line (which could easily have been outflanked and overtaken, if the Confederates had grasped the desperate military situation of their adversaries), the South would probably have won at Gettysburg, leading to potential victory in the war, a sundering of the United States, the balkanization of our continent, and the end (with markedly negative consequences for human history) of the world's most promising experiment in democracy. And if George Bailey had never been born (an alternative scenario that his guardian angel constructed for his consideration), the history of his town would have been equally sensible but altogether less pleasant for everyone actually loved by this apparently insignificant man. And so both the historical Mr. Chamberlain and the fictional Mr. Bailey (of America's most beloved movie) learned that one ostensibly small and meaningless life can make all the difference, sometimes for an entire world at a tipping point (in the admittedly grandiose and a bit extreme, but still not so utterly implausible, fable at the beginning of this paragraph), and more often for the few people whom we love and whom we yearn to serve as a source of comfort. The literati embrace contingency because no other theme so affirms the moral weight, and the practical importance, of each human life.

Thus, to end where this book began with Charles Darwin and his personal importance to our understanding of this grandest earthly enterprise, the tree of life, I must side with the literati and insist that my decision to focus this book on Darwin and the logic of his explanatory system for life's history and evolution's mechanism does not merely record an idiosyncratic or antiquarian indulgence. I will grant one point to my scientific colleagues and freely allow that if Charles Darwin had never been born, a well-prepared and waiting scientific world, abetted by a cultural context more than ready for such a reconstruction of nature, would still have promulgated and won general acceptance for evolution in the mid 19th century. At some point, the mechanism of natural selection would also have been formulated and eventually validated, perhaps by Wallace himself who might then have expanded his few pages of speculation, written during a malarial fit on Ternate, into the same kind of factual compendium that Darwin composed, and that guaranteed the triumph of this view of life.

So why fret and care that the actual version of the destined deed was done by an upper class English gentleman who had circumnavigated the globe as a vigorous youth, lost his dearest daughter and his waning faith at the same time, wrote the greatest treatise ever composed on the taxonomy of barnacles, and eventually grew a white beard, lived as a country squire just south of London, and never again traveled far enough even to cross the English Channel? We care for the same reason that we love okapis, delight in the fossil evidence of trilobites, and mourn the passage of the dodo. We care because the broad events that had to happen, happened to happen *in a certain particular way*. And something almost unspeakably holy—I don't know how else to say this—underlies our discovery and confirmation of the actual details that made our world and also, in realms of contingency, assured the minutiae of its construction in the manner we know, and not in any one of a trillion other ways, nearly all of which would not have included the evolution of a scribe to record the beauty, the cruelty, the fascination, and the mystery.

Yes, the Renaissance would have unfolded—indeed, Europe already bathed in its midst—if Michelangelo had never been born. But how much poorer would our world have been without the magnificent statue of Moses, furious and disconsolate as he holds the tablets of the law while his people dance about the golden calf, still presiding in the Church of San Pietro in Vincoli; and without the gigantic fresco of the Last Judgment, revealing all our blessed humanity in all our earthly sins, and still covering, in brilliant restoration, a full wall of the Sistine Chapel?

No difference truly separates science and art in this crucial respect. We only perceive a division because our disparate traditions lead us to focus upon different scales of the identity. The art historian looks right at Moses and knows the importance of its individuality. The scientist tends to gaze upon a world ready for evolution, and then discounts the centrality of a single, admittedly fascinating, individual named Charles Darwin. But if Darwin had never been born, we would have suffered the equivalent of a Renaissance without Moses or the Last Judgment—a biological revolution without the *Origin of Species;* without the invocation of Julia Pastrana, the bearded circus lady with two sets of teeth, to illustrate correlation of growth; without the Galapagos fauna to embody the principle of imperfection to prove the pathways of history; without pigeons to illustrate artificial selection; without barnacles to puncture half our pride with their dwarfed males upon the hermaphrodites.

Most of all, we would have experienced the same biological revolution without the stunning clarity, illustrated by wonderfully apposite metaphors, of a complex central logic so brilliantly formulated, and so bristling with implications extending nearly forever outward, at least well past our current reckoning. In this alternate world, we would probably be honoring a different and far less compelling founder by occasional visits to a statue in a musty pantheon, and not by constant dialogue with a man whose ideas live, breathe, challenge, taunt, and inspire us every day of our lives, more than a century after his bones came to rest on a cathedral floor at the foot of whatever persists in the material being of Isaac Newton.

We would be enjoying an evolutionary view of life, but not the specific grandeur of "this view of life." What can be more ennobling than a factual reality—the uniquely actualized result among innumerable potentials that did not obtain the most precious privilege of emergence into concrete existence? And what a stunning piece of good fortune, that this actuality came to us with all the grace, the moral weight, and the intellectual power of Darwin's particular struggles and insights, clothing the structure of his thought in that apotheosis of human achievement—wisdom, which the *Book of Proverbs,* citing the same icon that Darwin would borrow more than two millennia later, called Etz Chayim, the tree of life. "Length of days is in her right hand," for "she is a tree of life to them that lay hold upon her; and happy is every one that retaineth her."

# Bibliography

Adams, F. D. 1938. *The Birth and Development of the Geological Sciences*. London (reprint 1954, N.Y.: Dover).

Adams, Robert McC. 2000. Accelerated technological change in archaeology and ancient history. In G. M. Feinman and L. Manzanilla, eds., *Cultural Evolution: Contemporary Viewpoint*. N.Y.: Kluwer Publishers, pp. 95–118.

Agassiz, L. 1857. Essay on Classification. In *Contributions to the Natural History of the United States*, Vol. 1. Boston MA: Little Brown & Co.

——— 1874. Evolution and permanence of type. *Atlantic Monthly* 33 (January), pp. 92–101.

Ager, D. V. 1973. *The Nature of the Stratigraphic Record*. N.Y.: John Wiley.

——— 1976. The nature of the fossil record. *Proc.Geologists' Assoc.* 87:131–160.

——— 1983. Allopatric speciation—an example from the Mesozoic Brachiopoda. *Palaeontology* 26: 555–565.

Aguinaldo, A. M. A., J. M. Turbeville, L. S. Linford, M. C. Rivera, T. R. Garey, R. A. Raff, and J. A. Lake. 1997. Evidence for a clade of nematodes, arthropods and other moulting animals. *Nature* 387: 489–493.

Akam, M. 1998. *Hox* genes, homeosis and the evolution of segment identity: no need for hopeless monsters. *Int. Jour. Devptl. Biol.* 42: 445–451.

Akam, M., M. Averof, J. Castelli-Gair, R. Dawes, F. Falciani, and D. Ferrier. 1994. The evolving role of *Hox* genes in arthropods. *Development 1994 Supplement:* 209–215.

Albanesi, G. L., and C. R. Barnes. 2000. Subspeciation within a punctuated equilibrium evolutionary event: phylogenetic history of the Lower-Middle Ordovician *Paristodus originalis–P. horridus* complex (Conodonta). *Jour. Paleontology* 74: 492–502.

Alberch, P. 1982. Developmental constraints in evolutionary processes. In J. T. Bonner, ed., *Evolution and Development*. Berlin: Springer-Verlag, pp. 313–332.

Alberch, P., S. J. Gould, G. Oster, and D. Wake. 1979. Size and shape in ontogeny and phylogeny. *Paleobiology* 5: 296–317.

Alexander, G. 1962. *General Biology*. N.Y.: Thomas Y. Crowell.

Allaby, M. (ed.) 1985. *The Oxford Dictionary of Natural History*. N.Y.: Oxford Univ. Press.

Allee, W. C., A. E. Emerson, A. E. Park, O. Park, and K. P. Schmidt. 1949. *Principles of Animal Ecology*. Philadelphia PA: W. B. Saunders.

Allen, J. C., W. M. Schaffer, and D. Rosko. 1993. Chaos reduces species extinction by amplifying local population noise. *Nature* 364: 229–232.

Almada, V. C., and R. S. Santos. 1995. Parental care in the rocky intertidal: a case study of adaptation and exaptation in Mediterranean and Atlantic blennies. *Revs. In Fish Biology and Fisheries* 5: 23–37.

Alters, B. J., and W. F. McComas. 1994. Punctuated Equilibrium: the missing link in evolution education. *American Biol. Teacher* 56: 334–340.

Alvarez, L. W., W. Alvarez, F. Asaro, and H. V. Michel. 1980. Extraterrestrial cause for the Cretaceous-Tertiary extinction. *Science* 208: 1095–1108.

Andrews, H. 1971. *Turritella mortoni (Gastropoda) and biostratigraphy of the Aquia Formation (Paleocene) of Maryland and Virginia.* Cambridge MA: Ph.D. Diss. Harvard Univ.

Anonymous. 1969. What will happen to geology? *Nature* 221: 903.

Anstey, R. L., and J. F. Pachut. 1995. Phylogeny, diversity, history, and speciation in Paleozoic bryozoans. In D. H. Erwin and R. L. Anstey, eds., *New Approaches to Speciation in the Fossil Record.* N.Y.: Columbia Univ. Press, pp. 239–284.

Antonovics, J., and P. H. van Tienderen. 1991. Ontoecogenophyloconstraints? The chaos of constraint terminology. *Trends. Ecol. Evol.* 6: 166–168.

Appel, T. A. 1987. *The Cuvier-Geoffroy Debate. French Biology in the Decades before Darwin.* N.Y.: Oxford Univ. Press.

Arendt, D., U. Technau, and J. Wittbrodt. 2001. Evolution of the bilaterian larval foregut. *Nature* 409: 81–85.

Arnold, A. J. 1982. Hierarchical structure in evolutionary theory: applications in the Foraminiferida. Ph.D. Dissertation. Dept. of Geology. Harvard University.

Arnold, A. J. and K. Fristrup. 1982. The theory of evolution by natural selection: a hierarchical expansion. *Palaeobiology* 8: 113–129.

Arnold, A. J., D. C. Kelly, and W. C. Parker. 1995. Causality and Cope's Rule: evidence from the planktonic Foraminifera. *Jour. Paleontology* 69: 203–210.

Arnold, E. N. 1994. Investigating the origins of performance advantage: adaptation, exaptation and lineage effects. In *Phylogenetics and Ecology.* London: The Linnean Soc. of London.

Arthur, W., and M. Farrow. 1999. The pattern of variation in centipede segment number as an example of developmental constraint in evolution. *Jour. Theoret. Biol.* 200: 183–191.

Arthur, W., T. Jewett, and A. Panchen. 1999. Segments, limbs, homology, and co-option. *Evolution and Development* 1: 74–76.

Asimov, I. 1989. *Asimov's Chronology of Science and Discovery.* N.Y.: Harper & Row.

Averof, M., and M. Akam. 1995. *Hox* genes and the diversification of insect and crustacean body plans. *Nature* 376: 420–423.

Averof, M., and S. M. Cohen. 1997. Evolutionary origin of insect wings from ancestral gills. *Nature* 385: 627–630.

Averof, M., and N. H. Patel. 1997. Crustacean appendage evolution associated with changes in *Hox* gene expression. *Nature* 388: 682–686.

Avers, C. J. 1989. *Process and Pattern in Evolution.* N.Y.: Oxford Univ. Press.

Avise, J. C. 1977. Is evolution gradual or rectangular? Evidence from living fishes. *Proc. Natl. Acad. Sci. USA* 74: 5083–5087.

Ayala, F. J. 1982. Microevolution and macroevolution. In D. S. Bendall, ed., *Evolution From Molecules to Man.* Cambridge UK: Cambridge Univ. Press, pp. 387–402.

Ayala, F. J., and A. Rzhetsky. 1998. Origin of the metazoan phyla: molecular clocks confirm paleontological estimates. *Proc. Natl. Acad. Sci. USA* 95: 606–611.

Bahn, P. G., and J. Vertut. 1988. *Images of the Ice Age.* N.Y.: Facts on File.

Bailey, M. 1990. *Evolution: Opposing Viewpoints.* St. Paul MN: Greenhaven Press.

Bak, P., and K. Sneppen. 1993. Punctuated equilibrium and criticality in a simple model of evolution. *Physical Rev. Letters* 71: 4083–4086.

Baker, H. B. 1924. Land and freshwater mollusks of the Dutch Leeward Islands. *Occ. Pap. Mus. Zool. Univ. Mich.* 152: 1–158.

Baker, J. J. W., and G. E. Allen. 1968. *The Study of Biology*. Reading MA: Addison-Wesley.

Baker, V. R., and D. Nummedal. 1978. *The Channeled Scablands*. Washington, D.C.: National Aeronautics and Space Administration, Planetary Geology Program.

Bang, R., R. DeSalle, and W. Wheeler. 2000. Transformationalism, taxism, and developmental biology in systematics. *Syst. Biol.* 49: 19–27.

Banner, F. T., and F. M. D. Lowry. 1985. The stratigraphical record of planktonic foraminifera and its evolutionary implications. In J. C. W. Cope and P. W. Skelton, eds., Evolutionary Case Histories From the Fossil Record. *Special Papers Palaeontology* 33: 103–116.

Barnosky, A. D. 1987. Punctuated equilibrium and phyletic gradualism: Some facts from the Quaternary mammalian record. In Genoways, H. H., ed., *Current Mammalogy, Vol 11*: 109–147.

Barrett, P. H., P. J. Gautey, S. Herbert, D. Kohn, and S. Smith., eds. 1987. *Charles Darwin's Notebooks, 1836–1844*. Cambridge UK: Cambridge Univ. Press.

Bates, M. 1960. Ecology and evolution. In S. Tax, ed., *Evolution After Darwin*. Volume I, *The Evolution of Life*. Chicago IL: Univ. of Chicago Press, pp. 547–568.

Bateson, W. 1894. *Materials for the Study of Variation*. London: Macmillan.

———— 1909. Heredity and variation in modern lights. In A. C. Seward, ed., *Darwin and Modern Science*. Cambridge UK: Cambridge Univ. Press, pp. 85–101.

———— 1913. *Problems of Genetics*. New Haven CT: Yale Univ. Press.

———— 1928. *William Bateson, F. R. S. Naturalist*. His Essays and Addresses, Together With A Short Account of His Life, by Beatrice Bateson. Cambridge UK: Cambridge Univ. Press.

Beatty, J. 1988. Ecology and evolutionary biology in the war and postwar years: questions and comments. *Jour. History Biol.* 21: 245–263.

Beeman, R. W., K. S. Friesen, and R. E. Denell. 1992. Maternal-effect selfish genes in flour beetles. *Science* 256: 89–92.

Begg, C. B., and J. A. Berlin. 1988. Publication bias: a problem in interpreting medical data. *Jour. Roy. Statistical Soc.* 151: 419–463.

Bell, M. A., J. V. Baumgartner, and E. C. Olson. 1985. Patterns of temporal change in single morphological characters of a Miocene stickleback fish. *Paleobiology* 11: 258–271.

Bell, M. A., and P. Legendre. 1987. Multicharacter clustering in a sequence of fossil sticklebacks. *Syst. Zool.* 36: 52–61.

Bell, M. A. and T. R. Haglund. 1982. Fine-scale temporal variation of the Miocene stickleback *Gasterosteus doryssus*. *Paleobiology* 8: 282–292.

Benson, R. H. 1983. Biochemical stability and sudden change in the evolution of the deep-sea ostracode *Poseidonamicus*. *Paleobiology* 9: 398–413.

Benton, M. J. 1996. On the nonprevalence of competitive replacement in the evolution of tetrapods. In D. Jablonski, D. H. Erwin, and J. H. Lipps, eds., *Evolutionary Paleobiology*. Chicago IL: Univ. of Chicago Press, pp. 185–210.

Bergstrom, J. and R. Levi-Setti. 1978. Phenotypic variation in the Middle Cambrian trilobite *Paradoxides davidis* Salter at Manuels, SE Newfoundland. *Geologica et Palaeontologica* 12: 1–40.

Berry, M. S. 1982. *Time, Space, and Transition in Anasazi Prehistory*. Salt Lake City: Univ. of Utah Press.

Bethell, T. 1976. Darwin's mistake. *Harper's*, February.

Bier, E. 1997. Anti-neural-inhibition: a conserved mechanism for neural induction. *Cell* 89: 681–684.

Biological Sciences Curriculum Study, 1968. *Biological Science: Molecules to Man*. Boston MA: Houghton Mifflin.

———— 1973. *Biological Science: An Ecological Approach.* Chicago IL: Rand McNally.

Blackburn, D. G. 1995. Saltationist and punctuated equilibrium models for the evolution of viviparity and placentation. *J. Theoret. Biol.* 174: 199–216.

Blanc, M. 1982. Les théories de l'évolution aujourd'hui. *La Recherche,* January, pp. 26–40.

Bliss, R. B., G. E. Parker, and D. T. Gish. 1980. *Fossils: Key to the Present.* San Diego CA: CLP Publishers.

Bock, W. J. 1959. Preadaptation and multiple evolutionary pathways. *Evolution* 13: 194–211.

Böhm, H., V. Brinkmann, M. Drab, A. Henske, and T. V. Kurzchalia. 1997. Mammalian homologues of *C. elegans* PAR-1 are asymmetrically localized in epithelial cells and may influence their polarity. *Current Biology* 7: 603–606.

Bolker, J. A., and R. A. Raff. 1996. Developmental genetics and traditional homology. *Bioessays* 18: 489–494.

Bonner, J. T. 1962. *The Ideas of Biology.* N.Y.: Harper & Row.

Borgia, G. 1994. The scandals of San Marco. *Quart. Rev. Biol.* 69: 373–375.

Boucot, A. J. 1983. Does evolution take place in an ecological vacuum? *Jour. Paleontology* 57: 1–30.

Boulding, K. E. 1992. Punctuationalism in societal evolution. In A. Somit and S. A. Peterson, eds., *The Dynamics of Evolution: The Punctuated Equilibrium Debate in the Natural and Social Sciences.* Ithaca NY: Cornell Univ. Press, pp. 171–186.

Bowler, P. J. 1979. Theodor Eimer and orthogenesis: evolution by "definitely directed variation." *Jour. Hist. Med. Allied Sci.* 34: 40–73.

———— 1983. *The Eclipse of Darwinism: Anti-Darwinian Evolution Theories in the Decades around 1900.* Baltimore MD: Johns Hopkins Univ. Press.

———— 1988. *The Non-Darwinian Revolution.* Baltimore MD: Johns Hopkins Univ. Press.

Bown, T. M. 1979. Geology and mammalian paleontology of the Sand Creek facies, Lower Willwood Formation (Lower Eocene), Washakie County, Wyoming. *Mem. Geol. Surv. Wyo.* 2:1–151.

Boyajian, G., and G. & T. Lutz. 1992. Evolution of biological complexity and its relation to taxonomic longevity in the Ammonoidea. *Geology* 20: 983–986.

Boyle, R. 1688. *A Disquisition About the Final Causes of Natural Things.* London: John Taylor.

Brace, C. L. 1977. *Human Evolution.* N.Y.: MacMillan.

Brackman, A. C. 1980. *A Delicate Arrangement: The Strange Case of Charles Darwin and Alfred Russel Wallace.* London: Times Books.

Brandon, R. N. 1982. The levels of selection. *PSA 1982,* volume 1, Philosophy of Science Assoc., pp. 315–323.

———— 1988. Levels of selection: a hierarchy of interactors. In H. C. Plotkin, ed., *The Role of Behavior in Evolution.* Cambridge MA: MIT Press, pp. 51–71.

———— 1990. *Adaptation and Environment.* Princeton NJ: Princeton Univ. Press.

Brenner, S. 1999. Refuge of spandrels. *Current Biology,* p. 669.

Breton, G. 1996. Ponctualisme et gradualisme au sein d'une même lignée: réflexions sur la complexité et l'imprédictibilité des phénomènes évolutifs. *Geobios* 29: 125–130.

Brett, C. E., and G. C. Baird. 1995. Coordinated stasis and evolutionary ecology of Silurian to Middle Devonian faunas in the Appalachian Basin. In D. H. Erwin and R. L. Anstey, eds., *New Approaches to Speciation in the Fossil Record.* N.Y.: Columbia Univ. Press, pp. 285–317.

Brett, C. E., L. C. Ivany, and K. M. Schopf. 1995. Coordinated stasis: An overview. *Palaeogeol. Palaeoclimat. Palaeoecol.* 127: 1–20.

Breuil, H. 1906. *L'évolution de la peinture et de la gravure sur murailles dans les cavernes ornées de l'age de renne.* Paris: Congrès Préhistorique de France.

Briggs, D. E. G., R. A. Fortey, and M. A. Wills. 1992. Morphological disparity in the Cambrian. *Science* 256: 1670–1673.

Britten, R. J., and E. H. Davidson. 1971. Repetitive and non-repetitive DNA sequences and a speculation on the origins of evolutionary novelty. *Quart. Rev. Biol.* 46: 111–133.

Bromham, L., A. Rambaut, R. Fortey, A. Cooper, and D. Penny. 1998. Testing the Cambrian explosion hypothesis by using a molecular dating technique. *Proc. Natl. Acad. Sci. USA* 95: 12389–12393.

Brooks, W. K. 1883. *The Law of Heredity.* N.Y.: John Murphy & Co.

Brosius, J. 1999. Transmutation of tRNA over time. *Nature Genetics* 22: 8–9.

Brosius, J., and S. J. Gould. 1992. On "genomenclature": a comprehensive (and respectful) taxonomy for pseudogenes and other "junk DNA." *Proc. Natl. Acad. Sci. USA* 89: 10706–10710.

Brosius, J., and H. Tiedge. 1996. Reverse transcriptase: mediator of genomic plasticity. *Virus Genes* 11: 163–179.

Brown, J. H., and B. A. Maurer. 1986. Body size, ecological dominance, and Cope's rule. *Nature* 324: 248–250.

Browne, J. 1980. Darwin's botanical arithmetic and the principle of divergence, 1854–1858. *Jour. Hist. Biol.* 13: 53–89.

———— 1995. *Voyaging. Charles Darwin. A Biography.* Volume 1. N.Y.: Knopf.

Budd, A. F., and A. G. Coates. 1992. Non-progressive evolution in a clade of Cretaceous *Montastraea*-like corals. *Paleobiology* 18: 425–446.

Bull, J. J., I. J. Molineux, and J. H. Werren. 1992. Selfish Genes. *Science* 256: 65.

Burchfield, J. D. 1975. *Lord Kelvin and the Age of the Earth.* N. Y.: Science History Publications.

Burke, A. C., C. E. Nelson, B. A. Morgan, and C. Tabin. 1995. *Hox* genes and the evolution of vertebrate axial morphology. *Development* 121: 333–346.

Burkhardt, R. W. 1977. *The Spirit of System: Lamarck and Evolutionary Biology.* Cambridge MA: Harvard Univ. Press.

Busch, M. A., K. Bomblies, and D. Weigel. 1999. Activation of a floral homeotic gene in *Arabidopsis. Science* 285: 585–587.

Bush, G. L. 1982. What do we really know about speciation? In R. Milkman, ed., *Perspectives on Evolution.* Sunderland MA: Sinauer Associates, pp. 119–128.

Buss, D. M., M. G. Haselton, T. K. Shackelford, A. L. Bleske, and J. C. Wakefield. 1998. Adaptations, exaptations, and spandrels. *American Psychologist* 53: 533–548.

Buss, L. W. 1987. *The Evolution of Individuality.* Princeton NJ: Princeton Univ. Press.

Cain, A. J. 1979. Introduction to general discussion. *Proc. Roy. Soc. London Series B* 205: 599–604.

———— 1988. Evolution. *Jour. Evol. Biol.* 1: 185–194.

Cain, A. J., and J. D. Currey. 1963a. Area effects in *Cepaea. Philosophical Transactions Roy. Soc. London Series B* 246: 1–81.

———— 1963b. The causes of area effects. *Heredity* 18: 467–471.

Cain, A. J., and P. M. Sheppard. 1950. Selection in the polymorphic land snail *Cepaea nemoralis. Heredity* 4: 275–294.

———— 1952. The effects of natural selection on body color in the land snail *Cepaea nemoralis. Heredity* 6: 217–223.

———— 1954. Natural selection in *Cepaea. Genetics* 39: 89–116.

Cairns, J., J. Overbaugh, and S. Miller. 1988. The origin of mutants. *Nature* 335: 142–145.

Cairns-Smith, A. G. 1971. *The Life Puzzle.* Edinburgh: Oliver & Boyd.

Cameron, R. A. D., M. A. Carter, and M. A. Palles-Clark. 1980. *Cepaea* on Salisbury Plain: patterns of variation, landscape history, and habitat stability. *Biol. Jour. Linnean Soc. London* 14: 335–358.

Campbell, D. T. 1974. Downward causation in hierarchical organized biological systems. In F. Ayala and T. Dobzhansky, eds., *Studies in the Philosophy of Biology*. San Francisco: Univ. of California Press.

Carmines, E. G., and J. A. Stimson. 1989. *Issue Evolution: Race and the Transformation of American Politics*. Princeton NJ: Princeton Univ. Press.

Carrasco, A. E., W. McGinnis, W. J. Gehring, and E. M. De Robertis. 1984. Cloning of an *X. laevis* gene expressed during early embryogenesis coding for a peptide region homologous to *Drosophila* homeotic genes. *Cell* 37: 409–414.

Carroll, S. B. 1995. Homeotic genes and the evolution of arthropods and chordates. *Nature* 376: 479–485.

Carroll, S. B., J. Gates, D. Keys, S. W. Paddock, G. F. Panganiban, J. Selegue, and J. A. Williams. 1994. Pattern formation and eyespot determination in butterfly wings. *Science* 265: 109–114.

Carroll, S. B., S. D. Weatherbee, and J. A. Langeland. 1995. Homeotic genes and the regulation and evolution of insect wing number. *Nature* 375:58–61.

Carson, H. L. 1975. The genetics of speciation at the diploid level. *Amer. Naturalist* 109: 83–92.

Cartwright, P., J. Bowsher, and L. W. Buss. 1999. Expression of a Hox gene, *Cnox-2*, and the division of labor in a colonial hydroid. *Proc. Natl. Acad. Sci. USA* 96: 2183–2186.

Cassares, F., and R. S. Mann. 1998. Control of antennal versus leg development in *Drosophila*. *Nature* 392: 723–726.

Castle, W. E. 1916. Further studies of piebald rats and selection. *Carnegie Institute of Washington Publ.* No. 241, pp. 163–192.

———— 1919. Piebald rats and the theory of genes. *Proc. Natl. Acad. Sci. USA* 5: 126–130.

Catmull, J., D. C. Hayward, N. E. McIntyre, J. S. Reece-Hoyes, R. Mastro, P. Callaerts, E. E. Ball, and D. J. Miller. 1998. *Pax-6* origins—implications from the structure of two coral *Pax* genes. *Dev. Genes Evol.* 208: 352–356.

Catton, W. R. Jr. 1998. Darwin, Durkheim, and Mutualism. In *Advances in Human Ecology*, Volume 7. JAI Press Inc., pp. 89–138.

Cavalier-Smith, T. 1980. How selfish is DNA? *Nature* 285: 617–618.

Chaisson, E. 1988. *Universe: An Evolutionary Approach to Astronomy*. Englewood Cliffs NJ: Prentice Hall.

Chalfie, M. 1998. The worm revealed. *Nature* 396: 620–621.

Chaline J. 1982. *Modalités, rythmes, mécanismes de l'évolution biologique: gradualisme phylétique ou équilibres ponctués?* Colloque International. Paris: CNRS (Centre Nationale de Recherche Scientifique), 335 pp.

Chaline, J., and B. Laurin. 1986. Phyletic gradualism in a European Plio-Pleistocene *Mimomys* lineage (Arvicolidae, Rodentia). *Paleobiology* 12: 203–216.

[Chambers, R.] 1844. *Vestiges of the Natural History of Creation*. London: John Churchill (published anonymously).

Chatterjee, S. 1997. The beginning of avian flight. *Proc. Symposium Sponsored by Arizona State University*, pp. 311–335.

Chau, H. F. 1994. Scaling behavior of the punctuated equilibrium model of evolution. *Phys. Rev. E.*

Cheetham, A. H. 1986. Tempo of evolution in a Neogene bryozoan: rates of morphologic change within and across species boundaries. *Paleobiology* 12:190–202.

———— 1987. Tempo of evolution in a Neogene bryozoan: are trends in single morphologic characters misleading? *Paleobiology* 13: 286–296.

Cheetham, A. H., and J. B. C. Jackson. 1995. Process from pattern: tests for selection versus random change in punctuated bryozoan speciation. In D. H. Erwin and R. L. Anstey, eds., *New Approaches to Speciation in the Fossil Record.* N.Y.: Columbia Univ. Press, pp. 184–207.

Chen, J.-Y., G. D. Edgecombe, L. Ramsköld, and G.-Q. Zhou. 1995. Head segmentation in early Cambrian *Fuxianhuia:* implications for arthropod evolution. *Science* 268: 1339–1343.

Chen, J.-Y., P. Oliveri, C.-W. Li, G.-Q. Zhou, F. Gao, J. W. Hagadorn, K. J. Peterson, and E. H. Davidson. 2000. Precambrian animal diversity: putative phosphatized embryos from the Doushantuo Formation of China. *Proc. Natl. Acad. Sci. USA* 97: 4457–4462.

Clark, H. W. 1982. Evolutionist fights back. *Signs of the Times* (Seventh Day Adventist Press), Volume 109 (September).

Clark, R. B. 1964. *Dynamics in Metazoan Evolution: The Origin of the Coelom and Segments.* Oxford UK: Clarendon Press.

Coates, M. I., and J. A. Clack. 1990. Polydactyly in the earliest tetrapod limbs. *Nature* 347: 66- 69.

Coates, M. I., and M. J. Cohn. 1998. A common plan for dorsoventral patterning in Bilateria. *BioEssays* 20: 371–381.

Coddington, J. A. 1988. Cladistic tests of adaptational hypotheses. *Cladistics* 4: 3–22.

Coen, E. S., and E. M. Meyerowitz. 1991. The war of the whorls: genetic interactions controlling flower development. *Nature* 353: 31–37.

Cohen, S. M., G. Bronner, F. Kuttner, G. Jürgens, and H. Jäckle. 1989. *Distal-less* encodes a homeodomain protein required for limb development in *Drosophila. Nature* 338: 432–434.

Cohn, M. J., and C. Tickle. 1999. Developmental basis of limblessness and axial patterning in snakes. *Nature* 399: 474–479.

Colwell, R. K. 1981. Evolution of female-biased sex ratios: the essential role of group selection. *Nature* 290: 401–404.

Conway Morris, S. 1998. *The Crucible of Creation.* Oxford UK: Oxford Univ. Press.

Conway Morris, S., and S. J. Gould. 1998. Showdown on the Burgess Shale. (The Challenge by S. Conway Morris and The Reply by S. J. Gould.) *Nat. Hist.* 107 (Dec.–Jan): 48–55.

Coope, G. R. 1979. Late Cenozoic fossil Coleoptera. *Ann. Rev. Ecol. Syst.* 10: 247–267.

———— 1980. The invasion of Northern Europe during the Pleistocene by Mediterranean species of Coleoptera. In F. Di Castri et al., eds., *Biological Invasions in Europe and the Mediterranean Basin.* Dordrecht: Kluwer, pp. 203–215.

———— 1994. The response of insect faunas to glacial-interglacial climatic fluctuations. *Philos. Trans. Roy. Soc. London B,* 344: 19–26.

Cope, E. D. 1887. *The Origin of the Fittest.* N.Y.: D. Appleton & Co.

Cope, J. C. W., and P. W. Skelton (eds.) 1985. Evolutionary case histories from the fossil record. *Special Papers in Palaeontology* 33: 1–203.

Corsi, P. 1978. The importance of French transformist ideas for the second volume of Lyell's "Principles of Geology." *British Jour. Hist. Sci.* 11: 221–244.

———— 1988. *The Age of Lamarck: Evolutionary Theories in France, 1790–1830.* Berkeley CA: Univ. of California Press.

Courtillot, V. 1995. *La vie en catstrophes: Du hasard dans l'évolution des espèces.* Paris: Fayard.

Crampton, H. E. 1916. Studies on the variation, distribution, and evolution of the genus *Partula.* The species inhabiting Tahiti. *Carnegie Inst. Washington Publ.* 228: 1–311.

———— 1932. Studies on the variation, distribution, and evolution of the genus

*Partula.* The species inhabiting Moorea. *Carnegie Inst. Washington Publ.* 410: 1–335.

Cronin, H. 1991. *The Ant and the Peacock.* Cambridge UK: Cambridge Univ. Press.

Cronin, J. E., N. T. Boaz, C. B. Stringer, and Y. Rak. 1981. Tempo and mode in hominid evolution. *Nature* 292: 113–122.

Cronin, T. M. 1985. Speciation and stasis in marine Ostracoda: climatic modulation of evolution. *Science* 277: 60–62.

Crow, J. F. 1994. Advantages of sexual reproduction. *Dev. Genet.* 15: 205–213.

Curtis, H. 1962. *Biology.* N.Y. Worth.

Curtis, H., and N. S. Barnes. 1985. *Invitation to Biology.* N.Y.: Worth Publishers.

Cuvier, G. 1800–1805. *Leçons d'anatomie comparée* de G. Cuvier, recueillies et publiées sous ses yeux par C. Duméril. C. Duméril, éd., vol. 1–2; G.-L. Duvernoy, éd., vol. 3–5, Paris.

———— 1812. *Recherches sur les ossemens fossiles des quadrupèdes, où l'on rétablit les caractères de plusieurs espèces d'animaux que les révolutions du globe paroissent avoir détruites,* 4 vol. Paris: Déterville.

———— 1817. *Le Règne animal distribué d'après son organisation, pour servir de base à l'histoire naturelle des animaux et d'introduction à l'anatomie comparée,* 4 vol. Paris: Déterville.

———— 1818. *Essay on the Theory of the Earth* (Jameson translation). N.Y.: Kirk & Mercein.

———— 1832. Éloge de M. de Lamarck, lu à l'Académie royale des sciences de l'institut le 26 Novembre 1832. *Mémoires de l'Academie royale des sciences de l'institut de France.* 13: 1–30.

Cvekl, A., C. M. Sax, E. H. Bresnick, and J. Piatigorsky. 1994. A complex array of positive and negative elements regulates the chicken $\alpha$A-crystalline gene: involvement of Pax-6, USF, CREB and/or CREM, and $AP_1$ proteins. *Mol. Cell. Biol.* 14: 7363–7376.

Dacqué, E. 1921. *Vergleichende biologische Formenkunde der fossilen niederen Tiere.* Berlin: Bornträger.

Damuth, J. 1985. Selection among species: a formulation in terms of natural functional units. *Evolution* 39: 1132–1146.

Damuth, J., and J. L. Heisler. 1988. Alternative formulations of multilevel selection. *Biol. & Philos.* 3: 407–430.

Darling, L., and L. Darling. 1961. *The Science of Life.* Cleveland: World Publishing Co.

Darwin, C. 1842. *The Structure and Distribution of Coral Reefs. Being the First Part of the Geology of the Voyage of the Beagle.* London: Smith, Elder.

———— 1844. *Geological Observations on the Volcanic Islands Visited during the Voyage of H. M. S. "Beagle."* London: Smith, Elder.

———— 1846. *Geological Observations on South America.* London: Smith, Elder.

———— 1859. *On the Origin of Species by Means of Natural Selection, or Preservation of Favored Races in the Struggle for Life.* London: Murray.

———— 1862. *On the Various Contrivances by which British and Foreign Orchids are Fertilized by Insects, and on the Good Effects of Intercrossing.* London: Murray.

———— 1868. *The Variation of Animals and Plants under Domestication.* 2 vols. London: Murray.

———— 1871. *The Descent of Man, and Selection in Relation to Sex.* 2 vols. London: Murray.

———— 1872a. *The Expression of the Emotions in Man and Animals.* London: Murray.

———— 1872b. *The Origin of Species,* Sixth Edition. London: Murray.

———— 1875a. *Insectivorous Plants.* London: Murray.

———— 1875b. *The Movement and Habits of Climbing Plants.* London: Murray.

———— 1877. *The Different Forms of Flowers on Plants of the Same Species.* London: Murray.

———— 1880a. *The Power of Movement in Plants* (assisted by Francis Darwin). London: Murray.

———— 1880b. Sir Wyville Thomson and natural selection. *Nature* 23: 32.

———— 1881. *The Formation of Vegetable Mould, through the Action of Worms, with Observations on their Habitats.* London: Murray.

———— 1887. *The Life and Letters of Charles Darwin, Including an Autobiographical Chapter.* Edited by Francis Darwin. 3 vols. London: Murray.

———— 1903. *More Letters of Charles Darwin. A Record of his Work in a Series of Hitherto Unpublished Letters.* Edited by Francis Darwin and A. C. Seward. London: Murray.

———— 1960. Darwin's notebooks on transmutation of species, edited by Gavin De Beer. *Bull. British Mus.* (Nat. Hist.) *Historical Ser.* 2: 23–200.

———— 1987. *The Correspondence of Charles Darwin. Vol. 3: 1844–1846.* Cambridge UK: Cambridge Univ. Press.

Davidson, E. H., K. J. Peterson, and R. A. Cameron. 1995. Origin of bilaterian body plans: evolution of developmental regulatory mechanisms. *Science* 270: 1319–1325.

Davis, D. D. 1949. Comparative anatomy and the evolution of vertebrates. In G. L. Jepsen, E. Mayr, and G. G. Simpson, eds., *Genetics, Paleontology and Evolution.* Princeton NJ: Princeton Univ. Press, pp. 64–89.

———— 1964. The giant panda: a morphological study of evolutionary mechanisms. *Fieldiana Memoirs* (Zoology, Chicago Museum of Natural History) 3: 1–339.

Dawes, R., A. Dawson, F. Falciani, G. Tear, and M. Akam. 1994. *Dax,* a locust Hox gene related to *fushi-tarazu* but showing no pair-rule expression. *Development* 120: 1561–1572.

Dawkins, R. 1976. *The Selfish Gene.* N.Y.: Oxford Univ. Press.

———— 1978. Replication, selection and the extended phenotype. *Zeits. Tierpsychologie* 47: 61- 76.

———— 1982. *The Extended Phenotype.* N.Y.: Oxford Univ. Press.

———— 1986. *The Blind Watchmaker.* N.Y.: W. W. Norton.

———— 1996. *Climbing Mount Improbable.* N.Y.: W. W. Norton.

De Beer, G. 1930. *Embryology and Evolution.* Oxford UK: Clarendon Press.

Demus, O. 1984. *The Mosaics of San Marco in Venice.* Four volumes. Chicago IL: Univ. of Chicago Press.

Dennert, E. 1904. *At the Deathbed of Darwinism.* Burlington IA: German Literary Board.

Dennett, D. C. 1995. *Darwin's Dangerous Idea.* N.Y.: Simon & Schuster.

———— 1997. "Darwinian fundamentalism": an exchange. Letter to the Editors. *N.Y. Review of Books,* Aug. 14, 1997, pp. 64–65.

Den Tex, E. 1990. Punctuated equilibria between rival concepts of granite genesis in the late 18th, 19th, and early 20th centuries. *Geol. Jour.* 25: 215–219.

De Robertis, E. M. 1997. The ancestry of segmentation. *Nature* 387: 25–26.

De Robertis, E. M., G. Oliver, and C. V. E. Wright. 1990. Homeobox genes and the vertebrate body plan. *Scientific American,* July, pp. 46–52.

De Robertis, E. M., and Y. Sasai. 1996. A common plan for dorsoventral patterning in Bilateria. *Nature* 380: 37–40.

De Rosa, R., J. K. Grenier, T. Andreeva, C. E. Cook, A. Adouette, M. Akam, S. B. Carroll, and G. Balavoine. 1999. *Hox* genes in brachiopods and priapulids and protostome evolution. *Nature* 399: 772–776.

Desmond, A. 1982. *Archetypes and Ancestors: Palaeontology in Victorian London, 1850–1875.* London: Blond & Briggs.

Desmond, A., and J. Moore. 1991. *Darwin: The Life of a Tormented Evolutionist.* London: Penguin Books.

Deutsch, J., and H. Le Guyader. 1998. The neuronal zootype. An hypothesis. *C. R. Acad. Sci. Paris, Sciences de la vie/Life Sciences.* 321: 713–719.

Devillers, C., and J. Chaline. 1989. *La théorie de l'évolution.* Paris: Dunod.

De Vries, H. 1889. *Intracellulare Pangenesis.* Jena: Gustav Fisher.

—— 1901–1903. *Die Mutationstheorie.* Leipzig: Veit.

—— 1905. *Species and Varieties: Their Origin By Mutation.* Chicago IL: Open Court Publishing Co.

—— 1907a. *Plant Breeding.* Chicago IL: Open Court Publishing Co.

—— 1907b. Evolution and mutation. *The Monist* 17: 6–22.

—— 1909a. *The Mutation Theory,* 2 Vols. N.Y.: D. Appleton & Co.

—— 1909b. Variation. In A. C. Seward, ed., *Darwin and Modern Science.* Cambridge UK: Cambridge Univ. Press, pp. 66–84.

—— 1910. *Intracellular Pangenesis.* Chicago IL: Open Court Publishing Co.

—— 1915. The coefficient of variation in *Oenothera biennis* L. *Botanical Gazette* 49: 169–196.

—— 1922. Age and area and the mutation theory. In J. C. Willis, *Age and Area.* Cambridge UK: Cambridge Univ. Press, pp. 222–227.

De Vries, W. 1974. Caribbean land mollusks: notes on Cerionidae. *Studies of the Fauna of Curaçao and other Caribbean Islands* 45: 81–117.

Dobzhansky, Th. 1937. *Genetics and The Origin of Species.* N.Y.: Columbia Univ. Press.

—— 1940. Catastrophism versus evolution. *Science* 92: 356–358.

—— 1941. *Genetics and The Origin of Species.* Second Edition. N.Y.: Columbia Univ. Press.

—— 1958. *Genetics and The Origin of Species.* Third Edition. N.Y.: Columbia Univ. Press.

—— 1967. *The Biology of Ultimate Concern.* N.Y.: New American Library.

—— 1980. Morgan and his school in the 1930's. In E. Mayr and W. B. Provine, eds., *The Evolutionary Synthesis.* Cambridge MA: Harvard Univ. Press, pp. 445–452.

Dodson, E. O., and P. Dodson. 1990. *Evolution: Process and Product.* Boston MA: Prindle, Weber & Schmidt.

Dollé, P., A. Dierich, M. LeMeur, T. Schimmang, B. Schubaur, P. Chambon, and D. Duboule. 1993. Disruption of the *Hoxd-13* gene induces localized heterochrony leading to mice with neotenic limbs. *Cell* 75: 431–441.

Doolittle, W. F. 2000. Phylogenetic classification and the universal tree. *Science* 284: 2124–2129.

Doolittle, W. F., and C. Sapienza. 1980. Selfish genes, the phenotype paradigm and genome evolution. *Nature* 284: 601–607.

Dott, R. H., Jr., and D. R. Prothero. 1994. *Evolution of the Earth.* N.Y.: McGraw Hill.

Dover, G. A. 1982. A molecular drive through evolution. *Bioscience* 32: 526–533.

Duboule, D. 1992. The vertebrate limb: a model system to study the *Hox/Hom* gene network during development and evolution. *BioEssays* 14: 375–384.

Duboule, D., and A. S. Wilkins. 1998. The evolution of "bricolage." *Trends in Genetics.* 14: 54–59.

Eberhard, W. G. 1980. Evolutionary consequences of intracellular organelle competition. *Quart. Rev. Biol.* 55: 231–249.

—— 1990. Animal genitalia and female choice. *American Scientist* 78: 134–141.

Eble, G. T. 1999. On the dual nature of chance in evolutionary biology and paleobiology. *Paleobiology* 25: 75–87.

Ehrlich, P. L. 1986. *The Machinery of Nature.* N.Y.: Simon & Schuster.

Eimer, G. H. T. 1890. *Organic Evolution as the Result of the Inheritance of Acquired Characters According to the Laws of Organic Growth.* London: Macmillan.

—— 1897. *Orthogenese der Schmetterlinge. Ein Beweis bestimmt gerichter*

*Entwicklung und Ohnmacht der natürlicher Zuchtwahl bei der Artbildung.*
Leipzig: W. Engelemann.

Eiseley, L. 1958. *Darwin's Century.* N.Y.: Doubleday.

——— 1959. Charles Darwin, Edward Blyth and the theory of natural selection. In B. Glass, O. Temkin, and W. L. Straus, Jr., eds., *Forerunners of Darwin: 1745–1859.* Baltimore MD: Johns Hopkins Univ. Press.

——— 1979. *Darwin and the Mysterious Mr. X: New Light on the Evolutionists.* N.Y.

Eldredge, N. 1971. The allopatric model and phylogeny in Paleozoic invertebrates. *Evolution* 25: 156–167.

——— 1979. Alternative approaches to evolutionary theory. In J. H. Schwartz and H. B. Rollins, eds., Models and Methodologies in Evolutionary Theory. *Bull. Carnegie Mus. Nat. Hist.* 13: 7–19.

——— 1985a. *Unfinished Synthesis: Biological Hierarchies and Modern Evolutionary Thought.* N.Y.: Oxford Univ. Press.

——— 1985b. *Time Frames.* N.Y.: Simon & Schuster.

——— 1989. *Macroevolutionary Patterns and Evolutionary Dynamics: Species, Niches and Adaptive Peaks.* N.Y.: McGraw-Hill.

——— 1995. *Reinventing Darwin: The Great Debate at The High Table of Evolutionary Theory.* N.Y.: John Wiley.

——— 1999. *The Pattern of Evolution.* N.Y.: W. H. Freeman.

Eldredge, N., and S. J. Gould. 1972. Punctuated equilibria: an alternative to phyletic gradualism. In T. J. M. Schopf, ed., *Models in Paleobiology.* San Francisco: Freeman, Cooper & Co., pp. 82–115.

——— 1977. Evolutionary models and biostratigraphic strategies. In E. G. Kauffman and J. E. Hazel, eds., *Concepts and Methods of Biostratigraphy.* Stroudsburg PA: Dowden, Hutchinson & Ross, pp. 25–40.

——— 1988. Punctuated Equilibrium prevails. *Nature* 332: 211–212.

Eldredge, N., and M. Grene. 1992. *Interactions: The Biological Context of Social Systems.* N.Y.: Columbia Univ. Press.

Elena, S. F., V. S. Cooper, and R. E. Lenksi. 1996. Punctuated Evolution caused by selection of rare beneficial mutations. *Science* 272: 1802–1804.

Emerson, A. E. 1960. The evolution of adaptation in population systems. In S. Tax, ed., *Evolution After Darwin.* Volume I, *The Evolution of Life.* Chicago IL: Univ. of Chicago Press, pp. 307–348.

Emry, R. J. 1981. Additions to the mammalian fauna of the type Duchesnean, with comments on the status of the Duchesnean. *J. Paleontology* 55:563–570.

Erdtmann, B. D. 1986. Early Ordovician eustatic cycles and their bearing on punctuations in early nematophorid (planktic) graptolite evolution. In O. H. Walliser, ed., *Global Bio-Events: A Critical Approach.* Berlin: Springer Verlag, pp. 139–152.

Erwin, D. H., and R. L. Anstey. 1995. Speciation in the Fossil Record. In D. H. Erwin and R. L. Anstry, eds., *New Approaches to Speciation in the Fossil Record.* N.Y.: Columbia Univ. Press, pp. 11–38.

Falconer, H. 1868. *Palaeontological Memoirs and Notes* (C. Murchison, ed.) 2 volumes. London: Robert Hardwicke.

Fausto-Sterling, A. 1985. *Myths of Gender: Biological Theories About Women and Men.* N.Y.: Basic Books.

Fedonkin, M. A., and B. M. Waggoner. 1997. The late Precambrian fossil *Kimberella* is a mollusc-like bilaterian organism. *Nature* 388: 868–871.

Ferrier, D. E. K., C. Minguillon, P. W. H. Holland, and J. Garcia-Fernandez. 2000. The amphioxus *Hox* cluster: deuterostome posterior flexibility and *Hox14*. *Evolution and Development* 2: 284–293.

Finney, S. C. 1986. Heterochrony, punctuated equilibrium, and graptolite zonal boundaries. In Hughes, C. P. and R. B. Rickards, eds., *Palaeoecology and Biostratigraphy of Graptolites,* pp. 103–113.

Fisher, R. A. 1918. The correlation between relatives on the supposition of Mendelian inheritance. *Trans. Royal Soc. Edinburgh* 52: 399–433.

——— 1930. *The Genetical Theory of Natural Selection.* Oxford UK: Oxford Univ. Press.

——— 1958. *The Genetical Theory of Natural Selection.* Second Edition. N.Y.: Dover.

Fitch, D. H. 1997. Evolution of male tail development in rhabditid nematodes related to *Caenorhabditis elegans. Syst. Biol.* 46: 145–179.

Flynn, L. J. 1986. Species longevity, stasis, and stairsteps in rhizomyid rodents. *Contributions to Geology, Univ. of Wyoming, Special Paper* 3: 273–285.

Foote, M., and S. J. Gould 1992. Cambrian and recent morphological disparity. *Science* 258: 1816.

Ford, V. S., and L. D. Gottlieb. 1992. *Bicalyx* is a natural homeotic floral variant. *Nature* 358: 671–673.

Fortey, R. A. 1985. Gradualism and punctuated equilibrium as competing and complementary theories. *Special Papers in Palaeontology* 33: 17–28.

——— 1988. Seeing is believing: gradualism and punctuated equilibria in the fossil record. *Sci. Prog.,* 72: 1–19.

Franco, A. O. 1985. La teoria del equilibrio puntuado. Una alternativa al Neodarwinismo. *Ciencias,* UNAM, Mexico City, pp. 46–59.

François, V., and E. Bier. 1995. *Xenopus chordin* and *Drosophila short gastrulation* genes encode homologous proteins functioning in dorsal-ventral axis formation. *Cell* 80: 19–20.

François, V., M. Solloway, J. W. O'Neill, J. Emery, and E. Bier. 1994. Dorsal-ventral patterning of the *Drosophila* embryo depends on a putative negative growth factor encoded by the *short gastrulation* gene. *Genes and Development* 8: 2602–2616.

Frank, L. G. 1997. Evolution of genital masculinization: why do female hyaenas have such a large "penis"? *Trends Ecol. Evol.* 12: 58–62.

Frazzetta, T. H. 1975. *Complex Adaptations in Evolving Populations.* Sunderland MA: Sinauer Asssociates.

Frohnhöfer, H. G., and C. Nüsslein-Volhard. 1986. Organization of anterior pattern in the *Drosophila* embryo by the maternal gene *bicoid. Nature* 324: 120–125.

Fryer, G., P. H. Greenwood, and J. F. Peake. 1983. Punctuated equilibria, morphological stasis, and the paleontological documentation of speciation: a biological appraisal of a case history in an African lake. *Biol. Jour. Linnaean Soc.* 20: 195–205.

Futuyma, D. J. 1979. *Evolutionary Biology.* Sunderland MA: Sinauer.

——— 1986. *Evolutionary Biology.* Sunderland MA: Sinauer.

——— 1987. On the role of species in anagenesis. *Amer. Nat.* 130: 465–473.

——— 1988a. Macroevolutionary Consequences of Speciation: Inferences from Phytophagous Insects. In J. Endler and D. Otte, eds., *Speciation and its Consequences.*

——— 1988b. *Sturm* and *Drang* and the evolutionary synthesis. *Evolution* 42: 217–226.

Galton, F. 1869. *Hereditary Genius.* London: Macmillan.

——— 1884. *Hereditary Genius: An inquiry into its laws and consequences.* New and Revised Edition. N.Y.: D. Appleton & Co.

——— 1889. *Natural Inheritance.* London: Macmillan.

——— 1892. *Fingerprints.* London: Macmillan.

——— 1894. Discontinuity in evolution. *Mind* 11: 362–372.

Gans, C. 1987. Punctuated Equilibria and political science: a neontological view. *Politics and the Life Sciences* 5: 220–244.

Gans, C., and R. G. Northcutt. 1983. Neural crest and the origin of vertebrates: a new head. *Science* 220: 268–274.

Garstang, W. 1928. The morphology of the Tunicata and its bearings on the phylogeny of the Chordata. *Quart. Jour. Exp. Biol.* 5: 112–134.

Gaskell, W. H. 1908. *The Origin of Vertebrates.* London: Longmans, Green and Co.

Gasman, D. 1971. *The Scientific Origins of National Socialism.* London: Macdonald.

Geary, D. H. 1990. Patterns of evolutionary tempo and mode in the radiation of *Melanopsis* (Gastropoda: Melanopsidae). *Paleobiology* 16: 492–511.

———— 1995. The importance of gradual change in species-level transitions. In D. H. Erwin and R. L. Anstey, eds., *New Approaches to Speciation in the Fossil Record.* N.Y.: Columbia Univ. Press, pp. 67–86.

Gehring, W. J. 1996. The master control gene for morphogenesis and evolution of the eye. *Genes Cells* 1: 11–15.

———— 1998. *Master Control Genes in Development and Evolution.* New Haven CT: Yale Univ. Press.

Geoffroy Saint-Hilaire, E. 1818. *Philosophie anatomique.* Vol. 1, *Des organes respiratoires sous le rapport de la détermination et de l'identité de leurs pièces osseuses.* Paris: J.-B. Baillière.

———— 1822. Considérations générales sur la vertèbre. *Mémoires du Muséum national d'histoire naturelle* 9: 88–119.

———— 1830. *Principes de philosophie zoologique discutés en Mars 1830 au sein de l'Académie Royale des Sciences.* Paris: Pichon et Didier.

———— 1831. Sur des écrits de Goethe lui donnant les droits au titre de savant naturaliste. *Ann. sci. nat.* 22: 188–193.

Gerhart, J. 2000. Inversion of the chordate body axis: are there alternatives? *Proc. Natl. Acad. Sci. USA* 97: 4445–4448.

Gerhart, J., and M. Kirschner. 1997. *Cells, Embryos and Evolution.* Oxford UK: Blackwell Scientific.

Gersick, C. J. G. 1988. Time and transition in work teams: toward a new model of group development. *Acad. Management Jour.* 31: 9–41.

———— 1991. Revolutionary change theories: a multi-level exploration of the punctuated equilibrium paradigm. *Acad. Management Rev.,* Jan., pp. 10–35.

Ghiselin, M. T. 1969. *The Triumph of the Darwinian Method.* Berkeley CA: Univ. of California Press.

———— 1974a. *The Economy of Nature and the Evolution of Sex.* Berkeley CA: Univ. of California Press.

———— 1974b. A radical solution to the species problem. *Syst. Zool.* 23: 536–544.

———— 1987. Species concepts, individuality, and objectivity. *Biol. Philos.* 2: 127–144.

Gilinsky, N. L. 1981. Stabilizing species selection in the Archaeogastropoda. *Paleobiology* 7: 316–331.

———— 1986. Species selection as a causal process. *Evol. Biol.* 20: 248–273.

———— 1994. Volatility and the Phanerozoic decline of background extinction intensity. *Paleobiology* 20: 445–458.

Gillispie, C. C. 1959. Lamarck and Darwin in the History of Science. In B. Glass, O. Temkin, and W. L. Strauss, eds., *Forerunners of Darwin: 1745–1859.* Baltimore MD: Johns Hopkins Univ. Press, pp. 265–291.

Gilluly, J. 1949. Distribution of mountain building in geological time. *Bull. Geol. Soc. Amer.* 60: 561–590.

Gilluly, J., A. C. Waters, and A. O. Woodford. 1959. *Principles of Geology.* 2nd ed. 1968. San Francisco: W. H. Freeman.

Gingerich, P. D. 1974. Stratigraphic record of early Eocene *Hyopsodus* and the geometry of mammalian phylogeny. *Nature* 248: 107–109.

—— 1976. Paleontology and phylogeny: patterns of evolution at the species level in early Tertiary mammals. *Am. Jour. Sci.* 276: 1–28.

—— 1978. Evolutionary transition from the ammonite *Subprionocyclus* to *Reedsites* – punctuated or gradual? *Evolution* 32: 454–456.

—— 1980. Evolutionary patterns in early Cenozoic mammals. *Ann. Rev. Earth Planet. Sci.* 8: 407–424.

—— 1984a. Punctuated equilibria—where is the evidence? *Syst. Zool.* 33: 335–338.

—— 1984b. Darwin's gradualism and empiricism: discussion and reply. *Nature* 309: 116.

—— (with reply by F. H. T. Rhodes) 1985. Darwin's gradualism and empiricism. *Nature* 309: 116.

—— 1987. Evolution and the fossil record: patterns, rates, and processes. *Can. Jour. Zool.* 65: 1053–1060.

—— 1989. New earliest Wasatchian mammalian fauna from the Eocene of northwestern Wyoming. *Univ. Mich. Pap. Paleontology* 28: 1–27.

Glaubrecht, M. 1995. *Der lange Atem der Schöpfung. Was Darwin gern gewusst hätte.* Berlin: Rasch und Röhring.

Gleick, J. 1987. The pace of evolution: a fossil creature moves to center of debate. *New York Times,* December 22.

Glen, W. 1982. *The Road to Jaramillo.* Stanford CA: Stanford Univ. Press.

—— 1994. *The Mass Extinction Debates: How Science Works in a Crisis.* Stanford CA: Stanford Univ. Press.

Glennon, L. (ed.) 1995. *Our Times.* Atlanta GA: Turner Publishing.

Godinot, M. 1985. Evolutionary implications of morphological changes in Palaeogene primates. In J. C. W. Cope and P. W. Skelton, eds., Evolutionary Case Histories From The Fossil Record. *Special Papers in Palaeontology* 33, pp. 39–47.

Goethe, J. W. von. 1790. *Versuch der Metamorphose der Pflanzen zu Erklären.* Gotha: Etting.

—— 1831. Reflexions de Goethe sur les débats scientifiques de mars 1830 dans le sein de l'Academie des Sciences. *Ann. sci. nat.* 22:179–188.

—— 1832. Derniers pages de Goethe expliquant à l'Allemagne les sujets de philosophie naturelle controversées au sein de l'Acadamie des Sciences de Paris. *Rev. encyc.* 53: 563–573 and 54: 54–68.

Gold, T. 1999. *The Deep Hot Biosphere.* N.Y.: Copernicus.

Goldschmidt, R. 1933. Some aspects of evolution. *Science* 78: 539–547.

—— 1940. *The Material Basis of Evolution.* New Haven CT: Yale Univ. Press.

—— 1955. *Theoretical Genetics.* Berkeley CA: Univ. of California Press.

—— 1960. *In and Out of the Ivory Tower: The Autobiography of Richard B. Goldschmidt.* Seattle WA: Univ. of Washington Press.

Golub, R., and E. Brus. 1990. *The Almanac of Science and Technology.* N.Y.: Harcourt Brace Jovanovich.

Goodfriend, G. A., and S. J. Gould. 1996. Paleontology and chronology of two evolutionary transitions by hybridization in the Bahamian land snail *Cerion. Science* 274: 1894–1897.

Goodwin, B. 1994. *How the Leopard Changed Its Spots: The Evolution of Complexity.* N.Y.: Simon & Schuster.

Gould, J. L., and C. G. Gould. 1989. *Sexual Selection.* N.Y.: W. H. Freeman, Scientific American Library.

Gould, J. L., and W. T. Keeton. 1996. *Biological Science.* N.Y.: W. W. Norton.

Gould, S. J. 1965. Is uniformitarianism necessary? *Amer. Jour. Sci.* 263: 223–228.

———— 1966. Allometry and size in ontogeny and phylogeny. *Biol. Rev.* 41: 587–640.

———— 1967. Evolutionary patterns in pelycosaurian reptiles: a factor-analytic study. *Evolution* 21: 385–401.

———— 1969. An evolutionary microcosm: Pleistocene and Recent history of the land snail P. (*Poecilozonites*) in Bermuda. *Bull. Mus. Comp. Zool.* 138: 407–532.

———— 1970a. Evolutionary paleontology and the science of form. *Earth-Sci. Rev.* 6: 77–119.

———— 1970b. Dollo on Dollo's law: irreversibility and the status of evolutionary laws. *Jour. Hist. Biol.* 3: 189–212.

———— 1971a. The paleontology and evolution of *Cerion*, II: age and fauna of Indian shell middens on Curaçao amd Aruba. *Mus. Comp. Zool., Breviora* 372: 1–26.

———— 1971b. D'Arcy Thompson and the science of form. *New Literary Hist.* 2: 229–258.

———— 1971c. Precise but fortuitous convergence in Pleistocene land snails from Bermuda. *Jour. Paleont.* 45: 409–418.

———— 1972. Allometric fallacies and the evolution of *Gryphaea*: a new interpretation based on White's criterion of geometric similarity. In Th. Dobzhansky et al., eds., *Evolutionary Biology*, vol. 6, pp. 91–118.

———— 1974. The origin and function of "bizarre" structures: antler size and skull size in the "Irish Elk," *Megaloceros giganteus*. *Evolution* 28: 191–220.

———— 1976. In defense of the analog: a commentary to N. Hotton. In R. B. Masterton, E. Hodos, and H. Jerison, eds., *Evolution, Brain and Behavior*. Hillsdale N.J.: Lawrence Erlbaum Assoc., pp. 175–179.

———— 1977a. Eternal metaphors of paleontology. In A. Hallam, ed., *Patterns of Evolution*. Amsterdam: Elsevier Sci. Publ. Co., pp. 1–26.

———— 1977b. *Ontogeny and Phylogeny*. Cambridge MA: Harvard Univ. Press.

———— 1977c. *Ever Since Darwin*. N.Y.: W. W. Norton.

———— 1980a. The promise of paleobiology as a nomothetic, evolutionary discipline. *Paleobiology* 6: 96–118.

———— 1980b. Is a new and general theory of evolution emerging? *Paleobiology* 6: 119–130.

———— 1980c. The evolutionary biology of constraint. *Daedalus* 109: 39–52.

———— 1980d. *The Panda's Thumb*. N.Y.: W. W. Norton.

———— 1980e. G. G. Simpson, Paleontology, and the Modern Synthesis. In E. Mayr and W. B. Provine, eds., *The Evolutionary Synthesis*. Cambridge MA: Harvard Univ. Press, pp. 153–172.

———— (ed.) 1980f. *The Evolution of Gryphaea*. N.Y.: Arno Press.

———— 1981a. *The Mismeasure of Man*. N.Y.: W. W. Norton.

———— 1981b. The rise of Neo-Lamarckism in America. *Internatl. Colloquium on Lamarck*. Paris: Librairie Philosophique J. Vrin, pp. 81–91.

———— 1981c. The Titular Bishop of Titiopolis. *Nat. Hist.* 90 May: 20–24.

———— 1982a. The uses of heresy: an introduction to Richard Goldschmidt's "The Material Basis of Evolution," pp. xiii–xlii. New Haven CT and London: Yale Univ. Press.

———— 1982b. Darwinism and the expansion of evolutionary theory. *Science* 216: 380–387.

———— 1982c. The meaning of punctuated equilibrium and its role in validating a hierarchical approach to macroevolution. In R. Milkman, ed., *Perspectives on Evolution*. Sunderland MA: Sinauer Associates, pp. 83–104.

———— 1982d. Introduction to Th. Dobzhansky, "Genetics and the Origin of Species." In N. Eldredge and S. J. Gould, eds., *The Columbia Classics in Evolution Series*. N.Y.: Columbia Univ. Press, pp. xvii–xli.

—— 1983a. Unorthodoxies in the first formulation of natural selection. *Evolution* 37: 856–858.

—— 1983b. The hardening of the Modern Synthesis. In: Marjorie Grene, ed., *Dimensions of Darwinism*. Cambridge UK: Cambridge Univ. Press.

—— 1983c. Irrelevance, submission, and partnership: the changing role of palaeontology in Darwin's three centennials, and a modest proposal for macroevolution. In D. S. Bendall, ed., *Evolution from Molecules to Men*. Cambridge UK: Cambridge Univ. Press.

—— 1984a. Toward the vindication of punctuational change. In W. W. Berggren and J. A. Van Couvering, eds., *Catastrophes and Earth History*. Princeton NJ: Princeton Univ. Press, pp. 9–34.

—— 1984b. Covariance sets and ordered geographic variation in *Cerion* from Aruba, Bonaire, and Curaçao: a way of studying nonadaptation. *Syst. Zool.* 33: 217–237.

—— 1984c. Morphological channeling by structural constraint: convergence in styles of dwarfing and gigantism in *Cerion,* with a description of two new fossil species and a report on the discovery of the largest *Cerion*. *Paleobiology* 10: 172–194.

—— 1984d. A most ingenious paradox. *Nat. Hist.* 93, Dec.: pp. 20–29.

—— 1985a. The paradox of the first tier: an agenda for paleobiology. *Paleobiology* 11: 2–12.

—— 1985b. All the news that's fit to print and some opinions that aren't. *Discover,* November: 86–91.

—— 1985c. *The Flamingo's Smile*. N.Y.: W. W. Norton, 476 pp.

—— 1986. Evolution and the triumph of homology, or why history matters. *Amer. Scientist,* Jan.–Feb.: 60–69.

—— 1987a. Freudian Slip. *Nat. Hist.* 96 (Feb.): 14–21.

—— 1987b. *Time's Arrow, Time's Cycle*. Cambridge MA: Harvard Univ. Press.

—— 1987c. *An Urchin in the Storm*. N.Y.: W. W. Norton, 255 pp.

—— 1988a. The case of the creeping fox terrier clone. *Nat. Hist.* 97 (Jan.): 16–24.

—— 1988b. Trends as changes in variance: a new slant on progress and directionality in evolution (Presidential Address). *Jour. Paleont.* 62: 319–329.

—— 1988c. Prolonged stability in local populations of *Cerion agassizi* (Pleistocene-Recent) on Great Bahama Bank. *Paleobiology* 14: 1–18.

—— 1989a. A developmental constraint in *Cerion,* with comments on the definition and interpretation of constraint in evolution. *Evolution* 43: 516–539.

—— 1989b. Full of Hot Air. *Nat. Hist.* 98 (Oct.): 28–38.

—— 1989c. *Wonderful Life: The Burgess Shale and the Nature of History*. N.Y.: W. W. Norton, 347 pp.

—— 1989d. The wheel of fortune and the wedge of progress. *Nat. Hist.* 98 (March): 14–21.

—— 1989e. Punctuated equilibrium in fact and theory. *J. Social Biol. Struct.* 12: 117–136.

—— 1991a. The disparity of the Burgess Shale arthropod fauna and the limits of cladistic analysis: why we must strive to quantify morphospace. *Paleobiology* 17: 411–423.

—— 1991b. *Bully for Brontosaurus*. N.Y.: W. W. Norton, 540 pp.

—— 1991c. The smoking gun of eugenics. *Nat. Hist.* 100 (Dec.): 8–17.

—— 1992a. "Red in Tooth and Claw." *Nat. Hist.* 101 (Nov.): 14–23.

—— 1992b. Constraint and the square snail: life at the limits of a covariance set. The normal teratology of *Cerion disforme*. *Biological Jour. Linnean Soc.* 47: 407–437.

—— 1993a. A special fondness for beetles. *Nat. Hist.* 102 (Jan.): 4–12.

———— 1993b. The inexorable logic of the punctuational paradigm: Hugo de Vries on species selection. In *Evolutionary Patterns and Processes*. London: The Linnean Soc. of London, pp. 3–18.

———— 1993c. How to analyze Burgess Shale disparity—a reply to Ridley. *Paleobiology* 19: 522–523.

———— 1993d. *The Book of Life*. Preface, pp. 6–21. N.Y.: W. W. Norton (S. J. Gould general editor, 10 contributors).

———— 1993e. *Eight Little Piggies*. N.Y.: W. W. Norton.

———— 1994. Tempo and mode in the macroevolutionary reconstruction of Darwinism. *Proc. Natl. Acad. Sci. USA* 91: 6764–6771.

———— 1995. *Dinosaur in a Haystack*. N.Y.: Harmony Books.

———— 1996a. *Full House: The Spread of Excellence from Plato to Darwin*. N.Y.: Harmony Books.

———— 1996b. A Lesson from the Old Masters. *Nat. Hist.* 105 (Aug.): 16–22, 58–59.

———— 1997a. The taxonomy and geographic variation of *Cerion* on San Salvador (Bahama Islands). *Proceedings 8th Symposium on the Geology of the Bahamas and other Carbonate Regions,* James L. Carew, ed. San Salvador, Bahamas: Bahamian Field Station Ltd., pp. 73–91.

———— 1997b. Cope's rule as psychological artifact. *Nature* 385: 199–200.

———— 1997c. As the worm turns. *Nat. Hist.* (Feb.) 106: 24–27, 68–73.

———— 1997d. Darwinian Fundamentalism, part 1. *The New York Review of Books,* June 12, pp. 34–37. Evolution: The Pleasures of Pluralism, part 2. *The New York Review of Books,* June 26, pp. 47–52.

———— 1997e. The exaptive excellence of spandrels as a term and prototype. *Proc. Natl. Acad. Sci. USA* 94: 10750–10755.

———— 1997f. The paradox of the visibly irrelevant. *Nat. Hist.* 106 (Dec.): 12–18, 60–66.

———— 1998a. The Great Asymmetry. *Science* 279: 812–813.

———— 1998b. *Leonardo's Mountain of Clams and The Diet of Worms*. N.Y.: Harmony Books.

———— 1999a. A Darwinian gentleman at Marx's funeral. *Nat. Hist.* 108 (Sept.): 32–33, 56–66.

———— 1999b. *Rocks of Ages: Science and Religion in the Fullness of Life*. N.Y.: Ballantine Publ., 241 pp.

———— 2000a. What does the dreaded "E" word mean anyway? *Nat. Hist.* 109 (Feb.): 28–44.

———— 2000b. Beyond competition. *Paleobiology* 26: 1–6.

———— 2000c. Linnaeus's Luck? *Nat. Hist.* 109 (Sept.): 18–25, 66–76.

———— 2000d. A Tree Grows in Paris: Lamarck's division of Worms and Revision of Nature. In: S. J. Gould, *The Lying Stones of Marrakech: Penultimate Reflections in natural History*. N.Y.: Harmony Books, pp. 115–143.

———— 2000e. Of coiled oysters and big brains: how to rescue the terminology of heterochrony, now gone astray. *Evolution and Development* 2: 241–248.

———— 2001. Humbled by the Genome's Mysteries. *N.Y. Times* Op-Ed. Feb 19.

Gould, S. J., and C. B. Calloway. 1980. Clams and brachiopods—ships that pass in the night? *Paleobiology* 6: 383–396.

Gould, S. J., and N. Eldredge. 1971. Speciation and punctuated equilibria: an alternative to phyletic gradualism. G. S. A. Ann. Meeting, Washington, DC, *Abstracts with Programs,* pp. 584–585.

———— 1977. Punctuated equilibria: the tempo and mode of evolution reconsidered. *Paleobiology:* 3: 115–151.

———— 1983. Darwin's gradualism. *Systematic Zool.* 32: 444–445.

———— 1986. Punctuated equilibrium at the third stage. *Systematic Zool.* 35: 143–148.

—— 1988. Species selection: its range and power. *Nature* 334: 19.

—— 1993. Punctuated equilibrium comes of age. *Nature* 366: 223–227.

Gould, S. J., N. L. Gilinsky, and R. Z. German. 1987. Asymmetry of lineages and the direction of evolutionary time. *Science* 236: 1437–1441.

Gould, S. J., and R. C. Lewontin. 1979. The spandrels of San Marco and the Panglossian paradigm: a critique of the adaptationist programme. *Proc. R. Soc. Lond. B* 205: 581–598.

Gould, S. J., and E. A. Lloyd. 1999. Individuality and adaptation across levels of selection: how shall we name and generalize the unit of Darwinism? *Proc. Natl. Acad. Sci. USA* 96: 11904–11909.

Gould, S. J., D. M. Raup, J. J. Sepkoski, Jr., T. J. M. Schopf, and D. S. Simberloff. 1977. The shape of evolution: a comparison of real and random clades. *Paleobiology* 3: 23–40.

Gould, S. J., and S. Vrba, 1982. Exaptation—a missing term in the science of form. *Paleobiology* 8:4–15.

Gould, S. J., and D. S. Woodruff. 1986. Evolution and systematics of *Cerion* (Mollusca: Pulmonata) on New Providence Island: a radical revision. *Bull. Amer. Mus. Nat. Hist.* 182: 389–490.

—— 1987. Systematics and levels of covariation in *Cerion* from the Turks and Caicos. *Bull. Mus. Comp. Zool.* 151: 321–363.

—— 1990. History as a cause of area effects: an illustration from *Cerion* on Great Inagua, Bahamas. *Biol. Jour. Linn. Soc.* 40: 67–98.

Gould, S. J., N. D. Young, and Bill Kasson. 1985. The consequences of being different: sinistral coiling in *Cerion*. *Evolution* 39: 1364–1379.

Gorman, J. 1980. The tortoise or the hare? *Discover,* October.

Grant, P. 1986. *Ecology and Evolution of Darwin's Finches.* N.J.: Princeton Univ. Press.

Grant, V. 1983. The Synthetic Theory strikes back. *Biol. Zentralblatt* 102: 149–158.

Grantham, T. A. 1995. Hierarchical approaches to macroevolution: Recent work on species selection and the "effect hypothesis." *Ann. Rev. Ecol. Syst.* 26: 301–321.

Greiner, G. O. G. 1974. Environmental factors controlling the distribution of Recent benthic Foraminifera. *Breviora Mus. Comp. Zool. Harvard Univ.* Number 420.

Grenier, J. K., T. L. Garber, R. Warren, P. M. Whitington, and S. Carroll. 1997. Evolution of the entire arthropod *Hox* gene set predated the origin and radiation of the onychophoran/arthropod clade. *Current Biology* 7: 547–553.

Griffis, K., and D. J. Chapman. 1988. Survival of phytoplankton under prolonged darkness. Implications for the Cretaceous-Tertiary boundary darkness hypothesis. *Palaeogeog. Palaeoclimatol. Palaeoecol.* 67: 305–314.

Grine, F. E. 1993. Australopithecine taxonomy and phylogeny. In R. L. Ciochon and J. Fleagle, eds., *The Human Evolution Source Book.* Englewood Cliffs N.J.: Prentice Hall, pp. 145–175.

Gruber, H. E., and P. H. Barrett. 1974. *Darwin on Man: A Psychological Study of Scientific Creativity.* N.Y.: Dutton.

Haas, O., and G. G. Simpson. 1946. Analysis of some phylogenetic terms with attempts at redefinition. *Proc. Amer. Phil. Soc.* 90: 319–349.

Haeckel, E. 1866. *Generelle Morphologie der Organismen.* Berlin: Georg Reimer.

—— 1909. Charles Darwin as an anthropologist. In A. C. Seward, ed., *Darwin and Modern Science.* Cambridge UK: Cambridge Univ. Press, pp. 137–151.

Haim, A., G. Heth, H. Pratt, and E. Nevo. 1983. Photoperiodic effects of the thermoregulation in a "blind" subterranean mammal. *Jour. Experimental Biol.* 107: 59–64.

Halanych, K. M., J. D. Bacheller, A. M. A. Aguinaldo, S. M. Liva, D. M. Hillis, and J. A. Lake. 1995. Evidence of 18S ribosomal DNA that the lophophorates are protostome animals. *Science* 267: 1641–1643.

Haldane, J. B. S. 1932. *The Causes of Evolution.* London: Longmans Green.

Halder, G., P. Callaerts, and W. J. Gehring. 1995. Induction of ectopic eyes by targeted expression of the *eyeless* gene in *Drosophila. Science* 267: 1788–1792.

Hallam, A. 1968. Morphology, palaeoecology and evolution of the genus *Gryphaea* in the British Lias. *Phil. Trans. Roy. Soc. London* 254: 91–128.

——— 1978. How rare is phyletic gradualism and what is its evolutionary significance? Evidence from Jurassic bivalves. *Paleobiology* 4:16–25.

——— 1990. Biotic and abiotic factors in the evolution of early marine molluscs. In R. M. Ross and W. D. Allmon, eds., *Causes of Evolution: A Paleontological Perspective.* Chicago IL: Univ. of Chicago Press, pp. 249–269.

——— 1997. Speciation patterns and trends in the fossil record. *GEOBIOS* 30: 921–930.

Halstead, B. 1984. Neo-Darwinism rules. *New Scientist,* May 3, p. 40.

——— 1985. The Evolution debate continues. *Modern Geology* 9: 317–326.

Hamilton, W. D. 1964. The genetical evolution of social behavior. *Jour. Theoret. Biol.* 7: 1–52.

——— 1971. Selection of selfish and altruistic behavior in some extreme models. In J. E. Eisenberg and W. S. Dillon, eds., *Man and Beast. Comparative Social Behavior.* Washington DC: Smithsonian Inst. Press, pp. 57–91.

——— 1987. Discriminating nepotism: expectable, common, overlooked. In D. S. C. Fletcher and C. D. Michener, eds., *Kin Recognition in Animals.* N.Y.: John Wiley, pp. 417–437.

Hansen, T. A. 1978. Larval dispersal and species longevity in Lower Tertiary gastropods. *Science* 199: 885–887.

——— 1980. Influence of larval dispersal and geographic distribution on species longevity in neogastropods. *Paleobiology* 6: 193–207.

Hanson, N. R. 1961. *Patterns of Discovery.* Cambridge UK: Cambridge Univ. Press.

Harraway, D. 1989. *Primate Visions: Gender, Race and Nature in the World of Modern Science.* N.Y.: Routledge.

——— 1991. *Simians, Cyborgs and Women: The Reinvention of Nature.* London: Free Association Books.

Hatcher, J. B., O. C. Marsh, and R. S. Lull. 1907. The Ceratopsia. *Monographs U.S. Geol. Survey,* volume 49.

Heaton, T. H. 1993. The Ologocene rodent *Ischyromys* of the Great Plains: replacement mistaken for anagenesis. *Jour. Paleontology* 67: 297–308.

——— 1996. Ischyromyidae. In *The Terrestrial Eocene-Oligocene Transition in North America,* pp. 373–398.

Hempel, C. G. 1965. *Aspects of Scientific Explanation.* N.Y.: Free Press.

Hendriks, W., T. Leunissen, E. Nevo, H. Bloemendal, and W. W. de Jong. 1987. The lens protein alpha-A-crystallin of the blind mole rat *Spalax ehrenbergi:* evolutionary change and functional constraints. *Proc. Natl. Acad. Sci. USA* 84: 5320–5324.

Hilgendorf, F. 1866. *Planorbis multiformis* im Steinheimer Süsswasserkalk. Ein Beispiel von Gestaltsveränderung im Laufe der Zeit. Berlin: Weber.

Hofer, J., L. Turner, R. Hellens, M. Ambrose, P. Matthews, A. Michael, and N. Ellis. 1997. *Unifoliata* regulates leaf and flower morphogenesis in pea. *Current Biology* 7: 581–587.

Hoffman, P. F., A. J. Kaufman, G. P.. Halverson, and D. P. Schrag. 1998. A Neoproterozoic snowball earth. *Science* 281: 1342–1346.

Hoffman, A. 1989. *Arguments on Evolution.* N.Y.: Oxford.

Hogan, B. L. M. 1995. Upside-down ideas vindicated. *Nature* 376: 210–211.

Holland, L. Z., and N. D. Holland. 1998. Developmental gene expression in amphioxus: new insights into the evolutionary origin of vertebrate brain regions, neural crest, and rostrocaudal segmentation. *Amer. Zool.* 38: 647–658.

Holland, L. Z., M. Kene, N. A. Williams, and N. D. Holland. 1997. Sequence and embryonic expression of the amphioxus engrailed gene (*AmphiEn*): the metameric pattern of transcription resembles that of its segment-polarity homolog in *Drosophila. Development* 124: 1723–1732.

Holland, P. W. H., L. Z. Holland, N. A. Williams, and N. D. Holland. 1992. An amphioxus homeobox gene: sequence conservation, spatial expression during development and insight into vertebrate evolution. *Development* 116: 653–661.

Holley, S. A., P. D. Jackson, Y. Sasai, B. Lu, E. M. De Robertis, F. M. Hoffmann, and E. L. Ferguson. 1995. A conserved system for dorsal-ventral patterning in insects and vertebrates involving *sog* and *chordin. Nature* 376: 249–253.

Honma, T., and K. Goto. 2001. Complexes of MADS-box proteins are sufficient to convert leaves into floral organs. *Nature* 409: 525–529.

Hooykaas, R. 1963. *The Principle of Uniformity in Geology, Biology, and Theology.* Leiden: E. J. Brill.

Houston, A. I. 1997. Are the spandrels of San Marco really panglossian pendentives? *Trends Ecol. Evol.* 12: 125.

Howe, J. A. 1956. The Oligocene rodent *Ischyromys* in relationship to the paleosols of the Brule Formation. MS. Thesis, Univ. of Nebraska, 89 pp.

Howells, W. 1959. *Mankind in the Making.* Garden City NJ: Doubleday.

Hughes, N. C. 2000. The rocky road to Mendel's play. *Evol. and Develop.* 2: 63–66.

Hull, D. L. 1973. *Darwin and His Critics.* Cambridge MA: Harvard Univ. Press.

——— 1976. Are species really individuals? *Systematic Zool.* 25: 174–191.

——— 1980. Individuality and selection. *Ann. Rev. Ecol. Systematics* 11: 311–332.

——— 1984. Lamarck among the Anglos. In J.-B. Lamarck, *Zoological Philosophy.* Chicago IL: Univ. of Chicago Press, pp. xi–lxvi.

——— 1985. Darwinism as a historical entity. In D. Kohn, ed., *The Darwinian Heritage.* Princeton N.J.: Princeton Univ. Press, pp. 773–812.

——— 1988. *Science as a Process.* Chicago IL: Univ. of Chicago Press.

Hummelinck, P. W. 1940. Mollusks of the genera *Cerion* and *Tudora. Studies of the Fauna of Curaçao, Aruba, Bonaire and the Venezuelan Islands* 2: 43–82.

Hutchinson, G. E. 1948. In memoriam D'Arcy Wentworth Thompson. *Amer. Scientist* 36: 577–606.

Huxley, A. 1982. Address of the President. *Proc. Roy. Soc. London Series B* 214: 137–152.

Huxley, J. S. 1932. *Problems of Relative Growth.* London: Methuen & Co.

——— 1942. *Evolution, the modern synthesis.* London: Allen and Unwin.

——— 1953. *Evolution in Action.* London: Chatto & Windus.

——— 1960. The Evolutionary Vision. In S. Tax and C. Callender, eds., *Evolution After Darwin*, Volume III. *Issues in Evolution,* pp. 249–261.

Huxley, T. H. 1893. Evolution and Ethics. In *Evolution and Ethics and Other Essays.* Volume 9 (published 1894) of T. H. Huxley's Collected Essays. N.Y.: D. Appleton.

——— 1894 (reprint of address given in 1868). Presidential Address, Geological Society of London. In *Discourses, Biological and Geological Essays.* N.Y.: D. Appleton.

Hyatt, A. 1880. The genesis of the Tertiary species of *Planorbis* at Steinheim. *Anniversary Mem. Boston Soc. Nat. Hist.* (1830–1880), pp. 1–114.

——— 1889. Genesis of the Arietidae. *Bull. Mus. Comp. Zool. Harvard Univ.* 16: 1–238.

——— 1897a. Cycle in the life of the individual (ontogeny) and in the evolution of its own group (phylogeny). *Proc. Am. Acad. Arts Sci.* 32: 209–224.

——— 1897b. The influence of woman in the evolution of the human race. *Natural Science* 11: 89–93.

Hyde, W. T., T. J. Crowley, S. K. Baum, and W. R. Peltier. 2000. Neoproterozoic "snowball Earth" simulations with a coupled climate/ice-sheet model. *Nature* 405: 425–429.

Imbrie, J. 1957. The species problem with fossil animals. In E. Mayr, ed., *The Species Problem.* Am. Assoc. Adv. Sci. Pub. No. 50, pp. 125–153.

Ivany, L. C. 1996. Coordinated stasis or coordinated turnover? Exploring intrinsic *vs.* extrinsic controls on pattern. *Palaeogeog. Palaeoclimat. Palaeoecol.* 127:1–18.

Ivany, L. C., and K. M. Schof, eds., 1996. New Perspetives on Faunal Stability in the Fossil Record. *Special Issue of Palaeogeog. Palaeoclimatol. Palaeoecol.* volume 127, 359 pp.

Jablonski, D. 1986a. Larval ecology and macroevolution in marine invertebrates. *Bull. Marine Sci.* 39: 565–587.

———— 1986b. Background and mass extinctions: the alternation of macroevolutionary regimes. *Science* 231: 129–133.

———— 1987. Heritability at the species level: analysis of geographic ranges of Cretaceous mollusks. *Science* 238: 360–363.

———— 1996. Body size and macroevolution. In D. Jablonski, D. H. Erwin, and J. H. Lipps, eds., *Evolutionary Paleobiology.* Chicago IL: Univ. of Chicago Press, pp. 256–289.

———— 1997. Body-size evolution in Cretaceous molluscs and the status of Cope's rule. *Nature* 385: 250–252.

———— 1999. The future of the fossil record. *Science* 284: 2114–2116.

Jablonski, D., and D. J. Bottjer. 1983. Soft-substratum epifaunal suspension-feeding assemblages in the late Cretaceous: implications for the evolution of benthic communities. In M. J. Tevesz and P. L. McCall, eds., *Biotic Interactions in Recent and Fossil Benthic Communities.* N.Y.: Plenum, pp. 747–812.

Jablonski, D., S. Lidgard, and P. D. Taylor. 1997. Comparative ecology of bryozoan radiations: origin of novelties in cyclostomes and cheilostomes. *Palaios* 12: 505–523.

Jablonski, N. G., and G. Chaplin. 1999. Chimp cultural diversity. *Science* 285: 836–837.

Jackson, J. B. C., and A. H. Cheetham. 1990. Evolutionary significance of morphospecies: a test with Cheilostome Bryozoa. *Science* 248: 579–582.

———— 1994. Phylogeny reconstruction and the tempo of speciation in cheilostome Bryozoa. *Paleobiology* 20: 407–423.

———— 1999. Tempo and mode of speciation in the sea. *Trends Ecol. Evol.* 14: 72–77.

Jacobs, K., and L. Godfrey. 1982. Cerebral leaps and bounds: a punctuational perspective on hominid cranial capacity increase. *Man and His Origins* 21: 77–87.

Janzen, D. 1977. What are dandelions and aphids? *Amer. Nat.* 111: 586–589.

———— 1985. On ecological fitting. *Oikos* 45: 308–310.

Jenkin, H. C. F. 1867. "The Origin of Species." *North British Review* 46: 277–318.

Johanson, D., and M. Edey. 1981. *Lucy.* N.Y.: Simon & Schuster.

Johanson, D., and B. Edgar. 1996. *From Lucy to Language.* N.Y.: Simon & Schuster.

Johnson, A. L. A. 1985. The rate of evolutionary change in European Jurassic scallops. In J. C. W. Cope and P. W. Skelton, eds., *Evolutionary Case Histories From The Fossil Record. Special Papers in Palaeontology* 33: 91–102.

Johnson, J. G. 1975. Allopatric speciation in fossil brachiopods. *Jour. Paleontol.* 49: 646–661.

———— 1982. Occurrence of phyletic gradualism and punctuated equilibria through geological time. *Jour. Palaeontol.* 56: 1329–1331.

Jones, D. S., and S. J. Gould. 1999. Direct measurement of age in fossil *Gryphaea:* the solution to a classic problem in heterochrony. *Paleobiology* 25: 158–187.

Jones, K. C., and A. J. Gaudin. 1977. *Introductory Biology.* N.Y.: John Wiley.

Jukes, T. H. 1991. Early development of the neutral theory. *Perspectives Biol. Medicine* 34: 473–485.

Jürgens, G. 1997. Memorizing the floral ABC. *Nature* 386: 17.

Kammer, T. W., T. K. Baumiller, and W. I. Ausich. 1997. Species longevity as a function of niche breadth: Evidence from fossil crinoids. *Geology* 25: 219–222.

Kauffman, S. A. 1993. *The Origins of Order: Self-Organization and Selection in Evolution.* Oxford: Oxford Univ. Press.

Kazazian, H. H., Jr. 2000. L1 retrotransposons shape the mammalian genome. *Science* 289: 1152–1153.

Kelley, P. H. 1983. Evolutionary patterns of eight Chesapeake group molluscs: evidence for the model of punctuated equilibria. *Jour. Paleontol.* 57: 581–598.

——— 1984. Multivariate analysis of evolutionary patterns of seven Miocene Chesapeake Group molluscs. *Jour. Paleontol* 58: 1235–1250.

Kellogg, V. L. 1907. *Darwinism Today.* London: G. Bell & Sons.

——— 1917. *Headquarters Nights.* Boston.

Kemp, P. D., and M. D. Bertness. 1984. Snail shape and growth rates: evidence for plastic shell allometry in *Littorina littorea. Proc. Natl. Acad. Sci. USA* 81: 811–813.

Kerr, R. A. 1994. Between extinctions, evolutionary stasis. *Science* 266: 29.

——— 1995. Did Darwin get it all right? *Science* 267: 1421–1422.

Kessel, M., R. Balling, and P. Gruss. 1990. Variations of cervical vertebrae after expression of a *Hox-1.1* transgene in mice. *Cell* 61: 301–308.

Kessel, M., and P. Gruss. 1991. Murine developmental control genes. *Science* 249: 374–379.

Kettle, C., W. Arthur, T. Jowett, and A. Minelli. 1999. Homeotic transformation in a centipede. *Trends in Genetics* 15: 393.

Kilgour, F. G. 1998. *The Evolution of the Book.* N.Y.: Oxford Univ. Press.

Kimbel, W. H., D. C. Johanson, and Y. Rak. 1994. The first skull and other new discoveries of *Australopithecus afarensis* at Hadar, Ethiopia. *Nature* 368: 449–451.

Kimmel, C. B. 1996. Was *Urbilateria* segmented? *Trends in Genetics* 12: 329–331.

Kimura, M. 1968. Evolutionary rate at the molecular level. *Nature* 217: 624–626.

——— 1983. *The Neutral theory of Molecular Evolution.* Cambridge UK: Cambridge Univ. Press.

——— 1985. The neutral theory of molecular evolution. *New Scientist,* pp. 41–46.

——— 1991a. The neutral theory of molecular evolution: A review of recent evidence. *Japanese Journal of Genetics* 66: 367–386.

——— 1991b. Recent development of the neutral theory viewed from the Wrightian tradition of theoretical population genetics. *Proc. Natl. Acad. Sci. USA* 88: 5969–5973.

King, M.-C., and A. C. Wilson. 1975. Evolution at two levels in humans and chimpanzees. *Science* 188: 107–116.

Kingsolver, J. G., and M. A. R. Koehl. 1985. Aerodynamics, thermoregulation, and the evolution of insect wings: differential scaling and evolutionary change. *Evolution* 39: 488–504.

Kinsey, A. C. 1936. The origin of higher categories in *Cynips. Indiana Univ. Publ. Science Series* No. 4, 336 pp.

Kinsey, A. C., W. C. Pomeroy, C. E. Martin, and P. H. Gebhard. 1953. *Sexual Behavior in the Human Female.* Philadelphia PA: W. B. Saunders.

Kirschner, M., and J. Gerhart. 1998. Evolvability. *Proc. Natl. Acad. Sci. USA* 95: 8420–8427.

Kitchell, J. A., D. L. Clark, and A. M. Gombos, Jr. 1986. Biological selectivity of extinction: a link between background and mass extinction. *Palaios* 1: 504–511.

Kitcher, P. 1985. Darwin's achievements. In N. Rescher, ed., *Reason and Rationality in Natural Science.* N.J.: Univ. Press of America, pp. 127–189.

Knoll, A. H., and S. B. Carroll. 1999. Early animal evolution: emerging views from comparative biology and geology. *Science* 284: 2129–2137.

Kobayashi, M., H. Furuya, and P. W. H. Holland. 1999. Dicyemids are higher animals. *Nature:* 401: 762.

Koestler, A. 1971. *The Case of the Midwife Toad.* N.Y.: Random House.

Kohn, D. 1980. Theories to work by: rejected theories, reproduction, and Darwin's path to natural selection. *Studies Hist. Biol.* 4: 67–170.

———— 1981. On the origin of the principle of diversity. *Science* 213: 1105–1108.

———— 1985a. Darwin's principle of divergence as internal dialogue. In D. Kohn, ed., *The Darwinian Heritage.* Princeton N.J.: Princeton Univ. Press, pp. 245–257.

———— 1985b. *The Darwinian Heritage.* Princeton N.J.: Princeton Univ. Press.

Konner, M. 1986. Revolutionary biology. *The Sciences,* p. 608.

Korey, K. 1984. *The Essential Darwin.* Boston MA: Little, Brown.

Kottler, M. J. 1985. Charles Darwin and Alfred Russel Wallace: two decades of debate over natural selection. In D. Kohn, ed., *The Darwinian Heritage.* Princeton NJ: Princeton Univ. Press, pp. 367–432.

Kramer, E. M., and V. F. Irish. 1999. Evolution of genetic mechanisms controlling petal development. *Nature* 399: 144–148.

Kraus, D. 1983. *Concepts in Modern Biology.* N.Y.: Globe Books.

Krishtalka, L., and R. K. Stuckey. 1985. Revision of the Wind River faunas, early Eocene of central Wyoming. Part 7. Revision of *Diacodexis* (Mammalia, Artiodactyla). *Ann. Carnegie Mus.* 54:413–486.

———— 1986. Early Eocene artiodactyls from the San Juan Basin, New Mexico, and the Piceance Basin, Colorado. In K. M. Flanagan and J. A. Lillegraven, eds., *Vertebrates, Phylogeny and Paleontology. Univ. Wyo. Contrib. Geol. Spec. Pap.* 3: 183–197.

Kropotkin, P. A. 1902. *Mutual Aid: A Factor of Evolution.* N.Y.: McClure Phillips.

Kruuk, H. 1972. *The Spotted Hyena.* Chicago IL: Univ. of Chicago Press.

Kucera, M., and B. A. Malmgren. 1998. Differences between evolution of mean form and evolution of new morphotypes: an example from Late Cretaceous planktonic foraminifera. *Paleobiology* 24: 49–63.

Kuhn, T. S. 1962. *The Structure of Scientific Revolutions.* Chicago IL: Univ. of Chicago Press.

———— 1969. *Postscript to Second Edition of The Structure of Scientific Revolutions.* Chicago IL: Univ. of Chicago Press.

Lamarck, J.-B. 1801. *Système des Animaux sans vertèbres, ou Tableau général des classes, des ordres et des genres de ces animaux; . . . précédé du Discours d'ouverture du cours de zoologie donné dans le Muséum national d'Histoire naturelle, l'an VIII de la République, le 21 floréal.* Paris: Déterville.

———— 1802a. *Hydrogéologie ou recherches sur l'influence qu'ont les eaux sur la surface du globe terrestre.* Paris: chez l'auteur et Agasse, Maillard.

———— 1802b. *Recherches sur l'organisation des corps vivans.* Paris: Maillard.

———— 1809. *Philosophie zoologique, ou exposition des considérations relatives à l'histoire naturelle des animaux,* 2 vol. Paris: Dentu.

———— 1815–1822. *Histoire naturelle des animaux sans vertèbres, présentant les caractères généraux et particuliers de ces animaux, leur distribuiton, leurs classes, leurs familles, leurs genres, et la citation des principales espèces qui s'y rapportent.* 7 vols. Paris: Déterville.

Lampl, M., J. D. Veldhuis, and M. L. Johnson. 1992. Saltation and stasis: a model of human growth. *Science* 258: 801–803.

Lande, R. 1976. Natural selection and random genetic drift in phenotypic evolution. *Evolution* 30: 314–334.

———— 1986. The dynamics of peak shifts and the pattern of morphological evolution. *Paleobiology* 12: 343–354.

Lang, W. D. 1923. Evolution: a resultant. *Proc. Geologists' Assoc.* 34: 7–20.

Lankester, E. R. 1870. On the use of the term homology in modern zoology, and the distinction between homogenetic and homoplastic agreements. *Annals Mag. Nat. Hist.* 4th series, 6: 34–43.

Lawless, J. V. 1988. Punctuated equilibrium and paleohydrology. *Proc. New Zealand Geothermal Workshop* 10: 165–169.

Leakey, M. G., F. Spoor, F. H. Brown, P. N. Gathogo, C. Klarie, L. N. Leakey, and I. McDougall. 2001. New hominin genus from eastern Africa shows diverse middle Pliocene lineages. *Nature* 410: 433–440.

Le Conte, T. 1888. *Evolution: Its Nature, Its Evidences, and Its Relation to Religious Thought.* N.Y.: D. Appleton.

Lee, D. C., P. Gonzalez, and G. Wistow. 1994. $\xi$-Crystallin: a lens-specific promoter and the gene recruitment of an enzyme as a crystallin. *Jour. Molec. Biol.* 236: 669–678.

Lee, S. E., and D. K. Jacobs. 1999. Expression of Distal-less in molluscan eggs, embryos, and larvae. *Evolution and Development* 1: 172–179.

Leigh Jr., E. G. 1977. How does selection reconcile individual advantage with the good of the group. *Proc. Natl. Acad. Sci. USA* 74: 4542–4546.

———— 1991. Genes, bees and ecosystems: the evolution of a common interest among individuals. *Trends Ecol. Evol.* 6: 257–262.

Lemen, C. A., and P. W. Freeman. 1981. A test of macroevolutionary problems with neoontological data. *Paleobiology* 7: 311–315.

———— 1989. Testing macroevolutionary hypotheses with cladistic analysis: evidence against rectangular evolution. *Evolution* 43: 1538–1554.

Le Mouellic, H., Y. Lallemand, and P. Brulet. 1992. Homoeosis in the mouse induced by a null mutation in the *Hox-3.1* gene. *Cell* 69: 251–264.

Lenski, R. E., and M. Travisano. 1994. Dynamics of adaptation and diversification: a 10,000-generation experiment with bacterial populations. *Proc. Natl. Acad. Sci. USA* 91: 6808–6814.

Lerner, I. M. 1954. *Genetic Homeostasis.* N.Y.: John Wiley.

———— 1959. The concept of natural selection: a historical view. *Proc. Amer. Philosophical Soc.* (Special Volume for the Centennial of Darwin's *Origin of Species*.)

Leroi-Gourhan, A. 1967. *Treasures of Prehistoric Art.* N.Y.: H. N. Abrams.

Lester, T., and P. Bowler. 1995. *E. Ray Lankester and the Making of British Biology.* London: British Society for the History of Science.

Levin, H. L. 1991. *The Earth Through Time.* Fort Worth TX: W. B. Saunders.

Levine, D. 1991. Punctuated Equilibrium: the modernization of the proletarian family in the age of ascendant capitalism. *International Labor and Working Class History* No. 39.

Levinton, J. 1988. *Genetics, Paleontology, and Macroevolution.* Cambridge UK: Cambridge Univ. Press.

Lewin, R. 1980. Evolutionary theory under fire. *Science* 210: 883–887.

———— 1986. Punctuated Equilibrium is now old hat. *Science* 231:672–673.

———— 1996. Evolution's new heretics: a growing number of evolutionary biologists think that the interests of groups sometimes supersede those of individuals. *Natural History* 105 (1996): 12–17.

Lewis, E. B. 1978. A gene complex controlling segmentation in *Drosophila. Nature* 276: 565–570.

———— 1992. Clusters of master control genes regulate the development of higher organisms. *Jour. Amer. Medical Assoc.* 267: 1524–1531.

Lewontin, R. C. 1970. The units of selection. *Ann. Rev. Ecol. Systematics* 1: 1–18.

———— 1974. *The Genetic Basis of Evolutionary Change.* N.Y.: Columbia Univ. Press.

———— 1978. Adaptation. *Scientific American* 239: 156–169.

———— 2000. *The Triple Helix.* Cambridge MA: Harvard Univ. Press.

Lich, D. K. 1990. *Cosomys primus:* a case for stasis. *Paleobiology* 16: 384–395.

Lieberman, B. S. 1995. Phylogenetic trends and speciation: analyzing macroevolutionary processes and levels of selection. In D. H. Erwin and R. L. Anstey, eds., *New Approaches to Speciation in the Fossil Record.* N.Y.: Columbia Univ. Press, pp. 316–337.

Lieberman, B. S., W. D. Allmon, and N. Eldredge. 1993. Levels of selection and macroevolutionary patterns in the turritellid gastropods. *Paleobiology* 19: 205–215.

Lieberman, B. S., C. E. Brett, and N. Eldredge. 1994. Patterns and processes of stasis in two species lineages of brachiopods from the Middle Devonian of New York State. *Amer. Mus. Nat. Hist. Novitates* Number 3114.

—— 1995. A study of stasis and change in two species lineages from the Middle Devonian of New York state. Paleobiology 21: 15–27.

Lieberman, B. S. and S. Dudgeon. 1996. An evaluation of stabilizing selection as a mechanism for stasis. *Palaeogeog. Palaeoclimat. Palaeoecol.* 127: 229–238.

Lieberman, B. S. and E. S. Vrba. 1995. Hierarchy theory, selection, and sorting. *BioScience* 45: 394–398.

Liem, K. F. 1973. Evolutionary strategies and morphological innovations: cichlid pharyngeal jaws. *Systematic Zool.* 22: 425–441.

—— 1988. Form and function of lungs: the evolution of airbreathing mechanisms. *Amer. Zool.* 28: 739–759.

Limoges, C. 1968. Darwin, Milne-Edwards et le principe de divergence. *Actes XII Cong. Internat. Hist. Sci.* 8: 111–119.

Lindberg, D. R., and R. A. Dobbertsen. 1981. Umbilical brood protection and sexual dimorphism in the boreal trochid gastropod *Margarites vorticiferus* Dall. *Jour. Invert. Reprod.* 3: 347–355.

Lipps, J. H., and P. W. Signor. 1993. *Fossil Prokaryotes and Protists.* Boston MA: Blackwell Scientific Publishers.

Lister, A. M. 1993a. "Gradualistic" evolution: its interpretation in Quaternary large mammal species. *Quarternary International* 19: 77–84.

—— 1993b. Mammoths in miniature. *Nature* 362: 288–289.

—— 1994. The evolution of the giant deer, *Megaloceros giganteus* (Blumenbach). *Jour. Linnaean Soc. London* 112: 65–100.

—— 1996. Dwarfing in island elephants and deer: processes in relation to time and isolation. *Symp. Zool. Soc. Lond.* 69: 277–292.

Lloyd, E. A. 1988. *The Structure and Confirmation of Evolutionary Theory.* N.Y.: Greenwood Press.

Lloyd, E. A., and S. J. Gould. 1993. Species selection on variability. *Proc. Natl. Acad. Sci. USA* 90: 595–599.

Loch, Christoph H. 1999. A punctuated equilibrium model of technology diffusion. Abstract for Dynamics of Computation Group Meeting, Xerox Palo Alto Research Center.

Losos, J. B., K. I. Warheit, and T. W. Schoener. 1997. Adaptive differentiation following experimental island colonization in *Anolis* lizards. *Nature* 387: 70–73.

Lowe, C. J., and G. A. Wray. 1997. Radical alterations in the roles of homeobox genes during echinoderm evolution. *Nature* 389: 718–721.

Lufkin, T., M. Mark, C. P. Hart, P. Dollé, M. LeMeur, and P. Chambon. 1992. Homeotic transformation of the occipital bones of the skull by ectopic expression of a homeobox gene. *Nature* 359: 835–841.

Lumsden, A., and R. Krumlauf, 1996. Patterning the vertebrate neuraxis. *Science* 274: 1112

Lu Jizhuan. 1983. Is the Darwinian theory of evolution really wrong—a highlighted debate. *Ren Min Ri Bao* (The People's Daily), Beijing, March 21.

Lyell, C. 1830–1833. *The Principles of Geology; or, The Modern Changes of the Earth and Its Inhabitants as Illustrative of Geology,* 3 vol. London: Murray.

Lyne, J., and H. F. Howe. 1986. "Punctuated equilibria": rhetorical dynamics of a scientific controversy. *Quart. Jour. Speech* 72: 132–147.

MacFadden, B. J. 1986. Fossil horses from "eohippus" (*Hyracotherium*) to *Equus*: scaling laws and the evolution of body size. *Paleobiology* 12: 355–369.

MacFadden, B. J., and R. Hulbert, Jr. 1988. Explosive speciation at the base of the adaptive radiation of Miocene grazing horses. *Nature* 336: 466–468.

MacFadden, B. J., N. Solounias, and T. E. Cerling. 1999. Ancient diets, ecology, and extinction of 5-million-year-old horses from Florida. *Science* 283: 824–827.

MacGillavry, H. J. 1968. Modes of evolution mainly among marine invertebrates. *Bijdragen tot de Dierkunde* 38: 69–74.

MacLeod, N. 1991. Punctuated anagenesis and the importance of stratigraphy to paleobiology. *Paleobiology* 17: 167–188.

Maddux, J. 1994. Punctuated equilibrium by computer. *Nature* 371: 197.

Malmgren, B. A., W. A. Berggren, and G. P. Lohman. 1983. Evidence for punctuated gradualism in the Late Neogene *Globorotalia tumida* lineage of planktonic foraminifera. *Paleobiology* 9: 377–389.

Malmgren, B. A., and J. P. Kennett. 1981. Phyletic gradualism in a Late Cenozoic lineage: DSDP Site 284, Southwest Pacific. *Paleobiology* 7: 230–240.

Manak, J. R., and M. P. Scott. 1994. A class act: conservation of homeodomain protein functions. *Development (Suppl.)*: 61–71.

Marchant, S. 1916. *Alfred Russell Wallace, Letters and Reminiscences*, 2 volumes. London: Cassell.

Mark, R. 1996. Architecture and evolution. *American Scientist* 84: 383–389.

Markey, T. 1997. The problem of runic origins: where The Hobbit went to school. Preprint, Tucson, AZ.

Marshall, C. R. 1994. Confidence intervals on stratigraphic ranges: partial relaxation of the assumption of randomly distributed fossil horizons. *Paleobiology* 20: 459–469.

——— 1995. Distinguishing between sudden and gradual extinction in the fossil record: Predicting the position of the Cretaceous-Tertiary iridium anomaly using the ammonite fossil record on Seymour Island, Antarctica. *Geology* 23: 731–734.

Martinez, D. E., D. Bridge, L. M. Masuda-Nakagawa, and P. Cartwright. 1998. Cnidarian homeoboxes and the zootype. *Nature* 393: 748–749.

Matthew, P. 1831. *On Naval Timber and Arboriculture*. London: Longman.

Mayden, R. L. 1986. Speciose and depauperate phylads and tests of punctuated *vs.* gradual evolution: fact or artifact? *Syst. Zool.* 35: 591–602.

Maynard Smith, J. 1976. Group selection. *Quart. Rev. Biol.* 51: 277–283.

——— 1978. *The Evolution of Sex*. Cambridge UK: Cambridge Univ. Press.

——— 1983. The genetics of stasis and punctuation. *Ann. Rev. Genetics* 17: 11–25.

——— 1984. Palaeontology at the high table. *Nature* 309: 401–402.

——— 1987. How to model evolution. In T. Dupré, ed., *The Latest on the Best: Essays on Evolution and Optimality*. Cambridge MA: MIT Press, pp. 119–131.

——— 1988. Evolutionary progress and the levels of selection. In M. H. Nitecki, ed., *Evolutionary Progress*. Chicago IL: Univ. of Chicago Press, pp. 219–230.

Maynard Smith, J., R. Burian, S. Kauffman, P. Alberch, J. Campbell, B. Goodwin, R. Lande, D. Raup, and L. Wolpert. 1985. Developmental constraints and evolution. *Quart. Rev. Biol.* 60: 265–287.

Maynard Smith, J., and E. Szathmary. 1995. *The Major Transitions in Evolution*. N.Y.: W. H. Freeman.

Mayo, D., and N. Gilinsky. 1987. Models of group selection. *Philosophy of Science* 54: 515–538.

Mayr, E. 1942. *Systematics and the Origin of Species*. N.Y.: Columbia Univ. Press.

——— 1954. Change of genetic environment and evolution. In J. S. Huxley, A. C.

Hardy, and E. B. Ford, eds., *Evolution As A Process*. London: G. Allen & Unwin., pp. 157–180.

———— 1960. The emergence of evolutionary novelties. In S. Tax, ed., *Evolution after Darwin*. Volume I, *The Evolution of Life*. Chicago IL: Univ. of Chicago Press, pp. 349–380.

———— 1963. *Animal species and evolution*. Cambridge MA: Harvard Univ. Press.

———— 1964. *Introduction to facsimile edition of Darwin's Origin of Species,* first edition of 1859. Cambridge MA: Harvard Univ. Press.

———— 1972. Lamarck revisited. *Jour. Hist. Biol.* 5: 55–94.

———— 1974. Cladistic analysis or cladistic classification. *Zeits. Zool. Syst. Evolutionsforschung* 12: 94–128.

———— 1980. How I became a Darwinian. In E. Mayr and W. B. Provine, eds., *The Evolutionary Synthesis*. Cambridge MA: Harvard Univ. Press, pp. 413–423.

———— 1982a. Adaptation and selection. *Biologisches Zentralblatt* 101: 161–174.

———— 1982b. *The Growth of Biological Thought: Diversity, Evolution, and Inheritance*. Cambridge MA: Belknap Press of Harvard Univ. Press.

———— 1985. Darwin's five theories of evolution. In D. Kohn, ed., *The Darwinian Heritage*. Princeton N.J.: Princeton Univ. Press, pp. 755–772.

———— 1991. *One Long Argument: Charles Darwin and the Genesis of Modern Evolutionary Thought*. Cambridge MA: Harvard Univ. Press.

———— 1992. Speciational evolution and punctuated equilibria. In A. Somit, and S. A. Peterson, eds., *The Dynamics of Evolution*. Ithaca NY: Cornell Univ. Press, pp. 21–53.

Mayr, E., and W. B. Provine (eds.). 1980. *The Evolutionary Synthesis: Perspectives on the Unification of Biology*. Cambridge MA: Harvard Univ. Press.

Mazur, A. 1992. Periods and question marks in the punctuated evolution of human social behavior. In A. Somit and S. A. Peterson, eds., *The Dynamics of Evolution: The Puntuated Equilibrium Debate in the Natural and Social Sciences*. Ithaca NY: Cornell Univ. Press, pp. 221–234.

McAlester, A. L. 1962. Some comments on the species problem. *Jour. Paleontology* 36: 1377–1381.

McCauley, D. E., and M. J. Wade. 1980. Group selection: the genetic and demographic basis for the phenotypic differentiation of small populations of *Tribolium castaneum*. *Evolution* 34: 813–821.

McCosh, J., and G. Dickie. 1869. *Typical Forms and Special Ends in Creation*. N.Y.: Robert Carter & Bros.

McHenry, H. M. 1994. Hominid dualism. *Science* 265.

McKinney, F. K., S. Lidgard, J. J. Sepkoski Jr., and P. D. Taylor. 1998. Decoupled temporal patterns of evolutionary and ecology in two Post-Paleozoic clades. *Science* 281: 807–809.

McKinney, M. L. 1988. Classifying heterochrony: allometry, size, and time. In M. L. McKinney, ed., *Heterochrony in Evolution: A Multidisciplinary Approach*. N.Y.: Plenum Press, pp. 17–34.

———— 1999. Heterochrony: beyond words. *Paleobiology* 25: 149–153.

McKinney, M. L., and W. D. Allmon. 1995. Metapopulations and disturbance: from patch dynamics to biodiversity dynamics. In: D. H. Erwin and R. L. Anstrey, eds., *New Approaches to Speciation in the Fossil Record*. N.Y.: Columbia Univ. Press, pp. 123–183.

McKinney, M. L., and D. S. Jones. 1983. Oligopycoid echinoids and the biostratigraphy of the Ocala limestone of peninsular Florida. *Southeastern Geology* 23:21–30.

McKinney, M. L., and K. J. McNamara. 1991. *Heterochrony: The Evolution of Ontogeny*. N.Y.: Plenum Press.

McMenamin, M. A. S., and D. L. S. McMenamin. 1990. *The Emergence of Animals: The Cambrian Breakthrough.* N.Y.: Columbia Univ. Press.

McNamara, K. J. 1997. *Shapes of Time.* Baltimore MD: Johns Hopkins Univ. Press.

McPhee, J. 1980. *Basin and Range.* N.Y.: Farrar, Straus, and Giroux.

McPherson, J. M. 1988. *Battle Cry of Freedom: The Civil War Era.* N.Y.: Oxford Univ. Press.

McShea, D. W. 1993. Evolutionary change in the morphological complexity of the mammalian vertebral column. *Evolution* 47: 730–740.

——— 1994. Mechanisms of large-scale evolutionary trends. *Evolution* 48: 1747–1763.

McShea, D. W., B. Hallgrimsson, and P. D. Gingerich. 1995. Testing for evolutionary trends in non-hierarchical developmental complexity. *Abstracts Annual Meeting Geol. Soc. Amer.* A53-A54.

Medawar, P. B. 1967. D'Arcy Thompson and *Growth and Form.* In P. B. Medawar, *The Art of the Soluble.* London: MacMillan, pp. 21–35.

Meinke, D. W. 1992. A homoeotic mutant of *Arabidopsis thaliana* with leafy cotyledons. *Science* 258: 1647–1650.

Merton, R. K. 1965. *On the Shoulders of Giants.* N.Y.: Harcourt Brace.

Mettler, L. E., T. G. Gregg, and H. E. Schaffer. 1988. *Population Genetics and Evolution.* Englewood Cliffs NJ: Prentice Hall.

Meyer, C. J. A. 1878. Micrasters in the English chalk: two or more species? *Geol. Mag.* 5: 115–117.

Meyerowitz, E. M. 1999. Today we have naming of parts. *Nature* 402: 731–732.

Michaux, B. 1987. An analysis of allozymic characters of four species of New Zealand *Amalda* (Gastropoda: Olividae: Ancillinae). *New Zealand Jour. Zool.* 14: 359–366.

——— 1989. Morphological variation of species through time. *Biol. Jour. Linnaean Soc.* 38: 239–255.

Michener, C. D. 1949. Parallelisms in the evolution of the saturniid moths. *Evolution* 3: 129–141.

Mill, J. S. 1881. A System of Logic. Reprinted in J. S. Mill, 1950, *Philosophy of Scientific Method.* N.Y.: Hafner.

Miller, A. I. 1998. Biotic Transitions in Global Marine Diversity. *Science* 281:1157–1160.

Miller, D. J., and A. Miles. 1993. Homeobox genes and the zootype. *Nature* 365: 215–216.

Mindel, D. P., J. W. Sites, Jr., D. Grauer. 1989. Speciational evolution: a phylogenetic test with allozymes in *Sceloporus* (Reptilia). *Cladistics* 5: 49–61.

Minelli, A. 2000. Limbs and tail as evolutionarily diverging duplicates of the main body axis. *Evolution and Development.* 2: 157–165.

Mivart, St. G. 1871. *On the Genesis of Species.* London: Macmillan.

Mokyr, J. 1990. Punctuated equilibrium and technological progress. *Amer. Econ. Rev.* 80: 350–354.

Montgomery, M. K., and M. J. McFall-Ngai. 1992. The muscle-derived lens of a squid bioluminescent organ is biochemically convergent with the ocular lens: evidence for recruitment of aldehyde dehydrogenase as a predominant structural protein. *J. Biol. Chem.* 267: 20999–21003.

Mooi, R. 1990. Paedomorphosis, Aristotle's lantern, and the origin of sand dollars. *Paleobiology* 16: 25–48.

Moore, J., and P. Willmer. 1997. Convergent evolution in invertebrates. *Biol. Rev.* 72: 1–60.

Moore, R. C., and L. R. Laudon. 1943. Evolution and classification of Paleozoic crinoids. *Geol. Soc. Amer. Special Paper* 46: 1–153.

Morrell, V. 1996. Sizing up evolutionary radiations. *Science* 274: 1462–1463.

Moretti, F. 1996. *Modern Epic: The World System From Goethe to Garcia Marquez.* N.Y.: Verso.

Morgan, B. A., and C. Tabin. 1994. *Hox* genes and growth: early and late roles in limb bud morphogenesis. *Development 1994 Supplement:* 181–186.

Morgan, T. H. 1903. *Evolution and Adaptation.* N.Y.: Macmillan.

———— 1909. For Darwin. *Popular Science Monthly* 72: 367–380.

Morris, P. J. 1996. Testing patterns and causes of stability in the fossil record, with an example from the Pliocene Lusso Beds of Zaire. *Palaeogeog. Palaeoclimatol. Palaeoecol.* 127: 313–337.

Morris, P. J., L. C. Ivany, K. M. Schopf, and C. E. Brett. 1995. The challenge of paleoecological stasis: reassessing sources of evolutionary stability. *Proc. Natl. Acad. Sci. USA* 92: 11269–11273.

Mueller, B., and C. J. Engard. 1952. *Goethe's Botanical Writings.* Honolulu HI: Univ. of Hawaii Press.

Müller, M., E. von Weizäcker, and J. A. Campos-Ortega. 1996. Expression domains of a zebrafish homologue of the *Drosophila* pair-rule gene *hairy* correspond to primordia of alternating somites. *Development* 122: 2071–2078.

Nehm, R. H., and D. H. Geary. 1994. A gradual morphological transition during a rapid speciation event in marginellid gastropods (Neogene, Dominican Republic). *Jour. Paleontology* 68: 787–795.

Nei, M. 1987. *Molecular Evolutionary Genetics.* N.Y.: Columbia Univ. Press.

Nelson, G. E., G. G. Robinson, and G. Boolootian. 1967. *Fundamental Concepts of Biology.* N.Y.: John Wiley.

Neumann, C. J., and C. Nüsslein-Volhard. 2000. Patterning of the zebrafish retina by a wave of *Sonic Hedgehog* activity. *Science* 289: 2137–2139.

Newell, N. D. 1949. Phyletic size increase—an important trend illustrated by fossil invertebrates. *Evolution* 3: 103–124.

———— 1967. Revolutions in the history of life. *Geol. Soc. Amer. Special Paper* 89: 63–91.

Newman, C. M., J. E. Cohen, and C. Kipnis. 1985. Neo-Darwinian evolution implies punctuated equilibria. *Nature* 315: 400–401.

Nichols, D. J. 1982. Phyletic gradualism or punctuated equilibria: the evidence from fossil pollen of the Juglandaceae. *Palynology* 6: 288–289.

Nicholson, A. J. 1960. The role of population dynamics in natural selection. In S. Tax, ed., *Evolution After Darwin,* Volume I, *The Evolution of Life.* Chicago IL: Univ. of Chicago Press, pp. 477–521.

Nield, E. W., and V. C. T. Tucker. 1985. *Palaeontology: An Introduction.* Oxford UK: Pergamon Press.

Nietzsche, F. 1887. (1967 ed.) *On the Genealogy of Morals* (translated by W. Kaufmann). N.Y.: Vintage.

Nitecki, M. H. (ed.) 1988. *Evolutionary Progress.* Chicago IL: Univ. Chicago Press.

Ohno, S. 1970. *Evolution by Gene Duplication.* N.Y.: Springer Verlag.

Oken, L. 1806. *Beiträge zur Vergleichenden Zoologie, Anatomie und Physiologie.* Bamberg: J. A. Göbhardt.

———— 1809–1811. *Lehrbuch der Naturphilosophie.* Jena: F. Frommand.

———— 1847. *Elements of Physiophilosophy.* London: Ray Soc.

Olson, E. C. 1952. The evolution of a Permian vertebrate chronofauna. *Evolution* 6: 181–196.

———— 1960. Morphology, paleontology, and evolution. In S. Tax, ed., *Evolution After Darwin,* Volume I, *The Evolution of Life.* Chicago IL: Univ. of Chicago Press, pp. 523–545.

O'Neill, R. V., D. L. DeAngelis, J. B. Waide, T. H. F. Allen. 1986. *A Hierarchical Concept of Ecosystems.* Princeton N.J.: Princeton Univ. Press.

Orgel, L., and F. Crick. 1980. Selfish DNA: the ultimate parasite. *Nature* 284: 604–607.

Orwell, G. 1949, *1984*. N.Y.: Harcourt, Brace & Jovanovich.

Osborn, H. F. 1894. *From the Greeks to Darwin*. N.Y.: Macmillan.

——— 1902. Homology as a law of latent or potential homology. *Amer. Naturalist* 36: 259–271.

——— 1905. The ideas and terms of modern philosophical anatomy. *Science* 21: 959–961.

——— 1909. Life and works of Darwin. *Popular Science Monthly* 72: 315–343.

Ospovat, D. 1981. *The Development of Darwin's Theory: Natural History, Natural Theology and Natural Selection, 1838–1859*. Cambridge UK: Cambridge Univ. Press.

Owen, R. 1843. *Lectures on the Comparative Anatomy and Physiology of the Invertebrate Animals*. London: Longman, Brown, Green.

——— 1848. *On the Archetype and Homologies of the Vertebrate Skeleton*. London: John Van Voorst.

——— 1849. *On the Nature of Limbs*. London: John Van Voorst.

——— 1894. *The Life of Richard Owen*. London: Murray.

Paley, W. 1794. *View of the Evidences of Christianity*. London: R. Faulder.

——— 1802. *Natural Theology: or, Evidences of the Existence and Attributes of the Deity, Collected From the Appearances of Nature*. London: R. Faulder.

——— 1803. *Natural Theology. 5th Edition*. London: R. Faulder.

Panganiban, G., S. M. Irvine, C. Lowe, H. Roehl, L. S. Corley, B. Sherbon, J. K. Grenier, J. F. Fallon, J. Kimble, M. Walker, G. A. Wray, B. J. Swalla, M. Q. Martindale, and S. B. Carroll. 1997. The origin and evolution of animal appendages. *Proc. Natl. Acad. Sci. USA* 94: 5162–5166.

Panganabin, G., A. Sebring, I. Nagy, and S. Carroll. 1995. The development of crustacean limbs and the evolution of arthropods. *Science* 270: 1363–1366.

Papadopoulos, D., D. Schneider, J. Meier-Eiss, W. Arber, R. E. Lenski, and M. Blot. 1999. Genomic evolution during a 10,000-generation experiment with bacteria. *Proc. Natl. Acad. Sci. USA* 96: 3807–3812.

Parsons, P. A. 1993. Stress, extinctions and evolutionary change: from living organisms to fossils. *Biol. Rev.* 68: 313–333.

Patten, W. 1912. *The Evolution of the Vertebrates and Their Kin*. Philadelphia PA: P. Blakiston's Sons.

——— 1920. *The Grand Strategy of Evolution*. Boston MA: Gorham Press.

Patterson, C. 1988. Homology in classical and molecular biology. *Molec. Biol. Evol.* 5: 603–625.

Paul, C. R. C. 1985. The adequacy of the fossil record reconsidered. In J. C. W. Cope and P. W. Skelton, eds., *Evolutionary Case Histories From The Fossil Record. Special Papers in Palaeontology* 33: 7–15.

Pelaz, S., G. S. Ditta, E. Baumann, E. Wisman, and M. F. Yanofsky. 2000. B and C floral organ identity functions require *Sepallata MADS-box* genes. *Nature* 405: 200–203.

Pennisi, E. 1997. New developmental clock discovered. *Science* 278: 1564.

Pennisi, E., and W. Roush. 1997. Developing a new view of evolution. *Science* 277: 34–37.

Penny, D. 1983. Charles Darwin, gradualism and punctuated equilibrium. *Systematic Zool.* 32: 72–74.

——— 1985. Two hypotheses on Darwin's gradualism. *Systematic Zool.* 34: 201–205.

Peterson, K. J., R. A. Cameron, and E. H. Davidson. 1997. Set-aside cells in maximal indirect development: evolutionary and developmental significance. *BioEssays* 19: 623–631.

Peterson, K. J., and E. H. Davidson. 2000. Regulatory evolution and the origin of the bilaterians. *Proc. Natl. Acad. Sci. USA* 97:4430–4433.

Piatigorsky, J. 1992. Lens crystallins. *Jour. Biol. Chemistry* 267: 4277–4280.

——— 1993a. Gene sharing in the visual system. *Trans. Am. Ophthalmol. Soc.* 81: 283–297.

——— 1993b. Puzzle of crystallin diversity in eye lenses. *Developmental Dynamics* 196: 267–272.

Piatigorsky, J., and G. J. Wistow. 1989. Enzyme crystallins: gene sharing as an evolutionary strategy. *Cell* 57: 197–199.

——— 1991. The recruitment of crystallins: new functions precede gene duplication. *Science* 252: 1078–1079.

Piatigorsky, J., M. Kantorow, R. Gopal-Srivastava, and S. I. Tomarev. 1994. Recruitment of enzymes and stress proteins as lens crystallins. In B. Jansson, H. Jömvall, U. Rydberg, L. Terenius, and B. L. Vallee, eds., *Toward a Molecular Basis of Alcohol Use and Abuse.* Basel, Switzerland: Birkhäuser Verlag, pp. 241–250.

Pineda, D., J. Gonzalez, P. Callaerts, K. Ikeo, W. J. Gehring, and E. Salo. 2000. Searching for the prototypic eye genetic network: *Sine oculis* is essential for eye regeneration in planarians. *Proc. Natl. Acad. Sci. USA* 97: 4525–4529.

Plate, L. 1903. *Über die Bedeutung des Darwinischen Selectionsprinzip und Probleme der Artbildung.* Leipzig: Engelmann.

Playfair, J. 1802. *Illustrations of the Huttonian Theory of the Earth.* Edinburgh: William Creech.

Podos, J. 2001. Correlated evolution of morphology and vocal signal structure in Darwin's finches. *Nature* 409: 185–188.

Pollock, R. A., T. Sreenath, L. Ngo, and C. J. Bieberich. 1995. Gain of function mutations for paralogous *Hox* genes: implications for the evolution of *Hox* gene function. *Proc. Natl. Acad. Sci. USA* 92: 4492–4496.

Porter, R. 1976. Charles Lyell and the principles of the history of geology. *British Jour. Hist. Sci.* 9: 91–103.

Powers, T. P., J. Hogan, Z. Ke, K. Dymbrowski, X. Wang, F. H. Collins, and T. C. Kaufman. 2000. Characterization of the *Hox* cluster from the mosquito *Anopheles gambiae* (Diptera: Calicidae). *Evolution & Development.* 2: 311–325.

Price, G. R. 1970. Selection and covariance. *Nature* 227: 520–521.

——— . 1972. Extension of covariance selection mathematics. *Annals of Human Genetics* 35: 485–490.

Price, P. W. 1996. *Biological Evolution.* Fort Worth TX: W. B. Saunders.

Prothero, D. R., and T. H. Heaton. 1996. Faunal stability during the Early Oligocene climatic crash. *Palaeogeog. Palaeoclimatol. Palaeoecol.* 127: 257–283.

Prothero, D. R., and N. Shubin. 1983. Tempo and mode of speciation in Oligocene mammals. *GSA Absracts with Programs* 15: 665.

——— 1989. The evolution of Oligocene horses. In D. R. Prothero and R. M. Schoch, eds., *The Evolution of Perissodactyls.* Oxford UK: Oxford Univ. Press, pp. 142–175.

Provine, W. B. 1971. *Origins of Theoretical Population Genetics.* Chicago IL: Univ. of Chicago Press.

——— 1983. The development of Wright's theory of evolution: systematics, adaptation, and drift. In M. Grene, ed., *Dimensions of Darwinism.* Cambridge UK: Cambridge Univ. Press, pp. 43–70.

——— 1986. *Sewall Wright and Evolutionary Biology.* Chicago IL: Univ. of Chicago Press.

Queller, D. C. 1995. The spaniels of St. Marx and the Panglossian paradox: a critique of a rhetorical programme. *Quart. Rev. Biol.* 70: 485–489.

——— 2000. Pax argentinica. *Nature* 405: 519–520.

Quiring, R., U. Walldorf, U. Kloter, and W. J. Gehring. 1994. Homology of the *eye-*

*less* gene of *Drosophila* to the *Small eye* gene in mice and *Aniridia* in humans. *Science* 265: 785–789.

Raff, R. A. 1996. *The Shape of Life*. Chicago IL: Univ. of Chicago Press.

Rancourt, D. E., T. Tsuzuki, and M. R. Capecchi. 1995. Genetic interaction between *hoxb-5* and *hoxb-6* is revealed by nonallelic noncomplementation. *Genes and Development* 9:108–122.

Rand, D. A., and H. B. Wilson. 1993. Evolutionary catastrophes, punctuated equilibria and gradualism in ecosystem evolution. *Proc. R. Soc. Lond* 253: 137–141.

Rand, D. A., H. B. Wilson, and J. M. McGlade. 1993. Dynamics and evolution: evolutionary stable attractors, invasion exponents and phenotype dynamics. *Warwick Preprints* 53/92.

Raup, D. M. 1975. Taxonomic survivorship curves and Van Valen's Law. *Paleobiology* 1: 82–96.

——— 1983. On the early origins of major biologic groups. *Paleobiology* 9: 107–115.

——— 1985. Mathematical models of cladogenesis. *Paleobiology* 11: 42–52.

——— 1991a. *Extinction: Bad Genes or Bad Luck*. N.Y.: W. W. Norton.

——— 1991b. A kill curve for Phanerozoic marine species. *Paleobiology* 17: 37–46.

——— 1992. Large-body impact and extinction in the Phanerozoic. *Paleobiology* 18: 80–88.

——— 1996. Extinction models. In D. Jablonski, D. H. Erwin, and J. H. Lipps, eds., *Evolutionary Paleobiology*. Chicago IL: Univ. of Chicago Press, pp. 419–433.

Raup, D. M., and S. J. Gould. 1974. Stochastic simulation and evolution of morphology—towards a nomothetic paleontology. *Syst. Zool.* 23: 305–322.

Raup, D. M., S. J. Gould, T. J. M. Schopf, and D. S. Simberloff. 1973. Stochastic models of phylogeny and the evolution of diversity. *Jour. Geology* 81: 525–542.

Raup, D. M., and J. J. Sepkoski, Jr. 1984. Periodicity of extinctions in the geologic past. *Proc. Natl. Acad. Sci. USA* 81: 801–805.

Ray, J. 1691. *The Wisdom of God Manifested in the Works of the Creation*. London: W. Innys & R. Manby.

——— 1735. *The Wisdom of God Manifested in the Works of the Creation, 10th Edition*. London: W. Innys & R. Manby.

Ray, T. S. 1992. *Artificial Life*. In C. G. Langton, ed., Santa Fe Institute Studies in the Sciences of Complexity, *Proc. Vol. X*. Redwood City CA: Addison Wesley, pp. 371–408.

Reif, W. E. 1976. Die Erforschung des Steinheimer Beckens. In A. Akermann, ed., *75 Jahre Heimat-und Altertumsverein Heidenheim*, 1901–1976. Heidenheim, pp. 66–85.

——— 1983. Hilgendorf's (1866) dissertation on the Steinheim planorbids (Gastropoda; Miocene): the development of a phylogenetic research program for paleontology. *Paläont. Zeits.* 57: 7–20.

Rensberger, B. 1980. Recent studies spark revolution in interpretation of evolution. *New York Times*, November 4.

——— 1986. *How the World Works: A Guide to Science's Greatest Discoveries*. N.Y.: William Morrow.

Rensch, B. 1947. *Neuere Probleme der Abstammungslehre*. Stuttgart: F. Enke.

——— 1960. *Evolution Above the Species Level*. N.Y.: Columbia Univ. Press.

Retallak, G. J. 1993. Were the Ediacaran fossils lichens? *Paleobiology* 20: 523–544.

Reyment, R. A. 1975. Analysis of a generic level transition in Cretaceous ammonites. *Evolution* 28: 665–676.

——— 1982. Analysis of trans-specific evolution in Cretaceous ostracodes. *Paleobiology* 8: 293–306.

Reznick, D. N., F. H. Shaw, F. H. Rodd, and R. G. Shaw. 1997. Evolution of the rate of evolution in natural populations of guppies (*Poecilia reticulata*). *Science* 275: 1934–1937.

Rhodes, F. H. T. 1983. Gradualism, punctuated equilibrium, and the *Origin of Species*. *Nature* 305: 269–272.

Richards, R. C. 1992. *The Meaning of Evolution: The Morphological Construction and Ideological Reconstruction of Darwin's Theory*. Chicago IL: Univ. of Chicago Press.

Ridley, M. 1980. Evolution and gaps in the fossil record. *Nature* 286: 444–445.

———— 1993. *Evolution*. Boston MA: Blackwell.

Riedl, R. 1978. *Order in Living Organisms*. N.Y.: John Wiley & Sons.

Rightmire, G. P. 1981. Patterns in the evolution of *Homo erectus. Paleobiology* 7: 241–246.

———— 1985. The tempo of change in the evolution of mid-Pleistocene *Homo*. In E. Delson, ed., *Ancestors: The Hard Evidence*. N.Y.: Alan R. Liss, pp. 255–264.

———— 1986. Stasis in *Homo erectus* defended. *Paleobiology* 12: 324–325.

Roberts, J. 1981. Control mechanisms of Carboniferous brachiopod zones in eastern Australia. *Lethaia* 14:123–134.

Robison, R. A. 1975. Species diversity among agnostoid trilobites. *Fossils and Strata* 4: 219–226.

Robson, G. C., and O. W. Richards. 1936. *The Variations of Animals in Nature*. London: Longmans Green.

Roe, D. 1980. The handaxe makers. In A. Sherratt, ed., *The Cambridge Encyclopedia of Archaeology*. N.Y.: Columbia Univ. Press, pp. 71–78.

Romanes, G. J. 1900. *Darwin, and After Darwin,* Vol. 2. *Post-Darwinian Questions*. London: Longmans, Green & Co.

Ross, R. M. 1990. The evolution and biogeography of Neogene Micronesian ostracodes: the role of sea level, geography, and dispersal. Ph.D. Dissertation, Harvard University.

Ross, R. M., and W. D. Allmon (eds.). 1990. *Causes of Evolution: A Paleontological Perspective*. Chicago IL: Univ. of Chicago Press.

Roth, V. L. 1991. Homology and hierarchies: problems solved and unsolved. *Jour. Evol. Biol.* 4: 167–194.

Roux, W. 1881. *Der Kampf der Teile im Organismus*. Leipzig.

Roy, K. 1996. The roles of mass extinction and biotic interaction in large-scale replacements: a reexamination using the fossil record of stromboidean gastropods. *Paleobiology* 22: 436–452.

Rubinstein, E. 1995. Punctuated equilibrium in scientific publishing. *Science* 268:1415.

Rudwick, M. J. S. 1969. The strategy of Lyell's *Principles of Geology. Isis* 61: 5–33.

———— 1992. *Scenes From Deep Time*. Chicago IL: Univ. of Chicago Press.

Ruse, M. 1980. Charles Darwin and group selection. *Ann. Sci.* 37: 615–630.

———— 1992. Is the theory of punctuated equilibria a new paradigm? In A. Somit and S. A. Peterson, eds., *The Dynamics of Evolution: The Punctuated Equilibrium Debate in the Natural and Social Sciences*. Ithaca N.Y.: Cornell Univ. Press, pp. 139–167.

———— 1996. *Monad to Man: The Concept of Progress in Evolutionary Biology*. Cambridge MA: Harvard Univ. Press.

Russell, E. S. 1916. *Form and Function*. London: J. Murray.

Ryan, M. J. 2001. Food, song and speciation. *Nature* 409: 139–140.

Sacher, G. A. 1966. Dimensional analysis of factors governing longevity in mammals. *Proc. Int. Congr. Gerontology,* Vienna, p. 14.

Sadler, P. M. 1981. Sediment accumulation rates and the completeness of stratigraphic sections. *Jour. Geol.* 89: 569–584.

Saether, O. A. 1983. The canalized evolutionary potential: inconsistencies in phylogenetic reasoning. *Systematic Zool.* 32: 343–349.

Salvatori, N. 1984. Paleontologi a confronto sui tempi e i modi dell'evoluzione. *Airone,* December, p. 34.

Salvini-Plawen, L. V., and E. Mayr. 1977. On the evolution of photoreceptors and eyes. In: M. K. Hecht, W. C. Steere, and B. Wallace, eds., *Evolutionary Biology.* Vol. 10. N.Y.: Plenum, pp. 207–263.

Saranak, J., and K. W. Foster. 1997. Rhodopsin guides fungal phototaxis. *Nature* 387: 465–466.

Sasai, Y., B. Lu, H. Steinbeisser, D. Geissert, L. K. Gont, and E. M. De Robertis. 1994. *Xenopus chordin:* a novel dorsalizing factor activated by organizer-specific homeobox genes. *Cell* 79: 779–790.

Savage, C. H., and G. F. F. Lombard. 1983. *Sons of the Machine: Case Studies of Social Change in the Workplace.* Cambridge MA: MIT Press.

Schaeffer, B., M. K. Hecht, and N. Eldredge. 1972. Phylogeny and paleontology. In Th. Dobzhansky, M. K. Hecht, and W. C. Steere, *Evolutionary Biology,* Volume 6 (Festschrift for G. G. Simpson). N.Y.: Appleton-Century-Crofts, pp. 31–46.

Schankler, D. M. 1980. Faunal zonation of the Willwood Formation in the central Bighorn Basin, Wyoming. In P. D. Gingerich, ed., Early Cenozoic Paleontology and Stratigraphy of the Bighorn Basin, Wyoming. *Univ. Mich. Pap. Paleontology* 24: 99–114.

———— 1981. Local extinction and ecological re-entry of early Eocene mammals. *Nature* 293:135–138.

Schindel, D. E. 1982. Resolution analysis: a new approach to the gaps in the fossil record. *Paleobiology* 8: 340–353.

Schindewolf, O. H. 1963. Neokatastrophismus? *Zeits. Deutsch. Geol. Res.* 114: 430–435.

Schoch, R. M. 1984. Possible mechanisms for punctuated patterns in the fossil record. *GSA Abstracts with Programs* 16: 62.

Schoonover, L. M. 1941. A stratigraphic study of the molluscs of the Calvert and Choptank Formations of Southern Maryland. *Bull. Amer. Paleontol.* 25: 169–299.

Schopf, K. M. 1996. Coordinated stasis: biofacies revisited and the conceptual modeling of whole-fauna dynamics. *Palaeogeog. Palaeoclimat. Palaeoecol.* 127: 157–175.

Schwartz, J. H. 1999. *Sudden Origins: Fossils, Genes, and the Emergence of Species.* N.Y.: John Wiley and Sons.

Schweber, S. S. 1977. The origin of the Origin revisited. *Jour. History Biol.* 10: 229–316.

———— 1980. Darwin and the political economists: divergence of character. *Jour. Hist. Biol.* 13: 195–289.

———— 1985. The wider British context in Darwin's theorizing. In D. Kohn, ed., *The Darwinian Heritage.* Princeton NJ: Princeton Univ. Press., pp. 35–69.

———— 1988. The correspondence of the young Darwin. *Jour. Hist. Biol.* 21: 501–519.

Schwenk, K. 1995. A utilitarian approach to evolutionary constraint. *Zoology* 98: 251–262.

Scilla, A. 1670. *La vana speculazione disingannata dal senso.* Naples: A. Colicchia.

———— 1747. *De corporibus marinis lapidescentibus.* Rome: A. Rubeis.

Scott, W. B. 1891. On the osteology of *Mesohippus* and *Leptomeryx,* with observations on the modes and factors of evolution in the Mammalia. *Jour. Morphology* 5: 301–402.

———— 1896. Paleontology as a morphological discipline. *Biol. Lectures Marine Biol. Lab. Woods Hole* for 1895, pp. 43–61.

Scudo, F. M. 1985. Darwin, Darwinian theories and punctuated equilibria. *Systematic Zool.* 34: 239–242.

Seilacher, A. 1970. Arbeitskonzept zur Konstruktionsmorphologie. *Lethaia* 3: 393–396.

———— 1989. Vendozoa: organismic construction in the Proterozoic biosphere. *Lethaia* 22: 229–239.

Selander, R. K., S. Y. Yang, R. C. Lewontin, and W. E. Johnson. 1970. Genetic variation in the horseshoe crab (*Limulus polyphemus*), a phylogenetic "relic." *Evolution* 24: 402–419.

Selzer, J. (ed.) 1993. *Understanding Scientific Prose*. Madison WI: Univ. of Wisconsin Press.

Sepkoski, J. J. 1982. A compendium of fossil marine families. *Milwaukee Publ. Mus. Contrib. Biol. Geol.* 51: 1–25.

———— 1988. Alpha, beta, or gamma—where does all the diversity go? *Paleobiology* 14: 221–234.

———— 1991. Population biology models in macroevolution. In N. L. Gilinsky and P. W. Signor, eds., *Analytical Paleontology*. Paleont. Soc. Short Courses in Paleontology No. 4. Knoxville TN, pp. 136–156.

———— 1993. Ten years in the library: new data confirm paleontological patterns. *Paleobiology* 19: 43–51.

———— 1996. Competition in macroevolution: the double wedge revisited. In D. Jablonski, D. H. Erwin, and J. H. Lipps, eds., *Evolutionary Paleobiology*. Chicago IL: Univ. of Chicago Press, pp. 211–255.

———— 1997. Biodiversity: past, present, and future. *Jour. Paleontol.* 71: 533–539.

Sepkoski, J. J., F. K. McKinney, and S. Lidgard. 2000. Competitive displacement among post-Paleozoic cyclostome and cheilostome bryozoans. *Paleobiology* 26: 7–18.

Sequeiros, L. 1981. La evolucion biologica, teoria en crisis. *Razon y Fe*, December, pp. 586–593.

Seward, A. C. (ed.) 1909. *Darwin and Modern Science*. Essays in Commemoration of the Centenary of the Birth of Charles Darwin. Cambridge UK: Cambridge Univ. Press.

Shalizi, C. R. 1998. Scientific models: claiming and validating. *Santa Fe Institute Bulletin*, pp. 9–10.

Shapiro, E. A. 1978. Natural selection in a Miocene pectinid: a test of the punctuated equilibria model. *Geol Soc. Am. Abstracts Annual Meeting* 10: 490.

Shattuck, G. B. 1904. Geological and paleontological relations with a review of earlier investigations. *Maryland Geological Survey, Miocene Volume*.

Shaw, A. B. 1969. Adam and Eve, paleontology and the non-objective arts. *Jour. Paleontology* 43: 1085–1098.

Sheehan, P. M. 1996. A new look at ecological evolutionary units (EEUs). *Palaeogeog. Palaeoclimat. Palaeoecol.* 127: 21–32.

Sheehan, P. M., and T. A. Hansen. 1986. Detritus feeding as a buffer to extinction at the end of the Cretaceous. *Geology* 14: 868–870.

Sheehan, P. M., D. E. Fastovsky, R. G. Hoffman, C. D. Berghan, and D. L. Gabriel. 1991. Sudden extinction of the dinosaurs: Late Cretaceous, Upper Great Plains, USA. *Science* 254: 835–839.

Sheldon, P. R. 1987. Parallel gradualistic evolution of Ordovician trilobites. *Nature* 330: 561- 563.

———— 1990. Shaking up evolutionary patterns. *Nature* 345: 772.

———— 1996. Plus ça change—a model for stasis and evolution in different environments. *Palaeogeog. Palaeoclimatol. Palaeoecol.* 127: 209–227.

Sheng, G., E. Thouvenot, D. Schmucker, D. S. Wilson, and C. Desplan. 1997. Direct regulation of *rhodopsin 1* by *Pax-6/eyeless* in *Drosophila*: evidence for a conserved function in photoreceptors. *Genes and Development* 11: 1122–1131.

Shu, D., S. Conway Morris, X.-L. Zhang, L. Chen, Y. Li, and J. Han. 1999. A pipiscid-like fossil from the Lower Cambrian of south China. *Nature* 400: 746–749.

Shubin, N., C. Tabin, and S. Carroll. 1997. Fossils, genes and the evolution of animal limbs. *Nature* 388: 639–648.

Signor, P. W., and J. H. Lipps. 1982. Sampling bias, gradual extinction patterns, and catastrophes in the fossil record. *Geol. Soc. America Special Papers* 190: 291–296.

Simberloff, D. 1983. Competition theory, hypothesis testing, and other community ecology buzzwords. *Amer. Nat.* 122: 626.

———— 1984. The great god of competition. *The Sciences* 24: 16–22.

Simeone, A., D. Acampora, M. Gulisano, A. Stornaiuolo, and E. Boncinelli. 1992. Nested expression domains of four homeobox genes in developing rostral brain. *Nature* 358: 687–690.

Simpson, G. G. 1944. *Tempo and Mode in Evolution*. N.Y.: Columbia Univ. Press.

———— 1945. The principles of classification and a classification of mammals. *Bull. Amer. Mus. Nat. Hist.* 85: 1–350.

———— 1947. *The Meaning of Evolution*. N.Y.: Harcourt Brace & World.

———— 1949. Rates of evolution in animals. In G. L. Jepsen, E. Mayr, and G. G. Simpson, eds., *Genetics Paleontology and Evolution*. Princeton N.J.: Princeton Univ. Press, pp. 205–228.

———— 1953. *The Major Features of Evolution*. N.Y.: Columbia Univ. Press.

———— 1960. The History of Life. In S.Tax, ed., *Evolution After Darwin*, Volume I, *The Evolution of Life*. Chicago IL: Univ. of Chicago Press, pp. 117–180.

———— 1961a. Lamarck, Darwin and Butler. *Amer. Scholar* 30: 238–249.

———— 1961b. *Principles of Animal Taxonomy*. N.Y.: Columbia Univ. Press.

———— 1963. Historical science. In C. C. Albritton, Jr., ed., *The Fabric of Geology*. Reading MA: Addison-Wesley, pp. 24–48.

———— 1964. *This View of Life*. N.Y.: Harcourt, Brace & World.

Simpson, G. G., C. S. Pittendrigh, and L. H. Tiffany. 1957. *Life: An Introduction to Biology*. N.Y.: Harcourt Brace.

Slack, J. M. W. 1997. Our flexible friends. *Nature* 387: 866–867.

Slack, J. M. W., P. W. H. Holland, and C. F. Graham. 1993. The zootype and the phylotypic stage. *Nature* 361: 490–492.

Smit, A. F. A. 1999. Interspersed repeats and other momentos of transposable elements in mammalian genomes. *Current Opinion Genet. Develop* 9: 657–663.

Smith, A. B. and C. R. C. Paul. 1985. Variation in the irregular echinoid *Discoides* during the early Cenomanian. *Special Papers in Palaeontology* 33: 29–37.

Smith, C. G. 1994. Tempo and mode in deep-sea benthic ecology: punctuated equilibrium revisited. *Palaios* 9: 3–13.

Smith, J. P. 1898. Evolution of fossil Cephalopoda. In D. S. Jordan, ed., *Footnotes to Evolution*. N.Y.: D. Appleton.

Smocovitis, V. B. 1996. *Unifying Biology*. Princeton NJ: Princeton Univ. Press.

———— 2000. The 1959 Darwin centennial celebration in America. *Osiris* 14: 274–323.

Sneppen, K., P. Bak, H. Flyvbjerg, and M. H. Jensen. 1994. Evolution as a self-organized critical phenomenon. *Preprint*.

Sober, E. 1980. Holism, individualism, and the units of selection. *Philos. Sci. Assoc.* 2: 1–29.

———— 1984. *The Nature of Selection*. Cambridge MA: MIT Press.

———— 1993. *Philosophy of Biology*. Boulder CO: Westview Press.

Sober, E., and R. C. Lewontin. 1982. Artifact, cause and genic selection. *Philosophy of Science* 49: 157–180.

Sober, E., and D. S. Wilson. 1998. *Unto Others: The Evolution and Psychology of Unselfish Behavior*. Cambridge MA: Harvard Univ. Press.

Somit, A., and S. A. Peterson. 1992. *The Dynamics of Evolution: The Punctuated*

*Equilibrium Debate in the Natural and Social Sciences*. Ithaca N.Y.: Cornell Univ. Press.

Sorhannus, U. 1990. Punctuated morphological change in a Neogene diatom lineage: "local" evolution or migration? *Historical Biol*. 3: 241–247.

Spencer, H. 1893. The inadequacy of natural selection. *Contemp. Rev.* 64 (Feb. & March), pp. 153–166, 434–456.

——— 1893. A rejoinder to Professor Weismann. *Contemp. Rev.* 64 (Dec.), pp. 893–912.

Stanley, S. M. 1973. An explanation for Cope's rule. *Evolution* 27: 1–26.

——— 1975. A theory of evolution above the species level. *Proc. Natl. Acad. Sci. USA* 72: 646–650.

——— 1978. Chronospecies' longevities, the origin of genera, and the punctuated model of evolution. *Paleobiology* 4: 26–40.

——— 1979. *Macroevolution: Pattern and Process*. San Francisco CA: W. H. Freeman.

——— 1982. Macroevolution and the fossil record. *Evolution* 36: 460–473.

——— 1985. Rates of evolution. *Paleobiology* 11: 13–26.

——— 1987. The controversy over punctuational evolution: where do we stand? *GSA Abstracts with Programs* 198:54–855.

Stanley, S. M., and X. Yang. 1987. Approximate evolutionary stasis for bivalve morphology over millions of years: a multivariate, multilineage study. *Paleobiology* 13:113–139.

Stauffer, R. C. (ed.) 1975. *Charles Darwin's Natural Selection: Being the Second Part of His Big Species Book Written From 1856–1858*. Cambridge UK: Cambridge Univ. Press.

Stearns, S. C. 1986. Natural selection and fitness, adaptation and constraint. In D. M. Raup and D. Jablonski, eds., *Patterns and Processes in the History of Life*. Berlin: Springer Verlag, pp. 23–44.

Stebbins, G. L. 1950. *Variation and Evolution in Plants*. N.Y.: Columbia Univ. Press.

——— 1959. The role of hybridization in evolution. *Proc. Amer. Philosophical Soc.* Chapter 7 (Special Volume for the Centennial of the *Origin of Species*).

——— 1969. *The Basis of Progressive Evolution*. Chapel Hill NC: Univ. of North Carolina Press.

Stebbins, G. L., and F. J. Ayala. 1981a. Is a new evolutionary synthesis necessary? *Science* 213: 967–971.

——— 1981b. The evolution of Darwinism. *Scientific American*: 72–82.

Stenseth, N. C., and J. Maynard Smith. 1984. Coevolution in ecosystems: red queen evolution or stasis? *Evolution* 38: 870–880.

Stidd, B. M. 1985. Are punctuationists wrong about the Modern Synthesis? *Philosophy of Science* 52: 98–109.

Stokes, W. L. 1973. *Essentials of Earth History*. Englewood Cliffs NJ: Prentice-Hall.

Stomps, T. J. 1954. On the rediscovery of Mendel's work by Hugo de Vries. *Jour. Heredity* 45: 293–294.

Strathmann, R. R. 1978. The evolution and loss of feeding larval stages of marine invertebrates. *Evolution* 32: 894–906.

——— 1988. Larvae, phylogeny, and von Baer's law. In C. R. C. Paul and A. B. Smith, eds., *Echinoderm Phylogeny and Evolutionary Biology*. Oxford UK: Clarendon Press, pp. 51–68.

Struhl, G., K. Struhl, and P. M. Macdonald. 1989. The gradient morphogen *bicoid* is a concentration dependent transcriptional activator. *Cell* 57: 1259–1273.

Stuart, J. J., S. J. Brown, R. W. Beeman, and R. E. Denell. 1991. A deficiency of the homoeotic complex of the beetle *Tribolium*. *Nature* 350: 72–74.

Sulloway, F. J. 1979. Geographical isolation in Darwin's thinking. *Stud. Hist. Biol.* 3: 23–65.

——— 1982a. Darwin and his finches: the evolution of a legend. *Jour. Hist. Biol.* 15: 1–53.

——— 1982b. Darwin's conversion: the *Beagle* and its aftermath. *Jour. Hist. Biol.* 15: 325–396.

Sun, H., A. Rodin, Y. Zhou, D. P. Dickinson, D. E. Harper, D. Hewett-Emmett, and W.-H. Li. 1997. Evolution of paired domains: isolation and sequencing of jellyfish and hydra *Pax* genes related to *Pax-5* and *Pax-6*. *Proc. Natl. Acad. Sci. USA* 94: 5156–5161.

Swalla B., and W. Jeffery. 1996. Requirement of the *Manx* gene for expression of chordate features in a tailless ascidian larva. *Science* 274: 1205–1208.

Swisher, C. C., W. J. Rink, S. C. Anton, H. P. Schwarcz, G. H. Curtis, A. Suprijo, and Widiasmoro. 1996. Latest *Homo erectus* of Java. *Science* 274: 1870–1874.

Sylvester-Bradley, P. C. (ed.) 1956. *The Species Concept in Paleontology.* London: Systematics Assoc. Publication No. 2, 145 pp.

Symons, D. 1979. *The Evolution of Human Sexuality.* N.Y.: Oxford Univ. Press.

Szathmáry, E. 1991. Common interest and novel evolutionary units. *Trends Ecol. Evol.* 6: 407–408.

Tabin, C. J. 1992. Why we have (only) five fingers per hand: Hox genes and the evolution of paired limbs. *Development* 116: 289–296.

Tabin, C. J., S. B. Carroll, and G. Panganiban. 1999. Out on a limb: parallels in vertebrate and invertebrate limb patterning and the origin of appendages. *Amer. Zool.* 39: 650–663.

Tamarin, R. H. 1986. *Principles of Genetics.* Boston MA: Prindle, Weber & Schmidt.

Tand, C. M. and D. J. Bottjer. 1996. Long-term faunal stasis without evolutionary coordination: Jurrasic benthic marine paleocommunities, Western Interior, United States. *Geology* 24: 815–818.

Tattersall, I. 1998. *Becoming Human: Evolution and Human Uniqueness.* N.Y.: Harcourt Brace.

Tattersall, I., and J. Schwartz. 2000. *Extinct Humans.* Boulder CO: Westview Press.

Tax, S. (ed.) 1960. *Evolution After Darwin.* 3 vols. Chicago: Univ. of Chicago Press. Vol. 1: *The Evolution of Life;* vol. 2: *The Evolution of Man;* vol. 3: *Issues in Evolution,* edited with C. Callender.

Telford, M. J., and R. H. Thomas. 1998. Of mites and *zen:* expression studies in a chelicerate arthropod confirm *zen* is a divergent *Hox* gene. *Dev. Genes Evol.* 208: 591–594.

Telford, W. H., and D. Kennedy. 1965. *The Biology of Organisms.* N.Y.: John Wiley.

Templeton, A.R. 1979. The unit of selection in *Drosophilia mercatorum.* II. Genetic revolution and the origin of coadapted genomes in parthenogenetic strains. *Genetics* 92: 1265–1282.

——— 1982. Genetic architecture of speciation. In C. Barigozzi, ed., *Mechanisms of Speciation.* N.Y.: Alan R. Liss, pp. 105–121.

Tennyson, Alfred, Lord. 1850. *In Memoriam.* Boston MA: Ticknor, Reed & Fields.

Thain, M., and M. Hickman. 1990. *The Penguin Dictionary of Biology,* New Edition. N.Y.: Penguin Books.

Theissen, G., and H. Saedler. 2001. Floral quartets. *Nature* 409: 469–471.

Thompson, D'Arcy W. 1917. *On Growth and Form.* Cambridge UK: Cambridge Univ. Press.

Thompson, D'Arcy W. 1942. *On Growth and Form.* 2nd Edition. Cambridge UK: Cambridge Univ. Press.

Thompson, P. 1983. Tempo and mode in evolution: punctuated equilibria and the modern synthetic theory. *Philosophy of Science* 50: 432–452.

Thomson, W. (Lord Kelvin). 1866. The "doctrine of uniformity" in geology briefly refuted. *Proc. Royal Soc. Edinburgh* 5: 512–513.

——— 1868. On geological time; of geological dynamics. Lectures read in 1868 to the Glasgow Geological Society. Published in Volume 2 of *Popular Lectures and Addresses,* 3 Volumes, 1891–1894. London: MacMillan.

Thurow, L. C. 1996. *The Future of Capitalism.* N.Y.: Penguin Books.

Tickle, C. 1992. A tool for transgenesis. *Nature* 358: 188–189.

Tinbergen, N. 1960. Behaviour, systematics, and natural selection. In S. Tax, ed., *Evolution After Darwin,* Volume I, *The Evolution of Life.* Chicago IL: Univ. of Chicago Press, pp. 595–613.

Todes, D. P. 1988. Darwin's Malthusian metaphor and Russian evolutionary thought, 1859–1917. *Isis* 78: 537–551.

Tomarev, S. I., P. Callaerts, L. Kos, R. Zinovieva, G. Halder, W. Gehring, and J. Piatigorsky. 1997. Squid *Pax-6* and eye development. *Proc. Natl. Acad. Sci. USA* 94: 2421–2426.

Traub, J. 1995. Shake them bones. *New Yorker,* March 13, p. 60.

Trueman, A. E. 1922. The use of *Gryphaea* in the correlation of the Lower Lias. *Geol. Mag.* 49: 256–268.

——— 1940. The meaning of orthogenesis. *Trans. Geol. Soc. Glasgow* 20 (volume published 1945, article originally issued separately in 1940): 77–95.

Tsutsui, N. D., A. V. Suarez, D. A. Holway, and T. J. Case. 2000. Reduced genetic variation and the success of an invasive species. *Proc. Natl. Acad. Sci.* 97: 5948–5953.

Tudge, C. 1986. The evolution of evolution. *The Listener* Volume 115, June 19, pp. 11–12.

Turner, J. R. G. 1984. Why we need evolution by jerks. *New Scientist,* February 9.

——— 1986. The genetics of adaptive radiation: a neo-Darwinian theory of punctuational evolution. In D. M. Raup and D. Jablonski, eds., *Patterns and Processes in the History of Life.* Berlin: Springer Verlag.

Valdecasas, A. G., and D. V. Herreros. 1982. La teoria de la evolucion: los terminos de controversia. *Revista Universidad Complutense Madrid,* pp. 153–158.

Valentine, J. W. 1990. The macroevolution of clade shape. In R. M. Ross and W. D. Allmon, eds., *Causes of Evolution: A Paleontological Perspective.* Chicago IL: Univ. of Chicago Press, pp. 128–150.

Valentine, J. W., and A. G. Collins. 2000. The significance of moulting in ecdysozoan evolution. *Evolution and Development* 2: 152–156.

Van der Pas, P. W. 1970. The correspondence of Hugo de Vries and Charles Darwin. *Janus* 57: 173–213.

Van Valen, L. 1973. A new evolutionary law. *Evol. Theory* 1: 1–30.

Vermeij, G. J. 1977. The Mesozoic marine revolution: evidence from snails, predators, and grazers. *Paleobiology* 3: 245–258.

——— 1978. *Biogeography and Adaptation.* Cambridge MA: Harvard Univ. Press.

——— 1980. Gastropod growth rate, allometry, and adult size: environmental implications. In D. C. Rhoads and R. A. Lutz, eds., *Skeletal Growth in Organisms.* N.Y.: Plenum, pp. 379–394.

——— 1987. *Evolution and Escalation: An Ecological History of Life.* Princeton N.J.: Princeton Univ. Press.

Villee, C. A., W. F. Walker, Jr., and R. D. Barnes. 1989. *Biology.* Philadelphia PA: W. B. Saunders.

Vogel, K. 1983. *Macht die biologische Evolution Sprünge?* Wiesbaden: Franz Steiner.

Von Baer, K. E. 1828. *Über Entwickelungsgeschichte der Thiere.* Königsberg: Bornträger.

——— 1866. Über Prof. Nic. Wagner's Entdeckung von Larven die sich fortpflanzen, und über die Pädogenesis überhaupt. *Bull. Acad. Imp. Sciences St. Petersburg* 9: 63–137.

Vrba, E. S. 1980. Evolution, species and fossils: how does life evolve? *South African Jour. Sci.* 76: 61–84.

———— 1983. Macroevolutionary trends: new perspectives on the roles of adaptation and incidental effect. *Science* 221: 387–389.

———— 1984a. Evolutionary pattern and process in the sister-group Alcelaphini-Aepycerotini (Mammalia: Bovidae). In N. Eldredge and S. M. Stanley, eds., *Living Fossils*. N.Y.: Springer-Verlag, pp. 62–79.

———— 1984b. What is species selection? *Systematic Zool.* 33: 318–328.

———— (ed.) 1985a. *Species and Speciation*. Pretoria: Transvaal Museum.

———— 1985b. Environment and evolution: alternative causes of temporal distribution of evolutionary events. *South Afr. Jour. Sci.* 81: 229–236.

———— 1989. Levels of selection and sorting with special reference to the species level. *Oxford Surveys Evol. Biol.* 6:111–168.

Vrba, E. S., and N. Eldredge. 1984. Individuals, hierarchies and processes: towards a more complete evolutionary theory. *Paleobiology* 10: 146–171.

Vrba, E., and S. J. Gould. 1986. The hierarchical expansion of sorting and selection: Sorting and selection cannot be equated. *Paleobiology* 12: 217–228.

Waagen, W. 1869. Die Formenreihe des *Ammonites subradiatus. Paläont. Beiträge* 2: 179–256.

Waddington, C. H. 1960. Evolutionary Adaptation. In S. Tax, ed., *Evolution After Darwin*, Volume I, *The Evolution of Life*. Chicago IL: Univ. of Chicago Press, pp. 381–402.

———— 1967. Discussion. In P. S. Moorehead and M. M. Kaplan, eds., *Mathematical Challenges to the Neodarwinian Interpretation of Evolution*. Philadelphia PA: Wistar Institute.

Wade, M. J. 1978. A critical review of the model of group selection. *Quart. Rev. Biol.* 53: 101–114.

———— 1985. Soft selection, hard selection, kin selection, and group selection. *Amer. Naturalist* 125: 61–73.

Wade, M. J., and D. E. McCauley. 1980. Group selection: the phenotypic and genotypic differentiation of small populations. *Evolution* 34: 799–812.

Wagner, D., R. W. M. Sablowski, and E. M. Meyerowitz. 1999. Transcriptional activation of *APETALA1* by *LEAFY. Science* 285: 582–584.

Wagner, G. P. 1988. The influence of variation and developmental constraints on the rate of multivariate phenotypic evolution. *Jour. Evol. Biol.* 1: 45–66.

———— 1989. The biological homology concept. *Ann. Rev. Ecol. Systematics* 20: 51–69.

Wagner, P. J. 1995. Testing evolutionary constraint hypotheses with early Paleozoic gastropods. *Paleobiology* 21: 248–272.

———— 1996. Contrasting the underlying pattern of active trends in morphologic evolution. *Evolution* 50: 990–1007.

———— 1999. The utility of fossil data in phylogenetic analyses: a likelihood example using Ordovician-Silurian species of the Lophospiridae (Gastropoda: Murchisoniina). *Am. Malacol. Bull.* 15: 1–31.

Wagner, P. J., and D. H. Erwin. 1995. Phylogenetic patterns as tests of speciation models. In D. H. Erwin and R. L. Anstey, eds., *New Approaches to Speciation in the Fossil Record*. N.Y.: Columbia Univ. Press, pp. 87–122.

Wake, D. B. 1991. Homoplasy: the result of natural selection or evidence of design limitation. *Amer. Naturalist* 138: 543–567.

Wake, D. B., G. Roth, and M. H. Wake. 1983. On the problem of stasis in organismal evolution. *J. Theor. Biol.* 101: 211–224.

Waldrop, M. M. 1990. Spontaneous order, evolution, and life. *Science* 247: 1543–1545.

Wallace, A. R. 1870. The measurement of geological time. *Nature* 1: 399–401, 452–455.

———— 1892. The earth's age. *Nature* 47: 175.

———— 1893. The earth's age. *Nature* 47: 227.

———— 1895a. The age of the earth. *Nature* 51: 607.

———— 1895b. Uniformitarianism in geology. *Nature* 52: 4.

———— 1909. The origin of the theory of natural selection. *Popular Science Monthly* 72: 396–400.

Walther, C., and P. Gruss. 1991. *Pax-6,* a murine paired box gene, is expressed in the developing CNS. *Development* 113: 1435–1449.

Ward, P. 1992. *On Methuselah's Trail: Living Fossils and the Great Extinctions.* N.Y.: W. H. Freeman.

Warren, R., L. Nagy, J. Selegue, J. Gates, and S. Carroll. 1994. Evolution of homeotic gene regulation and function in flies and butterflies: testing the Lewis hypothesis. *Nature* 372: 458–461.

Warsh, D. 1990. What goes up sometimes levels off. *Boston Globe.*

———— 1992. Redeeming Karl Marx. *Boston Globe,* May 3.

Waters, J. A. 1981. Evolution of the carboniferous blastoid *Pentremites* Say: a case for phyletic gradualism or punctuated equilibrium? *GSA Abstracts with Programs* 13: 576.

Weatherbee, S. D., and S. B. Carroll. 1999. Selector genes and limb identity in arthropods vertebrates. *Cell* 97: 283–286.

Weatherbee, S. D., H. F. Nijhout, L. W. Grunert, G. Halder, R. Galant, J. Selegue, and S. Carroll. 1999. *Ultrabithorax* function in butterfly wings and the evolution of insect wing patterns. *Current Biology* 9: 109–115.

Wei, K. 1994. Allometric heterochrony in the Pliocene-Pleistocene planktic foraminiferal clade *Globoconella. Paleobiology* 20: 66–84.

Wei, K. Y., and J. P. Kennett. 1988. Phyletic gradualism and punctuated equilibrium in the late Neogene planktonic foraminiferal clade *Globoconella. Paleobiology* 14: 345–363.

Weigel, D., and E. M. Meyerowitz. 1994. The ABCs of floral homeotic genes. *Cell* 78: 203–209.

Weigel, D., and O. Nilsson. 1995. A developmental switch sufficient for flower initiation in diverse plants. *Nature* 377: 495–500.

Weiner, A. M., and N. Maizels. 1999. A deadly double life. *Science* 284: 63–64.

Weiss, H., and R. S. Bradley. 2001. What drives societal collapse? *Science* 291: 609–610.

Weiss, K. M. 1990. Duplication with variation: metameric logic in evolution from genes to morphology. *Yearbook of Physical Anthropology* 33: 1–23.

Weismann, A. 1891. *Amphimixis; oder, die Vermischung der Individuen.* Jena: G. Fischer.

———— 1893. Allsufficiency of natural selection. *Contemp. Rev.* 64: 309–338.

———— 1895. *Neue Gedanken zur Vererbungsfrage.* Jena: Gustav Fischer.

———— 1896. *On Germinal Selection.* Chicago IL: Open Court Publishing Co.

———— 1902. *Vorträge über Descendenztheorie.* Jena: Gustav Fischer.

———— 1903. *The Evolution Theory.* London: Edward Arnold.

———— 1909. The selection theory. In A. C. Seward, ed., *Darwin and Modern Science.* Cambridge UK: Cambridge Univ. Press, pp. 18–65.

Weller, J. M. 1961. The species problem. *Jour. Paleontology* 35: 1181–1192.

Wells, W. C. 1818. An account of a female of the white race of mankind, part of whose skin resembles that of a Negro. In W. C. Wells, *Two Essays* (his collected works, posthumously published). London: S. Constable & Co., pp. 425–439.

Werren, J. H. 1991. The paternal-sex-ratio chromostome of *Nasonia. Amer. Naturalist* 137: 392–402.

Werren, J. H., and L. W. Beukeboom. 1993. Population genetics of a parasitic chromosome: theoretical analysis of PSR in subdivided populations. *Amer. Naturalist* 142: 224–241.

Werren, J. H., U. Nur, and C. I. Wu. 1988. Selfish genetic elements. *Trends Ecol. Evol.* 3: 297–302.

Wessels, N. K., and J. L. Hopson. 1988. *Biology.* N.Y.: Random House.

West, R. M. 1979. Apparent prolonged evolutionary stasis in the middle Eocene hoofed mammal *Hyopsodus. Paleobiology* 5: 252–260.

West-Eberhard, M. J. 1986. Review of *Evolution Through Group Selection,* by V. C. Wynne-Edwards. *Jour. Genetics* 65: 213–217.

Westoll, T. S. 1949. On the evolution of the Dipnoi. In G. L. Jepsen, E. Mayr, and G. G. Simpson, eds., *Genetics, Paleontology and Evolution.* Princeton N.J.: Princeton Univ. Press.

Whatley, R. C. 1985. Evolution of the ostracods *Bradleya* and *Poseidonamicus* in the deep-sea Cainozoic of the south-west Pacific. *Special Papers in Palaeontology* 33: 103–116.

Wheeler, W. M. 1909. Predarwinian and postdarwinian biology. *Popular Science Monthly* 72: 381–385.

Whewell, W. 1837. *History of the Inductive Sciences.* London: John W. Parker.

———— 1869. *History of the Inductive Sciences, 3rd Edition.* N.Y.: D. Appleton.

White, M. T. D. 1945. *Animal Cytology and Evolution.* Cambridge UK: Cambridge Univ. Press.

White, T. D., and J. M. Harris. 1977. Suid evolution and correlation of African localities. *Science* 198: 13–21.

Whiteman, H. H. 1994. Evolution and facultative paedomorphosis in salamanders. *Quart. Rev. Biol.* 69: 205–221.

Whiten, A., and R. A. Byrne. 1988. Taking (Machiavellian) intelligence apart. In R. A. Byrne and A. Whiten, eds., *Machiavellian Intelligence.* Oxford UK: Oxford Univ. Press, pp. 50–55.

Whitman, C. O. 1919. *Orthogenetic Evolution in Pigeons.* Carnegie Inst. Washington Publications No. 257.

Wiester, J. L. 1980. *The Genesis Connection.*

Wiggins, V. D. 1986. Two punctuated equilibrium dinocyst events in the Upper Miocene of the Bering Sea. *AASP Contrib. Series* 17: 159–167.

Willey, A. 1911. *Convergence in Evolution.* London: Murray.

Williams, E. 1980. Evolution's believers need a church of their own. *St. Petersburg Times.*

Williams, G. C. 1966. *Adaptation and Natural Selection.* Oxford UK: Oxford Univ. Press.

———— 1975. *Sex and Evolution.* Princeton NJ: Princeton Univ. Press.

———— 1992. *Natural Selection: Domains, Levels and Challenges.* N.Y.: Oxford Univ. Press.

Williamson, P. G. 1981. Paleontological documentation of speciation in Cenozoic molluscs from Turkana Basin. *Nature* 252: 298–300.

———— 1985. Punctuated equilibrium, morphological stasis and the paleontological documentation of speciation: A reply to Fryer, Greenwood and Peake's critique of the Turkana Basin mollusk sequence. *Biol. Jour. Linn. Soc.* 26: 307–324.

———— 1987. Selection or constraint?: A proposal on the mechanism of stasis. In K. S. W. Campbell and M. F. Day, eds., *Rates of Evolution.* London: Allen & Unwin, pp. 129–142.

Willis, J. C. 1922. *Age and Area.* Cambridge UK: Cambridge Univ. Press.

Wilson, D. S. 1980. *The Natural Selection of Populations and Communities.* Menlo Park CA: Benjamin Cummings.

———— 1983. The group selection controversy: history and current status. *Ann. Rev. Ecol. Systematics* 14: 159–181.

———— 1989. Levels of selection: an aternative to individualism in biology and the human sciences. *Social Networks* 11:257–272.

Wilson, D. S., and R. K. Colwell. 1981. Evolution of sex ratio in structured demes. *Evolution* 35: 882–897.

Wilson, D. S., and E. Sober. 1989. Reviving the superorganism. *Jour. Theor. Biol.* 136: 337–356.

———— 1994. Reintroducing group selection to the human behavioral sciences. *Behavioral Brain Sci.* 17: 585–654.

Wilson, E. O., T. Eisner, W. R. Briggs, R. E. Dickerson, R. L. Metzenberg, R. D. O'Brien, M. Susman, and W. E. Boggs. 1973. *Life on Earth.* Stamford CT: Sinauer Assoc.

Wilson, L. G. 1970. *Sir Charles Lyell's Scientific Journals on the Species Question.* New Haven CT: Yale Univ. Press.

Wimsatt, W. 1981. Units of selection and the structure of the multi-level genome. *PSA 1980, vol. 2, Philosophy of Science Assoc.,* pp. 122–183.

Wistow, G. 1993. Lens crystallins: gene recruitment and evolutionary dynamism. *Trends Biochem. Sci.* 18: 301–306.

Wolfe, J. A. 1990. Long-term biotic effects of major climatic perturbations. *Fourth Int. Cong. Syst. Evol. Biol.* Abstract Volume.

Wollin, Andrew. 1996. A hierarchy-based approach to punctuated equilibrium: an alternative to thermodynamic self-organization in explaining complexity. Abstract, INFORMS National Meeting, Atlanta GA.

Wolpoff, M. H. 1984. Evolution in *Homo erectus:* the question of stasis. *Paleobiology* 10: 389–406.

Wray, G. A. 1995. Punctuated evolution of embryos. *Science* 267: 1115–1116.

Wray, G. A., and C. J. Lowe. 2000. Developmental regulatory genes and echinoderm evolution. *Systematic Biol.* 49: 151–174.

Wray, G. A., J. S. Levinton, L. H. Shapiro. 1996. Molecular evidence for deep Precambrian divergences among metazoan phyla. *Science* 274: 568–573.

Wright, R. 1994. *The Moral Animal.* N.Y.: Pantheon Books.

Wright, S. 1931. Evolution in Mendelian populations. *Genetics* 16: 97–159.

———— 1932. The roles of mutation, inbreeding, crossbreeding, and selection in evolution. *Proc. 6th Int. Cong. Genetics* 1: 356–366.

———— 1960. Physiological genetics, ecology of population, and natural selection. In S. Tax, ed., *Evolution After Darwin,* Volume I, *The Evolution of Life.* Chicago IL: Univ. of Chicago Press, pp. 429–475.

———— 1967. Comments on the preliminary working papers of Eden and Waddington. In P. S. Moorehead and M. M. Kaplan, eds., *Mathematical Challenges to the Neodarwinian Interpretation of Evolution.* Philadelphia PA: Wistar Institute Press: Monograph 5.

———— 1978. *Evolution and the Genetics of Populations.* Four Volumes. Chicago IL: Univ. of Chicago Press.

———— 1980. Genic and organismic selection. *Evolution* 34: 825–843.

———— 1982. Character change, speciation and the higher taxa. *Evolution* 36: 427–443.

Wynne-Edwards, V. C. 1962. *Animal Dispersion in Relation to Social Behavior.* Edinburgh: Oliver & Boyd.

Xiao, S., Y. Zhang, and A. H. Knoll. 1998. Three-dimensional preservation of algae and animal embryos in a Neoproterozoic phosphorite. *Nature* 391: 553–558.

Yoon, C. K. 1996. Bacteria seen to evolve in spurts. *New York Times,* June 25.

Yu, K., M. A. Sturtevant, B. Biehs, V. François, R. W. Padgett, R. K. Blackman, E. Bier. 1996. The *Drosophila decapentaplegic* and *short gastrulation* genes function antagonistically during adult wing vein development. *Development* 122: 4033–4044.

Ziegler, A. M. 1966. The Silurian brachiopod *Eocoelia hemisphaerica* (J. de C. Sowerby) and related species. *Palaeontology* 9: 523–543.

# Illustration Credits

| Figure | Source |
|--------|--------|
| 5.06 | From Gould, 1993e. |
| 5.07 | From Gould, 1993e. |
| 5.08 | From Gould, 1993e. |
| 5.09a | From De Vries, 1909a, Plate III. |
| 5.09b | From De Vries, 1909a, p. 655, fig. 149. |
| 5.10 | From De Vries, 1909a, Plate I. |
| 7.01 | From Fisher, 1930. |
| 7.02 | Drawing by Laszlo Meszoly. |
| 7.03 | From *Biological Science: An Ecological Approach,* BSCS Green Version, 3e. (Boulder, CO: Rand McNally and Company, 1973), p. 622, fig. 18.15. |
| Part II | From Filippo Buonanni, *Ricreatione dell'occhio e della mente nell'osservation delle Chiocciole* (Rome, 1681). |
| 8.01 | From Jere H. Lipps (ed.), *Fossil Prokaryotes and Protists* (London: Blackwell Science Ltd., 1993), p. 45, fig. 4.14. Reprinted by permission of Blackwell Science Ltd. |
| 8.02 | Drawing by Laszlo Meszoly. |
| 8.03 | From Thomas J. Schopf, *Models in Paleobiology* (New York: Freeman, Cooper & Company, 1972), p. 113, fig. 5–10. |
| 8.04 | Used with permission, from *Annual Review of Ecology & Systematics,* Vol. 26 © 1991 by Annual Reviews, www.AnnualReviews.org. |
| 8.05 | Drawing by Laszlo Meszoly. |
| 8.06 | Drawing by Laszlo Meszoly. |
| 8.07 | From Arnold, Kelly, and Parker, 1995, p. 204, fig. 1. |
| 8.08a | From Wagner, 1996, p. 999, fig. 7. |
| 8.08b | From Wagner, 1996, p. 1000, fig. 8. |
| 9.01 | Drawing by Laszlo Meszoly. |
| 9.02 | From author's collection. |
| 9.03a | Reprinted with permission from *Nature,* vol. 293, no. 5832, Oct. 8, 1981, p. 4, fig. 8. Copyright ©1981 Macmillian Magazines Limited. |
| 9.03b | Reprinted with permission from *Nature,* vol. 293, no. 5832, Oct. 8, 1981, p. 5, fig. 4. Copyright ©1981 Macmillian Magazines Limited. |
| 9.04 | From Goodfriend and Gould, 1996, p. 1896, fig. 3. Copyright ©1996 American Association for the Advancement of Science. |
| 9.05 | From R. C. Moore, C. G. Lalicker, and A. G. Fischer, *Invertebrate Fossils* (New York: The McGraw-Hill Companies, 1952), p. 33, fig. 1.14. Copyright © 1952 by The McGraw-Hill Companies. Reproduced with permission of The McGraw-Hill Companies. |
| 9.06 | From Simpson, 1944. |
| 9.07 | From R. C. Moore, C. G. Lalicker, and A. G. Fischer, *Invertebrate Fossils,* (New York: The McGraw-Hill Companies, 1952), p. 33, fig. 1.15. Copyright © 1952 by The McGraw-Hill Companies. Reproduced with permission of The McGraw-Hill Companies. |
| 9.08 | From Michaux, 1989. |
| 9.09 | Drawing by Laszlo Meszoly. |
| 9.10 | From Gould and Eldredge, 1989, p. 142. |
| 9.11 | From *Proc. Natl. Acad. Sci.,* USA, 91 (1994), p. 6811, fig. 5. Copyright ©1994 National Academy of Sciences, USA. Used by permission. |
| 9.12a | From Glenn L. Jepson, Ernst Mayr, and George Gaylord Simpson (eds.), *Genetics, Paleontology, and Evolution* (Princeton: Princeton University Press, 1949). Copyright ©1949 by Princeton University Press. Reprinted by permission of Princeton University Press. |

| Figure | Source |
|---|---|
| 9.12b | From Raup, David M. and Steven M. Stanley, *Principles of Paleontology* (San Francisco: W. H. Freeman and Co. 1971), p. 266, fig. 10–10a. |
| 9.13 | From Erwin and Anstey, 1995. Copyright ©1995 by Columbia University Press. Reproduced with permission of Columbia University Press in the format Textbook via Copyright Clearance Center. |
| 9.14 | From Smith and Paul, 1985, p. 35, fig. 4. Reprinted by permission of the Paleontological Association. |
| 9.15 | From Cronin, 1985, p. 61, fig. 1b. |
| 9.16 | From Kucera and Malmgran, 1998, p. 56, fig. 6. |
| 9.17 | From Kucera and Malmgran, 1998, p. 57, fig. 7. |
| 9.18 | From Cheetham, 1986, p. 196, fig. 5. |
| 9.19 | From H. M. McHenry, "Tempo and mode in human evolution, " *Proc. Natl. Acad. Sci.,* USA 91, 1994, p. 6781, fig. 1b. Copyright ©1994 National Academy of Sciences, USA. Used by permission. |
| 9.20 | From Heaton, 1993, p. 302, fig. 11. |
| 9.21 | From Heaton, 1993, p. 299, fig. 3. |
| 9.22 | From Heaton, 1993, p. 301, fig. 9. |
| 9.23 | From Kelley, 1984, p. 1247, fig. 11. |
| 9.24 | From Stanley and Yang, 1987, p. 124, fig. 8. |
| 9.25 | From Stanley and Yang, 1987, p. 132, fig. 16. |
| 9.26 | From Prothero and Heaton, 1996, p. 262, fig. 2. Copyright ©1996, with permission from Elsevier Science. |
| 9.27 | From Erwin and Anstey, 1995, p. 68, fig. 3.1. Copyright ©1995 by Columbia University Press. Reproduced with permission of Columbia University Press in the format Textbook via Copyright Clearance Center. |
| 9.28 | From Sheldon, 1996, p. 214, fig. 1. Copyright ©1996, with permission from Elsvier Science. |
| 9.29 | Adapted from illustrations by David Starwood, from "The Evolution of Life on the Earth" by Stephen Jay Gould. *Scientific American,* October 1994, p. 86. Copyright ©1994 by *Scientific American*. All rights reserved. |
| 9.30 | Adapted from "Universal Phylogenetic Tree in Rooted Form." Copyright ©1994 by Carl R. Woese. *Microbiological Reviews,* 58, 1994, pp. 1–9. Adapted with permission of the author. |
| 9.31 | From Gould, 1996a, p. 150, fig. 21. Copyright ©1996. Used by permission of Crown Publishers. |
| 9.32 | Adapted from Gould, 1988b. Copyright ©1988 by Stephen Jay Gould. Adapted with permission of *Journal of Paleontology.* |
| 9.33 | From "The Evolution of the Horse" by W. D. Matthew. Appeared in *Quarterly Review of Biology* 1926. Neg. no. 123823. Courtesy Department of Library Services, American Museum of Natural History. |
| 9.34 | From Lenski and Travisano, 1994, p. 6810, fig. 2. Copyright ©1994 National Academy of Sciences, USA. Used by permission. |
| 9.35 | From Blackburn, 1995, p. 203, fig. 1. Used by permission of Academic Press, London. |
| 9.36 | From Kilgour, 1998. Copyright ©1998 by Frederick Kilgour. Used by permission of Oxford University Press, Inc. |
| 9.37 | "Poorly punctuated equilibrium," *American Scientist,* May-June 1997, p. 225. Used by permission of Mark Heath. |
| 9.38 | Adapted from Bliss, Parker, and Gish, 1980. Drawing by Laszlo Meszoly. |

| Figure | Source |
|---|---|
| 9.39 | From Price, 1996. Copyright ©1996. Reprinted with permission of Brooks/Cole, an imprint of Wadsworth Group, a division of Thompson Learning. Fax 800–730–2215. |
| 10.01 | From Jones and Gould, 1999, p. 161, fig. 1. |
| 10.02 | From Jones and Gould, 1999, p. 173, fig. 7. |
| 10.03 | From Stephen Jay Gould, "Of coiled oysters and big brains," *Evolution & Development,* 2:5, September-October 2000, p. 248, fig. 3. Reprinted by permission of Blackwell Science Ltd. |
| 10.04 | From Stephen Jay Gould, "Of coiled oysters and big brains." *Evolution & Development,* 2:5, September-October 2000, p. 247, fig. 1. Reprinted by permission of Blackwell Science Ltd. |
| 10.05 | From Gould, 1980f, p. 199, fig. 23. |
| 10.06 | From Gould, 1980f, p. 120, fig. 24. |
| 10.07 | From Gould , 1989a, p. 521, fig. 2. |
| 10.08 | From Gould, 1984b, p. 219, fig. 1. |
| 10.09 | From Gould, 1984b, p. 219, fig. 3. |
| 10.10 | Drawing by Laszlo Meszoly. |
| 10.11 | Drawing by Laszlo Meszoly. |
| 10.12 | From Weigel and Mayerowitz, 1994, p. 204, fig. 1. Copyright ©1994, with permission from Elsevier Science. |
| 10.13 | From Osborn, 1905. |
| 10.14 | From Weigel and Mayerowitz, 1994, p. 204, fig. 2. Copyright ©1994, with permission from Elsevier Science. |
| 10.15 | From author's collection. |
| 10.16 | From Rudolf A. Raff, *The Shape of Life* (Chicago: University of Chicago Press, 1996), p. 343, fig. 10.2. Reprinted by permission of the publisher, The University of Chicago Press. |
| 10.17 | Reprinted with permission from *Nature,* vol. 358, August 20, 1992, p. 627. Copyright ©1992 Macmillan Magazines, Limited. |
| 10.18 | Reprinted with permission from Andrew Lumsden and Robb Krumlauf, "Patterning the vertebrate Neuraxis, " *Science,* vol. 274, Nov. 15, 1996, p. 1112, fig. 5. Copyright ©1996 American Association for the Advancement of Science. |
| 10.19 | From Gaskell, 1908. |
| 10.20 | From E. M. De Robertis and Y. Sasai. Reprinted with permission from *Nature,* volume 380. Copyright ©1996 by Macmillan Magazines Limited. |
| 10.21 | From Gehring, 1996, p. 14, figs. 3a and 3b. Reprinted by permission of Blackwell Science Ltd. |
| 10.22 | From author's collection. |
| 10.23 | From Tomarev, Callaerts, Kos, Zinovieva, Holder, Gehring, and Piatigorsky, 1997, p. 2425, fig. 4A. Copyright ©1997 National Academy of Sciences, USA. Used by permission. |
| 10.24 | Reprinted with permission from *Nature,* vol. 399, June 24, 1999, p. 775. Copyright ©1999 Macmillan Magazines Limited. |
| 10.25 | From Pennisi and Roush, 1997, p. 37. Copyright ©1997 American Association for the Advancement of Science. Reprinted with permission From Pennisi and Roush. |
| 10.26 | From Cartwright, Bowsher, and Buss, 1999, p. 2183, fig. 1. Copyright ©1999 National Academy of Sciences, USA. Used by permission. |
| 10.27 | Reprinted with permission from *Nature,* vol. 376, August 3, 1995, p. 423, fig. 4. Copyright ©1995 Macmillan Magazines Limited. |

| 10.28 | From Carrol, 1995, p. 61, fig. 4. Reprinted with permission from *Nature*. Copyright ©1995 Macmillan Magazines Limited. |
| 10.29 | From Gould, 1989c. |
| 10.30 | From Coates and Cohn, 1998, p. 374, fig. 2. Copyright ©1998 by John Wiley & Sons, Inc. Reprinted by permission of Wiley-Leiss, Inc., a subsidiary of John Wiley & Sons, Inc. |
| 10.31 | From Dobzhansky, 1937. |
| 11.01 | From Thompson, 1917. |
| 11.02 | From Gould, 1971b, p. 243, fig. 1. Copyright ©The University of Virginia. Reprinted by permission of the Johns Hopkins University Press. |
| 11.03 | From Thompson, 1917. |
| 11.04 | From Thompson, 1917. |
| 11.05 | From Gould, 1971b, p. 243, fig. 3. Copyright ©The University of Virginia. Reprinted by permission of the Johns Hopkins University Press. |
| 11.06 | From Thompson, 1917. |
| 11.07 | From Arnold, 1994, p. 140, fig. 4. Copyright ©1994. Used by permission of Academic Press Ltd., London. |
| 11.08 | From Otto Deumus, *The Mosaics of San Marco in Venice I: The Eleventh and Twelfth Centuries. Volume Two: Plates* (Chicago: The University of Chicago Press, 1984), Plate 1. Reprinted by permission of the publisher, The University of Chicago Press. |
| 11.09 | From *Proc. Natl. Acad. Sci.,* USA, 94 (1997), p. 10751, fig. 1. Copyright ©1997 National Academy of Sciences, USA. Used by permission. |
| 11.10 | From *Megaloceros* from Cougnac Cave, southwest France. Modified after Lorblancet et al., 1993, by A. M. Lister, 1994, *Zoological Journal of the Linnean Society,* London. Used by permission of Dr. A. M. Lister. |
| 12.01 | From David Jablonski, Douglas H. Erwin, and Jere H. Lipps, *Evolutionary Paleobiology* (Chicago: The University of Chicago Press, 1996), p. 422, fig. 16.1. Reprinted by permission of the publisher, The University of Chicago Press. |
| 12.02 | From David Jablonski, Douglas H. Erwin, and Jere H. Lipps, *Evolutionary Paleobiology* (Chicago: The University of Chicago Press, 1996), p. 423, fig. 16.2. Reprinted by permission of the publisher, The University of Chicago Press. |

# Index